The Fossil Record 2

The Fossil Record 2

Jointly sponsored by
the Palaeontological Association,
the Royal Society
and the Linnean Society

Edited by **M. J. Benton**
Reader, Department of Geology,
University of Bristol, UK

Charts produced by Rachael J. Walker

CHAPMAN & HALL
London · Glasgow · New York · Tokyo · Melbourne · Madras

Published by Chapman & Hall, 2–6 Boundary Row, London SE1 8HN

Chapman & Hall, 2–6 Boundary Row, London SE1 8HN, UK

Blackie Academic & Professional, Wester Cleddens Road, Bishopbriggs, Glasgow G64 2NZ, UK

Chapman & Hall Inc., 29 West 35th Street, New York NY10001, USA

Chapman & Hall Japan, Thomson Publishing Japan, Hirakawacho Nemoto Building, 6F, 1-7-11 Hirakawa-cho, Chiyoda-ku, Tokyo 102, Japan

Chapman & Hall Australia, Thomas Nelson Australia, 102 Dodds Street, South Melbourne, Victoria 3205, Australia

Chapman & Hall India, R. Seshadri, 32 Second Main Road, CIT East, Madras 600 035, India

---

First edition 1993

Typeset in 9/11 pt Palatino by Excel Typesetters Company
Printed in Great Britain by the University Press, Cambridge

ISBN 0 412 39380 8

A catalogue record for this book is available from the British Library

Library of Congress Cataloging-in-Publication data available

CONTENTS

ANIMALS: VERTEBRATES

PLANTS

CONTRIBUTORS

Dr R. J. Aldridge,
Department of Geology,
The University,
LEICESTER,
LE1 7RH, UK.

Dr J. G. Baldauf,
Ocean Drilling Program,
Texas A & M University,
Research Park,
1000 Discovery Drive,
COLLEGE STATION,
Texas 77845-9547, USA.

Dr M. J. Benton,
Department of Geology,
University of Bristol,
BRISTOL, BS8 1RJ, UK.

Dr I. D. Boomer,
Institute of Earth Studies,
The University College of Wales,
ABERYSTWYTH, SY23 3DB, UK.

Professor M. C. Boulter,
Division of Environmental Sciences,
University of East London,
Romford Road,
LONDON, E15 4LZ, UK.

Dr P. R. Bown,
Department of Geological Sciences,
University College London,
Gower Street,
LONDON, WC1E 6BT, UK.

Dr D. E. G. Briggs,
Department of Geology,
University of Bristol,
BRISTOL, BS8 1RJ, UK.

Dr C. H. C. Brunton,
Department of Palaeontology,
Natural History Museum,
Cromwell Road,
LONDON, SW7 5BD, UK.

Mr G. E. Budd,
Department of Earth Sciences,
Downing Street,
CAMBRIDGE, CB2 3EQ, UK.

Dr H. Cappetta,
Laboratoire de Paléontologie,
Université de Montpellier II,
Place Eugène Bataillon,
34095 MONTPELLIER CEDEX 5,
France.

Dr W. T. Chang,
Institute of Geology and
Palaeontology,
Academia Sinica,
Chi-Ming-Ssu,
NANKING,
People's Republic of China.

Dr C. J. Cleal,
Department of Botany,
National Museum of Wales,
Cathays Park,
CARDIFF, CF1 3NP, UK.

Dr L. R. M. Cocks,
Department of Palaeontology,
Natural History Museum,
Cromwell Road,
LONDON, SW7 5BD, UK.

Dr M. E. Collinson,
Department of Geology,
Royal Holloway and
Bedford New College,
Egham Hill,
EGHAM, TW20 0EX, UK.

Dr P. Copper,
Geology Department,
Laurentian University,
SUDBURY,
Ontario, Canada, P3E 2C6.

Professor W. T. Dean,
Department of Geology,
University of Wales,
PO Box 914,
CARDIFF, CF1 3YE, UK.

Dr F. Debrenne,
Institut de Paléontologie,
8 rue de Buffon,
75005 PARIS CEDEX 5,
France.

Dr K. J. Dorning,
Pallab Research,
58 Robertson Road,
SHEFFIELD, S6 5DX, UK.

Dr E. N. Doyle,
Department of Geology,
University of the West Indies,
Mona Campus,
KINGSTON 7, Jamaica.

Dr P. Doyle,
School of Earth Sciences,
University of Greenwich,
Walburgh House, Bigland Street,
LONDON, E1 2NG, UK.

Dr C. J. Duffin,
113 Shaldon Drive,
MORDEN,
SM4 4BQ, UK.

Dr G. D. Edgecombe,
Department of Invertebrates,
American Museum of Natural History,
Central Park West at 79th Street,
NEW YORK, NY 10024, USA.

Dr D. Edwards,
Department of Geology,
University of Wales,
PO Box 914,
CARDIFF, CF1 3YE, UK.

Dr D. H. Erwin,
Department of Paleobiology,
National Museum of Natural History,
WASHINGTON, DC 20560, USA.

Dr M. Feist,
Laboratoire de Paléontologie,
Université de Montpellier II,
Place Eugène Bataillon,
34095 MONTPELLIER CEDEX 5,
France.

Dr R. A. Fortey,
Department of Palaeontology,
Natural History Museum,
Cromwell Road,
LONDON, SW7 5BD, UK.

Dr A. S. Gale,
Department of Geology,
Imperial College,
Prince Consort Road,
LONDON, SW7 2BP, UK.

Dr L. T. Gallagher,
Department of Geological Sciences,
University College London,
Gower Street,
LONDON, WC1E 6BT, UK.

Professor B. G. Gardiner,
Division of Biosphere Sciences,
King's College London,
Campden Hill Road,
LONDON, W8 7AH, UK.

Dr P. Gilliland,
3 Monxton Road,
ANDOVER, SP10 3LY, UK.

Dr B. F. Glenister,
Department of Geology,
The University of Iowa,
IOWA CITY, Iowa 52242, USA.

Dr N. Grambast-Fessard,
Laboratoire de Paléobotanique et
Évolution des Vegetaux,
Université de Montpellier II,
Place Eugène Bataillon,
34095 MONTPELLIER CEDEX 5,
France.

Dr L. B. Halstead, (deceased),
Department of Geology,
Imperial College,
Prince Consort Road,
LONDON, SW7 2BP, UK.

Dr D. A. T. Harper,
Department of Geology,
University College,
GALWAY, Republic of Ireland.

Professor M. B. Hart,
Department of Geological Sciences,
University of Plymouth,
Drake Circus,
PLYMOUTH, PL4 8AA, UK.

Dr R. A. Hewitt,
Department of Geology,
McMaster University,
HAMILTON, Ontario,
L85 4M1, Canada.

Dr D. J. Holloway,
National Museum of Victoria,
285–321 Russell Street,
MELBOURNE,
Victoria, Australia 3000.

Dr P. L. Holmes,
Division of Environmental Sciences,
University of East London,
Romford Road,
LONDON, E15 4LZ, UK.

Professor M. R. House,
Department of Geology,
University of Southampton,
SOUTHAMPTON, SO9 5NH, UK.

Dr E. A. Jarzembowski,
Town Hall, Bartholomew Square,
BRIGHTON, BM1 1JA,
and
Postgraduate Research Institute for
Sedimentology,
University of Reading,
Whiteknights,
READING, RG6 2AB, UK.

Dr A. L. Jeffrey,
Department of Geology,
University College,
GALWAY, Republic of Ireland.

Dr A. H. King,
English Nature,
South West Region,
Roughmoor,
Bishop's Hull,
TAUNTON, Somerset, TA1 5AA, UK.

Dr J. Kullmann,
Paläontologisches Institut der
Universität,
Sigwartstrasse 10,
D-7400 TÜBINGEN 1, Germany.

Dr P. D. Lane,
Department of Geology,
University of Keele,
KEELE, ST5 5BG, UK.

Dr M. C. McKenna,
Department of Vertebrate
Paleontology,
American Museum of Natural History,
Central Park West at 79th Street,
NEW YORK, NY 10024, USA.

Dr A. R. Milner,
Department of Biology,
Birkbeck College,
Malet Street,
LONDON, WC1E 7HX, UK.

Dr J. R. Nudds,
The Manchester Museum,
The University,
MANCHESTER, M13 9PL, UK.

Dr A. W. Owen,
Department of Geology
and Applied Geology,
The University,
GLASGOW, GL2 8QQ, UK.

Dr E. F. Owen,
64 Collingwood Road,
HORSHAM,
RH12 2QW, UK.

Dr R. M. Owens,
Department of Geology,
National Museum of Wales,
CARDIFF, CF1 3NP, UK.

Dr K. N. Page,
Geology Section,
English Nature,
Northminster House,
PETERBOROUGH, PE1 1UA, UK.

Professor A. R. Palmer,
Geological Society of America,
3300 Penrose Place,
PO Box 9140,
BOULDER, CO 80301, USA.

Dr M. A. Parkes,
Department of Geology,
Trinity College,
DUBLIN 2, Republic of Ireland.

Dr C. Patterson,
Department of Palaeontology,
Natural History Museum,
Cromwell Road,
LONDON, SW7 5BD, UK.

Dr L. E. Popov, VSEGEI,
Srednii Prospekt 74,
199026 ST PETERSBURG, Russia.

Dr A. J. Powell,
Millennia Ltd., Unit 3,
Weyside Park,
Newman Lane,
ALTON, GU34 2PJ, UK.

Dr C. D. Prosser,
Geology Section,
English Nature,
Northminster House,
PETERBOROUGH, PE1 1UA, UK.

Dr R. B. Rickards,
Department of Earth Sciences,
Downing Street,
CAMBRIDGE, CB2 3EQ, UK.

Dr R. Riding,
Department of Geology,
University of Wales,
PO Box 914,
CARDIFF, CF1 3YE, UK.

Dr J. K. Rigby,
Department of Geology,
210 Page School,
Brigham Young University,
PROVO, Utah 84602, USA.

Dr M. Romano,
Earth Sciences Unit,
Beaumont Building,
University of Sheffield,
Brookhill,
SHEFFIELD, S3 7HF, UK.

Dr E. P. F. Rose,
Department of Geology,
Royal Holloway and
Bedford New College,
Egham Hill,
EGHAM, TW20 0EX, UK.

Mr A. J. Ross,
The Booth Museum of Natural History,
194 Dyke Road,
BRIGHTON, BN1 5AA, UK.

Dr A. W. A. Rushton,
British Geological Survey,
KEYWORTH,
NG12 5GG, UK. [contribution published by permission of the Director, British Geological Survey (NERC)]

Dr H.-P. Schultze,
Museum of Natural History,
Dyche Hall,
University of Kansas,
LAWRENCE,
Kansas 66045-2454, USA.

Dr P. A. Selden,
Department of Geology,
The University,
MANCHESTER, M13 9PL, UK.

Dr J. J. Sepkoski Jr,
Department of the Geophysical Sciences,
University of Chicago,
CHICAGO,
Illinois 60637, USA.

Dr G. D. Sevastopulo,
Department of Geology,
Trinity College,
DUBLIN 2, Republic of Ireland.

Dr J. H. Shergold,
Bureau of Mineral Resources,
Geology and Geophysics,
Box 378,
PO CANBERRA CITY ACT,
Australia 2601.

Dr M. J. Simms,
Department of Geology,
University of Bristol,
BRISTOL, BS8 1RJ, UK.
Currently at
Department of Geology,
National Museum of Wales,
Cathays Park,
CARDIFF, CF1 3NP, UK.

Dr David J. Siveter,
Department of Geology,
The University,
LEICESTER, LE1 7RH, UK.

Dr Derek J. Siveter,
Department of Earth Sciences,
Parks Road,
OXFORD, OX1 3PW, UK.

Dr P. Skelton,
Department of Earth Sciences,
The Open University,
Walton Hall,
MILTON KEYNES, MK7 6AA, UK.

Dr M. P. Smith,
School of Earth Sciences,
University of Birmingham,
Edgbaston,
BIRMINGHAM, B15 2TT, UK.

Dr R. K. Stucky,
Denver Museum of Natural History,
2001 Colorado Boulevard,
City Park,
DENVER,
Colorado 80205, USA.

Dr P. D. Taylor,
Department of Palaeontology,
Natural History Museum,
Cromwell Road,
LONDON, SW7 5BD, UK.

Dr T. N. Taylor,
Department of Plant Biology,
The Ohio State University,
COLUMBUS, Ohio 43210, USA.

Mr J. Todd,
Institute of Earth Studies,
University College of Wales,
ABERYSTWYTH, SY23 3DB, UK.

Mr S. Tracey,
Tertiary Research Group,
12 Bercta Road,
New Eltham,
LONDON, SE9 3TZ, UK.

Dr D. M. Unwin,
Department of Geology,
University of Bristol,
BRISTOL, BS8 1RJ, UK.

Ms R. J. Walker,
Department of Geology,
University of Bristol,
BRISTOL, BS8 1RJ, UK.

Dr Wang Yi-Gang,
Department of Geological Sciences,
University of British Columbia,
6339 Stores Road,
VANCOUVER,
BC, Canada, V6T 2B4.

Dr M. J. Weedon,
Department of Geology,
University of Bristol,
BRISTOL, BS8 1RJ, UK.

Professor R. C. Whatley,
Institute of Earth Studies,
The University College of Wales,
ABERYSTWYTH, SY23 3DB, UK.

Dr M. A. Whyte,
Earth Sciences Unit,
Beaumont Building,
University of Sheffield,
Brookhill,
SHEFFIELD, S3 7HF, UK.

Dr C. L. Williams,
Department of Geological Sciences,
University of Plymouth,
Drake Circus,
PLYMOUTH, PL4 8AA, UK.

Mr M. A. Wills,
Department of Geology,
University of Bristol,
BRISTOL, BS8 1RJ, UK.

Dr R. A. Wood,
Department of Earth Sciences,
Downing Street,
CAMBRIDGE, CB2 3EQ, UK.

Dr J. Zidek,
New Mexico Bureau of Mines,
Campus Station,
SOCORRO,
New Mexico 87801, USA.

ACKNOWLEDGEMENTS

I thank the Linnean Society (the NERC Taxonomic Publications Fund), the Royal Society, the Palaeontological Association and Chapman & Hall for financial assistance towards the completion of this volume. In addition, the Leverhulme Trust has supported much of the work that I have done on this database. Dr Simon Tull of Chapman & Hall was enthusiastic about the project from the start, as was his successor, Ms Ruth Cripwell. Helen Heyes and Andy Finch carried out the onerous task of guiding the volume to completion. Ms Rachael J. Walker contributed greatly to the usefulness of the volume by compiling and endlessly revising the stratigraphical range-chart diagrams (some of which passed through as many as ten revisions). Dr Glenn Storrs of the University of Bristol Department of Geology kindly aided in the final stages of the volume's editing. I must also thank those numerous authors who were willing to be cajoled and brow beaten into producing the text of the book, a task that without exception was much harder, and much more time-consuming, than any of them at first realized. All have vowed never again to become involved in such a gruesome project. Finally, I owe an enormous debt of gratitude to my wife Mary, and children Philippa and Donald, for their forbearance over the past two years while I have been deep in editing thousands of pages of incomprehensible (and, dare one say, dull) texts.

PREFACE

The present volume had its origin in 1987, when Michael Whyte and the editor approached the Palaeontological Association with the idea of an update of *The Fossil Record* published by the Geological Society of London in 1967. That volume had resulted from discussions between the Geological Society of London and the Palaeontological Association, and followed a meeting held in 1965. The most valuable part of the original publication had been the extensive documentation of families, and we decided to focus on that aspect, and not to include any analytical or commentary papers in the present volume, or to hold a meeting.

The 1967 *Fossil Record* was produced by nine editors and 125 contributors, and amounted to 827 pages. The 1993 edition was produced by one editor and 90 contributors, and amounts to 845 pages: a sure sign of increasing efficiency by the palaeontological community! Of the original 125 contributors, only eight have been involved in the present edition (P. Copper, W. T. Dean, B. G. Gardiner, L. B. Halstead (= L. B. H. Tarlo), M. R. House, C. Patterson, R. B. Rickards and A. W. A. Rushton). Of the 1967 contributors, 105 are listed with UK addresses, nine from the United States of America, four from Australia, three from France, two from the Republic of Ireland, and one each from Canada and The Netherlands (i.e. 84% of the authors were British). Comparative figures for the present edition are that 61 of the 90 contributors are based in the UK, 12 in the United States, four in each of France and the Republic of Ireland, two in each of Australia and Canada, and one in each of China, Germany, Jamaica, the former USSR and Sweden. The British contingent represents 68% of the total of authors. The rise in non-British authors from 16% in 1967 to 32% in 1992 could be interpreted as a laudable move to internationalize the project: equally, the fall from 84% to 68% could indicate the relative decline of palaeontology in Britain over the past 25 years (indeed, many of the British contributors to the present edition, 18 of the 61) are graduate students, postdoctoral scientists or essentially unemployed.

RATIONALE

To many palaeobiologists, of course, this kind of enterprise is highly suspect. The reasons for this view are not hard to find. For example, it will be possible for experts to criticize nearly every entry since the authors have had to make difficult decisions concerning which taxa to include in a family and which to exclude, how to deal with questionable and incomplete material, how to treat specimens of uncertain age, and how to divide up the families. However, the scope of this publication has allowed authors to comment on all of these kinds of complex issues. Hence, users of the data will be able to decide how to code the information, whether to include families represented by single species or not, how to deal with incomplete and poorly defined early records, how to interpret uncertain stratigraphical assignments and so on.

One of these problems may be insurmountable for many critics: the validity of families, or indeed of any other higher taxon. How are families to be determined and how are they to be rendered comparable between bacteria and mammals, or between trilobites and birds? There is no counter-argument other than practicality. Our view has been that, if it is worth studying large-scale evolutionary patterns, palaeobiogeographical distributions, and other macro-evolutionary phenomena, one has to have some raw data to work with. Better to have a 1993 database, shot through with errors as it may be, than to continue to use a 1967 listing *faute de mieux*. The critics might have been partially disarmed by a generic-level listing, or even a species-level listing, but these would have entailed other kinds of scientific problems, as well as the practical ones of finding authors with the stamina to complete the task, and a publisher with the generosity to deal with such a monster.

There have also been criticisms that the stratigraphical stage (or epoch for the Precambrian, Cambrian, Ordovician, Silurian, Carboniferous, Miocene and Pleistocene) is too crude and can be improved upon for many groups. While this is doubtless true for certain marine fossils used in biostratigraphy, it would have been impossible to go to substages or zones for most groups. Indeed, it was hard enough to achieve stage-level accuracy for many terrestrial groups! Hence, the family and the stage (or epoch, as noted) were chosen as the most appropriate working units for this volume. None the less, where possible, many authors have used stage-level terminology for the Palaeozoic and Cainozoic erathems.

DATA COMPILATION

The editor decided to follow broadly the chapter divisions used in *The Fossil Record* (1967), and to commission authors/editors who would oversee each major group. Each of these was to use their specialist knowledge of the phylum – or other major group – in question, to select and commission portions of the text, and then to compile the whole chapter, plugging gaps and providing an overview. The first letters inviting contributions went out in mid-1988, and several chapters were successfully allocated in this way.

As time went on, it became clear that it would not be possible to complete the book in such a simple fashion: in many cases, appropriate authors did not exist, or they had other commitments that prevented them from completing the work on time. Early in 1990, Chapman & Hall agreed to publish the book and, later that year, generous grants were received from the Linnean Society (administering the NERC Taxonomic Publications Grant), the Royal Society (a Scientific Publications Grant), and the Palaeontological Association. This money was used to pay for the completion of certain chapters and parts of chapters (1, 3–6, 8, 10, 11, 13–16, 18–21, 28, 29, 42, 45) that otherwise could not

Eonothem	Erathem	Sub-erathem, System, Sub-system		Series	Stage	Alternative stage Designations		
Phanerozoic	Cainozoic	Quaternary Q		Holocene	HOL			
				Pleistocene	PLE			
		Tertiary T	Neogene Ng	Pliocene PLI	Piacenzian PIA			
					Zanclian ZAN			
				Miocene UMI	Messinian MES			
					Tortonian TOR			
				Miocene MMI	Serravallian SRV			
					Langhian LAN			
				Miocene LMI	Burdigalian BUR			
					Aquitanian AQT			
			Palaeogene Pg	Oligocene OLI	Chattian CHT			
					Rupelian RUP			
				Eocene EOC	Priabonian PRB			
					Bartonian BRT			
					Lutetian LUT			
					Ypresian YPR			
				Palaeocene PAL	Thanetian THA			
					Danian DAN			
	Mesozoic	Cretaceous K		Senonian SEN	Maastrichtian MAA			
					Campanian CMP			
					Santonian SAN			
					Coniacian CON			
				Gallic GAL	Turonian TUR			
					Cenomanian CEN			
					Albian ALB			
					Aptian APT			
					Barremian BRM			
				Neocomian NEO	Hauterivian HAU			
					Valanginian VAL			
					Berriasian BER	Ryazanian RYA		
		Jurassic J		Malm MLM	Tithonian TTH	Portlandian POR		
					Kimmeridgian KIM			
					Oxfordian OXF			
				Dogger DOG	Callovian CAL			
					Bathonian BTH			
					Bajocian BAJ			
					Aalenian AAL			
				Lias LIA	Toarcian TOA			
					Pliensbachian PLB			
					Sinemurian SIN			
					Hettangian HET			
		Triassic Tr		Upper u	Rhaetian RHT			
					Norian NOR			
					Carnian CRN			
				Middle m	Ladinian LAD			
					Anisian ANS			
				Scythian SCY	Spathian SPA	Smithian SMI	Olenekian OLK	
					Nammalian NML	Dienerian DIE		
					Griesbachian GRI		Induan IND	
	Palaeozoic	Permian P		Zechstein ZEC	Changxingian CHX	Ochoan OCH	Tatarian TAT	Dorashamian DOR
					Longtanian LGT			Djulfian/Dzhulfian DZH
					Capitanian CAP	Guadalupian GUA	Kazanian KAZ	
					Wordian WOR			
					Ufimian UFI		Leonardian LEN	
				Rotliegendes ROT	Kungurian KUN	Roadian ROD		
					Artinskian ART			
					Sakmarian SAK		Wolfcampian WOL	
					Asselian ASS			

Fig. P.1 The geological time scale used in *The Fossil Record 2*, Permian to Recent.

Eonothem	Erathem	Sub-erathem, System	Sub-system	Series	Stage		Alternative stage Designations	
Phanerozoic	Palaeozoic	Carboniferous C	Pennsylvanian C(u)	Gzelian GZE	Noginskyian NOG	C	Stephanian STE	Silesian SLS
					Klazminskyian KLA	B		
				Kasimovian KAS	Dorogomilovskian DOR	B		
					Chamovnicheskian CHV	A		
					Krevyakinskian KRE		Cantabrian CTB	
				Moscovian MOS	Myachkovskian MYA	D	Westphalian WPH	
					Podolskian POD	C		
					Kashirskian KSK	B		
					Vereiskian VRK	A		
				Bashkirian BSK	Melekesskian MEL	A		
					Cheremshanskian CHE	A		
					Yeadonian YEA	C	Namurian NAM	
					Marsdenian MRD	B		
					Kinderscoutian KIN	B		
			Mississippian C(l)	Serpukhovian SPK	Alportian ALP	A		
					Chokierian CHO	A		
					Arnsbergian ARN	A		
					Pendleian PND	A		
				Visean VIS	Brigantian BRI		Dinantian DIN	
					Asbian ASB			
					Holkerian HLK			
					Arundian ARU			
					Chadian CHD			
				Tournaisian TOU	Ivorian IVO			
					Hastarian HAS			
		Devonian D		Upper u	Famennian FAM			
					Frasnian FRS			
				Middle m	Givetian GIV			
					Eifelian EIF			
				Lower l	Emsian EMS			
					Pragian PRA		Siegenian SIG	
					Lochkovian LOK		Gedinnian GED	
		Silurian S		Pridoli PRD				
				Ludlow LUD	Ludfordian LDF			
					Gorstian GOR			
				Wenlock WEN	Homerian HOM		Gleedonian GLE; Whitwellian WHI	
					Sheinwoodian SHE			
				Llandovery LLY	Telychian TEL			
					Aeronian AER		Fronian FRO; Idwian IDW	
					Rhuddanian RHU			
		Ordovician O	Bala BAL	Ashgill ASH	Hirnantian HIR			
					Rawtheyan RAW			
					Cautleyan CAU			
					Pusgillian PUS			
				Caradoc CRD	Onnian ONN			
					Actonian ACT			
					Marshbrookian MRB			
					Longvillian LON			
					Soudleyan SOU			
					Harnagian HAR			
					Costonian COS			
			Dyfed DFD	Llandeilo LLO	Late LLO3			
					Middle LLO2			
					Early LLO1			
				Llanvirn LLN	Late LLN2			
					Early LLN1			
			Canadian CND	Arenig ARG				
				Tremadoc TRE				

Fig. P.1 The geological time scale used in *The Fossil Record 2,* Ordovician to Carboniferous.

Eonothem	Erathem	Sub-erathem, System, Sub-system	Series	Stage	Alternative stage Designations
Phanerozoic	Palaeozoic	Cambrian €	Merioneth MER	Dolgellian DOL	
				Maentwrogian MNT	
			St David's STD	Menevian MEN	
				Solvan SOL	
			Caerfai/ Comley CRF	Lenian LEN	Toyonian TOY
				Atdabanian ATB	Botomian BOT
				Tommotian TOM	
Precambrian	Proterozoic	Sinian			
		Vendian V	Ediacara EDI	Poundian POU	
				Wonokan WON	
			Varanger VAR	Mortensnes MOR	
				Smalfjord SMA	
		Sturtian		STU	
		Riphean RIF		Karatau KAR	
				Yurmatin YUR	
				Burzyan BUZ	
		Animikean		ANI	
		Huronian		HUR	
	Archaen	Randian		RAN	
		Swazian		SWZ	
		Isuan		ISU	
		Hadean		HDE	
PЄ					

Fig. P.1 The geological time scale used in *The Fossil Record 2*, Archaean to Cambrian.

have been produced in time, and to assist with editorial costs.

Chapman & Hall paid for the production of the stratigraphical range charts, which were generated during 1991 and early 1992 from authors' texts by Ms Rachael Walker in Bristol, using the graphics software Canvas 2.1 on a Macintosh personal computer. The diagrams are on disc, and may be updated readily, or adapted for various uses.

STRATIGRAPHICAL FRAMEWORK AND OTHER STANDARDS USED

Authors were invited to use any stratigraphical scheme that they thought was appropriate, but to use those summarized in Harland *et al.* (1990) if they could. This was an attempt to standardize the stratigraphical periods and stages used, as well as the abbreviations, and of course involved no consideration of the exact ages in millions of years given by those authors. The relevant features of the stratigraphical scheme of Harland *et al.* (1990) are summarized in Fig. P.1, and some equivalent divisions of time used by some authors are also given. In addition, authors who used different schemes from the Harland *et al.* (1990) standard, have commented on this in their chapter introductions.

Other standards used in recording data are broadly as they were in the 1967 *Fossil Record* (see pp. 158–9 therein). The **First** and **Last** records of each family are given, based on published and unpublished data. Living families are indicated as **Extant**, although families with no fossil record are not always listed. For some groups, **Intervening** records are indicated, at stage level, to allow assessment of the gappiness (proportion of stages lacking fossils to stages with fossils) of the ranges quoted. Indeed, the measure of gappiness of intervening values can help to assess the likelihood of accuracy of the first and last records on a range bar, since error bars may be calculated (Strauss and Sadler, 1989).

An attempt was made to minimize the number of bibliographic references listed for each chapter, by referring to recent monographs and volumes of the *Treatise on Invertebrate Paleontology*, where available, for range records. Fuller documentation is presented where no such overview publications exist. Authors and dates of establishment of all taxa are also noted fairly completely, another great advance over the 1967 edition, but bibliographic data are **not** given for such authorships.

In the diagrams, all families, or family-equivalent taxa, are represented as noted by the author(s) of the chapters. Certain ranges are indicated by a solid line, and uncertain range terminations by a dashed line. No attempt is made in the charts to indicate gaps in the intervening range. Taxa with no fossil representatives are not shown on the charts.

In view of the shifting geography of eastern Europe and the former Soviet Union, the following terms are used throughout: 'former USSR', 'former Yugoslavia' and 'Germany'. Former Soviet regions revert to their former titles, e.g. 'Buryat SSR' becomes 'Buryatia'.

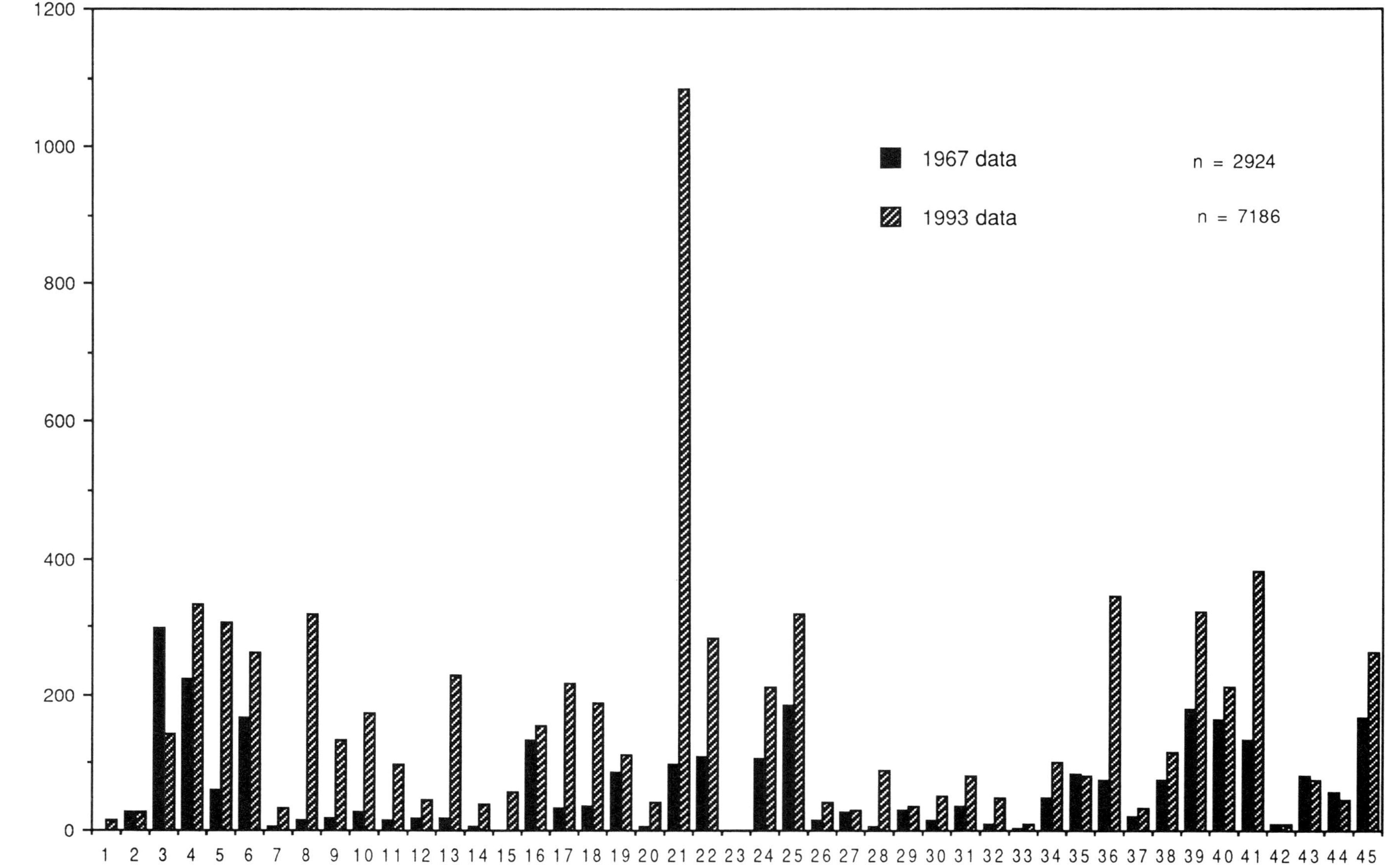

Fig. P.2 The number of families per chapter of *The Fossil Record 2*, compared to numbers in *The Fossil Record* (1967). Chapter 21 is Insecta.

	Number of families	
Group	1967	1993
1. Monera	0	14*
2. Fungi	28**	26
3. 'Algae'	298*	144
4. Protozoa	222	334
5. Porifera	59**	306
6. Coelenterata	168*	262
7. Mollusca: Amphineura, Monoplacophora	5**	32
8. Mollusca: Gastropoda	16**	318
9. Mollusca: Cephalopoda (Nautiloidea)	17**	133
10. Mollusca: Cephalopoda (pre-Jurassic Ammonoidea)	27**	174
11. Mollusca: Cephalopoda (post-Triassic Ammonoidea)	16**	97
12. Mollusca: Cephalopoda (Coleoidea)	18**	45
13. Mollusca: Rostroconchia, Scaphopoda, Bivalvia	18**	228
14. ?Mollusca incertae sedis	7**	40
15. Annelida	0	57
16. Arthropoda (Trilobita)	133*	154
17. Arthropoda (Aglaspidida, Chelicerata, Pycnogonida)	32**	217
18. Arthropoda (Crustacea, excluding Ostracoda)	36**	189
19. Arthropoda (Crustacea: Ostracoda)	86	113
20. Arthropoda (Euthycarcinoidea, Myriapoda)	6**	41
21. Arthropoda (Hexapoda: Insecta)	98**	1083
22. Brachiopoda	109*	282
23. Phoronida	0	1
24. Bryozoa	107*	212
25. Echinodermata	185*	319
26. Basal deuterostomes	14**	43
27. Graptolithina	28*	31
28. Problematica	6*	90
29. Miscellania	31*	35
30. Conodonta	16	52
31. Agnatha	36**	81
32. Placodermi	10**	49
33. Acanthodii	4**	9
34. Chondrichthyes	47*	100
35. Osteichthyes: basal actinopterygians	82	79
36. Osteichthyes: Teleostei	75**	345
37. Osteichthyes: Sarcopterygii	20	33
38. Amphibian-grade Tetrapoda	74*	115
39. Reptilia	178*	323
40. Aves	163	211
41. Mammalia	135**	381
42. Bryophyta	8*	9
43. Pteridophyta	81*	74
44. Gymnospermophyta	58*	44
45. Magnoliophyta ('Angiospermae')	167	261
TOTALS	2924	7186

Fig. P.3 Numbers of families recorded in the 1967 and 1993 editions of *The Fossil Record*. Key: *some taxa not divided to family level; **most taxa not divided to family level.

CHANGES SINCE 1967

Since 1967, a number of factors have combined to enhance the value of an updated second edition. Firstly, many more palaeobiologists than in 1967 are involved in research that requires accurate documentation of the fossil record, especially in the study of patterns of diversification, mass extinction, rates of evolution, clade shapes, completeness measures and phylogenetic bases of the data. Secondly, of course, much work has been done that will tend to change the nature of the family entries: systematic revisions of major groups, reassessments of numerous 'first' and 'last' taxa, discoveries of new fossils and revisions of stratigraphical schemes. All of these have resulted in a remarkable change in the database within 25 years: for example, Maxwell and Benton (1990) found that 416 out of 718 families of tetrapods (58%) listed in *The Fossil Record* (1967) had changed their durations in a 1987 compilation of data, and indeed most of these 416 changed families (57%) showed increased durations. Comparison of the independently compiled lists of marine animal families produced by Sepkoski (1982, 1992) shows similar large-scale changes in the database, here in the course of only ten years. It will be interesting to compare the 1967 and 1992 databases in similar ways in order to discover how much, and why, they have changed.

The Figs P.2 and P.3 indicate the numbers of families, or family-level equivalents identified for each major group in the 1967 and the 1992 editions of *The Fossil Record*. The overall increase in numbers of families listed, from 2924 to 7186, superficially reflects the effects of new finds and some taxonomic splitting in the intervening 25 years. However, much of the increase is a result of the fact that more groups in 1967 were covered at ordinal level than in the present volume. Also, of course, in many cases, families have been lost as a result of taxonomic revisions.

Hence, there has been a particular advance in the coverage of the sponges, molluscs, annelids, arthropods (especially insects, chelicerates and crustaceans), brachiopods, bryozoans, echinoderms, conodonts, vertebrates and angiosperms. Much of the increase in taxon numbers within these groups has been the result of the more consistent effort to identify families in 1992 than in 1967. However, for some groups, such as insects, chelicerates, teleosts and angiosperms, detailed documentation had not been attempted previously in the way presented here. The composition of family lists for certain groups has also been heavily affected by the introduction of a cladistic methodology. Classifications of vertebrates and of some major groups of sponges, gastropods, arthropods, echinoderms and angiosperms in the present work are wholly, or largely, cladistic. This should mean that most, or all, taxa listed in those chapters are monophyletic; further details are given in individual chapter introductions.

Features of *The Fossil Record 2* (1993) that represent advances over the 1967 version include, in summary:

1. consistent family-level coverage for all groups, except Monera;
2. consistent coverage to the stratigraphical stage level for most records, epoch level for most Precambrian, Cambrian, Ordovician, Silurian, Carboniferous, Miocene and Pliocene, records. For some groups, such as ammonoids, substage designations are given;
3. presentation of 'Intervening' data for many groups;
4. standardized presentation of details for 'First' and 'Last' records;
5. monophyletic, cladistically determined, families within many groups.

REFERENCES

Harland, W. B., Holland, C. H., House, M. R. *et al.* (eds) (1967) *The Fossil Record; a Symposium with Documentation.* Geological Society of London, London, 827 pp.

Harland, W. B., Armstrong, R. L., Cox, A. V. *et al.* (1990) *A Geologic Time Scale 1989.* Cambridge University Press, Cambridge, 263 pp.

Maxwell, W. D. and Benton, M. J. (1990) Historical tests of the absolute completeness of the fossil record of tetrapods. *Paleobiology*, **16**, 322–35.

Sepkoski, J. J. Jr (1982) A compendium of fossil marine families. *Milwaukee Public Museum Contributions in Biology and Geology*, **51**, 1–125.

Sepkoski, J. J. Jr (1992) A compendium of fossil marine animal families. 2nd edition. *Milwaukee Public Museum Contributions in Biology and Geology*, **83**, 1–156.

Strauss, D. and Sadler, P. M. (1989) Classical confidence intervals and bayesian probability estimates for ends of local taxon ranges. *Mathematical Geology*, **21**, 411–27.

Basal Groups

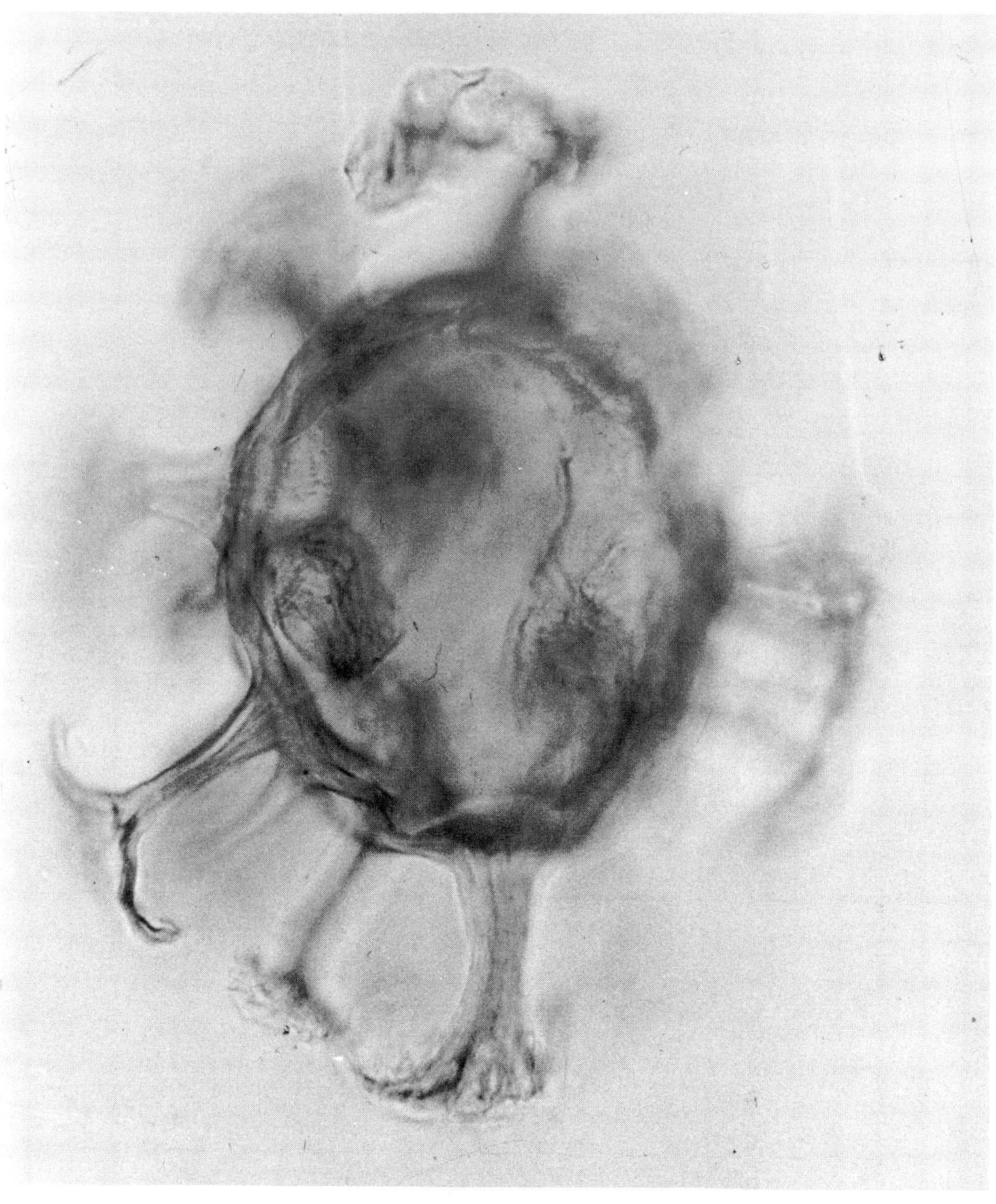

Cordosphaeridium cantharellum (Brosius) Gocht, 1969, the benthic resting cyst of a planktonic dinoflagellate from the Bisciaro Formation (Miocene, Burdigalian) of Marche, Italy. The youngest member of the *Cordosphaeridium* Complex. The central body measures about 50 μm across, and the specimen is viewed in mid-focus. Photograph courtesy of A. J. Powell.

1

MONERA (BACTERIA, BLUE-GREEN ALGAE)

D. Edwards

Perhaps the most momentous developments in palaeontology since the publication of the first *Fossil Record* relate to direct evidence for life in the Precambrian. More than three hundred publications describing micro-organisms from the Archaean and Proterozoic have now been published. The time interval has also seen a biological revolution in the acceptance of the fundamental division of living organisms at the cellular level into eukaryotes and prokaryotes. This has resulted in major changes in classification. Thus in 1967, fossil bacteria from Phanerozoic and only two Precambrian occurrences were merely listed below an introduction citing the problems relating to the compilation. The Schizophyceae (oxygen-producing blue-green algae) were included in the algae, whereas here they are reported under Cyanobacteria.

In the euphoria of the first phases of research into Precambrian biotas there was a tendency to assign the microfossils, particularly of putative cyanobacteria, uncritically to extant families, in some cases incorporating into their names those of extant genera. A somewhat more unbiased approach was employed on the earliest eukaryotes. However, increasing awareness among palaeontologists of the striking morphological resemblances between various groups of prokaryotes, e.g. between sulphur and iron bacteria and oscillatoracean cyanobacteria resulting from their simplicity of organization, plus the realization of the enormous metabolic versatility of the prokaryotes (Knoll and Bauld, 1989) and its role in their classification, have necessitated a more cautious approach. With the exception of the more highly differentiated cyanobacteria, the morphology of prokaryotes is of limited value in inferring affinity and physiology. Such caution has been combined with a considerably heightened sceptism towards claims of biogenicity (see e.g. terminology suggested by Hofmann, 1972), and the appreciation that, while the records documented below demonstrate the presence of a taxon at that time, because of sampling deficiencies it may well have existed earlier. Further, more detailed examination of assemblages and comparisons with modern cultures have shown that some taxa may be based on features produced by *post mortem* degradation of cells and on mucillage sheaths. The analyses presented here reflect these changes of attitude, and are based on re-evaluations of Schopf and Walter (1983) on Archaean records and Mendelson and Schopf (1992) from *c.* 3500 Proterozoic and earliest Cambrian occurrences gleaned from 316 papers published before August 1988. DE is exceedingly grateful to these authors for access to this information, and to Andrew Knoll for further advice. However it should be appreciated that this section has been compiled by a non-expert and readers are advised to consult the primary sources.

The earliest generally accepted evidence for life on this planet comes from carbonaceous stromatolitic sediments in the Warrawoona Group, North Pole Dome, Western Australia (3.3–3.5 Ga) and from cherts in the Onverwacht Group, Barberton Mountain Land, South Africa (3.5 Ga). The status of spheroidal and filamentous structures associated with the stromatolites was assessed by Schopf and Walter (1983) for the Australian records and more recently by Schopf and Packer (1987) on the discovery of new fossils. For the South African records, although the biogenicity of the spheroids can be debated (Knoll and Barghoorn, 1977; Schopf and Walter, 1983), abundant filaments (of two kinds) are far more compelling evidence for life in the early Archaean (Walsh and Lowe, 1985).

Kingdom MONERA Haeckel, 1866 (see Fig. 1.1)

Phylum BACTERIA *sensu lato*

Subphylum COCCOID FORMS P€. (HUR)–Rec. Mar./FW

First: 'coccoid microfossils' (Lanier, 1986), Monte Cristo Formation, Chuniespoort Group, Transvaal Supergroup, South Africa (*c.* 2330 Ma). **Extant**

Comments: Certain spheroids in the Archaean are assessed as dubiofossils by Schopf and Walter (1983), these authors finding no unequivocal evidence for coccoid micro-

The Fossil Record 2. Edited by M. J. Benton. Published in 1993 by Chapman & Hall, London. ISBN 0 412 39380 8

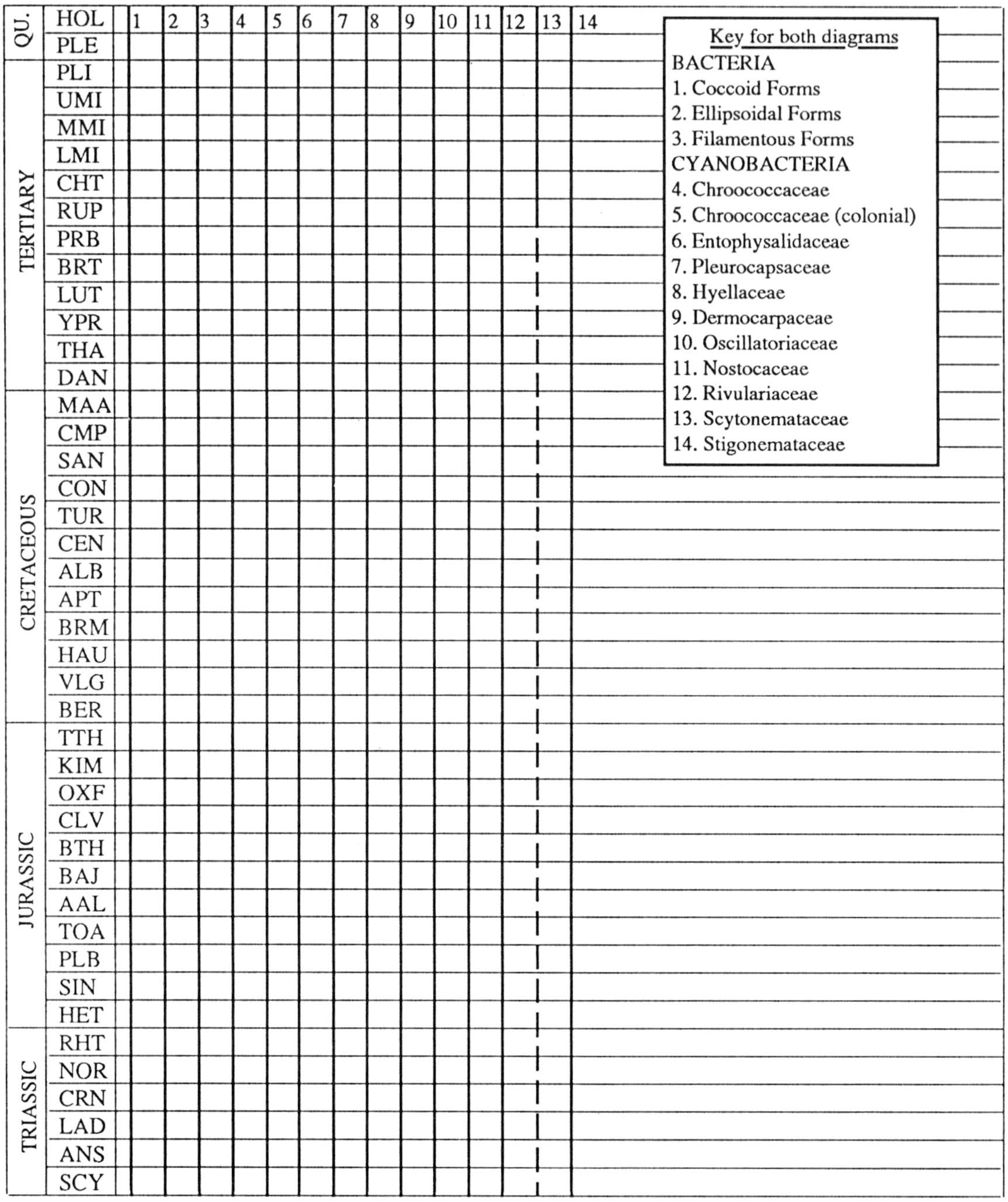

		1	2	3	4	5	6	7	8	9	10	11	12	13	14
QU.	HOL														
	PLE														
TERTIARY	PLI														
	UMI														
	MMI														
	LMI														
	CHT														
	RUP														
	PRB														
	BRT														
	LUT														
	YPR														
	THA														
	DAN														
CRETACEOUS	MAA														
	CMP														
	SAN														
	CON														
	TUR														
	CEN														
	ALB														
	APT														
	BRM														
	HAU														
	VLG														
	BER														
JURASSIC	TTH														
	KIM														
	OXF														
	CLV														
	BTH														
	BAJ														
	AAL														
	TOA														
	PLB														
	SIN														
	HET														
TRIASSIC	RHT														
	NOR														
	CRN														
	LAD														
	ANS														
	SCY														

Key for both diagrams
BACTERIA
1. Coccoid Forms
2. Ellipsoidal Forms
3. Filamentous Forms
CYANOBACTERIA
4. Chroococcaceae
5. Chroococcaceae (colonial)
6. Entophysalidaceae
7. Pleurocapsaceae
8. Hyellaceae
9. Dermocarpaceae
10. Oscillatoriaceae
11. Nostocaceae
12. Rivulariaceae
13. Scytonemataceae
14. Stigonemataceae

Fig. 1.1

organisms in that period. Their coccoidal dubiofossils (i.e. possibilities) include some of the smaller carbonaceous spheroids (cf. *Archaeosphaeroides barbertonensis*) described by Muir and Grant (1976) from carbonaceous cherts and shales in the Onverwacht Group (3.540 ± 0.030 Ga) and *A. barbertonensis* from the Fig Tree Group (?3.5–3.1 Ga). Spheres called *Isuasphaera isua* Pflug from the Isua Supracrustal Belt, southwestern Greenland (3.770 ± 0.042 Ga) and described as yeast-like micro-organisms (e.g. Pflug and Jaeschke-Boyer, 1979) are considered to be non-fossils being interpreted as 'metamorphically produced multiphase inclusions possibly containing organic fluids'.

Subphylum ELLIPSOIDAL FORMS
P€. (?HUR)–Rec. Mar./FW

First: ellipsoidal (rod-shaped) microfossils (Lanier, 1986), Monte Cristo Formation, South Africa (*c.* 2330 Ma). **Extant**
Comments: The earliest colonial ellipsoidal bacteria are described together with solitary examples as *Eosynechoccus moorei* by Hofmann (1976) in an extensive assemblage of prokaryotes from the Kasegalik Formation, Belcher Group, Canada. Rod-shaped structures (*Eobacterium isolatum*) from the Onverwacht Group (SA) examined ultrastructurally by Barghoorn and Schopf are now considered to be modern

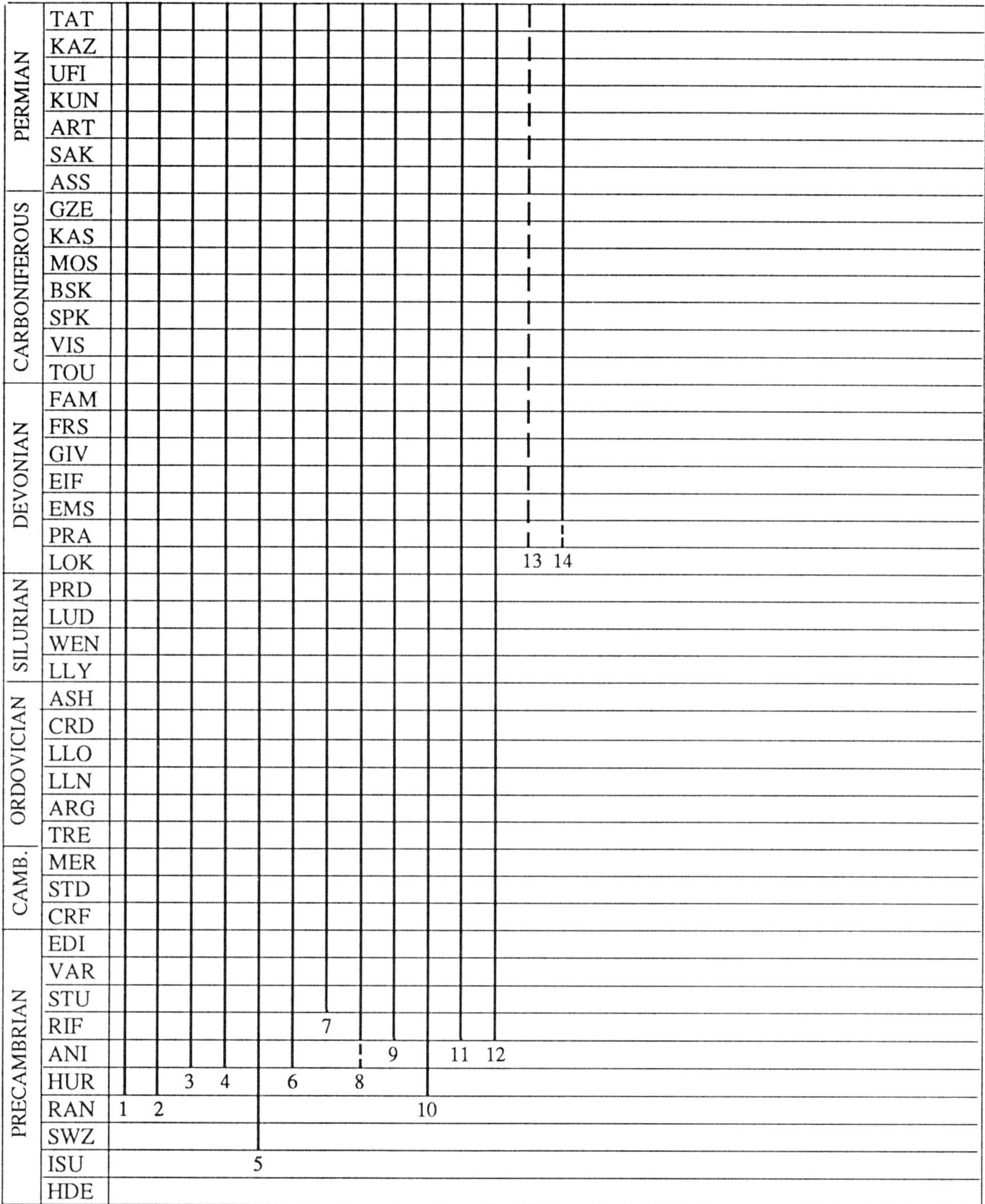

Fig. 1.1

contaminants of preparation and provide a good example of the pitfalls when researching into fossil bacteria.

Subphylum FILAMENTOUS FORMS
PЄ. (ANI)–Rec. Mar./FW

First: *Biocatenoides sphaerula* Schopf, 1968 (septate filamentous bacterium) Kasegalik Formation, Belcher Group, Canada (*c.* 2150 Ma) (Schopf, 1968; Hofmann, 1976).

Comments: The oldest unequivocal fossils accepted by Schopf and Walter (1983) and called 'filamentous Archean micro-organisms' occur in the carbonaceous stromatolitic cherts in the Warrawoona Group, North Pole Dome, Western Australia. They were originally described as 'filamentous fossil bacteria' by Awramik *et al.* (1983), but it is also possible that they are partially degraded cyanobacterial filaments (as indeed may be the first record!). More recently collected cherts of undoubted provenance (i.e. Apex Basalt, 3.3 Ga) contain clasts with sinuous unbranched filaments (*c.* 3.0 μm diameter, 30–40 μm long) composed of uniseriate more-or-less equant cells. These are compared with the trichomes of extinct and extant prokaryotes with modern analogues among the oscillatoraceans, beggiatoaceans and chloroflexaceans (Schopf and Packer, 1987). Groups of radiating filaments from the same locality (Awramik *et al.*, 1983) were assessed as dubiofossils, while similar structures (*Eoastrion*) in the

Gunflint Formation were accepted as prokaryotes, but of uncertain affinity (Mendelson and Schopf, 1992).

Phylum CYANOBACTERIA Stanier *et al.*, 1978

Order CHROOCOCCALES Wettstein, 1924

F. CHROOCOCCACEAE Nägeli, 1849 (solitary fossils) P€. (ANI)–Rec. Mar./FW

First: *Huroniospora* sp. and type 2 of solitary coccoid bacteria in Hofmann and Jackson (1969), the latter determined as cf. *Leptoteichos* by Mendelson and Schopf (1992), Belcher Group, Canada (*c.* 2150 Ma). **Extant**

F. CHROOCOCCACEAE (colonial) P€. (SWZ)–Rec. Mar./FW

First: Carbonaceous or iron-stained three-dimensionally preserved colonies of few to many sheath-enclosed spheroidal cells (Schopf and Packer, 1987), Towers Formation (3.4 Ga), Warrawoona Group. **Extant**

Comments: Globular colonies contain cells in two size ranges, *c.* 8 µm and *c.* 21 µm diameter. The latter are enclosed within a multilamellated sheath, and this, together with their size, is strongly suggestive of chroococcacean affinity. The oldest chroococcacean prefixed by '?' in Mendelson and Schopf is *Polyedrosphaeridium bullatum* Timofeev, 1966 from the Ikabijsk Formation, Siberia (*c.* 2200 Ma) (Timofeev *et al.*, 1976) and the oldest unequivocal Proterozoic example *Favososphaeridium bothnicum* Timofeev, 1966 from the Bothnia Formation, Siberia (*c.* 1770 Ma). The earliest cuboidal colonial coccoid cyanobacteria, were described as *Eucapsis*? in Licari *et al.* (1969) and Licari and Cloud (1972) from the Paradise Creek Formation, Australia (*c.* 1650 Ma).

F. ENTOPHYSALIDACEAE Geitler, 1925 (colonial) P€. (ANI)–Rec. Mar./FW

First: *Eoentophysalis belcherensis* Hofmann, 1976 Kasegalik Formation (*c.* 2150 Ma), Belcher Supergroup, Hudson Bay, Canada. **Extant**

Order PLEUROCAPSALES Geitler, 1925

F. PLEUROCAPSACEAE Geitler, 1925 P€. (STU)–Rec. Mar./FW

First: *Palaeopleurocapsa wopfneri* Knoll, 1975, Skillogallee Dolomite, Australia (770 Ma) (Knoll *et al.*, 1975). **Extant**

Comments: Older possible colonial coccoid pleurocapsaleans are *P. kelleri* Krylov and Sergeev, 1987, in the Satka Formation, Eurasia (*c.* 1550 Ma) or *Myxomorpha janecekii* Muir, 1976 in Oehler (1978) from the Nathan Group, Balbirini Dolomite (*c.* 1483 Ma), Northern Territory which is interpreted as a putative dermocarpacean or pleurocapsacean colonial coccoid cyanobacterium. *Pleurocapsa* (?) sp. from the Kasegalik Formation, Belcher Supergroup (*c.* 2150 Ma) (Hofmann, 1975) would be the oldest representative if the assignment is correct.

F. HYELLACEAE Borzi, 1914 P€. (ANI/BUZ)–Rec. Mar./FW

First: *Eohyella campbelliae* Zhang and Golubic, 1987, Dahongya Formation, Hebei Province, China (*c.* 1650 Ma) (Zhang and Golubic, 1987). **Extant**

Comments: Far better-preserved endoliths of coids (cf. *Hyella gigas*) were described by Green *et al.* (1988) from the Eleonore Bay Group, Greenland (*c.* 750 Ma).

F. DERMOCARPACEAE Geitler, 1925 P€. (YUR)–Rec. Mar./FW

First: *Polybessurus bipartitus* Fairchild, 1975, ex Green *et al.* (1987). Victor Bay Formation, Bylot Supergroup, Baffin Island, Canada (1260–1250 Ma). **Extant**

Comments: *P. bipartitus* was recorded from the *c.* 750 Ma Eleonore Bay Group by Green *et al.* (1987) and identified as a benthic probable pleurocapsalean, producing a unidirectional secretion of extracellular mucillage.

Order NOSTOCALES Geitler, 1925

F. OSCILLATORIACEAE (S. F. Gray) Dumortier ex Kirchner, 1898 P€. (HUR)–Rec. Mar./FW

First: *Siphonophycus transvaalense* Beukes *et al.* (1987), Gamohaan Formation, Ghaap Group, Transvaal Supergroup (2450–2250 Ma) (Klein *et al.*, 1987). **Extant**

Intervening: Throughout Precambrian, but rare in Phanerozoic.

Comments: Mainly recorded as tubular sheaths. Oscillatoriaceans are important mat-forming organisms in stromatolites. However, there is no direct unequivocal evidence that they were involved in the stromatolitic sedimentary structures recorded from the Archaean.

F. NOSTOCACEAE Dumortier ex Engler, 1892 P€. (KAR)–Rec. Mar./FW

First: *Anabaenidium johnsonii* Schopf, 1968, *Veteronostocale* Schopf and Blacic, 1971, Bitter Springs Formation, Australia (*c.* 850 Ma) are both considered possible nostocaceans by Mendelson and Schopf (1992) (Schopf, 1968; Schopf and Blacic, 1971). **Extant**

F. RIVULARIACEAE Kützing, ex Bornet and Flahault, 1887 P€. (KAR)–Rec. Mar./FW

First: *Caudiculophycus acuminatus* Schopf and Blacic, 1971 in Jankanskas (1982), Podinzer Formation, former USSR (*c.* 925 Ma) is considered to be a possible member. **Extant**

Comments: This species, together with *C. rivularioides* Schopf, 1968 was first described from the Bitter Springs Formation, and again both are considered possible rivulariaceans. Knoll (1981) described *C. rivularioides* as sheathless trichomes of oscillatorian cyanobacteria, emphasizing problems of *post-mortem* degradation in recognition of affinity.

F. SCYTONEMATACEAE Kützing, 1843, ex Bornet and Flahault, 1886 D. (?PRA)–Rec. FW/Mar.

First: *Rhyniella vermiformis* Croft and George (1959) Rhynie Chert, Scotland, UK. Data from Tappan (1980). **Extant**

Comment: Most other records are in the Tertiary (earliest in Oligocene *Epivalvia* and *Encrusta* (Daley, 1975). However, Precambrian *Palaeoscytonema* (Edhorn, 1973) is not accepted by Mendelson and Schopf (1992).

Order STIGONEMATALES Geitler, 1925

F. STIGONEMATACEAE Hassal, 1845 D. (?PRA)–Rec. FW/Mar.

First: *Langiella scourfieldi* Croft and George (1959); *Kidstoniella fritschi* Croft and George (1959), Rhynie Chert, Scotland. **Extant**

REFERENCES

Awramik, S. M., Schopf, J. W. and Walter, M. R. (1983) Filamentous fossil bacteria from the Archean of Western Australia. *Precambrian Research*, **20**, 357–74.

Croft, W. N. and George, E. A. (1959) Blue-green algae from the Middle Devonian of Rhynie, Aberdeenshire. *Bulletin of the British Museum (Natural History), Geology Section*, **3**, 339–53.

Daley, P. (1975) Shell encrusting algae from the Bembridge Marls (Lattorfian) of the Isle of Wight, Hampshire, England. *Revue de Micropaléontologie*, **17**, 15–22.

Edhorn, A.-S. (1973) Further investigations of fossils from the Animikie, Thunder Bay, Ontario. *Proceedings of the Geological Association of Canada*, **25**, 37–66.

Green, J. W., Knoll, A. H., Golubic, S. and Swett, K. (1987) Paleobiology of distinctive benthic microfossils from the Upper Proterozoic Limestone–Dolomite 'Series', East Greenland. *American Journal of Botany*, **74**, 928–40.

Green, J. W., Knoll, A. H. and Swett, K. (1988) Microfossils from oolites and pisolites of the upper Proterozoic Eleonore Bay Group, central East Greenland. *Journal of Paleontology*, **62**, 835–52.

Hofmann, H. J. (1972) Precambrian remains in Canada: fossils, dubiofossils and pseudofossils. *Twenty-fourth International Geological Congress. Section 1*, 20–30.

Hofmann, H. J. (1976) Precambrian microflora, Belcher Islands, Canada: significance and systematics. *Journal of Paleontology*, **50**, 1043–73.

Hofmann, H. J. and Jackson, G. D. (1969) Precambrian (Aphebian) microfossils from Belcher Islands, Hudson Bay, *Canadian Journal of Earth Sciences*, **6**, 1137–44.

Jankanskas, T. V. (1982) Riphean microfossils of the Southern Urals. In: *Stratotyp Rifeya: Paleontologiya i Paleomagnetizm*. (ed. B. M. Keller), Nauka, Moscow [in Russian], pp. 84–120.

Klein, C., Beukes, N. J. and Schopf, J. W. (1987) Filamentous microfossils in the early Proterozoic Transvaal Supergroup: their morphology, significance and palaeoenvironmental setting. *Precambrian Research*, **36**, 81–94.

Knoll, A. H. (1981) Paleoecology of Late Precambrian microbial assemblages. In: *Paleobotany, Paleoecology and Evolution* (ed. K. J. Niklas), Praeger, New York, pp. 17–54.

Knoll, A. H. and Barghoorn, E. S. (1977) Archaean microfossils showing cell divisions from the Swaziland System of South Africa. *Science*, **198**, 396–98.

Knoll, A. H. and Bauld, J. (1989) The evolution of ecological tolerance in prokaryotes. *Transactions of the Royal Society of Edinburgh: Earth Sciences*, **80**, 209–23.

Knoll, A. H., Barghoorn, E. S. and Golubic, S. (1975) *Palaeopleurocapsa wopfnerii* gen. et sp. nov. A late Precambrian alga and its modern counterpart. *Proceedings of the National Academy of Sciences, USA*, **72**, 2488–92.

Lanier, W. P. (1986) Approximate growth rates of early Proterozoic microstromatolites as deduced by biomass productivity. *Palaios*, **1**, 525–42.

Licari, G. R. and Cloud, P. (1972) Prokaryotic algae associated with Australian Proterozoic stromatolites. *Proceedings of the National Academy of Sciences USA*, **69**, 2500–4.

Licari, G. R., Cloud, P. and Smith, W. D. (1969) A new chroococcacean alga from the Proterozoic of Queensland. *Proceedings of the National Academy of Sciences USA*, **62**, 56–62.

Mendelson, C. V. and Schopf, J. W. (1992) Proterozoic and selected Early Cambrian microfossils and microfossil-like objects. In: *The Proterozoic Biosphere, A Multidisciplinary Study*, (eds J. W. Schopf and C. Klein), Cambridge University Press, New York, pp. 865–951.

Muir, M. D. and Grant, P. R. (1976) Micropalaeontological evidence from the Onverwacht Group, South Africa, In: *The Early History of the Earth* (ed. B. F. Windley), Wiley, New York, pp. 595–604.

Oehler, D. Z. (1978) Microflora of the middle Proterozoic Balbirini Dolomite (McArthur Group) of Australia, *Alcheringa*, **2**, 269–309.

Pflug, H. D. and Jaeschke-Boyer, H. (1979) Combined structural and chemical analyses of 3,800 Myr-old microfossils. *Nature*, **280**, 483–6.

Schopf, J. W. (1968) Microflora of the Bitter Springs Formation, late Precambrian, central Australia. *Journal of Paleontology*, **42**, 651–88.

Schopf, J. W. and Blacic, J. M. (1971) New microorganisms from the Bitter Springs Formation (late Precambrian) of north-central Amadeus Basin, Australia. *Journal of Paleontology*, **45**, 925–60.

Schopf, J. W. and Packer, B. M. (1987) Early Archean (3.3 billion to 3.5 billion-year-old) microfossils from Warrawoona Group, Australia. *Science*, **237**, 70–3.

Schopf, J. W. and Walter, M. R. (1983) Archean microfossils: new evidence of ancient microbes, in *Earth's Earliest Biosphere* (ed. J. W. Schopf), Princeton University Press, Princeton, pp. 214–39.

Tappan, H. (1980) *The Paleobiology of Plant Protists*. W. H. Freeman, San Francisco, 1028 pp.

Timofeev, B. V., Hermann, T. N. and Mikhaylova (1976) *Plant microfossils of the Precambrian, Cambrian and Ordovician*. Nauka, Leningrad.

Von Wettstein, R. (1924) *Handbuch der Systematischen Botanik* 3rd edn, Wien, Leipzig, 1017 pp.

Walsh, M. M. and Lowe, D. R. (1985) Filamentous microfossils from the 3,500-Myr-old Onverwacht Group, Barberton Mountain Land, South Africa. *Nature*, **314**, 530–2.

Zhang, Y. and Golubic, S. (1987) Endolithic microfossils (Cyanophyta) from Early Proterozoic stromatolites, Hebei, China. *Acta Micropalaeontologica Sinica*, **4**, 3–15.

2

FUNGI

T. N. Taylor

Although convincing examples of Precambrian fungi are not known, examples of the group can be documented throughout the rest of the geological record. Owing to the complexities of many fungal life histories, features used in taxonomy that cannot be demonstrated in fossils, and the often poor level of preservation, some of the earliest reports at the level of order and family were in error. By far the best record of the group is known from permineralized remains or epiphyllous types in which a sufficient suite of characters makes identification more reliable. In some instances, fungal spores have been useful in documenting some groups. New reports of fossil fungi are continually expanding the geological range of modern families, while at the same time demonstrating the existence of groups for which there are no modern analogues. In recent years, emphasis in palaeomycology has also centred on a variety of fungal interactions that can be demonstrated in the fossil record (Stubblefield and Taylor, 1988). Documenting interactions and fungal evolution will rely not only on past reports of fungi such as those listed by Tiffney and Barghoorn (1974), but also the discovery of additional forms from throughout the geological column (Taylor *et al.* 1992a,b,c). The distribution of fungi in time and space, based on fossil evidence, together with rapidly accumulating molecular data, will provide the continuing impetus to characterize more accurately the phylogeny of the group.

Kingdom FUNGI (see Fig. 2.1)

Division MASTIGIOMYCOTA

Form ***Class*** PALEOMASTIGOMYCETES

Order PALEOCHYTRIDALES D. (PRA) Terr.

First: *Milleromyces rhyniensis* (Taylor, Hass and Remy, 1992). (PRA). Together with *Lyonomyces pyriformis* and *Krispiromyces discoides* these aquatic fungi occur in the Rhynie Chert associated with the green alga *Palaeonitella*.

Class CHYTRIDIOMYCETES

Order CHYTRIDIALES

F. UNNAMED D. (PRA?)–Rec. Terr.

First: *Horneophyton lignieri* (Illman, 1984), Rhynie Chert bed, Aberdeenshire, Scotland, UK. This report is based on zoosporic fungal sporangia inside trilete spores of the taxon. Comparison is also made with oomycetes and hyphochytridiomycetes. **Extant**

Intervening: VRK.

F. OLPIDIACEAE T. (LUT/BRT)–Rec. Terr.

First: *Pleotrachelus askaulos* Bradley, 1967, Green River Formation, Wyoming, USA. Holocarpic zoosporangium with discharge tubes. **Extant**

F. PHLYCTIDIACEAE T. (LUT/BRT)–Rec. Terr.

First: *Entophlyctis willoughbyi* Bradley, 1967, Green River Formation, Wyoming, USA. Sporangia with zoospore cyst and germ tube. **Extant**

Class HYPHOCHYTRIDIOMYCETES

Order HYPHOCHYTRIALES C. (u.)–Rec. Terr.

First: ? Unnamed (Millay and Taylor, 1978), Breathitt Formation, Kentucky, USA. May also be included within other families of the Chytridiomycetes. **Extant**

Class OOMYCETES

Order SAPROLEGNIALES S. (GLE/GOR)–Rec. Terr.

First: *Palaeachlya perforans* (Duncan, 1876). Believed to have parasitized the coral *Goniophyllum pyramidale*. **Extant**
Intervening: SIG, C. (u.), T.

Order PERONOSPORALES C. (MOS)–Rec. Terr.

First: Unnamed (Stidd and Cosentino, 1975), Des Moines Series, Oskaloosa, Iowa, USA. Oogonia and disrupted tissues in the seed *Nucellangium* that are morphologically similar to the symptoms caused by the extant fungus *Albugo*. **Extant**

Intervening: RUP/CHT.

Division AMASTIGOMYCOTA

Class ZYGOMYCETES

Order MUCORALES

F. GLOMACEAE C.–Rec. Terr.

First: *Paleobasidiospora taugourdeauii* Locquin, 1983. This new family was recently proposed to include the large number of vesicles and chlamydospores found throughout

The Fossil Record 2. Edited by M. J. Benton. Published in 1993 by Chapman & Hall, London. ISBN 0 412 39380 8

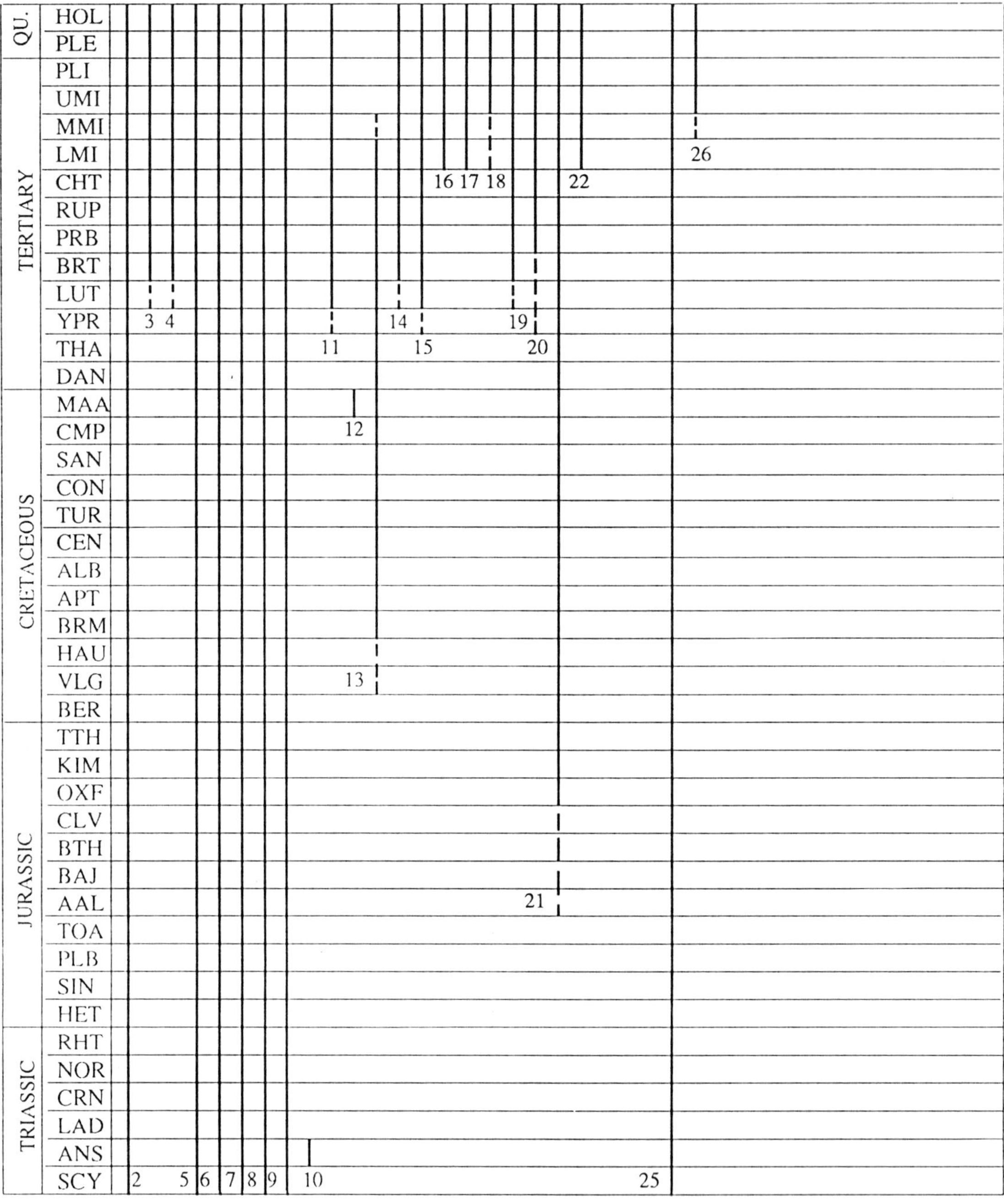

Fig. 2.1

the geological column that morphologically resemble the endogonaceous genus *Glomus* (Pirozynski and Daplé, 1989). Included in this family are the chlamydospores from the Lower Devonian Rhynie Chert (Kidston and Lang, 1921) described as *Palaeomyces*, and numerous Carboniferous representatives termed *Glomus*-like (Wagner and Taylor, 1982). It is not known whether these fungi were biotrophic symbiotic endophytes. The most convincing evidence of endophytic mycorrhizal fungi come from the Triassic in the form of arbuscles in root cells (Stubblefield *et al.*, 1987a,b). Also included in this family is the modern genus *Sclerocystis* which also dates from the Triassic (Stubblefield *et al.*, 1987a,b). Some of the taxa placed in this family have previously been included within the Mucoraceae (Tiffney and Barghoorn, 1974). **Extant**

F. ENDOGONACEAE C. (BSK)–Rec. Terr.

First: *Traquairia carruthersii* (Scott) Stubblefield and Taylor, 1983, Westphalian A equivalent, Dulesgate, Lancashire, England, UK. These structures which are termed sporocarps include several genera that have also been included with the ascomycetes (*Myocarpon*, *Coleocarpon*, *Sporocarpon*, *Dubiocarpon* (Stubblefield *et al.*, 1983)), but see Taylor and White (1989) for another interpretation. **Extant**
Intervening: ANS.

Class TRICHOMYCETES

Order ECCRINALES Tr. (ANS) Terr.

First and Last: Unnamed (White and Taylor, 1989), Fremouw Formation, central Transantarctic Mountains,

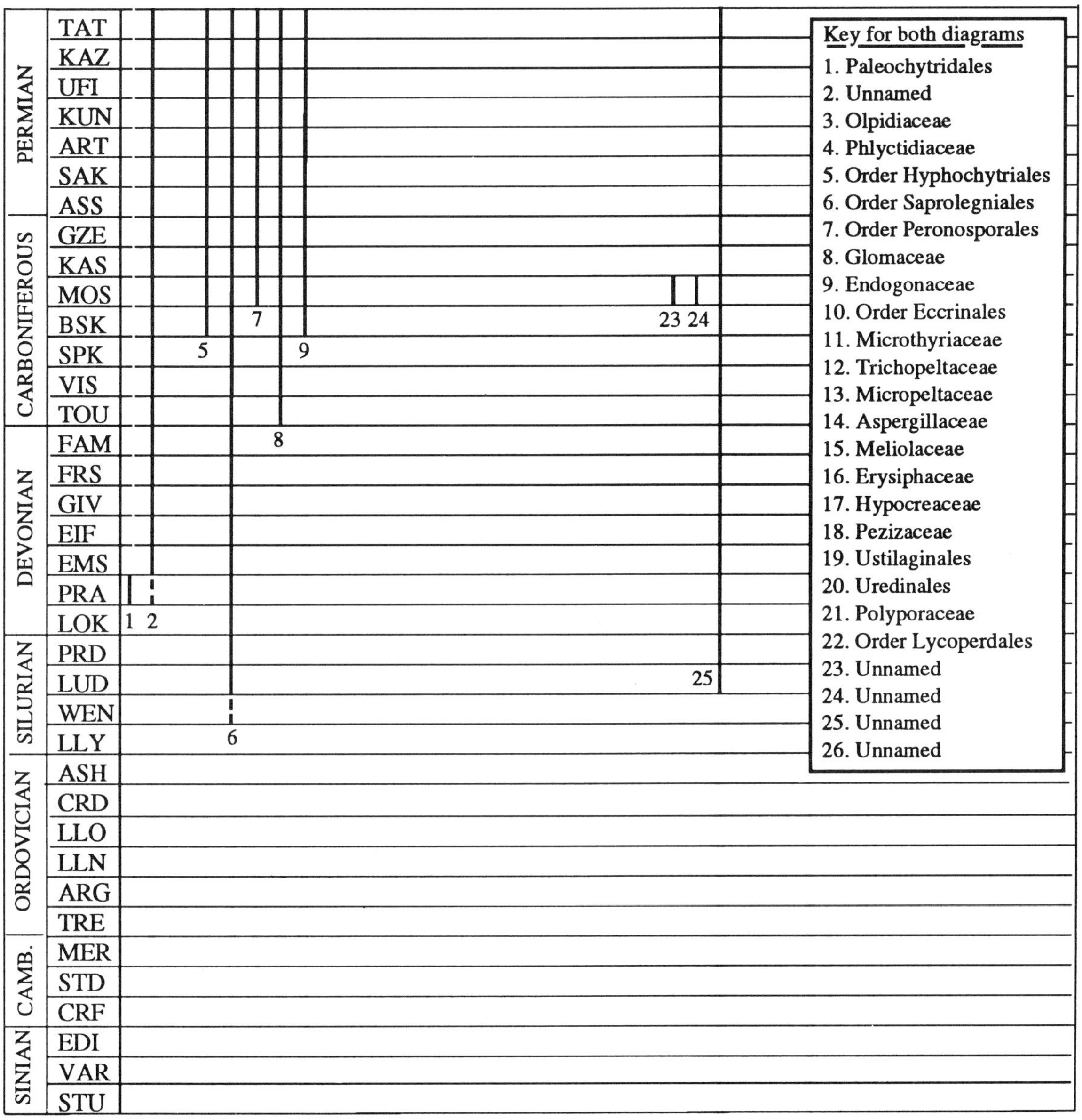
PERMIAN
TAT
KAZ
UFI
KUN
ART
SAK
ASS
CARBONIFEROUS
GZE
KAS
MOS
BSK
SPK
VIS
TOU
DEVONIAN
FAM
FRS
GIV
EIF
EMS
PRA
LOK
SILURIAN
PRD
LUD
WEN
LLY
ORDOVICIAN
ASH
CRD
LLO
LLN
ARG
TRE
CAMB.
MER
STD
CRF
SINIAN
EDI
VAR
STU
1 2
5
6
7
8
9
23 24
25
Key for both diagrams
1. Paleochytridales
2. Unnamed
3. Olpidiaceae
4. Phlyctidiaceae
5. Order Hyphochytriales
6. Order Saprolegniales
7. Order Peronosporales
8. Glomaceae
9. Endogonaceae
10. Order Eccrinales
11. Microthyriaceae
12. Trichopeltaceae
13. Micropeltaceae
14. Aspergillaceae
15. Meliolaceae
16. Erysiphaceae
17. Hypocreaceae
18. Pezizaceae
19. Ustilaginales
20. Uredinales
21. Polyporaceae
22. Order Lycoperdales
23. Unnamed
24. Unnamed
25. Unnamed
26. Unnamed

Fig. 2.1

Antarctica. Fragment of what is interpreted as arthropod cuticle containing numerous fungal thalli and spores.

Subdivision ASCOMYCOTINA

Class LOCULOASCOMYCETES

Order DOTHIDEALES

F. MICROTHYRIACEAE T. (YPR/LUT)–Rec. Terr.

First: Numerous genera of spores, hyphae and fructifications on various angiosperm leaves (Dilcher, 1965; Selkirk, 1974), Claiborne Formation, USA. **Extant**
Intervening: NG.

F. TRICHOPELTACEAE K. (MAA) Terr.

First: *Trichopeltinites* sp., Stroma (Sweet and Kalgutkar, 1989), Canada. Spores and hyphae assigned to this genus also reported (Dilcher, 1965; Selkirk, 1974). **Extant**

F. MICROPELTACEAE K. (VLG/BRM)–T. (AQT/LAN) Terr.

First: *Stomiopeltites cretacea* Alvin and Muir, 1970, Wealden, Isle of Wight, England, UK. Known from thyrothecia and, hyphae.
Last: *Dictyotopileos* sp., AQT–LAN1 (Selkirk, 1974).

Class PLECTOMYCETES

Order EUROTIALES

F. ASPERGILLACEAE T. (LUT/BRT)–Rec. Terr.

First: *Cryptocolax clarnensis* Scott, 1956, Clarno Formation, Oregon, USA. Cleistothecia bearing asci and ascospores in dicotyledenous wood. A second species, *C. parvular*, also reported. **Extant**

Class PYRENOMYCETES

Order MELIOLALES

F. MELIOLACEAE T. (YPR/LUT)–Rec. Terr.

First: *Meliolinites dilcherii* Daghlian, 1978, Rockdale Formation, Texas, USA. Known from hyphae, spores and perithecia. **Extant**
Intervening: T.

Order ERYSIPHALES

F. ERYSIPHACEAE T. (Ng.)–Rec. Terr.

First: ?*Erysiphites* (Pampaloni, 1902), formation unknown, Neogene. **Extant**

Order SPHAERIALES

F. HYPOCREACEAE T. (Ng.)–Rec. Terr.

First: *Polystigmites* sp. (Massalongo and Scarabelli, 1858–1859), details unknown. **Extant**

Class DISCOMYCETES

Order PEZIZALES

F. PEZIZACEAE T. (AQT/MES)–Rec. Terr.

First: *Pezizites* sp. (Ettingshausen, 1868), details unknown. **Extant**

Subdivision BASIDOMYCOTINA

Class TELIOMYCETES

F. USTILAGINALES T. (LUT/BRT)–Rec. Terr.

First: Unnamed (Currah and Stockey, 1991), British Columbia, Canada, based on spores that morphologically resemble teliospores of the extant genus *Microbotryum*. **Extant**

Order UREDINALES T. (YPR/PRB)–Rec. Terr.

First: *Puccinia* sp. (Wolf, 1969), Kentucky, USA. Based on isolated teliospores recovered from sediment.

Class HYMENOMYCETES

Order APHYLLOPHORALES

F. POLYPORACEAE J.(DOG/MLM)–Rec. Terr.

First: *Phelinites digiustoi* Singer and Archangelsky, 1958, Matilda Formation, Patagonia, Argentina.
Intervening: T. **Extant**

Class GASTEROMYCETES

Order LYCOPERDALES T. (Ng.)–Rec. Terr.

First: *Geasterites florissantensis* Cockerell, 1908, Colorado, USA.
Comments: Unnamed, branched and septate filaments with terminal and intercalary chlamydospores, from the Blackiston Formation (FAM), Indiana, USA, are associated with decay patterns similar to those produced by modern white-rot fungi. Similar symptoms and clamp connections are seen in Permian and Triassic woods from Antarctica (Stubblefield and Taylor, 1986).

F. UNNAMED C. (MOS) Terr.

First and Last: *Palaeancistrus martinii* Dennis, 1970, Carbondale Formation, Illinois, USA. Named for a mycelium with clamp connections and chlamydospores, and compared with extant genus *Panus*.

F. UNNAMED C. (MOS) Terr.

First and Last: *Palaeosclerotium pusillum* (Rothwell, 1972), Carbondale Formation, Illinois, USA. Cleistothecium-like structure with asci containing spores and hyphae with clamp connections (Dennis, 1976). Interpretations include: a fungus intermediate between ascomycete and basidiomycete; ascomycete parasitized by a basidiomycete; ascomycete closely related to the Eurotiales (Singer, 1977).

Form *subdivision* DEUTEROMYCOTINA

Class UNKNOWN

F. UNNAMED S. (GOR/LDF)–Rec. Terr.

First: Unnamed (Sherwood-Pike and Gray, 1985), Burgsvik Sandstone, Gotland, Sweden. Multicellate spores, some with scar suggestive of those found in conidial fungi with holoblastic development; branched hyphae that resemble a conidiophore also suggest Ascomycetes.
Intervening: Mesozoic, Cainozoic.

Form *Class* HYPHOMYCETES

F. UNNAMED T. (LAN2/MES)–Rec. Terr.

First: Unnamed (Sherwood-Pike, 1988), Clarkia Locality, Idaho and Oregon, USA. Large helicoid spores that are similar to the extant aquatic hyphomycetes *Helicoon pluriseptatum* and *Helicodendron giganteum*. Also numerous helicoid spores from the Upper Cretaceous might be included in this form class.

REFERENCES

Alvin, K. L. and Muir M. D. (1970) An epiphyllous fungus from the Lower Cretaceous. *Biological Journal of the Linnean Society*, **2**, 55–9.

Bradley, W. H. (1967) Two aquatic fungi (Chytridales) of Eocene age from the Green River Formation of Wyoming. *American Journal of Botany*, **54**, 577–82.

Cockerell, T. D. A. (1908) Descriptions of Tertiary plants II. *American Journal of Science*, **26**, 537–44.

Currah, R. S. and Stockey, R. A. (1991) A fossil smut fungus from the anthers of an Eocene angiosperm. *Nature*, **350**, 698–9.

Daghlian, C. P. (1978) A new melioloid fungus from the Early Eocene of Texas. *Palaeontology*, **21**, 171–6.

Dennis, R. L. (1970) A pennsylvanian basidiomycete mycelium with clamp connections. *Mycologia*, **62**, 578–84.

Dennis, R. L. (1976) *Palaeosclerotium*, a Pennsylvanian age fungus combining features of modern ascomycetes, and basidiomycetes. *Science*, **192**, 66–8.

Dilcher, D. L. (1965) Epiphyllous fungi from Eocene deposits in western Tennessee, U.S.A. *Palaeontographica, Abteilung, B*, **116**, 1–54.

Duncan, P. M. (1876) On some unicellular algae parasitic within Silurian and Tertiary corals, with a notice of their presence in *Calceola sandalina* and other fossils. *Quarterly Journal of the Geological Society of London*, **23**, 205–11.

Ettingshausen, C. (1868) Die fossile Flora der alteren Braunkohlenformation der Wetterau. *Sitzungsberichten der Mathematisch-Naturwissenschaften Classe der kaiserlische Akademie der Wissenschaften*, **60**, p. I.

Illman, W. I. (1984) Zoosporic fungal bodies in the spores of the Devonian fossil vascular plant, *Horneophyton*. *Mycologia*, **76**, 545–7.

Kidston, R. and Lang, W. H. (1921) Old Red Sandstone plants showing structure, from the Rhynie Chert Bed, Aberdeenshire,

Part V. The Thallophyta occurring in the peat bed. *Transactions of the Royal Society of Edinburgh*, **52**, 855–902.

Locquin, M. (1983) Nouvelles recherches sur le champignons fossiles. *108ème Congress Natural Sociétés Savantes, Grenoble, Fascicule 1*, **2**, 179–90.

Massalongo, A. and Scarabelli, G. (1858–1859) *Studi sulla Flora fossile e geologia stratigrafica del Sinigagliese*. Imola.

McLean, D. M. (1976) *Eocladopyxis peniculatum* Morgenroth, 1966, Early Tertiary ancestor of the modern dinoflagellate *Pyrodinium bahamense* Plate 1906. *Micropalaeontology*, **22**, 347–51.

Millay, M. A. and Taylor, T. N. (1978) Chytrid-like fossils of Pennsylvanian age. *Science*, **200**, 1147–9.

Pampaloni, L. (1902) I resti organici nel disolile di Melilli in Sicilia. *Paleontographica Italica*, **8**, 121–30.

Pirozynski, K. A. and Dalpé, Y. (1989) Geological history of the Glomaceae with particular reference to mycorrhizal symbiosis. *Symbiosis*, **7**, 1–36.

Rothwell, G. W. (1972) *Palaeosclerotium pusillum* gen. et sp. nov., a fossil eumycete from the Pennsylvanian of Illinois. *Canadian Journal of Botany*, **50**, 2353–6.

Scott, R. A. (1956) *Cryptocolax*, a new genus of fungus (Aspergillaceae) from the Eocene of Oregon. *American Journal of Botany*, **43**, 589–93.

Selkirk, D. R. (1974) Tertiary fossil fungi from Kiandra, New South Wales. *Proceedings of the Linnean Society of New South Wales*, **100**, 70–94.

Sherwood-Pike, M. A. (1988) Freshwater fungi: fossil record and paleoecological potential. *Palaeogeography, Palaeoclimatology and Palaeoecology*, **62**, 271–85.

Sherwood-Pike, M. A. and Gray, J. (1985) Silurian fungal remains: probable records of the Class Ascomycetes. *Lethaia*, **18**, 1–20.

Singer, R. (1977) An interpretation of *Palaeosclerotium*, *Mycologia*, **69**, 850–4.

Singer, R. and Archangelsky, S. (1958) A petrified basidiomycete from Patagonia. *American Journal of Botany*, **45**, 194–8.

Stidd, B. M. and Cosentino, K. (1975) *Albugo*-like oogonia from the American Carboniferous. *Science*, **190**, 1092–3.

Stubblefield, S. P. and Taylor, T. N. (1983) Studies of Paleozoic fungi. I. The structure and organization of *Traquairia* (Ascomycota). *American Journal of Botany*, **70**, 387–99.

Stubblefield, S. P. and Taylor, T. N. (1986) Wood decay in silicified gymnosperms from Antarctica. *Botanical Gazette*, **147**, 116–25.

Stubblefield, S. P. and Taylor, T. N. (1988) Tansley Review No. 12. Recent Advances in palaeomycology. *New Phytologist*, **108**, 3–25.

Stubblefield, S. P., Taylor, T. N., Miller, C. E. *et al.* (1983) Studies of Carboniferous fungi. II. The structure and organization of *Mycocarpon*, *Sporocarpon*, *Dubiocarpon* and *Coleocarpon* (Ascomycotina). *American Journal of Botany*, **70**, 1482–98.

Stubblefield, S. P., Taylor, T. N. and Seymour, R. L. (1987a) A possible endogonaceous fungus from the Triassic of Antarctica. *Mycologia*, **79**, 905–6.

Stubblefield, S. P., Taylor, T. N. and Trappe, J. M. (1987b) Vesicular–arbuscular mycorrhizae from the Triassic of Antarctica. *American Journal of Botany*, **74**, 1904–11.

Sweet, A. R. and Kalgutkar, R. M. (1989) *Trichopeltinites* Cookson from the latest Maastrichtian of Canada. *Contributions to Canadian Palaeontology, Geological Survey of Canada, Bulletin*, **396**, 223–7.

Taylor, T. N. and White, J. F. (1989) Fossil fungi (Endogonaceae) from the Triassic of Antarctica. *American Journal of Botany*, **76**, 389–96.

Taylor, T. N., Remy, W. and Hass, H. (1992a) Parasitism in a 400-million-year-old green alga. *Nature*, **357**, 493–4.

Taylor, T. N., Remy, W. and Hass, H. (1992b) Fungi from the Lower Devonian Rhynie Chert: Chytidiomycetes. *American Journal of Botany*, **79**, 1233–41.

Taylor, T. N., Hass, H. and Remy, W. (1992c) Devonian fungi: interactions with the green alga *Palaeonitella*. *Mycologia*, **84**, 901–10.

Tiffney, B. H. and Barghoorn, E. S. (1974) The fossil record of fungi. *Occasional Papers of the Farlow Herbarium of Cryptogamic Botany, Harvard University*, **7**, 1–42.

Wagner, C. A. and Taylor, T. N. (1982) Fungal chlamydospores from the Pennsylvanian of North America. *Review of Palaeobotany and Palynology*, **37**, 317–28.

White, J. F. and Taylor, T. N. (1989) A trichomycete-like fossil from the Triassic of Antarctica. *Mycologia*, **81**, 643–46.

Wolf, F. A. (1969) A rust and an alga in Eocene sediment from western Kentucky. *Journal of the Elisha Mitchell Society*, **85**, 57–8.

3

'ALGAE'

D. Edwards, J. G. Baldauf, P. R. Bown, K. J. Dorning, M. Feist, L. T. Gallagher, N. Grambast-Fessard, M. B. Hart, A. J. Powell and R. Riding

Recent major changes in the classification of protists involve the photosynthesizing, non-embryophyte, non-archegoniate, green plants previously grouped as the algae. In the 'five kingdoms' approach, they were included along with unicellular animals in the Protoctista or the Protista, but recent research based particularly on molecular phylogenies (e.g. rRNA) (Fernholm *et al.*, 1989) has prompted a reassessment by protozoologists (see Chapter 4) and the incorporation of certain groups, e.g. the dinoflagellates into the Protozoa, a subkingdom of Animalia. In that there as yet appears to be no consensus on the treatment of the former photosynthesizing protists, a more traditional scheme is adopted here, considering them as phyla within the eukaryotes. The classification is based on Round (1973, 1980), as modified from Engler (1954) and includes twelve phyla in all, subdivision being based on pigments present, type of food reserves, cell and flagellum ultrastructure, type of cell division and life cycles as well as morphology.

Recognition of these organisms in the fossil record, in the absence of biochemical and ultrastructural information and with a dearth of anatomical data, is often impossible. Exceptions are calcified forms and phytoplankton with resilient walls. Problems are compounded by convergence in evolutionary trends in organization (e.g. coccoid to filamentous forms) and in gross morphology in frondiose thalli. Thus it may be difficult to identify filaments even when preserved as permineralizations, and impossible to assign coalified compressions differentiated into holdfast, stipe and blade to reds, browns or even greens in the Palaeozoic (see e.g. Fry and Banks, 1955; Fry, 1983; Leary, 1986). Particularly intriguing are carbonaceous residues or impressions of presumed metaphytes in Precambrian rocks. The oldest structured carbonaceous filaments were described by Hofmann and Chen (1981) from the 1800 Ma Changcheng System, China, while ribbon-like structures in the 1300 Ma Belt Supergroup, Montana, were compared by Walter *et al.* (1976) with red, green and brown algae. It is, however, possible that they are aggregated sheaths of bacteria, an interpretation recently applied to the strap-shaped vendotaenids, common in the late Proterozoic (570–650 Ma) and considered to be allied to the Phaeophyta (Gnilovskaya, 1983). Vidal (1989) interprets them as the abandoned sheaths of sulphide-oxidizing organotrophic beggiatoacean bacteria. More compelling evidence for multicellular thallophytes derives from the longfengshaniids and tawuiids, widespread in the mid to late Proterozoic (1000 to 700 Ma) and splendidly illustrated in the Middle Proterozoic Little Dal macrobiota, Canada (Hofmann, 1985).

Coverage here is patchy, concentrating on unicellular forms with resilient walls (e.g. Dinophyta, Bacillariophyta, Haptophyta) or those with skeletons (e.g. Charophyta, calcified algae). Major calcified groups are indicated by*, and non-calcified members (if any) have usually not been included. For the remainder of the 'algae', the reader is referred to the original, *The Fossil Record*, although additional information on permineralized material of particularly early members of a group is given.

The systematic divisions used in this chapter cut across some commonly used micropalaeontological terms. Hence, the calcareous nannoplankton include the coccoliths (Phylum Haptophyta), some calcareous dinoflagellates (Phylum Dinophyta/Division Pyrrhophyta), and some *Incertae sedis* forms. Other well-known groups, such as the acritarchs, are of uncertain taxonomic position.

Contributions to this chapter were as follows: Chlorophyta (D. Edwards and R. Riding), Charophyta (M. Feist and N. Grambast-Fessard), Haptophyta (P. R. Bown and L. T. Gallagher), Bacillariophyta (J. G. Baldauf), Euglenophyta and Prasinophyta (M. B. Hart), Dinophyta/Pyrrhophyta (A. J. Powell), Acritarcha (K. J. Dorning), Rhodophyta (D. Edwards and R. Riding), *Incertae sedis* (P. R. Bown and L. T. Gallagher).

Acknowledgements – RR is indebted to the following colleagues for helpful advice: Juan Carlos Braga on corallines, Esmail Moussavian on corallines and peyssoneliaceans, and Andrey Zhuravlev on receptaculitids. KJD thanks S. G. Molyneux for assistance with the acritarchs.

The Fossil Record 2. Edited by M. J. Benton. Published in 1993 by Chapman & Hall, London. ISBN 0 412 39380 8

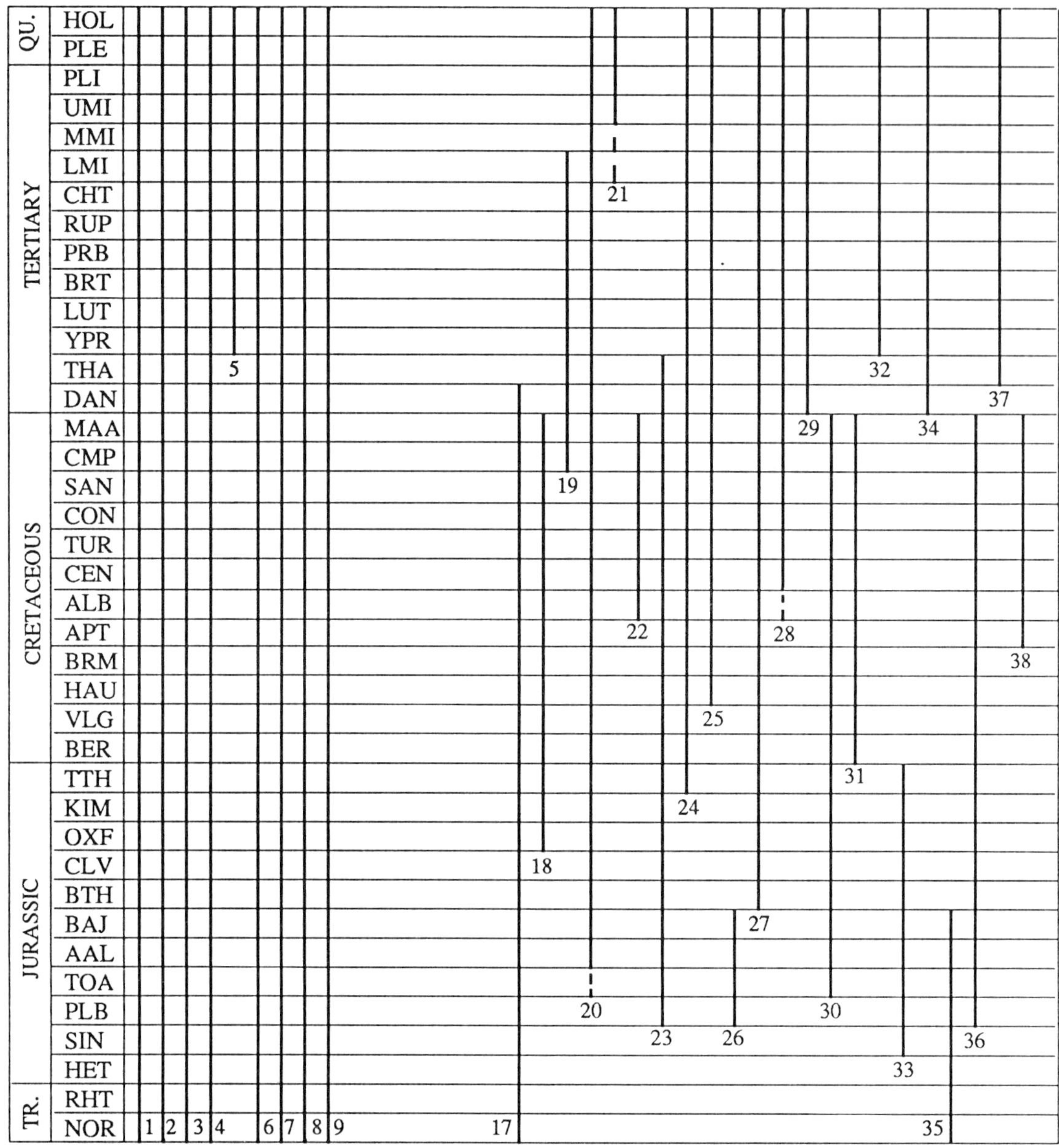

Fig. 3.1

Phylum CHLOROPHYTA Pascher, 1914
(see Fig. 3.1)

Class CHLOROPHYCEAE Kützing, 1833

Order CHLOROCOCCALES Marchand, 1985
orth. mut. Pascher 1915 P€. (RIF)–Rec. FW/Mar.

This group probably includes the first coccoid green algae, possibly even the earliest eukaryotes, but unequivocal identification in the Precambrian is unlikely to be achieved. Thus, for example, *Carosphaeroides pristina* and *C. tetras* from the 850 Ma Bitter Springs Formation, Australia, are described as either coccoid algae of chlorophycean affinity or cyanobacterial, as are the coeval *Latisphaera wrightii* Licari, 1978 from the Pahrump Group, Beck Spring Dolomite, California, *Glenobotrydion majorinum* Schopf and Blacic, from *c.* 740 Ma Min'yar Formation, former USSR, and an unnamed diverse assemblage of solitary coccoids from *c.* 750 Ma Chichkan Formation, former USSR (Schopf *et al.*, 1977). The last two records are also compared with rhodophytes, as are the putatively meiotically produced ?spores of *Ambiguaspora parvula* Volkova, 1976 from the 560 Ma Kotlin Formation, former USSR. The earliest botryococcoids are possibly the colonies described by Mendelson and Schopf (1992) as *Pila*-like colonial algae from the *c.* 615 Ma, Post-Spilitic Group, Czechoslovakia (Konzalova, 1973). Knoll *et al.* (1991) argued that there is reasonable, if not compelling, evidence to use the name *Myxococcoides chlorelloidae* for coccoids in the Neoproterozoic Draken Conglomerate Formation, Spitsbergen. **Extant**

Order ULOTRICHALES Borzi, 1895

F. ULOTRICHACEAE Kützing *orth. mut.* Haulk, 1883 D. (?PRA)–Rec. FW

First: *Mackiella rotundata* Edwards and Lyon, 1983; *Rhynchertia punctata* Edwards and Lyon, 1983, Rhynie Chert, Aberdeenshire, Scotland, UK (Edwards and Lyon, 1983).

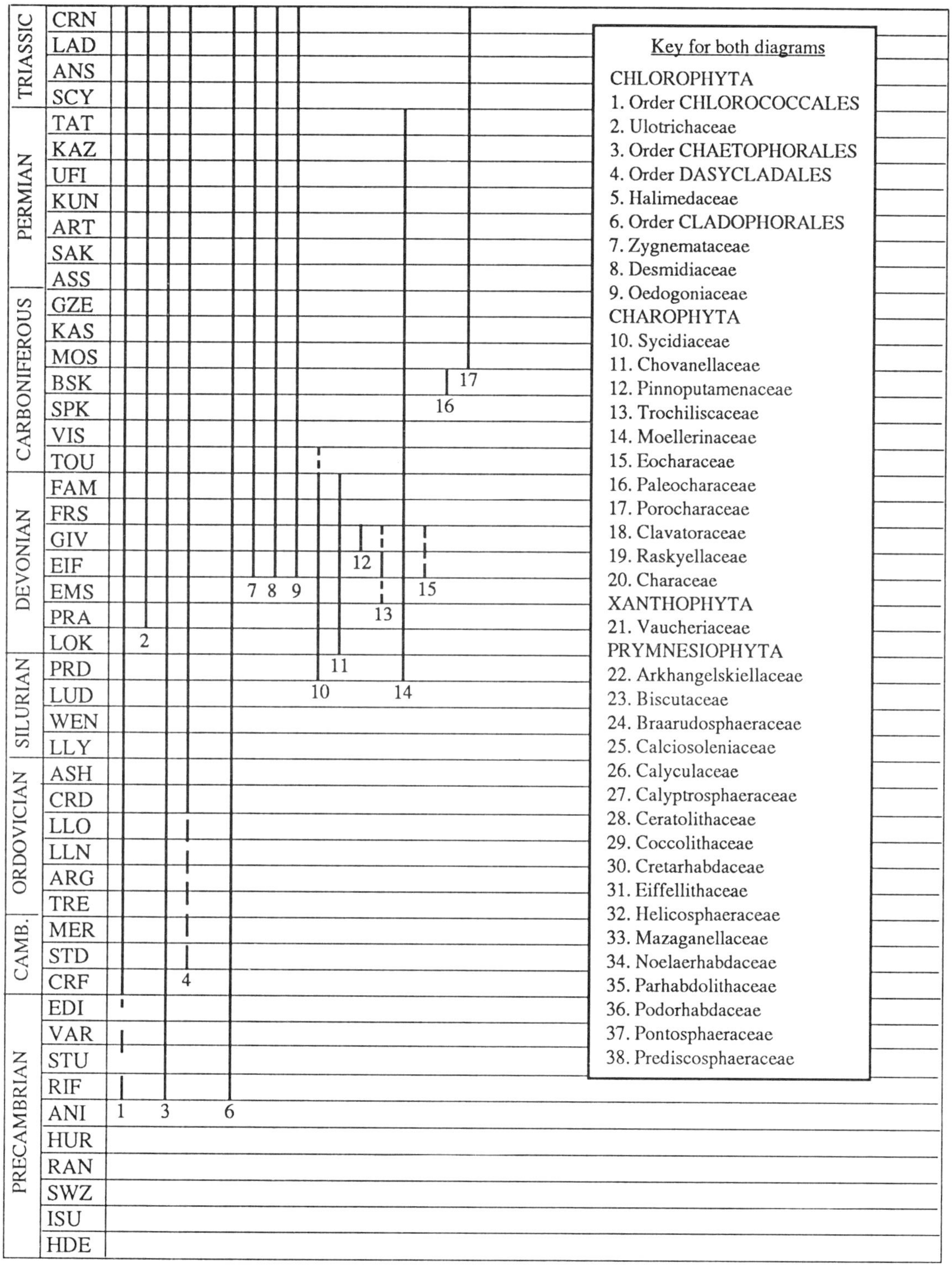

TRIASSIC: CRN, LAD, ANS, SCY
PERMIAN: TAT, KAZ, UFI, KUN, ART, SAK, ASS
CARBONIFEROUS: GZE, KAS, MOS, BSK, SPK, VIS, TOU
DEVONIAN: FAM, FRS, GIV, EIF, EMS, PRA, LOK
SILURIAN: PRD, LUD, WEN, LLY
ORDOVICIAN: ASH, CRD, LLO, LLN, ARG, TRE
CAMB.: MER, STD, CRF
PRECAMBRIAN: EDI, VAR, STU, RIF, ANI, HUR, RAN, SWZ, ISU, HDE

1 2 3 4 6 7 8 9 10 11 12 13 14 15 16 17

Key for both diagrams

CHLOROPHYTA
1. Order CHLOROCOCCALES
2. Ulotrichaceae
3. Order CHAETOPHORALES
4. Order DASYCLADALES
5. Halimedaceae
6. Order CLADOPHORALES
7. Zygnemataceae
8. Desmidiaceae
9. Oedogoniaceae

CHAROPHYTA
10. Sycidiaceae
11. Chovanellaceae
12. Pinnoputamenaceae
13. Trochiliscaceae
14. Moellerinaceae
15. Eocharaceae
16. Paleocharaceae
17. Porocharaceae
18. Clavatoraceae
19. Raskyellaceae
20. Characeae

XANTHOPHYTA
21. Vaucheriaceae

PRYMNESIOPHYTA
22. Arkhangelskiellaceae
23. Biscutaceae
24. Braarudosphaeraceae
25. Calciosoleniaceae
26. Calyculaceae
27. Calyptrosphaeraceae
28. Ceratolithaceae
29. Coccolithaceae
30. Cretarhabdaceae
31. Eiffellithaceae
32. Helicosphaeraceae
33. Mazaganellaceae
34. Noelaerhabdaceae
35. Parhabdolithaceae
36. Podorhabdaceae
37. Pontosphaeraceae
38. Prediscosphaeraceae

Fig. 3.1

Comment: Record is based on silicified filaments and attribution to the Ulotrichaceae on their general morphology, the typical chloroplasts not being preserved in the mat-like filaments. A similar reasoning was also used in naming filaments with relatively broad cells from an Upper Devonian chert as *Palaeogeminella folkii* Fairchild *et al.*, 1979. *Archaeoellipsoides conjunctivus* Zhang, 1985 from the *c.* 1325 Ma Wumishan Formation, China, is described as a 'broad septate (ulotrichalean?) filament with elongate cells' in Mendelson and Schopf (1992). **Extant**

Order CHAETOPHORALES West, 1904
P€. (RIF)–Rec. FW/Mar.

Butterfield *et al.* (1988) in a preliminary account of an Upper Riphean (700–800 Ma) biota from the Svanbergfjellet Formation, found a close structural analogue for some unnamed, repeatedly branched, non-cellular fibroids in the holdfasts of certain heterotrichous chaetophoraleans. **Extant**

Class BRYOPSIDOPHYCEAE Round, 1971

*Order DASYCLADALES Pascher, 1931
?Є. (STD)/O.(CRD)–Rec. Mar.

First: Possibly *Yakutina aciculata* (Korde, 1973), first recorded as *Siberiella aciculata* Korde, 1957 from the Middle Cambrian, Amga River, SE Siberia, former USSR (Korde 1957, p. 69, text-fig. 1). See Riding (1991, p. 327) for comments. Definitely *Rhabdoporella bacillum* Stolley, 1893, probably from the 'Leptaena Limestone', Upper Ordovician (also see Høeg, 1932, p. 75).
Comment: The above records refer to strongly calcified forms. Weakly calcified representatives include *Primicorallina trentonensis* Whitfield, 1894 from the Middle Ordovician (Trenton Limestone) of New York, USA; and *Archaeobatophora typa* Nitecki, 1976 from the Upper Ordovician (Richmondian) of Michigan, USA. **Extant**

*Order CAULERPALES Setchell, 1929

F. HALIMEDACEAE Hillis-Colinvaux, 1984
T. (Eoc./RUP)–Rec. Mar./FW

First: *Palaeoporella* sp. Ellenberger Group, Texas. Upper Cambrian/Lower Ordovician (Johnson, 1954, pp. 51 and 56, pl. 24). **Extant**

Order CLADOPHORALES West, 1904
PЄ. (KAR)–Rec. Mar./FW

First: In a preliminary account of a late Riphean (700–800 Ma) biota from the Svanbergfjellet Formation, Spitsbergen, Butterfield *et al.* (1988) compared some filamentous thalli of large cylindrical cells with distinctive thickened septal plates, and intercalary branching with cladophoralean greens (Ulvophyceae). **Extant**

Class ZYGNEMAPHYCEAE Round, 1971

Order ZYGNEMATALES Round, 1963

F. ZYGNEMATACEAE (Meneghini) Kützung *orth. mut.* Engler, 1898 D. (EIF)–Rec. FW/Mar.

First: *Paleodidymoprium didymum* Baschnagel, 1966, Onondaga Formation, New York State, USA (Baschnagel, 1966). **Extant**
Intervening: Eoc. (*Spirogyra*: Wyoming, USA [FR]).
Comment: Earliest record is based on filaments preserved in chert.

Order DESMIDIALES Round, 1963

F. DESMIDIACEAE Kützing, 1833, ex Ralfs (1845) *orth. mut.* Stizenbergen, 1860 D. (EIF)–Rec. FW/Mar.

First: *Paleoclosterium leptum* Baschnagel, 1966, Onondaga Formation, New York State, USA (Baschnagel, 1966). **Extant**
Comment: The record is based on single crescentic cells preserved in chert, which may be more realistically assigned to the acritarchs.

Class OEDOGONIOPHYCEAE Sarma, 1964

Order OEDOGONIALES Blackman and Tansley ex West, 1904

F. OEDOGONIACEAE (Thuret) de Bary, 1900
D. (EIF)–Rec. Mar./FW

First: *Paleoeodogonium micrum* Baschnagel, 1966, Onondaga Formation, New York State, USA. **Extant**
Comment: Record based on silicified filaments with arguably typical apical cells.

Phylum CHAROPHYTA

Class CHAROPSIDA

Order SYCIDIALES Mädler, 1952

F. SYCIDIACEAE Karpinsky, *in* Peck, 1934
S. (PRD)–C. (?TOU) ?Mar. FW

First: *Praesycidium siluricum* T. and A. Ishchenko, 1982, Shalsky Formation, Podolia, Ukraine, former USSR (Ishchenko and Ishchenko, 1982).
Last: *Sycidium foveatum* Peck, 1934 and *S. clathratum* Peck, 1934, Sylamore Formation, basal Mississippian, Missouri, USA.

F. CHOVANELLACEAE Grambast, 1962
D. (LOK–FAM) FW

First: *Chovanella* sp. Li and Cai, 1978, Xiaxishancun and Xitun Formations, southern China.
Last: *Chovanella kovaleri* Reitlinger and Jartzeva, 1958 and three other species; Khovansk Beds, Tula region, Ukraine, former USSR (Peck and Eyer, 1963a).

F. PINNOPUTAMENACEAE Z. Wang and Lu, 1980
D. (GIV) FW

First and Last: *Pinnoputamen yunnanensis* Wang and Lu, 1980, Xichong Formation, Yunnan, China (Wang and Lu, 1980).

Order TROCHILISCALES Mädler, 1952

F. TROCHILISCACEAE Karpinsky, *in* Peck, 1934 emend. Z. Wang and Lu, 1980 D. (EMS/GIV) FW

First: *Trochiliscus lipuensis* Wang *et al.*, 1980, Szepai Formation, Guangxi, China.
Last: *Trochiliscus zhanyensis* Wang and Lu, 1980, Xichong Formation, Yunnan, China (Wang and Lu, 1980).

F. MOELLERINACEAE Feist and Grambast-Fessard, 1990 S. (PRD)–P. (ZEC) ?Mar. FW

First: ?*Primochara calvata* Ishchenko and Saidakovski, 1975, Shalsky Formation, Podolia, Ukraine, former USSR.
Last: *Gemmichara sinensis* Z. Wang, 1984, Sunan Formation, Gansu, China (Wang, 1984).
Intervening: EIF/GIV–C. (l.).
Comment: *Primochara* is described as carbonaceous remains with an unknown number of dextral spiral cells, thus questionably assigned to the family (Feist and Grambast-Fessard, 1990).

Order CHARALES Lindley, 1836

F. EOCHARACEAE Grambast, 1959
D. (EIF/GIV) FW

First and Last: *Eochara wickendeni* Choquette, 1956, Elk Point Group, Alberta, Canada (Grambast, 1962).

F. PALEOCHARACEAE Pia, 1927 C. (BSK) FW

First and Last: *Paleochara acadica* Bell, 1922, lower part of Coal Measures, Nova Scotia, Canada (Peck and Eyer, 1963b).

F. POROCHARACEAE Grambast, 1962
C. (MOS)–T. (DAN) FW

First: *Stomochara moreyi* (Peck, 1934) Grambast, 1962, Lagonda Formation, Pennsylvanian, Missouri, USA, extending to Lower Permian (Peck and Eyer, 1963b).
Last: *Porochara* sp. (*Feistiella* sp.?), Willow Creek Formation, Member E, Alberta, Canada.
Comment: Porocharaceae were the main components of the Triassic and Jurassic charophyte floras. They gave rise to the three post-Palaeozoic families.

F. CLAVATORACEAE Pia, 1927 J. (OXF)–K. (MAA) FW

First: *Clavator pecki* Mädler, 1952, Vellerat Formation, northern Jura, France and Switzerland (Mojon, 1989).
Last: *Septorella ultima* Grambast, 1971, Rognacian, Provence, France.
Comment: The main development of the family occurred during the early Cretaceous (Grambast, 1974).

F. RASKYELLACEAE Grambast, 1957 K. (CMP)–T. (AQT) FW

First: *Saportanella maslovi* Grambast, 1962, Begudien, Provence, France.
Last: *Rantzieniella nitida* Grambast, 1962, Calcaire blanc de l'Agenais, Paulhiac, Aquitaine, France.
Comment: The main genus is *Raskyella* Grambast, 1954, Eocene (Grambast, 1967).

F. CHARACEAE Agardh, 1824 J. (?TOA)–Rec. FW

First: *Aclistochara* aff. *jonesi* Feist *et al.*, 1991, Gondwana Group, Kota Formation, Pranhita-Godavari Valley, India (Feist *et al.*, 1991). **Extant**
Comment: Characeae played a very unobtrusive part during the Jurassic and Lower Cretaceous (Grambast, 1974). The number of genera and species increased, after the regression of Porocharaceae and Clavatoraceae, during the late Cretaceous (CMP–MAA). From the Miocene onward, the family regressed to the present state, with only six genera, which include all the extant charophyte species.

Phylum XANTHOPHYTA Polyansby and Hollerbach, 1951

Order VAUCHERIALES Bohlin, 1901

F. VAUCHERIACEAE T. (Mio.)–Rec. FW

First: *Vaucheria antiqua* Ludwig, Miocene Braunköhle (from Tappan, 1980). **Extant**
Comment: *Botryococcus*, cited as the oldest genus in the first *Fossil Record*, is now considered to be a green alga. *Palaeovaucheria clavata* Hermann [German], 1981, Zl'merdak Formation, Bederysh Member, former USSR (*c.* 1000 Ma) is described as branched, septate *Vaucheria*-like, possibly eukaryotic filaments by Mendelson and Schopf (1992), while *Palaeosiphonella cloudii* Licari, 1978, Beck Spring Dolomite is interpreted as possibly a branching filamentous eukaryotic alga (as in original description) or a cyanobacterial sheath by Mendelson and Schopf. Butterfield *et al.* (1988) described some solitary filaments as 'resembling and possibly representing germinating zoospores of filamentous protists with modern analogues including the xanthophycean algae' from the ?700–800 Ma Svanbergfjellet Formation, Spitsbergen.

Phylum HAPTOPHYTA Hibberd, 1972

Division PRYMNESIOPHYTA Hibberd, 1976

Class PRYMNESIOPHYCEAE Hibberd, 1976

Order COCCOSPHAERALES Parke and Dixon, 1976

The Order Coccosphaerales includes all coccolith-bearing prymnesiophytes. First and last occurrences have not been referred to a particular locality and horizon due to the vast amount of biostratigraphical data which exists in the literature. In the late Mesozoic and Cainozoic, in particular, many of the stratigraphical ranges are based upon data gathered by the Deep Sea Drilling Project and Ocean Drilling Program.

F. ARKHANGELSKIELLACEAE Bukry, 1969 K. (ALB–MAA) Mar.

First: *Acaenolithus vimineus* Black, 1973 ??
Last: *Arkhangelskiella cymbiformis* Vekshina, 1959, Upper MAA.
Intervening: CEN–CMP.

F. BISCUTACEAE Black, 1971 J. (PLB)–T. (THA) Mar.

First: *Biscutum novum* (Goy, 1979), lower PLB.
Last: *Biscutum harrisonii* (Varol, 1989).
Intervening: TOA–DAN.

F. BRAARUDOSPHAERACEAE Deflandre, 1947 J. (TTH)–Rec. Mar.

First: *Braarudosphaera regularis* Black, 1973. **Extant**
Intervening: BER–HOL.

F. CALCIOSOLENIACEAE Kamptner, 1927 K. (HAU)–Rec. Mar.

First: *Scapholithus fossilis* Deflandre, 1954 **Extant**
Intervening: BRM–HOL.
Comments: Rare and sporadic throughout its range.

F. CALYCULACEAE Noel, 1973 J. (PLB–BAJ) Mar.

First: *Calyculus cribrum* Noel, 1973, lower PLB.
Last: *Calyculus cribrum* Noel, 1973 ??, lower BAJ.
Intervening: TOA, AAL.

F. CALYPTROSPHAERACEAE Boudreaux and Hay, 1969 J. (BTH)–Rec. Mar.

First: *Anfractus harrisonii* Medd, 1979, lower BTH. **Extant**
Intervening: CLV–HOL.
Comment: A strictly morphological taxonomic grouping which embraces coccolithophorids which secrete holococcoliths. Almost certainly includes taxa which also secrete heterococcoliths during non-motile phases and are then included in other family groupings (for further explanation, see e.g. Manton and Leedale, 1969; Wind and Wise, 1978).

F. CERATOLITHACEAE Norris, 1965 ?K. (ALB)–Rec. Mar.

First: Mesozoic *Ceratolithina hamata* Martini, 1967? Cainozoic *Amaurolithus primus* (Bukry and Percival, 1971) **Extant**
Intervening: CEN–MAA; TOR–HOL.
Comment: The Mesozoic representatives of this extant family are comparable to, but distinct from, the Cainozoic

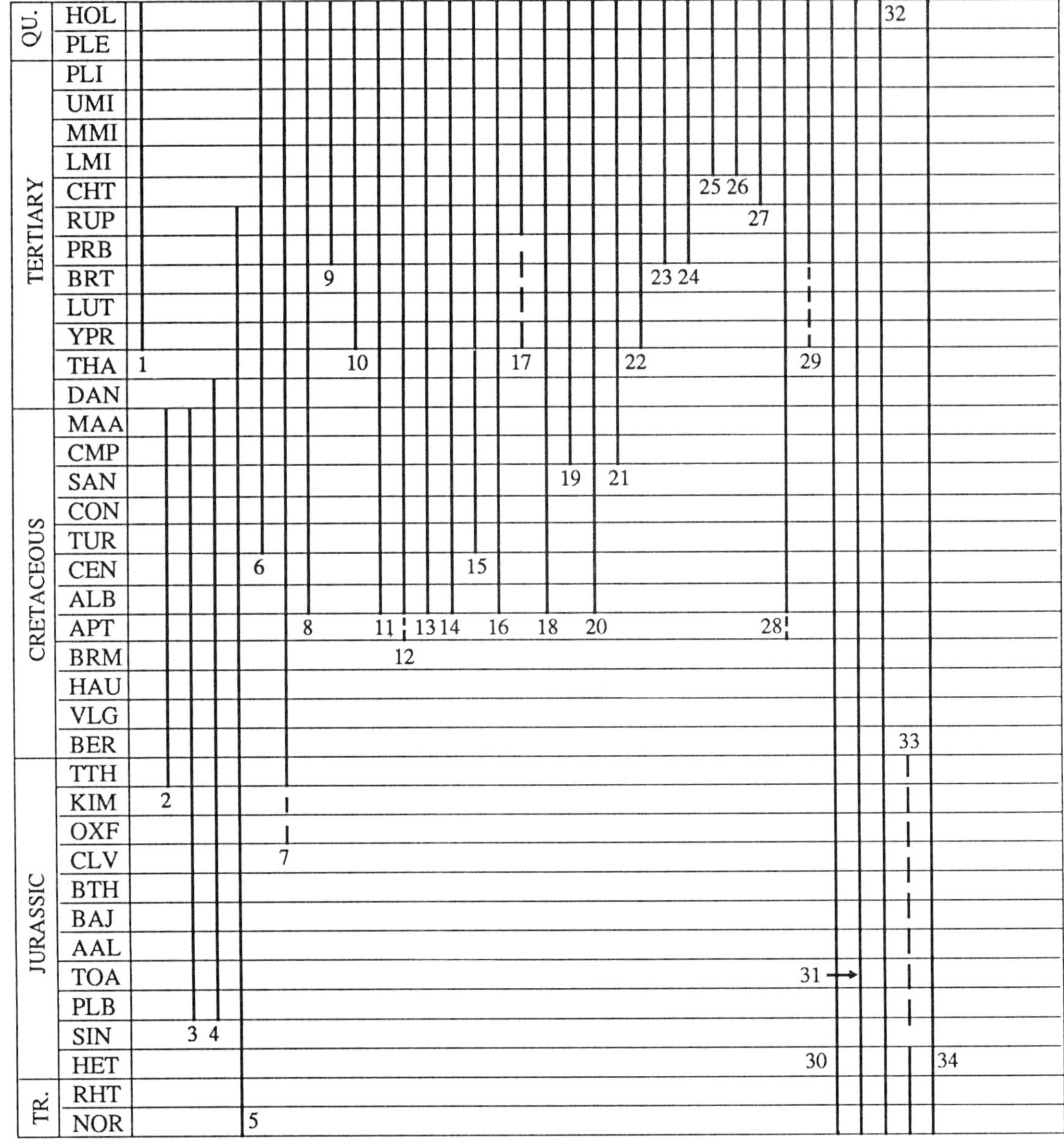

Fig. 3.2

forms. An extremely long stratigraphical interval separates the ranges of the two groups (Perch-Nielsen, 1985).

F. COCCOLITHACEAE Poche, 1913
T. (DAN)–Rec. Mar.

First: *Cruciplacolithus inseadus* Perch-Nielsen, 1969. **Extant**

Intervening: THA–HOL.

F. CRETARHABDACEAE Thierstein, 1973
J. (TOA)–K. (MAA) Mar.

First: *Retecapsa incompta* Bown and Cooper, 1989, upper TOA.

Last: *Cretarhabdus crenulatus* Bramlette and Martini, 1964, upper MAA.

Intervening: AAL–CMP.

F. EIFFELLITHACEAE Reinhardt, 1965
K. (BER–MAA) Mar.

First: *Eiffellithus primus* Applegate and Bergen, 1989.

Last: *Eiffellithus gorkae* Reinhardt, 1965.

Intervening: VLG–CMP.

F. HELICOSPHAERACEAE Black, 1971
T. (YPR)–Rec. Mar.

First: *Helicosphaera lophota* Bramlette and Sullivan, 1961. **Extant**

Intervening: LUT–HOL.

F. MAZAGANELLACEAE Bown, 1987
J. (SIN–TTH) Mar.

First: *Mazaganella pulla* Bown, 1987.

Last: *Triscutum* sp. indet.

Intervening: PLB–KIM.

F. NOELAERHABDACEAE Jerkovic, 1970
T. (DAN)–Rec. Mar.

First: *Futyanla petalosa* (Ellis and Lohmann, 1973). **Extant**

Intervening: THA–HOL.

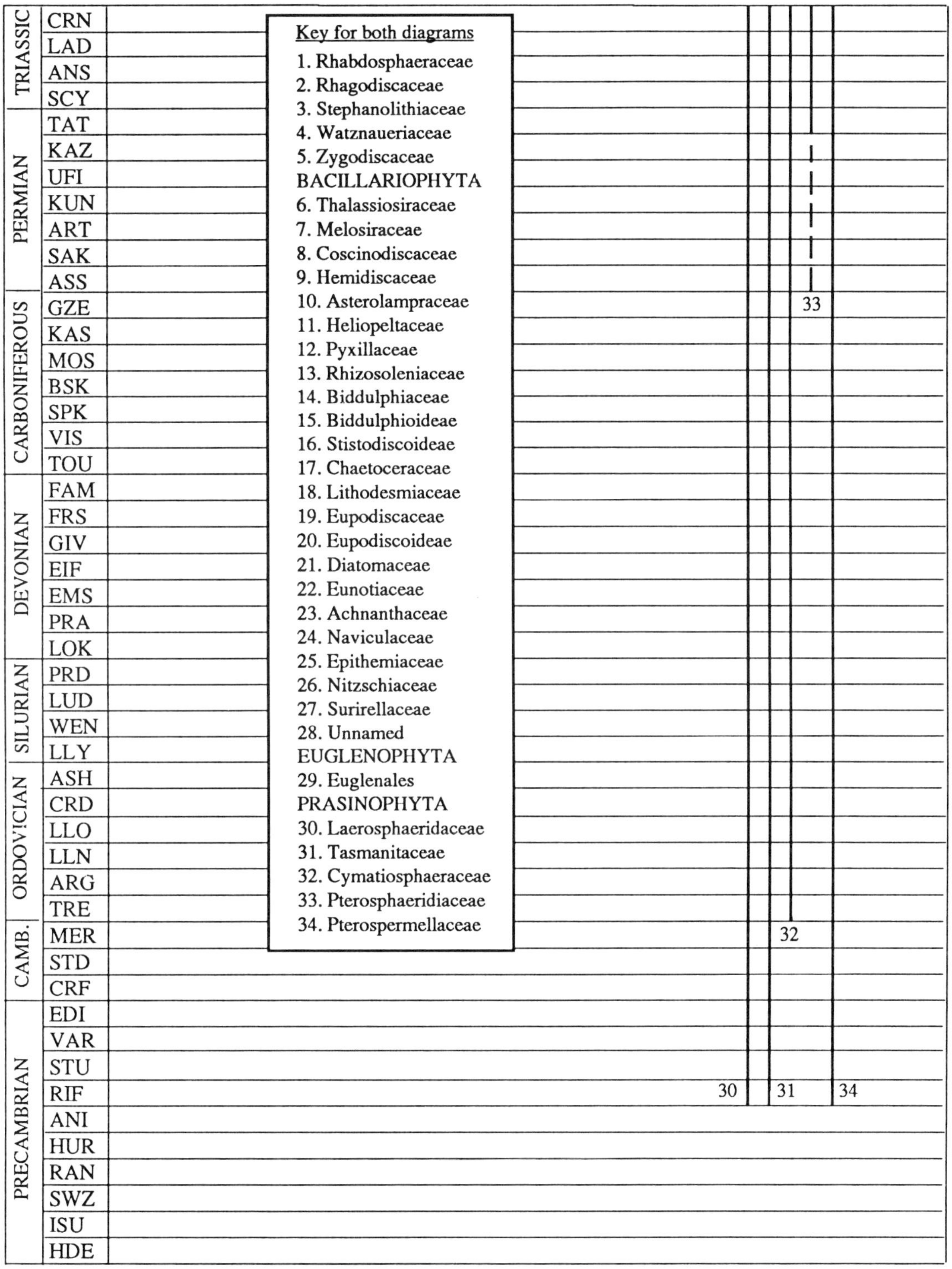

Period	Stage
TRIASSIC	CRN
	LAD
	ANS
	SCY
PERMIAN	TAT
	KAZ
	UFI
	KUN
	ART
	SAK
	ASS
CARBONIFEROUS	GZE
	KAS
	MOS
	BSK
	SPK
	VIS
	TOU
DEVONIAN	FAM
	FRS
	GIV
	EIF
	EMS
	PRA
	LOK
SILURIAN	PRD
	LUD
	WEN
	LLY
ORDOVICIAN	ASH
	CRD
	LLO
	LLN
	ARG
	TRE
CAMB.	MER
	STD
	CRF
PRECAMBRIAN	EDI
	VAR
	STU
	RIF
	ANI
	HUR
	RAN
	SWZ
	ISU
	HDE

Key for both diagrams

1. Rhabdosphaeraceae
2. Rhagodiscaceae
3. Stephanolithiaceae
4. Watznaueriaceae
5. Zygodiscaceae

BACILLARIOPHYTA

6. Thalassiosiraceae
7. Melosiraceae
8. Coscinodiscaceae
9. Hemidiscaceae
10. Asterolampraceae
11. Heliopeltaceae
12. Pyxillaceae
13. Rhizosoleniaceae
14. Biddulphiaceae
15. Biddulphioideae
16. Stistodiscoideae
17. Chaetoceraceae
18. Lithodesmiaceae
19. Eupodiscaceae
20. Eupodiscoideae
21. Diatomaceae
22. Eunotiaceae
23. Achnanthaceae
24. Naviculaceae
25. Epithemiaceae
26. Nitzschiaceae
27. Surirellaceae
28. Unnamed

EUGLENOPHYTA

29. Euglenales

PRASINOPHYTA

30. Laerosphaeridaceae
31. Tasmanitaceae
32. Cymatiosphaeraceae
33. Pterosphaeridiaceae
34. Pterospermellaceae

33 32 30 31 34

Fig. 3.2

F. PARHABDOLITHACEAE Bown, 1987
Tr. (NOR)–J. (BAJ) Mar.

First: *Crucirhabdus minutus* Jafar, 1983.
Last: *Diductius constans* Goy, 1979.
Intervening: RHT–AAL.

F. PODORHABDACEAE Noel, 1965
J. (PLB)–K. (MAA) Mar.

First: *Axopodorhabdus atavus* Grun *et al.*, 1974.
Last: *Teichorhabdus ethmos* Wind and Wise, *in* Wise and Wind, 1977.
Intervening: TOA–CMP.

F. PONTOSPHAERACEAE Lemmermann, 1908
T. (THA)–Rec. Mar.

First: *Pontosphaera plana* Bramlette and Sullivan, 1961.
Extant
Intervening: YPR–HOL.

Comments: The Upper Cretaceous species *Prolatipatella multicarinata* (Gartner, 1968) has very similar morphology to Upper Cainozoic forms and is considered by some to be the first representative of this family.

F. PREDISCOSPHAERACEAE Rood *et al.*, 1971 K. (APT–MAA) Mar.

First: *Prediscosphaera* cf. *P. stoveri* Perch-Nielsen, 1968, upper APT.
Last: *Prediscosphaera grandis* Perch-Nielsen, 1979, upper MAA.
Intervening: CEN–CMP.

F. RHABDOSPHAERACEAE Lemmermann, 1908 T. (YPR)–Rec. Mar.

First: *?Blackites spinosus* (Deflandre and Fert, 1954), upper YPR. **Extant**
Intervening: LUT–HOL.

F. RHAGODISCACEAE Hay, 1977 J. (TTH)–K. (MAA) Mar.

First: *Rhagodiscus asper* Stradner, 1963, upper TTH.
Last: *Rhagodiscus splendens* (Deflandre, 1953), upper MAA.
Intervening: BER–CMP.

F. STEPHANOLITHIACEAE Black, 1968 J. (PLB)–K. (MAA) Mar.

First: *Stradnerlithus clatriatus* (Rood *et al.* 1973)?
Last: *Rotelapillus crenulatus* (Stover, 1966), upper MAA.
Intervening: TOA–CMP.

F. WATZNAUERIACEAE Rood *et al.*, 1971 J. (PLB)–T. (DAN) Mar.

First: *Lotharingius primigenius* Bown, 1987.
Last: *Cyclagelosphaera alta* Perch-Nielsen, 1979.
Intervening: TOA–MAA.

F. ZYGODISCACEAE Hay and Mohler, 1967 Tr. (NOR)–T. (RUP) Mar.

First: *Archaeozygodiscus koessenensis* Bown, 1985, upper NOR.
Last: *Ismolithus rhenanus* Martini, 1973.
Intervening: RHT–PRB.

Phylum BACILLARIOPHYTA

Class BACILLARIOPHYCEAE (Diatoms)

The diatom classification of Simonsen (1979), used here, subdivides the diatoms into two orders, five suborders, and twenty-one families, based predominantly on the presence, position and arrangement of various processes. Recent studies of Aptian–Albian diatoms by Harwood and Gersonde (1990) and Gersonde and Harwood (1990), as well as others, indicate the need for revision of Simonsen's classification, but revision awaits completion of additional studies. The recently proposed classification of Round *et al.* (1990) is not used here as detailed discussions are required to evaluate the merit of this classification.

Families are used here as discussed in Simonsen (1979) with the following exceptions: the genus *Azpeitia* Peragallo emend. Sims is included in the family Hemidiscaceae; the genera *Crucidenticula* Akiba and Yanagisawa, and *Neodenticula* Akiba and Yanagisawa are included in the family Nitzschiaceae; and numerous new genera proposed by Harwood and Gersonde (1990) and Gersonde and Harwood (1990) are included here because these new genera comprise one of the earliest diatom floras documented to date, but these genera are not assigned to a specific family until further analysis can be completed.

F. THALASSIOSIRACEAE Lebour, 1930 emend. Hasle, 1973 K. (TUR)–Rec. Mar./FW

First: *Thalassiosira* Cleve (most likely *Thalassiosiropsis*), Turonian, Czechoslovakia (Wiesner, 1936) and France (Deflandre, 1941). **Extant**
Includes genera: *Aulacosira* Thwaites, *Bacterosira* Gran, *Cyclotella* Kutz, *Cymatodiscus* Hendey, *Cymatosira* Hendey, *Detonula* Schutt, *Landeria* Cleve, *Minidiscus* Hasle, *Planktoniella* Schutt, *Porosira* Jorgensen, *Skeletonema* Greville, *Stephanodiscus* Ehrenberg, *Thalassiosira* Cleve, *Thalassiosiropsis* Hasle and Syvertsen, and *Tryblioptychus* Hendey.

F. MELOSIRACEAE Kutzing, 1844 J. (u.)–Rec. Mar./FW

First: *Stephanopyxis parentes?*, Upper Jurassic, western Siberia, former USSR (Vekshina, 1960); *Pyxidicula bollensis?*, Jurassic (Rothpletz, 1896). **Extant**
Includes genera: *Annellus* Tempere, *Corethron* Castracane, *Druridgea* Donkin, *Endictya* Ehrenberg, *Hyalodiscus* Ehrenberg, *Leptocylindrus* Cleve, *Melosira* Agardh, *Parlaria* Heiberg, *Podosira* Ehrenberg, *Pyxidicula* Ehrenberg, *Stephanopyxis* Ehrenberg and *Trochosira* Kitton.

F. COSCINODISCACEAE Kutzing, 1844 K. (ALB)–Rec. Mar/FW

First: *Coscinodiscus* Ehrenberg, Albian clays of the Penza region, western Siberia, former USSR (Jouse, 1949). **Extant**
Includes genera: *Benethoras* Hanna, *Coscinodiscus* Ehrenberg, *Craspedodiscus* Ehrenberg, *Fenestrella* Greville, *Gossleriella* Schutt, *Isodiscus* Rattray, *Kozloviella* Jouse, *Palmeria* Greville, *Pomphodiscus* Barker and Meakin, *Porodiscus* Greville and *Pseudotriceratium* Gran.

F. HEMIDISCACEAE Hendy, 1937 emend. Simonsen, 1975 T. (PRB)–Rec. Mar./FW

First: *Azpeitia tuberculata* (Greville) Sims (as *Coscinodiscus tuberculatus*), Oamaru Formation, South Island, New Zealand, Upper Eocene (Fryxell *et al.*, 1986). *Azpeitia oligocenica* (Jouse) Sims recorded by Fenner (1977, 1981) as *Coscinodiscus oligocenicus* from Eocene and Oligocene sediments from the equatorial and South Atlantic (Sims *et al.*, 1989). **Extant**
Includes genera: *Actinocyclus* Ehrenberg, *Azpeitia* Peragallo emend. Sims, *Hemidiscus* Wallich and *Roperia* Grunow.

F. ASTEROLAMPRACEAE Smith, 1872 T. (YPR)–Rec. Mar./FW

First: *Bergonia primitiva* Gombos, uppermost Lower Eocene sediments, DSDP Site 384, western North Atlantic (Gombos, 1980). **Extant**
Includes genera: *Asterolampra* Ehrenberg, *Asteromphalus* Ehrenberg, *Bergonia* Tempere, *Brightwellia* Ralfs and *Rylandsia* Greville.

F. HELIOPELTACEAE Smith, 1872 K. (ALB)–Rec. Mar./FW

First: *Aulacodiscus* Debys, Albian clays, Penza region, western Siberia (Jouse, 1949). **Extant**
Includes genera: *Actinoptychus* Ehrenberg, *Aulacodiscus* Debys, *Glorioptychus* Hanna, *Lepidodiscus* Witt, *Polymyxus* Bailey, *Sturtiella* Simonsen and Schrader and *Wittia* Pantocsek.

F. PYXILLACEAE Schutt, 1896 K. (APT/ALB)–Rec. Mar./FW

First: *Gladius antiquus* Forti and Schulz and *Gladius perfectus* Gersonde and Harwood (and numerous other *Gladius*), Aptian–Albian sediments, Ocean Drilling Program (ODP) Site 693, Weddell Sea: *Gladius* sp., Albian clays, Penza region, western Siberia, former USSR (Jouse, 1949). **Extant**
Includes genera: *Gladius* Forti and Schulz, *Gyrodiscus* Witt, *Mastogloia* Thwaites, and *Pyxilla* Greville.

F. RHIZOSOLENIACEAE Petit, 1888 K. (ALB)–Rec. Mar./FW

First: *Dactyliosolen* sp., Lower Cretaceous, 'Phosphorite des Gaults' (Forti and Schulz, 1932); *Rhizosolenia* Ehrenberg, Campanian, western Siberia (Strelnikova, 1975); *Rhizosolenia cretacea* Hajos and Stradner, Upper Campanian, DSDP Site 275 (Hajos and Stradner, 1975) and of Seymour Island (Harwood, 1988).
Includes genera: *Rhizosolenia* Ehrenberg, *Dactyliosolen* Castracane and *Guinardia* Paragallo.

F. BIDDULPHIACEAE Kutzing, 1844

Subfamily HEMIAULOIDEAE Jouse *et al.*, 1949 K. (ALB)–Rec. Mar./FW

First: *Hemiaulus fragilis* Jouse, Albian clays, Penza region, western Siberia, former USSR (Jouse, 1949). **Extant**
Includes genera: *Attheya* West, *Baxteriopsis* Karsten, *Campylosira* Grunow, *Cerataulina* Peragallo, *Climacodium* Ehrenberg, *Cymatosira* Grun, *Eucampia* Ehrenberg, *Goniothecium* Ehrenberg, *Hemiaulus* Ehrenberg, *Monobrachia* Schrader, *Odonotropis* Gran, *Pseudorutilaria* Grove and Sturt, *Pseudostictodiscus* Grun, *Riedelia* Jouse and Scheschykova-Poretzkaja and *Trinacria* Heiberg.

Subfamily BIDDULPHIOIDEAE Schutt, 1896 K. (TUR)–Rec. Mar./FW

First: *Biddulphia* Gray, upper Turonian, near Usti and Labem, Czechoslovakia (Wiesner, 1936). **Extant**
Includes genera: *Anaulus* Ehrenberg, *Biddulphia* Gray, *Entogonia* Greville, *Eunotiopsis* Grun, *Eunotogramma* Weisse, *Isrhmia* Agardh, *Kittonia* Grove and Sturt, *Meretrosulus* Hanna, *Porperia* Bailey, *Terpsinoe* Ehrenberg, *Trigonium* Cleve.

Subfamily STICTODISCOIDEAE Simonsen K. (ALB)–Rec. Mar./FW

First: *Stictodiscus punctata* Jouse, Albian clays, Penza region, western Siberia (Jouse, 1949). **Extant**
Includes genera: *Anthodiscus* Grove and Sturt, *Arachonoidiscus* Deane, ex Pritch, *Chrysanthemodiscus* Mann, *Ethmodiscus* Castratance, *Pleurodiscus* Barker and Meakin, *Stictocyclus* Mann, and *Stictodiscus* Greville.

F. CHAETOCERACEAE Smith, 1872 T. (?Eoc.)–Rec. Mar./FW

First: *Chaetoceros clavigerium* Grunow, Eocene?, Franz Josef's Land and Ulyanovsk Province, former USSR. **Extant**
Includes genera: *Acanthoceras* Honigmann, *Bacteriastrum* Shadbolt, *Chaetoceros* Ehrenberg.

F. LITHODESMIACEAE H. and M. Peragallo, 1897–1908 K. (ALB)–Rec. Mar./FW

First: *Ditylum* sp., Lower Cretaceous, 'Phosphorite des Gaults' (Forti and Schultz, 1932). **Extant**
Includes genera: *Bellerochea* Van Heurck, *Ditylum* Bailey, *Lithodesium* Ehrenberg, *Neostreptotheca* Stosch, and *Strephotheca* Shrubsole.

F. EUPODISCACEAE Kutzing, 1849

Subfamily RUTILARIOIDEAE Pantocsek, 1889 K. (CMP)–Rec. Mar./FW

First: *Synedtocystis* Ralfs and *Rutilaria* Greville, Campanian, western Siberia (Strelnikova, 1975). **Extant**
Includes genera: *Rutilaria* Greville, *Synedetocystis* Ralfs, and *Syndetoneis* Grun.

Subfamily EUPODISCOIDEAE Kutzing, 1849 K. (ALB)–Rec. Mar./FW

First: *Triceratium schulzii* Jouse, Albian clays, Penza region, western Siberia (Jouse, 1949). **Extant**
Includes genera: *Actinodiscus* Greville, *Auliscus* Ehrenberg, *Cerataulus* Ehrenberg, *Corona* Lefebure and Cheneviere, *Craspedoporus* Greville, *Eupodiscus* Bailey, *Glyphodiscus* Greville, *Grovea* A. Schmidt, *Hendeya* Hanna, *Huttoniella* Karsten, *Odonetella* Agardh, *Pseoauliscus* Leuduger-Fortmorel, *Pseudocerataulus* Pant, *Rattrayella* De Toni, *Triceratium* Ehrenberg.

F. DIATOMACEAE Dumortier, 1823 K. (CMP)–Rec. Mar./FW

First: *Rhaphoneis* Ehrenberg and *Sceptroneis dimorpha* Strelnikova, Campanian, western Siberia (Strelnikova, 1975) and Arctic Ocean, CEASER 6 cores (Barron, 1985); *Rhaphoneis elliptica* Jouse and *Sceptroneis wittii* Jouse, and *Sceptroneis grunowii* Anissimova, upper Campanian, DSDP Sites 275, South Pacific (Hajos and Stradner, 1975). **Extant**
Includes genera: *Licmophora Agardh*, *Meridion* Agardh, *Opephora* Petit, *Plagiogramma* Greville, *Podocystis* Bailey, *Pseudodimerogramma* Schrader, *Rhabdonema* Kutz, *Rhaphoneis* Ehrenberg, *Sceptroneis* Ehrenberg, *Striatella* Agardh, *Subsilcea* Stosch and Reimann, *Synedra* Ehrenberg, *Synderosphenia* Peragallo, *Tabellaria* Ehrenberg, *Tetracyclus* Ralfs, *Thalassionema* Grun, *Thalassiothrix* Cleve and Grun, *Trachysphenia* Petit and *Tubularia* Brun.

F. PROTORAPHIDACEAE Simonsen, 1970 **Extant** Mar./FW

First: *Protoraphis* was described by Simonsen (1970) from the modern flora. Further studies are required to ascertain its chronological range.
Includes genera: *Protoraphis* Simonsen and *Pseudohimantidium* Hustedt and Krasske.

F. EUNOTIACEAE Kutzing, 1844 T. (Eoc.)–Rec. Mar./FW

First: *Eunotia striata* (Grove and Sturt) Grunow recorded as *Euodia striata* Grove and Sturt, Waiareka Volcanic Formation, Oamaru, South Island, New Zealand. **Extant**

Includes genera: *Actinella* Lewis, *Eunotia* Ehrenberg, *Peronia* Brebisson and Arnott, and *Semiorbis* Patrick.

F. ACHNANTHACEAE Kutzing, 1844 T. (PRB)–Rec. Mar./RW

First: *Cocconeis costata* Gregory and *Cocconeis greville* Smith (and others), Upper Eocene, Waiareka Volcanic Formation, Oamaru, South Island, New Zealand. **Extant**
Includes genera: *Achnanthes* Bory, *Anorthoneis* Grun, *Campyloneis* Grun and *Cocconeis* Ehrenberg.

F. NAVICULACEAE Kutzing, 1844 T. (PRB)–Rec. Mar./FW

First: *Navicula hennedyi* Smith and *Navicula pratexta* Ehrenberg, Upper Eocene, Ulyanovsk Province, former USSR (Witt, 1885). **Extant**
Includes genera: *Amphipleura* Kutz, *Amphiprora* Ehrenberg, *Amphora* Ehrenberg, *Anomoeoneis* Pfitzer, *Berkeleya* Greville, *Berkella* Ross and Sims, *Berbissonia* Grun, *Caloneis* Cleve, *Capartogramma* Kufferath, *Catenula* Mereschk, *Catillus* Hendey, *Cistula* Cleve, *Cymatoneis* Cleve, *Cymbella* Agardh, *Diatomella* Greville, *Dictyoneis* Cleve, *Didymosphenia* Schmidt, *Dimidiata* Hajos, *Diploneis* Ehrenberg, *Frickea* Heiden, *Frustulis* Rabh., *Gomphocalonies* Meister, *Gomphocymbella* Muller, *Gomphonema* Ehrenberg, *Gomphopleura* Reichelt, *Gyrosigme* Hassall, *Hassall* Simonsen, *Krasskella* Ross and Sims, *Mastogloia* Thwaites, *Navicula* Bory, *Neidiu* Pfitzer, *Oestrupia* Heiden, *Pachyneis* Simonsen, *Pinnularia* Ehrenberg, *Plagiotropis* Pfitzer, *Pleurosigma* Smith, *Pseudoamphiprora* Cleve, *Raphidodiscus* Smith, *Rhoicosphenia* Grun, *Rouxia* Brun and Heribaud, *Scoliopleura* Grun, *Scoliotropis* Cleve, *Stenoneis* Cleve, *Toxonidea* Donkin, and *Trachyneis* Cleve.

F. AURICULACEAE Hendey, 1964 **Extant** Mar./FW

First: Described from the Isle of Lesina. Further studies are required to determine its chronological range.
Includes genera: *Auricula* Castracane and *Hustedtia* Meister.

F. EPITHEMIACEAE Grunow, 1860 T. (AQT)–Rec. Mar./FW

First: *Epithema argus* Kutzing (and others), Lower Miocene (Aquitanian), Lozere, France (Lauby, 1910).
Includes genera: *Epithemia* Brebisson and *Rhopalodia* Muller.

F. NITZSCHIACEAE Grunow, 1860 T. (LMI)–Rec. Mar./FW

First: *Nitzschia maleinterpretaria* and several *Crucidenticula* species are common to abundant in the Lower Miocene sediment of the Pacific (Yanagisawa and Akiba, 1990; Barron, 1985). **Extant**
Includes genera: *Bacillaria* Gmelin, *Crucidenticula* Akiba and Yanagisawa, *Cylindrotheca* Rabh, *Cymatonitizschia* Hustedt, *Denticula* Kutz, *Denticulopsis* Simonsen emend. Akiba, *Gomphonitzschia* Grun, *Hantzschia* Grun, *Neodenticula* Akiba and Yanagisawa, *Nitzschia* Hassall, *Pseudoeunotia* Grun, and *Simonsenia* Lange-Bertalot.

F. SURIRELLACEAE Kutzing, 1844 T. (CHT)–Rec. Mar./FW

First: *Campylodiscus thuretii* Brebisson, *Surirella brunii* Heribaud, and *Surirella striatula* Turpin, Upper Oligocene, Puy-de-Mur, France (Lauby, 1910). **Extant**
Includes genera: *Campylodiscus* Ehrenberg, *Cymatopleura* Smith, *Hydrosilicon* Brun, *Stenopterobia* Breb. and *Surirella* Turpin.

F. UNNAMED K. (APT/ALB) Mar.

First and Last: *Amblypyrgus campanellus* Gersonde and Harwood, *Ancylopyrgus reticulatus* Gersonde and Harwood, *Archepyrgus melosiroides* Gersonde and Harwood, *Basilicostephanus ornatus* Gersonde and Harwood, *Bilingua rossii* Gersonde and Harwood, *Gladiopsis lagenoides* Gersonde and Harwood, *Kerkis bispinosa* Gersonde and Harwood, *Keagra bifaclcata* Gersonde and Harwood, *Microorbic convexus* Gersonde and Harwood, *Praethallssiosiropsis hasleae* Gersonde and Harwood, *Rhynchopyxis siphonoides* Gersonde and Harwood, *Trochus elegantulus* Gersonde and Harwood, *Calyptosporium carinatum* Gersonde and Harwood, *Cypellachaetes intricatus*, Gersonde and Harwood, *Hyalotrochuis imcompositus* Gersonde and Harwood, *Dasyangea dactylethra* Gersonde and Harwood (among others) from Aptian–Albian sediments, ODP Site 693, Weddell Sea (Gersonde and Harwood, 1990; Harwood and Gersonde, 1990).

Phylum EUGLENOPHYTA Pascher, 1931

One of the earliest classifications (Klebs, 1883) was based on nutritional characteristics, although this is now recognized as somewhat artificial. Leedale (1967) has proposed a classification based on the flagellar characteristics and other cytological features; this was adopted by Tappan (1980).

Order EUTREPTIALES Leedale, 1967

A Holocene group characterized by *Eutreptia* Perty, 1852, *Eutreptiella* de Cunha, 1913, and *Distigma* Ehrenberg, 1838. **Extant**

Order EUGLENALES Engler, 1898 T. (Eoc.)–Rec. FW

First: *Phacus* cf. *cordata* Huber, Eocene, Colorado. **Extant**
Comment: Bradley (1929) described fossil material from the Green River Formation of Colorado, but applied the name of an extant taxon, hence the 'cf.' in the above record.

Order RHABDOMONADALES Leedale, 1967

A monogeneric (*Rhabdomonas* Fresenius, 1858) group of Holocene euglenophytes. **Extant**

Order SPHENOMONADALES Leedale, 1967

A Holocene group characterized by *Anisonema* Dujardin, 1841, *Petalomonas* Stein, 1878 and *Sphenomonas* Stein, 1878. **Extant**

Order HETERONOMATALES Leedale, 1967

A Holocene group characterized by *Heteronema* Dujardin, 1841, *Peranema* Dujardin, 1841 and *Urceolus* Mereschkowsky, 1879. **Extant**

Order EUGLENAMORPHALES Leedale, 1967

A monogeneric (*Euglenamorpha* Wenrich, 1924) group of Holocene euglenophytes. **Extant**

Phylum PRASINOPHYTA Round, 1971

A distinctive group (first recognized as such by Chadefaud, 1950) of non-cellulosic green algae. They are closely related to the Chlorophyta, to which they are probably ancestral. Some have been included previously within the acritarchs or the dinoflagellates. One of the most recent summaries of the group is that of Tappan (1980).

Order PEDINOMONADALES

F. PEDINOMONADACEAE Korshikov, 1938

A monogeneric (*Pedinomonas* Korshikov, 1923) family of Holocene prasinophytes. **Extant**

Order PYRAMIMONADALES Chadefaud, 1950

F. PYRAMIMONADACEAE

A Holocene group characterized by *Asteromonas* Artari, 1913, *Chloraster* Ehrenberg, 1848, *Pyramimonas* Schmarda, 1850 and *Stephanoptera* Dangeard, 1910. **Extant**

F. HALOSPHAERACEAE

A Holocene group characterized by *Halosphaera* Schmitz, 1878 and *Hyalophysa* Cleve, 1900. **Extant**

F. LAEROSPHAERIDIACEAE Timofeev, 1956 *nom. corr.* Mädler, 1963 (=HALOSPHAERACEAE) P€. (RIF)–Rec. Mar.

First: A range of Riphean taxa, such as *Polyedrosphaeridium* Timofeev, 1966, *Protoleiosphaeridium* Timofeev, 1966, *Trachysphaeridium* Timofeev, 1966 and *Zonosphaeridium* Timofeev, 1969. **Extant**

Order PRASINOCLADALES

F. PRASINOCLADACEAE

A Holocene group characterized by *Platymonas* West, 1916, *Prasinochloris* Belcher, 1966, *Prasinocladus* Kuckuck, 1894 and *Tetraselmis* Stein, 1878. **Extant**

Order PTEROSPERMATALES Schiller, 1925

F. NEPHROSELMIDACEAE Pascher, 1913

A Holocene group characterized by *Bipedinomonas* Carter, 1937, *Micromonas* Manton and Parke, 1960, *Nephroselmis* Stein, 1878 and *Pseudoscourfieldia* Manton, 1975. **Extant**

F. PTEROSPERMATACEAE Lohmann, 1904

A Holocene group characterized by *Hexasterias* Cleve, 1900, *Pachysphaera* Ostenfeld, 1899, *Pterosperma* Pouchet, 1894 and *'Sphaeropsis'* Meunier, 1910 (*non* Saccardo, 1880). **Extant**

F. TASMANITACEAE Sommer, 1956, *nom. corr.* Tappan, 1980 P€. (RIF)–Rec. Mar.

Tappan (1980) quotes a range of Precambrian to Recent, and cites a list of typical genera. **Extant**

F. CYMATIOSPHAERACEAE Mädler, 1963 O.–?Qu. (HOL) Mar.

Tappan (1980) quotes a range of Ordovician to ?Holocene, and cites a list of typical genera. **?Extant**

F. PTEROSPHAERIDIACEAE Mädler, 1963 P.–J. Mar.

First: *Pterosphaeridia* Mädler, 1963, Permian.
Last: *Pterosphaeridia* Mädler, 1963, Jurassic.

F. PTEROSPERMELLACEAE Eisenack, 1972 P€. (RIF)–?Q. (HOL) Mar.

Tappan (1980) cites a range of Precambrian to ?Holocene, and lists typical genera. **?Extant**

Division PYRRHOPHYTA Pascher, 1914

Class DINOPHYCEAE Fritsch, 1929 (see Fig. 3.3)

This review concerns the stratigraphical record of dinoflagellates that have produced fossilizable cysts. To state the obvious, it does not include those dinoflagellates that do not, and have not, produced fossilizable cysts. This factor is stressed at the outset because the dinoflagellate fossil record is open to dangerous misinterpretation; it is 'an ambiguous record of dinoflagellate evolution' (Evitt, 1985, p. 276).

The Division Pyrrhophyta is divisible into three classes on the basis of nuclear structure. These are Dinophyceae, Syndiniophyceae and Oxyrrhidophyceae. The great majority of dinoflagellate taxa are contained within the Dinophyceae (see Fensome *et al.*, 1989).

Dinoflagellates are unicellular, aquatic, biflagellate algae. During the motile, vegetative, thecate, planktonic stage in the life cycle (contributing to biocoenoses), most dinoflagellates exist within the photic zone of surface waters. Various representatives of the class can be either autotrophic or heterotrophic. The distribution and productivity of dinoflagellates in a biocoenosis depends upon a number of factors: water temperature, salinity, nutrient levels, sunlight and the nature of water-mass movements and currents. As part of their life cycle, some dinoflagellate thecae encyst. As a result, benthic resting cysts (dinoflagellate cysts) are produced. These contribute to thanatocoenoses and stand a chance of fossilization, depending upon the nature of the cyst wall. Almost all fossil dinoflagellates in palynological preparations are sporopollenin (organic-walled) cysts. Fossil thecae (i.e. tests as opposed to cysts) are also known, as are calcareous and siliceous dinoflagellates. The term 'dinocyst' is a commonly used abbreviation for a dinoflagellate resting cyst.

Dinoflagellates have a fossil record stretching back to at least the Middle Triassic, and possibly to the late Silurian. However, because not all dinoflagellates produce fossilizable cysts, and presumably have not done so in the geological past, it is impossible to state with any confidence the age when the first dinoflagellates evolved. Unfortunately, potential preservability as a fossil is not an adaptation that favours evolution.

Since Sarjeant and Downie (1967) produced their review of the stratigraphical distribution of fossil dinoflagellates, there has been a considerable expansion in the levels of published taxonomic information concerning this class. In 1967, according to the data in Bujak and Williams (1979), there were just over 1000 validly described species assigned to 200 genera. By 1989 (Lentin and Williams, 1989) nearly 3000 species and over 500 genera were valid, representing increases of 200% and 150% respectively. A review of the suprageneric distribution of dinoflagellates is therefore long overdue. In addition, the unresolved question of the suprageneric classification of fossil dinoflagellates, and its relationship to the classification of modern dinoflagellates in particular, makes the present review problematic.

Earlier attempts at suprageneric classification have generally lost favour, because either they were developed before the realization that fossil dinoflagellates were cysts; they ignored existing classifications of modern dinoflagel-

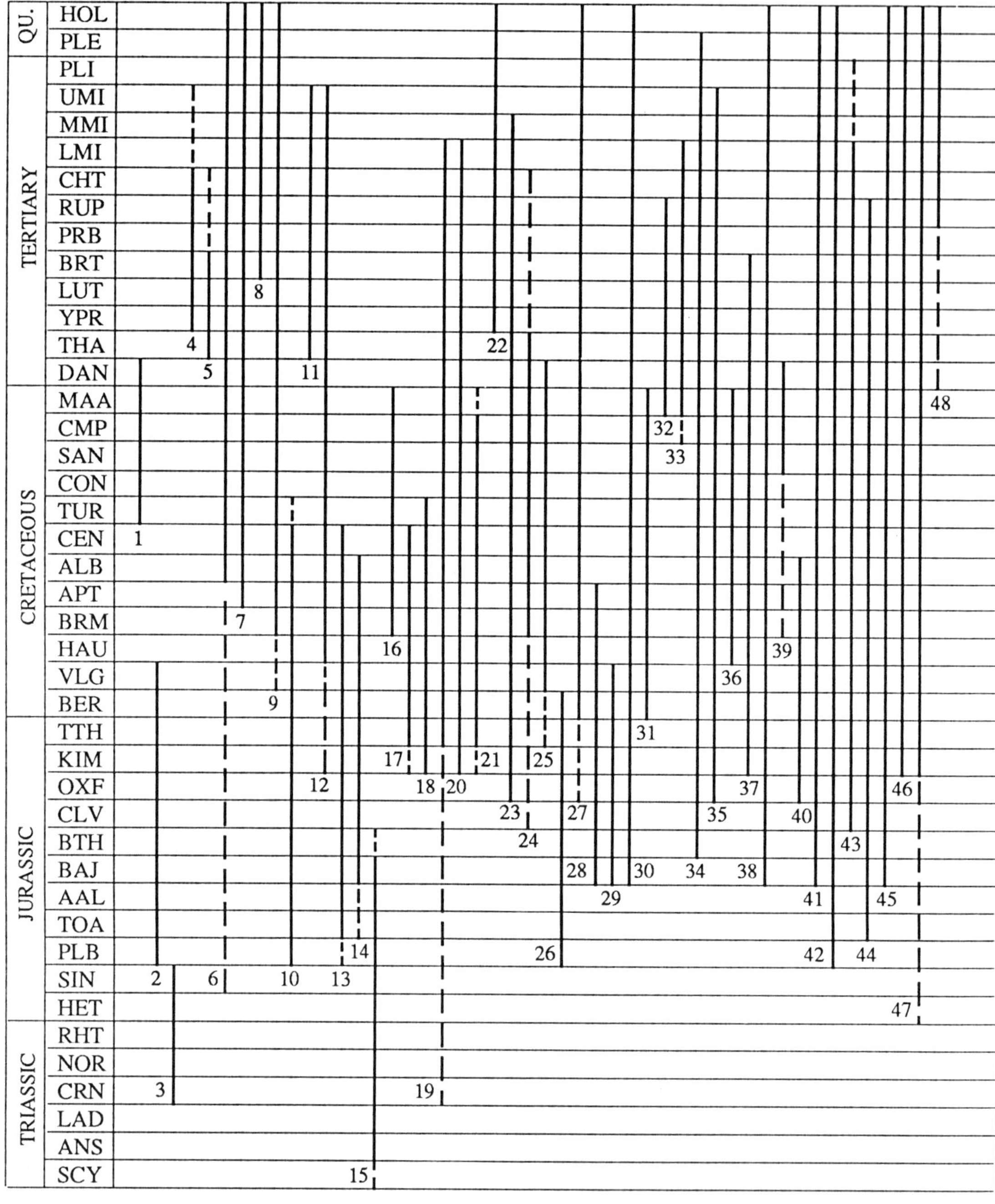

Fig. 3.3

lates; they did not incorporate morphological evidence from the thecae; they made insufficient allowance for the fact that not all dinoflagellates produce fossilizable cysts; or they overemphasized the importance of the archaeopyle (excystment aperture). Work is in progress (Fensome *et al.*, 1989) to formulate a comprehensive phylogenetic classification for fossil and living dinoflagellates.

In their review, Sarjeant and Downie (1967) subdivided the Class Dinophyceae on the grounds of general morphology (proximate, chorate, cavate, etc.), and then on archaeopyle style (precingular, apical, intercalary, etc.). I have adopted a generally similar approach here in that I have not employed a phylogenetic suprageneric scheme. Instead, I have used the phenetic groupings of Evitt (1985) for sporopollenin dinoflagellate cysts, together with additional categories for calcareous and siliceous forms. Evitt's scheme, in which paratabulation plays the major role in categorization, is not based upon biological grounds. Rather, the categories 'are intended principally as a convenience for organizing extensive observational or descriptive data that have been reduced to identifications at the generic and specific level *without* requiring a commitment to any formal suprageneric scheme' (Evitt, 1985, p. 174).

The generic and specific taxonomy of Lentin and Williams (1989) is applied without question, because their Index serves as a useful single reference point; it is the only source where all fossil dinoflagellates are treated, ostensibly in equal fashion. It would have been inappropriate to have strayed from their judgements for the purposes of the present review.

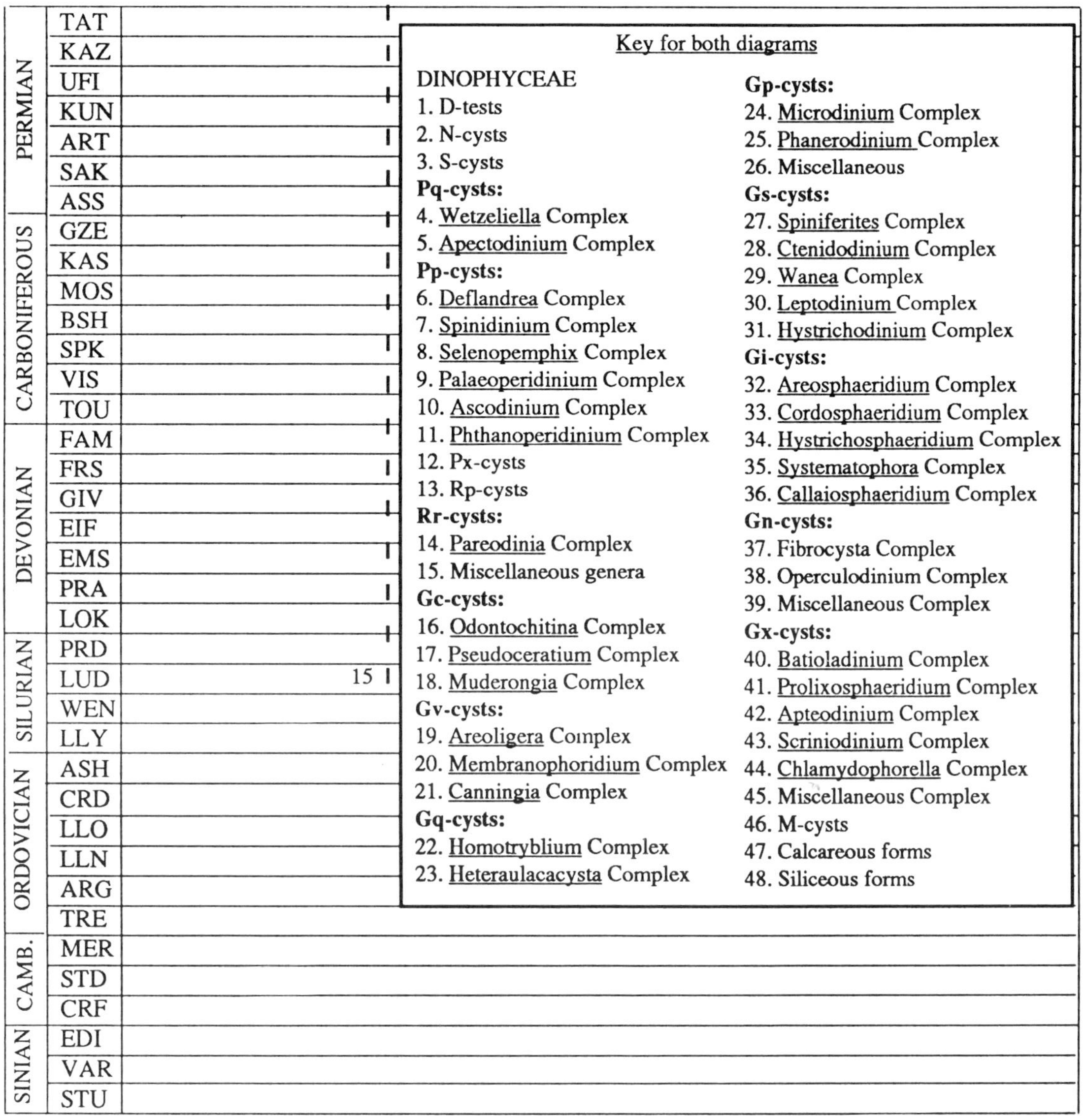

PERMIAN: TAT, KAZ, UFI, KUN, ART, SAK, ASS
CARBONIFEROUS: GZE, KAS, MOS, BSH, SPK, VIS, TOU
DEVONIAN: FAM, FRS, GIV, EIF, EMS, PRA, LOK
SILURIAN: PRD, LUD, WEN, LLY
ORDOVICIAN: ASH, CRD, LLO, LLN, ARG, TRE
CAMB.: MER, STD, CRF
SINIAN: EDI, VAR, STU

15

Key for both diagrams

DINOPHYCEAE
1. D-tests
2. N-cysts
3. S-cysts
Pq-cysts:
4. Wetzeliella Complex
5. Apectodinium Complex
Pp-cysts:
6. Deflandrea Complex
7. Spinidinium Complex
8. Selenopemphix Complex
9. Palaeoperidinium Complex
10. Ascodinium Complex
11. Phthanoperidinium Complex
12. Px-cysts
13. Rp-cysts
Rr-cysts:
14. Pareodinia Complex
15. Miscellaneous genera
Gc-cysts:
16. Odontochitina Complex
17. Pseudoceratium Complex
18. Muderongia Complex
Gv-cysts:
19. Areoligera Complex
20. Membranophoridium Complex
21. Canningia Complex
Gq-cysts:
22. Homotryblium Complex
23. Heteraulacacysta Complex
Gp-cysts:
24. Microdinium Complex
25. Phanerodinium Complex
26. Miscellaneous
Gs-cysts:
27. Spiniferites Complex
28. Ctenidodinium Complex
29. Wanea Complex
30. Leptodinium Complex
31. Hystrichodinium Complex
Gi-cysts:
32. Areosphaeridium Complex
33. Cordosphaeridium Complex
34. Hystrichosphaeridium Complex
35. Systematophora Complex
36. Callaiosphaeridium Complex
Gn-cysts:
37. Fibrocysta Complex
38. Operculodinium Complex
39. Miscellaneous Complex
Gx-cysts:
40. Batioladinium Complex
41. Prolixosphaeridium Complex
42. Apteodinium Complex
43. Scriniodinium Complex
44. Chlamydophorella Complex
45. Miscellaneous Complex
46. M-cysts
47. Calcareous forms
48. Siliceous forms

Fig. 3.3

Category of D-cysts K. (TUR)–T. (DAN) Mar.

First: *Dinogymnium cretaceum* (Deflandre) Evitt *et al.*, 1967, Silex du Turonien supérieur (sommet), Ruyaulcourt, Pas-de-Calais, France (Foucher, 1974).
Last: *Dinogymnium acuminatum* Evitt *et al.*, 1967, 'Boundary Clay', El Haria, Tunisia (Brinkhuis and Zachariasse, 1988).

Category of N-cysts J. (PLB)–K. (VLG) Mar.

First: *Nannoceratopsis senex* van Helden, 1977, Wilkie Point Formation, Emerald Island, North-west Territories, Canada (van Helden, 1977).
Last: *Nannoceratopsis* spp., Upper Barrow Group, Barrow-25 well, Carnarvon Basin, offshore western Australia (Helby *et al.*, 1987).
Comments: Helby *et al.* (1987) indicated that in offshore Australia, undifferentiated *Nannoceratopsis* species range inconsistently above the top of the Tithonian succession to within the lower Valanginian.

Category of S-cysts Tr. (CRN)–J. (SIN) Mar.

First: *Suessia swabiana* Morbey, 1975 emend. Below, 1987, unnamed unit, Sahul Shoals-1 well, Bonaparte Basin, offshore north-western Australia (Helby *et al.*, 1987).
Last: *Beaumontella langii* (Wall) emend. Below, 1987, Black Ven Marls, Dorset coast, England, UK (Wall, 1965).
Comments: Helby *et al.* (1987) indicated that in Australia, the last occurrence of *S. swabiana* coincides with that of *B. langii.*

Category of P-cysts

Category of Pq-cysts

Wetzeliella Complex T. (YPR–?TOR/MES) Mar.

First: *Wetzeliella astra* Denison *in* Costa *et al.*, 1978, lower London Clay, Bean, England, UK (Costa *et al.*, 1978).
Last: *Wetzeliella symmetrica* Weiler, 1956, Untere Graftenberger Schichten, Tönisberg Borehole, Krefeld, Germany (Benedek, 1972).
Comments: The records of *W. symmetrica* by Benedek (1972) include *W. gochtii* Costa and Downie, 1976 as 'W.

(*W.*) *symmetrica* Weiler var. Gocht, 1969'. There are Miocene records of the *Wetzeliella* Complex which may possibly be accounted for by reworking. For example, Harland (1979) recorded '*Wetzeliella* sp.' from DSDP Hole 400A (Bay of Biscay) in Upper Miocene sediments (TOR/MES), which if *in situ* would be the last true representation of the complex.

Apectodinium Complex T. (THA–BRT/?RUP/CHT) Mar.

First: *Apectodinium hyperacanthum* (Cookson and Eisenack) Lentin and Williams, 1977, Woolwich Beds, Charlton Brickpit, Erith and Upnor, England, UK (Costa and Downie, 1976).
Last: *Apectodinium homomorphum* (Deflandre and Cookson) Lentin and Williams, 1977, emend. Harland, 1979, middle Barton Beds, Barton, England, UK (Bujak *et al.*, 1980).
Comments: In their compiled range chart, Costa and Downie (1979) indicate that *A. homomorphum* ranges up to calcareous nannoplankton Biozone NP23 of Martini (1971). If verified, this would give an Oligocene (Rupelian/Chattian) last appearance for the *Apectodinium* Complex.

Category of Pp-cysts

Deflandrea Complex ?J. (SIN)/K. (ALB/CEN)–Rec. Mar.

First: *Isabelidinium glabrum* (Cookson and Eisenack) Lentin and Williams, 1977, subsurface Balcatta, Western Australia (Cookson and Eisenack, 1969). **Extant**
Comments: *Liasidium variabile* Drugg, 1978, originally described from Sinemurian material (Balingen, Swabia, Germany), would be the first representative if confidently assigned to the *Deflandrea* Complex. There are no unequivocal records of the complex between the Sinemurian and the Albian/Cenomanian. Species of *Leipokatium* Bradford, 1975 and *Omanodinium* Bradford, 1975 are known from Recent sediments.

Spinidinium Complex K. (APT)–Rec. Mar.

First: *Spinidinium denticulatum* Pothe de Baldis and Ramos, 1983, Rio Fosiles, Provincia de Santa Cruz, Argentina (Pothe de Baldis and Ramos, 1983). **Extant**
Comments: Some living species of *Protoperidinium* Bergh, 1882 are included within the *Spinidinium* Complex, e.g. *P. divaricatum* (Meunier) Parke and Dodge, 1976.

Selenopemphix Complex T. (BRT)–Rec. Mar.

First: *Selenopemphix nephroides* Benedek, 1972 emend. Bujak *in* Bujak *et al.*, 1980, lower Barton Beds, Barton, Hampshire, England, UK (Bujak *et al.*, 1980). **Extant**
Comments: Cysts of the extant species *Protoperidinium subinerme* (Paulsen) Loeblich III, 1970 are referable to *S. nephroides* (see Harland, 1982).

Palaeoperidinium Complex K. (?VLG/BRM)–Rec. Mar./FW

First: *Palaeoperidinium cretaceum* Pocock, 1962 emend. Davey, 1970, Cement Beds, Speeton Clay, Speeton, Yorkshire, England, UK (Duxbury, 1980). **Extant**
Comments: *Luxadinium? dabendorfense* (Alberti) Bujak and Davies, 1983 was described originally from subsurface Valanginian material from Dabendorf near Berlin, Germany. However, Bujak and Davies (1983) only questionably included this species in the genus *Luxadinium* Brideaux and McIntyre, 1975, and as a result it cannot be confidently referred to as the first representative of the *Palaeoperidinium* Complex. The genus *Palaeoperidinium* Deflandre, 1935, emend. Sarjeant, 1967 last appeared during Thanetian times (e.g. Holmehus Formation, Jutland, Denmark; Heilmann-Clausen, 1985). The genus *Saeptodinium* Harris, 1974 is known from Palaeogene and Neogene sediments, while *Protoperidinium limbatum* (Stokes) Lemmermann, 1900, and several other freshwater species which belong to the complex, are extant.

Ascodinium Complex J. (PLB)/K. (CEN/?TUR)–Rec. Mar.

First: *Ovoidinium waltonii* (Pocock) Lentin and Williams, 1976, Lower Member, Watrous Formation, Saskatchewan, Canada (Pocock, 1972).
Last: *Ovoidinium verrucosum* (Cookson and Hughes) Davey, 1970, Chalk Marl, Cement Works Pit, Barrington, Cambridge, England, UK (Cookson and Hughes, 1964).
Comments: The early stratigraphical record of the *Ascodinium* Complex is erratic. Pocock (1972) recorded *O. waltonii* from the PLB Lower Member of the Watrous Formation of Saskatchewan. There are no intervening occurrences between this record and the younger post-KIM Upper Member of the Vanguard Formation. The range of *O. waltonii* given by Jansonius (1986) is 'Toarcian?(–Bajocian?)'. Members of the complex are known to occur fairly regularly through the APT to CEN succession. Foucher (1974, 1979) recorded *Ascodinium* cf. *parvum* (Cookson and Eisenack) Cookson and Eisenack, 1960 from 'Silex du Turonien supérieur (sommet) de Ruyaulcourt', Pas-de-Calais, France, which would make it the last representative stratigraphically. However, there is a Cainozoic form, *Ovoidinium granulatum* Z.-C. Song *in* Z.-C. Song *et al.*, 1985, described from the Shelf Basin of the East China Sea (Donghai) region. Furthermore, Evitt (1985) indicated that the extant species *Protoperidinium claudicans* (Paulsen) Balech, 1974, *P. oblongum* (Aurivillius) Parke and Dodge, 1976 and *Peridinium wisconsinense* Eddy, 1930 are also attributable to the *Ascodinium* Complex.

Phthanoperidinium Complex T. (THA–TOR) Mar.

First: *Phthanoperidinium crenulatum* Heilmann-Clausen, 1985, Holmehus Formation, Jutland, Denmark (Heilmann-Clausen, 1985).
Last: *Phthanoperidinium lambdoideum* (Nagy) Eisenack and Kjellström, 1971, subsurface, Mecseck Mountains, Hungary (Nagy, 1966).
Comments: Nagy (1966) stated that *P. lambdoideum* was recovered originally from Pannonian sediments. A Tortonian age is indicated therefore for the last representation of the *Phthanoperidinium* Complex. Stover and Evitt (1978) considered *P. lambdoideum* to be a problematic species and only questionably allocated it to the genus *Phthanoperidinium* Drugg and Loeblich, 1967 (*sic.*). There are also stratigraphical difficulties because there are no known early or middle Miocene records of the genus. The next oldest species is *P. polytrix* Benedek, 1972, emend. Benedek and Sarjeant, 1981, which ranges into the Upper Oligocene (Untere Lintforter Schichten, Germany; Benedek, 1972).

Category of Px-cysts ?J. (?KIM)/K. (HAU)–T. (Mio.) Mar./FW

First: *Subtilisphaera terrula* (Davey) Lentin and Williams,

1976, emend. Harding, 1986, Mottled Clay, Division C1B, Speeton Clay, Yorkshire, England, UK (Duxbury, 1977).
Last: *Geiselodinium tyonekense* Engelhardt, 1976, Tyonek Formation, Cook Inlet, Alaska, USA (Engelhardt, 1976).
Comments: This catch-all category consists of those P-cysts which cannot be allocated to either the Pp-cysts or Pq-cysts. *Subtilisphaera? inaffecta* (Drugg) Bujak and Davies, 1983, is known to be restricted to the KIM. However, because this species is only questionably referred to *Subtilisphaera* Jain and Millepied, 1973, emend. Lentin and Williams, 1976, it cannot be said confidently to be the first representative of the genus. Apart from the KIM species *S.? paeminosa* (Drugg) Bujak and Davies, 1983, there are no other known records of the genus between the KIM and HAU. Both *Geiselodinium* Krutsch, 1962 and *Teneridinium* Krutsch, 1962, are freshwater representatives.

Category of R-cysts

Category of Rp-cysts J. (?PLB/TOA)–K. (ALB/CEN) Mar.

First: *Susadinium faustum* (Bjaerke) Lentin and Williams, 1985, Wilhelmøya Formation, Kong Karl's Land, Svalbard (Bjaerke, 1980).
Last: *Angustidinium acribes* (Davey and Verdier) Goodman and Evitt, 1981, Franciscan Elk Creek mélange and Great Valley sequence, Rice Valley, California (Goodman and Evitt, 1981).
Comments: Helby *et al.* (1987) indicated that their *Susadinium* sp. A first occurs within the PLB succession of Australia. There is an apparent gap in the stratigraphical distribution of Rp-cysts between the BTH (the last known *Susadinium* Dörhöfer and Davies, 1980) and the TTH (the first known *Angustidinium* Goodman and Evitt, 1981).

Category of Rr-cysts

Pareodinia Complex J. (?TOA/BAJ)–K. (ALB) Mar.

First: *Pareodinia ceratophora* Deflandre, 1947 emend. Gocht, 1970, Lincolnshire Limestone, Nettleton, Lincolnshire, England, UK (Riding, 1987).
Last: *Pareodinia psilata* Jain and Millepied, 1975, borehole BR-1, Senegal (Jain and Millepied, 1975).
Comments: Morbey (1975) indicated that *P. ceratophora* ranges into upper Lias (TOA) sediments of north-west Europe.

Miscellaneous genera of Rr-cysts ?S. (?LUD)/Tr. (ANS)–J. (BAJ/BTH) Mar.

First: *Sahulidinium ottii* Stover and Helby, 1987, unnamed unit, Sahul Shoals-1 Well, Bonaparte Basin, offshore north-western Australia (Helby *et al.*, 1987).
Last: *Valvaeodinium aquilonium* (Dörhöfer and Davies), emend. Below, 1987, Lower Savik Formation and Jaeger Member, Ellef Ringnes Island, Northwest Territories, Canada (Dörhöfer and Davies, 1980).
Comments: The problematic species *Arpylorus antiquus* Calandra, 1964, emend. Sarjeant, 1978 is known from Upper Silurian (LUD) subsurface sediments from Mecheguig, Tunisia (see discussion in Sarjeant, 1978 and Evitt, 1985 for details concerning the nature and assignation of this species). There are no unequivocal dinoflagellate cyst species known between the Upper Silurian and the Middle Triassic. Stover and Evitt (1978) discredited the Permian dinoflagellate cysts records of Tasch (1963), while those of Jansonius (1962) are believed to be the result of contamination by younger material (Evitt, 1985).

Category of G-cysts

Category of Gc-cysts

Odontochitina Complex K. (BRM-MAA) Mar.

First: *Odontochitina operculata* (O. Wetzel) Deflandre and Cookson, 1955, Cement Beds, Speeton Clay, Speeton, Yorkshire, England, UK (Duxbury, 1980).
Last: *Odontochitina costata* Alberti, 1961, emend. Clarke and Verdier, 1967, Lixhe 1 Member, Gulpen Formation, Halembaye, Belgium (Foucher, *in* Robaszynski *et al.*, 1985).

Pseudoceratium Complex J. (KIM/TTH)–K. (CEN) Mar.

First: *Pseudoceratium spitiense* Jain and Garg *in* Jain *et al.*, 1984, Spiti Shale (Formation), Malla Johar, Himalaya, India (Jain *et al.*, 1984).
Last: *Endoceratium ludbrookiae* (Cookson and Eisenack) Loeblich and Loeblich, 1966, emend. Morgan, 1980, Bathurst Island Formation, Bathurst Island, Northern Territory, Australia (Norvick and Berger, 1976).

Muderongia Complex J. (KIM)–K. (TUR) Mar.

First: *Muderongia simplex auct. non* Alberti, 1961, *rotunda* Zone, Warlingham Borehole, Surrey, England, UK (Gitmez and Sarjeant, 1972).
Last: *Muderongia perforata* Alberti, 1961, Pirna Borehole, Sachsen, Germany (Alberti, 1961).
Comments: *M. simplex sensu* Gitmez and Sarjeant (1972) is most usually referred to as *Muderongia* sp. A of Davey (1979).

Category of Gv-cysts

Areoligera Complex ?Tr. (u.)/J. (TTH)–T. (AQT) Mar.

First: *Circulodinium compta* (Davey) Helby, 1987, Sandringham Sands, North Runcton, Norfolk, England, UK (Davey, 1982).
Last: *Chiropteridium mespilanum* (Maier) Lentin and Williams, 1973, Atlantic Shelf Corehole 5/5B, Blake Plateau, offshore South Carolina, USA (Stover, 1977).
Comments: The oldest possible representative of the *Areoligera* Complex is the allegedly Upper Triassic *Cyclonephelium granulatum* (Horowitz) Stover and Evitt, 1978. However, there is doubt not only concerning the age of the sediments from which the type material was derived (see Conway and Cousminer, 1983) but also about its allocation to the genus *Cyclonephelium* Deflandre and Cookson, 1955. emend. Stover and Evitt, 1978. The type species of *Glaphyrocysta* Stover and Evitt, 1978, *G. retiintexta* (Cookson) Stover and Evitt, 1978, is the only representative of that genus within the *Areoligera* Complex; the other species belong to the *Areosphaeridium* Complex of Gi-cysts.

Membranophoridium Complex J. (KIM)–T. (AQT) Mar.

First: *Senoniasphaera jurassica* (Gitmez and Sarjeant) Lentin and Williams, 1976, *baylei* Zone, Staffin Bay, Skye, Scotland, UK (Gitmez and Sarjeant, 1972).
Last: *Membranophoridium aspinatum* Gerlach, 1961 ex Gocht, 1969, Atlantic Shelf Borehole 5/5B, Blake Plateau, offshore South Carolina, USA (Stover, 1977).

Canningia Complex J. (KIM/TTH)–K. (CMP/MAA) Mar.

First: *Canningia apiculata* Jain and Garg, *in* Jain *et al.*, 1984, Spiti Shale (Formation), Malla Johar, Himalaya, India (Jain *et al.*, 1984).
Last: *Canningia xinjiangensis* Chen *et al.*, 1988, Western Xinjiang, China (J.-X. Yu and W.-P. Zhang, 1980).
Comments: Lentin and Williams (1989) give a TUR to MAA age for the type *C. xinjiangensis*. Other members of the *Canningia* Complex are known to range at least to the CMP, e.g. *C. senonica* Clarke and Verdier, 1967 in the Paris Basin, France (Foucher, 1979).

Category of Gq-cysts

Homotryblium Complex T. (YPR)–Rec. Mar.

First: *Eocladopyxis peniculata* Morgenroth, 1966, London Clay, Whitecliff Bay, Isle of Wight, England, UK (Bujak *et al.*, 1980). **Extant**
Comments: Mclean (1976) considered *E. peniculata* to be an ancestor to the extant *Pyrodinium bahamense* Plate, 1906. This dinoflagellate, whose cysts are referred to as *Polysphaeridium zoharyii* (Rossignol) Bujak *et al.*, 1980, is a member of the *Homotryblium* Complex.

Heteraulacacysta Complex J. (OXF/KIM)/ K. (ALB)–T. (SRV) Mar.

First: *Dinopterygium bicuneatum* (Deflandre) Lentin and Williams 1981, Les Marnes de Villers-sur-Mer, Calvados, France (Deflandre, 1938).
Last: *Heteraulacacysta* cf. *campanula* Drugg and Loeblich, 1967, Cassinasco Formation, Langhe, Piemonte, Italy (Powell, 1986a).
Comments: There is an apparent gap in the stratigraphical record of the *Heteraulacacysta* Complex between the KIM and ALB. Gitmez and Sarjeant (1972) recorded *D. bicuneatum* ranging into the KIM *rotunda* Zone, while the first occurrence of *Dinopterygium cladoides* Deflandre, 1935 is not known until the ALB (e.g. Foucher, 1981).

Category of Gp-cysts

Microdinium Complex ?J. (?CLV)/K. (BRM/APT)– T. (THA/?Oli.) Mar.

First: *Fibradinium variculum* Stover and Helby, 1987, unnamed greensand unit, Warnbro Group, Houtman-1 Well, Western Australia (Stover and Helby, 1987).
Last: *Cladopyxidium saeptum* (Morgenroth) Stover and Evitt, 1978, Holmehus Formation, Viborg 1 borehole, Jutland, Denmark (Heilmann-Clausen, 1985).
Comments: Lentin and Williams (1989) questionably transferred *Phanerodinium follis* Below, 1987 and *P. diatretiforme* Below, 1987 to the genus *Fibradinium* Morgenroth, 1968. If confidently accepted within the genus, the *Microdinium* Complex would first appear within the CLV and range to the Oligocene. The type species of *Microdinium* Cookson and Eisenack, 1960, emend. Stover and Evitt, 1978, *M. ornatum* Cookson and Eisenack, 1960, alone is placed within the complex; the other species of *Microdinium* belong to the *Phanerodinium* Complex. *'Microdinium' veligerum* (Deflandre) Davey, 1969, another member of the complex, now lies within the monospecific genus *Rhiptocorys* Lejeune-Carpentier and Sarjeant, 1983.

Phanerodinium Complex J./K. (TTH/VLG)–T. (DAN) Mar.

First: *Druggidinium apicopaucicum* Habib, 1973, DSDP Site 105, Hatteras Abyssal Plain, offshore North Carolina, USA (Habib, 1973).
Last: *Glyphanodinium facetum* Drugg, 1964, Dos Palos Shale Member, Moreno Formation, Escarpado Canyon, California, USA (Drugg, 1964).
Comments: The contention of Below (1987) that *Microdinium* Cookson and Eisenack, 1960 emend. Stover and Evitt, 1978 (among others) is a junior synonym of *Phanerodinium* Deflandre, 1937, emend. Below, 1987 was rejected by Lentin and Williams (1989). *Fibradinium* Morgenroth, 1968, *Rhiptocorys* Lejeune-Carpentier and Sarjeant, 1983 and *Subtilidinium* Morgenroth, 1968, also considered to be junior synonyms of *Phanerodinium* by Below (1987), belong to the *Microdinium* Complex, along with the type species of *Microdinium*, *M. ornatum* Cookson and Eisenack, 1960.

Miscellaneous Complex of Gp-cysts J. (PLB)–K. (BER) Mar.

First: *Luehndea spinosa* Morgenroth, 1970, phosphatic nodules, blue-grey Liassic clays, Lühnde, Lower Saxony, Germany (Morgenroth, 1970).
Last: *Paragonyaulacysta capillosa* (Brideaux and Fisher) Stover and Evitt, 1978, Lower Sandstone Division, Buff Sandstone Unit, *Buchia volgensis* Zone, Martin Creek, Arlavik Range, District of Mackenzie, Canada (Brideaux and Fisher, 1976).
Comments: This is a group of species having partiform hypocysts but not necessarily other features in common. Evitt (1985, p. 232) intimated that *Paragonyaulacysta* Johnson and Hills, 1973, may be synonymous with *Carpathodinium* Drugg, 1978, and (p. 234) that *Lacrymodinium* Albert *et al.*, 1986 may be congeneric with *Pluriarvalium* Sarjeant, 1962.

Category of Gs-cysts

Spiniferites Complex ?J. (?OXF)/K. (BER)–Rec. Mar.

First: *Spiniferites ramosus* (Ehrenberg) Loeblich and Loeblich, 1966, Les Marnes de Villers-sur-Mer, Calvados, France (Deflander, 1938). **Extant**
Comments: Although Deflandre (1938, p. 186) recorded *S. ramosus*, as *Hystrichosphaera furcata* (Ehrenberg) O. Wetzel, 1933, from OXF sediments (and as also reported by Sarjeant, 1960 and 1962), it is questionable whether or not this represents the first occurrence of the genus. *Spiniferites*. Davey (1979, Fig. 6), for example, placed this event within the VLG. Duxbury (1977) recorded both *Achomosphaera? neptuni* (Eisenack) Davey and Williams, 1966 and *Avellodinium falsificum* Duxbury, 1977 from the Blue Beds, D6 division of the Speeton Clay (Speeton, Yorkshire, England, UK). Although both are only tentatively attributable to the *Spiniferites* Complex, they are the two next oldest records. Included within the *Spiniferites* Complex are some living species of *Gonyaulax spinifera* (Claparède and Lachmann) Diesing, 1886.

Ctenidodinium Complex J. (BAJ)–K. (APT) Mar.

First: *Ctenidodinium continuum* Gocht, 1970, Upper Inferior Oolite, *garantiana* Zone, Dorset, England, UK (Woollam and Riding, 1983).
Last: *Ctenidodinium elegantulum* Millioud, 1969, *Scapites*

Beds, Ferruginous Sands Series, Atherfield, Isle of Wight, England, UK (Duxbury, 1983).

Wanea Complex J. (BAJ)–K. (VLG) Mar.

First: *Energlynia acollaris* (Dodekova) Sarjeant, 1978, Upper Inferior Oolite, *garantiana* Zone, Dorset, England, UK (Woollam and Riding, 1983).
Last: *Isthmocystis distincta* Duxbury, 1979, Division D3D, Speeton Clay, Speeton, Yorkshire, England, UK (Duxbury, 1979).

Leptodinium Complex J. (BAJ)–Rec. Mar./FW

First: *Rhynchodiniopsis? regalis* (Gocht) Jan du Chêne *et al.*, 1985, *garantiana* Zone, southern England (Fenton and Fisher, 1978). **Extant**
Comments: Although *R.? regalis* is only provisionally accepted in the genus *Rhynchodiniopsis* Deflandre, 1935, emend. Jan du Chêne *et al.*, 1985, this does not preclude its allocation to the *Leptodinium* Complex. A possible older representative is the undescribed *Leptodinium* sp. 1 of Fenton and Fisher (1978) which first occurs within the *humphriesianum* Zone of eastern England. Living representatives of the complex include *Gonyaulax* spp. indet. which produce cysts referable to *Impagidinium* Stover and Evitt, 1978 and the freshwater species *Gonyaulax apiculata* (Penard) Entz, 1904.

Hystrichodinium Complex K. (BER–MAA) Mar.

First: *Hystrichodinium pulchrum* Deflandre, 1935, Coprolite Bed, Division E, Speeton Clay, Speeton, Yorkshire, England, UK (Duxbury, 1977).
Last: *Hystrichodinium pulchrum* Deflandre, 1935, northwest Europe (Foucher, 1979).
Comments: Foucher (1979) indicated that the last occurrence of *H. pulchrum* has been observed in MAA sediments from Belgium, The Netherlands, Denmark, and Sweden.

Category of Gi-cysts

Areosphaeridium Complex K. (MAA)–T. (RUP) Mar.

First: *Glaphyrocysta ordinata* (Williams and Downie) Stover and Evitt, 1978, White Chalk, borehole Tuba 13 and Kjølby Gaard, Denmark (Hansen, 1977).
Last: *Glaphyrocysta microfenestrata* (Bujak) Stover and Evitt, 1978, Upper Hamstead Beds, Hamstead Formation, Isle of Wight, England, UK (Liengjarern *et al.*, 1980).
Comments: The *Areosphaeridium* Complex excludes the type species of *Glaphyrocysta* Stover and Evitt, 1978, *G. retiintexta* (Cookson) Stover and Evitt, 1978, which is allocated to the *Areoligera* Complex of Gv-cysts.

Cordosphaeridium Complex K. (CMP/MAA)–T. (BUR) Mar.

First: *Cordosphaeridium senegalense* Jain and Millepied, 1975, subsurface, Borehole CM-1, offshore Senegal (Jain and Millepied, 1975).
Last: *Cordosphaeridium cantharellum* (Brosius) Gocht, 1969, Bisciaro Formation, Montebello d'Urbino, Marche, Italy (Biffi and Manum, 1988).

Hystrichosphaeridium Complex J. (BTH)–Q. (PLE) Mar.

First: *Adnatosphaeridium caulleryi* (Deflandre) Williams and Downie, 1969, Lower Fuller's Earth, *zigzag* Zone, vicinity of Bath, England, UK (Riding *et al.*, 1985).
Last: *Hystrichokolpoma rigaudiae* Deflandre and Cookson, 1955, Utahime Member (Ma1 Member), Saho Formation, Osaka Group, northern Utahime, Nara City, Japan (Matsuoka, 1976).
Comments: Matsuoka (1976, 1979) indicated that *H. rigaudiae* last occurs stratigraphically below the normal Jaramillo Event within the Matsuyama Reverse Epoch; an early Pleistocene age is thus indicated.

Systematophora Complex J. (OXF)–T. (MES) Mar.

First: *Systematophora areolata* Klement, 1960, Ancholme Clay Group, *cordatum* Zone, Nettleton Bottom borehole, Lincolnshire, England, UK (Riding, 1987).
Last: *Systematophora placacantha* (Deflandre and Cookson) Davey *et al.*, 1969 emend. May, 1980, Castellania Formation, Sant'Agata Fossili, Italy (Powell, 1986b).

Callaiosphaeridium Complex K. (HAU–MAA) Mar.

First: *Callaiosphaeridium asymmetricum* (Deflandre and Courteville) Davey and Williams, 1966, Main *speetonensis* Bed, C6 Division, Speeton Clay, Speeton, Yorkshire, England, UK (Duxbury, 1977).
Last: *Callaiosphaeridium asymmetricum* (Deflandre and Courteville) Davey and Williams, 1966, Hollviken Borehole No.1, Scania, Sweden (Kjellström, 1973).

Category of Gn-cysts

Fibrocysta Complex J. (KIM)–T. (BRT) Mar.

First: *Fibrocysta acornuta* Norris and Jux, 1984, southern England (Norris and Jux, 1984).
Last: *Fibrocysta vectensis* (Eaton) Stover and Evitt, 1978, Barton Beds, Whitecliff Bay, Isle of Wight, England, UK (Bujak *et al.*, 1980).

Operculodinium Complex J. (BAJ)–Rec. Mar.

First: *Cleistosphaeridium polytrichum* (Valensi) Davey *et al.*, 1969, *Truellei* Bed, Burton Limestone, *parkinsoni* Zone, Burton Bradstock, Dorset, England, UK (Fenton *et al.*, 1980). **Extant**
Comments: Woollam and Riding (1983) indicated that members of the *C. polytrichichum* Group do not range higher than the Jurassic, as least in England. There is, therefore, an apparent break in the stratigraphical distribution of the *Operculodinium* Complex between the TTH and HAU, when *Kiokansium polypes* (Cookson and Eisenack) Below, 1982 first occurs (e.g. Division C8 of the Speeton Clay, Speeton, Yorkshire, England, UK, Duxbury, 1977). The extant dinoflagellates *Gonyaulax grindleyi* Reinecke, 1967 and *G. polyhedra* Stein, 1883 produce cysts referable to *Operculodinium centrocarpum* (Deflandre and Cookson) Wall, 1967 and *Lingulodinium machaerophorum* (Deflandre and Cookson) Wall, 1967, respectively.

Miscellaneous Complex of Gn-cysts K. (BRM/SAN)–?T. (?DAN) Mar.

First: *Cauca parva* (Alberti) Davey and Verdier, 1971, Cement Beds, Speeton Clay, Speeton, Yorkshire, England, UK (Duxbury, 1980).
Last: *Stenopyxinium grassei* Deflandre, 1968, Cretaceous (Senonian) flint (Deflandre, 1968).
Comments: Members of the Miscellaneous Complex of Gn-cysts are characterized by spinose projections which are

apparently distributed according to the general outline of the cyst, but not evenly and not in association with paratabulation. *Cauca? velata* (W. Wetzel) Sarjeant, 1984 was described originally from a DAN flint from drift deposits near Kiel, Germany. However, Sarjeant (1984) only attributed it provisionally to the genus *Cauca* Davey and Verdier, 1971, and as a result in cannot be taken confidently as the last representative of the complex.

Category of Gx-cysts

Batioladinium Complex J. (u.)–K. (ALB) Mar.

First: *Batioladinium imbatodinense* (Vozzhennikova) Lentin and Williams, 1985, Verkhe-Imbatskoi, Well 1, western Siberia, former USSR (Vozzhennikova, 1967).
Last: *Batioladinium jaegeri* (Alberti) Brideaux, 1975, *inflatum* Zone, Vallentigny, Paris Basin, France (Davey and Verdier, 1971).

Prolixosphaeridium Complex J. (BAJ)–Rec. Mar.

First: *Ellipsoidictyum reticulatum* (Valensi) Lentin and Williams, 1977, Bajocian, Normandy, France (Valensi, 1953). **Extant**
Comments: There is an apparent stratigraphical break in the distribution of members of the *Prolixosphaeridium* Complex between the Middle Miocene, when *Distatodinium fusiforme* (Matsuoka) Bujak and Matsuoka, 1976 last appeared (e.g. Toyoda Formation, Fujiwara Group, Nara City, Japan; Matsuoka, 1974), and the late Pliocene. The extant species *Polykrikos schwartzii* Bütschli, 1873 has been reported from Upper Pliocene and younger sediments, offshore Peru (Powell *et al.*, 1990).

Apteodinium Complex J. (PLB)–Q. (HOL) Mar./FW

First: *Mendicodinium reticulatum* Morgenroth, 1970, Lias delta ('Amaltheen-Schichten'), Stichkanal Hildesheim, Lühnde, Germany (Morgenroth, 1970).
Last: *Muiradinium dorsispirale* (Churchill and Sarjeant, 1962) Harland and Sarjeant, 1970, Flandrian peats, south-western Australia (Churchill and Sarjeant, 1962).
Comments: Fossil (Flandrian) freshwater dinoflagellate cysts belonging to the *Apteodinium* Complex have been reported from Victoria and Western Australia (see Harland and Sarjeant, 1970). It is unclear whether or not these forms are extant. Aside from these occurrences, the next youngest representative of the complex is *Apteodinium mecsekense* (Nagy) Helenes, 1984, described from upper Pannonian (i.e. Upper Miocene) sediments from Hungary (Nagy, 1969).

Scriniodinium Complex J. (CLV)–T. (AQT/?PIA) Mar.

First: *Dingodinium harsveldtii* Herngreen *et al.*, 1984, Achterhoek, eastern Netherlands (Herngreen *et al.*, 1984).
Last: *Lophocysta sulcolimbata* Manum, 1979, DSDP Site 338, Vöring Plateau, Norwegian Sea (Manum, 1979).
Comments: Aside from *D. harsveldtii*, *Scriniodinium crystallinum* (Deflandre) Klement, 1960 is also known from CLV sediments (e.g. Woollam and Riding, 1983). Harland (1979) recorded '*Thalassiphora delicata*' *auct. non* Williams and Downie, 1966, emend. Eaton, 1976 in sediments as young as late Pliocene from the Bay of Biscay. According to Edwards (1984), these records are referable to *Invertocysta lacrymosa* Edwards, 1984. This record may represent the last appearance of the *Scriniodinium* Complex.

Chlamydophorella Complex J. (TOA)–T. (RUP) Mar.

First: *Scriniocassis weberi* Gocht, 1964, Ziegeleigrube Frommern, near Balingen, Württemberg, Germany (Gocht, 1964).
Last: *Samlandia chlamydophora* Eisenack, 1965, Walsumer Schichten, Tönisberg borehole, Krefeld, Germany (Benedek, 1972).

Miscellaneous Complex of Gx-cysts J. (BAJ)–Rec. Mar.

First: *Aldorfia aldorfensis* (Gocht) Stover and Evitt, 1978, Upper Inferior Oolite, garantiana Zone, Dorset (Woollam and Riding, 1983). **Extant**
Comments: Members of this varied catch-all complex do not fit into any of the other categories of Gx-cysts. Cysts referred to as *Tectatodinium pellitum* Wall, 1967 are produced by the extant *Gonyaulax* sp. indet.

Category of M-cysts J. (KIM)/K. (CEN/CON)–Rec. Mar.

First: *Cryptarchaeodinium calcaratum* Deflandre, 1939, emend. Gitmez, 1970, Schistes bitumineux d'Orbagnoux, Jura, France (Deflandre, 1939). **Extant**
Comments: The category of M-cysts is a catch-all grouping which includes cysts with discernibly non-gonyaulacoid paratabulations. There is a stratigraphical gap between the KIM, after which *Cryptarchaeodinium* Deflandre, 1939, emend. Sarjeant, 1984 is unknown, and the CEN, when *Caligodinium aceras* (Manum and Cookson) Lentin and Williams, 1973 first appeared (Hassel Formation, Ellef Ringnes Island, Northwest Territories, Canada; Manum and Cookson, 1964). The extant *Pyrophacus* Stein, 1883 produces cysts referable to *Tuberculodinium* Wall, 1967. Evitt (1985) indicated that the extant forms *Diplopsalis* Bergh, 1882 and *Zygabikodiniurn* Loeblich and Loeblich, 1970 are also referable to this category.

Category of calcareous dinoflagellates J. (HET/KIM)–Rec. Mar.

First: *Schizosphaerella punctulata* Deflandre and Dangeard, 1938, HET–KIM (Lentin and Williams, 1989). **Extant**
Comments: Kalin and Bernoulli (1984) suggested that microfossil forms assigned to *Schizosphaerella* Deflandre and Dangeard, 1938 may be dinoflagellates (and not calcareous nannoplankton). Extant calcareous dinoflagellate cysts include those referable to *Scrippsiella* Balech, 1959, ex Loeblich III, 1965. An alternative arrangement of this groups is to divide it into two families, the Schizosphaeraceae Deflandre, 1959, with a range from HET–OXF, and the Thoracosphaeraeceae Schiller, 1930, emend. Tangen, 1982, with a range of NOR–Rec.

Category of siliceous dinoflagellates T. (Pal.)–Rec. Mar.

First: *Peridinites parvulus* Lefevre, 1933, Lower Tertiary, Barbados, Caribbean (Lefevre, 1933).
Comments: Lefevre (1933) described a number of species of *Peridinites* Lefevre, 1933 from Lower Tertiary sediments of Barbados. Representatives of the genus *Actiniscus* Ehrenberg, 1840 are extant.

Group ('Phylum') ACRITARCHA Evitt, 1963 (see Fig. 3.4)

The acritarchs are organic-walled microfossils of unknown and uncertain affinity. They are unicellular or apparently unicellular, although they may be found in clusters. The acritarchs are an informal, polyphyletic, organic-walled microfossil group, conceived as a holding category for a varied collection of *incertae sedis*. It is anticipated that as the affinities of certain acritarchs become firmly established, transfers can be made to the appropriate biotic classes. Many acritarchs are considered to be cysts or temporary resting stages of marine planktonic algae, although some are recorded in apparently lacustrine and terrestrial environments. A few have clear affinities with the Prasinophyta and Chlorophyta (Tappan, 1980). Fensome *et al.* (1990) note forms with questionable dinoflagellate affinity. A few have morphological similarities with amoeboid cysts, copepod eggs and masuelloids. To date, little progress has been made with the possible affinities of most acritarchs. At present, published and unpublished records suggest that well over 10 000 acritarch species are preserved in the fossil record. Fensome *et al.* (1990, 1991) list over 7000 in an index. Acritarchs are regularly recorded in abundance in routine palynological preparations, particularly in marine Palaeozoic sediments. Owing to their abundance and diversity, they are of particular importance in biostratigraphical correlation and palaeoenvironmental studies.

Authors have varied considerably in the adoption of possible biological affinities. Herein, acritarch records of forms considered by some workers to have affinities with known biotic groups are cross-referenced. Downie *et al.* (1963) divided the acritarchs into informal morphological subgroups, and parts of the scheme have been used by some other workers. Subsequent observations have noted that there is gradation between the spherical to subspherical vesicles of some acanthomorph acritarchs and the subpolygonal vesicles of some three-dimensional polygonomorph acritarchs, with the polygonal outline produced as a result of crushing during sediment compaction. A few species of both acanthomorph and herkomorph acritarchs have been observed enclosed within an outer laevigate sphaeromorph-type vesicle. Downie (1973, 1984) reviewed the classification of the acritarchs, and noted groupings on the basis of overall morphology, wall type and excystment openings. The inclusion of the sphaeromorph acritarchs in the informal group Cryptarcha, Diver and Peat, 1979 has not been generally adopted. The transfer of non-marine algal cysts, such as *Chlorella*, to the anteturma Cryptosporites, Richardson *et al.*, 1984 by Strother (1991) on the basis of an ecological subdivision is not seen as appropriate. Alete spores may be distinguished from acritarchs in having significant variations in wall thickness and colour within an individual. Generally, at low geothermal alteration, cryptospores are noticably darker in overall coloration than acritarchs with equivalent wall thickness.

None of the subdivision schemes for the acritarchs is entirely satisfactory, at least in terms of possible biological affinities, and many workers currently list all acritarchs, algal cysts and colonial algae together alphabetically when describing palynological assemblages.

The first record of acritarchs has been taken where possible from reference sections with independent age evidence. The last record of acritarchs, in common with most palynomorphs, is difficult to establish with certainty. Acritarchs are readily recycled during the erosion of significantly older sediments, and may prove particularly resistant to degradation if subjected to geothermal alteration (Dorning, 1986). In modern sediments, it is often difficult to distinguish with certainty between modern cysts and recycled forms derived from low-thermal alteration Mesozoic and Cainozoic sediments.

Most acritarchs fall into three main morphological categories; acritarchs without flanges or processes, acritarchs with flanges but no processes, and acritarchs with processes, with or without flanges. Taxonomic references may be found in Fensome *et al.* (1990).

1. ACRITARCHS WITHOUT PROCESSES OR FLANGES

This category includes the subgroup Sphaeromorphitae Downie *et al.*, 1963, the sphaeromorph acritarchs, together with the *Navifusa* group of Downie (1973). This grouping is almost certainly polyphyletic, including forms possibly attributable to the Chlorophyta, Prasinophyta, Cyanophyta and other algal and faunal groups. *Moyeria* Thusu, 1973 is considered by Gray and Boucot (1989) to be a euglenoid cyst.

Subgroup SPHAEROMORPHITAE Downie *et al.*, 1963 ?P€. (RIF/ANI)–Rec. Mar.

First: Spaeromorphs, 20–200 μm in diameter, from the Chuanlinggou Formation (1800 Ma), Jixian, China, are compared with later Precambrian *Kildinosphaera*, *Leiosphaeridia*, and *Chuaria*, and are considered to be the cysts of marine planktonic algae, the first possibly prasinophycean (see above) (Z. Zhang, 1986). **Extant**

2. ACRITARCHS WITH FLANGES, BUT WITHOUT PROCESSES

This category includes most forms in the subgroups Herkomorphitae Downie *et al.*, 1963 and Pteromorphitae Downie *et al.*, 1963. Some of these forms, including *Cymatiosphaera* and *Pterospermopsis* are related to the Prasinophyta (Tappan, 1980).

Subgroup HERKOMORPHITAE Downie *et al.*, 1963 C. (CRF)–Rec. Mar.

First: *Cymatiosphaera postii* Jankauskas, Lower Cambrian, Olenellid (Potter, unpubl.; Downie, 1984). **Extant**

Subgroup PTEROMORPHITAE Downie *et al.*, 1963 O. (CRD)–Rec. Mar.

First: *Pterospermopsis* sp., Cheney Longville Flags, Marshbrook Quarry, Shropshire, England, UK (Dorning, unpubl.). **Extant**

3. ACRITARCHS WITH PROCESSES, WITH OR WITHOUT FLANGES

This category includes many forms in the subgroups Acanthomorphitae Downie *et al.*, 1963, Diacromorphitae Downie *et al.*, 1963, Netromorphitae Downie *et al.*, 1963, Oomorphitae Downie *et al.*, 1963, Polygonomorphitae Downie *et al.*, 1963 and Prismatomorphitae Downie *et al.*,

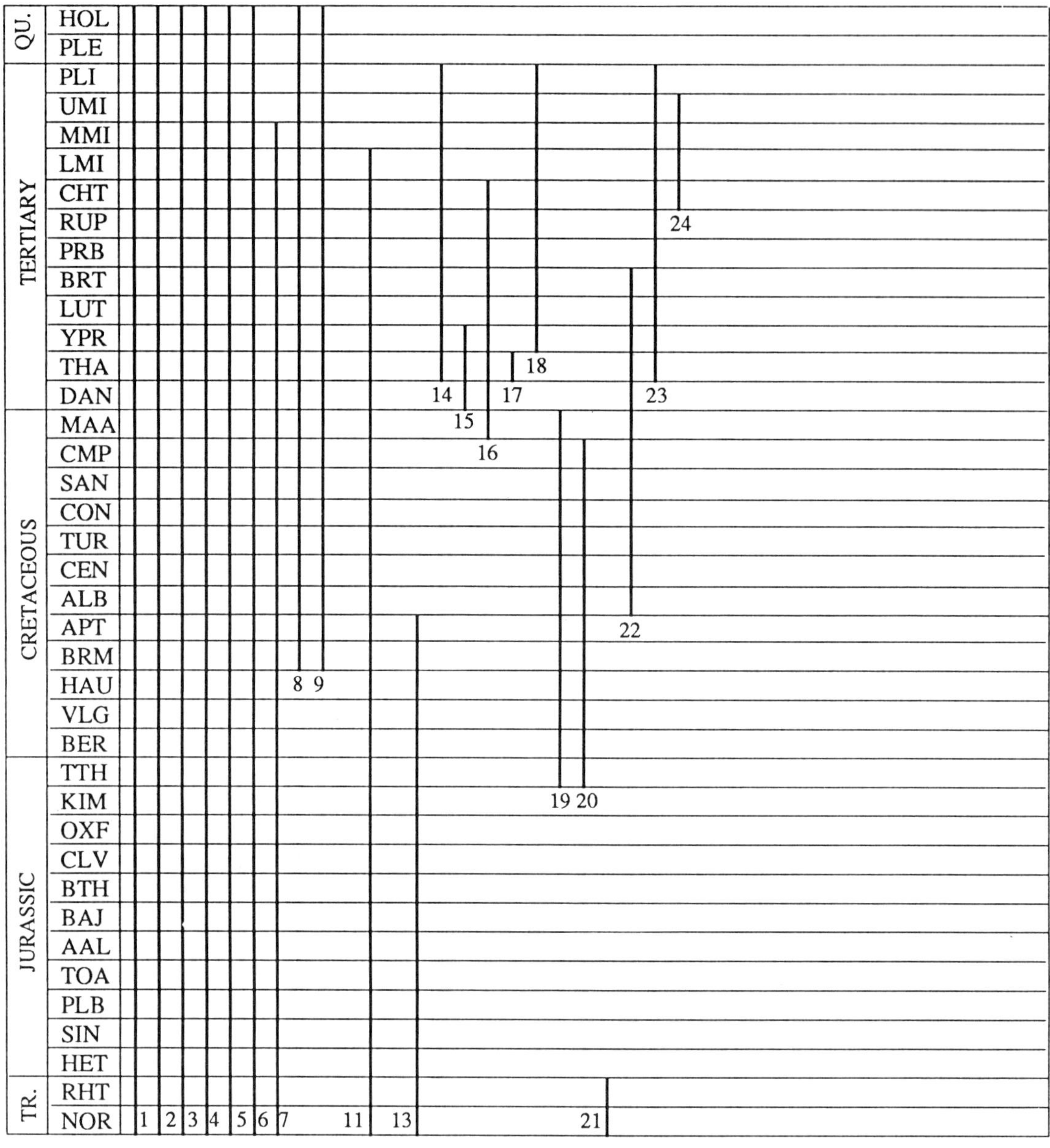

Fig. 3.4

1963. This group is almost certainly polyphyletic. Despite superficial resemblances of some to dinoflagellate cysts, the absence of consistent process-centred reflected tabulation and lack of stratigraphical continuity merely provides evidence of probable affinities of some within the marine microflora. The *Baltisphaeridium*, *Micrhystridium–Veryhachium*, *Leiofusa*, *Acanthodiacrodian* and *Visbysphaera* groups of Downie (1973) and the *Estiastra* and *Tunisphaeridium* groups of Dorning and Bell (1987) are in this category. A few of the very large forms may have affinities with the masuelloids (Order Muellerisphaerida Kozur, 1984).

Subgroup ACANTHOMORPHITAE Downie *et al.*, 1963 P€. (EDI)–Rec. Mar.

First: ?*Octaedryxium truncatum* Rudavskaya, 1973, Vendian, former USSR, or *Micrhystridium* spp. including *Micrhystridium tornatum* Volkova, 1968, uppermost Vendian–lowermost Cambrian.
Comment: Extant form is '*Baltisphaeridium*' from the Mediterranean (Rossignol, 1964).

Phylum RHODOPHYTA Wettstein, 1901

Class RHODOPHYCEAE Ruprecht, 1831

Subclass BANGIOPHYCIDAE de Toni, 1897, *orth. mut.* L. M. Newton, 1953

Order BANGIALES Engler, 1892

F. BANGIACEAE (Gray) Nägeli, 1847 P€.(YUR/STU)–Rec. Mar.

First: Unnamed multiseriate filaments preserved in chert, Hunting Formation, Somerset Island, Arctic Canada (1250–750 Ma: Butterfield *et al.*, 1990). **Extant**

F. PORPHYRIDACEAE Kylin, 1937 S. (?GOR)–Rec. Mar.

First: *Palaeoconchocelis starmachii* Campbell *et al.*, 1979 in Widowo core, Bielks Podlaski, eastern Poland, Lower Ludlow. Formation not given (Campbell *et al.*, 1979). **Extant**

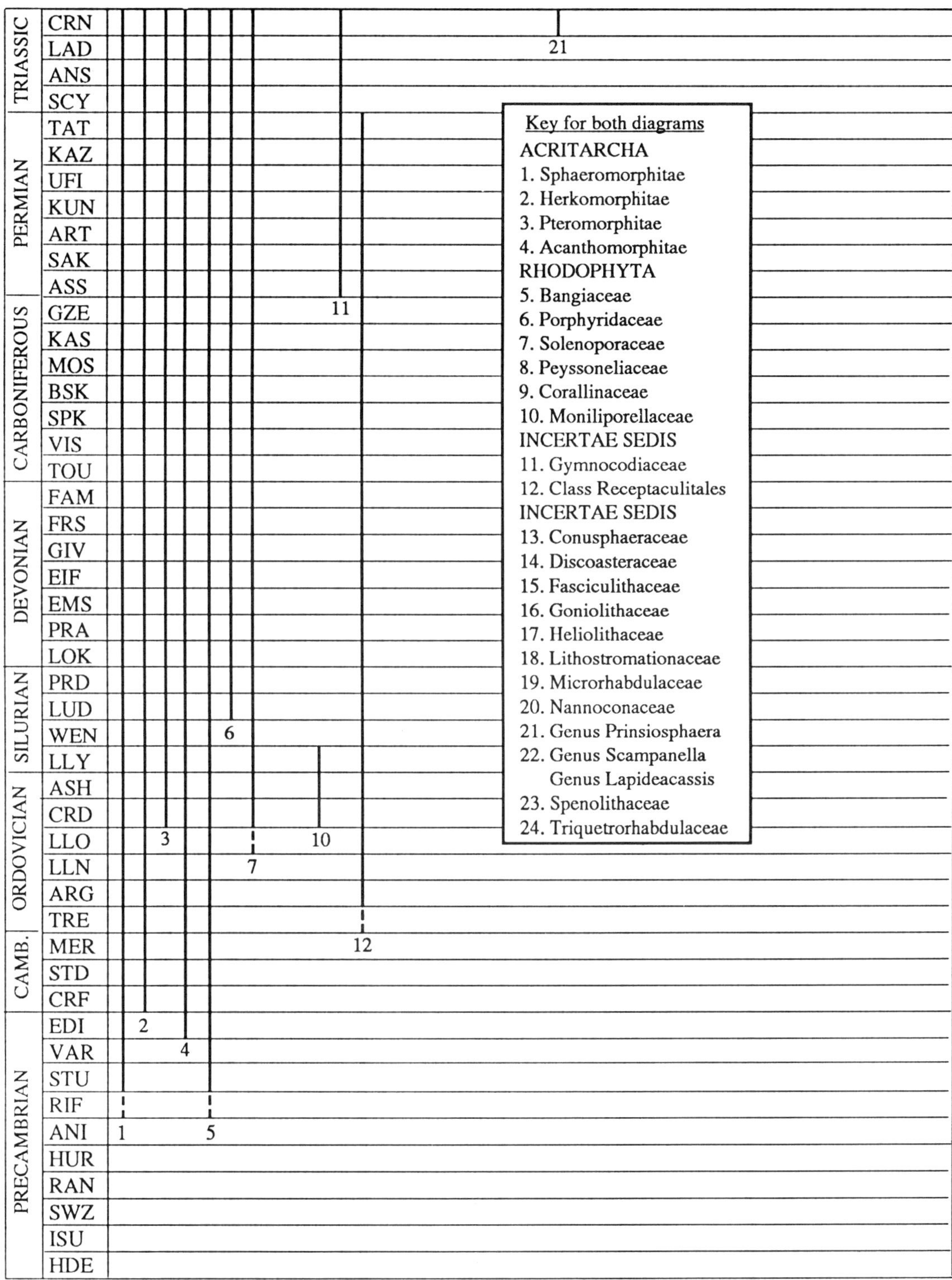

Fig. 3.4

Comment: Conchocelis stage of an endolithic microalga was compared with the modern bangiacean rhodophyte *Porphyra nereocystis*. Tappan (1980) cited several probable Precambrian cf. Porphyridaceae, the earliest being *Clonophycus* J. Oehler 1977 from the 1500–1600 Ma McArthur Group, although Oehler (1977) himself and Mendelson and Schopf consider it to be a colonial coccoid (?chroococcacean) cyanobacterium.

Subclass FLORIDEOPHYCIDAE (Lamouroux, 1813) Engler, 1892

Record is centred on calcified forms, there being insufficient characters to permit identification of foliose forms (e.g. Fry, 1983 for Ordovician problematica). However phosphatized *Thallophyca ramosa* Zhang, 1989 from the 680 Ma Doushantuo Formation, South China, shows remarkable fountain-type cellular preservation compared with that in modern florideophyceans (Y. Zhang, 1989).

* *Order* CRYPTONEMIALES Schmitz, *in* Engler, 1892

F. SOLENOPORACEAE Pia, 1927 O. (LLO/u.)–T. (MMI) Mar.

First: *Petrophyton* kiaeri Høeg, 1932, Kalstad Limestone, Upper or upper Middle Ordovician, Meldal, Trondheim area, Norway (Høeg, 1932, p. 82).
Last: *Neosolenopora patrini* Mastrorilli, 1955, Middle Miocene (Helvetian), Pavia area (Mastrorilli, 1955). Originally described as *Lithophyllum vinassai* by Patrini (1932). Also see Elliott (1965).
Comment: Unlike *Solenopora*, *Petrophyton* has well-developed cross-partitions and lacks septa-like projections, and is on this basis more likely to be an alga. The larger cells and absence of sporangia distinguish it from corallinaceans. See Johnson and Høeg (1961, pp. 22, 113).

* F. PEYSSONELIACEAE Denisot, 1968 K. (BRM)–Rec. Mar.

First: *Pseudolithothamnium album* Pfender, 1936, Lower Cretaceous (Barremian) (Moussavian, 1988).
Comment: Some Pennsylvanian phylloid algae have been compared with Squamariaceae (=Peyssoneliaceae) (Wray, 1977, p. 53). Also see Corallinaceae below.

* F. CORALLINACEAE (Lamouroux) Harvey, 1849 K. (BRM)–Rec. Mar.

First: *Sporolithon rude* Lemoine, 1925, Lower Cretaceous (Barremian), southern France.
Comment: Several coralline species appear for the first time in the Barremian. In addition to *S. rude*, these are *Kymalithon belgicum* Foslie, 1909 (Lemoine, 1970) and *Parakymalithon phylloideum* Bucur and Dragastan, 1985 (Moussavian, 1987; E. Moussavian pers. comm., 1992). Published reports of Jurassic (Johnson, 1961, p. 48; Johnson 1962, p. 19) and even Upper Triassic (Zankl, 1969; Elliott, 1979), corallinaceans appear to be misidentifications of *Thaumatoporella* (as Lemoine, 1970 suggests), or of solenoporaceans or *Marinella*. However, there are undescribed Upper Triassic red algae which resemble corallines and peyssoneliaceans (E. Moussavian pers. comm., 1992). The phylloid alga (see below) *Archaeolithophyllum* from the Pennsylvanian is likely a rhodophyte but is not a coralline.

Phylum ?RHODOPHYTA

* F. MONILIPORELLACEAE Gnilovskaya, 1972 O. (CRD)–S. (LLY) Mar.

First: *Contexta binata* Gnilovskaya, 1972 and *Ansoporella ansa* Gnilovskaya, 1972, Bestamakskaya Suite (upper part), (upper part of lower Caradoc), eastern Kazakhstan, Chingiz, right bank of the River Chagan, former USSR (Gnilovskaya, 1972, pp. 110–12, 117–19, and table p. 166).
Last: *Moniliporella halysitoides* Gnilovskaya, 1972, (Llandovery), eastern Kazakhstan, Chingiz, Kokaiykir Mountain, former USSR (Gnilovskaya, 1972, pp. 106–8). Chuvashov and Riding, 1984 (Fig. 1) show Moniliporellaceae ranging to end-Devonian, but do not give details.

INCERTAE SEDIS 1

* F. GYMNOCODIACEAE Elliott, 1955 P. (ASS)–T. (BUR) Mar.

First: Lower Permian *Permocalculus* cf. *P. tenellus* (Pia), Upper *Pseudoschwagerina* Limestone, Rattendorf Stage, Austria (Flügel, 1966, p. 16).
Last: *Permocalculus iagifuensis* Simmons and Johnston, 1991, Lower Miocene (Burdigalian), Darai Limestone Formation, Papua New Guinea (Simmons and Johnston, 1991).

* Class RECEPTACULITALES Sushkin, 1962 O. (TRE/ARG)–P. Mar.

First: *Calathella* Rauff, 1894: *Calathella* sp., Pogonip Group (TRE–ARG), Nevada, USA (Nitecki and Debrenne, 1979, pl. 4); and *Calathium* Billings 1865: *Calathium* sp., El Paso Group (ARG), West Texas, (Toomey, 1970, pp. 1323–4, Fig. 10).
Last: Unidentified genus, Permian, Sicily (Parona, 1933). Also see Rietschel and Nitecki (1984, p. 415).

INCERTAE SEDIS 2

The groups included here are generally extinct taxa which had calcite tests within the size limits of calcareous nannoplankton but with morphologies which are distinct from members of the Coccosphaerales.

F. CONUSPHAERACEAE (informal) Tr. (NOR)–K. (APT) Mar.

First: *Eoconusphaera zlambachensis* Jafar, 1983, Upper NOR.
Last: *Conusphaera rothii* (Thierstein, 1971), lower APT.
Intervening: RHT, SIN–lower TOA, U. KIM–BRM.

F. DISCOASTERACEAE Tan, 1927 T. (THA–PIA) Mar.

First: *Discoasteroides bramlettei* Bukry and Percival, 1971.
Last: *Discoaster brouweri* (Tan, 1927), upper PIA.
Intervening: YPR–ZAN.

F. FASCICULITHACEAE Hay and Mohler, 1967 T. (DAN–YPR) Mar.

First: *Fasciculithus magnus* Bukry and Percival, 1971.
Last: *Fasciculithus thomasii* Perch-Nielsen, 1971.
Intervening: THA.

F. GONIOLITHACEAE Deflandre, 1957 K. (MAA)–T. (CHT) Mar.

First: *Goniolithus fluckigeri* Deflandre, 1957.
Last: *Goniolithus fluckigeri* Deflandre, 1957.
Intervening: DAN–RUP.
Comments: Rare and sporadic stratigraphical distribution.

F. HELIOLITHACEAE Hay and Mohler, 1967 T. (THA) Mar.

First: *Heliolithus elegans* (Roth, 1973).
Last: *Heliolithus megastypus* (Bramlette and Sullivan, 1961).

F. LITHOSTROMATIONACEAE Deflandre, 1959 T. (YPR–ZAN) Mar.

First: *Trochoaster simplex* Klumpp, 1953.
Last: *Trochoaster deflandrei* (Stradner, 1959).
Intervening: BRT–MES.

F. MICRORHABDULACEAE Deflandre, 1963 J. (TTH)–K. (MAA) Mar.

First: *Pseudolithraphidites multibacillatus* Keupp, 1976?
Last: *Lithraphidites kennethii* Perch-Nielsen, 1984, upper MAA.

Intervening: BER–CMP.

F. NANNOCONACEAE Deflandre, 1959 J. (TTH)–K. (CMP) Mar.

First: *Nannoconus compressus* Bralower and Thierstein, *in* Bralower *et al.*, 1989, upper TTH.
Last: ?*Nannoconus elongatus* Bronnimann, 1955, upper CMP.
Intervening: BER–SAN.

Genus PRINSIOSPHAERA Jafar, 1983 Tr. (CRN–RHT) Mar.

First: *Prinsiosphaera triassica* Jafar, 1983.
Last: *Prinsiosphaera triassica* Jafar, 1983, upper RHT.
Intervening: NOR.

Genus SCAMPANELLA Forchheimer and Stradner, 1973 and *Genus* LAPIDEACASSIS Black, 1971 K. (ALB)–T. (BRT) Mar.

First: *Lapideacassis mariae* Black, 1971?
Last: *Scampanella bispinosa* Perch-Nielsen, 1977
Intervening: CEN–LUT.
Comments: Extremely rare and sporadic stratigraphical distribution.

F. SPHENOLITHACEAE Deflandre, 1952 T. (THA–PIA) Mar.

First: *Sphenolithus primus* Perch-Nielsen, 1971.
Last: *Sphenolithus neoabies* Bukry and Bramlette, 1969.
Intervening: YPR–ZAN.

F. TRIQUETRORHABDULACEAE Lipps, 1969 T. (CHT–MES) Mar.

First: *Triquetrorhabdulus carinatus* Martini, 1965.
Last: *Triquetrorhabdulus striatus* Muller, 1974.
Intervening: AQT–TOR.

REFERENCES

Note: References cited in the dinoflagellate section that are not given here may be found in Lentin and Williams (1989).

Baldauf, J.G. and Monjanel, A. L. (1989) An Oligocene diatom biostratigraphy for the Labrador Sea. *DSDP Site 112 and ODP Hole 105*, 323–47.

Barron, J. A. (1985) Diatom biostratigraphy of the CESAR 6 core, Alpha Ridge, in *Initial Geological Report on CESAR: The Canadian Expedition to Study the Alpha Ridge, Arctic Ocean* (eds H. R. Jackson, P. J. Mudie and S. M. Blasco), Geological Survey of Canada. Ottawa, Paper, **84–22**, 137–48.

Baschnagel, R. (1966) New fossil algae from the Middle Devonian of New York. *Transactions of the American Microscopical Society*, **85**, 297–302.

Biffi, U. and Manum, S. B. (1988) Late Eocene–Early Miocene dinoflagellate cyst stratigraphy from the Marche Region (Central Italy). *Bolletino della Società Paleontologica Italiana*, **27**, 163–212.

Bradley, W. H. (1929) Freshwater algae from the Green River Formation of Colorado. *Bulletin of the Torrey Botanical Club*, **56**, 421–8.

Brinkhuis, H. and Zachariasse, W. J. (1988) Dinoflagellate cysts, sea level changes and planktonic foraminifera across the Cretaceous–Tertiary boundary at El Haria, northwest Tunisia. *Marine Micropaleontology*, **13**, 153–91.

Bujak, J. P. and Williams, G. L. (1979) Dinoflagellate diversity through time. *Marine Micropaleontology*, **4**, 1–12.

Butterfield, N. J., Knoll, A. H. and Swett, K. (1988) Exceptional preservation of fossils in an Upper Proterozoic Shale. *Nature*, **334**, 424–7.

Butterfield, N. J., Knoll, A. H. and Swett, K. (1990) A bangiophyte red alga from the Proterozoic of arctic Canada. *Science*, **250**, 104–7.

Campbell, S., Kazmierczak, J. and Golubic, S. (1979) *Palaeoconchocelis starmachii* gen. n., sp. n., an endolithic rhodophyte (Bangiaceae) from the Silurian of Poland. *Acta Palaeontologica Polonica*, **24**, 405–8.

Chadefaud, M. (1950) Les cellules nageuses des algues dans l'embranchement des Chlorophycées. *Comptes Rendus Hebdomadaires des Séances de l'Académie des Sciences, Paris*, **231**, 989–90.

Chuvashov, B. and Riding, R. (1984) Principal floras of Palaeozoic calcareous algae. *Palaeontology*, **27**, 487–500.

Deflandre, G. (1941) Sur la présence de diatomées dans certains silex creaux Turoniens et sur un nouveau mode de fossilisation de ces organismes. *Comptes Rendus de l'Académie des Sciences, Paris*, **213**, 878–80.

Dorning, K. J. (1986) Organic microfossil geothermal alteration and interpretation of regional tectonic provinces. *Journal of the Geological Society*, **143**, 219–20.

Dorning, K. J. and Bell, D. G. (1987) The Silurian carbonate shelf microflora: acritarch distribution in the Much Wenlock Limestone Formation, in *Micropalaeontology of Carbonate Environments* (ed. M. B. Hart), Ellis Horwood, Chichester, pp. 266–87.

Downie, C. (1973) Observations on the nature of the acritarchs. *Palaeontology*, **16**, 239–59.

Downie, C. (1984) Acritarchs in British stratigraphy. *Geological Society Special Report*, **17**, 1–26.

Downie, C., Evitt, W. R. and Sarjeant, W. A. S. (1963) Dinoflagellates, hystrichospheres and the classification of the acritarchs. *Stanford University Publications (Geological Sciences)*, **7**, 1–16.

Edwards, D. S. and Lyon, A. G. (1983) Algae from the Rhynie Chert. *Botanical Journal of the Linnean Society*, **86**, 37–55.

Elliott, G. F. (1965) Tertiary solenoporacean algae and the reproductive structures of the solenoporaceae. *Palaeontology*, **7**, 695–702.

Elliott, G. F. (1979) Influences of organic reefs on the evolution of post-Palaeozoic algae. *Geological Magazine*, **116**, 375–83.

Engler, A. (1954) *Syllabus der Pflanzenfamilien*, 12th edn, Bd. 1, Gebrüder Borntraeger, Berlin.

Evitt, W. R. (1985) *Sporopollenin Dinoflagellate Cysts–their Morphology and Interpretation*. American Association of Stratigraphic Palynologists Foundation, Dallas, TX, xv + 333 pp.

Fairchild, T. R., Schopf, J. W. and Folk, R. L. (1973) Filamentous algal microfossils from the Caballos novaculite, Devonian of Texas. *Journal of Paleontology*, **47**, 946–52.

Feist, M. and Grambast-Fessard, N. (1990) The genus concept in Charophyta, from Palaeozoic to Recent. *Lecture Notes in Earth Sciences*, Springer Verlag (in press).

Feist, M., Bhatia, S. B. and Yadagiri, P. (1991) On the oldest representative of the family Characeae and its relationships with the Porocharaceae. *Bulletin de la Société Botanique de France*, **138** (1), 25–32.

Fenner, J. (1977) Cenozoic diatom biostratigraphy of the equatorial and southern Atlantic Ocean. *Initial Reports of the DSDP*, **39**, 491–624.

Fenner, J. (1981) *Diatoms in Eocene and Oligocene sediments off NW Africa–Their Stratigraphic and Paleographic Occurrences*. Dissertation zur Erlangung Doktorgrades MathNaturwiss., Fakultat Christian-Albrecht's-Universitat zu Kiel.

Fensome, R. A., Taylor, F. J. R., Sarjeant, W. A. S. *et al.* (1989) A classification of fossil and living dinoflagellates 44, in *Fourth International Conference on Modern and Fossil Dinoflagellates. Program and Abstracts*. Marine Biological Laboratory, Wood's Hole, Massachusetts.

Fensome, R. A., Williams, G. L., Barss, M. S. *et al.* (1990) Acritarchs and fossil prasinophytes: an index to genera, species and infraspecific taxa. *AASP Contributions Series*, **25**, 1–771.

Fensome, R. A., Williams, G. L., Barss, M. S. *et al.* (1991) Alphabetical listing of acritarch and fossil prasinophyte species. *AASP Contributions Series*, **26**, 1–111.

Fernholm, B., Bremer, K., Brundin, L. *et al.* (1989) *The Hierarchy of Life: Molecules and Morphology in Phylogenetic Analysis*. Excerpta Medica, Amsterdam, 499 pp.

Flügel, E. (1966) Algen aus dem Perm der Karnischen Alpen. *Carinthia II: Sonderheft*, **25**, 3–76.

Forti, A. and Schulz, P. (1932) Erste Mitteilung über Diatomeen aus dem hannoverschen Gault. *Beihefts zur Botanik, Centralblatt*, **50**, 241–6.

Foucher, J.-C. (1974) Microfossiles des silex Turonien supérieur du Ruyaulcourt (Pas-de-Calais). *Annales de Paléontologie (Invertébrés)*, **60**, 113–64.

Foucher, J.-C. (1979) Distribution stratigraphiques de kystes de dinoflagellés et des acritarches dans le Crétacé supérieur du Bassin de Paris et de l'Europe septentrionale. *Palaeontographica, Abteilung B*, **169**, 78–105.

Foucher, J.-C. (1981) Kystes de dinoflagellés du Crétacé Moyen Européen: proposition d'une échelle biostratigraphique pour le Domaine Nord-occidental. *Cretaceous Research*, **2**, 331–8.

Fry, W. L. (1983) An algal flora from the Upper Ordovician of the Lake Winnipeg Region, Manitoba, Canada. *Review of Palaeobotany and Palynology*, **39**, 313–41.

Fry, W. L. and Banks, H. P. (1955) Three new genera of algae from the Upper Devonian of New York. *Journal of Paleontology*, **29**, 37–44.

Fryxell, G., Sims, P. and Watkins, P. (1986) *Azpeitia* (Bacillariophyceae). Related genera and promorphology. *Systematic Botany Monographs*, **13**, 1–74.

Gartner, S. (1968) Coccoliths and related calcareous nannofossils from Upper Cretaceous deposits of Texas and Arkansas. *University of Kansas Paleontological Contributions*, **48**, 1–56.

Geroch, S. (1978) Lower Cretaceous diatoms in the Polish Carpathians. *Rocznik, Polskiego Towarzvstwa Geolgicznego*, **48**, 283–95.

Gersonde, R. and Harwood, D. (1990) Lower Cretaceous diatoms from ODP Leg 113 Site 693 (Weddell Sea). Part 1: Vegetative cells, in *Initial Reports ODP 113* P. Barker, J. Kennett *et al.*, pp. 365–402.

Gnilovskaya, M. B. (1972) *Calcareous Algae from the Middle and Late Ordovician of Eastern Kazakhstan*. USSR Acad. Sci., Inst. Precambrian Geology and Geochronology, Nauka, Leningrad, 196 pp. [in Russian].

Gnilovskaya, M. B. (1983) Vendotaenids, In *Upper Precambrian and Cambrian Palaeontology of the East European Platform* (eds A. Urbanek and A. Y. Rozanov), Publishing House Wychdwnictwa Geologiczne, Warsaw, pp. 45–56.

Gombos, A. M. Jr (1980) The early history of the diatom family Asterolampraceae. *Bacillaria*, **3**, 227–72.

Grambast, L. (1962) Classification de l'embranchement des Charophytes. *Naturalia Monspel.*, **14**, 63–86.

Grambast, L. (1967) Charophyta, in *The Fossil Record* (eds W. B. Harland *et al.*), Geological Society of London, pp. 216–17.

Grambast, L. (1974) Phylogeny of the Charophyta. *Taxon*, **23**, 463–81.

Gray, J. and Boucot, A. J. (1989) Is *Moyeria* a euglenoid? *Lethaia*, **22**, 447–56.

Hajos, M. and Stradner, H. (1975) Late Cretaceous Archaeomonadaceae, Diatomaceae, and Silicoflagellatae from the South Pacific Ocean, Deep Sea Drilling Project Leg 29, Site 275, In *Initial Reports DSDP 29*. (ed. J. P. Kennett, R. E. Houtz *et al.*), US Government Printing Office, Washington, pp. 913–1109.

Harwood, D. M. (1988) Upper Cretaceous and lower Paleocene diatom and silicoflagellate biostratigraphy of Seymour Island, eastern Antarctic Peninsula. *Geological Society of America Memoir*, **169**, 55–129.

Harwood, D. M. and Gersonde, R. (1990) Lower Cretaceous diatoms from ODP Leg 113 (Weddell Sea). Part 2: Resting Spores, Chrysophycean cysts, endoskeletal dinoflagellates, and notes on the origin of diatoms. In *Initial Reports ODP 113*, (eds P. Barker, J. Kennett *et al.*), pp. 402–14.

Helby, R., Morgan, R. and Partridge, A.D. (1987) A palynological zonation of the Australian Mesozoic, in *Studies in Australian Mesozoic Palynology* (ed. P. A. Jell), Association of Australian Palaeontologists, Sydney, pp. 1–94.

Høeg, O. (1932) Ordovician algae from the Trondheim area. *Skrifter utgitt av Det Norske Videnskaps-Akademi i Oslo. I Mat.–Naturv. Klasse*, **4**, 63–96.

Hofmann, H. J. (1985) The mid-Proterozoic Little Dal Macrobiota, Mackenzie Mountains, north-west Canada. *Palaeontology*, **28**, 331–54.

Hofmann, H. J. and Chen, J. (1981) Carbonaceous megafossils from the Precambrian (1800 Ma) near Jixian, northern China. *Canadian Journal of Earth Sciences*, **16**, 150–66.

Ishchenko, T. A. and Ishchenko, A. A. (1982) Novaia nakhodka kharophitov v verkhniem silure Podolii, in *Sistematika i Evolutsia Drevnik Rastenii Ukraini*. Nauk Dumka, Kiev, pp. 21–32.

Jansonius, J. (1962) Palynology of Permian and Triassic sediments, Peace River area, western Canada. *Palaeontographica, Abteilung B*, **110**, 35–98.

Johnson, J. H. (1954) An introduction to the study of rock-building algae and algal limestones. *Colorado School of Mines, Quarterly*, **49** (2), 117 pp.

Johnson, J. H. (1961) *Limestone-building Algae and Algal Limestones*. Colorado School of Mines, Boulder, 297 pp.

Johnson, J. H. (1962) The algal genus *Lithothamnium* and its fossil representatives. *Colorado School of Mines, Quarterly*, **57** (1), 111 pp.

Johnson, J. H. and Høeg, O. (1961) Studies of Ordovician algae. *Colorado School of Mines, Quarterly* **56** (2), 1–120.

Jouse, A. P. (1949) New upper Cretaceous diatoms and silicoflagellate from argillaceous sands along the Bol'shoi kitoi River, east slope of the northern Urals. *Botan. Mater. Otd. Spor. Rast. Bot. Inst. Akad. Nauk. SSSR*, **6**, 1–6.

Klebs, G. (1883) Über die Organisation einiger Flagellaten-Gruppen und ihre Beziehungen zu Algen und Infusorien. *Untersuchungen der Botanischer Institut, Tübingen*, **1**, 233–361.

Knoll, A. H., Swett, K. and Mark, J. (1991) Paleobiology of a Neoproterozoic tidal flat/lagoonal complex: the Draken Conglomerate Formation, Spitsbergen. *Journal of Paleontology*, **65**, 531–70.

Konzalova, M. (1973) Algal colony and rests of other microorganisms in the Bohemian Upper Proterozoic, *Vèstnik Ustrédiho ústavu Geologického*, **48**, 31–4.

Korde, K. B. (1957) New representatives of siphonous algae. *Mater. Principles Palaeontol.*, **1**, 67–75 [in Russian.]

Lauby, A. (1910) Recherches paléophytologiques dans le Massif Central. *Bulletin du Service de la Carte géologique de France*, **20**, 1–398.

Leary, R. L. (1986) Three new genera of fossil noncalcareous algae from Valmeyeran (Mississippian) strata of Illinois. *American Journal of Botany*, **73**, 369–75.

Leedale, G. F. (1967) *Euglenoid flagellates*. Prentice Hall, Englewood Cliffs, New Jersey, 242 pp.

Lemoine, M. (1970) Les algues floridées calcaires du Crétacé du Sud de la France. *Mémoires du Muséum Nationale d'Histoire Naturelle, Paris*, **7** (10), 127–240.

Lentin, J. K. and Williams, G. L. (1989) Fossil dinoflagellate index to genera and species 1989 Edition. *American Association of Stratigraphic Palynologists Contribution Series*, **20**, vi + 1–473.

Manton, I. and Leedale, G. F. (1969) Observations on the microanatomy of *Coccolithus pelagicus* and *Cricosphaera carterae*, with special reference to the origin and nature of coccoliths and scales. *Journal of the Marine Biological Association of the UK*, **49**, 1–16.

Martini, E. (1971) Standard Tertiary and Quaternary calcareous nannoplankton zonation, in *Proceedings of the 2nd Planktonic Conference, Roma, 1970* (ed. A. Farinacci), Tecnoscienza, Roma, 2, pp. 739–85.

Mastrorilli, V. I. (1955) Sui noduli fossiliferi di M. Vallassa (Appennino Pavese). *Atti del'Istituto Geologia Universita Pavia*, **6**, 61–74.

Mendelson, C. V. and Schopf, J. W. (1992) Proterozoic and selected Early Cambrian microfossils and microfossil-like objects, in *The Proterozoic Biosphere. A Multidisciplinary Study*. (eds J. W. Schopf and C. Klein), Cambridge University Press, New York, pp. 865–951.

Mojon, P. O. (1989) Charophytes et ostracodes laguno-lacustres du Jurassique de la Bourgogne (Bathonien) et du Jura septentrional franco-suisse (Oxfordien). *Revue de Paléobiologie, Volume Spéciale*, **3**, 1–18.

Morbey, S. J. (1975) Late Triassic and Early Jurassic subsurface palynostratigraphy in Northwestern Europe. *Palinologia, Número Extraordinaario*, **1**, 355–65.

Moussavian, R. (1987) *Parakymalithon*, eine neue Gattung der Corallinaceen (Rhodophyceen) aus der Unterkreide. *Facies*, **16**, 187–94.

Moussavian, R. (1988) Die Peyssonneliaceen (auct.: Squamariaceae; Rhodophyceae) der Kreide und des Paläogen der Ostalpen. *Mitteilungen der Bayerisches Staatssammlung für Paläontologie und Historische Geologie*, **28**, 89–124.

Nitecki, M. H. and Debrenne, F. (1979) The nature of radiocyathids and their relationship to receptaculitids and archaeocyathids. *Geobios*, **12**, 5–27.

Oehler, J. H. (1977) Microflora of the H. Y. C. Pyritic Shale Member of the Barney Creek Formation (McArthur Group), Middle Proterozoic of northern Australia. *Alcheringa*, **1**, 315–49.

Parona, C. F. (1933) Le spugne della fauna permiane di Palazzo Adriano (Bacino del Sosio) in Sicilia. *Memorie della Società Geologia Italiana*, **1**, 1–58.

Patrini, P. (1932) Su di un nuovo litofillo miocenico. *Rivista Italiana di Paleontologia*, **38**, 53–60.

Peck, R. E. and Eyer, J. A. (1963a) Representatives of *Chovanella*, a Devonian charophyte in North America. *Micropalaeontology*, **9**, 97–100.

Peck, R. E. and Eyer, J. A. (1963b) Pennsylvanian, Permian and Triassic Charophyta of North America. *Journal of Paleontology*, **37**, 835–44.

Perch-Nielsen, K. (1985) Mesozoic calcareous nannofossils, Cenozoic calcareous nannofossils, in *Plankton Stratigraphy* (eds H. M. Bolli, J. B. Saunders and K. Perch-Nielsen), Cambridge University Press, pp. 329–554.

Powell, A. J. (1986a) A dinoflagellate cyst biozonation for the late Oligocene to middle Miocene succession of the Langhe region, northwest Italy. *American Association of Stratigraphic Palynologists Contribution Series*, **17**, 105–27.

Powell, A. J. (1986b) The stratigraphic distribution of late Miocene dinoflagellate cysts from the Castellanian superstage stratotype, northwest Italy. *American Association of Stratigraphic Palynologists Contribution Series*, **17**, 129–49.

Powell, A. J., Dodge, J. D. and Lewis, J. (1990) Late Neogene to Pleistocene palynological facies of the Peruvian Continental Margin, Leg 112, in *Proceedings of the Ocean Drilling Program, Scientific Results*, E. Suess, R. Von Huene *et al.*), 112, Ocean Drilling Program, College Station, Texas, pp. 297–321.

Riding, R. (1991) Cambrian calcareous cyanobacteria and algae, in *Calcareous Algae and Stromatolites* (ed. R. Riding), Springer Verlag, Berlin, pp. 305–34.

Rietschel, S. and Nitecki, M. H. (1984) Ordovician receptaculitid algae from Burma. *Palaeontology*, **27**, 415–20.

Robaszynski, F., Bless, M. J. M., Felder, P. J. *et al.* (1985) The Campanian–Maastrichtian boundary in the chalky facies close to the type-Maastrichtian area. *Bulletin des Centres de Recherche Exploration–Production Elf-Aquitaine*, **9**, 1–113.

Rossignol, M. (1964) Hystrichosphères du Quaternaire en Méditerranée orientale, dans les sédiments Pléistocènes et les boues marines actuelles. *Revue de Micropaléontologie*, **7**, 83–99.

Rothpletz, A. (1896) Uber die Flysch Fucoiden und einige andere fossile Algen, sowie über liassische diatomeenführende Hornschwämme. *Zeitschrift des Deutsch geologische Gesellschaft*, **48**, 854-914.

Round, F. E. (1973) *The Biology of the Algae*, 2nd edn, Edward Arnold, London.

Round, F. E. (1980) The taxonomy of the Chlorophyta. II. *British Phycological Journal*, **6**, 235–264.

Round, F. E., Crawford, R. and Mann, D. (1990) *The Diatoms. Biology and Morphology of the Genera*. Cambridge University Press, Cambridge.

Sarjeant, W. A. S. and Downie, C. (1967) Class Dinophyceae Pascher, in *The Fossil Record–a symposium with documentation* (eds W. B. Harland *et al.*), Geological Society of London, London, pp. 195–207.

Schopf, J. W., Dolnik, T. A., Krylov, I. N. *et al.* (1977) Six new stromatolitic microbiotas from the Proterozoic of the Soviet Union. *Precambrian Research*, **4**, 269–84.

Simmons, M. D. and Johnston, M. J. (1991) *Permocalculus iagifuensis* sp. nov.: a new Miocene gymnocodiacean alga from Papua New Guinea. *Journal of Micropalaeontology*, **9**, 238–44.

Simonsen, R. (1970) Protoraphidaceae, eine neue Familie der Diatomeen. *Nova Hedwigia, Beiheften*, **31**, 383–413.

Simonsen, R. (1979) The diatom system: ideas on phylogeny. *Bacillaria*, **2**, 9–63.

Sims, P. A., Fryxell, G. A. and Baldauf, J. G. (1989) Critical examination of the diatom genus *Azpeitia*: species useful as stratigraphic markers for the Oligocene and Miocene Epochs. *Micropaleontology*, **35**, 293–307.

Stanier, R. Y., Sistrom, W. R., Hansen, T. A. *et al.* (1978) Proposal to place nomenclature of the Cyanobacteria (blue-green algae) under the rules of the Code of Nomenclature of Bacteria. *International Journal of Systematic Bacteriology*, **28**, 335–6.

Strelnikova, N. I. (1975) Diatoms of the Cretaceous Period, in *Third Symposium on Recent and Fossil Diatoms, Kiel. Nova Hedwigia*, **53**, (ed. R. Simonsen), pp. 311–21.

Strother, P. K. (1991) A classification schema for the cryptospores. *Palynology*, **15**, 219–36.

Tappan, H. (1980) *The Paleobiology of Plant Protists*. W. H. Freeman, San Francisco, 1028 pp.

Toomey, D. F. (1970) An unhurried look at a Lower Ordovician mound horizon, southern Franklin Mountains, West Texas. *Journal of Sedimentary Petrology*, **40**, 1318–34.

Vekshina, V. N. (1960) Diatomovie vodordsli vershneyurskich otozhenii zapadnpsibirskoi nizmennosti. *Trudi Sib. NAUK ISSL. in-ta geol. geofiz. i. miner. Ser.*, **8**, 160–2.

Vidal, G. (1989) Are late Proterozoic carbonaceous megafossils metaphytic algae or bacteria? *Lethaia*, **22**, 375–79.

Walter, M. S., Oehler, J. H. and Oehler, D. Z. (1976) Megascopic algae 1300 million years old from the Belt Supergroup, Montana: a reinterpretation of Walcott's *Helminthoidichnites*. *Journal of Paleontology*, **50**, 872–81.

Wang, Z. (1984) Two new charophyte genera from the Upper Permian and their bearing on the phylogeny and classification of Charales and Trochilischales. *Acta Micropalaeontologia Sinica*, **1**, 49–60 [in Chinese].

Wang, Z. and Lu, H. N. (1980) New discovery of Devonian charophytes from South China with special reference in classification and gyrogonite orientation of Trochiliscales and Sycidiales. *Acta Palaeontologia Sinica*, **19**, 190–200 [in Chinese].

Wiesner, H. (1936) Diatomées dans le Crétacé supérieur de la Bohèmie. *Annales de Protistologie*, **5**, 151–5.

Wind, F. H. and Wise, S. S. (1978) Mesozoic holococcoliths. *Geology*, **6**, 140–2.

Witt, O. (1885) Uber der Polierschiefer von Archangelsk-Kurojedowo im gouv. Simbrisk. *Verhandlungen, russisch-kaiserliche mineralogische Gesellschaft zu St. Petersburg. Ser II.* **22**, 137–77.

Woese, C. and Fox, G. (1977) Phylogenetic structure of the prokaryotic domain. *Proceedings of the National Academy of Sciences, USA*, **74**, 5088–90.

Wray, J. L. (1977) *Calcareous Algae*. Elsevier, Amsterdam, 185 pp.

Yanagisawa, Y. and Akiba, F. (1990) Taxonomy and phylogeny of the three marine diatom genera, *Crucidenticula*, *Denticulopsis*, and *Neodenticula*. *Bulletin of the Geological Survey of Japan*, **41**, 197–301.

Zankl, H. (1969) Der Hohe Goll: Aufbau und Lebensbild eines Dachsteinkalk-Riffes in der Obertrias der nordlichen Kalkalpen. *Senckenbergische Naturforschende Gesellschaft*, **519**, 1–123.

Zhang, Y. (1989) Multicellular thallophytes with differentiated tissues from the late Proterozoic phosphate rocks of south China. *Lethaia*, **22**, 113–92.

Zhang, Z. (1986) Clastic facies microfossils from the Chuanlinggou Formation (1800 Ma) near Jixian, North China. *Journal of Micropalaeontology*, **5**, 9–16.

NOTE ADDED IN PROOF

Hans and Runnegar (1992) extended the range of *Grypania*, a spirally coiled, carbonaceous cylindrical filament with transverse markings (1 mm wide and 90 mm long) from 1.800 Ma (Zhang, 1986) to 2.100 Ma and interpreted it as a non-calcified dasycladacean alga. Although such an affinity is equivocal, these fossils have been accepted as eukaryotic, with major implications for the composition of the coeval atmosphere previously considered anoxic (Riding, 1992).

Hans, T. M. and Runnegar, B. (1992) Megascopic eukaryotic algae from the 2.1-billion-year-old Negaunee Iron-Formation, Michigan. *Science*, **257**, 232–5.

Riding, R. (1992) The algal breath of life. *Nature*, **359**, 13–14.

Animals: Invertebrates

Sanctacaris uncata Briggs and Collins, 1988 from the Burgess Shale (Middle Cambrian) of Mount Stephen, British Columbia, Canada. The oldest chelicerate arthropod. The specimen is about 70 mm long. Photograph courtesy of D. E. G. Briggs; published with thanks to the Royal Ontario Museum.

4

PROTOZOA

M. B. Hart and C. L. Williams

The Protozoa embrace a wide and diverse group of organisms, regarded by many as a separate phylum and regarded by others as a taxonomic dustbin. Since the publication of the original *Fossil Record* in 1967, scanning electron microscopy and transmission electron microscopy have revolutionized the study of these minute organisms. This group, more than the majority of the groups represented in the fossil record, has received a great deal of attention from biologists and zoologists. It is therefore proposed that, for the purpose of this chapter, a biological classification should be adopted. The Protozoa are now regarded as a Subkingdom of the Kingdom Animalia – at least by the Society of Protozoologists, whose classification proposals were published in 1980 (Levine *et al.*, 1980). Unfortunately, while solving many of the micropalaeontological problems of the 'animal-based' workers, it separates out the botanically oriented forms. For micropalaeontologists, therefore, there are still problems, but at least many of the common groups (Foraminifera, Radiolaria, thecamoebians, dinoflagellates, etc.) can now be described in terms of their biological affinities.

The botanically related protistids have been dealt with in a separate chapter as, at the present time, this is probably the most logical grouping. The algae, dinoflagellates, acritarchs, diatoms, haptophytes and charophytes are to be found, therefore, in Chapter 3.

Kindom ANIMALIA

Subkingdom PROTOZOA

Phylum SARCOMASTIGOPHORA Honigberg and Balamuth, 1963

Subphylum SARCODINA Schmarda, 1871

Superclass RHIZOPODA von Siebold, 1845

Class LOBOSEA Carpenter, 1861 (see Fig. 4.1)

Subclass TESTACELOBOSIA De Saedeleer, 1934

Order ARCELLINIDA Kent, 1880

Superfamily ARCELLACEA Ehrenberg, 1832

F. ARCELLIDAE Ehrenberg, 1832 Qu. (PLE)–Rec. FW

First: *Arcella* Ehrenberg, 1832; monotypic; cosmopolitan. **Extant**

F. CENTROPYXIDAE Jung, 1942 K. (ALB)–?Rec. FW

First: Lower Albian: Genus *Marsipos* (may approximate to *Centropyxis*) Medioli *et al.*, 1990; Ruby Creek, Alberta, Canada. **?Extant**

F. TRIGONOPYXIDAE Loeblich and Tappan, 1964 T. (Mio.)–Rec. FW

First: *Cyclopyxis* Deflandre, 1829; Java; North America; South America; Africa; Europe. **Extant**

F. HYALOSPHENIIDAE Schulze, 1877 K. (ALB)–Rec. FW

First: Lower Albian: Genera *Ochros* Medioli *et al.*, 1990 and *Sacculus* Medioli *et al.*, 1990; Ruby Creek, Alberta, Canada. **Extant**

F. DIFFLUGIIDAE Wallich, 1864 C. (NAM)–Rec. FW

First: *Prantlitina* Vašiek and Ruíka, 1957; Czechoslovakia. **Extant**

Superfamily CRYPTODIFFLUGIACEA Jung, 1942

F. CRYPTODIFFLUGIIDAE Jung, 1942 Q. (PLE)–Rec. FW

First: *Cryptodifflugia* Penard, 1890; Europe. **Extant**

F. PHRYGANELLIDAE Jung, 1942 Q. (PLE)–Rec. FW

First: *Phryganella* Penard, 1902; India; western Europe; North America. **Extant**

Class FILOSEA Leidy, 1879

Order GROMIIDA Claparède and Lachmann, 1859

Superfamily GROMIACEA Reuss, 1862

F. GROMIIDAE Reuss, 1862 Q. (PLE)–Rec. FW

First: *Pseudodifflugia* Schlumberger, 1845; Europe. **Extant**

F. AMPHITREMATIDAE Poche, 1913 Q. (PLE)–Rec. FW

The Fossil Record 2. Edited by M. J. Benton. Published in 1993 by Chapman & Hall, London. ISBN 0 412 39380 8

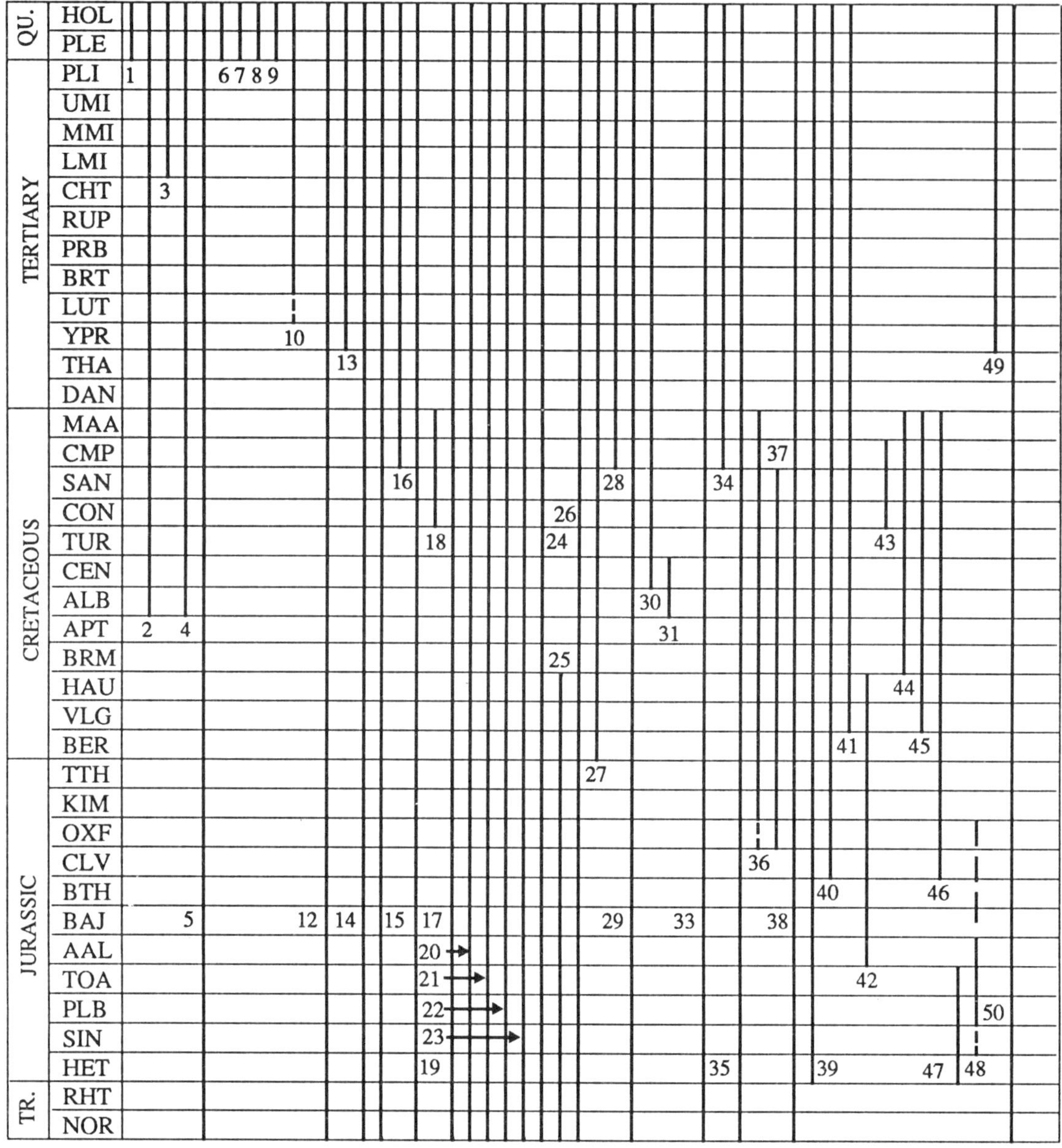

Fig. 4.1

First: *Amphitrema* Archer, 1867 (Europe) and *Archerella* Loeblich and Tappan, 1961 (Europe; North America). **Extant**

Superfamily EUGLYPHACEA Wallich, 1864

F. EUGLYPHIDAE Wallich, 1864
T. (LUT/BRT)–Rec. FW

First: *Euglypha* Dujardin, 1840; cosmopolitan. **Extant**

Class GRANULORETICULOSEA De Saedeleer, 1934

Order FORAMINIFERIDA Eichwald, 1830

Alcide d'Orbigny (1826) was the first author to use the name 'foraminifères' for this group of protozoans. He regarded them as a subdivision of the Cephalopoda and, as such, only placed them in an informal classificatory position. This, in the opinion of many workers, invalidated his use of the word 'foraminifera', at least in the higher levels of classification. Many of the species he described at the same time are still used as valid taxa. This would appear to be slightly inconsistent and there may be some justification for attributing the Foraminiferida to d'Orbigny. Current accepted usage, however, dictates that Eichwald (1830) be regarded as the origin of the Order. Range data are from Loeblich and Tappan (1964, 1988) and other sources.

Suborder ALLOGROMIINA Loeblich and Tappan, 1961

F. MAYLISORIIDAE Bykova, 1961 Є. (CRF)–S. FW

First: *Chitinodendron* Eisenack, 1938; USA; Estonia, former USSR; Germany.
Last: Silurian: same genus.

F. ALLOGROMIIDAE Rhumbler, 1904
O. (CRD)–Rec. FW

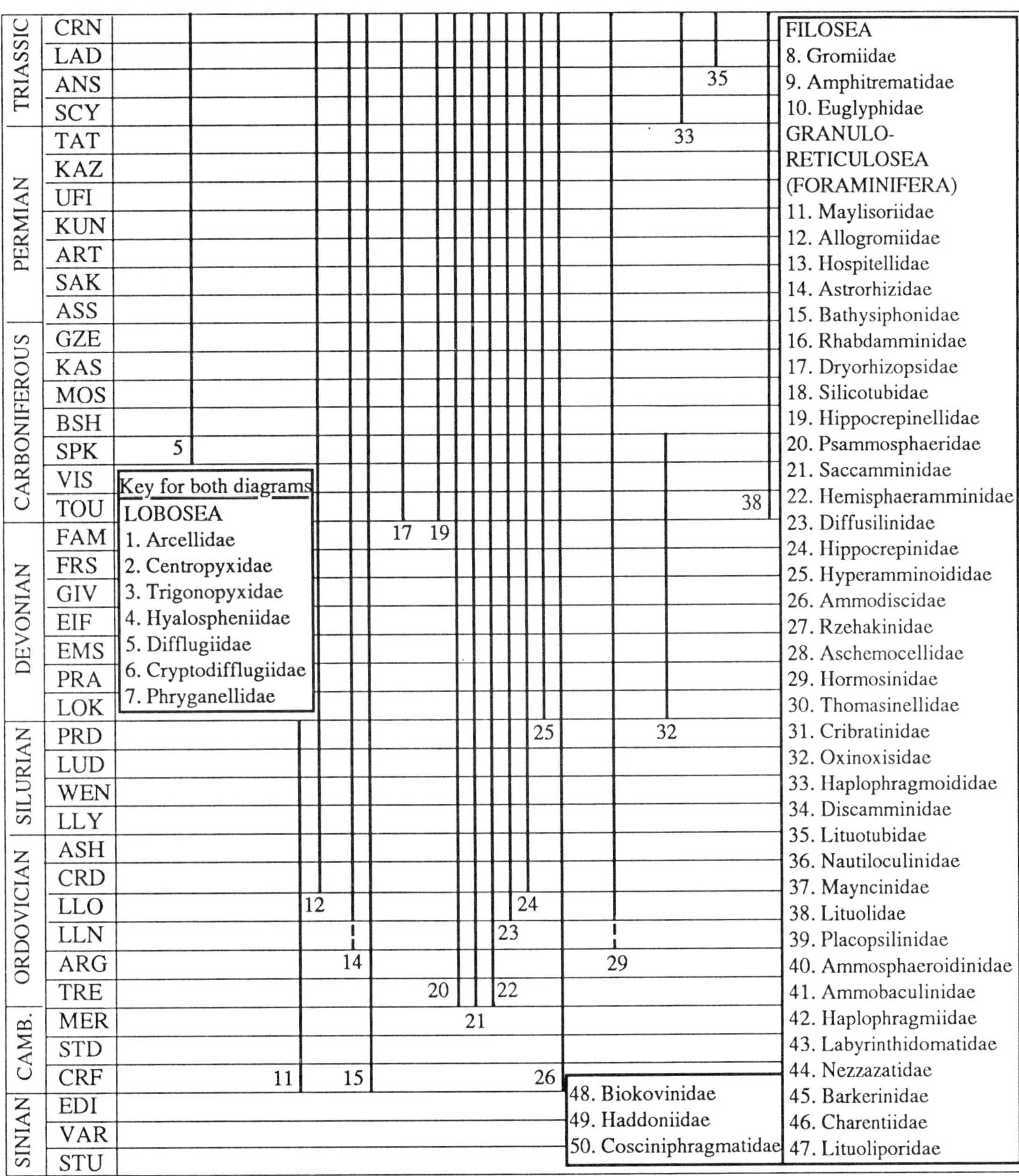

Fig. 4.1

First: *Chitinolagena* Bykova, 1961 and *Labyrinthochitinia* Bykova, 1961; Kazakhstan, former USSR. **Extant**

F. HOSPITELLIDAE Loeblich and Tappan, 1984
T. (Eoc.)–Rec. FW

First: *Thalamophaga* Rhumbler, 1911; France. **Extant**

Suborder TEXTULARIINA Delage and Hérouard, 1896

Superfamily ASTRORHIZACEA Brady, 1881

F. ASTRORHIZIDAE Brady, 1881
O. (LLN/LLO)–Rec. Mar.

First: *Astrorhiza* Sandahl, 1858; cosmopolitan. **Extant**

F. BATHYSIPHONIDAE Avnimelech, 1952
Є. (CRF)–Rec. Mar.

First: *Platysolenites* Eichwald, 1860; former USSR; Poland; Norway; USA: California. **Extant**

F. RHABDAMMINIDAE Brady, 1884
K. (CMP)–Rec. Mar.

First: *Psammatodendron* Norman, 1881; Romania. **Extant**

F. DRYORHIZOPSIDAE Loeblich and Tappan, 1984
C. (l.)–Rec. Mar.

First: *Dryorhizopsis* Henbest, 1963; USA: Texas. **Extant**

F. SILICOTUBIDAE Vyalov, 1968 K. (SEN) Mar.

First: *Silicotuba* Vyalov, 1966; Poland; Germany; Czechoslovakia.
Last: Upper Cretaceous (Senonian): same genus.

F. HIPPOCREPINELLIDAE Loeblich and Tappan, 1984 C.–Rec. Mar.

First: *Hippocrepinella* Heron-Allen and Earland, 1932; South Atlantic. **Extant**

F. PSAMMOSPHAERIDAE Haeckel, 1894 O.–Rec. Mar.

First: *Cellonina* Kristan-Tollmann, 1971 (Austria); *Psammosphaera* Schulze, 1875 (cosmopolitan); *Pseudastrorhiza* Eisenack, 1932 (Estonia, former USSR); *Raibosammina* Moreman, 1930 (USA: Oklahoma). **Extant**

F. SACCAMMINIDAE Brady, 1884 O.–Rec. Mar.

First: *Amphitremoida* Eisenack, 1938 and *Ordovicina* Eisenack, 1938; Baltic region. **Extant**

F. HEMISPHAERAMMINIDAE Loeblich and Tappan, 1961 O.–Rec. Mar.

First: *Tholosina* Rhumbler, 1895 (Atlantic; Antarctic; North America; Europe) and possibly *Nephrosphaera* Kristan-Tollmann, 1971 (Austria). **Extant**

F. DIFFUSILINIDAE Loeblich and Tappan, 1961 O. (LLN/LLO)–Rec. Mar.

First: Middle Ordovician (Trenton): Genus *Kerionammina* Moreman, 1933; USA: Oklahoma. **Extant**

Superfamily HIPPOCREPINACEA Rhumbler, 1895

F. HIPPOCREPINIDAE Rhumbler, 1895 O. (CRD)–Rec. Mar.

First: *Arenosiphon* Grubbs, 1939; Baltic region. **Extant**

F. HYPERAMMINOIDIDAE Loeblich and Tappan, 1984 D. (l.)–K. (NEO) Mar.

First: *Sansabaina* Loeblich and Tappan, 1984; USA.
Last: *Kechenotiske* Loeblich and Tappan, 1984; former USSR, western Siberia.

Superfamily AMMODISCACEA Reuss, 1862

F. AMMODISCIDAE Reuss, 1862 Є. (CRF)–Rec. Mar.

First: *Spirosolenites* Glaessner, 1979; Norway: Finmark. **Extant**

Superfamily RZEHAKINACEA Cushman, 1933

F. RZEHAKINIDAE Cushman, 1933 K. (l.)–Rec. Mar.

First: *Miliammina* Heron-Allen and Earland, 1930 (cosmopolitan) and *Rothina* Hanzliková, 1966 (Czechoslovakia: Carpathians). **Extant**

Superfamily HORMOSINACEA Haeckel, 1894

F. ASCHEMOCELLIDAE Vyalov, 1966 K. (CMP)–Rec. Mar.

First: *Aschemocella* Vyalov, 1966; Romania; former USSR. **Extant**

F. HORMOSINIDAE Haeckel, 1894 O. (LLN/LLO)–Rec. Mar.

First: *Reophax* de Montfort, 1808; cosmopolitan. **Extant**

F. THOMASINELLIDAE Loeblich and Tappan, 1984 K. (CEN)–Rec. Mar.

First: *Thomasinella* Schlumberger, 1893; Algeria; Tunisia; Egypt; India. **Extant**

F. CRIBRATINIDAE Loeblich and Tappan, 1964 K. (ALB–CEN) Mar.

First: *Cribratina* Sample, 1932; USA: Texas, Oklahoma.
Last: Upper Cretaceous (lower Cenomanian): same genus; USA.

Superfamily LITUOLACEA de Blainville, 1827

F. OXINOXISIDAE Vyalov, 1968 D. (l.)–C. (l.) Mar.

First: *Oxinoxis* Gutschick, 1962; USA: Montana, Illinois, Missouri.
Last: same genus; USA.

F. HAPLOPHRAGMOIDIDAE Maync, 1952 Tr. (SCY)–Rec. Mar.

First: *Ammosiphonia* He, 1977; China: Yunnan Province. **Extant**

F. DISCAMMINIDAE Mikhalevich, 1980 K. (CMP)–Rec. Mar.

First: *Ammoscalaria* Höglund, 1947; Asiatic former USSR; North and South Atlantic; North and South Pacific; Gulf of Mexico. **Extant**

F. LITUOTUBIDAE Loeblich and Tappan, 1984 Tr. (LAD)–Rec. Mar.

First: *Plagioraphe* Kristan-Tollmann, 1973; Late Ladinian, northern Alps; Germany; Bulgaria. **Extant**

F. NAUTILOCULINIDAE Loeblich and Tappan, 1985 J. (OXF/KIM)–K. (SEN) Mar.

First: *Nautiloculina* Mohler, 1938; former USSR; Egypt; Middle East; France; Switzerland; former Yugoslavia.
Last: *Murgeina* Bilotte and Decrouez, 1979; Italy; Greece; former Yugoslavia; Lebanon.

F. MAYNCINIDAE Loeblich and Tappan, 1985 J. (OXF)–K. (SAN) Mar.

First: *Flabellocyclolina* Gendrot, 1964; Israel; France.
Last: *Gendrotella*Maync, 1972 (France) and *Flabellocyclolina* Gendrot, 1964 (France; Israel).

F. LITUOLIDAE de Blainville, 1827 C. (l.)–Rec. Mar.

First: *Ammobaculites* Cushman, 1910; cosmopolitan. **Extant**

F. PLACOPSILINIDAE Rhumbler, 1913 J. (l.)–Rec. Mar.

First: *Subbdelloidina* Frentzen, 1944; France; Switzerland; Germany. **Extant**

Superfamily HAPLOPHRAGMIACEA Eimer and Fickert, 1899

F. AMMOSPHAEROIDINIDAE Cushman, 1927 J. (CLV)–Rec. Mar.

First: *Recurvoides* Earland, 1934; former USSR. **Extant**

F. AMMOBACULINIDAE Saidova, 1981 K. (VLG)–Rec. Mar.

First: *Bulbobaculites* Maync, 1952; Colombia; Germany; former USSR: Ukraine. **Extant**

F. HAPLOPHRAGMIIDAE Eimer and Fickert, 1899 J. (M.)–K. (HAU) Mar.

First: *Haplophragmium* Reuss, 1860; Europe.
Last: same genus.

F. LABYRINTHIDOMATIDAE Loeblich and Tappan, 1988 K. (CON–CMP) Mar.

First: *Labyrinthidoma* Adams, Knight and Hodgkinson, 1973; England, UK.
Last: *Bulbophragmium* Maync, 1952; Europe.

F. NEZZAZATIDAE Hamaoui and Saint-Marc, 1970 K. (BRM–MAA) Mar.

First: *Nezzazatinella* Darmoian, 1976; Iraq; Romania; France.
Last: *Antalyna* Farinacci and Köylüoglu, 1985; Turkey.

F. BARKERINIDAE Smout, 1956 K. (VLG–SEN) Mar.

First: *Barkerina* Frizzell and Schwartz, 1950; USA: Texas; Greece; Sardinia.
Last: same genus.

Superfamily BIOKOVINACEA Gusic, 1977

F. CHARENTIIDAE Loeblich and Tappan, 1985 J. (CLV)–K. (MAA) Mar.

First: *Praekaraisella* Kurbatov, 1972; Uzbekistan, former USSR.
Last: *Praepeneroplis* Hofker, 1952; The Netherlands; France.

F. LITUOLIPORIDAE Gusic and Velic, 1978 J. (l.) Mar.

First: *Lituolipora* Gusic and Velic, 1978; former Yugoslavia: Croatia.
Last: same genus.

F. BIOKOVINIDAE Gusic, 1977 J. (SIN/PLB–AAL/OXF) Mar.

First: *Biokovina* Gusic, 1977 (former Yugoslavia: Croatia) and *Bosniella* Gusic, 1977 (former Yugoslavia: Bosnia).
Last: M./L. Jurassic (L. Aalenian/M. Oxfordian): Genus *Chablaisia* Septfontaine, 1978; France; Switzerland.

Superfamily COSCINOPHRAGMATACEA Thalmann, 1951

F. HADDONIIDAE Saidova, 1981 T. (Eoc.)–Rec. Mar.

First: *Haddonia* Chapman, 1898; Cuba; Germany. **Extant**

F. COSCINOPHRAGMATIDAE Thalmann, 1951 Tr. (l.)–Rec. Mar.

First: *Alpinophragmium* Flügel, 1967; Germany. **Extant**

Superfamily CYCLOLINACEA Loeblich and Tappan, 1964

F. CYCLOLINIDAE Loeblich and Tappan, 1964 K. (VLG–CMP) Mar.

First: *Ammocycloloculina* Maync, 1958; lower VLG, France.
Last: *Ilerdorbis* Hottinger and Caus, 1982; Spain.

F. ORBITOPSELLIDAE Höttinger and Caus, 1982 J. (l.–KIM) Mar.

First: *Cyclorbitopsella* Cherchi, Schroeder and B. G. Zhang, 1984 (Tibet), *Labyrinthina* Weynschenk, 1951 (Spain; Morocco; former USSR; Italy; former Yugoslavia) and *Orbitopsella* Munier-Chalmas, 1902 (Italy; Cyprus; former Yugoslavia; Greece; Mallorca; Morocco; Iran; Oman; Arabia; China).
Last: *Labyrinthina* Weynschenk, 1951; as above.

Superfamily LOFTUSIACEA Brady, 1884

F. MESOENDOTHYRIDAE Voloshinova, 1958 Tr. (LAD)–J. (TTH) Mar.

First: *Mesoendothyra* Dain, 1958; Bulgaria; former Yugoslavia; former USSR: Ukraine.
Last: *Audienusina* Bernier, 1985; France.

F. HOTTINGERITIDAE Loeblich and Tappan, 1985 J. (OXF)–K. (BRM) Mar.

First: *Alveosepta* Höttinger, 1967; Switzerland; France; Portugal; Morocco.
Last: *Hottingerita* Loeblich and Tappan, 1985; Switzerland.

F. CYCLAMMINIDAE Marie, 1941 J. (l.)–Rec. Mar.

First: *Amijiella* Loeblich and Tappan, 1985; Iraq; Turkey; Italy; France; Switzerland; former Yugoslavia. **Extant**

F. ECOUGELLIDAE Loeblich and Tappan, 1985 K. (BRM–APT) Mar.

First: Genus *Ecougella* Arnaud-Vanneau, 1980; France.
Last: same genus.

F. SPIROCYCLINIDAE Munier-Chalmas, 1887 J. (l.)–T. (Eoc.) Mar.

First: *Haurania* Henson, 1948; (Morocco, Iraq, China) and *Streptocyclammina* Höttinger, 1967; (Morocco, Italy, former Yugoslavia)
Last: *Saudia* Henson, 1948; Saudi Arabia; Iraq; former Yugoslavia.

F. LOFTUSIIDAE Brady, 1884 K. (BRM–MAA) Mar.

First: *Praereticulinella* Deloffre and Hamaoui, 1970; Spain.
Last: *Loftusia* Brady, 1870 (Iran, Turkey, Sumatra) and *Reticulinella* Cuvillier, Bonnefous, Hamaoui and Tixier, 1970 (Algeria, Libya).

Superfamily SPIROPLECTAMMINACEA Cushman, 1927

F. SPIROPLECTAMMINIDAE Cushman, 1927 C.–Rec. Mar.

First: *Spiroplectammina* Cushman, 1927; cosmopolitan. **Extant**

F. TEXTULARIOPSIDAE Loeblich and Tappan, 1982 J. (PLB)–K. (MAA) Mar.

First: *Textulariopsis* Banner and Pereira, 1981; cosmopolitan.

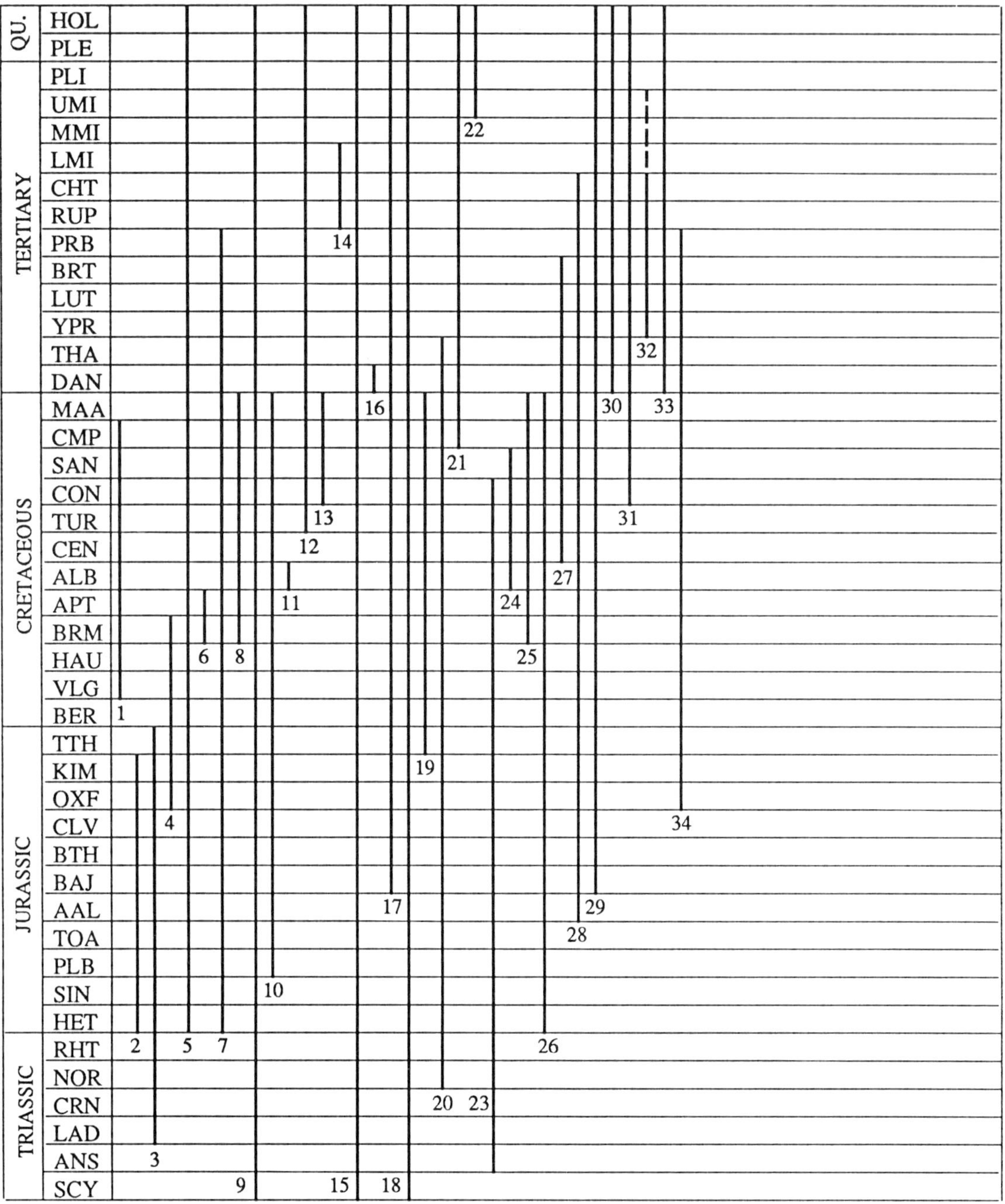

Fig. 4.2

Last: *Plectinella* Marie, 1956 (Egypt, France, Belgium, former USSR: Belorussia, Australia, USA: Texas) and *Pleurostomelloides* Majzon, 1943 (Hungary).

F. PLECTORECURVOIDIDAE Loeblich and Tappan, 1964 K. (ALB) Mar.

First and Last: *Plectorecurvoides* Noth, 1952; Austria; Czechoslovakia; former USSR; Eastern Atlantic.

F. NOURIIDAE Chapman and Parr, 1936 K. (TUR)–Rec. Mar.

First: *Abdullaevia* Suleymanov, 1965; former USSR: Uzbekistan, Kyzyl Kum. **Extant**

Superfamily PAVONITONACEA Loeblich and Tappan, 1961

F. MARIEITIDAE Loeblich and Tappan, 1986 K. (SEN) Mar.

First and Last: *Hensonia* Marie, 1954 (France, Spain) and *Marieita* Loeblich and Tappan, 1964 (France, Spain).

F. PAVONITINIDAE Loeblich and Tappan, 1961 T. (RUP–LMI) Mar.

First: *Pavonitina* Schubert, 1914; France; Poland; Austria; former Yugoslavia; Africa: offshore Cabinda (Angola).
Last: *Spiropsammia* Seiglie and Baker, 1984; Italy; Cameroon; Cabinda, Angola.

Superfamily TROCHAMMINACEA Schwager, 1877

F. TROCHAMMINIDAE Schwager, 1877 C. (u.)–Rec. Mar.

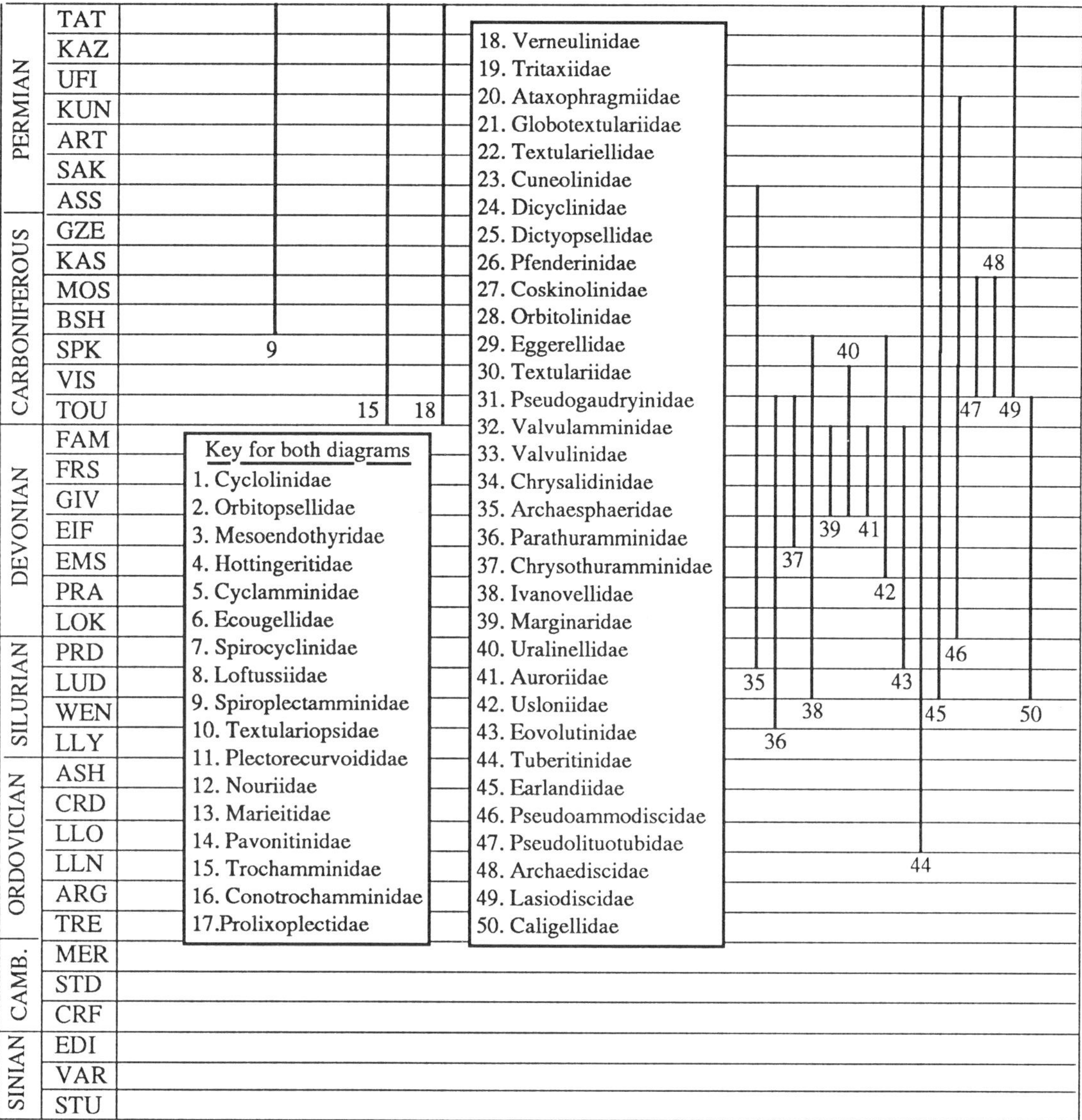

Fig. 4.2

First: *Trochammina* Parker and Jones, 1859; cosmopolitan. **Extant**

Superfamily VERNEUILINACEA Cushman, 1911

F. CONOTROCHAMMINIDAE Saidova, 1981 T. (DAN) Mar.

First and Last: *Conotrochammina* Finlay, 1940; New Zealand.

F. PROLIXOPLECTIDAE Loeblich and Tappan, 1985 J. (BAJ)–Rec. Mar.

First: *Riyadhella* Redmond, 1965; Saudi Arabia; western India. **Extant**

F. VERNEUILINIDAE Cushman, 1911 C.–Rec. Mar.

First: *Mooreinella* Cushman and Waters, 1928; USA: Texas; Australia. **Extant**

F. TRITAXIIDAE Plotnikova, 1979 J. (TTH)–K. Mar.

First: *Bitaxia* Plotnikova, 1978; former USSR: Crimea, Ukraine.

Last: *Tritaxia* Reuss, 1860; cosmopolitan.

Superfamily ATAXOPHRAGMIACEA Schwager, 1877

F. ATAXOPHRAGMIIDAE Schwager, 1877 Tr. (NOR)–T. (Pal.) Mar.

First: *Agglutisolena* Senowbari-Daryan, 1984 (Italy: Sicily) and *Kaeveria* Senowbari-Daryan, 1984 (Turkey, Italy: Sicily, Austria)

Last: *Arenobulimina* Cushman, 1927 (Europe, USA: Arkansas, Texas) and *Ataxophragmium* Reuss, 1860 (France, Germany, England, UK, Sweden, former USSR).

F. GLOBOTEXTULARIIDAE Cushman, 1927 K. (CMP)–Rec. Mar.

First: *Remesella* Vašíek, 1947; Czechoslovakia; Switzerland; Austria; Romania; former USSR: Crimea; New Zealand. **Extant**

F. TEXTULARIELLIDAE Grönhagen and Luterbacher, 1966 T. (UMI)–Rec. Mar.

First: *Guppyella* Brönnimann, 1951; West Indies; Costa Rica; Venezuela. **Extant**

F. CUNEOLINIDAE Saidova, 1981 Tr. (ANS)–K. (CON) Mar.

First: *Palaeolituonella* Bérczi-Makk, 1981; Hungary; Italy; former Yugoslavia; Austria; Bulgaria.
Last: *Cuneolina* d'Orbigny, 1839; China; USA; Europe.

F. DICYCLINIDAE Loeblich and Tappan, 1964 K. (ALB–SAN) Mar.

First: *Dicyclina* Munier-Chalmas, 1887; France; Spain; former Yugoslavia.
Last: same genus.

F. DICTYOPSELLIDAE Brönnimann *et al.*, 1983 K. (BRM–MAA) Mar.

First: *Conorbinella* Poroshina, 1976; France; former USSR: Azerbaijan SSR.
Last: *Dictyopsella* Munier-Chalmas, 1900; France; Spain.

F. PFENDERINIDAE Smout and Sugden, 1962 J. (l.)–K. (MAA) Mar.

First: *Pseudopfenderina* Höttinger, 1967 (Morocco) and *Kurnubia* Henson, 1948 (SW Asia, Morocco, former Yugoslavia, Crete).
Last: *Gyroconulina* Schroeder and Darmoian, 1977; Iraq.

F. COSKINOLINIDAE Moullade, 1965 K. (CEN)–T. (LUT/BRT) Mar.

First: *Pseudolituonella* Marie, 1955; France; Spain; Israel; Turkey.
Last: *Coleiconus* Höttinger and Drobne, 1980 (West Indies, USA: Florida), *Coskinolina* Stache, 1875 (former Yugoslavia, France) and *Coskinon* Höttinger and Drobne, 1980 (former Yugoslavia, Czechoslovakia, USA: Florida).

Superfamily ORBITOLINACEA Martin, 1890

F. ORBITOLINIDAE Martin, 1890 J. (m.)–T. (RUP/CHT) Mar.

First: *Gutnicella* Moullade, Haman and Huddleston, 1981 (Spain: Balearic Islands), *Kilianina* Pfender, 1933 (France) and *Meyendorffina* Aurouze and Bizon, 1958 (France).
Last: *Dictyoconus* Blanckenhorn, 1900; cosmopolitan.

Superfamily TEXTULARIACEA Ehrenberg, 1838

F. EGGERELLIDAE Cushman, 1937 J. (BAJ)–Rec. Mar.

First: *Pseudomarssonella* Redmond, 1965; Saudi Arabia; western India. **Extant**

F. TEXTULARIIDAE Ehrenberg, 1838 T. (Pal.)–Rec. Mar.

First: *Textularia* Defrance, 1824; cosmopolitan. **Extant**

F. PSEUDOGAUDRYINIDAE Leoblich and Tappan, 1985 K. (u.)–Rec. Mar.

First: *Clavulinopsis* Banner and Desai, 1985 (USA: Texas, Arkansas), *Pseudoclavulina* Cushman, 1936 (Mexico, West Indies: Trinidad, USA, England, UK, Denmark, Poland, Sweden, Germany, The Netherlands), *Pseudogaudryina* Cushman, 1936 (West Indies, USA: South Carolina, Australia, Gulf of Mexico, Atlantic, Germany) and *Valvoreussella* Hofker, 1957 (Czechoslovakia, Germany, The Netherlands). **Extant**

F. VALVULAMMINIDAE Loeblich and Tappan, 1986 T. (Eoc.–Oli./Mio.) Mar.

First: *Arenagula* Bourdon and Lys, 1955 (France, Greece, USA: Florida, Pacific: Marshall Islands), *Discorinopsis* Cole, 1941 (USA: Florida) and *Valvulammina* Cushman, 1933 (France, Cuba).
Last: *Arenagula* as above.

F. VALVULINIDAE Berthelin, 1880 T. (Pal.)–Rec. Mar.

First: *Clavulina* d'Orbigny, 1826; cosmopolitan. **Extant**

F. CHRYSALIDINIDAE Neagu, 1968 J. (OXF)–T. (PRB) Mar.

First: *Parurgonina* Cuvillier *et al.*, 1968; Italy; Greece; Switzerland; former Yugoslavia; Algeria.
Last: *Pfendericonus* Höttinger and Drobne, 1980 (former Yugoslavia) and *Pseudochrysalidina* Cole, 1941 (USA: Florida, former Yugoslavia, Somalia, Cuba, Dominican Republic, Jamaica).

Suborder FUSULININA Wedekind, 1937

Superfamily PARATHURAMMINACEA Bykova, 1955

F. ARCHAESPHAERIDAE Malakhova, 1956 S. (LUD/PRD)–P. (ASS) Mar.

First: *Arakaevella* Pronina, 1964 (former USSR) and *Eoammosphaeroides* Pronina, 1970 (former USSR).
Last: *Insolentitheca* Vachard, 1979; former USSR: southern Urals, Donbas, Kazakhstan; Belgium; France; Spain; Algeria; Morocco; Libya; Afghanistan; Japan; northern Thailand.

F. PARATHURAMMINIDAE Bykova, 1955 S. (WEN)–C. (Tou) Mar.

First: *Parathuramminites* Poyarkov, 1969; former USSR: Urals, Russian Platform.
Last: *Parathurammina* Suleymanov, 1945 (Europe, Asia) and *Irregularina* Bykova, 1955 (former USSR: Russian Platform, central Asia).

F. CHRYSOTHURAMMINIDAE Loeblich and Tappan, 1986 D. (EIF)–C. (Tou) Mar.

First: *Salpinogthurammina* Poyarkov, 1961; former USSR: Ukraine, Kazakh, Urals.
Last: *Chrysothurammina* Neumann, Pozaryska and Vachard, 1975; Poland; former USSR: Tomsk, western Urals, Tien Shan.

F. IVANOVELLIDAE Chuvashov and Yuferev, 1984 S. (LUD)–C. (l.) Mar.

First: *Elenella* Pronina, 1969 (former USSR: Urals, China) and *Ivanovella* Pronina, 1969 (former USSR: Urals).
Last: *Elenella* as above.

F. MARGINARIDAE Loeblich and Tappan, 1986 D. (GIV–FAM) Mar.

First: *Marginara* Petrova, 1984; former USSR.
Last: *Cordatella* Petrova, 1984 (former USSR) and *Turcmeniella* Miklukho-Maklay, 1965 (former USSR).

F. URALINELLIDAE Chuvashov *et al.*, 1984 D. (GIV)–C. (VIS) Mar.

First: *Uralinella* Bykova, 1952; former USSR: Urals, Bashkir, Tatar, Siberia, Tien Shan.
Last: *Sogdianina* Saltovskaya, 1973; former USSR: Tadzhikstan.

F. AURORIIDAE Loeblich and Tappan, 1986 D. (GIV–FAM) Mar.

First: *Apertauroria* Sabirov, 1984 (former USSR) and *Auroria* Poyarkov, 1969 (former USSR: Kirgiz).
Last: same genera.

F. USLONIIDAE Miklukho-Maklay, 1963 D. (EIF/GIV)–C. (l.) Mar.

First: *Bisphaera* Birina, 1948; former USSR.
Last: same genus.

F. EOVOLUTINIDAE Loeblich and Tappan, 1986 S. (PRD)–D. (FAM) Mar.

First: *Cribrohemisphaeroides* Pronina, 1980 (former USSR: Urals) and *Eovolutina* Antropov, 1950 (former USSR: Turkistan, Gissar, Russian Platform).
Last: *Eovolutina* as above.

F. TUBERITINIDAE Miklukho-Maklay, 1958 S. (LLY)–P. Mar.

First: *Illigata* Bykova, 1956; Lithuania, former USSR.
Last: *Tuberitina* Galloway and Harlton, 1928; cosmopolitan.

Superfamily EARLANDIACEA Cummings, 1955

F. EARLANDIIDAE Cummings, 1955 S. (LUD)–P. (ZEC) Mar.

First: *Earlandia* Plummer, 1930; USA: Texas; England, UK; former USSR; Poland.
Last: *Aeolisaccus* Elliott, 1958; Iraq; Arabia.

F. PSEUDOAMMODISCIDAE Conil and Lys, 1970 D.–P. (ROT) Mar.

First: *Brunsia* Mikhaylov, 1935 (former USSR, Belgium) and *Pseudoglomospira* Bykova, 1955 (former USSR: Urals, Russian Platform).
Last: *Brunsiella* Reytlinger, 1950; former USSR: Moscow Basin, Tatar, Kazan, Ukraine, Tien Shan.

F. PSEUDOLITUOTUBIDAE Conil and Longerstaey, 1980 C. (VIS–MOS) Mar.

First: *Pseudolituotuba* Vdovenko, 1971; western Europe; former USSR.
Last: same genus.

Superfamily ARCHAEDISCACEA Cushman, 1928

F. ARCHAEDISCIDAE Cushman, 1928 C. (VIS–MOS) Mar.

First: *Archaediscus* Brady, 1873 (Europe, Asia, North America), *Eosigmoilina* Ganelina, 1956 (former USSR, Czechoslovakia, North America), *Glomodiscus* Malakhova, 1973 (Belgium, former USSR), *Nudarchaediscus* Conil and Pirlet, 1974 (Belgium, England, UK, Germany, USSR, Iran, Egypt), *Planoarchaediscus* Miklukho-Maklay, 1956 (Iran, Egypt, Sinai, Belgium, former USSR), *Planospirodiscus* Sosipatrova, 1962 (former USSR, North America), *Tournarchaediscus* Conil and Pirlet, 1974 (France, Belgium), *Tubispirodiscus* Browne and Pohl, 1973 (North America), *Uralodiscus* Malakhova, 1973 (former USSR, Belgium, Iran), *Asteroarchaediscus* Miklukho-Maklay, 1956 (former USSR, North America, Iran?), *Neoarchaediscus* Miklukho-Maklay, 1956 (former USSR, Iran, North America), *Nodasperodiscus* Conil and Pirlet, 1974 (former USSR, western Europe, Iran, North America), *Nodosarchaediscus* Conil and Pirlet, 1974 (former USSR, western Europe, Iran, North America) and *Permodiscus* Dutkevich, 1948 (former USSR).
Last: *Archaediscus* and *Asteroarchaediscus* as above.

F. LASIODISCIDAE Reytlinger, 1956 C. (VIS)–P. Mar.

First: *Howchinia* Cushman, 1927; England, UK; Belgium; former USSR; Iran; USA: Alaska; Canada.
Last: *Glomotrocholina* Nikitina, 1977 (former USSR), *Lasiodiscus* Reichel, 1946 (former USSR, Cyprus, Greece) and *Lasiotrochus* Reichel, 1946 (Greece, former USSR: Azerbaijhan).

Superfamily MORAVAMMINACEA Pokorny, 1951

F. CALIGELLIDAE Reytlinger, 1959 S. (LUD)–C. (Tou) Mar.

First: *Glubokoevella* Pronina, 1970 (former USSR) and *Paracaligella* Lipina, 1955 (former USSR, USA).
Last: *Baituganella* Lipina, 1955 (former USSR) and *Paracaligella* as above.

F. MORAVAMMINIDAE Pokorny, 1951 D (GIV–FRS) Mar. (see Fig. 4.3)

First: *Kettnerammina* Pokorny, 1951 (Czechoslovakia), *Moravammina* Pokorny, 1951 (Czechoslovakia, former USSR) and *Saccorhina* Bykova, 1955 (former USSR).
Last: *Kettnerammina* and *Saccorhina* as above.

F. PARATIKHINELLIDAE Loeblich and Tappan, 1984 D. (GIV)–C. (TOU/VIS) Mar.

First: *Vasicekia* Pokorny, 1951; Czechoslovakia.
Last: *Paratikhinella* Reytlinger, 1954; former USSR; England, UK, Avonian.

Superfamily NODOSINELLACEA Rhumbler, 1895

F. EARLANDINITIDAE Loeblich and Tappan, 1984 D. (FRS)–C. (NAM) Mar.

First: *Tikhinella* Bykova, 1952; former USSR: Russian Platform; Canada.
Last: *Earlandinita* Cummings, 1955 (USA: Texas, Great Britain) and *Lugtonia* Cummings, 1955 (USA, UK).

F. NODOSINELLIDAE Rhumbler, 1895 S. (LUD)–P. Mar.

First: *Eolagena* Lipina, 1959; former USSR: Siberia.
Last: *Biparietata* Zolotova, 1980 (former USSR: Urals) and *Nodosinella* Brady, 1876 (England, UK).

Superfamily GEINITZINACEA Bozorgnia, 1973

F. GEINITZINIDAE Bozorgnia, 1973 D. (FRS)–P. (ZEC) Mar.

First: *Frondilina* Bykova, 1952; former USSR; USA: Alaska.
Last: *Lunucammina* Spandel, 1898; Europe; Asia; Australia; North America.

F. PACHYPHLOIIDAE Loeblich and Tappan, 1984 P. (ROT–ZEC) Mar.

First: *Pachyphloia* Lange, 1925; Malay Archipelago; Sumatra; former USSR; Iran; Turkey.
Last: *Maichelina* Sosnina, 1977 (former USSR) and *Robustopachyphloia* Lin, 1980 (China).

Superfamily COLANIELLACEA Fursenko, 1959

F. COLANIELLIDAE Fursenko, 1959 D. (u.)–P. (ZEC) Mar.

First: *Multiseptida* Bykova, 1952; former USSR: Russian Platform; Canada: Alberta; USA: Alaska.
Last: *Colaniella* Likharev, 1939 (Vietnam, China, Japan, former USSR, Greece, Turkey), *Cylindrocolaniella* Loeblich and Tappan, 1985 (former eastern USSR) and *Pseudowanganella* Sosnina, 1983 (former USSR).

Superfamily PTYCHOCLADIACEA Elias, 1950

F. PTYCHOCLADIIDAE Elias, 1950 D. (FRS)–C. (STE) Mar.

First: *Shuguria* Antropov, 1950; former USSR: Russian Platform.
Last: *Ptychocladia* Ulrich and Bassler, 1904; USA: Illinois, Nebraska, Oklahoma).

Superfamily PALAEOTEXTULARIACEA Galloway, 1933

F. SEMITEXTULARIIDAE Pokorny, 1956 D. (GIV)–C. (u.) Mar.

First: *Paratextularia* Pokorny, 1951; USA: Iowa; Czechoslovakia; Poland; former USSR: Russian Platform.
Last: *Koskinotextularia* Eickhoff, 1968; Germany; Belgium; France; USA: Oklahoma; Canada.

F. PALAEOTEXTULARIIDAE Galloway, 1933 C. (TOU)–P. Mar.

First: *Palaeotextularia* Schubert, 1921; cosmopolitan.
Last: *Climacammina* Brady, 1873 (cosmopolitan), *Cribrogenerina* Schubert, 1908 (Sumatra, former USSR, USA) and *Palaeobigenerina* Galloway, 1933 (cosmopolitan).

F. BISERIAMMINIDAE Chernysheva, 1941 C. (TOU)–P. (DZH) Mar.

First: *Biseriammina* Chernysheva, 1941; former USSR: Urals, Bashkirian.
Last: *Globivalvulina* Schubert, 1921 (cosmopolitan), *Paraglobivalvulina* Reytlinger, 1965 (former USSR, Turkey, Iran, India, Thailand, China), *Paraglobivalvulinoides* Zaninetti and Jenny-Deshusses, 1985 (Iran), *Dagmarita* Reytlinger, 1965 (former USSR, India, Iran, Turkey, China), *Paradagmarita* Lys, 1978 (Turkey, Iran, Oman, ?Afghanistan) and *Louisettita* Altiner and Brönnimann, 1980 (Turkey).

Superfamily TOURNAYELLACEA Dain, 1953

F. TOURNAYELLIDAE Dain, 1953 D. (FRS)–C. (NAM) Mar.

First: *Eotournayella* Lipina and Pronina, 1964; former USSR: Urals.
Last: *Chernobaculites* Conil and Lys, 1977; former USSR; Europe: Pyrenees.

F. PALAEOSPIROPLECTAMMINIDAE Loeblich and Tappan, 1984 D. (FAM)–C. (VIS) Mar.

First: *Palaeospiroplectammina* Lipina, 1965 (former USSR, western Europe, USA: Arizona) and *Rectochernyshinella* Lipina, 1960 (former USSR).
Last: *Eotextularia* Mamet, 1970 (former USSR, Belgium, Denmark), *Halenia* Conil, 1980 (Belgium), *Palaeospiroplectammina* as above and *Endospiroplectammina* Lipina, 1970 (former USSR).

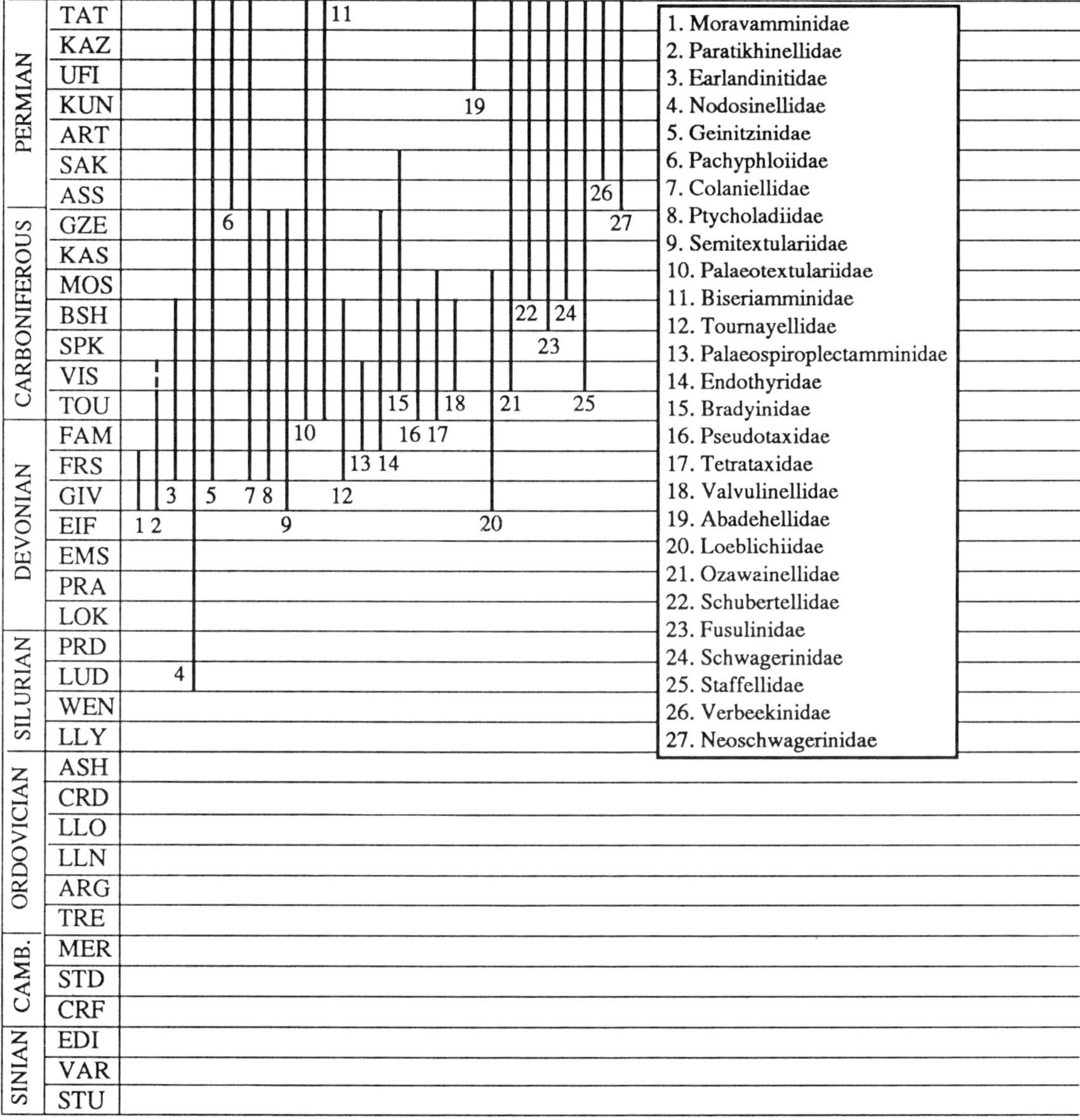
PERMIAN
CARBONIFEROUS
DEVONIAN
SILURIAN
ORDOVICIAN
CAMB.
SINIAN
TAT
KAZ
UFI
KUN
ART
SAK
ASS
GZE
KAS
MOS
BSH
SPK
VIS
TOU
FAM
FRS
GIV
EIF
EMS
PRA
LOK
PRD
LUD
WEN
LLY
ASH
CRD
LLO
LLN
ARG
TRE
MER
STD
CRF
EDI
VAR
STU
1. Moravamminidae
2. Paratikhinellidae
3. Earlandinitidae
4. Nodosinellidae
5. Geinitzinidae
6. Pachyphloiidae
7. Colaniellidae
8. Ptycholadiidae
9. Semitextulariidae
10. Palaeotextulariidae
11. Biseriamminidae
12. Tournayellidae
13. Palaeospiroplectamminidae
14. Endothyridae
15. Bradyinidae
16. Pseudotaxidae
17. Tetrataxidae
18. Valvulinellidae
19. Abadehellidae
20. Loeblichiidae
21. Ozawainellidae
22. Schubertellidae
23. Fusulinidae
24. Schwagerinidae
25. Staffellidae
26. Verbeekinidae
27. Neoschwagerinidae

Fig. 4.3

Superfamily ENDOTHYRACEA Brady, 1884

F. ENDOTHYRIDAE Brady, 1884
D. (FAM)–C. (STE) Mar.

First: *Klubovella* Lebedeva, 1956 (former USSR) and *Quasiendothyra* Rauzer-Chernousova, 1948 (former USSR, Turkey).
Last: *Endothyranella* Galloway and Harlton, 1930; former USSR; USA: Texas, Oklahoma, Indiana.

F. BRADYINIDAE Reytlinger, 1950
C. (VIS)–P. (SAK) Mar.

First: *Bradyina* von Möller, 1878; former USSR; Iran; Mongolia; China; UK: England, Scotland; Belgium; USA.
Last: *Pseudobradyina* Reytlinger, 1950; former USSR: Moscow Basin, central Asia; USA: Kansas.

Superfamily TETRATAXACEA Galloway, 1933

F. PSEUDOTAXIDAE Mamet, 1974
C. (TOU–NAM) Mar.

First: *Pseudotaxis* Mamet, 1974; Europe; Asia; North America; Australia.
Last: *Pseudotaxis* as above and *Vissariotaxis* Mamet, 1970 (former USSR, England, UK, Belgium, North Africa).

F. TETRATAXIDAE Galloway, 1933
C. (TOU–MOS) Mar.

First: *Tetrataxis* Ehrenberg, 1854; UK: England, Scotland; Belgium; Spain; Poland; Algeria; Iran; former USSR; China; Australia; USA.
Last: *Tetrataxis* as above and *Polytaxis* Cushman and Waters, 1928 (USA, Austria, Algeria).

F. VALVULINELLIDAE Loeblich and Tappan, 1984
C. (VIS–NAM) Mar.

First: *Valvulinella* Schubert, 1908; England, UK; Scotland, UK; Belgium; Germany; former Yugoslavia; Turkey; former USSR; North America.
Last: Same genus.

F. ABADEHELLIDAE Loeblich and Tappan, 1984
P. (ZEC) Mar.

First and Last: *Abadehella* Okimura and Ishi, 1975; Iran; India; Malaysia; Japan; Cambodia; former USSR.

Superfamily FUSULINACEA von Möller, 1878

F. LOEBLICHIIDAE Cummings, 1955
D. (GIV)–C. (MOS) Mar.

First: *Rhenothyra* Beckman, 1950; Germany.
Last: *Seminovella* Rauzer-Chernousova, 1951; former USSR.

F. OZAWAINELLIDAE Thompson and Foster, 1937
C. (VIS)–P. (ZEC) Mar.

First: *Millerella* Thompson, 1942 (North America, former USSR, North Africa, China, Japan), *Chomatomediocris* Vdovenko, 1973 (former USSR), *Eostaffella* Rauzer-Chernousova, 1948 (North America, Europe, Asia), *Mediocris* Rozovskaya, 1961 (USA, Canada), *Ninella* Malakhova, 1975 (former USSR), *Plectomediocris* Brazhnikova and Vdovenko, 1983 (former USSR) and *Pseudoendothyra* Mikhaylov, 1939 (former USSR, Mongolia, China, Japan, North America, Europe).
Last: *Eostaffeloides* Miklukho-Maklay, 1959 (former USSR), *Parareichelina* Miklukho-Maklay, 1959 (former USSR), *Pseudokahlerina* Sosnina, 1968 (former USSR), *Reichelina* Erk, 1942 (former USSR, China, SE Asia, Greece, Turkey, North America), *Sichotenella* Tumanskaya, 1953 (former USSR), *Kangvarella* Saurin, 1962 (Cambodia) and *Primoriina* Sosnina, 1981 (former USSR).

F. SCHUBERTELLIDAE Skinner, 1931
C. (MOS)–P. (ZEC) Mar.

First: *Eoschubertella* Thompson, 1937 (North and South America, Europe, Asia), *Fusiella* Lee and Chen, 1930 (former USSR, North America, China, Japan) and *Neofusulinella* Deprat, 1912 (Laos).
Last: *Neoschubertella* Saurin, 1962 (Cambodia, Laos), *Codonofusiella* Dunbar and Skinner, 1937 (USA, Canada, Japan, Cambodia, Pakistan, former Yugoslavia, Greece, Turkey, former USSR), *Dunbarula* Ciry, 1948 (former Yugoslavia, former USSR, China, Japan, North Africa, North America), *Gallowaiina* Chen, 1934 (former USSR, southern China), *Minojapanella* Fujimoto and Kanuma, 1953 (Japan, China, Inner Mongolia, Sumatra, former Yugoslavia, former USSR), *Nanlingella* Rui and Sheng, 1981 (southern China), *Palaeofusulina* Deprat, 1912 (Vietnam, China, Japan, Timor, former USSR, former Yugoslavia), *Paradoxiella* Skinner and Wilde, 1955 (USA, Japan), *Paradunbarula* Skinner, 1969 (Turkey, southern China, former USSR), *Parananlingella* Rui and Shen, 1981 (southern China), *Russiella* Miklukho-Maklay, 1957 (southern China, former USSR), *Tewoella* Sun, 1979 (northern China) and *Ziguiella* Lin, 1980 (China).

F. FUSULINIDAE von Möller, 1878
C. (BSH)–P. (ZEC) Mar.

First: *Verella* Dalmatskaya, 1951 (former USSR), *Eowedekindellina* Ektova, 1977 (former USSR).
Last: *Yangchienia* Lee, 1934; southern China; Japan; Korea; former USSR; former Yugoslavia; Sicily; Greece; Turkey; Afghanistan; Algeria.

F. SCHWAGERINIDAE Dunbar and Henbest, 1930
C. (MOS)–P. (ZEC) Mar.

First: *Moniparus* Rozovskaya, 1948; former USSR; China; Japan.
Last: *Darvasites* Miklukho-Maklay, 1959 (former USSR, China, Japan), *Nipponitella* Hanzawa, 1938 (Japan, former USSR) and *Rugososchwagerina* Miklukho-Maklay, 1959 (Sicily, Iran, Iraq, Afghanistan, China, former USSR).

F. STAFFELLIDAE Miklukho-Maklay, 1949
C. (VIS)–P. (TAT) Mar.

First: *Reitlingerina* Rauzer-Chernousova, 1985; Europe; Asia; Canada.
Last: *Eoverbeekina* Lee, 1934; former USSR; China; Japan; former Yugoslavia; USA; Belize; Guatemala; Mexico.

F. VERBEEKINIDAE Staff and Wedekind, 1910
P. (SAK)–P. (TAT) Mar.

First: *Brevaxina* Schenck and Thompson, 1940 (Laos, Japan, southern China, former USSR) and *Misellina* Schenck and Thompson, 1940 (former USSR, Japan, China, Laos, Turkey, Sumatra, former Yugoslavia).
Last: *Pseudodoliolina* Yabe and Hanzawa, 1932; Japan; Vietnam; China; former USSR; USA.

F. NEOSCHWAGERINIDAE Dunbar and Condra, 1927 P. (ASS–DZH) Mar.

First: *Shengella* Yang, 1985; China.
Last: *Yabeina* Deprat, 1914; Japan; China; Vietnam; Cambodia; New Zealand; former USSR; Canada; USA.

Suborder INVOLUTINA Hohenegger and Piller, 1977

F. INVOLUTINIDAE Butschli, 1880 P. (ROT)–K. (CEN) Mar. (see Fig. 4.4)

First: *Neohemigordius* Wang and Sun, 1973; China.
Last: *Hensonina* Moullade and Peybernes, 1974 (Qatar, northern Spain), *Involutina* Terquem, 1862 (Europe, Asia) and *Trocholina* Paalzow, 1922 (cosmopolitan).

F. HIRSUTOSPIRELLIDAE Zainetti *et al.*, 1985 Tr. (NOR) Mar.

First and Last: *Hirsutospirella* Zainetti *et al.*, 1985; former Yugoslavia; Sicily.

F. VENTROLAMINIDAE Weynshcenck, 1950 J. (BAJ)–K. (BER) Mar.

First: *Archaeosepta* Wernli, 1970; France; Sardinia.
Last: *Protopeneroplis* Weynschenck, 1950; Austria; Italy; France; former Yugoslavia; Switzerland; Israel; Turkey.

Suborder SPIRILLININA Hohenegger and Piller, 1975

F. SPIRILLINIDAE Reuss and Fritsch, 1861 Tr. (RHT)–Rec. Mar.

First: *Spirillina* Ehrenberg, 1843; cosmopolitan. **Extant**

F. PATELLINIDAE Rhumbler, 1906 K. (APT)–Rec. Mar.

First: *Hergotella* Ludbrook; 1966 (South Australia) and *Patellina* Williamson, 1858; cosmopolitan. **Extant**

Suborder CARTERININA Loeblich and Tappan, 1981

F. CARTERINIDAE Loeblich and Tappan, 1955 T. (PRB)–Rec. Mar.

First: *Carterina* Brady, 1884; Spain. **Extant**

Suborder MILIOLINA Delage and Hérouard, 1896

Superfamily SQUAMULINACEA Reuss and Fritsch, 1861

F. SQUAMULINIDAE Reuss and Fritsch, 1861 K. (MAA)–Rec. Mar.

First: *Brasiliella* Troelson, 1978; Brazil. **Extant**

Superfamily CORNUSPIRACEA Schultze, 1854

F. CORNUSPIRIDAE Schultze, 1854 C. (VIS)–Rec. Mar.

First: *Trepeilopsis* Cushman and Waters, 1928; North America; Europe; North Africa; Asia; Australia. **Extant**

F. HEMIGORDIOPSIDAE A. Nikitina, 1969 C. (VIS)–Rec. Mar.

First: *Hemigordius* Schubert, 1908; Germany; Czechoslovakia; Iran; China; USA. **Extant**

F. BAISALINIDAE Loeblich and Tappan, 1986 P. (ROT–ZEC) Mar.

First: *Septagathammina* Lin, 1984; China; Sumatra.
Last: *Nikitinella* Sosnina, 1983 (former USSR) and *Pseudobaisalina* Sosnina, 1983 (former USSR).

F. FISCHERINIDAE Millett, 1898 J. (BAJ)–Rec. Mar.

First: *Dolosella* Danich, 1969; former USSR; England, UK. **Extant**

F. NUBECULARIIDAE Jones, 1875 Tr. (ANS)–Rec. Mar.

First: *Gheorghianina* Loeblich and Tappan, 1986; Romania; Bulgaria. **Extant**

F. OPHTHALMIDIIDAE Wiesner, 1920 Tr. (ANS)–Rec. Mar.

First: *Eoophthalmidium* Langer, 1968; Turkey. **Extant**

F. DISCOSPIRINIDAE Wiesner, 1931 T. (MMI)–Rec. Mar.

First: *Discospirina* Munier-Chalmas, 1902; Atlantic; Mediterranean. **Extant**

Superfamily MILIOLACEA Ehrenberg, 1839

F. MILIOLECHINIDAE Zaninetti *et al.*, 1985 Tr. (NOR)–Rec. Mar.

First: *Miliolechina* Zaninetti *et al.*, 1985; former Yugoslavia. **Extant**

F. SPIROLOCULINIDAE Wiesner, 1920 J. (BAJ)–Rec. Mar.

First: *Palaeomiliolina* Antonova, 1959; former USSR. **Extant**

F. HAUERINIDAE Schwager, 1876 J.–Rec. Mar.

First: *Cycloforina* Luczkowska, 1972; cosmopolitan. **Extant**

F. MILIOLIDAE Ehrenberg, 1839 T. (LUT)–Rec. Mar.

First: *Neaguites* Andersen, 1984 (USA) and *Miliola* Lamarck, 1804 (France, Belgium, USA). **Extant**

F. RIVEROINIDAE Saidova, 1981 T. (RUP/CHT)–Rec. Mar.

First: *Pseudohauerina* Ponder, 1972; Atlantic; Pacific. **Extant**

F. AUSTROTRILLINIDAE Loeblich and Tappan, 1986 T. (LUT/BRT–LMI) Mar.

First: *Reticulogyra* Adams and Belford, 1979; New Guinea.
Last: *Austrotrillina* Parr, 1942; Pacific Islands; Australia; India; Sri Lanka; Malaysia; Sarawak; New Guinea; Somalia; Kenya; Tanzania; Libya; Iraq; Iran; Turkey; Greece; Spain.

Superfamily ALVEOLINACEA Ehrenberg, 1839

F. FABULARIIDAE Ehrenberg, 1839 K. (SEN)–T. (?RUP/CHT) Mar.

First: *Lacazina* Munier-Chalmas, 1882 (France, Spain, Israel) and *Periloculina* Munier-Chalmas and Schlumberger, 1885 (France).
Last: *Lacazinella* Crespin, 1962; New Guinea; Indonesia; Turkey.

F. RHAPYDIONINIDAE Keijzer, 1945 K. (CEN)–?Rec. Mar.

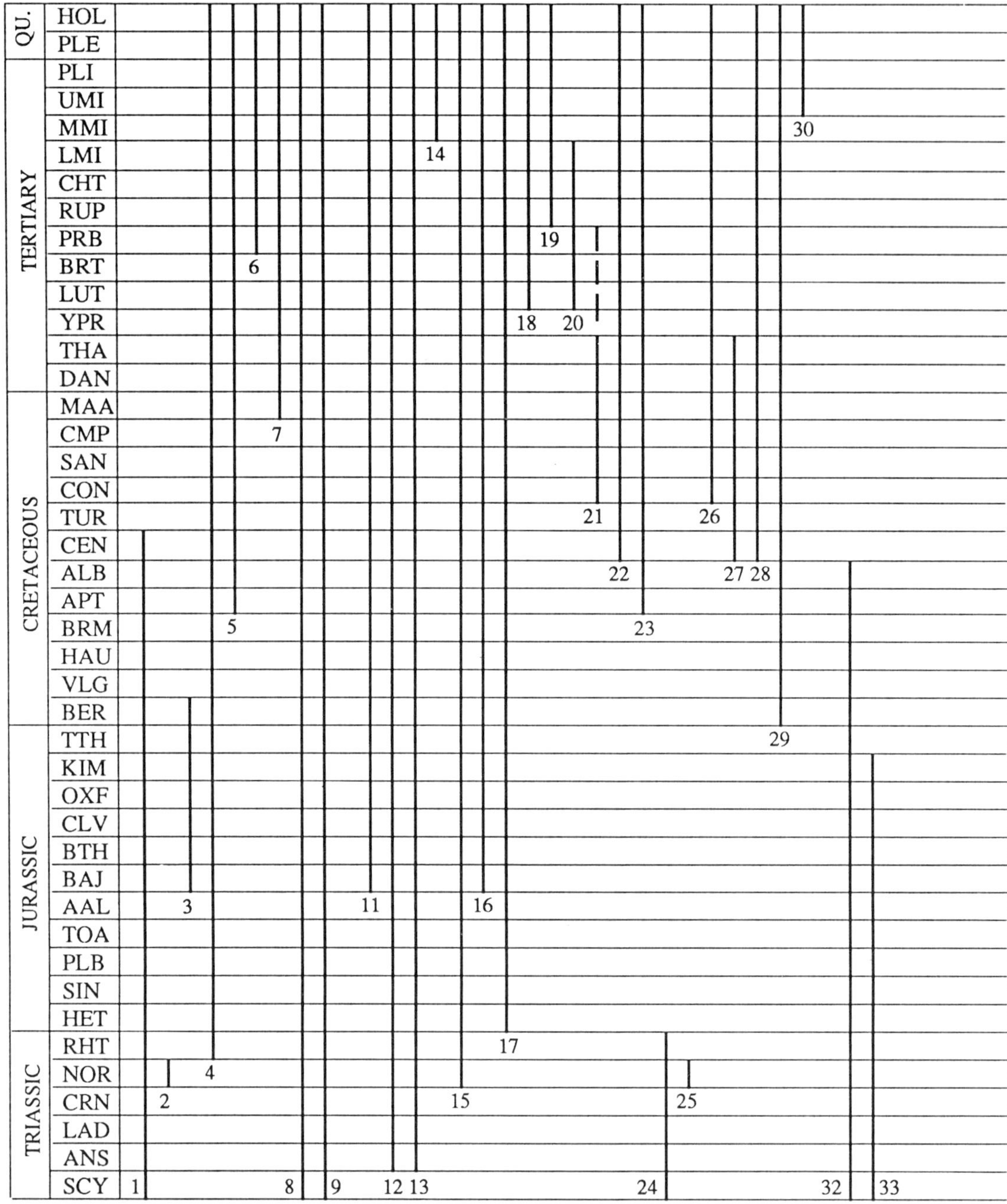

Fig. 4.4

First: *Pseudedomia* Henson, 1948; Qatar; Kuwait; Tunisia; Lebanon; Iraq; Israel; Italy; Portugal; former Yugoslavia; Greece. **?Extant**
Comment: The extant species may have been a reworked Cretaceous specimen.

F. ALVEOLINIDAE Ehrenberg, 1839
K. (APT)–Rec. Mar.

First: *Archaealveolina* Fourcade, 1980; Spain; Italy; Algeria. **Extant**

Superfamily SORITACEA Ehrenberg, 1839

F. MILIOLIPORIDAE Brönniman and Zaninetti, 1971
P. (DZL)–Tr. (RHT) Mar.

First: *Kamurana* Altiner and Zaninetti, 1977; Turkey; Bulgaria.
Last: *Galeanella* Kristan, 1958 (Austria, former Yugoslavia, Iran), *Orthotrinacria* Zaninetti *et al.*, 1985 (Turkey, Sicily, former Yugoslavia), *Cucurbita* Jablonsky, 1973 (Czechoslovakia, Sicily, Greece) and *Miliolipora* Brönnimann and Zaninetti, 1971 (Austria, Czechoslovakia, Iran).

F. SIPHONOFERIDAE Senowbari-Daryan and Zaninetti, 1986 Tr. (NOR) Mar.

First and Last: *Siphonofera* Senowbari-Daryan, 1983; Sicily.

F. PENEROPLIDAE Schultze, 1854
K. (CON)–Rec. Mar.

Key for both diagrams

1. Involutinidae	13. Ophthalmidiidae	24. Milioliporidae
2. Hirsutospirellidae	14. Discospirinidae	25. Siphonoferidae
3. Ventrolaminidae	15. Milioechinidae	26. Peneroplidae
4. Spirillinidae	16. Spiroloculinidae	27. Meandropsinidae
5. Patellinidae	17. Hauerinidae	28. Soritidae
6. Carterinidae	18. Milioidae	29. Keramosphaeridae
7. Squamulinidae	19. Riveroinidae	30. Silicoloculinidae
8. Cornuspiridae	20. Austrotrillinidae	31. Syzraniniidae
9. Hemigordiopsidae	21. Fabulariidae	32. Ichthyolariidae
10. Baisalinidae	22. Rhapydioninidae	33. Robuloididae
11. Fischerinidae	23. Alveolinidae	34. Partisaniidae
12. Nubeculariidae		

Fig. 4.4

First: *Vandenbroeckia* Marie, 1958; France. **Extant**

F. MEANDROPSINIDAE Henson, 1948 K. (CEN)–T. (THA) Mar.

First: *Broeckina* Munier-Chalmas, 1882 (France, Syria) and *Pastrikella* Cherchi, Radoičić and Schroeder, 1976 (former Yugoslavia, France).
Last: *Hottingerina* Drobne, 1975; former Yugoslavia.

F. SORITIDAE Ehrenberg, 1839 K. (CEN)–Rec. Mar.

First: *Edomia* Henson, 1948 (Iran, Israel) and *Pseudorhipidionina* de Castro, 1971 (Italy, Israel, Algeria). **Extant**

F. KERAMOSPHAERIDAE Brady, 1884 K. (BER)–Rec. Mar.

First: Genus *Pavlovecina* Loeblich and Tappan, 1988; France; Switzerland. **Extant**

Suborder SILICOLOCULININA Resig *et al.*, 1980

F. SILICOLOCULINIDAE Resig *et al.*, 1980 T. (UMI)–Rec. Mar.

First: *Miliammellus* Saidova and Burmistrova, 1978; Bering Sea; Antarctic; North and Central Pacific and Indian Oceans. **Extant**

Suborder LAGENINA Delage and Hérouard, 1896

Superfamily ROBULOIDACEA Reiss, 1963

F. SYZRANIIDAE Vachard, 1981 S. (PRD)–P. (DZH) Mar. (see Fig. 4.5)

First: *Tuborecta* Pronina, 1980; former USSR.
Last: *Rectostipulina* Jenny-Deshusses, 1985; Turkey; Cyprus; Greece; Afghanistan; Iran; India; former USSR: Armenia.

F. ICHTHYOLARIIDAE Loeblich and Tappan, 1986 P. (ART)–K. (ALB) Mar.

First: *Protonodosaria* Gerke, 1959; former USSR; China; Burma.

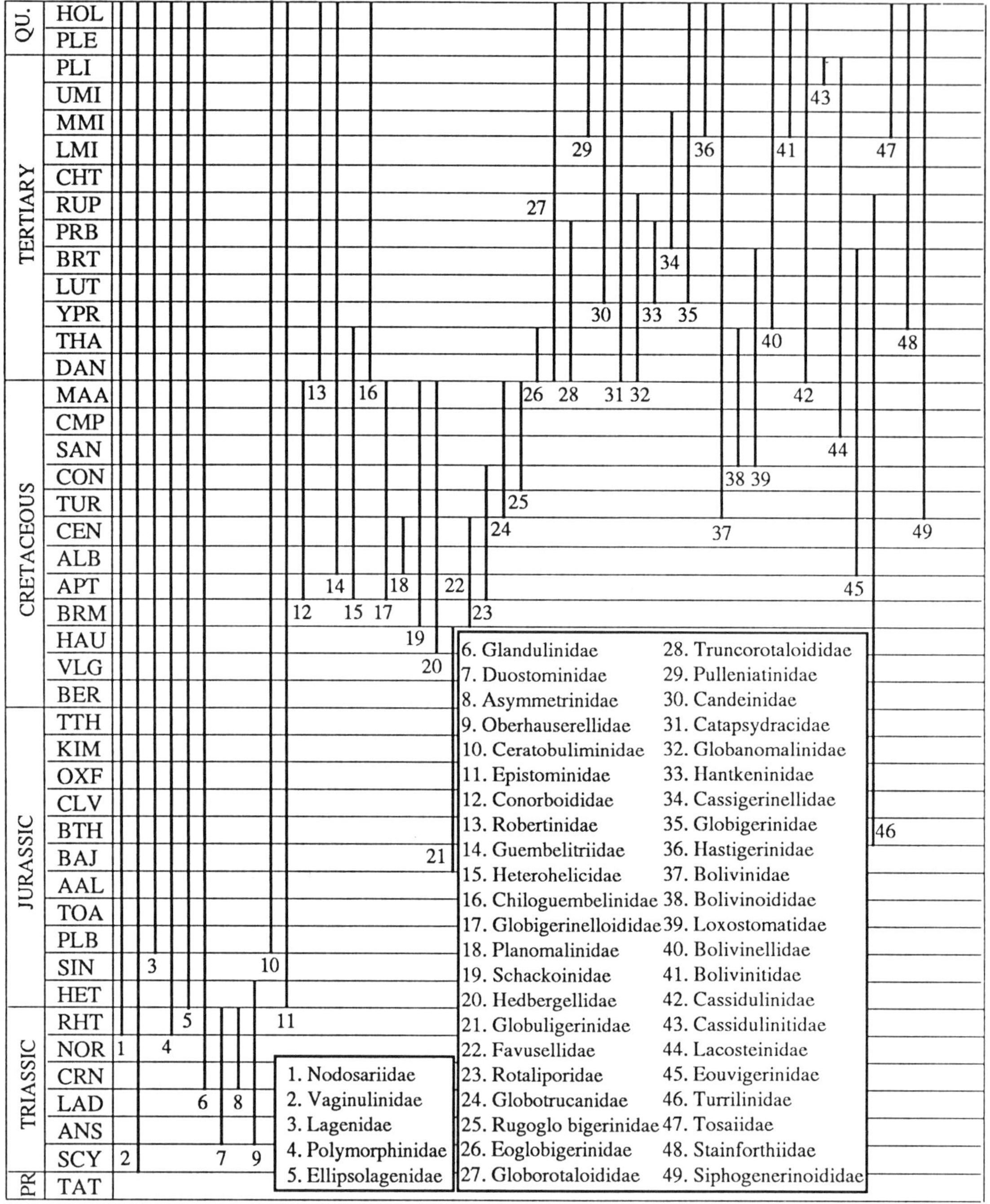

Fig. 4.5

Last: *Lingulonodosaria* Silvestri, 1903; cosmopolitan.

F. ROBULOIDIDAE Reiss, 1963 P. ('m.')–J. (KIM) Mar.

First: *Robuloides* Reichel, 1946; Greece; Italy; former Yugoslavia; Austria; former USSR; Sumatra.
Last: *Parinvolutina* Pelissié and Peybernes, 1982; France.

F. PARTISANIIDAE Loeblich and Tappan, 1984 P. (ROT–ZEC) Mar.

First: *Xintania* Lin, 1984; China.
Last: *Partisania* Sosnina, 1978; former USSR.

Superfamily NODOSARIACEA Ehrenberg, 1838

F. NODOSARIIDAE Ehrenberg, 1838 Tr. (RHT)–Rec. Mar.

First: *Berthelinella* Loeblich and Tappan, 1957; France; Austria; USA: Alaska. **Extant**

F. VAGINULINIDAE Reuss, 1860 Tr.–Rec. Mar.

First: *Lenticulina* Lamarck, 1804 (cosmopolitan) and *Vaginulinopsis* Silvestri, 1904 (cosmopolitan) **Extant**

F. LAGENIDAE Reuss, 1862 J. (PLB)–Rec. Mar.

First: *Reussoolina* Colom, 1956; cosmopolitan. **Extant**

F. POLYMORPHINIDAE d'Orbigny, 1839 Tr. (RHT)–Rec. Mar.

First: *Eoguttulina* Cushman and Ozawa, 1930 (Europe, North America, South America), *Pyrulinoides* Marie, 1941 (cosmopolitan) and *Sagoplecta* Tappan, 1951 (USA: Alaska). **Extant**

F. ELLIPSOLAGENIDAE A. Silvestri, 1923 J.–Rec. Mar.

First: *Oolina* d'Orbigny, 1839; cosmopolitan. **Extant**

F. GLANDULINIDAE Reuss, 1860 Tr. (CRN)–Rec. Mar.

First: *Glandulinoides* Hu, 1977; former USSR: Siberia; China: Yunnan. **Extant**

Suborder ROBERTININA Loeblich and Tappan, 1984

Superfamily DUOSTOMINACEA Brotzen, 1963

F. DUOSTOMINIDAE Brotzen, 1963 Tr. (ANS–RHT) Mar.

First: *Diplotremina* Kristan-Tollmann, 1960 (Austria, Hungary, Bulgaria, Poland, former Yugoslavia, China, USA), *Duostomina* Kristan-Tollmann, 1960 (Austria, Italy, former Yugoslavia, Bulgaria, Hungary, China) and *Variostoma* Kristan-Tollmann, 1960 (Austria, China, USA).
Last: *Duostomina* as above.

F. ASYMMETRINIDAE Brotzen, 1960 Tr. (CRN–RHT) Mar.

First: *Plagiostomella* Kristan-Tollmann, 1960; Austria.
Last: *Asymmetrina* Kristan-Tollmann, 1960; (Austria) and *Involvina* Kristan-Tollmann, 1960 (Austria).

F. OBERHAUSERELLIDAE Fuchs, 1970 Tr. (ANS)–J. (HET) Mar.

First: *Krikoumbilica* He, 1984 (China: Guizhou) and *Oberhauserella* Fuchs, 1967 (Austria, Bulgaria, Poland, China).
Last: *Oberhauserella* as above.

Superfamily CERATOBULIMINACEA Cushman, 1927

F. CERATOBULIMINIDAE Cushman, 1927 J. (PLB)–Rec. Mar.

First: *Reinholdella* Brotzen, 1948; Europe; former USSR: Siberia, Azerbaijan. **Extant**

F. EPISTOMINIDAE Wedekind, 1937 J. (l.)–Rec. Mar.

First: *Epistomina* Terquem, 1883 (cosmopolitan), *Garantella* Kaptarenko-Chernousova, 1956 (Europe, former USSR: Ukraine, Madagascar) and *Sublamarckella* Antonova, 1958 (former USSR: northern Caucasus). **Extant**

Superfamily CONORBOIDACEA Thalmann, 1952

F. CONORBOIDIDAE Thalmann, 1952 K. (APT–MAA) Mar.

First: *Stedumia* Bertram and Kemper, 1982; Germany.
Last: *Colomia* Cushman and Bermúdez, 1948; Germany; Austria; USA: California; Cuba; Australia: Victoria.

Superfamily ROBERTINACEA Reuss, 1850

F. ROBERTINIDAE Reuss, 1850 T. (DAN/THA)–Rec. Mar.

First: *Robertina* d'Orbigny, 1846; Europe; North America; New Zealand; Atlantic; Pacific; Arctic; Antarctic. **Extant**

Suborder GLOBIGERININA Delage and Hérouard, 1896

Superfamily HETEROHELICACEA Cushman, 1927

F. GUEMBELITRIIDAE Montanaro Gallitelli, 1957 K. (ALB)–Rec. Mar.

First: *Guembelitria* Cushman, 1933; cosmopolitan. **Extant**

F. HETEROHELICIDAE Cushman, 1927 K. (APT)–T. (DAN/THA) Mar.

First: *Spiroplecta* Ehrenberg, 1844; cosmopolitan.
Last: *Spiroplecta* as above and *Bifarina* Parker and Jones, 1872 (Europe, North America, Africa).

F. CHILOGUEMBELINIDAE Reiss, 1963 T. (DAN)–Rec. Mar.

First: *Chiloguembelina* Loeblich and Tappan, 1956; cosmopolitan. **Extant**

Superfamily PLANOMALINACEA Bolli *et al.*, 1957

F. GLOBIGERINELLOIDIDAE Longoria, 1974 K. (APT–MAA) Mar.

First: *Blowiella* Kretzschmar and Gorbachik, 1971 (cosmopolitan) and *Globigerinelloides* Cushmann and ten Dam, 1948 (cosmopolitan).
Last: *Globigerinelloides* as above.

F. PLANOMALINIDAE Bolli *et al.*, 1957 K. (ALB–CEN) Mar.

First and Last: *Planomalina* Loeblich and Tappan, 1946; Europe; North and East Africa; USA: Texas, California; Atlantic; Caribbean.

F. SCHACKOINIDAE Pokorny, 1958 K. (BRM–MAA) Mar.

First: *Leupoldina* Bolli, 1957; Trinidad; Mexico; France; Tunisia.
Last: *Schackoina* Thalmann, 1932; cosmopolitan.

Superfamily ROTALIPORACEA Sigal, 1958

F. HEDBERGELLIDAE Loeblich and Tappan, 1961 K. (HAU–MAA) Mar.

First and Last: *Hedbergella* Brönnimann and Brown, 1958; cosmopolitan.

F. GLOBULIGERINIDAE Loeblich and Tappan, 1984 J. (BAJ)–K. (HAU) Mar.

First: *Conoglobigerina* Morozova, 1961; former USSR: Caucasus, Crimea, Azerbaijan; Turkmenia; Poland; Bulgaria; Canada: Nova Scotia.
Last: *Globuligerina* Bignot and Guyader, 1971; former USSR: Lithuania, Komi, Turkmen, Dagestan, Caucasus, Azerbaijan, Crimea; Sweden; France; Germany.

F. FAVUSELLIDAE Longoria, 1974 K. (BRM–CEN) Mar.

First and Last: *Favusella* Michael, 1973; cosmopolitan.

F. ROTALIPORIDAE Sigal, 1958 K. (APT–CON) Mar.

First and Last: *Clavihedbergella* Banner and Blow, 1959; Europe; North America.

Superfamily GLOBOTRUNCANACEA Brotzen, 1942

F. GLOBOTRUNCANIDAE Brotzen, 1942
K. (TUR–MAA) Mar.

First: *Sigalitruncana* Korchagin, 1982 (cosmopolitan) and *Marginotruncana* Hofker, 1956 (cosmopolitan).
Last: *Gansserina* Caron *et al.*, 1984 (Trinidad, Spain, Turkey, Egypt, Tunisia, Mid-Pacific), *Globotruncana* Cushman, 1927 (cosmopolitan), *Kassabiana* Salaj and Solakius, 1984 (Egypt, Iraq, Tunisia, Pakistan), *Rugotruncana* Brönnimann and Brown, 1956 (Cuba, Trinidad), *Globotruncanella* Reiss, 1957 (cosmopolitan) and *Abathomphalus* Bolli *et al.*, 1957 (cosmopolitan).

F. RUGOGLOBIGERINIDAE Subbotina, 1959
K. (CON–MAA) Mar.

First: *Archaeoglobigerina* Pessagno, 1967; cosmopolitan.
Last: *Archaeoglobigerina* as above, *Bucherina* Brönnimann and Brown, 1956 (Cuba, North and South Atlantic, Egypt), *Klugerina* Brönnimann and Brown, 1956 (Trinidad, Tunisia) and *Trinitella* Brönnimann, 1952 (cosmopolitan).

Superfamily GLOBOROTALIACEA Cushman, 1927

F. EOGLOBIGERINIDAE Blow, 1979
T. (DAN–THA) Mar.

First: *Eoglobigerina* Morozova, 1959 (cosmopolitan), *Globoconusa* Khalilov, 1956 (cosmopolitan), *Parvularugoglobigerina* Hofker, 1978 (Italy, Germany, Spain, former USSR: Caspian Sea, Pacific, Atlantic: off Florida, Caribbean) and *Postrugoglobigerina* Salaj, 1986 (NW Tunisia).
Last: *Eoglobigerina* as above.

F. GLOBOROTALIIDAE Cushman, 1927
T. (DAN)–Rec. Mar.

First: *Planorotalites* Morozova, 1957; cosmopolitan. **Extant**

F. TRUNCOROTALOIDIDAE Loeblich and Tappan, 1961 T. (THA–PRB) Mar.

First: *Morozovella* McGowran, 1968; cosmopolitan.
Last: *Muricoglobigerina* Blow, 1979; cosmopolitan.

F. PULLENIATINIDAE Cushman, 1927
T. (MMI)–Rec. Mar.

First: *Globigerinopsis* Bolli, 1962; Venezuela; Dominican Republic. **Extant**

F. CANDEINIDAE Cushman, 1927
T. (LUT)–Rec. Mar.

First: *Tenuitella* Fleisher, 1974; cosmopolitan. **Extant**

F. CATAPSYDRACIDAE Bolli, Loeblich and Tappan, 1957 T. (DAN)–Rec. Mar.

First: *Subbotina* Brotzen and Pozaryska, 1961; cosmopolitan. **Extant**

Superfamily HANTKENINACEA Cushman, 1927

F. GLOBANOMALINIDAE Loeblich and Tappan, 1984 T. (DAN–RUP) Mar.

First and Last: *Globanomalina* Haque, 1956; cosmopolitan.

F. HANTKENINIDAE Cushman, 1927
T. (LUT)–T. (PRB) Mar.

First: *Aragonella* Thalmann, 1942 (cosmopolitan) and *Hantkenina* Cushman, 1924 (cosmopolitan).
Last: *Cribrohantkenina* Thalmann, 1942 (cosmopolitan) and *Hantkenina* as above.

F. CASSIGERINELLIDAE Bolli *et al.*, 1957
T. (PRB–MMI) Mar.

First and Last: *Cassigerinella* Pokorny, 1955; cosmopolitan.

Superfamily GLOBIGERINACEA Carpenter *et al.*, 1862

F. GLOBIGERINIDAE Carpenter *et al.*, 1862
T. (LUT)–Rec. Mar.

First: *Globigerinatheka* Brönnimann, 1952 (cosmopolitan), *Inordinatosphaera* Mohan and Soodan, 1967 (India), *Orbulinoides* Cordey, 1968 (cosmopolitan) and *Porticulasphaera* Bolli, Loeblich and Tappan, 1957 (cosmopolitan). **Extant**

F. HASTIGERINIDAE Bolli, Loeblich and Tappan, 1957 T. (MMI)–Rec. Mar.

First: *Hastigerinopsis* Saito and Thompson, 1976; cosmopolitan, tropical to warm subtropical. **Extant**

Suborder ROTALIINA Delage and Hérouard, 1896

Superfamily BOLIVINACEA Glaessner, 1937

F. BOLIVINIDAE Glaessner, 1937 K. (TUR)–Rec. Mar.

First: *Grimsdaleinella* Bolli, 1959; Trinidad. **Extant**

F. BOLIVINOIDIDAE Cushman, 1927
K. (SAN)–T. (DAN/THA) Mar.

First and Last: *Bolivinoides* Cushman, 1927; cosmopolitan.

Superfamily LOXOSTOMATACEA Loeblich and Tappan, 1962

F. LOXOSTOMATIDAE Loeblich and Tappan, 1962
K. (SAN)–T. (LUT/BRT) Mar.

First: *Loxostomum* Ehrenberg, 1854; England, UK; Republic of Ireland; USA: Arkansas, Texas.
Last: *Aragonia* Finlay, 1939 (North America, Caribbean, Italy, Morocco, New Zealand) and *Zeauvigerina* Finlay, 1939 (cosmopolitan).

F. BOLIVINELLIDAE Hayward, 1980
T. (Eoc.)–Rec. Mar.

First: *Bolivinella* Cushman, 1927; cosmopolitan, tropical to subtropical shallow water. **Extant**

Superfamily BOLIVINITACEA Cushman, 1927

F. BOLIVINITIDAE Cushman, 1927 T. (MMI)–Rec. Mar.

First: *Bolivinita* Cushman, 1927; Atlantic; Pacific; Indo-Pacific. **Extant**

Superfamily CASSIDULINACEA d'Orbigny, 1839

F. CASSIDULINIDAE d'Orbigny, 1839
T. (DAN/THA)–Rec. Mar.

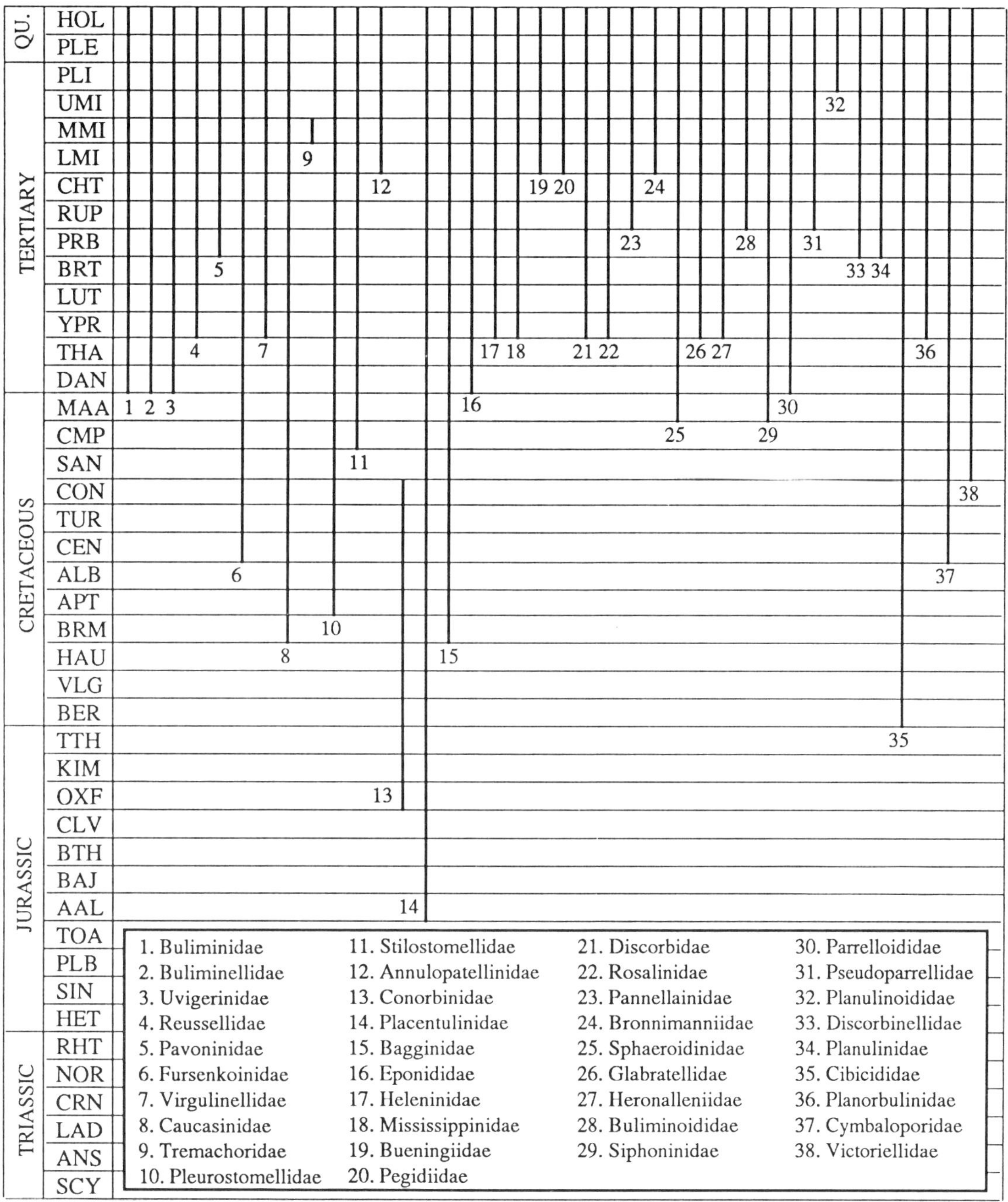

Fig. 4.6

First: *Islandiella* Nørvang, 1959; cosmopolitan. **Extant**

F. CASSIDULINITIDAE Saidova, 1981 T. (PLI) Mar.

First and Last: *Cassidulinita* Suzin, 1952; former USSR: northern Caucasus.

***Superfamily* EOUVIGERINACEA Cushman, 1927**

F. LACOSTEINIDAE Sigal, 1952 K. (CMP)–T. (PLI) Mar.

First: *Elhasaella* Hamam, 1976 (Jordan) and *Lacosteina* Marie, 1945 (Morocco; USA: Alaska, California).
Last: *Felsinella* Conato, 1964 (Italy) and *Spirobolivina* Hofker, 1956 (Ecuador, South Atlantic).

F. EOUVIGERINIDAE Cushman, 1927 K. (ALB)–T. (?LUT/BRT) Mar.

First and Last: *Eouvigerina* Cushman, 1926; cosmopolitan.

***Superfamily* TURRILINACEA Cushman, 1927**

F. TURRILINIDAE Cushman, 1927 J. (BTH)–T. (RUP/CHT) Mar.

First: *Praebulimina* Hofker, 1953; cosmopolitan.
Last: *Turrilina* Andreae, 1884; France; Denmark; The Netherlands; Poland.

F. TOSAIIDAE Saidova, 1981 T. (MMI)–Rec. Mar.

First: *Tosaia* Takayanagi, 1953; Japan; Egypt; Pacific, off Panama. **Extant**

F. STAINFORTHIIDAE Reiss, 1963 T. (Eoc.)–Rec. Mar.

First: *Hopkinsina* Howe and Wallace, 1932 (North America, Europe) and *Stainforthia* Hofker, 1956 (cosmopolitan). **Extant**

Superfamily BULIMINACEA Jones, 1875

F. SIPHOGENERINOIDIDAE Saidova, 1981 K. (TUR)–Rec. Mar.

First: *Siphogenerinoides* Cushman, 1927 (North and South America, Africa, Middle East) and *Orthokarstenia* Dietrich, 1935 (North and South America, Africa). **Extant**

F. BULIMINIDAE Jones, 1875 T. (DAN)–Rec. Mar. (see Fig. 4.6)

First: *Bulimina* d'Orbigny, 1826 (cosmopolitan) and *Globobulimina* Cushman, 1927 (cosmopolitan). **Extant**

F. BULIMINELLIDAE Hofker, 1951 T. (DAN)–Rec. Mar.

First: *Quadratobuliminella* de Klasz, 1953; Germany; France; USA: California. **Extant**

F. UVIGERINIDAE Haeckel, 1894 T. (DAN/THA)–Rec. Mar.

First: *Trifarina* Cushman, 1923; Atlantic; Pacific; New Zealand; Egypt. **Extant**

F. REUSSELLIDAE Cushman, 1933 T. (Eoc.)–Rec. Mar.

First: *Chrysalidinella* Schubert, 1908; Caribbean; Cuba; USA; Kerimba Archipelago; Pacific; Indonesia. **Extant**

F. PAVONINIDAE Eimer and Fickert, 1899 T. (PRB)–Rec. Mar.

First: *Finlayina* Hayward and Morgans, 1981; Mexico; New Zealand. **Extant**

Superfamily FURSENKOINACEA Loeblich and Tappan, 1961

F. FURSENKOINIDAE Loeblich and Tappan, 1961 K. (CEN)–Rec. Mar.

First: *Coryphostoma* Loeblich and Tappan, 1962; cosmopolitan. **Extant**

F. VIRGULINELLIDAE Loeblich and Tappan, 1984 T. (YPR)–Rec. Mar.

First: *Pseudobuliminella* de Klasz *et al.*, 1964; Gabon; Nigeria. **Extant**

Superfamily DELOSINACEA Parr, 1950

F. CAUCASINIDAE Bykova, 1959 K. (BRM)–Rec. Mar.

First: *Epistominitella* Poroshina, 1966; former USSR: Azerbaijan. **Extant**

F. TREMACHORIDAE Lipps and Lipps, 1967 T. (MMI) Mar.

First and Last: *Tremachora* Lipps and Lipps, 1967; USA: California.

Superfamily PLEUROSTOMELLACEA Reuss, 1860

F. PLEUROSTOMELLIDAE Reuss, 1860 K. (APT)–Rec. Mar.

First: *Pleurostomella* Reuss, 1860; cosmopolitan. **Extant**

Superfamily STILOSTOMELLACEA Finlay, 1947

F. STILOSTOMELLIDAE Finlay, 1947 K. (CMP)–Rec. Mar.

First: *Nodogenerina* Cushman, 1927; cosmopolitan. **Extant**

Superfamily ANNULOPATELLINACEA Loeblich and Tappan, 1964

F. ANNULOPATELLINIDAE Loeblich and Tappan, 1964 T. (Mio.)–Rec. Mar.

First: *Annulopatellina* Parr and Collins, 1930; Australia; Trinidad. **Extant**

Superfamily DISCORBACEA Ehrenberg, 1838

F. CONORBINIDAE Reiss, 1963 J. (OXF)–K. (CON) Mar.

First: *Topalodiscorbis* Neagu, 1970; Romania.
Last: *Conorbina* Brotzen, 1936; Europe; North America.

F. PLACENTULINIDAE Kaismova *et al.*, 1980 J. (AAL)–Rec. Mar.

First: *Trispirina* Danich, 1977; former USSR: Azerbaijan, Moldavia; France. **Extant**

F. BAGGINDAE Cushman, 1927 K. (BRM)–Rec. Mar.

First: *Serovaina* Sliter, 1968; cosmopolitan. **Extant**

F. EPONIDIDAE Hofker, 1951 T. (DAN/THA)–Rec. Mar.

First: *Rectoeponides* Cushman and Bermúdez, 1936; Cuba; Europe. **Extant**

F. HELENINIDAE Loeblich and Tappan, 1988 T. (YPR)–Rec. Mar.

First: *Hyderia* Haque, 1962; western Pakistan. **Extant**

F. MISSISSIPPINIDAE Saidova, 1981 T. (Eoc.)–Rec. Mar.

First: *Schlosserina* Hagn, 1954 (Europe) and *Stomatorbina* Doreen, 1948 (Indo-Pacific, Cuba, Australia). **Extant**

F. BUENINGIIDAE Saidova, 1981 T. (LMI)–Rec. Mar.

First: *Bueningia* Finlay, 1939; New Zealand; Indonesia; Bikini Atoll. **Extant**

F. PEGIDIIDAE Heron-Allen and Earland, 1928 T. (Mio.)–Rec. Mar.

First: *Pegidia* Heron-Allen and Earland, 1928; Indian Ocean; West Pacific. **Extant**

F. DISCORBIDAE Ehrenberg, 1838 T. (LUT/BRT)–Rec. Mar.

First: *Discorbis* Lamarck, 1804 (cosmopolitan) and *Trochulina* d'Orbigny, 1839 (cosmopolitan). **Extant**

F. ROSALINIDAE Reiss, 1963 T. (Eoc.)–Rec. Mar.

First: *Rosalina* d'Orbigny, 1826; cosmopolitan. **Extant**

F. PANNELAINIDAE Loeblich and Tappan, 1984 T. (RUP/CHT)–Rec. Mar.

First: *Pannellainia* Seiglie and Bermúdez, 1976; USA: Mississippi; NW Australia. **Extant**

F. BRONNIMANNIIDAE Loeblich and Tappan, 1984 T. (Mio.)–Rec. Mar.

First: *Bronnimannia* Bermúdez, 1952; Gulf of Mexico; Atlantic; Pacific. **Extant**

F. SPHAEROIDINIDAE Cushman, 1927 K. (MAA)–Rec. Mar.

First: *Pullenoides* Hofker, 1951; The Netherlands. **Extant**

Superfamily GLABRATELLACEA Loeblich and Tappan, 1964

F. GLABRATELLIDAE Loeblich and Tappan, 1964 T. (YPR)–Rec. Mar.

First: (Cuisian): Genus *Pseudoruttenia* Le Calvez, 1959; France; Belgium. **Extant**

F. HERONALLENIIDAE Loeblich and Tappan, 1986 T. (Eoc.)–Rec. Mar.

First: *Heronallenia* Chapman and Parr, 1931; cosmopolitan. **Extant**

F. BULIMINOIDIDAE Seiglie, 1970 T. (RUP/CHT)–Rec. Mar.

First: *Buliminoides* Cushman, 1911 (cosmopolitan on shallow-water reefs) and *Elongobula* Finlay, 1939 (New Zealand, USA: Alabama). **Extant**

Superfamily SIPHONINACEA Cushman, 1927

F. SIPHONINIDAE Cushman, 1927 K. (MAA)–Rec. Mar.

First: *Pulsiphonina* Brotzen, 1948; North America; Europe. **Extant**

Superfamily DISCORBINELLACEA Sigal, 1952

F. PARRELLOIDIDAE Hofker, 1956 T. (DAN/THA)–Rec. Mar.

First: *Cibicidoides* Thalmann, 1939 (cosmopolitan) and *Woodella* Haque, 1956 (Pakistan). **Extant**

F. PSEUDOPARRELLIDAE Voloshinova, 1952 T. (RUP/CHT)–Rec. Mar.

First: *Alabaminoides* Gudina and Saidova, 1967 (cosmopolitan) and *Pseudoparrella* Cushman and ten Dam, 1948 (cosmopolitan). **Extant**

F. PLANULINOIDIDAE Saidova, 1981 T. (PLI)–Rec. Mar.

First: *Planulinoides* Parr, 1941; Australia; Japan. **Extant**

F. DISCORBINELLIDAE Sigal, 1952 T. (PRB)–Rec. Mar.

First: *Biapertorbis* Pokorny, 1956, Czechoslovakia; Poland; former USSR: western Ukraine. **Extant**

Superfamily PLANORBULINACEA Schwager, 1877

F. PLANULINIDAE Bermúdez, 1952 T. (PRB)–Rec. Mar.

First: *Planulina* d'Orbigny, 1826; cosmopolitan. **Extant**

F. CIBICIDIDAE Cushman, 1927 K. (BER)–Rec. Mar.

First: *Epithemella* Sliter, 1968; Sweden; Germany; USA: California; former USSR: Ukraine, Crimea. **Extant**

F. PLANORBULINIDAE Schwager, 1877 T. (Eoc.)–Rec. Mar.

First: *Planorbulina* d'Orbigny, 1826 (cosmopolitan) and *Planorbulinella* Cushman, 1927 (Atlantic, Mediterranean, Pacific, Australia, New Zealand, Cuba, Mexico, USA: North Carolina. **Extant**

F. CYMBALOPORIDAE Cushman, 1927 K. (CEN)–Rec. Mar.

First: *Archaecyclus* Silvestri, 1908; Italy. **Extant**

F. VICTORIELLIDAE Chapman and Crespin, 1930 K. (SAN)–Rec. Mar.

First: *Haerella* Belford, 1960; Western Australia. **Extant**

Superfamily ACERVULINACEA Schultze, 1854

F. ACERVULINIDAE Schultze, 1854 T. (DAN/THA)–Rec. Mar. (see Fig. 4.7)

First: *Sphaerogypsina* Galloway, 1933; Western Europe; former USSR: Ukraine; Australia; New Guinea; Borneo; Caribbean; Jamaica; Costa Rica; San Domingo; Peru. **Extant**

F. HOMOTREMATIDAE Cushman, 1927 T. (Eoc.)–Rec. Mar.

First: *Sporadotrema* Hickson, 1911; cosmopolitan in warmer waters. **Extant**

Superfamily ASTERIGERINACEA d'Orbigny, 1839

F. EPISTOMARIIDAE Hofker, 1954 K. (SAN)–Rec. Mar.

First: *Nuttallinella* Belford, 1959; Western Australia; New Zealand. **Extant**

F. ALFREDINIDAE Singh and Kalia, 1972 T. (LUT)–Rec. Mar.

First: *Alfredina* Singh and Kalia, 1972; India. **Extant**

F. ASTERIGERINATIDAE Reiss, 1963 K. (CMP)–Rec. Mar.

First: *Eoeponidella* Wickenden, 1949; cosmopolitan. **Extant**

F. ASTERIGERINIDAE d'Orbigny, 1839 T. (YPR)–Rec. Mar.

First: *Asterigerina* d'Orbigny, 1839; cosmopolitan. **Extant**

F. AMPHISTEGINIDAE Cushman, 1927 T. (Eoc.)–Rec. Mar.

First: *Amphistegina* d'Orbigny, 1826; cosmopolitan. **Extant**

F. BORELOIDEDAE Reiss, 1963 T. (LUT/BRT–PRB) Mar.

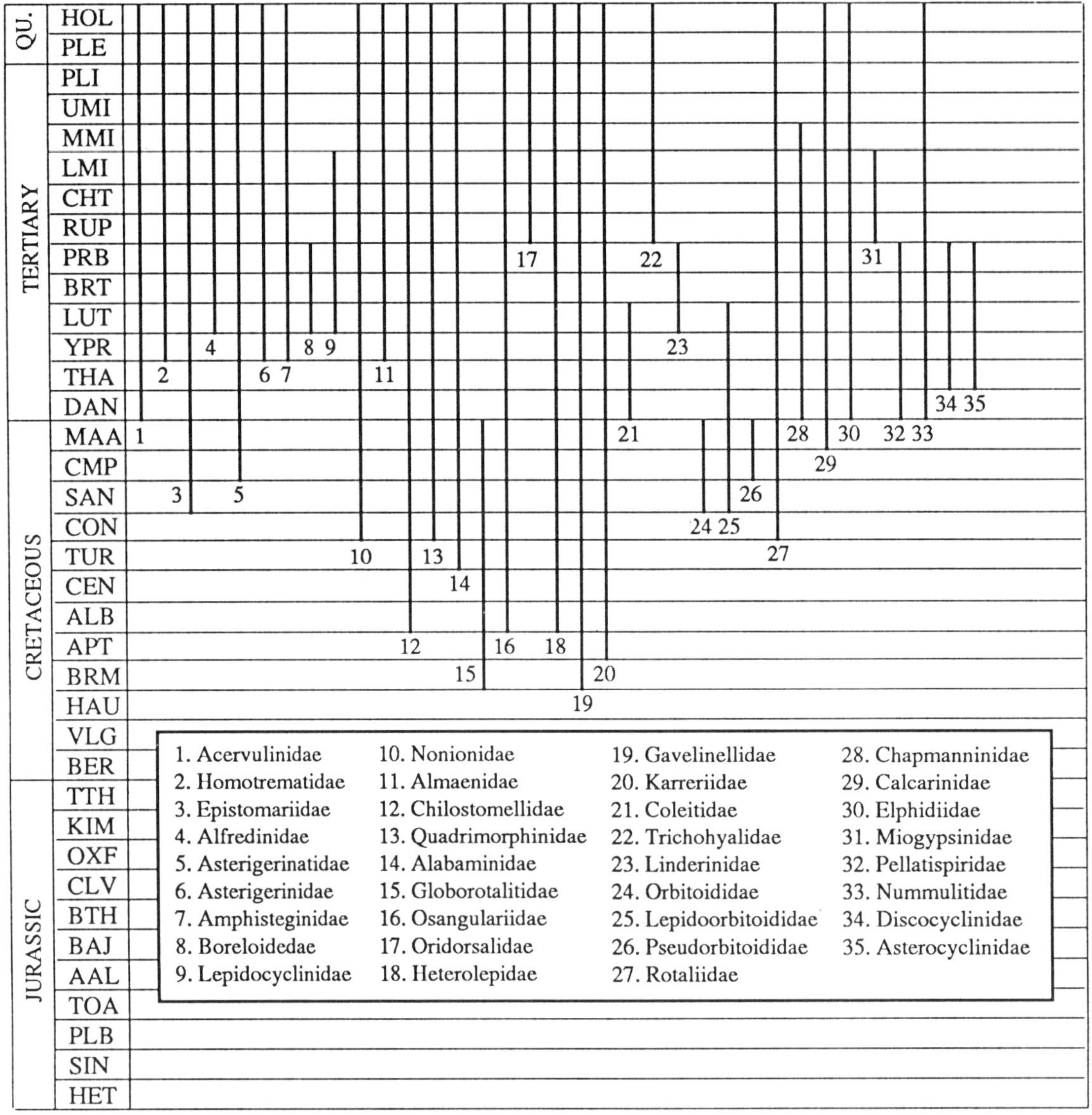

Fig. 4.7

First and Last: *Boreloides* Cole and Bermúdez, 1947 (Cuba, Pacific: Eniwetok Atoll, Marshall Islands) and *Eoconuloides* Cole and Bermúdez, 1944 (Cuba, Leeward Islands, Barbados). **Extant**

F. LEPIDOCYCLINIDAE Scheffen, 1932 T. (LUT/BRT–LMI) Mar.

First: *Eulinderina* Barker and Grimsdale, 1936 (Mexico), *Helicolepidina* Tobler, 1922 (North and South America), *Helicostegina* Barker and Grimsdale, 1936 (Trinidad, Mexico), *Nephrolepidina* Douvillé, 1911 (North and South America, North Africa, Europe, Indo-Pacific), *Caudriella* Haman and Huddleston, 1984 (Venezuela), *Lepidocyclina* Gümbel, 1870 (North and South America) and *Pseudolepidina* Barker and Grimsdale, 1937 (Mexico, Jamaica).
Last: *Nephrolepidina* as above.

Superfamily NONIONACEA Schultze, 1854

F. NONIONIDAE Schultze, 1854 K. (CON)–Rec. Mar.

First: *Nonionella* Cushman, 1926; cosmopolitan. **Extant**

F. ALMAENIDAE Myatlyuk, 1959 T. (YPR)–Rec. Mar.

First: *Ganella* Aurouze and Boulanger, 1954; France. **Extant**

Superfamily CHILOSTOMELLACEA Brady, 1881

F. CHILOSTOMELLIDAE Brady, 1881 K. (ALB)–Rec. Mar.

First: *Bagginoides* Podobina, 1975 (USA: South Dakota, Texas; Canada: Alberta; former USSR: western Siberia; Europe) and *Pallaimorphina* Tappan, 1957 (Greenland, USA: Alaska). **Extant**

F. QUADRIMORPHINIDAE Saidova, 1981 K. (CON)–Rec. Mar.

First: *Quadrimorphina* Finlay, 1939; cosmopolitan. **Extant**

F. ALABAMINIDAE Hofker, 1951 K. (TUR)–Rec. Mar.

First: *Valvalabamina* Reiss, 1963; cosmopolitan. **Extant**

F. GLOBOROTALITIDAE Loeblich and Tappan, 1984 K. (BRM–MAA) Mar.

First: *Conorotalites* Kaever, 1958; Europe.
Last: *Globorotalites* Brotzen, 1942; cosmopolitan.

F. OSANGULARIIDAE Loeblich and Tappan, 1964 K. (ALB)–Rec. Mar.

First: *Charltonina* Bermúdez, 1952 (Caribbean, Cuba, England, UK, Romania, Egypt, Australia) and *Osangularia* Brotzen, 1940 (cosmopolitan). **Extant**

F. ORIDORSALIDAE Loeblich and Tappan, 1984 T. (RUP/CHT)–Rec. Mar.

First: *Oridorsalis* Andersen, 1961; North America; Caribbean; Japan; Europe. **Extant**

F. HETEROLEPIDAE Gonzales-Donoso, 1969 K. (ALB)–Rec. Mar.

First: *Anomalinoides* Brotzen, 1942; cosmopolitan. **Extant**

F. GAVELINELLIDAE Hofker, 1956 K. (BRM)–Rec. Mar.

First: *Gavelinella* Brotzen, 1942; cosmopolitan. **Extant**

F. KARRERIIDAE Saidova, 1981 K. (APT)–Rec. Mar.

First: *Simionescella* Neagu, 1975; Romania. **Extant**

F. COLEITIDAE Loeblich and Tappan, 1984 T. (DAN–LUT) Mar.

First: *Coleites* Plummer, 1934; USA: Texas, Alabama, New Jersey, California; Guatemala; Cuba; Haiti; Sweden; Pakistan.
Last: (Lutetian): same genus.

F. TRICHOHYALIDAE Saidova, 1981 T. (RUP/CHT)–Rec. Mar.

First: *Buccella* Andersen, 1952; cosmopolitan. **Extant**

Superfamily ORBITOIDACEA Schwager, 1876

F. LINDERINIDAE Loeblich and Tappan, 1984 T. (LUT/BRT–PRB) Mar.

First: *Eoannularia* Cole and Bermúdez, 1944 (Cuba), *Epiannularia* Caudri, 1974 (Venezuela) and *Linderina* Schlumberger, 1893 (France, Spain, Romania, Turkey, Somalia, North America, Indonesia).
Last: *Linderina* as above.

F. ORBITOIDIDAE Schwager, 1876 K. (SAN–MAA) Mar.

First: *Orbitoides* d'Orbigny, 1848; Europe; North America; Caribbean; India.
Last: *Orbitoides* as above, *Pseudomphalocyclus* Meric, 1980 (Turkey), *Simplorbites* de Gregorio, 1882 (Sicily, France, Carpathians, Egypt), *Sivsella* Sirel and Gündüz, 1978 (Turkey, Greece) and *Omphalocyclus* Bronn, 1853 (France, Netherlands, Switzerland, Italy, Greece, former Yugoslavia, Romania, Turkey, Iran, Syria, Tunisia, India, Tibet, Cuba).

F. LEPIDOORBITOIDIDAE Vaughan, 1933 K. (SAN)–T. (LUT) Mar.

First: *Sirtina* Brönnimann and Wirz, 1962 (Iran, Libya, France) and *Praesiderolites* Wannier, 1983 (Spain, France).
Last: *Daviesina* Smout, 1954 (Qatar, Egypt, East Africa, Spain, France) and *Penoperculoides* Cole and Gravell, 1952 (Cuba).

Superfamily ROTALIACEA Ehrenberg, 1839

F. PSEUDORBITOIDIDAE M. G. Rutten, 1935 K. (CMP–MAA) Mar.

First: *Pseudorbitoides* Douvillé, 1922 (Jamaica, Cuba, Haiti, USA: Texas, Louisiana), *Rhabdorbitoides* Brönnimann, 1955 (Cuba), *Sulcorbitoides* Brönnimann, 1954 (Cuba, USA: Texas, Louisiana, Florida), *Asterorbis* Vaughan and Cole, 1932 (Cuba, Guatemala, Central Pacific, USA: Mississippi, Louisiana, Florida) and *Orbitocyclina* Vaughan, 1929 (Mexico, Cuba, USA: Florida, Louisiana).
Last: *Historbitoides* Brönnimann, 1956 (Cuba), *Pseudorbitoides* as above, *Vaughanina* Palmer, 1934 (Cuba, Venezuela, Guatemala, Mexico, USA: Florida), *Asterorbis* as above and *Orbitocyclina* as above.

F. ROTALIIDAE Ehrenberg, 1839 K. (CON)–Rec. Mar.

First: *Pararotalia* Le Calvez, 1949 (cosmopolitan), *Orbitokathina* Höttinger, 1966 (Spain) and *Rotalia* Lamarck, 1804 (cosmopolitan). **Extant**

F. CHAPMANINIDAE Thalmann, 1938 T. (DAN–MMI) Mar.

First: *Sherbornina* Chapman, 1922; cosmopolitan in warm water.
Last: *Chapmanina* Silvestri, 1931 (Italy, France, Spain, Greece, Romania) and *Sherbornina* as above.

F. CALCARINIDAE Schwager, 1876 K. (MAA)–Rec. Mar.

First: *Siderolites* Lamarck, 1801; Europe; Middle East; India. **Extant**

F. ELPHIDIIDAE Galloway, 1933 T. (DAN/THA)–Rec. Mar.

First: *Elphidiella* Cushman, 1936; cosmopolitan. **Extant**

F. MIOGYPSINIDAE Vaughan, 1928 T. (RUP/CHT–LMI) Mar.

First: *Miogypsinoides* Yabe and Hanzawa, 1928; Europe; North America; Indo-Pacific; India: Kutch.
Last: *Lepidosemicyclina* Rutten, 1911 (Borneo, Saipan, Australia, New Zealand, India, Japan), *Miogypsina* Sacco, 1893 (Europe, North and South America, Indo-Pacific), *Miogypsinita* Drooger, 1952 (Mexico, Trinidad, USA: Florida, Mississippi, Louisiana), *Miogypsinoides* as above and *Miolepidocyclina* Silvestri, 1907 (Mediterranean region, Mexico, Puerto Rico, Ecuador, Panama, East Indies, USA: California, Florida, Mississippi).

Superfamily NUMMULITACEA de Blainville, 1827

F. PELLATISPIRIDAE Hanzawa, 1937 T. (DAN/THA–PRB) Mar.

First: *Miscellanea* Pfender, 1935; Pakistan; Qatar; Saudi Arabia; Somalia; Nicaragua.
Last: *Biplanispira* Umbgrove, 1937 (Indo-Pacific, New Guinea, Saipan, eastern India, Andaman Islands) and *Pellatispira* Boussac, 1906 (Italy, Hungary, Pakistan, India, Japan, East Africa, Pacific, East Indies).

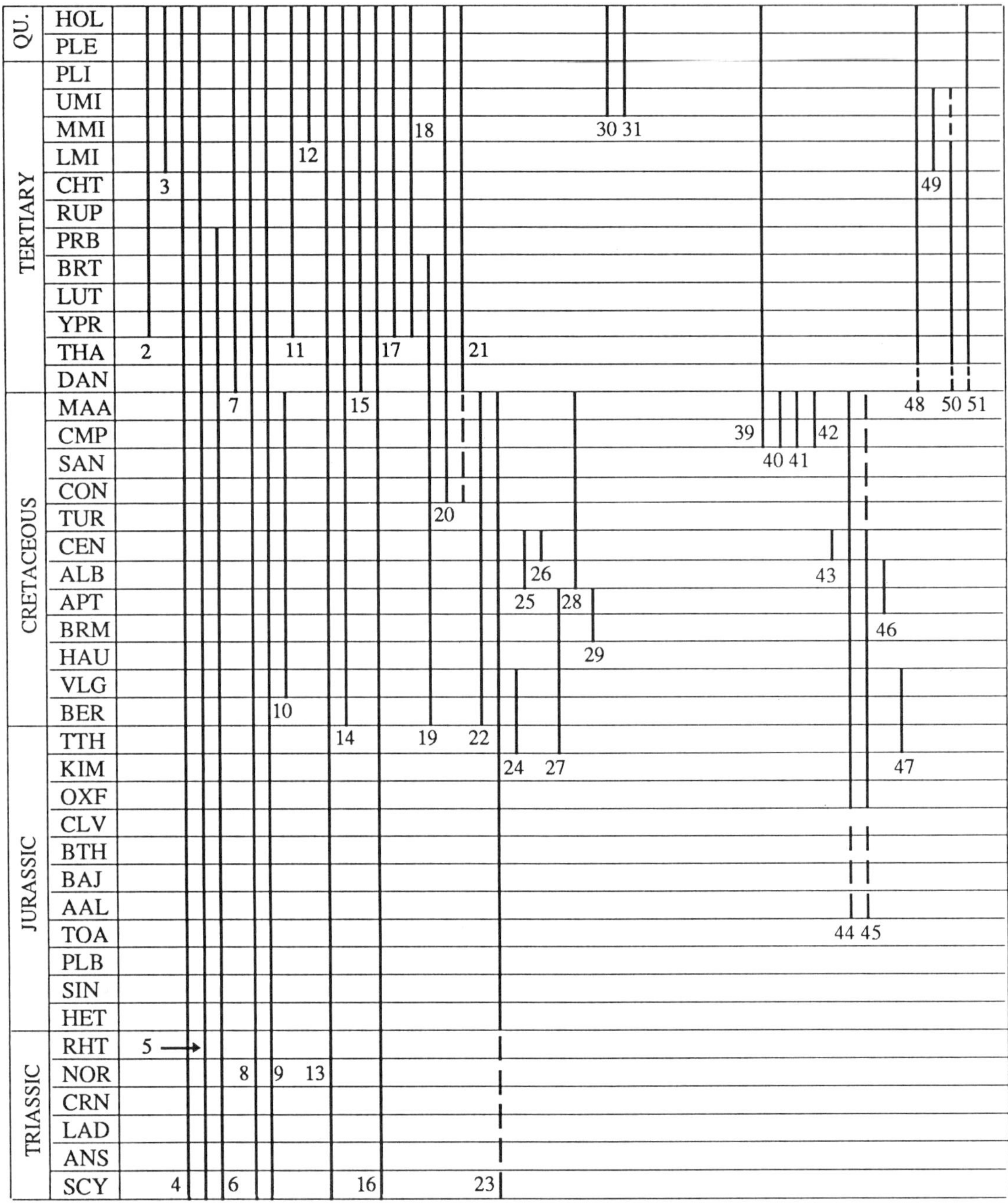

Fig. 4.8

F. NUMMULITIDAE de Blainville, 1827
T. (DAN/THA) Mar.

First: *Nummulites* Lamarck, 1801; tropical and subtropical cosmopolitan. **Extant**

F. DISCOCYCLINIDAE Galloway, 1928
T. (THA–PRB) Mar.

First: *Discocyclina* Gümbel, 1870; tropical and temperate cosmopolitan.
Last: *Actinocyclina* Gümbel, 1870 (Europe, N. America), *Asterophragmina* Rao, 1942 (Burma, Israel), *Discocyclina* as above and *Orbitoclypeus* Silvestri, 1907 (Europe).

F. ASTEROCYCLINIDAE Brönnimann, 1951
T. (THA–PRB) Mar.

First: *Asterocyclina* Gümbel, 1870; tropical and subtropical cosmopolitan.
Last: *Asterocyclina* as above and *Pseudophragmina* Douvillé, 1923 (North and South America).

Superclass ACTINOPODA Calkins, 1909

Class POLYCYSTINEA Ehrenberg, 1838

The classes Polycystinea and Phaeodarea are equivalent to the Radiolaria (Levine *et al.*, 1980). The classification of the radiolaria is in a state of flux, which was initiated when Haeckel's (1887) system was abandoned. In recent decades, most classifications have been based on the work of Riedel and Sanfilippo, 1977; Pessagno, 1977; Sanfilippo and Riedel, 1985; and Petrushevskaya, 1977 among others.

Key for both diagrams

ACTINOPODA
(RADIOLARIA)
POLYCYSTINEA
1. Entactiniidae
2. Orosphaeridae
3. Collosphaeridae
4. Actinommidae
5. Phacodiscidae
6. Coccodiscidae
7. Porodiscidae
8. Spongodiscidae
9. Hagiastriidae
10. Pseudoaulophacidae
11. Pyloniidae
12. Tholoniidae
13. Litheliidae
14. Plagoniidae
15. Acanthodesmiidae
16. Theoperidae
17. Carpioncaniidae
18. Pterocoryidae
19. Amphipyndacidae
20. Artostrobiidae
21. Cannobotryidae
22. Rotatormiidae
23. Archaeodictyomitridae
24. Parvingulidae
25. Pseudodityomitridae
26. Spongocapsidae
27. Syringocapsidae
28. Williriedellidae
29. Xitidae
PHAEODAREA
30. Challengeridae
31. Getticellidae
?RADIOLARIA
32. Albaillellidae
33. Palaeoscenidiidae
34. Anakrusidae
35. Inaniguttidae
36. Ceratoikiscidae
37. Haplentactiniidae
38. Pylentonemiidae
SILICOFLAGELLIDA
39. Dictychaceae
40. Vallacertaceae
41. Cornuaceae
42. Lyramulaceae
"CALCISPHERES"
43. Bonetocardiellidae
44. Stomiosphaeridae
45. Cadosinidae
46. Colomiellidae
47. Calpionellidae
EBRIOPHYCEAE
48. Hermesinaceae
49. Ditripodiaceae
50. Ammodochiaceae
51. Ebriaceae

PERMIAN: TAT, KAZ, UFI, KUN, ART, SAK, ASS
CARBONIFEROUS: GZE, KAS, MOS, BSK, SPK, VIS, TOU
DEVONIAN: FAM, FRS, GIV, EIF, EMS, PRA, LOK
SILURIAN: PRD, LUD, WEN, LLY
ORDOVICIAN: ASH, CRD, LLO, LLN, ARG, TRE
CAMB.: MER, STD, CRF
SINIAN: EDI, VAR, STU

1 4 8 9 13 32 33 34 35 36 37 38

Fig. 4.8

Range data are derived from these works and others, including Foreman (1973) and Campbell and Moore (1954).

Variations on these themes continue to be used, often with the addition of new families, as the phyletic relationships are elucidated. In some cases, family names are used without reference to their position in the overall classification and these are included under 'Radiolaria *Incertae sedis*'.

In general, this work uses Levine *et al.* (1980) down to Order, and Riedel and Sanfilippo (1977) and others for family designations. In view of the ongoing development of the classification, only family ranges are presented, together with some relevant reference(s).

Order SPUMELLARIDA Ehrenberg, 1875

F. ENTACTINIIDAE Riedel, 1967 S. (LLY)–C. Mar. (see Fig. 4.8)

Range: Lower Silurian to Carboniferous; Nazarov and Ormiston (1986).

F. OROSPHAERIDAE Haeckel, 1887 T. (Eoc.)–Rec. Mar.

Range: Eocene to Recent; Riedel and Sanfilippo, 1977. **Extant**

F. COLLOSPHAERIDAE Müller, 1858 T. (LMI)–Rec. Mar.

Range: Lower Miocene to Recent; Anderson (1983). **Extant**

F. ACTINOMMIDAE Haeckel, 1862; *sensu* Riedel, 1967b Pz./Tr.–Rec. Mar.

Range: Palaeozoic?/Triassic to Recent. **Extant**

F. PHACODISCIDAE Haeckel, 1881 Mz.–Rec. Mar.

Range: Mesozoic to Recent. **Extant**

F. COCCODISCIDAE Haeckel, 1862 Mz.–T. (Eoc.) Mar.

Range: Mesozoic to Eocene.

F. PORODISCIDAE Haeckel, 1881; *sensu* Kozlova in Petrushevskaya and Kozlova, 1972 Cz. Mar.

Range: Cainozoic; Riedel and Sanfilippo, 1977; Anderson (1983).

F. SPONGODISCIDAE Haeckel, 1862 emend. D.–Rec. Mar.

Range: Devonian to Recent. **Extant**

F. HAGIASTRIIDAE Pz.–Mz./Cz. Mar.

Range: Palaeozoic to Mesozoic/?Cainozoic; Pessagno (1977); Sanfilippo and Riedel (1985).
Comment: Split into two families by Baumgartner (1980); the Hagiastriidae and the Patulibracchiidae Pessagno, 1971.

F. PSEUDOAULOPHACIDAE Riedel, 1967 K. (VLG–MAA) Mar.

Range: Upper Valanginian to Maastrichtian; Pessagno (1977).

F. PYLONIIDAE Haeckel, 1881 T. (Eoc.)–Rec. Mar.

Range: Eocene to Recent; Anderson (1983). **Extant**

F. THOLONIIDAE Haeckel, 1887 T. (MMI)–Rec. Mar.

Range: Middle Miocene to Recent; Riedel and Sanfilippo (1977). **Extant**

F. LITHELIIDAE Haeckel, 1862 C.–Rec. Mar.

Range: Carboniferous to Recent. **Extant**

Order NASSELLARIDA Ehrenberg, 1875

Suborder SPYRIDA Ehrenberg, 1847 T. (THA)–Rec. Mar.

Range: Upper Palaeocene to Recent; Riedel and Sanfilippo (1977). **Extant**
Comment: Families not yet satisfactorily defined.

Suborder CYRTIDAE Haeckel, 1862 emend. Petrushevskaya, 1971

F. PLAGONIIDAE Haeckel, 1881 emend. K.–Rec. Mar.

Range: Cretaceous to Recent. **Extant**

F. ACANTHODESMIIDAE Hertwig, 1879 Cz. Mar.

Range: Cainozoic.
Comment: Also known as Trissocyclidae Goll, 1968.

F. THEOPERIDAE Haeckel, 1881, emend. Tr.–Rec. Mar.

Range: Triassic to Recent. **Extant**

F. CARPONCANIIDAE Haeckel, 1881, emend. T. (Eoc.)–Rec. Mar.

Range: Eocene to Recent. **Extant**

F. PTEROCORYIDAE Haeckel, 1881, emend. T. (Eoc.)–Rec. Mar.

Range: Eocene to Recent. **Extant**

Comment: = Pterocorythidae Riedel and Sanfilippo, 1977.

F. AMPHIPYNDACIDAE Riedel, 1967 K.–T. (LUT/BRT) Mar.

Range: Cretaceous to Middle Eocene; Riedel and Sanfilippo (1977).

F. ARTOSTROBIIDAE Riedel, 1967 K. (u.)–Rec. Mar.

Range: Upper Cretaceous to Recent; Sanfilippo and Riedel (1985). **Extant**

F. CANNOBOTRYIDAE Haeckel, 1881 emend. K. (u.)/T. (Pal./Eoc.)–Rec. Mar.

Range: Upper Cretaceous or Lower Palaeogene to Recent; Riedel and Sanfilippo (1977). **Extant**
Comment: = Cannobotrythidae; Riedel and Sanfilippo (1977).

F. ROTATORMIIDAE K. Mar.

Range: Cretaceous.

F. ARCHAEODICTYOMITRIDAE Pessagno, 1976 Tr./J.–K. Mar.

Range: Triassic/Jurassic to Cretaceous; Pessagno (1977); Sanfilippo and Riedel (1985).

F. PARVICINGULIDAE Pessagno, 1977a J. (TTH)–K. (VLG) Mar.

Range: Tithonian to Upper Valanginian; Sanfilippo and Riedel (1985).

F. PARVICINGULIDAE Pessagno, 1977a J. (TTH)–K. (VLG) Mar.

Range: Tithonian to Upper Valanginian; Sanfilippo and Riedel (1985).

F. SPONGOCAPSULIDAE Pessagno, 1977a K. (?CEN) Mar.

Range: Approx. Cenomanian; Sanfilippo and Riedel (1985).

F. SYRINGOCAPSIDAE Foreman, 1973 J. (TTH)–K. (APT) Mar.

Range: Tithonian to Aptian; Sanfilippo and Riedel (1985).

F. WILLIRIEDELLIDAE Dumitrica, 1970 K. (ALB–MAA) Mar.

Range: Albian to Maastrichtian; Sanfilippo and Riedel (1985).

F. XITIDAE Pessagno, 1977b K. (BRM–APT) Mar.

Range: Barremian to Aptian; Sanfilippo and Riedel (1985).

Class PHAEODAREA Haeckel, 1879

This class is very rare in the fossil record. Two families (Challengeridae and the Getticellidae) recognized by Dumitrica (1964 and 1965).

F. CHALLENGERIDAE T. (TOR)–Rec. Mar.

First: Specimens from the late TOR of Romanian eastern Predkaratje. **Extant**

F. GETTICELLIDAE T. (TOR)–Rec. Mar.

First: Specimens from the late TOR of Romanian eastern Predkaratje. **Extant**

RADIOLARIA *INCERTAE SEDIS*

F. ALBAILLELLIDAE Defandre, 1952b S.–C. Mar.

Range: Silurian to Carboniferous.

F. PALAEOSCENIDIIDAE Riedel, 1967 ?O./D.–C. Mar.

Range: ?Ordovician/Devonian to Carboniferous; Nazarov and Ormiston (1986).

F. ANAKRUSIDAE Nazarov, 1977 O. (CRD–ASH) Mar.

Range: Upper Caradoc to Ashgill (Ordovician); Webby and Blom (1986); Renz (1990).

F. INANIGUTTIDAE (Nazarov and Ormiston, 1984) O. (TRE/ARC)–S. (LUD/PRD) Mar.

Range: Lower Ordovician to Upper Silurian (at least); Nazarov and Ormiston (1986); Renz (1990).
Comment: = the informal group Palaeoactinommids (Holdsworth, 1977).

F. CERATOIKISCIDAE S. (WEN)–? Mar.

Range: Middle Silurian to ?; Nazarov and Ormiston (1986).

F. HAPLENTACTINIIDAE Nazarov, 1980 O. (TRE/ARG)–S./P. Mar.

Range: Lower Ordovician to Silurian–Permian; Nazarov and Ormiston (1986).

F. PYLENTONEMIIDAE Deflandre, 1963 O. Mar.

Range: Ordovician; Nazarov and Ormiston (1986).
Comment: Many other family names are to be seen in the literature; often survivors of the original Haeckelian classification.

Subphylum MASTIGOPHORA Diesing, 1866

Class PHYTOMASTIGOPHOREA Calkins, 1909

Order CRYPTOMONADIDA Senn, 1900

Extant

Order DINOFLAGELLIDA Butschli, 1885

These orders contain the normal members of the 'Division' DINOFLAGELLATA, which are included in Chapter 3.

Order PRYMNESIIDA Hibberd, 1976

This order contains the coccoliths and their relatives (see Chapter 3).

Order SILICOFLAGELLIDA Borgert, 1891

Tappan (1980) uses Order Vallacertales Glezer, 1966, and places the Silicoflagellida in synonymy.

F. DICTYOCHACEAE Wallich, 1865 K. (CMP)–Rec. Mar.

First: *Corbisema* Hanna, 1928 and *Dictyocha* Ehrenberg, 1837. **Extant**
Comment: This is based on data in Perch-Nielsen (1985), although Tappan (1980) does quote a range from early Cretaceous for the family.

F. VALLACERTACEAE Deflandre, 1950 K. (CMP–MAA) Mar.

First and Last: *Vallacerta* Hanna, 1928.

F. CORNUACEAE Gemeinhardt, 1930 K. (CMP–MAA) Mar.

First and Last: *Cornua* Schulz, 1928.

F. LYRAMULACEAE Tsumura, 1928 K. (CMP–MAA) Mar.

First and Last: *Lyramula* Hanna, 1928.

Phylum CILIOPHORA Doflein, 1901

Class POLYMENOPHOREA Jankowski, 1967

Subclass SPIROTRICHIA Butschli, 1889

Order OLIGOTRICHIDA Butschli, 1887

Suborder TINTINNINA Kofoid and Campbell, 1929

The Suborder Tintinnina are a group of organic-walled Recent ciliates. Their relationship with the calcareous-walled calpionellids is still under review. In accordance with normal practice, the latter group is placed in *Incertae sedis*.

Class *Incertae sedis* 'calcispheres'

Spherical/elliptical bodies with a simple, circular aperture which are commonly encountered in pelagic successions. Masters and Scott (1978), while acknowledging that they may be of algal affinity, left them as *Incertae sedis*. They classified them on the basis of their wall ultrastructure. The size, shape and orientation of the calcite prisms that form the wall are used to define the family-level divisions.

Subsequently, Keupp (1984) has suggested that these planktonic 'calcispheres' should be regarded as calcified dinoflagellates. This suggestion has been taken up by Willems (1985, 1988, 1990) who has included them within the Dinophyceae on the evidence of internal features (see also Chapter 3 herein). Unfortunately, Willems (1985, 1988, 1990) does not refer to Masters and Scott (1978), and so it is almost impossible to reconcile the two classifications. Tappan (1980) appears to have a mixture of classifications for these fossils. Masters and Scott (1978) recognized three families, as follows:

F. BONETOCARDIELLIDAE Masters and Scott, 1978 K. (CEN) Mar.

First and Last: *Bonetocardiella* Dufour, 1968.

F. STOMIOSPHAERIDAE Wanner, 1940, emend. Masters and Scott, 1978 J. (m.)–K. (MAA) Mar.

First: *Stomiosphaera* Wanner, 1940; Timor and cosmopolitan.
Last: *Inocardion* Masters and Scott, 1978 and *Pithonella* Lorenz, 1902, cosmopolitan.

F. CADOSINIDAE Wanner, 1940, emend. Masters and Scott, 1978 J. (m.)–K. (MAA) Mar.

First: *Cadosina* Wanner, 1940, Timor and cosmopolitan.
Last: *Cadosina* is described from the CEN; later forms may be MAA.

Superfamily CALPIONELLIDEA

Recent tintinnids possess an organic lorica, while fossil calpionellids have a calcareous test. Mineralized tests are completely unknown among ciliates in general, and tintinnids in particular.

F. COLOMIELLIDAE Bonet, 1956 K. (APT–ALB) Mar.

First and Last: *Colomiella* Bonet, 1956; Mexico; Cuba; Tunisia; Aquitaine Basin, France.

F. CALPIONELLIDAE Bonet, 1956 J. (TTH)–K. (VLG) Mar.

First: *Chitinoidella* Doben.
Last: *Tintinnopsella* Colom, 1948, top of lower VLG.

Phylum *INCERTAE SEDIS* 'EBRIDIANS'

Class EBRIOPHYCEAE Loeblich, 1970

Order EBRIALES Honigsberg *et al.*, 1964

F. HERMESINACEAE Hovasse, 1943 T. (Pal.)–Rec. Mar.

First: *Spongebria marthae* Deflandre, 1951. **Extant**

F. DITRIPODIACEAE Deflandre 1951 T. (Mio) Mar.

First and Last: *Ditripodium latum* Hovasse, 1932, Miocene, California, USA.

F. AMMODOCHIACEAE Deflandre, 1950 T. (Pal.–Mio.) Mar.

First and Last: *Ammodochium* Hovasse, 1932.

F. EBRIACEAE Lemmermann, 1901 T. (Pal.)–Rec. Mar.

First: *Falsebria ambigua* Deflandre, 1950. **Extant**

REFERENCES

Anderson, O. R. (1983) *Radiolaria*. Springer, New York.

Baumgartner, P. D. (1980) Late Jurassic Hagiastridae and Patulibracchiidae (Radiolaria) from the Argolis Peninsula (Peloponnesus, Greece). *Micropalaeontology*, **26**, 274–322.

Campbell, A. S. and Moore, R. C. (1954) Protista 3: Chiefly Radiolarians and Tintinnines, in *Treatise on Invertebrate Paleontology, Part D* (ed. R. C. Moore), Geological Society of America and University of Kansas Press, Boulder, Colorado and Lawrence, Kansas.

Dumitrica, P. (1964) Aspura prezentei unor radiolari din familia Challengeridae (Ord. Phaeodaria) in Tortonianul din Subcarpati. *Studii si Cercetari de geologie-geografie Filiala Cluj, Academia RPR*, **9**, 217–22.

Dumitrica, P. (1965) Sur le présence de Phéodaires dans le Tortonien des Subcarpathes roumaines. *Comptes rendus de l'Académie des Sciences, Paris*, **260**, 250–3.

Eichwald, C. E. von (1830) *Zoologia Specialis*, Vol. 2, D. E. Eichenwaldus, Vilnae, 323 pp.

Foreman, H. P. (1973) Radiolaria of Leg X, with systematics and ranges for the families Amphipyndacidae, Artostrobiidae and Theoperidae, in *Initial Reports of the Deep Sea Drilling Project*, 10 (eds J. L. Worzel *et al.*), US Government Printing Office, Washington.

Haeckel, E. (1887) Report on the Radiolaria collected by H. M. S. Challenger during the years 1873–1876. *Challenger Report, Zoology*, **18**, 1–1803.

Holdsworth, B. K. (1977) Paleozoic Radiolaria: stratigraphic distribution in Atlantic Borderlands, in *Stratigraphic Micropaleontology of Atlantic Basin and Borderlands* (ed. F. M. Swain), Elsevier, Amsterdam, pp. 167–84.

Keupp, H. (1984) Revision der kalkigen Dinoflagellaten-Zystem G. Deflandres, 1948. *Paläontologische Zeitschrift*, **58**, 9–31.

Levine, N. D., Corliss, J. O., Cox, F. E. G. *et al.* (1980) A newly revised classification of the Protozoa. *Journal of Protozoology*, **27**, 37–58.

Loeblich, A. R. Jr and Tappan, H. (1964) Sarcodina chiefly 'Thecamoebians' and Foraminiferida, in *Treatise on Invertebrate Paleontology. Part C, Protista 2* (ed. R. C. Moore), Geological Society of America and University of Kansas Press, Boulder, Colorado, and Lawrence, Kansas.

Loeblich, A. R. Jr and Tappan, H. (1988) *Foraminiferal Genera and their Classification*. Van Nostrand Reinhold Co., New York.

Masters, B. A. and Scott, R. W. (1978) Microstructure, affinities and systematics of Cretaceous calcispheres. *Micropaleontology*, **24**, 210–21.

Medioli, F. S., Scott, D. B., Collins, E. S. *et al.* (1990) Thecamoebians: present status and prospects for the future, in *Paleoecology, Biostratigraphy, Paleoceanography and Taxonomy of Agglutinated Foraminifera* (eds C. Hemleben, M. A. Kaminski, W. Kuhnt *et al.*), Kluwer Academic Publishers, London, pp. 813–39.

Nazarov, B. B. and Ormiston, A. R. (1986) Trends in the development of Paleozoic Radiolaria. *Marine Micropaleontology*, **11**, 3–32.

Orbigny, A. d' (1826) Tableau méthodique de la classe des Céphalopodes. *Annales des Sciences Naturelles*, **7**, 245–314.

Perch-Nielsen, K. (1985) Silicoflagellates, in *Plankton Stratigraphy* (eds H. M. Bolli, J. B. Saunders and K. Perch-Nielsen), Cambridge University Press, pp. 811–46.

Pessagno, E. A. (1977) Radiolaria in Mesozoic stratigraphy, in *Oceanic Micropaleontology* (ed. A. T. S. Ramsay), Academic Press, London, pp. 913–50.

Petrushevskaya, M. G. (1971) Spumellarian and nassellarian Radiolaria in the plankton and bottom sediments of the central Pacific, in *The Micropalaeontology of Oceans* (eds B. M. Funnell and W. R. Riedel), Cambridge University Press, Cambridge, pp. 309–17.

Renz, G. W. (1990) Ordovician Radiolaria from Nevada and Newfoundland – a comparison at the family level. *Marine Micropaleontology*, **15**, 393–402.

Riedel, W. R. and Sanfilippo, A. (1977) Cainozoic Radiolaria, in *Oceanic Micropaleontology* (ed. A. T. S. Ramsay), Vol. 2, Academic Press, London, pp. 847–912.

Sanfilippo, A. and Riedel, W. R. (1985) Morphological characters for a natural classification of Cenozoic Radiolaria, reflecting phylogenies. *Marine Micropaleontology*, **11**, 151–70.

Tappan, H. (1980) *The Paleobiology of Plant Protists*. W. H. Freeman, Los Angeles, 1028 pp.

Webby, B. and Blom, W. (1986) The first well preserved radiolarians from the Ordovician of Australia. *Journal of Paleontology*, **60**, 145–57.

Willems, H. (1985) *Tetramerosphaera lacrimula*, eine intern gefacherte Calcisphaere aus der Ober-Kreide. *Senckenbergiana Lethaea*, **66**, 177–201.

Willems, H. (1988) Kalkige Dinoflagellaten-Zystem aus der oberkretazischen Schreibkreide-Fazies N-Deutschlands (Coniac bis Maastricht). *Senckenbergiana Lethaea*, **68**, 433–77.

Willems, H. (1990) *Tetratropis*, eine neue Kalkdinoflagellaten-Gattun (Pithonelloideae) aus der Ober-Kreide von Lagerdorf (N-Deutschland). *Senckenbergiana Lethaea*, **70**, 239–57.

5

PORIFERA

J. K. Rigby, G. E. Budd, R. A. Wood and F. Debrenne

The palaeontological record of the Porifera is mainly that of marine forms and extends in broken fashion from the early Cambrian to the Recent. Unlike the record of many phyla, our knowledge of sponges, in the historic sense, is still largely based on major monographic works. For example, such papers would include those by Walcott (1920), treating Cambrian forms, Rigby and Webby (1988) on Ordovician ones, Hall and Clarke (1899) and Rigby (1986) on Devonian ones, Finks (1960) and Senowbari-Daryan (1990) on Permian ones, Schrammen (1910, 1912, 1924, 1936), Moret (1926), Lagneau-Hérengér (1962), and Hinde (1884, 1887–1893) on the Mesozoic faunas of Europe. The data are so limited that single studies still produce anomalies in any statistical treatments. Such papers show clearly in the 'first and last' occurrences compiled below. Assemblages of fossil sponges are generally localized, both in terms of geography and time. The record unduly emphasizes forms with calcareous skeletons or fused skeletons, in contrast to those with loose spicular skeletons or horny keratose skeletons, such as abound in modern seas, and which probably were equally common in ancient seas. The record is incomplete, and so the ranges given are generally only the minimum possible – all could be questioned.

For many years, the taxonomy of fossil Porifera appears to have been relatively stable, however, discoveries in the past two or three decades have prompted major revisions, many still under way. Spicule data have proved poriferan affinity for many stromatoporoids and chaetetids, and have also demonstrated that these formerly discrete systematic groups, together with the sphinctozoans, are polyphyletic. Hence, the higher systematic placing of sponges with an additional calcareous skeleton is still open to considerable interpretation. Archaeocyaths are also now considered as poriferans by several specialists (Debrenne *et al.*, 1990). Similarly, classifications of major groups within Demospongia and Hexactinellida are undergoing major revisions. Some of those revisions are reflected in the lists below, as noted, but many changes are not reflected in the classifications used, for they are only in manuscripts still awaiting publication in the revision of the *Treatise on Invertebrate Paleontology, Part E*.

Here, non-calcified poriferans and the heteractinids have been documented by J. Keith Rigby and calcified forms (archaeocyaths, stromatoporoids, sphinctozoans and chaetetids) by Graham Budd, Rachel Wood and Francoise Debrenne.

Class DEMOSPONGIA Sollas, 1875

Order KERATOSA Grant, 1861

F. VAUXIIDAE Walcott 1920 C. (STD) Mar.

First: *Vauxia magna* Rigby, 1980, from the *Glossopleura* Zone, Spence Tongue, Lead Bell Shale, Wasatch Mountains, near Brigham City, Utah, USA. A Lower Cambrian(?) *V. gracilenta* has been reported from the Buen Formation of north Greenland (Rigby, 1986).
Last: *Vauxia gracilenta* Walcott, 1920 and several other species of *Vauxia*, Burgess Shale, Stephen Formation near Field, British Columbia, Canada.

Order MONAXONIDA Sollas, 1883

F. SPONGILLIDAE Gray, 1867 K.–Rec. FW.

First: *Palaeospongilla chubutensis* Ott and Volkheimer, 1972, Chubut Group, near Cerro Condor, Patagonia, Argentina. **Extant**
Intervening: Mio.–PLE.

F. HAZELIIDAE de Laubenfels, 1955 C. (STD) Mar.

First: *Hazelia grandis* Walcott, 1920, *Ogygopsis* Shale of Stephen Formation, Mount Stephen, near Field, British Columbia, Canada.
Last: *Hazelia palmata* Walcott, 1920, and associated species, plus the associated *Falospongia falata* Rigby, 1986, *Crumillospongia frondosa* (Walcott, 1920) and *C. biporosa* Rigby, 1986 that co-occur, *Bathyuriscus–Elrathina* Zone, Burgess Shale, Stephen Formation, near Field, British Columbia, Canada.

The Fossil Record 2. Edited by M. J. Benton. Published in 1993 by Chapman & Hall, London. ISBN 0 412 39380 8

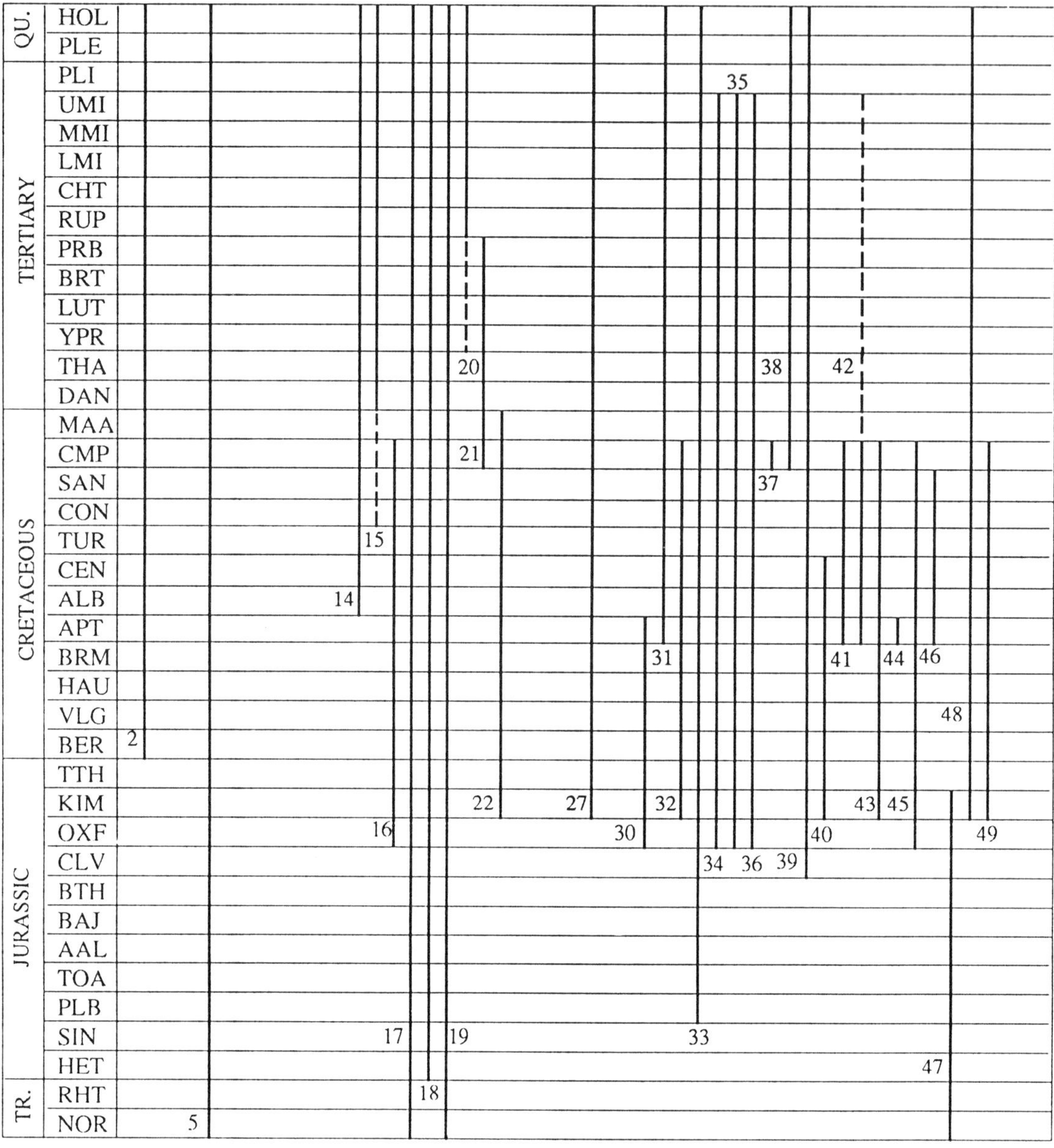

Fig. 5.1

F. HELIOSPONGIIDAE Finks, 1960 C. (WES C–D)–(ART) Mar.

First: *Coelocladia spinosa* Girty, 1908, Gaptank Formation, Desmoinesian, Glass Mountains, Texas, USA.

Last: *Heliospongia vokesi* King, 1943, Bone Spring Limestone, Leonardian, Sierra Diablo, north-west of Van Horn, Texas, USA.

F. CLIONIDAE Gray, 1867 €. (CRF)–Rec. Mar.

First: Unnamed boring sponge, Forteau Formation, near Port Amour, Labrador, Canada (Kobluk, 1981); *Clionolithes radicans* Clarke, 1908, Devonian, New York is perhaps the oldest named species. **Extant**

F. DYSTACTOSPONGIIDAE Miller, 1889 O. (LLN/LLO–ASH)/C. (WES)? Mar.

First: *Dystactospongia minor* Ulrich and Everett, 1890, Platteville Limestone, Dixon, Illinois, USA.

Last: *Dystactospongia madisonensis* Foerste, 1909, Saluda Limestone, near Madison, Indiana, USA. *Batospongia spicata* Miller, 1889 is questionably included in the family, from 'near the base of the Coal measures,' Seville, Illinois, USA.

F. LEPTOMITIDAE de Laubenfels 1955 €. (CRF–STD) Mar.

First: *Leptomitus teretiusculus*, *Leptomitella conica*, *Leptomitella confusa*, *Paraleptomitella dictyodroma*, and *P. globula*, all described by J.-Y. Chen *et al.*, 1989, Yuanshan Member, Chiungchussu Formation, near Chengjiang, Yunnan, China.

Last: *Letomitella metta* (Rigby, 1983), *Bolaspidella* Zone, Marjum Formation, House Range, Utah, USA.

F. HAMPTONIIDAE de Laubenfels, 1955 € (STD) Mar.

First and Last: *Hamptonia bowerbanki* Walcott, 1920,

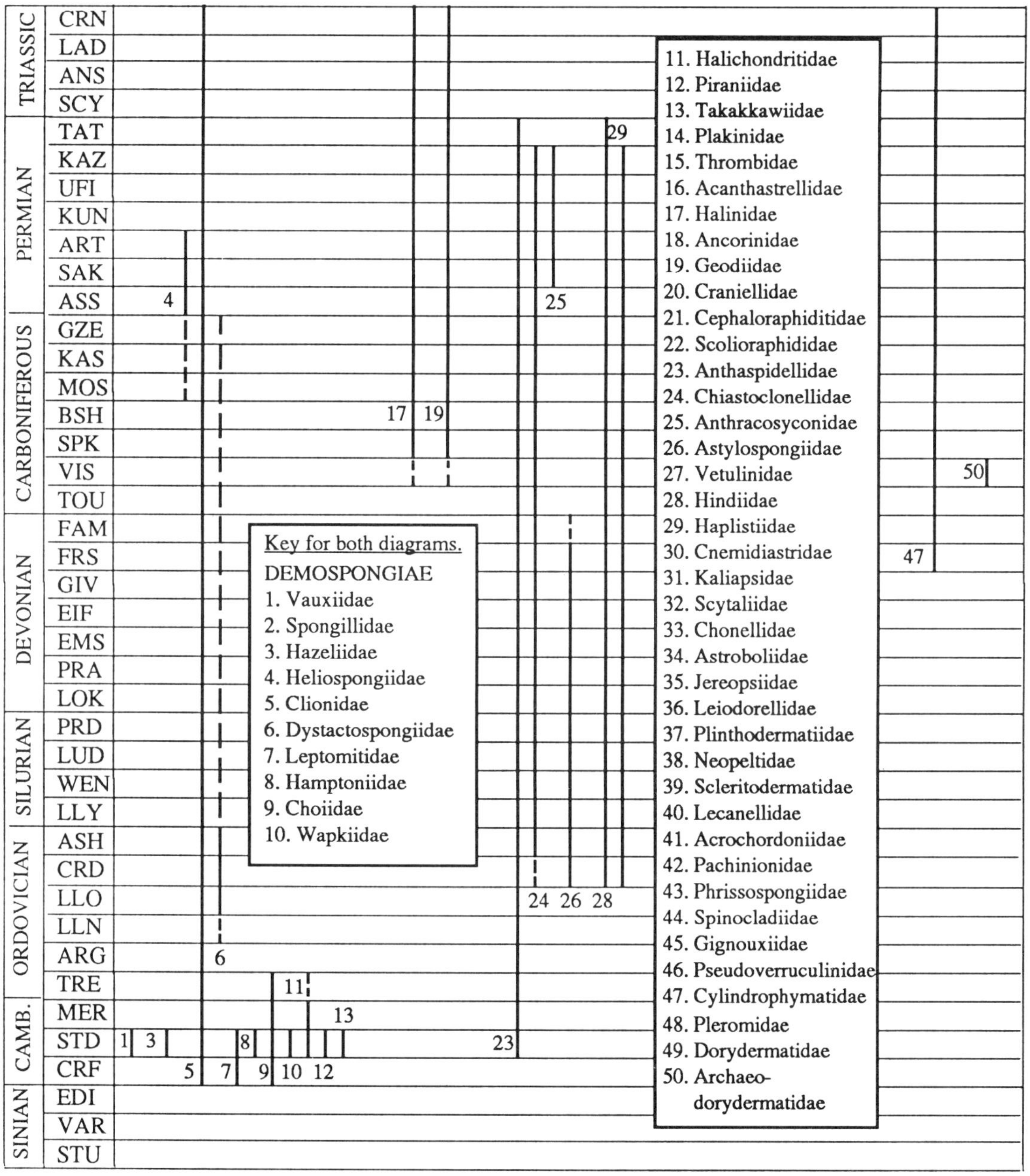

TRIASSIC
CRN
LAD
ANS
SCY
PERMIAN
TAT
KAZ
UFI
KUN
ART
SAK
ASS
CARBONIFEROUS
GZE
KAS
MOS
BSH
SPK
VIS
TOU
DEVONIAN
FAM
FRS
GIV
EIF
EMS
PRA
LOK
SILURIAN
PRD
LUD
WEN
LLY
ORDOVICIAN
ASH
CRD
LLO
LLN
ARG
TRE
CAMB.
MER
STD
CRF
SINIAN
EDI
VAR
STU
Key for both diagrams.
DEMOSPONGIAE
1. Vauxiidae
2. Spongillidae
3. Hazeliidae
4. Heliospongiidae
5. Clionidae
6. Dystactospongiidae
7. Leptomitidae
8. Hamptoniidae
9. Choiidae
10. Wapkiidae
11. Halichondritidae
12. Piraniidae
13. Takakkawiidae
14. Plakinidae
15. Thrombidae
16. Acanthastrellidae
17. Halinidae
18. Ancorinidae
19. Geodiidae
20. Craniellidae
21. Cephaloraphiditidae
22. Scolioraphididae
23. Anthaspidellidae
24. Chiastoclonellidae
25. Anthracosyconidae
26. Astylospongiidae
27. Vetulinidae
28. Hindiidae
29. Haplistiidae
30. Cnemidiastridae
31. Kaliapsidae
32. Scytaliidae
33. Chonellidae
34. Astroboliidae
35. Jereopsiidae
36. Leiodorellidae
37. Plinthodermatiidae
38. Neopeltidae
39. Scleritodermatidae
40. Lecanellidae
41. Acrochordoniidae
42. Pachinionidae
43. Phrissospongiidae
44. Spinocladiidae
45. Gignouxiidae
46. Pseudoverruculinidae
47. Cylindrophymatidae
48. Pleromidae
49. Dorydermatidae
50. Archaeo-dorydermatidae

Fig. 5.1

Bathyuriscus–Elrathina Zone, Burgess Shale, Stephen Formation, near Field, British Columbia, Canada.

F. CHOIIDAE de Laubenfels, 1955 €. (CRF–O.) (TRE?) Mar.

First: *Choia hindei* (Dawson, 1896), Buen Formation, north Greenland.
Last: *Choia hindei* (Dawson, 1896), black Levis(?) Shale, Tremadocian(?), Little Métis, south-eastern Quebec, Canada.
Intervening: STD.

F. WAPKIIDAE de Laubenfels, 1955 €. (STD) Mar.

First and Last: *Wapkia grandis* Walcott, 1920, *Bathyuriscus–Elrathina* Zone, Burgess Shale, Stephen Formation, near Field, British Columbia, Canada.

F. HALICHONDRITIDAE Rigby, 1986 €. (STD)–O. (TRE?) Mar.

First: *Halichondrites elissa* Walcott, 1920, *Bathyuriscus–Elrathina* Zone, Burgess Shale, Stephen Formation, near Field, British Columbia, Canada.
Last: *Halichondrites confusus* Dawson, 1889, black Levis(?) Shale, Tremadocian(?), Little Métis, south-eastern Quebec, Canada.

F. PIRANIIDAE de Laubenfels, 1955 €. (STD) Mar.

First: *Pirania muricata* Walcott, 1920, *Ogygopsis* Beds, Stephen Formation, Mt. Stephen, near Field, British Columbia, Canada.
Last: *Pirania muricata* Walcott, 1920 and *Moleculospina mammillata* (Walcott, 1920), *Bathyuriscus-Elrathina* zone, Burgess Shale, Stephen Formation, near Field, British Columbia, Canada.

F. TAKAKKAWIIDAE de Laubenfels, 1955
€. (STD) Mar.

First and Last: *Takakkawia lineata* Walcott, 1920, *Bathyuriscus-Elrathina* Zone, Burgess Shale, Stephen Formation, near Field, British Columbia, Canada.

Order CHORISTIDA Sollas, 1880

F. PLAKINIDAE Reid, 1968 K. (ALB)–Rec. Mar.

First: *Plakina?* sp. Hinde, 1885, Upper Greensand, southern England, UK. **Extant**
Intervening: Eoc.

F. THROMBIDAE Sollas, 1880 ?K (SEN)–Rec. Mar.

First: *Thrombus?* sp. or *Ortmannia* sp. Schrammen 1924, Upper Chalk (Mukronaten Kreide), Misberg, Germany. **Extant**
Intervening: Eoc.

F. ACANTHASTRELLIDAE Schrammen 1924
?J. (OXF)–K. (CMP) Mar.

First: *Acanthastrella*(?) sp., loose spicules only, Weissjura, southern Germany.
Last: *Acanthastrella panniculosa* Schrammen 1924, Upper Chalk (Mukronaten Kreide), Misburg, northern Germany.

F. HALINIDAE de Laubenfels, 1955
C. (VIS/NAM)–Rec. Mar.

First: *Pachastrella vetusta* Hinde, 1883, Lower Limestone, Dalry, Ayrshire and other localities, Scotland, and Republic of Ireland. **Extant**
Intervening: J. (l), APT, ALB, SEN, Eoc.

F. ANCORINIDAE Schmidt, 1870 J. (l)–Rec. Mar.

First: *Oppligera clavaeformis* (Oppliger, 1921), Austria. **Extant**
Intervening: ALB, TUR–SEN, Eoc.

F. GEODIIDAE Gray, 1867 C. (VIS/NAM)–Rec. Mar.

First: *Geodites antiquus* (Hinde, 1883), and *G. deformis* Hinde, 1887, Lower Limestone, Dalry, Ayrshire and other localities, Scotland, UK. **Extant**
Intervening: J. (l), APT, ALB, SEN, Eoc., Mio.

F. CRANIELLIDAE de Laubenfels, 1955
?T. (Eoc.?)–Rec. Mar.

First: *Craniella* sp., based on loose spicules only, Hinde and Holmes 1892, Plantagenet Beds, Western Australia. **Extant**

F. CEPHALORAPHIDITIDAE Reid, 1968
K. (CMP)–T. (Eoc.) Mar.

First: *Cephaloraphidites milleporatus* Schrammen, 1910, calcareous marls of Quadratenkreide at Oberg, Germany, co-occurring with *Megaloraphium auriforme* Schrammen, 1910 and *Polytretia seriatopora* Schrammen, 1910.
Last: *Ophiraphidites infundibuliformis* Schrammen, 1899, Castle Hayne Limestone, Eocene, North Carolina, USA.

F. SCOLIORAPHIDIDAE Zittel, 1879
J. (KIM)–K. (SEN) Mar.

First: *Helminthopyllum feifeli* Schrammen, 1936, Weissjura γ, Schwabtal, Germany.
Last: *Scolioraphis cerebriformis* Zittel, 1878, sandy marls, Granulatenkreide, Sudmerberg, Germany.

Order LITHISTIDA Schmidt, 1870

Suborder ORCHOCLADINA Rauff, 1893

F. ANTHASPIDELLIDAE Miller, 1889
€. (STD)–P. (TAT) Mar.

First: *Capsospongia undulata* (Walcott, 1920), Burgess Shale, Stephen Formation, British Columbia, Canada. If *Fieldospongia* (Walcott, 1920), from the Mount Whyte Formation, Rocky Mountains, British Columbia, belongs here it extends the range of the family to near the base of the Middle Cambrian. Dendroclones such as characterize the family have been reported from the lower Middle Cambrian Tindall Limestone, near Tipperary, Northern Australia.
Last: *Timorella permica* Gerth, 1929, Amarassi Formation, Timor.
Intervening: MER, TRE–LUD, GED–EIF, FRS–FAM, MOS, ART–KAZ. Family is major element in lower Palaeozoic faunas.

F. CHIASTOCLONELLIDAE Rauff, 1895
O. (CRD/ASH)–P. (KAZ) Mar.

First: *Syltispongia ingemariae* Van Kempen, 1990, erratics containing probably CRD/ASH age sponges, in sediments on Island of Sylt, north-west Germany. *Chiastoclonella headi* Rauff, 1894, Middle Silurian Brownsport Formation (WEN), Tennessee is the earliest species of the family with a certain provenance.
Last: *Actinocoelia maeandrina* Finks, 1960, San Andreas Formation Guadalupian, near Carlsbad, New Mexico, USA.
Intervening: WEN–LUD, FRS, ASS?–ART.

F. ANTHRACOSYCONIDAE Finks, 1960
P. (SAK–KAZ) Mar.

First: *Anthracosycon regulare* (King, 1943), Skinner Ranch Formation (Wolfcampian), Glass Mountains, Texas, USA.
Last: *Collatipora discreta* Finks, 1960, Rader Limestone Member, Upper Guadalupian Bell Canyon Formation, Guadalupe Mountains, Texas, USA.
Intervening: ART, KUN, UFI.

Suborder EUTAXICLADINA Rauff, 1893

F. ASTYLOSPONGIIDAE Zittel, 1877
O. (CRD)–D. (FRS/FAM?) Mar.

First: *Caryospongia parvulum* (Billings, 1861), Cobourg Limestone, Trenton Group, Ottawa, Ontario, Canada.
Last: *Scheielloides(?)* sp., Virgin Hills Formation (FAM); but *Scheielloides conica* Rigby, 1986 and *Sadleria pansa* Rigby, 1986 occur together in Sadler Limestone (FRS); all three in Devonian Canning Basin reef complexes of northern Western Australia.
Intervening: WEN, LUD, GED–EIF, FRS.

F. VETULINIDAE von Lendenfeld, 1904
K. (KIM)–Rec. Mar.

First: *Mastosia wetzleri* Zittel, 1878, Weissjura ε, extends up to Weissjura δ, Sozenhausen and Gussenstadt, Germany. **Extant**
Intervening: TUR–CMP.
Comment: Part of this family (Mastosiidae Schrammen, 1924) was placed in the Anomocladina by de Laubenfels (1953, p. E64), but clearly belongs in the Eutaxicladina for, as de Laubenfels noted, the skeleton 'is composed of sphaeroclones . . .'.

Suborder TRICRANOCLADINA Reid, 1968

F. HINDIIDAE Rauff, 1893 O. (CRD)–P. (TAT) Mar.

First: *Hindia sphaeroidalis* Duncan, 1879, and species of *Belubulaspongia, Palmatohindia, Arborohindia, Mamelohindia* and *Fenestrospongia*, all described by Rigby and Webby (1988), lower Malongulli Formation, Coppermine Greek breccias, New South Wales, Australia.
Last: *Scheiia tuberosa* Tschneryschew and Stepanov, 1916, Amarassi Formation (Dzhulfian), Timor.
Intervening: LLN, ASH, SIG–EMS, VIS–KAZ.

Suborder RHIZOMORINA Zittel, 1895

Comment: Family classification follows that of de Laubenfels (1955). The suborder is being significantly revised in the new version of the *Treatise on Invertebrate Paleontology*.

F. HAPLISTIIDAE de Laubenfels, 1955
O. (CRD)–P. (KAZ) Mar.

First: *Haplistion regularis* Rigby and Webby, 1988, and species of *Warrigalia, Taplowia, Lewinia* and *Boonderooia*, all described by Rigby and Webby (1988), lower Malongulli Formation, Coppermine Creek breccias, New South Wales, Australia.
Last: *Haplistion aeluroglossa* Finks, 1960, Word Formation, Glass Mountains, Texas, USA.
Intervening: LUD, EIF–FRS, TOU, VIS, MOS, ART, KUN.

F. CNEMIDIASTRIDAE Schrammen, 1924
J. (OXF)–K. (APT) Mar.

First: *Cnemidiastrum rimulosum* (Goldfuss, 1833), Weissjura γ with *C. granulosum* (Quenstedt, 1878); *C. stellatum* (Goldfuss, 1833) occurs first in Weissjura β–γ; all at Hossingen, Tieringen and nearby localities, southern Germany.
Last: *Lithostrobilis reticulatus* Lagneau-Hérengér, 1962, Aptian beds, Mas de Artis, Catalonia, Spain.
Intervening: KIM.

F. KALIAPSIDAE de Laubenfels, 1936
K. (APT)–Rec. Mar.

First: *Seliscothon phlyctioides* Moret, 1925, Aptian calcareous sandstone, Can Casanyas Castellet, Catalonia, Spain. **Extant**
Intervening: TUR–CMP, ?Eoc., Mio.

F. SCYTALIIDAE de Laubenfels, 1955
J. (KIM)–K. (CMP) Mar.

First: *Yrrhiza immunata* (Kolb, 1910), Weissjura ε, Sontheim, Germany.
Last: *Scytalia terebrata* (Phillips, 1835), Mukronatenkreide, Misburg area, north-western Germany.
Intervening: APT, CEN–CMP.

F. CHONELLIDAE Schrammen, 1924
J.–Q. (PLB)–Rec. Mar.

First: *Platychonia brodiei* Sollas, 1885, middle Lias, Ilminster, Somerset, England, UK. **Extant**
Intervening: BAJ, OXF–KIM, APT, CEN–CMP, Mio.

F. ASTROBOLIIDAE de Laubenfels, 1955
J. (OXF)–T. (Mio.) Mar.

First: *Cytoracia goldfussi* (Quenstedt, 1878), Weissjura α, Streitberg and other localities, north-western Germany.
Last: *Phlyctia expansa* Pomel, 1872, Miocene, Cartennian terrain, Djebel Djambeida, Algeria.
Intervening: APT, CEN–CMP.

F. JEREOPSIIDAE de Laubenfels, 1955
J. (OXF)–T. (Mio.) Mar.

First: *Hyalotragos patella* (Goldfuss, 1833), Weissjura α, Streitberg and other localities, Germany.
Last: *Jereopsis inaequalis* Pomel, 1872, and *Moretispongia pyriformis* (Pomel, 1872), Miocene, Cartennian terrain, Djebel Djambeida, Algeria.
Intervening: KIM, CEN–CMP.

F. LEIODORELLIDAE Schrammen, 1924
J. (OXF)–T. (Mio.) Mar.

First: *Leiodorella pustulosa* Schrammen, 1936, Weissjura α–ε, near Streitberg and Waldhausen, southern Germany. *Pyrgochonia acetabula* (Goldfuss, 1833) also in Weissjura α in southern Germany.
Last: *Scythophymia crassa* Pomel, 1872, and *Pleurophymia cotyle* Pomel, 1872, along with several species of *Verruculina*, Miocene, Algeria.
Intervening: KIM, APT, CEN–CMP.

F. PLINTHODERMATIIDAE de Laubenfels, 1955
K. (CMP) Mar.

First and Last: *Plinthodermatidium exile* Schrammen 1910, Mukronatenkreide, at Misberg, Germany.

F. NEOPELTIDAE Sollas, 1888 K. (CMP)–Rec. Mar.

First: *Trachynoton auricula* (Schrammen, 1912), Mukronatenkreide, Misburg, Germany. **Extant**

F. SCLERITODERMATIDAE Sollas, 1888
J. (CLV)–Rec. Mar.

First: *Azorica calloviensis* Moret, 1928, La Voulte-Sur-Rhône (Ardèche), France. **Extant**
Comment: *Azorica* was also questionably reported from the SAN of France. Cretaceous record of *Scleritoderma* cited by de Laubenfels (1955, p. E49) not found. Otherwise only Recent forms known.

F. LECANELLIDAE Schrammen, 1924
J. (KIM)–K. (CEN) Mar.

First: *Lecanella pateraeformis* Zittel, 1878, Weissjura γ, Hossingen and ε–ξ, Heuchstein and Gerstetten, Germany.
Last: *Regnardia lapparenti* Moret, 1926, Cenomanian, yellowish Chalk, Coulonges-les-Sablon, France.

Suborder DICRANOCLADINA Schrammen, 1924

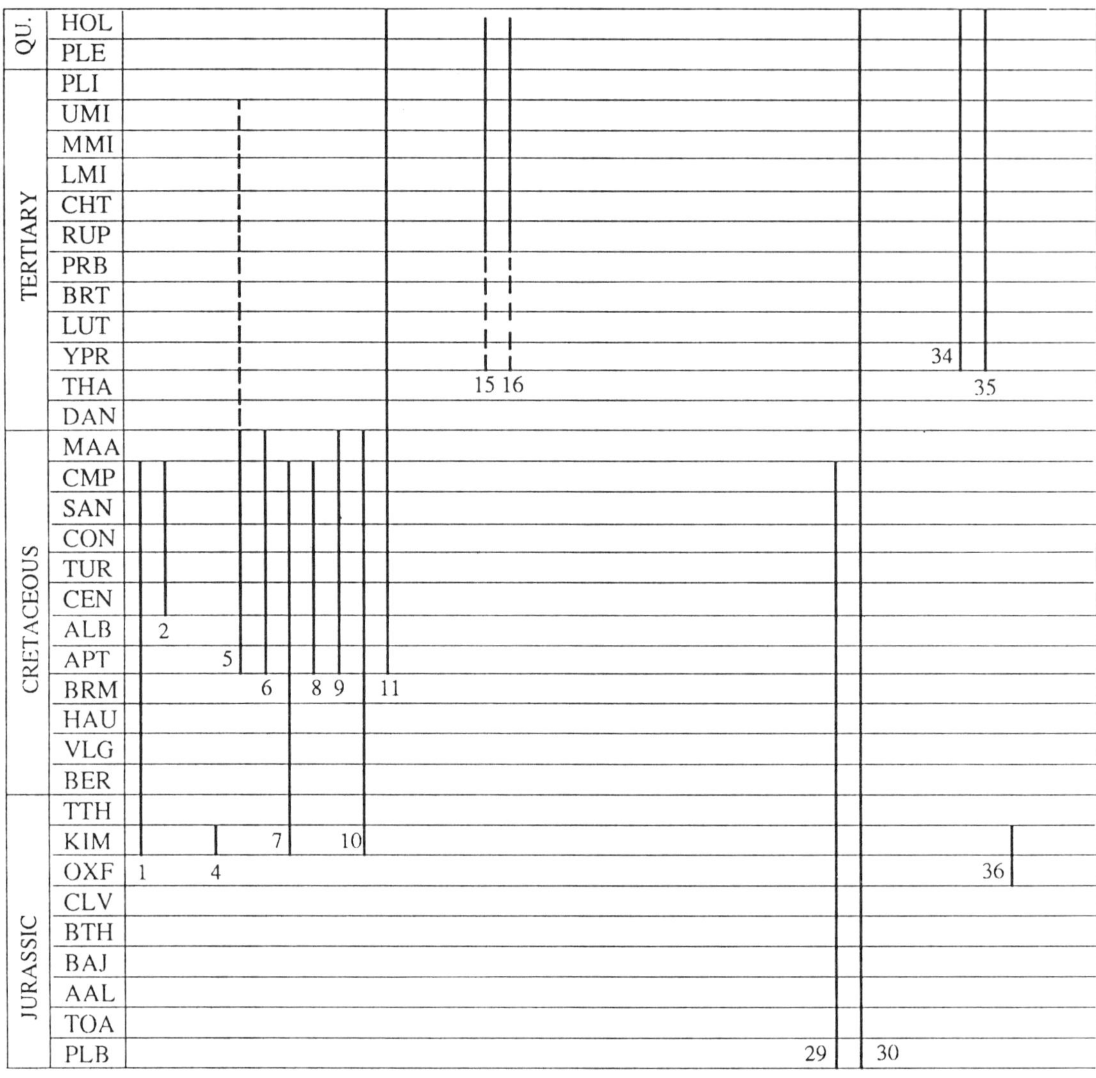

Fig. 5.2

Comment: Usage of the suborder follows that of Reid (1968, p. 23) and, with some modification, follows the general family divisions recognized by Lagneau-Hérenger (1962).

F. ACROCHORDONIIDAE Schrammen, 1910 K. (APT–CMP) Mar.

First: *Acrochordonia stellata* Lagneau-Hérenger, 1962, Aptian calcareous sandstone, Can Casanyas Castellet, Catalonia, Spain.
Last: *Acrochordonia ramosa* Schrammen, 1901 and *A. auricula* Schrammen, 1901, Quadratenkreide Olberg, Germany.
Intervening: SAN.

F. PACHINIONIDAE Schrammen, 1924 K. (APT–CMP)/T. (?Mio.) Mar.

First: *Pachinion scriptum* (Roemer, 1864) and *Pseudoverucculina globosa* Lagneau-Hérenger, 1962, Aptian calcareous sandstone, Can Casanyas Castellet, Catalonia, Spain. Co-occurring *Gilletia catalaunica*, *Pycnoclonella dactyliformis*, and *P. ramosa*, also named by Lagneau-Hérenger (1962) for specimens from Can Casanyas Castellet.
Last: *Pachinion scriptum* (Roemer, 1864), Mukronatenkreide, Misburg and Oberg, and other localities in northwestern Germany; and questionably as a species in the Miocene of Algeria.
Intervening: TUR–CMP.

F. PHRISSOSPONGIIDAE Lagneau-Hérenger, 1962 J. (KIM)–K. (CMP) Mar.

First: *Kyphoclonella multiformis* Kolb, 1910, Weissjura ε, Sontheim, Germany.
Last: *Schrammeniella scytaforme* (Schrammen, 1910), Mukronatenkreide, Misburg, Germany.
Intervening: APT, SAN, TUR–CMP.

F. SPINOCLADIIDAE Lagneau-Hérenger, 1962 K. (APT) Mar.

First and Last: *Spinocladia tubulata* Lagneau-Hérenger, 1962, calcareous sandstone beds, Can Casanyos Castellet, Catalonia, Spain.

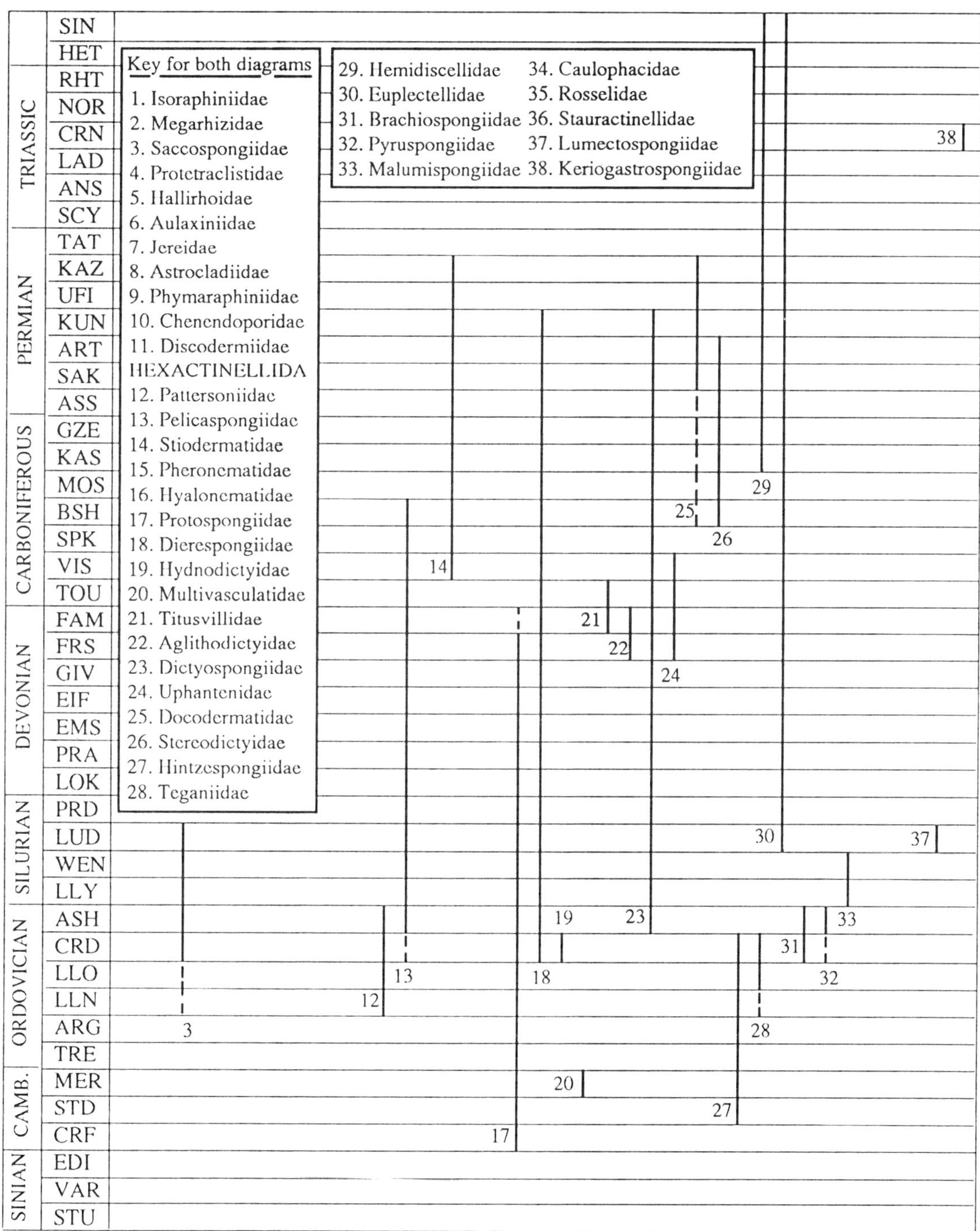

Fig. 5.2

F. GIGNOUXIIDAE de Laubenfels, 1955
J. (OXF)–K. (CMP) Mar.

First: *Dicranoclonella praecursor* Schrammen, 1936, Weissjura α, Streitberg, Germany.

Last: *Gignouxia niciensis* Moret, 1926, CMP, blue marl, Saint-Ardre, near Nice, France.

Intervening: KIM, APT, SAN–CMP.

Comment: This family, as used by de Laubenfels (1955, p. E61–2), included sponges now placed in other families (Lagneau-Hérenger, 1962, pp. 161–74). Only those genera not included by her elsewhere are included here. Systematics of the Dicranocladina will be further revised in the new version of the *Treatise on Invertebrate Paleontology*. The reported Cretaceous occurrence of a species of the living *Coscinospongia* is questioned; the family appears to have a Jurassic–Cretaceous range.

F. PSEUDOVERRUCULINIDAE de Laubenfels, 1955
K. (APT–SAN) Mar.

First: *Pseudoverruculina globosa* Lagneau-Hérenger, 1962, Aptian calcareous sandstone, Can Casanyas Castellet, Spain.

Last: *Pseudoverruculina niciensis* Moret, 1926, SAN, blue-grey marls, Nice, France.

Suborder DIDYMMORINA Rauff, 1893

F. CYLINDROPHYMATIDAE Schrammen, 1936 D. (FRS)–J. (KIM) Mar.

First: *Malinowskiella actinosum* Hurcewicz, 1985, upper FRS limestone, Kowala, Holy Cross Mountains, south-western Poland. The next record is *Melonella ovata* Sollas, 1883, Inferior Oolite, near Bristol, England, UK.
Last: *Cylindrophyma milleporata* (Goldfuss, 1833), Weissjura ξ, Gerstetten and other localities, southern Germany.
Intervening: BAJ, OXF.
Comment: The Palaeozoic *Heliospongia* and *Coelocladia* included in the family by de Laubenfels (1955, p. E64) are more properly included in the monactine sponges (Finks, 1960, pp. 40–52). The Eospongiidae, included as didymmorines, and as used by de Laubenfels (1955), included a mixture of genera that are principally included in the anthaspidellid Orchocladina, as that term was coined by Reid (1968, p. 23).

Suborder MEGAMORINA Zittel, 1878

F. PLEROMIDAE Sollas, 1888 J. (KIM)–Rec. Mar.

First: *Megalithistida foraminosa* Zittel, 1878, Weissjura ε and ζ, Heuchstetten and other localities, southern Germany, and Switzerland. **Extant**
Intervening: CEN–MAA.

F. DORYDERMATIDAE Moret, 1926 J. (KIM)–K. (CMP) Mar.

First: *Placonella perforata* Hinde, 1883, Upper Jurassic, Nattheim, Germany.
Last: *Doryderma ramosum* (Mantell, 1822), Upper Chalk, near Brighton and at Oare, Wiltshire, England, UK.
Intervening: ALB–CMP.

F. ARCHAEODORYDERMATIDAE Reid, 1968 C. (VIS) Mar.

First and Last: *Archaeodoryderma dalryense* (Hinde, 1883), Dockra Limestone, Scotland, UK.

F. HELOBRACHIIDAE Schrammen, 1910 K. (CMP) Mar. (see Fig. 5.2)

First and Last: *Helobrachium consecatum* Schrammen, 1910, Quadratenkreide, Oberg, Germany.

F. ISORAPHINIIDAE Schrammen, 1924 J. (KIM)–K. (CMP) Mar.

First: *Heloraphinia arborescens* Schrammen, 1936, Weissjura δ, Schwabtal, Germany,
Last: *Isoraphinia texta* (Roemer, 1864), Upper Chalk, Flamborough, Yorkshire, England, UK.
Intervening: APT, CEN–SEN.

F. MEGARHIZIDAE Schrammen, 1901 K. (CEN–CMP) Mar.

First: *Megarhiza colungensis* Moret, 1926, CEN, Upper Chalk, Coulonges-sur-Sablons, France.
Last: *Megarhiza dubia* Schrammen, 1901, calcareous marl, Quadratenkreide, Misburg and Oberg, Germany.

F. SACCOSPONGIIDAE Rigby and Dixon, 1979 O. (LLN/CAR)–S. (LUD) Mar.

First: *Saccospongia rudis* and *S. danvillensis* Ulrich, 1889, Trenton Beds, the former near Lexington and Frankfort, and the latter near Danville, Kentucky, USA.
Last: *Eochaunactis radiata* and co-occurring *Haplistionella garnieri* and *H. minitraba*, described by Rigby and Dixon (1979), Read Bay Formation, Garnier Bay, Somerset Island, Arctic Canada.
Intervening: CRD, ASH.

Suborder TETRACLADINA Zittel, 1878

F. PROTETRACLISTIDAE Schrammen, 1924 J. (KIM) Mar.

First and Last: *Protetraclis linki* Steinmann, 1881, Weissjura ε, Heuchstetten, Sontheim, Nattheim and Randen, Germany. *Rhizotetraclis plana*, *Sontheimia parasitica* and *S. perforata*, all named by Kolb (1910, pp. 206–9), co-occur in the Weissjura ε at Sontheim.

F. HALLIRHOIDAE de Laubenfels, 1955 K. (APT)–K. (MAA)/T. (Mio.) Mar.

First: *Siphonia pyriformis* (Goldfuss, 1833), Mas de Artis, Catalonia, Spain, in association with *S. königi* (Mantell, 1822), *Phymatella intumescens* (Roemer, 1864), and *Callopegma plana* Lagneau-Hérenger, 1962, Spain.
Last: *Siphonia königi* (Mantell, 1822) Upper Chalk, Flamborough, Yorkshire, and near Brighton, southern England, UK. The species has also been reported questionably from the Miocene of Italy.
Intervening: ALB, SAN, CMP.

F. AULAXINIIDAE de Laubenfels, 1955 K. (APT–MAA) Mar.

First: *Aulaxinia ventricosa* Schrammen, 1910, 1912, APT, Mas de Artis and Can Casanyas Castellet, Catalonia, Spain.
Last: *Aulaxinia sulcifera* (Roemer, 1864), Upper Chalk, England, UK and Mukronaten- and Quadratenkreide, Misburg, Oberg and other localities, north-western Germany, and in the Weiss Schreibkreide, Rügen, northern Germany.
Intervening: ALB, TUR–CMP.

F. JEREIDAE de Laubenfels, 1955 J. (OXF)–K. (CMP) Mar.

First: *Jereica* sp. Trammer, 1982, Jasna Góra Beds, Niegowonice, Poland. Described species *Jerea striata* Lagneau-Hérenger, 1962, and *Jerea excavata* (Goldfuss, 1833) from APT, Mas de Artis, Catalonia, Spain.
Last: *Placoscytus jereaeformis* (Schrammen, 1901), Mukronatenkriede, Misburg and Oberg, Germany.
Intervening: APT, CEN, TUR.

F. ASTROCLADIIDAE Schrammen, 1910 K. (APT–CMP) Mar.

First: *Ingentilotus ostreiformis* Lagneau-Hérenger, 1962, APT calcareous sandstone, Can Casanyas Castellet, Catalonia, Spain.
Last: *Myrmeciophytum verrucosum* Schrammen, 1910, calcareous marl of Quadratenkreide, Misburg, Oberg and Biewende, Germany.
Intervening: CEN–TUR.

F. PHYMARAPHINIIDAE Schrammen, 1910 K. (APT–MAA) Mar.

First: *Phymaraphinia plana* Lagneau-Hérenger, 1962, APT calcareous sandstone, Can Casanyas Castellet, Catalonia, Spain.
Last: (?)*Compsaspis cretacea* Sollas, 1880, MAA Trimmingham Chalk, England, UK, known only from isolated spicules.
Intervening: TUR-CMP.

F. CHENENDOPORIDAE Schrammen, 1910 J. (KIM)–K. (MAA) Mar.

First: *Tretoechus coniformis* Oppliger, 1915, Upper Jurassic, Switzerland.
Last: *Chenendopora fungiformis* Lamoroux 1821, Mukronatenkreide, Misburg, Germany, and *Turonia* cf. *T. cerebriformis* Schrammen, 1910, Weiss Schreibkriede, Rügen, northern Germany.
Intervening: APT, ALB, CEN, SAN.

F. DISCODERMIIDAE Schrammen, 1910 K. (APT)–Rec. Mar.

First: *Phyllodermia incrassata* (Goldfuss, 1833), APT, Spain. **Extant**
Intervening: ?ALB, CEN, SAN, CMP, Eoc., Mio.
Comment: Records of *Discoderma*, *Phylloderma* and *Cladodermia* begin together in Aptian deposits of Can Casanyas Castellet or Mas de Artis, Catalonia, Spain (Lagneau-Héringer, 1962, p. 134). Eight species of these genera were described from there, essentially as contemporaneous fossils. *P. incrassata* is considered to be characteristic.

Class HEXACTINELLIDA Schmidt, 1870

Comment: General classification follows that outlined by Finks (1983).

Subclass AMPHIDISCOPHORA Schulze, 1887

Order AMPHIDISCOSA Schrammen, 1924

F. PATTERSONIIDAE Miller, 1889 O. (LLN–ASH) Mar.

First: *Pattersonia aurita* (Beecher, 1889), Bigby Limestone, Benson Creek, Franklin County, Kentucky, USA.
Last: *Pattersonia tuberosa* (Beecher, 1889), Bellvue Limestone, Turners Station, Kentucky, USA.
Intervening: CRD.

F. PELICASPONGIIDAE Rigby 1970 O. (CRD/ASH)–C. (WES) Mar.

First: *Walliospongia gracilus*, *Liscombispongia nodosa*, *Wareembia concentrica* and *Kalimnospongia pertusa*, all described by Rigby and Webby (1988), co-occur in lower breccias of the Malongulli Formation, Cliefden Caves area, New South Wales, Australia.
Last: *Arakespongia mega* Rigby *et al.*, 1970, Wapanucka Limestone, near Hartshorne, Oklahoma, USA.
Intervening: EMS–FRS, VIS.

F. STIODERMATIDAE Finks, 1960 C. (VIS)–P. (KAZ) Mar.

First: *Stioderma smithii* (Young and Young, 1888), Lower Limestone, Dalry, Ayrshire, Scotland, UK.
Last: *Stioderma coscinum* Finks, 1960, sandstone tongue, Cherry Canyon Formation, Guadalupe Mountains, New Mexico, USA.
Intervening: NAM–WES.

F. PHERONEMATIDAE Gray, 1872 T. (?Eoc.)–Rec. Mar.

First: *Pheronema* sp. Hinde and Holmes 1891, Tertiary spiculite, near Oamaru, Otago, New Zealand. **Extant**

F. HYALONEMATIDAE Gray 1857 T. (?Eoc.)–Rec. Mar.

First: *Hyalonema* sp. Hinde and Holmes, 1892, Tertiary spiculite, near Oamaru, Otago, New Zealand. **Extant**

Order RETICULOSA Reid, 1958

Superfamily PROTOSPONGIOIDEA Hinde, 1887

F. PROTOSPONGIIDAE Hinde, 1887 Є. (CRF)–D. (FAM?) Mar.

First: *Protospongia fenestrata* Salter, 1864, as spicules; earliest intact skeletons are *P. fenestrata* Salter, 1864, *P. hicksi* Hinde, 1887 and *Diagoniella hindei* (Walcott, 1920), Burgess shale, Stephen Formation, near Field, British Columbia, Canada.
Last: *Gabelia pedunculus* Rigby and Murphy, 1987, unnamed Devonian shale, Roberts Mountains, Nevada, USA.
Intervening: STD, LLO, LLY–LUD, FRS.

Superfamily DIERESPONGIOIDEA Rigby and Gutschick, 1976

F. DIERESPONGIIDAE Rigby and Gutschick, 1976 O. (CRD)–P. (KUN) Mar.

First: *Dierospongia palla* Rigby and Gutschick, 1976, Pooleville Limestone Member, Bromide Formation, Criner Hills, Oklahoma, USA.
Last: *Polylophidium discus* Finks, 1960, Road Canyon Formation, Glass Mountains, Texas, USA.

F. HYDNODICTYIDAE Rigby 1971 O. (CRD) Mar.

First and Last: *Hydnodictya acantha* Rigby, 1971, Cat Head Member, Red River Formation, Lake Winnipeg, Manitoba, Canada.

F. MULTIVASCULATIDAE de Laubenfels, 1955 Є. (MER) Mar.

First and Last: *Multivasculatas ovatus* Howell and Van Houten, 1940, 'Gallatin' Formation, Bighorn Mountains, Wyoming, USA.

F. TITUSVILLIDAE Caster, 1939 D. (FRS/FAM)–C. (TOU) Mar.

First: *Protoarmstrongia ithacensis* Caster, 1941, Enfield Shale, Portage Formation, Ithaca, New York, USA.
Last: *Titusvillia drakei* Caster, 1939, Tidioute Shale, Titusville, or Corry Sandstone, McKean and Warren Counties, Pennsylvania, USA.
Comment: If *Annulispongia interrupta* Rigby and Moyle, 1959, is a titusvillid, it would be the first, but its poriferan nature is in question, Upper Manning Canyon Shale, near Ophir, Oquirrh Mountains, Utah, USA.

F. AGLITHODICTYIDAE Hall and Clarke, 1899 D. (FRS/FAM) Mar.

First and Last: *Aglithodictya numulina* Hall and Clarke,

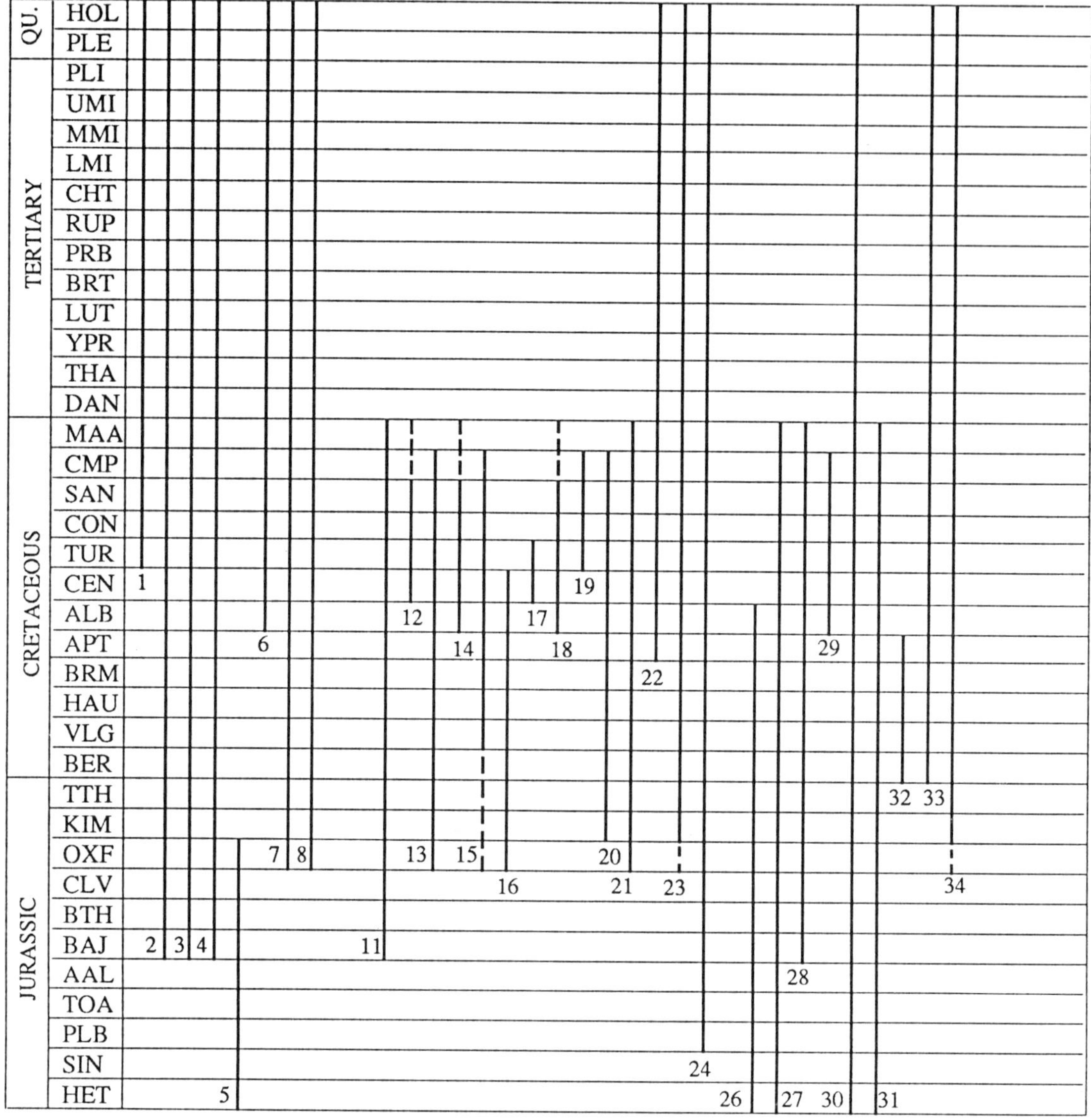

Fig. 5.3

1899, sandstones of Chemung Group, near Laurenceville, Pennsylvania, USA.

Superfamily DICTYOSPONGIOIDEA Hall and Clarke, 1899

F. DICTYOSPONGIIDAE Hall and Clarke, 1899
O. (ASH)–P. (KUN) Mar.

First: *Tiddalickia quadrata* Rigby and Webby, 1988, Sugarloaf Creek breccia, top of Malongulli Formation, Cliefden Caves area, New South Wales, Australia.
Last: *Microstaura doliolum* Finks, 1960, Road Canyon Formation, Glass Mountains, Texas, USA.
Intervening: ASH, LUD, SIG, FRS–FAM, TOU–NAM.

F. UPHANTENIDAE de Laubenfels, 1955
D. (FRS)–C. (VIS) Mar.

First: *Uphantenia chemungensis* Vanuxem, 1842, sandstone of Chemung Group, near Owego, New York, USA.
Last: *Physospongia dawsoni* Whitfield, 1881, calcareous shale of Keokuk Group, Crawfordsville, Indiana, USA.

F. DOCODERMATIDAE Finks, 1960
C. (?WES)–P. (KAZ) Mar.

First: *Docoderma* sp. Finks, 1960, spicules, Magdalena Formation, near Alamogordo, New Mexico, USA; oldest sponge with intact skeletons *Docoderma rigidum* Finks, 1960, which co-occurs with *Carphites plectus* Finks, 1960 and *Acanthocoryna stauroma* Finks, 1960 in the Road Canyon Formation, Glass Mountains, Texas, USA.
Last: *Carphites diabloensis* Finks, 1960, upper Bone Springs Limestone, Sierra Diablo, Texas, may be slightly younger than species listed above. Isolated spicules of *Docoderma* sp. Finks, 1960, Monos Formation, El Antimonio, Sonora, Mexico appear to be youngest evidence of the family (KAZ).
Intervening: ART, KUN.

F. STEREODICTYIDAE Finks, 1960
C. (WES)–P. (ART) Mar.

First: *Stereodictyum proteron* Rigby and Washburn, 1972, Diamond Peak Formation, Buck Mountain, near Eureka, Nevada, USA.

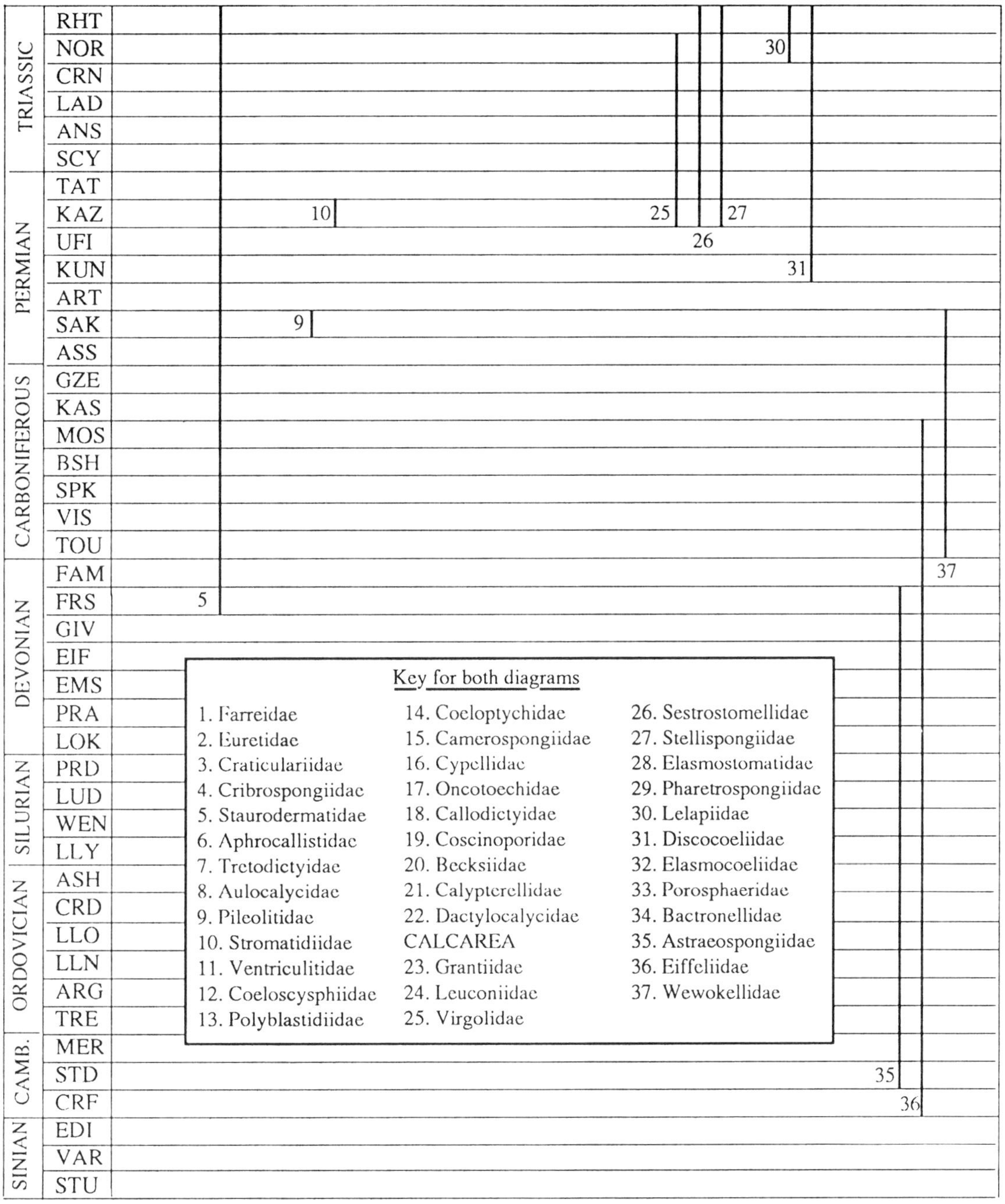
TRIASSIC
RHT
NOR
CRN
LAD
ANS
SCY
PERMIAN
TAT
KAZ
UFI
KUN
ART
SAK
ASS
CARBONIFEROUS
GZE
KAS
MOS
BSH
SPK
VIS
TOU
DEVONIAN
FAM
FRS
GIV
EIF
EMS
PRA
LOK
SILURIAN
PRD
LUD
WEN
LLY
ORDOVICIAN
ASH
CRD
LLO
LLN
ARG
TRE
CAMB.
MER
STD
CRF
SINIAN
EDI
VAR
STU
30
10
25
27
26
31
9
37
5
35
36
Key for both diagrams
1. Farreidae
2. Euretidae
3. Craticulariidae
4. Cribrospongiidae
5. Staurodermatidae
6. Aphrocallistidae
7. Tretodictyidae
8. Aulocalycidae
9. Pileolitidae
10. Stromatidiidae
11. Ventriculitidae
12. Coeloscysphiidae
13. Polyblastidiidae
14. Coeloptychidae
15. Camerospongiidae
16. Cypellidae
17. Oncotoechidae
18. Callodictyidae
19. Coscinoporidae
20. Becksiidae
21. Calypterellidae
22. Dactylocalycidae
CALCAREA
23. Grantiidae
24. Leuconiidae
25. Virgolidae
26. Sestrostomellidae
27. Stellispongiidae
28. Elasmostomatidae
29. Pharetrospongiidae
30. Lelapiidae
31. Discocoeliidae
32. Elasmocoeliidae
33. Porosphaeridae
34. Bactronellidae
35. Astraeospongiidae
36. Eiffeliidae
37. Wewokellidae

Fig. 5.3

Last: *Stereodictyum orthoplectum* Finks, 1960, Bone Springs Limestone, Guadalupe Mountains, Texas, USA.

Superfamily HINTZESPONGIOIDEA Finks, 1983

F. HINTZESPONGIIDAE Finks, 1983
€. (STD)–O. (CRD) Mar.

First: *Stephenospongia magnipora* Rigby, 1986, *Ogygopsis* Shale, Stephen Formation, near Field, British Columbia, Canada.
Last: *Cyathophycus reticulatus* Walcott, 1879, Utica Shale, near Trenton, New York, USA.
Intervening: LLO–CRD.

F. TEGANIIDAE de Laubenfels, 1955
O. (?LLN–CRD) Mar.

First: *Polyplectella mira* Ruedemann, 1925, Frankfort Shale, Six Mile Creek, New York, USA.
Last: *Teganium subsphaerica* (Walcott, 1879), Utica Shale, Holland Patent, New York, USA.

Order HEMIDISCOSA Schrammen, 1924

F. HEMIDISCELLIDAE Kling and Reif, 1969
C. (WES)–K. (CMP) Mar.

First: *Microhemidiscia ortmanni* Kling and Reif, 1969, Itarare Formation, Sausal de la Vuelta, north-eastern Uruguay.

Last: *Hemidiscella schrammeni* Reid, 1958, spicules figured by Schrammen, 1924, Upper Chalk (Quadratenkreide), Oberg, Germany.

Subclass HEXASTEROPHORA Schultz, 1887

Order LYSSACINOSA Zittel, 1877

Superfamily EUPLECTELLOIDEA Finks, 1960

F. EUPLECTELLIDAE Gray, 1867 S. (LUD)–Rec. Mar.

First: *Corticulospongia floccosa* Rigby and Chatterton, 1989, Cape Phillips Formation, Baillie-Hamilton Island, Arctic Canada. The next youngest known sponge is *Cypellospongia fimbriartis* Rigby and Gosney, 1983, Lower Triassic Thaynes Formation, Wasatch Mountains, Utah, USA.
Intervening: SCY, SIN, CEN–SEN. **Extant**

Superfamily BRACHIOSPONGIOIDEA Beecher, 1889

F. BRACHIOSPONGIIDAE Beecher, 1889 O. (CRD–ASH) Mar.

First: *Brachiospongia digitata* (Owen, 1858), Bigby Limestone, near Bright's Mill, Franklin County, Kentucky, USA.
Last: *Colpospongia lineata* Lamont, 1935, Sholeshook Limestone, Sholeshook, Haverford West, Wales, UK.

F. PYRUSPONGIIDAE Rigby, 1971 O. (CRD–?ASH) Mar.

First: *Pyruspongia ruga* Rigby, 1971, Cat Head Member, Red River Formation, Lake Winnipeg, Manitoba, Canada.
Last: *Wongaspongia minor* and *W. major*, both named by Rigby and Webby (1988), Sugarloaf Breccias, upper Malongulli Formation, Cliefden Caves area, New South Wales, Australia.

F. MALUMISPONGIIDAE Rigby, 1967 S. (LLY–WEN) Mar.

First: *Malumispongium hartnageli* (Clarke, 1924), La Vieille Formation, Black Cape Gaspé, Quebec, Canada.
Last: *Oncosella catinum* Rauff, 1894, Wenlock Limestone, Dudley, England, UK.

F. CAULOPHACIDAE Ijma, 1927 T. (Eoc)–Rec. Mar.

First: *Caulophacus* sp. Hinde and Holmes, 1891, spiculite Oamaru, Otago, New Zealand. **Extant**

F. ROSSELIDAE Schulze, 1887 T. (Eoc)–Rec. Mar.

First: Fossils known only as loose spicules. **Extant**

F. STAURACTINELLIDAE de Laubenfels, 1955 J. (OXF–KIM) Mar.

First and Last: *Stauractinella jurassica* Zittel, 1878, Weissjura α–γ; Streitberg and Hossinger, and other localities, southern Germany, and in Poland and France.

(?)F. LUMECTOSPONGIIDAE Rigby and Chatterton, 1989 S. (LUD) Mar.

First and Last: *Lumectospongia uncinata* Rigby and Chatterton, 1989, Cape Phillips Formation, Baillie-Hamilton Island, Arctic Canada.

(?)F. KERIOGASTROSPONGIIDAE Wu, 1989 Tr. CRN Mar.

First and Last: *Keriogastrospongia phialoides* Wu, 1989, co-occurs with several new genera and species, Hanwang Formation, near Anxian, Sichuan, China.

Order HEXACTINOSA Schrammen, 1903

Comment: Family usage generally follows that of Reid (1964).

Suborder CLAVULARIA Schultze, 1885

F. FARREIDAE Schulze, 1885 K. (TUR)–Rec. Mar. (see Fig 5.3)

First: *Farrea oakley* Reid, 1958, Chalk, southern England, UK. **Extant.**
Intervening: SAN, CMP.

Suborder SCOPULARIA Schultze, 1885

F. EURETIDAE Zittel, 1877 J. (BAJ)–Rec. Mar.

First: *Plectospyris elegans* Sollas, 1883 or *Mastodictyum whidborne* Sollas, 1883, Inferior Oolite, Burton Bradstock, southern England, UK. **Extant**
Intervening: OXF, K. (l.,u.), Oli., Mio.

F. CRATICULARIIDAE Rauf, 1893 J. (BAJ)–Rec. Mar.

First: *Craticularia clathrata* (Goldfuss, 1833) and *C. foliata* (Quenstedt, 1878), both from Inferior Oolite, Burton Bradstock, Dorset, southern England, UK. **Extant**
Intervening: OXF–KIM, MAA, Mio., ?Pli.

F. CRIBROSPONGIIDAE Roemer, 1864 J. (BAJ)–Rec. Mar.

First: *Cribrospongia sparsum* and *C. incertum*, both Hinde, 1893, Inferior Oolite, Burton Bradstock, Dorset, southern England, UK. **Extant**
Intervening: OXF–KIM, TUR, SEN.

F. STAURODERMATIDAE Zittel, 1877 D. (FRS)–J. (OXF) Mar.

First: *Paleostauronema transversallis*, co-occurring with *Poriferella formosum* and *Protremadictyon kainos*, all Hurcewicz, 1985, Upper FRS Limestone, Kowala, Holy Cross Mountains, south-western Poland. The next record is *Stauroderma explanatum* Hinde, 1893, Inferior Oolite, Burton Bradstock, Dorset, southern England, UK.
Last: *Stauroderma lochensis* (Quenstedt, 1858), Weissjura α–γ, Streitberg and other localities, southern Germany.
Intervening: BAJ.

F. APHROCALLISTIDAE Gray, 1867 K. (ALB)–Rec. Mar.

First: *Aphrocallistes verrucosus* and *A. macroporus*, both Lagneau-Hérenger, 1962, ALB, Gourdon, Maritime Alps, southern France. **Extant**
Intervening: MAA, Eoc.–Oli., Mio.

F. TRETODICTYIDAE Schulze, 1887 J. (OXF)–Rec. Mar.

First: *Psilocalyx nitidus* (Schrammen, 1936), Weissjura α, Streitberg, southern Germany. **Extant**

Intervening: TUR, SEN.

(?)F. AULOCALYCIDAE Ijima, 1927 J. (OXF)–Rec. Mar.

First: *Polygonatium sphaeroides* Schrammen, 1936, Weissjura α, Streitberg, southern Germany, questionably included in family (Reid, 1964). **Extant**

F. PILEOLITIDAE Finks, 1960 P. (SAK) Mar.

First and Last: *Pileolites baccatus* Finks, 1960, Skinner Ranch Formation, Glass Mountains, Texas, USA.

F. STROMATIDIIDAE Finks, 1960 P. (KAZ) Mar.

First and Last: *Stromatidium typicale* Girty, 1908, ranges from Getaway Limestone Member of Cherry Canyon Formation, to Pinery(?) and Rader Members, Bell Canyon Formation, Guadalupe Mountains, Texas, USA.

Order LYCHNISCOSA Schrammen, 1903

Comment: Family usage follows de Laubenfels (1955), although extensive revision of the order will be done by Reid in the revision of the *Treatise on Invertebrate Paleontology*.

F. VENTRICULITIDAE Smith, 1848 J. (BAJ)–K. (MAA) Mar.

First: *Calathiscus variolatus* Sollas, 1883, Inferior Oolite, Burton Bradstock, Dorset, southern England, UK.
Last: *Rhizopterion cervicornis* (Goldfuss, 1826–1833), Upper Chalk, Bromberg Halden, Germany. Several species of *Ventriculites* also co-occur in the Upper Chalk of England, along with several species of *Cephalites; V. radiatus* Mantell, 1822 occurs with *Rhizopoterion tubiforme* Schrammen, 1912 and *Leiostracosia angustata* (Roemer, 1841) in Weiss Schreibekreide, Rügen, northern Germany.
Intervening: OXF, VAL–ALB, TUR–SEN.

F. COELOSCYSPHIIDAE de Laubenfels, 1955 K. (CEN–SEN) Mar.

First: *Sestrocladia furcata* Hinde, 1884, Grey Chalk of Lower Chalk, Dover, England, UK.
Last: *Coeloscyphia sulcata* Tate, 1865, Upper Chalk, Island Magee, near Belfast, Northern Ireland.

F. POLYBLASTIDIIDAE Schrammen, 1912 J. (OXF)–K. (CMP) Mar.

First: *Phlyctaenium coniforme* Quenstedt, 1878, Weissjura γ–δ, Oberdigisheim, and other localities, southern Germany.
Last: *Polyblastidium luxurians* Zittel, 1877, Upper Chalk, Hanover, Germany. The coeval Upper Chalk species *P. tuberosum* and *P. racemosum* Toulmin-Smith, 1848, come from Kent and Sussex, England, UK.

F. COELOPTYCHIDAE Zittel 1877 K. (ALB–SEN) Mar.

First: *Coeloptychium* sp. Lagneau-Hérenger, 1962, ALB, Andon, southern France.
Last: *Coeloptychium agaricoides* Goldfuss, 1833, and several other species, Upper Chalk (Mukronatenkreide), near Misberg and other localities, north-western Germany.

F. CAMEROSPONGIIDAE Schrammen, 1912 J?/K. (OXF?/VLG)–K. (CMP) Mar.

First: *Toulminia jurassica* Hinde, 1884, Upper Jura, Randen, Switzerland, or *Polygonatium sphaeroides* Schrammen, 1936, Weissjura α, Streitberg, southern Germany. Genus questionably included in family by de Laubenfels (1955). Next youngest species *Camerospongia neocomiensis* Lagneau-Hérenger, 1947, VLG, Chateauneuf-de-Chabre, High Alps, France.
Last: *Camerospongia pervia* Schrammen, 1912, Quadratenkreide, Oberg, north-western Germany; several species of *Camerospongia* occur in the Upper Chalk of southern England, UK.
Intervening: VLG, ALB, TUR–SEN.

F. CYPELLIDAE Schrammen, 1936 J. (OXF)–K. (CEN) Mar.

First: *Cypelia prolifera* (Quenstedt, 1878) and other species of the genus; Weissjura α, Lochem and Heuberg, southern Germany.
Last: *Ophrystoma micrommatum* Roemer, 1864, Grey Chalk of Lower Chalk, Dover, England, UK.
Intervening: ALB, TUR.

F. ONCOTOECHIDAE Schrammen, 1912 K. (CEN–TUR) Mar.

First: *Onchotoechus cavernosus* Schrammen, 1912, Scaphitenpläner (Lower Chalk), southern Germany.
Last: *Onchotoechus subrutus* (Quenstedt, 1878), Cuvieripläner (Middle Chalk), Grossern Heere, southern Germany, and Chalk Marl (Lower Chalk), Isle of Wight, England, UK.

F. CALLODICTYIDAE Zittel, 1877 K. (ALB–?MAA) Mar.

First: *Callodictyon fragile* (Roemer, 1841), ALB, Escragnolles, southern France, and TUR Scaphitenpläner, Oppeln, Germany. *Sclerokalia cunningtoni* Hinde, 1884, ALB Greensands near Devizes, Wiltshire, England, UK, was questionably included in the family by de Laubenfels (1955). *Sporadiscina decheni* (Goldfuss, 1826) and *S. teutoniae* Schrammen, 1912 were also found in ALB, Gourdon, southern France.
Last: *Callodictyon infundibulum* (Zittel, 1877), Upper Chalk (Mukronatenkreide), Mizberg, Oberg and Ohlten, north-western Germany; *Marshallia tortuosa* (Roemer, 1864) and *Pleurope lacunosa* (Roemer, 1841), Mukronatenkreide, Misberg and/or Oberg. To these are added coeval *Porochonia simplex* (Toulmin-Smith, 1848), Upper Chalk, Sussex and Kent, and *Diplodictyon bayfieldi* Hinde, 1884, Upper Chalk, Norwich, southern England, UK.
Intervening: ALB, SEN.

F. COSCINOPORIDAE Zittel, 1877 K. (TURS–CMP) Mar.

First: *Cinclidella solitaria* (Schrammen, 1912) and *Coscinopora infundibuliformis* (Goldfuss, 1826), Cuvieripläner (Middle Chalk), Grosse Heere and Störmede, north-western Germany.
Last: *Balantonella elegans* (Schrammen, 1902), calcareous marls of Quadratenkreide, Misburg and Oberg, north-western Germany.

F. BECKSIIDAE Schrammen, 1912 J. (KIM)–K. (CMP) Mar.

First: *Ceriodictyon coniformis* Oppliger, 1907, Birmensdorferschichten, Mont Rivel, Jura, Switzerland.
Last: *Cyclostigma acinosa* Schrammen, 1902 and *C. mean-*

drina Schrammen, 1912, Mukronatenkreide, Misberg and Oberg, north-western Germany.
Intervening: APT, ALB, SEN.

F. CALYPTERELLIDAE Schrammen, 1912
J. (OXF)–K. (MAA) Mar.

First: *Coscinaulus micropora* Schrammen, 1936, Weissjura $\gamma-\delta$, Erkenbrechtsweiler, southern Germany.
Last: *Calypterella bertae* Schrammen, 1912 and *Saraphora armata* Schrammen, 1912, calcareous marls of the Quadratenkreide, Oberg, north-western Germany, and *Plectascus labrosus* (Toulmin-Smith, 1848), Weiss Schreibenkreide, Rügen, northern Germany.
Intervening: TUR, SEN.

F. DACTYLOCALYCIDAE Gray, 1867
K. (APT)–Rec. Mar.

First: *Exanthesis aptiensis* Lagneau-Hérenger, 1942 and *Moretiella elegans* Lagneau-Hérenger, 1962, APT, Can Casanyas Castellet, Catalonia, Spain. **Extant**
Intervening: ALB, CEN-SEN.

Class CALCAREA Bowerbank, 1864

Comment: Extensive revision of classification of the Calcarea, and fossils traditionally included in the class in the past, will be treated in the revision of *Treatise on Invertebrate Paleontology, Part E.*

Subclass CALCINEA Bidder, 1898

Comment: The classification here follows that of de Laubenfels (1955), with some modifications.

Order SOLENIDA de Laubenfels, 1955

F. CAMAROCLADIIDAE de Laubenfels, 1955
Rec. Mar.

Comment: Fossils listed in the family by de Laubenfels are probably trace fossils.

Order LABETIDA de Laubenfels, 1955

F. GRANTIIDAE Dendy, 1892
J. (?OXF/KIM)–Rec. Mar.

First: *Protosycon punctata* (Goldfuss, 1833), Weissjura, Streitberg, Germany. **Extant**

F. LEUCONIIDAE Vosmaer, 1886 J. (PLB)–Rec. Mar.

First: *Leuconia walfordia* (Hinde, 1883), Marlstone of middle Lias, Kings Sutton, Northampton, England, UK. **Extant**

Order PHARETRONIDA Zittel, 1878

F. VIRGOLIDAE Termier *et al.*, 1977
P. (KAZ)–Tr. (NOR) Mar.

First: *Virgola neptunia* (Girty, 1909), Capitan Limestone, Guadalupe Mountains, Texas, USA.
Last: *Reticulocoelia arborescens* Cuif, 1973, Antalya, Korkuteli, Turkey.
Intervening: CRN.

F. SESTROSTOMELLIDAE de Laubenfels, 1955
P. (KAZ)–K. (ALB) Mar.

First: *Precorynella crysanthemum* (Parona, 1933), Palazzo d'Adriano, Sicily.
Last: *Sestrostomella rugosa* Hinde, 1882, Upper Greensand, near Le Havre, France.
Intervening: CRN, OXF, KIM.

F. STELLISPONGIIDAE de Laubenfels, 1955
P. (KAZ)–K. (MAA) Mar.

First: *Stellispongia permica* Parona, 1933, Palazzo d'Adriano, Sicily; or *Stellispongia radiata* Rigby *et al.*, 1989, Maokou Formation, Xiangbo, Guangxi, China.
Last: *Synopella goldfussi* Hinde, 1883, Upper Chalk, Maastricht, The Netherlands.
Intervening: CRN, OXF–KIM.

F. ELASMOSTOMATIDAE de Laubenfels, 1955
J. (BAJ)–K. (MAA) Mar.

First: *Diaplectia auricula* Hinde, 1883, Inferior Oolite, near Cheltenham, England, UK.
Last: *Elasmostoma subpeziza* (D'Orbigny, 1847), Upper Chalk, Maastricht, The Netherlands.
Intervening: OXF–KIM.

F. PHARETROSPONGIIDAE de Laubenfels, 1955
K. (ALB–CMP) Mar.

First: *Pharetrospongia strahani* Sollas, 1877, Lower Greensand, near Cambridge, England, UK.
Last: *Pharetrospongia strahani* Sollas, 1877, Upper Chalk, Bromley, Kent, England, UK.

F. LELAPIIDAE Dendy and Row, 1913
Tr. (NOR)–Rec. Mar.

First: *Corynella penetrata* Quenstedt, 1878, St Cassian Formation, Dolomite Alps, Italy. **Extant**
Intervening: OXF–KIM, NEO, APT, MAA.

F. DISCOCOELIIDAE de Laubenfels, 1955
P. (KUN)–K. (MAA) Mar.

First: *Peronidella beipeiensis* Rigby *et al.*, 1989, Maokou Formation, Xiangbo, Guangxi, China.
Last: *Peronidella ocellata* (Hinde, 1883), Upper Chalk, Maastricht, The Netherlands.
Intervening: BAJ, OXF–KIM, BER, VLG.

F. ELASMOCOELIIDAE de Laubenfels, 1955
K. (NEO–APT) Mar.

First: *Elasmoierea orbiculata* (Roemer, 1864), Hils, Germany.
Last: *Elasmoierea farringdonensis* (Mantell, 1854), and associated species, Lower Greensand, Farringdon, Berkshire, England, UK.

F. POROSPHAERIDAE de Laubenfels, 1955
K. (Ng)–Rec. Mar.

First: *Porosphaerella subglobosa* Welter, 1910, Greensand, Essen, Germany. **Extant.**
Intervening: MAA.

F. BACTRONELLIDAE de Laubenfels, 1955
J. (OXF/KIM)–Rec. Mar.

First: *Bactronella pusillum* Hinde, 1883, Upper Jura probably from Thurnau, Bavaria, Germany. **Extant**
Intervening: PG.

Order HETERACTINIDA de Laubenfels, 1955

Classification, with minor modifications, follows Rigby (1983, 1986).

F. ASTRAEOSPONGIIDAE Miller, 1889 €. (STD)–D. (FRS) Mar.

First: *Jawonyia gurumal* Kruse, 1987, and *Wagima galbanyin* Kruse, 1987, lower parts of the lower Middle Cambrian, Tindall Limestone, near Tipperary, Northern Territory, Australia.
Last: *Astraeospongium*(?) spicule, Sadler Limestone, Sadler Ridge, reef tract, Canning Basin, Western Australia.
Intervening: STD–D. (m).

F. EIFFELIIDAE Rigby, 1986 €. (ATB)–C. (MOS) Mar.

First: *Eiffelia araniformis* (Mizzarzhevsky, 1981), Shabakty Formation, upper Atdabanian, Aktugay, former USSR, and coeval units from Europe, Mongolia, China and Australia.
Last: *Zangerlispongia richardsoni* Rigby and Nitecki, 1975, Carbondale Formation, Moscovian (Alleghenian), Fulton County, Illinois, USA.
Intervening: STD, O. (l., m.).

F. WEWOKELLIDAE King, 1943 C. (TOU)–P. (SAK) Mar.

First: *Asteractinella expansa* Hinde, 1888, lower part of Lower Limestone, Ayrshire, Scotland, UK, and from upper TOU limestones, Neufvilles-Soignes, Belgium.
Last: *Talpaspongia clavata* King, 1943, Talpa Member, Clyde Formation, Runnels County, central Texas, or top of Neal Ranch Formation, Wolfcampian, Glass Mountains, western Texas, USA.
Intervening: VIS, NAM, WES, STE.

'SPHINCTOZOANS', 'CHAETETIDS', 'STROMATOPOROIDS'

These three groups are maintained as separate entities here, in line with traditional views based on their skeletons. However, spicule and soft-tissue data suggest that all three are polyphyletic.

Class 'SPHINCTOZOANS'

Families and references pertaining to them are taken from Senowbari-Daryan (1990).

Order VERTICILLITIDA Termier and Termier, 1977

F. VERTICILLITIDAE Steinmann, 1882 P. (?ART)–K. (CMP) Mar.

First: *Stylothalamia permica* Senowbari-Daryan, 1990, Pietro di Salamone, Sicily.
Last: *Stylothalamia lehmanni* (Engeser and Neumann, 1986), Krappfeld-Gosau, Germany.
Intervening: KUN–SPA, HET–TOA, CON–SAN.

Order PERMOSPHINCTA Termier and Termier, 1974

F. SEBARGASIIDAE de Laubenfels, 1955 €. (?TOY)–Tr. (RHT) Mar.

First: ?*Amblysiphonella parvula* Pickett and Jell, 1986, New South Wales, Australia.
Last: *Amblysiphonella maxima* Senowbari-Daryan and Di Stefano, 1988, Sicily.
Intervening: ?PUS–?HIR, NAM–TAT, ANS–NOR.

F. COLOSPONGIIDAE Senowbari-Daryan, 1990 €. (TOY)–Tr. (RHT) Mar.

First: *Nucha naucum* Pickett and Jell, 1983, New South Wales.
Last: *Parauvanella* sp., Northern Calcareous Alps, Salzburg, Austria.
Intervening: ASH, NAM–TAT, ANS–NOR.

F. TEBAGATHALAMIIDAE Senowbari-Daryan and Rigby, 1988 P. (ART–TAT) Mar.

First and Last: *Tebagathalamia cylindrica* Senowbari-Daryan and Rigby, 1988, Tunisia.
Intervening: KUN–KAZ.

F. ANNAECOELIIDAE Senowbari-Daryan, 1978 Tr. (CRN–RHT) Mar.

First and Last: *Annaecoelia maxima* Senowbari-Daryan, 1978, former Yugoslavia.
Intervening: NOR.

F. CHEILOSPORITIIDAE Fischer, 1962 Tr. (NOR–RHT) Mar.

First and Last: *Cheilosporites tirolensis* Wähner, 1903, Northern Calcareous Alps, Italy.

F. SALZBURGIIDAE Senowbari-Daryan and Schäfer, 1979 P. (ASS)–Tr. (RHT) Mar.

First: *Salzburgia variabilis* Senowbari-Daryan and Schäfer, 1979, Sicily.
Last: *Salzburgia variabilis* Senowbari-Daryan and Schäfer, 1979, Northern Calcareous Alps, Italy.
Intervening: SAK.

F. CRIBROTHALAMIIDAE Senowbari-Daryan, 1990 Tr. (NOR–RHT) Mar.

First and Last: *Cribrothalamia gulloae* Senowbari-Daryan, 1990, Reef Carbonate, Madonie-Gebirge, Germany.

F. ANGULLONGIIDAE Webby and Rigby, 1985 O. (COS–HIR) Mar.

First and Last: *Angullongia vesica* Webby and Rigby, 1985, New South Wales, Australia.
Intervening: HAR–RAW.

F. PHRAGMOCOELIIDAE Ott, 1974 D. (EIF)–Tr. (CRN) Mar.

First: *Radiothalamos uniramosus* Picket and Rigby, 1983, Garra Formation, New South Wales, Australia.
Last: *Phragmocoelia endersi* Ott, 1974, Northern Calcareous Alps, Italy.
Intervening: GIV.

F. POLYSOLOTHIIDAE Seilacher, 1962 Tr. (CRN–RHT) Mar.

First and Last: *Fania astoma* (Seilacher, 1926), Luning Formation, Pilot and Cedar Mountains, Nevada, USA.
Intervening: NOR.

F. SOLENOLMIIDAE Engeser, 1986 D. (EIF)–Tr. (RHT) Mar.

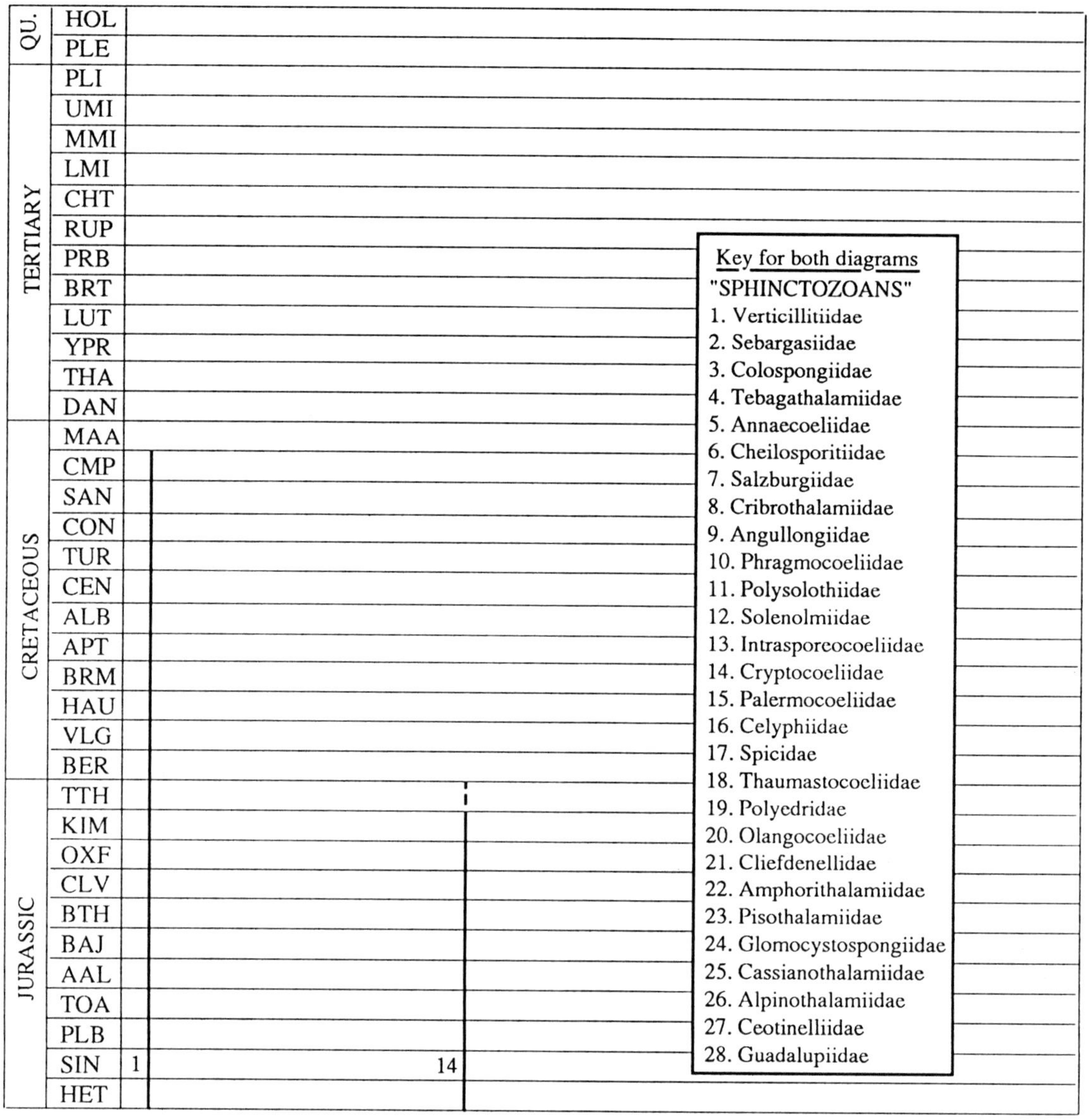

QU.: HOL, PLE
TERTIARY: PLI, UMI, MMI, LMI, CHT, RUP, PRB, BRT, LUT, YPR, THA, DAN
CRETACEOUS: MAA, CMP, SAN, CON, TUR, CEN, ALB, APT, BRM, HAU, VLG, BER
JURASSIC: TTH, KIM, OXF, CLV, BTH, BAJ, AAL, TOA, PLB, SIN, HET

1 14

Key for both diagrams
"SPHINCTOZOANS"
1. Verticillitiidae
2. Sebargasiidae
3. Colospongiidae
4. Tebagathalamiidae
5. Annaecoeliidae
6. Cheilosporitiidae
7. Salzburgiidae
8. Cribrothalamiidae
9. Angullongiidae
10. Phragmocoeliidae
11. Polysolothiidae
12. Solenolmiidae
13. Intrasporeocoeliidae
14. Cryptocoeliidae
15. Palermocoeliidae
16. Celyphiidae
17. Spicidae
18. Thaumastocoeliidae
19. Polyedridae
20. Olangocoeliidae
21. Cliefdenellidae
22. Amphorithalamiidae
23. Pisothalamiidae
24. Glomocystospongiidae
25. Cassianothalamiidae
26. Alpinothalamiidae
27. Ceotinelliidae
28. Guadalupiidae

Fig. 5.4

First: *Hormospongia labyrinthica* Rigby and Blodgett, 1983, western central Alaska.
Last: *Paradeningeria alpina* Senowbari-Daryan and Schäfer, 1979, Northern Calcareous Alps, Italy.
Intervening: GIV, ASS–TAT, ANS–NOR.

F. INTRASPOREOCOELIIDAE Fan and Zhang, 1985
P. (?ART–TAT) Mar.

First and Last: *Intrasporeocoelia hubeiennsis* Fan and Zhang, 1985, Pietro di Salamone, Sosio Valley, Sicily.
Intervening: KUN–KAZ.

F. CRYPTOCOELIIDAE Steinmann, 1882
S. (?GOR)–J. (?TTH) Mar.

First: *Rigbyspongia catenulata* De Freitas, 1987, Cornwallis Island, Arctic Canada.
Last: ?*Sphinctonella trestini* Hurcewicz, 1975, Poland. The next youngest is *Cryptocoelia zitteli* Steinmann, 1882, Mufara Formation, Sicily (RHT).
Intervening: ANS–RHT, OXF–KIM.

F. PALERMOCOELIIDAE Senowbari-Daryan, 1990
Tr. (NOR–RHT) Mar.

First and Last: *Palermocoelia tubifera* Senowbari-Daryan, 1990, Reef Carbonate, La Montagnola, Sicily.

F. CELYPHIIDAE Laubenfels, 1955
O. (?LLN)–Tr. (RHT) Mar.

First: *Porefieldia robusta* Rigby and Potter, 1986, north California, USA.
Last: *Celyphia submarginata* (Münster, 1841), Cassian Formation, Italy.
Intervening: ?LLO–HIR, ART–KUN, ANS–NOR.

F. SPICIDAE Termier and Termier, 1977
O. (?LLN)–Tr. (RHT) Mar.

First: *Cystothalamiella ducta* Rigby and Potter, 1986, California, USA.
Last: *Russospongia lupensis* (Senowbari-Daryan, 1980), Sicily.

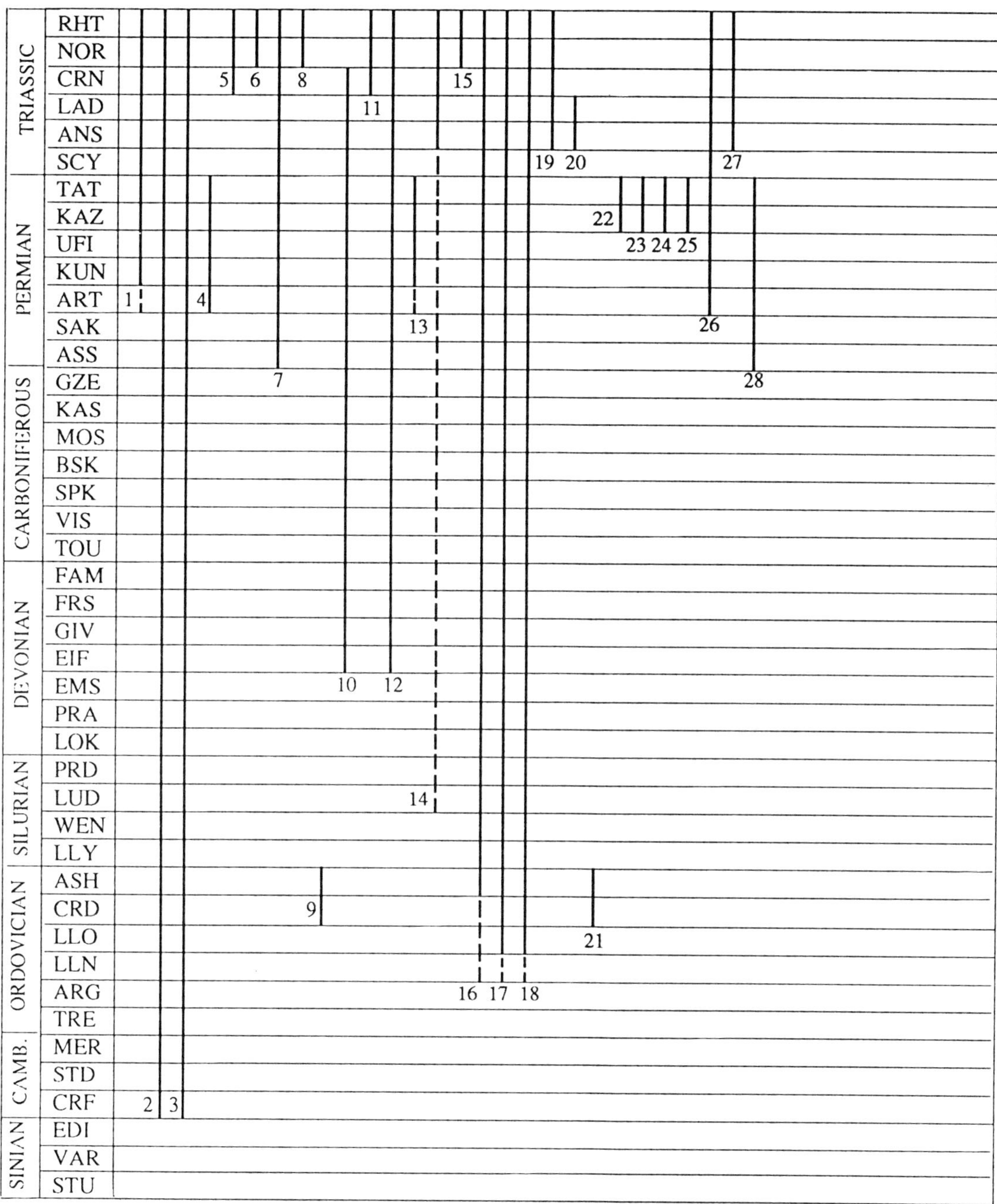

Fig. 5.4

Intervening: LLO–HIR, ?KRE–NOG, ART–KUN, ANS–NOR.

F. THAUMASTOCOELIIDAE Ott, 1967
O. (?LLN)–Tr. (RHT) Mar.

First: *Girtyocoelia canna* Rigby and Potter, 1986, eastern Klamath Mountains, northern California, USA.
Last: *Thaumastocoelia cassiana* Steinmann, 1882, Cassian Formation, Italy.
Intervening: LLO–HIR, ?ALP–TAT, CRN–NOR.

F. POLYEDRIDAE Termier and Termier, 1977
P. (ANS)–Tr. (RHT) Mar.

First: *Pseudoguadalupia alveolaris* Termier and Termier, 1977, Sosio Valley, Sicily.
Last: *Polyedra tebagensis* Termier and Termier, 1955, Djebel Tebaga, southern Tunisia.
Intervening: LAD–NOR.

F. OLANGOCOELIIDAE Bechstädt and Brandner, 1970 Tr. (ANS–LAD) Mar.

First and Last: *Olangocoelia otti* Bechstädt and Brandner, 1970, Dolomite Alps, Austria.

F. CLIEFDENELLIDAE Webby, 1969
O. (COS–HIR) Mar.

First and Last: *Cliefdenella etheridgei* Webby, 1969, New South Wales, Australia.
Intervening: HIR–RAW.

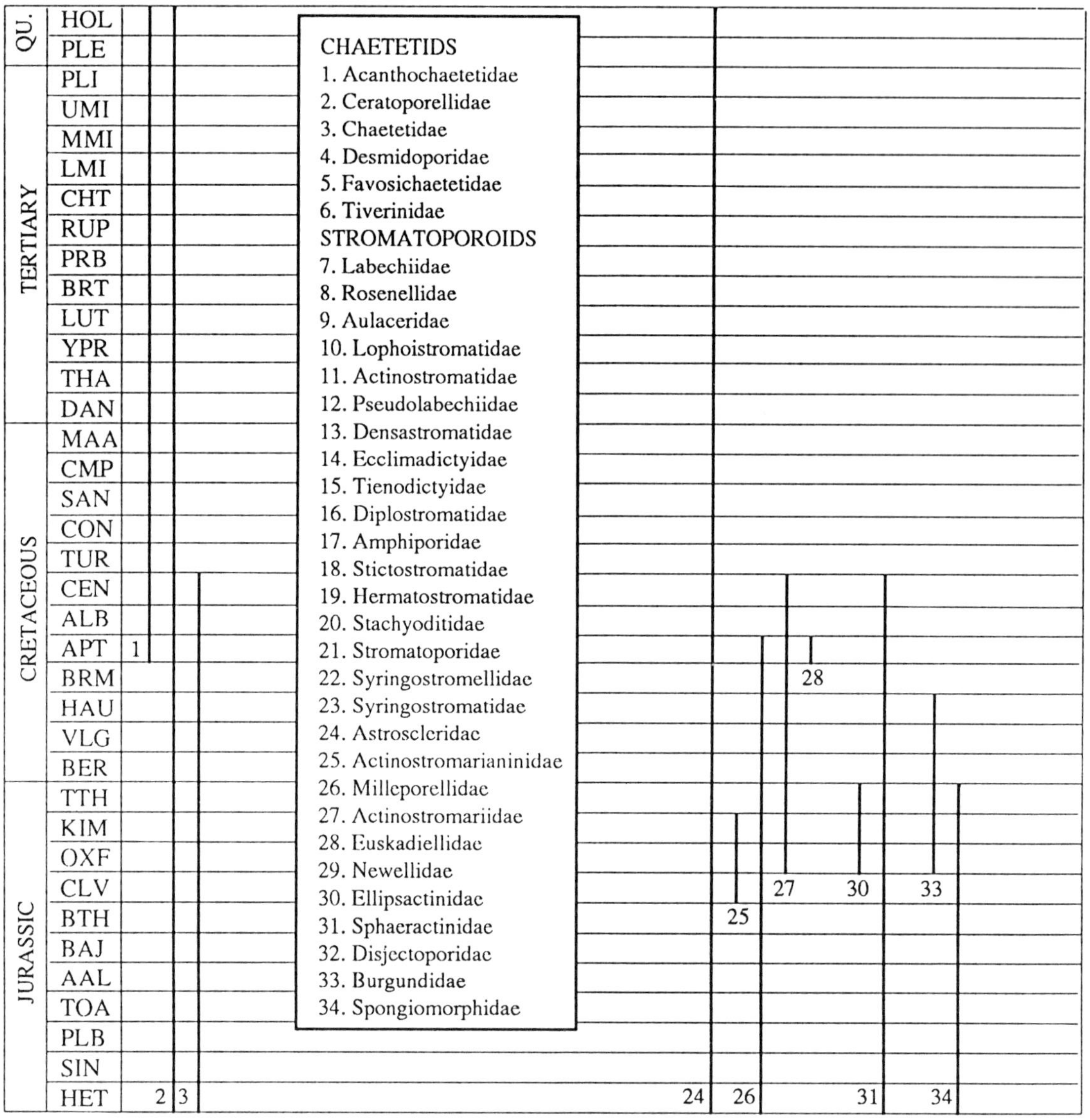

QU.: HOL, PLE
TERTIARY: PLI, UMI, MMI, LMI, CHT, RUP, PRB, BRT, LUT, YPR, THA, DAN
CRETACEOUS: MAA, CMP, SAN, CON, TUR, CEN, ALB, APT, BRM, HAU, VLG, BER
JURASSIC: TTH, KIM, OXF, CLV, BTH, BAJ, AAL, TOA, PLB, SIN, HET

CHAETETIDS
1. Acanthochaetetidae
2. Ceratoporellidae
3. Chaetetidae
4. Desmidoporidae
5. Favosichaetetidae
6. Tiverinidae
STROMATOPOROIDS
7. Labechiidae
8. Rosenellidae
9. Aulaceridae
10. Lophoistromatidae
11. Actinostromatidae
12. Pseudolabechiidae
13. Densastromatidae
14. Ecclimadictyidae
15. Tienodictyidae
16. Diplostromatidae
17. Amphiporidae
18. Stictostromatidae
19. Hermatostromatidae
20. Stachyoditidae
21. Stromatoporidae
22. Syringostromellidae
23. Syringostromatidae
24. Astroscleridae
25. Actinostromarianinidae
26. Milleporellidae
27. Actinostromariidae
28. Euskadiellidae
29. Newellidae
30. Ellipsactinidae
31. Sphaeractinidae
32. Disjectoporidae
33. Burgundidae
34. Spongiomorphidae

1 2 3 24 25 26 27 28 30 31 33 34

Fig. 5.5

F. AMPHORITHALAMIIDAE Senowbari-Daryan and Rigby, 1988 P. (KAZ–TAT) Mar.

First and Last: *Amphorithalamia cateniformis* Senowbari-Daryan and Rigby, 1988, Djebel Tebaga, Tunisia.

F. PISOTHALAMIIDAE Senowbari-Daryan and Rigby, 1988 P. (KAZ–TAT) Mar.

First and Last: *Pisothalamia spiculata* Senowbari-Daryan and Rigby, 1988, Djebel Tebaga, Tunisia.

F. GLOMOCYSTOSPONGIIDAE Rigby *et al.*, 1989 P. (KAZ–TAT) Mar.

First and Last: *Glomocystospongia gracilis* Rigby *et al.*, 1989, Djebel Tebaga, Tunisia.

Class DEMOSPONGIAE Sollas, 1875

Order HADROMERIDA Topent, 1928

F. CASSIANOTHALAMIIDAE Reitner, 1987 P. (KAZ–TAT) Mar.

First and Last: *Cassianothalamia zardinii* Reitner, 1987, Cassian Formation, Italy.

Order INCERTAE SEDIS

F. ALPINOTHALAMIIDAE Senowbari-Daryan, 1990 P. (ART)–Tr. (RHT) Mar.

First: *Uvanella irregularis* Ott, 1967, Guangxi, China.
Last: *Leinia schneebergensis* Senowbari-Daryan, 1990, Pantokrator Chalk, Greece.
Intervening: LAD–NOR.

F. CEOTINELLIIDAE Senowbari-Daryan, 1978 Tr. (ANS–RHT) Mar.

First and Last: *Coetinella mirunae* Pantic, 1975, Montenegro, former Yugoslavia.
Intervening: LAD–NOR.

Order GUADALUPIIDA Termier and Termier, 1977

F. GUADALUPIIDAE Termier and Termier, 1977 P. (ASS–TAT) Mar.

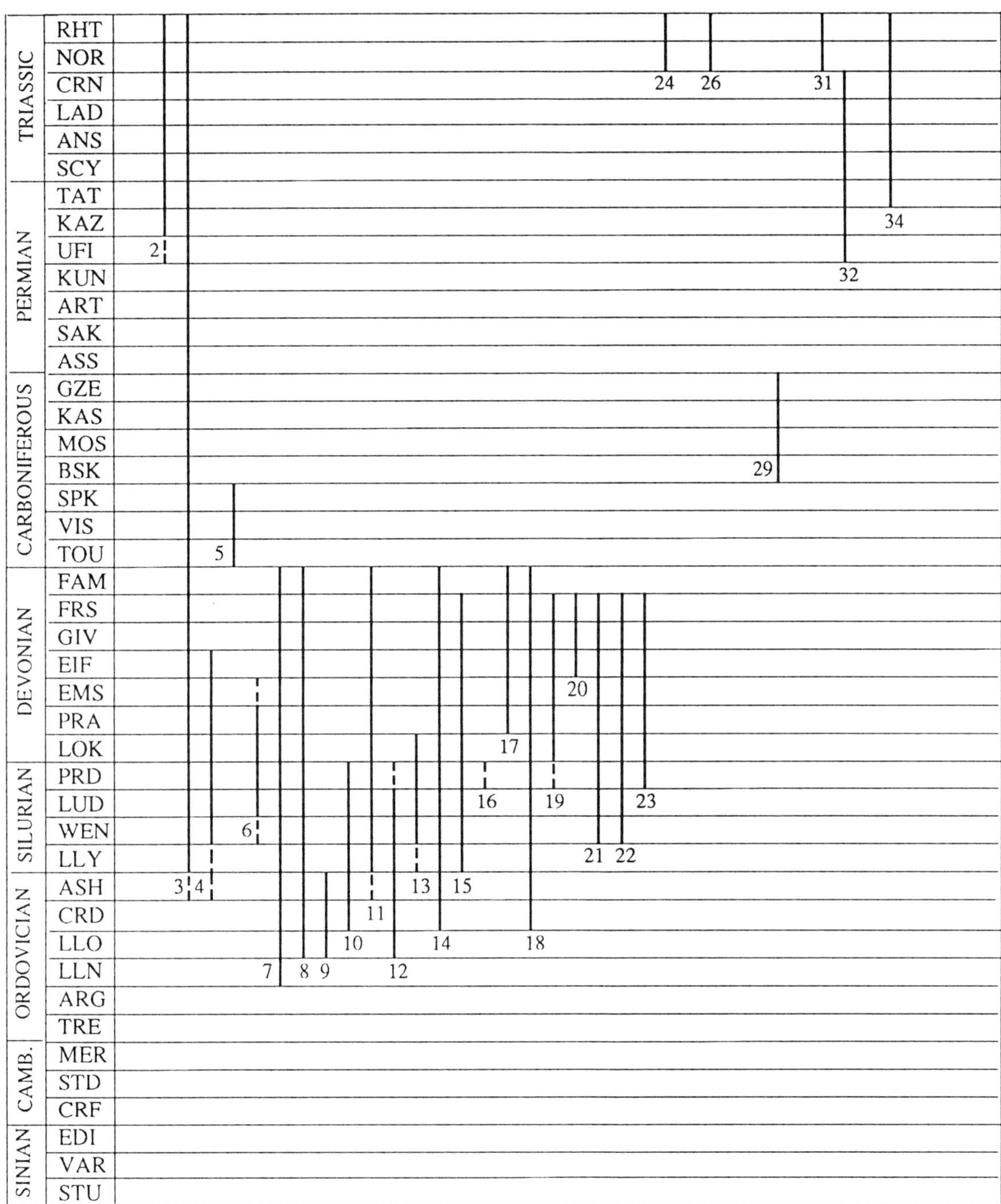

Fig. 5.5

First and Last: *Lemonea cylindrica* (Girty, 1908), Texas, USA.
Intervening: SAK–KAZ.

'CHAETETIDS'

Data are from Hill (1981) unless otherwise stated.

Class DEMOSPONGIAE Sollas, 1875

F. ACANTHOCHAETETIDAE Fischer, 1970
K. (APT)–Rec. Mar.

First: *Acanthochaetetes suenesi* Reitner and Engeser, 1983, Spain (Reitner and Engeser, 1983). **Extant**

F. CERATOPORELLIDAE Hickson, 1911
P. (?UFI)–Rec. Mar.

First: *Ceratoporella* sp., Tunisia. **Extant**

F. CHAETETIDAE Milne-Edwards and Haime, 1850
O. (?PUS)–K. (CEN) Mar.

First: *Chaetetella filiformis* Sokolov, 1962, former northeastern USSR.
Last: *Blastochaetetes irregularis* Wood and Reitner, 1987, Spain (Wood and Reitner, 1989).

F. DESMIDOPORIDAE Preobrazhenskiy, 1968
O. (?PUS)–D. (EIF) Mar.

First: *Schizolites floriformis* Preobrazhenskiy, 1968, former north-eastern USSR.
Last: *Desmidopora* sp., Tien Shan, former USSR.
Intervening: SHE–GLE.

F. FAVOSICHAETETIDAE Yang, 1978 C. (MIS) Mar.

First and Last: *Favosichaetetes multiporosus* Yang, 1978, China.

F. TIVERINIDAE Hill, 1981 S. (?SHE)–D. (?EMS) Mar.

First: *'Barrandeolites bowerbanki'* Sokolov and Prantl, 1965 (*nom. nud.*), Dudley, England, UK. Genus only summarily diagnosed.
Last: *Tiverina vermiculata* Sokolov and Tesakov, 1968, Podolia, former USSR.

PALAEOZOIC 'STROMATOPOROIDS'

Families and the references pertaining to them are from Stearn (1980); genera, unless otherwise noted, are from Lecompte, 1956, as amended by Stearn (1980).

Order LABECHIIDA Kühn, 1927

F. LABECHIIDAE Nicholson, 1879 O. (LLN)–D. (FAM) Mar.

First: *Stromatocerium rugosum* Hall, 1847, North America.
Last: *Stylostroma crassum* Gorsky, 1938, former USSR.
Intervening: LLO–FRS.

F. ROSENELLIDAE Yavorsky, 1973 O. (LLO)–D. (FAM) Mar.

First and Last: *Rosenella macrocystis* Nicholson, 1886, Gotland.
Intervening: COS–FRS.

F. AULACERIDAE Kühn, 1927 O. (LLO–HIR) Mar.

First: *Cryptophragmus antiquatum* Raymond, 1914, North America.
Last: *Aulacera undulata* (Billings, 1857), Indiana, USA.
Intervening: COS–RAW.

F. LOPHIOSTROMATIDAE Nestor, 1966 O. (COS)–S. (PRD) Mar.

First and Last: *Lophiostroma schmidti* (Nicholson, 1890), North America.
Intervening: HAR–LDF.

Order ACTINOSTROMATIDA Bogoyavlenskaya, 1969

F. ACTINOSTROMATIDAE Nicholson, 1886 O. (?PUS)–D. (FAM) Mar.

First: *Plectostroma* sp. Nestor, 1966, Estonia, former USSR (Nestor, 1966).
Last: *Atelodictyon fallax* Lecompte, 1951, Belgium.
Intervening: ?CAU–FRS.

F. PSEUDOLABECHIIDAE Bogoyavlenskaya, 1969 O. (LLO)–S. (?PRD) Mar.

First and Last: *Pseudolabechia granulata* Yabe and Siguyama, 1930, Europe.
Intervening: COS–LDF.

F. DENSASTROMATIDAE Bogoyavlenskaya, 1977 S. (?TEL)–D. (GED) Mar.

First: *Densastroma pexisum* Mori, 1968, Visby Beds, Gotland (Mori, 1968).
Last: *Araneosustroma* sp. Lessovaya, 1970, Uzbekistan, former USSR.
Intervening: SHE–PRD.

Order CLATHODICTYIDA Bogoyavlenskaya, 1969

F. ECCLIMADICTYIDAE Stearn, 1980 O. (COS)–D. (FAM) Mar.

First: ??*Pseudostylodictyon poshanense* Ozaki, 1938, China.
Last: *Stylodictyon columnaris* (Nicholson, 1875), former USSR.
Intervening: HAR–FRS.

F. TIENODICTYIDAE Bogoyavlenskaya, 1965 S. (?TEL)–D. (FRS) Mar.

First: *Intexodictyon olevi* Yavorsky, 1963, Estonia, former USSR (Stearn, 1969).
Last: *Hammatostroma nodosum* Stearn, 1961, Redwater Reef, Alberta, Canada (Stearn, 1969).
Intervening: SHE–GIV.

F. DIPLOSTROMATIDAE Stearn, 1980 S. (?PRD) Mar.

First and Last: *Diplostroma* sp., Nestor, 1966, Estonia, former USSR.

Order INCERTAE SEDIS

F. AMPHIPORIDAE Rukhin, 1938 D. (SIG–FAM) Mar.

First and Last: *Amphipora ramosa* (Phillips, 1841), England, UK.
Intervening: EMS–FRS.

Order STROMATOPORELLIDA Stearn, 1980

F. STICTOSTROMATIDAE Khalfina and Yavorsky, 1973 O. (?COS)–D. (FAM) Mar.

First: *Styloporella* sp., Siberia, former USSR (Stearn, 1980).
Last: *Stromatoporella granulata* (Nicholson, 1873), Belgium.
Intervening: HAR–FRS.

F. HERMATOSTROMATIDAE Nestor, 1964 S. (?PRD)–D. (FRS) Mar.

First and Last: *Hermatostroma schlüteri* Nicholson, 1886, Germany.
Intervening: GED–GIV.

F. STACHYODITIDAE Khromych, 1967 D. (EIF–FRS) Mar.

First: *Stachyodes verticillata* (M'Coy, 1851), Germany.
Last: *Stachyodes radiata* Lecompte, 1951, Belgium.
Intervening: GIV.

Order STROMATOPORIDA Stearn, 1980

F. STROMATOPORIDAE Winchell, 1867 S. (?SHE)–D. (FRS) Mar.

First: *Parallelostroma* sp. Nestor, 1966, Estonia, former USSR.
Last: *Stromatopora concentrica* Goldfuss, 1826, Belgium.
Intervening: WHI–GIV.

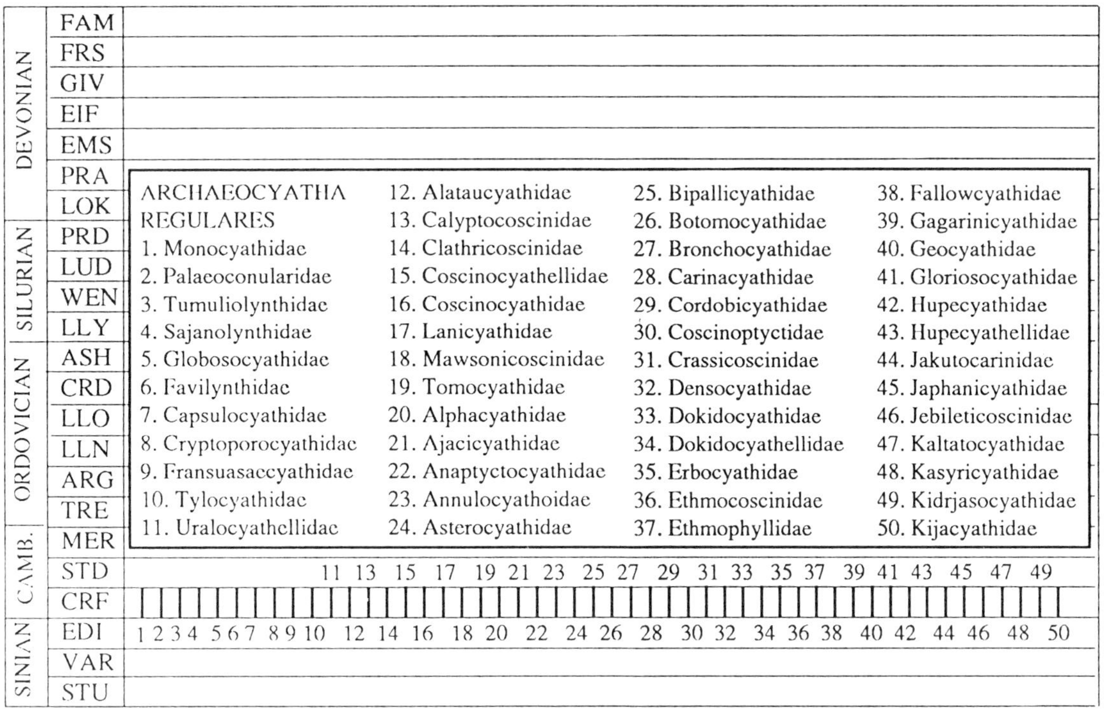

DEVONIAN: FAM, FRS, GIV, EIF, EMS, PRA, LOK
SILURIAN: PRD, LUD, WEN, LLY
ORDOVICIAN: ASH, CRD, LLO, LLN, ARG, TRE
CAMB.: MER, STD, CRF
SINIAN: EDI, VAR, STU

ARCHAEOCYATHA REGULARES
1. Monocyathidae
2. Palaeoconularidae
3. Tumuliolynthidae
4. Sajanolynthidae
5. Globosocyathidae
6. Favilynthidae
7. Capsulocyathidae
8. Cryptoporocyathidae
9. Fransuasaecyathidae
10. Tylocyathidae
11. Uralocyathellidae
12. Alataucyathidae
13. Calyptocoscinidae
14. Clathricoscinidae
15. Coscinocyathellidae
16. Coscinocyathidae
17. Lanicyathidae
18. Mawsonicoscinidae
19. Tomocyathidae
20. Alphacyathidae
21. Ajacicyathidae
22. Anaptyctocyathidae
23. Annulocyathoidae
24. Asterocyathidae
25. Bipallicyathidae
26. Botomocyathidae
27. Bronchocyathidae
28. Carinacyathidae
29. Cordobicyathidae
30. Coscinoptyctidae
31. Crassicoscinidae
32. Densocyathidae
33. Dokidocyathidae
34. Dokidocyathellidae
35. Erbocyathidae
36. Ethmocoscinidae
37. Ethmophyllidae
38. Fallowcyathidae
39. Gagarinicyathidae
40. Geocyathidae
41. Gloriosocyathidae
42. Hupecyathidae
43. Hupecyathellidae
44. Jakutocarinidae
45. Japhanicyathidae
46. Jebileticoscinidae
47. Kaltatocyathidae
48. Kasyricyathidae
49. Kidrjasocyathidae
50. Kijacyathidae

11 13 15 17 19 21 23 25 27 29 31 33 35 37 39 41 43 45 47 49

1 2 3 4 5 6 7 8 9 10 12 14 16 18 20 22 24 26 28 30 32 34 36 38 40 42 44 46 48 50

Fig. 5.6

F. SYRINGOSTROMELLIDAE Stearn, 1980
S. (?SHE)–D. (FRS) Mar.

First: *Syringostromella* sp. Nestor, 1966, Estonia, former USSR (Stearn, 1980).
Last: *Parallelopora paucicanaliculata* Lecompte, 1951, Belgium.
Intervening: S. (WHI)–D. (GIV)

F. SYRINGOSTROMATIDAE Lecompte, 1951
S. (PRD)–D. (FRS) Mar.

First: *Syringostroma centrotum* Girty, 1953, New York, USA.
Last: *Syringostroma densa* Nicholson, 1886, Belgium.
Intervening: GED–GIV.

MESOZOIC 'STROMATOPOROIDS'

Data are from Lecompte (1956) unless otherwise stated.

Class DEMOSPONGIAE Sollas, 1875

Order AXINELLIDA Lévi, 1956

F. ASTROSCLERIDAE Lister, 1900 Tr. (NOR)–Rec. Mar.

First: *Astrosclera* sp., Italy, Wood, 1987. **Extant**

F. ACTINOSTROMARIANINIDAE Wood, 1987
J. (CLV–KIM) Mar.

First and Last: *Actinostromarianina lecompti* Hudson, 1955, Alam Abayadh, North Yemen.
Intervening: OXF.

F. MILLEPORELLIDAE Yabe and Sugiyama, 1935
Tr. (NOR)–K. (APT) Mar.

First: *Murania* sp., Italy (Wood, 1987).
Last: *Murania lefeldi* Kazmierczak, 1974, Slovakian-Tatra Mountains, Poland (Kazmierczak, 1974).
Intervening: KIM.

F. ACTINOSTROMARIIDAE Hudson, 1955
J. (OXF)–K. (CEN) Mar.

First: *Actinostromaria* sp., Slovenia (Wood, 1987).
Last: *Actinostromaria stellata* Dehorne, 1915, L'Ile Madame, France.
Intervening: KIM, VLG–BRM.

Order HAPLOSCLERIDA Topent, 1928

F. EUSKADIELLIDAE Reitner, 1987 K. (APT) Mar.

First and Last: *Euzkadiella erenoensis* Reitner, 1987, Guiuzcoa Province, north Spain (Reitner, 1987).

F. NEWELLIDAE Wood *et al.*, 1989 C. (PEN) Mar.

First and Last: *Newellia mira* (Newell, 1935), Nevada, USA (Wood *et al.*, 1989).

Order *INCERTAE SEDIS*

F. ELLIPSACTINIDAE Poljak, 1936 J. (OXF–TTH) Mar.

First and Last: *Ellipsactinia ellipsoidea* Steinmann, 1878, Australia.
Intervening: KIM.

F. SPHAERACTINIDAE Kühn, 1927
Tr. (NOR)–K. (CEN) Mar.

First: *Lithopora* sp., Europe (Grubic, 1961).
Last: *Lithopora* sp., Tunisia (Grubic, 1961).
Intervening: ?SIN–TOA, OXF.

F. DISJECTOPORIDAE Tornquist, 1901
P. (UFI)–Tr. (CRN) Mar.

First: *Disjectopora japonica* Yabe and Sugiyama, 1935, Japan.
Last: *Balatonia kochi* Vinassa, 1908, Hungary.
Intervening: KAZ–LAD.

F. BURGUNDIDAE Dehorne, 1920
J. (OXF)–K. (HAU) Mar.

First: *Burgundia trinorchii* Dehorne, 1916, France.
Last: *Burgundia trinorchii* Dehorne, 1916, Japan (Wood, 1987).
Intervening: KIM–VLG.

F. SPONGIOMORPHIDAE Frech, 1890
P. (?TAT)–J. (TTH) Mar.

First: *Spongiomorpha* sp. Frech, 1890, Tunisia.
Last: *Heptastylis stromatoporoides* Frech, 1890, Japan.

Division ARCHAEOCYATHA Bornemann, 1884
(see Fig. 5.6)

Class REGULARES Vologdin, 1937

Information is taken from Debrenne *et al.* (1990b), unless otherwise stated.

Order MONOCYATHIDA Okulitch, 1935

F. MONOCYATHIDAE R. Bedford and W. R. Bedford, 1934 €. (TOM–BOT) Mar.

First: *Archaeolynthus porosus* (Bedford and W. R. Bedford, 1934), western Sayan, former USSR.
Last: *Kyarocyathus duplus* Kruse, 1982, Cymbric Vale Formation, New South Wales, Australia.
Intervening: ATB.

F. PALAEOCONULARIDAE Tchudinova, 1959
€. (ATB–BOT) Mar.

First: ?*Butakovicyathus butakovi* Zhuravleva, 1980, eastern Sayan, former USSR.
Last: *Palaeoconularia prima* Tchudinova, 1959, western Sayan, former USSR.

F. TUMULIOLYNTHIDAE Rozanov, 1966
€. (TOM–BOT) Mar.

First: *Tumuliolynthus* sp., Gorny Altay, former USSR.
Last: *Tumuliolynthus tubexternus* (Vologdin, 1932), Gorny Altay, former USSR.
Intervening: ATB.

F. SAJANOLYNTHIDAE Rozanov, 1973
€. (BOT) Mar.

First and Last: *Sajanolynthus desideratus* Vologdin and Kashina, 1972, eastern Sayan, former USSR.

F. GLOBOSOCYATHIDAE Okuneva, 1969
€. (ATB–BOT) Mar.

First: *Propriolynthus* sp., Far East of former USSR.
Last: *Melkanicyathus limitatus* Beljaeva, 1969, Far East of former USSR, Dzhagdy Ridge.

F. FAVILYNTHIDAE Debrenne, 1974 €. (BOT) Mar.

First and Last: *Favilynthus mellifer* (F. Bedford and W. R. Bedford, 1936), Ajax Limestone, South Australia.

Order CAPSULOCYATHIDA Zhuravleva, 1964

F. CAPSULOCYATHIDAE Zhuravleva, 1964
€. (ATB–BOT) Mar.

First: *Capsulocyathus subcallosus* Zhuravleva, 1964, eastern Sayan, former USSR.
Last: *Mirandocyathus artus* Beljaeva, 1974, Far East of former USSR, Dzhagdy Ridge.

F. CRYPTOPOROCYATHIDAE Zhuravleva, 1963
€. (TOM–ATB) Mar.

First: *Cryptoporocyathus junicanensis* Zhuravleva, 1960, north of Krasnoyask Region, Siberia, former USSR.
Last: *Korshunovicyathus melnikovi* Korshunov and Zhuravleva, 1967, northern Yakutia, Khara-Ulakh, former USSR.

F. FRANSUASAECYATHIDAE Debrenne, 1964
€. (ATB–BOT) Mar.

First: *Fransuasaecyathus subtumulatus* Zhuravleva, 1960, southern Yakutia, Lena River, former USSR.
Last: *Acanthopyrgus yukonensis* Handfield, 1967, western Canada, Yukon Province, MacKenzie Mountains.

F. TYLOCYATHIDAE Zhuravlev, 1988 €. (BOT) Mar.

First and Last: *Tylocyathus bullata* (Zhuravlev, 1961), eastern Sayan, former USSR.

F. URALOCYATHELLIDAE Zhuravleva, 1964
€. (BOT) Mar.

First and Last: *Rhabdolynthus conicus* Zhuravleva, 1960, southern Yakutia, Lena River, former USSR.

F. ALATAUCYATHIDAE Zhuravleva, 1955
€. (ATB–BOT) Mar.

First and Last: *Alataucyathus jaroschevitschi* Zhuravleva, 1955, Kuznetsky Alatau, former USSR.

F. CALYPTOCOSCINIDAE Debrenne, 1964
€. (BOT) Mar.

First and Last: *Calyptocoscinus cornucopiae* (Bornemann, 1887), Nebida Formation, Sardinia.

F. CLATHRICOSCINIDAE Rozanov, 1964
€. (ATB–BOT) Mar.

First: *Clathricoscinus* sp., Kuznetsky Alatau, former USSR.
Last: *Clathricoscinus infirmus* Zhuravleva, 1955, Kuznetsky Alatau, former USSR.

F. COSCINOCYATHELLIDAE Zhuravleva, 1956
€. (BOT) Mar.

First and Last: *Coscinocyathellus parvus* Vologdin, 1940, western Sayan, former USSR.

F. COSCINOCYATHIDAE Taylor, 1910
€. (ATB–BOT) Mar.

First: ?*Coscinocyathus* sp., Altay Sayan, former USSR.
Last: *Coscinocyathus dianthus* Bornemann, 1884, Nebiba Formation, Sardinia.

F. LANICYATHIDAE Debrenne *et al.*, 1989
€. (BOT) Mar.

First and Last: *Lanicyathus albus* Belkjaeva, 1975, Far East of former USSR, Dzhagdy Ridge.

F. MAWSONICOSCINIDAE Debrenne and Kruse, 1986 €. (BOT) Mar.

First and Last: *Mawsonicoscinus sigmoides* Debrenne and Kruse, 1986, Shackleton Limestone, Antarctica, Nimrod Glacier.

F. TOMOCYATHIDAE Debrenne *et al.*, 1989 €. (BOT) Mar.

First and Last: *Coscinocyathella nikitini* Vologdin, 1959, Kuznetsky Alatau, former USSR.

Order AJACICYATHIDA R. Bedford and J. Bedford, 1939

F. ALPHACYATHIDAE R. Bedford and J. Bedford, 1939 €. (BOT) Mar.

First and Last: *Alphacyathus simplex* (Taylor, 1910), Ajax Limestone, South Australia.

F. AJACICYATHIDAE R. Bedford and J. Bedford, 1939 €. (TOM–BOT) Mar.

First: *Nochoroicyathus mirabilis* Zhuravleva, 1951, southern Yakutia, Lena River, former USSR.
Last: *Ajacicyathus ajax* (Taylor, 1910), Ajax Limestone, South Australia.
Intervening: ATB.

F. ANAPTYCTOCYATHIDAE Debrenne, 1970 €. (BOT) Mar.

First and Last: *Anaptyctocyathus cribripora* (R. Bedford and J. Bedford, 1934), Ajax Limestone, South Australia.

F. ANNULOCYATHOIDAE Krasnopeeva, 1953 €. (BOT) Mar.

First and Last: *Annulocyathus pulcher* Vologdin, 1937, western Sayan, former USSR.

F. ASTEROCYATHIDAE Vologdin, 1956 €. (TOM–BOT) Mar.

First: *Erismacoscinus* sp., Siberian Platform, former USSR.
Last: *Rozanovicoscinus fonini* Debrenne, 1970, Ajax Limestone, Southern Limestone.
Intervening: ATB.

F. BIPALLICYATHIDAE Debrenne *et al.*, 1989 €. (ATB–BOT) Mar.

First and Last: *Bipallicyathus manifestus* Zhuravlev, 1982, Salaany Gol Formation, Western Mongolia, Khasagt–Khayrkhan Ridge.

F. BOTOMOCYATHIDAE Zhuravleva, 1955 €. (BOT) Mar.

First: *Botomocyathus zelenovi* Zhuravleva, 1955, southern Yakutia, Botoma River, former USSR.
Last: *Clathrithalamus mawsoni* Debrenne and Kruse, 1986, Shackleton Limestone, Antarctica, Nimrod Glacier.

F. BRONCHOCYATHIDAE R. Bedford and J. Bedford, 1936 €. (ATB–BOT) Mar.

First: *Thalamocyathus muchattensis* (Taylor, 1910), southern Yakutia, Lena River, former USSR.
Last: *Stillicidocyathus aulax* (Taylor, 1910), Ajax Limestone, South Australia.

F. CARINACYATHIDAE Krasnopeeva, 1953 €. (ATB–BOT) Mar.

First: *Carinacyathus* sp., Siberian Platform, former USSR.
Last: *Porocyathellus bouddi* Debrenne, 1977, Jbel Irhoud, Morocco.

F. CORDOBICYATHIDAE Perejon, 1975 €. (ATB) Mar.

First and Last: *Cordobicyathus deserti* Perejon, 1975, Cordoba, Spain.

F. COSCINOPTYCTIDAE Debrenne *et al.*, 1989 €. (BOT) Mar.

First and Last: *Coscinoptycta convoluta* Taylor, 1910, Ajax Limestone, South Australia.

F. CRASSICOSCINIDAE Debrenne *et al.*, 1988 €. (ATB–BOT) Mar.

First: *Crassicoscinus repandus* (Rozanov, 1964), Ajax Limestone, South Australia.
Last: *Coscinocyathellus vulgaris* Vologdin, 1940, Gorny Altay, former USSR.

F. DENSOCYATHIDAE Vologdin, 1937 €. (ATB–BOT) Mar.

First: *Heckericyathus heckeri* (Zhuravleva, 1960), southern Yakutia, Lena River, former USSR.
Last: *Diplocyathellus retezona* (Taylor, 1910), Ajax Limestone, South Australia.

F. DOKIDOCYATHIDAE R. Bedford and W. R. Bedford, 1936 €. (TOM–BOT) Mar.

First: *Dokidocyathus* sp., Siberian Platform, former USSR.
Last: *Dokidocyathus simplicissimus* Taylor, 1910, Ajax Limestone, South Australia.
Intervening: ATB.

F. DOKIDOCYATHELLIDAE Debrenne, 1964 €. (ATB–BOT) Mar.

First: *Dokidocyathella incognita* Zhuravleva, 1960, southern Yakutia, Lena River, former USSR.
Last: *Incurvocyathus voronovae* Rozanov, 1966, Tuva, former USSR.

F. ERBOCYATHIDAE Vologdinand Zhuravleva, 1956 €. (BOT–TOY) Mar.

First: *Ladaecyathus limbatus* (Zhuravleva, 1955), Kuznetsky Alatau, former USSR.
Last: *Erbocyathus heterovallum* (Vologdin, 1928), Kuznetsky Alatau, former USSR.

F. ETHMOCOSCINIDAE Zhuravleva, 1957 €. (BOT) Mar.

First and Last: *Ethmocoscinus papillipora* (R. Bedford and J. Bedford, 1934) Ajax Limestone, South Australia.

F. ETHMOPHYLLIDAE Okulitch, 1937 € (ATB–TOY) Mar.

First: *?Squamosocyathus taumatus* Zhuravleva, 1960, southern Yakutia, Lena River, former USSR.
Last: *Angaricyathus cyrenovi* Zhuravleva, 1965, northern Transbaikalia, former USSR.
Intervening: BOT.

1. Kisasacyathidae	11. Papillocyathidae	21. Rudanulidae	31. Tegerocyathidae
2. Kolbicyathidae	12. Peregrinicyathidae	22. Sajanocyathidae	32. Tercyathidae
3. Kordecyathidae	13. Piamaecyathellidae	23. Salairocyathidae	33. Tumulifungiidae
4. Kymbecyathidae	14. Polycoscinidae	24. Sanarkocyathidae	34. Tumulocoscinidae
5. Lenocyathidae	15. Porocoscinidae	25. Schumnyicyathidae	35. Tumulocyathidae
6. Leptosocyathidae	16. Pretiosocyathidae	26. Sigmocoscinidae	36. Vologdinocyathidae
7. Lunulacyathidae	17. Putapacyathidae	27. Sigmocyathidae	37. Wrighticyathidae
8. Mootwingeecyathidae	18. Rewardocyathidae	28. Soanicyathidae	38. Xestecyathidae
9. Marssocyathidae	19. Robertocyathidae	29. Sylviacoscinidae	39. Zhuravlevaecyathidae
10. Olgaecyathidae	20. Rozanovicyathidae	30. Tatijanaecyathidae	40. Zonacoscinidae

Fig. 5.7

F. FALLOCYATHIDAE Rozanov, 1969
€. (ATB–BOT) Mar.

First: *Fallocyathus dubius* Rozanov, 1969, southern Yakutia, Lena River, former USSR.
Last: *Yukonocyathus francesci* Handfield, 1971, western Canada, Yukon Province.

F. GAGARINICYATHIDAE Debrenne *et al.*, 1989
€. (ATB–BOT) Mar.

First and Last: *Gagarinicyathus ethmophylloides* Zhuravleva, 1969, north of Krasnoyarsk Region, Sukharikha River, former USSR.

F. GEOCYATHIDAE Debrenne, 1964
€. (ATB–BOT) Mar.

First and Last: *Geocyathus botomaensis* (Zhuravleva, 1955), southern Yakutia, Botoma River, former USSR.

F. GLORIOSOCYATHIDAE Rozanov, 1969
€. (ATB–BOT) Mar.

First: *Dupliporocyathus tumulosus* Jazmir, 1975, Transbaikalia, Vitim Ranges, former USSR.
Last: *Gloriosocyathus permultus* Rozanov, 1969, northern Yakutia, Olenek River Basin, former USSR.

F. HUPECYATHIDAE Debrenne *et al.*, 1990
€. (ATB) Mar.

First and Last: *Hupecyathus sphinctoides* Debrenne, 1964, Amouslekian, Anti Atlas, Morocco.

F. HUPECYATHELLIDAE Rozanov, 1973 €. (BOT) Mar.

First and Last: *Hupecyathellus chouberti* Rozanov, 1968, north of Krasnoyarsk Region, Sukharikha River, former USSR.

F. JAKUTOCARINIDAE Debrenne *et al.*, 1989
€. (ATB–BOT) Mar.

First: *Jakutocarinus jakutensis* Zhuravleva, 1960, southern Yakutia, Mukhatta River, former USSR.
Last: *Rossocyathella ninaekosti* Zhuravleva, 1960, southern Yakutia, Botoma River, former USSR.

F. JAPHANICYATHIDAE Rozanov, 1973
€. (ATB–BOT) Mar.

First and Last: *Japhanicyathus genurosus* Korshunov, 1969, southern Yakutia, Lena River, former USSR.

F. JEBILETICOSCINIDAE Debrenne *et al.*, 1989
€. (BOT) Mar.

First and Last: *Jebileticoscinus huvelini* Debrenne, 1977, Jbel Irhoud, Morocco.

F. KALTATOCYATHIDAE Rozanov, 1964
€. (ATB–BOT) Mar.

First and Last: *Kaltatocyathus kaschinae* Rozanov, 1964, eastern Sayan, former USSR.

F. KASYRICYATHIDAE Zhuravleva, 1961
€. (ATB–BOT) Mar.

First: ?*Agyrekocyathus malovi* Konjuschkov, 1967, north-eastern Kazakhstan, former USSR. Both familial assignment and age are questionable – may be Toyonian.
Last: *Kasyricyathus schirokovae* Zhuravleva, 1961, eastern Sayan, former USSR.

F. KIDRJASOCYATHIDAE Rozanov, 1964
€. (ATB–BOT) Mar.

First and Last: *Kidrjasocyathus uralensis* Rozanov, 1960, south Urals, former USSR.

F. KIJACYATHIDAE Zhuravleva, 1964 €. (ATB–BOT) Mar.

First: *Kijacyathus chomentovskii* Zhuravleva, 1969, Kuznetsky Alatau, former USSR.
Last: *Aporosocyathus mucroporus* Kruse, 1978, Cymbric Vale Formation, New South Wales, Australia.

F. KISASACYATHIDAE Konjuschkov, 1972 €. (BOT) Mar. (see Fig. 5.7)

First and Last: *Kisasacyathus microtumulatus* Konjuschkov, 1972, western Sayan, former USSR.

F. KOLBICYATHIDAE Debrenne *et al.*, 1988 €. (ATB–BOT) Mar.

First and Last: *Kolbicyathus kolbiensis* Zhuravlev, 1988, Kuznetsky Alatau, former USSR.

F. KORDECYATHIDAE Missarzhevsky, 1961 €. (BOT) Mar.

First and Last: *Kordecyathus shiveligensis* Missarzhevsky, 1961, Tuva, former USSR.

F. KYMBECYATHIDAE Debrenne *et al.*, 1989 €. (BOT) Mar.

First and Last: *Kymbecyathus avius* Debrenne and Kruse, 1986, Shackleton Limestone, Byrd Glacier, Antarctica.

F. LENOCYATHIDAE Zhuravleva, 1956 €. (ATB) Mar.

First and Last: *Lenocyathus lenaicus* Zhuravleva, 1955, southern Yakutia, Lena River, former USSR.

F. LEPTOSOCYATHIDAE Vologdin, 1961 €. (ATB–BOT) Mar.

First: *Tennericyathus malycanicus* Rozanov, 1969, southern Yakutia, Lena River, former USSR.
Last: *Ichnusocyathus ichnusae* (Meneghinin, 1881), Nebida Formation, Sardinia.

F. LUNULACYATHIDAE Debrenne, 1973 €. (BOT) Mar.

First and Last: *Lunulacyathus minimiporus* (R. Bedford and J. Bedford, 1934), Ajax Limestone, South Australia.

F. MOOTWINGEECYATHIDAE Kruse, 1982 €. (BOT) Mar.

First and Last: *Mootwingeecyathus mootwingeensis* Kruse, 1982, Cymbric Vale Formation, New South Wales, Australia.

F. MRASSUCYATHIDAE Vologdin, 1960 €. (ATB) Mar.

First and Last: *Membranacyathus repinae* Rozanov, 1960, Gornaya Shoriya, former USSR.

F. OLGAECYATHIDAE Borodina, 1974 €. (BOT) Mar.

First and Last: *Olgaecyathus fistulosus* Borodina, 1974, western Sayan, former USSR.

F. PAPILLOCYATHIDAE Rozanov, 1973 €. (ATB) Mar.

First and Last: *Papillocyathus vacuus* Rozanov, 1964, eastern Sayan, former USSR.

F. PEREGRINICYATHIDAE Zhuravleva, 1967 €. (BOT) Mar.

First and Last: *Peregrinicyathus dorotheae* Zhuravleva, 1967, Tuva, former USSR.

F. PIAMAECYATHELLIDAE Rozanov, 1973 €. (BOT) Mar.

First and Last: *Piamaecyathellus simplex* Rozanov, 1964, Gornay Altay, former USSR.

F. POLYCOSCINIDAE Debrenne, 1964 €. (ATB–BOT) Mar.

First: *Mennericyathus kundatus* (Rozanov, 1966), Kuznetsky Alatau, former USSR.
Last: *Polycoscinus contortus* (R. Bedford and J. Bedford, 1937), Ajax Limestone, South Australia.

F. POROCOSCINIDAE Debrenne, 1964 €. (ATB–BOT) Mar.

First: ?*Geniculicyathus varius* Debrenne, 1960, Anti Atlas, Morocco. The type is poorly preserved, but *G. amplus* from the same locality can be assigned to this family with more confidence.
Last: *Porocoscinus flexibilis* Debrenne, 1964, Nebida Formation, Sardinia.

F. PRETIOSOCYATHIDAE Rozanov, 1969 €. (ATB–BOT) Mar.

First and Last: *Pretiosocyathus subtilis* Rozanov, 1966, Kuznetsky Alatau, former USSR.

F. PUTAPACYATHIDAE R. Bedford and J. Bedford, 1939 €. (BOT) Mar.

First and Last: *Putapacyathus regularis* (R. Bedford and J. Bedford, 1936), Ajax Limestone, South Australia.

F. REWARDOCYATHIDAE Rozanov, 1973 €. (ATB–BOT) Mar.

First and Last: *Torosocyathus provisus* Kashina, 1972, eastern Sayan, former USSR.

F. ROBERTOCYATHIDAE Rozanov, 1969 €. (ATB–BOT) Mar.

First: *Robertocyathus* sp., Rozanov, 1969, Altay Sayan, former USSR.
Last: *Urcyathella tercyathoides* Zhuravleva, 1961, eastern Sayan, former USSR.

F. ROZANOVICYATHIDAE Korshunov, 1969 €. (ATB–BOT) Mar.

First: *Churanocyathus aculeatus* Sundukov, 1984, southern Yakutia, Lena River, former USSR.
Last: *Dentatocoscinus sektensis* (Korshunov and Zhuravleva, 1967), northern Yakutia, Khara-Ulakh, former USSR.

F. RUDANULIDAE Debrenne *et al.*, 1989 €. (BOT) Mar.

First: *Rudanulus petersi* (R. Bedford and J. Bedford), 1934, Ajax Limestone, South Australia.
Last: *Pilodicoscinus yuani* Debrenne and Jiang, 1989, Tsanglanpu Formation, Yunnan Province, China.

F. SAJANOCYATHIDAE Vologdin, 1956 €. (ATB–TOY) Mar.

First: *Chakassicyathus galinae* (Zhuravleva, 1967), Tuva, former USSR.
Last: *Siderocyathus duncanae* Debrenne *et al.*, 1990, Battle Mountains, Nevada (Debrenne *et al.*, 1990).
Intervening: BOT.

F. SALAIROCYATHIDAE Zhuravleva, 1956 Є. (ATB–BOT)

First: *Kotuyicoscinus minaevae* Sundukov, 1983, Anabar Region, former USSR.
Last: *Salairocyathus zenkovae* Vologdin, 1940, Salair, former USSR.

F. SANARKOCYATHIDAE Hill, 1972 Є. (ATB–BOT) Mar.

First: *Ringifungia vavilovi* Korshunov, 1969, southern Yakutia, Lena River, former USSR.
Last: *Sanarkocyathus mamaevi* Zhuravleva, 1963, south Urals, former USSR.

F. SCHUMNYICYATHIDAE Debrenne *et al.*, 1989 Є. (BOT) Mar.

First and Last: *Schumnyicyathus validus* Zhuravleva, 1968, north of the Krasnoyarsk Region, Sukharikha River, former USSR.

F. SIGMOCOSCINIDAE R. Bedford and J. Bedford, 1939 Є. (BOT) Mar.

First and Last: *Sigmocoscinus sigma* R. Bedford and J. Bedford, 1936, Ajax Limestone, South Australia.

F. SIGMOCYATHIDAE Krasnopeeva, 1953 Є. (BOT) Mar.

First and Last: *Sigmocyathus didymoteichus* (Taylor, 1910), Ajax Limestone, South Australia.

F. SOANICYATHIDAE Rozanov, 1964 Є. (ATB–BOT) Mar.

First: *Batschykicyathus angulosus* Zhuravlev, 1983, southern Yakutia, Lena River, former USSR.
Last: *Subtilocyathus subtilis* (Vologdin, 1932), Gorny Altay, former USSR.

F. SYLVIACOSCINIDAE Debrenne, Rozanov and Zhuravlev, 1989 Є. (BOT) Mar.

First and Last: *Sylviacoscinus sylvia* (R. Bedford and J. Bedford, 1937), Ajax Limestone, South Australia.

F. TATIJANAECYATHIDAE Korshunov, 1976 Є. (BOT) Mar.

First and Last: *Muchattocyathus sibiricus* Rozanov, 1976, southern Yakutia, Lena River, former USSR.

F. TEGEROCYATHIDAE Krasnopeeva, 1972 Є. (BOT–TOY) Mar.

First: *Krasnopeevaecyathus tyrgaensis* Rozanov, 1964, Gorny Altay, former USSR.
Last: *Tegerocyathus edelsteini* (Vologdin, 1931), Kuznetsky Alatau, former USSR.

F. TERCYATHIDAE Vologdin, 1937 Є. (BOT) Mar.

First and Last: *Tercyathellus capistarium* Borodina, 1974, western Sayan, former USSR.

F. TUMULIFUNGIIDAE Rozanov, 1973 Є. (ATB–TOY) Mar.

First: *Arturocyathus borisovi* Rozanov, 1973, Kuznetsky Alatau, former USSR.
Last: *Sclerocyathus* sp., Tuva, former USSR.
Intervening: BOT.

F. TUMULOCOSCINIDAE Zhuravleva, 1960 Є. (ATB–BOT) Mar.

First: *Tumulocoscinus atdabanensis* Zhuravleva, 1960, southern Yakutia, Lena River, former USSR.
Last: ?*Asterotumulus receptori* Kashina, 1964, eastern Sayan, former USSR.

F. TUMULOCYATHIDAE Krasnopeeva, 1953 Є. (TOM–BOT) Mar.

First: *Plicocyathus krasnyi* Vologdin, 1960, Far East of former USSR, Dzhagdy Ridge.
Last: *Tumulocyathus pustulatus* Vologdin, 1937, Salaany-Gol Formation, Western Mongolia, Khasagt–Kayrkhan Ridge.
Intervening: ATB.

F. VOLOGDINOCYATHIDAE Yaroshevitch, 1957 Є. (BOT–TOY) Mar.

First: *Syringocyathus aspectabilis* Vologdin, 1940, western Sayan, former USSR.
Last: *Vologdinocyathus erbiensis* Yaroshevitch, 1957, Kuznetsky Alatau, former USSR.

F. WRIGHTICYATHIDAE Kruse, 1978 Є. (BOT) Mar.

First and Last: *Wrighticyathus nexus* Kruse, 1978, Cymbric Vale Formation, New South Wales, Australia.

F. XESTECYATHIDAE Debrenne *et al.*, 1989 Є. (BOT) Mar.

First and Last: *Xestecyathus zigzag* Kruse, 1982, Cymbric Vale Formation, New South Wales, Australia.

F. ZHURAVLEVAECYATHIDAE Rozanov, 1973 Є. (BOT) Mar.

First and Last: *Zhuravlevaecyathus pulchellus* Rozanov, 1964, western Sayan, former USSR.

F. ZONACOSCINIDAE Debrenne, 1971 Є. (BOT) Mar.

First: *Orienticyathus mamontovi* Beljaeva, 1969, Far East of former USSR, Dzhagdy Ridge.
Last: *Zonacoscinus tumulosus* Debrenne, 1971, Nebida Formation, Sardinia.

Class IRREGULARES Vologdin, 1937 (see Fig. 5.8)

Comment: A major revision of Irregulares systematics has been completed recently (Debrenne and Zhuravlev, 1992), and this is the scheme of classification followed herein. This work is in press at the time of writing, and so its appearance before *The Fossil Record* cannot be guaranteed. Owing to this, it is deemed that **this section on Archaeocyathida, Class Irregulares is invalid for taxonomic purposes**.

F. LOCULICYATHIDAE Zhuravleva, 1955 Є. (ATB–?DOL) Mar.

First: *Okulichicyathus discoformis* (Zhuravleva, 1955), Lena River, Yakutia, former USSR.

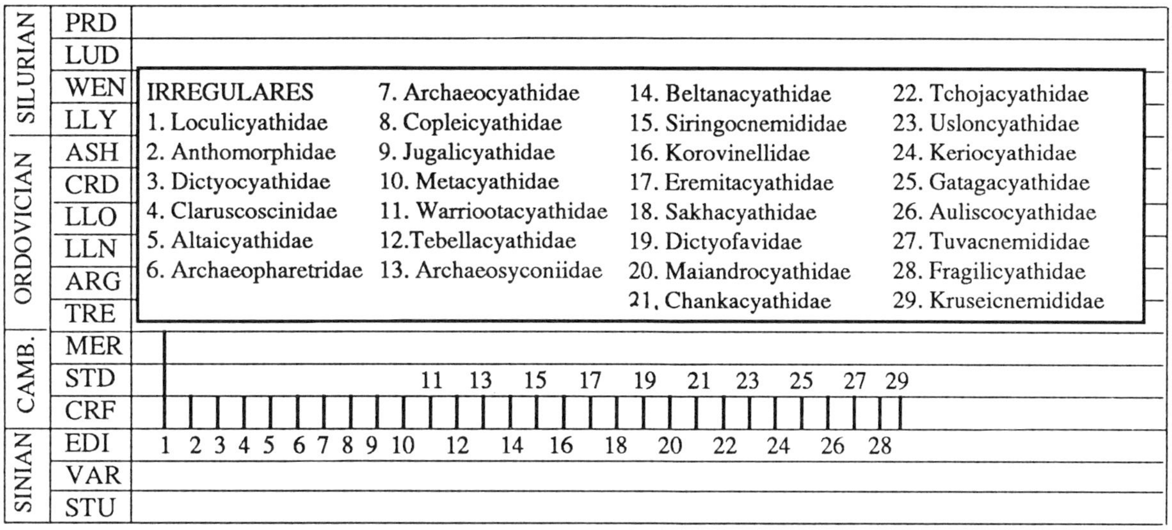

SILURIAN	PRD				
	LUD				
	WEN	IRREGULARES	7. Archaeocyathidae	14. Beltanacyathidae	22. Tchojacyathidae
	LLY	1. Loculicyathidae	8. Copleicyathidae	15. Siringocnemididae	23. Usloncyathidae
ORDOVICIAN	ASH	2. Anthomorphidae	9. Jugalicyathidae	16. Korovinellidae	24. Keriocyathidae
	CRD	3. Dictyocyathidae	10. Metacyathidae	17. Eremitacyathidae	25. Gatagacyathidae
	LLO	4. Claruscoscinidae	11. Warriootacyathidae	18. Sakhacyathidae	26. Auliscocyathidae
	LLN	5. Altaicyathidae	12.Tebellacyathidae	19. Dictyofavidae	27. Tuvacnemididae
	ARG	6. Archaeopharetridae	13. Archaeosyconiidae	20. Maiandrocyathidae	28. Fragilicyathidae
	TRE			21. Chankacyathidae	29. Kruseicnemididae
CAMB.	MER				
	STD	11 13 15 17 19 21 23 25 27 29			
	CRF				
SINIAN	EDI	1 2 3 4 5 6 7 8 9 10 12 14 16 18 20 22 24 26 28			
	VAR				
	STU				

Fig. 5.8

Last: ?*Antarcticocyathus webersi* Debrenne and Rozanov, *in* Debrenne *et al.*, 1984, Ellsworth Mountains, Antarctica.

F. ANTHOMORPHIDAE Okulitch, 1935 Є. (BOT) Mar.

First and Last: *Anthomorpha margarita* Bornemann, 1884, Sardinia.

F. DICTYOCYATHIDAE Taylor, 1910 Є. (TOM–TOY) Mar.

First: *Dictyocyathus translucidus* Zhuravleva, 1960, Lena River, Yakutia, former USSR.
Last: *Molybdocyathus juvenalis* Debrenne and Gangloff, 1990, Battle Mountain, Nevada, USA.
Intervening: ATB–BOT.

F. CLARUSCOSCINIDAE Debrenne and Zhuravlev, 1992 Є. (BOT–TOY) Mar.

First: *Stevocyathus elictus* Debrenne *et al.*, 1989, Caborca, Mexico.
Last: *Claruscocoscinus billingsi* (Vologdin, 1940), Altay Sayan, former USSR.

F. ALTAICYATHIDAE Debrenne and Zhuravlev, 1992 Є. (BOT) Mar.

First and Last: *Altaicyathus notabilis* Vologdin, 1932, Altay Mountains, former USSR.

F. ARCHAEOPHARETRIDAE Bedford and Bedford, 1936 Є. (ATB–BOT) Mar.

First: *Sphinctocyathus (Dictyosycon) gravis* Zhuravleva, 1960, Lena River, Yakutia, former USSR.
Last: *Protopharetra junensis* Zhuravleva, *in* Voronova *et al.*, 1987, Mackenzie Mountains, Canada.

F. ARCHAEOCYATHIDAE Hinde, 1889 Є. (BOT–TOY) Mar.

First: *Archaeocyathus solidus* (Vologdin, 1940), western Sayan, former USSR.
Last: *Archaeocyathus kusmini* Vologdin, 1932, Lena River, Yakutia, former USSR.

F. COPLEICYATHIDAE Bedford and Bedford, 1937 Є. (TOM–BOT) Mar.

First: *Spinosocyathus maslennikovae* Zhuravleva, 1960, Siberian Platform, former USSR.
Last: *Gabrielsocyathus gabrielsensis* (Okulitch, 1955), British Columbia, Canada.
Intervening: ATB.

F. JUGALICYATHIDAE Gravestock, 1984 Є. (BOT) Mar.

First: *Jugalicyathus tardus* Gravestock, 1984, Ajax Limestone, South Australia.
Last: *Alaskocoscinus tatondukenisis* Debrenne *et al.*, *in* Debrenne and Zhuravlev, 1990a,b, Alaska, USA.

F. METACYATHIDAE Bedford and Bedford, 1934 Є. (BOT–TOY) Mar.

First: *Metaldetes cylindricus* Taylor, 1910, South Australia.
Last: *Metaldetes profundus* (Billings, 1861), Labrador, Canada.

F. WARRIOOTACYATHIDAE Debrenne and Zhuravlev, 1992 Є. (ATB–BOT) Mar.

First: *Warriootacyathus wilkawillinensis* Gravestock, 1984, South Australia.
Last: *Warriootacyathus lucidus* Gravestock, 1984, South Australia.

F. TEBELLACYATHIDAE Fonin, 1963 Є. (BOT) Mar.

First and Last: *Taeniaecyathellus semenovi* Zhuravleva, 1960, western Sayan, former USSR.

F. ARCHAEOSYCONIIDAE Zhuravleva, 1949 Є. (BOT–TOY) Mar.

First: *Archaeosycon copulatus* (Debrenne and Gangloff, *in* Voronova *et al.*, 1987).
Last: *Archaeosycon billingsi* (Walcott, 1886), Labrador, Canada.

F. BELTANACYATHIDAE Debrenne, 1970 Є. (ATB–BOT) Mar.

First: *Beltanacyathus digitus* Gravestock, 1984, South Australia.
Last: *Beltanacyathus wirrialpensis* (Taylor, 1910), South Australia.

F. SYRINGOCNEMIDIDAE Taylor, 1910 €. (BOT) Mar.

First and Last: *Syriungocnema favus* Taylor, 1910, Ajax Limestone, South Australia.

F. KOROVINELLIDAE Khalfina, 1960 €. (BOT) Mar.

First: *Bicoscinus sdzuyi* Debrenne, 1977, Morocco.
Last: *Korovinella sajanica* (Yaworsky, 1932), western Sayan, former USSR.

F. EREMITACYATHIDAE Debrenne, *in* Zamarreno and Debrenne, 1975 €. (ATB) Mar.

First and Last: *Eremiticyathus fissus* Debrenne, 1975, Cordoba, Spain.

F. SAKHACYATHIDAE Debrenne and Zhuravlev, 1990 €. (TOM) Mar.

First and Last: *Sakhacyathus subartus* (Zhuravleva, 1960), Lena River, Siberian Platform, former USSR.

F. DICTYOFAVIDAE Debrenne and Zhuravlev, in press a) €. (BOT)

First: *Dictyofavus obtusus* Gravestock, 1984, South Australia.
Last: *Zunyicyathus grandus* (Yuan and Zhang, 1981), Chengkou Province, China.

F. MAIANDROCYATHIDAE Debrenne, 1970 €. (BOT) Mar.

First and Last: *Maiandrocyathus insigne* (R. Bedford and W. R. Bedford, 1936), Ajax Limestone, South Australia.

F. CHANKACYATHIDAE Yakovlev, 1959 €. (BOT) Mar.

First and Last: *Chankacyathus strachovi* Yakovlev, 1959, Primore, Far East of former USSR.

F. TCHOJACYATHIDAE Debrenne and Zhuravlev, in press b) €. (ATB) Mar.

First and Last: *Tchojacyathus validus* Rozanov, 1960, Altay, former USSR.

F. USLONCYATHIDAE Fonin, 1966, *in* Vologdin and Fonin, 1966 €. (ATB–BOT) Mar.

First and Last: *Usloncyathus miculus* Fonin, 1966, Transbaikalia, former USSR.

F. KERIOCYATHIDAE Debrenne and Gangloff, 1990 €. (ATB–BOT) Mar.

First and Last: *Keriocyathus arachnaius* Debrenne and Gangloff, *in* Debrenne *et al.*, 1990b, Battle Mountains, Nevada, USA.

F. GATAGACYATHIDAE Debrenne and Zhuravlev, in press a) €. (BOT) Mar.

First and Last: *Gatagacyathus mansyi* Debrenne and Zhuravlev, in press a), British Columbia, Canada.

F. AULISCOCYATHIDAE Debrenne and Zhuravlev, in press a) €. (BOT) Mar.

First and Last: *Auliscocyathus multifidus* (R. Bedford and W. R. Bedford, 1936), Ajax Limestone, South Australia.

F. TUVACNEMIDIDAE Debrenne and Zhuravlev, 1990 €. (BOT) Mar.

First and Last: *Tuvacnema tannuolensis* (Rodionova, 1967), eastern Tannu-Ola Ridge, Tuva, former USSR.

F. FRAGILICYATHIDAE Beljaeva *in* Beljaeva *et al.*, 1975 €. (BOT) Mar.

First and Last: *Fragilicyathus zhuravlevae* Beljaeva, 1969, Far East of former USSR.

F. KRUSEICNEMIDIDAE Debrenne and Zhuravlev, 1990 €. (BOT) Mar.

First and Last: *Kruseicnema gracilis* (Gordon, 1920), Antarctica.

REFERENCES

Chen, J.-Y., Hou, X.-G. and Lu, H.-Z. (1989) Lower Cambrian leptomitids (Demospongea), Chengjiang, Yunnan. *Acta Paleontologica Sinica*, **28**, 17–31.

Debrenne, F. (1974) Les Archéocyathes irreguliers d'Ajax Mine (Cambrien inf. Australie). *Bullétin de la Muséum national d'Histoire naturelle, Paris; Sciences de la Terre*, **33**, 185–258.

Debrenne, F. and Zhuravlev, A. (1990) New irregular archaeocyath taxa. *Geobios*, **23**, 299–305.

Debrenne, F. and Zhuravlev, A. (in press a) Les calicules, structure intervallaire de type chaetetide chez les archéocyathes irreguliers. *Geobios*.

Debrenne, F. and Zhuravlev, A. (1992) *Irregular Archaeocyaths*. Cahiers de Paléontologie, CNRS Editions Paris 212 pp.

Debrenne, F., Rozanov, A. Yu. and Webers, G.F. (1984) Upper Cambrian Archaeocyatha from Antarctica. *Geological Magazine*, **121**, 291–9.

Debrenne, F., Gandin, A. and Gangloff, R. A. (1990a) Analyse sédimentologique et paléontologique des calcaires organogènes du Cambrien Inférieur de Battle Mountain (Nevada, U.S.A.) *Annales de Paléontologie*, **76**, 73–119.

Debrenne, F., Rozanov, A. and Zhuravlev, A. (1990b) Regular Archaeocyaths. *Cahiers de Paléontologie, Éditions du Centre National de la Recherche Scientifique, Paris*, 1–218.

Finks, R. M. (1960) Late Paleozoic sponge faunas of the Texas Region. The siliceous sponges. *Bulletin of the American Museum of Natural History*, **120**, 1–160.

Finks, R. M. (1983) Fossil Hexactinellida, in *Sponges and Spongiomorphs*. Notes for a Short Course. University of Tennessee, Department of Geological Sciences. Studies in Geology, 7 (ed. T. W. Broadhead), Knoxville, Tennessee, pp. 101–13.

Gravestock, D. I. (1984) *Archaeocyatha from the Lower Part of the Lower Cambrian Carbonate Sequence in South Australia*. Association of Australian Palaeontologists, Sydney, Memoir, 2, 139 pp.

Grubic, A. (1961) Nouveau coup d'oeil retrospectif sur les problèmes de la stratigraphie de Sphaeractinides. *Institut des Recherches Géologiques et Géophysiques. Belgrad, Bullétin*, **19**, 159–79.

Hall, J. and Clarke, J. M. (1899) A memoir on the Paleozoic reticulate sponges constituting the family Dictyospongidae. *Memoir, New York State Museum*, **2**, 1–197.

Hill, D. (1981) Tabulata, in *Treatise on Invertebrate Paleontology. Part F. Coelenterata, Supplement 1: Rugosa and Tabulata*. (eds R. C. Moore and C. Teichert), Geological Society of America and University of Kansas Press, Boulder and Lawrence, Kansas, F430–F669.

Hinde, G. J. (1884) *Catalogue of the Fossil Sponges of the British Museum*. British Museum, London, 248 pp.

Hinde, G. J. (1887–1893) A monograph of the British fossil sponges. *Palaeontographical Society. London, [Monograph]*, **2**, 93–188.

Hinde, G. J. and Holmes, W. M. (1892) On the sponge remains in the Lower Tertiary strata near Oamaru, Otago, New Zealand. *Linnean Society Journal Zoology*, **24**, 177–262.

Kazmierczak, J. (1974) Lower Cretaceous sclerosponges from the

Slovakian Tatra Mountains. *Palaeontology*, **17**, 341–7.

Kobluk, D. R. (1981) Lower Cambrian cavity-dwelling endolithic boring sponges. *Canadian Journal of Earth Sciences*, **18**, 972–80.

Kolb, R. (1910) Die Kieselspongien des schwäbischen weissen Jura. *Palaeontographica*, **57**, 141–256.

Lagneau-Hérenger, L. (1962) Contribution à l'étude des Spongiaires silicieux du Crétacé inférieur. *Mémoires de la Société Géologique de France*, **95**, 1–252.

Laubenfels, M. W. de (1955) Part E, Archaeocyatha and Porifera, in *Treatise on Invertebrate Paleontology* (ed. R. C. Moore), Geological Society of America and University of Kansas Press, Boulder, Colorado, and Lawrence, Kansas, E21–E122.

Lecompte, M. (1956) Part E. Stromatoporoidea, in *Treatise on Invertebrate Paleontology, Part F. Coelenterata.* (ed. R. C. Moore), Geological Society of America and University of Kansas Press, Boulder, Colorado, and Lawrence, Kansas, F108–F144.

Moret, L. (1926) Contribution a l'étude des Spongiaires siliceux du Crétacé supérieur français. *Mémoires de la Société Géologique de France*, n.s., **5**, 1–308.

Moret, L. (1952) Embranchement des Spongiaires (Porifera, Spongiata), in *Traité de Paléontologie*, 1. (ed. J. Piveteau), Masson et Cie, Paris 333–74.

Mori, K. (1968) Stromatoporoids from the Silurian of Gotland. *Stockholm Contributions in Geology*, **19**, 1–100.

Nestor, H. (1966) Stromatoporoidei wenlocki i ludlowa Estonii. *Akademii Nauk Estonskoi SSR*, 1–87.

Reid, R. E. H. (1958) A monograph of the Upper Cretaceous Hexactinellida of Great Britain and Northern Ireland. Part I. *Palaeontographical Society, London [Monograph]*, i–xlvi.

Reid, R. E. H. (1964) The Upper Cretaceous Hexactinellida of Great Britain and Northern Ireland. *Palaeontographical Society, London [Monograph]*, xlix–cliv.

Reid, R. E. H. (1968) Microscleres in demosponge classification. *University of Kansas Paleontological Contributions, Paper* **35**, 1–10.

Reitner, J. (1987) *Euzkadiella erenoensis* n. gen. n. sp., ein Stromatopore mit spikularem aus dem Oberapt von Ereno (Prov. Guipuzcoa, Nordspanien) und die systematische Stellung der Stromatoporen. *Paläontologische Zeitschrift*, **61**, 203–22.

Reitner, J. and Engeser, T. (1983) Contributions to the systematics and the paleoecology of the family Acanthochaetetidae FISCHER, 1970 (Order Tabulospongida, Class Sclerospongiae). *Geobios*, **16**, 773–9.

Rigby, J. K. (1983) Heteractinida, in *Sponges and Spongiomorphs* (ed. T. W. Broadhead), Notes for a Short Course. University of Tennessee, Department of Geological Sciences, Studies in Geology, 7, pp. 70–89.

Rigby, J. K. (1986) Late Devonian sponges of Western Australia, *Geological Survey of Western Australia, Report*, **18**, 1–59.

Rigby, J. K. and Dixon, O. A. (1979) Sponge fauna of the Upper Silurian Read Bay Formation, Somerset Island, District of Franklin, Arctic Canada. *Journal of Paleontology*, **53**, 587–627.

Rigby, J. K. and Webby, B. D. (1988) Late Ordovician sponges from the Malongulli Formation of central New South Wales, Australia. *Palaeontographica Americana*, **56**, 1–147.

Schrammen, A. (1910, 1912) Die Kieselspongien der oberen Kreide von Nordwestdeutschland. I. Teil. Tetraxonia, Monaxonia und Silicea *incertae sedis* (1910); II. Teil. Triaxonia (Hexactinellida) (1912). *Palaeontographica, Supplement*, **5**, 1–385.

Schrammen, A. (1924) Die Kieselspongien der oberen Kreide von Nordwestdeutschland, III, und letzter Teil. *Monographien zur Geologie und Paläontologie, Berlin, Pt. 1*, **2**, 1–159.

Schrammen, A. (1936) Die Kieselspongien des oberen Jura von Suddeutschland. *Palaeontographica*, **84**, 1–194.

Senowbari-Daryan, B. (1990) Die Systematische Stellung der thalamiden Schwämme und ihr Bedeutung in der Erdegeschichte. *Münchner Geowissenschaftliche Abhandlungen, A*, **21**, 1–326.

Stearn, W. (1969) The stromatoporoid genera *Tienodictyon, Intexodictyon, Hammatostroma* and *Plexodictyon*. *Journal of Paleontology*, **43**, 753–66.

Stearn, W. (1980) Classification of the Paleozoic stromatoporoids. *Journal of Paleontology*, **54**, 881–902.

Walcott, C. D. (1920) Middle Cambrian Spongiae. *Cambrian Geology and Paleontology. Smithsonian Miscellaneous Collections*, **67**, 261–364.

Wood, R. A. (1987) Biology and revised systematics of some late Mesozoic stromatoporoids. *Special Papers in Palaeontology*, **37**, 1–89.

Wood, R. A. and Reitner, J. (1987) The upper Cretaceous "chaetitid" demosponge *Stromatoaxinella irregularis* n.g. (MICHELIN) and its systematic implications. *Neues Jahrbuch fur Geologie und Paläontologie, Monatshefte*, **1987**, 213–24.

Wood, R. A., Reitner, J. and West, R. (1989) Systematics and phylogenetic implications of the haplosclerid stromatoporoid *Newellia mira* nov. gen. *Lethaia*, **22**, 85–93.

6

COELENTERATA

J. R. Nudds and J. J. Sepkoski Jr

The stratigraphical information for the first and last occurrences of each family has been extracted mainly from Harland *et al.* (1967), Sepkoski (1982) and from the various volumes of the *Treatise on Invertebrate Paleontology* which cover the coelenterates (Bayer *et al.*, 1956; Glaessner, 1979; Hill, 1981). Evidence for the first occurrences of some of the very early members of the phylum is from Scrutton (1979, 1984). Bibliographical references to most of this information are given in these various works and are not repeated here. Additional references are included herein only when they offer newer data.

Acknowledgements – We are extremely grateful to Dr B. R. Rosen (Natural History Museum, London), Dr C. T. Scrutton (University of Durham) and Dr P. Wyse Jackson (Trinity College, Dublin) for their willing help in providing both literature references and specialist advice, often at very short notice. Sepkoski acknowledges partial support for research from NASA Grant NAGW-1693.

Class 'PETALONAMAE' Pflug, 1970a

Leaf-like structures, often with a median line (the 'petal organisms' of Pflug, 1970–1973); some appear to represent colonial organisms and many are considered to be pennatulaceans, although Pflug (1970–1973) considered that they represent a separate phylum. Classification mostly follows Glaessner (1979).

Order ERNIETTOMORPHA Pflug, 1972

Dickinsoniidae and Bomakellidae possibly belong in this order, but the former should more probably be assigned to the primitive phylum Proarticulata (Fedonkin, 1990), while the latter probably belongs to the Arthropoda (Fedonkin, 1990).

F. ERNIETTIDAE Pflug, 1972 V. (POU) Mar.

Glaessner (1979) declares most of these genera to be unrecognizable.
First and Last: Various genera, e.g. *Ernietta*, *Erniofossa*, *Ernionorma*, *Erniobeta*, *Erniograndis*, from the Nama Group, Namibia, SW Africa.

F. PTERIDINIIDAE Richter, 1955 V. (POU) Mar.

First and Last: *Pteridinium simplex* Gurich, 1933, Nama Group, Namibia, SW Africa.

F. UNCERTAIN V. (POU) Mar.

First and Last: *Namalia villiersiensis* Germs, 1968 and *Nasepia altae* Germs, 1972, Nama Group, Namibia, SW Africa; *Baikalina sessilis* Sokolov, 1972, upper Vendian, southern Siberia.

Order RANGEOMORPHA Pflug, 1972

The Sprigginidae possibly belongs in this order, but should more probably be assigned to the Polychaeta. It should be noted that Jenkins (1985) infers the Charniidae to be pennatulaceans and places the remaining Rangeomorpha within the Octocorallia.

F. CHARNIIDAE Glaessner, 1979 V. (POU) Mar.

First and Last: *Charnia masoni* Ford, 1958 and *Charniodiscus concentricus* Ford, 1958, Woodhouse and Bradgate Formation, Maplewell Group, Charnwood Forest, Leicestershire, England, UK and *Charniodiscus*, Pound Quartzite Formation, upper Adelaide Series, Ediacara, South Australia.

F. RANGEIDAE Glaessner, 1979 V. (POU) Mar.

First and Last: *Rangea schneiderhohni* Gurich, 1930, Nama Group, Namibia, SW Africa.

Class CYCLOZOA Fedonkin, 1983

Classification follows Fedonkin (1985), and includes numerous Ediacaran medusoids, such as: *Beltanelliformis*, *Cyclomedusa*, *Ediacaria*, *Kaisalia*, *Kimberia*, *Lorenzinites*, *Medusinites*, *Nimbia*, *Paliella*, *Pseudorhizostomites*, *Rugoconites*, *Tirasiana* and others. Glaessner (1979) describes these as 'medusae of uncertain affinities', while Harland *et al.* (1967) class them as 'medusae *incertae sedis*'.

? F. MAWSONITIDAE Sun, 1986 V. (POU) Mar.

Sun (1986) places this family in the Scyphozoa, questionably in the order Coronatida. However, the specimens appear to be preserved in hyporelief, suggesting that they are 'resting traces' rather than medusoids. We place the family questionably with the Cyclozoa.
First and Last: *Mawsonites spriggi* Glaessner and Wade, 1966, Ediacara Member, Rawnsley Quartzite, Pound Subgroup, Ediacara, South Australia.

F. UNCERTAIN V. (POU) Mar.

The Fossil Record 2. Edited by M. J. Benton. Published in 1993 by Chapman & Hall, London. ISBN 0 412 39380 8

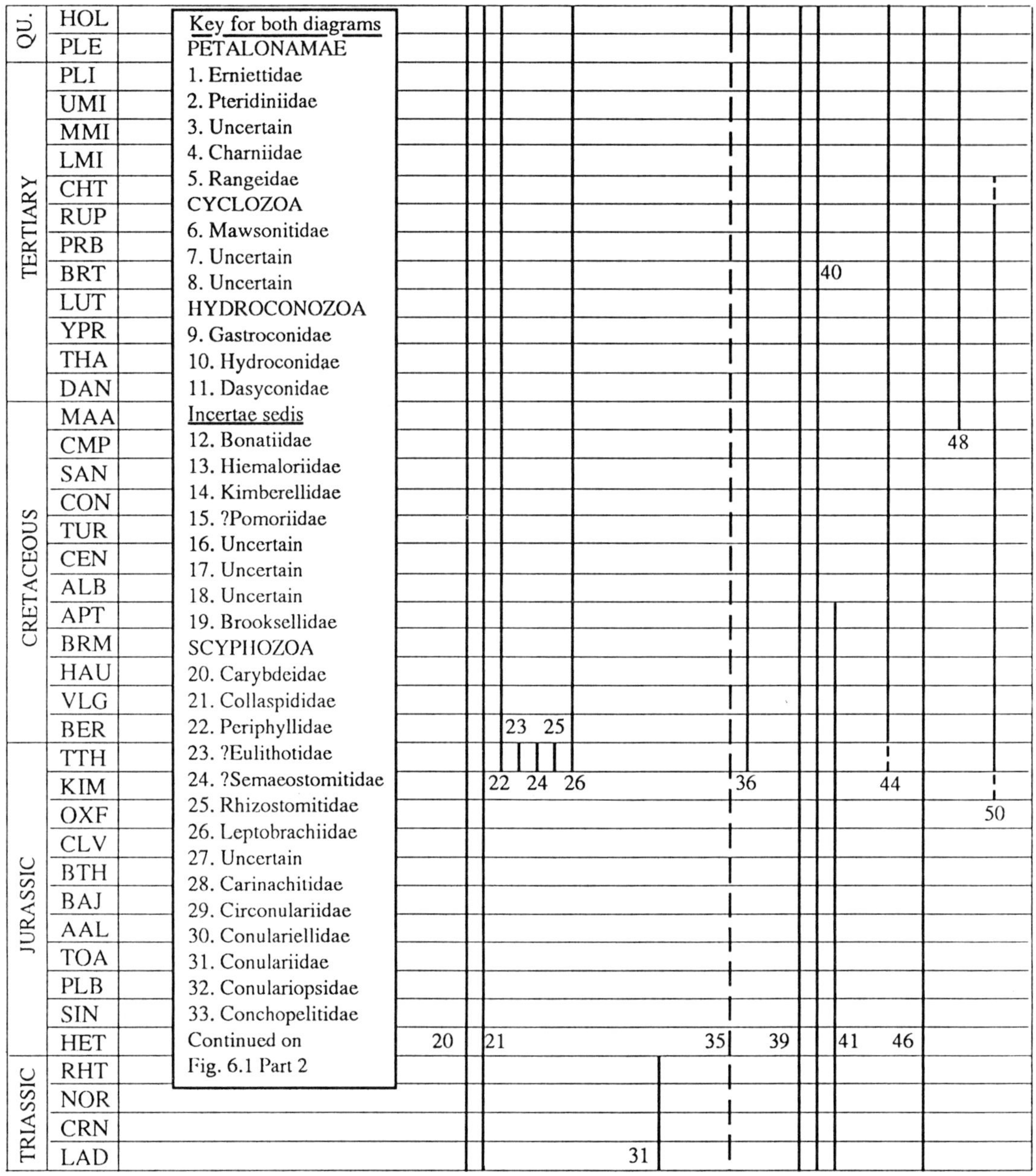

QU. HOL PLE
TERTIARY PLI UMI MMI LMI CHT RUP PRB BRT LUT YPR THA DAN
CRETACEOUS MAA CMP SAN CON TUR CEN ALB APT BRM HAU VLG BER
JURASSIC TTH KIM OXF CLV BTH BAJ AAL TOA PLB SIN HET
TRIASSIC RHT NOR CRN LAD

Key for both diagrams
PETALONAMAE
1. Erniettidae
2. Pteridiniidae
3. Uncertain
4. Charniidae
5. Rangeidae
CYCLOZOA
6. Mawsonitidae
7. Uncertain
8. Uncertain
HYDROCONOZOA
9. Gastroconidae
10. Hydroconidae
11. Dasyconidae
Incertae sedis
12. Bonatiidae
13. Hiemaloriidae
14. Kimberellidae
15. ?Pomoriidae
16. Uncertain
17. Uncertain
18. Uncertain
19. Brooksellidae
SCYPHOZOA
20. Carybdeidae
21. Collaspididae
22. Periphyllidae
23. ?Eulithotidae
24. ?Semaeostomitidae
25. Rhizostomitidae
26. Leptobrachiidae
27. Uncertain
28. Carinachitidae
29. Circonulariidae
30. Conulariellidae
31. Conulariidae
32. Conulariopsidae
33. Conchopelitidae
Continued on Fig. 6.1 Part 2

40 48 23 25 22 24 26 36 44 50 20 21 35 39 41 46 31

Fig. 6.1

First and Last: Various genera, including *Cyclomedusa*, Pound Quartzite Formation, upper Adelaide Series, Ediacara, South Australia.

F. UNCERTAIN V. (POU) Mar.

First and Last: Various genera, including *Medusinites*, Pound Quartzite Formation, upper Adelaide Series, Ediacara, South Australia.

Class HYDROCONOZOA Korde, 1963

Classification follows Korde (1963). The Tabulaconidae, included herein by Sepkoski and Kasting (in press), were referred to the Anthozoa by Debrenne *et al.* (1987).

F. GASTROCONIDAE Korde, 1963 €. (LEN) Mar.

First and Last: *Gastroconus venustus* Korde, 1963, Lower Cambrian, Bolshoy Shangan River, Tuva, former USSR.

F. HYDROCONIDAE Korde, 1963 €. (ATB–LEN) Mar.

First and Last: *Hydroconus mirabilis* Korde, 1963, Lower Cambrian, Bolshoy Shangan River, Tuva, former USSR.

F. DASYCONIDAE Korde, 1963 €. (LEN) Mar.

First and Last: *Dasyconus porosus* Korde, 1963, Lower Cambrian, Bolshoy Shangan River, Tuva, former USSR.

Class INCERTAE SEDIS

Informal classification follows Sepkoski (1982) and Sepkoski and Kasting (in press).

F. BONATIIDAE Fedonkin, 1985 V. (POU) Mar.

First and Last: *Bonata septata* Fedonkin, 1980, Bed 1, Valdai Series, mouth of Yelovyy Stream, Zimniy coast of White Sea, northern Russian Platform.

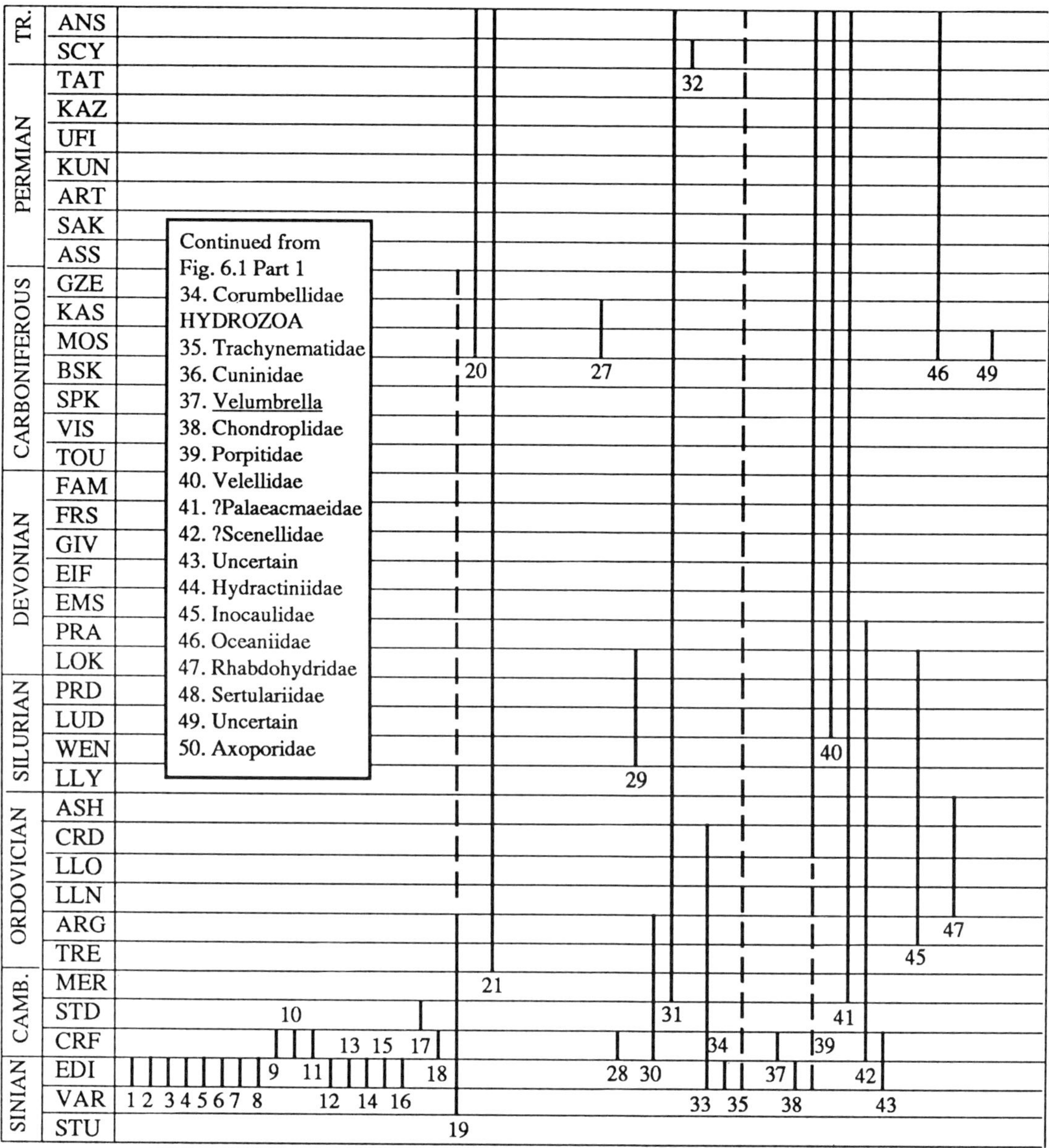

Fig. 6.1

F. HIEMALORIIDAE Fedonkin, 1985 V. (POU) Mar.

First and Last: *Hiemalora stellaris* Fedonkin, 1982 and *Evmiaksia aksionovi* Fedonkin, 1984, upper Vendian, former USSR.

F. KIMBERELLIDAE Wade, 1972 V. (POU) Mar.

First and Last: *Kimberella quadrata* Wade, 1972, Pound Quartzite Formation, upper Adelaide Series, Ediacara, South Australia.

F. ?POMORIIDAE Fedonkin, 1985 V. (POU) Mar.

First and Last: *Pomoria corolliformis* Fedonkin, 1980, Bed 11, Valdai Series, Medvezh'iy Stream, Zimniy coast of White Sea, north Russian Platform.

F. UNCERTAIN V. (POU) Mar.

First and Last: *Staurinidia crucicula* Fedonkin, 1985, upper Vendian, former USSR.

F. UNCERTAIN €. (MEN) Mar.

First and Last: *Cambromedusa furcula* Willoughby and Robison, 1979, *Bathyuriscus fimbriatus* Subzone, Wheeler Shale, upper Middle Cambrian, Wheeler Amphitheatre, House Range, Millard County, west-central Utah, USA.

F. UNCERTAN €. (LEN) Mar.

First and Last: *Rosellatana jamesi* Kobluk, 1984, *Bonnia–Olenellus* Zone, Rosella Formation, Atan Group, Lenian, east of Dease River, 70 km SW of fork with Liard River, near Yukon–British Columbia boundary, Cassiar Mountains, northern British Columbia, Canada.

Class INCERTAE SEDIS

Bayer *et al.* (1956) and Harland *et al.* (1967) referred the following to the Class Protomedusae, but we suggest suppression of this 'trash-can taxon'.

F. BROOKSELLIDAE Walcott, 1898 V. (VAR)–C. (u.)? Mar.

First: *Brooksella canyonensis* Bassler, 1941, Nankoweap Group, Algonkian, Grand Canyon, Arizona, USA. It is very doubtful that this is a medusoid; it may be a trace fossil or even a non-biogenic cluster of intraclasts.
Last: *Laotira* sp. Couyat and Fritel, 1912, Pennsylvanian, Sinai, Egypt? This is not substantiated by figures or descriptions. The next youngest species is *Duodecimedusina aegyptiaca* Avnimelech, 1966, Um Bogma Formation (?Namurian), Eastern Desert, Egypt.
Intervening: *Brooksella silurica* (von Huene, 1904), 'oberen rothen Orthocerenkalk', Ordovician (Arenig), Kargarde, Dalarna, Sweden. Walcott's *Brooksella* from the Middle Cambrian Conasauga Shale of Coosa Valley, Alabama, USA, are pseudofossils, being lobate concretions.

Class SCYPHOZOA Gotte, 1887

Classification follows Bayer *et al.* (1956) and Harland *et al.* (1967). Approximately 13 extant families without fossil representatives (including all in the Stauromedusida) are not listed.

Order CARYBDEIDA Claus, 1886

F. CARYBDEIDAE Gegenbaur, 1856 C. (MOS)–Rec. Mar.

First: *Anthracomedusa turnbulli* Johnson and Richardson, 1968, Upper Moscovian, Pit Eleven, Mazon Creek, Essex, Illinois, USA. **Extant**
Intervening: *Quadrimedusina quadrata* (Haeckel, 1869), Solnhofen Limestone (TTH), Bavaria, Germany.

Order CORONATIDA Vanhoffen, 1892

F. COLLASPIDIDAE Haeckel, 1880 O. (TRE)–Rec. Mar.

First: ?*'Camptostroma' germanicum* Hundt, 1939, Phycodes-Schichten, upper Tremadoc, eastern Thuringia, Germany. (Harland *et al.*, 1967, record *C. roddyi* from the Cambrian, but this is now referred to the Echinodermata.) **Extant**
Intervening: ?*Cannostomites multicirratus* Maas, 1906, Solnhofen Limestone (lower TTH), Bavaria, Germany; ?*Lorenzinia*, Cretaceous, Italy and Tertiary (Eocene) of Italy, Cyprus, Poland and Albania.

F. PERIPHYLLIDAE Claus, 1886 J. (TTH)–Rec. Mar.

First: *Epiphyllina distincta* (Maas, 1906), Solnhofen Limestone, Bavaria, Germany. **Extant**
Intervening: None.

Order SEMAEOSTOMATIDA L. Agassiz, 1862

F. ?EULITHOTIDAE Kieslinger, 1939 J. (TTH) Mar.

First and Last: *Eulithota fasciculata* Haeckel, 1869, Solnhofen Limestone, Bavaria, Germany.

F. ?SEMAEOSTOMITIDAE Harrington and Moore, 1956 J. (TTH) Mar.

First and Last: *Semaeostomites zitteli* Haeckel, 1869, Solnhofen Limestone, Bavaria, Germany.

Order LITHORHIZOSTOMATIDA von Ammon, 1886

F. RHIZOSTOMITIDAE Harrington and Moore, 1956 J. (TTH) Mar.

First and Last: *Rhizostomites admirandus* Haeckel, 1869, Solnhofen Limestone, Bavaria, Germany.

Order RHIZOSTOMATIDA Cuvier, 1799

F. LEPTOBRACHIIDAE Claus, 1883 J. (TTH)–Rec. Mar.

First: *Leptobrachites trigonobrachius* Haeckel, 1869, Solnhofen Limestone, Bavaria, Germany. **Extant**

F. UNCERTAIN C. (MOS/KAS) Mar.

First and Last: *Prothysanostoma eleanorae* Ossian, 1973, Wea Shale, Cherryville Formation, lower Stephanian, Limekiln Hollow, Pottawattamie County, western Iowa, USA.

Order ?CONULARIIDA Miller and Gurley, 1896

Considered by many to be unrelated to the coelenterates and often assigned to the Problematica.

F. CARINACHITIDAE Є. (TOM) Mar.

First and Last: *Quadrosiphogonuchites* and *Paranabarites*, Lower Cambrian, Meishucun, China.

F. CIRCONULARIIDAE Bischoff, 1978b S. (LLY)–D. (GED) Mar.

First: *Circonularia eosilurica* Bischoff, 1978b, *Garraconularia* 'n. sp. A.' Bischoff, 1978b and 'n.g. b, n. sp. b.' Bischoff, 1978b all from Panuara Group, lower Llandovery, southern bank of Cobblers Creek, 30 km SW of Orange, New South Wales, Australia.
Last: *Garraconularia multicostata* Bischoff, 1978a, basal Garra Formation, upper Lochkovian, New South Wales, Australia.

F. CONULARIELLIDAE Kiderlen, 1937 Є. (TOM)–O. (ARG) Mar.

First: *Arthrochites* and *Barbitositheca*, lower Meishucun, China.
Last: *Conulariella*, Lower Ordovician, Europe.

F. CONULARIIDAE Walcott, 1886 Є. (MER)–Tr. (RHT) Mar.

First: *Conularia*, Upper Cambrian, world-wide.
Last: *'Conularia' stromeri* Osswald, 1918, oberen Kossener Mergel, Tegernsee, Bavaria, Germany.

F. CONULARIOPSIDAE Sugiyama, 1942 Tr. (SCY) Mar.

First and Last: *Conulariopsis quadrata* Sugiyama, 1942, Triassic (Olenekian), Japan.

Order *INCERTAE SEDIS*

F. CONCHOPELTIDAE Moore and Harrington, 1956 V. (POU)–O. (CRD) Mar.

Considered by Oliver (1984) to be unrelated to the conularids: possibly scyphozoan or hydrozoan.
First: *Conomedusites lobatus* Glaessner and Wade, 1966, Pound Quartzite Formation, upper Adelaide Series, Ediacara, South Australia.
Last: *Conchopeltis alternata* Walcott, 1876, Trenton Limestone (upper third), Trentonian, Trenton Falls, New York, USA.

F. ?CORUMBELLIDAE Hahn *et al.*, 1982 V. (POU) Mar.

First and Last: *Corumbella werneri* Hahn *et al.*, 1982, Corumba Group, Corumba–Ladario, Mato Grosso, SW Brazil.

Class HYDROZOA Owen, 1843

Classification follows Bayer *et al.* (1956), Glaessner (1979) and Harland *et al.* (1967). Stromatoporoidea and Spongiomorphida are not listed as they are now considered to be Porifera (q.v.). An additional approximately 25 extant families without fossil representatives (including all in the Siphonophorida) are also not listed.

Order TRACHYLINIDA Haeckel, 1877

F. TRACHYNEMATIDAE Gegenbaur, 1856 V. (POU)?–Rec. Mar.

First: ?*Beltanella gilesi* Sprigg, 1947, Pound Quartzite Formation, upper Adelaide Series, Ediacara, South Australia. This taxon is considered to be doubtful and a possible synonym of *Ediacaria* (Cyclozoa) by Glaessner and Wade (1966). No other fossil records; numerous extant genera. **Extant**

F. CUNINIDAE Broch, 1929 J. (TTH)–Rec. Mar.

First: *Acalepha deperdita* Beyrich, 1849, Solnhofen Limestone, Bavaria, Germany. **Extant**

F. UNCERTAIN €. (LEN) Mar.

First and Last: *Velumbrella czarnockii* Stasinska, 1960, Lower Cambrian, Poland.

Order HYDROIDA Johnston, 1836

The first six families cited here belong to the suborder Chondrophorina, which has previously been included in the Siphonophorida (Bayer *et al.*, 1956; Harland *et al.*, 1967). It is now widely accepted that this suborder belongs to the Hydroida (Glaessner, 1979).

F. CHONDROPLIDAE Wade, 1971 V. (POU) Mar.

First and Last: *Ovatoscutum concentricum* Glaessner and Wade, 1966, Pound Quartzite Formation, upper Adelaide Series, Ediacara, South Australia. Fedonkin (1985, 1990) assigns this genus to the Cyclozoa.

F. PORPITIDAE Brandt, 1835 V. (POU)?–Rec. Mar.

First: *Eoporpita medusa* Wade, 1972, Pound Quartzite Formation, upper Adelaide Series, Ediacara, South Australia? Fedonkin (1990) assigns this genus to the Cyclozoa. The next oldest porpitid is *Discophyllum peltatum* Hall, 1847, from the Middle Ordovician (Caradoc), New York, USA. **Extant**

F. VELELLIDAE Brandt, 1835 S. (LUD)–Rec. Mar.

First: *Silurovelella casteria* Fisher, 1957, Vernon Shale, New York, USA (Yochelson *et al.*, 1983). **Extant**

F. ?PALAEACMAEIDAE Grabau and Shimer, 1909 €. (DOL)–K. (BRM) Mar.

Originally considered to be a monoplacophoran (Mollusca); placed in the Chondrophorina by Yochelson and Stanley (1981).

First: *Palaeacmaea typica* Hall and Whitfield, 1872, Potsdam Group, New York, USA.

Last: *Palaelophacmaea annulata* (Yokoyama, 1890), upper Ishido Formation, Sanchu area, Japan (Stanley and Kanie, 1985).

F. ?SCENELLIDAE Wenz, 1938 €. (TOM)–D. (PRA) Mar.

Many, but not necessarily all, species in this family were considered to be possible chondrophorines rather than monoplacophoran molluscs by Yochelson (1984).

First: *Scenella jijiapoensis* Chen, 1984, middle Meishucunian, Jijiapo, Yichang, Hubei, China and *Scenella micropora* He, 1984, Dengying Formation, middle Meishucunian, Szechwan, China.

Last: *Calloconus humilis* (Perner, 1903), Lower Devonian, Czechoslovakia.

F. UNCERTAIN V. (POU)–€. (TOM) Mar.

First: *Kullingia delicata* (Fedonkin, 1981), Redkino 'Series', Russian Platform, former USSR (Narbonne *et al.*, 1991).

Last: *Kullingia delicata* (Fedonkin, 1981), base of Member 2, Chapel Island Formation, Burin Peninsula, Newfoundland, Canada (Narbonne *et al.*, 1991).

F. HYDRACTINIIDAE Agassiz, 1862 J. (TTH)?–Rec. Mar.

First: *Hydractinia* sp. Records from Jurassic are tentative. Mainly from Lower Cretaceous and Lower Eocene onwards. **Extant**

F. INOCAULIDAE Ruedemann, 1947 O. (ARG)–D. (GED) Mar.

Considered to be graptolites by Ruedemann (1947), but assigned to the Hydroida by Mierzejewski (1986).

First: *Inocaulis* sp. Mierzejewski, 1986, Lower Ordovician, Podborowisko Borehole 'IG 1', Poland.

Last: *Inocaulis multiramous* Ruedemann, 1947, Manlius Limestone, Helderberg Group, Syracuse, New York, USA.

F. OCEANIIDAE Eschscholtz, 1829 C. (MOS/GZE)–Rec. Mar.

First: *Crucimedusina walcotti* (Barbour, 1914), Stephanian, Burlington Quarries, South Bend, Nebraska, USA. **Extant**

F. RHABDOHYDRIDAE Mierzejewski, 1986 O. (LLN)–O. (ASH) Mar.

The authors consider this to be a hydroid rather than a graptolite.

First: *Rhabdohydra tridens* Kozłowski, 1959, Ontikan Limestones, Oland, Sweden.

Last: *Rhadbohydra tridens* Kozłowski, 1959, limestone erratic, Zegrze, Poland.

F. SERTULARIIDAE Fleming, 1828 K. (MAA)–Rec. Mar.

First: *Hydrallmania graptolithiformis* Voigt, 1973, Horizon Mc, upper Maastrichtian, Verlassener Steinbruch, 'Van der Tombe', SW Seite des St Pietersberges, Maastricht, The Netherlands. **Extant**

F. UNCERTAIN C. (MOS) Mar.

First and Last: *Mazohydra megabertha* Schram and Nitecki, 1975, Francis Creek Shale, Westphalian C, upper Mos-

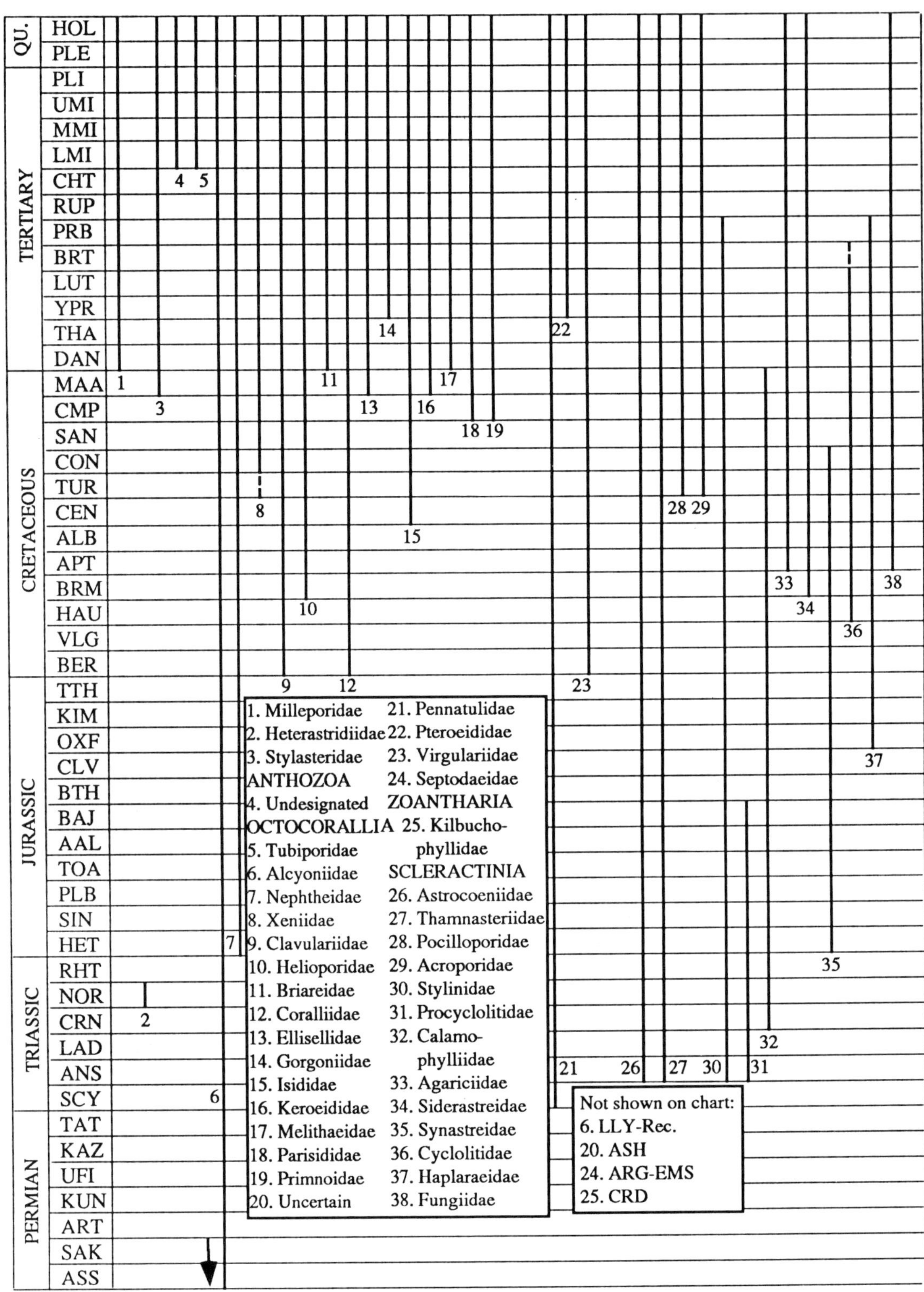

Fig. 6.2

covian, Pit Eleven, Mazon Creek, Will-Kankakee Counties, NE Illinois, USA.

Suborder CALYPTOBLASTINA Allman, 1871
€. (StD)–Rec. Mar.

First: *Archaeocryptolaria* Chapman, 1919, *Archaeolafoea* Chapman, 1919, *Protohalecium hallianum* Chapman and Thomas, 1936, *Sphenoecium* Chapman and Thomas, 1936, all from Dinesus–Hydroid Beds, Victoria, South Australia. Also *Archaeolafoea terranovaensis* Howell, 1963, Kelligrew Brook Formation, *Paradoxides davidis* Zone, Murphy's Cove, Little Lawn Harbor, Burin Peninsula, southern Newfoundland, Canada. **Extant**

Order MILLEPORINA Hickson, 1901

The Heterastridiidae were tentatively included in the

Hydroida by Bayer *et al.* (1956), and in the Sphaeractinida by Harland *et al.* (1967), but are herein assigned to the Milleporina.

F. AXOPORIDAE Boschma, 1951 J. (KIM/TTH)–T. (RUP/CHT) Mar.

First: *Subaxopora xizangensis* Deng, 1982, ?Lagongtang Group, Butou Village, near Dengqen, Tibet.
Last: *Axopora michelini* Duncan, 1866, Oligocene, England, UK.

F. MILLEPORIDAE Fleming, 1828 T. (DAN)–Rec. Mar. (see Fig. 6.2)

First: *Millepora parva* Nielsen, 1919, Chalk, Faxe Quarry, Denmark. **Extant**

F. HETERASTRIDIIDAE Frech, 1890 Tr. (NOR) Mar.

First and Last: *Heterastridium conglobatum* Reuss, 1865, Upper Triassic, Europe and Asia (Timor).

Order STYLASTERINA Hickson and England, 1905

F. STYLASTERIDAE Gray, 1847 K. (MAA)–Rec. Mar.

First: Unnamed stylasterines from White Chalk, Denmark (Floris, 1979). **Extant**

Class ANTHOZOA Ehrenberg, 1834

Subclass CERIANTIPATHARIA van Beneden, 1898

F. UNDESIGNATED T. (LMI)–Rec. Mar.

First: *Leiopathes glaberrima* Esper, 1792, Miocene, Italy. **Extant**

Subclass OCTOCORALLIA Haeckel, 1866

Classification mostly follows Bayer *et al.* (1956). Approximately 27 extant families without fossil representatives (including all in the Telestacea) are not listed.

Order STOLONIFERA Hickson, 1883

The Clavulariidae, included herein by Bayer *et al.* (1956), have been reassigned to the Coenothecalia by Bayer (1979).

F. TUBIPORIDAE Ehrenberg, 1828 T. (LMI)–Rec. Mar.

First: *Tubipora*, Miocene, North America (Sokolov, *in* Orlov, 1958–1964). **Extant**

Order ALCYONACEA Lamouroux, 1816

F. ALCYONIIDAE Lamouroux, 1812 S. (LLY)–Rec. Mar.

First: *Atractosella cataractaca* Bengtson, 1981, lower and upper Visby Beds and Hogklint Beds, upper Llandovery to lower Wenlock, Gotland, Sweden. **Extant**

F. NEPHTHEIDAE Gray, 1862 J. (HET)–Rec. Mar.

First: *Nephthea*, Lower Jurassic, Europe. **Extant**

F. XENIIDAE Ehrenberg 1828 K. (TUR)?–Rec. Mar.

First: *Nephtya*, upper Turonian, Bohemia, Czechoslovakia (Sokolov, *in* Orlov, 1958–1964). **Extant**

Order COENOTHECALIA Bourne, 1895

F. CLAVULARIIDAE Hickson, 1894 K. (BER)–Rec. Mar.

These were assigned to the Stolonifera by Bayer *et al.* (1956), but reassigned to the Coenothecalia by Bayer (1979).
First: Bayer (1981) lists the only definite fossil here as *Scyphopodium ingolfi* (Madsen), Upper Pleistocene, dredged sample off Oregon, USA. However, Bayer *et al.* (1956) list *Ephiphaxum auloporoides* Lonsdale, 1850, Cretaceous, England, UK. **Extant**

F. HELIOPORIDAE Moseley, 1876 K. (BRM)–Rec. Mar.

First: *Pseudopolytremacis hanagaensis* Kuz'micheva, 1975, Lower Cretaceous, Zeyva River, David-Bek village, Kafan District, Armenia, former USSR. **Extant**

Order GORGONACAE Lamouroux, 1816

F. BRIAREIDAE J. E. Gray, 1859 T. (DAN)–Rec. Mar.

First: *Kaluginella turkmenensis* Kuz'micheva, 1980, Chaaldzha Formation, western Kopet Dag, Turkmenia, former USSR. **Extant**

F. CORALLIIDAE Lamouroux, 1812 K. (BER)–Rec. Mar.

First: *Corallium*, Lower Cretaceous, California, USA. **Extant**

F. ELLISELLIDAE Gray, 1859 K. (MAA)–Rec. Mar.

First: *Nicella bursini* Kuz'micheva, 1980, lower Maastrichtian, Mount Besh-Kosh, Bakchisaray, SW Crimea, former USSR. **Extant**

F. GORGONIIDAE Lamouroux, 1812 T. (EOC)–Rec. Mar.

First: *Gorgonoid banmiti* Giammona and Stanton, 1980, Stone City Formation, Claiborne Group, Middle Eocene, south bank of Brazos River, west of College Station, Texas, USA. **Extant**

F. ISIDIDAE Lamouroux, 1812 K. (CEN)–Rec. Mar.

First: *Moltkia faveolata* (Reuss, 1865) and *Moltkia solida* (Stol.), Upper Cretaceous, Czechoslovakia (see Kuz'micheva, 1980). **Extant**

F. KEROEIDIDAE Kinoshita, 1910 K. (MAA)–Rec. Mar.

First: *Krimella klikusini* Kuz'micheva, 1980, lower Maastrichtian, Kuybyshevo, Kuybyshev District, SW Crimea, former USSR. **Extant**

F. MELITHAEIDAE Gray, 1870 T. (DAN)–Rec. Mar.

First: *Acabaria mangyshlakensis* Kuz'micheva, 1980 and *Melithaea* sp., Kuz'micheva, 1980, Lower Tertiary, Burlyu Well, Mangyshlak. **Extant**

F. PARISIDIDAE Aurivillius, 1931 K. (CMP)–Rec. Mar.

First: *Parisis steenstrupi* Nielsen, 1917, lower Campanian, Denmark (see Kuz'micheva, 1980). **Extant**

F. PRIMNOIDAE Gray, 1857 K. (CMP)–Rec. Mar.

First: *Primnoa costata* Nielsen, 1913, Upper Cretaceous, Poland (Malecki, 1982). **Extant**

F. UNCERTAIN O. (ASH) Mar.

First and Last: *Pragnellia arborescens* Leith, 1952, Upper Ordovician, Manitoba, Canada.

Order PENNATULACEA Verrill, 1865

F. PENNATULIDAE Ehrenberg, 1828 Tr.–Rec. Mar.

First: *Prographularia*, Triassic, Germany (Sokolov, *in* Orlov, 1958–1964). **Extant**

F. PTEROEIDIDAE Kolliker, 1880 T. (EOC.)–Rec. Mar.

First: *Pteroeides*, Tertiary, Sumatra. **Extant**

F. VIRGULARIIDAE Verrill, 1868 K. (BER)–Rec. Mar.

First: *Virgularia*, Cretaceous, world-wide. **Extant**

Order ?SEPTODAEARIA Bischoff, 1978a

?F. SEPTODAEIDAE, Bischoff 1978a O. (ARG)–D. (EMS) Mar.

Conti and Serpagli (1984) regard this taxon as a bryozoan.
First: *Septodaeum siluricum* Bischoff, 1978a, Lower Ordovician, Oland, Sweden.
Last: *Septodaeum siluricum* Bischoff, 1978a, lower Emsian, north of Capertee, New South Wales, Australia.

Subclass ZOANTHARIA de Blainville, 1830

A number of extant families without fossil representatives (including all in the Zoanthiniaria and Corallimorpharia) are not listed.

Order KILBUCHOPHYLLIDA Scrutton and Clarkson, 1991

All details from Scrutton and Clarkson, 1991.

F. KILBUCHOPHYLLIDAE Scrutton and Clarkson, 1991 O. (CRD) Mar.

First and Last: *Kilbuchophyllia discoidea* Scrutton and Clarkson, 1991, Kirkcolm Formation, middle Caradoc (Soudleyan–Actonian), Kilbucho, near Biggar, Southern Uplands, Scotland, UK.

Order SCLERACTINIA Bourne, 1900

The classification of the scleractinian corals has always been controversial and is presently the subject of major revision. In the absence of any strong alternatives, the classification herein follows Bayer *et al.* (1956), but no suprafamily grouping has been attempted.

F. ASTROCOENIIDAE Koby, 1890 Tr. (ANS)–Rec. Mar.

First: *Koilocoenia decipiens* (Laube, 1865), Middle Triassic, Europe. **Extant**

F. THAMNASTERIIDAE Vaughan and Wells, 1943 Tr. (ANS)–Rec. Mar.

First: *Thamnasteria*, Middle Triassic, Europe. **Extant**

F. POCILLOPORIDAE Gray, 1842 K. (TUR)–Rec. Mar.

First: *Madracis*, Upper Cretaceous, Europe, North America. **Extant**

F. ACROPORIDAE Verrill, 1902 K. (TUR)–Rec. Mar.

First: *Astreopora*, Upper Cretaceous, Europe. **Extant**

F. STYLINIDAE d'Orbigny, 1851 Tr. (ANS)–T. (PRB) Mar.

First: *Procyathophora furstenbergensis* (Eck, 1880), Middle Triassic, Germany.
Last: *Ewaldocoenia hawelkai* Oppenheim, 1921, Upper Eocene, Switzerland.

F. PROCYCLOLITIDAE Vaughan and Wells, 1943 Tr. (ANS)–J. (BAJ) Mar.

First: *Triadophyllum posthumum* Weissermel, 1925, Middle Triassic, Austria.
Last: *Phylloseris rugosa* Tomes, 1882, Middle Jurassic, England, UK.

F. CALAMOPHYLLIIDAE Vaughan and Wells, 1943 Tr. (CRN)–K. (MAA) Mar.

First: *Isastrea haueri* Cuif, 1975, Upper Triassic, no locality given (Roniewicz and Morycowa, 1989).
Last: *Calamophyllia sandbergeri* Felix, 1891, Upper Cretaceous, Mixteca Alta region, State of Oaxara, Mexico (Reyeros de Castillo, 1983).

F. AGARICIIDAE Gray, 1847 K. (APT)–Rec. Mar.

First: *Trochoseris*, Lower Cretaceous, Haldon and Atherfield, England, UK and *Brachyphyllia*, Lower Cretaceous, Spain. **Extant**

F. SIDERASTREIDAE Vaughan and Wells, 1943 K. (BRM)–Rec. Mar.

First: *Siderastrea senecta* Morycowa, 1971, Lower Cretaceous, Osojnica, former Yugoslavia (Turnsek and Buser, 1974). **Extant**

F. SYNASTREIDAE Alloiteau, 1952 J. (HET)–K. (CON) Mar.

First: *Proleptophyllia granulum* (de Fromentel and Ferry, 1866), Liassic, France.
Last: *Acrosmilia* and *Placoseris*, Upper Cretaceous, Europe, North America.

F. CYCLOLITIDAE d'Orbigny, 1851 K. (HAU)–T. (LUT/BRT) Mar.

First: *Funginella*, Lower Cretaceous, France.
Last: *Cyclolitopsis patera* d'Achiardi, 1867, Middle Eocene, Italy.

F. HAPLARAEIDAE Vaughan and Wells, 1943 J. (OXF)–T. (PRB) Mar.

First: *Diplaraea elegans* (Milaschevitsch, 1876), upper Oxfordian, reef near Col, southern Slovenia, former Yugoslavia (Turnsek, 1972).
Last: *Confusastraraea obsoleta* Gerth, 1921, Upper Eocene, Indonesia.

F. FUNGIIDAE Dana, 1846 K. (APT)–Rec. Mar.

First: *Cycloseris escosurae* Mallada, 1887, Lower Cretaceous, Spain. **Extant**

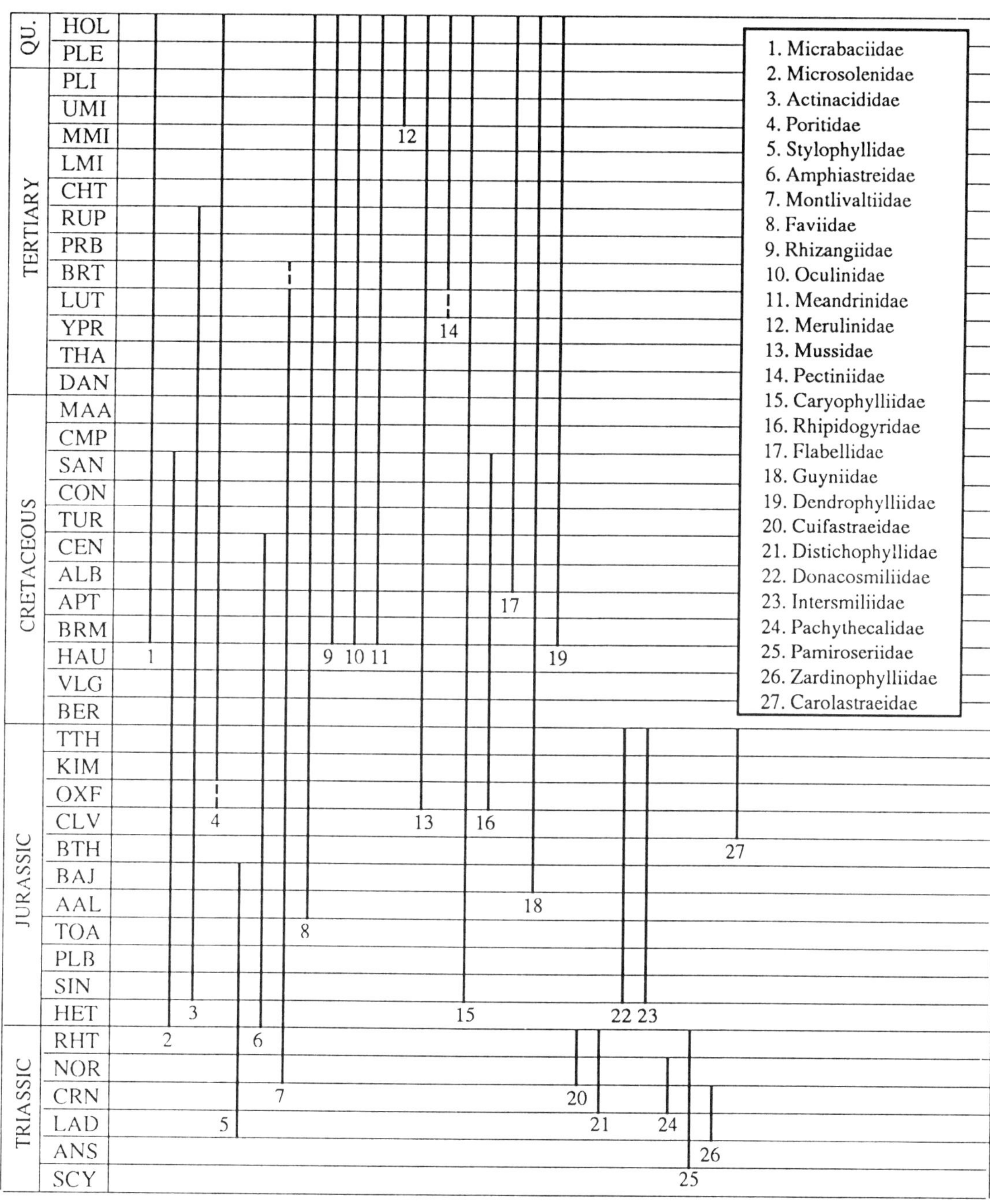

Fig. 6.3

F. MICRABACIIDAE Vaughan, 1905 K. (BRM)–Rec. Mar.

First: *Micrabacia beaumonti* Milne Edwards and Haime, 1851, Lower Cretaceous, France. **Extant**

F. MICROSOLENIDAE Koby, 1890 J. (HET)–K. (SAN) Mar.

First: *Chomatoseris*, Lias, Europe.
Last: *Gosaviaraea*, Upper Cretaceous, Austria.

F. ACTINACIDIDAE Vaughan and Wells, 1943 J. (SIN)–T. (RUP) Mar.

First: *Actinaraea*, Lower Jurassic, no locality given (Roniewicz and Morycowa, 1989).
Last: *Actinacis*, Lower Oligocene, Eurasia, North America, South America, West Indies, Africa.

F. PORITIDAE Gray, 1842 J. (OXF/KIM)–Rec. Mar.

First: *Goniopora*, Upper Jurassic, NW of former Yugoslavia (Turnsek, 1989). **Extant**

F. STYLOPHYLLIDAE Volz, 1896 Tr. (LAD)–J. (BAJ) Mar.

First: *Protoheterastrea leonardi* (Volz, 1896), Middle Triassic, Europe.
Last: *Lepidophyllia*, Middle Jurassic, UK, South America and *Heterasastrea*, Middle Jurassic, Europe.

F. AMPHIASTREIDAE Ogilvie, 1896 J. (HET)–K. (CEN) Mar.

First: *Discocoenia ruperti* (Duncan, 1867), *angulata* Zone, lower Liassic, Down Yatherly, Gloucestershire, England, UK (Negus, 1983).
Last: *Budaia travisensis* Wells, 1933, Upper Cretaceous, Texas, USA *Amphiastrea* and *Axosmilia*, Upper Cretaceous, Italy.

F. MONTLIVALTIIDAE Dietrich, 1926 Tr. (NOR)–T. (LUT/BRT) Mar.

First: *Palaeastraea*, Upper Triassic, Europe (Roniewicz and Morycowa, 1989).
Last: *Placosmilia*, Middle Eocene, Europe and Alabama, USA and *Elasmophyllia*, Middle Eocene, Europe, South America.

F. FAVIIDAE Gregory, 1900 J. (AAL)–Rec. Mar.

First: *Goniocora concinna* Tomes, 1882, Middle Jurassic, Jebel Bou Dahar, Morocco (Beauvais, 1986). **Extant**

F. RHIZANGIIDAE d'Orbigny, 1851 K. (BRM)–Rec. Mar.

First: *Arctangia nathorsti* (Lindstrom, 1900), Lower Cretaceous, King Charles Island, Arctic. **Extant**

F. OCULINIDAE Gray, 1847 K. (GAL)–Rec. Mar.

First: Harland *et al.* (1967) cite *Pseudogatheria hiraigensis* Eguchi, 1951 from the Aptian Miyako Group of Japan, but this genus is considered by Bayer *et al.* (1956) to belong to the Stylinidae. *Archohelia* occurs in the Middle Cretaceous of North America, Central America and the West Indies. **Extant**

F. MEANDRINIDAE Gray, 1847 K. (BRM)–Rec. Mar.

First: *Dendrogyra*, Lower Cretaceous, Mexico, Spain, Italy. **Extant**

F. MERULINIDAE Verrill, 1866 T. (UMI)–Rec. Mar.

First: *Merulina ampliata* (Ellis and Solander, 1786), Upper Miocene, Indonesia (Chevalier and Beauvais, 1987). **Extant**

F. MUSSIDAE Ortmann, 1890 J. (OXF)–Rec. Mar.

First: *Palaeomussa*, Rauracian, Europe. **Extant**

F. PECTINIIDAE Vaughan and Wells, 1943 T. (LUT/BRT)–Rec. Mar.

First: *Pectinia pseudomeandrites* d'Archiardi, 1867, Middle Eocene, Italy. **Extant**

F. CARYOPHYLLIIDAE Gray, 1847 J. (SIN)–Rec. Mar.

First: *Thecocyathus*, Lower Jurassic, Europe. **Extant**

F. RHIPIDOGYRIDAE Koby, 1904 J. (OXF)–K. (SAN) Mar.

First: *Acanthogyra micra* Eliasova, 1973, *Kologyra aldingeri* Geyer, 1955 and *Pruvostastrea labyrinthiformis* Alloiteau, 1957, all from lower Oxfordian, Algarve, southern Portugal (von Rosendahl, 1985).
Last: *Aplosmilia*, upper Santonian, Austria and France and *Fromentelligyra*, upper Santonian, France.

F. FLABELLIDAE Bourne, 1905 K. (ALB)–Rec. Mar.

First: *Adkinsella edwardsensis* Wells, 1933, Lower Cretaceous, Texas, USA. **Extant**

F. GUYNIIDAE Hickson, 1910 J. (BAJ)–Rec. Mar.

First: *Microsmilia* sp. Mariotti *et al.*, 1979, Middle Jurassic, Sasso di Pale, Foligno, Umbria, Italy. **Extant**

F. DENDROPHYLLIIDAE Gray, 1847 K. (BRM)–Rec. Mar.

First: *Palaeopsammia*, Lower Cretaceous, eastern Siberia (Turnsek and Mihajlovic, 1981). **Extant**

F. CUIFASTRAEIDAE Mel'nikova, 1983 Tr. (NOR–RHT) Mar.

First: *Cuifastraea granulata* Mel'nikova, 1983, Bor-Tepa Formation, upper Norian, Karakul'-Ashu Pass, SE Pamir, Tajikistan, *C. tenuiseptata* (Mel'nikova, 1967), Bor-Tepa Formation, upper Norian, Bor-Tepa Gorge, SE Pamir, Tajikstan and *Gillastraea delicata* Mel'nikova, 1983, Bor-Tepa Formation, upper Norian, between Khan-Yuly and Beik Gorges, SE Pamir, Tajikstan.
Last: *Cuifastraea incurva* Mel'nikova, 1983, Chichkautek Formation, Bezymyannyy Gorge, left bank of Karauldyn-Dana Valley, SE Pamir, Tajikistan.

F. DISTICHOPHYLLIDAE Cuif, 1976 Tr. (CRN–RHT) Mar.

First: *Distichophyllia*, Upper Triassic, no locality given (Roniewicz and Morycowa, 1989).
Last: *Distichophyllia*, Upper Triassic, no locality given (Roniewicz and Morycowa, 1989).

F. DONACOSMILIIDAE Krasnov, 1970 J. (SIN–TTH) Mar.

First: *Prodonacosmilia dronovi* Mel'nikova, *in* Mel'nikova and Roniewicz, 1976, Churumdinskaya Member, mouth of Djangi-davan saj, SE Pamir, Tajikistan.
Last: *Prodonacosmilia* sp. Mel'nikova and Roniewicz, 1976 and *Donacosmilia corallina* de Fromentel, 1861, both from Upper Jurassic, Stramberk, western Carpathians.

F. INTERSMILIIDAE Mel'nikova and Roniewicz, 1976 J. (SIN–TTH) Mar.

First: *Intersmilia djartyrabatica* Mel'nikova, *in* Mel'nikova and Roniewicz, 1976, Ghurumdinskaya Member, Djarty-Rabat Mountain, SE Pamir, Tajikistan.
Last: *Intersmilia*, Upper Jurassic, Eurasia (Mel'nikova and Roniewicz, 1976).

F. PACHYTHECALIDAE Cuif, 1975 Tr. (CRN–NOR) Mar.

First: *Pachysolenia mardjinaica* (Mel'nikova), Upper Triassic, Katta-Mardzhinay River, SE Pamir Range, Tajikstan (Il'ina, 1983).
Last: *Pachythecalis major* Cuif, 1975, *Pachydendron microthallos* Cuif, 1975 and *Pachysolenia cylindrica* Cuif, 1975, all from lower Norian, Alakir Kai, Turkey (Il'ina, 1983).

F. PAMIROSERIIDAE Melnikova, 1984 Tr. (ANS–RHT) Mar.

First: *Pamiroseris silesiaca* (Beyrich, 1852), Gorazdze Beds, lower Anisian, Upper Silesia, Poland (Malinowska, 1986).
Last: *Pamiroseris rectilamellosa* (Winkler, 1861), Sub-tatric

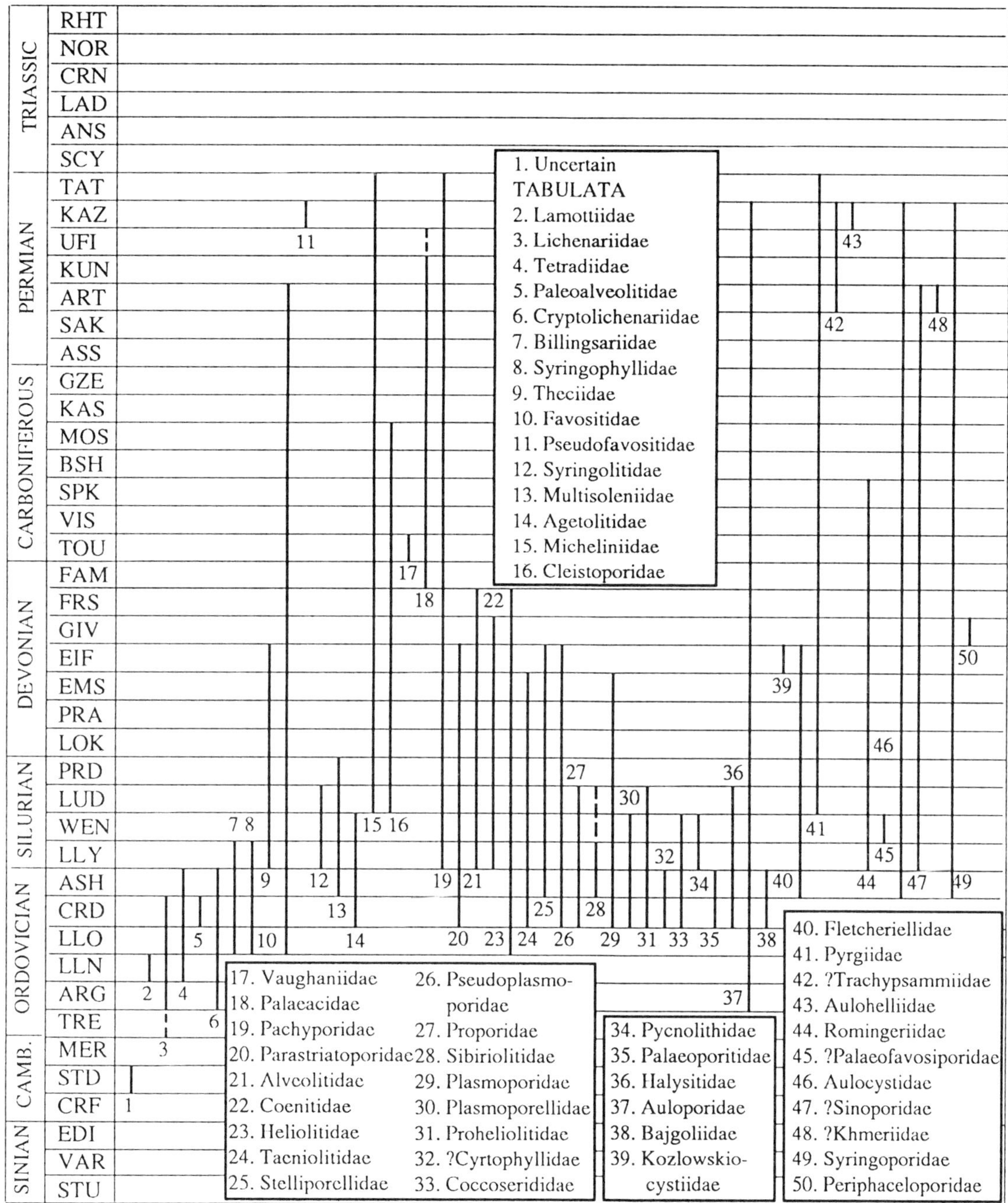

Fig. 6.4

Series, Upper Triassic, Tatra Mountains, Poland (Malinowska, 1986).

F. ZARDINOPHYLLIIDAE Montanaro Gallitelli, 1975 Tr. (LAD–CRN) Mar.

First: *Zardinophyllum zardinii* Montanaro Gallitelli, 1975, Cassiani Beds, Middle Triassic, near Cortina, northern Italy. **Last:** *Zardinophyllum zardinii* Montanaro Gallitelli, 1975, Cassiani Beds, Upper Triassic, near Cortina, northern Italy.

F. CAROLASTRAEIDAE Eliasova, 1976 J. (CLV)–J. (TTH) Mar.

First: *Carolastraea* sp. Mel'nikova and Roniewicz, 1976, Koltshakskaya Member, middle Callovian, Koltshak Mountain, Kuntej saj, SE Pamir, Tajikistan. **Last:** *Carolastraea fraji* Eliasova, 1976, Upper Jurassic, west Carpathians.

Order ACTINIARIA R. Hertwig, 1882

F. UNCERTAIN Є. (StD) Mar.

First and Last: *Mackenzia costalis* Walcott, 1911, Burgess Shale, Middle Cambrian, British Columbia, Canada.

Subclass TABULATA Milne Edwards and Haime, 1850

Classification follows Hill (1981), but excludes the Chaetetidae, Desmidoporidae, Tiverinidae and Acanthochaetetidae (Order Chaetetida), which are now assigned to the Porifera (q.v.). Other families referred to the Chaetetida by Hill (1981) are herein included with the Lichenariida and the Tetradiida, following Laub (1984).

Order LICHENARIIDA Sokolov, 1962

F. LAMOTTIIDAE Sokolov, 1950 O. (LLN) Mar.

First and Last: *Lamottia heroensis* Raymond, 1924, Day Point Limestone, Top of Lower Chazyan, upper Llanvirn, W. C. Hall's Pasture, 2 miles SW of South Hero, Vermont, USA.

F. LICHENARIIDAE Okulitch, 1936 O. (TRE?–CRD)

First: *Lichenaria cloudi* Bassler, 1950, upper Tremadoc, Texas, Missouri, USA. Laub (1984) doubts that these early forms are true *Lichenaria* and regards the earliest undisputed example to be *Lichenaria prima* Okulitch, 1936 from the Chazyan (Llanvirn) of Tennessee, USA.
Last: *Lichenaria typa* Winchell and Schuchert, 1895, Guttenberg Member, Decorah Formation, Blackriverian, upper Caradoc, St Paul, Minnesota, USA.

Order TETRADIIDA Okulitch, 1936

Several workers have expressed doubt that the tetradiids are true tabulate corals, and perhaps they should be considered as Problematica.

F. TETRADIIDAE Nicholson, 1879 O. (LLN–ASH) Mar.

First: *Phytopsis cellulosum* Hall, 1847, base of Middle Ordovician, New York, USA.
Last: *Tetradium*, upper Ashgill, North America.

F. PALEOALVEOLITIDAE Okulitch, 1935 O. (CRD) Mar.

First and Last: *Paleoalveolites*, lower Caradoc, Tennessee, Indiana, USA.

F. CRYPTOLICHENARIIDAE Sokolov, 1959 O. (ARG–ASH) Mar.

First: *Cryptolichenaria miranda* Sokolov, 1955, upper Arenig, River Moyero, northern Siberian Platform, former USSR.
Last: *Cryptolichenaria*, upper Ashgill, Estonia, former USSR, and *Porkunites amalloides* (Dybowski, 1873), upper Ashgill, Porkuni, eastern Estonia, former USSR.

Order SARCINULIDA Sokolov, 1950

F. BILLINGSARIIDAE Okulitch, 1936 O. (LLO)–S. (LLY) Mar.

First: *Billingsaria parva* (Billings, 1859), Valcour Limestone, Lake Champlain Region, New York and Vermont, USA; Mingan Formation, Quebec, Canada; Lenoir Limestone, Tennessee, USA.
Last: *Qianbeilites multitabulatus* Ge and Yu, 1974, lower Llandovery, Kweichow, China.

F. SYRINGOPHYLLIDAE Roemer, 1883 O. (LLO)–S. (LLY) Mar.

First: *Nyctopora vantuyli* Bassler, 1950, Valcour Limestone, Middleburg, Vermont, USA, and *Eofletcheria incerta* (Billings, 1859), Valcour Limestone, Sloop Island, New York, USA; Minegan Formation, Quebec and Aylmer Limestone, Ontario, Canada.
Last: *Reuschia aperta* Kiaer, 1930, lower Llandovery, Island of Strand, Bergen area, Norway.

F. THECIIDAE Milne Edwards and Haime, 1849 S. (LLY)–D. (EIF) Mar.

First: *Thecia swinderniana* (Goldfuss, 1829), upper Llandovery, widespread, UK, and *Romingerella major* Rominger, 1876, upper Clintonian, Indiana, USA.
Last: *Fossopora devonica* Leleshus, 1965, Middle Devonian, southern slope of Gissar Range, upper reaches of River Sorbukh, basin of River Kafirnigan, Kazakhstan, former USSR.

Order FAVOSITIDA Wedekind, 1937

F. FAVOSITIDAE Dana, 1846 O. (LLO)–P. (ART) Mar.

First: *Favosites ramulosus* Phillips, 1848, upper Llandeilo, Abberley, England, UK.
Last: *Sutherlandia*, lower Artinskian, Europe (Urals).

F. PSEUDOFAVOSITIDAE Sokolov, 1950 P. (KAZ) Mar.

First and Last: *Pseudofavosites stylifer* Gerth, 1921 and *Stylonites porosus* Gerth, 1921, upper Kazanian, Timor.

F. SYRINGOLITIDAE Waagen and Wentzel, 1886 S. (LLY–LUD) Mar.

First: *Syringolites huronensis* Hinde, 1879, Brassfield-Sexton Creek Limestone, middle Llandovery, near Hamburg and Cluster Park, western Illinois, USA.
Last: *Syringolites*, Middle Silurian, Gotland, Sweden, Estonia, former USSR.

F. MULTISOLENIIDAE Fritz, 1950 O. (ASH)–S. (PRD) Mar.

First: *Priscosolenia prisca* (Sokolov, 1951), Porkuni Stage, upper Ashgill, Estonia, former USSR.
Last: *Mesosolenia festiva* (Chernyshev, 1951), Upper Silurian, Kuzbas, left bank of River Chumysh, Mount Glyaden, Siberia, former USSR.

F. AGETOLITIDAE Kim, 1962 O. (CRD)–S. (WEN) Mar.

First: *Agetolites mirabilis* Sokolov, 1955, Upper Ordovician, SW foothills of Chingiz Range, Kazakhstan, former USSR, and *Agetolitella prima* Kim, 1962, Upper Ordovician, Zeravshan-Gissar Range, Tien Shan, China.
Last: *Somphopora daedalea* Lindstrom, 1883, Middle Silurian, Chan-Tien, Szechwan, China.

F. MICHELINIIDAE Waagen and Wentzel, 1886 S. (LUD)–P. (TAT) Mar.

First: *Pleurodictyum*, upper Ludlow, Kentucky, Tennessee, USA; New South Wales, Australia.
Last: *Michelinia*, lower Tatarian, Europe, Asia.

F. CLEISTOPORIDAE Easton, 1944 S. (LUD)–C. (MOS) Mar.

First: *Araiostrotion yohi* Guo, 1965, upper Ludlow, Asia (Inner Mongolia).
Last: *Donetzites milleporoides* Dampel, 1940, lower

Moscovian, Europe (Donbas), Asia (Iran, Vietnam, Kweichow).

F. VAUGHANIIDAE Lecompte, 1952 C. (IVO) Mar.

First and Last: *Vaughania cleistoporoides* Garwood, 1913, upper Tournaisian, UK.

F. PALAEACIDAE Roemer, 1883 D. (FAM)–P. (KUN/UFI) Mar.

First: *Palaeacis enorme* (Meek and Worthen, 1860), Louisiana Limestone and Saverton Shale, upper Famennian, eastern Missouri, USA.
Last: *Palaeacis regularis* Gerth, 1921 and *Palaeacis tubifer* Gerth, 1921, Middle Permian, Timor.

F. PACHYPORIDAE Gerth, 1921 S. (LLY)–P. (TAT) Mar.

First: *Striatopora flexuosa* Hall, 1851, Brassfield Formation, middle Llandovery, Cincinnati Arch region, Cincinnati, Ohio, USA (Laub, 1979).
Last: *Gertholites curvata* (Waagen and Wentzel, 1886), Salt Range, Pakistan; lower Tatarian, Timor, ?Yakutia, Australia.

F. PARASTRIATOPORIDAE Chudinova, 1959 O. (CRD)–D. (EIF) Mar.

First: *Kolymopora irjudiensis* Preobrazhenskiy, 1964, upper Caradoc, River Kolyma, NE of former USSR.
Last: *Fomichevia salairica* Dubatolov, 1959, Shanda Beds, Kuzbas, former USSR.

F. ALVEOLITIDAE Duncan, 1872 S. (LLY)–D. (FRS) Mar.

First: *Archypora tuvella* Chekhovich, 1975, lower part of upper Chergak subhorizon, western Tuva, Khondelen.
Last: Various genera, e.g. *Alveolites*, *Crassialveolites*, *Kitakamiia*, upper Frasnian, cosmopolitan.

F. COENITIDAE Sardeson, 1896 S. (LLY)–D. (GIV) Mar.

First: *Planocoenites*, upper Llandovery, Estonia, former USSR.
Last: *Planocoenites*, Middle Devonian, Asia and New South Wales, Australia.

Order HELIOLITIDA Frech, 1897

F. HELIOLITIDAE Lindstrom, 1876 O. (LLO)–D. (FRS) Mar.

First: *Heliolites inordinatus* (Lonsdale, 1839), Middle Ordovician, Robeston Walthen, Pembrokeshire, Wales, UK.
Last: *Heliolites porosus* Goldfuss, 1826, Upper Devonian, Torquay and Plymouth, England, UK; Ron, Annam, Indo-China.

F. TAENIOLITIDAE Lin and Chow, 1977 O. (CRD)–D. (l.) Mar.

First: *Wormsipora hirsutus* (Lindstrom, 1899), middle Caradoc, Tasmania.
Last: *Bogimbailites sytovae* Bondarenko, 1966, Nadaynasu Horizon, 5 km NE of ruins of Bogimba, Kazakhstan, former USSR.

F. STELLIPORELLIDAE Bondarenko, 1971 O. (ASH)–D. (EIF) Mar.

First: *Parastelliporella columella* Lin and Chow, 1977, Upper Ordovician, Kwangsi, China.
Last: *Podollites*, Middle Devonian, Kazakhstan, former USSR.

F. PSEUDOPLASMOPORIDAE Bondarenko, 1963 O. (CRD)–D. (EIF) Mar.

First: *Visbylites stella* (Lindstrom, 1899), Dulankar Horizon, upper Caradoc, Tarbagatau Range, Kazakhstan, former USSR.
Last: *Pachyhelioplasma kettnerovae* Kim, 1966, Middle Devonian, Zeravshan-Gissar Range, basin of River Kashkadari, Khodza-Kurgan Gully, Kazakhstan, former USSR.

F. PROPORIDAE Sokolov, 1949 O. (CRD)–S. (LUD) Mar.

First: *Propora tubulata* (Lonsdale, 1839), lower Caradoc, Bala Limestone, Gwynedd, Wales, UK.
Last: *Innapora incredula* (Chernova, *in* Kovalevskiy *et al.*, 1960), Dalyan Horizon, left side of River Isfara, north slopes of Turkestan Range, southern Tien Shan, China.

F. SIBIRIOLITIDAE Lin, 1977 O. (ASH)–?S. (Middle) Mar.

First: *Mongoliolites paradoxides* Bondarenko and Minzhin, 1977, lower Ashgill, southern foot of Khangay Range, Central Mongolia.
Last: *Sibiriolitella*, ?Middle Silurian, China.

F. PLASMOPORIDAE Sardeson, 1896 O. (CRD)–D. (l.) Mar.

First: *Plasmopora petalliformis* Lonsdale, 1839, Coniston Limestone, Coniston, England, UK; Dyfed, Wales, UK.
Last: *Squameolites* sp., Lower Devonian, Kazakhstan, former USSR.

F. PLASMOPORELLIDAE Kovalevskiy, 1964 O. (CRD)–S. (WEN) Mar.

First: *Plasmoporella*, middle Caradoc, New South Wales, Australia.
Last: *Camptolithus papillata* (Rominger, 1876), upper Wenlock, Michigan, USA.

F. PROHELIOLITIDAE Kiaer, 1899 O. (CRD)–S. (LUD) Mar.

First: *Kiaerolites kalstadensis* Bondarenko, 1977, upper Caradoc, Norway and *Protoheliolites norvegicus* Bondarenko, 1977, upper Caradoc, Stavnaestangen, Ringerike, Norway.
Last: *Avicenia aseptata* Leleshus, 1974, lower part of Dalyan Horizon, Southern Tien Shan, China.

F. ? CYRTOPHYLLIDAE Sokolov, 1950 O. (CRD–ASH) Mar.

First: *Cyrtophyllum densum* Lindstrom, 1882, middle Caradoc, Middle Tunguska River, Siberia, former USSR.
Last: *Karagemia altaica* Dzubo, 1960, upper Ashgill, right side of River Karagem, Altay Mountains, former USSR, and *Rhaphidophyllum constellatum* Lindstrom, 1882, upper Ashgill, middle Tunguska river, above last rapids before River Chuna, Siberia, former USSR.

F. COCCOSERIDIDAE Kiaer, 1899 O. (CRD)–S. (WEN) Mar.

First: *Coccoseris*, low in Fauna 1, lower mid-Mohawkian, New South Wales, Australia.
Last: *Acidolites* lower Wenlock, locality unknown.

F. PYCNOLITHIDAE Lindstrom, 1899
S. (LLY/WEN) Mar.

First and Last: *Pycnolithus bifidus* Lindstrom, 1899, upper Llandovery or lower Wenlock, lower Visby shore, Gotland, Sweden. [Not *in situ*.]

F. PALAEOPORITIDAE Kiaer, 1899
O. (CRD–ASH) Mar.

First: *Trochiscolithus micraster* (Lindstrom, 1899), middle Caradoc, Norway, Sweden and Estonia, former USSR.
Last: *Palaeoporites estonicus* Kiaer, 1899, Borkholm Beds, upper Ashgill, Borkholm, Estonia, former USSR; Upper *Chasmops* Beds, upper Ashgill, Norway.

F. HALYSITIDAE Milne Edwards and Haime, 1849
O. (CRD)–S. (LUD) Mar.

First: *Quepora quebecensis* (Lambe, 1899), Chaumont Limestone, Blackriveran (lower Caradoc), 2 miles south of Blue Point, Lake St John, Quebec, Canada.
Last: Various genera, e.g. *Halysites, Catenipora, Acanthohalysites, Cystihalysites, Falsicatenipora, Solenihalysites*, Upper Silurian, cosmopolitan.

Order AULOPORIDA Sokolov, 1947

F. AULOPORIDAE Milne Edwards and Haime, 1851
O. (ARG)–P. (KAZ) Mar.

First: *Aulopora*, Lower Ordovician, Europe (Baltic), Asia (Irkutsk).
Last: *Aulopora timorica* Gerth, 1922, Basleo Limestone, Timor.

F. BAJGOLIIDAE Hill, 1981 O. (CRD–ASH) Mar.

First: *Bajgolia altaica* Dzyubo, 1962, lower Caradoc, Tasmania, New South Wales, Australia.
Last: *Bajgolia altaica* Dzyubo, 1962, lower Ashgill, Baygol Creek, 1 km above junction with River Kayna, Altay, former USSR.

F. KOZLOWSKIOCYSTIIDAE Stasinska, 1969
D. (EIF) Mar.

First and Last: *Kozlowskiocystia polonica* (Stasinska, 1958), Middle Devonian, Grzegorzowice, Poland.

F. FLETCHERIELLIDAE Sokolov, 1965
O. (ASH)–D. (EIF) Mar.

First: *Eofletcheriella primitiva* Lin and Chow, 1977, Upper Ordovician, Kiangsi, China.
Last: *Pseudofletcheria fundibula* Chi, 1976, lower Middle Devonian, Dong Ujimqin Qi, NE Inner Mongolia.

F. PYRGIIDAE de Fromentel, 1861
S. (LUD)–P. (TAT) Mar.

First: *Bainbridgia typicalis* Ball, 1933, lower Ludlow, Missouri, USA.
Last: *Cladochonus*, lower Tatarian, cosmopolitan.

F. ?TRACHYPSAMMIIDAE Gerth, 1921
P. (ART–KAZ) Mar.

First: *Oculinella gerthi* Yakovlev, 1939, upper Artinskian, Oufimskoe Plateau, Donbas-Krasnoufimsk, Urals, former USSR.
Last: *Trachypsammia dendroides* Gerth, 1922, Basleo Limestone, upper Kazanian, Timor.

F. AULOHELIIDAE Sokolov, 1950
P. (KAZ) Mar.

First and Last: *Aulohelia irregularis* Gerth, 1921 and *Aulohelia laevis* Gerth, 1921, Basleo Limestone, upper Kazanian, Timor.

F. ROMINGERIIDAE Sokolov, 1950
S. (LLY)–C. (l.) Mar.

First: *Romingeria*, upper Llandovery, Estonia, former USSR.
Last: *Protopora cystoides* (Grabau, *in* Greene, 1901), upper Mississippian, Indiana, USA.

F. ?PALAEOFAVOSIPORIDAE Stasinska, 1976
S. (WEN) Mar.

First and Last: *Palaeofavosipora clausa* (Lindstrom, 1866), Middle Silurian, Gotland, Sweden.

F. AULOCYSTIDAE Sokolov, 1950
O. (ASH)–P. (KAZ) Mar.

First: *Adaverina acritos* Webby, 1977, Top of Malachi's Hill Beds, middle to upper Cincinnatian (Richmondian), New South Wales, Australia.
Last: *Pseudoromingeria kotoi* Yabe and Hayasaka, 1915, *Yabeina* Zone, upper Kazanian, Kinsyozan, Gifu Prefecture, Fuwa-gun, Japan.

F. ?SINOPORIDAE Sokolov, 1955
S. (LLY)–P. (ART) Mar.

First: *Sinopora*, Lower Silurian, Estonia, former USSR, and *Sinoporella fenggangensis* Kim and Yang, *in* Yang et al., 1978, Shiniulan Formation, Fenggang, Guizhou, Kweichow, China.
Last: *Sinopora dendroidea* (Yoh, *in* Yoh and Huang, 1932), Chihsia Limestone, lower Artinskian, Chi-lung-shan, near Ho-chou, SE Anhui, China.

F. ?KHMERIIDAE Montanaro-Gallitelli, 1954
P. (ART) Mar.

First and Last: *Khmeria problematica* Mansuy, 1914, lower Artinskian, Cambodia, Japan, Armenia, Sicily, Kazan, Donbas, Urals, Tunisia, etc.

F. SYRINGOPORIDAE de Fromentel, 1861
O. (ASH)–P. (KAZ) Mar.

First: *Syringopora*, upper Ashgill, Urals, former USSR.
Last: *Enigmalites lectus* Chudinova, 1975, upper Tastuba subhorizon, right bank of River Kosva, western slopes, central Urals, former USSR.

F. PERIPHACELOPORIDAE Hill, 1981 D. (GIV) Mar.

First and Last: *Periphacelopora exornata* Dethier and Pel, 1971, lower Givetian, Hampteau, Belgium.

F. TETRAPORELLIDAE Sokolov, 1950
O. (CRD)–P. (TAT) Mar.

First: *Labyrinthites chidlensis* Lambe, 1906, middle Caradoc, shore of west central Lake Manicouagan, Quebec, Canada.
Last: *Hayasakaia aequitabulata* (Huang, 1932), *H. elegantula* (Yabe and Hayasaka, 1915) and *H. lanchugensis* (Huang,

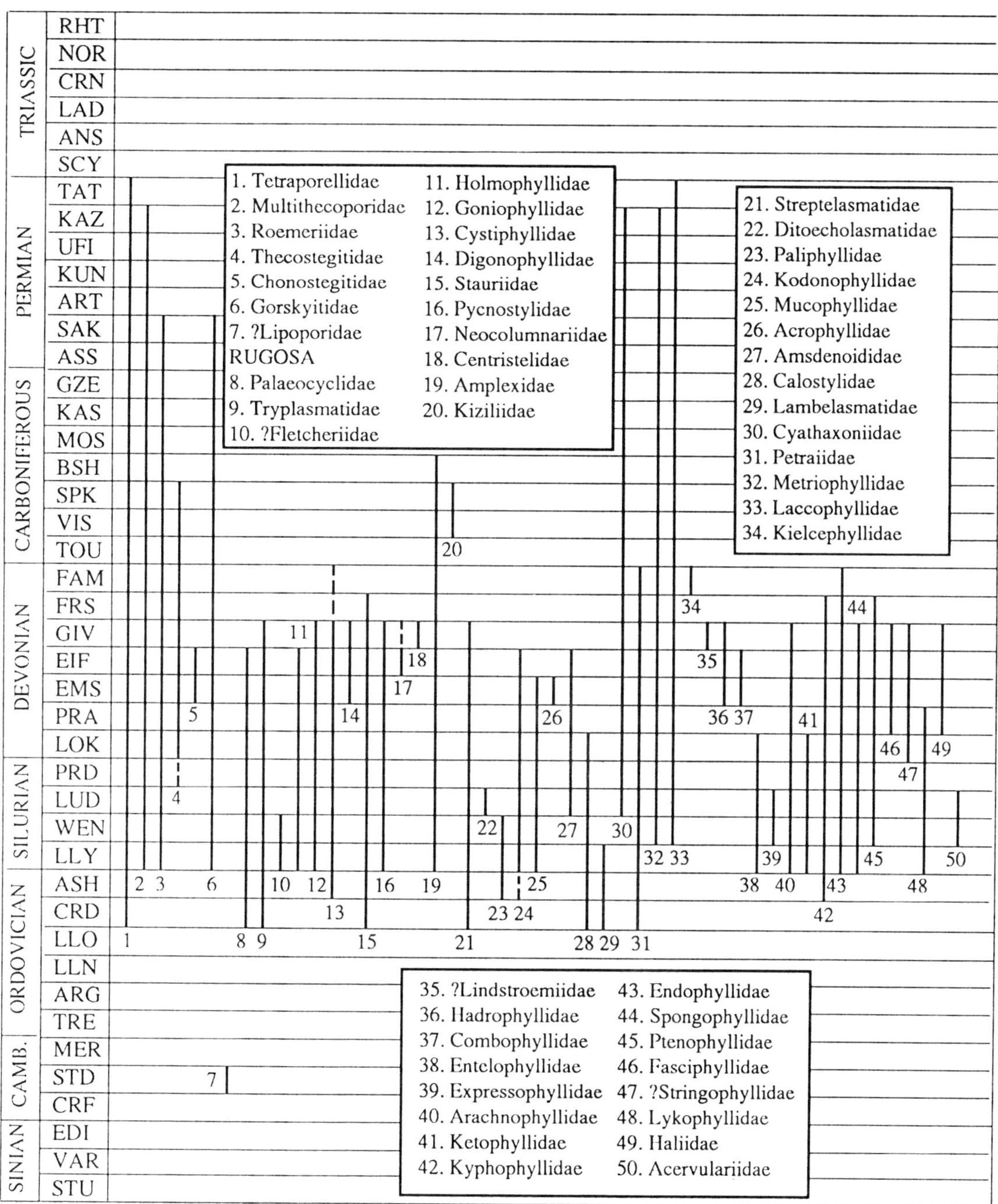

Fig. 6.5

1932), all from Wushan Limestone, lower Tatarian, Hupei and northern Szechwan, China.

F. MULTITHECOPORIDAE Sokolov, 1950 S. (LLY)–P. (KAZ) Mar.

First: *Multithecopora*, Lower Silurian, Norway.
Last: *Multithecopora*, Upper Permian, China, Japan, Iran, Afghanistan, former Yugoslavia; Urals, former USSR; Yukon, Canada.

F. ROEMERIIDAE Pocta, 1904 ?S. (LLY)–P. (SAK) Mar.

First: *Roemerolites*, ?Lower Silurian, Kweichow, China.
Last: *Bayhaium merriamorum* Langenheim and McCutcheon, 1959, McCloud Limestone, Lower Permian, Shasta County, California; Nevada, USA.

F. THECOSTEGITIDAE de Fromentel, 1861 S. (PRD)/D. (GED)–C. (SPK) Mar.

First: *Thecostegites*, Upper Silurian or Lower Devonian, Polar Urals, former USSR.
Last: *Duncanopora duncanae* Sando, 1975, Moffat Trail Limestone Member, ?lower Namurian, Wyoming, USA; ?lower Namurian, Idaho, Utah, USA.

F. CHONOSTEGITIDAE Lecompte, 1952 D. (EMS–EIF) Mar.

First and Last: *Chonostegites*, Lower–Middle Devonian, eastern North America.

F. GORSKYITIDAE Lin, 1963 S. (LLY)–P. (SAK) Mar.

First: *Meitanopora convexocystosa* Yang, 1973, Shiniulan Formation, Lower Silurian, Meitan, Guizhou, Kweichow, China.
Last: *Neosyringopora*, Lower Permian, Spitsbergen; Urals, former USSR; Devon Island, Nevada, USA.

Order UNCERTAIN

F. ?LIPOPORIDAE Jell and Jell, 1976 Є. (StD) Mar.

First and Last: *Lipopora lissa* Jell and Jell, 1976, Coonigan Formation, lower Middle Cambrian, western New South Wales, Australia.

Subclass RUGOSA Milne Edwards and Haime, 1850

Classification follows Hill (1981).

Order CYSTIPHYLLIDA Nicholson, 1889

F. PALAEOCYCLIDAE Dybowski, 1873 O. (CRD)–D. (EIF) Mar.

First: *Primitophyllum primum* Kaljo, 1956, lower Caradoc, Estonia, former USSR.
Last: *Bojocyclus bohemicus* Prantl, 1939, Hlubocepy Limestone, 'White Bed' at top, Quarry at Holyne, west of Prague, Czechoslovakia.

F. TRYPLASMATIDAE Etheridge, 1907 O. (CRD)–D. (GIV) Mar.

First: *Hillophyllum priscum* Webby, 1971, Cliefden Caves Limestone (lower part), lower Caradoc, Boonderoo, New South Wales, Australia.
Last: *Aphyllum*, Middle Devonian, Tajikistan, former USSR.

F. ?FLETCHERIIDAE Zittel, 1876 S. (LLY–WEN) Mar.

Stasinska (1967) suggests that this genus belongs instead to the Tabulata (?Auloporida).
First: *Parafletcheria dupliformis* Yang, 1973, Shiniulan Formation, Shimenkan, Shiqian, Guizhou, Kweichow, China.
Last: *Fletcheria tubifera* Milne Edwards and Haime, 1851, Middle Silurian, Gotland, Sweden.

F. HOLMOPHYLLIDAE Wang, 1947 S. (LLY)–D. (EIF) Mar.

First: *Holmophyllum*, Lower Silurian, New South Wales, Australia.
Last: *Holmophyllum uralicum* (Zhavoronkova, 1972), Middle Devonian, western slopes of South Urals, River Maly Ik, former USSR.

F. GONIOPHYLLIDAE Dybowski, 1873 S. (LLY)–D. (GIV) Mar.

First: *Goniophyllum pyramidiale* (Hisinger, 1831), upper Llandovery; UK; Gotland, Sweden; Ontario; Canada; Iowa, USA and *Araeopoma prismaticum* (Lindstrom, 1868), upper Llandovery, Gotland, Sweden.
Last: *Calceola sandalina* (Linnaeus, 1771), upper Givetian, Europe, Africa, Asia, Australia.

F. CYSTIPHYLLIDAE Milne Edwards and Haime, 1850 O. (ASH)–D. (FAM?) Mar.

First: *Cystiphyllum*, Upper Ordovician, cosmopolitan.
Last: *Cystiphyllum (Zonophyllum)*, ?Upper Devonian, Poland, western Australia. Definitely known from the Givetian – e.g. *C. (Z.) thomasi* (Taylor, 1951), upper Middle Devonian, Richmond Walk Quarry, Stonehouse, Plymouth, England, UK; *Cystiphylloides (Lythophyllum) exzentricum* (Borchers MS, *in* Wedekind, 1925), Berndorf, Germany; *Diplochone striata* Frech, 1886, upper *Stringocephalus* Beds, Rhineland, Germany.

F. DIGONOPHYLLIDAE Wedekind, 1923 D. (EMS–GIV) Mar.

First: *Digonophyllum* and *Mesophyllum*, both from Zlichovian, Western and Arctic Canada and Alaska (Oliver and Pedder, 1979).
Last: Various genera, e.g. *Digonophyllum (Digonophyllum), D. (Mochlophyllum), Mesophyllum (Mesophyllum), M. (Dialytophyllum), M. (Hemicosmophyllum), M. (Lekanophyllum), M. (Zonodigonophyllum)*, all from Givetian of Germany, etc.

Order STAURIIDA Verrill, 1865

F. STAURIIDAE Milne Edwards and Haime, 1850 O. (CRD)–D. (FRS) Mar.

First: *Favistina undulata* (Bassler, 1950), Platteville Limestone, Blackriverian, lower Caradoc, Beloit, Wisconsin, USA.
Last: *Columnaria*, Upper Devonian, Altai-Sayan, former USSR (Oliver and Pedder, 1979).

F. PYCNOSTYLIDAE Stumm, 1953 S. (LLY)–D. (GIV) Mar.

First: *Protopilophyllum*, upper Llandovery, Siberian Platform, former USSR.
Last: *Cyathopaedium paucitabulatum* (Schluter, 1880), *Stringocephalus* Limestone, Hebborn, Bergisch-Gladbach, Germany; *Depasophyllum adnetum* Grabau, 1936, bioherm in Four Mile Dam Formation, Traverse Group, Four Mile Dam, Thunder Bar River, Alpena County, Michigan, USA, and *Fletcherina simplex* (Yabe and Hayasaka, 1915), Middle Devonian, Queensland, Australia.

F. NEOCOLUMNARIIDAE Soshkina, 1949 D. (EIF–GIV?) Mar.

First: *Neocolumnaria vagranensis* Soshkina, 1949, Eifelian, Krasnaya shapochka no. 19, northern Urals, former USSR.
Last: *Neocolumnaria*, ?Givetian, NW Territories, Canada.

F. CENTRISTELIDAE Tsyganko, 1971 D. (GIV) Mar.

First and Last: *Centristela fasciculata* Tsyganko, 1967, Middle Devonian, Pay-Khoy, Belovskaya River, former USSR and *C. anavarensis* (Goryanov, *in* Bulvanker *et al.*, 1968), Middle Devonian, southern Fergana, Katran Range, former USSR.

F. AMPLEXIDAE Chapman, 1893 S. (LLY)–C. (BSH) Mar.

First: *Amplexoides severnensis* (Parks, 1915), lower Llandovery, Limestone rapids, Severn River, northern Ontario, Canada.
Last: *Gorskyella tschigariensis* (Fomichev, 1953), Donbas Limestone, C_23, Upper Bashkirian, right bank of Zheleznaya ravine, former USSR.

F. KIZILIIDAE Degtyarev, 1965 C. (VIS–SPK) Mar.

First: *Kizilia concavitabulata* Degtyarev, 1965, Kizilian Suite, upper Viséan, Kizil, southern Urals and *Melanophyllidium lativesiculosum* Kropacheva, 1966, lower part of Puma Suite, upper Viséan, southern Fergana, southern slopes of Katran Range, former USSR.
Last: *Kizilia*, upper Serpukhovian, Urals, former USSR (Semenoff-Tian-Chansky and Sutherland, 1982).

F. STREPTELASMATIDAE Nicholson, 1889 O. (CRD)–D. (GIV) Mar.

First: *Streptelasma corniculum* (Hall, 1847), *N. gracilis* Zone, Blackriver Formation, lower Caradoc, New York, Michigan, USA.
Last: *Altaiophyllum belgebaschicum* Ivaniya, 1955, upper Givetian, right tributary of River Chi, River Belgebash, Altay and *Xenocyathellus thedfordense* (Stewart, 1936), Hamilton-Arkona beds, upper Givetian, Dam on Aus Sables River at mouth of Rock Glen, Arkona, Ontario, Canada.

F. DITOECHOLASMATIDAE Sutherland, 1965 S. (LUD) Mar.

First and Last: *Ditoecholasma fanninganum* (Safford, 1869), Brownsport Formation, Tennessee, USA and *D. lawrencense* Sutherland, 1965, Henryhouse Formation, Oklahoma, USA.

F. PALIPHYLLIDAE Soshkina, 1955 O. (ASH)–S. (WEN) Mar.

First: Various genera, e.g. *Paliphyllum*, *Protocyathactis*, *Sumsarophyllum*, all from Upper Ordovician of Asia.
Last: *Neopaliphyllum soshkinae* Zheltonogova, 1961, Baskuskan Suite, left bank of River Baskuskan, Salair Mountains, former USSR.

F. KODONOPHYLLIDAE Wedekind, 1927 O. (ASH?)–D. (EIF) Mar.

First: *Kodonophyllum*, ?Upper Ordovician, Estonia, former USSR. *Schlotheimophyllum patellatus* (Schlotheim, 1820), definitely occurs in the upper Llandovery of the UK, Sweden and Norway.
Last: *Sinochlamydophyllum crassiseptatum* Guo, 1976, lower Middle Devonian, Inner Mongolia and *Zelophyllia tabulatum* (Soshkina, 1937), Middle Devonian, left bank of River Vagran, Urals, former USSR.

F. MUCOPHYLLIDAE Soshkina, 1947 S. (LLY)–D. (EMS) Mar.

First: *Kungejophyllum ajagusense* Sultanbekova, 1971, upper Llandovery, River Ayaguz, Chingiz Range, Kazakhstan, former USSR.
Last: *Briantia repleta* Barrois, 1889 and *Pseudamplexus ligeriensis* (Barrois, 1889), Calcaire d'Erbray, Chateau-Briant, France.

F. ACROPHYLLIDAE Stumm, 1949 D. (EMS) Mar.

First and Last: *Acrophyllum oneidaense* (Billings, 1859), Oonodaga Limestone, Ontario, Canada, and *Scenophyllum conigerum* (Rominger, 1876), Coral Zone in Jeffersonville Limestone, Falls of the Ohio, Indiana, USA.

F. AMSDENOIDIDAE Hill, 1981 S. (LUD)–D. (EIF) Mar.

First: *Amsdenoides acutiannulatum* (Amsden, 1949), Brownsport Formation, Tennessee, USA.
Last: *Multicarinophyllum multicarinatum* Spasskiy, 1965, Middle Devonian, River Kyzylagach, Dzhungarian Alatau, former USSR.

F. CALOSTYLIDAE Zittel, 1879 O. (CRD)–D. (GED) Mar.

First: Various genera, e.g. *Calostylis*, *Ningnanophyllum*, *Yohophyllum*, all from Middle Ordovician of Szechwan and Kweichow, China.
Last: *Calostylis*, Aynasu Horizon, Kazakhstan, former USSR.

F. LAMBELASMATIDAE Weyer, 1973 O. (CRD)–S. (LLY) Mar.

First: *Lambeophyllum profundum* (Conrad, 1843), Platteville Limestone, Black River Group, lower Caradoc, Mineral Point, Wisconsin, USA.
Last: *Prototryplasma oroniana* Ivanovskiy, 1963, upper Llandovery, basin of River Imanga, Norilsk district, Siberian Platform, former USSR.

F. CYATHAXONIIDAE Milne Edwards and Haime, 1850 S. (LUD)–P. (KAZ) Mar.

First: *Columnaxon angelae* Scrutton, 1971, Upper Silurian, Rio Aricagua section, Merida Andes, Venezuela.
Last: *Cyathaxonia*, Upper Permian, Cambodia.

F. PETRAIIDAE de Koninck, 1872 O. (CRD)–D. (FAM) Mar.

First: *Protozaphrentis minor* Yu, 1957, Middle Ordovician, Kuluk-Tag Range, Liu-Wang-shan, Sinkiang, China.
Last: *Petraiella kielcensis* Rozkowska, 1969, lower Famennian, Kielce, Poland.

F. METRIOPHYLLIDAE Hill, 1939 S. (WEN)–P. (KAZ) Mar.

First: *Duncanella borealis* Nicholson, 1874, 'Lower' Silurian, Waldron, Indiana, USA.
Last: *Asserculinia prima* Schouppe and Stacul, 1959, lower Kazanian, Basleo, Timor.

F. LACCOPHYLLIDAE Grabau, 1928 S. (WEN)–P. (TAT) Mar.

First: *Laccophyllum acuminatum* Simpson, 1900, lower Wenlock, Perry County, Tennessee, USA.
Last: *Amplexocarinia*, *Palaeofusulina* Zone, upper Tatarian, Salt Range, Timor (Flügel, 1970).

F. KIELCEPHYLLIDAE Rozkowska, 1969 D. (FAM) Mar.

First and Last: *Kielcephyllum cupulum* Rozkowska, 1969 and *Kozlowskinia flos* Rozkowska, 1969, both from lower Famennian, Kadzielnia, Holy Cross Mountains, Poland, and *Thecaxon rozkowskae* Weyer, 1978, lower Famennian, Germany.

F. ?LINDSTROEMIIDAE Pocta, 1902 D. (GIV) Mar.

First and Last: *Lindstroemia columnaris* Nicholson and Thomson, 1876, lower Givetian, North America.

F. HADROPHYLLIDAE Nicholson, 1889 D. (EMS-GIV) Mar.

First: *Hadrophyllum orbignyi* Milne Edwards and Haime, 1850, Jeffersonville Limestone, Indiana or Kentucky, USA; Speeds Limestone, Charlestown, Indiana, USA.
Last: *Microcyclus discus* Meek and Worthen, 1868, St Laurent Limestone, Hamilton Group, Grand Tower, northern end of Backbone Ridge, Illinois, USA.

F. COMBOPHYLLIDAE Weyer, 1975 D. (EMS–EIF) Mar.

First: *Combophyllum ibericum* Plusquellec, *in* Marin and Plusquellec, 1973, Lower Devonian, near Cabrero, Teruel, Spain.
Last: *Combophyllum osismorum* Milne Edwards and Haime, 1850, lower Eifelian, Brest roadstead, Le Fret, France.

F. ENTELOPHYLLIDAE Hill, 1940 S. (LLY)–D. (GED) Mar.

First: *Petrozium dewari* Smith, 1930, *Pentamerus* Beds, upper Llandovery, Morrell's Wood Brook, near Buildwas, UK.
Last: *Kysylagathophyllum michnevitchi* Kaplan, 1971, lower part of Pribalkhash Horizon, near Kyzyl-Agat, Kazakhstan, former USSR and *Scyphophyllum*, Lower Devonian, northern Urals, central Kazakhstan, former USSR.

F. EXPRESSOPHYLLIDAE Strelnikov, 1968 S. (WEN–LUD) Mar.

First: *Micula antiqua* Sytova, 1952, lower Wenlock, northern shore of Mikhailovskii Pond, western slope of Central Urals, former USSR and *Pseudopilophyllum moyerense* Ivanovskiy, 1963, lower Wenlock, River Moyero, Siberian Platform, former USSR.
Last: *Micula simplex* (Strelnikov, 1968) and *Contortophyllum tchernovi* Strelnikov, 1968, Durnayuiskiy Horizon, River Kozhim, Polar Urals, former USSR.

F. ARACHNOPHYLLIDAE Dybowski, 1873 S. (LLY)–D. (GIV) Mar.

First: *Arachnophyllum*, upper Llandovery, Europe, North America and *Angullophyllum warrisi* McLean, 1974, Lower Limestone Horizon, upper Llandovery, Cobblers' Creek, Angullong District, near Orange, central New South Wales, Australia.
Last: *Craterophyllum*, Onondaga Limestone, Kentucky, Indiana, USA; Ontario, Canada.

F. KETOPHYLLIDAE Lecompte, 1952 S. (LLY)–D. (GED) Mar.

First: *Dentilasma honorabilis* Ivanovskiy, 1962, upper Llandovery, River Mogokta, western Siberian Platform and *Heterolasma foerstei* Ehlers, 1919, Manistique Formation, ?upper Llandovery (or lower Wenlock), half a mile south of Gould City, Michigan, USA.
Last: *Chavsakia chavsakiensis* Lavrusevich, 1959, *Pholidophyllum* Beds, Khavsak Gully, Zeravshan-Gissar Range, Tajikistan, former USSR and *Nataliella poslavskajae* Sytova, *in* Sytova and Ulitina, 1966, upper part of Aynasu Horizon, Nurin Syncline, left bank of River Medine, central Kazakhstan, former USSR.

F. KYPHOPHYLLIDAE Wedekind, 1927 O. (ASH)–D. (FRS) Mar.

First: *Donacophyllum*, Upper Ordovician, Estonia, former USSR.
Last: Various genera, e.g. *Bouvierphyllum*, *Mikkwaphyllum*, *Wapitiphyllum*, *Parasmithiphyllum*, *Tarphyphyllum*, *Kakisaphyllum*, all from upper Frasnian, western Alberta, eastern British Columbia and southern District of Mackenzie, western Canada (McLean and Pedder, 1984).

F. ENDOPHYLLIDAE Torley, 1933 S. (LLY)–D. (FAM) Mar.

First: *Mictocystis endophylloides* Etheridge, 1908, Quarry Creek Limestone, upper Llandovery, central New South Wales, Australia.
Last: *Smithiphyllum*, Upper Devonian, Poland. May also occur in Carboniferous of Timan.

F. SPONGOPHYLLIDAE Dybowski, 1873 S. (LLY)–D. (GIV) Mar.

First: *Heterospongophyllum simplex* (He MS *in* Kong and Huang, 1978), Shiniulan Formation, Shiqian, Guizhou, Kweichow, China.
Last: *Neovepresiphyllum immersum* (Hill, 1942), Middle Devonian, Arthur's Creek, Burdekin Downs, Queensland, Australia.

F. PTENOPHYLLIDAE Wedekind, 1923 S. (WEN)–D. (FRS) Mar.

First: *Cymatelasma corniculum* Hill and Butler, 1936, Woolhope Limestone, Woolhope, near Hereford, UK.
Last: *Hankaxis tinocystis* (Frech, 1885), Iberger Kalk, upper Frasnian, Grund, Germany.

F. FASCIPHYLLIDAE Soshkina, 1954 D. (SIG–GIV) Mar.

First: *Fasciphyllum* sp., Pragian, western and Arctic Canada and Alaska, former USSR, eastern Australia and New Zealand (Oliver and Pedder, 1979).
Last: *Fasciphyllum conglomeratum* (Schluter, 1881), Middle Devonian, Eifel, Germany, and *Crista compacta* Tsyganko, 1971, Middle Devonian, River Nadota, Polar Urals, former USSR.

F. ?STRINGOPHYLLIDAE Wedekind, 1922 D. (GED–GIV) Mar.

First: *Rhegmaphyllum*, Lochkovian, Tien Shan, China and former USSR (Oliver and Pedder, 1979).
Last: *Stringophyllum*, *Parasociophyllum* and *Sociophyllum*, common in upper Givetian, Germany and Belgium.

F. LYKOPHYLLIDAE Wedekind, 1927 S. (LLY)–D. (SIG) Mar.

First: *Holophragma*, upper Llandovery, Gotland, Sweden; Ohio, USA; *Pseudophaulactis lykophylloides* Zaprudskaya, *in* Ivanovskiy, 1963, upper Llandovery, River Gorbiyachin, former USSR; *Rukhinia cuneata* Strelnikov, 1963, upper Llandovery, River Letney, western part of Siberian Platform and *Zeravschania prima* Lavrusevich, 1964, upper Llandovery, Mount Daurich, Zeravshan–Gissar Range, former USSR.
Last: *Camurophyllum camurum* Kravtsov, 1966, Lower, Zone of Valnevsk Horizon, Tsivolko Bay, Novaya Zemlya.

F. HALLIIDAE Chapman, 1893 D. (SIG–GIV) Mar.

First: *Kobeha walcotti* Merriam, 1974, Coral Zone B, Nevada

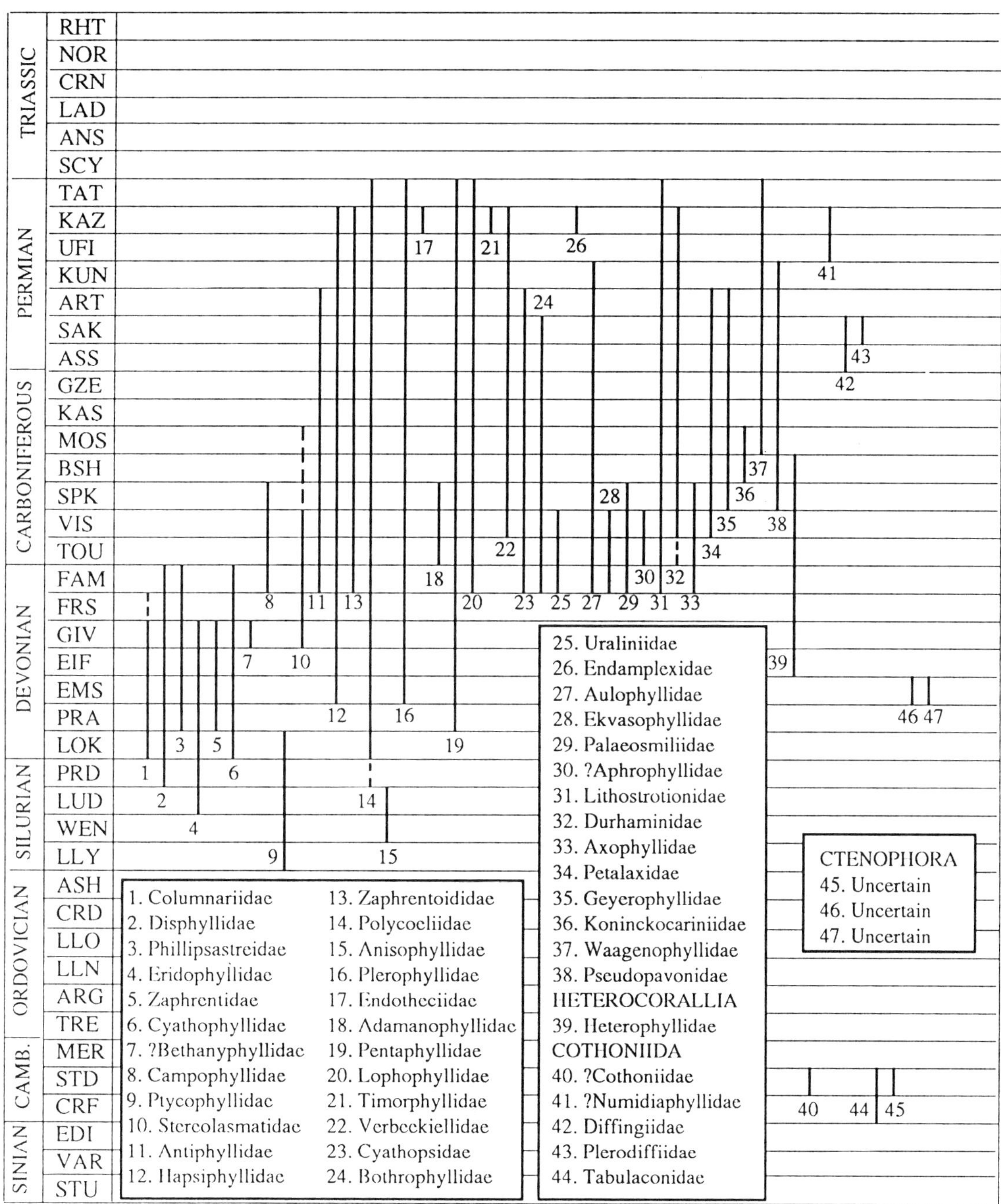

Fig. 6.6

Formation, southern Sulphur Springs Range, Nevada, USA.

Last: Various genera, e.g. *Hallia, Aulacophyllum, Odontophyllum*, all from Middle Devonian of North America and *Aspasmophyllum* from Middle Devonian of Germany.

F. ACERVULARIIDAE de Fromentel, 1861 S. (WEN-LUD) Mar.

First: *Acervularia* and *Diplophyllum*, Middle Silurian, Europe, North America.

Last: *Oliveria planotabulata* Sutherland, 1965, Henryhouse Formation, Lawrence Uplift, Oklahoma, USA.

F. COLUMNARIIDAE Nicholson, 1879 D. (GED–FRS?) Mar. (see Fig. 6.6)

First: *Circumtextiphyllum annulatum* Kaplan, 1971, lower part of Pribalkhash Horizon, near Kyzel-Agat, Kazakhstan, former USSR.

Last: *Columnaria sulcata* Goldfuss, 1826, ?lower Frasnian (or upper Givetian), Paffrather Mulde, near Bensberg, 10 miles east of Cologne, Germany.

F. DISPHYLLIDAE Hill, 1939 S. (PRD)–D. (FAM) Mar.

First: *Radiastraea*, Upper Silurian, Canadian Arctic.

Last: *Temnophyllum*, Upper Devonian, Urals, former USSR.

F. PHILLIPSASTREIDAE Hill, 1954 D. (SIG–FAM) Mar.

First: *Phillipsastrea*, Lower Devonian, Victoria, New South Wales, Australia; France.
Last: *Peneckiella* sp. Rozkowska, 1969 and *Phacellophyllum germanicum* Rozkowska, 1969, upper Famennian, Holy Cross Mountains, Poland (Sorauf and Pedder, 1986).

F. ERIDOPHYLLIDAE de Fromentel, 1861 S. (LUD)–D. (GIV) Mar.

First: *Capnophyllum hedlundi* Sutherland, 1965, Henryhouse Formation, Lawrence Uplift, Oklahoma, USA.
Last: *Cylindrophyllum elongatum* Simpson, 1900, Onondaga Limestone, upper Givetian, Clarksville, New York and *Asterobillingsia confluens* (Vanuxem, 1842), Onondaga Limestone, upper Givetian, quarry south of Chittenango, near Perryville, New York, USA.

F. ZAPHRENTIDAE Milne Edwards and Haime, 1850 D. (SIG–GIV) Mar.

First: *Heliophyllum*, Lower Devonian, Victoria, Australia.
Last: *Heliophyllum*, upper Givetian, eastern North America, Morocco, Spanish Sahara (NW Africa), Spain.

F. CYATHOPHYLLIDAE Dana, 1846 D. (GED–FAM) Mar.

First: *Radiophyllum*, Lower Devonian, Kuzbas, former USSR.
Last: *Commutatophyllum cincinnatum* Kaplan, 1971, *C. sulcifer* beds, northern Cis-Balkhash, Kazakhstan, former USSR.

F. ?BETHANYPHYLLIDAE Stumm, 1949 D. (GIV) Mar.

First and Last: *Bethanyphyllum* and *Tortophyllum*, Middle Devonian, North America, e.g. *B. robustum* (Hall, 1876), Hamilton Group, western New York, USA and *T. cysticum* (Winchell, 1866), Upper Blue Shale, Gravel Point Formation, Traverse Group, Bell Quarry, Michigan, USA.

F. CAMPOPHYLLIDAE Wedekind, 1992 D. (FAM)–C. (SPK) Mar.

First: *Campophyllum flexuosum* (Goldfuss, 1826), upper Famennian, and *Campophyllum cylindricum* (Onoprienko, 1979), lower Elergetkehyn Suite, upper Famennian, Omolon Massif, NE of former USSR (Sorauf and Pedder, 1986).
Last: *Campophyllum*, lower Serpukhovian, locality unknown.

F. PTYCHOPHYLLIDAE Dybowski, 1873 S. (LLY)–D. (GED) Mar.

First: *Ptychophyllum*, Lower Silurian, Michigan, USA; Ontario, Canada; Siberian Platform, Altay, Sayan, Tajikistan, former USSR; New South Wales, Queensland, Australia.
Last: *Implicophyllum vesiculosum* Sytova, *in* Sytova and Ulitina, 1966, Zone of *Nataliella poslavskajae*, Aynasu Horizon, south flank of Karaganda Basin, central Kazakhstan, former USSR.

F. STEREOLASMATIDAE Fomichev, 1953 D. (GIV)–C. (MOS?) Mar.

First: *Stereolasma rectum* (Hall, 1876), *Amplexiphyllum hamiltoniae* (Hall, 1876), *Lopholasma carinatum* Simpson, 1990 and *Stewartophyllum intermittens* (Hall, 1876), all from Hamilton Shale, New York, USA.
Last: *Lopholasma*, ?lower Moscovian (or upper Viséan), China.

F. ANTIPHYLLIDAE Ilina, 1970 D. (FAM)–P. (ART) Mar.

First: *Fasciculophyllum* aff. *rushianum* Rozkowska, 1969 and *Fasciculophyllum* (?) *dobroljubovae* Rozkowska, 1969, upper Famennian, Holy Cross Mountains, Poland and *Pseudoclaviphyllum*, upper Famennian, no locality given (Sorauf and Pedder, 1986).
Last: *Lytvolasma asymmetricum* Soshkina, 1925, Lower Permian, 6 km below Alexandrovskiy Works, River Lytva, central Urals, former USSR.

F. HAPSIPHYLLIDAE Grabau, 1928 D. (EMS)–P. (KAZ) Mar.

First: *Adradosia barroisi* Birenheide and Soto, 1977, Lower Devonian, Spain.
Last: *Duplophyllum zaphrentoides* Koker, 1924, upper Kazanian, Basleo, Timor.

F. ZAPHRENTOIDIDAE Schindewolf, 1938 D. (FAM)–P. (KAZ) Mar.

First: *Zaphrentoides* (?) *ecavatus* Hill, 1954, *Zaphrentoides* (?) sp. a Yang *et al.*, 1983 and *Amplexizaphrentis adyrensis* (Soshkina, 1960), upper Famennian, various localities (Sorauf and Pedder, 1986).
Last: *Basleophyllum indicum* (Koker, 1924), Upper Permian, Basleo, Timor.

F. POLYCOELIIDAE de Fromentel, 1861 D. (GED?)–P. (TAT) Mar.

First: *Amandaraia prima* Lavrusevich, 1968, Kunzhak Horizon, Lower Devonian (?or Upper Silurian), Shishkat ravine, north slope of Zeravshan Range, Tajikistan, former USSR.
Last: *Calophyllum donatianum* (King, 1848), Magnesian Limestone, lower Tatarian, Humbleton Hill, County Durham, UK; *Gerthia angusta* (Rothpletz, 1892), Upper Permian, Ajer Mati River, near Kupang, Timor; *Groenlandophyllum teicherti* (Flügel, 1973), *Productus* Limestone, Kap Stosch, eastern Greenland; *Hexalasma*, Upper Permian, Timor, NE of former USSR; *Tetralasma*, Upper Permian, Greenland; *Prosmilia cyathophylloides* (Gerth, 1921), Upper Permian, Basleo, Timor.

F. ANISOPHYLLIDAE Ivanovskiy, 1965 S. (WEN–LUD) Mar.

First and Last: *Anisophyllum agassizi* Milne Edwards and Haime, 1850, 30–45 feet (*c.* 11–16 m) above Dixon–Brownsport contact, Brownsport Formation, Blue Mount Glade, Perryville Quadrangle, Tennessee, USA.

F. PLEROPHYLLIDAE Koker, 1924 D. (EMS)–P. (TAT) Mar.

First: *Ufimia*, upper Emsian, Germany.
Last: *Plerophyllum radiciforme* Gerth, 1921 and *Pleramplexus similis* Schindewolf, 1940, lower Tatarian, Basleo and Oilmasi, Timor and *Barbarella stellaforma* (Flügel, 1972), Lower Jamul Formation, Kuk-e-Bagh-e-Vang, eastern Iran.

F. ENDOTHECIIDAE Schindewolf, 1942 P. (KAZ) Mar.

First and Last: *Endothecium apertum* Koker, 1924, upper Kazanian, Basleo, Timor.

F. ADAMANOPHYLLIDAE Vasilyuk, 1959
C. (TOU–SPK) Mar.

First: *Tachyphyllum artyshtense* Dobrolyubova, *in* Soshkina *et al.*, 1962, lower Terei Horizon, River Artyshta, Kuzbas, former USSR.
Last: *Adamanophyllum incertum* Vasilyuk, 1959, Namurian, right bank of River Berestovayya, opposite Obilnoe, Donbas, Ukraine, former USSR.

F. PENTAPHYLLIDAE Schindewolf, 1942
D. (SIG)–P. (TAT) Mar.

First: *Oligophyllum quinqueseptatum* Pocta, 1902, Dvorce Limestone, Lower Devonian, Dvorce, Czechoslovakia.
Last: *Pentamplexus*, *Palaeofusulina* Zone, upper Tatarian, Iran (Flügel, 1970).

F. LOPHOPHYLLIDAE Grabau, 1928
D. (FAM)–P. (TAT) Mar.

First: *Lophophyllum caninoides* Gorskiy, 1935, upper Famennian, Stroganova Gulf, southern island of Novaya Zemlya, former USSR (Sorauf and Pedder, 1986).
Last: *Lophophyllidium*, *Codonofusiella* Zone, lower Tartarian, North America (Flügel, 1970).

F. TIMORPHYLLIDAE Soshkina, 1941 P. (KAZ) Mar.

First and Last: *Timorphyllum wanneri* Gerth, 1921, upper Kazanian, Timor.

F. VERBEEKIELLIDAE Schouppe and Stacul, 1955
C. (CHD)–P. (KAZ) Mar.

First: *Cravenia rhytoides* Hudson, 1928, lower Viséan, Haw Crag Lower Quarry, Bell Busk, Yorkshire, England, UK.
Last: *Verbeekiella cristata* (Gerth, 1921), upper Kazanian, Basleo, Timor.

F. CYATHOPSIDAE Dybowski, 1873
D. (FAM)–P. (ART) Mar.

First: *Caninia cornucopiae* Michelin, 1840, Upper Devonian, no locality given (Sorauf and Pedder, 1986).
Last: *Paracaninia sinensis* Chi, 1937, upper Wumaling Series, Yungsin District, Wumahuitou, Kiangsi, China, *Arctophyllum intermedium* (Toula, 1875), lower Artinskian (? or Upper Carboniferous), Bellsund, Spitsbergen and *Fomichevella*, ?lower Artinskian, Urals, former USSR.

F. BOTHROPHYLLIDAE Fomichev, 1953
D. (FAM)–P. (SAK) Mar.

First: *Bothrophyllum* sp. a, *Bothrophyllum* sp. b and *Caninophyllum* sp. a, all Sorauf unpubl., Box Member, Percha Formation, upper Famennian, SW New Mexico, USA and *Caninophyllum adapertum* (Onoprienko, 1979), lower Elergetkehyn Suite, upper Famennian, Omolon Massif, NE USSR (Sorauf and Pedder, 1986).
Last: *Caninophyllum*, Lower Permian, Spitsbergen; Urals, former USSR; Carnic Alps and *Hornsundia lateseptata* Fedorowski, 1965, 5th coral limestone Horizon, upper Treskelodden Beds, Hornsund, Vestspitsbergen.

F. URALINIIDAE Dobrolyubova, 1962
D. (FAM)–C. (VIS) Mar.

First: *Uralinia megacystosa* Gorskiy, 1935 and *Enygmophyllum dubium* Gorskiy, 1935, upper Famennian, Stroganova Gulf, southern island of Novaya Zemlya, former USSR and *Cystophentis*, Arshakiachpyursky Horizon, upper Famennian, Southern Transcaucasus, former USSR (Sorauf and Pedder, 1986).
Last: *Bifossularia ussowi* (Gabuniya, 1919), Podyakov Horizon, upper Viséan, River Tom, Kuzbas, former USSR, *Keyserlingophyllum*, Viséan, Iran, Kuzbas, ?France and *Liardiphyllum hagei* Sutherland, 1954, middle Mississippian, Liard Range, North-west Territories, Canada.

F. ENDAMPLEXIDAE Schouppe and Stacul, 1959
P. (KAZ) Mar.

First and Last: *Endamplexus dentatus* Koker, 1924 and *Spineria diplochone* (Koker, 1924), upper Kazanian, Basleo, Timor.

F. AULOPHYLLIDAE Dybowski, 1873
D. (FAM)–P. (ROT) Mar.

First: Various genera from upper Famennian, various localities (Sorauf and Pedder, 1986).
Last: *Heintzella multiseptata* Fedorowski, 1967, Lower Permian, Treskelodden, Vestspitsbergen.

F. EKVASOPHYLLIDAE Hill, 1981
D. (FAM)–C. (VIS) Mar.

First: *Zaphriphyllum* sp. a, Sorauf unpubl., Box Member, Percha Formation, upper Famennian, SW New Mexico, USA (Sorauf and Pedder, 1986).
Last: *Ekvasophyllum inclinatum* Parks, 1951, *Faberophyllum occultum* Parks, 1951 and *Turbophyllum multiconum* Parks, 1951, all from Brazer Limestone, Meramecian, lower upper Mississippian, Wasatch Mountains, Utah, USA.

F. PALAEOSMILIIDAE Hill, 1940
D. (FAM)–C. (SPK) Mar.

First: *Palaeosmilia*, upper Famennian, Belgium, Germany, Western Australia.
Last: *Palaeosmilia*, ?Namurian, Europe, Asia, North Africa, *Palastraea*, Namurian, Kirghiz, former USSR.

F. ?APHROPHYLLIDAE Hill, 1973 C. (TOU–VIS) Mar.

First: *Naoides rangariensis* Pickett, 1967, Rangari Limestone, upper Tournaisian, Parish Rangari, County Nandewar, New South Wales, Australia.
Last: Various genera, e.g. *Aphrophyllum*, *Aphrophylloides*, *Coenaphrodia*, *Merlewoodia*, *Nothaphrophyllum*, *Symplectophyllum*, all from Viséan, New South Wales and Queensland, Australia.

F. LITHOSTROTIONIDAE D'Orbigny, 1852
D. (FAM)–P. (TAT) Mar.

First: *Stelechophyllum alferovi* (Gorskiy, 1935), *Stelechophyllum nalivkini* (Gorskiy, 1935) and *Stelechophyllum plativesiculosum* (Gorskiy, 1935), all from upper Famennian, Stroganova Gulf, southern island of Novaya Zemlya, former USSR (Sorauf and Pedder, 1986).
Last: *Yatsengia*, lower Tatarian, ?locality.

F. DURHAMINIDAE Minato and Kato, 1965
C. (TOU?)–P. (KAZ) Mar.

First: *Protolonsdaleiastraea atbassarica* Gorskiy, 1932, ?Tournaisian, Atbassarsky region, River Dzhezky, Kirghiz Steppe, Western Kazakhstan, former USSR.
Last: *Tanbaella izuruhense* (Sakaguchi and Yamagiwa, 1958), *Neoschwagerina* Zone, lower Kazanian, Osaka Prefecture, Takatsuki City, Izuruha-Shimojo, Japan.

F. AXOPHYLLIDAE Milne Edwards and Haime, 1851 D. (FAM)–C. (SPK) Mar.

First: *Axophylum* sp. a, Sorauf unpubl., Box Member, Percha Formation, upper Famennian, SW New Mexico, USA (Sorauf and Pedder, 1986).
Last: *Lonsdaleia*, II Member, Tagnana Formation, upper Serpukhovian, Bechar Basin, NW Sahara, Algeria (Semenoff-Tian-Chansky and Sutherland, 1982).

F. PETALAXIDAE Fomichev, 1953 C. (VIS)–P. (ART) Mar.

First: *Paralithostrotion podboriense* (Dobrolyubova, 1958), upper Viséan, Lyobytin district, Podbore, NW Russian Platform, former USSR.
Last: *Lytvophyllum tschernowi* (Soshkina, *in* Soshkina *et al.*, 1941), lower Artinskian, River Lytva, southern Urals, former USSR and ?*Lithostrotionella unica* (Yabe and Hayasaka, 1915), Lower Permian, Hui-tso-hsien, Kungshan, Yunnan, China.

F. GEYEROPHYLLIDAE Minato, 1955 C. (SPK)–P. (ART) Mar.

First: *Amygdalophylloides*, Namurian, Japan, Moscow Basin, Spain, Vietnam, China and *Darwasophyllum*, Namurian, Japan.
Last: *Amygdalophylloides*, Lower Permian, Austria, *Carinthiaphyllum kahleri* Heritsch, 1936, Lower *Pseudoschwagerina* Limestone, Lower Permian, Carnic Alps and *Lonsdaleoides boswelli* Heritsch, 1936, Lower *Schwagerina* Limestone, Lower Permian, Zollner See, Carnic Alps.

F. KONINCKOCARINIIDAE Dobrolyubova, 1962 C. (BSH–MOS) Mar.

First: *Koninckocarinia*, lower Bashkirian, Iowa, USA.
Last: *Koninckocarinia flexuosa* (Dobrolyubova, 1937), Podolsk Horizon, upper Moscovian, Shchurovo, Moscow Basin, former USSR.

F. WAAGENOPHYLLIDAE Wang, 1950 C. (MOS)–P. (TAT) Mar.

First: *Chielasma yui* (Chi, 1931), Laokanchai Limestone, Weiningian, Lipohsien, north of Piaochai, Kweichow, southern China and *Huangia chutsingensis* Chi, 1931, White Limestone, Weiningian, Chutsinghsien, Tungshan, Yunnan, southern China.
Last: Various genera, e.g. *Waagenophyllum*, *Ipciphyllum*, *Paraipciphyllum*, *Pavastehphyllum*, *Pseudohuangia*, *Wentzelella*, *Wentzelloides*, etc. from the lower Tatarian of Pakistan, Szechwan, Shensi, southern China, Japan, Vietnam, Cambodia, Iran, etc.

F. PSEUDOPAVONIDAE Yabe *et al.*, 1943 C. (SPK)–?P. (ROT) Mar.

First: Various genera, e.g. *Amygdalophyllidium*, *Hiroshimaphyllum*, *Omniphyllum*, *Ozakiphyllum*, all from the Namurian, Japan.
Last: *Pseudopavona*, Lower Permian, Japan.

Order ?HETEROCORALLIA Schindewolf, 1941

F. HETEROPHYLLIIDAE Dybowski, 1873 D. (EIF)–C. (BSH) Mar.

First: *Pseudopetraia devonica* Soshkina, 1951, Middle Devonian, near Pokrovsk Egorshin, eastern slopes of central Urals, former USSR.
Last: *Hexaphyllia*, upper Namurian, Europe, Asia, North America.

Order COTHONIIDA Oliver and Coates, 1987

F. ?COTHONIIDAE Jell and Jell, 1976 Є. (SOL) Mar.

First and Last: *Cothonion sympomatum* Jell and Jell, 1976, lower Middle Cambrian, New South Wales, Australia.

Order UNCERTAIN

F. ?NUMIDIAPHYLLIDAE Flugel, 1976 P. (UFI/KAZ) Mar.

First and Last: *Numidiaphyllum gillianum* Flügel, 1976, Upper Permian, Djebel Tebaga, Tunisia.

F. DIFFINGIIDAE Fedorowski, 1985 P. (ASS–SAK) Mar.

First: *Diffingia valida* Fedorowski, 1985, Neal Ranch Formation, lower Wolfcampian, western Texas, USA.
Last: Various species of *Diffingia* and *Turgidiffia*, Skinner Ranch Formation and Hess Formation, upper Wolfcampian, western Texas, USA.

F. PLERODIFFIIDAE Fedorowski, 1985 P. (SAK) Mar.

First and Last: *Plerodiffia eaglebuttensis* Fedorowski, 1985, Hueco Formation, upper Wolfcampian, western Texas, USA.

Subclass UNCERTAIN

F. TABULACONIDAE Debrenne *et al.*, 1987 Є. (LEN–SOL) Mar.

First: *Tabulaconus kordeae* Handfield, 1969, Sekwi Formation, Mackenzie Mountains, District of Mackenzie, North-west Territories, Canada.
Last: *Tabulaconus kordeae* Handfield, 1969, Atan Group, Cassiar Mountains, northern British Columbia, Canada.

Subphylum CTENOPHORA Eschscholtz, 1829

F. UNCERTAIN Є. (StD) Mar.

Collins *et al.* (1983) compare this taxon to ctenophores, but full description is needed (Briggs and Conway Morris, 1986).
First: *Fasciculus vesanus* Simonetta and Delle Cave, 1978, Stephen Formation, locality 9, Mount Stephen, British Columbia, Canada.
Last: *Fasciculus vesanus* Simonetta and Delle Cave, 1978, Burgess Shale, Stephen Formation, Mount Field, British Columbia, Canada.

F. UNCERTAIN D. (EMS) Mar.

First and Last: *Archeocydippida hunsreckiana* Stanley and Sturmer, 1987, Hunsruckschiefer, lower Emsian, Germany.

F. UNCERTAIN D. (EMS) Mar.

First and Last: *Paleoctenophora brasseli* Stanley and Sturmer, 1983, Hunsruckschiefer, lower Emsian, Germany.

REFERENCES

Bayer, F. M. (1979) The correct name of the helioporacean octocoral *Lithotelesto micropora* Bayer and Muzik. *Proceedings of the Biological Society of Washington*, **92**, 873–5.

Bayer, F. M. (1981) On some genera of stoloniferous octocorals

(Coelenterata: Anthozoa), with descriptions of new taxa. *Proceedings of the Biological Society of Washington*, **94**, 878–901.

Bayer, F. M., Boschma, H., Harrington, H. J. *et al.* (1956) *Treatise on Invertebrate Paleontology, Pt. F: Coelenterata*. Geological Society of America and University of Kansas Press, Boulder, Colorado, and Lawrence, Kansas, xx + 498 pp.

Beauvais, L. (1986) Monographie des Madreporaires du Jurassique inférieur du Maroc. *Palaeontographica, Abteilung A*, **194**, 1–68.

Bengtson, S. (1981) *Atractosella*, a Silurian alcyonacean octocoral. *Journal of Paleontology*, **55**, 281–94.

Bischoff, G. C. O. (1978a) *Septodaeum siluricum*, a representative of a new subclass Septodaearia of the Anthozoa, with partial preservation of soft parts. *Senckenbergiana Lethaea*, **59**, 229–73.

Bischoff, G. C. O. (1978b) Internal structures of conulariid tests and their functional significance, with special reference to Circonulariina n. suborder (Cnidaria, Scyphozoa). *Senckenbergiana Lethaea*, **59**, 275–327.

Briggs, D. E. G. and Conway Morris, S. (1986) Problematica from the Middle Cambrian Burgess Shale of British Columbia, in *Problematic Fossil Taxa* (eds A. Hoffman and M. H. Nitecki), Oxford University Press, Oxford, pp. 167–84.

Chen, P. (1984) [Discovery of Lower Cambrian small shelly fossils from Jijiapo, Yichang, West Hubei and its significance.] *Professional Papers of Stratigraphy and Palaeontology (Geological Publishing House, Beijing)*, **13**, 49–64 (in Chinese).

Chevalier, J.-P. and Beauvais, L. (1987) Order des Scleractiniaires. XI. – Sytématique, in *Cnidaires, Anthozoaires (Traité de Zoologie: Anatomie, Systematique, Biologie Vol. 3) Fascicule 3* (ed. P.-P. Grassé), Masson, Paris, pp. 679–753.

Collins, D., Briggs, D. E. G. and Conway Morris, S. (1983) New Burgess Shale fossil sites reveal Middle Cambrian faunal complex. *Science*, **222**, 163–7.

Conti, S. and Serpagli, E. (1984) A new interpretation of the anthozoan *Septodaeum* Bischoff 1978 as a bryozoan. *Bollettino della Societá paleontologica italiana*, **23**, 3–20.

Cuif, J.-P. (1975) Caractères morphologiques, microstructuraux et systématiques des *Pachythecalidae*, nouvelle famille de madréporaires Triasiques. *Geobios*, **8**, 157–80.

Debrenne, F. M., Gangloff, R. A. and Lafuste, J. G. (1987) *Tabulaconus* Handfield: microstructure and its implications in the taxonomy of primitive corals. *Journal of Paleontology*, **61**, 1–9.

Deng, Z. (1982) Mesozoic Milleporina and Tabulatomorphic corals from Xizang, in *Zhongguo Kexueyuan Qinzang Gaoyan Zonghe Kexue Koacha Dui. [The Series of the Comprehensive Scientific Expedition to the Qinghai–Xizang Plateau.] Xizang gushengwu [Palaeontology of Xizang]*. Beijing Scientific Press, pp. 184–8.

Eliasova, H. (1976) Nova celed' z podradu Amphiastraeina Alloiteau, 1952 (Hexacoralla, tithon CSSR). *Casopsis pro Mineralogii a Geologii*, **21**, xxx–xxx.

Fedonkin, M. A. (1980) New Precambrian Coelenterata in the north of the Russian Platform. *Paleontological Journal*, **14** (2), 7–15.

Fedonkin, M. A. (1983) Organicheskii mir Venda, in *Stratigrafiya. Paleontologiya. Itogi naukii techniki viniti an SSSR*, Vol. 12, Moscow, 128 pp.

Fedonkin, M. A. (1985) Sistematicheskoe opisanie Vendskikh Metazoa, in *Paleontologiya* (*Vendskaya Sistema* Vol. 1 (eds B. S. Sokolov and A. B. Ivanoskii), Moscow, pp. 70–106.

Fedonkin, M. A. (1990) Precambrian metazoans, in *Palaeobiology: a Synthesis* (eds D. E. G. Briggs and P. R. Crowther), Blackwell, Oxford, pp. 17–24.

Fedorowski, J. (1965) Diffingiina, a new suborder of the rugose corals from SW Texas. *Acta Palaeontologica Polonica*, **30**, 209–40.

Floris, S. (1979) Maastrichtian and Danian corals from Denmark, in *Cretaceous–Tertiary Boundary Events Symposium. 1. The Maastrichtian and Danian of Denmark* (eds T. Birkelund and R. G. Bromley), University of Copenhagen, Copenhagen, pp. 92–4.

Flügel, H. W. (1970) Die Entwicklung der rugosen Korallen im hohen Perm. *Verhandlungen der Geologischen Bundesanstalt*, **1**, 146–61.

Giammona, C. P. and Stanton, R. J. (1980) Octocorals from the Middle Eocene Stone City Formation, Texas. *Journal of Paleontology*, **54**, 71–80.

Glaessner, M. F. (1979) Precambrian, in *Treatise on Invertebrate Paleontology, Pt. A: Introduction*. Geological Society of America and University of Kansas Press, Boulder, Colorado and Lawrence, Kansas, pp. A79–A118.

Hahn, G., Hahn, R., Leonardos, O. H. *et al.* (1982) Körperlich erhaltene Scyphozoen-Reste aus dem Jungpräkambrium Brasiliens. *Geologica et Palaeontologica*, **16**, 1–18.

Handfield, R. C. (1969) Early Cambrian coral-like fossils from the northern Cordillera of western Canada. *Canadian Journal of Earth Sciences*, **6**, 782–5.

Harland, W. B., Holland, C. H., House, M. R. (1967) *The Fossil Record*. Geological Society of London, London, xi + 827 pp.

He, T.-G. (1984) [Discovery of *Lapworthella bella* assemblage from Lower Cambrian Meishucun Stage in Niuniuzhai, Leibo County, Sichuan Province.] *Professional Papers of Stratigraphy and Palaeontology (Geological Publishing House, Beijing)*, **13**, 23–34 (in Chinese).

Hill, D. (1981) *Treatise on Invertebrate Paleontology, Pt. F: Coelenterata: Supplement 1, Rugosa and Tabulata* (Vols 1 and 2). Geological Society of America and University of Kansas Press, Boulder, Colorado, and Lawrence, Kansas, x + 762 pp.

Il'ina, T. G. (1983) On the origin of the Scleractinia. *Paleontological Journal*, **1983** (1), 10–23.

Jenkins, R. J. F. (1985) The enigmatic Ediacaran (late Precambrian) genus *Rangea* and related forms. *Paleobiology*, **11**, 336–55.

Johnson, R. G. and Richardson, E. S. (1968) The Essex fauna and medusae. *Fieldiana: Geology*, **12**, 109–15.

Kobluk, D. R. (1984) A new compound skeletal organism from the Rosella Formation (Lower Cambrian), Atan Group, Cassiar Mountains, British Columbia. *Journal of Paleontology*, **58**, 703–8.

Korde, K. B. (1963) Hydroconozoa – novyi klass kishechnopolostnykh zhivotnykh. *Palaeontologicheskii Zhurnal*, **2**, 20–5.

Kuz'micheva, Ye. I. (1975) Systematic composition and development of the family Helioporidae (Octocoralla). *Paleontological Journal*, **1975** (3), 278–86.

Kuz'micheva, Ye. I. (1980) Fossil Gorgonida. *Paleontological Journal*, **14** (4), 3–12.

Laub, R. S. (1984) *Lichenaria* Winchell and Schuchert, 1895, *Lamottia* Raymond, 1924 and the early history of the tabulate corals. *Palaeontographica Americana*, **54**, 159–63.

Malecki, J. (1982) Bases of Upper Cretaceous octocorals from Poland. *Acta Palaeontologica Polonica*, **27**, 65–75.

Malinowska, L. (Editor). (1986) *Geology of Poland*, Vol. 3 *Atlas of Guide and Characteristic Fossils. Part 2a. Mesozoic, Triassic*, Wydawnictwa Geologiczne, Warsaw, 253 pp.

Mariotti, N., Nicosia, G., Pallini, G. *et al.* (1979) Coralli ed Ammoniti nel Bajociano del Sasso di Pale (Umbria). *Geologica Roma*, **18**, 225–51.

McLean, R. A. and Pedder, A. E. H. (1984) Frasnian rugose corals of Western Canada. *Palaeontographica, Abteilung A*, **185**, 1–38.

Mel'nikova, G. K. (1983) New Upper Triassic Scleractinia from the Pamir region. *Paleontological Journal*, **17**, 41–9.

Mel'nikova, G. K. and Roniewicz, E. (1976) Contribution to the systematics and phylogeny of Amphiastraeina (Scleractinia). *Acta Palaeontologica Polonica*, **21**, 97–114.

Mierzejewski, P. (1986) Ultrastructure, taxonomy and affinities of some Ordovician and Silurian organic microfossils. *Palaeontologia Polonica*, **47**, 129–220.

Montanaro Gallitelli, E. (1975) Hexanthiniaria – a new order of Zoantharia (Anthozoa, Coelenterata). *Bollettino della Societá Paleontologica Italiana*, **14**, 21–5.

Narbonne, G. M., Myrow, P., Landing, E. *et al.* (1991) A chondrophorine (medusoid hydrozoan) from the basal Cambrian (Placentian) of Newfoundland. *Journal of Paleontology*, **65**, 186–91.

Negus, P. E. (1983) Distribution of the British Jurassic corals. *Proceedings of the Geologists' Association*, **94**, 251–7.

Oliver, W. A. (1984) *Conchopeltis*: its affinities and significance. *Palaeontographica Americana*, **54**, 141–7.

Oliver, W. A. and Pedder, A. E. H. (1979) Rugose corals in Devonian stratigraphical correlation. *Special Papers in Palaeontology*, **23**, 233–48.

Orlov, Yu. A. (1958–1964) *Osnovy Paleontologii*. Moscow.

Ossian, C. R. (1973) New Pennsylvanian scyphomedusan from western Iowa. *Journal of Paleontology*, **47**, 990–5.

Pflug, H. D. (1970–1973) Zur Fauna der Nama-Schichten in Sudwest-Afrika. Parts I–IV. *Palaeontographica, Abteilung A*, **134**, 153–262 [1970a]; **135**, 198–231 [1970b]; **139**, 134–70 [1972]; **144**, 166–202 [1973].

Reyeros de Castillo, M. M. (1983) Algunas Formaciones cretacicas del Estada de Oaxara. *Paleontologia Mexicana*, **47**, 1–67.

Roniewicz, E. and Morycowa, E. (1989) Triassic Scleractinia and the Triassic/Liassic boundary. *Memoirs of the Association of Australasian Palaeontologists*, **8**, 347–54.

Ruedemann, R. (1947) Graptolites of North America. *Geological Society of America Memoir*, **19**, 1–652.

Schram, F. R. and Nitecki, M. H. (1975) Hydra from the Illinois Pennsylvanian. *Journal of Paleontology*, **49**, 549–51.

Scrutton, C. T. (1979) Early fossil cnidarians, in *The origin of Major Invertebrate Groups* (ed. M. R. House), Systematics Association Special Volume, 12. Academic Press, London and New York, pp. 161–207.

Scrutton, C. T. (1984) Origin and early evolution of tabulate corals. *Palaeontographica Americana*, **54**, 110–18.

Scrutton, C. T. and Clarkson, E. (1991) A new scleractinian-like coral from the Ordovician of the Southern Uplands, Scotland. *Palaeontology*, **34**, 179–94.

Semenoff-Tian-Chansky, P. and Sutherland, P. K. (1982) Coral distributions near the Middle Carboniferous boundary, in *Biostratigraphic Data for a Mid-Carboniferous Boundary*. (eds W. H. C. Ramsbottom, W. B. Saunders and B. Owens), IUGS Subcommission on Carboniferous Stratigraphy, 11, Leeds, pp. 134–44.

Sepkoski, J. J. Jr (1982) A compendium of marine fossil families. *Milwaukee Public Museum Contributions in Biology and Geology*, **51**, 1–125.

Sepkoski, J. J., Jr and Kasting, J. F. (in press) Models for Vendian–Cambrian diversity and for Proterozoic atmospheric and ocean chemistry, in *The Proterozoic Biosphere: a Multidisciplinary Study*, Vol. 2 (eds. J. W. Schopf and C. Klein), Cambridge University Press, Cambridge.

Simonetta, A. and Delle Cave, L. (1978) Notes on new and strange Burgess Shale fossils (Middle Cambrian of British Columbia). *Memorie della Societá toscona di Scienze naturali Serie A*, **85**, 45–9.

Sorauf, J. E. and Pedder, A. E. H. (1986) Late Devonian rugose corals and the Frasnian–Famennian crisis. *Canadian Journal of Earth Sciences* **23**, 1265–87.

Stanley, G. D., Jr and Kanie, Y. (1985) The first Mesozoic chondrophorine (medusoid hydrozoan), from the Lower Cretaceous of Japan. *Palaeontology*, **28**, 101–9.

Stanley, G. D. and Sturmer, W. (1983) The first fossil ctenophore from the Lower Devonian of West Germany. *Nature*, **303**, 518–20.

Stanley, G. D. and Sturmer, W. (1987) A new fossil ctenophore discovered by x-rays. *Nature*, **328**, 61–3.

Stasinska, A. (1967) Tabulata from Norway, Sweden and from the erratic boulders of Poland. *Palaeontologica Polonica*, **18**, 1–112.

Sun, W.-G. (1986) Late Precambrian scyphozoan medusa *Mawsonites randellensis* sp. nov. and its significance in the Ediacara metazoan assemblage, South Australia. *Alcheringa*, **10**, 169–81.

Turnsek, D. (1972) Upper Jurassic corals of southern Slovenia. *Slovenska Akademija Znanosti in Umetnosti, Razprave*, **15** (6), 1–121.

Turnsek, D. (1989) Diversifications of corals and coral reef associations in Mesozoic palaeogeographic units of northwestern Yugoslavia. *Memoirs of the Association of Australasian Palaeontologists*, **8**, 283–9.

Turnsek, D. and Buser, S. (1974) The Lower Cretaceous corals, hydrozoans, and chaetetids of Banjska Planota and Trnovski Gozd. *Slovenska Akademija Znanosti in Umetnosti, Razprave*, **17** (2), 1–44.

Turnsek, D. and Mihajlovic, M. (1981) Lower Cretaceous cnidarians from eastern Siberia. *Slovenska Akademija Znanosti in Unetnosti, Razprave*, **23** (1), 1–54.

Voigt, E. (1973) *Hydrallmania graptolithiformis* n. sp., eine durch Biomuration erhaltene Sertulariidae (Hydroz.) aus der Maastrichter Tuffkreide. *Paläontologische Zeitschrift*, **47**, 25–31.

Von Rosendahl, S. (1985) Die oberjurassische Korallenfazies von Algarve (Sudportugal). *Arbeiten aus dem Institut für Geologie und Paläontologie an der Universität Stuttgart*, **82**, 1–125.

Willoughby, R. H. and Robison, R. A. (1979) Medusoids from the Middle Cambrian of Utah. *Journal of Paleontology*, **53**, 494–500.

Yochelson, E. L. (1984) North American Middle Ordovician *Scenella* and *Macroscenella* as possible chondrophorine coelenterates. *Palaeontographica Americana*, **54**, 148–53.

Yochelson, E. L. and Stanley, G. D., Jr (1981) An early Ordovician patelliform gastropod, *Palaelophacmaea*, reinterpreted as a coelenterate. *Lethaia*, **14**, 323–30.

Yochelson, E. L., Stumer, W. and Stanley, G. D., Jr (1983) *Plectodiscus discoideus* (Rauff): a redescription of a chondrophorine from the Early Devonian Hunsruck Slate, West Germany. *Paläontologische Zeitschrift*, **57**, 39–68.

7

MOLLUSCA: AMPHINEURA AND 'MONOPLACOPHORA'

M. J. Benton and D. H. Erwin

The Amphineura and 'Monoplacophora' are of uncertain phylogenetic position within the Mollusca. The database on Amphineura is based on Smith (1960, 1973), Van Belle (1981) and Smith and Hoare (1987), and updated according to cited references. The familial revisions in Smith and Hoare (1987), in particular, gave quite different range data from previously published accounts.

Untorted univalved molluscs with multiple muscle scars have traditionally been placed in the Class Monoplacophora. Peel (1991a,b) has reviewed the various arguments concerning the class, and concluded that the class is best rejected completely in favour of two classes, the Class Tergomya Horný, 1965 and Helcionelloida Peel, 1991. Unfortunately, the assignment of many problematic Cambrian univalves is unknown, as are the precise limitations of each class. Peel's recommendations as to membership are followed here.

The Amphineura and Aplacophora were compiled by MJB and the 'Monoplacophora' by DHE.

The stratigraphical nomenclature of Harland *et al.* (1990) is used throughout this chapter.

Acknowledgements – MJB thanks Richard Hoare (Bowling Green, Ohio), David Jablonski (Chicago) and Ben McHenry (South Australian Museum) for their extensive and invaluable help in updating the coverage of Amphineura and Aplacophora.

Class AMPHINEURA von Ihering, 1876

Subclass POLYPLACOPHORA de Blainville, 1816

Superorder PALAEOLORICATA Bergenhayn, 1955

The Yangtzechitonidae, noted by W. Yu (1984) as possible palaeoloricate chitons, have been assigned to the Paracarinachitidae, an unassigned family of Problematica (see Chapter 28).

Order CHELODIDA Bergenhayn, 1943

F. CHELODIDAE Bergenhayn, 1943
O. (CRD)–D. (EIF/GIV) Mar.

First: *Eochelodes bergenhayni* Marek, 1962, lower part of Cernin Beds, central Bohemia, Czechoslovakia.
Last: *Probolaeum corrugatus* Sandberger and Sandberger, 1856, Stringocephalen Kalk, Hesse-Nassau, Germany.
Intervening: D. (l.).
Comment: The Chelodidae, according to Smith and Hoare (1987), excludes the numerous species of *Chelodes*, which are mainly placed in Mattheviidae.

F. MATTHEVIIDAE Walcott, 1886
€. (DOL)–S. (LUD/PRD) Mar.

First: *Matthevia variabilis* Walcott, 1885, Hoyt Limestone Member, Theresa Formation, ?upper Franconian and lower Trempealeauan, Saratoga Springs, New York, USA; *Matthevia walcotti* Runnegar *et al.*, 1979, Black Earth Member, St Lawrence Formation, lower Trempealeauan, Eikey Quarry, Wisconsin, USA (Runnegar *et al.*, 1979).
Last: *Chelodes bohemicus* (Barrande, 1867), Upper Silurian, Bohemia, Czechoslovakia, and *C. calceoloides* Etheridge, 1897, Series of Yass, Upper Silurian, King County, New South Wales, Australia.
Intervening: TRE-LLO, WEN.
Comment: A younger record may be *Chelodes? sarthacensis* (Oehlert, 1881) from the Lower Devonian of Sarthe, France, but its affinities are not clear (Smith and Hoare, 1987, p. 48).

F. GOTLANDOCHITONIDAE Bergenhayn, 1955
O. (TRE/ARG)–S. (WEN) Mar.

First: *Gotlandochiton hami* Smith, *in* Smith and Toomey, 1964, *Kindbladochiton arbucklensis* (Smith, *in* Smith and Toomey, 1964), and *Paleochiton kindbladensis* Smith, *in* Smith and Toomey, 1964, all Kindblade Formation, Arbuckle Mountains, Oklahoma, USA.
Last: *Gotlandochiton interplicatus* Bergenhayn, 1955, *G. birhomibalvis* Bergenhayn, 1955, *G. laterodepressus* Bergenhayn, 1955, and *G. troedssoni* Bergenhayn, 1955, Gotlandian, Gotland, Sweden.
Intervening: none.

F. PREACANTHOCHITONIDAE Bergenhayn, 1960
€. (DOL)–O. (TRE) Mar.

First: *Preacanthochiton cooperi* Bergenhayn, 1960 and *P. productus* Bergenhayn, 1960, Eminence Formation, Trempealeauan, Shannon County, Missouri, USA (Bergenhayn, 1960).
Last: *Preacanthochiton cooperi* Bergenhayn, 1960 and *P. depressus* Bergenhayn, 1960, Gasconade Formation, Madison

The Fossil Record 2. Edited by M. J. Benton. Published in 1993 by Chapman & Hall, London. ISBN 0 412 39380 8

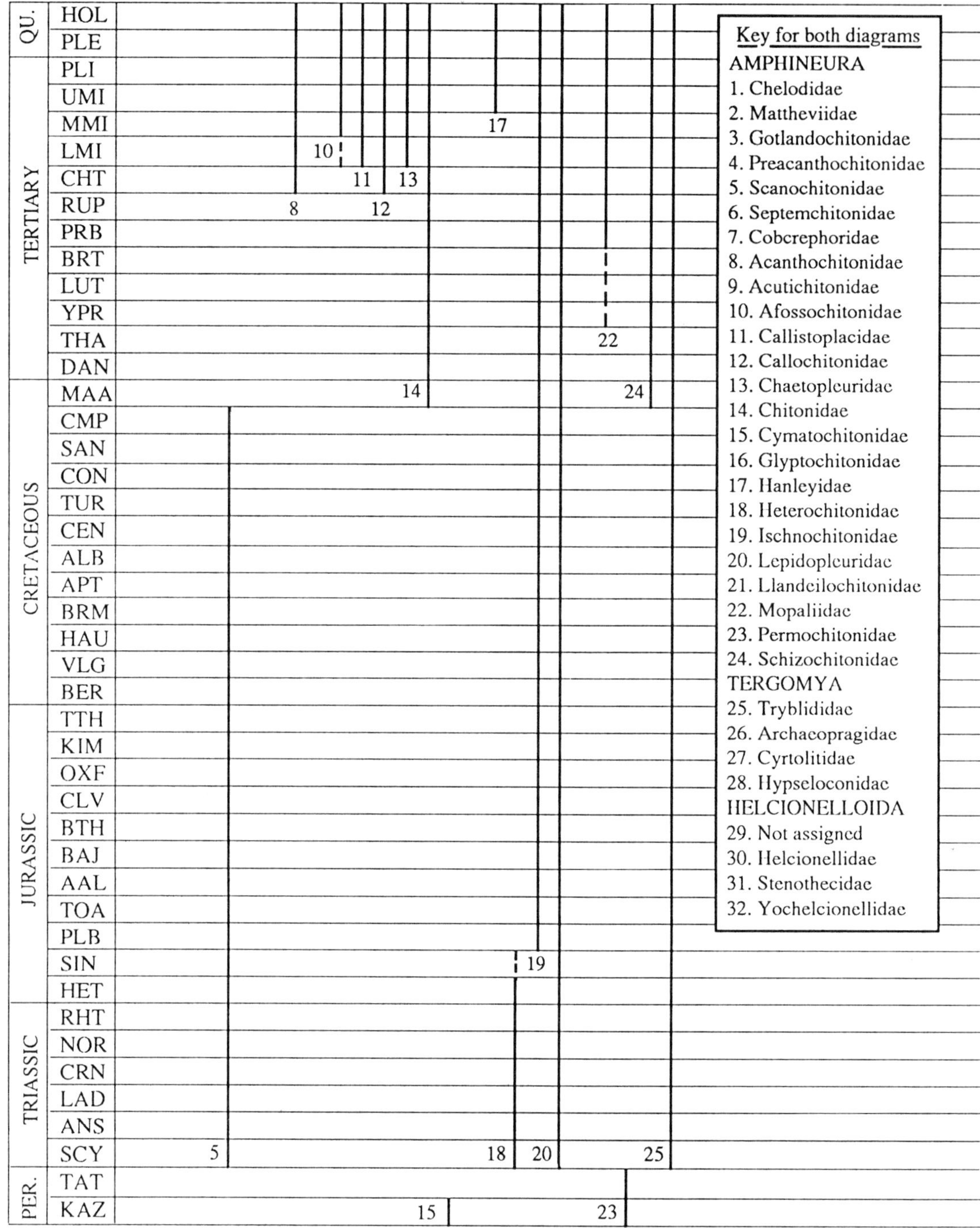

Fig. 7.1

County and Reynolds County respectively, Missouri, USA (Bergenhayn, 1960).

F. SCANOCHITONIDAE Bergenhayn, 1943
O. (TRE/ARG)–K. (CMP) Mar.

First: *Ivoechiton calathicolus* Smith, *in* Smith and Toomey, 1964 and *I. oklahomensis* Smith, *in* Smith and Toomey, 1964, both Kindblade Formation, Arbuckle Mountains, Oklahoma, USA.
Last: *Scanochiton jugatus* Bergenhayn, 1943, *Haeggochiton haeggi* Bergenhayn, 1943, *Ivoechiton levis* Bergenhayn, 1943, and *Olingechiton triangulatus* Bergenhayn, 1943, all *Mammillatus* Zone, Scania, Sweden.
Intervening: None.

F. SEPTEMCHITONIDAE Bergenhayn, 1955
O. (LLO/CRD–ASH) Mar.

First: *Solenocaris elongata* (Hadding, 1913), LLO/CRD, Aakirkeby, Sweden.
Last: *Septemchiton grayiae* (Woodward, 1885) and *S.? thraivensis* (Reed, 1911), both Starfish Bed, Drummuck Group, near Girvan, Ayrshire, Scotland, UK.
Intervening: CRD.

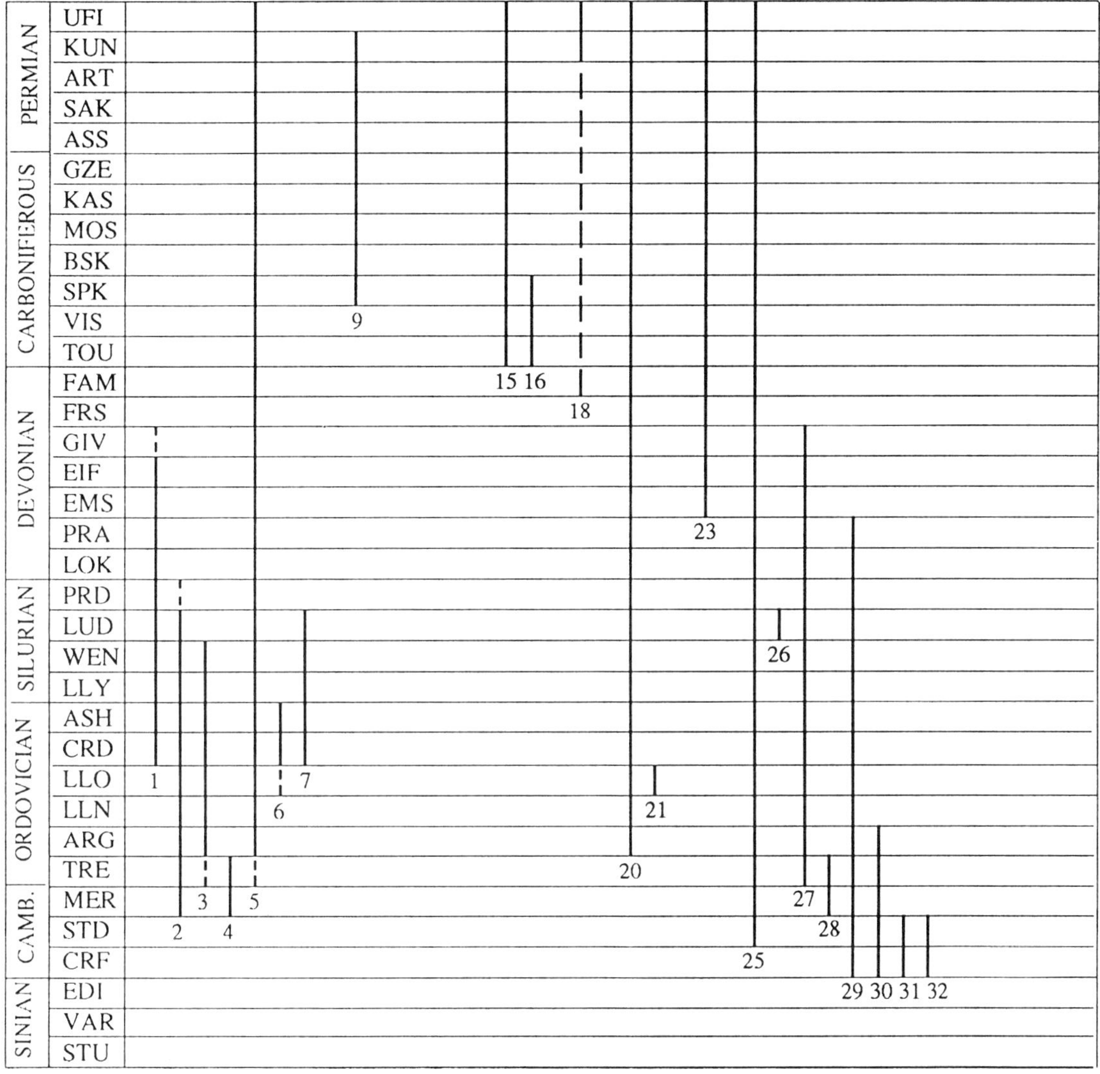

Fig. 7.1

Superorder PHOSPHATOLORICATA Bischoff, 1981

Order PHOSPHATOLORICATIDA Bischoff, 1981

F. COBCREPHORIDAE Bischoff, 1981
O. (CRD)–S. (LUD) Mar.

First: *Cobcrephora silurica* Bischoff, 1981, clasts in Cuga Burga Volcanics, Wellington area, New South Wales, Australia (Bischoff, 1981).
Last: *Cobcrephora silurica* Bischoff, 1981, Borenore Limestone, near Orange, and Kildrummie Formation, *crispus* Zone, LUD, Rockley, New South Wales, Australia (Bischoff, 1981).
Intervening: LLY, WEN.

Superorder NEOLORICATA Bergenhayn, 1955

Order LEPIDOPLEURIDA Thiele, 1910

F. ACANTHOCHITONIDAE Pilsbry, 1893
T. (CHT)–Rec. Mar.

First: *Acanthochiton (Notoplax) ashbyi* (Laws, 1932), Duntroonian, upper CHT, New Zealand (Beu and Maxwell, 1990). **Extant**

F. ACUTICHITONIDAE Hoare, Mapes, and Atwater, 1983 C. (SPK)–P. (KUN) Mar.

First: *Soleachiton soleaformis* (Etheridge, 1882) and *Acutichiton etheridgei* Smith and Hoare, 1987, both Main or Hurlet Limestone, Namurian A, Dalry, Ayrshire, Scotland, UK (Hoare *et al.*, 1983; Hoare and Smith, 1984), and *Elachychiton juxtaterminus* Hoare and Mapes, 1985, Imo Formation, Chesterian, Searcy County, Arkansas, USA.
Last: *Soleachiton yochelsoni* Hoare and Smith, 1984, Leonard and Road Canyon formations, upper Leonardian, West Texas, USA (Hoare and Smith, 1984).
Intervening: BSH–ART.

F. AFOSSOCHITONIDAE Ashby, 1925
T. (LMI/MMI)–Rec. Mar.

First and Last: *Afossochiton cudmorei* Ashby, 1925, Balcombian, Hamilton, Victoria, Australia. **Extant**
Comment: This family is synonymized with the Acanthochitonidae by Gowlett-Holmes (1987).

F. CALLISTOPLACIDAE Pilsbry, 1893
T. (LMI)–Rec. Mar.

First: *Callistochiton mafiaensis* Davis, 1954, Lower Miocene, Mafia Island, East Africa. **Extant**
Intervening: PLI, PLE.

F. CALLOCHITONIDAE Plate, 1899
T. (CHT)–Rec. Mar.

First: *Callochiton chattonensis* Ashby, 1929, Duntroonian, upper CHT, New Zealand (Beu and Maxwell, 1990). **Extant**

F. CHAETOPLEURIDAE Plate, 1899
T. (BUR)–Rec. Mar.

First: *Chaetopleura apiculata* (Say, 1834), Camp Roosevelt Shell Bed, Calvert Formation, upper Burdigalian, Plum Point, Maryland, USA. **Extant**

F. CHITONIDAE Rafinesque, 1815
K. (MAA)–Rec. Mar.

First: *Chiton (Chiton) cretaceus* Berry, 1940, Coon Creek Formation, lowermost MAA, Coon Creek, Tennessee, and Severn [formerly Monmouth] Formation, lower MAA, Brightseat, Maryland, USA; *Chiton (Chiton) berryi* Smith *et al.*, 1968, Ripley Formation, lower to middle MAA, Quitman County, Georgia, USA (Smith *et al.*, 1968). **Extant**
Comment: The ages of the Coon Creek and Ripley Formations have been revised to fall entirely within the MAA (Pojeta and Sohl, 1987), instead of being partially, or wholly, upper CMP as in Smith *et al.* (1968).

F. CHORIPLACIDAE Cotton and Weeding, 1939 **Extant**

F. CYMATOCHITONIDAE Sirenko and Starobogatov, 1977 C. (TOU)–P. (KAZ) Mar.

First: *Cymatochiton? kirkbyi* (de Koninck, 1883), Lower Scar Limestone, Settle, Yorkshire, England, UK.
Last: *Cymatochiton howseanus* (Kirkby, 1857) and *C. loftusianus* (King, 1850), upper Magnesian Limestone, Durham, England, UK.
Intervening: C. (u.), MOS, KUN, UFI.

F. GLYPTOCHITONIDAE Starobogatov and Sirenko, 1975 C. (TOU-SPK) Mar.

First: *Glyptochiton cordifer* (de Koninck, 1844), Tournai, Belgium.
Last: *Glyptochiton kirkbyanus* (Etheridge, 1882), *G. quadratus* (Etheridge, 1882), *G. subquadratus* (Kirkby and Young, 1867), and *G. youngianus* (Kirkby, *in* Young, 1865), all Main or Hurlet Limestone, Dalry, Ayrshire, Scotland, UK.
Intervening: None.

F. HANLEYIDAE T. (TOR)–Rec. Mar.

First: *Hanleya? multigranosa* (Reuss, 1860), Korytica Clays, lower Tortonian, Holy Cross Mountains, Poland and Bohemia, Czechoslovakia (Baluk, 1971). **Extant**

F. HETEROCHITONIDAE Van Belle, 1978
D. (FAM)/P. (KUN)–J. (HET/SIN) Mar.
(includes Ochmazochitonidae Hoare and Smith, 1984)

First: *Lobarochiton anomalus* (Rowley, 1908), Louisiana Limestone (FAM), Missouri, USA or *Ochmazochiton comptus* Hoare and Smith, 1984, Road Canyon Formation, upper Leonardian (KUN), West Texas, USA (Hoare and Smith, 1984).
Last: *Allochiton gemmellaroi* Fucini, 1912, Lias inferiore, Montagna di Casale, Sicily (Hoare and Smith, 1984).
Intervening: KUN.

F. ISCHNOCHITONIDAE Dall, 1889
J. (PLB)–Rec. Mar.

First: *Ischnochiton marloffsteinensis* Fiedel and Keupp, 1988, *gibbosus* Subzone, upper PLB, NE Erlangen, Franconia, Germany (Fiedel and Keupp, 1988). **Extant**

F. LEKISKOCHITONIDAE Smith and Hoare, 1987
P. (KUN–KAZ) Mar.

First: *Lekiskochiton fornicis* Hoare and Smith, 1984, Road Canyon Formation, uppermost Leonardian, Glass Mountains, West Texas, USA (Hoare and Smith, 1984).
Last: *Lekiskochiton? cordatus* (Kirkby, 1859), Upper Magnesian Limestone, Tunstall Hill, Durham, England, UK.
Intervening: None.

F. LEPIDOPLEURIDAE Pilsbry, 1892
O. (ARG)–Rec. Mar.

First: *Helminthochiton? aequivoca* Robson, 1913, Šárka and Malé Prilepy, Bohemia, Czechoslovakia. **Extant**
Intervening: LLY, D. (l., m.), TOU–KAS, SAK–KUN.

F. LLANDEILOCHITONIDAE Bergenhayn, 1955
O. (LLO) Mar.

First and Last: *Llandeilochiton ashbyi* Bergenhayn, 1955, Barr Group, Balclatchie, southern Scotland, UK.
Comment: The status of the single specimen is doubtful (Van Belle, 1981, p. 22; Smith and Hoare, 1987, p. 14).

F. MOPALIIDAE Dall, 1889 T. (Eoc.)–Rec. Mar.

First: *Plaxiphora (Plaxiphora) concentrica* Ashby and Torr, 1901, Eocene, Moorabool, Victoria, Australia. **Extant**

F. PERMOCHITONIDAE Sirenko and Starobogatov, 1977 D. (EMS)–P. Mar.

First: *Euleptochiton lebescontei* (Barrois, 1889), EMS, Loire Inférieure, France.
Last: *Permochiton australianus* Iredale and Hull, 1926, upper Marine Series, Permo-Carboniferous beds of Bundanoon, New South Wales, Australia (DeBrock *et al.*, 1984).
Intervening: SPK–KAS/GZE.

F. SCHIZOCHITONIDAE Dall, 1889
K. (MAA)–Rec. Mar.

First: *Aulacochiton praecursor* Smith *et al.*, 1968, San Germán Formation, lower MAA, Sabana Grande, Puerto Rico (Smith *et al.*, 1968). **Extant**

F. SCHIZOPLACIDAE Bergenhayn, 1955 **Extant**

F. SUBTERENOCHITONIDAE Bergenhayn, 1930
Extant

Subclass APLACOPHORA von Ihering, 1876

The only fossil aplacophoran recorded is an example suggested by Amelie Scheltema from the Mazon Creek locality, Illinois, USA (Francis Creek Shale, Carbondale Formation, upper Moscovian) (Lindberg, 1985, p. 236).

Order NEOMENIIDA Simroth, 1893

F. NEOMENIIDAE von Ihering, 1876 **Extant**

F. PRONEOMENIIDAE Simroth, 1893 **Extant**

F. LEPIDOMENIIDAE Pruvot, 1890 **Extant**

F. GYMNOMENIIDAE Odhner, 1921 **Extant**

Order CHAETODERMATIDA Simroth, 1893

F. CHAETODERMATIDAE von Ihering, 1876 **Extant**

'MONOPLACOPHORA'

Class TERGOMYA Horný, 1965

Traditionally, all untorted molluscs have been placed in the Class Monoplacophora. As discussed by Peel (1991a,b), so many different groups have been assigned to the class that the term is virtually meaningless. Consequently, Peel has replaced the class with two new classes, the Tergomya and the Helcionelloida, as followed here. The definition of these classes and constituent orders given here follows Peel (1991a,b) and Runnegar and Jell (1976), which should be consulted for details and references. In fact, neither the higher taxa nor the genera of Lower Palaeozoic cap-shaped molluscs are well understood.

Order TRYLIDIIDA Lemche, 1957

F. TRYBLIDIDAE Horný, 1965 €. (SOL)–Rec. Mar.

First: *Helcionopsis* sp. Runnegar and Jell, 1976, Coonigan Formation, New South Wales, Australia. **Extant**

F. ARCHAEOPRAGIDAE Horný, 1963 S. (LUD) Mar.

First and Last: *Archaeopraga pinnaeformis* (Perner, 1903) noted in Horný (1963), Přídolí Beds near Prague, Czechoslovakia.

Order CYRTONELLIDA Horný, 1963

F. CYRTOLITIDAE Miller, 1889 O. (TRE)–D. (GIV) Mar.

First: *Cyrtonellopsis huzzahensis* Yochelson, 1958, Gasconade Dolomite, near Steeleville, Missouri, USA (Yochelson, 1958).
Last: *Cyrtonella mitella* (Hall, 1879) noted in Rollins *et al.*, 1971, Marcellus Formation, New York, USA.

Order HYPSELOCONIDA Peel, 1991

F. HYPSELOCONIDAE Knight, 1956 €. (MER)–O. (TRE) Mar.

First: *Cambrioconus expansus* Stinchcomb, 1986, lower Eminence Formation, Missouri, USA (Stinchcomb, 1986).
Last: *Hypseloconus ozarkensis* Ulrich and Bridge, 1930, Gasconade Formation, Missouri, USA.

Class HELCIONELLOIDA Peel, 1990

F. not assigned (formerly SCENELLIDAE Wenz, 1938) €. (ATB)–D. (PRA) Mar.

Yochelson and Gil Cid (1984) suggested that at least some species assigned to *Scenella* may be chondrophorine floats. The position of the type species is unclear; if it is a chondrophore, then a new name must be established for the genera remaining in this group.
First: *Tanuella elata* Missarzhevskii, 1969, Anabar Massif, Sanashtykgolian Horizon, Tuva, former USSR.
Last: *Calloconus humilis* (Perner, 1903), Bande f, Koneprussy, Bohemia, Czechoslovakia.

F. HELCIONELLIDAE Wenz, 1938 €. (TOM)–O. (CND) Mar.

First: *Latouchella korobkovi* (Vostokova, 1962) and *L. memorabilis* Missarzhevskii, 1969, widespread among rocks of the Tommotian Stage, Siberia, former USSR.
Last: *Nyyella bialvi* Rozov, 1975 noted in Runnegar and Jell (1976), Siberian Platform, former USSR.

F. STENOTHECIDAE Runnegar and Jell, 1980 €. (TOM–STD) Mar.

First: *Anabarella plana* Vostokova, 1969, and *A. indecora* Missarzhevskii, 1969, Tommotian Stage of Anabar Massif, Uchuro-Maya region, Siberia, former USSR.
Last: *Mellopegama georginensis* Runnegar and Jell, 1976, Currant Bush Limestone, Queensland, Australia.

F. YOCHELCIONELLIDAE Runnegar and Jell, 1976 €. (?LEN–?MEN) Mar.

First: *Yochelcionella chinensis* Pei, 1985 noted in Bengston *et al.* (1990), Oraparinna Shale, central Flinders Range, South Australia.
Last: *Eotebenna viviannae* Peel, 1991, Andrarum Limestone, Bornholm, Denmark.

REFERENCES

Baluk, W. (1971) Lower Tortonian chitons from the Kortynica Clays, southern slopes of the Holy Cross Mountains. *Acta Geologica Polonica*, **21**, 449–72.

Bengtson, S., Conway Morris, S., Cooper, B. J. *et al.* (1990) Early Cambrian fossils from South Australia. *Association of Australian Palaeontologists, Memoir*, **9**, 1–364.

Bergenhayn, J. R. M. (1960) Cambrian and Ordovician loricates from North America. *Journal of Paleontology*, **34**, 168–78.

Beu, A. G. and Maxwell, P. A. (1990) Cenozoic Mollusca of New Zealand. *New Zealand Geological Survey Palaeontological Bulletin*, **58**, 1–518.

Bischoff, G. C. O. (1981) *Cobrecophora* n.g., representative of a new polyplacophoran order Phosphatoloricata with calcium phosphatic shells. *Senckenbergiana Lethaea*, **61**, 173–215.

Debrock, M. D., Hoare, R. D. and Mapes, R. H. (1984) Pennsylvanian (Desmoinesian) Polyplacophora (Mollusca) from Texas. *Journal of Paleontology*, **58**, 1117–35.

Fiedel, U. and Keupp, H. (1988) *Ischnochiton marloffsteinensis* n. sp., eine Polyplacophore aus dem fränkischen Lias. *Paläontologische Zeitschrift*, **62**, 49–58.

Gowlett-Holmes, K. L. (1987) The Suborder Choriplacina Starobogatov & Sirenko, 1975 with redescription of *Choriplax grayi* (H. Adams & Angas, 1864) (Mollusca: Polyplacophora). *Transactions of the Royal Society of South Australia*, **111**, 105–10.

Hoare, R. D. and Smith, A. G. (1984) Permian Polyplacophora (Mollusca) from West Texas. *Journal of Paleontology*, **58**, 82–103.

Hoare, R. D., Mapes, R. H. and Atwater, D. E. (1983) Pennsylvanian Polyplacophora (Mollusca) from Oklahoma and Texas. *Journal of Paleontology*, **57**, 992–1000.

Horný, R. J. (1963) *Archaeopraga*, a new problematic genus of monoplacophoran molluscs from the Silurian of Bohemia. *Journal of Paleontology*, **37**, 1071–3.

Lindberg, D. R. (1985) Aplacophorans, polyplacophorans, scaphopods: the lesser classes, in *Mollusca: Notes for a Short Course.* (eds D. J. Bottjer, C. S. Hickman and P. D. Ward), University of Tennessee, Knoxville, pp. 230–47.

Peel, J. S. (1991a) Functional morphology of the Class Helcionelloida nov., and the early evolution of the Mollusca, in *The Early Evolution of Metazoa and the Significance of Problematic Taxa.* (eds A. M. Simonetta and S. Conway Morris), Cambridge University Press, Cambridge, pp. 157–77.

Peel, J. S. (1991b) The functional morphology, evolution and systematics of Early Palaeozoic univalved molluscs. *Grønlands Geologiske Undersøgelse, Bulletin*, **161**, 1–116.

Pojeta, J. Jr and Sohl, N. F. (1987) *Ascaulocardium armatum* (Morton, 1833), new genus (Late Cretaceous): the ultimate variation on the bivalve paradigm. *Paleontological Society Memoir*, **24**, 1–77.

Rollins, H. B., Eldredge, N. and Spiller, J. (1971) Gastropoda and Monoplacophora of the Solsville Member (Middle Devonian, Marcellus Formation) in the Chenango Valley, New York State. *Bulletin of the American Museum of Natural History*, **144**, 131–70.

Runnegar, B. and Jell, P. A. (1976) Australian Middle Cambrian molluscs and their bearing on early molluscan evolution. *Alcheringa*, **1**, 109–38.

Runnegar, B., Pojeta, J., Taylor, M. E. *et al.* (1979) New species of the Cambrian and Ordovician chitons *Matthevia* and *Chelodes* from Wisconsin and Queensland: evidence for the early history of the polyplacophoran mollusks. *Journal of Paleontology*, **53**, 1374–94.

Smith, A. G. (1960) Amphineura, in *Treatise on Invertebrate Paleontology. Part I* (ed. R. C. Moore), Geological Society of America and University of Kansas Press, Boulder, Colorado and Lawrence, Kansas, pp. I41–I76.

Smith, A. G. (1973) Fossil chitons from the Mesozoic – a checklist and bibliography. *Occasional Papers of the California Academy of Sciences*, **103**, 1–30.

Smith, A. G. and Hoare, R. D. (1987) Paleozoic Polyplacophora: a checklist and bibliography. *Occasional Papers of the California Academy of Sciences*, **146**, 1–71.

Smith, A. G., Sohl, N. F. and Yochelson, E. L. (1968) New Upper Cretaceous Amphineura (Mollusca). *US Geological Survey Professional Paper*, **593G**, G1–G9.

Stinchcomb, B. (1986) New Monoplacophora (Mollusca) from Late Cambrian and Early Ordovician of Missouri. *Journal of Paleontology*, **60**, 606–26.

Van Belle, R. A. (1981) *Catalogue of Fossil Chitons*. Dr W. Backhuys, Rotterdam, 82 pp.

Yochelson, E. L. (1958) Some Lower Ordovician monoplacophoran mollusks from Missouri. *Journal of the Washington Academy of Sciences*, **48**, 8–14.

Yochelson, E. L. and Gil Cid, D. (1984) Reevaluation of the systematic position of *Scenella*. *Lethaia*, **17**, 331–40.

Yu, W. (1984) Early Cambrian molluscan faunas of Meishucun Stage with special reference to Precambrian–Cambrian boundary, in *Developments in Geoscience*. Science Press, Beijing, pp. 21–35.

8

MOLLUSCA: GASTROPODA

S. Tracey, J. A. Todd and D. H. Erwin

The higher-level classification of the Class Gastropoda is in a state of flux. The classification in the *Treatise on Invertebrate Paleontology* (Knight *et al.*, 1960) is incorrect; several superfamilies recognized there actually represent convergent grades rather than monophyletic clades. Several people, including one of us (DHE) are engaged in phylogenetic analyses of the Palaeozoic taxa in the hope of resolving these difficulties. Thus the classification outlined below is preliminary, but is the best current approximation. Moreover, since many Palaeozoic groups have not been restudied since the Treatise, it is generally far from clear whether a species is properly assigned to a genus or family. Hence, the first and last species listed below should be regarded as approximations only, and the data presented here should not be used for further analyses without additional study.

As a result of ongoing anatomical work on living species, the higher classification of caenogastropods and docoglossans has recently undergone a revolution and is still far from settled. The high degree of convergence of gross shell morphology between some distantly related groups has rendered difficult the placement of exclusively fossil taxa. Characters such as shell microstructure and protoconch morphology are now being used to attempt to construct a more robust phylogenetic system for fossil taxa.

We have based the classification of the caenogastropods, opisthobranchs and pulmonates on the schemes of Haszprunar (1988), Ponder and Warén (1988) and Vaught (1989), with a few small changes and additions due to recent work such as that of McLean (1990) and Bandel (1991, and other papers listed therein). Although several long-established neogastropod families were merged by Ponder and Warén (1988) owing to anatomical similarities, we have preferred to list all such families separately in view of their easily recognized shell characters.

Families known to be polyphyletic are marked with an asterisk (*). Where the superfamilial assignment of a family is uncertain it is marked with a question mark (?). If no source is listed for a record, the specimens are unpublished and housed in the collections of the United States National Museum (Palaeozoic groups) or the British Museum (Natural History) (other groups). DHE compiled the Palaeozoic data, while ST and JAT compiled the Mesozoic and Cainozoic records.

Acknowledgements – We are indebted to Ellis Yochelson, Smithsonian Institution, Peter Wagner, University of Chicago, Robert Blodgett, United States Geological Survey, and Noel Morris and John Cooper, Natural History Museum, London, for assistance in compiling these data.

Class GASTROPODA Cuvier, 1797

Subclass STREPTONEURA Spengel, 1881

Order EUOMPHALINA de Koninck, 1881

The Euomphalina include two distinct groups, the Lower Palaeozoic euomphalids, *sensu stricto*, and a number of Mesozoic genera previously assigned to the Euomphalina. Macluritids are considered as a separate Order. The hydrothermal vent family Neomphalidae have, on the basis of their soft-part anatomy, been considered to represent an extant superfamily of the Euomphalina. However, Batten (1984a) suggested that the shell structure of *Neomphalus* allies it to advanced archaeogastropods or to primitive caenogastropods. Warén and Bouchet (1989) considered the family to be closely related to the Peltospiridae, and it was considered by Haszprunar (1988) to be closely related to the Vetigastropoda.

Superfamily EUOMPHALOIDEA de Koninck, 1881

F. EUOMPHALIDAE de Koninck, 1881
O. (ARG)–P. (CAP)/?Tr. (NOR) Mar.

First: *Lytospira grocera* (Roemer, 1876), Serrite Limestone, New Mexico, USA (Flower, 1968).
Last: ??*Euomphalus cassianus* Koken, 1889, St Cassian Formation, Italy.
Comment: This form, as with other Triassic euomphalids, appears to be convergent with Palaeozoic members of the

The Fossil Record 2. Edited by M. J. Benton. Published in 1993 by Chapman & Hall, London. ISBN 0 412 39380 8

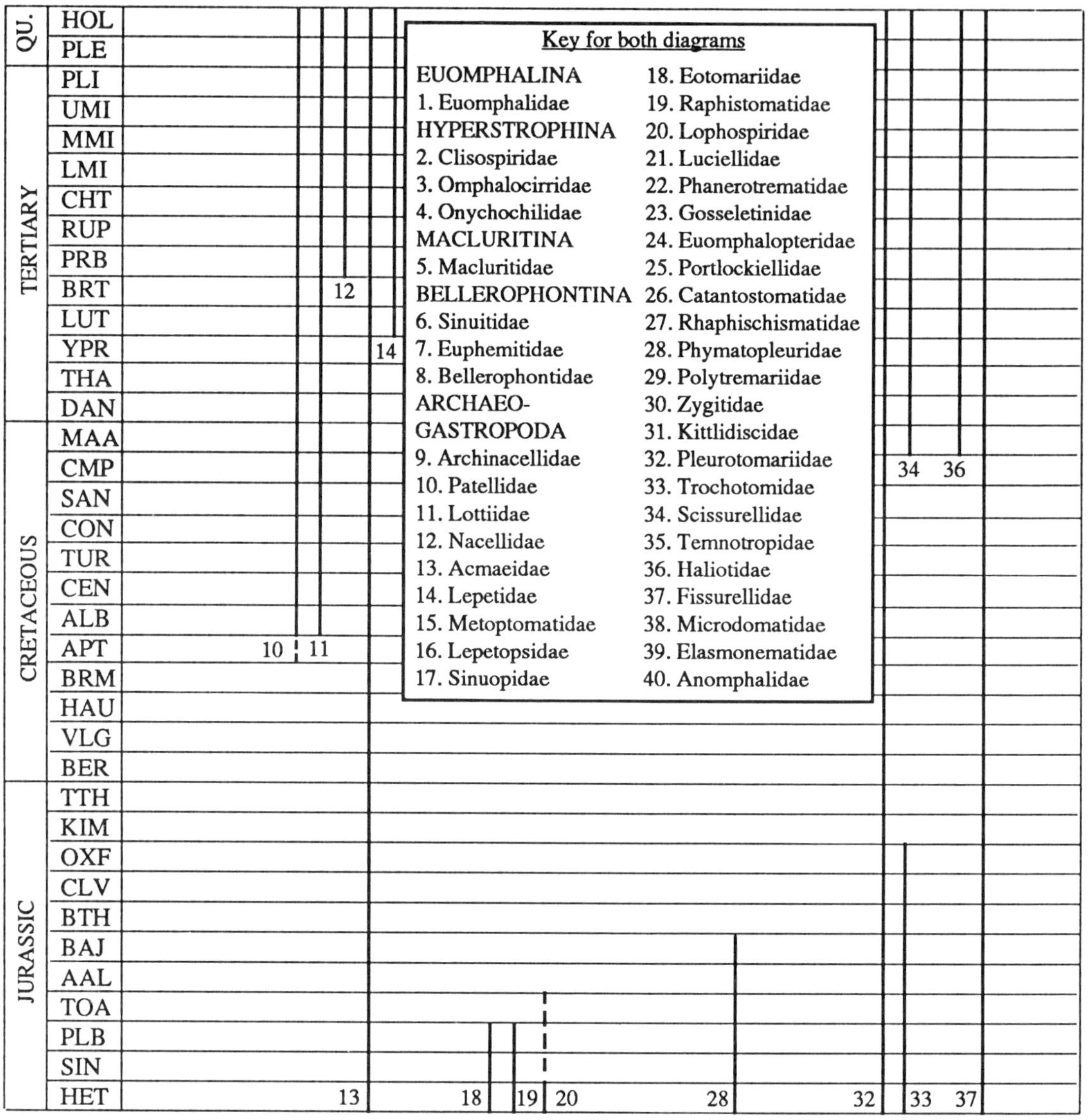
QU.
HOL
PLE
TERTIARY
PLI
UMI
MMI
LMI
CHT
RUP
PRB
BRT
LUT
YPR
THA
DAN
CRETACEOUS
MAA
CMP
SAN
CON
TUR
CEN
ALB
APT
BRM
HAU
VLG
BER
JURASSIC
TTH
KIM
OXF
CLV
BTH
BAJ
AAL
TOA
PLB
SIN
HET
10
11
12
13
14
18
19
20
28
32
33
37
34
36
Key for both diagrams
EUOMPHALINA
1. Euomphalidae
HYPERSTROPHINA
2. Clisospiridae
3. Omphalocirridae
4. Onychochilidae
MACLURITINA
5. Macluritidae
BELLEROPHONTINA
6. Sinuitidae
7. Euphemitidae
8. Bellerophontidae
ARCHAEO-
GASTROPODA
9. Archinacellidae
10. Patellidae
11. Lottiidae
12. Nacellidae
13. Acmaeidae
14. Lepetidae
15. Metoptomatidae
16. Lepetopsidae
17. Sinuopidae
18. Eotomariidae
19. Raphistomatidae
20. Lophospiridae
21. Luciellidae
22. Phanerotrematidae
23. Gosseletinidae
24. Euomphalopteridae
25. Portlockiellidae
26. Catantostomatidae
27. Rhaphischismatidae
28. Phymatopleuridae
29. Polytremariidae
30. Zygitidae
31. Kittlidiscidae
32. Pleurotomariidae
33. Trochotomidae
34. Scissurellidae
35. Temnotropidae
36. Haliotidae
37. Fissurellidae
38. Microdomatidae
39. Elasmonematidae
40. Anomphalidae

Fig. 8.1

order. The next youngest species is *Euomphalus levicarinatus* Yochelson, 1956 from the Bell Canyon Formation, Texas, USA.

Order HYPERSTROPHINA Linsley and Kier, 1984

Linsley and Kier (1984) erected the class Paragastropoda for a number of apparently untorted molluscs, which superficially appear to be similar to the Gastropoda, including the Euomphalina and several other groups. Dzik (1983) proposed the suborder Mimospirina to separate the Clisospiridae and the Onychochilidae from the Euomphalina. Linsley and Kier's more inclusive order is used here to include a variety of groups which are distinct from both the Euomphalina and Macluritina. The relationships between these three families are not clear, and the order may be polyphyletic.

F. CLISOSPIRIDAE S. A. Miller, 1887
O. (TRE)–D. (GIV) Mar.

First: *Clisospira curiosa* Billings, 1865, Beekmantown Limestone, Quebec, Canada.

Last: *Progalerus concoides* Holzapfel, 1895, Massenkalk, Frettermühle, Germany.

F. OMPHALOCIRRIDAE Linsley and Kier, 1984
D. (PRA–GIV) Mar.

First: *Liomphalus northi* (Etheridge, 1890), Lilydale Limestone, Victoria, Australia (Philip and Talent, 1959).

Last: *Omphalocirrus goldfussi* (Archiac and Verneuil, 1842), *Stringocephalus* Beds, Germany (Linsley and Kier, 1984).

F. ONYCHOCHILIDAE Koken, 1925
Є. (MER)–??C. (TOU) Mar.

First: *Kobayashiella circe* (Walcott, 1905), Chau-Mi-Tien Limestone, Shantung, China.

Last: *Onychochilus minutissimus* Yoo, 1988, Dangerfield Formation, New South Wales, Australia.

Comment: This species is considerably younger than the next youngest record of the family, and may not be correctly assigned. The next youngest record is *Sinistracirsa* sp. from Emsian-age rocks, Limestone Mountain, Alaska, USA (Blodgett *et al.*, 1988).

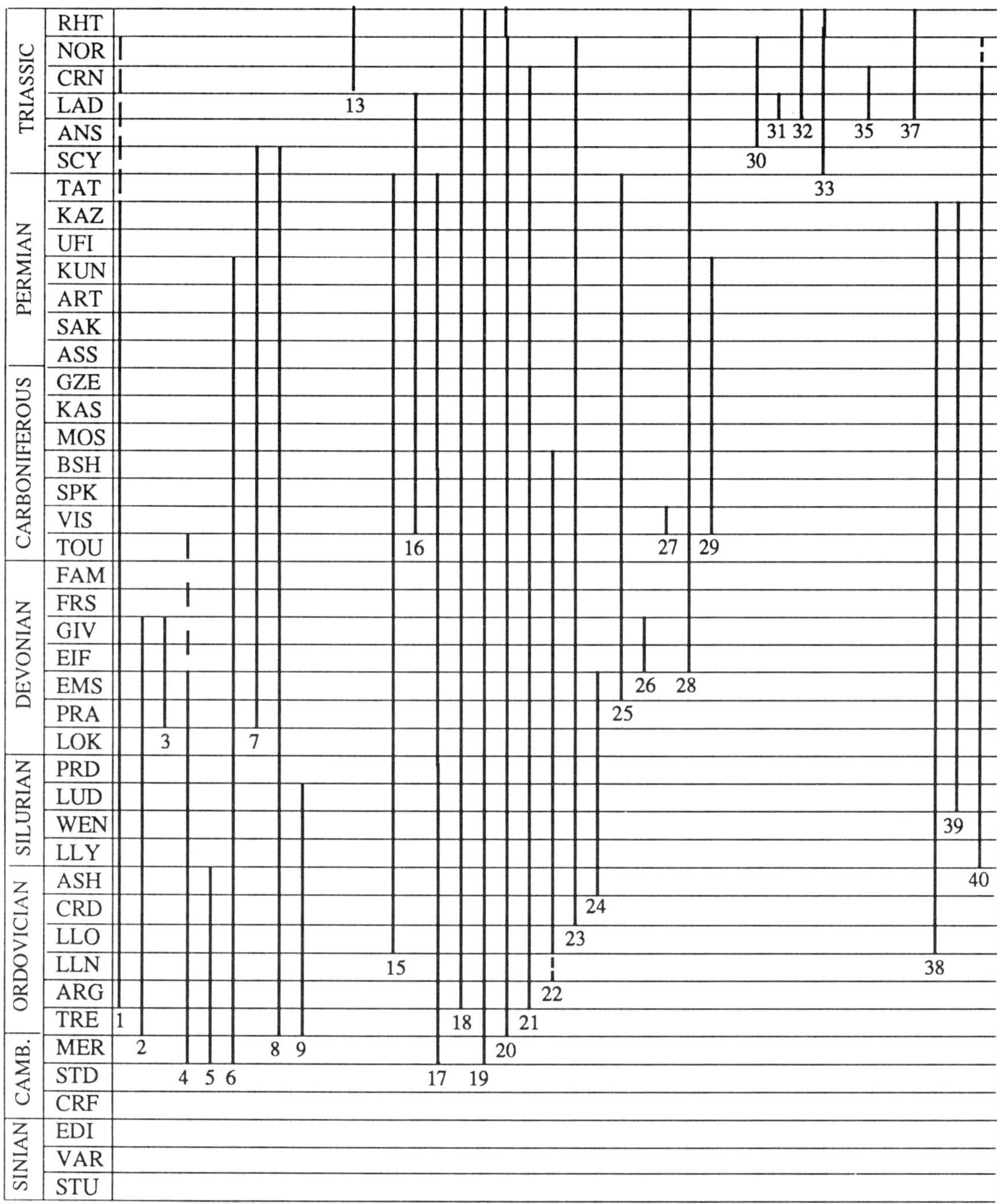

Fig. 8.1

Order MACLURITINA Cox and Knight, 1960

F. MACLURITIDAE Fischer, 1885
Є. (MER)–O. (ASH) Mar.

First: *Macluritella? walcotti* (Howell, 1946), ?Mendha Formation, Nevada, USA (Yochelson and Stinchcomb, 1987).
Last: *Maclurites manitobensis* (Whiteaves, 1887), Bighorn Group, Wyoming, USA (Rohr, 1979).

Order BELLEROPHONTINA Ulrich and Schofield, 1897

Superfamily BELLEROPHONTOIDEA M'Coy, 1851

F. SINUITIDAE Dall in Zittel-Eastman, 1913
Є. (MER)–P. (KUN) Mar.

First: *Sinuella minuta* Knight, 1947, ?Riley Formation, Texas, USA. The position of the locality from which these specimens were collected is in some doubt.
Last: *Sinuitina keytei* Yochelson, 1960, Leonard Formation, Texas, USA. Triassic specimens previously assigned to the family appear to be representatives of the Euphemitidae.

F. EUPHEMITIDAE Knight, 1956
D. (PRA)–Tr. (GRI) Mar.

First: ?*Paleuphemites petrboki* Horný, 1963, Dvorce Prokop Limestone, Czechoslovakia.

Last: *Euphemites* sp. Runnegar, 1969, *in* Yochelson and Yin, 1985, Woondun, Queensland, Australia.

F. BELLEROPHONTIDAE M'Coy, 1851
O. (TRE)–Tr. (SCY) Mar.

?First: *Eobucania mexicana* Yochelson, 1968, Tinu Formation, Oaxa, Mexico.
Last: *Retispira asiatica* (Wirth, 1936), Feixianguan Formation, Guizhou Province, China (Yochelson and Yin, 1985).

Order ARCHAEOGASTROPODA Thiele, 1925

Suborder DOCOGLOSSA Troschel, 1866

Superfamily PATELLOIDEA Rafinesque, 1815

F. ARCHINACELLIDAE Knight, 1956
O. (?TRE)–S. (LUD) Mar.

First: *Archinacella ?* cf. *elongata* (Cullison, 1944), Poulsen Cliff Formation, Greenland. The next youngest is *Floripatella rousseaui* Yochelson, 1988, Kanosh Shale, Utah, USA.
Last: *Guelphinacella canadense* (Whiteaves, 1884), Guelph Formation, Ontario, Canada.

F. PATELLIDAE Rafinesque, 1815
K. (APT/ALB)–Rec. Mar.

First: *Patella ? miyakoensis* Kase, 1984, Hiraiga Formation, Oshima Island, Miyako area, Japan.
Comment: Triassic records from the St Cassian Formation, Italy, and earlier, are problematic.

F. LOTTIIDAE Gray, 1840 K. (ALB)–Rec. Mar.

First: *Patelloida tenuistriata* (Michelin, 1838), Gault Clay, Folkestone and Dunton Green, Kent, England, UK (Akpan *et al.*, 1982; Lindberg, 1988). **Extant**

Superfamily NACELLOIDEA Thiele, 1929

F. NACELLIDAE Thiele, 1929 T. (PRB)–Rec. Mar.

First: *Cellana ampla* Lindberg and Hickman, 1986, Middle Member, Keasey Formation, Oregon, USA. **Extant**
Comment: This is the earliest definite occurrence. Others recorded from the Cretaceous of New Zealand and Australia are represented only by steinkerns and cannot be referred to this family with certainty (Lindberg and Hickman, 1986).

F. ACMAEIDAE Carpenter, 1857
Tr. (CRN)–Rec. Mar.

First: *Scurriopsis cycloides* Tichy, 1979, Raibl Group, Raibl, Italy. **Extant**

F. LEPETIDAE Dall, 1869 T. (LUT)–Rec. Mar.

First: *Lepeta ?boutillieri* (Cossman, 1888), Calcaire grossier, Parnes, Paris Basin, France (Dolin *et al.*, 1980). **Extant**

Suborder UNCERTAIN

F. METOPTOMATIDAE Wenz, 1938
O. (LLO)–P. (KUN) Mar.

First: *Palaeoscurria ordoviciana* Horný, 1961, Dobrotiva Beds, near Beroun, Czechoslovakia.
Last: *Metoptoma* Phillips, 1836, Guadalupian (Roadian), West Texas, USA.
Comment: The family is of uncertain affinities but is probably docoglossan. *Lepetopsis* Whitfield, 1882, is differentiated from *Metoptoma* by the muscle scar and position of the apex (Yochelson, 1960; McLean, 1990), and the Lepetopsidae are here retained as a separate family.

Suborder LEPETOPSINA McLean, 1990

Superfamily LEPETOPSOIDEA McLean, 1990

F. LEPETOPSIDAE McLean, 1990
C. (VIS)–Tr. (LAD) Mar.

First: *Lepetopsis levettei* White, 1882, Warsaw Formation, Spergen Hill, Salem, Indiana, USA.
Last: *Lepetopsis ? petricola* (Kittl, 1895), *L. campannaeformis* (Münster, 1841) and *L. costulata* (Münster, 1841), all St Cassian Formation, Italy (Yochelson, 1960).

F. NEOLEPETOPSIDAE McLean, 1990 **Extant** Mar.

This family was erected for extant species with similar shells to the Lepetopsidae and perhaps derived from them, but lacking radular mineralization, which was thought to have been lost (McLean, 1990).

Suborder VETIGASTROPODA Salvini-Plawen, 1980

Superfamily LEPETODRILOIDEA McLean, 1988

F. LEPETODRILIDAE McLean, 1988 **Extant** Mar.

F. GORGOLEPTIDAE McLean, 1988 **Extant** Mar.

Superfamily PLEUROTOMARIOIDEA Swainson, 1840

*F. SINUOPEIDAE Wenz, 1938
Є. (MER)–P (DZH) Mar.

First: *Sinuopea sweeti* Whitfield, 1882, Norwalk Sandstone, Wisconsin, USA.
Last: *Keenia* sp. Waterhouse, 1980, Mangarewa Formation, New Zealand (Waterhouse, 1980).

F. EOTOMARIIDAE Wenz, 1938
O. (ARG)–J. (PLB) Mar.

First: *Clathospira* sp., Smithville Formation, Arkansas, USA.
Last: *Ptychomphalus expansus* (J. Sowerby, 1821), Bakony Mountains, Hungary, and in Pliensbachian deposits throughout Europe and North Africa (Szabó, 1980).

*F. RAPHISTOMATIDAE Koken, 1896
Є. (MER)–J. (PLB) Mar.

First: *Schizopea typica* (Ulrich and Bridge, 1931), Eminence Dolomite, Missouri, USA.
Last: *Sisenna subturrita* (Deslongchamps, 1849), Calvados, France and southern Germany, and *S. turrita* (Deslongchamps, 1849), Bakony Mountains, Hungary (Szabó, 1980).

F. LOPHOSPIRIDAE Wenz, 1938
O. (TRE)–J. (LIA) Mar.

First: *Lophospira perangulata* (Hall, 1847), upper Durness Limestone, Scotland, UK (Donald, 1902).
Last: *Worthenia zhongshanensis* Pan, 1982, Lower Jurassic, south-west China.
Comment: The Lophospiridae are considered here to

include most of the Trochonematidae, a polyphyletic assemblage of pleurotomariids, most of which are closely related to the Lophospiridae.

*F. LUCIELLIDAE Knight, 1956 O. (ARG)–Tr. (CRN) Mar.

First: *Rhombella umbilicata* (Ulrich and Bridge, *in* Drake and Bridge, 1932), Deadwood Formation, USA.
Last: *Luciella infrasinuata* Koken, 1897, Ober-Röthelstein, Hallstatt, Austria (Ferenc, 1961).

F. PHANEROTREMATIDAE Knight, 1956 O. (LLN/LLO)–C. (BSK) Mar.

First: *Brachytomaria baltica* (Verneuil, 1845), *Orthoceras* Limestone, Estonia, former USSR (Koken, 1925).
Last: *Phanerotrema ornatum* Sayre, 1930, Drum Limestone, Kansas, USA (Forney and Nitecki, 1976).

F. GOSSELETINIDAE Wenz, 1938 O. (CRD)–Tr. (NOR) Mar.

First: *Cataschisma typa* Branson, 1909, Auburn Chert, Missouri, USA (Forney and Nitecki, 1976).
Last: *Gosseletina* sp., Seven Devils Formation, Idaho, USA (specimens at American Museum of Natural History, New York).

F. EUOMPHALOPTERIDAE Koken, 1896 O. (ASH)–D. (EMS) Mar.

First: *Euomphalopterus* sp. Rohr and Blodgett, 1985, Lone Mountain, Alaska, USA.
Last: *Euomphalopterus* sp. Blodgett *et al.*, 1988, Limestone Mountain, Alaska, USA.

F. PORTLOCKIELLIDAE Batten, 1956 D. (EMS)–P. (DZH) Mar.

First: ??*Agniesella aratula* Perner, 1903, southern Urals, former USSR.
Last: *Shansiella* sp. Batten, 1973, Chhidru Formation, Salt Range, Pakistan.

F. CATANTOSTOMATIDAE Wenz, 1938 D. (EIF–GIV) Mar.

First and Last: *Catantostoma clathratum* Sandberger, 1842, *Stringocephalus* Beds, Germany.

F. RHAPHISCHISMATIDAE Knight, 1956 C. (VIS) Mar.

First and Last: *Raphischisma planorbiformis* de Koninck, 1881 (Knight, 1956).

F. PHYMATOPLEURIDAE Batten, 1956 D. (EIF)–T. (DAN) Mar.

First: *Dictyomaria capillaria* (Conrad, 1842), Marcellus Formation, New York, USA (Rollins *et al.*, 1971).
Last: *Leptomaria sublevis* Traub, 1979, Palaeocene, north of Salzburg, Austria.

F. POLYTREMARIIDAE Wenz, 1938 C. (VIS)–P. (KUN) Mar.

First: *Polytremaria catenata* (de Koninck, 1843), 'Asse 6', Visé, Belgium.
Last: *Plocostoma josephina*, *P. neumayri* and *P. piazzi* Gemmellaro, 1890, Sosio Beds, Italy.

F. ZYGITIDAE Cox, 1960 Tr. (ANS–NOR) Mar.

First: *Zygites elegans* Yin and Yochelson, 1983, Qingyan Formation, Qingyan, Guizhou Province, China.
Last: *Zygites marmorea* (Koken, 1897), Hallstätten Kalke, Hallstatt, Austria.

F. KITTLIDISCIDAE Cox, 1960 Tr. (LAD) Mar.

First and Last: *Kittlidiscus bronni* (Klipstein, 1845), St Cassian Formation, southern Tyrol, Italy (Knight *et al.*, 1960).

F. PLEUROTOMARIIDAE Swainson, 1840 Tr. (LAD)–Rec. Mar.

First: *Stuorella subconcava* (Münster, 1841), Ladinian, southern Tyrol, Italy (Knight *et al.*, 1960). **Extant**

F. TROCHOTOMIDAE Cox, 1960 Tr. (SCY)–J. (OXF) Mar.

First: *Trochotoma* ? *orbita* Pan, 1982, Lower Triassic, southwest China.
Last: *Discotoma amata* (d'Orbigny, 1850), Rauracian?, France (Knight *et al.*, 1960).

Superfamily SCISSURELLOIDEA Gray, 1847

F. SCISSURELLIDAE Gray, 1847 K. (MAA)–Rec. Mar.

First: *Scissurella* sp. (Sohl, 1987) Maastrichtian, Puerto Rico.
Next oldest: *Scissurella annulata* Ravn, 1933, Calcaire de Faxe, Faxe, Denmark. **Extant**

Superfamily HALIOTOIDEA Rafinesque, 1815

F. TEMNOTROPIDAE Cox, 1960 Tr. (LAD–CRN) Mar.

First: *Temnotropis carinata* (Münster, 1841), St Cassian Formation, Cortina d'Ampezzo, Italy.
Last: *Temnotropis carinata* (Münster, 1841) and *T. bicarinata* Laube, 1869, Balaton Mountains, Hungary (Ferenc, 1961).

F. HALIOTIDAE Rafinesque, 1815 K. (MAA)–Rec. Mar.

First: *Haliotis* sp. (Sohl, 1987), Maastrichtian, Puerto Rico. **Extant**

Superfamily FISSURELLOIDEA Fleming, 1822

F. FISSURELLIDAE Fleming, 1822 Tr. (LAD)–Rec. Mar.

First: *Emarginula muensteri* Pictet, 1860?, *E. zardini* (Garavello-Spaetti) and *E. cristata* Zardini, 1978, all St Cassian Formation, Cortina d'Ampezzo, Italy (Zardini, 1978). **Extant**

F. CYPEOSECTIDAE McLean, 1989 **Extant** Mar.

Superfamily TROCHOIDEA Rafinesque, 1815

F. MICRODOMATIDAE Wenz, 1938 O. (LLO)–P. (WOR) Mar.

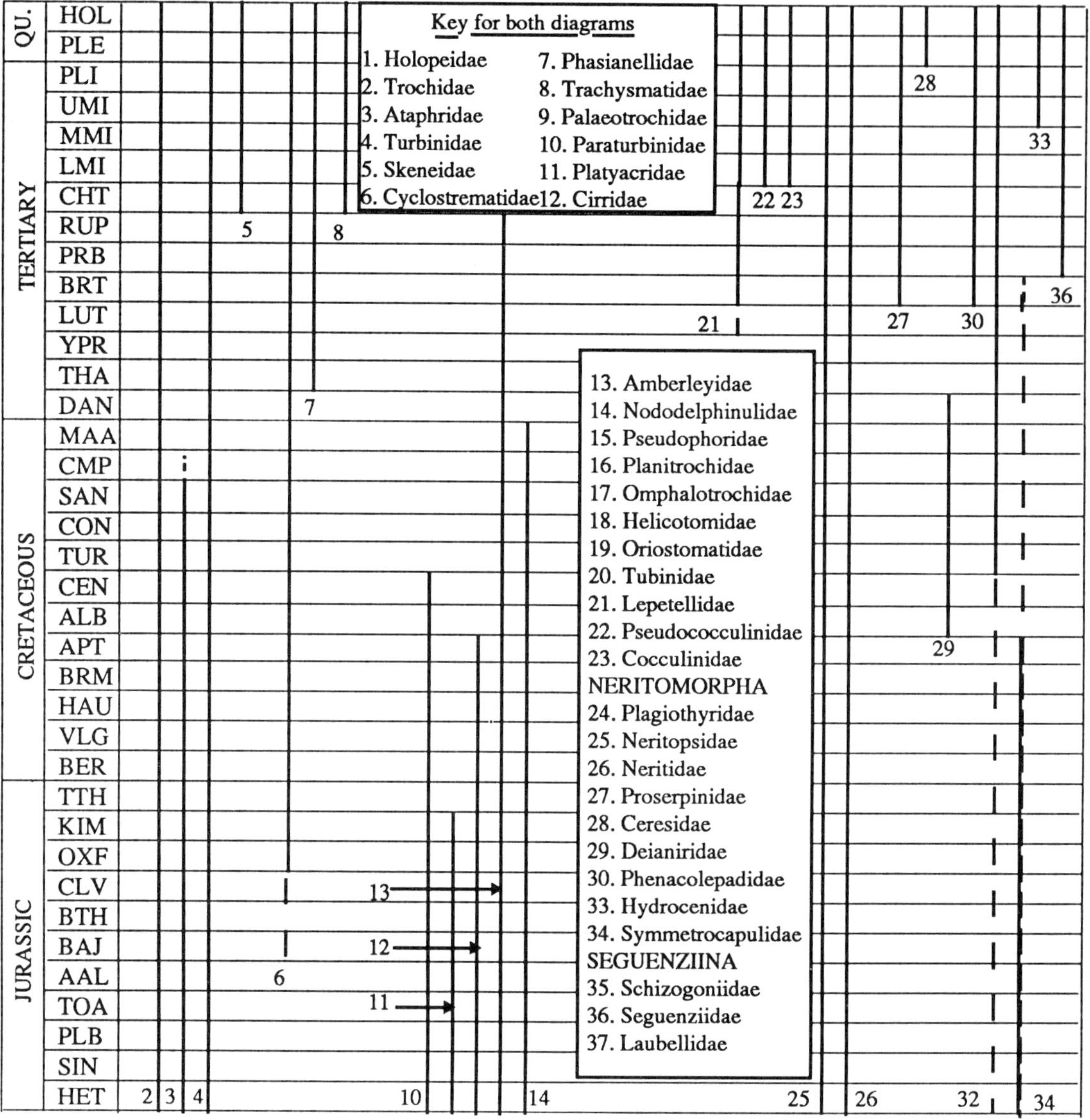

Fig. 8.2

First: *Daidia cerithioides* (Salter, 1859), Leray-Rockland Beds, Allumette Island, Ottawa River, Canada.
Last: *Microdoma variegata*, *M. nodosa* and *Glyptospira*? sp. Batten, 1979, 1985, H. S. Lee Mine No. 8, Perak, Malaysia.

F. ELASMONEMATIDAE Knight, 1956 S. (LUD)–P. (WOR) Mar.

First: *Elasmonema*? sp., Racine Dolomite, specimens at Field Museum, Chicago, listed as *Callonema? elevatum* Wing *nomen nudum*.
Last: *Anematina* sp. Batten, 1979, H. S. Lee Mine No. 8, Perak, Malaysia.

F. ANOMPHALIDAE Wenz, 1938 S. (LLY)–Tr. (?NOR) Mar.

First: *Grantlandispira christiei* Peel, 1984, Offley Island Formation, Kap Tyson, Greenland. Wenz (1939) claimed a Middle Ordovician record for *Pycnomphalus*, but it is unclear to which species he referred.
Last: *Anomphalus helicoides* Münster, 1841 and *A. biconcavus* Haas, 1953, Pucará Group, Cerro de Pasco area, Peru (Haas, 1953).

F. PELYCIDIIDAE Ponder, 1983 **Extant** Mar.

F. HOLOPEIDAE Wenz, 1938 O. (ARG)–P. (WOR) Mar. (see Fig. 8.2)

First: *Straparollina pelagica* Billings, 1865, Quebec Group, Newfoundland, Canada.
Last: *Yunnania meridionalis* Mansuy, 1914, H. S. Lee Mine No. 8, Perak, Malaysia (Batten, 1979), and *Yunnania* sp. Glass Mountains, Texas, USA.
Comment: There are more uncertain records of *Cinclidonema* and *Rhabdotocochlis* from Western Guizhou Province, China (Wang and Xi, 1980). Family uncertain.

F. TROCHIDAE Rafinesque, 1815 Tr. (LAD)–Rec. Mar.

First: *Diplochilus bistriatus* (Münster, 1841), *Pseudoclanculus cassianus* (Wissmann, *in* Münster, 1841), *P. nodosus* (Münster,

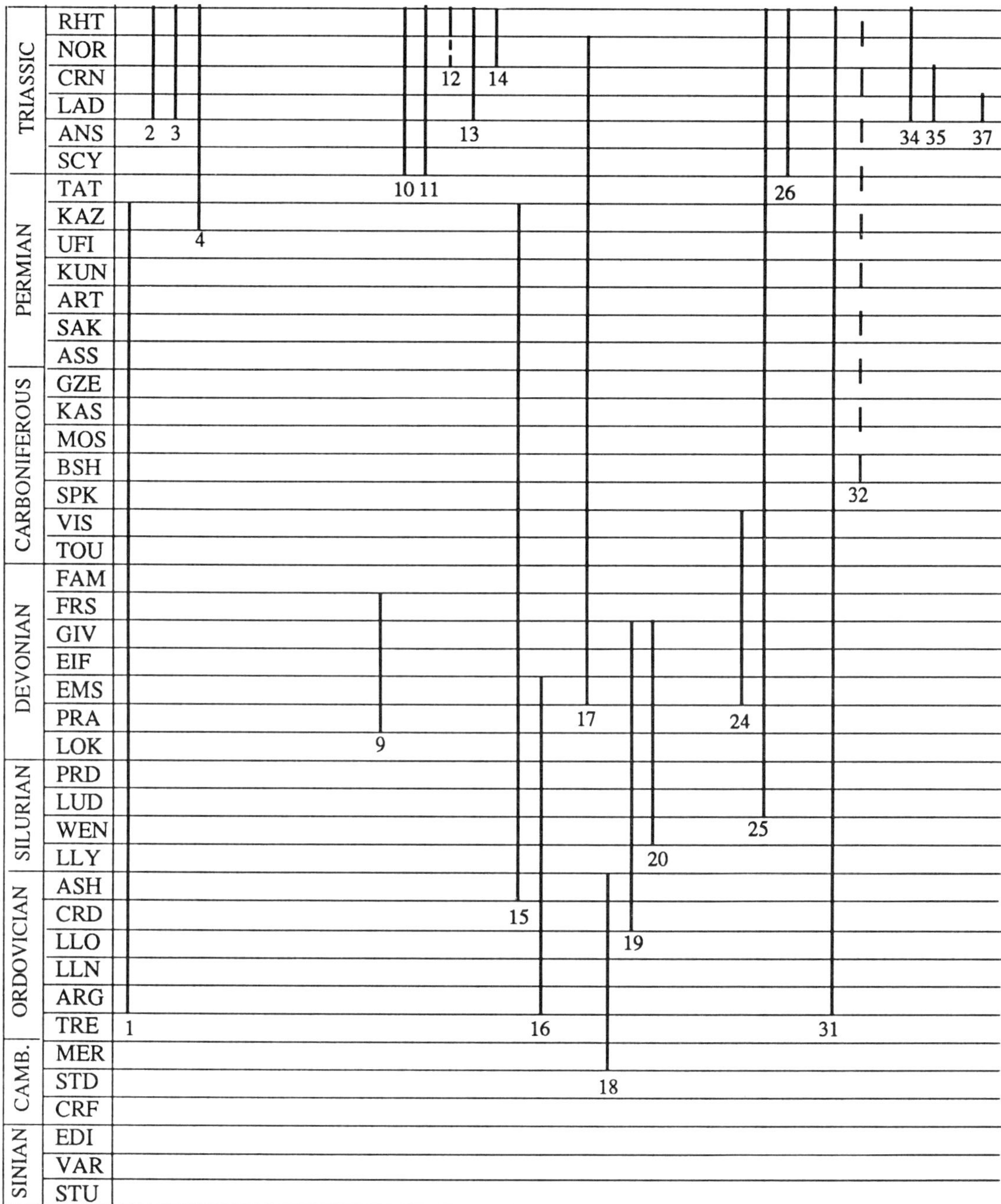

Fig. 8.2

1841), *Solarioconulus nudus* (Münster, 1841), all from the St Cassian Formation, southern Tyrol, Italy (Ferenc, 1961). **Extant**

Comment: There do not seem to be any valid Permian records for the family.

F. ATAPHRIDAE Cossmann, 1918 Tr. (LAD)–K. (SAN/CMP) Mar.

First: *Cirsostylus glandulus* (Laube, 1869), Ladinian, southern Tyrol (Knight *et al.*, 1960).

Last: *Ataphrus compactus* Gabb, Chico Formation, Texas Flat, California, USA (Wenz, 1938).

F. TURBINDAE Rafinesque, 1815 P. (WOR)–Rec. Mar.

First: *Eucycloscala asiatica* Batten, 1985, H. S. Lee Mine No. 8, Perak, Malaysia. **Extant**

F. SKENEIDAE Clark 1851 T. (CHT)–Rec. Mar.

First: *Skenea andersoni* and *S. radiostriata* R. Janssen, 1978, Kassel Meeressandes, Glimmerode, Niederhessen, Germany. **Extant**

F. CYCLOSTREMATIDAE Fischer, 1885 J. (?BAJ/OXF)–Rec. Mar.

First: ? *Ataphropsis pygmaeus* Conti and Fischer, 1982, Case Canepine, Spoleto, Italy.

Next oldest: *Teinostomopsis saharae* Chavan, 1954, Upper Rauracian, France (Knight *et al.*, 1960). **Extant**

F. PHASIANELLIDAE Swainson, 1840 T. (THA)–Rec. Mar.

First: *Tricolia laubrierei* (Cossmann, 1888), Sables de Chalons-sur-Vesle, Chenay (Marne), France (Cossmann and Pissarro, 1913). **Extant**

F. TRACHYSMATIDAE Thiele, 1910
T. (CHT)–Rec. Mar.

First: *Trachysma* sp. (Moisescu, 1982), Merisor, Hateg Basin, Romania. **Extant**

F. TROCHACLIDIDAE Thiele, 1928 **Extant** Mar.

Superfamily PLATYCERATOIDEA Hall, 1859

F. PLATYCERATIDAE Hall, 1859
O. (ARG)–P. (CAP)/? Tr. (NOR) Mar.

First: *Cyclonema montrealense* Billings, 1865?, Ottawa Formation, St Lawrence Valley, Canada.
Last: *Platyceras (Orthonychia) bowsheri* Yochelson, 1956, Cherry Canyon Formation, Texas, USA.
Comment: The subgenus *Orthonychia* is widely distributed in approximately coeval sediments.

F. PARATURBINIDAE Cossmann, 1916
Tr. (NML)–K. (CEN) Mar.

First: *Chartronella unicostata* and *C.? pagina* Batten and Stokes, 1980, Sinbad Member, Moenkopi Formation, Windowblind Butte, Utah, USA.
Last: *Paraturbo kumasoana* (Matsumoto, 1938), Goshonoura Group, Goshonoura Island, Kumamoto area, Japan (Hayami and Kase, 1977).

Superfamily AMBERLEYOIDEA Wenz, 1938

F. PLATYACRIDAE Wenz, 1938
Tr. (SCY)–J. (KIM) Mar.

First: *Lepidotrochus perfectus* Pan, 1982, Scythian, southwest China.
Last: *Platyacra (Asperilla) longispina* (Rolle), Kimmeridgian, Nattheim, Austria (Brosamlen, *in* Wenz, 1938).

F. CIRRIDAE Cossmann, 1916
Tr. (?NOR)–K. (APT) Mar.

First: *Sororcula gracilis* and *S. costata* Haas, 1953, Pucará Group, Cerro de Pasco area, Peru.
Last: *Shikamacirrus nipponicus* Kase, 1984, Tanohata Formation, Koikorobe, Miyako, Japan.

F. AMBERLEYIDAE Wenz, 1938
Tr. (LAD)–T. (RUP) Mar.

First: *Eunemopsis epaphus* (Laube, 1869), Marmolada Mountains, Italy (Ferenc, 1961).
Last: *Eucyclus bundensis* (von Koenen, 1892), Lattorfian, Brandhorst, Germany (Cossmann, 1915).

F. NODODELPHINULIDAE Cox, 1960
Tr. (NOR)–K. (SEN) Mar.

First: *Metriomphalus textorius* (Broili, 1907), Pachycardientuffe, Southern Alps, Italy (Ferenc, 1961).
Last: *Hanaispira tabulata* (Cossmann, 1903), Senonian, France, or *Trochacanthus tuberculatocinctus* (Münster, *in* Goldfuss, 1844), Senonian, Germany (Knight *et al.*, 1960).

?***Superfamily*** PSEUDOPHOROIDEA
Miller, 1889

Ponder and Warén (1988) placed these families in the Xenophoroidea, but with little justification.

?F. PSEUDOPHORIDAE Miller, 1889
O. (ASH)–P. (WOR) Mar.

First: *Pseudophorus* sp. Rohr and Blodgett, 1985, Lone Mountain, Alaska, USA.
Last: *Sallya linsa* and *S. striata* Yochelson, 1956, Cherry Canyon Formation, Texas, USA and *S. terendaka* Batten, 1979, H. S. Lee Mine No. 8, Perak, Malaysia.

F. PLANITROCHIDAE Knight, 1956
O. (ARG)–D. (EMS) Mar.

First: *Raphistomina inflata* Cullison, 1944, Theodosia Formation, Missouri, USA.
Last: *Planitrochus* sp., Kasaan Island, Alaska, USA (collections of R. Blodgett, USGS).

Suborder UNCERTAIN

Superfamily MURCHISONIOIDEA Koken, 1896

F. MURCHISONIIDAE Koken, 1896
O. (ARG)–Tr. (NOR) Mar.

First: *Hormotoma artemsia* Billings, 1865 Smithville Formation, Arkansas and Beekmantown Formation, New York State, USA. *Hormotoma* sp. Fortey and Peel, 1990 occurs in the Poulsen Cliff Formation, Lower Ordovician (?TRE) of northern Greenland. *Gascondia putilla* (Sardeson, 1896), Gasconade Dolomite, Missouri, has been considered to be a murchisoniid, but the assignment is doubtful.
Last: *Vestilia klipsteini* Koken, 1897, Sandling, Hallstatt, Austria.

F. PLETHOSPIRIDAE Wenz, 1938
O. (TRE/ARG)–Tr. (NOR) Mar.

First: *Plethospira ? floweri* Fortey and Peel, 1990, Poulsen Cliff Formation, northern Greenland.
Last: *Wortheniopsis margarethae* (Kittl, 1895), Hallstatt, Austria (Ferenc, 1961).

F. CROSSOSTOMATIDAE Cox, 1960
Tr. (LAD)–J. (BTH) Mar.

First: *Palaeocollonia laevigata* (Münster, 1841), St Cassian Formation, southern Tyrol, Italy (Ferenc, 1961).
Last: *Crossostoma discoideum* Morris and Lycett, 1850, Great Oolite, Gloucestershire, England, UK, and *C. nudum* Fisher, 1969, Calcaire pseudo-oolithique sup., La Hérie, Hirson, Ardennes, France.

F. ACANTHONEMATIDAE Wenz, 1938
D. (EIF)–Tr. (ANS/LAD) Mar.

First: *Acanthonema newberryi* (Meek, 1871), Upper Monroan, Ohio. Wenz (1938) records the genus from the Upper Silurian, but no such species can be identified.
Last: *Orthonema paedice* Erwin, 1988, Cathedral Mountain Formation, Texas, USA, and *O. striatonodosum* Chronic, 1952, Bone Spring Limestone, Kaibab Formation, Walnut Canyon, Arizona, USA (Erwin, 1988).
Comments: *Acanthonema* and *Orthonema* were assigned by Knight (1934) and Knight *et al.* (1960) to the Turritellidae. Acanthonematids have an orthocline outer lip with a slight inflection ab- and adapically which corresponds with a spiral angulation or carina. The central part of the whorl is rather smooth (Knight, 1934). In contrast, turritellids have a generally well-developed lateral sinus, its vertex not dependent on the position of spiral ornament and an overall orthocline to strongly prosocline outer lip. The present family is tentatively placed in the Murchisonioidea.

Suborder UNCERTAIN

? F. OMPHALOTROCHIDAE Knight, 1945 D. (EMS)–Tr. (NOR) Mar.

First: *Bassotrochus angulatus* Tassell, 1977, Bell Point Limestone, Victoria, Australia.
Last: *Discotropis* sp., Seven Devil's Formation, Idaho, USA.

Suborder UNCERTAIN

F. HELICOTOMIDAE Wenz, 1938 €. (MER)–O. (ASH) Mar.

First: *Prohelicotoma uniangularia* (Hall, 1847), Smith Basin Limestone, New York, USA.
Last: ? *Helicotoma patula* Lamont, 1946, lower Drummuck Group, Scotland, UK.
Comment: Wenz (1938) and the *Treatise* (Knight *et al.*, 1960) included two unrelated groups in the Helicotomidae. Here the family is restricted to the Lower Palaeozoic forms.

Suborder UNCERTAIN

Superfamily ORIOSTOMATOIDEA Wenz, 1938

F. ORIOSTOMATIDAE Wenz, 1938 O. (CRD)–D. (GIV) Mar.

First: *Oriostoma bromidensis* Rohr and Johns, 1990, Bromide Formation, Oklahoma, USA.
Last: *Oriostoma* aff. *gerbaulti* Oehlert (Blodgett and Johnson, 1993), Dehay Formation, Nevada, USA.

F. TUBINIDAE Knight, 1956 S. (WEN)–D. (GIV) Mar.

First: *Semitubina sakoi* Kase, 1986, Yokokurayama Formation, Shikoku, Japan.
Last: *Semitubina* Linsley, 1979.

Suborder COCCULINIFORMIA Haszprunar, 1987

Superfamily LEPETELLOIDEA Dall, 1882

F. LEPETELLIDAE Dall, 1882 T. (LUT/BRT)–Rec. Mar.

First: *Sablea minuta* Allen, 1970, Middle Eocene, Red River, Grant Parish, Louisiana, USA. **Extant**

F. PSEUDOCOCCULINIDAE Hickman, 1983 T. (BUR)–Rec. Mar.

First: *Kaiparapelta singularis* Marshall, 1986 and *Notocrater maxwelli* Marshall, 1986, Otaian, Pakaurangi Point, Kaipara, New Zealand (Beu and Maxwell, 1990). **Extant**

F. PYROPELTIDAE McLean and Haszprunar, 1987 **Extant** Mar.

F. OSTEOPELTIDAE Marshall, 1987 **Extant** Mar.

F. COCCULINELLIDAE Moskalev, 1971 **Extant** Mar.

F. ADDISONIIDAE Dall, 1882 **Extant** Mar.

F. CHORISTELLIDAE Bouchet and Warén, 1979 **Extant** Mar.

F. BATHYPHYTOPHILIDAE Moskalev, 1978 **Extant** Mar.

Superfamily COCCULINOIDEA Dall, 1882

F. COCCULINIDAE Dall, 1882 T. (BUR)–Rec. Mar.

First: *Cocculina pristina*, *Coccopigya otaiana* and *C. komitica* Marshall, 1986, Otaian, Pakaurangi point, Kaipara, New Zealand (Beu and Maxwell, 1990). **Extant**

F. BATHYSCIADIIDAE Dautzenberg and Fischer, 1990 **Extant** Mar.

Suborder NERITIMORPHA Golikov and Starobogatov, 1975

Superfamily NERITOIDEA Rafinesque, 1815

F. PLAGIOTHYRIDAE Knight, 1956 D. (EMS)–C. (VIS) Mar.

First: *Plagiothyra* sp. Blodgett *et al.*, 1988, Lowther Island, Canadian Arctic.
Last: *Littorinides solida* (de Koninck, 1843) 'Asse 6', Visé, Belgium, and *Littorinides* sp. Kase, 1988, Hikoroichi Formation, Kitakami Mountains, Japan.

F. NERITOPSIDAE Gray, 1847 S. (LUD)–Rec. Mar.

First: *Naticopsis transversa* (Lindstrom, 1884), Hogklint Beds, Gotland. **Extant**

Superfamily PALAEOTROCHOIDEA Knight, 1956

F. PALAEOTROCHIDAE Knight, 1956 D. (PRA–FRS) Mar.

First: *Palaeotrochus* sp. Linsley, 1979.
Last: *Turbonopsis apachiensis* Day and Bues, 1982, Martin Formation, Arizona, USA.

F. NERITIDAE Rafinesque, 1815 Tr. (NML)–Rec. Mar./Brackish/FW

First: *Neritaria costata* Batten and Stokes, 1980, Sinbad Member, Moenkopi Formation, Windowblind Butte, Utah, USA. **Extant**

F. PROSERPINIDAE Gray, 1847 T. (BRT)–Rec. Terr.

First: *Proserpina* sp., Creechbarrow Limestone Formation, East Creech, Dorset, England, UK (Preece, 1980). **Extant**

F. CERESIDAE Thompson, 1980? (PLE)–Rec. Terr.

First: *Linidiella (Staffola) derbyi* Dall, 1905, Pleistocene, Brazil (Knight *et al.*, 1960). **Extant**

F. DEIANIRIDAE Wenz, 1938 K. (ALB)–T. (DAN) Mar.

First: *Deianira* sp. 1 (Kollmann, 1980), Losenstein Formation, Losenstein, Austria.
Last: *Deianira bicarinata* (Zekeli, 1852) Danian, Austria (Knight *et al.*, 1960).

F. PHENACOLEPADIDAE Thiele, 1929 T. (BRT)–Rec. Mar.

First: *Plesiothyreus parmophoroides* Cossmann, 1885, Auvers-sur-Oise (Val d'Oise), France (Cossmann and Pissarro, 1911). **Extant**

Superfamily HELICINOIDEA Latreille, 1825

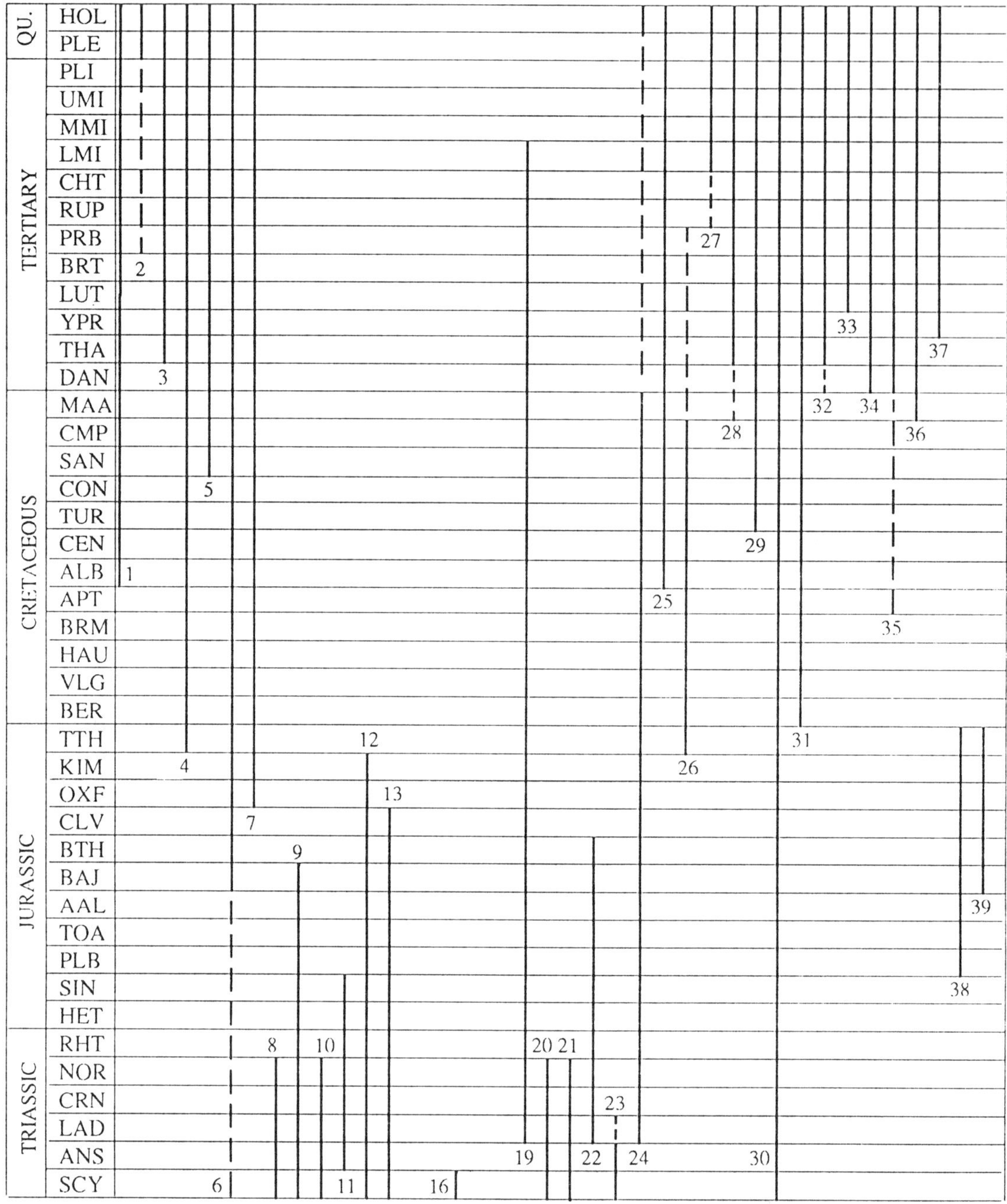

Fig. 8.3

F. HELICINIDAE Latreille, 1825
?C. (BSH)/K. (TUR)–Rec. Terr.

First: *?Dawsonella meeki* (Bradley, 1872), Carbondale Formation, Herrin Coal Member (lower Westphalian C), Petty's Ford, Little Vermilion River, Georgetown, Illinois, USA (Solem and Yochelson, 1979). **Extant**
Next oldest: *Dimorphoptychia* Sandberger, 1871, Turonian, Europe (noted by Wenz, 1938).

Superfamily HYDROCENOIDEA Troschel, 1856

F. HYDROCENIDAE Troschel, 1856
T. (TOR/MES)–Rec. Terr.

First: *Hydrocena troili* Schlickum, 1978, Upper Miocene, Germany. **Extant**

Superfamily TITISCANIOIDEA Bergh, 1890

F. TITISCANIIDAE Bergh, 1890 **Extant** Mar.

Superfamily SYMMETROCAPULOIDEA Wenz, 1938

F. SYMMETROCAPULIDAE Wenz, 1938
Tr. (LAD)–K. (APT)/?T. (LUT/BRT) Mar.

First: *Phryx bilateralis* (Blaschke, 1905) St Cassian Formation, southern Tyrol, Italy (Knight *et al.*, 1960).
Last certain: *Symmetrocapulus hanaii* Kase, 1984, Tanohata and Hiraiga Formations, Miyako area, NE Japan.
Last: *?Symmetrocapulus* sp. (Squires, 1989), Tejon Formation, Tehachapi Mountains, Great Valley, California, USA.
Comment: The morphology of the protoconch of some symmetrocapulids supports their placement within the

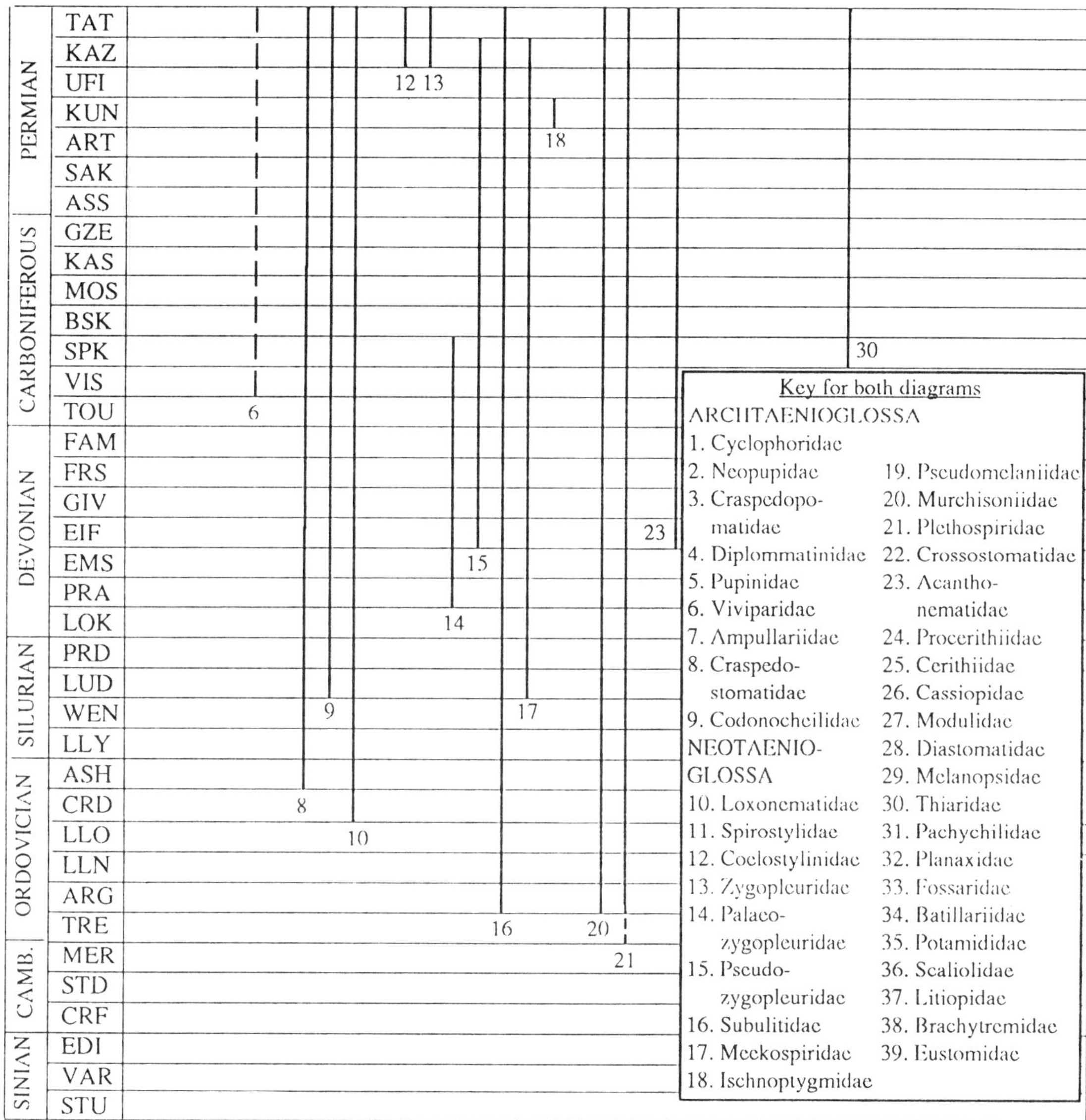

Fig. 8.3

Neritomorpha, most closely related to the following superfamilies (McLean, 1988).

Superfamily PELTOSPIROIDEA McLean, 1989

F. PELTOSPIRIDAE McLean, 1989 **Extant** Mar.

The Triassic genus *Phryx* Blaschke, 1905 resembles this family but not enough is known of its shell characters to permit the assignment (McLean, 1989). Following Knight *et al.* (1960), it is here retained in the Symmetrocapulidae.

Superfamily NEOMPHALOIDEA McLean, 1981

F. NEOMPHALIDAE McLean, 1981 **Extant** Mar.

Suborder SEGUENZIINA Salvini-Plawen and Haszprunar, 1987

F. SCHIZOGONIIDAE Cox, 1960 Tr. (LAD–CRN) Mar.

First: *Schizogonium scalare* (Münster, 1841), Ladinian, S. Tyrol, Italy (Ferenc, 1961).

Last: *Pseudoschizogonium turriculatum* Kutassy, 1937, Carnian, Hungary (Knight *et al.*, 1960).

F. SEGUENZIIDAE Verrill, 1894 T. (PRB)–Rec. Mar.

First: *Calliobasis eos* Marshall, 1983, Runangan Stage, New Zealand (Beu and Maxwell, 1990). **Extant**

F. LAUBELLIDAE Cox, 1960 Tr. (LAD) Mar.

First and Last: *Laubella delicata* (Laube, 1868) St. Cassian Formation, Stuoresmergel, St. Cassian, Italy (Wenz, 1938)

Suborder ARCHTAENIOGLOSSA Haller, 1892 (see Fig. 8.3)

Superfamily CYCLOPHOROIDEA Gray, 1847

F. CYCLOPHORIDAE Gray, 1847 K. (ALB)–Rec. Terr./FW

First: *Pseudarinia* sp., Bear River Formation, Bear River,

near Beartown, Wyoming, USA (Fürsich and Kauffman, 1984).
Intervening: BAR–PLE **Extant**
Comment: The Carboniferous Dendropupinae, formerly included here, have been shown to be buliminid pulmonates (Solem and Yochelson, 1979).

F. NEOCYCLOTIDAE Kobelt and Moellendorff, 1897
Extant Terr.

F. NEOPUPIDAE Kobelt, 1902
?T. (PRB/RUP)/Q. (PLE)–Rec. Terr.

First: ?*Megalostoma*? sp. (Poinar and Roth, 1991), preserved in amber, Upper Eocene–Upper Oligocene, Dominican Republic. An indeterminate shell with operculum, perhaps, represents this or a similar genus (Poinar and Roth, 1991).
Next oldest: A few species of the endemic genera *Hainesia* L. Pfeiffer, 1856 and *Acroptychia* Crosse and Fischer, 1877 occur in the Pleistocene of Madagascar (Wenz, 1938). **Extant**

F. CRASPEDOPOMATIDAE Kobelt and Moellendorff, 1897 T. (THA)–Rec. Terr.

First: *Craspedopoma conoideum* (Boissy, 1848), and *C. insuetum* (Deshayes, 1863), both Thanetian, Chenay (Marne), France (Cossmann, 1888). **Extant**

F. DIPLOMMATINIDAE Pfeiffer, 1856
J. (TTH)–Rec. Terr.

First: *Diplommatoptychia conulus* Maillard, 1885, Source de l'Ain, Nozeroy, Jura, France (Wenz, 1938).
Intervening: EOC. **Extant**

F. PUPINIDAE H. and A. Adams, 1855
K. (SAN)–Rec. Terr.

First: *Cyclomastoma pachygaster* Hrubesch, 1965, Gosau Group, Glanegg near Salzburg, Austria (Hrubesch, 1965a), and *Rognacia abbreviata* (Matheron, 1842), Senonian, Rognac, Les Pennes (Bouches-du-Rhône), France (Wenz, 1938). **Extant**

Superfamily AMPULLARIOIDEA Gray, 1824

F. VIVIPARIDAE Gray, 1847
?C. (VIS)/J. (BAJ)–Rec. FW/Brackish

First: ?*Bernicia praecursor* Cox, 1927, Scremerston Coal Group, Viséan D2, Scremerston, Northumberland, England, UK (Solem and Yochelson, 1979). **Extant**
Next oldest: *Viviparus langtonensis* (Hudleston, 1896), Inferior Oolite, Langton Bridge, England, UK (Cossmann, 1921), or *V. wangjianshanensis* Guo *et al.*, 1982, Middle Jurassic, north-west China.

F. AMPULLARIIDAE Gray, 1824 J. (MLM)–Rec. FW/Semi-terr.

First: *Ampullaria nipponica* (Kobayashi and Suzuki, 1937), Itoshiro Subgroup, Tetori Group, Furukawa area, central Japan (Hayami and Kase, 1977). **Extant**

Suborder UNCERTAIN

Superfamily UNCERTAIN

Ponder and Warén (1988) assigned these two families to the Suborder Archtaenioglossa. The two groups are very poorly known, at least in the Palaeozoic.

*F. CRASPEDOSTOMATIDAE Wenz, 1938
O. (ASH)–Tr. (NOR) Mar.

First: *Umbonellina infrasilurica* Koken, 1925, Bornholm, Estonia, former USSR, based on a single specimen, whereabouts unknown.
Next oldest: *Bucanospira expansa* Ulrich and Schofield, 1897, Niagara Group, Tennessee, USA.
Last: *Brochidium vareasense* Haas, 1953, Pucará Group, Cerro de Pasco area, Peru.

*F. CODONOCHEILIDAE Miller, 1889
S. (LUD)–J. (BAJ) Mar.

First: *Codonocheilus striatum* Whiteaves, 1884, Guelph Dolomite, Ontario, Canada.
Last: *Ventricaria* ? *vesicula* Szabó, 1983, Bakonybél, Somhegy, Bakony Mountains, Hungary.

Order APOGASTROPODA Salvini-Plawen and Haszprunar, 1987

Suborder CAENOGASTROPODA Cox, 1959

Section CERITHIIMORPHA Golikov and Starobogotov, 1975

Superfamily LOXONEMATOIDEA Koken, 1889

This superfamily probably represents the stem-group of the Cerithioidea and hence the Caenogastropoda (Haszprunar, 1988). Bandel (1991) has dismembered this superfamily and has placed the last four families below in a new superfamily, Zygopleuroidea, which he considers to belong to the Ptenoglossa.

F. LOXONEMATIDAE Koken, 1889
O. (CRD)–Tr. (NOR) Mar.

First: *Loxonema murrayana* Salter, 1859, Ottawa Group, Canada (Wilson, 1951).
Last: *Polygyrina elegans* (Hörnes, 1855), Gastropodenschicht, Sandling, and *Loxonema pagoda* Koken, 1897, Sommeraukogel, both Hallstatt, Austria (Koken, 1897).

F. SPIROSTYLIDAE Cossmann, 1909
Tr. (ANS)–J. (SIN) Mar.

First: *Spirostylus cf. subcolumnaris* (Münster, 1841), Qingyan Formation, Qingyang, Guizhou Province, China (Yochelson and Yin, 1983).
Last: *Climacina catharinae* and *C. mariae* Gemmellaro, 1878, Montagna del Casale, Palermo, Sicily, Italy (Wenz, 1938).

F. COELOSTYLINIDAE Cossmann, 1909
P. (WOR)–J. (KIM) Mar.

First: *Omphaloptychia paleozoica* and *O. cingulata* Batten, 1985, near Perak, Malaysia.
Last: *Bourguetia saemanni* (Oppel, 1856), *Rhactorhynchia inconstans* Bed, Black Head, Dorset, England, UK (Brookfield, 1978) and Hereri Shales, Hereri, NE Kenya (Cox, 1965).

F. ZYGOPLEURIDAE Wenz, 1938
P. (WOR)–J. (CLV) Mar.

First: *Allocosmia* ? *multicostata* and *Raha* ? *yabei* (Hayasaka, 1943), Akasaka Limestone, Kinshozan, Gifu area, Japan (Hayami and Kase, 1977).
Last: *Zygopleura mysis* (d'Orbigny, 1850), Marault, Oiron, Nantua and Pisieux (Sarthe), France, and *Z. cotteaui*

Cossmann, 1913, Saintpuits (Yonne), France (Cossmann, 1913).

F. PALAEOZYGOPLEURIDAE Horny, 1955 D. (PRA)–C. (l.) Mar.

First: *Palaeozygopleura alinae* (Perner, 1903), Dvorce Limestone, Czechoslovakia.
Last: *Palaeozygopleura venusta* and *P. welleri* Thein and Nitecki, 1974, Lower Okaw Group, Illinois, USA.

F. PSEUDOZYGOPLEURIDAE Knight, 1930 D. (EIF)–P. (WOR) Mar.

First: *Alaskozygopleura crassicosta* Blodgett, 1992, 'Cascaden Ridge' unit, east-central Alaska, USA.
Last: *Pseudozygopleura pleurozyga*, *P. obliqua*, *P. convexa* and *P. lirata* Batten, 1985, H. S. Lee Mine No. 8, Kampar, Perak, Malaysia.
Comment: Wang and Xi (1980) recorded two species of *Pseudozygopleura* from the Wuchiapinian Stage (DZH), China, but the information is insufficient.

F. ABYSSOCHRYSIDAE Tomlin, 1927 **Extant** Mar.

No fossil record as such, but Houbrick (1979) has provided convincing evidence that the modern deep-sea genus *Abyssochrysos* Tomlin, 1927 is a living representative of the Zygopleuridae/Pseudozygopleuridae.

Superfamily SUBULITOIDEA Lindström, 1884

F. SUBULITIDAE Lindström, 1884 O. (ARG)–Tr. (SCY) Mar.

First: *Fusispira* sp., Smithville Formation, Arkansas, USA.
Last: *Strobeus paludinaeformis* (Hall, 1858), Moenkopi Formation, Utah, USA (noted in Batten and Stokes, 1986).

F. MEEKOSPIRIDAE Knight, 1956 S. (LUD)–P. (WOR) Mar.

First: *Auriptygma fortior* Perner, 1903, Band e, Lochkov, Czechoslovakia.
Last: *Meekospira melanoides* and *Meekospira ligoni* Batten, 1985, at H. S. Lee Mine No. 8, Kampar, Perak, Malaysia.
Comment: *Meekospira* sp. Ishii and Murata, 1974, from the upper Sisophon Limestone, Cambodia, may be a representative of the genus, and may be younger, but neither is clear.

F. ISCHNOPTYGMIDAE Erwin, 1988 P. (KUN) Mar.

First and Last: *Ischnoptygma archensis* Erwin, 1988, Road Canyon Formation, Texas, USA.

Superfamily PSEUDOMELANIOIDEA Fischer, 1885

F. PSEUDOMELANIIDAE Fischer, 1885 Tr. (LAD)–T. (BUR) Mar.

First: *Oonia subtortilis* (Münster, 1841), Marmolatakalke, Marmolada Mountains, Italy.
Last: *Bayania purpusilla* (Degranges-Touzin, 1895), Burdigalian, Dax, France (Cossmann, 1909).
Comment: A poorly known and probably polyphyletic group of elongate, often rather featureless shells. Cenomanian *Liocium* and typically Palaeogene *Bayania* seem to be close relatives, but their relationship to other genera assigned here is problematic. Included here are the Trajanellidae Pčhelintsev, 1853, K. (APT–SEN) which do not show enough distinguishing features to be recognized with confidence.

Superfamily CERITHIOIDEA Férussac, 1819

F. PROCERITHIIDAE Cossmann, 1906 Tr. (LAD)–K. (MAA)–? Rec. Mar.

First: *Paracerithium subcerithiforme* (Kittl, 1895), Marmolatakalke, Marmolata, Italy.
Last: *Nudivagus simplicus* Wade, 1917, Ripley Formation, Tennessee, USA (Sohl, 1960). **Extant**
Comment: By the Norian the family was abundant, species-rich and diversified (Haas, 1953). Various Tertiary species placed in the Cerithiidae, e.g. *Batillona amara* Finlay, 1927 (Oligocene, New Zealand), have the shell morphology of the Procerithiidae. Also probably cofamilial with these is *Argyropeza* Melvill and Standen, 1901 (e.g. *A. schepmaniana* Melville, 1912) living in the Indo-Pacific (Houbrick, 1980). Cossmann (1906) classed his genera *Metacerithium* and *Nerineopsis* as procerithiids, but these seem to be Cerithiidae/Potamididae and Cerithiopsidae, respectively. *Spanionema* from the mid-Devonian of England, placed by Knight *et al.* (1960) in this family almost certainly does not belong here.

F. CERITHIIDAE Férussac, 1819 K. (ALB)–Rec. Mar.

? First: *Exechocirsus* aff. *subpustulosus* Pčhelintsev, 1953, and *Ageria costata* (J. de C. Sowerby, 1827), Upper Greensand, Haldon and Blackdown, Devon, England, UK (Abbass, 1973). **Extant**
Comment: This family may be polyphyletic, or may have been derived from the Eustomidae rather than the homeomorphic Procerithiidae (Abbass, 1973). *Bittium* Leach, *in* Gray, 1847 may be one of the earliest cerithiids but a first occurrence is difficult to trace owing to convergence with the procerithiids. The above taxa can be fairly confidently assigned to the Cerithiidae.

F. CASSIOPIDAE Kollmann, 1979 J. (TTH)–K. (CMP), ?T. (Eoc.) Mar./Brackish

First: *Paraglauconia* sp. nov. (Cleevely and Morris, 1988), *Indotrigonia danielli* horizon, Sana'a, Yemen.
Last: *Cassiope cf. kefersteinii* (Münster, *in* Goldfuss, 1844). Aachen Greensand, Aachen and Ronheide, Germany (Holzapfel, 1888).
Comment: No undoubted species are known from the Tertiary. *Pseudoglauconia* Douville, 1921 from the Eocene of Negritos, Piura, Brazil, was included doubtfully by Cleevely and Morris (1988).

F. MODULIDAE Fischer, 1885 T. (Oli./Mio.)–Rec. Mar.

First: *Modulus turbinatus* (Heilprin, 1887), *Orthaulax* Bed, Ballast Point, Tampa Bay, Florida, USA (Dall, 1915). **Extant**
Comment: The Maastrichtian genus, *Turbinopsis* Conrad, 1860 from the Ripley Formation of Tennessee, USA, has been included in the Modulidae, but Sohl (1964a) considered it to belong to the Trichotropidae.

F. DIASTOMATIDAE Cossmann, 1895 ?K. (MAA)/T. (THA)–Rec. Mar.

First: ?*Diastoma arcotense* (Stoliczka, 1867), Arivalur Group, south India. **Extant**
Next oldest: *Diastoma multispiratum* Cossmann, 1881, Sables de Bracheux, Abbecourt near Beauvais, France (Cossmann, 1889).
Comment: Houbrick (1981) restricted the family to the typical genus, including one living representative.

F. MELANOPSIDAE H. and A. Adams, 1854 K. (TUR)–Rec. FW/Brackish

First: *Megalonoda reussi* (Hörnes, 1856), Turonian, Styrie (Kollmann, 1984). **Extant**
Comment: *Melanopsis* today lives in fresh water, but from the Palaeocene to at least the late Eocene, many species seem to have inhabited brackish and shallow-marine environments.

F. THIARIDAE Troschel, 1857 C. (SPK)–Rec. FW/Brackish

First: ?*Carbonispira scotica* Yen, 1949, Namurian A, Top Hosie Limestone, Scotland, UK (Solem and Yochelson, 1979).
Intervening: MLM, TTH, SEN–PLE **Extant**
Comment: Considered likely to be polyphyletic by Houbrick (1988).

F. PACHYCHILIDAE Troschel, 1857 K. (BER)–Rec. FW

First: *Pachychilus (Pachychiloides) manselli* (de Loriol, 1866) and *P. (P.) attenuatus* (J. de C. Sowerby, 1836), *Corbula* Beds, Middle Purbeck Beds, Durlston Bay, Dorset, England, UK (Arkell, 1941). **Extant**

F. PLANAXIDAE Gray, 1850 T. (DAN/THA)–Rec. Mar.

First: *Planaxis africana* Adegoke, 1977, Ewekoro Formation, Ewekoro, Nigeria. **Extant**
Comment: The Albian species '*P.*' *simplex* Mahmoud, 1956 from Egypt is unlikely to belong to this family.

F. FOSSARIDAE Troschel, 1861 T. (LUT)–Rec. Mar.

First: *Fossarus* sp. Middle Eocene, Selsey Formation, Bracklesham, Sussex, England, UK (but not '*Fossarus*' *dixoni* Wrigley, 1942, which is a trochacean from an earlier horizon at the same locality: S. Tracey, unpublished data), and ?*Zeradina obliquicostata* (Marshall and Murdoch, 1920), Bortonian, New Zealand (Beu and Maxwell, 1990). **Extant**
Comment: The shells of this family are to some extent convergent with the Vanikoridae, which is where the affinities of most Tertiary genera, formerly regarded as fossarid, will probably prove to lie.

F. BATILLARIIDAE Thiele, 1929 T. (DAN)–Rec. Brackish

First: *Batillaria (Vicinocerithium) inopinata* (Deshayes, 1864), Calcaire de Mons, Mons, Belgium (Glibert, 1973). **Extant**

F. POTAMIDIDAE H. and A. Adams, 1854 ?K (APT/ALB)–Rec. Brackish/Semi-terr.

First: *Echinobathra vicina* (Verneuil and Lorrière, 1868) APT, Utrillas, Spain (Kollmann, 1979), *Pyrazus ? scalariformis* and *P.*? sp. (Nagao, 1934), Hiraiga Formation (APT), Hiraiga, Miyako, Japan, have a similar form and ornament to *Pyrazus*, but this assignation cannot be confirmed as details of the aperture and body whorl are missing (Kase, 1984). **Extant**
Comment: The differentiation of potamidid from other cerithiform gastropods is problematic in the Mesozoic. Albian cerithiids such as *Metacerithium trimonile* (Michelin, 1838) from the Gault of Folkestone, Kent, England, UK (Abbass, 1973), have the shape and ornament of three beaded cords that is found in various Tertiary potamidids (*Ptychopotamides*, *Eotympanotonus*) and cerithiids (*Serratocerithium*).

F. SCALIOLIDAE Iredale and McMichael, 1962 K. (MAA)–Rec. Mar.

First: *Springvaleia*? sp. Sohl, *in* Boucot, 1990, Owl Creek Formation, Ripley County, Mississippi, USA (Boucot, 1990). **Extant**

F. LITIOPIDAE Gray, 1847 T. (YPR)–Rec. Mar. (pelagic)

First: *Litiopa* (2 spp. undescribed), London Clay Formation, Hampstead Tunnel, London, England, UK (Newton, 1891).
Next oldest: *Litiopa acuminata* (Baudon, 1856?), Cuisian, Paris Basin, France (Cossmann and Pissarro, 1911). **Extant**

F. BRACHYTREMIDAE Cossmann, 1906 J. (PLB–TTH) Mar.

First: *Brachytrema labiosum* Deslongchamps, 1866, Pliensbachian, May, France.
Last: *Brachytrema superbum* Zittel, 1873, *B. (Petersia) victrix* and *B. (P.) curtum* (Zittel, 1873), all Tithonian, Stramberg, and *B. (P.) costatum* (Gemmellaro, 1870), Tithonian, Sicily, Italy, and *B. lamberti* Cossmann, 1913, Auxerre, France (Cossmann, 1906, 1913). **Extant**

F. EUSTOMIDAE Cossmann, 1906 J. (BAJ–TTH) Mar.

First: *Diatinostoma euterpe* (d'Orbigny, 1852), Arthis, Calvados, France.
Last: *Ditretus nodosostriatus* (Peters, 1865) and *D. ?colloti* Cossmann, 1913, Portlandian, Murles, Hérault, France (Cossmann, 1913).

F. FAXIIDAE Ravn, 1933 T. (DAN) Mar. (see Fig. 8.4)

First and Last: *Faxia macrostoma* Ravn, 1933, Calcaire de Faxe, Denmark (Wenz, 1939).

F. OBTORTIONIDAE Thiele, 1925 K. (MAA)–Rec. Mar.

First: *Sandbergeria antecedens* and *S. crispicans* (Stoliczka, 1867), Arivalur Group, southern India (Cossmann, 1906; Wenz, 1939). **Extant**

F. SILIQUARIIDAE Anton, 1838 ?Tr. (LAD)/? K. (MAA)/T. (YPR)–Rec. Mar.

First: ? *Siliquaria triadica* Kittl, 1892, St Cassian Formation, Austria (Cossmann, 1912; Wenz, 1939).
Next oldest: *Siliquaria (Agathirses) lima* (Lamarck, 1818) and *S. (Pyxipoma) gracilis* (Deshayes, 1861), Cuisian, Vregny, Paris Basin, France (Cossmann and Pissarro, 1910).
Intervening: MAA of Egypt (Quaas, 1902, noted by K. Bandel in MS) **Extant**

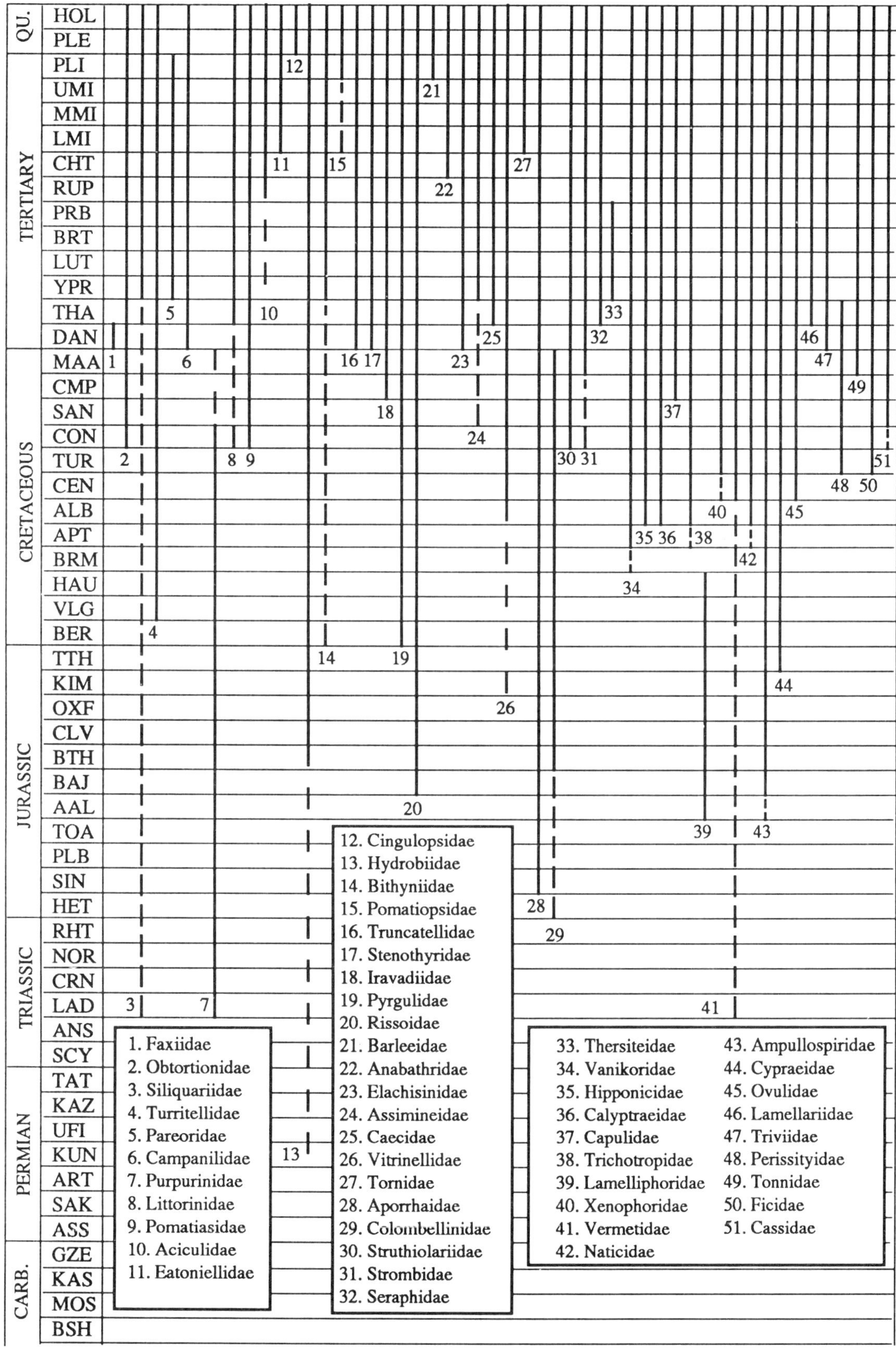

QU.
TERTIARY
CRETACEOUS
JURASSIC
TRIASSIC
PERMIAN
CARB.
HOL
PLE
PLI
UMI
MMI
LMI
CHT
RUP
PRB
BRT
LUT
YPR
THA
DAN
MAA
CMP
SAN
CON
TUR
CEN
ALB
APT
BRM
HAU
VLG
BER
TTH
KIM
OXF
CLV
BTH
BAJ
AAL
TOA
PLB
SIN
HET
RHT
NOR
CRN
LAD
ANS
SCY
TAT
KAZ
UFI
KUN
ART
SAK
ASS
GZE
KAS
MOS
BSH
1. Faxiidae
2. Obtortionidae
3. Siliquariidae
4. Turritellidae
5. Pareoridae
6. Campanilidae
7. Purpurinidae
8. Littorinidae
9. Pomatiasidae
10. Aciculidae
11. Eatoniellidae
12. Cingulopsidae
13. Hydrobiidae
14. Bithyniidae
15. Pomatiopsidae
16. Truncatellidae
17. Stenothyridae
18. Iravadiidae
19. Pyrgulidae
20. Rissoidae
21. Barleeidae
22. Anabathridae
23. Elachisinidae
24. Assimineidae
25. Caecidae
26. Vitrinellidae
27. Tornidae
28. Aporrhaidae
29. Colombellinidae
30. Struthiolariidae
31. Strombidae
32. Seraphidae
33. Thersiteidae
34. Vanikoridae
35. Hipponicidae
36. Calyptraeidae
37. Capulidae
38. Trichotropidae
39. Lamelliphoridae
40. Xenophoridae
41. Vermetidae
42. Naticidae
43. Ampullospiridae
44. Cypraeidae
45. Ovulidae
46. Lamellariidae
47. Triviidae
48. Perissityidae
49. Tonnidae
50. Ficidae
51. Cassidae

Fig. 8.4

F. TURRITELLIDAE Lovén, 1847 K. (VLG/APT)–Rec. Mar.

First: *Torquesia ? dupiniana* (Orbigny, 1842), Valanginian, Marolles, France. **Extant**
Next oldest: *Torquesia tamra* (Abbass, 1962), Crackers, Atherfield Clay, Atherfield, Isle of Wight, England, UK.
Comments: Phylogenetic analysis of the Cerithioidea (Houbrick, 1988) has shown the Turritellidae to be one of the most highly derived families. This supports the view that Palaeozoic genera sometimes placed in this family (and herein assigned to the Acanthonematidae) are probably turritellid homeomorphs. Supposed Triassic and Jurassic species often have heterostrophic apices preserved and can be referred to the Heterostropha.

F. PAREORIDAE Finlay and Marwick, 1937 T. (YPR–ZAN) Mar.

First: *Pareora* sp. Mangaorapan, White's Creek, North Canterbury, New Zealand (Beu and Maxwell, 1990).
Last: *Pareora striolata* (Hutton, 1885) Opoitian, New Zealand (Beu and Maxwell, 1990).
Comment: Maintained as a separate family by Beu and Maxwell, but included in the Turritellidae by Ponder and Warén (1988).

Superfamily CAMPANILOIDEA Douvillé, 1904

F. CAMPANILIDAE Douvillé, 1904 K. (CEN)–Rec. Mar.

First: *Procampanile* sp. (Pan, 1990), lower Member, Kukebai Formation, South Xinjiang, China. **Extant**

Section NEOTAENIOGLOSSA Haller, 1892

Superfamily LITTORINOIDEA Gray, 1840

F. PURPURINIDAE Zittel, 1895 Tr. (LAD)–K. (SEN) Mar.

First: *Angularia subpleurotomaria* (Münster, 1841) Stuoresmergel, St Cassian, Italy, and *Aristerostrophia gracilis* Broili, 1907, Ladinian, southern Alps (Ferenc, 1961).
Last: *Purpurina subcaucasica* Dzhalilov, 1977, Upper Cretaceous, Turkmenistan, former USSR.

F. LITTORINIDAE Gray, 1840 ?K (SEN)/T. (THA)–Rec. Mar./Semi-terr.

First: ?*Lemniscolittorina berryi* (Wade, 1926), Ripley Formation, Mississippi, USA (Sohl, 1960). In form and ornament this is not close to later littorinids, and may be unrelated (Reid, 1989). The protoconch of a species from the Campanian Coffee Sand Formation of Mississippi, referred to this genus by D. T. Dockery (pers. comm.), shows it to belong to the Mathildidae. **Extant**
Next oldest: *Melaraphe rissoides* (Deshayes, 1861) and *M. mausseneti* (Cossmann, 1907), both Sables de Chalons-sur-Vesle, Chenay (Marne), France (Cossmann and Pissarro, 1913), and *Littorina* ? sp. (Woods and Saul, 1986), Sepultura Formation, Baja California, Mexico (Reid, 1989).
Comments: Wenz's (1938) Cretaceous records of *Littorinopsis* from Europe and East Africa were not corroborated by Reid (1989). *Lacunina* Kittl, 1899 and species referred to *Lacuna* Turton, 1827 from the Triassic St Cassian fauna from Italy do not appear to be littorinids and are of uncertain systematic position.

F. POMATIASIDAE Gray, 1852 K. (CON)–Rec. Terr.

First: *Anapomatias astrongylum* Hrubesch, 1965, Gosau Group, Windisch Garsten, Austria (Hrubesch, 1965a). **Extant**

F. ANNULARIIDAE Henderson and Bartsch, 1920 Extant Terr.

F. ACICULIDAE Woodward, 1854 T. (?Eoc.)/(CHT)–Rec. Terr.

First: *Acicula (Platyla)* sp. was indicated by Wenz (1939) from the Eocene of Europe, but the record could not be located.
Next oldest: *Pseudotruncatella microceras* (A. Braun, 1843), Chattian, Landschneckenkalk, Hochheim-Flörsheim, Hessen-Nassau (Wenz, 1939). **Extant**

F. SKENEOPSIDAE Iredale, 1915 Extant Mar.

Superfamily CINGULOPSOIDEA Fretter and Patil, 1958

F. EATONIELLIDAE Ponder, 1968 T. (BUR)–Rec. Mar.

First: *Eatoniella (Dardanula) praecursor* (Laws, 1939), *E. (D.) sedicula* and *E. (D.) subexcavata* (Laws, 1941), all Otaian, New Zealand (Beu and Maxwell, 1990). **Extant**

F. CINGULOPSIDAE Fretter and Patil, 1958 Q. (PLE)–Rec. Mar.

First: *Eatonina (Otatara) subflavescens* Iredale, 1915, Pleistocene, New Zealand (Beu and Maxwell, 1990). **Extant**

F. RASTODENTIDAE Ponder, 1966 Extant Mar.

Superfamily RISSOOIDEA Gray, 1847

F. HYDROBIIDAE Troschel, 1857 ?P. (KUN)/J. (DOG)–Rec. FW/Brackish

First: ?*Hydrobia gondwanica* Cox, 1953, Ecca Beds, Karroo, Zimbabwe. Doubtfully hydrobiid, described as having a slightly heterostrophic protoconch (Cox, 1953).
Next oldest: *Pseudamnicola acuta, Amnicola kushuixaensis* and *A. shuidonggouensis* Pan, 1982, Middle Jurassic, China (Guo *et al.* , 1982). **Extant**

F. BITHYNIIDAE Gray, 1857 J. (MLM)–Rec. FW

First: *Bithynia haizhouensis* Yu, 1982, Fuxin Formation, western Liaoning, China. **Extant**
Comment: The calcified opercula, characteristic of this family, first occur in the lower Ypresian Reading Formation, Harefield, Middlesex, England (Cooper, 1976).

F. HYDROCOCCIDAE Thiele, 1928 Extant Brackish

F. POMATIOPSIDAE Stimpson, 1865 T. (Mio.)–Rec. FW

First: *Prosothenia* spp. (Neumayr, 1869), Dalmatian Miocene, former Yugoslavia (noted in Davis, 1979). **Extant**
Comment: Ponder (1985) implied a possible Cretaceous occurrence of the family.

F. TRUNCATELLIDAE Gray, 1840 T. (DAN)–Rec. Brackish/Semi-terr.

First: *Truncatella minor* Briart and Cornet, 1887, Calcaire de Mons, Mons, Belgium (Glibert, 1973). **Extant**

F. STENOTHYRIDAE Fischer, 1885 T. (DAN)–Rec. FW/Brackish

First: *Stenothyra pupiformis* (Briart and Cornet, 1887), Calcaire de Mons, Mons, Belgium (Glibert, 1973). **Extant**

F. FALSICINGULIDAE Slavoshevskaya, 1975 **Extant** Mar.

F. IRAVADIIDAE Thiele, 1928 K. (CMP)–Rec. Brackish/Mar.

First: *Nozeba* sp. Coffee Sand Formation, Friendship near Chapelville, Lee County, Mississippi, USA (D. T. Dockery, pers. comm.) **Extant**

F. PYRGULIDAE Brusina, 1881 K. (NEO)–Rec. FW

First: *Micromelania katoensis* and *Itomelania basicordata* (Suzuki, 1943), Naktong Group, Kyongsang-namdo and K.-bukdo, South Korea (Hayami and Kase, 1977). **Extant**
Intervening: ?DAN, Mio.–PLI.

F. RISSOIDAE Gray, 1847 J. (BAJ)–Rec. Mar./Brackish

First: *Rissocerithium nicosiai* Conti and Fischer, 1981, *Zebinostoma nicolisi* and *Z. turrita* (Parona, 1894), and *Trochoturboella tethysiana* Conti and Fischer, 1982, all Case Canepine, Spoleto, Italy (Conti and Fischer, 1982). **Extant**

F. BARLEEIDAE Gray, 1857 T. (PIA)–Rec. Mar.

First: *Fictonoba similis* (Laws, 1950), lower Piacenzian, New Zealand (Beu and Maxwell, 1990). **Extant**

F. EMBLANDIDAE Ponder, 1985 **Extant** Mar.

F. ANABATHRIDAE Coan, 1964 T. (CHT)–Rec. Mar./Brackish.

First: *Anabathron (Scrobs) chattonensis* (Laws, 1950), Landon Creek, North Otago, New Zealand (Beu and Maxwell, 1990). **Extant**

F. EPIGRIDAE Ponder, 1985 **Extant** Mar.

F. ELACHISINIDAE Ponder, 1985 T. (DAN)–Rec. Mar.

First: *Dissochilus lineatus* (Briart and Cornet, 1887), Calcaire de Mons, Mons, Belgium (Glibert, 1973). **Extant**

F. ASSIMINEIDAE H. and A. Adams, 1856 ?K. (SAN)/T. (YPR)–Rec. Brackish/Semi-terr.

First: ?*Turbacmella europaea* Hrubesch, 1965, Gosau Group, Glanegg near Salzburg, Austria (Hrubesch, 1965a).
Next oldest: *Assiminea stenochora* Cossmann, 1888, Sparnacian, Mont-Bernon near Epernay, France (Cossmann, 1888). **Extant**
Intervening: LUT–PLE.

F. CAECIDAE Gray, 1850 T. (THA)–Rec. Mar.

First: *Watsonia novallacense* (Cossmann, 1907), Sables de Bracheux, Noailles near Beauvais, France (Cossmann, 1912). **Extant**

F. VITRINELLIDAE Bush, 1897 ?J. (KIM)/K. (CEN)–Rec. Mar.

First: ?'*Teinostoma*' *valfinense* Loriol, 1886?, Calcaires blancs, Valfin, France (Cossmann, 1918).
Next oldest: *Cenomanella archiaciana* (Orbigny, 1847), Cenomanian, Le Mans (Sarthe), France (Wenz, 1938). **Extant**
Comment: *Teinostoma* spp., showing little difference from living forms, were widespread by the late Cretaceous.

F. TORNIDAE Sacco, 1896 T. (BUR)–Rec. Mar.

First: *Tornus trigonostoma* (Basterot, 1825) and *T. quadrifasciatus* (Grateloup, 1832), Laag van Miste, Winterswijk–Miste, The Netherlands (Janssen, 1984) and *Naricava huttoni* (Marwick, 1924), Altonian, New Zealand (Beu and Maxwell, 1990). **Extant**
Comments: The Tornidae are difficult to identify in the Tertiary as the shells are convergent with those of Vitrinellidae and related families. *Adeorbis bicarinatus* and *A. spirorbis* (Lamarck, 1806) of the French Ypresian resemble some tornids in shape but are connected to typical *Adeorbis* (Vitrinellidae) by a range of intermediate forms.

Superfamily STROMBOIDEA Rafinesque, 1815

F. APORRHAIDAE Gray, 1850 J. (SIN)–Rec. Mar.

First: *Anchura (Pietteia) hudlestoni* (Wilson, 1887), lower Lias, Bristol, Avon, England, UK (Cossmann, 1904). **Extant**
Comments: Wenz (1940) doubtfully suggested a Rhaetian age for *Spinigera* Orbigny, 1850, but the species referred to is not certain. The family is considered to include the Harpagodidae Pčhelintsev, 1963.

F. COLOMBELLINIDAE Fischer, 1884 J. (?LIA/BTH)–K. (MAA) Mar.

First: ?*Alariopsis clathrata* Gemmellaro, 1878, Lias, Sicily, Italy.
Next oldest: *Columbellaria bathonica* Cossmann, 1913, St Gaultier, France.
Last: *Colombellina americana* (Wade, 1926), Ripley Formation, Coon Creek, Tennessee and Mississippi, USA (Sohl, 1960).
Comment: This family is perhaps the stem group of the Cypraeoidea and Tonnoidea (Taylor and Morris, 1988).

F. STRUTHIOLARIIDAE (Gabb, 1868) Fischer, 1884 K. (SEN)–Rec. Mar.

First: *Conchothyra parasitica* Hutton, 1877, Waipara Formation, Waipara River, New Zealand (Beu and Maxwell, 1990). **Extant**

F. STROMBIDAE Rafinesque, 1815 K. (?CON/MAA)–Rec. Mar.

First: ?*Dientomochilus stueri* Cossmann, 1904, Condat (Lot-et-Garonne), France (Cossmann, 1904). Based on an internal mould missing its lip, and therefore open to some doubt.
Next oldest: '*Calyptraphorus*' *palliatus* (Forbes, 1848), Trichinopoly and Arivalur Groups, south India (Stoliczka, 1868). **Extant**
Comment: The genus *Pugnellus*, dating from the Santonian, is now considered an aporrhaid (Popenoe, 1983).

F. SERAPHIDAE Gray, 1853 T. (THA)–Rec. Mar.

First: *Seraphs sopitus* (Solander, *in* Brander, 1766), l'Aude or l'Hérault, southern France (noted in Jung, 1974). **Extant**

F. THERSITEIDAE Savorin, 1914 T. (Eoc.) Mar.

First and Last: *Thersitea gracilis* (Coquand, 1862), Tunis, Tunisia, and *T. ponderosa* (Coquand, 1862), Tébessa, Constantine Province, Algeria (Wenz, 1943).

Superfamily **VANIKOROIDEA Gray, 1840**

F. VANIKORIDAE Gray, 1840 K. (BRM/APT)–Rec. Mar.

First: *Vanikoropsis decussata* (Deshayes, 1842), Kimigahama Formation, Choshi area, Japan (Hayami and Kase, 1977). **Extant**

F. HIPPONICIDAE Troschel, 1861 K. (ALB)–Rec. Mar.

First: *'Hipponix'* cf. *dixoni* Deshayes, Cambridge Greensand, Cambridge, England, UK. **Extant**

Superfamily **CALYPTRAEOIDEA Lamarck, 1809**

F. CALYPTRAEIDAE Lamarck, 1809 K. (ALB)–Rec. Mar.

First: *Galericulus altus* (Seeley, 1861), Cambridge Greensand, Cambridge, England, UK (Wenz, 1938). **Extant**

F. CAPULIDAE Fleming, 1822 K. (CMP)–Rec. Mar.

First: *Capulus verus* (Böhm, 1885), Aachen Greensand, Vaals, The Netherlands and Königsthor, Germany (Holzapfel, 1888). **Extant**
Comments: This appears to be the first recognizable capulid. Most earlier records of capuliform gastropods are of uncertain affinities, including those figured by Zardini (1978) from the Triassic St Cassian Formation.

F. TRICHOTROPIDAE H. and A. Adams, 1854 K. (?APT/ALB)–Rec. Mar.

First: ?*Atresius cornuelianus* (Orbigny, 1843), Aptian, Vaucluse, France. **Extant**
Next oldest: *Atresius lallierianus* (Orbigny, 1843), Gault Yonne, Sainte Florentin, France.

Superfamily **XENOPHOROIDEA Troschel, 1852**

F. LAMELLIPHORIDAE Korobkov, 1955 J. (AAL)–K. (NEO) Mar.

First: *Lamelliphorus supraliasinus* (Vacek, 1886), Aalenian, Tuscany, Calabria and Veneto, Italy.
Last: *Lamelliphorus ? tortilis* (Peron, 1900), Neocomian, Yonne, France (Ponder, 1983).
Comment: The shell morphology resembles that of xenophorids but lacks the attachment of foreign objects. Their systematic position is, however, still uncertain.

F. XENOPHORIDAE Troschel, 1852 K. (?CEN/TUR)–Rec. Mar.

First: ?*Xenophora* ? sp. (noted in Stephenson, 1952), Woodbine Formation, Texas, USA.
Next oldest: *Xenophora simpsoni* Stanton, 1893, Turonian, Colorado, USA, and *Xenophora grasi* Roman and Mazeran, 1920, Turonian, France (Ponder, 1983). **Extant**

Superfamily **VERMETOIDEA Rafinesque, 1815**

F. VERMETIDAE Rafinesque, 1815 ?Tr. (LAD)–Rec. Mar.

First: ?*Provermicularia circumcarinata* (Stoppani), Esinokalk, Ca'nova am Monte Croce, Esino, Italy (Kittl, 1899 noted in Wenz, 1939), and ?*Pseudobrochidium germanicum* Grupe, 1907, lower Keuper, Hannover, Germany (Wenz, 1939). **Extant**
Next oldest: *Burtinella damesi* (Noetling, 1885), Cenomanian, Germany (Cossmann, 1912; Wenz, 1939).
Comment: The family, as presently understood, is probably a polyphyletic group featuring lax coiling. Convergence makes it difficult to identify Mesozoic vermetids from the adult shells.

Superfamily **NATICOIDEA Forbes, 1838**

F. NATICIDAE Forbes, 1838 K. (APT/ALB)–Rec. Mar.

First: *Euspira* sp., Hiraiga Formation, Miyako area, NE Japan (Kase, 1984). **Extant**
Comments: This would appear to be one of the earliest records of an unequivocal naticid. Shells of *Gyrodes* spp. from the upper Albian Blackdown Greensand of Devon, UK, occur in association with drill holes, as made by living naticids (Taylor *et al.*, 1983).

F. AMPULLOSPIRIDAE Cox, 1930 J. (AAL/BAJ)–Rec. Mar.

First: *Ampullospira adducta* (Phillips, 1829), Dogger, Yorkshire coast, England, UK (Hudleston, 1892). **Extant**
Comments: The simple morphology of the naticiform shell is seen to recur in distantly related groups. Earlier Mesozoic forms are of uncertain affinities. Bandel (1988b) has shown the larval shell morphology of genera such as *Amauropsis* to be neritoidean in character and he considers that there are no confirmed naticoids from the Triassic. Kase (1990) briefly notes that the sole surviving species, *Cernina fluctuata* (G. B. Sowerby, 1825) is an archtaenioglossan. Pending a fuller investigation, the family is retained in the Naticoidea.

Superfamily **CYPRAEOIDEA Rafinesque, 1815**

F. CYPRAEIDAE Rafinesque, 1815 J. (TTH)–Rec. Mar.

First: *Palaeocypraea tithonica* and *Bernaya gemmellaroi* Stefano, 1882, Termini Imerese, Sicily, Italy (Schilder and Schilder, 1971). **Extant**

F. OVULIDAE Fleming, 1822 K. (CEN)–Rec. Mar.

First: *Eocypraea pilulosa* (Stoliczka, 1867), Cenomanian, India (Schilder and Schilder, 1971). **Extant**

Superfamily **LAMELLARIOIDEA d'Orbigny, 1841**

F. LAMELLARIIDAE d'Orbigny, 1841 T. (THA)–Rec. Mar.

First: *Lamellaria inopinata* Cossmann, 1907 Jonchery (Aisne), France (Cossmann and Pissarro, 1910). **Extant**

F. TRIVIIDAE Troschel, 1863 T. (DAN)–Rec. Mar.

First: *Johnstrupia faxensis* Ravn, 1933, Calcaires à Coralliaires, Faxe, Denmark (Schilder and Schilder, 1971). **Extant**

F. PSEUDOSACCULIDAE Wenz, 1940 **Extant** Mar.

Superfamily TONNOIDEA Suter, 1913

F. PERISSITYIDAE Popenoe and Saul, 1987 K. (TUR)–T. (THA) Mar.

First: *Pseudocymia aurora* Popenoe and Saul, 1987, Frazier Siltstone, Salt Creek, Shasta County, California, USA.
Last: *Perissitys stewarti* (Zinsmeister, 1983), Lower Santa Susana Formation, Simi Hills, Ventura County, California, USA (Popenoe and Saul, 1987).

F. TONNIDAE Suter, 1913 K. (MAA)–Rec. Mar.

First: *Protodolium speighti* (Trechmann), Maastrichtian, Waipara, New Zealand (Wenz, 1941). **Extant**

F. FICIDAE Meek, 1864 ?K. (CMP)/T. (DAN)–Rec. Mar.

First: ?*Protopirula capensis* (Rennie, 1931), Umzamba Beds, Pondoland, South Africa (Wenz, 1939).
Next: *Priscoficus bicarinatus* (Briart and Cornet, 1870), Calcaire de Mons, Mons, Belgium (Glibert, 1973) and *P. obtusus* (Marshall, 1917), Wangaloa Formation, Wangaloa, New Zealand (Beu and Maxwell, 1990). **Extant**

F. CASSIDAE Latreille, 1825 K. (CON/SAN)–Rec. Mar.

First: *Pseudogaleodea tricarinata* Nagao, 1932, Kawakami-Mine, Toyohara-gun, Sakhalin, former USSR (Hayami and Kase, 1977). **Extant**

F. RANELLIDAE Gray, 1854 K. (BRM)–Rec. Mar. (see Fig. 8.5)

First: *Eoranella kiliani* Sayn, 1932, Urgonian sables calcaires, Bogaris, near Barcelona, Spain. **Extant**
Comment: Considered by Sayn (1932) to have many ranellid features, this species also bears a strong resemblance to later Muricidae.

F. BURSIDAE Thiele, 1925 ?K. (APT)/T. (DAN/THA)–Rec. Mar.

First: ?*Hanaibursa aequilana* (Parona, 1909), Hiraiga Formation, Miyako area, Japan (Kase, 1984). The species is tentatively assigned to the Bursidae.
Next oldest: *Bursa saundersi* Adegoke, 1977, Ewekoro Formation, Ewekoro, Nigeria. **Extant**

F. LAUBIERINIDAE Warén and Bouchet, 1990 **Extant** Mar.

Superfamily CARINARIOIDEA Blainville, 1818

F. ATLANTIDAE Rang, 1829 T. (DAN)–Rec. Mar. (pelagic)

First: *Eoatlanta* sp. nov. (Rosenkrantz, 1960), Calcaire de Faxe, Denmark. Recorded by Ravn (1933) as *E. spiruloides* (Lamarck, 1824). **Extant**

F. CARINARIIDAE Blainville, 1818 J. (TOA)–Rec. Mar. (pelagic)

First: *Coelodiscus minutus* (Zieten, 1832) and *C. fluegeli* Bandel and Hemleben, 1987, *Posidonia* Shales, Franconia and Swabia, Germany (Bandel and Hemleben, 1987). **Extant**

F. FIROLIDAE Rang, 1829 J. (TOA)–Rec. Mar. (pelagic)

First: *Pterotrachea liassica* Bandel and Hemleben, 1987, *Posidonia* Shales, Germany (Bandel and Hemleben, 1987). **Extant**

F. PTEROSOMATIDAE Rang, 1829 **Extant** Mar. (pelagic)

Section PTENOGLOSSA Gray, 1853

Superfamily TRIPHOROIDEA Gray, 1847

F. CERITHIOPSIDAE H. and A. Adams, 1853 K. (VLG)–Rec. Mar.

First: *Uchauxia wisei* Abbass, 1973, Claxby Ironstone, Nettleton, Lincolnshire, UK (Abbass, 1973). **Extant**

F. TRIFORIDAE Jousseaume, 1884 K. (APT)–Rec. Mar.

First: *Orthochetus hantoniensis* Abbass, 1973, Lower Greensand, Atherfield, Isle of Wight, England, UK (Abbass, 1973). **Extant**

F. SHERBORNIIDAE Iredale, 1917 **Extant** Mar.

F. TRIPHORIDAE Gray, 1847 ?K. (MAA)/T. (THA)–Rec. Mar.

First: ?*Triphora cincta* (Kaunhowen, 1898), Chalk, Maastricht, The Netherlands (Cossmann, 1906).
Next oldest: *Triphora staadti* (Cossmann, 1907), Sables de Chalons-sur-Vesle, Chenay (Marne), France. **Extant**

Superfamily JANTHINOIDEA Lamarck, 1812

F. EPITONIIDAE S. S. Berry, 1910 (1812) J. (BTH)–Rec. Mar.

First: *Proacirsa inornata* (Terquem and Jourdy, 1869) Fuller's Earth, Clapes, Lorraine, France (Cossmann, 1912). **Extant**

F. JANTHINIDAE Lamarck, 1812 T. (MES)–Rec. Mar. (pelagic)

First: *Hartungia typica* Bronn, 1861, Messinian, East Cape, New Zealand; Santa Maria Islands and Azores (Beu and Maxwell, 1990). **Extant**
Comment: Not *Janthina cimbrica* Sorgenfrei, 1958 (Miocene, The Netherlands) which was shown to be a juvenile cypraeid (Janssen, 1984).

Suborder UNCERTAIN

Superfamily EULIMOIDEA Troschel, 1853

F. ACLIDIDAE G. O. Sars, 1878 T. (DAN)–Rec. Mar.

First: *Graphis formosa* (Briart and Cornet, 1873), Calcaire de Mons, Mons, Belgium (Glibert, 1973). **Extant**

F. EULIMIDAE Troschel, 1853 K. (?CEN/MAA)–Rec. Mar.

First: *Eulima persimplica* and *E. laevigata* Wade, 1926, Ripley Formation, Tennessee, USA (Sohl, 1964a). **Extant**
Comments: These are among the earliest described species, although Sohl (1967) reported unspecified eulimids from the Cenomanian to Maastrichtian of New Mexico, Colorado and Wyoming, USA. Kier (1981) described an upper Albian spatangoid echinoid from Texas showing

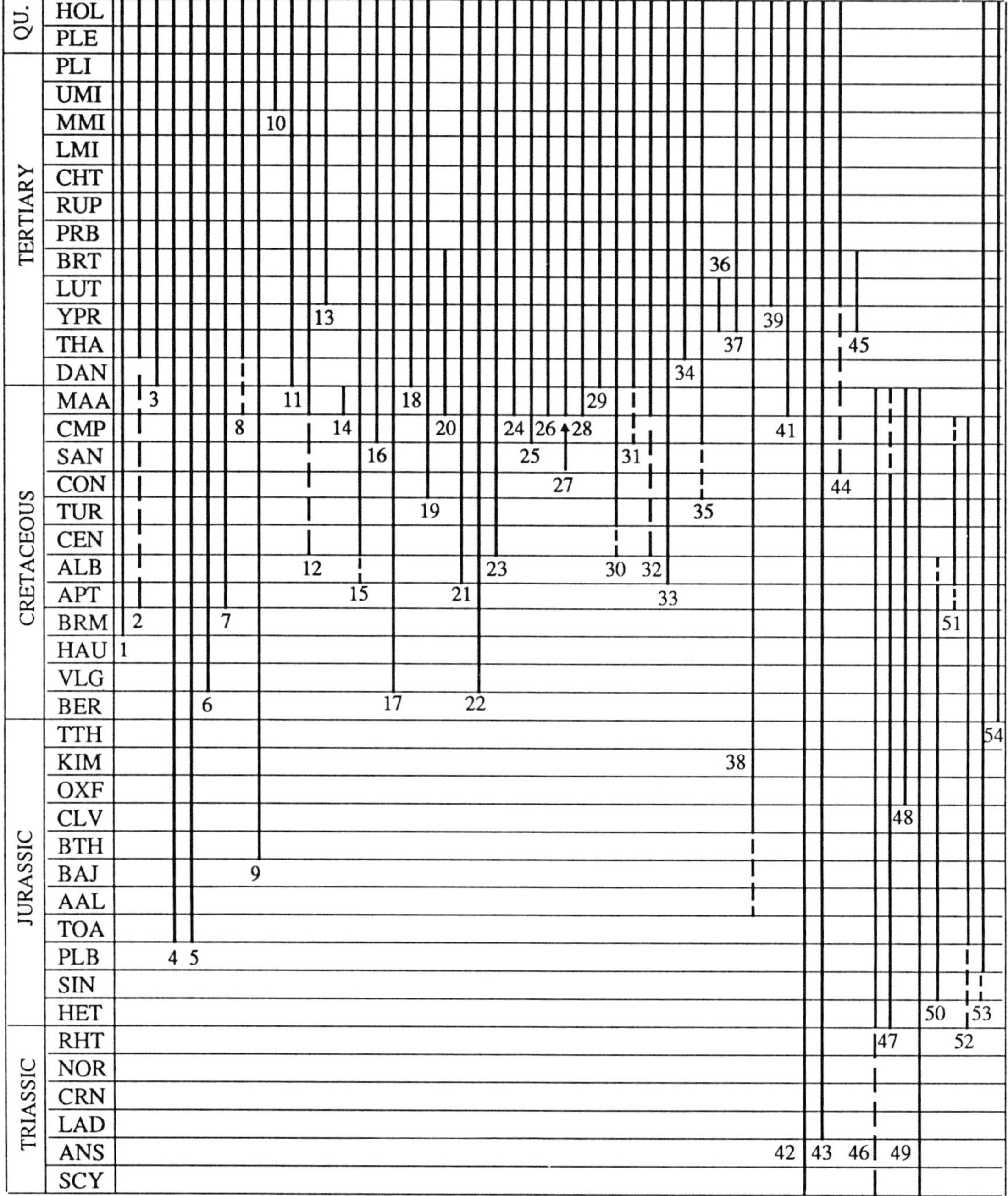

Fig. 8.5

the characteristic ambulacral 'perforation' of a parasitic eulimid.

F. STILIFERIDAE H. and A. Adams, 1853 T. (LUT)–Rec. Mar.

First: *Semistylifer pellucidus* (Deshayes, 1824), Calcaire Grossier, Parnes, France (Cossmann, 1888). **Extant**
Comments: The earliest stiliferid is listed here, although the group is now included in the Eulimidae (Warén, 1983). Also included in the Eulimidae by Warén (1983) are the families Thycidae Thiele, 1931, Asterophilidae Thiele, 1925, Paedophoropodidae Ivanov, 1933, and Entoconchidae Gill, 1871, all Recent, marine and without fossil records.

Section UNCERTAIN

Superfamily UNCERTAIN

F. WEEKSIIDAE Sohl, 1960 K. (MAA) Mar.

First: *Weeksia lubbocki* (Stephenson, 1941), Nacotch Sand, Navarro Group, Texas, USA (Sohl, 1960).
Last: *Weeksia deplanata* (Johnson, 1905), Prairie Bluff Chalk, Alabama and Mississippi, USA (Sohl, 1960).
Comments: Sohl included the Triassic genera *Discohelix* Dunker, 1848 and *Amphitomaria* Koken, 1897 in his family of supposed archaeogastropods. Bandel (1988a) showed *Discohelix* to be a trochoidean, *Amphitomaria* a hetero-

Fig. 8.5

strophan and placed *Weeksia* in the Neogastropoda, although Batten (1984b) considered it a mesogastropod.

Section NEOGASTROPODA Thiele, 1929

Superfamily MURICOIDEA da Costa, 1776

F. MURICIDAE da Costa, 1776 K. (?ALB/CEN)–Rec. Mar.

First: ?*Paramorea lineata* (J. de C. Sowerby, 1836), Blackdown Greensand, Blackdown, Devon, England, UK (Taylor *et al.*, 1983). **Extant**
Next oldest: *Poirieria* (?*Paziella*) *cenomae* Garvie, 1991, Cenomanian, Saxony, eastern Germany.
Comments: The relationships of pre-Tertiary forms are unclear. If the Moreinae Conrad, 1860 are muricids (see Sohl, 1964a) then the above record is the oldest. The many species of supposed Rapaninae from the Albian Blackdown Greensand of England, UK (Taylor *et al.*, 1983) may be rapaninid homeomorphs and not closely related (N. J. Morris, pers. comm.).

F. CORALLIOPHILIDAE Chenu, 1859 K. (CMP)–Rec. Mar.

First: *Lowenstamia funiculus* Sohl, 1964a, Coffee Sand Formation, near Ratliff, Lee Co., Mississippi, USA. **Extant**
Comment: *Lowenstamia* Sohl, 1964 and the contemporaneous *Sargana* Stephenson, 1923 are possibly neotaenioglossan muricoid homeomorphs.

F. BUCCINIDAE Rafinesque, 1815 K. (VLG)–Rec. Mar./Brackish

First: *'Buccinum' incertum* Orbigny, *in* Murchison, 1845,

Speeton Clay, Speeton, North Yorkshire, England, UK (N. J. Morris, pers. comm.). **Extant**

F. COLUMBELLIDAE Swainson, 1840 T. (DAN)–Rec. Mar.

First: *Mitrella (Columbellopsis) edmondi* (Briart and Cornet, 1870), Calcaire de Mons, Mons, Belgium (Glibert, 1973). **Extant**

F. NASSARIIDAE Iredale, 1916 K. (CON)–Rec. Mar.

First: *Buccinopsis* sp. nov. Sohl, 1964b, Eutaw Formation, Alabama, USA. **Extant**
Comment: *Buccinopsis* Conrad, 1857 is considered to represent the first recognizable nassariid (Nuttall, *in* Taylor *et al.*, 1980).

F. STREPSIDURIDAE Cossmann, 1901 K. (MAA)–T. (BRT) Mar.

First: *Hydrotribulus nodosus* Wade, 1916, Ripley Formation, Mississippi, USA (Sohl, 1964a).
Last: *Strepsidura turgida* (Solander, *in* Brander, 1766) Barton Clay Formation, Barton, Hampshire, England, UK.
Comment: Three other species from Oligocene strata, formerly assigned to *Strepsidura* Swainson, 1840, have all proved to belong to the Nassariidae (Nuttall and Cooper, 1973).

F. MELONGENIDAE Gill, 1867 K. (ALB)–Rec. Mar.

First: *Tantunia clathrata* (J. de C. Sowerby), Blackdown Greensand, Blackdown, Devon, England, UK (Taylor *et al.*, 1983). **Extant**

F. FASCIOLARIIDAE Gray, 1853 K. (VLG)–Rec. Mar.

First: *'Fusus' valanginiensis* Pictet and Campiche, 1872?, Auberson, St Croix, Switzerland. **Extant**
Comment: Such early fusiform gastropods may belong to a stem group of the paraphyletic Fasciolariidae and the Turridae (N. J. Morris, pers. comm.).

F. VOLUTIDAE Rafinesque, 1815 K. (CEN)–Rec. Mar.

First: *Carota pendula* Stephenson, 1952, Lewisville Member, Woodbine Formation, Texas, USA. **Extant**

F. HARPIDAE Bronn, 1849 K. (MAA)–Rec. Mar.

First: *Eoharpa sinuosa* Stephenson, 1955, Owl Creek Formation, Missouri and Mississippi, USA (Sohl, 1964a). **Extant**

F. VASIDAE H. and A. Adams, 1853 K. (CMP)–Rec. Mar.

First: *'Tudicla' monheimi* (Müller, 1851), Aachen Greensand, Vaals, The Netherlands (Holzapfel, 1888). **Extant**
Comment: By the Maastrichtian the family was represented by about ten genera in the USA alone (Sohl, 1964a; Saul, 1988).

F. COLUMBARIIDAE Dall, 1904 K. (MAA)–Rec. Mar.

First: *Columbarium heberti* (Briart and Cornet, 1880), Tuffeau de Maastricht, Limburg, The Netherlands (Darragh, 1969). **Extant**
Comment: The columbariids are now generally considered a subfamily of the Vasidae.

F. PSEUDOLIVIDAE Fischer, 1884 K. (MAA)–Rec. Mar.

First: *Popenoeum subcostatum* (Stoliczka, 1868), Arivalur Group, Ninnyoor area, south India (Squires *et al.*, 1989), and *Ptychosyca inornata* Gabb, 1876, Ripley Formation, Mississippi, USA (Sohl, 1964a). **Extant**

F. OLIVIDAE Latreille, 1825 K. (MAA)–Rec. Mar.

First: *Tanimasanoria japonica* (Kase, 1990), Azenotami Mudstone Member, lower MAA, Mutsuo Formation, Osaka, Japan. **Extant**

F. MARGINELLIDAE Fleming, 1828 T. (DAN)–Rec. Mar.

First: *Gibberula* sp. nov. (Glibert, 1973), Calcaire de Mons, Mons, Belgium. **Extant**

F. MITRIDAE Swainson, 1831 K. (?CEN/TUR)–Rec. Mar.

First: *Imbricaria (Sohlia) conoidea* (Matheron, 1843), Port de Figuiéres, near Marseilles, France (Cernohorsky, 1970). **Extant**

F. VOLUTOMITRIDAE Gray, 1854 ?K. (CMP)/T. (DAN)–Rec. Mar.

First: ?*Conomitra* sp. (Sohl, 1964b), Coffee Sand Formation, Mississippi, USA. This is possibly referable to the Volutidae (Cernohorsky, 1970).
Next oldest: *Conomitra glabra* (Ravn, 1933), Calcaire de Faxe, Nez, Faxe, Denmark. **Extant**

F. COSTELLARIIDAE MacDonald, 1860 K. (?CEN/MAA)–Rec. Mar.

First: ?*Mesorhytis decorosa* (Stephenson, 1952), Woodbine Formation, Texas, USA.
Next oldest: *Mesorhytis ripleyana* (Wade, 1926), Ripley Formation, Coon Creek, Tennessee, USA (Cernohorsky, 1970). **Extant**

Superfamily CANCELLARIOIDEA Gray, 1853

F. CANCELLARIIDAE Gray, 1853 K. (ALB)–Rec. Mar.

First: *Palaeocancellaria hoelleitenensis* Kollmann, 1976, Losenstein Formation, Losenstein, Austria. **Extant**
Comment: Ponder (1973) suggests that the cancellariids represent the sister group of all other living Neogastropoda. Their recognition earlier in the Cretaceous is difficult as the diagnostic shell characters were not fully developed.

Superfamily CONOIDEA Fleming, 1822

F. CONIDAE Fleming, 1822 T. (THA)–Rec. Mar.

First: *Hemiconus bicoronatus* (Melleville, 1843), Sables de Chalons-sur-Vesle, Jonchery-sur-Vesle (Marne), France (Cossmann, 1913). **Extant**
Comment: All records of Conidae from the Cretaceous

appear to be based on doubtful forms or are mistakenly dated.

F. TURRIDAE Swainson, 1840 K (?CON/SAN/CMP)–Rec. Mar.

First: ?'*Pleurotoma*' *subfusiformis* d'Orbigny, 1831, Gosau Group, Austria (Stoliczka, 1867).
Next oldest: *Struthiolariopsis ferrieri* (Philippi, 1887), Quiriquina, Chile, and *Beretra preclara* Sohl, 1964b, Coffee Sand Formation, Mississippi, USA (Sohl, 1964b). **Extant**

F. SPEIGHTIIDAE Powell, 1942 T. (YPR–LUT) Mar.

First: *Andicula occidentalis* (Woods, 1922), Negritos of Peru (Powell, 1966).
Last: *Speightia spinosa* (Suter, 1917), Bortonian, Waihao River, South Island, New Zealand (Powell, 1966).
Comment: Although allowed familial status by some authors, some if not all of the speightiids may prove to belong in the Fasciolariidae.

F. TEREBRIDAE Mörch, 1852 T. (YPR)–Rec. Mar.

First: *Hastula (Terebrellina) plicatula* (Lamarck, 1805), Blackheath Beds, Abbey Wood, London, England, UK (Wrigley, 1942). **Extant**
Comment: Wenz (1943) doubtfully extended the range of the family into the Cretaceous but this appears to be unsubstantiated.

Suborder HETEROSTROPHA Fischer, 1885

Superfamily VALVATOIDEA Gray, 1840 and Thompson, 1840

F. VALVATIDAE Gray, 1840 and Thompson, 1840 J. (DOG)–Rec. FW

First: *Amplovalvata antiqua* and *A. obliqua* Pan, 1982, Middle Jurassic, China (Guo *et al.*, 1982). **Extant**

F. CORNIROSTRIDAE Ponder, 1990 **Extant** Shallow mar.

F. ORBITESTELLIDAE Iredale, 1917 T. (LUT)–Rec. Mar.

First: *Orbitestella (Omalogyrina) plicatella* (Cossmann, 1888) middle Lutetian, Villiers-St-Frédéric (Yvelines), France (J. Le Renard, unpublished data). **Extant**

Superfamily PYRAMIDELLOIDEA Gray, 1840

F. STREPTACIDIDAE Knight, 1931 D. (GIV)–P. (LEN) Mar.

First: *Streptacis* sp. Linsley, 1967, Anderdon Limestone, Michigan, USA.
Last: *Streptacis piercei* Erwin, 1988, Bone Spring Formation, Guadelupe Mountains, Texas, USA.

F. PYRAMIDELLIDAE Gray, 1840 K. (MAA)–Rec. Mar.

First: *Creonella triplicata* Wade, 1917, *C. subangulata* and *C. turretiforma* Sohl, 1964, all Ripley Formation, Mississippi, USA (Sohl, 1964a). **Extant**

F. AMATHINIDAE Ponder, 1988 **Extant** Mar.

Superfamily ARCHITECTONICOIDEA Gray, 1850

F. MATHILDIDAE Dall, 1889 P. (GUA)–Rec. Mar.

First: *Promathilda spirocostata* Batten and Stokes, 1986, Moenkopi Formation, Utah, USA. **Extant**

F. ARCHITECTONICIDAE Gray, 1850 Tr. (LAD)–Rec. Mar.

First: *Rinaldoconchus ampezzanus* (Zardini, 1980) and *Amphitomaria cassiana* (Koken, 1889), St Cassian Formation, Cortina d'Ampezzo, Italy (Bandel, 1988a). **Extant**

Superfamily RISSOELLOIDEA Gray, 1850

F. RISSOELLIDAE Gray, 1850 **Extant** Mar.

Superfamily OMALOGYROIDEA G. O. Sars, 1878

F. OMALOGYRIDAE G. O. Sars, 1878 ?K. (SAN)/T. (LUT)–Rec. Mar./Brackish

First: ?*Omalogyra* sp. (Bandel, 1988), Amman Formation, Jordan.
Next oldest: *Omalogyra cf. atomus* (Philippi, 1841), Thionville-sur-Opton (Yvelines), France (J. Le Renard, unpublished data). **Extant**
Comment: The family is linked to the Architectonicidae by the intermediate genera, *Amphitomaria* Koken, 1897 and *Neamphitomaria* Bandel, 1988a, from the late Cretaceous of the eastern USA, and the Tertiary, respectively (Bandel, 1988a).

Superfamily GLACIDORBOIDEA Ponder, 1986

F. GLACIDORBIDAE Ponder, 1986 **Extant** FW

Suborder and Superfamily UNCERTAIN

F. OMALAXIDAE Wenz, 1938 T. (YPR–BRT) Mar.

First: *Omalaxis* sp. (Bandel, 1988), Marnes de la Tuilerie de Gan (Pyrenées Atlantiques), France, and *O. deshayesi* (Michaud, *in* Deshayes, 1863) and *O. laudunensis* (Defrance, 1828), Sables de Cuise, Cuise-la-Motte (Oise), France (Gougerot and Le Renard, 1980).
Last: *Omalaxis disjunctus* (Lamarck, 1804) and *O. cresnensis* (Morlet, 1888), Marinesian, Le Quoniam (Val-d'Oise), France (Gougerot and Le Renard, 1980).
Comment: The placement of this Eocene family is problematic.

Superfamily NERINEOIDEA Zittel, 1873

Pčhelintsev (1965) considered the three families below to be superfamilies containing fifteen families. This scheme has not won general acceptance, and a more traditional system is used here.

F. NERINEIDAE Zittel, 1873 ?P. (ART)/J. (HET)–K. (MAA) Mar.

First: ?*Prodiozoptyxis permiana* Batten, 1985, Colina Limestone, Arizona, USA.
Last: *Diozoptyxis marrotiana* (d'Orbigny, 1842), Font-Barrade, near Bergerac, France (Cossmann, 1896).
Comment: The genus *Prodiozoptyxis* is assigned to the Nerineidae with question, based largely on the similarities

to the Cretaceous genus *Diozoptyxis*, which according to Houbrick (1981), may not be a nerineid.

F. NERINELLIDAE Pčhelintsev, 1960 J. (HET)–K. (SEN) Mar.

First: *Nerinella norigliensis* (Tausch), HET, Alpes méridionelles, France (Cossmann, 1896).
Last: *Multiptyxis gissarensis* (Pčhelintsev, 1960), Senonian, former 'Soviet Central Asia' (Pčhelintsev, 1965).

F. ITIERIIDAE Cossmann, 1896 J. (OXF)–K. (MAA) Mar.

First: *Itieria cabanetiana* (Orbigny, 1841), Oyonnax, near Nantua (Ain), France.
Last: *Itruvia carinata* (Reuss), Calcaire à Hippurites, Czechoslovakia (Cossmann, 1896; Kollmann and Sohl, 1980).

Subclass EUTHYNEURA Spengel, 1881

Superorder OPISTHOBRANCHIA Milne-Edwards, 1848

Order CEPHALASPIDEA Fischer, 1883

Superfamily CYLINDROBULLINOIDEA Wenz, 1947

F. CYLINDROBULLINIDAE Wenz, 1847 C. (TOU)–K. (MAA) Mar.

First: *Girtyspira* sp. Yoo, 1988, Namoi Formation, New South Wales, Australia.
Last: *Acteonina obesa* Stoliczka, 1868, Arivalur Formation, southern India (Cossmann, 1895).

F. CERITELLIDAE Cossmann, 1895 J. (SIN)–K. (APT/ALB) Mar.

First: *Ceritella exilis* (Martens), Semur (Cossmann, 1895).
Last: *Pseudonerinea stantoni* and *P. sturtoni* Allison, 1955, Punta China, Baja California, Mexico.

Superfamily ACTEONELLOIDEA Zilch, 1959

F. ACTEONELLIDAE Zilch, 1959 K. (APT/ALB–SAN/CMP) Mar.

First: *Acteonella baconica* Benköne-Czabalay, 1965, upper Aptian/Albian, Hungary (Dzhalilov, 1972).
Last: *Acteonella laevis* (J. de C. Sowerby, 1835), *A. elongata* and *A. (Sogdianella) zekelii* Kollmann, 1965, all Gosau Group, Austria (Dzhalilov, 1972).

F. TROCHACTAEONIDAE Akopyan, 1963 J. (LIA)–K. (MAA) Mar.

First: *Cylindrites acutus* (J. de C. Sowerby, 1824), Lias, Europe (Akopyan, 1972).
Last: *'Trochactaeon' truncatus* and *T. minutus* (Stoliczka, 1868) Arivalur Formation, southern India (Stoliczka, 1868).
Comment: Zilch's (1959) record of *Trochactaeon* from the Danian could not be substantiated.

Superfamily ACTEONOIDEA Orbigny, 1842

F. ACTEONIDAE Orbigny, 1842 J. (?SIN/PLB)–Rec. Mar.

First: ?*Tornatellaea heberti* (Piette), lower Lias, France (Cossmann, 1895).
Next oldest: *Tornatellaea fontis* (Duméril, 1806?), Charmouthian, eastern France (Cossmann, 1895). **Extant**

Superfamily RINGICULOIDEA Philippi, 1853

F. RINGICULIDAE Philippi, 1853 K. (NEO)–Rec. Mar.

First: *Cinulia globulosa* (Deshayes), and *Eriptycha ringens* (Orbigny, 1842), and *Ringinella albensis* (Orbigny, 1842), all Neocomian, Marolles, France (Cossmann, 1895; Zilch, 1959). **Extant**

Order TECTIBRANCHIA

Superfamily BULLOIDEA Rafinesque, 1815

The bullomorph families are distinguished on anatomical criteria. Of the shells of Tertiary forms, only those which closely resemble living species can be assigned fairly confidently to their respective families. The affinities of pre-Campanian forms are particularly problematic.

F. HYDATINIDAE Pilsbry, 1895 J. (?SIN/BAJ)–Rec. Mar. (see Fig. 8.6)

First: ?*Palaeohydatina flouesti* (Deslongchamps), Sinemurian, ?Geneva, Switzerland (Cossmann, 1895).
Next oldest: *Palaeohydatina* sp. (Cossmann, 1895), Bajocian, near Nancy, France. **Extant**

F. BULLINIDAE ?Rudman, 1971 **Extant** Mar.

F. CYLICHNIDAE A. Adams, 1850 K. (?SAN/MAA)–Rec. Mar.

First: ?*Roxania ovoides* (d'Archiac, 1854) and ?*Cylichna palassoni* (d'Archiac, 1854), both Santonian, Corbières, Bain de Rennes, France.
Next oldest: *Cylichna secalina* Schumard, 1861 and *C. intermissa* Sohl, 1964, both Ripley Formation, Mississippi, USA (Sohl, 1964a). **Extant**

F. TORNATINIDAE Fischer, 1887 J. (KIM)–Rec. Mar.

First: *Acteocina* sp. nov. (Cossmann, 1895), Rauracian, Glos, France. **Extant**
Intervening: TTH, MAA–PLI.

F. PHILINIDAE Gray, 1850 T. (YPR)–Rec. Mar.

First: *Megistostoma gabbianum* (Stoliczka, 1868), Llajas Formation, Simi Valley area, southern California, USA (Squires, 1984). **Extant**

F. AGLAJIDAE Pilsbry, 1895 **Extant** Mar.

F. GASTROPTERIDAE Swainson, 1840 **Extant** Mar.

F. DIAPHANIDAE Odhner, 1922 ?J. (BAJ)/?K. (CEN)/T. (THA)–Rec. Mar.

First: ?*Diaphana* sp. nov. (Cossmann, 1895), Bajocian, France.
Next oldest: ?*Diaphana truncata* (Stanton), Cenomanian, Colorado, USA (Cossmann, 1895).
Next oldest: *Diaphana moloti* (Cossmann, 1907), Sables de Chalons-sur-Vesle, Jonchery (Marne), France. **Extant**

F. NOTODIAPHANIDAE ?Vaught, 1989 **Extant** Mar.

F. BULLIDAE Rafinesque, 1815 J. (PLB)–Rec. Mar.

First: *Bulla liasina* Deslongchamps, 1869?, Charmouthian, Normandy, France (Cossmann, 1895). **Extant**

Comment: Although usually placed here, the North American Upper Cretaceous genus, *Bullopsis* Conrad, 1858 contains globular shells with two strong columellar pleats, and is unlikely to belong in this family.

F. HAMINOEIDAE Pilsbry, 1895 K. (MAA)–Rec. Mar./Brackish

First: *'Cylichna' incisa* Stephenson, 1941, Ripley Formation, Mississippi, USA (Sohl, 1964a). **Extant**
Comments: This species appears to be congeneric with a number of Tertiary species, variously assigned to *Roxania* Gray, 1847, *Mnestia* H. and A. Adams, 1854, or *Atys* Montfort, 1810. The fossil forms more closely resemble *Aliculastrum* Pilsbry, 1896, and are better placed in this family.

F. RETUSIDAE Thiele, 1926 J. (BTH)–Rec. Mar./Brackish

First: *Retusa* sp. nov. (Cossmann, 1895), Boulonnais, France. **Extant**
Intervening: CLV–PLI.

Superfamily RUNCINOIDEA Gray, 1857

F. RUNCINIDAE Gray, 1857 **Extant** Mar.

F. ILDICIDAE Burn, 1963 **Extant** Mar.

Superfamily HEDYLOPSOIDEA Bergh, 1896

This comprises the families Acochlidiidae Küthe, 1935, Hedylopsidae Bergh, 1896, Platyhedylidae, ?Tantulidae and Microhedylidae Odhner, 1938, which have no shell. There is no fossil record of this superfamily, although calcareous spicules occur in the mantle of some groups and theoretically could be detected in micropalaeontological samples. **Extant**

Superfamily PHILINOGLOSSOIDEA

This comprises the Philinoglossidae in which the shell is absent, and the Plusculidae with a vestigial shell. No fossil record. **Extant**

Order SACOGLOSSA Ihering, 1876

Superfamily JULIOIDEA E. A. Smith, 1885

F. CYLINDROBULLIDAE Thiele, 1931 **Extant** Mar.

F. VOLVATELLIDAE Pilsbry, 1895 **Extant** Mar.

F. JULIIDAE E. A. Smith, 1885 T. (YPR)–Rec. Mar.

First: *Berthelinia elegans elegans* Crosse, and *Gougerotia orthodonta* Le Renard, 1980, Sables de Pierrefonds, Mercin (Aisne), France (Le Renard, 1983). **Extant**

F. OXYNOIDAE H. and A. Adams, 1854 **Extant** Mar.

F. GASCOIGNELLIDAE Jensen, 1983 **Extant** Mar.

F. CALIPHYLLIDAE Clark, 1982 **Extant** Mar.

F. LIMAPONTIIDAE Gray, 1847 **Extant** Mar.

F. ELYSIIDAE H. and A. Adams, 1854 **Extant** Mar.

F. BOSELLIIDAE Marcus, 1982 **Extant** Mar.

Order APLYSIOMORPHA

Superfamily APLYSIOIDEA Rafinesque, 1815

F. APLYSIIDAE Rafinesque, 1815 T. (BUR)–Rec. Mar.

First: *Dolabella aldrichi* Dall, 1890, Chipola Formation, Alum Bluff, western Florida, USA (Dall, 1915). **Extant**

F. NOTARCHIDAE Eales, 1925 **Extant** Mar.

Order ANASPIDEA Fischer, 1883

F. AKERIDAE Pilsbry, 1893 J. (CLV)–Rec. Brackish/Mar.

First: *Akera mediojurensis* Cossmann, 1896, Callovian, France, and *A. tanganyicensis* Cox, 1965, Mandawa–Mahokondo Series, SE Tanzania. **Extant**

Order NOTASPIDEA Fischer, 1883

Superfamily TYLODINOIDEA Gray, 1847

F. UMBRACULIDAE Dall, 1889 T. (YPR)–Rec. Mar.

First: *Umbraculum laudunense* (Melleville, 1843), Sables de Pierrefonds, Laon, France (Cossmann and Pissarro, 1913), or *U. sylvaerupis* (Harris) and *Eosinica elevata* (Aldrich), Bashi Marl Member, Hachetigbee Formation, Woods Bluff, Alabama, USA (Jablonski and Bottjer, 1990). **Extant**

F. TYLODINIDAE Gray, 1847 T. (PLI)–Rec. Mar.

First: *Tylodina (Tylodinella) rafinesquei* Philippi, 1844?, Pliocene, Sicily, Italy (Cossmann, 1895). **Extant**

Superfamily PLEUROBRANCHOIDEA Menke, 1828

F. PLEUROBRANCHIDAE Menke, 1828 **Extant** Mar.

Order PTEROPODA Cuvier, 1804

Suborder EUTHECOSOMATA Meisenheimer, 1905

Superfamily LIMACINOIDEA Blainville, 1823

F. LIMACINIDAE Blainville, 1823 T. (THA)–Rec. Mar. (pelagic)

First: *Limacina mercinensis* (Watelet and Lefèvre, 1885), Tuscahoma Sand Formation, Bear Creek Marls Member, Bear Creek, Alabama, USA (S. Tracey, unpublished data). **Extant**

F. CUVIERINIDAE Gray, 1840 T. (YPR)–Rec. Mar. (pelagic)

First: *Euchilotheca ganensis* Curry, 1981 and *Camptoceratops priscus* (Godwin-Austen, 1882), marnes de la tuilerie de Gan (Pyrenées Atlantiques), France (Curry, 1981). **Extant**

Suborder PSEUDOTHECOSOMATA Meisenheimer, 1905

F. PERACLIDAE C. W. Johnson, 1915 **Extant** Mar. (pelagic)

F. CYMBULIIDAE Gray, 1840 **Extant** Mar. (pelagic)

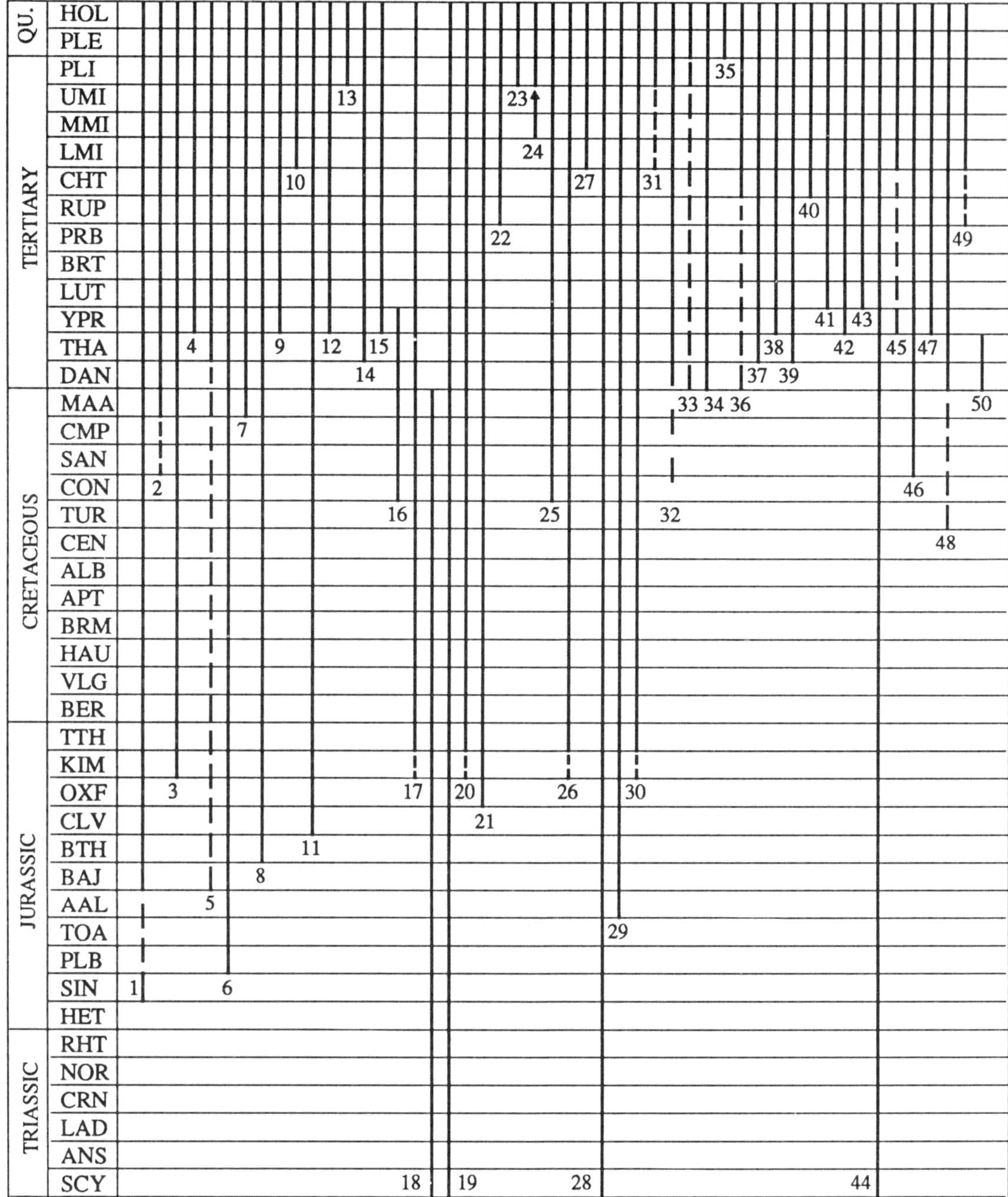

Fig. 8.6

F. DESMOPTERIDAE Chun, 1889 **Extant**
Mar. (pelagic)

Order GYMNOSOMATA Blainville, 1824

Comprising the families Pneumodermatidae Latreille, 1825, Notobranchaeidae Pelseneer, 1886, Cliopsidae Costa, 1873, Thliptodontidae Kwietniewski, 1902, Hydromylidae Pruvot-Fol, 1942, Laginiopsidae and Clionidae Rafinesque, 1815, which have no shell in the adult phase. Protoconchs referable to the Gymnosomata have been found in Pleistocene samples from the northeast Atlantic (Van der Spoel and Diester-Haass, 1976). **Extant**

Order NUDIBRANCHIA Cuvier, 1817

This order contains 66 families of opisthobranchs without shells, and there are no reliable fossil records. **Extant**

Superorder GYMNOMORPHA

Order SYSTELLOMMATOPHORA

Superfamily ONCHIDIOIDEA Gray, 1824

F. ONCHIDIIDAE Gray, 1824 **Extant** Mar.

Order SOLEOLIFERA

This comprises the families Veronicellidae Gray, 1840, Rhodopidae and Rathouisiidae Sarasin, 1899 which have no shells and are not recorded as fossils. **Extant**

Superorder PULMONATA Cuvier, 1817

Order ARCHAEOPULMONATA Morton, 1955

F. ACROREIDAE Wenz, 1923 K. (SEN)–T. (YPR)
Probably brackish

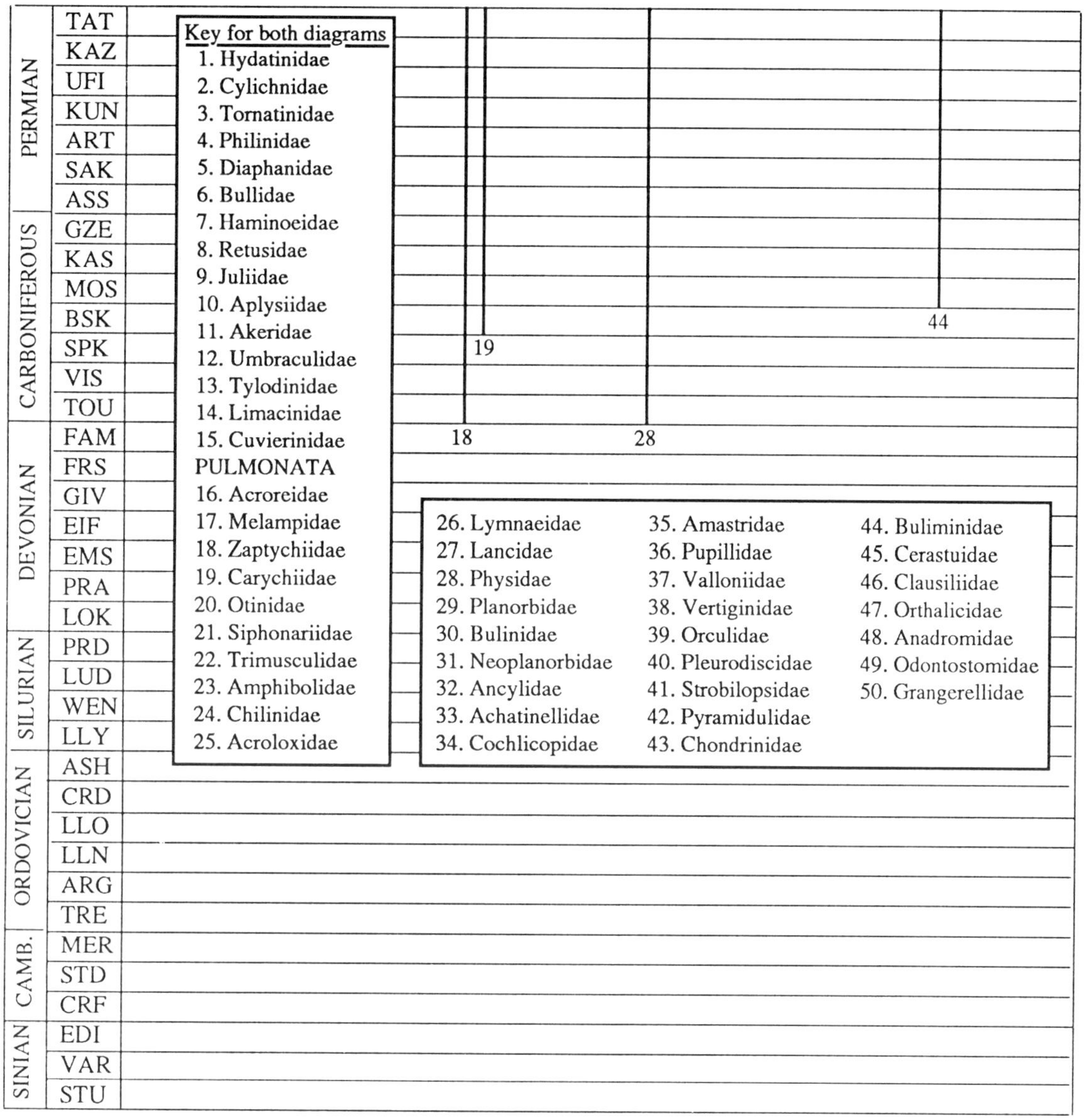

PERMIAN: TAT, KAZ, UFI, KUN, ART, SAK, ASS
CARBONIFEROUS: GZE, KAS, MOS, BSK, SPK, VIS, TOU
DEVONIAN: FAM, FRS, GIV, EIF, EMS, PRA, LOK
SILURIAN: PRD, LUD, WEN, LLY
ORDOVICIAN: ASH, CRD, LLO, LLN, ARG, TRE
CAMB.: MER, STD, CRF
SINIAN: EDI, VAR, STU

19 44 18 28

Key for both diagrams
1. Hydatinidae
2. Cylichnidae
3. Tornatinidae
4. Philinidae
5. Diaphanidae
6. Bullidae
7. Haminoeidae
8. Retusidae
9. Juliidae
10. Aplysiidae
11. Akeridae
12. Umbraculidae
13. Tylodinidae
14. Limacinidae
15. Cuvierinidae
PULMONATA
16. Acroreidae
17. Melampidae
18. Zaptychiidae
19. Carychiidae
20. Otinidae
21. Siphonariidae
22. Trimusculidae
23. Amphibolidae
24. Chilinidae
25. Acroloxidae

26. Lymnaeidae
27. Lancidae
28. Physidae
29. Planorbidae
30. Bulinidae
31. Neoplanorbidae
32. Ancylidae
33. Achatinellidae
34. Cochlicopidae
35. Amastridae
36. Pupillidae
37. Valloniidae
38. Vertiginidae
39. Orculidae
40. Pleurodiscidae
41. Strobilopsidae
42. Pyramidulidae
43. Chondrinidae
44. Buliminidae
45. Cerastuidae
46. Clausiliidae
47. Orthalicidae
48. Anadromidae
49. Odontostomidae
50. Grangerellidae

Fig. 8.6

First: *Acroreia (Vasculum) obliqua* (White, 1889), Laramie Group, California, USA.
Last: *Acroreia baylei* (Cossmann, 1885), Sables d'Hérouval, Hérouval, Paris Basin, France (Cossmann, 1895; Zilch, 1959).

F. MELAMPIDAE Stimpson, 1851 (ELLOBIIDAE H. and A. Adams, 1855) J. (KIM/TTH)–Rec. Brackish/Semi-terr.

First: *Mesauriculastra accelerata* (White, 1886), Morrison Formation, Canyon City, Colorado, USA, *M. morrisonensis* and *M. spiralis* Yen, 1952, Morrison Formation, USA (Yen, 1952). **Extant**

F. ZAPTYCHIIDAE Zilch, 1959 C. (l.)–K. (SEN) FW

First: *Zaptychius carbonarius* Walcott, 1883, Eureka District, Nevada, USA (Henderson, 1935).
Last: *Tortacella wyomingensis* Henderson, 1935, Laramie Formation, Bear River Valley, Wyoming, USA.
Comment: Generally considered to be ancestral to the Melampidae and often included in that family.

F. CARYCHIIDAE Jeffreys, 1830 C. (BSH)–Rec. Terr./Semi-aquatic/FW

First: *Anthracopupa britannica* Cox, 1926, Westphalian A, northern England and southern Scotland (Solem and Yochelson, 1979). **Extant**

F. OTINIDAE H. and A. Adams, 1855 J. (KIM/TTH)–Rec. Mar./Semi-terr.

First: ?*Limnopsis jurassica* Yen, 1952, Morrison Formation, Wyoming, USA (Yen, 1952). **Extant**
Comment: This monotypic genus is based on a dubious internal mould and is unlikely to be connected to this family. There are no intervening occurrences.

Order BASOMMATOPHORA Keferstein, 1864

Superfamily SIPHONARIOIDEA Gray, 1840

F. SIPHONARIIDAE Gray, 1840 J. (OXF)–Rec. Littoral mar./Brackish

First: *Berlieria ledonica* de Loriol, Chatillon-sur-Ain, France (Zilch, 1959). **Extant**

F. TRIMUSCULIDAE Zilch, 1959 T. (Oli.)–Rec. Mar.

First: *Trimusculus* sp. Oligocene, Europe (noted in Zilch, 1959). **Extant**

Superfamily AMPHIBOLOIDEA H. and A. Adams, 1855

F. AMPHIBOLIDAE H. and A. Adams, 1855 T. (PIA)–Rec. Brackish

First: *Salinator neozelanica* Laws, 1950, Waipipian, New Zealand (Beu and Maxwell, 1990). **Extant**

Superfamily CHILINOIDEA H. and A. Adams, 1855

F. CHILINIDAE H. and A. Adams, 1855 T. (PIA)–Rec. FW

First: *Chilina* sp. Upper Pliocene, Argentina (noted in Zilch, 1959). **Extant**

F. LATIIDAE Hannibal, 1912 **Extant** FW

Superfamily ACROLOXOIDEA Thiele, 1931

F. ACROLOXIDAE Thiele, 1931 K. (SEN)–Rec. FW

First: *Pseudancylastrum* sp., Upper Cretaceous, Europe (noted in Zilch, 1959). **Extant**
Comment: *Acroloxus radiatulus* Whiteaves, 1885 from the Paskapoo Formation of Alberta, Canada, was referred doubtfully to *Ferrissia* Walker, 1903 by Henderson (1935).

Superfamily LYMNAEOIDEA Rafinesque, 1815

F. LYMNAEIDAE Rafinesque, 1815 J. (KIM/TTH)–Rec. FW

First: *Lymnaea hopii* (Robinson, 1915), Painted Desert Beds, NE Arizona and *Galba accelerata*, *G. ativuncula* and *G. consortis* (White, 1886), Morrison Formation, north of Canyon City, Colorado, USA (Henderson, 1935). *Galba* Schrank, 1803 is widespread in the 'Purbeckian' (BER) (Zilch, 1959; Pan, 1983). **Extant**

F. LANCIDAE Hannibal, 1914 T. (Mio.)–Rec. FW

First: *Lanx undulatus* (Meek, 1870), Fossil Hill, Kawsoh Mountains, Nevada, USA. **Extant**
Comment: The genus *Zalophancylus* (*Z. morani* Hannibal, 1912) from the Pliocene Idaho Lake Beds of Oregon, included as a subgenus of *Lanx* by Zilch (1959), proved to be based on a fragment of fish vertebra (Henderson, 1935).

Superfamily PHYSOIDEA Fitzinger, 1833

F. PHYSIDAE Fitzinger, 1833 C.(l.)–Rec. FW

First: *Physa prisca* Walcott, 1883, Eureka district, Nevada, USA (Henderson, 1935). **Extant**

Superfamily PLANORBOIDEA Rafinesque, 1815

F. PLANORBIDAE Rafinesque, 1815 J. (DOG)–Rec. FW

First: *Anisopsis calculus* (Sandberger, 1875), Cajac, (Dép. Lot), France (Zilch, 1959). **Extant**

F. BULINIDAE Oken, 1815 J. (KIM/TTH)–Rec. FW

First: *Graptophysa spiralis* Yen and Reeside, 1946, Morrison Formation, Wyoming, USA (Zilch, 1959). **Extant**

F. PROTANCYLIDAE B. Walker, 1923 **Extant** FW

F. PATELLOPLANORBIDAE ?Hubendick, 1957 **Extant** FW

F. NEOPLANORBIDAE Hannibal, 1912 T. (?Mio./PLI)–Rec. FW

First: *Amphigyra dalli* (White, 1883), Idaho Formation (and also ?Miocene), Idaho, USA (Henderson, 1935). **Extant**

F. ANCYLIDAE Rafinesque, 1815 ?K. (SEN)/T. (Pal.)–Rec. FW

First: ?*Ferrissia ?minuta* (Meek and Hayden, 1876), Laramie Group of Fort Union, North Dakota, USA. Based on small, distorted specimens which cannot be safely assigned to this family (Henderson, 1935).
Next oldest: *Palaeancylus radiatus* Yen, 1948, Fort Union Formation, southern Montana, USA (noted by Cooper, 1979), and *Ancylus* sp. (Cooper, 1979), Ardtun Leaf Beds, Isle of Mull, Scotland, UK. **Extant**

Order STYLOMMATOPHORA Schmitt, 1855

Superfamily ACHATINELLOIDEA Gulick, 1873

F. ACHATINELLIDAE Gulick, 1873 ?T. (DAN)/Q. (PLE)–Rec. Terr.

First: ?*Protornatellina isoclina* (White, 1895), Bear River Formation, 20 miles north of Cokeville, Wyoming, USA (Henderson, 1935; Zilch, 1959).
Next oldest: *Tornatellides (Waimea) rudicostatus* (Ancey), and *Tornatellaria* spp., Pleistocene, Hamakua, Hawaii (Zilch, 1959). **Extant**

Superfamily COCHLICOPOIDEA Pilsbry, 1900

F. COCHLICOPIDAE Pilsbry, 1900 T. (Pal.)–Rec. Terr.

First: *Azeca* Fleming, 1828 is indicated from the Palaeocene of Europe and USA and *Cochlicopa* Risso, 1826 from Europe by Zilch (1959). **Extant**

F. AMASTRIDAE Pilsbry, 1911 Q. (PLE)–Rec. Terr.

First: *Carelia* H. and A. Adams, 1855, *Amastra* H. and A. Adams, 1855, *Planamastra* Pilsbry, 1911 and *Cyclamastra* Pilsbry and Vanatta, 1905; numerous species in the Pleistocene of Hawaiian Islands (Zilch, 1959). **Extant**

Superfamily PUPILLOIDEA Turton, 1831

The earliest record of the Pupilloidea (family uncertain) is *Pupa bigsbii* Dawson, 1880, Carboniferous (Westphalian B), Group XV, Cumberland Group, Joggins, Nova Scotia, Canada (Solem and Yochelson, 1979).

F. PUPILLIDAE Turton, 1831 T. (Pal.)–Rec. Terr.

First: *Gastrocopta*? sp. (La Rocque, 1960), Flagstaff Formation, Utah, USA and *Albertanella minuta* Russell, 1931, Saunders Formation, McLeod River, Alberta, Canada and Flagstaff Formation, central Utah, USA (La Rocque, 1960). **Extant**

F. VALLONIIDAE Pilsbry, 1900 T. (THA)–Rec. Terr.

First: *Acanthinula dumasi* (Boissy, 1848), Calcaire de Rilly, Rilly, France (Cossmann and Pissarro, 1913), and *Agallospira* *multispiralis* and *Shanghuspira costata* Yü, 1977, Palaeocene, South China. **Extant**

F. VERTIGINIDAE Stimpson, 1851 T. (YPR)–Rec. Terr.

First: *Vertigo oviformis* Michaud *in* Deshayes, 1863 and *V. interferens* (Deshayes, 1863), both Sparnacian, Grauves, France (Cossmann and Pissarro, 1913). **Extant**

F. ORCULIDAE Pilsbry, 1918 T. (THA)–Rec. Terr.

First: *Orcula plateaui* (Cossmann, 1889), Sables de Chalons, Chenay (Marne), France (Cossmann and Pissarro, 1913). **Extant**

F. PLEURODISCIDAE Wenz, 1923 T. (?CHT/PLI)–Rec. Terr.

First: *Pleurodiscus* sp. Chattian, Europe (noted by Zilch, 1959).
Next oldest: *Pleurodiscus falkneri* Schlickum, 1978, Pliocene, Hungary. **Extant**

F. STROBILOPSIDAE Pilsbry, 1918 T. (LUT)–Rec. Terr.

First: *Paleostrobilops menardi* (Brongniart, 1810), Lutetian, Paris Basin, France (Poinar and Roth, 1991). **Extant**

F. PYRAMIDULIDAE Kennard and Woodward, 1914 T. (Eoc.)–Rec. Terr.

First: *Pyramidula* Fitzinger, 1833 is indicated from the Eocene of Europe by Zilch (1959). **Extant**

F. CHONDRINIDAE Steenburg, 1925 T. (LUT)–Rec. Terr.

First: *Abida* Turton, 1831 is indicated from the Middle Eocene of Europe by Zilch (1959). **Extant**

Superfamily BULIMINOIDEA Clessin, 1879

F. BULIMINIDAE Clessin, 1879 [ENIDAE Woodward, 1903] C. (MOS)–Rec. Terr.

First: *Dendropupa vestuta* (Dawson, 1855), group VIII, Cumberland Group, Joggins, Nova Scotia, Canada, and *D. grandaeva* (Dawson, 1880), *D. primaeva* (Matthew, 1895), both Fern Ledges beds, Little River Group, St John, New Brunswick, Canada, all Westphalian B. **Extant**
Comment: Solem and Yochelson (1979) gave reasons for including *Dendropupa* in this family, rather than the Cyclophoridae where they were formerly placed.

F. CERASTUIDAE Wenz, 1923 T. (?YPR/LMI)–Rec. Terr.

First: ?*Procerastus dautzenbergi* (Cossmann, 1907), Sparnacian, Grauves, France (Cossmann and Pissarro, 1913).
Next oldest: *Cerastua* sp. Lower Miocene, East Africa (noted by Zilch, 1959). **Extant**
Comment: The genus *Procerastus* was included with some doubt in the Anadromidae by Nordsieck (1986).

Superfamily CLAUSILIOIDEA Mörch, 1864

F. CLAUSILIIDAE Mörch, 1864 K. (SAN)–Rec. Terr.

First: *Dextrospira minutula* Hrubesch, 1965a, Gosau Group, Glanegg near Salzburg, Austria. **Extant**

Superfamily PARTULOIDEA Pilsbry, 1900

F. PARTULIDAE Pilsbry, 1900 **Extant** Terr.

Superfamily ORTHALICOIDEA Albers-Martens, 1860

F. ORTHALICIDAE Albers-Martens, 1860 T. (Eoc.)–Rec. Terr.

First: *Palaeobulimulus eocenicus* Parodiz, 1949, Mallin Blanco, Chubut, Patagonia, Argentina (Zilch, 1959). **Extant**

F. ANADROMIDAE Zilch, 1959 K. (?TUR/SAN)–T. (DAN) Terr.

First: ?*Conobulimus fuggeri* (Tausch) and ?*Juvavina juvaviensis* (Tausch), both Turonian, Algen near Salzburg (Zilch, 1959).
Next oldest: *Lychnus sanchezi* (Vidal, 1874), Gosau Group, Glanegg near Salzburg, Austria (Hrubesch, 1965).
Last: *Lychnus ellipticus* Matheron, 1832, Les Baux (Bouches-du-Rhone), southern France, *Anadromus proboscideus* (Matheron), Peynier (Bouches-du-Rhône), southern France, and *Vidaliella darderi* (Vidal), Danian, Selva, Mallorca, Spain (Zilch, 1959). **Extant**

F. ODONTOSTOMIDAE Pilsbry and Vanatta, 1898 T. (Oli./Mio.)–Rec. Terr.

First: *Hyperaulax americanus* (Heilprin, 1887) *s.s.* (and nine other species/subspecies, *Silex* Beds, Ballast Point near Tampa, Florida, USA (Dall, 1915; Henderson, 1935). **Extant**

F. GRANGERELLIDAE, Russell, 1931 T. (Pal.) Terr.

First: *Grangerella macleodensis* (Russell, 1929), upper Saunders Formation, west side of McLeod River, Alberta, Canada (Henderson, 1935).

F. AMPHIBULIMIDAE Crosse and Fischer, 1873 **Extant** Terr.

F. UROCOPTIDAE Pilsbry and Vanatta, 1898 T. (DAN/THA)–Rec. Terr.

First: *Holospira* cf. *leidyi* (Meek, 1872), Flagstaff Formation, central Utah, USA (La Rocque, 1960). **Extant**

F. CERIONIDAE Pilsbry, 1901 T. (Oli./Mio.)–Rec. Terr.

First: *Eostrophia anodonta* (Dall, 1890), Silex Beds, Ballast Point nr. Tampa, Florida, USA (Dall, 1915; Henderson, 1935).

F. MEGASPIRIDAE Pilsbry, 1904 ?K. (TUR)/T. (THA)–Rec. Terr.

?First: *Palaeostoa exarata* (Michaud), Sables de Chalons, Chalons-sur-Vesle near Reims (Marne), France (Cossmann, 1889). **Extant**
Comment: Nordsieck (1987) indicated a late Cretaceous origin for *Palaeostoa* Andreae, 1884 (removed to the Palaeostoidae H. Nordsieck, 1986) and a Turonian origin for *Ptychicula* Tausch, which is possibly cofamilial.

F. COELOCIIDAE H. Nordsieck, 1986 **Extant** Terr.

Superfamily ACHATINOIDEA Swainson, 1840

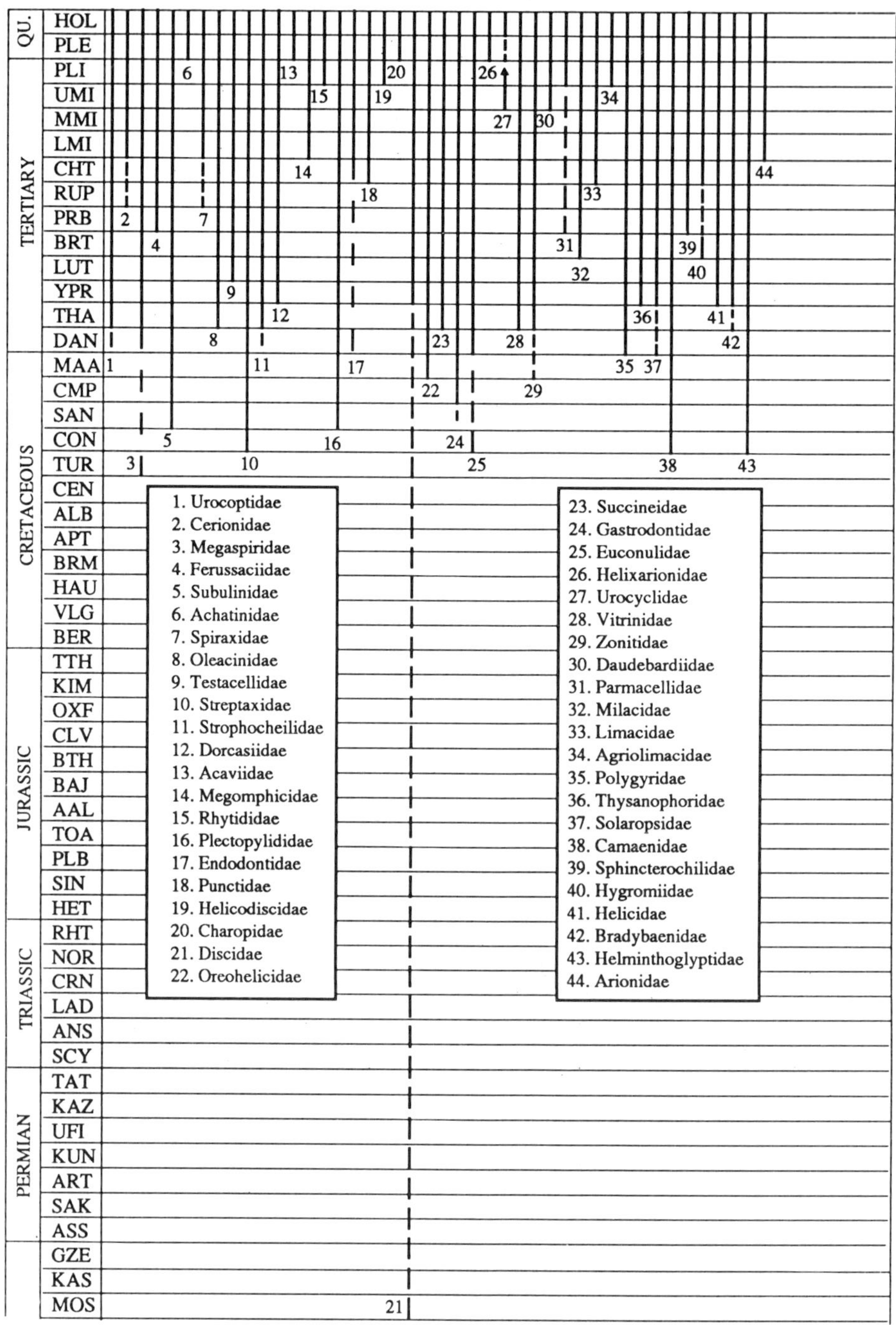

Fig. 8.7

F. FERUSSACIIDAE Bourguignat, 1883 T. (PRB)–Rec. Terr.

First: *Coilostele* sp., Upper Eocene, southern Europe (noted by Zilch, 1959). **Extant**

F. SUBULINIDAE Thiele, 1931 K. (SAN)–Rec. Terr.

First: *Cylindrellina permaxima* Hrubesch, 1965a, Gosau Group, Glanegg near Salzburg, Austria. **Extant**

F. ACHATINIDAE Swainson, 1840 Q. (PLE)–Rec. Terr.

First: *Achatina (Lissachatina)* sp., PLE, tropical Africa (noted in Zilch, 1959). **Extant**

F. COELIAXIDAE Pilsbry, 1907 **Extant** Terr.

F. THYROPHORELLIDAE Thiele, 1926 **Extant** Terr.

Superfamily AILLYOIDEA Baker, 1930

F. AILLYIDAE Baker, 1930 **Extant** Terr.

Superfamily OLEACINOIDEA H. and A. Adams, 1855

F. SPIRAXIDAE Baker, 1939 T. (?PRB/RUP/Oli./Mio.)–Rec. Terr.

First: ?*Spiraxis* sp. (Poinar and Roth, 1991), Upper Eocene/Upper Oligocene, Dominican Republic.

Next oldest: *Spiraxis tampae* Dall, 1915, Silex Beds, Ballast Point near Tampa, Florida, USA (Dall, 1915; Henderson, 1935). **Extant**

F. OLEACINIDAE H. and A. Adams, 1855 T. (THA)–Rec. Terr.

First: *Palaeoglandina cordieri* (Deshayes, 1863), Thanetian, Cramant, France (Zilch, 1959). **Extant**

F. TESTACELLIDAE Gray, *in* Turton, 1840 T. (LUT)–Rec. Terr.

First: *Parmacellina vitrinaeformis* Sandberger, 1872, Buchsweiler, Unter Elsass, Germany (Zilch, 1959). **Extant**

Superfamily STREPTAXOIDEA Gray, 1860

F. STREPTAXIDAE Gray, 1806 K. (CON)–Rec. Terr.

First: *Eoplicadomus austriaca* Hrubesch, 1965b, Gosau Group, Windisch Garsten, Austria. **Extant**

Superfamily STROPHOCHEILOIDEA Thiele, 1926

F. STROPHOCHEILIDAE Thiele, 1926 T. (DAN/THA)–Rec. Terr.

First: *Strophocheilus* Spix, 1827, Patagonia, Brazil (noted by Zilch, 1959). **Extant**

F. DORCASIIDAE Thiele, 1926 T. (Eoc.)–Rec. Terr.

First: *Dorcasia antiqua* Wenz, 1926, marly sandstone at Chalcedontafelberg, west of the old Lüderitz fields, Namibia (Connolly, 1939). **Extant**

Superfamily ACAVOIDEA Pilsbry, 1895

F. ACAVIDAE Pilsbry, 1895 Q. (PLE)–Rec. Terr.

First: Numerous species of the endemic genera, *Ampelita* Beck, 1837, *Helicophanta* Férussac, 1821, and *Clavator* Martens, *in* Albers, 1860, occur in the Pleistocene of Madagascar (Zilch, 1959). **Extant**

F. CARYODIDAE Thiele, 1926 **Extant** Terr.

F. MACROCYCLIDAE Thiele, 1926 **Extant** Terr.

F. MEGOMPHICIDAE Baker, 1930 T. (Mio.)–Rec. Terr.

First: *Polygyrella lunata* (Conrad, 1870), John Day beds, eastern Oregon, USA (Henderson, 1935). **Extant**

Superfamily RHYTIDOIDEA Pilsbry, 1895

F. RHYTIDIDAE Pilsbry, 1895 ?T. (PLI)–Rec. Terr.

First: *Rhytida* Albers, 1860, Neogene of Tasmania (noted by Zilch, 1959). **Extant**

F. CHLAMYDEPHORIDAE Cockerell, 1935 **Extant** Terr.

F. SYSTROPHIIDAE Thiele, 1926 [SCOLODONTIDAE Baker, 1925] **Extant** Terr.

F. HAPLOTREMATIDAE Baker, 1925 **Extant** Terr.

Superfamily PLECTOPYLIDOIDEA Moellendorff, 1900

F. SCULPTARIIDAE ?Vaught, 1989 **Extant** Terr.

F. PLECTOPYLIDIDAE Moellendorff, 1900 K. (SAN)–Rec. Terr.

First: *Proterocorilla europaea* Hrubesch, 1965a, Gosau Group, Glanegg near Salzburg, Austria. **Extant**
Comment: No intervening occurrences are recorded. *Proterocorilla* Hrubesch, 1965 was one of the original genera included in the Upper Cretaceous Family Anostomopsidae H. Nordsieck, 1986.

Superfamily PUNCTOIDEA Morse, 1864

F. ENDODONTIDAE Pilsbry, 1895 T. (Pal./Oli.)–Rec. Terr.

First: *Hebeispira hebeiensis* Youlou, 1978, Lower Tertiary, Bohai, China.
Next oldest: *Cookeconcha subpacificus* (Ladd, 1958), Lower Miocene, Bikini Atoll, Pacific Ocean (Solem, 1977). **Extant**

F. PUNCTIDAE Morse, 1864 T. (CHT)–Rec. Terr.

First: *Punctum* Morse, 1864, Lower Oligocene, Europe (noted by Zilch, 1959). **Extant**

F. HELICODISCIDAE Baker, 1927 T. (PLI)–Rec. Terr.

First: *Helicodiscus singleyanus* (Pilsbry, 1890), Rexroad Formation, XI Member, Seward County, south-west Kansas, USA (Taylor, 1960). **Extant**

F. CHAROPIDAE Hutton, 1884 Q. (PLE)–Rec. Terr.

First: The endemic subgenera of *Ptychodon*, *Helenoconcha* Pilsbry 1892 and *Pseudohelenoconcha* Germain, 1932 occur in the Pleistocene of St Helena (Zilch, 1959). **Extant**

F. OTOCONCHIDAE Cockerell, 1893 **Extant** Terr.

F. DISCIDAE Thiele, 1931 ?C. (KSK)/T. (DAN/THA)–Rec. Terr.

First: ?*Protodiscus priscus* (Carpenter, 1867), Group VIII (Westphalian B), Cumberland Group, Joggins, Nova Scotia, Canada (Solem and Yochelson, 1979). **Extant**
Next oldest: *Discus* cf. *ralstonensis* (Cockerell, 1914), Flagstaff Formation, central Utah, USA (La Rocque, 1960).

F. OREOHELICIDAE Pilsbry, 1939 K. (MAA)–Rec. Terr.

First: *Oreohelix angulifera*, and *O. obtusata* (Whiteaves, 1885), both St Mary Cretaceous, Pincher Creek and Old Man River, Alberta, Canada (Henderson, 1935). **Extant**

F. OOPELTIDAE Cockerell, 1891 **Extant** Terr.

Superfamily SUCCINEOIDEA Beck, 1837

F. SUCCINEIDAE Beck, 1837 T. (THA)–Rec. Semi-aquatic (FW)

First: *Succinea boissyi* (Deshayes, 1863), Calcaire de Rilly, Rilly, France. **Extant**

Superfamily ATHORACOPHOROIDEA Fischer, 1883

F. ATHORACOPHORIDAE Fischer, 1883 **Extant** Terr.

Shell reduced or absent. No fossil record.

Superfamily SAGDIDOIDEA Pilsbry, 1895

F. SAGDIDIDAE Pilsbry, 1895 **Extant** Terr.

Superfamily GASTRODONTOIDEA Tryon, 1866

F. GASTRODONTIDAE Tryon, 1866 K. (SAN/CMP)–Rec. Terr.

First: *Ventridens lens* (Gabb, 1864), Chico Formation, Texas Flat, Placer County, California, USA (Henderson, 1935). **Extant**

Superfamily HELIXARIONOIDEA Godwin-Austen, 1888

F. EUCONULIDAE Baker, 1928 ?K (SEN)/T. (Pal.)–Rec. Terr./Semi-aquatic

First: *Euconulus* Reinhardt, 1883 is indicated by Zilch (1959) from the Middle Palaeocene of Europe and doubtfully from the Upper Cretaceous. **Extant**

F. HELIXARIONIDAE Godwin-Austen, 1888 Q. (PLE)–Rec. Terr.

First: The mostly endemic genera, *Ctenophila* Ancey, 1882, *Ctenoglypta* Ancey, 1904, *Pseudocaelatura* Wenz, 1947, and *Erepta* Albers, 1850 occur in the Pleistocene of Mauritius and Réunion (Zilch, 1959). **Extant**

F. ARIOPHANTIDAE Godwin-Austen, 1888 **Extant** Terr.

F. TROCHOMORPHIDAE Moellendorff, 1890 **Extant** Terr.

F. UROCYCLIDAE Simroth, 1889 Q. (?PLE)–Rec. Terr.

First: *Hamya revoili* Bourguignat, 1885, occurs in Neogene deposits in Portugal and East Africa (Zilch, 1959). **Extant**

F. CYSTOPELTIDAE ?Vaught, 1989 **Extant** Terr.

Superfamily VITRINOIDEA Fitzinger, 1833

F. VITRINIDAE Fitzinger, 1833 T. (THA)–Rec. Terr.

First: *Provitrina rillyensis* (Boissy, 1848), Calcaire de Rilly, Rilly near Reims, France (Zilch, 1959). **Extant**

F. ZONITIDAE Mörch, 1864 ?K. (MAA)/T. (THA)–Rec. Terr.

First: ?*Vitrea obliqua* Meek and Hayden, 1857, Judith River Beds, Judith River, Montana, USA (Henderson, 1935; Zilch, 1959).
Next oldest: *Grandipatula (Grandipatula) hemisphaerica* (Michaud), and *G. (Sphaerozonites) oppenheimi* (Pfeffer, 1929), both Calcaire de Rilly, Rilly, near Reims, France. **Extant**

F. DAUDEBARDIIDAE Kobelt, 1906 T. (VMI)–Rec. Terr.

First: *Daudebardia* sp., Upper Miocene, Europe (noted by Zilch, 1959). **Extant**

F. PARMACELLIDAE Gray, 1860 T. (?PRB/PIA)–Rec. Terr.

First: ?*Parmacella* Cuvier, 1804, Upper Pliocene, Europe and doubtfully in the Upper Eocene (noted by Zilch, 1959). **Extant**

F. MILACIDAE Ellis, 1926 T. (BRT)–Rec. Terr.

First: *Milax* cf. *latus* (Edwards, 1852), Creechbarrow Limestone Formation, East Creech, Dorset, England, UK (Preece, 1980). **Extant**

Superfamily LIMACOIDEA Rafinesque, 1815

The flattened internal shells of these and the milacid slugs can often be recognized at generic level in Pleistocene and younger deposits, but the allocation of older examples is highly tentative.

F. LIMACIDAE Rafinesque, 1815 T. (CHT)–Rec. Terr.

First: *Limax*? sp. Shells probably referable to the family occur in the Oligocene of Europe (Zilch, 1959). **Extant**

F. AGRIOLIMACIDAE Wagner, 1935 T. (PLI)–Rec. Terr.

First: *Deroceras aenigma* Leonard, 1950, lower Rexroad Formation, High Plains, Kansas, USA (Taylor, 1960). **Extant**
Comment: Some authors treat this as a subfamily of the Limacidae.

Superfamily TRIGONOCHLAMYDOIDEA Hesse, 1882

F. TRIGONOCHLAMYDIDAE Hesse, 1882 **Extant** Terr.

Superfamily POLYGYROIDEA Pilsbry, 1895

F. POLYGYRIDAE Pilsbry, 1895 T. (DAN/THA)–Rec. Terr.

First: *Zhenjiangospira dignata* Yü, 1977, Palaeocene, South China, and *'Mesodon' rutherfordi* (Russell, 1929), Saunders Formation, McLeod River, Alberta, Canada (Henderson, 1935; Zilch, 1959). **Extant**

F. THYSANOPHORIDAE Pilsbry, 1926 T. (Eoc.)–Rec. Terr.

First: *Microphysula oxyaenae* (Cockerell, 1914), Sand Coulee Beds, Park County, Wyoming, USA (Henderson, 1935). **Extant**

Superfamily CAMAENOIDEA Pilsbry, 1895

F. SOLAROPSIDAE H. Nordsieck, 1986 ?T. (Pal./Eoc.)–Rec. Terr.

First: ?*Hodopoeus crassus* Pilsbry and Cockerell, 1945, New Mexico or Texas, USA. **Extant**
Comment: Based on an internal mould. No intervening fossil occurrences are recorded.

F. CAMAENIDAE Pilsbry, 1895 K. (SEN)–Rec. Terr.

First: *Kanabohelix kanabensis* (White, 1876), Point of Rocks Group, Upper Kanab, Utah, USA (Solem, 1978). **Extant**

Superfamily HELICOIDEA Rafinesque, 1815

F. SPHINCTEROCHILIDAE T. (PRB)–Rec. Terr.

First: *Dentellocaracolus damnata* (Brongniart, 1810), and *D. (Prothelidomus) acrochordon* (Oppenheim, 1895), Upper Eocene, Vicentin, Ronca and Monte Altissimo, Vicenza, Italy (Zilch, 1959). **Extant**
Intervening: ?RUP–CHT.

F. HYGROMIIDAE Tryon, 1866 T. (?BRT/CHT)–Rec. Terr.

First: ?*Archygromia durbani* (Edwards, 1852), Creechbarrow Limestone Formation, East Creech, Dorset, England, UK (Preece, 1980). **Extant**
Next oldest: *Pseudoxerotricha subconspurcata* (Sandberger, 1873), and *Hemistenotrema quadrisinuosa* (O. Boettger, 1897), Upper Oligocene, Hochheim–Flörsheim, Hessen (Zilch, 1959).

F. HELICIDAE Rafinesque, 1815 T. (YPR)–Rec. Terr.

First: *Loganiopharynx rarus* (Boissy, 1840), Sparnacian, Mont-de-Bernon near Epernay (Marne), France (Zilch, 1959) or *Nanhaispira eversilabia* Yu and Zhang, 1982, Palaeocene or Eocene, Sanshui Basin, Guangdong, China.
Intervening: ?BRT, CHT–HOL.

F. EULOTIDAE Moellendorff, 1898 [BRADYBAENIDAE Pilsbry, 1934] T. (THA/YPR)–Rec. Terr.

First: *Coneulota diarti* (Cossmann, 1907), Sparnacian, Grauves (Marne), France (Cossmann and Pissarro, 1913; Zilch, 1959). **Extant**

F. HELMINTHOGLYPTIDAE Pilsbry, 1939 K. (SEN)–Rec. Terr.

First: *Mesoglypterpes sagensis* Yen, 1952, Mill Creek, Wind River Mountains, Wyoming, USA. **Extant**
Intervening: DAN–CHT.

F. XANTHONYCHIDAE Strebel and Pfeffer, 1880 **Extant** Terr.

Superfamily ARIONOIDEA Gray, *in* Turton, 1840

F. PHILOMYCIDAE Gray, 1847 **Extant** Terr.

F. ARIONIDAE Gray *in* Turton, 1840 T. (Mio.)–Rec. Terr.

First: *Craterarion pachyostracon* Taylor, 1954, Barstow Formation, San Bernadino Co., California, USA and *Geomalacus indifferens* (O. Boettger, 1911?), Landschneckenmergel, Frankfurt am Main, Germany (Zilch, 1959). **Extant**

REFERENCES

Omitted references to sources prior to 1941 can be found in Knight (1941) and Knight *et al.* (1960).

Abbass, H. L. (1962) The English Cretaceous Turritellidae and Mathildidae (Gastropoda). *Bulletin of the British Museum (Natural History), Geology*, **7**, 173–96.

Abbass, H. L. (1973) Some British Cretaceous gastropods belonging to the families Procerithiidae, Cerithiidae and Cerithiopsidae (Cerithiacea). *Bulletin of the British Museum (Natural History), Geology Series*, **23**, 103–75.

Adegoke, O. S. (1977) Stratigraphy and palaeontology of the Ewekoro Formation (Paleocene) of southwestern Nigeria. *Bulletin of American Paleontology*, **71** (295), 1–379.

Akopyan, V. T. (1972) Systematics of Late Cretaceous trochactaeonids (Gastropoda). *Paleontological Journal*, **6**, 1–12.

Akpan, E. T., Farrow, G. E. and Morris, N. (1982) Limpet grazing on Cretaceous algal-bored ammonites. *Palaeontology*, **25**, 361–7.

Allen, J. E. (1970) New species of Eocene Mollusca from the Gulf Coast. *Tulane Studies in Geology and Paleontology*, **8**, 69–78.

Allison, E. C. (1955) Middle Cretaceous Gastropoda from Punta China, Baja California, Mexico. *Journal of Paleontology*, **29**, 400–32.

Arkell, W. J. (1941) The Gastropods of the Purbeck Beds. *Quarterly Journal of the Geological Society of London*, **97**, 79–128.

Bandel, K. (1988a) Repräsentieren die Euomphaloidea eine natürliche Einheit der Gastropoden? *Mitteilungen aus dem Geologisch–Paläontologischen Institut der Universität Hamburg*, **67**, 1–33.

Bandel, K. (1988b) Early ontogenetic shell and shell structure as aid to unravel gastropod phylogeny and evolution, in *Prosobranch Phylogeny. Proceedings of a Symposium Held at the 9th International Malacological Congress, Edinburgh* (ed. W. F. Ponder), Malacological Review, Supplement, 4, Ann Arbor, Michigan, pp. 267–72.

Bandel, K. (1991) Über triassische 'Loxonematoidea' und ihre Beziehungen zu rezenten und paläozoischen Schnecken. *Paläontologische Zeitschrift*, **65**, 239–68.

Bandel, K. and Hemleben, C. (1987) Jurassic heteropods and their modern counterparts (planktonic Gastropoda, Mollusca). *Neues Jahrbuch für Geologie und Paläontologie, Abhandlung*, **174**, 1–22.

Batten, R. L. (1973) The vicissitudes of the gastropods during the interval of Guadalupian–Ladinian time, in *The Permian and Triassic Systems and their Mutual Boundary* (ed. A. Logan), Canadian Society of Petroleum Geologists, Memoir, **2**, pp. 597–607.

Batten, R. L. (1979) Permian gastropods from Perak, Malaysia. Part 2. The trochids, patellids and neritids. *American Museum Novitates*, **2829**, 1–40.

Batten, R. L. (1984a) Shell structure of the Galapagos Rift Limpet *Neomphalus fretterae* McLean, 1981, with notes on muscle scars and insertions. *American Museum Novitates*, **2276**, 1–13.

Batten, R. L. (1984b) The calcitic wall in the Paleozoic families Euomphalidae and Platyceratidae (Archeogastropoda). *Journal of Paleontology*, **58**, 1186–92.

Batten, R. L. and Stokes, W. L. (1986) Early Triassic gastropods from the Sinbad Member of the Moenkopi Formation, San Rafel Swell, Utah. *American Museum Novitates*, **2864**, 1–33.

Beu, A. G. and Maxwell, P. A. (1990) Cenozoic Mollusca of New Zealand. *New Zealand Geological Survey Palaeontological Bulletin*, **58**, 1–518.

Blodgett, R. B. (1988) *Wisconsinella*, a new raphistomatid gastropod genus from the Middle Devonian of Wisconsin. *Journal of Paleontology*, **62**, 442–4.

Blodgett, R. B. (1993) Taxonomy and palaeobiogeographic affinities of an early Middle Devonian (Eifelian) gastropod faunule from the Livengood Quadrangle, East-Central Alaska. *Palaeontographica Abt.* A, **221**, 125–68.

Blodgett, R. B. and Johnson, J. G. (1992) Early Middle Devonian (Eifelian) gastropods of Central Nevada. *Palaeontographic Abt. A*, **222**, 85–139.

Blodgett, R. B., Rohr, D. M. and Boucot, A. J. (1988) Lower Devonian gastropod biogeography of the Western Hemisphere, in *Devonian of the World* (eds N. J. McMillan, A. F. Embry and D. J. Glass). Canadian Society of Petroleum Geologists, Memoir, 3, pp. 281–94.

Boucot, A. J. (1990) Carrier Shells, in *Evolutionary Paleobiology of Behaviour and Coevolution*. Elsevier, Amsterdam, pp. 463–7.

Brookfield, M. E. (1978) The lithostratigraphy of the upper Oxfordian and lower Kimmeridgian Beds of South Dorset, England. *Proceedings of the Geologists' Association*, **89**, 1–32.

Cernohorsky, W. O. (1970) Systematics of the families Mitridae and Volutomitridae (Mollusca: Gastropoda). *Bulletin of the Auckland Institute and Museum*, **8**, 1–190.

Cleevely, R. J. and Morris, N. J. (1988) Taxonomy and ecology of Cretaceous Cassiopidae (Mesogastropoda). *Bulletin of the British Museum (Natural History), Geology Series*, **44**, 233–91.

Connolly, M. (1939) A Monographic Survey of South African Non-marine Mollusca. *Annals of the South African Museum*, **33**, 1–660.

Conti, M. A. and Fischer, J. C. (1982) La faune à gastropodes du Jurassique Moyen de Case Canepine (Umbria, Italie).

Systématique, paléobiogéographie, paléoécologie. *Geologica Romana*, **21**, 125–83.

Cooper, J. (1976) Report of Field Meeting to Harefield, Middlesex, 14.III.1976. *Tertiary Research*, **1**, 31–4.

Cooper, J. (1979) Lower Tertiary fresh-water Mollusca from Mull, Argyllshire. *Tertiary Research*, **2**, 69–74.

Cossmann, M. (1888–1889) Catalogue illustré des coquilles fossiles de l'Eocène des environs de Paris. 3–4. *Annales de la Société Malacologique de Belgique*, **23**, 1–328, **24**, 1–385.

Cossmann, M. (1895–1925) *Essais de Paléoconchologie comparée*. Livr. 1–13, Paris.

Cossmann, M. (1913) Contribution à la paléontologie française des terrains Jurassique: III Cerithiacea et Loxonematacea. *Mémoires de la Société Géologique de France, Paléontologie*, **46**, 145–245.

Cossmann, M. and Pissarro, G. (1907–1913) *Iconographie Complète des Coquilles Fossiles de l'Eocène des Environs de Paris*, Vol. 2, Paris.

Cox, L. R. (1953) Gastropoda from the Karroo Beds of Southern Rhodesia. *Geological Magazine*, **90**, 201–7.

Cox, L. R. (1965) Jurassic Bivalvia and Gastropoda from Tanganyika and Kenya. *Bulletin of the British Museum (Natural History), Geology Series, Supplement*, **1**, 1–213.

Curry, D. (1981) Ptéropodes Eocènes de la tuilerie de Gan (Pyrénées- Atlantiques) et de quelques autres localités du SW de la France. *Cahiers de Micropaléontologie*, **4**, 35–44.

Dall, W. H. (1915) A monograph of the molluscan fauna of the *Orthaulax pugnax* Zone of the Oligocene of Tampa, Florida, *Bulletin of the United States National Museum*, **90**, 1–173.

Darragh, T. A. (1969) A revision of the Family Columbariidae (Mollusca: Gastropoda). *Proceedings of the Royal Society of Victoria*, **83**, 63–119.

Davis, G. M. (1979) The origin and evolution of the gastropod Family Pomatiopsidae, with emphasis on the Mekong River Triliculinae. *Academy of Natural Sciences of Philadelphia, Monograph*, **20**, 1–120.

Day, J. E. and Beus, S. S. (1982) *Turbonopsis apachiensis*, a new species of archaeogastropod from the Devonian (Frasnian) of Arizona. *Journal of Paleontology*, **56**, 1119–23.

Dolin, C., Dolin, L. and Le Renard, J. (1980) Inventaire systématique des mollusques de l'Auversien à 'facies charrié' de Baron (Oise), et remarques paléontologiques. *Bulletin de l'Information des Géologues du Bassin de Paris*, **17** (2), 26–48.

Donald, J. (1902) On some of the Proterozoic Gastropoda which have been referred to *Murchisonia* and *Pleurotmaria*, with descriptions of new subgenera and species. *Quarterly Journal of the Geological Society of London*, **58**, 313–39.

Dzhalilov, M. R. (1972) Systematics of acteonellids (Gastropoda). *Paleontological Journal*, **6**, 13–19.

Dzhalilov, M. R. (1977) [*Cretaceous gastropods of S.E. of Central Asia.*] Donish, Dushanbe, 1–202 (in Russian).

Dzik, J. (1983) Larval development and relationships of *Mimospira* – a presumably hyperstrophic Ordovician gastropod. *Geologiska Föreningens i Stockholm Förhandlingar*, **104**, 231–9.

Erwin, D. H. (1988) Permian Gastropoda of the southwestern United States: Subulitacea. *Journal of Paleontology*, **62**, 56–69.

Ferenc, G. (1961) Regionale Metachronozönose der Alpinen Triadischen Gastropoden. *Jahrbuch der Ungarischen Geologischen Anstatt*, **49**, 21–8.

Flower, R. H. (1968) Fossils from the Fort Ann Formation. *New Mexico Bureau of Mines and Mineral Resources Memoir*, **22**, 1–68.

Forney, G. G. and Nitecki, M. H. (1976) Type fossil Mollusca (Hyolitha, Polyplacophora, Scaphopoda, Monoplacophora and Gastropoda) in Field Museum. *Fieldiana, Geology*, **35**, 1–24.

Fortey, R. A. and Peel, J. S. (1990) Early Ordovician trilobites and molluscs from the Poulsen Cliff Formation, Washington Land, western North Greenland. *Bulletin Geological Society of Denmark*, **38**, 11–32.

Fürsich, F. T. and Kauffman, E. G. (1984) Palaeoecology of marginal marine sedimentary cycles in the Albian Bear River Formation of south-western Wyoming. *Palaeontology*, **27**, 501–36.

Garvie, C. L. (1991) Two new species of Muricinae from the Cretaceous and Paleocene of the Gulf coastal plain, with comments on the genus *Odontopolys* Gabb, 1860. *Tulane Studies in Geology and Paleontology*, **24**, 87–92.

Glibert, M. (1973) Revision des Gastropoda du Danien et du Montien de la Belgique, 1. Les Gastropoda du Calcaire de Mons. *Insitut Royal des Sciences Naturelles de Belgique, Mémoire*, **173**, 1–116.

Golikov, A. N. and Starobogatov, Y. I. (1975) Systematics of prosobranch gastropods. Malacologia, **15**, 185–232.

Gougerot, L. and Le Renard, J. (1980) Clefs de détermination des petites espèces de gastéropodes de l'Eocène du Bassin Parisien. 13. Le genre *Omalaxis*. *Cahiers des Naturalistes, Paris*, **36**, 1–7.

Guo, F., Yü, W. and Pan, H.-Z. (1982) Phylum Mollusca. Class Gastropoda, in *Xian Institute of Geology and Mineral Resources. Palaeontological Atlas of Northwest China. Part 3. Mesozoic and Cenozoic*. Geological Publishing House, Beijing, pp. 1–181 (in Chinese), pp. 28–52.

Haas, A. (1953) Mesozoic invertebrate faunas of Peru, Parts I and II. *Bulletin of the American Museum of Natural History*, **101**, 1–328.

Haszprunar, G. (1988) On the origin and evolution of major gastropod groups, with special reference to the Streptoneura. *Journal of Molluscan Studies*, **54**, 367–440.

Hayami, I. and Kase, T. (1977) A systematic survey of the Paleozoic and Mesozoic Gastropoda and Paleozoic Bivalvia from Japan. *Bulletin of the University Museum, The University of Tokyo*, **13**, 1–156.

Henderson, J. (1935) Fossil non-marine Mollusca of North America. *Geological Society of America, Special Paper*, **3**, 1–313.

Holzapfel, E. (1888) Die Mollusken der Aachener Kreide. *Palaeontographica*, **34**, 29–180.

Horný, R. (1961) Lower Paleozoic Monoplacophora and patellid Gastropoda (Mollusca) of Bohemia. *Sbornik Ústředniho Ústavu geologického Serie paléontologique*, **28**, 1–69.

Horný, R. (1963) Lower Paleozoic Bellerophontina (Gastropoda) of Bohemia. *Sbornik Geologickych Ved*, **30**, 57–161.

Houbrick, R. S. (1979) Classification and systematic relationships of the Abyssochrysidae, a relict family of bathyal snails (Prosobranchia: Gastropoda). *Smithsonian Contributions to Zoology*, **290**, 1–21.

Houbrick, R. S. (1980) Review of the deep-sea genus *Argyropeza* (Gastropoda: Prosobranchia: Cerithiidae. *Smithsonian Contr. Zool.*, **321**, 1–20.

Houbrick, R. S. (1981) Anatomy of *Diastoma melanoides* (Reeve, 1849) with remarks on the systematic position of the family Diastomatidae (Prosobranchia: Gastropoda). *Proceedings of the Biological Society of Washington*, **94**, 598–621.

Houbrick, R. S. (1988) Cerithioidean phylogeny, in *Prosobranch Phylogeny: Proceedings of a Symposium held at the 9th International Malacological Congress, Edinburgh* (ed. W. F. Ponder), Malacological Review, Supplement 4. Ann Arbor, Michigan, pp. 88–128.

Hrubesch, K. (1965a) Die santone Gosau–Landschneckenfauna von Glanegg bei Salzburg, Österreich. *Mitteilungen der Bayerische Staatssammlung für Paläontologie und historische Geologie*, **5**, 83–120.

Hrubesch, K. (1965b) Gosau–Landschnecken des Coniac von Unterlaussa bei Windisch Garsten, Oberösterreich. *Mitteilungen der Bayerische Staatssammlung für Paläontologie und historische Geologie*, **5**, 121–6.

Hudleston, W. H. (1887–1896) A monograph of the British Jurassic Gasteropoda. *Palaeontographical Society (Monograph)*, **40–50**, 1–514.

Ishii, K. I. and Murata, M. (1974) *Khumerspira*, a new genus of Bellerophontidae, and some Middle Permian gastropods from Cambodia. *Journal of Geosciences, Japan*, **17**, 73–86.

Jablonski, D. and Bottjer, D. J. (1990) The origin and diversification of major groups: environmental patterns and macroevolutionary lags, in *Major Evolutionary Radiations* (eds P. D. Taylor and G. P. Larwood). Systematics Association Special Volume 42,

Clarendon Press, Oxford, pp. 17–57.

Janssen, A. W. (1984) *Mollusken uit het Mioceen van Winterswijk-Miste. Een Inventarisatie, met Beschrijvingen en Afbeeldingen van alle Aangetroffen Soorten*. Koninklijke Nederlandse Natuurhistorische Vereniging, Amsterdam, 452 pp.

Janssen, R. (1978) Die Scaphopoden und Gastropoden des Kasseler Meeressandes von Glimmerode (Niederhessen). *Geologisches Jahrbuch (A)*, **41**, 3–195.

Jung, P. (1974) A revision of the family Seraphsidae (Gastropoda: Strombacea). *Palaeontographica Americana*, **8** (47), 1–72.

Kase, T. (1984) *Early Cretaceous Marine and Brackish-water Gastropoda from Japan*. National Science Museum, Tokyo, 199 pp.

Kase, T. (1986) Mode of life of the Silurian uncoiled gastropod *Semitubina sakoi* n. sp. from Japan. *Lethaia*, **19**, 327–37.

Kase, T. (1990) Late Cretaceous gastropods from the Izumi Group of southwest Japan. *Journal of Paleontology*, **64**, 563–78.

Kier, P. M. (1981) A bored Cretaceous echinoid. *Journal of Paleontology*, **55**, 656–9.

Kittl, E. (1895) Die triadischen Gastropoden der Marmolata und verwandter Fundstellen in der weissen Riffkalken Südtirols. *Jahrbuch der Kaiserlich–Königlichen Geologischen Reichsanstalt*, **44**, 99–182.

Knight, J. B. (1934) The gastropods of the St Louis, Missouri, Pennsylvanian outlier: VIII. The Turritellidae. *Journal of Paleontology*, **8**, 443–7.

Knight, J. B. (1941) Paleozoic gastropod genotypes. *Special Papers of the Geological Society of America*, **32**, 1–510.

Knight, J. B. (1956) New families of gastropods. *Washington Academy of Sciences Journal*, **46**, 41–2.

Knight, J. B., Cox, L. R., Keen, A. M., *et al.* (1960) Mollusca – Gastropoda; systematic descriptions (with supplement by Knight, J. B., Batten, R. L., Yochelson, E. L. and Cox, L. R. Paleozoic and some Mesozoic Caenogastropoda and Opisthobranchia), in *Treatise on Invertebrate Paleontology, 1 (Mollusca 1)*, (ed. R. C. Moore), Geological Society of America and University of Kansas Press, Boulder, Colorado and Lawrence, Kansas, pp. 1171–351.

Koken, E. (1897) Die Gastropoden der Trias um Hallstatt. *Abhandlungen der Kaiserlich–Königlichen Geologischen Reichsanstalt*, **17**, 135–225.

Koken, E. (1925) *Die Gastropoden des baltisten Untersilurs* (Ed. J. Perner) Memoires de l'Academie des Sciences de Russie, 8 ser. Classe Physico-Mathematique, **37**(1), Leningrad.

Kollmann, H. A. (1976) Gastropoden aus den Losensteiner Schichten der Umgebung von Losenstein (Oberösterreich) 1 Teil, Euthyneura und Prosobranchia 1 (Neogastropoda). *Annalen des Naturhistorischen Museums, Wien*, **80**, 163–207.

Kollmann, H. A. (1979) Gastropoden aus den Losensteiner Schichten der Umgebung von Losenstein (Oberösterreich), 3 Teil: Cerithiacea (Mesogastropoda). *Annalen Naturhistorischen Museums*, Wein, **82**, 11–51.

Kollmann, H. A. (1984) *Megalonoda* n. gen. (Melanopsidae, Gastropoda) Oberkreide der nordlichen Kalkalpen (Osterreich). *Annalen des Natuurhistorischen Museums, Wien*, **86**, 55–62.

Kollmann, H. A. and Sohl, N. F. (1980) Western Hemisphere Cretaceous Itieriidae gastropods. *US Geological Survey Professional Paper*, **1125A**, 1–50.

La Rocque, A. (1960) Molluscan faunas of the Flagstaff Formation of central Utah. *Memoirs of the Geological Society of America*, **78** 1–100.

Le Renard, J. (1983) Mise en évidence d'algueraies à *Caulerpa* par les Juliidae (Gastéropodes à 2 valves: Sacoglossa) dans l'Eocène du Bassin de Paris. *Geobios*, **16**, 39–51.

Lindberg, D. R. (1988) The Patellogastropoda, in *Prosobranch Phylogeny: Proceedings of a Symposium Held at the 9th International Malacological Congress, Edinburgh* (ed. W. F. Ponder), Malacological Review, Supplement 4, Ann Arbor, Michigan, pp. 35–63.

Lindberg D. R. and Hickman, C. S. (1986) A new anomalous giant limpet from the Oregon Eocene (Mollusca: Patellida). *Journal of Paleontology*, **60**, 661–8.

Linsley, R. M. (1968) Gastropods of the Middle Devonian Anderdon Limestone. *Bulletin of American Paleontology*, **54** (244), 329–465.

Linsley, R. M. (1979) Gastropods of the Devonian. *Special Papers in Palaeontology*, **23**, 249–54.

Linsley, R. M. and Kier, W. M. (1984) The Paragastropoda: a proposal for a new class of Paleozoic Mollusca. *Malacologia*, **25**, 241–54.

McLean, L. H. (1988) New archaeogastropod limpets from hydrothermal vents, Superfamily Leptodrilacea. 1. Systematic descriptions. *Philosophical Transactions of the Royal Society of London B*, **319**, 1–32.

McLean, L. H. (1989) New archaeogastropod limpets from hydrothermal vents: new Family Peltospiridae, new Superfamily Peltospiracea. *Zoologica Scripta*, **18**, 49–66.

McLean, L. H. (1990) Neolepetopsidae, a new docoglossate limpet family from hydrothermal vents and its relevance to patellogastropod evolution. *Journal of Zoology*, **222**, 485–528.

Moisescu, V. (1982) Contributions à la connaissance de la faune de mollusques Oligocènes de calcaire ducaquicole de Merisor (Bassin de Hateg). *Dări de Seamă ale Sedintelor Institutul Geologic al României*, **67**, 99–107.

Newton, R. B. (1891) *Systematic List of the Frederick E. Edwards collection of British Oligocene and Eocene Mollusca in the British Museum (Natural History)*. British Museum (Natural History), London, 365 pp.

Nordsieck, H. (1986) The System of the Stylommatophora (Gastropoda) with special regard to the systematic position of the Clausiliidae, 2. Importance of the shell and distribution. *Archiv für Molluskenkunde*, **117**, Appendix, 93–116.

Nuttall, C. P. and Cooper, J. (1973) A Review of some English Palaeogene Nassariidae, formerly referred to *Cominella*. *Bulletin of the British Museum (Natural History) Geology Series*, **23**, 179–219.

Pan, H.-Z. (1982) Triassic marine fossil gastropods from S.W. China. *Bulletin of the Nanjing Institute of Geology and Palaeontology*, **4**, 153–188 [in Chinese with English summary].

Pan, H.-Z. (1990) Late Cretaceous gastropod dominated communities of the Western Tarim Basin, Southern Xinjiang, China. *Lethaia*, **23**, 273–89.

Pĉhelintsev, V. F. (1965) Mesozoic Murchisoniata of the Crimean Highlands. *A. P. Karpinsky Memorial Museum Geology, Akademiya Nauk, SSR Monograph*, **8**, 1–216 [in Russian].

Phillip, G. M. and Talent, J. A. (1959) The gastropod genera *Liomphalus* Chapman and *Scalaetrochus* Etheridge. *Journal of Paleontology*, **33**, 50–4.

Poinar, G. O., Jr and Roth, B. (1991) Terrestrial snails (Gastropoda) in Dominican amber. *The Veliger*, **34**, 253–8.

Ponder, W. F. (1973) The origin and evolution of the Neogastropoda. *Malacologia*, **12**, 295–338.

Ponder, W. F. (1983) A revision of the Recent Xenophoridae of the World and of the Australian fossil species (Mollusca: Gastropoda), with an appendix by W. F. Ponder and J. Cooper. *Australian Museum Memoir*, **17**, 1–126.

Ponder, W. F. (1985) A review of the genera of the Rissoidae (Mollusca: Mesogastropoda: Rissoacea). *Records of the Australian Museum, Supplement*, **4**, 1–221.

Ponder, W. F. (1988) The truncatelloidan (= rissoacean) radiation – a preliminary phylogeny, in *Prosobranch Phylogeny: Proceedings of a Symposium Held at the 9th International Malacological Congress, Edinburgh* (ed. W. F. Ponder), Malacological Review, Supplement, 4, Ann Arbor, Michigan, pp. 129–66.

Ponder, W. F. and Warén, A. (1988) Classification of the Caenogastropoda and Heterostropha – a list of the family-group names and higher taxa, in *Prosobranch Phylogeny: Proceedings of a Symposium held at the 9th International Malacological Congress, Edinburgh* (ed. W. F. Ponder), Malacological Review, Supplement, 4, Ann Arbor, Michigan, pp. 288–326.

Popenoe, W. P. (1983) Cretaceous Aporrhaidae from California:

Aporrhainae and Arrhoginae. *Journal of Paleontology*, **57**, 742–65.

Popenoe, W. P. and Saul, L. R. (1987) Evolution and classification of the Late Cretaceous – Early Tertiary gastropod *Perissitys*. *Natural History Museum of Los Angeles County, Contributions in Science*, **380**, 1–37.

Powell, A. W. B. (1966) The molluscan families Speightiidae and Turridae. *Bulletin of the Auckland Institute and Museum*, **5**, 1–184.

Preece, R. (1980) The Mollusca of the Creechbarrow Limestone Formation (Eocene) of Creechbarrow Hill, Dorset. *Tertiary Research*, **2**, 169–80.

Ravn, J. P. J. (1933) Études sur les pélécypodes et gastropodes daniens du Calcaire de Faxe. *Mémoires de l'Académie Royale des Sciences et des Lettres de Danemark (Section des Sciences). 9th. ser.*, **5**, 55–170.

Reid, D. G. (1989) The comparative morphology, phylogeny and evolution of the gastropod Family Littorinidae. *Philosophical Transactions of the Royal Society of London*, **B324**, 1–110.

Rohr, D. M. (1979) Geographic distribution of the Ordovician gastropod *Maclurites*, in *Historical Biogeography, Plate Tectonics and the Changing Environment* (eds J. Gray and A. J. Boucot), Oregon State Univ. Press, Corvallis, pp. 45–52.

Rohr, D. M. and Blodgett, R. B. (1985) Upper Ordovician Gastropoda from west-central Alaska. *Journal of Paleontology*, **59**, 667–73.

Rohr, D. M. and Johns, R. A. (1990) First occurrence of *Oriostoma* (Gastropoda) from the Middle Ordovician. *Journal of Paleontology*, **64**, 732–5.

Rollins, H. B., Eldredge, N. and Spiller, J. (1971) Gastropoda and Monoplacophora of the Solsville Member (Middle Devonian, Marcellus Formation) in the Chenango Valley, New York State. *Bulletin of the American Museum of Natural History*, **144**, 131–70.

Rosenkrantz, A. (1960) Danian Mollusca from Denmark, in *Report of the International Geological Congress, XXI Session, Norden 1960. Part V: The Cretaceous – Tertiary Boundary*, Det Berlingske Bogtrykkeri, Copenhagen, pp. 193–8.

Saul, L. R. (1988) Latest Cretaceous and Early Tertiary Tudiclidae and Melongenidae (Gastropoda) from the Pacific Slope of North America. *Journal of Paleontology*, **62**, 880–9.

Sayn, G. (1932) Description de la faune de l'Urgonien de Barcelonne (Drome). *Travaux du Laboratoire de Geologie de la Faculté des Sciences de Lyon, Mémoire* 15, **18**, 1–70.

Schilder, M. and Schilder, F. A. (1971) A catalogue of living and fossil cowries, Taxonomy and bibliography of Triviacea and Cypraeacea (Gastropoda, Prosobranchia). *Institut Royal des Sciences Naturelles de Belgique, Mémoires, 2nd série*, **85**, 1–246.

Schlickum, W. R. (1978) Zur oberpannonen Mollusken Fauna von Öcs. 1. *Archiv für Molluskenkunde*, **108**, 245–61.

Schlickum, W. R. (1979) Die Gattung *Hydrocena* im europäischen Tertiär (Neritacea: Hydrocenidae). *Archiv für Molluskenkunde*, **110**, 71–3.

Sohl, N. F. (1960) Archeogastropoda, Mesogastropoda and Stratigraphy of the Ripley, Owl Creek and Prairie Bluff Formations. *US Geological Survey Professional Paper*, **331-A**, 1–151.

Sohl, N. F. (1964a) Neogastropoda, Opisthobranchia and Basommatophora from the Ripley, Owl Creek, and Prairie Bluff Formations. *US Geological Survey Professional Paper*, **331-B**, 153–344.

Sohl, N. F. (1964b) Gastropods from the Coffee Sand (Upper Cretaceous) of Mississippi. *US Geological Survey Professional Paper*, **331-C**, 345–94.

Solem, A. (1977) Fossil endodontid land snails from Midway Atoll. *Journal of Paleontology*, **51**, 902–11.

Solem, A. (1978) Cretaceous and Early Tertiary camaenid land snails from western North America (Mollusca: Pulmonata). *Journal of Paleontology*, **52**, 581–9.

Solem, A. and Yochelson, E. L. (1979) North American Paleozoic land snails, with a summary of other Paleozoic nonmarine snails. *US Geological Survey Professional Paper*, **1072**, 1–39.

Squires, R. L. (1984) Megapaleontology of the Eocene Llajas Formation, Simi Valley, California. *Contributions in Science*, Los Angeles, **350**, 1–76.

Squires, R. L. (1987) Cretaceous gastropods: contrasts between tethys and the temperate provinces. *Journal of Paleontology*, **61**, 1085–111.

Squires, R. L. (1989) First Tertiary occurrence of a rare patelliform gastropod (Archaeogastropoda: Symmetrocapulidae), Eocene Tejon Formation, Tenachapi Mountains, California. *The Veliger*, **32**, 406–8.

Squires, R. L., Zinsmeister, W. J. and Paredes-Mejia, L. M. (1989) *Popenoeum*, a new pseudolivine gastropod genus widespread and most diversified during the Paleocene. *Journal of Paleontology*, **63**, 38–47.

Stephenson, L. W. (1952) The larger invertebrate fossils of the Woodbine Formation (Cenomanian) of Texas. *US Geological Survey Professional Paper*, **242**, 1–226.

Stoliczka, F. (1867–1868) Cretaceous Fauna of Southern India. Volume II, Gastropoda. *Palaeontographica Indica*, **2** (ser. 5) pts 1–10, 497 pp.

Szabó, J. (1980) Lower and Middle Jurassic Gastropods from the Bakony Mountains (Hungary). Part 2. Pleurotomariacea (Archaeogastropoda). *Annales Historico – Naturales Musei Nationalis Hungarici*, **72**, 1–50.

Szabó, J. (1983) Lower and Middle Jurassic Gastropods from the Bakony Mountains (Hungary). Part 5. Supplement to Archaeogastropoda; Caenogastropoda. *Annales Historico – Naturales Musei Nationalis Hungarici*, **75**, 27–46.

Tassell, C. B. (1977) Gastropods from the Early Devonian Bell Point Limestone, Cape Liptrap Peninsula, Victoria. *Memoir Natural History Museum of Victoria*, **38**, 19–32.

Taylor, D. W. (1960) Late Cenozoic molluscan faunas from the High Plains. *US Geological Survey Professional Paper*, **337**, 1–94.

Taylor, D. W. and Morris, N. J. (1988) Relationships of mesogastropods, in *Prosobranch Phylogeny: proceedings of a Symposium held at the 9th International Malacological Congress, Edinburgh* (ed. W. F. Ponder), Malacalogical Review, Supplement, **4**, Ann Arbor, Michigan, pp. 167–79.

Taylor, D. W., Cleevely, R. J. and Morris, N. J. (1983) Predatory gastropods and their activities in the Blackdown Greensand (Albian) of England. *Palaeontology*, **26**, 521–53.

Thein, M. and Nitecki, M. H. (1974) Chesternian (Upper Mississippian) Gastropoda of the Illinois Basin. *Fieldiana, Geology*, **34**, 1–238.

Tichy, G. (1979) Gastropoden und Scaphopoden aus der Raibler Gruppe (Karn) von Raibl (Cave de Predil), Italien. *Verhandlungen der Geologischen Bundesanstatt, Wien*, **1979**, 443–61.

Traub, F. (1979) Weitere Paleozän Gastropoden aus den Helvetikum des Haunsberges nördlich von Salzburg. *Mitteilungen des Bayerischen Staatssammlung für Paläontologic und Hiztorische Geologie*, **19**, 93–123.

Van der Spoel, S. and Diester-Haass, L. (1976) First record of fossil gymnostomatous protoconchae (Pteropoda, Gastropoda). *Bulletin, Zoologisch Museum, Universiteit van Amsterdam*, **5**, 85–8.

Vaught, K. C. (1989) *A Classification of the Living Mollusca*. Melbourne, Florida, 195 pp.

Wang, H. and Xi, Y. (1980) *Stratigraphy and Palaeontology of the Late Permian Coal Measure of Western Guizhou and Eastern Yunnan*. Science Press, Beijing, pp. 195–232.

Warén, A. (1983) A generic revision of the Family Eulimidae (Gastropoda, Prosobranchia). *The Journal of Molluscan Studies, Supplement*, **13**, 1–96.

Warén, A. and Bouchet, P. (1989) New gastropods from East Pacific hydrothermal vents. *Zoologica Scripta*, **18**, 67–102.

Warén, A. and Bouchet, P. (1990) Laubierinidae and Pisanianurinae (Ranellidae), two new deep-sea taxa of the Tonnoidea (Gastropoda: Prosobranchia). *The Veliger*, **33**, 56–102.

Waterhouse, J. B. (1980) Scaphopod, gastropod, and rostroconch species from the Permian of New Zealand. *Journal of the Royal Society of New Zealand*, **10**, 195–214.

Wenz, W. (1938–1944) Gastropoda, Teil 1: Allgemeiner Teil und

Prosobranchia, in *Handbuch der Paläozoologie*, 6, (ed. O. H. Schindewolf), Borntraeger, Zehlendorf, Berlin, pp. 1–1639.

Wilson, A. E. (1951) Gastropoda and Conularida of the Ottawa Formation of the Ottowa–St Lawrence Lowland. *Geologic Survey of Canada Bulletin*, **17**, 1–149.

Wrigley, A. (1942) English Eocene *Hastula* with remarks on the coloration of the Terebridae. *Proceedings of the Malacological Society*, **25**, 17–24.

Yen, T.-C. (1952) Molluscan fauna of the Morrison Formation. *US Geological Survey Professional Paper*, **233-B**, 21–51.

Yochelson, E. L. (1956) Permian gastropoda of the southwestern United States. 1. Euomphalacea, Trochonematacea, Pseudophoracea, Anomphalacea, Craspedostomatacea, and Platyceratacea. *Bulletin of the American Museum of Natural History*, **110**, 177–275.

Yochelson, E. L. (1960) Permian Gastropoda of the Southwestern United States. 3. Bellerophontacea and Patellacea. *Bulletin of the American Museum of Natural History*, **119**, 205–94.

Yochelson, E. L. (1968) Tremadocian mollusks from the Nochixtlán Region, Oaxaca, Mexico. *Journal of Paleontology*, **42**, 801–3.

Yochelson, E. L. (1988) A new genus of Patellacea (Gastropoda) from the Middle Ordovician of Utah: the oldest known example of the superfamily. *New Mexico Bureau of Mines and Mineral Resources Memoir*, **44**, 195–200.

Yochelson, E. L. and Stinchcomb, B. L. (1987) Recognition of *Macluritella* (Gastropoda) from the Upper Cambrian of Missouri and Nevada. *Journal of Paleontology*, **61**, 56–61.

Yochelson, E. L. and Yin, H.-F. (1983) Middle Triassic Gastropoda from Qingyan, Ghizhou Province, China: 3 – Euomphalacea and Loxonematacea. *Journal of Paleontology*, **57**, 1098–127.

Yochelson, E. L. and Yin, H.-F. (1985) Redescription of *Bellerophon asiaticus* Wirth (Early Triassic: Gastropoda) from China, and a survey of Triassic Bellerophontacea. *Journal of Paleontology*, **59**, 1305–19.

Yoo, E. K. (1988) Early Carboniferous Mollusca from Gundy, Upper Hunter, New South Wales. *Records of Australian Museum*, **40**, 233–64.

Youlou (Lu, W., Mao X.-L., Chen Z.-G., *et al.*) (1978) *Early Tertiary Gastropod Fossils from the Coastal Region of Bohai*. Science Press, Beijing, 157 pp [in Chinese].

Yü, W. (1977) Cretaceous and Early Tertiary non-marine gastropods from South China with their stratigraphical significance. *Acta Palaeontologica Sinica*, **16**, 191–213 [in Chinese with English summary].

Yu, X. (1982) Freshwater gastropods (Mollusca) of the Upper Jurassic Fuxin Formation, western Liaoning. *Bulletin of the Shenyang Institute of Geology and Mineral Resources*, **4**, 195–207 [in Chinese].

Zardini, R. (1978) *Fossili Cassiani (Trias Medio–Superiore); Atlante dei Gasteropodi di S. cassiano Raccolti Nella Regione Dolomitico Altorno a Cortina d'Ampezzo*, Edizioni Ghedina, Cortina d'Ampezzo, Italy, 58 pp.

Zilch, A. (1959–1960) Gastropoda, Teil 2: Euthyneura, in *Handbuch der Paläozoologie*, **6**, (ed. W. Wenz), Gebrüder Borntraeger, Berlin–Nikolassee, pp. 1–834.

9

MOLLUSCA: CEPHALOPODA (NAUTILOIDEA)

A. H. King

The term Nautiloidea is used here in a broad sense, incorporating the subclasses Endoceratoidea, Actinoceratoidea, Orthoceratoidea and Nautiloidea as perceived by Teichert (1967). Classification at order level is relatively stable; taxonomic complications arise in recognizing the greater diversity of nautiloids in comparison with ammonoids and reconciling this within a classification at higher level (Holland, 1979, 1987). The scheme adopted below essentially follows Teichert and Moore (*in* Teichert *et al.*, 1964) to order level, variations from this are noted in the text; the treatment of Cambrian orders follows Chen and Teichert (1983).

The enigmatic genera *Salterella* (=*Volborthella*) and *Vologdinella* have been regarded as taxa 'doubtfully classifiable' within the nautiloids and treated as a separate order Volborthellida (Teichert *in* Teichert *et al.*, 1964). The present author follows Yochelson (1977, 1981), and assigns these genera to the extinct phylum Agmata.

No detailed cladistic analysis has been carried out on nautiloids; comments concerning the mono- or paraphyletic nature of some orders are mainly the author's own views.

Order PLECTRONOCERIDA Flower, 1964 (see Fig. 9.1)

F. PLECTRONOCERATIDAE Kobayashi, 1935 €. (DOL) Mar.

First: *Plectronoceras cambria* (Walcott, 1905), *Ptychaspis–Tsinania* Zone, lower Fengshan Formation, Shandong Province, China; *P. liaoningense* Kobayashi, 1935 from Liaodong Peninsula and *P. huaibeiense* Chen and Qi 1979, from Anhui Province, China, are coeval (Chen and Teichert, 1983).
Last: *Eodiaphragmoceras sinense* Chen and Qi, 1979, *Jiagouceras cordatum* Chen and Tsou, 1979, *Lunanoceras changshanense* Chen and Qi, 1979, *L. precordium* Chen and Qi, 1979, *Rectseptoceras eccentricum* Tsou and Chen, 1979 and *Paraplectronoceras* spp., *Sinoeremoceras* zone (=*Acaroceras–Eburnoceras* Zone), upper Fengshan Formation, Shandong and Anhui Provinces, China. *Paleoceras mutabile* Flower, 1954 and *P. undulatum* Flower, 1964, San Saba Limestone, central Texas, USA, are of similar age (Chen and Teichert, 1983).

F. BALKOCERATIDAE Flower, 1964 €. (DOL) Mar.

First and Last: *Theskeloceras benxiense* Chen and Teichert, 1983, *T. subrectum* Chen and Teichert, 1983, upper Fengshan Formation, Liaoning Province, China; *Balkoceras gracile* Flower, 1964, San Saba Limestone, central Texas, USA (Chen and Teichert, 1983).

Order ELLESMEROCERIDA Flower, *in* Flower and Kummel, 1950

Suborder Ellesmerocerina Flower, 1964

F. ELLESMEROCERATIDAE Kobayashi, 1934 €. (DOL)–O. (CRD) Mar.

First: *Eburnoceras pissinum* Chen and Qi, 1980, upper *Quadraticephalus* Zone, lower Fengshan Formation, Anhui Province, China (Chen and Teichert, 1983).
Last: *Oelandoceras* sp. Lai, 1965, Baota (=Pagoda) Limestone Formation, Ningkiang, Shaanxi (=Shensi) Province, China (Lai, 1965).
Intervening: TRE–LLO.

F. ACAROCERATIDAE Chen *et al.*, 1979 €. (DOL) Mar.

First: *Acaroceras primordium* Chen and Qi, 1982, lower *Quadraticephalus* Zone, lower Fengshan Formation, Hanjia, Suxian County, Anhui Province, China (Chen and Teichert, 1983).
Last: *Weishanhuceras rarum* Chen and Qi, 1979, numerous species assigned to *Acaroceras* (Chen and Teichert, 1983), upper Fengshan and Siyangshan formations, Liaoning, Shandong, Anhui, Shanxi, Zhejiang Provinces and Nei Monggol, China (Chen and Teichert, 1983).

F. HUAIHECERATIDAE Zou and Chen, 1979 €. (DOL) Mar.

First: *Huaiheceras ? longicollum* Zou and Chen, 1979, upper *Quadraticephalus* Zone, lower Fengshan Formation, Suxian County, Anhui Province, China (Chen and Teichert, 1983).
Last: *Huaihecerina elegans* Chen and Teichert, 1983, *Zhuibanoceras conicum* Chen and Qi, 1981, numerous species assigned to *Huaiheceras*, upper Fengshan and Siyangshan formations, Liaoning, Shandong, Anhui and Zhejiang Provinces, China (Chen and Teichert, 1983).

The Fossil Record 2. Edited by M. J. Benton. Published in 1993 by Chapman & Hall, London. ISBN 0 412 39380 8

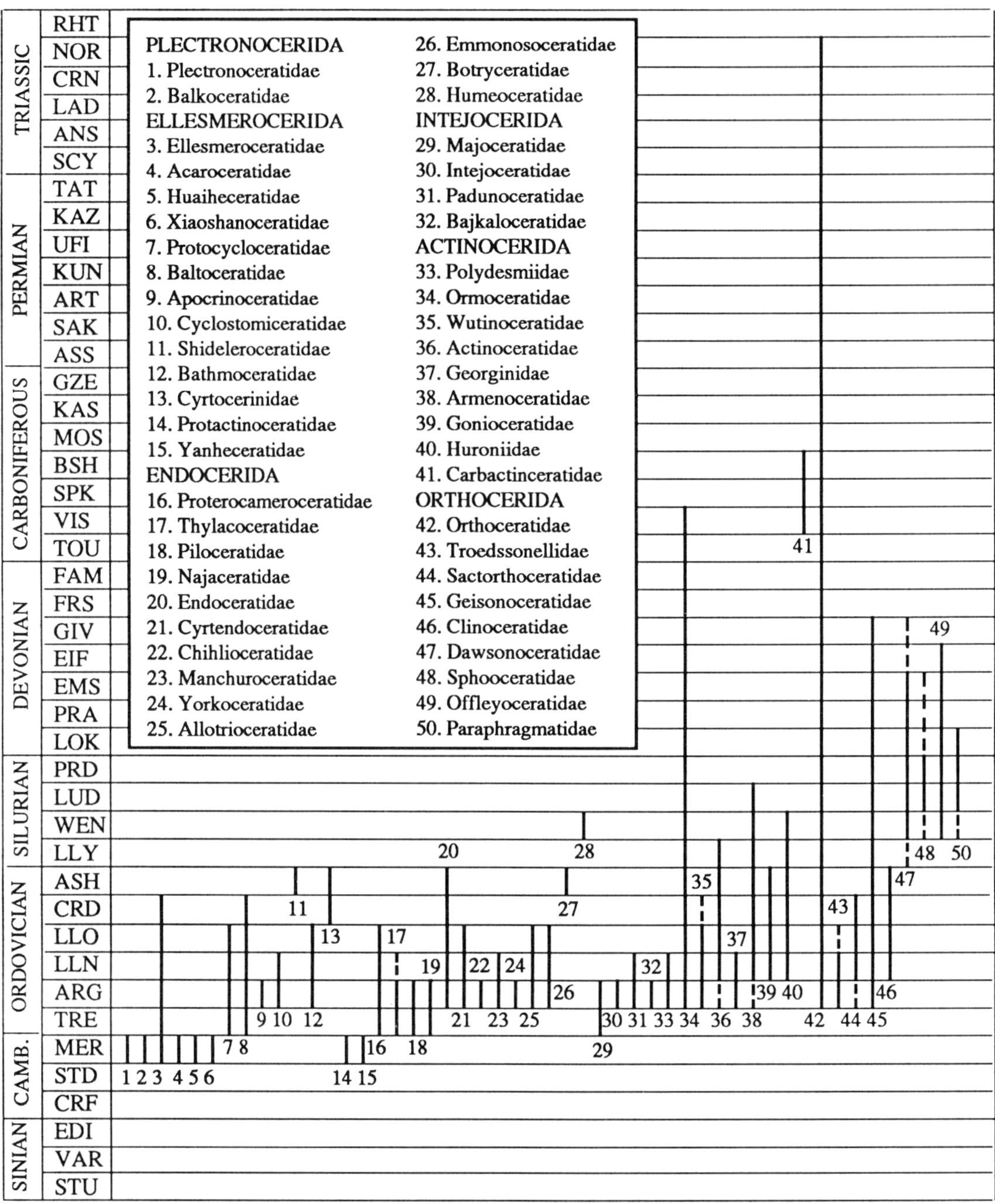

Fig. 9.1

F. XIAOSHANOCERATIDAE Chen and Teichert, 1983 Є. (DOL) Mar.

First and Last: *Xiaoshanoceras jini* Chen and Teichert, 1983, *X. subcirculare* Chen and Teichert, 1983, *Acaroceras* Zone, upper Siyangshan Formation, Xiaoshigai, Zhejiang Province, China (Chen and Teichert, 1983).

F. PROTOCYCLOCERATIDAE Kobayashi, 1935 O. (TRE–LLO$_3$) Mar.

First: *Walcottoceras obliquum* Ulrich, Foerste, Miller, and Unklesbay, 1944, lower Canadian (Gasconadian), Iowa, USA (Ulrich *et al.*, 1944).
Last: *Protocycloceras deprati* Reed, 1917, *P. wongi* (Yü, 1930), Shihtzupu Formation, Hunan Province, China (Sheng, 1980).
Intervening: ARG–LLN.

F. BALTOCERATIDAE Kobayashi, 1935 O. (TRE–CRD) Mar.

First: *Microbaltoceras minore* Flower, 1964, Tanyard Formation, Gillespie County, Texas, USA. This species is based on fragmentary material and may be referable to the Ellesmeroceratidae. The next oldest species is *Rioceras expansum* Flower, 1964, Cooks Formation, El Paso Group, Rhodes Canyon, New Mexico, USA (Flower, 1964a).
Last: *Cartersoceras noveboracense* Flower, 1952, Amsterdam

Limestone, New York, USA, *C. shideleri* Flower, 1964, Carter's Limestone, Tennessee, USA (Flower, 1964a).
Intervening: ARG–LLO.

F. APOCRINOCERATIDAE Flower in Flower and Teichert, 1957 O. (ARG) Mar.

First: *Apocrinoceras talboti* Teichert and Glenister, 1954, Emanuel Formation ('horizon 3'), Emanuel Creek, Kimberley Division, Western Australia (Teichert and Glenister, 1954).
Last: *Glenisteroceras obscurum* Flower *in* Flower and Teichert, 1957, Fort Cassin Formation, Champlain Valley, New York, USA (Flower, 1964a).

F. CYCLOSTOMICERATIDAE Foerste, 1925 O. (ARG–LLN_1) Mar.

First: *Cyclostomiceras cassinense* (Whitfield, 1886), *C. minimum* (Whitfield, 1886), Fort Cassin Formation, Champlain Valley, New York, USA (Flower, 1964a).
Last: *Pictetoceras eichwaldi* (De Verneuil, 1845), Kundan Stage (Aluojan Substage), Ural Mountains and Estonia, former USSR (Mutvei and Stumbur, 1971).

F. SHIDELEROCERATIDAE Flower, 1946 O. (RAW) Mar.

First and Last: *Shideleroceras sinuatum* Flower, 1946, *S. simplex* Flower, 1946, lower Whitewater Formation, Cincinnati; *S. gracile* Flower, 1946, Saluda Formation, Cincinnati, USA, all Richmondian (Flower, 1964a).

Suborder Cyrtocerinina Flower, 1964

F. BATHMOCERATIDAE Holm, 1899 O. (ARG–LLO) Mar.

First: *Eothinoceras americanum* Ulrich, Foerste, Miller and Unklesbay, 1944, Rochdale Limestone, New York, USA; *E. maitlandicum* Teichert and Glenister, 1954, Emanuel Formation, Emanuel Creek, Kimberley Division, Western Australia; *E. marchense* Balashov, 1960, Chunskiy Stage, Siberian Platform, former USSR (Flower, 1964a).
Last: *Bathmoceras norvegicum* Sweet, 1958, Cephalopod Shale, Helgøya, Nes-Hamar District, Norway (Sweet, 1958).
Intervening: LLN.

F. CYRTOCERINIDAE Flower, 1946 O. (HAR/SOU–RAW) Mar.

First: *Cyrtocerina* sp. Flower, 1952, Lowville Formation, Ottawa, Canada (Flower, 1952a).
Last: *Cyrtocerina madisonensis* (Miller, 1894), *C. patella* Flower, 1943, *C. modesta* Flower, 1943, upper Whitewater Formation, Madison, Indiana, USA; *C ? carinifera* Flower, 1946, Saluda Formation, Oxford, Ohio, USA (Flower, 1964a).

Order PROTACTINOCERIDA Chen and Qi, 1979

F. PROTACTINOCERATIDAE Chen and Qi, 1979 Є. (DOL) Mar.

First: *Wanwanoceras exiguum* Chen and Qi, 1982, lower *Quadraticephalus* Zone, lower Fengshan Formation, Hanjia, Suxian County, Anhui Province, China (Chen and Teichert, 1983).
Last: Numerous species assigned to *Wanwanoceras*, *Sinoeremoceras*, *Protactinoceras*, *Physalactinoceras*, *Benxioceras* and *Mastoceras*, upper Fengshan Formation, Liaoning, Shandong and Anhui Provinces, China (Chen and Teichert, 1983).

Order YANHECERIDA Chen and Qi, 1979

F. YANHECERATIDAE Chen and Qi, 1979 Є. (DOL) Mar.

First and Last: *Yanheceras anhuiense* Chen and Qi, 1979, *Y. endogastrum* Chen, 1979, *Y. longiconum* Chen and Qi, 1979, *Aetheloxoceras suxianense* Chen and Qi, 1979, *Archendoceras conipartitum* Chen and Qi, 1979, *Oonendoceras sinicum* Chen and Qi, 1982, upper Fengshan Formation, Shandong and Anhui Provinces, China (Chen and Teichert, 1983).

Order ENDOCERIDA Teichert, 1933

The earliest endocerids (Proterocameroceratidae) are commonly regarded to have arisen from longiconic ellesmerocerids (Ellesmeroceratidae) via the development of endocones from siphonal diaphragms (Flower, 1976c). The Piloceratidae may also be more simply derived from curved breviconic ellesmeroceratids than from the Proterocameroceratidae as believed by Flower (1976b). Such a phylogeny implies that endocones originated more than once from different sources within the Ellesmeroceratidae, and consequently the Endocerida are polyphyletic.

F. PROTEROCAMEROCERATIDAE Kobayashi, 1937 O. (TRE–LLO) Mar.

First: *Proendoceras annuliferum* (Flower, 1941a), Roubidoux Formation, New York, USA; Flower (1964a) noted 'straight endocerid siphuncles, probably *Proendoceras*' from the Cooks Formation (first endoceroid zone), El Paso Group, New Mexico, USA (Flower, 1941a, 1964a).
Last: *Lamottoceras ruedemanni* Flower, 1955a, *L. nodosum* Flower, 1955a, *L. franklini* Flower, 1958, Valcour Formation, Champlain Valley, New York, USA (Flower, 1955a, 1958).
Intervening: ARG–LLN.

F. THYLACOCERATIDAE Teichert and Glenister, 1954 O. (TRE–ARG/LLN_1) Mar.

First: *Talassoceras kumyschtagense* Balashov, 1960, upper Tremadoc, Khrebet Talasskiy, Kirgizia, former USSR (Balashov, 1960).
Last: *Thylacoceras yangtzeense* (Yü, 1930), Dawan Formation, Hubei Province, China (Lai, 1965).
Intervening: ARG.

F. PILOCERATIDAE Miller, 1889 O. (TRE–ARG_1) Mar.

First: *Piloceras* and *Bisonoceras* (*B. corniforme* Flower, 1964, *Bisonoceras* spp. Flower, 1964), Victorio Hills Formation (first piloceroid zone), El Paso Group, New Mexico, USA (Flower, 1964a, 1964b); also *Piloceras tuvense* Balashov, 1968 and *Allopiloceras sevierense* Ulrich *et al.* 1943, Malyy Karatau Mountains, Kazakhstan, former USSR (Balashov, 1968).
Last: *Cassinoceras explanator* (Whitfield, 1886), Fort Cassin Formation, Vermont, USA, and St George Group, western Newfoundland (Ulrich and Foerste, 1936; Flower, 1978). Other species assigned to *Cassinoceras*, *Piloceras* and *Allopiloceras* are known from western Newfoundland, north-western Australia and northern China (Teichert and Glenister, 1954; Flower, 1978; Sheng, 1980); these are all

of approximate Cassinian age but cannot be separated stratigraphically.

F. NAJACERATIDAE Flower, 1976 O. (TRE–ARG_2) Mar.

First: *Pronajaceras yichangense* Lai and Xu, 1983, *P. hubeiense* Lai and Xu, 1987, *P. eccentricum* Lai and Xu, 1987, upper Fenxiang Formation, Hubei Province, China (Lai and Xu, 1987).
Last: *Najaceras triangulum* Flower, 1971, *N. bilobatum* Flower, 1971, *N. chevroniferum* Flower, 1971, Oil Creek Formation, Oklahoma, USA (Flower, 1971).

F. ENDOCERATIDAE Hyatt, 1883 O. (ARG_1–CAU/RAW) Mar.

First: Stumbur (1962) reported *Endoceras* from the basal Arenig Leetse Stage, but provided no illustration or description. *Dideroceras leetsense* Balashov, 1968, Leztskiy Beds, Estonia, former USSR, is poorly known and *Allocotoceras insigne* Teichert and Glenister, 1953, upper Canadian, Adamsfield, Tasmania, is known only from isolated endosiphuncles (Balashov, 1968; Teichert and Glenister, 1953). The earliest undoubted endoceratid is *Dideroceras glauconiticum* (Heinrichson, 1935), 'Dikari Beds', lower Volkhovian Stage, Tiskre, Estonia, former USSR (Heinrichson, 1935).
Last: *Rossicoceras priguense* Balashov, 1968, Pirgu Stage, Estonia, former USSR; *Foerstellites faberi* (Foerste, 1930), Saluda Formation, Indiana, USA (Kobayashi, 1940; Balashov, 1968).
Intervening: LLN–CRD.

F. CYRTENDOCERATIDAE Hyatt, *in* Zittel, 1900 O. (ARG–LLO) Mar.

First: *Cyrtendoceras carnegiei* Teichert and Glenister, 1954, Emanuel Formation, Emanuel Creek, Kimberley Division, Western Australia (Teichert and Glenister, 1954).
Last: *Cyrtendoceras hircus* (Holm, 1892), Folkeslunda Formation, Lasnamaegian Stage, Öland, Sweden; *C. schmidti* (Holm, 1892), *Echinosphaerites* Limestone, Lasnamaegian/Uhakuan Stage, Estonia, former USSR (Holm, 1892); also *Cyrtendoceras* n. sp. Sweet, 1958, Cephalopod Shale, Nes-Hamar District, Norway (Sweet, 1958).
Intervening: LLN.

F. CHIHLIOCERATIDAE Grabau, 1922 O. (ARG) Mar.

First and Last: *Chihlioceras nathani* Grabau, 1922, *C. chingwangtaoense* Grabau, 1922, Peilintze Limestone (=Hunghuayuan Formation), Hubei Province, China (Grabau, 1922).

F. MANCHUROCERATIDAE Kobayashi, 1935 O. (ARG_1–LLN_1)

Genera assigned to the Coreanoceratidae (Chen, 1976; Lai and Xu, 1987) are typically based on incomplete, often recrystallized endosiphuncles. The taxonomic importance of endosiphuncular structures is currently being reassessed (MS in preparation). Consequently, *Coreanoceras* and allied forms are here provisionally included within the Manchuroceratidae pending further study.
First: *Coreanoceroides variabile* Qi, 1980, *Chaohuceras contractum* Qi, 1980, and numerous species assigned to *Manchuroceras* and *Coreanoceras* from the lower Arenig of East Asia (including North China, Nei Monggol, Korea, southern Thailand), Tasmania and Texas, USA (Chen, 1976; Zou, 1981; Stait and Burrett, 1984).
Last: *Manchuroceras asiaticum* Balashov, 1962, Siberian Platform, former USSR (Balashov, 1962); *Chaohuceras* ? sp. Stait and Burrett, 1984, upper Thung Song Formation, Satun Province, Thailand (Stait and Burrett, 1984).

F. YORKOCERATIDAE Flower, 1968 O. (ARG_1) Mar.

First and Last: *Yorkoceras discordium* Flower, 1968, *Sewardoceras tellerense* Flower, 1968, *Telleroceras undulatum* Flower 1968, upper Canadian, Seward Peninsula, Alaska (Flower, 1968e).

F. ALLOTRIOCERATIDAE Flower, 1955 O. (ARG_1–LLO) Mar.

First: Several species assigned to *Williamsoceras* and *Cacheoceras* from the Garden City and Juab Formations (lower Whiterockian), Utah, USA (Flower, 1968b, 1976b).
Last: *Mirabiloceras multitubulatum* Flower, 1955, Valcour Formation, Lake Champlain, New York, USA (Flower, 1955a).
Intervening: LLN.

F. EMMONSOCERATIDAE Flower, 1958 O. (ARG–LLO) Mar.

First: *Juaboceras braithwaiti* Flower, 1968, upper Juab Formation, western Utah, USA. Flower (1968b) tentatively placed this species within the Manchuroceratidae, but the large size and endosiphuncle form of *Juaboceras* are regarded here as more characteristic of the Emmonsoceratidae.
Last: *Emmonsoceras aristos* (Flower, 1955), Valcour Formation, Lake Champlain, New York, USA (Flower, 1955b).
Intervening: LLN.

F. BOTRYCERATIDAE Flower, 1968 O. (PUS) Mar.

First and Last: *Botryceras enigma* Flower, 1968, Second Value Formation, New Mexico, USA (Flower, 1968d).

?F. HUMEOCERATIDAE Teichert, 1964 S. (SHE/WHI) Mar.

Flower (1968c) regarded *Humeoceras* as a member of the Piloceratidae despite the large stratigraphical interval between known occurrences. The present author prefers to follow Teichert (*in* Teichert *et al.*, 1964) and to regard *Humeoceras* as a possible homeomorph after *Piloceras*.
First: *Humeoceras tardum* Flower, 1968, Severn River Formation, Rivière Malouin, James Bay Lowland, Canada (Flower, 1968c).
Last: *Humeoceras durdeni* Flower, 1968, Ekwan River Formation, Rapides des Papillons, Harricana River, James Bay Lowland, Canada; *H. unguloideum* Foerste, *in* Hume, 1925, Lockport Dolomite, Ontario, Canada is approximately coeval (Foerste, *in* Hume, 1925; Flower, 1968c).

Order INTEJOCERIDA Balashov, 1960

Relationships between genera assigned to this order are unclear and familial distinctions blurred. Teichert (*in* Teichert *et al.*, 1964) recognized the Endocerida and Intejocerida within the Endoceratoidea. Flower (1976b) concluded that *Intejoceras* and *Bajkaloceras* were not related to the endocerids but may be derived from the Baltoceratidae (Ellesmerocerida). Flower (1976b) and Gil (1988) placed *Padunoceras*, *Evencoceras* and *Rossoceras* in the Padunoceratidae and retained them in the Endocerida. Zhuravleva (1978) described the Bajkaloceratida as comprising the Bajkaloceratidae, Offleyoceratidae and Sichuanoceratidae. The last two families are recognized here as belonging in the Orthocerida and the taxonomic scheme used below follows Balashov (1968) and Crick (1988).

F. MAJOCERATIDAE Zhuravleva, 1964 O. (TRE/ARG) Mar.

First and Last: *Majoceras jakutense* Zhuravleva, 1964, Yakutia, Maiya River, Siberian Platform, former USSR (Zhuravleva, 1964).

F. INTEJOCERATIDAE Balashov, 1960 O. (ARG) Mar.

First and Last: *Intejoceras angarense* Balashov, 1969, Chunskiy Stage, Angara River Basin, Siberian Platform, former USSR (Balashov, 1960).

F. PADUNOCERATIDAE Balashov, 1960 O. (ARG–LLN) Mar.

First: *Rossoceras lamelliferum* Flower, 1964, *R. dentiferum* Flower, 1968, upper Garden City Formation, and *R. circulatum* Gil, 1988, *Evencoceras raymondi* Gil, 1988, Juab Formation from the early Whiterockian, Utah, USA (Flower, 1968b; Gil, 1988); *E. angarense* Balashov, 1960, *E. rozhkovense* Balashov, 1960, Chunskiy Stage, Angara River Basin, Siberian Platform, former USSR are approximately coeval (Balashov, 1960).
Last: *Padunoceras rugosaeforme* Balashov, 1960, Krivolutskiy Stage, Angara River Basin, Siberian Platform, former USSR (Balashov, 1960).

F. BAJKALOCERATIDAE Balashov, 1962 O. (ARG) Mar.

First and Last: *Bajkaloceras angarense* Balashov, 1962, *B. centrale* Balashov, 1960, *B. rozhkovense* Balashov, 1962, Chunskiy Stage, Angara River basin, Siberian Platform, former USSR (Balashov, 1962).

Order ACTINOCERIDA Teichert, 1933

F. POLYDESMIIDAE Kobayashi, 1940 O. (ARG_1–LLN) Mar.

First: *Ordosoceras contractum* Chen, 1976, lower Chiatsun Formation (Zone 3), Mt. Jolmolungma, Himalayas (Sheng, 1980). Several other species assigned to *Ordosoceras* and *Polydesmia* occur in Arenig-aged limestones (Santaokan, Chuotzeshan, Pelanchuang, lower Machiakou formations) in Nei Monggol, Hebei and Shandong Provinces, China (Chen, 1976; Sheng, 1980; Lai, 1989).
Last: *Polydesmia watanabei* Kobayashi, 1940, *P. shimamurai* Kobayashi, 1940, *P. peshanense* Kobayashi, 1940, Maruyama Beds, Shandong Province, China and Korea; *P. elegans* (Endo, 1932), post-Wuting limestones, Manchuria, China, is approximately coeval (Flower, 1976a).

F. ORMOCERATIDAE Saemann, 1853 O. (ARG_1)–C. (VIS) Mar.

First: *Ormoceras nyalamense* Chen, 1976, lower Chiatsun Formation (Zone 4), Mt. Jolmolungma, Himalayas (Sheng, 1980). Sheng also records undescribed species of *Ormoceras* from lower in the Chiatsun Formation (Zone 3) and from the Santaokan Formation of Nei Monggol.
Last: *Mstikhinoceras mirabile* Shimanskiy, 1961, Viséan, Moscow Basin, former USSR (Shimanskiy, 1961).
Intervening: LLN–WEN/LUD, D. (l., m.).

F. WUTINOCERATIDAE Shimizu and Obata, 1936 emend. Flower, 1968 O. (ARG_1–LLN, LLO/CRD ?) Mar.

First: *Wutinoceras* sp., Santaokan Formation, Nei Monggol, China (Sheng, 1980).
Last: *Wutinoceras aigawaense* (Endo, 1932), Ssuyen Limestone, southern Manchuria, China (Endo, 1932). The structure of the siphonal canal in this species may indicate assignment elsewhere. The next youngest forms are numerous species assigned to *Wutinoceras*, *Cyrtonybyoceras* and *Adamsoceras* from the Antelope Valley Formation, Toquima Range, Nevada, the Table Head Limestone of Newfoundland, the Kundan–Aserian stages of Sweden and Estonia and other Whiterockian equivalents (Flower, 1957, 1968a, 1976a; Mutvei, 1964).

F. ACTINOCERATIDAE Saemann, 1853 O. (ARG/LLN_1)–S. (LLY) Mar.

First: *Metactinoceras boreale* Balashov, 1962, *Actinoceras sp. nov.*, Machiakou Limestone (middle beds), Shandong Province, China (Chen, 1976).
Last: *Actinoceras*, Glenbower Beds, Cavan, New South Wales, Australia (Teichert and Glenister, 1952).
Intervening: LLN–ASH.

F. GEORGINIDAE Wade, 1977 O. (ARG_2–LLN) Mar.

First: *Georgina taylori* Wade, 1977, *G. linda* Wade, 1977, *G. andersonorum* Wade, 1977, *G. beuteli* Wade, 1977, upper Coolibah Formation, Georgina Basin, Queensland and Northern Territory, Australia (Wade, 1977).
Last: *Georgina dwyeri* Wade, 1977, upper Nora Formation, Georgina Basin, Australia (Wade, 1977); *Georgina* sp. (Stait and Burrett, 1984), Tha Manao Formation, Kanchanaburi Province, Thailand and *G. kongurensis* Chen and Wang, 1983, Kuweixi Formation, Mount Kongur, Xinjiang Province, China are approximately coeval (Chen and Wang, 1983; Stait and Burrett, 1984).

F. ARMENOCERATIDAE Troedsson, 1926 O. (ARG_2/LLN_1)–S. (LUD) Mar.

First: *Armenoceras* cf. *tani* (Grabau, 1922), Chuotzeshan Limestone, Nei Monggol, China (Sheng, 1980).
Last: *Armenoceras pseudoimbricatum* (Barrande, 1874), *A. kiaeri* Teichert 1934, Hemse Group, Gotland, Sweden (Mutvei, 1964).
Intervening: LLN–WEN.

F. GONIOCERATIDAE Hyatt, 1894 O. (LLN–RAW) Mar.

First: *Hoeloceras yimengshanense* Chen and Liu, 1976, Machiakou Limestone, Shandong Province, China (Chen and Liu, 1976).
Last: *Lambeoceras richmondense* Foerste, 1935, Lower Whitewater and Saluda formations (late Richmondian), Cincinnati, USA (Flower, 1957).
Intervening: LLO–CRD.

F. HURONIIDAE Foerste and Teichert, 1930 O. (LLN)–S. (WEN) Mar.

First: *Discoactinoceras wuyangshanense* Chen and Liu, 1976, *D. platyventrum* Chen and Liu, 1976, Machiakou Limestone, Shandong Province, China (Chen and Liu, 1976); *D. multiplexum* Kobayashi, 1927, Middle Ordovician, southern Manchuria, is approximately coeval (Kobayashi, 1927); *Climacoceras wuyangshanense* Chen, 1976, Machiakou Limestone, Shandong Province, China, is based on an isolated siphuncle barely referable to the Huroniidae (Chen, 1976).
Last: *Huronia bigsbyi* Stokes, 1824; *H. paulodilata* Foerste, 1925; *Huroniella inflecta* (Parks, 1915), Niagaran, Michigan, USA; *Huronia vertebralis* Stokes, 1824, Middle Silurian, Drummond Island, Lake Huron, Canada (Foerste, 1925; Shimer and Shrock, 1944). These taxa cannot be separated stratigraphically until revised.
Intervening: ASH–LLY.

F. CARBACTINOCERATIDAE Schindewolf, 1943 C. (VIS–BSH) Mar.

First: *Carbactinoceras torleyi* Schindewolf, 1943, Viséan, Germany (Schindewolf, 1943).
Last: *Rayonnoceras solidiforme* Croneis, 1926, *R. fayettevillensis* Croneis, 1926, *R. bassleri* Foerste and Teichert, 1930, *R. girtyi* Foerste and Teichert, 1930, Fayetteville Formation (and equivalents), Mississippian, of Oklahoma, Arkansas and California, USA (Gordon, 1964); *Rayonnoceras* is also recorded from the Namurian (E_2 zone) of Scotland (Turner, 1951) and the Moscow Basin, former USSR (Shimanskiy, 1961).

Order ORTHOCERIDA Kuhn, 1940

Derived from the Baltoceratidae (Ellesmerocerida) during Arenig (Cassinian) times. Hook and Flower (1976, 1977) regarded the Orthocerida (=Michelinoceratida of Flower, *in* Flower and Kummel, 1950) as diphyletic; baltoceratids with 'siphonal rods' giving rise to the Troedssonellidae and vacuosiphonate baltoceratids leading to the Michelinoceratidae (*sensu* Hook and Flower). Many of these early orthocerids are based solely on thin sections of fragmentary conchs.

Superfamily ORTHOCERATACEAE M'Coy, 1844

F. ORTHOCERATIDAE M'Coy, 1844 O. (ARG)–Tr. (NOR) Mar.

First: *Michelinoceras primum* Flower, 1962, Scenic Drive Formation, El Paso, Texas, USA (Flower, 1962b). Several other orthocerids (*Michelinoceras floridaense* Hook and Flower, 1977, *M. melleni* Hook and Flower, 1977, *M.* spp. Hook and Flower, 1977, *Wardoceras orygoforme* Hook and Flower, 1977) have been recorded from the lower Arenig Florida Mountains Formation and Wahwah Formation of Texas and western Utah, USA (Hook and Flower, 1977); all these forms possess relatively broad siphuncles with extensive siphonal deposits. Such features are not known in *Michelinoceras sensu stricto* (Ristedt, 1968; Serpagli and Gnoli, 1977) and assignment to this genus is suspect. Further study may ultimately separate these early American taxa into a new family which probably includes Baltic Volkhovian forms described under *'Orthoceras' nilssoni* (Boll, 1857) and *'O.' wahlenbergi* (Boll, 1857) by Dzik (1984).
Last: *'Orthoceras'* spp., 'Zone of *Trachyceras archelaus*', Italian Alps (Mojsisovics, 1882), Halorites Limestone, Himalayan Mountains (Mojsisovics, 1899) and 'Karnisch–Norische Trias' (Bülow, 1915).
Intervening: LLN–LUD, D. (m.), C. (l.), P.

F. TROEDSSONELLIDAE Kobayashi, 1935 O. (ARG_1–LLN_2/LLO_1) Mar.

First: *Tajaroceras wardae* Hook and Flower, 1976, Wahwah Formation, Ibex area, Utah, USA (Hook and Flower, 1976).
Last: *Troedssonella endoceroides* (Troedsson, 1932), Folkeslunda Limestone Formation, Öland, Sweden (Troedsson, 1932); *T. huanense* Lai and Tsi, 1977, and *T.* cf. *endoceroides* (Troedsson, 1932), Shihtzupu Formation, Hunan Province, China are approximately coeval (Lai and Tsi, 1977).

F. SACTORTHOCERATIDAE Flower, 1946 O. (ARG_2?, LLN–CRD) Mar.

First: *Sactorthoceras* spp. nov. (Chen, 1976), upper Ningkuoan (Machiakou Limestone), Shandong Province, China. According to Chen (1976) this occurrence is of late Arenig (*Didymograptus hirundo* Biozone) age. However, associated cephalopods (including *Dideroceras wennanense* Chen and Liu, 1976, *Discoactinoceras wuyangshanense* Chen and Liu, 1976, *Stolbovoceras boreale* Balashov, 1962), support an early Llanvirn age (Chen, 1976; Chen and Liu, 1976; Sheng, 1980).
Last: *Centroonoceras josephianum* (Foerste, 1932), Lourdes Formation, Port au Port Peninsula, western Newfoundland (Stait, 1988).
Intervening: LLO.

F. GEISONOCERATIDAE Zhuravleva, 1959 O. (ARG_2)–D. (GIV) Mar.

First: *Geisonoceras* sp., erratic boulder of Volkhovian age, Rozewie, Poland (Dzik, 1984).
Last: *Striacoceras typum* (Saemann, 1854), Cherry Valley Limestone, New York, USA (Flower, 1936); *Temperoceras caucasium* Zhuravleva, 1978, Zhivetskii Stage, Nakhichevan, Armenia, former USSR (Zhuravleva, 1978).
Intervening: LLV–EIF.

F. CLINOCERATIDAE Flower, 1946 O. (LLN–PUS) Mar.

First: *Clinoceras maskei* (Dewitz, 1879), 'Red Orthoceras Limestone of Biii–Ci Stufe' (Kundan–Aserian stages), erratic boulder, DDR (Neben and Krueger, 1971).
Last: *Whiteavesites winnipegense* (Whiteaves, 1892), Red River Formation, southern Manitoba, Canada (Foerste, 1929).
Intervening: LLO–CRD.

F. DAWSONOCERATIDAE Flower, 1962 S. (LLY?/WEN)–D. (l./m.?) Mar.

First: *Dawsonoceras tenuilineatum* Savage, 1927, Lower Silurian of North America is poorly known (Flower, 1962b); several species of *Dawsonoceras* (including the type *D. annulatum* Sowerby, 1818) are recorded from the Middle Silurian of North America, Europe and Gotland (Flower, 1962b).
Last: *Dawsonoceras americanum* Foord, 1888, Lower Devonian, Michigan, USA and Ontario, Canada (Shimer and Shrock, 1944). *Arazdajanites mamedovi* Zhuravleva, 1978, Middle Devonian, Mt. Arazdayai, former USSR, is

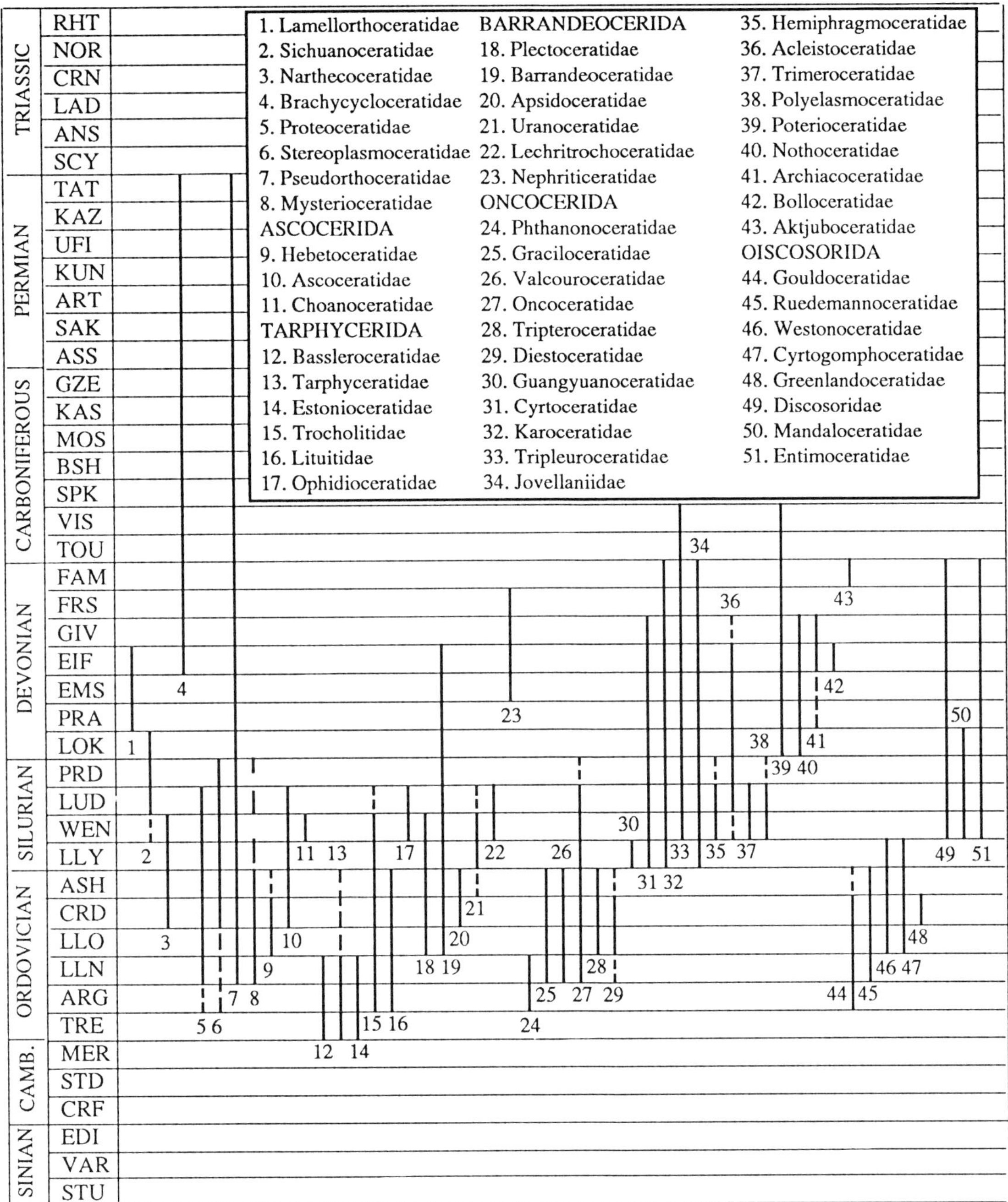

Fig. 9.2

represented only by a small conch fragment which may belong in the Geisonoceratidae (Zhuravleva, 1978).
Intervening: LUD.

F. SPHOOCERATIDAE Flower, 1962
S. (WEN/LUD), D. (l.)? Mar.

First and Last: *Sphooceras truncatum* (Barrande, 1868), Middle Silurian, Czechoslovakia (Barrande, 1868). *Sphooceras* ? sp., Lower Devonian, south-western Sardinia (Gnoli, 1982) is based on an incomplete conch in which the decollation features are imperfectly known, and therefore its generic assignment is tentative. No other records are known.

F. OFFLEYOCERATIDAE Flower, 1962
S. (WEN ?)–D. (EIF) Mar.

Flower (1962b) erected the Offleyoceratidae for orthocerids with relatively large siphuncles and holochoanitic septal necks. The present author follows Zhuravleva (1978) in regarding the Folioceratidae as synonymous.
First: *Offleyoceras arcticum* (Foord, 1888), Middle Silurian (Wenlock?), Kennedy Channel, north-west Greenland (Flower, 1962b).
Last: *Folioceras segmentum* Collins, 1969, Eids Formation, northern Canada; *F. sulmenevense* (Foerste, 1925), Sulmeneva Fjord, Novaya Zemlya, former Arctic USSR (Zhuravleva, 1978).

Intervening: GED.

F. PARAPHRAGMITIDAE Flower, *in* Flower and Kummel, 1950 S. (WEN/LUD)–D. (GED) Mar.

First: Several species assigned to *Paraphragmites*, *Calocyrtoceras*, *Gaspocyrrtoceras* and *Lyecoceras*, Middle Silurian of Canada, Sweden and Czechoslovakia (Barrande, 1866; Flower, 1943a; Mutvei, 1957).
Last: *Arterioceras concavum* Zhuravleva, 1978, 'Borshchovskiy horizon', Podolia, former USSR (Zhuravleva, 1978).

F. LAMELLORTHOCERATIDAE Teichert, 1961 D. (SIG–EIF) Mar. (see Fig. 9.2)

First: *Lamellorthoceras gracile* Termier and Termier, 1950, Siegenian, Morocco (Termier and Termier, 1950).
Last: *Lamellorthoceras vermiculare* Termier and Termier, 1950, Morocco; *Arthrophyllum crassus* Roemer, 1843, Harz Mountains, Germany; *Gorgonoceras visendum* Zhuravleva, 1961, Sverdlovsk District, former USSR; *Coralloceras coralliforme* Le Maitre, 1950, North Africa (Zhuravleva, 1978).
Intervening: EMS.

F. SICHUANOCERATIDAE Zhuravleva, 1978 S. (WEN/LUD)–D. (GED) Mar.

Zhuravleva (1978) distinguished this family from the Offleyoceratidae by the former's shorter septal necks. However, both families contain forms with similar 'segmented' siphuncles and future work may ultimately synonymize all these Middle Silurian–Lower Devonian taxa within the Offleyoceratidae.
First: *Jangziceras yinkiangense* Lai, 1964, Middle Silurian, Tsingki, Guizhou Province, China (Lai, 1964); *Neosichuanoceras columinum* Chen and Liu, 1974, and numerous species assigned to *Sichuanoceras* are recorded from the Middle Silurian of Shaanxi, Gansu and Ningxia Provinces, China (Chen and Liu, 1974; Lai, 1982).
Last: *Jangziceras cherkesovae* Zhuravleva, 1978, 'Vaigachskiy horizon' Vaigach, former Arctic USSR (Zhuravleva, 1978).

F. NARTHECOCERATIDAE Flower, 1958 O. (CRD)–S. (WEN) Mar.

First: *Tasmanoceras zeehanense* Teichert and Glenister, 1952, Smelters Limestone Formation, Zeehan, western Tasmania (Teichert and Glenister, 1952).
Last: *Narthecoceras subannulatum* Flower, 1968, *N. brevicameratum* Flower, 1968, *N. exile* Flower, 1968, *Donacoceras timiskamingense* Foerste *in* Hume, 1925, *D. arundineum* Foerste *in* Hume, 1925, *D. mutabile* Flower, 1968, *D. leve* Flower, 1968, *D. humei* Flower, 1968, Ekwan River Formation, Rapides des Papillons, Harricana River, James Bay Lowland, Canada (Flower, 1968c). Some of the above species are also recorded from coeval beds (Thornloe Limestone) in the Lake Timiskaming Region, Ontario, Canada (Foerste, *in* Hume, 1925).
Intervening: ASH.

F. BRACHYCYCLOCERATIDAE Furnish *et al.*, 1964 D. (EIF)–P. Mar.

First: *Pythonoceras boreum* Zhuravleva, 1978, upper reaches of Pechora River, Komi, former USSR (Zhuravleva, 1978).
Last: *Brachycycloceras*, Permian, Western Australia (Teichert and Glenister, 1952).
Intervening: SPK–MOS/KAS.

Superfamily PSEUDORTHOCERATACEAE Flower and Caster, 1935

F. PROTEOCERATIDAE Flower, 1962 O. (ARG_2/LLN_1)–S. (LUD) Mar.

First: *Gangshanoceras jurongense* Zou, 1988, *G. densum* Zou, 1988, Dawan Formation, Jurong, Jiangsu Province, China (Zou, 1988).
Last: *Cyrtactinoceras rebelle* (Barrande, 1866), 'Stage e2', Bubovitz, Czechoslovakia (Barrande, 1866).
Intervening: LLO-ASH.

F. STEREOPLASMOCERATIDAE Kobayashi, 1934 O. (ARG_2–CRD)/S. ? Mar.

First: *Stereoplasmoceras*, lower Machiakou Limestone (*Didymograptus hirundo* Biozone), Hebei Province, China (Sheng, 1980).
Last: *Ningkiangoceras centrale* Lai, 1965, Lojoping Formation, Ningkiang, Shaanxi Province, China (Lai, 1965). The siphonal deposits are of uncertain structure in this Middle Silurian species which may equally well be referable to the Pseudorthoceratidae or Proteoceratidae. The youngest taxon assignable to the Stereoplasmoceratidae is *Badouceras pyriforme* Chen and Liu, 1976, Badou Formation, Shandong Province, China (Chen and Liu, 1976).
Intervening: LLO.

F. PSEUDORTHOCERATIDAE Flower and Caster, 1935 O. (LLN)–P. (ZEC) Mar.

First: *Paradnatoceras modestum* Chen, 1987, Xainza, northern Xizang Province, China (Chen, 1987).
Last: *Shikhanoceras sphaerophorum* Shimanskiy, 1954, Lower Permian, southern Urals, former USSR (Shimanskiy, 1954).
Intervening: LLO–C. (u.).

F. MYSTERIOCERATIDAE Sweet, 1964 O. (LLN–LLO)/S. ?

First: *Mysterioceras tunguskense* Balashov, 1962, lower Krivolushkiy Stage, Podkamennaya Tunguska River, Siberian Platform, former USSR (Balashov, 1962).
Last: *Mysterioceras shengi* Lai, 1965, Nancheng Shale Formation, Liangshan, Shensi Province, China (Lai, 1965). Siphonal deposits are poorly known in this species and assignment to *Mysterioceras* is uncertain. The next youngest record is *M. australe* Teichert and Glenister, 1953, Gordon Group (Cashion's Creek Limestone and equivalents) of Tasmania (Stait and Flower, 1985).

Order ASCOCERIDA Kuhn, 1949

An apparently monophyletic group of nautiloids which exhibit truncation of conch with septal and apertural modification at maturity; derived from the Orthocerida (Clinoceratidae) (Flower, 1976c).

F. HEBETOCERATIDAE Flower, 1941 O. (LLO–CRD/ASH) Mar.

First: *Montyoceras arcuatum* Flower, 1941, *M. curviseptatum* Flower, 1941, *M. titaniforme* Flower, 1941, *M. tuba* Flower, 1963, *Hebetoceras mirandum* Flower, 1941, *Ecdyceras sinuiferum* Flower, 1941, Valcour and Crown Point Formations, New York, USA (Flower 1941b, 1963).
Last: *Ecdyceras foerstei* Flower, 1962, Arnheim Beds, Labanon, Kentucky, USA (Flower 1962a).

F. ASCOCERATIDAE Kuhn, 1949
O. (CRD)–S. (LUD) Mar.

First: *Redpathoceras clarki* Flower, 1963, Leray Limestone, Joliette Ridge, Quebec, Canada (Flower, 1963).
Last: *Ascoceras* cf. *gradatum* Lindström, 1890, Sundre Limestone Formation, Gotland, Sweden (Lindström 1890); *Ascoceras, Aphragmites* and *Glossoceras* are all recorded from 'Stage e2' (Budňanian), Czechoslovakia (Barrande, 1865).
Intervening: ASH, WEN.

F. CHOANOCERATIDAE Miller, 1932
S. (WEN) Mar.

First and Last: *Choanoceras mutabile* Lindström, 1890, 'Stratum h', Gotland, Sweden (Lindström, 1890).

Order TARPHYCERIDA Flower, *in* Flower and Kummel, 1950

Probably a monophyletic order, derived from the Ellesmerocerida via bassleroceratid-like forms (Teichert, 1967; Flower, 1976c). The Bassleroceratidae retain connecting rings of ellesmerocerid type but lack siphonal diaphragms. Following Flower (1976c), the family is assigned to the Tarphycerida.

F. BASSLEROCERATIDAE Ulrich *et al.*, 1944
O. (TRE–LLN) Mar.

First: *Anguloceras ovatum* Unklesbay and Young, 1956, *A. depressum* Unklesbay and Young, 1956, *A. rotundum* Unklesbay and Young, 1956, *Bassleroceras* cf. *bridgei* Ulrich, Foerste, Miller, and Unklesbay, 1944, Chepultepec–Stonehenge, Strasburg, Virginia, USA (Unklesbay and Young, 1956).
Last: *Bassleroceras xintaiense* Chen, 1976, Machiakou Limestone, Shandong Province, China (Chen, 1976).
Intervening: ARG.

F. TARPHYCERATIDAE Hyatt, 1894
O. (TRE–ARG_2/LLN_1, LLO ?, ASH ?) Mar.

First: *Campbelloceras*, Victorio Hills Formation (first piloceroid zone), El Paso Group, New Mexico, USA (Flower, 1964a).
Last: *Tarphyceras* ? *morkokense* Balashov, 1962, *T* ? *excentricum* Balashov, 1962, Upper Ordovician, Morkoka River, Siberian Platform, former USSR (Balashov, 1962); Flower (1984) regarded these generic assignments as questionable. He previously (1976c) reported a new undescribed genus from the Chazyan of New York. The youngest undoubted member of the family is *Centrotarphyceras* sp., Neichiashanian portion of Shuichuankou Formation, Ningxia Province, China (Lai, 1982).
Intervening: ARG.

F. ESTONIOCERATIDAE Hyatt *in* Zittel, 1900
O. (TRE–LLN) Mar.

First: *Aphetoceras*, Victorio Hills Formation (first piloceroid zone), El Paso Group, New Mexico, USA (Flower, 1964a).
Last: *Pakrioceras holmi* Stumbur and Mutvei, 1983, Aserian Stage, Pakri Islands, northern Estonia, former USSR (Stumbur and Mutvei, 1983).
Intervening: ARG.

F. TROCHOLITIDAE Chapman, 1857
O. (ARG_1)–S. (WEN/LUD) Mar.

First: *Trocholitoceras walcotti* Hyatt, 1894, *Beekmanoceras priscum* Ruedemann, 1906, *Curtoceras eatoni* (Whitfield, 1886), *C. cassinense* (Whitfield, 1886), *C. internastriatum* (Whitfield, 1886), Fort Cassin Formation (and equivalents), New York and Vermont, USA (Ruedemann, 1906); several species assigned to *Hardmanoceras*, Western Australia and East Asia, are approximately coeval (Stait and Burrett, 1984).
Last: *Graftonoceras graftonense* (Meek and Worthen, 1870), Niagaran Series, Ohio, USA (Foerste, 1925).
Intervening: LLN–ASH.

F. LITUITIDAE Phillips, 1848
O. (ARG_2–ASH) Mar.

Considerable morphological evidence (including form of cameral deposits and 'muscle scars') suggests that this family may belong in the Orthocerida. The family is conventionally retained within the Tarphycerida here, pending results from the present author's revision of Baltoscandian lituitids (MS in preparation).
First: *Rhynchorthoceras* aff. *beyrichi* (Remelé, 1880), erratic boulder, Volkhovian Stage, Opatów, Poland (Dzik, 1984); *Pseudoancistroceras hubeiense* Lai and Xu, 1987, middle Dawan Formation, Hubei Province, China, is approximately coeval (Lai and Xu, 1987).
Last: *Tyrioceras warburgae* Frey, 1982, Boda Formation, Dalarna, Sweden; *T. kjerulfi* Strand, 1934, '*Trinucleus* limestone', Frognøya, Oslo area, Norway (Strand, 1934; Frey, 1982).
Intervening: LLN–CRD.

F. OPHIDIOCERATIDAE Hyatt, 1894
S. (WEN–LUD) Mar.

First: *Ophioceras simplex* Barrande, 1865, *O. rudens* Barrande, 1865, Wenlock, Czechoslovakia (Barrande, 1865); *O. reticulatum* Angelin (*in* Angelin and Lindström, 1880), Gotland, Sweden (Angelin and Lindström, 1880).
Last: *Ophiceras rota* Lindström, 1890, Ludlow, Gotland, Sweden (Lindström, 1890).

Order BARRANDEOCERIDA Flower, *in* Flower and Kummel, 1950

Originally separated from the Tarphycerida on the basis of their thin, homogeneous connecting rings. Flower (1984) briefly presented new evidence indicating that the Barrandeocerida was polyphyletic and suggested the order should be abolished and forms previously assigned to it be included within the Tarphycerida. The Barrandeocerida is temporarily retained here pending further study.

F. PLECTOCERATIDAE Hyatt *in* Zittel, 1900
O. (LLO)–S. (WEN) Mar.

First: *Plectoceras jason* (Billings, 1859), *Avilionella multicaneratum* (Ruedemann, 1906); Chazy Limestone, Champlain Valley, New York, USA and Canada (Ruedemann, 1906; Flower, 1984).
Last: *Laureloceras cumingsi* Flower, 1943, Laurel Limestone Formation, Indiana, USA (Flower, 1943b).
Intervening: CRD.

F. BARRANDEOCERATIDAE Foerste, 1925
O. (LLO)–D. (EIF) Mar.

First: *Barrandeoceras natator* (Billings, 1859), Chazy Limestone, Champlain Valley, New York, USA, Ontario and Quebec, Canada (Ruedemann, 1906; Flower, 1984).

Last: *Haydenoceras acutum* Flower, 1949, Middle Devonian, Nevada, USA (Flower, 1949).
Intervening: CRD-ASH, WEN.

F. APSIDOCERATIDAE Hyatt, 1884 O. (CRD–ASH) Mar.

First: *Apsidoceras montrealense* Flower, 1943, upper Trenton Limestone, Isle Jesus, Quebec, Canada; *Fremontoceras jewetti* Flower, 1947, Sherman Falls Limestone, New York, USA (Flower, 1943c, 1947).
Last: *Charactoceras triangulum* Frey, 1982, *C. kallholnense* Frey, 1982, *C. suecicum* Frey, 1982, *C. raettvikense* Frey, 1982, *Bodoceras torticoni* Frey, 1982, Boda Limestone Formation, Dalarna, Sweden, (Frey, 1982); *Charactoceras baeri* (Meek and Worthen, 1865), Richmondian, Ohio, USA (Foerste, 1924); *Deckeroceras adaense* Foerste, 1935, Richmondian, Oklahoma, USA (Foerste, 1935).

F. URANOCERATIDAE Hyatt, *in* Zittel, 1900 (O. (ASH)?/S. (WEN/LUD) Mar.

First and Last: Records of *Uranoceras* from the Upper Ordovician (Ashgill) of Sweden (Sweet *in* Teichert *et al.*, 1964) are probably referable to *Charactoceras* of the Apsidoceratidae (Frey, 1982). Undoubted members of the Uranoceratidae are species assigned to *Uranoceras*, *Cumingsoceras* and *Jolietoceras* from the Middle Silurian of Europe and the USA (Foerste, 1925); without further revision these cannot be separated stratigraphically.

F. LECHRITROCHOCERATIDAE Flower, *in* Flower and Kummel, 1950 S. (WEN–LUD/PRD) Mar.

First: *Lechritrochoceras desplainense* Hall, 1868, Racine Member, Wisconsin, USA; *L. placidum* (Barrande, 1865), Butovitz, Czechoslovakia; *Trochodictyoceras slocomi* Foerste, 1926, Racine Member, Stony Island, Illinois, USA: *Leurotrochoceras aeneas* (Hall, 1868), Racine Member, Iowa, USA (Foerste, 1926).
Last: *Catyrephoceras giganteum* (Blake, 1882), early Ludlow, Leintwardine, UK; *Peismoceras optatum* (Barrande, 1865) and *Systrephoceras arietinum* (Barrande, 1865), 'Stage e' at Lochkov and Kozorz, Czechoslovakia respectively, may be coeval or slightly younger (Foerste, 1926).

F. NEPHRITICERATIDAE Hyatt, 1894 D. (EMS–FRS) Mar.

First: *Sphyradoceras clio* (Hall, 1861), Schoharie Formation, New York, USA (Zhuravleva, 1974).
Last: *Baeopleuroceras incipiens* Williams, *in* Cooper and Williams, 1935, Tully Limestone, New York, USA; *Triplooceras* is also reported from the Tully Limestone of New York (Flower, 1945).

Order ONCOCERIDA Flower, *in* Flower and Kummel, 1950

The Oncocerida were thought to have developed from bassleroceratid-like forms via the Graciloceratidae (Flower, 1976c). The discovery of older oncocerids (Phthanoncoceratidae) from Sweden and Spitsbergen (Evans and King, 1990) indicates the order may have arisen from at least two independent ellesmerocerid lineages and is therefore polyphyletic.

F. PHTHANONCOCERATIDAE Evans and King, 1990 O. (ARG–LLN_1) Mar.

This family possesses primary siphonal diaphragms and thickened connecting rings typical of the Ellesmerocerida, but exhibits an exogastric curvature and narrow siphuncle characteristic of the Oncocerida. Evans and King (1990) regarded conch form (combined with stratigraphical arguments) as sufficient evidence for assignment of the Phthanoncoceratidae to the Oncocerida.
First: *Valhalloceras floweri* Evans and King, 1990, Olenidsletta Member, Valhallfonna Formation, North Ny Friesland, Spitsbergen (Evans and King, 1990).
Last: *Phthanoncoceras oelandense* Evans and King, 1990, 'glauconitic, grey Vaginatum Limestone', early Kundan Stage, Hälludden, Öland, Sweden (Evans and King, 1990).

F. GRACILOCERATIDAE Flower *in* Flower and Kummel, 1990 O. (LLN–ASH) Mar.

First: *Leonardoceras parvum* Flower, 1968, Antelope Valley Limestone Formation, Nevada, USA (Flower, 1968b).
Last: *Ringoceras praecurvum* Strand, 1934, Lyckholm-Stufe, Norway (Strand, 1934).
Intervening: LLN–CRD.

F. VALCOUROCERATIDAE Flower, 1945 O. (LLN–ASH) Mar.

First: *Hemibeloitoceras ellipsoidale* Balashov, 1962, *H. ellipsoidale nujense* Balashov, 1962, Krivolutskiy Stage, Stolbovoy and Nyui Rivers, Siberian Platform, former USSR (Balashov, 1962).
Last: *Kindleoceras reversatum* Foerste, 1924, *Manitoulinoceras lysander* (Billings, 1865), Ontario, Canada (Foerste, 1924); *Augustoceras? molense* Stait, 1982, Den Member, Chudleigh Limestone, northern Tasmania (Stait, 1982).
Intervening: LLO–CRD.

F. ONCOCERATIDAE Hyatt, 1884 O. (LLN)–S. (LUD/PRD) Mar.

First: *Oonoceras* sp. and undescribed oncoceratids, mid-Kundan Stage, Öland, Sweden (Dzik, 1984); *Richardsonoceras tangyaense* Lai and Xu, 1987, upper Dawan Formation, Hubei Province, China, is approximately equivalent. Flower (1976c) reported undescribed Whiterockian oncoceratids from Newfoundland and Spitsbergen.
Last: *Oocerina lentigradium* (Barrande, 1866), 'Stage e', Lochkov, Czechoslovakia; *O. stygiale* (Barrande, 1877), 'Stage e', Dvoretz, Czechoslovakia; *O. strangulatum* (Barrande, 1877), 'Stage e2', Dvoretz, Czechoslovakia (Foerste, 1926); *Paroocerina podolskensis* Zhuravleva, 1961, Upper Silurian, Podolia, former USSR (Zhuravleva, 1961).
Intervening: LLO–WEN.

F. TRIPTEROCERATIDAE Flower, 1941 O. (LLO–ASH) Mar.

Some taxa placed within this family exhibit extensive cameral deposits of a type usually associated with the Orthocerida (Flower, 1962b).
First: *Allumettoceras mjoesense* Sweet, 1958, Cephalopod Shale, Helgøya, Nes-Hamar district, Norway (Sweet, 1958).
Last: *Tripteroceras xiphias* (Billings, 1857), Richmondian, Anticosti Island, Canada (Foerste, 1926).
Intervening: CRD.

F. DIESTOCERATIDAE Foerste, 1926
O. (LLN/LLO ?, CRD–ASH)

First: *Xainzanoceras xainzaense* Chen, 1987, Xungmei Formation, northern Xizang province, China (Chen, 1987). Chen questioned the familial assignment of this taxon, whose conch form alone indicates placement elsewhere. Earliest undoubted representatives of the Diestoceratidae are *Diestoceras lavalense* Flower, 1952, *D. sinclairi* Flower, 1952, Terrebonne Formation, Quebec, Canada; *D. sycon* Flower, 1952, Black River Limestone, Poland, New York and *Danoceras inutile* Flower, 1952, Gull River Formation, Ontario, Canada (Flower, 1952a).
Last: *Diestoceras indianense* (Miller and Faber, 1894), lower Whitewater Formation, Oxford, Ohio and Saluda Formation, Versailles, Indiana, USA; *D. scalare* Foerste, 1921, Richmond Formation, Anticosti Island, Canada (Foerste, 1926). *Lyckholmoceras estoniae* Teichert, 1930, Lyckholm-Stufe, Estonia and *L. graciliforme* Lai, 1982, Peikuoshan Formation, Shaanxi Province, China, are approximately coeval (Teichert, 1930; Lai, 1982).

F. GUANGYUANOCERATIDAE Lai and Zhu, 1985
S. (LLY) Mar.

First and Last: *Guangyuanoceras depressum* Lai and Zhu, 1985, *G. planodorsum* Lai and Zhu, 1985, *Guangyuanoceroides sichuanense* Lai and Zhu, 1985, Lower Silurian, Shangsi region of Guangyuan, Sichuan Province, China (Lai and Zhu, 1985).

F. CYRTOCERATIDAE Chapman, 1857
S. (LLY)–D. (GIV) Mar.

First: *Blakeoceras llandoveri* (Blake, 1882), lower Llandovery, Craig-yr-Wyddon, Wales, UK (Foerste, 1926).
Last: *Cyrthoceratites depressus* (Bronn, 1835), *C. lineatus* (D'Archaic and De Verneuil, 1842), *C. alatus* (Holzapfel, 1895), Middle Devonian, Germany (Zhuravleva, 1974).
Intervening: LVD, EIF.

F. KAROCERATIDAE Teichert, 1939
S. (LLY)–D. (FAM) Mar.

First: *Osbornoceras swinnertoni* Foerste, 1936, Lower Silurian, Ohio, USA (Foerste, 1936).
Last: *Geitonoceras lucidum* Zhuravleva, 1974, Aktyubinsk Region, Kirigziya Steppe, Kazakhstan, former USSR (Zhuravleva, 1974).
Intervening: WEN/LUD, D. (l.).

F. TRIPLEUROCERATIDAE Foerste, 1926
S. (WEN)–C. (VIS) Mar.

First: *Tripleuroceras robsoni* Whiteaves, 1906, Niagaran, Stonewall, Manitoba, Canada (Foerste, 1926).
Last: *Psiaoceras hesperis* (Eichwald, 1860), Viséan, Kalouga, former USSR (Shimanskiy, 1957).
Intervening: EMS–GIV.

F. JOVELLANIIDAE Foord, 1888
S. (LLY)–D. (FAM) Mar.

First: *Mixosiphonoceras sichuanense* Lai and Zhu, 1985, *M. subglobum* Lai and Zhu, 1985, *M. simplex* Lai and Zhu, 1985, *M. discum* Lai and Zhu, 1985, *M. subcirculare* Lai and Zhu, 1985, Lower Silurian, Shangsi, Szechwan Province, China (Lai and Zhu, 1985).
Last: *Agrioceras gregarium* Zhuravleva, 1974, *Corysoceras karatauense* Zhuravleva, 1974, *Mimolychnoceras zolkinae* Zhuravleva, 1974, *Almaloceras obaeratum* Zhuravleva, 1974, *Lynchoceras occultum* Zhuravleva, 1974, *Mnemoceras galithkyi* Zhuravleva, 1974, Khrebet Karatauo, Kazakhstan, former USSR (Zhuravleva, 1974).

F. HEMIPHRAGMOCERATIDAE Foerste, 1926
S. (WEN–LUD/PRD) Mar.

First and Last: Numerous species assigned to *Hemiphragmoceras*, *Conradoceras*, *Tetrameroceras* and *Hexameroceras*, Viscocilka, Hinter-Kopanina, Dvoretz, Lochkov, Czechoslovakia (Barrande, 1867).

F. ACLEISTOCERATIDAE Flower, *in* Flower and Kummel, 1950 S. (WEN/LUD)–D. (EIF/GIV) Mar.

First: Numerous species assigned to *Amphicyrtoceras*, *Anomeioceras*, *Austinoceras*, *Byronoceras*, *Chadwickoceras*, *Crateroceras*, *Eocyrtoceras*, *Euryrizoceras*, *Galtoceras*, *Hercocyrtoceras*, *Perioidanoceras*, *Rhomboceras*, *Slocomoceras*, *Streptoceras*, and *Worthenoceras*, Middle Silurian, Illinois, Ohio, New York, USA and Ontario, Quebec, Canada (Foerste, 1924, 1930, 1934).
Last: *Acleistoceras olla* (Saemann, 1854), *A. eximium* (Hall, 1888), *A. mitra* (Hall, 1888), Middle Devonian, Columbus, Ohio, USA (Foerste, 1926); *Paracleistoceras devonicans* (Barrande, 1865), *Poteriocerina lumbosum* (Barrande, 1877), *Gonatocyrtoceras heteroclytum* (Barrande, 1866), *G. postscripti* (Barrande, 1866), 'Stage g3', Middle Devonian, Hlubocep, Czechoslovakia (Barrande, 1865–77).

F. TRIMEROCERATIDAE Hyatt, *in* Zittel, 1900
S. (WEN/LUD) Mar.

First and Last: Several species assigned to *Trimeroceras*, *Clathroceras*, *Eotrimeroceras*, *Inversoceras*, *Pentameroceras* and *Stenogomphoceras*, Middle Silurian of Europe (Sweden, Czechoslovakia), North America and Canada (Barrande, 1865; Foerste, 1928, 1930).

F. POLYELASMOCERATIDAE Shimanskiy, 1956
S. (WEN–LUD/PRD) Mar.

First: *Danaoceras danai* (Barrande, 1866), *D. insociale* (Barrande, 1866), 'Stage e', Kozorz, Czechoslovakia (Foerste, 1926).
Last: *Cyclopites cuclops* (Venyukov, 1886), *Evlanoceras evlanensis* (Nalivkin, 1947), Siberian Platform, former USSR (Zhuravleva, 1972); *Codoceras indomitum* (Barrande, 1866), 'Stage e', Lochkov, Czechoslovakia (Foerste, 1926).
Intervening: EMS–GIV.

F. POTERIOCERATIDAE Foord, 1888
D. (GED)–C. (VIS) Mar.

First: *Cyrtogomphoceras lunatus* (Hall, 1879), *C. metula* (Hall, 1879), upper Helderbergian, New York (Zhuravleva, 1972).
Last: *Poterioceras fusiforme* (Sowerby, 1829), *P. latiseptatum* (Foord, 1898), Viséan, Dublin, Republic of Ireland (Foerste, 1926); *Welleroceras liratum* (Miller and Furnish, 1938), lower Mississippian, Missouri, USA (Miller and Furnish, 1938); *Argocheilus ? chinense* Shimanskiy, 1957, Lower Carboniferous, China (Shimanskiy, 1957).
Intervening: D. (m., u.).

F. NOTHOCERATIDAE Fischer, 1882
D. (l.–GIV) Mar.

First: *Lorieroceras lorieri* (Barrande, 1874), Lower Devonian, Courtoisières, Sarthe, France (Foerste, 1926).

Last: *Nothoceras bohemicum* Barrande, 1867, Horizon G_3, Hlubocep, Czechoslovakia (Foerste, 1926); *Oligoceras russanovi* (Kuzmin, 1965), Eifelian, Novaya Zemlya, former Arctic USSR (Foerste, 1926).

F. ARCHIACOCERATIDAE Teichert, 1939 D. (SIG–EIF ?/GIV) Mar.

First and Last: *Archiacoceras subventricosus* (D'Archaic and De Verneuil, 1842), Givetian, Rhenish Schiefergebirge, Germany (Crick and Teichert, 1979). *Archiacoceras rarum* Kuzmin, 1966, Givetian, Novaya Zemlya, former Arctic USSR lacks the diagnostic actinosiphonate lamellae and therefore its affinities are uncertain. Sweet (*in* Teichert *et al.*, 1964) recorded *Cyrtoceratites flexuosus* (Schlotheim, 1820) from the Middle Devonian of Germany, but noted *Cyrtoceratites* as possibly being a senior synonym of *Archiacoceras*. Teichert *et al.* (1979) reported *Cyrtoceratites* to range from the Siegenian to the Givetian.

F. BOLLOCERATIDAE Zhuravleva, 1962 D. (EIF) Mar.

First and Last: Numerous species assigned to *Bolloceras*, *Metaphragmoceras* and *Paraconradoceras*, Eifelian, Czechoslovakia and New York, USA (Zhuravleva, 1974).

F. AKTJUBOCERATIDAE Zhuravleva, 1972 D. (FAM) Mar.

First: *Irinites editus* Zhuravleva, 1972, *Atopoceras vodoresovi* Zhuravleva, 1972, *Aktjubocheilus anaticula* Zhuravleva, 1972, *A. imbellus* Zhuravleva, 1972, *A. verbosus* Zhuravleva, 1972, *A. longus* Zhuravleva, 1972, middle Famennian, Aktyubinsk and Chelyabinsk regions, former USSR (Zhuravleva, 1972).
Last: *Kijoceras clarum* Zhuravleva, 1972, upper Famennian, Aktyubinsk region, former USSR (Zhuravleva, 1972).

Order DISCOSORIDA Flower, *in* Flower and Kummel, 1950

Probably monophyletic; believed to be directly derived from the Plectronocerida (Flower and Teichert, 1957; Flower, 1964a) although an origin in the Ellesmerocerida may be equally plausible.

F. GOULDOCERATIDAE Stait, 1984 O. (ARG_2–CRD/ASH) Mar.

First: *Madiganella magna* Teichert and Glenister, 1952, Horn Valley Siltstone, Amadeus Basin, central Australia (Teichert and Glenister, 1952).
Last: *Gouldoceras synchonema* Stait, 1980, *G. obliquum* (Teichert and Glenister, 1953), *G. benjaminense* Stait, 1984, Benjamin Limestone, western Tasmania (Stait, 1984). Stait and Burrett (1987) reported undescribed gouldoceratids from the upper Benjamin Limestone which are of early Ashgill (Pusgillian) age.

F. RUEDEMANNOCERATIDAE Flower, 1940 O. (LLN–ASH) Mar.

First: Small doubtfully assigned fragment of *Ruedemannoceras*? sp., Badger Flat Limestone, Inyo County, California, USA (Flower, 1968b); the approximately coeval taxon *Elkanoceras pluto* (Billings, 1865), from the Table Head Limestone, Newfoundland is better known (Flower, 1971).
Last: *Taoqupoceras peculare* Lai, 1982, Peikuoshan Formation, Shaanxi Province, China (Lai, 1982).
Intervening: LLO–CRD.

F. WESTONOCERATIDAE Teichert, 1933 O. (LLO)–S (LLY) Mar.

First: *Sinclairoceras haha* Flower, 1952, Chazyan, Ste. Anne de Chicoutimi, Quebec, Canada (Flower, 1952a).
Last: *Glyptodendron eatonense* Claypole, 1878, Lower Silurian, Ohio, USA (Flower, *in* Flower and Teichert, 1957).
Intervening: CRD–ASH.

F. CYRTOGOMPHOCERATIDAE Flower, 1940 O. (LLO)–S. (LLY) Mar.

First: *Strandoceras strandi* Sweet, 1958, Cephalopod Shale, Helgøya, Norway (Sweet, 1958).
Last: *Konglungenoceras norvegicum* Sweet, 1959, middle Stricklandian Series, Konglungen, Oslo, Norway (Sweet, 1959).
Intervening: CRD–ASH.

?F. GREENLANDOCERATIDAE Shimizu and Obata, 1935 O. (CRD) Mar.

The form of the siphonal deposits in *Greenlandoceras* are poorly known and assignment of the genus to the Orthocerida may be correct.
First and Last: *Greenlandoceras striatum* (Troedsson, 1926), *G. lineatum* (Troedsson, 1926), Cape Calhoun Series, Cape Calhoun, north Greenland (Troedsson, 1926).

F. DISCOSORIDAE Miller, 1889 S. (WEN)–D. (FAM) Mar.

First: Several species assigned to *Discosorus*, *Endodiscosorus*, *Kayoceras*, and *Stokesoceras*, Middel Silurian, USA (New York, Ohio, Iowa, Michigan), Canada and Europe (Flower and Teichert, 1957).
Last: *Alpenoceras*? *robustum* (Schindewolf, 1944), Kellerwald, Germany, is poorly known from only two specimens but assignment to *Alpenoceras* seems likely (Flower, *in* Flower and Teichert, 1957).
Intervening: LUD, GIV.

F. MANDALOCERATIDAE Flower, *in* Flower and Kummel, 1957 S. (WEN)–D. (GED) Mar.

First: Several species assigned to the genera *Mandaloceras*, *Cinctoceras*, *Ovocerina*, *Pseudogomphoceras*, *Umbeloceras* and *Vespoceras*, all from the Middle Silurian of North America (Ohio, Illinois, New York), Canada (Quebec), England, UK and Czechoslovakia (Shimer and Shrock, 1944; Flower, *in* Flower and Teichert, 1957).
Last: *Mandaloceras emaciatum* (Barrande, 1866), Czechoslovakia (Zhuravleva, 1972).
Intervening: LUD–PRD.

F. ENTIMOCERATIDAE Zhuravleva, 1972 S. (WEN)–D. (FAM) Mar.

First: *Gonatocyrtoceras inflatum* Foerste, 1930, Niagaran, Illinois, USA (Foerste, 1930).
Last: *Selenoceras onerosum* Zhuravleva, 1972, *Pantoioceras mutum* Zhuravlava, 1972, *Lysagoroceras* ?*separatum* Zhuravleva, 1972, Levigitovyi Horizon, Aktyubinsk region, former USSR (Zhuravleva, 1972).
Intervening: GED?, SIG?, EMS–FRS.

?F. MESOCERATIDAE Hyatt, 1884 S. (WEN/LUD) Mar. (see Fig. 9.3)

Mesoceras is only known from body chambers and recognition of the family is uncertain.
First and Last: *Mesoceras bohemicum* Barrande, 1877, Middle Silurian, Czechoslovakia (Barrande, 1877).

F. LOWOCERATIDAE Flower, 1940
S. (WEN/LUD) Mar.

First and Last: *Lowoceras southamptonense* Foerste and Savage, 1927, *Tuyloceras percurvatum* Foerste and Savage, 1927, Hudson Bay, Canada (Foerste and Savage, 1927).

F. PHRAGMOCERATIDAE Miller, 1877
S. (WEN/LUD)–D. (GED) Mar.

First: Several species assigned to *Phragmoceras*, Endoplectoceras, *Protophragmoceras* and *Sthenoceras*, Czechoslovakia and Gotland, Sweden (Hedström, 1917; Zhuravleva, 1972).
Last: *Protophragmoceras nonnullum* Zhuravleva, 1972, *Endoplectoceras podolicum* Zhuravleva, 1972, Khudkovtsy, Melbintsy–Podolbskoi, former USSR; *Sthenoceras aduncum* (Barrande, 1866), 'Stage g1', Czechoslovakia (Zhuravleva, 1972).
Intervening: PRD.

F. NAEDYCERATIDAE Shimanskiy, 1956
S. (WEN/LUD)–D. (FAM) Mar.

First: *Oxygonioceras oxynotum* (Barrande, 1865), 'Stage e', Kozorz, Czechoslovakia (Foerste, 1926).
Last: *Mitroceras? intactum* Zhuravleva, 1972, Khrevet Karatan, Kazakhstan, former USSR (Zhuravleva, 1972).
Intervening: SIG–FRS.

F. UKHTOCERATIDAE Zhuravleva, 1972
S. (LUD/PRD)–D. (FAM) Mar.

First: *Turoceras schnyrevae* Zhuravleva, 1959, Upper Silurian, Sverdovsk region, former USSR (Zhuravleva, 1972).
Last: *Ropaloceras implicatum* Zhuravleva, 1972, *R? illicitum* Zhuravleva, 1972, *Nipageroceras riphaeum* Zhuravleva, 1972, *Metrioceras desertum* Zhuravleva, 1972, Levigitovyi Horizon, Aktyubinsk region, former USSR (Zhuravleva, 1972).
Intervening: GED–FRS.

F. BREVICOCERATIDAE Flower, 1941
D. (GED–FAM) Mar.

First: *Xenoceras oncoceroides* Flower, 1951, Helderbergian, New York, USA (Flower, 1951).
Last: *Aipetoceras lebedjanicum* Zhuravleva, 1972, Lebedyanskie Beds, upper Famennian, Lipetsk region, former USSR (Zhuravleva, 1972).
Intervening: EIF–FRS.

F. TAXYCERATIDAE Zhuravleva, 1972
D. (GED–FAM) Mar.

First: *Brodekoceras dnestrovense* Balashov *in* Balashov and Kiselev, 1968, Borshchovskiy Horizon, Podolia, former USSR (Zhuravleva, 1972).
Last: *Pachtoceras asiaticum* Zhuravleva, 1972, Prolobitovyi Horizon, Karagandinsk region, former USSR (Zhuravleva, 1972).
Intervening: EMS–FRS.

F. MECYNOCERATIDAE Zhuravleva, 1972
D. (l.–FAM) Mar.

First: *Laumontoceras laumonti* (Barrande, 1866), Lower Devonian, Nehou, Manche, France (Foerste, 1926).
Last: *Mecynoceras rex* (Pacht, 1856), *Paramecynoceras fixum* Zhuravleva, 1972, Novaya Zemlya, former Arctic USSR; *Laumontoceras improvisum* Zhuravleva, 1972, Shushakovsk region, former USSR (Zhuravleva, 1972).

F. DEVONOCHELIDAE Zhuravleva, 1972
D. (EMS–FAM) Mar.

First: *Platyconoceras? kuzmini* Zhuravleva, 1972, Novaya Zemlya, former Arctic USSR (Zhuravleva, 1972).
Last: *Pelagoceras lautum* Zhuravleva, 1972, *P. mendicum* Zhuravleva, 1972, Prolobitovyi Horizon, Aktyubinsk region, former USSR (Zhuravleva, 1972).
Intervening: EIF–FRS.

Order NAUTILIDA Agassiz, 1847

Most, if not all, Devonian to Triassic nautiloids are apparently derived from the Rutoceratidae (Tainocerataceae). Early forms have cyrtochoanitic siphuncles with actinosiphonate deposits interpreted as relict features retained from their ancestors in the Oncocerida (Kummel *in* Teichert *et al.*, 1964).

Superfamily TAINOCERATACEAE Hyatt, 1883

F. RUTOCERATIDAE Hyatt, 1884
D. (SIG)–C. (BSH) Mar.

First: *Ptenoceras alatum* (Barrande, 1865), *Trochoceras davidsoni* Barrande, 1865, 'Stage f2', Upper Koněprusy Limestone, Pragian, Czechoslovakia (Foerste, 1926).
Last: *Aphractus adempta* Shimanskiy, 1967, upper Namurian, southern Urals, former USSR (Shimanskiy, 1967).
Intervening: EMS–FRS, VIS.

F. TETRAGONOCERATIDAE Flower, 1945
D. (EIF–GIV) Mar.

First: *Nassauoceras subtuberculatum* (Sandberger and Sandberger, 1852), *Orthoceras*-Schiefer, Rhineland, Germany; *Wellsoceras columbiense* (Whitfield, 1882), Colombus Limestone, Ohio, USA (Zhuravleva, 1974).
Last: *Tetragonoceras gracile* Whiteaves, 1891, Winnipegosis Dolomite, Manitoba, Canada (Whiteaves, 1891).

F. TAINOCERATIDAE Hyatt, 1883
C. (VIS)–Tr. (NOR) Mar.

First: *Gzheloceras antiquum* Shimanskiy, 1967, *G. striatum* Shimanskiy, 1967, *Celox erratica* Shimanskiy, 1967, Viséan, Kazakhstan, former USSR; *Tylonautilus nodiferus* (Armstrong, 1866), Viséan, Clydesdale, Scotland, UK; *T. ornatissimus* (Tzwetaev, 1898), Viséan, Tul'skaya Province, former USSR (Shimanskiy, 1967).
Last: *Enoploceras ausseeanus* Diener, 1919, *E. lepsiusiformis* Diener, 1919, *E. lepsiusi* Mojsisovics, 1902, 'Karnisch-Norische Misch-fauna', Hallstätter Kalke, Eastern Alps; *Phloioceras welteri* Kieslinger, 1924, Norian, Timor; *Germanonautilus kyotanii* Nakazawa, 1959, Nariwa Group, Lapan (Diener, 1919; Kieslinger, 1924; Nakazawa, 1959).
Intervening: C. (u.)–P. (u.), SCY–CRN.

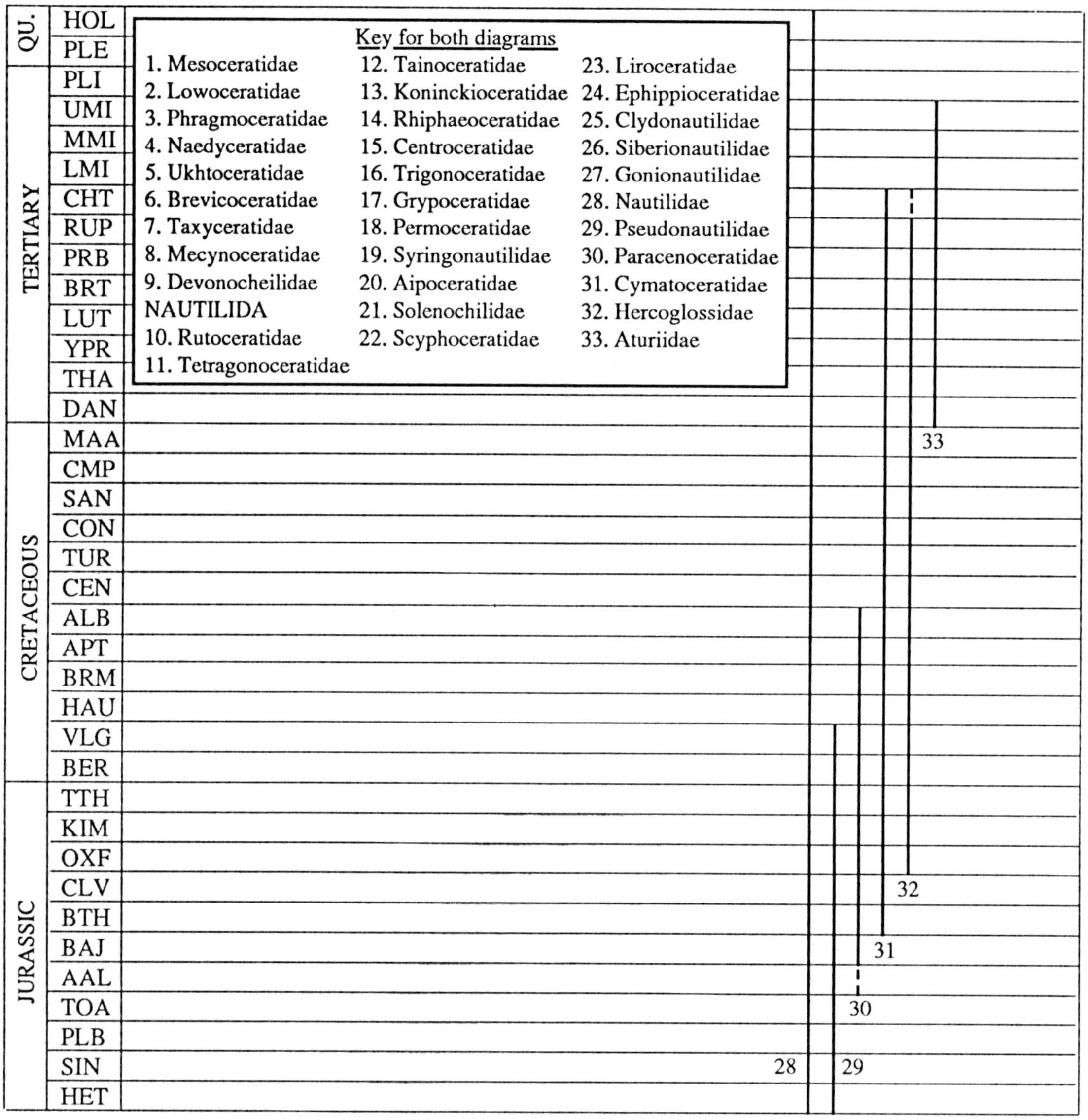

Key for both diagrams

1. Mesoceratidae
2. Lowoceratidae
3. Phragmoceratidae
4. Naedyceratidae
5. Ukhtoceratidae
6. Brevicoceratidae
7. Taxyceratidae
8. Mecynoceratidae
9. Devonocheilidae

NAUTILIDA

10. Rutoceratidae
11. Tetragonoceratidae
12. Tainoceratidae
13. Koninckioceratidae
14. Rhiphaeoceratidae
15. Centroceratidae
16. Trigonoceratidae
17. Grypoceratidae
18. Permoceratidae
19. Syringonautilidae
20. Aipoceratidae
21. Solenochilidae
22. Scyphoceratidae
23. Liroceratidae
24. Ephippioceratidae
25. Clydonautilidae
26. Siberionautilidae
27. Gonionautilidae
28. Nautilidae
29. Pseudonautilidae
30. Paracenoceratidae
31. Cymatoceratidae
32. Hercoglossidae
33. Aturiidae

Fig. 9.3

F. KONINCKIOCERATIDAE Hyatt, *in* Zittel, 1990 C. (VIS)–P. (ART) Mar.

First: *Millkoninckioceras konincki* (Miller and Kemp, 1947), Viséan, Belgium; *Lophoceras rossicum* Shimanskiy, 1957, *L. regulus* (Eichwald, 1857), Viséan?, Lower Carboniferous; *L. pentagonum* (Sowerby, 1819), Viséan?, Lower Carboniferous, England, UK; *Planetoceras retardatum* Hyatt, 1893, Viséan, Belgium; *Subvestinautilus crassimarginatus* (Foord, 1900), Viséan, Republic of Ireland; *Temnocheilus coronatus* (M'Coy, 1844), Viséan, Republic of Ireland (Shimanskiy, 1967).
Last: *Foordiceras goliathum* (Waagen, 1879), *Productus* Limestone, Lower Permian, Salt Range, Pakistan (Waagen, 1879); *Knightoceras kempae* Miller and Youngquist, 1949, lower Lueder's Formation, Texas, USA (Miller and Youngquist, 1949); *Kummeloceras sibiricum* Shimanskiy, 1967, Verkhoyansk, former USSR (Shimanskiy, 1967).
Intervening: C. (u.).

F. RHIPHAEOCERATIDAE Ruzhentsev and Shimanskiy, 1954 P. (ZEC) Mar.

First and Last: *Pararhiphaeoceras tastubense* (Krugov, 1928), *Rhiphaeonautilus curticostatus* Ruzhentsev and Shimanskiy, 1954; *Sholakoceras bisulcatum* Ruzhentsev and Shimanskiy, 1954, Lower Permian, southern Urals, former USSR (Ruzhentsev and Shimanskiy, 1954).

Superfamily TRIGONOCERATACEAE Hyatt, 1884

F. CENTROCERATIDAE Hyatt, *in* Zittel, 1900 D. (EIF)–P. (ZEC) Mar.

First: Undescribed species of *Centroceras*, Pine Point Limestone, Great Slave Lake region, Canada; Columbus Limestone, Ohio and Jeffersonville Limestone, Indiana, USA (Flower, 1952b).
Last: *Phacoceras*, Lower Permian, Western Australia (Teichert and Glenister, 1952).
Intervening: GIV, FAM, VIS–SPK, MOS.

F. TRIGONOCERATIDAE Hyatt, 1884 C. (VIS)–P. (ZEC) Mar.

First: Numerous species assigned to *Trigonoceras*, *Aphelaeceras*, *Chouteauceras*, *Diodoceras*, *Discitoceras*, *Epi-*

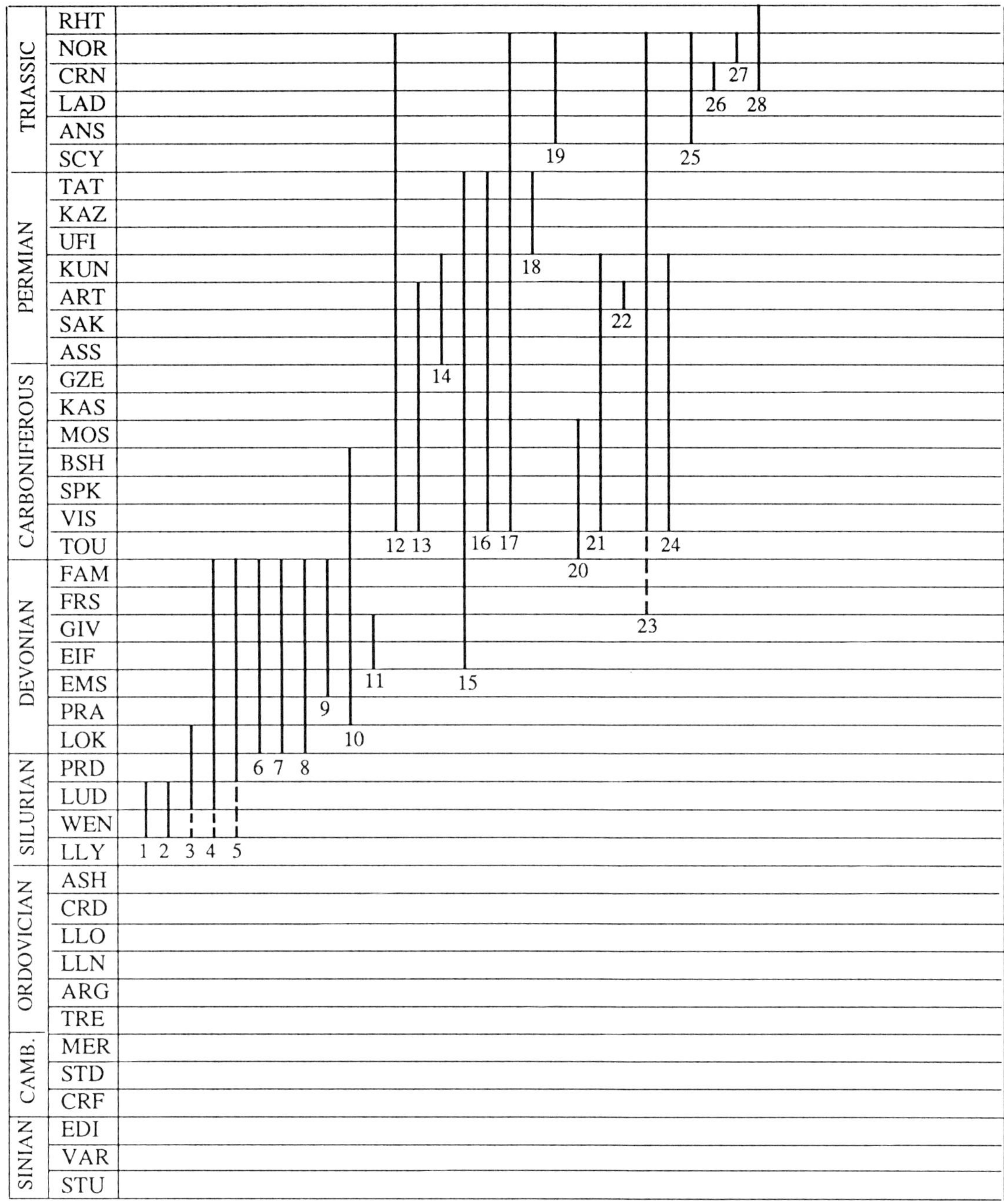

Fig. 9.3

stroboceras, Leuroceras, Lispoceras, Maccoyoceras, Mesochasmoceras, Pararineceras, Rinecaris, Stroboceras, Subclymenia, and *Thrincoceras,* Lower Carboniferous of England, UK, Belgium, Germany, Canada, USA, and Republic of Ireland (Furnish and Glenister, *in* Teichert *et al.*, 1964). It is difficult to relate the age of Viséan, Lower Carboniferous or Mississipian records.

Last: *Apogonoceras remotum* Ruzhentsev and Shimanskiy, 1954, Lower Permian, southern Urals, former USSR (Ruzhentsev and Shimanskiy, 1954).

Intervening: SPK–BSK.

F. GRYPOCERATIDAE Hyatt, *in*, Zittel, 1990 C. (VIS)–Tr. (NOR) Mar.

First: *Epidomatoceras maccoyi* Turner, 1954, Viséan, Republic of Ireland; *E. aemulum* Shimanskiy, 1967, Viséan, Kazakhstan, former USSR; *E ?doohylensae* (Foord, 1900), Viséan, Republic of Ireland (Shimanskiy, 1967).

Last: *Grypoceras mesodicum* (Quenstedt, 1845), Norian, Alps (Mojsisovics, 1873).

Intervening: SPK–C. (u.), ZEC, SCY.

F. PERMOCERATIDAE Miller and Collinson, 1953 P. (ZEC) Mar.

First and Last: *Permoceras bitauniensis* (Haniel, 1915), Lower Permian, Timor, Indonesia (Miller and Collinson, 1953).

F. SYRINGONAUTILIDAE Mojsisovics, 1902 Tr. (ANS–NOR)

First: *Syringonautilus lilianus* Mojsisovics, 1882, Anisian,

Alps (Mojsisovics, 1882); *Syringoceras*, Anisian, Nevada and California, USA (Kummel, 1953).
Last: *Clymenonautilus ehrlichi* (Mojsisovics, 1873), *Juvavionautilus heterophyllus* (Hauer, 1849), *Oxynautilus acutus* (Hauer, 1846), Norian, Alps (Mojsisovics, 1873, 1902); *Syringoceras*, Norian, Timor, Indonesia (Kummel, 1953).
Intervening: CRN.

Superfamily AIPOCERATACEAE Hyatt, 1883

F. AIPOCERATIDAE Hyatt, 1883 C. (TOU–MOS) Mar.

First: *Aipoceras gibberosum* (De Koninck 1880), Tournaisian, Belgium; *A. compressum* (Foord, 1900), Tournaisian, Republic of Ireland; *A. easleyense* Miller and Furnish, 1938, *A. oweni* Miller and Furnish, 1938, *A. pinhookense* Miller and Furnish, 1938, Chouteau Limestone, Kinderhookian, Missouri, USA (Miller and Furnish, 1938); *Asymptoceras crassilabrum* (Foord, 1900), Tournaisian, Republic of Ireland (Foord, 1900).
Last: *Librovitschiceras atuberculatus* (Tzwetaev, 1888), Moscow, former USSR (Shimanskiy, 1967).
Intervening: VIS–SPK.

F. SOLENOCHILIDAE Hyatt, 1893 C. (VIS)–P. (ZEC) Mar.

First: *Acanthonautilus bispinosus* Foord, 1896, Viséan, Republic of Ireland; *A.* sp. Shimanskiy, 1967, Viséan, European former USSR (Shimanskiy, 1967).
Last: *Solenochilus kempae* Miller and Youngquist, 1949, lower Lueder's Formation, Texas, USA (Miller and Youngquist, 1949); *Solenocheilus ?auriculus* Chao, 1954, is from the Yangsinian Formation of Tanchiashan, Hunan Province, China (Chao, 1954), but its siphuncular structure is unknown and generic assignment is tentative. If confirmed this would represent a younger record than *S. kempae*.
Intervening: C. (u.).

F. SCYPHOCERATIDAE Ruzhentsev and Shimanskiy, 1954 P. (ART) Mar.

First and Last: *Scyphoceras dionysi* Ruzhentsev and Shimanskiy, 1954, *S. ellipticum* Ruzhentsev and Shimanskiy, 1954, *S. angultum* Ruzhentsev and Shimanskiy, 1954, *Dentoceras magnum* Ruzhentsev and Shimanskiy, 1954, *D. latum* Ruzhentsev and Shimanskiy, 1954, *Venatoroceras verae* Ruzhentsev and Shimanskiy, 1954, Artinskian, southern Urals, former USSR (Ruzhentsev and Shimanskiy, 1954); *Mariceras* sp., Leonardian, Arizona, USA (Furnish and Glenister, *in* Teichert *et al.*, 1964) is approximately coeval.

Superfamily CLYDONAUTILACEAE Hyatt, *in* Zittel, 1990

F. LIROCERATIDAE Miller and Youngquist, 1949 D. (FRS)?/C. (VIS)–Tr. (NOR) Mar.

First: *Potoceras dubium* Hyatt, 1894; Hyatt (1894) thought the single type specimen to be of Devonian age, Kummel (1953) suggested an early Carboniferous (Viséan) age. Kummel (*in* Teichert *et al.*, 1964) noted that the specimen was thought probably to have come from the Frasnian Iberger Kalk of Germany or Viséan of Belgium. If a Viséan age is correct then the taxon becomes inseparable from: *Liroceras fornicatum* Shimanskiy, 1967, Viséan, former European USSR; *L. hyatti* Miller *et al.* 1933, Viséan, Belgium; *L. praelunense* Shimanskiy, 1967, Viséan, Severnyye Uvaly, former USSR; *Bistrialites bistrialis* Phillips, 1836, Viséan?, England, UK (Shimanskiy, 1967).
Last: *Indonautilus kraffti* (Mojsisovics, 1902), Norian, Himalayas; *Paranautilus simonyi* (Hauer, 1849), Norian, Alps (Mojsisovics, 1902).
Intervening: SPK–C. (u.), P., ANS–CRN.

F. EPHIPPIOCERATIDAE Miller and Youngquist, 1949 C. (VIS)–P. (ZEC) Mar.

First: *Ephippioceras bilobatum* (Sowerby, 1840), Lower Carboniferous, Scotland, UK; *E. spirale* Ramsbottom and Moore, 1961, Viséan, Republic of Ireland (Ramsbottom and Moore, 1961).
Last: *Ephippioceras hunanense* Chao, 1954, *E. involutum* Chao, 1954, Lower Permian, Hunan Province, China (Chao, 1954); *E. inexpectans* Miller and Youngquist, 1949, Lower Permian, Texas, USA (Miller and Youngquist, 1949).
Intervening: SPK–MOS.

F. CLYDONAUTILIDAE Hyatt, *in* Zittel, 1900 Tr. (ANS–NOR) Mar.

First: *Styrionautilus*, Ansian, Nevada, USA (Kummel, 1953).
Last: *Clydonautilus noricus* (Mojsisovics, 1873), Norian, Alps (Mojsisovics, 1873); *Callaionautilus turgidus* Kieslinger, 1924, *Cosmonautilus*, Upper Triassic, Timor, Indonesia, and *Proclydonautilus griesbachi* (Mojsisovics, 1896), Upper Triassic, India are probably coeval (Kummel, 1953).

F. SIBERIONAUTILIDAE Popov, 1951 Tr. (CRN) Mar.

First and Last: *Siberionautilus multilobatus* Popov, 1951, Carnian, Siberia, former USSR (Popov, 1951).

F. GONIONAUTILIDAE Kummel, *in* Flower and Kummel, 1950 Tr. (NOR) Mar.

First and Last: *Gonionautilus securis* (Von Dittmar, 1866), Norian, Alps (Mojsisovics, 1902).

Superfamily NAUTILACEAE De Blainville, 1825

F. NAUTILIDAE De Blainville, 1825 Tr. (CRN)–Rec. Mar.

First: *Cenoceras trechmanni* Kummel, 1953, Carnian, Hokonui Hills, New Zealand (Kummel, 1956).

F. PSEUDONAUTILIDAE Shimanskiy and Erlanger, 1955 J. (l.)–K. (VLG) Mar.

First: *Pseudoaganides kochi* (Prinz, 1906), Lias of Austria and Hungary (Kummel, 1956).
Last: *Pseudonautilus aturoides* (Pictet, 1867), lower Valanginian, France (Kummel, 1956).
Intervening: BAJ–OXF, TTH, BER.

F. PARACENOCERATIDAE Spath, 1927 J. (AAL/BAJ)–K. (ALB) Mar.

First: *Somalinautilus fuscus* (Crick, 1898), England, UK (Kummel, 1956).
Last: *Paracenoceras rhodani* (Roux, 1848), Switzerland (Kummel, 1956).
Intervening: BTH–TTH.

F. CYMATOCERATIDAE Spath, 1927

J. (BTH)–T. (CHT) Mar.

First: *Cymatoceras julii* (D'Orbigny, 1850), Belfort, France; *Procymatoceras subtruncatum* (Morris and Lycett, 1850), *P. baberi* (Morris and Lycett, 1850), Gloucestershire, England, UK (Kummel, 1956).
Last: *Cymatoceras tsukushiense* (Kobayashi, 1954), Oligocene, northern Kyushu, Japan (Kummel, 1956).
Intervening: CLV–J. (u.), HAU, MAA.

F. HERCOGLOSSIDAE Spath, 1927

J. (u.)–T. (RUP/CHT) Mar.

First: *Cimomia turcicus* (Krumbeck, 1905), upper Jurassic, Libya (Kummel, 1956).
Last: *Cimomia blakei* (Avnimelech, 1947), Oligocene, Palestine; *Deltoidonautilus bakeri* Teichert, 1947, Oligocene, Victoria, Australia (Kummel, 1956).
Intervening: K. (u.) DAN–PRB.

F. ATURIIDAE Chapman, 1857

T. (DAN/THA–Mio.) Mar.

First: *Aturia praeziczac* Oppenheim, 1903, Palaeocene, Thebes, Egypt (Kummel, 1956).
Last: Numerous species assigned to *Aturia* (including *A. aturi* (Basterot, 1825), *A. angustata* Conrad, 1866, *A. coxi* Miller, 1947), from the Miocene of France, USA, Venezuela, Zanzibar, Angola, Italy and Japan (Kummel, 1956).
Intervening: T. (Eoc.–Oli.).

REFERENCES

Angelin, N. P. and Lindström, G. (1880) *Fragmenta Silurica e dono Caroli Henrici Wegelin*, Holmiae, Stockholm, 60 pp.

Balashov, Z. G. (1960) Novye Nautiloidei Ordovika SSSR. Novye vidy drevneishikh rastenii i bespozvonochykh. *Trudy VSEGEI*, **2**, 123–45 [in Russian].

Balashov, Z. G. (1962) Nautiloidei Ordovika Sibirskoi platformi. *Izdatelstvo Leningradskogo Universiteta*, 204 pp. [in Russian].

Balashov, Z. G. (1968) Endoceratoidei Ordovika SSSR. *Izdatelstvo Leningradskogo Universiteta*, 277 pp. [in Russian].

Barrande, J. (1865–77) *Systême Silurien du Centre de la Bohême*, Première Partie: Recherches paléontologiques, Vol. 2, Classe des Mollusques, Ordre des Céphalopodes: pt. 1, xxxvi + 712 pp. (1867); pt. 2, xi + 263 pp. (1870); pt. 3, xxiv + 804 pp. (1874); pts 4–5, lx + 742 pp., xx + pp. 743–1505 (1877); pls 1–107 (1865); pls 108–244 (1866); pls 245–350 (1868): pls 351–460 (1870): Supplement, pt. 1, viii + 297 pp. (1877); pt. 2, pls 461–544 (1877). Published by the author, Prague and Paris.

Bülow, E. (1915) Orthoceren und belemnitiden der Trias von Timor. *Paläontologie von Timor*, **4** (7), 1–72.

Chao, K. K. (1954) Permian cephalopods from Tanchiashan, Hunan. *Acta Palaeontologica Sinica*, **2**, 1–58.

Chen, J.-Y. (1976) Advances in the Ordovician stratigraphy of North China with a brief description of nautiloid fossils. *Acta Palaeontologica Sinica* **15** (1), 55–76 [in Chinese with English abstract].

Chen, J.-Y. and Liu, G.-W. (1974) Ordovician and Silurian Nautiloidea, in *Handbook of Geology and Palaeontology of Southwest China*. Nanjing [in Chinese], pp. 138–94.

Chen, J.-Y. and Liu, G.-W. (1976) Description of selected Ordovician fossils (Cephalopoda), in Ordovician Biostratigraphy and Palaeozoology of China. *Memoirs of the Nanjing Institute of Geology and Palaeontology, Academia Sinica* 7 (ed. Y.-H. Lu, C.-L. Chu and Y.-Y. Chien), Science Press, Nanjing, pp. 42–53 [in Chinese with English index].

Chen, J.-Y. and Teichert, C. (1983) Cambrian Cephalopoda of China. *Palaeontographica, Abteilung A*, **181**, 1–102.

Chen, T.-E. (1987) Ordovician nautiloids from Xainza, northern Xizang. *Bulletin of the Nanjing Institute of Geology and Palaeontology, Academia Sinica* **11** (11), 133–91 [in Chinese with English abstract and descriptions of new genera and species].

Chen, T.-E. and Wang, M. Q. (1983) Discovery of Ordovician actinoceroids from the Mt. Kongur area, Southwest Xinjiang. *Contributions to the Geology of the Quinghai–Xizang (Tibet) Platform*, **2**, 81–9, [in Chinese with English summary].

Crick, R. E. (1988) Buoyancy regulation and macroevolution in nautiloid cephalopods. *Senckenbergiana Lethaea*, **69**, 13–42.

Crick, R. E. and Teichert, C. (1979) Siphuncular structures in the Devonian nautiloid *Archiacoceras* from the Eifel of West Germany. *Palaeontology*, **22** (4), 745–64.

Diener, C. (1919) Nachträge zur Kenntnis der Nautiloideenfauna der Hallstätter Kalke. *Denkschrifter des Akadamie der Wissenschaften Wien*, **96**, 751–78.

Dzik, J. (1984) Phylogeny of the Nautiloidea. *Palaeontologia Polonica*, **45**, 219 pp.

Endo, R. (1932) The Canadian and Ordovician of southern Manchuria. *United States National Museum, Bulletin*, **164**, 1–152.

Evans, D. H. and King, A. H. (1990) The affinities of early oncocerid nautiloids from the Lower Ordovician of Spitsbergen and Sweden. *Palaeontology*, **33**, 623–30.

Flower, R. H. (1936) Cherry Valley cephalopods. *Bulletin of American Paleontology*, **22**, 1–96.

Flower, R. H. (1941a) Notes on the structure and phylogeny of eurysiphonate cephalopods. *Palaeontographica Americana*, **3** (13), 1–56.

Flower, R. H. (1941b) Development of the Mixochoanites. *Journal of Paleontology*, **15**, 523–48.

Flower, R. H. (1943a) Cephalopods from the Silurian of Arisaig, Nova Scotia. *Journal of Paleontology*, **17**, 248–57.

Flower, R. H. (1943b) Studies of Paleozoic Nautiloidea, VI. Some Silurian cyrtoconic cephalopods from Indiana with notes on stratigraphic problems. *Bulletin of American Paleontology*, **28** (109), 83–101.

Flower, R. H. (1943c) *Apsidoceras* in the Trenton of Montreal. *Journal of Paleontology*, **17**, 258–63.

Flower, R. H. (1945) Classification of Devonian cephalopods. *American Midland Naturalist*, **33**, 675–724.

Flower, R. H. (1947) New Ordovician nautiloids from New York. *Journal of Paleontology*, **21**, 429–33.

Flower, R. H. (1949) New genera of Devonian nautiloids. *Journal of Paleontology*, **23**, 74–80.

Flower, R. H. (1951) A Helderbergian cyrtoconic cephalopod. *Wagner Free Institute*, **26** (1), 1–7.

Flower, R. H. (1952a) New Ordovician cephalopods from eastern North America. *Journal of Paleontology*, **26**, 24–59.

Flower, R. H. (1952b) The ontogeny of *Centroceras* with remarks on the phylogeny of the Centroceratidae. *Journal of Paleontology*, **26**, 519–28.

Flower, R. H. (1955a) Status of Endoceroid classification. *Journal of Paleontology*, **29**, 329–37.

Flower, R. H. (1955b) New Chazyan orthocones. *Journal of Paleontology*, **29**, 857–67.

Flower, R. H. (1957) Studies of the Actinoceratida. I. The Ordovician development of the Actinoceratida, with notes on actinoceroid morphology and Ordovician stratigraphy. *New Mexico Bureau of Mines and Mineral Resources, Memoir*, **2**, 3–62.

Flower, R. H. (1958) Some Chazyan and Mohawkian Endoceratida. *Journal of Paleontology*, **32**, 433–58.

Flower, R. H. (1962a) The phragmocone of *Ecdyceras*. *New Mexico Bureau of Mines and Mineral Resources, Memoir*, **9**, 1–28.

Flower, R. H. (1962b) Notes on the Michelinoceratida. *New Mexico Bureau of Mines and Mineral Resources, Memoir*, **10** (II), 21–44.

Flower, R. H. (1963) New Ordovician Ascoceratida. *Journal of Paleontology*, **37**, 69–85.

Flower, R. H. (1964a) The Nautiloid order Ellesmeroceratida. *New Mexico Bureau of Mines and Mineral Resources, Memoir*, **12**, 1–234.

Flower, R. H. (1964b) Nautiloid shell morphology. *New Mexico*

Bureau of Mines and Mineral Resources, Memoir, **13**, 1–79.

Flower, R. H. (1968a) The first great expansion of the actinoceroids. *New Mexico Bureau of Mines and Mineral Resources, Memoir*, **19** (I), 1–16.

Flower, R. H. (1968b) Some additional Whiterock cephalopods. *New Mexico Bureau of Mines and Mineral Resources, Memoir*, **19** (II), 17–53.

Flower, R. H. (1968c) Silurian cephalopods of James Bay Lowland, with a revision of the Narthecoceratidae. *Geological Survey of Canada, Bulletin*, **164**, 1–88.

Flower, R. H. (1968d) *Botryoceras*, a remarkable nautiloid from the Second Value of New Mexico. *New Mexico Bureau of Mines and Mineral Resources, Memoir*, **21** (I), 3–5.

Flower, R. H. (1968e) Endoceroids from the Canadian of Alaska. *New Mexico Bureau of Mines and Mineral Resources, Memoir*, **21** (III), 8–17.

Flower, R. H. (1971) Cephalopods of the Whiterock Stage, in *Paleozoic Perspectives. A Paleontological Tribute to G. Arthur Cooper* (ed. J. T., Jr Dutro), Smithsonian Contributions to Paleobiology, 3, Smithsonian Institution Press, Washington, DC, pp. 101–11.

Flower, R. H. (1976a) New American Wutinoceratidae with a review of actinocerid occurrences in the eastern hemisphere. *New Mexico Bureau of Mines and Mineral Resources, Memoir*, **28** (I), 5–12.

Flower, R. H. (1976b) Some Whiterock and Chazy Endocerids. *New Mexico Bureau of Mines and Mineral Resources, Memoir*, **28** (II), 13–39.

Flower, R. H. (1976c) Ordovician cephalopod faunas and their rəle in correlation, in *The Ordovician System. Proceedings of a Palaeontological Association Symposium* (ed. M. G. Bassett), University of Wales Press and National Museum of Wales, Cardiff, pp. 523–52.

Flower, R. H. (1978) St. George and Table Head cephalopod zonation in western Newfoundland. *Geological Survey of Canada (Current Research, Part A), Paper*, **78-1A**, 217–24.

Flower, R. H. (1984) *Bodeiceras*, a new Mohawkian oxycone, with revision of the older Barrandeoceratida and discussion of the status of the order. *Journal of Paleontology*, **58**, 1372–9.

Flower, R. H. and Kummel, B. (1950) A classification of the Nautiloidea. *Journal of Paleontology*, **24**, 604–16.

Flower, R. H. and Teichert, C. (1957) The cephalopod order Discosorida. *University of Kansas Paleontological Contributions, Mollusca, Article* **6**, 1–114.

Foerste, A. F. (1924) Notes on American Paleozoic cephalopods. *Denison University Bulletin, Journal of the Scientific Laboratories*, **20**, 193–268.

Foerste, A. F. (1925) Notes on cephalopod genera; chiefly coiled Silurian forms. *Denison University Bulletin, Journal of the Scientific Laboratories*, **21**, 1–70.

Foerste, A. F. (1926) Actinosiphonate, trochoceroid, and other cephalopods. *Denison University Bulletin, Journal of the Scientific Laboratories*, **21**, 285–384.

Foerste, A. F. (1928) A restudy of some of the Ordovician and Silurian cephalopods described by Hall. *Denison University Bulletin, Journal of the Scientific Laboratories*, **23**, 173–230.

Foerste, A. F. (1929) The cephalopods of the Red River formation of southern Manitoba. *Denison University Bulletin, Journal of the Scientific Laboratories*, **24**, 129–235.

Foerste, A. F. (1930) Port Byron and other Silurian cephalopods. *Denison University Bulletin, Journal of the Scientific Laboratories*, **25**, 1–24.

Foerste, A. F. (1934) Silurian cyrtoconic cephalopods from Ohio, Ontario and other areas. *Denison University Bulletin, Journal of the Scientific Laboratories*, **29**, 107–94.

Foerste, A. F. (1935) Bighorn and related cephalopods. *Denison University Bulletin, Journal of the Scientific Laboratories*, **30**, 1–96.

Foerste, A. F. (1936) Several new Silurian cephalopods and crinoids, chiefly from Ohio and Hudson Bay. *Ohio Journal of Science*, **36**, 261–72.

Foerste, A. F. and Savage, T. E. (1927) Ordovician and Silurian cephalopods of the Hudson Bay area. *Denison University Bulletin, Journal of the Scientific Laboratories*, **22**, 1–107.

Foord, A. H. (1897–1903) Monograph of the Carboniferous Cephalopoda of Ireland. *Monograph of the Palaeontographical Society of London*, 234 pp. (Pt. 1, pp. 1–22 (1897); Pt. 2, pp. 23–48 (1898); Pt. 3, pp. 49–126 (1900); Pt. 4, pp. 127–146 (1901); Pt. 5, pp. 147–234 (1903)).

Frey, M. W. (1982) Upper Ordovician (Harjuan) nautiloid cephalopods from the Boda Limestone of Sweden. *Journal of Paleontology*, **56**, 1274–92.

Gil, A. V. (1988) Whiterock (lower Middle Ordovician) cephalopod fauna from the Ibex area, Millard County, western Utah. *New Mexico Bureau of Mines and Mineral Resources, Memoir*, **44**, 27–59.

Gnoli, M. (1982) Lower Devonian orthocone cephalopods from Iglesiente and Sulcis regions (southwestern Sardinia). *Bollettino della Società Paleontologica Italiana*, **21**, 73–98.

Gordon, M. (1964) Carboniferous cephalopods of Arkansas. *US Geological Survey, Professional Paper*, **460**, 1–322.

Grabau, A. W. (1922) Ordovician fossils of North China. *Palaeontologia Sinica Series B*, **1** (1), 3–95.

Hedström, T. (1917) Über die Gattung *Phragmoceras* in der Obersilurformation Gotlands. *Sveriges Geologiska Undersökning, ser. C: A*, **15**, 1–35.

Heinrichson, T. (1935) Über *Endoceras glauconiticum* n. sp. aus dem Glaukonitkalk Bii Estlands. *Publications of the Geological Institution of the University of Tartu*, **42**, 3–6.

Holland, C. H. (1979) Early Cephalopoda, in *The Origin of Major Invertebrate Groups*. (ed. M. R. House), Systematics Association, Special Publication, **12**, pp. 367–78.

Holland, C. H. (1987) The nautiloid cephalopods; a strange success. *Journal of the Geological Society of London*, **144**, 1–15.

Holm, G. (1892) Om tvenne *Gyroceras*-formigt böjda *Endoceras*-arter. *Geologiska Föreningens i Stockholm Förhandlingar*, **14** (2), 125–37.

Hook, S. C. and Flower, R. H. (1976) *Tajaroceras* and the origin of the Troedssonellidae. *Journal of Paleontology*, **50** (2), 293–300.

Hook, S. C. and Flower, R. H. (1977) Late Canadian (Zones J, K) Cephalopod Faunas from Southwestern United States. *New Mexico Bureau of Mines and Mineral Resources, Memoir*, **32**, 102 pp.

Hume, G. S. (1925) The Palaeozoic outlier of Lake Timiskaming, Ontario and Quebec. *Geological Survey of Canada, Memoir*, **145**, 129 pp.

Hyatt, A. (1894) Phylogeny of an acquired characteristic. *Proceedings of the American Philosophical Society*, **32**, 349–647.

Kieslinger, A. (1924) Die Nautiloideen der mittleren und oberen Trias von Timor [incl. Nachtrag], *Mijnwezen Nederland, Oost.-Indië, Jaarbuch*, **51**, (1922), 53–124, Nachtrag, 127–45.

Kobayashi, T. (1927) Ordovician fossils from Korea and South Manchuria. *Japanese Journal of Geology and Geography*, **5** (4), 173–212.

Kobayashi, T. (1940) Nomenclatural note on *Foerstella*. *Journal of the Geological Society of Japan*, **47**, p. 261.

Kummel, B. (1953) American Triassic coiled nautiloids. *US Geological Survey, Professional Paper*, **250**, 1–104.

Kummel, B. (1956) Post-Triassic nautiloid genera. *Bulletin of the Harvard Museum of Comparative Zoology*, **7**, 324–484.

Lai, C.-G. (1964) *Jangziceras*, a new Silurian nautiloid genus. *Acta Palaeontologica Sinica*, **12** (1), 124–8 [in Chinese with English summary].

Lai, C.-G. (1965) Ordovician and Silurian cephalopods from Hanzhung and Ningkiang of Shensi. *Acta Palaeontologica Sinica*, **13** (2), 308–42 [in Chinese with English summary and descriptions of new species].

Lai, C.-G. (1982) Cephalopoda. in *Palaeontological Atlas of North West China (Shaanxi, Gansu and Ningxia Provinces), 1 Precambrian and Early Palaeozoic*, (ed. Xi'an Institute of Geology and Mineral Resources), Geological Publishing House, Beijing, China, pp. 189–208.

Lai, C.-G. (1989) Biogeography of the Ordovician cephalopods from China. *Journal of Southeast Asian Earth Sciences*, **3**, 125–30.

Lai, C.-G. and Tsi, S.-P. (1977) Ordovician cephalopods from northwest Hunan. Collection of Palaeontology and stratigraphy papers. *Academia Geologica Sinica*, **3**, 1–73 [in Chinese].

Lai, C.-G. and Xu, G.-H. (1987) Cephalopoda, in *Biostratigraphy of the Yangtze Gorge Area (2) Early Palaeozoic Era*. Xi'an Institute of Geology and Mineral Resources, Geological Publishing House, Beijing, China [in Chinese], pp. 245–93.

Lai, C.-G. and Zhu, K.-Y. (1985) Oncocerid cephalopods from the Lower Silurian of Guangyuan, Sichuan, in *Professional Papers of Stratigraphy and Palaeontology*, 15 (Chinese Academy of Geological Sciences), Geological Publishing House, Beijing, China, pp. 40–72.

Lindström, G. (1890) The Ascoceratidae and the Lituitidae of the Upper Silurian formation of Gotland. *Kungliga Svenska Vetenskaps-akadamie, Handlingar*, **23** (12), 54 pp.

Miller, A. K. and Collinson, C. (1953) An aberrant nautiloid of the Timor Permian. *Journal of Paleontology*, **27**, 293–5.

Miller, A. K. and Furnish, W. M. (1938) Lower Mississippian nautiloid cephalopods of Missouri, in *Stratigraphy and Paleontology of the Lower Mississipian of Missouri* (ed. E. B. Branson), Studies of Missouri University 13, pp. 140–78.

Miller, A. K. and Youngquist, W. L. (1949) American Permian nautiloids. *Geological Society of America, Memoir*, **63**, 218 pp.

Mojsisovics, E. (1873–1902) Das Gebirge um Hallstätter, Abt. 1, Die Cephalopoden der Hallstätter kalke. *Abhandlungen geologische Reichsanst. Wein*, **6**, 356 pp. [No. 1, pp. 1–82, (1873); No. 2, pp. 83–174, (1875); suppl., pp. 175–356, (1902)].

Mojsisovics, E. (1882) Die Cephalopoden der mediterranean Triasprovinz. *K. K. geologische Reichsanst. Wein. Abhandlungen*, **10**, 322 pp.

Mojsisovics, E. (1899) Upper Triassic Cephalopoda faunae Himálaya. *Memoirs of the Geological Survey of India, Palaeontologia Indica, ser. 15*, **3** (1), 1–157.

Mutvei, H. (1957) On the relations of the principal muscles to the shell in *Nautilus* and some fossil nautiloids. *Arkiv för Mineralogi och Geologi*, **2** (10), 219–59.

Mutvei, H. (1964) On the secondary internal calcareous lining of the wall of the siphonal tube in certain fossil 'nautiloid' cephalopods. *Arkiv för Zoologi, ser. 2*, **16** (21), 375–424.

Mutvei, H. and Stumbur, H. (1971) Remarks on the genus *Pictetoceras* (Cephalopoda: Ellesmerocerida). *Bulletin of the Geological Institutions of the University of Uppsala N. S.*, **2** (13), 117–22.

Nakazawa, K. (1959) Two cephalopod species from the Norian Nariwa Group in Okayama Prefecture, West Japan. *Japanese Journal of Geology and Geography*, **30**, 127–33.

Neben, W. and Krueger, H. H. (1971) Fossilien ordovicischer geschiebe. *Staringia*, **1**, 55 pp.

Popov, Y. N. (1951) Slozhnoe rasshcheplenie suturniykh liniy u Nautiloidea. *Akademiya Nauk SSSR, Doklady new ser.*, **78**, 765–7.

Ramsbottom, W. H. C. and Moore, E. W. J. (1961) Coiled nautiloids from the Viséan of Ireland. *Liverpool and Manchester. Geological Journal*, **2**, 630–644.

Ristedt, H. (1968) Zur revision der Orthoceratidae. *Akademie der Wissenschaften und der Literatur, Abhandlungen der Math-naturwissenschaften. Klasse Jahrbuch 1968*, **4**, 213–97.

Ruedemann, R. (1906) Cephalopoda of the Beekmantown and Chazy formations of the Champlain basin. *New York State Museum, Bulletin*, **90**, 393–605.

Ruzhentsev, V. E. and Shimanskiy, V. N. (1954) Nizhnepermskie svernutie i sognutie nautiloidei yuzhnovo Urala. *Trudy Paleontologicheskogo Instituta Akademiya Nauk SSSR*, **50**, 152 pp.

Schindewolf, O. H. (1943) Über das Apikalende der Actinoceren (Cephalopoden). *Reichsamt Bodenforschrift, Jahrbuch 1941*, **62**, 207–47.

Serpagli, E. and Gnoli, M. (1977) Upper Silurian cephalopods from south-western Sardinia. *Bollettino della Società Paleontologica Italiana*, **16** (2), 153–96.

Sheng, S.-F. (1980) The Ordovician System in China, Correlation Chart and Explanatory Notes. *International Union of Geological Sciences, Publication* **1**, 7 pp.

Shimanskiy, V. N. (1954) Pryamye nautiloidei i baktritoidei Sakmarskogo i Artinskogo yarusov Yuzhnogo Urala. *Trudy Paleontologicheskogo Instituta, Akademiya Nauk SSSR*, **44**, 156 pp. [in Russian].

Shimanskiy, V. N. (1957) Kamennougolnye Oncoceratida. *Akademiya Nauk SSSR, Doklady*, **112** (3), 530–32 [in Russian].

Shimanskiy, V. N. (1961) K evolyutsii Kamennougolnykh aktinoseratoidei. *Paleontologischeskiy Zhurnal*, **3**, 33–40 [in Russian].

Shimanskiy, V. N. (1967) Kamennougolnykh Nautilida. *Trudy Paleontologischeskogo Instituta Akademiya Nauk SSSR*, **115**, 258 pp. [in Russian].

Shimer, H. W. and Shrock, R. R. (1944) Index fossils of North America. *Massachusetts Institute of Technology, Technology Press Publication*, 837 pp.

Stait, B. (1982) Ordovician Oncoceratida (Nautiloidea) from Tasmania, Australia. *Neues Jahrbuch für Geologie und Paläontologie, Monatshefte* **10**, 607–18.

Stait, B. (1984) Ordovician nautiloids of Tasmania, Australia – Gouldoceratidae fam. nov. (Discosorida). *Proceedings of the Royal Society of Victoria*, **96**, 187–207.

Stait, B. (1988) Nautiloids of the Lourdes Formation (Middle Ordovician), Port au Port Peninsula, western Newfoundland. *New Mexico Bureau of Mines and Mineral Resources, Memoir*, **44**, 61–77.

Stait, B. and Burrett, C. F. (1984) Ordovician nautiloid faunas of Central and Southern Thailand. *Geological Magazine*, **121** (2), 115–24.

Stait, B. and Burrett, C. F. (1987) Biogeography of Australian and Southeast Asian Ordovician nautiloids. in *Gondwana Six; Stratigraphy, Sedimentology and Paleontology*. (ed. G. D. McKenzie), Geophysical Monograph, 41, pp. 21–8.

Stait, B. and Flower, R. H. (1985) Michelinoceratida (Nautiloidea) from the Ordovician of Tasmania, Australia. *Journal of Paleontology*, **59** (1), 149–59.

Strand, T. (1934) The Upper Ordovician cephalopods of the Oslo area. *Norsk Geologisk Tidsskrift*, **14** (1), 1–117.

Stumbur, H. (1962) Rasprostranenie nautiloidei v Ordoviki Estonii (s opisaniem necotorye novye rodov). *Trudy Instituta Geologii, Akademiya Nauk Est. SSSR*, **10**, 131–48 [in Russian].

Stumbur, H. and Mutvei, H. (1983) A new Middle Ordovician torticonic nautiloid. *Geologiska Föreningens i Stockholm Förhandlingar*, **105** (1), 43–7.

Sweet, W. C. (1958) The Middle Ordovician of the Oslo Region, Norway 10. Nautiloid Cephalopods. *Norsk Geologisk Tidsskrift*, **38** (1), 178 pp.

Sweet, W. C. (1959) Muscle-attachment impressions in some Paleozoic nautiloid cephalopods. *Journal of Paleontology*, **33**, 293–304.

Teichert, C. (1930) Die Cephalopoden-Fauna der Lyckholm-Stufe des Ostbaltikums. *Paläontologische Zeitschrift*, **12**, 264–312.

Teichert, C. (1967) Major features of Cephalopod Evolution, in *Essays in Paleontology and Stratigraphy, R. C. Moore, Commemorative volume*. (eds C. Teichert and E. L. Yochelson), University of Kansas, Special Publications, 2, pp. 162–210.

Teichert, C. and Glenister, B. F. (1952) Fossil nautiloid faunas from Australia. *Journal of Paleontology*, **26** (5), 730–52.

Teichert, C., Glenister, B. F. (1953) Ordovician and Silurian cephalopods from Tasmania, Australia. *Bulletins of American Paleontology*, **34** (144), 66 pp.

Teichert, C. and Glenister, B. F. (1954) Early Ordovician cephalopod fauna from Northwestern Australia. *Bulletins of American Paleontology*, **35** (150), 113 pp.

Teichert, C., Glenister, B. F. and Crick, R. (1979) Biostratigraphy of Devonian nautiloid cephalopods. *Special Paper in Palaeontology*, **23**, 259–62.

Teichert, C., Kummel, B., Sweet, W. C. *et al.* (1964) Cephalopoda – general features, Endoceratoidea, Actinoceratoidea, Nautiloidea, Bactritoidea, in *Treatise on Invertebrate Paleontology*,

Part K. (ed. R. C. Moore), University of Kansas Press, Boulder, Colorado, and Lowrence, Kansas.

Termier, G. and Termier, H. (1950) Invertébrés l'Ère primaire, Fascule 3, Mollusques. *Paléontologie Marocaine*, **2**, 116 pp.

Troedsson, G. T. (1926) On the Middle and Upper Ordovician faunas of Northern Greenland, I. Cephalopods. *Meddelelser om Grønland*, **71**, 157 pp.

Troedsson, G. T. (1932) Studies on Baltic fossil cephalopods. II Vertically striated or fluted Orthoceracones in the Orthoceras Limestone. *Lunds Universitets Årsskrift, N. F. 2*, **28** (6), 38 pp.

Turner, J. S. (1951) On the Carboniferous nautiloids: *Orthocera gigantea* J. Sowerby and allied forms. *Transactions of the Royal Society of Edinburgh*, **62**, 169–90.

Ulrich, E. O. and Foerste, A. F. (1936) New genera of Ozarkian and Canadian cephalopods. *Denison University Bulletin, Journal of the Scientific Laboratories*, **30**, 259–90.

Ulrich, E. O., Foerste, A. F., Miller, A. K. and Unklesbay, A. G. (1944) Ozarkian and Canadian cephalopods. Part III. Longicones and Summary. *Geological Society of America, Special Paper*, **58**, 137 pp.

Unklesbay, A. G. and Young, R. S. (1956) Early Ordovician nautiloids from Virginia. *Journal of Paleontology*, **30**, 481–91.

Waagen, W. H. (1879) Salt Range fossils: Productus limestone fossils. *Memoirs of the Geological Survey of India, Palaeontologia Indica, ser.*, **13** (1), 72 pp.

Wade, M. (1977) Georginidae, a new family of actinoceratid cephalopods, Ordovician, Australia. *Memoirs of Queensland Museum*, **18** (1), 1–15.

Whiteaves, J. F. (1891) Description of some new or previously unrecorded species of fossils from the Devonian rocks of Manitoba. *Transactions of the Royal Society of Canada*, **8** (14), 93–110.

Yochelson, E. L. (1977) Agmata, a proposed extinct phylum of early Cambrian age. *Journal of Paleontology*, **51**, 437–54.

Yochelson, E. L. (1981) A survey of *Salterella* (Phylum Agmata), in *Short Papers for the Second International Symposium on the Cambrian System 1981.* (ed. M. E. Taylor), Open File Report, US Geological Survey, 81–743, pp. 244–8.

Zhuravleva, F. A. (1961) Nekotorye Paleozoyskiye nautiloidei Podolii. *Paleontologicheskiy Zhurnal*, **4**, 55–9 [in Russian].

Zhuravleva, F. A. (1964) Novye Ordovikskie i Siluriiskie tsefalopody Sibirskoi platformi. *Paleontologicheskiy Zhurnal*, **4**, 87–100 [in Russian].

Zhuravleva, F. A. (1972) Devonskie nautiloidei, Otryad Discosorida. *Trudy Paleontologicheskogo Instituta, Akademiya Nauk SSSR*, **134**, 320 pp. [in Russian].

Zhuravleva, F. A. (1974) Devonskie nautiloidei, Otryady Oncoceratida, Tarphyceratida, Nautilida. *Trudy Paleontologicheskogo Instituta, Akademiya Nauk SSSR*, **142**, 159 pp. [in Russian].

Zhuravleva, F. A. (1978) Devonskie orthotserody, Nadotryad Orthoceratoidea. *Trudy Paleontologicheskogo Instituta, Akademiya Nauk SSSR*, **148**, 223 pp. [in Russian].

Zou, X.-P. (1981) Early Ordovician nautiloids from Qingshuihe, nei Monggol (Inner Mongolia) and Pianguan, Shanxi Province. *Acta Palaeontologica Sinica*, **20** (4), 353–62 [in Chinese with English summary].

Zou, X.-P. (1988) Ordovician nautiloid fauna from Lunshan, Jurong, Jiangsu. *Acta Palaeontologica Sinica*, **27** (3), 309–30 [in Chinese with English summary and description of new species].

10

MOLLUSCA: CEPHALOPODA (PRE-JURASSIC AMMONOIDEA)

R. A. Hewitt, J. Kullmann, M. R. House, B. F. Glenister and Wang Yi-Gang

This chapter was complied by several authors, each working on a geological period: MRH on the Devonian ammonoids, JK on the Carboniferous, BFG on the Permian, and RAH and WYG on the Triassic. The typical ammonoids of the Jurassic and Cretaceous are treated in the next chapter. The listings for Permian ammonoids are based largely on MSS for the revision of Volume L, Mollusca 4 of the *Treatise on Invertebrate Paleontology*, prepared by B. F. Glenister, W. M. Furnish and Z.-Z. Zhou. The Triassic ammonoid classification is based on Shevyrev (1986), which should probably be analysed by equating his superfamilies with the cited families of Palaeozoic ammonoids. Some of the Shevyrev families are placed in synonymy and are indicated by upper-case letters in the body of the text.

The subdivision of the Devonian follows recent recommendations by the Subcommission on Devonian stratigraphy (Kirchgasser and House, 1981). The stratigraphical subdivisions of the Permo-Carboniferous follow the standard stage names set out at the beginning of the volume, but several schemes are mixed in order that the most appropriate terms may be used for each case. However, note that the Lower Carboniferous zones cannot be applied for ammonoid zonation (Kullman *et al.*, 1991).

Russian stage names for the Lower Permian (Asselian, Sakmarian, Artinskian) are legitimate time-rock terms that serve effectively for international reference. However, those recommended (e.g. Harland *et al.*, 1982) for the higher Permian (Kungurian, Ufimian, Kazanian, Tatarian) are primarily ecological entities with limited chronological significance. Serviceable post-Artinskian references (Roadian, Wordian and Capitanian = Guadalupian) are available in objective stratigraphical succession in the North American south-west, and are supplemented by the highest Permian sequences in Transcaucasia (Dzhulfian, Dorashamian). The present chapter uses the succession Asselian (ASS), Sakmarian (SAK), Artinskian (ART), Roadian (ROD) as Lower Permian, and Guadalupian (GUA) (Wordian (WOR) and Capitanian (CAP)), Dzhulfian (DZH) and Dorashamian (DOR) (= Changxingian (CHX)) as Upper Permian. The authors are mindful, however, that Roadian through Capitanian (roughly Kungurian to early Kazanian) will probably be segregated in future as a Middle Permian Series (Glenister *et al.*, in press), and that the presently defined base of the boreal Triassic in Canada (GRI) may well be defined eventually to correspond approximately with the base of the Asian Dorashamian (Sheng *et al.*, 1984).

The Triassic has been divided into 39 zonal units, each of which is composed of no more than two subzones (a, older; b, younger half of zone). They are numbered in chronological order within each of the stages cited in this volume, and their totals per stage are as follows: GRI (4), NML (4), SPA (3), ANS (9), LAD (5), CRN (5), NOR (6), RHT (3). The RHT symbol denotes the upper Norian substage defined by the base of the *Cordilleranus* Zone (Tozer, 1984). The other zonal units were derived from table 2 of Tozer (1984) by addition of the *Mulleri* Zone (ANS, 1) of Bucher (1989) and recognition of informal upper (ANS, 5) and lower (ANS, 4) *Ismidicum* Zones (Fantini Sestini, 1988). The three main Triassic ammonoid extinctions extended through two subzones, and are situated in the Ellesmerian Substage (GRI, 3–4), the *Austriacum* Zone (CRN, 2), and the final *Marshi* Zone (RHT, 3).

The Fossil Record 2. Edited by M. J. Benton. Published in 1993 by Chapman & Hall, London. ISBN 0 412 39380 8

Subclass AMMONOIDEA Zittel, 1984

Order ANARCESTIDA Miller and Furnish, 1954

Suborder BACTRITINA Miller and Furnish, 1954

It is generally agreed that this group of orthocones with egg-shaped protoconchs and marginal siphuncles is ancestral to the Ammonoidea. But it is possible that similar egg-shaped protoconchs may have arisen several times among orthoconic nautiloids, and deciding which, if any, of the known early forms with this feature is likely to be the ancestor, is a speculative and rather idle occupation. Few show the poorly developed protoconch apparatus within the protoconch that characterizes the coiled Ammonoidea, a feature which they share with belemnoids and spirulids, and preservation of such characters is rare. A further complication is that the Bactritina are thought to be ancestral to some Coleoidea (Belemnoidea) and many palaeontologists would prefer to place the Bactritina (or Bactritida) in the Nautiloidea to avoid having coleoids derived from the Ammonoidea, and to avoid having the Ammonoidea range back into the early Palaeozoic if certain bactritid-like forms are included within the suborder. Bactritids became abundant in the Emsian, where the first coiled goniatites appeared, and there are intermediates between the straight bactritids and the earliest coiled ammonoids. It seems most reasonable to presume these are the earliest known, until such time as relationships can be established with earlier forms.

F. BACTRITIDAE Hyatt, 1884 D. (EMS)–Tr. (CRN) Mar. (see Fig. 10.1)

First: *Bactrites* n. sp. B. (Erben, 1960). Schönauer Kalk, Kellerwald, lower Emsian.
Last: *Dillerites shastensis* Gordon, 1966, Hosselkus Limestone, Triassic (CRN 4), California, USA.
Intervening: Fair later Devonian, Carboniferous and Permian record.
Comment: This assignment excludes the Ordovician *Eobactrites* Schindewolf, 1932, the very poor Silurian specimen alleged to be *Bactrites* sp. figured by Termier and Termier, 1950 and excludes also the *Bactrites* recorded from the *nilsonni* Zone, Silurian, by Ristedt (1981).

F. BOJOBACTRITIDAE Horny, 1956 D. (EMS) Mar.

First and Last: *Pseudobactrites bicarinatus* (=*Bojobactrites ammonitans*) Horny, 1956, Zlichov Limestone, upper Emsian, Czechoslovakia.

F. PARABACTRITITIDAE Shimanskiy, 1951 C. (VIS)–P. (ART) Mar.

First: *Angustobactrites saundersi* Mapes, 1979, Imo Formation, Chester, Mississippian, Arkansas, USA.
Last: *Parabactrites rhuzhencevi* and other spp., Shimanskiy, 1962, Artinskian, Permian, southern Urals, former USSR.

F. SINUOBACTRITIDAE Mapes, 1979 C. (BSH–KAS) Mar.

First: *Sinuobactrites morrowanensis* Mapes, 1979, Gene Autry Formation, Morrowian, Upper Carboniferous, Oklahoma, USA.
Last: *Dilatobactrites missouriensis* Mapes, 1979, Eudora Shale, upper Missourian, Upper Carboniferous, Kansas, USA.

Suborder AGONIATITINA Ruzhentsev, 1957

Superfamily MIMOCERATACEAE Steinmann, 1890

F. MIMOSPHINCTIDAE Erben, 1953 D. (EMS–EIF), Mar

First: *Anetoceras mattei* Feist, 1970. D. (EMS), *dehiscens* conodont Zone, Montagne Noire, France.
Last: *Kokenia obliquecostata* Holzapfel, 1895, Odershauser Kalk, D. (U. EIF), Bonzel, Germany.
Comment: This classification groups the cyrtoconic to partly gyroconic genera *Metabactrites* and *Kokenia* with the anetoceratids rather than the bactritids as was recommended by Dzik (1984).

F. MIMOCERATIDAE Steinmann, 1890, D. (EMS), Mar.

First: *Gyroceratites laevis* Chlupáč and Turek, 1980, Daleje Shale, lower Emsian, Czechoslovakia.
Last: *Gyroceratites gracilis* Chlupáč and Turek, 1980, Trebetov Limestone, upper Emsian, Czechoslovakia.

Superfamily AGONIATITACEAE Holzapfel, 1899

F. MINAGONIATITIDAE Miller, 1938 D. (EMS)–D. (EIF) Mar.

This family comprises advolute and convolute agoniatitaceans with perforate first whorls.
First: *Mimagoniatites falcistria* Chlupáč and Turek, 1983, lower Emsian, Germany.
Last: *Mimagoniatites bohemicus* Chlupáč and Turek, 1983, lower Eifelian, Czechoslovakia.

F. PARENTITIDAE Bogoslovskiy, 1980 D. (EMS) Mar.

First and Last: *Kimoceras lentiforme* and *Dillerites* sp., Zlichovian, Emsian, former USSR.

F. PINACITIDAE Schindewolf, 1933 (emend.) D. (EIF) Mar.

First and Last: *Pinacites jugleri* Chlupáč and Turek, 1980, Eifelian, Czechoslovakia, Morocco, etc.

F. AGONIATITIDAE Holzapfel, 1899 D. (EMS–GIV) Mar.

This family comprises involute agoniatiaceans with imperforate first whorls.
First: *Paraphyllites tabuloides* Chlupáč and Turek, 1983, Trebetov Limestone, upper Emsian, Czechoslovakia.
Last: *Agoniatites* sp. House, 1963, *terebratum* Zone, mid-Givetian, Trevose Slates, north Cornwall. Similar records in Germany and Morocco.

Superfamily AUGURITACEAE Bogoslovski, 1961

F. AUGURITIDAE Bogoslovski, 1961 D. (EMS) Mar.

First and Last: *Celaeceras praematurum* Chlupáč and Turek, 1980, upper Zlichovian or possibly lower Dalejan, Emsian, Czechoslovakia.

Suborder ANARCESTINA Steinmann, 1890

Superfamily ANARCESTACEAE Steinmann, 1890

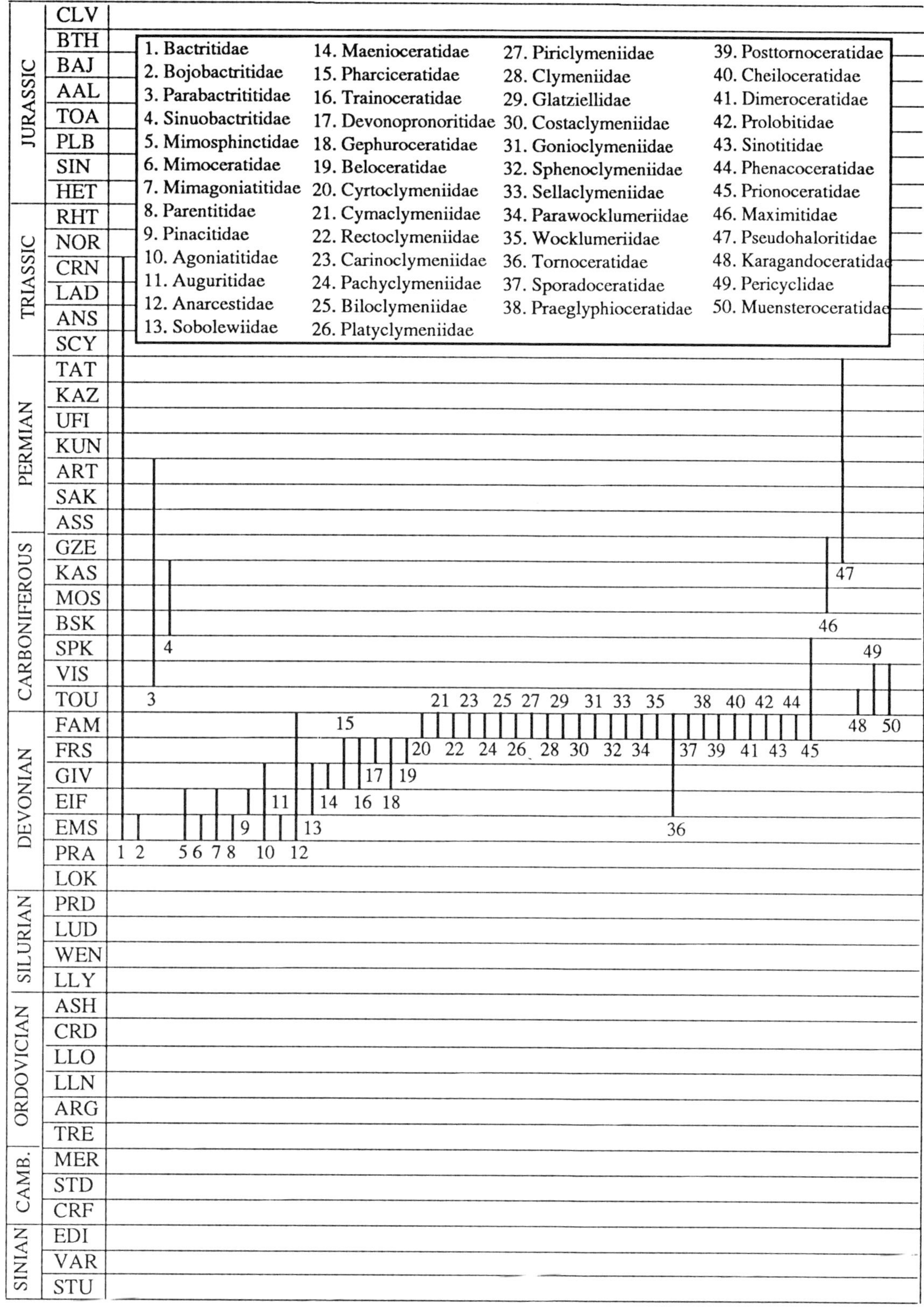

Fig. 10.1

F. ANARCESTIDAE Steinmann, 1890 D. (EMS–FAM) Mar.

First: *Latanarcestes boreus* Bogoslovskiy, 1972, *regularissimus* Zone, lower Dalejan, Emsian, Novaya Zemlya, former USSR.

Last: *Archoceras paeckelmanni* Schindewolf, 1937b, lower Famennian IIα, Nehdenschiefer, Germany.

Intervening: GIV, FRS.

F. SOBOLEWIIDAE House, 1989 D. (EIF–GIV) Mar.

First: *Sobolewia rotella* Petter, 1959, middle Eifelian, Morocco.

Last: *Sobolewia nuciformis* Petter, 1959, middle Givetian, *terebratum* Zone, Morocco.

Superfamily PHARCICERATACEAE Hyatt, 1900

F. MAENIOCERATIDAE Bogoslovskiy, 1958 D. (GIV) Mar.

First: *Maenioceras molarium* House, 1963, *molarium* Zone, lower Givetian, Wolborough Limestone, Devon, UK.
Last: *Maenioceras crassum* Bensaïd, 1974, *tridens* Zone, Givetian, Oued Mzerreb, Morocco.

F. PHARCICERATIDAE Hyatt, 1900 D. (GIV–FRS) Mar.

First: *Pharciceras amplexum* House, 1962, Tully Limestone, Mid. *varcus* conodont Zone, New York State, USA.
Last: *Petteroceras feisti* House *et al.*, 1985, lowest Frasnian (Iα), *lower asymmetricus* conodont Zone, Montagne Noire.

F. TRAINOCERATIDAE Hyatt, 1884 D. (GIV–FRS) Mar.

First: *Tamarites subitus* Bogoslovskiy, 1958, upper Givetian, Tarbagatae Mountains, Kazakhstan, former USSR.
Last: *Schindewolfoceras chemungense* House and Kirchgasser, 1993, Frasnian, Rhinestreet Shale, New York State, USA (House and Kirchgasser, 1993).

F. DEVONOPRONORITIDAE Bogoslovskiy, 1958 D. (FRS) Mar.

First and Last: *Devonopronorites ruzhencevi* Bogoslovskiy, 1969, Frasnian, Rudnyi Altai, former USSR.

Superfamily GEPHUROCERATACEAE Frech 1897

F. GEPHUROCERATIDAE D. (GIV–FRS) Mar.

First: *Pseudoprobeloceras nebechense* Bensaïd, 1974, *lunulicosta* Zone, Givetian (Iα), Morocco, now late Givetian on redefinition of base of Frasnian.
Last: *Crickites holzapfeli* Becker *et al.*, 1989, *holzapfeli* Zone, upper Frasnian (Iδ), Montagne Noire, France.

F. BELOCERATIDAE Hyatt, 1884 D. (FRS) Mar.

First: *Probeloceras lutheri* House and Kirchgasser, 1993, Cashaqua Shale, New York State, USA.
Last: *Beloceras tenuistriatum* Feist *et al.*, 1989, *holzapfeli* Zone, upper Frasnian (Iδ), Montagne Noire, France.

Order CLYMENIIDA Wedekind, 1914

These extraordinary dorsal-siphuncled ammonoids are confined to the Famennian. Their diversity will confuse taxon tots. The classification mostly follows one recently proposed by Korn (1992) and acknowledgement is made of the unpublished thesis of Price (1982). The ranges against the detailed Famennian zonation given here is preliminary.

Suborder CLYMENIINA Hyatt, 1884

Superfamily CYRTOCLYMENIACEAE Hyatt, 1884

F. CYRTOCLYMENIIDAE Hyatt, 1884 D. (FAM) Mar.

First: *Cyrtoclymenia involuta* Wedekind, 1908, *delphinus* Zone, Famennian (IIIβ), Enkeberg, Germany.
Last: *Cyrtoclymenia angustiseptata* Schindewolf, 1937a, *sphaeroides* Zone, Famennian (VI), Hönnetal, Germany.

F. CYMACLYMENIIDAE Hyatt, 1884 D. (FAM) Mar.

First: *Genuclymenia borni* Schindewolf, 1923, *sandbergeri* Zone, Famennian (IIIα), Kirch-Gattendorf, Germany.
Last: *Cymaclymenia evoluta* (= *euryomphala*) Schindewolf, 1937a, *euryomphala* Zone, Famennian (VI), Hönnetal, Germany.

F. RECTOCLYMENIIDAE Schindewolf, 1923 D. (FAM) Mar.

First: *Rectoclymenia roemeri* Wedekind, 1908, *delphinus* Zone, Famennian (IIIβ), Enkeberg, Germany.
Last: *Falciclymenia falcifera* auctt. (Price, 1982, gen. *nov.* E), Brügge, 1973, *annulata* Zone/*Clymenia* Stufe boundary, Thuringia, Germany.

F. CARINOCLYMENIIDAE Bogoslovskiy, 1975 D. (FAM) Mar.

First: *Carinoclymenia beuelense* Lange, 1929, *annulata* Zone, Famennian (IV), Enkeberg, Germany.
Last: *Pinacoclymenia inexpectata* Bogoslovskiy, 1975, *acuticostata* Zone, *Clymenia* Stufe, Famennian (V), Urals, former USSR.

F. PACHYCLYMENIIDAE Korn, 1992 (T. S. Price, 1982) D. (FAM) Mar.

First: *Uraloclymenia nodosa* Bogoslovskiy *in* Simakov (ed.), 1985, upper *annulata* Zone, Famennian (IV), Urals, former USSR.
Last: *Pachyclymenia intermedia* Bogoslovskiy *in* Simakov (ed.) 1985, upper *Clymenia* Stufe, Famennian (V), Urals, former USSR.

F. BILOCLYMENIIDAE Bogoslovskiy, 1955 D. (FAM) Mar.

First: *Borkinia kozlowskii* Czarnocki, 1989, upper *Platyclymenia* Stufe or *Clymenia* Stufe, Famennian (VI or V), Holy Cross Mountains, Poland.
Last: *Dimeroclymenia semicostata* Czarnocki, 1989, *paradoxa* Zone, Famennian (VI), Holy Cross Mountains, Poland.

Superfamily CLYMENIACEAE Edwards, 1849

F. PLATYCLYMENIIDAE Wedekind, 1914 D. (FAM) Mar.

First: *Platyclymenia brevicosta* and others, Wedekind, 1908, *delphinus* Zone, Famennian (IIIβ), Enkeberg, Germany.
Last: *Platyclymenia (Spinoclymenia) aculeata* Bogoslovskiy, *in* Simakov *et al.* 1985, *ornata* Zone, Famennian (V), Mugodzhar, former USSR.

F. PIRICLYMENIIDAE Korn, 1992 (T. S. Price, 1982) D. (FAM) Mar.

First: *Sulcoclymenia sulcata* Schindewolf, 1971, *delphinus* Zone, Famennian (IIIβ), Ebersdorf, Poland (Schindewolf, 1972).
Last: *Piriclymenia piriforme* Schmidt, 1924, *piriforme* Zone, Famennian (V), Frankenberg Germany.

F. CLYMENIIDAE Edwards, 1849 D. (FAM) Mar.

First: *Genuclymenia dunkeri* Lange, 1929 (see Price, 1982), *annulata* Zone, Famennian (IV), Gattendorf, Germany.
Last: *Kosmoclymenia (Lissoclymenia) wocklumeri* Korn and Price, 1987, upper *paradoxa* Zone, Famennian (VI), Germany.

F. GLATZIELLIDAE Schindewolf, 1928
D. (FAM) Mar.

First: *Soliclymenia paradoxa* Schindewolf, 1937a, *subarmata* Zone, Famennian (Vβ), Hönnetal, Germany.
Last: *Postglatziella carinata* Schindewolf, 1937a, *sphaeroides* Zone, Famennian (VI), Hönnetal, Germany.

Superfamily GONIOCLYMENIACEAE Hyatt, 1884

F. COSTACLYMENIIDAE Schindewolf, 1920
D. (FAM) Mar.

First: *Costaclymenia multicostata* Bogoslovskiy, 1981, *annulata* Zone, Famennian (Vα), Urals, former USSR.
Last: *Costaclymenia kiliani* Wedekind, 1914, *acuticostata* Zone, Famennian (Vβ), Dasberg, Germany.

F. GONIOCLYMENIIDAE Hyatt, 1884
D. (FAM) Mar.

First: *Gonioclymenia (Gonioclymenia) speciosa* Price, 1982, *Clymenia* Stufe, Famennian (V), Germany.
Last: *Gonioclymenia (Finiclymenia) wocklumensis* Price and Korn, 1989, upper *paradoxa* Zone, Famennian (VI), Schübelhammer, Germany.

F. SPHENOCLYMENIIDAE Korn, 1992 (T. S. Price, 1982) D. (FAM) Mar.

First: *Sphenoclymenia* sp. nov. Bogoslovskiy, *in* Simakov (ed.) 1985, *acuticostata* Zone, Famennian (V), Urals, former USSR.
Last: *Kalloclymenia brevispina* Schindewolf, 1937a, upper *subarmata* Zone, Famennian (VI), Oberrödinghausen, Germany.

F. SELLACLYMENIIDAE Schindewolf, 1923
D. (FAM) Mar.

First: *Sellaclymenia torleyi* Wedekind, 1914, *annulata* Zone, Famennian (IV), Hoevel, Germany.
Last: *Sellaclymenia plana* Schindewolf, 1923, *Wocklumeria* Stufe, Famennian (VI), Germany.

Superfamily WOCKLUMERIACEAE
Schindewolf, 1937

F. PARAWOCKLUMERIIDAE Schindewolf, 1937
D. (FAM) Mar.

First: *Parawocklumeria paprothae* Clausen *et al.*, 1989, lower *paradoxa* Zone (= *endogona* Zone, Famennian (VI), Müssenberg, Germany.
Last: *Parawocklumeria paradoxa* Schindewolf, 1937a, *sphaeroides* Zone, Famennian (VI), Hönnetal, Germany.

F. WOCKLUMERIIDAE Schindewolf, 1937
D. (FAM) Mar.

First and Last: *Wocklumeria sphaeroides* Schindewolf, 1937a, *sphaeroides* Zone, Famennian (VI), Hönnetal, Germany.

Order GONIATITIDA Hyatt, 1884

Goniatite ranges are based on Bogolovskiy (1971a,b), Wedekind (1917 (1918)), and other sources.

Suborder TORNOCERATINA Wedekind, 1917

Superfamily TORNOCERATACEAE Arthaber, 1911

F. TORNOCERATIDAE Arthaber, 1911
D. (EIF–FAM) Mar.

First: *Parodiceras brachystoma* Petter, 1959, *jugleri* Zone, Eifelian, Algeria.
Last: *Lobotornoceras* (= *Falcitornoceras*) sp. Bartzsch and Weyer, 1986, *sphaeroides* Zone, upper Famennian (VI), Thüringia, Germany.

Superfamily PRAEGLYPHIOCERATACEAE
Ruzhencev, 1957

F. SPORADOCERATIDAE Miller and Furnish, 1957
D. (FAM) Mar.

First: *Maeneceras biferum* and others, Schindewolf, 1923, *Cheiloceras* Stufe, Famennian (IIβ), Germany (also Australia, Morocco, etc.).
Last: *Sporadoceras orbiculare* Schindewolf, 1937a, *sphaeroides* Zone, upper Famennian (VI), Germany.

F. PRAEGLYPHIOCERATIDAE Ruzhentsev, 1957
D. (FAM) Mar.

First: *Lagowites schindewolfi* Matern, 1931, Becker (1990), *Cheiloceras* Stufe, Famennian (IIβ), Germany.
Last: *Praeglyphioceras pseudosphaericum* Lange, 1929, *Platyclymenia* Stufe, Famennian (IV), Germany.

F. POSTTORNOCERATIDAE Bogoslovskiy, 1962
D. (FAM) Mar.

First: *Posttornoceras balvei* Lange, 1929, *annulata* Zone, Famennian (IV), Germany.
Last: *Discoclymenia cucullata* Schindewolf, 1937a, *sphaeroides* Zone, upper Famennian (VI), Hönnetal, Germany.

Superfamily DIMEROCERATACEAE Hyatt, 1884

F. CHEILOCERATIDAE Frech, 1897
D. (FAM) Mar.

First: *Cheiloceras verneuili* Buggisch and Clausen, 1972, upper *Pa. triangularis* conodont Zone, upper Famennian (IIα), Erfoud, Morocco.
Last: *Raymondiceras simplex* House, 1962, *annulata* Zone, Famennian (IV), Three Forks Shale, Montana, USA.

F. DIMEROCERATIDAE Hyatt, 1884
D. (FAM) Mar.

First: Gen. nov. aff. *Dimeroceras petterae* Becker, 1990, *Cheiloceras* Stufe, lower Famennian (IIα), Enkeberg, Germany.
Last: *Dimeroceras mamilliferum* Wedekind, 1908, *delphinus* Zone, Famennian (III), Enkeberg, Germany.

F. PROLOBITIDAE Wedekind, 1913
D. (FAM) Mar.

First: *Prolobites insulcatus* Lange, 1929, *Cheiloceras* Stufe, lower Famennian (IIy), Germany.
Last: *Renites kiensis* Bogoslovskiy, 1969, *Clymenia* Stufe, Famennian (V), Urals, former USSR.

F. SINOTITIDAE Chang, 1960 D. (FAM) Mar.

First: *Sinotites aktubensis* Bogoslovskiy, 1971a, *Cheiloceras* Stufe, Famennian (II), Urals, former USSR.
Last: *Sunites suni* Chang, 1960, Famennian (?III), Great Khingan, China.

?F. PHENACOCERATIDAE Frech, 1902
D. (FAM) Mar.

First: *Clymenoceras insolatum* Schindewolf, 1937b, Famennian (IIIβ), Enkeberg, Germany.
Last: *Cycloclymenia planorbiforme* Schindewolf, 1937a, upper Famennian (IV), Germany.

Superfamily PRIONOCERATACEAE Hyatt, 1884

F. PRIONOCERATIDAE D. (FAM)–C. (ARN) Mar.

First: *Prionoceras divisum* (Münster, 1832), *P. frechi* (Wedekind, 1913), *Platyclymenia* Zone, Rheinisches Schiefergebirge, Germany (Schindewolf, 1923).
Last: *Irinoceras stevanovici* Kullmann, 1962, *Eumorphoceras* Zone, Serbia, former Yugoslavia (Kullmann, 1962).
Comment: Only one group of simple quinquelobate ammonoids, *Imitoceras s.l.*, crossed the Devonian/Carboniferous boundary. At the beginning of the Carboniferous, it produced several distinct evolutionary lineages.

F. MAXIMITIDAE Ruzhentsev, 1960 C. (MOS–GZE) Mar.

First: *Maximites alexanderi* Nassichuk, 1975, Hare Fiord Formation, Atokan, Ellesmere Island, Arctic Archipelago, Canada.
Last: *Maximites cherokeensis* (Miller and Owen, 1939), Ochelata Group, Missourian, Oklahoma, USA (Frest *et al.*, 1981).
Comment: The family is monotypic, and it is characterized by a bifid ventral lobe that is unique within the Cheiloceratаceae (Frest *et al.*, 1981).

F. PSEUDOHALORITIDAE Ruzhentsev, 1957 C. (GZE)–P. (DOR) Mar.

First: *Neoaganides grahamensis* Plummer and Scott, 1937, Missourian, Iowa–Texas, Ohio, USA.
Last: *Neoaganides* sp. 2 (Frest *et al.*, 1981), Ali Bashi Formation, Kuh-e-Ali Bashi, Iran.
Comment: The family is characterized by the position of the siphuncle at maturity, ranging from subcentral to within the dorsal septal flexure. Three component subfamilies are based on the advent and extent of sutural serration, 'goniatitic', 'ceratitic', or 'ammonitic'.

Suborder GONIATITINA Hyatt, 1884

Superfamily KARAGANDOCERATACEAE Librovich, 1957

F. KARAGANDOCERATIDAE Librovich, 1957 C. (TOU) Mar.

First: *Voehringerites peracutus* (Vöhringer, 1960), *Gattendorfia* Zone, Rheinisches Schiefergebirge, Germany (Manger, 1971).
Last: *Karagandoceras* sp., Luton Formation, upper Tournaisian, New South Wales, Australia (Campbell *et al.*, 1983).
Comment: These are the oldest goniatitids with a subdivided ventral lobe; this is an artificial grouping, the relationship is not known.

Superfamily PERICYCLACEAE Hyatt, 1900

F. PERICYCLIDAE Hyatt, 1900 C. (TOU–VIS) Mar.

First: *Goniocyclus blairi* (Miller and Gurley, 1896), Caballero Formation, upper Kinderhookian series, New Mexico, USA (Gordon, 1986).
Last: *Ammonellipsites kochi* (Holzapfel), Erdbacher Kalk, *Pericyclus* Zone (lower VIS), Rheinisches Schiefergebirge, Germany, and other species of pericyclids.
Comment: The mostly prominently sculptured Pericyclidae were strictly limited to the *Pericyclus* Zone (upper TOU and lower VIS) and ended rather abruptly.

F. MUENSTEROCERATIDAE Librovich, 1957 C. (TOU–VIS) Mar.

First: *Muensteroceras ?medium* Miller and Collinson, 1951, Northview Formation, upper Kinderhookian Series, Missouri, USA (Gordon, 1986).
Last: *Bollandites castletonense* (Bisat, 1924), *Beyrichoceras* Zone, England, UK.
Comment: The family comprises forms with smooth shell surface. It is not clear if the first muensteroceratid form is *Intoceras osagense* (Swallow, 1860), Chouteau Limestone, upper Kinderhookian series, Missouri, USA (see below, Girtyoceratidae). If the peculiar genus *Cluthoceras* Currie belongs in the Muensterocaeratidae, the last species would be: *Cluthoceras truemani* Currie, 1954, *Eumorphoceras* Zone, Scotland, UK.

Superfamily NOMISMOCERATACEAE Librovich, 1957

F. NOMISMOCERATIDAE Librovich, 1957 C. (VIS–MRD) Mar. (see Fig. 10.2)

First: *Pseudonomismoceras spiratissimum* (Holzapfel, 1889), Erdbacher Kalk, *Pericyclus* Zone (lower VIS), Rheinisches Schiefergebirge, Germany.
Last: *Baschkirites vasilkovskyi* (Ruzhentsev and Bogoslovskaya, 1978), upper *Reticuloceras* Zone, Central Asia.
Comment: This group is characterized at the beginning by a thin-discoidal and vermicular conch form which is smooth and later with simple or divaricate ribs.

Superfamily DIMORPHOCERATACEAE Hyatt, 1884

F. DIMORPHOCERATIDAE Hyatt, 1884 C. (VIS–YEA) Mar.

First: *Dimorphoceras gilbertsoni* (Phillips, 1836), *Goniatites* Zone, upper VIS, England, UK.
Last: *Metadimorphoceras subdivisum* Manger and Quinn, 1972, lower *Gastrioceras* Zone, Arkansas, USA (Manger, 1988).
Comment: This is the first goniatitid group with denticulate or subdivided ventral or lateral lobes.

F. BERKHOCERATIDAE Librovich, 1957 C. (VIS–ARN) Mar.

First: *Kazakhoceras hawkinsi* (Moore, 1930), *Goniatites* Zone, upper VIS, Republic of Ireland.
Last: *Kazakhoceras hawkinsi* (Moore, 1930), *Eumorphoceras* Zone, England, Spain, former USSR.
Comment: Only one genus known, no descendants.

F. ANTHRACOCERATIDAE Plummer and Scott, 1937 C. (ARN–MOS) Mar.

First: *Anthracoceras paucilobum* (Phillips, 1836), *Eumorphoceras* Zone, England, UK (Bisat, 1934).
Last: *Anthracoceratites vanderbeckei* (Ludwig, 1863), Westphalian B, England, UK (Ramsbottom, 1970).

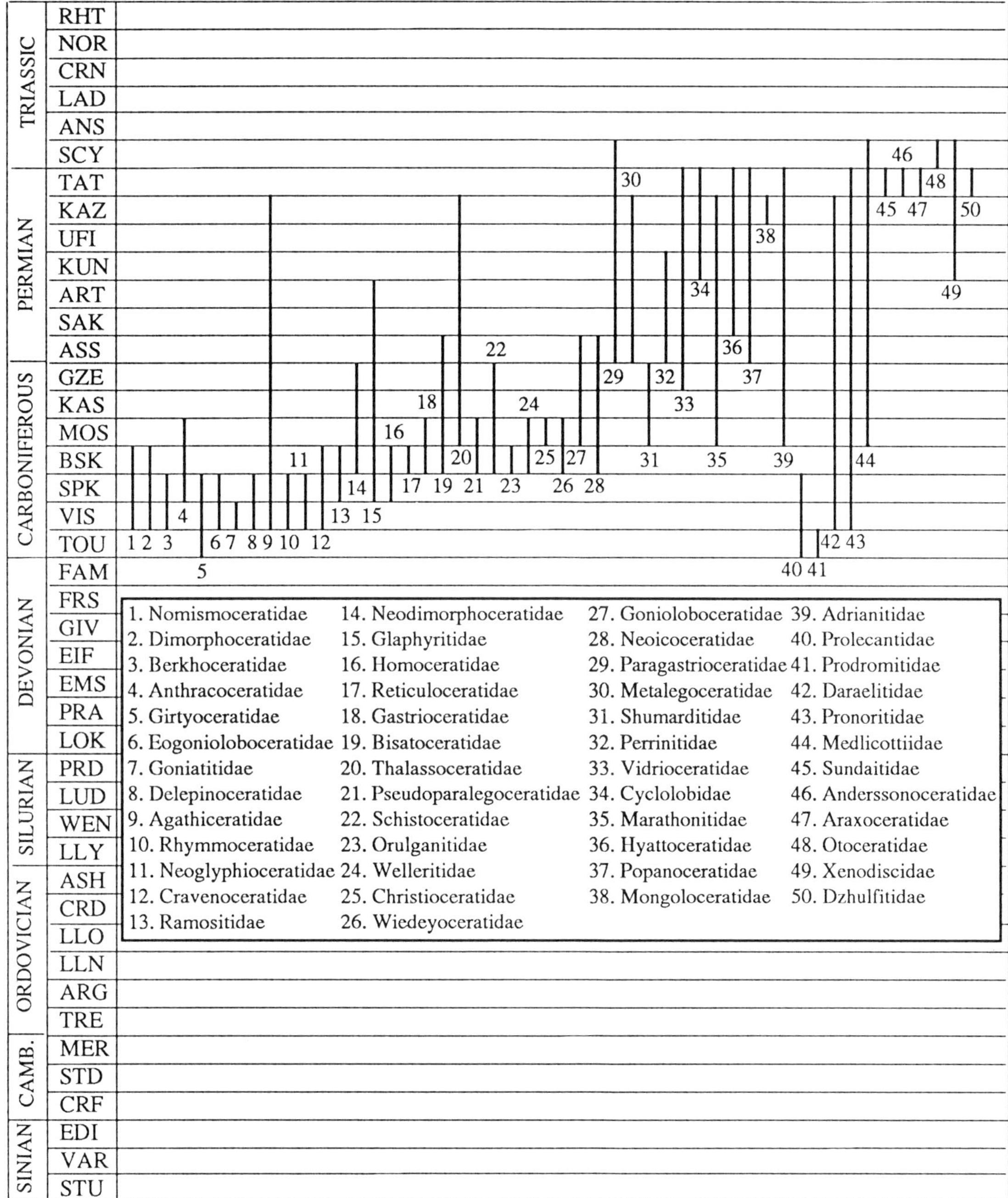

Fig. 10.2

F. GIRTYOCERATIDAE Wedekind, 1918 C. (TOU–ARN) Mar.

First: *Intoceras osagense* (Swallow, 1860), upper Kinderhookian, Missouri, USA.

Last: *Peytonoceras ornatum* Saunders, 1966, Chesterian Series, Arkansas, USA.

Comment: If the assignment of the subfamily Intoceratinae to Muensteroceratidae is correct, then the first appearance of this family would be: *Winchelloceras allei* (Winchell, 1862), Marshall Sandstone, Osagean Series, Michigan, USA.

F. EOGONIOLOBOCERATIDAE Ruzhentsev and Bogoslovskaya, 1978 C. (VIS–ARN) Mar.

First: *Eogonioloboceras asiaticum* (Librovich, 1940), *Goniatites* Zone, upper VIS, Kazakhstan, former USSR.

Last: *Arcanoceras burmai* (Miller and Downs, 1950), *Eumorphoceras* Zone, southern Urals, former USSR.

Superfamily GONIATITACEAE de Haan, 1825

F. GONIATITIDAE de Haan, 1825 C. (VIS) Mar.

First: *Goniatites hudsoni* (Bisat, 1934), Brigantian, *Goniatites* Zone, upper VIS, England, UK.

Last: *Lusitanoceras granosum* (Portlock, 1843), *Goniatites* Zone, upper VIS, Republic of Ireland.

Comment: Older species of *Goniatites* doubtful.

F. DELEPINOCERATIDAE Ruzhentsev, 1957 C. (VIS–ARN) Mar.

First: *Platygoniatites molaris* Ruzhentsev, 1956, *Goniatites* Zone, upper VIS, southern Urals, former USSR.

Last: *Delepinoceras bressoni* Ruzhentsev, 1958, *Eumorphoceras* Zone, southern Urals, former USSR.

F. AGATHICERATIDAE Arthaber, 1911 C. (VIS)–P. (WOR) Mar.

First: *Dombarites poststriatus* (Brüning, 1923), *Goniatites* Zone, Rheinisches Schiefergebirge, Germany.
Last: *Agathiceras suessi* Gemmellaro, 1888, Sosio beds, Sicily, Italy.
Comment: *Dombarites* seems to occur at the same time in many regions. The families Delepinoceratidae and Agathiceratidae show for the first time in the history of the ammonoids an increase of lateral lobes through trifurcation. The latter family persisted until late Permian times without any significant change in shell form.

Superfamily NEOGLYPHIOCERATACEAE Plummer and Scott, 1937

F. RHYMMOCERATIDAE Ruzhentsev and Bogoslovskaya, 1971 C. (VIS–ARN) Mar.

First: *Ophilyroceras tersum* Ruzhentsev and Bogoslovskaya, 1971, *Goniatites* Zone, southern Urals, former USSR.
Last: *Rhymmoceras gracilentum* Ruzhentsev, 1958, *Eumorphoceras* Zone, Serbia, former Yugoslavia.

F. NEOGLYPHIOCERATIDAE Plummer and Scott, 1937 C. (VIS–PND) Mar.

First: *Neoglyphioceras spirale* (Phillips, 1841), *Goniatites* Zone, upper VIS, England, UK.
Last: *Neoglyphioceras litvinovichae* Ruzhentsev and Bogoslovskaya, 1971, *Eumorphoceras* Zone, Kazahkstan, former USSR.
Comment: Numerous short-lived species of Neoglyphioceratins and related genera became extinct at about the same time.

F. CRAVENOCERATIDAE Ruzhentsev, 1957 C. (VIS–KIN) Mar.

First: *Lyrogoniatites eisenbergi* (Ruprecht, 1937), *Goniatites* Zone, Rheinisches Schiefergebirge, Germany.
Last: *Aenigmaticeras rhipaeum* Ruzhentsev and Bogoslovskaya, 1978, *Reticuloceras* Zone, southern Urals, former USSR.

Superfamily NEODIMORPHOCERATACEAE Furnish and Knapp, 1966

F. RAMOSITIDAE Ruzhentsev and Bogoslovskaya, 1971 C. (ARN–YEA) Mar.

First: *Cravenoceratoides edalensis* (Bisat, 1928), *Eumorphoceras* Zone, England, UK.
Last: *Ramosites praesagus* (Ruzhentsev and Bogoslovskaya, 1978), *Gastrioceras* Zone, Middle Asia.

F. NEODIMORPHOCERATIDAE Furnish and Knapp, 1966 C. (BSK–GZE) Mar.

First: *Cymoceras miseri* McCaleb, 1964, Bloyd, upper Morrowan Series, Arkansas, USA.
Last: *Neodimorphoceras (Neodimorphoceras) texanum* (Smith, 1903), Graham Formation, Virgilian series, Texas, USA.
Comment: Tendency of subdividing ventral portion of suture.

Superfamily GASTRIOCERATACEAE Hyatt, 1884

F. GLAPHYRITIDAE Ruzhentsev and Bogoslovskaya, 1971 C. (ARN)–P. (ART) Mar.

First: Various species of *Glaphyrites* and related genera in Europe, Asia and North America.
Last: *Neoshumardites triceps* (Ruzhentsev, 1936), Aktastinian Stage, southern Urals, former USSR.
Comment: This family is closely linked with the Cravenoceratidae and the Gastrioceratidae.

F. HOMOCERATIDAE Spath, 1934 C. (CHO–YEA) Mar.

First: *Homoceras subglobosum* (Bisat, 1924), *Homoceras* Zone, England, UK.
Last: *Umbetoceras aravanense* (Ruzhentsev and Bogoslovskaya, 1978), Middle Asia.

F. RETICULOCERATIDAE Librovich, 1957 C. (KIN–BSH) Mar.

First: *Reticuloceras compressum* (Bisat and Hudson, 1943), *Reticuloceras* Zone, England, UK.
Last: *Melvilloceras sabinense* Nassichuk, 1975, upper Morrowan Series, Melville Island, Arctic Canada.

F. GASTRIOCERATIDAE Hyatt, 1884 C. (MRD–MOS) Mar.

First: *Gastrioceras sigma* Wright, 1926, upper *Reticuloceras* Zone, England, UK.
Last: *Gastrioceras glenisteri* Nassichuk, 1975, Ellesmere Island, Arctic Canada.
Comment: The majority of gastrioceratids appeared for the first time in the Yeadonian Stage, with various species of *Cancelloceras*.

Superfamily THALASSOCERATACEAE Hyatt, 1900

F. BISATOCERATIDAE Miller and Furnish, 1957 C. (YEA)–P. (ASS) Mar.

First: *Bisatoceras hoeni* Nassichuk, 1975, Axel Heiberg Island, lower Morrowan Series, Arctic Canada.
Last: *Neoglaphyrites satrus* (Maximova, 1940), southern Urals, former USSR (Nassichuk, 1975).

F. THALASSOCERATIDAE Hyatt, 1900 C. (MOS)–P. (WOR) Mar.

First: *Eothalassoceras inexpectans* (Miller and Owen, 1937), lower Missourian Series, Oklahoma, USA.
Last: *Epithalassoceras ruzencevi* Miller and Furnish, 1940, Coahuila, Mexico (Miller and Furnish, 1940).
Comment: Tendency of increasing digitation of the lobes, but no increase of sutural elements.

Superfamily SCHISTOCERATACEAE Schmidt, 1929

F. PSEUDOPARALEGOCERATIDAE Librovich, 1957 C. (BSH–MOS) Mar.

First: *Phaneroceras compressum* (Hyatt, 1891), upper Morrowan Series, Oklahoma, USA (McCaleb, 1968).
Last: *Pseudoparalegoceras tzwetaevae* Ruzhentsev, 1951, Podolskian Stage, Kazakhstan, former USSR.

F. SCHISTOCERATIDAE Schmidt, 1929 C. (BSH–GZE) Mar.

First: *Branneroceras branneri* (Smith, 1896), upper Morrowan Series, Arkansas, USA.

Last: *Schistoceras uralense* Ruzhentsev, 1939, Orenburgian Stage, southern Urals, former USSR.
Comment: Additional suture elements in umbilical area.

F. ORULGANITIDAE Ruzhentsev, 1965 C. (BSH) Mar.

First: *Yakutoceras aldanicum* Popov, 1965, Omolon Massif, eastern Siberia, former USSR.
Last: *Kayutoceras triangulare* Ruzhentsev and Ganelin, 1971, Omolon Massif, eastern Siberia, former USSR (Ruzhentsev and Ganelin, 1971).
Comment: These are triangularly coiled ammonoids, restricted to the upper Bashkirian strata of eastern Siberia. This endemic group is thought to be derived from the Schistoceratidae.

F. WELLERITIDAE Plummer and Scott, 1937 C. (BSH–MOS) Mar.

First: *Axinolobus modulus* Gordon, 1961, upper Morrowan Series, Arkansas, USA.
Last: *Wellerites mohri* Plummer and Scott, 1937, Desmoinisian Series, Texas, USA.

F. CHRISTIOCERATIDAE Nassichuk and Furnish, 1965 C. (MOS) Mar.

First and Last: *Christioceras trifurcatum* Nassichuk and Furnish, 1965, Atokan, Arctic Canada and Texas, USA.
Comment: This family is based on *Christioceras* only, with its distinctive trifurcation of all external lobes, probably a descendant of the Schistoceratidae.

Superfamily GONIOLOBOCERATACEAE Spath, 1934

F. WIEDEYOCERATIDAE Ruzhentsev and Bogoslovskaya, 1978 C. (BSH–MOS) Mar.

First: *Donetzoceras donetzense* (Librovich, 1939), C. (BSH), Donbass, former USSR.
Last: *Wiedeyoceras sanctijohanis* Wiedey, 1929, C. (MOS), Desmoinesian Series, Iowa, USA (Furnish and Spinosa, 1968).
Comment: If the peculiar form *Pennoceras seamani* Miller and Unklesbay, 1942 from the Missourian Series of Pennsylvania, USA belongs here, the range of the family ends later (KAS). This genus differs from the Wiedeyoceratidae in its straight, prominent ribbing.

F. GONIOLOBOCERATIDAE Spath, 1934 C. (MOS)–P. (ASS) Mar.

First: *Gonioloboceratoides curvatus* Nassichuk, 1975, Atokan Series, Ellesmere Island, Arctic Canada.
Last: *Mescalites discoidalis* (Böse, 1920), New Mexico, USA (Furnish and Glenister, 1971).

Superfamily NEOICOCERATACEAE Hyatt, 1900

F. NEOICOCERATIDAE Hyatt, 1900 C. (BSK)–P. (ASS) Mar.

First: *Neoicoceras elkhornensis* (Miller and Gurley, 1896), upper Morrowan, Kentucky, USA.
Last: *Eoasianites subhanieli* Ruzhentsev, 1933, southern Urals, former USSR (Furnish and Knapp, 1966).
Comment: This family comprises only three genera, and is probably an artificial grouping of evolute goniatitids.

F. PARAGASTRIOCERATIDAE Ruzhentsev, 1951 P. (ASS)–Tr. (GRI) Mar.

First: *Svetlandoceras* spp., widespread, but documented best from the Urals, former USSR (Ruzhentsev, 1951) and Western Australia (Glenister *et al.*, 1990).
Last: *Pseudogastrioceras* sp., Bed ACT 31 with *Metophiceras* (GRI, 1–2), Meishan, Zheijiang Province and elsewhere, China (Wang, 1984).
Comments: Paragastrioceratids in which growth lines form a ventral salient (Paragastrioceratinae) are restricted to the Lower Permian (ASS–ART, Baigendzhinian Substage), whereas those with a shallow to deep hyponomic sinus (Pseudogastrioceratinae) characterize the Upper Permian (Mikesh *et al.*, 1988).

F. METALEGOCERATIDAE Plummer and Scott, 1937 P. (ASS–WOR) Mar.

First: *Juresanites* spp., locally abundant in the Boreal Realm, but documented best from the Urals, former USSR (Ruzhentsev, 1952), Western Australia and Timor (Glenister *et al.*, 1973).
Last: *Clinolobus telleri* (Gemmellaro, 1887), Sosio Limestone, Sicily, Italy.
Comments: Metalegoceratinae are the most abundant representatives of the family, and characterize the Lower Permian (ASS–ART, Baigendzhinian Substage). Other subfamilies extend through the Lower Permian (ROD). The monotypic Clinolobinae are recognized herein as the youngest Metalegoceratidae, extending as rare elements into the Upper Permian (WOR).

Superfamily SHUMARDITACEAE Plummer and Scott, 1937

F. SHUMARDITIDAE Plummer and Scott, 1937 C. (?MOS–GZE) Mar.

First: *Aktubites trifidus* Ruzhentsev, 1955, ?MOS, southern Urals, former USSR.
Last: *Shumardites* spp., abundant in Orenburgian of the Urals, former USSR (Ruzhentsev, 1950) and Virgilian of Texas, USA (Miller and Downs, 1950).

F. PERRINITIDAE Miller and Furnish, 1940 P. (ASS–ROD) Mar.

First: *Properrinites bakeri* (Plummer and Scott, 1937), Wolfcampian, Texas, New Mexico, Kansas, USA (Tharalson, 1984).
Last: *Perrinites vidriensis* Böse, 1919, Roadian, Texas, Oklahoma, New Mexico, Arizona, USA; Coahuila, Mexico; Venezuela; Colombia; Afghanistan.
Comment: Identity of the shumarditid ancestor of *Properrinites* is uncertain (Tharaldson, 1984). Perrinitids characterize Tethyan faunas, but are virtually absent from the Boreal Realm.

Superfamily CYCLOLOBACEAE Zittel, 1895

This superfamily originated from the marathonitacean family Marathonitidae, and diversified in the Permian to range through that system. Their rapid evolution (especially in the late Permian) and general abundance make them one of the biostratigraphically most useful groups of Permian ammonoids.

F. VIDRIOCERATIDAE Plummer and Scott, 1937 C. (GZE)–P. (DOR) Mar.

First: *Vidrioceras* spp. (the only Carboniferous representative), Virgilian, Texas and Kansas, USA (Plummer and Scott, 1937); Orenburgian, southern Urals, former USSR (Ruzhentsev, 1950).
Last: *Stacheoceras* spp., widespread in uppermost Permian (DOR).
Comment: Total number of lobes increased during phylogenesis from twenty to more than fifty, but denticulation remained confined to lobe bases.

F. CYCLOLOBIDAE Zittel, 1895 P. (ROD–DOR) Mar.

First: The family were derived in ROD from Vidrioceratidae, and diversified rapidly before the late Permian.
Last: *Changsingoceras* spp., DOR, Zheijiang, Szechwan, Anhui, Shanxi and Hunan Provinces, China (Zhao *et al.*, 1978).
Comment: During phylogenesis, the prongs of the ventral lobe became strongly expanded, the number of external lobes across the flanks to umbilical shoulders increased from three to 12 pairs, denticulation of lobes extended almost to the crest of saddles, and the sutural traces became strongly arched.

Superfamily MARATHONITACEAE Ruzhentsev, 1938

F. MARATHONITIDAE Ruzhentsev, 1938 C. (MOS)–P. (WOR) Mar.

First: *'Marathonites'* sp. Chatelain, 1984, Smithwick Shale, Atokan, Mid-continent (Texas), USA.
Last: *Pseudovidrioceras pygmaeum* (Gemmellaro, 1887), Sosio limestone, WOD, Italy (Sicily) (Glenister and Furnish, 1988a).
Comment: The origin of the Marathonitidae in the Atokan is probable, and definite occurrences commence in the overlying Desmoinesian. The family represents a dominant ammonoid stock from the GZE (Missourian) through the ART (Lower Permian).

F. HYATTOCERATIDAE Miller and Furnish, 1957 P. (?SAK–DZH) Mar.

First: *Leeites leei* (Glenister and Furnish, 1987), Riepetown Formation, Nevada, USA.
Last: *Hyattoceras subgeinitzi* Haniel, 1915, Amarassi Beds, Indonesia (Timor) (Glenister and Furnish, 1987). Comparable forms are known from correlative strata, Beihua Mts (Yangtze), southern China (Zhao *et al.*, 1978).
Comment: Common occurrence confined to WOD.

Superfamily POPANOCERATACEAE Hyatt, 1900

F. POPANOCERATIDAE Hyatt, 1900 P. (ASS–DZH) Mar.

First: *Protopopanoceras sublahuseni* (Gerassimov, 1937), platform adjacent to southern Urals, former USSR (Ruzhentsev, 1951).
Last: *Epitauroceras soewarnoi* Glenister and Furnish 1988b, Amarassi Beds, Indonesia (Timor).
Comment: This ancestral family originated in the ASS with one rare, diminutive species. Size, abundance and diversity increased to a maximum in the 'middle' Permian (ART through WOD). Thereafter, the group declined to eventual extinction ending with a rare paedomorphic relic in the basal DZH Stage, substantially before the end of the Permian (Glenister and Furnish, 1988b).

F. MONGOLOCERTIDAE Ruzhentsev and Bogoslovskaya, 1978 P. (GUA:WOR, ?CAP) Mar.

First: *Mongoloceras omanicum* Glenister and Furnish, 1988b, base of Hamrat Dura Group, WOD, Ba'ad area, Oman.
Last: *?Angrenoceras langcuoense* Sheng, 1988, Langcuo Formation, ?CAP, Angren District, Xizang, China.
Comment: This family is rare; morphology and stratigraphical occurrence are both poorly known. Its early Permian history, after presumed divergence from primitive popanoceratids, is unknown (Glenister and Furnish, 1988b).

Superfamily ADRIANITACEAE Schindewolf, 1931

F. ADRIANITIDAE Schindewolf, 1931 C. (MOS)–P. (DZH) Mar.

First: *Dunbarites n. sp.* Chatelain, 1984, Desmoinesian, Wewoka Formation, Oklahoma, USA.
Last: *Epadrianites timorensis* (Boehm, 1908) and *E. involutus* (Haniel, 1915), Amarassi Beds, Indonesia (Timor).
Comment: A gradation in conch form, sculpture and sutural complexity exists between virtually all adrianitids, and this major complex constitutes the Adrianitinae (Upper Carboniferous, GZE (Missourian)–Upper Permian (DZH)). Rare monotypic extremes in conch form are recognized as the primitive Dunbaritinae (Upper Carboniferous, MOS–GZE (Desmoinesian–Virgilian)) and advanced Hoddmanniinae (Upper Permian, WOR), and rounded lobes with parallel sides characterize the Texoceratinae (Lower Permian, ROD). The family is normally rare, with abundance and diversity greatest in GUA (WOR). Superficial homeomorphs exist in some associated taxa, especially Agathiceratidae, but represent a distinctive separate lineage.

Order PROLECANITIDA Miller and Furnish, 1954

Superfamily PROLECANITACEAE Hyatt, 1884

F. PROLECANITIDAE Hyatt, 1884 C. (TOU–ARN) Mar.

First: *Eocanites nodosus* (Schmidt, 1925), *Gattendorfia* Zone, Rheinisches Schiefergebirge, Germany.
Last: *Metacanites chancharensis* (Ruzhentsev, 1949), *Eumorphoceras* Zone, Kazakhstan, former USSR.
Comment: Tendency of increase of lateral suture elements.

F. PRODROMITIDAE Arthaber, 1911 C. (TOU) Mar.

First: *Eoprodromites kinderhooki* Work *et al.*, 1988, Kinderhookian Series, Missouri, USA (Work *et al.*, 1988).
Last: *Acrocanites tornacensis* Delepine, 1940, *Pericyclus* Zone, Belgium (Weyer, 1972).
Comment: This family comprises multilobate forms, some of them exhibiting serrate lateral lobes.

F. DARAELITIDAE Tchernov, 1907 C. (VIS)–P. (WOR) Mar.

First: *Praedaraelites culmiensis* (Kobold), *Goniatites* Zone, upper VIS, Rheinisches Schiefergebirge, Germany.

Last: *Daraelites meeki* Gemmelaro, 1888, Sosio Beds, Sicily, Italy.
Comment: Digitate lateral suture elements.

Superfamily MEDLICOTTIACEAE Karpinsky, 1889

F. PRONORITIDAE Frech, 1901 C. (VIS)–P. (DZH) Mar. (includes the SHIKHANITIDAE Ruzhentsev, 1951)

First: *Pronorites cyclolobus* (Phillips, 1836), *Goniatites* Zone, upper VIS, England (Kullmann, 1963).
Last: gen. and sp. *nov.*, Glenister *et al.*, in prep., Amarassi Beds, Timor, Indonesia. This genus is interpreted as a terminal pronoritid paedomorph, characterized by small size, sutural simplification and low abundance, and conforming to the pattern displayed by many family-level Upper Palaeozoic ammonoid taxa (Glenister and Furnish, 1988a,b).
Comment: Ancestral medlicottaceans characterized by moderately evolute conch and relatively simple sutures.

F. MEDLICOTTIIDAE Karpinsky, 1889 C. (KAS)–Tr.(NML) Mar.

First: *Prouddenites* n. sp., Chatelain, 1984, Wewoka Formation, Desmoinesian Series, Oklahoma, USA; next oldest is *Proudennites primus* Miller, 1930, Missourian Series, Missouri, USA.
Last: *Episageceras noetlingi* Haniel, 1915, with *Anasibirites* (NML, 4), Timor (Shevyrev, 1986); *Latisageceras* Ruzhentsev, 1956 is known from GRI of Pakistan (Salt Range) and possibly Indonesia (Timor), with *Episageceras* spp.
Comments: This family comprises advanced medlicottaceans characterized by involute conch and complex sutures. Each suture comprises a total of 13 to approximately 30 lobes. The ventral prong of the broad, primary, lateral lobe was transformed during phylembryogenesis into a progressively more complexly subdivided ventrolateral saddle. The family comprises Uddenitinae (MOS (Desmoinesian)–ART (Baigendzhinian Substage)), Medlicottiinae (GZE (Virgilian)–DZH), Sicanitinae (ASS–WOR), Propinacoceratinae (ASS–DZH), and Episageceratinae (DZH–NML). An expanded Episageceratidae Ruzhentsev, 1956 (ASS–NML, 4) was recognized as an alternative by Shevyrev (1986).

F. SUNDAITIDAE Ruzhentsev, 1957 P. (DZH) Mar.

First and Last: This is a monotypic family, with undoubted representatives known only from the Amarassi Beds, Timor, Indonesia.

Order CERATITIDA Hyatt, 1884

Suborder OTOCERATINA Shevyrev and Ermakova, 1979

The affinities of the Permian 'Ceratitida' were reviewed by Spinosa *et al.* (1975), Zhao *et al.* (1978), Shevyrev and Yermakova (1978), Shevyrev (1986) and Zakharov (1988). If the base of the Triassic continues to be defined as the base of the *Otoceras woodwardi* ammonoid zone, then the Otoceratina survived the late Permian interval of extinction to extend into the earliest Triassic (GRI).

F. ANDERSSONOCERATIDAE Ruzhentsev, 1959 P. (DZH) Mar.

First and Last: *Anderssonoceras anfuense* Grabau, 1924, *Pericarinoceras robustum* Zhao and Liang, 1966, and other species, 20-m-thick shale and calcareous sandstone unit 2b, *Anderssonoceras–Protoceras* zone, lower Loping Series, Jiangxi, China (Zhao *et al.*, 1978).

F. ARAXOCERATIDAE Ruzhentsev, 1959 P. (DZH) Mar.

First: *Protoceras complanatum* Zhao *et al.*, 1978 and other species, with *Araxoceras*, 20-m-thick shale and sandstone unit 2b, *Anderssonoceras–Protoceras* Zone, lower Loping Series, Jiangxi, China (Zhao *et al.*, 1978).
Last: *Sanyangites tricarinatus* Zhao *et al.*, 1978, *Sanyangites* Zone, upper Loping Series, unit 2d 100 m above 2b, Jiangxi, China (Zhao *et al.*, 1978).

F. OTOCERATIDAE Hyatt, 1900 Tr. (GRI) Mar.

First: *Otoceras concavum* Tozer, 1967, 10 m above base, Blind Fjord Group (GRI, 1), Axel Heiberg Island, and also Wordie Creek Formation, East Greenland (Tozer, 1965; Trümpy, 1969).
Last: *Anotoceras nala* (Diener, 1897), Bed JSQ 7b, Kangshare Formation, *Sakuntala* Zone (GRI, 3), Selong, southern Tibet (Wang and He, 1980).
Comment: There is little difference between this family and the two ancestral families postulated by Zhao *et al.* (1978). All or part of the overlying CHX stage is probably contemporaneous with the boreal facies which traditionally defined the base of the Triassic (Wang, 1984).

Suborder PARACELTITINA Shevyrev, 1968

F. XENODISCIDAE Frech, 1902 P. (ROD)–?Tr. (GRI) Mar.

First: *Paraceltites elegans* Girty 1908, lower Permian (Roadian), Cutoff Formation, USA (Texas, New Mexico), Spinosa *et al.* (1975).
Last: *Xenodiscus chaotianus* Zhao *et al.*, 1978, upper Chaotian Member, Dalong Formation (CHX = ?GRI), Chaotian, Szechwan, China. *Tapashanites changxingense* Zhao *et al.*, 1978, lower Baojing Member, Changxing Formation, Meishan, Zhejiang Province (Zhao *et al.*, 1978).
Comment: Tozer (1981a), unlike Shevyrev (1986), has maintained the formal link between this family and their descendants the Kashmiritidae.

F. DZHULFITIDAE Shevyrev, 1965 P. (DOR) Mar. (see Fig. 10.3)

First and Last: *Dzhulfites nodosus* Shevyrev, 1965, *Paratirolites trapezoidalis* Shevyrev, 1965, *Abichites stoyanowi* (Kiparisova, 1947), Dzhulfa region, Dorasham Formation (LGT), former USSR (Shevyrev, 1986).

F. PSEUDOTIROLITIDAE Zhao, 1965 ?P. (CHX)–?Tr. (GRI) Mar.

First: *Pseudotirolites dapuense* (Zhao *et al.*, 1978), lower Baojing Member, Changxing Formation (CHX = ?LGT), Meishan, Zhejiang Province, China.
Last: ?*Pleuronodosus* sp., top of Dalong Formation (ACT 915–4, below GRI, 3, Wang, 1984). *Rotodiscoceras asiaticum* Zhao and Liang, 1966, upper Meishan Member, Changxing Formation (CHX = GRI), Meishan, China (Zhao *et al.*, 1978).

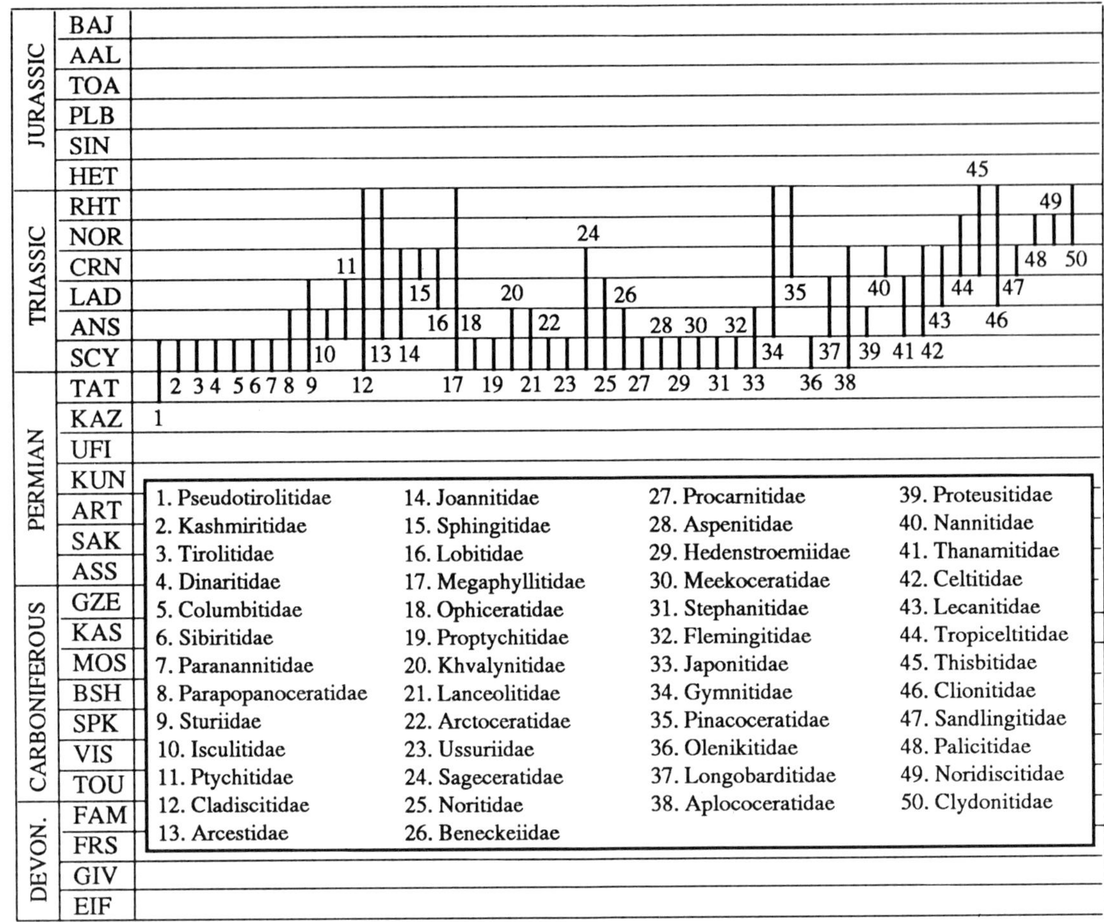

Fig. 10.3

F. KASHMIRITIDAE Spath, 1930 Tr. (GRI–SPA) Mar.

First: *Metophiceras trivale* Spath, 1930, Wordie Creek Formation, *Concavum* Zone (GRI, 1), East Greenland (Trümpy, 1969).
Last: *Hemilecanites* cf. *paradiscus* Kummel, 1969, *Welteri* Beds, Prida Formation (SPA, 3b), Star Creek Canyon, Nevada (Bucher, 1989).

F. TIROLITIDAE Mojsisovics, 1882 Tr. (NML–SPA) Mar.

First: *Tirolites injucundus* Krafft and Diener, 1909, Spiti (NML, 3–4?), Nepal (Wang, 1985). *T. cassianus* (Quenstedt, 1849) is SPA, 1 index (Krystyn, 1974).
Last: *Tirolites carniolicus* Mojsisovics, 1882 and *T. mangyshlakensis* (Shevyrev, 1968), 10 m below Gutensteinerdolomit, Campill Beds, Muć (SPA, 2), former Yugoslavia (Krystyn, 1974).

F. DINARITIDAE Mojsisovics, 1882 Tr. (SPA) Mar.

First: *Dorikranites bogdoanus* (Buch, 1831) and *D. discoides* (Astakhova, 1960), *Dorikranites* Beds (SPA, 1), Mangyshlak, eastern Caspian Sea. *Dinarites* cf. *dalmatinus* (Hauer, 1865), 20 m up the Campill Beds (SPA, 1), Muć, former Yugoslavia (Krystyn, 1974; Dagys *et al.*, 1979).
Last: *Stacheites concavus* Shevyrev, 1968, *Stacheites* Beds (SPA, 3?), Mangyshlak. *Dorikranites* sp., Unit C (SPA, 3?), Kialingkiang Group, Weiyuan, Szechwan Province, China (Wang, 1985).
Comment: Shevyrev (1986) separated a monogeneric Dorikranitidae Astachova, 1960, but indicated that it is most closely related to this family. Tozer (1981a) subordinated it to the Tirolitidae.

F. COLUMBITIDAE Spath, 1934 Tr. (SPA) Mar.

First: *Columbites parisianus* Hyatt and Smith, 1905, middle shale unit, Thaynes Formation (SPA, 1), Nevada (Silberling and Tozer, 1968).
Last: *Epiceltitoides epiceltitoides* Guex, 1978, Bed 21, *Pakistanum* Zone (SPA, 2), Landu, Salt Range, Pakistan (Guex, 1978).

F. SIBIRITIDAE Mojsisovics, 1896 Tr. (NML–SPA) Mar.

First: *Kazakhstanites pilatoides* Guex, 1978, Bed 6d at Zaluch, *Pluriformis* Zone (NML, 4), Salt Range, Pakistan. *Parasibirites grambergi* (Popov, 1961), lower *Spiniplicatus* Zone (SPA, 1), Lena–Anabar region, former USSR (Dagys *et al.*, 1979, p. 38; Guex, 1978).
Last: *Kazakhstanites dolnapensis* Shevyrev, 1968, Mangyshlak (SPA, 2?) and *Sibirites eichwaldi* (Keyserling, 1845), upper *Spiniplicatus* Zone (SPA, 2), Lena–Anabar region, former USSR (Dagys *et al.*, 1979, p. 39).
Comment: Shevyrev (1986) accepts that the monogeneric Kazakhstanitidae Shevyrev, 1986 are most closely related to

this family and restricts the range to his *Parisianus* and *McKelvei* Zones (SPA, 2–3?).

Suborder ARCESTINA Hyatt, 1884, *s.l.*

F. PARANANNITIDAE Spath, 1930 Tr. (NML–SPA) Mar.

First: Dwarf *Melagathiceras crassus* (Tozer, 1961), *Juvenites canadensis* Tozer, 1961 and *Paranannites spathi* (Frebold, 1930), 0.6-m-thick *Meekoceras bed of Romunderi* Zone (NML, 3), Blind Fiord Group, Ellesmere Island (Tozer, 1965, 1984).
Last: *Isculitoides* sp. A of Bucher (1989), upper *Haugi* Zone (SPA, 3a), Prida Formation, northern Humboldt Range, Nevada, USA.
Comment: The dwarf and cosmopolitan Melagathiceratidae Tozer, 1971 are sometimes associated with one of the largest Triassic ammonoids, the 550-mm-diameter *Hedenstroemia*, and has the same limited stratigraphical range (NML, 3). Shevyrev (1986) interprets it as the ancestral form of this family. Actually it has the same first record and may be considered as a minor initial variant of a stock which produced the Arcestina *s.s.*

F. PARAPOPANOCERATIDAE Tozer, 1971 Tr. (SPA–ANS) Mar.

First: *Prosphingites czekanowski* Mosisovics, 1886, 8-m-thick upper *Spiniplicatus* Zone (SPA, 2?), Lena–Anabar region, former USSR (Dagys *et al.*, 1979, p. 49; Arkadiev and Vavilov, 1984). A more morphological definition of the family yields a slightly later first record: *Stenopopanoceras karangatiense* (Popov, 1968) and *S. mirabile* Popov, 1961, *Evolutus* Subzone (ANS, 1), Lena–Anabar, Siberia, former USSR.
Last: *Amphipopanoceras dzeginense* (Voinova, 1961), *Humboldtensis* Zone (ANS, 8), Kolyma Basin, Siberia (Dagys *et al.*, 1979, p. 72).

F. STURIIDAE Kiparisova, 1958 Tr. (SPA–LAD) Mar.

First: *Ziyunites ziyunensis* Wang, 1978, 0.2-m-thick Ziyun limestone (SPA, 2 or condensed), Guizhou Province, China. *Ziyunites asseretoi* Fantini Sestini, 1981 and *Sturia* sp., Bed T329 Chios, Greece (SPA, 3-ANS, 2?); *Psilosturia* and *Eosturia* in rocks of similar age elsewhere (Wang, 1985; Shevyrev, 1986; Bucher, 1989).
Last: *Sturia karpinskyi* Mojsisovics, 1902; the CRN of Alps record is not verified by modern stratigraphy (Tozer, 1981a). *Sturia semiarata* Mojsisovics, 1882, Unit 5 with LAD, 4 *Doleriticus* fauna, upper Laibuxi Formation, Tulong, Tibet (Wang and He, 1980, upper '*Protrachyceras–Joannites* bed').

F. ISCULITIDAE Spath, 1951 Tr. (ANS) Mar.

First: *Isculites meeki* (Hyatt and Smith, 1905), *Caurus* Zone (ANS, 2), Prida Formation, Humboldt Range, Nevada, USA (Silberling and Nichols, 1982).
Last: *Isculites asseretoi* Fantini Sestini, 1988, upper *Ismidicum* Zone (ANS, 5), Nodular Limestone of Gebze, Turkey (Fantini Sestini, 1988).

F. PTYCHITIDAE Mojsisovics, 1882 Tr. (ANS–LAD) Mar.

First: *Malletoptychites kotschetkovi* Popov, 1961, *Decipiens* Subzone (ANS, 5), Lena–Anabar Region, former USSR (Dagys *et al.*, 1979, p. 40).
Last: *Aristoptychites kolymaensis* (Kiparisova, 1937), *Mcconnelli* Subzone (LAD 4–5), Omulevsk region (Dagys *et al.*, 1979, p. 93).

F. CLADISCITIDAE Zittel, 1884 Tr. (SPA–RHT) Mar.

First: *Procladiscites simplex* Wang, 1978, 0.2-m-thick Ziyun Limestone (SPA, 2 or later), Guizhou Province. *Procladiscites* sp. of *Procladiscites* beds and Bed T329 Chios, Greece (Wang, 1985; Bucher, 1989 imply SPA).
Last: *Cladiscites* sp. and *Procladiscites* sp., middle member, Gabbs Formation (RHT 3a), New York Canyon, Nevada (Silberling and Tozer 1968, Tozer, 1980). *Cladiscites* sp., *Marshi* Zone (RHT 3), Koessen Beds, Austria (Krystyn, 1973).

F. ARCESTIDAE Mojsisovics, 1875 Tr. (ANS–RHT) Mar.

First: *Proarcestes* from *Osmani* Zone (ANS, 3) implied by Shevyrev (1986). *Proarcestes bramantel* (Mojsisovics, 1869), upper *Ismidicum* Zone (ANS, 5), Nodular Limestone of Gebze, Turkey (Fantini Sestini, 1988).
Last: *Arcestes* sp., upper member, Gabbs Formation (RHT 3b), Nevada. *A. rhaeticus*, Clark, 1887 and *A. tenuis* Pompeckj, 1895, Koessen Beds (RHT, 3b?), Austria. *Stenarcestes polysphinctus* (Mojsisovics, 1875), Koessen Beds (RHT, 3a?), Austria (Silberling and Tozer, 1968; Krystyn, 1973; Tozer, 1980).

F. JOANNITIDAE Mojsisovics, 1882 Tr. (ANS–CRN) Mar.

First: *Joannites compressus* Wang and He, 1976, Bed JSB 83b, *Rugifer* Zone (ANS, 7), Laibuxi Formation, Tulong, southern Tibet (Wang and He, 1980).
Last: *Joannites cymbiformis* Wulfen, 1793, *Austriacum* Subzone (CRN, 2a), Feuerkogel, Austria (Krystyn, 1973) and the *Sirenites nanseni* Zone of China.

F. SPHINGITIDAE Arthaber, 1911 Tr. (CRN) Mar.

First and Last: *Sphingites coangustatus* (Hauer, 1860) and spp., Bed 70/78 in lower *Anatropites Bereich*, Feuerkogel (CRN, 5), Austria (Krystyn, 1973). All the typical species of the genus are described as CRN in the old literature and it is not clear why this age was revised to LAD–?CRN by Tozer (1981a) and LAD, 3–CRN, 1 by Shevyrev (1986).

F. LOBITIDAE Mojsisovics, 1875 Tr. (LAD–CRN) Mar.

First: *Lobites paceanus* McLearn, 1937, *Meginae* Zone (LAD, 3), Peace River, British Columbia, Canada (Tozer, 1981b).
Last: *Coroceras* cf. *suessi* (Mojsisovics, 1875), *Sirenites nanseni* Zone (CRN, 2), Kupreanof Island, Alaska, USA (Silberling and Tozer, 1968).

F. MEGAPHYLLITIDAE Mojsisovics, 1896 (SPA–RHT) Mar.

First: *Megaphyllites sandalinus* Mojsisovics, 1882, *Qinghaiensis* Subzone (ANS, 1 of Wang, SPA, 3 of Bucher who rejects the ANS taxa of Wang), Naocangjiango Formation, Qinghai Province, China. *M. chiosensis* Fantini Sestini, 1981, Bed T329, Chios, Greece (Wang, 1985).
Last: *Megaphyllites robustus* Wiedmann, 1973, Zlambach Beds, Grünbachgraben, Austria. *M. boehmi* Pompeckji, 1895 and *M. insectus* Mojsisovics, 1873, Koessen Beds and White Crinoidal Limestone of Steinbergkogel, Austria. Shevyrev

(1986) implies that *Megaphyllites* died out later than the index of the *Marshi* Zone (RHT 3b).

Comment: Shevyrev (1986) and other authors have regarded this family and the Lobitidae as separate problematic suborders with an affinity with the Arcestina. Previous comparisons with *Nathorsites* and *Parapopanoceras* are plausibly rejected (Arkadiev and Vavilov, 1984). The derivation of the general morphology of *Megaphyllites* from *Procarnites* (Wang, 1985) is opposed by Shevyrev (1986), who favours an origin in the early Paranannitidae.

The next suborder (Sageceratina) was independently derived from the Medlicottiaceae by Zacharov (1988), but is united here with the Ceratitida as suggested by Shevyrev (1986).

Suborder SAGECERATINA Shevyrev, 1983 *s.l.*

F. OPHICERATIDAE Arthaber, 1911 Tr. (GRI–SPA) Mar.

First: *Ophiceras (O.)* aff. *transitorum* Spath, 1930 and *O. (Lytophiceras) spathi* Trümpy, 1969, *Subdemissum* Zone (GRI, 2b), Kumait, East Greenland. *O. (L.) sukuntala* Diener, 1897, Bed 52b at Guryul, lower *Woodwardi* Zone (GRI 2?), Kashmir (Bando, 1981).

Last: *Nordophiceras schmidti* Mojsisovics, 1886 and *N.* sp., upper *Spiniplicatus* Zone (SPA, 2?), Lena–Anabar region (Dagys *et al.*, 1979, p. 49).

F. PROPTYCHITIDAE Waagen, 1895 Tr. (GRI–SPA) Mar.

First: *Proptychites* sp. of Bando (1981), Bed 56 at Guryul, middle *Woodwardi* Zone (GRI, 2?), Kashmir. *P. strigatus* Tozer, 1961 and *P. rosenkrantzi* Spath, 1930, 30 m above *Boreale* Zone, Blind Fiord Group (GRI, 4), Griesbach Creek, Axel Heiberg Island (Tozer, 1965).

Last: *Proptychitoides decipiens* Spath, 1930, *Subcolumbites* Beds, Kčira (SPA, 2), Albania. *Paranoritoides paranoritoides* Guex, 1978, Bed 22 at Nammal (SPA, 1), Salt Range, Pakistan (Guex, 1978).

Comment: This family links the Sageceratina *sensu* Shevyrev (1986) with the ancestral Ophioceratidae in his phylogenetic interpretation.

F. KHVALYNITIDAE Shevyrev, 1968 Tr. (SPA–ANS) Mar.

First: *Khvalynites mangyshlakensis* Shevyrev, 1968, *Parisianus* Zone (SPA, 2), Mangyshlak, Transcaucasia (Shevyrev, 1986).

Last: *Ismidites marmarensis* Arthaber, 1914, upper *Ismidicum* Zone (ANS, 5), Nodular Limestone of Gebze, Turkey (Fantini Sestini, 1988).

F. LANCEOLITIDAE Spath, 1934 Tr. (NML–ANS) Mar.

First: *Lanceolites compactus* Hyatt and Smith, 1905, *Gracilitatus* Zone (NML, 3), California, USA (Silberling and Tozer, 1968).

Last: *Metadagnoceras youngi* Bucher, 1989, *Mulleri* Zone (ANS, 1), Coyote Canyon, Nevada, USA (Bucher, 1989). *Lanceolites* sp., Campill Beds (SPA, 2), Bulgaria (Wang, 1985).

Comment: The septal sutures of *Metadagnoceras* and *Dagnoceras* link them to the more involute and compressed Lanceolitidae (NML, 3-SPA, 3) of Shevyrev (1986).

F. ARCTOCERATIDAE Arthaber, 1911 Tr. (NML) Mar.

First and Last: *Arctoceras polare* (Mojsisovics, 1886), Fish Beds, Sassendal (NML, 3), Spitsbergen. *A. mushbachanum* (White, 1879), USA and Siberia, also correlated with *Romunderi* Zone (Dagys *et al.*, 1979).

F. USSURIIDAE Spath, 1930 Tr. (NML–SPA) Mar.

First: *Metussuria waageni* (Hyatt and Smith, 1905), *Gracilitatus* Zone (NML, 3), Idaho and Utah, USA (Silberling and Tozer, 1968).

Last: *Parussuria latilobata* Zhao, 1959, *Columbites asymmetricus* Zone (SPA, 1), Linglo, Guangxi Province, China (Wang and He, 1980). Shevyrev (1986) only accepted NML, 3.

F. SAGECERATIDAE Hyatt, 1884 Tr. (NML–CRN) Mar.

First: *Pseudosageceras multilobatum* Noetling, 1905, Lower Limestone and Ceratite Marls, Chideru (NML, 1?), Salt Range, Pakistan. Modern records are later and include *Sverdrupi* Zone (NML, 2), Blind Fiord Group, Axel Heiberg Island (Tozer, 1965) and Bed JSB 15 at Tulong (NML, 3) in southern Tibet (Wang and He, 1980).

Last: *Sageceras haidingeri* (Hauer, 1847), Bed 70/6 at Feuerkogel, *Austriacum* Subzone (CRN, 2a), Austria (Krystyn, 1973).

F. NORITIDAE Karpinsky, 1889 Tr. (SPA–LAD) Mar.

First: *Metahedenstroemia kastriotoe* (Arthaber, 1911) and ?*Beatites berthoe* Arthaber, 1911, *Subcolumbites* Beds (SPA, 2), Kčira, Albania.

Last: *Norites subcarinatus* (Hauer, 1887), *Avisianum* Zone (ANS, 9), Zgorigrad, Bulgaria, and allies ranging up to Bed 162 of Grenzbituminenzone (LAD, 1), Monte San Giorgio, Switzerland (Rieber, 1973; Parnes, 1986). *Neoclypites desertorum* Johnson, 1941, New Pass, Nevada and *Neoclypites*? *perigrinus* Müller, 1973, Grenzdolomit, Reisdorf (CRN, 1), were accepted as isolated records by Shevyrev (1986). They are reinterpreted as a later offshoot of the *Sageceras* here.

F. BENECKEIIDAE Waagen, 1895 Tr. (SPA–ANS) Mar.

First: *Beneckia buchi* (Alberti, 1849), *Magnus* Zone and underlying platform facies (SPA–ANS, 2) briefly noted by Wang (1985). *Beneckia tenuis* (Seebach, 1857), Lower *Röth*, Jena, Germany (ANS, 4? of Parnes, 1986).

Last: *Beneckeia levantina* Parnes, 1962, middle Gevanim Formation, Makhlesh Ramon, Israel (ANS, 6–7? of Parnes, 1986).

Comment: This rather unsatisfactory monogeneric family could be a polyphyletic group of the Sageceratidae in a marginal marine facies.

F. PROCARNITIDAE Zhao, 1959 Tr. (SPA) Mar.

First: *Procarnites oxynostus* Zhao, 1959, *Columbites asymmetricus* Zone (SPA, 1), Linglo, Guangxi Province, China (Wang and He, 1980).

Last: *Neopopanoceras haugi* (Hyatt and Smith, 1905), upper *Haugi* Zone (SPA, 3a), Union Wash and Prida Formations, Nevada (Bucher, 1989).

F. ASPENITIDAE Spath, 1934 Tr. (NML) Mar.

First and Last: *Aspenites acutus* Hyatt and Smith, 1905, *Gracilitatus* Zone (NML, 3), Aspen Mountains, Idaho, USA and elsewhere (Silberling and Tozer, 1968).
Comment: The tentative extension of the range to SPA by Shevyrev (1986) was based on *Beatites*: here transferred to the Noritidae.

F. HEDENSTROEMIIDAE Waagen, 1895 Tr. (NML) Mar.

First: ?*Clypeoceras crassum* (Krafft, 1909), Bed JSB 10 of Kamshare Formation, *Psilogyrus* Zone (NML, 1), Tulong, southern Tibet (Wang and He, 1980; Shevyrev, 1986).
Last: *Hedenstroemia hedenstroemi* (Keyserling, 1845), *Hedenstroemi* Zone (NML, 3), Verkhoyansk (Dagys *et al.*, 1970, p. 27). Tozer (1981a) plausibly restricted the range of the family.

Suborder PINACOCERATINA Waagen, 1895 *s.l.*

F. MEEKOCERATIDAE Waagen, 1895 Tr. (NML) Mar.

First: *Gyronites plicatilis* (Waagen, 1895), Bed 6b, upper Lower Ceratite Limestone, defines base of NML, 1, Nammal, Pakistan (Guex, 1978).
Last: ?*Proavites hueffeli* Arthaber, 1896, Reiflingkalk, Stavljan (ANS, 6), Bosnia, former Yugoslavia (Shevyrev, 1986). *Wasatchites spinatus* Guex, 1978, Bed 20, Upper Ceratite Limestone, Nammal (NML, 4), Pakistan (Guex, 1978).
Comment: The Meekoceratidae *s.s.* (NML, 1–3) appear to continue as the Prionitidae Hyatt, 1900 (NML, 3–4) and closely allied Inyoitidae Spath, 1934 (NML, 3), rather than the stratigraphically isolated *Proavites*. Shevyrev (1986) derived *Inyoites* from the Prionitidae, but it could be a variant of the frequently associated *Anasibirites*.

F. STEPHANITIDAE Arthaber, 1896 Tr. (NML) Mar.

First and Last: *Stephanites superbus* Waagen, 1895, Upper Ceratite Limestone, Childroo, and *S. corona* Waagen, 1895 with *Wasatchites* in Bed 7 at Zaluch (NML, 4), Upper Ceratite Limestone, Salt Range, Pakistan (Guex, 1978). *Amphistephanites parisensis* (Zacharov, 1968), NML, 4 of Shevyrev (1986).

F. FLEMINGITIDAE Hyatt, 1900 Tr. (NML–SPA) Mar.

First: *Flemingites rohilla* Diener, 1897, Bed JSB10 at Tulong, Kangshare Formation, *Psilogyrus* Zone (NML, 1), sourthern Tibet (Wang and He, 1980).
Last: *Preflorianites* sp. (cf. *P. intermedius* Tozer, 1965), upper *Haugi* Zone (SPA, 3a), Prida Formation, Nevada, USA (Bucher, 1989).

F. JAPONITIDAE Tozer, 1971 Tr. (SPA–ANS) Mar.

First: *Eogymnites arthaberi* (Diener, 1915), *Subcolumbites* Beds (SPA, 2) Kćira, Albania. *Japonites* and *Aegeiceras* from SPA, 2-ANS, 2 (Wang, 1985).
Last: *Tropigymnites* sp., *Paraceratites cricki* Beds (ANS, 7), Favret Formation, Nevada, USA (Silberling and Nichols, 1982). Shevyrev (1986) tentatively cites *Bukowskiites colvini* Diener, 1907 from ANS, 9, Himalayas.

F. GYMNITIDAE Waagen, 1895 Tr. (ANS–RHT) Mar.

First: *Gymnites billingsi* Bucher, 1989, *Mulleri* Zone (ANS, 1), Prida Formation, Humboldt Range, Nevada, USA (Bucher, 1989).
Last: *Placites* sp., middle member, Gabbs Formation (RHT, 3a), Nevada and Sutton Formation, British Columbia, Canada (Silberling and Tozer, 1968; Tozer, 1980).

F. PINACOCERATIDAE Mojsisovics, 1879 Tr. (CRN–RHT) Mar.

First: *Pompeckjites layeri* (Hauer, 1847), basal bed 68/58 Feuerkogel, *Aon* Subzone (CRN, 1a), Austria (Krystyn, 1973). Tozer (1981a) cites LAD–CRN.
Last: *Pinacoceras* sp., middle member, Gabbs Formation (RHT, 3a), *Crickmayi* Zone, Nevada (Silberling and Tozer, 1968). *P. metternichii* (Hauer, 1846), highest bed 68/55 or Rotkalk, Sommeraukogel, *Suessi* Zone (RHT, 1–2), Austria (Krystyn, 1973; Tozer, 1980).

Suborder CERATITINA Hyatt, 1884

The phylogenetic interpretation of Shevyrev (1986) is modified by the linkage of the Paraceratitinae Silberling, 1962 to *Eoprotrachyceras*, and changes made by Krystyn (1982) and Krystyn and Wiedmann (1985) to late Triassic systematics. The Phylloceratida and other orders reviewed by Page (this vol.) were derived from the Flemingitidae (above) in Shevyrev (1986) and appear to be allied to the Ceratitina.

Superfamily CLYDONITACEAE Mojsisovics, 1879 *s.l.*

F. OLENIKITIDAE Tozer, 1971 Tr. (SPA) Mar.

First: *Hyrcanites nodosus* Shevyrev, 1968, Mangÿshlak, Transcaucasia, former USSR, *Harti* Zone (SPA, 1) of Shevyrev (1986). *Olenikites spiniplicatus* (Mojsisovics, 1886), lower *Spiniplicatus* Zone (SPA, 1), Lena–Anabar region, former USSR (Dagys *et al.*, 1979).
Last: *Olenekites* sp. indet., upper *Haugi* Zone (SPA, 3a), Nevada (Bucher, 1989). *O. canadensis* Tozer, 1961, 72 m up Blind Fjord Group, *Subrobustus* Zone (SPA, 3a), Ellesmere Island (Tozer, 1965).

F. LONGOBARDITIDAE Spath, 1951 Tr. (ANS–LAD) Mar.

First: *Grambergia* sp., *Mulleri* Zone (ANS, 1), Prida Formation, Nevada (Bucher, 1989). *G. taimyrensis* Popov, 1961, *Evolutus* Subzone (ANS, 1), Lena–Anabar region, former USSR (Dagys *et al.*, 1979, p. 39).
Last: *Nathorstites concentricus* (Oeberg, 1877) (=*N. gibbosus* Stolley, 1911), *Tenuis* Zone (LAD, 5 or CRN, 1), Spitsbergen and Yana–Kolyma region, Siberia, former USSR (Dagys *et al.*, 1979, p. 81; Tozer, 1981b; Arkadiev and Vavilov, 1984).
Comment: A combination of ontogenetic and stratigraphical work has produced an arbitrary boundary between this family and the Nathorstitidae Spath, 1951, at the base of strata bearing *Indigirites krugi* Popov, 1961 (LAD, 3a of Siberia). Since there is no indication of lineage bifurcation, the two so-called families can be combined, although this was not advocated by Arkadiev and Vavilov (1984, 1989).

F. APLOCOCERATIDAE Spath, 1951 Tr. (SPA–CRN) Mar.

First: *Karangatites multicameratus* (Smith, 1914), upper *Haugi* Zone (SPA, 3a), Prida Formation, Nevada, USA (Bucher, 1989).
Last: *Epiceratites elevatus* (Dittmar, 1866), Halstatt Limestone, Austria, CRN, 4–5 according to Tozer (1981a) and Shevyrev (1986).
Comment: The relatively limited and stratigraphically correlated intraspecific variation of *Aplococeras* species was investigated in a pioneer statistical study by S. Bubnoff (Spath, 1951) and Silberling and Nichols (1982). Thus the family is likely to a monophyletic group, rather than polyphyletic dwarfs produced by the ecophenotypic starvation of macromorphs.

F. PROTEUSITIDAE Spath, 1951 Tr. (ANS) Mar.

First: *Tropigastrites lahontanus* Smith, 1914, base of *Rotelliformis* Zone (ANS, 7), Humboldt Range, Nevada, USA (Silberling and Nichols, 1982). *Proteusites kellneri* Hauer, 1887, *Trinodosus* Zone (ANS, 7), Han Bulog, Bosnia, former Yugoslavia (Spath, 1951).
Last: *Tozerites polygyratus* (Smith, 1914), upper *Nevadites* Beds, *Occidentalis* Zone (ANS, 9), Prida Formation, Humboldt Range, Nevada, USA (Silberling and Nichols, 1982).
Comment: The *Tropigastrites–Tozerites* lineage of Nevada was associated with similar but less regularly modified *Aplococeras*, and seems unlikely to be conspecific with this potentially ancestral, long-ranging morphology.

F. NANNITIDAE Mojsisovics, 1884 Tr. (CRN) Mar.

First and Last: *Nannites spurius* (Münster, 1848), *Aonoides* Zone (CRN, 1), Cassian Formation, Italy (Shevyrev, 1986).
Comment: Spath (1951) suggested that this rare dwarf morphology was derived from the local aplococeratids, while O. H. Schindewolf showed that the goniatitic septal suture was somewhat different to the associated dwarf Lecanitidae (Shevyrev, 1986). Shevyrev indicates a close phylogenetic position to the latter family and the longer-ranging Celtitidae. Tozer (1981a) rejects the Celtitidae, while maintaining the separate status of the three dwarf families from the Cassian Formation.

F. THANAMITIDAE Tozer, 1971 Tr. (ANS–LAD) Mar.

First: *Thanamites ?contractus* (Smith, 1914), *Nevadites humboltensis* Beds, *Occidentalis* Zone (ANS, 9), Humboldt Range, Nevada, USA (Silberling and Nichols, 1982). Shevyrev (1986) cited no records before the *Archelaus* Zone (LAD, 4?).
Last: *Drumoceras minor* Wang and He, 1976, Bed JSB 83b, *Protrachyceras–Joannites* Zone, unit 3 (LAD, 4), Tulong, southern Tibet (Wang and He, 1980). *Thanamites bitterni* (Mosisovics, 1882), Lower Pachycardienstuffe, Tschipitbach, Austria (Urlichs, 1977), appears contemporaneous with *Drumoceras* and *Thanamites*, *Meginae* Zone (LAD, 3), Canada (Tozer, 1981b). Shevyrev (1986) implies a range into his *Regoledanus* Zone (LAD, 5).

F. CELTITIDAE Mojsisovics, 1893 Tr. (ANS–CRN) Mar.

First: ?*Celtites* cf. *fumagalli* Stabile, 1861, Beds 83–136, *Reitzi* Zone (ANS, 9) Grenzbituminenzone, Monte San Giorgio, Switzerland (Rieber, 1973). *Celtites epolensis* Mojsisovics, 1882, Lower Pachycardienstuffe, Tschipitbach (LAD, 3?), Austria (Urlichs, 1977). *Indoceltites trigonalis* (Diener, 1908), Bed JSB 11 with *Doleriticus* fauna (LAD, 4), Tulong, southern Tibet (Wang and He, 1980).
Last: *Orthoceltites buchii* (Klipstein, 1843), Cassian Formation, Italy, *Aonoides* Zone (CRN, 1) of Shevyrev (1986).

F. LECANITIDAE Hyatt, 1900 Tr. (LAD–CRN) Mar.

First: *Lecanites glaucus* (Münster, 1834), *Pachycardienstuffe* with *Maclearnoceras* (LAD, 4), Seiser Alm, Austria (Ulrichs, 1977). Both Tozer (1981a) and Shevyrev (1986), doubt or overlook LAD records.
Last: *Lecanites glaucus* (Münster, 1834) and *Badiotites eryx* (Münster, 1834), *Aonoides* Zone (CRN, 1), Cassian Formation, Italy (Shevyrev, 1986). *Lecanites trauthi* Johnson, 1941, *Desatoyense* Zone (CRN, 1), New Pass, Nevada, USA (Spath, 1951; Silberling and Tozer, 1968).
Comment: The contemporaneous and monogeneric Badiotitidae Hyatt, 1900 (CRN, 1) was kept separate by Tozer (1981a) and Shevyrev (1986), although Spath (1951) remarked 'that *Lecanites* and *Badiotites* are not separable into two distinct families'.

F. TROPICELTITIDAE Spath, 1951 Tr. (CRN–NOR) Mar.

First: *Tornquistites obolinus* (Dittmar, 1866), Bed VII Feuerkogel and elsewhere, *Subbullatus* Zone (CRN, 4), Hallstatt Limestone, Austria (Krystyn, 1973). Tozer (1981a) indicates first occurrence in *Dilleri* Zone (CRN, 3).
Last: *Tropiceltites columbianus* (McLearn, 1940), *Kerri* Zone (NOR, 1b), Pardonet Hill, British Columbia, Canada (Tozer, 1981a, 1984). *Tropiceltites rotundus* Mojsisovics, 1893, sample B16, *Jandianus* Zone (NOR, 1), Hallstatt Limestone, Austria (Krystyn, 1973).

F. THISBITIDAE Spath, 1951 Tr. (CRN–RHT) Mar.

First: *Thisbites dawsoni* (McLearn, 1940), *Macrolobatus* Zone (CRN, 5), Pardonet Formation, Brown Hill, British Columbia, Canada (Tozer 1981a, 1984). *Thisbites agricolae* Mojsisovics, 1893, Bed IV Feuerkogel, Obere *Anatropites*-Bereich (CRN, 5b), Hallstatt Limestone, Austria (Krystyn, 1973).
Last: ?*Glyphidites docens* Mojsisovics, 1893, Limestone block of *Cordilleranus* Zone (RHT, 1), Aliambata, Timor (Tozer, 1980). *Scheutzites bifunicarnatus* Tatzreiter, 1985, Hangend Rotkalk, *Hogarti* Subzone (NOR, 5b), Schneckenkogel, Austria (Tatzreiter, 1985).

F. CLIONITITIDAE Arabu, 1932 Tr. (LAD–RHT) Mar.

First: *Clionitites* sp., *Sutherlandi*/*Regoledanum* Zones (LAD, 5), *Roten Bankkalk*, Epidauros, Greece (Krystyn and Mariolakos, 1975). The occurrence in the *Maclearni* Zone (LAD, 4) is cited by E. T. Tozer in Dagys *et al.* (1979). *Clionitites rarecostatus* (Parnes, 1962), Matsoq Nahal Ramon, *Sirenitiforme* Zone (LAD, 5 of Parnes, 1986), Israel. *Clionitites barwick* (Johnston, 1941), *Desatoyense* Zone (CRN, 1), New Pass, Nevada (Johnston, 1941).
Last: *Steinmannites* sp., mixed *Suessi* Zone fissure fauna with *Cladiscites* (RHT, 1–2), Millibrunnkogel, Austria (Tozer, 1980) and elsewhere (Krystyn, 1973).

F. SANDLINGITIDAE Tozer, 1971 Tr. (CRN) Mar.

First: *Traskites merriami* (Hyatt and Smith, 1905), *Dilleri* Zone (CRN, 3), California, USA (Silberling and Tozer, 1968).
Last: *Sandlingites oribasus* (Dittmar, 1866) Hallstatt Limestone, Austria. *Sandlingites* sp., *Subbullatus* Zone (CRN, 4), Saltzkammergut area (Krystyn, 1973), and a generic range of CRN, 3–5 (Tozer, 1981a) indicates the age of this species.

F. PALICITIDAE Krystyn, 1982 Tr. (NOR) Mar.

First: ?*Pterotoceras* (=?*Anolcites*, LAD, 5) included by Shevyrev (1986), perhaps in error. *Palicites mojsisovicsi* Gemmellaro, 1904, Bed 78b, upper *Jandianus* Zone (NOR, 1b), Thinigaon Formation, Jomson, Nepal (Krystyn, 1982). *Mojsisovicsites* sp., basal Bed III Feuerkogal, lower *Jandianus* Zone (NOR, 1a), Austria (Krystyn, 1973). *M. kerri* (McLearn, 1930), *Kerri* Zone (NOR, 1b), Brown and Pardonet Hills, British Columbia, Canada (Tozer, 1981b).
Last: *'Pterotoceras' caurinum* McLearn, 1939, *Magnus* Zone (NOR, 3), Pardonet Formation, Brown Hill, British Columbia, Canada (McLearn, 1960; Tozer, 1984).

F. NORIDISCITIDAE Spath, 1951 Tr. (NOR) Mar.

First and Last: *Noridiscites viator* (Mojsisovics, 1893) Hallstatt Limestone, Austria. *Nairites armenius* Kiparisova and Azarian, 1963 and *Nairites laevis* Kiparisova and Azurian, 1963, Armenia, former USSR. All from *Bicrenatus* Zone (NOR, 4) according to Shevyrev (1986).

F. CLYDONITIDAE Mojsisovics, 1879 Tr. (NOR–RHT) Mar.

First: *Parathisbites baunensis* Tatzreiter, 1980, Bed 12, *Watsoni* Subzone (NOR, 5a), Limestone block A, Baun, Timor. *Leislingites pseudoarchibaldi* Tatzreiter, 1980, Bed 16, base of *Macer* Zone (NOR, 6a), block A, Baum, and also ?Lower *Suessi* Zone (RHT, 1), Timor (Tatzreiter, 1980).
Last: *Choristoceras marshi* Hauer, 1865, middle *Suessi* (RHT, 1) to upper *Marshi* Zones (RHT, 3b), Koessen Beds, Austria (Krystyn and Wiedmann, 1986). The last occurrence of the Thetiditidae Tozer, 1971 is *Pseudothetidites praemarshi* Krystyn and Wiedmann, 1986, of Bed 18, highest *Macer* Zone (NOR, 6b), block A, Baum, Timor. The last occurrence of the CLYDONITIDAE *s.s.* is *Leislingites* of the lower *Suessi* Zone (RHT, 1) cited by Tatzreiter (1980) and Shevyrev (1986).
Comment: Tatzreiter (1980) combined this family with the Thetiditidae Tozer, 1971 and Shevyrev (1986) derived the one directly from the other. Krystyn and Wiedmann (1986) have undermined the Choristocerataceae Hyatt, 1900 (RHT) by deriving *Choristoceras* from the last *Pseudothetidites*. They have a similar morphology, but are separated by a small stratigraphical gap (most of RHT, 1). Shevyrev (1986) suggested that *Choristoceras* and the early heteromorphic members of the Choristocerataceae were derived from *Helictites* (NOR, 6) of the Metasibiritidae. An expanded Clydonitidae is an alternative source of the heteromorphs and appears to be the last of the Ceratitida in both Nevada (Silberling and Tozer, 1968) and elsewhere (Tozer, 1980).

F. METASIBIRITIDAE Spath, 1951 Tr. (NOR–RHT) Mar. (see Fig. 10.4)

First: ?*Helictites involutus* Wang and He, 1976, Bed Fd VI–5, *Socius* Zone (NOR, 4), Longjiang, southern Tibet (locality in Wang and He, 1980). *Helicites subgeniculatus* Mojsisovics, 1893 and *H. decorus* McLearn, 1940, *Columbianus* Zone (NOR, 5–6), Pardonet Formation, Black Bear Ridge, British Columbia, Canada (Tozer, 1981a, 1984).
Last: *Metasibirites* cf. *spiniscens* (Hauer, 1855), Bed 20, block F, *Reticulatus* Subzone (RHT, 2) Baum, Timor, and beds of equivalent age at Kotel, Bulgaria (Tozer, 1980; Krystyn and Wiedmann 1986).

F. LISSONITIDAE Tatzreiter, 1985 Tr. (RHT) Mar.

First and Last: *Lissonites canadensis* Tozer, 1979, Upper *Cordilleranus* Zone (RHT 1b), Pardonet Formation, Ne-Parle-Pas Rapids and elsewhere, British Columbia, Canada. *'Tozerites'* (junior homonymn) *hernsteini* Tatzreiter, 1985 and *Psamateiceras saxicastelli* Tatzreiter, 1985, Grey Hallstatt Limestone, upper Suessi Zone (RHT, 2), Hernstein, Austria. Also Kotel, Bulgaria (Tatzreiter, 1985; Tozer, 1982).

F. RHABDOCERATIDAE Tozer, 1979 Tr. (RHT) Mar.

First: *Rhabdoceras boreale* Afitsky, 1965, *Ochotica* Zone (RHT, 1), Siberia (Dagys *et al.*, 1979, p. 153). *R. suessi* Hauer, 1860, lower *Cordilleranus* Zone (RHT 1a), Mt. Ludington, USA (Silberling and Tozer, 1968).
Last: *Rhabdoceras* (?) sp., Zlambach Marls, Grunbachgraben, Austria (Wiedmann, 1977). *Rhabdoceras suessi* Hauer, 1860, lower *Marshi* Zone (RHT 3a), Koessen Beds and crinoidal limestone of Steinbergkogel, Austria (Krystyn, 1973; Tozer, 1980).

F. CYCLOCELTITIDAE Tozer, 1979 Tr. (RHT) Mar.

First and Last: *Cycloceltites corneus* Kollarova-Andrusova, 1973, *Stuerzenbaumi* Subzone (RHT 3a), Bleskoy Pramen, Czechoslovakia. *C. arduini* Mojsisovics, 1882, middle member, Gabbs Formation (RHT, 3a), Nevada, and other contemporaneous records (Tozer, 1980).

F. COCHLOCERATIDAE Hyatt, 1900 Tr. (RHT) Mar.

First and Last: *Cochloceras fisheri* Hauer, 1860 and *C. canaliculatum* Hauer, 1860, Grey Hallstatt Limestone, *Reticularis* Subzone (RHT, 2), Steinbergkogel, Austria. *C. fisheri*, Lower member, Gabbs Formation, *Amoebum* Zone (RHT, 2), Nevada, USA, and other contemporaneous localities (Tozer, 1980).
Comment: This helicoid spiral was derived from the relatively straight *Peripleurites* (RHT, 1–2) of the Rhabdoceratidae (Shevyrev, 1986), and should perhaps be united with it.

F. DANUBITIDAE Spath, 1951 Tr. (SPA–ANS) Mar.

First: *Eodanubites xingyuanensis* Wang, 1978 and *E. costulatus* Wang, 1978, 0.2 m thick Ziyun Limestone (SPA, 2-? ANS, 1), Guizhou Province (Wang, 1985). *Paradanubites depressus* Fantini Sestini, 1981, Bed T329, Chios (SPA, 3-ANS, 2?), Greece (Wang, 1985). Interpreted as ANS, 1–2 by Shevyrev (1986).
Last: *Ticinites polymorphus* Rieber, 1973, Bed 50, *Polymorphus* Zone (ANS, 8), Grenzbitumenzone, Monte San Giorgio, Switzerland (Rieber, 1973).
Comment: This family was given an important position in the superfamily by Tozer (1981a) and Shevyrev (1986), but has been classified with the Ceratitaceae in the past, and at least some genera might still be placed there (Spath, 1951).

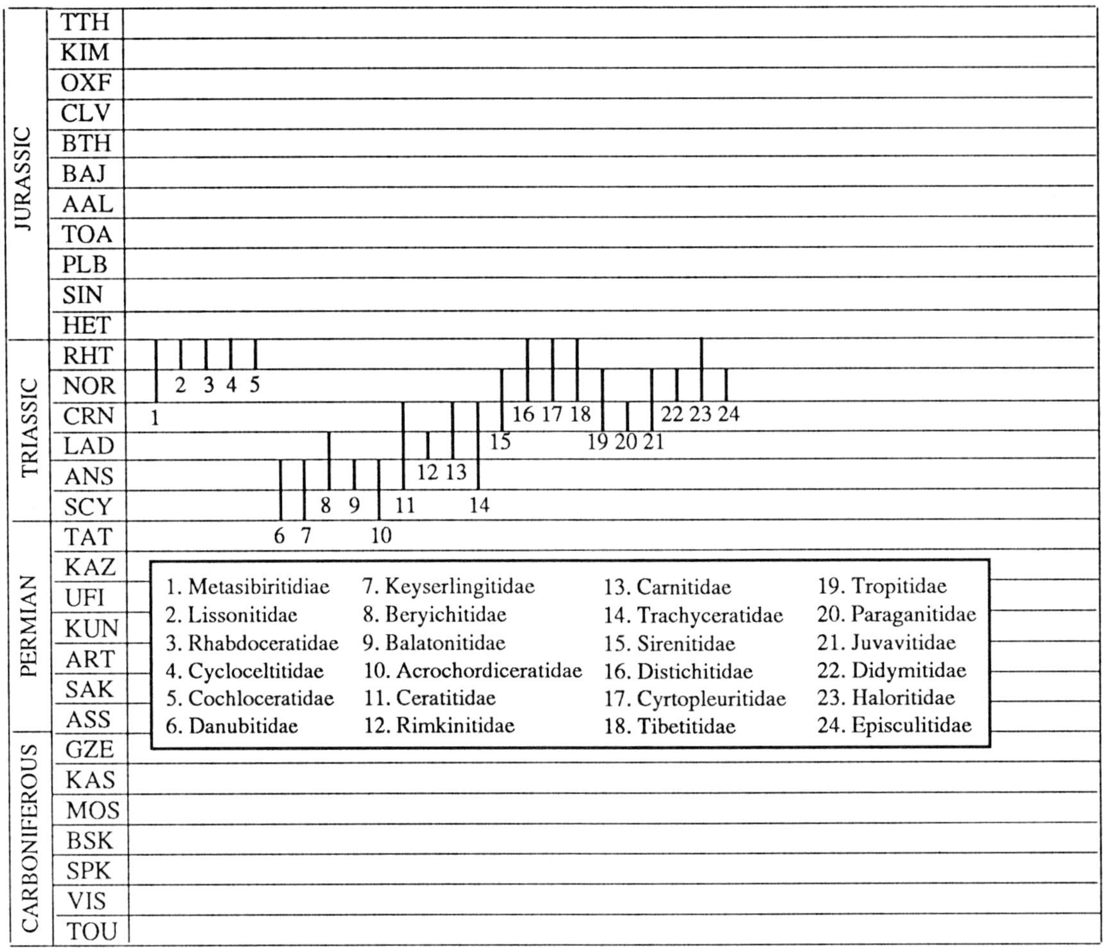

Fig. 10.4

Superfamily CERATITACEAE Mojsisovics, 1879 *s.l.*

F. KEYSERLINGITIDAE Zacharov, 1970 Tr. (SPA–ANS) Mar.

First: *Olenekoceras middendorffi* (Keyserling, 1845), middle *Spiniplicatus* Zone (SPA, 1 of Wang, 1985), Lena–Anabar region, former USSR (Dagys *et al.*, 1979, p. 30).
Last: *Keyserlingites dieneri* (Mojsisovics, 1903), *Meridianus* Subzone (ANS, 1 or 3), Naocanjianguo, Qinghai Province, China (Wang, 1985; Bucher, 1989). *Silberlingites mulleri* Bucher, 1989 and *S. tregoi* Bucher, 1989, *Mulleri* Zone (ANS, 1), Coyote Canyon, northern Humboldt Range, Nevada, USA (Bucher, 1989).

F. BEYRICHITIDAE Spath, 1934 Tr. (ANS–LAD) Mar.

First: Dimorphic *Nicomedites osmani* Toula, 1896, base of *Osmani* Zone and Nodular Limestone of Gebze (ANS, 3), Turkey (Fantini Sestini, 1988).
Last: *Frechites johnstoni* Silberling and Nichols, 1982, *Subaspersum* Zone (LAD, 1), Humboldt Range, Nevada, USA (Silberling and Nichols, 1982).
Comment: The inclusion of *Serpianites antecedens* (Beyrich, 1958) in this family (Shevyrev, 1986), might suggest that *'Paraceratites' binodosus* (Hauer, 1850) should be transferred here, and not regarded as the ancestral form of the Trachyceratidae (e.g. Siberling and Nichols, 1982; Urlichs and Mundlos, 1985; Tozer, 1981b; Fantini Sestini, 1988). However this *'Paraceratites'* shows a larger number of saddles than *Serpianities*, and the Nevada Berichitidae. Sheyrev (1986) interprets this family as a separate and ultimately infertile lineage.

F. BALATONITIDAE Spath, 1951 Tr. (ANS) Mar.

First: *Platycuccoceras bonaevistae* (Hyatt and Smith, 1905), lower *Hyatti* Zone (ANS, 3), Prida Formation, Nevada, USA (Silberling and Nichols, 1982).
Last: *Balatonites balatonicus* Mojsisovics, 1873 and ?*Reiflingites eugeniae* Arthaber, 1896, 0.15-m-thick Deqen limestone, *Trinodosus* Zone (ANS, 7), Chaqupu Formation, 67 km NW of Lhasa, Tibet (Wang and He, 1980).

F. ACROCHORDICERATIDAE Arthaber, 1911 Tr. (SPA–ANS) Mar.

First: *Eoacrochordiceras ziyunense* Wang, 1978 and *E. evolutum* Wang, 1978, 0.2-m-thick Ziyun Limestone (SPA, 2 or later), Guizhou Province (Wang, 1985). *Paraacrochordiceras pandya* (Diener, 1895), Bed T329 (SPA, 3-ANS, 2?), Chios, Greece (Fantini Sestini, *in* Wang, 1985). *Paraacrochordiceras* sp. and *Paraacrochordiceras silberlingi* Bucher, 1989, *Welteri* and *Guexi* Beds (SPA, 3b), northern Humboldt Range, Nevada, USA (Bucher, 1989).
Last: *Acrochordiceras haueri* Arthaber, 1911, *Trinodosus* Zone (ANS, 7), Nodular Limestone of Gebze, Turkey (Fantini Sestini, 1988).

F. CERATITIDAE Mojsisovics, 1879 Tr. (ANS–CRN) Mar.

First: ??*'Hungarites' yatesi* Hyatt and Smith, 1905, *Haugi* Zone (SPA, 3a), California and Nevada, USA (Spath, 1951; Bucher, 1989). ?*'Paraceratites' binodosus* (Hauer, 1850), *Balatonicus* Zone (ANS, 6), Nodular Limestone of Gebze, Turkey and Dont, Italy (Fantini Sestini, 1988). ?*Paraceratitoides brotzeni* (Avinimelech, 1956), Lower Sahoronim Formation, Har Gavanum, Israel (ANS, 7 of Parnes, 1986). ?*Hungarites* sp., 0.15 m Deqen Limestone, *Trinodosus* Zone (ANS, 7), Tibet (Wang and He, 1980). *Progonoceratites flexuosus* (Philippi, 1901), Trochitenbank 2, upper Muschelkalk (LAD, 1?), Baden (Urlichs and Mundlos, 1985; Tozer, 1981b). *Progonoceratites nanjiangensis* Zhao and Wang, 1974, Badong Formation (?ANS–LAD), SW China (Wang and He, 1980). *Progonoceratites poseidon* Tozer, 1967, *Poseidon* Zone (LAD, 2), British Columbia (Tozer, 1981b).
Last: ?*Alloceratites schmidi* (Zimmerman, 1883), *Grenzdolomit* (CRN, 1?), Sülzenbrucken, Germany (Müller, 1973). ?*Perrinoceras novaditus* Johnston, 1941, *Desatoyense* Zone (CRN, 1), South Canyon, Nevada, USA (Johnston, 1941). *Discoceratites dorsoplanus* (Philippi, 1901), upper Ceratite Beds, upper Muschelkalk (LAD), Germany (Parnes, 1986).
Comment: The family and order are defined by the rediscovered holotype of *Ceratites nodosa* Bruguiere, 1792, which has the morphology of *Doloceratites robustus* Riedel, 1918 from the *Spinosus* Zone, lower Ceratite Beds, Upper Muschelkalk (LAD, 2?), Germany (Rieber and Tozer, 1986). It is reasonable to define the Ceratitidae *s.s.* by the succession of morphologies in the poorly dated Upper Muschelkalk and the more cosmopolitan occurrences of *Progonoceratites* noted above. Both the CRN species listed above have more affinity with the Hungaritidae Waagen, 1895 defined by Shevyrev (1986). Thus the Ceratitidae *s.s.* would be confined to the LAD, and the nebulous and Middle Eastern Hungaritidae would have a longer range (ANS–CRN). Parnes (1986) rejected the reinterpretation of his ANS genera by Tozer (1981a) and Shevyrev (1986), while Silberling and Nichols (1982) demonstrated that the Paraceratitinae and Beyrichitidae lost any resemblance to *Progonoceratites* by ANS, 8–9. The plausible hypothesis of ceratitization in marginal marine embayments (Urlichs and Mundlos, 1985), therefore must have involved the Hungarititidae and the closely related Rimkinitidae if it took place in Germany.

It is proposed that the Ceratitidae and the Hungaritidae be united as a separate and highly variable group, which appeared in the *Balatonensis* Zone (ANS, 6) and persisted in marginal marine facies (Israel, China, Germany), without loss of lobes, until it became widespread in the *Poseidon* (LAD, 2) and *Desatoyense* (CRN, 1) Zones. The ancestral form seems likely to have been derived from the early Beyrichitidae.

F. RIMKINITIDAE Wang and He, 1976 Tr. (LAD) Mar.

First and Last: *Rimkinites nitiensis* (Mojsisovics, 1896) and other species, Bed JSB-11, correlated with the *Maclearni* Zone (LAD, 4), Laibuxi Formation, Tulong, southern Tibet (Wang and He, 1980).
Comment: This family provides an ammonitic link between *Hungarites* and *Carnites*, with a range confined to the LAD (Spath, 1951; Shevyrev, 1986).

F. CARNITIDAE Arthaber, 1911 Tr. (LAD–CRN) Mar.

First: *Pseudocarnites arthaberi* Simionescu, 1913, Agighol, Romania (LAD, 4–5 of Shevyrev, 1986).
Last: *Carnites multilobatus* Diener, 1908, Bed JSB 33b, *Acutus* Zone (CRN, 5), Tulong, southern Tibet (Wang and He, 1980), and contemporaneous *Klamathites*, California, USA (Silberling and Tozer, 1968).

F. TRACHYCERATIDAE Haug, 1894 Tr. (ANS–CRN) Mar.

First: *Paraceratites burckhardti* (Smith, 1914), base *Rotelliformis* Zone (ANS, 7), Prida Formation, Nevada, USA. Alternative definitions are favoured by Tozer (1981a): *Eoprotrachyceras dunni* (Smith, 1914), base of *Subaspersum* Zone (LAD, 1), Nevada, USA (Silberling and Nichols, 1982) defines first Arpaditidae Hyatt, 1900; *Anolcites doleriticum* (Mojsisovics, 1869), middle Pachycardienstuffen (LAD, 4), Seiser Alm, Italy (Urlichs, 1977) defines the first Trachyceratidae *s.s.*
Last: *Trachysagenites herbichi* (Mojsisovics, 1893), lower *Subbullatus* Zone (CRN, 4a), Feuerkogel and elsewhere, Austria (Krystyn, 1973).
Comment: The distinction made by Silberling and Nichols (1982) between *Nevadites* (supposed Ceratitidae) and *Eoprotrachyceras*, appears to be an arbitrary stratigraphical convention, which previously confused Rieber (1973). Since there is hardly anything in common between the Arpaditidae Hyatt, 1900 of Tozer (1981a) and that of Shevyrev (1986), it may be wise to unite it with this expanded Trachyceratitidae. The origin of both families in the Danubitidae (Shevyrev, 1986) is rejected in favour of a linkage with the Acrochordiceratidae via the Paraceratitinae Silberling, 1962.

F. SIRENITIDAE Tozer, 1971 Tr. (CRN–NOR) Mar.

First: *Diplosirenites raineri* (Mojsisovics, 1893), generic-level identification of Krystyn (1973), Bed 70/66, *Aonoides* Subzone (CRN, 1b), Feuerkogal, Austria. Also the same subzone in Nepal (Krystyn, 1982).
Last: *Wangoceras pax* (Tozer, 1980), *Malayites bococki* bed, *Dawsoni* Zone (NOR, 2), Pardonet Formation, Pardonet Hill, British Columbia, Canada (Tozer, 1981a).

F. DISTICHITIDAE Diener, 1920 Tr. (NOR–RHT) Mar.

First: *Heraclites robustus* (Hauer, 1855) and *Ectolcites pseudoaries* (Hauer, 1849): generic identifications by Krystyn (1973), *Brecrenatus* lager (NOR, 4), Feuerkogel, Austria.
Last: *Sagenites reticulatus* (Hauer, 1849), *Reticulatus* Subzone (RHT, 2) Grey Hallstatt Limestone, Steinbergkogel, Austria (Tozer, 1980, 1984).

F. CYRTOPLEURITIDAE Diener, 1925 Tr. (NOR–RHT) Mar.

First: *Lipuites totiae* Jeannet, 1959, Bed 78b upper Thinigaon Formation, upper *Jandianus* Zone (NOR, 1b), Jomson, Nepal (Krystyn, 1982).
Last: *Pseudosirenites* sp., Limestone of Aliambata, *Cordilleranus* Zone (RHT, 1), Timor (Tozer, 1980). Krystyn (1982) limits family to NOR.

F. TIBETITIDAE Hyatt, 1900 Tr. (NOR–RHT) Mar.

First: *Nodotibetites nodosus* Zhao and Wang, 1973, Bed JSB 34b of Dasalong Formation, *Nodosus* Zone (NOR 1a), Tulong, S. Tibet (Wang and He, 1980; Krystyn, 1982).
Last: *Paratibetites welteringi* (Krumbeck, 1913), *Cordilleranus* Zone (RHT, 1), Fogi Beds, Buru (Tozer, 1980). *Paratibetites bertrandi* Mojsisovics, 1896, Bed 113 at Jomson and *Halorites* Limestone at Bambanag, upper *Columbianus* Zone (NOR, 6), Nepal and India (Krystyn, 1982).

Superfamily TROPITACEAE Mojsisovics, 1875

This superfamily was defined by Krystyn (1982) as a branch of the CRN Trachyceratidae. It is separated here to preserve the nomenclature of the Ceratitaceae Mojsisovics, 1879 and to emphasize that Shevyrev (1986) derived these families from the early Tropiceltitidae.

F. TROPITIDAE Mojsisovics, 1875 Tr. (CRN–NOR) Mar.

First: *Gymnotropites dinarus* (Diener, 1916), Bed VII Feuerkogel, *Dilleri* Zone (CRN, 3), Austria (Krystyn, 1973).
Last: *Euisculites heimi* (Mojsisovics, 1893) is implied to range up to the *Kerri* Zone (NOR, 1b) at Feuerkogel, Austria (Krystyn, 1980; Tozer, 1981a). *E. bittneri* (Gemmellaro, 1904) from Tibet (CRN, 5b) is reviewed by Krystyn (1982).

F. PARAGANITIDAE Wang and He MSS Tr. (CRN) Mar.

First and Last: New family for the generally unclassified *Paraganides californicus* Hyatt and Smith, 1905, California, and similar Chinese new genus. Family ranges through CRN.

F. JUVAVITIDAE Tozer, 1971 Tr. (CRN–NOR) Mar.

First: *Projuvavites crasseplicatus* (Mojsisovics, 1893), Bed VII Feuerkogel, base of *Subbullatus* Zone (CRN, 4), Austria (Krystyn, 1980).
Last: *Indojuvavites angulatus* (Diener, 1908), Bed 106, top of *Magnus* Zone (NOR, 3b), Tarap Formation, Jomson, Nepal (Krystyn, 1982).

F. DIDYMITIDAE Haug, 1894 Tr. (NOR) Mar.

First and Last: *Didymites simplex* Wang and He, 1976, Bed Fdt 21, *Socius* Zone (NOR, 4), East Hill of Longjiang, Tibet (Wang and He, 1980). *D. subglobus* Mojsisovics, 1875, sample 68/96, *Bicrenatus* Zone (NOR, 4), Sommeraukogel, Austria (Krystyn, 1973).

F. HALORITIDAE Mojsisovics, 1893 Tr. (NOR–RHT) Mar.

First: *Parajuvavites* cf. *buddhaicus* Mojsisovics, 1896, *Columbianus* Zone (NOR, 5–6), Sikanni Chief River, British Columbia, Canada (McLearn, 1960; Tozer, 1981a).
Last: *Halorites* sp., Grey Hallstatt Limestone, *Reticulatus* Subzone (RHT, 2), Steinbergkogel, Austria (Krystyn, 1973; Tozer, 1980).

F. EPISCULITIDAE Spath, 1951 Tr. (NOR) Mar.

First and Last: *Episculites descresens* (Hauer, 1855), Hallstatt Limestone, Austria and other species from British Columbia, Nevada and Timor, all within the *Columbianus* Zone (NOR, 5–6) (McLearn, 1960; Silberling and Tozer, 1968; Tozer, 1981a; Shevyrev, 1986).

REFERENCES

Arkadiev, V. V. and Vavilov, M. N. (1984) Middle Triassic Parapopanoceratidae and Nathorstitidae (Ammonoidea) of boreal regions: internal structure, ontogeny and phylogenetic pattern. *Geobios*, **17**, 397–415.

Arkadiev, V. V. and Vavilov, M. N. (1989) Anisian–Ladinian boundary in the boreal region based on Ammonoidea. *Palaeontographica, Abteilung A*, **207**, 49–78.

Bando, Y. (1981) Lower Triassic ammonoids from Guryul Ravine and the spur three kilometres north of Barus. *Palaeontologica Indica*, **46**, 135–77.

Bartszch, K. and Weyer, D. (1986) Biostratigraphie der Devon/Karbon-Grenze im Bohlen-Profil bei Saalfeld. *Zeitschrift für Geologischen Wissenschaft*, **14**, 147–52.

Becker, R. T. (1990) Stratigraphische Gliederung und Ammonoideen-Fauna im Nehdenium (Oberdevon II) von Europa und Nord-Afrika. Unpublished Dissertation, University of Bochum, 464 pp.

Becker, R. T., Feist, R., Flajs, G., House, M. R. *et al.* (1989) Frasnian–Famennian extinction events in the Devonian at Coumiac, southern France. *Comptes Rendus de l'Académie des Sciences, Paris, Série II*, **309**, 259–66.

Bensaïd, M. (1974) Études sur les Goniatites à la limite du Devonien moyen et supérieur, du Sud Marocain. *Notes de la Service Géologique Marocaine*, **36**, 81–140.

Bisat, W. S. (1934) The goniatites of the *Beyrichoceras* Zone in the North of England. *Proceedings of the Yorkshire Geological Society*, **22**, 80–309.

Bogoslovskiy, B. I. (1958) Devonskie Ammonoidei Rudnogo Altaya. *Trudy Paleontologicheskogo Instituta*, **64**, 1–155.

Bogoslovskiy, B. I. (1969) Devonskie Ammonoidei, I. Agoniatity. *Trudy Paleontologicheskogo Instituta*, **124**, 1–341.

Bogoslovskiy, B. I. (1971a) Devonskie Ammonoidei, II. Goniatity. *Trudy Paleontologicheskogo Instituta*, **127**, 1–228.

Bogoslovskiy, B. I. (1971b) Novoe rannedevonskie golovonogie Novoya Zemlya. *Paleontologicheskiy Zhurnal*, **1972**, 44–51.

Bogoslovskiy, B. I. (1975) Novoe semeystvo klymeniy. *Paleontologicheskiy Zhurnal*, **1975**, 35–41.

Bogoslovskiy, B. I. (1981) Devonskie Ammonoidei, III. Climenii. *Trudy Paleontologicheskogo Instituta*, **191**, 1–123.

Brügge, N. (1973) Zur stratigraphischen Einstufen des Oberdevon-Profil 'Alte Heerstrasse' bei Schleiz, Bezirk Gera. *Zeitschrift für Geologischen Wissenschaft*, **1**, 319–27.

Bucher, H. (1989) Lower Anisian ammonoids from the northern Humboldt Range (Northwestern Nevada, U.S.A.) and their bearing on the Lower–Middle Triassic boundary. *Eclogae Geologicae Helvetiae*, **82**, 945–1002.

Buggisch, W. and Clausen, C.-D. (1972) Conodonten- und Goniatiten-Faunen aus dem oberen Frasnium und unteren Famennian Marokjkos (Tafilalt, Anti Atlas). *Neues Jahrbuch für Geologie und Paläontologie, Abhandlungen*, **141**, 137–67.

Campbell, K. S. W., Brown, D. A. and Coleman, A. R. (1983) Ammonoids and correlation of the Lower Carboniferous rocks of eastern Australia. *Alcheringa*, **7**, 75–123.

Chang, A. C. (1960) New late Upper Devonian faunas of the Great Khingan and its biological significance. *Acta Paleontologica Sinica*, **8**, 180–92.

Chlupáč, I. and Turek, V. (1980) Devonian goniatites from the Barrandian area, Czechoslovakia. *Rozpravy Ustredniho Ustavu Geologického*, **46**, 1–159.

Czarnocki, J. (1989) Klimenie Gór Swietokrzyskich. *Prace Panstwowego Instytutu Geologicznego*, **127**, 1–91.

Dagys, A. S., Arkhipov, Yu. V. and Bychkov, Yu. M. (1979) Stratigraphy of the Triassic system of the North-eastern Asia. *Trudy Instituta Geologii i Geofizikii, Sibirskoe Otdelenie Akademii Nauk SSSR, Moscow*, **447**, 1–243 [in Russian].

Dzik, J. (1984) Phylogeny of the Nautiloidea. *Paleontologica Polonica*, **45**, 1–219.

Erben, H. K. (1960) Primitive Ammonoidea aus dem Unterdevon

Nordfrankreichs und Deutschlands. *Neues Jahrbuch für Geologie und Paläeontologie, Abhandlungen*, **110**, 1–128.

Fantini Sestini, N. (1988) Anisian ammonites from Gebze area (Kokaeli Peninsula, Turkey). *Rivista Italiana di Paleontologia e Stratigraphia*, **94**, 35–80.

Feist, F. (1970) Présence d'*Anetoceras (Erbeonoceras) mattei* n.sp. (Ammonoidéa primitive) dans le Dévonien Inférieur de la Montagne Noire. *Comptes Rendus de l'Académie des Sciences, Paris, Série II*, **270**, 290–3.

Frest, T. J., Glenister, B. F. and Furnish, W. M. (1981) Pennsylvanian–Permian cheiloceratacean ammonoid families Maximitidae and Pseudohaloritidae. *Paleontological Society Memoir*, **11**, 1–46.

Furnish, W. M. and Glenister, B. F. (1971) Permian Gonioloboceratidae (Ammonoidea), in *Paleozoic Perspectives: a Paleontological Tribute to G. Arthur Cooper* (ed. J. T. Dutro), Smithsonian Contributions to Paleobiology, **3**, pp. 301–12.

Furnish, W. M. and Knapp, W. D. (1966) Lower Pennsylvanian fauna from eastern Kentucky: Part 1, Ammonoids. *Journal of Paleontology*, **40**, 296–308.

Furnish, W. M. and Spinosa, C. (1968) Historic Pennsylvanian ammonoids from Iowa. *Iowa Academy of Science*, **73**, 253–9.

Glenister, B. F. and Furnish, W. M. (1987) New Permian representatives of ammonoid superfamilies Marathonitaceae and Cyclolobaceae. *Journal of Paleontology*, **61**, 982–98.

Glenister, B. F. and Furnish, W. M. (1988a) Terminal progenesis in Late Paleozoic ammonoid families, in *Cephalopods – Present and Past* (eds J. Wiedmann and J. Kullmann), Schweitzerbart'sche Verlagsbuchhandlung, Stuttgart, pp. 51–66.

Glenister, B. F. and Furnish, W. M. (1988b) Patterns in stratigraphic distribution of Popanocerataceae, Permian ammonoids. *Senckenbergiana Lethaea*, **69**, 43–71.

Glenister, B. F., Windle, D. L. and Furnish, W. M. (1973) Australasian Metalegoceratidae (Lower Permian ammonoids). *Journal of Paleontology*, **47**, 1031–43.

Glenister, B. F., Baker, C., Furnish, W. M. *et al*. (1990) Additional Early Permian ammonoid cephalopods from Western Australia. *Journal of Paleontology*, **64**, 392–9.

Glenister, B. F., Boyd, D. W., Furnich, W. M., Harris, M. T. *et al.* (in press) The Guadalupian: proposed international standard for a Middle Permian Series, in *Permian System of the World: Proceedings. Earth Sciences and Resources Institute* (ed. A. E. M. Nairn), Columbia, South Carolina.

Gordon, M. Jr (1966) An Upper Triassic bactritoid cephalopod from California. *Journal of Paleontology*, **40**, 1220–2.

Gordon, M. Jr (1986) Late Kinderhookian (Early Mississippian) ammonoids of western United States. *Journal of Paleontology*, **60**, *Memoir*, **19**, 1–36.

Guex, J. (1978) Le Trias inférieur des Salt Ranges (Pakistan): problèmes biochronologiques. *Eclogae Geologicae Helvetiae*, **71**, 105–41.

Harland, W. B., Cox, A. V., Llewellyn, P. G. *et al*. (1982) *A Geologic Time Scale*. Cambridge University Press, Cambridge, 131 pp.

House, M. R. (1962) Observations on the ammonoid succession of the North American Devonian. *Journal of Paleontology*, **36**, 247–84.

House, M. R. (1963) Devonian ammonoid successions and facies in Devon and Cornwall. *Quarterly Journal of the Geological Society of London*, **119**, 1–27.

House, M. R. and Kirchgasser, W. T. (1993) Devonian goniatite biostratigraphy and facies movements in the Frasnian of eastern North America. Geological Society Special Publication, **70** (in press).

House, M. R., Kirchgasser, W. T., Price, J. D. *et al.* (1985) Goniatites from Frasnian (Upper Devonian) and adjacent strata in the Montagne Noire. *Hercynica*, **1**, 1–21.

Johnston, F. N. (1941) Trias at New Pass, Nevada (new lower Karnic ammonoids). *Journal of Paleontology*, **15**, 447–91.

Kirchgasser, W. T. and House, M. R. (1981) Upper Devonian biostratigraphy, in *Devonian biostratigraphy of New York* (eds W. A. Oliver, Jr and G. Klapper), Subcommission on Devonian Stratigraphy, IUGS, Washington, pp. 39–55.

Korn, D. (1992) Relationship between shell form, septal construction and suture line in clymeniid cephalopods (Ammonoidea; Upper Devonian). *Neues Jahrbuch für Geologie und Paläontologie, Abhandlungen*, **185**, 115–30.

Korn, D. and Price, J. (1987) Taxonomy and Phylogeny of the Kosmoclymeniinae subfam. nov. (Cephalopoda, Ammonoidea, Clymeniida). *Courier Forschungsinstitut Senckenberg*, **92**, 5–75.

Krystyn, L. (1973) Zur Ammoniten- und Conodonten-Stratigraphie der Hallstatter Obertrias (Salzkammergut, Österreich). *Verhandlungen der Geologischen Bundesanstalt, A*, **1973**, 113–53.

Krystyn, L. (1974) Die *Tirolites*-Fauna der untertriassischen Werfener Schichten Europas und ihre stratigraphische Bedeutung. *Sitzungsberichte der Mathematisch–Naturwissenschaftlichen Klasse der Österreichischen Akademie der Wissenschaften, Abteilung 1*, **183**, 29–50.

Krystyn, L. (1980) Triassic conodont localities of the Saltzkammergut region. *Abhandlungen der Geologischen Bundesanstalt, Wien*, **35**, 61–98.

Krystyn, L. (1982) Obertriassische Ammonoideen aus dem zentralnepalesischen Himalaya (Gebiet vom Jomsom). *Abhandlungen der Geologischen Bundesanstalt, Wien*, **36**, 1–63.

Krystyn, L. and Mariolakos, I. (1975) Stratigraphie und Tektonik der Hallstätter-kalk-scholle von Epidauros. *Sitzungsberichte der Mathematisch–Naturwissenschaftlichen Klasse der Österreichischen Akademie der Wissenschaften, Abteilung 1*, **184**, 181–95.

Krystyn, L. and Wiedmann, J. (1986) Ein *Choristoceras*-Vorlaufer (Ceratitina, Ammonoidea) aus dem Nor von Timor. *Neues Jahrbuch für Geologie und Paläontologie, Monatshefte*, **1986**, 27–37.

Kullmann, J. (1962) Die Goniatiten des Unterkarbons im Kantabrischen Gebirge (Nordspanien). II. Paläontologie der U. O. Prolecanitina Miller and Furnish. Die Altersstellung der Faunen. *Neues Jahrbuch für Geologie und Paläontologie, Abhandlungen*, **116**, 269–324.

Kullmann, J. (1963) II. Paläontologischer Teil, *Namirski kat Druzetica i Njegora Gonijatitska Fauna* (ed. P. Stevanovic and J. Kullmann), Glasnik Prirodnachkog Muzeja Beograd, Serija A, 16/17, pp. 71–112.

Kullmann, J., Korn, D. and Weyier, D. (1991) Ammonoid zonation of the Lower Carboniferous subsystem. *Courier Forschungsinstitut Senckenberg*, **130**, 127–31.

Lange, W. (1929) Zur Kenntnis des Oberdevons am Enkeberg und bei Balve (Sauerland). *Abhandlung der Preußischen Geologischen Landesanstalt, Neues Folge*, **119**, 1–132.

Manger, W. L. (1971) The Mississippian ammonoids *Karagandoceras* and *Kazakhstania* from Ohio. *Journal of Paleontology*, **45**, 33–9.

Manger, W. L. (1988) Phylogeny of the Carboniferous ammonoid family Dimorphoceratidae, in *Cephalopods – Present and Past* (eds J. Wiedmann and J. Kullmann), Schweitzerbart'sche Verlagsbuchhandlung, Stuttgart, pp. 29–42.

Mapes, R. H. (1979) Carboniferous and Permian Bactritoidea (Cephalopoda) in North America. *University of Kansas Paleontological Contributions*, **64**, 1–75.

Matern, H. (1931) Das Oberdevon der Dill-Mulde. *Abhandlungen der Preußischen Geologischen Landesanstalt, Neues Folge*, **134**, 1–139.

McCaleb, J. A. (1968) Lower Pennsylvanian ammonoids from the Bloyd Formation of Arkansas and Oklahoma. *Special Paper of the Geological Society of America*, **96**, 1–91.

McLearn, F. H. (1960) Ammonoid faunas of the Upper Triassic Pardonet Formation, Peace River Foothills, British Columbia. *Memoir, Geological Survey of Canada*, **311**, 1–118.

Mikesh, D. L., Glenister, B. F. and Furnish, W. M. (1988) *Stenolobulites* n. gen., Early Permian ancestor of predominantly Late Permian paragastrioceratid subfamily Pseudogastrioceratinae. *The University of Kansas Paleontological Contributions, Paper*, **123**, 1–19.

Miller, A. K. and Downs, H. R. (1950) Ammonoids of the Pennsylvanian Finis Shale of Texas. *Journal of Paleontology*, **24**, 185–218.

Miller, A. K. and Furnish, W. M. (1940) Permian ammonoids of the Guadalupe Mountain region and adjacent areas. *Special Paper of the Geological Society of America*, **26**, 1–242.

Müller, A. H. (1973) Über Ammonoidea (Cephalopoda) aus der Grenzdolomit-region des germanischen Unterkeupers. *Zeitschrift für Geologische Wissenschaften, Berlin*, **1**, 935–45.

Nassichuk, W. W. (1975) Carboniferous ammonoids and stratigraphy in the Canadian Arctic Archipelago. *Bulletin of the Geological Society of Canada*, **237**, 1–240.

Paeckelmann, W. (1924) Das Devon und Karbon der Umgebung von Balve in Westfalen. *Jahrbuch der Preußischen Geologischen Landesanstalt*, **41**, 51–97.

Parnes, A. (1986) Middle Triassic cephalopods from the Negev (Israel) and Sinai (Egypt). *Bulletin, Geological Survey of Israel*, **79**, 1–59.

Petter, G. (1959) Goniatites dévoniennes du Sahara. *Publications du Sérvice de la Carte Géologique de l'Algérie, N.S., Paléontologie, Mémoires*, **2**, 1–371.

Plummer, F. R. and Scott, G. (1937) Upper Paleozoic ammonites in Texas. *The University of Texas Bulletin*, **3701** (1), 1–516.

Price, J. D. (1982) Some Famennian (Upper Devonian) Ammonoids from North Western Europe. Unpublished PhD Thesis, University of Hull, 555 pp.

Ramsbottom, W. H. (1970) Some British Carboniferous goniatites of the family Anthracoceratidae. *Bulletin of the Geological Survey of Great Britain*, **32**, 53–60.

Rieber, H. (1973) Cephalopoden aus der Grenzbitumen Zone (Mittlere Trias) des Monte San Giorgio (Kanton Tessin, Schweiz). *Schweizerische Paläontologische Abhandlungen*, **73**, 1–96.

Rieber, H. and Tozer, E. T. (1986) Discovery of the original specimen of *Ammonites nodosa* Bruguière 1789, type species of *Ceratites* De Haan 1825 (Ammonoidea, Triassic). *Eclogae Geologicae Helvetiae*, **79**, 827–34.

Ristedt, H. (1981) Bactriten aus dem Obersilur Böhmens. *Mitteilungen des Geologischen–Paläontologischen Institut, Universität Hamburg*, **51**, 23–6.

Ruzhentsev, V. E. (1950) Upper Carboniferous ammonoids from the Urals. *Akademiya Nauk SSSR, Paleontologicheskogo Instituta, Trudy*, **29**, 1–223.

Ruzhentsev, V. E. (1951) Lower Permian ammonoids of the southern Urals, I. Ammonoids of the Sakmarian Stage. *Akademiya Nauk SSSR, Paleontologicheskogo Instituta, Trudy*, **33**, 1–186.

Ruzhentsev, V. E. (1952) Biostratigraphy of the Sakmarian Stage in the Aktyubinsk region of Kazakhstan, USSR. *Akademiya Nauk SSSR, Paleontologicheskogo Instituta, Trudy*, **42**, 1–90.

Ruzhentsev, V. E. and Ganelin, V. G. (1971) Rukovodiashchie y srednekamennougol'nye ammonoidei na Omolomskom massive. *Paleontologicheskii Zhurnal*, **1971** (1), 49–61.

Schindewolf, O. H. (1923) Beiträge zur Kenntnis der Paläozoicums in Oberfranken, Ostthüringen und dem Sachsischen Vogtlande. I. Stratigraphie und Ammoneenfauna des Oberdevons von Hof a. S. *Neues Jahrbuch für Mineralogie, Geologie und Paläontologie, Beilage-Band*, **49**, 250–358, 393–509.

Schindewolf, O. H. (1932) Zur Stammesgeschichte der Ammoniteen. *Palaeontologische Zeitschrift*, **14**, 164–81.

Schindewolf, O. H. (1937a) Zur Stratigraphie und Paläontologie der Wocklumer Schichten. *Abhandlungen der Preußischen Geologisches Landesamt, Neue Folge*, **178**, 1–132.

Schindewolf, O. H. (1937b) Zwei neue bemerkenswerte Goniatiten-Gattungen des rheinischen Oberdevons. *Jahrbuch der Preußischen Geologisches Landesanstalt, Neue Folge*, **58**, 242–55.

Schindewolf, O. H. (1972) Über Clymenien und andere Cephalopoden. *Abhandlungen der Akademie der Wissenschaft und Literatur, Mainz, Mathematische–Naturwissenschaftliche Klasse*, **1971** (3), 55–141.

Schmidt, H. (1924) Zwei Cephalopodenfaunen an der Devon-Carbongrenze im Sauerland. *Jahrbuch der Preußischen Geologischen Landesanstalt*, **44**, 98–171.

Sheng, J.-Z., Chen, C.-Z., Wang, Y.-G., *et al.* (1984) Permian–Triassic boundary in middle and eastern Tethys. *Journal of the Faculty of Science, Hokkaido University, Series IV*, **21**, 133–81.

Shevyrev, A. A. (1986) Triassic Ammonoidea. *Trudy Paleontologicheskogo Instituta, Leningrad*, **217**, 1–184.

Shevyrev, A. A. and Yermakova, S. P. (1978) On the systematics of the ceratitoids. *Paleontologischeskii Zhurnal*, **1979** (1), 52–8.

Shimanskiy, V. N. (1951) K voprosu ob evoliutsii verlynepaleozoiskikh prynamkh golovonogikh. *Doklady Akademia Nauk*, **79**, 867–70.

Shimanskiy, V. N. (1962) Nadotrjad Bactritoidea. Bactritoidei, in *Osnovy Paleontologii, 5, Molluski Golovonogie*, 1 (ed. Y. A. Orlov), Nauka, Moscow, pp. 229–39.

Silberling, N. and Nichols, K. (1982) Middle Triassic molluscan-fossils of biostratigraphic significance from the Humboldt Range, Northwestern Nevada. *Professional Paper, United States Geological Survey*, **1207**, 1–77.

Silberling, N. and Tozer, E. T. (1968) Biostratigraphic classification of the marine Triassic in North America. *Special Papers, Geological Society of America*, **110**, 1–63.

Simakov, K. V. (ed.) (1985) Biostratigrafiya pogranichnykh Devonu i Karbona. *Severo-Bostochnyi Kompleksnyu Nauchno-Issledovatel'skii Institut, Vypusk*, **9**, 1–55.

Spath, L. F. (1951) Catalogue of the Fossil Cephalopoda in the British Museum (Natural History). Part V. The Ammonoidea of the Trias (II). British Museum, London, 228 pp.

Spinosa, C., Furnish, W. M. and Glenister, B. F. (1975) The Xenodiscidae, Permian ceratitoid ammonoids. *Journal of Paleontology*, **49**, 239–83.

Tatzreiter, F. (1980) Neue trachyostrake Ammonoideen aus dem Nor (Alaun 2) der Tethys. *Verhandlungen der Geologischen Bundesanstalt*, **1980**, 123–59.

Tatzreiter, F. (1985) Zur Kenntnis der obertriadischen (Nor; Alaun, Sevat) trachyostraken Ammonoideen. *Jahrbuch der Geologischen Bundesanstalt Wien*, **128**, 219–26.

Termier, H. and Termier, G. (1950) Paléontologie marocaine. II Invertébrés de l'Ère Primaire. Fascicule 3, Mollusques. *Service Géologique Protectorat de la République Française Marocaine, Notes et Mémoires*, **78**, 1–246.

Tharalson, D. B. (1984) Revision of the Early Permian ammonoid family Perrinitidae. *Journal of Paleonotology*, **58**, 804–33.

Tozer, E. T. (1965) Lower Triassic stages and ammonoid zones in Arctic Canada. *Paper, Geological Survey of Canada*, **65-12**, 1–14.

Tozer, E. T. (1980) Latest Triassic (Upper Norian) ammonoid and Monotis faunas and correlations. *Rivista Italiana di Paleontologia e Stratigraphia*, **85**, 843–76.

Tozer, E. T. (1981a) Triassic Ammonoidea: classification, evolution and relationship with Permian and Jurassic forms, in *The Ammonoidea* (eds M. R. House and J. R. Senior), Systematics Association, London, pp. 65–100.

Tozer, E. T. (1981b) Triassic Ammonoidea: geographic and stratigraphic distribution, in *The Ammonoidea* (eds M. R. House and J. R. Senior), Systematics Association, London, pp. 397–431.

Tozer, E. T. (1982) Late Triassic (Upper Norian) and earliest Jurassic (Hettangian) rocks and ammonoid faunas, Halfway River and Pine Pass map areas, British Columbia. *Paper, Geological Survey of Canada*, **82-1A**, 385–91.

Tozer, E. T. (1984) The Trias and its ammonoids: The evolution of a time scale. *Miscellaneous Reports, Geological Survey of Canada*, **35**, 1–171.

Trümpy, R. (1969) Lower Triassic ammonites from Jameson Land (East Greenland). *Meddelelser om Grønland*, **168**, 77–116.

Urlichs, M. (1977) Zur Altersstellung der Pachycardienstufe und der Unteren Cassianer Schichten in den Dolomiten (Italien). *Mitteilungen der Bayerischen Staatssammlung für Paläontologie und historische Geologie*, **17**, 15–25.

Urlichs, M. and Mundlos, R. (1985) Immigration of cephalopods into the Germanic Muschelkalk Basin and its influence on their suture, in *Sedimentary and Evolutionary Cycles* (eds U. Bayer and A. Seilacher), Lecture Notes in Earth Sciences, 1, Springer,

Stuttgart, pp. 221–36.

Wang, Y. G. (1984) Earliest Triassic ammonoid fauna from Jiangsu and Zhejiang and its bearing on the definition of the Permian–Triassic boundary. *Acta Palaeontologica Sinica*, **23**, 257–70.

Wang, Y. G. (1985) Remarks on the Scythian–Anisian boundary. *Rivista Italiana di Paleontologia e Statigraphia*, **90**, 515–44.

Wang, Y. G. and He, G.-X. (1980) Triassic ammonoid sequence of China. *Rivista Italiana di Paleontologia e Stratigraphia*, **85**, 1207–20.

Wedekind, R. (1908) Die Cephalopodenfauna des hoheren Oberdevon am Enkeberge. *Neues Jahrbuch für Mineralogic, Geologie und Paläontologie*, **26**, 565–634.

Wedekind, R. (1913) Die Goniatitenkalke des unteren Oberdevon von Martenberg bei Adorf. *Sitzungsberichte der Gesellschaft Naturforschender Freunde zu Berlin*, **1**, 1–77.

Wedekind, R. (1914) Monographie der Clymenien des Rheinischen Gebirges. *Abhandlungen der Könige Gesellschaft der Wissenschaften zu Göttingen, Mathematische–Physikalische Klasse, Neue Folge*, **10**, 1–73.

Wedekind, R. (1917 (1918)) Die Genera der Palaeoammonoidea (Goniatiten). *Palaeontographica*, **62**, 85–184.

Weyer, D. (1972) Trilobiten und Ammonoideen aus der *Entogonites nasutus*-Zone (Unterkarbon) des Buchenbergsattels (Elbingeroder Komplex, Harz). Teil 2. *Geologie*, **21**, 318–49.

Wiedmann, J. (1977) On the significance of ammonite nuclei from sieve residues. *Annales des Mines et de la Géologie Tunis*, **28**, 135–61.

Zacharov, Yu. D. (1988) Parallelism and ontogenetic acceleration in ammonoid evolution, in *Cephalopods – Present and Past* (eds J. Wiedmann and J. Kullamann), Schweizerbart'sche Buchhandlung, Stuttgart, pp. 191–206.

Zhao, J.-K., Liang, X. L. and Zheng, Z. G. (1978) Late Permian cephalopods of South China. *Palaeontologica Sinica*, **154**, 1–194 [in Chinese].

11

MOLLUSCA: CEPHALOPODA (AMMONOIDEA: PHYLLOCERATINA, LYTOCERATINA, AMMONITINA AND ANCYLOCERATINA)

K. N. Page

The classification followed here is essentially that of Donovan *et al.* (1981) and Wright (1981). When known, bio- and chronostratigraphical details are quoted, in a form which follows conventional usage, i.e. for Triassic, Jurassic and early Cretaceous (Berriasian to Barremian) zones, non-italicized specific names of indices are used (e.g. Herveyi Zone); for mid to late Cretaceous (Aptian to Maastrichtian) zones, specific names are italicized (e.g. *ultimus* Zone).

Only selected examples of early and late records are quoted, and it should be noted that for many, alternative localities and nominal taxa could be cited, frequently in regions widely separated geographically.

Faunal provincialism can create problems of correlation between such regions, and commonly precludes the accurate stratigraphical separation of different records. In addition, the frequent use of the first occurrence of genera to correlate the base of zones can lead to a spurious simultaneity of early occurrences in different provinces.

Continually improving zonal schemes and interregional correlations will eventually resolve some of these problems, and it should be emphasized therefore that no list of first and last ammonite family occurrences can ever be in any way definitive.

Contributions from J. H. Callomon, D. T. Donovan and P. F. Rawson (University College, London), W. J. Kennedy (University of Oxford) and R. A. Hewitt (Leigh-on-Sea) are indicated by [JHC], [DTD], [PFR], [WJK] and [RAH] respectively.

Acknowledgements – J. H. Callomon, D. T. Donovan, W. J. Kennedy and R. A. Hewitt read and suggested alterations to sections of an early draft and provided additional records, and P. F. Rawson assisted with the tracing of further records.

Order AMMONOIDEA Zittel, 1884

Suborder PHYLLOCERATINA Zittel, 1884
(see Fig. 11.1)

Superfamily PHYLLOCERATACEAE Zittel, 1884

F. USSURITIDAE Hyatt, 1900 Tr. (SCY–RHT) Mar.

First: *Burijites skorochodi* (Burijzharnikovi); Prinoye, former USSR (Shevyrev, 1986). Upper Nammalian. [RAH]
Last: *Eopsiloceras planorboides* (Gümbel); Schicht alpha 5, Alps (Wiedmann *et al.*, 1979, p. 138). Marshi Zone, Stuerzenbaumi Subzone (RHT 3a).

F. DISCOPHYLLITIDAE Spath, 1927 Tr. (NML)–(RHT) Mar.

First: *Rhacophyllites zitteli* (Mojsisovics); *Lubites* Beds, Feuerkogel, Austria (Shevyrev, 1986, Spath, 1934). Aonoides Zone (CRN 1). [RAH]
Last: *Rhacophyllites neojurensis* (Quenstedt); Kössener Schichten, Austria (Krystyn, 1973). ?Upper Marshi Zone (RHT 3b). [RAH]

F. PHYLLOCERATIDAE Zittel, 1884 J. (HET)–K. (MAA) Mar.

First: *?Calliphylloceras psilomorphum* (Neumayr); eastern Karwendelgebirge north of Innsbruck, Austria (Lange, 1952). Calliphyllum Zone (Planorbis Zone of sub-boreal zonation).
Last: Phylloceratid indet.; Upper Unit 12 (1 m below Cretaceous–Tertiary boundary, Zumaya, Spain (Wiedmann, 1988, p. 136) [RAH]. *Neophylloceras velledaeforme* (Schlüter); terminal Maastrichtian hardground, Stevns Klint, Denmark (Birkelund, 1979). *Belemnitella casimirovensis* Zone.

F. JURAPHYLLITIDAE Arkell, 1950 J. (HET–TOA) Mar.

First: *Schistophylloceras aulonotum* (Herbich); eastern Karwendelgebirge north of Innsbruck, Austria (Lange, 1952). Calliphyllum Zone (Planorbis Zone). *Nevadaphyllites compressus* Guex; Bed Z-5, New York Canyon, Gabbs Valley Range, Nevada, USA (Guex, 1980, pp. 129, 135; pl. 1, fig. 7). Planorbis Zone, Planorbis Subzone.
Last: *Meneghiniceras lariense* (Meneghini); Grey Shales,

The Fossil Record 2. Edited by M. J. Benton. Published in 1993 by Chapman & Hall, London. ISBN 0 412 39380 8

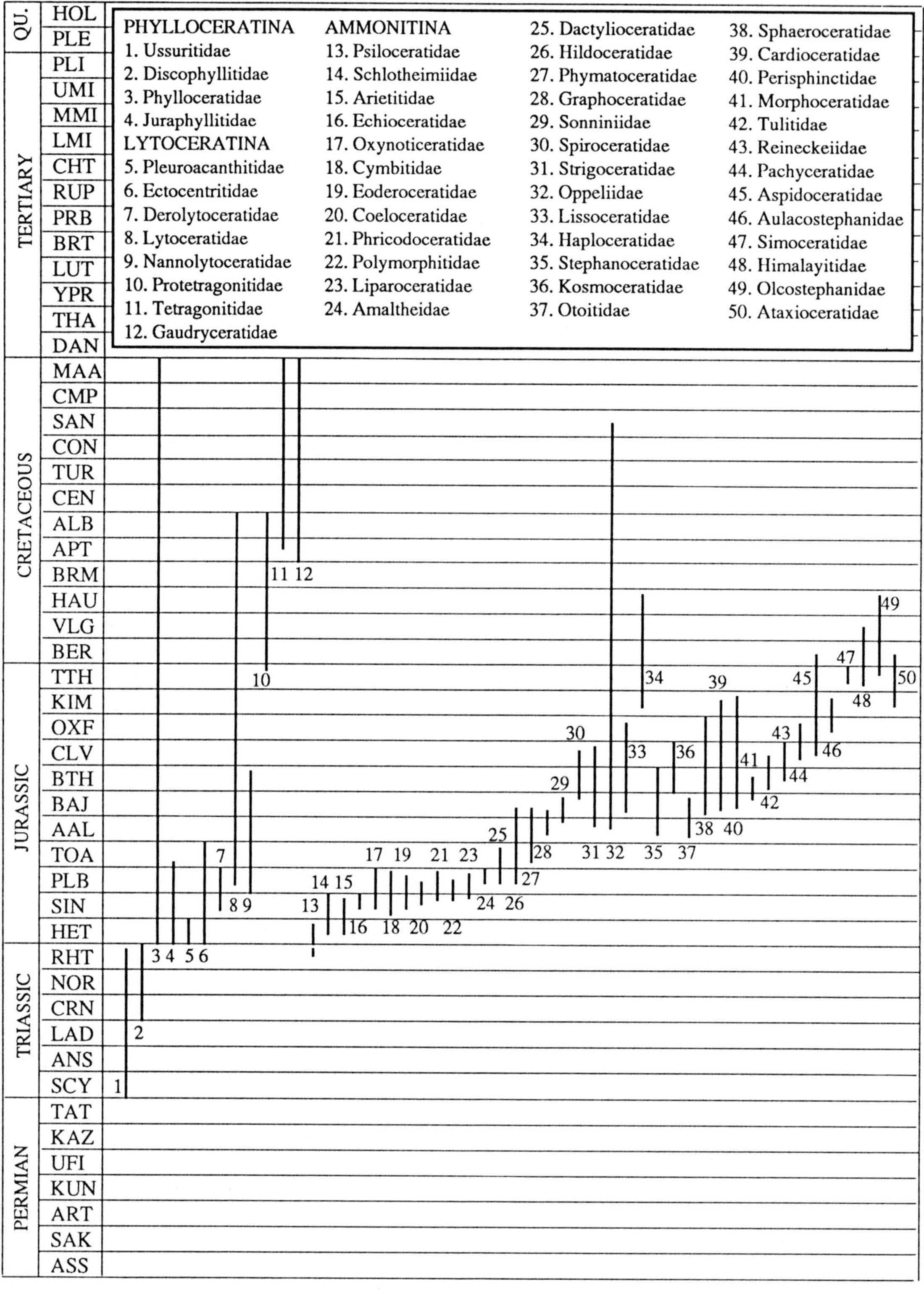

Fig. 11.1

near Whitby, North Yorkshire, England, UK (Howarth, 1976; Donovan and Howarth, 1982, p. 439). Tenuicostatum Zone, Semicelatum Subzone. *Juraphyllites libertus* (Gemmellaro); Rappel, Morocco (Benshili, 1989, pp. 43, 154). Polymorphum Zone, ?Mirabile Subzone (equivalent, in part, to the Tenuicostatum Zone of sub-boreal zonation).

Suborder LYTOCERATINA Hyatt, 1889

Superfamily LYTOCERATACEAE Neumayr, 1875

F. PLEUROACANTHITIDAE Hyatt, 1900

J. (HET) Mar.

First: *Analytoceras articulatus* (Sowerby); Austria (Wähner, 1894, pl. 3, fig. 3; pl. 7, figs 1–5; pl. 8, fig. 1; pl. 9, figs 1–2; Roman, 1938, p. 31, fig. 3–23). Lower Hettangian.

Last: *Pleuroacanthites biformis* (J. de C. Sowerby); Italy (Canavari, 1882), Hettangian.

F. ECTOCENTRITIDAE Spath, 1926
J. (HET–TOA) Mar.

First: *Ectocentrites* cf. *petersi* (Hauer); South America (von Hillebrandt, 1981; Riccardi *et al.*, 1990, p. 95). Reissi Zone.
Last: *Holcolytoceras*; Oranie, Algeria (Elmi and Caloo-Fortier, 1985). Middle Aalensis Zone (Levesquei Zone, Aalensis Subzone of sub-boreal region).

F. DEROLYTOCERATIDAE Spath, 1927
J. (?SIN–PLB) Mar.

First: *Derolytoceras haueri* (Stur); Hierlatz, Austria (Rosenberg, 1909; Roman, 1938, p. 32). ?Sinemurian (according to Roman).
Last: *Derolytoceras tortum* (Quenstedt); Swabia, Germany (Rosenberg, 1909, p. 259, figs 29a–d, 30; Roman, 1938, p. 32, fig. 3–24). Upper Pliensbachian.

F. LYTOCERATIDAE Neumayr, 1875
J. (PLB)–K. (ALB) Mar.

First: *Lytoceras interlineatum* (Buckman); Radstock, Avon, England, UK (Buckman, 1921, pl. 204A, B). Ibex Zone, Valdani Subzone.
Last: *Ammonoceratites ezoense* Yabe; Japan; (Matsumoto, 1954; discussion in Kennedy and Klinger, 1978). Upper Albian. [WJK]

F. NANNOLYTOCERATIDAE Spath, 1927
J. (PLB–BTH) Mar.

First: *Audaxlytoceras* sp.; Breggia River section near Chiasso, Lombardi Alps, Italy (Wiedenmayer, 1980, p. 39). Jamesoni Zone.
Last: *Nannolytoceras* sp. nov.; Bas Autan, near Digne, Basses-Alpes, SE France (Sturani, 1966, p. 23, pl. 3, fig. 8). Progracilis Zone.

Superfamily TETRAGONITACEAE Hyatt, 1900

F. PROTETRAGONITIDAE Spath, 1923
J. (TTH)–K. (ALB) Mar.

First: *Protetragonites quadrisulcatum* (d'Orbigny); Djebel Oust, northern Tunisia (Memmi and Salaj, 1975). Upper Tithonian, 'Jacobi Zone' (lower part, = Microacanthum Zone?).
Last: *Hemitetragonites strangulatus* (d'Orbigny), *H.* cf. *crebrisulcatus* Uhlig; east of Vohimaranitra, Madagascar (Collignon, 1963, pl. 244, figs 1048–1050). Albian.

F. TETRAGONITIDAE Hyatt, 1900
K. (APT–MAA) Mar.

First: *Jauberticeras jauberti* (d'Orbigny); Lesches-en-Diois, Drôme, SE France (Moullade, 1966, fig. 16). Lower 'Gargasien' (Upper Aptian), *guettardi* Zone.
Last: *Saghalinites* sp.; terminal Maastrichtian hardground, Stevns Klint, Denmark (Birkelund, 1979). *Belemnitella casimirovensis* Zone.

F. GAUDRYCERATIDAE Spath, 1927
K. (APT–MAA) Mar.

First: *Eogaudryceras numidum* (Coquand) Sayn; Gargasien, SE France (Jacob, 1908; Casey, 1960, pp. 7–8). Aptian.
Last: *Zelandites varuna* Forbes; terminal Maastrichtian, Lopez de Bertodano Formation, Seymour Island, Antarctica (Macellari, 1986). *ultimus* Zone.

Suborder AMMONITINA Hyatt, 1889

Superfamily PSILOCERATACEAE Hyatt, 1867

F. PSILOCERATIDAE Hyatt, 1867
?Tr. (RHT)–J. (HET) Mar.

First: Psiloceratid sp. indet.; Westbury Formation, Penarth Group, Chipping Sodbury, Avon, England, UK (Donovan *et al.*, 1989). This record is significantly earlier than the first occurrence of *Psiloceras* ex grp *planorbis* (J. de C. Sowerby) in the British succession in the overlying Lower Lias Group. Following Torrens and Getty (*in* Cope *et al.*, 1980), the Penarth Group is considered to be of Triassic (Rhaetian) age.
Last: *Caloceras leptoptychum* (Lange); Lower Lias Group, Burton Row Borehole, Somerset, England, UK (Ivimey-Cook and Donovan, *in* Whitaker and Green, 1983). Liasicus Zone, upper part of Laqueus Subzone.

F. SCHLOTHEIMIIDAE Spath, 1923
J. (HET–SIN) Mar.

First: *Waehneroceras* spp. including *W. prometheus* (Reynès); Bed 43, Lower Lias Group, West Somerset coast, England, UK (Ivimey-Cook and Donovan, *in* Whitaker and Green, 1983). Liasicus Zone, basal Portlocki Subzone.
Last: *Angulaticeras* [*Pseudoschlotheimia*] *densilobata* (Pompeckj); (Pompeckj, 1893; Spath, 1924, p. 194). ?Raricostatum Zone. [DTD]

Superfamily ARIETITACEA Hyatt, 1875

F. ARIETITIDAE Hyatt, 1875 J. (HET–SIN) Mar.

First: *Alsatites* [*Proarietites*] *laqueolus* (Schloenbach); Lias, southern Germany (Lange, 1931). 'Zone des *Saxoceras schroederi*' (alpha 1d), (Liasicus Zone, Portlocki Subzone; Donovan *in* Dean *et al.*, 1961, p. 446).
Last: *Eparietites denotatus* (Simpson); Lower Lias Group, Robin Hood's Bay, North Yorkshire, England, UK (Simpson, 1855, p. 76; Buckman, 1912, pl. 67A, B). Obtusum Zone, Denotatus Subzone.
Comment: *Eparietites* evolved directly into *Oxynoticeras* (family Oxynoticeratidae); the disappearance of the Arietitidae is therefore a pseudo-extinction.

F. ECHIOCERATIDAE Buckman, 1913
J. (SIN) Mar.

First: *Epophioceras* ['*Caenisites*'] *pseudobonnardi* (Spath); Bed 75, Shales-with-Beef, Lyme Regis, Dorset, England, UK (Lang *et al.*, 1923, p. 76); Turneri Zone, Birchi Subzone.
Last: *Paltechioceras* ex grp. *aplanatum* (Hyatt); Lower Lias Group, Robin Hood's Bay, North Yorkshire, England, UK (Getty, 1973, p. 20). Raricostatum Zone, upper Aplanatum Subzone.
Comments: The Echioceratidae are direct descendants of the Arietitid genus *Epophioceras* (Getty, 1973). Inclusion of *Epophioceras* in the Echioceratidae is therefore preferable on phylogenetic grounds. Features characteristic of *Epophioceras* occur already in '*Caenisites*' *pseudobonnardi* (Spath).

F. OXYNOTICERATIDAE Hyatt, 1875
J. (SIN–PLB) Mar.

First: *Oxynoticeras simpsoni* (Simpson); Lower Lias Group, Robin Hood's Bay, North Yorkshire, England, UK

(Simpson, 1843, pp. 37, 38; Buckman, 1912, pls 66A, B). Oxynotum Zone, basal Simpsoni Subzone.
Last: *Fanninoceras disciforme* von Hillebrandt; Chile, Argentina, Peru (von Hillebrandt, 1981; 1984). Disciforme Zone (top Pliensbachian). [DTD]

Superfamily CYMBITACEAE Buckman, 1919

F. CYMBITIDAE Buckman, 1919 J. (SIN–PLB) Mar.

First: *Cymbites laevigatus* (J. de C. Sowerby); Lower Lias Group, Burton Row Borehole, Somerset, England, UK (Ivimey-Cook and Donovan, *in* Whitaker and Green, 1983, p. 127). Bucklandi Zone, Rotiforme Subzone.
Last: *Cymbites centriglobus* (Oppel); Bed 2, Eype Clay, Middle Lias, near Eype, Dorset, England, UK (Howarth, 1957, p. 196, pl. 17, figs 3a–d). Margaritatus Zone, Stokesi Subzone.
Comment: A group of dwarf, globose ammonites. Diagnostic features are few and a polyphyletic origin has often been postulated (e.g. *in* Arkell *et al.*, 1957, p. L240). However, an early sutural ontogeny distinct from contemporary Ammonitina may indicate a single origin (Schindewolf, 1962; Donovan *et al.*, 1981, p. 109).

Superfamily EODEROCERATACEAE Spath, 1929

F. EODEROCERATIDAE Spath, 1929 J. (SIN–PLB) Mar.

First: *Microderoceras* sp.; Bed 74g, Shales-with-Beef, Lower Lias Group, Lyme Regis–Charmouth district, Dorset, England, UK (Lang *et al.*, 1923). Turneri Zone, basal Birchi Subzone. A record from the Semicostatum Zone (Walliser, 1956) is considered to be unconfirmed by Donovan *et al.* (1967, p. 453).
Last: *Metaderoceras mouterdei* (Frebold); western Canada and USA (Smith *et al.*, 1988, p. 1511). Lowest Kunae Zone, probably equivalent to upper Davoei Zone of European succession.

F. COELOCERATIDAE Haug, 1910 J. (SIN–PLB) Mar.

First: *Tetraspidoceras* aff. *birchiades* (Rosenberg); Gola del Fiume Bosse, Appennino Marchiagiano, Umbria, Italy (Ferretti, 1975, p. 177, pl. 23; Donovan, 1990a, p. 259). Oxynotum Zone, Oxynotum Subzone. [DTD]
Last: *Coeloceras pettos* (Quenstedt); Bed 118b, Belemnite Marls, Lower Lias Group, Charmouth-Seatown district, Dorset coast, England, UK (Lang *et al.*, 1928, p. 192). Jamesoni Zone, Jamesoni Subzone.

F. PHRICODOCERATIDAE Spath, 1938 J. (SIN–PLB) Mar.

First: *Epideroceras exhaeredatum* S. Buckman; Bed 103, Black Ven Marls, Lower Lias Group, Charmouth, Dorset, England, UK (Lang *et al.*, 1926, p. 155). Raricostatum Zone, Raricostatoides Subzone.
Last: *Phricoderoceras subtaylori* (Krumbeck); Arzo, Breggia, etc. Lombardi Alps (Wiedenmayer, 1980, pp. 50–51). Margaritatus Zone, Subnodosus Subzone.

F. POLYMORPHITIDAE Haug, 1887 J. (SIN–PLB) Mar.

First: *Leptonotoceras suessi* (Hauer); Lower Lias Group, Stowell Park Borehole, Gloucestershire, England, UK (Spath, 1956, p. 149). Raricostatum Zone, ?Densinodulum Subzone.
Last: *Acanthopleuroceras lepidum* Tutcher and Trueman (=*alisiense* auctt., *stahli* auct. non Oppel); Portugal (Dommergues and Mouterde, 1981). Ibex Zone, Luridum Subzone [JHC].

F. LIPAROCERATIDAE Hyatt, 1867 J. (SIN–PLB) Mar.

First: *Vicininodiceras simplicicosta* (Trueman); fine-grained sandstone band in Pabba Shale, Allt Fearns, Isle of Raasay, Hebrides, Scotland, UK (Donovan, 1990b). Raricostatum Zone, topmost Aplanatum Subzone. [DTD]
Last: *Liparoceras (Becheiceras) nautiliforme* (J. Buckman); Thorncombe Sands, Middle Lias, near Eype, Dorset, England, UK (Howarth, 1957, p. 196). Margaritatus Zone, Subnodosus Subzone.

F. AMALTHEIDAE Hyatt, 1867 J. (PLB) Mar.

First: *Amaltheus* ex grp *stokesi* (J. Sowerby); Bed 131 Green Ammonite Beds, Lower Lias Group, Seatown district, Dorset, England, UK (Lang, 1936, p. 431). Margaritatus Zone, Stokesi Subzone.
Last: *Pleuroceras hawskerense* (Young and Bird); Bed 25, Kettleness, North Yorkshire, England, UK (Howarth, 1955, pp. 156, 164). Spinatum Zone, Hawskerense Subzone.
Comment: Descendants of the *Beaniceras–Aegoceras–Oistoceras* lineage (Liparoceratidae), originating in the Ibex Zone.

F. DACTYLIOCERATIDAE Hyatt, 1867 J. (PLB–TOA) Mar.

First: *Reynesocoeloceras praeincertum* Dommergues and Mouterde; Peniche, Portugal (Dommergues *et al.*, 1983). Ibex Zone, Luridum Subzone.
Last: *Catacoeloceras confectum* Buckman, *C. dumortieri* (Mauberge); Cotswolds Sands, Gloucestershire, England, UK (Buckman, 1889). Variabilis Zone.

Superfamily HILDOCERATACEAE Hyatt, 1867

F. HILDOCERATIDAE Hyatt, 1867 J. (PLB–BAJ) Mar.

First: *Protogrammoceras mellahense* Dubar; 3 m below *Tropidoceras stahli* Bed, south-east of Gourama and also west of Ziz, Morocco (Dubar, 1961). Ibex Zone.
Last: *Vacekia intermedia* Imlay [Macroconch]/*Asthenoceras delicatum* Imlay [Microconch]; Oregon and south Alaska, USA (Imlay, 1973). Lower Bajocian, ?Ovalis Zone. [JHC]
Comment: A complex multibranching family comprising many separate lineages.

F. PHYMATOCERATIDAE Hyatt, 1867 J. (TOA–BAJ) Mar.

First: *Phymatoceras* ex grp *lilli* Hauer; *Hildoceras bifrons* Horizon (D), central-west France (Gabilly *in* Gabilly *et al.*, 1971, p. 612). Bifrons Zone, ?Fibulatum Subzone.
Last: *Fissilobiceras fissilobatum* (Waagen), Sandford Lane, near Sherbourne, Dorset, England, UK (Callomon and Chandler, 1990, p. 97). Laeviuscula Zone, Trigonalis Subzone.

F. GRAPHOCERATIDAE Buckman, 1905 J. (AAL–BAJ) Mar.

First: *Ludwigia crassa* (Horn), *L. haugi* Douvillé; *opalinoides* Horizon, SE French Jura (Contini, 1969). Murchisonae Zone, Haugi Subzone.
Last: *Hyperlioceras subtectum* (Buckman); Horizon Bj 3,

near Broadwinsor, Dorset, England, UK (Callomon and Chandler, 1990, p. 96). Upper Discites Zone. [JHC]

F. SONNINIIDAE Buckman, 1892 J. (AAL–BAJ) Mar.

First: *Euhoploceras* cf. *crassiforme* (S. Buckman); basal part of Bradford Abbas Bed, Inferior Oolite, Bradford Abbas, Dorset, England, UK (Parsons, 1974, p. 171). Concavum Zone, Concavum Subzone.
Last: *Dorsetensia regrediens* Haug; Bed 4c, Oborne Road Stone, Inferior Oolite, Oborne, Dorset, England, UK (Parsons, 1976, p. 130). Humphriesianum Zone, Humphriesianum Subzone.

Superfamily SPIROCERATACEAE Hyatt, 1900

F. SPIROCERATIDAE Hyatt 1900 J. (BAJ–CLV) Mar.

First: *Spiroceras baculatum* (Quenstedt); Bed 6d, Cadomensis Beds, Inferior Oolite, Oborne, Dorset, England, UK (Parsons, 1976, pp. 126, 129). Subfurcatum Zone, Baculata Subzone.
Last: *Parapatoceras* sp. [*Spiroceras calloviense* Morris]; Vesaignes, Haute-Marne, France (Corroy, 1932, p. 60). ?Jason Zone.
Comment: As discussed by Callomon (*in* Donovan *et al.*, 1981, p. 130), the Spirocerataceae are stratigraphically isolated from later heteromorphs assigned to the Suborder Ancyloceratina. Important sutural differences may also suggest a separate derivation, with the Spirocerataceae apparently developing from the Hildoceratacean genus *Tmetoceras* (Callomon, loc. cit.).

Superfamily HAPLOCERATACEAE Zittel, 1884

F. STRIGOCERATIDAE Buckman, 1924 J. (AAL–CLV) Mar.

First: *Praestrigites praenuntius* S. Buckman; Hornpark Ironshot Beds, Inferior Oolite, Beaminster, Dorset, England, UK (Buckman, 1924, pl. 466; Callomon *in* Donovan *et al.*, 1981, pp. 120, 144). Murchisonae Zone, Bradfordensis Subzone.
Last: *Phlycticeras pustulatum* (Reinecke); Beds 8c–d, Sengenthal, Franconia, Germany (Callomon *et al.*, 1987). Upper Coronatum Zone. [JHC]

F. OPPELIIDAE Douvillé, 1890 J. (AAL)–K. (SAN) Mar.

First: *Bradfordia liomphala* S. Buckman; basal part of Bradford Abbas Bed, Inferior Oolite, Bradford Abbas, Dorset, England, UK (Parsons, 1974, p. 171). Concavum Zone, Concavum Subzone.
Last: *Binneyites rugosus* Cobban; Kelvin Member, southwest of Shelby, Montana, USA (Cobban, 1961). *vermiformis* Zone.

F. LISSOCERATIDAE Douvillé, 1885 J. (BAJ–OXF) Mar.

First: *Lissoceras* cf. *semicostulatum* S. Buckman; Bed 9, Elton Farm Limestone, Inferior Oolite, Dundry, Avon, England, UK (Parsons, 1979, p. 141). Laeviuscula Zone, Laeviuscula Subzone.
Last: *Lissoceras (Lissoceratoides) erato* (d'Orbigny); Schilli Horizon, France (Enay *et al.*, 1971, p. 640). Tranversarium Zone, Parandieri Subzone.

F. HAPLOCERATIDAE Zittel, 1884 J. (KIM)–K. (HAU) Mar.

First: *Haploceras subclimatum* (Fontannes); southern Germany (White Jura E) and Ardèche, France (Berckhemer and Hölder, 1959, p. 106). Beckeri Zone (Upper Kimmeridgian). [JHC]
Last: *Haploceras (Neolissoceras) grasianum* (d'Orbigny); Beds 1–4, Angles, Basses-Alpes, SE France (Busnardo, 1965, p. 105, table 1). Sayni Zone.

Superfamily STEPHANOCERATACEAE Neumayr, 1875

F. STEPHANOCERATIDAE Neumayr, 1875 J. (AAL–CLV) Mar.

First: *Stephanoceras (Abbasitoides) modestum* (Vacek); Inferior Oolite, Dorset, England, UK (Callomon and Chandler, 1990, p. 94). Murchisonae Zone, Murchisonae Subzone. [JHC]
Last: *Cadomites altispinosum* Dietl and Herold; *quenstedti* Horizon, Macrocephalen-Oolith, SW Germany (Dietl and Herold, 1986; Callomon *et al.*, 1989). Herveyi Zone, Keppleri Subzone.

F. KOSMOCERATIDAE Haug, 1887 J. (BTH–CLV) Mar.

First: *Kepplerites costidersus* (Imlay), Rierdon Formation, Montana, USA (Callomon, 1984, pp. 153, 159; Westermann, 1956). Bathonian. [JHC]
Last: *Kosmoceras (K.) spinosum* (J. de C. Sowerby); Lamberti Bed, Oxford Clay, Woodham, Buckinghamshire, England, UK (Arkell, 1939, p. 138). Lamberti Zone, Lamberti Subzone.

F. OTOITIDAE Mascke, 1907 J. (AAL–BAJ) Mar.

First: *Abbasites abbas* S. Buckman; Paving Bed, Inferior Oolite, Bradford Abbas, Dorset, England, UK (Buckman, 1912, pl. 236). Murchisonae Zone, Bradfordensis Subzone.
Last: *Zemistephanus richardsoni* (Whiteaves); Queen Charlotte Islands, British Columbia, Canada; southern Alaska, USA (Hall and Westermann, 1980, p. 20). Richardsoni Subzone (=upper Sauzei Zone?). [JHC]

F. SPHAEROCERATIDAE Buckman, 1930 J. (BAJ–OXF) Mar.

First: *Sphaeroceras [Chondroceras] sp. nov.*; Bed 3 (Spissa Bed), Corton Denham Member, Inferior Oolite, Oborne, Dorset, England, UK (Parsons, 1976, p. 132). Laeviuscula Zone, Laeviuscula Subzone.
Last: *Epimayites subtumidus* (Waagen), *E. transiens* (Waagen), etc.; Kantcote Sandstone, west of Kantcote, Cutch, India (Spath, 1928). Upper Oxfordian.

F. CARDIOCERATIDAE Siemiradzki, 1891 J. (BAJ–KIM) Mar.

First: *Cranocephalites borealis* (Spath); East Greenland (Callomon, 1985, pp. 63, 64). Boreal Bathonian, Borealis Zone (equivalent to upper Bajocian of sub-boreal region).
Last: *Amoeboceras (Nannocardioceras) anglicum* (Salfeld); Kimmeridge Clay, Dorset, Yorkshire, widespread in the UK (Callomon, 1985, p. 74). Lower Autissiodorensis Zone.

Superfamily PERISPHINCTACEAE Steinmann, 1890

F. PERISPHINCTIDAE Steinmann, 1890 J. (BAJ–KIM) Mar.

First: *Leptosphinctes (L.) chadonensis* Pavia; Chandon, SE France (Pavia, 1983, p. 163). Humphriesianum Zone, Blagdeni Subzone. [JHC]
Last: *Mesosimoceras zullianum* (Parona); Bed 177, Chateauneuf d'Oze, SE France (Atrops, 1982, p. 305). Divisum Zone.
Comment: The Perisphinctaceae have always created classification problems, largely due to wide intraspecific variation which compounds the problems caused by the close homeomorphism between stratigraphically and geographically unrelated forms. An attempt at a primarily phylogenetic classification by Callomon (*in* Donovan *et al.*, 1981, p. 123) 'revealed a family tree of almost innumerable trunks and branches of all lengths and thicknesses'.

F. MORPHOCERATIDAE Hyatt, 1900 J. (BAJ–BTH) Mar.

First: *Dimorphinites* sp.; Astarte Bed, Inferior Oolite, near Bridport, Dorset, England, UK (Senior *et al.*, 1970, p. 166). Garantiana Zone, Acris Subzone.
Last: *Asphinctes tenuiplicatus* (Brauns); Swabia, Germany (Hahn, *in* Buck *et al.*, 1966, p. 34; Torrens, 1971, p. 586). Zigzag Zone, Yeovilensis Subzone (=Tenuiplicatus Subzone of Hahn, 1971).

F. TULITIDAE Buckman, 1921 J. (BTH–CLV) Mar.

First: *Bullatimorphites latecentratus* (Quenstedt); Fuscus Bank, Swabia, Germany (Hahn, *in* Torrens, 1971, p. 58). Zigzag Zone, Yeovilensis Subzone.
Last: *Kheraiceras (Bomburites) bombur* (Oppel); Swabia, Germany (Calloman *et al.*, 1989, p. 11). Koenigi Zone, Curtilobus Subzone, *subcostarius* Horizon. [JHC]

F. REINECKEIIDAE Hyatt, 1900 J. (BTH–CLV) Mar.

First: *Neuqueniceras* ex grp *steinmanni* (Stehn); Andean Province, eastern Pacific (Riccardi *et al.*, 1989). Upper Bathonian, Steinmanni Zone. [JHC]
Last: *Reineckeia (Collotia)* cf. *angustilobata* (Brasil) Lamberti Bed, Oxford Clay, Woodham, Buckinghamshire, England, UK (Arkell, 1939, pp. 185–7). Lamberti Zone, Lamberti Subzone.

F. PACHYCERATIDAE Buckman, 1918 J. (CLV–OXF) Mar.

First: *Erymnoceras* ex grp *coronatum* (Bruguière); Bed 14, Lower Oxford Clay, Peterborough, England, UK (Callomon, *in* Sylvester-Bradley and Ford, 1968, p. 279). Coronatum Zone, Obductum Subzone.
Last: *Tornquistes liesbergense* (De Loriol) and *T.* spp.; Swiss and French Jura (Thierry and Charpy, 1982). Plicatilis Zone, Antecedens Subzone.

F. ASPIDOCERATIDAE Zittel, 1895 J. (CLV–BER) Mar.

First: *Binatisphinctes comptoni* (Pratt); Comptoni Bed, Lower Oxford Clay, East Midlands, England, UK (Callomon, *in* Sylvester-Bradley and Ford, 1968). Coronatum Zone, Grossouvrei Subzone.
Last: *Aspidoceras neobürgense* (Oppel) [=*A. rogozicense?* (Zeuschner)]; southern Spain (Cecca, 1985, p. 111). Jacobi Zone. [JHC]

F. AULACOSTEPHANIDAE Spath, 1924 J. (OXF–KIM) Mar.

First: *Decipia decipiens* (J. Sowerby), *D. lintonensis* Arkell; Bed 12, Ampthill Clay, Knapwell, Cambridgeshire, England, UK (Hancock, 1954, Gallois and Cox, 1977, pp. 218, 219). Tenuiserratum Zone, Blakei Subzone.
Last: *Aulacostephanus autissiodorensis* (Cotteau); Kimmeridge Clay, Dorset, widespread in England, UK, also occurs in France (Cope, 1968). Autissiodorensis Zone. [JHC]

F. SIMOCERATIDAE Spath, 1924 J. (TTH) Mar.

First: *Aulasimoceras auberti* (Pervinquière); Betic Cordilleras, southern Spain (Enay and Geyssant, 1975, p. 41). Lower Tithonian, upper part of Hybonotum Zone.
Last: *Cordubiceras gleminatum* Oloriz and Tavera [=Simoceratinae gen. et sp. nov.]; Betic Cordilleras, southern Spain (Enay and Geyssant, 1975, p. 45). Upper Tithonian, Microacanthum Zone. [JHC]

F. HIMALAYITIDAE Spath, 1925 J. (TTH)–K. (VLG) Mar.

First: *Aulacosphinctes* ex grp *linoptychus* Uhlig, *A.* ex grp. *spitiensis* Uhlig, *A.* cf. *rectefurcatus* (Zittel), *'Microacanthoceras' ponti* (Fallto and Termier); Betic Cordilleras, southern Spain (Enay and Geyssant, 1975, p. 44). Middle Tithonian, Ponti Zone.
Last: *Himalayites (?) nieri* (Pictet); SE France (Le Hégarat, 1971, pp. 144, 145). Roubardi Zone, Pertransiens Subzone.

F. OLCOSTEPHANIDAE Pavlow, 1892 J. (TTM)–K. (HAU) Mar.

First: *Spiticeras (Proniceras) pronum* (Oppel); SE France (Le Hégarat, 1971, p. 232). Transitorius Zone.
Last: *Olcostephanus (Jeannoticeras) jeannoti* (d'Orbigny), *O. (O.) variegatus* Paquier; SE France (Bullot, 1990). Base of Nodosoplicatum Zone. [PFR]

F. ATAXIOCERATIDAE Buckman, 1921 J. (KIM)–K. (BER) Mar.

First: *'Orthospinctes' virgulatus* (Quenstedt) *laufensis* (Siemiradzki); *'O' suevicus* Siemiradzki, SE France (Atrops, 1982, p. 48). Bimammatum Zone.
Last: *Parapallasiceras bochianensis* (Mazenot), *P. busnardoi* Le Hégarat; SE France (Le Hégarat, 1971, pp. 44–8). Grandis Zone.

F. NEOCOMITIDAE Salfeld, 1921 J. (TTH)–K. (HAU) Mar. (see Fig. 11.2)

First: *Berriasella* spp. including *B. subeudichotoma* Nikolov; Titcha Formation, Ticha Gorge, near Preslav, Bulgaria (Nikolov, 1982, pp. 20, 22). Transitorius Zone, Micracanthus Subzone.
Last: *'Cruasiceras' cruasense* (Torcapel); SE France (Thieuloy, 1977, pp. 435–6). Nodosoplicatum Zone. [PFR]

F. OOSTERELLIDAE Breistroffer, 1940 K. (VLG–HAU) Mar.

First: *Oosterella cultrata* (d'Orbigny); Agadir Basin, Morocco (Wiedmann *et al.*, 1978); 'Middle' Valanginian. *O.* aff. *cultrata* (d'Orbigny); Varlheide, northern Germany (Kemper *et al.*, 1981, p. 302). Crassus Zone. [PFR]

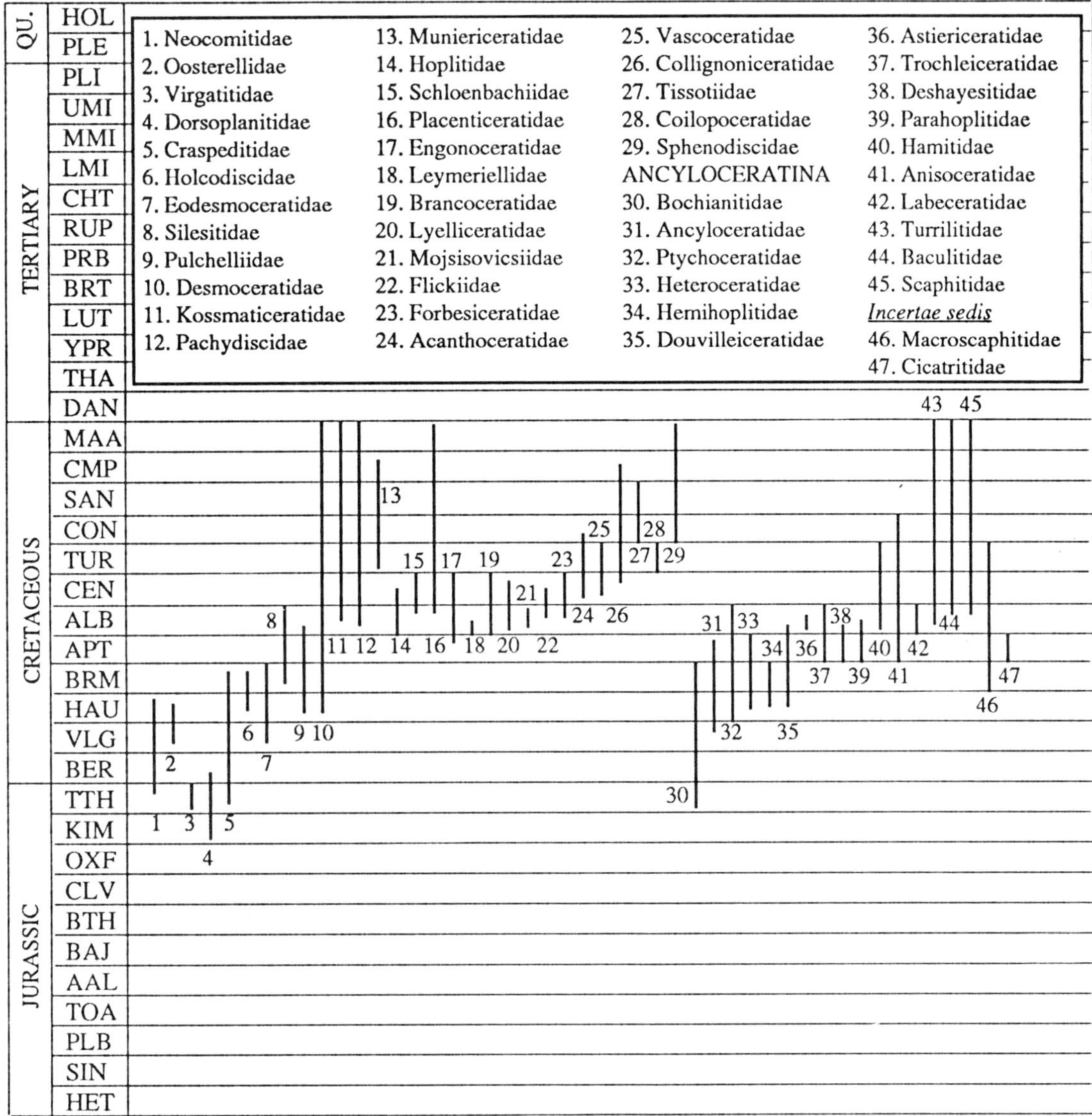

Fig. 11.2

Last: *Oosterella cultrata* (d'Orbigny); Kraptchéné, Mikhailovgrad region, Bulgaria (Nikolov, 1965). Lower Hauterivian, Radiatus Zone.

F. VIRGATITIDAE Spath, 1923 J. (TTH) Mar.

First: *Ilowaiskya klimovi* Ilovaisky and Florensky; Ural River Basin, former USSR (Mikhailov, 1964). Basal Lower Volgian. [JHC]

Last: *Epivirgatites nikitini* (Mich.), *E. bipliciformis* (Nikitin); Bed 13 (J_3V_2–nk), middle Volga, near Gorodische, Ulianovks, former USSR (Gerasimov *et al.*, 1971). Volgian, Nikitini Zone.

F. DORSOPLANITIDAE Arkell, 1950 J. (KIM)–K. (BER) Mar.

First: *Pectinatites (Propectinatites) websteri* Cope; Kimmeridge Clay, Dorset, England, UK (Cope, 1968, p. 2). Upper Autissiodorensis Zone.

Last: *Chetaites sibiricus*; former USSR (Gerasimov *et al.*, 1971, fig. 2). Ryazanian, Rjasanensis Zone.

F. CRASPEDITIDAE Spath, 1924 J. (TTH)–K. (BRM) Mar.

First: *Kachpurites* sp.; former USSR (Gerasimov *et al.*, 1971, p. 351). Volgian, Virgatus Zone, Rosanovi Subzone.

Last: *Simbirskites (Craspedodiscus)* cf. *juddi* Rawson; lower Division B, Bed 5e (LB5E), Speeton Clay Formation, Speeton, North Yorkshire, England, UK (Rawson and Mutterlöse, 1983). Variabilis Zone.

F. HOLCODISCIDAE Spath, 1924 K. (HAU–BRM) Mar.

First: *Holcodiscus intermedius* (d'Orbigny); La Charce, Drôme, SE France (Moullade, 1966, p. 161, fig. 8). Jeanneti–castellanensis Zone.

Last: *Holcodiscus* sp. juv.; Bed 116, Angles, Basse-Alpes, SE France (Busnardo, 1965, p. 106, table 1; Moullade, 1966, fig. 11). Lower Barremian, *compressissima* Horizon.

Superfamily DESMOCERATACEAE Zittel, 1895

F. EODESMOCERATIDAE Wright, 1955 K. (VLG–BRM) Mar.

First: *Eodesmoceras celestini* (Pictet and Campiche); Sainte-Croix, Switzerland. *E. haughtoni* Spath; Uitenhage, South Africa (Spath, 1930, p. 141; Howarth and Wright, *in* Donovan *et al.*, 1967, p. 455). Upper Valanginian.
Last: *Barremites cassidoides* (Uhlig); top of 'Formation K', Montclus, Hautes-Alpes, SE France (Moullade, 1966, p. 202, fig. 13). Topmost Barremian.

F. SILESITIDAE Hyatt, 1900 K. (BRM–ALB) Mar.

First: *Silesites* ex grp *vulpes* (Coquand), *S.* cf. *tenuis* Kar.; Bed 97, Angles, Basse-Alpes, SE France (Busnardo, 1965, p. 106). Lower Barremian, *compressissima horizon.*
Last: *Neosilesites madagascariensis* Collignon; west of Ambarimaninga, Madagascar (Collignon, 1963). Middle Albian, *Lemuroceras spathi* and *Brancoceras besairiei* Zone.

F. PULCHELLIIDAE Hyatt, 1903 K. (HAU–ALB) Mar.

First: *Psilotissotia* cf. *favrei* (Ooster); Bed 3, Crans, Ardèche, France (Sayn and Roman, 1905, p. 634; Donovan *et al.*, 1967, p. 455). Sayni Zone.
Last: *Trochleiceras balearense* Fallot and Termier; Komihevritra, Madagascar (Collignon, 1950, p. 488; Donovan *et al.*, 1967, p. 455). *mammilatum* Zone.

F. DESMOCERATIDAE Zittel, 1895 K. (HAU–MAA) Mar.

First: *Abrytusites julianyi* Honnorat-Bastide; Rottier, Drôme, SE France (Thieuloy, 1972). Nodosiplicatum Zone. [PRF]
Last: *Kitchinites laurae* Macellari; Lopez de Bertodano Formation, Seymour Island, Antarctica (Macellari, 1986). Terminal Maastrichtian, *ultimus* Zone.

F. KOSSMATICERATIDAE Spath, 1922 K. (ALB–MAA) Mar.

First: *Hulenites reesidei* (Anderson); California (Anderson, 1938). ?Upper Albian.
Last: *Maorites densicostatus* Kilian and Reboul; Lopez de Bertodano Formation, Seymour Island, Antarctica (Macellari, 1986). Terminal Maastrichtian, *ultimus* Zone.

F. PACHYDISCIDAE Spath, 1922 K. (ALB–MAA) Mar.

First: *Eopachydiscus marcianus* (Shumard); Duck Creek Limestone, north Texas, USA (Kennedy *et al.*, 1983). *marcianus* Zone.
Last: *Pachydiscus* aff. *colligatus* (von Binkhorst); terminal Maastrichtian hardground, Stevns Klint, Denmark (Birkelund, 1979); *Belemnitella casimirovensis* Zone. *P. ultimus* Macellari; Lopez de Bertodano Formation, Seymour Island, Antarctica (Macellari, 1986); terminal Maastrichtian, *ultimus* Zone.

F. MUNIERICERATIDAE Wright, 1952 K. (TUR–CMP) Mar.

First: *Tragodesmoceras bassi* Morrow, *T. socorroense* Cobban and Hook; Colorado and New Mexico, USA (Cobban, 1984). *nodosoides* Zone. [WJK]
Last: *Pseudoschloenbachia (Termierella) lenticularis* Collignon; Madagascar (Collignon, 1969). Upper Lower Campanian. [WJK]

Superfamily HOPLITACEAE H. Douvillé, 1890

F. HOPLITIDAE H. Douvillé, 1890 K. (ALB–CEN) Mar.

First: *Farnhamia farnhamensis* Casey, and *F.* spp.; '*Farnhamia* band', Folkestone Beds, Farnham, Surrey, England, UK (Casey, 1965, pp. 463–71). Lower Albian, *tardefurcata* Zone, *farnhamensis* Subzone.
Last: *Hyphoplites curvatus aransionensis* (Hébert and Munier-Chalmas). Lower Chalk, southern England, UK (Kennedy and Wright, 1984, p. 74). *dixoni* Zone.

F. SCHLOENBACHIIDAE Parona and Bonnarelli, 1897 K. (ALB–CEN) Mar.

First: *Saltericeras salteri* (Sharpe); Akkup, Tuarkyr, Transcaspia, former USSR (Marinowski, 1983, p. 15). *dispar* Zone. [WJK]
Last: *Schloenbachia lymense* Spath and *S.* spp.; Lower Chalk, southern England, UK (Kennedy, 1971, p. 103). *naviculare* Zone.

F. PLACENTICERATIDAE Hyatt, 1900 K. (ALB–MAA) Mar.

First: *Semenovites iphitus* (Spath), *S. gracilis* (Spath); Upper Greensand, Blackdown, Devon, England, UK (Spath, 1926, p. 183; 1927, p. 187; Casey, 1965, p. 461). *inflatum* Zone, *varicosum* Subzone.
Last: *Hoplitoplacenticeras lasfresnayanum* (d'Orbigny); Calcaire à *Baculites*, Cotentin Peninsula, Manche, France (Kennedy, 1986a). Upper Maastrichtian.
Comment: Casey (1965, p. 461) considers the genus *Semenovites* Glasunova, 1960 to be the 'root stock' of Albian placenticeratids such as *Hengestites* Casey, 1960.

F. ENGONOCERATIDAE Hyatt, 1900 K. (APT–CEN) Mar.

First: *Knemiceras* sp.; Apon Formation, Venezuela (Rod and Maync, 1954, p. 267, Casey, 1978, p. 585). Upper Aptian.
Last: *Metengonoceras acutum* Hyatt, Britton Formation, north-central Texas (Kennedy, 1988, p. 34). Upper Cenomanian, *gracile* Zone.

Superfamily ACANTHOCERATACEAE Hyatt, 1900

F. LEYMERIELLIDAE Breistroffer, 1951 K. (ALB) Mar.

First: *Proleymeriella schrammeni* (Jacob); northern Germany (Brinkmann, 1937; Casey, 1979, p. 593). Lower Albian, *tardefurcata* Zone, *schrammeni* Subzone.
Last: *Leymeriella (L.) tardefurcata* (d'Orbigny), *L. (L.) densicostata* Spath, *L. (Neoleymeriella) consueta* Casey, *L. (N.) regularis* (Bruguière), *L. (N.) intermedia* Spath, *L. (N.) pseudoregularis* Seitz, *L. (N.) diabolus* Casey, *L. (N.) renascens* Seitz; Band II, Woburn Sands–Gault Clay junction beds, Arnolds Pit, Billington Crossing, Leighton Buzzard, Bedfordshire, England, UK (Casey, 1978, pp. 592–620). Lower Albian, *tardefurcata* Zone, *regularis* Subzone.

F. BRANCOCERATIDAE Spath, 1933 K. (ALB–CEN) Mar.

First: *Brancoceras senequieri* (d'Orbigny); French Alps (Jacob, 1907, p. 314, Spath, 1931, p. 346). *tardefurcata* Zone.
Last: *Euhystrichoceras baylissi* Kennedy and Wright; Divison C, Cenomanian Limestone, Shapwick Grange, Devon, England, UK (Kennedy and Wright, 1981). Upper Cenomanian.

F. LYELLICERATIDAE Spath, 1921 K. (ALB–CEN) Mar.

First: *Tegoceras gladiator* (Bayle), *T. miles* Casey; main *mammillatum* Bed, Folkestone Beds, Folkestone, Kent, England, UK (Casey, 1978, pp. 624–7). *mammillatum* Zone, *floridum–raulinianus* subzones.
Last: *Stoliczkaia (Lamnegella) chancellori* Wright and Kennedy; Cenomanian Limestone, Shapwick Grange, Devon, England, UK (Wright and Kennedy, 1981, p. 78). *mantelli* Zone, *saxbii* Subzone or *dixoni* Zone. [WJK]

F. MOJSISOVICSIIDAE Hyatt, 1903 K. (ALB) Mar.

First: *Oxytropidoceras alticarinatum* (Spath); main *mammillatum* Bed, Folkestone Beds, Folkestone, Kent, England, UK (Casey, 1978, p. 632). *mammillatum* Zone, *floridum–raulinianus* subzones.
Last: *Diploceras pseudaon* Spath, *D. bouchardianum* (d'Orbigny); Bed IX, Upper Gault, Folkestone, Kent, England, UK (Spath, 1931, pp. 374, 378). *inflatum* Zone, *orbignyi* Subzone.

F. FLICKIIDAE Adkins, 1928 K. (ALB–CEN) Mar.

First: *Salaziceras salazacensis* (Hébert and Munier-Chalmas), 'Craie à Fossiles de Rouen', near Salazac, Gard, France (Wright and Kennedy, 1979). *dispar* Zone (lower part, = *gardonica* Subzone *auctt.*).
Last: *Adkinsia bosquensis* Adkins; Del Rio Clay, Texas, USA (Wright and Kennedy, 1979). Lower Cenomanian.

F. FORBESICERATIDAE Wright, 1952 K. (ALB–CEN) Mar.

First: *Paradolphia prisca* Casey. Cambridge Greensand, Cambridge, England, UK (Casey, 1965, p. 461). *dispar* Zone.
Last: *Forbesiceras bicarinatum* Szász; Sables à *Catopygus obtusus*, Briollay, Maine-et-Loire, France (Wright and Kennedy, 1984, p. 96). *geslinianum* Zone.

F. ACANTHOCERATIDAE Hyatt, 1900 K. (CEN–CON) Mar.

First: *Graysonites adkinsi* Young; Texas, USA (Young, 1958). Lower Cenomanian. [WJK]
Last: *Buchiceras bilobatum* Hyatt; Otusco, Peru (Brùggen, 1910; Kennedy *et al.*, 1980). Lower Coniacian.

F. VASCOCERATIDAE Spath, 1925 K. (CEN–TUR) Mar.

First: *Vascoceras diartianum* (d'Orbigny); New Mexico, USA (Cobban *et al.*, 1989, p. 47). *masbyense* Zone. [WJK]
Last: *Hourquia hataii* Matsumoto; Hokkaido, Japan (Matsumoto, 1973, pp. 315–18). Uppermost Turonian or Lower Coniacian. [WJK]

F. COLLIGNONICERATIDAE Wright and Wright, 1951 K. (CEN–CMP) Mar.

First: *Cibolaites* sp. cf. *molenacri* Cobban and Hook; Devon, England, UK; Aube, France (Kennedy *et al.*, 1986, p. 209). *juddii* Zone. [WJK]
Last: *Texanites campaniensis* (Grossouvre); Aquitaine, France (Séronie-Vivien, 1972, pp. 141, 142). Middle Campanian.

F. TISSOTIIDAE Hyatt, 1900 K. (CON–SAN) Mar.

First: *Metatissotia nodosa* Hyatt; Torsac, Charente, France; *M. desmoulinsi* (de Grossouvre) and *M.? nanclusi* (de Grossouvre); Périgueux, Dordogne, France (Kennedy, 1984). *petrocoriensis* Zone.
Last: *Tissotia steinmanni* Lissón; Célandin Formation, northern Peru (Benavides-Caceres, 1956, p. 31). *baltai* Zone. [WJK]

F. COILOPOCERATIDAE Hyatt, 1893 K. (TUR) Mar.

First: *Hoplitoides*; Algerian Sahara (Armand *et al.*, 1981). Lower Turonian, Zone VI.
Last: *Coilopoceras springeri* Hyatt, *C. stephani* Renz, *C. laraense* Renz, etc.; Santa Rosa Formation, etc. Venezuela (Renz, 1982, p. 101). Upper Turonian.

F. SPHENODISCIDAE Hyatt, 1900 K. (CON–MAA) Mar.

First: *Lenticeras andii* (Gabb.); top of La Luna Formation, Quebrade Chiriria, Lobaterita, Venezuela (Renz, 1982, p. 32). Lower Coniacian.
Last: *Sphenodiscus binkhorsti* (Böhm); Vistula Valley, Poland (Machaloka and Walaszczyki, 1988, pp. 67–70). Uppermost Maastrichtian, *Belemnitella casimirovensis* Zone. [WJK]

Suborder ANCYLOCERATINA Wiedmann, 1960

Superfamily ANCYLOCERATACEAE Gill, 1871

F. BOCHIANITIDAE Spath, 1922 J. (TTH)–K. (BRM) Mar.

First: *Protancyloceras* sp.; Betic Cordilleras, southern Spain (Enay and Geyssant, 1975, p. 43). Lower Tithonian, Darwini Zone.
Last: *Kabylites* ? sp.; Medjez Ska region, Algeria (Busnardo and David, 1957, p. 74). Barremian.

F. ANCYLOCERATIDAE Gill, 1871 K. (VLG–APT) Mar.

First: *Crioceratites* sp.; Orpièrre, Hautes-Alpes, SE France (Moullade, 1966, fig. 4). Upper Valanginian, '*Lyticoceras* sans *Crioceratites* Zone'.
Last: *Ammonitoceras sowerbyi* Casey; Upper *Crioceras* Beds, Lower Greensand, Isle of Wight, England, UK (Casey, 1960, p. 64). *martinioides* Zone.

F. PTYCHOCERATIDAE Gill, 1871 K. (HAU–ALB) Mar.

First: *Euptychoceras curnieri* Thieuloy; *Peregrinella peregrina* Beds, Cunier, Drôme, SE France (Thieuloy, 1972, pp. 44–5). ?Lower Hauterivian. [PFR]
Last: *Ptychoceras closteroide* Etheridge, Point Charles Beds,

Darwin Formation, Point Charles, Northern Territory, Australia (Whitehouse, 1928). Upper Albian.

F. HETEROCERATIDAE Hyatt, 1900 K. (HAU–APT) Mar.

First: *Moutoniceras annulare* (d'Orbigny); Basses-Alpes, SE France (Thomel, 1964, p. 65). Top of Duvali Zone.
Last: *Heteroceras (?Argvethites) vohimaranitraensis* Collignon; Vohimaranitra, Madagascar (Collignon, 1962, pl. 221, fig. 962). Upper Aptian, *Epicheloniceras tscherischeuri* Zone.

F. HEMIHOPLITIDAE Spath, 1924 K. (HAU–BRM) Mar.

First: *Pseudothurmannia angulicostata* (d'Orbigny); Bed 42, near Angles, Basses-Alpes, SE France (Busnardo, 1965, p. 107; 1970b). Angulicostata Zone.
Last: *Hemihoplites feraudi* (d'Orbigny), *H. solieri* Matheron and *H.* spp.; Bed 165, near Angles, Basses-Alpes, SE France (Busnardo, 1965, p. 107; Moullade, 1966, fig. 11). Seranonis Zone.

Superfamily DOUVILLEICERATACEAE Parona and Bonarelli, 1897

F. DOUVILLEICERATIDAE Parona and Bonarelli, 1897 K. (HAU–ALB) Mar.

First: *Paraspiticeras* gr. *percevali* (Uhlig); Angles, SE France (Busnardo and Vermeulen, 1986, p. 458). Upper Hauterivian.
Last: *Douvilleiceras clementianum* (d'Orbigny); Lower Gault, Aube, Paris Basin, France (Casey, 1962, p. 263). *dentatus* Zone, *benettianus* Subzone.

F. ASTIERICERATIDAE Breistroffer, 1953 K. (ALB) Mar.

?First: *Astiericeras* sp. [= '*Scaphites (Eoscaphites)* nov. sp. of Colleté *et al.*, 1982, pl. 25, fig. 6]; Courcelles, Aube, France (Kennedy, 1986b). *dentatus* Zone, *lyelli* Subzone.
?Last: *Astiericeras astierianum* (d'Orbigny); phosphorite deposits, Escragnolles, Gourdon, etc., SE France (Wiedmann, 1965a; Kennedy, 1986b). *dentatus* Zone, *lyelli* or *spathi* subzone.
Comments: The affinities and age of *A. asterianum* and the monotypic family Astiericeratidae are discussed by Kennedy (1986b).

F. TROCHLEICERATIDAE Breistroffer, 1952 K. (APT–ALB) Mar.

First: *Trochleiceras balearense* (Fallot); Sierras of Alicante and Murcia, southern Spain (Wiedmann, 1965a, table V). Aptian.
Last: *Trochleiceras magneti* (Collignon); Madagascar (Collignon, 1950). Lower Albian.

Superfamily DESHAYESITACEAE Stoyanow, 1949

F. DESHAYESITIDAE Stoyanow, 1949 K. (APT–ALB) Mar.

First: *Prodeshayesites fissicostatus* (Phillips); Top B. beds, Speeton Clay Formation, Speeton, North Yorkshire, England, UK (Casey, 1964, p. 358). Basal Aptian, *fissicostatus* Zone, *bodeni* Subzone.
Last: '*Dufrenoyia*' spp.; Apulo anticline, Cundinamarca, Colombia (Bürgl, 1955, p. 13; Casey, 1964, p. 377). Basal Albian?

F. PARAHOPLITIDAE Spath, 1922 K. (APT–ALB) Mar.

First: *Parahoplites weissi* (Neumayr and Uhlig); 'Couches de la Carrière à ciment', Cassis, near La Bédoule, Bouches du Rhône, France (Fabre-Taxy *et al.*, 1965, fig. 8, p. 189). Lower Aptian (basal Bédoulian), 'Zone I'.
Last: *Hypacanthohoplites* [?cf.] *milletianus* (d'Orbigny); 'main *mammillatum* bed', Folkestone Beds, Folkestone, England, UK (Casey, 1965, p. 436). Lower Albian, *mammillatum* Zone.

Superfamily TURRILITACEAE Gill, 1871

F. HAMITIDAE Gill, 1871 K. (ALB–TUR) Mar.

First: *Hamites pseudattenuatus* Casey; Lower Greensand, West Dereham, Norfolk, England, UK (Casey, 1961, p. 96). *mammillatum* Zone, *kitchini* Subzone.
Last: *Metaptychoceras smithi* (Woods); Chalk Rock, Cuckhamsley, England, UK (Wright, 1979, p. 284). *neptuni* Zone. [WJK]

F. ANISOCERATIDAE Hyatt, 1900 K. (APT–CON) Mar.

First: *Metahamites parcetuberculatus* Collignon; west of Ambanjabe, Madagascar (Collignon, 1962, pl. 229). *Aconeceras niscus* and *Melchiorites melchioris* Zone.
Last: *Phlycticrioceras trinodosus* (Geinitz); Craie de Villedieu, La Ribochère, Loir-et-Cher, France (Kennedy, 1984). Upper Coniacian, *serratomarginatus* Zone.

F. LABECERATIDAE Spath, 1925 K. (ALB) Mar.

First: *Labeceras crassetuberculatum* Klinger; Zululand, Africa (Klinger, 1989). Lower Albian V.
Last: *Labeceras besairiei* Collignon; Zululand, Africa (Klinger, 1989). Upper Albian V.
Comment: Klinger (1989, p. 190) discusses the dimorphic pairing of *Labeceras* [m] and *Myloceras* [M]. He suggests that the family Labeceratidae be reduced to a subfamily Labeceratinae containing a single dimorphic genus. Inclusion in the Ancyloceratidae is a possibility, but stratigraphical confirmation is lacking.

F. TURRILITIDAE Gill, 1871 K. (ALB–MAA) Mar.

First: *Turrilitoides densicostatus* Pass.; Madagascar (Collignon, 1965b, p. 307). *acuticarinatum* and *jacobi* Zone (Middle Albian).
Last: *Diplomoceras* sp., *?Phylloptychoceras* sp; terminal Maastrichtian hardground, Stevns Klint, Denmark (Birkelund, 1979); *Belemnitella casimivovensis* Zone. *Diplomoceras lambi* Spath; Lopez de Bertodano Formation, Seymour Island, Antarctica. (Macellari, 1986); terminal Maastrichtian, *ultimus* Zone.

F. BACULITIDAE Meek, 1876 K. (ALB–MAA) Mar.

First: *Lechites gaudini* (Pictet and Campiche), *L.* sp.; SE France (Moullade, 1966, p. 240). *inflatum* Zone.
Last: *Baculites vertebralis* Lamark, *B. valognensis* Boehn;

terminal Maastrichtian hardground, Stevns Klint, Denmark (Birkelund, 1979). *Belemnitella casimirovensis* Zone.

Superfamily SCAPHITACEAE Meek, 1876

F. SCAPHITIDAE Meek, 1876 K. (ALB–MAA) Mar.

First: *Eoscaphites circularis* (J. de C. Sowerby), *E. subcircularis* (Spath); Bed X, Gault Clay, Kent, England, UK (Wiedmann, 1965b, p. 404). *varicosum* Zone, *inflatum* Subzone.
Last: *Hoploscaphites constrictus constrictus* (J. Sowerby), *H. constrictus* (Sow.) *crassus* Lopuski; terminal Maastrichtian hardground, Stevns Klint, Denmark (Birkelund, 1979). *Belemnitella casimirovensis* Zone.

Superfamily INCERTAE SEDIS

F. MACROSCAPHITIDAE Hyatt, 1900 K. (BRM–TUR) Mar.

First: *Macroscaphites yvani* (Puzos); Silesia (Kilian, 1892). Barremian.
Last: *Macroscaphites* sp.; Uchaux, France (Roman and Mazeran, 1913; Roman, 1938, p. 39). Turonian.
Comment: Included in the Lytoceratina by Wright (*in* Arkell *et al.*, 1957, p. L204), although subsequently only provisionally (Wright, 1981).

F. CICATRITIDAE Spath, 1927 K. (APT) Mar.

First and Last: *Cicatrites abichi* (Anthula); Caucasus, etc. (Anthula, 1899). Lower to Upper Aptian.
Comment: Only one species known, apparently showing affinities to *Costidiscus* (Macroscaphitidae) (Wright *in* Arkell *et al.*, 1957, p. L205).

REFERENCES

Anderson, F. M. (1938) Lower Cretaceous deposits in California and Oregon. *Geological Society of America Special Paper*, **16**, 339 pp.

Anthula, D. J. (1899) Über die Kreidefossilien des Kaukasus. *Beiträge zur Paläontologie und Geologie Österreich – Ungarns und des Orients, Band*, **2**, 55–102.

Arkell, W. J. (1939) The ammonite succession at the Woodham Brick Company's pit, Akeman Street Station, Buckinghamshire, and its bearing on the classification of the Oxford Clay. *Quarterly Journal of the Geological Society of London*, **95**, 135–222.

Arkell, W. J., Kummel, B. and Wright, C. W. (1957) Mesozoic Ammonoidea, in *Treatise on Invertebrate Palaeontology. Part L, Mollusca*, **4**, *Cephalopoda, Ammonoidea* (ed. R. C. Moore), Geological Society of America and University of Kansas Press, New York and Lawrence, Kansas.

Armand, B., Collignon, M. and Roman, J. (1981) Étude stratigraphique et paléontologique du Crétacé supérieur et Paléocène du Tinhert – W. et Tademait – E. (Sahara algérian). *Documents de la Laboratoire de Géologie, Faculté des Sciences, Lyon*, **6**, 15–173.

Atrops, F. (1982) La sous-famille des Ataxioceratinae (Ammonitina) dans le Kimmeridgian inférieur du sud-est de la France. *Documents de la Laboratoire de Géologie, Faculté des Sciences, Lyon*, **83**, 371 pp.

Benavides-Cáceres, V. E. (1956) Cretaceous system in northern Peru. *Bulletin of the American Museum of Natural History*, **108**, 359–493.

Benshili, K. (1989) Lias–Dogger du Moyen-Atlas Plissé (Maroc). Sedimentologie, biostratigraphie et évolution paléogeographique. *Documents de la Laboratoire de Géologie, Faculté des Sciences, Lyon*, **106**, 284 pp.

Berckhemer, F. and Hölder, H. (1959) Ammoniten aus dem Oberen Weissen Jura Süddeutschlands. *Beiträge zur Geologische Jahrbuch*, **35**, 135 pp.

Birkelund, T. (1979) The last Maastrichtian ammonites, in *Cretaceous–Tertiary Boundary Events Symposium, The Maastrichtian and Danian of Denmark.* (eds T. Birkelund and R. G. Bromley), University of Copenhagen, Copenhagen, pp. 51–7.

Brinkmann, R. (1937) Biostratigraphie des Leymeriellenstammes nebst Bemerkungen zür Paläogeographie des Nordvestdeutschen Alb. *Mitteilungen aus dem Geologischen Staatsinstitut in Hamburg*, 1–18.

Brùggen, H. (1910) Die Fauna des unteren Senons von nord Péru. *Neues Jahrbuch für Geologie und Paläontologie*, Beilage Band, **30**, 717–88.

Buck, E., Hahn, W. and Schädel, K. (1966) Zur Stratigraphie des Bajociums und Bathoniums (Dogger delta epsilon) der Schwäbischen Alb. *Jahrbuch des Geologisches Landesamt Baden-Württembergs*, **8**, 23–46.

Buckman, S. S. (1889) On the Cotteswold, Midford and Yeovil Sands, and the division between the Lias and the Oolite. *Quarterly Journal of the Geological Society of London*, **45**, 440–73.

Buckman, S. S. (1909–1930) *Yorkshire Type Ammonites*, **1–2**, continued as *Type Ammonites*, **3–7**. Published by the author, London and Thame.

Bullot, L. (1990) Evolution des Olcostephaninae (Ammonitina, Cephalopoda) dans le context palaeo-biogeographique du Crétacé Inférieur (Valanginian–Hauterivian) du Sud-Est de la France. Unpublished Diploma Thesis, Université de Bourgogne (Dijon).

Bürgl, H. (1955) El Anticlinal de Apulo. *Instituto Geologico Nacional Colombia Bolettino Geologia*, **3** (2), 2–22.

Busnardo, R. (1965) Le stratotype du Barrémien, in *Colloque sur le Crétacé Inférieur, Lyon, 1963*, pp. 161–9. *Memoires du Bureau des Recherches Géologique et Minières*, **34**.

Busnardo, R. (1970) Les *Pseudothurmannia* (Ammonoidea de l'Hauterivian Supérieur de la Montagne de Lure. *Documents de la Laboratoire de Géologie, Faculté des Sciences, Lyon*, **37**, 133–44.

Busnardo, R. and David, C. (1957) Contribution à l'étude des faunes d'ammonoïdes de Medjez Sfa (est Constantinois). *Bulletin de la Service de la Carte Géologique de l'Algérie* (Nouvelle Série), **13** (for 1956), 67–123.

Busnardo, R. and Vermeulen, J. (1986) La limite Hauterivien–Barrémien dans la région stratotypique d'Anglès (Sud-Est de la France). *Comptes rendus de l'Académie des Sciences, Paris*, **302**, Sér. II, 457–9.

Callomon, J. H. (1984) A review of the biostratigraphy of the post-Lower Bajocian ammonites of western and northern North America. *Geological Association of Canada Special Paper*, **27**, 143–74.

Callomon, J. H. (1985) The evolution of the Jurassic ammonite family Cardioceratidae. *Special Papers in Palaeontology*, **33**, 49–90.

Callomon, J. H. and Chandler, R. B. (1990) A review of the ammonite horizons of the Aalenian and Lower Bajocian Stages in the Middle Jurassic of Southern England, in *Proceedings of Meeting on Bajocian Stratigraphy.* (eds G. Crest and G. Pavia), *Memorie Descrittive della Carta Geologica d'Italia*, **40**, pp. 85–111.

Callomon, J. H., Dietl, G., Galacz, A. *et al.* (1987) Zur stratigraphie der mittel- und unteren Oberjuras in Sengenthal bei Neumarket/ Opf (Fränkische Alb). *Stuttgarter Beiträge zür Naturkunde*, Serie B., **132**, 1–53.

Callomon, J. H., Dietl, G. and Niederhöfer, H.-J. (1989) Die Ammonitenfaunen-Horizonte im Grenzbereich Bathonium/ Callovium des Schwäbischen Juras und deren Korrelation mit W-Frankreich und England. *Stuttgarter Beiträge zür Naturkunde*, Serie B., **148**, 1–13.

Canavari, M. (1882) Beiträge zur Fauna des unteren Lias von Spezia. *Palaeontographica*, **29**, 122–92.

Casey, R. (1960–1980) A monograph of the Ammonoidea of the Lower Greensand. *Palaeontographical Society [Monograph]*, 660 pp.

Cecca, A. (1985) Les Aspidoceratiformes en Europa (Ammonitina Fam. Aspidoceratidae Physodoceratinae). Unpublished PhD Thesis, Universidad de Granada, 413 pp.

Chao, K. (1959) Lower Triassic ammonoids from western Kwangsi, China. *Palaeontologia Sinensis* (B), **145**, 1–355.

Cobban, W. A. (1961) The ammonite family Binneyitidae in the western interior of the United States. *Journal of Paleontology*, **35**, 737–58.

Cobban, W. A. (1984) Mid-Cretaceous ammonite zones, Western Interior, United States. *Bulletin of the Geological Society of Denmark*, **33**, 71–89.

Cobban, W. A., Hook, S. C. and Kennedy, W. J. (1989) Upper Cretaceous rocks and ammonite faunas of southwest New Mexico. *Memoir of the New Mexico Bureau of Mining and Mineralogical Research*, **45**, 1–137.

Colleté, C., Destombs, P., Fricot, C. *et al.* (1982) *Les Fossiles de l'Albien de l'Aube*. Association Géologique Auboise, Troyes, 100 pp.

Collignon, M. (1950) Recherches sur les faunes albiennes de Madagascar III. L'Albien de Komihevitra. *Annals géologiques de Madagascar*, **17**, 19–54.

Collignon, M. (1962) *Atlas de Fossiles Caractéristiques de Madagascar (Ammonites) VIII, Berriasien, Valanginien, Hauterivien, Barrémien*. Service Géologique Tananarive, Madagascar, 96 pp.

Collignon, M. (1963) *Atlas des Fossiles Caractéristiques de Madagascar. X, Albien*. Service Géologique, Tananarive, Madagascar, 184 pp.

Collignon, M. (1965a) *Atlas des Fossiles Caractéristiques de Madagascar (Ammonites) – XII, Turonien*. Service Géologique, Tananarive, Madagascar, 82 pp.

Collignon, M. (1965b) L'Albien à Madagascar, ses subdivisions comparées a celles de l'Europe occidentale; essai de chronostratigraphie aussi générale que possible. Colloque sur le Crétacé inférieur, Lyon, 1963. *Mémoire du Bureau des Recherches Géologiques et Miniéres*, **34**, 303–10.

Collignon, M. (1969) *Atlas des Fossiles Caractéristiques de Madagascar (Ammonites) – XV, Campanien inférieur*. Service Géologique, Tananarive, Madagascar, 216 pp.

Contini, D. (1969) Les Graphoceratidae de Jura franc–comtois. *Annales de Science de l'Université Bésançon*, (**3**), *Géologie*, **7**, 95 pp.

Cope, J. C. W. (1968) *Propectinatites*, a new Lower Kimmeridgian ammonite genus. *Palaeontology*, **11**, 16–18.

Cope, J. C. W., Getty, T. A., Howarth, M. K. *et al.* (1980) A correlation of Jurassic rocks in the British Isles. 1. Introduction and Lower Jurassic. *Special Report of the Geological Society of London*, **14**, 73 pp.

Corroy, G. (1932) Le Callovien de la Bordure orientale du Bassin de Paris. *Mémoire du Service de la Carte Géologique Détaillé de la France*, 337 pp.

Dean, W. T., Donovan, D. T. and Howarth, M. K. (1961) The Liassic ammonite zones and subzones of the North-west European Province. *Bulletin of the British Museum (Natural History) Geology Series*, **4**, 435–505.

Dietl, G. and Herold, G. (1986) Erstfund von *Cadomites* (Ammonoidea) in Unter-Callovium (Mittl Jura) von Südwest-Deutschland. *Stuttgarter Beiträge zür Naturkunde, Serie B*, **120**, 1–9.

Dommergues, J. L. and Mouterde, R. (1981) Les acanthopleuroceratines portuguais et leurs relations avec les formes subboréales. *Ciências da Terra (UNL)*, **6**, 77–120.

Dommergues, J. L., Ferreti, A., Géczy, B. *et al.* (1983) Éléments de corrélation entre faunes d'ammonites mésogéennes (Hongrie, Italie) et subboréales (France, Portugal) au Carixien et au Domérien inférieur. *Geobios*, **16**, 471–99.

Donovan, D. T. (1990a) Sinemurian and Pliensbachian ammonite faunas of central Italy. *Atti del Secondo Convegno Internazionale Fossili, Evoluzione, Ambiente 1987, Pergola 1987*, 253–62.

Donovan, D. T. (1990b) The late Sinemurian ammonite genus *Vicininodiceras* Trueman. *Cahiers de l'Université Catholique de Lyon, Séries Science*, **4**, 29–37.

Donovan, D. T. and Howarth, M. K. (1982) A rare lytoceratid ammonite from the Lower Lias of Radstock. *Palaeontology*, **25**, 439–42.

Donovan, D. T., Hodson, F. and Howarth, M. K. (1967) Mollusca: Cephalopoda (Ammonoidea), in *The Fossil Record*. (eds W. B. Harland *et al.*), Geological Society of London, pp. 445–60.

Donovan, D. T., Callomon, J. H. and Howarth, M. K. (1981) Classification of the Jurassic Ammonitina, in *The Ammonoidea*. (eds M. R. House and J. R. Senior), Systematics Association Special Volume, **18**, Academic Press, London and New York, pp. 101–55.

Donovan, D. T., Curtis, M. T. and Curtis, S. A. (1989) A psiloceratid ammonite from the supposed Triassic Penarth Group of Avon, England. *Palaeontology*, **32**, 231–5.

Dubar, G. (1961) Les Hildoceratidae du Domérien des Pyrénées et l'apparition de cette famille au Pliensbachien inférieur en Afrique du nord. *Annales des Services d'Information Géologique du Bureau Recherches Géologiques, Geophysique et Minières*, **1961**, 245–57.

Elmi, S. and Caloo-Fortier, B. (1985) Éléments essentiels des peuplements d'Ammonites du Toarcien terminal–Aaléniens en Oranie (Algérie occidentale). *Cahiers de l'Institut Catholique de Lyon*, **14**, 43–53.

Enay, R. and Geyssant, J. R. (1975) Faunas tithoniques des chaînes bétiques (Espagne méridionale). Colloque sur la limite Jurassique–Crétacé. Lyon, Neuchâtel 1973. *Mémoires du Bureau de Recherches Géologiques et Minières, France*, **86**, 39–55.

Enay, R., Tintant, H. and Cariou, E. (1971) Les faunes Oxfordiennes d'Europe méridionale, essai de zonation. Colloque du Jurassique, Luxembourg, 1967. *Mémoire du Bureau de Recherches Géologiques et Minières, France*, **75**, 635–64.

Fabre-Taxy, F., Moullade, M. and Thomel, G. (1965) Le Bédoulien dans sa region type, La Bédoule-Cassis (B. du. R.). Colloque sur le Crétacé inférieur, Lyon, 1963. *Mémoires du Bureau de Recherches Géologiques et Minières, France*, **34**, 1, 73–199.

Ferretti, A. (1975) Ricerche biostratigrafiche sul Sinemuriano, Pliensbachiano nella Gola del F. Basso (Appennino Marchiagiano). *Rivista Italiana di Paleontologia e Stratigrafia*, **81**, 161–94.

Gabilly, J., Elmi, S., Mattei, J. *et al.* (1971) L'étage Toarcien zones et sous-zones d'ammonites. *Colloque du Jurassique Luxembourg 1967. Mémoires du Bureau de Recherches Géologiques et Minières, France*, **75**, 605–34.

Gallois, R. W. and Cox, B. M. (1977) The stratigraphy of the Middle and Upper Oxfordian sediments of Fenland. *Proceedings of the Geologists' Association*, **88**, 207–28.

Gerasimov, P. A., Kuznetzowa, K. I. and Mikhailov, N. P. (1971) Volgian Stage and its zonal subdivision. *Colloque du Jurassique Luxembourg 1967. Mémoires du Bureau de Recherches Géologiques et Minières, France*, **75**, 347–55.

Getty, T. A. (1973) A revision of the generic classification of the family Echioceratidae (Cephalopoda, Ammonoidea) (Lower Jurassic). *Paleontological Contributions, Kansas University*, **63**, 1–32.

Glasunova, A. E. (1960) On a new Albian genus of ammonites from Transcaspia. *Informats Sbornik*, **35**, VSEGEI, Leningrad [in Russian].

Guex, J. (1980) Remarques prélimaires sur la distribution stratigraphique des ammonites hettangiennes du New York Canyon (Gabbs Valley Range, Nevada). *Bulletin de Géologie, Lausanne*, **250**, 127–40.

Hall, R. L. and Westermann, G. E. G. (1980) Lower Bajocian (Jurassic) cephalopod faunas from western Canada and proposed assemblage zones for the Lower Bajocian of North America. *Palaeontographica Americana*, **9** (52), 1–93.

Hancock, J. M. (1954) A new Ampthill Clay fauna from Knapwell, Cambridgeshire. *Geological Magazine*, **91**, 249–51.

Hégarat, G. Le (1971) Le Berriasien du sud-est de la France.

Documents de la Laboratoire de Géologie, Faculté des Sciences, Lyon, **43**, 537 pp.

Howarth, M. K. (1955) Domerian of the Yorkshire coast. *Quarterly Journal of the Geological Society of London*, **30**, 147–75.

Howarth, M. K. (1957) The Middle Lias of the Dorset coast. *Quarterly Journal of the Geological Society of London*, **113**, 185–204.

Howarth, M. K. (1976) An occurrence of the Tethyan ammonite *Meneghinoceras* in the Upper Lias of the Yorkshire Coast. *Palaeontology*, **19**, 773–77.

Imlay, R. W. (1973) Middle Jurassic (Bajocian) ammonites from eastern Oregon. *Professional Paper of the United States Geological Survey*, **418-B**, 1–126.

Jacob, C. (1907) Études paléontologiques et stratigraphiques sur la partie moyenne des terrains crétacés dans les alpes françaises. *Travaux du Laboratoire de Géologie de la Faculté des Sciences de l'Université de Grenoble*, **I**, 280–590.

Jacob, C. (1908) Études sur quelques ammonites du Crétacé moyen. *Mémoires de la Société Géologique de France, Paléontologie*, **38**, (1907), 1–64.

Kemper, E., Rawson, P. F. and Thieuloy, J.-P. (1981) Ammonites of Tethyan ancestry in the early Lower Cretaceous of North-west Europe. *Palaeontology*, **24**, 251–311.

Kennedy, W. J. (1971) Cenomanian ammonites from southern England. *Special Papers in Palaeontology*, **8**, 1–133.

Kennedy, W. J. (1984) Systematic palaeontology and stratigraphic distribution of the ammonite faunas of the French Coniacian. *Special Papers in Palaeontology*, **31**, 1–160.

Kennedy, W. J. (1986a) The ammonite fauna of the Calcaire à *Baculites* (Upper Maastrichtian) of the Cotentin Peninsula (Manche, France). *Palaeontology*, **29**, 25–83.

Kennedy, W. J. (1986b) Observations on *Astiericeras asterianum* (d'Orbigny 1842) (Cretaceous Ammonoidea). *Geological Magazine*, **123**, 507–13.

Kennedy, W. J. (1988) Late Cenomanian and Turonian ammonite faunas from north-east and central Texas. *Special Papers in Palaeontology*, **39**, 1–131.

Kennedy, W. J. and Klinger, M. C. (1978) Cretaceous faunas from Zululand and Natal, South Africa: The Ammonite Family Lytoceratidae Neumayr, 1875. *Annals of the South African Museum*, **74**, 257–333.

Kennedy, W. J. and Wright, C. W. (1979) Vascoceratid ammonites from the type Turonian. *Palaeontology*, **22**, 665–83.

Kennedy, W. J. and Wright, C. W. (1981) *Euhystrichoceras* and *Algericeras*, the last Mortoniceratine ammonites. *Palaeontology*, **24**, 417–35.

Kennedy, W. J. and Wright, C. W. (1984) The Cretaceous ammonite *Ammonites requienianus d'Orbigny, 1841. Palaeontology*, **27**, 281–93.

Kennedy, W. J., Wright, C. W. and Hancock, J. M. (1980) Origin, evolution and systematics of the Cretaceous ammonite *Spathites*. *Palaeontology*, **23**, 821–32.

Kennedy, W. J., Wright, C. W. and Chancellor, G. R. (1983) The Cretaceous ammonite *Eopachydiscus* and the origin of the Pachydiscidae. *Palaeontology*, **26**, 655–62.

Kennedy, W. J., Amedro, F. and Colette, C. (1986) Late Cenomanian and Turonian ammonites from Ardennes, Aube and Yonne, east Paris Basin (France). *Neues Jahrbuch für Geologie und Paläontologie Abhandlungen*, **172**, 193–217.

Kilian, W. (1892) Céphalopodes nouveaux ou peu connus de la période Secondaire, B. *Macroscaphites yvani* Puz. mut. *striatisulcata* d'Orb. *Travaux du Laboratoire de Géologie de la Faculté des Sciences de l'Université de Grenoble*, **3**, 2–19.

Klinger, H. C. (1989) The ammonite subfamily Labeceratinae Spath 1925. Systematics, Phylogeny, Dimorphism and Distribution. *Annals of the South African Museum*, **98**, 189–219.

Krystyn, L. (1973) Zur Ammoniten and Conodonten-Stratigraphie der Hallstaten Obertrias (Salz Kammergut Östlericht). *Verhandlungen des geologische Bundesanstalt*, **1973**, 113–41.

Lang, W. D. (1936) The Green Ammonite Beds of the Dorset Lias. *Quarterly Journal of the Geological Society of London*, **92**, 423–37.

Lang, W. D., Spath, L. F. and Richardson, W. A. (1923) Shales-with-'Beef', a sequence in the Lower Lias of the Dorset coast. *Quarterly Journal of the Geological Society of London*, **79**, 47–99.

Lang, W. D., Spath, L. F. and Richardson, W. A. (1926) The Black Marl of Black Ven and Stonebarrow, in the Lias of the Dorset coast. *Quarterly Journal of the Geological Society of London*, **82**, 144–87.

Lang, W. D., Spath, L. F. and Richardson, W. A. (1928) The Belemnite Marls of Charmouth, a series in the Lias of the Dorset coast. *Quarterly Journal of the Geological Society of London*, **84**, 179–257.

Lange, W. (1931) Die biostratigraphischen Zonen des Lias alpha und Vollraths petrographische Leithorizonte. *Zentralblatt für Mineralogie, Geologie und Paläontologie*, **1931** (B), 349–72.

Lange, W. (1952) Der Untere Lias Fonsjoch (östliches Karwendelgebirge) und seine Ammoniten-fauna. *Palaeontographica, Abteilung A*, **102**, 49–162.

Macellari, C. (1986) Late Campanian–Maastrichtian ammonite fauna from Seymour Island, Antarctic Peninsula. *Memoirs of the Paleontological Society*, **18**, 1–55.

Machalski, M. and Walaszczyki, I. (1988) The youngest (uppermost Maastrichtian) ammonites in the middle Vistule valley, Central Poland. *Bulletin of the Polish Academy of Sciences, Earth Sciences*, **36**, 67–70.

Marinowski, R. (1983) Upper Albian and Cenomanian Ammonites from some sections of the Mangyshlak and Tuarkyr regions. Transcaspia, Soviet Union. *Neues Jahrbuch für Geologie und Paläontologie, Monatshefte*, **1983**, 156–80.

Matsumoto, T. (1954) *The Cretaceous System in the Japanese Islands*. Japanese Society for the Promotion of Science, Tokyo.

Matsumoto, T. (1960) Upper Cretaceous ammonites of California. Part 3. *Memoirs of the Faculty of Science, Kyushu University (D), Special Volume*, **2**, 1–204.

Matsumoto, T. (1973) Vascoceratid ammonites from the Turonian of Hokkaido. *Transactions and Proceedings of the Palaeontological Society, Japan, N.S.*, **89**, 27–41.

Memmi, L. and Salaj, J. (1975) Le Berriasien de Tunisie. Succession de faunes d'Ammonites, de Foraminifères et de Tintinnoïdiens. Colloque sur la limite Jurassique–Crétacé, Lyon Neuchâtel 1975. *Mémoires du Bureau de Recherches Géologiques et Miniéres, France*, **86**, 58–67.

Mikhailov, N. P. (1964) Boreal Late Jurassic (Lower Volgian) ammonites (Virgatosphinctinae). *Akademiya Nauk SSSR Geologischeskii Institut*, **107**, 1–88 [in Russian].

Moullade, M. (1966) Étude stratigraphique et micropaléontologique du Crétacé inférieur de la fosse vocontienne. *Documents de la Laboratoire de Géologie, Faculté des Sciences, Lyon*, **15**, 369 pp.

Nikolov, T. G. (1965) Étages, sous-étages et zones d'ammonites du Crétacé inférieur en Bulgarie du nord. Colloque sur le Crétacé inférieur, Lyon 1963. *Mémoires du Bureau de Recherches Géologiques et Miniéres, France*, **34**, 803–17.

Nikolov, T. G. (1982) *Les Ammonites de la Famille Berriasellidae Spath 1922; Tithonique Supérieur–Berriasien*. Académie Bulgare des Sciences, Université de Sofia, 251 pp.

Parsons, C. F. (1974) The *sauzei* and 'so-called' *sowerbyi* Zones of the Lower Bajocian. *Newsletters in Stratigraphy*, **3**, 153–80.

Parsons, C. F. (1976) A stratigraphical revision of the *humphriesianum–subfurcatum* Zone rocks (Bajocian Stage, Middle Jurassic) of southern England. *Newsletters in Stratigraphy*, **5**, 114–42.

Parsons, C. F. (1979) A stratigraphic revision of the Inferior Oolite of Dundry Hill, Bristol. *Proceedings of the Geologists' Association*, **90**, 133–51.

Pavia, G. (1983) Ammoniti e biostratigrafia del Baiociano inferiore di Digne (Francia, SE, Dip. Alpes Haute Provence). *Monografia del Museo regale Scienze Naturali, Torino, Italia*, 254 pp.

Pompeckj, J. F.(1893) Beiträge zu einer revision der Ammoniten des Schwäbischen Jura. *Jahresheft des Vereins für Vaterlandische*

Naturkunde in Württemberg, **49**, 151–248.
Rawson, P. F. and Mutterlöse, J. (1983) Stratigraphy of the Lower B and basal Cement Beds (Barremian) of the Speeton Clay, Yorkshire, England. *Proceedings of the Geologists' Association*, **94**, 133–46.
Renz, O. (1982) *The Cretaceous Ammonites of Venezuela*. Birkhäuser Verlag, Basel, 132 pp.
Reyment, R. A. (1955) The Cretaceous Ammonoidea of Southern Nigeria and the Southern Cameroons. *Bulletin of the Geological Survey of Nigeria*, **25**, 1–112.
Riccardi, A. C., Westermann, G. E. G. and Elmi, S. (1989) The Bathonian–Callovian ammonite zones of the Argentine Chilean Andes. *Geobios*, **22**, 553–97.
Riccardi, A. C., Damborenea, S. E. and Manceñido, M. O. (1990) Lower Jurassic of South America and Antarctic Peninsula. *Newsletters on Stratigraphy*, **21**, 75–103.
Rieber, H. (1963) Ammoniten und Stratigraphie des Braunjura der Schwäbischen Alb. *Palaeontographica, Abteilung A*, **122**, 1–89.
Rod, E. and Maync, W. (1954) Revision of Lower Cretaceous stratigraphy of Venezuela. *Bulletin of the American Association of Petroleum Geologists*, **38**, 193–283.
Roman, F. (1938) *Les Ammonites Jurassiques et Crétacées*. Masson, Paris, 554 pp.
Roman, F. and Mazeran, (1913) Faune du bassin d'Uchaux et de ses dépendences. *Archives du Muséum d'histoire Naturelle de Lyon*, **12**, 1–137.
Rosenberg, P. (1909) Die liassische Cephalopoden Fauna der Kratzalpe in Hagengebirge. *Beiträge zur Paläontologie und Geologie Österreich Ungarns und des Orients*, **22**, 1–259.
Sayn, G. and Roman, F. (1905) L'Hauterivien et le Barrémien de la rive droite du Rhône et du Bas-Languedoc. *Bulletin de la Societé Géologique de France*, **4**, 605–40.
Schindewolf, O. H. (1962) Studien zur Stammesgeschichte der Ammoniten. *Abhandlungen der Akademie der Wissenschaften zu Mainz, mathematisch-naturwissenschaftliche Klasse*, **8**, 429–572.
Senior, J. R., Parsons, C. F. and Torrens, H. S. (1970) New sections in the Inferior Oolite of south Dorset. *Proceedings of the Dorset Natural History and Archaeological Society*, **91**, 114–19.
Séronie-Vivien, M. (1972) Contribution à l'étude du seronien en Aquitaine Septentrionale, in *Les Stratotypes Français, 3*. Centre National de la Recherche Scientifique, Paris, 195 pp.
Shevyrev, N. (1986) Triassic ammonoids. *Trudy Palaeontologicheskogo Instituta Moscow*, **217**, 181–84 [in Russian].
Simpson, M. (1843) *A Monograph of the Ammonites of the Yorkshire Lias*. London, 60 pp.
Simpson, M. (1855) *The Fossils of the Yorkshire Lias; described from Nature*. London and Whitby, 149 pp.
Smith, P. L., Tipper, H. W., Taylor, D. G. *et al.* (1988) An ammonite zonation for the Lower Jurassic of Canada and the United States: the Pliensbachian. *Canadian Journal of Earth Sciences*, **25**, 1503–23.
Spath, L. F. (1923–1943) Ammonoidea of the Gault. *Palaeontographical Society [Monograph]*, 787 pp.
Spath, L. F. (1924) The Ammonites of the Blue Lias. *Proceedings of the Geologists' Association*, **35**, 186–217.
Spath, L. F. (1930) On the Cephalopoda of the Uitenhage Beds. *Annals of the South African Museum*, **28**, 131–58.
Spath, L. F. (1934) *Catalogue of the Fossil Cephalopoda in the British Museum (Natural History). Part IV. The Ammonoidea of the Trias (I)*. British Museum (Natural History), London, 521 pp.
Spath, L. F. (1956) The Liassic ammonite faunas of the Stowell Park Borehole. *Bulletin of the Geological Survey of Great Britain*, **11**, 140–64.
Sturani, C. (1966) Ammonites and stratigraphy of the Bathonian in the Digne–Barrême area. *Bolletino della Società Paleontologica Italiana*, **5**, 3–57.
Sylvester-Bradley, P. C. and Ford, T. D. (eds) (1968) *Geology of the East Midlands*. Leicester University Press, Leicester, 400 pp.
Thierry, J. and Charpy, N. (1982) Le genre *Tornquistes* (Ammonitina, Pachyceratidae) à l'Oxfordien inférieur et moyen en Europe occidentale. *Geobios*, **15**, 619–77.
Thieuloy, J.-P. (1972) Biostratigraphie des lentilles à Peregrinelles (Brachiopodes) de l'Hauterivien de Rottier (Drôme, France). *Geobios*, **5**, 5–53.
Thieuloy, J.-P. (1977) Les ammonites boréales des formations néocomiennes du Sud-Est français (Province submediterranéenne). *Geobios*, **10**, 395–61.
Thomel, G. (1964) Contribution à la connaissance des céphalopodes Crétacés du sud-est de la France. Note sur les ammonites déroulées du Crétacé inférieur vocontien. *Mémoires de la Societé Géologique de France*, **101**, 1–80.
Torrens, H. (1971) New names for microconch ammonite genera from the Middle Bathonian (Jurassic) of Europe and their macroconch counterparts. *Bollettino della Società Paleontologica Italiana*, **9**, 136–46.
von Hillebrandt, A. (1981) Faunas de ammonites del Liásico inferior y medio (Hettangino hasta Pliensbachiano) de América del Sur (excluyendo Argentina). *Ciencias Sedimentarias del Jurásico y Cretácio de América del Sur*, **2**, 499–538.
von Hillebrandt, A. (1984) The faunal relations of the Lower Jurassic ammonites of South America. *Proceedings of the 2nd International Symposium on the Jurassic System, Erlangen 1984*, **III**, 716–29. Geological Survey of Denmark, Copenhagen.
Wähner, F. (1894) Beiträge zur Kenntnis der tieferen Zonen des unteren Lias in dem nordöstlichen Alpen. *Beiträge zur Paläontologie und Geologie Österreich Ungarns und des Orients, Band*, **2–11**, 291 pp.
Walliser, O. H. (1956) Chronologie des Lias alpha zwischen Fildern und Klettgau. *Neues Jahrbuch für Mineralogie, Geologie und Paläontologie, Abhandlungen*, **103**, 181–222.
Westermann, G. E. G. (1956) Phylogenie des Stephanocerataceae and Perisphinctaceae des Dogger. *Neues Jahrbuch für Geologie und Paläontologie, Abhandlungen*, **103**, 233–79.
Whitaker, A. and Green, G. W. (1983) Geology of the country around Weston-super-Mare. *Memoirs of the Geological Survey of Great Britain*, 147 pp.
Whitehouse, F. W. (1928) Additions to the Cretaceous ammonite fauna of Eastern Australia. Part 2. *Memoirs of the Queensland Museum*, **8**, 200–6.
Wiedenmayer, F. (1980) *Die Ammoniten der mediterraneen Provinz im Pliensbachian und unteren Toarcian aufgrund neuer Untersuchungen im Generosa-Becken (Lombardische Alpen)*. Birkhäuser Verlag, Basel, 263 pp.
Wiedmann, J. (1965a) Sur le possibilité à une subdivision et des corrélations du Crétacé inférieur ibérique. Colloque sur le Crétacé inférieur. *Memoires du Bureau de Recherches Géologiques et Minières, France*, **34**, 821–3.
Wiedmann, J. (1965b) Origin, limits and systematic position of *Scaphites*. *Palaeontology*, **8**, 397–453.
Wiedmann, J. (1988) Ammonoid extinction and the 'Cretaceous–Tertiary boundary event', in *Cephalopods Past and Present* (eds J. Wiedmann and J. Kullmann), Schweizerbart'sche, Stuttgart, 117–40.
Wiedmann, J., Butt, A. and Einsele, G. (1978) Vergleich von marokkanischen Kreideküstenaufschlüssen und Tiefebohrung (DSDP): Stratigraphie, Paläoenvironment und Subsidenz an einen passiven Kontinentalrand. *Geologische Rundschau*, **67**, 454–508.
Wiedmann, J., Fabricus, F., Krystyn, C. *et al.* (1979) Über Umfang und Stellung des Rhaet. *Newsletters in Stratigraphy*, **8**, 133–52.
Wright, C. W. (1979) The ammonites of the English Chalk Rock (Upper Turonian). *Bulletin of the British Museum (Natural History), Geology Series*, **31**, 281–332.
Wright, C. W. (1981) Cretaceous Ammonoidea, in *The Ammonoidea* (eds M. R. House and J. R. Senior), Systematics Association Special Volume, **18**, Academic Press, London and New York, pp. 157–75.
Wright, C. W. and Kennedy, W. J. (1979) Origin and evolution of

the Cretaceous micromorph ammonite Family Flickiidae. *Palaeontology*, **22**, 685–704.

Wright, C. W. and Kennedy, W. J. (1981) The Ammonoidea of the Plenus Marls and the Middle Chalk. *Palaeontographical Society [Monograph]*, 148 pp.

Wright, C. W. and Kennedy, W. J. (1984) The Ammonoidea of the Lower Chalk. *Palaeontographical Society [Monograph]*, **1**, 1–126.

Young, K. (1958) *Graysonites*, a Cretaceous ammonite in Texas. *Journal of Paleontology*, **32**, 171–82.

12

MOLLUSCA: CEPHALOPODA (COLEOIDEA)

P. Doyle

The higher classification of the Coleoidea is presently in a state of flux; the scheme used below largely follows Jeletzky (1966), ranking major groups as orders. However, a departure is the separation of the Spirulida from the Sepiida at order level, which largely follows Donovan (1977) and Engeser and Bandel (1988). No attempt has been made to distinguish paraphyletic from monophyletic families, due to the lack of detailed cladistic treatment of the coleoids at family level. The Coleoidea are wholly marine.

Order Status UNCERTAIN

??F. PROTOAULACOCERATIDAE Bandel *et al.*, 1983 D. (EMS) Mar. (see Fig. 12.1)

First and Last: *Protoaulacoceras longirostris* Bandel, Reitner, and Stürmer, 1983, Hunsrückschiefer, Kaisergrube, Hunsrück, West Germany (Bandel *et al.*, 1983).

Comment: This taxon, possibly the earliest of the Aulacocerida, is a doubtful coleoid, given its subcentral siphuncle and late ontogenetic development of the 'rostrum'. Some authorities believe that the 'rostrum' of this taxon could actually be a fish spine (J. Dzik, pers. comm., 1991).

Order AULACOCERIDA Stolley, 1919

F. AULACOCERATIDAE Mojsisovics, 1882 D. (EIF?)–Tr. (RHT) Mar.

First: *Aulacoceras*? sp. indet., Couvain, Belgium. This record is based on a very poor specimen figured by de Koninck (1843) and cannot be considered to be reliable. The next oldest forms are *Paleoconus bakeri* Flower and Gordon, 1959, *Hematites barbarae* Flower and Gordon, 1959, *H. Burbankensis* Flower and Gordon, 1959, *Bactrimimus ulrichi* Flower and Gordon, 1959 and *H. girtyi* Flower and Gordon, 1959, Mississippian, Fayetteville Shale, Arkansas and Chainman Shale, Utah, USA (Flower and Gordon, 1959).

Last: *Prographularia triadica* Frech, 1890, *Austroteuthis kuehni* Jeletzky and Zapfe, 1967, Zamblach Marl, Fischerweise near Aussee, Austria (Jeletzky and Zapfe, 1967).

Intervening: P. (u.), ANS–NOR.

F. PALAEOBELEMNOPSIDAE Chen, 1982 P. (ZEC) Mar.

First and Last: *Palaeobelemnopsis sinensis* Chen, 1982, Upper Permian, Dalong Formation, Hubei and Zhejiang Provinces, China (Chen and Sun, 1982).

F. XIPHOTEUTHIDIDAE Naef, 1922 Tr. (SCY)–J. (OXF?) Mar.

First: *Moisisovicsteuthis*? sp., Scythian, Olena River, northeast Siberia (Diener, 1915).

Last: *Atractites*? *argoviense* Dreyfuss, 1957, Vis Valley, Hérault, France. This species is based on phragmocones only and cannot be assigned definitely to the Aulacocerida (Dreyfuss, 1957). The next youngest record is *Atractites idunense* (Meneghini, 1867), *A. inflatum* (Stoppanini, 1857) and *A. stoppanii* (Meneghini, 1867), Toarcian, Rosso Amonitico, Lombardy, Italy (Meneghini, 1867).

Intervening: ANS–RHT, SIN–TOA.

F. CHITINOTEUTHIDIDAE Müller-Stoll, 1936 J. (PLB) Mar.

First: *Chitinoteuthis* sp. indet., Lower Pliensbachian, Kirchheim, Württemberg, Germany (Müller-Stoll, 1936).

Last: *Chitinoteuthis decidua* Müller-Stoll, 1936, *C. crassicristata* Müller-Stoll, 1936, Lower Depressus Beds, Württemberg, Germany (Müller-Stoll, 1936).

Order PHRAGMOTEUTHIDA Jeletzky, 1965

F. PHRAGMOTEUTHIDIDAE Mojsisovics, 1881 P. (TAT)–J. (TOA) Mar.

First: *Permoteuthis groenlandica* Rosenkrantz, 1946, Foldvik Creek Formation, Posidonia Shale Member, Clavering Island, East Greenland (Rosenkrantz, 1946).

Last: *Phragmoteuthis conocauda* (Quenstedt, 1849), *Hildoceras bifrons* Zone, Posidonienschiefer, Dotternhausen to Aalen, Württemberg, Germany (Riegraf *et al.*, 1984).

Intervening: ANS, €RN, NOR, PLB.

Order BELEMNITIDA Zittel, 1895

The age of the oldest Belemnitida is open to some discussion. Phylogenetically, derivation of the Belemnitida from the Bactritida (Jeletzky, 1966) would support Palaeozoic belemnites, while derivation from the Aulacocerida or Phragmoteuthida (Engeser and Bandel, 1988) would allow only for Mesozoic belemnites. Palaeozoic records are equivocal: *Eobelemnites caneyensis* Flower is a true belemnite phragmocone, reputed to be from the Caney Shale (Mississippian), Oklahoma, USA (Flower, 1945), and the coleoid *Jeletzkya douglassae* Johnson and Richardson from the middle Pennsylvanian of Illinois is considered by Gordon and Jeletzky (*in* Gordon, 1971) to be a true belemnite. As the former is represented by a single museum specimen with indifferent provenance, it can be discounted, with the

The Fossil Record 2. Edited by M. J. Benton. Published in 1993 by Chapman & Hall, London. ISBN 0 412 39380 8

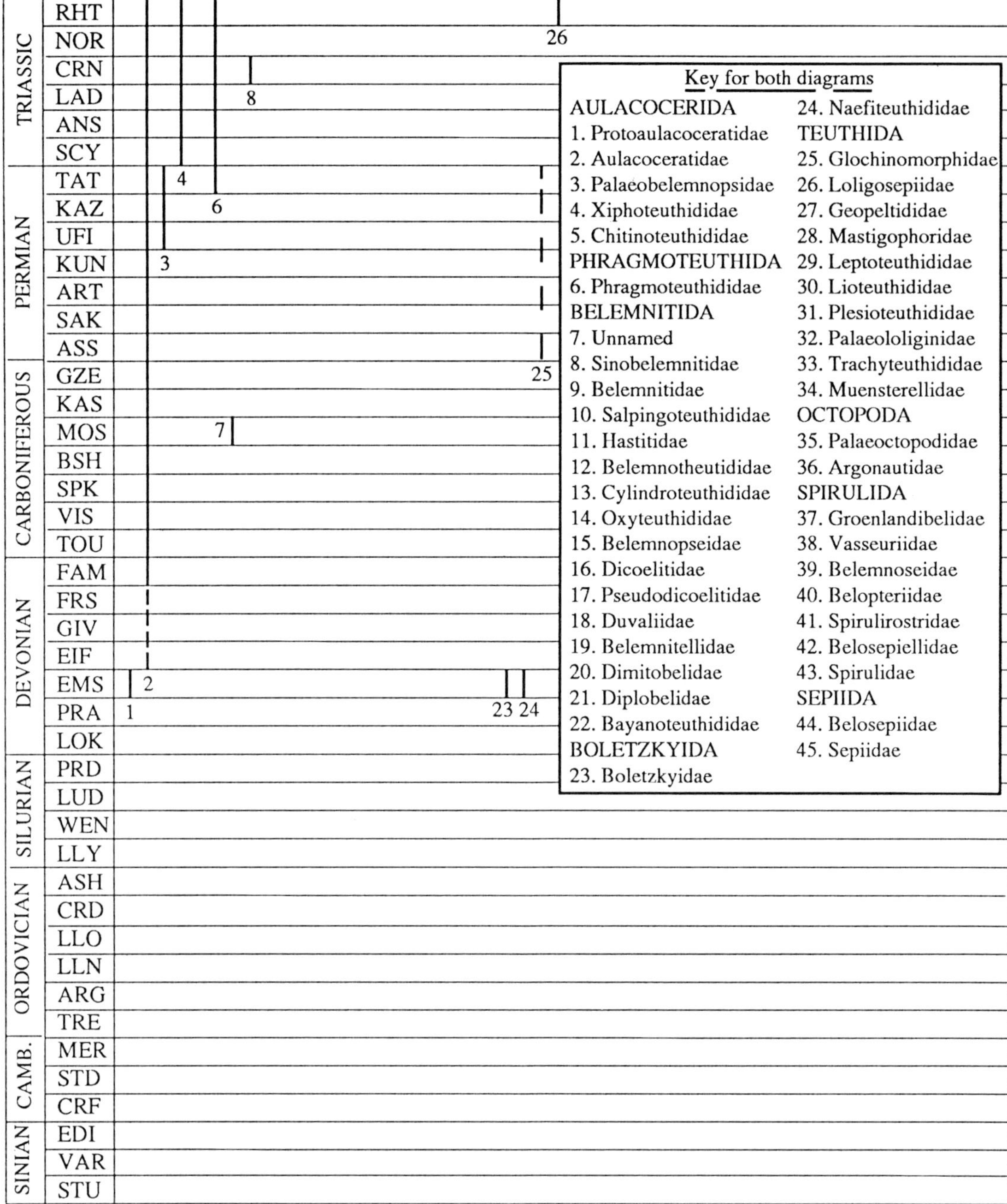

Fig. 12.1

reidentification of the only other specimen (Flower and Gordon, 1959) as an orthoconic nautiloid (Gordon, 1966). *Jeletzkya*, although an undoubted Carboniferous coleoid, is perhaps even more problematic. On the basis of associated gladius-like structures, Saunders and Richardson (1979) reaffirmed Johnson and Richardson's assignment to the Teuthida, but these structures are now known to be fish scales (Riccardi and Sabattini, 1985). Therefore, on the basis of its ten equal arms with paired arm hooks *Jeletzkya* could be placed tentatively with the Belemnitida. Other alternatives are that it is an aulacocerid, or a phragmoteuthid, both of which are considered to have possessed ten arms, although arm hooks are confirmed only in phragmoteuthids.

Suborder Status UNCERTAIN

??F. UNNAMED C. (MOS) Mar.

First and Last: *Jeletzkya douglassae* Johnson and Richardson, 1968, Middle Pennsylvanian (Desmoinesean) Francis Creek Shale, Western Springs, Illinois, USA (Saunders and Richardson, 1979).

Comments: Considered by Jeletzky (*in* Gordon, 1971) to be a belemnite, and by Johnson and Richardson (1968) and Saunders and Richardson (1979) to be a teuthid.

Suborder Status UNCERTAIN

F. SINOBELEMNITIDAE Zhu and Bian, 1984 Tr. (CRN) Mar.

First and Last: *Sinobelemnites cornutus* Zhu and Bian, 1984, *S. typica* Zhu and Bian, 1984, *S. elongata* Zhu and Bian, 1984,

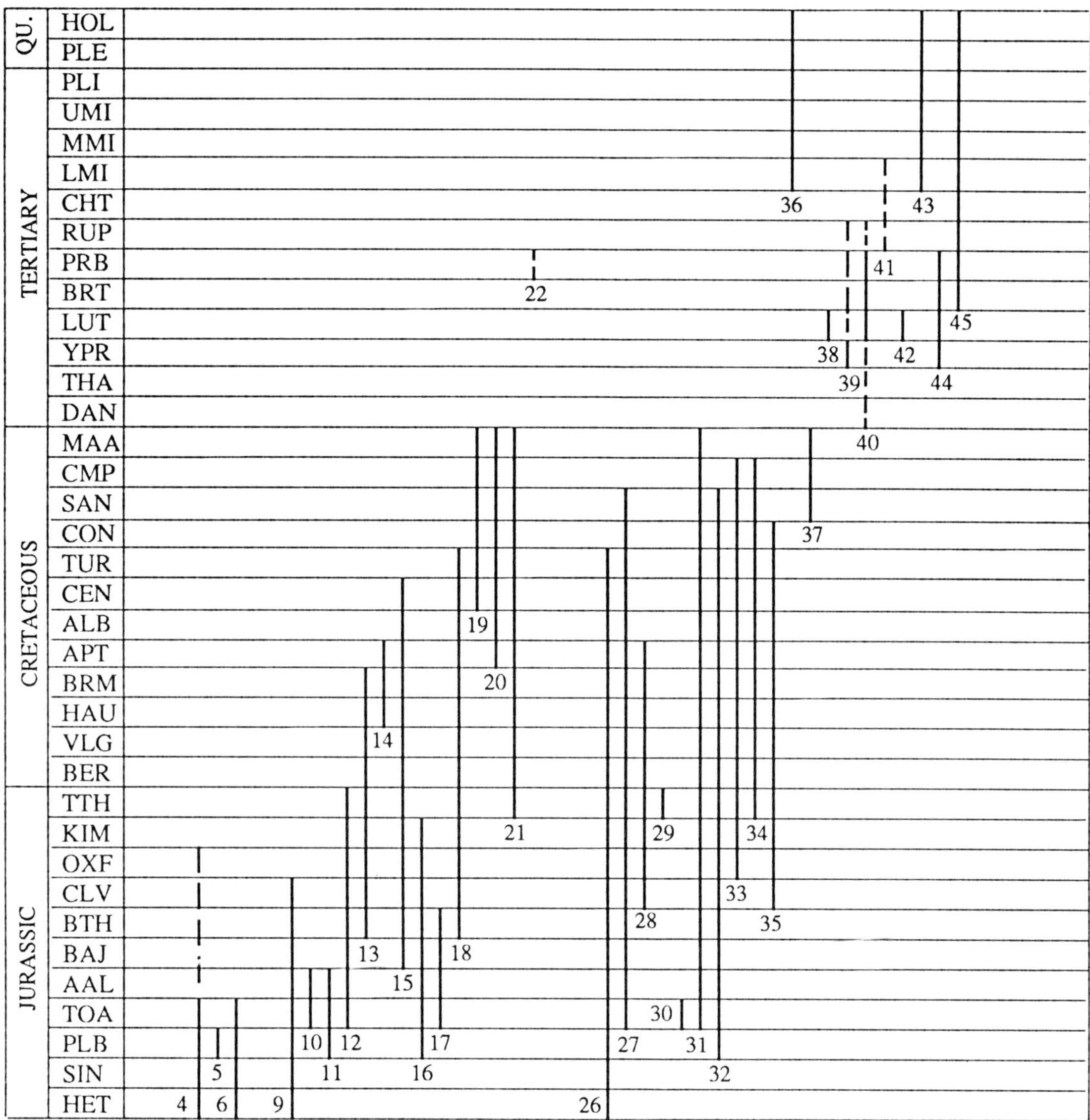

Fig. 12.1

S. maantangensis Zhu and Bian, 1984, *Sichuanobelus longmenshanensis* Zhu and Bian, 1984 and *S. yangi* Zhu and Bian, 1984, Carnian, Maantang of Jiangyu District, Sichuan Province, China (Zhu and Bian, 1984).

Comments: The belemnoids figured by Zhu and Bian (1984) are, as far as can be ascertained from the illustrations and descriptions, true belemnites. However, they are unusual in being closer in morphology to Middle and Upper Jurassic Belemnopseidae than to the earlier Jurassic Belemnitidae. This poses significant phylogenetic problems, and further clarification of stratigraphical data is necessary.

Suborder BELEMNITINA Zittel, 1895

F. BELEMNITIDAE d'Orbigny, 1845 J. (HET–CLV) Mar.

First: *Schwegleria feifeli* (Schwegler, 1939), *S. praecox* (Schwegler, 1939), *S. psilonoti* (Schwegler, 1939), Psilonotentone, Steinenberg near Nürtlingen, Württemberg, Germany (Riegraf, 1980).

Last: *Paramegateuthis ishmensis* Gustomesov, 1960, *P. timanensis* Gustomesov, 1960, *Arcticoceras ishmae* Zone, Pechora River Basin, Siberia, former USSR (Nal'nyaeva, 1983).

Intervening: SIN–BTH.

F. SALPINGOTEUTHIDIDAE Doyle, 1992 J. (TOA–AAL) Mar.

First: *Salpingoteuthis tessoniana* (d'Orbigny, 1842), Toarcian, Amaye-sur-Orne, Calvados, France (Doyle, 1992).

Last: *Salpingoteuthis hartmanni* Lissajous, 1927, Aalenian of Vénède, Lozère, France (Lissajous, 1927).

F. HASTITIDAE Naef, 1922 J. (PLB–AAL) Mar.

First: *Hastites clavatus* (Schlotheim, 1820), *Uptonia jamesoni* Zone, Reutlingen, Württemberg, Germany (Riegraf, 1980).

Last: *Rhabdobelus (Neoclavibelus) neumarktensis* (Oppel, 1856), Middle Jurassic Alpha, Metzingen, Württemberg, Germany (Riegraf, 1980). All post-Aalenian records of *Hastites* are more properly referred to *Hibolithes* (Belemnopseidae).

Intervening: TOA.

F. BELEMNOTHEUTIDIDAE Zittel, 1885
J. (TOA–TTH) Mar.

First: *Chondroteuthis wunnenbergi* Böde, 1933, Posidonienschiefer, Hondelage, near Braunschweig, Lower Saxony, Germany (Jeletzky, 1966).
Last: *Belemnotheutis mayri* Engeser and Reitner, 1981, Solnhofen Limestone, Solnhofen, Bavaria, Germany (Engeser and Reitner, 1981).
Intervening: CLV, KIM.

F. CYLINDROTEUTHIDIDAE Stolley, 1919
J. (BTH)–K. (BRM) Mar.

First: *Cylindroteuthis spathi* Saks and Nal'nyaeva, 1964, Vardekløft Formation, Jameson Land, East Greenland (Saks and Nal'nyaeva, 1964).
Last: *Acroteuthis (Boreioteuthis)* spp. Aklavik Range, northern Richardson Mountains, North-West Territories, Canada (Doyle and Kelly, 1988).
Intervening: CLV–HAU.

F. OXYTEUTHIDIDAE Stolley, 1919
K. (HAU–APT) Mar.

First: *Praeoxyteuthis jasikofiana* (Lahusen, 1874), Bed VIIA, Speeton Clay, Speeton, North Yorkshire, England, UK (Mutterlose, 1983).
Last: *Oxyteuthis borealis* Frebold, 1935, Upper Aptian, Store Koldewey, East Greenland (Frebold, 1935).
Intervening: BRM.

Suborder BELEMNOPSEINA Jeletzky, 1965

F. BELEMNOPSEIDAE Naef, 1922
J. (BAJ)–K. (CEN) Mar.

First: *Belemnopsis canaliculata* (Schlotheim, 1820), *Stephanoceras humphriesianum* Zone, Gammelshausen, Württemberg, Germany (Riegraf, 1980). Earlier *Belemnopsis* possibly occur in the *Ludwidgia murchisonae* Zone (Aalenian) of Normandy, France (Riegraf, 1980, p. 172).
Last: *Neohibolites ultimus* (d'Orbigny, 1845), *Mantelliceras mantelli* Zone, Glauconitic Marl and Lower Chalk, south-east England, UK (Donovan and Hancock, 1967).
Intervening: BTH–APT.

F. DICOELITIDAE Saks and Nal'nyaeva, 1967
J. (PLB–KIM) Mar.

First: *Conodicoelites riliensis* Gaković and Stoyanova-Vergilova, 1978, Domerian, Rilje, Zalomka River, eastern Hercegovina, former Yugoslavia (Gaković and Stoyanova-Vergilova, 1978).
Last: *Conodicoelites flemingi* Stevens, 1965, *C. orakaensis* Stevens, 1965, Heterian (Kimmeridgian), Oraka Sandstone, Totara Peninsula, Kawhia Harbour, New Zealand (Stevens, 1965).
Intervening: TOA, BAJ–OXF.

F. PSEUDODICOELITIDAE Saks and Nal'nyaeva, 1967 J. (TOA–BTH) Mar.

First: *Lenobelus lenensis* Gustomesov, 1966, *Dactylioceras commune* Zone, Toarcian, Siberia (Saks and Nal'nyaeva, 1975).
Last: *Pseudodicoelites* sp., *Arctocephalites elegans* Zone, Bathonian, Siberia (Saks and Nal'nyaeva, 1975).
Intervening: AAL, BAJ.

F. DUVALIIDAE Pavlow, 1914
J. (BTH)–K. (TUR) Mar.

First: *Rhopaloteuthis gillieroni* (Mayer, 1866), Couches de Klaus, Fribourg, Switzerland; Trzebionka, Poland (Małecki, 1984).
Last: *Duvalia rafarae* Combémorel, 1988, Marnes supra-basaltiques du Turonian, Antantiloky, Madagascar (Combémorel, 1988).
Intervening: CLV, OXF, TTH–HAU, APT.

F. BELEMNITELLIDAE Pavlow, 1914
K. (CEN–MAA) Mar.

First: *Actinocamax (Praeactinocamax) primus* Arkhangelsky, 1912, *Mantelliceras mantelli* Zone, Russian Platform, former USSR; Glauconite Sands, Antrim, Northern Ireland, UK (Christensen, 1974).
Last: *Fusiteuthis polonica* Kongiel, 1962, *Belemnella casimirovensis* Zone, middle Vistula Valley, central Poland (Kongiel, 1962).
Intervening: TUR–CMP.

F. DIMITOBELIDAE Whitehouse, 1924
K. (APT–MAA) Mar.

First: *Peratobelus australis* (Phillips, 1867), Lower Aptian, Minimi Member, Blythesdale Formation, Roma-Mitchell area, Surat Basin, Queensland, Australia (Day, 1969). A possible early dimitobelid, *Patagonibelus enigmaticus*, was described by Aguirre-Urreta and Doyle (1989) from the Rio Mayer Formation (VAL–HAU) of Patagonia.
Last: *Dimitobelus (Dimitocamax) hectori* Stevens, 1965, Haumurian (Maastrichtian), Amuri Bluff, New Zealand (Stevens, 1965).
Intervening: ALB–CMP.

Suborder DIPLOBELINA Jeletzky, 1965

F. DIPLOBELIDAE Naef, 1926
J. (TTH)–K. (MAA) Mar.

First: *Diplobelus belemnitoides* (Zittel, 1868), Tithonian, Stramberg, Czechoslovakia (Jeletzky, 1981); *Quiricobelus italicus* Combémorel and Mariotti, 1986, Tithonian, Aptychus Limestones, Serra San Quirico, Ancona Province, Italy (Combémorel and Marriotti, 1986).
Last: Unnamed, Maastrichtian of the Pacific Coast of Canada (Jeletzky, 1981). The next youngest species is *Conoteuthis syriacea* (Roger, 1944) from the Cenomanian of the Lebanon (Jeletzky, 1981).
Intervening: BER–ALB.

Order Status UNCERTAIN

?F. UNNAMED K. (ALB) Mar.

First and Last: *Belospirula compressa* Ayyasami and Jaganndha Rao, 1987, Grey shales of Dalmiapuram below the Uttattur Group (Albian), Tiruchirapalli District, Tamil Nadu, southern India (Ayyasami and Jaganndha Rao, 1987).
Comments: This taxon is claimed by the authors to be an 'intermediate form' between belemnoids and sepiids. Although it possesses a rostrum which is somewhat similar to that of *Spirulirostra* (Sepiida), it is clearly a belemnoid, Further work is needed to clarify its position.

Order Status UNCERTAIN

?F. BAYANOTEUTHIDIDAE Naef, 1922
T. (PRB?) Mar.

First and Last: *Bayanoteuthis rugifer* (Schloenbach, 1868), Bremier, Sables de Beauchamp, France (Munier-Chalmas, 1872).
Comments: The true nature of *Bayanoteuthis* is yet to be determined, with its phylogenetic and taxonomic position unclear. Riegraf (1991), following earlier authors (e.g. Branco, 1885), considers that this form is actually a pennatulacean coral.

Order BOLETZKYIDA Bandel *et al.*, 1983

Ancestral 'teuthids', combining phragmocone with gladius; grouped by Bandel *et al.* (1983) in a superorder, Palaeoteuthomorpha. Additional work is required to confirm the status of this group.

F. BOLETZKYIDAE Bandel *et al.*, 1983
D. (EMS) Mar.

First and Last: *Boletzkya longa* Bandel *et al.*, 1983, *B. hunsrueckensis* Bandel *et al.*, 1983, Hunsrückschiefer, Kaisergrube, Hunsrück, Germany.

F. NAEFITEUTHIDIDAE Bandel *et al.*, 1983
D. (EMS) Mar.

First and Last: *Naefiteuthis breviphagmoconus* Bandel *et al.*, 1983, Hunsrückschiefer, Kaisergrube, Hunsrück, Germany.

Order TEUTHIDA Naef, 1922

The more familiar nominal order Teuthida Naef, 1922 is employed here in place of Vampyromorpha Robson, 1930, used for these fossils by Engeser and Bandel (1988) and Engeser (1988b), based on a variety of evidence discussed by Bandel and Leich (1986). Although the gladius of the living genus *Vampyroteuthis* resembles those of fossil genera *Loligosepia* and *Teudopsis* (Loligosepiidae), other forms included in the Vampyromorpha by these authors (such as *Plesioteuthis*) have gladii which are closer to those of modern teuthids (Donovan and Toll, 1988). Clearly, further work is needed to clarify the position of these fossils. Engeser (1988b) has exhaustively reviewed all fossil teuthids.

Suborder Status UNCERTAIN

??F. GLOCHINOMORPHIDAE Gordon, 1971
P. Mar.

First and Last: *Glochinomorpha stifeli* Gordon, 1971, Meade Peak Phosphatic Shale Member, Phosphoria Formation, Box Elder County, Utah, USA (Gordon, 1971).
Comments: This taxon has doubtful status within the Teuthida, and requires much further study.

Suborder LOLIGOSEPIINA Jeletzky, 1965

F. LOLIGOSEPIIDAE Regteren Altena, 1949
Tr. (RHT)–K. (TUR) Mar.

First: *Loligosepia neidersachensis* Reitner, 1978, Kössener Schichten, Zollstation Griesen near Garmisch-Partenkirchen, Germany.
Last: *Neololigosepia vinarensis* (Fritsch, 1910), Weissenberger Schichten, Vinar near Hohenmauth, Czechoslovakia.
Intervening: SIN, TOA, TTH, BRM.

F. GEOPELTIDIDAE Regteren Altena, 1949
J. (TOA)–K. (SAN) Mar.

First: *Geopeltis simplex* (Voltz, 1830), Posidonienschiefer, Ohmden, Württemberg, Germany.
Last: *Geopeltis* sp., Fish Beds, Sahel-Alma, Lebanon.
Intervening: TTH.

F. MASTIGOPHORIDAE Engeser and Reitner, 1985
J. (CLV)–K. (APT) Mar.

First: *Mastigophora brevipinnis* Owen, 1856, Oxford Clay, Chippenham, Wiltshire, England, UK.
Last: *Mastigophora stuehmeri* Engeser and Reitner, 1985, Lower Aptian of Heligoland, Schleswig-Holstein, Germany.
Intervening: TTH.

F. LEPTOTHEUTHIDIDAE Naef, 1921
J. (TTH) Mar.

First and Last: *Leptotheuthis gigas* Meyer, 1834, Solnhofen Limestone, Solnhofen, Bavaria, Germany.

F. LIOTEUTHIDIDAE Naef, 1922
J. (TOA) Mar.

First and Last: *Lioteuthis problematica* Naef, 1922, Posidonienschiefer, Ohmden, Württemberg, Germany.

Suborder PROTOTEUTHINA Naef, 1921

F. PLESIOTEUTHIDIDAE Naef, 1921
J. (TOA)–K. (MAA) Mar.

First: *Paraplesioteuthis hastata* (Münster, 1843); *P. sagitata* (Münster, 1843), Posidonienschiefer, Holzmaden, Württemberg, Germany.
Last: *Maioteuthis maestrichtensis* (Binkhorst van den Binkhorst, 1862), Maastrichtian of Fauquemont, Limbourg, Holland.
Intervening: CLV, TTH, BRM, APT, SAN, CMP.

Suborder MESOTEUTHINA Naef, 1921

F. PALAEOLOLIGINIDAE Naef, 1922
J. (PLB)–K. (SAN) Mar.

First: *Teudopsis glevensis* (Smithe, 1877), Marlstone Rock Bed, Churchdown, Gloucestershire, England, UK.
Last: *Parateudopsis libanotica* (Naef, 1922), Fish Beds, Sahel-Alma, Lebanon.
Intervening: TOA, CLV, TTH.

F. TRACHYTEUTHIDIDAE Naef, 1922
J. (OXF)–K. (CMB) Mar.

First: *Trachyteuthis palmeri* (Schevill, 1950), Oxfordian of Jagua Vieja, Vinâles Region, Cuba.
Last: *Actinosepia canadensis* Whiteaves, 1897, Bearpaw Shale, mouth of Swift Current Creek, southern Saskatchewan, Canada.
Intervening: KIM, TTH, APT, CEN, TUR, SAN.

Suborder KELAENINA Starobogatov, 1983

F. MUENSTERELLIDAE Roger, 1952
J. (TTH)–K. (CMB) Mar.

First: *Muensterella scutellaris* (Münster, 1843), *Calaenoteuthis incerta* Naef, 1922, Solnhofen Limestone, Solnhofen, Bavaria, Germany.
Last: *Tusoteuthis longa* Logan, 1898, upper Smoky Hill Member, Niobrara Formation, Logan County, Kansas, USA (Nicholls and Isaak, 1987).
Intervening: CON, SAN.

Order OCTOPODA Leach, 1818

F. PALAEOCTOPODIDAE Dollo, 1912 J. (CLV)–K. (CON) Mar.

First: *Proteroctopus ribeti* Fischer and Riou, 1982, *Bositra buchi* and *Ophiopinna elegans* beds, Boissine, Voulte-sur-Rhône, Ardeche, France (Fischer and Riou, 1982). Engeser (1988a) considers that this record is not a true octopod, with *Palaeoctopus* the sole representative of the family.
Last: *Palaeoctopus newboldi* (Woodward, 1896), Fish Beds of Sahel-Alma, Lebanon (Roger, 1944).

F. ARGONAUTIDAE Naef, 1912 T. (Oli)–Rec. Mar.

First: *Obinautilus pulchra* Kobayashi, 1954, Oligocene, Nichinan Formation, between Obi and Aburatsu, Province Hyuga, Miyazaki Prefecture, Japan (Kobayashi, 1954). Originally described as a nautiloid (Noda *et al.*, 1986).
Intervening: T. (MMI, PLI). **Extant**

Order SPIRULIDA Stolley, 1919

Separation of Spirulida from Sepiida follows Donovan (1977) and Engeser and Bandel (1988).

F. GROENLANDIBELIDAE Jeletzky, 1966 K. (SAN–MAA) Mar.

First: *Naefia neogaeia* Wetzel, 1930, Upper Yezo Group (SAN–CMP), Sakasa-gawa, Haboro area, Hokkaido, Japan (Hewitt *et al.*, 1991); Ariyalur Group (CMP–MAA), Pondicherry, southern India (Doyle, 1986).
Last: *Groenlandibelus rosenkrantzi* (Birkelund, 1956), Upper Maastrichtian, Nûgssuaq, West Greenland (Jeletzky, 1966).
Intervening: K. (MAA).

F. VASSEURIIDAE Naef, 1921 T. (LUT) Mar.

First and Last: *Vasseuria occidentalis* Munier-Chalmas, 1880, Lutetian, Bois-Gouët, Loire-Inférieure, France (Curry, 1955); *Styracoteuthis orientalis* Crick, 1905, (?Middle) Eocene, Sharkeeyab, Oman (Crick, 1905).

F. BELEMNOSEIDAE Wiltshire, 1869 T. (YPR–?RUP) Mar.

First: *Belemnosis anomala* (Sowerby, 1829), Eocene, London Clay, Highgate, London, England, UK (Naef, 1922).
Last: *Spirulirostrella szainochae* (Vojcik, 1903), Lower Oligocene of Przemysl, Galicia, Poland (Naef, 1922).
Intervening: ?LUT/?BRT.

F. BELOPTERIIDAE Naef, 1922 T. (?THA–?RUP) Mar.

First: *Belocurta yahavensis* Avinmelech, 1958, Taquiya Marls, Ein Yahav, Araba Valley, Israel (Avinmelech, 1958).
Last: *Beloptera postera* Koenen, 1894, Lower Oligocene, Westeregeln, northern Germany (von Bülow-Trummer, 1920).
Intervening: ?YPR, LUT.

F. SPIRULIROSTRIDAE Naef, 1921 T. (?RUP–?BUR) Mar.

First: *Spirulirostridium obtusum* Naef, 1922, Lower Oligocene Cement Marls, Häring near Kufstein, Tyrol, Austria (Naef, 1922).
Last: *Amerirostra americana* (Berry, 1922), Miocene, Vera Cruz, Mexico (Jeletzky, 1969); *Spirulirostrina lovisatoi* Canavari, 1892, Miocene, Gagliari, Sardinia (Naef, 1922).
Intervening: ?RUP.

F. BELOSEPIELLIDAE Naef, 1921 T. (LUT) Mar.

First and Last: *Belosepiella cossmani* Alessandri, 1905, *B. parisiensis* Alessandri, 1905, Calcaire Grossier, Chaussy and Trye, Paris Basin, France (Curry, 1955).

F. SPIRULIDAE d'Orbigny, 1826 T. (BUR)–Rec. Mar.

First: *Spirula* sp., Waitakere Group (Lower Miocene), Hukatere Peninsula, Kaipara Harbour, New Zealand (Hayward, 1976); *Spirula mizunamiensis* Tomida and Itoigawa, 1981, Nataki Conglomerate, Oldwawara Formation, Mitzunami Group (Miocene), Okuna, Toki-chô, Mitzunami City, Gifu Prefecture, Japan (Tomida and Itoigawa, 1981). **Extant**

Order SEPIIDA Zittel, 1895

Donovan (1977) considers that *Trachyteuthis* (=*Voltzia*) (Trachyteuthididae) is an ancestral sepiid from the Upper Jurassic. Most other authors consider it to be purely homeomorphic, and a teuthid, and it is treated here as such (above).

F. BELOSEPIIDAE Naef, 1921 T. (YPR–PRB) Mar.

First and Last: *Belosepiella cossmani* Alessandri, 1905, *B. parisiensis* Alessandri, 1905, Calcaire Grossier, Chaussy and Trye, Paris Basin, France (Curry, 1955).

F. SEPIIDAE Keferstein, 1866 T. (BRT)–Rec. Mar.

First: *Archaeosepia naefi* Szörenyi, 1933, Bartonian, Hungary (Roger, 1952). **Extant**

REFERENCES

Please note that purely taxonomic references are not included.

Aguirre-Urreta, M. and Doyle, P. (1989) A problematical belemnite from the Lower Cretaceous of the Austral Basin, Patagonia, Argentina. *Neues Jahrbuch für Geologie und Paläontologie. Monatshefte*, **1989**, 345–55.

Avinmelech, M. (1958) A new belemnoid genus from the Paleocene of Israel. *Bulletin of the Research Council of Israel*, **7G**, 61–5.

Ayyasami, K. and Jaganndha Rao, B. R. (1987) New belemnoid from the Cretaceous of South India. *Geological Survey of India, Special Publications*, **11**, 410–12.

Bandel, K. and Leich, H. (1986) Jurassic Vampyromorpha (dibranchiate cephalopods). *Neues Jahrbuch für Geologie und Paläontologie, Monatshefte*, **1986**, 129–48.

Bandel, K., Reitner, J. and Stürmer, W. (1983) Coleoids from the Lower Devonian Black Slate ('Hunsrück-Schiefer') of the Hunsrück (West Germany). *Neues Jahrbuch für Geologie und Paläontologie, Abhandlungen*, **165**, 397–417.

Branco, W. (1885) Über einige neue Arten von *Graphularia* und über tertiäre Belemniten. *Zeitschrift der Deutschen Geologischen Gesellschaft*, **37**, 422–32.

Bülow-Trummer, E. von. (1920) *Fossilium Catalogus. I: Animalia, Pars 11 Cephalopoda dibranchiata*, Junk, Berlin, 271 pp.

Chen, T. E. and Sun, Z. H. (1982) Discovery of Permian belemnoids in South China, with comments on the origin of Coleoidea. *Acta Paleontologica Sinica*, **21**, 181–90 [in Chinese, English summary].

Christensen, W. K. (1974) Morphometric analysis of *Actinocamax plenus* from England. *Bulletin of the Geological Survey of Denmark*, **23**, 1–26.

Combémorel, R. (1988) Les bélemnites de Madagascar. *Documents des Laboratoires de Géologie Lyon*, **104**, 1–237.

Combémorel, R. and Mariotti, N. (1986) Les bélemnites de la carrière de Serra San Quirico (Province d'Ancona, Apennin

Central, Italie) et la paléobiogéographie des bélemnites de la Téthys méditerranéene au Tithonique inférieur. *Geobios*, **19**, 299–321.

Crick, G. C. (1905) On a dibranchiate cephalopod. *Styracoteuthis orientalis*, n. gen. & n. sp., from the Eocene of Arabia. *Proceedings of the Malacological Society of London*, **6**, 274–78.

Curry, D. (1955) The occurrence of the dibranchiate cephalopods *Vasseuria* and *Belosepiella* in the English Eocene, with notes on their structure. *Proceedings of the Malacological Society of London*, **31**, 111–22.

Day, R. W. (1969) The Lower Cretaceous of the Great Artesian Basin, in *Stratigraphy and Palaeontology: Essays in Honour of Dorothy Hill*, (ed. K. S. W. Compbell). ANUP, Canberra, pp. 140–73.

Diener, C. (1915) *Fossilium Catalogus: I Animalia, Pars 8 Cephalopoda triadica*, (ed. F. Frech), Junk, Berlin, pp. 1–360.

Donovan, D. T. (1977) Evolution of the dibranchiate cephalopoda. *Symposia of the Zoological Society of London*, **38**, 15–48.

Donovan, D. T. and Hancock, J. M. (1967) Mollusca: Cephalopoda (Coleoidea), in *The Fossil Record* (eds W. B. Harland, C. H. Holland, M. R. House *et al.*), Geological Society of London, London, pp. 461–7.

Donovan, D. T. and Toll, R. B. (1988) The gladius in coleoid (Cephalopoda) evolution, in *Paleontology and Neontology of Cephalopods (The Mollusca Vol. 12.)*, (eds M. R. Clarke and E. R. Trueman), Academic Press, San Diego, pp. 89–101.

Doyle, P. (1986) *Naefia* (Coleoidea) from the late Cretaceous of southern India. *Bulletin of the British Museum (Natural History)*, Geology, **40**, 133–9.

Doyle, P. (1990) The British Toarcian (Lower Jurassic) belemnites. *Palaeontographical Society (Monographs)*, Part 1, 49 pp.

Doyle, P. (1992) The British Toarcian (Lower Jurassic) belemnites. Part 2. *Palaeontographical Society (Monographs)*, 29 pp.

Doyle, P. and Kelly, S. R. A. (1988) The Jurassic and Cretaceous belemnites of Kong Karls Land, Svalbard. *Norske Polarinstitut Skrifter*, **189**, 1–78.

Dreyfuss, M. (1957) Présence d'un céphalopode de la familie des Aulacocératidés dans l'Argovien des environs de Gorniès (Hérault). *Bulletin de la Société Géologique de France, série 6*, **7**, 61–4.

Engeser, T. (1988a) Fossil 'Octopods' – A critical review, in *Paleontology and Neontology of Cephalopods (The Mollusca, Vol. 12)*, (eds M. R. Clarke and E. R. Trueman), Academic Press, San Diego, pp. 81–7.

Engeser, T. (1988b) *Fossilium Catalogus I: Animalia. Pars 130 Vampyromorpha ('Fossile Teuthiden')*, Kugler, Amsterdam, 167 pp.

Engeser, T. (1990) Phylogeny of the fossil Cephalopoda (Mollusca). *Berliner geowissenschaftlichen Abhandlungen*, **124**, 123–91.

Engeser, T. and Bandel, K. (1988) Phylogenetic classification of coleoid cephalopods, in *Cephalopods – Present and Past*, (eds J. Wiedmann and J. Kullmann), Schweizerbart, Stuttgart, pp. 105–15.

Engeser, T. and Reitner, J. (1981) Beiträge zur Systematik von phragmokontragenden Coleoiden aus dem Untertithonium (Malm zeta, 'Solnhofener Plattenkalk') von Solnhofen und Eichstätt (Bayern). *Neues Jahrbuch für Geologie und Paläontologie. Monatshefte*, **1981**, 527–45.

Fischer, J. C. and Riou, B. (1982) Le plus ancien Octopode connu (Cephalopoda, Dibranchiata): *Proteroctopus ribeti* nov. gen., nov. sp., du Callovien de l'Ardèche (France). *Comptes Rendus de l'Académie des Sciences. Paris, Série II*, **295**, 277–80.

Flower, R. H. (1945) A belemnite from a Mississippian boulder of the Caney Shale. *Journal of Paleontology*, **19**, 490–503.

Flower, R. H. and Gordon, M. (1959) More Mississippian belemnites. *Journal of Paleontology*, **33**, 809–42.

Frebold, H. (1935) Marines Aptien von der Koldeway Insel (nördliches Ostgrönland). *Meddelelser om Grønland*, **95**, 1–112.

Gaković, M. and Stoyanova-Vergilova, M. (1978) *Conodicoelites riliensis* spec. nov. (Belemnitida) from the Domerian in the Dinarides. *Annales Géologiques de la Péninsule Balkanique*, **42**, 413–20 [in Bulgarian, English summary].

Gordon, M. (1966) Classification of Mississippian coleoid cephalopods. *Journal of Paleontology*, **40**, 449–52.

Gordon, M. (1971) Primitive squid gladii from the Permian of Utah. *United States Geological Survey Professional Paper*, **750-C**, 34–8.

Hayward, B. W. (1976) *Spirula* (Sepioidea: Cephalopoda) from the Lower Miocene of Kaipara Harbour, New Zealand. *New Zealand Journal of Geology and Geophysics*, **29**, 145–7.

Hewitt, R. A., Yoshiike, T. and Westermann, G. E. G. (1991) Shell microstucture and ecology of the Cretaceous coleoid cephalopod *Naefia* from the Santonian of Japan. *Cretaceous Research*, **12**, 47–54.

Jeletzky, J. A. (1966) Comparative morphology, phylogeny, and classification of fossil coleoids. *University of Kansas Paleontological Contributions*, Mollusca. Art, **7**, 1–162.

Jeletzky, J. A. (1969) New or poorly understood Tertiary sepiids from southeastern United States and Mexico. *University of Kansas Paleontological Contributions*, **41**, 1–39.

Jeletzky, J. A. (1981) Lower Cretaceous diplobelinid belemnites from the Anglo-Paris Basin. *Palaeontology*, **24**, 115–45.

Jeletzky, J. A. and Zapfe, H. (1967) Coleoid and orthocerid cephalopods of the Rhaetian Zlambach Marl from the Fischerwiese near Aussee, Styria (Austria). *Annalen des Naturhistorischen Museums, Wien*, **71**, 69–106.

Johnson, R. G. and Richardson, E. S. (1968) Ten-armed fossil cephalopod from the Pennsylvanian in Illinois. *Science*, **159**, 526–8.

Kobayashi, T. (1954) A new Palaeogene paracenoceratoid from southern Kyushu in Japan. *Japanese Journal of Geology and Geography*, **24**, 181–4.

Kongiel, R. (1962) On belemnites from Maestrichtian, Campanian and Santonian sediments from the middle Vistula Valley (Central Poland). *Prace Muzeum Ziemi*, **5**, 3–158.

Koninck, L. G. de. (1843) Notice sur une coquille fossile de terrains anciens de Belgique. *Bulletin de l'Académie Royale des Sciences, des Lettres et des Beaux-Arts de Belgique*, **10**, 207–8.

Lissajous, M. (1927) Déscription de quelques nouvelles espèces de bélemnites jurassiques. *Travaux de Laboratoire de Géologique de la Faculté des Sciences de Université de Lyon*, **7** *(Supplément)*, 1–43.

Małecki, J. (1984) Belemnites of the genus *Rhopaloteuthis* Lissajous, 1915 from the Lower and Middle Oxfordian in the vicinities of Cracow. *Bulletin of the Polish Academy of Sciences, Earth Sciences*, **32**, 53–63.

Meneghini, J. (1867) Monographie des fossiles du calcaire rouge ammonitique (Lias supérieur) de Lombardie et de l'Appennin central, in *Paléontologie Lombarde*, 4 série, (A. Stoppani), Milan, 242 pp.

Meyer, J. C. (1989) Un nouveau cephalopode coleoide dans le Paléocene inférieur de Vigny (95): *Ceratisepia elongata*, n. gen., n. sp. *SAGA Information*, **94**, 30–51.

Müller-Stoll, H. (1936) Beitrage zur Anatomie der Belemnoidea. *Nova Acta Leopoldina. New Series*, **4**, 159–226.

Munier-Chalmas, E. (1872) Note sur les nouveaux genres *Belopterina* et *Bayanoteuthis*. *Bulletin de la Société Géologique de France, 2 série*, **29**, 530–1.

Mutterlose, J. (1983) Phylogenie und Biostratigraphie der unterfamilie Oxyteuthinae (Belemnitida) aus dem Barrême (Unter-Kreide) NW-Europas. *Palaeontographica, Abteilung A* **180**, 1–90.

Naef, A. (1922) *Die Fossilien Tintenfische. Eine Paläozoologisches Monographie*, Fischer, Jena, 322 pp.

Nal'nyaeva, T. I. (1983) Biostratigraphic and biogeographic analysis of associations of belemnites of the Upper Jurassic and Neocomian in the Pechora Basin. *Trudy̆ Instituta Geologii i Geofiziki. Sibirskoe Otdelenie*, **528**, 113–21 [in Russian].

Newton, R. B. and Harris, G. F. (1894) A revision of the British Eocene Cephalopoda. *Proceedings of the Malacological Society of London*, **1**, 119–31.

Nicholls, E. L. and Isaak, H. (1987) Stratigraphic and taxonomic significance of *Tusoteuthis longa* Logan (Coleoidae, Teuthida) from the Pembina Member, Pierre Shale (Campanian), of Manitoba. *Journal of Paleontology*, **61**, 727–37.

Noda, H., Ogasawara, K. and Nomura, R. (1986) Systematic and palaeobiogeographic studies on the Miocene argonautid *'Nautilus' izumoensis, Science Reports of the Institute of Geosciences, University of Tsukuba*, **B7**, 14–42.

Riccardi, A. C. and Sabattini, N. (1985) Supposed coleoid remains reinterpreted as fish scales. *Neues Jahrbuch für Geologie und Paläontologie, Monatshefte*, **1985**, 700–5.

Riegraf, W. (1980) Revision der Belemniten des Schwäbischen Jura, teil 7. *Palaeontographica, Abteilung A*, **169**, 128–206.

Riegraf, W. (1991) Triassic belemnoids (Cephalopoda, Coleoidea) formerly described as corals, versus Tertiary corals formerly described as belemnites. *Fossil Cnidaria*, **20**, 40–5.

Riegraf, W., Werner, G. and Lörcher, F. (1984) *Der Posidonienschiefer. Biostratigraphie, Fauna und Fazies des südwestdeutschen Untertoarciums (Lias ε)*, Enke, Stuttgart, 195 pp.

Roger, J. (1944) Phylogénie des céphalopodes octopodes: *Palaeoctopus newboldi* (Sowerby, 1846) Woodward. *Bulletin de la Société Géologique de France, 5 série*, **14**, 83–98.

Roger, J. (1952) Sous-classe des Debranchiata, in *Traite de Paleontologie*. (ed. J. Piveteau) vol. 2, Masson, Paris, pp. 689–755.

Rozenkrantz, A. (1946) Krogbarnende Cephalopoder fra Østgrønlands Perm. *Meddelelser Dansk Geologisk Forening*, **11**, 160–1.

Saks, V. N. and Nal'nyaeva, T. I. (1964) *Upper Jurassic and Lower Cretaceous belemnites of the northern USSR. The genera* Cylindroteuthis *and* Lagonibelus. Nauka, Leningrad, 260 pp. [in Russian].

Saks, V. N. and Nal'nyaeva, T. I. (1975) *Lower and Middle Jurassic belemnites of the northern USSR. Megateuthinae and Pseudodicoelitinae*. Nauka, Leningrad, 185 pp. [in Russian].

Saunders, W. B. and Richardson, E. S. (1979) Middle Pennsylvanian (Desmoinesean) cephalopoda of the Mazon Creek Fauna, northeastern Illinois, in *Mazon Creek Fossils*. (ed. M. H. Nitecki), Academic Press, San Diego. pp. 333–59.

Stevens, G. R. (1965) The Jurassic and Cretaceous belemnites of New Zealand and a review of the Jurassic and Cretaceous belemnites of the Indo-Pacific region. *New Zealand Geological Survey Paleontological Bulletin*, **36**, 1–283.

Stinnesbeck, W. (1986) Zu den faunistischen und palökologischen Verhältnissen in der Quiriquina Formation (Maastrichtian) Zentral-Chiles. *Palaeontographica, Abteilung A*, **194**, 99–237.

Tomida, S. and Itoigawa, J. (1981) *Spirula mizunamiensis*, a new fossil Sepiida from the Miocene Mitzunami Group, central Japan. *Bulletin of the Mizunami Fossil Museum*, **8**, 21–4.

Zhu, K. Y. and Bian, Z. X. (1984) Sinobelemnitidae, a new family of Belemnitida from the Upper Triassic of Longmenshan, Sichuan. *Acta Palaeontologica Sinica*, **23**, 300–319 [in Chinese, English summary].

13

MOLLUSCA: ROSTROCONCHIA, SCAPHOPODA AND BIVALVIA

P. W. Skelton and M. J. Benton

Three molluscan classes, the extinct Rostroconchia and the extant Scaphopoda and Bivalvia were united by Runnegar and Pojeta (1974) in the Subphylum Diasoma ('through-body'), characterized by a primitively bilobate shell associated with a relatively straight gut running from near the anterior arch to the posterior arch of the shell aperture. The rostroconchs are regarded as having evolved as a shallow-burrowing offshoot of early monoplacophorans, in turn giving rise to both the scaphopods and the bivalves. However, Peel (1991) expressed an opposing view, that the Diasoma is a diphyletic group, since he considers that the exogastric and endogastric molluscs had separate origins. Hence, the Bivalvia and Rostroconchia would be entirely separate.

In the rostroconchs, the two lateral lobes of the shell are separated ventrally by a narrow gape, which became occluded in some advanced forms, leaving only anterior and posterior gapes. Continuous fracturing and repair along the dorsal midline of the shell accompanied the forcing apart of the opposing shell lobes as they grew ventrally, although, in certain advanced forms, the midline seems to have become flexible. In early forms, the two lobes are joined internally by a transverse dorsal ridge (*pegma*). Most rostroconchs are believed to have been deposit feeders, though a few may have been suspension feeders. The systematics of the class is comprehensively reviewed by Pojeta and Runnegar (1976) and Pojeta (1985), from whom the data given here are derived. Rostroconchs were of moderate diversity in shallow Palaeozoic seas, but declined in the Permian, becoming extinct by its close.

The adult shell of scaphopods is fused ventrally, so forming a conical tube, open at both ends, and extended by growth mainly at the posterior end. They are sluggish shallow burrowers in marine sediments, capturing microscopic prey with anterior feeding tentacles. Their diversity has never been high. See Pojeta (1985) and Palmer (1974) for discussion.

The bivalves are diasome molluscs in which the shell consists of two matching lateral valves, united by a dorsal horny ligament, and which can be drawn together by a pair of adductor muscles (reduced to one in some) against the opening counter-force of the ligament.

Primitively, this arrangement has permitted active penetration of sediments, aided by the blade-shaped foot, associated with deposit- and suspension-feeding. Subsequent adaptations have included attachment to surfaces either by means of an organic byssus or by cementation of one of the valves, boring, swimming and the adoption of miniaturized, commensal habits, and, in a few genera, predation. Today, bivalves are among the most common of benthic invertebrates, especially on marine shelves, although there are also many species in fresh water and in abyssal (including rift-vent) habitats.

The fossil record of the class is world-wide and spans the Phanerozoic. In spite of the abundance of bivalve fossils (only a handful of living families are unrepresented), and an extensive literature on their lower-level taxonomy, much uncertainty remains over their higher-level relationships. The relatively small number of taxonomically useful shell characters (e.g. form and arrangement of the teeth on the valve hinges, type of ligament insertion, disposition of muscle scars and the general shape of the shell) are mostly of a simple nature, and frequently show convergence. Cladistic analysis is therefore difficult and has been conducted on only a limited scale to date; the monophyletic versus paraphyletic status of many families is unknown. Also, the taxonomic diversity of families is very uneven: several monogeneric families exist, but most have from a few up to a couple of dozen genera, while two families (Unionidae and Veneridae) have around a hundred genera each.

The precision of stratigraphical information available for bivalve taxa often leaves much to be desired. With a few notable exceptions (e.g. Coal Measures forms, inoceramids, buchiids), the

The Fossil Record 2. Edited by M. J. Benton. Published in 1993 by Chapman & Hall, London. ISBN 0 412 39380 8

bivalves themselves are little used for correlation, and many records are from strata which are difficult to date in any case. In most instances, stage-level resolution is all that can be established.

The indispensible sources for taxonomic and stratigraphical information are the three '*Treatise*' volumes (Cox *et al.*, 1969, 1971). These formed the basis for the compilation of marine families given by Sepkoski (1982). A full systematic and bibliographic (although not stratigraphical) catalogue of genera is given by Vokes (1980). Photographic illustrations of a number of representative taxa, with guidance on identification, may be found in Skelton (1985a). Several papers in Yonge and Thompson (1978) make important revisions of taxonomic groupings and of ranges across the class, while Pojeta and Runnegar (1985) and Pojeta (1985) provide updated information concerning its early Palaeozoic record. Carter (1990) provides updated coverage of three of the subclasses (Palaeotaxodonta, Pteriomorphia and Isofilibranchia), although a number of the ranges given have not been revised from the *Treatise* and are thus inconsistent with other data included in his discussions of the taxa. In the list which follows, we have used these sources, together with a number of more specialized works.

Acknowledgements – We have also drawn extensively on personal communications and corrections from C. Babin, Lyons, M. Brasier, Oxford, J. C. W. Cope, Cardiff, J. A. Crame, Cambridge, S. E. Damborenea, La Plata, T. A. Darragh, Melbourne, A. V. Dhondt, Brussels, E. M. Harper, Cambridge, I. Hayami, Japan, A. L. A. Johnson, Derby, S. R. A. Kelly, Cambridge, J. Kříž, Prague, N. J. Morris, London, S. P. Tunnicliff, Keyworth, J. B. Waterhouse, Nelson, New Zealand, T. Yancey, Texas A & M and Yin Honggu, Hubei. Their help is gratefully acknowledged, although they cannot be held responsible for any errors herein.

EDITOR'S NOTE

This chapter was compiled in several stages. PWS provided the outline classification and stratigraphical range information, as well as more detailed documentation of bivalve families from Palaeotaxodonta to Ostreoida inclusive plus Hippuritoida. MJB filled the gaps, providing documentation of rostroconchs and scaphopods, and other bivalve families from Lucinoida to the end of the listing, and augmenting the coverage earlier in the chapter. We warn potential users of this database of its patchy quality: some parts are good, where specialists have helped us, but large parts of the listing were compiled at speed, and without expert commentary.

Class ROSTROCONCHIA Pojeta *et al.*, 1972

The classification and basic information on rostroconchs are taken from Pojeta and Runnegar (1976), except where otherwise indicated. The oldest rostroconch is *Watsonella crosbyi* Grabau, 1900, Placentian (pre-Tom), Burin Peninsula, Newfoundland, Canada (Landing, 1989).

Order RIBEIRIOIDA Kobayashi, 1933

F. RIBEIRIIDAE Kobayashi, 1933
Є. (TOM)–O. (ASH) Mar.

First: *Heraultipegma* sp., TOM, Tiktirikteekh, River Lena, Siberia, former USSR (Matthews and Missarzhevsky, 1975); *H. varensalense* (Cobbold, 1935), Georgien, near St Geniès de Varensal, Hérault, France; *Watsonella crosbyi* Grabau, 1990, Lower Cambrian boulders, Cohasset, Massachusetts, USA.
Last: *Ribeiria* sp., ASH, Bohemia, Czechoslovakia.

F. TECHNOPHORIDAE Miller, 1889
Є. (ATB)–S. (LLY) Mar.

First: 'technophorid', ATB, Nevada, USA (Pojeta, 1985).
Last: *Technophorus*, LLY, China (Pojeta, 1985).
Comments: The next oldest technophorid is *Oepikila cambrica* (Runnegar and Pojeta, 1974), Georgina Limestone, Idamean (MNT), western Queensland, Australia, and technophorids are otherwise indicated as having died out in the latest Ordovician.

Order ISCHYRINIOIDA Pojeta and Runnegar, 1976

F. ISCHYRINIIDAE Kobayashi, 1933
Є. (DOL)–O. (ASH) Mar.

First: *Pseudotechnophorus typicalis* Kobayashi, 1933 and *Eoischyrinia billingsi* Kobayashi, 1933, both Wanwankou Dolomite, Wanwanian, southern Manchuria.
Last: *Ischyrinia winchelli* Billings, 1866, English Head and Ellis Bay formations (Richmondian), Anticosti Island, Quebec, Canada; *I. schmidti* Teichert, 1930, Lyckholm-Stufe, Estonia, former USSR.
Comments: The Wanwankou Dolomite was originally dated as TRE, but is reclassified as Fengshanian, or latest Cambrian.

Order CONOCARDIOIDA Neumayr, 1891

F. EOPTERIIDAE Miller, 1889
Є. (DOL)–O. (CRD) Mar.

First: *Eopteria flora* Kobayashi, 1933, Wanwankou Dolomite, Wanwanian, Fengshanian, southern Manchuria.
Last: *Eopteria conocardiformis* Pojeta and Runnegar, 1976, Little Oak Limestone, Porterfieldian, Pelham, Alabama, USA and High Bridge Group, Wildernessian, Highbridge, Kentucky, USA.
Comments: A possible eopteriid is noted from the ASH of Queensland, Australia (Shergold, 1976).

F. CONOCARDIIDAE Miller, 1889
D. (PRA)–P. (SAK) Mar.

First: *Conocardium attenuatum* (Conrad, 1842), Schoharic Formation, New York State, USA.
Last: *Arceodomus langenheimi* (Wilson, 1970), McCloud Limestone, Shasta County, California, USA.

F. BRANSONIIDAE Pojeta and Runnegar, 1976
O. (TRE)–P. (KAZ) Mar.

First: *Bransonia chapronierei* Pojeta *et al.* 1977, Ninmaroo Formation (Datsonian, lower TRE), eastern Georgina Basin, western Queensland, Australia (Pojeta *et al.*, 1977).
Last: *Bransonia projecta* Waterhouse, 1987, Flat Top Formation, Bowen Basin, Queensland, Australia (Waterhouse, 1987).

F. HIPPOCARDIIDAE Pojeta and Runnegar, 1976 O. (LLO)–C. (MOS) Mar.

First: *Hippocardia calcis* (Baily, 1860), ?LLO, Tipperary, Republic of Ireland; *H? diptera* (Salter, 1851), LLO, Scotland, UK.
Last: *Hippocardia* and *Pseudobigalea* noted by Pojeta (1985).

Class SCAPHOPODA Bronn, 1862

The classification is taken from Palmer (1974), and the taxonomic ranges from Ludbrook (1960), except where otherwise stated.

Order DENTALIOIDA Palmer, 1974

F. LAEVIDENTALIIDAE Palmer 1974 O. (CRD)–Rec. Mar.

First: *Rhytiodentalium kentuckyensis* Pojeta and Runnegar, 1979, middle Lexington Limestone (Shermanian), central Kentucky, USA (Pojeta and Runnegar, 1979). **Extant**
Comments: Another early record, *Plagioglypta undulatum* (Münster, 1844) from the Lower Ordovician of the former USSR (Ludbrook, 1960) is not confirmed. The next oldest specimens of *Plagioglypta* are late Devonian in age (Ludbrook, 1960).

F. DENTALIIDAE Gray, 1834 D. (PRA)–Rec. Mar.

First: *Prodentalium martini* (Whitfield, 1882), upper Helderberg Limestone, Dublin, Ohio, USA. **Extant**
Comments: The Family Dentaliidae was redefined by Palmer (1974) to include only '*Dentalium sensu lato*' and *Prodentalium*. Other dentalioids were assigned by him to the Laevidentaliidae.

Order SIPHONODENTALIOIDA Palmer, 1974

F. CADULIDAE Grant and Gale, 1931 K. (l.)–Rec. Mar.

First: *Cadulus ovulum* (Philippi, 1844), Lower Cretaceous, Europe, North America, Greenland. **Extant**

F. SIPHONODENTALIIDAE Simroth, 1894 P. (SAK)–Rec. Mar.

First: *Calstevenus arcturus* Yancey, 1973, Riepetown Formation, Arcturus Group, White Pine County, Nevada, USA (Yancey, 1973). **Extant**

Class BIVALVIA Linné, 1758

Details of taxa are taken from Cox *et al.* (1969, 1971) unless otherwise noted.

Subclass PALAEOTAXODONTA Korobkov, 1954

Order NUCULOIDA, Dall, 1889

F. THORALIIDAE Morris, 1980 O. (ARG) Mar.

First and Last: *Thoralia languedociana* (Thoral, 1935), lower ARG, Languedoc, southern France (Morris, 1980).

F. PRAENUCULIDAE McAlester 1969 Ꞓ. (TOM)–D. (EMS) Mar.

First: *Pojetaia runnegari* Jell, 1980, lower Parara Limestone, upper TOM, Yorke Peninsula, South Australia, and China (Pojeta, 1988).
Last: *Deceptrix carinata* Fuchs, 1919, Hunsrückschiefer and Koblenzschichten, Lorelei region, Germany.

F. NUCULIDAE Gray, 1824 D. (GIV)–Rec. Mar.

First: *Nuculoidea opima* (Hall, 1843) and *N. deceptriformis* Bailey, 1983, both Hamilton Group, New York State, USA (Carter, 1990).
Comments: The diagnostic internal resilium is unknown in any Ordovician forms (Pojeta and Runnegar, 1985), so extension of the range of this genus into the Ordovician, cited by Cox *et al.* (1969) and Carter (1990, p. 145), may be discounted.

F. PRISTIGLOMIDAE Sanders and Allen, 1973 **Extant** Mar.

F. MALLETIIDAE Adams and Adams, 1858 O. (TRE)–Rec. Mar.

First: *Ctenodonta iruyensis* (Harrington, 1938), black shales with *Parabolibnella*, Lower TRE, La Rioja, and *C. famatinensis* Harrington, 1938, shales with *Kainella*, lower TRE, Iruya, both northern Argentina (Pojeta, 1988).
Comments: The family 'Ctenodontidae' (see Cox *et al.*, 1969) is suppressed by Pojeta (1988), and incorporated provisionally in Malletiidae (among which other Ordovician genera are known, however).

F. POLIDEVCIIDAE Kumpera *et al.*, 1960 D. (GIV)–K. (MAA) Mar.

First: *Phestia* Chernyshev, 1951, Ukraine, former USSR.
Last: *Nuculana (Nuculana) producta* (Nilsson, 1827), and other species, Middle Vistula Valley, central Poland, and elsewhere (Abdel-Gawad, 1986).
Comment: Next youngest is *Ryderia* Wilton, 1830, Jurassic (Carter, 1990).

F. NUCULANIDAE Adams and Adams, 1858 K.–Rec. Mar.

First: *Ezonuculana mactraeformis* Nagao, 1938, Japan (Carter, 1990). **Extant**

F. NEILONELLIDAE Allen, 1978 T. (LUT)–Rec. Mar.

First: *Pseudotindaria* (?) *ferrari* (Fleming, 1950), Bortonian, Kaipara Harbour, Northland, New Zealand (Beu and Maxwell, 1990).

F. YOLDIIDAE Allen, 1985 K. (MAA)–Rec. Mar.

First: *Yoldia (Yoldia) hyperborea* Torell, 1859, Fox Hills Formation, South Dakota, USA (Speden, 1970). **Extant**

F. SILICULIDAE Allen and Sanders, 1973 **Extant** Mar.

F. LAMETILIDAE Allen and Sanders, 1973 **Extant** Mar.

F. TINDARIIDAE Scarlato and Starobogatov, 1971 T. (PLI?)–Rec. Mar.

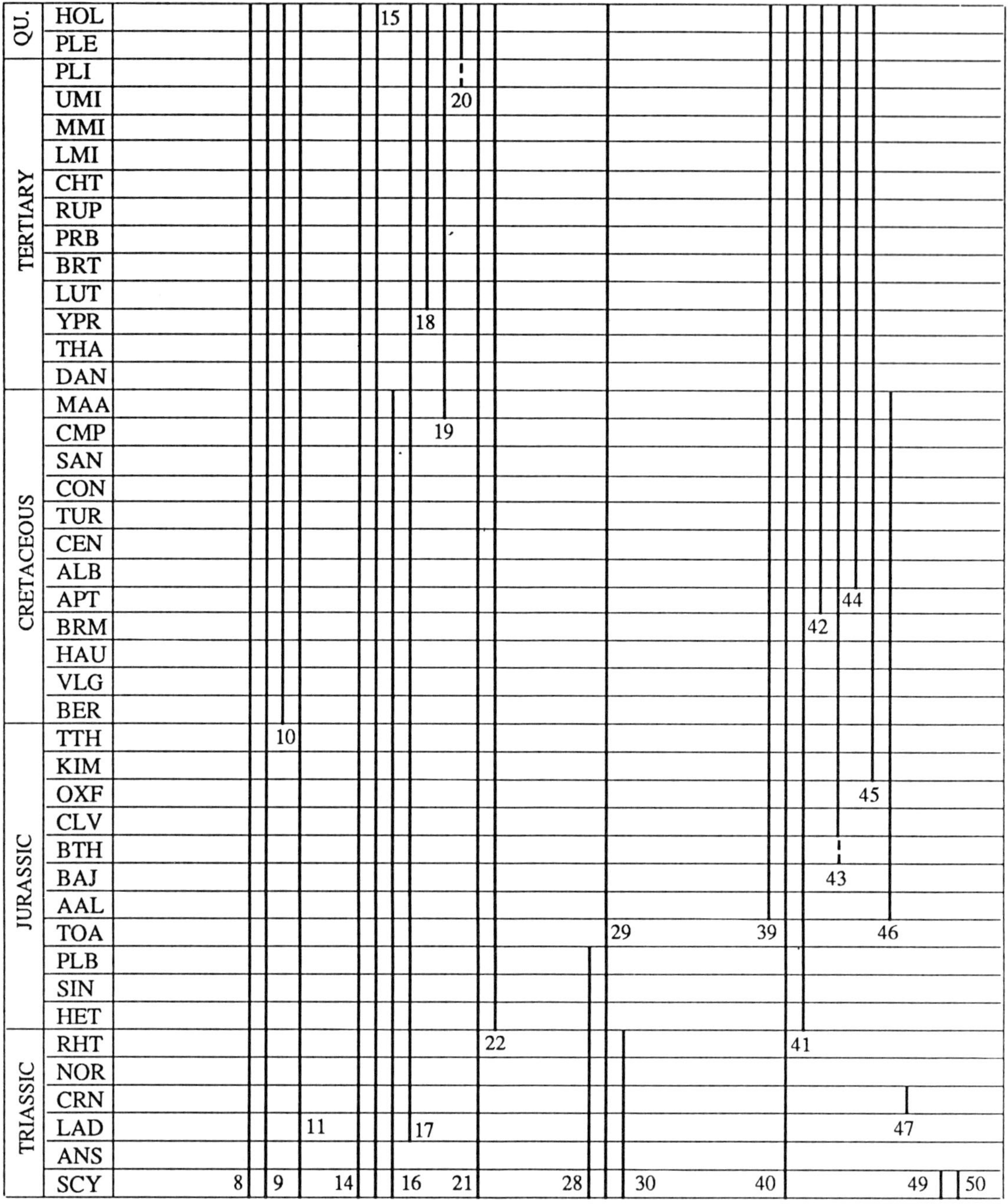

Fig. 13.1

First: *Tindaria arata* Bellardi, 1875, lower PLI, Piemonte and Liguria, Italy. **Extant**
Comments: The fossil record of these deep-water forms is very poor (N. J. Morris, pers. comm., 1992). See also Allen (1978).

Order SOLEMYOIDA Dall, 1889

F. SOLEMYIDAE Adams and Adams, 1857 (1840)
O. (LLN)–Rec. Mar.

First: *Psiloconcha senecta* (Sardeson, 1896), St Peter Sandstone, St Paul, Minnesota, USA; *Dystactella ordovicicus* (Pojeta and Runnegar, 1985) and *D. aedilis* (Eichwald, 1856), Kukruse Stage, Estonia, former USSR (Pojeta, 1988). **Extant**

F. NUCINELLIDAE Vokes, 1956
J. (HET)–Rec. Mar.

First: *Nucinella liasina* (Bistram), near Lake Lugano, Switzerland (Pojeta, 1988). **Extant**
Comments: The Nucinellidae is included in the Manzanellidae by many authors (e.g. Beu and Maxwell, 1990).

F. MANZANELLIDAE Chronic, 1952
P. (LEN) Mar.

First and Last: *Manzanella elliptica* Girty, 1909, Yeso Formation, New Mexico and Kaibab Formation, Arizona, USA, both Leonardian (Pojeta, 1988).
Comments: Placement of the family in the Limopsacea Dall, 1895, *in* Cox *et al.* (1969) is rejected (see Pojeta, 1988).

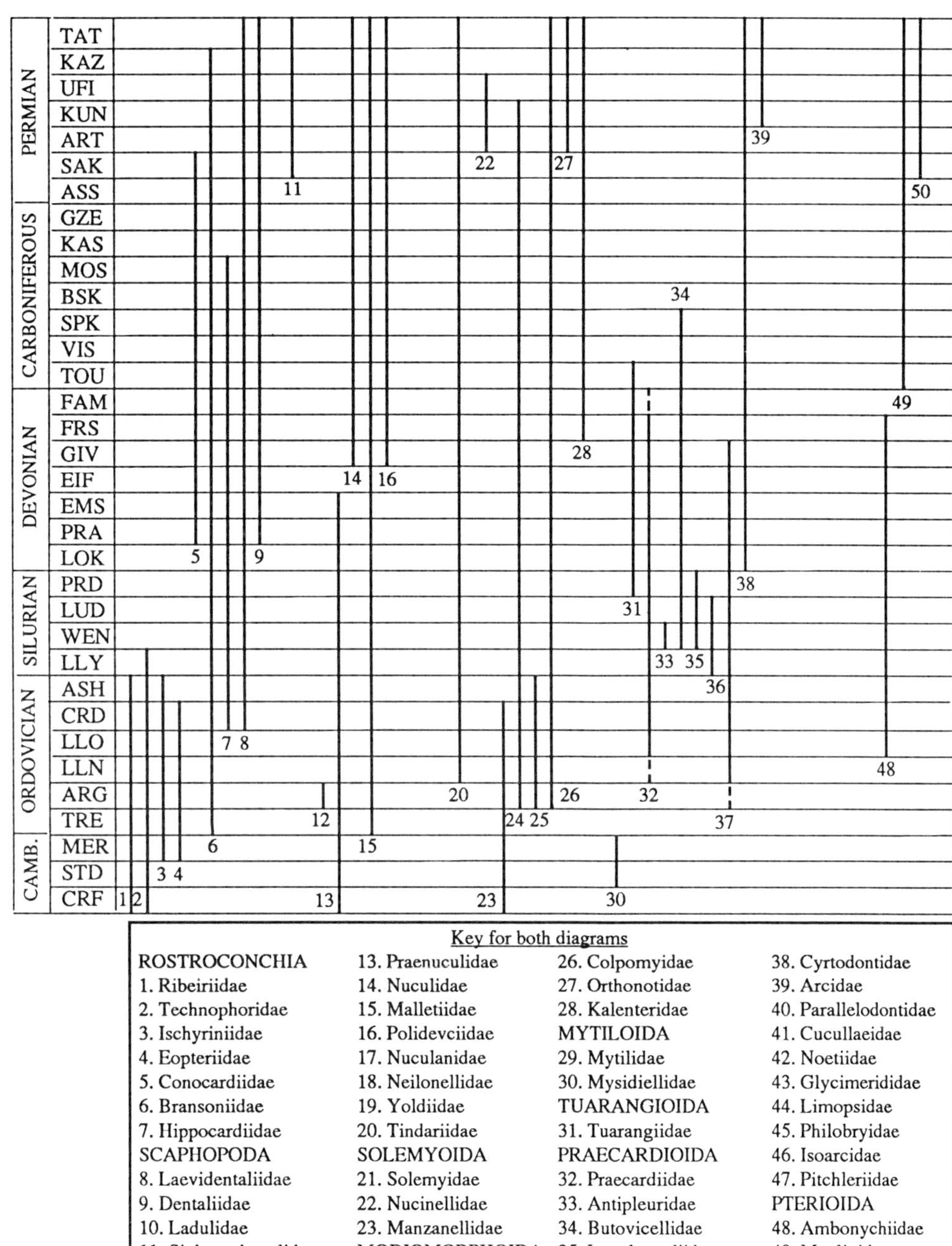

Fig. 13.1

Subclass ISOFILIBRANCHIA Iredale, 1939

Order MODIOMORPHOIDA Newell, 1969

F. FORDILLIDAE Pojeta, 1975
€. (TOM)–O. (CRD) Mar.

First: Fordilla sp., TOM, Khara-Ulakh, Lena River, Siberia, Russia (Jermak, 1986 *in* Rozanov and Zhoravlev, 1992). The next oldest fossil is *Fordilla troyensis* (Barrande, 1881), *Hyolithes* Limestone (lower ATB), Leicestershire, England, as well as Siberia, former USSR and New York, USA (Brasier, 1984).

Last: *Neofordilla elegans* Krasilova, 1977, middle Dolborian Beds, Cherlechine River, Siberia, former USSR (Pojeta and Runnegar, 1985).

Comments: *Fordilla* was tentatively assigned to the Isofilibranchia by Pojeta (1975) on the basis of similarity in its external morphology to later forms. It is not clear whether the feeding gill had evolved before the Ordovician (J. C. W. Cope, pers. comm., 1992).

F. MODIOMORPHIDAE Miller, 1877
(= Modiolopsidae Fischer, 1887)
O. (ARG)–P. (ROT) Mar.

First: *Modiolopsis?* sp., Couches du Foulon, Croix de-Roquebrun, Montagne Noire, France (Babin, 1982a).

Last: *Goniophora, Goniophorina*, Lower Permian, southwestern USA.

Comments: *Glyptarca primaeva* Hicks, 1873, Ogof Hên

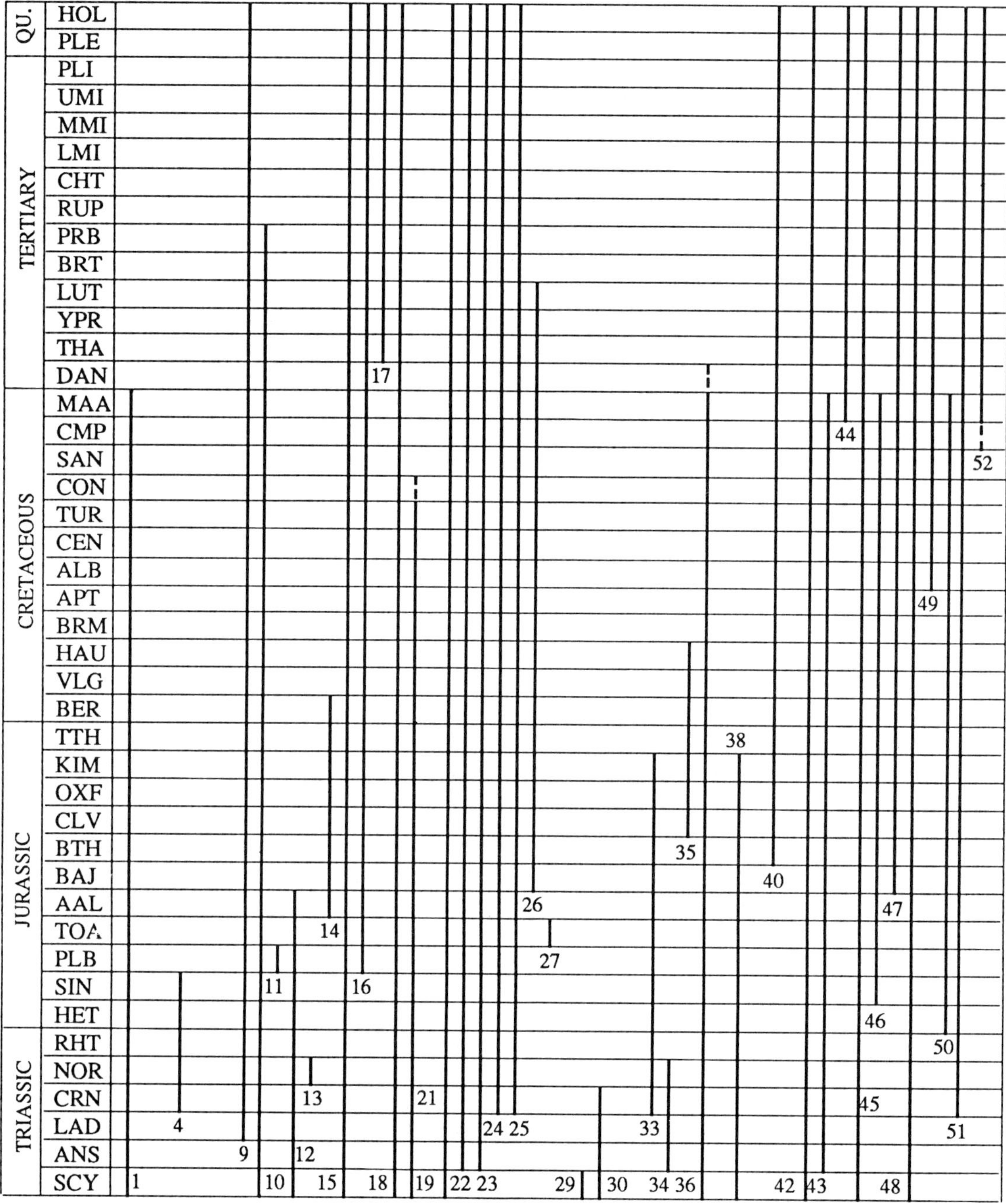

Fig. 13.2

Formation (ARG), Pembrokeshire, Wales, UK (R. M. Carter, 1971) is a poorly defined form, and its assignment is unclear (J. C. W. Cope, pers. comm., 1992). Records of *Modiomorpha* in the Permian are probably not valid (T. Yancey, pers. comm., 1992).

F. COLPOMYIDAE Pojeta and Gilbert-Tomlinson, 1977 O. (ARG–PUS) Mar.

First: *Colpantyx woolleyi* Pojeta and Gilbert-Tomlinson, 1977 and *Xestoconcha kraciukae* Pojeta and Gilbert-Tomlinson, 1977, both Pacoota Sandstone, Northern Territory, Australia (Pojeta and Gilbert-Tomlinson, 1977).
Last: Colpomyid, Maysvillian, USA (Pojeta and Gilbert-Tomlinson, 1977).

F. ORTHONOTIDAE Miller, 1877 O. (ARG)–P. (ZEC) Mar.

First: ?*Cymatonota* sp., Couches du Foulon, Croix de Roquebrun, Montagne Noire, France (Babin, 1982a).
Last: *Solenomorpha minor* (M'Coy, 1844), (Runnegar, 1974).

F. KALENTERIDAE Marwick, 1953 P. (LEN)–J. (PLB) Mar.

First: *Rimmyjimima arcula* Chronic, 1952, Kaibab Formation, Walnut Canyon, Arizona, USA.
Last: *Kalentera flemingi* Marwick, 1953, upper Uraroan (PLB/TOA?), Awakino, Auckland, New Zealand. A similar species extends up to Early PLB in western Argentina (S. Damborenea, pers. comm., 1992).

Order MYTILOIDA Férussac, 1822

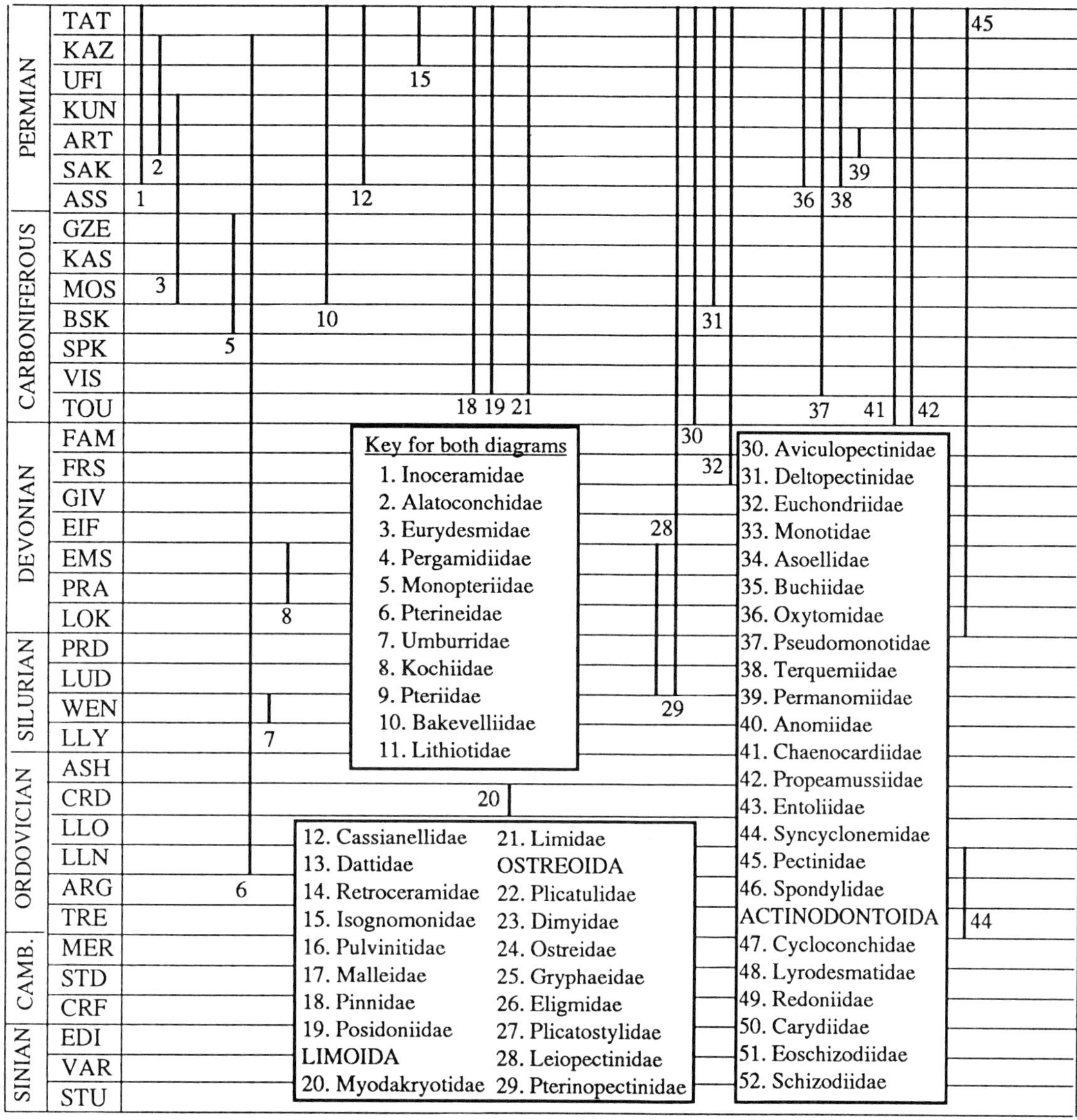

Fig. 13.2

F. MYTILIDAE Rafinesque, 1815
D. (FRS)–Rec. Mar.

First: ?*Mytilops precedens* (Hall, 1870), Chemung Group, New York State, USA. **Extant**

F. MYSIDIELLIDAE Cox, 1964
Tr. (SCY–RHT) Mar.

First: *Protopis triptycha* Kittl, 1904, Werfen Beds, Austrian Alps.

Last: *Mysidiella orientalis* (Bittner, 1891), Anatolia; *M. cordillerana* Newton, 1987, NOR, Wallowa Terrane, Oregon, USA (Carter, 1990); *M. aequilateralis* (Stoppani, 1865), Zlambach Beds, Austria (Hallam, 1981).

Subclass PTERIOMORPHIA Beurlen, 1944

Order TUARANGIOIDA MacKinnon, 1982

F. TUARANGIIDAE MacKinnon, 1982
Є. (MEN–MNT) Mar.

First: *Tuarangia paparua* MacKinnon, 1982, Tasman Formation, New Zealand; *T. gravgaerdensis* Berg-Madsen, 1987, Andarum Limestone, Bornholm, Denmark.

Last: *Tuarangia* sp., Miedzyzdroye, Poland (Berg-Madsen, 1987).

Order PRAECARDIOIDA Newell, 1965

F. PRAECARDIIDAE Hörnes, 1884
S. (PRD)–C. (HAS) Mar.

First: *Praecardium primulum* Barrande, 1881, E_2, Middle Silurian, Bohemia, Czechoslovakia.

Last: *Dexiobia whitei* Winchell, 1863, Kinderhookian, Iowa, USA.

Comments: *Eopteria* Billings, 1865, included in the family by Cox *et al.* (1969), is a rostroconch.

F. ANTIPLEURIDAE Neumayr, 1891
O. (LLN/LLO)–D. (FRS/FAM) Mar.

First: *Shanina vlastoides* Reed, 1915, upper Naugkangyi Beds, North Shan States, India.

Last: *Hercynella beyrichi* Kayser, 1878, Harz Mountains, Germany.

F. BUTOVICELLIDAE Kříž, 1965 S. (WEN) Mar.

First and Last: *Butovicella migrans* (Barrande, 1881), Motol Formation, *C. lundgreni* Zone, Bohemia, Czechoslovakia.

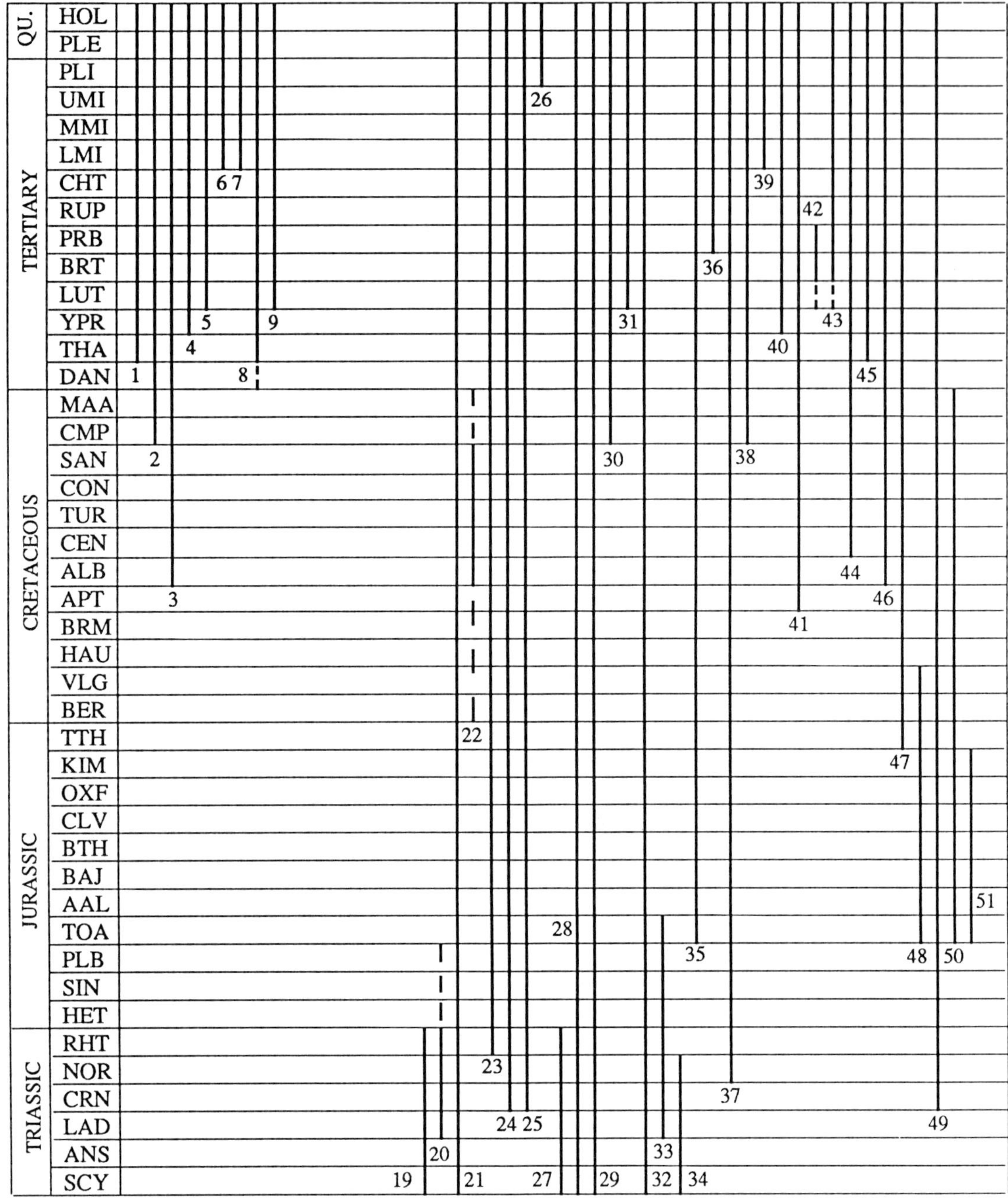

Fig. 13.3

F. LUNULACARDIIDAE Fischer, 1887 S. (WEN)–C. (l.) Mar.

First: *Patrocardia* sp., WEN, Carnic Alps, Austria (J. Křìž, pers. comm., 1992).

Last: *Lunulacardium (Lunulacardium) semistriatum* Münster, 1840, Europe, North America.

Comments: All the early taxa are poorly understood, and the bivalve status of some is even questionable. *Euchasma* Billings, 1865, included here by Cox *et al.* (1969), is a rostroconch (Pojeta and Runnegar, 1976). In addition, *Maminka* has been reassigned to the Antipleuridae (Carter, 1990, p. 204).

Order ARCOIDA Stoliczka, 1871

F. CARDIOLIDAE Fischer, 1886 S. (WEN–PRD) Mar.

First: *Cardiobeleba lavina* Kříž, 1979, lower WEN, Carnic Alps, Austria.

Last: *Cardiolinka concubina* Kříž, 1979, uppermost PRD, Bohemia, Czechoslovakia.

Comments: Devonian and Carboniferous records, noted in Cox *et al.* (1969, p. N245) are not accepted by J. Kříž (pers. comm., 1992).

F. SLAVIDAE Kříž, 1982 S. (LLY–LUD) Mar.

First: *Slava semirugata* (Portlock, 1843), Pomeroy, Co. Tyrone, Northern Ireland.

Last: *Slavinka oforata* (Barrande, 1881), LUD, Bohemia, Czechoslovakia.

F. CYRTODONTIDAE Ulrich, 1894 O. (ARG?)–D. (GIV) Mar.

First: *Cyrtodontula hadzeli* Pojeta and Gilbert-Tomlinson,

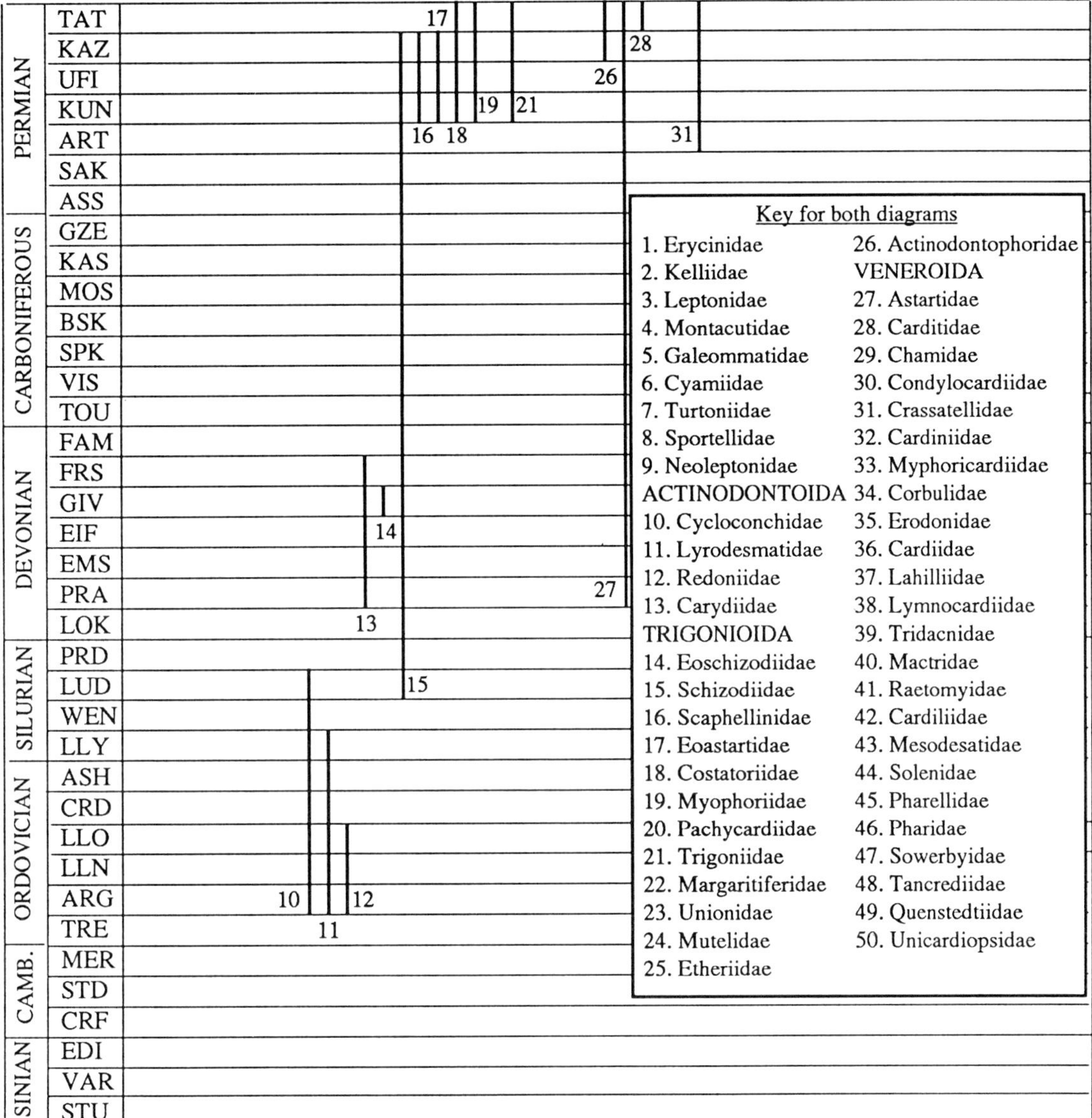

Fig. 13.3

1977 and *Pharcidoconcha raupi* Pojeta and Gilbert-Tomlinson, 1977, both Pacoota Sandstone, Amadeus Basin, southern Northern Territory, Australia (Pojeta and Gilbert-Tomlinson, 1977).
Last: *Ptychodesma knappianum* Hall and Whitfield, 1872, Hamilton Group, New York State, USA; *Pycinodesma* Kirk, 1927, Alaska, USA (Pojeta and Runnegar, 1985).
Comments: *Pharcidoconcha parallelus* (Hsu, 1948) has been noted from the ?TRE of China, but these beds are probably ARG in age (J. C. W. Cope, pers. comm., 1992). *?Cyrtodonta oboloidea* (Hicks, 1873), Ogof Hên Formation, Pembrokeshire, Wales, UK (Carter, 1971) is poorly defined.

F. ARCIDAE Lamarck, 1809 J. (AAL)–Rec. Mar.

First: *Arca (Eonavicula) minuta* J. de C. Sowerby, 1824, various formations, England, UK, France, Germany (Hallam, 1976). **Extant**

F. PARALLELODONTIDAE Dall, 1898
P. (KUN)–Rec.? Mar.

First: *Parallelodon capillatus* Waterhouse, 1987, Brae Formation, Bowen Basin, Queensland, Australia (Waterhouse, 1987). **Extant?**
Comments: Pojeta (1971) regards the poorly known *'Parallelodon' antiquus* Barrois, 1891, from the Grès Armoricain, France (ARG), as misplaced in this family; likewise the *Glyptarca* Hicks, 1873, cited by Cox *et al.* (1969, p. N256). Devonian *'Macrodus'* from the Rhine Devonian may belong here.

F. CUCULLAEIDAE Stewart, 1930
J. (HET)–Rec. Mar.

First: *Cucullaea (Idonearca) mabuchii* Hayami, 1958, *Alsatites* Bed, Shizukawa Group, NE Japan (Damborenea, 1987). **Extant**

F. NOETIIDAE Stewart, 1930 K. (APT)–Rec. Mar.

First: *Noetia (Incanopsis) palestina* (Whitfield, 1891), Lebanon. **Extant**

F. GLYCYMERIDIDAE Newton, 1922
J. (?BTH)–Rec. Mar.

First: *'Limopsis' minima* (J. de C. Sowerby, 1825), BTH, England, UK (Hallam, 1976). **Extant**
Comment: This taxon has been transferred from the Limopsidae (Oliver, 1981).

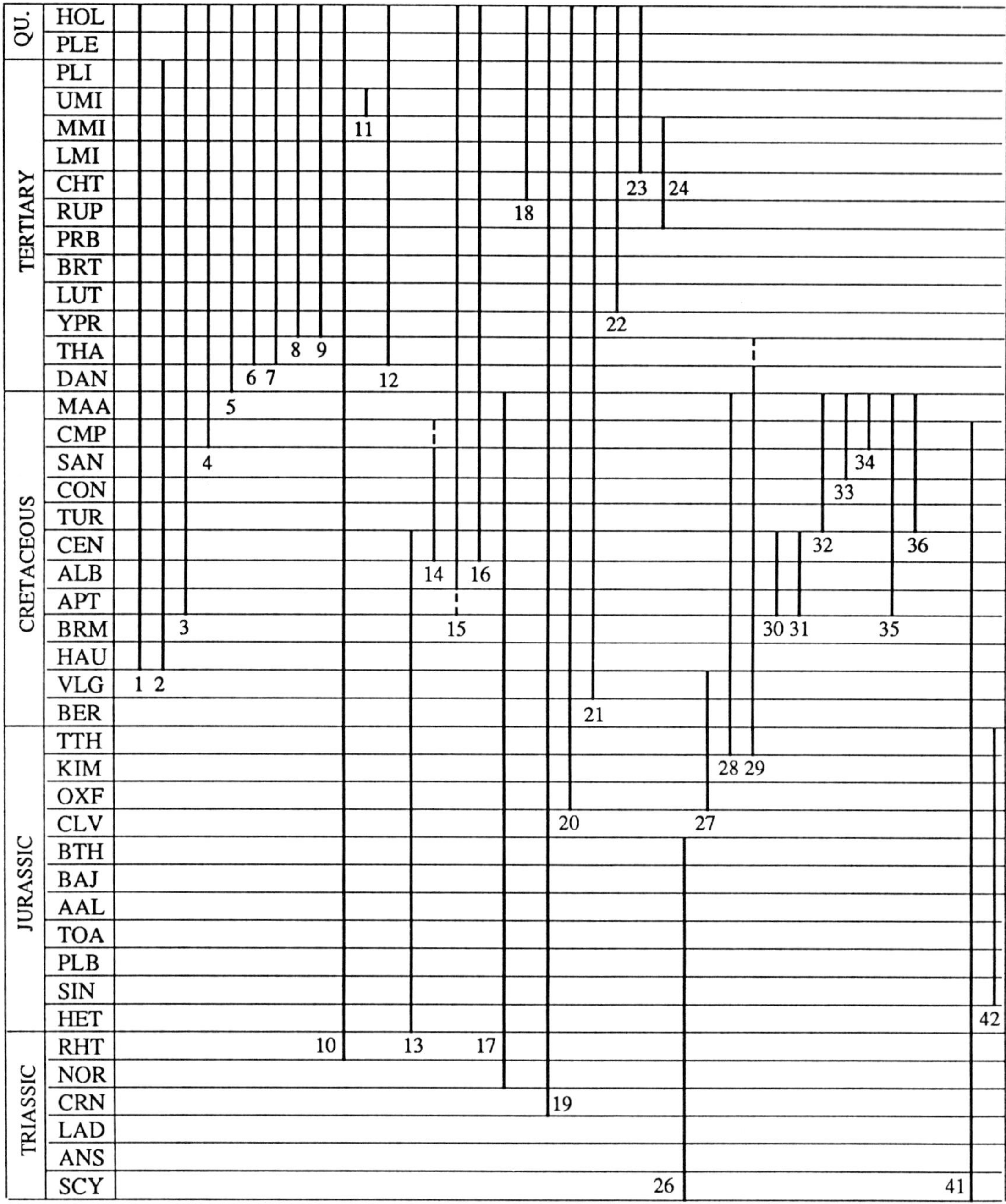

Fig. 13.4

F. LIMOPSIDAE Dall, 1895 K. (ALB)–Rec. Mar.

First: *Limopsis albiensis* (Woods 1899), UK (Carter, 1990); *L. delicatulus* Kauffman, 1976, DSDP hole 317A, Manihiki Plateau, South Pacific (Kauffman, 1976; Oliver, 1981). **Extant**

Comments: *Hoferia* Bittner, 1894 and *Pitchleria* Bittner, 1894 are rejected from the family (Oliver, 1981); their affinities are unknown. Likewise, Jurassic species previously assigned to *Limopsis* Sassi, 1827, are now considered to be glycymeridids (see Carter, 1990, p. 194).

F. PHILOBRYIDAE Bernard, 1897 J. (KIM)–Rec. Mar.

First: Philobryid (P. Wignall, pers. comm., 1992). **Extant**

F. ISOARCIDAE Keen, 1969 J. (AAL)–K. (MAA) Mar.

First: *Isoarca subspirata* (Münster, 1837), Germany (Hallam, 1976).

Last: *Isoarca* sp. The latest BMNH specimens are CEN, but there may be a MAA record (N. J. Morris, pers. comm., 1992).

Comments: The family has been transferred to the Arcoida from the Palaeotaxodonta: Keen's assertion (*in* Cox *et al.*, 1969, p. N241) that the shell is nacreous (implying placement of the family with palaeotaxodonts) is not borne out by inspection of specimens of type species in the British Museum (Natural History), London (N. J. Morris, pers. comm., 1992).

F. PITCHLERIDAE Scarlato and Starobogatov, 1979 Tr. (CRN) Mar.

First and Last: *Pitchleria auingeri* (Laube, 1865) and *Hoferia*

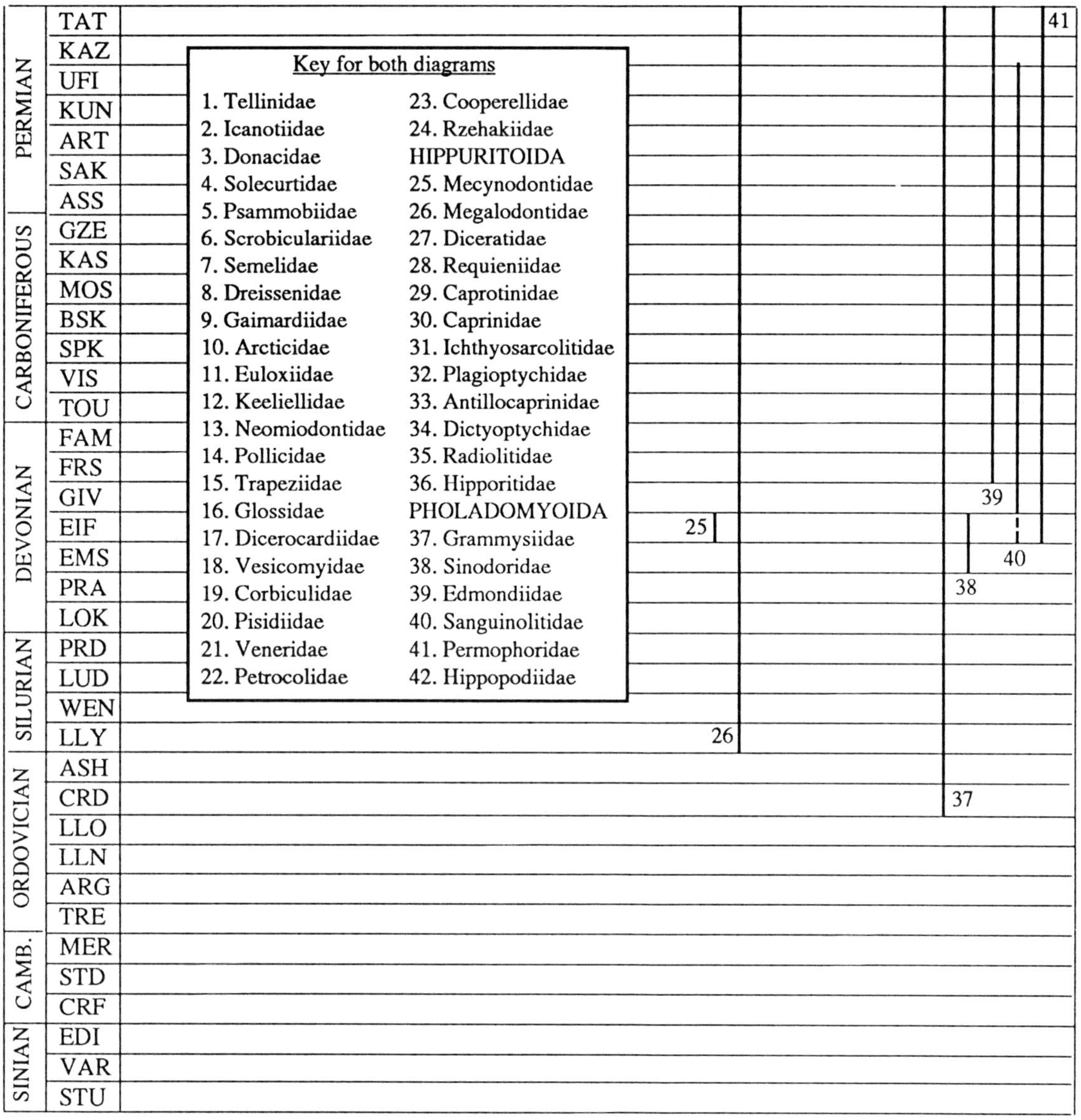

Fig. 13.4

duplicata (Münster, 1838), both Cassian Formation, Italian Alps (Carter, 1990, p. 196).

Order PTERIOIDA Newell, 1965

F. AMBONYCHIIDAE Miller, 1877 O. (LLO)–D. (FRS) Mar.

First: *Cleionychia lamellosa* (Whitfield, 1882), Chazy Formation, New York State, USA; *Pteronychia haupti* Pojeta and Gilbert-Tomlinson, 1977, Stairway Sandstone (LLN/LLO), Amadeus Basin, southern Northern Territory, Australia (Pojeta and Runnegar, 1985).
Last: *Mytilarca chemungensis* (Conrad, 1842), Chemung Group (FRS), New York State, USA.
Comments: *Mytilarca* has been noted doubtfully from the lower Mississippian of Europe and North America.

F. MYALINIDAE Frech, 1891 C. (TOU)–Tr. (SCY) Mar./Brackish

First: *Myalina (Myalina) goldfussiana* De Koninck, 1842, Cementstone Group, Calciferous Sandstone Series, Berwickshire, Scotland, UK, and other species (Gray, 1988).
Last: *Promyalina groenlandica* (Newell, 1955) and *Myalinella meeki* (Dunbar, 1924), both SCY, East Greenland.
Comments: Two Jurassic taxa, *Pachymytilus* Zittel, 1881 and *Pseudopachymytilus* Krumbeck, 1923, and one Devonian form, *Boiomytilus* Růžička and Prantl, 1961, were assigned to the Myalinidae by Cox *et al.* (1969, pp. N289–N295), but these are all disregarded here.

F. ATOMODESMIDAE Waterhouse, 1976 P. (SAK)–Tr. (SCY) Mar.

First: *Aphanaia tivertonensis* Waterhouse, 1979, Tiverton Formation, Bowen Basin, Queensland, Australia (Waterhouse, 1979a).
Last: *Atomodesma variabile* Wanner, 1922, Timor; Panjang Formation, Nepal (possibly derived, J. B. Waterhouse, pers. comm., 1992).

F. INOCERAMIDAE Zittel, 1881 P. (SAK)–K. (MAA) Mar.

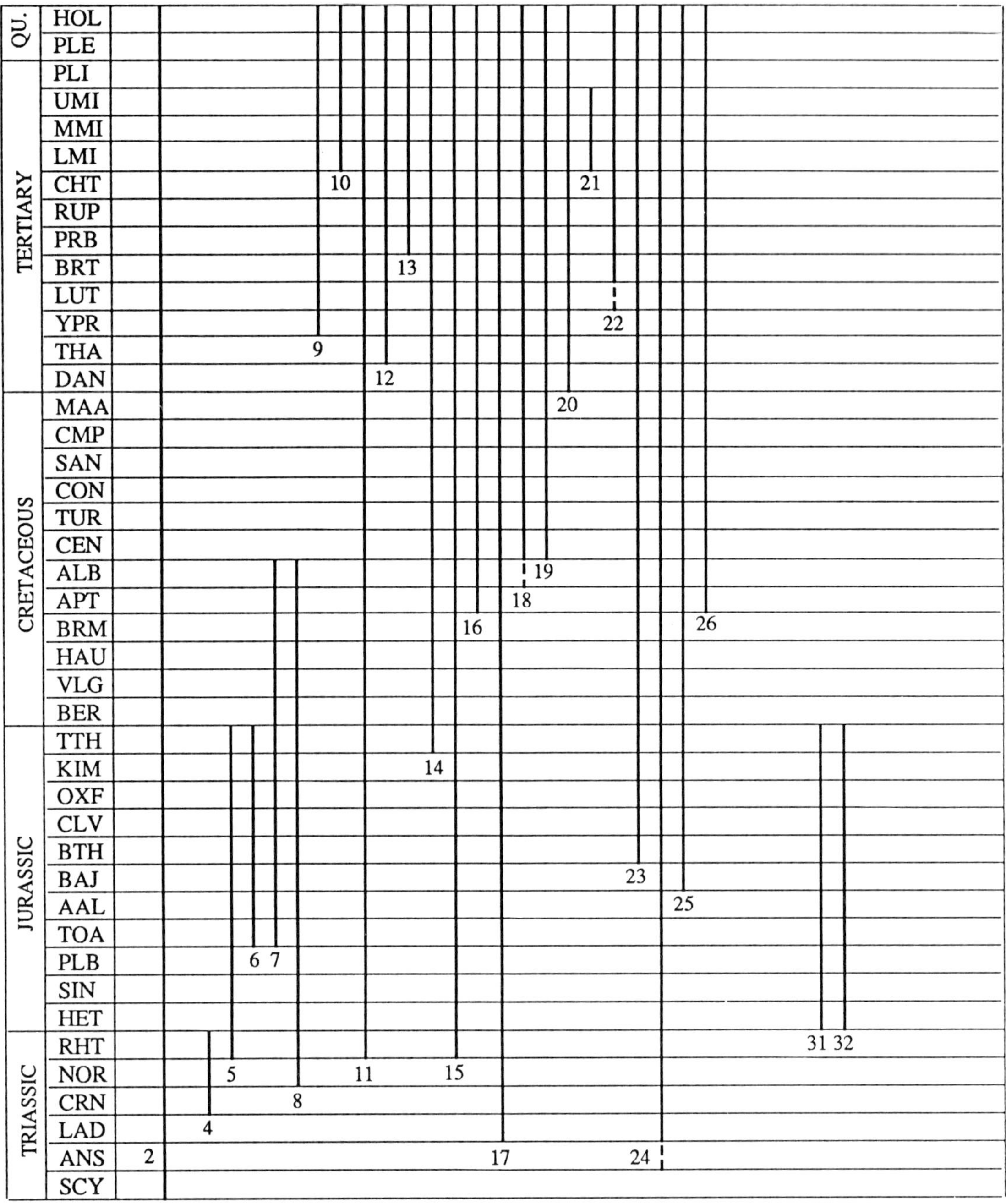

Fig. 13.5

First: *Permoceramus brownei* Waterhouse, 1970, Muree Formation, Sydney Basin, New South Wales, Australia (Waterhouse, 1970).
Last: *Tenuipteria argentea* (Conrad, 1858), Owl Creek Formation, Montana, USA; uppermost MAA, Middle Vistula Valley, Poland (Abdel-Gawad, 1986); and upper MAA, Sint Pietersberg, The Netherlands.
Comments: It is not clear whether *Tenuipteria* is a true inoceramid; it occurs in the MAA after the extinction of *Inoceramus* species. See Ward *et al.* (1991) and Dhondt (1992).

F. ALATOCONCHIDAE Termier *et al.*, 1973
P. (ART–CAP) Mar.

First: *Shikamaia (Tanchintongia) perakensis* (Runnegar and Gobbett, 1975) and *Saikraconcha (Dereconcha) kamparensis* Yancey and Boyd, 1983, both *Pseudofusulina ambigua* Zone, Malaysia.
Last: *Shikamaia (Alatoconcha) vampyra* (Termier *et al.*, 1973), Afghanistan, *Saikraconcha ogulineci* (Kochansky-Devidé, 1978), former Yugoslavia, and *S. tunisiensis* Yancey and Boyd, 1983, Tunisia; all in the *Neoschwagerina* Assemblage Zone (Yancey and Boyd, 1983).

F. EURYDESMIDAE Reed, 1932
C. (MOS)–P. (KUN) Mar.

First: *Eurydesma cordatum truncatum* Waterhouse, 1987, Fairyland Formation, Bowen Basin, Queensland, Australia, and *Eurydesma* sp., Gonzalez, 1972, Westphalian, Patagonia (Waterhouse, 1982a).

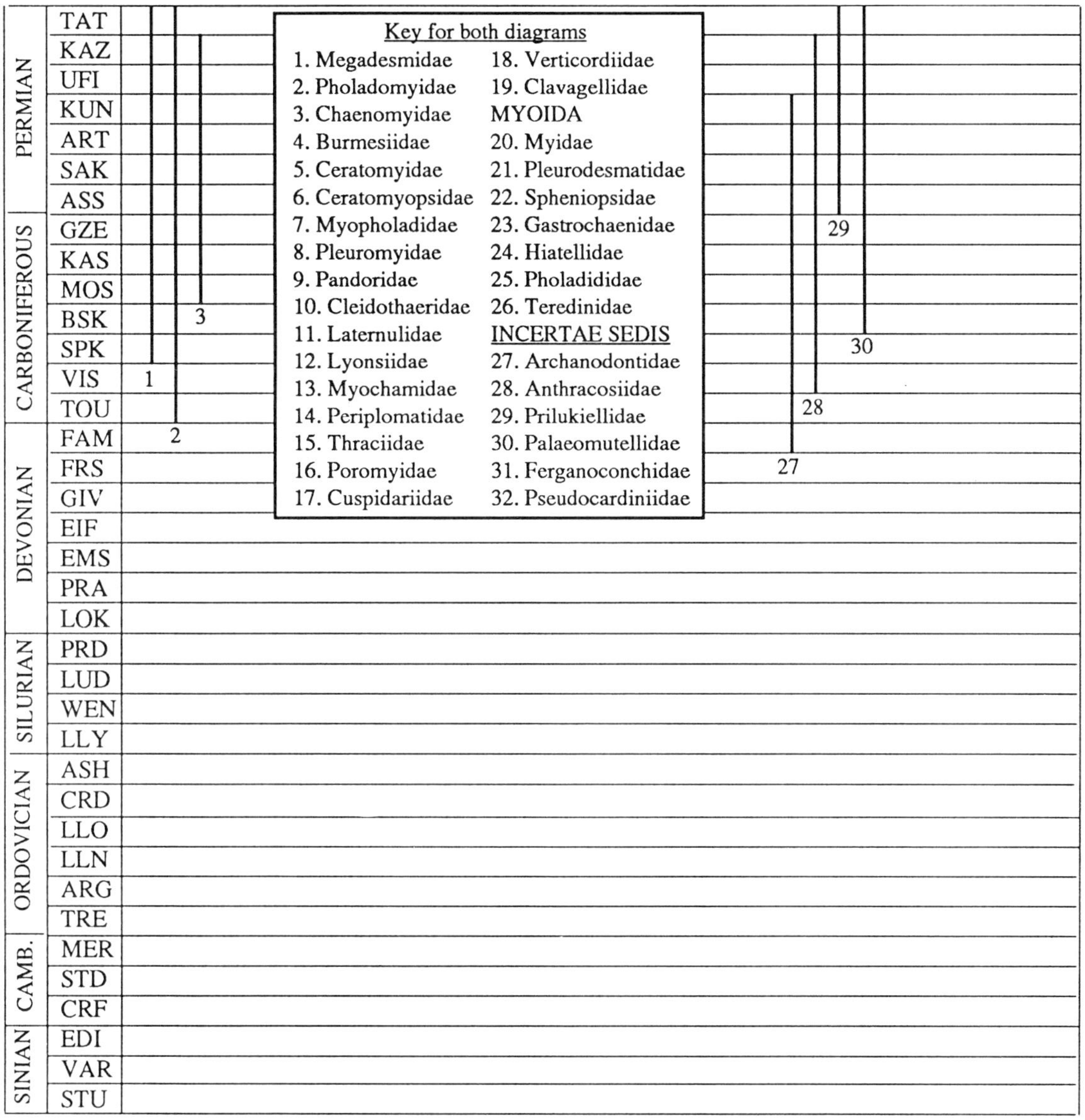
PERMIAN
TAT
KAZ
UFI
KUN
ART
SAK
ASS
CARBONIFEROUS
GZE
KAS
MOS
BSK
SPK
VIS
TOU
DEVONIAN
FAM
FRS
GIV
EIF
EMS
PRA
LOK
SILURIAN
PRD
LUD
WEN
LLY
ORDOVICIAN
ASH
CRD
LLO
LLN
ARG
TRE
CAMB.
MER
STD
CRF
SINIAN
EDI
VAR
STU
1
2
3
27
28
29
30
Key for both diagrams
1. Megadesmidae
2. Pholadomyidae
3. Chaenomyidae
4. Burmesiidae
5. Ceratomyidae
6. Ceratomyopsidae
7. Myopholadidae
8. Pleuromyidae
9. Pandoridae
10. Cleidothaeridae
11. Laternulidae
12. Lyonsiidae
13. Myochamidae
14. Periplomatidae
15. Thraciidae
16. Poromyidae
17. Cuspidariidae
18. Verticordiidae
19. Clavagellidae
MYOIDA
20. Myidae
21. Pleurodesmatidae
22. Spheniopsidae
23. Gastrochaenidae
24. Hiatellidae
25. Pholadididae
26. Teredinidae
INCERTAE SEDIS
27. Archanodontidae
28. Anthracosiidae
29. Prilukiellidae
30. Palaeomutellidae
31. Ferganoconchidae
32. Pseudocardiniidae

Fig. 13.5

Last: *Eurydesma alisulcatum* Waterhouse, 1987, Ulladulla Formation, Sydney Basin, New South Wales, Australia (Waterhouse, 1982a).

F. PERGAMIDIIDAE Cox, 1969
Tr. (CRN)–J. (SIN) Mar.

First: *Oretia coxi* Marwick, 1953, Oretian, North and South Islands, New Zealand; *Manticula problematica* (Zittel, 1864), Otamitan, North and South Islands, New Zealand and upper CRN, Baie de Saint-Vincent, New Caledonia (Waterhouse, 1979b).
Last: *Semuridia dorsetensis* (Cox, 1926), Lias, Dorset, England, UK (Hallam, 1987).
Comment: A possible later pergamidiid, *Manticula* sp., is known from the Cretaceous (BER) of South Shetland Island, Antarctica (J. A. Crame, pers. comm., 1992).

F. MONOPTERIIDAE Newell, 1969
C. (BSH–GZE) Mar.

First: *Monopteria longispina* Meek and Worthen, 1866, Desmoinesian, Appalachian Basin, USA.
Last: *Monopteria* sp., Virgilian, Jacksboro, Texas, USA.
Comment: Cox *et al.* (1969, p. N297) note '?L. Perm.' material from the USA, but this has not been substantiated.

F. PTERINEIDAE Miller, 1877
O. (LLN)–P. (KAZ) Mar.

First: *Denticelox turtuosa* (Tate, 1896), Stairway Sandstone, Amadeus Basin, southern Northern Territory, Australia (Pojeta and Gilbert-Tomlinson, 1977).
Last: *Leptodesma* sp., Tenjinnoki Formation, Miyagi, Japan (Nakazawa and Newell, 1968).

F. UMBURRIDAE Johnston, 1991 S. (WEN) Mar.

First and Last: *Umburra cinefacta* Johnston, 1991, Walker Volcanics, Fairlight Station, Canberra, Australia (Johnston, 1991).

F. KOCHIIDAE Maillieux, 1931
D. (PRA–EMS) Mar.

First: *Kochia capuliformis* (Koch, 1881), Siegener Grauwacke, Rheinland, Germany.

Last: *Kochia capuliformis* (Koch, 1881) and *K. alata* Maurer, 1902, Grès de Mormont, Ardennes, Belgium.

F. PTERIIDAE Gray, 1847 Tr. (LAD)–Rec. Mar.

First: *Arcavicula cassiana* (Bittner, 1895), ?Pachycardientuffe, Cassian Formation, eastern Alps. **Extant**

F. BAKEVELLIIDAE King, 1850 C. (MOS)–T. (PRB) Mar.

First: *Bakevellia* sp., Buckhorn Asphalt, Desmoinesian Stage, south-central Oklahoma, USA (T. Yancey, pers. comm., 1992).
Last: *Aviculoperna aviculina* (Deshayes, 1864), Paris Basin, France.
Comment: The next oldest well-known record is *Bakevellia binneyi* (Brown, 1841), Lower Magnesian Limestone (UFI), north-east England, UK.

F. LITHIOTIDAE Reis, 1903 J. (PLB) Mar.

First and Last: *Lithiotis problematica* Gümbel, 1874, widespread in lower Lias on southern Tethyan platforms, and *Cochlearites loppianus* (Tausch, 1890), upper Calcari Grigi, Verona, north Italy (Chinzei, 1982).
Comment: Could be included in Plicatostylidae (S. Damborenea, pers. comm., 1992).

F. CASSIANELLIDAE Ichikawa, 1958 P. (SAK)–J. (AAL) Mar.

First: *Cassianella crassispinosus* Chronic, 1949, Copacabana Group, Peru.
Last: Cassianellid (N. J. Morris, pers. comm., 1992).
Comments: If the Permian records are not accepted, the oldest cassianellid would be *Septihoernesia johannisaustriae* (Klipstein, 1845), south Tyrol, eastern Alps; *S. subglobosa* (Credner), Muschelkalk, Germany (both LAD).

F. DATTIDAE Healey 1908 Tr. (NOR) Mar.

First and Last: *Datta oscillaris* Healey, 1908, ?NOR (N. J. Morris, pers. comm., 1992), Burma.
Comments: This family is of dubious status, being based on a single internal mould of a left valve (Cox *et al.*, 1969, p. N314). The age is given as RHT by the latter authors.

F. RETROCERAMIDAE Pergament *in* Koschelkina, 1971 J. (AAL)–K. (BER) Mar.

First: *Retroceramus subtilis, P. macklintocki* Zone, lower AAL, Priokhotye, Siberia, former USSR; *R. levis* Koskelkina, 1969, ?TOA and AAL, and four other AAL species, former USSR (Damborenea, 1990).
Last: *Retroceramus everesti* (Oppel, 1865), lower Fossil Bluff Formation, Callisto Cliffs and Tombaugh Cliffs, Alexander Island, Antarctica (Crame, 1982); *R. foliiformis* Pokhialainen, 1969, BER, Anadyr-Koryak, Far East of former USSR.

F. ISOGNOMONIDAE Woodring, 1925 P. (KAZ)–Rec. Mar.

First: *Waagenoperna (Permoperna) hayamii* Nakazawa and Newell, 1968, Tenjinnoki Formation, NE Japan. **Extant**
Comment: *Waagenoperna lateplanata* (Waagen, 1907), supposedly from the SAK of Japan, is an Upper Triassic species (I. Hayami, pers. comm., 1992).

F. PULVINITIDAE Stephenson, 1941 J. (PLB)–Rec. Mar.

First: *Pulvinites (Hypotrema) liasicus* Damborenea, 1987, *Fanninoceras* Zone, upper PLB, southern Mendoza and Neuquén Provinces, Argentina (Damborenea, 1987). **Extant**

F. MALLEIDAE Lamarck, 1819 T. (THA)–Rec. Mar.

First: *Vulsella* sp., THA, France. **Extant**
Comments: Waller (1978) removed several of the genera cited by Cox *et al.* (1969, pp. 326–32) from this family, leaving *Vulsella* as the earliest genus, of which the earliest undoubted record is Upper Palaeocene (N. J. Morris, pers. comm., 1992). A confirmed record is *V. angusta* Deshayes, 1858, Auversian (PRB), France.

F. PINNIDAE Leach, 1819 C. (VIS)–Rec. Mar.

First: *Pinna costata* Phillips, 1848, Visé, Belgium; *Pinna spatula* M'Coy, 1853, Visé, Belgium, Central France, Harz Mountains, Germany. **Extant**
Comments: The family is allied with the Pteriacea (Pojeta, 1978).

F. POSIDONIIDAE Frech, 1909 C. (VIS)–K. (CON?) Mar.

First: *Posidonia becheri* Bronn, 1828, Lower Carboniferous, Germany.
Last: *Didymotis variabilis* Gerhardt, 1897, lower CON, Colombia.
Comments: Affinities of the family are reviewed by Carter (1990, p. 211). Inclusion of *Didymotis* in the family is questionable (N. J. Morris, pers. comm., 1992); the next youngest taxon is *Pseudodidymotis lamberti* Gillet, 1924, NEO, France.

Order LIMOIDA Rafinesque, 1815

F. MYODAKRYOTIDAE Tunnicliff, 1987 O. (CRD) Mar.

First and Last: *Myodakryotus deigryn* Tunnicliff, 1987, Allt-Tair-ffynon Beds, Llanfyllin, Powys, and Gelli Grin Formation, Bala, Cwm Eigiau Formation, Roman Bridge, and Cwm Rhiwarth Siltstone Formation, Cwm Rhiwarth, Gwynedd (all Longvillian age), North Wales and specimens from North America (Tunnicliff, 1987).

F. LIMIDAE Rafinesque, 1815 C. (VIS)–Rec. Mar.

First: *Palaeolima simplex* (Phillips, 1836), Harz Mountains, Germany, eastern Alps, Austria, Yorkshire, England, UK, and other species. **Extant**
Comments: Ordovician material from North America referred to *Prolobella* by Pojeta and Runnegar (1985) is referred to Myodakryotidae by Tunnicliff (1987).

Order OSTREOIDA Férussac, 1822 (Waller, 1978)

F. PLICATULIDAE Watson, 1930 Tr. (ANS)–Rec. Mar.

First: *Plicatula (Eoplicatula) imago* Bittner, 1895 and *P. (E.) filifera* Bittner, 1895, north Italy (Yin, 1985; Carter, 1990, pp. 221–6). **Extant**

F. DIMYIDAE Fischer, 1886 Tr. (ANS)–Rec. Mar.

First: *Dimyodon* sp. (Yin, 1985); *Proostrea* sp., China (N. J. Morris, pers. comm., 1992). **Extant**
Comments: The next oldest dimyids are *Atreta intustriata* (Emmerich, 1853), early RHT of Europe, and *Dimyodon schlumbergeri* Munier-Chalmas, *in* Fischer, 1886, BTH, France.

F. OSTREIDAE Rafinesque, 1815 Tr. (CRN)–Rec. Mar./Brackish

First: *'Lopha' calceoformis* (Broili, 1904) and other species, CRN of Tethyan and Pacific realms (Stenzel, 1971, pp. N1157–N1158; Carter, 1990, p. 229). **Extant**
Comments: Older material of ?*'Lopha'* has been reported from the Upper Permian of Japan (Nakazawa and Newell, 1968), so the range may be greater.

F. GRYPHAEIDAE Vialov, 1936 Tr. (CRN)–Rec. Mar.

First: *Gryphaea arcuataeformis* Kiparisova, 1936, Korbodon River, eastern Siberia, former USSR, and other species from Arctic regions listed by Stenzel (1971, pp. N1097–N1100); also CRN of British Columbia, Canada and Nevada and Oregon, USA (A. L. A. Johnson, pers. comm., 1992). **Extant**

F. ELIGMIDAE Gill, 1871 J. (BAJ)–T. (LUT) Mar.

First: *Eligmus rollandi* (Douvillé, 1907), East Africa (Hallam, 1977).
Last: *Euphenax jamaicensis* (Trechmann, 1923), LUT, Jamaica, North Africa, Pakistan; *Nyadina (Exputens) llajasensis* Clark, 1934, LUT, California, USA, Jamaica; *Pseudoheligmus morgani* Douvillé, 1904, LUT, France.
Comments: *Eligmus* was included in the Malleidae by Cox *et al.* (1969, p. N329), but this was rejected by Waller (1978), and the family is recognized as distinct, with the inclusion of the 'Chondrodontidae' (N. J. Morris, pers. comm., 1992). Cox (1965) noted *E. rollandi* from the Muddo Erri Limestone (CLV), NE Kenya.

F. PLICATOSTYLIDAE Lupher and Packard, 1930 J. (TOA) Mar.

First and Last: *Plicatostylus gregarius* Lupher and Packard, 1930, Robertson Formation, Oregon, USA; also Peru and Chile.
Comments: Probably an ostreacean, but not a rudist (*pace* Cox *et al.*, 1969, p. N866).

F. LEIOPECTINIDAE Krasilova, 1959 S. (LUD)–D. (l.) Mar.

First: *Rhombopteria mira* (Barrande, 1881), stage e2, lower LUD, Bohemia, Czechoslovakia.
Last: *Leiopecten rectangularis* Khalfin, 1940, Pribalkhash, Siberia, former USSR.

F. PTERINOPECTINIDAE Newell, 1938 S. (LUD)–Tr. (SPA) Mar.

First: *Pterinopecten (Pterinopecten) undosus* Hall, 1883, LUD, cosmopolitan.
Last: *Claraia griesbachi* (Bittner, 1900), SPA, northern Pacific, Timor; *Periclaraia* sp., China (Yin, 1985).
Comments: Carter (1990, pp. 239–40), following other authors, includes *Claraia* Bittner, 1901 and *Pseudoclaraia* here. If these are not pterinopectinids, the youngest form is *Pterinopectinella welleri* Newell, 1938, Leonardian, Kansas, USA.

F. AVICULOPECTINIDAE Meek and Hayden, 1864 C. (TOU)–Tr. (CRN) Mar.

First: *Aviculopecten ? affinis* de Koninck, 1885, Tournai, Belgium; *A. grandocostatus* White, 1862, Kinderhookian, Iowa and Missouri, USA, and other species.
Last: *Eumorphotis telleri* (Bittner, 1898), CRN, western Tethys, Arctic (Hallam, 1981).
Comments: *Otapiria* Marwick, 1935 (TTH) has been transferred to the Monotidae, and *Claraia* Bittner, 1901 may belong to the Pterinopectinidae (q.v.).

F. DELTOPECTINIDAE Dickins, 1957 C. (MOS)–P. (CHX) Mar.

First: *Orbiculopecten parma* Gonzalez, 1978, lower Tepuel Group, *Levipustula* Zone, Tecka, Chubut, Patagonia, Argentina (Carter, 1990, pp. 241, 243).
Last: *Corrugopecten altoprimus* Waterhouse, 1982, Mangarewa Formation, South Island, New Zealand and Flat Top Formation, Bowen Basin, Queensland, Australia (both KAZ); *Corrugopecten* sp., upper Dorashamian–Changxingian of Pig Valley Formation, Nelson, New Zealand (Waterhouse, 1982b).

F. EUCHONDRIIDAE Newell, 1938 D. (FRS)–P. (ZEC). Mar.

First: *Crenipecten crenulatus* Hall, 1883, New York State, USA.
Last: *Euchondria*, Upper Permian, North America, Europe, Japan.

F. MONOTIDAE Fischer, 1887 Tr. (CRN)–J. (TTH) Mar.

First: *Otapiria dubia* (Ichikawa, 1954), Kochigatani Group, Tokushima Prefecture, Shikoku, Japan (Damborenea, 1987).
Last: *Otapiria masoni* Marwick, 1953, Puaroan, TTH, Port Waikato, North Island, New Zealand.
Comments: *Otapiria* has been transferred to this family from the Aviculopectinidae (Carter, 1990, pp. 245–8). The KIM and TTH records of *Otapiria* is doubtful (J. A. Crame, pers. comm., 1992). *Monotis boreas* Öberg, 1877, SCY, Dickson Land, Spitsbergen is only doubtfully regarded as a member of this family (S. Damborenea, pers. comm., 1992).

F. ASOELLIDAE Begg and Campbell, 1985 Tr. (ANS–NOR) Mar.

First: *Etalia johnstoni* Begg and Campbell, 1985, Etalian Stage, Murihiku Supergroup, Nelson and Southland, South Island, New Zealand (Begg and Campbell, 1985).
Last: *Asoella confertoradiata* (Tokuyama, 1959) and *A. laevigata* (Tokuyama, 1959), both Atsu and Mine Series, west Japan.
Comment: *Asoella* was placed in the Aviculopectinae by Cox *et al.* (1969, p. N337). It was transferred to the Buchiacea (= Monotoidea) by Waller (1978), and linked with *Etalia* in a new family by Begg and Campbell (1985). The latter authors very tentatively include *Chlamys kotakiensis* Takai and Hayami, 1957 from the Lias of Japan and China in the family (a pectinid, I. Hayami, pers. comm., 1992).

F. BUCHIIDAE Cox, 1953 J. (CLV)–K. (HAU) Mar.

First: *Praebuchia anabarensis* Zakharov, 1981, *Cadoceras emelianzevi* Zone, Siberia, former USSR.
Last: *Buchia crassicollis* (Keyserling, 1846), *Simbirskites decheni* Zone, central Russian Platform (S. R. A. Kelly, pers. comm., 1992).
Comments: The range of this family is given as CRN to

CEN by Cox *et al.* (1969), pp. N374–N377), while Carter (1990, pp. 248–9) indicates possible earlier members from the mid Permian (*Glendella dickinsi* Runnegar, 1970) and Middle Triassic (*Sichuania marwicki* Waterhouse, 1980). These, as well as other late Triassic, and early to middle Jurassic taxa are not accepted as buchiids by J. A. Crame or S. R. A. Kelly (pers. comm., 1992).

F. OXYTOMIDAE Ichikawa, 1958
P. (SAK)–T. (?DAN) Mar.

First: *Cyrtorostra limitans* Waterhouse, 1987, Elvinia Formation, Bowen Basin, Queensland, Australia (Waterhouse, 1987).
Last: *Hypoxytoma* sp., MAA or ?DAN, Ocean Point, north Alaska, USA (J. A. Crame, pers. comm., 1992, based on the work of L. Marincovich).

F. PSEUDOMONOTIDAE Newell, 1938
C. (VIS)–P. (CHX) Mar.

First: *Pachypteria nobilissima* de Koninck, 1851, Visé Limestone, Visé, Belgium, Derbyshire and Yorkshire, England, UK, ?Iowa, USA (Newell and Boyd, 1970).
Last: *Pseudomonotis (Pseudomonotis) speluncaria* (von Schlotheim, 1816), Zechstein, Germany.
Comments: If *Claraia* Bittner, 1901 is included (Newell and Boyd, 1970), the range is extended to SCY, but this genus is placed in the Pterinopectinidae here.

F. TERQUEMIIDAE Cox, 1964 P. (SAK)–J. (KIM)
Mar.

First: *Palaeowaagia cooperi* Newell and Boyd, 1970, Neal Ranch Formation, Glass Mountains, west Texas, USA (Newell and Boyd, 1970).
Last: *Terquemia ostreiformis* (d'Orbigny), Jura, France.

F. PERMANOMIIDAE Carter, 1990 P. (ART) Mar.

First and Last: *Permanomia texana*, 1970, Leonardian, west Texas, USA (Newell and Boyd, 1970).
Comments: Carter (1990, p. 253) erected the family Permanomiidae since *Permanomia* has a duplivincular ligament, and thus could no longer be regarded as an anomiid.

F. ANOMIIDAE Rafinesque, 1815 J. (BTH)–Rec.
Mar.

First: *Eonomia timida* Fürsich and Palmer, 1982, *Oppelia (Oxycerites) aspidoides* and *Clydoniceras discus* Zones, upper BTH, Normandy, France and Bradford Clay, Wiltshire, England, UK (Fürsich and Palmer, 1982). **Extant**
Comments: E. M. Harper (pers. comm., 1992) notes that *Eonomia* was not cemented, as first thought, so *Permanomia* is not precluded as the earliest anomiid (*pace* Carter, 1990).

F. CHAENOCARDIIDAE Miller, 1889
(= Streblochondriidae Newell, 1938)
C. (l.)–P. (LGT) Mar.

First: *Streblochondria* sp., Mississippian, USA (Newell and Boyd, 1985).
Last: *Guizhoupecten wangi* Chen, 1962, Wujiaping Stage, Ziyun, Guizhou, China.
Comments: *Streblochondria sculptilis* (Miller, 1891) is noted from the upper Pennsylvanian to Lower Permian by Newell and Boyd (1985), who also state that the genus is known from the Lower Carboniferous. *Guizhoupecten* was transferred from the Aviculopectinidae to this family by Newell and Boyd (1985).

F. PROPEAMUSSIIDAE Tucker Abbott, 1954
C. (IVO)–Rec. Mar.

First: *Pernopecten limaformis* (White and Whitfield, 1862), Burlington Beds, Osagean, Burlington, Iowa, USA. **Extant**
Comments: Waller (1978, p. 362) notes the oldest definite representatives of this family as SCY, but regards *Pernopecten* as an ancestral propeamussiid. *Pernopecten* was placed in the polyphyletic family 'Entoliidae' by Cox *et al.* (1969, p. N347), and in the family Pernopectinidae Nevesskaya *et al.*, 1971 by Carter (1990, p. 259).

F. ENTOLIIDAE von Teppner, 1922
Tr. (ANS)–K. (MAA) Mar.

First: *Entolium kellneri* (Kittl, 1903), Buloger-Kalk, Haliluci, near Sarajevo, Bosnia former Yugoslavia; *E. discites* (von Schlotheim, 1820), Muschelkalk, Germany (Johnson, 1984).
Last: *Entolium* sp., Miria Formation, Western Australia; *E. seymourensis* Zinsmeister and Macellari, 1988 and *E. sadleri* Zinsmeister and Macellari, 1988, López de Bertodano Formation, Seymour Island, Antarctica (Zinsmeister and Macellari, 1988); *E. membranaceum* (Nilsson, 1827), CMP, Sweden and uppermost MAA, Maastricht, The Netherlands (A. V. Dhondt, pers. comm., 1992).
Comments: The family ranges back to the Triassic, with the exclusion of *Pernopecten* and its assignment to the Propeamussiidae (q.v.). Allasinaz (1972) noted *Entolium discites* from the SCY, but provides no supporting evidence (A. L. A. Johnson, pers. comm., 1992). Yin (1985) gives the range of *Entolium* back to LGT, or even GUA, but this is not accepted here.

F. SYNCYCLONEMIDAE Waller, 1978
K. (MAA)–Rec. Mar.

First: *Syncyclonema simplicia* (Conrad, 1860) and *S. halli* (Gabb, 1891), both from various localities and formations in North America, such as Ripley Formation, Alabama and Coon Creek Formation, Tennessee, USA. **Extant**

F. PECTINIDAE Rafinesque, 1815 P. (CHX)–Rec.
Mar.

First: *Hunanopecten exilis* Zhang, 1977 and *H. qujiangensis* Zhang, 1977, both Dalong Formation, South China (Yin, 1985). **Extant**

F. NEITHEIDAE Sobetzky, 1960 J. (SIN)–K. (MAA)
Mar.

First: *Weyla (Lywca) unca* (Philippi, 1899), several localities in central Chile and western Argentina (Damborenea, 1987).
Last: *Neithea alpina* (d'Orbigny, 1847), *N. regularis* (Schlotheim, 1813), and many other species, MAA, cosmopolitan (Dhondt, 1973).

F. SPONDYLIDAE Gray, 1826 J. (BAJ)–Rec. Mar.

First: *Spondylus consobrinus* Deslongchamps, Ste. Hônorine, Bayeaux, France (E. M. Harper, pers. comm., 1992). **Extant**

Subclass LUCINATA Pojeta, 1978

Order LUCINOIDA Dall, 1889

F. BABINKIDAE Horný, 1960 O. (TRE–LLN)
Mar.

First: *Babinka prima* Barrande, 1881, Schistes de Saint-Chinian inférieur, Fédou, Minervois occidental, France (Babin, 1982a).
Last: *Babinka prima* Barrande, 1881, various formations, Spain; Sárka Beds, Bohemia, Czechoslovakia (Babin and Gutiérrez-Marco, 1991).

F. LUCINIDAE Fleming, 1828 S. (WEN)–Rec. Mar.

First: *Illionia canadensis* Billings, 1875, Canada; *I. prisca* (Hisinger, 1837), Sweden. **Extant**

F. THYASIRIDAE Dall, 1901 K. (ALB)–Rec. Mar.

First: *Thyasira (Thyasira)* sp. **Extant**
Comments: The Middle Triassic form, *Storthodon liscaviensis* Giebel, 1856, included in this family by Cox *et al.* (1969, p. N511), is excluded here.

F. MACTROMYIDAE Cox, 1929 J. (HET)–Rec. Mar.

First: *Mactromyopsis (Mactromyella) inflata* (Thévenin, 1909), Charmouthian, France, England, UK. **Extant**
Comments: Palaeozoic taxa included in this family by Cox *et al.* (1969, pp. N511–N513), such as *Montanaria devonicus* (Beushausen, 1884) from the Devonian of Europe, *Paracyclas elliptica* Hall, 1843 from the Devonian of North America and Europe, *Palaeolucina carbonaria* Chao, 1927 from the Carboniferous of China, and *Plesiocyprinella carinata* Holdhaus, 1918 from the Permian of South America, are excluded here.

F. FIMBRIIDAE Nicol, 1950 Tr. (CRN)–Rec. Mar.

First: *Schafhaeutlia ovatum* (Schafhaeutl, 1871) and *S. mellingi* (Hauer, 1857), Europe, South America and Sakawa, Japan. **Extant**
Comments: The Carboniferous species *Scaldia lambotteana* de Ryckholt, 1871, included in this family by Cox *et al.* (1969, p. N514), is transferred to the Edmondiidae here (N. J. Morris, pers. comm., 1992).

F. UNGULINIDAE Adams and Adams, 1857 K. (?CMP)–Rec. Mar.

First: *Brachymeris alta* Conrad, 1875, North Carolina, USA; *Felaniella (Zemysia)* sp. **Extant**

F. ERYCINIDAE Deshayes, 1850 T. (DAN/THA)–Rec. Mar.

First: *Erycina (Erycina) pellucida* Lamarck, 1805 and *Semierycina (Erycinopsis) semipecten* (Cossmann, 1887), both Palaeocene, France. **Extant**

F. KELLIIDAE Forbes and Hanley, 1878 K. (CMP)–Rec. Mar.

First: *Rhectomyax undulatus* (Gabb, 1864), Ten Mile Member, Chico Formation, lower CMP, California, USA. **Extant**

F. LEPTONIDAE Gray, 1847 K. (ALB)–Rec. Mar.

First: *Lepton (Lepton)* sp. **Extant**

F. MONTACUTIDAE Clark, 1855 T. (YPR)–Rec. Mar.

First: *Montacuta herberti* Aldrich, 1921, Bashí Member, Hatchetigbee Formation, Upper Wilcox Group, Clarke County, Alabama, USA; *Mysella minuta* (Aldrich, 1921), Bells Landing Member, Tuscahoma Formation, Middle Wilcox Group, Monroe County, Alabama, USA (Palmer and Brann, 1965). **Extant**
Comment: Slightly younger forms are *Laseoneara radiata* (Deshayes, 1824), *Laubriereia emarginata* (Deshayes, 1860), and *Namnetia discoides* Cossmann, 1905, all LUT, Paris Basin, France.

F. GALEOMMATIDAE Gray, 1840 T. (LUT)–Rec. Mar.

First: *Scintilla clarkeana* Aldrich, 1897, Bashi Member, Hatchetigbee Formation, Upper Wilcox Group, Clarke County, Alabama, USA (Palmer and Brann, 1965). **Extant**
Comments: These two taxa are indicated as uncertain by Cox *et al.* (1969, p. N536), *Passya* since it may not be a galeommatid, and *Scintilla* since the Eocene record is uncertain.

F. CHLAMYDOCONCHIDAE Dall, 1899 **Extant** Mar.

F. CYAMIIDAE Philippi, 1845 T. (LAN)–Rec. Mar.

First: *Dicranodesma calvertensis* (Glenn, *in* Dall, 1900), LAN, Calvert Cliffs, Maryland, USA. **Extant**

F. TURTONIIDAE Clark, 1855 T. (Mio.)–Rec. Mar.

First: *Turtonia* sp. **Extant**

F. SPORTELLIDAE Dall, 1899 T. (?DAN/THA)–Rec. Mar.

First: *Sportella (Sportella) dubia* (Deshayes, 1824), *Anisodonta complanatum* Deshayes, 1858, and *Hindsiella arcuata* (Lamarck, 1807), all Palaeocene, France; *Cerullia* sp.; *S. subaequilateralis* Cossmann, 1908, and other species, Calcaire de Mons (THA), Mons, Belgium (Glibert and Van de Poel, 1973). **Extant**
Comments: Cox *et al.* (1969, p. N541) include *Vokesella inopinata* Chavan, 1952, from the Jurassic of France, in this family, but it is excluded here.

F. NEOLEPTONIDAE Thiele, 1934 T. (LUT)–Rec. Mar.

First: *Goodalliopsis terminalis* (Deshayes, 1858), LUT, Paris Basin, France (Glibert and Van de Poel, 1967). **Extant**
Comments: If this record is not confirmed, the next oldest are a Duntroonian (CHT) occurrence of *Neolepton* in New Zealand, and *Puyseguria crenulifera* Maxwell, 1969, Waitakian (LMI), New Zealand (Beu and Maxwell, 1990).

Subclass HETEROCONCHIA Hertwig, 1895

Order ACTINODONTOIDA Douvillé, 1912

F. CYCLOCONCHIDAE Ulrich, 1894 O. (ARG)–S. (LUD) Mar./FW

First: *Copidens browni* Pojeta and Gilbert-Tomlinson, 1977, Nora Formation (ARG/LLN), Toko Range, western Queensland, Australia; *Actinodonta ramseyensis* (Hicks, 1873), Ogof Hên Formation, Pembrokeshire, Wales, UK (Carter, 1971); *Actinodonta* sp., Armoricain Sandstone, Brittany, France (Pojeta, 1971).

Last: *Actinodonta cuneata* Phillips, 1848, lower LUD, Malvern Hills, England, UK (Stubblefield, 1938).

F. LYRODESMATIDAE Ulrich, 1894 O. (ARG)–S. (LLY) Mar.

First: *Tromelinodonta armoricana* (De Tromelin and Lebesconte, 1876), middle ARG, Pont-Réan, Massif Armoricain, France (Babin, 1982b).
Last: *Lyrodesma* sp., Bransfield Formation, Adams County, Ohio, USA (Harrison and Harrison, 1975).

F. REDONIIDAE Babin, 1966 O. (ARG–LLO) Mar.

First: *Redonia michelae* Babin, 1982, lower ARG, La Maurerie, Montagne Noire, France.
Last: *Redonia deshayesiana* Rouault, 1851, upper LLO, Almaden, Spain (Babin and Gutiérrez-Marco, 1991).
Comments: The family Redoniidae was subsumed under Permophoridae by Cox *et al.* (1969, p. N543), but is treated as independent here.

F. CARYDIIDAE Haffer, 1959 D. (PRA–FRS) Mar.

First: *Carydium gregarium* Beushausen, 1895, Singhofen, Germany.
Last: *Carydium concentricum* Spriestersbach, 1915, Lenneschiefer, upper EIF, Germany; also known from FRS of Spain and Armorican Massif, France (C. Babin, pers. comm., 1992).

Order TRIGONIOIDA Dall, 1889

F. EOSCHIZODIIDAE Newell and Boyd, 1975 D. (GIV) Mar.

First and Last: *Eoschizodus truncatus* (Goldfuss, 1837), *Stringocephalus* Zone, Rhineland, Germany (Newell and Boyd, 1975).
Comments: Cox *et al.* (1969, p. N473) note Permian examples of this genus.

F. SCHIZODIIDAE Newell and Boyd, 1975 S. (LUD)–P. (KAZ) Mar.

First: *Schizodus* sp., Elwy Group, *Cucullograptus scanicus* Zone, lower LUD, Conway, Wales, UK (Newell and Boyd, 1975).
Last: *Schizodus obscurus* (Sowerby, 1821), Lower Magnesian Limestone, north-east England, UK; Zechstein, Germany, Kazanian and equivalents of former USSR (Newell and Boyd, 1975).

F. SCAPHELLINIDAE Newell and Ciriacks, 1962 P. (KUN–KAZ) Mar.

First: *Scaphellina concinnus* (Branson, 1930), Kaibab Formation, Arizona and Grandeur Member, Park City Formation, both upper LEN, Wyoming, USA (Newell and Boyd, 1975).
Last: *Scaphellina concinnus* (Branson, 1930), Ervay Member, Park City Formation, Wyoming, and Shedhorn Sandstone and Franson Member, all lower GUA, Montana, USA (Newell and Boyd, 1975).

F. EOASTARTIDAE Newell and Boyd, 1975 P. (KUN–KAZ) Mar.

First: *Eoastarte subcircularis* Ciriacks, 1963, Grandeur Member, Park City Formation, Wyoming, Cathedral Mountain Formation and Road Canyon Formation, Texas, USA (all upper LEN), as well as other genera and species (Newell and Boyd, 1975).
Last: *Kaibabella (Flattopia) axinia* Waterhouse, 1987, Flat Top Formation, Bowen Group, Queensland, Australia (Waterhouse, 1987).

F. COSTATORIIDAE Newell and Boyd, 1975 P. (KUN–TAT) Mar.

First: *Procostatoria cooperi* Newell and Boyd, 1975, Road Canyon Formation (upper LEN), Texas and Franson Member (lower GUA), Park City Formation, Wyoming, USA (Newell and Boyd, 1975).
Last: *Costatoria costata* (Zenker, 1833), Dzhulfian, Japan.
Comments: Cox *et al.* (1969, p. N473) note Triassic examples of *Costatoria*, from the Muschelkalk of Germany and the Upper Triassic of the southern Tyrol, but these are excluded here.

F. MYOPHORIIDAE Bronn, 1849 P. (KUN)–Tr. (RHT) Mar.

First: *Paraschizodus elongatus* Newell and Boyd, 1975, Grandeur Member, Park City Formation and lower Goose Egg Formation (both upper LEN), Wyoming, USA (Newell and Boyd, 1975).
Last: *Lyriomyophoria elongatus* (Moore, 1861), Penarth Group, England, UK, *Myophoria inflata* (Emmerich), Zlambach and Kössen Beds, Austria (Hallam, 1981), and species of *Neoschizodus*.
Comments: Cox *et al.* (1969, pp. N472–N476) give a much wider interpretation to this family, including in it members of the Costatoriidae, Eoschizodidae, and Schizodidae, all of which were excluded by Newell and Boyd (1975). Other taxa included in the family by Cox *et al.* (1969) extended its range from Devonian to (?) Lower Jurassic, but these genera (*Cytherodon, Hefteria, Liotrigonia, Rhenania, Toechomya*) are also omitted from the Myophoriidae here. Newell and Boyd (1975) indicate a range to Upper Ctretaceous with *Nipponotrigonia*.

F. PACHYCARDIIDAE Cox, 1961 Tr. (LAD)–J. (?PLB) Mar./Brackish

First: *Pachycardia rugosa* Hauer, 1857, Muschelkalk, Germany; *Trigonodus sandbergeri* (Stoliczka, 1871), Lettenkohle, Germany.
Last: *Cardinioides varidus* Hayami, 1957, Kuruma Group, central Japan.
Comments: This family is problematic, being indicated with a '?' by Cox *et al.* (1969, p. N467), and being synonymized with Trigonodídae by Vokes (1980, p. 95). The Permian record of *Kidodia* noted by Cox *et al.* (1969, pp. N467–N468) is not accepted here. If the Jurassic records are not accepted, latest forms are CRN and NOR of the Alpine region and Oregon, USA.

F. TRIGONIIDAE Lamarck, 1819 P. (KUN)–Rec. Mar.

First: *Lyroschizodus orbicularis* Newell and Boyd, 1975, Cathedral Mountain and Road Canyon formations, upper LEN, Texas, USA (Newell and Boyd, 1975). **Extant**
Comments: Cox *et al.* (1969, pp. N476–N488) give the oldest as Middle Triassic (ANS): *Trigonia (Trigonia) sulcata* (Hermann, 1781). Some authors (e.g. Cooper, 1991) split this large family into several, each smaller family, corresponding to one of the subfamilies assumed here.

F. TRIGONIOIDIDAE Cox, 1952 K. (l.–SEN) FW

First: *Trigonioides kodairai* Kobayashi and Suzuki, 1936, Naktong–Wakino Series, Japan.
Last: *Trigonioides kobayashi* (Hoffet, 1937), SEN, Laos.

F. MARGARITIFERIDAE Haas, 1940 Tr. (RHT)–Rec. FW

First: *Margaritifera* sp., RHT, central Iran (Gray, 1988, p. 51). **Extant**

F. UNIONIDAE Fleming, 1828 Tr. (CRN)–Rec. FW

First: *Unio subplanatus* Simpson, 1896 and six other species; *Diplodon gregori* Reeside, 1927, *D. haroldi* Reeside, 1927, *Antediplodon dockumensis* (Simpson, 1896), and two other species, Chinle Formation and Dockum Group of SW United States, and Pekin Formation, Tuckahoe Formation, Stockton Formation, Lockatong Formation, Wolfville Formation, Newark Supergroup of eastern North America (Good, 1989). **Extant**

F. MUTELIDAE Swainson, 1840 Tr. (CRN)–Rec. FW

First: *Mycetopoda dilicula* Pilsbry, 1921, Newark Supergroup, York County, Pennsylvania, USA (Good, 1989). **Extant**

F. ETHERIIDAE Swainson, 1840 T. (PLI)–Rec. FW

First: *Etheria semilunata* Lamarck, 1807, Africa/Indian Ocean.

F. ACTINODONTOPHORIDAE Newell, 1969 P. (KAZ)–Tr. (RHT) FW

First: *Actinodontophora katsurensis* Ichikawa, 1951, Katsura Formation, Shikoku, Japan.
Last: *Palaeopharus elongatus* Moore, 1861, Penarth Group, England, UK (Hallam, 1981).

Order VENEROIDA Adams and Adams, 1856

F. ASTARTIDAE d'Orbigny, 1844 D. (PRA)–Rec. Mar.

First: *Eodon* aff. *bicostula* (Krantz, 1857), Eifel, Germany; *Prosocoelus (Prosocoelus) priscus* (Roemer, 1843) and *P. (Tripleura) pesanseris* Zeil and Wirtgen, 1851, both Spiriferensandstein, NW Harz, Germany (Morris, 1978, p. 264). **Extant**
Comments: Cox *et al.* (1969, p. N566) doubtfully include the Middle Ordovician *Matheria tenera* Billings, 1858 in this family, but it is excluded here, being possibly a cyrtodontid.

F. CARDITIDAE Fleming, 1828 P. (TAT)–Rec. Mar.

First: *Gujocardita oviformis* Nakazawa and Newell, 1968, Gujo Formation, Japan (Nakazawa and Newell, 1968). **Extant**
Comments: Cox *et al.* (1969, p. N554) include the latest Silurian (Downtonian) *Carditomantea spinata* Quenstedt, 1929 in this family, but it is excluded here.

F. CHAMIDAE Blainville, 1825 K. (CMP)–Rec. Mar.

First: *Chama haueri* Zittel, 1865 and *C. detrita* Zittel, 1865, Gosau Beds, Austria (Kennedy *et al.* 1970). **Extant**

F. CONDYLOCARDIIDAE Bernard, 1897 T. (LUT)–Rec. Mar.

First: *Condylocardia atomus* (Deshayes, 1858), LUT, Paris Basin, France (Glibert and Van de Poel, 1970). **Extant**
Comment: There are abundant condylocardiids in the BRT of France, North America, and New Zealand: *Micromeris (Micromeris) minutissima* (Lea, 1833), *Cuna monroensis* (Meyer, 1887), and *C. parva* (Lea, 1833), all Gosport Sand, uppermost Claiborne Group, Monroe County, Alabama, USA (Palmer and Brann, 1965); *Condylocuna subaequilateralis* (Maxwell, 1966), Kaiatan, New Zealand (Beu and Maxwell, 1990). **Extant**

F. CRASSATELLIDAE Férussac, 1822 P. (ART)–Rec. Mar.

First: *Oriocrassatella stokesi* Etheridge, 1907, Norton Greywacke, Kennedy Range, Western Australia (T. A. Darragh, pers. comm., 1992). **Extant**
Comments: Cox *et al.* (1969, pp. N573–N578) include the Devonian to Carboniferous *Cypricardella* Hall, 1858 in this family, but it is excluded here.

F. CARDINIIDAE Zittel, 1881 Tr. (LAD)–J. (TOA) Mar.

First: *Balantioselena gairi* Speden, 1962, Torlesse Group, North Otago, South Island, New Zealand.
Last: *Cardinia concinna* (Sowerby, 1817), *C. attenuata* Stutchbury, 1842, and *Nidarica slatteri* (Wilson and Crick, 1889), all Domerian, lower TOA, Yorkshire, England, UK.
Comments: Cox *et al.* (1969, pp. N578–N580) give the range of this family as Ordovician to Recent, by including *Cypricardinia* Hall, 1859 (Ordovician–Permian) and *Tellidorella* Berry, 1963 (Miocene–Recent), both of which are excluded here. Cox *et al.* (1969, p. N579) also note doubtful BAJ material of *Cardinia*.

F. MYOPHORICARDIIDAE Chavan, *in* Vokes, 1967 Tr. (SCY–NOR) Mar.

First: *Pseudocorbula gregaria* (Münster, 1837).
Last: *Myophoriopsis* sp.

F. CORBULIDAE Lamarck, 1818 J. (TOA)–Rec. Mar./FW

First: *Corbula didimtuensis* Cox, 1965, Didimtu Beds, Didimtu Hill, NE Kenya (Cox, 1965). **Extant**
Comments: The oldest record in Cox *et al.* (1969, pp. N692–N698) is *Corbulomima*: Upper Jurassic.

F. ERODONIDAE Winckworth, 1932 T. (PRB)–Rec. Mar.

First: *Potamomya plana* (J. Sowerby, 1814), Headon Beds, Hampshire and Isle of Wight, England, UK (Glibert and Van de Poel, 1966b). **Extant**

F. CARDIIDAE Lamarck, 1809 Tr. (NOR)–Rec. Mar.

First: *Protocardia* sp., NOR, Tethys; *Septocardia typica* Hall and Whitfield, 1877, NOR, North and South America (Hallam, 1981). **Extant**

F. LAHILLIIDAE Finlay and Marwick, 1937 K. (CMP)–Rec. Mar.

First: *Lahillia* n. sp. aff. *L. angulata* (Philippi, 1887), Piripauan, Haumuri Bluff, New Zealand (Beu and Maxwell, 1990); *L. larseni* Sharman and Newton, 1897, upper CMP

to lower MAA, James Ross Basin, western Antarctica (Zinsmeister and Macellari, 1988). **Extant**
Comment: Ages of early records are uncertain. Also *L. larsoni* (Sharman and Newton, 1897), López de Bertodano Formation (MAA), Seymour Island, Antarctica (Zinsmeister and Macellari, 1988).

F. LYMNOCARDIIDAE Stoliczka, 1870 T. (LMI)–Rec. Brackish

First: *Eoprosodacna (Eoprosodacna) kartlicum* Davidaschvilli, 1934 and *E. (Succuridacna) goriense* (Davidaschvilli, 1934), both south Caucasus, Georgia, former USSR. **Extant**

F. TRIDACNIDAE Lamarck, 1819 T. (YPR)–Rec. Mar.

First: *Avicularium aviculare* (Lamarck, 1805), YPR, Paris Basin, France (Glibert and Van de Poel, 1970). **Extant**
Comments: A doubtful Late Cretaceous record of *Tridacna (Chametrachea) crocea* (Lamarck, 1819), noted for this family by Cox *et al.* (1969, p. N594) is not included here.

F. MACTRIDAE Lamarck, 1809 K. (APT)–Rec. Mar.

First: *Nelltia elliptica* (Whitfield, 1891), APT, Olive Mountain, Syria. **Extant**

F. RAETOMYIDAE Newton, 1919 T. (LUT/BRT–PRB) Mar.

First: *Amotapus arbolensis* (Woods, 1922), Middle Eocene, Peru.
Last: *Raetomya schweinfurthi* (Mayer-Eymar, 1887), PRB, Nigeria.

F. ANATELLIDAE Gray, 1853 **Extant** Mar.

F. CARDILIIDAE Fischer, 1887 T. (LUT/BRT)–Rec. Mar.

First: *Hemicyclonosta michelini* Deshayes, 1850, Middle Eocene, France. **Extant**
Comments: This early record is doubtful, according to Cox *et al.* (1969, p. N608). The next oldest record is *Cardilona bensoni* Marwick, 1943 from the Oligocene of New Zealand.

F. MESODESMATIDAE Gray, 1839 K. (CEN)–Rec. Mar.

First: *Mesodesmatid.* **Extant**
Comments: The oldest records in Cox *et al.* 1969, pp. N608–N610 are Eocene in age (*Mactropsis, Myadesma*).

F. SOLENIDAE Lamarck, 1809 T. (THA)–Rec. Mar.

First: *Siliqua*? sp., Mexia Member, Wills Point Formation, Upper Midway Group, Williamson County, Texas, USA (Palmer and Brann, 1965). **Extant**
Comments: Oldest records in Cox *et al.*, 1969, pp. N610–N611 are Lower Eocene (*Solen (Plectosolen)*).

F. PHARELLIDAE Adams and Adams, 1856 (= Cultellidae Davies, 1935) K. (ALB)–Rec. Mar.

First: *Leptosolen biplicata* (Conrad, 1858), North America/eastern Europe? **Extant**

F. PHARIDAE Adams and Adams, 1856 J. (TTH)–Rec. Mar.

First: *Senis petschorae* (Keyserling, 1846), Spilsby Sandstone erratics, middle Volgian, Norfolk, England, UK (S. R. A. Kelly, pers. comm., 1992). **Extant**
Comments: This family is not listed in Cox *et al.* (1969), being included in their Solecurtidae (pp. N637–639). The oldest pharid noted here might be a solecurtid (S. R. A. Kelly, pers. comm., 1992).

F. SOWERBYIDAE Cox, 1929 J. (TOA)–K. (VLG) Mar.

First: *Sowerbya triangularis* (Phillips, 1848), *thouarense* Zone, upper TOA, Europe.
Last: *Sowerbya* sp., basal Claxby Ironstone, early VLG, Lincolnshire, England, UK (S. R. A. Kelly, pers. comm., 1992).
Comments: The Upper Triassic *Rhaetidia zittellii* Bittner, 1895 was included doubtfully in this family by Cox *et al.* (1969, p. N639), but is excluded here (see also Jablonski and Bottjer, 1990, pp. 58–9).

F. TANCREDIIDAE Meek, 1864 Tr. (CRN)–Rec. Mar.

First: *Sakawanella triadica* Ichikawa, 1950, *Oxytoma–Mytilus* Bed, CRN, lower Kochigatani Group, Tosa Province, Kochi Prefecture, Shikoku Island, Japan (Jablonski and Bottjer, 1990). **Extant**

F. QUENSTEDTIIDAE Cox, 1929 J. (TOA)–K. (MAA) Mar.

First: *Quenstedtia laevigata* (Phillips, 1848), Cephalopod Bed, upper TOA, Gloucestershire, England, UK (Jablonski and Bottjer, 1990).
Last: Quenstedtiid.
Comments: The youngest representatives in Cox *et al.* (1969, pp. N634–N635) are middle Jurassic in age: *Quenstedtia*, England, UK.

F. UNICARDIOPSIDAE Vokes, 1967 J. (TOA–KIM) Mar.

First: *Unicardiopsis incertum* (Phillips, 1848), Blea Wyck Sands, upper TOA, Yorkshire, England, UK (Jablonski and Bottjer, 1990).
Last: *Unicardiopsis aceste* (d'Orbigny, 1850), Sequanian, France.

F. TELLINIDAE de Blainville, 1814 K. (HAU)–Rec. Mar.

First: *Linearia subhercynica* Maas, 1895, Hilskonglomerat, Hannover, Germany; *L. subconcentrica* (d'Orbigny), Calcaire à Spatangues, Paris Basin, France, both lower HAU (Jablonski and Bottjer, 1990). **Extant**

F. ICANOTIIDAE Casey, 1961 K. (HAU)–T. (PLI) Mar.

First: *Scittila* cf. *S. nasuta* Casey, 1961, Marne d'Hauterive, lower HAU, Neuchâtel, Switzerland (Jablonski and Bottjer, 1990).
Last: Icanotiid (N. J. Morris and C. P. Nuttall, pers. comm., 1992).
Comments: The youngest Icanotiid in Cox *et al.* (1969, p. N635) is MAA: *Icanotia*.

F. DONACIDAE Fleming, 1828 K. (APT)–Rec. Mar.

First: *Notodonax (Protodonax) minutissimus* (Whitfield, 1891), Abeih Formation, lower APT, 'Olive locality', Abeih, Lebanon (Jablonski and Bottjer, 1990). **Extant**

F. SOLECURTIDAE d'Orbigny, 1846
K. (CMP)–Rec. Mar.

First: *Protagelus albertinus* (d'Orbigny, 1846), Silakkudi Formation, Ariyular Group, lower CMP, Tamil Nadu State, India (Jablonski and Bottjer, 1990). **Extant**
Comments: The family might extend back to the TTH, if specimens of *Senis*, assigned to the Pharidae (q.v.) here, are misidentified.

F. PSAMMOBIIDAE Fleming, 1828 T. (DAN)–Rec. Mar.

First: *Gari (Gobraeus) debilis* (Deshayes, 1855), Unnamed unit, Luzankova, northern Ukraine, former USSR; Sochaczew Beds, Bochotnica, central Poland; *Echinanthus carinatus* Zone, eastern flank of Caspian Sea; *Adansonella duponti* (Cossmann, 1886), Calcaire grossier de Mons, NP2–NP3 (THA), Mons, Belgium (Jablonski and Bottjer, 1990). **Extant**
Comments: Cox *et al.* (1969, p. N633) include the Upper Cretaceous *Rhectomyax undulata* (Gabb, 1864) in this family, but it is excluded here.

F. SCROBICULARIIDAE H. and A. Adams, 1856
T. (THA)–Rec. Mar.

First: *Scrobiculabra condamini* (Morris, 1854), Woolwich Shell Beds, upper THA (NP9), Thames Valley, England, UK (Jablonski and Bottjer, 1990). **Extant**

F. SEMELIDAE Stoliczka, 1870 T. (THA)–Rec. Mar.

First: *Semele langdoniana* Aldrich, 1921, Bells Landing Member, Tuscahmoa Formation, upper THA (NP9), Bells Landing, Alabama, USA (Jablonski and Bottjer, 1990). **Extant**

F. DREISSENIDAE Gray, 1840 T. (Eoc.)–Rec. FW/Brackish

First: *Dreissena (Dreissena) polymorpha* (Pallas, 1771), Paratethys. **Extant**

F. GAIMARDIIDAE Hedley, 1916 T. (YPR)–Rec. Mar.

First: Gaimardiid. **Extant**
Comments: The oldest form in Cox *et al.* (1969, p. N644) is Miocene: *Kidderia*.

F. ARCTICIDAE Newton, 1891 Tr. (RHT)–Rec. Mar.

First: *Isocyprina (Eotrapezium) alpinus* (Winkler), Kössen and Zlambach Beds, Austria; *I. (E.) ewaldi* (Bomemann), Penarth Group, England, UK, *Plesiocyprina gaudryi* Fischer, 1887, RHT, Europe (Hallam, 1981). **Extant**

F. BERNARDINIDAE Keen, 1963 **Extant** Mar.

F. EULOXIDAE Gardner, 1943 T. (UMI) Mar.

First and Last: *Euloxa latisulcata* (Conrad, 1840), UMI, Virginia, USA: *Cabralista schmitzi* (Böhm, 1899), UMI, Salvages Island, East Atlantic.

F. KELLIELLIDAE Fischer, 1887 T. (THA)–Rec. Mar.

First: *Kelliella ? aldrichi* Gardner, 1935 and *K. ? evansi* Gardner, 1935, both Kincaid Formation, Texas, USA (Palmer and Brann, 1965). **Extant**

F. NEOMIODONTIDAE Casey, 1955
J. (HET)–K. (CEN) FW/Brackish

First: *Eomiodon lunulatus* (Yokoyama, 1904), Niranohama Formation, Shizukawa Group, NE Japan.
Last: *Eomiodon matsubasensis* Tamura, 1977, Mifune Group, Kyushu, western Japan.

F. POLLICIDAE Stephenson, 1953 K. (CEN–?CMP) Mar.

First: *Pollex obesus* Stephenson, 1953, Woodbine Formation, Texas, USA.
Last: *Neritra polliciformis* Stephenson, 1954, ?CMP, ?New Jersey, USA.

F. TRAPEZIIDAE Lamy, 1920 K. (?APT)–Rec. Mar.

First: *Corbiculopsis birdi* Whitfield, 1891, ?APT, Syria. **Extant**
Comments: The hinge of *Corbiculopsis* is imperfectly known (Cox *et al.*, 1969, pp. N655–N656). If it is not a trapeziid, the next oldest record is *Pseudopleurophorus rochi* Chavan, 1954 from the Upper Cretaceous of Chad.

F. GLOSSIDAE Gray 1847 K. (CEN)–Rec. Mar.

First: Glossid (N. J. Morris, pers. comm., 1992). **Extant**
Comments: The oldest record in Cox *et al.*, 1969, pp. N657–N658 is Palaeocene: *Glossus (Meiocardia)*.

F. DICEROCARDIIDAE Kutassy, 1934
Tr. (NOR)–K. (MAA) Mar.

First: *Dicerocardium jani* Stoliczka, 1871, NOR, northern Italy; *D. curionii* Stoppani, 1865, NOR, Sicily, Italy; *D. dolomiticum* (Loretz), NOR, northern Italy (Hallam, 1981).
Last: *Agelasina plenodonta* Riedel, 1932, MAA, West Africa.
Comment: Probably a polyphyletic assemblage.

F. VESICOMYIDAE Dall, 1908 T. (CHT)–Rec. Mar.

First: *Hubertschenckia ezoensis* (Yokoyama, 1890), Hokkaido, Japan; *Pleurophopsis unioides* Palmer, 1919, CHT, West Indies, Central America, NW South America. **Extant**

F. CORBICULIDAE Gray, 1847 Tr. (CRN)–Rec. Mar./FW

First: ?Corbiculid, lacustrine beds, Durham, North Carolina, USA (Olsen *et al.*, 1989, pp. 30, 32). **Extant**

F. PISIDIIDAE Gray, 1857 J. (MLM)–Rec. FW

First: Pisidiid, Muling Formation, Jixi Group, north-east China (Gray, 1988, p. 54). **Extant**
Comments: Cox *et al.* (1969, pp. N669–N670) note another doubtful late Jurassic record of *Sphaerium (Sphaerium) corneum* (Linnaeus, 1758).

F. VENERIDAE Rafinesque, 1815 K. (VLG)–Rec. Mar.

First: *'Venus' vendoperana* (Leymerie, 1842), VLG, eastern France, Switzerland (Dhondt and Dieni, 1988). **Extant**

F. PETRICOLIDAE Deshayes, 1839 T. (LUT)–Rec. Mar.

First: *Petricola ? novaegyptica* Whitfield, 1885, Manasquan Formation, Ocean County, New Jersey, USA (Palmer and Brann, 1965). **Extant**

F. COOPERELLIDAE Dall, 1990 T. (LMI)–Rec. Mar.

First: *Lajonkairia rupestris* (Brocchi, 1814), AQT, Lariey, Bordelais, France (Glibert and Van de Poel, 1966a). **Extant**

F. GLAUCONOMIDAE Gray, 1853 **Extant** Mar.

F. RZEHAKIIDAE Korobkov, 1954 T. (RUP–MMI) Brackish

First: *Ergenica cimlanica* (Popov, 1959), Solenovskiy Horizon, 'Middle Oligocene', Turkmenia, Kazakhstan, Ukraine, former USSR; *Urbnisia lata* Goncharova, 1981, Solenovskiy Horizon, 'Middle Oligocene', Georgia and northern Caucasus, former USSR (Goncharova, 1981).
Last: *Rzehakia dubiosa dubiosa* (Hörnes, 1859), Middle Tarkhanian, Georgia, former USSR (Goncharova, 1981).

Order HIPPURITOIDA Newell, 1965

F. MECYNODONTIDAE Haffer, 1959 D. (EIF) Mar.

First and Last: *Mecynodon carinatus* (Goldfuss, 1837) and *M. oblongus* (Goldfuss, 1837), *Stringocephalus* Limestone, Germany and Onondagan, North America.

F. MEGALODONTIDAE Morris and Lycett, 1853 S. (LLY)–J. (BTH) Mar.

First: Megalodonts, Canadian Arctic (Freitas *et al.*, in press).
Last: *Pachyrisma grande* Morris and Lycett, 1853, Great Oolite, Cheltenham, England, UK (Hallam, 1976).
Comments: Cox *et al.* (1969, pp. N743–N749) give an extended range for this family, by including Middle Silurian (WEN) *Megalomoidea canadensis* (Hall, 1852), regarded here as a modiomorphid (N. J. Morris, pers. comm., 1992), and *Pachyrismella* and *Pterocardia* (OXF), which are transferred to the Cardiidae (N. J. Morris, pers. comm., 1992).

F. DICERATIDAE Dall, 1895 J. (OXF)–K. (VLG) Mar.

First: *Diceras arietinum* Lamarck, 1805, 'Hauts-de-Meuse' Coral Rag, Middle OXF, France (Enay and Boullier, 1981).
Last: *Valletia tombecki* Munier Chalmas, 1873, Salève, Haute-Savoie, France (Skelton, 1985b).
Intervening: KIM–BER.

F. REQUIENIIDAE Douvillé, 1914 J. (TTH)–K. (MAA) Mar.

First: *Matheronia salevensis* Joukowsky and Favre, 1913, Salève, Haute-Savoie, France (Skelton, 1985b).
Last: *Apricardia sicoris* Astre, 1932, Calcaire nankin, north pre-Pyrenees, France (Bilotte, 1985).
Intervening: BER–CMP.

F. CAPROTINIDAE Gray, 1848 (emend. Skelton, 1978 J. (TTH)–?T. (THA) Mar.

First: *Monopleura crimica* Yanin, 1975, Crimea, former USSR (Skelton, 1985b).
Last: *Monopleura moroi* (Vidal, 1877), Marnes d'Auzas, north pre-Pyrenees, France (Bilotte, 1985).
Intervening: BER–CEN, SAN, CMP.
Comment: The dating (THA) and affinities of *Paramonopleura* (see Cox *et al.*, 1969, p. N783) remain to be confirmed. *Valletia* is here transferred to the Diceratidae.

F. CAPRINIDAE d'Orbigny, 1850 (emend, Skelton, 1978) K. (APT–CEN) Mar.

First: *Praecaprina varians* Paquier, 1905, Urgonian, Vercors, France (Skelton, 1985b).
Last: *Caprinula* sp., Calcaires bioclastiques A2, Akros Massif, Greece (Mermighis *et al.*, 1991).
Intervening: ALB.

F. ICHTHYOSARCOLITIDAE Douvillé, 1887 K. (APT–CEN) Mar.

First: *Ichthyosarcolites* sp., Urgonian facies, Bulgaria (Skelton, 1991).
Last: *Ichthyosarcolites* sp., terminal CEN, Old World Tethys (Philip and Airaud-Crumière, 1991).
Intervening: None.

F. PLAGIOPTYCHIDAE Douvillé, 1888 K. (TUR–MAA) Mar.

First: *Plagioptychus paradoxus* Matheron, 1843, *P. arnaudi* Douvillé, 1888, and *P. toucasianus* Matheron, 1843, all TUR, Bouche-du-Rhône, France.
Last: *Plagioptychus* spp., several species, *Titanosarcolites* Limestone, Jamaica (Chubb, 1971).
Intervening: CON–CMP.

F. ANTILLOCAPRINIDAE MacGillavry, 1937 K. (SAN–MAA) Mar.

First: *Antillocaprina* sp., Peters Hill Limestone, Jamaica (PWS, pers. obs., 1992).
Last: *Antillocaprina occidentalis* (Whitfield, 1897), *Titanosarcolites* Limestone, Jamaica (Chubb, 1971).

F. DICTYOPTYCHIDAE Skelton, *fam. nov.* (ex Trechmannellinae MacGillavry, 1937, *nom. van.*) K. (CMP–MAA) Mar.

First: *Eodictyoptychus arumaensis* Skelton and El-Asa'ad, 1992, upper Khanasir Limestone Member, Aruma Formation, central Saudi Arabia (Skelton and El-Asa'ad, 1992).
Last: *Dictyoptychus morgani* (Douvillé, 1904), *Loftusia persica* Beds, Bakhtiari Mountains, Iran.

F. RADIOLITIDAE Gray, 1848 K. (APT–MAA) Mar.

First: *Eoradiolites* sp., Shuaiba Formation, Oman (Skelton, 1985b).
Last: *Praeradiolites* spp., several species, Marnes d'Auzas, French Pre-Pyrenees (Bilotte, 1985).
Intervening: ALB–CMP.

F. HIPPURITIDAE Gray, 1848 K. (TUR–MAA) Mar.

First: *Vaccinites fontalbensis* Philip, 1983, basal TUR, Provence, France (Philip and Airaud-Crumière, 1991).
Last: *Hippuritella castroi* (Vidal, 1878), and other species, Marnes d'Auzas, French Pre-Pyrenees (Bilotte, 1985).
Intervening: CON–CMP.

Subclass ANOMALODESMATA Dall, 1889

Order PHOLADOMYOIDA Newell, 1965

F. GRAMMYSIIDAE, S. A. Miller, 1877 O. (CRD)–P. (ZEC) Mar.

First: *Cuneamya* sp., Trentonian, Watertown, New York, USA (Pojeta, 1978).

Last: *Prothyris (Prothyris) elegans* Meek, 1871, ZEC, western Europe and North America; *Solenomorpha minor* (M'Coy, 1844), ZEC, Europe.
Comments: Cox *et al.* (1969, pp. N819–N823) doubtfully include a Lower Ordovician (TRE) form, *Davidia ornata* Hicks, 1873 in this family.

F. SINODORIDAE Pojeta and Zhang, 1984 D. (EMS–EIF) Mar.

First: *Sinodora concava* Pojeta and Zhang, 1984, upper Ertang Formation, Wuxuan County, Guangxi, China; *S. bisulcata* Pojeta and Zhang, 1984, Sipai Formation, Yongfu County, Guangxi, China (Pojeta and Zhang, 1984).
Last: *Sinodora semiglabra* Pojeta and Zhang, 1984, Yingtang Formation, Xiangzhou County, Guangxi, China (Pojeta and Zhang, 1984).

F. EDMONDIIDAE, King, 1850 D. (FRS)–P. (TAT) Mar.

First: *Edmondia aequimarginalis* (Winchell, 1862), New York, USA; *E. bodana* (Roemer, 1860), Pilton Beds, Devon, England, UK, and Harz Mountains, Germany.
Last: *Edmondia* sp., Gujo Formation, upper TAT, Kyoto, Japan (Nakazawa and Newell, 1968); *Pachymyonia elata* Popov, 1958, Verchoyan, KAZ, Siberia, former USSR.

F. SANGUINOLITIDAE S. A. Miller, 1877 D. (EIF/GIV)–P. (UFI) Mar.

First: *Sanguinolites* sp. (Morris, 1978).
Last: *Sanguinolites kamiyassensis* Nakazawa and Newell, 1968, Shegejizawa Member, lower Kanokura Formation, Miyagi, Japan (Nakazawa and Newell, 1968).

F. PERMOPHORIDAE van de Poel, 1959 D. (EIF)–K. (CMP) Mar.

First: *Pleurodapis multicincta* Clarke, 1913, São Domingos Member, middle EIF, Ponta Grossa and Santa Cruz, Paraná Basin, Brazil.
Last: Permophorid.
Comments: Cox *et al.* (1969, p. N546) doubtfully include the Ordovician *Redonia* in this family, but it is now placed in the Redoniidae, a family of actinodontoid heteroconchs.

F. HIPPOPODIIDAE Cox, 1965 J. (SIN–TTH) Mar.

First: *Hippopodium ponderosum* Sowerby, 1819, Lias, Dorset, England, UK.
Last: *Hippopodium quenstedti* (Dietrich, 1933), *Nerinella* Bed and *'Trigonia smeei'* Bed, Tendaguru, Tanzania (Cox, 1965).
Comments: Cox *et al.* (1969, p. N583) include the Devonian *Tusayana cibola* Stoyonow, 1948 in this family, but note that this 'is very dubious'.

F. MEGADESMIDAE Vokes, 1967 C. (SPK)–P. (TAT) Mar.

First: *Unklesbayella geinitzi* (Meek, 1867), Putnam Hill Shale, Vanport Shale, lower Mercer Shale, Ames Shale, Ohio and West Virginia, USA.
Last: *Pyramus planus* Nakazawa and Newell, 1968, Gujo Formation, upper TAT, Kyoto, Japan (Nakazawa and Newell, 1968).

F. PHOLADOMYIDAE Gray, 1847 C. (HAS)–Rec. Mar.

First: *Grammysia blairi* Miller, 1891, and other species, Kinderhookian, Missouri, Iowa, USA. **Extant**
Comment: Includes Margaritariidae Runnegar, 1974.

F. CHAENOMYIDAE Waterhouse, 1966 C. (MOS)–P. (KAZ) Mar.

First: *Chaenomya leavenworthensis* Meek and Hayden, 1866, MOS, USA.
Last: *Vacunella oblonga* Waterhouse, 1969, Flat Top, Ingelara, and Blenheim Formations, Bowen Basin, Queensland, Australia, and *V. similis* Lyutkevich and Lobanova, 1960, KAZ, Taimyr Peninsula, former USSR.

F. BURMESIIDAE Healey, 1908 Tr. (CRN–RHT) Mar.

First: *Burmesia latouchii* Healey, 1980, Asia (Jordan–Burma–Indonesia–Japan).
Last: *Burmesia latouchii* Healey, 1908 and *Prolaria sollasi* Healey, 1908, both Burma, Indonesia, and *P. armenica* Robinson, Armenia, former USSR.
Comments: Cox *et al.* (1969, p. N838) include early Jurassic records of *Burmesia*, but these are not accepted here (I. Hayami, pers. comm., 1992).

F. CERATOMYIDAE Arkell, 1934 Tr. (RHT)–J. (TTH) Mar.

First: *Pteromya crowcombeia* Moore, 1861, lower Penarth Group, England, UK; Kössen Beds, Austria.
Last: *Ceratomya excentrica* (Roemer, 1836), Dakacha Limestones, Melka Dakacha, NE Kenya (Cox, 1965).
Comments: Cox *et al.* (1969, pp. N839–N841) doubtfully include the Miocene *Ceromyella miotaurina* Sacco, 1901 in this family, but this is not accepted here.

F. CERATOMYOPSIDAE Cox, 1964 J. (TOA–TTH) Mar.

First: *Ceratomyopsis* sp., Europe.
Last: *Ceratomyopsis helvetica* (de Loriol, 1875), TTH, France.

F. MYOPHOLADIDAE Cox, 1964 J. (TOA)–K. (ALB) Mar.

First: *Myopholas minor* Hayami, 1972, Lo-Duc Bed, near Saigon, Vietnam.
Last: *Myopholas ledouxi* Douvillé, 1907, upper ALB, France.

F. PLEUROMYIDAE Dall, 1900 Tr. (CRN)–K. (ALB) Mar.

First: *Pleuromya nipponica* Kobayashi and Ichikawa, 1951, Kochigatani Group, Sakawa and Sakamoto areas, Japan (Hayami, 1975).
Last: *Pleuromya* sp.

F. PANDORIDAE Rafinesque, 1815 T. (YPR)–Rec. Mar.

First: *Pandora dilatata* Deshayes, 1858, YPR, Paris Basin, France (Glibert and Van de Poel, 1966a). **Extant**

F. CLEIDOTHAERIDAE Hedley, 1918 T. (LMI)–Rec. Mar.

First: *Cleidothaerus albidus* (Lamarck, 1819), Waitakian, Chatton Formation, Brydon, New Zealand (Beu and Maxwell, 1990) and Fossil Bluff, Wynyard, Tasmania, Australia. **Extant**

F. LATERNULIDAE Hedley, 1918 Tr. (RHT)–Rec. Mar.

First: *Cercomya (Cercomya) praecursor* (Quenstedt), Penarth Group, England, UK (Hallam, 1981). **Extant**

F. LYONSIIDAE Fischer, 1887 T. (THA)–Rec. Mar.

First: *Lyonsia subplicata* (d'Orbigny, 1850), THA, Paris Basin, France (Glibert and Van de Poel, 1966a). **Extant**
Comments: Cox *et al.* (1969, pp. N845–N847) give the oldest as Eocene: *Lyonsia (Lyonsia).*

F. MYOCHAMIDAE Bronn, 1862 T. (PRB)–Rec. Mar.

First: *Myadora lamellata*, Browns Creek Clay, Johanna, Victoria and Blanche Point Marl, Aldinga, South Australia (Darragh, 1985). **Extant**

F. PERIPLOMATIDAE Dall, 1895 J. (TTH)–Rec. Mar.

First: Periplomatid. **Extant**
Comments: Cox *et al.* (1969, pp. N849–N850) give the oldest as *Periploma (Periploma) subgracile* Whitfield, Fox Hills Formation (MAA), South Dakota, USA.

F. THRACIIDAE Stoliczka, 1870 Tr. (RHT)–Rec. Mar.

First: Thraciid. **Extant**
Comments: Cox *et al.* (1969, pp. N850–N852) give the oldest as Jurassic: *Thracia (Thracia).*

F. POROMYIDAE Dall, 1886 K. (APT)–Rec. Mar.

First: *Poromya* sp., DSDP borehole 317A, Manihiki Plateau, South Pacific (Kauffman, 1976). **Extant**

F. CUSPIDARIIDAE Dall, 1886 Tr. (LAD)–Rec. Mar.

First: *Cuspidaria triassica* (Stoppani, 1865), Austria (Morris, 1967, p. 475). **Extant**

F. VERTICORDIIDAE Stoliczka, 1871 K. (?ALB)–Rec. Mar.

First: *Verticordia* sp. and ?*Manihikia erecta* Kauffman, 1976, DSDP hole 317A, Manihiki Plateau, South Pacific (Kauffman, 1976). **Extant**

F. CLAVAGELLIDAE d'Orbigny, 1843 K. (CEN)–Rec. Mar.

First: *Clavagella cenomanensis* d'Orbigny, 1850, CEN, Le Mans, France. **Extant**

Order MYOIDA Stoliczka, 1870

F. MYIDAE Lamarck, 1809 T. (DAN)–Rec. Mar.

First: *Sphenia leptomorpha* Cossmann, 1891, THA, Paris Basin, France (Glibert and Van de Poel, 1966a). **Extant**

F. PLEURODESMATIDAE Cossmann and Peyrot, 1909 T. (LMI/UMI). Mar.

First: *Pleurodesma moulinsi* (Potiez and Michaud, 1844), BUR, Bordelais, France (Glibert and Van de Poel, 1966a).
Last: *Pleurodesma mayeri* Hörnes, 1859 and *P. moulinsi* (Potiez and Michaud, 1844), Tortonian, Vienna Basin, Austria (Glibert and Van de Poel, 1966a).

F. SPHENIOPSIDAE Gardner, 1928 T. (LUT/BRT)–Rec. Mar.

First: *Spheniopsis scalaris* (Braun, 1851), ?Europe. **Extant**

F. GASTROCHAENIDAE Gray, 1840 J. (BTH)–Rec. Mar.

First: *Gastrochaena moreana* Buvignier, Europe (Hallam, 1976). **Extant**

F. HIATELLIDAE Gray, 1824 Tr. (ANS/LAD)–Rec. Mar.

First: *Panopea (Panopea) glycimeris* (Born, 1778), Europe. **Extant**
Comments: Cox *et al.* (1969, pp. N700–N702) include the Permian *Roxoa intrigans* (Mendes, 1944) in this family, but this species is excluded here.

F. PHOLADIDAE Lamarck, 1809 J. (BAJ)–Rec. Mar.

First: *Martesia (Particoma) australis* Moore, 1870, Western Australia (Kelly, 1988). **Extant**
Comments: Cox *et al.* (1969, pp. N711–N712) note a doubtful *Martesia* from the Carboniferous, but this is excluded from the family, following Kelly (1988). The ichnogenus *Teredolites*, probably formed by wood-boring bivalves, is known from the PLB.

F. TEREDINIDAE Rafinesque, 1815 K. (APT)–Rec. Mar.

First: Teredinid. **Extant**
Comments: Cox *et al.* (1969, pp. N740–N741) note several Cretaceous taxa: *Terebrimya*, CEN, Texas, USA; *Teredolites*, Europe; *Turnus*, North America. *Teredolites* is a trace fossil, possibly assignable to the Pholadidae (q.v.) and *Turnus* is a pholadid (S. R. A. Kelly, pers. comm., 1992).

Subclass and ***Order*** UNCERTAIN

F. ARCHANODONTIDAE Weir, 1969 D. (FAM)–P. (ROT) FW

First: *Archanodon jukesi* (Forbes, 1853), Kiltorcan Formation, Upper Old Red Sandstone, Kilkenny, Republic of Ireland.
Last: *Neamnigenia beljanini* Khalfin, 1950 and *N. longa* Betekhtina, 1966, Siberia, former USSR.

F. ANTHRACOSIIDAE Amalitsky, 1892 C. (VIS)–P. (KAZ) FW

First: *Anthracomya ovalis* (Dawson, 1860) and *Carbonicola bradorica* (Dawson, 1868), Nova Scotia, Canada, and *Naiadites fragilis* (Meek and Worthen, 1866), Keokuk, Osagean, Illinois, USA.
Last: *Palaenodonta castor* (Eichwald, 1856) and others, Oka-Volga Basin, former USSR, Beaufort Series, South Africa, East Africa, Burma, Norway.
Comments: The next oldest in Cox *et al.*, 1969, pp. N406–N407 is *Carbonicola* from the SPK (Namurian).

F. PRILUKIELLIDAE Starabogatov, 1970 (= Microdontidae) P. (ROT–ZEC) FW

First: *Amnigeniella kumsassiana* (Ragozin, 1960), Upper Carboniferous–Lower Permian, western Siberia, former USSR (Cox *et al.*, 1969, p. N860).
Last: *Microdontella problematica* Lebedev, 1944, Kolchugino Series, Kuznetsk Basin, Siberia, former USSR.

F. PALAEOMUTELIDAE Weir, *in* Vokes, 1967 C. (u.)–P. (ZEC) FW

First: *Angarodon kumsassiensis* Ragozin, 1935, western Siberia, former USSR.
Last: *Palaeomutela (Palaeomutela) verneuilli* Amalitsky, 1892, *P. (P.) keyserlingi*, *P. (Oligodon) geinitzi* Amalitzky, 1892, and *P. (O.) zitteli*, all ZEC, former USSR.

F. FERGANOCONCHIDAE Martinson, 1956 J. (l.–u) FW

First: *Ferganoconcha sibirica* Chernyshev, 1937, Siberia, former USSR.
Last: *Ferganoconcha sibirica* Chernyshev, 1937, Siberia, former USSR.

F. PSEUDOCARDINIIDAE Martinson, 1961 J. (l.–u.) FW

First: *Tutuella chachlovi* Ragozin, 1938 and *Utschamiella tungussica* Ragozin, 1938, both Lias, Tungusska Basin, Siberia, former USSR.
Last: *Kija tjazhinensis* Lebedev, 1958 and *K. kibetenensis* Lebedev, 1958, both Upper Jurassic, Chulimo-Yenisei Basin, Central Asia.

REFERENCES

Abdel-Gawad, G. I. (1986) Maastrichtian non-cephalopod mollusks (Scaphopoda, Gastropoda and Bivalvia) of the Middle Vistula Valley, central Poland. *Acta Geologica Polonica*, **36**, 69–224.

Allasinaz, A. (1972) Revisione dei Pettinidi triassici. *Rivista Italiana di Paleontologia e Stratigrafia*, **78**, 189–428.

Allen, J. A. (1978) Evolution of the deep-sea protobranch bivalves. *Philosophical Transactions of the Royal Society, Series B*, **284**, 387–401.

Babin, C. (1982a) Mollusques bivalves et rostroconches, in *Brachiopodes (Articulés), Mollusques (Bivalves, Rostroconches, Monoplacophores, Gastropodes) de l'Ordovicien Inférieur (Tremadocien –Arenigien) de la Montagne Noire (France Meridionale)*, (eds C. Babin, R. Courtessole, M. Melou *et al.*), Mémoire de la Société des Études Scientifiques de l'Aude, Carcassonne, France, pp. 37–49.

Babin, C. (1982b) *Tromelinodonta* nov. gen., bivalve lyrodesmatide (Mollusca) de l'Arenigien (Ordovicien inférieur). *Geobios*, **15**, 423–7.

Babin, C. and Gutierrez-Marco, J.-C. (1991) Middle Ordovician bivalves from Spain and their phyletic and palaeogeographic significance. *Palaeontology*, **34**, 109–43.

Begg, J. H. and Campbell, H. J. (1985) *Etalia*, a new Middle Triassic (Anisian) bivalve from New Zealand, and its relationship with other pteriomorphs. *New Zealand Journal of Geology and Geophysics*, **28**, 725–41.

Berg-Madsen, V. (1987) *Tuarangia* from Bornholm (Denmark) and similarities in Baltoscandian and Australasian late Middle Cambrian faunas. *Alcheringa*, **11**, 245–259.

Beu, A. G. and Maxwell, P. A. (1990) Cenozoic Mollusca of New Zealand. *New Zealand Geological Survey Palaeontological Bulletin*, **58**, 1–518.

Bilotte, M. (1985) Le Crétacé supérieur des plâtes-formes Est-Pyrenéennes. *Strata (Actes du Laboratoire Géologique et Sedimentologique, Université Paul-Sabatier, Toulouse), Mémoire*, **5**, 1–438.

Brasier, M. D. (1984) Microfossils and small shelly fossils from the Lower Cambrian *Hyolithes* Limestone at Nuneaton, English Midlands. *Geological Magazine*, **121**, 229–253.

Carter, J. G. (ed.) (1990) *Skeletal Biomineralization; Patterns, Processes and Evolutionary Trends*, Vol. 1, Van Nostrand Reinhold, New York, 832 pp.

Carter, R. M. (1971) Revision of Arenig Bivalvia from Ramsey Island, Pembrokeshire. *Palaeontology*, **14**, 259–61.

Chinzei, K. (1982) Morphological and structural adaptations to soft substrates in the Early Jurassic monomyarians *Lithiotis* and *Cochlearites*. *Lethaia*, **15**, 179–97.

Chubb, L. J. (1971) Rudists of Jamaica. *Palaeontographica Americana*, **45**, 159–257.

Cooper, M. R. (1991) Lower Cretaceous Trigonioida (Mollusca, Bivalvia) from the Algoa Basin, with a revised classification of the order. *Annals of the South African* Museum, **100**, 1–52.

Cox, L. R. (1965) Jurassic Bivalvia and Gastropoda from Tanganyika and Kenya. *Bulletin of the British Museum (Natural History), Geology, Supplement*, **1**, 1–213.

Cox, L. R., Newell, N. D., Boyd, D. W. *et al.* (1969, 1971) *Treatise on Invertebrate Paleontology. Part N. Mollusca 6 Bivalvia*, Vols 1, 2 and 3. Geological Society of America and University of Kansas Press, Boulder, Colorado, and Lawrence, Kansas, 952 pp.

Crame, J. A. (1982) Late Jurassic inoceramid bivalves from the Antarctic Peninsula and their stratigraphic use. *Palaeontology*, **25**, 555–603.

Damborenea, S. D. (1987) Early Jurassic Bivalvia of Argentina. *Palaeontographica, Abteilung A*, **199**, 23–111, 113–216.

Damborenea, S. D. (1990) Middle Jurassic inoceramids from Argentina. *Journal of Paleontology*, **64**, 736–59.

Darragh, T. A. (1985) Molluscan biogeography and biostratigraphy of the Tertiary of southeastern Australia. *Alcheringa*, **9**, 83–116.

Dhondt, A. V. (1973) Systematic revision of the subfamily Neitheinae (Pectinidae, Bivalvia, Mollusca) of the European Cretaceous. *Institut Royal des Sciences Naturelles de Belgique, Mémoires*, **176**, 1–101.

Dhondt, A. V. (1992) Cretaceous inoceramid biogeography: a review. *Palaeogeography, Palaeoclimatology, Palaeoecology*, **92**, 217–32.

Dhondt, A. V. and Dieni, I. (1988) Early Cretaceous bivalves of eastern Sardinia. *Memorie di Scienze Geologiche (Padova)*, **40**, 1–97.

Enay, R. and Boullier, A. (1981) L'âge du complexe récifal des Côtes de Meuse entre Verdun et Commercy et la stratigraphie de l'Oxfordien dans l'est du Bassin de Paris. *Geobios*, **14**, 727–71.

Freitas, T. A. de, Brunton, F. and Bernecker, T. (in press) Silurian megalodont bivalves of the Canadian Arctic and Australia: paleoecology and evolutionary significance.

Fürsich, F. T. and Palmer, T. J. (1982) The first true anomiid? *Paleontology*, **25**, 897–903.

Glibert, M. and Van de Poel, L. (1966a) Les bivalves fossiles du Cénozoïque étranger des collections de l'Institut Royal des Sciences Naturelles de Belgique. III. *Institut Royal des Sciences Naturelles de Belgique, Mémoires, Deuxième Série*, **81**, 1–82.

Glibert, M. and Van de Poel, L. (1966b) Les bivalves fossiles du Cénozoïque étranger des collections de l'Institut Royal des Sciences Naturelles de Belgique. IV. *Institut Royal des Sciences Naturelles de Belgique, Mémoires, Deuxième Série*, **82**, 1–108.

Glibert, M. and Van de Poel, L. (1967) Les bivalves fossiles du Cénozoïque étranger des collections de l'Institut Royal des Sciences Naturelles de Belgique. V. *Institut Royal des Sciences Naturelles de Belgique, Mémoires, Deuxième Série*, **83**, 1–152.

Glibert, M. and Van de Poel, L. (1970) Les bivalves fossiles du Cénozoïque étranger des collections de l'Institut Royal des Sciences Naturelles de Belgique. VI. *Institut Royal des Sciences Naturelles de Belgique, Mémoires, Deuxième Série*, **84**, 1–185.

Glibert, M. and Van de Poel, L. (1973) Les Bivalvia du Danien et du Montien de la Belgique. *Institut Royal des Sciences Naturelles de Belgique, Mémoire*, **175**, 1–89.

Goncharova, I. A. (1981) On the composition and possible origin of the family Rzekakiidae (Bivalvia). *Palaeontological Journal*, **2**, 12–22.

Good, S. C. (1989) Nonmarine Mollusca in the Upper Triassic Chinle Formation and related strata of the western interior: systematics and distribution, in *Dawn of the Age of Dinosaurs in the American Southwest* (ed. S. G. Lucas and A. P. Hunt), New Mexico Museum of Natural History, Albuquerque, pp. 233–48.

Gray, J. (1988) Evolution of the freshwater ecosystem: the fossil record. *Palaeogeography, Palaeoclimatology, Palaeoecology*, **62**, 1–214.

Hallam, A. (1976) Stratigraphic distribution and ecology of European Jurassic bivalves. *Lethaia*, **9**, 245–59.

Hallam, A. (1977) Jurassic bivalve biogeography. *Paleobiology*, **3**, 58–73.

Hallam, A. (1981) The end-Triassic bivalve extinction event. *Palaeogeography, Palaeoclimatology, Palaeoecology*, **35**, 1–44.

Hallam, A. (1987) Radiations and extinctions in relation to environmental change in the marine Lower Jurassic of northwest Europe. *Paleobiology*, **13**, 152–68.

Harrison, W. B. III and Harrison, L. K. (1975) A Maquoteka-like molluscan community in the Bransfield Formation (Early Silurian) of Adams County, Ohio. *Bulletins of American Paleontology*, **67**, 193–234.

Hayami, I. (1975) A systematic survey of the Mesozoic Bivalvia from Japan. *The University Museum, Tokyo, Bulletin*, **10**, 1–249.

Jablonski, D. and Bottjer, D. J. (1990) Onshore-offshore trends in marine invertebrate evolution, in *Causes of Evolution; a Paleontological Perspective* (ed. R. M. Allmon and W. D. Allmon), Chicago University Press, Chicago, pp. 21–75.

Johnson, A. L. A. (1984) The palaeobiology of the bivalve families Pectinidae and Propeamussiidae in the Jurassic of Europe. *Zitteliana*, **11**, 1–235.

Johnston, P. A. (1991) Systematics and ontogeny of a new bivalve, *Umburra cinefacta*, from the Silurian of Australia. *Alcheringa*, **15**, 293–319.

Kauffman, E. G. (1976) Deep-sea Cretaceous macrofossils: hole 317A, Manihiki Plateau. *Initial Reports of the Deep-Sea Drilling Project*, **33**, 503–35.

Kelly, S. R. A. (1988) Cretaceous wood-boring bivalves from western Antarctica with a review of the Mesozoic Pholadidae. *Palaeontology*, **31**, 341–72.

Kennedy, W. J., Morris, N. J. and Taylor, J. D. (1970) The shell structure, mineralogy, and relationships of the Chamacea (Bivalvia). *Palaeontology*, **13**, 379–413.

Landing, E. (1989) Paleoecology and distribution of the Early Cambrian rostroconch *Watsonella crosbyi* Grabau. *Journal of Paleontology*, **63**, 566–73.

Ludbrook, N. H. (1960) Scaphopoda, in *Treatise on Invertebrate Paleontology, Part I. Mollusca 1* (ed. R. C. Moore and C. W. Pitrat), Geological Society of America and University of Kansas Press, Boulder, Colorado, and Lawrence, Kansas, pp. I37–I41.

Matthews, S. C. and Missarzhevsky, V. V. (1975) Small shelly fossils of late Precambrian and early Cambrian age: a review of recent work. *Journal of the Geological Society of London*, **131**, 289–304.

Mermighis, A., Philip, J. and Tronchetti, G. (1991) Séquences et cortèges de dépôts de plate-forme carbonatée au passage Cénomanien-Turonien dans les Héllénides internes (Péloponnèse, Grèce). *Bulletin de la Société Géologique de France* **160**, 544–52.

Morris, N. J. (1967) Mollusca: Scaphopoda and Bivalvia, in *The Fossil Record* (ed. Harland *et al.*). Geological Society of London, pp. 469–77.

Morris, N. J. (1978) The infaunal descendants of the Cycloconchidae: an outline of the evolutionary history and taxonomy of the Heteroconchia, superfamilies Cycloconchacea to Chamacea. *Philosophical Transactions of the Royal Society, Series B*, **284**, 259–75.

Morris, N. J. (1980) A new Lower Ordovician bivalve family, the Thoraliidae (?Nuculoida), interpreted as actinodont deposit feeders. *Bulletin of the British Museum (Natural History), Geology*, **34**, 265–72.

Nakazawa, K. and Newell. N. D. (1968) Permian bivalves of Japan. *Memoirs of the Faculty of Science, Kyoto University, Series of Geology and Mineralogy*, **35**, 1–108.

Newell, N. D. and Boyd, D. W. (1970) Oyster-like Permian Bivalvia. *Bulletin of the American Museum of Natural History*, **143**, 217–82.

Newell, N. D. and Boyd, D. W. (1975) Parallel evolution in early trigoniacean bivalves. *Bulletin of the American Museum of Natural History*, **154**, 53–162.

Newell, N. D. and Boyd, D. W. (1985) Permian scallops of the pectinacean family Streblochondriidae. *American Museum Novitates*, **2831**, 1–13.

Oliver, P. G. (1981) The functional morphology and evolution of Recent Limopsidae (Bivalvia, Arcoidea). *Malacologia*, **21**, 61–93.

Olsen, P. E., Schlische, R. W. and Gore, P. J. W. (1989) Tectonic, depositional, and paleoecological history of Early Mesozoic rift basins, eastern North America. *28th International Geological Congress Field Trip Guidebook*, **T351**, 1–174.

Palmer, C. P. (1974) A supraspecific classification of the scaphopod Mollusca. *Veliger*. **17**, 115–23.

Palmer, K. V. W. and Brann, D. C. (1965) Catalogue of the Paleocene and Eocene Mollusca of the southern and eastern United States. Part I. Pelecypoda, Amphineura, Pteropoda, Scaphopoda, and Cephalopoda. *Bulletin of American Paleontology*, **48**, 1–466.

Peel, J. S. (1991) Functional morphology of the Class Helcionelloida nov., and the early evolution of the Mollusca, in *The Early Evolution of Metazoa and the Significance of Problematic Taxa* (eds A. M. Simonetta and S. Conway Morris), Cambridge University Press, Cambridge, pp. 157–77.

Philip, J. M. and Airaud-Crumière, C. (1991) The demise of the rudist-bearing carbonate platforms at the Cenomanian-Turonian boundary: a global control. *Coral Reefs*, **10**, 115–25.

Pojeta, J. Jr (1971) Review of Ordovician pelecypods. *United States Geological Survey Professional Paper*, **695**, 1–46.

Pojeta, J. Jr (1975) *Fordilla troyensis* Barrande and early pelecypod phylogeny. *Bulletins of American Paleontology*, **67**, 363–79.

Pojeta, J. Jr (1978) The origin and early taxonomic diversification of pelecypods. *Philosophical Transactions of the Royal Society, Series B*, **284**, 225–46.

Pojeta, J. Jr (1985) Evolutionary history of diasome mollusks, in *Mollusks–Notes for a Short Course* (eds C. S. Hickman and P. D. Ward), University of Tennessee Department of Geological Sciences Studies in Geology, **13**, pp. 102–21.

Pojeta, J. Jr (1988) The origin and Paleozoic diversification of solemyid pelecypods. *New Mexico Bureau of Mines and Mineral Resources Memoir*, **44**, 201–71.

Pojeta, J. Jr and Gilbert-Tomlinson, J. (1977) Australian Ordovician pelecypod molluscs. *Bulletin of the Australian Bureau of Mineral Resources, Geology, and Geophysics*, **174**, 1–64.

Pojeta, J. Jr and Runnegar, B. (1976) The paleontology of rostroconch mollusks and the early history of the Phylum Mollusca. *US Geological Survey Professional Paper*, **968**, 1–88 pp.

Pojeta, J. Jr and Runnegar, B. (1979) *Rhytiodentalium kentuckensis*, a new genus and new species of Ordovician scaphopod, and the early history of scaphopod molluscs. *Journal of Paleontology*, **53**, 530–41.

Pojeta, J. Jr and Runnegar, B. (1985) The early evolution of diasome molluscs, in *The Mollusca, 10, Evolution of Mollusca* (eds E. R. Trueman and M. R. Clarke), Academic Press, New York.

Pojeta, J. Jr and Zhang, R. (1984) *Sinodora* n. gen. – a Chinese Devonian homeomorph of Cenozoic pandoracean bivalves. *Journal of Paleontology*, **58**, 1010–25.

Pojeta, J. Jr, Gilbert-Tomlinson, J. and Shergold, J. H. (1977) Cambrian and Ordovician rostroconch molluscs. *Bulletin of the Australian Bureau of Mineral Resources, Geology and Geophysics*, **171**, 1–54.

Rozanov, A. Yu. and Zhuravlev, A. Yu. (1992) The Lower Cambrian fossil record of the Soviet Union, in *Origin and Evolution of the Metazoa* (eds J. H. Lipps and P. W. Signor). Plenum, New York, pp. 205–82.

Runnegar, B. (1974) Evolutionary history of the bivalve subclass Anomalodesmata. *Journal of Palaeontology*, **48**, 904–39.

Runnegar, B. and Pojeta, J. Jr (1984) Molluscan phylogeny: the paleontological viewpoint. *Science*, **186**, 311–17.

Sepkoski, J. J. Jr (1982) A compendium of fossil marine families. *Milwaukee Public Museum Contributions in Biology and Geology*, **51**, 1–125.

Shergold, J. H. (1976) Biostratigraphical synopsis: Eastern Georgina

Basin, Qld. *Record of the Australian Bureau of Mineral Resources, Geology and Geophysics*, **1975/69**, 1–50.

Skelton, P. W. (1985a) Chapters 6.4 Bivalvia and 6.5 Rostroconchia, in *Atlas of Invertebrate Macrofossils* (ed. J. W. Murray), Longman, London, pp. 81–101.

Skelton, P. W. (1985b) Preadaptation and evolutionary innovation in rudist bivalves, in *Evolutionary Case Histories from the Fossil Record. Special Papers in Palaeontology*, 33 (ed. J. C. W. Cope and P. W. Skelton), Palaeontological Association, London, pp. 159–73.

Skelton, P. W. (1991) Morphogenetic versus environmental cues for adaptive radiations, in *Constructional Morphology and Evolution* (eds N. Schmid-Kittler and K. Vogel), Springer Verlag, Stuttgart, pp. 375–88.

Skelton, P. W. and El-Asa'ad, G. M. A. (1992) A new canaliculate rudist bivalve from the Aruma Formation of central Saudi Arabia. *Geologica Romana*, **28**, 105–17.

Speden, I. G. (1970) The type Fox Hills Formation, Cretaceous (Maestrichtican), South Dakota. *Bulletin of the Peabody Museum, Yale University*, **33**, 1–222.

Stenzel, H. B. (1971) Oysters, in *Treatise on Invertebrate Paleontology. Part N. Mollusca 6 Bivalvia*, Vol. 3, Geological Society of America and University of Kansas Press, Boulder, Colorado, and Lawrence, Kansas, pp. N954–N1224.

Stubblefield, C. J. (1938) Types and figured specimens in Phillips and Salter's Palaeontological Appendix to John Phillips' Memoir on the Malvern Hills compared with the Palaeozoic Districts of Abberley, etc. *Summary of Progress of the Geological Survey for 1936*, pp. 27–51.

Tunnicliff, S. P. (1987) Caradocian bivalve molluscs from Wales. *Palaeontology*, **30**, 677–90.

Vokes, H. E. (1980) *Genera of the Bivalvia. A Systematic and Bibliographic Catalogue (revised and updated)*. Paleontological Research Institution, Ithaca, New York, 307 pp.

Waller, T. R. (1978) Morphology, morphoclines and a new classification of the Pteriomorphia (Mollusca: Bivalvia). *Philosophical Transactions of the Royal Society London, B*, **284**, 345–65.

Ward, P. D., Kennedy, W. J., Macleod, K. G. *et al.* (1991) Ammonite and inoceramid bivalve extinction patterns in Cretaceous/Tertiary boundary sections of the Biscay region (southwestern France, northern Spain). *Geology*, **19**, 1181–4.

Waterhouse, J. B. (1970) *Permoceramus*, a new inoceramid bivalve from the Permian of eastern Australia. *New Zealand Journal of Geology and Geophysics*, **13**, 760–6.

Waterhouse, J. B. (1979a) New members of the Atomodesmatidae (Bivalvia) from the Permian of Australia and New Zealand. *Papers of the Department of Geology, University of Queensland*, **9**, 1–22.

Waterhouse, J. B. (1979b) The Upper Triassic bivalve *Oretia* Marwick, 1953. *New Zealand Journal of Geology and Geophysics*, **22**, 621–5.

Waterhouse, J. B. (1982a) Palacoecology and evolution of the Permian bivalve genus *Eurydesma* Morris. *Recent Researches in Geology*, **9**, 1–19.

Waterhouse, J. B. (1982b) Permian Pectinaceae and Limaceae (Bivalvia) from New Zealand. *New Zealand Geological Survey Palaeontological Bulletin*, **49**, 1–75.

Waterhouse, J. B. (1987) Late Palaeozoic Mollusca and correlations from the south-east Bowen Basin, east Australia. *Palaeontographica, Abteilung A*, **198**, 129–233.

Yancey, T. E. (1973) A new genus of Permian siphonodentalid scaphopods, and its bearing on the origin of the Siphonodentalidae. *Journal of Paleontology*, **47**, 1062–4.

Yancey, T. E. and Boyd, D. W. (1983) Revision of the Alatoconchidae: a remarkable family of Permian bivalves. *Palaeontology*, **26**, 497–520.

Yin, H.-F. (1985) Bivalves ner the Permian-Triassic boundary in South China. *Journal of Paleontology*, **59**, 572–600.

Yonge, C. M. and Thompson, T. E. (eds) (1978) Evolutionary systematics of bivalve molluscs. *Philosophical Transactions of the Royal Society of London, B*, **284**, 199–436.

Zinsmeister, W. J. and Macellari, C. E. (1988) Bivalvia (Mollusca) from Seymour Island, Antarctic Peninsula. *Memoir of the Geological Society of America*, **169**, 253–84.

14

?MOLLUSCA *INCERTAE SEDIS*

M. A. Wills

This chapter contains a number of groups possessing molluscan characteristics, but which cannot be assigned to the phylum with certainty, either because they exhibit additional characteristics outside the accepted molluscan scope, or because they also show affinities with some other group, most frequently the annelids.

Acknowledgements – I thank Dr. J. W. Cowie for generously supplying references. This chapter was researched during the tenure of a University of Bristol Postgraduate Scholarship.

Class TENTACULITOIDEA Ljashenko, 1957 (= CRICONARIDA Fisher, 1962)

Tentaculitoids are best known from the Palaeozoic. They are first recorded in the Lower Ordovician, although they remain fairly uncommon until the Devonian. Maximum diversity is attained in the Middle Devonian, but the class becomes less prominent by the end of this period. Weedon (1990, 1991) considers that certain vermiform 'gastropods' should be included within the Tentaculitoidea, establishing the order 'Microconchida' to accommodate them. This extends the record of the class up into the Lower Triassic. Additionally, he erects another new order for the enigmatic Devonian genus *Trypanopora*. Members of the class are united by the possession of gradually tapering, narrow shells, with transverse rings, ringlets and striae.

Order TENTACULITIDA Ljashenko, 1955 (see Fig. 14.1)

Tentaculites all possess minute, conical, calcareous shells with a circular cross-section. These may be composed of a number of layers, and are occasionally divided by septa. The group is most often assigned to the Mollusca, although affinities are also perceived with the annelids. Tentaculitids probably lived as nektobenthos and plankton in shallow seas, and their shells are often a significant constituent of Ordovician and Devonian limestones world-wide. The classification is taken mainly from Larsson (1979).

F. VOLYNITIDAE Ljashenko, 1959 S. (GLE)–D. (FRS) Mar.

First: *Volynites muldiensis* Larsson, 1979, and *Volynites scalpratus* Larsson, 1979, Mulde Beds, Mulde region of Gotland, Sweden.
Last: *Seretites bolonica* Farsan, 1983, and *Tripartites ferquensis* Farsan, 1983, both lower FRS of Ferques, Boulonnais, northern France.

F. TENTACULITIDAE Walcott, 1886 O. (TRE)–D. (FRS) Mar.

First: *Tentaculites lowdoni* Fisher and Young, 1955, Chepultepec Limestone, lowermost Ordovician, Virginia, USA.
Last: *Tentaculites donensis* Ljashenko, 1959 (*nomen nudum*, 1955), middle FRS Semiluk Beds, central Russian Platform, former USSR. *T. semilukianus*, Ljashenko, 1959 (*nomen nudum*, 1953), Semiluk Beds, central Russian Platform, former USSR. *T. spiculus*, Hall, 1876, 'Chemung' facies (exact position within FRS uncertain), south of Ithaca and south of Cortland, New York, USA.
Intervening: LUD.

F. GOTLANDELLITIDAE Larsson, 1979 S. (LLY–WEN) Mar.

First: *Gotlandellites visbyensis* Larsson, 1979, lower Visby Beds, Gotland, Sweden.
Last: *Gotlandellites areolatus* Larsson, 1979, upper Visby Beds, Gotland, Sweden.

F. ROSSIITIDAE Ljashenko, 1959 S. (WEN)–D. (FRS) Mar.

First: *Dicricoconus acutalis*, *D. clintianus*, *D. gutnicus*, *D. svarvariensis* and *D. valliensis* Larsson, 1979, Slite Beds, Gotland, Sweden.
Last: *Dicricoconus* sp., FRS of Russia. *Dicricoconus mesodevonicus* (Ljashenko, 1954), Middle Devonian, former USSR (Larsson, 1979).

Order HOMOCTENIDA Bouček, 1964

F. HOMOCTENIDAE Ljashenko, 1955 D. (EIF–FRS) Mar.

First: *Homoctenus hanusi* Bouček, 1964, Daleje Shales, lowermost EIF, Hlubocepy, Bohemia, Czechoslovakia.
Last: *Polycylindrites nalivkini* Ljashenko, 1954, lower Voronezh Beds, upper FRS, central Russian Platform, former USSR (Ljashenko, 1959).

Order DACRYCONARIDA Fisher, 1962

These are small criconarids, with a pronounced, tear-drop-shaped embryonal bulb at the apex of the shell. The shell broadens faster towards the aperture than in the tentaculitids, and the wall may be appreciably thicker, although none have internal septa. Their greatest abundance is achieved in the Upper Silurian and Lower Devonian of

The Fossil Record 2. Edited by M. J. Benton. Published in 1993 by Chapman & Hall, London. ISBN 0 412 39380 8

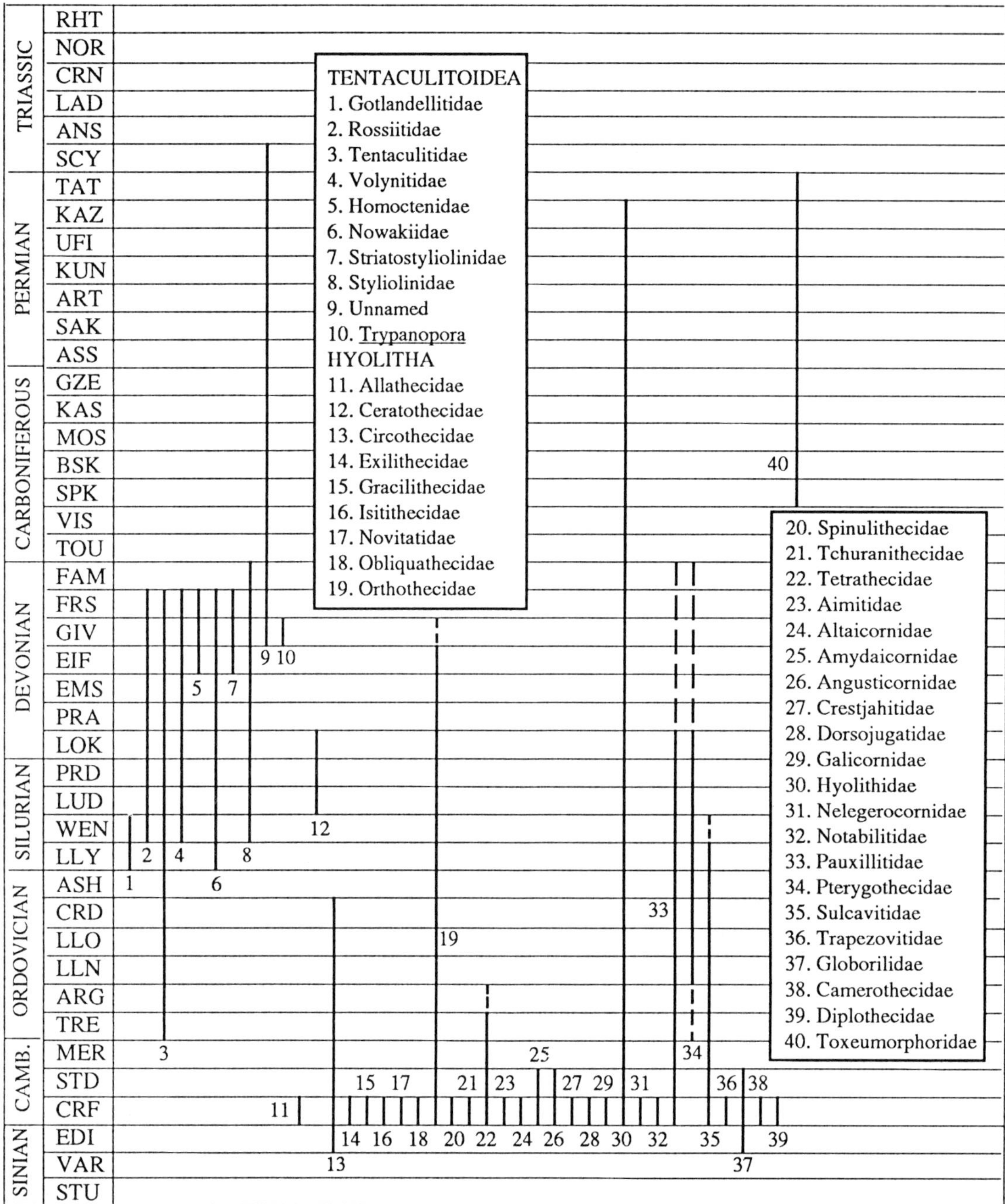

Fig. 14.1

Australia, and Upper Devonian of North America. Their classification is taken mainly from Bouček (1964).

F. NOWAKIIDAE Bouček and Prantl, 1960 S. (LLY)–D. (FRS) Mar.

First: *Nowakia brevis*, Tunnicliff, 1983, Limehill Beds at Lime Hill, Pomeroy, Co. Tyrone, Republic of Ireland.
Last: *Crassilina timanica* Ljashenko, 1957, Khvorostan Beds (early FRS), Timan, former USSR. *Viriatella spatiosa* Ljashenko, 1959, Khvorostan Beds, Tatar, former USSR (Ljashenko, 1959).
Comment: The position of this stratigraphical interval is in doubt, owing to debate over the position of the Silurian–Devonian boundary in continental Europe.

F. STRIATOSTYLIOLINIDAE Bouček, 1964 D. (EIF/GIV–FRS) Mar.

First: *Styliola strialula* Novak, 1882, Daleje Formation, Czechoslovakia (Fisher, 1962).
Last: *Striatostyliona striata* (Richter, 1854), *Undastriatostyliolina cirvimarginata*, *U. concavimarginata*, *U. crassa* and *U. globicellata* Farsan, 1983, Beaulieu Formation, lower FRS of Ferques, Boulonnais, northern France (Farsan, 1983).

F. STYLIOLINIDAE Grabau, 1912 S. (WEN)–D. (FAM) Mar.

First: *Styliolina* cf. *laevis* Richter, 1854, *Monograptus riccartonensis* Zone, Zebingyi Beds, near Maymyo, Burma

(Richter, 1854). '*Styliola*' sp., Rochester Formation, Maryland, USA (Swartz, 1913).
Last: *Styliolina* sp., Gowanda Formation (lower FAM), western New York, USA (Clarke, 1904).

Order MICROCONCHIDA Weedon, 1991

These are forms with helically coiled, attached shells, three-layered in the larva.

F. UNNAMED D. (GIV)–Tr. (GRI) Mar.

First: Vermiform 'gastropods' from the Middle Devonian of northern France (Weedon, 1990).
Last: Vermiform 'gastropods' from the Lower Triassic (Weedon, 1990).

Order TRYPANOPORIDA Weedon, 1991

Small, helically coiled shells with the initial whorls attached, and later whorls regularly or irregularly uncoiled. Larval shell calcitic with simple microlamellar microstructure. Septa branched and anastomosing.

F. UNNAMED D. (GIV) Mar.

First and Last: *Trypanopora* sp. Mistiaen and Poncet, 1983, Griset (lower) Member of the Blacourt Formation, Banc Noir Quarry, near Ferques, Boulonnais region, northern France. Also *Torquaysalpinx* sp. (Weedon, 1991).

Class HYOLITHA Marek, 1963 (= CALYPTOMATIDA Fisher, 1962)

Hyoliths originated in the Lower Cambrian, were very abundant up to the Ordovician and are last represented in the Permian. All are bilaterally symmetrical, mollusc-like, marine animals, with thin, conical, calcareous shells, sometimes internally divided into chambers at the apex. The ventral surface is flattened, and projected anteriorly into a tongue. The aperture of the shell is covered by an oval, curved operculum, with sculpture on its inner side which may have supported the soft parts. Long, recurved, tubular appendages lead back along the shell from the oral region in some forms. The system of classification adopted here primarily follows Missarzhevsky (1969), with some additional families.

Order ORTHOTHECIDA Marek, 1963 Mar.

F. ALLATHECIDAE Missarzhevsky, 1969 €. (TOM) Mar.

First: *Egdethica aldanica* Missarzhevsky, 1969, lower Tommotian, *Aj. sunnaginicus–T. licis* Zone and *D. regularis* Zone, Aldan River, Siberia, former USSR.
Next oldest: *Allatheca corrugata* Missarzhevsky, 1969, Tommotian, *D. regularis* Zone, middle reaches of Lena and Aldan Rivers, Siberia, former USSR. *A. concinna* Missarzhevsky, 1969, Tommotian, *D. regularis* Zone, Kotui River, Ary-Mas-Yuryakh River Mouth, former USSR. *Allatheca* sp., non-trilobite Zone of UK and Irish Republic (Cowie *et al.*, 1972; Sokolov and Zhuravleva, 1983).
Last: *Majatheca tumefacta* Missarzhevsky, 1969, upper part of Tommotian, *D. lenaicus–M. tumefacta* Zone, Churgan Village, middle reaches of River Lena, former USSR.

F. CERATOTHECIDAE Fischer, 1962 S. (LUD)–D. (LOK) Mar.

First: *Ceratotheca abundica* (Barrande, 1867), Czechoslovakia and England, UK.
Last: *Ceratotheca abundica* (Barrande, 1867), England, UK.

F. CIRCOTHECIDAE Missarzhevsky, 1969 V. (N–DA)–O. (CRD) Mar.

First: *Circotheca longiconica* Qian, 1978, lower Meischucunian *Anabarites–Circotheca–Protohertzina* Zone, eastern Yunnan, China (Brasier, 1989).
Next oldest: *Crossbitheca arcuaria* Missarzhevsky, 1974, *Aldanocyathus sunnaginicus–T. licis* Zone, Tommotian Stage, Siberian Platform, former USSR. *Laratheca nana* Missarzhevsky, 1969, *Aj. sunnaginicus–T. licis* Zone, Dvortsy Rocks, Aldan River, Siberian Platform, former USSR. *Circotheca kuteinikovi* Missarzhevsky, 1969, *Aj. sunnaginicus–T. licis* Zone, north-western slope of Anabar Massif; *Turcutheca cotuiensis* (Sysoiev, 1959b), *Aj. sunnaginicus–T. licis* Zone, north-western slope of Anabar Massif, Aldan River; *T. cotuiensis* (Sysoiev, 1959b), lower half of the Tommotian on the Siberian Platform, Russia, former USSR.
Last: Unnamed species, Europe, Siberia and North America (Missarzhevsky, 1969).

F. ISITITHECIDAE Sysoiev, 1968 €. (TOM–LEN) Mar.

First: *Isititheca lenae* Sysoiev, 1968, *Dorsojugatus sedecostatus* Zone, Tommotian Stage, Siberian Platform, former USSR (Sysoiev, 1972).
Last: *Plicitheca sulcata* Sysoiev, 1968, *Grandicornus validus* and *Lenatheca triconcava* Zones, Atdabanian Stage, and *Inflaticornus stratigus* Zone, Botomian Stage, Siberian Platform (Sysoiev, 1972).

F. ORTHOTHECIDAE Sysoiev, 1958 €. (TOM)–D. (EIF/GIV) Mar.

First: *Trapezotheca bicostata* Missarzhevsky, 1969, *D. regularis* and *D. lenaicus–M. tumefacta* Zones, Siberian Platform, former USSR.
Last: *Orthotheca intermedia* Novak, 1886, Middle Devonian of Czechoslovakia.

F. TETRATHECIDAE Sysoiev, 1968 €. (TOM)–O. (TRE/ARG) Mar.

First: *Tetratheca hexagona* Sysoiev, 1968, Tommotian Stage, Siberian Platform, former USSR (Sokolov and Zhuravleva, 1983).
Last: Unnamed species, Lower Ordovician (Sepkoski, 1982).
Intervening: ATB, LEN.

F. EXILITHECIDAE Sysoiev, 1968 €. (TOM–LEN) Mar.

First: *Exilitheca multa* Sysoiev, 1968, *A. sunnaginicus* and *D. regularis* Zones, Tommotian Stage. *Exilitheca ancestralis* and *E. oblongata* Sysoiev, 1968, *D. regularis* Zone, Lower Cambrian, TOM, Siberian Platform, former USSR, and Mongolia (Sokolov and Zhuravleva, 1983).
Last: *Holmitheca obvia* Sysoiev, 1968, *H. zhuravlevae* Sysoiev, 1972, and *Micatheca stupenda* Sysoiev, 1968, *B. asiaticus* Zone, Lower Cambrian, Botomian Stage, Siberian Platform (Sokolov and Zhuravleva, 1983).
Other late spp.: *Lenatheca dolosa* Sysoiev, 1972, *L. pyramidata* (Sysoiev, 1968), and *L. triconcava* (Sysoiev, 1968), Lower Cambrian, Atdabanian and Botomian Stages, Siberian Platform; *L. groenlandica* (Poulson, 1932), Lower Cambrian, Atdabanian and Botomian Stages, Siberian

Platform (Sokolov and Zhuravleva, 1983). *Holmitheca ulterior* (Sysoiev, 1968), Lower Cambrian, Atdabanian Stage, Siberian Platform, former USSR (Sokolov and Zhuravleva, 1983).

F. GRACILITHECIDAE Sysoiev, 1968 Є. (LEN) Mar.

First and Last: *Gracilitheca ternata* Sysoiev, 1968, *B. micmacciformis–Eribella* Zone, Lower Cambrian, Botomian Stage, Siberian Platform, former USSR (Sokolov and Zhuravleva, 1983).

F. NOVITATIDAE Sysoiev, 1968 Є. (ATB–LEN) Mar.

First: *Novitatus oblongus* (Meshkova, 1974), *R. zegebarti* and *P. pinus* Zones, Atdabanian Stage, Siberian Platform, former USSR (Sokolov and Zhuravleva, 1983).
Last: *Novitatus lermontovae* and *N. tarynicus* Sysoiev, 1968, *N. incompletus* (Meshkova, 1974), *B. micmacciformis* Zone, Lower Cambrian, Botomian Stage, Siberian Platform, former USSR (Sokolov and Zhuravleva, 1983).

F. OBLIQUATHECIDAE Sysoiev, 1968 Є. (TOM–LEN) Mar.

First: *Obliquatheca bicostata* (Missarzhevsky, 1969), and *O. aldanica* (Sysoiev, 1960), Lower Cambrian, Tommotian and Atdabanian Stages, Siberian Platform (Sokolov and Zhuravleva, 1983).
Last: *Obliquatheca pulchella* Valkov, 1975, *B. asiaticus* Zone, Lower Cambrian, Botomian Stage, Siberian Platform, former USSR (Sokolov and Zhuravleva, 1983). *Obliquatheca acostae* Sysoiev, 1968, and *O. inermis* Sysoiev, 1968, Lower Cambrian, Atdabanian and Botomian Stages, Siberian Platform, former USSR (Sokolov and Zhuravleva, 1983).

F. SPINULITHECIDAE Sysoiev, 1968 Є. (TOM) Mar.

First: *Spinulitheca billingsi* (Sysoiev, 1962) and *S. kuteinikovi* (Missarzhevsky, 1969), *A. sunnaginicus* Zone, Lower Cambrian, lower Tommotian Stage, Siberian Platform. *Ladatheca annae* (Sysoiev, 1959), *A. sunnaginicus* Zone, Lower Cambrian, lower Tommotian Stage, Siberian Platform, former USSR and M. Karatau, Mongolia (Sokolov and Zhuravleva, 1983).
Last: *Ladatheca sysoievi* (Meshkova, 1974), *D. regularis* Zone, Tommotian Stage, Siberian Platform, former USSR (Sokolov and Zhuravleva, 1983).

F. TCHURANITHECIDAE Sysoiev, 1968 Є. (TOM–ATB) Mar.

First: *Turcutheca crasseochila* (Sysoiev, 1962), *A. sunnaginicus* Zone, Lower Cambrian, lower Tommotian Stage, Siberian Platform, former USSR and M. Karatau, Mongolia. *Ovalitheca rasa* Sysoiev, 1968, *D. regularis* and *D. lenaicus–T. primigenius* Zones, *Tchuranitheca simplicis* and *T. sinuata* Sysoiev, 1968, *D. regularis* Zone, *Uniformitheca rhombiformis* (Sysoiev, 1963), *D. regularis* Zone, Lower Cambrian, Tommotian Stage, Siberian Platform, former USSR (Sokolov and Zhuravleva, 1983).
Last: *Uniformitheca jasmiri* (Sysoiev, 1959), *D. lenaicus–T. primigenius* and *R. zegebarti* Zones, Lower Cambrian, Tommotian and Atdabanian Stages, Siberian Platform, former USSR (Sokolov and Zhuravleva, 1983).

Order HYOLITHIDA Matthew, 1889

These are bilaterally symmetrical, pyramidal shells, with a conical embryonic chamber not separated from the rest of the shell. The operculum has one or two pairs of bilaterally symmetrical muscle scars.

F. AIMITIDAE Sysoiev, 1968 Є. (TOM–ATB) Mar.

First and Last: *Oxytus sagittalis* Sysoiev, 1968, *D. lenaicus–T. primegenius* Zone, upper Tommotian Stage, to *R. zegebarti* Zone, Atdabanian Stage, Siberian Platform, former USSR (Sokolov and Zhuravleva, 1983).

F. ALTAICORNIDAE Sysoiev, 1970 Є. (LEN) Mar.

First and Last: *Erraticornus debilis*, *E. cordeae*, *Insignicornus rectus*, *Nitoricornus pictus*, *N. subtilis* and *N. vegetus* Sysoiev, 1973, *B. asiaticus* Zone, Botomian Stage, Siberian Platform, former USSR (Sokolov and Zhuravleva, 1983).

F. AMYDAICORNIDAE Valkov, 1975 Є. (ATB–STD) Mar.

First: *Kuonamkicornus gracilis* Valkov, 1975, *Kuonamkicornus tenuis* Valkov, 1975, *Galicornus lenaicus* Zone, Lenian Stage, Siberian Platform, former USSR.
Last: *Amydaicornus fortis*, *A. mirabilis*, *A. modicus* and *Anabaricornus jucundus* Valkov, 1970, *Amydaicornus fortis* Zone, Amginian Stage, Middle Cambrian, Siberian Platform, former USSR. *Anabaricornus nanus* Valkov, 1970, *Linevitus mitralis* Zone, Amginian Stage, Middle Cambrian, Siberian Platform, former USSR (Valkov, 1975).

F. ANGUSTICORNIDAE Sysoiev, 1968 Є. (ATB–SOL) Mar.

First: *Angusticornus acutangulus* Sysoiev, 1968, *N. kokoulini*, *F. lermontovae*, and *B. micmacciformis–Erbiella* Zones, *Firmicornus obliteratus*, Sysoiev, 1968, *F. lermontovae* and *B. micmacciformis–Erbiella* Zones, Lower Cambrian, Atdabanian and Botomian Stages (Sokolov and Zhuravleva, 1983).
Next oldest: *A. reflexus* Sysoiev, 1968, *B. micmacciformis–Erbiella* Zone, Botomian Stage, Siberian Platform, former USSR (Sokolov and Zhuravleva, 1983).
Last: *Ketemecornus ermakovi* and *K. licitus*, Sysoiev, 1974, *B. ketemensis* Zone, Toyonian Stage, Siberian Platform, former USSR. *Ketemecornus viduus* (Sysoiev, 1968), *B. micmacciformis–Erbiella* and *B. ketemensis* Zones, Botomian and Toyonian Stages, Siberian Platform, former USSR (Sokolov and Zhuravleva, 1983).

F. CRESTJAHITIDAE Sysoiev, 1968 Є. (TOM–LEN) Mar.

First: *Jacuticornus tenuistrigatus* (Sysoiev, 1962), *D. regularis* Zone, *Burithes distortus* (Sysoiev, 1962), *D. regularis* and *D. lenaicus–T. primigenius* Zones, *Crestjahitus compressus* Sysoiev, 1968, *D. regularis* and *D. lenaicus–T. primigenius* Zones, Tommotian Stage, Siberian Platform, former USSR (Sokolov and Zhuravleva, 1983).
Last: *Borealicornus depsilis* Sysoiev, 1968, *B. micmacciformis* Zone, *Burithes elongatus* Missarzhevsky, 1969, *F. lermontovae*, *P. anabarus* and *B. micmacciformis–Erbiella* Zones, Atdabanian and Botomian Stages, Siberian Platform, former USSR (Sokolov and Zhuravleva, 1983).

F. DORSOJUGATIDAE Sysoiev, 1968 €. (TOM–LEN) Mar.

First: *Dorsojugatus sedecostatus* (Sysoiev, 1962), *D. regularis* and *D. lenaicus–T. primigenius* Zones, Tommotian Stage, Siberian Platform, former USSR (Sokolov and Zhuravleva, 1983).
Last: *Dorsojugatus multicostatus* (Sysoiev, 1968), *R. zegebarti* Zone, Atdabanian Stage, Siberian Platform, former USSR (Sokolov and Zhuravleva, 1983).

F. GALICORNIDAE Valkov, 1975 €. (ATB–LEN) Mar.

First: *Galicornus anabarus* Valkov, 1975, Atdabanian Stage, Siberian Platform, former USSR.
Last: *Galicornus lenaicus* Valkov, 1975, *B. micmacciformis–Erbiella* Zone, Botomian Stage, Siberian Platform, former USSR.

F. HYOLITHIDAE Nicholson, 1872 €. (TOM)–P. (KAZ) Mar.

First: *Korlithes bilabiatus* Missarzhevsky, 1969, *D. regularis* Zone (*L. tortuosa* Subzone), Tommotian Stage, north-western slope of Anabar Massif (Fomich River), middle reaches of River Lena, Chekurovka Village, Siberian Platform, former USSR.
Last: *Hyolithes kirkbyi* (as *Theca? kirkbyi* Howse, 1857), 'Magnesian Limestone Series', Durham, England, UK. Also Permian, upper Guadalupian, *Timorites* Zone (Downie *et al.*, 1967).

F. NELEGEROCORNIDAE Meshkova, 1974 €. (TOM) Mar.

First and Last: *Jacutolituus fusiformis* and *Microcornus simus* Missarzhevsky, 1974, *D. regularis* Zone, Tommotian Stage, Siberian Platform, former USSR (Sokolov and Zhuravleva, 1983).

F. NOTABILITIDAE Sysoiev, 1968 €. (TOM–LEN) Mar.

First: *Notabilitus costatus* and *Oblisicornus tetraconcavus* Sysoiev, 1968, *D. regularis* Zone, Tommotian Stage. *N. orientalis, N. simplex, Oblisicornus compositus* and *O. dupliconcavus* Sysoiev, 1968, *D. regularis* and *D. lenaicus–T. primigenius* Zone, Tommotian Stage, Siberian Platform, former USSR (Sokolov and Zhuravleva, 1983).
Last: *Doliutus brevis* Meshkova, 1974, *B. micmacciformis–Erbiella* Zone, Botomian Stage, Siberian Platform, former USSR (Sokolov and Zhuravleva, 1983).

F. PAUXILLITIDAE Marek, 1967 emended €. (CRF)–D. Mar.

First: *Neopauxilites zlatarskii* Malinky, 1989, Lower Cambrian of Newfoundland, Canada.
Last: *Recilites* sp. Marek, 1967, Devonian of south central Europe (Malinky, 1989).

F. PTERYGOTHECIDAE Sysoiev, 1958 O. (CND/DFD)–D. Mar.

First: *Virgulaxonaria elegans* Yin, 1937, Lower Ordovician, China (Fisher, 1962).
Last: *Pterygotheca barrandei* Novak, 1891, Devonian of Czechoslovakia.

F. SULCAVITIDAE Sysoiev, 1957 €. (TOM)–S. (LLY/WEN) Mar.

First: *Yacutolituus fusiformis* Missarzhevsky, 1974, and *Microcornus simus* Missarzhevsky, 1974, *Dokidocyathus regularis* Zone, Tommotian Stage, Siberian Platform, former USSR; *Linevitus distortus* Sysoiev, 1962, Tommotian Stage, Aldan River, Siberia, former USSR (Brasier, 1983).
Last: ?*Linevitus* sp., Silurian, Norway, Sweden, former USSR (Fisher, 1962).
Intervening: ATB, O.

F. TRAPEZOVITIDAE Valkov, 1975, €. (TOM–LEN) Mar.

First: *Tuoidachithes figuratus* Missarzhevsky, 1969, *T. costulatus* Missarzhevsky, 1969, *D. regularis* Zone, Tommotian Stage, Siberian Platform, former USSR (Valkov, 1975).
Last: *Trapezovitus latus* Valkov, 1975, Botomian Stage, Siberian Platform; *T. sinscus* Sysoiev, 1958, Atdabanian and Botomian Stages, Siberian Platform, former USSR (Sokolov and Zhuravleva, 1983).

Order GLOBORILIDA Sysoiev, 1957

These are bilaterally symmetrical, curved shells, with curvature increasing towards the apex, and a globular embryonic chamber. The cross-section of the main shell is circular to subtriangular. A very low, conical operculum is subcircular to subquadrate in outline. No external ornamentation is visible.

F. GLOBORILIDAE Sysoiev, 1958 V. (N–DA)–€. (STD) Mar.

First: *Wyattia reedensis* Taylor, 1966, Nemakit–Daldyn Horizon, Reed Dolomite, White–Inyo Mountain Area, Inyo County, California, USA.
Last: *Globorilus globiger* (Saito, 1936) Korea. *Globeringa ?mantoensis* Walcott, ?1905, lower Fouchouan or Hsuchuan Stage at Yenchuang, Shantung, Hwangho Basin, South Korea (Kobayashi, 1956).
Comment: The assignment of *Wyattia* to the globorilids by Taylor (1966) is questionable.

Order CAMEROTHECIDA Sysoiev, 1957

These shells are bilaterally symmetrical, with an oval cross-section. The embryonic stage is tubular and parallel walled. Small chambers are separated by imperforate partitions in this portion. The angle of divergence increases in adult stages, although the sides become almost parallel near the aperture. An operculum is unknown.

F. CAMEROTHECIDAE Sysoiev, 1958 €. (LEN) Mar.

First and Last: *Camerotheca gracilis* Matthew, 1885, North America (Fisher, 1962).

F. DIPLOTHECIDAE Sysoiev, 1958 €. (LEN) Mar.

First and Last: *Diplotheca acadia* Matthew, 1885, Canada (Fisher, 1962).

Order TOXEUMORPHORIDA Shimansky, 1962

F. TOXEUMORPHORIDAE ?Shimansky, 1962 C. (SPK)–P. (TAT) Mar.

First: *Toxeumorphora dombarense* Shimansky, 1963, Namurian, southern Urals, former USSR (Shimansky and Barskov, 1970). Also, *Toxeumorphora* sp., TOU/VIS Stages, Kozhim River, northern Urals – an older specimen, but taxonomic assignment is in doubt (Shimansky and Barskov, 1970).
Last: *Macrotheca wynnei* Waagen 1880, U. Productus Limestone, Salt Range, Pakistan; *Macrotheca almgreeni* Peel and Yochelson, 1984, Foldvik Greek Formation (Upper Permian) at Kap Stosch, northern East Greenland (Peel and Yochelson, 1984).

REFERENCES

Barrande, J. (1867) *Système Silurien du Centre de la Bohême. Ière Partie: Recherches Paléontologiques, 3, Classe des Mollusques. Ordre des Ptéropodes*. Prague and Paris, 179 pp.

Bouček, B. (1964) *The Tentaculites of Bohemia*. Czechoslovak Academy of Science, Prague, 215 pp.

Brasier, M. D. (1983) Microfossils and small shelly fossils from the Lower Cambrian *Hyolithes* Limestone at Nuneaton, English Midlands. *Geological Magazine*, **121**, 229–53.

Brasier, M. D. (1989) Towards a biostratigraphy of the earliest skeletal biotas, in *The Precambrian–Cambrian Boundary* (eds J. W. cowie and M.D. Brasier), Clarendon Press, Oxford, pp. 117–65.

Clarke, J. M. (1904) Naples fauna in western New York. *New York State Museum Memoirs*, **6**, 1–454.

Cowie, J. W., Rushton, A. W. A. and Stubblefield, C. J. (1972) A correlation of Cambrian rocks in the British Isles. *Special Report of the Geological Society of London*, **2**, 1–42.

Downie, C., Fisher, D. W., Goldring, R. *et al.* (1967) Miscellania, in *The Fossil Record* (eds W. B. Harland, C. H. Holland, M. R. House *et al.*), Geological Society of London, London, pp. 613–26.

Farsan, N. M. (1983) Tentaculites du Frasnian inférieur de Ferques (Boulonnais, Nord de la France). *Palaeontographica, Abteilung A*, **182**, 26–43.

Fisher, D. W. (1962) Small conoidal shells of uncertain affinities, in *Treatise on Invertebrate Paleontology. Part W* (ed. C. Teichert), Geological Society of America and University of Kansas Press, Boulder, Colorado and Lawrence, Kansas, pp. W98–143.

Fisher, D. W. and Young, R. S. (1955) The oldest known tentaculitid from the chepultec limestone (Canadian) of Virginia. *Journal of Paleontology*, **29**, 871–5.

Kobayashi, T. (1956) The Cambro-Ordovician formations and faunas of South Korea. Part X. Stratigraphy of the Chosen Group in Korea and South Manchuria. Section C. The Cambrian of Eastern Asia and other parts of the continent. *Journal of the Faculty of Science, University of Tokyo, Section II*, **16**, 209–301.

Larsson, K. (1979) Silurian tentaculitids from Gotland and Scania. *Fossils and Strata*, **11**, 1–180.

Ljashenko, G. P. (1959) [*Devonian Coniconchia in the Central District of the Russian Platform.*] VNIGNI, Moscow, 220 pp. [in Russian].

Malinky, J. M. (1989) New early Paleozoic Hyolithida and Orthothecida (Hyolitha) from North America. *Journal of Paleontology*, **63**, 302–19.

Missarzhevsky, V. V. (1969) [Description of hyolithids, gastropods, hyolithelminthes, camenides and forms of an obscure taxonomic position], in *The Tommotian Stage and the Cambrian Lower Boundary Problem* (eds A. V. Pieve, K. I. Kuznetsova, V. V. Menner *et al.*), Trudy Geologecheskii Institut, Nauka, Moscow **206** [in Russian. English translation, US Department of the Interior, 1981], pp. 127–205.

Missarzhevsky, V. V. (1974) Novye dannye o drevneishikh okamenelostyakh rannego kembriya sibirskoi platformi, in *Biostratigrafiya i Paleontologiya Nizhnego Kembriya Evropy i Severnoy Asii* (eds I. T. Zhuravleva and A. Yu. Rozanov), Nauka, Moscow, pp. 179–89.

Mistiaen, B. and Poncet, J. (1983) Stromatolithes, serpulides et *Trypanopora* (vers?) associes dans de petits biohermes Givetiens du Boulonnais (France). *Palaeogeography, Palaeoclimatology, Palaeoecology*, **41**, 125–38.

Novak, O. (1886) Zur Kenntnis der Fauna der Etage F-f1 in der Paläozoïschen Schichtengruppe Böhmens. *Königliche böhmische Gesellschaft Wissenschaften, Sitzungsberichten*, **1886**, 660–85.

Novak, O. (1891) Revision der Palaeozoischen Hyolithiden Böhmens. *Königliche böhmische Gesellschaft Wissenschaften, Abhandlungen*, **7**, 1–18.

Peel, J. S. and Yochelson, E. L. (1984) Permian Toxeumorphorida from Greenland: an appraisal of the molluscan class. *Lethaia*, **17**, 211–22.

Richter, R. (1854) Thuringische Tentacoliten. *Deutsche geologische Gesellschaft, Zeitschrift*, **6**, 275–90.

Sepkoski, J. J. Jr (1982) A compendium of fossil marine families. *Milwaukee Public Museum Contributions in Biology and Geology*, **51**, 1–212.

Shimansky, V. N. and Barskov, I. S. (1970) New data on the Order Toxeumorphorida. *Paleontological Journal*, **1970**, 430–4.

Sokolov, B. S. and Zhuravleva, I. T. (1983) Lower Cambrian Stage subdivision of the Siberia. Atlas of fossils. *Transactions of the Institute of Geology and Geophysics*, **558**, 1–216 [in Russian].

Swartz, C. K. (1913) *Correlation of Silurian Formations in Maryland with those of Other Areas*. Maryland Geological Survey, 749 pp.

Sysoiev, V. A. (1972) *Biostratigrafiya i Khiolity Ortotetsimorfy Nizhnego Kembriya Sibirskoi Platformy*. Nauka, Moscow, 152 pp. [in Russian].

Taylor, M. E. (1966) Precambrian mollusc-like fossils from Inyo County, California. *Science*, **153**, 198–201.

Tunnicliff, S. P. (1983) The oldest known nowakiid (Tentaculitoidea). *Palaeontology*, **26**, 851–4.

Valkov, A. K. (1975) *Biostratigrafiya i khiolity Kembriya severovostoka Sibirskoi Platformi*. Nauka, Moscow, 137 pp. [in Russian].

Weedon, M. J. (1990) Shell structure and affinity of vermiform 'gastropods'. *Lethaia*, **23**, 297–309.

Weedon, M. J. (1991) Microstructure and affinity of the enigmatic Devonian tubular fossil *Trypanopora*. *Lethaia*, **24**, 227–34.

15

ANNELIDA

M. A. Wills

The soft-bodied annelids would not be expected to leave an extensive or reliable fossil record. Very few specimens have been preserved intact, most remains being chaetae, scolecodonts, tubes, burrows and castings. Some secreted tubes are very characteristic and can be used reliably to identify the organism that produced them. The calcareous tubes of serpulids, for example, are unmistakable, and known from Palaeozoic and younger rocks. Most burrows and casts, however, are of little value in this respect.

Scolecodonts are thought to be chitinized jaw apparatuses, similar to those found in a number of extant errant polychaetes. They are common in sedimentary rocks of all ages from the Ordovician to the Recent. Their classification is confused, since many eunicids, to which at least some of the fossils have been referred, have five or more jaw pairs. Only in a limited number of cases have complete jaw sets been found, the relationships of dissociated scolecodonts being inferred from these (Clark, 1969). Much elegant work has been conducted by Kozłowski (1956), Kielan-Jaworowska (1962, 1966), and Szaniawski (1968, 1974), on Ordovician and Silurian erratic material from eastern Europe.

Acknowledgement – This chapter was researched during the tenure of a University of Bristol Postgraduate Scholarship.

Phylum ANNELIDA Lamarck, 1809

Class POLYCHAETA Grube, 1850 (see Fig. 15.1)

Polychaetes can be assigned to about 87 families without difficulty, but apart from grouping some of them together in what may loosely be regarded as orders, it has not proved possible to devise an entirely satisfactory hierarchical classification (Clarke, 1978). The classification adopted here is modified from Clark (1969), with additions of extinct families from Kielan-Jaworowska (1966). Approximately 44 extant families with no known fossil records are not listed. (*Indicates possible relationship to an unlisted living family.)

Order AMPHINOMORPHA

F. AMPHINOMIDAE Savigny, 1818 C. (MOS)–Rec. Mar

First: *Raphidiophorus hystix* Thompson, 1979, Pennsylvanian Essex fauna, Francis Creek Shale, Mazon Creek, northern Illinois, USA. **Extant**

Order EUNICEMORPHA

F. ARCHAEOPRIONIDAE Mierzejewski and Mierzejewska, 1975 O. (CRD) Mar.

First and Last: *Archaeoprion quadricristatus* Mierzejewski and Mierzejewska, 1975, erratic calcareous pebble, *Amorphognathus superbus* Zone, near Orzechowo, Koszalin Province, Poland.

F. ATRAKTOPRIONIDAE Kielan-Jaworowska, 1966 O. (LLO)–Tr. (ANS) Mar.

First: *Atraktoprion cornutus* Kielan-Jaworowska, 1962, *A. mirabilis*, *A. robustus*, and two other species of *Atraktoprion*, Kielan-Jaworowska, 1966, all from erratic limestone boulders, Mochty Province of Warsaw, Poland. *Xanthoprion erraticus* Kielan-Jaworowska, 1966, erratic limestone boulder, Zakrocym Province of Warsaw, Poland.
Last: *Atraktoprion anatinus* Zawidzka, 1975, lower Muschelkalk, southern Poland.
Intervening: FRS, ZEC.

F. DORVILLEIDAE Chamberlin, 1919 J. (CLV)–Rec. Mar.

First: *Ophryotrocha lukowensis* Szaniawski, 1974, Poland. **Extant**

F. KALLOPRIONIDAE Kielan-Jaworowska, 1966 O. (LLO)–P. (UFI/WOR) Mar.

First: *Kalloprion ovalis* Kielan-Jaworowska, 1962, Kukruse or Idavere Stage of the Estonian sequence, *K. triangularis*, plus two other species of Kalloprion, Kielan-Jaworowska, 1966, Keila Stage of the Estonian seguence, all from the Mochty Province of Warsaw. *Leptoprion artus* Kielan-Jaworowska, 1966, Mochty Province, *L. polonicus* Kielan-Jaworowska, 1966, Kukruse Stage of the Estonian sequence, Wyszogród–Zakrocym Province of Warsaw, Poland.
Last: *Eunicites sp. indet.* Tasch and Stude, 1965, lower Zechstein, Poland (Kielan-Jaworowska, 1966).

The Fossil Record 2. Edited by M. J. Benton. Published in 1993 by Chapman & Hall, London. ISBN 0 412 39380 8

Fig. 15.1

Comment: Kielan-Jaworowska (1966) considers *Eunicites* to be a bona fide kalloprionid.

F. KIELANOPRIONIDAE Szaniawski, 1968
D. (FRS)–Tr. (LAD) Mar.

First: *Kielanoprion elleri* Szaniawski and Wrona, 1973, Frasnian Limestone, Opole Lubelskie Borehole, right bank of Vistula River, south of Lublin, south-east Poland.
Last: *Kielanoprion longidentatus* Zawidzka, 1975, upper Muschelkalk, Wierchlesie, southern Poland. *K. oertlii* Zawidzka, 1975, lower Muschelkalk, Silesia-Cracow and Fore-Sudetic monoclines, southern Poland.
Intervening: D.

F. LUMBRINEREIDAE Malmgren, 1867
C. (SPK)–Rec. Mar.

First: *Phiops aciculorum* Schram, 1979, uppermost Mississippian, Bear Gulch Limestone, Central Montana, USA.

F. MOCHTYELLIDAE Kielan-Jaworowska, 1966
O. (LLO/CRD)–P. (ZEC) Mar.

First: *Mochtyella cristata* Kielan-Jaworowska, 1966, Kukruse or Idavere Stage of Estonian sequence, Mochty Province of Warsaw, Poland.
Last: *Oxyprion compressus* Szaniawski, 1968, Zechstein, second cyclothem, main dolomite horizon, Pomerania, Poland.

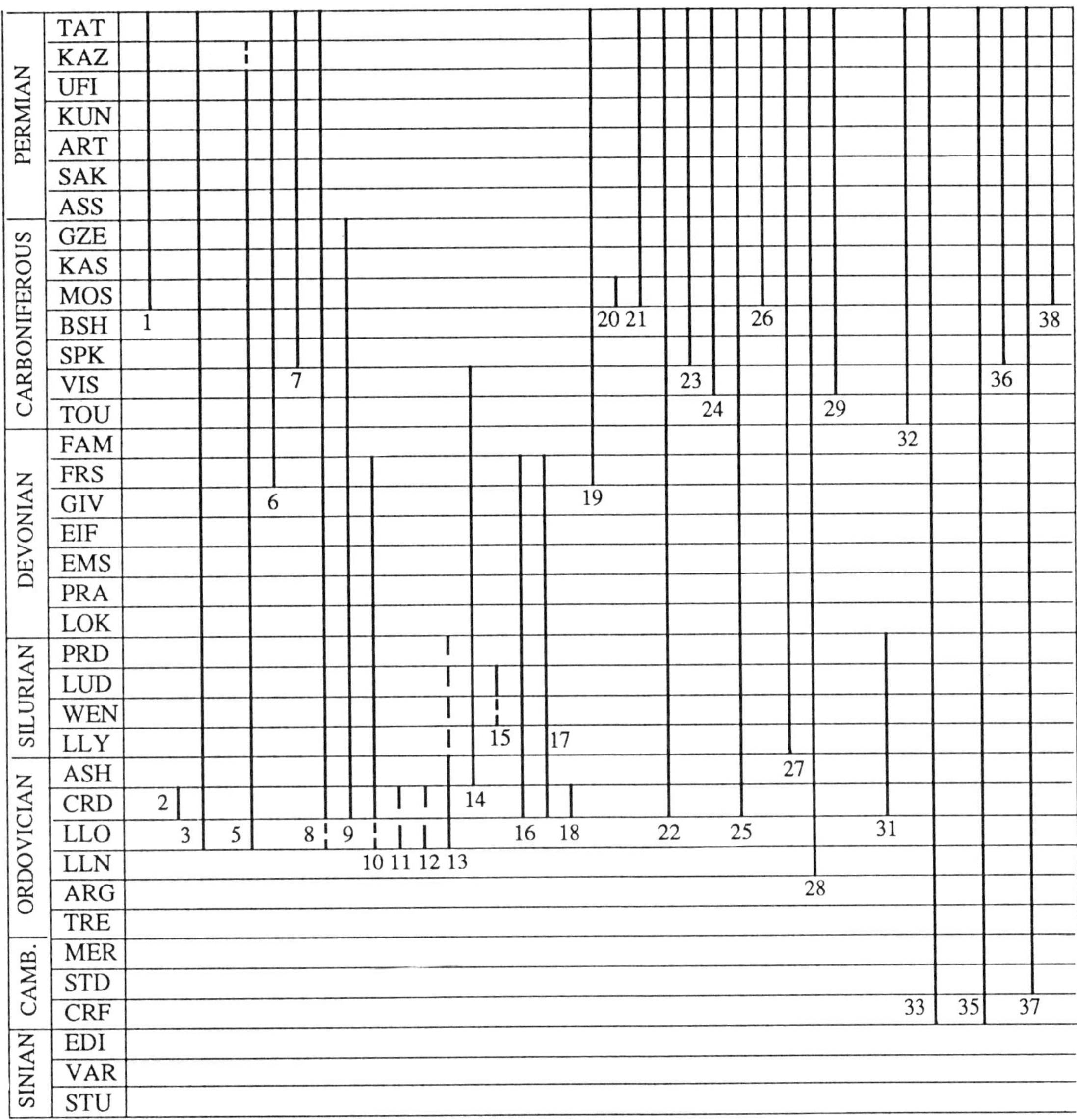

Fig. 15.1

F. PAULINITIDAE Lange, 1947 O. (CRD)–C. Mar.

First: *Nereidavus angulostus* Eller, 1945, Trenton Series, Ontario, Canada. *Elmhurstia nododentata* Potter, Middle Ordovician (Liberty Formation), Cincinnati, USA (Kielan-Jaworowska, 1966).

Last: *Paulinites* spp., Silurian through Carboniferous of North America, South America and Europe (Kielan-Jaworowska, 1966).

Next youngest: *Paulinites paranaensis* Lange, 1947, Ponta Grossa Formation, Middle Devonian, Brazil (Howell, 1962; Kielan-Jaworowska, 1966).

Comment: Both the genus *Nereidavus* and the species *Elmhurstia nododentata* are placed within the Paulinitidae by Kielan-Jaworowska (1966). She also indicates that the family may possibly be recorded as recently as the Permian.

F. POLYCHAETASPIDAE Kielan-Jaworowska, 1966 O. (??LLO)–D. (FRS) Mar.

First: *Polychaetaspis wyszogrodensis* Kozłowski, 1956, ?Middle Ordovician erratic boulder, Wyszgrod Province of Warsaw; *P. tuberculatus* Kielan-Jaworowska, 1966, *P. gadomskae* Kielan-Jaworowska, 1966, and *P.* sp. a, ?Middle Ordovician erratic boulder, Mochty Province of Warsaw; *P. warkae* Kozłowski, 1956, Middle Ordovican erratic boulder, ?Uhaku Stage of Estonian sequence, Warka Province of Warsaw; *P. varsoviensis* Kielan-Jaworowska, 1966, Middle Ordovician erratic boulder, ?Kukruse or Idavere Stage of Estonian sequence, Mochty Province of Warsaw; '*Polychaetaspis*' *incisus*, Kielan-Jaworowska, 1966, ?Middle Ordovician erratic boulder, Mochty Province of Warsaw; *Kozlowskiprion brevialatus*, ?Middle Ordovician erratic boulder, Mochty Province of Warsaw, Poland (all in Kielan-Jaworowska, 1966).

Last: *Polychaetaspis hindei* and *Polychaetaspis* sp. a. Szaniawski and Wrona, 1973, FRS, Opole Lubelskie Borehole, SE Poland.

Intervening: Throughout S.

F. POLYCHAETURIDAE Kozłowski, 1956 O. (LLO/CRD) Mar.

First and Last: *Polychaetura gracilis* Kozłowski, 1956

and *Polychaetura* sp. a, Kielan-Jaworowska, 1966, ?Kukruse or Idavere Stage of the Estonian sequence, Mochty Province of Warsaw, Poland (Kielan-Jaworowska, 1966).
Comment: Specimens assigned to *Polychaetura gracilis* are thought to belong to three or four separate species.

F. RAMPHOPRIONIDAE Kielan-Jaworowska, 1966 O. (LLO/CRD–CRD) Mar.

First: *Ramphoprion elongatus, R. urbaneki* and *Ramphoprion* sp. b Kielan-Jaworowska, 1966, ?Middle Ordovician; *Ramphoprion* sp. a, *R.* sp. c, and *R.* sp. d Kielan-Jaworowska, 1966, ?Kukruse Stage of the Estonian sequence, Mochty Province of Warsaw, Poland.
Last: *Ramphoprion* sp. Gries, 1944, basal Liberty Formation, Ohio, USA (Kielan-Jaworowska, 1966).

F. RHYTIPRIONIDAE Kielan-Jaworowska, 1966 O. (LLO)–S. Mar.

First: *Rhytioprion magnus* Kielan-Jaworowska, 1966, Ordovician, Mochty Province of Warsaw, Poland.
Last: *Rhytioprion* sp. uncertain, Kielan-Jaworowska, 1966, ?Silurian of the Baltic region, Poland.

*F. SKALENOPRIONIDAE Kielan-Jaworowska, 1962 O. (ASH)–C. (VIS) Mar.

First: *Skalenoprion alatus* Kielan-Jaworowska, 1962, and *Skalenoprion* spp. a, b and c Kielan-Jaworowska, 1966, Ashgillian erratic boulder, Baltic region, Poland (Kielan-Jaworowska, 1966).
Last: *Arabellites hamiltonensis* Stauffer, 1939, lower Mississippian, Lake Erie district, USA (Kielan-Jaworowska, 1962).
Intervening: D.
Comment: The family may possibly occur as recently as the Permian (Kielan-Jaworowska, 1966).

F. SYMMETROPRIONIDAE Kielan-Jaworowska, 1966 S. (?WEN–GOR) Mar.

First: *Symetroprion reduplicatus* Kielan-Jaworowska, 1966, ?Wenlockian erratic boulder, Mochty Province of Warsaw, Poland; *Symmetroprion* sp. a Kielan-Jaworowska, 1966, ?Wenlockian erratic boulder, Debina, near Ustka, Baltic coast, Poland.
Last: *Symetroprion reduplicatus* Kielan-Jaworowska, 1966, lower Ludlow erratic boulder, Mochty Province of Warsaw, Poland.

F. TETRAPRIONIDAE Kielan-Jaworowska, 1966 O. (CRD)–D. (FRS) Mar.

First: *Tetraprion pozaryskae* Kielan-Jaworowska, 1966, erratic boulder, Mochty Province of Warsaw and Baltic region, Poland.
Last: *Tetraprion* sp. Szaniawski and Wrona, 1973, Frasnian strata from Opole Lubelskie borehole, SE Poland (Szaniawski and Wrona, 1973).
Intervening: *Tetraprion* sp. Szaniawksi, 1970, Silurian erratic boulders throughout Poland (Szaniawski and Wrona, 1973).

F. XIANIOPRIONIDAE Kielan-Jaworowska, 1962 O. (LLO/CRD)–D. (FRS) Mar.

First: *Xianioprion borealis* Kielan-Jaworowska, 1962, erratic boulder, Kukruse or Idavere Stage of Estonian sequence, Mochty Province of Warsaw, Poland.
Last: *Xianioprion walliseri* and *Processoprion longiprocessus* Szaniawski and Wrona, 1973, upper FRS, Opole Lubelskie Borehole, SE Poland.

F. UNNAMED O. (CRD) Mar.

First and Last: *Trentonia shegiriana* Pickerill and Forbes, 1978, Trenton Limestone of Quebec City area, Canada.

Order PHYLLODOCEMORPHA

F. APHRODITIDAE Savigny, 1818 D. (FRS)–Rec. Mar.

First: *Protonympha salicifolia* Clarke, 1903, Upper Devonian, Portage Group, New York, USA. **Extant**

F. FOSSUNDECIMIDAE Thompson, 1979 C. (MOS) Mar.

First and Last: *Fossundecima konecniorum* Thompson, 1979, Pennsylvanian Essex fauna, Francis Creek Shale, Mazon Creek, northern Illinois, USA.

F. HESIONIDAE Malmgren, 1867 C. (MOS)–Rec. Mar.

First: *Rutellifrons wolfforum* Thompson, 1979, Pennsylvanian Essex fauna, Francis Creek Shale, Mazon Creek, northern Illinois, USA. **Extant**

F. GLYCERIDAE Grube, 1850 O. (BAL)–Rec. Mar.

First: *Glycerites sulcatus* Hinde, 1879, Upper Ordovician, Cincinnatian, North America (Howell, 1962). **Extant**

F. GONIADIDAE Kinberg, 1866b C. (SPK)–Rec. Mar.

First: *Carbosesostris megaliphagon* Schram, 1979, uppermost Mississippian, Bear Gulch Limestone, Central Montana, USA. Genus and species unknown, Mazon Creek, northern Illinois, USA. **Extant**

F. NEPHTYIDAE Grube, 1850 C. (VIS)–Rec. Mar.

First: *Astreptoscolex anasillosus* Thompson, 1979, uppermost Mississippian, Bear Gulch Limestone, central Montana, USA (Schram, 1979). **Extant**

F. NEREIDAE Savigny, 1820 O. (CRD)–Rec. Mar.

First: *Pronereites primus* Stauffer, 1933, Middle Ordovician, upper Glenwood Beds, Washington Avenue Bridge, Minnesota, USA; *Dinoscolites mirabilis* Stauffer, 1933, Middle Ordovician, basal Platteville Limestone, Johnson St. Quarry, Minnesota, USA; *Ungulites tridentatus* Stauffer, 1933, Middle Ordovician, basal Platteville Limestone, Johnson St. Quarry, Minnesota; *Ungulites tridentatus,* Stauffer, 1933, basal Decorah Shale, Guttenberg and Fillmore, Iowa; *Paleonereites cervicornis* Stauffer, 1933, Middle Ordovician, lower Decorah Shale, University of Minnesota; *Ungulites bicuspidatus* Stauffer, 1933, Middle Ordovician, Decorah Shale, Minnesota, USA. **Extant**

F. PALMYRIDAE Kinberg, 1858 C. (MOS)–Rec. Mar.

First: Unnamed species, Pennsylvanian Essex Fauna, Francis Creek Shale, Mazon Creek, northern Illinois, USA (Thompson, 1979). **Extant**

F. PHYLLODOCIDAE Grube, 1850 and Williams, 1851 S.–Rec. Mar.

First: *Palaeochaeta devonica* Clarke, 1903, North America and Europe (Czechoslovakia), (Howell, 1962). **Extant**
Intervening: MOS.

F. SIGALIONIDAE Kinberg, 1855 O. (CRD)–Rec. Mar.

First: *Thalenessites lobatus* Stauffer, 1933, Middle Ordovician, Decorah Shale, Minnesota, USA (Howell, 1962). **Extant**

F. TOMOPTERIDAE Grube, 1848 C. (VIS)–Rec. Mar.

First: *Eotomopteris aldridgei* Briggs and Clarkson, 1987, Lower Carboniferous, Granton, Edinburgh, Scotland, UK. **Extant**
Comments: Glaessner (1976) somewhat controversially places *Spriggina floundersi* within the Tomopteridae, which would extend the record of the family back into the Tommotian (see 'Sprigginidae' in Chapter 28 on 'Problematica', this volume, section 1).

Order SPIOMORPHA Carus, 1863

F. CIRRATULIDAE Carus, 1863 T. (Mio).–Rec. Mar.

First: *Dodecaceria concharum* Örsted, 1843, cosmopolitan (Howell, 1962). **Extant**

F. KEILORITIDAE Allan, 1927 O. (CRD)–S. Mar.

First: *Keilorites coriacea* (Phillips, 1848), Caradoc beds, Hill Side Formation, Abberly District, England, UK (Phillips and Salter, 1848).
Last: *Keilorites squamosa* (Phillips, 1848), Silurian (Yeringian) strata, junction of the Woori Yallock and Yarra, Victoria, Australia (Allan, 1910); *Keilorites crassituba* (Chapman, 1910), Silurian (Melbournian) strata, Yarra Improvement Works, South Yarra, Victoria, Australia (all *in* Allan, 1927). **Extant**

F. SABELLARIIDAE Johnston, 1865 C.–Rec. Mar.

First: *Sabellaria* Lamarck, 1818, cosmopolitan genus (Howell, 1962). **Extant**
Comments: Produces tubes of sand which typically occur in clusters on the ocean floor.

F. SPIONIDAE Grube, 1850, Sars, 1861 Є. (TOM)–Rec. Mar.

First: ?*Scolelepsis* sp. Glaessner, 1958, Emu Bay Shale, Lower Cambrian of Kangaroo Island, South Australia (Glaessner, 1976). **Extant**

Order DRILOMORPHA

F. ARENICOLIDAE Audouin and Edwards, 1833 Tr. (RHT)–Rec. Mar.

First: *Archarenicola rhaetica* Horwood, 1912, Rhaetic black shales, Glen Parva, Leicestershire, England, UK. **Extant**

F. OPHELIIDAE Malmgren, 1867 Є. (LEN)–Rec. Mar.

First: Genus and sp. uncertain, Emu Bay Shale, Lower Cambrian of Kangaroo Island, South Australia (Glaessner, 1976). **Extant**

Order TEREBELLOMORPHA

F. PECTINARIIDAE Quatrefages, 1865 = ?AMPHICTENIDAE Malmgren, 1867 C. (SPK)–Rec. Mar.

First: *Crininicaminus haneyensis* Ettensohn, 1981, Newman Limestone, Haney Member, Chesterian, north of Stanton, east-central Kentucky, USA. **Extant**

F. TEREBELLIDAE Grube, 1850 Є. (STD)–Rec. Mar.

First: *Terebellites franklini* Howell, 1943, Clouds Rapids Formation, Middle Cambrian, Newfoundland, Canada (Howell, 1962). **Extant**

Order FLABELLIGERIMORPHA

F. FLABELLIGERIDAE Saint-Joseph, 1894 C. (MOS)–Rec. Mar.

First: Undescribed sp., Pennsylvanian Essex Fauna, Francis Creek Shale, Mazon Creek, northern Illinois, USA (Thompson, 1979). **Extant**

Order SERPULIMORPHA

F. SABELLIDAE Malmgren, 1867 T. (DAN)–Rec. Mar. (see Fig. 15.2)

First: *Spirographites ellipticus* ?Astre, 1937, Garumnian of Saldés (and throughout Europe), (Howell, 1962). **Extant**

F. SERPULIDAE Burmeister, 1837 O. (LLN)–Rec. most Mar., some FW

First: *Serpularia crenata* Münster, 1840, Orthoceratite Limestone, southern Germany (Howell, 1962). **Extant**
Intervening: C.

Order INCERTAE SEDIS

F. BURGESSOCHAETIDAE Conway-Morris, 1979 Є. (MEN) Mar.

First and Last: *Burgessochaeta setigera* (Walcott, 1911), Phyllopod bed, Walcott Quarry, Middle Cambrian Burgess Shale, British Columbia, Canada (Conway Morris, 1979).

F. CANADIIDAE Walcott, 1911 Є. (MEN) Mar.

First and Last: *Canadia spinosa* Walcott, 1911, Phyllopod Bed, Walcott Quarry, Middle Cambrian Burgess Shale, British Columbia, Canada (Conway Morris, 1979).

F. INSILICORYPHIDAE Conway Morris 1979 Є. (MEN) Mar.

First and Last: *Insilicorypha psygma* Conway Morris, 1979, Middle Cambrian Burgess Shale, British Columbia, Canada (Conway Morris, 1979).

F. PERONOCHAETIDAE Conway Morris, 1979 Є. (MEN) Mar.

First and Last: *Peronochaeta dubia* (Walcott, 1911), Middle Cambrian Burgess Shale, British Columbia, Canada (Conway Morris, 1979).

F. STEPHENOSCOLECIDAE Conway Morris, 1979 Є. (MEN) Mar.

First and Last: *Stephenoscolex argutus* Conway Morris, 1979, Middle Cambrian Burgess Shale, British Columbia, Canada (Conway Morris, 1979).

F. UNNAMED O. (ASH) Mar.

First and Last: *Eopolychaetus albaniensis* Ruedmann, 1901,

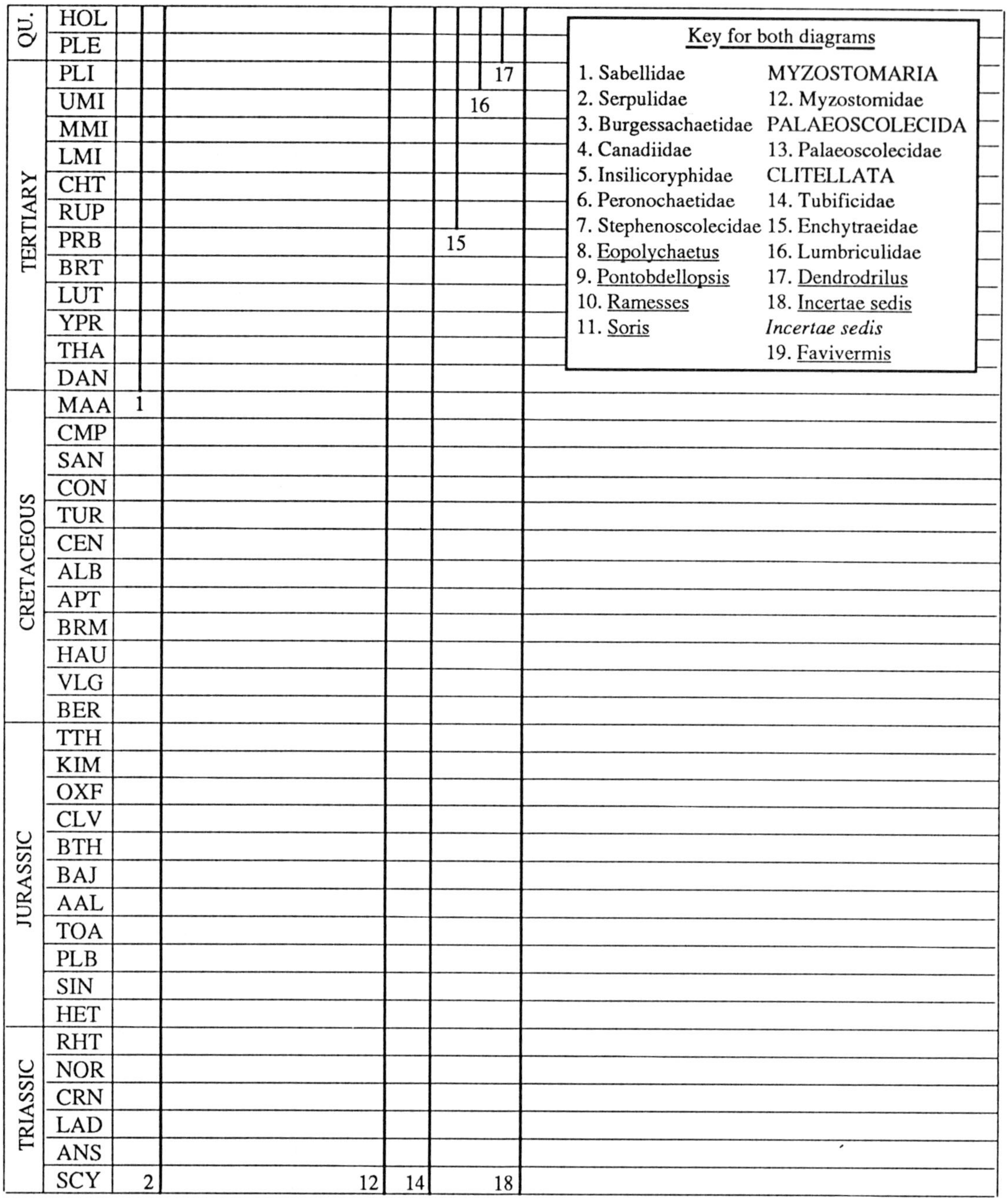

QU.: HOL, PLE

TERTIARY: PLI, UMI, MMI, LMI, CHT, RUP, PRB, BRT, LUT, YPR, THA, DAN

CRETACEOUS: MAA, CMP, SAN, CON, TUR, CEN, ALB, APT, BRM, HAU, VLG, BER

JURASSIC: TTH, KIM, OXF, CLV, BTH, BAJ, AAL, TOA, PLB, SIN, HET

TRIASSIC: RHT, NOR, CRN, LAD, ANS, SCY

1 17 16 15 2 12 14 18

Key for both diagrams

1. Sabellidae	MYZOSTOMARIA
2. Serpulidae	12. Myzostomidae
3. Burgessachaetidae	PALAEOSCOLECIDA
4. Canadiidae	13. Palaeoscolecidae
5. Insilicoryphidae	CLITELLATA
6. Peronochaetidae	14. Tubificidae
7. Stephenoscolecidae	15. Enchytraeidae
8. Eopolychaetus	16. Lumbriculidae
9. Pontobdellopsis	17. Dendrodrilus
10. Ramesses	18. Incertae sedis
11. Soris	*Incertae sedis*
	19. Favivermis

Fig. 15.2

Hudson River Beds near Albany, New York, USA (Fisher, 1962).

F. UNNAMED O. (ASH) Mar.

First and Last: *Pontobdellopsis cometa* Ruedmann, 1901, Hudson River Beds near Albany, New York, USA (Fisher, 1962).

F. UNNAMED C. (SPK) Mar.

First and Last: *Ramesses magnus* Schram, 1979, uppermost Mississippian, Bear Gulch Limestone, central Montana, USA.

F. UNNAMED C. (SPK) Mar.

First and Last: *Soris labiosus* Schram, 1979, uppermost Mississippian, Bear Gulch Limestone, central Montana, USA.

Class MYZOSTOMARIA

The Myzostomaria may be closely related to the polychaetes. They are parasitic on echinoderms, particularly crinoids, often forming gall-like cysts. There are seven extant families, of which only one has a fossil record.

F. MYZOSTOMIDAE Graff, 1884 O.–Rec. Mar.

First: *Myzostomites clarkei* Clarke, 1921, a cosmopolitan species from the Ordovician. **Extant**

Class PALAEOSCOLECIDA Conway Morris and Robison, 1986

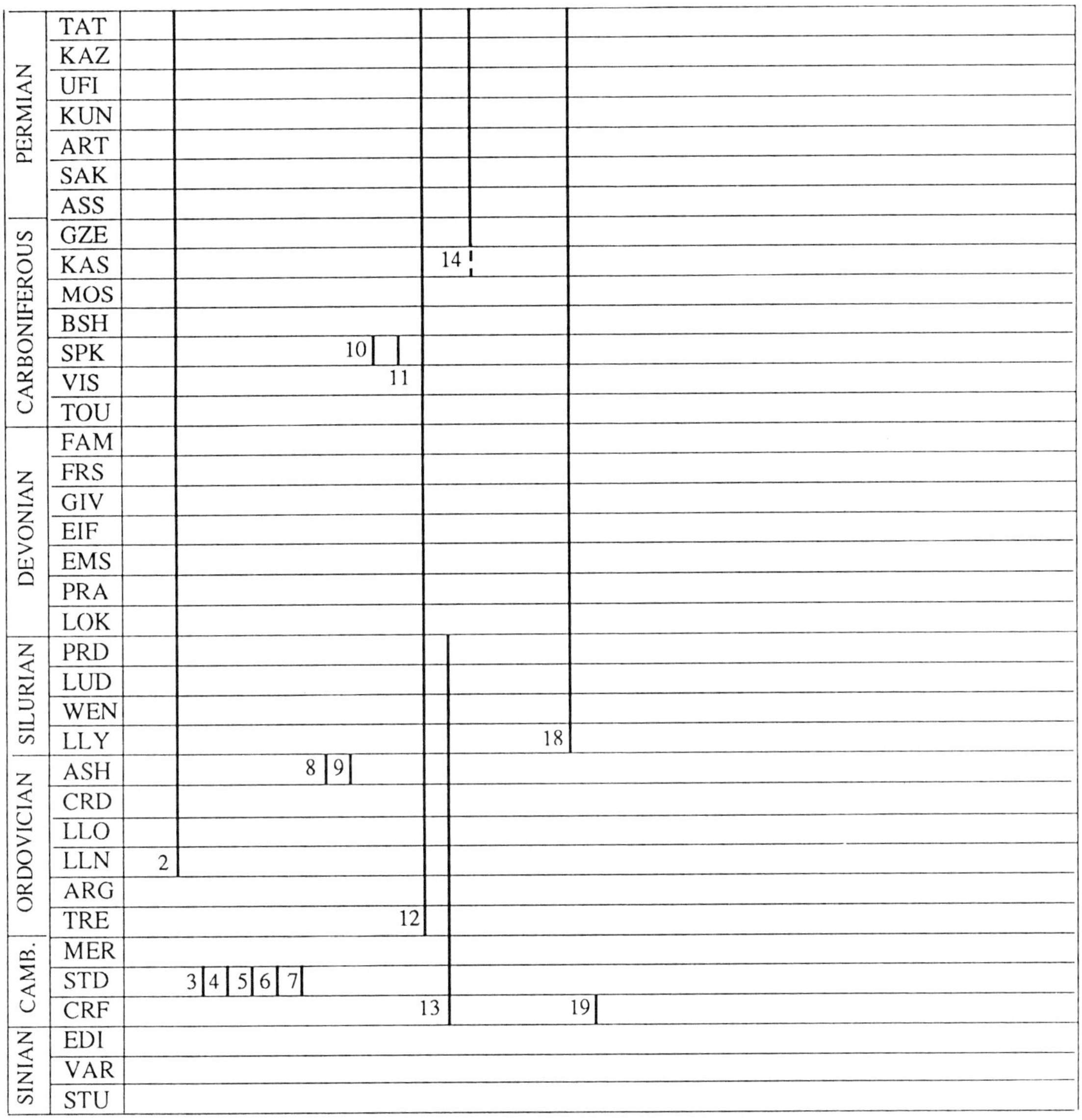

Fig. 15.2

F. PALAEOSCOLECIDAE Whittard, 1953
Є. (ATB)–S. (u) Mar.

First: *Palaeoscoleca sinensis* Hou and Sun, 1988, Chiungchussu Formation, Yunnan Province, China.
Last: *Palaeoscolex ruedmanni* Roy and Croneis, 1931, upper Lockport Shale (Silurian), Illinois, USA (Conway Morris, 1977).

Class CLITELLATA

Subclass OLIGOCHAETA Grube, 1850

Fossil oligochaetes are rare and disputed. *Protoscolex*, sometimes referred to this subclass, is known from marine deposits from the Upper Ordovician to the Upper Silurian. *P. batheri* from the Silurian of New York (Lockport) inhabited shallow lagoons which may have been brackish. The genus is also referred to the Miskoa (Polychaeta), (Roger, 1959).

Order PLESIOTHECA Michaelsen, 1930

F. TUBIFICIDAE Vejdovsky
C. (GZE/KAS)–Rec. Mar./FW

First: *Pronaidites carbonarius* Kusta, 1888, Rakovnik, Bohemia, Czechoslovakia (Howell, 1962). **Extant**

Order PROSOTHECA Michaelsen, 1930

F. ENCHYTRAEIDAE Vejdovsky, 1879
T. (RUP)–Rec. Terr.

First: *Enchytraeus sepultus* Menge, 1866, Oligocene Baltic amber (Conway Morris *et al.*, 1982). **Extant**

Order PROSPORA

F. LUMBRICULIDAE T. (PLI)–Rec. Terr./FW

First: *Lumbriculus sp.* Strauss, 1970, Upper Pliocene lake sediments of Willershausen, western Germany (Conway Morris *et al.*, 1982). **Extant**

Order INCERTAE SEDIS

F. INCERTAE SEDIS Q. (PLE)–Rec. FW

First: *Dendrodrilus rubidus* Schwert, 1979, Quaternary

lacustrine sequence in Kitchener, southern Ontario, Canada (Schwert, 1979). **Extant**

Subclass HIRUDINEA d'Orbigny and Lafresnaye, 1837

There are three families of leeches, of which one has a fossil record.

F. INCERTAE SEDIS S. (LLY)–Rec. Mar.

First: Unnamed species, Lower Silurian, Waukesha County, near Milwaukee, USA (Mikulik *et al.*, 1985). **Extant**

Class INCERTAE SEDIS

Order INCERTAE SEDIS

F. INCERTAE SEDIS Є. (ATB) Mar.

First and Last: *Facivermis yunnanicus* Hou and Chen, 1989, Chiungchussu Formation, Yunnan Province, China.

REFERENCES

Allan, R. S. (1927) *Keilorites* (a new generic name for a Silurian annelid from Australia). *Geological Magazine*, **64**, p. 240.

Briggs, D. E. G. and Clarkson, E. N. K. (1987) The first tomopterid, a polychaete from the Carboniferous of Scotland. *Lethaia*, **20**, 257–62.

Clarke, J. M. (1903) Some Devonic worms. *New York State Museum, Bulletin*, **69**, 1234–8.

Clarke, J. M. (1921) *Organic Dependence and Disease; Their Origin and Significance*. Yale University Press, New Haven, 114 pp.

Clarke, R. B. (1969) Systematics and phylogeny: Annelida, Echiura, Sipuncula, in *Chemical Zoology IV, Annelida, Echiura, Sipuncula* (eds M. Florkin and T. Scheer), Academic Press, New York and London, pp. 1–68.

Clarke, R. B. (1978) Composition and relationships, in *Physiology of the Annelids* (ed. P. J. Mill), Academic Press, New York and London, pp. 1–32.

Conway Morris, S. (1979) Middle Cambrian polychaetes from the Burgess Shale of British Columbia. *Philosophical Transactions of the Royal Society of London, Series*, **B285**, 227–74.

Conway Morris, S. (1987) Fossil priapulid worms, in *Special Papers in Palaeontology*, **20**, Palaeontological Association of London, 75 pp.

Conway Morris, S., Pickerill, R. K. and Harland, T. L. (1982) A possible annelid from the Trenton Limestone (Ordovician) of Quebec, with a review of fossil oligochaetes and other annulate worms. *Canadian Journal of Earth Sciences*, **19**, 2150–7.

Ettensohn, F. R. (1981) *Crininicaminus haneyensis*, a new agglutinated worm tube from the Chesterian of east-central Kentucky. *Journal of Paleontology*, **55**, 479–82.

Fisher, D. W. (1962) Small conoidal shells of uncertain affinities, in *Treatise on Invertebrate Paleontology. Part W.* (ed. C. Teichert), Geological Society of America and University of Kansas Press, Boulder, Colorado, and Lawrence, Kansas, pp. W98–W143.

Glaessner, M. F. (1958) New fossils from the base of the Cambrian in South Australia. *Transactions of the Royal Society of Southern Australia*, **81**, 185–8.

Glaessner, M. F. (1976) Early Phanerozoic annelid worms and their geological and biological significance. *Journal of the Geological Society of London*, **132**, 259–75.

Horwood, A. R. (1912) On *Archarenicola rhaetica* sp. nov. *Geological Magazine*, (**5**), 9, p. 395.

Hou, X.-G. and Chen, J.-Y. (1989) Early Cambrian tentacled worm-like animals (*Facivermis* gen. nov.) from Chengjiang, Yunnan. *Acta Palaeontologica Sinica*, **28**, 32–41 [in Chinese with an English abstract].

Hou, X.-G. and Sun, W.-G. (1988) Discovery of Chengjiang fauna at Meishucan, Jinning, Yunnan. *Acta Palaeontologica Sinica*, **27**, 1–12 [in Chinese with an English abstract].

Howell, B. F. (1962) Worms, in *Treatise on Invertebrate Paleontology. Part W*, (ed. C. Teichert), Geological Society of America and University of Kansas Press, Lawrence, Kansas, pp. W144–W177.

Kielan-Jaworowska, Z. (1962) New Ordovician genera of polychaete jaw apparatuses [Nowe rodzaje ordowickich szczekowych wieloszczetów] (Annelida, Polychaeta). *Palaeontologia Polonica*, **7**, 291–332.

Kielan-Jaworowska, Z. (1966) Polychaete jaw apparatuses from the Ordovician and Silurian of Poland and a comparison with modern forms. *Palaeontologia Polonica*, **16**, 1–152.

Kozłowski, R. (1956) Sur quelques appareils masticateurs des annélides polychètes ordoviciens. *Acta Palaeontologica Polonica*, **1**, 165–210.

Mierzejewski, P. and Mierzejewska, G. (1975) Zenognath type of polychaete jaw apparatuses. *Acta Palaeontologica Polonica*, **20**, 437–42.

Mikulic, D. G., Briggs, D. E. G. and Kluessendorf, J. (1985) A Silurian soft-bodied biota. *Science*, **228**, 715–17.

Phillips, J. and Salter, J. W. (1848) Palaeontological appendix to Prof. John Philip's memoir on the Malvern Hills compared with the Palaeozoic districts of Abberly, Woolhope, May Hill, Tortworth, and Usk. *Memoirs of the Geological Survey of Great Britain and of the Museum of Practical Geology*, **2**, 331–86.

Pickerill, R. K. and Forbes, W. H. (1978) A trace fossil preserving its producer (*Trentonia shegirianâ*) from the Trenton Limestone of the Quebec City area. *Canadian Journal of Earth Sciences*, **15**, 659–64.

Roger, J. (1959) Annelida, in *Traité de Zoologie* Vol. 5, fasc. 1 (ed. P. Grassé), Masson, Paris, pp. 687–713.

Schram, F. R. (1979b) Worms of the Mississipian Bear Gulch Limestone of Central Montana, USA. *Transactions of the San Diego Natural History Society*, **19**, 107–20.

Schwert, D. P. (1979) Description and significance of a fossil earthworm (Oligochaeta: Lumbricidae) cocoon from post-glacial sediments in southern Ontario. *Canadian Journal of Zoology*, **57**, 1402–5.

Stauffer, C. R. (1933) Middle Ordovician Polychaeta from Minnesota. *Bulletin of the Geological Society of America*, **44**, 1173–218.

Szaniawski, H. (1968) Three new polychaete jaw apparatuses from the Upper Permian of Poland. *Acta Palaeontologica Polonica*, **13**, 255–80.

Szaniawski, H. (1974) Some Mesozoic scolecodonts congeneric with Recent forms. *Acta Palaeontologica Polonica*, **19**, 179–95.

Szaniawski, II. and Wrona, R. M. (1973) Polychaete jaw apparatuses and scolecodonts from the Upper Devonian of Poland. *Acta Palaeontologica Polonica*, **18**, 223–67.

Tasch, P. and Stude, J. R. (1965) A scolecodont natural assemblage from the Kansas Permian. *Transactions of the Kansas Academy of Science*, **67**, p. 4.

Thompson, K. S. (1979) Errant polychaetes (Annelida) from the Pennsylvanian Essex fauna of northern Illinois. *Palaeontographica, Abteilung A*, **163**, 166–99.

Zawidzka, K. (1975) Polychaete remains and their stratigraphic distribution in the Muschelkalk of southern Poland. *Acta Geologica Polonica*, **25**, 259–74.

16

ARTHROPODA (TRILOBITA)

M. Romano, W. T. Chang, W. T. Dean, G. D. Edgecombe, R. A. Fortey, D. J. Holloway, P. D. Lane, A. W. Owen, R. M. Owens, A. R. Palmer, A. W. A. Rushton, J. H. Shergold, Derek J. Siveter and M. A. Whyte

The classification of the Trilobita above family level has always been contentious. Since the 1959 Treatise, there have been several attempts to review classification (Bergström, 1973; Fortey, 1990), and new high-level taxa have been introduced, such as the Order Proetida Fortey and Owens, 1975. Some of the new concepts have passed into general use, but the classification is not stable, or fully resolved phylogenetically. In this work, the classification mostly follows Fortey (1990, 1991), and should be similar to that which will be used in the revision of the *Treatise on Invertebrate Paleontology, Part O*, currently in progress.

Class TRILOBITA Walch, 1771

Order REDLICHIIDA Richter, 1933 (See Fig. 16.1)

Comments: Redlichiids and olenellids are treated as separate orders in some classifications. Here they are combined in the Redlichiida, as a paraphyletic group including most of the primitive families having more than two or three thoracic segments in the holaspis.

Suborder OLENELLINA Walcott, 1890

Superfamily OLENELLOIDEA Walcott, 1890

F. OLENELLIDAE Walcott, 1890 Є. (CRF) Mar.

First: *Olenellus truemani* Walcott, 1913, Sekwi Formation (Botomian), north-western Canada (Fritz, 1972).
Last: *Olenellus gilberti* Meek, 1974, and at least four other olenellids, Pioche and Carrara Formations (Toyonian), south-western USA (Palmer and Halley, 1979).

F. HOLMIIDAE Hupé, 1953 Є. (CRF) Mar.

First: *Schmidtiellus mickwitzi* (Schmidt, 1888), Lukati Formation (Atdabanian), Estonia, former USSR (Mens *et al.*, 1990).
Last: *Elliptocephala asaphoides* Emmons, 1844, West Castleton Formation (?Botomian), USA (Theokritoff, 1984). Intercontinental correlation of later Early Cambrian beds is very uncertain. This species seems to be the youngest holmiid with reasonably good stratigraphical control.

Superfamily FALLOTASPOIDEA

F. FALLOTASPIDAE Hupé, 1953 Є. (CRF) Mar.

Frist: *Profallotaspis jakutensis* Repina, 1965, Pestrotsvet Formation (Atdabanian), south-eastern Siberian Platform, former USSR (Khomentovskiy and Repina, 1965).
Last: *Parafallotaspis grata* Fritz, 1972, Sekwi Formation (Atdabanian), north-western Canada (Fritz, 1972)

F. ARCHAEASPIDAE Repina, 1979 Є. (CRF) Mar.

First: *Archaeaspis hupei* Repina, 1965, Pestrotsvet Formation (Atdabanian), south-eastern Siberian Platform, former USSR (Khomentovskiy and Repina, 1965).
Last: *Bradyfallotaspis patula* Fritz, 1972, Sekwi Formation (Atdabanian), north-western Canada (Fritz, 1972). Based on correlations suggested by Repina *in* Spizharski *et al.* (1986), this species seems to be younger than any Siberian archaeaspids.

F. DAGUINASPIDAE Hupé, 1953 Є. (CRF) Mar.

First: *Choubertella spinosa* Hupé, 1953, Amouslek Formation (Atdabanian), Morocco (Hupé, 1953).
Last: *Daguinaspis ambroggii* Hupé and Abadie, 1950, Amouslek Formation (Atdabanian), Morocco (Hupé, 1953).

Superfamily NEVADIOIDEA Hupé, 1953

F. NEVADIIDAE Hupé, 1953 Є. (CRF) Mar.

First: *Nevadella (Paranevadella) subgroenlandica* (Repina, 1965), Pestrotsvet Formation (Atdabanian), south-eastern Siberian Platform, former USSR (Khomentovskiy and Repina, 1965).
Last: *Nevadella (Nevadella) perfecta* (Walcott, 1913), Mural Formation (Botomian), south-western Canada (Fritz, 1992).

F. JUDOMIIDAE Repina, 1979 Є. (CRF) Mar.

First: *Judomia mattajensis* Lazarenko, 1962, Tyuser Formation (Atdabanian), north-eastern Siberian Platform, former USSR (Repina *et al.*, 1974).
Last: *Judomiella heba* Lazarenko, 1962, Perekhod Formation (Atdabanian), south-eastern Siberian Platform, former USSR (Repina, 1979).
Comments: In addition to the references cited above, considerable information on olenellid distribution on the Siberian Platform, former USSR is provided by Astashkin *et al.* (1991).

Suborder REDLICHIINA Harrington, 1959

Superfamily EMUELLOIDEA Pocock, 1970

The Fossil Record 2. Edited by M. J. Benton. Published in 1993 by Chapman & Hall, London. ISBN 0 412 39380 8

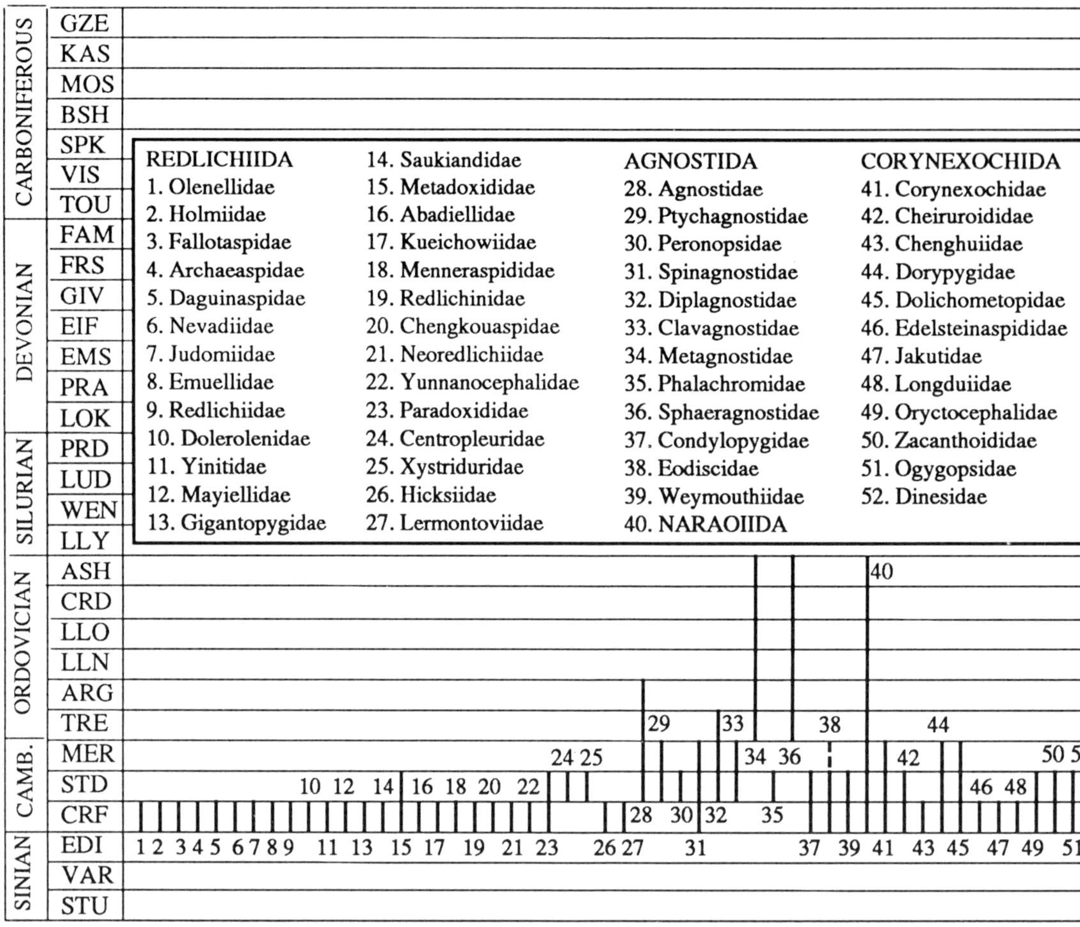

CARBONIFEROUS: GZE, KAS, MOS, BSH, SPK, VIS, TOU
DEVONIAN: FAM, FRS, GIV, EIF, EMS, PRA, LOK
SILURIAN: PRD, LUD, WEN, LLY
ORDOVICIAN: ASH, CRD, LLO, LLN, ARG, TRE
CAMB.: MER, STD, CRF
SINIAN: EDI, VAR, STU

REDLICHIIDA
1. Olenellidae
2. Holmiidae
3. Fallotaspidae
4. Archaeaspidae
5. Daguinaspidae
6. Nevadiidae
7. Judomiidae
8. Emuellidae
9. Redlichiidae
10. Dolerolenidae
11. Yinitidae
12. Mayiellidae
13. Gigantopygidae
14. Saukiandidae
15. Metadoxididae
16. Abadiellidae
17. Kueichowiidae
18. Menneraspididae
19. Redlichinidae
20. Chengkouaspidae
21. Neoredlichiidae
22. Yunnanocephalidae
23. Paradoxididae
24. Centropleuridae
25. Xystriduridae
26. Hicksiidae
27. Lermontoviidae

AGNOSTIDA
28. Agnostidae
29. Ptychagnostidae
30. Peronopsidae
31. Spinagnostidae
32. Diplagnostidae
33. Clavagnostidae
34. Metagnostidae
35. Phalachromidae
36. Sphaeragnostidae
37. Condylopygidae
38. Eodiscidae
39. Weymouthiidae
40. NARAOIIDA

CORYNEXOCHIDA
41. Corynexochidae
42. Cheiruroididae
43. Chenghuiidae
44. Dorypygidae
45. Dolichometopidae
46. Edelsteinaspididae
47. Jakutidae
48. Longduiidae
49. Oryctocephalidae
50. Zacanthoididae
51. Ogygopsidae
52. Dinesidae

40
29 33 38 44
24 25 34 36 42 50 52
10 12 14 16 18 20 22 46 48
28 30 32 35
1 2 3 4 5 6 7 8 9 11 13 15 17 19 21 23 26 27 31 37 39 41 43 45 47 49 51

Fig. 16.1

F. EMUELLIDAE Pocock, 1970 €. (late CRF) Mar.

First: *Emuella polymera* Pocock, 1970, lower part of White Point Conglomerate, South Australia (Pocock, 1970).
Last: *Emuella dalgarnoi* Pocock, 1970, Emu Bay Shale, South Australia (Pocock, 1970).

Superfamily REDLICHOIDEA Poulsen, 1927

F. REDLICHIIDAE Poulsen, 1927 €. (early–late CRF) Mar.

First: *Wutingaspis tingi* Kobayashi, 1935, lower Chiungchussuan Stage, eastern Yunnan, south-western China (Chang *et al.*, 1980).
Last: *Redlichia nobilis* Walcott, 1913, top of the Manto Formation, Shandong, North China (Walcott, 1913). The *Redlichia* faunas of northern Australia, including *Redlichia chinensis* Walcott, are of early middle Cambrian age (Ordian Stage) (Öpik, 1958, p. 11).

F. DOLEROLENIDAE Kobayashi, 1951 €. (early–early middle CRF) Mar.

First: *Dolerolenus zoppii* (Meneghini), Punta Manna Formation, Sardinia, Italy; *Dolerolenus (Malungia) laevigata* Lu, 1975, upper Chiungchussuan Stage, eastern Yunnan, SW China (Chang *et al.*, 1980; Pillola, 1989).
Last: *Paramalungia lubrica* Chang, upper Tsanglangpuan Stage, eastern Yunnan, SW China (Chang *et al.*, 1980).

F. YINITIDAE Hupé, 1953 €. (CRF) Mar.

First: *Yinites typicalis* Lu, 1975, Minghsingssu Formation, northern Guizhou, SW China (Chang *et al.*, 1980).
Last: *Drepanuroides latilimbatus* Chang, 1966, *Drepanuroides* Zone, eastern Yunnan, SW China (Chang *et al.*, 1980).

F. MAYIELLIDAE Chang, 1966 €. (CRF) Mar.

First and Last: *Mayiella tuberculata* Chang, 1966, upper Tsanglangpuan Stage, eastern Yunnan, SW China (Chang *et al.*, 1980).

F. GIGANTOPYGIDAE Harrington, 1959 €. (CRF) Mar.

First and Last: Lower Cambrian: *Gigantopygous papillatus* Hupé, 1953, *Longianda* and *Gigantopygous* Zone, Morocco (Hupé, 1953).

F. SAUKIANDIDAE Hupé, 1953 €. (CRF) Mar.

First and Last: *Saukianda andalusiae* Richter and Richter, 1940, Spain and Morocco (Richter and Richter, 1940; Hupé, 1953).

F. METADOXIDIDAE Whitehouse, 1939 €. (CRF–early STD) Mar. (=ANADOXIDIDAE Nicosia and Rasetti, 1970)

First: *Metadoxides torosus* Meneghini, lower Punta Manna Formation, Sardinia, Italy (Rasetti, 1972; Pillola, 1989).
Last: *Onaraspis somniurna* Öpik, 1967, Ordian Stage, Northern Territory, Australia (Öpik, 1967).

F. ABADIELLIDAE Hupé, 1953
Є. (early CRF) Mar.

First: *Parabadiella huoi* Chang, lower Chiungchussuan Stage, southern Shaanxi and eastern Yunnan, SW China (Chang *et al.*, 1980).
Last: *Guangyuanaspis modaoyaensish* Chang and Qian, upper Chiungchussuan Stage, northern Sichuan, SW China (Chang *et al.*, 1980).

F. KUEICHOWIIDAE Lu, 1965 Є. (CRF) Mar.

First and Last: *Kueichowia liui* Lu, 1965, Tsanglangpuan Stage, northern Guizhou, SW China (Chang *et al.*, 1980).

F. MENNERASPIDIDAE Pokrovskya, 1959
Є. (CRF) Mar.

First and Last: *Menneraspis striatus* Pokrovskya, 1959, Tojohnian Stage, Tuva, former USSR.

F. REDLICHINIDAE Chang and Lin, 1980
Є. (CRF) Mar.

First and Last: *Redlichina vologdini* Lermontova, Tojohnian Stage, Sayan-Altai, former USSR.

F. CHENGKOUASPIDAE Chang and Lin, 1980
Є. (CRF) Mar.

First: *Pseudoresserops oculatus* Repina, 1965, Atdabanian Stage, Siberia, former USSR.
Last: *Chengkouaspis longioculus* Chang and Lin, 1980, Yingzuiyan Formation (upper Tsanglangpuan Stage), northern Szechwan, SW China (Chang *et al.*, 1980).

F. NEOREDLICHIIDAE Hupé 1953 Є. (CRF) Mar.

?F. YUNNANOCEPHALIDAE Hupé, 1953
Є. (CRF) Mar.

First and Last: *Yunnanocephalus*, SW China.

Superfamily PARADOXIDOIDEA Hawle and Corda, 1847

F. PARADOXIDIDAE Hawle and Corda, 1847
Є. (CRF–STD) Mar.

First: *Anabaraceps*, *Anabaraspis*, Toyonian Stage, former USSR (Yakutia). *Paradoxides s.l.* (including *Acadoparadoxides*) appears in the early Middle Cambrian in Europe, North Africa and Scandinavia, where a discordance is often present at the Lower/Middle Cambrian boundary. According to Geyer (1990) the first occurrence of *Paradoxides* (*s.l.*) in Morocco is older than in Scandinavia.
Last: *Paradoxides forchhammeri* Angelin, 1854, upper Middle Cambrian, ?*Lejopyge laevigata* Zone, Scandinavia.

F. CENTROPLEURIDAE Angelin, 1854
Є. (middle–late STD) Mar.

First: *Clarella*, *P. hicksii* Zone, eastern Canada.
Last: *Centropleura*, *P. forchhammeri* 'Stage', Scandinavia, China and Australia.

F. XYSTRIDURIDAE Whitehouse, 1939
Є. (early STD) Mar.

First and Last: *Xystridura* (including subgenera), Australia and SW China (Hainan).

?F. HICKSIIDAE Hupé, 1953 Є. (CRF) Mar.

First and Last: *Hicksia*, Portugal and Spain.

F. LERMONTOVIIDAE Suvorora, 1956
Є. (CRF) Mar.

Order AGNOSTIDA Salter, 1864

Eodiscina and Agnostina are regarded as belonging to separate clades by some authorities (e.g. Shergold, 1991). The peculiarities of Agnostina are taken to indicate that they are derived 'separately' from the rest of the trilobites, and the several similarities to Eodiscina are considered to be a result of convergence. Fortey (1990) enumerated a number of characters shared between Agnostina and Eodiscina, and observed that many of the characters of Eodiscina were plesiomorphic. For this reason, Eodiscina and Agnostina are included as a clade here.

Suborder AGNOSTINA

Comments: Classification of Agnostida: from Shergold *et al.* 1990; genera *incertae familiae* omitted.

Superfamily AGNOSTOIDEA M'Coy, 1849

F. AGNOSTIDAE M'Coy, 1849 Є. (STD)–O. (ARG) Mar. (= MICRAGNOSTIDAE Howell, 1935b; = GLYPTAGNOSTIDAE Whitehouse, 1936; = HASTAGNOSTIDAE Howell, 1937; = RUDAGNOSTIDAE Lermontova, 1951)

First: *Agnostus (Agnostus) pisiformis* Brongniart, *L. laevigata* Zone, Sweden (Westergård, 1946).
Last: *Micragnostus serus* Fortey, 1980, *D. bifidus* Zone, Spitsbergen (Fortey, 1980).
Comments: Includes Agnostinae M'Coy, 1849; Ammagnostinae Öpik, 1967; Glyptagnostinae Whitehouse, 1936.

F. PTYCHAGNOSTIDAE Kobayashi, 1939
Є. (STD–MER) Mar. (= Triplagnostinae Kobayashi, 1939; = Lejopyginae Harrington, *in* Kobayashi, 1939; = Tomagnostinae Kobayashi, 1939; = CANOTAGNOSTIDAE Rusconi, 1951)

First: *Pentagnostus praecurrens* (Westergård, 1946), *Paradoxides pinus* Zone, Sweden (Westergård, 1946).
Last: *Lejopyge laevigata* (Dalman), *Acmarhachis quasivespa* Zone (= *U. Cedaria* Zone), Queensland, Tasmania (Laurie, 1989).
Comments: Last occurrence depends on correlation between the pre-*Glyptagnostus stolidotus* Zones of Kazakhstan (former USSR) and Australia: if *Kormagnostus simplex* post-dates *Acmarhachis quasivespa*, then *Lejopvge armata* (Linnarsson, 1869) is the youngest taxon.

F. PERONOPSIDAE Westergård, 1936
Є. (STD) Mar.

First: *Peronopsis cuneifera* (Barrande), *E. pusillus* Zone (Upper *P. oelandicus* Zone), Bohemia, Czechoslovakia (Šnajdr, 1983).
Last: *Diplorrhina quadrata* (Tullberg), *Jincella brachymetopa* Zone (*Paradoxides forchhammeri* Zone), Sweden (Westergård, 1946).
Comments: First occurrence depends on correlation of *Paradoxides* faunas across Europe.

F. SPINAGNOSTIDAE Howell, 1935a
Є. (CRF–MER) Mar. (= QUADRAGNOSTIDAE

Howell, 1935a *sensu* Öpik, 1961;
= CYCLOPAGNOSTIDAE Howell, 1937)

First: *Eoagnostus roddyi* Resser and Howell, 1938, Upper *Olenellus* Zone, Pennsylvania (Resser and Howell, 1938).
Last: *Peratagnostus* sp. cf. *P. nobilis* Öpik, 1961, *Rhaptagnostus apsis*/*Wentsuia iota* Zone, Queensland (Lower *Conaspis* Zone), (Shergold, 1980).
Comments: Includes Spinagnostinae Howell (1935b); Cyclopagnostinae Howell (1937); Hypagnostinae Ivshin (1953); Euagnostinae Öpik (1979); Doryagnostinae Shergold *et al.* (1990).

F. DIPLAGNOSTIDAE Whitehouse, 1936
Є. (STD)–O. (TRE) Mar.

First: *Diplagnostus? abbatiae* Rushton, 1979, *Paradoxides aurora* Zone (=*T. gibbus* Zone), England, UK (Rushton, 1979).
Last: *Neoagnostus aspidoides* Kobayashi, 1955, *Symphysurina* Zone, British Columbia, Canada (Kobayashi, 1955).
Comments: Includes Diplagnostinae Whitehouse, 1936; Oidalagnostinae Öpik, 1967; Pseudagnostinae Whitehouse, 1936.

F. CLAVAGNOSTIDAE Howell, 1937
Є. (STD–MER)
Mar. (= ACANTHAGNOSTIDAE Qian, 1982)

First: *Clavagnostus repandus* (Westergård, 1946), *Jincella brachymetopa* Zone (*Paradoxides forchhammeri* Zone), Sweden (Westergård, 1946).
Last: *Aspidagnostus rugosus* Palmer, 1962, *G. reticulatus* Zone, Nevada (Palmer, 1962).
Comments: Includes Clavagnostinae Howell, 1937; Aspidagnostinae Pokrovskaya, 1960.

F. METAGNOSTIDAE Jaekel, 1909 O. (TRE–ASH)
Mar. (= TRINODIDAE Howell, 1935b;
= GERAGNOSTIDAE Howell, 1935b;
= ARTHRORHACHIDAE Raymond, 1913)

First: *Anglagnostus? lacaunensis* (Capèra *et al.*, 1978), lower Tremadoc, Montagne Noire, France (Capèra *et al.*, 1978).
Last: *Arthrorhachis tarda* (Barrande), upper Ashgill, *Dalmanitina* Beds, Poland (Kielan, 1959).
Comments: Species of *Corrugatagnostus*, *Dividuagnostus* and *Geragnostus* are also reported from the Ashgill.

Superfamiliae* *incertae sedis

F. PHALACHROMIDAE Hawle and Corda, 1847
Є. (STD) Mar. (= PLATAGNOSTIDAE Howell, 1935b)

First: *Phalacroma bibullatum* (Barrande), *E. pusillus* Zone (upper *P. oelandicus* Zone), Bohemia, Czechoslovakia (Šnajdr, 1958).
Last: *Phalacroma calva* Pokrovskaya, upper *P. davidis* Zone (= *P. punctuosus* Zone), southern Siberia, former USSR (Rozova *in* Lisogar *et al.*, 1988).
Comments: This family also contains *Dignagnostus* Hairullina, 1975 and *Lisogoragnostus* Rozova, 1988.

F. SPHAERAGNOSTIDAE Kobayashi, 1939
O. (TRE–ASH) Mar.

First: ?*Sphaeragnostus* sp., Kendyktas, Kazakhstan, former USSR (Lisogor, 1961).
Last: *Sphaeragnostus cingulatus* (Olin), *Staurocephalus clavifrons* Zone, Middle Ashgill, Poland, Skåne, Bornholm (Ahlberg, 1989).
Comments: *Sphaeragnostus* in the Tremadoc of Kazakhstan cannot be definitely confirmed. The next youngest species, of Llandeilo age, is *S. similis* (Barrande) from Czechoslovakia. Ashgillian species occurring in Scandinavia and Bornholm have been synonymized by Ahlberg (1989). *S. gaspensis* Cooper and Kindle (1936) occurs in the Ashgillian of Quebec, and is also reported from the Rawtheyan (middle Ashgill), *D. anceps* Zone of Wales (Thomas *et al.*, 1984).

Superfamily CONDYLOPYGOIDEA Raymond, 1913

F. CONDYLOPYGIDAE Raymond, 1913
Є. (CRF–STD) Mar.

First: *Condylopyge amitina* Rushton, 1966, Upper *Protolenus* Zone, England, UK (Rushton, 1966).
Last: *Pleuroctenium bifurcatum* (Illing), *P. punctuosus* Zone, England, UK (*Paradoxides davidis* Zone, Newfoundland) (Rushton, 1979).
Comments: *Condylopyge spinigera* Westergård, 1944 is also reported from the *Ptychagnostus punctuosus* Zone at Andrarum, Sweden.

Suborder EODISCINA

Superfamily EODISCOIDEA Raymond, 1913

Families from Jell (1975).

F. EODISCIDAE Raymond, 1913
Є. (CRF–basal MER?) Mar.

First: *Tsunyidiscus liangshanensis* Chang, 1988, Chingchussu Formation, NW Yangtze Platform, China, or *Hupeidiscus orientalis* (Chang, 1988), Chungchussuian Stage, Juimenchong Formation, Hunan, and Hsuijingtuo Formation, Hubei, China (Chang, 1988).
Last: *Opsidiscus* spp., *Lejopyge laevigata* Zone, including *O. depolitus* Romanenko, Middle–Upper Cambrian passage beds (Jago, 1972).
Intervening: Most of the Lower, all of the Middle Cambrian, cosmopolitan.

F. WEYMOUTHIIDAE (*sensu* Jell, 1975)
Є. (CRF–low STD) Mar.

First: *Serrodiscus bellimarginatus* Shaler and Foerste, *Callavia broeggeri* Zone, Branchian Series, North Atlantic Province (Massachusetts, USA, Newfoundland; England, UK), (Rushton, 1966).
Last: *Cobboldites? simplex* Cobbold, 1931 Quarry Ridge Grit (local basal Middle Cambrian), Shropshire, England, UK (Cobbold, 1931).
Intervening: Almost exclusively Lower Cambrian.

Order NARAOIIDA Є. (CRF)–O. (ASH) Mar.

Naraoiida are an uncalcified sister group to the rest of the Trilobita, with which they may be formally included.
First: *Naraoia* sp. from China (R. A. Fortey, pers. comm.).
Last: Undescribed species cf. *Tarricoyia*, Soom Shale, South Africa (R. A. Fortey, pers. comm.).

Order CORYNEXOCHIDA Kobayashi, 1935

Suborder CORYNEXOCHINA

Families from Suvorova (1964) and Zhang *et al.* (1980).

F. CORYNEXOCHIDAE Angelin, 1854
Є. (CRF–low MER) Mar.

First: *Bonnaspis* spp., Atdabanian, Siberian Platform, former USSR (Suvorova, 1964).
Last: *Corynexochus plumula* Whitehouse, Idamean, Australia, Kazakhstan, former USSR (Öpik, 1967).

F. CHEIRUROIDIDAE Chang, 1963
Є. (CRF–low STD) Mar.

First: *Hunanocephalus ovalis* Lee, Shuijingtuo Formation (lower Tsanglangpuian), SW China (Chang, 1988).
Last: *Cheiruroides arcticus* Chernysheva, 1962, Amgan, Siberian Platform, former USSR (Suvorova, 1964).

F. CHENGHUIIDAE *in* Zhang *et al.*, 1980
Є. (middle CRF) Mar.

First and Last: *Chengkouia* and *Xuigiella* spp., Bianmachong Formation (middle Tsanglangpuian), SW China (Zhang *et al.*, 1980).

F. DORYPYGIDAE Kobayashi, 1935 (including OGYGOPSIDAE Rasetti, 1951)
Є. (CRF–early MER) Mar.

First: *Kootenia* sp., black shale below Balang Formation (early or mid-Tsanglangpuian), Hunan, China (Chang, 1988).
Last: *Olenoidestranans* Öpik (1967, p. 174), Mindyallan, Australia; or *Dorypyge* sp., Mila Formation (basal MER), Iran (Fortey and Rushton, 1976).

F. DOLICHOMETOPIDAE Walcott, 1916
Є. (late CRF–early MER) Mar.

First: *Hoffetella* spp., Lungwangmiao Formation, SW China (Zhang *et al.*, 1980).
Last: *Hemirhodon* spp., Dresbachian, Vermont (Raymond, 1937).

F. EDELSTEINASPIDIDAE Hupé, 1953
Є. (CRF) Mar.

First: *Edelsteinaspis gracilis* Lermontova, Botomian, Siberian Platform, former USSR (Suvorova, 1964).
Last: *Edelsteinaspis ornata* Lermontova, Toyonian, Siberian Platform, former USSR (Suvorova, 1964).

F. JAKUTIDAE Suvorova, 1959 Є. (CRF) Mar.

First: *Malykania gribovae* Suvorova, 1960, *M. grandis* Suvorova, early Botomian, Siberian Platform, former USSR (Suvorova, 1960).
Last: *Jakutus amplus* Egorova, 1976, Toyonian, Siberian Platform, former USSR (Egorova *et al.*, 1976).

F. LONGDUIIDAE Zhang and Qian, *in* Zhang *et al.*, 1980 Є. (mid CRF) Mar.

First and Last: *Longduia* spp., Tsanglangpu Formation (middle Tsanglangpuian), SW China (Zhang *et al.*, 1980), and Punta Manna Formation (mid CRF), Sardinia (Pillola, 1990).

F. ORYCTOCEPHALIDAE Beecher, 1897
Є. (late CRF–STD) Mar.

First: *Arthricocephalus chauveaui* Bergeron, Balang Formation, upper Tsanglangpuian, Hunan, China (Chang, 1988).
Last: *Tonkinella kobayashi* Resser, Machari Formation (high STD, *Tonkinella* Zone), Neietsu, Korea (Chang, 1988).

F. ZACANTHOIDIDAE Swinnerton, 1915
Є. (CRF–STD) Mar.

First: species of *Zacanthopsis*, *Zacanthopsina* and *Stephenaspis*, upper *Olenellus* Zone, Nevada, USA (Palmer, 1964) or *Chuchiaspis* spp., Shilungtung Formation (Lunwangmiaoan), SW China (Chang, 1988).
Last: *Zacanthoides* sp., Marjum Formation (*Bolaspidella* Zone, *contracta* Subzone), House Range, Utah, USA (Robison, 1964).

F. OGYGOPSIDAE Rasetti, 1951
Є. (CRF–STD) Mar.

First: *Ogygopsis batis* (Walcott, 1889), upper *Olenellus* Zone, Nevada, USA (Palmer, 1964).
Last: *Ogygopsis*, USA.

F. DINESIDAE Lermontova, 1940
Є. (CRF–STD) Mar.

No additional information since *Treatise O* (1959).

Suborder SCUTELLUINA Hupé, 1953

F. STYGINIDAE Vogdes, 1890
(= SCUTELLUIDAE O. (ARG)–D. (FRS) Mar.
(See Fig. 16.2)

First: *Raymondaspis limbata* (Angelin), Scandinavia.
Last: *Scutellum*, *Scabriscutellum*, Germany, Czechoslovakia.

F. ILLAENIDAE Hawle and Corda, 1847
O. (TRE)–S. (LLY/?WEN) Mar.

Comments: Range taken from Lane and Thomas (1983).

F. PHILLIPSINELLIDAE Whittington, 1950
O. (ARG–ASH) Mar.

First: *Phillipsinella matutina* Dean, 1973, Volkhov Stage, Sobova Formation, Taurus Mountains, Turkey (Dean, 1973).
Last: *Phillipsinella parabola s.l.* (Barrande, 1846), Hirnantian, Côte de la Surprise Member, Percé, Quebec (Lespérance, 1988).

F. TSINANIIDAE Kobayashi, 1933 Є. (MER) Mar.

First and Last: *Tsinania (T.)* spp. and *Tsinania (Dictyites)* spp., Chaumitien Limestone, Shantung, China (Kobayashi, 1933) and Chatsworth Limestone, Western Queensland, Australia (Shergold, 1975). Australian records of later late Cambrian age; *T. (T.)* cf. *nomas* Shergold of 'latest late Cambrian' age in Thailand.

Suborder LEIOSTEGIINA Bradley, 1925

F. LEIOSTEGIIDAE Bradley, 1925
Є. (STD)–O. (?CRD) Mar.

Comments: Fortey and Shergold (1984) extended the concept of this family to include Ordovician genera which had previously been referred to other families. These authors also regarded the Eucalymenidae (Lu, 1975) as synonymous.

F. PAGODIIDAE Kobayashi, 1935
Є. (STD)–O. (TRE/ARG) Mar.

F. KAOLISHANIIDAE Kobayashi, 1935
Є. (MER) Mar.

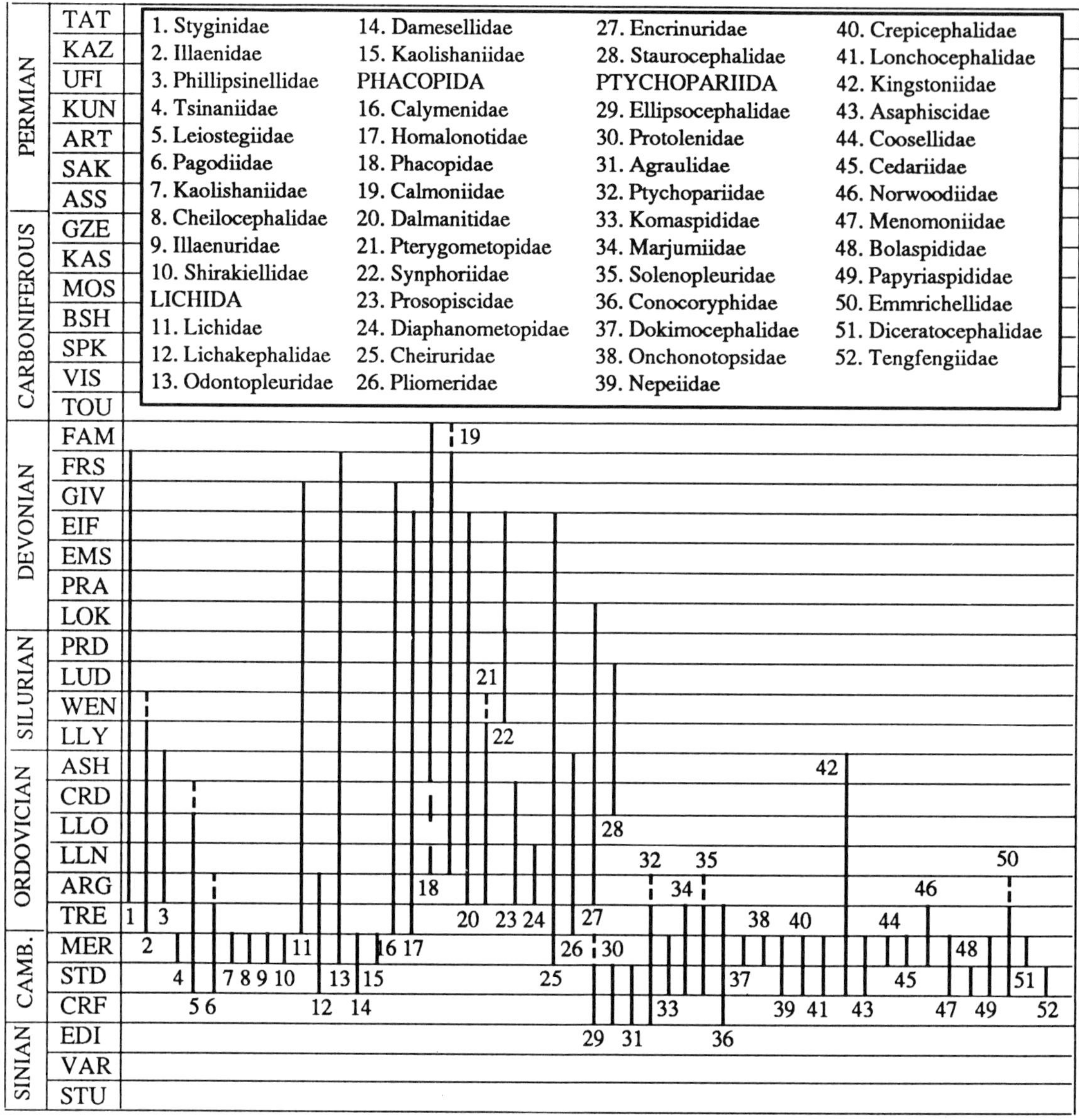

Fig. 16.2

F. CHEILOCEPHALIDAE Shaw (*sensu* Palmer, 1965), 1956 Є. (MER) Mar.

F. ILLAENURIDAE Vogdes, 1890 Є. (MER) Mar.

F. SHIRAKIELLIDAE Hupé, 1953 Є. (MER) Mar.

Order LICHIDA Moore, 1959

Superfamily LICHOIDEA (*sensu* Fortey, 1991)

F. LICHIDAE Hawle and Corda, 1847 O. (TRE)–D. (GIV) Mar.

First: *Metopolichas? klouceki* (Ruzicka, 1926), Czechoslovakia.
Last: *Radiolichas aranea* (Holzapfel, 1895), Massenkalk Limestone, Germany.
Comments: 'A. T. Thomas reported: In the first edition of *The Fossil Record*, Tripp listed *Craspedarges wilcanniae* as the last record. The (lost) type material of that species was collected from erratic boulders of the Amphitheatre Group of New South Wales, Australia. So far as I can gather, the Amphitheatre Group is of D_1, Pragian age.' (See also Thomas and Holloway, 1988.)

F. LICHAKEPHALIDAE Tripp, 1957 Є. (STD)–O. (ARG) Mar.

First: *Eoacidaspis*? sp., western Siberia, former USSR (*E. salairica* Poletaeva, 1964 occurs in the Upper Cambrian of the same area).
Last: *Lichakephalina schilikta* Antcygin, *in* Varganov, 1973, Urals, former USSR.

Superfamily ODONTOPLEUROIDEA (*sensu* Fortey, 1991)

F. ODONTOPLEURIDAE Burmeister, 1843 Є. (MER)–D. (FRS) Mar.

First: *Acidaspides praecurrens* Lermontova, 1951, Kazakhstan, former USSR.
Last: *Radiaspis radiata* (Goldfuss), Germany.
Comments: Bruton (1983a) assigned *Belovia calva* of late middle Cambrian age (eastern Siberia) to the

Eoacidaspididae of the Odontopleuroidea. This species has subsequently been assigned tentatively to *Eoacidaspis* of Lichakephalidae (Thomas and Holloway, 1988).

Superfamily DAMESELLOIDEA Kobayashi, 1935

F. DAMESELLIDAE Kobayashi, 1935
€. (STD)–€. (MER) Mar.

F. KAOLISHANIIDAE Kobayashi, 1935
€. (MER) Mar.

Order PHACOPIDA Salter, 1864

Suborder CALYMENINA Swinnerton, 1915

F. CALYMENIDAE Milne Edwards, 1840
O. (TRE)–D. (GIV) Mar.

First: *Pharostomina* species, including *P. oepiki* Sdzuy, 1955 and *P. ferentaria* Sdzuy, 1955, Leimitz Schiefer, Bavaria, Germany; *P. mexicana* Robison and Pantoja-Alor, 1968, Lower Tiflu Formation, Mexico; *P. trapezoidalis quaesita* Přibyl and Vaněk, 1958, *Kainella meridionalis* Zone, Bolivia; *P.* cf. *trapezoidalis* of Fortey and Owens, *in* Owens *et al.*, 1981, *Clonograptus tenellus* Zone shales, Wales, UK.
Last: *Gravicalymene zhenzishanensis* Nan, 1980, Hetai Formation, north-east China.
Intervening: ARG–EIF
Comments: All the above *Pharostomina* species occur in strata of known early Tremadoc age. The genus also occurs in the upper Tremadoc of Argentina, and it may occur in the lower Tremadoc of Czechoslovakia. Other, unnamed calymenid species occur in 'Middle' Devonian (?GIV) strata.

F. HOMALONOTIDAE Chapman, 1890
O. (TRE)–D. (?GIV) Mar.

First: *Bavarilla hofensis* (Barrande), Germany (Sdzuy, 1955).
Last: *Dipleura dekayi* Green, North and South America (Kozłowski, 1923).

Suborder PHACOPINA Struve, *in* Moore, 1959

Within the Phacopina, GDE recognizes a Superfamily Acastacea Delo, 1935 comprising a plesion 'Kloucekiinae' Destombes, 1972 (O. (LLN–ASH)), F. Calmoniidae Delo, 1935 (S. (PRD)–D. (FAM?)) and F. Acastidae Delo, 1935 (S. (LLY)–D. (FRS)). The plesion is regarded as a paraphyletic grade of Acastacea *s.l.*, within which *Baniaspis* Destombes, 1972 is most closely related to the Calmoniidae and Acastidae. The classification adopted by DJH follows more closely that used in the first edition of *The Fossil Record*, and five families are recognized (Phacopidae, Calmoniidae, Dalmanitidae, Pterygometopidae, Synphoriidae). Pending the publication of the new edition of the Treatise, the DJH classification is followed.

F. PHACOPIDAE Hawle and Corda, 1847
O. (LLN?/LLO)–D. (FAM) Mar.

First: *Morgatia primitiva* Hammann, 1972, Postolonnec Formation, France (Henry, 1980, p. 166). This species may belong to *Kloucekia* (Calmoniidae) rather than to *Morgatia*. The next oldest species are *Morgatia zguidensis* (Destombes, 1972), Morocco (LLN?/LLO) and *M. hupei* (Nion and Henry, 1966), France, Portugal and Spain (LLO), (Henry, 1980, p. 167–8).
Last: Several species of *Phacops s.l.*, *Phacops (Omegops)*, *Dianops* and *Cryphops*, England, UK, France, Morocco, Belgium, Germany, Poland, former USSR (Hahn and Hahn, 1975, pp. 21–6; Struve, 1976; Thomas *et al.*, 1984, pp. 62–3).
Intervening: ASH–FRS.

F. CALMONIIDAE Delo, 1935
O. (LLN)–D. (FRS/FAM?) Mar.

First: *Kloucekia drevermanni drevermanni* Hammann, 1972, Spain (Hammann, 1974, p. 79).
Last: Several species of *Neocalmonia (Neocalmonia)*, *N. (Bradocryphaeus)*, *N. (Heliopyge)* and *N. (Quadratispina)* are known from the FRS of Spain, England, UK, France, Belgium, Germany and Afghanistan (Hahn and Hahn, 1975, pp. 27–8; Morris, 1988, p. 37). Of these, *N. (Bradocryphaeus) occidentalis* (Whidborne, 1897) occurs in the Morte Slates, England, UK, which may range into the lower FAM (Morris, 1988).
Intervening: LLO–GIV.
Comments: The family Calmoniidae is here considered to comprise Calmoniinae, Acastinae, 'Acastavinae' and Asteropyginae of *Treatise O* usage, the last two being transferred from the Dalmanitidae. This is consistent with the views of Eldredge and Branisa (1980, p. 191) who regarded Acastinae, Acastavinae and Asteropyginae as the sister group of the Malvinokaffric calmoniids.

F. DALMANITIDAE Vogdes, 1935
O. (ARG)–D. (EIF) Mar.

First: *Ormathops borni* (Dean, 1966), *D. extensus* Biozone, from the Couches du Landeyran, Montagne Noire, France.
Last: *Dalmanites patacamayaensis* Kozłowski, 1923, Sicasica Formation?, Bolivia (EIF, precise horizon and age uncertain; Wolfart, 1968, p. 70); *Odontochile* aff. *carinata* Maksimova (1968, p. 108), Kazakh Horizon (lower EIF), Kazakhstan; *Odontochile (Reussia) kailensis* Maksimova (*in* Modzalevskaya, 1969, p. 148), Imatchinskaya Suite (lower EIF), eastern Siberia, former USSR.
Intervening: LLN–EMS.

F. PTERYGOMETOPIDAE Reed, 1905
O. (ARG)–S. (LLY/WEN?) Mar.

First: *Pterygometopus sclerops* (Dalman, 1827), *Expansus* Limestone, Sweden (Whittington, 1950, p. 538); *P. borni* (Dean, 1966), Couches du Landeyran, Montagne Noire, France, and strata of similar age in Morocco (Destombes, 1972, p. 27) (this species possibly belongs in *Ormathops*–see 'Dalmanitidae' above); *P. bredensis* Weber, 1948, Bredy Horizon, southern Urals, former USSR (Ancigin, 1970, p. 14).
Last: *Podowrinella straitonensis* (Lamont, 1965), Knockgardner Formation, Scotland, UK (Howells, 1982, p. 63).
Intervening: LLN–ASH.

F. SYNPHORIIDAE Delo, 1935
S. (WEN)–D. (EIF) Mar.

First: Several species of *Lygdozoon* from the St Clair Limestone, Arkansas, and strata of similar age in Texas, Oklahoma, Illinois and Indiana, USA (Holloway, 1981, p. 717); *Delops nobilis marri* Rickards, 1965, Brathay Flags and Lower Coldwell Beds, England, UK (Holloway, 1981; Thomas *et al.*, 1984, fig. 23, p. 52).

Last: *Trypaulites calypso* (Hall, 1861), Grand Tower Limestone, Illinois and Dundee Limestone, Michigan, USA (Lespérance, 1975, p. 113); *Coronura marylandicus* (Prosser and Kindle, 1913), Romney Formation, Maryland, USA (Lespérance and Bourque, 1971, fig. 2).
Intervening: LUD, GED–EIF.

F. PROSOPISCIDAE Fortey and Shergold, 1984
O. (ARG–CRD) Mar.

First: *Prosopiscus praecox* Fortey and Shergold, 1983, lower part of Nora Formation, central Australia, mid–late Arenig age. *Prosopiscus* sp. A also occurs with *P. praecox*. *P. latus* is known from the upper Arenig of SW China.
Last: *Prosopiscus* spp., Asia.
Intervening: LLN.
Comments: *Prosopiscus* is the only genus of the family, and had previously been placed within the F. Cheiruridae or F. Encrinuridae.

Suborder CHEIRURINA Harrington and Leanza, 1957

F. DIAPHANOMETOPIDAE Jaanusson, 1959
O. (ARG–LLN) Mar.

First: *Gyrometopus lineatus* (Angelin, 1854), *Megistaspis planilimbata* Zone, Västergotland, Oltorp, Sweden.
Last: *Diaphanometopus volborthi* Schmidt, 1881, Baltoscandia.
Comments: Jaanusson (1975) upgraded Diaphanometopinae to include a new genus, *Gyrometopus*, which he regarded was an immediate ancestor of the Phacopina. *Diaphanometopus* (a monotypic genus) is poorly known, but the family was considered by Jaanusson (1975, p. 215) as most likely to belong to the Cheiruroidea.

F. CHEIRURIDAE Salter, 1864
€. (MER)–D. (EIF) Mar.

First: *Eocheirurus* spp., former USSR.
Last: *Crotalocephalus sternbergi* (Boeck), Czechoslovakia.

F. PLIOMERIDAE Raymond, 1913
O. (TRE–ASH) Mar.

First: *Parapilekia anxia* Sdzuy, 1955, Leimitz-Schiefer, Germany.
Last: *Placoparia (Hawleia) prantli* Kielan, 1959, *S. clavifrons* Zone, Poland.

F. ENCRINURIDAE Angelin, 1854
O. (ARG)–D. (GED) Mar.

First: *Cybelurus sokoliensis* Bursky, 1970, top Sokoliy Horizon, Pai-Khoya, southern Novaya Zemlya, former USSR (see Fortey, 1980, p. 100). *Lyrapyge ebriosus* Fortey, 1980 from the middle Olenidsletta Member of the Valhallfonna Formation (*I. gibberulus* Zone), Ny Friesland, Spitsbergen, may be the next oldest.
Last: *Encrinurus* sp. Bourque and Lespérance, 1977, West Point Formation, Madeleine River area, Gaspé, Canada. *Encrinurus (Pacificurus)* cf. *robustus* (Mitchell, 1924) from the Wallace Shale, central New South Wales, could extend into the GED (Strusz, 1980, p. 35). *Encrinurus mariannae* Maximova, 1975 may occur in the lower GED Kokbaytal Horizon in Kazakhstan (Strusz, 1980, pp. 52, 56).
Intervening: ARG2–PRD.

F. STAUROCEPHALIDAE Prantl and Přibyl, 1948
O. (CRD)–S. (LUD) Mar. (*sensu* Tomczykowa, 1987)

First: *Staurocephalus pilafrons* Owen and Bruton, 1980, Onnian Stage, upper Solvang Formation, central Oslo Region, Norway.
Last: *Staurocephalus mitchelli* Chatterton and Campbell, 1980, Ludfordian Stage, Black Bog Shale, near Bowning, New South Wales, Australia; *Staurocephalus azuella* Šnajdr, 1980, Kopanina Formation, Bohemia, Czechoslovakia.
Intervening: PUS–HIRN, SHE–HOM.

Order PTYCHOPARIIDA Swinnerton, 1915

This group is difficult to define, and is here understood as a paraphyletic group including those families in which the hypostome is not attached to the doublure (nata t condition).

Suborder PTYCHOPARIINA Richter, 1933

Superfamily ELLIPSOCEPHALOIDEA Matthew, 1887

F. ELLIPSOCEPHALIDAE Matthew, 1887
€. (CRF–STD/MER) Mar.

First: Genera include *Ellipsocephalus*, *Antatlasia*, *Lusatiops* and *Strenuaeva* from Europe, North Africa (Geyer, 1990b, Bani-Stufe, *Antatlasia hollardi* Zone), Middle East, eastern Canada and Australia.
Last: ?*Manchurocephalus*, *L. laevigata* Zone/*A. pisiformis* Zone, NE China.

F. PROTOLENIDAE R. and E. Richter, 1948
€. (late CRF–early STD) Mar.

First: Several genera and subgenera, including *H. (Hamatolenus)*, *H. (Myopsolenus)*, *Protolenus* and *Pseudolenus*, Europe, North Africa and eastern Canada.
Last: *H. (Hamatolenus)*, *H. (Lotzeia)* and *H. (Myopsolenus)*, North Africa and Spain.
Comments: The above ages are based on the conventional view that the Protolenid–Strenuellid Zone of England and Wales, UK (Cowie *et al.*, 1972; Bassett *et al.*, 1976), and the *Protolenus* Zone of eastern Newfoundland (Hutchinson, 1962) are of late early Cambrian age. On the basis of more recent work in Morocco, Geyer (1990a,b) considered these units to be of early middle Cambrian age as they correspond with strata in which *Paradoxides (s.l.)* occurs.

F. AGRAULIDAE Raymond, 1913
€. (CRF–STD) Mar.

Superfamily PTYCHOPARIOIDEA Matthew, 1887

F. PTYCHOPARIIDAE Matthew, 1887
€. (CRF)–O. (TRE/ARG) Mar.

F. KOMASPIDIDAE Kobayashi, 1935
€. (STD/MER) Mar.

First: *Komaspis typa* and *K. convexus* Kobayashi, 1935, *Olenoides* Zone. South Korea (Kobayashi, 1935, p. 141; Öpik, 1963, p. 95).
Last: *Irvingella nuneatonensis* (Sharman, 1886), top of the *Olenus cataractes* Zone, Nuneaton, central England, UK; *I. major* Ulrich and Resser, *in* Walcott, 1924, *Elvinia* Zone, USA, also *Ivshinagnostus ivshini* Zone, Kazakhstan, former USSR (Rushton, 1983, p. 113).

F. MARJUMIIDAE Kobayashi, 1935
€. (STD)–O. (TRE) Mar.

F. SOLENOPLEURIDAE Angelin, 1854
€. (STD)–O. (TRE/ARG) Mar.

F. CONOCORYPHIDAE Angelin, 1854
€. (CRF)–O. (TRE) Mar.

F. DOKIMOCEPHALIDAE Kobayashi, 1935
€. (MER) Mar.

F. ONCHONOTOPSIDAE Rasetti, 1946
€. (MER) Mar.

F. NEPEIIDAE Whitehouse, 1939
€. (STD–MER) Mar.

F. CREPICEPHALIDAE Kobayashi, 1935
€. (STD–MER) Mar.

F. LONCHOCEPHALIDAE Hupé, 1953
€. (STD–MER) Mar.

F. KINGSTONIIDAE Kobayashi, 1933
€. (STD)–O. (ASH) (including F. SHUMARDIIDAE) Mar.

F. ASAPHISCIDAE Raymond, 1924
€. (STD–MER) Mar.

F. COOSELLIDAE Palmer, 1954 €. (MER) Mar.

F. CEDARIIDAE Raymond, 1937 €. (MER) Mar.

F. NORWOODIIDAE Walcott, 1916
€. (MER)–O. (TRE) Mar.

F. MENOMONIIDAE Walcott, 1916
€. (STD–MER) Mar.

F. BOLASPIDIDAE Howell, 1959 €. (STD) Mar.

F. PAPYRIASPIDIDAE Whitehouse, 1939
€. (STD–MER) Mar.

F. EMMRICHELLIDAE Kobayashi, 1935
€. (STD)–O. (TRE/ARG) Mar.

F. DICERATOCEPHALIDAE Lu, 1954
€. (MER) Mar.

F. TENGFENGIIDAE Chang, 1963 €. (STD) Mar.

Comments: East Asia and western Canada.

F. LISANIIDAE Chang, 1963 €. (STD–MER) Mar.

Comments: upper Middle Cambrian to lower Upper Cambrian, China.

F. INOUYIIDAE Chang, 1963 €. (STD–MER) Mar.

Comments: lower Middle Cambrian to lower Upper Cambrian, East Asia.

F. WUANIIDAE Chang and Yuan, 1981
€. (STD) Mar.

Comments: Hsuchuangian Stage, NE China.

F. LORENZELLIDAE Chang, 1963 €. (STD) Mar.

Comments: Hsuchuangian Stage, China.

F. PROASAPHISCIDAE Chang, 1963
€. (STD) Mar.

Comments: Hsuchuangian and Changhian Stages, Asia.

F. IGNOTOGREGATIDAE Zhang Wentang and Jell, 1987 €. (STD) Mar.

Comments: Hsuchuangian and lower Changhian Stages of North and NE China.

F. HOLANSHANIIDAE Chang, 1963
€. (STD) Mar.

Comments: lower Middle Cambrian, Hsuchuang Formation, North China.

F. MAPANIIDAE Chang, 1963 €. (STD) Mar.

Comments: Changhsia Formation, Changhian Stage; North and southern NE China, Australia.

Suborder OLENINA Burmeister, 1843

F. OLENIDAE Burmeister, 1843
€. (MER)–O. (ASH) Mar.

First: *Olenus alpha* Henningsmoen, 1957, *A. pisiformis* Zone, Norway.
Last: *Triarthrus* spp. listed by Ludvigsen and Tuffnell (1983), eastern USA, Scandinavia, SW China. Ludvigsen and Tuffnell cite (1983) a *Triarthrus* sp. from the Whitby Formation of southern Ontario, Canada, as the last olenid.

Suborder HARPINA Whittington, 1959

F. HARPETIDAE Hawle and Corda, 1847
O. (TRE)–D. (FRS) Mar.

First: *Scotoharpes lauriei* Jell and Stait, 1985, Florentine Valley, Tasmania, Australia.
Last: *Harpes neogracilis* R. Richter and E. Richter, 1926, Europe.

F. HARPIDIDAE Whittington, 1950
€. (MER)–O. (LLN) Mar. (see Fig. 16.3)

First: *Harpidoides explicatus* Apollonov, Batyrbay Section, Kazakhstan, former USSR.
Last: *Harpides atlanticus* Billings, Table Head, Newfoundland, Canada (Whittington, 1965).

F. ENTOMASPIDIDAE Ulrich, *in* Bridge, 1930
€. (MER)–O. (TRE) Mar.

First: *Entomaspis radiatus* Ulrich, *in* Bridge, 1930, USA.
Last: *Hypothetica rawi* Ross, 1951, USA.

Order ASAPHIDA Salter, 1864 emend. Fortey, 1991

Superfamily ANOMOCAROIDEA Poulsen, 1927

F. ANOMOCARIDAE Poulsen, 1927
€. (STD–MER) Mar.

F. PTEROCEPHALIIDAE Kobayashi, 1935
€. (MER) Mar.

Comments: This classification follows that of Fortey and Chatterton (1988) who downgraded the Housiidae to include it within the Pterocephaliidae. The family ranges from the *Aphelaspis* Zone to the *Elvinia* Zone.

F. PARABOLINOIDIDAE Lochman, 1956
€. (MER) Mar.

Comments: *Conaspis* Zone to *Ptychaspis–Prosaukia* Zone.

F. DIKELOKEPHALINIDAE Kobayashi, 1935
€. (MER)–O. (LLN/?LLO) Mar.

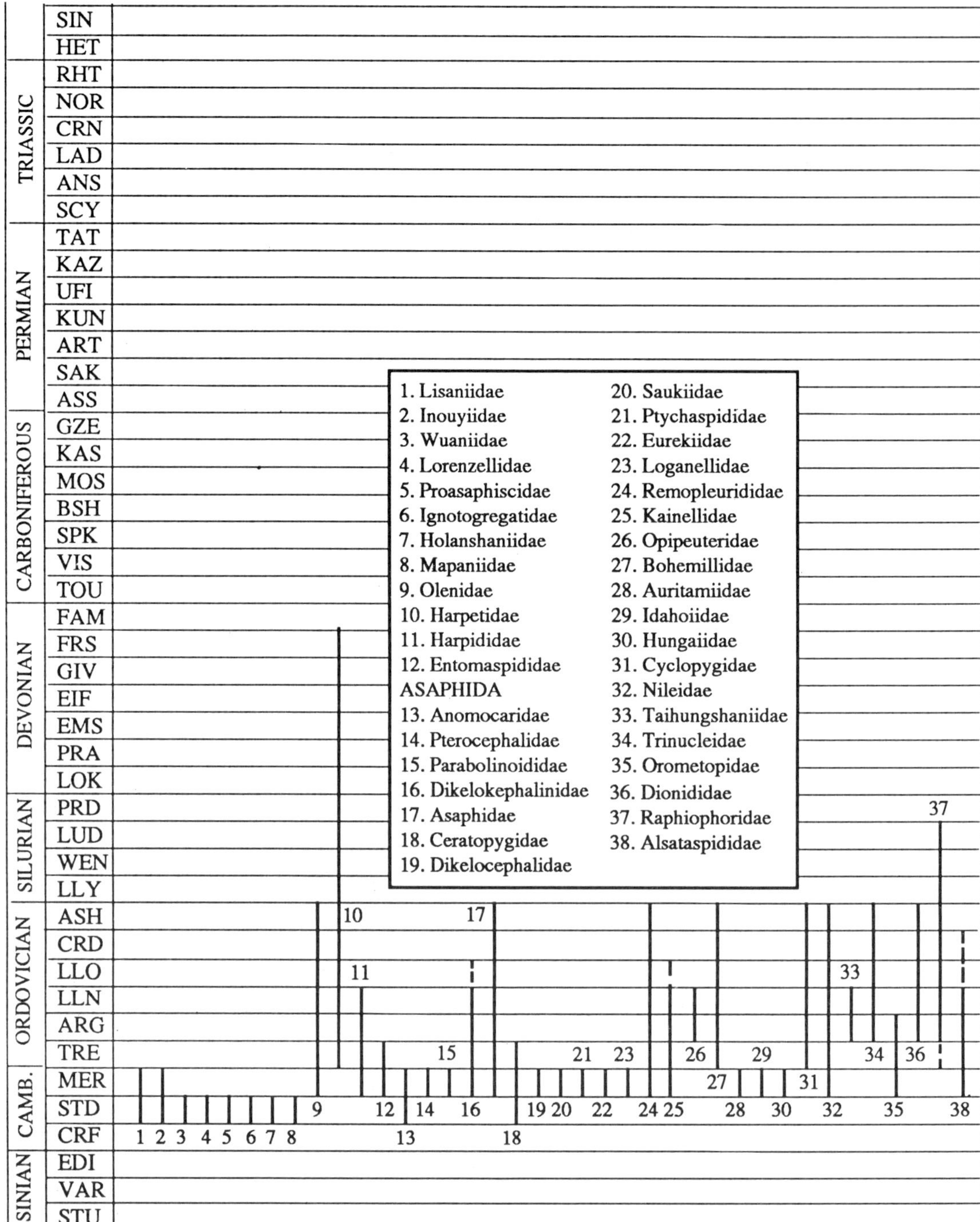

Fig. 16.3

First: *Nomadinis pristinus* Öpik, 1967, Mindyallan, Queensland, Australia.
Last: *Hungioides bohemicus* (Perner, 1918), LLN, Europe, Argentina, China and Australia. Lu (1975, p. 372) tentatively records the species from the LLO of China.

Superfamily ASAPHOIDEA Burmeister, 1843

F. ASAPHIDAE Burmeister, 1843
Є. (MER)–O. (ASH) Mar.

First: *Griphasaphus griphus* Öpik, 1967, Mindyallan, Queensland, Australia.
Last: *Ectenaspis beckeri* (Slocum), USA. Owen (1986, p. 237) lists four genera from the Hirnantian Stage.

F. CERATOPYGIDAE Linnarsson, 1869
Є. (STD)–O. (TRE) Mar.,
includes MACROPYGIDAE Kobayashi, 1953

First: *Proceratopyge* spp. (Westergård, 1948), Sweden, Australia.
Last: *Ceratopyge forficula* Sars, *Ceratopyge* Limestone, Scandinavia.

Superfamily DIKELOCEPHALOIDEA Miller, 1889 emend. Ludvigsen and Westrop, 1983

F. DIKELOCEPHALIDAE Miller, 1889
Є. (MER) Mar.

Comments: *Ptychaspis–Prosaukia* Zone to *Saukia* Zone.

F. SAUKIIDAE Ulrich and Resser, 1930
Є. (MER) Mar.

Comments: ?lower Upper Cambrian in SW China, *Ptychaspis–Prosaukia* Zone to *Saukia* Zone in USA.

F. PTYCHASPIDIDAE Raymond, 1924
Є. (MER) Mar.

Comments: *Conaspis* Zone to *Saukia* Zone in USA and East Asia.

F. EUREKIIDAE Hupé, 1953 Є. (MER) Mar.

Comments: *Ptychaspis–Prosaukia* Zone, and *Saukia* Zone, USA.

F. LOGANELLIDAE Rasetti, 1959 Є. (MER) Mar.

Superfamily REMOPLEURIDOIDEA Hawle and Corda, 1847

F. REMOPLEURIDIDAE Hawle and Corda, 1847
Є. (MER)–O. (ASH) Mar.

First: *Richardsonella megalops* (Billings, 1860), Canada.
Last: *Remopleurides* sp. C from the upper Langara Formation, Hirnantian, of the Oslo region, Norway (Owen, 1981, p. 16).

F. KAINELLIDAE Ulrich and Resser, 1930
Є. (MER)–O. (LLN/LLO) Mar.

F. OPIPEUTERIDAE Fortey, 1974
O. (ARG–LLN) Mar.

First: *Opipeuter inconnivus* Fortey, 1974, Olenidsletta Member, Valhallfonna Formation, Spitsbergen.
Last: *Cremastoglottos occipitalis* Whittard, 1961, Hope Shales, Shropshire, England, UK.

F. BOHEMILLIDAE Barrande, 1872
O.(TRE–ASH) Mar.

First: *Bohemilla (Fenniops) sabulon* Fortey and Owens, 1987, *Bergamia rushtoni* Zone, Pontyfenni Formation, upper Arenig, South Wales, UK. An older, upper Tremadoc form is reported by Fortey and Owens (1987, p. 128).
Last: *Bohemilla scotica scotica*, Mill Formation, Whitehouse Subgroup, Girvan of Pusgillian age (J. K. Ingham, pers. comm.); also recorded from Co. Clare, Republic of Ireland (Whittard, 1952). *Bohemilla* is present in the Králův Dvůr Formation of Czechoslovakia.

F. AURITAMIIDAE Öpik, 1967 Є. (MER) Mar.

F. IDAHOIIDAE Lochman, 1956 Є. (MER) Mar.

Comments: *Elvinia* Zone to *Saukia* Zone, USA.

F. HUNGAIIDAE Raymond, 1924 Є. (MER) Mar.

Superfamily CYCLOPYGOIDEA Raymond, 1925

F. CYCLOPYGIDAE Raymond, 1925
O. (TRE–ASH) Mar.

First: *Prospectatrix genatenta* (Stubblefield, 1927), *Shumardia pusilla* Zone, Shineton Shales, Shropshire, England, UK.
Last: *Cyclopyge*, Hirnantian Stage, Pomeroy, North Ireland. Several other genera from Europe, including *Ellipsotaphrus, Microparia s.s., M. (Degamella), Psilacella* and *Symphysops* (Marek, 1961; Whittard, 1952), are of Ashgill age.

F. NILEIDAE Angelin, 1854
Є. (MER)–O. (ASH) Mar.

First: *Platypeltoides wimani* Troedsson, 1931, ?uppermost Upper Cambrian, Batyrbai Section, Shabakty Formation, Kazakhstan, former USSR.
Last: *Elongatinileus convexus* Ji, Pagoda Formation, China.

F. TAIHUNGSHANIIDAE Sun, 1931
O. (ARG–LLN) Mar.

First: *Tungtzuella szechuanensis* Sheng, 1958, *Acanthograptus–Tungtzuella* Zone, Fenhsiang Formation, China (Lu, 1975, p. 350).
Last: *Omeipsis huangi* (Sun, 1931), Meitan Formation, China (Kobayashi, 1951, p. 16; Lu, 1975, p. 346) or *Taihungshania multisegmentata* Sheng, 1958, upper Meitan Formation, China.

Superfamily TRINUCLEIOIDEA Hawle and Corda, 1847

F. TRINUCLEIDAE Hawle and Corda, 1847
O. (ARG–ASH) Mar.

First: *Myttonia* cf. *fearnsidesi* Whittington, 1966 of Fortey and Owens (1987), lower part of Moridunian Stage (*Merlinia selwynii* Zone) in South Wales, UK. *Myttonia* spp., *Anebolithus simplicior* (Whittard, 1966) and *Hanchungolithus* sp. occur in Moridunian strata elsewhere in the Anglo-Welsh area (Fortey and Owens, 1987, pp. 203–4).
Last: *Cryptolithus portageeinsis* Lespérance, 1988, Hirnantian, Côte de la Surprise Member, White Head Formation, near Percé, Quebec, Canada. Undetermined trinucleids are also recorded from the Hirnantian of the Anglo-Welsh area (Lespérance, 1988, p. 369).
Intervening: LLN–CRD.

F. OROMETOPIDAE Hupé, 1955
Є. (MER)–O. (ARG) Mar.
(*sensu* Fortey and Owens 1991 *non* Fortey and Shergold 1984; Fortey and Chatterton, 1988)

First: *Araiopleura* spp. and *Jegorovaia* spp., uppermost Cambrian of (*inter alia*) Wales and China (see Fortey and Owens, 1991, pp. 451, 453).
Last: *Araiopleura reticulata* Harrington and Leanza, 1957, Arenig of Argentina.
Intervening: TRE.

F. DIONIDIDAE Gurich, 1908
O. (ARG–ASH) Mar.

First: *Dionidella*? sp. indet. 1 of Fortey and Owens (1987, p. 222), Pontyfenni Formation (lower Fennian Stage, *Stapeleyella abyfrons* Zone = *c. Isograptus gibberulus* Zone).
Last: *Dionide* spp., Rawtheyan of Scandinavia, UK, Bohemia and Poland (Ingham, 1974, pp. 63–4; Owen, 1981, pp. 36–7).
Intervening: LLN–CRD.

F. RAPHIOPHORIDAE Angelin, 1854
O. (TRE/ARG)–S. (LUD) Mar.
(*sensu* Fortey, 1975, = Endymioniidae Raymond, 1930)

First: *Ampyx* cf. *pater* Holm, 1882 of Tjernvik (1956), *Megistaspis armata* Zone (lower Hunneberg Stage), Sjurberg, Dalarna, Sweden.
Last: *Raphiophorus parvulus* (Forbes, 1848), Gorstonian

strata, Ludlow district, Welsh borderland, UK; basal Ludlow Series, NW Morocco (Thomas, 1978, p. 55).
Intervening: ARG–ASH, LLY.

F. ALSATASPIDIDAE Turner, 1940 Є. (MER)–O. (LLN/?CRD) Mar. (*sensu* Fortey and Owens 1991, = HAPLOPLEURIDAE Harrington and Leanza, 1957)

First: *Species of Calycinoidia* Lu and Chien, 1978 and *Hermosella* Lu and Chien, 1978 from the Upper Cambrian of Guizhou, China were tentatively assigned to the Alsataspididae by Fortey and Owens (1991, p. 453). Most members of the family are Tremadoc and Arenig in age, including species of *Falanaspis*, *Haplopleura*, *Nambeetella* and *Seleneceme* from the Anglo-Welsh area, Scandinavia, China, South America and Australia.
Last: *Seleneceme acuticaudata* (Hicks, 1875), lower LLN, Wales and Welsh borderlands, UK; *Seleneceme* spp., probable lower LLN, Texas and the Appalachians, USA (see Whittard, 1960, pp. 117–18). If *Yumenaspis* Chang and Fan, 1960 is correctly ascribed to the Alsataspididae, the range of the family extends to the Upper Caradoc with *Yumenaspis* sp. of Owen *et al.* (1986) from the Raheen Formation of Co. Waterford, Republic of Ireland.
Intervening: TRE–ARG, ?LLO–CRD.

F. LIOSTRACINIDAE Raymond, 1937 Є. (?STD–MER) Mar. (see Fig. 16.4)

F. RHYSSOMETOPIDAE

F. MONKASPIDIDAE (?=CHELIDONOCEPHALIDAE)

Suborder UNCERTAIN

F. ITYOPHORIDAE Warburg 1925 O. (CRD–ASH) Mar.

First: *Frognaspis stoermeri* Nikolaisen, 1965, Pusgillian, Hogberg Member, Solvang Formation, Frognøya, Ringerike, Norway.
Last: *Ityophorus undulatus* Warburg, 1925, Cautleyan/Rawtheyan, Boda Limestone, Kallholn, Dalarna, Sweden.

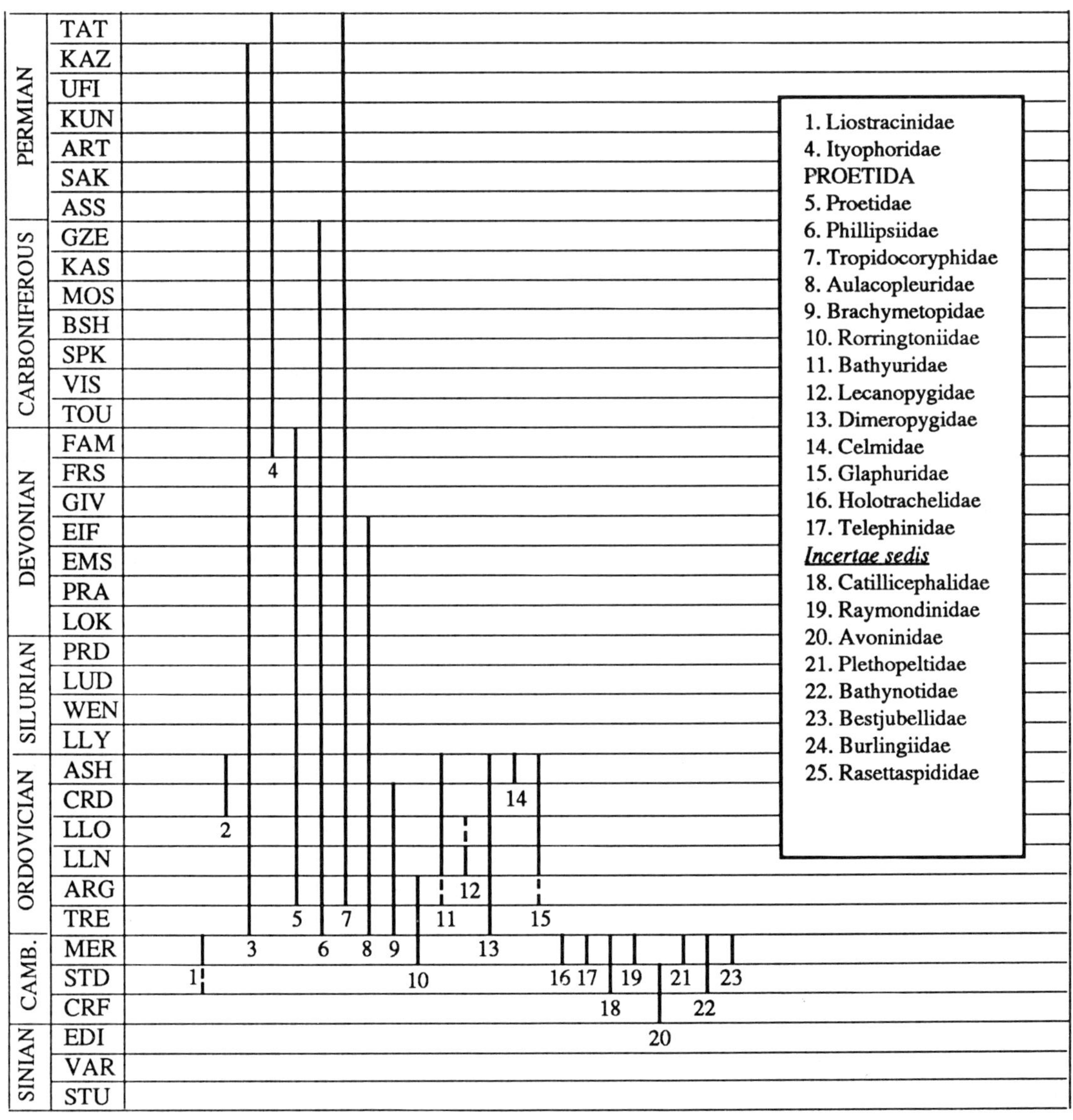

Fig. 16.4

Order PROETIDA Fortey and Owens, 1975

Superfamily PROETOIDEA Salter, 1864

F. PROETIDAE Salter, 1864 O. (TRE)–P. (KAZ) Mar.

First: *Proetus* (*s.l.*) *owensi* Tripp, 1980, Caradoc, Harnagian–Soudleyan, Balclatchie Conglomerate, Girvan District, Scotland, UK; *Cyphoproetus facetus* Tripp, 1954, Caradoc, approximately Longvillian–Actonian, Craighead (Kiln) Mudstone, Girvan district, Scotland, UK; *C. wilsonae* (Sinclair, 1944) Caradoc, lower Trenton Group, Ontario, Canada (Tripp, 1954, 1980; Ludvigsen, 1979).
Last: *Neogriffithides siculus* (Gemmellaro, 1892), Sosio, Sicily; (*Kathwaia capitorosa* Grant, 1966, upper Guadalupian–lower Dzhulfian, Middle *Productus* Limestone, Salt Range, Pakistan, if this is a proetid and not a phillipsiid (Owens, 1983).
Intervening: CRD, Westphalian, ART.

F. PHILLIPSIIDAE Oehlert, 1886 D. (FAM)–P. (TAT) Mar.

First: Include *Waribole dunhevidensis* (Thomas, 1909), FAM, Launceston district, Cornwall, England, UK; *W. beulensis* (Richter and Richter, 1926), FAM Rheinisches Schiefergebirge, Germany; *W. familiaris* (Alberti, 1975), FAM, Morocco. (*Cyrtosymbole* spp. such as *C. escoti* (von Koenen, 1886), FAM, southern France and *C. (Franconicabole) dillensis* (Drevermann, 1901), FAM Rheinisches Schiefergebirge, Germany appear earlier, but may be proetids, not phillipsiids), (Osmolska, 1962).
Last: Include *Acropyge weggeni* Hahn and Hahn, 1981, uppermost Upper Permian, Iran; *Ditomopyge fatmil* Grant, 1966, upper Guadalupian–lower Dzhulfian, Middle *Productus* Limestone, Salt Range, Pakistan; *Pseudophillipsia obtusicaudata* (Kayser, 1883), TAT, Yangtze Province, China (Owens, 1983).
Intervening: Carboniferous–Permian (KAZ)

F. TROPIDOCORYPHIDAE Přibyl, 1946 O. (ARG)–D. (FAM) Mar.

First: *Decoroproetus*? sp. of Fortey and Owens, 1975, Tourmakeady Limestone, Mayo, Republic of Ireland (Fortey and Owens, 1975).
Last: *Constantina pulchra* Cisne, 1970, Conewango Stage, Pennsylvania, USA (Cisne, 1970).
Intervening: Silurian–Devonian (FRS).

Superfamily AULACOPLEUROIDEA Angelin, 1854

F. AULACOPLEURIDAE Angelin, 1854 O. (TRE)–C. (MOS) Mar.

First: *Otarion (Aulacopleura) szechuanica* Lu, 1975, Panho Formation, Sichuan, China (Thomas and Owens, 1978).
Last: *Namuropyge sinica* Hahn *et al.*, 1989, Westphalian A, Dala Formation, Guangxi, China (Hahn *et al.*, 1989).
Intervening: ARG–FRS, C.(l.)

F. BRACHYMETOPIDAE Prantl and Přibyl, 1950 O. (ARG)–P. (TAT) Mar.

First: *Oenonella paulula* Fortey, 1980, Valhallan, Spitsbergen (another, similar species occurs in the Meiklejohn Bioherm, Nevada, USA), (Owens, *in* Owens and Hammann, 1990).
Last: *Cheiropyge himalayensis* Diener, 1897, Chitichun Limestone, Himalayas, NE India (Owens, 1983).
Intervening: CRD–ASH, WEN, LOK–KAZ.

F. RORRINGTONIIDAE Owens, 1990 O. (TRE)–D. (EIF) Mar.

First: *Protarchegonus moroffi* Sdzuy, 1955, Leimitz-Schiefer, Bavaria, Germany (Owens, *in* Owens and Hammann, 1990).
Last: *Aulacopleurina peltata peltata* (Novak, 1890), *Acanthopyge* and Chotec Limestones, Barrandian area, Czechoslovakia; *A. peltata glabra* Rietschel, 1964, *A. peltata schmidti* Rietschel, 1964, both EIF, Rheinisches Schiefergeberge, Germany; *Eodrevermannia rara* Přibyl and Plas, 1954, highest EMS, Trebotov Limestone, Barrandian area, Czechoslovakia (Rietschel, 1964; Šnajdr, 1980).
Intervening: LLN–ASH, WEN, PRA–EIF.

Superfamily BATHYUROIDEA Walcott, 1886

F. BATHYURIDAE Walcott, 1886 O. (TRE–CRD) Mar.

First: *Peltabellia*, Garden City Formation, Utah, USA (Whittington and Ross, *in* Whittington, 1953), Nevada, USA, and Greenland (Fortey and Peel, 1990).
Last: *Bathyurus*, *Raymondites*, eastern USA (Whittington, 1953).

F. LECANOPYGIDAE Lochman, 1953 Є. (MER)–O. (ARG) Mar.

First: *Lecanopyge*, *Platydiamesus*, *Rasettia* and *Resseraspis*, especially in eastern USA (Lochman, 1953).
Last: *Benthamaspis*, western Newfoundland, Canada (Fortey, 1979).

F. DIMEROPYGIDAE Hupé, 1953 O. (?ARG–ASH) Mar.

First: *Dimeropygiella* spp., Pogonip Group, western USA (Hintze, 1953).
Last: *Dimeropye* recorded from the Hirnantian of Mackenzie Mountains, NW Canada (Chatterton and Ludvigsen, 1983; Owen, 1986). *Toernquistia nicholsoni* Reed, 1896, Keisley Limestone, England, UK; *Dimeropyge* and *Toernquistia nicholsoni*, Chair of Kildare Limestone, Republic of Ireland.

F. CELMIDAE Jaanusson, 1956 O. (ARG/?LLN) Mar.

First and Last: *Celmus granulatus* Angelin, 1854, Lysaker Member (*Asaphus* Shale), upper ARG (Fennian), Oslo, Norway. The species also occurs in Sweden (Jaanusson, 1956; Bruton, 1983b).

?F. GLAPHURIDAE Hupé, 1953 O. (TRE–ASH) Mar.

First: *Glaphurus alimbeticus* Balashova, 1961, former USSR (Balashova, 1961, p. 137).
Last: *Glaphurella harknessi* (Reed, 1896), Keisley Limestone, Westmorland, England, UK; Chair of Kildare Limestone, Republic of Ireland (Dean, 1971, p. 44). Other Ashgill occurrences are known.

F. HOLOTRACHELIDAE Warburg, 1925 O. (ASH) Mar.

First and Last: *Holotrachelus* is widely distributed in Britain, Scandinavia, eastern USA and the former USSR. *H.* cf. *punctillosus* (Törnquist, 1884) is recorded from the lower

Sorbakken Formation, Ringerike, Norway, of Cautleyan age; *Holotrachelus* is known from the Hirnantian of Siberia (Owen, 1981, 1986; Chugaeva, 1983) and the Kildare Limestone of Eire (Dean, 1971) (middle Cautleyan to Hirnantian; J. K. Ingham, pers. comm.).

F. TELEPHINIDAE Marek, 1952
O. (?ARG–ASH) Mar.

First: *Goniophrys prima* Ross, 1951, Garden City Formation, Utah, USA, close to Tremadoc/Arenig boundary.
Last: *Telephina* cf. *linnarssoni* (Ulrich, 1930), Chair of Kildare Limestone, Republic of Ireland (Dean, 1971, p. 46) of middle Cautleyan to lower Hirnantian age. *Telephina* spp., Europe.

POLYPHYLETIC GROUPS IN NEED OF FURTHER ANALYSIS

F. CATILLICEPHALIDAE Raymond, 1938
€. (MER) Mar.

F. RAYMONDINIDAE Clark, 1924 €. (MER) Mar.

F. AVONINIDAE Lochman, 1936
€. (STD–MER) Mar.

F. PLETHOPELTIDAE Raymond, 1925
€. (MER) Mar.

Order INCERTAE ORDINIS

F. BATHYNOTIDAE Hupé, 1953
€. (CRF–STD) Mar.

First: *Bathynotus holopyga* (Hall), Parker Shale (*Olenellus* Zone), Vermont, USA (Whittington, 1988).
Last: *Bathynotellus kielcensis* (Bednarczyk, 1970), lower Middle Cambrian (*Paradoxides oelandicus* Zone), Holy Cross Mountains, Poland (*Jakutus kielcensis* Bednarczyk is here transferred to *Bathynotellus*).
Comments: Of unknown origin (Fortey, 1991); eastern USA, Arctic Eurasia.

F. BESTJUBELLIDAE Ivshin, 1983 €. (MER) Mar.

First and Last: *Bestjubella munificus* Ivshin, 1983, Kuyandinian Stage, Seletinian Horizon with *Irvingella*, Seleta River Basin District, Kazakhstan, former USSR.
Comments: Ivshin (1983) tentatively included the family under Odontopleuracea, and noted the close similarity to Eoacidaspididae.

F. BURLINGIIDAE Walcott, 1908
€ (STD–MER) Mar.

First: *Burlingia laevis* Westergård, 1936, *Paradoxides pinus* Zone, Sweden.
Last: *Schmalenseeia amphionura* Moberg, 1903, *Agnostus pisiformis* Zone, Sweden (Westergård, 1947).

F. RASETTASPIDIDAE Ivshin, 1983
€ (MER) Mar.

First and Last: *Rasettaspis francoi* Ivshin, 1983, Kuyandinian Stage, upper Seletinian Horizon, Seleta River Right Bank Basin District, central Kazakhstan, former USSR.

REFERENCES

Ahlberg, P. (1989) Agnostid trilobites from the Upper Ordovician of Sweden and Bornholm, Denmark. *Bulletin of the Geological Society of Denmark*, **37**, 213–26.

Ancigin, N. Ya. (1970) Trilobity semeistva Pterygometopidae iz ordovikskikh otlozhenii Urala, in *Materialy po Paleontologi Urala* (eds M. G. Breivel' and G. N. Papulov), Akademiya Nauk SSSR, Ural'skii filial, Institut geologii i geokhimii, Sverdlovsk, pp. 13–42.

Astashkin, V. A., Pegel', T. V., Repina, L. N. *et al.* (1991) The Cambrian System of the Siberian Platform. *International Union of Geological Sciences Publication*, **27**, 133 pp.

Balashova, E. A. (1961) Tremadoc trilobites of the Aktubiask district. *Trudy Geol. Inst. Leningrad*, **18**, 102–45.

Bassett, M. G., Owens, R. M. and Rushton, A. W. A. (1976) Lower Cambrian fossils from the Hell's Mouth Grits, St Tudwal's Peninsula, North Wales. *Journal of the Geological Society, London*, **132**, 623–44.

Bednarczyk, W. (1970) Trilobites of the Lower *Paradoxides oelandicus* Stage from the Brzechow area in the western part of the Swietokryskie Mts. *Bullétin de l'Académie Polonaise des Sciences, Série Geologique*, **18**, 29–35.

Bergström, J. (1973) Organization, life, and systematics of trilobites. *Fossils and Strata*, **2**, 1–69.

Bourque, P.-A. and Lespérance, P. J. (1977) The Silurian–Devonian boundary in north-eastern Gaspé Peninsula, in *The Silurian–Devonian Boundary* (ed. A. Martinsson), IUGS Series A, No. 5, pp. 245–55.

Bruton, D. L. (1983a) Cambrian origins of the odontopleurid trilobites. *Palaeontology*, **26**, 875–85.

Bruton, D. L. (1983b) The morphology of *Celmus* (Trilobita) and its classification, in *Trilobites and Other Early Arthropods: Papers in Honour of Professor H. B. Whittington, F.R.S.* (eds. D. E. G. Briggs and P. D. Lane), Special Papers in Palaeontology, **30**, 213–19.

Capèra, J. C., Courtessole, R. and Pillet, J. (1978) Contribution à l'étude de l'Ordovicien inférieur de la Montagne Noire. Biostratigraphie et révision des Agnostida. *Extrait des Annales de la Société géologique du Nord*, **98**, 67–88.

Chang, W. T. (1963) A classification of the lower and middle Cambrian trilobites from north and northeast China, with descriptions of new families and new genera. *Acta Palaeontologica Sinica*, **11**, 475–87 [English summary].

Chang W. T. (1988) The Cambrian System in eastern Asia: correlation chart and explanatory notes. *International Union of Geological Sciences*, **24**, 81 pp.

Chang, W. T., Lu, Y. H., Zhu, Z. L. *et al.* (1980) Cambrian trilobite fauna of southwestern China. *Palaeontologia Sinica*, **159** (N. S., B, no. 16), Science Press, Beijing [in Chinese with English summary], 497 pp.

Cisne, J. L. (1970) *Constantina pulchra*: an unusual proetid trilobite from the Devonian of Pennsylvania. *Journal of Paleontology*, **44**, 522–3.

Cobbold, E. S. (1931) Additional fossils from the Cambrian rocks of Comely, Shropshire. *Quarterly Journal of the Geological Society, London*, **87**, 459–512.

Cooper, G. A. and Kindle, C. H. (1936) New brachiopods and trilobites from the Upper Ordovician of Percé, Quebec. *Journal of Paleontology*, **10**, 348–72.

Cowie, J. W., Rushton, A. W. A. and Stubblefield, C. J. (1972) A correlation of Cambrian rocks in the British Isles. *Geological Society of London Special Report*, **2**, 1–40.

Dean, W. T. (1966) The Lower Ordovician stratigraphy and trilobites of the Landeyran Valley and the neighbouring district of the Montagne Noire, south-western France. *Bulletin of the British Museum (Natural History), Geology Series*, **12**, 245–353.

Dean, W.T. (1971) The trilobites of the chair of Kildare Limestone (Upper Ordovician) of eastern Ireland. *Monograph of the Palaeontographical Society, London*, **125**, 1–60.

Dean, W. T. (1973) The Lower Palaeozoic stratigraphy and faunas of the Taurus Mountains near Beysehir, Turkey, II. The trilobites of the Sobova Formation (Lower Ordovician). *Bulletin of the British Museum (Natural History), Geology Series*, **24**, 279–348.

Destombes, J. (1972) Les trilobites du sous-ordre des Phacopina de l'Ordovicien de l'Anti-Atlas (Maroc). *Notes et Mémoires du Service géologique du Maroc*, **240**, 112 pp.

Egorova, L. I., Shabanov, Yu. Ya., Rozanov, A. Yu. *et al.* (1976) [Elankan and Kuonamian facial stratotype of the Lower–Middle Cambrian Boundary]. *Trudy sibirskogo nauchno-issledovatel'skogo Instituta Geologii, Geofiziki i mineral'nogo Syrya (SNIIGGIMS)*, **211**, 167 pp.

Eldredge, N. and Branisa, L. (1980) Calmoniid trilobites of the Lower Devonian *Scaphiocoelia* zone of Bolivia, with remarks on related species. *Bulletin of the American Museum of Natural History*, **165**, 181–290.

Fortey, R. A. (1974) A new pelagic trilobite from the Ordovician of Spitsbergen, Ireland and Utah. *Palaeontology*, **17**, 111–24.

Fortey, R. A. (1975) The Ordovician trilobites of Spitsbergen. II. Asaphidae, Nileidae, Raphiophoridae and Telephinidae of the Valhallfonna Formation. *Norsk Polarinstitut Skrifter*, **162**, 1–125.

Fortey, R. A. (1979) Early Ordovician trilobites from the Catoche Formation (St George Group) western Newfoundland. *Bulletin of the Geological Survey of Canada*, **321**, 61–114.

Fortey, R. A. (1980) The Ordovician trilobites of Spitsbergen. III. Remaining trilobites of the Valhallfonna Formation. *Norsk Polarinstitut Skrifter*, **171**, 1–163.

Fortey, R. A. (1990) Ontogeny, hypostome attachment and trilobite classification. *Palaeontology*, **33**, 529–76.

Fortey, R. A. (1991) Revision of Trilobite Treatise, in *The Trilobite Papers 3* (ed. R. Ludvigsen), Denman Institute for Research on Trilobites, pp. 4–8.

Fortey, R. A. and Chatterton, B. D. E. (1988) Classification of the trilobite suborder Asaphina. *Palaeontology*, **31**, 165–222.

Fortey, R. A. and Owens, R. M. (1975) Proetida – a new order of Trilobites. *Fossils and Strata*, **4**, 227–39.

Fortey, R. A. and Owens, R. M. (1987) The Arenig Series in South Wales. *Bulletin of the British Museum (Natural History), Geology Series*, **41**, 69–307.

Fortey, R. A. and Owens, R. M. (1991) A trilobite fauna from the highest Shineton Shales in Shropshire, and the correlation of the latest Tremadoc. *Geological Magazine*, **128**, 437–64.

Fortey, R. A. and Peel, J. S. (1990) Early Ordovician trilobites and molluscs from the Poulsen Cliff Formation, Washington Land, western North Greenland. *Bulletin of the Geological Society of Denmark*, **38**, 11–32.

Fortey, R. A. and Rushton, A. W. A. (1976) *Chelidonocephalus* trilobite fauna from the Cambrian of Iran. *Bulletin of the British Museum (Natural History) Geology Series*, **27**, 321–40.

Fortey, R. A. and Shergold, J. H. (1984) Early Ordovician trilobites, Nora Formation, Central Australia. *Palaeontology*, **27**, 315–66.

Fritz, W. H. (1972) Lower Cambrian trilobites from the Sekwi Formation type section, Mackenzie Mountains, northwestern Canada. *Canadian Geological Survey Bulletin*, **212**, pp. 1–90.

Fritz, W. H. (1992) Walcott's Lower Cambrian olenellid trilobite collection 61k, Mount Robson Area, Rocky Mountains, Canada. *Geological Survey of Canada Bulletin* (in press).

Geyer, G. (1990a) Revised Lower to lower Middle Cambrian biostratigraphy of Morocco. *Newsletters on Stratigraphy*, **22**, 53–70.

Geyer, G. (1990b) Die marokkanischen Ellipsocephalidae (Trilobita: Redlichiida). *Beringeria*, **3**, 363 pp.

Hahn, G. and Hahn, R. (1975) Die Trilobiten des Ober-Devon, Karbon und Perm. *Leitfossilien*, **1**, 1–127.

Hahn, G., Hahn, R. and Yuan, J.-L. (1989) Trilobites from the Upper Carboniferous (Westphalian A) of S. China (N. Guangxi). *Geologica et Palaeontologica*, **23**, 113–203.

Hammann, W. (1972) Neue propare Trilobiten aus dem Ordovizium Spaniens. *Senckenbergiana Lethaea*, **53**, 371–81.

Hammann, W. (1974) Phacopina und Cheirurina (Trilobita) aus dem Ordovizium von Spanien. *Senckenbergiana Lethaea*, **55**, 1–151.

Hawle, I. and Corda, A. J. C. (1847) Prodrom einer Monographie der böhmischen Trilobiten. *Abhandlungen der Königliche Böhmischen Gesellschaft Wissenschaften*, **5**, 121–292.

Henningsmoen, G. (1957) The trilobite family Olenidae. *Skrifter utgitt av det Norske Videnskaps-Akademi i Oslo, for 1957. 1. Matematisk–Naturvidenskapelig Klasse*. **1**, 303 pp.

Henry, J.-L. (1980) Trilobites ordoviciens du Massif Armoricain. *Mémoires de la Société géologique et minéralogique de Bretagne*, **22**, 1–250.

Holloway, D. J. (1981) Silurian dalmanitacean trilobites from North America, and the origins of the Dalmanitinae and Synphoriinae. *Palaeontology*, **24**, 695–731.

Howell, B. F. (1935a) New Middle Cambrian agnostian trilobites from Vermont. *Journal of Paleontology*, **9**, 218–21.

Howell, B. F. (1935b) Cambrian and Ordovician trilobites from Hérault, southern France. *Journal of Paleontology*, **9**, 222–38.

Howell, B. F. (1937) Cambrian *Centropleura vermontensis* fauna from northwestern Vermont. *Geological Society of America Bulletin*, **48**, 1147–1210.

Howells, Y. (1982) Scottish Silurian Trilobites. *Monograph of the Palaeontographical Society, London*, 1–76.

Hupé, P. (1953) Contribution a l'étude du Cambrian inférieur et du Precambrien III de l'Anti-Atlas marocain. *Service Geologique du Maroc (Rabat) Notes et Mémoires*, **103**, 402 pp.

Hutchinson, R. D. (1962) Cambrian stratigraphy and trilobite faunas of southeastern Newfoundland. *Geological Society of Canada, Bulletin*, **88**, 1–156.

Ingham, J. K. (1974) The upper Ordovician trilobites from the Cautley and Dent districts of Westmorland and Yorkshire. *Palaeontographical Society (Monograph)*, **2**, 59–87.

Ivshin, N. K. (1953) *Middle Cambrian trilobites of Kazakhstan. Part 1. Fauna of the Boschchekul Gorizont*. Institut Geologicheskikh Nauk, Akademiya Nauk, Alma-Ata, 226 pp.

Ivshin, N. K. (1983) *Upper Cambrian Trilobites of Kazakhstan. Part II. Seletinian Horizon of the Kuyandinian Stage of Central Kazakhstan*. Institut Geologicheskikh Nauk, Akademiya Nauk, Alma-Ata, 432 pp.

Jaanusson, V. (1956) On the trilobite genus *Celmus* Angelin, 1854. *Bulletin of the Geological Institutions of the University of Uppsala*, **36**, 35–49.

Jaanusson, V. (1975) Evolutionary processes leading to the trilobite suborder Phacopina. *Fossils and Strata*, **4**, 209–18.

Jaekel, O. (1909) Über die Agnostiden. *Zeitschrift der Deutschen Geologischen Gesellschaft*, **61**, 380–400.

Jago, J. B. (1972) Two new Cambrian trilobites from Tasmania. *Palaeontology*, **15**, 226–37.

Jell, P. A. (1975) Australian Middle Cambrian Eodiscoids with a review of the Superfamily. *Palaeontographica, Abteilung A*, **150**, 1–97.

Khomentovskiy, V. V. and Repina, L. N. (1965) *Lower Cambrian of the Stratotype Section of Siberia*, Nauka, Moscow [in Russian], 200 pp.

Kielan, Z. (1959) Upper Ordovician trilobites from Poland and some related forms from Bohemia and Scandinavia. *Palaeontologia Polonica*, **11**, 198 pp.

Kobayashi, T. (1933) Upper Cambrian of the Wuhutsui Basin, Liaotung, with special reference to the limit of the Chaumitien (or Upper Cambrian) of eastern Asia, and its subdivision. *Jap. J. Geol. Geogr.*, **11**, 51–155.

Kobayashi, T. (1935) The Cambro-Ordovician formations and faunas of South Chosen. Palaeontology. Part III. The Cambrian faunas of South Chosen with a special study on the Cambrian trilobite genera and families. *Journal of the Faculty of Science. Imperial University of Tokyo. Section II*, **4**, 49–344.

Kobayashi, T. (1939) On the agnostids (Part 1). *Journal of the Faculty of Science. Imperial University of Tokyo. Section II*, **5** (5), 69–198.

Kobayashi, T. (1951) On the Ordovician trilobites in Central China. *Journal of the Faculty of Science, University of Tokyo, Section II*, **8** (1), 1–87.

Kobayashi, T. (1955) The Ordovician fossils from the McKay Group in British Columbia, western Canada, with a note on the Early Ordovician palaeogeography. *Journal of the Faculty of Science. University of Tokyo. Section II*, **9** (3), 355–493.

Kozlowski, R. (1923) Faune dévonienne de Bolivie. *Annales de Paléontologie*, **12**, 1–112.

Lane, P. D. and Thomas, A. T. (1983) A review of the trilobite Suborder Scutelluina, in *Trilobites and Other Early Arthropods – Papers in Honour of Professor H. B. Whittington, F.R.S.* (eds D. E. G. Briggs and P. D. Lane), Special papers in Palaeontology, **30**, 141–60.

Laurie, J. R. (1989) Revision of species of *Goniagnostus* Howell and *Lejopyge* Corda from Australia (Agnostida, Cambrian). *Alcheringa*, **13**, 175–91.

Lermontova, E. V. (1951) *Upper Cambrian Trilobites and Brachiopods from Boschchekul (Northeastern Kazakhstan).* Vsesoyuzniy Nauchno-Issledovatelskiy Geologicheskiy Institut (VSEGEI), Gosudarstvennoe Izdatelstvo, Moscow, 49 pp.

Lespérance, P. J. (1975) Stratigraphy and paleontology of the Synphoriidae (Lower and Middle Devonian dalmanitacean trilobites). *Journal of Paleontology*, **49**, 91–137.

Lésperance, P. J. (1988) Trilobites. *Bulletin of the British Museum (Natural History) Geology Series*, **43**, 357–76.

Lespérance, P. J. and Bourque, P.-A. (1971) The Synphoriidae: an evolutionary pattern of Lower and Middle Devonian trilobites. *Journal of Paleontology*, **45**, 182–208.

Linnarsson, J. G. O. (1869) Om Vestergötlands Cambriska och Siluriska aflagringar. *Kongliga Svenska Vetenskaps – Akademiens Handlingar*, **8** (2), 1–89.

Lisogor, K. A. (1961) Trilobites of the Tremodoc and adjacent deposits of Kendyktas. *Akademiya Nauk SSSR. Trudy Geologicheskiy Instituta*, **18**, 55–92.

Lisogor, K. A., Rozov, S. N. and Rozova, A. V. (1988) Correlation of the Middle Cambrian deposits of Lesser Karatau and the Siberian Platform by trilobites. *Akademiya Nauk SSSR. Sibirskoe Otdelenie. Trudy Institut Geologii i Geofiziki, Novosibirsk*, **720**, 54–82.

Lu, Y.-H. (1975) Ordovician trilobite faunas of central and south-west China. *Paleontologica sinica (N.S.)* (B) **11**, 1–463.

Ludvigsen, R. (1979) Fossils of Ontario. Part 1: The trilobites, in *Royal Ontario Museum Life Sciences Miscellaneous Publications*, Toronto, 96 pp.

Ludvigsen, R. and Tuffnell, P. A. (1983) A revision of the Ordovician olenid trilobite *Triarthus* Green. *Geological Magazine*, **120**, 567–77.

Maksimova, Z. A. (1968) Srednepaleozoiskie trilobity tsentral'nogo Kazakhstana. *Trudy Vsesoyuznogo Nauchno-Issledovatel'skogo Geologicheskogo Instituta (VSEGEI)*, **165**, 1–208.

Marek, L. (1961) The trilobite family Cyclopygidae Raymond in the Ordovician of Bohemia. *Rozpravy Ústředniho Ustavu Geologického*, **28**, 85 pp.

Marek, L. (1966) Nadceled Bohemillacea. *Cas. Narodnčho Muzea v Praze*, **135**, 145–53.

M'Coy, F. (1849) On the classification of some British fossil Crustacea with notices of some new forms in the University collection at Cambridge. *Annals and Magazine of Natural History, Series 2*, **4**, 161–79, 330–5, 392–414.

Mens, K., Bergström, J. and Lendzion, K. (1990) The Cambrian System on the East European Platform. *International Union of Geological Sciences, Publication*, **25**, 1–73.

Modzalevskaya, E. A. (1969) (ed.) *Polevoi Atlas Siluriiskoi, Devonskoi i Rannekamennougol'noi Fauny Dal'nego Vostoka*. Izdatel'stvo 'Nedra', Moscow, 327 pp.

Morris, S. F. (1988) A review of British trilobites, including a synoptic revision of Salter's monograph. *Monograph of the Palaeontographical Society, London* (publication no. **574**, part of volume **140** for 1986), 1–316.

Nan, J. (1980) Trilobita, in *Palaeontological Atlas of Northeast China, Part 1, Palaeozoic* (ed. Shenyang Institute of Geology and Mineral Resources), Geological Publishing House, Peking, pp. 484–519.

Öpik, A. A. (1958) The Cambrian trilobite *Redlichia*: Organization and generic concept. *Bulletin of the Bureau of Mineral Resources, Geology and Geophysics of Australia*, **42**, 1–38.

Öpik, A. A. (1961) The geology and palaeontology of the head-waters of the Burke River, Queensland. *Bulletin of the Bureau of Mineral Resources, Geology and Geophysics of Australia*, **53**, 249 pp.

Öpik, A. A. (1963) Early Upper Cambrian fossils from Queensland. *Bulletin of the Bureau of Mineral Resources, Geology and Geophysics of Australia*, **64**, 133 pp.

Öpik, A. A. (1967) The Mindyallan fauna of north-western Queensland. *Bulletin of the Bureau of Mineral Resources, Geology and Geophysics of Australia*, **74** (1), 404 pp., (2), 167 pp.

Öpik, A. A. (1979) Middle Cambrian agnostids: systematics and biostratigraphy. *Bulletin of the Bureau of Mineral Resources, Geology and Geophysics of Australia*, **172** (1), 188 pp., (2), 67 pls.

Osmolska, H. (1962) Famennian and Lower Carboniferous Cyrtosymbolinae (Trilobites) from the Holy Cross Mountains, Poland. *Acta Palaeontologica Polonica*, **7**, 53–222.

Owen, A. W. (1981) The Ashgill trilobites of the Oslo Region, Norway. *Palaeontographica Abteilung A*, **175**, 1–88.

Owen, A. W. (1986) The uppermost Ordovician (Hirnantian) trilobites of Girvan, SW Scotland with a review of coeval trilobite faunas. *Transactions of the Royal Society of Edinburgh: Earth Sciences*, **77**, 231–59.

Owen, A. W. and Bruton, D. L. (1980) Late Caradoc–early Ashgill trilobites of the central Oslo Region, Norway. *Paleontological Contributions of the University of Oslo*, **245**, 1–63.

Owen, A. W., Tripp, R. P. and Morris, S. F. (1986) The trilobite fauna of the Raheen Formation (upper Caradoc), Co. Waterford, Ireland. *Bulletin of the British Museum (Natural History) Geology Series*, **40**, 91–122.

Owens, R. M. (1983) A review of Permian trilobite genera, in *Trilobites and Other Early Arthropods: Papers in Honour of Professor H. B. Whittington, F. R. S.* (eds D. E. G. Briggs and P. D. Lane), Special Papers in Palaeontology, **30**, pp. 15–41.

Owens, R. M., Fortey, R. A., Cope, J. C. W. *et al.* (1981) Tremadoc faunas from the Carmarthen district, South Wales. *Geological Magazine*, **119**, 1–112.

Owens, R. M. and Hammann, W. (1990) Proetide trilobites from the Cystoid Limestone (Ashgill) of NW Spain, and the supra-generic classification of related forms. *Palaeontologisches Zeitschrift*, **64**, 221–44.

Palmer, A. R. (1962) *Glyptagnostus* and associated trilobites in the United States. *US Geological Survey Professional Paper*, **374F**, 49 pp.

Palmer, A. R. (1964) An unusual Lower Cambrian triolobite fauna from Nevada. *US Geological Survey Professional Paper*, **483F**, 13 pp.

Palmer, A. R. and Halley, R. B. (1979) Physical stratigraphy and trilobite biostratigraphy of the Carrara Formation (Lower and Middle Cambrian) in the southern Great Basin. *US Geological Survey Professional Paper*, **1047**, 131 pp.

Pillola, G. L. (1989) Trilobites du Cambrien Inférieur du S. W. de la Sardaigne, Italie. Thèse, Université de Rennes I, 190 pp.

Pillola, G. L. (1990) Lithologie et trilobites du Cambrien inférieur du SW de la Sardaigne (Italie): implications paléobiogeo-graphiques. *Comptes rendues de l'Académie des Sciences*, **310**, série 2, 321–8.

Pocock, K. J. (1970) The Emuellidae, a new family of trilobites from the Lower Cambrian of South Australia. *Palaeontology*, **13**, 522–62.

Pokrovskaya, N. V. (1958) Middle Cambrian agnostids of Yakutia, part 1. *Akademiya Nauk SSSR, Trudy Geologicheskogo Instituta*, **16**, 96 pp.

Pokrovskaya, N. V. (1960) Order Miomera, in *Osnovy Paleontologii: Chlenistonogie–Trilobitoobraznye i Rakoobraznye* (ed. N. E. Chernysheva), Gosudarstvennoe Nauchno-Tekhicheskoe Izdatelstvo, Moscow, pp. 54–61.

Qian, Y.-Y. (1982) Ontogeny of *Pseudagnostus benxiensis* sp. nov. (Trilobita). *Acta Palaeontologica Sinica*, **21**, 632–44.

Rasetti, F. (1972) Cambrian faunas of Sardinia. *Atti dell'Accademie*

Nazionale Lincei, Memorie, S. 8, **11**, ser. II^a, n. 1, 1–98.

Raymond, P. E. (1913) Some changes in the names of genera of trilobites. *The Ottawa Naturalist*, **26** (11), 137–42.

Raymond, P. E. (1937) Upper Cambrian and lower Ordovician Trilobita and Ostracoda from Vermont. *Bulletin of the Geological Society of America*, **48**, 1079–145.

Repina, L. N. (1979) Dependence of morphologic features on habitat conditions in trilobites and evaluation of their significance for the systematics of the superfamily Olenelloidea. *Trudy Institute of Geology and Geophysics, Siberian Section, Academy of Sciences USSR*, **431**, 11–30.

Repina, L. N., Lazarenko, N. P., Meshkova, N. P. *et al.* (1974) *Biostratigraphy and Fauna of the Lower Cambrian of Kharaulakh (Tuora-Sis Ridge)*. 'Nauka', Moscow, 299 pp. [in Russian].

Resser, C. E. and Howell, B. F. (1938) Lower Cambrian *Olenellus* Zone of the Appalachians. *Bulletin of the Geological Society of America*, **49**, 195–248.

Richter, R. and Richter, E. (1940) Studien im Paläozoikum der Mittelmeer-Länder, 5, Die Saukianda-Stufe von Andalusien, eine fremade Fauna im europaischen Ober-Kambrium. *Senckenberg naturforschend Gesellschaft (Frankfurt a. M.), Abhandlungen*, **450**, 1–81.

Rietschel, S. (1964) *Aulacopleurina* (Trilobita) aus der südlichen Lahnmulde (Rheinisches Schiefergebirge). *Senckenbergiana Lethaea*, **45**, 135–49.

Robison, R. A. (1964) Late Middle Cambrian faunas from western Utah. *Journal of Paleontology*, **38**, 510–66.

Ross, R. J. (1951) Stratigraphy of the Garden City Formation in Northeastern Utah, and its trilobite faunas. *Bulletin of the Peabody Museum of Natural History, Yale University*, **6**, 161 pp.

Rusconi, C. (1951) Mas trilobitas Cambricos de San Isidro, Cerro Pelado y Canota. *Revista del Museo Historia Natural de Mendoza*, **5**, 3–30.

Rushton, A. W. A. (1966) The Cambrian trilobites from the Purley shales of Warwickshire. *Palaeontolographical Society Monograph*, **120** [for 1966], 1–55.

Rushton, A. W. A. (1979) A review of the Middle Cambrian Agnostida from the Abbey Shales, England. *Alcheringa*, **3**, 43–61.

Rushton, A. W. A. (1983) Trilobites from the Upper Cambrian *Olenus* Zone in central England, in *Trilobites and other Early Arthropods – Papers in Honour of Professor H. B. Whittington, F.R.S.* (eds D. E. G. Briggs and P. D. Lane), Special Papers in Palaeontology, **30**, pp. 107–39.

Salter, J. W. (1864) A monograph of the British trilobites from the Cambrian, Silurian and Devonian formations. *Palaeontographical Society Monograph*, **16**, 1–80.

Sdzuy, K. (1955) Die Fauna der Leimitz-Schiefer (Tremadoc). *Abhandlungen von der Senckenbergischen Naturforschenden Gesellschaft*, **492**, 1–74.

Shergold, J. H. (1975) Late Cambrian and Early Ordovician Trilobites from the Burke River Structural Belt, Western Queensland, Australia. *Bureau of Mineral Resources of Australia, Bulletin*, **153**, v. 1, 251 pp., v. 2, 58 pls.

Shergold, J. H. (1980) Late Cambrian trilobites from the Chatsworth Limestone, western Queensland. *Bulletin of the Bureau of Mineral Resources, Geology and Geophysics of Australia*, **186**, 111 pp.

Shergold, J. H. (1991) Protaspis and early meraspid growth stages of the eodiscoid trilobite *Pagetia ocellata* Jell, and their implications for classification. *Alcheringa*, **15**, 65–86.

Shergold, J. H., Laurie, J. R. and Sun X.-W. (1990) Classification and review of the trilobite order Agnostida Salter, 1864: an Australian perspective. *Bureau of Mineral Resources of Australia. Report*, **296**, 92 pp.

Šnajdr, M. (1958) Trilobiti českeho středniho Kambria. *Rozpravy Ústředniho Ustavu Geologického*, **24**, 280 pp. [English summary, pp. 237–80].

Šnajdr, M. (1980) Bohemian Silurian and Devonian Proetidae (Trilobita). *Rozpravy Ústředniho Ustavu Geologického*, **45**, 324 pp.

Šnajdr, M. (1983) Revision of the trilobite type material of I. Hawle and A. J. C. Corda, 1847. *Sbornik Národního Muzea v Praze. Acta Musei Nationalis Pragae*, **39** (B) (3), 129–212.

Spizharski, T. N., Zhuravleva, I. T., Repina, L. N. *et al.* (1986) The Stage Scale of the Cambrian System. *Geological Magazine*, **123**, 387–92.

Strusz, D. L. (1980) The Encrinuridae and related trilobite families with a description of Silurian species from southeastern Australia. *Palaeontographica Abteilung A*, **168**, 1–68.

Struve, W. (1976) *Phacops (Omegops)* n. sg. (Trilobita; Ober-Devon). *Senckenbergiana Lethaea*, **56**, 429–51.

Suvorova, N. P. (1960) [Cambrian trilobites of the eastern Siberian Platform. 2. Olenellidae–Granulariidae]. *Trudy paleontologiskogo Instituta [Moscow]*, **84**, 1–238.

Suvorova, N. P. (1964) [Corynexochid trilobites and their historical development]. *Trudy paleontologiskogo Instituta [Moscow]*, **103**, 319 pp.

Theokritoff, G. (1984) Early Cambrian biogeography in the North Atlantic region. *Lethaia*, **18**, 283–93.

Thomas, A. T. (1978) British Wenlock Trilobites. *Monograph of the Palaeontographical Society, London*, **132**, 1–56.

Thomas, A. T. and Holloway, D. J. (1988) Classification and phylogeny of the trilobite order Lichida. *Philosophical Transactions of the Royal Society of London, Series B*, **321**, 179–262.

Thomas, A. T. and Owens, R. M. (1978) A review of the trilobite family Aulacopleuridae. *Palaeontology*, **21**, 65–81.

Thomas, A. T., Owens, R. M. and Rushton, A. W. A. (1984) Trilobites in British stratigraphy. *Geological Society of London, Special Report*, **16**, 78 pp.

Tjernvik, T. E. (1956) On the early Ordovician of Sweden. *Bulletin of the Geological Institution of Uppsala*, **36**, 107–284.

Tomczykowa, E. (1987) [Taxonomy of Staurocephalidae Prantl et Přibyl, 1947 in connection with morphology of the anterior cranidial border.] Biuletyn Instytutu Geologicznego (Warszawa), **354**, 183–96 [in Polish with English summary].

Tripp, R. P. (1954) Caradocian trilobites from the mudstones at Craighead Quarry, near Girvan, Ayrshire. *Transactions of the Royal Society of Edinburgh: Earth Sciences*, **62**, 655–93.

Tripp, R. P. (1980) Trilobites from the Ordovician Balclatchie and lower Ardwell groups of the Girvan district, Scotland. *Transactions of the Royal Society of Edinburgh: Earth Sciences*, **71**, 123–45.

Walcott, C. D. (1913) The Cambrian faunas of China. *Carnegie Inst. (Washington) Publication 54, Research in China*, **3**, 1–276.

Westergård, A. H. (1936) *Paradoxides oelandicus* Beds of Öland. *Sveriges Geologiska Undersökning, Series C*, **394**, 66 pp.

Westergård, A. H. (1944) Borrinngar Genom Skånes Alunskiffer 1941–42. *Sveriges Geologiska Undersökning, Series C*, **459**, 45 pp.

Westergård, A. H. (1946) Agnostidea of the Middle Cambrian of Sweden. *Sveriges Geologiska Undersökning, Series C*, **477**, 141 pp.

Westergård, A. H. (1947) Supplementary notes on the Upper Cambrian trilobites of Sweden. *Sveriges Geologiska Undersökning, Series C*, **489**, 1–34.

Westergård, A. H. (1948) Non-agnostidean trilobites of the Middle Cambrian of Sweden. *Sveriges Geologiska Undersökning, Avhandlingar och uppsatser, Series C*, **498**, 32 pp.

Whitehouse, F. W. (1936) The Cambrian faunas of northeastern Australia. Parts 1 and 2. *Memoirs of the Queensland Museum*, **11**, 59–112.

Whittard, W. F. (1952) Cyclopygid trilobites from Girvan and a note on *Bohemilla*. *Bulletin of the British Museum of Natural History, Geology*, **1**, 305–24.

Whittard, W. F. (1960) The Ordovician trilobites of the Shelve Inlier. *Palaeontographical Society [Monograph]*, **4**, 117–62.

Whittard, W. F. (1961) The Ordovician trilobites of the Shelve Inlier. *Palaeontographical Society [Monograph]*, **5**, 163–96.

Whittington, H. B. (1950) Sixteen Ordovician genotype trilobites. *Journal of Paleontology*, **24**, 531–65.

Whittington, H. B. (1953) North American Bathyuridae and Leiostegiidae (Trilobita). *Journal of Paleontology*, **27**, 647–78.

Whittington, H. B. (1965) Trilobites of the Ordovician Table Head

Formation, Western Newfoundland. *Bulletin of the Museum of Comparative Zoology, Harvard*, **132**, 275–442.
Whittington, H. B. (1988) Hypostomes and ventral cephalic sutures in Cambrian trilobites. *Palaeontology*, **31**, 577–609.
Wolfart, R. (1968) Die Trilobiten aus dem Devon Boliviens und ihre Bedeutung für Stratigraphie und Tiergeographie. *Beihefte zum Geologischen Jahrbuch*, **74**, 5–201.
Zhang, W.-T., Lu, Y.-H., Zhu, Z.-L. *et al.* (1980) Cambrian trilobite faunas of southwestern China. *Palaeontologia Sinica*, **159**, *New Series*, no. 16, 497 pp.

17

ARTHROPODA (AGLASPIDIDA, PYCNOGONIDA AND CHELICERATA)

P. A. Selden

Class AGLASPIDIDA Walcott, 1911 (see Fig. 17.1)

Aglaspidida were removed from Chelicerata by Briggs *et al.* (1979).

F. PALEOMERIDAE Størmer, 1955 Є. (SOL) Mar.

First and Last (monotypic family): *Paleomerus hamiltoni* Størmer, 1955, Kinnekulle, Sweden.
Comments: This animal was described in detail by Størmer (1956), in a paper communicated to the *Kunglige Svenska Akademien* in 1955 but not published until 5 December 1956; meanwhile, Størmer had included the new family, genus and species in the *Treatise*, published in 1955, so the latter publication included the first description of the new taxa. Bergström (1971) has expressed doubts that this family belongs in the Aglaspidida.

F. AGLASPIDIDAE Miller, 1877 Є. (MEN–DOL) Mar.

First: *Beckwithia typa* Resser, 1931, probably upper Weeks Formation, Weeks Canyon, Utah, USA. This form was reviewed by Hesselbo (1989), who removed it from (thereby suppressing) the monotypic family Beckwithiidae.
Last: *Aglaspis spinifer* Raasch, 1939, *A. simplex* Raasch, 1939, *Aglaspella granulifera* Raasch, 1939, *A. eatoni* Whitfield, 1880, *Glypharthrus thomasi* (Walter, 1924), *Aglaspoides sculptilis* Raasch, 1939, *Uarthrus instabilis* Raasch, 1939, *Cyclopites vulgaris* (Raasch, 1939), *Craspedops modesta* Raasch, 1939, *Setaspis spinulosa* Raasch, 1939, Lodi Shale, St Lawrence Formation, Point Jude, Richland County, Wisconsin, USA.

F. STRABOPIDAE Gerhardt, 1932 Є. (DOL)–O. (PUS) Mar.

First: *Strabops thacheri* Beecher, 1901, Potosi Dolomite, St Lawrence Formation, St François County, Missouri, USA.
Last: *Neostrabops martini* Caster and Macke, 1952, Maysville Formation, Ohio, USA.
Comment: Bergström (1971) doubted that this family belonged in the Aglaspidida.

F. SINAGLASPIDAE Hong and Niu, 1981 P. (ASS) Mar.

First and Last (monotypic family): *Sinaglaspis xiashanensis* Hong and Niu, 1981, lower Shanxi Formation, Gancaoshan, Xiangning County, Shanxi Province, China.

Class PYCNOGONIDA Latreille, 1810

The classification of Bergström *et al.* (1980) is used here.

First?: A larval form (D), more comparable with pycnogonids than any other arthropod group, was described by Müller and Walossek (1986) from Upper Cambrian Orsten of Sweden, Є. (MNT).

Order PALAEOISOPODA Bergström *et al.*, 1980

F. PALAEOISOPODIDAE Dubinin, 1957 D. (PRA) Mar.

First and Last (monotypic family and order): *Palaeoisopus problematicus* Broili, 1928, Hunsrückschiefer, Bundenbach, Germany.

Order PALAEOPANTOPODA Broili, 1930

F. PALAEOPANTOPODIDAE Hedgpeth, 1955 D. (PRA) Mar.

First and Last (monotypic family and order): *Palaeopantopus maucheri* Broili, 1929, Hunsrückschiefer, Bundenbach, Germany.

Order PANTOPODA Gerstaecker, 1863 D. (PRA)–Rec. Mar.

First: *Palaeothea devonica* Bergström *et al.*, 1980, Hunsrückschiefer, Bundenbach, Germany. **Extant**
Comment: Bergström *et al.* (1980) declined to place this form in a family; it has *incertae sedis* status as the only known fossil in the order Pantopoda.

Phylum CHELICERATA Heymons, 1901

Sanctacaris Є. (MEN) Mar.

First: *Sanctacaris uncata* Briggs and Collins, 1988, Burgess Shale, British Columbia, Canada. Plesion: plesiomorphic sister taxon to all other Chelicerata.
Comment: The traditional division of the Chelicerata into primarily marine Merostomata and primarily terrestrial Arachnida is untenable. Most authors consider Xiphosura to be a sister group to all other Chelicerata (except *Sanctacaris*) and not to Eurypterida alone. Many authors place Scorpionida as a sister group to Eurypterida, while others consider Eurypterida to be the sister group of all other chelicerates except Xiphosura and *Sanctacaris* (i.e. Arachnida). Recently, a novel phylogeny was put forward by Shultz (1990). See Selden (1990) for a discussion.

A particular problem with the terrestrial arachnid groups, such as pseudoscorpions, spiders and mites, is that the majority of fossils are known from Tertiary ambers, the dating of which is insecure. Apart from the difficulty of dating the amber pieces, many specimens which come to be

The Fossil Record 2. Edited by M. J. Benton. Published in 1993 by Chapman & Hall, London. ISBN 0 412 39380 8

QU.: HOL, PLE
TERTIARY: PLI, UMI, MMI, LMI, CHT, RUP, PRB, BRT, LUT, YPR, THA, DAN
CRETACEOUS: MAA, CMP, SAN, CON, TUR, CEN, ALB, APT, BRM, HAU, VLG, BER
JURASSIC: TTH, KIM, OXF, CLV, BTH, BAJ, AAL, TOA, PLB, SIN, HET
TRIASSIC: RHT, NOR, CRN, LAD, ANS, SCY

23 24

7 22 25 54 56

Key for both diagrams

AGLASPIDIDA	27. Hughmilleriidae
1. Paleomeridae	28. Carcinosomatidae
2. Aglaspididae	29. Adelophthalmidae
3. Strabopidae	30. Mixopteridae
4. Sinaglaspidae	31. Lanarkopteridae
PYCNOGONIDA	32. Megalograptidae
5. Palaeoisopodidae	33. Eurypteridae
6. Palaeopantopodidae	34. Dolichopteridae
7. Pantopoda	35. Erieopteridae
CHELICERATA	36. Stylonuridae
XIPHOSURA	37. Drepanopteridae
8. Sanctacaris	38. Parastylonuridae
9. Eolimulidae	39. Laurieopteridae
10. Chasmataspidae	40. Kokomopteridae
11. Diploaspididae	41. Hardieopteridae
12. Heteroaspididae	42. Brachyopterellidae
13. Weinberginidae	43. Rhenopteridae
14. Bunodidae	44. Mycteropidae
15. Pseudoniscidae	45. Woodwardopteridae
16. Elleriidae	46. Pterygotidae
17. Bellinuridae	47. Jaekelopteridae
18. Euproopidae	48. Hibbertopteridae
19. Liomesaspidae	49. Cyrtoctenidae
20. Rolfeiidae	SCORPIONIDA
21. Moravuridae	50. Protoscorpiones
22. Paleolimulidae	51. Proscorpioidea
23. Austrolimulidae	52. Archaeoctonoidea
24. Heterolimulidae	53. Praearcturus
25. Limulidae	54. Mesoscorpionina
EURYPTERIDA	55. Paleosterni
26. Slimoniidae	56. Orthosterni

Fig. 17.1

described have an unspecified provenance, thus rendering their dating by fossils in associated sediments impossible. Dominican amber occurs at numerous sites, spanning an interval of possibly 25–40 Ma BP (Lambert *et al.*, 1985), but probably the majority are at the younger end of that range. There is no consensus regarding the relative ages among Tertiary ambers, and correlation is understandably difficult. For the purposes of this work, Baltic amber is taken to be T. (RUP) in age, and Dominican, Mexican and most other Tertiary ambers with arachnids are placed as T. (CHT) in age.

Class XIPHOSURA Latreille, 1802

Recent ideas on the phylogeny and classification of the class Xiphosura were discussed by Selden and Siveter (1987), and their classification is followed here.

F. EOLIMULIDAE Bergström, 1968
Є. (LEN) Mar.

First and Last (monotypic family): *Eolimulus alatus* (Moberg, 1892), Ekerum, Öland, Sweden (Bergström, 1968).

Comment: The carapace only is known, so the xiphosuran identity of this animal is uncertain; if confirmed, this would be the oldest known xiphosuran.

Order CHASMATASPIDIDA Caster and Brooks, 1956

A large arthropod from the Devonian (GIV/FRS?) of

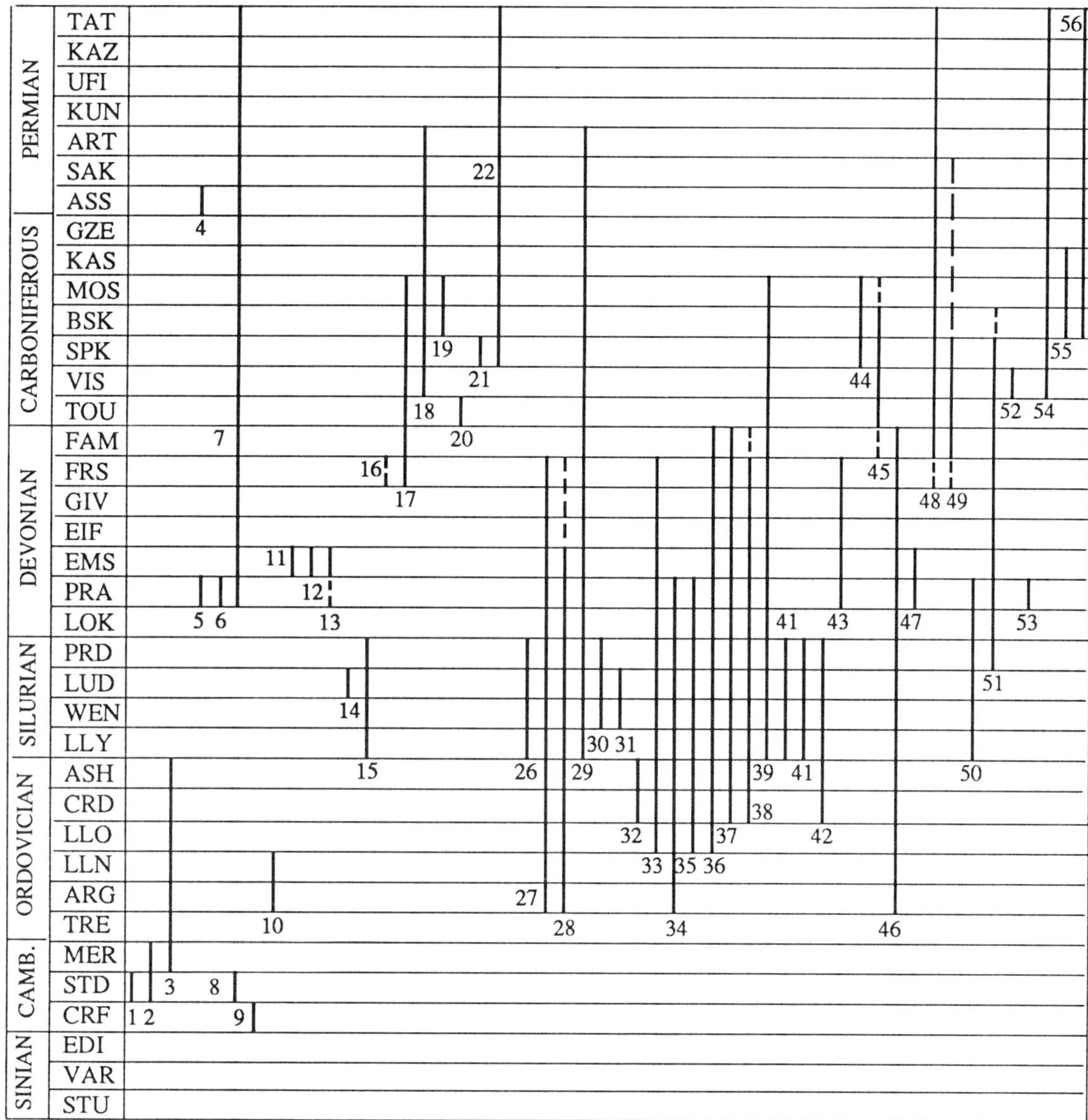

Fig. 17.1

Vietnam was referred to as Chasmataspidida? gen. *et* sp. indet. by Janvier *et al.* (1989). Study of the fossil by P. A. Selden (unpublished) suggests that it is a carcinosomatid eurypterid.

F. CHASMATASPIDAE Caster and Brooks, 1956 O. (ARG/LLN1) Mar.

First and Last (monotypic family): *Chasmataspis laurencii* Caster and Brooks, 1956, Douglas Dam, Tennessee, USA.

F. DIPLOASPIDIDAE Størmer, 1972 D. (EMS) Mar. ?FW

First and Last (monotypic family): *Diploaspis casteri* Størmer, 1972, Alken-an-der-Mosel, Germany.

F. HETEROASPIDIDAE Størmer, 1972 D. (EMS) Mar. ?FW

First and Last (monotypic family): *Heteroaspis novojilovi* Størmer, 1972, Alken-an-der-Mosel, Germany.

Order XIPHOSURIDA Latreille, 1802

Suborder SYNZIPHOSURINA Packard, 1886

A possible synziphosurine (unnamed), from the Brandon Bridge Formation, Waukesha County, Wisconsin, S. (TEL), was described briefly by Mikulic *et al.* (1985).

F. WEINBERGINIDAE Richter and Richter, 1929 D. (PRA/EMS–PRA) Mar.

First: *Legrandella lombardii* Eldredge, 1974, Rumicorral, Cochabamba Province, Bolivia.

Last: *Weinbergina opitzi* Richter and Richter, 1929, Hunsrückschiefer, Bundenbach, Germany.

F. BUNODIDAE Packard, 1886 S. (GOR–LDF) Mar.

First: *Bunodes lunula* Eichwald, 1854, *Eurypterus* Dolomite, Saaremaa, Estonia, former USSR.

Last: *Limuloides limuloides* Salter, *in* Woodward, 1865, Leintwardine Group, Leintwardine, Shropshire, England, UK.

Suborder LIMULINA Richter and Richter, 1929

Infraorder PSEUDONISCINA Eldredge, 1974

Superfamily PSEUDONISCOIDEA Packard, 1886

F. PSEUDONISCIDAE Packard, 1886 S. (TEL–PRD) Mar.

First: *Cyamocephalus loganensis* Currie, 1927, Patrick Burn Formation, Logan Water, Lesmahagow, Scotland, UK.

Last: *Pseudoniscus clarkei* Ruedemann, 1916, and ?*Bunaia woodwardi* Clarke, 1919, Bertie Waterlime, New York, USA.
Intervening: GOR

Infraorder LIMULICINA Richter and Richter, 1929

Superfamily BELLINUROIDEA Zittel and Eastman, 1913

Most authors (e.g. Eldredge, 1974; Fisher, 1982; Raymond, 1944; Selden and Siveter, 1987) agree that this group is the paraphyletic stem group which gave rise to the Euproopoidea and the Limuloidea. Eldredge (1974) put the 'primitive' bellinuroids (e.g. *Bellinuroopsis rossicus* Chernyshev, 1933, *Bellinurus bellulus* König, 1851) in an unnamed family within the Pseudoniscoidea, but later work, shown in the phylogenetic diagrams of Fisher (1982, fig. 1; 1984, fig. 2) and Selden and Siveter (1987, fig. 2) depict their relationships more clearly.
Intervening (species not included in families below): D. (?FRS, *Bellinuroopsis rossicus* Chernyshev, 1933, top horizons of Devonian section, right bank of River Don, near Lebedian, former USSR ?FAM, *'Paleolimulus?' randalli* (Beecher, 1902), Chemung Sandstone, Pennsylvania, USA.

F. ELLERIIDAE Raymond, 1944 D. (?FRS) Mar./?FW

First and Last (monotypic family): *Elleria morani* (Eller, 1938), Salamanca Sandstone, North Warren, Pennsylvania, USA.
Comment: Elleriidae was placed in Bellinuroidea by Bergström (1975) and Fisher (1982), but Siveter and Selden (1987) considered this placement unsupportable on the basis of the morphological evidence.

F. BELLINURIDAE Zittel and Eastman, 1913 D.(?GIV)–C. (KSK) Mar./FW

First: *Bellinurus carteri* Eller, 1940, lower Cattaraugus Beds, Bradford, Pennsylvania, USA.
Last: *Bellinurus trechmanni* Woodward, 1918, Clauxheugh, County Durham, England, UK.
Intervening: FRS, FAM, FAM/HAS, IVO, CHE, MEL, VRK.

Superfamily EUPROOPOIDEA Eller, 1938

F. EUPROOPIDAE Eller, 1938 C. (CHD/BRI)–P. (ART) Mar./FW

First: *Euproops thompsoni* Raymond, 1944, Windsor Group, Cumberland County, Nova Scotia, Canada.
Last: *Anacontium carpenteri* Raymond, 1944, *A. brevis* Raymond, 1944, Wellington Formation, Noble County, Oklahoma, USA.
Intervening: ALP, YEA–CHE, CHE, MEL, VRK, KSK, POD, MYA, MYA/KRE, KLA.

F. LIOMESASPIDAE Raymond, 1944 C. (CHE–MYA) Mar./FW

First: *Pringlia birtwelli* (Woodward, 1872), ?Soapstone Bed, above Lower Mountain Mine, Cornfield Pit, Padiham, Lancashire, England, UK.
Last: *Prolimulus woodwardi* Frič, 1899, Nýřany, Czechoslovakia.

Superfamily LIMULOIDEA Zittel, 1885

F. ROLFEIIDAE Selden and Siveter, 1987 C. (IVO) Mar.

First and Last (monotypic family): *Rolfeia fouldenensis* Waterston, 1985, Foulden, Berwickshire, Scotland, UK.

F. MORAVURIDAE Přibyl, 1967 C. (PND) Mar.

First and Last: *Moravurus rehori* Přibyl, 1967, Kyjovice Formation, Tichá, Czechoslovakia; *Xaniopyramis linseyi* Siveter and Selden, 1987, above Great Limestone, Weardale, England, UK.

F. PALEOLIMULIDAE Raymond, 1944 C. (ALP)–J. (HET) Mar./FW

First: *Paleolimulus? longispinus* Schram, 1979, Bear Gulch Limestone, Fergus County, Montana, USA.
Last: *Limulitella* cf. *bronni* Schimper, 1850, Helmstedt, Germany.
Intervening: KRE–NOG, MYA, ASS, SAK, ART, GRI–SPA, ANS, NOR, RHT.

F. AUSTROLIMULIDAE Riek, 1968 Tr. (LAD) FW

First and Last (monotypic family): *Austrolimulus fletcheri* Riek, 1955, Hawkesbury Sandstone, Brookvale, New South Wales, Australia.

F. HETEROLIMULIDAE Vía Boada and Villalta, 1966 Tr. (LAD) Mar.

First and Last (monotypic family): *Heterolimulus gadeai* Vía Boada and Villalta, 1966, Canteras de Montral-Alcover, Tarragona Province, Spain.

F. LIMULIDAE Zittel, 1885 Tr. (SPA)–Rec. Mar./FW

First: *Psammolimulus gottingensis* Lange, 1922, Göttingen, Germany.
Intervening: LAD, HET–TOA, OXF/KIM, TTH, APT, CEN, CMP, MAA, BUR. **Extant**

Class EURYPTERIDA Burmeister, 1843

The classification of Tollerton (1989) is used here. A number of genera were excluded from the Order Eurypterida by Tollerton, but their removal has no overall effect on the ranges given below. Tollerton (1989) also excluded the hibbertopteroids, which had been returned to the Order Eurypterida by Waterston *et al.* (1985); here, they are included in the Order Cyrtoctenida, a taxon originally proposed by Størmer and Waterston (1968) for these unusual forms. Eurypterids occur mainly in marginal marine facies, lacking good marine, stratigraphically useful, fossils; therefore, the dating of many of the horizons is under constant review.

Order EURYPTERIDA Burmeister, 1843

Suborder EURYPTERINA Burmeister, 1843

Superfamily SLIMONIOIDEA Novojilov, 1962

F. SLIMONIIDAE Novojilov, 1962 S. (TEL–PRD) Mar./?FW

First: *Slimonia dubia* Laurie, 1899, Reservoir Formation,

Gutterford Burn, Pentland Hills, Scotland, UK; *Slimonia acuminata* (Salter, 1856), Patrick Burn and Kip Burn Formations, Lesmahagow, Scotland, UK.
Last: *Salteropterus abbreviatus* (Kjellesvig-Waering, 1951), Temeside Shale Formation, Perton, Herefordshire, England, UK.
Intervening: LDF
Comment: A *Slimonia* specimen from Lochkov, Czechoslovakia (Prantl and Přibyl, 1948), D. (GED) was studied by Kjellesvig-Waering (1973) who doubted the identification.

Superfamily HUGHMILLERIOIDEA Kjellesvig-Waering, 1951

F. HUGHMILLERIIDAE Kjellesvig-Waering, 1951 O. (ARG)–D. (FRS) Mar.

First: *Waeringopterus? priscus* (Ruedemann, 1942) Deepkill Formation, Mt. Merino, Hudson, New York, USA.
Last: *Grossopterus? inexpectans* (Ruedemann, 1919), Oneonta Formation, Gilboa, New York, USA.
Intervening: COS–ONN, GED, SIG.

F. CARCINOSOMATIDAE Størmer, 1934 O. (ARG)–D. (EMS, possibly GIV/FRS?) Mar.

First: *Eocarcinosoma ruedemanni* (Flower, 1945), Deepkill Formation, Rensselaer County, New York, USA.
Last: *Carcinosoma* sp., Alken-an-der-Mosel, Germany (Størmer, 1974).
Intervening: COS–ONN, HIR, RHU–GLE, FRO, TEL, SHE, GLE, GOR, LDF, PRD, GED.
Comment: A probable carcinosomatid, referred to under Chasmataspidida above, was reported by Janvier *et al.* (1989) from the Grey Devonian of Dô Son, Haïphong, Vietnam, which these authors considered may be D. (GIV/FRS).

F. ADELOPHTHALMIDAE Tollerton, 1989 S. (RHU/FRO)–P. (ART) Mar./FW

First: *Parahughmilleria maria* (Clarke, 1907), Tuscarora Formation, Swatara Gap, Pennsylvania, USA.
Last: *Adelophthalmus sellardsi* (Dunbar, 1924) Wellington Formation, Dickinson County, Kansas, and Red Rock, Oklahoma, USA.
Intervening: TEL, SHE, LDF, PRD, LOK, PRA, EMS, FAM, HLK, MRD, CHE, MEL, VRK, KSK, POD, MYA, NOG, ASS, SAK.

Superfamily MIXOPTEROIDEA Caster and Kjellesvig-Waering, 1955

F. MIXOPTERIDAE Caster and Kjellesvig-Waering, 1955 S. (SHE/GLE–PRD) Mar.

First: *Mixopterus?* sp., Lower Sintan Formation, Hsin Tan, Hubei, China (Chang, 1957).
Last: *Mixopterus multispinosus* (Clarke and Ruedemann, 1912), Vernon Shales, New York, USA; *M. kiaeri* Størmer, 1934, Sundvolle Formation, Ringerike, Norway; *M.* sp., Perton, Herefordshire, England, UK (Kjellesvig-Waering, 1951).
Intervening: LDF.

F. LANARKOPTERIDAE Tollerton, 1989 S. (GLE/GOR) Mar.

First and Last (monotypic family): *Lanarkopterus dolichoschelus* Ritchie, 1968, Fish Beds of Hagshaw Hills and Lesmahagow, Scotland, UK.

Superfamily MEGALOGRAPTOIDEA Caster and Kjellesvig-Waering, 1955

F. MEGALOGRAPTIDAE Caster and Kjellesvig-Waering, 1955 O. (COS/ONN–HIR) Mar.

First: *Echinognathus clevelandi* Walcott, 1882, Utica Shales, Oneida County, New York, USA; *Megalograptus alveolatus* Caster and Kjellesvig-Waering, 1964, Martinsburg Formation, Walker Mountain, Virginia, USA.
Last: *Megalograptus ohioensis* Caster and Kjellesvig-Waering, 1964, Elkhorn Formation, Manchester, Ohio, USA.
Intervening: CAU, RAW.

Superfamily EURYPTEROIDEA Burmeister, 1843

F. EURYPTERIDAE Burmeister, 1843 O. (LLO)–D. (FRS) Mar./FW

First: *Eurypterus? decipiens* Ruedemann, 1942, Normanskill Grit, Albany County, New York, USA.
Last: *Eurypterus?* sp. Bergisch Gladbach, Germany (Jux, 1967).
Intervening: COS–ONN, TEL, SHE, WHI, GLE, GOR, LDF, PRD, LOK, EMS.

F. DOLICHOPTERIDAE Kjellesvig-Waering and Størmer, 1952 O. (ARG)–D. (SIG) Mar.

First: *Dolichopterus antiquus* Ruedemann, 1942, Deepkill Formation, Merino, Hudson, New York, USA.
Last: *Strobilopterus princetonii* Ruedemann, 1935, Beartooth Butte, Wyoming, USA.
Intervening: COS–ONN, FRO–LDF, SHE, GOR, PRD, LOK.

F. ERIEOPTERIDAE Tollerton, 1989 O. (LLO)–D. (SIG) Mar./?FW

First: *Erieopterus chadwicki* Clarke and Ruedemann, 1912, Normanskill Shale, Catskill, New York, USA.
Last: *Erieopterus statzii* Størmer, 1936, upper Siegener Sandstone, Wahnbachtal, Siegburg, Germany; *Erieopterus latus* Ruedemann, 1935, Beartooth Butte, Wyoming, USA.
Intervening: MRB, RHU, GOR, PRD, LOK.

Superfamily STYLONUROIDEA Diener, 1924) (=DREPANOPTEROIDEA Kjellesvig-Waering, 1966)

F. STYLONURIDAE Diener, 1924 O. (LLO)–D. (FAM) Mar./?FW

First: *Stylonurella? modestus* (Clarke and Ruedemann, 1912), Normanskill Shale, Catskill, New York, USA.
Last: *Stylonurus? shaffneri* Willard, 1933, Canadaway Formation, Galeton, Pennsylvania, and *Stylonurella? arnoldi* (Ehlers, 1935), Chadakoin Formation, Bush Hill, Pennsylvania, USA.
Intervening: TEL, SHE, WHI, GLE, GOR, LDF, LOK, EMS, ?EIF.

F. DREPANOPTERIDAE Kjellesvig-Waering, 1966 O. (COS/ONN)–D. (FAM) Mar./?FW

First: *Drepanopterus? ruedemanni* (O'Connell, 1916), Schenectady Formation, Schenectady, New York, USA.
Last: *Drepanopterus abonensis* Simpson, 1951, Portishead, Somerset, England, UK.

Intervening: TEL, PRD, GED, EMS.

F. PARASTYLONURIDAE Waterston, 1979 O. (HAR)–D. (?FAM) Mar./?FW

First: *Brachyopterus stubblefieldi* Størmer, 1951, Bausley House Shales, Abberley, Montgomeryshire, Wales, UK.
Last: *Parastylonurus? beecheri* (Hall, 1884), ?Conewango Formation, Warren, Pennslyvania, USA.
Intervening: COS–ONN, TEL, GOR, LDF, PRD.

F. LAURIEOPTERIDAE Kjellesvig-Waering, 1966 S. (TEL)–C. (MYA) Mar./FW

First: *Laurieopterus elegans* (Laurie, 1899), Reservoir Formation, Gutterford Burn, Pentland Hills, Scotland, UK.
Last: *Mazonipterus cyclophthalmus* Kjellesvig-Waering, 1963, Francis Creek Shale, Mazon Creek, Illinois, USA.
Intervening: FRO–LDF, GED, LOK/PRA, FRS, FRS/FAM.

Superfamily KOKOMOPTEROIDEA Kjellesvig-Waering, 1966

F. KOKOMOPTERIDAE Kjellesvig-Waering, 1966 S. (TEL–PRD) Mar./?FW

First: *Lamontopterus knoxae* (Lamont, 1955), Reservoir Formation, Gutterford Burn, Pentland Hills, Scotland, UK.
Last: *Kokomopterus longicaudatus* (Clarke and Ruedemann, 1912), Kokomo Formation, Kokomo, Indiana, USA.

F. HARDIEOPTERIDAE Tollerton, 1989 S. (RHU/FRO–PRD) Mar.

First: *Hardieopterus myops* (Clarke, 1907), Tuscarora Formation, Swatara Gap, Pennsylvania; Shawangunk Formation, Delaware Water Gap, Pennsylvania, USA.
Last: *Hardieopterus megalops* (Salter, 1859), Downton Castle Sandstone Formation, Ludlow, Shropshire, England, UK.
Intervening: RHU–GLE, TEL, FRO–LDF.

Superfamily BRACHYOPTERELLOIDEA Tollerton, 1989

F. BRACHYOPTERELLIDAE Tollerton, 1989 O. (COS/ONN)–S. (PRD) Mar.

First: *Brachyopterella? magna* (Clarke and Ruedemann, 1912), Schenectady Formation, Schenectady, New York, USA.
Last: *Brachyopterella pentagonalis* (Størmer, 1934), Sundvolle Formation, Ringerike, Norway.
Intervening: GOR.

Superfamily RHENOPTEROIDEA Størmer, 1951

F. RHENOPTERIDAE Størmer, 1951 D. (SIG–FRS) Mar./FW

First: *Rhenopterus tuberculatus* Størmer, 1936, Overath, Eifel, Germany.
Last: *Rhenopterus?* sp., Gogo Formation, Kimberley, Australia (Rolfe, 1966).
Intervening: EMS, GIV.

Superfamily MYCTEROPOIDEA Cope, 1886

F. MYCTEROPIDAE Cope, 1886 C. (PND–MYA) FW

First: *Mycterops? blairi* Waterston, 1968, below Johnstone Shell Bed, Limestone Coal Group, Midlothian, Scotland, UK.
Last: *Mycterops* fragments, Francis Creek Shale, Mazon Creek, Illinois, USA (Kjellesvig-Waering, 1963).
Intervening: MEL.

F. WOODWARDOPTERIDAE Kjellesvig-Waering, 1959 D. (FAM)/C. (HAS)–C. (MEL/MYA) FW

First: *Woodwardopterus scabrosus* (Woodward, 1887), Kiltorcan, Republic of Ireland.
Last: *Vernonopterus minutisculptus* (Peach, 1907), Coal Measures of Airdrie, Scotland, UK.

Suborder PTERYGOTINA Caster and Kjellesvig-Waering, 1964

Superfamily PTERYGOTOIDEA Clarke and Ruedemann, 1912

F. PTERYGOTIDAE Clarke and Ruedemann, 1912 O. (ARG)–D. (FAM) Mar.

First: *Pterygotus? deepkillensis* Ruedemann, 1934, Deepkill Formation, Deepkill, New York, USA.
Last: *Pterygotus (Pterygotus) montanensis* Ruedemann, 1935b, Three Forks Shale, Montana, USA.
Intervening: LLO, COS–ONN, RHU–GLE, RHU–FRO, FRO–LDF, TEL, SHE, GLE, GOR, LDF, PRD, GED, SIG, EMS, EIF, GIV.

F. JAEKELOPTERIDAE Størmer, 1974 D. (PRA–EMS) Mar.

First: *Jaekelopterus rhenaniae* (Jaekel, 1914), Overath, Eifel, Germany.
Last: *Jaekelopterus rhenaniae* (Jaekel, 1914), Alken-an-der-Mosel, Germany.

Order CYRTOCTENIDA Størmer and Waterston, 1968

Although highly specialized forms, cyrtoctenids may be derived from primitive eurypterids (Waterston *et al.*, 1985).

Superfamily HIBBERTOPTEROIDEA Kjellesvig-Waering, 1959

F. HIBBERTOPTERIDAE Kjellesvig-Waering, 1959 D. (?FRS)–P. (?KAZ) Mar./FW

First: *Hibbertopterus? sewardi* (Strand, 1928), Wagon Drift Formation, Witteberg Group, Grahamstown, South Africa.
Last: *Campylocephalus oculatus* (Kutorga, 1838), Copper-bearing Sandstones, Dourassoff, Urals, former USSR; *Hibbertopterus permianus* Ponomarenko, 1985, Intinskaya Suite, Inta, Komi, former USSR.
Intervening: FAM/HAS, ASB, BRI, PND–MEL, ARN, MEL, UFI–TAT.

F. CYRTOCTENIDAE Waterston *et al.*, 1985 D. (?FRS)–P. (?SAK) Mar./FW

First: *Dunsopterus? wrightianus* (Dawson, 1881), Portage Sandstone, Italy, Yates County, New York, USA.
Last: *Hastimima whitei* White, 1908, Rio Bonito Formation, Minas, Santa Catarina, Brazil, and Tatui Formation, Sao Paolo, Brazil.
Intervening: FAM, IVO, HLK, ASB, BRI, ARN.

Class SCORPIONIDA Latreille, 1810

A major new classification of Scorpionida was proposed by Kjellesvig-Waering (1986), but this posthumous publication was marred by compilation defects, and updating is required in the light of more recent work by Jeram (1990) and Stockwell (1989). The classification of Stockwell is used here without comment (it has yet to be published formally), because it conveys current understanding of the classification of scorpions in the light of unpublished work and is likely to become more widely adopted in the future than that of Kjellesvig-Waering (1986). The monotypic families and superfamilies of Kjellesvig-Waering (1986) are not used here.

A number of fossil scorpions cannot be assigned to categories of lower rank within the Scorpionida, and most of these were listed as Scorpionida *incertae sedis* by Stockwell (1989). In addition, *Hubeiscorpio gracilitarsis* Walossek *et al.*, 1990, from the Devonian of China is added to the list, and *Tiphoscorpio hueberi* Kjellesvig-Waering, 1986 is removed from Scorpionida (Selden and Shear, 1992). They range through the following stages: S. (PRD), D. (FRS), C. (HLK, MEL, KSK, POD, MYA, DOR), Tr. (ANS).

Order PROTOSCORPIONES Petrunkevitch, 1949 S. (TEL)–D. (PRA) Mar./Terr.?

First: *Dolichophonus loudonensis* (Laurie, 1899), Reservoir Formation, Gutterford Burn, Pentland Hills, Scotland, UK.
Last: *Palaeoscorpius devonicus* Lehmann, 1944, Hunsrückschiefer, Bundenbach, Germany.
Intervening: SHE, GOR/LDF.

Order PALAEOSCORPIONES Stockwell, 1989

Superfamily PROSCORPIOIDEA Scudder, 1885 S. (PRD)–C. (ALP/CHE) Mar./?FW/Terr.

First: *Proscorpius osborni* (Whitfield, 1885), *Archaeophonus eurypteroides* Kjellesvig-Waering, 1966, *Stoermeroscorpio delicatus* Kjellesvig-Waering, 1986, Bertie Waterlime, New York, USA.
Last: *Labriscorpio alliedensis* Leary, 1980, Milan, Illinois, USA.
Intervening: EMS.

Superfamily ARCHAEOCTONOIDEA Petrunkevitch, 1949 C. (HLK) ?Terr.

First and Last: *Archaeoctonus glaber* (Peach, 1883), *Pseudoarchaeoctonus denticulatus* (Kjellesvig-Waering, 1986), *Loboarchaeoctonus squamosus* (Kjellesvig-Waering, 1986), Glencartholm Volcanic Beds, Langholm, Scotland, UK.

Order SCORPIONES Hemprich and Ehrenberg, 1829 *Praearcturus* D. (PRA) Terr.

First: *Praearcturus gigas* Woodward, 1871, Rowlestone, Herefordshire, England, UK (placed as Scorpiones *incertae sedis* by Stockwell (1989)).
Intervening: many fossil scorpions cannot be assigned to categories of lower rank within the Scorpiones, and were listed as Scorpiones *incertae sedis* by Stockwell (1989). They range through the following stages: PRA, HLK, BRI, ALP, CHE, KSK, MEL, MYA, KLA, ANS, TOA.

Suborder MESOSCORPIONINA Stockwell, 1989 C. (HLK)–J. (MET/TOA) Terr.

First: *Phoxiscorpio peachi* Kjellesvig-Waering, 1986, Dalmeny, Edinburgh, Scotland, UK.
Last: *Mesophonus? maculatus* Brauer *et al.*, 1889, Ust-Balei, Irkutsk, Siberia, former USSR.
Intervening: CHE, MEL, VRK, KSK, MYA, KLA, ANS.

Suborder NEOSCORPIONINA Thorell and Lindström, 1885

Infraorder PALEOSTERNI Stockwell, 1989 C. (MEL–DOR) Terr.

First: *Allobuthiscorpius major* (Wills, 1960), Kilburn Coal, Trowell Colliery, Nottinghamshire, England, UK.
Last: *Buthiscorpius pescei* Vachon and Heyler, 1985, shales above First Blanzy-Montceau Coal, Montceau-les-Mines, France.
Intervening: VRK, KSK.

Infraorder ORTHOSTERNI Pocock, 1911 (see Fig. 17.2) C. (MEL)–Rec. Terr.

First: *Compsoscorpius elegans* Petrunkevitch, 1949, Coseley, Staffordshire, England, UK. **Extant**
Intervening: MYA, and others assigned to families below.

Superfamily BUTHOIDEA Simon, 1880

F. BUTHIDAE Simon, 1880 T. (RUP)–Rec. Terr.

First: *Tityus eogenus* Menge, 1854, Baltic amber. **Extant**
Intervening: CHT.

Superfamily SCORPIONOIDEA Peters, 1861

Araripescorpio

First: *Araripescorpio ligabuei* Campos, 1986, Santana Formation K. (APT), Chapado de Araripe, Brazil.

F. SCORPIONIDAE Peters, 1861 T. (RUP)–Rec. Terr.

First: *Scorpio schweiggeri* Holl, 1829, Baltic amber. **Extant**
Intervening: TOR/MES.

Class PSEUDOSCORPIONES Latreille, 1817

Intervening: undescribed pseudoscorpions were reported from Lebanese amber, K. (BER–APT), by Whalley (1980).

F. DRACOCHELIDAE Schawaller *et al.*, 1991 D. (GIV) Terr.

First and Last (monotypic family): *Dracochela deprehendor* Schawaller *et al.*, 1991 (protonymph and tritonymph), Gilboa Mudstones, New York, USA (Shear *et al.*, 1989; Schawaller *et al.*, 1991).
Comment: Only protonymph and tritonymph are known, which, although modern in many aspects, cannot be assigned with confidence to extant taxa because both diagnostic characters in the fossils and cladistic assessment of extant forms are lacking (Shear *et al.*, 1989; Schawaller *et al.*, 1991).

Suborder CHTHONIINA Beier, 1932

Superfamily CHTHONIOIDEA Beier, 1932

F. CHTHONIIDAE Hansen, 1894 T. (RUP)–Rec. Terr.

First: *Chthonius mengei* Beier, 1937, Baltic amber. **Extant**
Intervening: CHT.

F. DITHIDAE Chamberlin, 1931 T. (RUP)–Rec. Terr.

First: *Chelignathus kochii* (Menge, 1854), Baltic amber. **Extant**

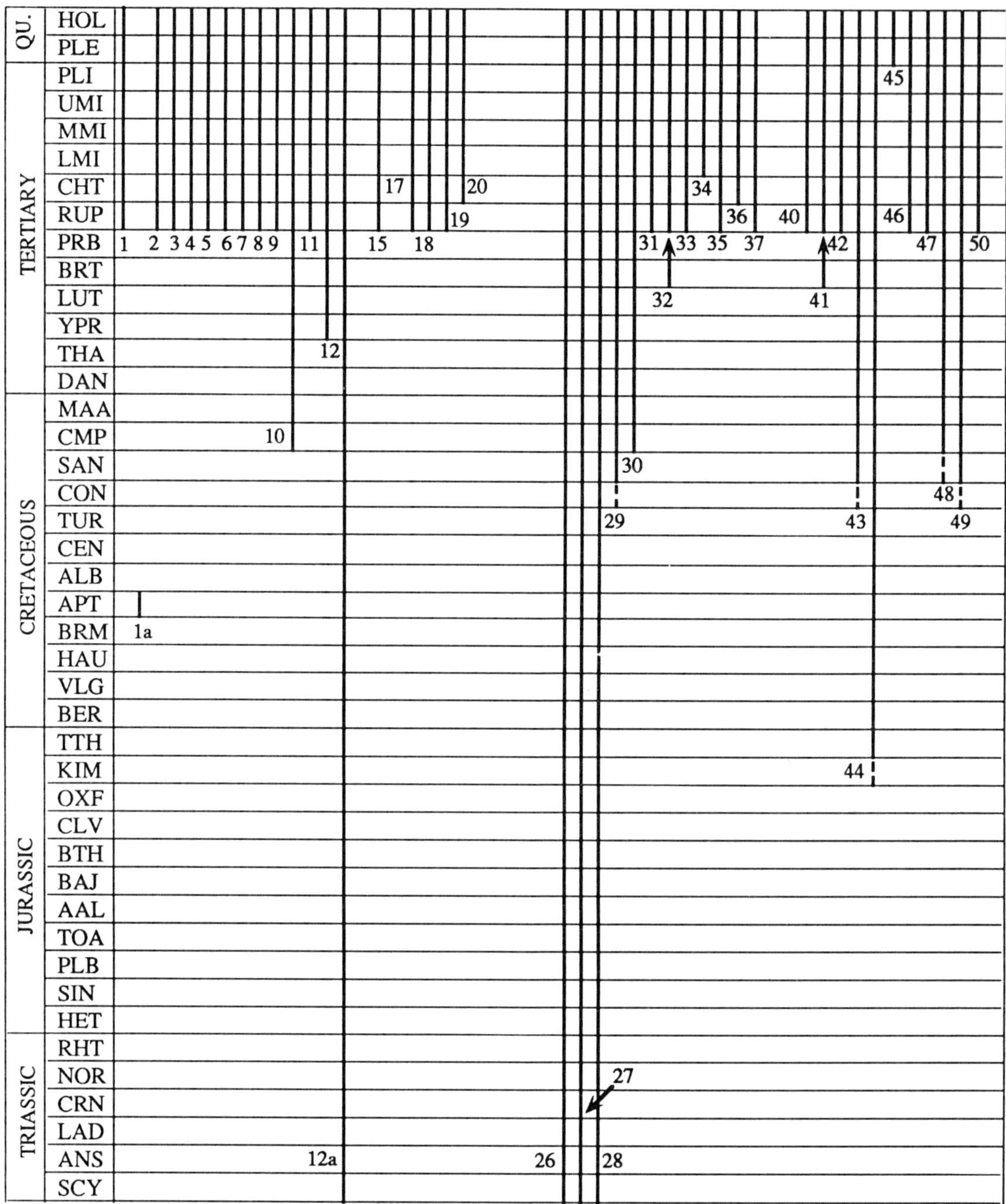

Fig. 17.2

Suborder NEOBISIINA Beier, 1932

Superfamily NEOBISIOIDEA Beier, 1932

F. NEOBISIIDAE Chamberlin, 1930 T. (RUP)–Rec. Terr.

First: *Neobisium rathkii* (Koch and Berendt, 1854), *N. exstinctum* Beier, 1937, *Roncus succineus* Beier, 1955, Baltic amber. **Extant**
Intervening: CHT, AQT.

Superfamily GARYPOIDEA Beier, 1932

F. OLPIIDAE Chamberlin, 1930 T. (RUP)–Rec. Terr.

First: *Garypinus electri* Beier, 1937, Baltic amber. **Extant**

F. GARYPIDAE Hansen, 1894 T. (RUP)–Rec. Terr.

First: *Geogarypus macrodactylus* Beier, 1937, *G? major* Beier, 1937, Baltic amber. **Extant**
Intervening: AQT.

Superfamily FEAELLOIDEA Beier, 1932

F. PSEUDOGARYPIDAE Chamberlin, 1923 T. (RUP)–Rec. Terr.

First: *Pseudogarypus hemprichii* Beier, 1937, *P. extensus* Beier, 1937, *P. minor* Beier, 1947, Baltic amber. **Extant**

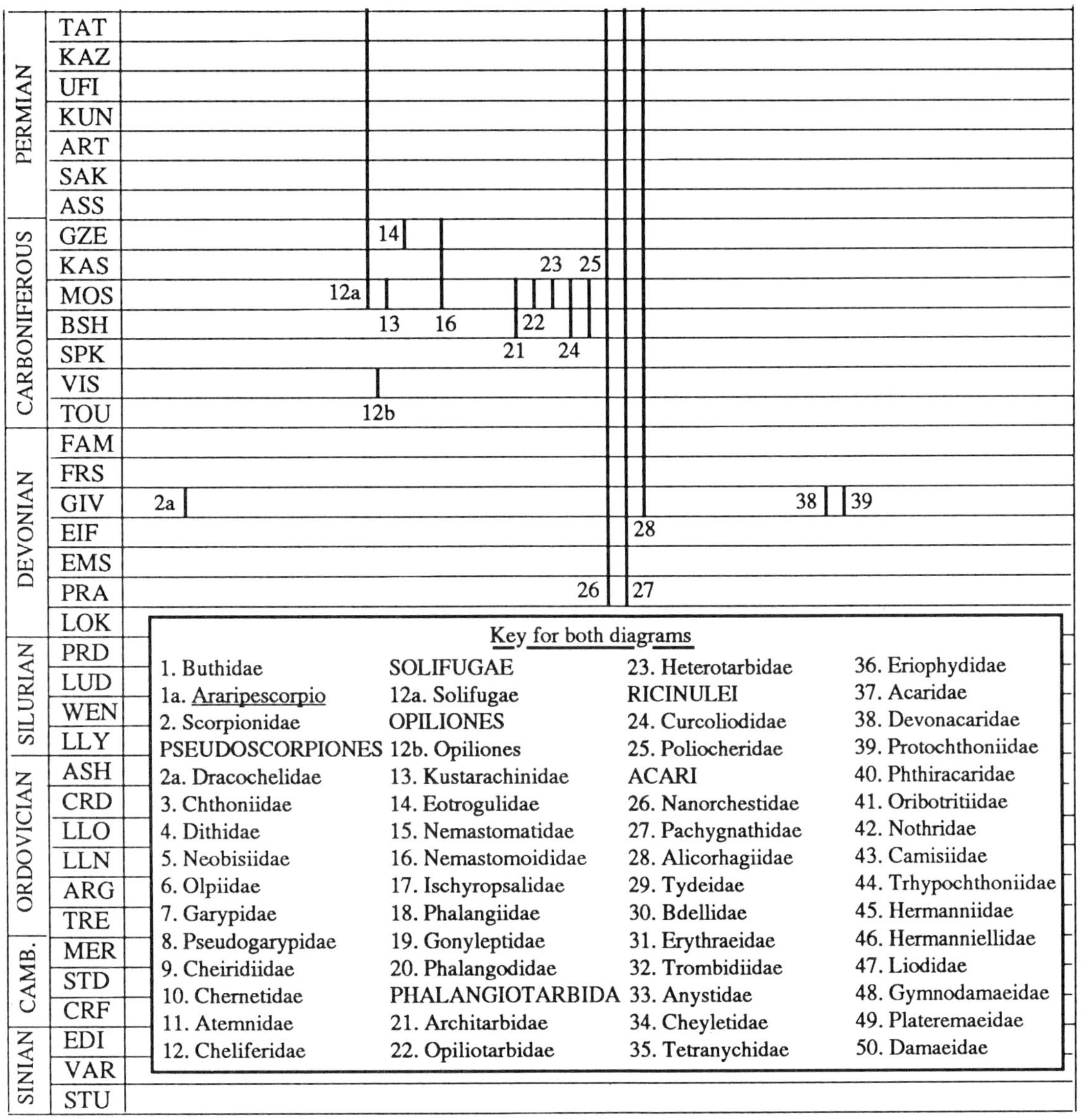
PERMIAN
TAT
KAZ
UFI
KUN
ART
SAK
ASS
CARBONIFEROUS
GZE
KAS
MOS
BSH
SPK
VIS
TOU
DEVONIAN
FAM
FRS
GIV
EIF
EMS
PRA
LOK
SILURIAN
PRD
LUD
WEN
LLY
ORDOVICIAN
ASH
CRD
LLO
LLN
ARG
TRE
CAMB.
MER
STD
CRF
SINIAN
EDI
VAR
STU
14
23 25
12a
13 16
22
21 24
12b
2a
38 39
28
26 27
Key for both diagrams
1. Buthidae
1a. Araripescorpio
2. Scorpionidae
PSEUDOSCORPIONES
2a. Dracochelidae
3. Chthoniidae
4. Dithidae
5. Neobisiidae
6. Olpiidae
7. Garypidae
8. Pseudogarypidae
9. Cheiridiidae
10. Chernetidae
11. Atemnidae
12. Cheliferidae
SOLIFUGAE
12a. Solifugae
OPILIONES
12b. Opiliones
13. Kustarachinidae
14. Eotrogulidae
15. Nemastomatidae
16. Nemastomoididae
17. Ischyropsalidae
18. Phalangiidae
19. Gonyleptidae
20. Phalangodidae
PHALANGIOTARBIDA
21. Architarbidae
22. Opiliotarbidae
23. Heterotarbidae
RICINULEI
24. Curcoliodidae
25. Poliocheridae
ACARI
26. Nanorchestidae
27. Pachygnathidae
28. Alicorhagiidae
29. Tydeidae
30. Bdellidae
31. Erythraeidae
32. Trombidiidae
33. Anystidae
34. Cheyletidae
35. Tetranychidae
36. Eriophydidae
37. Acaridae
38. Devonacaridae
39. Protochthoniidae
40. Phthiracaridae
41. Oribotritiidae
42. Nothridae
43. Camisiidae
44. Trhypochthoniidae
45. Hermanniidae
46. Hermanniellidae
47. Liodidae
48. Gymnodamaeidae
49. Plateremaeidae
50. Damaeidae

Fig. 17.2

Suborder CHELIFERIINA Hagen, 1879

Superfamily CHEIRIDIOIDEA Chamberlin, 1931

F. CHEIRIDIIDAE Chamberlin, 1924 T. (RUP)–Rec. Terr.

First: *Cheiridium hartmanni* (Menge, 1854b), Baltic amber. **Extant**
Intervening: CHT.

Superfamily CHELIFEROIDEA Chamberlin, 1931

F. CHERNETIDAE Menge, 1854b K. (CMP)–Rec. Terr.

First: Unnamed deutonymph, Manitoban amber, Canada (Schawaller, 1991). **Extant**
Intervening: RUP, CHT.

F. ATEMNIDAE Chamberlin, 1931 T. (RUP)–Rec. Terr.

First: *Progonatemnus succineus* Beier, 1955, Baltic amber. **Extant**

F. CHELIFERIDAE Stecker, 1874 T. (YPR)–Rec. Terr.

First: *Trachychelifer liaoningense* Hong, 1983, main coal bed of Guchengzi Formation, Liaoning Province, China. **Extant**
Intervening: RUP, CHT.

Class SOLIFUGAE Sundevall, 1833

C. (MYA)–Rec. Terr.

First: *Protosolpuga carbonaria* Petrunkevitch, 1949, Francis Creek Shale, Mazon Creek, Illinois, USA.
Intervening: CHT.

Class OPILIONES Sundevall, 1833

Unnamed C. (BRI) Terr.

First: Unnamed specimen, East Kirkton Limestone, near Bathgate, Scotland, UK (Wood *et al.*, 1985).
Intervening: Unnamed opilionid Koonwarra Fossil Bed, Lower Gippsland, Victoria, Australia (BRM/APT) (Jell and Duncan, 1986).

F. KUSTARACHNIDAE Petrunkevitch, 1913 C. (MYA) Terr.

First and Last (monotypic family): *Kustarachne tenuipes* Scudder, 1890, *K. conica* Petrunkevitch, 1913, *K. extincta* (Melander, 1903), Francis Creek Shale, Mazon Creek, Illinois, USA.
Comment: The order Kustarachnida was erected by Petrunkevitch (1913) for *Kustarachne*. The specimens have been reassessed by Beall (1986) who considered them to belong to Opiliones, an assignment with which other arachnologists who have seen the specimens (P. A. Selden, W. A. Shear) concur.

Suborder PALPATORES Thorell, 1876

Superfamily TROGULOIDEA Sundevall, 1833

F. EOTROGULIDAE Petrunkevitch, 1955 C. (KLA/NOG) Terr.

First and Last (monotypic family): *Eotrogulus fayoli* Thévenin, 1901, Commentry, France.

F. NEMASTOMATIDAE Simon, 1879 T. (RUP)–Rec. Terr.

First: *Nemastoma denticulatum* Koch and Berendt, 1854, *N. tuberculatum* Koch and Berendt, 1854, *N. clavigerum* Menge, 1854, *N. succineum* Röwer, 1939, Baltic amber. **Extant**

F. NEMASTOMOIDIDAE Petrunkevitch, 1955 C. (MYA)–(KLA/NOG) Terr.

First: *Nemastomoides longipes* (Petrunkevitch, 1913) and *N. depressus* (Petrunkevitch, 1913), Francis Creek Shale, Mazon Creek, Illinois, USA.
Last: *Nemastomoides elaveris* Thévenin, 1901, Commentry, France.

F. ISCHYROPSALIDAE Simon, 1879 T. (RUP)–Rec. Terr.

First: *Sabacon bachofeni* Röwer, 1939, Baltic amber. **Extant**

Superfamily PHALANGIOIDEA Thorell, 1876

F. PHALANGIIDAE Thorell, 1876 T. (RUP)–Rec. Terr.

First: *Caddo dentipalpus* (Koch and Berendt, 1854), *Cheiromachus coriaceus* Menge, 1854, *Dicranopalpus ramiger* (Koch and Berendt, 1854), *D. corniger* Menge, 1854, *D. palmnickensis* Röwer, 1939, *Liobunum sarapum* Menge, 1854, *L. inclusum* Röwer, 1939, *Opilio ovalis* Koch and Berendt, 1854, Baltic amber. **Extant**
Intervening: CHT.

Suborder LANIATORES Thorell, 1876

F. GONYLEPTIDAE Sundevall, 1833 T. (RUP)–Rec. Terr.

First: *Gonyleptes nemastomoides* Koch and Berendt, 1854, Baltic amber. **Extant**

F. PHALANGODIDAE Simon, 1879 T. (CHT)–Rec. Terr.

First: *Pellobunus proavus* Cokendolpher, 1987, *Philacarus hispaniolensis* Cokendolpher and Poinar, 1992, *Kimula?* sp., Dominican amber. **Extant**

Class PHALANGIOTARBIDA Petrunkevitch, 1949

Kjellesvig-Waering left a posthumous monograph on this group in which he drastically reduced in number the genera and species erected by Petrunkevitch (1913–1949). This work was being readied for publication but has yet to appear in print. Thus, the classification used is that of Petrunkevitch (1955).

F. ARCHITARBIDAE Karsch 1882 C. (MEL–MYA) Terr.

First: *Goniotarbus tuberculatus* (Pocock, 1911), *G. angulatus* (Pocock, 1911), *Mesotarbus intermedius* Petrunkevitch, 1949, *M. hindi* (Pocock), *M. angustus* (Pocock, 1911), *M. eggintoni* (Pocock, 1911), *Leptotarbus torpedo* (Pocock, 1911), Coseley, Dudley, England, UK.
Last: *Orthotarbus nyranensis* Petrunkevitch, 1953, Nýřany, Czechoslovakia. *Geratarbus bohemicus* Petrunkevitch, 1953 may also come from this horizon in the Coal Measures of Nýřany.
Intervening: VRK.

F. OPILIOTARBIDAE Petrunkevitch, 1945 C. (MYA) Terr.

First and Last (monotypic family): *Opiliotarbus elongatus* (Scudder, 1890), Francis Creek Shale, Mazon Creek, Illinois, USA.
Comment: Petrunkevitch (1953) considered that *Opiliotarbus kliveri* Waterlot, 1934, C. (KSK/POD), did not belong in this family.

F. HETEROTARBIDAE Petrunkevitch, 1913 C. (MYA) Terr.

First and Last (monotypic family): *Heterotarbus ovatus* Petrunkevitch, 1913, Francis Creek Shale, Mazon Creek, Illinois, USA.

Class RICINULEI Thorell, 1892

Suborder NEORICINULEI Selden, 1992

Comment: No fossil record. **Extant**

Suborder PALAEORICINULEI Selden, 1992

F. CURCULIOIDIDAE Cockerell, 1916 C. (MRD–MYA) Terr.

First: *Curculioides adompha* Brauckmann, 1987, Hagen-Vorhalle, Ruhr, Germany.
Last: *Curculioides scaber* (Scudder, 1890), *Amarixys sulcata* (Melander, 1903), *A. gracilis* (Petrunkevitch, 1945a). Francis Creek Shale, Mazon Creek, Illinois, USA.
Intervening: MEL.

F. POLIOCHERIDAE Scudder, 1884 C. (MEL–MYA) Terr.

First: *Terpsicroton alticeps* (Pocock, 1911), Coseley, Dudley, England, UK.
Last: *Poliochera punctulata* Scudder, 1884, *P. glabra* Petrunkevitch, 1913, Francis Creek Shale, Mazon Creek, Illinois, USA.

Class ACARI Latreille, 1802

A review of the classification of the mites in relation to the fossil record was given by Bernini (1986). Acari are divided fundamentally into two major divisions: Actinotrichida and Anactinotrichida, which some authors (e.g. van der Hammen, 1989) considered to be separate arachnid lineages whose common features (small size, modes of life) are due to convergence. For convenience, they are treated as a whole here; the classification of Lindquist (1984) is followed, with the alternative names used by van der Hammen and others in parentheses.
Intervening: Undescribed mites were reported from Lebanese amber, K. (BER–APT), by Whalley (1980).

Order ACTINOTRICHIDA Oudemans, 1931 (ACARIFORMES Zachvatkin, 1952)

Suborder ACTINEDIDA van der Hammen, 1968 (PROSTIGMATA Kramer, 1877)

F. NANORCHESTIDAE Grandjean, 1937
D. (PRA)–Rec. Terr.

First: *Protospeleorchestes pseudoprotacarus* Dubinin, 1962, Rhynie Chert, Aberdeenshire, Scotland, UK. **Extant**

F. PACHYGNATHIDAE Kramer, 1877
D. (PRA)–Rec. Terr.

First: *Protacarus crani* Hirst, 1923, *Pseudoprotacarus scoticus* Dubinin, 1962, *Palaeotydeus devonicus* Dubinin, 1962, *Paraprotacarus hirsti* Dubinin, 1962, Rhynie Chert, Aberdeenshire, Scotland, UK. **Extant**
Comment: These mites were scattered across three families by Dubinin (1962), but Kethley (*in* Kethley *et al.*, 1989 and Norton *et al.*, 1989) suggested the need for restudy of this material, and considered them all to belong in the Pachygnathidae.

F. ALICORHAGIIDAE Grandjean, 1939
D. (GIV)–Rec. Terr.

First: *Archaeacarus dubinini* Kethley and Norton, 1989, Gilboa mudstones, Gilboa, New York, USA. **Extant**

F. TYDEIDAE Kramer, 1877
K. (CON/SAN)–Rec. Terr.

First: Possible tydeid larva, amber of Taimyr Peninsula, former USSR (Bulanova-Zakhavatkina, 1974). **Extant**

F. BDELLIDAE Dugès, 1834 K. (CMP)–Rec. Terr.

First: *Bdella vetusta* Ewing, 1937, Manitoban amber, Canada. **Extant**
Intervening: RUP.

F. ERYTHRAEIDAE Oudemans, 1902
T. (RUP)–Rec. Terr.

First: *Erythraeus foveolatus* Koch and Berendt, 1854, *E. longipes* Koch and Berendt, 1854, *E. illustris* Koch and Berendt, 1854, *E. incertus* Koch and Berendt, 1854, *E. hirsutissimus* Koch and Berendt, 1854, *E. raripilus* Koch and Berendt, 1854, *E. lagopus* Koch and Berendt, 1854, *E. proavus* Koch and Berendt, 1854, *Arythaena troguloides* Menge, 1854, Baltic amber. **Extant**

F. TROMBIDIIDAE Leach, 1815
T. (RUP)–Rec. Terr.

First: *Trombidium clavipes* Koch and Berendt, 1854, *T. saccatum* Koch and Berendt, 1854, *T. scrobiculatum* Menge, 1854, *T. heterotrichum* Menge, 1854, *T. crassipes* Menge, 1854, *T. granulatum* Menge, 1854, Baltic amber. **Extant**

F. ANYSTIDAE Oudemans, 1902
T. (RUP)–Rec. Terr.

First: *Anystis venustula* Koch and Berendt, 1854, Baltic amber. **Extant**

F. CHEYLETIDAE Leach, 1815
T. (AQT)–Rec. Terr.

First: *Cheyletus burmiticus* (Cockerell, 1917), Burmese amber. **Extant**

F. TETRANYCHIDAE Donnadieu, 1875
T. (RUP)–Rec. Terr.

First: *Tetranychus gibbus* Koch and Berendt, 1854, Baltic amber. **Extant**

F. ERIOPHYIDAE Nalepa, 1898
T. (CHT)–Rec. Terr. (phytoparasites)

First: *Eriophyes daphnogene* Ambrus and Hably, 1979, Hungary. **Extant**

Suborder ACARIDIDA Latreille, 1802 (ASTIGMATA Canestrini, 1891)

F. ACARIDAE Ewing and Nesbitt, 1942
T. (RUP)–Rec, Terr., parasitic

First: *Acarus rhombeus* Koch and Berendt, 1854, Baltic amber. **Extant**
Intervening: CHT, TOR.

F. LISTROPHORIDAE Canestrini, 1892
T. (CHT)–Rec. Terr. (ectoparasitic)

First: unnamed listrophorid, Dominican amber (Poinar, 1988). **Extant**

Suborder ORIBATIDA Michael, 1884 (CRYPTOSTIGMATA Canestrini, 1891)

The vast majority of known fossil mites belong to this suborder. These animals are characterized by a hard cuticle which allows their preservation, often in large numbers, in certain sedimentary environments. Many are known from Quaternary deposits (especially freshwater forms), but for the most part these have not been included here; the families to which they belong generally have a pre-Quaternary history.

Division ENARTHRONOTA Grandjean, 1947

F. DEVONACARIDAE Norton, 1988
D. (GIV) Terr.

First and Last (monotypic family): *Devonacarus sellnicki* Norton, 1988, Gilboa mudstones, New York, USA.

F. PROTOCHTHONIIDAE Norton, 1988
D. (GIV) Terr.

First and Last (monotypic family): *Protochthonius gilboa* Norton, 1988, Gilboa mudstones, New York, USA.

Division MIXONOMATA Grandjean, 1969

Superfamily PHTHIRACAROIDEA Perty, 1841

F. PHTHIRACARIDAE Perty, 1841
T. (RUP)–Rec. Terr.

First: *Hoploderma multipunctatum* Sellnick, 1918, Baltic amber. **Extant**

Superfamily EUPHTHIRACAROIDEA Jacot, 1930

F. ORIBOTRITIIDAE Grandjean, 1954
T. (RUP)–Rec. Terr.

First: *Oribotritia translucida* Sellnick, 1931, *O. pyropus* (Sellnick, 1918), Baltic amber. **Extant**

Division DESMONOMATA Woolley, 1973

Superfamily CROTONIOIDEA Thorell, 1876

F. NOTHRIDAE Berlese, 1896
T. (RUP)–Rec. Terr.

First: *Nothrus illautus* Sellnick, 1918, Baltic amber. **Extant**

F. CAMISIIDAE Oudemans, 1900
K. (CON/SAN)–Rec. Terr.

First: *Eocamisia sukatshevae* Bulanova-Zakhavatkina, 1974, amber of Taimyr Peninsula, former USSR. **Extant**
Intervening: RUP, PLE/HOL, HOL.

F. TRHYPOCHTHONIIDAE Willmann, 1931
J. (OXF/TTH)–Rec. Terr.

First: *Palaeochthonius krasilovi* Krivolutsky and Ryabinin, 1976, *Juracarus serratus* Krivolutsky and Ryabinin, 1976, Burea River, Far East of former USSR. **Extant**
Intervening: RUP.

Superfamily HERMANNIOIDEA Sellnick, 1928

F. HERMANNIIDAE Sellnick, 1928
Q. (PLE)–Rec. Terr.

First: *Hermannia gigantea* Sitnikova, 1975, Anabar River, Siberia, former USSR. **Extant**
Intervening: HOL.

Division BRACHYPYLINA Hull, 1918

Superfamily HERMANNIELLOIDEA Grandjean, 1934

F. HERMANNIELLIDAE Grandjean, 1934
T. (RUP)–Rec. Terr.

First: *Hermanniella concamerata* Sellnick, 1931, *H. tuberculata* Sellnick, 1918, Baltic amber. **Extant**

Superfamily LIODOIDEA Grandjean, 1934

F. LIODIDAE Grandjean, 1934
T. (RUP)–Rec. Terr.

First: *Liodes quadriscutatus* Sellnick, 1918, *Platyliodes ensigerus* (Sellnick, 1918), *Embolocarus pergratus* Sellnick, 1918, Baltic amber. **Extant**

Superfamily PLATEREMAEOIDEA Trägårdh, 1931

F. GYMNODAMAEIDAE Grandjean, 1954
K. (SAN/CMP)–Rec. Terr.

First: Unnamed gymnodamaeid, Canadian amber (McAlpine and Martin, 1963). **Extant**
Intervening: RUP.

F. PLATEREMAEIDAE Trägårdh, 1931
K. (CON/SAN)–Rec. Terr.

First: *Rasnitsynella punctulata* Bulanova-Zachvatina, 1974, Taimyr Peninsula, former USSR. **Extant**

F. DAMAEIDAE Berlese, 1896
T. (RUP)–Rec. Terr.

First: *Damaeus? genadensis* Sellnick, 1931, Baltic amber. **Extant**
Intervening: CHT, TOR, ?PLE/HOL.

Superfamily CEPHEOIDEA Berlese, 1896

F. CEPHEIDAE Berlese, 1896
T. (RUP)–Rec. Terr. (see Fig.17.3)

First: *Cepheus implicatus* (Sellnick, 1918), Baltic amber. **Extant**

Superfamily EREMAEOIDEA Sellnick, 1928

F. EREMAEIDAE Sellnick, 1928
T. (RUP)–Rec. Terr.

First: *Eremaeus oblongus* C. L. Koch, 1836, Baltic amber. **Extant**
Intervening: CHT, ?PLE/HOL.

Superfamily GUSTAVIOIDEA Oudemans, 1900

F. XENILLIDAE Woolley and Higgins, 1966
T. (RUP)–Rec. Terr.

First: *Xenillus tegeocraniformis* (Sellnick, 1918), Baltic amber. **Extant**

F. ASTEGISTIDAE Balogh, 1961
J. (OXF–TTH)–Rec. Terr.

First: *Cultroribula jurassica* Krivolutsky and Ryabinin, 1976, Burea River, Far East of former USSR. **Extant**
Intervening: RUP.

F. METRIOPPIIDAE Balogh, 1943
T. (RUP)–Rec. Terr.

First: *Ceratoppia bipilis* (Hermann, 1804), Baltic amber. **Extant**
Intervening: PIA, PLE, HOL.

Superfamily CARABODOIDEA C. L. Koch, 1837

F. CARABODIDAE C. L. Koch, 1837
T. (RUP)–Rec. Terr.

First: *Carabodes gerberi* Sellnick, 1931, *C. dissonus* Sellnick, 1931, *C. coriaceus* C. L. Koch, 1836, *C. labyrinthicus* (Michael, 1879), *Plategeocranus sulcatus* (Karsch, 1884), *Scutoribates perornatus* Sellnick, 1918, *Odontocepheus?* sp., Baltic amber. **Extant**
Intervening: TOR.

F. OTOCEPHEIDEA Balogh, 1961
T. (RUP)–Rec. Terr.

First: *Otocepheus niger* Sellnick, 1931, *O. praesignis* Sellnick, 1931, Baltic amber. **Extant**

Superfamily TECTOCEPHEOIDEA Grandjean, 1954

F. TECTOCEPHEIDAE Grandjean, 1954
T. (RUP)–Rec. Terr.

First: *Tectocepheus similis* Sellnick, 1931, Baltic amber. **Extant**

Superfamily OPPIOIDEA Grandjean, 1951
T. (RUP)–Rec. Terr.

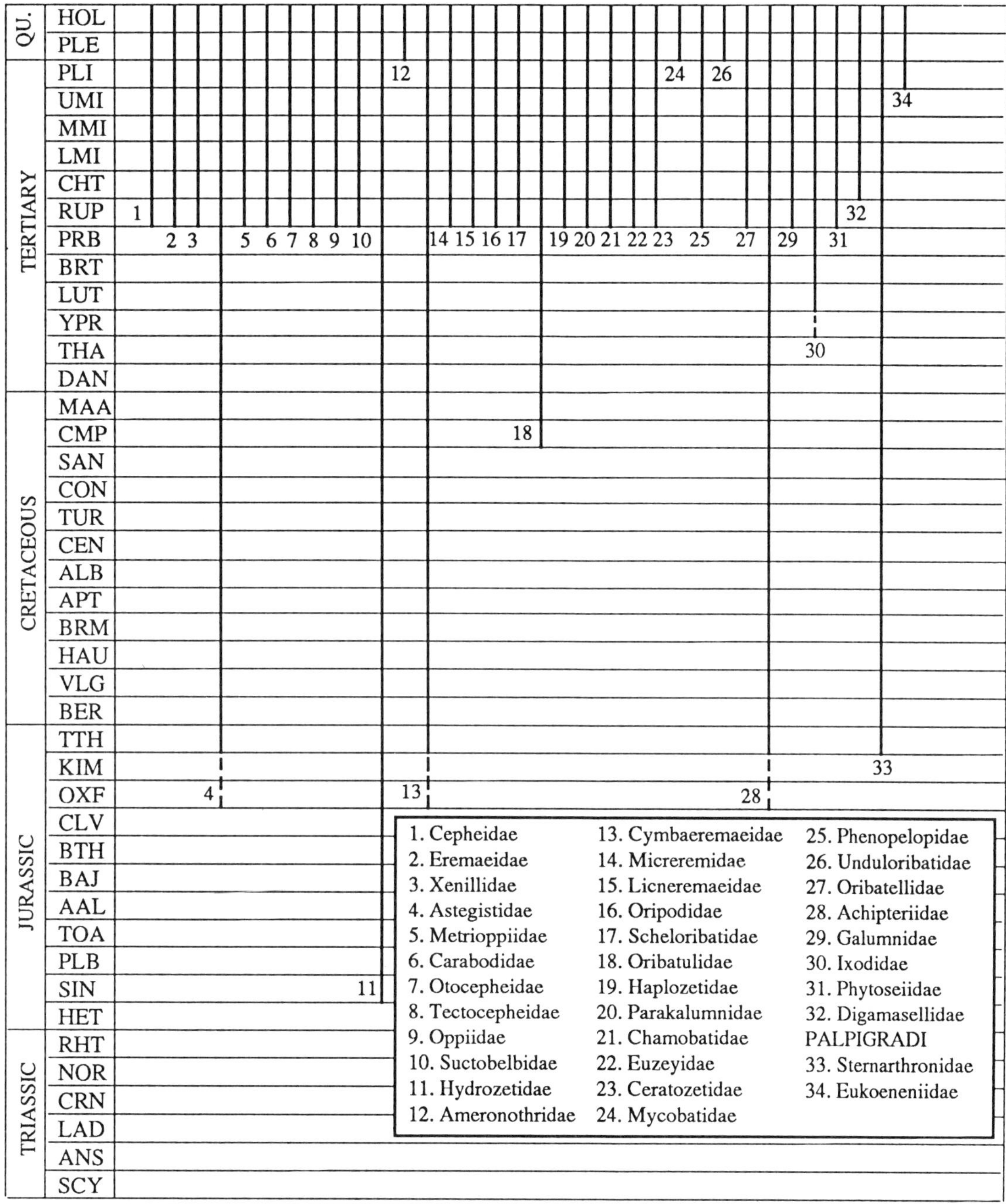

Fig. 17.3

F. OPPIIDAE Grandjean, 1951
T. (RUP)–Rec. Terr.

First: *Oppia curvicornum* (Sellnick, 1931), Baltic amber. **Extant**

Intervening: CHT, TOR.

F. SUCTOBELBIDAE Jacot, 1938
T. (RUP)–Rec. Terr.

First: *Suctobelbella subtrigona* (Oudemans, 1900), Baltic amber. **Extant**

Superfamily HYDROZETOIDEA Grandjean, 1954

F. HYDROZETIDAE Grandjean, 1954
J. (SIN)–Rec. Terr. (FW)

First: *Hydrozetes* sp., borehole, Döshult Formation, Höganäs, Skåne, Sweden (Sivhed and Wallwork, 1978). **Extant**

Intervening: AAL–TTH, CHT, PIA, PLE, HOL.

Superfamily AMERONOTHROIDEA Willmann, 1931

F. AMERONOTHRIDAE Willmann, 1931
Q. (PLE)–Rec. Terr.

First: *Ameronothrus lineatus* (Thorell, 1871), Anabar River, Siberia, former USSR. **Extant**

Intervening: HOL.

Superfamily CYMBAEREMAEOIDEA Sellnick, 1928

F. CYMBAEREMAEIDAE Sellnick, 1928
J. (OXF–TTH)–Rec. Terr.

First: *Jureremus foveolatus* Krivolutsky and Ryabinin, 1976, Burea River, Far East of former USSR. **Extant**
Intervening: RUP, CHT.

F. MICREREMIDAE Grandjean, 1954
T. (RUP)–Rec. Terr.

First: *Micreremus scrobiculatus* Sellnick, 1931, *M. reticulatus* Sellnick, 1931, Baltic amber. **Extant**

***Superfamily* LICNEREMAEOIDEA Grandjean, 1931**

F. LICNEREMAEIDAE Grandjean, 1931
T. (RUP)–Rec. Terr.

First: *Licneremaeus fritschi* Sellnick, 1931, Baltic amber. **Extant**

***Superfamily* ORIPODOIDEA Jacot, 1925**

F. ORIPODIDAE Jacot, 1925
T. (RUP)–Rec. Terr.

First: *Oripoda baltica* Sellnick, 1931, Baltic amber. **Extant**
Intervening: CHT.

F. SCHELORIBATIDAE Grandjean, 1933
T. (RUP)–Rec. Terr.

First: *Scheloribates areatus* Sellnick, 1931, *S. setatus* Sellnick, 1931, Baltic amber. **Extant**

F. ORIBATULIDAE Thor, 1929
K. (CMP)–Rec. Terr.

First: Unnamed oribatulid, Manitoban amber, Canada (McAlpine and Martin, 1963). **Extant**
Intervening: YPR–PRB, RUP, CHT, PLE.

F. HAPLOZETIDAE Grandjean, 1936
T. (RUP)–Rec. Terr.

First: *Protoribates longipilis* Sellnick, 1931, Baltic amber. **Extant**

F. PARAKALUMNIDAE Grandjean, 1936
T. (RUP)–Rec. Terr.

First: *Neoribates borussicus* Sellnick, 1931, Baltic amber. **Extant**

***Superfamily* CERATOZETOIDEA Jacot, 1925**

F. CHAMOBATIDAE Grandjean, 1954
T. (RUP)–Rec. Terr.

First: *Chamobates difficilis* Sellnick, 1931, Baltic amber. **Extant**

F. EUZETIDAE Grandjean, 1954
T. (RUP)–Rec. Terr.

First: *Euzetes convexulus* (Koch and Berendt, 1854), Baltic amber. **Extant**

F. CERATOZETIDAE Jacot, 1925
T. (RUP)–Rec. Terr.

First: *Melanozetes foderatus* Sellnick, 1931, *M. mollicomus* (C. L. Koch, 1839), *Sphaerozetes convexulus* (Koch and Berendt, 1854), *S. primus* Sellnick, 1931, Baltic amber. **Extant**
Intervening: PLE, HOL.

F. MYCOBATIDAE Grandjean, 1954
Q. (PLE)–Rec. Terr.

First: *Mycobates* sp., Anabar River, former USSR (Golosova *et al.*, 1985). **Extant**
Intervening: HOL.

***Superfamily* PHENOPELOPOIDEA Petrunkevitch, 1955**

F. PHENOPELOPIDAE Petrunkevitch, 1955
T. (RUP)–Rec. Terr.

First: *Phenopelops punctulatus* (Sellnick, 1931), *Notaspis* sp., Baltic amber. **Extant**

F. UNDULORIBATIDAE Kunst, 1971
Q. (PLE)–Rec. Terr.

First: *Scutozetes lanceolatus* (Hammer, 1952), Anabar River, Siberia, former USSR. **Extant**
Intervening: HOL.

***Superfamily* ORIBATELLOIDEA Jacot, 1925**

F. ORIBATELLIDAE Jacot, 1925
T. (RUP)–Rec. Terr.

First: *Oribatella mirabilis* Sellnick, 1931, *Tectoribates parvus* Sellnick, 1931, Baltic amber. **Extant**

***Superfamily* ACHIPTERIOIDEA Thor, 1929**

F. ACHIPTERIIDAE Thor, 1929
J. (OXF–TTH)–Rec. Terr.

First: *Achipteria? obscura* Krivolutsky and Ryabinin, 1976, Far East of former USSR. **Extant**
Intervening: PLE, HOL.

***Superfamily* GALUMNOIDEA Jacot, 1925**

F. GALUMNIDAE Jacot, 1925
T. (RUP)–Rec. Terr.

First: *Galumna clavata* Sellnick, 1931, *Galumna diversa* Sellnick, 1931, Baltic amber. **Extant**

***Order* ANACTINOTRICHIDA Oudemans, 1931**

***Suborder* OPILIOACARIDA With, 1902 (NOTOSTIGMATA With, 1904) Terr.**

No fossil record. **Extant**

***Suborder* HOLOTHYRIDA Reuter, 1909 (TETRASTIGMATA Evans *et al.*, 1961) Terr.**

No fossil record. **Extant**

***Suborder* IXODIDA Leach, 1815 (METASTIGMATA Canestrini, 1891)**

F. IXODIDAE Leach, 1815
T. (YPR/LUT)–Rec. Terr. (ectoparasites)

First: *Ixodes tertiarius* Scudder, 1890, Green River Formation, Wyoming, USA. **Extant**
Intervening: PLE.

***Suborder* GAMASIDA Leach, 1815 (MESOSTIGMATA Canestrini, 1891)**

F. PHYTOSEIIDAE Berlese, 1916
T. (RUP)–Rec. Terr.

First: *Seius bdelloides* Koch and Berendt, 1854, Baltic amber. **Extant**

F. DIGAMASELLIDAE Evans, 1956
T. (CHT)–Rec. Terr.

First: *Dendrolaelaps fossilis* Hirschmann, 1971, Chiapas amber, Mexico. **Extant**

Class PALPIGRADI Thorell, 1888

F. STERNARTHRONIDAE Haase, 1890
J. (TTH) Terr.

First and Last (monotypic family): *Sternarthron zitteli* Haase, 1890, Lithographic Limestone, Solnhofen, Germany.
Comment: These fossils require restudy to check their identification as palpigrades (Rowland and Sissom, 1980).

F. EUKOENENIIDAE Petrunkevitch, 1955
T. (?ZAN/PIA)–Rec. Terr./?Mar.

First: *Paleokoenenia mordax* Rowland and Sissom, 1980, 'Onyx Marble', Bonner Quarry, Ashfork, Arizona, USA. **Extant**

Class HAPTOPODIDA Pocock, 1911 (see Fig. 17.4)

F. PLESIOSIRONIDAE Pocock, 1911
C. (CHE)–(MEL) Terr.

First: *Plesiosiro madeleyi* Pocock, 1911, Sparth Bottoms, Rochdale, England, UK.
Last: *Plesiosiro madeleyi* Pocock, 1911, Coseley, Dudley, England, UK.

Class ANTHRACOMARTIDA Karsch, 1882

F. ANTHRACOMARTIDAE Haase, 1890
C. (CHE–KLA/NOG) Terr.

First: *Cryptomartus?* sp., Kohlscheid-Schichten, Grube Carolus Magnus, Palenberg bei Aachen, Germany (Guthörl, 1964).
Last: *Pleomartus palatinus* (Ammon, 1901), Breitenbach-Schichten, Grube Steinbach bei Brücken, Pfalz, Germany.
Intervening: MEL, VRK, KSK, POD, MYA, DOR.

Class PULMONATA Firstman, 1973

Intervening: *Ecchosis pulchribothrium* Selden and Shear, 1991, D. (GIV), was placed as Pulmonata *incertae sedis*.

Order TRIGONOTARBIDA Petrunkevitch, 1949
S. (PRD)–C. (DOR) Terr.

First: Unnamed trigonotarbid from Ludlow Bone Bed Member, Ludford Lane, Ludlow, Shropshire, England, UK (Jeram *et al.*, 1990).
Intervening: D. (EMS), *Alkenia mirabilis* Størmer, 1970, Alken-an-der-Mosel, Germany. This was removed from Palaeocharinidae by Shear *et al.* (1987).

F. PALAEOCHARINIDAE Hirst, 1923
D. (PRA–GIV) Terr.

First: *Palaeocharinus rhyniensis* Hirst, 1923, *P. scourfieldi* Hirst, 1923, *P. calmani* Hirst, 1923, P. *kidstoni* Hirst, 1923, *P. hornei* Hirst, 1923, Rhynie Chert, Aberdeenshire, Scotland, UK.
Last: *Gilboarachne griersoni* Shear *et al.*, 1987, *Gelasinotarbus reticulatus* Shear *et al.*, 1987, *G. bonamoae* Shear, *et al.*, 1987, *G. bifidus* Shear *et al.*, 1987, *G. heptops* Shear *et al.*, 1987, *Aculeatarbus depressus* Shear *et al.*, 1987, Gilboa Mudstones, Gilboa, New York, USA.

F. ANTHRACOSIRONIDAE Pocock, 1903b
C. (MEL–KSK) Terr.

First: *Anthracosiro woodwardi* Pocock, 1903, *A. fritschi* Pocock, 1903b, Shipley Clay Pit, Ilkeston, England, UK and Coseley, Dudley, England, UK.
Last: *Anthracosiro woodwardi* Pocock, 1903, Crawcrook, Ryton-on-Tyne, England, UK.
Intervening: VRK.

F. EOPHRYNIDAE Karsch, 1882
C. (HAS–MYA) Terr.

First: *Pocononia whitei* (Ewing, 1930), lower part of Pocono Formation, Allegheny, Virginia, USA.
Last: *Hemiphrynus longipes* Frič, 1901, *H. hofmanni* Frič, 1901, Nýřany, Czechoslovakia; *Gondwanarachne argentinensis* Pinto and Hünicken, 1980, Bajo de Véliz Formation, San Luis Province, Argentina.
Intervening: PND, ARN, MRD, YEA, CHE, MEL, VRK, KSK, POD.

F. TRIGONOTARBIDAE Petrunkevitch, 1949
D. (EMS)–C. (MEL) Terr.

First: *Archaeomartus levis* Størmer, 1970, and *A. tuberculatus* Størmer, 1970, Alken-an-der-Mosel, Germany.
Last: *Trigonotarbus johnsoni* Pocock, 1911, Coseley, Dudley, England, UK.
Intervening: MRD.

F. APHANTOMARTIDAE Petrunkevitch, 1945
C. (PND/ALP–KRE) Terr.

First: *Aphantomartus?* sp., Namurian A, Upper Silesian Coal Basin, Czechoslovakia (Přibyl, 1960).
Last: *Aphantomartus areolatus* Pocock, 1911, Prado Formation, borehole, Cerezal, Léon Province, Spain.
Intervening: CHE, MEL, VRK, KSK, POD, MYA.

Order ARANEAE Clerck, 1757

The classification used here is essentially the consensus scheme given in Shear (1986). The vast majority of families are first known from Baltic amber; many of the original specimens described by Koch and Berendt (1854) and Menge (*in* Koch and Berendt, 1854) have been lost or were poorly described so that available specimens cannot be matched to the descriptions (Petrunkevitch, 1958). Consequently, only those early specimens which were studied by Petrunkevitch have been included in the already lengthy lists of 'firsts'. It should be noted, too, that Petrunkevitch's work itself is in need of review, in the light of new ideas on classification, and the realization that some supposed Baltic amber is now thought to be pieces of comparatively Recent copal.

Attercopus D. (PRA) Terr./FW

First: *Attercopus fimbriunguis* (Shear, *et al.*, 1987), Gilboa Mudstones, New York, USA. Plesion: plesiomorphic sister taxon of all other spiders.
Intervening: Undescribed spiders were reported from Lebanese amber, K. (BER–APT), by Whalley (1980).

Suborder LIPHISTIOMORPHAE Pocock, 1892

F. ARTHROLYCOSIDAE Frič, 1904
C. (MEL–MYA) Terr.

First: *Eocteniza silvicola* Pocock, 1911, Coseley, Dudley, England, UK.

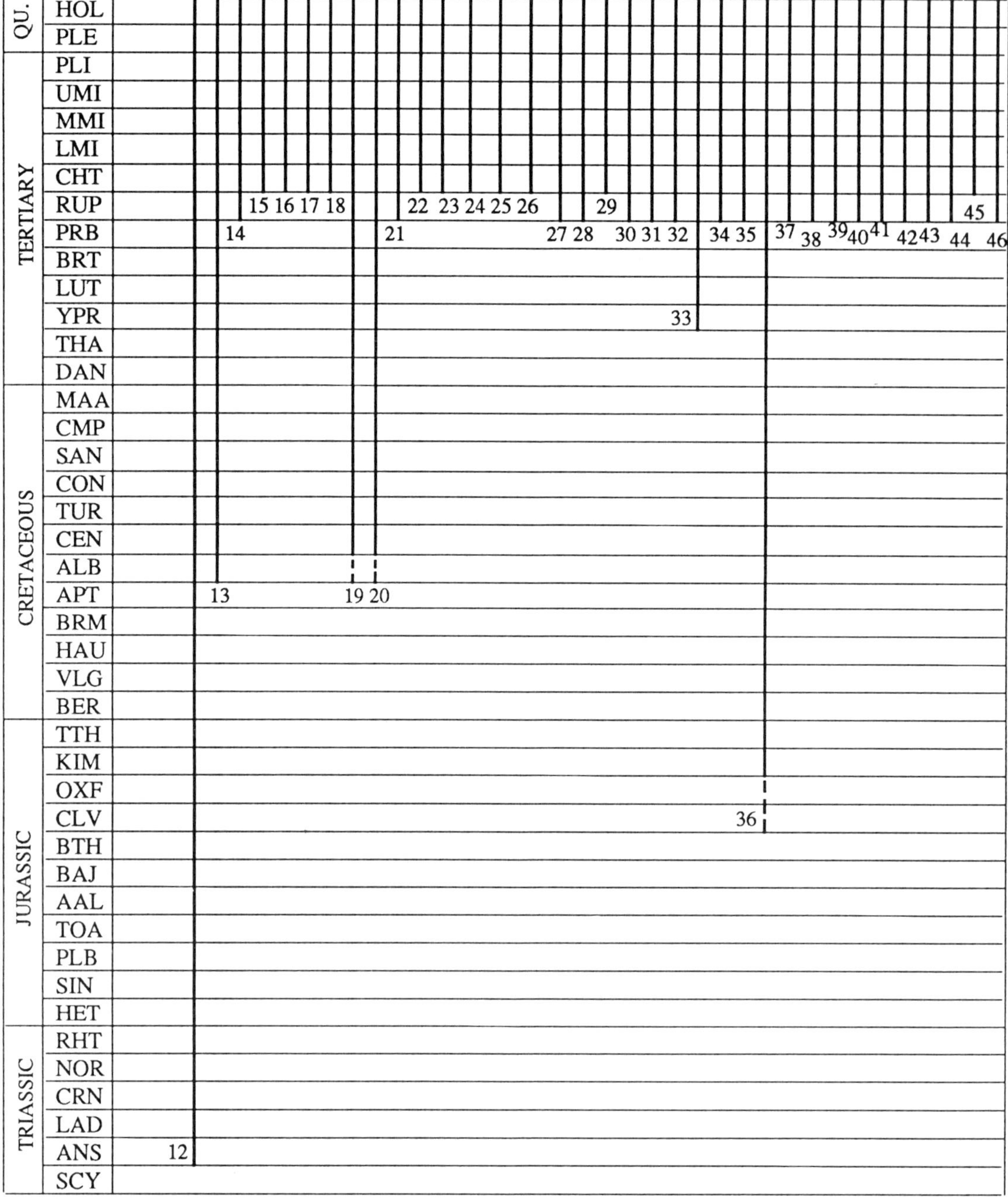

Fig. 17.4

Last: *Arthrolycosa antiqua* Harger, 1874, *A. danielsi* Petrunkevitch, 1913, Francis Creek Shale, Mazon Creek, Illinois, USA.
Intervening: C. (POD).

F. ARTHROMYGALIDAE Petrunkevitch, 1923 C. (MEL–KRE) Terr.

First: *Protocteniza britannica* Petrunkevitch, 1949, Coseley, Dudley, England, UK.
Last: *Protolycosa cebennensis* Laurentiaux-Vieira and Laurentiaux, 1963, couche Le Pin, La Grand'Combe, Cévennes, France.
Intervening: POD, MYA.

Suborder **OPISTHOTHELAE Pocock, 1892**

Infraorder **MYGALOMORPHAE Pocock, 1892**

The monotypic family Megarachnidae was erected for the giant Carboniferous fossil *Megarachne servinei* Hünicken, 1980, but there is doubt about the araneid (or even arachnid) nature of the fossil (Eskov and Zonshtein, 1990), which has no preserved spinnerets.

F. HEXATHELIDAE Simon, 1892 Tr. (ANS)–Rec. Terr.

First: *Rosamygale grauvogeli* Selden and Gall, 1992, Grès à Voltzia, Vosges, France. **Extant**

F. MECICOBOTHRIIDAE Holmberg, 1882 K. (?ALB)–Rec. Terr.

First: *Cretohexura coylei* Eskov and Zonshtein, 1990, Semyon Creek, Elizovo, near Chita, Transbaikalia, former

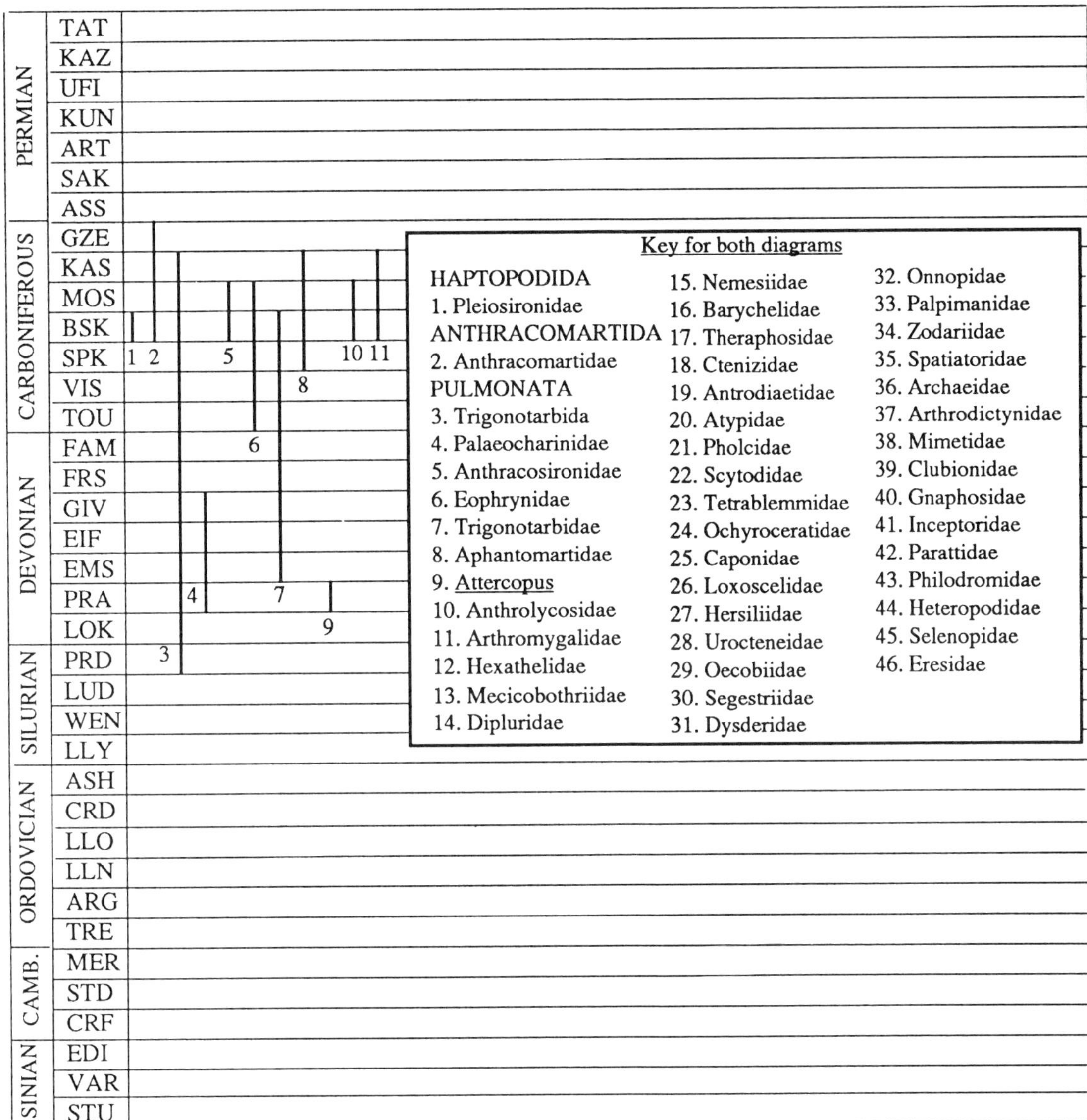

Fig. 17.4

USSR; *Cretomegahexura platnicki* Eskov and Zonshtein, 1990, Bon-Tsagan Lake, central Mongolia. **Extant**

F. DIPLURIDAE Simon, 1889
T. (RUP)–Rec. Terr.

First: *Clostes priscus* Menge, 1869, Baltic amber. **Extant**
Intervening: CHT.

F. NEMESIIDAE Simon, 1889
T. (CHT)–Rec. Terr.

First: Unnamed pycnotheline, Dominican amber (Schawaller, 1981). **Extant**

F. BARYCHELIDAE Simon, 1889
T. (CHT)–Rec. Terr.

First: *Plasistops hispaniolensis* Wunderlich, 1988, Dominican amber. **Extant**

F. THERAPHOSIDAE Thorell, 1869
T. (CHT)–Rec. Terr.

First: *Ischnocolinopsis acutus* Wunderlich, 1988, Dominican amber. **Extant**

F. CTENIZIDAE Thorell, 1887
T. (CHT)–Rec. Terr.

First: *Bolostromus destructus* Wunderlich, 1988, Dominican amber. **Extant**

F. ANTRODIAETIDAE Gertsch, 1940
K. (?ALB)–Rec. Terr.

First: *Cretacattyma raveni* Eskov and Zonshtein, 1990, Bon-Tsagan Lake, central Mongolia. **Extant**

F. ATYPIDAE Thorell, 1870 K. (?ALB)–Rec. Terr.

First: *Ambiortiphagus ponomarenkoi* Eskov and Zonshtein, 1990, Bon-Tsagan Lake, central Mongolia. **Extant**
Intervening: RUP.

Infraorder ARANEOMORPHAE Smith, 1902

The records of araneomorph spiders from Palaeozoic strata do not hold up under close scrutiny (Shear *et al.*, 1989; Selden *et al.*, 1991). These are: *Archaeometa devonica* Størmer, 1976, D. (EMS); *Archaeometa nephilina* Pocock, 1911,

Arachnometa tuberculata Petrunkevitch, 1949, C. (KSK); and *Eopholcus pedatus* Frič, 1904, C. (POD).

Superfamily PHOLCOIDEA Simon, 1874

F. PHOLCIDAE Simon, 1874 T. (RUP)–Rec. Terr.

First: *Micropholcus heteropus* Petrunkevitch, 1942, Baltic amber. **Extant**
Intervening: CHT.

F. SCYTODIDAE Blackwall, 1864 T. (CHT)–Rec. Terr.

First: *Scytodes piliformis* Wunderlich, 1988. *S. stridulans* Wunderlich, 1988, *S. planithorax* Wunderlich, 1988, Dominican amber. **Extant**

F. TETRABLEMMIDAE O. Pickard-Cambridge, 1873 T. (CHT)–Rec. Terr.

First: *Monoblemma*(?) *spinosum* Wunderlich, 1988, Dominican amber. **Extant**

F. OCHYROCERATIDAE Fage, 1912 T. (CHT)–Rec. Terr.

First: *Arachnolithus pygmaeus* Wunderlich, 1988, Dominican amber. **Extant**

F. CAPONIDAE Simon, 1887 T. (CHT)–Rec. Terr.

First: *Nops lobatus* Wunderlich, 1988, *N. segmentatus* Wunderlich, 1988, Dominican amber. **Extant**

F. LOXOSCELIDAE Gertsch, 1949 T. (CHT)–Rec. Terr.

First: *Loxosceles deformis* Wunderlich, 1988, *L. defecta* Wunderlich, 1988, Dominican amber. **Extant**

Superfamily HERSILIOIDEA Thorell, 1869

F. HERSILIIDAE Thorell, 1869 T. (RUP)–Rec. Terr.

First: *Hersilia miranda* Koch and Berendt, 1854, *Gerdia myura* Menge, 1854, Baltic amber. **Extant**
Intervening: CHT.

F. UROCTEIDAE Simon, 1875 T. (RUP)–Rec. Terr.

First: *Paruroctea blauvelti* Petrunkevitch, 1942, Baltic amber. **Extant**

F. OECOBIIDAE Blackwall, 1862 T. (CHT)–Rec. Terr.

First: Unnamed oecobiid, Dominican amber (Schawaller, 1981). **Extant**
Intervening: CRD.

Superfamily DYSDEROIDEA C. L. Koch, 1837

F. SEGESTRIIDAE Petrunkevitch, 1933 T. (RUP)–Rec. Terr.

First: *Segestria succinei* Berland, 1939, *S. cylindrica* Koch and Berendt, 1854, *S. cristata* Menge, 1854, *S. elongata* Koch and Berendt, 1854, *S. exarata* Menge, 1854, *S. nana* Koch and Berendt, 1854; *S. pusilla* Menge, 1854, *S. undulata* Menge, 1854; *S. tomentosa* Koch and Berendt, 1854, *S. elongata* Koch and Berendt, 1854, *S. plicata* Petrunkevitch, 1950, Baltic amber. **Extant**
Intervening: CHT.

F. DYSDERIDAE C. L. Koch, 1837 T. (RUP)–Rec. Terr.

First: *Dysdera glabrata* Menge, 1854, *D. hippopodium* Menge, 1854, *D. scobiculata* Menge, 1854, *D. tersa* Koch and Berendt, 1854, *Dasumia subita* Petrunkevitch, 1950, *Harpactes extinctus* Petrunkevitch, 1950, *Thereola petiolata* Koch and Berendt, 1854, Baltic amber. **Extant**
Intervening: CHT, ZAN/PIA.

F. OONOPIDAE Simon, 1892 T. (RUP)–Rec. Terr.

First: *Orchestina baltica* Petrunkevitch, 1942, Baltic amber. **Extant**
Intervening: CHT.

Superfamily PALPIMANOIDEA Forster and Platnick, 1984

F. PALPIMANIDAE Thorell, 1870 T. (YPR)–Rec. Terr.

First: *Protochersis spinosus* Gourret, 1886, Aix-en-Provence, France. **Extant**
Intervening: CHT.

F. ZODARIIDAE Simon, 1892 T. (RUP)–Rec. Terr.

First: *Anniculus balticus* Petrunkevitch, 1942, *Eocydrele mortua* Petrunkevitch, 1958, Baltic amber. **Extant**

F. SPATIATORIDAE Petrunkevitch, 1942 T. (RUP) Terr.

First and Last: *Spatiator praeceps* Petrunkevitch, 1942, *Adorator brevipes* Petrunkevitch, 1942, *A. samlandicus* Petrunkevitch, 1942, Baltic amber.

F. ARCHAEIDAE Koch and Berendt, 1854 (CLV–KIM)–Rec. Terr.

First: *Jurarchaea zherikhini* Eskov, 1987, Karatau Mountains, Kazakhstan, former USSR. **Extant**
Intervening: RUP.

F. ARTHRODICTYNIDAE Petrunkevitch, 1942 T. (RUP) Terr.

First and Last (monotypic family): *Arthrodictyna segmentata* Petrunkevitch, 1942, Baltic amber.

F. MIMETIDAE Simon, 1895 T. (RUP)–Rec. Terr.

First: *Ero permunda* Petrunkevitch, 1942, *E. aberrans* Petrunkevitch, 1958, *E. carboneana* Petrunkevitch, 1942, Baltic amber. **Extant**
Intervening: MES.

Superfamily CLUBIONOIDEA Simon, 1895

F. CLUBIONIDAE Simon, 1895 T. (RUP)–Rec. Terr.

First: *Ablator triguttatus* Koch and Berendt, 1854, *A. lanatus* Petrunkevitch, 1958, *Abliguritor niger* Petrunkevitch, 1942, *A. felix* Petrunkevitch, 1958, *A. plumosus* Petrunkevitch, 1942, *Machilla setosa* Petrunkevitch, 1958, *Eomazax pulcher* Petrunkevitch, 1958, *Phrurolithus ipseni* Petrunkevitch, 1958, *P. extinctus* Petrunkevitch, 1958, *P. fossilis* Petrunkevitch, 1958, *Eodeter magnificus* Petrunkevitch, 1958, *Cryptoplanus? paradoxus* Petrunkevitch, 1958, *Concursator nudipes* Petrunkevitch, 1958, Baltic amber. **Extant**
Intervening: CHT, MES.

F. GNAPHOSIDAE Pocock, 1898
T. (RUP)–Rec. Terr.

First: *Eomactator mactatus* Petrunkevitch, 1958, *Captrix lineata* Koch and Berendt, 1854, Baltic amber. **Extant**
Intervening: CHT.

F. INCEPTORIDAE Petrunkevitch, 1942
T. (RUP) Terr.

First and Last (monogeneric family): *Inceptor aculeatus* Petrunkevitch, 1942, *I. dubius* Petrunkevitch, 1942, Baltic amber.

F. PARATTIDAE Petrunkevitch, 1922
T. (RUP) Terr.

First and Last (monogeneric family): *Parattus evocatus* Scudder, 1890, *P. latitatus* Scudder, 1890, *P. oculatus* Scudder, 1890, *P. resurrectus* Scudder, 1890, Florissant Shales, Colorado.

Superfamily PHILODROMOIDEA Thorell, 1970

F. PHILODROMIDAE Thorell, 1970
T. (RUP)–Rec. Terr.

First: *Syphax asper* Petrunkevitch, 1942, *S. crassipes* Petrunkevitch, 1942, *S. fuliginosus* Koch and Berendt, 1854, *S. gracilis* Koch and Berendt, 1854, *S. hirtus* Menge, 1854, *S. megacephalus* Koch and Berendt, 1854, *Eothanatus diritatis* Petrunkevitch, 1942, Baltic amber. **Extant**
Intervening: CHT.

F. HETEROPODIDAE Thorell, 1873
T. (RUP)–Rec. Terr.

First: *Adulatrix fusca* Petrunkevitch, 1942, *A. decumana* (Koch and Berendt, 1854), *A. fusca* Petrunkevitch, 1942, *A. rufa* Petrunkevitch, 1942, *A. parva* Petrunkevitch, 1942, *Zachria restincta* Petrunkevitch, 1958, *Z. peculiata* Petrunkevitch, 1942, *Z. desiderabilis* Petrunkevitch, 1942, *Caduceator minutus* Petrunkevitch, 1942, *C. quadrimaculatus* Petrunkevitch, 1942, *Collacteus captivus* Petrunkevitch, 1942, *Eostaianus succini* Petrunkevitch, 1942, *Eostasina aculeata* Petrunkevitch, 1942, *Eoprychia succini* Petrunkevitch, 1958, Baltic amber. **Extant**
Intervening: CHT.

F. SELENOPIDAE Simon, 1897
T. (CHT)–Rec. Terr.

First: *Selenops beynai* Schawaller, 1984, Dominican amber. **Extant**

Superfamily ERESOIDEA C. L. Koch, 1837

F. ERESIDAE C. L. Koch, 1837
T. (RUP)–Rec. Terr.

First: *Eresus curtipes* Koch and Berendt, 1854, *E. monachus* Koch and Berendt, 1854, Baltic amber. **Extant**

Superfamily DICTYNOIDEA Simon, 1874

F. DICTYNIDAE Simon, 1874 T. (YPR)–Rec. Terr. (see Fig. 17.5)

First: *Sinodictyna fushunensis* Hong, 1982, Guchengzi Formation, Fushun Coalfield, Liaoning Province, China. **Extant**
Intervening: RUP.

F. ARGYRONETIDAE Menge, 1869b
T. (BUR)–Rec. (FW)

First: *Elvina antiqua* (von Heyden, 1859) Brown Coal of Siebengebirge, Germany; *Argyroneta longipes* Heer, 1865, Switzerland. **Extant**
Comment: von Heyden placed *Elvina antiqua* in the Argyronetidae on the basis of the freshwater nature of the enclosing sediment; Petrunkevitch (1946) doubted this family assignation.

F. ANYPHAENIDAE Bertkau, 1878
T. (RUP)–Rec. Terr.

First: *Anyphaena fuscata* Koch and Berendt, 1854, Baltic amber. **Extant**
Intervening: CHT.

F. HAHNIIDAE Bertkau, 1878
T. (RUP)–Rec. Terr.

First: *Eohahnia succini* Petrunkevitch, 1958, Baltic amber. **Extant**

Superfamily AGELENOIDEA Simon, 1898

F. AGELENIDAE Simon, 1898
T. (RUP)–Rec. Terr.

First: *Eocryphoeca gracilipes* (Koch and Berendt, 1854), *E. distincta* (Petrunkevitch, 1942), *E. fossilis* (Petrunkevitch, 1942), *Myro extinctus* Petrunkevitch, 1942, *M. hirsutus* Petrunkevitch, 1942, *Mastigusa acuminata* Menge, 1854, *M. Modesta* Wonderlich, 1986 Baltic amber. **Extant**
Intervening: CHT, TOR/MES, MES.

F. AMAUROBIIDAE L. Koch, 1868
T. (RUP)–Rec. Terr.

First: *Amaurobius succini* Petrunkevitch, 1942, *Auximus succini* Petrunkevitch, 1942, *A. fossilis* Petrunkevitch, 1942, Baltic amber. **Extant**
Intervening: CHT.

F. INSECUTORIDAE Petrunkevitch, 1942
T. (RUP) Terr.

First and Last (monogeneric family): *Insecutor aculeatus* Petrunkevitch, 1942, *I. mandibulatus* Petrunkevitch, 1942, *I. rufus* Petrunkevitch, 1942, Baltic amber.

Superfamily LYCOSOIDEA Sundevall, 1833

F. LYCOSIDAE Sundevall, 1833
T. (RUP)–Rec. Terr.

First: *Lycosa florissanti* Petrunkevitch, 1922, Florissant Shales, Colorado, USA. **Extant**
Intervening: TOR/MES.

F. PSECHRIDAE Simon, 1892
T. (RUP)–Rec. Terr.

First: *Eomatachia latifrons* Petrunkevitch, 1942, Baltic amber. **Extant**

F. CTENIDAE T. (CHT)–Rec. Terr.

First: *Nanoctenus longipes* Wunderlich, 1988, Dominican amber. **Extant**

F. OXYOPIDAE Thorell, 1870 T. (RUP)–Rec. Terr.

First: *Oxyopes succini* Petrunkevitch, 1958, Baltic amber. **Extant**
Intervening: CHT.

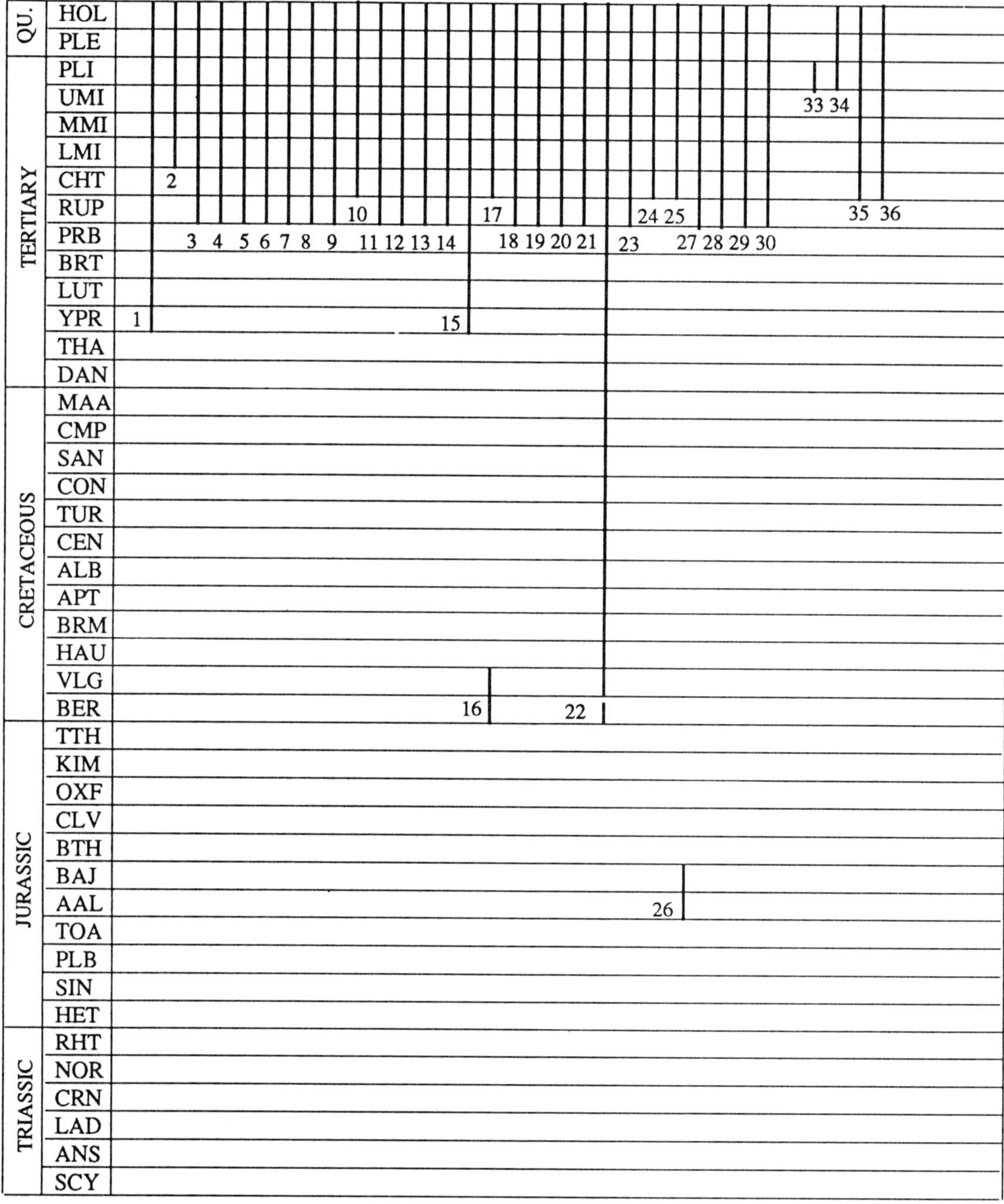

Fig. 17.5

F. ZOROPSIDAE Simon, 1892 T. (RUP)–Rec. Terr.

First: *Adamator succineus* Petrunkevitch, 1942, Baltic amber. **Extant**

F. PISAURIDAE Simon, 1897 T. (RUP)–Rec. Terr.

First: *Esuritor spinipes* Petrunkevitch, 1942, *E. aculeatus* Petrunkevitch, 1958, *Eopisaurella? valdespinosa* Petrunkevitch, 1958, Baltic amber. **Extant**
Intervening: CHT.

Superfamily THOMISOIDEA Sundevall, 1833

F. THOMISIDAE Sundevall, 1833 T. (RUP)–Rec. Terr.

First: *Facundia clara* Petrunkevitch, 1942, *Fiducia tenuipes* Petrunkevitch, 1942, *Filiolella argentata* (Petrunkevitch, 1942), *Medela baltica* Petrunkevitch, 1942, *Misumena samlandica* Petrunkevitch, 1942, Baltic amber. **Extant**
Intervening: CHT, MES, ZAN/PIA, HOL.

Superfamily SALTICOIDEA F. Pickard-Cambridge, 1900

F. SALTICIDAE F. Pickard-Cambridge, 1900 T. (YPR)–Rec. Terr.

First: *Attoides eresiformis* Brongniart, 1877, Aix-en-Provence, France. **Extant**
Intervening: RUP, CHT, TOR/MES.

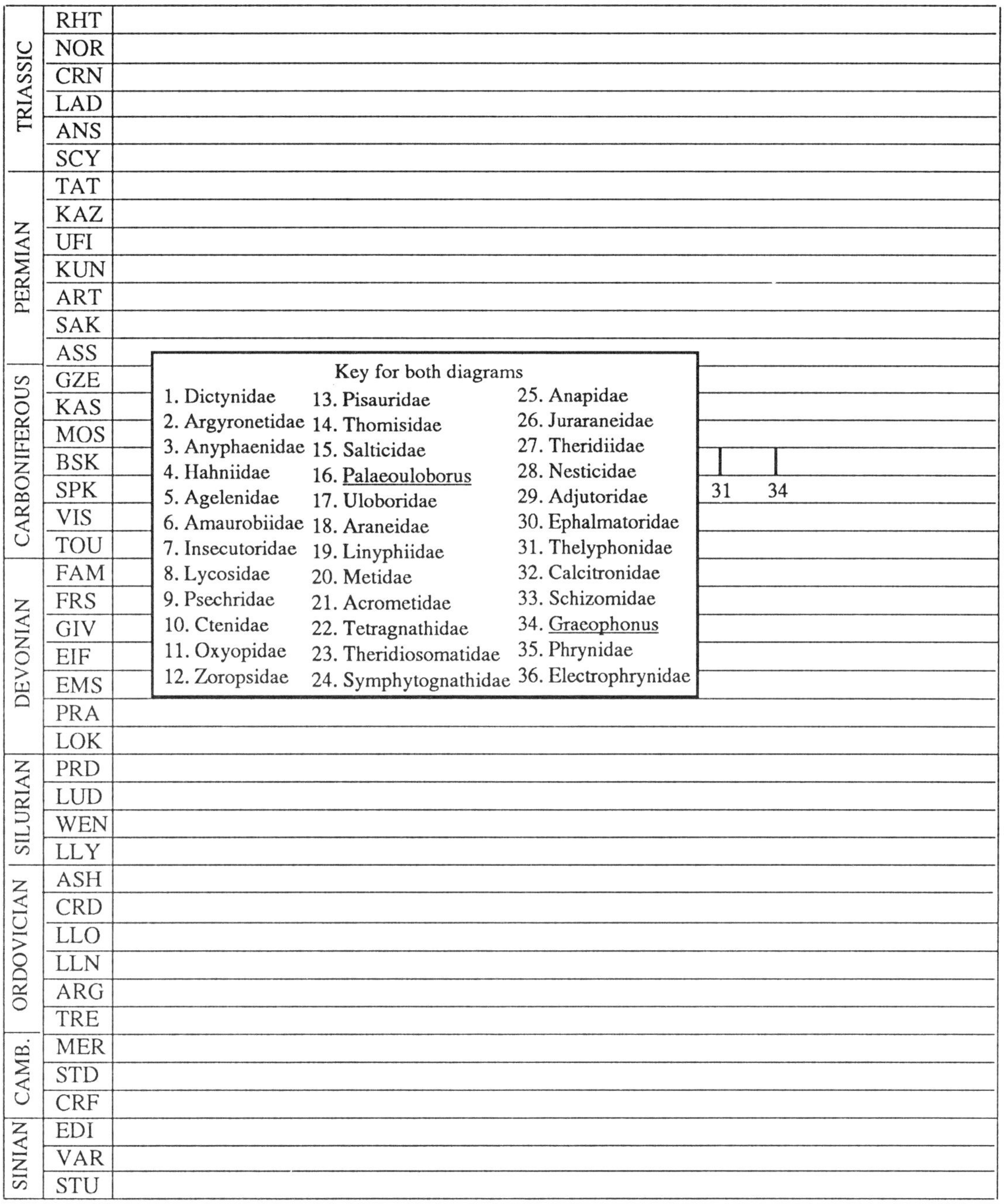

TRIASSIC: RHT, NOR, CRN, LAD, ANS, SCY
PERMIAN: TAT, KAZ, UFI, KUN, ART, SAK, ASS
CARBONIFEROUS: GZE, KAS, MOS, BSK, SPK, VIS, TOU
DEVONIAN: FAM, FRS, GIV, EIF, EMS, PRA, LOK
SILURIAN: PRD, LUD, WEN, LLY
ORDOVICIAN: ASH, CRD, LLO, LLN, ARG, TRE
CAMB.: MER, STD, CRF
SINIAN: EDI, VAR, STU

31 34

Key for both diagrams

1. Dictynidae	13. Pisauridae	25. Anapidae
2. Argyronetidae	14. Thomisidae	26. Juraraneidae
3. Anyphaenidae	15. Salticidae	27. Theridiidae
4. Hahniidae	16. Palaeouloborus	28. Nesticidae
5. Agelenidae	17. Uloboridae	29. Adjutoridae
6. Amaurobiidae	18. Araneidae	30. Ephalmatoridae
7. Insecutoridae	19. Linyphiidae	31. Thelyphonidae
8. Lycosidae	20. Metidae	32. Calcitronidae
9. Psechridae	21. Acrometidae	33. Schizomidae
10. Ctenidae	22. Tetragnathidae	34. Graeophonus
11. Oxyopidae	23. Theridiosomatidae	35. Phrynidae
12. Zoropsidae	24. Symphytognathidae	36. Electrophrynidae

Fig. 17.5

Superfamily DEINOPOIDEA C. L. Koch, 1851

Palaeouloborus K. (BER/VLG) Terr.

First: *Palaeouloborus lacasae* Selden, 1990, lithographic limestone, Sierra de Montsech, Lérida Province, Spain.
Comment: Although not assigned to a family, this fossil bears a greater similarity to the Uloboridae than to any other extant spider family.

F. ULOBORIDAE Simon, 1892 T. (CHT)–Rec. Terr.

First: *Miagrammopes* sp., Dominican amber (Schawaller, 1982; Wunderlich, 1988). **Extant**

Superfamily ARANEOIDEA Latreille, 1806

Intervening: *Cretaraneus vilaltae* Selden, 1990*a*, K. (BER/VAL), lithographic limestone, Sierra de Montsech, Lérida Province, Spain, was referred to Araneoidea but without a family assignation; it probably belongs in Theridiidae or Linyphiidae.

F. ARANEIDAE Leach, 1819 T. (LUT)–Rec. Terr.

First: unnamed Araneidae, oil shales, Grube Messel, Darmstadt, Germany (Wunderlich, 1986).
Intervening: RUP, CHT, TOR/MES, MES, ZAN/PIA. **Extant**

F. LINYPHIIDAE Dahl, 1913 T. (RUP)–Rec. Terr.

First: *Impulsor neglectus* Petrunkevitch, 1942, *I. mutilus*

Petrunkevitch, 1958, *Liticen setosus* Petrunkevitch, 1942, *Mystagogus glaber* Petrunkevitch, 1942, *Custodela cheiracantha* (Koch and Berendt, 1854), *Eopopino longipes* Petrunkevitch, 1942, *Linyphia oblonga* Koch and Berendt, 1854, *Malleator niger* Petrunkevitch, 1942, *Meditrina circumvallata* Petrunkevitch, 1942, *Mystagogus dubius* Petrunkevitch, 1942, *M. glaber* Petrunkevitch, 1942, *Obnisus tenuipes* Petrunkevitch, 1942, Baltic amber. **Extant**
Intervening: CHT, MES.

F. METIDAE Simon, 1892 T. (RUP)–Rec. Terr.

First: *Memoratrix rydei* Petrunkevitch, 1942, *Eometa samlandica* Petrunkevitch, 1958, *E. longipes* Petrunkevitch, 1958, *E. aberrans* Petrunkevitch, 1958, *E. robusta* Petrunkevitch, 1958, *Priscometa tenuipes* Petrunkevitch, 1958, Baltic amber. **Extant**
Intervening: CHT.

F. ACROMETIDAE Wunderlich, 1979 T. (RUP) Terr.

First: *Acrometa cristata* Petrunkevitch, 1942, *A. samlandica* (Petrunkevitch, 1942), *A. minutum* (Petrunkevitch, 1942), *A. robustum* (Petrunkevitch, 1942), *E. succini* Petrunkevitch, 1942, *A. setosus* (Petrunkevitch, 1942), *Pseudoacrometa gracilipes* Wunderlich, 1986, *Anandrus inermis* (Petrunkevitch, 1942), *A. redemptus* (Petrunkevitch, 1958), *A. quaesitus* (Petrunkevitch, 1958), *Cornuanandrus maior* Wunderlich, 1986, Baltic amber.

F. TETRAGNATHIDAE Menge, 1866 K. (BER/VAL)–Rec. Terr.

First: *Macryphantes cowdeni* Selden, 1990*a*, lithographic limestone, Sierra de Montsech, Lérida Province, Spain. **Extant**
Intervening: CHT.

F. THERIDIOSOMATIDAE Simon, 1895 T. (RUP)–Rec. Terr.

First: *Cyclososoma succini* Petrunkevitch, 1958, Baltic amber. **Extant**
Intervening: CHT.

F. SYMPHYTOGNATHIDAE Hickman, 1931 T. (CHT)–Rec. Terr.

First: Unnamed symphytognathid, Dominican amber (Schawaller, 1981). **Extant**

F. ANAPIDAE T. (CHT)–Rec. Terr.

First: *Palaeoanapis nana* Wunderlich, 1988, Dominican amber. **Extant**

F. JURARANEIDAE Eskov, 1984 (AAL/BAJ) Terr.

First and Last (monotypic family): *Juraraneus rasnitsyni* Eskov, 1984, Mukhor-Shibir, near Novospasskoye, Buryat, Siberia, former USSR.

F. THERIDIIDAE Sundevall, 1833 T. (RUP)–Rec. Terr.

First: *Theridion simplex* Koch and Berendt, 1854, *T. alutaceum* Koch and Berendt, 1854, *T. detersum* Koch and Berendt, 1854, *T. granulatum* Koch and Berendt, 1854, *T. hirtum* Koch and Berendt, 1854, *T. ovale* Koch and Berendt, 1854, *T. ovatum* Koch and Berendt, 1854, *Flegia longimana* Koch and Berendt, 1854, *F. succini* Petrunkevitch, 1942, *Mysmena succini* (Petrunkevitch, 1942), *Municeps minutus* Petrunkevitch, 1942, *M. pulcher* (Petrunkevitch, 1942), *Mictodipoena stridula* (Petrunkevitch, 1958), *Dipoena infulata* (Koch and Berendt, 1854), *Eomysmena moritura* Petrunkevitch, 1942, *E. bassleri* (Petrunkevitch, 1942), *E. baltica* (Petrunkevitch, 1946), *E. stridens* Petrunkevitch, 1958, *Nactodipoena dunbari* Petrunkevitch, 1942, *Steatoda succini* Petrunkevitch, 1942, *Eodipoena oculata* Petrunkevitch, 1942, *E. consulta* Petrunkevitch, 1958, *E. germanica* Petrunkevitch, 1958, *E. kaestneri* Petrunkevitch, 1958, *E. nielseni* Petrunkevitch, 1958, *E. regalis* Petrunkevitch, 1958, *Nanomysmena gracilis* Petrunkevitch, 1958, *N. aculeata* Petrunkevitch, 1958, *N. munita* Petrunkevitch, 1958, *Astodipoena crassa* Petrunkevitch, 1958, *Lithyphantes anticus* Berland, 1939, Baltic amber. **Extant**
Intervening: CHT, MES.

F. NESTICIDAE T. (RUP)–Rec. Terr.

First: *Eopopino inopinatus* Wunderlich, 1986, *E. rarus* Wunderlich, 1986, *E. longipes* Petrunkevitch, 1942, *Balticonesticus flexuosus* Wunderlich, 1986, *Heteronesticus magnoparacymbialis* Wunderlich, 1986, Baltic amber. **Extant**
Intervening: CHT.

F. ADJUTORIDAE Petrunkevitch, 1942 T. (RUP) Terr.

First and Last: *Adjutor mirabilis* Petrunkevitch, 1942, *A. deformis* Petrunkevitch, 1958, *Ajunctor similis* Petrunkevitch, 1942, *Admissor aculeatus* Petrunkevitch, 1942, Baltic amber.

F. EPHALMATORIDAE Petrunkevitch, 1958 T. (RUP) Terr.

First and Last (monotypic family): *Ephalmator eximius* Petrunkevitch, 1958, *E. fossilis* Petrunkevitch, 1958, Baltic amber.

Order UROPYGI Thorell, 1882

F. THELYPHONIDAE Lucas, 1835 C. (MRD)–Rec. Terr.

First: *Prothelyphonus naufragus* Brauckmann and Koch, 1983, Hagen-Vorhalle, Ruhr, Germany. **Extant**
Intervening: CHE, MEL, KSK, MYA, ?ZAN/PIA.

Order SCHIZOMIDA Petrunkevitch, 1945

F. CALCITRONIDAE Petrunkevitch, 1945 T. (?ZAN/PIA) Terr.

First and Last: *Calcitro fischeri* Petrunkevitch, 1945, *Onychothelyphonus bonneri* Pierce, 1950, 'Onyx Marble', Bonner Quarry, Ashfork, Arizona.

F. SCHIZOMIDAE Chamberlin, 1922 T. (?ZAN/PIA)–Rec. Terr.

First: *Calcoschizomus latisternum* Pierce, 1951, 'Onyx Marble', Bonner Quarry, Ashfork, Arizona, USA. **Extant**

Order AMBLYPYGI Thorell, 1882

First: *Graeophonus anglicus* Pocock, 1911, C. (MEL) Coseley, near Dudley, England, UK.
Intervening: KSK, MYA.

Suborder APULVILLATA Quintero, 1986

F. PHRYNIDAE Wood, 1863 T. (CHT)–Rec. Terr.

First: *Phrynus resinae* Schawaller, 1979, Dominican amber. **Extant**

Suborder PULVILLATA Quintero, 1986

F. ELECTROPHRYNIDAE Petrunkevitch, 1971 T. (CHT) Terr.

First and Last (monotypic family): *Electrophrynus mirus* Petrunkevitch, 1971, Chiapas amber, Mexico.

REFERENCES

Beall, B. S. (1986) Reinterpretation of the Kustarachnida. (Abstract). *American Arachnology*, **34**, 4.

Bergström, J. (1968) *Eolimulus*, a Lower Cambrian xiphosurid from Sweden. *Geologiska Föreningens i Stockholm Förhandlingar*, **90**, 489–503.

Bergström, J. (1971) *Paleomerus*–merostome or merostomoid. *Lethaia*, **4**, 393–401.

Bergström, J. (1975) Functional morphology and evolution of xiphosurids. *Fossils and Strata*, **4**, 291–305.

Bergström, J., Stürmer, W. and Winter, G. (1980) *Palaeoisopus*, *Palaeopantopus* and *Palaeothea*, pycnogonid arthropods from the Lower Devonian Hunsrück Slate, West Germany. *Paläontologische Zeitschrift*, **54**, 7–54.

Bernini, F. (1986) Current ideas on the phylogeny and the adaptive radiations of Acarida. *Bolletino di Zoologia*, **53**, 279–313.

Briggs, D. E. G., Bruton, D. L. and Whittington, H. B. (1979) Appendages of the arthropod *Aglaspis spinifer* (Upper Cambrian, Wisconsin) and their significance. *Palaeontology*, **22**, 167–80.

Bulanova-Zachvatina, E. M. (1974) New genera of oribatid mites from the Upper Cretaceous of Taimyr. *Palaeontologicheskii Zhurnal*, **2**, 141–4.

Caster, K. E. and Brooks, H. K. (1956) New fossils from the Canadian-Chazyan (Ordovician) hiatus in Tennessee. *Bulletins of American Paleontology*, **36** (157), 153–99.

Chang, A. C. (1957) On the discovery of the Wenlockian *Eurypterus*-fauna from South China. *Acta Palaeontologica Sinica*, **5**, 439–50.

Dubinin, V. B. (1962) Class Acaromorpha: mites or gnathosomic chelicerate arthropods, in *Fundamentals of Paleontology* (ed. B. B. Rodendorf), Academy of Sciences of the USSR, Moscow. [in Russian], pp. 447–73.

Eldredge, N. (1974) Revision of the suborder Synziphosurina (Chelicerata, Merostomata), with remarks on merostome phylogeny. *American Museum Novitates*, **2543**, 1–41.

Eskov, K. and Zonshtein, S. (1990) First Mesozoic mygalomorph spiders from the Lower Cretaceous of Siberia and Mongolia, with notes on the system and evolution of the infraorder Mygalomorphae (Chelicerata: Araneae). *Neues Jahrbuch für Geologie und Paläontologie, Abhandlungen*, **178**, 325–68.

Fisher, D. C. (1982) Phylogenetic and macroevolutionary patterns within the Xiphosurida. *Proceedings of the Third North American Paleontological Convention*, **1**, 175–80.

Fisher, D. C. (1984) The Xiphosura: archetypes of bradytely?, in *Living Fossils* (eds N. Eldredge and S. M. Stanley). Springer-Verlag, Berlin, pp. 196–212.

Golosova, L. D., Druk, Y. A., Karppinen, E. *et al.* (1985) Subfossil oribatid mites (Acarina, Oribatei) of northern Siberia. *Annales Entomologici Fennici*, **51**, 3–18.

Guthörl, P. (1964) Zur Arthropoden-Fauna des Karbons und Perms. 20. Neue Arachniden-Funde (Anthracom.) aus dem Westfal A des Aachener Karbons. *Paläontologische Zeitschrift*, **38**, 98–103.

Hammen, L. van der (1989) *An Introduction to Comparative Arachnology*. SPB Academic Publishing, The Hague, x + 576pp.

Hesselbo, S. P. (1989) The aglaspidid arthropod *Beckwithia* from the Cambrian of Utah and Wisconsin. *Journal of Paleontology*, **63**, 636–42.

Janvier, P., Thanh, T. D. and Gerrienne, P. (1989) Les placodermes, arthropodes et lycophytes des Grès Dévoniens de Dô Son (Haïphong, Viêt Nam). *Geobios*, **22**, 625–39.

Jell, P. A. and Duncan, P. M. (1986) Invertebrates, mainly insects, from the freshwater, Lower Cretaceous, Koonwarra Fossil Bed (Korumburra Group), South Gippsland, Victoria. *Memoirs of the Association of Australasian Palaeontologists*, **3**, 311–405.

Jeram, A. J. (1990) The Micropalaeontology of Palaeozoic Scorpions. Unpublished PhD Thesis, University of Manchester.

Jeram, A. J., Selden, P. A. and Edwards, D. (1990) Land animals in the Silurian: arachnids and myriapods from Shropshire, England. *Science*, **250**, 658–61.

Jux, U. (1967) Eurypteriden im Oberen Plattenkalk von Bergisch Gladbach (O. Devon, Rheinisches Schiefergebirge). *Neues Jahrbuch für Mineralogie, Geologie und Paläontologie, Monatshefte*, **6**, 350–60.

Kethley, J. B., Norton, R. A., Bonamo, P. M. *et al.* (1989) A terrestrial alicorhagiid mite (Acari: Acariformes) from the Devonian of New York. *Micropaleontology*, **35**, 367–73.

Kjellesvig-Waering, E. N. (1951) Downtonian (Silurian) Eurypterida from Perton, near Stoke Edith, Herefordshire. *Geological Magazine*, **88**, 1–24.

Kjellesvig-Waering, E. N. (1963) Pennsylvanian invertebrates of the Mazon Creek area, Illinois: Eurypterida. *Fieldiana (Geology)*, **12**, 85–106.

Kjellesvig-Waering, E. N. (1973) A new Silurian *Slimonia* (Eurypterida) from Bolivia. *Journal of Paleontology*, **47**, 549–550.

Kjellesvig-Waering, E. N. (1986) A restudy of the fossil Scorpionida of the world. *Palaeontographica Americana*, **55**, 1–287.

Koch, C. L. and Berendt, G. C. (1854) *Die im Bernstein Befindlichen Crustaceen, Myriapoden, Arachniden und Apteren Vorwelt*. Berlin. 124 pp. [published posthumously with footnotes by A. Menge].

Lambert, J. B., Frye, J. S. and Poinar, G. O. (1985) Amber from the Dominican Republic: analysis by nuclear magnetic resonance spectroscopy. *Archaeometry*, **27**, 43–51.

Lindquist, E. E. (1984) Current theories on the evolution of major groups of Acari and on their relationships with other classification, in *Acarology VI*, Vol. 1 (eds D. A. Griffiths and C. E. Bowman), Ellis Horwood, Chichester, pp. 28–62.

McAlpine, J. F. and Martin, J. E. H. (1963) Canadian amber – a paleontological treasure-chest. *The Canadian Entomologist*, **101**, 819–38.

Mikulic, D. G., Briggs, D. E. G. and Kluessendorf, J. (1985) A new exceptionally preserved biota from the Lower Silurian of Wisconsin, USA. *Philosophical Transactions of the Royal Society of London, Series B*, **311**, 75–85.

Müller, K. J. and Walossek, D. (1986) Arthropod larvae from the Upper Cambrian of Sweden. *Transactions of the Royal Society of Edinburgh: Earth Sciences*, **77**, 157–79

Norton, R. A., Bonamo, P. M., Grierson, J. D. *et al.* (1989) Fossil mites from the Devonian of New York State. in *Systematics and Taxonomy of Acari* (Progress in Acarology, Vol. 1) (eds G. P. Channabasavanna and C. A. Viraktamath), Oxford and IBH Publishing Co. Pvt. Ltd, New Delhi, Bombay, Calcutta, pp. 271–7.

Petrunkevitch, A. (1913) A monograph of the terrestrial Palaeozoic Arachnida of North America. *Transactions of the Connecticut Academy of Arts and Sciences*, **18**, 1–137.

Petrunkevitch, A. (1945) Palaeozoic Arachnida of Illinois. An inquiry into their evolutionary trends. *Illinois State Museum, Scientific Papers*, **3** (2), 1–72.

Petrunkevitch, A. (1946) Fossil spiders in the collection of the American Museum of Natural History. *American Museum Novitates*, **1328**, 1–36.

Petrunkevitch, A. (1949) A study of Palaeozoic Arachnida. *Transactions of the Connecticut Academy of Arts and Sciences*, **37**, 69–315.

Petrunkevitch, A. (1953) Palaeozoic and Mesozoic Arachnida of Europe. *Geological Society of America, Memoir*, **53**, xi + 1–128pp.

Petrunkevitch, A. (1955) Arachnida, in *Treatise on Invertebrate Paleontology. Part P. Arthropoda 2* (ed. R. C. Moore), Geological Society of America and University of Kansas Press, Boulder, Colorado, and Lawrence, Kansas, pp. p42–p162.

Petrunkevitch, A. (1958) Amber spiders in European collections.

Transactions of the Connecticut Academy of Arts and Sciences, **41**, 97–400.

Poinar, G. O. (1988) Hair in Dominican amber: Evidence for Tertiary land animals in the Antilles. *Experientia*, **44**, 88–9.

Prantl, F. and Pribyl, A. (1948) Revision of the Bohemian Silurian eurypterids. *Rozpravy Statniho Geologickeho Ustavu Republiky Ceskoslovenske, Svezek*, **10**, 1–111.

Pribyl, A. (1960) Nové poznatky o svrchokarbonské sladkovodní kontinentální fauně z ostravsko-karvinské oblasti. *Rozpravy Ceskoslovenské Akademie Ved. Praha. Matematickych a Přírodnich Věd*, **70**, 1–71.

Raymond, P. E. (1944) Late Paleozoic xiphosurans. *Bulletin of the Museum of Comparative Zoology, Harvard University*, **94**, 476–508.

Rolfe, W. D. I. (1966) Phyllocarid crustacean fauna of European aspect from the Devonian of Western Australia. *Nature*, **209**, 192.

Rowland, J. M. and Sissom, W. D. (1980) Report on a fossil palpigrade from the Tertiary of Arizona, and a review of the morphology and systematics of the order (Arachnida: Palpigradida). *Journal of Arachnology*, **8**, 69–86.

Schawaller, W. (1981) Übersicht über Spinnen-Familien im Dominicanischen Bernstein und anderen tertiären Harzen (Stuttgarter Bernsteinsammlung: Arachnida, Araneae). *Stuttgarter Beiträge zur Naturkunde, Serie B*, **77**, 1–10.

Schawaller, W. (1991) The first Mesozoic pseudoscorpion, from the Cretaceous Canadian amber. *Palaeontology*, **34**, 971–6.

Schawaller, W., Shear, W. A. and Bonamo, P. (1991) The first Paleozoic pseudoscorpions (Arachnida, Pseudoscorpionida). *American Museum Novitates*, **3009**, 1–17.

Selden, P. A. (1990) Fossil history of the arachnids. *Newsletter of the British Arachnological Society*, **58**, 4–6.

Selden, P. A. and Shear, W. A. (1992) A myriapod identity for the Devonian 'scorpion' *Tiphoscorpio hueberi*. *Bericht der naturwissenschaftlich-Medezinischen Vereins in Innsbruck, Supplement 10*, 35–6.

Selden, P. A. and Siveter, D. J. (1987) The origin of the limuloids. *Lethaea*, **20**, 383–92.

Selden, P. A., Shear, W. A. and Bonamo, P. M. (1991) A spider and other arachnids from the Devonian of New York, and reinterpretations of Devonian Araneae. *Palaeontology*, **34**, 241–81.

Shear, W. A. (1986) Taxonomic glossary, in *Spiders – Webs, Behavior, and Evolution* (ed. W. A. Shear), Stanford Press, Stanford, California, pp. 401–32.

Shear, W. A., Selden, P. A., Rolfe, W. D. I. *et al.* (1987) New terrestrial arachnids from the Devonian of Gilboa, New York (Arachnida, Trigonotarbida). *American Museum Novitates*, **2901**, 1–74.

Shear, W. A., Schawaller, W. and Bonamo, P. M. (1989) Record of Palaeozoic pseudoscorpions. *Nature*, **341**, 527–9.

Shultz, J. W. (1990) Evolutionary morphology and phylogeny of Arachnida. *Cladistics*, **6**, 1–38.

Siveter, D. J. and Selden, P. A. (1987) A new, giant xiphosurid from the lower Namurian of Weardale, County Durham. *Proceedings of the Yorkshire Geological Society*, **46**, 153–68.

Sivhed, U. and Wallwork, J. A. (1978) An early Jurassic oribatid mite from southern Sweden. *Geologiska Föreningens i Stockholm Förhandlingar*, **100**, 65–70.

Stockwell, S. A. (1989) Revision of the Phylogeny and Higher Classification of Scorpions (Chelicerata). Unpublished PhD Thesis, University of California, Berkeley.

Størmer, L. (1955) Merostomata, in *Treatise on Invertebrate Paleontology. Part P. Arthropoda 2* (ed. R. C. Moore), Geological Society of America and University of Kansas Press, Boulder, Colorado, and Lawrence, Kansas, pp. p4–p41.

Størmer, L. (1974) Arthropods from the Lower Devonian (Lower Emsion) of Alken-an-der-Mosel, Germany. Part 4: Eurypterida, Drepanopteridae, and other groups. *Senckenbergiana Lethaea*, **54**, 359–451.

Størmer, L. (1956) A Lower Cambrian merostome from Sweden. *Arkiv för Zoologi, Serie 2*, **9**, 507–514.

Størmer, L. and Waterston, C. D. (1968) *Cyrtoctenus* gen. nov., a large, late Palaeozoic arthropod with pectinate appendages. *Transactions of the Royal Society of Edinburgh*, **68**, 63–104.

Tollerton, V. P. (1989) Morphology, taxonomy, and classification of the order Eurypterida Burmeister, 1843. *Journal of Paleontology*, **63**, 642–57.

Waterston, C. D., Oelefsen, B. W. and Oosthuizen, R. D. F. (1985) *Cyrtoctenus wittebergensis* sp. nov. (Chelicerata: Eurypterida), a large sweep-feeder from the Carboniferous of South Africa. *Transactions of the Royal Society of Edinburgh: Earth Sciences*, **76**, 339–58.

Whalley, P. E. S. (1980) Neuroptera (Insecta) in amber from the Lower Cretaceous of Lebanon. *Bulletin of the British Museum (Natural History), Geology Series*, **33**, 157–64.

Wood, S. P., Panchen, A. L. and Smithson, T. R. (1985) A terrestrial fauna from the Scottish Lower Carboniferous. *Nature*, **314**, 355–6.

Wunderlich, J. (1988) Die fossilen Spinnen in Dominikanischen Bernstein. *Beitrage für Araneologie*, **2**, 1–378.

18

ARTHROPODA (CRUSTACEA EXCLUDING OSTRACODA)

D. E. G. Briggs, M. J. Weedon and M. A. Whyte

This chapter includes all crustaceans except the ostracodes, which are treated in Chapter 19.

Acknowledgements – E. A. Bousfield, R. M. Feldmann, S. F. Morris, W. A. Newman, W. D. I. Rolfe and F. R. Schram generously supplied information and comments.

STEM LINEAGE CRUSTACEANS

This category (Walossek and Müller, 1990) accommodates those Upper Cambrian arthropods from the Orsten of Sweden which are interpreted as descendants of early offshoots from the stem-lineage of Crustacea. The most primitive of these stem lineage crustaceans is *Henningsmoenia scutula* Walossek and Müller, 1990; the most advanced is *Martinssonia elongata* Müller and Walossek, 1986 (Walossek and Müller, 1990). Only two of the stem lineage crustaceans have been grouped in a higher taxon, currently of unspecified rank.

Taxon CAMBROPACHYCOPIDAE Walossek and Müller, 1990 Є. (MNT) Mar. (see Fig. 18.1)

First and Last: *Cambropachycope clarksoni* Walossek and Müller, 1990 and *Goticaris longispinosa* Walossek and Müller, 1990, *Agnostus pisiformis* Zone, Alum Shale Formation, Västergötland, Sweden.

Subphylum CRUSTACEA Pennant, 1777 (excluding Ostracoda)

The classification used here follows that of Schram (1986), with minor additions and modifications. This does not necessarily imply agreement with all his taxonomic conclusions. Families without a fossil record are not included.

Class UNNAMED

Order SKARACARIDA

F. SKARAIDAE Є. (MNT) Mar.

First and Last: *Skara anulata* Müller, 1983, and *S. minuta* Müller and Walossek, 1985, *Agnostus pisiformis* Zone, Alum Shale Formation, Västergötland, Sweden.

Class REMIPEDIA Yager, 1981

The class includes two orders, the living Nectiopoda and the fossil Enantiopoda (Schram, 1986).

Order ENANTIOPODA Birshtein, 1960

F. TESNUSOCARIDIDAE Brooks, 1955 C. (VIS–MOS) Mar.

First: *Tesnusocaris goldichi* Brooks, 1955, Tesnus Formation, Brewster County, Texas, USA (Emerson and Schram, 1991).

Last: *Cryptocaris hootchi* Schram, 1974, Francis Creek Shale, Carbondale Formation, NE Illinois, USA (Emerson and Schram, 1991).

Class MALACOSTRACA Latreille, 1806

Subclass HOPLOCARIDA Calman, 1904

Order AESCHRONECTIDA Schram, 1969

F. KALLIDECTHIDAE Schram, 1969 C. (BSK–KAS/GZE) Mar.

First: *Kallidechtes eageri* Schram, 1979a, Bude, Cornwall, England, UK.

Last: *Kallidechtes richardsoni* Schram, 1969, Francis Creek Shale, Carbondale Formation, NE Illinois, USA.

F. AENIGMACARIDIDAE Schram and Horner, 1978 C. (VIS–KAS/GZE) Mar.

First: *Joanellia elegans* (Peach, 1883), Cementstone Group, Redesdale, Northumberland (Schram, 1979a).

Last: *Aenigmacaris minima* Schram and Schram, 1979, Madera Formation, New Mexico, USA.

F. ARATIDECTHIDAE Schram, 1979 C. (VIS–MOS) Mar.

First: *Crangopsis eskdalensis* (Peach, 1882), Cementstone Group, Courceyan, River Tweed, Berwickshire, UK.

Last: *Aratidecthes johnsoni* Schram, 1969, Logan Quarry Shale, Garrard Quarry, Indiana, USA.

Comment: Schram (1979b) recorded *Crangopsis* from a number of UK localities of this age.

Order PALAEOSTOMATOPODA Brooks, 1962

F. PERIMECTURIDAE C. (TOU–SPK) Mar.

First: *Archaeocaris graffhami* Brooks, 1962, Hastarian, Nevada, USA (Schram, 1979b).

Last: *Perimecturus rapax* Schram and Horner, 1978, Bear Gulch Limestone, central Montana, USA.

Comment: Schram (1980) noted the resemblance of *Eopterum devonicum* Rohdendorf, 1961 and *Eopteridium striatum* Rohdendorf, 1970, from the Upper Devonian (?Famennian) of the Timan Peninsula and the Ukraine,

The Fossil Record 2. Edited by M. J. Benton. Published in 1993 by Chapman & Hall, London. ISBN 0 412 39380 8

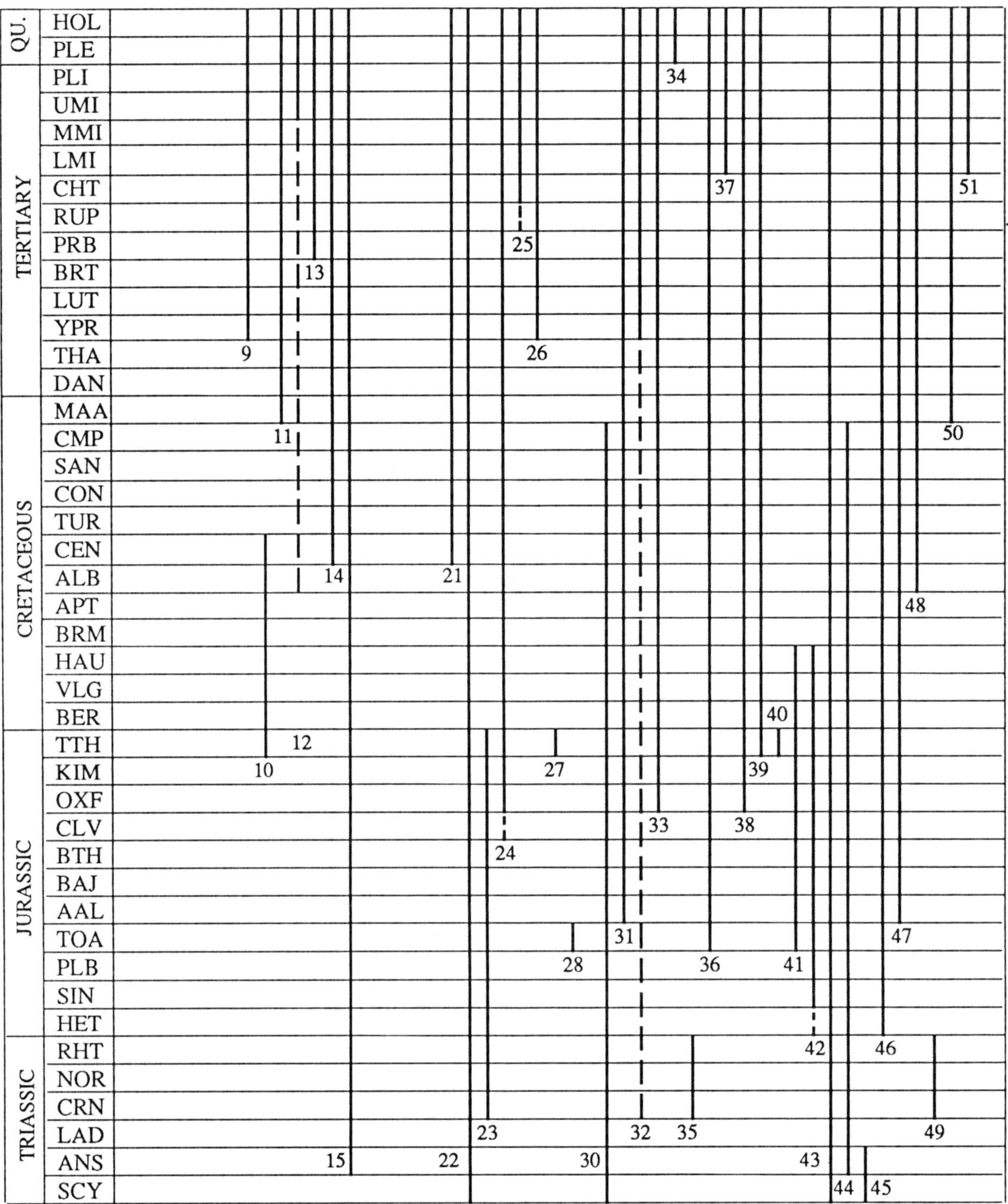

Fig. 18.1

respectively, to *Perimecturus*. Schram and Horner (1978) described *Bairdops beargulchensis* from the Bear Gulch Limestone, but Factor and Feldmann (1985) synonymized the material with *Tyrannophontes theridion*. Schram (pers. comm.) disagrees.

Order STOMATOPODA Latreille, 1817

Suborder ARCHAEOSTOMATOPODEA Schram, 1969

F. TYRANNOPHONTIDAE Schram, 1969
C. (SPK–KAS) Mar.

First: ?'*Perimecturus*' *pattoni* Peach, 1908, Top Hosie Limestone, East Kilbride, Scotland, UK.

Last: *Gorgonophontes peleron* Schram, 1984b, Dennis Formation, Nebraska, USA.

Comment: Although unsure about its identity, Schram (1979a) considered the specimen of ?'*Perimecturus*' *pattoni* likely to be a tyrannophontid. Other early occurrences of the family were reviewed by Schram (1984b).

Suborder UNIPELTATA Latreille, 1825

Superfamily BATHYSQUILLOIDEA Manning, 1967

F. BATHYSQUILLIDAE Manning, 1967
T. (YPR)–Rec. Mar.

First: *Bathysquilla wetherelli* (Woodward, 1879), London Clay, Kent, UK (Quayle, 1987). **Extant**

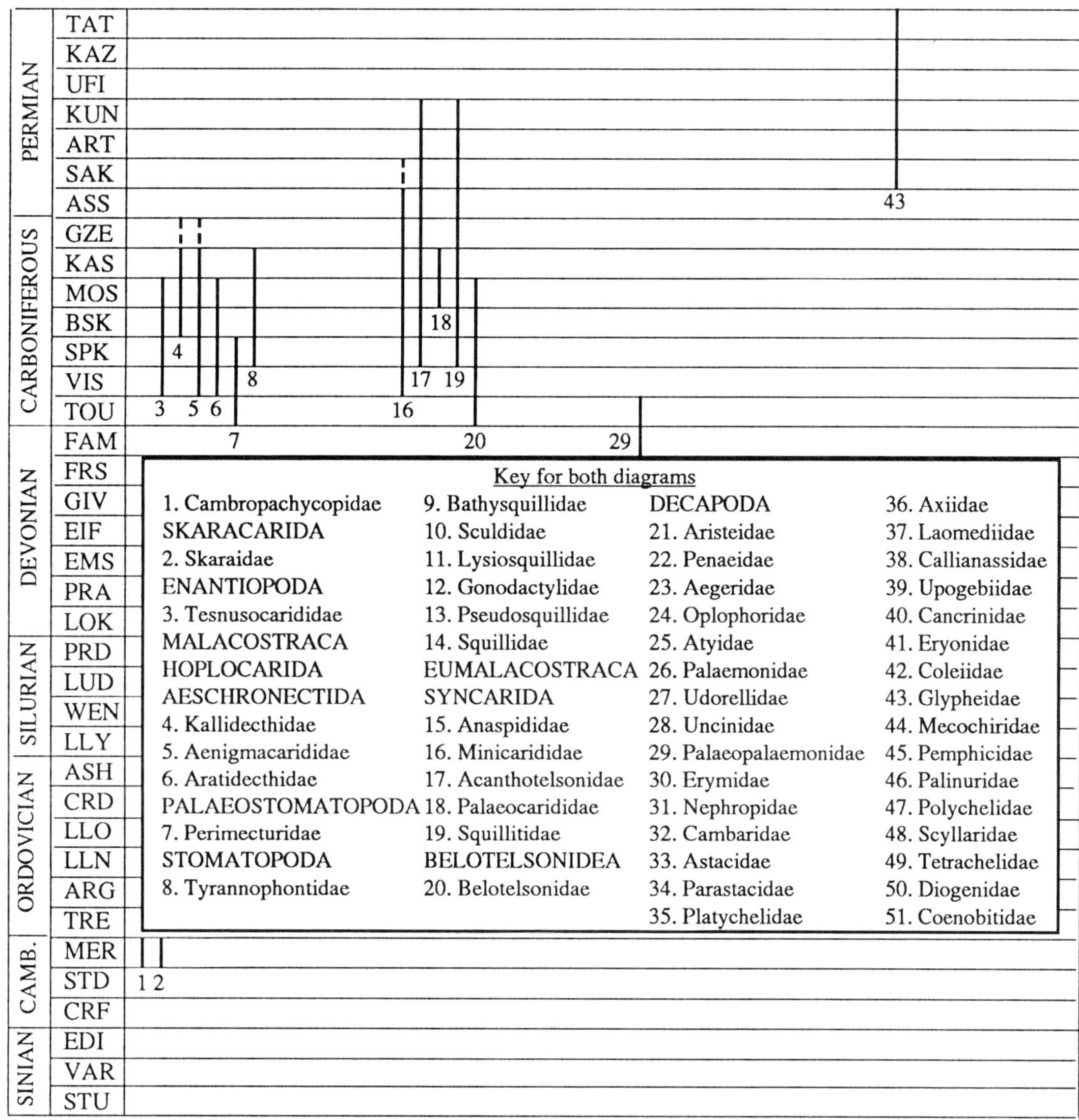

Fig. 18.1

F. SCULDIDAE Dames, 1866 J. (TTH)–K. (CEN) Mar.

First: *Sculda pennata* Münster, 1840, Solnhofen Limestone, Bavaria, Germany.
Last: *Sculda syriaca* Dames, 1886; *Pseudosculda laevis* Schlüter, 1872, Lebanon (see Roger, 1946).
Comment: *Clausocaris* (*Clausia*) and *Pseuderichthus* which were formerly assigned to this family are now recognized as Thylacocephala (see Arduini and Pinna, 1989).

Superfamily LYSIOSQUILLOIDEA Giesbrecht, 1910

F. LYSIOSQUILLIDAE Giesbrecht, 1910 K. (MAA)–Rec. Mar.

First: *Lysiosquilla nkporoensis* Förster, 1982, Nkporo Shale, Nigeria. **Extant**

Superfamily GONODACTYLOIDEA Giesbrecht, 1910

F. GONODACTYLIDAE Giesbrecht, 1910 (?K. (GAL)) T. (UMI)–Rec. Mar.

First: *Paleosquilla brevicoxa* Schram, 1968, Albian–Cenomanian, Simiti Formation, Antioquia, Columbia. **Extant**
Comment: Schram (1968, p. 1299) considered that this may be the earliest gonodactylid. Otherwise this family ranges from the Upper Miocene.

F. PSEUDOSQUILLIDAE Manning, 1977 T. (PRB)–Rec. Mar.

First: ?*Pseudosquilla wulfi* Förster, 1982, Gehlberg-Schichten, north-western Germany. **Extant**

Superfamily SQUILLOIDEA Latreille, 1803

F. SQUILLIDAE Latreille, 1803 K. (GUL)–Rec. Mar.

First: *Chloridella angolia* Berry, 1932, Upper Cretaceous, Angola (Holthuis and Manning, 1969, p. 541). **Extant**

Subclass EUMALACOSTRACA Grobben, 1892

Order SYNCARIDA Packard, 1885

Suborder ANASPIDACEA Calman, 1904

Schram (1986) was inconsistent in his spelling of this taxon giving both Anaspidacea (p. 542) and Anaspidinea (p. 85).

F. ANASPIDIDAE Thompson, 1893 Tr. (ANS)–Rec. FW

First: *Anaspidites antiquus* (Chilton, 1929) Triassic, Hawkesbury Sandstone, New South Wales, Australia. **Extant**

Order **PALAEOCARIDACEA Brooks, 1962**

F. MINICARIDIDAE Schram, 1984 C. (VIS)–P. (ASS/SAK) FW–?Mar.

First: *Minicaris* sp., Briggs and Clarkson, 1991, Granton Shrimp Bed, Holkerian–Asbian, Edinburgh, Scotland, UK.
Last: *Erythrogaulos carrizoensis* Schram, 1984a, Red Tanks Member, Madera Formation, Lower Permian, USA.
Comment: *Minicaris* was first described (*Minicaris brandi* Schram, 1979) from below the Pumpherston Shell Bed, Lower Oil Shale Group, VIS (?Asbian), England, UK.

F. ACANTHOTELSONIDAE Meek and Worthen, 1865 C. (SPK)–P. (ROT) FW

First: *Palaeosyncaris dakotensis* Brooks, 1962, Heath Shale, Madera Formation, USA.
Last: *Uronectes fimbriatus* (Jordan, 1947), Rotliegendes, Saarbrücken, Germany.

F. PALAEOCARIDIDAE Meek and Worthen, 1865 C. (MOS–KAS) FW

First: *Palaeocaris retractata* Calman, 1932, Ten Foot Ironstone Measures, lower Similis–Pulchra Zone, Worcestershire, England, UK.
Last: *Palaeocaris secretanae* Schram, 1984, First Blanzy-Montceau Coal, Montceau-les-Mines, France.

F. SQUILLITIDAE Schram and Schram, 1974 C. (SPK)–P. (ROT) FW

First: *Squillites spinosa* Scott, 1938, Heath Shale, Big Snowy Group, Montana, USA.
Last: *Nectotelson krejcii* (Fritsch, 1875), Rotliegendes, near Autun, France (Schram, 1984a).
Comment: Schram (1984a) attributed the Gasköhle of the Humboldt Mine at Nyrany, Bohemia, Czechoslovakia, where *N. krejcii* also occurs, to the Lower Permian, although it is Moscovian (Westphalian D) in age (Gray, 1988).

Order **BELOTELSONIDEA Schram, 1981**

F. BELOTELSONIDAE Schram, 1974 C. (TOU–MOS) Mar.

First: *Belotelson traquairi* (Peach, 1882), ?Hastarian, Cementstone Group, Berwickshire, Scotland, UK (Locality 9 of Schram, 1979a; the Foulden locality, Briggs and Clarkson, 1985, is similar in age).
Last: *Belotelson magister* (Packard, 1886), Francis Creek Shale, Mazon Creek, NE Illinois, USA.

Order **EUPHAUSIACEA Dana, 1852**

There are no well-substantiated fossil taxa, although possible Carboniferous forms have been alluded to in the literature (Dzik and Jazdzewski, 1978; Schram, 1981; see discussion in Schram, 1986). **Extant**

Order **DECAPODA Latreille, 1803**

Suborder **DENDROBRANCHIATA Bate, 1888**

Superfamily **PENAEOIDEA Rafinesque, 1815**

F. ARISTEIDAE Alcock, 1901 K. (CEN)–Rec. Mar.

First: *Benthesicymus libanensis* (Brocchi, 1875), Sahel Alma, Lebanon (see Glaessner, 1945; Roger, 1946). **Extant**

F. PENAEIDAE Bate, 1881 ?Tr. (SCY)–Rec. Mar.

First: *Antrimpos* Münster, 1839 and *Bombur* Münster, 1839, Triassic, Europe. **Extant**
Comment: Burkenroad (1963, p. 9) recognized *Antrimpos atavus* as Lower Triassic. Glaessner (1969) recorded *Antrimpos* from the Permo-Triassic and *Bombur* from the Upper Triassic. Schram (1986, p. 252) recognized no Triassic penaeids.

F. AEGERIDAE Münster, 1839 Tr. (CRN)–J. (TTH) Mar.

First: *Aeger straeleni* Glaessner, 1929, Carnian Shales, Austria.
Last: ?*Aeger tipularius* (Schlotheim, 1832), Solnhofen Limestone, Bavaria, Germany.
Comment: This family status of this taxon was advocated by Burkenroad (1963) and supported by Schram (1986). The taxonomic placement of an older species, *A. lehmanni* from the Middle Triassic of Thuringia, Germany is uncertain (Förster and Crane, 1984).

Suborder **EUKYPHIDA Boas, 1880**

Infra-order **CARIDEA Dana, 1852**

Superfamily **OPLOPHOROIDEA Dana, 1852**

F. OPLOPHORIDAE Dana, 1852 J. (?CLV)–Rec. Mar.

First: *Oplophorus* H. Milne Edwards, 1837, ?Upper Jurassic, central Africa (Glaessner, 1969). **Extant**

F. ATYIDAE de Haan, 1849 T. (RUP/CHT)–Rec. Mar.

First: *Caridina nitida* A. Milne Edwards, 1879, Oligocene, France (Glaessner, 1969). **Extant**

Superfamily **PALAEMONOIDEA Rafinesque, 1815**

F. PALAEMONIDAE Rafinesque, 1815 T. (YPR)–Rec. Mar./FW

First: *Bechleja rostrata* Feldmann *et al.*, 1981, Green River Formation, Wyoming, USA. **Extant**

Superfamily ***INCERTAE SEDIS***

F. UDORELLIDAE Van Straelen, 1924 J. (TTH) Mar.

First and Last: *Udorella agassizi* Oppel, 1862, Tithonian, Solnhofen Limestone, Bavaria, Germany.
Comment: Felgenhauer and Abele (1983) pointed out similarities between this taxon and the Infra-order Procarididea (Schram, 1986, p. 259).

Suborder **EUZYGIDA Burkenroad, 1981**

Infra-order **UNCINIDEA Beurlen, 1930**

F. UNCINIDAE, Beurlen, 1928 J. (TOA) Mar.

First and Last: *Uncina posidoniae*, Liassic, lower Toarcian, Holzmaden, Baden-Württemberg, Germany.
Comment: Schram (1986, p. 284) did not recognize this family, but states that '*Uncina* bears some sister-group relationship to the living stenopodideans'.

Suborder REPTANTIA Boas, 1880

F. PALAEOPALAEMONIDAE Rafinesque, 1815 D. (FAM)–C. (TOU) Mar.

First: *Palaeopalaemon newberryi* Whitfield, 1880, Ohio, New York and Iowa, USA (Schram *et al.*, 1978).
Last: ?*Palaeopalaemon newberryi* Whitfield, 1880, New Providence Formation, Boyle County, Kentucky, USA (Schram *et al.*, 1978, p. 1376, express some doubt regarding the identity of the specimen upon which this record is based).

Infra-order ASTACIDEA Latreille, 1803

F. ERYMIDAE Van Straelen, 1924 Tr. (SCY)–K. (CMP) Mar.

First: *Clytiopsis argentoratensis* Glaessner, 1929, Lower Triassic, Alsace (Glaessner, 1969).
Last: *Enoploclytia* sp. Beikirch and Feldmann, 1980, Austin Formation, Travis County, Texas, USA.
Comment: *Protoclytiopsis antiqua* Birshtein, 1958, uppermost Permian, Siberia, was placed in the Erymidae by Schram (1980), but he subsequently (1986) declared that it 'seems to be a glypheid'. There are uncertain Palaeocene records of *Enoploclytia* (Glaessner, 1969).

F. NEPHROPIDAE Dana, 1852 J. (DOG)–Rec. Mar.

First: *Palaeophoberus suevicus* (Quenstedt, 1867), Middle Jurassic, southern Germany (Glaessner, 1969).
Comment: There is some doubt about the placement of *P. suevicus* in this family (R. M. Feldmann, pers. comm.).

F. CAMBARIDAE Hobbs, 1942 ?Tr (u.)/T. (YPR)–Rec. FW

First: *Cambarus primaevus* Packard, 1880, YPR, North America (Glaessner, 1969).
Comment: Undescribed material from the Upper Triassic Chinle Formation of the Colorado Plateau may represent this family (Hasiotis and Mitchell, 1989). **Extant**

F. ASTACIDAE Latreille, 1803 J. (MLM)–Rec. Mar.

First: *Astacus licenti* van Straelen, 1928 and *A. spinirostrius* Imaizumi, 1938, Asia (eastern Mongolia – China). **Extant**
Comment: Glaessner (1969) expressed some doubt about the age of the first occurrence.

F. PARASTACIDAE Huxley, 1879 Q. (PLE)–Rec. Mar.

First: *Astacopsis*?, Pleistocene, SE Australia (Glaessner, 1969).
Comment: Sepkoski (1982) recorded the first occurrence as Eocene, but cited Glaessner (1969) as source.

F. PLATYCHELIDAE Glaessner, 1969 Tr. (u.) Mar.

First and Last: *Platychela trauthi* Glaessner, 1931, Upper Triassic, Australia (Glaessner, 1969), or *Platypleon nevadense* Van Straelen, 1936, Upper Triassic, Nevada, USA (Glaessner, 1969).

Infra-order THALASSINIDEA Latreille, 1831

F. AXIIDAE Huxley, 1879 J. (TOA)–Rec. Mar.

First: *Magila? bonjouri* (Etallon, 1861), Lower Jurassic, Europe (Glaessner, 1969). **Extant**

F. LAOMEDIIDAE Borradaile, 1903 T. (Mio.)–Rec. Mar.

First: *Jaxea kuemeli* Bachmayer, Miocene, Australia (Glaessner, 1969). **Extant**

F. CALLIANASSIDAE Dana, 1852 J. (MLM)–Rec. Mar.

First: *Protocallianassa* Beurlen, 1930 (Glaessner, 1969). **Extant**
Comment: Includes *Protocallianassa archiaci* (A. Milne Edwards) upper Senonian, The Netherlands (Glaessner, 1969).

F. UPOGEBIIDAE Borradaile, 1903 J. (TTH)–Rec. Mar.

First: *Upogebia dura* Moericke, 1889, Tithonian, Europe (Glaessner, 1969). **Extant**
Comment: The *U. clypeata* from the Bathonian of England was referred to the Mecochiridae by Woods (1928).

Infra-order PALINURA Latreille, 1803

F. CANCRINIDAE Beurlen, 1930 J. (TTH) Mar.

?First and Last: *Cancrinos claviger* Münster, 1839, Solnhofen Limestone, Bavaria, Germany (Glaessner, 1969).

F. ERYONIDAE De Haan, 1841 J. (TOA)–K. (NEO) Mar.

First: ?*Proeryon hartmanni* (Von Meyer, 1835), Liassic, lower Toarcian, Holzmaden, Baden-Württemberg, Germany.
Last: *Eryon neocomiensis* Woodward, 1881, Lower Cretaceous, Europe (Glaessner, 1969).
Comment: Rathbun (1919) assigned fragments from the Lower Miocene of the West Indies to this family.

F. COLEIIDAE Van Straelen, 1924 J. (HET/SIN)–K. (NEO) Mar.

This family was recognized by Glaessner (1969), (and by Sepkoski, 1982) but was not listed by Schram (1986). Unfortunately, Schram did not indicate where he would place the two included genera *Coleia* and *Hellerocaris*.
First: *Coleia antiqua*, lower Lias, England, UK (Glaessner, 1969).
Last: *Coleia* sp., Lower Cretaceous, India (Glaessner, 1969).

F. GLYPHEIDAE Winckler, 1883 P. (SAK)–Rec. Mar.

First: *Palaeopemphix* Gemmellaro, 1890 (three species from the Lower Permian Sosio Limestone of Sicily). *Protoclytiopsis antiqua* Birshtein, 1958, uppermost Permian, Siberia, was placed in the Erymidae by Schram (1980), but he subsequently (1986) declared that it 'seems to be a glypheid'. Otherwise the earliest records are of *Litogaster* ?Lower and Middle Triassic of Germany (Glaessner, 1969). **Extant**

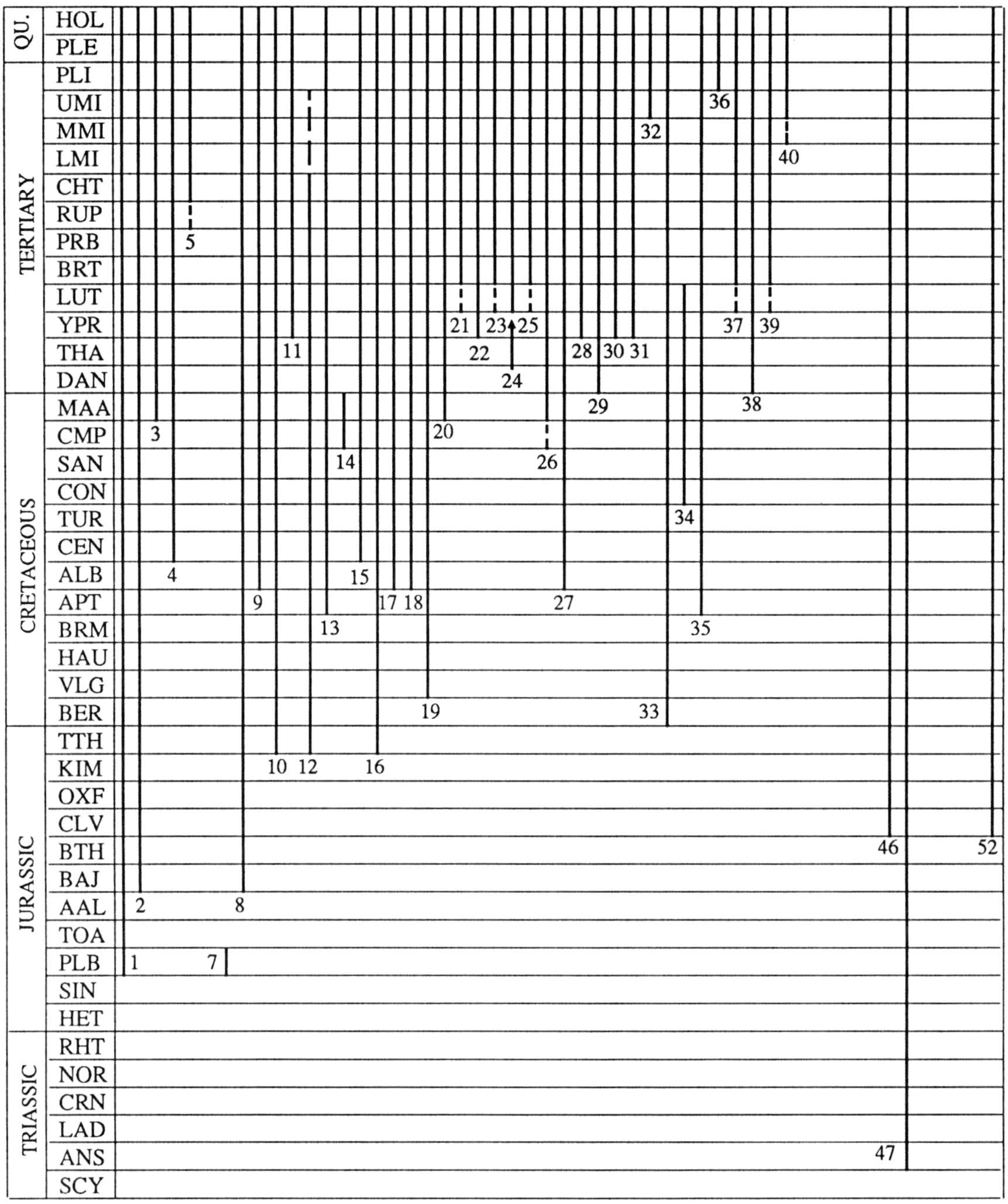

Fig. 18.2

Comment: Glaessner (1969, p. 436) considered that the *Palaeopemphix* material needs re-examination before assignment to the order Decapoda.

F. MECOCHIRIDAE Van Straelen, 1925 Tr. (m.)–K (CMP) Mar.

First: *Pseudoglyphea* Oppel, 1861, Triassic (Glaessner, 1969).

Last: *Meyeria* M'Coy, 1849, Campanian, cosmopolitan (Glaessner, 1969); Genus indet. mecochirid, Austin Formation, lower CMP, Travis County, Texas, USA (Beikirch and Feldmann, 1980).

F. PEMPHICIDAE Van Straelen, 1928 Tr. (SPA/ANS) Mar.

First and Last: *Pemphix* von Meyer, 1840, including *Pemphix sueurii* (Desmarest, 1822), or *Pseudopemphix* Wüst, 1903, including *Pseudopemphix albertii* (von Meyer, 1840), both Europe (Glaessner, 1969).

F. PALINURIDAE Latreille, 1803 J. (l.)–Rec. Mar.

First: *Palinurina* sp. Woodward, 1867, Lower Jurassic, England, UK (Glaessner, 1969). **Extant**

F. POLYCHELIDAE Wood-Mason, 1874 J. (DOG)–Rec. Mar.

First: *Willemoesiocaris ovalis* Van Straelen, 1925, Middle Jurassic, France (Glaessner, 1969). **Extant**

F. SCYLLARIDAE Latreille, 1825 K. (ALB)–Rec. Mar.

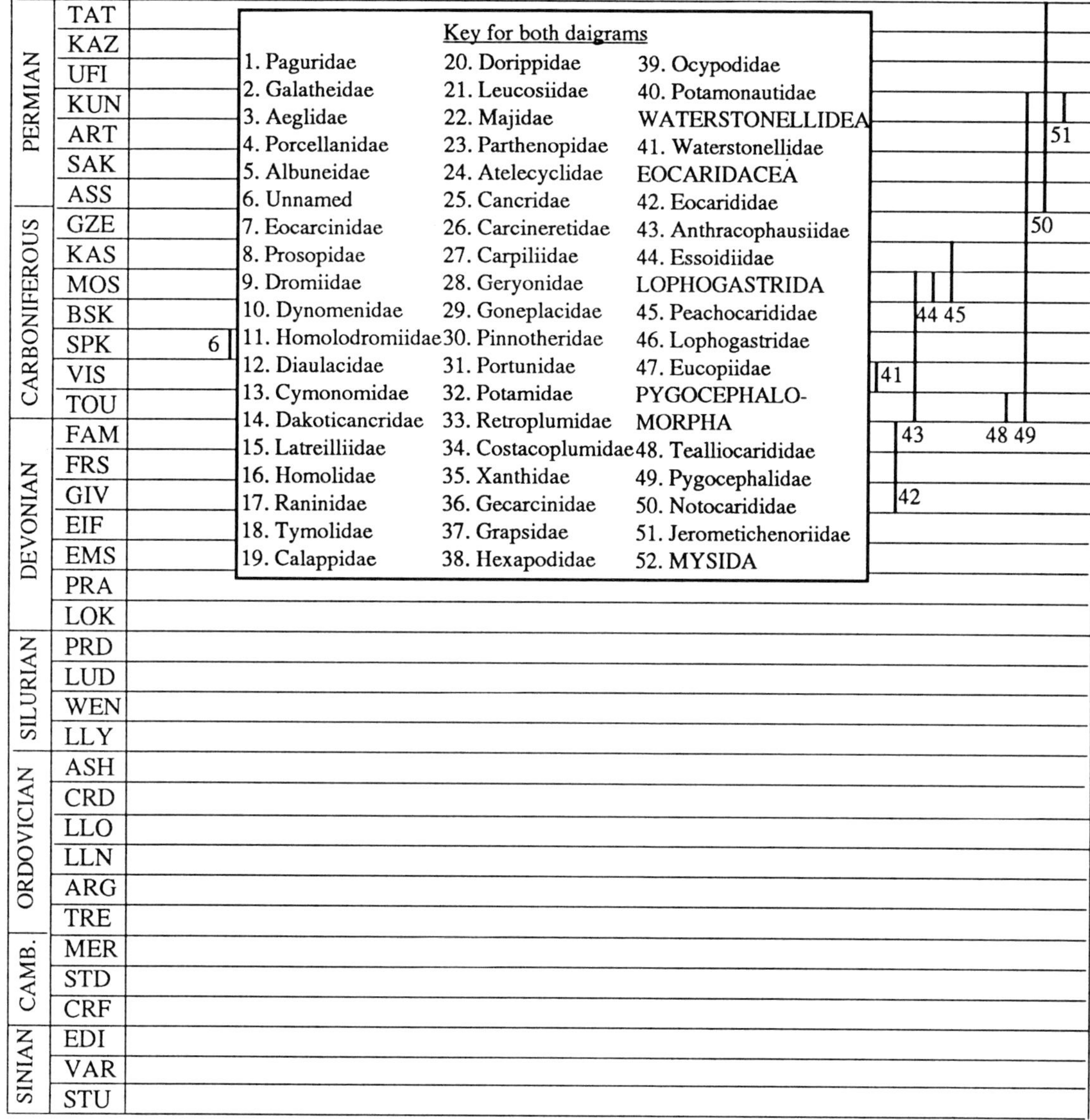

Fig. 18.2

First: *Scyllarides punctatus* Woods, 1925 and *Scyllarella gardneri* Woods, 1925, Gault, England, UK. **Extant**

F. TETRACHELIDAE Beurlen, 1930 Tr (u.) Mar.

First and Last: *Tetrachela raiblana* (Bronn, 1858), Upper Triassic, southern Europe (Glaessner, 1969).

Infra-order ANOMALA Boas, 1880

F. DIOGENIDAE Ortmann, 1892 K. (MAA)–Rec. Mar.

First: *Paguristes whitteni* Bishop, 1983, Coon Creek Formation, Mississippi, USA. **Extant**

F. COENOBITIDAE Dana, 1851 ?T. (Mio.)–Rec. Mar./Terr.

First: *Coenobita*? sp. Martin, 1883, ?Lower Miocene, Java. **Extant**

F. PAGURIDAE Latreille, 1803 J. (PLB)–Rec. Mar. (see Fig. 18.2)

First: *Palaeopagurus deslongchampsi* Van Straelen, 1925, Lias, France (Glaessner, 1969). **Extant**

F. GALATHEIDAE Samouelle, 1819 J. (BAJ)–Rec. Mar.

First: *Palaeomundiopsis moutieri* Van Straelen, 1925 (Glaessner, 1969). **Extant**

F. AEGLIDAE Dana, 1852 K. (MAA)–Rec. Mar.

First: *Haumuriaegla glaessneri* Feldmann, 1984, North Canterbury, New Zealand. **Extant**

F. PORCELLANIDAE Haworth, 1825 K. (CEN)–Rec. Mar.

First: *Porcellana antiqua*, A. Milne Edwards, 1862, Cenomanian, France (Glaessner, 1969). **Extant**

F. ALBUNEIDAE Stimpson, 1858 T. (RUP/CHT)–Rec. Mar.

First: *Blepharipoda occidentalis* Randall, 1840, Oligocene, North America (Glaessner, 1969). **Extant**

Infra-order BRACHYURA Latreille, 1803

Section DROMIACEA De Haan, 1833

F. UNNAMED ?C. (SPK) Mar.

First and Last: The earliest fossil referred to the Brachyura, albeit tentatively, is *Imocaris tuberculata* Schram and Mapes, 1984, Chesterian, Imo Formation, Arkansas, USA.

F. EOCARCINIDAE Withers, 1932 J. (PLB) Mar.

First and Last: *Eocarcinus praecursor* Withers, 1932, lower Lias, England, UK (Glaessner, 1969).

F. PROSOPIDAE von Meyer, 1860 J. (BAJ)–Rec. Mar.

First: *Prosopon* von Meyer, 1835, *Pithonoton (Pithonoton)* von Meyer, 1842, or *Coelopus* Étallon, 1861, all mid-Jurassic. **Extant**

Comment: Stratigraphy from Glaessner (1969).

F. DROMIIDAE De Haan, 1833 K. (ALB)–Rec. Mar.

First: *Mesodromilites glaber* Woodward, 1898, Albian (Gault), Folkestone, Kent, England, UK (Wright and Collins, 1972). **Extant**

F. DYNOMENIDAE Ortmann, 1892 J. (TTH)–Rec. Mar.

First: *Cyclothyreus reussi* (Gemmellaro, 1870), Tithonian, Moravia, Czechoslovakia (Glaessner, 1969). **Extant**

F. HOMOLODROMIIDAE Alcock, 1899 T. (Eoc.)–Rec. Mar.

First: *Homolodromia chaneyi* Feldmann and Wilson, 1988, La Meseta Formation, Seymour Island, Antarctica. **Extant**

F. DIAULACIDAE Wright and Collins, 1972 J. (TTH)–T. (?Oli./?Mio.) Mar.

First: *Diaulax* sp. Glaessner, 1931, Portlandian.
Last: *Diaulax* Bell, 1863, Oligocene.
Comment: Wright and Collins (1972, p. 56) reported the possibility of Miocene specimens.

Section ARCHAEOBRACHYURA Guinot, 1977

F. CYMONOMIDAE Bouvier, 1897 K. (APT)–Rec. Mar.

First: *Glaessneria latteri* Wright and Collins, 1972, lower Aptian, Binscombe, Surrey, England, UK. **Extant**

F. DAKOTICANCRIDAE Rathbun, 1917 K. (CMP–MAA) Mar.

First: *Dakoticancer* Rathbun, 1917, including *Dakoticancer overanus* Rathbun, 1917, lower CMP, Mississippi, USA; *Avitelmessus* Rathbun, 1923, including *Avitelmessus grapsoideus* Rathbun, 1923, upper Senonian, Tennessee, USA; *Tetracarcinus* Weller, 1905, including *Tetracarcinus subquadratus* Weller, 1905, lower CMP, New Jersey to Wyoming, USA.
Last: *Dakoticancer australis* Rathbun, 1935 and *Tetracarcinus subquadratus* Weller, 1905, Coon Creek Formation, Mississippi, USA (Bishop, 1983).

F. LATREILLIIDAE Alcock, 1899 K. (CEN)–Rec. Mar.

First: *Heeia villersensis* (Hée, 1924), Villers-sur-Mer, France, and Wilmington, Devon, England, UK (Wright and Collins, 1972). **Extant**

F. HOMOLIDAE De Haan, 1839 J. (TTH)–Rec. Mar.

First: *Gastrodorus neuhausensis* (von Meyer, 1864), Upper Jurassic, Germany (Glaessner, 1969). **Extant**
Comment: Some authors (notably Glaessner, 1969) regarded the Latreilliidae and the Homolidae as a single family.

F. RANINIDAE De Haan, 1839 K. (ALB)–Rec. Mar.

First: *Notopocorystes (N.) stokesi praecox* (Mantell, 1844), lower Albian (*D. mammilatum* Zone), Westerham, Kent and Oxted, Surrey, England, UK; *Notopocorystes (Eucorystes) oxtedensis* Wright and Collins, 1972, lower Albian (*D. mammilatum* Zone), Oxted, Surrey, England, UK (Wright and Collins, 1972). **Extant**

F. TYMOLIDAE Alcock, 1896 K. (ALB)–Rec. Mar.

First: *Torynomma* Woods, 1953, Albian including *Torynomma quadrata* Woods, 1953, upper Albian, Australia (Glaessner, 1969). **Extant**

Section EUBRACHYURA de St Laurent, 1980

Subsection HETEROTREMATA Guinot, 1977

F. CALAPPIDAE De Haan, 1833 K. (VLG)–Rec. Mar.

First: *'Gebia' controversa* Tribolet, 1874, Switzerland (Förster, 1968).
Comment: This record is based on claws. The earliest completely known species is *Paranecrocarcinus hexagonalis* Van Straelen, 1936, Hauterivian, Migraine (near Auxerre), Yonne, France (Wright and Collins, 1972). **Extant**

F. DORIPPIDAE MacLeay, 1838 K. (MAA)–Rec. Mar.

First: *Sodakus tatankayotankaensis* Bishop, 1978, Mobridge Member, Pierre Shale, South Dakota, USA. **Extant**
Comment: *Orithopsis bonneyi* Carter, 1872, Albian, England, UK, was referred to this family (e.g. Glaessner, 1969), but Wright and Collins (1972, p. 66–8) questioned the validity of this species, and synonymized it with *Necrocarcinus tricarinatus* Bell, 1863 (Calappidae).

F. LEUCOSIIDAE Samouelle, 1819 T. (LUT/BRT)–Rec. Mar.

First: *Typilobus trispinosus* Lörenthey, 1909, Egypt (Glaessner, 1969). **Extant**

F. MAJIDAE Samouelle, 1819 T. (YPR)–Rec. Mar.

First: *Mithracia libinoides*, Bell, 1858, Lower Eocene, England, UK (Glaessner, 1969). **Extant**
Comment: Glaessner (1969, p. 502) reported fragmentary claws of *Stenocionops* Desmarest, 1823, from the ?Upper Cretaceous, Arkansas, USA.

F. PARTHENOPIDAE MacLeay, 1838 T. (LUT/BRT)–Rec. Mar.

First: *Parthenope* Weber, 1795, Middle Eocene (Glaessner, 1969). **Extant**

F. ATELECYCLIDAE Ortmann, 1893 T. (LUT)–Rec. Mar.

First: *Montezumella fraasi* Lörenthey, 1907, Mokattam, Egypt. **Extant**

F. CANCRIDAE Latreille, 1819
T. (LUT/BRT)–Rec. Mar.

First: *Lobocarcinus* Reuss, 1857, e.g. *Lobocarcinus paulino-wurtembergensis* (von Meyer, 1847), Middle Eocene, Egypt. **Extant**

F. CARCINERETIDAE Beurlen, 1930
K. (APT–MAA) Mar.

First: *Withersella crepitans* (Wright and Collins, 1972), Crackers Bed, lower Aptian (*callidiscus* Subzone), Isle of Wight, England, UK.
Last: *Ophthalmoplax* Rathbun, 1935, including *Ophthalmoplax stephensoni* Rathbun, 1935, Texas, USA.

F. CARPILIIDAE Ortmann, 1894
K. (ALB)–Rec. Mar.

First: *Caloxanthus americanus* Rathbun, 1935, uppermost Albian, Texas, USA (Wright and Collins, 1972). **Extant**

F. GERYONIDAE Colosi, 1923
T. (YPR)–Rec. Mar.

First: *Coeloma (Litoricola) dentata* Woodward, 1873, *C. (L.) glabra* Woodward, 1873, and *C. bicarinatum* Rann, 1903, Lower Eocene, southern England, UK. **Extant**

F. GONEPLACIDAE MacLeay, 1838
T. (DAN)–Rec. Mar.

First: The following taxa occur in the Lower Palaeocene: *Branchioplax* Rathbun, 1916, *Martinezicancer* Van Straelen, 1939, *Tehuacana* Stenzel, 1949 (Glaessner, 1969). **Extant**
Comment: A Maastrichtian record of *Goniocypoda* H. Woodward, 1867, is doubtful (Glaessner, 1969). This genus may belong to the Hexapodidae.

F. PINNOTHERIDAE De Haan, 1833
T. (YPR)–Rec. Mar.

First: *Pinnixa* White, 1846, including *Pinnixa eocenica* Rathbun, 1926, Lower Eocene, Washington, USA (Glaessner, 1969). Also *Pinnixa (Palaeopinnixa) pornata* Collins and Morris, 1976, Lower or Middle Eocene, Scotland Beds, Spa, Barbados. **Extant**

F. PORTUNIDAE Rafinesque, 1815
T. (YPR)–Rec. Mar.

First: *Rhachiosoma* Woodward, 1871, Lower Eocene, England, UK, including *Rhachiosoma bispinosum* Woodward, 1871, Lower Eocene, southern England, UK (Glaessner, 1969). **Extant**
Comment: Glaessner (1969) records other Eocene portunids, but none of these is specifically from the Lower Eocene.

F. POTAMIDAE Ortmann, 1896
T. (UMI)–Rec. Mar.

First: *Archithelphusa punctulatus* (Heer, 1865), Oeningen, Switzerland and *Pseudopotamon speciosum* (Capellini, 1874), Italy. **Extant**

F. RETROPLUMIDAE Gill, 1894 K.–Rec. Mar.

First: *Archaeopus antennatus* Rathbun, 1908, ?Lower–Upper Cretaceous, California, USA. **Extant**

F. COSTACOPLUMIDAE St Laurent, 1989
K. (CON)–T. (LUT) Mar.

First: *Costacopluma concava* Collins and Morris, 1975, Arugu Limestone, Nigeria.
Last: *Retrocypoda almelai* Via, 1959, Spain (de Saint Laurent, 1989). **Extant**

F. XANTHIDAE MacLeay, 1838
K. (APT)–Rec. Mar.

First: *Xanthosia jacksoni* Wright and Collins, 1972, and *Actaeopsis wiltshirei* Carter, 1898, Crackers Bed, lower Aptian (*Deshayesites callidiscus* Subzone), Isle of Wight, England, UK. **Extant**

Subsection THORACOTREMATA Guinot, 1977
T. Mar.

F. GECARCINIDAE MacLeay, 1838
T. (PLI)–Rec. Mar.

First: *Cardisoma* Latreille, 1825, Pliocene, Fiji (Glaessner, 1969). **Extant**

F. GRAPSIDAE MacLeay, 1838
T. (LUT/BRT)–Rec. Mar.

First: *Varuna* H. Milne Edwards, 1830, ?Middle Eocene, Central America, and *Palaeograpsus* Bittner, 1875, Middle Eocene, Egypt, Europe (Glaessner, 1969). **Extant**

F. HEXAPODIDAE Miers, 1886
T. (Pal.)–Rec. Mar.

First: *Goniocypoda tessieri* Remy, 1954, West Africa. **Extant**

F. OCYPODIDAE Rafinesque, 1815
T. (LUT/BRT)–Rec. Mar.

First: *Loerentheya* Beurlen, 1929, including *Loerentheya carinata* Beurlen, 1929, Middle Eocene, Hungary. **Extant**

F. POTAMONAUTIDAE Bott, 1970
T. (MIO. (m./u.))–Rec. Mar.

First: *Potamonautes (Lirrangopotamonautes) tugenensis* Morris, 1976, Ngorora Formation, Kenya. **Extant**

Order WATERSTONELLIDEA Schram, 1981

F. WATERSTONELLIDAE Schram, 1979
C. (VIS) Mar.

First and Last: *Waterstonella grantonensis* Schram, 1979a, Lower Oil Shale Group, Calciferous Sandstone Series, Granton, Scotland, UK (Briggs *et al.*, 1991).

Order EOCARIDACEA Brooks, 1962

Schram (1986, p. 104) described this as 'a sort of catchall order, whose membership has slowly shrunk through the years as forms become better understood and assigned elsewhere'.

F. EOCARIDIDAE Brooks, 1962
D. (GIV–FAM) Mar.

First: *Eocaris oervigi* (Brooks, 1962a), Middle Devonian, Rhenish Massif, Germany.
Last: *Devonocaris destinezi* (Van Straelen, 1943), Belgium.

F. ANTHRACOPHAUSIIDAE Brooks, 1962
C. (TOU–MOS) Mar.

First: *Anthracophausia dunsiana* Peach, 1908, Cementstone

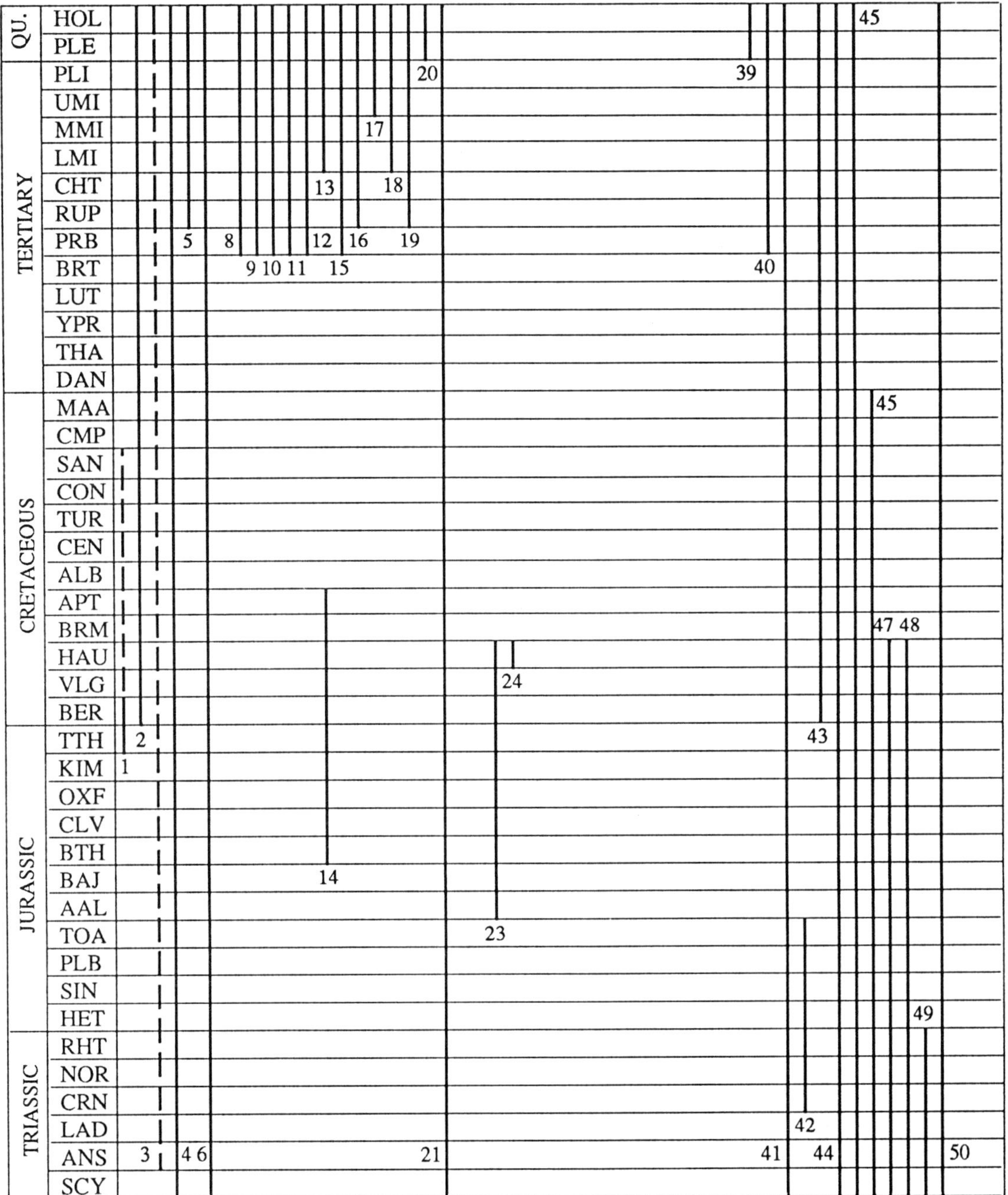

Fig. 18.3

Group, Chattlehope Burn, Redesdale, Northumberland, England, UK (Schram, 1979a).
Last: *Anthracophausia ingelsorum* Brooks, 1962, Francis Creek Shale, Mazon Creek, NE Illinois, USA.
Comment: Schram (1986, p. 105) suggested that *Anthracophausia* might be a fossorial euphausiacean.

F. ESSOIDIIDAE Schram, 1974 C. (MOS) Mar.

First and Last: *Essoidia epiceron* Schram, 1974, Francis Creek Shale, Mazon Creek, NE Illinois, USA.

Order LOPHOGASTRIDA Boas, 1883

F. PEACHOCARIDIDAE Schram, 1986
C. (MOS–KAS) Mar.

First: *Peachocaris strongi* (Brooks, 1962), Francis Creek Shale, Mazon Creek, NE Illinois, USA.
Last: *Peachocaris acanthouraea* Schram, 1984b, Missourian, Hushpuckney Shale, Crescent, Iowa, USA.

F. LOPHOGASTRIDAE Sars, 1857
J. (CLV)–Rec. Mar.

First: *Lophogaster voultensis*, Secretan and Riou, 1986, La Voulte-sur-Rhône, Ardeche, France.
Comments: Schram (1986, p. 124) discussed *Dollocaris ingens* Van Straelen, 1923 and *Kilianicaris lerichei* Van Straelen, 1923 from the same locality under this taxon, but remarks that they 'are nothing but poorly preserved carapaces'. They are now identified as belonging in the Thylacocephala (see Arduini and Pinna, 1989).

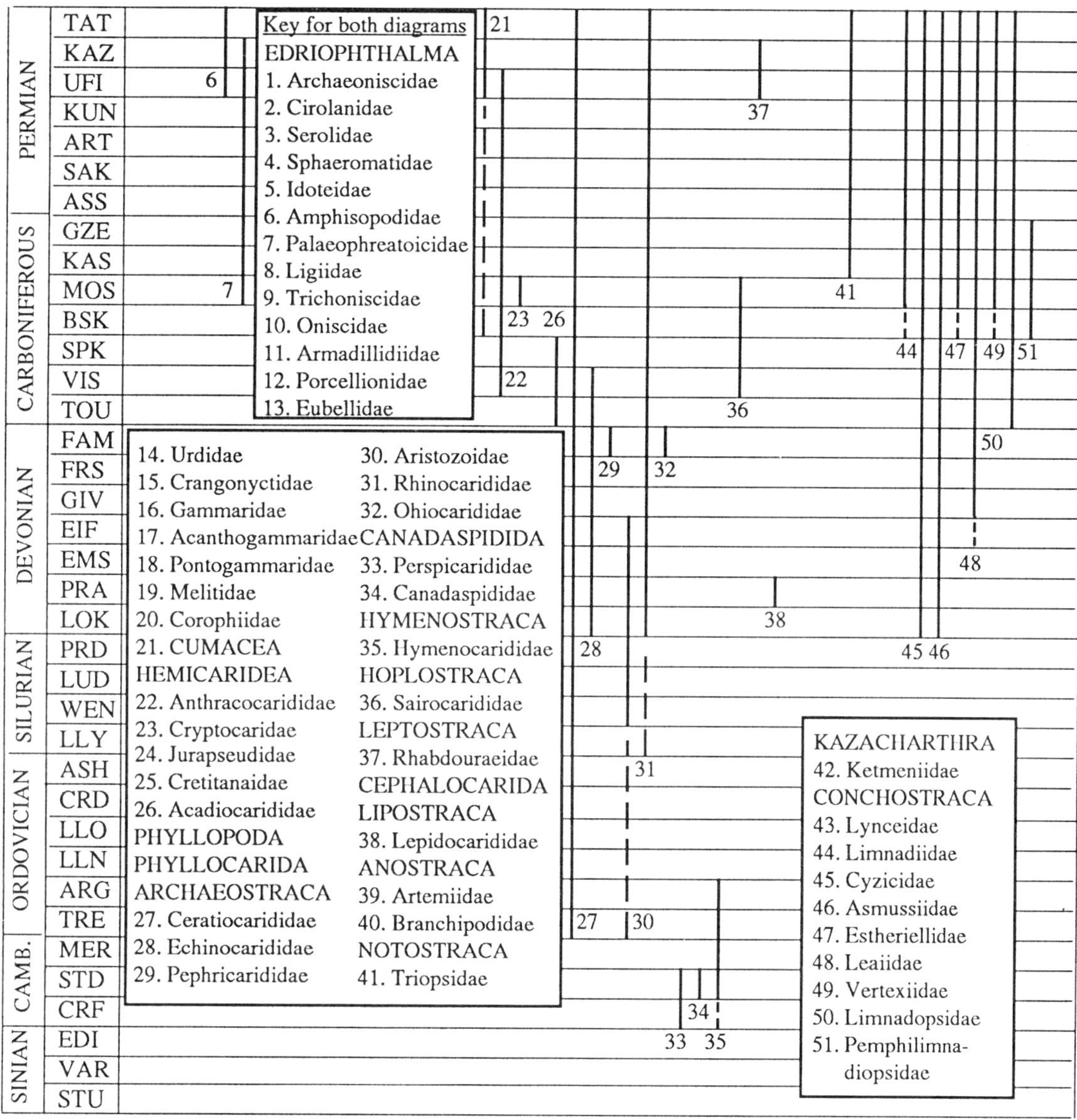

Fig. 18.3

F. EUCOPIIDAE Dana, 1852 Tr. (ANS)–Rec. Mar/FW

First: *Schimperella beneckei* and *Schimperella kessleri* Bill, 1914, Triassic, Grès à Voltzia, Alsace, France. **Extant**
Comment: Schram (1986, p. 124) discussed alternative assignments for this genus. The two species may be synonymous. Otherwise the first occurrence is *Eucopia praecursor* Secretan and Riou, 1986, Callovian, La Voulte-sur-Rhône, France.

Order PYGOCEPHALOMORPHA Beurlen, 1930

F. TEALLIOCARIDIDAE Brooks, 1962 C. (TOU) Mar./FW

First: *Pseudogalathea macconochiei* (Etheridge, 1879), Cementstone Group, East Lothian, Scotland, UK (Schram, 1979a).
Last: *Tealliocaris*, Namurian, Bearsden, Glasgow, Scotland, UK (N. D. L. Clark, in prep.).
Comment: Schram (1980, p. 580) recorded a specimen 'reminiscent of *Tealliocaris* and the pygocephalomorph mysidaceans' from the Upper Devonian near Moscow.

F. PYGOCEPHALIDAE Brooks, 1962 C. (TOU)–P. (ROT) Mar./FW

First: *Pseudotealliocaris palincsari* Schram, 1988, Kinderhookian, Lower Pocono Formation.
Last: *Mamayocaris jepseni* Brooks, 1962, from the Lower Permian of Oklahoma, USA.

F. NOTOCARIDIDAE Brooks, 1962 P. (ROT–ZEC) FW

First: *Notocaris tapscotti* (Woods, 1923), Lower Permian, South Africa.
Last: *Paulocaris pachoecoi* Clarke, 1920, Upper Permian, Brazil.
Comment: *Paulocaris* may be a pygocephalid (Briggs and Clarkson, 1990). These Permian taxa are in need of restudy.

F. JEROMETICHENORIIDAE Schram, 1978 P. (KUN) FW

First and Last: *Jerometchenoria grandis* Schram, 1978, Kungur Stage, Irensky Horizon, Komai Republic, former USSR.

Order MYSIDA Boas, 1883 J. (CLV)–Rec. Mar./FW

First: *Siriella antiqua* Secretan and Riou, 1986, La Voulte-sur-Rhône, Ardeche, France.
Comment: Possible mysidan fossils are rare (Schram, 1986). *Elder ungulatus* Münster, 1839 and *Francocaris grimmi* Broili, 1917, Solnhofen Limestone, Bavaria, Germany, may belong here, but Schram (1986, p. 124) regarded them as 'too poorly preserved to permit an unqualified assignment'. **Extant**

Order EDRIOPHTHALMA Leach, 1815 (see Fig. 18.3)

Suborder ISOPODA Latreille, 1817

Infra-order FLABELLIFERA Sars, 1882

A number of flabelliferan isopods have been described from the fossil record and remain unassigned to families (e.g. Hessler, 1969).

F. ARCHAEONISCIDAE Haack, 1918 J. (TTH)–K. (?SEN) Mar./FW

First: *Archaeoniscus brodiei* Milne Edwards, 1843, occurs in the Purbeck Beds of England, UK, the Serpulit of northwestern Germany and in the Abu Ballas Formation of Egypt (Barthel and Boettcher, 1978).
Last: *Archaeoniscus texanus* Wieder and Feldmann, 1992, ?Austin Chalk, Texas.

F. CIROLANIDAE Dana, 1853 K. (NEO)–Rec. Mar.

First: *Cirolana enigma* Wieder and Feldmann, 1992, Lakota Formation, South Dakota, USA.
Comment: A revision of species referred to *Palaega* might confirm older occurrences (Wieder and Feldmann, 1992).

F. SEROLIDAE Dana, 1853 ?Tr. (ANS)–Rec. Mar.

First: *Anhelkocephalon handlirschi* Bill, 1914, Triassic, Grès à Voltzia, Alsace, France, may belong here (Schwebel *et al.*, 1983). **Extant**
Comment: Hessler (1969) recognized no fossil serolids. Schram (1986, p. 151) noted the possibility of some Mesozoic records such as *Anhelkocephalon handlirschi.*

F. SPHAEROMATIDAE Milne Edwards, 1840 Tr.–Rec. Mar.

First: *Isopodites triasinus* (Picard, 1858), Triassic, Germany. **Extant**

Infra-order VALVIFERA Sars, 1882

F. IDOTEIDAE Milne Edwards, 1840 T. (RUP)–Rec. Mar.

First: *Proidotea haugi* Racovitza and Sevastos, 1910, Lower Oligocene–Middle Oligocene, Romania. **Extant**
Comment: *Anhelkocephalon handlirschi* Bill, 1914, Triassic, Grès à Voltzia, Alsace, France shows some similarity to the Idoteidae as well as to the Serolidae (Schwebel *et al.*, 1983).

Infra-order PHREATOICIDEA Stebbing, 1893

F. AMPHISOPODIDAE Nicholls, 1943 P. (GUA)–Rec. Mar.

First: *Protamphisopus reichelti* Glaessner and Malzahn, 1962, Upper Permian of Germany (Schram, 1986, p. 151). **Extant**

F. PALAEOPHREATOICIDAE Birshtein, 1962 C. (MOS)–P. (GUA) Mar.

First: *Hesslerella shermani* Schram, 1970, Francis Creek Shale, Mazon Creek, NE Illinois, USA.
Last: *Palaeophreatoicus sojanensis* Birshtein, 1962, Kazanian, Iva Gora, Soyama River, former USSR, or *Palaeocrangon problematicus* (von Schlotheim, 1820), Permian, Germany, and ?England, UK.
Comment: The relative ages of these Permian occurrences are not clear. The taxa were reviewed by Schram (1980).

Infra-order ONISCIDEA Latreille, 1803

Section DIPLOCHETA Vandel, 1957

F. LIGIIDAE Brandt, 1883 T. (PRB)–Rec. Terr.

First: *Ligidium splendidum* Strouhal, 1940, Baltic amber (Keilbach, 1982). **Extant**

Section SYNOCHETA Legrand, 1946

Superfamily TRICHONISCOIDEA Sars, 1899

F. TRICHONISCIDAE Sars, 1899 T. (PRB)–Rec. Terr.

First: *Trichoniscus* Brandt, 1833, Upper Eocene (Hessler, 1969). **Extant**

Section CRINOCHETA Legrand, 1946

Superfamily ONISCOIDEA Dana, 1852

F. ONISCIDAE Brandt, 1851 T. (PRB)–Rec. Terr.

First: *Oniscus* Linné, 1758, Upper Eocene, Europe (Hessler, 1969). *Oniscus convexus* Koch and Berendt, 1854 occurs in the Baltic amber (Keilbach, 1982). **Extant**

Superfamily ARMADILLOIDEA Verhoeff, 1917

F. ARMADILLIDIIDAE, Brandt, 1833 T. (PRB)–Rec. Terr.

First: *Protosphaeoniscus tertiarius* Schmalfuss, 1980, Baltic amber. **Extant**

F. PORCELLIONIDAE Verhoeff, 1918 T. (PRB)–Rec. Terr.

First: *Porcellio* Latreille, 1804, including three species from Baltic amber (Keilbach, 1982). **Extant**

F. EUBELIDAE Budde-Lund, 1899 T. (MIO. (l.))–Rec. Terr.

First: *Eubelum rusingaense* Morris, 1979, Lower Miocene, Rusinga Island, Lake Victoria, Kenya.

Infra-order UNCERTAIN

F. URDIDAE Kunth, 1870 J. (BTH)–K (APT) Mar.

First: *Urda* Münster, 1840 (Hessler, 1969; range from Sepkoski, 1982).

Suborder AMPHIPODA Latreille, 1816

The fossil record of the Amphipoda is reviewed in Bousfield (1982) and Karaman (1984).

Infra-order GAMMARIDEA Latreille, 1803

Superfamily CRANGONYCTOIDEA Bousfield, 1977

F. CRANGONYCTIDAE Bousfield, 1973 T. (PRB)–Rec. Terr./FW

First: *Palaeogammarus balticus* Lucks, 1928, *P. danicus* Just, 1974, and *P. sambiensis* Zaddach, 1864, all Upper Eocene, Bernstein Formation, Baltic amber. **Extant**

Superfamily GAMMAROIDEA Leach, 1814

F. GAMMARIDAE Leach, 1814
T. (RUP)–Rec. Mar./FW

First: *Jubeogammarus alsaticus* (Van Straelen, 1924), *Gammarus* sp., and *Condiciogammarus retzi* (Maikovsky, 1941), all Lower Oligocene, France (Karaman, 1984).
Comment: The familial placement of *Codiciogammarus* is uncertain (Karaman, 1984).

F. ACANTHOGAMMARIDAE Garjej, 1901
T. (MES)–Rec. FW

First: *Hellenis saltatorius* Petumnikov, 1914, Lower Miocene, Baku region, Caspian Basin or e.g. *Andrussovia bogacevi* Derzhavin, 1927, Middle Miocene, Caucasus (Karaman, 1984).
Comment: The familial placement of *Hellenis* is uncertain (Bousfield, 1982). **Extant**

F. PONTOGAMMARIDAE Bousfield, 1977
T. (Mio.)–Rec. Mar.

First: *Praegmelina andrussovi* Derzhavin, 1927, and *P. archangelskii* Derzhavin, 1927, Middle Miocene, Grozny, Caucasus (Karaman, 1984). **Extant**

Superfamily HADZIOIDEA Karaman, 1932

F. MELITIDAE Bousfield, 1973
T. (RUP)–Rec. Mar./FW

First: *Alsacomelita semipalmata* Karaman, 1984, Lower Oligocene, Alsace, France. **Extant**

Superfamily COROPHIOIDEA Dana, 1849

F. COROPHIIDAE Dana, 1849
Q. (PLE)–Rec. Mar.

First: *Corophium volutator*, Quaternary, England, UK (Bousfield, 1982).
Comment: This record is based on a trace fossil. **Extant**

Order HEMICARIDEA Schram, 1981

Suborder CUMACEA Kröyer, 1846
C. (u.)/P. (GUA)–Rec. Mar.

First: Schram (1986, p. 214) recorded undescribed primitive cumaceans from the Pennsylvanian of North America. Two species of *Ophthalmdiastylis* occur in the Zechstein of Germany (Malzahn, 1972). **Extant**

Suborder TANAIDACEA Dana, 1853

Infra-order ANTHRACOCARIDOMORPHA Sieg, 1980

F. ANTHRACOCARIDIDAE Schram, 1979
C. (VIS)–P. (UFI)? Tr. (RHT) Mar.

First: *Anthracocaris scotica* (Peach, 1882), Glencartholm, Scotland, UK, Calciferous Sandstone Series. (This taxon also occurs in the coeval Granton shrimp-bed, Edinburgh (Briggs *et al.*, 1991).)
Last: *Opthalmapseudes rhenanus* (Malzahn, 1957), Kamp-Lintfort, Germany, Zechstein 1, Lower Permian.
Comment: Schram *et al.* (1986) pointed out that *Opthalmapseudes* sp. (Végh and Bachmayer, 1965) from the RHT of Hungary, eventually may be determined to be more akin to *Jurapseudes*.

F. CRYPTOCARIDAE Sieg, 1980 C. (MOS) Mar.

First and Last: *Eucryptocaris asherorum* Schram, 1989, Francis Creek Shale, Mazon Creek, NE Illinois, USA.
Comment: This name was designated for some of the material formerly referred to *Cryptocaris hootchi* (see Schram, 1989).

Infra-order APSEUDOMORPHA Sieg, 1980

Superfamily JURAPSEUDOIDEA Schram *et al.*, 1986

F. JURAPSEUDIDAE Schram *et al.*, 1986
J. (DOG)–K. (HAU) Mar.

First: *Jurapseudes friedericianus* (Malzahn, 1965) Middle Jurassic (Dogger, *lineatum* Zone), Germany, and *J. acutirostris* (Sachariewa-Kowatschewa and Bachmayer, 1965), Middle Jurassic (Dogger), Germany.
Comments: The relative stratigraphical levels of these two species are uncertain (see Schram *et al.*, 1986).
Last: *Carlclausus emersoni* Schram *et al.*, 1986, lower HAU, Hannover, Germany.

Infra-order TANAIDOMORPHA Sieg, 1980

Superfamily CRETITANAOIDEA Schram *et al.*, 1986

F. CRETITANAIDAE Schram *et al.*, 1986
K. (HAU) Mar.

First and Last: *Cretitanais giganteus* (Malzahn, 1979), Lower HAU, Hannover, Germany.

Suborder SPELAEOGRIPHACEA Gordon, 1957

F. ACADIOCARIDIDAE Schram, 1974 C. (l.) Mar.

First and Last: *Acadiocaris novascotica* (Copeland, 1957), Mississippian, Canada.

Class PHYLLOPODA Latreille, 1825

Subclass PHYLLOCARIDA Packard, 1879

Order ARCHAEOSTRACA Claus, 1888

Suborder CERATIOCARINA Clarke, 1900

F. CERATIOCARIDIDAE Salter, 1860
O. (TRE)–P. (ZEC) Mar.

First: *C.? scanica* (Moberg and Segerberg, 1906), Sweden, and *C.* spp., Yukon (additional less well-constrained records (Rolfe, pers. comm.) are listed in Bate *et al.*, 1967).
Last: *Ceratiocaris* M'Coy, 1849, ?Upper Permian, cosmopolitan (Rolfe, 1969).

F. ECHINOCARIDIDAE Clarke, 1900
D. (LOK)–C. (VIS) Mar.

First: *Montecaris antecedens* Chlupáč, 1960 (Rolfe and Edwards, 1979).
Last: *Echinocaris* Whitfield, 1880, lower Mississippian (Rolfe, 1969).

F. PEPHRICARIDIDAE van Straelen, 1933
D. (FAM) Mar.

First and Last: *Pephricaris horripilata* Clarke, 1898, Upper

Devonian, Chemung, Alfred, Allegany County, New York, USA (Rolfe, 1969).

F. ARISTOZOIDAE Gürich, 1929
O. (?l)/S. (WEN)–D. (EIF) Mar.

First: *Aristozoe* Barrande, 1872, ?Lower Ordovician, Upper Silurian, Europe, Asia, Canada.
Last: *Aristozoe porcula*, Richter, upper Eifelian (Rolfe and Edwards, 1979).

Suborder RHINOCARINA Clarke, 1900

F. RHINOCARIDIDAE Hall and Clarke, 1888
?S./D. (l.)–P. (ZEC) Mar.

First: *Dithyrocaris* Scouler, *in* Portlock, 1843, ?Silurian, ?Lower Devonian (Rolfe, 1969), or *Nahecaris stuertzi* Jaekel, 1921, Lower Devonian, Germany (Rolfe, 1969).
Last: *Dithyrocaris* Scouler, *in* Portlock, 1843, ?Upper Permian (Rolfe, 1969).

F. OHIOCARIDIDAE Rolfe, 1962 D. (FAM) Mar.

First and Last: *Ohiocaris wycoffi* Rolfe, 1962, Chagrin Shale, Porter Creek, Cuyahoga County, Ohio, USA.

Order CANADASPIDIDA Novzhilov, 1960

F. PERSPICARIDIDAE Briggs, 1978
Є. (CRF–STD) Mar.

First: *Perspicaris*? sp., Hou, 1987, Lower Cambrian, *Nevadella* Zone, Chiungchussu Formation, Yunnan Province, China.
Last: *Perspicaris? ellipsopelta* Robison and Richards, 1981, Middle Cambrian, *Peronopsis punctuosus* Zone Robison and Richards, 1981, Marjum Formation, Red Wash, House Range, Utah, USA.

F. CANADASPIDIDAE Novozhilov (*in* Orlov), 1960
Є. (STD) Mar.

First: *Canadaspis* cf. *C. perfecta*, *Peronopsis bonnerensis* Zone, Spence Tongue of the Lead Bell Shale, Antimony Canyon, Utah, USA (Robison and Richards, 1981).
Last: *Canadaspis perfecta* Walcott, 1912, *Ptychagnostus praecurrens* Zone, Stephen Formation, Burgess Shale, Field, British Columbia, Canada.

Order HYMENOSTRACA Rolfe, 1969

F. HYMENOCARIDIDAE Haeckel, 1896
Є. (CRF/STD)–O. (ARG) Mar.

First: *Hymenocaris* Salter, 1853, Middle (?Lower) Cambrian, Europe, North America, Australia, New Zealand.
Last: *Hymenocaris vermicauda* Salter, 1853, Lower Ordovician, Wales, UK.
Comment: This genus has been used as a catch-all and is in need of revision. Most of the occurrences other than *H. vermicauda* from the UK are based on carapaces alone and their assignment must be regarded as uncertain.

Order HOPLOSTRACA Schram, 1973

F. SAIROCARIDIDAE C. (VIS–MOS) Mar.

First: *Sairocaris elongata* (Peach, 1882), Calciferous Sandstone Series, Glencartholm, Scotland, UK.
Last: *Kellibrooksia macrogaster* Schram, 1973, Moscovian, Francis Creek Shale, Mazon Creek, NE Illinois, USA.

Order LEPTOSTRACA Claus, 1880

F. RHABDOURAEIDAE Schram and Malzahn, 1984
P. (GUA) Mar.

First and Last: *Rhabdouraea bentzi* (Malzahn, 1958), Zechstein 1, Niederrhein, Germany.

Subclass CEPHALOCARIDA Sanders, 1955

Order BRACHYPODA Birshtein, 1960

Comment: *Dala*, from the Upper Cambrian of Sweden, was once thought to be related to the Cephalocarida, but this view is no longer held (Müller, 1983). **Extant**

Order LIPOSTRACA Scourfield, 1926

F. LEPIDOCARIDIDAE Scourfield, 1926
D. (PRA) FW

First and Last: *Lepidocaris rhyniensis* Scourfield, 1926, Rhynie Chert, Aberdeenshire, Scotland, UK.

Subclass SARSOSTRACA Tasch, 1969

Order ANOSTRACA Sars, 1867

F. ARTEMIIDAE Grochowski, 1896
Q. (PLE)–Rec. FW

? First: *Artemia* Leach, 1819, Pleistocene (Tasch, 1969). **Extant**

F. BRANCHIPODIDAE Simon, 1886
?T. (PRB)–Rec. FW

?First: *Branchipodites vectensis* Woodward, 1877, Eocene, Bembridge Limestone (latest PRB), Isle of Wight, England, UK. **Extant**
Comment: Schram (1986, p. 375) argued that the Palaeozoic records of Anostraca (Tasch, 1969) are unconvicing, although he illustrated (p. 376) a possible example from the Upper Silurian Kokomo Limestone (LUD/PRD) of Indiana. An undetermined anostracan has been described from the ?BRM/APT Koonwarra Fossil Bed, Victoria, Australia (Jell and Duncan, 1986). Schram (1986) regarded the undescribed anostracan from the Miocene Barstow Formation of California (Palmer, 1957) as representing a new family.

Subclass CALMANOSTRACA Tasch, 1969

Order NOTOSTRACA Sars, 1867

F. TRIOPSIDAE Keilhack, 1910 C. (STE)–Rec. FW

First: *Triops ornatus* (Goldenberg, 1873), Stephanian, Saarbrücken, Germany (Tasch, 1969). **Extant**
Comment: Schram (1986, p. 360) remarked that fossil notostracans 'are few, incompletely known, and essentially indistinguishable from the modern types'.

Order KAZACHARTHRA Nozohilov, 1957

F. KETMENIIDAE Novozhilov, 1957 Tr. (u.)–J. (l.)

First: *Almatium gusevi* Chernyshev, 1940, *Jeanrogerium sornayi* Novozhilov, 1959, and *Panacanthocaris ketmenia* Novozhilov, 1957, Upper Triassic Huangshanjie Formation, Turpan Basin, China (McKenzie *et al.*, 1991).
Last: *Ketmenia schultzi* Chernyshev, 1940, *Almatium gusevi* Chernyshev, 1940, *Iliella spinosa* Chernyshev, 1940, *Jeanrogerium sornayi* Novozhilov, 1959, *Kungeja tchakabaevi* Novozhilov, 1957, *Kysyltamia tchiiliensis* Novozhilov, 1957 and *Panacanthocaris ketmenia* Novozhilov, 1957, all Lias,

Ketmen Mountains, Kazakhstan, former USSR (Tasch, 1969).

Order CONCHOSTRACA Sars, 1867

Suborder LAEVICAUDATA Linder, 1945 K. (l.)–Rec.

This suborder is incorrectly called 'Laeviscaudata' in Tasch (1969), and 'Laevicauda' in Schram (1986) (see Fryer, 1987).

F. LYNCEIDAE Stebbing, 1902 K.(l.)–Rec. FW

First: *Lynceus stschukini* Chernyshev, 1940, Lower Cretaceous, Transbaikal, former USSR (Tasch, 1969). **Extant**

Suborder SPINICAUDATA Linder, 1945

This suborder is called Spinicauda in Schram (1986).

Superfamily LIMNADIOIDEA Baird, 1849

F. LIMNADIIDAE Baird, 1849 C. (BSK/MOS)–Rec. FW

First: *?Limnestheria ardra*, Wright, 1920, Westphalian A, Kilkenny Coal Measures, Republic of Ireland (Tasch, 1969).

Superfamily CYZICOIDEA Baird, 1849

F. CYZICIDAE Stebbing, 1910 D. (l.)–Rec. FW

First: *Cyzicus* Audouin, 1837, Lower Devonian, cosmopolitan (Tasch, 1969).

F. ASMUSSIIDAE Kobayashi, 1954 D. (l.)–K. (u.) FW

First: *Asmussia membranacea*, Pacht, 1849, Devonian, Livonia (Tasch, 1969).
Last: *Asmussia* Pacht, 1849, Upper Cretaceous (Tasch, 1969).

Superfamily ESTHERIELLOIDEA Kobayashi, 1954

F. ESTHERIELLIDAE Kobayashi, 1954 C. (BSK/MOS)–K. (NEO) FW

First: *Anomalonema* Raymond, 1946, Pennsylvanian, Europe and North America (Tasch, 1969).
Comment: Tasch (1969, p. 157) listed two subgenera from the Upper Carboniferous, but their relative ages are not clear.
Last: *Estheriella camerouni* Defretin, 1953 and *Afrograpta tricostata* (Defretin, 1953), Lower Cretaceous (Wealden), north Cameroon (Tasch, 1969).

Superfamily LEAIOIDEA Raymond, 1946

F. LEAIIDAE Raymond, 1946 D. (EIF/GIV)–K (NEO) FW

First: *Praeleaia quadricarinata* Lyutkevich, 1929, Middle Devonian, confluence of the Ruia and Pliusa Rivers, Estonia, former USSR, and *Pteroleaia canadensis*, Copeland, 1962, Middle Devonian, Melville Island, Arctic Canada (Tasch, 1969).
Last: *Japanoleaia rectangula* (Yokahama, 1894), Neocomian, Yuasa, Japan (Tasch, 1969).

Superfamily VERTEXIOIDEA Kobayashi, 1954

F. VERTEXIIDAE Kobayashi, 1954 C. (BSK/MOS)–Tr (u.) FW

First: *Cornia*, Lyutkevich, 1937, Westphalian, former USSR (Tasch, 1969).
Last: *Echinestheria marimbensis* Marlière, 1950, Upper Triassic, West Africa (Tasch, 1969).
Comment: Tasch (1969) recorded the range of this family as Lower Carboniferous–Upper Triassic, but listed no genus with a range older than Upper Carboniferous (Westphalian); Schram (1986) followed Tasch (1969).

F. LIMNADOPSIDAE Tasch, 1961 C (l.)–Rec. FW

First: *Palaeolimnadiopsis* Raymond, 1946, Lower Carboniferous (Tasch 1969). **Extant**
Comments: Tasch (1969, p. 162) regarded *Belgolimnadiopsis* Novozhilov, 1958 from the Lower Devonian of Belgium, as 'inadequately documented; doubtful'.

F. PEMPHILIMNADIOPSIDAE Tasch, 1961 C. (u.) FW

First and Last: *Pemphilimnadiopsis ortoni* (Clarke, 1900), Pennsylvanian, Carollton, Ohio, USA (Tasch, 1969).

F. IPSILONIIDAE Novozhilov, 1958 D.–K. (ALB/APT) FW (see Fig. 18.4)

First: *Ipsilonia auriculata* Novozhilov, 1953, Devonian, Koura Region, north of Caucasus mountains (Tasch, 1969).
Last: *Aculestheria novojilovi* Cardosa, 1963, Lower Cretaceous, Bahia Series, Brazil (Tasch, 1969).

Order CLADOCERA Latreille, 1829

Suborder EUCLADOCERA Erikson, 1932

Superfamily DAPHNOIDEA Straus, 1820 (=Anomopoda)

F. DAPHNIIDAE Straus, 1820 K. (NEO)–Rec. FW

First: Ephippia from the early Cretaceous of Mount Ukha, Mongolia, are attributed to this family (Smirnov, 1992).
Comment: Daphniid carapaces occur in the ?BRM/APT Koonwarra Fossil Bed, Victoria, Australia (Jell and Duncan, 1986). Other occurrences of daphniid ephippia are reviewed by Fryer (1991). **Extant**

F. PROCHYDORIDAE Smirnov, 1992 K. (NEO) FW

First and Last: *Prochydorus rotundus* Smirnov, 1992, *Archeoxus mirabilis* Smirnov, 1992, and *A. ventrosus* Smirnov, 1992, early Cretaceous of Mount Ukha, Mongolia.

F. MOINIDAE Goulden, 1968 ?K (NEO)–Rec. FW

?First: Ephippia from the early Cretaceous of Mount Ukha, Mongolia, are attributed to this family (Smirnov, 1992). **Extant**
Comment: Fryer (1991, p. 164) considered that the separation of this family from the Daphniidae 'seems scarcely justified'.

Class MAXILLOPODA Latreille, 1817

?Order PENTASTOMIDA Rudolphi, 1819 Terr.

The affinities and status of the parasitic pentastomids are uncertain. Some authorities assign them to a separate phylum, but there is evidence that they are highly modified crustaceans (Grygier, 1983; Abele *et al.*, 1989). An unnamed marine form resembling recent parasitic Pentastomida has

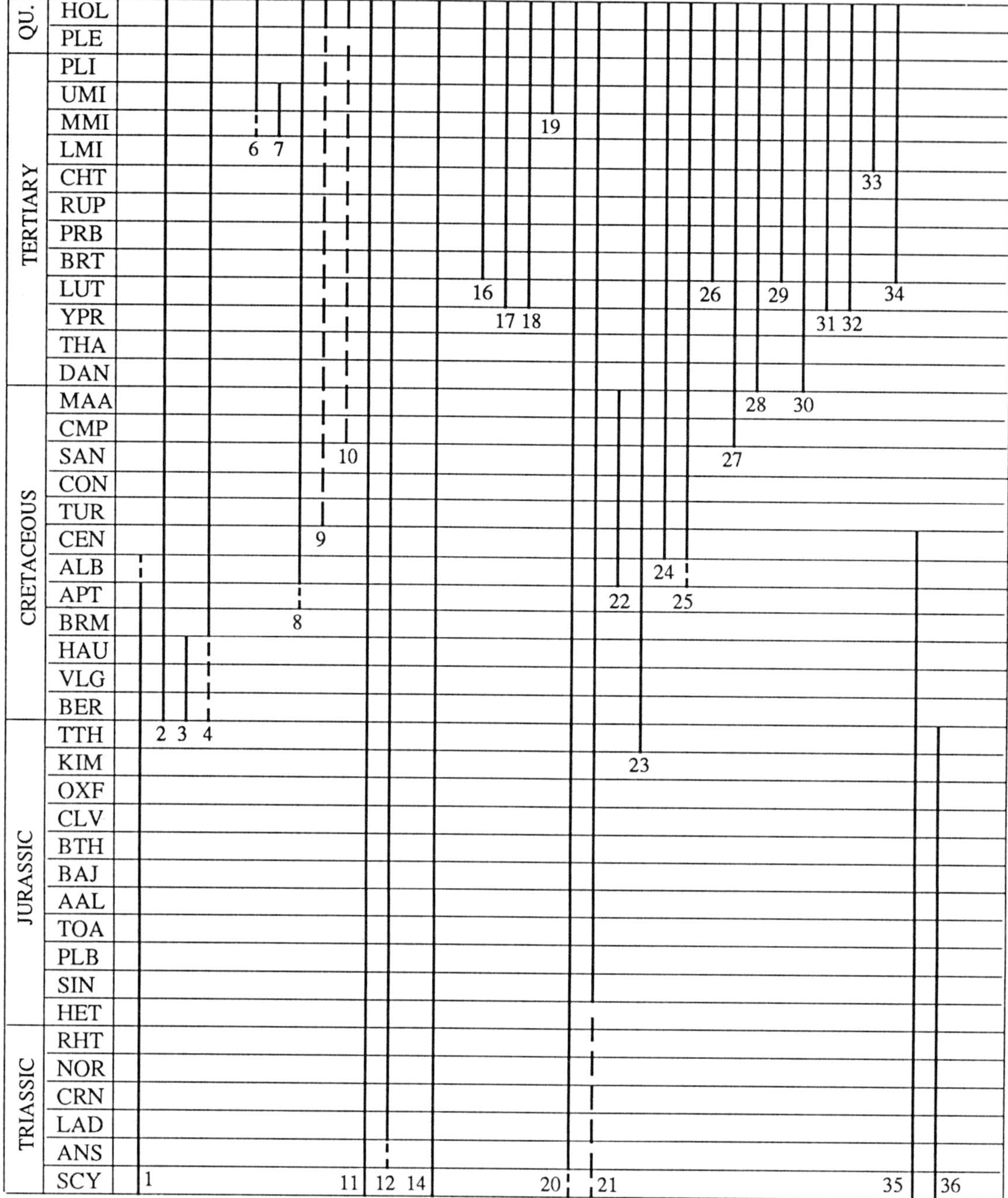

Fig. 18.4

been described from the Tremadoc of Öland, Sweden (Andres, 1989).

Order ORSTENOCARIDA, Müller and Walossek, 1988

F. BREDOCARIDIDAE, Müller and Walossek, 1988 Є. (DOL) Mar.

First and Last: *Bredocaris admirabilis* Müller, 1983, *Peltura* Zone, Alum Shale Formation, Västergötland, Sweden.

Subclass COPEPODA Milne Edwards, 1840

Order HARPACTICOIDA Sars, 1903

Suborder OLIGOARTHA Lang, 1948

Infra-order PODOGENNONTA Lang, 1948

Superfamily CLETODOIDEA Lang, 1948

F. CLETODIDAE T. Scott, 1904 T. (MMI–UMI)–Rec. FW

First: *Cletocamptus* Schmankevitsch, 1875, Middle–Upper Miocene, Mojave Desert, California, USA (Palmer, 1960).
Extant

Order MONSTRILLOIDA Sars, 1903

Order CYCLOPOIDA Burmeister, 1834

F. UNNAMED T. (MMI–UMI) FW

First: Genus undet., Middle–Upper Miocene, Mojave Desert, California, USA (Palmer, 1960).

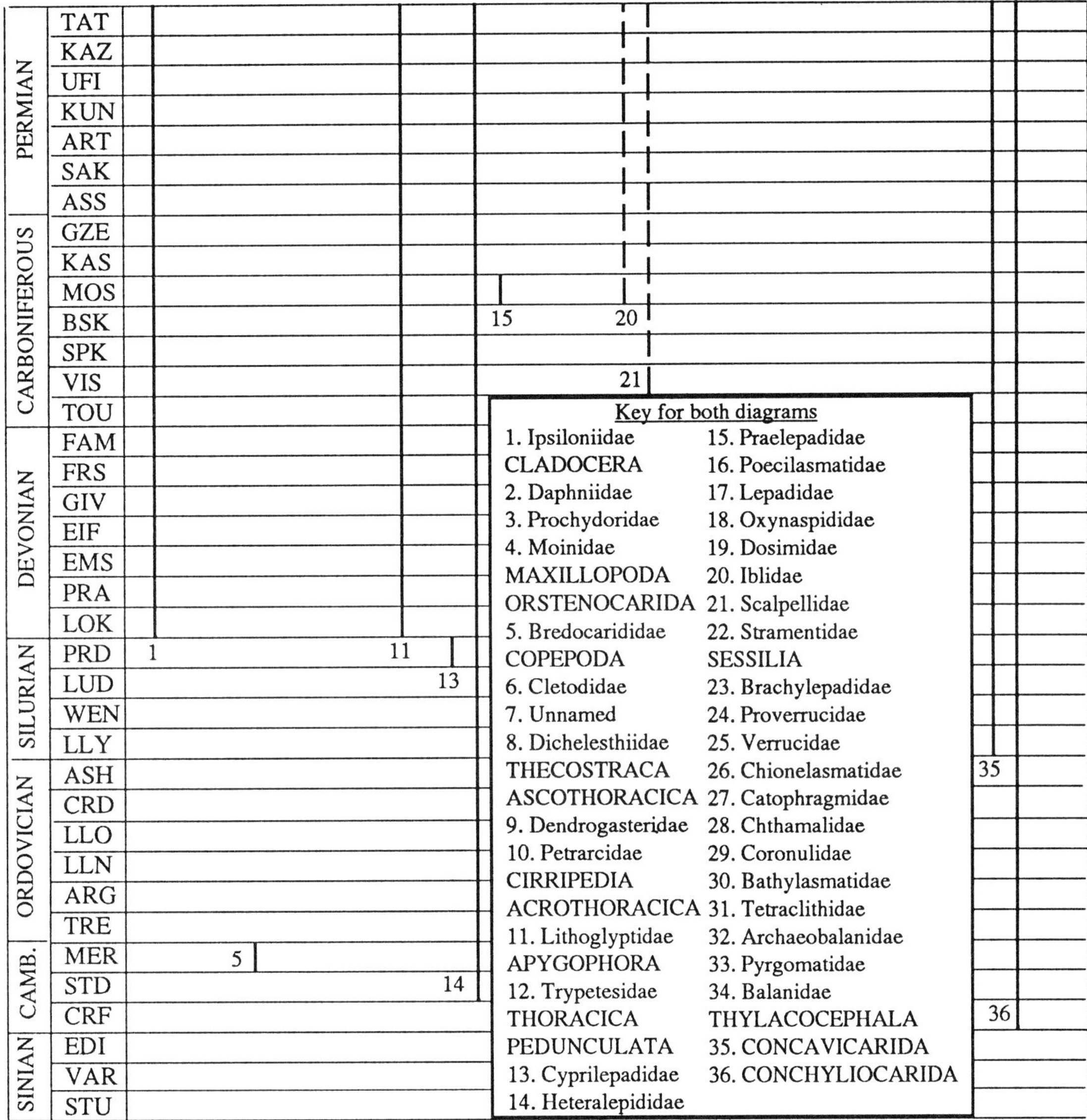

Fig. 18.4

Order SIPHONOSTOMATOIDA Thorell, 1859

F. DICHELESTHIIDAE Dana, 1853 K. (?APT)–Rec. Mar.

First: *Kabatarina pattersoni* Cressey and Boxshall, 1989, Santana Formation, Serro do Araripe, Ceara, Brazil.
Comment: These parasitic copepods were found in the gills of the teleost fish *Cladocyclus gardneri* Agassiz.

Class THECOSTRACA Gruvel, 1905

Since the publication of *The Fossil Record* (Harland *et al.*, 1967) and of the *Treatise* Vol. R (Newman *et al.*, 1969), there have been a number of reviews of the systematics of some, or all, of the groups which constitute the cirripedes in the widest sense of the term. These include the classifications of Bowman and Abele, 1982; Tomlinson, 1969; Newman and Ross, 1976; Zevina, 1980; Zullo, 1983; Buckeridge, 1983; Schram, 1982, 1986; and Newman, 1987. Where they overlap, the various classifications use broadly similar lists of families and differ principally in the range of forms included within the Cirripedia, and the taxonomic status accorded to this and to other groups within the higher-order classification. Unless otherwise stated the classification used here follows that of Schram (1986) with modifications based on Newman (1987). The Order Apoda Darwin, 1854, which had an uncertain status in earlier classifications, is no longer recognized (see Newman, 1982, p. 200).

Subclass FASCETOTECTA Grygier, 1984 Mar.

'The position of the enigmatic Y-larvae (Fascetotecta) is uncertain' (Schram, 1986, p. 539). They have no fossil record. **Extant**

Subclass ASCOTHORACICA Lacaze-Duthiers, 1880

No body fossils have been attributed to the Ascothoracica but some trace fossils attributed to the group have been recorded from the Upper Cretaceous. Many characters of the Order are interpreted as primitive and may indicate that the origins of the group lie far back in geological time (Schram, 1986; Newman, 1982, 1987; see Grygier, 1990 on similarity of nauplii to those of Cirripedia).

Order SYNAGOGOIDA Wagin, 1976

F. SYNAGOGIDAE Gruvel, 1905 Rec. Mar.

F. DENDROGASTERIDAE Gruvel, 1905 ?K. (TUR)–Rec. Mar.

Comments: Unnamed conical borings in the tests of the echinoid *Echinocorys* were considered by Madsen and Wolff (1965) to be identical to the borings of the extant genus *Ulophysema*. The material came from the Upper Cretaceous (TUR–MAA) of north Jutland. There is no other fossil record.

Order LAUROIDIDA Wagin, 1976

F. LAURIDAE Gruvel, 1905 Rec. Mar.

F. CTENOSCULIDAE Thiele, 1925 Rec. Mar.

F. PETRARCIDAE Gruvel, 1905 ?K. (CMP)–Rec. Mar.

Comments: Gall-like deformations in Cretaceous octocorals have been interpreted as the cysts of ascothoracic barnacles and could perhaps very tentatively be assigned to this family. The traces known as *Endosacculus* have been recorded from the Upper Cretaceous (CMP–MAA), (Voigt, 1959, 1967) but there is no other known fossil record.

Subclass RHIZOCEPHALA F. Müller, 1862

The wholly parasitic rhizocephalans have no known fossil record. The classification and evolutionary relationships within the group, which includes seven families in two orders, are not clear and there are a number of genera of uncertain status since their lack of a kentrogon stage has yet to be properly confirmed. Relationships with other cirripedes *sensu lato* were discussed by Newman (1982, 1987) and could imply that the group has had a long history but, as noted by Schram (1986), there is also the possibility that they are polyphyletic. Two genera are exclusively fresh water in their occurrence.

Subclass CIRRIPEDIA Burmeister, 1834

Superorder ACROTHORACICA Gruvel, 1905

The fossil record of the Acrothoracica is dominated by trace fossils which have been linked with the superorder, and in some cases with families on the basis of burrow morphology. Burrow characters, including the presence or absence of calcareous cement deposits, have also been used to recognize two ichnofamilies (Zapfellidae and Rogerellidae). More recently Grygier and Newman (1985) compared some described traces with extant forms and showed that in some cases a calcareous rostral plate, which is an apparently primitive feature of some species in this superorder, may be preserved; as they pointed out, this preservation of body parts has considerable nomenclatural significance. The palaeontology of the group undoubtedly needs extensive review in the light of their discoveries. A chitinous body fossil has also been recorded from the Upper Cretaceous (Turner, 1973).

Order PYGOPHORA Berndt, 1907

F. LITHOGLYPTIDAE Aurivillius, 1892 D. (LOK)–Rec. Mar.

First: *Zapfella* sp. in platyceratid gastropods, Lower Devonian, ?New Scotland Formation, Athens, New York (Baird *et al.*, 1990). **Extant**

Order APYGOPHORA Berndt, 1907

F. TRYPETESIDAE Stebbing, 1910 ?Tr. (ANS)–Rec. Mar.

Comments: Tomlinson's (1969) recognition of Carboniferous trypetesids has been shown to be erroneous by Seilacher (1969) who considered that trypetesid burrowing strategies evolved in the early Mesozoic and illustrated burrows, with lateral expansions, from the Trochitenkalk of Crailsheim in Germany.

Superorder THORACICA Darwin, 1854

Order PEDUNCULATA Lamarck, 1818

Suborder CYPRILEPADOMORPHA Newman *et al.*, 1969

F. CYPRILEPADIDAE Newman *et al.*, 1969 S. (PRD) Mar.

First and Last: *Cyprilepas holmi* Wills, 1962, Schicht K, Oesel, Estonia, former USSR.

Suborder HETERALEPADOMORPHA Newman, 1987

F. HETERALEPADIDAE Nilsson-Cantell, 1921 Є. (STD)–Rec. Mar.

Comments: *Priscansermarinus barnetti* Collins and Rudkin, 1981, from the Middle Cambrian Burgess Shale of British Columbia, has been compared with the heteralepadids. Briggs (1983) urged caution in the interpretation of this fossil. The family has no other fossil record. The suborder includes five other extant families.

Suborder PRAELEPADOMORPHA Tschernyshew, 1930

F. PRAELEPADIDAE Tschernyshew, 1930 C. (MOS) Mar.

First: *Praelepas jaworskii* Tschernyshew, 1930, Donetz and Kusnetz Basins, former USSR.

Comments: Two other species have also been recorded from rocks of a similar age in the former USSR. There is no other fossil record.

Suborder LEPADOMORPHA Pilsbry, 1916

The name Lepadomorpha is used by Newman (1987) and here in a more restricted sense than in most previous classifications.

F. POECILASMATIDAE Annandale, 1909 T. (BRT)–Rec. Mar.

First: *Trilasmis (Poecilasma) curryi* Withers, 1953, Middle Barton Beds, Barton, Hampshire, England, UK. **Extant**

F. LEPADIDAE Darwin, 1851 T. (LUT)–Rec. Mar.

First: *Lepas stenzeli* Withers, 1953, Claiborne Group, Texas, USA. **Extant**

F. OXYNASPIDIDAE Pilsbry, 1907 T. (LUT)–Rec. Mar.

First: *Oxynaspis eocenica* (Withers, 1953), Upper Bracklesham Beds, Selsey Bill, Sussex, England, UK (Withers, 1953). **Extant**

F. DOSIMIDAE Memmi, 1983 T. (UMI)–Rec. Mar.

First: *Dosima latiscutis* Zullo, 1973, Puente Formation, Los Angeles, California, USA (Weisbord, 1980). **Extant**

Comments: The family is here recognized and separated from the Lepadidae (in contrast to Schram, 1986). However, Memmi's (1983) transfer of the Dosimidae and Oxynaspididae to the Scalpelloidea (= Scalpellomorpha herein) is not followed.

Suborder IBLOMORPHA Leach, 1825

F. IBLIDAE Leach, 1825 ?C. (MOS)/Tr. (ANS)–Rec. Mar.

Comments: *Praelepas damrowi* Schram, 1975 from the Mazon Creek fauna has been reinterpreted, placed in the new genus *Illilepas*, and suggested to have affinities with the iblids (Schram, 1986). The genus *Eolepas*, (Tr. (ANS, RHT)–J. (HET, PLB, AAL–BTH, TTH)), (Withers, 1928) shows clearer affinities with the iblids (Whyte, in prep.). A fossil species of *Ibla* has recently been found in the Upper Jurassic (TTH) of Wiltshire (Whyte, pers. obs.). There is no other fossil record. **Extant**

Suborder SCALPELLOMORPHA Pilsbry, 1916

F. SCALPELLIDAE Pilsbry, 1916 ?C. (VIS)/J. (SIN)–Rec. Mar.

Comments: *Pabulum spathiforme* Whyte, 1976 from the upper Brigantian (C. (VIS)) of Yorkshire, England, UK, and Fife, Scotland, UK, may not be a barnacle. The next record of a scalpellid is ?*Neolepas augurata* Buckeridge and Grant-Mackie (1985), upper Aratauran to lower Ururoan J. (SIN/PLB), New Caledonia. If the acrothoracic rostral plate is derived from a scalpellid ancestor, then a pre-early Devonian divergence is implied for the two groups. **Extant**

F. STRAMENTIDAE Withers, 1920 K. (ALB–MAA) Mar.

First: *Stramentum syriacum* (Dames), 1874, ?Albian, Lebanon (Withers, 1935).

Last: *Loriculina fosteri* Buckeridge, 1983, Gingin Chalk, Molecape Hill, Western Australia.

Order SESSILIA Lamarck, 1818

Suborder BRACHYLEPADOMORPHA Withers, 1923

F. BRACHYLEPADIDAE Woodward, 1923 J. (TTH)–Rec. Mar.

First: *Pycnolepas tithonica* (Withers, 1935), Red and White Limestone, Tithonian, Stramberg, Moravia, Czechoslovakia (Withers, 1935). **Extant**

Comments: Prior to reports of a living form (Newman, *in* Kuhns, 1990), this family was thought to have become extinct in the middle Miocene.

Suborder VERRUCOMORPHA Pilsbry, 1916

The familial classification here follows Newman *in* Newman and Hessler, 1989.

F. NEOVERRUCIDAE Newman, 1989 Rec. Mar.

Comment: Some fragmentary fossils currently assigned to the Brachylepadomorpha might yet prove to be neoverrucids. **Extant**

F. PROVERRUCIDAE Newman, 1989 K. (CEN)–Rec. Mar.

First: *Proverruca nodosa* (Withers, 1935), Chalk detritus (probably Chalk Marl), Charing, Kent, England, UK. **Extant**

F. VERRUCIDAE Darwin, 1854 K. (ALB/CEN)–Rec. Mar.

First: *Verruca withersi* Schram and Newman, 1980, Albian–Cenomanian, Simiti Formation, Antioquia, Columbia. **Extant**

Suborder BALANOMORPHA Pilsbry, 1916

Superfamily CHIONELASMATOIDEA Buckeridge, 1983

F. CHIONELASMATIDAE Buckeridge, 1983 T. (BRT)–Rec. Mar.

First: *Chionelasmus darwini* (Pilsbry), Kaiatan T. (BRT), New Zealand (Buckeridge, 1983).

Comment: Fossil forms may be more closely related to *Eochionelasmus* than to *Chionelasmus* (Yamaguchi and Newman, 1990). **Extant**

Superfamily CHTHAMALOIDEA Darwin, 1854

F. CATOPHRAGMIDAE Utinomi, 1968 K. (CMP)–Rec. Mar.

First: *Pachydiadema cretaceum* (Withers), Ifö, Sweden (Withers, 1935). **Extant**

F. CHTHAMALIDAE Darwin, 1854 T. (DAN)–Rec. Mar.

First: *Pachylasma veteranum* Buckeridge, 1983, Matanginui Limestone (?Teurian–Mangaorapan), Chatham Islands, Pacific. **Extant**

Superfamily CORONULOIDEA Newman and Ross, 1976

F. CORONULIDAE Leach, 1817 T. (BRT)–Rec. Mar.

First: *Emersonius cybosyrinx* Ross, 1967, Upper Eocene, Florida, USA (Newman and Ross, 1976). **Extant**

F. BATHYLASMATIDAE Newman and Ross, 1971 T. (DAN)–Rec. Mar.

First: *Bathylasma rangatira* Buckeridge, 1983, Red Bluff Tuff (Teurian–Mangaorapan), Chatham Islands, Pacific. **Extant**

F. TETRACLITIDAE Gruvel, 1903 T. (LUT)–Rec. Mar.

First: *Eopopella eosimplex* Buckeridge, 1983, Port Willunga Beds (Aldingan), South Australia. **Extant**

Superfamily BALANOIDEA Leach, 1817

F. ARCHAEOBALANIDAE Newman and Ross, 1976 T. (LUT)–Rec. Mar.

First: *Solidobalanus* spp., McBean Formation, Claibornian, Georgia, USA (Zullo, 1984). **Extant**

F. PYRGOMATIDAE Gray, 1925 T. (LMI)–Rec. Mar.

First: *Ceratoconcha jungi* Newman and Ladd, 1974, Lower Miocene, Jamaica (Newman and Ross, 1976). **Extant**

F. BALANIDAE Leach, 1817 T. (BRT)–Rec. Mar.

First: *Megabalanus tintinnabulum* (Linneaus), 1767, Hungary (Davadie, 1963). **Extant**

Class THYLACOCEPHALA Pinna, Arduini *et al.*, 1982

The Thylacocephala are generally regarded as a separate extinct class of crustaceans. Their nature, assigned taxa and higher taxonomy are reviewed by Rolfe (1985), Arduini and Pinna (1989) and Schram (1990) respectively. A familial taxonomy has yet to be attempted; hence the ranges are given for orders.

Order CONCAVICARIDA Briggs and Rolfe, 1983 S. (LLY)–K. (CEN) Mar.

First: Unnamed species from the Lower Silurian Brandon Bridge Formation of Wisconsin, USA (Mikulic *et al.*, 1985a,b).
Last: *Protozoe damesi* Roger, 1946, *P. hilgendorfi* Dames, 1886, *Pseuderichthus cretaceus* Dames, 1886, Lebanon.

Order CONCHYLIOCARIDA Secretan, 1983 ?€ (CRF)–J. (TTH) Mar.

First: *Silesicaris nasuta* Gürich, 1926, Lower Cambrian, Poland.
Last: *Clausocaris lithographica* (Oppenheim, 1888), Solnhofen Limestone, Bavaria, Germany.
Comment: The assignment of *Silesicaris* here (Arduini and Pinna, 1989; Schram, 1990) is based on carapace alone, and may not be reliable.

REFERENCES

Abele, L. G., Kim, W. and Felgenhauer, B. E. (1989) Molecular evidence for inclusion of the phylum Pentastomida in the Crustacea. *Molecular Biology and Evolution*, **6**, 685–91.

Andres, D. (1989) Phosphatisierte fossilien aus dem unteren Ordoviz von Südschweden. *Berliner geowissenschaften Abhandlungen*, **106**, 9–19.

Arduini, P. and Pinna, G. (1989) *I Tilacocefali: una Nuova Classe di Crostacei Fossili*. Museo Civico di Storia Naturale di Milano, 35 pp.

Baird, G. C., Brett, C. E. and Tomlinson, J. T. (1990) Host-specific acrothoracid barnacles on Middle Devonian platyceratid gastropods. *Historical Biology*, **4**, 221–44.

Barthel, K. W. and Boettcher, R. (1978) Abu Ballas Formation (Tithonian/Berriasian; Southwestern Desert, Egypt) a significant lithostratigrahic unit of the former 'Nubian Series'. *Mitteilungen des Bayerische Staatssammlung für Paläontologie und Historische Geologie*, **18**, 153–66.

Bate, R. H., Collins, J. S. H., Robinson, J. E. *et al.* (1967) Arthropoda: Crustacea, in *The Fossil Record* (eds W. B. Harland, C. H. Holland, M. R. House *et al.*), Geological Society of London, pp. 535–63.

Beikirch, D. W. and Feldmann, R. M. (1980) Decapod crustaceans from the Pflugerville Member, Austin Formation (Late Cretaceous: Campanian) of Texas. *Journal of Paleontology*, **54**, 309–24.

Bishop, G. A. (1983) Fossil decapod Crustacea from the Late Cretaceous Coon Creek Formation, Union County, Mississippi. *Journal of Crustacean Biology*, **3**, 417–30.

Bousfield, E. L. (1982) Amphipoda, in *McGraw-Hill Yearbook of Science and Technology*. McGraw-Hill, New York, pp. 96–100.

Bowman, T. E. and Abele, L. G. (1982) Classification of the Recent Crustacea, in *The Biology of Crustacea*, Vol. 1 (ed. L. G. Abele), Academic Press, New York, pp. 1–27.

Briggs, D. E. G. (1983) Affinities and early evolution of the Crustacea: the evidence of the Cambrian fossils. *Crustacean Issues*, **1**, 1–22.

Briggs, D. E. G. and Clarkson, E. N. K. (1985) Malacostracan Crustacea from the Dinantian of Foulden, Berwickshire, Scotland. *Transactions of the Royal Society of Edinburgh: Earth Sciences*, **76**, 35–40.

Briggs, D. E. G. and Clarkson, E. N. K. (1990) The late Palaeozoic radiation of malacostracan crustaceans, in *Major Evolutionary Radiations* (eds P. D. Taylor and G. P. Larwood), Systematics Association Special Volume 42, Clarendon Press, Oxford, pp. 165–86.

Briggs, D. E. G. and Rolfe, W. D. I. (1983) New Concavicarida (new order: ? Crustacea) from the Upper Devonian of Gogo, Western Australia, and the palaeoecology and affinities of the group. *Special Papers in Palaeontology*, **30**, 249–76.

Briggs, D. E. G., Clark, N. D. L. and Clarkson, E. N. K. (1991) The Granton 'shrimp-bed', Edinburgh – a Lower Carboniferous Konservat-Lagerstätte. *Transactions of the Royal Society of Edinburgh: Earth Sciences*, **82**, 65–85.

Buckeridge, J. S. (1983) Fossil barnacles (Cirripedia: Thoracica) of New Zealand and Australia. *New Zealand Geological Survey Paleontological Bulletin*, **50**, 1–151.

Buckeridge, J. S. and Grant-Mackie, J. A. (1985) A new scalpellid barnacle (Cirripedia: Thoracica) from the Lower Jurassic of New Caledonia. *Géologie de la France*, **1**, 15–18.

Burkenroad, M. D. (1963) The evolution of the Eucarida (Crustacea, Eumalacostraca) in relation to the fossil record. *Tulane Studies in Geology*, **2**, 3–16.

Collins, D. H. and Rudkin, D. M. (1981) *Priscansermarinus barnetti*, a probable lepadomorph barnacle from the Middle Cambrian Burgess Shale of British Columbia. *Journal of Paleontology*, **55**, 1006–15.

Cressy, R. and Patterson, C. (1973) Fossil parasitic copepods from a Lower Cretaceous fish. *Science*, **180**, 1283–5.

Davadie, C. (1963) *Études des Balanes d'Europe et d'Afrique. Systématique et structure des Balanes fossiles d'Europe et d'Afrique*. Éditions du Centre National de la Recherche Scientifique, Paris, 146 pp.

Dzik, J. and Jazdzewski, K. (1978) The euphausiid species of the Antarctic region. *Polskie Archiwum Hydrobiologii*, **25**, 589–605.

Emerson, M. J. and Schram, F. R. (1991) Remipedia, Part 2: Paleontology. *Proceedings of the San Diego Society of Natural History*, **7**, 1–52.

Factor, D. F. and Feldmann, R. M. (1985) Systematics and paleoecology of malacostracan arthropods in the Bear Gulch Limestone (Namurian) of Central Montana. *Annals of Carnegie Museum*, **54**, 319–56.

Felgenhauer, B. E. and Abele, L. G. (1983) Phylogenetic relationships among shrimp-like decapods. *Crustacean Issues*, **1**, 291–311.

Förster, R. (1968) *Paranecrocarcinus libanoticus* n.sp. (Decapoda) und die Entwicklung der Calappidae in der Kreide. *Mitteilungen des Bayerische Staatssammlung für Paläontologie und Historische Geologie*, **8**, 167–95.

Förster, R. and Crane, M. D. (1984) A new species of the penaeid shrimp *Aeger* Münster (Crustacea, Decapoda) from the Upper Triassic of Somerset, England. *Neues Jahrbuch für Geologie und Paläontologie, Monatshefte*, **1984**, 455–62.

Fryer, G. (1987) A new classification of the branchiopod Crustacea. *Zoological Journal of the Linnean Society*, **91**, 357–83.

Fryer, G. (1991) A daphniid ephippium (Branchiopoda: Anomopoda) of Cretaceous age. *Zoological Journal of the Linnean Society*, **102**, 163–7.

Glaessner, M. F. (1945) Cretaceous Crustacea from Mount Lebanon, Syria. *Annals and Magazine of Natural History*, **12**, ser. 11, 694–707.

Glaessner, M. F. (1969) Decapoda, in *Treatise on Invertebrate Paleontology, Part R, Arthropoda, 4 (2)* (ed. R. C. Moore), Geological

Society of America and University of Kansas Press, Boulder, Colorado and Lawrence, Kansas, pp. R399–R533.

Glaessner, M. F. and Malzahn, E. (1962) Neue Crustaceen aus dem niederrheinischen Zechstein. *Fortschritte Geologie von Rheinland und Westfalen, Krefeld*, **6**, 245–64.

Gray, J. (1988) Evolution of the freshwater ecosystem: the fossil record. *Palaeogeography, Palaeoclimatology, Palaeoecology*, **62**, 1–214.

Grygier, M. J. (1983) Ascothoracida and the unity of Maxillopoda, *Crustacean Issues*, **1**, 73–104.

Grygier, M. J. (1990) Early planktonic nauplii of *Baccalaureus* and *Zibrowia* (Crustacea; Ascothoracida) from Okinawa, Japan. *Galaxea*, **8**, 321–37.

Grygier, M. J. and Newman, W. A. (1985) Motility and calcareous parts in extant and fossil Acrothoracica (Crustacea: Cirripedia), based primarily upon new species burrowing in the deep-sea scleractinian coral *Enallopsammia*. *Transactions of the San Diego Society of Natural History*, **21**, 1–22.

Harland, W. B., Holland, C. H., House, M. R. *et al.* (eds) (1967) *The Fossil Record*. Geological Society of London, London, 827 pp.

Hasiotis, S. T. and Mitchell, C. E. (1989) Lungfish burrows in the Upper Triassic Chinle and Dolores formations, Colorado Plateau – discussion: new evidence suggests origin by a burrowing decapod crustacean. *Journal of Sedimentary Petrology*, **59**, 871–5.

Hessler, R. R. (1969) Peracarida, in *Treatise on Invertebrate Paleontology, Part R, Arthropoda, 4 (1)* (ed. R. C. Moore), Geological Society of America and University of Kansas Press, Boulder, Colorado, and Lawrence, Kansas, R360–R393.

Holthuis, L. B. and Manning, R. B. (1969) Stomatopoda, in *Treatise on Invertebrate Paleontology, Part R, Arthropoda, 4 (2)* (ed. R. C. Moore), Geological Society of America and University of Kansas Press, Boulder, Colorado, and Lawrence, Kansas, pp. R535–52.

Jell, P. A. and Duncan, P. M. (1986) Invertebrates, mainly insects, from the freshwater, Lower Cretaceous, Koonwarra Fossil Bed (Korumburra Group), South Gippsland, Victoria. *Memoirs of the Association of Australasian Palaeontologists*, **3**, 111–205.

Karaman, G. S. (1984) Critical remarks to the fossil Amphipoda with description of some new taxa. *Poljoprivreda i Sumarstvo*, **30**, 87–104.

Keilbach, R. (1982) Bibliographie und Liste der Arten tierischer Einschlusse in fossilen Harzen sowie ihrer Aufbewahrungsorte, Teil 1. *Deutsche Entomologische Zeitschrift*, **29**, 129–286.

Kuhns, K. K. (ed.) (1990) *Scripps Institution of Oceanography Annual Report*. University of California, San Diego.

Madsen, N. and Wolff, T. (1965) Evidence of the occurrence of Ascothoracica (parasitic cirripedes) in Upper Cretaceous. *Meddelanden fra Dansk Geologiske Forendlingen*, **15**, 556–8.

Malzahn, E. (1972) Cumaceenfunde aus dem neiderrheinischen Zechstein. *Geologische Jahrbuch*, **90**, 441–62.

McKenzie, K. G., Chen, Pei-ji and Majoran, S. (1991) *Almatium gusevi* (Chernishev 1940): redescription, shield-shapes; and speculations on reproductive mode (Branchiopoda, Kazacharthra. *Paläontologische Zeitschrift*, **65**, 305–17.

Memmi, M. (1983) New concepts on the origin and position of pelagic barnacles in the system of the Suborder Lepadomorpha (Cirripedia, Thoracica). *Doklady Akademii Nauk SSR*, **273**, 1271–75 [in Russian].

Mikulic, D. G., Briggs, D. E. G. and Kluessendorf, J. (1985a) A Silurian soft-bodied biota. *Science*, **228**, 715–7.

Mikulic, D. G., Briggs, D. E. G. and Kluessendorf, J. (1985b) A new exceptionally preserved biota from the Lower Silurian of Wisconsin, USA. *Philosophical Transactions of the Royal Society of London*, **B311**, 75–84.

Müller, K. J. (1983) Crustacea with preserved soft parts from the Upper Cambrian of Sweden. *Lethaia*, **16**, 93–109.

Newman, W. A. (1982) Cirripedia, in *The Biology of Crustacea*, Vol. 1 (ed. L. G. Abele), Academic Press, New York, pp. 197–221.

Newman, W. A. (1987) Evolution of cirripedes and their major groups. *Crustacean Issues*, **5**, 3–42.

Newman, W. A. and Hessler, R. R. (1989) A new abyssal hydrothermal verrucomorphan (Cirripedia; Sessilia): The most primitive living sessile barnacle. *Transactions of the San Diego Society of Natural History*, **21**, 259–73.

Newman, W. A. and Ross, A. (1976) Revision of the balanomorph barnacles; including a catalog of the species. *San Diego Natural History Museum Memoir*, **9**, 1–108.

Newman, W. A., Zullo, V. A. and Withers, T. H. (1969) Cirripedia, in *Treatise on Invertebrate Paleontology, Part R, Arthropoda, 4 (1)* (ed. R. C. Moore), Geological Society of America and University of Kansas Press, Boulder, Colorado, and Lawrence, Kansas, R206–R295.

Palmer, A. R. (1957) Miocene arthropods from the Mojave Desert, California. *United States Geological Society, Professional Paper*, **294G**, 235–77.

Palmer, A. R. (1960) Miocene copepods from the Mojave Desert, California. *Journal of Paleontology*, **34**, 447–52.

Quayle, W. J. (1987) English Eocene Crustacea (lobsters and stomatopod). *Palaeontology*, **30**, 581–612.

Rathbun, M. J. (1919) West India Tertiary decapod crustaceans. *Publications of the Carnegie Institute, Washington*, **291**, 157–82.

Robison, R. A. and Richards, B. C. (1981) Larger bivalve arthropods from the Middle Cambrian of Utah. *University of Kansas Paleontological Contributions*, **106**, 1–28.

Roger, W. D. I. (1946) Les invertébrés des couches à poissons du Crétacé Superieur du Liban. Étude scientifiques de la Mission C. Arambourg en Syrie et en Iran (1938–1939). I, *Société Géologique de France, Mémoires, new series*, **23**, 1–92.

Rolfe, W. D. I. (1969) Phyllocarida, in *Treatise on Invertebrate Paleontology, Part R, Arthropoda, 4 (1)* (ed. R. C. Moore), Geological Society of America and University of Kansas Press, Boulder, Colorado, and Lawrence, Kansas, R296–R331.

Rolfe, W. D. I. (1985) Form and function in Thylacocephala, Conchyliocarida and Concavicarida (?Crustacea): a problem of interpretation. *Transactions of the Royal Society of Edinbrugh*, **76**, 391–9.

Rolfe, W. D. I. and Edwards, V. A. (1979) Devonian Arthropoda (Trilobita and Ostracoda excluded). *Special Papers in Palaeontology*, **23**, 325–9.

Saint Laurent, M. de (1989) La nouvelle superfamille des Retroplumoidea Gill, 1894 (Decapoda, Brachyura): systématique, affinités et évolution, in *Résultats des Campagnes Musorstom*, 5 (ed. J. Forest), Mémoires du Museum Nationelle d'Histoire Naturelle, A, **144**, pp. 103–79.

Schram, F. R. (1979a) British Carboniferous Malacostraca. *Fieldiana: Geology*, **40**, 1–129.

Schram, F. R. (1979b) The genus *Archaeocaris*, and a general review of the Palaeostomatopoda (Hoplocarida: Malacostraca). *Transactions of the San Diego Society of Natural History*, **19**, 57–66.

Schram, F. R. (1980) Miscellaneous Late Paleozoic Malacostraca of the Soviet Union. *Journal of Paleontology*, **54**, 542–7.

Schram, F. R. (1981) On the classification of the Eumalacostraca. *Journal of Crustacean Biology*, **1**, 1–10.

Schram, F. R. (1982) The fossil record and evolution of Crustacea, in *The Biology of Crustacea*, Vol. 1 (ed. L. G. Abele), Academic Press, New York, pp. 93–147.

Schram, F. R. (1984a) Fossil Syncarida. *Transactions of the San Diego Society Natural History*, **20**, 189–246.

Schram, F. R. (1984b) Upper Pennsylvanian arthropods from black shales of Iowa and Nebraska. *Journal of Paleontology*, **58**, 197–209.

Schram, F. R. (1986) *Crustacea*. Oxford University Press, Oxford, 606 pp.

Schram, F. R. (1989) Designation of a new name and type for the Mazon Creek (Pennsylvanian, Francis Creek Shale) tanaidacean. *Journal of Paleontology*, **63**, 536.

Schram, F. R. (1990) On Mazon Creek Thylacocephala. *Proceedings of the San Diego Society Natural History*, **3**, 1–16.

Schram, F. R. and Horner, J. (1978) Crustacea of the Mississippian Bear Gulch Limestone of central Montana. *Journal of Paleontology*, **52**, 394–406.

Schram, F. R. and Newman, W. A. (1980) *Verruca withersi* n.sp.

(Crustacea: Cirripedia) from the middle of the Cretaceous of Columbia. *Journal of Paleontology*, **54**, 229–33.

Schram, F. R., Feldmann, R. M. and Copeland, M. J. (1978) The Late Devonian Palaeopalaemonidae and the earliest decapod crustaceans. *Journal of Paleontology*, **52**, 1375–87.

Schram, F. R., Sieg, J. and Malzahn, E. (1986) Fossil Tanaidacea. *Transactions of the San Diego Society of Natural History*, **21**, 127–44.

Schwebel, L., Gall, J. C. and Grauvogel, L. (1983) Faune du Buntsandstein. V. Description du néotype d'*Anhelkocephalon handlirschi* Bill (1914), crustacé isopode du Trias Inférieur des Vosges (France). *Annales de Paléontologie*, **69**, 307–16.

Seilacher, A. (1969) Paleoecology of boring barnacles. *American Zoologist*, **9**, 705–19.

Sepkoski, J. J., Jr (1982) A compendium of fossil marine families. *Milwaukee Public Museum Contributions in Biology and Geology*, **51**, 125 pp.

Smirnov, N. N. (1992) Mesozoic Anomopoda (Crustacea) from Mongolia. *Zoological Journal of the Linnean Society*, **104**, 97–116.

Tasch, P. (1969) Branchiopoda, in *Treatise on Invertebrate Paleontology, Part R, Arthropoda, 4 (1)* (ed. R. C. Moore), Geological Society of America and University of Kansas Press, Boulder, Colorado and Lawrence, Kansas, R128–91.

Tomlinson, J. T. (1969) The burrowing barnacles (Cirripedia: Order Acrothoracica). *United States National Museum Bulletin*, **296**, 1–162.

Tschernischew, B. I. (1930) Cirripeden aus dem Bassin des Donez und von Kusnetsk. *Zoologischer Anzeiger*, **92**, 26–8.

Turner, R. F. (1973) Occurrence and implications of fossilized burrowing barnacles. *Geological Society of America Abstracts with Programs*, **5**, 230–1.

Voigt, E. (1959) *Endosacculus moltkiae* n.g. n.sp., ein vermutlicher fossiler Ascothoracide (Entomostr.) als Cystenbildner bei der Oktokoralle *Moltkia minuta*. *Paläontologische Zeitschrift*, **33**, 211–23.

Voigt, E. (1967) Ein vermutlicher Ascothoracide (*Endosacculus* (?) *najdini* n.sp.) als Bewohner einer kretazischen *Isis* aus der USSR. *Paläontologische Zeitschrift*, **41**, 86–90.

Walossek, D. and Müller, K. J. (1990) Upper Cambrian stem-lineage crustaceans and their bearing upon the monophyletic origin of Crustacea and the position of *Agnostus*. *Lethaia*, **23**, 409–27.

Weisbord, N. E. (1980) Fossil Lepadomorph, Brachylepadomorph and Verrucomorph Barnacles (Cirripedia) of the Americas. *Bulletin of American Paleontology*, **78**, 117–212.

Whyte, M. A. (1976) A Carboniferous pedunculate barnacle. *Proceedings of the Yorkshire Geological Society*, **41**, 1–12.

Wieder, R. W. and Feldmann, R. M. (1992) Mesozoic and Cenozoic isopods of North America. *Journal of Paleontology*, **66**, 958–72.

Wills, L. J. (1962) A pedunculate cirripede from the Upper Silurian of Oesel, Esthonia. *Nature*, **194**, p. 567.

Withers, T. H. (1928) *Catalogue of fossil Cirripedia in the Department of Geology, I. Triassic and Jurassic*. British Museum (Natural History), London, 154 pp.

Withers, T. H. (1935) *Catalogue of Fossil Cirripedia in the Department of Geology, II. Cretaceous*. British Museum (Natural History), London, 534 pp.

Withers, T. H. (1953) *Catalogue of fossil Cirripedia in the Department of Geology, III. Tertiary*. British Museum (Natural History), London, 396 pp.

Woods, H. (1928) A monograph of the fossil macrurous Crustacea of England. *Palaeontographical Society Monograph*, **82**, 73–88.

Wright, C. W. and Collins, J. S. H. (1972) British Cretaceous Crabs. *Palaeontographical Society Monograph*, **126**, (533), 114 pp.

Yamaguchi, T. and Newman, W. A. (1990) A new and primitive barnacle (Cirripedia: Balanomorpha) from the North Fiji Basin abyssal hydrothermal field, and its evolutionary implications. *Pacific Science*, **44**, 135–55.

Zevina, G. B. (1980) A new classification of Lepadomorpha (Cirripedia). *Zoologicheskii Zhurnal Akademie Nauk SSSR*, **54**, 689–98 [in Russian].

Zullo, V. A. (1983) Cirripedia, in *Synopsis and Classification of Living Organisms*, Vol. 2 (ed. S. P. Parker), McGraw-Hill Book Company, New York, 1232 pp.

Zullo, V. A. (1984) New genera and species of balanoid barnacles from the Oligocene and Miocene of North Carolina. *Journal of Paleontology*, **58**, 1312–38.

19

ARTHROPODA (CRUSTACEA: OSTRACODA)

R. C. Whatley, David J. Siveter and I. D. Boomer

The classification of the Ostracoda is in a state of flux; a revision of the *Treatise on Invertebrate Paleontology*, Part Q, is presently being undertaken under the co-ordination of the two senior contributors. Some minor families are excluded herein and not all higher taxonomic relationships have been elucidated prior to the publication of this revision. The classification herein is based on Moore (1961), Hartmann and Puri (1974) and the present contributors' observations and those of unspecified Part Q revision authors. Classification schemes adopted by colleagues in the former USSR are in many cases radically different from that outlined herein; for example, see the '*Treatises*' on Cainozoic (Nikolaeva *et al.*, 1989) and Palaeozoic (Abushik *et al.*, 1990) ostracods, respectively (the Mesozoic volume is in press). In particular, Russian colleagues often recognize many more families of ostracods than we identify herein.

The Ostracoda are regarded as a distinct Class of the Crustacea, distinguished as they are by their bivalved carapace which is pierced by pores and into which the entire animal can be withdrawn when the carapace is closed.

Acknowledgements – We are indebted to the following colleagues for supplying data on various ostracod groups: Dr A. F. Abushik (St Petersburg), Dr F. J. Adamczak (Stockholm), Dr G. Becker (Frankfurt), Dr J. M. Berdan (Washington), Dr C. P. Dewey (Mississippi), Dr H. Groos-Uffenorde (Göttingen), Dr I. Hinz (Bonn), Dr P. J. Jones (Canberra), Professor R. F. Lundin (Tempe), Professor K. J. Müller (Bonn), Dr L. Sarv (Tallin), Dr R. E. L. Schallreuter (Hamburg), Dr I. G. Sohn (Washington), Dr J. M. C. Vannier (Lyon), Dr M. Williams (Leicester; see Williams, 1990). DJS thanks NATO for its support through its collaborative research programme.

Class OSTRACODA Latreille, 1802

Order BRADORIIDA Raymond, 1935

The taxonomy of this heterogeneous (almost certainly polyphyletic) group is particularly contentious; many taxa which have been assigned to the group may belong outside the Ostracoda. The groupings and ranges given below are very conservative by some standards; in particular, large numbers of Chinese taxa (genera) have yet to be satisfactorily assessed and assigned with any degree of confidence to their respective families.

F. BRADORIIDAE Matthew, 1902 Є. (CRF–DOL) Mar. (see Fig. 19.1)

First: e.g. *Bradoria scrutator* Matthew, 1899, CRF, Nova Scotia, Canada.
Last: *Waldoria buchholzi* Gründel, 1981, *Leptoplastus paucisegmentatus* Zone, boulders, Baltic provenance.
Intervening: STD.

F. HIPPONICHARIONIDAE Sylvester-Bradley, 1961 Є. (CRF–MNT) Mar.

First: e.g. *Hipponicharion eos* Matthew, 1886, St John Group, New Brunswick, Canada.
Last: *Cyclotron cambricum* Gründel, 1981, *Olenus* zones, boulders, Baltic provenance.
Intervening: STD.

F. CAMBRIIDAE Li, 1975 Є. (ATB–?LEN) Mar.

First: e.g. *Cambria sibirica* Neckaja and Ivanova, 1956, ATB, eastern Siberia, former USSR.
Last: All species are of Lower Cambrian age.

F. KUNMINGELLIDAE Huo and Shu, 1985 Є. (CRF) Mar.

First: e.g. *Kunmingella maxima* Huo, 1956, CRF (Chiungchussu Formation), Yunnan Province, China.
Last: All species are of early Cambrian age.

F. HAOIIDAE Shu, 1990 Є. (CRF)–O. (TRE) Mar.

First: *Uskarella prisca* Koneva, 1978, CRF, Kazakhstan.
Last: *Septadella jackmanae* Stubblefield, 1933, TRE, Gloucestershire, England, UK.
Intervening: Rare and discontinuous.

F. SVEALUTIDAE Öpik, 1968 Є. (MEN–MER) Mar.

First: e.g. *Leperditia primordialis* Linnarsson, 1869, Olenidskiffer, *Leiopyge laevigata* Zone, Vestergotland, Sweden.
Last: e.g. *Anabarochilina ventriangulosa* Abushik, 1960, MER, eastern Siberia, former USSR.

F. COMPTALUTIDAE Öpik, 1968 Є. (STD) Mar.

First: *Aristaluta gutta* Öpik, 1961, Devoncourt Limestone, Boomerangian, Queensland, Australia.

The Fossil Record 2. Edited by M. J. Benton. Published in 1993 by Chapman & Hall, London. ISBN 0 412 39380 8

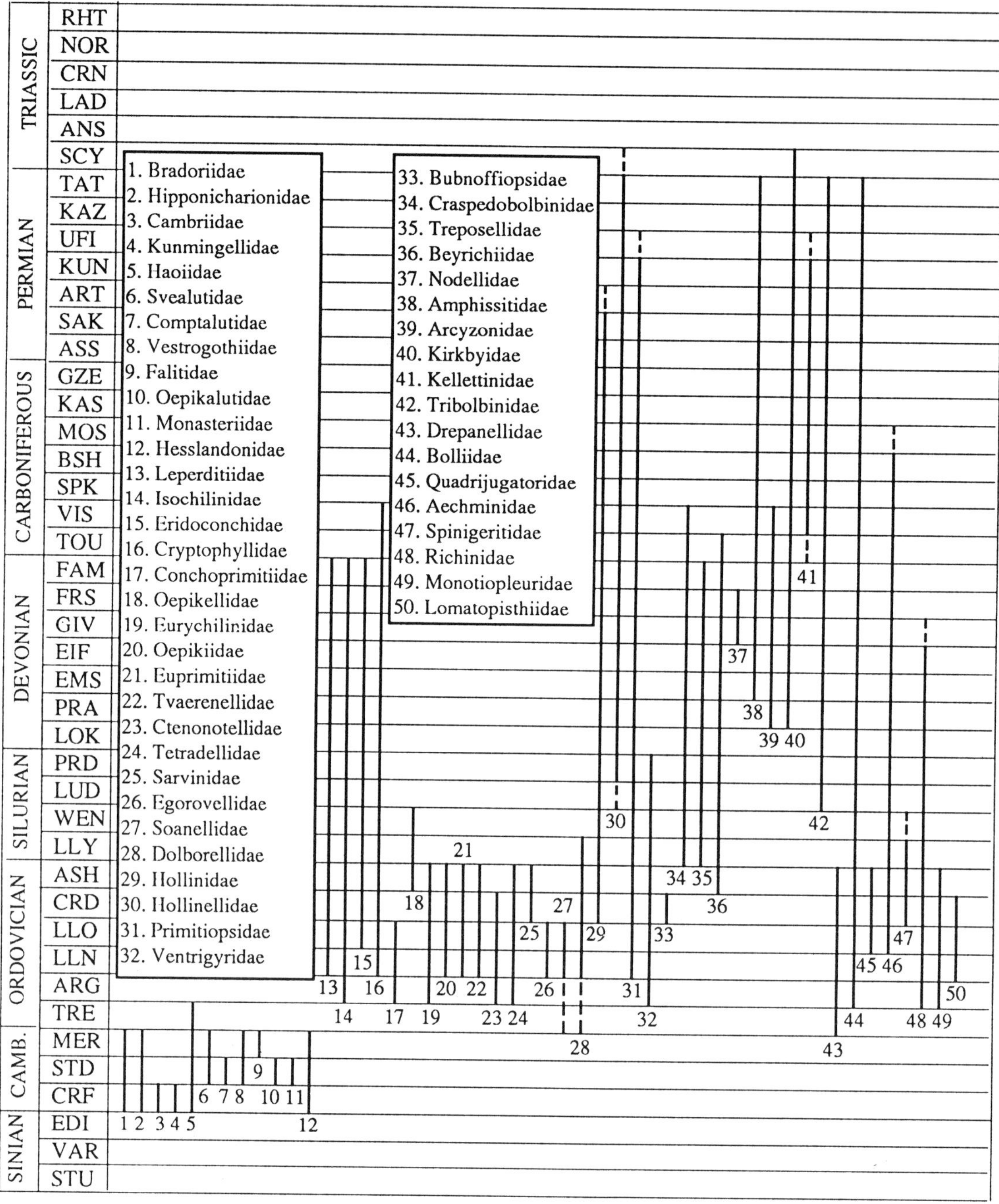

Fig. 19.1

Last: All species are of Middle Cambrian age.

Order PHOSPHATOCOPIDA Müller, 1964

Suborder VESTROGOTHIINA Müller, 1982

F. VESTROGOTHIIDAE Kozur, 1974 Є. (SOL/MEN–DOL) Mar.

First: *Vestrogothia longispinosa* Kozur, 1974, erratic boulder, STD, Sellin, Rügen, Germany.
Last: *Vestrogothia spinata* Müller, 1964, *Peltura minor* Zone, Sweden.
Intervening: MNT.

F. FALITIDAE Müller, 1964 Є. (MNT–DOL) Mar.

First: e.g. *Falites fala* Müller, 1964, *Agnostus* and *Olenus* zones, Sweden.
Last: *Falites angustiduplicata* Müller, 1964, *Peltura scarabaeoides* Zone, Sweden.

F. OEPIKALUTIDAE Jones and McKenzie, 1980 Є. (SOL) Mar.

First: e.g. *Zepaera rete* Fleming, 1973, Beetle Creek Formation, Templetonian 'stage', Duchess area, Queensland, Australia.
Last: e.g. *Oepikaluta dissuta* Jones and Mckenzie, 1980, Current Bush Limestone, Undillan 'stage', Georgina Basin, Australia.

F. MONASTERIIDAE Jones and McKenzie, 1980 Є. (SOL) Mar.

First and Last: *Monasterium oepiki* Fleming, 1973, Beetle

Creek Formation, Templetonian 'stage', Duchess area, Queensland, Australia.

Suborder HESSLANDONINA Müller, 1982

F. HESSLANDONIDAE Müller, 1964
€. (LEN–MNT) Mar.

First: *Hesslandona*? n. sp. of Hinz, 1987, *Strenuella* Limestone, lower Comley Limestones, Shropshire, England, UK.
Last: e.g. *Hesslandona unisulcata* Müller, 1982, *Agnostus pisiformis* Zone, Västergötland, Sweden.
Intervening: ?SOL, MEN.

Order LEPERDITICOPIDA Scott, 1961

F. LEPERDITIIDAE Jones, 1856
?O. (LLN_1)–D. (FAM) Mar.

First: *Eoleperditia ambigua* Berdan, 1976, Kanosh Shale, Pogonip Group, Chazyan, Utah, USA.
Last: Probably *Herrmannina fameniana* Schevtsov, 1971, upper FAM, Tataria, near Perm, former USSR. Abushik (1979) also records *Moelleritia* from the Upper Devonian of the Siberian Platform, former USSR.
Intervening: LLN_2–FRS.
Comments: The records from the ?Upper Cambrian of the USA of supposed leperditiids (Frederickson, 1946; Palmer, 1954) are not yet confirmed by modern studies, and neither is the record of 'Lower' (= Middle) Ordovician leperditiids from Greenland (Poulsen, 1937). Similarly, the records of leperditiids from the basal part of the Ordovician (Arenig ?) of Oklahoma (e.g. Harris, 1957, 1960) need to be re-appraised before they can be used with confidence in modern stratigraphy.

F. ISOCHILINIDAE Swartz, 1949
O. (?ARG)–D. (u.) Mar.

First: *Tirisochilina juabaria* Berdan, 1976, Juab Limestone, Pogonip Group, Chazyan, Utah, USA.
Last: Abushik (1979) says that *Hogmochilina* occurs in the Upper Devonian of Spitsbergen. This refers to *H. elliptica* (Solle, 1935).
Intervening: ?LLN_1–D_2.

Order ERIDOSTRACODA Adamczak, 1961

The taxonomic position and rank of this group is particularly controversial. Some opinions hold that the 'Eridostraca' may be an extinct group of marine branchiopods; others, that they are a part of the ostracod group Palaeocopida.

F. ERIDOCONCHIDAE Henningsmoen, 1953
O. (LLO)–D. (FAM) Mar.

First: *Eridoconcha simpsoni* Harris, 1931, Tulip Creek Formation, Oklahoma, USA.
Last: *Eridoconcha* sp. of Jones, 1968, Gumhole Formation, Fairfield Group, northern Canning Basin, Western Australia.
Intervening: COS–FRS.

F. CRYPTOPHYLLIDAE Adamczak, 1961
O. (LLN_1)–C. (BRI) Mar.

First: *Cryptophyllus magnus* (Harris, 1931), Oil Creek Formation, Oklahoma, USA.
Last: *Cryptophyllus* sp. 1 of Bless and Massa, 1982, upper M'rar Formation, Viséan, Rhadamès Basin, western Libya.
Intervening: LLN_2–ASB.

F. CONCHOPRIMITIIDAE Henningsmoen, 1953
O. (ARG–LLO_3) Mar.

This family is tentatively referred to the Eridostracoda; its systematic position is problematic.
First: *Conchoprimitia gammae* Öpik, 1935, Toila Formation, Ontik Regional 'series', Estonia, former USSR.
Last: *Conchoprimitia tolli* (Bonnema, 1909), Kukruse Regional 'stage', middle part of the Viru 'series', Estonia, former USSR.
Intervening: LLN_1–LLO_2.

Order PALAEOCOPIDA Henningsmoen, 1953

Superfamily EURYCHILINACEA Ulrich and Bassler, 1923

F. OEPIKELLIDAE Jaanusson, 1957
O. (HIR)–S. (WHI) Mar.

First: *Platybolbina plana* (Krause, 1889), top (Porkuni) part of the Harju 'series', Baltoscandia.
Last: *Platybolbina lunulifera* Henningsmoen, 1954, Steinsfjorden Formation, Ringerike district, Norway.
Intervening: RHU–SHE.

F. EURYCHILINIDAE Ulrich and Bassler, 1923
O. (ARG–HIR) Mar.

First: *Laccochilina estonula* (Öpik, 1935), middle part of Ontik 'series', Baltoscandia.
Last: *Uhakiella jonesiana* (Schmidt, 1941), upper (Porkuni) part of Harju 'series', Baltoscandia.
Intervening: LLN_1–RAW.

F. OEPIKIIDAE Jaanusson, 1957
O. (LLN_1–HIR) Mar.

First: *Oepikium novum* Sarv, 1959, Kunda 'stage', Baltoscandia.
Last: *Oepikium porkuniensis* Henningsmoen, 1954, upper (Porkuni) part of Harju 'series', Baltoscandia.
Intervening: LLN_2–RAW.

Superfamily HOLLINACEA Swartz, 1936

F. EUPRIMITIIDAE Hessland, 1949
O. (LLN_1–HIR) Mar.

First: *Steinfurtia macroreticulata* (Hessland, 1949), Kunda 'stage', Baltoscandia.
Last: *Gryphiswaldensia visbya* Schallreuter, 1969, upper (Porkuni) part of Harju 'series', Baltoscandia.
Intervening: LLN_2–RAW.

F. TVAERENELLIDAE Jaanusson, 1957
O. (LLN_1–HIR) Mar.

First: *Tvaerenella modesta* (Sarv, 1959), Kunda 'stage', Baltoscandia.
Last: *Eoaquapulex frequens* (Steusloff, 1894), upper (Porkuni)
part of Harjuan 'series', Baltoscandia.
Intervening: LLN_2–RAW.

F. CTENONOTELLIDAE Schmidt, 1941
O. (ARG–ACT) Mar.

First: *Tallinnellina palmata* (Krause, 1889), middle part of Ontik 'series', Baltoscandia.

Last: *Steusloffia neglecta* Sarv, 1959, upper (Rakvere) part of Viru 'series', Baltoscandia.
Intervening: LLN_1–MRB.

F. TETRADELLIDAE Swartz, 1936
O. (ARG–HIR) Mar.

First: *Glossomorphites bocki* (Öpik, 1935), middle part of Ontik 'series', Baltoscandia.
Last: *Tetradella plicatula* (Krause, 1892), top (Porkuni) part of the Harju 'series', Baltoscandia.
Intervening: LLN_1–RAW.

F. SARVINIDAE Schallreuter, 1966
O. (SOU–RAW) Mar.

First: *Distobolbina pinna* Schallreuter, 1964, middle (Idavere/Johvi) part of Viru 'series', Baltoscandia.
Last: *Anticostiella ellisensis* Copeland, 1973, Ellis Bay Formation, Anticosti Island, Canada.
Intervening: LON–CAU.

F. EGOROVELLIDAE Schallreuter, 1966
O. (LLN/LLO) Mar.

First: *Egorovella defecta* Ivanova, 1959, lower Volcinsk Formation, Middle Ordovician, Siberia, former USSR.
Last: *Egorovella poricostata* Kanygin, 1965, upper Kalycanskian Formation, Middle Ordovician, Siberia, former USSR.

F. SOANELLIDAE Kanygin, 1977 O. (l./m.) Mar.

First: *Fuscinullina pectinata* Kanygin, 1967, Sienskian Formation, Lower Ordovician, Cher Mountains, Siberia, former USSR.
Last: *Sibiretella furcata* Kanygin, 1967, Kalycanskian Formation, Middle Ordovician, Cher Mountains, Siberia, former USSR.

F. DOLBORELLIDAE Melnikova, 1976
O. (l./m.)–S. (?TEL) Mar.

First: *Planiprimites solitus* Kanygin, 1967, Sienskaya Formation, Lower Ordovician, Cher Mountains, Siberia, former USSR.
Last: *Costaegera hastata* Abushik, 1960, upper LLY, Siberia, former USSR.
Intervening: O. (u.)–?FRO.

F. HOLLINIDAE Swartz, 1936
O. (?LON)–P. (SAK/ART) Mar.

First: *Aloculatia hartmanni* Schallreuter, 1976, Skagen Limestone, Baltoscandia.
Last: *Tetrasacculus ?timorensis* Bless, 1987, Lower Permian, Timor.
Intervening: ?MRB–ASS.

F. HOLLINELLIDAE Bless and Jordon, 1971
S. (LDF/PRD)–Tr. (?GRI) Mar.

First: *Hollinnella originalis* Lundin, 1965, Henryhouse Formation, Oklahoma, USA.
Last: *Hollinnella* sp. of Bless and Jordon 1972, lowermost Triassic, Australia.
Intervening: GED–?TAT.

Superfamily PRIMITIOPSACEA Swartz, 1936

F. PRIMITIOPSIDAE Swartz, 1936
O. (LLN_2)–P. (KUN/UFI) Mar.

First: *Anisocyamus multiperforata* (Harris, 1957), Mclish Formation, Oklahoma, USA.
Last: *Coryellina indicata* Sohn, 1954, Leonard Formation or Word Formation, Texas, USA.
Intervening: LLO_1–ART.

F. VENTRIGYRIDAE Gründel, 1977
O. (ARG)–S. (PRD) Mar.

First: *Ventrigyrus sulcatus* (Kanygin, 1965), Sakkyryskaya Svita (lower part), NE Siberia, former USSR.
Last: *Zenkopsis enormis* (Zenkova, 1975), Demidske Beds, Ural Mountains, former USSR.
Intervening: LLN_1–LDF.

F. BUBNOFFIOPSIDAE Schallreuter, 1964
O. (SOU–ONN) Mar.

First: *Bubnoffiopsis bubnoffi* Schallreuter, 1964, middle (Idavere/Johvi) part of Viru 'series', Baltoscandia.
Last: *Lembitites incognitus* (Sidaraviciene, 1975), upper (Nabala) part of Viru 'series', Baltoscandia.
Intervening: LON–ACT.

Superfamily BEYRICHIACEA Matthew, 1886

F. CRASPEDOBOLBINIDAE Martinsson, 1962
S. (RHU)–C. (TOU/VIS) Mar.

First: *Bolbineossia pineaulti* Copeland, 1974, Becscie Formation, Anticosti Island, Canada.
Last: *Copelandella novascotia* (Jones and Kirkby, 1884), Horton Bluff Formation, lower Mississippian, Nova Scotia, Canada (Bless and Jordan, 1971).
Comments: Other Lower Carboniferous, possibly amphitoxotidine (craspedobolbinid) taxa are *Malnina spinosa* Jones, 1989, Bonaparte Basin, Australia and *Armilla sibirica* Bushmina, 1975, Kolymian Basin, former USSR.
Intervening: IDW–D. (u.).

F. TREPOSELLIDAE Henningsmoen, 1954
S. (RHU)–D. (FAM) Mar.

First: *Loutriella jupiterensis* Copeland, 1974, Becscie Formation, Anticosti Island, Canada.
Last: *Parabouckekius martinssoni* Jones, 1985 and *Katatona romei* Jones, 1985, Button Beds, 'Strunian', Bonaparte Basin, Western Australia.
Intervening: IDW–FRS.

F. BEYRICHIIDAE Matthew, 1886
O. (ASH)–C. (IVO) Mar.

First: *Fallaticella schaeferi* Schallreuter, 1984, Öjlemyrflint erratic boulder, ASH, Gotland, Sweden (Schallreuter, 1989). This is the only known pre-Silurian beyrichiacean species.
Last: *Pseudoleperditia tuberculifera* Schneider, 1956 (*sensu* Ivanova *et al.*, 1975; see Jones, 1989, p. 29), Kizel Horizon, upper Tournaisian, south Urals, former USSR.
Comments: Inclusion in this family of a number of possibly (i.e. very diffusely) dimorphic, more *Ochesaarina*-like and possibly paraparchitid-like Russian genera (e.g. see Tschigova, 1977 and Abushik *et al.*, 1990) would extend its stratigraphical range to younger horizons in the Carboniferous; however, based on evidence from literature sources alone, the taxonomic position of such forms remains uncertain.
Intervening: RHU–HAS.

Superfamily NODELLACEA Becker, 1968

F. NODELLIDAE Zaspelova, 1952 D. (GIV–FRS) Mar.

First: *Nodella faceata* Rozhdestvenskaya, 1972 (*sensu* Coen, 1985), Formation de Fromelennes, Ardennes region, Belgium.
Last: *Nodella lefrevei* Becker, 1971, Schistes à *Minatothyris maureri*, Dinant Basin, Belgium.

Superfamily KIRKBYACEA Ulrich and Bassler, 1906

F. AMPHISSITIDAE Knight, 1928 D. (EMS)–P. (ZEC) Mar.

First: *Vitissites comtei* Becker, 1981, Lavid Formation, northern Spain.
Last: *Neoamphissites costatus* Becker and Wang, 1992, Wuchiaping Formation, Szechwan Province, China.
Intervening: EIF–ROT.

F. ARCYZONIDAE Kesling, 1961 D. (SIG)–C. (?ARU) Mar.

First: *Chironiptrum bilinearis* Copeland, 1977, Delorme Formation, District of McKenzie, NW Canada.
Last: *Reticestus* sp. 1 of Becker, 1975, Hochwipfel Formation (Cu IIb/g), Carnic Alps, southern central Europe.
Intervening: EMS–?CHD.

F. KIRKBYIDAE Ulrich and Bassler, 1906 D. (SIG)–Tr. (SCY) Mar.

First: *Villozona? aspera* (Polenova, 1974), Valnevskii Horizon, Novaya Zemlya, former Arctic USSR.
Last: *Carinaknightina carinata* Sohn, 1970, Mianwali Formation, west Pakistan.
Intervening: EMS–ZEC.

F. KELLETTINIDAE Sohn, 1954 C. (IVO/CHD)–P. (KUN/UFI) Mar.

First: *Kindlella* sp. 1 of Becker and Wang, 1992, Bojiwan Formation, Guizhou Province, China.
Last: e.g. *Kellettina vidriensis* Hamilton, 1942, Leonard Formation or Word Formation, Glass Mountains, Texas, USA.
Intervening: ARU–ART.

Superfamily TRIBOLBINACEA Sohn, 1978

F. TRIBOLBINIDAE Sohn, 1978 S. (LDF)–P. (ZEC) Mar.

First: *?Kolmodinia grandis* (Kolmodin, 1869), Eke Beds, Gotland, Sweden.
Last: *Tribolbina doescheri* Sohn, 1978, upper Permian, Idra Island, Greece.
Intervening: C. (l., u.), ROT.

Superfamily DREPANELLACEA Ulrich and Bassler, 1923

Many authors hold that the six 'drepanellacean' families listed below are not palaeocopes; rather, that they constitute a major separate taxonomic group, the Order Binodicopa Schallreuter, 1972, containing two superfamilies, the Drepanellacea (families Drepanellidae, Bolliidae and Quadrijugatoridae) and Aechminacea (families Aechminidae, Spinigeritidae and Richinidae).

F. DREPANELLIDAE Ulrich and Bassler, 1923 O. (TRE–HIR) Mar.

First: *Pilla ?sinensis* (Hou, 1953), Tremadoc Series, Liaotung, China.
Last: *Duplexibollia duplex* (Krause, 1892), upper (Porkuni) part of Harju 'series', Baltoscandia.
Intervening: ARG–RAW.

F. BOLLIIDAE Boucek, 1936 O. (ARG)–P. (TAT) Mar.

First: *Klimphores kuemperi* Schallreuter, 1992, middle part of Ontik 'series', Baltoscandia.
Last: *Neoulrichia pulchra* Kozur, 1983, lower Dzhulfian, Buekk Mountains, Hungary.
Intervening: LLN_1–KAZ.

F. QUADRIJUGATORIDAE Kesling and Hussey, 1953 O. (LLO–HIR) Mar.

First: *Quadrijugator marcoi* Vannier, 1986, upper part of Formation de Postolonnec, Armorican Massif, France.
Last: *Harpabollia harparum* (Troedsson, 1918), upper (Porkuni) part of Harju 'series', Baltoscandia.
Intervening: ?COS–RAW.

F. AECHMINIDAE Boucek, 1936 O. (LLO)–C. (?VRK) Mar.

First: *Aechmina ?ventadorni* Vannier, 1986, Formation de Traveusot, Armorican Massif, France.
Last: *Mammoides mamillata* Bradfield, 1935, Deese Formation?, middle Pennsylvanian, Oklahoma, USA.
Intervening: ?COS–?MEL.

F. SPINIGERITIDAE Schallreuter, 1980 O. (CRD)–S. (?SHE) Mar.

First: *Eocytherella nioni* Vannier, 1986, Formation du Pont-de-Caen, Armorican Massif, France.
Last: *Susus sus* Schallreuter, 1991, 'Ostracodenkalk', Wenlock Series, Lindener Mark, near Giessen, Hesse, Germany.
Intervening: ?PUS–?TEL.

F. RICHINIDAE Scott, 1961 (=Circulinidae Neckaja, 1966) O. (ARG)–D. (?GIV) Mar.

First: *Rivillina schallreuteri* Vannier, 1983, Pissot Formation, Les Tanneres, Armorican Massif, France.
Last: *Richina selenicristata* Stover, 1956, Windom Shale Member, Moscow Formation, Hamilton Group, New York, USA.
Intervening: LLN_1–?EIF.

Order PALAEOCOPIDA?

Superfamily BARYCHILINACEA Ulrich, 1894

F. MONOTIOPLEURIDAE Guber and Jaanusson, 1964 O. (ARG–ASH) Mar.

First: *Primitiella ?brevisulcata* Hessland, 1949, Lower Grey *Orthoceras* Limestone, Siljan District, Sweden.
Last: *Monotiopleura auriculata* Guber and Jaanusson, 1964, Whitewater Formation, Richmond Group, Ohio, USA.
Intervening: LLN_1–?PUS.
Comment: This family has traditionally been assigned to the Kloedenellacea (=Platycopina); in his revision of that superfamily Adamczak (1991) maintains that the monotiopleurids (and the platycopes) are palaeocopes.

F. LOMATOPISTHIIDAE Guber and Jaanusson, 1964 O. (LLN_2–CRD) Mar.

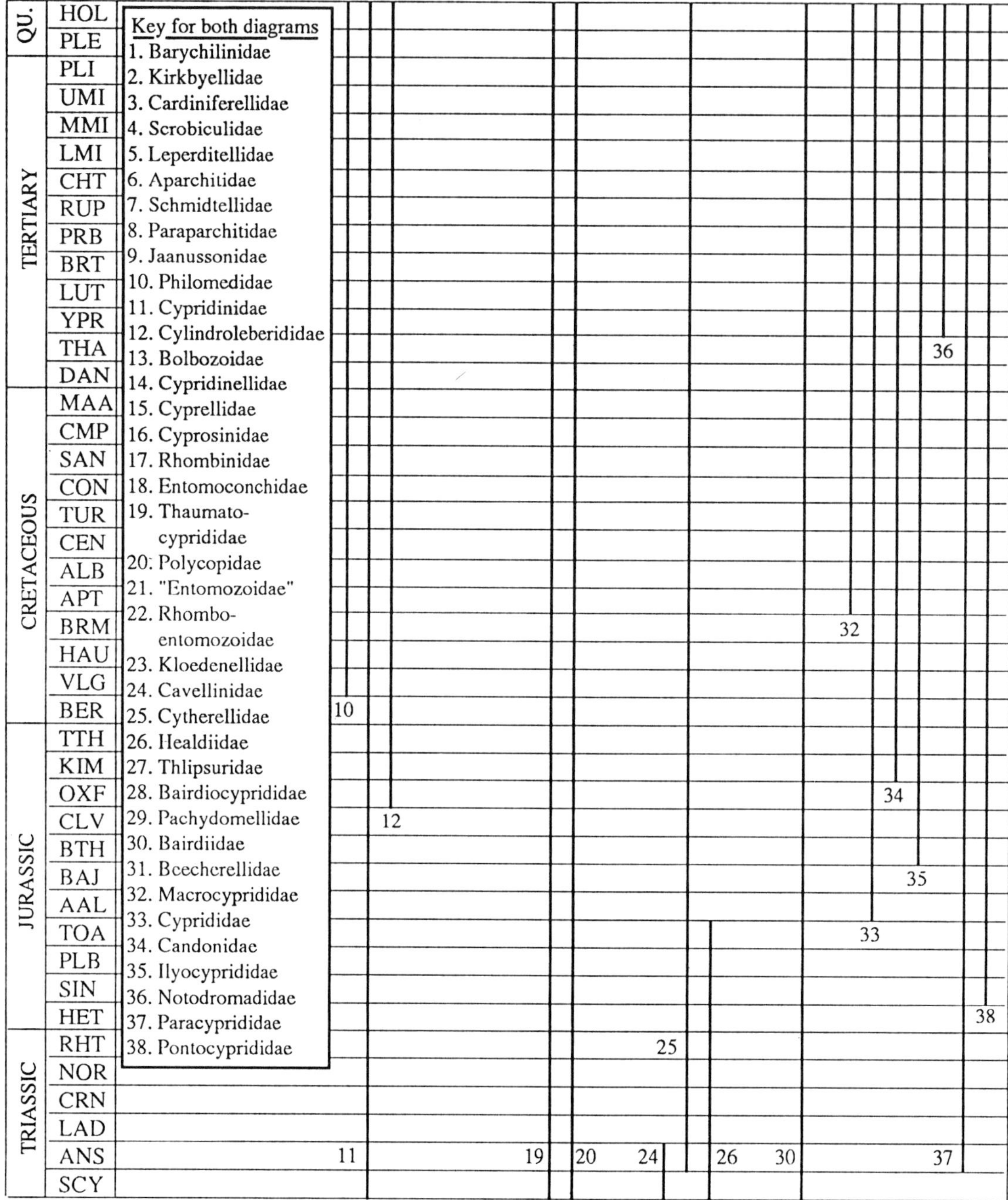

Fig. 19.2

First: *Lomatopisthia mclishi* (Harris, 1957), Mclish Formation, Oklahoma, USA.
Last: *Lomatopisthia auricula* (Harris, 1957), Pooleville Member, Bromide Formation, upper CRD, Oklahoma, USA.
Intervening: LLO_1–lower CRD.
Comment: Adamczak (1991) considers that this family is synonymous with the Monotiopleuridae.

F. BARYCHILINIDAE Ulrich, 1894
S. (FRO)–P. (ART) Mar. (Fig. 19.2)

First: *Neckajatia modesta* (Neckaja, 1958), Svencionys Formation, Lithuania, former USSR.
Last: *Ellipsella obliqua* Coryell and Rogatz, 1932, Arroyo Formation, Texas, USA.
Intervening: IDW–SAK.

Superfamily UNCERTAIN

F. KIRKBYELLIDAE Sohn, 1961
S. (WEN/LUD)–C. (?KSK) Mar.

First: *Psilokirkbyella prominens* Becker and Wang, 1992, Bateaobao Formation, Inner Mongolia, China.
Last: *Kirkbyella (Berdanella)* sp. A of Becker 1978, Escalada Formation, Westphalian C, north Spain.
Intervening: PRD–?VRK.

F. CARDINIFERELLIDAE Sohn, 1953
C. (?PND–ALP) Mar.

First: *Cardiniferella bowsheri* Sohn, 1953, Helms Formation, Chesterian, Texas, USA.
Last: *Cardiniferella ringwoodensis* (Harris and Jobe, 1956), Manning Zone, Chesterian, Oklahoma, USA.

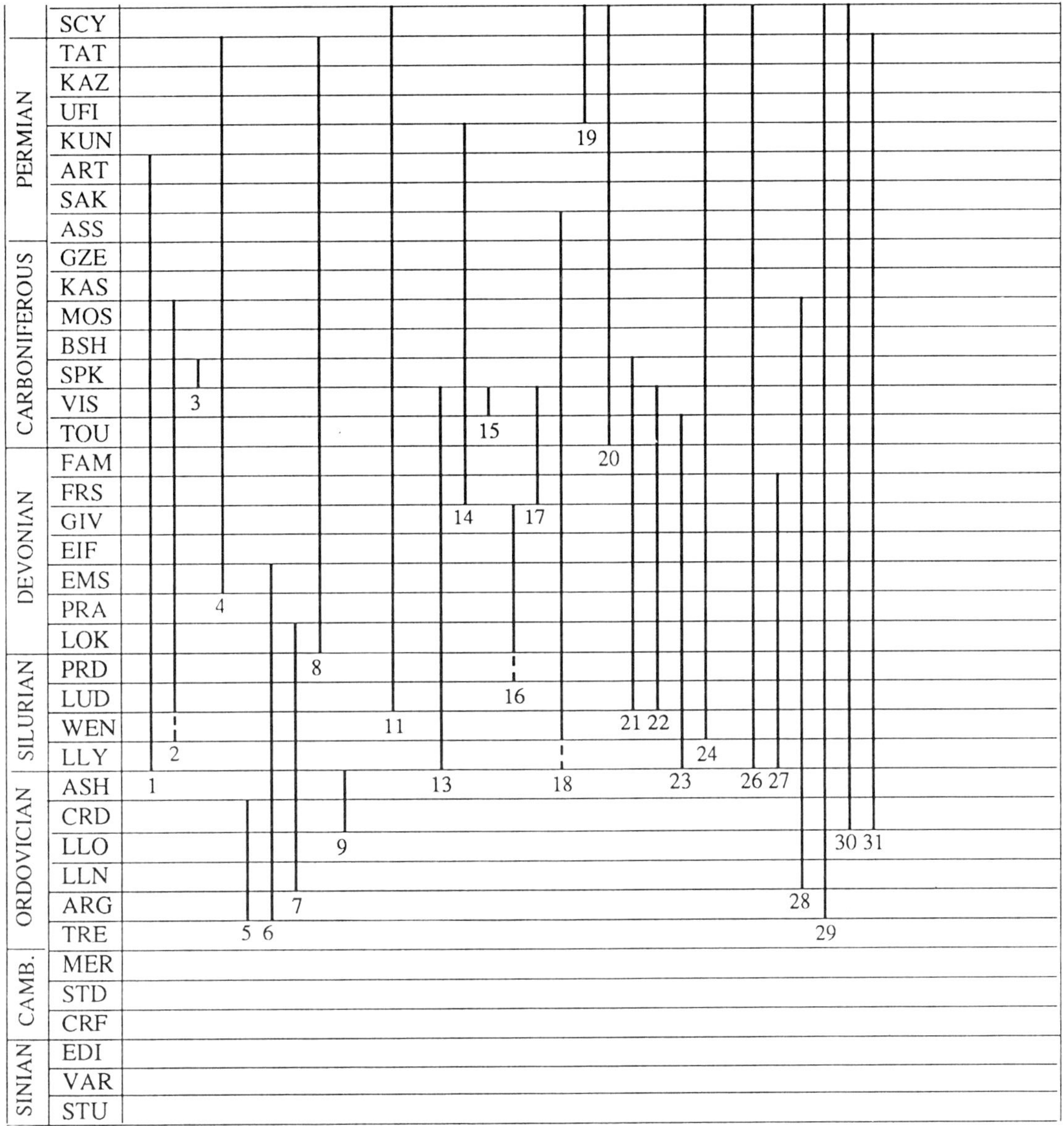

Fig. 19.2

F. SCROBICULIDAE Posner, 1951
D. (EMS)–P. (TAT) Mar.

First: *Doraclatum* sp. 3 of Feist and Groos-Uffenorde 1979, Calcaires à polypiers siliceux, Montagne Noire, France.
Last: *Placidea lutkevichi* (Spizharsky, 1939), Tatarian, Russian Platform, former USSR.
Intervening: EIF–KAZ.

F. LEPERDITELLIDAE Ulrich and Bassler, 1906
O. (ARG–CRD) Mar.

First: *Leperditella valida* Harris, 1957, Joins Formation, Oklahoma, USA.
Last: *Leperditella tumida* (Ulrich, 1892), Pooleville Member, Bromide Formation, upper CRD, Oklahoma, USA.
Intervening: LLN_1–lower CRD.
Comment: This family is herein tentatively referred to the Palaeocopida; its systematic position is problematic.

Order LEIOCOPA Schallreuter, 1973

Superfamily APARCHITACEA Jones, 1901

F. APARCHITIDAE Jones, 1901
O. (ARG)–D. (EMS) Mar.

First: *Longidorsa rectelloides* Schallreuter, 1985, Lower Ordovician (BIIIβ; ?upper ARG), Baltoscandia.
Last: *Brevidorsa devonica* Reynolds, 1978, *Receptaculites* Limestone, Murrumbidgee Group, Taemus, New South Wales, Australia.
Intervening: LLN_1–SIG.

F. SCHMIDTELLIDAE Neckaja, 1966
O. (LLN_1)–D. (GED) Mar.

First: *Punctoschmidtella ?umbopunctata* (Harris, 1957), Oil Creek Formation, Oklahoma, USA.
Last: *Paraschmidtella dorsopunctata* Swartz, 1936, Shriver Formation, Pennsylvania, USA.
Intervening: LLN_2–PRD.

Superfamily PARAPARCHITACEA Scott, 1959

F. PARAPARCHITIDAE Scott, 1959 D.–P. Mar.

The systematics of this group is in an extremely confused

state. The family occurs in the Devonian, Carboniferous (Mississippian and Pennsylvanian) and Permian, but at present it is not possible to give a more exact range.
Intervening: C.

F. JAANUSSONIDAE Schallreuter, 1971
O. (?CRD–HIR) Mar.

First: *Kayina hybosa* Harris, 1957, Mountain Lake Member, Bromide Formation, ?lower CRD, Oklahoma, USA.
Last: *Jaanussonia unicerata* Schallreuter, 1971, upper (Porkuni) part of Harju 'series', Baltoscandia.
Intervening: upper CRD–RAW.

Order MYODOCOPIDA Sars, 1866

Suborder MYODOCOPINA Sars, 1866

F. RUTIDERMATIDAE Brady and Norman, 1896
Mar. **Extant**

F. SARSIELLIDAE Brady and Norman, 1896
Mar. **Extant**

F. PHILOMEDIDAE Müller, 1912
K. (VLG)–Rec. Mar.

First: *Philomedes donzei* Neale, 1976, VLG, Vocontian Trough, Provence, France. **Extant**
Intervening: Very discontinuous, HAU–HOL.

F. CYPRIDINIDAE Baird, 1850
S. (?GOR)–Rec. Mar.

First: 'Cypridinid' gen. *et* sp. nov. A of Siveter *et al.*, 1987, La Lande Murée Formation, Mayenne, Laval area, Brittany, France. **Extant**
Intervening: Discontinous, LDF–HOL.

F. CYLINDROLEBERIDIDAE Müller, 1906
J. (u.)–Rec. Mar.

First: *Cycloleberis* sp. of Dzik, 1978, Upper Jurassic, Volga Region, former USSR. **Extant**
Intervening: Rare, K.–HOL.

F. BOLBOZOIDAE Boucek, 1936
S. (TEL)–C. (TOU/VIS) Mar.

First: *Entomozoe tuberosa* (Jones, 1861), Wether Law Linn Formation, North Esk Group, Pentland Hills, Scotland, UK. This species is currently the type species of the type genus of the Family Entomozoidae (Siveter and Vannier, 1990); an application to the ICZN is being prepared to resolve this taxonomic/nomenclatural problem.
Last: *Sulcuna lepus* Jones and Kirkby, 1874, Dinantian Limestone, Little Island, near Cork, Republic of Ireland.
Intervening: Discontinuous, SHE–D. (u.).
Comments: This family probably has closer affinity with the Myodocopina than the Halocypridina (Siveter *et al.*, 1987).

F. CYPRIDINELLIDAE Sylvester-Bradley, 1961
D. (FRS)–P. (ROT) Mar.

First: *Cypridella oertlii* Becker and Bless, 1987, lower Kellwasser Limestone, Germany.
Last: *Cypridella nasuta* Glebovskaja, 1939, Lower Permian, northern Urals, former USSR.
Intervening: Discontinuous, FAM–C. (u.).
Comment: Some representatives show close similarities to bolbozoids.

F. CYPRELLIDAE Sylvester-Bradley, 1961
C. (?VIS) Mar.

First and Last: Poorly known and rare group; needs revision. All known species appear to be of probable Viséan age; for example, *Cyprella chrysalidea* De Koninck, 1841, Lower Carboniferous (?VIS) Limestone, Visé, Belgium.

F. CYPROSINIDAE Whidborne, 1890
S. (?PRD)–D. (GIV) Mar.

First: Possibly *Cyprosina* sp. of Chapman, 1904, Silurian, Victoria, Australia.
Last: *Cyprosina whidbornei* Jones, 1881, Torquay Limestone, near Torquay, Devon, England, UK.
Intervening: None?
Comment: Very rare. Taxonomic position uncertain.

F. RHOMBINIDAE Sylvester-Bradley, 1951
D. (FRS)–C. (TOU/VIS) Mar.

First: *Palaeophilomedes ?neuvillensis* Casier, 1988, 'Montagne Shales', Belgium.
Last: *Palaeophilomedes bairdiana* (Jones and Kirkby, 1874), Dinantian Limestone, Little Island, near Cork, Republic of Ireland.
Intervening: Very rare?
Comment: Poorly known group; taxonomic position uncertain.

Suborder HALOCYPRIDINA Dana, 1852

F. ENTOMOCONCHIDAE Brady, 1868
S. (?TEL/SHE)–P. (ASS) Mar.

First: *Elpezoe ?borealis* Copeland, 1964, Allen Bay Formation, Canyon Fiord Region, Ellesmere Island, District of Franklin, Canada.
Last: *Elpezoe orbiculata* Kotschetkova and Gusseva, 1972, ASS, Bashkiria, Urals, former USSR.
Intervening: Very discontinuous, ?WHI–NOG.

F. THAUMATOCYPRIDIDAE Müller, 1906
P. (UFI/KAZ)–Rec. Mar.

First: *Thaumatomma piscifrons* Kornicker and Sohn, 1976, Limestone of early late Permian age, Idhra, Greece. **Extant**
Intervening: Very discontinuous, SCY–HOL.
Comment: Carboniferous records dubious.

F. HALOCYPRIDIDAE Dana, 1852 Mar. **Extant**

Comment: All supposed fossil records of halocyprids (such as *Conchoecia*) belong to other myodocopid genera.

Suborder CLADOCOPINA Sars, 1866

F. POLYCOPIDAE SARS, 1866
C. (HAS)–Rec. Mar.

First: *'Discoidella' spaerula* (Gründel, 1961), Gattendorfia 'stage', Thuringia, Germany. **Extant**
Intervening: Discontinuous, IVO–HOL.

F. 'ENTOMOZOIDAE' Pribyl, 1950
S. (GOR)–C. (ARN) Mar.

First: *Richteria migrans* (Barrande, 1872), Kopanina Formation, Bohemia, Czechoslovakia.
Last: *Truyolsina truyolsi* Becker and Bless, 1975, Entomozoen-Schiefer, E2a Chronozone, Cantabrian Mountains, north Spain. Wilkinson and Riley (1990) record 'entomozoids' from the slightly younger (but, again,

Arnsbergian) *Cravenoceratiodes nitidus* Marine band (E2b2 Chronozone), Namurian, from near Littledale, Lancashire, England, UK, but the material is too poorly preserved for further identification.
Intervening: LDF–PND.
Comment: Possibly belong to the Cladocopina or Halocypridina. This family requires a new name (see note, under Family Bolbozoidae, on the type species of *Entomozoe*).

F. RHOMBOENTOMOZOIDAE Gründel, 1962
S. (GOR)–C. (IVO/CHD) Mar.

First: *Rhomboentomozoe rhomboidea* (Barrande, 1872), Kopanina Formation, Bohemia, Czechoslovakia.
Last: *Franklinella mempeli* (Kummerow, 1939), highest TOU or lowest VIS, Germany.
Intervening: Discontinuous, LDF–HAS/IVO.
Comment: Poorly known group occurring in the same facies as 'entomozoids'; taxonomic position uncertain.

Order PODOCOPIDA Müller, 1894

Suborder PLATYCOPINA Sars, 1866

Some authors (e.g. Becker, 1990; Adamczak, 1991) consider that the Suborder Platycopina have (via the monotiopleurids) a closer affinity to the palaeocopes rather than to the Podocopida.

Superfamily KLOEDENELLACEA Ulrich and Bassler, 1908

F. KLOEDENELLIDAE Ulrich and Bassler, 1908
S. (TEL)–C. (HAS) Mar.

First: *Nyhamnella musculimonstrans* Adamczak, 1966, Visby Beds, Gotland, Sweden.
Last: *Dizygopleura mehli* Morey, 1935, Bushberg Formation, Kinderhookian 'series', lower Mississippian, Missouri, USA.
Intervening: SHE–FAM.

F. CAVELLINIDAE Egorov, 1950
S. (SHE)–Tr. (ANS) Mar.

First: *Gotlandella cornuta* (Krause, 1891), Jaani Regional 'stage', Estonia, former USSR.
Last: *Reubenella avnimelechi* Sohn, 1968, Geranim Formation, Makhtesh Ramon, Israel.
Intervening: WHI–SPA.
Comment: Adamczak (1991) regards this taxon as a subfamily of the Cytherellidae.

F. CYTHERELLIDAE Sars, 1866
Tr. (ANS)–Rec. Mar.

First: *Cytherella jenensis* Kozur, 1968, lower Muschelkalk, eastern Germany. **Extant**
Intervening: LAD–HOL.

Suborder METACOPINA Sylvester-Bradley, 1961

Superfamily THLIPSURACEA Ulrich, 1894

F. HEALDIIDAE Harlton, 1933
S. (FRO)–J. (TOA) Mar.

First: *Cyrtocypris inornata* Copeland, 1974, Jupiter Formation, Anticosti Island, Quebec, Canada.
Last: *Ogmoconcha convexa* Boomer, 1991, Mochras Borehole, near Harlech, North Wales, UK.
Intervening: TEL–PLB.

F. THLIPSURIDAE Ulrich, 1894
S. (RHU)–D. (FRS) Mar.

First: *Conbathella inornata* Copeland, 1974, Becscie Formation, Anticosti Island, Quebec, Canada.
Last: Possibly *Polyzygia neodevonica* (Matern, 1929) and other FRS thlipsurids, Dinant Basin, Belgium (Becker, 1971).
Intervening: IDW–GIV.
Comment: No undoubted Ordovician or Carboniferous members of this family are known.

Suborder PODOCOPINA Sars, 1866

Superfamily BAIRDIOCYPRIDACEA Shaver, 1961

F. BAIRDIOCYPRIDIDAE Shaver, 1961
O. (LLN/LLO)–C. (?KSK) Mar.

First: *Arcuaria sineclivula* Neckaja, 1958, Middle Ordovician, north-western part of Russian Platform.
Last: *Praepilatina homosibirica* Becker, 1991, Escalada Formation, northern Spain.
Intervening: O. (u.)/S.–?VRK.

F. PACHYDOMELLIDAE Berdan and Sohn, 1961
O. (ARG)–Tr. (SCY) Mar.

First: *Hesslandites ellipsiformis* (Hessland, 1949), (and other '*Bythocypris*' species), Lower Grey *Orthoceras* Limestone, Siljan District, Sweden.
Last: *Microcheilinella* sp. of Sohn, 1970, Mianwali Formation, Pakistan.
Intervening: LLN_1–P. (u.).

Superfamily BAIRDIACEA Sars, 1888

F. BAIRDIIDAE Sars, 1888 O. (CRD)–Rec. Mar.

First: *Bairdiacypris incurvata* Kraft, 1962, Edinburg Formation, lower/middle CRD, Shenandoah Valley, Virginia, USA. **Extant**
Intervening: O. (upper CRD)–HOL.

F. BEECHERELLIDAE Ulrich, 1894
O. (CRD)–P. (ZEC) Mar.

First: e.g. *Platyrhomboides virginiensis* Kraft, 1962, Edinburg Formation, lower/middle CRD, Shenandoah Valley, Virginia, USA.
Last: *Acanthoscapha blessi* Kozur, 1989, upper Abadehian, Lower Permian, Buekk Mountains, Hungary.
Intervening: O. (upper CRD)–P. (ROT).

Superfamily CYPRIDACEA Baird, 1845

F. MACROCYPRIDIDAE Müller, 1912
K. (APT)–Rec. Mar.

First: *Macrosarisa exquisita* (Kaye, 1964), APT, UK. **Extant**
Intervening: ALB–HOL.
Comment: There are certainly earlier occurrences in the Jurassic but of too poor preservation to be certain of genus. Palaeozoic records are not valid.

F. CYPRIDIDAE Baird, 1845 J. (m.)–Rec. FW.

First: *Djungarica ?yunnanensis* Ye *et al.*, 1977, Upper Member of Hepingxiang Formation, Middle Jurassic, Jiangju of Changxin, Yulong County, Yunnan Province, China. **Extant**
Intervening: J. (u.)–HOL.

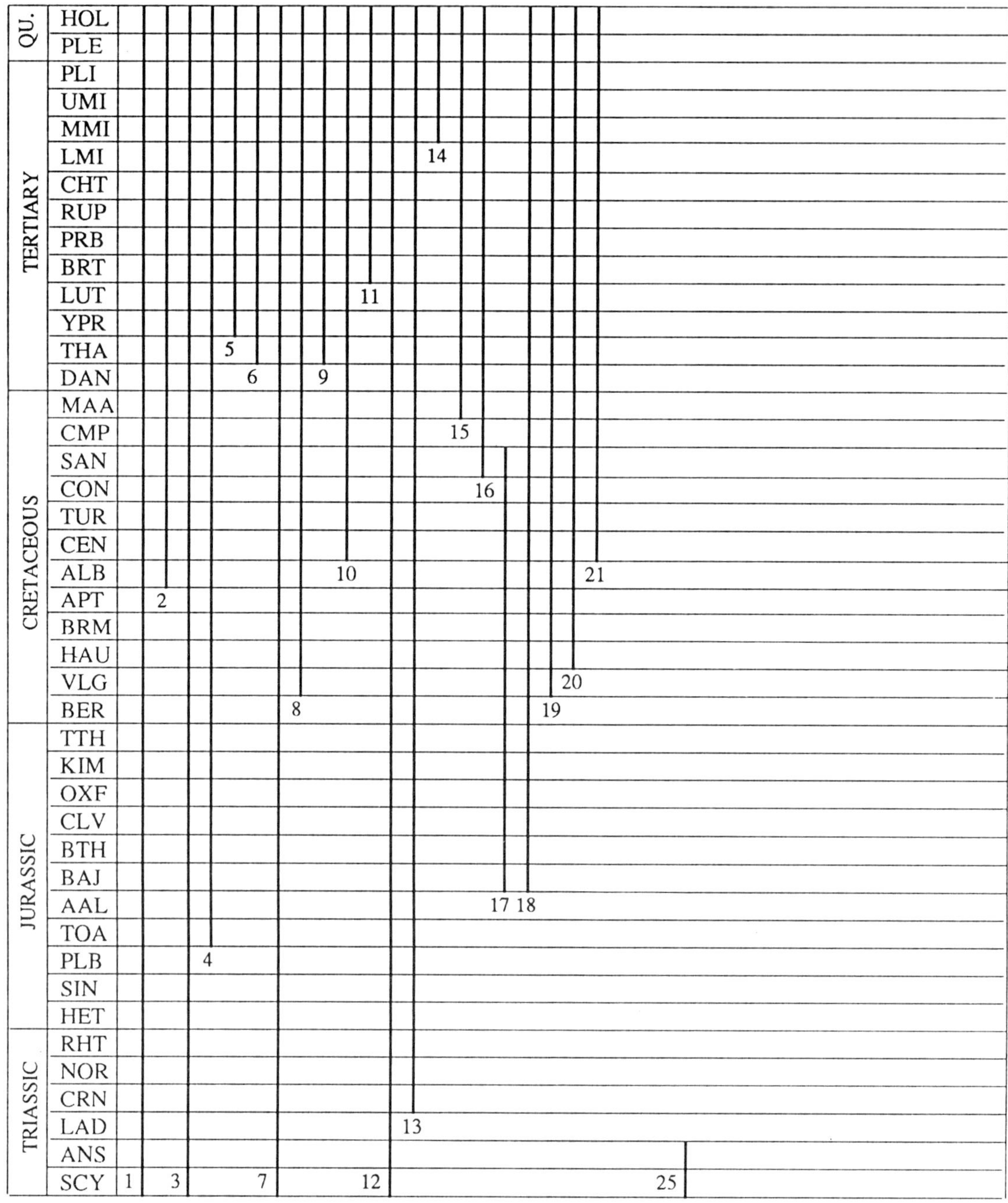

Fig. 19.3

F. CANDONIDAE Kaufmann, 1990
J. (KIM)–Rec. FW.

First: *Cetacella inermis* Martin, 1958, borehole, Thoren, north of Hannover, Germany. **Extant**
Intervening: TTH–HOL.

F. ILYOCYPRIDIDAE Kaufmann, 1990
J. (BTH)–Rec. FW.

First: *'Cyprideis'* sp. of Oertli, 1957, upper BTH, Route Nationale 42, 14.5 km east of Boulogne, France. **Extant**
Intervening: CLV–HOL.

F. NOTODROMADIDAE Kaufmann, 1990
T. (Eoc.)–Rec. FW.

First: *Cyprois ephraimensis* Swain, 1964, Colton Formation, Eocene, Utah, USA. **Extant**
Intervening: Discontinuous to HOL.

F. PARACYPRIDIDAE Sars, 1923
Tr. (ANS)–Rec. Mar.

First: *Triassocypris pusilla* (Kozur, 1968), Germany. **Extant**
Intervening: LAD–HOL.

F. PONTOCYPRIDIDAE Müller, 1894
J. (SIN)–Rec. Mar.

First: *Liasina lanceolata* (Apostolecsu, 1959), upper *raricostatum* Zone, uppermost SIN (*Lophodentina striata* ostracod zone of Boomer, 1991), Mochras Borehole, near Harlech, North Wales, UK. **Extant**

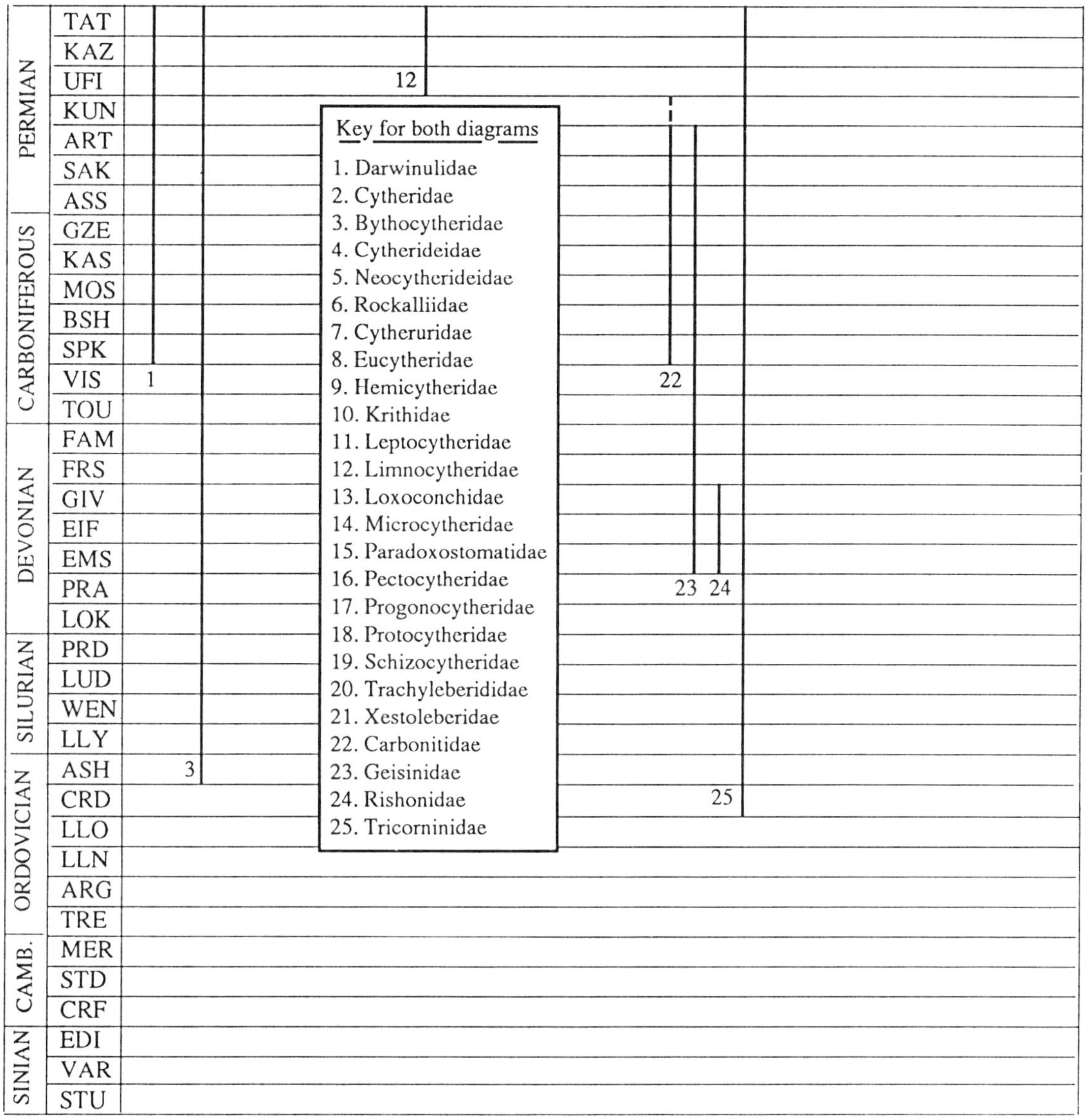

PERMIAN	TAT							
	KAZ							
	UFI			12				
	KUN							
	ART							
	SAK							
	ASS							
CARBONIFEROUS	GZE							
	KAS							
	MOS							
	BSH							
	SPK							
	VIS	1			22			
	TOU							
DEVONIAN	FAM							
	FRS							
	GIV							
	EIF							
	EMS							
	PRA					23	24	
	LOK							
SILURIAN	PRD							
	LUD							
	WEN							
	LLY							
ORDOVICIAN	ASH		3					
	CRD							25
	LLO							
	LLN							
	ARG							
	TRE							
CAMB.	MER							
	STD							
	CRF							
SINIAN	EDI							
	VAR							
	STU							

Key for both diagrams

1. Darwinulidae
2. Cytheridae
3. Bythocytheridae
4. Cytherideidae
5. Neocytherideidae
6. Rockalliidae
7. Cytheruridae
8. Eucytheridae
9. Hemicytheridae
10. Krithidae
11. Leptocytheridae
12. Limnocytheridae
13. Loxoconchidae
14. Microcytheridae
15. Paradoxostomatidae
16. Pectocytheridae
17. Progonocytheridae
18. Protocytheridae
19. Schizocytheridae
20. Trachyleberididae
21. Xestoleberidae
22. Carbonitidae
23. Geisinidae
24. Rishonidae
25. Tricorninidae

Fig. 19.3

Intervening: PLB–HOL.

Superfamily DARWINULACEA Brady and Norman, 1889

F. DARWINULIDAE Brady and Norman, 1889
C. (SPK)–Rec. FW/Brackish (see Fig. 19.3)

First: *Darwinula* sp. of Sohn, 1985, Bluestone Formation, West Virginia, USA. **Extant**
Intervening: C. (u.)–HOL.

Superfamily CYTHERACEA Baird, 1850

F. CYTHERIDAE Baird, 1850 K. (ALB)–Rec. Mar.

First: *Saida nettgauensis* Gründel, 1966, lower ALB, Nettgau Borehole 6/59, Saxony, Germany. **Extant**
Intervening: Discontinuous to HOL.

F. BYTHOCYTHERIDAE Sars, 1926
O. (HIR)–Rec. Mar.

First: *Sylthere vonhachi* Schallreuter, 1978, top (Porkuni) part of the Harju 'series', Baltoscandia. **Extant**
Intervening: RHU–HOL.

F. CYTHERIDEIDAE Sars, 1925
J. (TOA)–Rec. Mar./Brackish/FW

First: *Praeschuleridea pseudokinkellinella* Bate and Coleman, 1975, middle TOA, Empingham, Leicestershire, England, UK. **Extant**
Intervening: AAL–HOL.
Comment: Although such Carboniferous genera as *Basslerella* Kellett, 1935 are considered by some to belong to this family, this is not accepted here. Rather, they are regarded as Cytheracea *incertae sedis*. Family includes Schulerideidae Mandelstam, and Cuneocytheridae Mandelstam.

F. NEOCYTHERIDEIDAE Puri, 1957
T. (Eoc.)–Rec. Mar./Brackish

First: *Cushmanidea serangodes* Krutak, 1961, Yazoo Clay, Alabama, USA. **Extant**
Intervening: Oli.–HOL.

F. ROCKALLIIDAE Whatley *et al.*, 1982 T. (THA)–Rec. Mar.

First: *Arcacythere eocenica* (Whatley *et al.*, 1982), Waipawan Stage, DSDP Site 207, Lord Howe Rise, off New Zealand. **Extant**
Intervening: YPR–HOL.

F. CYTHERURIDAE Müller, 1894 Tr. (SCY)–Rec. Mar./Brackish

This family is in the process of being revised, and a number of Triassic cytheracean families are being accommodated within it. At present, it is not possible to give a first species for the family, but there are a number of SCY candidates. **Extant**
Intervening: ANS–HOL.

F. ENTOCYTHERIDAE Hoff, 1942 FW/Commensal **Extant**

North American, subterranean and commensal on other Crustacea. Not known fossil.

F. EUCYTHERIDAE Puri, 1954 K. (VLG)–Rec. Mar.

First: *Stravia crossata* Neale, Bed D2E, Speeton Clay, East Yorkshire, England, UK. **Extant**
Intervening: HAU–HOL.

F. HEMICYTHERIDAE Puri, 1953 T. (THA)–Rec. Mar./Brackish

First: *Leguminocythereis* sp. cf. *L. arachidnoides* Tambareau, 1972, Esperaza, France. **Extant**
Intervening: YPR–HOL.

F. KLIELLIDAE Schäfer, 1954 FW/Subterranean **Extant**

F. KRITHIDAE Mandelstam, 1958 K. (CEN)–Rec. Mar.

First: *Krithe* sp. 1 of Colin, 1973, upper CEN, Fournet, Dordogne, France. **Extant**
Intervening: TUR–HOL.

F. LEPTOCYTHERIDAE Hanai, 1957 T. (BRT)–Rec. Mar./Brackish

First: *Callistocythere kaiata* (Hornibrook, 1952), Kaiatan and Runungan stages, Jackson's Paddock, New Zealand. **Extant**
Intervening: PRB–HOL.

F. LIMNOCYTHERIDAE Klie, 1938 P. (ZEC)–Rec. FW/Brackish

First: ?*Permiana oblonga* Schneider, 1947, Volga Region, former USSR. **Extant**
Intervening: GRI–HOL.
Comment: In the *Treatise* revision this family will probably be expanded to include a number of families described from brackish and freshwater environments in the late Permian and Triassic.

F. LOXOCONCHIDAE Sars, 1925 Tr. (CRN)–Rec. Mar./Brackish/FW

First: *Gemmanella gracilis* (Bütler and Gründel, 1963), lower Keuper, Thuringer Becken, Germany. **Extant**
Intervening: CLV–HOL.

F. MICROCYTHERIDAE Klie, 1938 T. (LAN_1)–Rec. Mar.

First: *Microcythere moresiana* Stephenson, 1935, Middle Miocene, Louisiana, USA. **Extant**
Intervening: LAN_2–HOL.

F. PARADOXOSTOMATIDAE Brady and Norman, 1889 K. (MAA)–Rec. Mar./Brackish

First: *Paracytherois ?praegracilis* Herrig, 1964, upper MAA, eastern Germany. **Extant**
Intervening: DAN–HOL.

F. PECTOCYTHERIDAE Hanai, 1957 K. (SAN)–Rec. Mar.

First: *Praemunseyella ornata* Bate, 1972, Toolonga Calcilutite, Carnarvon Basin, Western Australia. **Extant**
Intervening: CMP–HOL.

F. PROGONOCYTHERIDAE Sylvester-Bradley, 1948 J. (BAJ)–K. (SAN) Mar.

First: *Acanthocythere (Protoacanthocythere) faveolata* Bate, 1963, Kirton Shale, Kirton Lindsey, Lincolnshire, England, UK.
Last: *Majungaella annula* Bate, 1972, Lower Toolonga Calcilutite, Carnarvon Basin, Western Australia.
Intervening: BTH–APT.

F. PROTOCYTHERIDAE Ljubimova, 1956 J. (BAJ)–Rec. Mar.

First: *Pleurocythere striata* Ljubimova, 1956, upper BAJ, Cherkasskoye, Donetsk Region, Ukraine, former USSR. **Extant**
Intervening: BTH–MAA, AQT–HOL.
Comment: The family was thought to have become extinct during the Cretaceous, but it is now known that *Abyssocythereis sulcatoperforata*, from Tertiary–Recent deep oceans, is a protocytherid.

F. PSAMMOCYTHERIDAE Klie, 1938 **Extant** Mar./Interstitial

F. SCHIZOCYTHERIDAE Howe, 1961 K. (VLG)–Rec. Mar.

First: *Sondagella valanginiana* Valicenti and Stephens, 1984, Sundays River Formation, Coega Brickworks, Algoa Basin, South Africa. **Extant**
Intervening: HAU–HOL.

F. TERRESTRICYTHERIDAE Schornikov, 1980 Terr. **Extant**

Not known fossil. From saline soils and the littoral around Vladivostok, former USSR. A single specimen from Bristol Channel littoral, England, UK.

F. TRACHYLEBERIDIDAE Sylvester-Bradley, 1948 K. (HAU)–Rec. Mar.

First: *Cythereis senckenbergi* Triebel, 1940, Eschershausen, Hils, north-western Germany. **Extant**
Intervening: BRM–HOL.

F. XESTOLEBERIDAE Sars, 1928 K. (CEN)–Rec. Mar./Brackish

First: *Xestoleberis burnetti* Weaver, 1982, lower CEN, Glauconitic Marl, southern England, UK. **Extant**

Intervening: TUR–HOL.

Order PODOCOPIDA

Suborder and Superfamily UNCERTAIN

F. CARBONITIDAE Swain, 1976 C. (SPK)–P. (?SAK/KUN) Non-mar.

First: *Carbonita* n. sp. of Sohn, 1985, Bluestone Formation, West Virginia, USA.
Last: *Gutschickia convexa* Tasch, 1963, Leonardian, Kansas, USA.
Intervening: KIN–?ASS.

F. GEISINIDAE Sohn, 1961 D. (EMS)–P. (ART) Mar.

First: *Neonyhamnella suboblonga* Wang, 1983, Sipai Formation, Guangxi, China.
Last: *Knoxiella simplex* Xie, 1983, Maokou Formation, Szechwan, China.
Intervening: Discontinuous, EIF–SAK.

F. RISHONIDAE Sohn, 1960 D. (EMS–GIV) Mar.

First: *Rishona sarcinula* Reynolds, 1978, *Receptaculites* Limestone, New South Wales, Australia.
Last: *Rishona* sp. of Braun, 1966, Hume Formation, Norman Wells area, western Canada.
Intervening: EIF.

F. TRICORNINIDAE Blumenstengel, 1965 O. (ONN)–Tr. (ANS) Mar.

First: *Margoplanitia brevispina* Knüpfer, 1968, Gräsentaler 'series', Thuringia, Germany.
Last: *Nagyella longispina* Kozur, 1970, ANS, Hungary.
Intervening: PUS–SPA.
Comments: Some researchers (e.g. Schallreuter) regard this family as Metacopina, other authors (e.g. Becker) assign it to the Cytheracea (Podocopina) or even (e.g. Becker and Bless, 1990) to the Drepanellacea (Palaeocopida).

REFERENCES

All relevant references **except** those listed below are to be found in Kempf's *Index and Bibliography of Non-marine* (1980) *and Marine Ostracoda* (1984–1988).

Abushik, A. F. (1979) Order Leperditicopida, in *Taxonomy, Biostratigraphy and Distribution of Ostracodes* (ed. N. Krstic), Proceedings of the 7th International Symposium on Ostracoda, Serbian Geological Society, Belgrade, pp. 29–34.

Abushik, A. F., Gusseva, E. A., Ivanova, V. A. *et al.* (1990) *Practical Manual on the microfauna of the USSR*, Vol. 4, *Paleozoic Ostracoda*. All-Union Geological Research Institute, Mineralogy Geology USSR, Nedra, Leningrad, 356 pp.

Adamczak, F. J. (1991) Kloedenellids: morphology and relation to non-myodocopide ostracodes. *Journal of Paleontology*, **65**, 255–67.

Baird, W. (1845) Arrangement of the British Entomostraca, with a list of species, particularly noticing those which have as yet been discovered within the bounds of the club. *Transactions of the Berwickshire Naturalists' Club*, **2**(13), 145–58.

Becker, G. (1971) Paleoecology of Middle Devonian ostracods from the Eifel region, Germany, in *Paléoécologie des Ostracodes. Bulletin du Centre des Recherches Pau-SNPA* (ed. H. J. Oetli), **5** (Supplement), 801–16.

Becker, G. (1990) On the morphology and taxonomy of Palaeozoic Ostracoda. With critical remarks on the significance of carapace structures. *Senckenbergiana Lethaea*, **70**, 147–69.

Becker, G. (1991) Shallow water ostracods from the upper Westphalian of Asturias (Cantabrian Mountains, N. Spain); Part 2, Podocopida. *Senckenbergiana Lethaea*, **71**(5), 383–425.

Becker, G. and Bless, M. J. M. (1987) Cypridinellidae (Ostracoda) from the Upper Devonian of Hassia (lower Kellwasser Limestone; Lahn–Dill Region and eastern Sauerland, Rechserheinisches Schiefergebirge). *Geologische Jahrbuch, Hessen*, **115**, 29–56.

Becker, G. and Bless, M. J. M. (1990) Biotope indicative features in Palaeozoic ostracods: a global phenomenon, in *Ostracoda and Global Events* (eds R. Whatley and C. Maybury), British Micropalaeontological Society Series, Chapman and Hall, London, pp. 421–36.

Becker, G. and Wang, S. (1992) Kirkbyacea and Bairdiacea (Ostracoda) from the Palaeozoic of China. *Palaeontographica, Abteilung A*, in press.

Bless, J. M. (1987) Lower Permian ostracods from Timor (Indonesia). *Proceedings Konincklijke Nederlandse Akademie van Wetenschappen, ser. B*, **90**, 1–30.

Bless, M. J. M. and Jordon, H. (1971) Classification of palaeocopid ostracodes belonging to the families Ctenoloculinidae, Hollinidae and Hollinellidae, in *Paléoécologie des Ostracodes. Bulletin du Centre des Recherches Pau-SNPA* (ed. H. J. Oertli), **5** (Supplement), 869–90.

Boomer, I. (1991) Lower Jurassic ostracod biozonation of the Mochras Borehole. *Journal of Micropalaeontology*, **9**, 205–18.

Braun, W. (1966) Stratigraphy and Microfauna of Middle and Upper Devonian formations, Norman Wells area, north-west Territories, Canada. *Neues Jahrbuch für Geologie und Paläontologie, Abhandlungen*, **125**, 247–64.

Bütler, G. and Gründel, J. (1963) Die Ostracoden des Unteren Keupers im Bereich des Thüringer Beckens. *Freiberger Forschungshefte, Serie C*, **164**, 33–92.

Casier, J.-G. (1988) Présence de Cypridinacea (Ostracodes) dans la partie supérieure du Frasnien du Bassin de Dinant. *Bullétin de l'Institut Royale des Sciences Naturelles de Belgique (Sciences de la Terre)*, **58**, 89–94.

Dzik, J. (1978) A myodocopid ostracod with preserved appendages from the upper Jurassic of the Volga River region, USSR. *Neues Jahrbuch für Geologie und Paläontologie, Monatshefte*, **1978**, 393–9.

Frederickson, E. A. (1946) A Cambrian ostracode from Oklahoma. *Journal Paleontology*, **20**, 578.

Gründel, J. (1977) Bemerkungen zur Taxonomie und Phylogenie der Primitiopsacea Swartz, 1936. *Zeitschrift für Geologischen Wissenschaften*, **5**, 1223–33.

Harris, R. W. (1957) Ostracode from the Simpson Group. *Bull. Oklahoma Geol. Surv.*, **75**, 333 pp.

Harris, R. W. (1960) Review of the systematics and Recent research of primitiopsia Ostracoda. *Oklahoma Geol. Notes*, **20**, 176–82.

Hartmann, G. and Puri, H. S. (1974) Summary of neontological and palaeontological classification of Ostracoda. *Mitteilungen der Hamburgsche Zoologische Museum und Institut*, **70**, 7–73.

Hinz, I. (1987) The Lower Cambrian microfauna of Comley and Rushton, Shropshire, England. *Palaeontographica, Abteilung A*, **198**, 41–100.

Hoff, C. C. (1942) The ostracods of Illinois. Their biology and taxonomy. *Illinois Biological Monographs*, **19**, 5–196.

Huo, S. and Shu, D. (1985) *Cambrian Bradoriida of south China*. Northwest University Press, Xian, 251 pp.

Ivanova, N. O. *et al.* (1975) Ostracoda, in *Palaeontological Atlas of Carboniferous Rocks of the Urals* (ed. D. L. Stepanov), Trudy VINIIGI, **383**, pp. 131–45.

Jones, P. J. (1989) Lower Carboniferous Ostracoda (Beyrichicopida and Kirkbyocopa) from the Bonoparte Basin, northwestern Australia. *Bulletin of the Bureau of Mineral Research, Geology and Geophysics*, **228**, 1–97.

Kauffmann, A. (1900) Cypriden und Darwinuliden der Schweiz. *Revue Suisse de Zoologie*, **8**, 209–423.

Kempf, E. K. (1980) Index and Bibliography of Non-marine Ostracoda: Parts 1 (Index A; 188 pp), 2 (Index B; 180 pp), 3 (Index C; 204 pp), 4 (Bibliography A; 186 pp), *Sonderverlagerungen des Geologisches Institut Universitäts Köln*, **35**, **36**, **37**, **38**.

Kempf, E. K. (1984–1988) Index and Bibliography of Marine Ostracoda: Parts 1 (Index A: 762 pp, 1986), 2 (Index B; 712 pp, 1986), 3 (Index C; 774 pp, 1987), 4 (Bibliography A; 454 pp, 1988), *Sonderverlagerungen des Geologisches Institut Universitäts Köln*, **50**, **51**, **52**, **53**.

Klie, W. (1938) Krebstiere oder Crustacea III: Ostracoda Muschelkrebse. *Die Tierwelt Deutschlands*, **34**, 1–230.

Kozur, H. (1989) New ostracode species from the upper Middle Carboniferous (upper Moscovian) and middle and upper Permian of the Buekk Mountains, north Hungary. *Geologische–paläontologische Mitteilungen Innsbruck, spec. vol.* **2** (for 1985), 1–145.

Martin, G. P. C. (1958) *Cetacella*, eine neue Ostracoden-Gattung aus dem Kimmeridge Nordwestdeutschlands. *Paläontologische Zeitschrift*, **32**, 190–6.

Moore, R. C. (ed) (1961) *Treatise on Invertebrate Paleontology, Part Q, Arthropoda 3, Ostracoda*. Geological Society of America and University of Kansas Press, Boulder, Colorado, and Lawrence, Kansas, 442 pp.

Nikolaeva, I. A., Pavlovskaya, V. I., Kovalenko, A. L. *et al.* (1989) *Practical Manual on the microfauna of the USSR*, Vol. 3, *Cenozoic Ostracoda*. All-Union Geological Research Institute, Mineralogy Geology, USSR, Nedra, Leningrad, 235 pp.

Oertli, H. J. (1957) Ostracoden als Salzgehalts-Indikatoren im obern Bathonien des Boulonnais. *Eclogae Geologicae Helvetiae*, **50**, 279–83.

Palmer, A. R. (1954) The faunas of the Riley Formation in central Texas. *Journal of Paleontology*, **28**, 709–86.

Poulsen, C. (1937) On the Lower Ordovician faunas of east Greenland. *Meddelelsar om Grünland*, **119**, 72 pp.

Puri, H. S. (1957) Notes on the subfamily Cytherideinae Puri, 1952. Postscript notes on the ostracod subfamily Brachycytherinae. *Journal of the Washington Academy of Sciences*, **47**, 305–8.

Schäfer, H. W. (1954) Grundwasser Ostracoden aus Griechenland. *Archiv für Hydrobiologie*, **40** (= Festband August Thienemann) (4), 847–66.

Schallreuter, R. E. L. (1989) The oldest known beyrichian ostracode. *Geschiebekunde Aktuell*, **5**, 17–20.

Schallreuter, R. E. L. (1991) Microfossils from the Ostracodenkalk (Silurian) of the Lindener Mark near Giessen (Hesse). *Neues Jahrbuch für Geologie und Paläontologie, Monatshefte*, **1991**, 105–18.

Schallreuter, R. E. L. (1993) Ostrakoden aus ordovizischen Geschieben Westalens 2. *Geologie und Paläontologie Westfalens*, in press.

Schneider, G. F. and Mandelstam, M. T. (1947) Otryad Ostracoda – Rakovincha tye raki, in *Atlas of the Guide Forms of the Fossil Faunas of the USSR*, Vol. 7, (*Triasovaya Systema*). Leningrad and Moscow, pp. 179–85.

Schornikov, E. I. (1980) Ostracoda in terrestrial biotopes. *Zoologiskaya Zhurnal*, **59**, 1306–19.

Shu, D. (1990) *Cambrian and Lower Ordovician Bradoriida from Zhejiang, Hunan and Shaanxi Provinces*. Northwest University Press, Xian, 95 pp.

Siveter, D. J. and Vannier, J. M. C. (1990) The Silurian myodocope ostracode *Entomozoe* from the Pentland hills, Scotland: its taxonomic, ecological and phylogenetic significance and the affinity of the bolbozoid myodocopes. *Transactions of the Royal Society of Edinburgh: Earth Sciences*, **81**, 45–67.

Siveter, D. J., Vannier, J. M. C. and Palmer, D. (1987) Silurian myodocopid ostracodes: their depositional environments and the origin of their shell microstructures. *Palaeontology*, **30**, 783–813.

Stubblefield, C. J. (1933) Notes on the fossils, in On the Occurrence of Tremadoc Shales in the Tortworth Inlier, Gloucestershire (ed. S. Smith), *Quarterly Journal of the Geological Society of London*, **89**, pp. 357–87.

Swain, F. M. (1964) Early Tertiary freshwater Ostracoda from Colorado, Nevada and Utah and their stratigraphic distribution. *Journal of Paleontology*, **38**, 265–80.

Swain, F. M. (1976) Evolutionary development of cypridopsid Ostracoda. *Abhandlungen Verhand der Natur Vereins Hamburg (NF)* **18/19** *(suppl.)*, pp. 103–18.

Tasch, P. (1963) Paleolimnology, Part 3, Marion and Dickinson counties, Kansas, with additional sections in Harvey and Sedgwick Counties: stratigraphy and biota. *Journal of Paleontology*, **37**, 1233–251.

Tschigova, Z. A. (1977) Stratigraphy and Correlation of Oil- and Gas-bearing Deposits in the Devonian and Carboniferous from the European part of the USSR and Neighbouring Regions, Nedra, Moscow, 263 pp.

Vannier, J. (1986) Ostracodes Binodicopa de L'Ordovicien (Arenig–Caradoc) Ibero-Armoricain. *Palaeontographica, Abteilung A*, **193**, 77–143.

Wilkinson, I. and Riley, N. M. (1990) Namurian entomozoacean Ostracoda, in *Ostracoda and Global Events* (eds R. Whatley and C. Maybury), British Micropalaeontological Society Series, Chapman and Hall, London, pp. 161–72.

Williams, M. (1990) Ostracoda (Arthropoda) of the Middle Ordovician Simpson Group, Oklahoma, U.S.A. Unpublished PhD Thesis, University of Leicester, 204 pp.

Ye, C.-H., Gou, Y.-X., Hou, Y.-T. *et al.* (1977) Mesozoic and Cainozoic ostracod fauna from Yunnan, in *Mesozoic Fossils from Yunnan, China*. Science Press, Beijing, 153–309.

Zenkova, G. G. (1975) Some Silurian ostracodes from the eastern slope of the Urals, in *Materials (Data) on the Paleontology of the Urals*, Academy of Sciences of the USSR, Urals Fil., Inst. Geol., Geochem. Min. Geol. RSFSR, Ural. Territ. Geol. Upravdene, Sverdlovsk, pp. 86–95.

20

ARTHROPODA (EUTHYCARCINOIDEA AND MYRIAPODA)

A. J. Ross and D. E. G. Briggs

Superclass EUTHYCARCINOIDEA Gall and Grauvogel, 1964 (see Fig. 20.1)

The euthycarcinoids are generally considered to be a primitive group of uniramians (Schram and Rolfe, 1982). They were previously assigned to the crustaceans (Gall and Grauvogel, 1964), merostomoids (Schram, 1971) and trilobitoids (Starobogatov, 1988). The classification followed here is that of Schram and Rolfe (1982). An undescribed euthycarcinoid has been recorded from the Upper Silurian of Western Australia (K. J. McNamara and N. H. Trewin, in preparation).

F. EUTHYCARCINIDAE Handlirsch, 1914 C. (WES D)–Tr. (ANS/LAD) ?Mar.–FW

First: *Kottixerxes gloriosus* Schram, 1971, and *Smithixerxes juliarum* Schram and Rolfe, 1982, Carbondale Formation, Mazon Creek, Illinois, USA.
Last: *Synaustrus brookvalensis* Riek, 1964, Hawkesbury Sandstone, New South Wales, Australia.

F. SOTTYXERXIDAE Schram and Rolfe, 1982 C. (WES D–STE B) ?Mar./FW

First: *Sottyxerxes pieckoae* Schram and Rolfe, 1982, Carbondale Formation, Mazon Creek, Illinois, USA.
Last: *Sottyxerxes multiplex* Schram and Rolfe, 1982, Montceau-les-Mines, France.

Superclass MYRIAPODA Latreille, 1796

For general comments see Chapter 21 (Superclass Hexapoda), paragraph two. The earliest undoubted myriapods occur in the Upper Silurian of the UK. A possible marine myriapod-like uniramian occurs in the Lower Silurian Brandon Bridge Formation of Wisconsin, USA (Mikulic *et al.*, 1985a,b). An earlier, more equivocal example (*Cambropodus gracilis*) was described from the Middle Cambrian Wheeler Formation of Utah, USA (Robison, 1990).

The classification and the authors of higher taxa are taken mainly from Hoffman (1969), supplemented with later papers.

ACKNOWLEDGEMENTS

J. Hannibal, P. A. Selden, W. D. I. Rolfe and E. A. Jarzembowski kindly supplied information and comments.

Class/Order KAMPECARIDA

F. UNNAMED S. (PRD)–D. (LOK) ?FW

First: e.g. undescribed kampecarid figured by Jeram *et al.* (1990) (family uncertain), Downton Castle Sandstone Formation, Ludlow, Shropshire, England, UK.
Last: e.g. *Kampecaris forfarensis* Peach, 1882 (family uncertain), Arbuthnott Group and Carmyllie Group, Angus and Kincardineshire, Scotland, UK (Almond, 1985a).

Class DIPLOPODA Gervais, 1844

F. UNNAMED S. (PRD) Terr.

First: Unnamed taxon described and figured by Almond (1985a) (order/family uncertain), Cowie Formation, Cowie, Scotland, UK.

Order ARCHIPOLYPODA Scudder, 1882

F. UNNAMED D. (LOK) Terr.

First: *Archidesmus macnicoli* Peach, 1882 (family uncertain), Arbuthnott Group, Carmyllie, Angus, Scotland, UK (Almond, 1985a).
Comment: This order was transferred to the Diplopoda by Burke (1979).

F. EUPHOBERIIDAE Scudder, 1882 C. (WES B–STE B) Terr.

First: e.g. *Myriacantherpestes ferox* (Salter, 1863), Middle Coal Measures, England, UK (Burke, 1979).
Last: Unnamed genus and species described and figured by Hannibal and Feldmann (1988), Hamilton, Kansas, USA.

Order POLYXENIDA Chamberlin and Hoffman, 1958

F. SYNXENIDAE Silvestri, 1923 T. (PRB)–Rec. Terr.

First: *Phryssonotus hystrix* (Menge, 1854), Baltic amber (Hoffman, 1969). **Extant**

F. POLYXENIDAE Gray, 1842 T. (PRB)–Rec. Terr.

First: e.g. *Polyxenus conformis* Koch and Berendt, 1854, Baltic amber (Keilbach, 1982). **Extant**

Order GLOMERIDA Cook, 1895

The Fossil Record 2. Edited by M. J. Benton. Published in 1993 by Chapman & Hall, London. ISBN 0 412 39380 8

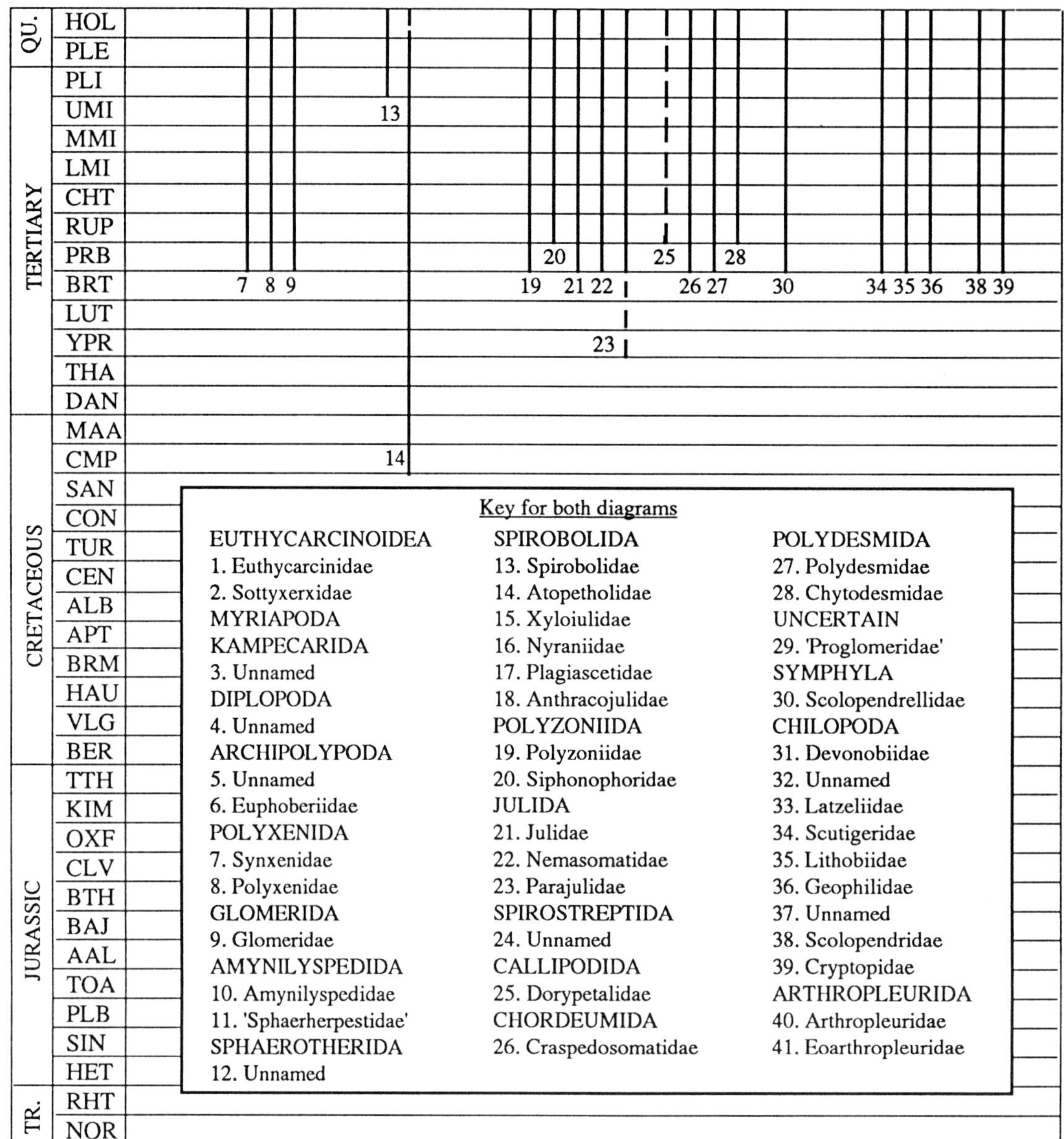

Fig. 20.1

F. GLOMERIDAE (GLOMERIDIDAE) Leach, 1815 T. (PRB)–Rec. Terr.

First: *Glomeris denticulata* Menge, 1854, Baltic amber (Keilbach, 1982). **Extant**

Order AMYNILYSPEDIDA Hoffman, 1969

F. AMYNILYSPEDIDAE Hoffman, 1969 (ACROGLOMERIDAE Fritsch, 1899) C. (WES D–STE B) Terr.

First: e.g. *Amynilyspes wortheni* Scudder, 1882, Carbondale Formation, Mazon Creek, Illinois, USA (Hannibal and Feldmann, 1981).
Last: *Amynilyspes typicus* Fritsch, 1899, Saarland, Germany (Forster, 1973).

F. 'SPHAERHERPESTIDAE' Fritsch, 1899 C. (WES D) Terr.

First and Last: *Glomeropsis ovalis* Fritsch, 1899, Gasköhle Formation, Nýřany, Czechoslovakia (Hoffman, 1969).
Comment: Hoffman (1969) assigned *Glomeropsis* to the Amnilyspedidae. Hannibal and Feldmann (1981) provided no definitive statement regarding its affinities, but rejected it from the Amnilyspedidae.

Order SPHAEROTHERIIDA Brandt, 1833

F. UNNAMED C. (WES D) Terr.

First: Unnamed taxon described and figured by Hannibal and Feldmann (1981) (family *incertae sedis*), Carbondale Formation, Mazon Creek, Illinois, USA.
Comment: The order is extant, but the assignment of this specimen is uncertain.

Order SPIROBOLIDA Cook, 1895

F. SPIROBOLIDAE Bollman, 1893 T. (PLI)–Rec. Terr.

First: *Spirobolus* (?) sp. Tiegs, 1956, Australian amber (Keilbach, 1982). **Extant**

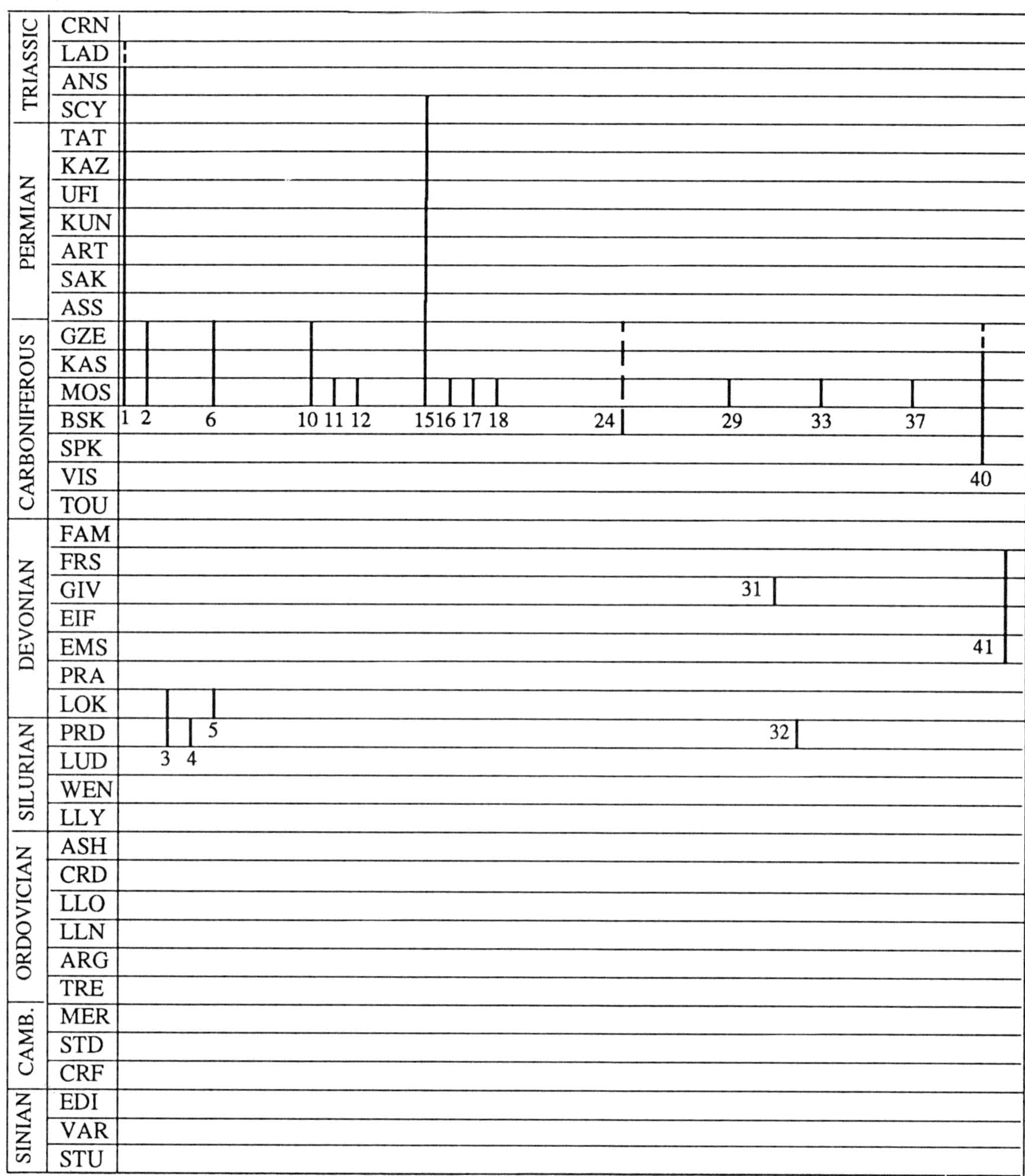

Fig. 20.1

F. ATOPETHOLIDAE Chamberlin, 1918 ?K. (CMP)–Rec. Terr.

First: *Gobiulus sabulosus* Dzik, 1975, and *Gobiulus* sp. Dzik, 1975, Barun Goyot Formation, Gobi Desert, Mongolia. **Extant**

Comment: Dzik (1975) tentatively assigned *Gobiulus* to this family.

F. XYLOIULIDAE Cook, 1895 C. (WES B)–Tr. (SCY) Terr.

First: *Xyloiulus sigillariae* (Dawson, 1860), Joggins, Nova Scotia, Canada (Hoffman, 1969).

Last: *Tomiulus angulatus*, Martynov, 1936, Malcevskaya Formation, Tom' River, Siberia, former USSR (Dzik, 1981).

F. NYRANIIDAE Hoffman, 1969 C. (WES D) Terr.

First and Last: *Nyranius costulatus* (Fritsch, 1883) and *N. tabulatus* (Fritsch, 1899), Gasköhle Formation, Nýřany, Czechoslovakia (Hoffman, 1963).

F. PLAGIASCETIDAE Hoffman, 1969 C. (WES D) Terr.

First and Last: *Plagiascetus lateralis* Hoffman, 1963, Allegheny Series, Linton Mine, Jefferson County, Ohio, USA.

F. ANTHRACOJULIDAE Hoffman, 1969 C. (WES D) Terr.

First and Last: *Anthracojulus pictus* Fritsch, 1899, Gasköhle Formation, Nýřany, Czechoslovakia (Hoffman, 1969).

Comment: Secretan (1980) noted that this family may be

represented in the Stephanian at Montceau-les-Mines, France.

Order POLYZONIIDA Cook, 1895

F. POLYZONIIDAE (POLYZONIDAE) Gervais, 1844 T. (PRB)–Rec. Terr.

First: *Polyzonium* sp. Bachofen-Echt, 1942, Baltic amber (Keilbach, 1982). **Extant**

F. SIPHONOPHORIDAE T. (RUP/CHT)–Rec. Terr.

First: *Siphonocybe* sp. Shear, 1981, Dominican amber, Dominican Republic. **Extant**

Order JULIDA Brandt, 1833

F. JULIDAE Meinert, 1868 T. (PRB)–Rec. Terr.

First: *Julus laevigatus* Koch and Berendt, 1854, Baltic amber (Keilbach, 1982). **Extant**

F. NEMASOMATIDAE Bollman, 1893 T. (PRB)–Rec. Terr.

First: *Blaniulus* sp. Menge, 1856, Baltic amber. **Extant**

F. PARAJULIDAE Bollman, 1893 T. (Eoc.?)–Rec. Terr.

First: '*Julus telluster*' Scudder, 1878, Green River Shales, Colorado, USA (Hoffman, 1969). **Extant**

Order SPIROSTREPTIDA Cook, 1895

F. UNNAMED C.(u.)–Rec. Terr.

First: *Archicambala dawsoni* (Scudder, 1868).
Comment: Doubtful record of this extant order (Hoffman, 1969).

Order CALLIPODIDA Bollman, 1893

F. DORYPETALIDAE Verhoeff, 1900 T. ?(RUP/CHT)–Rec. Terr.

First: *Protosylvestria sculpta* Handschin, 1944, France (Hoffman, 1969). **Extant**

Order CHORDEUMIDA Cook, 1895

F. CRASPEDOSOMATIDAE Cook, 1895 (CRASPEDOSOMIDAE Verhoeff, 1899) T. (PRB)–Rec. Terr.

First: *Craspedosoma angulatum* Koch and Berendt, 1854 and *C. affine* Koch and Berendt, 1854, Baltic amber (Keilbach, 1982). **Extant**

Order POLYDESMIDA Pocock, 1887

F. POLYDESMIDAE Leach, 1815 T. (PRB)–Rec. Terr.

First: *Polydesmus* sp. Menge, 1856, Baltic amber. **Extant**

F. CHYTODESMIDAE T. (RUP/CHT)–Rec. Terr.

First: *Docodesmus brodzinskyi* Shear, 1981, Dominican amber, Dominican Republic. **Extant**

Order UNCERTAIN

F. 'PROGLOMERIDAE' Fritsch, 1899 C. (WES D) Terr.

First and Last: *Archiscudderia paupera* Fritsch, 1899, Gasköhle Formation, Nýřany, Czechoslovakia (Hoffman, 1969).
Comment: Hoffman (1969) assigned *Archiscudderia* to the Amnilyspedidae. Hannibal and Feldmann (1981) provided no definitive statement regarding its affinities, but rejected it from the Amnilyspedidae.

Class PAUROPODA Lubbock, 1866

No fossil record.

Class SYMPHYLA Ryder, 1880

Order SCOLOPENDRELLIDA Hoffman, 1969

F. SCOLOPENDRELLIDAE Newport, 1845 T. (PRB)–Rec. Terr.

First: *Scolopendrella* sp. Bachofen-Echt, 1942, Baltic amber (Keilbach, 1982). **Extant**

Class CHILOPODA Latreille, 1817

Order DEVONOBIOMORPHA Shear and Bonamo, 1988 D. (GIV) Terr.

F. DEVONOBIIDAE Shear and Bonamo, 1988 D. (GIV) Terr.

First and Last: *Devonobius delta* Shear and Bonamo, 1988, Erian Series, Gilboa, New York, USA.

Order SCUTIGEROMORPHA (SCUTIGERIDA) Pocock, 1895

F. UNNAMED S. (PRD) Terr.

First: Undescribed ?scutigeromorph figured by Jeram *et al.* (1990) (family uncertain), Downton Castle Sandstone Formation, Ludlow, Shropshire, England, UK.
Comment: Doubtful ordinal record.

F. LATZELIIDAE Mundel, 1979 (GERASCUTIGERIDAE Scudder, 1890) C. (WES D) Terr.

First and Last: *Latzelia primordialis* Scudder, 1890, Carbondale Formation, Mazon Creek, Illinois, USA (Mundel, 1979).

F. SCUTIGERIDAE Newport, 1844 T. (PRB)–Rec. Terr.

First: *Scutigera illigeri* (Koch and Berendt, 1854) and *S. leachii* (Koch and Berendt, 1854), Baltic amber (Keilbach, 1982). **Extant**

Order LITHOBIOMORPHA (LITHOBIIDA) Pocock, 1895

F. LITHOBIIDAE Leach, 1814 T. (PRB)–Rec. Terr.

First: e.g. *Lithobius maxillosus* Koch and Berendt, 1854, Baltic amber (Keilbach, 1982). **Extant**

Order GEOPHILOMORPHA (GEOPHILIDA) Pocock, 1895

Comment: *Calciphilus abboti* Chamberlin, 1949 (family uncertain) is Cainozoic and not Cretaceous as stated in Hoffman (1969); see Shear and Bonamo (1988).

F. GEOPHILIDAE Newport, 1844 T. (PRB)–Rec. Terr.

First: *Geophilus* sp. Bachofen-Echt, 1942, Baltic amber (Hoffman, 1969). **Extant**

Order SCOLOPENDROMORPHA (SCOLOPENDRIDA) Pocock, 1865

F. UNNAMED C. (WES D) Terr.

First: *Mazoscolopendra richardsoni* Mundel, 1979, Carbondale Formation, Mazon Creek, Illinois, USA.

F. SCOLOPENDRIDAE Newport, 1844 T. (PRB)–Rec. Terr.

First: *Scolopendra proavita* Menge, 1854, Baltic amber (Keilbach, 1982). **Extant**

F. CRYPTOPIDAE (CRYPTOPSIDAE) T. (PRB)–Rec. Terr.

First: *Cryptops* sp. Bachofen-Echt, 1942, Baltic amber (Keilbach, 1982). **Extant**

Class/Order ARTHROPLEURIDA Waterlot, 1934

F. ARTHROPLEURIDAE Scudder, 1885 C. (NAM A–STE) Terr.

First: *Diplichnites cuithensis* Briggs *et al.*, 1979, Limestone Coal Group, Isle of Arran, Scotland, UK. This record is based on a trackway attributed to *Arthropleura*.
Last: *Arthropleura* sp. Almond (1985b), Montceau-les-Mines, France, and *Arthropleura armata* Jordan, 1854, Massif Central, France (Hahn *et al.*, 1986).

F. EOARTHROPLEURIDAE Størmer, 1976 D. (EMS–FRS) Terr.

First: *Eoarthropleura devonica* Størmer, 1976, Nellenköpfchen-Schichten, Alken an der Mosel, Germany.
Last: *'Tiphoscorpio' hueberi*, Onteora Formation, New York, USA (Selden and Shear, 1992).

REFERENCES

Almond, J. E. (1985a) The Silurian-Devonian fossil record of the Myriapoda. *Philosophical Transactions of the Royal Society of London, Series B*, **309**, 227–37.

Almond, J. E. (1985b) Les Arthropleurides du Stéphanien de Montceau-les-Mines, France. *Bulletin de la Société d'Histoire Naturelle, Autun*, **115**, 59–60.

Briggs, D. E. G., Rolfe, W. D. I. and Brannan, J. (1979) A giant myriapod trail from the Namurian of Arran, Scotland. *Palaeontology*, **22**, 273–91

Burke, J. J. (1979) A new millipede genus, *Myriacantherpestes* (Diplopoda, Archipolypoda) and a new species, *Myriacantherpestes bradebirksi*, from the English Coal Measures. *Kirtlandia*, **30**, 1–24.

Dzik, J. (1975) Spiroboloid millipeds from the Late Cretaceous of the Gobi Desert, Mongolia. *Palaeontologia Polonica*, **33**, 17–24.

Dzik, J. (1981) An early Triassic millipede from Siberia and its evolutionary significance. *Neues Jahrbuch für Geologie und Paläontologie, Monatshefte*, **1981**, 395–404.

Forster, R. (1973) Ein Diplopoden-Fund aus dem Oberkarbon des Saarlandes. *Neues Jahrbuch für Geologie und Paläontologie, Monatshefte*, **1973**, 67–71.

Gall, J.-C. and Grauvogel, L. (1964) Un arthropode peu connu le genre *Euthycarcinus* Handlirsch. *Annales de Paléontologie (Invertébrés)*, **1964**, 1–18.

Hahn, G., Hahn, R. and Brauckmann, C. (1986) Zur Kenntnis von *Arthropleura* (Myriapoda; Ober-Karbon). *Geologica et Palaeontologica*, **20**, 125–37.

Hannibal, J. T. and Feldmann, R. M. (1981) Systematics and functional morphology of oniscomorph millipedes (Arthropoda: Diplopoda) from the Carboniferous of North America. *Journal of Paleontology*, **55**, 730–46.

Hannibal, J. T. and Feldmann, R. M. (1988) Millipeds from late Paleozoic limestones at Hamilton, Kansas, in Regional Geology and Paleontology of Upper paleozoic Hamilton Quarry Area (eds G. Mapes and R. Mapes), *Kansas Geological Survey Guidebook* Series, 6, pp. 125–31.

Hoffman, R. L. (1963) New genera and species of Upper Paleozoic Diplopoda. *Journal of Paleontology*, **37**, 167–74.

Hoffman, R. L. (1969) Myriapoda, exclusive of Insecta, in *Treatise on Invertebrate Paleontology, Part R, Arthropoda 4 (2)* (ed. R. C. Moore), University of Kansas Press and Geological Society of America, Lawrence, Kansas, and Boulder, Colorado, 651 pp.

Jeram, A. J., Selden, P. A. and Edwards, D. (1990) Land animals in the Silurian: arachnids and myriapods from Shropshire, England. *Science*, **250**, 658–61.

Keilbach, R. (1982) Bibliographie und Liste der Arten tierischer Einschlüsse in fossilen Harzen sowie ihrer Aufbewahrungsorte, Teil 1. *Deutsche Entomologische Zeitschrift*, **29**, 129–286.

Mikulic, D. G., Briggs, D. E. G. and Kluessendorf, J. (1985a) A Silurian soft-bodied biota. *Science*, **228**, 715–17.

Mikulic, D. G., Briggs, D. E. G. and Kluessendorf, J. (1985b) A new exceptionally preserved biota from the Lower Silurian of Wisconsin, U.S.A. *Philosophical Transactions of the Royal Society of London, Series B*, **311**, 75–84.

Mundel, P. (1979) The centipedes (Chilopoda) of the Mazon Creek, in *Mazon Creek Fossils* (ed. M. H. Nitecki), Academic Press, New York, pp. 361–78.

Robison, R. A. (1990) Earliest-known uniramous arthropod. *Nature*, **343**, 163–4.

Schram, F. R. (1971) A strange arthropod from the Mazon Creek of Illinois and the Trans Permo-Triassic Merostomoidea (Trilobitoidea). *Fieldiana (Geology)*, **20**, 85–102.

Schram, F. R. and Rolfe, W. D. I. (1982) New euthycarcinoid arthropods from the Upper Pennsylvanian of France and Illinois. *Journal of Paleontology*, **56**, 1434–50.

Secretan, S. (1980) Les arthropodes du Stéphanien de Monteceau-les-Mines. *Bulletin de la Société d'Histoire Naturelle, Autun*, **94**, 23–35.

Selden, P. A. and Shear, W. A. (1992) A myriapod identity for the Devonian "Scorpion" *Tiphoscorpio hueberi*. *Bericht des Naturwissenschaftlich-Medizinishen Vereins in Innsbruck, Supplement* **10**, 35–6.

Shear, W. A. (1981) Two fossil millipeds from the Dominican amber (Diplopoda: Chytodesmidae, Siphonophoridae). *Myriapodologica*, **1**, 51–4.

Shear, W. A. and Bonamo, P. M. (1988) Devonobiomorpha, a new order of centipeds (Chilopoda) from the Middle Devonian of Gilboa, New York State, USA, and the phylogeny of centiped orders. *American Museum Novitates*, **2927**, 1–30.

Starobogatov, Ya. I. (1988) On the system of euthycarcinids (Arthropoda, Trilobitoides). *Byulleten Geologii*, **3**, 65–73 (in Russian).

Størmer, L. (1976) Arthropods from the Lower Devonian (Lower Emsian) of Alken an der Mosel, Germany. Part 5: Myriapoda and additional forms, with general remarks on fauna and problems regarding in invasion of land by arthropods. *Senckenbergiana Lethaea*, **57**, 87–183.

21

ARTHROPODA (HEXAPODA; INSECTA)

A. J. Ross and E. A. Jarzembowski

In the first edition of *The Fossil Record*, Crowson *et al.* (1967) only included the stratigraphical ranges of insect superfamilies and higher taxa; in this work we have attempted a listing of all families which have a pre-Quaternary fossil record, for the latter, see Buckland and Coope (1991). For ease of reference, the families are listed alphabetically within their orders. Alternative names, spelling variations and groups included are given in parentheses.

Period, epoch and stage abbreviations follow Harland *et al.* (1982); however, we have used the European nomenclature for the Upper Carboniferous. We have attempted to give ranges of families and orders to stage level, but this has often proved impossible because of the absence of accompanying detailed stratigraphical information. Thus ages of Permian localities follow Wootton (1981). For C.I.S. locality names we have used the English transliteration unless an alternative is better known, e.g. Transbaikalia instead of Zabaikale. Where there is some uncertainty as to the exact age of species, e.g. those in Baltic amber, we have made a decision as indicated below. For first and last occurrences, an example is given where there is more than one species of the same age. We have distinguished between the author of a species and a reference in which the species is mentioned by giving the latter as '*in*'. We have also used '*in*' for references where specimens are figured but undescribed.

The insect part of the *Treatise* by F. M. Carpenter appeared while this work was in press. However, it is based on published work up to 1983, and therefore we have surveyed the literature in detail from January 1984 up to December 1991, although some of the ranges in Dmitriev and Zherikhin (1988) may be based on unpublished material. The ranges given here should be supplemented by Carpenter (1992) which gives the authors of higher taxa documented before 1984 (we have only included the authors of families described after 1983) and much more information besides on the occurrence of fossil insects.

The *Treatise* arrangement of classes is followed; however, Kukalová-Peck (1987a) unites Collembola and Protura in the Parainsecta and includes Diplura in the Insecta *s.s.* A phylogenetic classification of these and other extant hexapods is discussed by Kristensen (1991). For common names of orders see Jarzembowski (1990b). The earliest hexapod is as for Collembola. The Class/Order Protura is Recent only. Most insects are terrestrial, but Wootton (1988) gives a useful overview of the geological history of aquatic insect groups. 'Terr.' has been used below in the broadest sense, including aerial or freshwater adults or larvae/nymphs.

To keep this list up to date please send your reprints on fossil insects to AJR.

Acknowledgements – Many thanks to all those who have commented on the manuscript and sent us reprints, particularly: Dr A. P. Rasnitsyn, Dr Y. A. Popov, Dr D. E. Shcherbakov, Professor Y. Hong, Dr J. Zhang, Professor R. G. Martins-Neto, Dr A. V. Gorokhov, Dr U. Spahr, Dr V. A. Blagoderov, Dr A. G. Ponomarenko, Dr L. N. Pritykina, Dr N. D. Sinichenkova, Dr I. D. Sukacheva and Dr V. V. Zherikhin. This is PRIS contribution No. 228 for E.A.J.

ORDER INDEX

The Fossil Record 2. Edited by M. J. Benton. Published in 1993 by Chapman & Hall, London. ISBN 0 412 39380 8

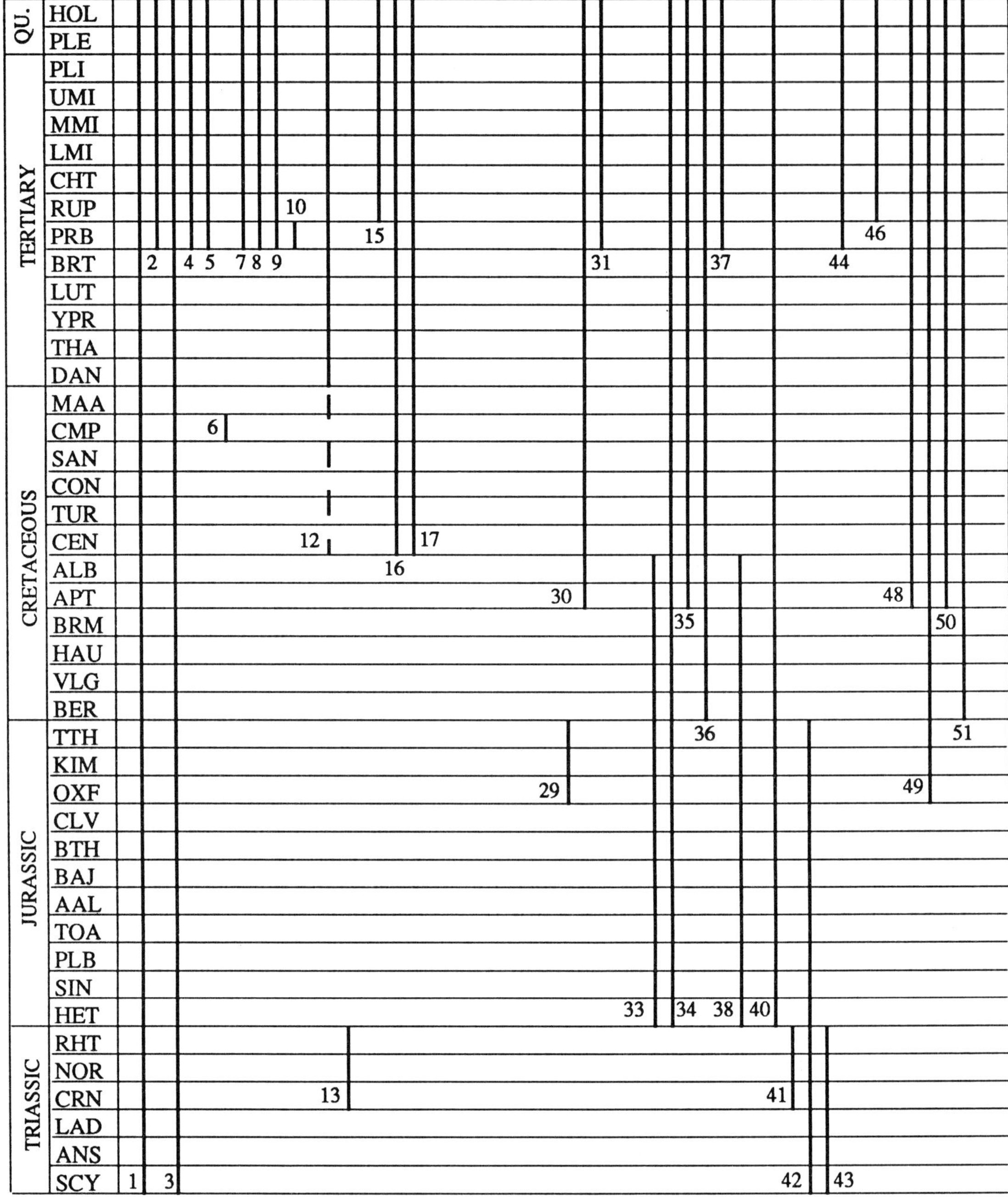

Fig. 21.1

Superclass HEXAPODA (Insects *sensu lato*)

Class/Order COLLEMBOLA D. (SIG)–Rec.

The Miocene family Paleosminthuridae belongs in the Hymenoptera: Formicidae, see Najt (1987). First as for Isotomidae.

F. ENTOMOBRYIDAE P. (ROT)–Rec. Terr.

Extant

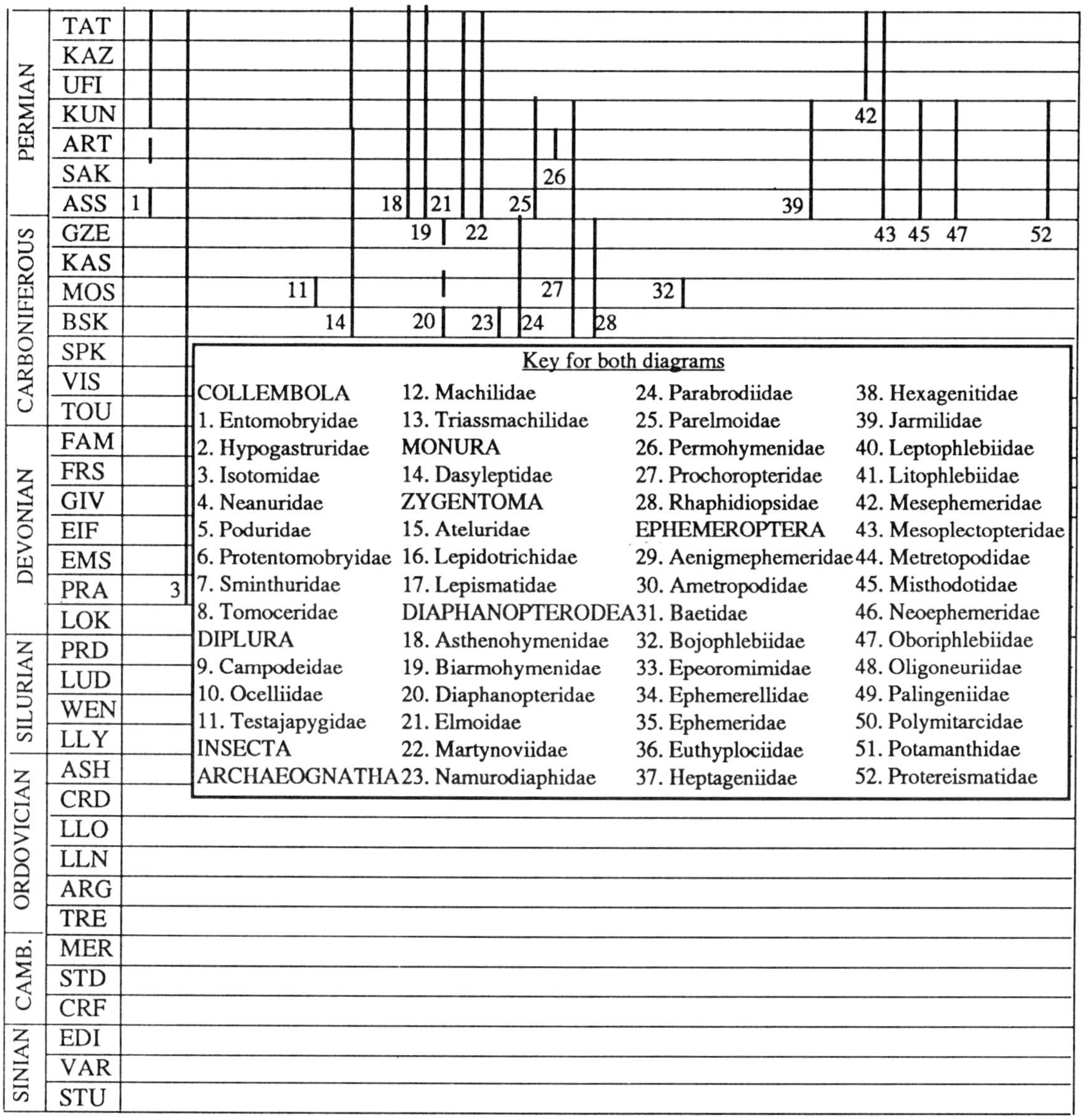

PERMIAN: TAT, KAZ, UFI, KUN, ART, SAK, ASS
CARBONIFEROUS: GZE, KAS, MOS, BSK, SPK, VIS, TOU
DEVONIAN: FAM, FRS, GIV, EIF, EMS, PRA, LOK
SILURIAN: PRD, LUD, WEN, LLY
ORDOVICIAN: ASH, CRD, LLO, LLN, ARG, TRE
CAMB.: MER, STD, CRF
SINIAN: EDI, VAR, STU

Key for both diagrams

COLLEMBOLA
1. Entomobryidae
2. Hypogastruridae
3. Isotomidae
4. Neanuridae
5. Poduridae
6. Protentomobryidae
7. Sminthuridae
8. Tomoceridae
DIPLURA
9. Campodeidae
10. Ocelliidae
11. Testajapygidae
INSECTA
ARCHAEOGNATHA
12. Machilidae
13. Triassmachilidae
MONURA
14. Dasyleptidae
ZYGENTOMA
15. Ateluridae
16. Lepidotrichidae
17. Lepismatidae
DIAPHANOPTERODEA
18. Asthenohymenidae
19. Biarmohymenidae
20. Diaphanopteridae
21. Elmoidae
22. Martynoviidae
23. Namurodiaphidae
24. Parabrodiidae
25. Parelmoidae
26. Permohymenidae
27. Prochoropteridae
28. Rhaphidiopsidae
EPHEMEROPTERA
29. Aenigmephemeridae
30. Ametropodidae
31. Baetidae
32. Bojophlebiidae
33. Epeoromimidae
34. Ephemerellidae
35. Ephemeridae
36. Euthyplociidae
37. Heptageniidae
38. Hexagenitidae
39. Jarmilidae
40. Leptophlebiidae
41. Litophlebiidae
42. Mesephemeridae
43. Mesoplectopteridae
44. Metretopodidae
45. Misthodotidae
46. Neoephemeridae
47. Oboriphlebiidae
48. Oligoneuriidae
49. Palingeniidae
50. Polymitarcidae
51. Potamanthidae
52. Protereismatidae

Fig. 21.1

F. HYPOGASTRURIDAE T. (PRB)–Rec. Terr.

First: e.g. *Hypogastrura* (*Ceratophysella*) sp., *in* Lawrence (1985), Baltic amber, Gdańsk, Poland. **Extant**

F. ISOTOMIDAE D. (SIG)–Rec. Terr.

First: *Rhyniella praecursor*, *in* Greenslade and Whalley (1986), Rhynie Chert, Rhynie, Scotland, UK. **Extant**

F. NEANURIDAE T. (PRB)–Rec. Terr.

First: *Pseudachorutes* sp., *in* Lawrence (1985), Baltic amber, Gdańsk, Poland. **Extant**

F. PODURIDAE T. (PRB)–Rec. Terr.

First: e.g. *Podura fuscata*, *in* Spahr (1990), Baltic amber. **Extant**

F. PROTENTOMOBRYIDAE K. (CMP) Terr.

First and Last: *Protentomobrya walkeri in* Spahr (1990), Canadian amber, Canada.

F. SMINTHURIDAE T. (PRB)–Rec. Terr.

First: e.g. *Sminthurinus* sp., *in* Lawrence (1985), Baltic amber, Gdańsk, Poland. **Extant**

F. TOMOCERIDAE T. (PRB)–Rec. Terr.

First: e.g. *Tomocerus* cf. *minor*, *in* Lawrence (1985), Baltic amber, Gdańsk, Poland. **Extant**

Class/Order DIPLURA C. (WES D)–Rec. Terr.

First as for Testajapygidae.

F. CAMPODEIDAE T. (PRB)–Rec. Terr.

First: *Campodea darwinii in* Spahr (1990), Baltic amber. **Extant**

F. OCELLIIDAE Spahr, 1990 T. (PRB) Terr.

First and Last: *Ocellia articulicornis*, *in* Spahr (1990), Baltic amber.

F. TESTAJAPYGIDAE Kukalová-Peck, 1987 C. (WES D) Terr.

First and Last: *Testajapyx thomasi* Kukalová-Peck (1987), Carbondale Formation, Mazon Creek, Illinois, USA.

Class INSECTA (SCARABAEODA; insects *sensu stricto*) D. (EMS?)–Rec.

First as for Archaeognatha. The Lower Devonian *Rhyniognatha hirsti* in Kukalová-Peck (1991) is a possible myriapod.

Subclass APTERYGOTA *s.s.* (LEPISMATONA; primitively wingless insects; THYSANURA *s.l.* of older classifications) D. (EMS?)–Rec.

First as for Archaeognatha.

Order ARCHAEOGNATHA (MACHILIDA *pars*) D. (EMS?)–Rec. Terr.

First: *Gaspea palaeoentognathae* (*nomen nudum*) Labandeira *et al.* (1988), Battery Point Formation, Gaspé Peninsula, Canada. This species has not been placed in a family. Jeram *et al.* (1990) do not consider that this species is a fossil hexapod. **Extant**

F. MACHILIDAE K (u.?)–Rec. Terr.

F. TRIASSOMACHILIDAE Tr. (u.)

Kukalová-Peck (1991) considers *Triassomachilis* to be a mayfly nymph.

Order MONURA (MACHILIDA *pars*) C. (u.)–P. Terr.

Only one recognized family.

F. DASYLEPTIDAE C. (u.)–P. Terr.

First: e.g. ?*Dasyleptus* sp. Kukalová-Peck (1985), Carbondale Formation, Mazon Creek, Illinois, USA.
Last: e.g. *Leoidodasypus sharoui*, *in* Labandeira and Beall (1990), Kansas, USA.

Order ZYGENTOMA (THYSANURA *s.s.*; LEPISMATIDA) C. (WES D)–Rec. Terr.

First: *Ramsdelepidion schusteri* Kukalová-Peck (1987), Carbondale Formation, Mazon Creek, Illinois, USA. Kukalová-Peck (1987) did not place this species in a family. **Extant**

F. ATELURIDAE T. (Oli.)–Rec. Terr.

Extant

F. LEPIDOTRICHIDAE (LEPIDOTHRICIDAE) K. (u.)–Rec. Terr.

Extant

F. LEPISMATIDAE K (u.)–Rec. Terr.

Extant

Subclass PTERYGOTA (SCARABAEONA; winged insects) C. (NAM A)–Rec. Terr.

The Order Archaeoptera is based on crustacean rather than insect remains (Rodendorf, 1972). First as for Protorthoptera.

Cohort PALAEOPTERA C. (NAM B)–Rec. Terr.

First as for Megasecoptera.

Order DIAPHANOPTERODEA (DIAPHANOPTERIDA) C. (NAM B)–P. Terr.

First as for Namurodiaphidae.

F. ASTHENOHYMENIDAE (DOTERIDAE) P. Terr.

e.g. *Doter minor*, *in* Hubbard (1987), Kansas, USA.

F. BIARMOHYMENIDAE P. Terr.

F. DIAPHANOPTERIDAE C. (u.) Terr.

F. ELMOIDAE P. Terr.

F. MARTYNOVIIDAE P. Terr.

F. NAMURODIAPHIDAE Kukalová-Peck and Brauckmann, 1990 C. (NAM B) Terr.

First and Last: *Namurodiapha sippelorum* Kukalová-Peck and Brauckmann (1990), Vorhalle Beds, Hagen-Vorhalle, Germany.

F. PARABRODIIDAE C. (u.) Terr.

F. PARELMOIDAE P. (ROT) Terr.

F. PERMOHYMENIDAE P. (ART) Terr.

First and Last: *Permohymen schucherti*, *in* Kukalová-Peck and Brauckmann (1990), Kansas, USA.

F. PROCHOROPTERIDAE C. (u.)–P. (ROT) Terr.

First: e.g. *Prochoroptera calopteryx*, *in* Kukalová-Peck and Brauckmann (1990), Carbondale Formation, Mazon Creek, Illinois, USA.

F. RHAPHIDIOPSIDAE C. (u.) Terr.

Order EPHEMEROPTERA (EPHEMERIDA, PLECTOPTERA) C. (WES C)–Rec. Terr.

Data taken from Hubbard (1987), unless stated otherwise. First as for Bojophlebiidae. The Triassomachilidae may prove to belong here (see Order Archaeognatha).

F. AENIGMEPHEMERIDAE J. (u.) Terr.

First and Last: *Aenigmephemera demoulini*, Karatau, Kazakhstan, former USSR.

F. AMETROPODIDAE K. (APT)–Rec. Terr.

Extant

F. BAETIDAE T. (PRB)–Rec. Terr.

First: e.g. *Baetis gigantea*, Baltic amber. Sinichenkova (1985a, 1989) included the Jurassic genus *Mesobaetis* in the Siphlonuridae, but Hubbard (1987) placed it in the Baetidae without discussion. Here, we have followed Sinichenkova (1985a, 1989). **Extant**

F. BOJOPHLEBIIDAE Kukalová-Peck, 1985 C. (WES C) Terr.

First and Last: *Bojophlebia prokopi*, Whetstone Horizon, Bohemia, Czechoslovakia. Klyuge (1989), however, considers that the 'nymph' of this species may belong in another order, 'probably Thysanura'.

F. EPEOROMIMIDAE (EPEOROMIDIDAE) J. (l.)–K. (l.) Terr.

First: e.g. *Epeoromimus kazlauskasi*, western Siberia, former USSR.
Last: e.g. *Epeoromimus cretaceus*, Transbaikalia, former USSR.

F. EPHEMERELLIDAE J. (l.)–Rec. Terr.

First: e.g. *Clephemera clava* Lin (1986), south China. The systematic position of this species is doubtful. **Extant**

F. EPHEMERIDAE (ICHTHYBOTIDAE) K. (APT)–Rec. Terr.

First: e.g. *Australiphemera revelata* McCafferty (1990), Santana Formation, Ceará, Brazil. The Upper Jurassic *'Ephemera' deposita* Weyenbergh does not belong to this family (McCafferty, 1990). **Extant**

F. EUTHYPLOCIIDAE K. (l.)–Rec. Terr.

First: e.g. *Pristiplocia rupestris* McCafferty (1990), Santana Formation, Ceará, Brazil. **Extant**

F. HEPTAGENIIDAE (ECDYURIDAE, ECDYONURIDAE) T. (PRB)–Rec. Terr.

First: e.g. *Heptagenia (Kageronia) fuscogrisea, in* Klyuge (1986), Baltic amber. **Extant**

F. HEXAGENITIDAE (PAEDEPHEMERIDAE, STENODICRANIDAE) J. (l.)–K. (l.) Terr.

First: e.g. *Siberiogenites angustatus*, Transbaikalia, former USSR.
Last: e.g. *Protoligoneuria limai, in* McCafferty (1990), Santana Formation, Ceará, Brazil.

F. JARMILIDAE P. (ROT) Terr.

First and Last: *Jarmila elongata*, Czechoslovakia.

F. LEPTOPHLEBIIDAE J. (l.)–Rec. Terr.

First: e.g. *Mesoneta antiqua*, Transbaikalia, former USSR. **Extant**

F. LITOPHLEBIIDAE (XENOPHLEBIIDAE) Tr. (u.) Terr.

First and Last: *Litophlebia optata*, Molteno Formation, Bird's River, South Africa. Hubbard (1987) believes this is a megasecopteran.

F. MESEPHEMERIDAE (PALINGENIOPSIDAE) P. (ZEC)–J. (TTH) Terr.

First: *Palingeniopsis praecox*, former USSR.
Last: e.g. *Mesephemera lithophila*, Lithographic Limestone, Solnhofen, Germany.

F. MESOPLECTOPTERIDAE P.–Tr. Terr.

e.g. *Mesoplectopteron longipes*, western Europe.

F. METRETOPODIDAE T. (PRB)–Rec. Terr.

First: e.g. *Metretopus henningseni*, Baltic amber. **Extant**

F. MISTHODOTIDAE (EUDOTERIDAE) P. (ROT) Terr.

e.g. *Misthodotes obtusus*, Wellington Formation, Kansas, USA.

F. NEOEPHEMERIDAE T. (Oli.)–Rec. Terr.

First: *Potamanthellus rubiensis*, Ruby River Basin, Montana, USA. **Extant**

F. OBORIPHLEBIIDAE P. (ROT) Terr.

e.g. *Oboriphlebia moravica*, Czechoslovakia.

F. OLIGONEURIIDAE (ISONYCHIIDAE) K. (APT)–Rec. Terr.

First: *Colocrus indivicum* McCafferty (1990), Santana Formation, Ceará, Brazil. McCafferty (1990) transferred the previously supposed oldest species *Protoligoneuria limai* to the Hexagenitidae. **Extant**

F. PALINGENIIDAE J. (u.)–Rec. Terr.

First: *Mesogenesia petersae*, Transbaikalia, former USSR. **Extant**

F. POLYMITARCIDAE (POLYMITARCYIDAE) K. (APT)–Rec. Terr.

First: *Caririnympha mandibulata* Martins-Neto and Caldas (1990), Santana Formation, Ceará, Brazil. **Extant**

F. POTAMANTHIDAE K. (l.)–Rec. Terr.

First: e.g. Potamanthidae (?) sp. 1 McCafferty (1990), Santana Formation, Ceará, Brazil. **Extant**

F. PROTEREISMATIDAE P. (ROT) Terr.

e.g. *Protereisma permianum*, Wellington Formation, Kansas, USA.

F. SIPHLONURIDAE J. (l.)–Rec. Terr.

First: e.g. *Mogzonurus elevatus*, Transbaikalia, former USSR. **Extant**

F. SYNTONOPTERIDAE C. (WES D) Terr.

e.g. *Syntonoptera schucherti, in* Carpenter (1988), Carbondale Formation, Mazon Creek, Illinois, USA. Carpenter (1988), however, preferred to place this family in Order Uncertain because important body structures, such as the mouthparts, are not yet known.

F. TOREPHEMERIDAE Sinichenkova, 1989 J. (u.)–K. (l.) Terr.

First: *Archaeobehningia edmundsi, in* Sinichenkova (1989), Udinskaya Formation, Transbaikalia, former USSR. Sinichenkova (1989) transferred this genus from the Behningiidae thus reverting the latter family back to Recent only.
Last: *Torephemera longipes* Sinichenkova (1989), Tsagantsabskaya Formation, Khutel-Khara, Mongolia.

F. TRIPLOSOBIDAE C. (STE) Terr.

First and Last: *Triplosoba pulchella*, Commentry, France.

Order MEGASECOPTERA (MISCHOPTERIDA) C. (NAM B)–P. Terr.

First as for Brodiopteridae, although *Xenoptera riojanaensis* ('Xenopteridae') may be older. The Upper Triassic Litophlebiidae (Ephemeroptera) may prove to belong to this order.

F. ALECTONEURIDAE P. Terr.

F. ANCHINEURIDAE C. (u.) Terr.

F. ANCOPTERIDAE P. Terr.

F. ARCIONEURIDAE P. Terr.

F. ASPIDOHYMENIDAE P. Terr.

F. ASPIDOTHORACIDAE C. (WES D–STE) Terr.

First: *Aspidothorax aestalis* Brauckmann (1991), Piesberg, Osnabrück, Germany.
Last: *Aspidothorax triangularis, in* Brauckmann (1991), Commentry, France.

F. BARDOHYMENIDAE C. (NAM B)–P. (KUN) Terr.

First: *Sylvohymen peckae* Brauckmann (1988b), Vorhalle Beds, Hagen-Vorhalle, Germany.

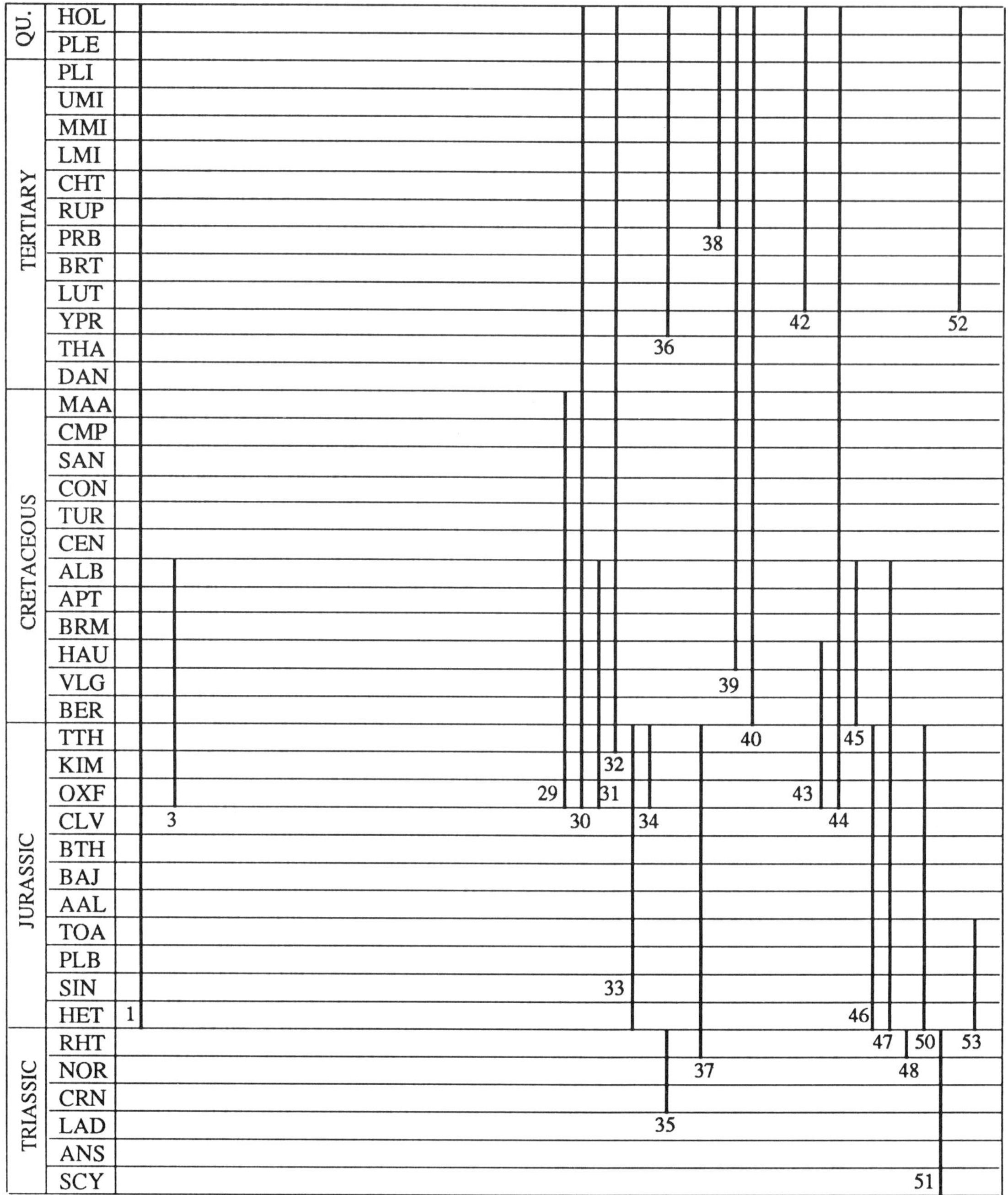

Fig. 21.2

Last: e.g. *Bardohymen magnipennifer, in* Brauckmann (1988b), Urals, former USSR.

F. BRODIIDAE C. (u.) Terr.

F. BRODIOPTERIDAE C. (NAM B–WES A) Terr.

First: *Brodioptera stricklani* Nelson and Tidwell (1988), Manning Canyon Shale Formation, Utah, USA.
Last: *Brodioptera cumberlandensis, in* Nelson and Tidwell (1988), USA.

F. CARBONOPTERIDAE C. (WES D)

First and Last: *Carbonoptera furcaradii, in* Brauckmann (1991), Neuenkirchen, Saarland, Germany.

F. CAULOPTERIDAE P. Terr.

F. CORYDALOIDIDAE C. (u.) Terr.

F. ENGISOPTERIDAE P. Terr.

F. FORIRIIDAE C. (u.) Terr.

F. FRANKENHOLZIIDAE C. (WES D)

First and Last: *Frankenholzia culmanni, in* Brauckmann (1991), Grube Frankenholz, Saarland, Germany.

F. HANIDAE P. Terr.

F. ISCHNOPTILIDAE C. (u.) Terr.

F. MISCHOPTERIDAE C. (u.) Terr.

e.g. *Mischoptera douglassi, in* Shear and Kukalová-Peck (1990), Carbondale Formation, Mazon Creek, Illinois, USA.

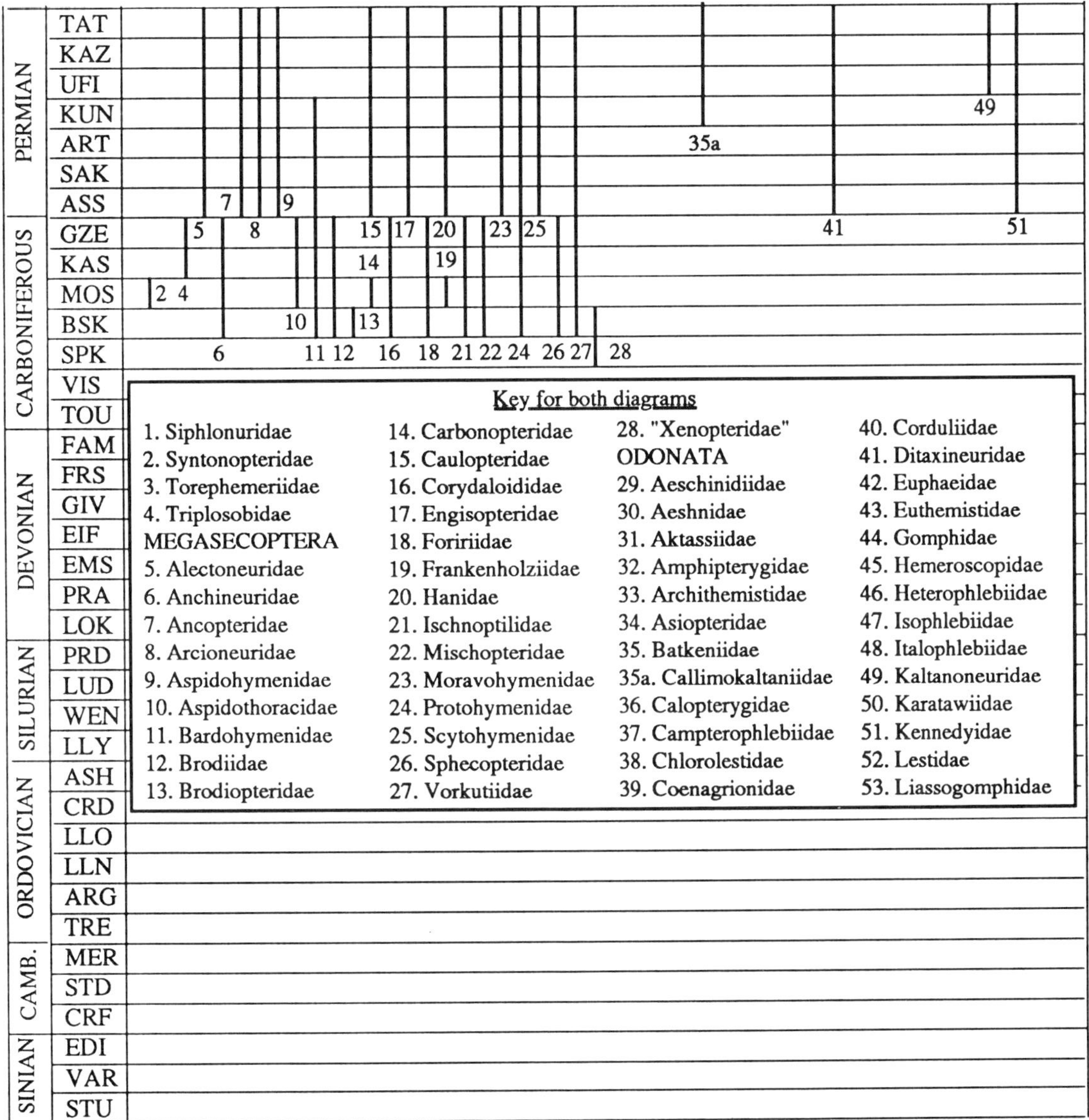

Fig. 21.2

F. MORAVOHYMENIDAE P. Terr.

F. PROTOHYMENIDAE C. (u.)–P. Terr.

First: *Sunohymen xishanensis* Hong (1985c), Shanxi Formation, Xishan, Shanxi, China.

F. SCYTOHYMENIDAE P. Terr.

F. SPHECOPTERIDAE C. (u.) Terr.

F. VORKUTIIDAE C. (u.)–P. Terr.

F. 'XENOPTERIDAE' Pinto, 1986 C. (NAM) Terr.

First and Last: *Xenoptera riojaensis* Pinto (1986), Malanzan Formation, Malanzan, Argentina. It is uncertain as to whether this formation is of Lower or Upper Carboniferous age. This family name is a junior homonym of Xenopteridae Riek (Orthoptera).

Order ODONATA (LIBELLULIDA)
P. (ROT)–Rec. Terr.

Earliest families: Ditaxineuridae and Kennedyidae, although some authors include the Protodonata in this order.

F. AESCHNIDIIDAE J. (u.)–K. (u.) Terr.

First: e.g. *Hebeiaeschnidia fengningensis*, *in* Hong (1985d), Hebei, China.

Last: *Aeschnidopsis flindersiensis*, *in* Rozefelds (1985), Queensland, Australia.

F. AESHNIDAE (AESCHNIDAE) J. (u.)–Rec.
Terr.

Extant

F. AKTASSIIDAE J. (u.)–K. (l.) Terr.

F. AMPHIPTERYGIDAE (STELEOPTERIDAE)
J. (TTH)–Rec. Terr.

First: *Steleopteron deichmuelleri*, *in* Ponomarenko (1985b), Lithographic Limestone, Solnhofen, Germany. **Extant**

F. ARCHITHEMISTIDAE J. (l.)–J. (u.) Terr.

First: e.g. *Dorsettia laeta* Whalley (1985), Lower Lias, Charmouth, Dorset, England, UK.

F. ASIOPTERIDAE J. (u.) Terr.

F. BATKENIIDAE Tr. (u.) Terr.

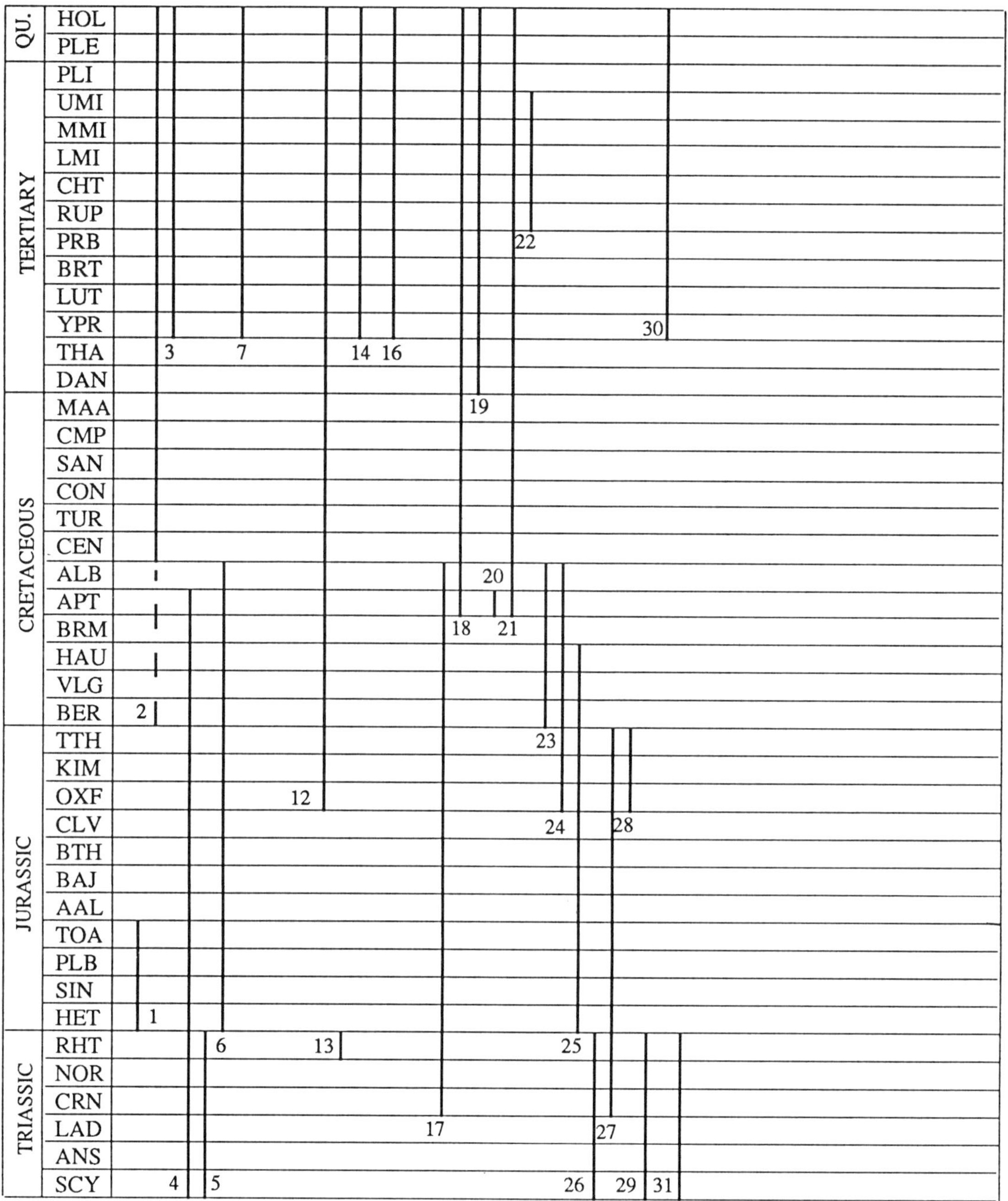

Fig. 21.3

F. CALLIMOKALTANIIDAE P. (ZEC) Terr.

F. CALOPTERYGIDAE T. (Eoc)–Rec. Terr.

Extant

F. CAMPTEROPHLEBIIDAE Tr. (RHT)–J. (u.) Terr.

First: *Samarura* sp. Rozefelds (1985), Aberdare Conglomerate, Ipswich, Australia.
Last: *Sibirioneura amurica, in* Pritykina (1985), Priamurya, former USSR.

F. CHLOROLESTIDAE T. (Oli.)–Rec. Terr.

Extant

F. COENAGRIONIDAE (COENAGRIIDAE) K. (HAU)–Rec. Terr.

First: *Cretacoenagrion alleni* Jarzembowski (1990a), Lower Weald Clay, Capel, Surrey, England, UK. Carle and Wighton (1990), however, think this species might belong in the Megapodagrionidae. **Extant**

F. CORDULIIDAE K. (l.)–Rec. Terr.

First: *Eocordulia cretacea* Pritykina (1986), Gurvaneren Formation, Myangad, Mongolia. **Extant**

F. DITAXINEURIDAE P. Terr.

F. EUPHAEIDAE (EPALLAGIDAE) T. (LUT/BRT)–Rec. Terr.

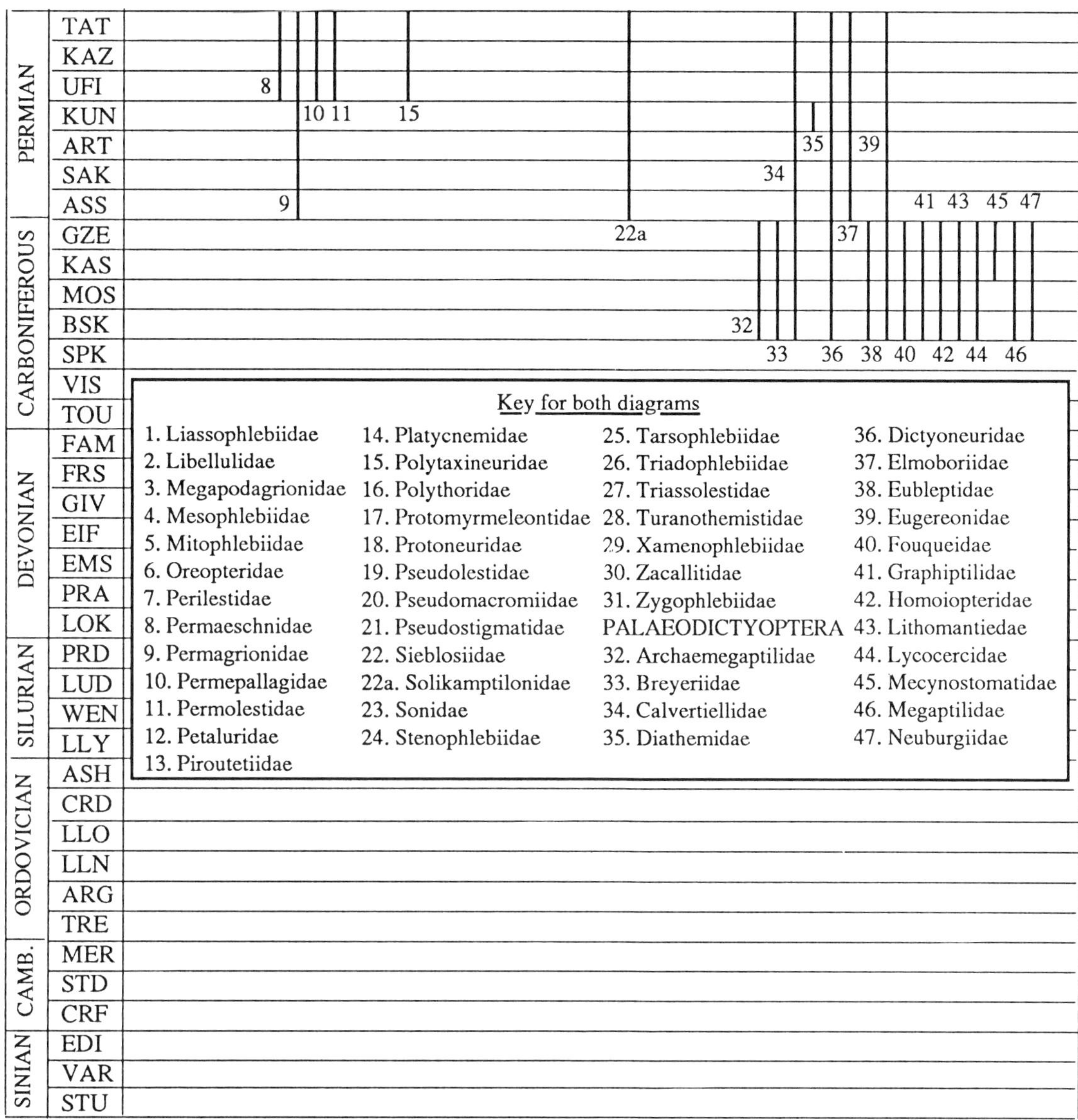

Fig. 21.3

First: e.g. *Epallagites avus*, *in* Nel (1988b), Green River Formation, Colorado, USA. **Extant**

F. EUTHEMISTIDAE J. (u.)–K. (HAU) Terr.

First: e.g. *Euthemis multinervosa*, *in* Jarzembowski (1990a), Karatau, Kazakhstan, former USSR.
Last: *Euthemis* sp. Jarzembowski (1990a), Lower Weald Clay, Capel, Surrey, England, UK. Pritykina (pers. comm.) considers that this species belongs in the Oreopteridae.

F. GOMPHIDAE (GOMPHINIDAE, PROTOLINDENIIDAE) J. (u.)–Rec. Terr.

First: e.g. *Sinogomphus taushanensis*, *in* Hong (1985d), north China. **Extant**

F. HEMEROSCOPIDAE K. (l.) Terr.

F. HETEROPHLEBIIDAE J. (l.–u.) Terr.

First: e.g. *Heterophlebia* sp. Whalley (1985), Lower Lias, Charmouth, Dorset, England, UK.

F. ISOPHLEBIIDAE J. (l.)–K. (l.) Terr.

First: *Dinosamarura tugnuica*, *in* Pritykina (1985), Transbaikalia, former USSR.

F. ITALOPHLEBIIDAE Whalley, 1986a Tr. (RHT) Terr.

First and Last: *Italophlebia gervasuttii* Whalley (1986a), Argilliti di Riva di Solto Formation, Bergamo, Italy.

F. KALTANONEURIDAE P. (ZEC) Terr.

F. KARATAWIIDAE J. (l.–u.) Terr.

First: e.g. *Karatawia sibirica*, *in* Pritykina (1985), Ust-Balei, Siberia, former USSR.
Last: e.g. *Karatawia turanica*, *in* Pritykina (1985), Karatau, Kazakhstan, former USSR.

F. KENNEDYIDAE P. (ROT)–Tr. Terr.

F. LESTIDAE T. (LUT/BRT)–Rec. Terr.

First: *Petrolestes hendersoni*, *in* Nel (1986), Green River Formation, Ute Trail, Colorado, USA. **Extant**

F. LIASSOGOMPHIDAE J. (l.) Terr.

F. LIASSOPHLEBIIDAE J. (l.) Terr.

e.g. *Liassophlebia pseudomagnifica* Whalley (1985), Lower Lias, Charmouth, Dorset, England, UK.

F. LIBELLULIDAE
K. (?BER)–Rec. Terr.

First: *Condalia woottoni* Whalley and Jarzembowski (1985), Lithographic Limestone, Montsech, Spain. This species is a libelluloid, but may not belong to this family. **Extant**

F. MEGAPODAGRIONIDAE
(MEGAPODOGRIONIDAE) T. (Eoc.)–Rec.
Terr.

Extant

F. MESOPHLEBIIDAE Tr.–K. (APT) Terr.

First: e.g. *Mesophlebia antinodalis, in* Rozefelds (1985), Coal Measures, Ipswich, Australia.
Last: *Peraphlebia tetrastichia* Jell and Duncan (1986), Koonwarra Fossil Bed, south Gippsland, Australia.

F. MITOPHLEBIIDAE Tr. Terr.

F. OREOPTERIDAE J. (l.)–K. (l.) Terr.

F. PERILESTIDAE T. (Eoc.)–Rec. Terr.

Extant

F. PERMAESCHNIDAE P. (ZEC) Terr.

e.g. *Gondvanoptilon brasiliense, in* Martins-Neto (1987a), Irati Formation, São Paulo, Brazil.

F. PERMAGRIONIDAE P. Terr.

F. PERMEPALLAGIDAE P. (ZEC) Terr.

F. PERMOLESTIDAE F. (ZEC) Terr.

F. PETALURIDAE J. (u.)–Rec. Terr.

First: e.g. *Mesuropetala koehleri, in* Ponomarenko (1985b), Lithographic Limestone, Solnhofen, Germany. **Extant**

F. PIROUTETIIDAE Nel, 1989 Tr. (RHT) Terr.

First and Last: *Piroutetia liasina, in* Nel (1989), Lower Lias, Mouchard, France.

F. PLATYCNEMIDAE (PLATYCNEMIDIDAE)
T. (Eoc.)–Rec. Terr.

Extant

F. POLYTAXINEURIDAE P. (ZEC) Terr.

e.g. *Polytaxineura stanleyi* in Rozefelds (1985), Belmont, New South Wales, Australia.

F. POLYTHORIDAE T. (Eoc)–Rec. Terr.

Extant

F. PROTOMYRMELEONTIDAE
(PROTOMYRMELEONIDAE,
TRIASSAGRIONIDAE) Tr (u.)–K. (l.) Terr.

First: e.g. *Terskeja paula, in* Zessin (1991), south Fergana, former USSR.

F. PROTONEURIDAE K. (APT)–Rec. Terr.

First: *Eoprotoneura hyperstigma* Carle and Wighton (1990), Santana Formation, Ceará, Brazil. **Extant**

F. PSEUDOLESTIDAE T. (Pal.)–Rec. Terr.

Extant

F. PSEUDOMACROMIIDAE Carle and Wighton,
1990 K. (APT) Terr.

First and Last: *'Pseudomacromia' sensibilis* Carle and Wighton (1990), Santana Formation, Ceará, Brazil. This genus name is a junior homonym of *Pseudomacromia* Kirby.

F. PSEUDOSTIGMATIDAE K. (APT)–Rec. Terr.

First: *Euarchistigma atrophium* Carle and Wighton (1990), Santana Formation, Ceará, Brazil. **Extant**

F. SIEBLOSIIDAE T. (Oli.–UMI) Terr.

First: e.g. *Stenolestes fischeri* Nel (1986), Malvezy, Aude, France.
Last: *Stenolestes hispanicus* Nel (1991), Diatomites, Bellver-en-Cerdaña, Spain.

F. SOLIKAMPTILONIDAE P. Terr.

F. SONIDAE Pritykina, 1986 K. (l.) Terr.

First and Last: *Sona nectes* Pritykina (1986), Gurvaneren Formation, Myangad, Mongolia.

F. STENOPHLEBIIDAE (STENOPHLEBIDAE)
J. (u.)–K. (l.) Terr.

First: e.g. *Stenophlebia latreillei, in* Ponomarenko (1985b), Lithographic Limestone, Solnhofen, Germany.
Last: e.g. *Sinostenophlebia zhanjiakouensis* Hong (1985d), north China.

F. TARSOPHLEBIIDAE J. (l.)–K. (HAU) Terr.

Last: *Tarsophlebia*? sp. Jarzembowski (1990a), Lower Weald Clay, Capel, Surrey, England, UK.

F. TRIADOPHLEBIIDAE Tr. Terr.

F. TRIASSOLESTIDAE (PROGONOPHLEBIIDAE)
Tr. (u.)–J. Terr.

First: e.g. *Triassolestes epiophlebioides, in* Rozefelds (1985), Blackstone Formation, Denmark Hill, Australia.

F. TURANOTHEMISTIDAE J. (u.) Terr.

F. XAMENOPHLEBIIDAE Tr. Terr.

F. ZACALLITIDAE T. (Eoc.)–Rec. Terr.

First: *Zacallites balli, in* Nel (1988b), Colorado, USA. **Extant**

F. ZYGOPHLEBIIDAE Tr. Terr.

Order PALAEODICTYOPTERA (DICTYONEURIDA,
PERMOTHEMISTIDA, EUBLEPTIDODEA,
ARCHODONATA) C. (NAM B)–P. (ZEC) Terr.

Oldest species from the Hagen Beds of Germany, see Dictyoneuridae and Graphiptilidae.

F. ARCHAEMEGAPTILIDAE C. (u.) Terr.

F. BREYERIIDAE C. (NAM B–STE) Terr.

First: *Jugobreyeria sippelorum* Brauckmann *et al.*, 1985, Vorhalle Beds, Hagen-Vorhalle, Germany.
Last: e.g. *Breyeria boulei, in* Brauckmann *et al.*, 1985, Commentry, France.

F. CALVERTIELLIDAE C. (u.)–P. Terr.

F. DIATHEMIDAE P. (KUN) Terr.

e.g. *Diathemidia monstruosa, in* Riek and Kukalová-Peck (1984), Koshelovo Formation, Suksun, Perm, former USSR.

F. DICTYONEURIDAE C. (NAM B)–P. Terr.

First: e.g. *Schmidtopteron adictyon*, *in* Brauckmann (1988a), Hagen Beds, Schmiedestrasse, Germany.

F. ELMOBORIIDAE P. Terr.

F. EUBLEPTIDAE C. (u.) Terr.

F. EUGEREONIDAE C. (u.)–P. Terr.

Last: e.g. *Eugereon boeckingi* in Shear and Kukalová-Peck (1990), Germany.

F. FOUQUEIDAE C. (u.) Terr.

F. GRAPHIPTILIDAE (PATTEISKYIDAE) C. (NAM B–STE) Terr.

First: *Patteiskya bouckaerti*, *in* Brauckmann (1988a), Hagen Beds, Schmiedestrasse, Germany.
Last: e.g. *Graphiptilus heeri*, *in* Brauckmann *et al.* (1985), Commentry, France.

F. HOMOIOPTERIDAE C. (NAM B–STE) Terr.

First: *Homoioptera vorhallensis*, *in* Brauckmann *et al.* (1985), Vorhalle Beds, Hagen-Vorhalle, Germany.
Last: e.g. *Homoioptera gigantea*, *in* Riek and Kukalová-Peck (1984), Commentry, France.

F. LITHOMANTEIDAE (LITHOMANTIDAE) C. (NAM B–STE) Terr.

First: *Lithomantis varius* Brauckmann *et al.*, 1985, Vorhalle Beds, Hagen-Vorhalle, Germany.
Last: *Macroptera fariai*, *in* Brauckmann *et al.*, 1985, Alto do Pejao, Portugal.

F. LYCOCERCIDAE C. (u.) Terr.

e.g. *Lycocerus goldenbergi* in Shear and Kukalová-Peck (1990), Commentry, France.

F. MECYNOSTOMATIDAE C. (STE) Terr.

First and Last: *Meynostomata* (= *Mecynostoma*) *dohrni* in Shear and Kukalová-Peck (1990), Commentry, France.

F. MEGAPTILIDAE C. (u.) Terr.

F. NEUBURGIIDAE C. (u.) Terr.

e.g. *Palaeoneura qiligouensis* Hong (1985c), Shanxi Formation, Xishan, Shanxi, China.

F. PERMONEURIDAE P. (ROT) Terr.

F. PERMOTHEMISTIDAE P. (ROT–KAZ) Terr.

Last: *Permothemis libelluloides*, *in* Riek and Kukalová-Peck (1984), Iva-Gora Beds, Archangelsk, former USSR.

F. PSYCHROPTILIDAE C. (u.) Terr.

F. RECTINEURIDAE C. (u.) Terr.

F. SPILAPTERIDAE C. (NAM B)–P. (KUN) Terr.

First: *Homaloneura ligeia* Brauckmann (1986), Vorhalle Beds, Hagen-Vorhalle, Germany.
Last: e.g. *Paradunbaria pectinata*, *in* Brauckmann (1986), Urals, former USSR.

F. STRAELENIELLIDAE Laurentiaux-Vieira and Laurentiaux, 1986a C. (NAM C–WES A) Terr.

First: e.g. *Straeleniella namurensis* Laurentiaux-Vieira and Laurentiaux (1986a), grey-black schists, Charbonnages d'Armercoeur, Belgium.

F. TCHIRKOVAEIDAE C. (u.) Terr.

Order PROTODONATA (ODONATA: MEGANISOPTERA) C. (NAM B)–Tr. (u.) Terr.

Oldest species from the Vorhalle Beds of Germany, see Meganeuridae and Erasipteridae, although those of Eugeropteridae may be older. Last as for Triadotypidae. Data are taken from Brauckmann and Zessin (1989).

F. ERASIPTERIDAE C. (NAM B–WES D) Terr.

First: *e.g. Erasipteroides valentini*, Vorhalle Beds, Hagen-Vorhalle, Germany.
Last: *Erasipterella piesbergensis*, Piesberg, Osnabrück, Germany.

F. EUGEROPTERIDAE Riek, 1984 C. (NAM) Terr.

e.g. *Eugeropteron lunatum*, Malanzan Formation, Malanzan, Argentina. It is uncertain as to whether this formation is of Lower or Upper Carboniferous age.

F. MEGANEURIDAE C. (NAM B)–P. (KAZ) Terr.

First: *Namurotypus sippeli*, Vorhalle Beds, Hagen-Vorhalle, Germany.
Last: e.g. *Arctotypus sinuatus*, Iva-Gora, Archangelsk, former USSR.

F. PARALOGIDAE C. (WES A)–P. (ART) Terr.

First: *Oligotypus britannicus* (*nomen nudum*), Lower Coal Measures, Staffordshire, England, UK.
Last: e.g. *Oligotypus tillyardi*, Wellington Formation, Elmo, Kansas, USA.

F. TRIADOTYPIDAE Tr. (SCY–u.) Terr.

First: *Triadotypus guillaumei*, Bunter Sandstone, France.
Last: e.g. *Triadotypus sogdianus*, Sogd, Fergana, former USSR.

Cohort NEOPTERA C. (NAM A)–Rec. Terr.

First as for Protorthoptera.

Superorder POLYNEOPTERA (GRYLLONES; Orthopteroid orders) C. (NAM A)–Rec. Terr.

First as for Protorthoptera.

Order BLATTODEA (BLATTIDA, BLATTIDEA, BLATTARIAE, BLATTOIDEA) C. (WES A)–Rec. Terr.

We have followed Schneider's (1984) classification of the Palaeozoic Blattodea; Durden (1984) gave an alternative classification. The Eocene Parallelophoridae consists of the anal fields of other cockroach families (Lutz, 1984).

F. ARCHIMYLACRIDAE C. (WES A)–J. (l.) Terr.

First: e.g. *Miroblatta costalis* Laurentiaux-Vieira and Laurentiaux (1987), passage beds, Charbonnages de Rieu-du-Coeur, Belgium.

F. ARCHOBLATTINIDAE C. (WES–STE) Terr.

F. BLABERIDAE T. (PRB)–Rec. Terr.

Extant

F. BLATTELLIDAE (PHYLLODROMIIDAE) K. (u.)–Rec. Terr.

F. BLATTIDAE K. (l.)–Rec. Terr.

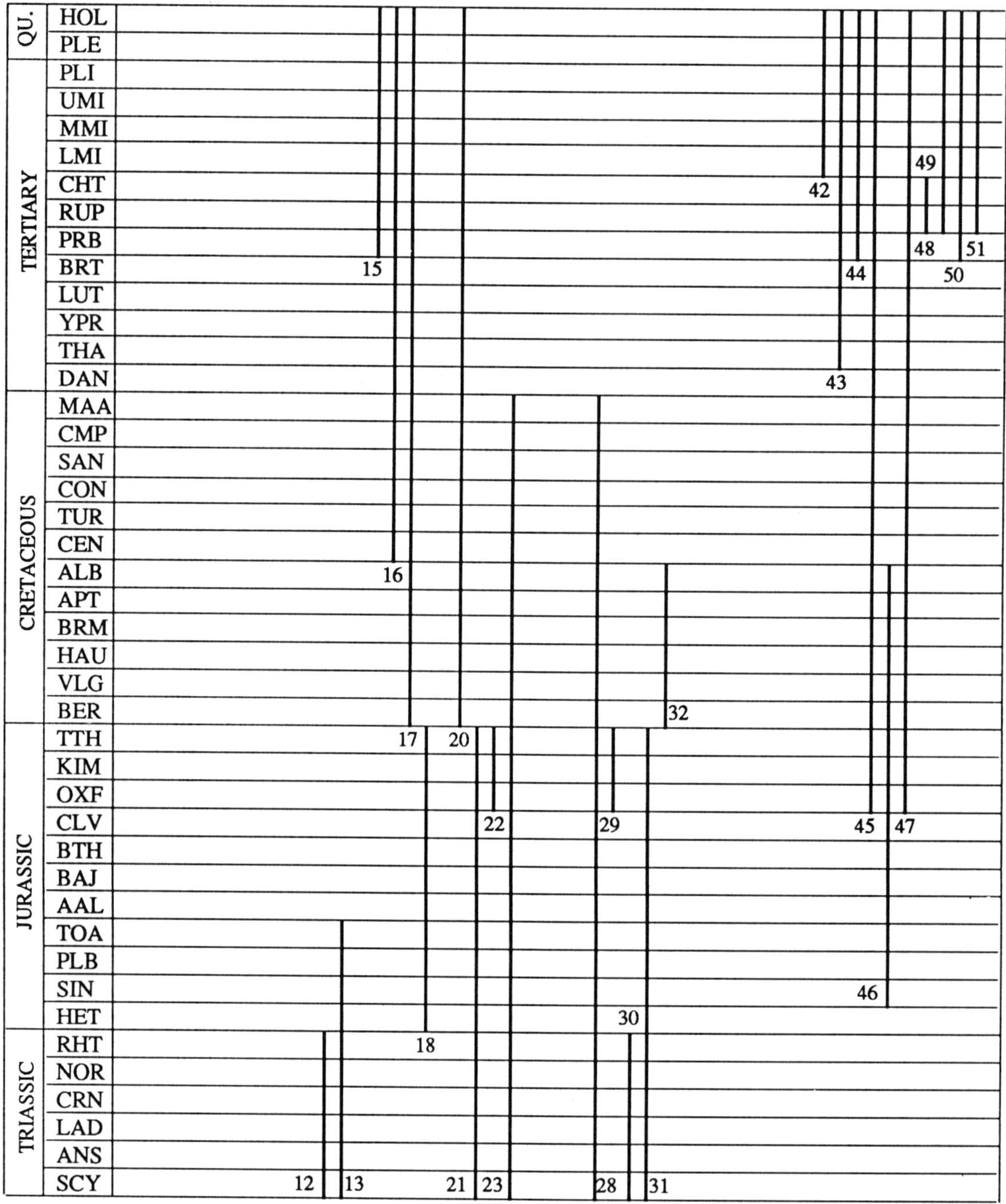

Fig. 21.4

First: e.g. *Methana*? sp. Jell and Duncan (1986), Koonwarra Fossil Bed, south Gippsland, Australia. **Extant**

F. BLATTULIDAE J. (l.–u.) Terr.

First: e.g. *Mesoblattula sincera* Lin (1986), south China.

F. COMPSOBLATTIDAE C. (STE)–P. (ROT) Terr.

Last: e.g. *Compsoblatta frankei, in* Schneider (1984), Thüringer, Germany.

F. CORYDIIDAE (POLYPHAGIDAE) K. (l.)–Rec. Terr.

Extant

F. DIECHOBLATTINIDAE P. (ROT)–J. (u.) Terr.

F. LATIBLATTIDAE J. (u.) Terr.

F. MESOBLATTINIDAE C. (STE)–K. (u.) Terr.

F. MYLACRIDAE C. WES–P. (ZEC) Terr.

First: e.g. figured but undescribed *in* Jarzembowski (1989a), Farrington Formation, Writhlington, Avon, England, UK.

F. NECYMYLACRIDAE C. (WES B–STE) Terr.

First: e.g. *Necymylacris handlirschi, in* Schneider (1984), Mercer group, Brookville, USA.

Last: e.g. *Necymylacris* sp. *in* Schneider (1984), former USSR.

F. PHYLOBLATTIDAE C. (WES)–P. Terr.

Last: e.g. *Phyloblatta*? sp. Brauckmann and Willmann (1990), Weiterstadt 1 borehole, Rhine Valley, Germany.

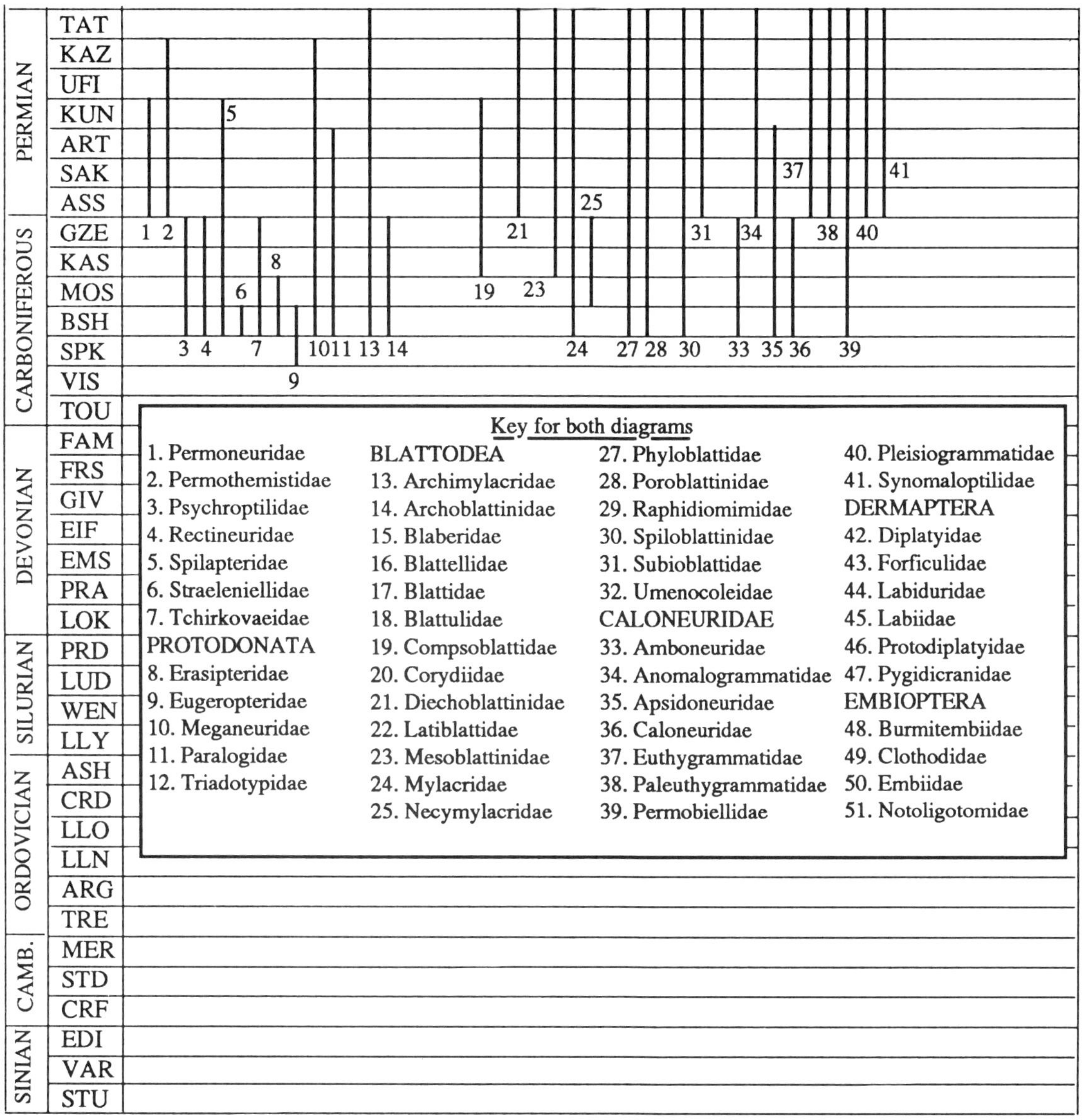
PERMIAN
TAT
KAZ
UFI
KUN
ART
SAK
ASS
CARBONIFEROUS
GZE
KAS
MOS
BSH
SPK
VIS
TOU
DEVONIAN
FAM
FRS
GIV
EIF
EMS
PRA
LOK
SILURIAN
PRD
LUD
WEN
LLY
ORDOVICIAN
ASH
CRD
LLO
LLN
ARG
TRE
CAMB.
MER
STD
CRF
SINIAN
EDI
VAR
STU
1 2
3 4
5
6
7
8
9
10 11 13 14
19 23
21
24
25
27 28 30
31
33
34
35 36
37
38
39
40
41
Key for both diagrams
1. Permoneuridae
2. Permothemistidae
3. Psychroptilidae
4. Rectineuridae
5. Spilapteridae
6. Straeleniellidae
7. Tchirkovaeidae
PROTODONATA
8. Erasipteridae
9. Eugeropteridae
10. Meganeuridae
11. Paralogidae
12. Triadotypidae
BLATTODEA
13. Archimylacridae
14. Archoblattinidae
15. Blaberidae
16. Blattellidae
17. Blattidae
18. Blattulidae
19. Compsoblattidae
20. Corydiidae
21. Diechoblattinidae
22. Latiblattidae
23. Mesoblattinidae
24. Mylacridae
25. Necymylacridae
27. Phyloblattidae
28. Poroblattinidae
29. Raphidiomimidae
30. Spiloblattinidae
31. Subioblattidae
32. Umenocoleidae
CALONEURIDAE
33. Amboneuridae
34. Anomalogrammatidae
35. Apsidoneuridae
36. Caloneuridae
37. Euthygrammatidae
38. Paleuthygrammatidae
39. Permobiellidae
40. Pleisiogrammatidae
41. Synomaloptilidae
DERMAPTERA
42. Diplatyidae
43. Forficulidae
44. Labiduridae
45. Labiidae
46. Protodiplatyidae
47. Pygidicranidae
EMBIOPTERA
48. Burmitembiidae
49. Clothodidae
50. Embiidae
51. Notoligotomidae

Fig. 21.4

F. POROBLATTINIDAE C. (WES D)–K. (u.) Terr.

First: *Poroblatta duffienxi, in* Schneider (1984), Lens, France.

F. RAPHIDIOMIMIDAE J. (u.) Terr.

F. SPILOBLATTINIDAE C. (WES)–Tr. (u.) Terr.

First: e.g. *Kinklidoblatta morini, in* Schneider (1984), Nord-Pas-de-Calais, France.

F. SUBIOBLATTIDAE P. (ROT)–J. (u.) Terr.

Last: *Subioblatta karatavica, in* Schneider (1984), Karatau, Kazakhstan, former USSR.

F. UMENOCOLEIDAE K. (l.) Terr.

We have followed Ponomarenko (pers. comm.) in considering that this family belongs to this order.

Order CALONEURODEA (CALONEURIDA) C. (u.)–P. Terr.

We do not consider that the Lower Cretaceous family Mesogrammatidae Hong, 1985d belongs in this order, because the only known specimen does not show the ordinal characters as given by Burnham (1984). This family probably belongs in the Orthoptera.

F. AMBONEURIDAE C. (u.) Terr.

F. ANOMALOGRAMMATIDAE P. Terr.

F. APSIDONEURIDAE C. (u.)–P. (ART) Terr.

First: e.g. *Apsidoneura sottyi* Burnham (1984), Montceau-les-Mines, France.

Last: *Apsidoneura flexa, in* Burnham (1984), Wellington Formation, Elmo, Kansas, USA.

F. CALONEURIDAE C. (u.) Terr.

First and Last: *Caloneura dawsoni, in* Burnham (1984), Commentry, France.

F. EUTHYGRAMMATIDAE P. Terr.

F. PALEUTHYGRAMMATIDAE P. Terr.

e.g. *Paleuthygramma tenuis* in Kukalová-Peck (1991), Urals, former USSR.

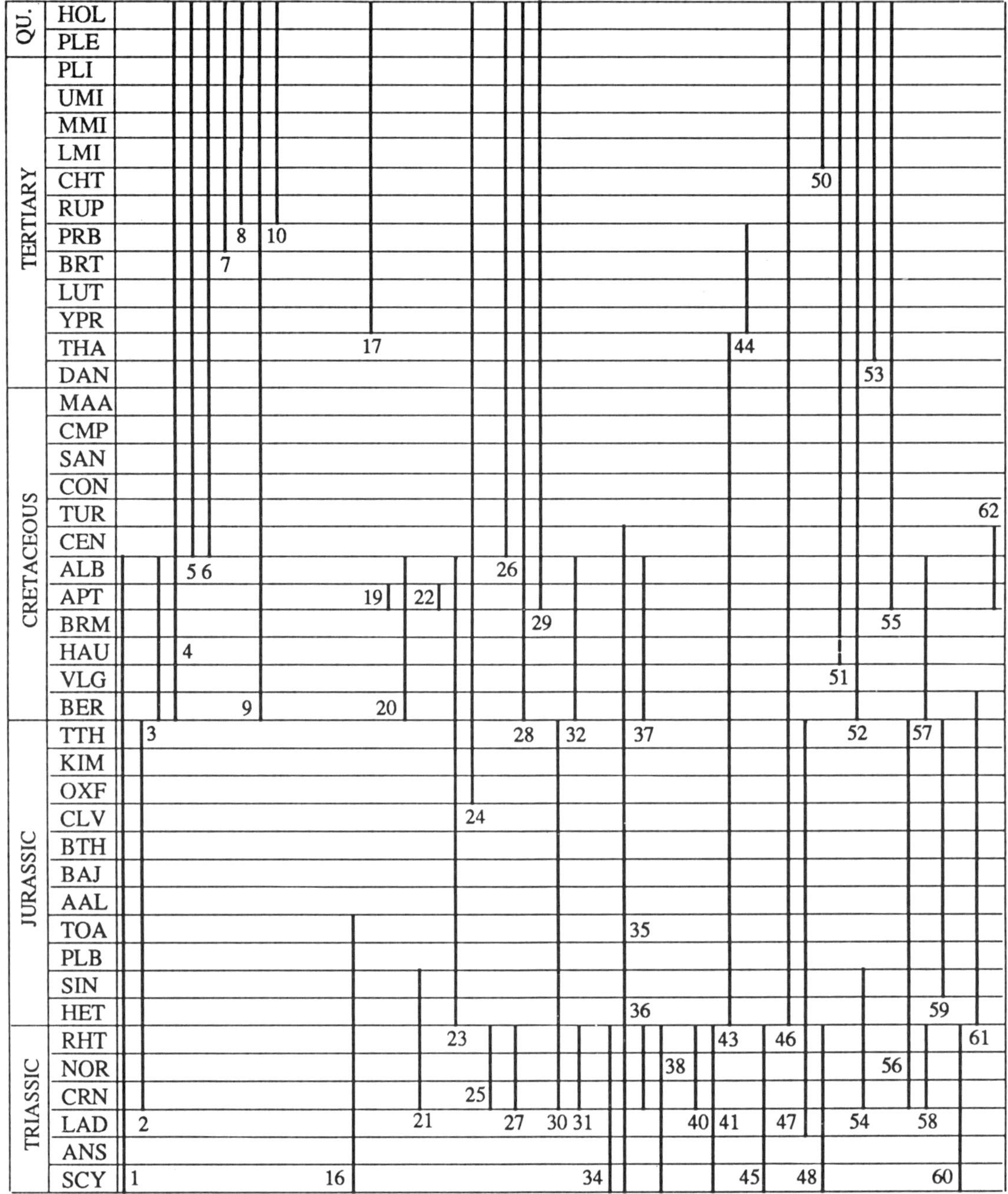

Fig. 21.5

F. PERMOBIELLIDAE C. (u.)–P. Terr.

F. PLEISIOGRAMMATIDAE P. Terr.

F. SYNOMALOPTILIDAE P. Terr.

e.g. *Synomaloptila longipes* in Kukalová-Peck (1991), Urals, former USSR.

Order DERMAPTERA (FORFICULIDA)
J. (SIN)–Rec. Terr.

First as for Protodiplatyidae.

F. DIPLATYIDAE T. (LMI)–Rec. Terr.

First: *Diplatys (Syndiplatys) protoflavicollis* Sakai and Fujiyama (1989), Seki, Sado Island, Japan. **Extant**

F. FORFICULIDAE T. (THA)–Rec. Terr.

First: *Forficula paleocaenica* Willmann (1990), Mo-clay, Knuden, Denmark. **Extant**

F. LABIDURIDAE T. (PRB)–Rec. Terr.

Extant

F. LABIIDAE J. (u.)–Rec. Terr.

First: e.g. *Semenoviola obliquotruncata* in Popham (1990), Turkestan, former USSR. **Extant**

F. PROTODIPLATYIDAE (PROTODIPLATIDAE)
J. (SIN)–K. (l.) Terr.

First: *Brevicula gradus* Whalley (1985), Lower Lias, Charmouth, Dorset, England, UK.

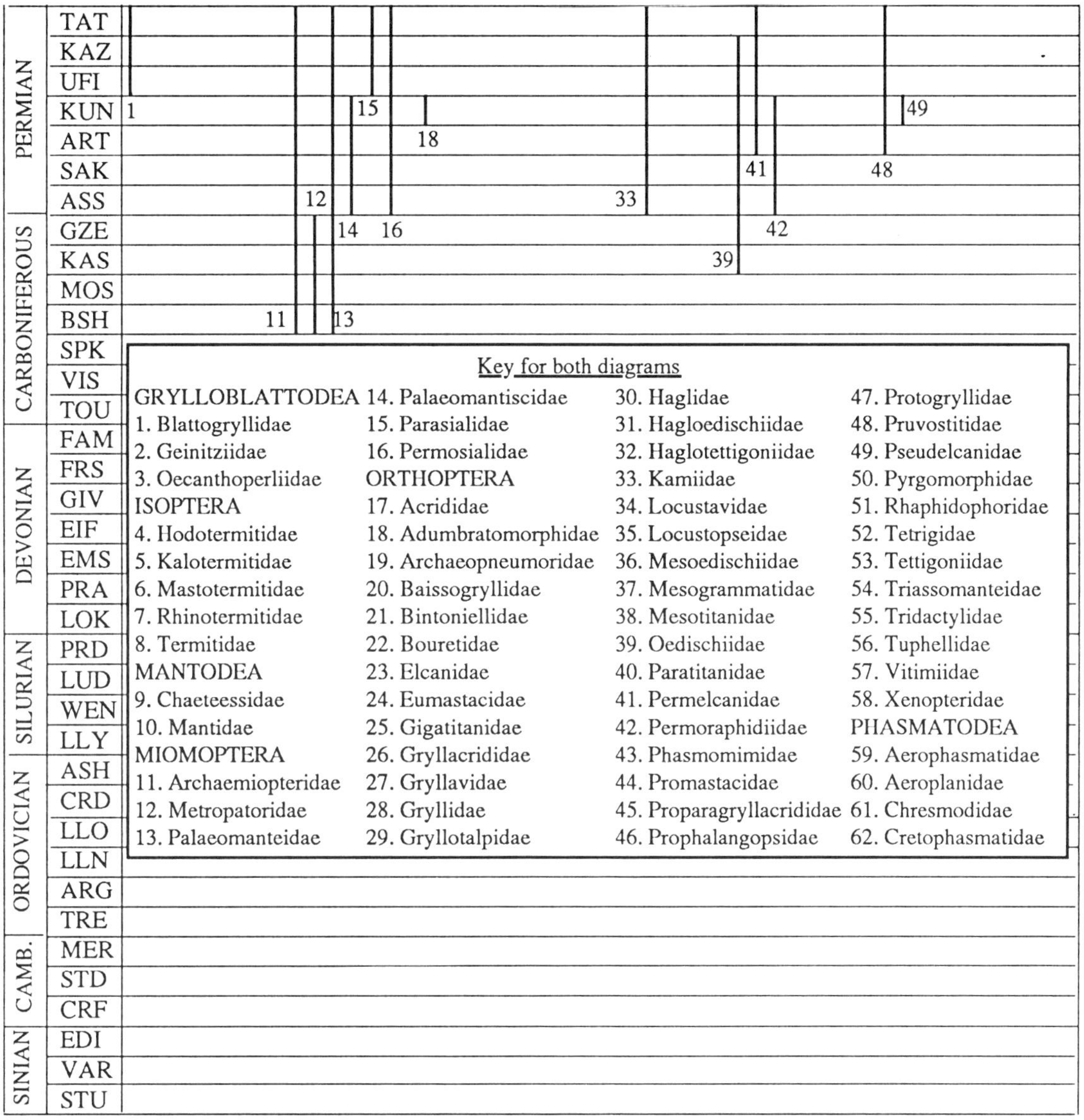

PERMIAN: TAT, KAZ, UFI, KUN, ART, SAK, ASS
CARBONIFEROUS: GZE, KAS, MOS, BSH, SPK, VIS, TOU
DEVONIAN: FAM, FRS, GIV, EIF, EMS, PRA, LOK
SILURIAN: PRD, LUD, WEN, LLY
ORDOVICIAN: ASH, CRD, LLO, LLN, ARG, TRE
CAMB.: MER, STD, CRF
SINIAN: EDI, VAR, STU

1 15 49 18 41 48 12 33 14 16 42 39 11 13

Key for both diagrams

GRYLLOBLATTODEA	14. Palaeomantiscidae	30. Haglidae	47. Protogryllidae
1. Blattogryllidae	15. Parasialidae	31. Hagloedischiidae	48. Pruvostitidae
2. Geinitziidae	16. Permosialidae	32. Haglotettigoniidae	49. Pseudelcanidae
3. Oecanthoperliidae	ORTHOPTERA	33. Kamiidae	50. Pyrgomorphidae
ISOPTERA	17. Acrididae	34. Locustavidae	51. Rhaphidophoridae
4. Hodotermitidae	18. Adumbratomorphidae	35. Locustopseidae	52. Tetrigidae
5. Kalotermitidae	19. Archaeopneumoridae	36. Mesoedischiidae	53. Tettigoniidae
6. Mastotermitidae	20. Baissogryllidae	37. Mesogrammatidae	54. Triassomanteidae
7. Rhinotermitidae	21. Bintoniellidae	38. Mesotitanidae	55. Tridactylidae
8. Termitidae	22. Bouretidae	39. Oedischiidae	56. Tuphellidae
MANTODEA	23. Elcanidae	40. Paratitanidae	57. Vitimiidae
9. Chaeteessidae	24. Eumastacidae	41. Permelcanidae	58. Xenopteridae
10. Mantidae	25. Gigatitanidae	42. Permoraphidiidae	PHASMATODEA
MIOMOPTERA	26. Gryllacrididae	43. Phasmomimidae	59. Aerophasmatidae
11. Archaemiopteridae	27. Gryllavidae	44. Promastacidae	60. Aeroplanidae
12. Metropatoridae	28. Gryllidae	45. Proparagryllacrididae	61. Chresmodidae
13. Palaeomanteidae	29. Gryllotalpidae	46. Prophalangopsidae	62. Cretophasmatidae

Fig. 21.5

Last: *Protodiplatys mongoliensis* Vishniakova (1986), Gurvaneren Formation, Mongolia.

F. PYGIDICRANIDAE (PYGIDIOCRANIDAE) J. (u.)–Rec. Terr.

Extant

Order EMBIOPTERA (EMBIIDA) P. (ROT)–Rec. Terr.

First: Figured but undescribed in Kukalová-Peck (1991), Urals, former USSR.

F. BURMITEMBIIDAE T. (Oli.) Terr.

F. CLOTHODIDAE T. (Oli.)–Rec. Terr.

Extant

F. EMBIIDAE T. (PRB)–Rec. Terr.

Extant

F. NOTOLIGOTOMIDAE T. (Oli.)–Rec. Terr.

Extant

Order GRYLLOBLATTODEA (GRYLLOBLATTIDA *pars*) P. (ZEC)–Rec. Terr. (see Fig. 21.5)

Some authors regard this order as Recent only; however, Storozhenko (1988, 1991) would also include a number of families which we have retained in the Protorthoptera.

F. BLATTOGRYLLIDAE P. (ZEC)–K. (l.) Terr.

First: *Protoblattogryllus zajsanicus* Storozhenko (1990b), Akkolka Formation, Karaungur, Kazakhstan, former USSR. **Last:** *Parablattogryllus obscurus* Storozhenko (1988), Zaza Formation, Baisa, Buryat ASSR, former USSR.

F. GEINITZIIDAE Tr. (u.)–J. (u.) Terr.

First: e.g. *Fletchitzia picturata*, *in* Martins-Neto (1991), Molteno Formation, Natal, South Africa.

F. OECANTHOPERLIDAE Storozhenko, 1988 K. (l.) Terr.

First and Last: *Oecanthoperla sibirica* Storozhenko (1988), Zaza Formation, Baisa, Buryat ASSR, former USSR.

Order ISOPTERA (TERMITIDA) K. (BER)–Rec. Terr.

First as for Hodotermitidae.

F. HODOTERMITIDAE K. (BER)–Rec. Terr.

First: *Meiatermes bertrani* Lacasa-Ruiz and Martínez-Delclòs (1986), Lithographic Limestone, Montsech, Spain. **Extant**

F. KALOTERMITIDAE (CALOTERMITIDAE) K. (u.)–Rec. Terr.

First: Termite nest in Rohr *et al*. (1986), Javelina Formation, Big Bend National Park, Texas, USA. The Lower Cretaceous *Hebeitermes* (= *Hopeitermes*) *weichangensis* in Hong (1985d) is a cockroach hindwing (Blattodea). **Extant**

F. MASTOTERMITIDAE K. (u.)–Rec. Terr.

Extant

F. RHINOTERMITIDAE T. (PRB)–Rec. Terr.

Extant

F. TERMITIDAE T. (Oli.)–Rec. Terr.

Extant

Order MANTODEA (MANTIDA, MANTEODEA) K. (l.)–Rec. Terr.

For new fossil families and species, see Gratshev and Zherikhin (in press).

F. CHAETEESSIDAE (CHAETEESSIIDAE) K. (l.)–Rec. Terr.

Extant

F. MANTIDAE (MANTEIDAE) T. (Oli.)–Rec. Terr.

Extant

Order MIOMOPTERA (PALAEOMANTEIDA) C. (u.)–J. (l.) Terr.

Kukalová-Peck (1991) considers this order belongs to the Oligoneoptera.

F. ARCHAEMIOPTERIDAE C. (u.)–P. Terr.

F. METROPATORIDAE C. (u.) Terr.

F. PALAEOMANTEIDAE C. (u.)–P. Terr.

F. PALAEOMANTISCIDAE P. (ROT) Terr.

F. PARASIALIDAE P. (ZEC) Terr.

F. PERMOSIALIDAE (PERMOSIALIDIDAE, EPIMASTACIDAE) P.–J. (l.) Terr.

Order ORTHOPTERA (GRYLLIDA, TITANOPTERA) C. (STE)–Rec. Terr.

First as for Oedischiidae.

F. ACRIDIDAE T. (Eoc)–Rec. Terr.

Extant

F. ADUMBRATOMORPHIDAE Gorokhov, 1987c P. (KUN) Terr.

Extant

First and Last: *Adumbratomorpha tettigonioides* Gorokhov (1987c), Chekarda, former USSR.

F. ARCHAEOPNEUMORIDAE (*nomen nudum*) Martins-Neto, 1987b K. (APT) Terr.

First and Last: *Archaeopneumora cretacea* (*nomen nudum*) Martins-Neto (1987b), Santana Formation, Ceará, Brazil.

F. BAISSOGRYLLIDAE Gorokhov, 1985 K. (l.) Terr.

e.g. *Baissogryllus sharovi*, *in* Gorokhov (1985), Zaza Formation, Baisa, Buryat ASSR, former USSR.

F. BINTONIELLIDAE Tr. (u.)–J. (SIN) Terr.

First: e.g. *Oshiella crassa* Gorokhov (1987b), Madygen, former USSR.
Last: *Bintoniella brodiei* in Martins-Neto (1991), Lower Lias, England, UK.

F. BOURETIDAE (*nomen nudum*) Martins-Neto, 1987b K. (APT) Terr.

First and Last: *Bouretia elegans* (*nomen nudum*) Martins-Neto (1987b), Santana Formation, Ceará, Brazil.

F. ELCANIDAE J. (l.)–K. (l.) Terr.

First: e.g. *Elcana liasina*, *in* Zessin (1987), Lower Lias, Strensham, Worcestershire, England, UK.
Last: e.g. *Eubaisselcana sharovi* Gorokhov (1986a), Gurvaneren Formation, Mongolia.

F. EUMASTACIDAE J. (u.)–Rec. Terr.

Extant

F. GIGATITANIDAE Tr. (u.) Terr.

F. GRYLLACRIDIDAE (GRYLLACRIDAE) K. (u.)–Rec. Terr.

The Triassic *Xenogryllacris reductus* and Jurassic *Jurassobatea gryllacroides* listed in Martins-Neto (1991) do not belong in this family. **Extant**

F. GRYLLAVIDAE Gorokhov, 1986b Tr. (u.) Terr.

e.g. *Paragryllavus curvatus* Gorokhov (1986b), Madygen, former USSR.

F. GRYLLIDAE (ENEOPTERIDAE, OECANTHIDAE, MYRMECOPHILIDAE, TRIGONIDIIDAE) K. (l.)–Rec. Terr.

First: e.g. *Gryllospeculum mongolicum* Gorokhov (1985), Bon-Tsagan, Mongolia. **Extant**

F. GRYLLOTALPIDAE K. (APT)–Rec. Terr.

First: e.g. *Palaeoscapteriscops cretacea* Martins-Neto (1991), Santana Formation, Ceará, Brazil. **Extant**

F. HAGLIDAE Tr. (u.)–J. (u.) Terr.

First: e.g. *Hagloptera intermedia* Gorokhov (1986b), Madygen, former USSR.

F. HAGLOEDISCHIIDAE Gorokhov, 1986b Tr. (u.) Terr.

First and Last: *Hagloedischia primitiva* Gorokhov (1986b), Madygen, former USSR.

F. HAGLOTETTIGONIIDAE Gorokhov, 1988b K. (l.) Terr.

First and Last: *Haglotettigonia egregia* Gorokhov (1988b), River Vitim, former USSR.

F. KAMIIDAE P. (ROT–ZEC) Terr.

F. LOCUSTAVIDAE Tr. Terr.

F. LOCUSTOPSEIDAE (LOCUSTOPSIDAE) Tr. (SCY)–K. (CEN) Terr.

First: *Praelocustopsis mirabilis*, Siberia, former USSR.
Last: *Zeunerella arborea*, *in* Ansorge (1991), Kzyl Dzhar, Kazakhstan, former USSR.

F. MESOEDISCHIIDAE Gorokhov, 1987b Tr. (u.) Terr.

e.g. *Mesoedischia kirgizica* Gorokhov (1987b), Madygen, former USSR.

F. MESOGRAMMATIDAE Hong, 1985d K. (l.) Terr.

First and Last: *Mesogramma divaricata* Hong (1985d), north China. See Caloneurodea.

F. MESOTITANIDAE (CLATROTITANIDAE) Tr. Terr.

e.g. *Clatrotitan scollyi* in Kukalová-Peck (1991), Hawkesbury Sandstone, Sydney, New South Wales, Australia.

F. OEDISCHIIDAE (TCHOLMANVISSIIDAE, ANELCANIDAE, PARELCANIDAE) C. (STE)–P. (KAZ) Terr.

First: e.g. *Oedischia williamsoni* in Kukalová-Peck (1991), Commentry, France.
Last: e.g. *Mezenoedischia maculosa* Gorokhov (1987a), Iva-Gora, Archangelsk, former USSR.
The Carboniferous species of *Plesioidischia*, *in* Jarzembowski (1988) probably belong in the Protorthoptera.

F. PARATITANIDAE Tr. (u.) Terr.

F. PERMELCANIDAE P. (ART)–Tr. (u.) Terr.

First: *Promartynovia venicosta*, *in* Zessin (1987), Wellington Formation, Elmo, Kansas, USA.
Last: e.g. *Meselcana permelcanoides* Gorokhov (1989), Madygen, former USSR.

F. PERMORAPHIDIIDAE P. (ROT) Terr.

e.g. *Permoraphidia americana*, *in* Oswald (1990), Wellington Formation, Elmo, Kansas, USA.

F. PHASMOMIMIDAE J. (l.)–T. (THA) Terr.

First: *Paraphasmomima sharovi* Zherikhin (1985a), Iya, Irkutsk Basin, former USSR.
Last: e.g. *Promastacoides albertae*, *in* Martins-Neto (1991), Paskapoo Formation, Alberta, Canada.

F. PROMASTACIDAE T. (Eoc.) Terr.

Gorokhov (1988a) transferred the Palaeocene genus *Promastacoides* to the Phasmomimidae.

F. PROPARAGRYLLACRIDIDAE Tr. Terr.

e.g. *Paraferganía sharovi* Gorokhov (1987c), Madygen, former USSR.

F. PROPHALANGOPSIDAE J. (l.)–Rec. Terr.

First: e.g. *Protaboilus praedictus* Gorokhov (1988c), Kizyl-Kiya, Kirgizia, former USSR. **Extant**

F. PROTOGRYLLIDAE Tr. (u.)–J. (u.) Terr.

First: *Protogryllus stormbergensis*, *in* Martins-Neto (1991), Stormberg Series, South Africa.
Last: e.g. *Falsispeculum karatavicus*, *in* Gorokhov (1985), Karatau, Kazakhstan, former USSR.

F. PRUVOSTITIDAE (TETTAVIDAE) P. (ART)–Tr. (u.) Terr.

First: *Paroedischia recta*, Wellington Formation, Elmo, Kansas, USA.
Last: e.g. *Provitimia pectinata*, *in* Martins-Neto (1991), Madygen Formation, Madygen, former USSR.

F. PSEUDELCANIDAE Gorokhov, 1987b P. (KUN) Terr.

e.g. *Pseudelcana permiana* Gorokhov (1987b), Chekarda, former USSR.

F. PYRGOMORPHIDAE T. (Mio)–Rec. Terr.

Extant

F. RHÁPHIDOPHORIDAE K. (?HAU)–Rec. Terr.

First: Figured but undescribed in Martínez-Delclòs (1989a), Las Hoyas, Cuenca, Spain. **Extant**

F. TETRIGIDAE K. (l.)–Rec. Terr.

Extant

F. TETTIGONIIDAE (LOCUSTIDAE, CONOCEPHALIDAE) T. (THA)–Rec. Terr.

First: *Pseudotettigonia amoena*, *in* Martins-Neto (1991), Moclay, Demmark. The Jurassic *Termitidium ignotum* listed in Martins-Neto (1991) probably belongs in the Prophalangopsidae. **Extant**

F. TRIASSOMANTEIDAE (TRIASSOMANTIDAE) Tr. (u.)–J. (SIN) Terr.

First: e.g. *Triassomanteodes madygenicus*, *in* Martins-Neto (1991) Madygen Formation, Madygen, former USSR.
Last: *Orichalcum ornatum* Whalley (1985), Lower Lias, Charmouth, Dorset, England, UK.

F. TRIDACTYLIDAE K. (APT)–Rec. Terr.

First: e.g. *Cratodactylus ferreirai* Martins-Neto (1990), Santana Formation, Ceará, Brazil. **Extant**

F. TUPHELLIDAE Gorokhov, 1988c Tr. (u.)–J. (u.) Terr.

First: e.g. *Tuphella rasnitzyni* Gorokhov (1986b), Madygen, former USSR.

F. VITIMIIDAE K. (l.) Terr.

e.g. *Deinovitimia insoluta* Gorokhov (1989), Zaza Formation, Baisa, Buryat ASSR, former USSR. Gorokhov (1987a) transferred the Triassic genus *Provitimia* to the Pruvostitidae.

F. XENOPTERIDAE Tr. (u.) Terr.

e.g. *Xenoferganella pini* Gorokhov (1989), Madygen, former USSR.

Order PHASMATODEA (PHASMATIDA, PHASMIDA) Tr. (SCY)–Rec. Terr.

F. AEROPHASMATIDAE J. (SIN–u.) Terr.

First: *Durnovaria parallela* Whalley (1985), Lower Lias, Charmouth, Dorset, England, UK.

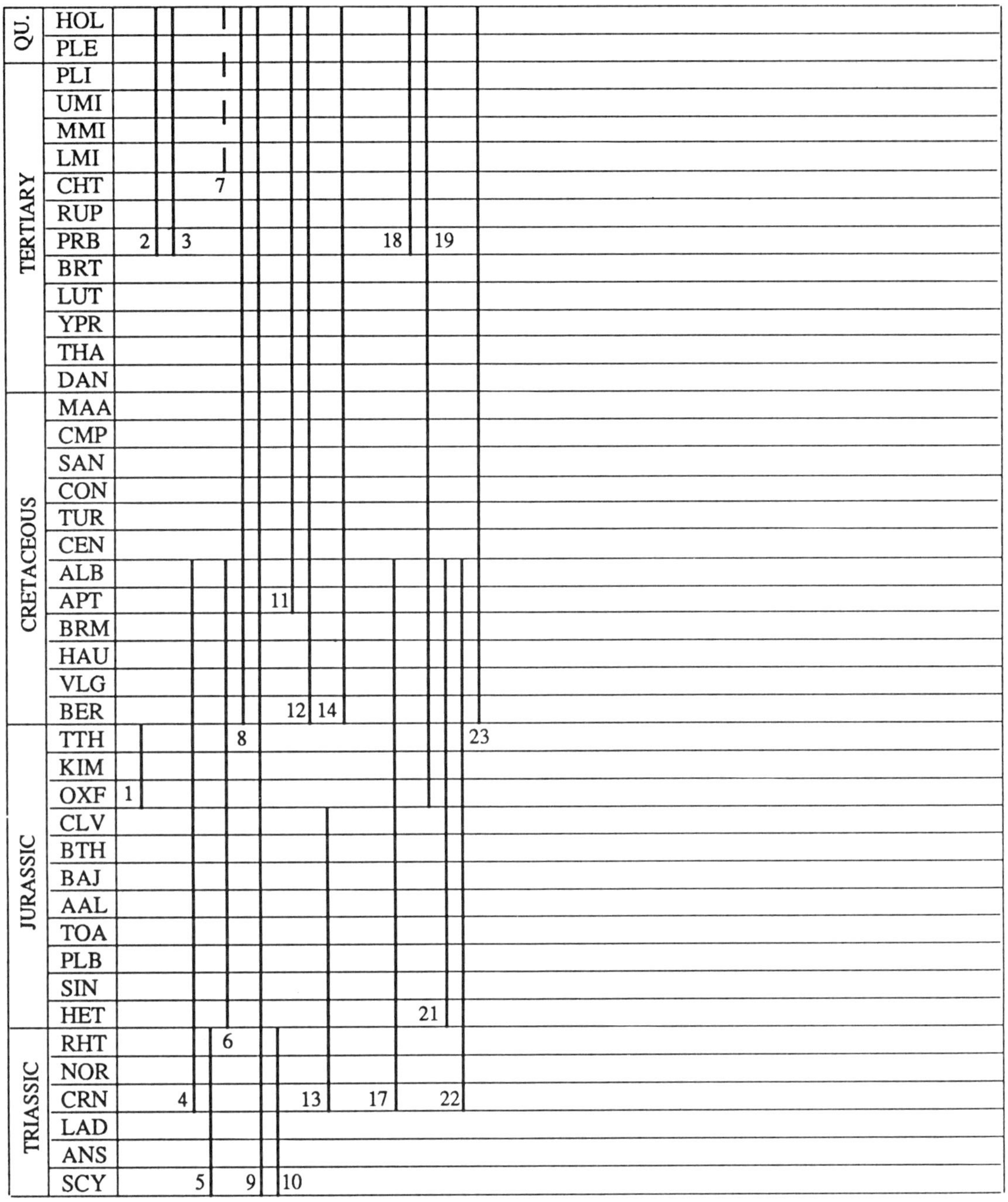

Fig. 21.6

F. AEROPLANIDAE Tr. Terr.

e.g. *Aeroplana mirabilis*, *in* Rozefelds (1985), Australia.

F. CHRESMODIDAE J.–K. (BER) Terr.

Last: *Chresmoda aquatica* Martinez-Delclòs (1989b), Lithographic Limestone, Montsech, Spain.
Some authors regard this family as belonging in other orders or as Order *incertae sedis*.

F. CRETOPHASMATIDAE K. (APT–CEN) Terr.

First: *Cretophasma araripensis* Martins-Neto (1989), Santana Formation, Ceará, Brazil.
Last: e.g. *Cretophasma raggei*, *in* Martins-Neto (1989), Kzyl-Dzhar, Karatau, Kazakhstan, former USSR.

F. NECROPHASMATIDAE J. (u.) Terr. (Fig. 21.6)

F. PHASMATIDAE T. (PRB)–Rec Terr.

Extant

F. PHYLLIIDAE (PHYLLIDAE) T. (PRB)–Rec. Terr.

Extant

F. PROCHRESMODIDAE Tr. (u.)–K. (l.) Terr.

F. XIPHOPTERIDAE Tr. Terr.

Order PLECOPTERA (PERLIDA, PERLARIA) P. (KUN)–Rec. Terr.

Oldest species from Chekarda (former USSR), see Palaeonemouridae, Perlopseidae and Tshekardoperlidae. Most data are taken from Sinichenkova (1987).

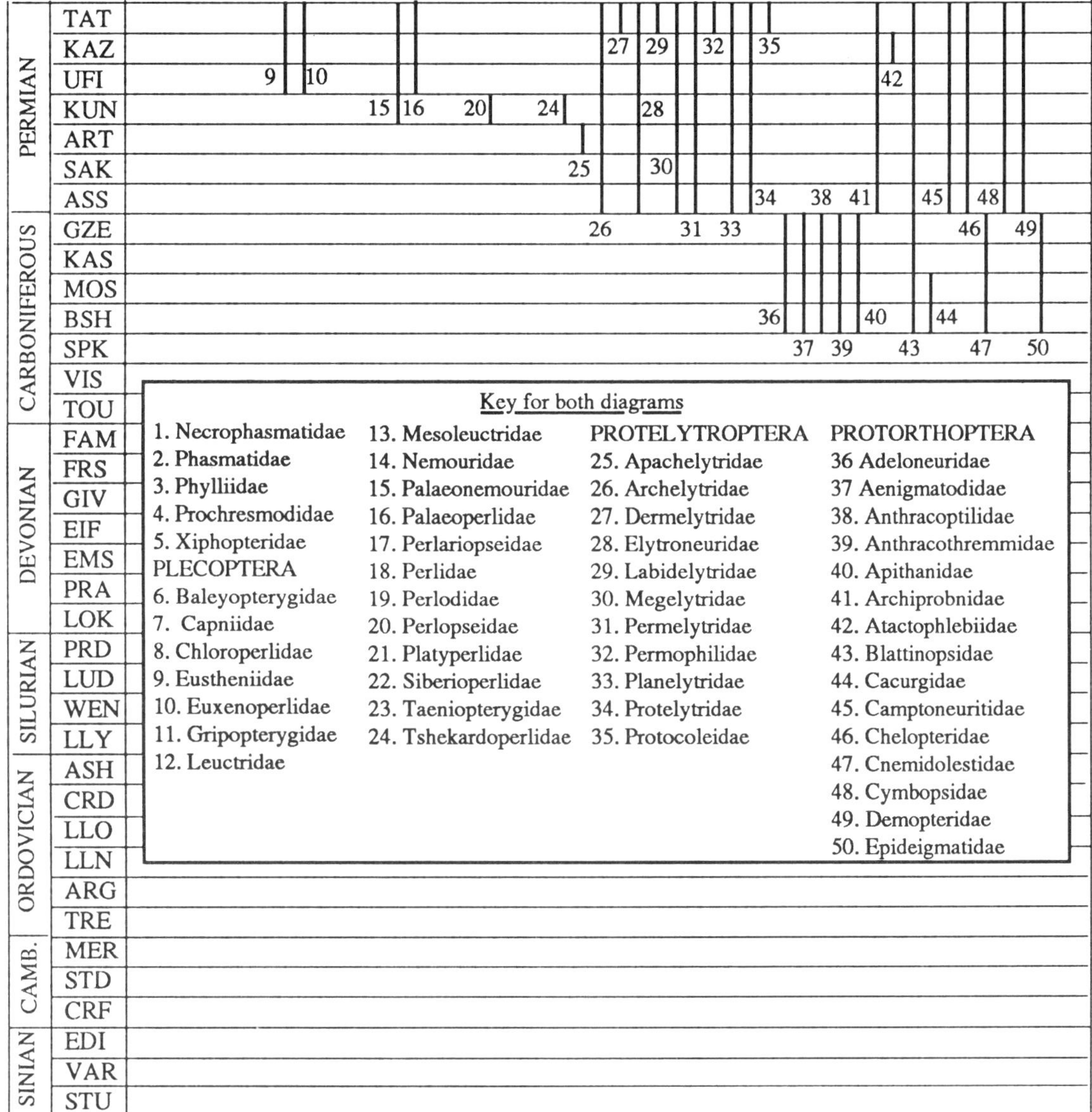

Fig. 21.6

F. BALEYOPTERYGIDAE Sinichenkova, 1985b J. (l.)–K. (l.) Terr.

First: e.g. *Baleyopteryx altera*, Kuznetsk Basin, Mongolia.
Last: e.g. *Baissoleuctra irinae*, Transbaikalia, former USSR.

F. CAPNIIDAE T.? (Mio.)–Rec. Terr.

Extant

F. CHLOROPERLIDAE K. (l.)–Rec. Terr.

First: e.g. *Dipsoperla serpentis*, Transbaikalia, former USSR. **Extant**

F. EUSTHENIIDAE P. (ZEC)–Rec. Terr.

First: *Stenoperlidium permianum*, Australia. **Extant**

F. EUXENOPERLIDAE P. (ZEC)–Tr. (u.) Terr.

First: e.g. *Euxenoperla simplex*, Middle Beaufort Series, Natal, South Africa.
Last: e.g. *Gondwanoperlidium argentinarum*, Potrevillos Formation, Mendoza, Argentina.

F. GRIPOPTERYGIDAE K. (APT)–Rec. Terr.

First: *Eodinotoperla duncanae* Jell and Duncan (1986), Koonwarra Fossil Bed, south Gippsland, Australia. **Extant**

F. LEUCTRIDAE (LEUCTRIIDAE) K. (l.)–Rec. Terr.

First: *Lycoleuctra lupina*, Transbaikalia, former USSR. **Extant**

F. MESOLEUCTRIDAE Tr. (u.)–J. (m.) Terr.

e.g. *Mesoleuctra gracilis*, Mongolia.

F. NEMOURIDAE K. (l.)–Rec. Terr.

First: *Nemourisca diligens*, Transbaikalia, former USSR. **Extant**

F. PALAEONEMOURIDAE Sinichenkova, 1987 P. (KUN–ZEC) Terr.

First: e.g. *Uralonympha varica*, Chekarda, former USSR.
Last: e.g. *Palaeonemoura clara*, Kuznetsk Basin, Mongolia.

F. PALAEOPERLIDAE P. (ZEC) Terr.

e.g. *Palaeoperla exacta*, Kuznetsk Basin, Mongolia.

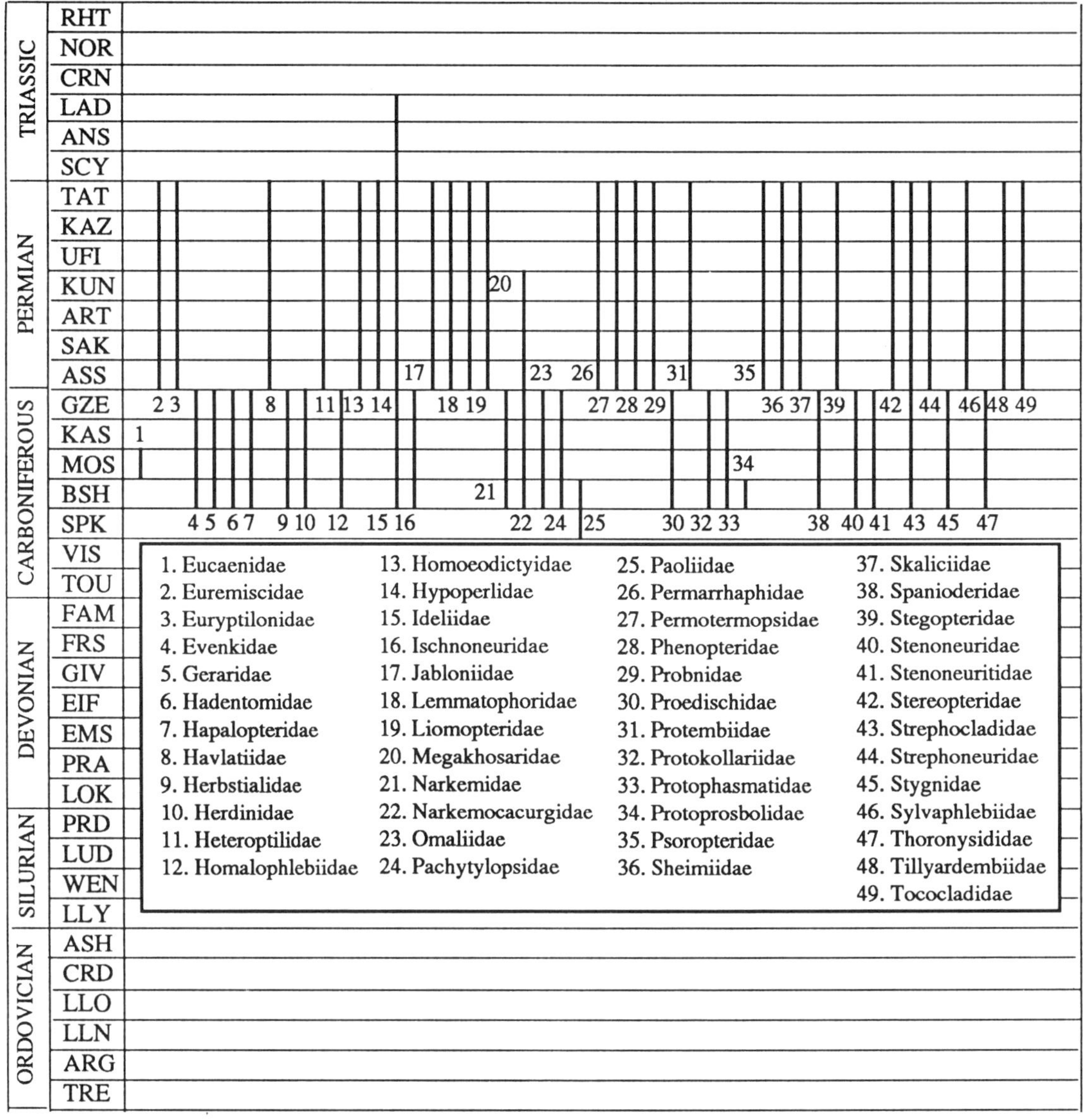

Fig. 21.7

F. PERLARIOPSEIDAE Sinichenkova, 1985b Tr. (u.)–K. (l.) Terr.

First: e.g. *Fritaniopsis brevicaulis*, former USSR.
Last: e.g. *Accretonemoura radiata*, Mongolia.

F. PERLIDAE T. (PRB)–Rec. Terr.

First: *Perla prisca*, Baltic amber. **Extant**

F. PERLODIDAE J. (u.)–Rec. Terr.

First: *Derancheperla collaris* Sinichenkova (1990b), Arhangay Aymag, Mongolia. **Extant**

F. PERLOPSEIDAE P. (KUN) Terr.

e.g. *Perlopsis filicornis*, Chekarda, former USSR.

F. PLATYPERLIDAE J. (l.)–K. (l.) Terr.

First: e.g. *Platyperla platypoda*, former USSR.
Last: *Platyperla parricidalis* Sinichenkova (1990a), Polosatik, Transbaikalia, former USSR.

F. SIBERIOPERLIDAE Tr. (u.)–K. (l.) Terr.

Last: e.g. *Flexoperla flexuosa*, Transbaikalia, former USSR.

F. TAENIOPTERYGIDAE K. (l.)–Rec. Terr.

First: e.g. *Gurvanopteryx effeta*, Mongolia. **Extant**

F. TSHEKARDOPERLIDAE Sinichenkova, 1987 P. (KUN) Terr.

e.g. *Tshekardoperla expulsa*, Chekarda, former USSR.

Order PROTELYTROPTERA (PROTELYTRIDA) P. Terr.

F. APACHELYTRIDAE P. (ART) Terr.

First and Last: Apachelytron transversum, *in* Shear and Kukalová-Peck (1990), Obora, Moravia, Czechoslovakia.

F. ARCHELYTRIDAE P. Terr.

F. DERMELYTRIDAE P. (TAT) Terr.

F. ELYTRONEURIDAE P. Terr.

F. LABIDELYTRIDAE (STENELYTRIDAE) Kukalová-Peck, 1988 P. (TAT) Terr.

e.g. *Labidelytron enervatum* in Kukalová-Peck (1988), Belmont, New South Wales, Australia.

F. MEGELYTRIDAE P. Terr.

F. PERMELYTRIDAE P. Terr.

F. PERMOPHILIDAE P. (TAT) Terr.

F. PLANELYTRIDAE P. Terr.

F. PROTELYTRIDAE P. Terr.

e.g. *Protelytron permianum*, *in* Jarzembowski (1990b), Wellington Formation, Elmo, Kansas, USA.

F. PROTOCOLEIDAE P. (TAT) Terr.

e.g. *Phyllelytron petalon* in Kukalová-Peck (1991), Belmont, New South Wales, Australia.

Order PROTORTHOPTERA *s.l.* (STHAROPODINA, PARAPLECOPTERA, GERARIDA, BLATTINOPSODEA, HYPOPERLIDA, GRYLLOBLATTIDA *pars*)
C. (NAM A)–Tr. Terr.

We have followed the traditional broad definition of this order *sensu* F. M. Carpenter. See Rodendorf and Rasnitsyn (1980) for an alternative classification. First as for Paoliidae.

F. ADELONEURIDAE C. (u.) Terr.

F. AENIGMATODIDAE C. (u.) Terr.

F. ANTHRACOPTILIDAE C. (u.) Terr.

F. ANTHRACOTHREMMIDAE C. (u.) Terr.

F. APITHANIDAE C. (u.) Terr.

F. ARCHIPROBNIDAE (ARCHIPROBNISIDAE) P. Terr.

F. ATACTOPHLEBIIDAE P. (KAZ) Terr.

e.g. *Atactophlebia termitoides*, *in* Storozhenko (1990a), Kama River, former USSR.

F. BLATTINOPSIDAE C. (u.)–P. Terr.

F. CACURGIDAE C. (NAM B–WES D) Terr.

First: e.g. *Kochopteron hoffmannorum*, *in* Brauckmann *et al.* (1985), Vorhalle Beds, Hagen-Vorhalle, Germany.
Last: e.g. *Cacurgus spilopterus*, *in* Brauckmann *et al.* (1985), Carbondale Formation, Mazon Creek, Illinois, USA.

F. CAMPTONEURITIDAE P. Terr.

F. CHELOPTERIDAE P. Terr.

F. CNEMIDOLESTIDAE C. (u.) Terr.

F. CYMBOPSIDAE P. Terr.

F. DEMOPTERIDAE P. Terr.

F. EPIDEIGMATIDAE C. (u.) Terr.

F. EUCAENIDAE C. (WES D) Terr.

First and Last: *Eucaenus ovalis*, *in* Baird *et al.* (1985), Carbondale Formation, Mazon Creek, Illinois, USA.

F. EUREMISCIDAE P. Terr.

F. EURYPTILONIDAE P. Terr.

F. EVENKIDAE C. (u.) Terr.

F. GERARIDAE C. (u.) Terr.

e.g. *Gerarus danielsi* in Jarzembowski (1990b), Carbondale Formation, Mazon Creek, Illinois, USA.

F. HADENTOMIDAE C. (u.) Terr.

F. HAPALOPTERIDAE C. (u.) Terr.

F. HAVLATIIDAE P. Terr.

F. HERBSTIALIDAE C. (u.) Terr.

F. HERDINIDAE C. (u.) Terr.

e.g. *Herdinia mirificus*, *in* Shear and Kukalová-Peck (1990), Carbondale Formation, Mazon Creek, Illinois, USA.

F. HETEROPTILIDAE P. Terr.

F. HOMALOPHLEBIIDAE C. (u.) Terr.

F. HOMOEODICTYIDAE P. Terr.

F. HYPOPERLIDAE P. Terr.

F. IDELIIDAE C. (u.)–Tr. (m.) Terr.

F. ISCHNONEURIDAE C. (u.) Terr.

F. JABLONIIDAE P. Terr.

F. LEMMATOPHORIDAE P. (ROT–ZEC) Terr.

First: *Lemmatophora typa* in Kukalová-Peck (1991), Wellington Formation, Elmo, Kansas, USA.
Last: *Karaungirella minuta* Storozhenko (1991), Karaungur, Kazakhstan, former USSR.

F. LIOMOPTERIDAE P. Terr.

e.g. *Liomopterites germanicus* Brauckmann and Willmann (1990), Weiterstadt 1 borehole, Rhine Valley, Germany.

F. MEGAKHOSARIDAE P. Terr.

e.g. *Megakhosarodes zajsanicus*, *in* Storozhenko (1991), Karaungur, Kazakhstan, former USSR.

F. NARKEMIDAE C. (u.) Terr.

e.g. *Narkeminopsis eddi*, *in* Jarzembowski (1988), Kilmersdon Colliery, Radstock, Avon, England, UK.

F. NARKEMOCACURGIDAE C. (u.)–P. (ZEC) Terr.

First: e.g. *Narkemina rodendorfi*, *in* Martins-Neto (1987a), Boituva, Brazil.

F. OMALIIDAE C. (u.) Terr.

F. PACHYTYLOPSIDAE C. (u.) Terr.

F. PAOLIIDAE C. (NAM A–WES A) Terr.

First: *Ampeliptera limburgica*, *in* Brauckmann (1988a), Gulpen, south Limburg, The Netherlands.
Last: e.g. *Zdenekia occidentalis* Laurentiaux-Vieira and Laurentiaux (1986b), micaceous schists, Charbonnages de Ressaix, Belgium.

F. PERMARRHAPHIDAE P. Terr.

F. PERMOTERMOPSIDAE P. Terr.

F. PHENOPTERIDAE P. Terr.

F. PROBNIDAE P. Terr.

F. PROEDISCHIDAE C. (u.) Terr.

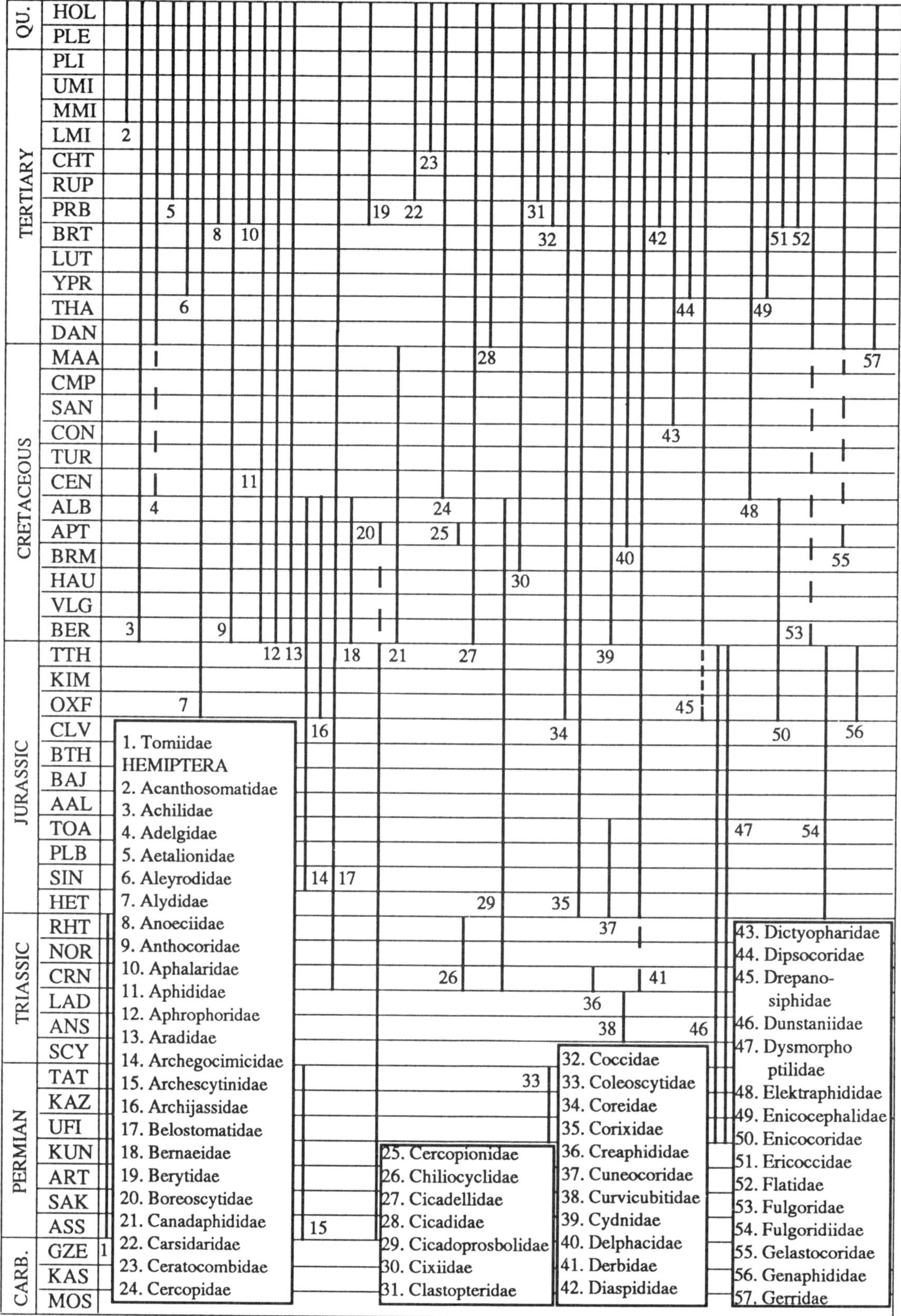

Fig. 21.8

First and Last: *Proedischia mezzalirai, in* Martins-Neto (1987a), Boituva, Brazil.

F. PROTEMBIIDAE P. Terr.

F. PROTOKOLLARIIDAE C. (u.) Terr.

F. PROTOPHASMATIDAE C. (u.) Terr.

F. PROTOPROSBOLIDAE C. (NAM B) Terr.

First and Last: *Protoprosbole straeleni, in* Brauckmann (1988a), Charleroi Coal Basin, Belgium.

F. PSOROPTERIDAE P. Terr.

F. SHEIMIIDAE P. Terr.

F. SKALICIIDAE P. Terr.

F. SPANIODERIDAE C. (u.) Terr.

F. STEGOPTERIDAE P. Terr.

F. STENONEURIDAE C. (u.) Terr.

F. STENONEURITIDAE C. (u.) Terr.

F. STEREOPTERIDAE P. Terr.

F. STREPHOCLADIDAE C. (u.)–P. Terr.

F. STREPHONEURIDAE P. Terr.

F. STYGNIDAE C. (u.) Terr.

F. SYLVAPHLEBIIDAE P. Terr.

F. THORONYSIDIDAE C. (u.) Terr.

F. TILLYARDEMBIIDAE P. Terr.

F. TOCOCLADIDAE P. Terr.

F. TOMIIDAE P.–Tr. Terr.

Superorder PARANEOPTERA (Hemipteroid orders) P. (ROT)–Rec. Terr.

Order HEMIPTERA (CIMICIDA) P. (ROT)–Rec. Terr.

Earliest families: Archescytinidae, Boreoscytidae, Ingruidae and Prosbolopseidae.

F. ACANTHOSOMATIDAE T. (MMI)–Rec. Terr.

First: e.g. *Acanthosoma* sp. Fujiyama (1987), Abura, Hokkaido, Japan. **Extant**

F. ACHILIDAE K. (l.)–Rec. Terr.

First: e.g. *Acixiites immodesta* Hamilton (1990), Santana Formation, Ceará, Brazil. **Extant**

F. ADELGIDAE K. (u.?)–Rec. Terr.

Extant

F. AETALIONIDAE T. (Oli.)–Rec. Terr.

Extant

F. ALEYRODIDAE T. (Eoc.)–Rec. Terr.

The earlier records in Boucot (1990) are extremely doubtful. **Extant**

F. ALYDIDAE J. (u.)–Rec. Terr.

Extant

F. ANOECIIDAE T. (PRB)–Rec. Terr.

First: *Berendtaphis cimicoides*, *in* Heie (1985), Baltic amber. **Extant**

F. ANTHOCORIDAE K. (l.)–Rec. Terr.

First: e.g. *Eoanthocoris cretaceus* Popov (1990), Turga Formation, Transbaikalia, former USSR. **Extant**

F. APHALARIDAE T. (PRB)–Rec. Terr.

First: *Paleopsylloides oligocaenica*, *in* Bekker-Migdisova (1985b), Baltic amber. **Extant**

F. APHIDIDAE K. (l.)–Rec. Terr.

First: e.g. *Sunaphis shandongensis* Hong and Wang (1990), Laiyang Formation, Shandong, China. **Extant**

F. APHROPHORIDAE K. (l.)–Rec. Terr.

F. ARADIDAE K. (l.)–Rec. Terr.

First: *Aradus nicholasi* Popov (1989), Bon-Tsagan, Mongolia. **Extant**

F. ARCHEGOCIMICIDAE (ARCHAEGOCIMICIDAE, EONABIDAE, DIATILLIDAE) J. (SIN)–K. (l.) Terr.

First: gen. *et* sp. indet. Whalley (1985), Lower Lias, Charmouth, Dorset, England, UK.
Last: e.g. *Sondalia kovalevi* Popov (1988b), Godymboyskaya Formation, Onokhoy, Chita, former USSR.

F. ARCHESCYTINIDAE P. (ROT–ZEC) Terr.

First: e.g. *Archescytina permiana*, *in* Wootton and Betts (1986), Wellington Formation, Elmo, Kansas, USA.
Last: e.g. *Protopincombea obscura*, *in* Bekker-Migdisova (1985a), New South Wales, Australia.

F. ARCHIJASSIDAE J. (u.)–K. (l.) Terr.

Last: *Archijassus? plurinervis* Zhang (1985), Laiyang Formation, Laiyang, Shandong, China. We have followed Hong and Wang (1990) in regarding this formation as Lower Cretaceous in age. Shcherbakov (pers. comm.) considers that this species does not belong in this family and that this family belongs in the Hylicellidae.

F. BELOSTOMATIDAE Tr. (u.)–Rec. Terr.

Extant

F. BERNAEIDAE K. (l.) Terr.

F. BERYTIDAE T. (PRB)–Rec. Terr.

Extant

F. BOREOSCYTIDAE P. (ROT)–K. ?(APT). Terr.

Last: *Megaleurodes megocellata* Hamilton (1990), Santana Formation, Ceará, Brazil. Shcherbakov (pers. comm.) considers that this species does not belong in this family.

F. CANADAPHIDIDAE K. (l.–u.) Terr.

First: *Nuuraphis gemma* Vengerek (1991), Bon-Tsagan, Mongolia.
Last: e.g. *Canadaphis carpenteri*, *in* Heie (1987), Canadian amber, Cedar Lake, Manitoba, Canada.

F. CARSIDARIDAE T. (RUP)–Rec. Terr.

First: e.g. *Carsidarina hooleyi*, *in* Bekker-Migdisova (1985b), Bembridge Marls, Isle of Wight, England, UK. **Extant**

F. CERATOCOMBIDAE T. (Mio.)–Rec. Terr.

Extant

F. CERCOPIDAE K. (u.)–Rec. Terr.

Extant

F. CERCOPIONIDAE Hamilton, 1990 K. (APT) Terr.

First and Last: *Cercopion reticulata* Hamilton (1990), Santana Formation, Ceará, Brazil.

F. CHILIOCYCLIDAE Tr. (u.) Terr.

F. CICADELLIDAE (JASSIDAE, SPINIDAE, JASCOPIDAE, APHRODIDAE, COELIDIIDAE, EUSCELIDAE, IASSIDAE, MACROPSIDAE, TETTIGELLIDAE) K. (l.)–Rec. Terr.

First: e.g. *Mesoccus lutarius* Zhang (1985), Laiyang Formation, Laiyang, Shandong, China. We have followed Hong and Wang (1990) in regarding this formation as Lower Cretaceous in age. **Extant**

F. CICADIDAE (TIBICINIDAE) T. (Pal.)–Rec. Terr.

Extant

F. CICADOPROSBOLIDAE Tr. (u.)–K. (l.) Terr.

Last: e.g. *Architettix compacta* Hamilton (1990), Santana Formation, Ceará, Brazil.

F. CIXIIDAE K. (BRM)–Rec. Terr.

First: *Cixius petrinus*, *in* Martins-Neto (1988b), Upper Weald Clay, South Godstone, Surrey, England, UK.
Comment: The Jurassic genus *Mesocixiella* is a junior synonym of *Cycloscytina*, which belongs in the Hylicellidae (Shcherbakov, 1988b). **Extant**

F. CLASTOPTERIDAE T. (Oli)–Rec. Terr.

Extant

F. COCCIDAE T. (PRB)–Rec. Terr.

Extant

F. COLEOSCYTIDAE P. (ZEC) Terr.

F. COREIDAE (CORIZIDAE, RHOPALIDAE) J. (u.)–Rec. Terr.

First: e.g. *Hebeicoris xinboensis* Hong (1985d), north China. **Extant**

F. CORIXIDAE J. (l.)–Rec. Terr.

First: e.g. *Venacorixa xiangzhongensis* Lin (1986), south China. **Extant**

F. CREAPHIDIDAE Shcherbakov and Vengerek, 1991 Tr. (CRN) Terr.

First and Last: *Creaphis theodora* Shcherbakov and Vengerek (1991), Madygen Formation, Dzhailou-Tcho, South Fergana, former USSR.

F. CUNEOCORIDAE J. (l.) Terr.

F. CURVICUBITIDAE Hong, 1984b Tr. (m.) Terr.

First and Last: *Curvicubitus triassicus* Hong (1984b), Tongchuan Formation, Jinshuoguan, Shaanxi, China. Kozlov (1988) transferred this family from the Lepidoptera.

F. CYDNIDAE K. (l.)–Rec. Terr.

First: e.g. *Clavicoris cretaceus* Popov (1986), Gurvaneren Formation, Mongolia. **Extant**

F. DELPHACIDAE (ARAEOPIDAE) K. (APT)–Rec. Terr.

Extant

F. DERBIDAE Tr. (u.?)–Rec. Terr.

First: *Sanctipaulus mendesi*, *in* Martins-Neto (1987a), Santa Maria Formation, Rio Grande do Sul, Brazil. Shcherbakov (pers. comm.) considers that this species does not belong to this family. **Extant**

F. DIASPIDIDAE T. (PRB)–Rec. Terr.

Extant

F. DICTYOPHARIDAE K. (SAN)–Rec. Terr.

First: *Netutela annunciator*, *in* Martins-Neto (1988a), Siberian amber, Taimyr, former USSR. **Extant**

F. DIPSOCORIDAE T. (Eoc.)–Rec. Terr.

Extant

F. DREPANOSIPHIDAE (CALLAPHIDIDAE) J. (u.?)–Rec. Terr.

First: *Jurocallis longipes*, *in* Heie (1987), Karabastau Formation, Karatau, Kazakhstan, former USSR. **Extant**

F. DUNSTANIIDAE P. (ZEC)–J. (u.) Terr.

F. DYSMORPHOPTILIDAE (EOSCARTERELLIDAE, DISMORPHOPTILIDAE) P. (ZEC)–J. (u.) Terr.

F. ELEKTRAPHIDIDAE K. (u.)–T. (PIA) Terr.

First: e.g. *Antonaphis brachycera*, *in* Heie (1987), Siberian amber, Taimyr, former USSR.
Last: *Schizoneurites* sp., *in* Heie (1985), Willershausen, Germany.

F. ENICOCEPHALIDAE T. (YPR)–Rec. Terr.

First: Enicocephalid juv. indet. Jarzembowski (1986), London Clay, Isle of Sheppey, Kent, England, UK. **Extant**

F. ENICOCORIDAE (XISHANIDAE, MESOLYGAEIDAE) J. (u.)–K. (l.) Terr.

e.g. *Mesolygaeus laiyangensis*, *in* Zhang (1991b), China and Mongolia.
Comment: The name Enicocoridae has date priority over Xishanidae and Mesolygaeidae; the latter was used erroneously by Hong and Wang (1990) and Zhang (1991b).

F. ERIOCOCCIDAE T. (PRB)–Rec. Terr.

First: e.g. *Kuenowicoccus pietrzeniukae* Koteja (1988b), Baltic amber. **Extant**

F. FLATIDAE T. (PRB)–Rec. Terr.

Extant

F. FULGORIDAE K.? (BER)–Rec. Terr.

First: Figured but undescribed in Gomez Pallerola (1986), Lithographic Limestone, Montsech, Spain. Shcherbakov (pers. comm.) considers that these specimens do not belong in this family. **Extant**

F. FULGORIDIIDAE J. (l.–u.) Terr.

First: e.g. *Valvifulgoria tiantungensis* Lin (1986), Guangxi, China.
Comments: Zhang (1989) transferred the Miocene species in Hong (1985b) to the genus *Limois* in the Fulgoridae. Shcherbakov (pers. comm.) considers that this family belongs in the Cixiidae.

F. GELASTOCORIDAE K.? (APT)–Rec. Terr.

First: Gelastocorid indet. Jell and Duncan (1986), Koonwarra Fossil Bed, south Gippsland, Australia. Popov

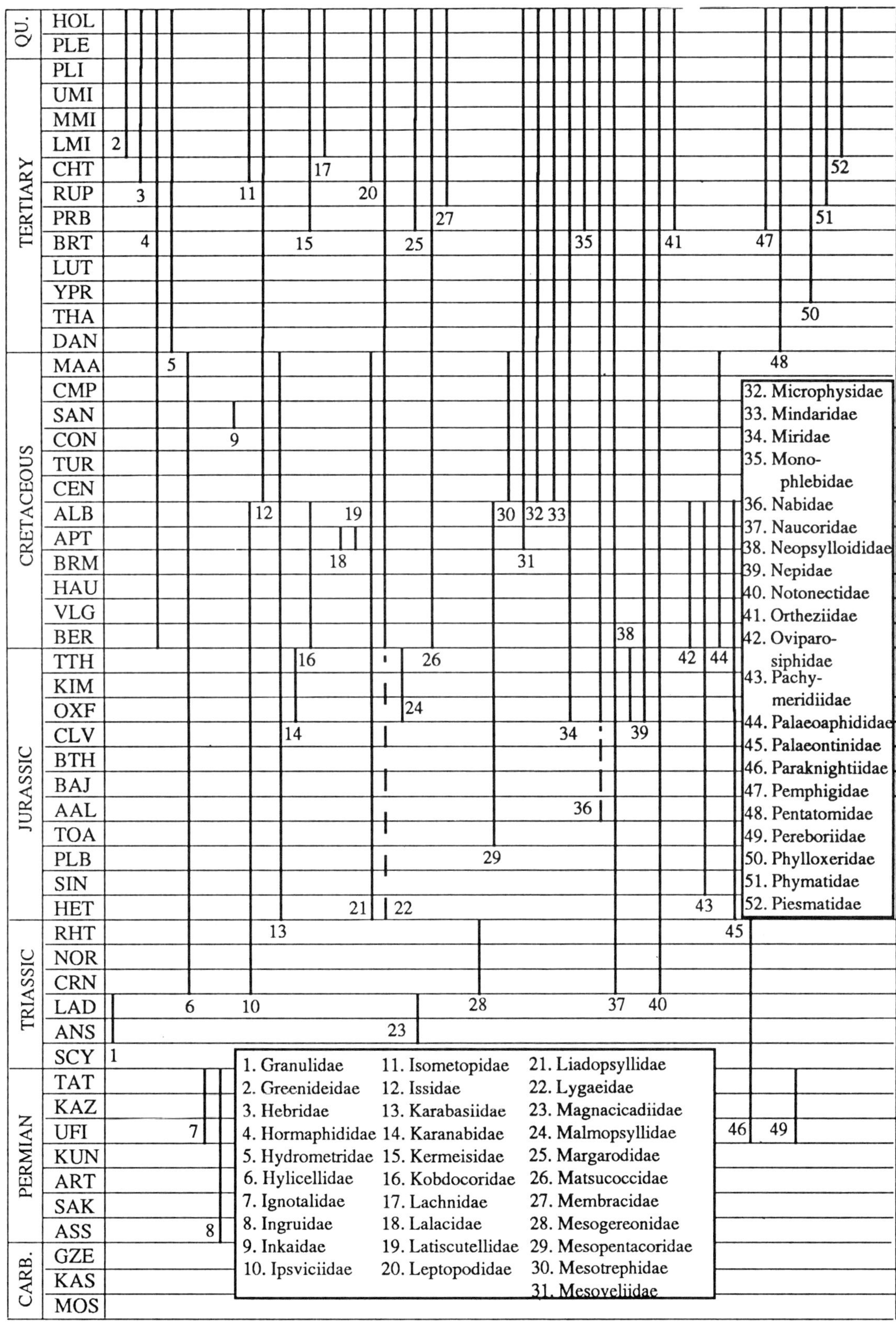
QU.
TERTIARY
CRETACEOUS
JURASSIC
TRIASSIC
PERMIAN
CARB.
HOL
PLE
PLI
UMI
MMI
LMI
CHT
RUP
PRB
BRT
LUT
YPR
THA
DAN
MAA
CMP
SAN
CON
TUR
CEN
ALB
APT
BRM
HAU
VLG
BER
TTH
KIM
OXF
CLV
BTH
BAJ
AAL
TOA
PLB
SIN
HET
RHT
NOR
CRN
LAD
ANS
SCY
TAT
KAZ
UFI
KUN
ART
SAK
ASS
GZE
KAS
MOS
1. Granulidae
2. Greenideidae
3. Hebridae
4. Hormaphididae
5. Hydrometridae
6. Hylicellidae
7. Ignotalidae
8. Ingruidae
9. Inkaidae
10. Ipsviciidae
11. Isometopidae
12. Issidae
13. Karabasiidae
14. Karanabidae
15. Kermeisidae
16. Kobdocoridae
17. Lachnidae
18. Lalacidae
19. Latiscutellidae
20. Leptopodidae
21. Liadopsyllidae
22. Lygaeidae
23. Magnacicadiidae
24. Malmopsyllidae
25. Margarodidae
26. Matsucoccidae
27. Membracidae
28. Mesogereonidae
29. Mesopentacoridae
30. Mesotrephidae
31. Mesoveliidae
32. Microphysidae
33. Mindaridae
34. Miridae
35. Mono-phlebidae
36. Nabidae
37. Naucoridae
38. Neopsylloididae
39. Nepidae
40. Notonectidae
41. Ortheziidae
42. Oviparo-siphidae
43. Pachy-meridiidae
44. Palaeoaphididae
45. Palaeontinidae
46. Paraknightiidae
47. Pemphigidae
48. Pentatomidae
49. Pereboriidae
50. Phylloxeridae
51. Phymatidae
52. Piesmatidae

Fig. 21.9

(pers. comm.) considers that these specimens belong in the Naucoridae. **Extant**

F. GENAPHIDIDAE J. (u.) Terr.

e.g. *Genaphis valdensis*, *in* Heie (1987), Lulworth Formation, Vale of Wardour, England, UK.

F. GERRIDAE T. (Pal.)–Rec. Terr.

Extant

F. GRANULIDAE Tr. m. Terr.

F. GREENIDEIDAE T. (LMI)–Rec. Terr.

First: e.g. *'Aphis' macrostyla*, *in* Heie (1987), Radoboj, Croatia. **Extant**

F. HEBRIDAE T. (CHT)–Rec. Terr.

First: ?*Hebrus* sp., *in* Spahr (1988), Mexican amber, Chiapas, Mexico. **Extant**

F. HORMAPHIDIDAE K. (l.)–Rec. Terr.

First: e.g. *Petiolaphis laiyangensis* Hong and Wang (1990), Laiyang Formation, Shandong, China. **Extant**

F. HYDROMETRIDAE T. (Pal.)–Rec. Terr.

Extant

F. HYLICELLIDAE Tr. (u.)–K. (u.) Terr.

F. IGNOTALIDAE (IGNATOLIDAE) P. (ZEC) Terr.

F. INGRUIDAE P. (ROT–ZEC) Terr.

First: e.g. *Scytoneurella major* in Shcherbakov (1984), Chekarda, former USSR.

F. INKAIDAE Koteja, 1989 K. (SAN) Terr.

First and Last: *ınka minuta* Koteja (1989), Siberian amber, Taimyr, former USSR.

F. IPSVICIIDAE Tr. (u.)–K. (l.) Terr.

First: *Ipsvicia jonesi* in Kukalová-Peck (1991), Australia. *Apheloscyta xiangdongensis* Lin (1986) was originally placed in this family; however, this genus belongs in the Scytinopteridae, see Shcherbakov (1984).

F. ISOMETOPIDAE T. (CHT)–Rec. Terr.

Extant

F. ISSIDAE K. (u.)–Rec. Terr.

Extant

F. KARABASIIDAE Popov, 1985 J. (l.)–K. (u.) Terr.

First: e.g. *Tegulicicada plana* in Popov (1989), Shiti Formation, Guangxi, China.

F. KARANABIDAE J. (u.) Terr.

Popov (pers. comm.) considers that this family belongs in the Mesovellidae.

F. KERMESIDAE T. (PRB)–Rec. Terr.

First: *Sucinikermes kulickae* Koteja (1988c), Baltic amber, Gdańsk, Poland. **Extant**

F. KOBDOCORIDAE Popov, 1986 K. (l.) Terr.

First and Last: *Kobdocoris aradinus* Popov (1986), Gurvaneren Formation, Mongolia.

F. LACHNIDAE T. (Mio.)–Rec. Terr.

First: e.g. *Cinara limnogena* Zhang (1989), Shanwang Formation, Shanwang, China. **Extant**

F. LALACIDAE Hamilton, 1990 K. (APT) Terr.

e.g. *Lalax mutabilis* Hamilton (1990), Santana Formation, Ceará, Brazil.

Comment: Shcherbakov (pers. comm.) considers that this family belongs in the Cixiidae.

F. LATISCUTELLIDAE K. (APT) Terr.

First and Last: *Latiscutella santosi*, *in* Martins-Neto (1987a), Codó Formation, Maranhão, Brazil.

Comment: Popov (pers. comm.) considers that this family belongs in the Cydnidae.

F. LEPTOPODIDAE T. (CHT)–Rec. Terr.

First: *Leptosalda chiapensis*, *in* Spahr (1988), Mexican amber, Chiapas, Mexico. **Extant**

F. LIADOPSYLLIDAE (LITHENTOMIDAE, ASIENTOMIDAE) J. (l.)–K. (u.) Terr.

First: *Liadopsylla geinitzi*, *in* Bekker-Migdisova (1985b), Upper Lias, Dobbertin, Germany.

F. LYGAEIDAE J.?–Rec. Terr.

First: *Leipolygaeus similis* Lin (1985) Hanshan, Anhui, China. The position of this species is doubtful. Popov (1989) transferred the Lower Jurassic *Hunanilarva micra* Lin (1986) to the Karabasiidae. **Extant**

F. MAGNACICADIIDAE Tr. (m.) Terr.

F. MALMOPSYLLIDAE Bekker-Migdisova, 1985b J. (u.) Terr.

First and Last: *Malmopsylla karatavica* Bekker-Migdisova (1985b), Karabastau Formation, Karatau, Kazakhstan, former USSR.

F. MARGARODIDAE T. (PRB)–Rec. Terr.

First: Figured in Koteja (1990), Baltic amber. **Extant**

F. MATSUCOCCIDAE K. (l.)–Rec. Terr.

First: e.g. *Eomatsucoccus sukachevae* Koteja (1988a), Zaza Formation, Baisa, Buryat ASSR, former USSR. **Extant**

F. MEMBRACIDAE T. (Oli.)–Rec. Terr.

First: Figured in Schlee (1990), Dominican amber. Popov (1989) transferred the Lower Jurassic *Tegulicicada plana* Lin (1986) to the Karabasiidae. **Extant**

F. MESOGEREONIDAE Tr. (u.) Terr.

e.g. *Mesogereon superbum* in Kukalová-Peck (1991), Queensland, Australia.

F. MESOPENTACORIDAE J. (TOA)–K. (l.) Terr.

Last: e.g. *Corienta transbaicalica* Popov (1990), Turga Formation, Transbaikalia, former USSR.

F. MESOTREPHIDAE K. (u.) Terr.

F. MESOVELIIDAE K. (APT)–Rec. Terr.

First: *Duncanovelia extensa* Jell and Duncan (1986), Koonwarra Fossil Bed, south Gippsland, Australia. **Extant**

F. MICROPHYSIDAE K. (u.)–Rec. Terr.

Extant

F. MINDARIDAE K. (u.)–Rec. Terr.

First: *Nordaphis sukatchevae, in* Heie (1987), Siberian amber, Taimyr, former USSR. **Extant**

F. MIRIDAE J. (u.)–Rec. Terr.

Extant

F. MONOPHLEBIDAE T. (PRB)–Rec. Terr.

First: *Monophlebus irregularis, in* Koteja (1990), Baltic amber. **Extant**

F. NABIDAE J. (m.?)–Rec. Terr.

First: *Sinanabis brevipes* Zhang (1986b), Hebei, China.
Comment: The position of this species is doubtful. Popov (pers. comm.) considers that it belongs in the Mesovellidae. **Extant**

F. NAUCORIDAE (APHLEBOCORIDAE, APOPNIDAE, ATOPOSITIDAE) Tr. (u.)–Rec. Terr.

Extant

F. NEOPSYLLOIDIDAE Bekker-Migdisova, 1985b J. (u.) Terr.

e.g. *Neopsylloides turutanovae* Bekker-Migdisova (1985b), Karatau, Kazakhstan, former USSR.

F. NEPIDAE J. (u.)–Rec. Terr.

Extant

F. NOTONECTIDAE Tr. (u.)–Rec. Terr.

Extant

F. ORTHEZIIDAE T. (PRB)–Rec. Terr.

First: e.g. *Protorthezia aurea* Koteja (1987), Baltic amber, Poland. **Extant**

F. OVIPAROSIPHIDAE K. (l.) Terr.

e.g. *Oviparosiphum jakovlevi, in* Heie (1987), Mongolia. This family may be synonymous with the Drepanosiphidae, see Jarzembowski (1989b).

F. PACHYMERIDIIDAE (PSYCHROCORIDAE, HYPOCIMICIDAE, SISYROCORIDAE) J. (SIN)–K. (l.) Terr.

First: e.g. gen. et sp. *nov.* 3A Whalley 1985, Lower Lias, Charmouth, Dorset, England, UK.
Last: e.g. *Pachycoridium letum* Popov, 1986, Gurvaneren Formation, Mongolia.

F. PALAEOAPHIDIDAE K. (l.–u.) Terr.

First: e.g. *Caudaphis spinalis* Zhang *et al.* (1989), Laiyang Formation, Shandong, China. We have followed Hong and Wang (1990) in regarding this formation as early Cretaceous in age.
Last: e.g. *Palaeoaphis archimedia, in* Heie (1985), Canadian amber, Cedar Lake, Manitoba, Canada.

F. PALAEONTINIDAE (PALEONTINIDAE) J. (l.)–K. (l.) Terr.

First: e.g. *Asiocossus costalis, in* Gomez Pallerola (1984), Dzhil Formation, Issyk-Kul, Kirgizia, former USSR.
Last: e.g. figured but undescribed *in* Jarzembowski (1984), Lower Weald Clay, Capel, Surrey, England, UK.

F. PARAKNIGHTIIDAE P. (ZEC)–Tr. (u.) Terr.

F. PEMPHIGIDAE T. (PRB)–Rec. Terr.

First: e.g. *Germaraphis dryoides, in* Heie (1985), Baltic amber. Heie (1985) listed the Upper Cretaceous *Palaeoforda taymyrensis* as systematic position unknown. **Extant**

F. PENTATOMIDAE (DINIDORIDAE) T. (Pal.)–Rec. Terr.

Extant

F. PEREBORIIDAE (PEREBORIDAE) P. (ZEC) Terr.

e.g. *Scytophara extensa, in* Shcherbakov (1984), Kargala, former USSR.

F. PHYLLOXERIDAE T. (Eoc.)–Rec. Terr.

Extant

F. PHYMATIDAE T. (RUP)–Rec. Terr.

Extant

F. PIESMATIDAE (PIESMIDAE) T. (Mio.)–Rec. Terr.

Extant

F. PINCOMBEIDAE (PINCOMBAEIDAE) P. (ZEC)–Tr. (u.)

First: e.g. *Pincombea mirabilis, in* Bekker-Migdisova (1985b), Belmont, New South Wales, Australia.
Last: *Madygenopsyllidium djailautshoense* Bekker-Migdisova (1985b), Madygen Formation, Fergana, former USSR.

F. PITYOCOCCIDAE K. (CMP)–Rec. Terr.

First: *Electrococcus canadensis, in* Koteja (1989), Canadian amber, Cedar Lake, Manitoba, Canada. **Extant**

F. PRICECORIDAE K. (APT) Terr.

First and Last: *Pricecoris beckerae, in* Martins-Neto (1987a), Codó Formation, Maranhão, Brazil.
Comment: Popov (pers. comm.) considers that this family belongs in the Cydnidae.

F. PROBASCANIIDAE (PROBASCANIONIDAE) J. (l.) Terr.

e.g. *Probascanion megacephalum, in* Ponomarenko and Schultz (1988), Upper Lias, Dobbertin, Germany.

F. PROCERCOPIDAE J. (l.)–K. (u.) Terr.

F. PROGONOCIMICIDAE (ACTINOSCYTINIDAE, CICADOCORIDAE, ACTINESCYTINIDAE, EOCIMICIDAE) P. (ZEC)–K. (u.) Terr.

First: e.g. *Actinoscytina belmontensis, in* Wootton and Betts (1986), Warner's Bay, New South Wales, Australia.

F. PROPREOCORIDAE J. (l.) Terr.

Comment: Popov (pers. comm.) considers that this family belongs in the Cydnidae.

F. PROSBOLECICADIDAE Pinto, 1987b P. (ZEC) Terr.

First and Last: *Prosbolecicada gondwanica* Pinto (1987b), Irati Formation, Rio Grande do Sul, Brazil.
Comment: Shcherbakov (pers. comm.) considers that this family belongs in the Prosbolidae.

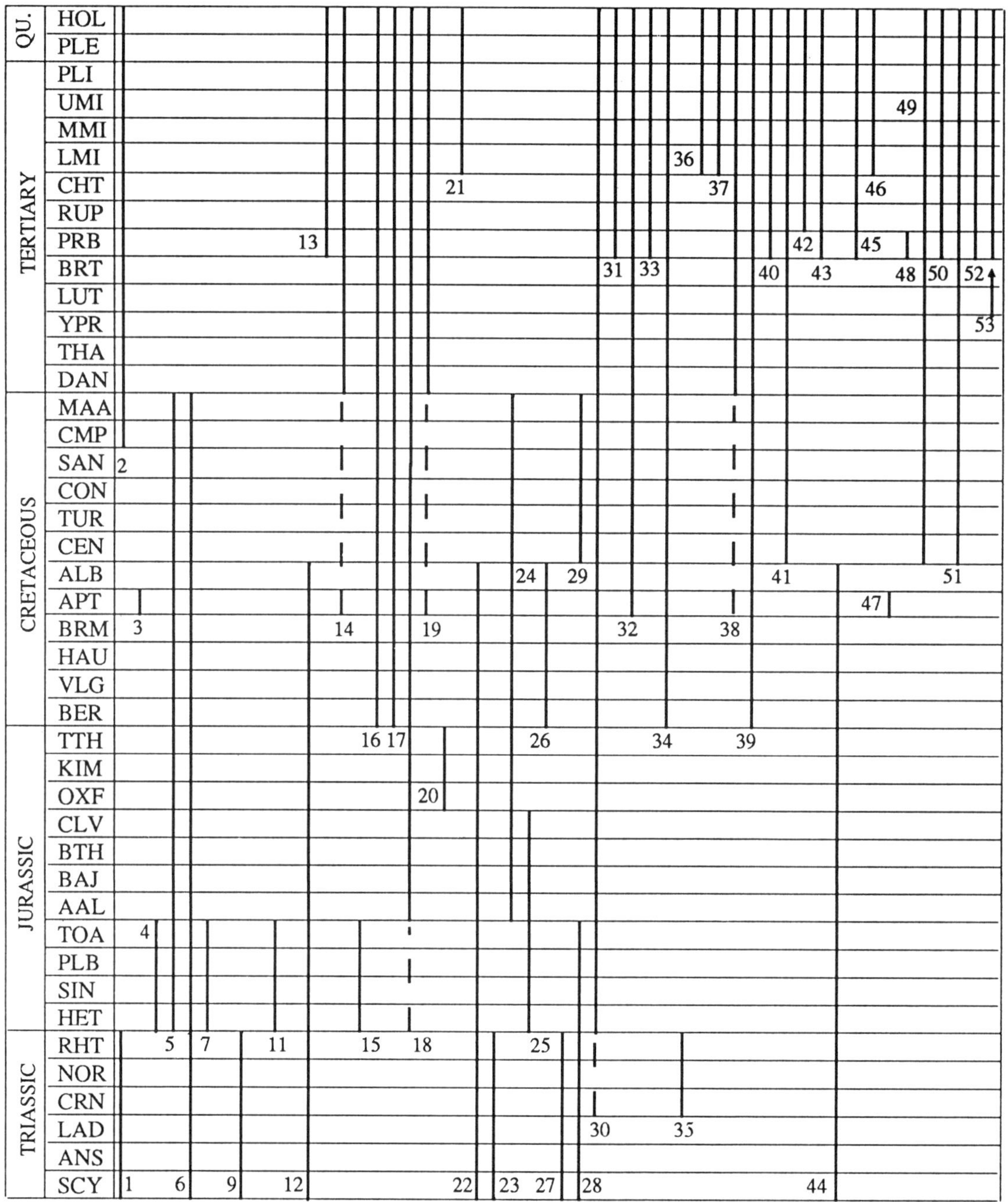

Fig. 21.10

F. PROSBOLIDAE (CICADOPSYLLIDAE, PERMOGLYPHIDAE, PERMOCICADOPSIDAE, SOJANONEURIDAE) P. (ZEC)–Tr. (u.) Terr.

First: e.g. *Permocicada integra* in Shcherbakov (1984), Iva-Gora Beds, Archangelsk, former USSR.
Last: e.g. *Sinisbole juvensis* Lin (1986), Hunan, China.
Comment: Shcherbakov (pers. comm.) considers that this species belongs in the Hylicellidae.

F. PROSBOLOPSEIDAE (PROSBOLOPSIDAE, IVAIIDAE, MUNDIDAE) P. (ROT–ZEC) Terr.

Last: e.g. *Prosboloneura kondomensis*, *in* Shcherbakov (1984), Kuznetsk Formation, Kaltan, Kuznetsk Basin, former USSR.

F. PROTOCORIDAE J. (l.) Terr.

e.g. *Pallicoris firmis* Lin (1986), Guangxi, China.
Comment: Popov (pers. comm.) considers that this species belongs in the Pachymeridiidae.

F. PROTOPSYLLIDIIDAE (EOPSYLLIDIIDAE, PERMOPSYLLIDAE, PERMAPHIDOPSEIDAE) P. (ZEC)–K. (l.) Terr.

First: e.g. *Permopsyllidium mitchelli*, *in* Bekker-Migdisova (1985b), New South Wales, Australia.
Last: *Aphidulum stenoptilium* Shcherbakov (1988a), Bon-Tsagan, Mongolia.

F. PSEUDOCOCCIDAE T. (PRB)–Rec. Terr.

Extant

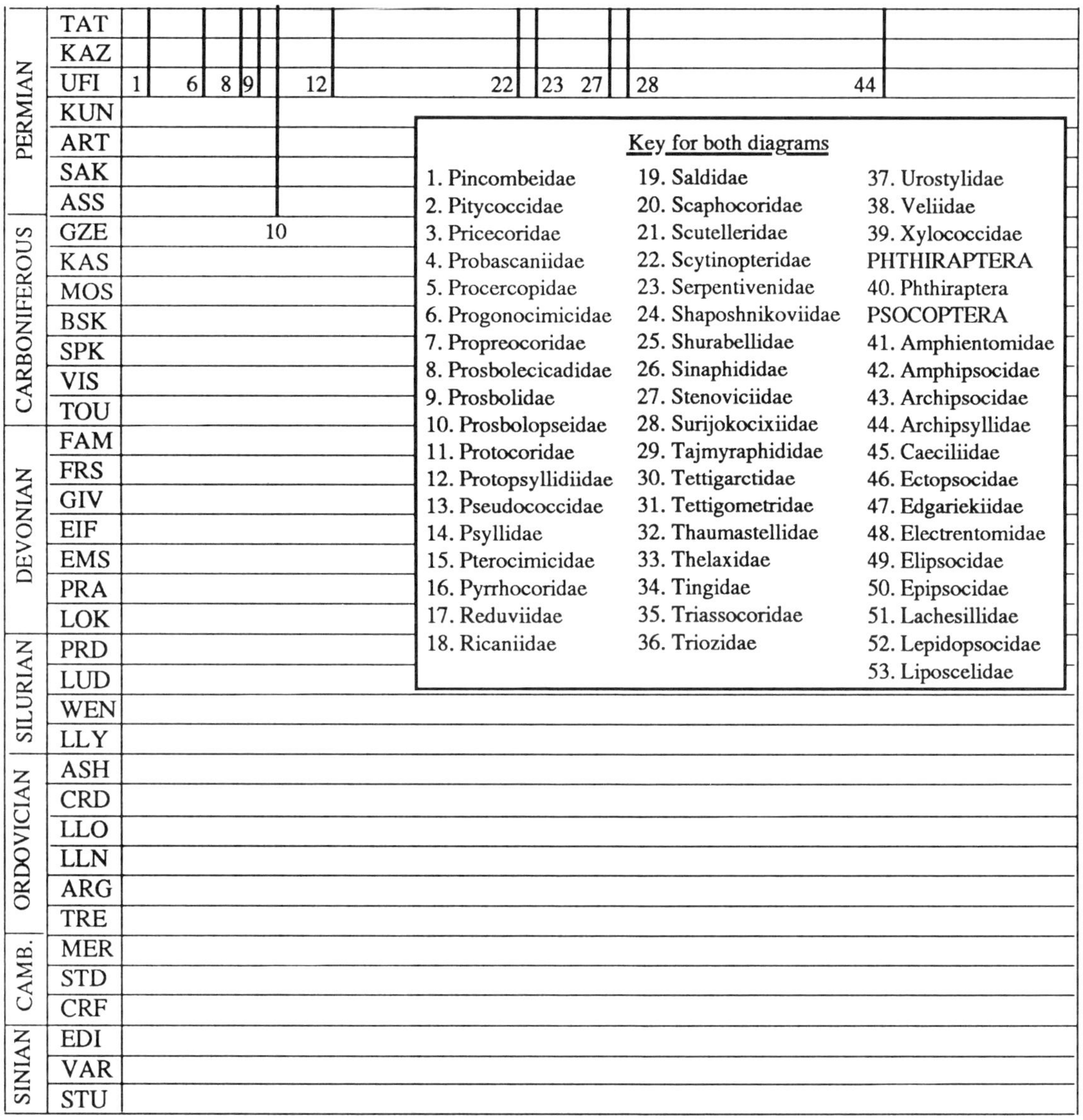

Period	Stage	
PERMIAN	TAT	
	KAZ	
	UFI	1 6 8 9 12 22 23 27 28 44
	KUN	
	ART	
	SAK	
	ASS	
CARBONIFEROUS	GZE	10
	KAS	
	MOS	
	BSK	
	SPK	
	VIS	
	TOU	
DEVONIAN	FAM	
	FRS	
	GIV	
	EIF	
	EMS	
	PRA	
	LOK	
SILURIAN	PRD	
	LUD	
	WEN	
	LLY	
ORDOVICIAN	ASH	
	CRD	
	LLO	
	LLN	
	ARG	
	TRE	
CAMB.	MER	
	STD	
	CRF	
SINIAN	EDI	
	VAR	
	STU	

Key for both diagrams

1. Pincombeidae
2. Pitycoccidae
3. Pricecoridae
4. Probascaniidae
5. Procercopidae
6. Progonocimicidae
7. Propreocoridae
8. Prosbolecicadidae
9. Prosbolidae
10. Prosbolopseidae
11. Protocoridae
12. Protopsyllidiidae
13. Pseudococcidae
14. Psyllidae
15. Pterocimicidae
16. Pyrrhocoridae
17. Reduviidae
18. Ricaniidae
19. Saldidae
20. Scaphocoridae
21. Scutelleridae
22. Scytinopteridae
23. Serpentivenidae
24. Shaposhnikoviidae
25. Shurabellidae
26. Sinaphididae
27. Stenoviciidae
28. Surijokocixiidae
29. Tajmyraphididae
30. Tettigarctidae
31. Tettigometridae
32. Thaumastellidae
33. Thelaxidae
34. Tingidae
35. Triassocoridae
36. Triozidae
37. Urostylidae
38. Veliidae
39. Xylococcidae

PHTHIRAPTERA

40. Phthiraptera

PSOCOPTERA

41. Amphientomidae
42. Amphipsocidae
43. Archipsocidae
44. Archipsyllidae
45. Caeciliidae
46. Ectopsocidae
47. Edgariekiidae
48. Electrentomidae
49. Elipsocidae
50. Epipsocidae
51. Lachesillidae
52. Lepidopsocidae
53. Liposcelidae

Fig. 21.10

F. PSYLLIDAE K.? (APT)–Rec. Terr.

First: Psyllid indet. Jell and Duncan (1986), Koonwarra Fossil Bed, south Gippsland, Australia.
Comment: Shcherbakov (pers. comm.) considers that this specimen belongs in the Protopsyllidae. **Extant**

F. PTEROCIMICIDAE J. (l.) Terr.

F. PYRRHOCORIDAE (LARGIDAE) K. (l.)–Rec. Terr.

First: *Mesopyrrhocoris fasciata* Hong and Wang (1990), Laiyang Formation, Shandong, China. **Extant**

F. REDUVIIDAE K. (l.)–Rec. Terr.

First: e.g. *Liaoxia longa* Hong (1987), Jiufutang Formation, Kezuo, China.
Comment: Popov (pers. comm.) considers that this species does not show structures attributable to this family. **Extant**

F. RICANIIDAE J. (l.?)–Rec. Terr.

First: *Qiyangiricania cesta* Lin (1986), Hunan, China.
Comment: Shcherbakov (pers. comm.) considers that this species does not belong in this family. **Extant**

F. SALDIDAE K.? (APT)–Rec. Terr.

First: Figured but undescribed *in* Grimaldi and Maisey (1990), Santana Formation, Ceará, Brazil.
Comment: Popov (pers. comm.) considers that this specimen was incorrectly identified. **Extant**

F. SCAPHOCORIDAE J. (u.) Terr.

F. SCUTELLERIDAE T. (Mio.)–Rec. Terr. **Extant**

F. SCYTINOPTERIDAE (SEYTINOPTERIDAE) P. (ZEC)–K. (l.) Terr.

First: e.g. *Scytinoptera sibirica*, *in* Shcherbakov (1984), Kuznetsk Formation, Kaltan, Kuznetsk Basin, former USSR.
Last: e.g. *Sunoscytinopteris lushangfenensis* Hong (1985d), north China.

Fig. 21.11

Comment: Shcherbakov (pers. comm.) considers that this species belongs in the Procercopidae.

F. SERPENTIVENIDAE (SERPENIVENIDAE) Shcherbakov, 1984 P. (ZEC)–Tr. (u.) Terr.

Last: e.g. *Serpentivena tigrina* Shcherbakov (1984), Madygen Formation, southern Fergana, former USSR.

F. SHAPOSHNIKOVIIDAE J. (u.)–K. (u.) Terr.

First: *Tinaphis sibirica* Vengerek (1989), Itatskaya Formation, Yenisey River, Siberia, former USSR.

Last: *Shaposhnikovia electri in* Heie (1987), Siberian amber, Taimyr, former USSR.

F. SHURABELLIDAE J. (l.–m.). Terr.

First: e.g. *Shurania sibirica* Popov (1985), Kuznetsk Basin, former USSR.

Last: e.g. *Shuragobia altaica* Popov (1988a), Bakhar Formation, Gobi Altai, Mongolia.

F. SINAPHIDIDAE Zhang *et al.*, 1989 K. (l.) Terr.

e.g. *Sinaphidum epichare* Zhang *et al.* (1989), Laiyang Formation, Shandong, China. We have followed Hong and Wang (1990) in regarding this formation as Lower Cretaceous in age. Vengerek (pers. comm.) considers that this family belongs in the Oviparosiphidae.

F. STENOVICIIDAE P. (ZEC)–Tr. (u.) Terr.

F. SURIJOKOCIXIIDAE P. (ZEC)–J. (l.) Terr.

F. TAJMYRAPHIDIDAE (TAYMYRAPHIDIDAE) K. (u.) Terr.

e.g. *Tajmyraphis beckermigdisovae*, *in* Heie (1987), Siberian amber, Taimyr, former USSR.

F. TETTIGARCTIDAE Tr. (u.?)–Rec. Terr.

First: e.g. *Quadrisbole vieta* Lin (1986), Hunan, China.
Comment: Shcherbakov (pers. comm.) considers that this species and *Lacunisbole ligonis* Lin (1986) do not belong in this family. **Extant**

F. TETTIGOMETRIDAE T. (PRB)–Rec. Terr.

Extant

F. THAUMASTELLIDAE K. (APT)–Rec. Terr.

Extant

F. THELAXIDAE T. (PRB)–Rec. Terr.

First: *Palaeothelaxes setosa*, *in* Heie (1987), Baltic amber. **Extant**

F. TINGIDAE K. (l.)–Rec. Terr.

First: e.g. *Sinaldocader drakei* Popov (1989), Bon-Tsagan, Mongolia. **Extant**

F. TRIASSOCORIDAE Tr. (u.) Terr.

F. TRIOZIDAE T. (Mio.)–Rec. Terr.

First: e.g. *Trioza similis*, *in* Bekker-Migdisova (1985b), Stavropol, former USSR. **Extant**

F. UROSTYLIDAE T. (Mio.)–Rec. Terr.

First: e.g. *Urochela pardalina* Zhang (1989), Shanwang Formation, Shanwang, Shandong, China. **Extant**

F. VELIIDAE K.? (APT)–Rec. Terr.

First: Veliid indet. Jell and Duncan (1986), Koonwarra Fossil Bed, south Gippsland, Australia.
Comment: Popov (pers. comm.) considers that this specimen belongs in the Mesovellidae. **Extant**

F. XYLOCOCCIDAE (XYLOCCIDAE) K. (l.)–Rec. Terr.

First: *Baisococcus victoriae*, *in* Koteja (1989), Zaza Formation, Baisa, Buryat ASSR, former USSR. **Extant**

***Order* PHTHIRAPTERA (ANOPLURA, MALLOPHAGA) T. (PRB)–Rec. Terr.**

First: Eggs in Boucot (1990), Baltic amber. These eggs cannot be determined to family level. **Extant**
Comment: There are no pre-Quaternary family records of this order.

***Order* PSOCOPTERA (PSOCIDA, CORRODENTIA) P. (ROT)–Rec. Terr.**

F. AMPHIENTOMIDAE K. (u.)–Rec. Terr.

Extant

F. AMPHIPSOCIDAE T. (Oli.)–Rec. Terr.

Extant

F. ARCHIPSOCIDAE T. (PRB)–Rec. Terr.

Extant

F. ARCHIPSYLLIDAE P. (ZEC)–K. (l.) Terr.

F. CAECILIIDAE T. (PRB)–Rec. Terr.

Extant

F. ECTOPSOCIDAE T. (Mio.)–Rec. Terr.

Extant

F. EDGARIEKIIDAE Jell and Duncan, 1986 K. (APT) Terr.

First and Last: *Edgariekia una* Jell and Duncan (1986), Koonwarra Fossil Bed, south Gippsland, Australia.
Comment: Zherikhin (pers. comm.) considers that this family belongs in the Lophioneuridae.

F. ELECTRENTOMIDAE T. (PRB) Terr.

F. ELIPSOCIDAE K. (u.)–Rec. Terr.

Extant

F. EPIPSOCIDAE T. (PRB)–Rec. Terr.

Extant

F. LACHESILLIDAE K. (u.)–Rec. Terr.

Extant

F. LEPIDOPSOCIDAE T. (PRB)–Rec. Terr.

Extant

F. LIPOSCELIDAE T. (PRB)–Rec. Terr.

Extant

F. LOPHIONEURIDAE P. (ROT)–K. (u.) Terr.

Comment: Some authors consider that this family does not belong in this order.

F. MARTYNOPSOCIDAE P. Terr.

F. MESOPSOCIDAE T. (Oli.)–Rec. Terr.

Extant

F. MYOPSOCIDAE T. (Mio.)–Rec. Terr.

Extant

F. PACHYTROCTIDAE T. (PRB)–Rec. Terr.

Extant

F. PERIPSOCIDAE T. (Oli.)–Rec. Terr.

Extant

F. PERMOPSOCIDAE P. Terr.

F. PHILOTARSIDAE T. (PRB)–Rec. Terr.

Extant

F. POLYPSOCIDAE T. (Oli.)–Rec. Terr.

Extant

F. PSEUDOCAECILIIDAE T. (PRB)–Rec. Terr.

Extant

F. PSOCIDAE T. (PRB)–Rec. Terr.

Extant

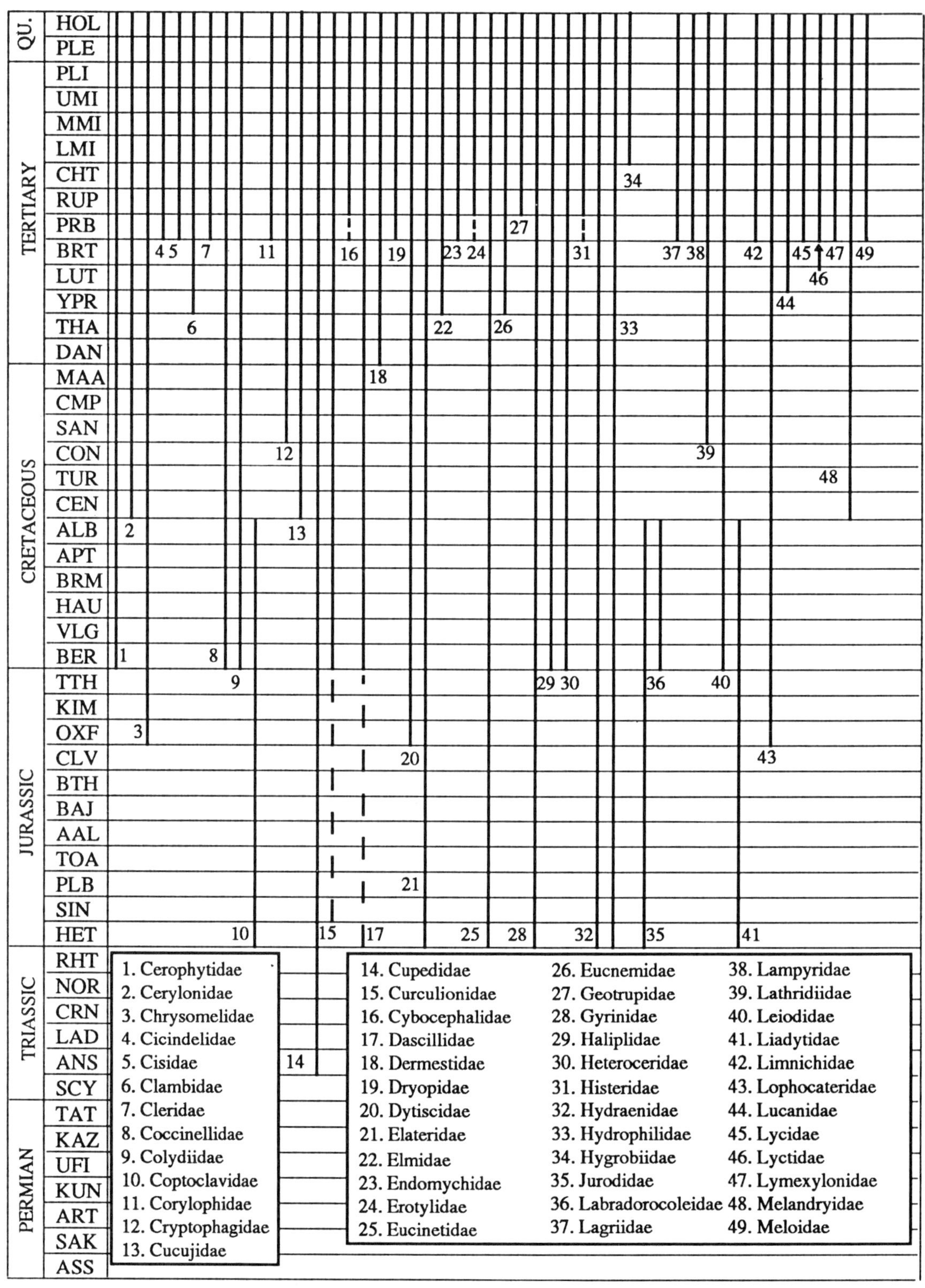

Fig. 21.12

F. PSOCIDIIDAE P. Terr.

F. PSYLLIPSOCIDAE K. (u.)–Rec. Terr.
Extant

F. SPHAEROPSOCIDAE K. (u.)–Rec. Terr.
Extant

F. SURIJOKOPSOCIDAE P. Terr.

F. TRICHOPSOCIDAE T. (PRB)–Rec. Terr.
Extant

F. TROGIIDAE K. (u.)–Rec. Terr.
Extant

F. ZYGOPSOCIDAE P. Terr.

Order THYSANOPTERA (THRIPIDA)
P. (ROT)–Rec. Terr.

F. AEOLOTHRIPIDAE (AEOTHRIPIDAE) K. (u.)–Rec. Terr.

Extant

F. HETEROTHRIPIDAE J. (u.)–Rec. Terr.

Extant

F. KARATAOTHRIPIDAE J. (u.) Terr.

F. LIASSOTHRIPIDAE J. (u.) Terr.

F. MEROTHRIPIDAE T. (PRB)–Rec. Terr.

Extant

F. PALAEOTHRIPIDAE T. (Eoc.) Terr.

F. PERMOTHRIPIDAE P. (ROT) Terr.

F. PHLAEOTHRIPIDAE T. (PRB)–Rec. Terr.

Extant

F. THRIPIDAE K. (u.)–Rec. Terr.

Extant

Order ZORAPTERA T. (Oli.)–Rec. Terr.

Only one fossil record.

F. ZOROTYPIDAE T. (Oli.)–Rec. Terr.

First: *Zorotypus palaeus* Poinar (1988), Dominican amber, Cordillera Septentrional, Dominican Republic. **Extant**

Superorder OLIGONEOPTERA (Holometabola, Endopterygota, Scarabaeiformes) C. (u.?)–Rec. Terr.

First as for Raphidioptera.

Order COLEOPTERA (SCARABAEIDA) P. (ROT)–Rec. Terr.

Lists of families that occur in amber are given in Hieke and Pietrzeniuk (1984) and Wunderlich (1986). Some Mesozoic species are given in Arnoldi, Zherikhin, Nikritin and Ponomarenko (1991).

F. ACANTHOCNEMIDAE (ACANTHOCUEMIDAE) J. (l.?)–Rec. Terr.

First: *Artinama qinghuoensis* Lin (1986), south China.
Comment: Ponomarenko (pers. comm.) considers that this species belongs in the Elateridae. **Extant**

F. ADEMOSYNIDAE Tr. (u.)–K. (l.) Terr.

First: e.g. *Ademosynoides juxta* Lin (1986), south China.

F. ADERIDAE (CIRCAEIDAE) T. (PRB)–Rec. Terr.

Extant

F. ANOBIIDAE K. (l.)–Rec. Terr.

Extant

F. ANTHICIDAE T. (Pal.)–Rec. Terr.

Extant

F. ANTHRIBIDAE K. (l.)–Rec. Terr.

Extant

F. APIONIDAE K. (l.)–Rec. Terr.

Extant

F. ARTEMATOPIDAE T. (?PRB)–Rec. Terr.

Extant

F. ASIOCOLEIDAE P. Terr.

F. ASPIDIPHORIDAE T. (PRB)–Rec. Terr.

Extant

F. ATTELABIDAE K. (l.)–Rec. Terr.

Extant

F. BOSTRYCHIDAE T. (YPR)–Rec. Terr.

Extant

F. BRENTHIDAE (BRENTIDAE) T. (Oli.)–Rec. Terr.

Extant

F. BRUCHIDAE J. (m.?)–Rec. Terr.

First: *Mesolaria longala* Zhang (1986b), northern Hebei, China.
Comment: Ponomarenko (pers. comm.) considers that this species does not show any structures attributable to this family. **Extant**

F. BUPRESTIDAE (ELECTRAPATIDAE) J. (m.)–Rec. Terr.

Extant

F. BYRRHIDAE J. (l.)–Rec. Terr.

Extant

F. BYTURIDAE T. (?PRB)–Rec. Terr.

Extant

F. CANTHARIDAE K. (APT)–Rec. Terr.

First: Cantharid indet. Jell and Duncan (1986), Koonwarra Fossil Bed, south Gippsland, Australia. **Extant**

F. CARABIDAE (PAUSSIDAE) J. (SIN)–Rec. Terr.

First: e.g. figured but undescribed, *in* Whalley (1985), lower Lias, Charmouth, Dorset, England, UK. **Extant**

F. CATINIIDAE (CANITIDAE) Tr. (u.)–K. (l.) Terr.

F. CATOPIDAE T. (PRB)–Rec. Terr.

Extant

F. CERAMBYCIDAE K. (l.)–Rec. Terr.

Extant

F. CEROPHYTIDAE K. (l.)–Rec. Terr.

Extant

F. CERYLONIDAE K. (u.)–Rec. Terr.

Extant

F. CHRYSOMELIDAE J. (u.)–Rec. Terr.

Extant

F. CICINDELIDAE T. (PRB)–Rec. Terr.

Extant

F. CISIDAE T. (PRB)–Rec. Terr.

Extant

F. CLAMBIDAE T. (Eoc.)–Rec. Terr.

Extant

F. CLERIDAE T. (PRB)–Rec. Terr.

Extant

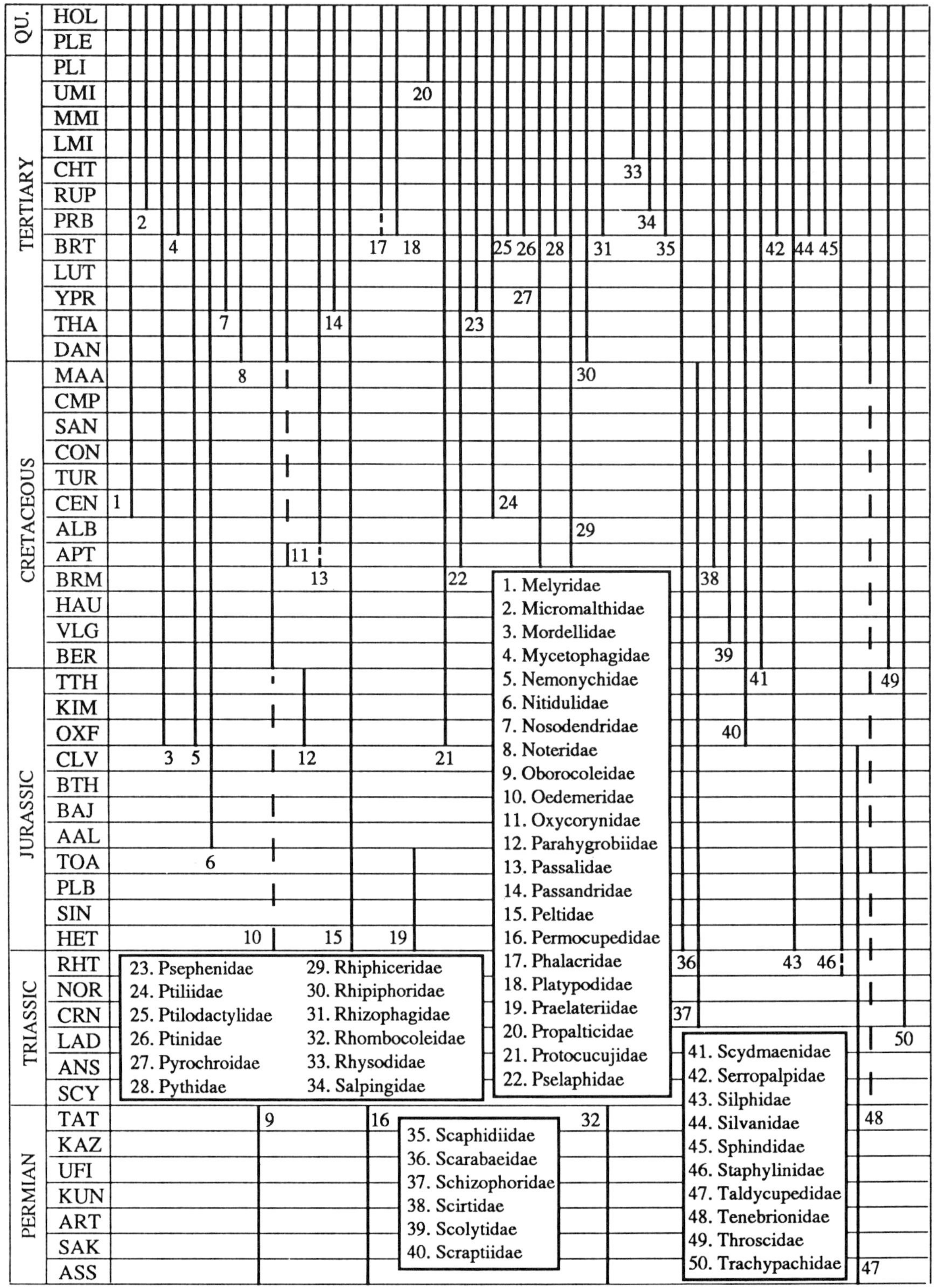

Fig. 21.13

F. COCCINELLIDAE K. (l.)–Rec. Terr.

Extant

F. COLYDIIDAE K. (l.)–Rec. Terr.

Extant

F. COPTOCLAVIDAE J. (l.)–K. (l.) Terr.

First: e.g. *Timarchopsis sainshandensis* Ponomarenko (1987), Khamarkhoburinskaya Formation, Tushilga, Mongolia.

Last: e.g. *Bolbonectes intermedius* Ponomarenko (1987), Byankinskaya Formation, Bolboy, former USSR.

F. CORYLOPHIDAE (ORTHOPERIDAE) T. (PRB)–Rec. Terr.

Extant

F. CRYPTOPHAGIDAE K. (SAN)–Rec. Terr.

First: *Nganasania khetica*, Siberian amber, Taimyr, former USSR. **Extant**

F. CUCUJIDAE K. (u.)–Rec. Terr.

Extant

F. CUPEDIDAE (CUPESIDAE, TRIADOCUPEDIDAE) Tr. (m.)–Rec. Terr.

First: *Chengdecupes shiluoense* Hong (1985d), north China. **Extant**

F. CURCULIONIDAE J. ?(SIN)–Rec. Terr.

First: gen. *et* sp. indet. Whalley (1985), Lower Lias, Charmouth, Dorset, England, UK.
Comment: Zherikhin (pers. comm.) considers that this specimen does not belong in this family. **Extant**

F. CYBOCEPHALIDAE T. (?PRB)–Rec. Terr.

Extant

F. DASCILLIDAE J.?–Rec. Terr.

Ponomarenko (pers. comm.) considers that the Mesozoic records of this family are doubtful. **Extant**

F. DERMESTIDAE T. (Pal.)–Rec. Terr.

Extant

F. DRYOPIDAE T. (PRB)–Rec. Terr.

Extant

F. DYTISCIDAE J. (u.)–Rec. Terr.

First: e.g. *Palaeodytes gutta* Ponomarenko (1987), Karabastau Formation, Mikhailovka, Kazakhstan, former USSR. **Extant**

F. ELATERIDAE J. (SIN)–Rec. Terr.

First: e.g. *Elaterophanes regius* Whalley (1985), Lower Lias, Charmouth, Dorset, England, UK. **Extant**

F. ELMIDAE T. (Eoc.)–Rec. Terr.

Extant

F. ENDOMYCHIDAE T. (PRB)–Rec. Terr.

Extant

F. EROTYLIDAE T. (?PRB)–Rec. Terr.

Extant

F. EUCINETIDAE J. (l.)–Rec. Terr.

Extant

F. EUCNEMIDAE T. (YPR)–Rec. Terr.

Extant

F. GEOTRUPIDAE T. (Oli.)–Rec. Terr.

Extant

F. GYRINIDAE J. (l.)–Rec. Terr.

First: e.g. *Mesogyrus sibiricus* Ponomarenko (1985a), former USSR. **Extant**

F. HALIPLIDAE K. (l.)–Rec. Terr.

Extant

F. HETEROCERIDAE K (l.)–Rec. Terr.

First: *Heterocerites kobdoensis* Ponomarenko (1986a), Gurvaneren Formation, Mongolia. **Extant**

F. HISTERIDAE T. (?PRB)–Rec. Terr.

Extant

F. HYDRAENIDAE J. (l.)–Rec. Terr.

First: e.g. *Ochtebiites minor* Ponomarenko (1985a), Kuznetsk Basin, former USSR. **Extant**

F. HYDROPHILIDAE J. (l.)–Rec. Terr.

First: e.g. *Hydrobiites crassus* Ponomarenko (1985a), Ustbalei, Irkutsk Basin, Siberia, former USSR. **Extant**

F. HYGROBIIDAE T. (Mio.)–Rec. Terr.

Extant

F. JURODIDAE Ponomarenko, 1985a J. (l.)–K. (l.) Terr.

First: *Jurodes ignoramus* Ponomarenko (1985a), Transbaikalia, former USSR.
Last: e.g. *Jurodes minor* Ponomarenko (1990), Transbaikalia, former USSR.
Comment: Crowson (1985) preferred to treat this genus as Adephaga *incertae sedis*.

F. LABRADOROCOLEIDAE K. (l.) Terr.

F. LAGRIIDAE T. (PRB)–Rec. Terr.

Extant

F. LAMPYRIDAE T. (PRB)–Rec. Terr.

Extant

F. LATHRIDIIDAE K. (SAN)–Rec. Terr.

First: *Succinimontia infleta*, Siberian amber, Taimyr, former USSR. **Extant**

F. LEIODIDAE (LIODIDAE) T. (PRB?)–Rec. Terr.

First: *Nyujwa zherichini* Perkovsky (1990), Zaza Formation, Baisa, Buryat ASSR, former USSR. **Extant**

F. LIADYTIDAE (LYADYTIDAE) J. (l.)–K. (l.) Terr.

First: e.g. *Liadytes major* Ponomarenko (1985a), Ust-balei, Irkutsk Basin, Siberia, former USSR.
Last: e.g. *Liadytes dajensis* Ponomarenko (1987), Glushkovskaya Formation, Daya River, former USSR.

F. LIMNICHIDAE T. (PRB)–Rec. Terr.

Extant

F. LOPHOCATERIDAE J. (u.)–Rec. Terr.

Extant

F. LUCANIDAE T. (LUT/BRT)–Rec. Terr.

First: Figured but undescribed in Lutz (1988), Messel Formation, Grube Messel, Germany. **Extant**

F. LYCIDAE T. (PRB)–Rec. Terr.

Extant

F. LYCTIDAE T. (PRB)–Rec. Terr.

Extant

F. LYMEXYLONIDAE (LYMEXYLIDAE, LYMEXILIDAE) T. (PRB)–Rec. Terr.

Extant

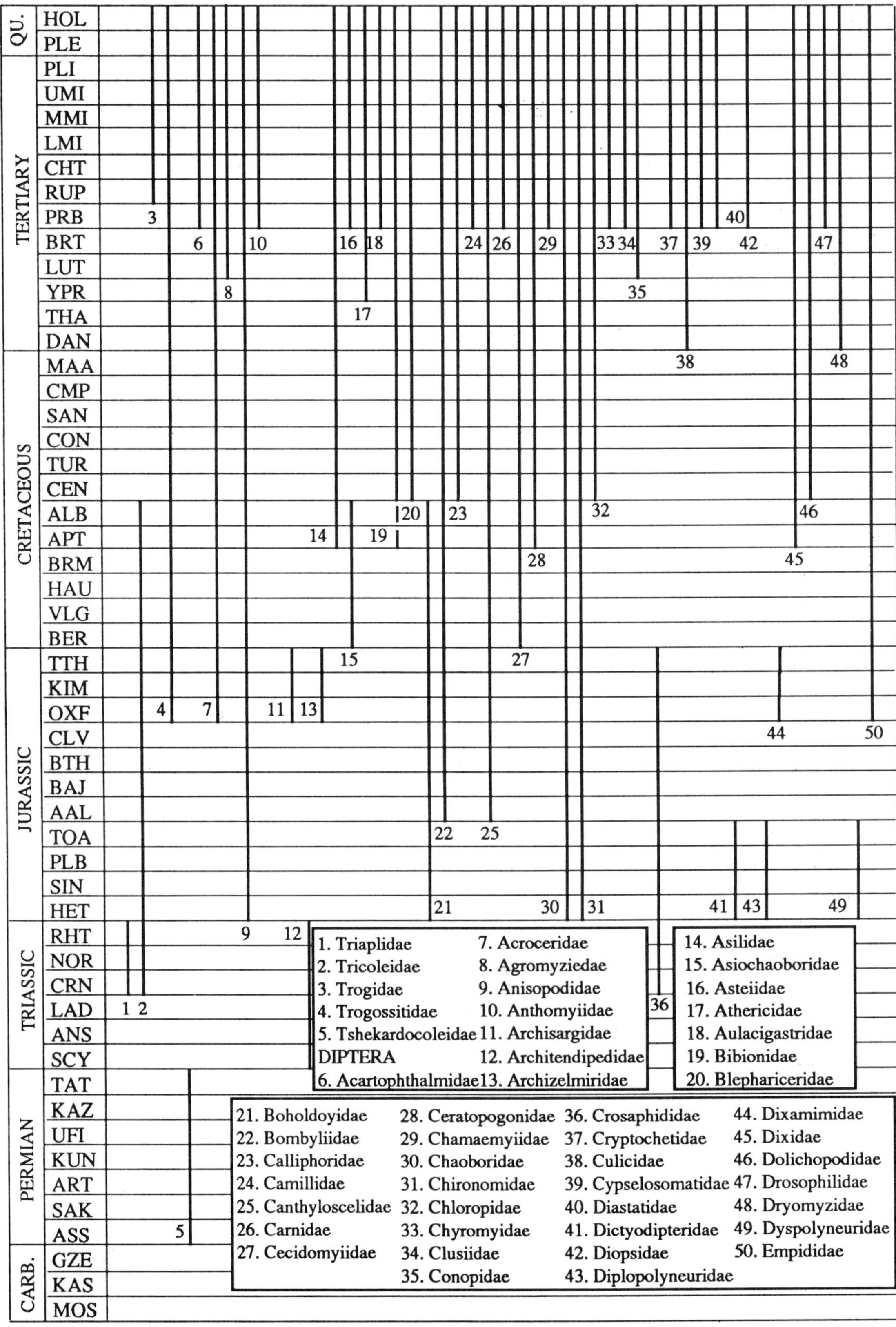
QU.
HOL
PLE
TERTIARY
PLI
UMI
MMI
LMI
CHT
RUP
PRB
BRT
LUT
YPR
THA
DAN
CRETACEOUS
MAA
CMP
SAN
CON
TUR
CEN
ALB
APT
BRM
HAU
VLG
BER
JURASSIC
TTH
KIM
OXF
CLV
BTH
BAJ
AAL
TOA
PLB
SIN
HET
TRIASSIC
RHT
NOR
CRN
LAD
ANS
SCY
PERMIAN
TAT
KAZ
UFI
KUN
ART
SAK
ASS
CARB.
GZE
KAS
MOS
1. Triaplidae
2. Tricoleidae
3. Trogidae
4. Trogossitidae
5. Tshekardocoleidae
DIPTERA
6. Acartophthalmidae
7. Acroceridae
8. Agromyziedae
9. Anisopodidae
10. Anthomyiidae
11. Archisargidae
12. Architendipedidae
13. Archizelmiridae
14. Asilidae
15. Asiochaoboridae
16. Asteiidae
17. Athericidae
18. Aulacigastridae
19. Bibionidae
20. Blephariceridae
21. Boholdoyidae
22. Bombyliidae
23. Calliphoridae
24. Camillidae
25. Canthyloscelidae
26. Carnidae
27. Cecidomyiidae
28. Ceratopogonidae
29. Chamaemyiidae
30. Chaoboridae
31. Chironomidae
32. Chloropidae
33. Chyromyidae
34. Clusiidae
35. Conopidae
36. Crosaphididae
37. Cryptochetidae
38. Culicidae
39. Cypselosomatidae
40. Diastatidae
41. Dictyodipteridae
42. Diopsidae
43. Diplopolyneuridae
44. Dixamimidae
45. Dixidae
46. Dolichopodidae
47. Drosophilidae
48. Dryomyzidae
49. Dyspolyneuridae
50. Empididae

Fig. 21.14

F. MELANDRYIDAE K. (u.)–Rec. Terr.

Extant

F. MELOIDAE T. (PRB)–Rec. Terr.

Extant

F. MELYRIDAE (MALACHIIDAE) K. (u.)–Rec. Terr.

Extant

F. MICROMALTHIDAE T. (Oli.)–Rec. Terr.

Extant

F. MORDELLIDAE J. (u.)–Rec. Terr.

Extant

F. MYCETOPHAGIDAE T. (PRB)–Rec. Terr.

Extant

F. NEMONYCHIDAE (EOBELIDAE) J. (u.)–Rec. Terr.

First: e.g. *Eobelus longipes*, Karabastau Formation, Galkini, Kazakhstan, former USSR. **Extant**

F. NITIDULIDAE J. (m.)–Rec. Terr.

First: e.g. *Sinonitidulina luanpingensis*, *in* Hong (1985d), north China.
Comment: Ponomarenko (pers. comm.) considers that this species does not show structures attributable to this family. **Extant**

F. NOSODENDRIDAE T. (Eoc.)–Rec. Terr.

Extant

F. NOTERIDAE T. (Pal.)–Rec. Terr.

Extant

F. OBOROCOLEIDAE P. Terr.

F. OEDEMERIDAE J.?–Rec. Terr.

Comment: Ponomarenko (pers. comm.) considers that the Mesozoic records of this family are doubtful. **Extant**

F. OXYCORYNIDAE K.? (APT)–Rec. Terr.

Extant

F. PARAHYGROBIIDAE J. (u.). Terr.

First and Last: *Parahygrobia natans*, *in* Zherikhin (1985b), Uda, Transbaikalia, former USSR.

F. PASSALIDAE K. (?APT)–Rec. Terr.

Extant

F. PASSANDRIDAE T. (Eoc.)–Rec. Terr.

Extant

F. PELTIDAE (OSTOMIDAE, OSTOMATIDAE) J.–Rec. Terr.

First: e.g. *Anhuistoma hyla* Lin (1985), Hanshan Formation, Hanshan, China.
Comment: Ponomarenko (pers. comm.) considers that this species does not show structures attributable to this family. **Extant**

F. PERMOCUPEDIDAE P. Terr.

e.g. *Kaltanicupes ponomarenkoi* Pinto (1987a), Irati Formation, Porto Alegre-Uruguaiana, Rio Grande do Sul, Brazil.

F. PHALACRIDAE T. (?PRB)–Rec. Terr.

Extant

F. PLATYPODIDAE T. (PRB)–Rec. Terr.

Extant

F. PRAELATERIDAE (PRAELATERIIDAE) J. (l.) Terr.

F. PROPALTICIDAE T. (Pli)–Rec. Terr.

Extant

F. PROTOCUCUJIDAE J. (u.)–Rec. Terr.

Extant

F. PSELAPHIDAE K. (APT)–Rec. Terr.

First: Pselaphid indet. Jell and Duncan (1986), Koonwarra Fossil Bed, south Gippsland, Australia. **Extant**

F. PSEPHENIDAE T. (Eoc.)–Rec. Terr.

Extant

F. PTILIIDAE (PTILIDAE) K. (u.)–Rec. Terr.

Extant

F. PTILODACTYLIDAE T. (PRB)–Rec. Terr.

Extant

F. PTINIDAE T. (PRB)–Rec. Terr.

Extant

F. PYROCHROIDAE K. (APT)–Rec. Terr.

First: *Cretaceimelittommoides cearensis* (*nomen nudum*) Vulkano and Pereira (1987), Santana Formation, Ceará, Brazil. **Extant**

F. PYTHIDAE T. (PRB)–Rec. Terr.

Extant

F. RHIPICERIDAE (SANDALIDAE) K. (APT)–Rec. Terr.

Extant

F. RHIPIPHORIDAE T. (Pal.)–Rec. Terr.

Extant

F. RHIZOPHAGIDAE T. (PRB)–Rec. Terr.

Extant

F. RHOMBOCOLEIDAE P. Terr.

F. RHYSODIDAE T. (Mio.)–Rec. Terr.

Extant

F. SALPINGIDAE T. (Oli.)–Rec. Terr.

Extant

F. SCAPHIDIIDAE T. (PRB)–Rec. Terr.

Extant

F. SCARABAEIDAE (MELOLONTHIDAE, MELONTHIDAE, RUTELIDAE, APHODIIDAE, HYBOSORIDAE, CETONIIDAE) J. (l.)–Rec. Terr.

Extant

F. SCHIZOPHORIDAE (SCHIZOCOLEIDAE) Tr. (u.)–K. (u.) Terr.

First: e.g. *Shijingocoleus margacrispus* Lin and Mou (1989), Xiaoping Formation, Guangzhou, China.

F. SCIRTIDAE (HELODIDAE) K. (APT)–Rec. Terr.

First: Helodid indet. Jell and Duncan (1986), Koonwarra Fossil Bed, south Gippsland, Australia. **Extant**

F. SCOLYTIDAE K. (VLG)–Rec. Terr.

First: *Paleoscolytus sussexensis* Jarzembowski (1990c), Wadhurst Clay, Crowborough, Sussex, England, UK. **Extant**

F. SCRAPTIIDAE (SCRAPTIDAE, SCARAPTIIDAE) J. (u.)–Rec. Terr.

Extant

F. SCYDMAENIDAE K. (l.)–Rec. Terr.

Extant

F. SERROPALPIDAE T. (PRB)–Rec. Terr.

Extant

F. SILPHIDAE J. (l.)–Rec. Terr.

First: e.g. *Mercata festira* Lin (1986), Guangxi, China. **Extant**

F. SILVANIDAE T. (PRB)–Rec. Terr.

Extant

F. SPHINDIDAE T. (PRB)–Rec. Terr.

Extant

F. STAPHYLINIDAE Tr.? (RHT)–Rec. Terr.

First: Figured but undescribed, *in* Gore (1988), Newark Supergroup, Culpeper Basin, Virginia, USA. **Extant**

F. TALDYCUPEDIDAE (TALDYCUPIDAE) P.–J. (m.) Terr.

Last: *Yuxianocoleus hebeiense* Hong (1985a), Xiahuayuan Formation, Yuxian, Hebei, China.

F. TENEBRIONIDAE (ALLECULIDAE) Tr.?–Rec. Terr.

Comment: Zherikhin (pers. comm.) considers that the Mesozoic records of this family are doubtful. **Extant**

F. THROSCIDAE (TRIXAGIDAE) K. (l.)–Rec. Terr.

Extant

F. TRACHYPACHIDAE (TRACHYPACHEIDAE, TRACHYPACHYIDAE) Tr. (u.)–Rec. Terr.

First: *Sogdodromeus altus*, Madygen Formation, Madygen, Batken, former USSR. **Extant**

F. TRIAPLIDAE Tr. (u.). Terr.

e.g. *Triaplus macroplatus*, Madygen Formation, south Fergana, former USSR. **Extant**

F. TRICOLEIDAE Tr. (u.)–K. (l.) Terr.

F. TROGIDAE T. (Oli.)–Rec. Terr.

Extant

F. TROGOSSITIDAE (TROGOSITIDAE) J. (u.)–Rec. Terr.

Extant

F. TSHEKARDOCOLEIDAE P. Terr.

e.g. *Votocoleus submissus* in Kukalová-Peck (1991), Moravia, Czechoslovakia. **Extant**

Order DIPTERA (MUSCIDA) P. (ZEC)–Rec. Terr.

We have followed the classification of extant families given in McAlpine (1981, 1987, 1989). Oldest families: Permotipulidae and Permotanyderidae. Grimaldi (1990) gives a list of the Cretaceous genera. Amber records taken from Spahr (1985) unless stated otherwise.

F. ACARTOPHTHALMIDAE T. (PRB)–Rec. Terr.

First: *Acartophthalmites tertiaria*, Baltic amber. **Extant**

F. ACROCERIDAE J. (u.)–Rec. Terr.

Extant

F. AGROMYZIDAE T. (LUT/BRT)–Rec. Terr.

First: Cambian mines, *in* Boucot (1990) **Extant**

F. ANISOPODIDAE (PROTOLBIOGASTRIDAE, MYCETOBIIDAE, RHYPHIDAE, OLBIOGASTRIDAE) J. (l.)–Rec. Terr.

Extant

F. ANTHOMYIIDAE T. (PRB)–Rec. Terr.

First: *Anthomyia* sp., Baltic amber. **Extant**

F. ARCHISARGIDAE J. (u.) Terr.

F. ARCHITENDIPEDIDAE Tr. Terr.

F. ARCHIZELMIRIDAE J. (u.). Terr.

F. ASILIDAE K. (APT)–Rec. Terr.

First: e.g. *Araripogon axelrodi* Grimaldi (1990), Santana Formation, Ceará, Brazil. **Extant**

F. ASIOCHAOBORIDAE Hong and Wang, 1990 K. (l.) Terr.

e.g. *Asiochaoborus tenuous* Hong and Wang (1990), Laiyang Formation, Shandong, China.

F. ASTEIIDAE T. (PRB)–Rec. Terr.

First: *Succiniasteia carpenteri*, *in* Spahr (1989), Baltic amber. **Extant**

F. ATHERICIDAE T. (Eoc.)–Rec. Terr.

Extant

F. AULACIGASTRIDAE T. (PRB)–Rec. Terr.

First: *Protaulacigaster electrica*, Baltic amber. **Extant**

F. BIBIONIDAE (PLECIIDAE, PENTHETRIIDAE, HESPERINIDAE) K. (?APT)–Rec. Terr.

First: Described but unnamed in Grimaldi (1990), Santana Formation, Ceará, Brazil. **Extant**

F. BLEPHARICERIDAE (BLEPHAROCERIDAE) K. (CEN)–Rec. Terr.

Extant

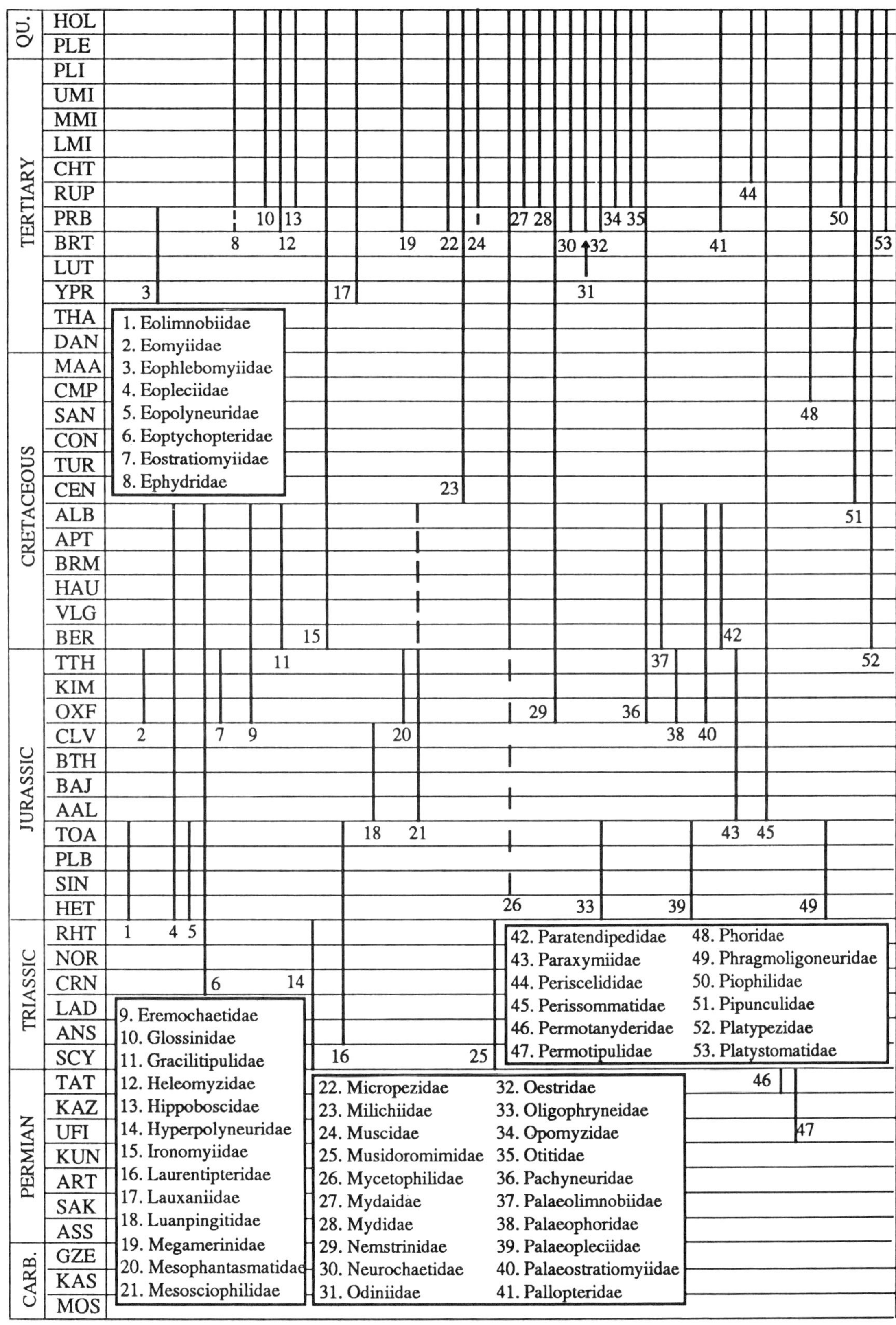
QU.
TERTIARY
CRETACEOUS
JURASSIC
TRIASSIC
PERMIAN
CARB.
HOL
PLE
PLI
UMI
MMI
LMI
CHT
RUP
PRB
BRT
LUT
YPR
THA
DAN
MAA
CMP
SAN
CON
TUR
CEN
ALB
APT
BRM
HAU
VLG
BER
TTH
KIM
OXF
CLV
BTH
BAJ
AAL
TOA
PLB
SIN
HET
RHT
NOR
CRN
LAD
ANS
SCY
TAT
KAZ
UFI
KUN
ART
SAK
ASS
GZE
KAS
MOS
1. Eolimnobiidae
2. Eomyiidae
3. Eophlebomyiidae
4. Eopleciidae
5. Eopolyneuridae
6. Eoptychopteridae
7. Eostratiomyiidae
8. Ephydridae
9. Eremochaetidae
10. Glossinidae
11. Gracilitipulidae
12. Heleomyzidae
13. Hippoboscidae
14. Hyperpolyneuridae
15. Ironomyiidae
16. Laurentipteridae
17. Lauxaniidae
18. Luanpingitidae
19. Megamerinidae
20. Mesophantasmatidae
21. Mesosciophilidae
22. Micropezidae
23. Milichiidae
24. Muscidae
25. Musidoromimidae
26. Mycetophilidae
27. Mydaidae
28. Mydidae
29. Nemstrinidae
30. Neurochaetidae
31. Odiniidae
32. Oestridae
33. Oligophryneidae
34. Opomyzidae
35. Otitidae
36. Pachyneuridae
37. Palaeolimnobiidae
38. Palaeophoridae
39. Palaeopleciidae
40. Palaeostratiomyiidae
41. Pallopteridae
42. Paratendipedidae
43. Paraxymiidae
44. Periscelididae
45. Perissommatidae
46. Permotanyderidae
47. Permotipulidae
48. Phoridae
49. Phragmoligoneuridae
50. Piophilidae
51. Pipunculidae
52. Platypezidae
53. Platystomatidae

Fig. 21.15

F. BOHOLDOYIDAE Kovalev, 1985 J. (l.)–K. (l.) Terr.

First: *Boholdoya alata* Kovalev, *in* Kalugina and Kovalev (1985), Transbaikalia, former USSR.
Last: e.g. ?*Boholdoya thoracica* Kovalev (1990), Turga, former USSR.

F. BOMBYLIIDAE (MYTHICOMYIIDAE, SYSTROPODIDAE, PHTHIRIIDAE, USIIDAE) J (m.)–Rec. Terr.

First: *Palaeoplatypygus zaitzevi*, *in* Zaytsev (1986), Siberia, former USSR. **Extant**

F. CALLIPHORIDAE (SARCOPHAGIDAE) K. (u.)–Rec. Terr.

First: *Cretaphormia fowleri*, *in* Grimaldi (1990), Edmonton Formation, Alberta, Canada. **Extant**

F. CAMILLIDAE T. (PRB)–Rec. Terr.

First: *Protocamilla succini*, Baltic amber. **Extant**

F. CANTHYLOSCELIDAE (HYPEROSCELIDIDAE) J. (m.)–Rec. Terr.

First: *Prohyperoscelis jurassicus* Kovalev, *in* Kalugina and Kovalev (1985), former USSR. **Extant**

F. CARNIDAE T. (PRB)–Rec. Terr.

First: *Meoneurites enigmaticus*, Baltic amber. **Extant**

F. CECIDOMYIIDAE (LESTREMIIDAE) K. (l.)–Rec. Terr.

First: ?*Catotricha mesozoica* Kovalev (1990), Daya, Transbaikalia, former USSR. **Extant**

F. CERATOPOGONIDAE (LEPTOCONOPIDAE) K. (APT)–Rec. Terr.

F. CHAMAEMYIIDAE T. (PRB)–Rec. Terr.

First: *Procremifania electrica*, Baltic amber. **Extant**

F. CHAOBORIDAE (MESOTENDIPEDIDAE, CHIRONOMAPTERIDAE) J. (l.)–Rec. Terr.

First: e.g. *Praechaoborus tugnuicus* Kalugina, *in* Kalugina and Kovalev (1985), Transbaikalia, former USSR. **Extant**

F. CHIRONOMIDAE (TENDIPEDIDAE, PROTOBIBIONIDAE) J. (l.)–Rec. Terr.

First: e.g. *Oryctochlus vulcanus* Kalugina, *in* Kalugina and Kovalev (1985), Transbaikalia, former USSR. **Extant**

F. CHLOROPIDAE K. (u.)–Rec. Terr.

Extant

F. CHYROMYIDAE (CHYROMYIIDAE) T. (PRB)–Rec. Terr.

First: *Gephyromyiella electrica*, Baltic amber. **Extant**

F. CLUSIIDAE T. (PRB)–Rec. Terr.

First: *Electroclusiodes meunieri*, Baltic amber. **Extant**

F. CONOPIDAE J. (LUT/BRT)–Rec. Terr.

First: *Poliomyia recta* in McAlpine (1989), Green River, Wyoming, USA. **Extant**

F. CROSAPHIDIDAE Kovalev, 1983 Tr. (u.)–J. (u.) Terr.

First: e.g. *Crosaphis anomala*, *in* Kovalev (1983), Ipswich Series, Mt. Crosby, Australia.

F. CRYPTOCHETIDAE (CRYPTOCHAETIDAE) T. (PRB)–Rec. Terr.

First: *Phanerochaetum tuxeni*, Baltic amber. **Extant**

F. CULICIDAE T. (Pal.)–Rec. Terr.

Extant

F. CYPSELOSOMATIDAE T. (PRB)–Rec. Terr.

First: *Cypselosomatites succini*, Baltic amber. **Extant**

F. DIASTATIDAE T. (PRB)–Rec. Terr.

First: e.g. *Pareuthychaeta electrica*, Baltic amber. **Extant**

F. DICTYODIPTERIDAE J. (l.) Terr.

F. DIOPSIDAE T. (PRB)–Rec. Terr.

First: e.g. *Prosphyracephala succini*, Baltic amber. **Extant**

F. DIPLOPOLYNEURIDAE J. (l.) Terr.

F. DIXAMIMIDAE J. (u.) Terr.

F. DIXIDAE K. (APT)–Rec. Terr.

First: Dixid indet. Jell and Duncan (1986), Koonwarra Fossil Bed, south Gippsland, Australia. **Extant**

F. DOLICHOPODIDAE K. (u.)–Rec. Terr.

First: e.g. *Retinitus nervosus*, *in* Grimaldi (1990), Siberian amber, former USSR. **Extant**

F. DROSOPHILIDAE T. (PRB)–Rec. Terr.

First: *Electrophortica succini in* Grimaldi (1987), Baltic amber. **Extant**

F. DRYOMYZIDAE T. (Pal.)–Rec. Terr.

Extant

F. DYSPOLYNEURIDAE J. (l.) Terr.

F. EMPIDIDAE (HYBOTIDAE) J. (u.)–Rec. Terr.

Extant

F. EOLIMNOBIIDAE (EOLIMUOBIIDAE, EOLIMBIIDAE) J. (l.) Terr.

e.g. *Chorolimnobia ostera* Lin (1986), Hunan, China.

F. EOMYIIDAE J. (u.) Terr.

F. EOPHLEBOMYIIDAE T. (Eoc.) Terr.

F. EOPLECIIDAE J. (TOA)–K. (l). Terr.

First: *Eoplecia primitiva*, *in* Kovalev (1990), Upper Lias, Germany.
Last: e.g. *Eomycetophila asymmetrica* Kovalev (1990), Daya, Transbaikalia, former USSR.

F. EOPOLYNEURIDAE J. (l.) Terr.

F. EOPTYCHOPTERIDAE Tr. (u.)–K. (l.) Terr.

Last: e.g. *Eoptychopterina baisica* Kalugina (1989), Zaza Formation, Baisa, Buryat ASSR, former USSR.

F. EOSTRATIOMYIIDAE J. (u.) Terr.

F. EPHYDRIDAE T. (?PRB)–Rec. Terr.

First: *Ephydra* sp., Baltic amber. **Extant**

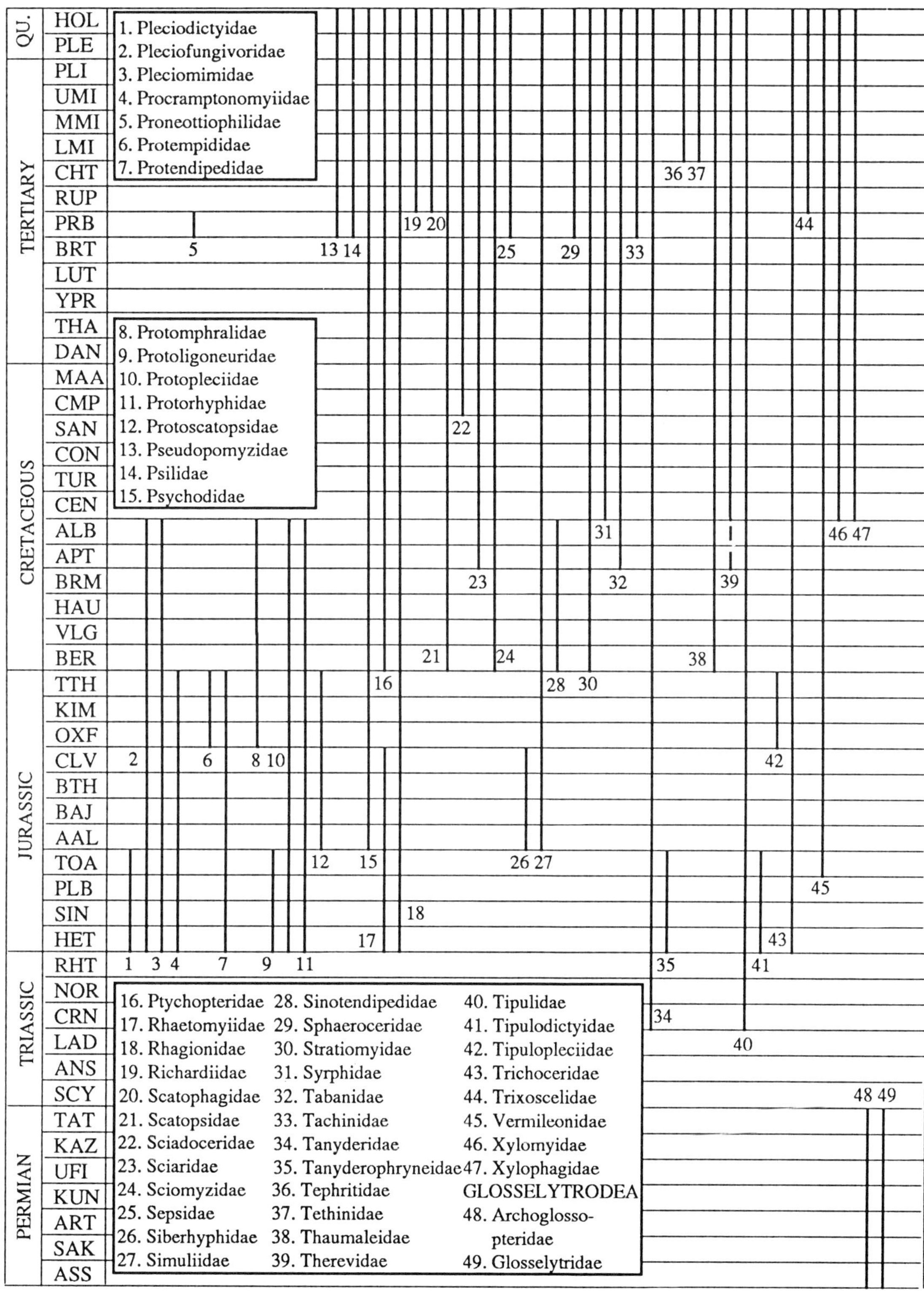

Fig. 21.16

F. EREMOCHAETIDAE (BREMOCHAETIDAE) J. (u.)–K. (l.) Terr.

First: e.g. *Pareremochaetus minor*, *in* Kovalev (1989a), Kazakhstan, former USSR.
Last: e.g. *Eremochaetosoma mongolicum* Kovalev (1986), Gurvaneren Formation, Mongolia.

F. GLOSSINIDAE T. (Oli.)–Rec. Terr.

Extant

F. GRACILITIPULIDAE Hong and Wang, 1990 K. (l.) Terr.

First and Last: *Gracilitipula asiatica* Hong and Wang (1990), Laiyang Formation, Shandong, China.

F. HELEOMYZIDAE (HELOMYZIDAE) T. (PRB)–Rec. Terr.

First: e.g. *Chaetohelomyza electrica*, Baltic amber. **Extant**

F. HIPPOBOSCIDAE T. (Oli.)–Rec. Terr.

Extant

F. HYPERPOLYNEURIDAE Tr. Terr.

F. IRONOMYIIDAE K. (l.)–Rec. Terr.

Extant

F. LAURENTIPTERIDAE (PSEUDODIPTERIDAE) Tr. (m.)–J. (l.) Terr.

First: *Laurentiptera gallica*, *in* Wootton and Ennos (1989), Vosges, France.
Last: *Ijapsyche sibirica*, *in* Wootton and Ennos (1989), former USSR.

F. LAUXANIIDAE T. (Eoc.)–Rec. Terr.

First: e.g. *Tryaneoides ellipticus*, Chinese amber, Fushun coalfield, China. **Extant**

F. LUANPINGITIDAE Zhang, 1986a J. (m.) Terr.

First and Last: *Luanpingites flavus* Zhang (1986a), Xiahuayuan Formation, Luanping, China.

F. MEGAMERINIDAE T. (PRB)–Rec. Terr.

First: *Palaeotanypeza spinosa*, Baltic amber. **Extant**

F. MESOPHANTASMATIDAE J. (u.) Terr.

F. MESOSCIOPHILIDAE (ALLACTONEURIDAE pars, FUNGIVORITIDAE pars) J. (m.)–K. (l.?) Terr.

First: e.g. *Mesosciophilina irinae* Kovalev, *in* Kalugina and Kovalev (1985), former USSR.
Last: *Mesosciophilites indefinites* Kovalev (1986), Gurvaneren Formation, Mongolia.

F. MICROPEZIDAE (CALOBATIDAE) T. (PRB)–Rec. Terr.

First: e.g. *Electrobata tertiaria*, *in* McAlpine (1990), Baltic amber. **Extant**

F. MILICHIIDAE (PHYLLOMYZIDAE) K. (u.)–Rec. Terr.

Extant

F. MUSCIDAE T. (?PRB)–Rec. Terr.

First: e.g. *'Musca' longipes*, Baltic amber. **Extant**

F. MUSIDOROMIMIDAE Tr. Terr.

F. MYCETOPHILIDAE (FUNGIVORIDAE, BOLITOPHILIDAE, ARACHNOCAMPIDAE, LYGISTORRHINIDAE, KEROPLATIDAE, DITOMYIIDAE, MANOTIDAE, MACROCERIDAE, DIADOCIDIIDAE) J.? (SIN)–Rec. Terr.

First: *Prodocidia spectra* Whalley (1985), Lower Lias, Charmouth, Dorset, England, UK. Blagoderov (pers. comm.) considers that this species does not belong in this family. **Extant**

F. MYDAIDAE T. (Oli.)–Rec. Terr.

Extant

F. MYDIDAE T. (Oli.)–Rec. Terr.

Extant

F. NEMESTRINIDAE J. (u.)–Rec. Terr.

Extant

F. NEUROCHAETIDAE T. (PRB)–Rec. Terr.

First: e.g. *Anthoclusia gephyrea*, *in* Spahr (1989), Baltic amber. **Extant**

F. ODINIIDAE T. (PRB)–Rec. Terr.

First: *Protodinia electrica*, Baltic amber. **Extant**

F. OESTRIDAE T. (PRB)–Rec. Terr.

First: *Novoberendtia baltica*, Baltic amber. **Extant**

F. OLIGOPHRYNEIDAE J. (l.) Terr.

F. OPOMYZIDAE T. (Oli.)–Rec. Terr.

Extant

F. OTITIDAE (ULIDIIDAE) T. (Oli.)–Rec. Terr.

Extant

F. PACHYNEURIDAE (CRAMPTONOMYIIDAE) J. (u.)–Rec. Terr.

Extant

F. PALAEOLIMNOBIIDAE Zhang *et al.*, 1986 K. (l.)

e.g. *Palaeolimnobia laiyangensis* Zhang *et al.* (1986), Laiyang Formation, Laiyang, Shandong, China. We have followed Hong and Wang (1990) in regarding this formation as Lower Cretaceous in age.

F. PALAEOPHORIDAE J. (u.) Terr.

F. PALAEOPLECIIDAE J. (l.) Terr.

F. PALAEOSTRATIOMYIIDAE (PALAEOSTRIOMYIIDAE) J. (u.)–K. (l.) Terr.

Last: e.g. *Stratiomyopsis robusta* Hong and Wang (1990), Laiyang Formation, Shandong, China.

F. PALLOPTERIDAE T. (PRB)–Rec. Terr.

First: e.g. *Pallopterites electrica*, Baltic amber. **Extant**

F. PARATENDIPEDIDAE Hong and Wang, 1990 K. (l.) Terr.

e.g. *Paratendipes laiyangensis* Hong and Wang (1990), Laiyang Formation, Shandong, China.

F. PARAXYMYIIDAE J. (m.–u.) Terr.

F. PERISCELIDIDAE (PERISCELIDAE) T. (CHT)–Rec. Terr.

First: *Periscelis annectens*, *in* McAlpine (1989), Mexican amber, Chiapas, Mexico. **Extant**

F. PERISSOMMATIDAE J. (m.)–Rec. Terr.

First: *Palaeoperissomma collessi* Kovalev, *in* Kalugina and Kovalev (1985), former USSR. **Extant**

F. PERMOTANYDERIDAE P. (TAT) Terr.

e.g. *Permotanyderus ableptus*, *in* Willmann (1989b), Warner's Bay, New South Wales, Australia.

F. PERMOTIPULIDAE P. (ZEC). Terr.

e.g. *Permotipula patricia*, *in* Willmann (1989a), Warner's Bay, New South Wales, Australia.

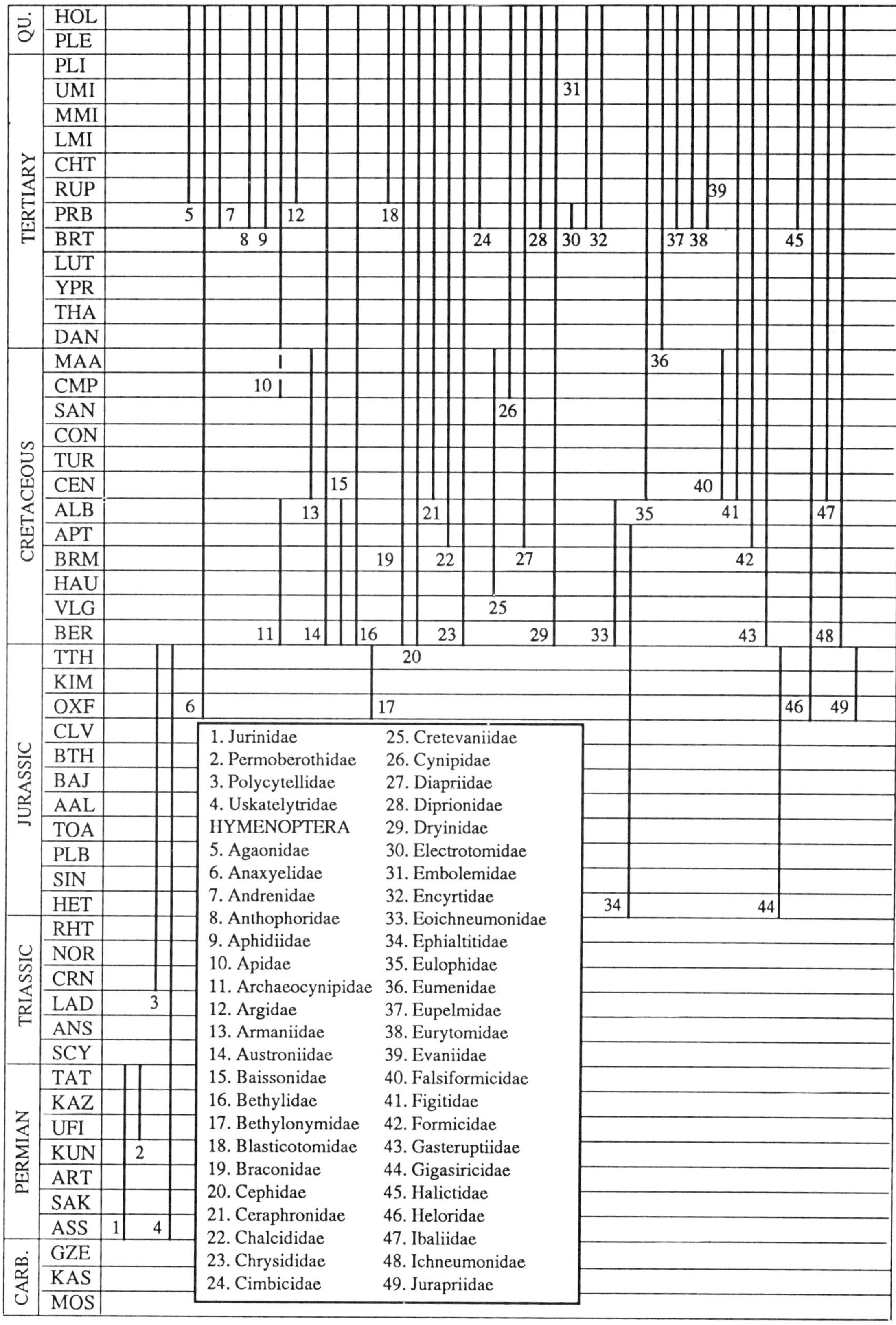
QU.
TERTIARY
CRETACEOUS
JURASSIC
TRIASSIC
PERMIAN
CARB.
HOL
PLE
PLI
UMI
MMI
LMI
CHT
RUP
PRB
BRT
LUT
YPR
THA
DAN
MAA
CMP
SAN
CON
TUR
CEN
ALB
APT
BRM
HAU
VLG
BER
TTH
KIM
OXF
CLV
BTH
BAJ
AAL
TOA
PLB
SIN
HET
RHT
NOR
CRN
LAD
ANS
SCY
TAT
KAZ
UFI
KUN
ART
SAK
ASS
GZE
KAS
MOS
1. Jurinidae
2. Permoberothidae
3. Polycytellidae
4. Uskatelytridae
HYMENOPTERA
5. Agaonidae
6. Anaxyelidae
7. Andrenidae
8. Anthophoridae
9. Aphidiidae
10. Apidae
11. Archaeocynipidae
12. Argidae
13. Armaniidae
14. Austroniidae
15. Baissonidae
16. Bethylidae
17. Bethylonymidae
18. Blasticotomidae
19. Braconidae
20. Cephidae
21. Ceraphronidae
22. Chalcididae
23. Chrysididae
24. Cimbicidae
25. Cretevaniidae
26. Cynipidae
27. Diapriidae
28. Diprionidae
29. Dryinidae
30. Electrotomidae
31. Embolemidae
32. Encyrtidae
33. Eoichneumonidae
34. Ephialtitidae
35. Eulophidae
36. Eumenidae
37. Eupelmidae
38. Eurytomidae
39. Evaniidae
40. Falsiformicidae
41. Figitidae
42. Formicidae
43. Gasteruptiidae
44. Gigasiricidae
45. Halictidae
46. Heloridae
47. Ibaliidae
48. Ichneumonidae
49. Jurapriidae

Fig. 21.17

F. PHORIDAE K. (CMP)–Rec. Terr.

First: *Metopina goeleti* Grimaldi (1989), New Jersey amber, Kinkora, New Jersey, USA. **Extant**

F. PHRAGMOLIGONEURIDAE J. (l.) Terr.

F. PIOPHILIDAE T. (Oli.)–Rec. Terr.

Extant

F. PIPUNCULIDAE K. (u.)–Rec. Terr.

Extant

F. PLATYPEZIDAE (OPETIIDAE) K. (l.)–Rec. Terr.

First: e.g. *Palaeopetia laiyangensis* Zhang (1987), Laiyang Formation, Laiyang, Shandong, China. We have followed Hong and Wang (1990) in regarding this formation as Lower Cretaceous in age. **Extant**

F. PLATYSTOMATIDAE T. (PRB)–Rec. Terr.

First: *Scholastes foordi*, English amber, Norfolk, England, UK. **Extant**

F. PLECIODICTYIDAE J. (l.) Terr.

F. PLECIOFUNGIVORIDAE (PLICIOFUNGIVORIDAE, ALLACTONEURIDAE pars, FUNGIVORITIDAE pars) J. (l.)–K. (l.) Terr.

First: e.g. *Archihesperinus phryneoides*, *in* Kovalev (1987), Issyk-Kul, Kirgizia, former USSR.
Last: e.g. *Bryanka lepida* Kovalev (1990), Daya, Transbaikalia, former USSR.

F. PLECIOMIMIDAE (SINEMEDIIDAE) J. (l.)–K. (l.) Terr.

Last: e.g. *Pleciomimella perbella* Zhang *et al*. (1986), Laiyang Formation, Shandong, China.

F. PROCRAMPTONOMYIIDAE Kovalev, 1985 J. (l.–u.) Terr.

First: e.g. *Procramptonomyia sibirica* Kovalev, *in* Kalugina and Kovalev (1985), Transbaikalia, former USSR.

F. PRONEOTTIOPHILIDAE T. (PRB) Terr.

First and Last: *Proneottiophilum extinctum*, Baltic amber.

F. PROTEMPIDIDAE J. (u.) Terr.

F. PROTENDIPEDIDAE J. Terr.

e.g. *Protendipes huabeiensis* Zhang (1986b), northern Hebei, China.

F. PROTOMPHRALIDAE J. (u.)–K. (l.) Terr.

Last: *Mesomphrale asiatica* Hong and Wang (1990), Laiyang Formation, Shandong, China.

F. PROTOLIGONEURIDAE J. (l.) Terr.

F. PROTOPLECIIDAE J. (l.)–K. (l.) Terr.

First: e.g. *Archipleciomima obtusipennis*, *in* Kalugina and Kovalev (1985), former USSR.

F. PROTORHYPHIDAE J. (l.)–K. (l.) Terr.

First: e.g. *Protorhyphus ovisimilis*, *in* Kalugina and Kovalev (1985), Upper Lias, Dobbertin, Germany.
Last: *Protorhyphus major* Kovalev (1990), Daya, Transbaikalia, former USSR.

F. PROTOSCATOPSIDAE J. (m.–u.) Terr.

First: *Mesoscatopse rohdendorfi* Kovalev, *in* Kalugina and Kovalev (1985), former USSR.

F. PSEUDOPOMYZIDAE T. (PRB)–Rec. Terr.

First: *Eopseudopomyza kuehnei*, Baltic amber. **Extant**

F. PSILIDAE T. (PRB)–Rec. Terr.

First: *Electrochyliza succini*, Baltic amber. **Extant**

F. PSYCHODIDAE (PHLEBOTOMIDAE) J. (m.?)–Rec. Terr.

First: *Eopericoma zherichini* Kalugina, *in* Kalugina and Kovalev (1985), Transbaikalia, former USSR. **Extant**

F. PTYCHOPTERIDAE K. (l.)–Rec. Terr.

First: *Ptychoptera mesozoica* Kalugina (1989), Zaza Formation, Baisa, Buryat ASSR, former USSR. **Extant**

F. RHAETOMYIIDAE (RHAETOMYIDAE) J. (l.–m.) Terr.

F. RHAGIONIDAE (RHAGIONEMPIDIDAE) J. (l.)–Rec. Terr.

First: e.g. *Palaeobrachyceron nagatomii*, *in* Zherikhin (1985b), Abashev Formation, Tom River, Kererovo, former USSR. **Extant**

F. RICHARDIIDAE T. (Oli.)–Rec. Terr.

Extant

F. SCATOPHAGIDAE (SCATHOPHAGIDAE) T. (Oli.)–Rec. Terr.

Extant

F. SCATOPSIDAE K. (l.)–Rec. Terr.

Extant

F. SCIADOCERIDAE K. (CMP)–Rec. Terr.

First: e.g. *Sciadophora bostoni*, *in* Grimaldi (1990), Canadian amber, Medicine Hat, Canada. **Extant**

F. SCIARIDAE (SCIAROIDAE) K. (APT)–Rec. Terr.

Extant

F. SCIOMYZIDAE K. (BER)–Rec. Terr.

First: Dipterous larva: species 3, Whalley and Jarzembowski (1985), Lithographic Limestone, Montsech, Spain. **Extant**

F. SEPSIDAE T. (PRB)–Rec. Terr.

First: *Protorygma electricum*, Baltic amber. **Extant**

F. SIBERHYPHIDAE Kovalev, 1985 J. (m.) Terr.

First and Last: *Siberhyphus lebedevi* Kovalev, *in* Kalugina and Kovalev (1985), former USSR.

F. SIMULIIDAE J. (m.)–Rec. Terr.

First: *Simulimima grandis*, *in* Crosskey (1991), Ichetuisk Formation, Transbaikalia, former USSR. **Extant**

F. SINOTENDIPEDIDAE Hong and Wang, 1990 K. (l.) Terr.

First and Last: *Sinotendipes tuanwangensis* Hong and Wang (1990), Laiyang Formation, Shandong, China.

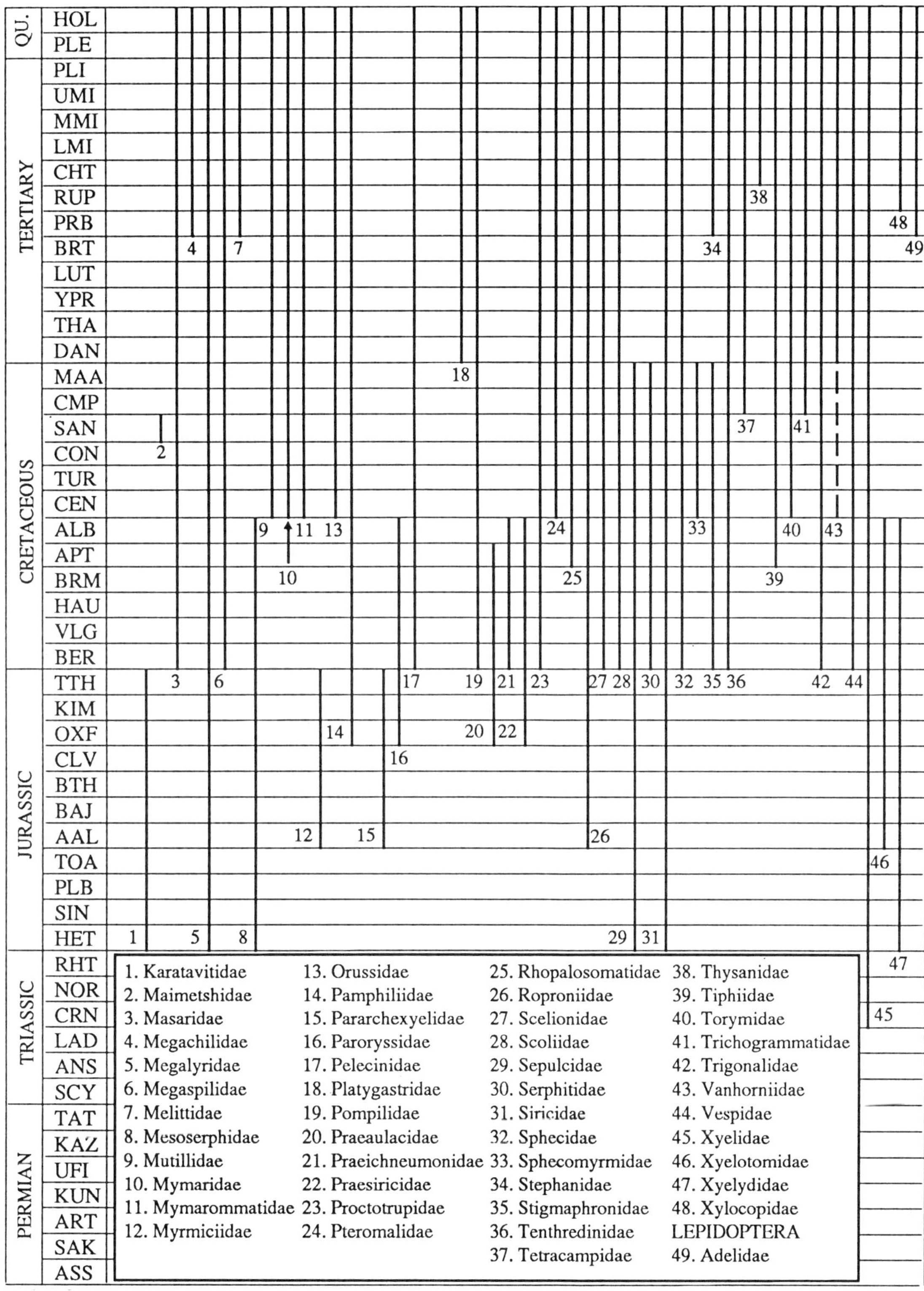

Fig. 21.18

F. SPHAEROCERIDAE T. (PRB)–Rec. Terr.

First: *Copromyza* sp., Baltic amber. **Extant**

F. STRATIOMYIDAE (STRATIOMYIIDAE, STRATIOMYRIIDAE) K. (BER)–Rec. Terr.

First: e.g. Dipterous larva: species 1, Whalley and Jarzembowski (1985), Lithographic Limestone, Montsech, Spain. **Extant**

F. SYRPHIDAE K. (u.)–Rec. Terr.

Extant

F. TABANIDAE K. (APT)–Rec. Terr.

Extant

F. TACHINIDAE T. (PRB)–Rec. Terr.

First: e.g. *Electrotachina smithii*, Baltic amber. **Extant**

F. TANYDERIDAE Tr. (u.)–Rec. Terr.

Extant

F. TANYDEROPHRYNEIDAE J. (l.) Terr.

F. TEPHRITIDAE T. (Mio.)–Rec. Terr.

Extant

F. TETHINIDAE T. (Mio.)–Rec. Terr.

Extant

F. THAUMALEIDAE (THAUMALAEIDAE) K. (l.)–Rec. Terr.

First: *Mesothaumalea fossilis* Kovalev (1989b), Baley Series, Dai, Transbaikalia, former USSR. **Extant**

F. THEREVIDAE K. (?APT)–Rec. Terr.

First: Described but unnamed, *in* Grimaldi (1990), Santana Formation, Ceará, Brazil. **Extant**

F. TIPULIDAE (LIMONIIDAE, ARCHITIPULIDAE, EOASILIDAE, CYLINDROTOMIDAE) Tr. (u.)–Rec. Terr.

Extant

F. TIPULODICTYIDAE J. (l.) Terr.

F. TIPULOPLECIIDAE J. (u.) Terr.

F. TRICHOCERIDAE J. (l.)–Rec. Terr.

First: e.g. *Eotrichocera christinae* Kalugina, *in* Kalugina and Kovalev (1985), Transbaikalia, former USSR. **Extant**

F. TRIXOSCELIDAE (TRIXOSCELIDIDAE) T. (Oli.)–Rec. Terr.

Extant

F. VERMILEONIDAE (PROTOBRACHYCERONTIDAE) J. (TOA)–Rec. Terr.

First: *Protobrachyceron liasinum*, *in* Ponomarenko and Schultz (1988), Upper Lias, Dobbertin, Germany. **Extant**

F. XYLOMYIDAE K. (u.)–Rec. Terr.

Extant

F. XYLOPHAGIDAE (RACHICERIDAE, COENOMYIIDAE) K. (u.)–Rec. Terr.

Extant

Order GLOSSELYTRODEA (JURINIDA) P.–J. (u.) Terr.

Last as for Polycytellidae.

F. ARCHOGLOSSOPTERIDAE P. Terr.

F. GLOSSELYTRIDAE P. Terr.

F. JURINIDAE P. Terr.

F. PERMOBEROTHIDAE P. (u.) Terr.

F. POLYCYTELLIDAE Tr. (u.)–J. (u.) Terr.

Last: *Mongolojurina altaica* Ponomarenko (1988), Gobi Altai, Mongolia.

F. UŠKATELYTRIDAE P.–J. Terr.

Order HYMENOPTERA (VESPIDA) Tr. (u.)–Rec. Terr.

Many ranges were taken from Rasnitsyn (1988). A list of Cretaceous genera (some with a revised position) is given by Rasnitsyn in Darling and Sharkey (1990). Earliest family: Xyelidae. Amber records taken from Spahr (1987) unless stated otherwise.

F. AGAONIDAE (AGAONTIDAE) T. (Oli.)–Rec. Terr.

F. ANAXYELIDAE J. (u.)–Rec. Terr.

First: e.g. *Xaxexis longhuaensis* (Hong, 1985d) **comb. nov.**, north China. New generic name from Pagliano and Scaramozzino (1990).

Comment: Rasnitsyn (pers. comm.) considers that this species belongs in the Xyelidae. **Extant**

F. ANDRENIDAE T. (PRB)–Rec. Terr.

First: *Andrena wrisleyi*, Baltic amber. **Extant**

F. ANTHOPHORIDAE T. (PRB)–Rec. Terr.

Extant

F. APHIDIIDAE T. (PRB)–Rec. Terr.

First: e.g. *Praeaphidius macrophthalmus*, Baltic amber. **Extant**

F. APIDAE (BOMBIDAE) K. ?(CMP)–Rec. Terr.

First: *Trigona prisca* Michener and Grimaldi (1988), New Jersey amber, Kinkora New Jersey, USA. Rasnitsyn and Michener (1991) doubt the age of this species.

Comment: Rasnitsyn, *in* Darling and Sharkey (1990) transferred the Lower Cretaceous *Palaeapis beiboziensis* Hong (1984a) to the Sphecidae. **Extant**

F. ARCHAEOCYNIPIDAE Rasnitsyn and Kovalev, 1988 K. (l.) Terr.

e.g. *Archaeocynips villosa* Rasnitsyn and Kovalev (1988), Zaza Formation, Baisa, Buryat ASSR, former USSR.

F. ARGIDAE T. (Oli.)–Rec. Terr.

Extant

F. ARMANIIDAE K. (u.) Terr.

e.g. *Armania robusta*, *in* Wilson (1987), Alskaya Formation, Magadan, former USSR.

F. AUSTRONIIDAE (TRUPOCHALCIDIIDAE) K. (l.)–Rec. Terr.

Extant

F. BAISSODIDAE K. (l.) Terr.

e.g. ?*Baissodes longus* Rasnitsyn (1986), Gurvaneren Formation, Mongolia.

F. BETHYLIDAE K. (l.)–Rec. Terr.

First: *Cretobethylellus lucidus* Rasnitsyn (1990), Pavlovka, former USSR. **Extant**

F. BETHYLONYMIDAE J. (u.) Terr.

e.g. *Bethylonymus sibiricus* in Zherikhin (1985b), Uda, Transbaikalia, former USSR.

F. BLASTICOTOMIDAE T. (Oli.)–Rec. Terr.

Extant

F. BRACONIDAE (BRANCONIDAE) K. (l.)–Rec. Terr.

First: *Eobraconus inopinatus*, *in* Rasnitsyn (1985), Mongolia. **Extant**

F. CEPHIDAE K. (l.)–Rec. Terr.

First: e.g. *Mesocephus sibiricus*, Zaza Formation, Baisa, Buryat ASSR, former USSR.
Comment: Pagliano and Scaramozzino (1990) transferred the Upper Jurassic *Cephenopsis mirabilis* Hong (1985c) to *incertae sedis*. **Extant**

F. CERAPHRONIDAE K. (u.)–Rec. Terr.

Extant

F. CHALCIDIDAE (CHALCIDAE) K. (APT)–Rec. Terr.

Extant

F. CHRYSIDIDAE K. (l.)–Rec. Terr.

First: *Dahurochrysis veta* Rasnitsyn (1990), Turga, former USSR. **Extant**

F. CIMBICIDAE T. (PRB)–Rec. Terr.

Extant

F. CRETEVANIIDAE K. (HAU–u.) Terr.

First: Figured but undescribed, *in* Jarzembowski (1984), Lower Weald Clay, Capel, Surrey, England, UK.
Last: e.g. *Cretevania minor*, Siberian amber, Taimyr, former USSR.

F. CYNIPIDAE K. (CMP)–Rec. Terr.

First: *Protimaspis costalis*, Canadian amber, Cedar Lake, Manitoba, Canada. **Extant**

F. DIAPRIIDAE K. (APT)–Rec. Terr.

First: *Cretacoformica explicata* Jell and Duncan (1986), Koonwarra Fossil Bed, south Gippsland, Australia. **Extant**

F. DIPRIONIDAE T. (PRB)–Rec. Terr.

Extant

F. DRYINIDAE K. (l.)–Rec. Terr.

Extant

F. ELECTROTOMIDAE T. (PRB) Terr.

First and Last: *Electrotoma succini*, Baltic amber.

F. EMBOLEMIDAE T. (PRB)–Rec. Terr.

First: e.g. *Ampulicomorpha succinalis*, Baltic amber. **Extant**

F. ENCYRTIDAE T. (PRB)–Rec. Terr.

First: *Propelma rohdendorfi*, Baltic amber. **Extant**

F. EOICHNEUMONIDAE Jell and Duncan, 1986 K. (l.) Terr.

e.g. *Eoichneumon duncanae* Jell and Duncan (1986), Koonwarra Fossil Bed, south Gippsland, Australia.

F. EPHIALTITIDAE (KARATAIDAE) J. (l.)–K. (APT) Terr.

First: e.g. *Sippelipterus liasinus* Zessin (1985), Upper Lias, Tongrube, Germany.
Last: *Karataus kourios* Sharkey, *in* Darling and Sharkey (1990), Santana Formation, Ceará, Brazil.

F. EULOPHIDAE (APHELIDAE) K. (u.)–Rec. Terr.

Extant

F. EUMENIDAE T. (Pal.)–Rec. Terr.

Extant

F. EUPELMIDAE T. (PRB)–Rec. Terr.

Extant

F. EURYTOMIDAE T. (PRB)–Rec. Terr.

Extant

F. EVANIIDAE T. (PRB)–Rec. Terr.

First: e.g. *Evania (Parevania) brevis*, Baltic amber. **Extant**

F. FALSIFORMICIDAE K. (u.) Terr.

e.g. *Falsiformica cretacea*, Siberian amber, former USSR.

F. FIGITIDAE K. (u.)–Rec. Terr.

Extant

F. FORMICIDAE (DOLICHODERIDAE, PALEOSMINTHURIDAE) K. (APT)–Rec. Terr.

First: *Cariridris bipetiolata* Brandão and Martins-Neto, *in* Brandão *et al.* (1989), Santana Formation, Ceará, Brazil. **Extant**

F. GASTERUPTIIDAE (AULACIDAE, KOTUJELLIDAE) K. (l.)–Rec. Terr.

First: e.g. *Manlaya laevinota* Rasnitsyn (1986), Gurvaneren Formation, Mongolia. **Extant**

F. GIGASIRICIDAE J. (l.–u.) Terr.

F. HALICTIDAE T. (PRB)–Rec. Terr.

Extant

F. HELORIDAE J. (u.)–Rec. Terr.

Extant

F. IBALIIDAE K. (u.)–Rec. Terr.

Extant

F. ICHNEUMONIDAE K. (l.)–Rec. Terr.

First: e.g. *Tanychora petriolata*, *in* Hong (1988), Sahai Formation, Kezuo, China; Zaza Formation, Baisa, Buryat ASSR, former USSR. **Extant**

F. JURAPRIIDAE J. (u.) Terr.

First and Last: *Jurapria sibirica*, *in* Zherikhin (1985b), Uda, Transbaikalia, former USSR.

F. KARATAVITIDAE J. (l.–u.) Terr.

F. MAIMETSHIDAE K. (SAN)

First and Last: *Maimetsha arctica*, Siberian amber, Taimyr, former USSR.

F. MASARIDAE K. (l.)–Rec. Terr.

Extant

F. MEGACHILIDAE T. (PRB)–Rec. Terr.

Extant

F. MEGALYRIDAE J. (l.)–Rec. Terr.

Extant

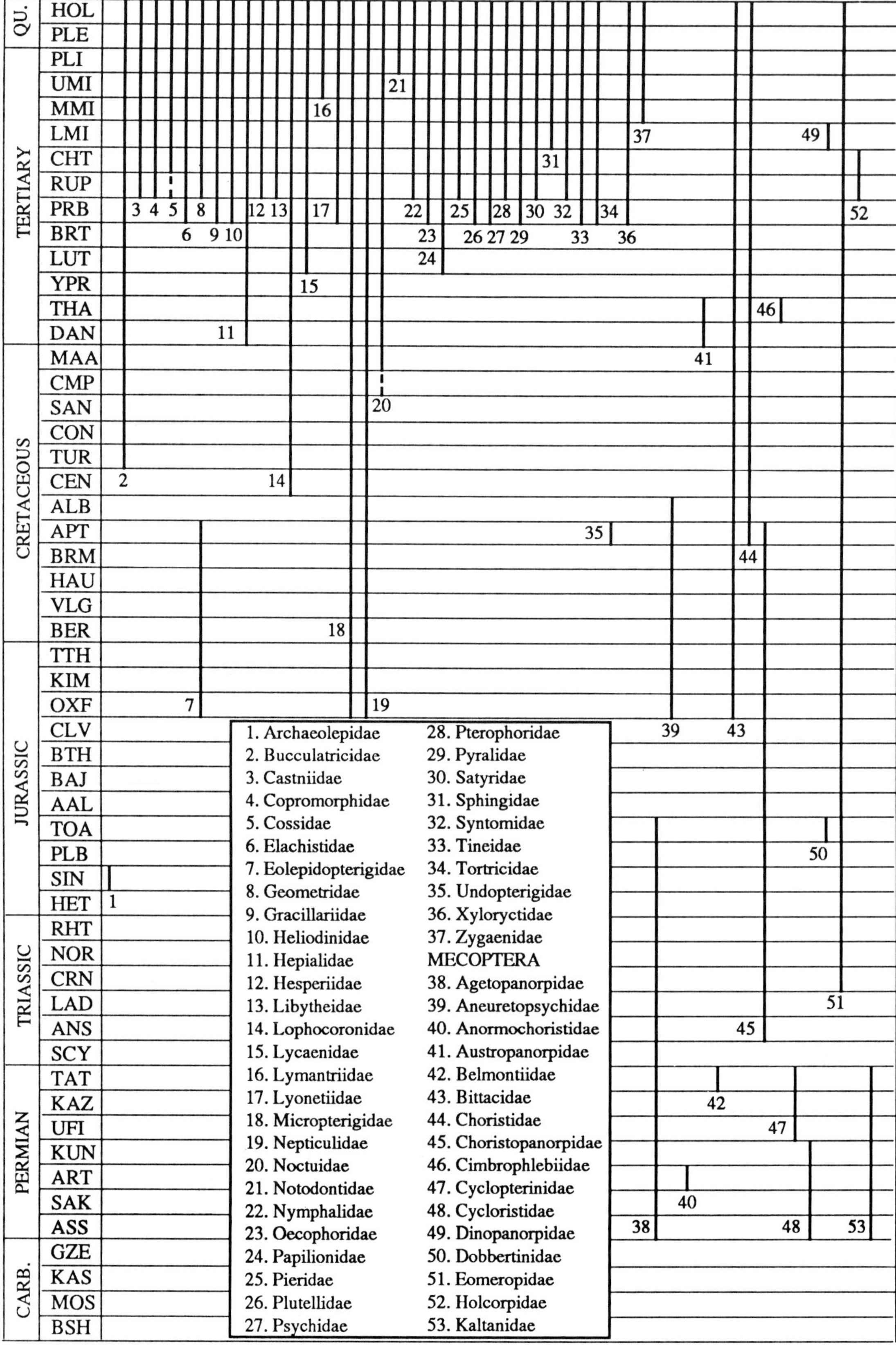
QU.
HOL
PLE
TERTIARY
PLI
UMI
MMI
LMI
CHT
RUP
PRB
BRT
LUT
YPR
THA
DAN
CRETACEOUS
MAA
CMP
SAN
CON
TUR
CEN
ALB
APT
BRM
HAU
VLG
BER
JURASSIC
TTH
KIM
OXF
CLV
BTH
BAJ
AAL
TOA
PLB
SIN
HET
TRIASSIC
RHT
NOR
CRN
LAD
ANS
SCY
PERMIAN
TAT
KAZ
UFI
KUN
ART
SAK
ASS
CARB.
GZE
KAS
MOS
BSH
1. Archaeolepidae
2. Bucculatricidae
3. Castniidae
4. Copromorphidae
5. Cossidae
6. Elachistidae
7. Eolepidopterigidae
8. Geometridae
9. Gracillariidae
10. Heliodinidae
11. Hepialidae
12. Hesperiidae
13. Libytheidae
14. Lophocoronidae
15. Lycaenidae
16. Lymantriidae
17. Lyonetiidae
18. Micropterigidae
19. Nepticulidae
20. Noctuidae
21. Notodontidae
22. Nymphalidae
23. Oecophoridae
24. Papilionidae
25. Pieridae
26. Plutellidae
27. Psychidae
28. Pterophoridae
29. Pyralidae
30. Satyridae
31. Sphingidae
32. Syntomidae
33. Tineidae
34. Tortricidae
35. Undopterigidae
36. Xyloryctidae
37. Zygaenidae
MECOPTERA
38. Agetopanorpidae
39. Aneuretopsychidae
40. Anormochoristidae
41. Austropanorpidae
42. Belmontiidae
43. Bittacidae
44. Choristidae
45. Choristopanorpidae
46. Cimbrophlebiidae
47. Cyclopterinidae
48. Cycloristidae
49. Dinopanorpidae
50. Dobbertinidae
51. Eomeropidae
52. Holcorpidae
53. Kaltanidae

Fig. 21.19

F. MEGASPILIDAE K. (l.)–Rec. Terr.

Extant

F. MELITTIDAE (CTENOPLECTRIDAE) T. (PRB)–Rec. Terr.

First: e.g. *Ctenoplectrella dentata*, Baltic amber. **Extant**

F. MESOSERPHIDAE J. (l.)–K. (l.) Terr.

Last: e.g. described but unnamed *in* Darling and Sharkey (1990), Santana Formation, Ceará, Brazil.

F. MUTILLIDAE (CRETAVIDAE) K. (u.)–Rec. Terr.

First: *Cretavus sibiricus*, *in* Manley and Poinar (1991), Siberia, USSR. Rasnitsyn in Darling and Sharkey (1990) transferred the Lower Cretaceous *Mesomutilla aptera* Zhang (1985) to Vespina (Apocrita) *incertae sedis*. **Extant**

F. MYMARIDAE K. (u.)–Rec. Terr.

First: e.g. *Triadomerus bulbosus*, Canadian amber, Medicine Hat, Canada. **Extant**

F. MYMAROMMATIDAE K. (u.)–Rec. Terr.

First: e.g. *Archaeromma minutissima*, Canadian amber, Alberta, Canada. **Extant**

F. MYRMICIIDAE (PSEUDOSIRICIDAE) J. (m.–u.) Terr.

F. ORUSSIDAE K. (u.)–Rec. Terr.

First: *Mesorussus taimyrensis*, Siberian amber, Taimyr, former USSR. **Extant**

F. PAMPHILIIDAE J. (u.)–Rec. Terr.

First: *Juralyda udensis*, *in* Zherikhin (1985b), Uda, Transbaikalia, former USSR. **Extant**

F. PARARCHEXYELIDAE J. (m.–u.) Terr.

F. PARORYSSIDAE (PARORYSIDAE) J. (u.)–K. (l.) Terr.

F. PELECINIDAE (PELECINOPTERIDAE) K. (l.)–Rec. Terr.

First: *Iscopinus baissieus*, River Vitim, former USSR. **Extant**

F. PLATYGASTRIDAE T. (Pal.)–Rec. Terr.

Extant

F. POMPILIDAE K. (l.)–Rec. Terr.

First: *Pompilopterus ciliatus*, Zaza Formation, Baisa, Buryat ASSR, former USSR. **Extant**

F. PRAEAULACIDAE (ANOMOPTERELLIDAE) J. (u.)–K. (APT) Terr.

Last: *Westratia nana* Jell and Duncan (1986), Koonwarra Fossil Bed, south Gippsland, Australia.

F. PRAEICHNEUMONIDAE K. (l.) Terr.

e.g. *Praeichneumon townesi*, Mongolia.

F. PRAESIRICIDAE J. (u.)–K. (l.) Terr.

Last: e.g. *Xyelydontes sculpturatus*, Mongolia.

F. PROCTOTRUPIDAE K. (l.)–Rec. Terr.

First: e.g. *Rasnitsynia huadongensis* (Zhang, 1985) **comb. nov.**, Laiyang Formation, Shandong, China. This species was originally described by Zhang (1985) as *Oligoneuroides huadongensis*; Pagliano and Scaramozzino (1990) proposed two replacement names for *Oligoneuroides* Zhang. We have adopted the first alphabetically. We have followed Hong and Wang (1990) in regarding this formation as Lower Cretaceous in age. **Extant**

F. PTEROMALIDAE (ORMYRIDAE, CLEONYMIDAE, PERILAMPIDAE) K. (u.)–Rec. Terr.

Extant

F. RHOPALOSOMATIDAE K. (APT)–Rec. Terr.

First: *Mesorhopalosoma cearae* Darling *in* Darling and Sharkey (1990), Santana Formation, Ceará, Brazil. **Extant**

F. ROPRONIIDAE J. (m.)–Rec. Terr.

Extant

F. SCELIONIDAE K. (l.)–Rec. Terr.

Extant

F. SCOLIIDAE K. (l.)–Rec. Terr.

Extant

F. SEPULCIDAE (PARAPAMPHILIIDAE) J. (l.)–K. (u.) Terr.

F. SERPHITIDAE K. (l.–u.) Terr.

Last: e.g. *Aposerphites solox*, Siberian amber, former USSR.

F. SIRICIDAE (SINOSIRICIDAE) J. (l.)–Rec. Terr.

Extant

F. SPHECIDAE (PEMPHREDONIDAE, TRYPOXYLIDAE, AMPULICIDAE, LARRIDAE, CRABONIDAE, PHILANTHIDAE) K. (l.)–Rec. Terr.

First: e.g. *Mataeosphex venulosus* Zhang (1985), Laiyang Formation, Shandong, China. We have followed Hong and Wang (1990) in regarding this formation as Lower Cretaceous in age. **Extant**

F. SPHECOMYRMIDAE K. (u.) Terr.

e.g. *Baikuris mandibularis* Dlussky (1987), Siberian amber, Taimyr, former USSR.

Comment: Wilson (1987), however, regards this family as synonymous with the Formicidae.

F. STEPHANIDAE T. (PRB)–Rec. Terr.

First: e.g. *Electrostephanus brevicornis*, Baltic amber. **Extant**

F. STIGMAPHRONIDAE K. (l.–u.) Terr.

First: e.g. *Aphrostigmon vitimense* Rasnitsyn (1991), Zaza Formation, Baisa, Buryat ASSR, former USSR.

Last: e.g. *Stigmaphron orphne*, Siberian amber, Taimyr, former USSR.

F. TENTHREDINIDAE K. (l.)–Rec. Terr.

First: *Palaeathalia laiyangensis* Zhang (1985), Laiyang Formation, Shandong, China. We have followed Hong and Wang (1990) in regarding this formation as Lower Cretaceous in age. **Extant**

F. TETRACAMPIDAE K. (CMP)–Rec. Terr.

First: e.g. *Baeomorpha dubitata*, Canadian amber, Cedar Lake, Manitoba, Canada. **Extant**

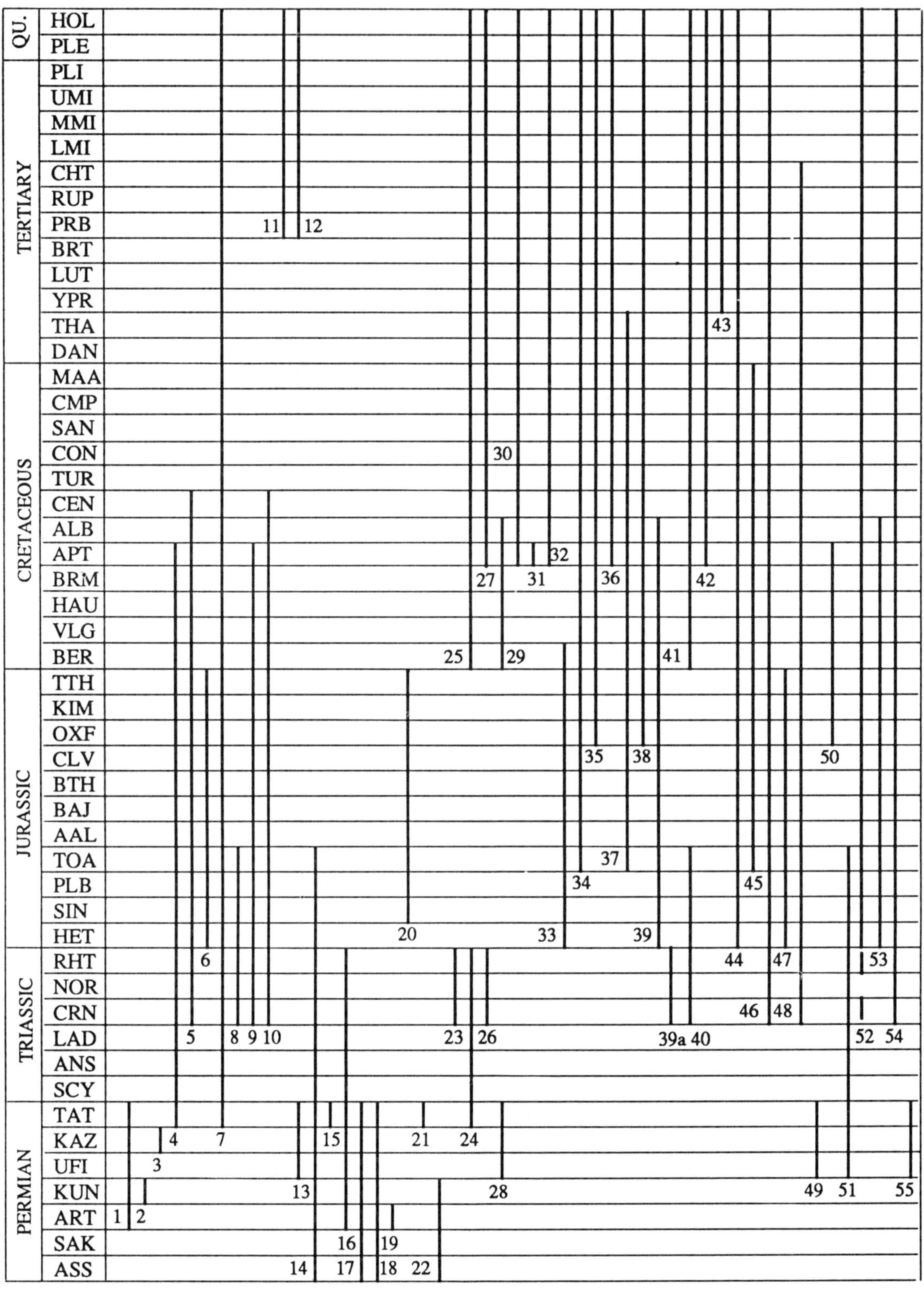

Fig. 21.20

F. THYSANIDAE T. (CHT)–Rec. Terr.

Extant

F. TIPHIIDAE (TIPHIDAE, METHOCIDAE) K. (APT)–Rec. Terr.

First: *Architiphia rasnitsyni* Darling, *in* Darling and Sharkey (1990), Santana Formation, Ceará, Brazil. **Extant**

F. TORYMIDAE K. (u.)–Rec. Terr.

Extant

F. TRICHOGRAMMATIDAE K. (CMP)–Rec.

First: *Enneagmus pristinus*, Canadian amber, Cedar Lake, Manitoba, Canada. **Extant**

F. TRIGONALIDAE (ICHNEUMOMIMIDAE) K. (l.)–Rec. Terr.

1. Lithopanorpidae
2. Marimerobiidae
3. Martynopanorpidae
4. Mesopanorpodidae
5. Mesopsychidae
6. Muchoriidae
7. Nannochoristidae
8. Neorthophlebiidae
9. Neoparachoristidae
10. Orthophlebiidae
11. Panorpidae
12. Panorpodidae
13. Permocentropidae
14. Permochoristidae
15. Permomeropidae
16. Permopanorpidae
17. Petrochoristidae
18. Protomeropidae
19. Protopanorpidae
20. Pseudopolycentropidae
21. Robinjohniidae
22. Tomiochoristidae
23. Triassochoristidae
24. Xenochoristidae

MEGALOPTERA

25. Corydalidae
26. Euchanliodidae
27. Sialidae
28. Tychtodelopteridae

NEUROPTERA

29. Allopteridae
30. Ascalaphidae
31. Babinskaiidae
32. Berothidae
33. Brongniartiellidae
34. Chrysopidae
35. Coniopterygidae
36. Hemerobiidae
37. Kalligrammatidae
38. Mantispidae
39. Mesithonidae
40. Mesopolystoechotidae
41. Myrmeleontidae
42. Nemopteridae
43. Neurorthidae
44. Nymphidae
45. Nymphitidae
46. Osmylidae
47. Osmylitidae
48. Osmylopsychopidae
40. Palaemerobiidae
50. Panfiloviidae
51. Permithonidae
52. Polystoechotidae
53. Prohemerobiidae
54. Psychopsidae
55. Sialidopsidae

Fig. 21.20

First: e.g. *Darbigonalus capitatus* Rasnitsyn (1986), Gurvaneren Formation, Mongolia. **Extant**

F. VANHORNIIDAE K. (u.?)–Rec. Terr.

Extant

F. VESPIDAE K. (l.)–Rec. Terr.

First: e.g. *Priorvespa bullata* Carpenter and Rasnitsyn (1990), Zaza Formation, Baisa, Buryat ASSR, former USSR. **Extant**

F. XYELIDAE Tr. (u.)–Rec. Terr.

First: e.g. Figured but undescribed in Kukalová-Peck (1991), Australia. **Extant**

F. XYELOTOMIDAE J. (m.)–K. (l.) Terr.

Last: e.g. ?*Undatoma undurgensis* Rasnitsyn (1990), Undurga, former USSR.

F. XYELYDIDAE (XYELIDIDAE) J. (l.)–K. (l.) Terr.

Last: *Sinoprolyda meileyingensis* Hong (1987), Jiufutang Formation, Kezuo, China.

F. XYLOCOPIDAE T. (Oli.)–Rec. Terr.

Extant

Order LEPIDOPTERA (PAPILIONIDA) J. (SIN)–Rec. Terr.

Data taken from Kozlov (1988). For reviews on fossil Lepidoptera see Whalley (1986b) and Shields (1988). First as for Archaeolepidae. More families are listed in Spahr (1989), but Kozlov (1988) regards the included records as *incertae sedis* or unsubstantiated.

F. ADELIDAE T. (PRB)–Rec.

First: e.g. *Adela kuznetzovi*, Baltic amber. **Extant**

F. ARCHAEOLEPIDAE Whalley, 1985 J. (SIN) Terr.

First and Last: *Archaeolepis mane*, Lower Lias, Charmouth, Dorset, England, UK. Kozlov (1988) placed this species in Papilionida *incertae sedis*.

F. BUCCULATRICIDAE K. (TUR)–Rec. Terr.

First: *Bucculatrix platani*, Karatau, Kazakhstan, former USSR. **Extant**

F. CASTNIIDAE T. (RUP)–Rec. Terr.

First: *Dominickus castnioides* Tindale (1985), Florissant, Colorado, USA. **Extant**

F. COPROMORPHIDAE T. (RUP)–Rec. Terr.

First: *Copromorpha fossilis*, Bembridge Marls, Isle of Wight, England, UK. **Extant**

F. COSSIDAE T. (?RUP)–Rec. Terr.

First: *Gurnetia durranti*, Bembridge Marls, Isle of Wight, England, UK. **Extant**

F. ELACHISTIDAE T. (PRB)–Rec. Terr.

First: e.g. *Elachistites inclusus*, Baltic amber. **Extant**

F. EOLEPIDOPTERIGIDAE (EOLEPIDOPTERYGIDAE) J. (u.)–K. (APT). Terr.

First: e.g. *Eolepidopterix jurassica*, Transbaikalia, former USSR.

Last: e.g. *Daiopterix rasnitsyni*, Daya River, Transbaikalia, former USSR.

F. GEOMETRIDAE T. (RUP)–Rec. Terr.

First: e.g. *Geometridites larentiiformis*, Bembridge Marls, Isle of Wight, England, UK. **Extant**

F. GRACILLARIIDAE T. (PRB)–Rec. Terr.

First: e.g. *Gracillariites lithuanicus*, Baltic amber. **Extant**

F. HELIODINIDAE T. (PRB)–Rec. Terr.

First: *Baltonides roeselliformis*, *in* Spahr (1989), Baltic amber. **Extant**

F. HEPIALIDAE T. (Pal.)–Rec. Terr.

First: *Prohepialus incertus*, Menat, France. **Extant**

F. HESPERIIDAE T. (RUP)–Rec. Terr.

First: *Pamphilites abdita*, Aix-en-Provence, France. **Extant**

F. LIBYTHEIDAE T. (RUP)–Rec. Terr.

First: e.g. *Prolibythea vagabunda*, Florissant, Colorado, USA. **Extant**

F. LOPHOCORONIDAE K. (u.)–Rec. Terr.

Extant

F. LYCAENIDAE T. (LUT/BRT)–Rec. Terr.

First: *Rhiodinella nympha*, Colorado, USA. **Extant**

F. LYMANTRIIDAE T. (MES)–Rec. Terr.

First: Figured but undescribed, *in* Cavallo and Galletti (1987), gypsiferous marls, Alba, Italy. **Extant**

F. LYONETIIDAE T. (PRB)–Rec. Terr.

First: *Prolyonetia cockerelli*, Baltic amber. **Extant**

F. MICROPTERIGIDAE (MICROPTERYGIDAE) J. (u.)–Rec. Terr.

First: *Auliepterix mirabilis* Kozlov (1989), Kazakhstan, former USSR. **Extant**

F. NEPTICULIDAE J. (u.)–Rec. Terr.

Extant

F. NOCTUIDAE K. (?CMP)–Rec. Terr.

First: *Noctuites* sp., amber, USA.
Comment: This record is doubtful, see Boucot (1990). **Extant**

F. NOTODONTIDAE T. (Pli)–Rec. Terr.

First: *Cerurites wagneri*, Willershausen, Germany. **Extant**

F. NYMPHALIDAE (DANAIDAE) T. (RUP)–Rec. Terr.

First: e.g. *Neorinopsis sepulta*, Aix-en-Provence, France. **Extant**

F. OECOPHORIDAE (SYMMOCIDAE) T. (PRB)–Rec. Terr.

First: e.g. *Borkhausenites bachofeni*, Baltic amber. **Extant**

F. PAPILIONIDAE T. (LUT/BRT)–Rec. Terr.

First: e.g. *Praepapilio colorado*, Colorado, USA. **Extant**

F. PIERIDAE T. (RUP)–Rec. Terr.

First: *Coliates proserpina*, Aix-en-Provence, France. **Extant**

F. PLUTELLIDAE (PROLYONETIIDAE) T. (PRB)–Rec. Terr.

First: e.g. *Epinomeuta truncatipennella*, Baltic amber. **Extant**

F. PSYCHIDAE T. (PRB)–Rec. Terr.

First: e.g. *Psychites pristinella*, Baltic amber. **Extant**

F. PTEROPHORIDAE T. (RUP)–Rec. Terr.

First: *Pterophorus oligocenicus* Bigot *et al.* (1986), Aix-en-Provence, France. **Extant**

F. PYRALIDAE T. (PRB)–Rec. Terr.

First: *Glendotricha olgae*, Baltic amber. **Extant**

F. SATYRIDAE T. (RUP)–Rec. Terr.

First: *Lethites reynesii*, Aix-en-Provence, France. **Extant**

F. SPHINGIDAE T. (Mio.)–Rec. Terr.

First: e.g. Gen. et sp. indet. Zhang (1989), Shanwang Formation, Shanwang, Shandong, China. **Extant**

F. SYNTOMIDAE T. (Oli.)–Rec. Terr.

First: *Oligamatites martynovi*, Kazakhstan, former USSR. **Extant**

F. TINEIDAE T. (PRB)–Rec. Terr.

First: e.g. *Monopibaltia ignitella*, Baltic amber. **Extant**

F. TORTRICIDAE T. (PRB)–Rec. Terr.

First: e.g. *Electresia zalesskii*, Baltic amber. **Extant**

F. UNDOPTERIGIDAE Kozlov, 1988 K. (APT) Terr.

e.g. *Undopterix sukatshevae*, Transbaikalia, former USSR.

F. XYLORYCTIDAE T. (PRB)–Rec. Terr.

First: *Oegocniites borisjaki*, Baltic amber. **Extant**

F. ZYGAENIDAE T. (MMI)–Rec. Terr.

First: *Zygaena? turolensis* Fernandez-Rubio, Peñalver and Martínez-Delclòs (1991), ritmitas oil shales, Rubielos de Mora, Spain. **Extant**

Order MECOPTERA (PANORPIDA) P. (ROT)–Rec. Terr.

We have followed the traditional classification including stemgroups Antliophora (Diptera plus Siphonaptera plus Mecoptera) and Mecopteroidea (Amphiesmenoptera (Trichoptera plus Lepidoptera) plus Antliophora) in the Mecoptera; i.e. Mecoptera as given here (*s.l.*) is paraphyletic with respect to Amphiesmenoptera, Diptera and Siphonaptera. Most data are taken from Willman (1978, 1989b). Families marked with an asterisk are considered not to belong to the Mecoptera (*s.l.*) (Willmann, 1989b).

F. AGETOPANORPIDAE (CHORISTOPSYCHIDAE, TYCHTOPSYCHIDAE) P. (ROT)–J. (l.) Terr.

First: e.g. *Agetopanorpa maculata*, Wellington Formation, Elmo, Kansas, USA.
Last: *Choristopsyche tenuinervis*, Shurab, Tadzhikistan, former USSR.

F. ANEURETOPSYCHIDAE Rasnitsyn and Kozlov, 1990 J. (u.)–K. (l.) Terr.

First: e.g. *Aneuretopsyche rostrata* Rasnitsyn and Kozlov (1990), Mikhailovka, former USSR.
Last: *Aneuretopsyche vitimensis* Rasnitsyn and Kozlov (1990), Zaza Formation, Baisa, Buryat ASSR, former USSR.

F. ANORMOCHORISTIDAE* P. (ART) Terr.

First and Last: *Anormochorista oligoclada*, Wellington Formation, Elmo, Kansas, USA.

F. AUSTROPANORPIDAE T. (Pal.) Terr.

First and Last: *Austropanorpa australis*, Redbank Plains Series, Redbank Plains, Queensland, Australia.

F. BELMONTIIDAE P. (TAT) Terr.

e.g. *Belmontia mitchelli*, Belmont Beds, Belmont, New South Wales, Australia.

F. BITTACIDAE J. (u.)–Rec. Terr.

First: *Probittacus avitus*, Karatau, Kazakhstan, former USSR. **Extant**

F. CHORISTIDAE K. (APT)–Rec. Terr.

First: *Cretacochorista parva* Jell and Duncan (1986), Koonwarra Fossil Bed, south Gippsland, Australia. **Extant**

F. CHORISTOPANORPIDAE Willmann, 1989b Tr. (m.)–K. (APT) Terr.

First: *Choristopanorpa bifasciata*, Beacon Hill, New South Wales, Australia.
Last: *Choristopanorpa drinnani* Jell and Duncan (1986), Koonwarra Fossil Bed, south Gippsland, Australia.

F. CIMBROPHLEBIIDAE T. (THA) Terr.

First and Last: *Cimbrophlebia bittaciformis*, Mo-clay, north Jutland, Denmark.

F. CYCLOPTERINIDAE* (CYCLOPTERIDAE) Carpenter, 1987 P. (ZEC) Terr.

First and Last: *Cyclopterina autumnalis in* Carpenter (1987), Iljinsk Formation, Kuznetsk Basin, former USSR.

F. CYCLORISTIDAE* P. (ROT) Terr.

First and Last: *Cyclorista convexicosta*, Kaltan, Kuznetsk Basin, former USSR.

F. DINOPANORPIDAE T. (LMI) Terr.

First and Last: *Dinopanorpa megarche*, former USSR.

F. DOBBERTINIIDAE* J. (TOA) Terr.

First and Last: *Dobbertinia reticulata*, Upper Lias, Dobbertin, Germany.

F. EOMEROPIDAE (NOTIOTHAUMIDAE) Tr. (u.)–Rec. Terr.

First: e.g. *Pronotiothauma neuropteroides*, Madygen Formation, Madygen, former USSR. **Extant**

F. HOLCORPIDAE T. (Oli.) Terr.

First and Last: *Holcorpa maculosa*, Florissant, Colorado, USA.

F. KALTANIDAE* P. (ROT–ZEC) Terr.

First: e.g. *Altajopanorpa kaltanica*, Kuznetsk Basin, former USSR.
Last: e.g. *Altajopanorpa iljinskiensis*, Iljinsk Formation, Kuznetsk Basin, former USSR.
The Lower Cretaceous *Cretacechorista qilianshanensis* Hong *et al.* (1989) does not agree with the family definition as given by Martynova (1962). Zherikhin (pers. comm.) considers that it belongs in the Blattodea: Raphidiomimidae.

F. LITHOPANORPIDAE P. (ART–ZEC) Terr.

First: *Lithopanorpa pusilla*, Wellington Formation, Elmo, Kansas, USA.
Last: *Lithopanorpa kuznetskiensis*, Gramoteino Formation, Kuznetsk Basin, former USSR.

F. MARIMEROBIIDAE* P. (KUN) Terr.

First and Last: *Marimerobius splendens*, Chekarda, former USSR.

F. MARTYNOPANORPIDAE Willmann, 1989b P. (KAZ) Terr.

e.g. *Martynopanorpa angustata*, Archangelsk, former USSR.

F. MESOPANORPODIDAE P. (TAT)–K. (APT) Terr.

First: e.g. *Mesopanorpodes belmontensis*, Warner's Bay, New South Wales, Australia.
Last: *Prochoristella leongatha* Jell and Duncan (1986), Koonwarra Fossil Bed, south Gippsland, Australia.

F. MESOPSYCHIDAE Tr. (u.)–K. (l.) Terr.

First: *Mesopsyche triareolata*, Ipswich Beds, Ipswich, Australia.
Last: *Undisca dobrokhotovae* Sukacheva (1990a), Transbaikalia, former USSR.

F. MUCHORIIDAE Willmann, 1989b J. Terr.

First and Last: *Muchoria reducta*, Transbaikalia, former USSR.

F. NANNOCHORISTIDAE P. (TAT)–Rec. Terr.

First: e.g. *Nannochoristella reducta*, Warner's Bay, New South Wales, Australia. **Extant**

F. NEORTHOPHLEBIIDAE Tr. (u.)–J. (TOA) Terr.

First: e.g. *Archebittacus exilis*, Ipswich Series, Mt. Crosby, Australia.
Last: e.g. *Neorthophlebia maculipennis*, Upper Lias, Dobbertin, Germany.

F. NEOPARACHORISTIDAE Tr. (u.)–K. (APT) Terr.

First: e.g. *Neoparachorista perkinsi*, Ipswich Series, Mt. Crosby, Queensland, Australia.
Last: *Neoparachorista clarkae* Jell and Duncan (1986), Koonwarra Fossil Bed, south Gippsland, Australia.

F. ORTHOPHLEBIIDAE Tr. (u.)–K. (l.) Terr.

First: e.g. *Orthophlebia curta*, Sogjuty, former USSR.
Last: e.g. *Mesopanorpa* sp. Jarzembowski (1984), Lower Weald Clay, Capel, Surrey, England, UK.

F. PANORPIDAE T. (PRB)–Rec. Terr.

First: e.g. *Panorpa mortua*, Baltic amber. **Extant**

F. PANORPODIDAE T. (PRB)–Rec. Terr.

First: e.g. *Panorpodes brevicauda*, Baltic amber. **Extant**

F. PERMOCENTROPIDAE P. (ZEC) Terr.

First and Last: *Permocentropus philopotamoides* Iva-Gora, Archangelsk, former USSR.

F. PERMOCHORISTIDAE (MESOCHORISTIDAE) P. (ROT)–J. (l.) Terr.

First: e.g. *Mesochorista javorskyi*, Kuznetsk Basin, former USSR.
Last: *Liassochorista anglicana*, Lower Lias, Brown's Wood, Warwickshire, England, UK.

F. PERMOMEROPIDAE* P. (TAT) Terr.

e.g. *Permomerope nanus*, Warner's Bay, New South Wales, Australia.

F. PERMOPANORPIDAE P. (ART)–Tr. (u.) Terr.

First: e.g. *Permopanorpa inaequalis*, Wellington Formation, Elmo, Kansas, USA.
Last: e.g. *Neopermopanorpa mesembria*, Ipswich Series, Mt. Crosby, Queensland, Australia.

F. PETROCHORISTIDAE Willmann, 1989b P. (ROT–ZEC) Terr.

First: e.g. *Petrochorista prona*, Kaltan, Kuznetsk Basin, former USSR.
Last: e.g. *Petrochorista elegantula*, Tichjie Gory, former USSR.

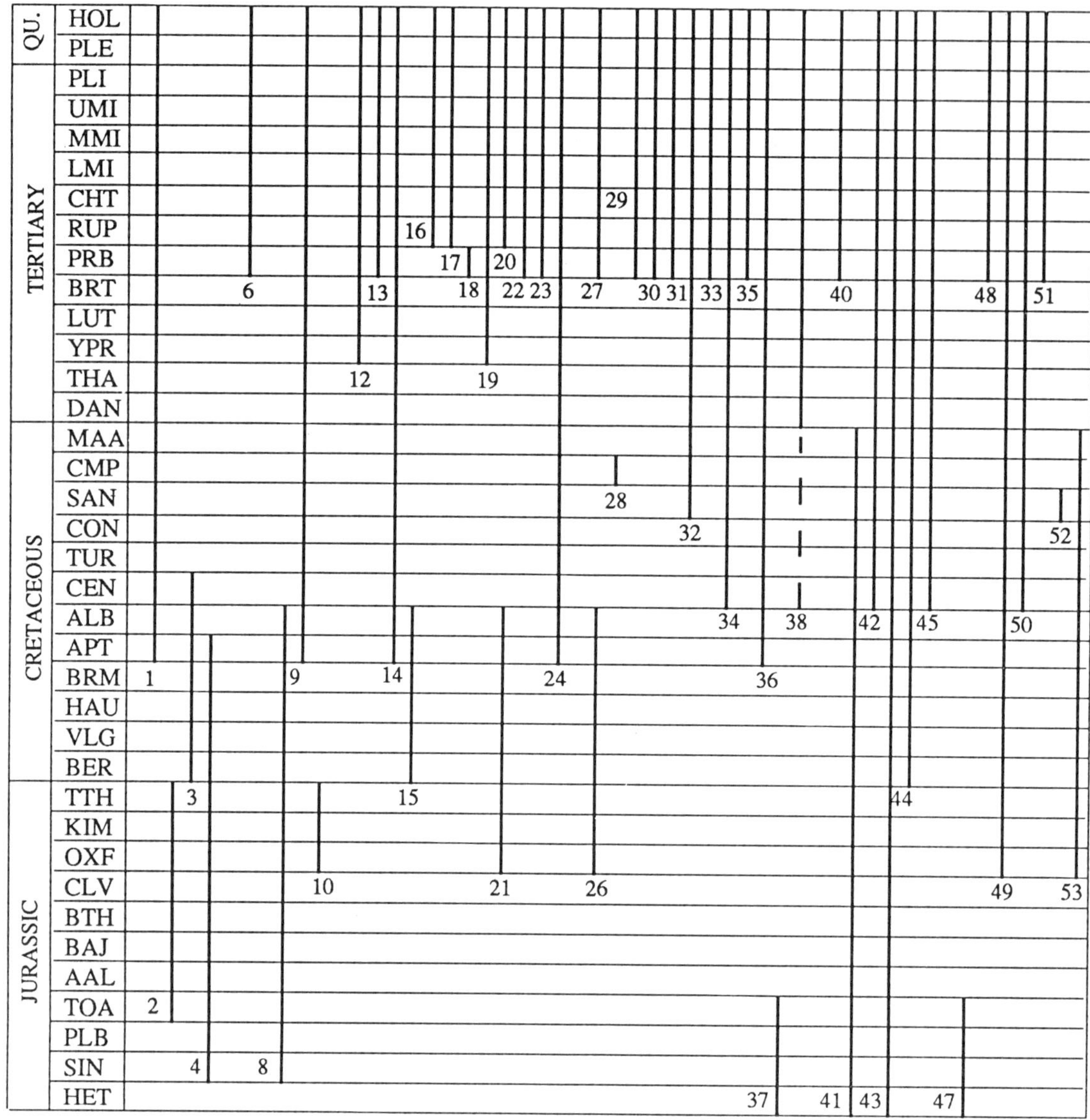

Fig. 21.21

F. PROTOMEROPIDAE* (PLATYCHORISTIDAE) P. (ASS–ZEC) Terr.

First: *Pseudomerope gallei* Kukalová-Peck and Willmann (1990), Říčany, Czechoslovakia.

F. PROTOPANORPIDAE P. (ART) Terr.

First and Last: *Protopanorpa permiana*, Wellington Formation, Elmo, Kansas, USA.

F. PSEUDOPOLYCENTROPIDAE (PSEUDOPOLYCENTROPIDIDAE, PSEUDOPOLYCENTROPODIDAE) J. (SIN–u.) Terr.

First: *Pseudopolycentropus prolatipennis* Whalley (1985), Lower Lias, Charmouth, Dorset, England, UK.
Last: *Pseudopolycentropus latipennis*, Karatau, Kazakhstan, former USSR.

F. ROBINJOHNIIDAE P. (TAT) Terr.

First and Last: *Robinjohnia tillyardi*, Warner's Bay, New South Wales, Australia.

F. TOMIOCHORISTIDAE* P. (ROT) Terr.

e.g. *Tomiochorista minuta*, Kaltan, Kuznetsk Basin, former USSR.

F. TRIASSOCHORISTIDAE Willmann, 1989b Tr. (u.) Terr.

First and Last: *Triassochorista nana*, Ipswich Series, Mt. Crosby, Queensland, Australia.

F. XENOCHORISTIDAE P. (TAT)–T. (u.) Terr.

First: *Xenochorista sobrina*, Warner's Bay, New South Wales, Australia.
Last: *Xenochoristella hillae*, Ipswich Series, Mt. Crosby, Queensland, Australia.

Order MEGALOPTERA (CORYDALIDA) P. (ZEC)–Rec. Terr.

F. CORYDALIDAE K. (l.)–Rec. Terr.

Extant

F. EUCHAULIODIDAE Tr. (u.) Terr.

1. Sisyridae
2. Solenoptilidae
RAPHIDIOPTERA
3. Alloraphidiidae
4. Baissopteridae
5. Fatjanopteridae
6. Inocelliidae
7. Letopalopteridae
8. Mesoraphidiidae
9. Raphidiidae
10. Sinoraphidiidae
11. Sojanoraphidiidae
SIPHONAPTERA
12. Ctenophthalmidae
13. Hystrichopsyllidae
14. Pulicidae
15. Saurophthiridae
STREPSIPTERA
16. Bohartillidae
17. Elenchidae
18. Mengeidae
19. Myrmecolacidae
TRICHOPTERA
20. Atopsychopsidae
21. Baissoferidae
22. Beraeidae
23. Brachycentridae
24. Calamoceratidae
25. Cladochoristidae
26. Dysoneuridae
27. Ecnomidae
28. Electralbertidae
29. Glossosomatidae
30. Goeridae
31. Helicopsychidae
32. Hydrobiosidae
33. Hydropsychidae
34. Hydroptilidae
35. Lepidostomatidae
36. Leptoceridae
37. Liassophilidae
38. Limnephilidae
39. Microptysmatidae
40. Molannidae
41. Necrotauliidae
42. Odontoceridae
43. Philopotamidae
44. Phryganeidae
45. Polycentropodidae
46. Prorhyacophilidae
47. Prosepididontidae
48. Psychomyiidae
49. Rhyacophilidae
50. Sericostomatidae
51. Stenopsychidae
52. Taymyrelectronidae
53. Vitimotauliidae

Fig. 21.21

First and Last: *Euchauliodes distinctus* in Kukalová-Peck (1991), South Africa.

F. SIALIDAE K. (APT)–Rec. Terr.

Extant

F. TYCHTODELOPTERIDAE P. (ZEC) Terr.

Order NEUROPTERA (PLANIPENNIA, MYRMELEONTIDA) P. (ZEC)–Rec. Terr.

F. ALLOPTERIDAE Zhang, 1991a K. (l.) Terr.

First and Last: *Allopterus luianus* Zhang (1991a), Laiyang Formation, Laiyang, Shandong, China. We have followed Hong and Wang (1990) in regarding this formation as Lower Cretaceous in age.

F. ASCALAPHIDAE K. (APT)–Rec. Terr.

First: *Cratopteryx robertosantosi* Martins-Neto and Vulcano (1989), Santana Formation, Ceará, Brazil. **Extant**

F. BABINSKAIIDAE Martins-Neto and Vulcano, 1989 K. (APT) Terr.

e.g. *Babinskaia pulchra* Martins-Neto and Vulcano (1989), Santana Formation, Ceará, Brazil.

F. BEROTHIDAE K. (APT)–Rec. Terr.

First: e.g. *Araripeberotha fairchildi* Martins-Neto and Vulcano (1990), Santana Formation, Ceará, Brazil. **Extant**

F. BRONGNIARTIELLIDAE J. (l.)–K. (BER) Terr.

First: e.g. *Actinophlebia intermixta*, *in* Whalley (1988), Upper Lias, Gloucestershire, England, UK.
Last: e.g. *Pterinoblattina pluma*, *in* Whalley (1988), Durlston Formation, Durlston Bay, England, UK.

F. CHRYSOPIDAE (MESOCHRYSOPIDAE, MESOCHRYSOPSIDAE) J. (TOA)–Rec. Terr.

First: *Liassochrysa stigmatica* Ansorge and Schlüter (1990), Upper Lias, Dobbertin, Germany. **Extant**

F. CONIOPTERYGIDAE J. (u.)–Rec. Terr.

Extant

F. HEMEROBIIDAE K. (APT)–Rec. Terr.

Extant

F. KALLIGRAMMATIDAE J. (TOA)–T. (Pal.) Terr.

First: *Paractinophlebia curtisii*, *in* Whalley (1988), Upper Lias, Alderton, Gloucestershire, England, UK.

F. MANTISPIDAE J. (u.)–Rec. Terr.

First: *Promantispa similis*, *in* Nel (1988a), Karatau, Kazakhstan, former USSR. **Extant**

F. MESITHONIDAE J. (l.)–K. (l.) Terr.

First: e.g. *Sibithone dichotoma* Ponomarenko (1984), Ichetuy Formation, Buryat ASSR, former USSR.

F. MESOBEROTHIDAE (PROBEROTHIDAE) Tr. (u.) Terr.

F. MESOPOLYSTOECHOTIDAE Tr. (u.)–J. (l.) Terr.

First: *Megapolystoechus magnificus*, *in* Whalley (1988), Worcestershire and Gloucestershire, England, UK.

F. MYRMELEONTIDAE (MYRMELEONIDAE) K. (l.)–Rec. Terr.

First: e.g. *Liaoximyia sinica* Hong (1988), Sahai Formation, Kezuo, China. **Extant**

F. NEMOPTERIDAE K. (APT)–Rec. Terr.

First: e.g. *Roesleria exotica* Martins-Neto and Vulcano (1989), Santana Formation, Ceará, Brazil. **Extant**

F. NEURORTHIDAE T. (Eoc.)–Rec. Terr.

Extant

F. NYMPHIDAE J.–Rec. Terr.

First: e.g. *Mesonymphes rohdendorfi*, *in* Lambkin (1988), Karatau, former USSR. **Extant**

F. NYMPHITIDAE J. (TOA)–K. (u.) Terr.

First: *Epigambria longipennis*, *in* Lambkin (1988), Upper Lias, Dobbertin, Germany.

F. OSMYLIDAE Tr. (u.)–Rec. Terr.

First: e.g. *Sogjuta speciosa*, *in* Lambkin (1988), Issyk-Kul, Kirgizia, former USSR. **Extant**

F. OSMYLITIDAE J. (l.–TTH) Terr.

First: *Idiastogyia fatisca* Lin (1986), Guangdong, China.
Last: *Osmylites protogaeus*, *in* Lambkin (1988), Lithographic Limestone, Eichstätt, Germany.

F. OSMYLOPSYCHOPIDAE (OSMYLOPSYCHOPSIDAE) J. (u.)–T. (Pg.) Terr.

F. PALAEMEROBIIDAE P. (ZEC) Terr.

F. PANFILOVIIDAE (GRAMMOSMYLIDAE) Makarkin, 1990 J. (u.)–K. (APT) Terr.

First: *Panfilovia acuminata*, *in* Makarkin (1990), Karatau, former USSR.

F. PERMITHONIDAE (ARCHEOSMYLIDAE, PERMOPSYCHOPIDAE) P. (ZEC)–J. (TOA) Terr.

First: e.g. *Permipsythone panfilovi*, *in* Martins-Neto (1987a), Irati Formation, Porto Alegre-Uruguaiana, Rio Grande do Sul, Brazil.
Last: e.g. *Archeosmylus complexus* Whalley (1988), Gloucestershire, England, UK.

F. POLYSTOECHOTIDAE Tr. ?(CRN)–Rec. Terr.

First: *Lithosmylidia lineata*, *in* Lambkin (1988), Mt. Crosby Formation, Mt. Crosby, Queensland, Australia. **Extant**

F. PROHEMEROBIIDAE J. (l.)–K. (l.) Terr.

First: e.g. *Prohemerobius oshinensis* Ponomarenko (1984), Zhargalanskaya Formation, Kobdos Aymak, Mongolia.

F. PSYCHOPSIDAE Tr. (u.)–Rec. Terr.

First: e.g. *Triassopsychops superba*, *in* Schlüter (1986), Ipswich, Queensland, Australia. **Extant**

F. SIALIDOPSIDAE P. (ZEC) Terr.

F. SISYRIDAE K. (APT)–Rec. Terr.

Extant

F. SOLENOPTILIDAE J. (TOA–u.) Terr.

First: *Solenoptilon kochi*, *in* Lambkin (1988), Upper Lias, Germany.
Last: *Solenoptilon martynovi*, *in* Lambkin (1988), Kazakhstan, former USSR.

Order RAPHIDIOPTERA (RAPHIDIIDA) C. (u.?)–Rec. Terr.

Most data are taken from Oswald (1990). First as for Fatjanopteridae, although the ordinal position of this family is doubtful. First definite raphidiopteran as for Sojanoraphidiidae.

F. ALLORAPHIDIIDAE K. (l.–CEN) Terr.

First: e.g. *Archeraphidia yakowlewi* Ponomarenko (1988), Mongolia.
Last: *Alloraphidia dorfi*, Redmond Formation, Knob Lake District, Canada.

F. BAISSOPTERIDAE (BAISSORAPHIDIIDAE) J. (SIN)–K. (l.) Terr.

First: *Priscaenigma obtusa*, *in* Whalley (1988), Lower Lias, Charmouth, Dorset, England, UK.
Last: e.g. *Baissoptera longissima* Ponomarenko (1988), Bon-Tsagan, Mongolia.

F. FATJANOPTERIDAE C. (u.) Terr.

First and Last: *Fatjanoptera mnemonica*, Tunguska Basin, former USSR.
The ordinal placement of this family is doubtful.

F. INOCELLIIDAE T. (PRB)–Rec. Terr.

First: e.g. *Inocellia peculiaris*, Baltic amber. **Extant**

F. LETOPALOPTERIDAE P. (ZEC) Terr.

e.g. *Letopaloptera albardiana*, former USSR.
The ordinal placement of this family is doubtful.

F. MESORAPHIDIIDAE (MESORAPHIIDIDAE) J. (SIN)–K. (l.) Terr.

First: *Mesoraphidia confusa*, *in* Whalley (1988), Lower Lias, Charmouth, Dorset, England, UK.
Last: e.g. *Mesoraphidia* sp. Whalley (1988), Lower Weald Clay, Capel, Surrey, England, UK.

F. RAPHIDIIDAE K. (APT)–Rec. Terr.

First: *Raphidia brasiliensis* Nel *et al.* (1990), Santana Formation, Ceára, Brazil. **Extant**

F. SINORAPHIDIIDAE J. (u.) Terr.

First and Last: *Sinoraphidia viridis*, Gansu, China.

F. SOJANORAPHIDIIDAE P. (ZEC) Terr.

First and Last: *Sojanoraphidia rossica*, former USSR.

Order SIPHONAPTERA (PULICIDA) K. (l.)–Rec. Terr.

Boucot (1990) discusses the Cretaceous records of this order.

F. CTENOPHTHALMIDAE T. (Eoc.)–Rec. Terr.

Extant

F. HYSTRICHOPSYLLIDAE T. (PRB)–Rec. Terr.

Extant

F. PULICIDAE K. (APT)–Rec. Terr.

First: e.g. Pulicid indet. 1, Jell and Duncan (1986), Koonwarra Fossil Bed, south Gippsland, Australia. **Extant**

F. SAUROPHTHIRIDAE Ponomarenko, 1986b K. (l.) Terr.

e.g. *Saurophthiroides mongolicus* Ponomarenko (1986b), Gurvaneren Formation, Mongolia. The position of this family is doubtful.

Order STREPSIPTERA (STYLOPIDA) T. (Eoc.)–Rec. Terr.

Data was taken from Kathirithamby (1989). First as for Myrmecolacidae.

F. BOHARTILLIDAE T. (Oli.)–Rec. Terr.

Extant

F. ELENCHIDAE T. (Oli.)–Rec. Terr.

Extant

F. MENGEIDAE T. (PRB) Terr.

e.g. *Mengea tertiaria*, Baltic amber.

F. MYRMECOLACIDAE T. (Eoc.)–Rec. Terr.

First: *Stichotrema eocaenicum*, Brown Coal, Halle-an-der-Saale, Germany. **Extant**

Order TRICHOPTERA (PHRYGANAEIDA, PHRYGANEIDA) P. (ART)–Rec. Terr.

We have followed the traditional classification, including stem group Amphiesmenoptera (Trichoptera plus Lepidoptera), i.e. Trichoptera as given here (*s.l.*) are paraphyletic with respect to Lepidoptera. Ranges taken from Kulicka and Sukacheva (1990). First as for Micro-ptysmatidae. Baltic amber records taken from Spahr (1989).

F. ATOPSYCHIDAE T. (Oli.)–Rec. Terr.

Extant

F. BAISSOFERIDAE J. (u.)–K. (l.) Terr.

F. BERAEIDAE T. (PRB)–Rec. Terr.

First: *Bereodes pectinatus*, Baltic amber. **Extant**

F. BRACHYCENTRIDAE T. (PRB)–Rec. Terr.

First: *Brachycentrus labialis*, Baltic amber. **Extant**

F. CALAMOCERATIDAE K. (APT)–Rec. Terr.

First: e.g. Calamoceratid pupa indet. Jell and Duncan (1986), Koonwarra Fossil Bed, south Gippsland, Australia. **Extant**

F. CLADOCHORISTIDAE P. (u.)–Tr. (u.) Terr.

First: *Cladochorista belmontensis*, *in* Willmann (1989b), Belmont, New South Wales, Australia.
Last: *Cladochoristella bryani*, *in* Willmann (1989b), Mt Crosby, Queensland, Australia.

F. DYSONEURIDAE (DISONEURIDAE) J. (u.)–K. (l.) Terr.

e.g. *Utania remissa* Sukacheva (1990b), Transbaikalia, former USSR.

F. ECNOMIDAE T. (PRB)–Rec. Terr.

First: e.g. *Archaeotinodes angusta*, Baltic amber. **Extant**

F. ELECTRALBERTIDAE K. (CMP) Terr.

First and Last: *Electralberta creiacica*, *in* Botosaneanu and Wichard (1984), Canadian amber, Medicine Hat, Canada.

F. GLOSSOSOMATIDAE T. (PRB)–Rec. Terr.

First: *Electragapetus scitulus*, Baltic amber. **Extant**

F. GOERIDAE T. (PRB)–Rec. Terr.

First: e.g. *Goera gracilicornis*, Baltic amber. **Extant**

F. HELICOPSYCHIDAE T. (PRB)–Rec. Terr.

First: e.g. *Electrohelicopsyche taeniata*, Baltic amber. **Extant**

F. HYDROBIOSIDAE K. (SAN)–Rec. Terr.

First: *Palaeohydrobiosis siberambra*, *in* Botosaneanu and Wichard (1984), Siberian amber, Taimyr, former USSR. **Extant**

F. HYDROPSYCHIDAE T. (PRB)–Rec. Terr.

First: e.g. *Hydropsyche viduata*, Baltic amber. **Extant**

F. HYDROPTILIDAE K. (u.)–Rec. Terr.

Extant

F. LEPIDOSTOMATIDAE T. (PRB)–Rec. Terr.

First: e.g. *Palaeocrunoecia crenata*, Baltic amber. **Extant**

F. LEPTOCERIDAE K. (APT)–Rec. Terr.

First: e.g. *Oecetis*? sp. Jell and Duncan (1986), Koonwarra Fossil Bed, south Gippsland, Australia. **Extant**

F. LIASSOPHILIDAE Tr. (SCY)–J. (l.) Terr.

F. LIMNEPHILIDAE K.? (CEN)–Rec. Terr.

First: Cases in Shields (1988), Windrow Formation, New

Ulm, Minnesota, USA. Sukacheva (pers. comm.) considers that the position of these specimens is doubtful. **Extant**

F. MICROPTYSMATIDAE P. (ART–ZEC) Terr.

First: *Microptysmella moravica* Kukalová-Peck and Willmann (1990), Bavcov Beds, Obora, Czechoslovakia.
Last: e.g. *Microptysmodes uralicus*, *in* Willmann (1989b), Tatar ASSR, former USSR.

F. MOLANNIDAE T. (PRB)–Rec. Terr.

First: e.g. *Molanna crassicornis*, Baltic amber. **Extant**

F. NECROTAULIIDAE (NECROTAULIDAE) Tr. (u.)–K. (u.) Terr.

F. ODONTOCERIDAE (ODONTOCERATIDAE) K. (u.)–Rec. Terr.

Extant

F. PHILOPOTAMIDAE Tr. (u.)–Rec. Terr.

Extant

F. PHRYGANEIDAE (PHRYGANAEIDAE) K. (l.)–Rec. Terr.

Extant

F. POLYCENTROPODIDAE K. (u.)–Rec. Terr.

First: e.g. *Archaeopolycentra zherikhini*, *in* Spahr (1989), Siberian amber, Taimyr, former USSR. **Extant**

F. PRORHYACOPHILIDAE Tr. (u.) Terr.

F. PROSEPIDIDONTIDAE J. (l.) Terr.

F. PSYCHOMYIIDAE (PSYCHOMYIDAE) T. (PRB)–Rec. Terr.

First: e.g. *Lype sericea*, Baltic amber. **Extant**

F. RHYACOPHILIDAE J. (u.)–Rec. Terr.

First: ?*Rhyacophila* sp. Sukacheva (1985), Mogzon, Transbaikalia, former USSR. **Extant**

F. SERICOSTOMATIDAE K. (u.)–Rec. Terr.

Extant

F. STENOPSYCHIDAE T. (PRB)–Rec. Terr.

First: *Stenopsyche imitata*, Baltic amber. **Extant**

F. TAYMYRELECTRONIDAE K. (SAN) Terr.

First and Last: *Taymyrelectron sukatshevae*, *in* Botosaneanu and Wichard (1984), Siberian amber, Taimyr, former USSR. **Extant**

F. VITIMOTAULIIDAE J. (u.)–K. (u.) Terr.

REFERENCES

Ansorge, J. (1991) *Locustopsis africanus* n.sp., (Saltatoria, Caelifera) aus der Unterkreide SW-Ägyptens. *Neues Jahrbuch für Geologie und Paläontologie Monatshefte*, 1991, 205–12.

Ansorge, J. and Schlüter, T. (1990) The earliest chrysopid: *Liassochrysa stigmatica* n.g., n.sp. from the Lower Jurassic of Dobbertin, Germany. *Neuroptera International*, **6**, 87–93.

Arnold, L. V., Zherikhin, V. V., Nikitin, L. M. and Ponomarenko, A. G. (1991) *Mesozoic Coleoptera*, Oxonian Press, New Delhi. [Translation of *Mesozoiskie Zhestokrylye*, Nauka, Moscow (1977)].

Baird, G. C., Shabica, C. W., Anderson, J. L. *et al.* (1985) Biota of a Pennsylvanian muddy coast: habitats within the mazonian delta complex, Northeast Illinois. *Journal of Paleontology*, **59**, 253–81.

Bekker-Migdisova, E. E. (1985a) Permian isopteran [sic] genera, *Protopincombea* and *Propatrix*. *Paleontological Journal*, **19** (1), 85–9.

Bekker-Migdisova, E. E. (1985b) Iskopaemye nasekomye psillomorfy. [The fossil psyllomorph insects.] *Trudy Paleontologicheskogo Instituta*, **206**, 92 pp. [in Russian].

Bigot, L., Nel, A. and Nel, J. (1986) Description de la première espèce fossile connue de Ptérophore (Lepidoptera Pterophoridae). *Alexanor*, **14**, 283–8.

Botosaneanu, L. and Wichard, W. (1984) Upper Cretaceous amber Trichoptera. *Proceedings of the 4th International Symposium on Trichoptera*, South Carolina, 1983, pp. 43–8.

Boucot, A. J. (1990) *Evolutionary Paleobiology of Behavior and Coevolution*. Elsevier, Amsterdam, 725 pp.

Brandão, C. R. F., Martins-Neto, R. G. and Vulcano, M. A. (1989) The earliest known fossil ant (first Southern Hemisphere Mesozoic record) (Hymenoptera: Formicidae: Myrmeciinae). *Psyche*, **96**, 195–208.

Brauckmann, C. (1986) Eine neue Spilapteriden-Art aus dem Namurium B von Hagen-Vorhalle (Insecta: Palaeodictyoptera; Ober-Karbon; West-Deutschland). *Dortmunder Beiträge zur Landeskunde*, **20**, 57–64.

Brauckmann, C. (1988a) Hagen-Vorhalle, a new important Namurian Insecta-bearing locality (Upper Carboniferous; F R Germany). *Entomologica Generalis*, **14**, 73–9.

Brauckmann, C. (1988b) Zwei neue Insekten (Odonata, Megasecoptera) aus dem Namurium von Hagen-Vorhalle (West-Deutschland). *Dortmunder Beiträge zur Landeskunde*, **22**, 91–101.

Brauckmann, C. (1991) Ein neuer insekten-rest (Megasecoptera) aus dem Ober-Karbon von Osnabrück. *Osnabrücker Naturwissenschafliche Mitteilungen*, **17**, 25–32.

Brauckmann, C. and Willmann, R. (1990) Insekten aus dem Permo-Silesium der Bohrung Weiterstadt 1 (Blattodea, 'Protorthoptera'; Oberrheinische Tiefebene, SW-Deutschland. *Neues Jahrbuch für Geologie und Paläontologie Monatshefte*, 1990, 470–8.

Brauckmann, C. and Zessin, W. (1989) Neue Meganeuridae aus dem Namurium von Hagen-Vorhalle (BRD) und die Phylogenie der Meganisoptera (Insecta, Odonota). *Deutsche Entomologische Zeitschrift*, **36**, 177–215.

Brauckmann, C., Koch, L. and Kemper, M. (1985) Spinnentiere (Arachnida) und Insekten aus den Vorhalle-Schichten (Namurium B; Ober-Karbon) von Hagen-Vorhalle (West-Deutschland). *Geologie und Paläontologie in Westfalen*, **3**, 131 pp.

Buckland, P. C. and Coope, G. R. (1991) *A bibliography and literature review of Quaternary entomology*, J. R. Collis, Sheffield, 85 pp.

Burnham, L. (1984) Les insectes du Carbonifère Supérieur de Montceau-les-Mines I. – L'Ordre des Caloneurodea. *Annales de Paléontologie*, **70**, 167–80.

Carle, F. L. and Wighton, D. C. (1990) Chapter 3. Odonata, in *Insects from the Santana Formation, Lower Cretaceous, of Brazil*. (ed. D. A. Grimaldi), *Bulletin of the American Museum of Natural History*, **195**, pp. 51–68.

Carpenter, F. M. (1986) Substitute names for the extinct genera *Cycloptera* Martynova (Mecoptera) and *Parelcana* Carpenter (Orthoptera). *Psyche*, **93**, 375–6.

Carpenter, F. M. (1987) Review of the extinct Family Syntonopteridae (Order Uncertain). *Psyche*, **94**, 373–88.

Carpenter, F. M. (1992) Hexapoda. *Treatise on Invertebrate Paleontology, Part R, Arthropoda 4 (3, 4)*. Geological Society of America and University of Kansas, Boulder, Colorado, and Lawrence, Kansas, 655 pp.

Carpenter, J. M. and Rasnitsyn, A. P. (1990) Mesozoic Vespidae. *Psyche*, **97**, 1–20.

Cavallo, O. and Galletti, P. A. (1987) Studi di Carlo Sturani su Odonati e altri insetti fossili del Messiniano albese (Piemonte) con descrizione di *Oryctodiplax gypsorum* n. gen. n. sp. (Odonata, Libellulidae). *Bollettino della Società Paleontologica Italiana*, **26**, 151–76.

Crosskey, R. W. (1991) The fossil pupa *Simulimima* and the evidence it provides for the Jurassic origin of the Simuliidae (Diptera). *Systematic Entomology*, **16**, 401–6.

Crowson, R. A. (1985) Report on a Russian treatise about Jurassic insects of Siberia and Mongolia. *Entomologia Generalis*, **11**, 99–101.

Crowson, R. A., Smart, J. and Wootton, R. J. (1967) Class Insecta, in *The Fossil Record* (eds W. B. Harland, C. H. Holland, M. R. House *et al.*), Geological Society of London, pp. 508–28.

Darling, C. D. and Sharkey, M. J. (1990) Chapter 7. Order Hymenoptera. *Insects from the Santana Formation, Lower Cretaceous, of Brazil.* (ed. D. A. Grimaldi), *Bulletin of the American Museum of Natural History*, 195, pp. 123–53.

Dlussky, G. M. (1987) New Formicoidea (Hymenoptera) of the Upper Cretaceous. *Paleontological Journal*, **21** (1), 146–50.

Dmitriev, V. U. and Zherikhin, V. V. (1988) Izmeneniya raznoobraziya semeistv nasekomykh po dannym metoda nakoplennykh poyavlenii, in *Melovoi Biotsenoticheskii Krizis i Evolyutsiya nasekomykh. [The Cretaceous Biocenotic crisis and Evolution of Insects]* (ed. A. G. Ponomarenko), Nauka, Moscow, pp. 208–15 [in Russian].

Durden, C. J. (1984) Age zonation of the Early Pennsylvanian using fossil insects. *Oklahoma Geological Survey Bulletin*, **136**, 175–83.

Fernandez-Rubio, F., Peñalver, E. and Martínez-Delclòs, X. (1991) *Zygaena? turolensis*, una nueva especie de Lepidoptera Zygaenidae del Mioceno de Rubielos de Mora (Teruel). Descripción y filogenia. *Estudios del Museo de Ciencias Naturales de Alava*, **6**, 77–93.

Fujiyama, I. (1987) Middle Miocene insect fauna of Abura, Hokkaido, Japan, with notes on the occurrence of Cenozoic fossil insects in the Oshima Peninsula, Hokkaido. *Memoirs of the National Science Museum Tokyo*, **20**, 37–44 [in Japanese, English summary].

Gomez Pallerola, J. E. (1984) Nuevos Paleontínidos del yacimiento Infracretácico de la 'Pedrera de Meiá' (Lérida). *Boletin Geológico y Minero*, **1984**, 301–9.

Gomez Pallerola, J. E. (1986) Nuevos insectos fósiles de las calizas litográficas del Cretácico Inferior del Montsech (Lérida). *Boletin Geológico y Minero*, **1986**, 717–36.

Gore, P. J. W. (1988) Paleoecology and sedimentology of a Late Triassic lake, Culpeper Basin, Virginia, U.S.A. *Palaeogeography, Palaeoclimatology, Palaeoecology*, **62**, 593–608.

Gorokhov, A. V. (1985) Mesozoic crickets (Orthoptera, Grylloidea) of Asia. *Paleontological Journal*, **19** (2), 56–66.

Gorokhov, A. V. (1986a) Pryamokrylye. Gryllida (= Orthoptera). [Grasshoppers. Gryllida (= Orthoptera)], in *Nasekomye v Rannemelovykh Ekosistemakh Zapadnoi Mongolii. [Insects in the Early Cretaceous ecosystems of West Mongolia]*. Transactions of the Joint Soviet–Mongolian Palaeontological Expedition, 28, pp. 171–4 [in Russian].

Gorokhov, A. V. (1986b) Triasovye pryamokrylye nadsemeistva Hagloidea (Orthoptera). [Triassic grasshoppers of the superfamily Hagloidea (Orthoptera)], in *Sistematika Faunistika i Ekologiya Ortopteroidnykh nasekomykh. [Systematics, Faunistics and Ecology of Orthopteroid Insects]*. Trudy Zoologicheskogo Instituta, **143**, pp. 65–100 [in Russian].

Gorokhov, A. V. (1987a) Permian Orthoptera of the Infraorder Oedischiidea (Ensifera). *Paleontological Journal*, **21** (1), 72–85.

Gorokhov, A. V. (1987b) Novye iskopaemye pryamokryle semeistv Bintoniellidae, Mesoedischiidae fam. n. i Pseudelcanidae fam. n. (Orthoptera, Ensifera) iz permskikh i triasovykh otlozhenii SSSR. [New Fossil Orthopterans of the Families Bintoniellidae, Mesoedischiidae fam. n. and Pseudelcanidae fam. n. (Orthoptera, Ensifera) from Permian and Triassic deposits of the USSR.] *Vestnik Zoologii*, **1987** (1), pp. 18–23 [in Russian].

Gorokhov, A. V. (1987c) Novye iskopaemye pryamokryle semeistv Adumbratomorphidae fam. n., Pruvostitidae i Proparagryllacrididae (Orthoptera, Ensifera) iz permskikh i triasovykh otlozhenii SSSR. [New Fossil Orthopterans of the Families Adumbratomorphidae fam. n., Pruvostitidae and Proparagryllacrididae (Orthoptera, Ensifera) from Perm and Triassic Deposits of the USSR.] *Vestnik Zoologii*, **1987** (4), 20–8 [in Russian].

Gorokhov, A. V. (1988a) O klassifikatsii iskopaemykh pryamokrylykh nadsemeistva Phasmomimoidea (Orthoptera) s opisannem novykh taksonov. [On the classification of fossil grasshoppers of the superfamily Phasmomimoidea (Orthoptera) with descriptions of new taxa.] *Trudy Zoologicheskogo Instituta*, **178**, 32–44 [in Russian].

Gorokhov, A. V. (1988b) Klassifikatsiya i filogeniya kuznetchikovykh (Gryllida = Orthoptera, Tettigonioidea). [Classification and phylogeny of the Long-horned Grasshoppers (Gryllida = Orthoptera, Tettigonioidea).] in *Melovoi Biotsenoticheskii Krizis i Evolyutsiya Nasekomykh. [The Cretaceous Biocenotic Crisis and Evolution of Insects.]* (ed. A. G. Ponomarenko), Nauka, Moscow [in Russian], pp. 145–90.

Gorokhov, A. V. (1988c) Pryamokrylye nadsemeistva Hagloidea (Orthoptera) nizhnei i srednei jury. [Grasshoppers of the superfamily Hagloidea (Orthoptera) from the Lower and Middle Jurassic.] *Paleontologicheskii Zhurnal*, **1988** (2), 54–66 [in Russian].

Gorokhov, A. V. (1989) Novye taksony pryamokrylykh semeistv Bintoniellidae, Xenopteridae, Permelcanidae, Elcanidae, i Vitimiidae (Orthoptera, Ensifera) iz Mezozoya Azii. [New taxa of the Orthopteran Families Bintoniellidae, Xenopteridae, Permelcanidae, Elcanidae and Vitimiidae (Orthoptera, Ensifera) from the Mesozoic of Asia.] *Vestnik Zoologii*, **1989** (4), 20–7 [in Russian].

Gratshev, V. G. and Zherikhin, V. V. (in press) [New fossil mantids (Insects, Mantida).] *Paleontologicheskii Zhurnal.*

Greenslade, P. and Whalley, P. E. S. (1986) The systematic position of *Rhyniella praecursor* Hirst and Maulik (Collembola). The earliest known hexapod, in *2nd International Seminar on Apterygota* (ed. R. Dallai), Siena, Italy, 1986, pp. 319–23.

Grimaldi, D. A. (1987) Amber fossil Drosophilidae (Diptera), with particular reference to the Hispaniolan taxa. *American Museum Novitates*, **2880**, 1–23.

Grimaldi, D. A. (1989) The genus *Metopina* (Diptera: Phoridae) from Cretaceous and Tertiary ambers. *Journal of the New York Entomological Society*, **97** (1), 65–72.

Grimaldi, D. A. (1990) Chapter 9. Diptera, in *Insects from the Santana Formation, Lower Cretaceous, of Brazil* (ed. D. A. Grimaldi), Bulletin of the American Museum of Natural History, 195, pp. 164–83.

Grimaldi, D. A. and Maisey, J. (1990) Introduction, in *Insects from the Santana Formation, Lower Cretaceous, of Brazil* (ed. D. A. Grimaldi) Bulletin of the American Museum of Natural History, 195, pp. 5–14.

Hamilton, K. G. A. (1990) Chapter 6. Homoptera, in *Insects from the Santana Formation, Lower Cretaceous, of Brazil* (ed. D. A. Grimaldi), Bulletin of the American Museum of Natural History, 195, pp. 82–122.

Harland, W. B., Cox, A. V., Llewellyn, P. G. *et al.* (1982) *A Geologic Time Scale*. Cambridge University Press, 131 pp.

Heie, O. E. (1985) Fossil aphids. A catalogue of fossil aphids, with comments on systematics and evolution, in *Evolution and Biosystematics of Aphids*. Proceedings of the International Aphidological Symposium, Jablonna, 1981, pp. 101–34.

Heie, O. E. (1987) Palaeontology and phylogeny, in *Aphids: their Biology, Natural Enemies and Control, Volume A.* (eds A. K. Minks and P. Harrewijn), Elsevier, Amsterdam, pp. 367–91.

Hieke, F. and Pietrzeniuk, E. (1984) Die Bernstein-Käfer des Museums für Naturkunde, Berlin (Insecta, Coleoptera). *Mitteilungen aus dem Zoologischen Museum in Berlin*, **60**, 297–326.

Hong, Y. (1984a) New fossil insects of Laiyang Group from Laiyang Basin, Shandong Province. *Professional Papers of Stratigraphy and Palaeontology*, **11**, 31–41 [in Chinese, English summary].

Hong, Y. (1984b) Curvicubitidae fam. nov. (Lepidoptera? Insecta) from Middle Triassic of Shaanxi. *Acta Palaeontologica Sinica*, **2** (6), 782–85 [in Chinese, English summary].

Hong, Y. (1985a) New fossil insects of Xiahuayuan Formation in Yuxian County, Hebei Province. *Bulletin Tianjin Institute*, **13**, 131–7 [in Chinese, English summary].

Hong, Y. (1985b) *Fossil Insects, Scorpionids and Araneids in the Diatom[ite]s of Shanwang*. Geological Publishing House, Beijing China, 80 pp. [in Chinese, English summary].

Hong, Y. (1985c) New fossil genera and species of Shanxi Formation in Xishan of Taiyuan. *Entomotaxonomia*, **7** (2), 83–91 [in Chinese, English summary].

Hong, Y. (1985d) Insecta, in *Palaeontological Atlas of North China II. Mesozoic Volume*. Geological Publishing House, Beijing, pp. 128–85 [in Chinese].

Hong, Y. (1987) The study of Early Cretaceous insects of 'Kezuo', West Liaoning. *Professional Papers in Stratigraphy and Palaeontology*, **18**, 76–86 [in Chinese, English summary].

Hong, Y. (1988) Early Cretaceous Orthoptera, Neuroptera, Hymenoptera (Insecta) of Kezuo in West Liaoning Province. *Entomotaxonomia*, **10** (1/2), 119–30 [in Chinese, English summary].

Hong, Y. and Wang, W. (1990) Fossil insects from the Laiyang Basin, Shandong Province, in *The Stratigraphy and Palaeontology of Laiyang Basin, Shandong Province*. Shandong Bureau of Geology and Mineral Resources, pp. 44–189 [in Chinese, English summary].

Hong, Y., Yan, D. and Wang, D. (1989) Discovery of Early Cretaceous *Cretacechorista* gen. nov. Insecta: Mecoptera from Jiuquan Basin, Gansu Province. *Memoirs of Beijing Natural History Museum*, **44**, 1–9 [in Chinese, English summary].

Hubbard, M. D. (1987) Ephemeroptera. *Fossilium Catalogus 1: Animalia*, **129**, 1–99.

Jarzembowski, E. A. (1984) Early Cretaceous insects from Southern England. *Modern Geology*, **9**, 71–93.

Jarzembowski, E. A. (1986) A fossil enicocephalid bug (Insecta: Hemiptera) from the London Clay (early Eocene) of the Isle of Sheppey, southern England. *Tertiary Research*, **8** (1), 1–5.

Jarzembowski, E. A. (1988) Prospecting for early insects. *Journal of the Open University Geological Society*, **9** (1), 34–40.

Jarzembowski, E. A. (1989a) Writhlington Geological Nature Reserve. *Proceedings of the Geologists' Association*, **100**, 219–34.

Jarzembowski, E. A. (1989b) A fossil aphid (Insecta: Hemiptera) from the Early Cretaceous of southern England. *Cretaceous Research*, **10**, 239–48.

Jarzembowski, E. A. (1990a) Early Cretaceous Zygopteroids of Southern England, with the description of *Cretacoenagrion alleni* gen. nov., spec. nov. (Zygoptera: Coenagrionidae; 'Anisozygoptera': Tarsophlebiidae, Euthemistidae). *Odonatologica*, **19** (1), 27–37.

Jarzembowski, E. A. (1990b) A century plus of fossil insects. *Proceedings of the Geologists Association*, **100**, 433–49.

Jarzembowski, E. A. (1990c) A boring beetle from the Wealden of the Weald. In *Evolutionary Paleobiology of Behavior and Coevolution* (ed. A. J. Boucot), Elsevier, Amsterdam, pp. 373–6.

Jell, P. A. and Duncan, P. M. (1986) Invertebrates, mainly insects, from the freshwater, Lower Cretaceous, Koonwarra Fossil Bed (Korumburra Group), South Gippsland, Victoria. *Memoir of the Association of Australasian Palaeontologists*, **3**, 111–205.

Jeram, A. J., Selden, P. A. and Edwards, D. (1990) Land animals in the Silurian: arachnids and myriapods from Shropshire, England. *Science*, **250**, 658–61.

Kalugina, N. S. (1989) Novye psikhodomorfnye dvukrylye nasekomye Mezozoya Sibiri (Diptera: Eoptychopteridae, Ptychopteridae). [New psychodomorph dipterous insects from the Mesozoic of Siberia (Diptera: Eoptychopteridae, Ptychopteridae).] *Paleontologicheskii Zhurnal*, **1989** (1), 65–77 [in Russian].

Kalugina, N. S. and Kovalev, V. G. (1985) *Dvukrylye Nasekomye Yury Sibiri. [Two-Winged Insects from the Jurassic of Siberia.]* Nauka, Moscow, 197 pp. [in Russian].

Kathirithamby, J. (1989) Review of the Order Strepsiptera. *Systematic Entomology*, **14**, 41–92.

Klyuge, N. Y. (1986) A recent may fly species (Ephemeroptera, Heptageniidae) in Baltic amber. *Paleontological Journal*, **20** (2), 106–7.

Klyuge, N. Y. (1989) Vopros o gomologii trakheinykh zhabr: paranotalnykh vyrostov lichinok nodenok i krylev nasekomykh v sryazi s sistematikoi i filogennei otryada podenok (Ephemeroptera). [A question of the homology of the tracheal gills and paranotal processi of mayfly larvae and wings of the insects with reference to the taxonomy and phylogeny of the Order Ephemeroptera.] *Chteniya Pamyati Nikolaya Aleksandrovicha Kholodkovskogo*, **1988**, 48–77 [in Russian].

Koteja, J. (1987) *Protorthezia aurea* gen. et sp. n. (Homoptera, Coccinea, Ortheziidae) from Baltic amber. *Polskie Pismo Entomologiczne*, **57**, 241–9.

Koteja, J. (1988a) *Eomatsucoccus* gen. n. (Homoptera, Coccinea) from Siberian Lower Cretaceous deposits. *Annales Zoologici*, **42** (4), 141–63.

Koteja, J. (1988b) Two new eriococcids from Baltic amber (Homoptera, Coccinea). *Deutsche Entomologische Zeitschrift*, **35**, 405–16.

Koteja, J. (1988c) *Sucinikermes kulickae* gen. et sp. n. (Homoptera, Coccinea) from Baltic amber. *Polskie Pismo Entomologiczne*, **58**, 525–35.

Koteja, J. (1989) *Inka minuta* gen. et sp. n. (Homoptera, Coccinea) from Upper Cretaceous Taymyrian amber. *Annales Zoologici*, **43** (5), 77–101.

Koteja, J. (1990) Paleontology, in *Armoured Scale Insects: Their Biology, Natural Enemies and Control, Volume A* . Elsevier, Amsterdam, pp. 149–63.

Kovalev, V. G. (1983) A new family of fossil Diptera from Australian Triassic deposits and its presumed descendants (Diptera: Crosaphididae, fam. nov., Mycetobiidae). *Entomological Review*, **62** (4), 130–6.

Kovalev, V. G. (1986) Infraotryady Bibionomorpha i Asilomorpha. [Infraorders Bibionomorpha and Asilomorpha], in *Nasekomye v Rannemelovykh Ekosistemakh Zapadnoi Mongolii. [Insects in the Early Cretaceous Ecosystems of Western Mongolia.]* (ed. A. P. Rasnitsyn), Transactions of the Joint Soviet–Mongolian Palaeontological Expedition, **28**, pp. 125–54 [in Russian].

Kovalev, V. G. (1987) The Mesozoic Mycetophiloid Diptera of the Family Pleciofungivoridae. *Paleontological Journal*, **21** (2), 67–79.

Kovalev, V. G. (1989a) Bremochaetidae [sic], the Mesozoic family of brachycerous dipterans. *Paleontological Journal*, **23** (2), 100–5.

Kovalev, V. G. (1989b) Geological history and systematic position of the Thaumaleidae (Diptera). *Entomological Review*, **69** (4), 1990, 121–31.

Kovalev, V. G. (1990) Dvukrylye. Muscida. [Dipterans. Muscida], in Pozdnemezozoiskie nasekomye vostochnogo Zabaikalya. [Late Mesozoic insects of Eastern Transbaikalia.] *Trudy Paleontologicheskogo Instituta*, **239**, pp. 123–77 [in Russian].

Kozlov, M. A. (1988) Paleontologiya cheshuekrylykh i voprosy filogenii otryada Papilionida. [Palaeontology of moths and butterflies and the problem of the phylogeny of the Order Papilionida.], in *Melovoi Biotsenoticheskii Krizis i evolyutsiya nasekomykh. [The Cretaceous Biocenotic Crisis and Evolution of Insects]* (ed. A. G. Ponomarenko), Nauka, Moscow, pp. 16–69 [in Russian].

Kozlov, M. A. (1989) New Upper Jurassic and Lower Cretaceous Lepidoptera (Papilionida). *Paleontological Journal*, **23** (4), 34–9.

Kristensen, N. P. (1991) Phylogeny of extant hexapods, *in The insects of Australia* (ed. I. D. Naumann), 2nd edn, volume 1, Melbourne University Press, Victoria, pp. 125–40.

Kukalová-Peck, J. (1985) Ephemeroid wing venation based upon new gigantic Carboniferous mayflies and basic morphology, phylogeny and metamorphosis of pterygote insects (Insecta, Ephemerida). *Canadian Journal of Zoology*, **63**, 933–55.

Kukalová-Peck, J. (1987) New Carboniferous Diplura, Monura, and Thysanura, the hexapod ground plan, and the role of thoracic side lobes in the origin of wings (Insecta). *Canadian Journal of Geology*, **65**, 2327–45.

Kukalová-Peck, J. (1988) A substitute name for the extinct genus *Stenelytron* Kukalová (Protelytroptera). *Psyche*, **94**, p. 339.

Kukalová-Peck, J. (1991) Fossil history and the evolution of hexapod structures, in *The Insects of Australia* (ed. I. D. Naumann), 2nd edn, Vol. 1. Melbourne University Press, Victoria, pp. 141–79.

Kukalová-Peck, J. and Brauckmann, C. (1990) Wing folding in pterygote insects, and the oldest Diaphanopterodea from the early Late Carboniferous of West Germany. *Canadian Journal of Zoology*, **68**, 1104–11.

Kukalová-Peck, J. and Willmann, R. (1990) Lower Permian 'mecopteroid-like' insects from central Europe (Insecta, Endopterygota). *Canadian Journal of Earth Sciences*, **27**, 459–68.

Kulicka, R. and Sukacheva, I. D. (1990) Rodziny kopalnych Trichoptera mezozoiku i kenozoiku. [The families of Fossil Trichoptera of the Mesozoic and Caenozoic Eras.] *Prace Muzeum Ziemi*, **41**, 65–75. [in Polish].

Labandeira, C. C. and Beall, B. S. (1990) Arthropod terrestriality, in *Arthropod Paleobiology* (ed. S. J. Culver), Short Courses in Paleontology, **3**, pp. 214–56.

Labandeira, C. C., Beall, B. S. and Hueber, F. M. (1988) Early insect diversification: evidence from a Lower Devonian bristletail from Quebec. *Science*, **242**, 913–16.

Lacasa-Ruiz, A. and Martínez-Delclòs, X. (1986) *Meiatermes, Nuevo Genero Fosil de Insecto Isoptero (Hodotermitidae) de las calizas Neocomiensis del Montsec (Provincia de Lérida, España)*. Institut d'Estudis Ilerdencs, Lleida, 65 pp.

Lambkin, K. J. (1988) A re-examination of *Lithosmylidia* Riek from the Triassic of Queensland with notes on Mesozoic 'Osmylid-like' fossil Neuroptera (Insecta: Neuroptera). *Memoirs of the Queensland Museum*, **25**, 445–58.

Laurentiaux-Vieira, F. and Laurentiaux, D. (1986a) Paléodictyoptère nouveau du Namurien belge. *Annales de la Société Géologique du Nord*, **105**, 187–93.

Laurentiaux-Vieira, F. and Laurentiaux, D. (1986b) Présence du genre *Zdenedkia* Kuk. (Protorthoptères Paoliides) dans le Westphalien inférieur de Belgique. *Annales de la Société Géologique du Nord*, **105**, 195–201.

Laurentiaux-Vieira, F. and Laurentiaux, D. (1987) Un remarquable Archimylacride du Westphalien inférieur belge. Ancienneté du dimorphisme sexuel des Blattes. *Annales de la Société Géologique du Nord*, **106**, 37–47.

Lawrence, P. N. (1985) Ten species of Collembola from Baltic Amber. *Prace Muzeum Ziemi*, **37**, 101–4.

Lin, Q. (1985) Insect fossils from the Hanshan Formation at Hanshan County, Anhui Province. *Acta Palaeontologica Sinica*, **24**, 300–11 [in Chinese, English summary].

Lin, Q. (1986) Early Mesozoic fossil insects from South China. *Palaeontologia Sinica, B*, **170** (21), 1–112 [in Chinese, English summary].

Lin, Q. and Mou, C. (1989) On insects from Upper Triassic Xiaoping Formation, Guangzhou, China. *Acta Palaeontologica Sinica*, **28**, 598–603 [in Chinese, English summary].

Lutz, H. (1984) Parallelophoridae – isolierte Analfelder eozäner Schaben (Insecta: Blattodea). *Paläontologische Zeitschrift*, **58**, 145–7.

Lutz, H. (1988) Riesenameisen und andere Raritaten – die Insektenfauna, in *Messel – Ein Schaufenster in die Geschichte der Erde und des Lebens* (ed. S. Schaal), Senckenburg-Büch, **64**, pp. 55–67.

Makarkin, V. N. (1990) Novye nazvaniya yurskikh setchatokrylykh. [New names for Jurassic lacewings.] *Paleontologicheskii Zhurnal*, **1990** (1), p. 120 [in Russian].

Manley, D. G. and Poinar, G. O. Jr (1991) A new species of fossil *Dasymutilla* (Hymenoptera: Mutillidae) from Dominican amber. *Pan-Pacific Entomologist*, **67**, 200–5.

Martínez-Delclòs, X. (1989a) Insectos del Cretácio Inferior de Las Hoyas (Cuenca), in *La Fauna del Pasado en Cuenca*. Actas del 1 Curso de Paleontología, Instituto 'Juan de Valdes', Cuenca, pp. 51–82.

Martínez-Delclòs, X. (1989b) *Chresmoda aquatica* n. sp. insecto Chresmodidae de Cretácio Inferior de la Sierra del Montsec (Lleida, Espãna). *Revista Espãnola de Paleontología*, **4**, 67–74.

Martins-Neto, R. G. (1987a) A paleoentomofauna Brasileira: estágio atual do conhecimento, in *Anais do X Congresso Brasileiro de Paleontologia*, Rio de Janeiro, 1987, pp. 567–91.

Martins-Neto, R. G. (1987b) Descrição de três novos gêneros e três novas espécies de Orthoptera (Insecta, Acridoidea) da Formação Santana, Bacia do Araripe (Cretáceo Inferior) Nordeste do Brasil, representando três famílias, sendo que duas novas: Archaeopneumoridae nov. fam. e Bouretidae nov. fam. *Anais da Academia Brasileira de Ciencias*, **59**, p. 444.

Martins-Neto, R. G. (1988a) A new fossil insect (Homoptera, Cixiidae) from the Santana Formation (Lower Cretaceous), Araripe Basin, Northeast Brasil. *Interciencia*, **13**, 313–16.

Martins-Neto, R. G. (1988b) A new genus and species of Cixiidae (Homoptera, Fulgoroidea) from the Santana Formation (Lower Cretaceous) Araripe Basin, Northeast Brazil. *Acta Geologica Leopoldensia*, No. **26**, **11**, 7–14.

Martins-Neto, R. G. (1989) Primeiro registro de Phasmatodea (Insecta: Orthopteromorpha) na Formação Santana, Bacia do Araripe (Cretáceo Inferior), Nordeste do Brasil (1). *Acta Geologica Leopoldensia*, **28** (12), 91–104.

Martins-Neto, R. G. (1990) Um novo gênero e duas novas espécies de Tridactylidae (Insecta, Orthopteridea) na Formação Santana (Cretáceo Inferior do Nordeste do Brasil). *Anais da Academia Brasileira de Ciencas*, **62**, 51–9.

Martins-Neto, R. G. (1991) Sistémática dos Ensifera (Insecta, Orthopteroidea) da Formação Santana, Cretáceo Inferior do Nordeste do Brasil. *Acta Geologica Leopoldensia*, **32** (14), 3–162.

Martins-Neto, R. G. and Caldas, E. B. (1990) Efèmeras escavadoras (Insecta, Ephemeroptera, Ephemeroidea) na Formação Santana Cretáceo Inferior), Bacia do Araripe Nordeste do Brasil: des crição de três novas gêneros e três novas espéçies (ninfas). *Actas do I Simpósio sobre a Bacia do Araripe e Basias Interiores do Nordeste*, Crato, Brazil, 1990, pp. 265–75.

Martins-Neto, R. G. and Vulcano, M. A. (1989) Neurópteros (Insecta, Planipennia) da Formação Santana (Cretáceo Inferior), Bacia do Araripe, Nordeste do Brasil. II. Superfamília Myrmeleontoidea. *Revista Brasileira de Entomologia*, **33**, 367–402.

Martins-Neto, R. G. and Vulcano, M. A. (1990) Neurópteros (Insecta, Planipennia) da Formação Santana (Cretáceo Inferior) Bacia do Araripe, Nordeste do Brasil. III. Superfamília Mantispoidea. *Revista Brasileira de Entomologia*, **34**, 619–25.

Martynova, O. M. (1962) Otryad Mecoptera. Skorpionnitsy. [Order Mecoptera. Scorpionflies], in *Osnovy Paleontologii: Chlenistonogie. Trakheinye i Khelitserovye*. (ed. B. B. Rodendorf), Akademii Nauk, Moscow, pp. 283–94.

McAlpine, D. K. (1990) A new apterous micropezid fly (Diptera: Schizophora) from western Australia. *Systematic Entomology*, **15**, 81–6.

McAlpine, J. F. (ed.) (1981) *Manual of Nearctic Diptera*, Vol. 1. Monograph of the Biosystematics Research Centre, Ottawa, **27**, 674 pp.

McAlpine, J. F. (1987) *Manual of Nearctic Diptera*, Vol. 2. *Monograph of the Biosystematics Research Centre*, Ottawa, **28**, pp. 675–1332.

McAlpine, J. F. (ed.) (1989) *Manual of Nearctic Diptera Volume 3*. Monograph of the Biosystematics Research Centre, 32, pp. 1333–581.

McCafferty, W. P. (1990) Chapter 2. Ephemeroptera, in *Insects from the Santana Formation, Lower Cretaceous, of Brazil* (ed. D. A. Grimaldi), Bulletin of the American Museum of Natural History, **195**, pp. 20–50.

Michener, C. D. and Grimaldi, D. A. (1988) A *Trigona* from Late Cretaceous amber of New Jersey (Hymenoptera: Apidae: Meliponinae). *American Museum Novitates*, **2917**, 1–10.

Najt, J. (1987) Le Collembole fossile *Paleosminthurus juliae* est un Hyménoptère. *Revue Française d'Entomologie*, **9** (4), 152–4.

Nel, A. (1986) Révision du genre cénozoïque *Stenolestes* Scudder, 1895; description de deux espèces nouvelles (Insecta, Odonata, Lestidae). *Bulletin du Muséum National d'Histoire Naturelle, Paris, Série 4*, **8** (C), 447–61.

Nel, A. (1988a) Deux nouveaux Mantispidae (Planipennia) fossiles de 1'Oligocène du sud-est de la France. *Neuroptera International*, **5**, 103–9.

Nel, A. (1988b) Parazacallitinae, nouvelle sous-famille et premier Epallagidae de l'Oligocène Européen (Odonata, Zygoptera).

Bulletin du Muséum National d'Histoire Naturelle, Paris, Série 4, **10** (C), 175–9.

Nel, A. (1989) *Piroutetia liasina* Meunier, 1907, insecte du Lias de France, espèce-type des Piroutetiidae nov. fam. *Bulletin du Museum National d'Histoire Naturelle, Paris, Ser.* 4, **11** (C), 15–19.

Nel, A. (1991) Description de quelques Sieblosiidae fossiles nouveaux (Odonata, Zygoptera, Lestoidea). *Nouvelle Revue d'Entomologie*, **8**, 367–75.

Nel, A., Séméria, Y. and Martins-Neto, R. G. (1990) Un Raphidioptera fossile du Crétacé Inférieur du Brésil (Neuropteroidea). *Neuroptera International*, **6**, 27–37.

Nelson, C. R. and Tidwell, W. D. (1988) *Brodioptera stricklani* n. sp. (Megasecoptera: Brodiopteridae), a new fossil insect from the Upper Manning Canyon Shale Formation, Utah (Lowermost Namurian B). *Psyche*, **94**, 309–16.

Oswald, J. D. (1990) Chapter 8. Raphidioptera, in *Insects from the Santana Formation, Lower Cretaceous, of Brazil* (ed. D. A. Grimaldi), Bulletin of the American Museum of Natural History, **195**, pp. 154–63.

Pagliano, G. and Scaramozzino, P. (1990) Elenco dei generi di Hymenoptera del mondo. *Memorie della Società Entomologica Italiana*, **68**, 1–210.

Perkovsky, Y. E. (1990) First discovery of Cretaceous insects of the Family Leiodidae (Coloptera). *Paleontological Journal*, **24** (4), 116–18.

Pinto, I. D. (1986) Carboniferous insects from Argentina III – Familia Xenopteridae Pinto, nov. Ordo Megasecoptera. *Pesquisas*, **18**, 23–9.

Pinto, I. D. (1987a) Permian insects from Paraná Basin, South Brazil IV Coleoptera. *Pesquisas*, **19**, 5–12.

Pinto, I. D. (1987b) Permian insects from Paraná Basin, South Brazil IV Homoptera – 2 – Cicadidea. *Pesquisas*, **19**, 13–22.

Poinar, G. O. Jr (1988) *Zorotypus palaeus*, new species, a fossil Zoraptera (Insecta) in Dominican amber. *Journal of the New York Entomological Society*, **96**, 253–9.

Ponomarenko, A. G. (1984) Neuroptera from the Jurassic in Eastern Asia. *Paleontological Journal*, **19** (3), 59–69.

Ponomarenko, A. G. (1985a) Zhestkokrylye iz yury Sibiri i zapadnoi Mongolii [Beetles from the Jurassic of Siberia and western Mongolia], in *Yurskie nasekomye Sibiri i Mongolii.* [*Jurassic insects of Siberia and Mongolia.*] (ed. A. P. Rasitsyn), Trudy Paleontologicheskogo Instituta, **211**, pp. 47–87 [in Russian].

Ponomarenko, A. G. (1985b) Fossil insects from the Tithonian 'Solnhofener Plattenkalke' in the Museum of Natural History, Vienna. *Annalen des Naturhistorischen Museums Wien*, **87** (A), 135–44.

Ponomarenko, A. G. (1986a) Zhestkokrylye. Scarabaeida (= Coleoptera). [Beetles. Scarabaeida (= Coleoptera)], in *Nasekomye v rannemelovykh Ekosistemakh Zapadnoi Mongolii.* [*Insects in the Early Cretaceous ecosystems of Western* Mongolia.] (ed. A. P. Rasnitsyn), Transactions of the Joint Soviet – Mongolian Palaeontological Expedition, **28**, pp. 84–100 [in Russian].

Ponomarenko, A. G. (1986b) Scarabaeiformes Incertae Sedis, in *Nasekomye v Rannemelovykh Ekosistemakh Zapadnio Mongolii.* [*Insects in the Early Cretaceous ecosystems of western* Mongolia.] (ed. A. P. Rasnitsyn), Transactions of the Joint Soviet – Mongolian Palaeontological Expedition, **28**, pp. 110–12 [in Russian].

Ponomarenko, A. G. (1987) New Mesozoic water beetles (Insecta, Coleoptera) from Asia. *Paleontological Journal*, **21** (2), 79–92.

Ponomarenko, A. G. (1988) Novye Mezozoiskie Nasekomye. [New Mesozoic Insects], in *Novye Iskopaemye Bespozvonochnye Mongolii*, [*New Fossil Invertebrates from Mongolia.*] (ed. A. Y. Rosanov), Transactions of the Joint Soviet – Mongolian Palaeontological Expedition, **33**, pp. 71–80 [in Russian].

Ponomarenko, A. G. (1990) Zhuki. Scarabaeida. [Beetles. Scarabaeida], in *Pozdnemezozoiskie Nasekomye Vostochnogo Zabaikalya.* [*Late Mesozoic Insects of Eastern Transbaikalia.*] (ed. A. P. Rasnitsyn), Trudy Paleontologicheskogo Instituta, **239**, pp. 39–87 [in Russian].

Ponomarenko, A. G. and Schultz, O. (1988) Typen der geologisch-paläontologischen abteilung: fossile insekten. *Kataloge der Wissenschaftlichen Sammlung des Naturhistorischen Museums in Wien*, **6**, *Paläozoologie (1)*, 1–39.

Popham, E. J. (1990) Chapter 4. Dermaptera, in *Insects from the Santana Formation, Lower Cretaceous, of Brazil.* (ed. D. A. Grimaldi), Bulletin of the American Museum of Natural History, **195**, pp. 69–75.

Popov, Y. A. (1985) Yurskie klopy i peloridiinovye yuzhnoi Sibiri i zapadnoi Mongolii. [Jurassic heteropteran bugs and new peloridiids from southern Siberia and western Mongolia], in *Yurskie Nasekomye Sibiri i Mongolii.* [*Jurassic Insects of Siberia and Mongolia.*] (ed. A. P. Rasnitsyn), Trudy Paleontologicheskogo Instituta, **211**, pp.28–47 [in Russian].

Popov, Y. A. (1986) Peloridiinovye i klopy Peloridiina (= Coleorrhyncha) et Cimicina (= Heteroptera). [New peloridiids and heteropteran bugs Peloridiina (= Coleorrhyncha et Cimicina (= Heteroptera), in *Nasekomye v Rannemelovykh Ekosistemakh Zapadnoi Mongolii. [Insects in the Early Cretaceous Ecosystems of West Mongolia].* Transactions of the Joint Soviet – Mongolian Palaeontological Expedition, **28**, pp. 50–83 [in Russian].

Popov, Y. A. (1988a) Novye Mesozoiskie klopy greblyaki (Corixidae, Shurabellidae). [New Mesozoic water bugs (Corixidae, Shurabellidae)] in *Novye Iskopaemye Bespozvonochnye Mongolii.* [New fossil *Invertebrates from Mongolia.*] (ed. A. Y. Rosanov), Transactions of the Joint Soviet – Mongolian Palaeontological Expedition, **33**, pp. 63–71 [in Russian.]

Popov, Y. A. (1988b) New Mesozoic peloridiinians and true bugs (Hemiptera: Coleorrhyncha and Heteroptera) from eastern Transbaikalia. *Paleontological Journal*, **22** (4), 64–73.

Popov, Y. A. (1989) New fossil Hemiptera (Heteroptera + Coleorrhyncha) from the mesozoic of Mongolia. *Neues Jahrbuch für Geologie und Paläontologie, Monatshefte*, **1989**, 166–81.

Popov, Y. A. (1990) Klopy. Cimicina. [Bugs. Cimicina], in *Pozdnemezozoiskie Nasekomye Vostochnogo Zabaikalya.* [*Late Mesozoic Insects of Eastern Transbaikalia.*] (ed. A. P. Rasnitsyn), *Trudy Paleontologicheskogo Instituta*, **239**, pp. 20–39 [in Russian].

Pritykina, L. N. (1985) Yurskie strekozy (Libellulida = Odonata) Sibiri i Zapadnoi Mongolii. [Jurassic dragonflies (Libellulida = Odonata) of Siberia and Western Mongolia], in *Yurskie Nasekomye Sibiri i Mongolii. [Jurassic Insects of Siberia and Mongolia.]* (ed. A. P. Rasnitsyn), Trudy Paleontologicheskogo Instituta, **211**, pp. 120–38 [in Russian.]

Pritykina, L. N. (1986) Two new dragonflies from the Lower Cretaceous deposits of West Mongolia (Anisoptera: Sonidae fam. nov., Corduliidae). *Odonatologica*, **15**, 169–84.

Rasnitsyn, A. P. (1985) *Eobraconus*, a substitute name for *Eobracon* Rasnitsyn (Hymenoptera, Braconidae). *Psyche*, **92** (1), p. 163.

Rasnitsyn, A. P. (1986) Pereponchatokrylye. Vespida (= Hymenoptera). [Wasps. Vespida (= Hymenoptera)], in *Nasekomye v Rannemelovykh Ekosistemakh Zapadnoi Mongolii. [Insects in the Early Cretaceous Ecosystems of Western Mongolia.] Transactions of the Joint Soviet – Mongolian Palaeontological Expedition*, **28**, pp. 154–64 [in Russian].

Rasnitsyn, A. P. (1988) An outline of evolution of the hymenopterous insects (Order Vespida). *Oriental Insects*, **22**, 115–45.

Rasnitsyn, A. P. (1990) Pereponchatokrylye. Vespida. [Hymenopterans. Vespida], in *Pozdnemezozoiskie Nasekomye Vostochnogo Zabaikalya. [Late Mesozoic Insects of Eastern Transbaikalia.]. Trudy* Paleontologicheskogo Instituta, **239**, pp. 177–205 [in Russian].

Rasnitsyn, A. P. (1991) Rannemelovye Predstaviteli Evaniomorfnykh Pereponchatokrylykh Nasekomykh Semeistv Stigmaphronidae i Cretevaniidae i Podsemeistva Kotujellitinae (Gasteruptiidae). [Early Cretaceous examples of Evanioid Hymenopterans of the insect families Stigmaphronidae and Cretevaniidae and subfamily Kotujellitinae (Gasteruptiidae).] Paleontologicheskii Zhurnal, **1991** (4), 128–32. [in Russian].

Rasnitsyn, A. P. and Kovalev, O. V. (1988) Drevneiskie orekhotrovki iz rannego mela Zabaikalya (Hymenoptera, Cynipoidea, Archaeocynipidae fam. n.). [Gall wasps from the Early Cretaceous of Transbaikalia (Hymenoptera, Cynipoidea,

Archaeocynipidae fam. n.).] *Vestnik Zoologii*, **1988** (1), 18–21 [in Russian].

Rasnitsyn, A. P. and Kozlov, M. V. (1990) Novaya gruppa iskopaemykh nasekomykh: skorpionnitsa s adaptsiyami tsikad i babochek. [A new group of fossil insects: Scorpionflies with the adaptations of bugs and butterflies.] *Doklady Akademii Nauk SSSR*, **310**, 973–6 [in Russian].

Rasnitsyn, A. P. and Michener, C. D. (1991) Miocene fossil bumble bee from the Soviet Far East with comments on the chronology and distribution of fossil bees (Hymenoptera: Apidae). *Annals of the Entomological Society of America*, **84**, 583–9.

Riek, E. F and Kukalová-Peck, J. (1984) A new interpretation of dragonfly wing venation based upon early Upper Carboniferous fossils from Argentina (Insecta: Odonatoidea) and basic character states in pterygote wings. *Canadian Journal of Zoology*, **62**, 1150–66.

Rodendorf, B. B. (1972) Devonian eopterids were not insects but eumalacostraceans (Crustacea). *Entomological Review*, **51**, 58–9.

Rodendorf, B. B. and Rasnitsyn, A. P. (eds) (1980) Istoricheskoe razvitie klassa nasekomykh. [Historical development of the Class Insecta]. *Trudy Paleontologicheskogo Instituta*, **175**, 269 pp. [in Russian].

Rohr, D. M., Boucot, A. J., Miller, J. and Abbott, M. (1986) Oldest termite nest from the Upper Cretaceous of west Texas. *Geology*, **14**, 87–8.

Rozefelds, A. C. (1985) A fossil zygopteran nymph (Insecta: Odonata) from the Late Triassic Aberdare Conglomerate, South-east Queensland. *Proceedings of the Royal Society of Queensland*, **96**, 25–32.

Sakai, S. and Fujiyama, I. (1989) New dermapteran fossil from Sado Islands, Japan with description of a new species. (Dermaptera, Diplatyidae). *Special Bulletin of Daito Bunka University*, **38**, 3102–3.

Schlee, D. (1990) Das Bernstein-Kabinett. *Stuttgarter Beiträge zur Naturkunde*, C, **28**, 1–100.

Schlüter, T. (1986) The fossil Planipennia–a review, in *Proceedings of the 2nd International Symposium on Neuropterology*, (eds J. Gepp, H. Aspöck and H. Hölzel), Hamburg, pp. 103–11.

Schneider, J. (1984) Die Blattodea (Insecta) des Paläozoikums Teil II: Morphogenese der Flugelstrukturen und Phylogenie. *Freiberger Forschungenshefte, C*, **391**, 5–34.

Shcherbakov, D. Y. (1984) Systematics and phylogeny of Permian Cicadomorpha (Cimicida and Cicadina). *Paleontological Journal*, **18** (2), 87–97.

Shcherbakov, D. Y. (1988a) Novye Mezozoiskie Ravnokrylye. [New Mesozoic Homoptera], in *Novye Iskopaemye Bespozvonochnye Mongolii. [New Fossil Invertebrates from Mongolia.]* (ed. A. Y. Rosanov), Transactions of the Joint Soviet–Mongolian Palaeontological Expedition, **33**, pp. 60–3.

Shcherbakov, D. Y. (1988b) New cicadas (Cicadina) from the later Mesozoic of Transbaikalia. *Paleontological Journal*, **22** (4), 52–63.

Shcherbakov, D. Y. and Vengerek, P. (1991) Creaphididae, a new and the oldest aphid family from the Triassic of middle Asia. *Psyche*, **98** (1), 81–5.

Shear, W. A. and Kukalová-Peck, J. (1990) The ecology of Paleozoic terrestrial arthropods: the fossil evidence. *Canadian Journal of Zoology*, **68**, 1807–34.

Shields, O. (1988) Mesozoic history and neontology of Lepidoptera in relation to Trichoptera, Mecoptera, and Angiosperms. *Journal of Paleontology*, **62**, 251–8.

Sinichenkova, N. D. (1985a) Yurskie podenki (Ephemerida–Ephemeroptera) yuzhnoi Sibiri i zapadnoi Mongolii. [Jurassic Mayflies (Ephemerida–Ephemeroptera) of southern Siberia and western Mongolia], in *Yurskie Nasekomye Sibiri i Mongolii. [Jurassic Insects of Siberia and Mongolia] (ed. A. P. Rasnitsyn)*, Trudy Paleontologicheskogo Instituta, **211**, pp. 11-23 [in Russian].

Sinichenkova, N. D. (1985b) Yurskie vesnyanki yuzhnoi Sibiri prilegayushchikh teritorii (Perlida = Plecoptera). [Jurassic stoneflies of southern Siberia and adjacent territories (Perlida = Plecoptera)], in *Yurskie Nasekomye Sibiri i Mongolii. [Jurassic Insects of Siberia and Mongolia] (ed. A. P. Rasnitsyn)*, Trudy Paleontologicheskogo Instituta, **211**, pp. 48–71 [in Russian].

Sinichenkova, N. D. (1987) Istoricheskoe razvitie vesnyanok. [Historical development of Stoneflies]. *Trudy Paleontologicheskogo Instituta*, **221**, 1–142 [in Russian].

Sinichenkova, N. D. (1989) New Mesozoic mayflies (Ephemerida) from Mongolia. *Paleontological Journal*, **23** (3), 26–37.

Sinichenkova, N. D. (1990a) Vesnyanki. Perlida. [Stoneflies. Perlida]. In *Pozdnemezozoiskie Nasekomye Vostochnogo Zabaikalya. [Late Mesozoic Insects of Eastern Transbaikalia.]* (ed. A. P. Rasnitsyn), Trudy Paleontologicheskogo Instituta, **239**, pp. 207–10 [in Russian].

Sinichenkova, N. D. (1990b) Novye Mezozoiskie vesnyanki Azii. [New Mesozoic stoneflies from Asia.] *Paleontologicheskii Zhurnal*, **1990** (3), 63–70 [in Russian].

Spahr, U. (1985) Ergänzungen und Berichtigungen zu R. KEILBACHS Bibliographie und Liste der Bernsteinfossilien–Ordnung Diptera. *Stuttgarter Beiträge zur Naturkunde, B*, **111**, 1–146.

Spahr, U. (1987) Ergänzungen und Berichtigungen zu R. KEILBACHs Bibliographie und Liste der Bernsteinfossilien – Ordnung Hymenoptera. *Stuttgarter Beiträge zur Naturkunde, B*, **127**, 1–121.

Spahr, U. (1988) Ergänzungen und Berichtigungen zu R. KEILBACHs Bibliographie und Liste der Bernsteinfossilien Überordnung Hemipteroidea. *Stuttgarter Beiträge zur Naturkunde, B*, **144**, 1–60.

Spahr, U. (1989) Ergänzungen und Berichtigungen zu R. KEILBACHs Bibliographie und Liste der Bernsteinfossilien – Überordnung Mecopteroidea. *Stuttgarter Beiträgē zur Naturkunde, B*, **157**, 1–87.

Spahr, U. (1990) Ergänzungen und Berichtigungen zu R. KEILBACHs Bibliographie und Liste der Bernsteinfossilien – "Apterygota". *Stuttgarter Beiträge zur Naturkunde, B*, **166**, 1–23.

Storozhenko, S. Y. (1988) New and little-known Mesozoic Grylloblattids. *Paleontological Journal*, **22** (4), 45–52.

Storozhenko, S. Y. (1990a) Permian fossil insects of North-East Europe: revision of the family Atactophlebiidae (Ins. Gerarida, Atactophlebiidae). *Deutsche Entomologische Zeitschrift*, **37**, 407–12.

Storozhenko, S. Y. (1990b) New Permian and Mesozoic insects (Insecta, Grylloblattida: Blattogryllidae, Geinitziidae) from Asia. *Paleontological Journal*, **24** (4), 53–61.

Storozhenko, S. Y. (1991) Grilloblattidovye nasekomye verzhnei permi vostochnogo kazakhstana. [Grylloblattidan insects from the Upper Permian of eastern Kazakhstan.] *Paleontologicheskii Zhurnal*, **1991** (2), 110–14 [in Russian].

Sukacheva, I. D. (1985) Yurskie rucheiniki yuzhnoi Sibiri. [Jurassic caddisflies of Southern Siberia], *in Yurskie Nasekomye Sibiri i Mongolii. [Jurassic Insects of Siberia and Mongolia.]* (ed. A. P. Rasnitsyn), Trudy Paleontologicheskogo Instituta, **211**, pp. 115–19 [in Russian].

Sukacheva, I. D. (1990a) Skorpionnitsy. Panorpida. [Scorpionflies. Panorpida.], in *Pozdnemezozoiskie Nasekomye Vostochnogo Zabaikalya. [Late Mesozoic Insects of Eastern Transbaikalia.]* (ed. A. P. Rasnitsyn), *Trudy Paleontologicheskogo Instituta*, **239**, pp. 88–94 [in Russian].

Sukacheva, I. D. (1990b) Rucheiniki. Phryganeida. [Caddisflies. Phryganeida], in *Pozdnemezozoiskie Nasekomye Bostochnogo Zabaikalya. [Late Mesozoic Insects of Eastern Transbaikalia]*. Trudy Paleontologicheskogo Instituta, **239**, pp. 94–122 [in Russian].

Tindale, N. B. (1985) A butterfly-moth (Lepidoptera, Castniidae) from the Oligocene Shales of Florissant, Colorado. *Journal of Research on the Lepidoptera*, **24** (1), 31–40.

Vengerek (Wegierek), P. (1989) New species of Mesozoic aphids (Shaposhnikoviidae, Homoptera). *Paleontological Journal*, **23** (4), 40–9.

Vengerek (Wegierek), P. (1991) O melovykh tlyakh semeistva Canadaphididae (Hemiptera, Aphidomorpha). [Cretaceous aphids of the Family Canadaphididae (Hemiptera, Aphidomorpha).] *Paleontologicheskii Zhurnal*, **1991** (2), 114–15 [in Russian].

Vishniakova, V. N. (1986) Uhovertki. Forficulida (=Dermaptera).

[Earwigs. Forficulida (=Dermaptera)], *in Nasekomye Rannemelovykh Ekosistemakh Zapadnoi Mongolii. [Insects in the Early Cretaceous Ecosystems of West Mongolia.] Transactions of the Joint Soviet–Mongolian Palaeontological Expedition*, **28**, p. 171 [in Russian].

Vulcano, M. A. and Pereira, F. S. (1987) Entomofauna fóssil da Chapada do Araripe, Ceará, Brasil *Cretaceimelittommoides cearensis* gen. nov. sp. nov. (Coleoptera: Pyrochroidae). *X Congresso Brasileiro de Paleontologia – Resumos das Comunicações*, **37**.

Whalley, P. E. S. (1985) The systematics and palaeogeography of the Lower Jurassic insects of Dorset, England. *Bulletin of the British Museum (Natural History)*, Geology **39**, 107–89.

Whalley, P. E. S. (1986a) Insects from the Italian Upper Trias. *Rivista del Museo Civico di Scienze Naturali 'Enrico Caffi'*, **10**, 51–60.

Whalley, P. E. S. (1986b) A review of current fossil evidence of Lepidoptera in the Mesozoic. *Biological Journal of the Linnean Society*, **28**, 253–71.

Whalley, P. E. S. (1988) Mesozoic Neuroptera and Raphidioptera (Insecta) in Britain. *Bulletin of the British Museum (Natural History), Geology*, **44**, 45–63.

Whalley, P. E. S. and Jarzembowski, E. A. (1985) Fossil insects from the Lithographic Limestone of Montsech (Late Jurassic–Early Cretaceous), Lérida Province, Spain. *Bulletin of the British Museum (Natural History), Geology*, **38**, 381–412.

Willmann, R. (1978) Mecoptera (Insecta, Holometabola). *Fossilium Catalogus 1: Animalia*, **124**, 1–139.

Willmann, R. (1989a) Rediscovered: *Permotipula patricia*, the oldest known fly. *Naturwissenschaften*, **76**, 375–7.

Willmann, R. (1989b) Evolution und phylogenetisches System der Mecoptera (Insecta: Holometabola). *Abhandlungen der Senckenbergischen Naturforschenden Gesellschaft*, **544**, 1–153.

Willmann, R. (1990) Insekten aus der Fur-Formation von Dänemark (Moler, ob. Paleozän/unt. Eozän?) 2. Dermaptera. *Meyniana*, **42**, 15–23.

Wilson, E. O. (1987) The earliest known ants: an analysis of the Cretaceous species and an inference concerning their social organization. *Paleobiology*, **13**, 44–53.

Wootton, R. J. (1981) Palaeozoic insects. *Annual Review of Entomology*, **26**, 319–44.

Wootton, R. J. (1988) The historical ecology of aquatic insects: an overview. *Palaeogeography, Palaeoclimatology, Palaeoecology*, **62**, 477–92.

Wootton, R. J. and Betts, C. R. (1986) Homology and function in the wings of Heteroptera. *Systematic Entomology*, **11**, 389–400.

Wootton, R. J. and Ennos, A. R. (1989) The implications of function on the origin and homologies of the dipterous wing. *Systematic Entomology*, **14**, 507–20.

Wunderlich, J. (1986) Liste der vom Baltischen und Dominikanischen Bernstein bekannten Familien fossiler Käfer (Coleoptera). *Entomologische Zeitschrift*, **96** (20), 298–301.

Zaytsev, V. F. (1986) New species of Cretaceous fossil bee flies and a review of paleontological data on the Bombyliidae (Diptera). *Entomological Review*, **66** (3), 1987, 150–9.

Zessin, W. (1985) Neue oberliassische Apocrita und die Phylogenie der Hymenoptera (Insecta, Hymenoptera). *Deutsche Entomologische Zeitschrift*, **32** (1–3), 129–42.

Zessin, W. (1987) Variabilität, Merkmalswandel und Phylogenie der Elcanidae im Jungpaläozoikum und Mesozoikum und die Phylogenie der Ensifera (Orthopteroida, Ensifera). *Deutsche Entomologische Zeitschrift*, **34** (1–3), 1–76.

Zessin, W. (1991) Die Phylogenie der Protomyrmeleontidae unter einbeziehung neuer oberliassischer funde (Odonata: Archizygoptera sens. nov.). *Odonatologica*, **20** (1), 97–125.

Zhang, J. (1985) New data on the Mesozoic fossil insects from Laiyang in Shandong. *Geology of Shandong*, **1** (2), 23–39 [in Chinese, English summary].

Zhang, J. (1986a) Luanpingitidae – a new fossil insect family. *Acta Palaeontologica Sinica*, **25** (1), 49–54 [in Chinese, English summary].

Zhang, J. (1986b) Some fossil insects from the Jurassic of Northern Hebei, China, in *The Paleontology and Stratigraphy of Shandong*. Paleontological Society of Shandong [in Chinese, English summary], pp. 74–81.

Zhang, J. (1987) Four new genera of Platypezidae. *Acta Palaeontologica Sinica*, **26**, 595–603 [in Chinese, English summary].

Zhang, J. (1989) *Fossil Insects from Shanwang, Shandong, China*. Shandong Science and Technology Publishing House, Jenan, China, 459 pp. [in Chinese, English summary].

Zhang, J. (1991a) A new family of Neuroptera (Insecta) from the Late Mesozoic of Shandong, China. *Science in China*, B, **34**, 1105–111.

Zhang, J. (1991b) Going further into late Mesozoic mesolygaeids (Heteroptera, Insecta). *Acta Palaeontologica Sinica*, **30**, 679–704 [in Chinese, English summary].

Zhang, J., Zhang, S., Lui, D. *et al.* (1986) Fossil insects (Diptera, Nematocera) of Laiyang Basin in Shandong Province. *Geology of Shandong*, **2** (1), 14–39 [in Chinese, English summary].

Zhang, J., Zhang, S., Hou, F. *et al.* (1989) Late Jurassic aphids (Homoptera, Insecta) from Shandong Province, China. *Geology of Shandong*, **5** (1), 28–46 [in Chinese, English summary].

Zherikhin, V. V. (1985a) Yurskie pryamokrylye yuzhnoi Sibiri i zapadnoi Mongolii (Gryllida = Orthoptera). [Jurassic grasshoppers of southern Siberia and western Mongolia (Gryllida = Orthoptera)], in *Yurskie Nasekomye Sibiri i Mongolii. [Jurassic Insects of Siberia and Mongolia.]* (ed. A. P. Rasnitsyn), Trudy Paleontologicheskogo Instituta, **211**, pp. 171–84.

Zherikhin, V. V. (1985b) Nasekomye. [Insects.], in *Jurskie Kontinentalnye Biotsenozy Yuzhnoi Sibiri i Sopredelnykh Territorii. [Jurassic Continental Biocoenoses in Southern Siberia and Adjacent Territories.]* (ed. A. P. Rasnitsyn), Trudy Paleontologicheskogo Instituta, **213**, pp. 100–31.

22

BRACHIOPODA

D. A. T. Harper, C. H. C. Brunton, L. R. M. Cocks, P. Copper, E. N. Doyle,
A. L. Jeffrey, E. F. Owen, M. A. Parkes, L. E. Popov and C. D. Prosser

The initial review of the first and last familial occurrences within the phylum (Ager *et al.*, 1967) post-dated publication of the first edition of the brachiopod part of the *Treatise on Invertebrate Paleontology* (Williams *et al.*, 1965). Range coverage was uneven. Data for nine superfamilies of inarticulate and three superfamilies of rhynchonellide were provided with limited familial information, while only superfamily data were reported for the orthides and strophomenides. Fuller information was listed for 30 families of spire-bearing brachiopod (including two families of thecideides) and 14 families of terebratulides. Moreover, since publication of the *Treatise*, there has been an accelerated proliferation of taxa (see Grant, 1980) together with a move towards cladistic classification within the group (Carlson, 1989; Holmer, 1991c). Consequently, data now are presented here for over 275 families across over 50 superfamilies. The split of the phylum into three classes, the Lingulata, Inarticulata and Articulata, is followed (Gorjansky and Popov, 1986).

Since the *Treatise*, there has been no comparable review of the phylum; moreover, few of the main groups within the phylum have been revised, as a whole, to account for new data. Accordingly, some contributors have not included data pertaining to post-*Treatise* families, while others have attempted to include as many verifiable families as possible. Most authors have presented data within accepted taxonomic structures; however, in a few cases, families have been listed alphabetically or in order of stratigraphical appearance.

There are many unresolved problems of correlation relevant to both well-known and less well-known faunas and their horizons, which, together with the uneven precision of taxonomic structure across the phylum, have presented major problems. Nevertheless, the project is finite and the information is presented within inevitable stratigraphical and systematic constraints. The *Treatise* (Williams *et al.*, 1965) remains the main source reference for the phylum while Doescher (1981 and updates) provides important generic lists and bibliographic data.

Acknowledgements – DATH and MAP thank Drs L. E. Holmer and L. E. Popov for their detailed comments on the non-articulate groups and Professor A. D. Wright for his comments on the Orthida. DATH thanks Professor P. G. Baker for detailed comments and new information on the Thecideida, Professor N. M. Savage for comments on the Palaeozoic Rhynchonelloidea and Dr R. E. Grant for detailed comments on the Stenoscismatoidea; DATH and ALJ thank Dr C. H. C. Brunton for his comments on the Spiriferida. CHCB thanks Dr P. R. Racheboeuf for helpful comments on the chonetid section, and Dr R. E. Grant for comments on the productid section. CDP is grateful to Drs D. E. Lee and E. F. Owen for their comments on the post-Palaeozoic Rhynchonellida. Professor Rong Jia-yu commented on parts of the entire manuscript. DATH thanks Professor D. L. Bruton for facilities at the Paleontologisk Museum, Oslo during final compilation and editing of the chapter, and Rex Doescher for access to his Brachiopod Bibliographic File in the Department of Paleobiology, Smithsonian Institution, Washington DC.

Phylum BRACHIOPODA Duméril, 1806

THE NON-ARTICULATE BRACHIOPODS

This informal grouping includes the classes Lingulata and Inarticulata of Gorjansky and Popov (1985).

Class LINGULATA Gorjansky and Popov, 1985

Order LINGULIDA Waagen, 1885

Superfamily LINGULOIDEA Menke, 1828

The classification of the Linguloidea is difficult, compounded by numerous generic assignments based only on outline and shape and not on internal morphology. Future revision may require substantial changes of some familial assignments and range data.

F. LINGULIDAE Menke, 1828 D. (u)–Rec. Mar.

First: Not positively known, since many pre-Devonian species are poorly documented, particularly the interiors

The Fossil Record 2. Edited by M. J. Benton. Published in 1993 by Chapman & Hall, London. ISBN 0 412 39380 8

QU.	HOL			
	PLE			
TERTIARY	PLI			
	UMI			
	MMI			
	LMI			
	CHT			
	RUP			
	PRB			
	BRT			
	LUT			
	YPR			
	THA			
	DAN			
CRETACEOUS	MAA			
	CMP			
	SAN			
	CON			
	TUR			
	CEN			
	ALB			
	APT			
	BRM			
	HAU			
	VLG			
	BER			
JURASSIC	TTH			
	KIM			
	OXF			
	CLV			
	BTH			
	BAJ			
	AAL			
	TOA			
	PLB			
	SIN	1	14	22
	HET			

Key for both diagrams

LINGULIDA	CHILEIDA
1. Lingulidae	27. Chileidae
2. Obolidae	NAUKATIDA
3. Elkaniidae	28. Naukatidae
4. Lingulasmatidae	ORTHIDA
5. Lingulellotretidae	29. Billingsellidae
6. Zhanatellidae	30. Nisusiidae
7. Andobolidae	31. Protorthidae
8. Paterulidae	32. Eoorthidae
ACROTRETIDA	33. Hesperonomiidae
9. Acrotretidae	34. Orthidiellidae
10. Curticidae	35. Toxorthidae
11. Acrothelidae	36. Orthidae
12. Botsfordiidae	37. Hesperorthidae
13. Trematidae	38. Plaesiomyidae
14. Discinidae	39. Cremnorthidae
15. Siphonotretidae	40. Leioridae
16. Eoconulidae	41. Alimbellidae
PATERINIDA	42. Bohemiellidae
17. Paterinidae	43. Ranorthidae
CRANIOPSIDA	44. Nanorthidae
18. Craniopsidae	45. Euorthisinidae
TRIMERELLIDA	46. Poramborthidae
19. Trimerellidae	47. Finkelnburgiidae
20. Ussuniidae	48. Plectorthidae
21. Adensuidae	49. Skenidiidae
CRANIIDA	ENTELETIDA
22. Craniidae	50. Enteletidae
KUTORGINIDAE	51. Draboviidae
23. Kutorginidae	52. Chrustenoporidae
24. Yorkiidae	53. Schizophoriidae
25. Agrekiidae	54. Proschizophoriidae
OBOLELLIDA	55. Linoporellidae
26. Obolellidae	

Fig. 22.1

which are needed to confirm assignments; for example *Lingula* sp. from the Upper Whitehouse of the Girvan district, SW Scotland (Harper, 1984) may be better assigned to the pseudolingulinids (Holmer, 1991b). The first is probably *Apsilingula parkesensis* Williams, 1977, Mandagery Sandstone, New South Wales, Australia (Williams, 1977).

Extant

F. OBOLIDAE King, 1846 €. (CRF)–C. (BSK/MOS) Mar.

First: There are many cosmopolitan occurrences of *Lingulella*: precise relative ages of the majority of older species being in doubt and not better constrained than lower Cambrian. The first is probably *Lingulella linguata* Pelman, 1977, Tiuser Formation and *L. variabilis* Pelman, 1977, Erkekty Formation of the Siberian Platform, former USSR (Pelman, 1977).

Last: *Trigonoglossa scotica* (Davidson, 1860), Harden Beds (Middle Border Group), Black Metals Marine Band (Limestone Coal Group) and Calmy Limestone (Upper Limestone Group), Midland Valley of Scotland (Graham, 1970). *Lachrymula pringlei* Graham, 1970, Skipsey's Marine Band, Midland Valley of Scotland, is only doubtfully referred to the obolids.

F. ELKANIIDAE Walcott and Schuchert, 1908 €. (MER)–O. (ASH) Mar.

First: Probably *Elkania hamburgensis* (Walcott, 1884), Upper Cambrian, Nevada, USA (Ulrich and Cooper, 1938) or possibly *Broeggeria salteri* (Holl, 1865), Upper Cambrian, Malvern Hills, England, UK (Cocks, 1978).

Last: *Tilasia rugosa* Holmer, 1991, Boda Limestone, Siljan district, Dalarna, Sweden (Holmer, 1991a).

F. LINGULASMATIDAE Winchell and Schuchert, 1893 O. (LLN–ASH) Mar.

First: *Lingulasma crassum* (Eichwald, 1829), Kunda rocks, east Baltic (Popov and Pushkin, 1986).

Last: Possibly *Lingulasma schucherti* Ulrich, 1889, Richmond Formation, Illinois, USA (Ulrich, 1889).

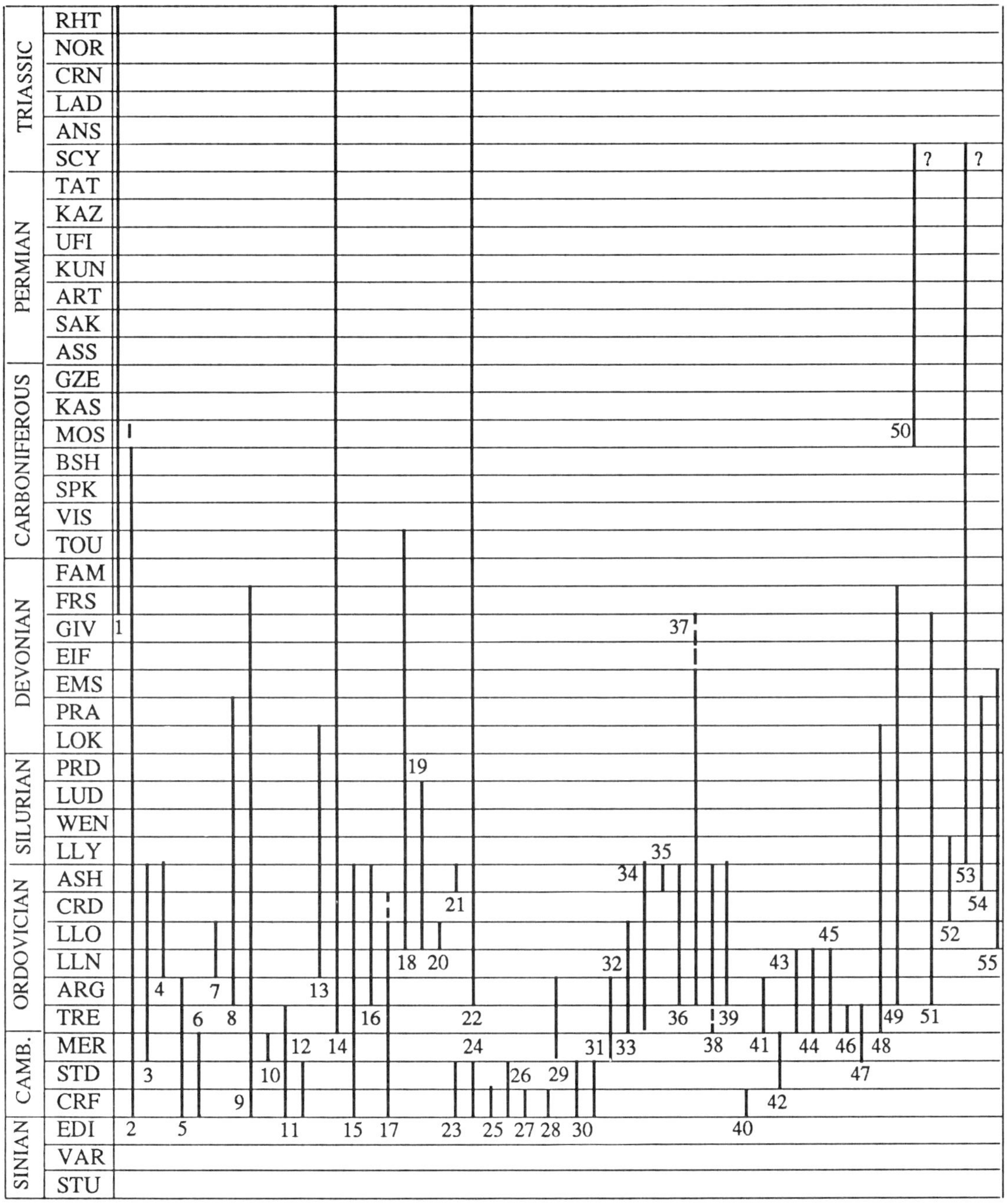

Fig. 22.1

F. LINGULELLOTRETIDAE Koneva, 1983
Є. (CRF)–O. (ARG) Mar.

First: *Lingulellotreta ergalievi* Koneva, 1983, Lower Cambrian, Malyi Karatau, southern Kazakhstan, former USSR (Gorjansky and Koneva, 1983).
Last: *Mirilingula* aff. *mutabilis* Popov, 1983, Lower Ordovician, Malyi Karatau, southern Kazakhstan (Koneva and Popov, 1983).

F. ZHANATELLIDAE Koneva, 1986
Є. (CRF–MER) Mar.

First: *Kyrshabaktella belli* (Pelman, 1977), Lower Cambrian, Siberian Platform (Koneva, 1986).
Last: *Zhanatella rotunda* Koneva, 1986, Selety Horizon, Kazakhstan (Koneva, 1986).

F. ANDOBOLIDAE Kozłowski, 1930 O. (?LLN/LLO) Mar.

First and Last: *Andobolus jackowskii* Kozłowski, 1930. This family is monotypic and apparently restricted to the Ordovician of Bolivia, South America (Kozłowski, 1930).

F. PATERULIDAE Cooper, 1956
O. (ARG)–D. (PRA) Mar.

First: *Paterula* sp., Volkhov rocks, east Baltic (Gorjansky, 1969).
Last: Possibly *Paterula nana* Wolfart, 1969, 'Unterdevonischen' black shales, northern Thailand (Jaeger *et al.*, 1969). [MAP]

Order ACROTRETIDA Kuhn, 1949

Suborder ACROTRETIDINA Kuhn, 1949

Superfamily ACROTRETOIDEA Schuchert, 1893

F. ACROTRETIDAE Schuchert, 1893 €. (ATB)–D. (FRS) Mar.

First: *Linnarssonia* sp., uppermost Atdabanian, Sinjaja River, Siberian Platform, former USSR (Rozanov and Sokolev, 1984).
Last: *Opisconidion arcticon* Ludvigsen, 1974, Michelle Formation, northern Yukon is the last described species (Ludvigsen, 1974), but there are records of torynelasmatines from the Adorf Stufe (FRS) of Germany (Langer, 1971).

F. EOCONULIDAE Rowell, 1965 O. (ARG–ASH) Mar.

First: Although the genus has been reported from the upper Cambrian (Popov, *in* Holmer, 1989) the earliest records to date are from Lower Ordovician rocks, for example, *Eoconulus cryptomyus* Gorjansky, 1969, Volkhov rocks, west Estonia, former USSR (Biernat, 1973).
Last: *Eoconulus* sp., Kildare Limestone Formation, County Kildare, Republic of Ireland (McClean, 1988).

F. CURTICIDAE Walcott and Schuchert, 1908 €. (MER) Mar.

First: *Curticia minuta* Bell, 1941, Holm Dal Formation, Greenland (Zell and Rowell, 1988) and Dresbachian, Montana, USA (Rowell and Bell, 1961).
Last: *Curticia elegantula* Walcott, 1905, Franconian, Minnesota, USA (Rowell and Bell, 1961).

F. ACROTHELIDAE Walcott and Schuchert, 1908 €. (CRF)–O. (TRE) Mar.

First: A number of species of *Acrothele* are known from rocks of early Cambrian age but are poorly constrained; better constrained is *Spinulothele dubia* (Walcott, 1912), Poleta Formation, Nevada, USA (Rowell, 1977).
Last: There are a number of TRE species of *Orbithele* recorded, for example, *O. contraria* (Barrande, 1868) from Czechoslovakia and *O. bicornis* Biernat, 1973 from the Holy Cross Mountains (Biernat, 1973); definitely *Orbithele undulosa* (Barrande, 1868), Klabava Formation, Bohemia, Czechoslovakia (Mergl, 1981).

F. BOTSFORDIIDAE Schindewolf, 1955 €. (ATB–STD) Mar.

First: *Botsfordia caelata* (Hall, 1847), uppermost Atdabanian, Sinjaja River, Siberian Platform, former USSR (Rozanov and Sokolev, 1984).
Last: *B. epigona* Mergl, 1988, lower Middle Cambrian, Morocco (Mergl, 1988).

Superfamily DISCINOIDEA Gray, 1840

F. DISCINIDAE Gray, 1840 O. (TRE)–Rec. Mar.

First: Possibly *Orbiculoidea? subovalis* Biernat, 1973, TRE, Wysoczki, Holy Cross Mountains (Biernat, 1973). **Extant**

F. TREMATIDAE Schuchert, 1893 O. (LLN)–D. (LOK) Mar.

First: *Schizocrania salopiensis* Williams, 1974, upper Llanvirn, Shelve Inlier, Shropshire (Williams, 1974).
Last: Possibly *Schizocrania? helderbergensis* (Hall, 1892), lower Helderberg Group, eastern USA.

Superfamily SIPHONOTRETOIDEA Kutorga, 1848

F. SIPHONOTRETIDAE Kutorga, 1848 €. (LEN)–O. (RAW) Mar.

First: *Dysoristus belli* Pelman, 1977, Lenian, Siberian Platform may be the oldest (Pelman, 1977); definitely *Schizambon reticulatus* MacKinnon, 1976, Elandinskii Horizon, Altai Mountains (Aksarina and Pelman, 1978).
Last: *Multispinula drummuckensis* Harper, 1984, South Threave Formation, Girvan district, SW Scotland, UK (Harper, 1984).

Order PATERINIDA Rowell, 1965

Superfamily PATERINOIDEA Schuchert, 1893

F. PATERINIDAE Schuchert, 1893 €. (TOM)–O. (LLO/CRD) Mar.

First: *Aldanotreta sunnaginensis* Pelman, 1977, Tommotian, Siberian Platform (Pelman, 1977).
Last: *Dictyonites fredriki* Holmer, 1989, Furudal Limestone, Dalarna, Sweden and *D. perforata* Cooper, 1956, Pratt Ferry Formation, Alabama, southern Appalachians, USA, are approximately coeval; the tentative record of the genus from the Dalby Limestone, Dalarna, Sweden may be the latest occurrence (Holmer, 1989).

Class INARTICULATA Huxley, 1869

Order CRANIOPSIDA Gorjansky and Popov, 1985

Superfamily CRANIOPSOIDEA Williams, 1963

F. CRANIOPSIDAE Williams, 1963 O. (LLO)–C. (TOU) Mar.

Comment: This family was elevated to a separate Order of the Class Inarticulata by Gorjansky and Popov (1985).
First: *Pseudopholidops pusilla* (Eichwald, 1829), Kukruse rocks, East Baltic (Popov and Pushkin, 1986).
Last: Possibly *Craniops* cf. *hamiltoniae* (Hall, 1860), Bedford Shales, Bedford, Ohio, USA.

Order TRIMERELLIDA Gorjansky and Popov, 1985

This group was given ordinal status by Gorjansky and Popov (1985).

Superfamily TRIMERELLOIDEA Davidson and King, 1872

F. TRIMERELLIDAE Davidson and King, 1872 O. (LLO)–S. (LUD) Mar.

First: *Palaeotrimerella superba* Nikitin and Popov, 1984, Bestamak Formation, Bestamak, Chingiz Range, Kazakhstan, former USSR (Nikitin and Popov, 1984).
Last: The group declined within the Ludlow; nevertheless *Trimerella lindstroemi* Dall, 1870, Klinteberg Beds, Gotland together with *T. ohioensis* Meek, 1871, Niagara Group, New York, USA, and *T. grandis* Billings, 1862, Guelph Group, Ontario, Canada and others are locally common during this epoch.

F. USSUNIIDAE Nikitin and Popov, 1984 O. (LLO) Mar.

First and Last: *Ussunia incredibilis* Nikitin and Popov, 1984, Bestamak Formation, Bestamak, Chingiz Range, Kazakhstan, former USSR (Nikitin and Popov, 1984).

F. ADENSUIDAE Popov and Rukavishnikova, 1986 O. (PUS) Mar.

First and Last: *Adenus monstratum* Popov and Rukavishnikova, 1986, Dulankara Formation, Chu-Ili Range, former Kazakhstan (Popov and Rukavishnikova, 1986).

Order CRANIIDA Waagen, 1885

Gorjansky and Popov (1985) have given this group ordinal status.

Superfamily CRANIOIDEA Menke, 1828

F. CRANIIDAE Menke, 1828 O. (ARG)–Rec. Mar.

First: Cambrian records, for example, *Philhedra columbiana* (Walcott, 1908), Middle Cambrian of British Columbia inadequately known; *Pseudocrania petropolitana* (Pander, 1830), Volkhov rocks, east Baltic may be oldest (Gorjansky, 1969). [DATH] **Extant**

POSITION AND STATUS UNCERTAIN

The following four orders probably have more in common with the Articulata than the non-articulate groups.

Order KUTORGINIDA Kuhn, 1949

Gorjansky and Popov (1985) tentatively parked the kutorginides between their non-articulates and the articulates.

Superfamily KUTORGINOIDEA Schuchert, 1893

F. KUTORGINIDAE Schuchert, 1893 €. (TOM–STD) Mar.

First: *Khasagtina primaria* Ushatinskaya, 1987, uppermost Tommotian, Salany-Gol Rivulet, west Mongolia (Ushatinskaya, 1987) and possibly *Kutorgina peculiaris* (Tate), upper Tommotian or lower Atdabanian, South Australia (Daily, 1956) or *K.? anglica* Cobbold and Pocock, 1934, Malvern Quartzite, England, UK (Rushton, 1974).
Last: There are a number of poorly constrained Middle Cambrian *Kutorgina* species, but *Kutorgina amzassica* Aksarina, 1978, Mundybashskii Horizon, Altai Mountains (Aksarina and Pelman, 1978) may be youngest.

F. YORKIIDAE Rowell, 1962 €. (CRF–STD) Mar.

First: Possibly *Yorkia wanneri* Walcott, 1897, Lower Cambrian, Emigsville and York, Pennsylvania, USA (Rowell, 1962).
Last: *Hauperia tasmani* MacKinnon, 1983, Hauperi Group, Nelson, New Zealand (MacKinnon, 1983).

F. AGYREKIIDAE Koneva, 1979 €. (BOT) Mar.

First and Last: *Agyrekia alta* Koneva, 1978, Bajanaul Horizon, Agyrek Mountains, Kazakhstan, former USSR (Koneva, 1978).

Order OBOLELLIDA Rowell, 1965

Gorjansky and Popov (1975) classified the calcareous-shelled obolellides with the articulates.

Superfamily OBOLELLOIDEA Walcott and Schuchert, 1908

F. OBOLELLIDAE Walcott and Schuchert, 1908 €. (ATB–SOL) Mar.

First: *Obolella chromatica* Billings, 1861, lower Atdabanian, Siberia Platform, former USSR (Pelman, 1977).
Last: Probably *Alisina sibirica* (Aksarina, 1978), Mundybashskii Horizon, Altai Mountains (Aksarina, 1978) or species of *Trematobolus*, for example, *T. simplex* (Vogel, 1962), Murero Formation, Zaragoza, Spain.

Order CHILEIDA Popov and Tikhonov, 1990

Popov and Tikhonov (1990) questionably assigned this order to the Inarticulata.

Superfamily CHILEIDOIDEA Popov and Tikhonov, 1990

F. CHILEIDAE Popov and Tikhonov, 1990 €. (BOT–LEN) Mar.

First: *Chile mirabilis* Popov and Tikhonov, 1990, Botomian rocks, Alai Range, south Kirgizia, former USSR (Popov and Tikhonov, 1990).
Last: *Acareorthis jelli* Roberts, 1990, Coonigan Formation, New South Wales, Australia (Roberts and Jell, 1990).

Order NAUKATIDA Popov and Tikhonov, 1990

Popov and Tikhonov (1990) assigned this order to the Articulata.

Superfamily NAUKATOIDEA Popov and Tikhonov, 1990

F. NAUKATIDAE Popov and Tikhonov, 1990 €. (BOT–LEN) Mar.

First: *Naukat proprium* Popov and Tikhonov, 1990 and *Oina rotunda* Popov and Tikhonov, 1990, Botomian, Alai Range, south Kirgizia (Popov and Tikhonov, 1990).
Last: *Bynguanoia perplexa* Roberts, 1990, Coonigan Formation, New South Wales, Australia (Roberts and Jell, 1990). [LEP and DATH]

Class ARTICULATA Huxley, 1869

Order ORTHIDA Schuchert and Cooper, 1932

Suborder ORTHIDINA Schuchert and Cooper, 1932

Superfamily BILLINGSELLOIDEA Schuchert, 1893

F. BILLINGSELLIDAE Schuchert, 1893 €. (STD)–O. (ARG) Mar.

First: Possibly *Billingsella destombesi* Mergl, 1983, Morocco (Mergl, 1983).
Last: Probably *Eosostrematorthis sinenesis* Wang (1955) from the Liangchiashan Limestone, Liaoning, northern China.

Superfamily NISUSIOIDEA Andreeva, 1987

F. NISUSIIDAE Walcott and Schuchert, 1908 €. (CRF–STD) Mar.

First: *Nisusia festinata* (Billings, 1861), Virginia, USA (Cooper, 1936).
Last: *Nisusia sulcata* Rowell and Caruso, 1985, Middle Cambrian, Utah, USA (Rowell and Caruso, 1985) or *Nisusia deissi* Bell, 1965, Middle Cambrian, Montana, USA (Bell, 1965).

Superfamily PROTORTHOIDEA Cooper, 1976

This family was raised to a superfamily by Cooper (1976) who described the earliest species of three new genera (*Glyptoria, Israelaria, Psiloria*) from rocks of early Cambrian age.

F. PROTORTHIDAE Schuchert and Cooper, 1931
Є. (CRF–STD) Mar.

First: *Psiloria dayi* Cooper, 1976, Burj Formation, Dead Sea, Jordan (Cooper, 1976).
Last: Possibly *Jamesella perpasta* (Pompeckj, 1896), Jince Formation, Bohemia, Czechoslovakia (Havlíček, 1977).

Superfamily ORTHOIDEA Woodward, 1852

F. EOORTHIDAE Walcott, 1908
Є. (MER)–O. (ARG) Mar.

First: Possibly *Diraphora venzlaffi* Wolfart, 1974, Lower Cambrian, Iran (Wolfart, 1974).
Last: *Jivinella slaviki* (Kloucek, 1915), Klabava Formation, Bohemia (Havlíček, 1977).

F. HESPERONOMIIDAE Ulrich and Cooper, 1936
O. (TRE–LLO) Mar.

First: Ulrich and Cooper (1936) recorded four species of *Hesperonomia*, for example, *H. planidorsalis*, Sarbach Formation (Lower Ordovician, middle TRE), Alberta, Canada.
Last: Possibly *Hesperonomia orientalis* (Su, 1976), Middle Ordovician, Inner Mongolia (Nei *et al.*, 1976).

F. ORTHIDIELLIDAE Ulrich and Cooper, 1936
O. (TRE–LLN) Mar.

First: Probably *Orthidium gemmiculum* (Billings, 1865), Upper Canadian rocks, Quebec, Canada (Cooper, 1956).
Last: Probably *Orthidium bellulum* Ulrich and Cooper, 1938, Upper Pogonip Group, Nevada, USA (Cooper, 1956).
Comment: *Portranella angulocostellata* Wright, 1964, Portrane Limestone, Portrane, eastern part of Irish Republic (Wright, 1964) has been assigned to the Enteletida (Hiller, 1980).

F. TOXORTHIDAE Rong, 1984 O. (HIR) Mar.

First and Last: *Toxorthis mirabilis* Rong, 1979, Kuanyinchiao Beds, Yichang, western Hubei and Changning, southern Szechwan, China and *T. proetus* morphs A and B Temple, 1968, Hirnantian, Keisley, Westmorland, England, UK (Rong, 1984).

F. ORTHIDAE Woodward, 1852 O. (ARG)–(RAW) Mar.

Since Havlíček (1977) separated the Ranorthidae, Nanorthidae and Bohemiellidae from the Orthidae, this family is now probably restricted to the Ordovician.
First: Many genera are currently no better constrained in age than early Ordovician, but see, for example, *Paralenorthis alata* (Sowerby, 1839), lower ARG, South Wales (Bates, 1969).
Last: *Nicolella asteroidea* (Reed, 1917), South Threave Formation (RAW), Girvan, Scotland, UK (Harper, 1984); *Toxorthis proteus* Temple, 1968, from Keisley is HIR (late ASH) in age, but was only questionably assigned to the Orthidae (Temple, 1968) and is now reassigned to the Toxorthidae.

F. HESPERORTHIDAE Schuchert and Cooper, 1931
O. (ARG)–D. (EMS) or (?EIF/GIV) Mar.

This family includes the Dolerorthinae, Hesperorthinae and Glyptorthinae as subfamilies, although some authors have raised them to family level.
First: *Lomatorthis mimula* Williams and Curry, 1985, Tourmakeady Limestone, south Mayo, Republic of Ireland (Williams and Curry, 1985).
Last: *Ptychopleurella reviviscens* Havlíček, 1977, Zlíchov Limestone, Bohemia, Czechoslovakia (Havlíček, 1977) or *Dolerorthis bumtungensis* Gupta, 1973, Middle Devonian (Gupta, 1973) – possibly spurious.

F. PLAESIOMYIDAE Schuchert, 1913
O. (TRE/ARG–ASH) Mar.

Several different genera could possibly be the first, and the last occurrences in this family, with few better constrained than either Lower Ordovician or Upper Ordovician.
First: Possibly *Valcourea intercarinata* (Ulrich and Cooper, 1938), Pogonip Group, Nevada, USA (Cooper, 1956).
Last: Possibly *Plaesiomys* aff. *porcata* (M'Coy, 1846), High Mains Formation, Girvan, SW Scotland, UK (Harper, 1984).

F. CREMNORTHIDAE Williams, 1963
O. (ARG–CAU) Mar.

First: Possibly *Phragmorthis mucronata* Williams and Curry, 1985, Tourmakeady Limestone, Mayo, Republic of Ireland (Williams and Curry, 1985).
Last: *Septorthis engurensis* Hints, 1973, Middle Ordovician, west Latvia, former USSR (Hints, 1973) or possibly *Cremnorthis* sp. B, Killey Bridge Formation, Pomeroy, Northern Ireland (Mitchell, 1977).

F. LEIORIDAE Cooper, 1976 Є. (CRF) Mar.

First and Last: Probably *Leioria bentori* Cooper, 1976, Nimra Formation, southern Negev, Israel (Cooper, 1976).

F. ALIMBELLIDAE Andreeva, 1960 emended Williams, 1974 O. (TRE–ARG) Mar.

First: *Alimbella armata* Andreeva, 1960, TRE, Urals, former USSR (Andreeva, 1960).
Last: *Astraborthis uniplicata* Williams, 1974, Shelve district, England, UK (Williams, 1974).

F. BOHEMIELLIDAE Havlíček, 1977
Є. (STD–MER) Mar.

First: Many species of the genera assigned to this family by Havlíček (1977) first occurred in the middle Cambrian. Stratigraphical records are too imprecise to name the first: as an example, *Bohemiella romingeri* (Barrande, 1879), Jince Formation, Bohemia, Czechoslovakia (Havlíček, 1977).
Last: *Shiragia biloba* Kobayashi, 1935, Chuangia Zone, South Korea (Kobayashi, 1935).

F. RANORTHIDAE Havlíček, 1977
O. (TRE–LLN) Mar.

First: *Ranorthis prima* Havlíček, 1977, lower TRE, Bohemia, Czechoslovakia (Havlíček, 1977).
Last: *Eodalmanella socialis* (Barrande, 1879), Šárka Formation, Bohemia, Czechoslovakia (Havlíček, 1977).

F. NANORTHIDAE Havlíček, 1977
O. (TRE–LLN) Mar.

First: *Nanorthis hamburgensis* (Walcott, 1884), Nevada and Colorado, USA (Ulrich and Cooper, 1938).
Last: *Trondorthis strandi* Neuman, 1974, Hølonda Limestone, Trondheim Region, Norway, or *Trondorthis bifurcatus* (Cooper, 1956), Antelope Valley Limestone, Nevada, USA (Neuman and Bruton, 1974).

F. EUORTHISINIDAE Havlíček, 1977 O. (TRE–LLN) Mar.

First: *Notorthisina notoconcha* Havlíček, 1977, Bolivia (Havlíček and Branisa, 1980).
Last: *Euorthisina minor* Havlíček, 1951, Bohemia, Czechoslovakia (Havlíček, 1951).

F. PORAMBORTHIDAE Havlíček, 1951 O. (TRE) Mar.

First: *Poramborthis grimmi* (Barrande, 1879), or *Poramborthis anomala* Havlíček, 1977, both Třenice Formation (lower TRE), Bohemia, Czechoslovakia (Havlíček, 1977).
Last: *Poramborthis klouceki* Havlíček, 1977, Mílina Formation (upper TRE), Bohemia, Czechoslovakia (Havlíček, 1977).

Superfamily PLECTORTHOIDEA Schuchert and Le Vene, 1929

According to Havlíček (1977), the families Plectorthidae, Finkelnburgiidae and Skenidiidae are members of this superfamily, previously treated as separate families within the superfamily Orthoidea of the Treatise.

F. FINKELNBURGIIDAE Schuchert and Cooper, 1931 €. (MER)–O. (TRE) Mar.

The stratigraphical records of genera belonging to this family are not well enough constrained to determine first and last occurrences with certainty.
First: *Finkelnburgia buttsi* Ulrich and Cooper, 1938, Upper Cambrian, Virginia, USA (Ulrich and Cooper, 1938).
Last: *Finkelnburgia arbucklensiformis* Severgina, 1985, Lower Ordovician, Gorny Altai (Severgina, 1985).

F. PLECTORTHIDAE Schuchert and Le Vene, 1929 O. (TRE)–D. (LOK) Mar.

First: *Plectorthis simplex* Havlíček, 1971, Upper TRE, Morocco (Havlíček, 1971a).
Last: Amsden (1968) notes several species of *Orthostrophia* of LOK age, for example, *Orthostrophia strophomenoides* (Hall, 1857), Helderberg Group, New York, USA (Schuchert and Cooper, 1932).

F. SKENIDIIDAE Kozłowski, 1929 O. (ARG)–D. (FRS) Mar.

First: Probably *Protoskenidioides minor* Xu and Liu, 1984 from the Meitan Formation, northern Guizhou, SW China or *Protoskenidioides revelata* Williams, 1974 from the Mytton Flags, Shelve district, England, UK (Williams, 1974).
Last: *Skenidium asellatum* Veevers, 1959, Sadler Formation, Fitzroy Basin, Western Australia (Veevers, 1959). [MAP]

THE PUNCTATE ORTHIDES

This diverse group, probably derived polyphyletically from the non-punctate orthides, is considered in terms of two superfamilies following Havlíček (1977).

Order ENTELETIDA Waagen, 1884

Superfamily ENTELETOIDEA Waagen, 1884

F. ENTELETIDAE Waagen, 1884 C. (MOS)–Tr. (GRI) Mar.

First: *Enteletes lamarcki* (Fischer von Waldheim, 1825), Westphalian, Spain, Austria, and the former USSR (Martinez-Chacon, 1979).
Last: *Enteletes dzhargrensis* Sokolskaja, 1965, Induan, Transcaucasus (Grant, 1970).

F. DRABOVIIDAE Havlíček, 1950 O. (ARG)–D. (GIV) Mar.

First: *Nocturniella nocturna* (Barrande, 1879) and *N. bachori* Havlíček, 1977, Klabava Formation, Zbiroh and Klabav, respectively, Czechoslovakia.
Last: *Spenophragmus nana* Imbrie, 1959, GIV, Michigan, USA.

F. CHRUSTENOPORIDAE Baarli, 1988 O. (CRD)–S. (LLY) Mar.

First: *Chrustenopora imbricata* Havlíček, 1968, Letná Formation, Chrustenice, Czechoslovakia (Havlíček, 1977).
Last: Probably *Jercia rongi* Baarli, 1988, Solvik Formation, Asker, southern Norway (Baarli, 1988).

F. SCHIZOPHORIIDAE Schuchert and Le Vene, 1929 S. (LLY)–Tr. (GRI) Mar.

First: *Schizophoria senecta* Hall and Clarke, 1892, Clinton Group, New York, USA (Schuchert and Cooper, 1932).
Last: *Orthotichia parva* Sokolskaja, 1965, Induan, Transcaucasus (Grant, 1970).

F. PROSCHIZOPHORIIDAE Boucot, Gauri, and Johnson, 1966 O. (ASH)–D. (SIG) Mar.

First: Possibly *Villicundella mozetici* (Levy and Nullo, 1974), ASH, San Juan Province, Argentina (Levy and Nullo, 1974).
Last: *Proschizophoria personata* (Zeiler, 1857), LOK and SIG, Europe (Maillieux, 1912).

F. LINOPORELLIDAE Schuchert and Cooper, 1931 O. (LLO)–D. (EMS) Mar.

First: *Salopia turgida* (M'Coy, 1851), Ffairfach Group, LLO and coeval strata, Builth, Wales, UK (Lockley and Williams, 1981).
Last: *Cycladigera cycladigerens* Havlíček, 1971, Zlíchov Limestone, Prague, Czechoslovakia (Havlíček, 1977).

Superfamily DALMANELLOIDEA Schuchert, 1911

F. PAURORTHIDAE Öpik, 1933 O. (ARG–ASH) Mar. (see Fig. 22.1)

First: *Paurorthina resima* Rubel, 1961, Latorp Stage, Estonia former USSR (Rubel, 1961).
Last: *Paurorthis gnoliana* Havlíček and Kriz, 1987, uppermost Beroun, Sardinia (Havlíček *et al.*, 1987).

F. TYRONELLIDAE Mitchell, 1977 O. (CAU) Mar.

First and Last: *Tyronella killeyensis* Mitchell, 1977, Killey Bridge Formation (Cautleyan), Pomeroy, Northern Ireland (Mitchell, 1977) and *Tyronella quebecensis* Sheehan and Lespérance, 1979, Whitehead Formation, Quebec, Canada.

F. DALMANELLIDAE Schuchert, 1913 O. (ARG)–D. (FRS) Mar.

First: *Dalmanella elementaria* Williams, 1974, Mytton Flags, Shelve (Williams, 1974).
Last: *Cariniferella carinata* (Hall, 1843), near cosmopolitan distribution in rocks of FRS age.

F. DICOELOSIIDAE Cloud, 1948 O. (PUS)–D. (FRS) Mar.

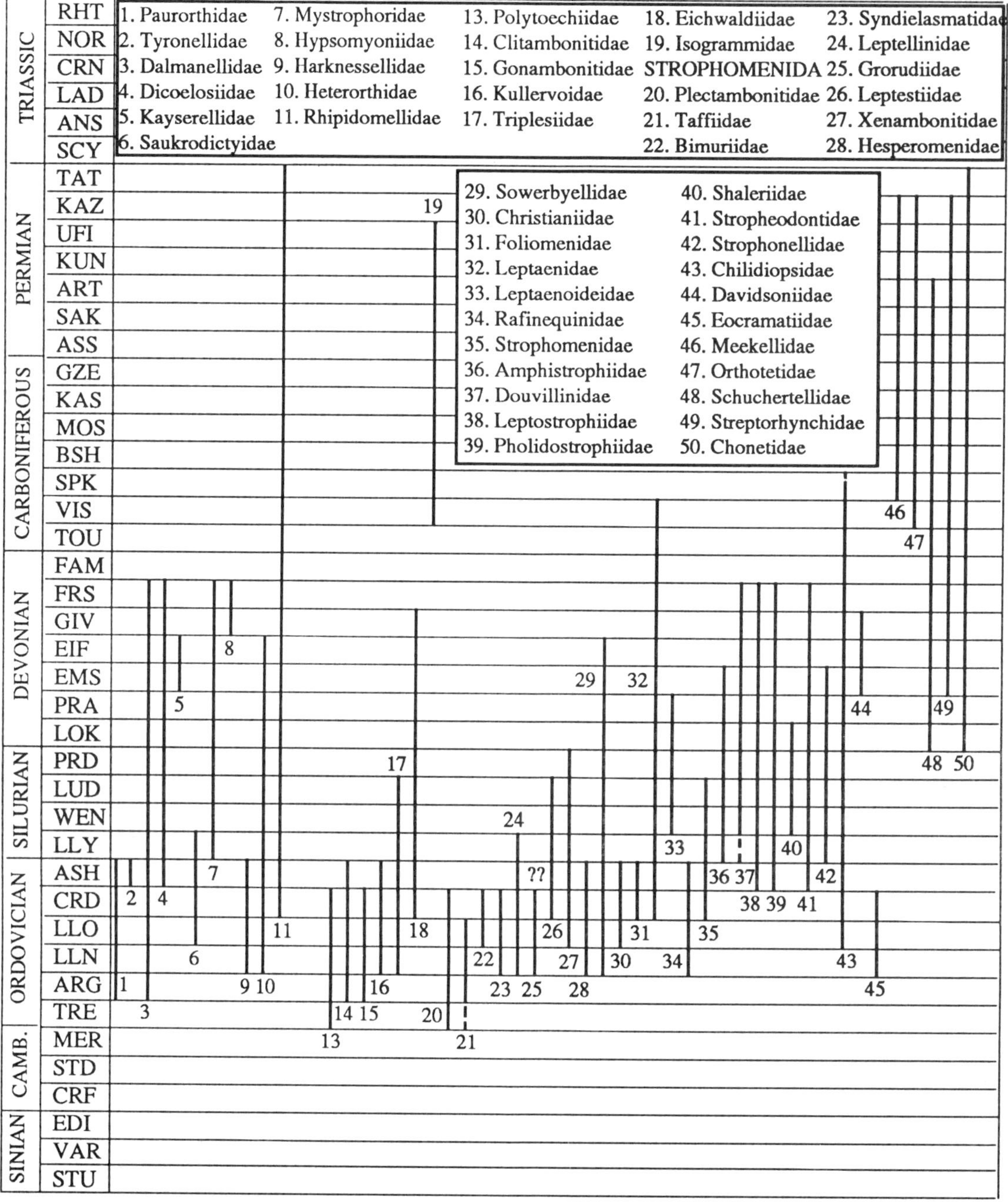

Fig. 22.2

First: *Dicoelosia* cf. *lata* Wright, 1964, and *D.* cf. *indentata* (Cooper, 1930), Mill Formation (Pusgillian), Whitehouse Group, Girvan, Scotland, UK (Harper, 1989).
Last: *Teichertina fitzroyensis* Veevers, 1959, Sadler Formation, Fitzroy Basin, Australia (Veevers, 1959).

F. KAYSERELLIDAE Wright, 1965 D. (EMS–EIF) Mar.

First: *Kayserella costulata* Lenz, 1977, Lower Devonian, Yukon, USA (Lenz, 1977).
Last: *Kayserella lepida* (Schnur, 1853), EIF, the Eifel, Germany and Moravia, Czechoslovakia (Havlíček, 1977).

F. SAUKRODICTYIDAE Wright, 1964 O. (LLO)–S. (LLY) Mar.

First: *Saukrodictya porosa* (Havlíček, 1977), Libeň and Letná formations, Czechoslovakia (Havlíček, 1977).
Last: *Saukrodictya arcana* Havlíček, 1977, Želkovice Formation, Hýskov, Czechoslovakia (Havlíček, 1977).

F. MYSTROPHORIDAE Schuchert and Cooper, 1931 S. (RHU)–D. (FRS) Mar.

First: Probably *Visbyella* sp. nov. Temple, 1987, lower LLY, Wales, UK (Temple, 1987).
Last: *Biernatium emanuelensis* (Veevers, 1959), FRS, Fitzroy Basin, Australia (Havlíček, 1977).

F. HYPSOMYONIIDAE Wright, 1965 D. (u.) Mar.

First and Last: *Hypsomyonia stainbrooki* Cooper, 1955, 'Independence Shale', Iowa, USA (Cooper, 1955).

F. HARKNESSELLIDAE Bancroft, 1928
O. (LLN–ASH) Mar.

First: *Horderleyella convexa* Williams, 1949, upper LLN Grits and Ashes near Llangadog, Wales, UK (Lockley and Williams, 1981).
Last: *Reuschella* sp., Portrane Limestone, Co. Dublin, Republic of Ireland (Wright, 1964) and *R.* sp., Dolhir Formation, Glyn Ceiriog, North Wales (Hiller, 1980); *Reuschella inexpectata* Temple, 1968, Hirnantian rocks overlying the Keisley Limestone, Keisley, Cumbria, England, UK (Temple, 1968) has been assigned to *Dysprosorthis* (Rong, 1984).

F. HETERORTHIDAE Schuchert and Cooper, 1931
O. (LLN)–D. (EIF) Mar.

First: *Tissintia prototypa* (Williams, 1949), *Didymograptus bifidus* Shales and equivalent units at Builth, Llandeilo and Shelve, Wales, UK (Lockley and Williams, 1981).
Last: *Platyorthis opercularis* (Verneuil, 1845), EIF, Brittany, Germany and the Ural Mountains (Harper *et al.*, 1969).

F. RHIPIDOMELLIDAE Schuchert, 1913
O. (CRD)–P. (TAT) Mar.

First: *Mendacella borrbyensis* Hints, 1976, Vormsi rocks, Estonia, former USSR (Hints, 1990).
Last: Although common in the Lower Permian, *Rhipidomella* is rare in the higher parts of the system; *Rhipidomella* sp. from the middle Productus Limestone of the Salt Range may be youngest (Grant, 1970).

Suborder CLITAMBONITIDINA Öpik, 1934

Superfamily CLITAMBONITOIDEA Winchell and Schuchert, 1893

F. POLYTOECHIIDAE Öpik, 1934 O. (TRE–CRD) Mar.

First: *Protambonites soror* (Barrande, 1879) and *Protambonites kolihai* (Havlíček, 1977), Holoubkov and Březany, respectively, Třenice Formation, Czechoslovakia (Havlíček, 1977).
Last: *Eremotoechia cloudi* Cooper, 1956, Arline Formation, Blount, Appalachians, USA (Cooper, 1956).

F. CLITAMBONITIDAE Winchell and Schuchert, 1893 O. (ARG–HIR) Mar.

First: *Apomatella ingrica* (Pahlen, 1877), Volkhov Stage, Estonia and Russia, former USSR (Hints, 1990).
Last: *Vellamo silurica* Öpik, 1934 and *Ilmarinia ponderosa* Öpik, 1934, Porkuni Stage, Estonia, former USSR (Öpik, 1934).

Superfamily GONAMBONITOIDEA Schuchert and Cooper, 1931

F. GONAMBONITIDAE Schuchert and Cooper, 1931
O. (ARG–CRD) Mar.

First: *Antigonambonites planus* (Pander, 1830), Latorp Stage, Estonia and Russia, former USSR (Öpik, 1934).
Last: *Estlandia pyron* (Eichwald, 1840) *silicificata* Öpik, 1934, Keila rocks, Estonia (Hints, 1990).

F. KULLERVOIDAE Öpik, 1934 O. (LLN–ASH) Mar.

First: Possibly *Kullervo* sp., Ffairfach Group, Mid-Wales, UK (Lockley and Williams, 1981) or *Kullervo lacunata* Öpik, 1934 and *Kullervo* sp. nov., Uhaku Stage, Estonia, former USSR (Rõõmusoks, 1970).
Last: *Kullervo complectens* (Wiman, 1907), 'West Baltic Leptaena Limestone', Hulterstad, Öland, Sweden and *K?* sp., Langøyene and Langara formations, Oslo Region, Norway (Cocks, 1982).

Suborder TRIPLESIIDINA Moore, 1952

Superfamily TRIPLESIOIDEA Schuchert, 1913

F. TRIPLESIIDAE Schuchert, 1913
O. (LLN)–S. (LUD) Mar.

First: *Onychoplecia kindlei* Cooper, 1956, Table Head Formation, Newfoundland, Canada (Cooper, 1956).
Last: *Plectotreta lindstroemi* Ulrich and Cooper, 1936a, Hemse Beds, Gotland (Bassett and Cocks, 1974). [DATH]

Order UNCERTAIN

Suborder DICTYONELLIDINA Cooper, 1956

Superfamily EICHWALDIOIDEA Schuchert, 1893

F. EICHWALDIIDAE Schuchert, 1893
O. (CRD)–D. (GIV) Mar.

First: *Eichwaldia subtrigonalis* Billings, 1858, Rockland Formation, Ontario, Canada (Billings, 1858).
Last: *Dictyonella* sp., Middle Devonian (GIV), northern Balkhash area, former USSR (Kaplun, 1967).

F. ISOGRAMMIDAE Schuchert and LeVene, 1929
C. (ASB)–P. (ART/UFI) Mar.

First: Several species of *Isogramma* are known from rocks of Viséan age, for example, *I. germanica* Paeckelmann, 1930, Lower Carboniferous, Austria, Germany, Lower Moscow Basin, and the Ukraine, former USSR; in Britain *Isogramma* sp. A, Dinwoodie Beds, southern Scotland, UK is the oldest (Brand, 1970).
Last: *Isogramma lobatum* Cooper and Grant, 1975, Cathedral Mountain Formation, west Texas, USA (Cooper and Grant, 1975). [DATH]

Order STROPHOMENIDA Öpik, 1934

Suborder STROPHOMENIDINA Öpik, 1934

This group is similar to that in the *Treatise* (A. Williams, *in* Williams *et al.*, 1965) except that the Thecospiridae and the Bactryniidae have been removed to the Spiriferida.

Superfamily PLECTAMBONITOIDEA Jones, 1928

The classification within this superfamily follows the recent revision by Cocks and Rong (1989).

F. PLECTAMBONITIDAE Jones, 1928
O. (TRE–CRD) Mar.

First: *Plectella uncinata* (Pander, 1830), B_1 Beds with *Cyrtometopus*, Maekula, Estonia, former USSR (Öpik, 1933).
Last: *Isophragma princeps* Popov, 1980, Erkebidaiski Horizon, Chelinograd, Kazakhstan, former USSR (Popov, 1980).

F. TAFFIIDAE Schuchert and Cooper, 1931
O. (TRE?–LLO) Mar.

First: May be *Leptella? exigua* Clark, 1917, *Shumardia* Zone of Beekmantown Group (TRE), Levis, Quebec, Canada (Cooper, 1956); but definitely Lower ARG, for example, *Schedophyla striata* (Xu *et al.*, 1974), lower Meitan Formation, Guizhou, China (Xu *et al.*, 1974).

Last: *Ahtiella lirata* Öpik, 1932, C_1 Beds, Tsitri, Estonia, former USSR (Öpik, 1932).

F. BIMURIIDAE Cooper, 1956 O. (LLO–CRD) Mar.

First: *Bimuria superba* Ulrich and Cooper, 1942, Arline Formation, Tennessee, USA (Cooper, 1956).
Last: *Bimuria youngiana* (Davidson, 1871), Craighead Limestone, Girvan, Scotland, UK (Williams, 1962).

F. SYNDIELASMATIDAE Cooper, 1956 O. (LLN–CRD) Mar.

First: *Syndielasma biseptatum* Cooper, 1956, upper Pogonip Group, Nevada, USA (Cooper, 1956).
Last: *Sowerbyites mongolicus* Rozman, 1981, Bairimski Beds, Mongolia (Rozman, 1981).

F. LEPTELLINIDAE Ulrich and Cooper, 1936 O. (LLN)–S. (LLY) Mar.

First: *Leptellina (Leptellina) occidentalis* Ulrich and Cooper, 1938, upper Pogonip Group, Nevada, USA (Ulrich and Cooper, 1938).
Last: *Leptellina (Merciella) vesper* (Lamont and Gilbert, 1945), Wych Formation, Malvern Hills, Worcestershire, England, UK (Lamont and Gilbert, 1945).

F. GRORUDIIDAE Cocks and Rong, 1989 O. (LLN–CRD) Mar.

First: *Tetraodontella? aquiloides* Fu, 1975, Xiliangsi Formation, Shaanxi, China (Fu, 1975).
Last: *Grorudia glabrata* Spjeldnæs, 1957, Arnestad Formation, Oslo, Norway (Spjeldnæs, 1957).

F. LEPTESTIIDAE Öpik, 1933 O. (CRD)–S. (LUD) Mar.

First: *Bilobia virginiensis* Cooper, 1956, Edinburg Formation, Virginia, USA (Cooper, 1956).
Last: *Leangella segmentum* (Lindström, 1861), Elton Beds, Shropshire, England, UK (L. R. M. Cocks *nov.*).

F. XENAMBONITIDAE Cooper, 1956 O. (LLO)–S. (PRD) Mar.

First: *Xenambonites undosus* Cooper, 1956, Pratt Ferry Formation, Alabama, USA (Cooper, 1956).
Last: *Jonesea mariaformis* (Lenz, 1977), Road River Formation, Yukon, Canada (Lenz, 1977).

F. HESPEROMENIDAE Cooper, 1956 O. (LLN–ASH) Mar.

First: *Hesperomena leptellinoidea* Cooper, 1956, Antelope Valley Limestone, Nevada, USA (Cooper, 1956).
Last: A number of species of *Kassinella* occur within the middle Ashgill (Cocks and Rong, 1989) including *Kassinella (Kassinella) moneta* (Barrande, 1879), Králův Dvůr Formation, Kosov, Czechoslovakia (Havlíček, 1967).

F. SOWERBYELLIDAE Öpik, 1930 O. (LLN)–D. (EIF) Mar.

First: *Sowerbyella antiqua* Jones, 1928, Ffairfach Group, Llandeilo, Wales, UK (Williams *et al.*, 1981).
Last: *Plectodonta (Dalejodiscus) comitans* (Barrande, 1879), Daleje Shales, Prague, Czechoslovakia (Havlíček, 1967).

Superfamily STROPHOMENOIDEA King, 1846

The classification within this superfamily is unsatisfactory and needs revision. As pointed out by Cocks (1978), since denticles appeared polyphyletically in four different strophomenide stocks, then the superfamily Stropheodontoidea used by some authors for these brachiopods should not be employed. The classification here generally follows Williams (*in* Williams *et al.*, 1965), but has been somewhat modified according to, e.g. Harper and Boucot (1978), Havlíček (1967). For ease of reference the families without denticles are listed first (Christianiidae to Strophomenidae) and these are followed by those with denticles (Amphistrophiidae to Strophonellidae).

F. CHRISTIANIIDAE Williams, 1953 O. (LLO–ASH) Mar.

First: *Christiania sulcata* Williams, 1962, Stinchar Limestone, Girvan, Scotland, UK (Williams, 1962).
Last: *Christiania* aff. *tenuicincta* (M'Coy, 1846), Langøyene Formation, Oslo, Norway (Cocks, 1982).

F. FOLIOMENIDAE Williams, 1965 O. (CRD–ASH) Mar.

First: *Foliomena exigua* Harper, 1989, upper Whitehouse Group, Girvan, Scotland, UK (Harper, 1989).
Last: *Foliomena*? sp. Tommarp Beds, Västergötland, Sweden (Cocks and Rong, 1988).

F. LEPTAENIDAE Hall and Clarke, 1893 O. (CRD)–C. (VIS) Mar.

First: *Bellimurina tenuicorrugata* (Reed, 1917), Balclatchie Mudstones, Girvan, Scotland, UK (Williams, 1962).
Last: *Leptagonia smithi* Brand, 1972, Calmy Limestone (Arnsbergian), Dalry, Ayrshire, Scotland, UK (Brand, 1972).

F. LEPTAENOIDEIDAE Williams, 1953 S. (WEN)–D. (PRA) Mar.

First: *Leptaenoidea gotlandica* Hedström, 1917, Upper Visby Beds, Gotland, Sweden (Bassett and Cocks, 1974).
Last: *Leptaenisca concava* (Hall, 1857), Haragan Formation, Oklahoma, USA (Amsden, 1958).

F. RAFINESQUINIDAE Schuchert, 1893 O. (LLN–ASH) Mar.

First: *Macrocoelia llandeiloensis* (Davidson, 1871), Lower Flags and Grits of Ffairfach Group, Llandeilo, Wales, UK (Williams *et al.*, 1981).
Last: *Drummuckina donax* (Reed, 1917), Lady Burn Starfish Beds, Girvan, Scotland, UK (Cocks, 1978).

F. STROPHOMENIDAE King, 1846 O. (CRD)–S. (LUD) Mar.

First: *Strophomena deficiens* Reed, 1917, Balclatchie Mudstones, Girvan, Scotland, UK (Williams, 1962).
Last: *Katastrophomena antiquata* (J. de C. Sowerby, 1839), Elton Beds, Wenlock Edge, Shropshire, England, UK (Cocks, 1968).

F. AMPHISTROPHIIDAE Harper and Boucot, 1978 S. (LLY)–D. (EMS) Mar.

First: *Eoamphistrophia whittardi* (Cocks, 1967), Purple Shales, Hughley, Shropshire, England, UK (Cocks, 1967).
Last: *Devonamphistrophia alveata* (Hall, 1863), Schoharie Grit, New York, USA (Harper and Boucot, 1978).

F. DOUVILLINIDAE Caster, 1939
S. (LLY?)–D. (FRS) Mar.

First: *Mesodouvillina (Mesodouviella)* sp., French River Formation, Arisaig, Nova Scotia, Canada (Harper and Boucot, 1978).
Last: *Douvillina arcuata* (Hall, 1857), Lime Creek Beds, Rockford, Iowa, USA (Harper and Boucot, 1978).

F. LEPTOSTROPHIIDAE Caster, 1939
O. (ASH)–D. (FRS) Mar.

First: *Aphanomena schmalenseei* Bergström, 1968, Dalmanitina Beds, Västergötland, Sweden (Bergström, 1968).
Last: *Nervostrophia* spp., Chemung Group, New York, USA (Harper and Boucot, 1978).

F. PHOLIDOSTROPHIIDAE Stainbrook, 1943
O. (ASH)–D. (FRS) (including Lissostrophiidae)
Mar.

First: *Eopholidostrophia matutinum* (Lamont, 1935), Lower Drummuck Group, Girvan, Scotland, UK (Cocks, 1978).
Last: *Pholidostrophia (Pholidostrophia) lepida* (Hall, 1857), 'Chemung' Group, Rock Island, Illinois, USA (Harper and Boucot, 1978).

F. SHALERIIDAE Williams, 1965
S. (WEN)–D. (LOK) Mar.

First: *Shaleriella delicata* Harper and Boucot, 1978, Slite Marl, Gotland, Sweden (Harper and Boucot, 1978).
Last: *Shaleria rigida* (de Koninck, 1876), Grès de Gdoumont, Gedinne, Belgium (Harper and Boucot, 1978).

F. STROPHEODONTIDAE Caster, 1939
O. (ASH)–D. (FRS) (including Eostropheodontidae)
Mar.

First: *Eostropheodonta* sp., Slade and Redhill Mudstone Formation, Haverfordwest, Dyfed, Wales, UK (Cocks and Price, 1975).
Last: *Strophonelloides reversa* (Calvin, 1878), Hackberry Group, Rockford, Iowa, USA (Harper and Boucot, 1978).

F. STROPHONELLIDAE Caster, 1939
S. (LLY)–D. (EMS) Mar.

First: *Eostrophonella eothen* (Bancroft, 1949), Gasworks Mudstone, Haverfordwest, Dyfed, Wales, UK (Williams, 1951).
Last: *Strophonella (Quasistrophonella) bohemica* (Barrande, 1848), Koněprusy Limestone, Prague, Czechoslovakia (Havlíček, 1967). [LRMC]

Superfamily DAVIDSONIOIDEA King, 1850

In general, the classification scheme of Havlíček (1967) is followed, with the possible addition of the Eocramatiidae. Relationships between the various families are rather obscure, and so they are listed here in alphabetical order.

F. CHILIDIOPSIDAE Boucot, 1959
O. (LLO)–C. (SPK) Mar.

First: *Gacella insolita* Williams, 1962, Stinchar Limestone, Girvan, Scotland, UK (Williams, 1962).
Last: *Pulsia mosquensis* Ivanov, 1925, SPK, Moscow Basin, former USSR (Sarytcheva and Sokolskaya, 1952).

F. DAVIDSONIIDAE King, 1850 D. (EMS–GIV)
Mar.

First: *Biconostrophia fragilis* (Barrande, 1879), Koněprusy Limestone, Prague, Czechoslovakia (Havlíček, 1967).
Last: *Davidsonia verneuili* Bouchard-Chantreaux, 1849, GIV Limestones, Čelechovice na Hané, Moravia, Czechoslovakia (Havlíček, 1967).

?F. EOCRAMATIIDAE Williams, 1974
O. (LLN–CRD) Mar.

First: *Eocramatia dissimulata* Williams, 1974, Hope Shales, Shelve Inlier, Shropshire, England, UK (Williams, 1974).
Last: *Neocramatia diffidentia* Harper, 1989, Upper Whitehouse Group, Girvan, Scotland, UK (Harper, 1989).
Comment: It is uncertain whether or not these genera, which have a bilobed cardinal process, but whose shell structure is unknown, should be assigned to the Strophomenoidea, Orthoidea or Davidsonioidea, but the latter seems most probable.

F. MEEKELLIDAE Stehli, 1954 C. (SPK)–P. (KAZ)
Mar.

First: *Orthotetina thomasi* (Sarytcheva and Sokolskaya, 1952), SPK, Tarussky, Moscow Basin, former USSR (Sarytcheva and Sokolskaya, 1952).
Last: *Geyerella americana* Girty, 1909, Capitan Formation (early KAZ, in age), Texas, USA (Cooper and Grant, 1974).

F. ORTHOTETIDAE Waagen, 1884
C. (VIS)–P. (KAZ) Mar.

First: *Orthotetes keokuk* (Hall, 1858), Keokuk Limestone (Chadian), Keokuk, Iowa, USA (Weller, 1914).
Last: *Orthotetes bisulcata* Waterhouse, 1966, Nangung Limestone (Kalabaghian), Nangung, Nepal (Waterhouse, 1966).

F. SCHUCHERTELLIDAE Williams, 1953
D. (LOK)–P. (ART) Mar.

First: *Areostrophia interjecta* (Barrande, 1879), Lockhov Limestone, Lockhov, Prague, Czechoslovakia (Havlíček, 1967).
Last: *Goniarina permiana* (Stehli, 1954), Bone Spring Formation, Texas, USA (Cooper and Grant, 1974).

F. STREPTORHYNCHIDAE Stehli, 1954
D. (EMS)–P. (KAZ) Mar.

First: *Aesopomum aesopeum* (Barrande, 1879), Koněprusy Limestone, Koněprusy, Czechoslovakia (Havlíček, 1967).
Last: *Tropidelasma gregarium* Girty, 1909, Capitan Formation, Texas, USA (Cooper and Grant, 1974).
[CHCB and LRMC]

Suborder CHONETIDINA Muir-Wood, 1955

There has been no complete revision of the suborder since the *Treatise* (1965), so those families are followed here, with the addition of the Rugosochonetidae, Strophochonetidae and Anopliidae, elevated from their subfamilies in the *Treatise*.

Superfamily CHONETOIDEA Bronn, 1862

F. CHONETIDAE Bronn, 1862 D. (LOK)–P. (LGT)
Mar.

First: *Dawsonelloides canadensis* (Billings, 1874), New Scotland Beds, northern Appalachians, USA (Boucot and Harper, 1968).

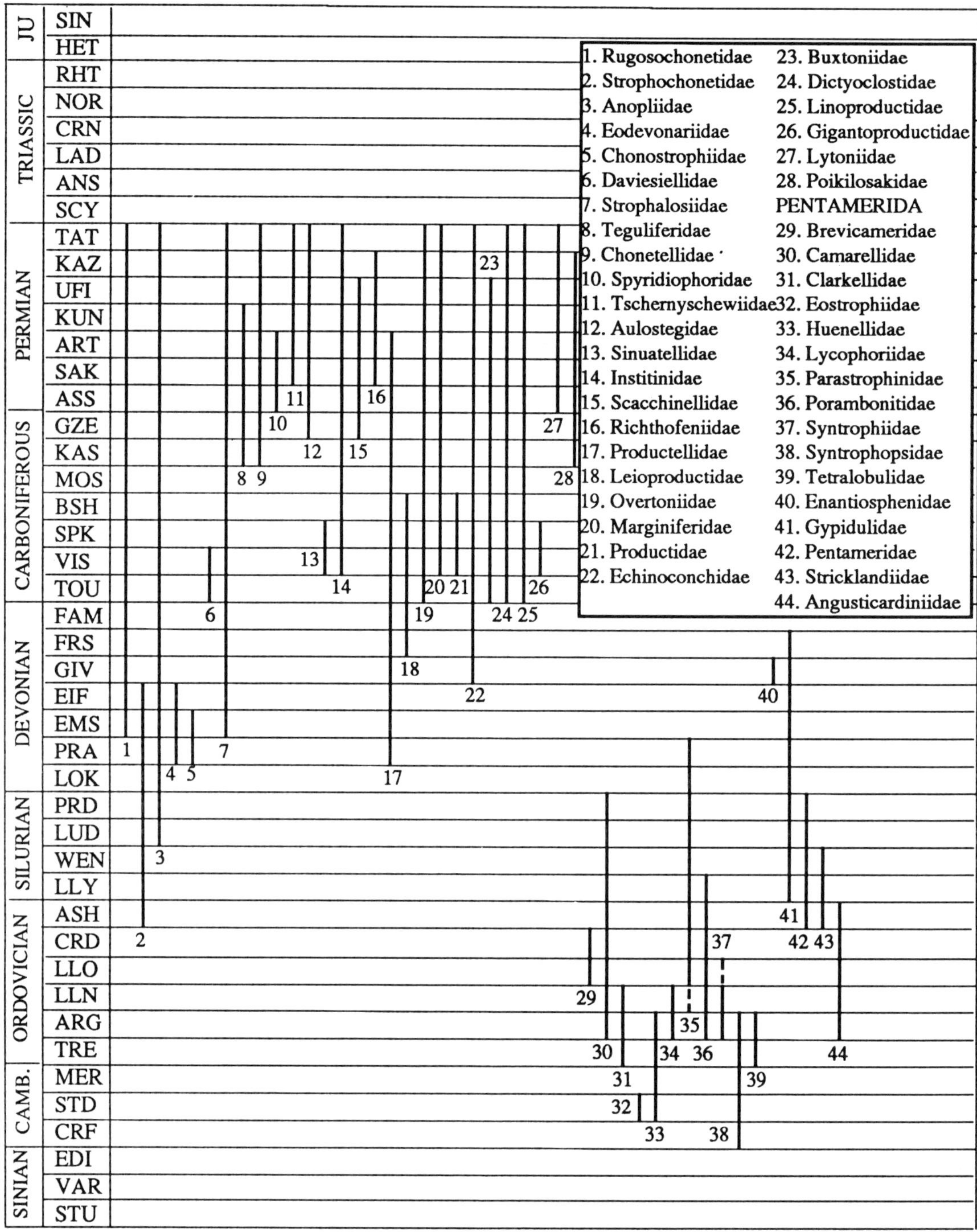

Fig. 22.3

Last: *Quinquenella glabra* Waterhouse, 1975, Chhidru Formation, Nepal.

Comment: The rugosochonetids, strophochonetids and anopliids of the *Treatise* are removed from here.

F. RUGOSOCHONETIDAE Muir-Wood, 1962 D. (EMS)–P. (CHX) Mar.

First: *Dagnachonetes (Luanquella) alcaldei* Racheboeuf, 1981, lower Moniello Formation, NW Spain.

Last: *Rugaria* aff. *nisalensis* Waterhouse, 1978, Senja Formation, Nepal.

F. STROPHOCHONETIDAE Muir-Wood, 1962 O. (ASH)–D. (EIF) Mar.

First: *Archaeochonetes primigenius* (Twenhofel, 1914), Vaureal Formation, Anticosti Island, Canada (Racheboeuf and Copper, 1986).

Last: *Chlupacina longispina* Havlíček and Racheboeuf, 1979, Chotec Limestone, lower *Pinacites juglen* Zone, Bohemia, Czechoslovakia (Racheboeuf, 1981).

F. ANOPLIIDAE Muir-Wood, 1962 S. (LUD)–P. (CHX) Mar.

First: *Eoplicanoplia collicula* (Foerste, 1909), Salina Formation, Indiana, USA (Boucot and Harper, 1968).

Last: *Glabrichonetes kuwaensis* Waterhouse, 1978, Senja Formation, Nepal.

F. EODEVONARIIDAE Sokolskaya, 1960 D. (PRA–EIF) Mar.

First: *Davoustia davousti* (Oehlert, 1887), St Cénére Formation, Laval, France (Racheboeuf, 1976).
Last: *Eodevonaria elymencheri* Boucot and Harper, 1968, Hazard Limestone, Pennsylvania, USA.

F. CHONOSTROPHIIDAE Muir-Wood, 1962 D. (PRA–EMS) Mar.

First: *Chonostrophiella complanata* (Hall, 1857), Tarratine Formation, Maine, USA (Boucot and Amsden, 1964).
Last: *Chonostrophiella cynthiae* Racheboeuf, 1987, Upper Blue Fjord Formation, Arctic Canada. Or, if proved, *Chonostrophia reversa* (Whitefield, 1802), Delaware Formation, Ohio, USA, possibly EIF.

F. DAVIESIELLIDAE Sokolskaya, 1960 C. (TOU–ASB) Mar.

First: *Delepinea* sp., Burrington, near Bristol, England, UK and *Delepinea carinata* Garwood, 1916, *Michelinia grandis* Zone, Cumbria, England, UK (Muir-Wood, 1962).
Last: *Daviesiella llangollensis* Davidson, 1862, Lower Brown Limestone, Llangollen, North Wales, UK (Muir-Wood, 1962).
Comment: *Megachonetes* Sokolskaya, 1950 was included in this family in the *Treatise*, with a range starting in the mid-Devonian and extending to the upper Carboniferous (MOS), but it is thought to be better assigned to the Rugosochonetidae. [CHCB]

Suborder PRODUCTIDINA Waagen, 1883

The 24 families included here are classified in four superfamilies, but unlike the *Treatise* (1965), the aulostegids are here treated as a superfamily, within which several previous strophalosioidean families now reside, as well as the Institinidae from the Productoidea. The Richthofenioidea and the Lyttonioidea remain unsatisfactory. Since 1965, other families have been proposed and provide good groupings, such as McKellar's (1970) use of the Leioproductidae and Sentosiidae: the latter for genera here retained within the Echinoconchidae. While authors such as Sarytcheva (1963), and Cooper and Grant (1975), have added families, the lack of any complete revision of the Productidina makes their inclusion here difficult.

Superfamily STROPHALOSIOIDEA Schuchert, 1913

F. STROPHALOSIIDAE Schuchert, 1913 D. (EMS)–P. (CHX) Mar.

First: *Ralia primigenia* Lazarev, 1987, *Oculipora angulata* Zone, Gobi Alti, southern Mongolia.
Last: *Marginalosia kalikotei* (Waterhouse, 1975), Senja Formation, Nepal.
Comment: The Upper Permian *Cyrtalosia circinata* Termier and Termier, 1970 from Cambodia has no interarea and is thought to be an aulostegid.

F. TEGULIFERIDAE Muir-Wood and Cooper, 1960 C. (KAS)–P. (KUN) Mar.

First: *Teguliferina armata* (Girty, 1908), Hertha Limestone, Missouri, USA (Dunbar and Condra, 1932).
Last: *Acritosia magnifica* Cooper and Grant, 1975, Road Canyon Formation, Texas, USA.

F. CHONETELLIDAE Likharew, 1960 C. (KAS)–P. (CHX) Mar.

First: *Chonetella dunbari* Newell, 1934, Missourian, Oklahoma, USA. There is some doubt about the assignment of this species.
Last: *Chonetella nasuta* Waagen, 1884, upper Productus Limestone, Chhidru Formation, Salt Range, Pakistan (Waterhouse, 1978).

Superfamily AULOSTEGOIDEA Muir-Wood and Cooper, 1960

F. SPYRIDIOPHORIDAE Muir-Wood and Cooper, 1960 P. (ASS–ART) Mar.

First: *Spyridiophora distincta* Cooper and Stehli, 1955, Neal Ranch, Texas, USA (Cooper and Grant, 1975).
Last: *Spyridiophora reticulata* (R. E. King, 1931), low Cathedral Mountain, Texas, USA (Cooper and Grant, 1975).

F. TSCHERNYSCHEWIIDAE Muir-Wood and Cooper, 1960 P. (SAK–CHX) Mar.

First: *Tschernyschewia inexpectans* Cooper and Grant, 1975, Taylor Ranch, Texas, USA.
Last: *T. typica* or *T. yakowlewi* Stoyanow, 1910, Djulfa, Armenia.
Comment: In North America, members of the family are restricted to the Lower Permian, while in Eurasia they are mainly late Permian in age.

F. AULOSTEGIDAE Muir-Wood and Cooper, 1960 C. (GZE)–P. (CHX) Mar.

First: *Limbella costellata* Cooper and Grant, 1975, Uddenites Shale, Texas, USA.
Last: *Megasteges nepalensis* Waterhouse, 1975, Senja Formation, Nepal.

F. SINUATELLIDAE Muir-Wood and Cooper, 1960 C. (ASB–ARN) Mar.

First: *Sinuatella sinuata* (de Koninck, 1851), Visé, Belgium and England, UK (Brunton and Mundy, 1988).
Last: *S. johnsoni* Brunton and Mundy, 1988, lower Namurian, Northumberland, NE England, UK.
Comment: Demanet (1958) recorded the species from the Ivorian of Tournai, but the identification is doubted.

F. INSTITINIDAE Muir-Wood and Cooper, 1960 C. (ASB)–P. (LGT) Mar.

First: *Rugicostella nystianus* (de Koninck, 1842), B2 Malham Formation, North Yorkshire, England, UK (Brunton and Mundy, 1988), Arundian of Visé, Belgium (?Pirlet, 1967).
Last: *Institina plicatiliformis* (Fredericks, 1932), Uralian, northern Russia, former USSR (Muir-Wood and Cooper, 1960).

F. SCACCHINELLIDAE Licharew, 1928 C. (GZE)–P. (UFI) Mar.

First: *Scacchinella primitiva* Cooper and Grant, 1975, Gaptank Formation, Texas, USA.
Last: *Scacchinella variabilis* Gemmellaro, 1897, Sosio Limestone, Sicily (Rudwick and Cowen, 1968).
Comment: The age of the Sosio Limestone remains in doubt.

Superfamily RICHTHOFENIOIDEA Waagen, 1885

F. RICHTHOFENIIDAE Waagen, 1885 P. (SAK–CAP) Mar.

First: *Hercosia uddeni* (Bose, 1916), Skinner Ranch Formation, Texas, USA (Cooper and Grant, 1975).
Last: *Sestropoma cribriferum* Cooper and Grant, 1975, Capitan Formation, Texas, USA.

Superfamily PRODUCTOIDEA Gray, 1840

F. PRODUCTELLIDAE Schuchert and Le Vene, 1929 D. (PRA)–P. (ART) Mar.

First: *Eoproductella menakovae* Rzhonsnitskaja, 1980, Pandzhrutskiy Horizon, central Asia.
Last: *Stictozoster leptus* Grant, 1976, ?Rat Buri Limestone, Thailand.
Comment: This is one of the longer-ranging families, and it is likely that some of the younger genera should reside elsewhere.

F. LEIOPRODUCTIDAE Muir-Wood and Cooper, 1960 D. (FRS)–C. (KIN) Mar.

First: *Devonoproductus intermedius* Cooper and Dutro, 1982, Sly Gap Formation, New Mexico, USA.
Last: *Productina pectinoides* (Phillips, 1836), Valdeteja Formation, northern Spain (Prins, 1968).
Comment: Here the probable Lower Permian genus *Jakutoproductus* Kashirtsev, 1959 is considered as an overtoniid.

F. OVERTONIIDAE Muir-Wood and Cooper, 1960 C. (HAS)–P. (CHX) Mar.

First: *Overtonia borodencovensis* (Tolmatchow, 1924), Taidon Formation, Kuznetsk, former USSR (Sarytcheva, 1963).
Last: *Dorashamia abichi* Sarytcheva, 1965, Djulfian, Transcaucasus, former USSR and *Krotovia arcuata* Waterhouse, 1978, Baisalian, Nepal.
Comment: This large *Treatise* (1965) family probably should not include genera such as the Upper Devonian *Laminatia* or *Sentosia*, which should reside in the Echinoconchiidae.

F. MARGINIFERIDAE Stehli, 1954 C. (CHD)–P. (CHX) Mar.

First: *Eomarginifera* sp. Shephard-Thorn, 1963, Waulsortian Limestone, Limerick, Republic of Ireland.
Last: *Spinomarginifera* sp. Grant, 1970, Mianwali Formation, Salt Range, Pakistan.
Comment: Perhaps the earliest-named species is *Eomarginifera* cf. *longispina* (J. Sowerby, 1814) known from the Chadian of Derbyshire, England, UK, while *Spinomarginifera* cf. *helica* (Abich), recorded by Nakazawa *et al.* (1975) from the upper TAT of Nepal, is perhaps the youngest.

F. PRODUCTIDAE Gray, 1840 C. (HLK–CHE) Mar.

First: *Productus garwoodi* Muir-Wood, 1928, S2 Bryozoa Beds, Westmorland (Cumbria), England, UK.
Last: *Productus carbonarius* de Koninck, 1842, Quarterburn Marine Band, Westmorland (Cumbria), England, UK (Calver, *in* Owens and Burgess, 1965).

F. ECHINOCONCHIDAE Stehli, 1954 D. (GIV)–P. (LGT) Mar.

First: *Caucasiproductus gretchishnikovae* Lazarev, 1987, *Indospirifer pseudowilliamsi* Zone, Transcaucasus, former USSR.
Last: *Waagenoconcha gangeticus* (Diener, 1897), Nangung Formation, Nepal (Waterhouse, 1978).
Comment: Lazarev's species was placed in the Sentosiidae McKellar 1970, within the Echinoconchoidea *sensu* Lazarev, 1987.

F. BUXTONIIDAE Muir-Wood and Cooper, 1960 C. (HAS)–P. (UFI) Mar.

First: *Marginatia vaughani* (Muir-Wood, 1928), K Zone, Avon Gorge, SW England, UK.
Last: *Vediproductus vediensis* Sarytcheva, 1965, Gnishisky Horizon, Transcaucasus, former USSR.
Comment: Whidborne (1897) recorded in his Devonian faunas *Productus scabriculus* Martin, 1809 from the Pilton Beds, but this also could be early TOU, although the identification is doubtful.

F. DICTYOCLOSTIDAE Stehli, 1954 C. (IVO)–P. (LGT) Mar.

First: *Dictyoclostus bristolensis* (Muir-Wood, 1928), Z2, Avon Gorge, Bristol, England, UK and Taidon, Kuznetsk Basin, USSR (Sarytcheva, 1963).
Last: *Araxilevis intermedius* (Abich, 1878), Djulfian, Transcaucasus, former USSR, and Iran (Sarytcheva, 1965).
Comment: If *Tolmatchoffia* Frederiks, 1933 is accepted in this family there are species from the Kuznetsk Basin, former USSR, of early to mid TOU age.

F. LINOPRODUCTIDAE Stehli, 1954 C. (HAS)–P. (CHX) Mar.

First: *'Cancrinella' panderi* (Auerbach, 1862), Cita, Donetz Basin, former USSR (Aizenverg, 1966) and *Ovatia laevicosta* (White, 1860), Taidon, Kuznetsk Basin, former USSR (Sarytcheva, 1963).
Last: *Linoproductus superba* Reed, 1944, Senja Formation, Nepal (Waterhouse, 1978).
Comment: *O. laevicosta* reported by Nalivkin (1979) from the Fammenian of the Urals has yet to be confirmed.

F. GIGANTOPRODUCTIDAE Muir-Wood and Cooper, 1960 C. (HLK–CHO) Mar.

First: *Linoprotonia ashfellensis* Ferguson, 1971, Ashfell Limestone, northern England, UK.
Last: *Titanaria horreitensis* Legrand-Blain, 1987, mid Tagnana, Algerian Sahara. Also Arnsbergian, *Latiproductus latipriscus* (Sarytcheva, 1928), SPK, Moscow Basin, former USSR. Also Newton Limestone, Northumberland, England, UK (Pattison, 1981).
Comment: *Bagrasia* Nalivkin, 1960 was doubtfully included here in the *Treatise* (1965) from Etroeungt beds in the former USSR, but is considered more likely to be a linoproductid.

Superfamily LYTTONIOIDEA Waagen, 1883

F. LYTTONIIDAE Waagen, 1883 P. (ASS–LGT) Mar.

First: *Keyserlingina filicis* (Keyserling, 1891) and *K. schellwieni* Tschernyschew, 1902, Schwagerina Limestone, Urals, former USSR.
Last: *Oldhamia transcaucasica* (Stoyanov, 1915), Djulfian, Transcaucasus, former USSR (Sarytcheva, 1965) and *Lyttonia*

sp., Kathwai Formation, Djulfian, Pakistan (Grant, 1970).
Comment: *Keyserlingina plana* Ivanova, 1936 of KAS age is now thought to belong in *Poikilosakos* Watson, 1917.

F. POIKILOSAKIDAE Williams, 1953
C. (KAS)–P. (CAP) Mar.

First: *Poikilosakos plana* (Ivanova, 1936), Moscow Basin, former USSR (Sarytcheva, 1952).
Last: *Cardinocrania indica* Waagen, 1885, Chhidru Formation, (Cephalopod Bed), Salt Range, Pakistan.
Comment: *Poikilosakos petaloides* Watson, 1917, from the Plattsmouth Limestone, USA (Dunbar and Condra, 1932) is slightly younger than *P. plana*.

F. SPINOLYTTONIIDAE Williams, 1965: invalid

Comment: This family has only one genus and, since the substrate, not the specimens, seems to be spinose, the family is invalid. [CHCB]

Order PENTAMERIDA Schuchert and Cooper, 1931

Superfamily PORAMBONITOIDEA Davidson, 1853

F. BREVICAMERIDAE Cooper, 1956
O. (LLO–CRD) (includes Parallelasmatidae) Mar.

First: *Brevicamera camerata* Cooper, 1956, Pratt Ferry Formation, Alabama, USA.
Last: *Vaga sinualis* Sapelnikov and Rukavishnikova, 1973, upper CRD, Chu-ili Mountains, Kazakhstan, former USSR (Sapelnikov and Rukavishnikova, 1973).

F. CAMARELLIDAE Hall and Clarke, 1894
O. (ARG)–S. (PRD) Mar.

First: *Idiostrophia perfecta* Ulrich and Cooper, 1936b, Mystic Conglomerate, Quebec, Canada.
Last: *Bleshidium triste* Havlíček, 1990, Přídolí Formation, Beroun, Czechoslovakia (Havlíček and Štorch, 1990).

F. CLARKELLIDAE Schuchert and Cooper, 1931
O. (TRE–LLN) Mar.

First: *Syntrophina altaica* Severgina, 1985, TRE, Gorny Altai, former USSR (Severgina, 1985).
Last: *Yangzteella poloi* (Martelli, 1901), Kuniutan Formation, Hubei Province, China.

F. EOSTROPHIIDAE Ulrich and Cooper, 1936
Є. (STD) Mar.

First and Last: *Cambrotrophia cambria* (Walcott, 1908), Middle Cambrian, Utah, USA.

F. HUENELLIDAE Schuchert and Cooper, 1931
Є. (STD)–O. (ARG) Mar.

First: *Huenella abnormis* (Walcott, 1912), Deep Creek Limestone, Wyoming, USA (Walcott, 1912).
Last: *Rectotrophia globularis* Bates, 1968, Treiorwerth Formation, Trefor, Anglesey, Wales, UK (Bates, 1968).

F. LYCOPHORIIDAE Schuchert and Cooper, 1931
O. (ARG–LLN) Mar.

First and Last: *Lycophoria nucella* (Dalman, 1828), Lower Ordovician, Baltoscandia (Schuchert and Cooper, 1931).

F. PARASTROPHINIDAE Ulrich and Cooper, 1938
O. (LLN/LLO)–D. (PRA) Mar.

First: *Parastrophina bilobata* Cooper, 1956, Pratt Ferry Formation, Alabama, USA (Cooper, 1956).
Last: *Anastrophia verneuili* (Hall, 1857), Helderberg Formation, New York, USA.

F. PORAMBONITIDAE Davidson, 1853
O. (ARG)–S. (LLY) Mar.

First: *Porambonites*? *umbonatus* Cooper, 1956, Pogonip Group, Nevada, USA.
Last: Definite *Porambonites* sp., Dolhir Formation, Glyn Ceiriog, North Wales, UK (Hiller, 1980), or possibly *Noetlingia tscheffkini* (de Verneuil, 1845), unrevised Lower Silurian, Urals, former USSR (*fide* Biernat *in* Williams *et al.*, 1965).

F. SYNTROPHIIDAE Schuchert, 1896
O. (ARG–LLN/LLO) Mar.

First: *Syntrophia lateralis* (Whitfield, 1886), middle ARG, Cassin Formation, Vermont, USA (Biernat *in* Williams *et al.*, 1965).
Last: *Xenelasmella perplexa* Liu *et al.*, 1984, middle Ordovician, Kunlun and Altun Mountains, China (Liu *et al.*, 1984) or *Syntrophia gigantea* Ross, 1987, Newfoundland, Canada (Ross and James, 1987).

F. SYNTROPHOPSIDAE Ulrich and Cooper, 1936
Є. (CRF)–O. (ARG) Mar.

First: *Tcharella amgensis* Andreeva, 1987, *Kooteniella* Zone, Amga River, Siberian Platform, former USSR (Andreeva, 1987).
Last: *Rhysostrophia nevadensis* Ulrich and Cooper, 1936b, upper Pogonip Group, Nevada, USA.

F. TETRALOBULIDAE Ulrich and Cooper, 1936
O. (TRE–ARG) Mar.

First: *Tetralobula*? *nundina* (Walcott, 1905), lower Pogonip Group, Nevada, USA (Ulrich and Cooper, 1938).
Last: *Tetralobula huanghuaensis* Wang, 1975, ARG, Yichang, south-west China.

Superfamily PENTAMEROIDEA M'Coy, 1844

F. ENANTIOSPHENIDAE Torley, 1934
D. (GIV) Mar.

First and Last: *Enantiosphen vicaryi* (Davidson, 1882), GIV, Germany.

F. GYPIDULIDAE Schuchert and Le Vene, 1929
S. (LLY)–D. (FRS) Mar.

First: There are a number of records of the Gypidulidae from the lower WEN and an unconfirmed report from the upper LLY of the Hudson Bay Lowlands of Canada.
Last: There are a number of records of the genus *Gypidula* from the Upper Devonian (FRS); these include, *Gypidula sublubrica* Zhao, 1977, Hupeh Province, China, together with *G. mimica* Cooper and Dutro, 1982, *G. stainbrooki* Cooper and Dutro, 1982 and *G. subcarinata* Cooper and Dutro, 1982, New Mexico, USA (Cooper and Dutro, 1982).

F. PENTAMERIDAE M'Coy, 1846
O. (ASH)–S. (PRD) Mar.

First: The genera *Eoconchidium*, *Holorhynchus*, *Proconchidium* and *Tscherkidium* are common in the middle Ashgill; they are represented by a number of species, for example, *Proconchidium muensteri* (St. Joseph, 1938), ASH, Ringerike, Norway.

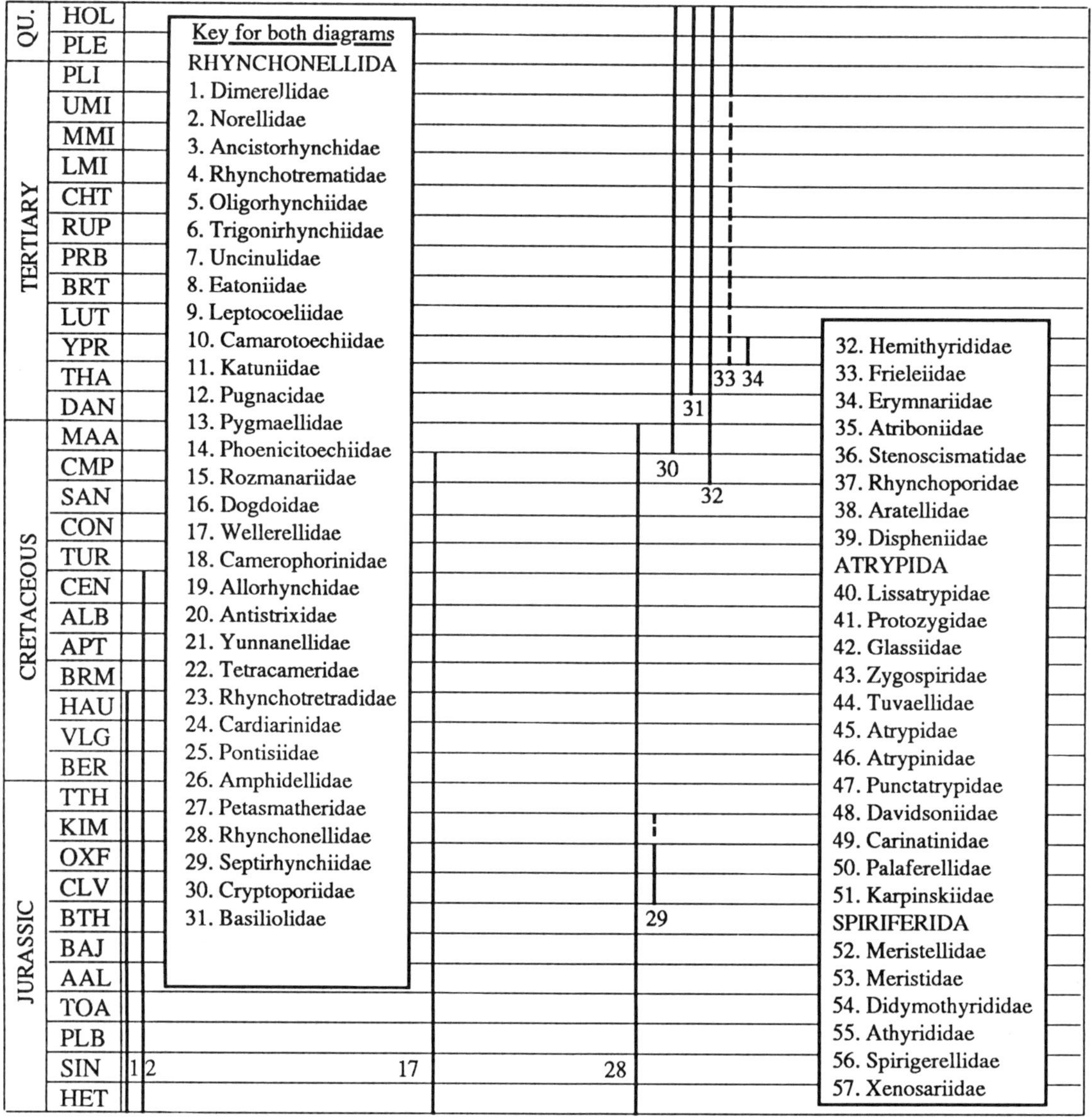

Fig. 22.4

Last: *Bisulcata indianensis* Boucot and Johnson, 1979, Kenneth Limestone Member, Salina Formation, Logansport, Indiana, USA.

F. STRICKLANDIIDAE Schuchert and Cooper, 1931 O. (ASH)–S. (WEN) Mar.

First: *Prostricklandia prisca* Rukavishnikova, 1973, Kazakhstan, former USSR (Sapelnikov, 1977); the systematic position of *Prostricklandia* requires revision.
Last: *Costistricklandia* sp. of Bassett (1977) from *linnarssoni* Zone age beds (Sheinwoodian), near Llandeilo, Wales, UK. [LRMC and END]

PRECISE POSITION UNCERTAIN

F. ANGUSTICARDINIIDAE Schuchert and Cooper, 1931 O. (ARG–ASH) Mar.

This group is better located within the porambonitides than the orthides where it possibly gave rise to the rhynchonellides (Jaanusson, 1971).
First: *Angusticardinia recta* (Pander, 1830) and *A. striata* (Pander, 1830), Latorp Stage, Estonia, former USSR (Rubel, 1961).
Last: *Apatorthis ultima* Öpik, 1933, Pirgu Stage, Estonia, former USSR (Öpik, 1933). [DATH]

Order RHYNCHONELLIDA Kuhn, 1949

The classification of Mesozoic and Cainozoic rhynchonellides used here is largely based on that proposed by Ager (1965) in the *Treatise*, but has been modified to include some of the less-tentative ideas put forward by Ager *et al.* (1972).

Superfamily DIMERELLOIDEA Buckman, 1918

F. DIMERELLIDAE Buckman, 1918 Tr. (CRN)–K. (HAU) Mar.

First: *Halorelloidea rectifrons* (Bittner, 1884), CRN, Plesivec, Slovakia, Eastern Alps, Czechoslovakia (Ager, 1968) is an early species of numerous poorly dated species of *Halorella* and *Halorelloidea* known from the CRN and NOR worldwide (Ager, 1968). It is not certain which species of these two genera is the first of the Dimerellidae.
Last: *Peregrinella multicarinata* (Lamarck, 1819), HAU, Drôme, south-west France (Biernat, 1957).

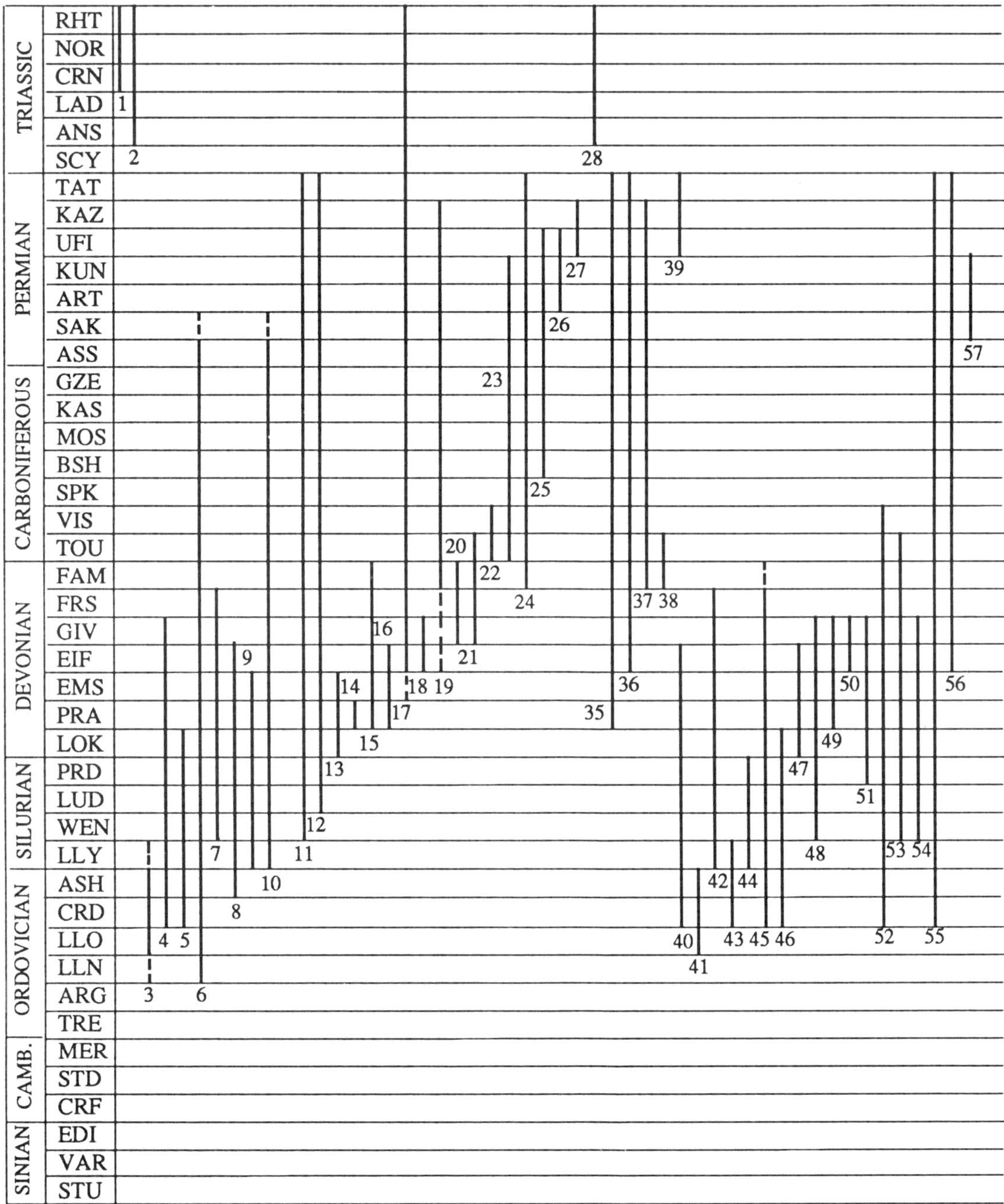

Fig. 22.4

F. NORELLIDAE Ager, 1959 Tr. (ANS)–K. (CEN) Mar.

First: *Norella refactifrons* Bittner, 1890, Schreyeralm Schichten, Upper ANS, northern Austrian Alps (Ager *et al.*, 1972).

Last: *Monticlarella jefferiesi* Owen, 1968, Plenus Marl, Yorkshire, England, UK (Owen, 1968).

Superfamily RHYNCHONELLOIDEA Gray, 1848

No overall consensus on the classification of the Palaeozoic families of this group has emerged since publication of the *Treatise*; this together with the establishment of a large number of new Middle Palaeozoic genera without clear familial attachments has limited the precision of this part of the project. Palaeozoic members of the superfamily have been listed mainly in order of stratigraphical appearance. Tentative proposals by Ager *et al.* (1972) to extend the families Pugnacidae and Erymnariidae into the Mesozoic are not followed here; there is currently insufficient evidence to warrant adoption of these proposals.

F. ANCISTORHYNCHIDAE Cooper, 1956 O. (LLN/LLO)–?S. (LLY) Mar.

First: Probably *Ancistorhynchia? perplexa* Cooper, 1956, McLish Formation, Oklahoma, USA, *A.? vacua* Cooper, 1956, Mingan Formation, Quebec, Canada (Cooper, 1956) or possibly *A. modesta* Popov, 1981, Baygard Formation and *A. prisca* Popov, 1981, Sarybidaik Formation, Kazakhstan, former USSR (Nikiforova and Popov, 1981).

Last: *Kritorhynchia seclusa* Rong and Yang, 1981, Xiangshuyuan Formation, NE Guizhou, China (Rong and Yang, 1981). The probable reassignment of this genus to the atrypides (J.-Y. Rong, unpublished data) may restrict this family to the Ordovician.

F. TRIGONIRHYNCHIIDAE McLaren, 1965 O. (LLN)–P. (ASS/SAK) Mar.

First: *Rostricellula triangularis* Williams, 1949, Upper LLN Ashes and Ffairfach Group, Llangadog, Wales, UK (Lockley and Williams, 1981).
Last: *Tricoria hirpex* Cooper and Grant, 1976, Skinner Ranch Formation, west Texas (Cooper and Grant, 1976).

F. RHYNCHOTREMATIDAE Schuchert, 1913 O. (CRD)–D. (GIV) Mar. (includes Lepidocyclidae)

First: *Rhynchotrema kentuckiense* Fenton and Fenton, 1922, Trenton Formation, Kentucky, USA, or *R. wisconsinense* Fenton and Fenton, 1922, Decorah Formation, Minnesota and Wisconsin, USA (Cooper, 1956).
Last: Possibly *Callipleura nobilis* (Hall, 1860), Hamilton Group, New York, USA.

F. OLIGORHYNCHIIDAE Cooper, 1956 O. (CRD)–D. (LOK) Mar.

First: *Oligorhynchia subplana gibbosa* Cooper, 1956, Lincolnshire Formation, Tennessee, USA or *O. conybeare* (Reed, 1917), Balclatchie Conglomerate, Girvan district, Scotland, UK (Williams, 1962).
Last: If *Rhynchotreta* is considered a member of this family, then *R. gansuensis* Fu, 1983, Putongyan Formation, NW China (Fu, 1983).

F. INNAECHIIDAE Baranov, 1980 O. (ASH)–D. (EIF) Mar.

First: *Lepidocycloides baikiticus* Nikiforova, 1961, Upper Ordovician, Siberia, former USSR.
Last: *Alekseevaella salagaensis* (Rzonsnickaja, 1967), Ural'tun Formation, Omulevskij Mountains, NE of former USSR.

F. EATONIIDAE Schmidt, 1965 O. (HIR)–D. (EIF) Mar.

First: *Clarkeia* sp., upper Bani Sandstone Formation, Morocco (Destombes *et al.*, 1985).
Last: Although there are records of *Costellirostra, Eucharitina, Pegmarhynchia* and *Tanerhynchia* species from the EMS, *Clarkeia bublitschekoi* Kaplun, 1968, EIF, North Pribalkhash, former USSR may be youngest (Kaplun, 1968).

F. LEPTOCOELIIDAE Boucot and Gill, 1956 S. (LLY)–D. (EMS) Mar.

First: Probably *Eocoelia hemisphaerica* (J. de C. Sowerby, 1839) from middle Llandovery localities in Wales and the Welsh Borderlands, UK (Cocks, 1978).
Last: Species of *Australocoelia, Leptocoelia* and *Pacificocoelia* occur in EMS rocks in a number of parts of the world; *Pacificocoelia infrequens* (Johnson, 1970) from EMS rocks in Nevada is one of the last (Boucot, 1975).

F. CAMAROTOECHIIDAE Schuchert and Le Vene, 1929 S. (LLY)–P. (ASS/SAK) Mar.

First: *Fenestrirostra primaeva* Jin, 1989, Merimack Formation, Anticosti Island, Canada (Jin *et al.*, 1990).
Last: *Paranorella aquilonia* Cooper and Grant, 1976, Bell Canyon Formation, west Texas (Cooper and Grant, 1976).

F. UNCINULIDAE Rzhonsnitskaya, 1956 S. (WEN)–D. (FRS) Mar.

First: Possibly *Hostimex hostimensis* Havlícek, 1982, Motol Formation, Hostim, Czechoslovakia (Havlíček, 1982b).
Last: Although several records of *Uncinunellina* from Upper Permian rocks, for example, *U. theobaldi* (Waagen, 1884), Salt Range, Pakistan, were considered youngest, Grant (1976) has transferred *Uncinunellina* to the Wellerellidae. Youngest may be *Hypothyridina cuboides* (J. de C. Sowerby, 1840), Upper Devonian, south Devon, England, UK.

F. KATUNIIDAE Xu and Yao, 1984 S. (WEN)–P. (TAT) Mar.

First: *Sulcatina sulcata* (Cooper, 1942), Waldron Shale, USA.
Last: *Terebratuloidea davidsoni* Waagen, 1883, Wargal Limestone, Salt Range, Pakistan (Grant, 1976).

F. PUGNACIDAE Rzhonsnitskaya, 1956 S. (LUD)–P. (ZEC) Mar.

First: *Xeniopugnax modicus* (Barrande, 1847) and *X. lynx* (Barrande, 1879) from the Kopanina Formation, Bohemia, Czechoslovakia (Havlíček, 1982a).
Last: Possibly *Pugnoides tardivensis* Waterhouse, 1982, Permian, New Zealand (Waterhouse, 1982).

F. PYGMAELLIDAE Baranov, 1977 D. (LOK–EMS) Mar.

First: Probably *Pygmaella pygmaea* Baranov, 1977, Sagyrian Horizon, in the NE of former USSR.
Last: *Pygmaella orbiculata* Baranov, 1977, Nelichenian Horizon, in the NE of former USSR (Baranov, 1977).

F. PHOENICITOECHIIDAE Havlíček, 1990 D. (PRA) Mar.

First and Last: This small family consists of three genera mainly occurring in the Koněprusy Limestone and coeval strata, Koněprusy area, Czechoslovakia; *Phoenicitoechia phoenix* (Barrande, 1847) and *Kotysex simulans* (Barrande, 1879) are representative (Havlíček, 1990).

F. ROZMANARIIDAE Havlíček, 1982 D. (PRA–FAM) Mar.

First: *Rackirhynchia lacerata* (Barrande, 1879) from the Koněprusy Limestone, Koněprusy, Czechoslovakia (Havlíček, 1990).
Last: Probably *Hadyrhyncha hadyensis* Havlíček, 1979 from the Hady Limestone, near Brno, Czechoslovakia (Havlíček, 1979).

F. DOGDOIDAE Baranov, 1982 D. (PRA–EIF) Mar.

First: Probably *Corvinopugnax bimbax* Havlíček, 1990 from the Vinařice Limestone, Koněprusy, Czechoslovakia.
Last: *Selennjachia abaimovae* Baranov, 1982, Seymchan Formation, Selennyakh Range, former USSR and *Tatjanica trigona* Baranov, 1982, Seymchan Formation, Ulakhan-Sis Range, former USSR.

F. WELLERELLIDAE Likharev *in* Rzhonsnitskaya, 1956 D. (EMS/EIF)–K. (CMP) Mar.

First: Possibly *Yakukijaella dubatolovi* Baranov, 1977, Nelichenian Horizon, Selennyakh Range, in the NE of former USSR (Baranov, 1977) or *Tetratomia amanshauseri* (Dahmer, 1942) from the EIF of the Rhineland (Havlíček, 1961).
Last: *Orbirhynchia bella* Pettitt, 1954 and *O. granum* Pettitt, 1954, both from the *pilula* or *quadrata* zones, Upper Chalk, southern England, UK (Pettitt, 1954).

F. CAMEROPHORINIDAE Rzhonsnitskaya, 1958 D. (EIF–GIV) Mar.

First: *Camarophorina pachyderma* (Quenstedt, 1871), EIF, Germany (Schmidt, 1941).
Last: *Camarophorina bijugata* (Schnur, 1851), GIV, Moravia (Schmidt, 1941).

F. ALLORHYNCHIIDAE Cooper and Grant, 1976 D. (EIF/FAM)–P. (GUA) Mar.

First: Probably *Yanetechia excelsa* Baranov, 1980 and *Y. limata* Baranov, 1980, Middle Devonian, in the NE of former USSR (Baranov, 1980) or possibly *Allorhynchoides kirki* Savage, Erberlein and Churkin, 1978, Refugio Formation, SE Alaska, USA (Savage *et al.*, 1978).
Last: Several allorhynchiids occur high in the Bell Canyon Formation and coeval units in west Texas, USA (Cooper and Grant, 1976a), for example, *Allorhynchus venustulum* Cooper and Grant, 1976, *Ptilotorhynchus delicatum* Cooper and Grant, 1976, *Deltarina magnicostata* Cooper and Grant, 1976, and *Fascicosta longaeva* (Girty, 1909).

F. ANTISTRIXIDAE Johnson, 1972 D. (GIV–FAM) Mar.

First: Probably *Antistrixia invicta* Johnson, 1972, GIV, Nevada, USA (Johnson, 1972).
Last: Possibly *Kindleina suemezensis* Savage, Erberlein and Churkin, 1978, Port Refugio Formation, SE Alaska, USA (Savage *et al.*, 1978).

F. YUNNANELLIDAE Rzhonsnitskaya, 1959 D. (GIV)–C. (TOU) Mar.

First: Possibly *Schnurella schnuri* (de Verneuil, 1840), Gerolstein, the Eifel, Germany (Schmidt, *in* Williams *et al.*, 1965).
Last: Possibly *Paraphorhynchus elongatum* Weller, 1914, lower Mississippian, Mississipi Valley Basin, USA (Weller, 1914).

F. TETRACAMERIDAE Licharew *in* Rzhonsnitskaya, 1956 C. (TOU–VIS) Mar.

First: *Rotaia subtrigona* (Meek and Worthen, 1860), lower Mississippian, Illinois, USA (Carter, 1990).
Last: *Tetracamara subcuneata* (Hall, 1858), middle Mississippian, Indiana, USA (Carter, 1990).

F. RHYNCHOTRETRADIDAE Licharew *in* Rzhonsnitskaya, 1956 C. (TOU)–P. (ROT) Mar.

First and Last: Possibly *Goniophoria monstrosa* Yanishevskiy, 1910, Lower Carboniferous to Lower Permian, south Urals, former USSR (Licharew, 1957).

F. CARDIARINIDAE Cooper, 1956 D. (FAM)–P. (ZEC) Mar.

First: *Loborina lobata* Balinski, 1982, Upper Devonian, Cracow Region, Poland (Grant, 1988).
Last: *Lambdarina iota* Grant, 1988, Upper Permian, Hydra Island, Greece (Grant, 1988).

F. PONTISIIDAE Cooper and Grant, 1976 C. (BSK)–P. (GUA) Mar.

First: *Pontisia* spp. have been reported from Pennsylvanian rocks in North America, for example, the Gaptank Formation of west Texas (Cooper and Grant, 1972); however, *Pontisia leonica* (Martinez-Chacon, 1979) from the Cantabrian Mountains, Spain, may be oldest.
Last: A number of pontisiids are present in the Bell Canyon Formation, west Texas, USA (Cooper and Grant, 1976a), for example, *Aphaurosia scutata* Cooper and Grant, 1976; *A. rotundata* Cooper and Grant, 1976; *Anteridocus bicostatus* Cooper and Grant, 1976; *A. swallovianus* (Shumard, 1860) and *Lirellaria costellata* Cooper and Grant, 1976.

F. AMPHIDELLIDAE Cooper and Grant, 1976 P. (ART–UFI) Mar.

First: *Amphipella arcaria* Cooper and Grant, 1976, base of Cathedral Mountain Formation, west Texas, USA (Cooper, and Grant, 1976a).
Last: *A. attenuata* Cooper and Grant, 1976, Road Canyon Formation, west Texas, USA (Cooper and Grant, 1976a).

F. PETASMATHERIDAE Cooper and Grant, 1976 P. (GUA) Mar.

First: *Iotina minuta* Cooper and Grant, 1976, Bone Spring Limestone, west Texas, USA (Cooper and Grant, 1976a).
Last: *Petasmatherus opulus* Cooper and Grant, 1976, Word and Cherry Canyon Formations, west Texas, USA (Cooper and Grant, 1976a).

F. RHYNCHONELLIDAE Gray, 1848 Tr. (ANS)–K. (MAA) Mar.

First: *Piarorhynchia trinodosi* (Bittner, 1890), Werfener Schiefer and Muschelkalk of numerous localities in the Alps (Bittner, 1890; Ager *et al.*, 1972).
Last: Three species of *Cretirhynchia* recorded by Pettitt (1950, 1954) from the *lanceolata* Zone, Upper Chalk, southern England, UK. *C. triminghamensis* Pettitt, 1950, *C. magna* Pettitt, 1950 and *C. limbata* (Schlotheim, 1813).

F. SEPTIRHYNCHIIDAE Muir-Wood and Cooper, 1951 J. (CLV-KIM?) Mar.

First: *Septirhynchia? budulcaensis* (Steffanini, 1932), CLV, Tunisia (Mancenido and Walley, 1979).
Last: Possibly *Septirhynchia azaisi* (Cottreau, 1924), Daghani Shales, Ida Kabeita, Somalia (Muir-Wood, 1935); but definitely *Septirhynchia hirschi* Feldman, 1987, Masajid Formation, Middle–Upper CLV, Gebel El-Maghara, northern Sinai (Feldman, 1987).

F. CRYPTOPORIIDAE, Muir-Wood, 1955 K. (MAA)–Rec. Mar.

First: *Cryptopora antiqua* Bitner and Pisera, 1979, MAA Chalk facies of Mielnik, eastern Poland (Bitner and Pisera, 1979). The second oldest is a record of *Cryptopora*, possibly *C. parvillima* (Sacco, 1902), Palaeocene (DAN), Crimea, former USSR (Popiel-Barczyk, 1980). **Extant**

F. BASILIOLIDAE, Cooper, 1959 T. (THA)–Rec. Mar.

First: *Probolarina chathamensis* Lee, 1978, Upper Palaeocene–lower Eocene, Chatham Islands, New Zealand (Lee, 1978). Other early species include *Eohemithyris alexi* Hertlein and Grant, 1944, Domengine Formation (Eocene),

California, USA (Cooper, 1959) and *Probolarina brevirostris* Cooper, 1988, Castle Hayne Formation (Eocene), North Carolina, USA (Cooper, 1988). The earliest record of the genus *Aetheia* which is arguably attributable to this family, is from the Eocene of Otago, New Zealand (Lee, 1978). **Extant**

F. HEMITHYRIDIDAE, Rzhonsnitzkaya, 1956 K. (CMP)–Rec. Mar.

First: *Protegulorhynchia meridionalis* Owen, 1980, Lower CMP, James Ross Island, Antarctica (Owen, 1980); next earliest is *Tegulorhynchia boongeroodaensis* McNamara, 1983, Boongerood Greensand (Lower Palaeocene–Lower Eocene), Western Australia (McNamara, 1983). **Extant**

F. FRIELEIIDAE, Cooper, 1959 T. (Eoc.?, PLI)–Rec. Mar.

First: May be an unnamed species of *Hispanirhynchia*, Eocene, Habana Province, Cuba (Cooper, 1959), but definitely *Sphenarina sicula* (Seguenza, 1870), Pliocene, Sicily (Davidson, 1870). **Extant**

F. ERYMNARIIDAE, Cooper, 1959 T. (YPR) Mar.

First: Probably *Erymnaria polymorpha* (Massalongo, 1850), Lower Eocene, Spilecco, Verona, Italy (Cooper, 1959).
Last: May be *Erymnaria cubensis* Cooper, 1959, Eocene, Matanzas Province, Cuba (Cooper, 1959). Poor dating and correlation make exact determination of first and last species impossible. [DATH and CDP]

***Superfamily* STENOSCISMATOIDEA Oehlert, 1887**

F. ATRIBONIIDAE Grant, 1965 D. (PRA)–P. (ZEC) Mar.

First: *Proatribonium alticum* Gratsianova, 1967, Lower Devonian, Altai Range, Russia, former USSR.
Last: *Ussuricamara majchensis* Koczyrkevicz, 1969, upper Permian, southern Primoria, Russia, former USSR.

F. STENOSCISMATIDAE Oehlert, 1887 D. (EIF)–P. (CHX) Mar.

First: *Coledium rhomboidale* (Hall and Clarke, 1894) from the Logansport Limestone, Indiana (Grant, 1965).
Last: Species of *Stenoscisma* are widespread in uppermost Permian rocks; a distinctive species from Dzhulfian rocks of Armenia may be youngest (Grant, 1970) or more probably unnamed species from the Changxing Formation, China, or from the Episkopi Limestone, Greece (Dr R. E. Grant, pers. comm.).

***Superfamily* RHYNCHOPOROIDEA Muir-Wood, 1955**

F. RHYNCHOPORIDAE Muir-Wood, 1955 D. (FAM)–P. (KAZ) Mar.

First: Probably *Rhynchopora? morini* (Drot, 1964), FAM, Morocco (Brunton, 1971).
Last: *Rhynchopora geinitziana* (Verneuil, 1845), Upper Permian, north of the Russian Platform, former USSR (Erlanger, 1981).

F. ARATELLIDAE Erlanger, 1986 D. (FAM)–C. (TOU) Mar.

First: *Aratella dichotomians* (Abrahamians, 1954), upper FAM, Transcaucasus, former USSR (Erlanger, 1986).
Last: *Aratella dichotomians* (Abrahamians, 1954) or *A. aratica* (Abrahamians, 1957), lower TOU, Transcaucasus, former USSR (Erlanger, 1986).

***Superfamily* POSITION UNCERTAIN**

F. DISPHENIIDAE Grant, 1988 P. (ZEC)

The placement of this aberrant group requires clarification; it may be a rhynchonellide or terebratulide.
First and Last: *Disphenia myiodes* Grant, 1988 from the Upper Permian rocks of Hydra Island, Greece (Grant, 1988). [DATH]

***Order* ATRYPIDA Rzhonsnitskaya, 1960 (see Fig. 22.4)**

***Suborder* LISSATRYPIDINA Twenhofel, 1914**

***Superfamily* LISSATRYPOIDEA Twenhofel, 1914**

F. LISSATRYPIDAE Twenhofel, 1914 O. (CRD)–D. (EIF) Mar.

First: *Webbyspira principalis* Percival, 1991, Dunhill Bluff Limestone, lower CRD, Australia (Percival, 1991).
Last: *Holynatrypa crucifera* Havlíček, 1973, Třebotov Limestone, lower EIF, Bohemia, Czechoslovakia.

F. PROTOZYGIDAE Copper, 1986 O. (LLO–ASH) Mar.

First: *Manespira* sp., Crown Point Formation, New York, USA (Copper, 1986).
Last: *Idiospira taoqupoensis* Fu, 1982, upper Beiguoshan Formation, China (Copper, 1986).

F. GLASSIIDAE Schuchert and Le Vene, 1928 S. (LLY)–D. (FRS) Mar.

First: Possibly *Glassia minuta* Rybnikova, 1967, upper LLY, Latvia (Copper, 1986).
Last: *Peratos beyrichi* (Kayser, 1872), 'Rotheisenstein', Brilon, Germany or possibly *P. drevermanni* Maillieux, 1936, Schistes de Matagne, Belgium (Copper, 1986).

***Suborder* ZYGOSPIRIDINA Waagen, 1883**

***Superfamily* ZYGOSPIROIDEA Waagen, 1883**

F. ZYGOSPIRIDAE Waagen, 1883 O. (CRD)–S. (LLY) Mar.

First: *Anazyga matutina* (Cooper, 1956), Little Oak Formation, Lower CRD, Alabama, USA
Last: *Pentlandella tenuistriata* Rubel, 1970, Adavere Formation, upper LLY, Estonia, former USSR (Copper, 1977), *Zygatrypa exigua* (Lindström, 1861), Lower Visby Beds, upper LLY, Gotland, Sweden.

F. TUVAELLIDAE Alikhova, 1960 S. (LLY–PRD) Mar.

First: *Tuvaella rackovskii* Chernyshev, 1937, Elegest Formation, Altai, former USSR.
Last: *Tuvaella gigantea buchtarmaensis* Kulkov and Kozlov, 1978, Sazhaev Formation, Altai, former USSR.

***Suborder* ATRYPIDINA Boucot, Johnson and Staton, 1964**

***Superfamily* ATRYPOIDEA Gill, 1871**

F. ATRYPIDAE Gill, 1871 O. (CRD)–D. (FRS/?FAM) Mar.

First: *Protatrypa malmoeyensis* Boucot *et al.*, 1964, lower LLY, Norway or *Sypharatrypa honora* Copper, 1982, Manitoulin Formation, Ontario, Canada.
Last: *Spinatrypa frequens* (Weller, 1907), Glen Park Limestone, Missouri, USA.

F. ATRYPINIDAE McEwen, 1939
O. (CRD)–D. (LOK) Mar.

First: *Sulcatospira parva* (Rukavishnikova, 1970), Dulankara Formation, Kazakhstan, former USSR.
Last: *Atrypina prosimpsoni* Johnson *et al.*, 1973, Fauna F, Nevada, USA.

Superfamily PUNCTATRYPOIDEA Rzhonsnitskaya, 1960

F. PUNCTATRYPIDAE Rzhonsnitskaya, 1960
D. (LOK–EIF) Mar.

First: *Punctatrypa tumidula* (Khodalevich, 1951), Saum Formation, eastern Urals, former USSR.
Last: *Punctatrypa siehli* Johnson, 1975, Greifensteinerkalk, Germany.

Suborder DAVIDSONIIDEA King, 1850

Superfamily DAVIDSONIOIDEA King, 1850

F. DAVIDSONIIDAE King, 1850
S. (WEN)–D. (GIV) Mar.

First: *Gracianella praecrista* Johnson, Boucot and Murphy, 1976 from the upper Wenlock of Nevada, USA.
Last: *Davidsonia antelopensis* Johnson and Trojan, 1982, Interval 23, upper GIV, Nevada, USA.

F. CARINATINIDAE Rzhonsnitskaya, 1960
D. (PRA–GIV) Mar.

First: *Carinatina comata* (Barrande, 1847), Koněprusy Limestone, Prague, Czechoslovakia; *Carinatina comatoidea* Alekseeva and Kulkov, 1970, Malobachat Limestone, Kuznetsk Basin, former USSR.
Last: *Carinatina signifera* (Schnur, 1853), Gerolstein, Eifel, Germany.

Superfamily PALAFERELLOIDEA Spriestersbach, 1942

F. PALAFERELLIDAE Spriestersbach, 1942
D. (EIF–GIV) Mar.

First: *Gruenewaldtia* sp., Wellersbach Horizon, lower EIF, Eifel, Germany.
Last: *Gruenewaldtia sibirica* Ivanova, 1956, Safonov Horizon, upper GIV, Kuznetsk Basin, former USSR (Ivanova, 1962).

F. KARPINSKIIDAE Poulsen, 1943
S. (PRD)–D. (GIV) Mar.

First: *Tectatrypa tectiformis* (Chernyshev, 1893), Severural Horizon, PRD, eastern Urals, former USSR or *Eokarpinskia nalivikini* (Nikiforova, 1937), Svetloroz Formation, Kazakhstan, former USSR.
Last: *Vagrania (Desatrypa) desquamata* (Sowerby, 1840), upper GIV, Devon, England, UK, or *Vagrania* (*Desatrypa*) *globosa* (Leidhold, 1928), Massenkalk, Germany. [PC]

Order SPIRIFERIDA Waagen, 1883

Suborder ATHYRIDIDINA Boucot, Johnson and Staton, 1964

Superfamily MERISTELLOIDEA Waagen, 1883

F. MERISTELLIDAE Waagen, 1883
O. (CRD)–C. (VIS) Mar.

First: Possibly *Hyattidina*? *sulcata* Williams, 1962, Kiln Mudstones, Girvan district, SW Scotland, UK (Williams, 1962).
Last: Several species of *Meristina* have been reported from Lower Carboniferous rocks, for example, *M. roemeri* Weller, 1914, Mississippi Basin, USA.

F. MERISTIDAE Hall and Clarke, 1895
S. (WEN)–C. (TOU) Mar.

First: Species of *Merista*, for example, *M. typa* Hall, 1859, are common in rocks of WEN age in the Western Hemisphere (Boucot *et al.*, 1969) or *Diacamaropsis parva* (Thomas, 1926), St Clair Limestone, Arkansas, USA (Amsden and Barrick, 1988).
Last: Probably *Camarophorella mutabilis* Hyde, 1953, Logan Formation, Ohio, USA (Hyde, 1953) or *Merista maccullochensis* Carter, 1971, lower Mississippian, Iowa, USA (Carter, 1971).

Superfamily ATHYRIDIDOIDEA Davidson, 1881

F. DIDYMOTHYRIDIDAE Modzalevskaya, 1977
S. (WEN)–D. (GIV) Mar.

First: *Glassina laeviuscula* (Sowerby, 1839), WEN, Welsh Borderland, UK (Cocks, 1978) or *G. usitata* Modzalevskaya, 1979, Mukshinian Horizon, Podolia, former USSR (Modzalevskaya, 1979).
Last: *Buchanathyris westoni* Talent, 1956, Buchan Caves Limestone, Victoria, Australia (Talent, 1956).

F. ATHYRIDIDAE Davidson, 1881
O. (CRD)–P. (ZEC) Mar.

First: Possibly *Apheathyris guyuanensis* Fu, 1982, middle Ordovician, NW China (Fu, 1982).
Last: Many species of *Cleiothyridina* are reported from the upper Permian, for example, *C. capillata* (Waagen, 1883), upper part of Chhidru Formation, west Pakistan (Grant, 1970).

F. SPIRIGERELLIDAE Grunt, 1965
D. (EIF)–P. (TAT) Mar.

First: *Meristospira michiganense* Grabau, 1910, Amherstberg Dolomite, Michigan, USA.
Last: *Spirigerella* spp., Mianwali Formation, Salt Range and Trans-Indus Ranges, west Pakistan (Grant, 1970).

F. XENOSARIIDAE Cooper and Grant, 1976
P. (GUA) Mar.

First and Last: *Xenosaria exotica* Cooper and Grant, 1976, Bell Canyon Formation, west Texas, USA (Cooper and Grant, 1976a).

F. DIPLOSPIRELLIDAE Schuchert, 1894
Tr. (SMI/SPA–RHT) Mar. (see Fig. 22.5)

First: *Sprigerellina pygmaea* Dagis, 1974, Olenekian, Mangyslak, Primoria, former USSR (Grunt, 1986).
Last: Several diplospirellid taxa are known from rocks of Rhaetian age, for example, *Oxycolpella oxycolps* (Emmrich, 1855) from the Alps, Carpathians, Crimea and the Caucasus (Dagis, 1962).

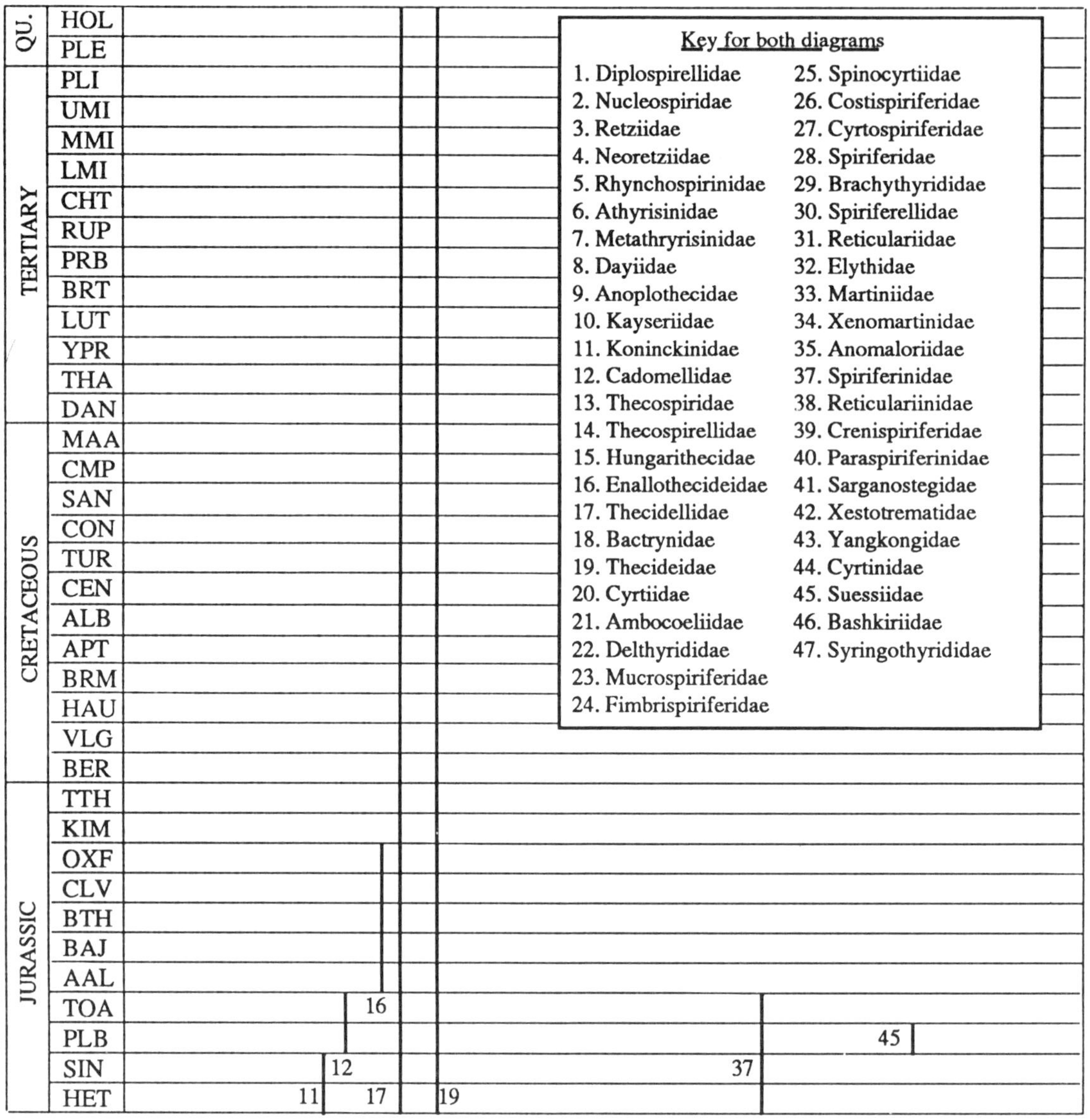

QU.: HOL, PLE
TERTIARY: PLI, UMI, MMI, LMI, CHT, RUP, PRB, BRT, LUT, YPR, THA, DAN
CRETACEOUS: MAA, CMP, SAN, CON, TUR, CEN, ALB, APT, BRM, HAU, VLG, BER
JURASSIC: TTH, KIM, OXF, CLV, BTH, BAJ, AAL, TOA, PLB, SIN, HET

16 12 45 37 11 17 19

Key for both diagrams

1. Diplospirellidae	25. Spinocyrtiidae
2. Nucleospiridae	26. Costispiriferidae
3. Retziidae	27. Cyrtospiriferidae
4. Neoretziidae	28. Spiriferidae
5. Rhynchospirinidae	29. Brachythyrididae
6. Athyrisinidae	30. Spiriferellidae
7. Metathryrisinidae	31. Reticulariidae
8. Dayiidae	32. Elythidae
9. Anoplothecidae	33. Martiniidae
10. Kayseriidae	34. Xenomartinidae
11. Koninckinidae	35. Anomaloriidae
12. Cadomellidae	37. Spiriferinidae
13. Thecospiridae	38. Reticulariinidae
14. Thecospirellidae	39. Crenispiriferidae
15. Hungarithecidae	40. Paraspiriferinidae
16. Enallothecideidae	41. Sarganostegidae
17. Thecidellidae	42. Xestotrematidae
18. Bactrynidae	43. Yangkongidae
19. Thecideidae	44. Cyrtinidae
20. Cyrtiidae	45. Suessiidae
21. Ambocoeliidae	46. Bashkiriidae
22. Delthyrididae	47. Syringothyrididae
23. Mucrospiriferidae	
24. Fimbrispiriferidae	

Fig. 22.5

Superfamily NUCLEOSPIROIDEA Davidson, 1881

F. NUCLEOSPIRIDAE Davidson, 1881
S. (WEN)–Tr. (RHT)

First: *Nucleospira pisiformis* Hall, 1859, upper Clinton Group, New York, USA (Dale, 1953).
Last: *Nucleospira cunctata* Cooper and Grant, 1976, Cathedral Mountain Formation, Texas, USA (Cooper and Grant, 1976a) or possibly *Amphitomella hemisphaeroidica* (Klipstein, 1843), RHT, Eastern Alps (Grunt, 1989).

Suborder RETZIIDINA Boucot, Johnson and Staton, 1964

Superfamily RETZIOIDEA Waagen, 1883

F. RETZIIDAE Waagen, 1883 D. (LOK–FRS) Mar.

First: *Leptospira costata* (Hall, 1859), New Scotland Formation, New York, USA (Boucot *et al.*, 1969).
Last: *Plectospira sexplicata* (White and Whitfield, 1862), Chemung Group, New York and Iowa, USA (Grunt, 1989).

F. NEORETZIIDAE Dagis, 1962
D. (FAM)–Tr. (RHT) Mar.

First: Possibly *Eumetria subtrigonalis* (Stainbrook, 1947), Percha Shale, New Mexico and Arizona, USA (Grunt, 1989).
Last: *Neoretzia superbescens* (Bittner, 1890), RHT, Crimea and the Caucasus, former USSR (Dagis, 1962).

F. RHYNCHOSPIRINIDAE Schuchert and Le Vene, 1929 S. (LLY)–D. (FAM) Mar.

First: Possibly *Homoeospirella*? sp., LLY, Haverfordwest, Wales, UK (Temple, 1987).
Last: *Rhynchospirina* cf. *haidingeri* Barrande, 1879, Hlubocepy Limestone, Czechoslovakia (Havlíček, 1956).

Superfamily ATHYRISINOIDEA Grabau, 1931

F. ATHYRISINIDAE Grabau, 1931
S. (PRD)–Tr. (NOR) Mar.

First: Possibly *Squamathyris glacialis* Modzalevskaya, 1981, Skala rocks, Podolia (Grunt, 1989).
Last: Probably *Misolia noetlingii* (Bittner, 1890), Sumra Formation, Arabia (Hudson and Jefferies, 1961).

F. METATHRYRISINIDAE Wang *et al.*, 1981
S. (LLY)–D. (LOK) Mar.

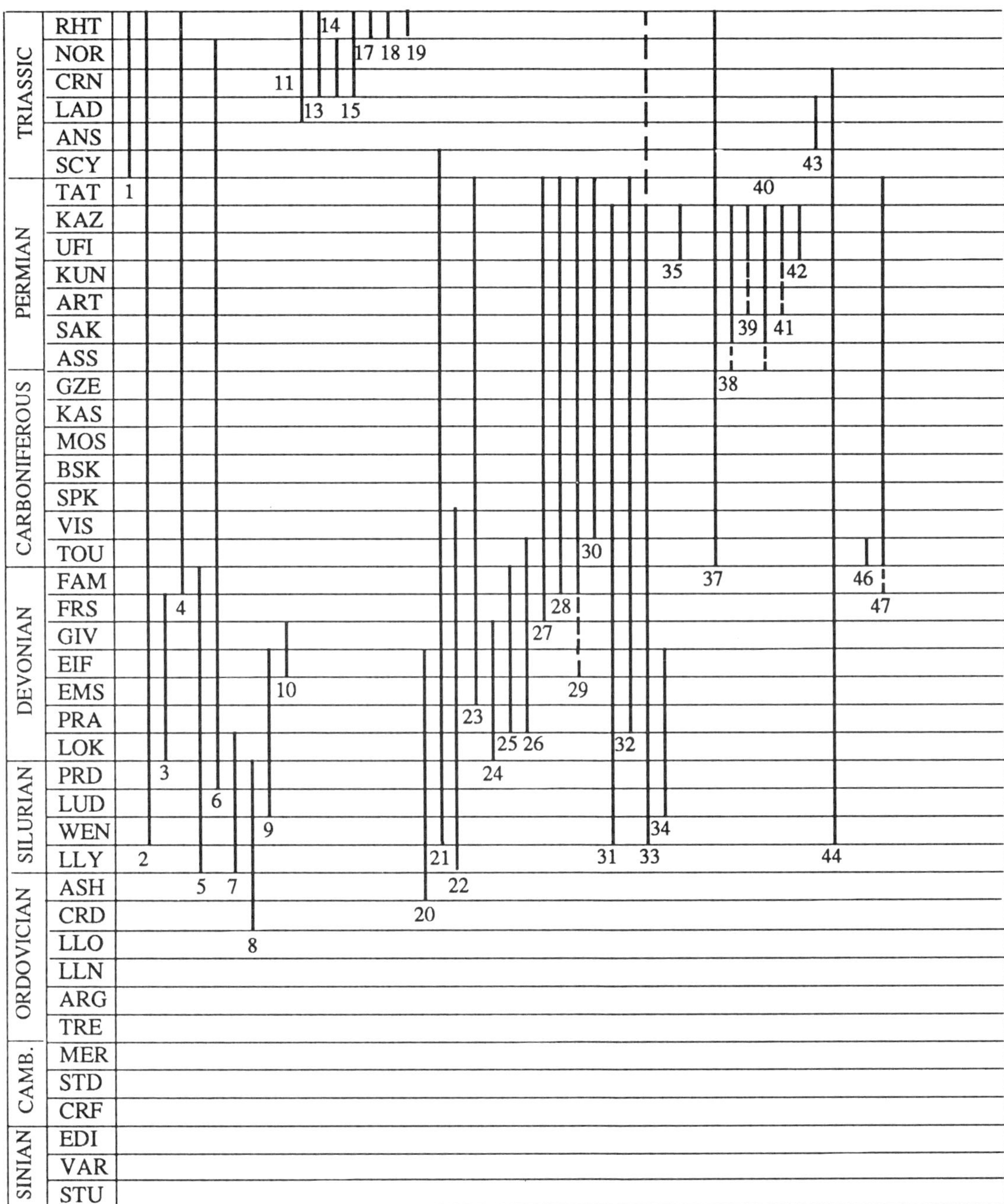

Fig. 22.5

First: Possibly *Metathyrisina merita* Rong and Yang, 1981, Leijiatun Formation, NE Guizhou, China (Rong and Yang, 1981).
Last: *Molongia elegans* Mitchell, 1921, Mt. Ida Beds, Victoria, Australia (Boucot *et al.*, 1969); in addition, several species of *Molongia* are also reported from LOK rocks in the western Qinling Mts, NW China.

Suborder DAYIIDINA Waagen, 1883

Superfamily DAYIOIDEA Waagen, 1883

F. DAYIIDAE Waagen, 1883
O. (CRD)–S. (PRD) Mar.

First: Possibly *Cyclospira? longa* Cooper, 1956, basal Ardwell Group, Girvan, SW Scotland, UK (Williams, 1962).
Last: *Dayia bohemica* Bouček, 1940, Upper Silurian rocks, Estonia and Podolia, former USSR (Rubel, 1977).

F. ANOPLOTHECIDAE Schuchert, 1894
S. (LUD)–D. (EIF) Mar.

First: *Coelospira saffordi* Foerste, 1903, Brownsport Formation, Tennessee, USA (Boucot and Johnson, 1967).
Last: *Coelospira camilla* Hall, 1867, Bois Blanc Formation, Lancaster, USA (Boucot and Johnson, 1967).

F. KAYSERIIDAE Boucot, Johnson and Staton, 1964
D. (EIF–GIV) Mar.

First: *Kayseria lens* (Phillips, 1841), Middle Devonian, SW England, UK.

Last: *Kayseria procera* Ficner and Havlíček, 1978, Middle Devonian, Čelechovic, Moravia, Czechoslovakia (Ficner and Havlíček, 1978).

Suborder KONINCKINIDINA Davidson, 1853

Superfamily KONINCKINOIDEA Davidson, 1853

F. KONINCKINIDAE Davidson, 1853 Tr. (LAD)–J. (SIN) Mar.

First: Bittner (1890) described a suite of koninckinids from the St Cassian Beds of the Italian Dolomites including *Koninckina amoena* Bittner, 1890 and *Koninckinella triassina* Bittner, 1890.
Last: Several koninckinids, including *Koninckodonta geyeri* Bittner, 1893 and *K. eberhardi* Bittner, 1893, are known from the middle Liassic of Corfu, Greece (Renz, 1932).

Superfamily CADOMELLOIDEA Schuchert, 1893

F. CADOMELLIDAE Schuchert, 1893 J. (PLB–TOA) Mar.

First: *Cadomella davidsoni* (Eudes Deslongchamps, 1854) and *C. moorei* (Davidson, 1876), Liassic, Somerset, England, UK (Davidson, 1876).
Last: *Cadomella davidsoni* (Eudes Deslongchamps, 1854), Liassic, Normandy, France (Davidson, 1876).

Suborder THECIDEIDINA Elliott, 1958

Baker's (1990) detailed revision of the group and its assignment to the Spiriferida are followed herein.

Superfamily THECOSPIROIDEA Bittner, 1890

F. THECOSPIRIDAE Bittner, 1890 Tr. (CRN–RHT) Mar.

First: *Thecospira semseyi* Bittner, 1890, CRN, Hungary (Dagis, 1972).
Last: *Thecospira haidingeri* (Suess, 1854), RHT, Northern Alps and Carpathians (Benigni and Ferliga, 1989).

F. THECOSPIRELLIDAE Dagis, 1972 Tr. (CRN–NOR) Mar.

First: Probably *Thecospirella loczyi* (Bittner, 1912), CRN, Lienz area, Southern Alps (Bittner, 1912).
Last: *Bitternella bittneri* Dagis, 1974, NOR, south-eastern Pamir (Dagis, 1974).

F. HUNGARITHECIDAE Dagis, 1972 Tr. (CRN–RHT) Mar.

First: *Hungaritheca andreaei* (Bittner, 1890), CRN, Hungary (Dagis, 1972).
Last: *Pamirotheca aulacothyridiformis* Dagis, 1974, RHT, south-eastern Pamir (Dagis, 1974).

Superfamily THECIDEOIDEA Gray, 1840

F. ENALLOTHECIDEIDAE Baker, 1983 J. (AAL–OXF) Mar.

First and Last: *Enallothecidea pygmaea* (Moore, 1861) from the Cotswold area of southern England, UK (Baker, 1983).

F. THECIDELLIDAE Elliott, 1958 Tr. (RHT)–Rec. Mar.

First: *Moorellina prima* Elliott, 1953, RHT, Hirtenberg, Austria (Elliott, 1953). **Extant**

F. BACTRYNIIDAE Williams, 1965 Tr. (RHT) Mar.

First and Last: *Bactrynium bicarinatum* Emmrich, 1855, Kossener Schichten of Austria (Rudwick, 1968).

F. THECIDEIDAE Gray, 1840 Tr. (RHT)–Rec. Mar.

First: *Davidsonella rhaetica* (Zugmayer, 1880) from the Rhaetian rocks of Waldegg, Austria (Zugmayer, 1880).
Extant
[DATH]

Suborder SPIRIFERIDINA Waagen, 1883

Superfamily CYRTIOIDEA Fredricks, 1919 (1924)

F. CYRTIIDAE Fredericks, 1919 (1924) O. (ASH)–D. (EIF) Mar.

First: *Iliella minima* Rukavischinikova, 1980, Choparskij Horizon, northern part of the Dulankar Mountains, Kazakhstan, former USSR (Appollonov *et al.*, 1980).
Last: Probably *Pinguispirifer infirmus* (Barrande, 1879), EIF, Czechoslovakia (Havlíček, 1971b).

F. AMBOCOELIIDAE George, 1931 S. (WEN)–Tr. (SCY) Mar.

First: Boucot (1975) reported a species of *Plicoplasia*, Cape Phillips Formation, Canadian Arctic; otherwise *Ambothyris praecox* Kozłowski, 1929, Ukraine or possibly *Ambocoelia operculifera* Havlíček, 1959, Bohemia, Czechoslovakia (Havlíček, 1959).
Last: *Paracrurithyris pigmaea* Liao, 1981, lower Lower Triassic, South China (Zhao, 1981).

Superfamily SPIRIFEROIDEA King, 1846

F. DELTHYRIDIDAE Waagen, 1883 S. (LLY)–C. (VIS) Mar.

First: Probably *Howellella anglicus* Lamont and Gilbert, 1945, Pentamerus Beds, Worcestershire, England, UK (Lamont and Gilbert, 1945).
Last: Possibly *Tylothyris clarksvillensis* (Winchell, 1865) from middle Mississippian rocks from the midcontinent of North America and elsewhere (Carter, 1990).

F. MUCROSPIRIFERIDAE Pitrat, 1965 D. (EMS)–P. (ZEC) Mar.

First: *Mucrospirifer mucronatus* (Conrad, 1841), Romney Formation, Maryland, USA (Vokes, 1957).
Last: *Pteroplecta laminatus* Waterhouse, 1978, northwestern Nepal (Waterhouse, 1978).

F. FIMBRISPIRIFERIDAE Pitrat, 1965 D. (LOK-GIV) Mar.

First: *Fimbrispirifer charybdis* (Barrande, 1879), lower Devonian, Bohemia, Czechoslovakia (Havlíček, 1959).
Last: *Fimbrispirifer venustus* (Hall, 1860), Hamilton Formation, New York, USA (Cooper, 1944).

F. SPINOCYRTIIDAE Ivanova, 1959 D. (PRA–FAM) Mar.

First: *Spinocyrtia affinis* (Fuchs, 1929), lower Siegenian, Belgium (Vandercammen, 1963).
Last: *Spinocyrtia struniana* (Gosselet, 1879), Couches d'Etroeungt, France (Vandercammen, 1956).

F. COSTISPIRIFERIDAE Termier and Termier, 1949
D. (PRA)–C. (TOU) Mar.

First: *Costispirifer arenosus* (Conrad, 1841), Oriskany Group, Maryland, USA (Vokes, 1957).
Last: Possibly *Eudoxina* species, for example *E. subrotunda* from the English River Sandstone, Iowa, USA (Carter, 1990).

F. CYRTOSPIRIFERIDAE Termier and Termier, 1949
D. (FRS)–P. (ZEC) Mar.

First: Possibly *Eodmitria supradisjunctus* (Obrutchew, 1917), FRS, Russian Platform and *E. s. boloniensis* Brice, 1982, FRS, northern France (Brice, 1982).
Last: *Tipispirifer oppilatus* Grant, 1976, Ko Muk, Thailand (Grant, 1976).

F. SPIRIFERIDAE King, 1846
(includes Neospiriferidae Termier, 1975)
D. (FAM)–P. (ZEC) Mar.

First: May be *Palaeospirifer karagatschicus* (Sverbilova, 1963), Upper Devonian, Kazakhstan, former USSR (Martynova and Sverbilova, 1968). But if *Glyptospirifer* belongs to this family, then *Glyptospirifer chui* Grabau, 1931 (Hou and Xian, 1975) and *Glyptospirifer chui cyrtinoides* Xian, 1978, lower–middle Devonian, China (Hou, 1979) may be the oldest.
Last: *Neospirifer chivatschensis* Zavodowskii, 1968, *N.* (?) *srjatkovi* Zavodowskii, 1968, *N. tricostatus* Zavodowskii, 1968, from the Omolona Basin, former USSR (Zavodowskii, 1968).

F. BRACHYTHYRIDIDAE Fredericks, 1919 (1924)
D. (EIF/FAM)–P. (ZEC) Mar.

First: Possibly *Brachythyris? talicensis* Khodalevich, 1959, Bauxite deposits, Urals, former USSR (Khodalevich and Breivel, 1959); otherwise *Brachythyris bisbeensis* Stainbrook, 1947 and *B. putilla* Stainbrook, 1947, Percha Shale, New Mexico and Arizona, USA (Stainbrook, 1947) and also *B. medioplicata* Martynova and Sverbilova, 1968, from the Upper Devonian, Kazakhstan, former USSR.
Last: *Purdonella oligsangonsis* Termier *et al.*, 1974, Wardak, central Afghanistan (Termier *et al.*, 1974).

F. SPIRIFERELLIDAE Termier, Termier, Lapparent and Marin, 1975 C. (VIS)–P. (TAT) Mar.

First: Probably *Spiriferella neglecta* (Hall, 1858) from the Keokuk Limestone, Mississippi Valley Basin (Weller, 1914).
Last: There are a number of upper Permian species of *Spiriferella*, for example, *Spiriferella* spp. from the Hawtel Limestone and the Stephens Formation, New Zealand (Waterhouse, 1968).

Superfamily RETICULARIOIDEA Waagen, 1883

F. RETICULARIIDAE Waagen, 1883
S. (WEN)–P. (KAZ) Mar.

First: Probably *Reticulariopsis silurica* Strusz, 1982 or *Vadum coppinsense* Strusz, 1982, Australian Capital Territory (Strusz, 1982).
Last: *Ambikella furca* Waterhouse, 1968, Stephens Formation, Maitai Group, Nelson, New Zealand.

F. ELYTHIDAE Fredericks, 1924
D. (PRA)–P. (ZEC) Mar.

First: *Elita* (*Elytha*) *saffordi* Hall, 1859, Oriskany Group, Pennsylvania, USA (Cleaves, 1939).
Last: *Spinomartinia spinosa* Waterhouse, 1968, Arthurton and Kiriwao groups, New Zealand (Waterhouse, 1968).

F. MARTINIIDAE Waagen, 1883
S. (WEN)–P. (KAZ/RHT) Mar.

First: *Tenellodermis praetenellus* Havlíček, 1971, Liteň Formation, Kozlupy, Lištice and Sedlec, Czechoslovakia (Havlíček, 1971b).
Last: *Notospirifer excelsus* Waterhouse, 1968, *N. microspinosus* Waterhouse, 1968, Arthurton Group, New Zealand (Waterhouse, 1968) or if *Mentzelia* belongs to the family, the last are Triassic (RHT), *Mentzelia kawhiana* Trechmann, 1918, *M.* cf. *ampla* Bittner, 1890, Hokonui Hills, Kawhia, New Zealand (Marwick, 1953).

F. XENOMARTINIIDA Havlíček, 1957
S. (LUD)–D. (EIF) Mar.

First: *Proreticularia carens* (Barrande, 1879), basal LUD, Czechoslovakia (Havlíček, 1950).
Last: *Obesaria obesa* (Barrande, 1848), EIF, Czechoslovakia (Havlíček, 1959).

F. ANOMALORIIDAE Cooper and Grant, 1976
P. (GUA) Mar.

First and Last: *Anomaloria anomala* Cooper and Grant, 1969, Bell Canyon Formation, west Texas, USA (Cooper and Grant, 1969).

Suborder SPIRIFERINIDINA King, 1846

This suborder includes the punctate spiriferides probably derived polyphyletically from the impunctate spiriferide group.

Superfamily SPIRIFERINOIDEA Davidson, 1884

F. SPIRIFERINIDAE Davidson, 1884
C. (TOU)–J. (LIA) Mar.

First: Possibly *Spiriferina paratransversa* Minato, 1952, base of TOU, Japan (Minato, 1952), or *Spiropunctifera tulnsis* Ivanova, 1971, Lower Carboniferous, Gorenskaya, Moscow Basin, former USSR (Ivanova, 1971).
Last: *Callospiriferina tumidus* (von Buch, 1836), Liassic, Morocco (Rousselle, 1977).

F. RETICULARIINIDAE Cooper and Grant, 1976
P. (ASS/SAK–GUA) Mar.

First: *Reticulariina hueconiana* Cooper and Grant, 1976, Hueco Canyon Formation, west Texas, USA (Cooper and Grant, 1976b).
Last: *Reticulariina phoxa* Cooper and Grant, 1976, Bell Canyon Formation, west Texas, USA (Cooper and Grant, 1976b).

F. CRENISPIRIFERIDAE Cooper and Grant, 1976
P. (ART/UFI–GUA) Mar.

First: *Crenispirifer sagus* Cooper and Grant, 1976, Bone Spring Formation, west Texas, USA (Cooper and Grant, 1976b).
Last: *Crenispirifer jubatus* Cooper and Grant, 1976, Bell Canyon Formation, west Texas, USA (Cooper and Grant, 1976b).

F. PARASPIRIFERINIDAE Cooper and Grant, 1979
P. (ASS/SAK–GUA) Mar.

First: *Paraspiriferina amoena* Cooper and Grant, 1976, Neal

Ranch Formation, west Texas, USA (Cooper and Grant, 1976b).
Last: *Paraspiriferina setulosa* Cooper and Grant, 1976, Word Formation, west Texas, USA (Cooper and Grant, 1976b).

F. SARGANOSTEGIDAE Cooper and Grant, 1976 P. (ART/UFI–GUA) Mar.

First: *Sarganostega prisca* Cooper and Grant, 1976, Bone Spring Formation, west Texas, USA (Cooper and Grant, 1976b).
Last: *Sarganostega pressa* Cooper and Grant, 1976, Bell Canyon Formation, west Texas, USA (Cooper and Grant, 1976b).

F. XESTOTREMATIDAE Cooper and Grant, 1976 P. (GUA) Mar.

First: Probably *Xestotrema pulchrum* (Meek, 1860), Park City and Phosphoria formations, Rocky Mountains, USA (Cooper and Grant, 1976b).
Last: *Arionthia lamaria* Cooper and Grant, 1976, Carlsbad, Capitan and Bell Canyon formations, west Texas (Cooper and Grant, 1976b).

F. YANGKONGIDAE Xu and Liu, 1983 Tr. (ANS/LAD) Mar.

First and Last: *Yangkongia planofolda*, *Y. qieermaensis*, and *Y. zhihemaensis* Xu and Liu, 1983, from the Middle Triassic rocks of the South Qilian Mountains, China (Xu and Liu, 1983).

Superfamily SUESSIOIDEA Waagen, 1883

F. CYRTINIDAE Fredericks, 1912 S. (WEN)–Tr. (CRN) (includes Laballidae Dagis, 1965) Mar.

First: *Cyrtina extensa* Bolton, 1957, Amabel Formation, Ontario, Canada (Bolton, 1957).
Last: Possibly *Psioidea australis* (Trechmann, 1918), *P. nelsonensis* (Trechmann, 1918), and *P. conjucta* (Hector, 1879), Balfour Group, Hokonui Hills, New Zealand (Marwick, 1953) or perhaps *Flabellocyrtia flabellum* Chorowicz and Termier, 1975, Svilaja, former Yugoslavia (Chorowicz and Termier, 1975).

F. SUESSIIDAE Waagen, 1883 J. (PLB) Mar.

First and Last: *Suessia costata* Eudes Deslongchamps, 1854, Liassic (Dacque, 1933) and *Suessia liasiana* Eudes Deslongchamps, 1854, Domerian, Morocco (Dubar, 1948).

F. BASHKIRIIDAE Nalivkin, 1979 C. (TOU) Mar.

First and Last: *Bashkiria gemma* Nalivkin, 1979, Lower Carboniferous, Urals, former USSR.

Superfamily SYRINGOTHYROIDEA Massa *et al.*, 1975

F. SYRINGOTHYRIDIDAE Fredericks, 1926 (includes Septothyringothyridae Massa *et al.*, 1975) D. (FAM)/C. (TOU)–P. (ZEC) Mar.

First: *Syringothyris spissus* Glenister, 1955, Moogooree Limestone, Western Australia (Glenister, 1955).
Last: *Paeckelmanella teneicostata* Termier, 1975, Wardak, central Afghanistan (Termier *et al.*, 1974) or *Licharewia kaninensis* Kulikov and Stepanov, 1975, Kanin Peninsula, former USSR (Stepanov *et al.*, 1975). [ALJ]

Order TEREBRATULIDA Waagen, 1883 (see Fig. 22.6)

Suborder CENTRONELLIDINA Stehli, 1965

Superfamily STRINGOCEPHALOIDEA King, 1850

F. CENTRONELLIDAE Waagen, 1882 D. (LOK)–P. (TAT) Mar.

First: *Nanothyris reeseidei* Cloud, 1942, Keyser Formation, USA (Cloud, 1942). Boucot (1957) regards *Nanothyris* as diagnostic of the Lower Devonian.
Last: *Notothyris warthi* Waagen, 1882, upper Productus Limestone, Pakistan, and *Cryptocanthia compacta* White and St John, 1867, Permian, Caucasus, former USSR (Waagen, 1882; Licharew, 1936).

F. RHIPIDOTHYRIDAE Cloud, 1942 D. (LOK)–C. (VIS) Mar.

First: *Prorensselaeria nylanderi* Raymond, 1923, Helderberg Formation, Maine, USA (Cloud, 1942). *'Waldheimia' mawei* Davidson, 1881, WEN, England, UK, is a spiriferid (Cloud, 1942). There are no terebratulids in the British succession below the Ludlow Bone Bed.
Last: *Girtyella intermedia* Weller, 1911, Chester Formation, Illinois, USA (Cooper, 1944).

F. STRINGOCEPHALIDAE King, 1850 D. (GIV) Mar.

First: *Subrensselandia claypoli* (Hall, 1891), Hamilton Formation, Pennsylvania, USA.
Last: *Stringocephalus burtini* Defrance, 1825, Europe and North America. Several other genera and species extend to the top of the GIV (Cloud, 1942; Cooper, 1944).

F. MEGANTERIDAE Schuchert and Le Vene, 1929 D. (PRA–EMS) Mar.

First and Last: *Meganteris suessi* Drevermann, 1901, PRA–EMS of Europe (Cloud, 1942).

Suborder TEREBRATULIDINA Waagen, 1883

Superfamily DIELASMATOIDEA Schuchert, 1913

F. DIELASMATIDAE Schuchert, 1913 *nom. transl.* Schuchert and Le Vene, 1929 S. (LUD)/D. (LOK)–P. (TAT) Mar.

First: *Brachyzyga pentameroides* Kozłowski, 1929 and *Podolella rensselaeroides* Kozłowski, 1929, Borszczow Formation, Poland. Earliest of *Cryptonella* series (Cryptonellinae and Cranaeninae) is *Cryptonella melonica* (Barrande, 1847), Upper Koněprusy Limestone, Bohemia, Czechoslovakia (Weller, 1911; Cloud, 1942). The *Cryptonella* series should probably be separated from the Dielasmatinae proper; they are ubiquitous in the Carboniferous and Permian.
Last: *Dielasma breviplicatum* Waagen, 1882 and *Hemiptychina himalayensis* Davidson, 1862, upper Productus Limestone, Pakistan (Waagen, 1882) and *Dielasma elongatum* (Schlotheim, 1816), Magnesian Limestone, and Upper Zechstein, western Europe; these are all late members of the Dielasmatinae. Of the *Cryptonella* series, last is *Heterelasma schumardianum* Girty, 1908, Permian, west Texas, USA (Cooper, 1944).

F. TEREBRATULIDAE Gray, 1840 Tr. (LAD)–Rec. Mar.

First: *'Terebratula' 'suborbicularis'* Munster, 1841, *'T'. capsella* Bittner, 1890, St Cassian Beds, Italian Alps (Bittner, 1890). Also *Plectoconcha aequiplicata* (Gabb, 1864), Triassic, Nevada, USA (Cooper, 1944). **Extant**
Comment: *'Terebratula' laricimontana* Bittner, 1890, Muschelkalk, central Europe, is earlier, but is inadequately known (Bittner, 1890).

F. ORTHOTOMIDAE Muir-Wood, 1936 J. (PLB) Mar.

First: *Orthotoma liasina* (Friren, 1896), *margaritatus* Zone, western Europe (Muir-Wood, 1936).
Last: *Orthotoma spinati* Rau, 1905, *spinatum* Zone, western Europe.

F. CHENIOTHYRIDIDAE Muir-Wood, 1965 J. (BAJ) Mar.

First and Last: *Cheniothyris morieri* (Eudes Deslongchamps, 1852), Europe, mainly England and France (Buckman, 1918).

F. DICTYOTHYRIDIDAE Makridin, 1964, *nom. transl.* J. (BTH–KIM) Mar.

First: *Dictyothyris coarctata* Parkinson, 1811, Jurassic, England, UK (Buckman, 1918).
Last: *Dictyothyris badensis* (Rollier, 1918), Jurassic, Europe (Makridin, 1964).

F. TEGULITHYRIDIDAE Muir-Wood, 1965 J. (CLV) Mar.

First and Last: *Tegulithyris bentleyi* (Davidson, 1851) Europe, mainly England, UK.

F. PYGOPIDAE Muir-Wood, 1965 J. (BAJ)–K. (NEO) Mar.

First: *Linguithyris bifida* (Rothpletz, 1886), Jurassic, western Europe (Buckman, 1914).
Last: *Pygites diphyoides* (d'Orbigny, 1849), Europe, mainly France and Switzerland, the Arctic and North Africa.

F. DYSCOLIIDAE Fischer and Oehlert, 1891 J. (CLV)–Rec. Mar.

First: *?Trigonithyris eruduwensis* Muir-Wood, 1935, Somalia, Africa. **Extant**

F. DIENOPIDAE Cooper, 1983 J. (CLV) Mar.

First and Last: *Dienope trigeri* (Eudes Deslongchamps, 1856), Europe, mainly France.

F. HESPERITHYRIDIDAE Cooper, 1983 J. (PLB) Mar.

First and Last: *Hesperithyris sinuosa* (Dubar, 1942), Domerian, Morocco.

F. MUIRWOODELLIDAE Tchorszhevsky, 1974 J. (BTH) Mar.

First and Last: *Muirwoodella muirwoodae* Tchorszhevsky, 1974, Transcarpathians and former USSR.

F. TCHEGEMITHYRIDIDAE Tchorszhevsky, 1972 J. (PLB–OXF) Mar.

First: *Viligothyris viligaensis* Dagis, 1968, northern Siberia, former USSR.
Last: *Bejrutella bejrutica* Tchorszhevsky, 1972, Lebanon and Syria.

Superfamily CANCELLOTHYRIDOIDEA Thomson, 1926

F. CANCELLOTHYRIDIDAE Thomson, 1926 K. (APT/ALB)–Rec. Mar.

First: *Terebratulina elongata* Davidson, 1874, Greensand, England, UK. **Extant**

Suborder TEREBRATELLIDINA Muir-Wood, 1955

Superfamily ZEILLERIOIDEA Allan, 1940

F. ZELLERIIDAE Allan, 1940 Tr. (CRN)–K. (ALB) Mar.

First: *'Terebratula' julica* Bittner, 1890, *'T.' paronica* Tommasi, 1887, and *'T.' woehrmanniana* Bittner, 1890, Carditaschichten, and *Camerothyris ramsaurri* Suess, 1855, the Alps (Bittner, 1890).
Last: *Modestella modesta* Owen, 1963, *tardefurcata* Zone, England, UK (Owen, 1963).

F. EUDESIIDAE Muir-Wood, 1965 J. (l–m) Mar.

First: *Apothyris* sp., Marrat Formation, Saudi Arabia (Cooper, 1989).
Last: *Apothyris abberans* Cooper, 1989, Jurassic, Saudi Arabia (Cooper, 1989).

Superfamily TEREBRATELLOIDEA King, 1850

F. DALLINIDAE Beecher, 1895 Tr. (NOR/RHT)–Rec. Mar.

First: *Eodallina peruviana* Elliott, 1959, Cerro de Pasco, Peru. **Extant**

F. MEGATHYRIDIDAE Dall, 1870 K. (CEN)–Rec. Mar.

First: *Argyrotheca megatrema* (J. de C. Sowerby, 1836), England, UK (Owen, 1988). **Extant**

F. KINGENIDAE Elliott, *nom. transl.* Owen 1970 K. (BRM–MAA) Mar.

First: *Zittelina orbis* (Quenstedt, 1858), White Jura, Westphalia, Germany (Owen, 1980).
Last: *Kingeneila kongieli* Popiel-Barczyk, 1968, Poland (Popiel-Barczyk, 1968).

F. LAQUEIDAE Thomson, 1927 K. (ALB)–Rec. Mar.

First: *Waconella wacoensis* (Roemer, 1852), Comanchean, Duck Creek Formation, Texas, USA (Owen, 1970). **Extant**

F. TEREBRATELLIDAE King, 1850 K. (ALB)–Rec. Mar.

First: *Australiarcula artesiana* Elliott, 1960, Oodnadata, South Australia. **Extant**

F. PLATIDIIDAE Thomson, 1927 T. (Eoc.)–Rec. Mar.

First: *Platidia cretacea* Weller, 1907, New Jersey, USA. **Extant**

F. KRAUSSINIDAE Allan, 1940 T. (Mio.)–Rec. Mar.

First: *Megerlia oblita* (Michelotti, 1839), north Italy. **Extant**

[EFO]

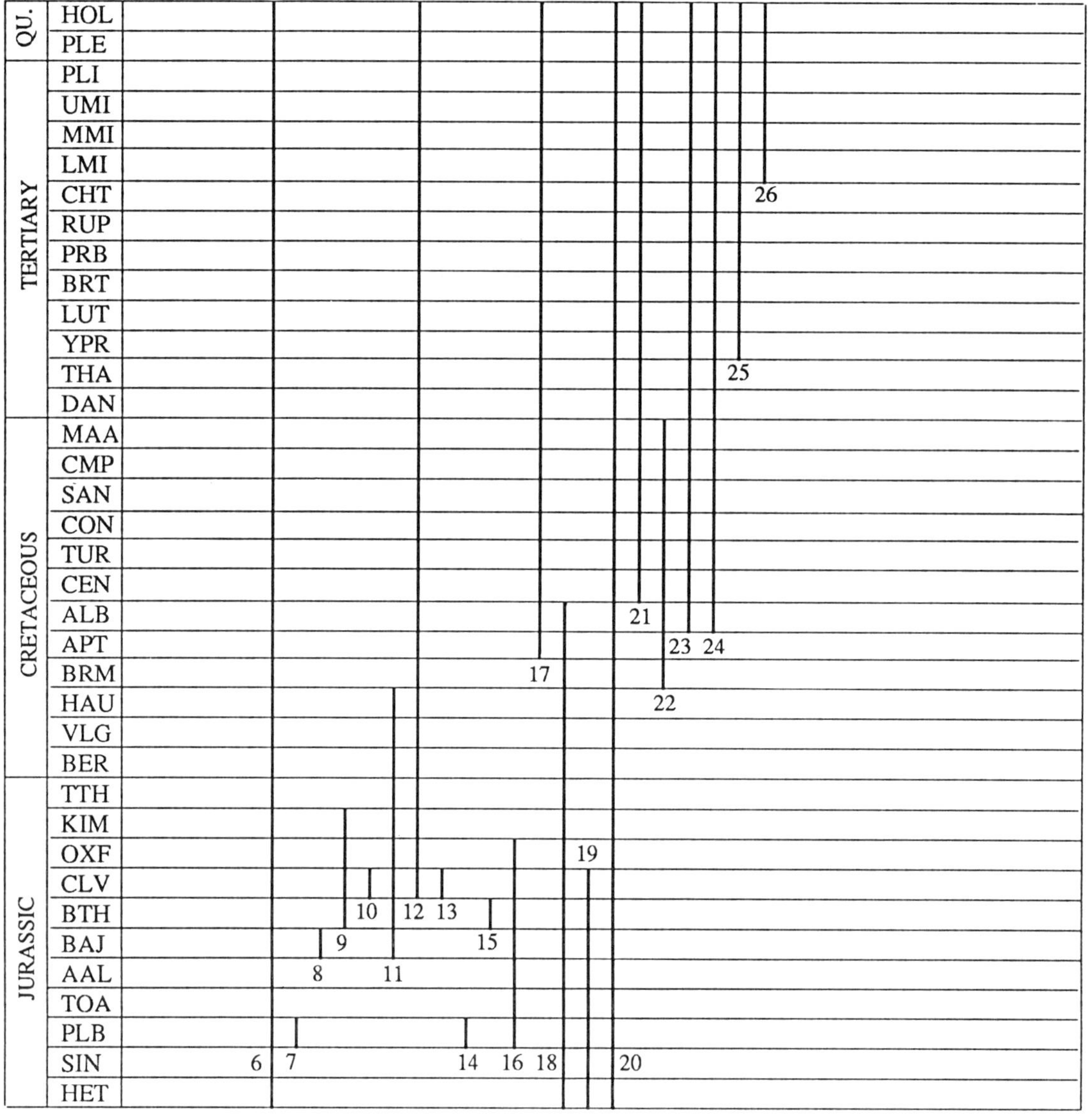

Fig. 22.6

PRECISE POSITION UNCERTAIN

F. TROPIDOLEPTIDAE Schuchert, 1896
D. (LOK–FRS) Mar.

According to Jaanusson (1971) this group possesses cyrtomatodont teeth which, together with the presence of a loop, indicate a close affinity with the terebratulides, where it is better accommodated.

First: *Tropidoleptus* sp., LOK, Annapolis Valley, Nova Scotia, Canada (Boucot *et al.*, 1969).

Last: *Tropidoleptus carinatus* (Conrad, 1839), Chemung Group, Chemung, New York, USA (Isaacson and Perry, 1977). [DATH]

REFERENCES

Ager, D. V. (1965) Mesozoic and Cenozoic Rhynchonellacea, in *Treatise on Invertebrate Paleontology, Part H. Brachiopoda* (ed. R. C. Moore), Geological Society of America and University of Kansas Press, Boulder, Colorado, and Lawrence, Kansas, pp. H597–H625.

Ager, D. V. (1968) The supposedly ubiquitous Tethyan brachiopod *Halorella* and its relations. *Journal of the Palaeontological Society of India*, **5–9**, (1960–1964), 54–70.

Ager, D. V., Copper, P., Dunlop, G. M. *et al.* (1967) Brachiopoda, in *The Fossil Record* (ed. W. B. Harland, C. H. Holland, M. R. House *et al.*) Geological Society of London, pp. 397–421.

Ager, D. V., Childs, A. and Pearson, D. A. B. (1972) The evolution of the Mesozoic Rhynchonellida. *Geobios*, **5**, 157–234.

Aizenverg, D. E. (1966) The fauna of the lowest part of Tournaisian (Zone Cl^{t}a) in the Donetz Basin. *Akademiia Nauk Ukrainskoi SSR, Institut Geologicheskikh Nauk, Kiev*, 1–128 [in Russian].

Aksarina, N. A. and Pelman, Y. L. (1978) [Cambrian brachiopods and bivalve molluscs of Siberia]. *Akademiia Nauk SSSR, Sibirskoe Otdelenie, Institut Geologii i Geofiziki Trudy*, **362**, 1–147 [in Russian].

Alekseeva, R. E. and Kulkov, N. I. (1970) in [*Stratigraphy and Brachiopods of the Lower Devonian of NE Salair*] (eds R. E. Alekseeva, R. T. Gratsianova, and E. A. Elkin *et al.*) *Akademiia Nauk SSSR, Sibirskoe Otdelenie, Institut Geologii i Geofiziki Trudy*, **72**, 1–188 [in Russian].

Amsden, T. W. (1958) Stratigraphy and paleontology of the Hunton Group in the Arbuckle Mountain region. Part II. Haragan articulate brachiopods. *Bulletin of the Oklahoma Geological Survey*, **78**, 1–144.

Amsden, T. W. (1968) Articulate brachiopods of the St. Clair

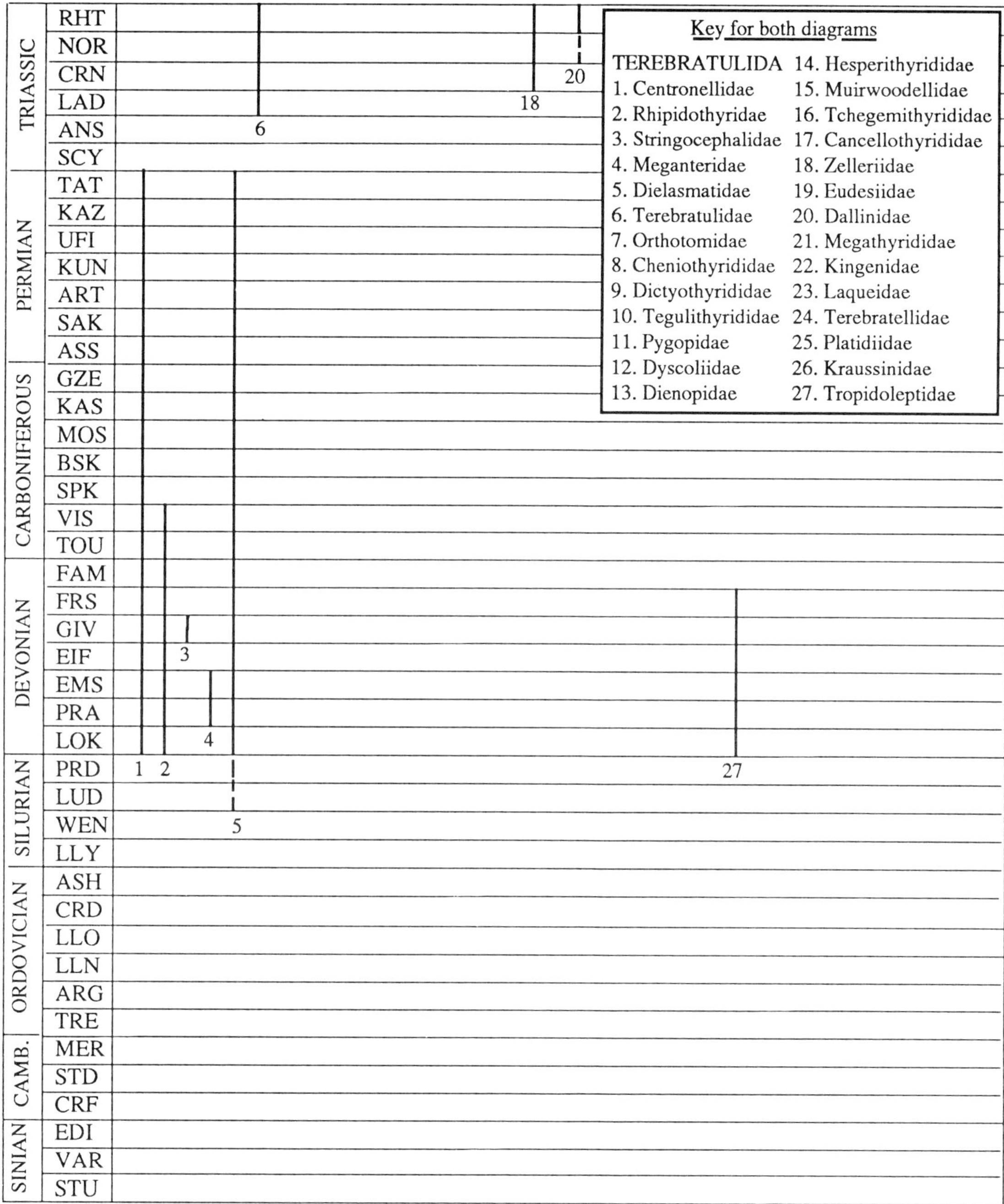

Fig. 22.6

Limestone (Silurian), Arkansas, and the Clarita Formation (Silurian), Oklahoma. *Paleontological Society Memoir*, **1**, 1–99.

Amsden, T. W. and Barrick, J. E. (1988) Late Ordovician–Early Silurian strata in the central United States and the Hirnantian Stage. *Bulletin of the Oklahoma Geological Survey*, **139**, 1–154.

Andreeva, O. N. (1960) in [*New Species of Ancient Plants and Invertebrates of the USSR*.] (ed. B. P. Morkowski), Moscow, 288 pp [in Russian].

Andreeva, O. N. (1987) [Cambrian articulate brachiopods]. *Paleontologicheskii Zhurnal*, **1987**, 31–40 [in Russian].

Appollonov, M. K., Bandaletov, S. M. and Nikitin, I. F. (eds) (1980) [*The Ordovician–Silurian Boundary in Kazakhstan*.] Nauka, Alma-Alta, 232 pp. [in Russian].

Baarli, B. G. (1988) The Llandovery enteletacean brachiopods of the central Oslo Region, Norway. *Palaeontology*, **31**, 1101–29.

Baker, P. G. (1983) The diminutive thecideidine brachiopod *Enallothecidea pygmaea* (Moore) from the Middle Jurassic of England. *Palaeontology*, **26**, 663–9.

Baker, P. G. (1990) The classification, origin and phylogeny of thecideidine brachiopods. *Palaeontology*, **33**, 175–91.

Baranov, V. V. (1977) [New early Devonian rhynchonellids in the northeast USSR]. *Paleontologiskii Zhurnal*, **1977**, 75–82 [in Russian].

Baranov, V. V. (1980) [Crural morphology and new rhynchonellid taxa]. *Paleontologiskii Zhurnal*, **1980**, 75–90 [in Russian].

Baranov, V. V. (1982) [New Devonian rhynchonellids and athyrids from eastern Yakutia.] *Paleontologiskii Zhurnal*, **1982**, 41–50 [in Russian].

Barrande, J. (1847) Über die Brachiopoden der silurischen Schichten von Böhmen. *Naturwissenschaft Abhandlungen*, **1**, 257–475.

Bassett, M. G. (1977) The articulate brachiopods from the Wenlock Series of the Welsh Borderland and South Wales, Part 4.

Monograph of the Palaeontographical Society, **547**, 123–76.
Bassett, M. G. and Cocks, L. R. M. (1974) A review of the Silurian brachiopods from Gotland. *Fossils and Strata*, **3**, 1–56.
Bates, D. E. B. (1968) The Lower Palaeozoic brachiopod and trilobite faunas of Anglesey. *Bulletin of the British Museum (Natural History) Geology*, **16**, 125–99.
Bates, D. E. B. (1969) Some early Arenig brachiopods and trilobites from Wales. *Bulletin of the British Museum (Natural History) Geology*, **18**, 1–28.
Bell, W. C. (1965) Cambrian Brachiopoda from Montana. *Journal of Paleontology*, **15**, 193–255.
Benigni, C. and Ferliga, C. (1989) Carnian Thecospiridae (Brachiopoda) from San Cassiano Formation (Cortina d'Ampezzo, Italy). *Rivista Italiana di Paleontologia e Stratigraphia*, **94**, 515–60.
Bergström, J. (1968) Upper Ordovician brachiopods from Västergotland, Sweden. *Geologica & Palaeontologica*, **2**, 1–21.
Biernat, G. (1957) On *Peregrinella multicarinata* (Lamark) (Brachiopoda). *Acta Palaeontologica Polonica*, **12**, 19–50.
Biernat, G. (1973) Ordovician inarticulate brachiopods from Poland and Estonia. *Palaeontologica Polonica*, **28**, 1–120.
Billings, E. (1858) Report for the year 1857 of E. Billings Esq., palaeontologist, in *Canada, Geological Survey, Report of Progress for the year 1857*, pp. 147–92.
Bittner, A. (1890) Brachiopoden der Alpinen Trias. *Geologische Reichsanstalt Abhandlungen*, **14**, 1–325.
Bittner, A. (1912) Brachiopoden aus der Trias des Kokonyer Waldes, in *Resultate der Wissenschaftlichen Erforschung des Baltonsees. Palaeontologischer Anhang*, **1**, Vienna, pp. 3–59.
Bittner, M. A. and Pisera, A. (1979) Brachiopods from the upper Cretaceous Chalk of Mielnik (Eastern Poland). *Acta Geologica Polonica*, **29**, 67–88.
Bolton, T. G. (1957) The Silurian stratigraphy and palaeontology of the Niagra Escarpment in Ontario. *Memoir of the Geological Survey of Canada*, **289**, 1–145.
Boucot, A. J. (1957) Position of the North Atlantic Siluro-Devonian boundary. *Bulletin of the Geological Society of America*, **68**, 170–2.
Boucot, A. J. (1975) *Evolution and Extinction Rate Controls*. Elsevier, New York, 426 pp.
Boucot, A. J. and Amsden, T. W. (1964) *Chonostrophiella*, a new genus of chonostrophid brachiopod. *Journal of Paleontology*, **38**, 881–4.
Boucot, A. J. and Harper, C. W. (1968) Silurian and Lower Middle Devonian Chonetacea. *Journal of Paleontology*, **42**, 143–76.
Boucot, A. J. and Johnson, J. G. (1967) Silurian and Upper Ordovician atrypids of the genera *Plectatrypa* and *Eospirigerina*. *Norsk Geologisk Tidsskrift*, **47**, 79–101.
Boucot, A. J. and Johnson, J. G. (1979) Pentamerinae (Silurian Brachiopoda). *Palaeontographica, Abteilung A*, **163**, 87–129.
Boucot, A. J., Johnson, J. G. and Talent, J. A. (1969) Early Devonian brachiopod zoogeography. *Special Paper Geological Society of America*, **119**, 1–106.
Brand, P. J. (1970) British Carboniferous Isogrammidae. *Bulletin of the Geological Survey of Great Britain*, **31**, 67–79.
Brand, P. J. (1972) Some British Carboniferous species of the brachiopod genus *Leptagonia* McCoy. *Bulletin Geological Survey GB*, **39**, 57–79.
Brice, D. (1982) *Eodmitria*, genre nouveau de Brachiopode–Cytospiriferidae du Frasnien inférieur et moyen. *Geobios*, **15**, 575–81.
Brunton, C. H. C. (1971) An endopunctate rhynchonellid brachiopod from the Viséan of Belgium and Britain. *Palaeontology*, **14**, 95–106.
Brunton, C. H. C. and Mundy, D. J. C. (1988) Strophalosiacean and Aulostegacean productoids (Brachiopoda) from the Craven Reef Belt (late Viséan) of North Yorkshire. *Proceedings of the Yorkshire Geological Society*, **47**, 55–8.
Buckman, S. S. (1914) *Genera of Some Jurassic Brachiopoda*. Wesley, London, 2 pp.
Buckman, S. S. (1918) The Brachiopoda of the Namyau Beds, Northern Shan States, Burma. *Palaeontologica Indica, New Series*, **3**, 1–299.
Carlson, S. J. (1989) The articulate brachiopod hinge mechanism: morphological and functional variation. *Paleobiology*, **15**, 364–86.
Carter, J. L. (1971) New early Mississippian silicified brachiopods from Central Iowa. *Smithsonian Contributions to Paleobiology*, **3**, 245–55.
Carter, J. L. (1990) Subdivision of the Lower Carboniferous in North America by means of articulate brachiopod generic ranges. *Courier Forschungen Institut Senckenberg*, **130**, 145–55.
Chernyshev, T. N. (1893) Die Fauna des unteren Devon am Ostabhange des Ural. *Trudy Geologischeskago Komiteta Leningrad*, **4**, 1–221.
Chernyshev, T. N. (1902) Die Obercarbonischen Brachiopoden den des Ural und des Timan. *Comité Géologique St Petersbourg, Mémoires*, **16**, 1–749.
Chernyshev, T. N. (1937) Silurian brachiopods of Mongolia and Tuva. *Trudy Mongolia Komiteta*, **29**, 1–94.
Chorowicz, J. and Termier, G. (1975) Une faunule silificée Nouvelle dans le Trias moyen de la Svilaja (Yougoslavie). *Annales Société Géologique du Nord*, **95**, 231–42.
Cleaves, A. B. (1939) in *The Devonian of Pennsylvania* (ed. B. Willard), *Bulletin of the Pennsylvanian Geological Survey*, G. **19**, 92–130.
Cloud, P. (1942) Terebratuloid Brachiopoda of the Silurian and Devonian. *Special Paper of the Geological Society of America*, **38**, 1–182.
Cocks, L. R. M. (1967) Llandovery stropheodontids from the Welsh borderland. *Palaeontology*, **10**, 245–65.
Cocks, L. R. M. (1968) Some strophomenacean brachiopods from the British Lower Silorian. *Bulletin British Museum of Natural History (Geology)*, **15**, 283–324.
Cocks, L. R. M. (1978) A review of British Lower Palaeozoic brachiopods, including a synoptic revision of Davidson's Monograph. *Palaeontographical Society Monograph*, **549**, 256 pp.
Cocks, L. R. M. (1982) The commoner brachiopods of the latest Ordovician of the Oslo–Asker district, Norway. *Palaeontology*, **25**, 755–81.
Cocks, L. R. M. and Price, D. (1975) The biostratigraphy of the upper Ordovician and lower Silurian of southwest Dyfed, with comments on the *Hirnantia* fauna. *Palaeontology*, **18**, 703–24.
Cocks, L. R. M. and Rong, J.-Y. (1988) A review of the late Ordovician *Foliomena* brachiopod fauna with new data from China, Wales and Poland. *Palaeontology*, **31**, 53–67.
Cocks, L. R. M. and Rong, J.-Y. (1989) Classification and review of the brachiopod superfamily Plectambonitacea. *Bulletin of the British Museum (Natural History) Geology*, **45**, 77–163.
Cooper, G. A. (1936) New Cambrian brachiopods from Alaska. *Journal of Paleontology*, **10**, 210–14.
Cooper, G. A. (1942) New genera of North American brachiopods. *Journal of the Washington Academy of Sciences*, **32**, 228–35.
Cooper, G. A. (1944) Phylum Brachiopoda, in *Index Fossils of North America* (ed. H. W. Shimer and R. R. Shrock), Wiley and Sons, New York, pp. 277–365.
Cooper, G. A. (1955) New genera of middle Paleozoic brachiopods. *Journal of Paleontology*, **29**, 45–63.
Cooper, G. A. (1956) Chazyan and related brachiopods. *Smithsonian Miscellaneous Collections*, **127**, 1–1245.
Cooper, G. A. (1959) Genera of Tertiary and Recent rhynchonelloid brachiopods. *Smithsonian Miscellaneous Collections*, **139**, 1–90.
Cooper, G. A. (1976) Lower Cambrian brachiopods from the Rift Valley (Israel and Jordan). *Journal of Paleontology*, **50**, 269–89.
Cooper, G. A. (1988) Some Tertiary brachiopods of the East Coast of the United States. *Smithsonian Contributions to Paleobiology*, **64**, 1–45.
Cooper, G. A. (1989) Jurassic brachiopods of Saudi Arabia. *Smithsonian Contributions to Paleobiology*, **65**, 1–138.
Cooper, G. A. and Grant, R. E. (1974) Permian brachiopods of west Texas, 2. *Smithsonian Contributions to Paleobiology*, **14**, 233–794.
Cooper, G. A. and Dutro, J. T. (1982) Devonian brachiopods of New Mexico. *Bulletin of American Paleontology*, **82/83**, (315), 1–215.

Cooper, G. A. and Grant, R. E. (1969) New Permian brachiopods from West Texas. *Smithsonian Contributions to Palaeobiology*, **1**, 1–20.

Cooper, G. A. and Grant, R. E. (1972) Permian brachiopods of west Texas, 1. *Smithsonian Contributions to Paleobiology*, **14**, 1–231.

Cooper, G. A. and Grant, R. E. (1975) Permian brachiopods of west Texas, 3. *Smithsonian Contributions to Paleobiology*, **19**, 795–1921.

Cooper, G. A. and Grant, R. E. (1976a) Permian brachiopods of west Texas, 4 (Part 1: text, Part 2: plates). *Smithsonian Contributions to Palaeobiology*, **21**, 1923–2607 [published in 2 separate volumes].

Cooper, G. A. and Grant, R. E. (1976b) Permian brachiopods of west Texas, 5. *Smithsonian Contributions to Palaeobiology*, **21**, 2609–3159.

Copper, P. (1977) *Zygospira* and some related Ordovician and Silurian atrypoid brachiopods. *Palaeontology*, **20**, 295–335.

Copper, P. (1982) Early Silurian atrypoids from Manitoulin Island and Bruce peninsula, Ontario. *Journal of Paleontology*, **56**, 68–702.

Copper, P. (1986) Evolution of the earliest smooth spire-bearing atrypoids (Brachiopoda: Lissatrypidae, Ordovician–Silurian). *Palaeontology*, **29**, 827–66.

Dacqué, E. (1933) Wirbellose des Jura. 1. *Leitfossilien*, **7**, 147–72.

Dagis, A. S. (1965) [*Triassic Brachiopods of Siberia.*] Akademii Nauk SSSR, Moscow, 186 pp. [in Russian].

Dagis, A. S. (1968) [Jurassic and early Cretaceous brachiopods of Northern Siberia.] *Akademiia Nauk SSSR, Sibirskoe Otdelenie, Institut Geologii i Geofiziki Trudy*, **41**, 1–167 [in Russian].

Dagis, A. S. (1972) [Ultrastructure of thecospirid shells and their position in brachiopod systematics]. *Paleontologicheskii Zhurnal*, **1972**, 87–98 [in Russian].

Dagis, A. S. (1974) [Triassic brachiopods.] *Akademia Nauk SSR, Sibirskoe Otdelenie, Institut Geologii Geofiziki Trudy*, **214**, 1–386 [in Russian].

Daily, B. (1956) Cambrian in South Australia, in *El Sistema Cambrico su Palaeogeographiá y el problema de su base*, 2. XX International Geological Congress, Mexico, pp. 91–147.

Dale, N. L. (1953) Geology and mineral resources of the Oriskany Quadrangle (Rane Quadrangle). *Bulletin New York State Museum*, **345**, 1–197.

Davidson, T. (1851–1852) The Oolitic and Liassic Brachiopoda. A monograph of the British fossil Brachiopoda I. *Monograph of the Palaeontographical Society*, 1–100; appendix to volume I, 1–30.

Davidson, T. (1870) On Italian Tertiary Brachiopoda. *Geological Magazine*, **7**, 359–408, 460–6.

Davidson, T. (1874) A monograph of the British fossil Brachiopoda IV, supplement to the Recent, Tertiary and Cretaceous species. *Monograph of the Palaeontographical Society*, **122**, 1–77.

Davidson, T. (1876–1878) A monograph of the British fossil Brachiopoda, volume 4, part 2, Jurassic-Triassic supplement. *Monograph of the Palaeontographical Society*, **135**, 73–242.

Fu Li-Pu (1975) Brachiopods, in *Stratigraphy of the Early Palaeozoic in the West Part of Dabashan Mountain* (eds Li Yao-xi, Song Li-sheng, Zhou Zhi-qiang *et al.*), Beijing, pp. 103–19, pls 1–70.

Deslongchamps, E. (1856) Catalogue des brachiopodes de Montreuil Bellay. *Bulletin de la Société Linnéenne de Normandie*, **8**, 161–297.

Destombes, J., Hollard, H. and Willefert, S. (1985) Lower Palaeozoic rocks of Morocco, in *Lower Palaeozoic of North-Western and West Central Africa* (ed. C. H. Holland), John Wiley, Chichester, pp. 91–336.

Doescher, R. A. (1981) Living and fossil brachiopod genera 1775–1979 lists and bibliography. *Smithsonian Contributions to Paleobiology*, **42**, 1–238.

Dubar, G. (1942) Études paléontologiques sur le Lias du Maroc. Brachiopodes, Térébratules et Zeilléries multiplissés. *Notes et Mémoires du Service Géologique du Maroc*, **57**, 1–103.

Dubar, G. (1948) La faune domerienne du Jebel Bou-Dahar pres de Beni-Taggite. *Notes et Mémoires du Service Géologique du Maroc*, **68**, 1–248.

Dunbar, C. and Condra, G. E. (1932) Brachiopoda of the Pennsylvanian system in Nebraska. *Bulletin of the Geological Survey of Nebraska*, **5**, 1–377.

Elliott, G. E. (1953) The classification of the thecidean brachiopods. *Annals and Magazine of Natural History*, (12), **6**, 693–701.

Elliott, G. E. (1959) Six new genera of Mesozoic Brachiopoda. *Geological Magazine*, **96**, 146–8.

Elliott, G. E. (1960) A new Mesozoic terebratelloid brachiopod. *Proceedings of the Geologists' Association*, **71**, 25–30.

Erlanger, O. A. (1981) [A contribution to revision of the genus *Rhynchopora*]. *Paleontologicheskii Zhurnal*, **1981**, 88–94 [in Russian].

Erlanger, O. A. (1986) [On the systematics of Rhynchoporacea (Brachiopoda)]. *Paleontologicheskii Zhurnal*, **1986**, 52–9 [in Russian].

Feldman, H. R. (1987) A new species of the Jurassic (Callovian) brachiopod *Septirhynchia* from Northern Sinai. *Journal of Paleontology*, **61**, 1156–72.

Ferguson, J. (1971) *Linoprotonia*, a new genus of Lower Carboniferous productoid. *Proceedings of the Yorkshire Geological Society*, **38**, 549–64.

Ficner, F. and Havlíček, V. (1978) Middle Devonian brachiopods from Celechovice, Moravia. *Sborník Geologických Věd Paleontologie*, **21**, 49–106.

Fu, L.-P. (1982) [Phylum Brachiopoda], in *Dizhi Kuangehan Bu Xian Dizhi Kuangchan Yanjiusuo.* [Xian Institute of Geology and Mineral Resources] *Xibei diqu gu shengwu. tuce: Shaan-Gan Ning fence [Palaeontological Atlas of Northwest China. Shaanxi–Gansu–Ninxia Volume. Part 1. Precambrian and Early Palaeozoic*]. Geological Publishing House, Peking, [in Chinese], pp. 95–179.

Fu, L.-P. (1983) [Early Devonian Rhynchonellida (Brachiopoda) from west Qinling]. *Acta Paleontologica Sinica*, **22**, 542–50 [in Chinese with English summary].

Glenister, B. F. (1955) Devonian and Carboniferous spiriferids from the North-West Basin, Western Australia. *Journal of the Royal Society of Western Australia*, **39**, 46–71.

Gorjansky, V. Yu. (1969) [*Inarticulate brachiopods of the Cambro-Ordovician rocks of the northwest Russian platform.*] NEDRA, Leningrad, 127 pp. [in Russian].

Gorjansky, V. Yu. and Koneva, S. P. (1983) [Lower Cambrian inarticulate brachiopods of the Maly Karatau Range (southern Kazakhstan)]. *Akademiia Nauk SSSR, Sibirskoe Otdelenie, Institut Geologii i Geofiziki Trudy*, **541**, 128–38 [in Russian].

Gorjansky, V. Yu. and Popov, L. E. (1985) [The morphology, systematic position and origin of inarticulate brachiopods with carbonate shells]. *Paleontologicheskii Zhurnal*, **1985**, 3–13 [in Russian].

Gorjansky, V. Yu. and Popov, L. E. (1986) On the origin and systematic position of the calcareous-shelled inarticulate brachiopods. *Lethaia*, **19**, 233–40.

Graham, D. K. (1970) Scottish Carboniferous Lingulacea. *Bulletin of the Geological Survey of Great Britain*, **31**, 139–84.

Grant, R. E. (1965) The brachiopod superfamily Stenoscismatacea. *Smithsonian Miscellaneous Collections*, **148**, 1–192.

Grant, R. E. (1970) Brachiopods from Permian–Triassic boundary beds and age of the Chhidru Formation, West Pakistan. *University of Kansas Special Publication*, **4**, 117–51.

Grant, R. E. (1976) Permian brachiopods from southern Thailand. *Journal of Paleontology*, **50** (Supplement 2), 1–269.

Grant, R. E. (1980) The human face of the brachiopod. *Journal of Paleontology*, **54**, 499–507.

Grant, R. E. (1988) The family Cardiarinidae (Late Paleozoic rhynchonellid Brachiopoda). *Senckenbergiana Lethaea*, **69**, 121–35.

Gratsionova, R. T. (1967) [*Brakhiopody i Stratigrafiia Nizhnego Devona Gorgnogo Altaia*]. Akademiia Nauk SSSR Sibirskoe Otdelenie Institut Geologii i Geofiziki, Moscow, 117 pp. [in Russian].

Grunt, T. A. (1986) [The systematics of brachiopods of the order Athyridida]. *Akademiia Nauk SSSR Paleontologicheskii Institut Trudy*, **215**, 1–187 [in Russian].

Grunt, T. A. (1989) [Order Athyridida (evolutionary morphology and phylogeny).] *Akademiia Nauk SSSR Paleontologicheskii Institut Trudy*, **238**, 1–139. [in Russian].

Gupta, V. J. (1973) Additional fossils from the Muth Quartzite of Kashmir, India. *Research Bulletin Punjab University*, **24**, 77–93.

Hall, J. (1860) Contributions to paleontology. *New York State Cabinet of Natural History Annual Report*, **13**, 53–125.

Hall, J. (1891) Preliminary notice of *Newberria*, a new genus of brachiopods, with remarks on its relations to *Rensselaeria* and *Amphigenia*. *Annual Report of the New York State Geologist*, **10**, 97–3.

Harper, D. A. T. (1984) Brachiopods from the Upper Ardmillan succession (Ordovician) of the Girvan District, Scotland, Part 1. *Monograph of the Palaeontographical Society*, **564**, 1–78.

Harper, D. A. T. (1989) Brachiopods from the Upper Ardmillan succession (Ordovician) of the Girvan District, Scotland, Part 2. *Monograph of the Palaeontographical Society*, **579**, 79–128.

Harper, C. W. and Boucet, A. J. (1978) The Stropheodontacea, Parts I–III. *Palaeontographica, Series A*, **161**, 55–175; **162**, 1–80.

Harper, C. W., Boucot, A. J. and Walmsley, V. G. (1969) The rhipidomellid brachiopod subfamilies Heterorthinae and Platyorthinae (new). *Journal of Paleontology*, **43**, 74–92.

Havlíček, V. (1950) [The Ordovician Brachiopoda from Bohemia]. *Ústředniho Ústavu Geologického Rozpravy*, **13**, 1–72 [in Czech].

Havlíček, V. (1951) [A paleontological study of the Devonian Celechovice; brachiopods (Pentameracea, Rhynchonellacea, Spiriferacea)]. *Ústředniho Ústavu Geologického Sborník*, **18**, 1–20 [in Czech].

Havlíček, V. (1956) [The brachiopods of the Branik and Hlubocepy limestones in the immediate vicinity of Prague]. *Ústředniho Ústavu Geologického Sborník*, **22**, 535–650 [in Czech].

Havlíček, V. (1959) The Spiriferidae of the Silurian and Devonian of Bohemia. *Ústředniho Ústavu Geologického Rozpravy*, **25**, 1–275.

Havlíček, V. (1961) Rhynchonelloidea des böhmischen altere Palaözoikums (Brachiopoda). *Ústředniho Ústavu Geologického Rozpravy*, **27**, 1–211.

Havlíček, V. (1967) Brachiopoda of the suborder Strophomenidina in Czechoslovakia. *Ústředniho Ústavu Geologického Rozpravy*, **33**, 1–235.

Havlíček, V. (1971a) Brachiopodes de l'Ordovicien du Maroc. *Notes et Mémoires du Service Géologique du Maroc*, **230**, 1–132.

Havlíček, V. (1971b) Non-costate and weakly costate Spiriferidina (Brachiopoda) in the Silurian and Lower Devonian of Bohemia. *Sborník Geologických Věd Praha Paleontologie*, **14**, 7–34.

Havlíček, V. (1973) New brachiopod genera in the Devonian of Bohemia. *Ústředniho Ústavu Geologického Věstník*, **48**, 337–40.

Havlíček, V. (1977) Brachiopods of the order Orthida in Czechoslovakia. *Vydal Ústředni Ústav Geologícky, Praha*, **44**, 1–328.

Havlíček, V. (1979) Upper Devonian and lower Tournaisian hynchonellida in Czechoslovakia. *Ústředniho Ústavu Geologického Věstník*, **54**, 87–101.

Havlíček, V. (1982a) New Pugnacidae and Plectorhynchellidae (Brachiopoda) in the Silurian and Devonian rocks of Bohemia. *Ústředniho Ústavu Geologického Věstník*, **57**, 111–14.

Havlíček, V. (1982b) New genera of rhynchonellid and camerellid brachiopods in the Silurian of Bohemia. *Ústředniho Ústavu Geologického Věstník*, **57**, 365–72.

Havlíček, V. (1990) New Lower Devonian (Pragian) rhynchonellid brachiopods in the Koneprusy area (Czechoslovakia). *Ústředniho Ústavu Geologického Věstník*, **65**, 211–21.

Havlíček, V. and Branisa, L. (1980) Ordovician brachiopods of Bolivia (Succession of assemblages, climate control, affinity to Anglo-French and Bohemian provinces). *Rozpravy Ceskoslovenske Akademie Věd*, **90**, 1–54.

Havlíček, V. and Štorch, P. (1990) Silurian brachiopods and benthic communities in the Prague Basin (Czechoslovakia). *Ústředniho Ústavu Geologického Rozpravy*, **48**, 1–275.

Havlíček, V., Kriz, J. and Sepagli, E. (1987) Upper Ordovician brachiopod assemblages of the Carnic Alps, Middle Carinthia and Sardinia. *Bolletino della Societa Paleontologico Italiana*, **25**, 277–311.

Hiller, N. (1980) Ashgill Brachiopoda from the Glyn Ceiriog District, North Wales. *Bulletin of the British Museum (Natural History) Geology*, **34**, 109–216.

Hints, L. (1973) New Orthacean brachiopods from the Middle Ordovician of the East Baltic area and Sweden. *Eesti NSV Teaduste Akadeemia Toimetised*, **22**, 248–56.

Hints, L. (1990) Ordovician articulate brachiopods, in *Field Meeting Estonia 1990, An Excursion Guidebook* (eds D. Kaljo and H. Nestor), Estonian Academy of Sciences, Tallinn, pp. 58–61.

Holmer, L. E. (1989) Middle Ordovician phosphatic inarticulate brachiopods from Västergötland and Dalarna, Sweden. *Fossils and Strata*, **26**, 1–172.

Holmer, L. E. (1991a) The taxonomy and shell characteristics of a new elkaniid brachiopod from the Ashgill of Sweden. *Palaeontology*, **34**, 195–204.

Holmer, L. E. (1991b) The systematic position of *Pseudolingula* Mickwitz and related lingulacean brachiopods, in *Brachiopods through Time* (eds D. I. Mackinnon, D. E. Lee and J. D. Campbell), Balkema, Rotterdam, pp. 3–14.

Holmer, L. E. (1991c) Phyletic relationships within the Brachiopoda. *Geologiska Föreningens i Stockholm Förhandlinger*, **113**, 84–6.

Hou, H.-F. (1979) Devonian stratigraphy of south China, in *Institute of Geology and Mineral Resources of the Chinese Academy of Geological Sciences, Symposium on the Devonian System of South China*. Geological Press, Peking, pp. 214–30.

Hou, H.-F. and Xian, S.-Y. (1975) [The lower and middle brachiopods from Kwangsi and Kueichow]. Dicenig Gushengwi Luwenzhi (*Professional Paper in Stratigraphical Palaeontology*), **1**, 1–85, illustr. [In Chinese with English summary] (summary separate publication, 1–2).

Hudson, R. G. S. and Jefferies, R. P. S. (1961) Upper Triassic brachiopods and lamellibranchs from the Oman Peninsula, Arabia. *Palaeontology*, **4**, 1–41.

Hyde, J. E. (1953) Mississippian formations of Central and Southern Ohio. *Bulletin of the Geological Survey of Ohio*, **51**, 1–355.

Isaacson, P. E. and Perry, D. G. (1977) Biogeography and morphological conservatism of *Tropidoleptus* (Brachiopoda, Orthida) during the Devonian. *Journal of Paleontology*, **51**, 1108–22.

Ivanova, E. A. (1962) [Ecology and development of Silurian and Devonian brachiopods of the Kutznetz, Minussinsk and Tuvinsk Basins]. *Akademii Nauk SSSR Trudy Palaeontologicheskikh Institut*, **83**, 1–150 [in Russian].

Ivanova, E. A. (1971) [On a new spiriferid genus.] *Paleontologicheskii Zhurnal*, **1971**, 120–3 [in Russian].

Jaanusson, V. (1971) Evolution of the brachiopod hinge. *Smithsonian Contributions to Paleobiology*, **3**, 33–46.

Jaeger, H., Stein, V. and Wolfart, R. (1969) Fauna (Graptolithen, Brachiopoden) der unterdevonischen Schwarzschiefer Nord-Thailands. *Neues Jahrbuch für Geologie und Paläontologie, Abhandlungen*, **133**, 171–90.

Jin, J., Caldwell, W. G. E. and Norford, B. S. (1990) Rhynchonellid brachiopods from the Upper Ordovician–Lower Silurian Beaverfoot and Nonda Formations of the Rocky Mountains, British Columbia. *Bulletin of the Geological Survey of Canada*, **396**, 21–59.

Johnson, J. G. (1972) The *Antistrix* brachiopod faunule from the Middle Devonian of Central Nevada. *Journal of Paleontology*, **46**, 120–4.

Johnson, J. G. (1975) Late Early Devonian brachiopods from the Disappointment Bay Formation, Lowther Island, Arctic Canada. *Journal of Paleontology*, **49**, 947–78.

Johnson, J. G. and Trojan, W. R. (1982) The *Tecnocyrtina* brachiopod fauna (?Upper Devonian) of central Nevada. *Geologica et Palaeontologica*, **16**, 119–49.

Kaplun, L. I. (1967) A discovery of the genus *Eichwaldia* in the middle Devonian. *Paleontologicheskii Zhurnal*, **1967**, 127–8.

Kaplun, L. I. (1968) [A new middle Devonian species of Uncinulidae from Kazakhstan.] in [*New species of Fossils Plants and Invertebrates from the USSR.*] (ed. B. P. Markovskii), NEDRA, Moscow [in Russian], pp. 115–16.

Khodalevich, A. N. (1951) [Lower Devonian and Eifelian

brachiopods from the Ivdel and Serov areas of the Sverdlovsk region]. *Sverdlovsk Gornogo Institut Trudy*, **18**, 1–107 [in Russian].

Khodalevich, A. N. and Breivel, M. G. (1959) [*Brachiopods and corals from the Eifelian bauxite-producing deposits of the eastern slope of the Central and Northern Urals*] Urals Geological Administration, Moscow, 282 pp. [in Russian].

Kobayashi, T. (1935) The Cambro-Ordovician formations and faunas of South Chosen: part 3. Cambrian faunas of South Chosen. *Tokyo Imperial University Faculty of Science Journal, Section 3*, **4**, 49–344.

Koczyrkevicz, B. V. (1969) K Sistematike Nadsemeistve Steno-scismatacea Oehlert, 1887 (1883) (Brachiopoda), in *Iskopaemaia Fauna i Flora Dal'nego Vostoka I*. Academiia Nauk SSSR, Vladivostok, pp. 7–16.

Koneva, S. P. (1978) [*Stenothecoidea and inarticulate brachiopods of the Lower Cambrian and lowest strata of Middle Cambrian in central Kazakhstan*]. Nauka, Alma Ata, 124 pp. [in Russian].

Koneva, S. P. (1986) A new family of Cambrian inarticulate brachiopods. *Paleontologicheskii Zhurnal*, **1986**, 28–35.

Koneva, S. P. and Popov, L. E. (1983) [Certain new lingulids from the Upper Cambrian and Lower Ordovician of Maly Karatau], in *The Lower Palaeozoic Stratigraphy and Palaeontology of Kazakhstan* (eds M. K. Appollonov, S. M. Bandetaletov and N. K. Ivshin), Nauka, Alma Ata, 176 pp. [in Russian], pp. 112–24.

Kozłowski, R. (1930) [*Andobolus* gen. nov. i kilka innych ramienionogow bezzawiasowych z ordowiku Boliwji.] *Sprawozdania Polskiego Instytuta Geologicznego Warszawa*, **6**, 293–313 [in Polish].

Kulkov, N. P. and Kozlov, M. S. (1978) [The fauna and bio-stratigraphy of Upper Ordovician and Silurian of the Altai-Sayan region: stratigraphy and brachiopods of the Silurian ore-bearing Altai]. *Akademiia Nauk SSSR, Sibirskoe Otdelenie, Institut Geologii i Geofiziki Trudy*, **405**, 57–84 [in Russian].

Lamont, A. and Gilbert, D. L. F. (1945) Upper Llandovery Brachiopoda from Coneygore Coppice and Old Storridge Common, near Alfrick, Worcestershire. *Annals and Magazine of Natural History*, **12**, 641–82.

Langer, W. (1971) Acrotretidae (Brachiopoda) im Devon des Sauer-und Bergischen Landes. *Decheniana*, **123**, 328–9.

Lazarev, S. S. (1987) The origin and systematic position of the main groups of productids (brachiopods). *Paleontological Journal*, **4**, 41–52.

Lee, D. E. (1978) Cenozoic and Recent rhynchonellide brachiopods of New Zealand: Genus *Aetheia*. *Journal of the Royal Society of New Zealand*, **8**, 93–113.

Legrand-Blain, M. (1987) Les Gigantoproductidae (brachiopodes) Namuriens du Sahara Algerien. *Bulletin de la Société Belge de Géologie*, **96**, 159–94.

Leidhold, C. (1928) Beitrag zur Kenntnis der Fauna des rheinischen Stringocephalenkalkes, insbesondere seiner Brachiopodenfauna. *Koniglich Preussichen Geologischen Landesanstalt, Abhandlungen*, **109**, 1–99.

Lenz, A. C. (1977) Upper Silurian and Lower Devonian brachiopods of Royal Creek, Yukon Canada, Part I (Orthida, Strophomenida, Pentamerida, Rhynchonellida). *Palaeontographica, Abteilung A*, **159**, 37–109.

Levy, R. and Nullo, F. (1974) La Fauna del Ordovicio (Ashgilliano) de Villicun San Juan, Argentina (Brachiopoda). *Ameghiniana*, **11**, 173–94.

Li, L. and Gu, F. (1980) [Carboniferous and Permian Brachiopoda], in *Shenyang Dizhi Kuangchan Yanjiusuo. Dongbei. Diqu gushengwu tuce. [Palaeontological Atlas of Northeast China]*. Vol. 1. *Palaeozoic*. Geological Publishing House, Peking, 686 pp. [in Chinese], pp. 327–428.

Licharew, B. K. (1936) Über einigen oberpaläozoische Brachiopoden. *Problemy Paleontologii*, **1**, 263–71.

Licharew, B. K. (1957) [On the genus *Goniophoria* Yanisch. and other related genera.] *Vsesoyuznoe Paleontologicheskoe Obshchestvo Ezhegodnik*, **16**, 134–41 [in Russian].

Liu, D.-Y., Zhang, Z.-X. and Di, Q.-L. (1984) Middle Ordovician brachiopods from Mts. Kunlun and Altun. *Acta Paleontologica Sinica*, **23**, 155–69.

Lockley, M. G. and Williams, A. (1981) Lower Ordovician Brachiopoda from mid and southwest Wales. *Bulletin of the British Museum (Natural History) Geology*, **35**, 1–78.

Ludvigsen, R. (1974) A new Devonian acrotretid (Brachiopoda, Inarticulata) with unique protegular ultrastructure. *Neues Jahrbuch für Geologie und Paläontologie, Monatshefte*, **1974**, 133–48.

MacKinnon, D. I. (1983) A Late Middle Cambrian orthide–kutorginide brachiopod fauna fron northwest Nelson, New Zealand. *New Zealand Journal of Geology and Geophysics*, **26**, 97–102.

Maillieux, E. (1912) Apparition de deux formes siegeniennes dans les schistes de Mondrepuits. *Bulletin Société Belge de Géologie de Paléontologie et d'Hydrologie*, **25**, 176–80.

Makridin, V. P. (1964) [*Jurassic Brachiopoda from the Russian Platform and Certain Adjacent Districts*.] Kharkov Gosudarsttvennyi Universitet Izdatel, Moscow, 339 pp. [in Russian].

Mancenido, M. O. and Whalley, C. D. (1979) Functional morphology and ontogenetic variation in the Callovian brachiopod *Septirhynchia* from Tunisia. *Palaeontology*, **22**, 317–37.

Martinez-Chacon, M. L. (1979) Braquiopodos carboniferos de la Cordillera Cantabrica (Orthida, Strophomenida y Rhynchonellida). *Memoria del Instituto Geologico y Minero de España*, **96**, 1–291.

Martynova, M. V. and Sverbilova, T. V. (1968) [First members of the Spiriferidae from the Devonian of Kazakhstan.] *Paleontologicheskii Zhurnal*, **1968**, 26–31 [in Russian].

Marwick, J. (1953) Divisions and faunas of the Hokonui System (Triassic and Jurassic). *Palaeontological Bulletin, Wellington*, **21**, 5–141.

McClean, A. E. (1988) Epithelial moulds from some Upper Ordovician acrotretide brachiopods of Ireland. *Lethaia*, **21**, 43–50.

McKellar, R. G. (1970) The Devonian productoid brachiopod faunas of Queensland. *Geological Survey of Queensland*, Publication **342**, 1–40.

McNamara, K. J. (1983) The earliest *Tegulorhynchia* (Brachiopoda: Rhynchonellida) and its evolutionary significance. *Journal of Paleontology*, **57**, 461–73.

Mergl, M. (1981) The genus *Orbithele* (Brachiopoda, Inarticulata) from the Lower Ordovician of Bohemia and Morocco. *Věstník Ústředniho Ústavu Geologického*, **56**, 287–92.

Mergl, M. (1983) New brachiopods (Cambrian-Ordovician) from Algeria and Morocco (Mediterranean Province). *Casopis Mineraloggii a Geologii, Praha*, **28**, 337–48.

Mergl, M. (1988) Inarticulate brachiopods of early Middle Cambrian age from the High Atlas, Morocco. *Ústředniho Ústavu Geologického Věstník*, **63**, 291–5.

Michelotti, G. (1839) Brevi cenni di alcuni resti delle classi Brachiopodi ed acefali trovati fossili in Italia. *Annali delle Regno Lombardo–Veneto*, **9**, 119–38, 157–73.

Minato, M. (1952) A further note on the Lower Carboniferous fossils of the Kitakami Mountainland, N.E. Japan. *Journal of the Faculty of Science Hokkaido University*, (4), **8**, 136–74.

Mitchell, W. I. (1977) The Ordovician Brachiopoda from Pomeroy, Co. Tyrone. *Monograph of the Palaeontographical Society*, **545**, 1–138.

Modzalevskaya, T. L. (1979) [Systematics of Paleozoic Athyrididae]. *Paleontologicheskii Zhurnal*, **1979**, 48–63 [in Russian].

Muir-Wood, H. M. (1928) British Carboniferous Producti II; *Productus* (s.s.), *semireticulatus* and *longispinus* groups. *Memoirs of the Geological Survey of Great Britain, Palaeontology*, **3**, 1–217.

Muir-Wood, H. M. (1935) Jurassic Brachiopoda, in *The Geology and Palaeontology of British Somaliland Part II, The Mesozoic Palaeontology of British Somaliland* (ed. W. A. Macfadyan *et al.*), London, pp. 75–147.

Muir-Wood, H. M. (1962) *On the Morphology and Classification of the Brachiopod Suborder Chonetoidea*. British Museum (Natural History), London, 132 pp.

Muir-Wood, H. M. and Cooper, G. A. (1960) Morphology, classification and life habits of the Productoidea. *Geological Society of America, Memoir*, **81**, 1–447.

Nalivkin, D. V. (1979) [Brachiopods of the Tournaisian stage of the Urals]. Nauka, Leningrad, 247 pp. [in Russian].

Nakazawa, K. *et al.* (1975) The Upper Permian and Lower Triassic in Kashmir, India. *Memoirs of the Faculty of Science, Kyoto University, Geology and Mineralogy Series*, **42**, 1–106.

Nei, M.-G., Qu, Z.-Z., Ju, D.-Z. *et al.* (1976) *Huabei Diqu Gushengwu Tuce.* Vol. 1. (*Fossils of Inner Mongolia*). Dizhi Chuban She, Peking, 502 pp. [in Chinese].

Neuman, R. B. and Bruton, D. L. (1974) Early middle Ordovician fossils from the Hølonda area, Trondheim Region, Norway. *Norsk Geologisk Tidsskrift*, **54**, 69–115.

Newell, N. D. (1934) Some mid-Pennsylvanian invertebrates from Kansas and Oklahoma. *Journal of Paleontology*, **8**, 422–32.

Nikiforova, O. I. (1937) [Upper Silurian Brachiopoda of the Central Asiatic part of the U.S.S.R.] *Akademiia Nauk SSSR, Paleontologicheskii Institut, Monographii po Paleontologii SSRR*, **35**, 1–94 [in Russian].

Nikiforova, O. I. (1961) [Ordovician and Silurian stratigraphy of the Siberian platform and its palaeontological basis (Brachiopoda)]. *Vsesoiuznyi Nauchno-Issledovatelskii Geologicheskii Institut (VSEGEI) Trudy*, **56**, 1–412 [in Russian].

Nikiforova, O. I. and Popov, L. E. (1981) [New data on Ordovician rhynchonellids of Kazakhstan and Soviet central Asia]. *Paleontologicheskii Zhurnal*, **1981**, 54–67 [in Russian].

Nikitin, I. F. and Popov, L. E. (1984) [Brachiopods of the Bestamak and Sargaldak suites (Middle Ordovician).] in [*Brachiopod Biostratigraphy of the Middle and Upper Ordovician of the Chingiz Range*] (ed. 126–66), Nauka, Alma Ata, 196 pp. [in Russian].

Öpik, A. A. (1932) Über die Plectellinen. *Acta et Commentationes Universitatis Tartuensis*, **23**, 1–85.

Öpik, A. A. (1933) Über Plectamboniten. *Acta et Commentationes Universitatis Tartuensis*, **24**, 1–79.

Öpik, A. A. (1934) Über Klitamboniten. *Acta et Commentationes Universitatis Tartuensis*, **26**, 1–239.

Owen, E. F. (1963) The brachiopod genus *Modestella* in the Lower Cretaceous of Great Britain. *Annals and Magazine of Natural History*, **61**, 199–203.

Owen, E. F. (1968) A further study of some Cretaceous rhynchonellid brachiopods. *Bulletin of the Indian Geological Association*, **1**, 17–32.

Owen, E. F. (1970) A revision of the subfamily Kingeninae Elliott. *Bulletin of the British Museum (Natural History) Geology*, **19**, 29–83.

Owen, E. F. (1980) Tertiary and Cretaceous brachiopods from Seymour, Cockburn and James Ross Islands, Antarctica. *Bulletin of the British Museum (Natural History) Geology*, **33**, 123–45.

Owen, E. F. (1988) Cenomanian brachiopods from the Lower Chalk of Britain and northern Europe. *Bulletin of the British Museum (Natural History) Geology*, **44**, 65–175.

Owens, B. and Burgess, I. C. (1965) The stratigraphy and palynology of the Upper Carboniferous outlier of Stainmore, Westmorland. *Bulletin of the Geological Survey of Great Britain*, **23**, 17–44.

Pattison, J. (1981) The stratigraphical distribution of gigantoproductoid brachiopods in the Viséan and Namurian rocks of some areas in northern England. *Institute of Geological Sciences, Report*, **81/9**, 1–30.

Pelman, Yu. L. (1977) [Early and Middle Cambrian inarticulate brachiopods from the Siberian platform.] *Akademiia Nauk SSSR, Sibirskoe Otdelenie, Institut Geologii i Geofiziki Trudy*, **316**, 1–168 [in Russian].

Percival, I. G. (1991) Late Ordovician articulate brachiopods from central New South Wales. *Memoirs of the Association of Australasian Palaeontologists*, **11**, 107–77.

Pettitt, N. E. (1950) Rhynchonellidae of the British Chalk, Part 1. *Monograph of the Palaeontographical Society*, **450**, 1–26.

Pettitt, N. E. (1954) Rhynchonellidae of the British Chalk, Part 2. *Monograph of the Palaeontographical Society*, **466**, 27–52.

Pirlet, H. (1967) Nouvelle interpretation des carrieres de Richelle: le Viséen de Visé. *Bulletin de la Société belge de Géologie*, **90**, 299–328.

Popiel-Barczyk, E. (1968) Upper Cretaceous terebratulids (Brachiopoda) from the Middle Vistula Gorge. *Prace Muzeum Ziemi*, **12**, 3–86.

Popiel-Barczyk, E. (1980) Brachiopod genus *Cryptopora* Jeffreys, from the Miocene deposits of the Lublin Upland. *Acta Geologica Polonica*, **30**, 111–20.

Popov, L. E. (1980) New strophomenoids from the Middle Ordovician of northern Kazakhstan. *Novie Vidi Drevnich Rastenii i Bespozbonochnich SSSR*, **5**, 54–7.

Popov, L. E. and Pushkin, V. I. (1986) [Ordovician inarticulate brachiopods from the southern pre-Baltics.] in [*New and Little-known species of Fossil Animals and Plants from Byelorussia*.] 'Nauka Tekhnika', Minsk, pp. 11–22.

Popov, L. E. and Rukavishnikova, T. B. (1986) New family of giant inarticulate brachiopods from the Upper Ordovician of southern Kazakhstan. *Paleontologicheskii Zhurnal*, **1986**, 56–60.

Popov, L. E. and Tikhanov, Yu. A. (1990) Early Cambrian brachiopods from Southern Kirgizia. *Paleontologicheskii Zhurnal*, **1990**, 33–45.

Prins, C. F. W. (1968) Carboniferous Productidina and Chonetidina of the Cantabrian Mountains (NW Spain): Systematics, stratigraphy and palaeoecology. *Leidse Geologische Mededelingen*, **43**, 1–126.

Racheboeuf, P. R. (1976) Chonetacea (Brachiopodes) du Dévonien Inférieur du Bassin de Laval (Massif Amoricain). *Palaeontographica, Abteilung A*, **52**, 14–89.

Racheboeuf, P. R. (1981) Chonetacés (Brachiopodes) Siluriens et Dévoniens du Sud-ouest de L'Europe. *Mémoires de la Société Géologique et Mineralogique de Bretagne*, **27**, 1–294.

Racheboeuf, P. R. (1987) Upper Lower and Lower Middle Devonian brachiopods from Bathurst, Devon and Ellesmere Islands, Canadian Arctic Archipelago. *Geological Survey of Canada, Bulletin*, **375**, 1–29.

Racheboeuf, P. R. and Copper, P. (1986) The oldest chonetacean brachiopods (Ordovician-Silurian, Anticosti Island, Quebec). *Canadian Journal of Earth Sciences*, **23**, 1297–308.

Rau, K. (1905) Die Brachiopoden des mittleren Lias Schwäbens mit Ausschluss der Spiriferinen. *Geologische und Paläontogische Abhandlungen*, **10**, 263–355.

Renz, C. (1932) Brachiopoden des sudschweizerischen und west griechischen Lias. *Schweizerische Paläontologische Gesellschaft Abhandlungen*, **52**, 1–61.

Roberts, J. and Jell, P. A. (1990) Early Middle Cambrian (Ordian) brachiopods of the Coonigan Formation, western New South Wales. *Alcheringa*, **14**, 257–309.

Rong, J.-Y. (1984) Brachiopods of latest Ordovician in the Yichang district, western Hubei, central China. *Stratigraphy and Palaeontology of Systemic Boundaries in China, Ordovician–Silurian*, **1**, 11–176.

Rong, J.-Y. and Yang, X.-C. (1981) Middle and late early Silurian brachiopod faunas in southwest China. *Memoirs of Nanjiing Institute of Geology and Palaeontology Academia Sinica*, **13**, 163–270.

Rõõmusoks, A. (1970) [*Stratigraphy of the Viruan Series (middle Ordovician) in northern Estonia.*] Tartu Riikliku Ülikool, Tallinn, 346 pp. [in Russian].

Ross, R. J. Jr and James, N. P. (1987) Brachiopod biostratigraphy of the middle Ordovician Cow Head and Table Head Groups, Western Newfoundland. *Canadian Journal of Earth Sciences*, **24**, 70–95.

Rouselle, L. (1977) Spiriferines du lias moyen et supérieur au Maroc (Rides Prerifaunes: Moyen Atlas) et en Espagne (Chaine Celtiberique orientale). *Notes et Mémoires du Service Géologique du Maroc*, **268**, 153–75.

Rowell, A. J. (1962) The genera of the brachiopod superfamilies Obolellacea and Siphonotretacea. *Journal of Paleontology*, **36**, 136–52.

Rowell, A. J. (1977) Early Cambrian brachiopods from the Southwestern Great Basin of California and Nevada. *Journal of*

Paleontology, **51**, 68–85.

Rowell, A. J. and Bell, C. (1961) The inarticulate brachiopod *Curticia* Walcott. *Journal of Paleontology*, **35**, 927–31.

Rowell, A. J. and Caruso, N. E. (1985) The evolutionary significance of *Nisusia sulcata*, an early articulate brachiopod. *Journal of Paleontology*, **59**, 1227–42.

Rozanov, A. Yu. and Sokolov, B. S. (eds) (1984) [*Lower Cambrian Stage Subdivision*]. Akademiia Nauk SSSR, Geologii, Geofiziki i Geokhimii, Moscow, 184 pp. [in Russian].

Rozman, K. S. (1981) Middle and Upper Ordovician brachiopods of Mongolia. *Trudy Akademia Nauk SSSR Geologicheskii Institut*, **354**, 117–76.

Rubel, M. (1961) [Lower Ordovician brachiopods of the superfamilies Orthacea, Dalmanellacea, and Syntrophiacea of the East Baltic.] *Eesti NSV Teaduste Akadeemia Geoloogia Instituudi Uurimused Trudy*, **6**, 141–226 [in Russian].

Rubel, M. (1977) [Revision of Silurian Dayiacea (Brach.) from the northeast Baltic.] *Eesti NSV Teaduste Akadeemia Toimetised (Keemia-Geoloogia)*, **26**, 211–20 [in Russian].

Rudwick, M. J. S. (1968) The feeding mechanisms and affinities of the Triassic brachiopods *Thecospira* Zugmayer and *Bactrynium* Emmrich. *Palaeontology*, **11**, 329–60.

Rudwick, M. J. S. and Cowen, R. (1968) The functional morphology of some aberrant strophomenide brachiopods from the Permian of Sicily. *Bolletino della Società Paleontologica Italiana*, **6**, 113–76.

Rushton, A. W. A. (1974) The Cambrian of Wales and England, in *The Cambrian of the British Isles, Norden and Spitzbergen* (ed. C. H. Holland), Wiley, London, pp. 43–121.

Sapelnikov, V. P. (1977) [Systematics and phylogeny of the superfamily Stricklandiacea (Brachiopoda).] *Akademiia Nauk SSSR Uralskii Nauchnyi Tsentr Instituta Geologii i Geokhimii Trudy*, **129**, 3–19 [in Russian].

Sapelnikov, V. P. and Rukavishnikova, T. B. (1973) [Two new genera of early Pentameracea (Brachiopoda) from southern Kazakhstan.] *Paleontologicheskij Zhurnal*, **1973**, 32–8 [in Russian].

Sarytcheva, T. G. (1928) [The Productidae of the group *Productus giganteus* Martin from the Visean of Moscow]. *Akademiia Nauk SSSR Insitut Geologicheskikh Nauk Trudy*, **1**, 1–71 [in Russian].

Sarytcheva, T. G. (1952) in [A description of the Palaeozoic Brachiopoda of the Moscow Basin.] (eds T. G. Sarytcheva and A. N. Sokolskaya), *Akademiia Nauk SSSR Paleontologicheskii Institut Trudy*, **38**, 1–307 [in Russian].

Sarytcheva, T. G. (1963) in [*Carboniferous Brachiopods and Palaeogeography of the Kuznetsk Basin*.] (eds T. G. Sarytcheva, A. N. Sokolskaya, G. A. Besnosova *et al.*), *Akademiia Nauk SSSR Paleontologicheskii Institut Trudy*, **95**, 1–547 [in Russian].

Sarytcheva, T. G. (1965) in [The development and successions of marine organisms at the Palaeozoic Mesozoic boundary.] (eds B. F. Ruzhentsev and T. G. Sarytcheva), *Akademiia Nauk SSSR Paleontologicheskii Institut Trudy*, **108**, 1–431 [in Russian].

Sarytcheva, T. G. and Sokolskaya, A. N. (1952) [Index of Palaeozoic Brachiopods in the Moscow Basin]. *Akademiya Nauk SSR, Paleontologicheskii Institut, Trudy*, **38**, pp. 1–307, pls 1–71.

Savage, N. M., Eberlein, G. D. and Churkin, M. Jr (1978) Upper Devonian brachiopods from the Port Refugio Formation, Suemez Island, southeastern Alaska. *Journal of Paleontology*, **52**, 370–93.

Schmidt, H. (1941) Die mitteldevonischen Rhynchonelliden der Eifel. *Senckenbergische Naturforschende Gesellschaft Abhandlungen*, **459**, 1–79.

Schnur, J. (1853) Zusammenstellung und Beschriebung sämtlicher im Übergangsgebirge der Eifel vorkommenden Brachiopoden nebst Abbildung derselben. *Palaeontographica*, **4**, 169–254.

Schuchert, C. and Cooper, G. A. (1931) Synopsis of the brachiopod genera of the suborders Orthoida and Pentameroidea, with notes on the Telotremata. *American Journal of Science*, **22**, 241–51.

Schuchert, C. and Cooper, G. A. (1932) Brachiopod genera of the suborders Orthoidea and Pentameroidea. *Yale University Peabody Museum Natural History Memoir*, **4**, 1–270.

Severgina, L. G. (1985) [Lower Ordovician brachiopods of the Gornoi Altai, Kuznetz Alatai, Shori hill and Salair.] *Trudy Akademia Nauk SSSR Sibirskoe Otdalenie*, **584**, 34–53 [in Russian].

Sheehan, P. M. and Lespérance, P. J. (1979) Late Ordovician (Ashgillian) brachiopods from the Percé region of Quebec. *Journal of Paleontology*, **53**, 950–67.

Shephard-Thorn, E. R. (1963) The Carboniferous Limestone succession in north-west County Limerick, Ireland. *Proceedings of the Royal Irish Academy*, **62B**, 267–94.

Sowerby, J. de C. (1839) Shells of the Lower Silurian rocks, in *The Silurian System* (ed. R. I. Murchison). Murray, London, pp. 634–44.

Sowerby, J. de C. (1840) Explanation of the plates and woodcuts. *Transactions of the Geological Society of London*, **5**, 633–703.

Spjeldnæs, N. (1957) The Middle Ordovician of the Oslo Region, Norway 8. Brachiopoda of the suborder Strophomenida. *Norsk Geologisk Tidsskrift*, **37**, 1–214.

St Joseph, J. K. S. (1938) The Pentameracea of the Oslo Region. *Norsk Geologisk Tidsskrift*, **17**, 225–336.

Stainbrook, M. A. (1947) Brachiopoda of the Percha shale of New Mexico and Arizona. *Journal of Paleontology*, **21**, 297–328.

Stepanov, D. L., Kulikov, M. V. and Sultanaev, A. A. (1975) [The stratigraphy and brachiopods of the upper Permian deposits of the Kanin Peninsula.] *Leningradskogo Universiteta (Geologia i Geografiia) Vestnik*, **6**, 51–65 [in Russian].

Strusz, D. L. (1982) Wenlock brachiopods from Canberra, Australia. *Alcheringa*, **6**, 105–42.

Su, Y.-Z. (1980) [Cambrian–Devonian Brachiopoda], in *Shenyang Dizhi Kuangchan Yanjiusuo. Dongbei Diqu gushengwu tuce. [Paleontological Atlas of Northeast China]* Vol. 1. *Paleozoic*. Geological Publishing House, Peking, pp. 254–327 [in Chinese].

Talent, J. A. (1956) Devonian brachiopods and pelecypods of the Buchan Caves Limestone, Victoria. *Proceedings of the Royal Society of Victoria*, **68**, 1–56.

Tchorszevsky, E. S. (1972) [A new family of Jurassic terebratulid from the Peninsk slope (Zakarpat) Zone.] *Kharkovskogo Universiteta (Geologia) Vestnik*, **72**, 62–5 [in Russian].

Tchorszevsky, E. S. (1974) [New data on the internal structure of shells and systematics of the superfamilies Terebratuloidea [sic] Gray, 1840 etc.]. *Kharkovskogo Universiteta (Geologia) Vestnik*, **108**, 42–58 [in Russian].

Temple, J. T. (1968) The Lower Llandovery (Silurian) brachiopods from Keisley, Westmorland. *Monograph of the Palaeontographical Society*, **521**, 1–58.

Temple, J. T. (1987) Early Llandovery brachiopods of Wales. *Monograph of the Palaeontographical Society*, **572**, 1–137.

Termier, G., Termier, H., de Lapparent, A. F. *et al.* (1974) Monographie du Permo-Carbonifere de Wardak (Afghanistan Central). *Documents des Laboratoires de Géologie de la Faculté des Sciences de Lyon*, **2**, 1–167.

Ulrich, E. O. (1889) On *Lingulasma*, a new genus, and eight new species of *Lingula* and *Trematis*. *American Geologist*, **3**, 377–91.

Ulrich, E. O. and Cooper, G. A. (1936a) New Silurian brachiopods of the family Triplesiidae. *Journal of Paleontology*, **10**, 331–47.

Ulrich, E. O. and Cooper, G. A. (1936b) New genera and species of Ozarkian and Canadian brachiopods. *Journal of Paleontology*, **10**, 616–31.

Ulrich, E. O. and Cooper, G. A. (1938) Ozarkian and Canadian Brachiopoda. *Geological Society of America Special Paper*, **13**, 1–323.

Ushatinskaya, G. T. (1987) [Unusual inarticulate brachiopods from the Lower Cambrian of Mongolia.] *Paleontologicheskii Zhurnal*, **1987**, 62–8 [in Russian].

Vandercammen, A. (1956) Revision de *Spinocyrtia struniana*. *Bulletin Institut Royal des Sciences Naturelles de Belgique*, **32** (59), 1–9.

Vandercammen, A. (1959) Essai d'étude statistique des *Cyrtospirifer* du Frasnian de la Belgique. *Mémoires de l'Institut Royal des Sciences Naturelles de Belgique*, **145**, 1–175.

Vandercammen, A. (1963) Spiriferidae du Devonian de la Belgique. *Mémoires de l'Institut Royal des Sciences Naturelles de Belgique*, **150**, 1–177.

Veevers, J. J. (1959) Devonian brachiopods from the Fitzroy Basin,

Western Australia. *Australian Bureau of Mineral Resources, Geology and Geophysics, Bulletin*, **45**, 1–220.

Vogel, K. (1962) Muscheln mit Schlösszähnen aus dem spanischen Kambrium und ihre Bedeutung für die Evolution der Lamellibranchiaten. *Akademie der Wissenschaft und Literatur Mainz, Mathematische–Naturwissenschaftliche Klasse*, **4**, 193–244.

Vokes, H. E. (1957) Geography and geology of Maryland. *Bulletin of the Maryland Department of Geology and Mines*, **19**, 1–243.

Waagen, W. H. (1882–1885) Salt Range fossils. part 4, Brachiopoda. *Memoirs of the Geological Survey of India, Series* **13**, 329–770.

Walcott, C. D. (1912) Cambrian Brachiopoda. *US Geological Survey Monograph*, **1**, 1–872.

Wang, Y. and Zhu, R.-F. (1979) [Beiliuan (Middle Middle Devonian) brachiopods from south Guizhou and central Guangxi.] *Palaeontologica Sinica*, **15**, 1–95 [in Chinese with English summary].

Waterhouse, J. B. (1966) Lower Carboniferous and Upper Permian brachiopods from Nepal. *Jahrbuch Geologische Bundestalt*, **12**, 5–99.

Waterhouse, J. B. (1968) The classifications and descriptions of Permian Spiriferida (Brachiopods) from New Zealand. *Palaeontographica, Abteilung A*, **129**, 1–94.

Waterhouse, J. B. (1975) New Permian and Triassic brachiopod taxa. *Papers of the Department of Geology, University of Queensland*, **7**, 1–23.

Waterhouse, J. B. (1978) Permian Brachiopoda and Mollusca from north-west Nepal. *Palaeontographica, Abteilung A*, **160**, 1–175.

Waterhouse, J. B. (1982) New Zealand Permian brachiopod systematics, zonation and palaeoecology. *Wellington Palaeontological Bulletin*, **48**, 1–155.

Weller, S. (1907) Kinderhook faunal studies. IV. The fauna of the Glen Park Limestone. *Academy of Science of St. Louis, Transactions*, **15**, 259–64.

Weller, S. (1911) Genera of Mississippian loop-bearing Brachiopoda. *Journal of Geology*, **19**, 439–48.

Weller, S. (1914) The Mississippian Brachiopoda of the Mississippi valley basin. *Monograph of the Illinois Geological Survey*, **1**, 1–508.

Whidbourne, G. F. (1897) A monograph of the Devonian fauna of the South of England. *Palaeontographical Society Monograph*, **47**, v. 3, part 2. *The fauna of the Marwood and Pilton Beds of North Devon and Somerset*, London, pp. 113–78, pls xvii–xxi.

Williams, A. (1951) Llandovery brachiopods from Wales with special reference to the Llandovery district. *Q. J. Geol. Soc. London*, **107**, 85–136.

Williams, A. (1962) The Barr and Lower Ardmillan Series (Caradoc) of the Girvan District, South West Ayrshire, with descriptions of the Brachiopoda. *Memoir of the Geological Society of London*, **3**, 1–267.

Williams, A. (1974) Ordovician Brachiopoda from the Shelve district, Shropshire. *Bulletin of the British Museum (Natural History), Geology, Supplement*, **11**, 1–163.

Williams, A. and Curry, G. B. (1985) Lower Ordovician Brachiopoda from the Tourmakeady Limestone, Co. Mayo, Ireland. *Bulletin of the British Museum (Natural History) Geology*, **38**, 183–269.

Williams, A. *et al.* (1965) Part H. Brachiopoda, in *Treatise on Invertebrate Paleontology* (eds R. C. Moore, A. J. Powell and H. M. Muir-Wood), Geological Society of America and University of Kansas Press, Boulder, Colorado, and Lawrence, Kansas, 927 pp.

Williams, A. J. (1977) Insight into lingulid evolution from the Late Devonian. *Alcheringa*, **1**, 401–6.

Williams, A., Lockley, M. G. and Hirst, J. M. (1981) Benthic palaeocommunities represented in the Ffairfach Group and coeval Ordovician successions of Wales. *Palaeontology*, **24**, 661–94.

Wolfart, R. (1974) Die Fauna (Brachiopoda, Mollusca, Trilobita) aus dem Unter-Kambrium von Kerman, Sudost-Iran. *Geologisches Jahrbuch*, **B8**, 5–70.

Wright, A. D. (1964) The fauna of the Portrane Limestone, II. *Bulletin of the British Museum (Natural History) Geology*, **9**, 157–256.

Xu, G.-R. and Liu, G.-C. (1983) Brachiopods, in [*Triassic of the South Qilian Mountains*.] (eds Z.-Y. Yang, H.-F. Yin, G.-H. Xu *et al.*), Peking Geological Publishing House [in Chinese with English summary], pp. 84–128.

Xu, H.-K. Rong, J.-Y. and Liu, D.-Y. (1974) [Ordovician brachiopods], in *Handbook of Stratigraphy and Palaeontology in Southwest China*. Nanjing Institute of Geology and Palaeontology, Academia Sinica, Nanjing [in Chinese], pp. 144–54.

Zavodowskii, V. M. (1968) [New species of Permian Spiriferida from north-east USSR], in [*New Species of Fossils Plants and Invertebrates from the USSR*] (ed. B. P. Markovskii), NEDRA, Moscow [in Russian], pp. 149–60.

Zell, M. G. and Rowell, A. J. (1988) Brachiopods of the Holm Dal Formation (Late Middle Cambrian), central north Greenland. *Meddelelser Grønland Geologiska*, **20**, 119–44.

Zhao, J.-K., Sheng, J.-Z., Yao, Z.-Q. *et al.* (1981) [The Changhsingian and Permian-Triassic boundary of south China]. *Bulletin Nanjing Institute of Geology and Palaeontology*, **2**, 1–112 [in Chinese with English summary].

Zugmayer, H. (1880) Ueber rhätische Brachiopoden. *Sitzungsberichte der Kaiserlich-Königlichen Geologischen Reichsanstalt Jahrbuch*, **30**, 149–56.

23

PHORONIDA

P. D. Taylor

The Phoronida are a phylum of tube-dwelling, vermiform lophophorates regarded as close relatives of brachiopods and bryozoans. Although cosmopolitan in their present-day distribution, only about 10 or 11 species distributed between two genera (*Phoronis*, *Phoronopsis*) are recognized (see Emig, 1982, 1985). Phoronids have no mineralized parts and are devoid of a body fossil record. However, living phoronids either burrow into soft sediment or bore into hard substrates, and various fossil burrows and borings have been attributed, with varying degrees of confidence, to the activities of phoronids. The range is plotted on the first bryozoan chart (Fig. 24.1) (see Chapter 24).

No family-level classification exists.

Phylum PHORONIDA Hatschek, 1888
??P€ (RIP)/?D. (SIG)–Rec. Mar.

First: ??*Skolithos*, Riphean, North Australia (Glaessner, 1969); ?'Formengruppe B' boring of Jux and Strauch (1965), Siegenian, Germany.

Comments: *Skolithos* Haldemann, 1840, a simple straight burrow, may be partly attributable to the Phoronida but, as Häntzschel (1975) emphasizes, this ichnogenus requires revision. Various borings can be attributed with more certainty to the Phoronida (see Voigt, 1975). While the earliest example may be an unnamed Devonian boring, the post-Palaeozoic ichnogenus *Talpina* von Hagenow, 1840 has been shown by Voigt (1975) to resemble closely the borings of modern *Phoronis ovalis*.

REFERENCES

Emig, C. C. (1982) The biology of Phoronida. *Advances in Marine Biology*, **19**, 1–89.

Emig, C. C. (1985) Phylogenetic systematics in Phoronida (Lophophorata). *Zeitschrift für Zoologische Systematik und Evolutionsforschung*, **23** (3), 184–93.

Glaessner, M. F. (1969) Trace fossils from the Precambrian and basal Cambrian. *Lethaia*, **2**, 369–94.

Häntzschel, W. (1975) Trace fossils and problematica. 2nd edition, in *Treatise on Invertebrate Paleontology, Part W, Supplement 1* (eds R. C. Moore and C. Teichert), Geological Society of America and University of Kansas, Boulder, Colorado, and Lawrence, Kansas, pp. W1–W269.

Jux, U. and Strauch, F. (1965) Angebohrte Spiriferen-Klappen; ein Hinweis auf palökologische Zusammenhänge. *Senckenbergiana Lethaea*, **46**, 89–125.

Voigt, E. (1975) Tunnelbaue rezenter und fossiler Phoronidea. *Paläontologisches Zeitschrift*, **49**, 135–67.

The Fossil Record 2. Edited by M. J. Benton. Published in 1993 by Chapman & Hall, London. ISBN 0 412 39380 8

24

BRYOZOA

P. D. Taylor

The Bryozoa are an exclusively colonial phylum of coelomate metazoans, closely related to brachiopods and phoronid worms. Most present-day bryozoans are marine and the majority of these secrete mineralized skeletons of calcium carbonate, usually calcite. As sessile members of the benthos, bryozoans have a good fossilization potential and are represented in the fossil record by an estimated 20 000 or more species. Unfortunately, the taxonomy and stratigraphical distribution of fossil bryozoans are both poorly known. This compilation must therefore be treated as very provisional; it is no more than a starting point for subsequent improvement. Major problems inherent in the compilation include:

1. Uncertain taxonomic status of families. Relatively few bryozoan families are defined rigorously as monophyletic clades using autapomorphies. A great many are likely to be paraphyletic or even polyphyletic groupings.
2. Subjectivity of taxonomic rank. There is little or no consistency in what constitutes a family, either within or between orders of bryozoans.
3. Stratigraphical range uncertainty. Many fossil bryozoan faunas have yet to be adequately described, and so family ranges are likely to be underestimated. Furthermore, difficulties exist in dating some faunas, especially for the Upper Palaeozoic.

Ryland (1982) provides the most up-to-date family-level classification of living bryozoans. There is no equivalent work for extinct families; the original bryozoan *Treatise* (Bassler, 1953) has been revised only in part (Boardman *et al.*, 1983).

The general lack of comprehensive monographic literature (cf. faunal studies) means that it is impossible to supply information about the presence/absence of families in intervening stages between their first and last occurrences.

Acknowledgements – I thank the following for their assistance with this compilation: D. P. Gordon (Wellington), E. Voigt (Hamburg), F. K. McKinney (Boone), C. J. Buttler (Cardiff), A. H. Cheetham (Washington) and S. Lidgard (Chicago).

Class PHYLACTOLAEMATA Allman, 1856

This is a small class of exclusively freshwater bryozoans which lack mineralized skeletons. Only one phylactolaemate body fossil has been reported – *Plumatellites proliferus* Fric, 1901 from the Cenomanian of Bohemia – but this indistinct fossil shows no certain phylactolaemate features (Mundy *et al.*, 1981). However, all modern phylactolaemates produce chitinous resting bodies called statoblasts which are resistant and should have a reasonable fossilization potential. Until recently, the oldest statoblasts recorded were from the 'late Tertiary' of Arctic Canada. Apparent statoblasts have now been described by Vinogradov (1985) from the Jurassic of the former USSR, and by Jell and Duncan (1986) from the Cretaceous of Australia. If phylactolaemates are a primitive bryozoan class, as is often believed, finds of statoblasts from older rocks might be anticipated.

F. FREDERICELLIDAE Hyatt, 1868 Rec.

F. PLUMATELLIDAE Allman, 1856
J. (l./m.)–Rec. Mar. (see Fig. 24.1)

First: *Plumatella mongoliensis* Vinogradov, 1985, Lower or Middle Jurassic, Jargalang Group, Jargalang, Mongolia; *P. sibiriensis* Vinogradov, 1985, *P.* sp., Lower or Middle Jurassic, Ichetuya Group, Novospasskoe settlement, Mukhor–Shibirskii region, Buryat, former USSR; *P. angaraensis* Vinogradov, 1985, *P. sedimentata* Vinogradov, 1985, Lower or Middle Jurassic, Cheremkovka Group, Irkutsk District, Olonkovsk region, Ust-Balei, former USSR. **Extant**

F. LOPHOPODIDAE Rogick, 1935 **Extant**

F. CRISTATELLIDAE Allman, 1856
T. (ZAN)–Rec. Mar.

First: *Cristatella mucedo* Cuvier, 1798, Beaufort Formation, Arctic Canada (see Kuc, 1973; Matthews *et al.*, 1990). **Extant**

Order CTENOSTOMATA Busk, 1852

The Fossil Record 2. Edited by M. J. Benton. Published in 1993 by Chapman & Hall, London. ISBN 0 412 39380 8

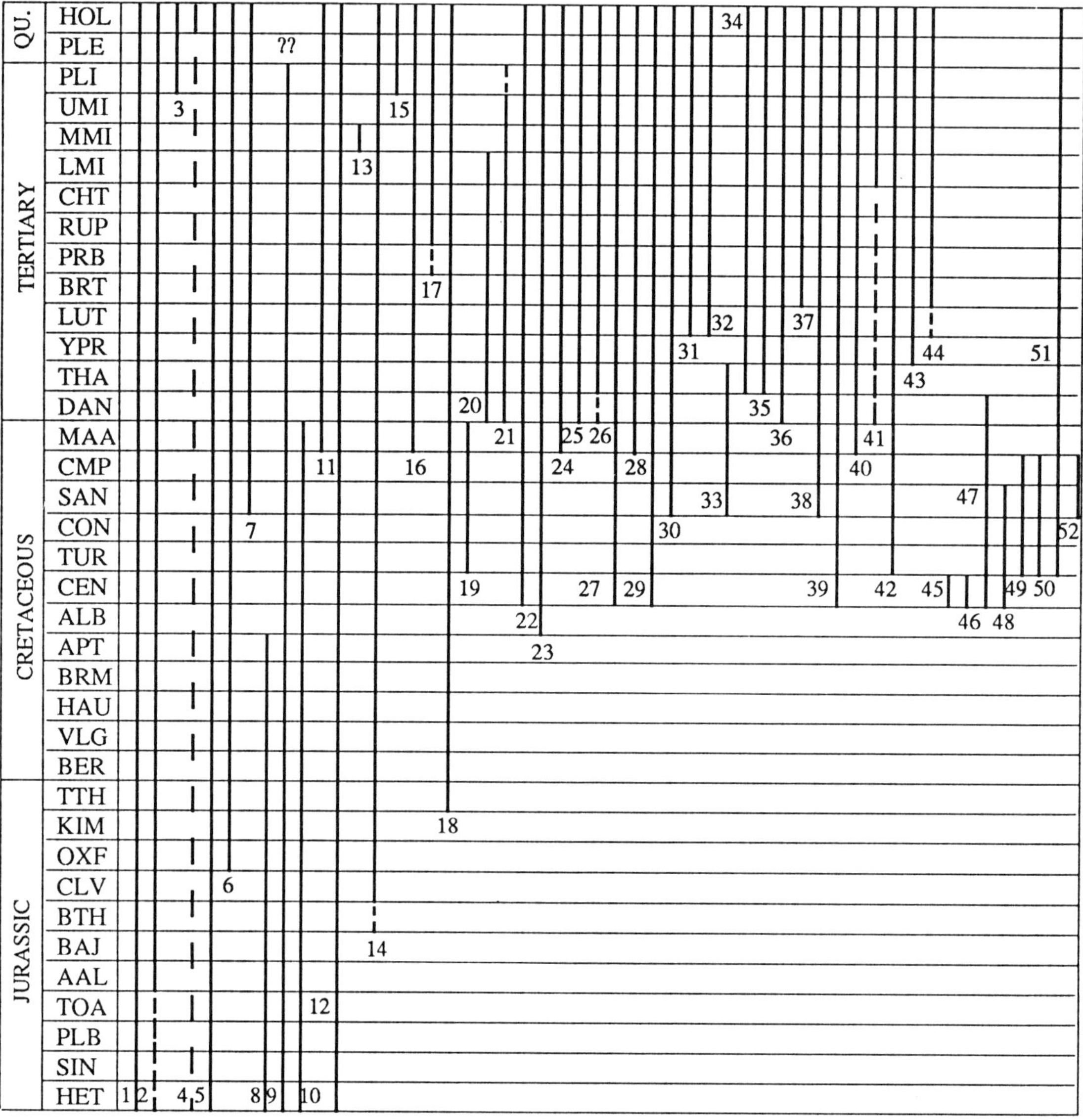

Fig. 24.1

Ctenostomes are a primitive paraphyletic group of gymnolaemate bryozoans which lack mineralized skeletons. Most modern ctenostomes are marine, but some are inhabitants of brackish water or fresh water (see d'Hondt, 1983 and Hayward, 1985 for identification guides to modern taxa). The majority of families are unrepresented in the fossil record. However, preservation of ctenostomes as fossils can occur in two ways: some species are represented by the borings that they excavate in calcareous substrata (see Pohowsky, 1978); non-boring species can be preserved as natural moulds on the underside of oysters, serpulid worms or other organisms which have overgrown them ('bioimmuration', see Taylor, 1990). Unfortunately, a dual nomenclature has developed for boring ctenostomes; certain palaeontologists regard them as ichnofossils, whereas others treat them as body fossils because the borings accurately reproduce the external shapes of the zooids.

F. BENEDENIPORIDAE Delage and Hérouard, 1897 **Extant** Mar.

F. HISLOPIIDAE Jullien, 1885 **Extant** FW

F. ALCYONIDIIDAE Hinck, 1880 ?C. (u.)–Rec. Mar.

A possible alcyonidiid from the Pennsylvanian of NW Pennsylvania, USA awaits description (R. J. Cuffey, pers. comm., 1990). **Extant**

F. CLAVOPORIDAE Soule, 1953 **Extant** Mar.

F. FLUSTRELLIDRIDAE Bassler, 1953 **Extant** Mar.

F. LOBIANCOPORIDAE Delage and Hérouard, 1897 **Extant** Mar.

F. PHERUSELLIDAE Soule, 1953 **Extant** Mar.

F. HAYWARDOZOONTIDAE d'Hondt, 1983 **Extant** Mar.

F. PALUDICELLIDAE Allman, 1844 **Extant** FW

F. ARACHNIDIIDAE Hincks, 1880 Tr. (LAD)–Rec. Mar./FW

Fig. 24.1

First: Arachnidiid, Upper Muschelkalk, Künzelsau-Garnberg, Germany (Todd and Hagdorn, in press). **Extant**

F. VICTORELLIDAE Hincks, 1880 **Extant** Mar./FW

F. PANOLICELLIDAE Jebram, 1985 **Extant** Mar.

F. SUNDANELLIDAE Jebram, 1973 **Extant** Mar.

F. MONOBRYOZOONTIDAE Remane, 1936 **Extant** Mar.

F. AETHOZOONTIDAE d'Hondt, 1983 **Extant** Mar.

F. PACHYZOONTIDAE d'Hondt, 1983 **Extant** Mar.

F. VALKERIIDAE Hincks, 1880 **Extant** Mar.

F. MIMOSELLIDAE Hincks, 1877 **Extant** Mar.

F. TRITICELLIDAE Sars, 1874 **Extant** Mar.

F. BATHYALOZOONTIDAE d' Hondt, 1975 **Extant** Mar.

F. FARELLIDAE d'Hondt, 1983 **Extant** Mar.

F. AEVERRILLIIDAE Jebram, 1973 **Extant** Mar.

F. HYPOPHORELLIDAE Prenant and Bobin, 1956 **Extant** Mar.

F. HARMERIELLIDAE d'Hondt, 1983 **Extant** Mar.

F. VESICULARIIDAE Hincks, 1880 **Extant** Mar.

F. BUSKIIDAE Hincks, 1880 J. (OXF)–Rec. Mar.

First: *Buskia nigribovis* Todd, 1993, Oxfordian, Villers-sur-Mer, Normandy, France, and Waterperry, Oxfordshire, England, UK (see Todd, in press).

F. IMMERGENTIIDAE Silén, 1946 K. (SAN)–Rec. Mar.

First: *Immergentia cruciata* (Mägdefrau, 1937), Eisenerz-konglomerat des Mittelsantons, Lengede-Broistedt, Peine, Niedersachsen, Germany.
Comments: Monogeneric. **Extant**

F. ROPALONARIIDAE Nickles and Bassler, 1990 O. (ASH)–K. (APT) Mar.

First: *Ropalonaria venosa* Ulrich, 1879, Richmond Group, Ohio and Indiana, USA.
Last: *Ropalonaria? bassleri* (Voigt and Soule, 1973), Faringdon Sponge Gravel, Faringdon, Oxfordshire, England, UK.
Comments: Monogeneric.

F. ORBIGNYOPORIDAE Pohowsky, 1978 S. (?SHE/?WHI/?GLE)–T. (PIA) Mar.

First: *Orbignyopora? capillaris* (Dollfus, 1877), Rochester Limestone, Lockport, New York State, USA.
Last: *O. archiaci* (Fischer, 1866), ?Astian, Italy (Pohowsky, 1978, p. 59).
Comments: Monogeneric.

F. VOIGTELLIDAE Pohowsky, 1978 P.–K. (MAA) Mar.

First: *Voigtella? timorensis* (Bassler, 1929), Permian, Noil Boewan, Timor.
Last: *V. regalis* Pohowsky, 1978, upper Maastrichtian, Trans-Caspian Province, former USSR (Pohowsky, 1978, p. 69); *V. secunda* Pohowsky, 1978, Ripley Formation, Alabama and Tennessee, USA (see Pohowsky, 1978, p. 75).
Comments: Monogeneric.

F. TEREBRIPORIDAE d'Orbigny, 1847 K. (MAA)–Rec. Mar.

First: *Marcusopora ripleyensis* Pohowsky, 1978, Ripley Formation, Coon Creek, Tennessee, USA (Pohowsky, 1978, p. 119). **Extant**

F. PENETRANTIIDAE Silén, 1946 Tr. (RHT)–Rec. Mar.

First: *Haimeina michelini* (Terquem, 1855), Kössener Schichten, Marmorgraben bei Mittenwald, Germany (Pohowsky, 1978, p. 90). **Extant**

F. COOKOBRYOZOONIDAE Pohowsky, 1978 T. (?LAN–?LAN/?SRV) Mar.

First: *Cookobryozoon lagaaiji* Pohowsky, 1978, Olcese Sand Member, Temblor Formation, Barkers Ranch, Kern Co., California, USA (Pohowsky, 1978, p. 93).
Last: *C. lagaaiji* Pohowsky, 1978, Kalimnan, Muddy Creek, Victoria, Australia.
Comments: Monospecific.

F. SPATHIPORIDAE Pohowsky, 1978 J. (?BTH/?CLV)–Rec. Mar.

First: *Spathipora cheethami* Pohowsky, 1978, Cornbrash, southern England, UK (Pohowsky, 1978, p. 103). **Extant**
Comment: Monogeneric.

Order CHEILOSTOMATA Busk, 1852

Cheilostomes are a seemingly monophyletic order of gymnolaemate bryozoans, which are mostly marine but also include a few brackish water species. All cheilostomes have mineralized skeletons of calcite or, less commonly, aragonite or a calcite:aragonite mix. In some families, however, mineralization is slight.

Cheilostomes are traditionally divided into two sub-orders: Anasca and Ascophora. These are best regarded as organizational grades. Anascans are a primitive and paraphyletic group from which ascophorans have probably been derived polyphyletically. Detailed phylogenetic relationships are poorly understood throughout the order, and classification is in a state of flux (see Gordon, 1989). Hence the validity and ranges of many families must be viewed as highly tentative.

Suborder ANASCA Levinsen, 1909

Anascans are a paraphyletic grouping of primitive cheilostomes without an ascus or equivalent structure. They are variously related to the more advanced ascophoran grade cheilostomes.

F. AETEIDAE Smitt, 1867 T. (ZAN)–Rec. Mar.

First: *Aetea truncata* (Landsborough, 1852), Lower Pliocene, Crete. **Extant**
Comments: Monogeneric. Pre-Pliocene records of this lightly mineralized family are erroneous or very doubtful; the only certain fossil examples are those preserved by bio-immuration (Voigt, 1983). However, the family is often regarded as primitive among cheilostomes, suggesting that older examples may be forthcoming.

F. SCRUPARIIDAE Busk, 1852 K. (MAA)–Rec. Mar.

First: *Scruparia* sp., upper MAA Chalktuff, Maastricht, The Netherlands (Voigt, 1985, p. 335). **Extant**

F. EUCRATEIDAE Johnston, 1838 **Extant** Mar.

F. LABIOSTOMELLIDAE Silén, 1942 **Extant** Mar.

F. MEMBRANIPORIDAE Busk, 1854 T. (??PRB)–Rec. Mar.

First: *Biflustra savartii texturata* (Reuss, 1848), PRB, Vicenza Province, Italy (Braga and Barbin, 1988). **Extant**
Comments: The Membraniporidae, as here understood, are distinguished by the presence of a twinned ancestrula. As the ancestrula is seldom preserved or described in fossil material, ascertaining the range of the family is difficult. *Membranipora s.s.* is a lightly calcified epiphyte of algae and has no certain fossil record; most of the numerous fossil species assigned to *Membranipora* belong in families such as the Calloporidae.

F. ELECTRIDAE Stach, 1937 J. (TTH)–Rec. Mar.

First: *Pyriporopsis portlandensis* Pohowsky, 1973, Portland Beds, southern England, UK. **Extant**
Comments: A primitive paraphyletic family of malacostegan cheilostomes.

F. TENDRIDAE Vigneaux, 1949 **Extant** Mar.

Monogeneric for *Tendra* Nordman, 1839.

F. FUSICELLARIIDAE Canu, 1990 K. (TUR–MAA) Mar.

First: *Fusicellaria pulchella* d'Orbigny, 1851, TUR, Sarthe, France (Voigt, 1981).
Last: *Encicellaria hofkeri* Keij, 1969, upper MAA, St Pietersberg, Maastricht, The Netherlands.

F. BICORNIFERIDAE Keij, 1977 T. (DAN–BUR) Mar.

First: '*Bicornifera* sp.', Dano-Montian, Limburg, The Netherlands (Keij, 1977).
Last: *Bifissurinella triangularis* Poignang and Ubaldo, 1974, Burdigalian, SW France (see Keij, 1977).

F. SKYLONIIDAE Sandberg, 1963 T. (DAN–?MES/?ZAN) Mar.

First: *Calvina kalloensis* Willems, 1972, Sonja Lens of the Agatdalen Formation (late DAN–early THA), Agatdalen, Nûgssuaq, West Greenland; Lower–Middle Palaeocene, central and northern Poland (Szczechura, 1990).
Last: '*Skylonia* sp. A', Klasaman Formation (late Miocene or Pliocene in age), Waileh, Salaweti Island, New Guinea (Keij, 1973).

F. QUADRICELLARIIDAE Gordon, 1984 K. (CEN)–Rec. Mar.

First: *Cellarinidra clavata* (d'Orbigny, 1851), CEN, Le Mans, Sarthe, France. **Extant**

F. FLUSTRIDAE Fleming, 1828 **Extant** Mar.

Supposed fossil records of *Flustra* (e.g. Busk, 1859) may be the basal calcitic parts of bimineralic cheilostomes in which the aragonitic parts of the skeleton have suffered dissolution.

F. CALLOPORIDAE Norman, 1903 K. (ALB)–Rec. Mar.

First: *Wilbertopora mutabilis* Cheetham, 1954, Kiamichi Formation (*Adkinisites bravoensis* Zone), Texas, USA; *Marginaria* sp., Hunstanton Red Chalk (probably *Mortoniceras inflatum* Zone), Norfolk, England, UK (Taylor, 1988, p. 53).
Comments: A primitive paraphyletic family of ovicellate cheilostomes. **Extant**

F. CHAPERIIDAE Jullien, 1888 K. (MAA)–Rec. Mar.

First: *Hagenowinella vaginata* (von Hagenow, 1851), MAA, Maastricht, The Netherlands. **Extant**

F. HIANTOPORIDAE Gregory, 1893 T. (DAN)–Rec. Mar.

First: *Hiantopora tripora* Canu, 1911, Rocanéen, Roca, Argentina. **Extant**
Comments: *Tremogasterina* Canu, 1911 is reassigned to the ascophoran family Arachnopusiidae (e.g. Gordon, 1989).

F. CUPULADRIIDAE Lagaaij, 1952 T. (?DAN/?THA)–Rec. Mar.

First: *Cupuladria ovalis* Gorodiski and Balavoine, 1962, 'Paléocène', Tambacounda, Senegal (Cook and Chimonides, 1983, p. 571). **Extant**
Comments: Includes Discoporellidae Baluk and Radwanski, 1984.

F. MICROPORIDAE Gray, 1848 K. (CEN)–Rec. Mar.

First: *Stichomicropora oceani* (d'Orbigny, 1852), Lower CEN, Le Mans and Lamnay, Sarthe, France (Taylor, 1988, fig. 6d); *Stichomicropora* sp., lower CEN, Mülheim-Ruhr, Germany (Voigt, 1991). **Extant**

F. PORICELLARIIDAE Harmer, 1926 K. (MAA)–Rec. Mar.

First: *Poricellaria* sp., upper MAA, Westmoreland Parish, Jamaica (Cheetham, 1968a). **Extant**
Comments: Monogeneric.

F. PSEUDOPORICELLARIIDAE d'Hondt, 1987 **Extant** Mar.

F. ONYCHOCELLIDAE Jullien, 1881 K. (CEN)–Rec. Mar.

First: *Onychocella* spp., lower CEN, Lamnay, Sarthe, France (Taylor, 1988, fig. 6c); lower CEN, Mülheim-Ruhr, Germany. **Extant**

F. SELENARIIDAE Busk, 1854 K. (SAN)–Rec. Mar.

First: *Lunulites plana* d'Orbigny, 1852, Craie de Villedieu, Villedieu, France (Nowicki, 1986). **Extant**
Comments: Commonly called Lunulitidae Lagaaij, 1952.

F. STEGINOPORELLIDAE Hincks, 1884 T. (LUT)–Rec. Mar.

First: *Steginoporella cylindrica* Ziko, 1985, *S. delicata* Ziko, 1985, *S. obtusa* Ziko, 1985, Qarara Formation, Gabal Qarara, Maghagha, Egypt. **Extant**

F. THALAMOPORELLIDAE Levinsen, 1902 T. (LUT)–Rec. Mar.

First: *Thalamoporella bifoliata* Ziko, 1985, Qarara Formation, Gabal Qarara, Maghagha, Egypt; *T. aegyptiaca* Ziko, 1985, Samalut Formation, El Fashn Formation, Gabal Qarara, Maghagha, Egypt. **Extant**
Comments: In the absence of a polypide tube, porous cryptocyst, and spicules it is not possible to confirm the assignment to this family of Coniacian–Maastrichtian species of *Woodipora* as provisionally argued by Schubert (1986).

F. COSCINOPLEURIDAE Canu, 1913 K. (SAN)–T. (THA) Mar.

First: *Acoscinopleura* (?) *vindocinensis* (Filliozat, 1908), SAN (*testudinarius* Zone), Vendôme, France (Voigt, 1956).
Last: *Coscinopleura digitata* (Morton, 1834), Vincentown Limesand, New Jersey, USA.

F. MACROPORIDAE Uttley, 1949 T. (THA)–Rec. Mar.

First: *Macropora aquia* Canu and Bassler, 1920, Bryozoan Bed at base of Aquia Formation, Upper Marlboro, Maryland, USA. **Extant**
Comments: Monogeneric. *Monoporella exsculpta* (Marsson, 1887) from the MAA, has calcified opercula and a minutely porous frontal wall, suggesting that it may be a macroporid.

F. SETOSELLIDAE Levinsen, 1909 T. (THA)–Rec. Mar.

First: *Setosinella prolifica* Canu and Bassler, 1933, Vincentown Limesand, New Jersey, USA. **Extant**

F. SETOSELLINIDAE Hayward and Cook, 1979 T. (DAN)–Rec. Mar.

First: *Setosellina houzeaui* (Meunier and Pergens, 1886), Calcaire de Mons, Mons, Belgium (Voigt, 1987). **Extant**

F. CHLIDONIIDAE Busk, 1884 T. (BRT)–Rec. Mar.

First: *Cothurnicella* sp., BRT, Biarritz, France (Debourle, 1975). **Extant**
Comments: Otherwise known as Cothurnicellidae Bassler, 1935.

F. ALYSIDIIDAE Levinsen, 1909 **Extant** Mar.

Catenariopsis morningtoniensis Maplestone, 1899, from the Oligocene (Balcombian) of Victoria, Australia, is now regarded as a chaperiid (D. P. Gordon, pers. comm.).

F. CELLARIIDAE Hincks, 1880 K. (SAN)–Rec. Mar.

First: *Escharicellaria polymorpha* Voigt, 1924, Granulatensenon, Gr. Bülten, Germany (Voigt, 1991). **Extant**

F. ASPIDOSTOMATIDAE Jullien, 1888 K. (CEN)–Rec. Mar.

First: *Euritina eurita* (d'Orbigny, 1851), CEN, France (Voigt, 1981, table 1). **Extant**

F. CABEREIDAE Busk, 1852 K. (MAA)–Rec. Mar.

First: *Eoscrupocellaria cretae* (Marsson, 1887), Lower MAA, Rügen, Germany (see Voigt, 1991). **Extant**
Comments: Equivalent to Scrupocellariidae Levinsen, 1909.

F. EUOPLOZOIDAE Harmer, 1926 **Extant** Mar.

F. EPISTOMIIDAE Gregory, 1903 T. (?YPR/AQT/BUR/LAN 1)–Rec. Mar.

First: 'Epistomiid? gen. and sp. indet.', YPR, DSDP Site 308, Koko Seamount, Pacific Ocean (Cheetham, 1975); *Synnotum* sp., Tuban Formation (Lower Miocene, Tertiary e5), Prupuh, East Java. **Extant**

F. FARCIMINARIIDAE Busk, 1884 **Extant** Mar.

F. QUADRICELLARIIDAE Gordon, 1984 K. (TUR)–Rec. Mar.

First: *Quadricellaria oblonga* d'Orbigny, 1851, Turonian, Sainte-Maure, Indre-et-Loire, France. **Extant**

F. BUGULIDAE Gray, 1848 **Extant** Mar.

F. BICELLARIELLIDAE Levinsen, 1909 T. (YPR)–Rec. Mar.

First: 'Bicellariellidae? gen. et sp. indet.', DSDP Site 246, Indian Ocean (Labracherie, 1975). **Extant**

F. DESMACYSTIDAE Dick and Ross, 1988 **Extant** Mar.

F. BEANIIDAE Canu and Bassler, 1927 T. (?LUT/?BRT)–Rec. Mar.

First: *Beania bermudezi* Lagaaij, 1968, Loma Candella Formation (Middle Eocene), Loma Candela, Pinar del Rio Province, Cuba. **Extant**

F. OTOPORIDAE Lang, 1916 K. (CEN) Mar.

First and Last: *Anaptopora disjuncta* Lang, 1916, *Anotopora inaurita* Lang, 1916, *Otopora auricula* Lang, 1916, Chalk Marl, Cambridge, England, UK.

F. CTENOPORIDAE Lang, 1916 K. (CEN) Mar.

First and Last: *Ctenopora pecten* Lang, 1916, Chalk Marl, Cambridge, England, UK.
Comments: Monospecific.

F. ANDRIOPORIDAE Lang, 1916 K. (CEN)–T. (DAN) Mar.

First: *Kankopora kankensis* Lang, 1916, Korycaner Schichten, Kaňk, Czechoslovakia.
Last: *Auchenopora guttur* Lang, 1916, Danian, Faxe, Denmark.

F. CALPIDOPORIDAE Lang, 1916 K. (CEN–SAN) Mar.

First: *Calpidopora diota* Lang, 1916, Korycaner Schichten, Kaňk, Czechoslovakia.
Last: *Graptopora scripta* Lang, 1916, SAN, Coulommiers, Loir-et-Cher, France.

F. THORACOPORIDAE Lang, 1916 K. (TUR–CMP) Mar.

First: *Thoracopora costata* Lang, 1916, *planus* Zone, Norfolk, England, UK.
Last: *T. monastica* (Brydone, 1909), CMP, Norfolk, England, UK.

F. LAGYNOPORIDAE Lang, 1916 K. (TUR–CMP) Mar.

First: *Lagynopora pediculus* (Reuss, 1874), Ober-Pläner, Strehlen, Dresden, Germany.
Last: *Leptocheilopora vulnerata* (Brydone, 1914), *mucronata* Zone, Norfolk, England, UK; *L. regularis* (d'Orbigny, 1851), Craie à *Baculites*, Sainte Colombe, Manche, France; *L. magna* Lang, 1916, *mucronata* Zone, Lüneburg, Germany.

F. PELMATOPORIDAE Lang, 1916 K. (TUR)–Rec. Mar.

First: *Sandalopora gallica* Lang, 1916, Turonian, Lavardin, Loir-et-Cher, France. **Extant**

F. TARACTOPORIDAE Lang, 1916 K. (CON–CMP) Mar.

First: *Taractopora confusa* Lang, 1916, *cortestudinarium* Zone, southern England, UK.
Last: *T. ernsti* Voigt and Schneemilch, 1986, lower CMP, Lägerdorf, Germany.

F. DISHELOPORIDAE Lang, 1916 K. (CON–SAN) Mar. (see Fig. 24.2)

First: *Hystricopora horrida* Lang, 1916, CON, Fécamp, Seine Inférieure, France.
Last: *Dishelopora claviceps* (Brydone, 1910), *Marsupites* Zone, England, UK.

F. RHACHEOPORIDAE Lang, 1916 K. (CON–MAA) Mar.

First: *Rhacheopora larvalis* Lang, 1916, CON, Fécamp, Seine Inférieure, France.

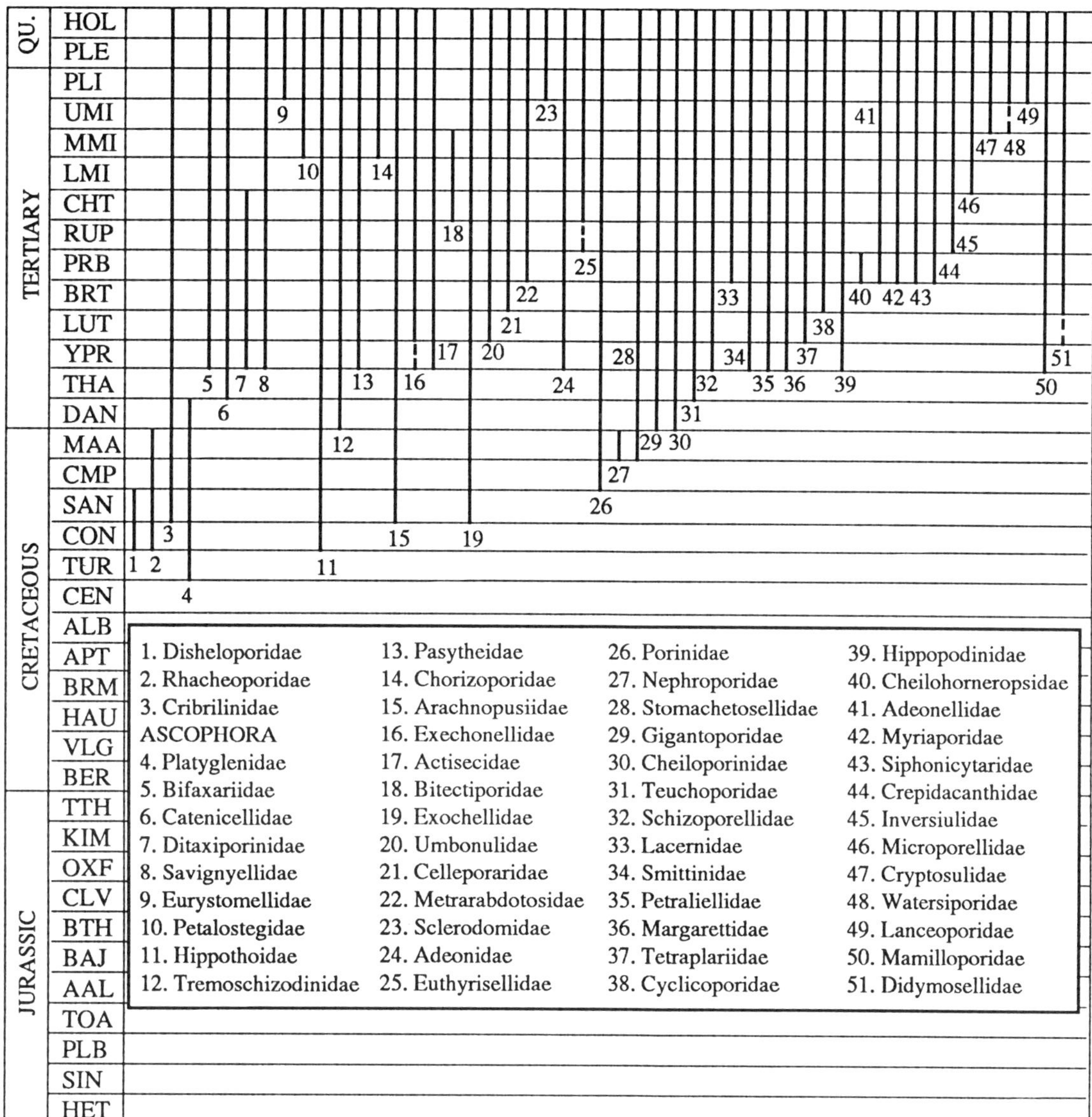

Fig. 24.2

Last: *Geisopora protecta* Lang, 1916, MAA, Rügen, Germany.

F. CRIBRILINIDAE Hincks, 1880 K. (SAN)–Rec. Mar.

First: *Mumiella mumia* (d'Orbigny, 1851), SAN, Sainte-Colombe, Manche, France. **Extant**

Comments: The relationship between the extant Cribrilinidae and the various Cretaceous/Palaeocene families of cribrimorphs listed above (Otoporidae to Rhacheoporidae) is unclear and awaits further study. In the meantime, the fossil families are retained as they reflect the high species diversity undoubtedly attained by cribrimorphs in the late Cretaceous.

Suborder ASCOPHORA Levinsen, 1909

Ascophorans are generally regarded as a polyphyletic, grade grouping of cheilostomes with calcified frontal shields covering a space (ascus or equivalent structure) into which sea-water enters as part of the hydrodynamic mechanism of tentacle protrusion. Although progress in distinguishing separate groups within the Ascophora is beginning to be made (e.g. Voigt, 1991; Gordon, 1989), the relationships between these groups and their anascan sister groups are still poorly understood. Therefore, the Ascophora are retained as a taxonomic entity for the present. Families are arranged according to the scheme adopted by Gordon (1989 and pers. comm., 1990), except that the cribrimorphs *s.s.* are retained within the Anasca.

As most ascophorans are relatively heavily calcified, they have a good preservation potential. However, some taxa have aragonitic skeletons prone to diagenetic dissolution, while others inhabit the deep sea and are seldom or never found fossil.

F. EUTHYROIDIDAE Levinsen, 1909 **Extant** Mar.

Monogeneric for *Euthyroides* Harmer, 1903.

F. PLATYGLENIDAE Marsson, 1887 K. (TUR)–T. (DAN) Mar.

First: *Platyglena culveriana* (Brydone, 1930), Chalk (*planus* Zone), Isle of Wight, England, UK; *P. altonensis* (Brydone,

1930), Chalk (*planus* Zone), Hampshire, England, UK (see Voigt, 1991).
Last: *P. ocellata* Marsson, 1887, DAN, Germany.

F. BIFAXARIIDAE Busk, 1884 T. (YPR)–Rec. Mar.

First: *'Bifaxaria* sp. 1', *'Bifaxaria* sp. 2', Lower Eocene, DSDP Site 246, Indian Ocean (Labracherie, 1975; see also Gordon, 1988). **Extant**

F. CATENICELLIDAE Busk, 1852 T. (THA)–Rec. Mar.

First: 'Vittaticellid new genus', Upper Palaeocene, DSDP Site 117, Rockall, North Atlantic (Cheetham and Håkansson, 1972). **Extant**

F. DITAXIPORINIDAE Cheetham, 1963 T. (YPR–CHT) Mar.

First: *Ditaxiporina labiatum* Canu, 1910, Auversian, Biarritz, France.
Last: *D. bifenestrata* Cheetham, 1963, Vicksburgian, Chicksasawhay Limestone, St Stephens, Alabama, USA.

F. SAVIGNYELLIDAE Levinsen, 1909 T. (YPR)–Rec. Mar.

First: *Savignyella*? sp., Lower Eocene, DSDP Site 246, Indian Ocean (Labracherie, 1975). **Extant**

F. EURYSTOMELLIDAE Levinsen, 1909 T. (PIA)–Rec. Mar.

First: *Eurystomella foraminigera* (Hincks, 1883), Mangapanian, Waipukurau, Hawkes Bay, New Zealand (Brown, 1952, p. 286). **Extant**
Comments: Monogeneric for *Eurystomella* Levinsen, 1909.

F. PETALOSTEGIDAE Gordon, 1984 T. (LAN)–Rec. Mar.

First: *Petalostegus tenuis* (Maplestone, 1899), *Chelidozoum vespertilio* (MacGillivray, 1895), Balcombian, Muddy Creek, Victoria, Australia (Gordon and d'Hondt, 1991). **Extant**

F. HIPPOTHOIDAE Busk, 1859 K. (CON)–Rec. Mar.

First: *Boreasina nowickii* Voigt, 1991, CON, St Christophe, Indre et Loire, France (see Voigt, 1991). **Extant**

F. TREMOSCHIZODINIDAE Vigneaux, 1949 T. (DAN)–Rec. Mar.

First: *Diplotresis europaea* Berthelsen, 1962, DAN, Denmark. **Extant**

F. PASYTHEIDAE Davis, 1934 T. (YPR)–Rec. Mar.

First: *Dittosaria wetherellii* Busk, 1866, London Clay, SE England, UK (Davis, 1934); *D.* sp., Lower Eocene, DSDP Site 246, Indian Ocean (Labracherie, 1975). **Extant**

F. CHORIZOPORIDAE Vigneaux, 1949 T. (SRV)–Rec. Mar.

First: *Chorizopora brongniartii* (Audouin, 1828), Badenian, Baden, Austria (see Vavra, 1977, p. 107); Tortonian, Gieraszowice, Poland (Malecki, 1962).
Comments: Monogeneric for *Chorizopora* Hincks, 1879. **Extant**

F. ARACHNOPUSIIDAE Jullien, 1888 K. (SAN)–Rec. Mar.

First: *Staurosteginopora irregularis* (d'Orbigny, 1852), Santonian, Villedieu, Loir et Cher, France (Voigt, 1991). **Extant**

F. EXECHONELLIDAE Harmer, 1957 T. (?YPR/?LUT)–Rec. Mar.

First: *'Exechonella*? sp. 1', *'Exechonella*? sp. 2', Lower Eocene, DSDP Site 246, Indian Ocean (Labracherie, 1975); *Exechonella* sp., Selsey Formation (Upper Bracklesham Beds), Sussex, England, UK (Cheetham, 1966). **Extant**

F. ACTISECIDAE Harmer, 1957 T. (YPR)–Rec. Mar.

First: 'Actisecidae? n. gen. 1 n. sp.', 'Actisecidae? n. gen. 2 n. sp.', Lower Eocene, DSDP Site 246, Indian Ocean (Labracherie, 1975). **Extant**
Comments: Monogeneric for *Actisecos* Canu and Bassler, 1927.

F. BITECTIPORIDAE MacGillivray, 1895 T. (CHT–SRV) Mar.

First: *Bitectipora lineata* MacGillivray, 1895, Janjukian, Gellibrand, Victoria, Australia.
Last: *Bitectipora lineata* MacGillivray, 1895, Bairnsdalian, Bairnsdale, Victoria, Australia.

F. EXOCHELLIDAE Bassler, 1935 K. (SAN)–Rec. Mar.

First: *'Cryptostomella*-like forms', SAN, ?locality (Voigt, 1991). **Extant**

F. UMBONULIDAE Canu, 1904 T. (LUT)–Rec. Mar.

First: *Umbonula leda* (d'Orbigny, 1851), LUT, Parnes, France (Canu, 1907–10). **Extant**
Comments: The slightly older *U. calcariformis* Gregory, 1893, from the British YPR, is probably not an umbonulid.

F. CELLEPORARIDAE Harmer, 1957 T. (BRT)–Rec. Mar.

First: *Celleporaria granulosa* (Canu and Bassler, 1920), Claibornian, Cook Mountain Formation, Caldwell Co., Texas, USA. **Extant**

F. METRARABDOTOSIDAE Vigneaux, 1949 T. (PRB)–Rec. Mar.

First: *Metrarabdotos micropora* (Gabb and Horn, 1862), Upper Eocene, Shubuta Clay, Clarke Co., Mississippi, USA (Cheetham, 1968b). **Extant**

F. SCLERODOMIDAE Levinsen, 1909 T. (PIA)–Rec. Mar.

First: *Cellarinella ?njegovanae* Rogick, 1956, *C.* sp., Scallop Hill Formation, Brown Peninsula, McMurdo Sound, Antarctica (Buckeridge, 1989). **Extant**
Comments: An Upper Oligocene–Lower Miocene species, described as *Porella operculata* Canu and Bassler, 1935, from Victoria, Australia, may perhaps belong to this family (Winston, 1983).

F. DHONDTISCIDAE Gordon, 1989 **Extant** Mar.

Comments: Monogeneric for *Dhondtiscus* Gordon, 1989.

F. HINCKSIPORIDAE Powell, 1968 **Extant** Mar.

Comments: Monospecific for *Hincksipora spinulifera* (Hincks, 1889).

F. PSEUDOLEPRALIIDAE Silén, 1942 **Extant** Mar.

Comments: Monospecific for *Pseudolepralia ellisinae* Silén, 1942.

F. ADEONIDAE Jullien, 1903 T. (YPR)–Rec. Mar.

First: *Adeonellopsis wetherelli* Gregory, 1893, London Clay, SE England, UK; *Bracebridgia*? sp., Lower Eocene, DSDP Site 246, Indian Ocean (Labracherie, 1975). **Extant**

F. CHLIDONIOPSIDAE Harmer, 1957 **Extant** Mar.

Comments: Monogeneric for *Chidoniopsis* Harmer, 1957.

F. EUTHYRISELLIDAE Bassler, 1953 T. (?RUP/?CHT)–Rec. Mar.

First: *Quadriscutella burlingtoniensis* (Waters, 1882), *Q. punctata* Bock and Cook, in press, Oligocene, Victoria, Australia. **Extant**

F. PORINIDAE d'Orbigny, 1852 K. (CMP)–Rec. Mar.

First: *Porina aftonensis* Brydone, 1930, *quadratus* Zone Chalk, Hampshire, England, UK. **Extant**
Comments: Includes Spiroporinidae Harmer, 1957.

F. NEPHROPORIDAE Marsson, 1887 K. (MAA) Mar.

First and Last: *Nephropora elegans* Marsson, 1887, MAA, Rügen, Germany.

F. STOMACHETOSELLIDAE Canu and Bassler, 1917 K. (MAA)–Rec. Mar.

First: *Taenioporella articulata* (Voigt, 1930), lower MAA, Rügen, Germany (Voigt, 1987, p. 90). **Extant**

F. GIGANTOPORIDAE Bassler, 1935 T. (DAN)–Rec. Mar.

First: *Tessaradoma rossica* Favorskaja, 1970, DAN, Crimea, former USSR. **Extant**

F. CHEILOPORINIDAE Bassler, 1936 T. (DAN)–Rec. Mar.

First: *Cianotremella gigantea* Canu, 1911, Roca Formation, Argentina (Voigt, 1985, p. 334). **Extant**

F. TEUCHOPORIDAE Neviani, 1895 T. (THA)–Rec. Mar.

First: *Perigastrella exserta* (Gabb and Horn, 1862), Vincentown Limesand, New Jersey, USA. **Extant**
Comments: Equivalent to the Phylactellidae Canu and Bassler, 1917.

F. SCHIZOPORELLIDAE Jullien, 1883 T. (YPR)–Rec. Mar.

First: 'Schizoporellidae gen. et sp. indet.', Lower Eocene, DSDP Site 246, Indian Ocean (Labracherie, 1975). **Extant**
Comments: *Schizoporella carinata* Hagenow from the Campanian is a hippothoid of the genus *Boreasina* (Voigt and Hillmer, 1984), and *Systenostoma* Marsson from the CMP–MAA is not a schizoporellid, according to Voigt (1985).

F. LACERNIDAE Jullien, 1888 T. (PRB)–Rec. Mar.

First: *Cribellopora* sp., Upper Eocene, Waiareka Tuff, Alma, North Otago, New Zealand (Parker and Gordon, 1992). **Extant**

F. SMITTINIDAE Levinsen, 1909 T. (YPR)–Rec. Mar.

First: 'Smittinid n. gen. ? n. sp.', Ypresian, DSDP Site 308, Koko Seamount, Pacific (Cheetham, 1975); *Porella* sp., 'Smittinid n. gen.? aff. Smittinid n. gen.? n. sp. Cheetham', 'Smittinidae n. gen.? n. sp.', DSDP Site 246, Indian Ocean (Labracherie, 1975). **Extant**

F. PETRALIELLIDAE Harmer, 1957 T. (YPR)–Rec. Mar.

First: *Hippopetraliella*? sp., Lower Eocene, DSDP Site 246, Indian Ocean (Labracherie, 1975). **Extant**

F. MARGARETTIDAE Harmer, 1957 T. (YPR)–Rec. Mar.

First: *Tubucella contorta* (Canu, 1910), Lower Eocene, North Aquitaine, France; '*Tubucella* n. sp. 1', '*Tubucella* n. sp. 2', YPR, DSDP Site 308, Koko Seamount, Pacific (Cheetham, 1975); *Margaretta* sp., Lower Eocene, DSDP Site 246, Indian Ocean (Labracherie, 1975). **Extant**

F. TETRAPLARIIDAE Harmer, 1957 T. (LUT)–Rec. Mar.

First: *Tetraplaria turgida* Tewari and Srivastava, 1967, Kirthar Stage, Sche, Kutch, India. **Extant**
Comments: Monogeneric.

F. CYCLICOPORIDAE Hincks, 1884 T. (BRT)–Rec. Mar.

First: *Cyclicopora fissurata* Canu and Bassler, 1920, *C. laticella* Canu and Bassler, 1920, Castle Hayne Limestone, Wilmington, North Carolina, USA. **Extant**
Comments: *Taenioporina arachnoides* (Goldfuss, 1826) from the MAA, tentatively assigned to this family by Larwood *et al.* (1967), is not a cyclicoporid (Voigt, 1985, p. 334) but is currently unplaced.

F. HIPPOPODINIDAE Levinsen, 1909 T. (YPR)–Rec. Mar.

First: *Hippoporina* sp., Lower Eocene, DSDP Site 246, Indian Ocean (Labracherie, 1975). **Extant**

F. CHEILOHORNEROPSIDAE Annoscia, Braga and Finotti, 1984 T. (PRB) Mar.

First and Last: *Cheilohorneropsis roveretana* Annoscia *et al.* 1984, PRB, Valle di Gresta, Italy.
Comments: Monospecific. Contrary to the original family description, *Semihaswellia* Canu and Bassler, 1917, and *Tremotoichos* Canu and Bassler, 1917, appear not to belong in this family (D. P. Gordon, pers. comm.).

F. ADEONELLIDAE Gregory, 1893 T. (PRB)–Rec. Mar.

First: *Adeonella syringopora* (Reuss, 1848), PRB, Vicenza, Italy (Braga and Barbin, 1988). **Extant**

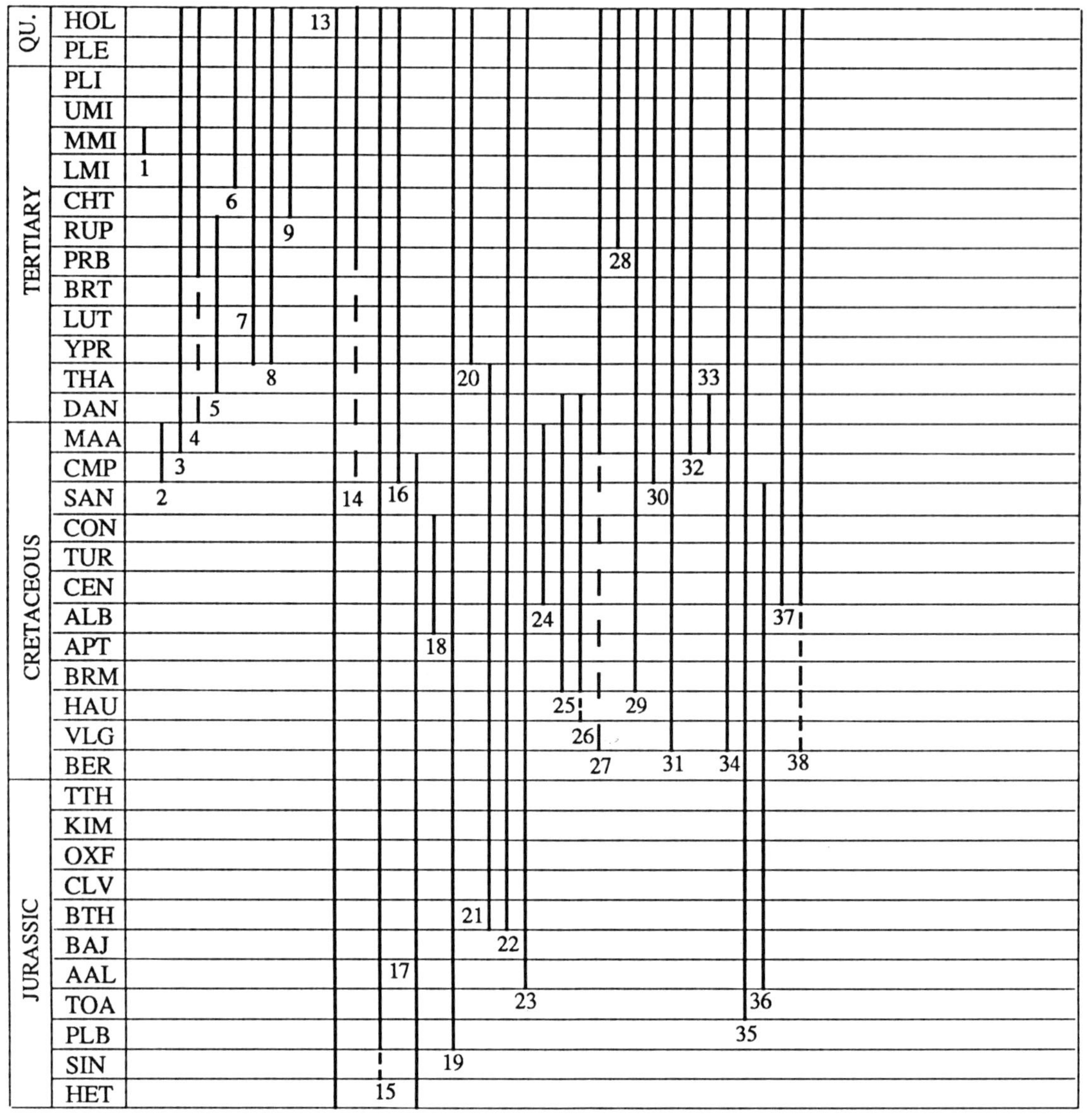

Fig. 24.3

F. MYRIAPORIDAE Gray, 1849
T. (PRB)–Rec. Mar.

First: *Myriopora* sp., Upper Eocene, San Marino (Annoscia, 1968). **Extant**

F. SIPHONICYTARIDAE Harmer, 1957
T. (PRB)–Rec. Mar.

First: *Tubitrabecularia clypeata* (Waters, 1881), Upper Eocene, Eua, Tonga (Cheetham, 1972). **Extant**

F. CREPIDACANTHIDAE Levinsen, 1909
T. (PRB)–Rec. Mar.

First: *Schizobathysella semilunata* Canu and Bassler, 1920, *S. saccifera* Canu and Bassler, 1920, Jacksonian, Wilmington, North Carolina, USA. **Extant**

F. INVERSIULIDAE Vigneaux, 1949
T. (RUP)–Rec. Mar.

First: *Inversiula airensis* Maplestone, 1910, *I. quadricornis* Maplestone, 1910, 'Aire Coastal Beds', Victoria, Australia. **Extant**

Comments: Monogeneric for *Inversiula* Jullien, 1888.

F. MICROPORELLIDAE Hincks, 1879
T. (AQT)–Rec. Mar.

First: *Microporella hyadesi* (Jullien, 1888), Bryozoan Bed (Otaian) overlying Takaka Limestone, Tarakohe Quarry, Nelson, New Zealand (Brown, 1952), (the specific identity of this *Microporella* has been questioned by Gordon, 1984, p. 102). **Extant**

F. CRYPTOSULIDAE Vigneaux, 1949
T. (TOR)–Rec. Mar.

First: *Cryptosula pallasiana* (Moll, 1803), Tortonian, Europe (Ghiurca, 1975). **Extant**

F. WATERSIPORIDAE Vigneaux, 1949
T. (?MES)–Rec. Mar.

First: *Watersipora* (?) sp., Messinian, Algeria (Moissette, 1988). **Extant**

Comments: Possible earlier records of this family are either doubtful or require confirmation.

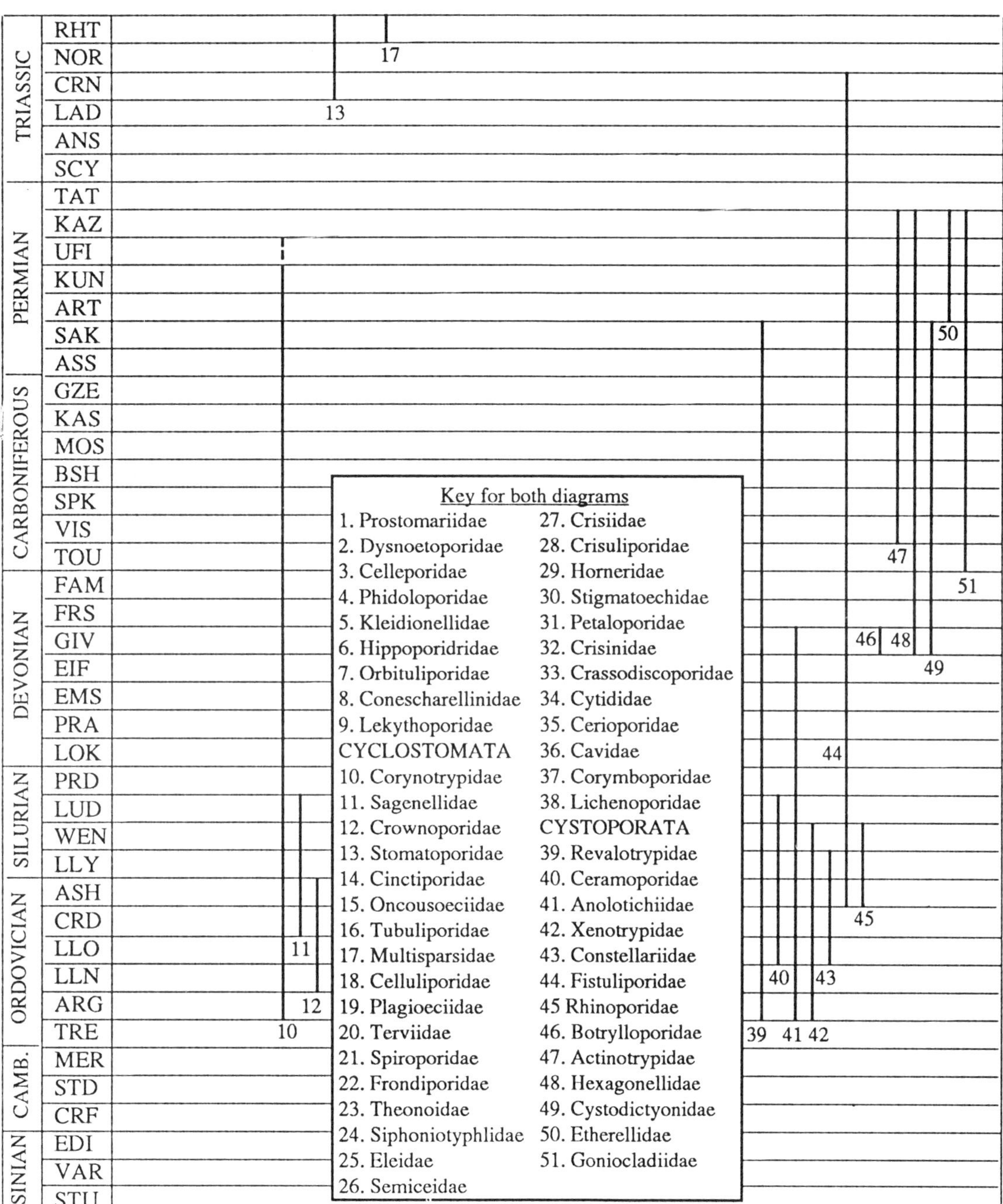

Fig. 24.3

F. LANCEOPORIDAE Harmer, 1957
T. (?ZAN/?PIA)–Rec. Mar.

First: *Lanceopora flabellata* (Livingstone, 1902), [probably Jemmys Point Formation, Kalimnan Stage], Jimmys Point, Reeves River, Victoria, Australia.

Comments: According to Voigt (1985, p. 334), *Bathystomella* Strand from the MAA is not related to *Parmularia* (=*Lanceopora*).
Extant

F. PETRALIIDAE Levinsen, 1909 **Extant** Mar.

Comments: Monogeneric for *Petralia* MacGillivray, 1869.

F. MAWATARIIDAE Gordon, 1990 **Extant** Mar.

Comments: Monogeneric for *Mawatarius* Gordon, 1990.

F. VICIDAE Gordon, 1988 **Extant** Mar.

Comments: Monogeneric for *Vix* Gordon, 1988.

F. EMINOOECIIDAE Hayward and Thorpe, 1988 **Extant** Mar.

F. CALWELLIIDAE MacGillivray, 1887 **Extant** Mar.

F. MAMILLOPORIDAE Canu and Bassler, 1927
T. (YPR)–Rec. Mar.

First: *Stenosipora* aff. *unirostris* Canu and Bassler, 1929, Lower Eocene, North Aquitaine, France (Labracherie, 1971).
Extant

F. DIDYMOSELLIDAE Brown, 1952
T. (?LUT/?BRT)–Rec. Mar.

First: *Didymosella* sp., Middle Eocene, North Aquitaine, France (Labracherie, 1971). **Extant**

F. PROSTOMARIIDAE MacGillivray, 1895 T. (LAN) Mar. (see Fig. 24.3)

First and Last: *Prostomaria gibbericollis* MacGillivray, 1895, Balcombian, Victoria, Australia (Gordon, 1990).

F. URCEOLIPORIDAE Bassler, 1936 **Extant** Mar.

F. DYSNOETOPORIDAE Voigt, 1971 K. (CMP–MAA) Mar.

First: *Dysnoetopora demissa* (White, 1879), 'Mesaverde' Formation and Lewis Shale, Wyoming; Pierre Shale, Colorado, USA (Toots and Cutler, 1962).
Last: *Dysnoetopora celleporoides* Canu and Bassler, 1926, Ripley Formation, Tennessee, USA.
Comments: Monogeneric.

F. CELLEPORIDAE Busk, 1852 K. (MAA)–Rec. Mar.

First: *'Cellepora' agglomerata* von Hagenow *in* Geinitz, 1846, Lower MAA Chalk, Baltic and England, UK (Voigt, 1985). **Extant**

F. PHIDOLOPORIDAE Gabb and Horn, 1862 T. (?DAN/?PRB)–Rec. Mar.

First: *Psilosecos angustidens* (Levinsen, 1925), Upper Danian, Denmark (Berthelsen, 1962); or *Reteporellina*? sp., Upper Eocene, Eua, Tonga (Cheetham, 1972). **Extant**
Comments: Otherwise known as Reteporidae Smitt, 1868, or Sertellidae Jullien and Calvet, 1903. Unlike typical members of this family, which have branches with autozooidal apertures opening on one side only, *Psilosecos* has apertures opening on both sides of the branches. Therefore, assignment to the Phidoloporidae is somewhat tentative (Voigt, 1985, p. 334). If *Psilosecos* is not a phidoloporid, the earliest known representative of this family may be from the Upper Eocene (PRB).

F. KLEIDIONELLIDAE Vigneaux, 1949 T. (THA–RUP) Mar.

First: *Hoplocheilina osculifera* (Reuss, 1872), Vincentown Limesand, New Jersey and Delaware, USA (Voigt, 1985, p. 335).
Last: *Kleidionella verrucosa* Canu and Bassler, 1920, Marianna Limestone, Alabama, USA.

F. HIPPOPORIDRIDAE Vigneaux, 1949 T. (BUR)–Rec. Mar.

First: *Hippoporidra edax* (Busk, 1859), BUR, France. **Extant**

F. ORBITULIPORIDAE Canu and Bassler, 1923 T. (YPR)–Rec. Mar.

First: *Atactoporidra globata* Labracherie, 1961, Lower Eocene, Marcheprime, Gironde, France; *Batopora stoliczkai* Reuss, 1867, Lower Eocene, Bordeaux, France (Labracherie, 1971). **Extant**

F. CONESCHARELLINIDAE Levinsen, 1909 T. (YPR)–Rec. Mar.

First: *Conescharellinopsis vigneauxi* Labracherie, 1975, Lower Eocene, Baloze, Aquitaine, France. **Extant**

F. LEKYTHOPORIDAE Levinsen, 1909 T. (CHT)–Rec. Mar.

First: *Lekythopora hystrix* MacGillivray, 1883, Janjukian, Victoria, Australia. **Extant**

Order CYCLOSTOMATA Busk, 1852

This exclusively marine order of stenolaemate bryozoans is probably paraphyletic. However, there is an acute need for systematic revision, not only to establish the status of the order and its relationship with other orders of stenolaemates, but also to define more regorously its constituent families. Many of the families listed below are of highly dubious value, and their ranges are very tentative.

Replacement of the ordinal name by Tubuliporata Johnston, 1847, to avoid homonymy with the fish order Cyclostomata Duméril, 1806, as in the revised *Treatise* (Boardman *et al.*, 1983), is considered to be unnecessary and potentially misleading. Such a change is not obligatory under the *Rules of Zoological Nomenclature* and could lead to confusion with the suborder Tubuloporina. Furthermore, the name Cyclostomata has fallen into disuse among vertebrate systematists.

F. CORYNOTRYPIDAE Dzik, 1981 O. (ARG)–P. (?UFI) Mar.

First: *Wolinella baltica* Dzik, 1981, Middle Volkhov Stage, Baltic (Dzik, 1981).
Last: *Corynotrypa voigtiana* (King, 1850), Middle Magnesian Limestone, Sunderland, England, UK and Zechstein, Pössneck, Germany (Taylor, 1985); *Lagenosypho permianus* Spandel, 1898, Zechstein, Germany (Langer, 1980).
Comments: There exist uncertainties about the age of the Zechstein deposits containing the youngest corynotrypids.

F. SAGENELLIDAE Brood, 1975 O. (?CRD)–S. (?LUD) Mar.

First: *Sagenella minnesotensis* (Ulrich, 1886), Black River, St Paul, Minnesota, USA.
Last: *Sagenella* sp. A, Hamra Beds, Vamlingbo, Gotland, Sweden (Brood, 1975).
Comments: Ranges of this family are highly tentative, pending detailed investigations of Palaeozoic cyclostomes.

F. CROWNOPORIDAE Ross, 1967 O. (LLN–ASH) Mar.

First: *Kukersella borealis* (Bassler, 1911), Orthoceras Limestone, Port Kunda, Estonia, former USSR.
Last: *K. borealis* (Bassler, 1911), Slade and Redhill Beds (Upper Rawtheyan), Dyfed, Wales, UK (Buttler, 1989).
Comments: Homonymous with Kukersellidae Brood, 1975.

F. STOMATOPORIDAE Pergens and Meunier, 1886 Tr. (CRN)–Rec. Mar.

First: *Stomatopora* sp., S. Cassiano Formation, Eastern Dolomites, Italy (Bizzarini and Braga, 1981). **Extant**
Comments: Palaeozoic records of supposed *Stomatopora* need re-examination; at least some are corynotrypids or crownoporids.

F. CINCTIPORIDAE Boardman, McKinney and Taylor, 1992 ?K. (??CMP, ?MAA), T. (RUP)–Rec. Mar.

First: ?*Cinctipora* sp., Maastrichtian (?upper Campanian), Need's Camp, Cape Province, South Africa. *Semicinctipora amplexus* Boardman *et al.* 1992, *Attinopora campbelli* Boardman *et al.* 1992, Whaingaroan, McDonald Limestone, Oamaru, New Zealand. **Extant**
Comments: The small size of the zooids in *Cinctipora* sp. from the South African Cretaceous call into question its assignment to this family, which is otherwise characterized by gigantic zooids and is endemic to New Zealand (Boardman *et al.*, 1992).

F. ONCOUSOECIIDAE Canu, 1918 J. (?SIN)–Rec. Mar.

First: ?*Oncousoecia* sp., Blue Lias (*bucklandi* Zone), Dunraven Bay, Glamorgan, Wales, UK (P. D. Taylor, unpublished). **Extant**
Comments: Carnian forms, described by Bizzarini and Braga (1985) as belonging to the Diastoporidae and Oncousoeciidae, may represent earlier records of this family, but more morphological details are required. Includes Annectocymidae Hayward and Ryland, 1985.

F. TUBULIPORIDAE Johnston, 1838 K. (?CMP)–Rec. Mar.

First: *Tubulipora suberecta* Brood, 1972, lower CMP, Ifö, Sweden. **Extant**
Comments: The generic composition and range of this family are highly uncertain. The range given here is based on the oldest species belonging to the type genus of the family. If *Idmonea* Lamouroux, 1821 is included in the Tubuliporidae (e.g. Brood, 1972), family range is extended down into the Aalenian.

F. MULTISPARSIDAE Bassler, 1935 Tr. (RHT)–K. (CMP) Mar.

First: *Reptomultisparsa hybensis* (Prantl, 1938), Hybe Beds, Hybe, Czechoslovakia (Taylor and Míchalik, 1991).
Last: *Heterohaplooecia monticulifera* Voigt and Viaud, 1983, lower CMP, Les Guignardières en Soullans, Vendée, France.

F. CELLULIPORIDAE Buge and Voigt, 1972 K. (ALB–CON) Mar.

First: *Cellulipora ornata* d'Orbigny, 1850, Upper Greensand, Devon, England, UK.
Last: *Cellulipora* (?) *rugosa* (d'Orbigny, 1853), CON, Villedieu, Loir-et-Cher, France.
Comments: Monogeneric. Contrary to Buge and Voigt (1972), *Berenicea spissa* Gregory, 1899, is not regarded as belonging to this family, but is a plagioeciid, *Mesonopora spissa* (see Pitt and Taylor, 1990).

F. PLAGIOECIIDAE Canu, 1918 J. (PLB)–Rec. Mar.

First: *Mesenteripora wrighti* Haime, 1854, Middle Lias Marlstone (*spinatum* Zone), King's Sutton, Northamptonshire, England, UK (Walter and Powell, 1973). **Extant**
Comments: Synonymous with the Diastoporidae Busk, 1859 of some authors (e.g. Brood, 1972; Hayward and Ryland, 1985). The name Diastoporidae is best avoided because of its common usage for tubuloporines lacking brooding zooids (e.g. Bassler, 1953).

F. TERVIIDAE Canu and Bassler, 1920 T. (YPR)–Rec. Mar.

First: *Lagonoecia lamellifera* Canu and Bassler, 1920, Bashi Formation, Woods Bluff, Alabama, USA. **Extant**

F. SPIROPORIDAE Voigt, 1968 J. (BTH)–T. (THA) Mar.

First: *Spiropora elegans* Lamouroux, 1821, BTH, Normandy, France.
Last: *S. verticillata* (Goldfuss, 1826), THA, Pont Labou, Pau, France (Voigt and Flor, 1970).
Comments: Monogeneric.

F. FRONDIPORIDAE Busk, 1859 J. (BTH)–Rec. Mar.

First: *Fasciculipora waltoni* Haime, 1854, upper BTH, Bath, England, UK. **Extant**
Comments: As here understood, this family includes also the Fasciculiporidae Walter, 1970.

F. THEONOIDAE Busk, 1859 J. (AAL)–Rec. Mar.

First: *Theonoa diplopora* (Branco, 1879), Lower Inferior Oolite, Gloucestershire, England, UK. **Extant**

F. SIPHONIOTYPHLIDAE Voigt, 1967 K. (CEN–MAA) Mar.

First: *Siphoniotyphlus tenuis* (von Hagenow, 1840), *Clinopora cenomanensis* Hillmer, 1971, Lower Cenomanian, Hannover, Germany (see Hillmer, 1971).
Last: *Clinopora lineata* (Beissel, 1865), upper MAA (*mayaroensis* Zone), Limhamn, Sweden (Brood, 1972, p. 268).

F. ELEIDAE d'Orbigny, 1852 K. (BRM)–T. (DAN) Mar.

First: *Meliceritites semiclausa* (Michelin, 1846) *sensu* Walter *et al.* 1975, lower BRM, Fontaine-Graillère Marls, South-Vercors, France.
Last: *Meliceritella steenstrupi* (Pergens and Meunier, 1887), *M. armata* (Levinsen, 1912), DAN, Baltic.
Comments: Homonymous with Melicerititidae Pergens, 1890, eleids are a well-defined, monophyletic family. The characteristic operculate zooids are absent in a supposed eleid from the Bathonian, *Cyclocites primogenitum* Canu and Bassler, 1922.

F. SEMICEIDAE Buge, 1952 K. (?HAU/?BRM)–T. (DAN) Mar.

First: *Poriceata ardescensis* Walter, 1983, upper HAU or lower BRM, Vaulion, Switzerland.
Last: *Filicea danica* Viskova, 1968, DAN, Crimea, Inkerman, former USSR.
Comments: *Cinctipora elegans* Hutton, 1873, a Recent species sometimes assigned to *Filicea*, has been reassigned to the Cinctiporidae.

F. CRISIIDAE Johnston, 1847 K. (??VLG/MAA)–Rec. Mar.

First: ??*Filicrisia* (?) *noeomiensis* Voigt and Walter, 1991, ??*Crisia* (?) *nozeroyensis* Voigt and Walter, 1991, Marnes grises à bryozoaires, Nozeroy, Jura, France and Switzerland; *Crisia* sp., *Crisidia inopinata* Lagaaij, 1975, upper MAA, Curfs Quarry, Maastricht, The Netherlands.
Comments: The diagonistic articulated colony form has still to be established in the material described by Voigt and Walter (1991). **Extant**

F. CRISULIPORIDAE Buge, 1979 T. (RUP)–Rec. Mar.

First: *Crisulipora prominens* Canu and Bassler, 1920, *C. rugosodorsalis* Canu and Bassler, 1920, *C. flabellata* Canu and Bassler, 1920, *C. grandipora* Canu and Bassler, 1920, Marianna Limestone, Alabama, USA. **Extant**

F. HORNERIDAE Smith, 1867 K. (BRM)–Rec. Mar.

First: *Siphodictyum gracile* Lonsdale, 1849, lower BRM, Fontaine-Graillère Marls, South-Vercors and Crupies, France (Walter *et al.*, 1975). **Extant**

F. STIGMATOECHIDAE Brood, 1972 K. (CMP)–?Rec Mar.

First: *Stigmatoechos punctatus* Marsson, 1887, upper ?CMP, Denmark and Sweden (Brood, 1972, p. 370). **Extant**
Comments: Putative Recent examples (e.g. *S. violacea* (Sars)) require confirmation.

F. PETALOPORIDAE Gregory, 1899 K. (VLG)–Rec. Mar.

First: *Petalopora rugosa* (d'Orbigny, 1853), lower VLG, Arzier, Jura, France (see Walter, 1972). **Extant**
Comments: Although often regarded as extinct, the Recent *Calvetia dissimilis* Borg, 1944 (for which Borg created a new family, the Calvetiidae) is probably a petaloporid.

F. CRISINIDAE d'Orbigny, 1853 ?K. (MAA)–Rec. Mar.

First: *Crisidmonea tripora* (Canu and Bassler, 1926), Ripley Formation, Coon Creek, Tennessee, USA. **Extant**
Comments: There is considerable confusion about this family arising from the existence of numerous taxa with superficially similar colony forms, but for which the details of the brooding zooids, crucial to ascertaining true affinities, have yet to be described; and problems concerning the type species of the type genus which, as figured by Voigt (1984, pl. 7, figs 6 and 7), is a tubuloporine and not a cancellate cyclostome. *Crisidmonea tripora* may be the oldest species with cancellate brooding zooids matching the concept of the Crisinidae as here understood.

F. CRASSODISCOPORIDAE Brood, 1972 K. (MAA)–T. (DAN) Mar.

First and Last: *Crassodiscopora alcicornis* (Levinsen, 1925), upper MAA (*mayaroensis* Zone)–DAN (*vexillifera* Zone), Denmark and Sweden (Brood, 1972, p. 391).
Comments: Monospecific.

F. CYTIDIDAE d'Orbigny, 1854 K. (VLG)–Rec. Mar.

First: *Chartecytis compressa* Canu and Bassler, 1926, *Voigticytis campicheana* (d'Orbigny, 1853), Lower VLG, SE France. **Extant**
Comments: The definition, composition and stratigraphical range of this family are problematic.

F. CERIOPORIDAE Busk, 1859 J. (TOA)–Rec. Mar.

First: *Heteropora tipperi* Henderson and Perry, 1981, lower TOA, north-central British Columbia, Canada. **Extant**
Comments: Synonymous with Heteroporidae Waters, 1880. Pending systematic revision, Tretocycloeciidae Canu, 1919, Leiosoeciidae Canu and Bassler, 1920, Canuellidae Borg, 1944, and Densiporidae Borg, 1944 are here included within the Cerioporidae. Putative cerioporids from the Triassic are calcified demosponges (Engeser and Taylor, 1989).

F. CAVIDAE d'Orbigny, 1854 J. (AAL)–K. (SAN) Mar.

First: *'Ceriocava corymbosa'* (Lamouroux, 1821), Lower Inferior Oolite, Gloucestershire, England, UK.
Last: *Ceriocava incrustata* (Roemer, 1840), SAN, Germany.
Comments: Possibly polyphyletic.

F. CORYMBOPORIDAE Smitt, 1866 K. (CEN)–Rec. Mar.

First: *Marssoniella cenomana* Voigt, 1974, *Amphimarssoniella klaumanni* Voigt, 1974, lower CEN, Mülheim/Ruhr, Germany. **Extant**
Comments: Taken to include the Fungellidae Kluge, 1955.

F. LICHENOPORIDAE Smitt, 1866 K. (??VLG/?CEN)–Rec. Mar.

First: ??*'Lichenopora' interradiata* Walter, 1989, *'L' neocomiensis* (d'Orbigny, 1853), upper VLG, Marnes à bryozoaires, Ste-Croix, Switzerland; ?*Lichenopora pedunculata* Voigt, 1989, lower CEN, Mülheim-Broich, Westfalia, Germany. **Extant**
Comments: Includes the Disporellidae Borg, 1944. For certain assignment to the Lichenoporidae, brood chambers consisting of an interior wall must be demonstrated, rarely possible in fossil material; therefore, the stratigraphical range is very tentative. An interior-walled brood chamber figured by Voigt (1989, pl. 6, fig. 6) may be the earliest known.

Order CYSTOPORATA Astrova, 1964

This order was created for some Palaeozoic stenolaemates previously included (e.g. Bassler, 1953) in the Cyclostomata. The bulk of the Cystoporata are likely to constitute a monophyletic grouping which is characterized by the presence of lunaria and/or vesiculose calcification between the zooids. However, there are doubts about the cystoporate affinities of certain genera, and also over the exact boundary with the Trepostomata. The order was revised by Utgaard (*in* Boardman *et al.*, 1983) for the *Treatise*, whose classification is employed here.

F. REVALOTRYPIDAE Goryunova, 1986 O. (ARG)–P. (SAK) Mar.

First: *Revalotrypa eugeniae* Goryunova, 1988, Latorpskiy Horizon, Estonia, former USSR.
Last: *Metelipora monstrata* Trizna, 1950, Sterlitamak Horizon (late SAK), Urals, former USSR.

F. CERAMOPORIDAE Ulrich, 1882 O. (LLO)–S. (LUD) Mar.

First: *Ceramoporella* sp., Chazyan, Day Point Limestone, New York State, USA (Ross, 1984, fig. 2).
Last: *Ceramopora perforata* Hennig, 1908, Hamra-Sundre Beds, Gotland, Sweden.
Comments: The large diameter of the 'zooids' in *Ceramopora? unapensis*, described by Ross (1966) from the Arenig Kindblade Formation of Oklahoma, throws doubt on the affinities of this potential early ceramoporid. The

systematic position of *Ganiella* Yaroshinskaya, *in* Astrova and Yaroshinskaya, 1968, a possible Lower Devonian ceramoporid, is highly doubtful.

F. ANOLOTICHIIDAE Utgaard, 1968
O. (ARG)–D. (GIV) Mar.

First: *Profistulipora arctica* Astrova, 1965, Nelidov Stage, Novaya Zemlya, former USSR (Nekhorosheva, 1974).
Last: *Altshedata belgebaschensis* (Nekhoroshev, 1948), GIV, Altai Mts, former USSR (Utgaard, *in* Boardman *et al.*, 1983, p. 370).

F. XENOTRYPIDAE Utgaard, 1983
O. (ARG)–S. (?SHE/?WHI/?GLE) Mar.

First: *Xenotrypa primaeva* (Bassler, 1911), Volkhov Stage, Estonia, former USSR.
Last: *Hennigopora florida* (Hall, 1852), Rochester Shale, New York State, USA.

F. CONSTELLARIIDAE Ulrich, 1896
O. (LLO)–S. (LLY) Mar.

First: *Constellaria islensis* Ross, 1963, Chazy Formation, Vermont and New York State, USA.
Last: *Constellaria* sp., 'lower Silurian of Siberia' (Ross, 1963, p. 53).

F. FISTULIPORIDAE Ulrich, 1882
O. (ASH)–Tr. (CRN) Mar.

First: *Fistulipora* sp., Slade and Redhill Beds, upper Rawtheyan, Whitland, Dyfed, Wales, UK (Buttler, 1991).
Last: *Cystitrypa cassiana* Schäfer and Fois, 1987, southern Cassiano Formation, Cortina d'Ampezzo, Dolomites, Italy.

F. RHINOPORIDAE Miller, 1889
O. (ASH)–S. (?SHE/?WHI/?GLE) Mar.

First: *Lichenalia* cf. *concentrica* Hall, 1852, Slade and Redhill Beds, upper Rawtheyan, Whitland, Dyfed, Wales, UK (Buttler, 1991).
Last: *Lichenalia concentrica* Hall, 1852, Rochester Shale, New York State, USA.

F. BOTRYLLOPORIDAE Miller, 1889
D. (GIV) Mar.

First and Last: *Botryllopora socialis* Nicholson, 1874, Hamilton Group, Ontario, Canada; New York State and Michigan, USA.
Comments: Monospecific.

F. ACTINOTRYPIDAE Simpson, 1897
C. (CHD)–P. (KAZ) Mar.

First: *Actinotrypa peculiaris* (Rominger, 1866), Keokuk Group, Missouri and Illinois, USA (Horowitz, 1968).
Last: *Epiactinotrypa flosculosa* Kiseleva, 1973, *E. incognita* Kiseleva, 1982, Chandalezy Suite, *Parafusulina stricta* Zone, Primor'e, former USSR (Kiseleva, 1982).

F. HEXAGONELLIDAE Crockford, 1947
D. (GIV)–P. (KAZ) Mar.

First: *Prismopora triquetra* Hall and Simpson, 1887, Hamilton Group, Indiana, USA.
Last: *Hexagonella ramosa* Waagen and Wentzel, 1886, Murgabian, *Colaniella parva* Zone, Primor'e, former USSR (Kiseleva, 1982).

F. CYSTODICTYONIDAE Ulrich, 1884
D. (GIV)–P. (SAK) Mar.

First: *Acrogenia prolifera* Hall, 1883, *Ptilocella parallela* (Hall and Simpson, 1887), *Semiopora bistigmata* Hall, 1883, *Stictocella sinuosa* (Hall, 1883), *Taeniopora exigua* Nicholson, 1874, *T. occidentalis* Ulrich, 1890, *T. penniformis* Nicholson, 1874, *T. recubans* (Hall and Simpson, 1887), *T. subcarinata* (Hall, 1883), *Thamnotrypa divaricata* (Hall, 1883), Hamilton Group, north-eastern USA and Ontario, Canada.
Last: *Filiramoporina kretaphilia* Fry and Cuffey, 1976, Wreford Limestone, Kansas, USA.

F. ETHERELLIDAE Crockford, 1957
P. (ART–KAZ) Mar.

First: *Etherella porosa* Crockford, 1957, *E. minor* Crockford, 1957, *E. irregularis* Crockford, 1957, *Liguloclema typicalis* Crockford, 1957, Noonkanbah Formation, Fitzroy Basin, Western Australia.
Last: *Etherella crassa* Kiseleva, 1973, Chandalezy Suite, *Parafusulina stricta* Zone, Primor'e, former USSR (Kiseleva, 1982).

F. GONIOCLADIIDAE Waagen and Pichl, 1885
C. (TOU)–P. (KAZ) Mar.

First: *Goniocladiella kasakhstanica* Nekhoroshev, 1953, TOU, Kazakhstan, former USSR.
Last: *Goniocladia timorensis* Bassler, 1929, Chandalezy Suite, *Parafusulina stricta* Zone, Primor'e, former USSR (Kiseleva, 1982).

Order TREPOSTOMATA Ulrich, 1882 (see Fig. 24.4)

This predominantly Palaeozoic stenolaemate order has no certain autapomorphies and is likely to be paraphyletic, although some major subgroups within the order (e.g. Suborder Halloporoidea) are undoubtedly monophyletic. R. S. Boardman is currently revising the trepostomes for the *Treatise*. Pending this publication, the family-level classification used here is largely derived from the work of Astrova (1978).

The oldest known bryozoans are two Tremadocian trepostome genera from the Yangtze Valley, China which will be described in a forthcoming publication by Hu and Spjeldnaes. Their family assignment is currently unknown.

F. ESTHONIOPORIDAE Vinassa de Regny, 1921
O. (ARG–CRD) Mar.

First: *Esthoniopora lessnikovae* (Modzalevskaya, 1953), BI, Leningrad Oblast', former USSR.
Last: *Esthoniopora subsphaerica* (Bassler, 1911), E, Estonia, former USSR.

F. ORBIPORIDAE Astrova, 1978
O. (ARG)–D. (GIV) Mar.

First: *Orbipora* sp. of Taylor and Cope (1987), Ogof Hên Formation, Bolahaul Member, Llangynog, Wales, UK.
Last: *Chondraulus densus* Duncan, 1939, *C. granosus* Duncan, 1939, and *C. petoskeyensis* Duncan, 1939, Traverse Group, Michigan, USA.

F. DITTOPORIDAE Vinassa de Regny, 1921
O. (ARG–ASH) Mar.

First: *Dittopora annulata* (Eichwald, 1860), *D. sokolovi* Modzalevskaya, 1953, *D. ramosa* Modzalevskaya, 1953, and *D. clavaeformis* Dybowski, 1877, BII, Estonia, former USSR.

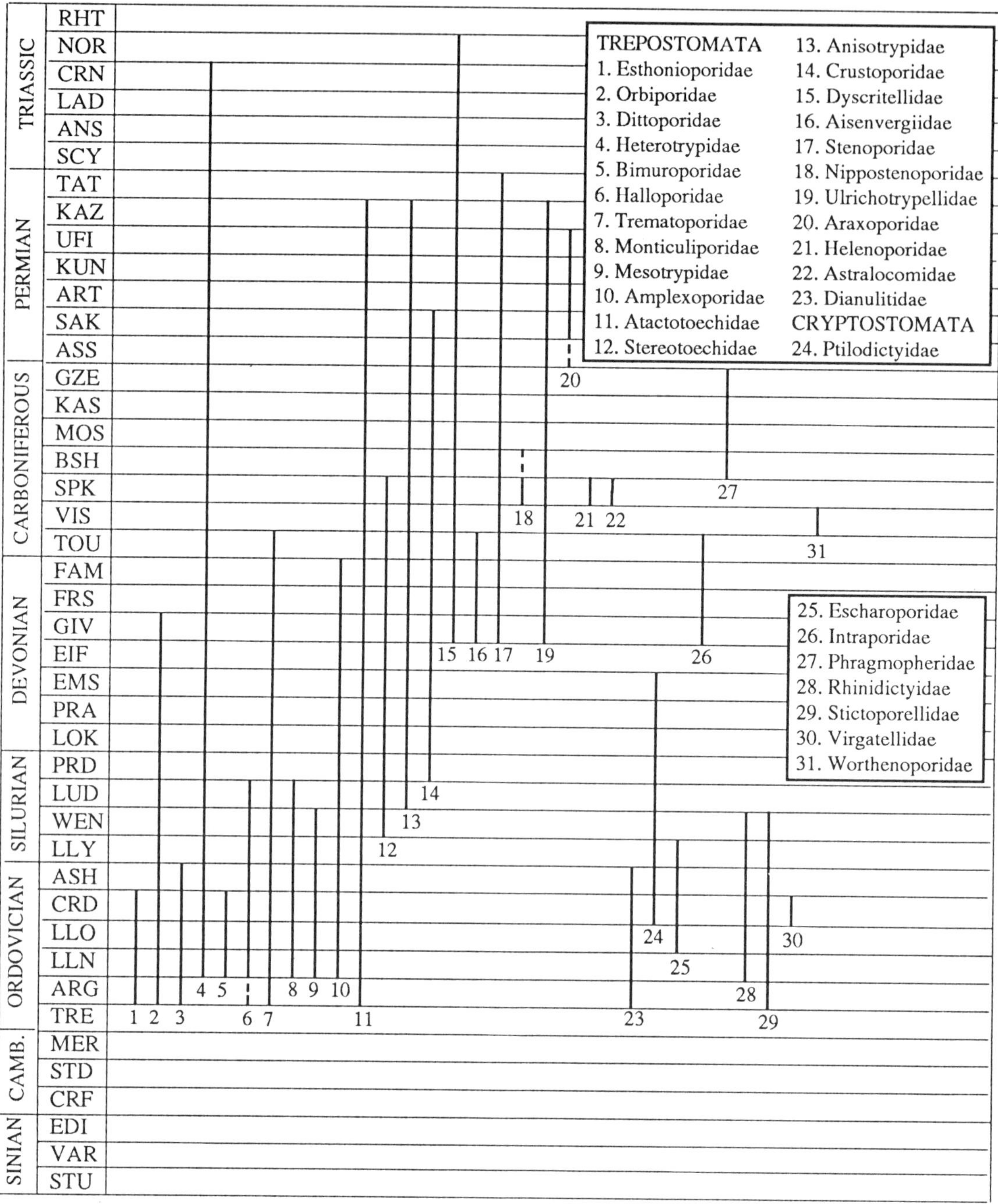

Fig. 24.4

Last: *Hemiphragma imperfectum* (Ulrich, 1890), and *H. whitfieldi* (James, 1875), Cincinnatian, USA.

F. HETEROTRYPIDAE Ulrich, 1890 O. (LLN)–Tr. (CRN) Mar.

First: *Stigmatella indenta* (Bassler, 1911), *S. inflecta* Bassler, 1911, and *Lioclemella spinea* (Bassler, 1911), BIII, Estonia, former USSR.

Last: *Zozariella stellata* Schäfer and Fois, 1987, Zozar Formation, Zanskar Region, West Himalaya, North India.

Comments: *Z. stellata* is assigned to this family with reservation by Schäfer and Fois (1987). The next youngest species is *Phragmotrypa ordinata* Schäfer and Fois (1987) from the lower to middle ANS of Nevada, USA.

F. BIMUROPORIDAE Key, 1990 O. (LLN–CRD) Mar.

First: *Champlainopora chazyensis* (Ross, 1963), Chazy, Day Point Limestone, New York State, USA.

Last: *Bimuropora winchelli* (Ulrich, 1886), Trenton, Guttenberg Formation, Iowa, USA.

F. HALLOPORIDAE Bassler, 1911 O. (?ARG/?LLN)–S. (LUD) Mar.

First: *Diplotrypa pusilla* Astrova, 1965, top of Lower–base of Middle Ordovician, Vaigach Island, Circumpolar Urals, former USSR.

Last: *Diplotrypa franklini* Bolton, 1966, LUD, Canadian Arctic Islands; *Hallopora elegantula* (Hall, 1852), LUD, Welsh

Borderlands, UK; *H. wajgatshensis* Astrova, 1965, LUD, Vaigach Island, former USSR.

F. TREMATOPORIDAE Miller, 1889 O. (ARG)–C. (?HAS/?IVO) Mar.

First: *Nicholsonella nelidovi* Nekhoroshëva, 1965, Nelidovo Horizon, Novaya Zemlya, former USSR; *N. genuina* Astrova, 1965, O. (l.), Novaya Zemlya, former USSR.
Last: *Neotrematopora tabulata* (Nekhoroshëv, 1956), TOU, Kazakhstan, former USSR.

F. MONTICULIPORIDAE Nicholson, 1881 O. (LLN)–S. (LUD) Mar.

First: *Homotrypa instabilis* Ulrich, 1886, CI, Estonia, former USSR.
Last: *Prasopora gotlandica* Hennig, 1908, Hemse Beds, Hemse, Gotland, Sweden.

F. MESOTRYPIDAE Astrova, 1965 O. (LLN)–S. (?SHE/?WHI/?GLE) Mar.

First: *Mesotrypa bystrowi* Modzalevskaya, 1953, CI, Estonia, former USSR; *M. torosa*, Modzalevskaya, 1953, CI, Leningrad Oblast', former USSR.
Last: *Mesotrypa suprasilurica* Hennig, 1908, Högklint Beds (b), Visby, Gotland, Sweden.

F. AMPLEXOPORIDAE Miller, 1889 O. (LLN)–D. (FAM) Mar.

First: *Monotrypa helenae* Modzalevskaya, 1953, BIII, Leningrad Oblast', former USSR.
Last: *Monotrypa hsui* Yang, 1950, Hsikuangshan Formation, Central Hunan, China (Yang *et al.*, 1988).

F. ATACTOTOECHIDAE Duncan, 1939 O. (ARG)–P. (KAZ) Mar.

First: *Cyphotrypa antiqua* (Modzalevskaya, 1953), BII, Leningrad Oblast', former USSR.
Last: *Permopora kapitzai* Romanchuk, 1967, Upper Permian, Khabarovsk region, former USSR; *Neoeridotrypella pulchra* Morozova, 1970, KAZ, Russian Platform, former USSR.

F. STEREOTOECHIDAE Yang, Hu and Xia, 1988 S. (?SHE/?WHI/?GLE)–C. (SPK) Mar.

First: *Eostenopora peculiaris* (Bassler, 1906), Rochester Shale, New York State, USA.
Last: *E. tumida* (Girty, 1911), Chesterian, Arkansas, USA.

F. ANISOTRYPIDAE Dunaeva and Morozova, 1967 S. (LUD)–P. (KAZ) Mar.

First: *Anisotrypa proavus* Astrova, 1970, S2, Skal'sk Horizon, Podolia, former USSR.
Last: *Anisotrypella borealis* Morozova, 1967, Lower Kazanian Substage, Russian Platform, former USSR.

F. CRUSTOPORIDAE Dunaeva and Morozova, 1967 S. (PRD)–P. (SAK) Mar.

First: *Callocladia kaugatumensis* Astrova, 1970, Upper Silurian, Kaugatumas Horizon, Estonia, former USSR.
Last: *Tabuliporella permiana* Baranova, 1960, SAK, northern Urals, former USSR.

F. DYSCRITELLIDAE Dunaeva and Morozova, 1967 D. (GIV)–Tr. (NOR) Mar.

First: *Dyscritella devonica* Volkova, 1968, GIV, Gornyi Altai, former USSR.
Last: *Dyscritella agischevi* Nekhoroshev, 1949, NOR, Khabarovsk region, former USSR; *Pseudobatostomella morbosa* Morozova, 1969, NOR, Pamir, former USSR; *P. maorica* (Wilckens, 1927), Wairoa, Nelson, New Zealand: *Paralioclema formosum* Morozova, 1969, NOR, Caucasus, former USSR.

F. AISENVERGIIDAE Dunaeva, 1964 D. (GIV)–C. (HAS) Mar.

First: *Polycylindricus asphinctus* Boardman, 1960, *P. clausus* Boardman, 1960, *P. devonicus* (Ulrich, 1890), Hamilton Group, USA.
Last: *Aisenvergia cylindrica* Dunaeva, 1964, *Volnovachia distincta* Dunaeva, 1964, Hastarian, Donbas, former USSR.

F. STENOPORIDAE Waagen and Wentzel, 1886 D. (GIV)–P. (CHX) Mar.

First: *Dyoidophragma typicale* Duncan, 1939, *D. serratum* Duncan, 1939, *D. polymorphum* Boardman, 1960, Traverse and Hamilton Groups, USA.
Last: *Stenopora* sp., Changxing Formation, South China (Yang and Lu, 1981).

F. NIPPOSTENOPORIDAE Xia, 1987 C. (SPK–?SPK/?BSK) Mar.

First: *Nippostenopora tabulata* Dunaeva, 1963, Steshevsky Horizon, early SPK, Donbas, former USSR.
Last: *Nippostenopora elegantula* Sakagami, 1960, *Millerella* Zone (late SPK or early BSH in age), Fukuji, Hida Massif, Japan.
Comments: The exact age of species from China assigned to the upper Lower Carboniferous by Xia (1987) is not clear.

F. ULRICHOTRYPELLIDAE Romanchuk, 1968 D. (GIV)–P. (KAZ) Mar.

First: *Petalotrypa compressa* Ulrich, 1890, *P. delicata* Ulrich, 1890, Hamilton Group, USA.
Last: *Hinganella clara* Morozova, 1970, Guadalupian Stage, Mongolia; *H. hinganensis* Romanchuk, 1967, Osakhta Group, Khabarovsk, former USSR; *H. sincera* Romanchuk, *in* Morozova, 1970, Osakhta Group, Khabarovsk, former USSR; *Ulrichotrypella wanneri* (Bassler, 1929), Barabash Group, Primor'e Region, former USSR; *U. prima* Romanchuk, 1967, Osakhta Group, Khabarovsk, former USSR; *U. oculata* Morozova, 1970, Kazanian Stage, Russian Platform, former USSR.

F. ARAXOPORIDAE Morozova, 1970 P. (?ASS/?SAK–UFI) Mar.

First: *Araxopora variana* (Yang, 1958), Chihsia Formation (lower Lower Permian), Yangtze region, China (Yang and Lu, 1979).
Last: *A. araxensis* (Nikiforova, 1933), *A. bifoliata* Morozova, 1965, *A. macrocava* Morozova, 1965, *A. spinata* Morozova, 1965, *A. minax* (Morozova, 1970), Gnishik Horizon, Transcaucasia, former USSR.
Comments: Monogeneric.

F. HELENOPORIDAE Ross, 1988 C. (SPK) Mar.

First and Last: *Helenopora duncanae* Ross, 1988, upper Mississippian, western interior of the USA.
Comments: Monospecific.

F. ASTRALOCOMIDAE Ross, 1988 C. (SPK) Mar.

First and Last: *Astralochoma helenae* Ross, 1988, upper Mississippian, western interior of the USA.
Comments: Monospecific.

F. DIANULITIDAE Vinassa, 1920 O. (ARG–ASH) Mar

First: *Dianulites fastigiatus* Eichwald, 1829, *D. petropolitana* Dybowski, 1877, *D. glauconiticus* Mannil, 1959, *D. janishewskyi* Modzalevskaya, 1953, *D. hexaporites* (Pander, 1830), *D. multimesoporicus* Modzalevskaya, 1953, Volkhov Stage, Estonia, former USSR.
Last: *D. globularis* Bassler, 1928, Ellis Bay Formation (Zone 4), Anticosti Island, Canada.
Comments: This family, here regarded as monogeneric, is of uncertain affinities; some authors have placed it in the Trepostomata, others in the Cystoporata.

Order CRYPTOSTOMATA Vine, 1883

This Palaeozoic stenolaemate order comprises bryozoans with restricted linear or planar loci of zooidal budding (Blake, 1980). Following removal of the Fenestrata as a separate order, three suborders of cryptostomes are recognized: Ptilodictyina Astrova and Morozova, 1956 (so-called 'bifoliate cryptostomes'), Rhabdomesina Astrova and Morozova, 1956, and Timanodictyina Morozova, 1966. The first two suborders have been revised for the *Treatise* (Boardman *et al.*, 1983) by Karklins and Blake respectively; the timanodictyines are a small, poorly known suborder. The phylogenetic relationships and exact taxonomic status of the three suborders have yet to be worked out.

Suborder PTILODICTYINA Astrova and Morozova, 1956

Family classification follows Karklins (*in* Boardman *et al.*, 1983). It should be noted that Karklins left several genera unassigned to families; these genera are therefore not taken into account in the range data that follow.

F. PTILODICTYIDAE Zittel, 1880 O. (CRD)–D. (EMS) Mar.

First: *Phaenopora stubblefieldi* Ross, 1962, Hoar Edge Group, Costian, Shropshire, England, UK.
Last: *Ensiphragma mirabilis* Astrova, *in* Astrova and Yaroshinskaya, 1968, Kireyev Stratum, Lower Emsian, *dehiscens* Conodont Biozone, Altai Mountains, former USSR.
Comments: The exact age of species of *Phaenopora* described by Nekhoroshëva (1966) from the Ordovician of Taimyr, former USSR is unclear, but they may possibly antedate *P. stubblefieldi*.

F. ESCHAROPORIDAE Karklins, 1983 O. (LLO)–S. (LLY) Mar.

First: *Chazydictya chazyensis* Ross, 1963, Chazy Limestone, Vermont, USA.
Last: *Graptodictya minuta* Kopajevich, 1975, *G. sulcata* Kopajevich, 1975, LLY, Estonia, former USSR.

F. INTRAPORIDAE Simpson, 1897 D. (GIV)–C. (?HAS/?IVO) Mar.

First: *Intrapora puteolata* Hall, 1883, Jeffersonville Limestone, Kentucky and Indiana, USA; *Coscinella elegantula* Hall, 1887, Hamilton Group, Ontario, Canada.
Last: *Intrapora texera* Troitskaya, 1975, Kassin Horizon (TOU), central Kazakhstan, former USSR.

F. PHRAGMOPHERIDAE Goryunova, 1969 C. (u.) Mar.

First and Last: *Phragmophera eximia* Goryunova, 1969, Upper Carboniferous, central Urals, former USSR.
Comments: This family is known only from one species occurring in one locality (Mal'tsevka on the Kos'va River) which has been dated no more precisely than 'Upper Carboniferous' (Goryunova, 1969).

F. RHINIDICTYIDAE Ulrich, 1893 O. (LLN)–S. (?SHE/?WHI/?GLE) Mar.

First: *Phyllodictya crystalaria* Hinds, 1970, Lehman Formation, Utah, USA.
Last: *Trigonodictya* sp., WEN, ?locality (Karklins, 1983, p. 21).

F. STICTOPORELLIDAE Nickles and Bassler, 1900 O. (ARG)–S. (?SHE/?WHI/?GLE) Mar.

First: *Stictoporellina gracilis* (Eichwald, 1840), Volkhov Stage, Estonia, former USSR.
Last: *Stictoporella asiatica* Astrova, 1957, WEN, Tuvinskaq, former USSR.

F. VIRGATELLIDAE Astrova, 1965 O. (CRD) Mar.

First and Last: *Pseudopachydictya multicapillaris* (Astrova, 1955), *Virgatella bifoliata* Astrova, 1955, Mangazeya Stage, Siberia, former USSR.

F. WORTHENOPORIDAE Ulrich, 1893 C. (CHD–ASB) Mar.

First: *Worthenopora spinosa* Ulrich, 1890, Keokuk Formation, Illinois Basin, USA.
Last: *W. castletonense* Owen, 1966, Reef limestone, Castleton, Derbyshire, England, UK.
Comments: This monogeneric family is of uncertain affinity but seems most likely to be related to the ptilodictyines (Hageman and Snyder, 1987).

Suborder RHABDOMESINA Astrova and Morozova, 1956

Rhabdomesines are a suborder of cryptostomes characteristically forming narrow-branched, dendroid colonies. Family-level classification adopted here is after Blake (*in* Boardman *et al.*, 1983), with modifications from Goryunova (1985).

F. GOLDFUSSITRYPIDAE Goryunova, 1985 O. (ARG)–P. (ART) Mar. (see Fig. 24.5)

First: *Goldfussitrypa abnormalis* Goryunova, 1985, *G. electa* Goryunova, 1985, Volkhov Stage, Estonia, former USSR.
Last: *Nicklesopora lepida* Nikiforova, 1939, ART, Bashkiria, former USSR.

F. MAYCHELLINIDAE Goryunova, 1985 C. (?BSH/?MOS/?KAS/?GZE)–P. (UFI) Mar.

First: *Maychellina aliena* Goryunova and Morozova, 1979, Middle–Upper Carboniferous, Mongolia.
Last: *M. edita* (Morozova, 1970), *M. nervosa* (Morozova, 1970), Omolonskii Horizon, NE of former USSR.
Comments: Monogeneric.

F. MAYCHELLIDAE Goryunova, 1985 P. (?WOR/?CAP) Mar.

First: *Maychella rhomboidea* Yang and Lu, 1983, Maokou Formation, KAZ, China.
Last: *M. tuberculata* Morozova, 1970, Osakhta Group, Soviet Far East; Chandalaz Group, Primor'e, former USSR.
Comments: The stratigraphical distribution of this monogeneric family is very uncertain.

F. ARTHROSTYLIDAE Ulrich, 1882
O. (LLO)–P. (CAP) Mar.

First: *Arthroclema vescum* Goryunova, 1985, LLO, Estonia, former USSR; *Ulrichostylus* sp., *Cuneatopora* sp., LLO, Llandeilo, Wales, UK (Buttler, 1988).
Last: *Permoheloclema merum* Ozhgibesov, 1983, KAZ, former Arctic USSR.

F. RHABDOMESIDAE Vine, 1884
S. (?SHE/?WHI/?GLE)–P. (CHX) Mar.

First: *Orthopora casualis* Goryunova, 1985, Rochester Shale, New York State, USA.
Last: *Rhabdomeson* sp., Changxing Formation, Xizang Province, China (Yang and Lu, 1981).

F. RHOMBOPORIDAE Simpson, 1895
D. (EIF)–P. (CAP) Mar.

First: *Saffordotaxis yukonensis* Astrova, 1972, EIF, Arctic Canada (originally assigned to *Rhombopora* but transferred to *Saffordotaxis* by Goryunova, 1985, p. 115).
Last: *Rhombopora ornata* Shishova, 1964, KAZ, Kirovsk Oblast', former USSR.

F. BACTROPORIDAE Simpson, 1897
D. (GIV) Mar.

First and Last: *Bactropora granistriata* (Hall, 1881), *B. simplex* (Ulrich, 1886), Hamilton Group, New York State, USA.

F. NIKIFOROVELLIDAE Goryunova, 1975
D. (FAM)–Tr. (CRN) Mar.

First: *Nikiforovella amazarica* Nekhoroshev, 1948, FAM, Altai Mts, former USSR; *N. nitida* Troitskaya, 1979, FAM, Kazakhstan, former USSR.
Last: *Tebitopora orientalis* Zhao-xun, 1984, Zozar Formation, West Himalaya, India.
Comments: *T. orientalis,* originally described as a trepostome, has been assigned tentatively to this family by Schäfer and Fois (1987). The next youngest nikiforovellid is *Pinegopora delicatula* Shisova, 1965 from the Kazanian of Arkhangel'sk Province, former USSR.

F. NUDYMIELLIDAE Goryunova, 1985
C. (BSH) Mar.

First and Last: *Nudymiella singula* Morozova, 1981, Middle Carboniferous, Severo-Vostoka, former USSR.

F. PSEUDOASCOPORIDAE Goryunova, 1985
D. (FAM) Mar.

First and Last: *Pseudascopora valentinae* Goryunova, 1985, FAM, Azerbaijan, former USSR.

F. HYPHASMOPORIDAE Vine, 1886
D. (FAM)–P. (TAT) Mar.

First: *Ipmorella tobolensis* Goryunova, 1985, FAM, Kazakhstan, former USSR.
Last: *Streblotrypa parva* Morozova, 1965, Dzhulfian, Dzhul'fa Gorge, Transcaucasia, former USSR.

Suborder TIMANODICTYINA Morozova, 1966

This exclusively Upper Palaeozoic suborder is characterized by dendroid or bifoliate branches with thick exozonal walls containing small styles ('capillaries' of Russian workers). Although regarded as cryptostomes, timanodictyines were not treated in the cryptostome section of the bryozoan *Treatise* (Boardman *et al.*, 1983).

F. TIMANODICTYIDAE Morozova, 1966
P. (SAK–KAZ) Mar.

First: *Timanodictya dichotoma* (Stuckenberg, 1895), SAK, northern Timan and Malozemelsk Tundra, Urals, former USSR; Ellesmere Island, Canada.
Last: *Timanotrypa borealis* Morozova, 1970, KAZ, Pinega River, Arkhangel'sk Oblast', former USSR.

F. GIRTYOPORIDAE Morozova, 1966
C. (VIS)–P. (TAT) Mar.

First: *Morozovapora akiyoshiensis* Sakagami and Sugimura, 1978, *Nagatophyllum satoi* Zone of the Akiyoshi Limestone, Mizuta, Japan.
Last: *Girtyoporina crassa* Morozova, 1970, 'Lyudyanza (?) Horizon', basin of the Khuanikhezda River, Primor'e Region, former USSR.

Order FENESTRATA Elias and Condra, 1957

This order of Palaeozoic stenolaemates was created for taxa previously assigned to the Cryptostomata (e.g. Bassler, 1953). Fenestrates are likely to be a monophyletic group closely related to cryptostomes (e.g. Blake, 1980), but possessing the apomorphic character of apertures opening along one side of the branches only. Many fenestrates have planar colonies with a net-like or pinnate form. The order is currently being revised by F. K. McKinney for the *Treatise*. Family-level classification adopted here is after Lavrentjeva (1985) and Morozova (1987).

F. ENALLOPORIDAE Miller, 1889
O. (ARG)–S. (LLY) Mar.

First: *Alwynopora orodamnus* Taylor and Curry, 1985, Tourmakeady Limestone, County Mayo, Republic of Ireland.
Last: *Pushkinella acanthoporoides* (Pushkin, 1976), LLY, Byelorussia, former USSR.

F. RALFINIDAE Lavrentjeva, 1985
O. (CRD) Mar.

First and Last: *Ralfina aluverensis* (Männil, 1958), *Ralfinella plana* (Männil, 1958), CRD, Estonia, former USSR.

F. PHYLLOPORINIDAE Ulrich, 1890
O. (LLO)–S. (WEN) Mar.

First: *Phylloporina punctata* (Bekker, 1921), *P. fragilis* Lavrentjeva, 1985, *Pseudohornera bifida* (Eichwald, 1855), LLO, former USSR.
Last: *Pseudohornera diffusa* (Hall, 1852), Rochester Shale, New York State, USA, and Ontario, Canada.

F. SARDESONINIDAE Lavrentjeva, 1985
O. (LLO)–O. (CRD) Mar.

First: *Sardesonina maxima* (Toots, 1952), Kuckers Shale, Estonia, former USSR.
Last: *Sardesonina corticosa* (Ulrich, 1886), Black River, Minnesota, USA.
Comments: Monogeneric.

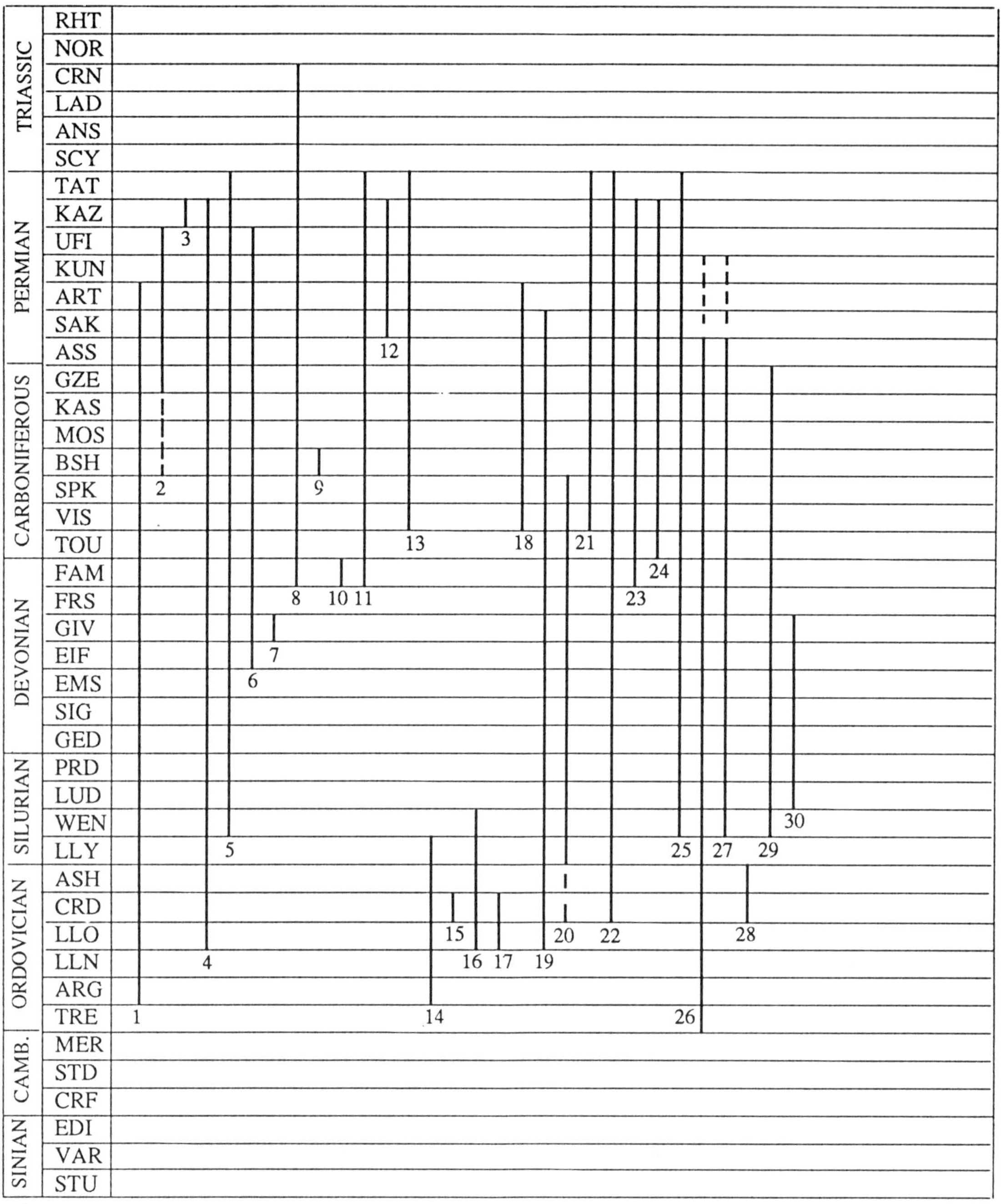

Fig. 24.5

F. CHAINODICTYONIDAE Nickles and Bassler, 1900 C. (VIS)–P. (ART) Mar.

First: *Chainodictyon aktasicum* Plamenskaja, 1964, Kazakhstan, former USSR. *C. undata* (M'Coy, 1844), Carboniferous Limestone, Republic of Ireland.

Last: *Chainodictyon lucidum* Goryunova, 1970, ART, Pamirs, former USSR.

F. CHASMATOPORIDAE Schulga-Nesterenko, 1955 O. (LLO)–P. (SAK) Mar.

First: *Esthonioporina quadrata* (Bekker, 1921), LLO, Estonia, former USSR.

Last: *Bashkirella nikiforovae* Schulga-Nesterenko, 1952, *B. ornata* Nikiforova, 1939, *B. operculata* Schulga–Nesterenko, 1952, Bashkiria, former USSR.

F. SEMICOSCINIUMIDAE Morozova, 1988 O. (?CRD/?ASH)–C. (SPK) Mar.

First: *Eosemicoscinium* sp., Upper Ordovician (see Morozova, 1987, p. 82).

Last: *'Neoreteporina'* sp., Namurian A, American Shelf (Ross, 1981, text-fig. 4).

F. SEPTOPORIDAE Morozova, 1962 C. (VIS)–P. (TAT) Mar.

First: *Septopora* sp., Lower Carboniferous (Morozova, 1987, fig. 5).

Last: *Synocladia rigida* Morozova, 1965, Dzhulfian, Dzhul'fa Gorge, Transcaucasia, former USSR.

F. FENESTELLIDAE King, 1850 O. (CRD)–P. (CHX) Mar.

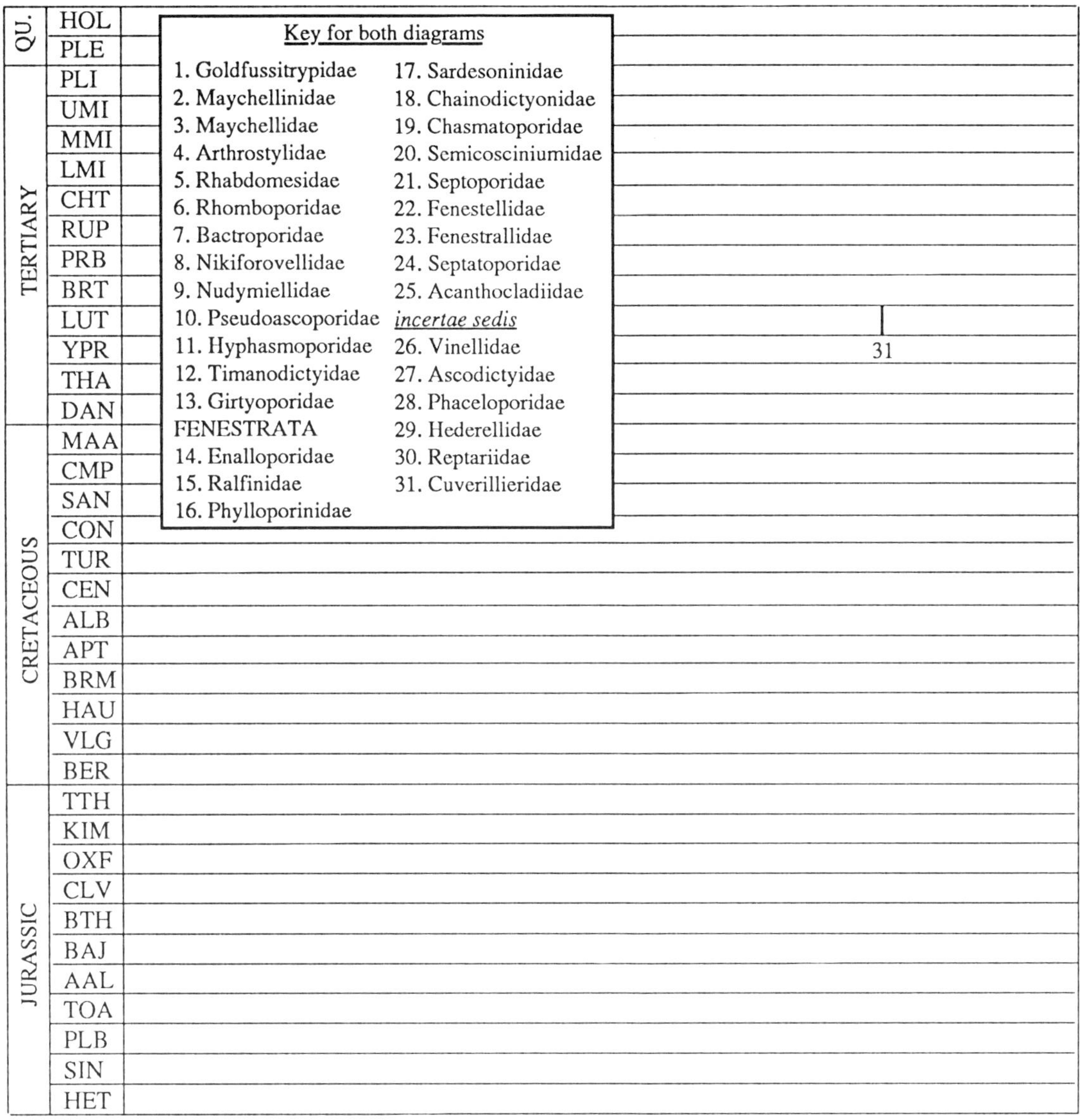

QU.	HOL
	PLE
TERTIARY	PLI
	UMI
	MMI
	LMI
	CHT
	RUP
	PRB
	BRT
	LUT
	YPR
	THA
	DAN
CRETACEOUS	MAA
	CMP
	SAN
	CON
	TUR
	CEN
	ALB
	APT
	BRM
	HAU
	VLG
	BER
JURASSIC	TTH
	KIM
	OXF
	CLV
	BTH
	BAJ
	AAL
	TOA
	PLB
	SIN
	HET

Key for both diagrams

1. Goldfussitrypidae
2. Maychellinidae
3. Maychellidae
4. Arthrostylidae
5. Rhabdomesidae
6. Rhomboporidae
7. Bactroporidae
8. Nikiforovellidae
9. Nudymiellidae
10. Pseudoascoporidae
11. Hyphasmoporidae
12. Timanodictyidae
13. Girtyoporidae

FENESTRATA

14. Enalloporidae
15. Ralfinidae
16. Phylloporinidae
17. Sardesoninidae
18. Chainodictyonidae
19. Chasmatoporidae
20. Semicosciniumidae
21. Septoporidae
22. Fenestellidae
23. Fenestrallidae
24. Septatoporidae
25. Acanthocladiidae

incertae sedis

26. Vinellidae
27. Ascodictyidae
28. Phaceloporidae
29. Hederellidae
30. Reptariidae
31. Cuverillieridae

31

Fig. 24.5

First: *Moorephylloporina typica* Bassler, 1952, Edinburg Formation, Virginia, USA.
Last: *Fenestella sinopermiana* Yang and Lu, upper Changhsing Limestone, South China (Zhao *et al.*, 1981).
Comments: Supposed Triassic records of fenestellids are probably erroneous.

F. FENESTRALLIDAE Morozova, 1963
D. (FAM)–P. (KAZ) Mar.

First: *Fenestralia* sp., ? locality (Morozova, 1987, fig. 5).
Last: *Parafenestralia arborescens* (Nechaev, 1893), *P. longa* Morozova, 1970, *Triznella viatkensis* (Nikiforova, 1945), *T. permiana* (Nikiforova, 1945), *T. formosa* (Morozova, 1963), lower KAZ, Russian Platform, former USSR.

F. SEPTATOPORIDAE Engel, 1975
C. (IVO)–P. (KAZ) Mar.

First: *Septatopora acarinata* (Crockford, 1947), *S. nodosa* Engel, 1975, upper Tournaisian (*Schellwienella* cf. *burlingtonensis* Zone), Raglan, New South Wales, Australia.
Last: *Septatopora* sp., KAZ, Australia (Ross, 1978, p. 351).
Comments: This monogeneric family is retained with reservation for fenestellid-like forms with septate apertures.

F. ACANTHOCLADIIDAE Zittel, 1880
S. (?SHE/?WHI/?GLE)–P. (CHX) Mar.

First: *Polyporella intermedia* (Shrubsole, 1880), Wenlock Limestone, Dudley, England, UK.
Last: *Polypora* sp., Changxing Formation, South China (Yang and Lu, 1981).
Comments: 'Induan' sediments containing acanthocladiids were formerly regarded as early Triassic in age, but are now thought to be late Permian (Rostovtsev and Azaryan, 1973).

INCERTAE SEDIS

The following families, except for the Cuvillieridae, were included in the bryozoan *Treatise* by Bassler (1953), but are either of unknown affinities within the phylum, or do not belong to the Bryozoa.

F. VINELLIDAE Ulrich and Bassler, 1904
O. (TRE)–P. (?ASS/?SAK/?ART/?KUN) Mar.

First: *Marcusodictyon priscum* (Bassler, 1911), Ungulite Sandstone, Estonia, former USSR (see Taylor, 1984).
Last: *Condranema parvula* (Condra and Elias, 1944), *C. magna* (Condra and Elias, 1944), Lower Permian, USA.
Comments: A heterogeneous grouping of thread-like encrusters, assigned traditionally to the soft-bodied order Ctenostomata with little justification. Most, if not all, are not bryozoans. At least some Mesozoic vinellids are foraminifera, but the bulk of the family are Palaeozoic species of problematic affinity (e.g. Taylor, 1984).

F. ASCODICTYIDAE Miller, 1889
S. (?SHE/?WHI/?GLE)–P. (?ASS/?SAK/?ART/?KUN) Mar.

First: *Ascodictyon siluriense* Vine, 1892, Wenlock Shale, Shropshire, England, UK; Rochester Shale, New York State, USA; Waldron Shale, Indiana, USA; *A. filiforme* Vine, 1882, Wenlock Shale, Shropshire, England, UK (Ulrich and Bassler, 1903).
Last: *A. nebraskensis* Condra and Elias, 1944, Big Blue Series, Nebraska, USA.
Comments: As with the Vinellidae, these encrusting, thread-like, supposed ctenostomes are unlikely to be bryozoans.

F. PHACELOPORIDAE Miller, 1889
O. (CRD–ASH) Mar.

First: *Phacelopora pertenuis* Ulrich, 1890, Trenton, Kentucky, USA.
Last: *Phacelopora* sp., Öjlemyrflint, Sylt, Germany (Hillmer and Schallreuter, 1987).
Comments: A monogeneric family, generally regarded as belonging to the Cyclostomata but in need of re-evaluation.

F. HEDERELLIDAE Kiepura, 1973
S. (?SHE/?WHI/?GLE)–C. (GZE) Mar.

First: *Hederella siluriana* Bassler, 1939, 'probably Klintehamn', Gotland, Sweden.
Last: *H. carbonaria* Condra and Elias, 1944, Missouri Series, Dewey Limestone, Oklahoma, USA.
Comments: Together with the Reptariidae, this family forms the Suborder Hederelloidea Bassler, 1939, which has been generally assigned to the Order Cyclostomata (cf. Dzik, 1981, who tentatively assigns them to the Phylactolaemata). However, the hederelloids, in great need of revision, are thought by some not to be bryozoans, but more closely related to auloporid tabulate corals.

F. REPTARIIDAE Simpson, 1897
S. (LUD)–D. (GIV) Mar.

First: *Reptaria steiningeri* (Barrande, 1868), Budnany Beds, Bohemia, Czechoslovakia (Prantl, 1938).
Last: *R. cloudi* Bassler, 1939, Tully Limestone, West Brook Member, New York State, USA.

F. CUVERILLIERIDAE Annoscia, 1965
T. (LUT) Mar.

First and Last: *Cuvilliera egyptiense* Pfender, 1934, upper LUT, Egypt.
Comments: This monospecific family was assigned to the anascan cheilostomes by Annoscia (1965), but the bryozoan affinities of *C. egyptiense* are dubious.

REFERENCES

Annoscia, E. (1965) Revisione del genere *Cuvilliera* Briozoo del Luteziano Superiore d'Egitto. *Memoire degli Istituti di Geologia e Mineralogia dell'Università di Padova*, **25**, 1–9.

Annoscia, E. (1968) Briozoi. *Palaeontographia Italica*, 397 pp.

Astrova, G. G. (1978) The history of development, system and phylogeny of the Bryozoa order Trepostomata. *Trudy Paleontologicheskogo Instituta*, **169**, 1–240 [in Russian].

Bassler, R. S. (1953) Bryozoa, in *Treatise on Invertebrate Paleontology. Part G* (ed. R. C. Moore), Geological Society of America and University of Kansas, Lawrence, pp. G1–G253.

Berthelsen, O. (1962) Cheilostome Bryozoa in the Danian deposits of East Denmark. *Danmarks Geologiske Undersogelse*, II. Raekke, **83**, 290 pp.

Bizzarini, F. and Braga, G. (1981) Prima segnalazione del genre *Stomatopora* (Bryozoa Cyclostomata) nel Trias superiore delle Dolomiti orientali. *Lavori Società Venezianza di Scienze Naturali*, **6**, 135–44.

Bizzarini, F. and Braga, G. (1985) *Braiesopora voigti* n. gen. n. sp. (cyclostome bryozoan) in the S. Cassiano Formation in the Eastern Alps (Italy), in *Bryozoa: Ordovician to Recent* (ed. C. Nielsen and G. P. Larwood), Olsen and Olsen, Fredensborg, pp. 25–33.

Blake, D. B. (1980) Homeomorphy in Paleozoic bryozoans: a search for explanations. *Paleobiology*, **6**, 451–65.

Boardman, R. S., Cheetham, A. H., Blake, D. B. *et al.* (1983) Bryozoa (revised). Volume 1, in *Treatise on Invertebrate Paleontology. Part G* (ed. R. C. Moore and R. A. Robison), Geological Society of America and University of Kansas, Boulder, Colorado, and Lawrence, Kansas, pp. 1–625.

Boardman, R. S., McKinney, F. K. and Taylor, P. D. (1992) Morphology, anatomy, and systematics of the Cinctiporidae, new family (Bryozoa: Stenolaemata). *Smithsonian Contributions to Paleobiology*, **70**, 1–81.

Braga, G. and Barbin, V. (1988) Les Bryozoaires du Priabonien stratotypique (Province Vicenza, Italie). *Revue de Paléobiologie*, **7**, 495–556.

Brood, K. (1972) Cyclostomatous Bryozoa from the Upper Cretaceous and Danian in Scandanavia. *Stockholm Contributions in Geology*, **26**, 1–464.

Brood, K. (1975) Cyclostomatous Bryozoa from the Silurian of Gotland. *Stockholm Contributions in Geology*, **28**, 45–119.

Brown, D. A. (1952) *The Tertiary Cheilostomatous Polyzoa of New Zealand*. British Museum (Natural History), London, 405 pp.

Buckeridge, J. S. (1989) Marine invertebrates from late Cainozoic deposits in the McMurdo Sound region, Antarctica. *Journal of the Royal Society of New Zealand*, **19**, 333–42.

Buge, E. and Voigt, E. (1972) Les *Cellulipora* (Bryozoa, Cyclostomata) du Cénomanian francais et la famille des Celluliporidae. *Geobios*, **5**, 121–50.

Busk, G. (1859) A monograph of the fossil Polyzoa of the Crag. *Palaeontographical Society Monograph*, **11** (49), 136 pp.

Buttler, C. J. (1989) New information on the morphology and skeletal ultrastructure of the Ordovician cyclostome bryozoan *Kukersella* Toots, 1952. *Paläontologische Zeitschrift*, **63**, 215–27.

Buttler, C. J. (1991) A new upper Ordovician bryozoan fauna from the Slade and Redhill Beds, South Wales. *Palaeontology*, **34**, 77–108.

Canu, F. (1907–10) Bryozoaires des terrains tertiaires des environs de Paris. *Annales de Paléontologie*, **2**, 57–88, 137–60; **3**, 61–104; **4**, 29–68; **5**, 89–112.

Cheetham, A. H. (1966) Cheilostomatous Polyzoa from the Upper Bracklesham Beds (Eocene) of Sussex. *Bulletin of the British Museum (Natural History)*, Geology Series, **13**, 1–115.

Cheetham, A. H. (1968a) Evolution of zooecial asymmetry and origin of poricellariid cheilostomes. *Atti della Società Italiana di Scienze Naturali e del Museo Civico di Storia Naturale di Milano*, **108**, 185–94.

Cheetham, A. H. (1968b) Morphology and systematics of the bryozoan genus *Metrarabdotos*. *Smithsonian Miscellaneous Collections*, **153** (1), 121 pp.

Cheetham, A. H. (1972) Cheilostome Bryozoa of late Eocene age from Eua, Tonga. *Professional Paper. United States Geological Survey*, **640-E**, v + 26 pp.

Cheetham, A. H. (1975) Preliminary report on early Eocene cheilostome bryozoans from Site 308–Leg 32, Deep Sea Drilling Project. *Initial Reports of the Deep Sea Drilling Project*, **32**, 835–51.

Cheetham, A. H. and Håkansson, E. (1972) Preliminary report on Bryozoa (Site 117). *Initial Reports of the Deep Sea Drilling Project*, **12**, 432–41.

Cook, P. L. and Chimonides, P. J. (1983) A short history of the lunulite Bryozoa. *Bulletin of Marine Science*, **33**, 566–81.

Davis, A. G. (1934) English Lutetian Polyzoa. *Proceedings of the Geologists' Association*, **45**, 205–45.

Debourle, A. (1975) New Bryozoa of south-west Aquitaine (France). *Documents des Laboratoires de Géologie de la Faculté des Sciences de Lyon, Hors Série*, **3**, 535–8.

Dzik, J. (1981) Evolutionary relationships of the early Palaeozoic 'cyclostomatous' Bryozoa. *Palaeontology*, **24**, 827–61.

Engeser, T. and Taylor, P. D. (1989) Supposed Triassic bryozoans in the Klipstein Collection from the Italian Dolomites redescribed as calcified demosponges. *Bulletin of the British Museum (Natural History), Geology*, **45**, 39–55.

Ghiurca, V. (1975) Les Bryozoaires Néogènes de la Paratethys. *Documents des Laboratoires de Géologie de la Faculté des Sciences de Lyon, Hors Série*, **3**, 497–518.

Gordon, D. P. (1984) The marine fauna of New Zealand: Bryozoa: Gymnolaemata from the Kermadec Ridge. *New Zealand Oceanographic Institute Memoir*, **91**, 1–198.

Gordon, D. P. (1988) The bryozoan families Sclerodomidae, Bifaxariidae, and Urceoliporidae and a novel type of frontal wall. *New Zealand Journal of Zoology*, **15**, 249–90.

Gordon, D. P. (1989) The marine fauna of New Zealand: Bryozoa: Gymnolaemata (Cheilostomida Ascophorina) from the western South Island continental shelf and slope. *New Zealand Oceanographic Institute Memoir*, **97**, 1–158.

Gordon, D. P. (1990) The Tertiary bryozoan family Prostomariidae – morphology and relationships. *Memoirs of the Museum of Victoria*, **50**, 467–72.

Gordon, D. P. and d'Hondt, J.-L. (1991) Bryozoa: The miocene to recent family Petalostegidae. Systematics, affinities, biogeography, in *Résultats des Campagnes* Musorstom, Vol. 8 (ed. A. Crosnier). *Mémoires de Museum National d'Histoire Naturelle, Paris,* (A), **151**, 91–123.

Goryunova, R. V. (1969) New Upper Carboniferous bryozoan genus from the Central Urals. *Paleontologicheskii Zhurnal*, **1969**, 129–31 [in Russian].

Goryunova, R. V. (1985) [Morphology systematics and phylogeny of Bryozoa (Order Rhabdomesida)]. *Trudy Paleontological Institute Nauk SSSR*, **208**, 1–152 [in Russian].

Hageman, S. J. and Snyder, E. M. (1987) *Worthenopora* (Bryozoa, Stenolaemata): morphology and stratigraphic usefulness of a cryptostome (Ptilodictyina) with a cheilostome appearance. *Geological Society of America Abstracts with Programs*, **19** (4), p. 202.

Hayward, P. J. (1985) Ctenostome bryozoans. *Synopses of the British Fauna*, **33**, 1–169.

Hayward, P. J. and Ryland, J. S. (1985) Cyclostome bryozoans. *Synopses of the British Fauna*, **34**, 1–147.

Hillmer, G. (1971) Bryozoen aus dem Alb und Cenoman von Hannover. *Beihefte zu den Berichten den Naturhistorische Gesellschaft zu Hannover*, **7**, 49–67.

Hillmer, G. and Schallreuter, R. (1987) Ordovician bryozoans from erratic boulders of northern Germany and Sweden, in *Bryozoa: Present and Past* (ed. J. R. P. Ross), Western Washington University, Bellingham, 333 pp.

d'Hondt, J.-L. (1983) Tabular keys for identification of the Recent ctenostomatous Bryozoa. *Mémoires de l'Institut Océanographique, Monaco*, **14**, 1–134.

Horowitz, A. S. (1968) The ectoproct (bryozoan) genus *Actinotrypa* Ulrich. *Journal of Paleontology*, **42**, 356–73.

Jell, P. A. and Duncan, P. M. (1986) Invertebrates, mainly insects, from the freshwater, Lower Cretaceous, Koonwarra Fossil Bed (Korumburra Group), South Gippsland, Victoria. *Memoir of the Association of Australasian Palaeontologists*, **3**, 111–205.

Karklins, O. L. (1983) Ptilodictyoid Cryptostomata Bryozoa from the Middle and Upper Ordovician rocks of central Kentucky. *Journal of Paleontology*, **57** (1, supplement 2), 1–31.

Keij, A. J. (1973) The bryozoan genus *Skylonia* Thomas (Cheilostomata). *Bulletin of the British Museum (Natural History), Geology Series*, **24**, 217–33.

Keij, A. J. (1977) The Tertiary bryozoan genera *Bicornifera* and *Bifissurinella* (Cheilostomata, Anasca). *Proceedings, Koninklijke Nederlandse Akademie van Wetenschappen, Series B*, **80**, 229–41.

Kiseleva, A. V. (1982) *Late Permian Bryozoa from the S. Primor'e region*. Paleontologicheskii Institut, Moscow, 128 pp [in Russian].

Kuc, M. (1973) Fossil statoblasts of *Cristatella mucedo* Cuvier in the Beaufort Formation and in interglacial and postglacial deposits of the Canadian Arctic. *Geological Survey Papers, Mines and Geology Branch, Canada*, **73** (28), 1–12.

Labracherie, M. (1971) Évolution générale des assemblages de Bryozoaires dans l'Éocène du bassin nord-aquitain. *Compte Rendu Sommaire des Séances de la Société Géologique de France*, **21**, 388–9.

Labracherie, M. (1975) Description des Bryozoaires Cheilostomes d'âge Eocèné inférieur du site 246 (croisière 25, Deep Sea Drilling Project). *Bullétin de l'Institut de Geologie du Bassin d'Aquitaine*, **18**, 149–202.

Langer, W. (1980) Über *Lagenosypho* Spändel (Bryozoa) und einige andere karbonatische Mikrofossilien aus den westdeutschen Devon. *Munstersche Forschungen zur Geologie und Paläontologie*, **52**, 97–118.

Larwood, G. P., Medd, A. W., Owen, D. E. *et al*. (1967) Bryozoa, in *The Fossil Record* (ed. W. B. Harland, C. H. Holland, M. R. House *et al*.), Geological Society of London, London, pp. 379–95.

Lavrentjeva, V. D. (1985) Bryozoa Suborder Phylloporina. *Trudy Paleontologicheskogo Instituta*, **214**, 1–101 [in Russian].

Malecki, J. (1962) Bryozoa from the Eocene of the Central Carpathians between Grybów and Dukla. *Polska Akademia Nauk Prace Geologiczne*, **16**, 1–158 [in Polish, with an English summary].

Matthews, J. V., Ovenden, L. E. and Fyles, J. G. (1990) Plant and insect fossils from the late Tertiary Beaufort Formation on Prince Patrick Island, N. W. T., in *Canada's Missing Dimension* (ed. C. R. Harington), Canadian Museum of Nature, Ottawa, pp. 105–39.

Moissette, P. (1988) Faunes de bryozoaires du Messinien d'Algerie Occidentale. *Documents des Laboratoires de Géologie Lyon*, **102**, 1–351.

Morozova, I. P. (1987) The morphogeny, systematics, and colonial integration of bryozoans of the Order Fenestrida. *Trudy Paleontologicheskogo Instituta*, **222**, 70–88.

Mundy, S. P., Taylor, P. D. and Thorpe, J. P. (1981) A reinterpretation of phylactolaemate phylogeny, in *Recent and Fossil Bryozoa* (ed. G. P. Larwood and C. Nielsen), Olsen and Olsen, Fredensborg, pp. 185–90.

Nekhoroshëva, L. V. (1966) Ordovician Ptilodictyidae of Taimyr. *Ucenye Zapiski, Nauchno-Issledovatel'skii Institut Geologii Arktiki*, **13**, 22–37 [in Russian].

Nekhoroshëva, L. V. (1974) Ordovician Bryozoa of the Soviet Arctic. in *The Ordovician System* (ed. M. G. Bassett), University of Wales Press and the National Museum of Wales, Cardiff, pp. 575–82.

Nowicki, M. J. (1989) The biostratigraphy of Coniacian-Santonian cheilostome Bryozoa of the North Aquitaine Basin. Unpublished PhD thesis, King's College, University of London, 283 pp.

Parker, S. A. and Gordon, D. P. (1992) A new genus of the bryozoan superfamily Schizoporelloidea, with remarks on the validity of the family Lacernidae Jullien, 1888. *Records of the South Australian Museum*, **26**, 67–71.

Pitt, L. J. and Taylor, P. D. (1990) Cretaceous Bryozoa from the Faringdon Sponge Gravel (Aptian) of Oxfordshire. *Bulletin of the British Museum (Natural History), Geology*, **46**, 61–152.

Pohowsky, R. A. (1978) The boring ctenostomate Bryozoa: taxonomy and paleobiology based on cavities in calcareous substrata. *Bulletin of American Paleontology*, **73**, 1–192.

Prantl, F. (1938) Revision of the Bohemian Paleozoic Reptariidae. (Bryozoa). *Sborník Národního Musea v Praze*, **1B** (6), Geologia et Paleontologia No. **2**, 73–84.

Ross, J. R. P. (1963) *Constellaria* from the Chazyan (Ordovician), Isle la Motte, Vermont. *Journal of Paleontology*, **37**, 51–6.

Ross, J. R. P. (1966) Early Ordovician ectoproct from Oklahoma. *Oklahoma Geology Notes*, **26**, 218–24.

Ross, J. R. P. (1978) Biogeography of Permian ectoproct Bryozoa. *Palaeontology*, **21**, 341–56.

Ross, J. R. P. (1981) Biogeography of Carboniferous ectoproct Bryozoa. *Palaeontology*, **24**, 313–41.

Ross, J. R. P. (1984) Palaeoecology of Ordovician Bryozoa, in *Aspects of the Ordovician System* (ed. D. L. Bruton), Palaeontological Contributions from the University of Oslo, No. 295, Universitetsforlaget, pp. 141–8.

Rostovtsev, K. O. and Azaryan, N. R. (1973) The Permian–Triassic boundary in Transcaucasia. *Memoirs of the Canadian Society of Petroleum Geologists*, **2**, 89–99.

Ryland, J. S. (1982) Bryozoa, in *Synopsis and Classification of Living Organisms*, 2 Vols (ed. S. P. Parker), McGraw-Hill, New York, pp. 743–69.

Schäfer, P. and Fois, E. (1987) Systematics and evolution of Triassic Bryozoa. *Geologica et Palaeontologica*, **21**, 173–225.

Schubert, T. (1986) Parallele Merkmalsentwicklung der Bryozoen-Arten von *Woodipora* Jullien 1888 im Coniacium bis Maastrichtium NW-Europas. *Geologisches Jahrbuch*, **A98**, 1–83.

Szczechura, J. (1990) Marginal bryozoan *Calvina* from the Paleocene of Poland. *Acta Palaeontologica Polonica*, **35**, 41–8.

Taylor, P. D. (1984) *Marcusodictyon* Bassler from the Lower Ordovician of Estonia: not the earliest bryozoan but a phosphatic problematicum. *Alcheringa*, **8**, 177–86.

Taylor, P. D. (1985) Carboniferous and Permian species of the cyclostome bryozoan *Corynotrypa* Bassler, 1911 and their clonal propagation. *Bulletin of the British Museum (Natural History), Geology*, **38**, 359–72.

Taylor, P. D. (1988) Major radiation of cheilostome bryozoans: triggered by the evolution of a new larval type? *Historical Biology*, **1**, 45–64.

Taylor, P. D. (1990) Bioimmured ctenostomes from the Jurassic and the origin of the cheilostome Bryozoa. *Palaeontology*, **33**, 19–34.

Taylor, P. D. and Cope, J. C. W. (1987) A trepostome bryozoan from the Lower Arenig of south Wales: implications of the oldest described bryozoan. *Geological Magazine*, **124**, 367–71.

Taylor, P. D. and Michalík, J. (1991) Cyclostome bryozoans from the late Triassic (Rhaetian) of the West Carpathians, Czechoslovakia. *Neues Jahrbuch für Geologie und Palaontologie, Abhandlungen*, **182**, 285–302.

Todd, J. A. (in press) The role of bioimmuration in the exceptional preservation of fossil ctenostomes, including a new Jurassic species of *Buskia*. *Proceedings of the 9th Conference of the International Bryozoology Association* (eds J. S. Ryland, P. J. Hayward and P. D. Taylor), Olsen and Olsen, Fredensberg.

Todd, J. A. and Hagdorn, H. (in press) First record of Muschelkalk Bryozoa: the earliest ctenosome body fossils, in *Muschelkalk, Schöntaler Symposium 1991* (eds H. Hagdorn and A. Seilacher) Goldschneck Verlag, Stuttgart.

Toots, H. and Cutler, J. F. (1962) Bryozoa from the 'Mesaverde' Formation (Upper Cretaceous) of southeastern Wyoming. *Journal of Paleontology*, **36**, 81–6.

Ulrich, E. O. and Bassler, R. S. (1903) A revision of the Paleozoic Bryozoa. Part 1. – On genera and species of Ctenostomata. *Smithsonian Miscellaneous Collections*, **45**, 256–94.

Vavra, N. (1977) Bryozoa tertiaria. *Catalogus Fossilium Austriae*, **5b/3**, 1–210.

Vinogradov, A. V. (1985) Bryozoa, in *The Jurassic Continental Biocoenoses of Southern Siberia and Adjacent Areas* (ed. A. P. Rasnitsyn), Trudy Paleontologicheskogo Instituta, **213**, 85–7 [in Russian].

Voigt, E. (1956) Untersuchungen über *Coscinopleura* Marss. (Bryoz. foss.) und verwandte Gattungen. *Mitteilungen aus dem Geologischen Staatinstitut in Hamburg*, **25**, 26–75.

Voigt, E. (1981) Répartition et utilisation stratigraphique des Bryozoaires du Crétacé Moyen (Aptien-Coniacien). *Cretaceous Research*, **2**, 439–62.

Voigt, E. (1983) Fossilerhaltung durch Bioimmuration: *Aetea* (Lamouroux) (Bryozoa Cheilostomata) aus dem Mittelmeerraum (Pliozän und rezent). *Facies*, **9**, 285–310.

Voigt, E. (1984) Die Genera *Reteporidea* d'Orbigny, 1849 und *Crisidmonea* Marsson (Bryozoa Cyclostomata) in der Maastrichter Tuffkreide (Oberes Maastrichtium) nebst Bemerkungen über *Polyascosoecia* Canu & Bassler und andere ähnliche Gattungen. *Mitteilungen Geologisch–Paläontologisches Institut der Universität Hamburg*, **56**, 385–412.

Voigt, E. (1985) The Bryozoa of the Cretaceous–Tertiary boundary, in *Bryozoa: Ordovician to Recent* (eds C. Nielsen and G. P. Larwood), Olsen and Olsen, Fredensborg, pp. 329–42.

Voigt, E. (1987) Die Bryozoen des klassischen Dano-Montiens von Mons (Belgien). *Mémoires pour servir à l'Explication des Cartes Géologiques et Minières de la Belgique*, **17**, 1–161.

Voigt, E. (1989) Neue cyclostome Bryozoen aus dem Untercenomanium von Mülheim-Broich (Westfalen). *Münstersche Forschungen zur Geologie und Paläontologie*, **69**, 87–113.

Voigt, E. (1991) Mono- or polyphyletic evolution of cheilostomatous bryozoan divisions? *Bulletin de la Société des Sciences Naturelles de l'Ouest de la France, Mémoire* HS, **1**, 502–22.

Voigt, E. and Flor, F. D. (1970) Homöomorphien bei fossilen cyclostomen Bryozoen, dargestellt am Beispiel der Gattung *Spiropora* Lamouroux 1821. *Mitteilungen Geologisch–Paläontologisches Institut der Universität Hamburg*, **39**, 7–96.

Voigt, E. and Hillmer, G. (1984) Oberkretazische Hippothoidae (Bryozoa Cheilostomata) aus dem Campanium von Schweden und dem Maastrichtium der Niederlande. *Mitteilungen Geologisch –Paläontologisches Institut der Universität Hamburg*, **54**, 169–208.

Voigt, E. and Walter, B. (1991) De possibles Crisiidae (Bryozoa–Articulata) dans le Néocomien du Jura Franco-Suisse. *Geobios*, **24**, 41–6.

Walter, B. (1972) Les bryozoaires neocomiens du Jura Suisse et Francais. *Geobios*, **5**, 277–354.

Walter, B. and Powell, H. P. (1973) Exceptional preservation in cyclostome Bryozoa from the Middle Lias of Northamptonshire. *Palaeontology*, **16**, 219–20.

Walter, B., Arnaud-Vanneau, A., Arnaud, H. *et al.* (1975) Les bryozoaires Barrémo-Aptiens du sud-est de la France. Gisements et paléoécologie, biostratigraphie. *Geobios*, **8**, 83–117.

Winston, J. E. (1983) Patterns of growth, reproduction and mortality in bryozoans from the Ross Sea, Antarctica. *Bulletin of Marine Science*, **33**, 688–702.

Xia, F.-S. (1987) Nippostenoporidae, a new family of Trepostomida (Bryozoa) from the late early Carboniferous of Zhanyi, eastern Yunnan. *Acta Micropalaeontologica Sinica*, **4**, 375–85 [in Chinese, with English summary].

Yang, J.-Z. and Lu, L.-H. (1979) *Araxopora* (Bryozoa) from the Lower Permian of southwest China. *Acta Palaeontologica Sinica*, **18**, 347–59 [in Chinese, with English abstract].

Yang, J.-Z. and Lu, L.-H. (1981) Geographical and stratigraphical distributions of Permian Bryozoa in China, in *Recent and Fossil Bryozoa* (eds G. P. Larwood and C. Nielsen), Olsen and Olsen, Fredensborg, pp. 305–7.

Yang, J.-Z., Hu, Z. X. and Xia, F.-S. (1988) Bryozoans from late Devonian and early Carboniferous of central Hunan. *Palaeontologica Sinica, New Series B*, **174** (23), 1–197 [in Chinese, with English summary].

Zhao, J.-K., Sheng, J.-Z., Yao, Z.-Q. *et al.* (1981) The Changhsingian and Permian–Triassic boundary of South China. *Bulletin of the Nanjing Institute of Geology and Palaeontology*, **2**, 1–85 [in Chinese, with English summary].

25

ECHINODERMATA

M. J. Simms, A. S. Gale, P. Gilliland, E. P. F. Rose and G. D. Sevastopulo

The overall classification of echinoderms used in this compilation is that of Smith (1984a) and Paul and Smith (1984). Details of the classification schemes used for particular groups are found at the start of the relevant sections. An understanding of the phylogeny of many echinoderm groups is far from clear. Of the five extant classes, echinoids are perhaps the best documented, both in terms of their representation in the fossil record and also how recently their phylogeny has been interpreted using cladistic methodology. Both asteroids and holothuroids have also been the subject of recent phylogenetic analyses. However, knowledge of the stratigraphical distribution of both groups has been hindered by their poor fossil record and, particularly in the case of the holothurians, problems inherent in identifying disarticulated material. Ophiuroids also have a rather poor fossil record, a problem further compounded by the inadequacy of the few classification schemes currently available. Crinoids represent another major area of uncertainty in echinoderm phylogenetics. Although they have a comparatively good fossil record, current classification schemes in no way can be considered to reflect their phylogeny on even the coarsest scale, and it is probable that only a minority of taxa below the level of order will, on further investigation, prove to be monophyletic.

Our understanding of the various extinct Palaeozoic echinoderm groups has benefited in particular from recent work by Andrew Smith and Chris Paul, although considerable areas of uncertainty still exist. The blastoids are one such group, which, although having been the subject of recent revision (Breimer and Macurda, 1972; Macurda, 1983; Horowitz *et al.*, 1986) are still essentially classified in terms of 'grade groups'.

It is evident from the following compilation that, although a great deal of recent work has been undertaken to elucidate echinoderm phylogenies, much still remains to be done and, in many cases, the data on first and last occurrences, compiled below, should not be considered to approximate in any way to real instances of origination or extinction. Perhaps the most valuable aspect of this exercise, therefore, is to highlight those areas most in need of further work. Some families of characteristically deep-water echinoids have been noted as such.

Sections are attributable to the contributors as follows: M. J. Simms – cystoids, blastoids (with GDS), crinoids (excluding disparids), edrioasteroids, ophiocistioids, ophiuroids, miscellaneous Lower Palaeozoic plesions; A. S. Gale – asteroids; P. Gilliland – holothurians (excluding ophiocistioids); E. P. F. Rose – echinoids; G. D. Sevastopulo – blastoids (with MJS), disparid crinoids.

Acknowledgements – We thank Hans Hagdorn, Clare Milsom, Chris Paul and Andrew B. Smith for assistance with certain sections.

Phylum ECHINODERMATA Bruguière, 1791

Plesion (F.) HELICOPLACIDAE Durham, 1963 Є. (CRF) Mar. (see Fig. 25.1)

First and Last: *Helicoplacus gilberti* Durham and Caster, 1963, *H. curtisi* Durham and Caster, 1963, *H. everndeni* Durham, 1967, *H. nelsoni* (Durham, 1967) and *Polyplacus kilmeri* Durham, 1967, *Nevadella* Zone, California, Nevada and Alberta.

Plesion (Genus) *CAMPTOSTROMA* Ruedemann, 1933 Є. (CRF) Mar.

First and Last: *Camptostroma roddyi* Ruedemann, 1933, Kinzer Formation, Pennsylvania, USA.

Comment: Paul and Smith (1984) regard *Camptostroma* as the latest common ancestor of pelmatozoans and eleutherozoans.

Subphylum PELMATOZOA Leuckart, 1848

Plesion (F.) LEPIDOCYSTIDAE Durham, 1967 Є. (CRF) Mar.

First and Last: *Lepidocystis wanneri* Foerste, 1938 and *Kinzercystis durhami* Sprinkle, 1973, *Olenellus* Zone, Kinzer Formation, Pennsylvania, USA.

Comment: Paul and Smith consider the lepidocystids to lie close to the common ancestry of both crinoids and cystoids.

The Fossil Record 2. Edited by M. J. Benton. Published in 1993 by Chapman & Hall, London. ISBN 0 412 39380 8

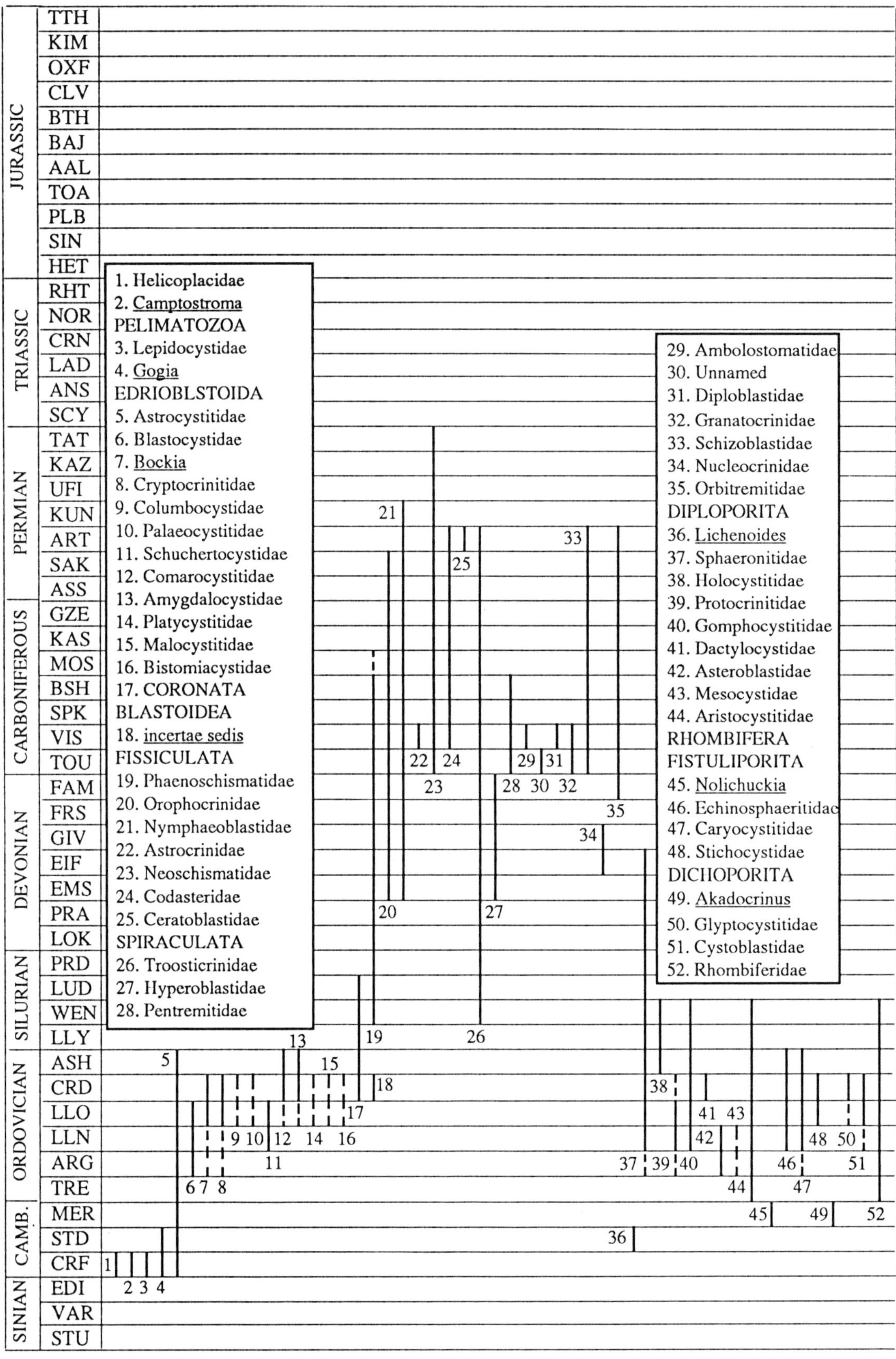
JURASSIC
TTH
KIM
OXF
CLV
BTH
BAJ
AAL
TOA
PLB
SIN
HET
TRIASSIC
RHT
NOR
CRN
LAD
ANS
SCY
PERMIAN
TAT
KAZ
UFI
KUN
ART
SAK
ASS
CARBONIFEROUS
GZE
KAS
MOS
BSH
SPK
VIS
TOU
DEVONIAN
FAM
FRS
GIV
EIF
EMS
PRA
LOK
SILURIAN
PRD
LUD
WEN
LLY
ORDOVICIAN
ASH
CRD
LLO
LLN
ARG
TRE
CAMB.
MER
STD
CRF
SINIAN
EDI
VAR
STU
1. Helicoplacidae
2. Camptostroma
PELIMATOZOA
3. Lepidocystidae
4. Gogia
EDRIOBLSTOIDA
5. Astrocystitidae
6. Blastocystidae
7. Bockia
8. Cryptocrinitidae
9. Columbocystidae
10. Palaeocystitidae
11. Schuchertocystidae
12. Comarocystitidae
13. Amygdalocystidae
14. Platycystitidae
15. Malocystitidae
16. Bistomiacystidae
17. CORONATA
BLASTOIDEA
18. incertae sedis
FISSICULATA
19. Phaenoschismatidae
20. Orophocrinidae
21. Nymphaeoblastidae
22. Astrocrinidae
23. Neoschismatidae
24. Codasteridae
25. Ceratoblastidae
SPIRACULATA
26. Troosticrinidae
27. Hyperoblastidae
28. Pentremitidae
29. Ambolostomatidae
30. Unnamed
31. Diploblastidae
32. Granatocrinidae
33. Schizoblastidae
34. Nucleocrinidae
35. Orbitremitidae
DIPLOPORITA
36. Lichenoides
37. Sphaeronitidae
38. Holocystitidae
39. Protocrinitidae
40. Gomphocystitidae
41. Dactylocystidae
42. Asteroblastidae
43. Mesocystidae
44. Aristocystitidae
RHOMBIFERA
FISTULIPORITA
45. Nolichuckia
46. Echinosphaeritidae
47. Caryocystitidae
48. Stichocystidae
DICHOPORITA
49. Akadocrinus
50. Glyptocystitidae
51. Cystoblastidae
52. Rhombiferidae

Fig. 25.1

Plesion (Superclass) CYSTOIDEA Von Buch, 1846

The overall classification of cystoids adopted here is based on Paul (1988) and Paul and Smith (1984).

Plesion (Genus) *GOGIA* Walcott, 1917
Є. (CRF–STD) Mar.

First: *Gogia (Alanisicystis) andalusiae* Ubaghs and Vizcaïno, 1990, upper part of Marianian Stage, Sierra Morena Oriental, Andalusia, Spain.
Last: *Gogia gondi* Ubaghs, 1987, upper Middle Cambrian, *Paradoxides mediterraneus* Zone, Montagne Noire, France.
Comment: Paul (1988) considers *Gogia* to represent the latest common ancestor of the main cystoid groups.

Order EDRIOBLASTOIDA Fay, 1962

Plesion (F.) ASTROCYSTITIDAE Bassler, 1938
Є. (MER)–O. (HIR) Mar.

First: *Cambroblastus enubilatus* Smith and Jell, 1990, Median Upper Cambrian (Franconian), Chatsworth Limestone, western Queensland, Australia.
Last: Undescribed deltoid plates, Boda Limestone, Sweden (C. R. C. Paul cited in Smith and Jell, 1990).
Comment: The relationship of edrioblastoids to other cystoids remains obscure (Paul, 1988).

Plesion (F.) BLASTOCYSTIDAE Jaekel, 1918
('Parablastoids') O. (ARG1–LLO) Mar.

First: *Blastoidocrinus antecedens* Paul and Cope, 1982, Bolahaul Member, Ogof Hen Formation, Llangynog, Dyfed, South Wales, UK.
Last: *Blastoidocrinus carchariaedens* Billings, 1859, Aylmer Formation, Montreal, Quebec, Canada and Valcour Limestone, New York, USA.

Order UNNAMED HIGHER TAXON COMPRISING CRYPTOCRINOIDS, PARACRINOIDS, CORONATES AND BLASTOIDS

Paul (1988) considers each of these groups to be monophyletic and to belong to a larger monophyletic group with *Bockia* as its latest common ancestor.

Plesion (Genus) *BOCKIA* Gekker, 1938
O. (ARG2?/LLN–L. CRD) Mar.

First: *Bockia*? sp. A, *Asaphus* Marl, Asker, Oslo region, Sweden (Bockelie, 1981). The earliest definite record is *Bockia mirabilis* Bockelie, 1981, *Ogygiocaris* Shale, Regnstrand, Oslo region, Sweden.
Last: *Bockia neglecta* Gekker, 1938, lower Caradoc, eastern Baltic.

'*Superfamily* CRYPTOCRINIDA' Bassler, 1938

There has been little recent work on this 'superfamily'.

F. CRYPTOCRINITIDAE Bassler, 1938
O. (ARG2?/LLN/LLO–CRD) Mar.

First: *Cryptocrinites ? similis* Bockelie, 1981, *Asaphus expansus* Shale, Vaekerø Farm, Oslo, Sweden. The earliest definite record is *Cryptocrinites laevis* (Pander, 1830), from the *Echinosphaerites* Limestone Pulkova, near Leningrad, former USSR.
Last: *Foerstecystis obliqua* Bassler, 1950, upper Llandeilo/ lower Caradoc, Tennessee, USA.

F. COLUMBOCYSTIDAE Bassler, 1950
O. (LLO3/CRD) Mar.

First and Last: *Columbocystis typica* Bassler, 1950 and *Springerocystis longicollis* Bassler, 1950, upper Llandeilo/ lower Caradoc, Benbolt Formation, Tennessee, USA.

F. PALAEOCYSTITIDAE Ubaghs, 1968
O. (LLO–LLO3/CRD) Mar.

First: *Palaeocystites tenuiradiatus* (Hall, 1847) and *P. dawsoni* Billings, 1858, Llandeilo, New York, USA and Canada.
Last: *Bromidocystis sinclairi* Sprinkle, 1982, Pooleville Member, Bromide Formation, Oklahoma, USA.

F. SCHUCHERTOCYSTIDAE Bassler, 1950
O. (LLN/LLO) Mar.

First and Last: *Schuchertocystis radiata* Bassler, 1950, Middle Ordovician, Tennessee, USA.

'*Superfamily* PARACRINOIDA' Regnéll, 1945

F. COMAROCYSTITIDAE Bather, 1899
O. (LLO3/CRD–RAW) Mar.

First: *Sinclairocystis praedicta* Bassler, 1950, *Bromidocystis* Bed, Bromide Formation, Oklahoma, USA.
Last: *Comarocystites* or *Sinclairocystis* sp., Starfish Bed, Threave Glen, Girvan, Scotland, UK (Paul, 1966).

F. AMYGDALOCYSTIDAE Jaekel, 1900
O. (LLO3/CRD–CRD/ASH) Mar.

First: *Oklahomacystis bibrachiatus* Parsley, 1982 and *O. tribrachiatus* (Bassler, 1943), Upper Echinoderm Zone, Bromide Formation, Oklahoma, USA.
Last: *Achradocystites grewingkii* Von Volborth, 1870, Upper Ordovician, Estonia, former USSR.

F. PLATYCYSTITIDAE Parsley and Mintz, 1975
O. (LLO3/CRD) Mar.

First: *Platycystites levatus* Bassler, 1943, Lower Echinoderm Zone, Bromide Formation, Oklahoma, USA.
Last: *Platycystites cristatus* Bassler, 1943, Upper Echinoderm Zone, Bromide Formation, Oklahoma, USA.

F. MALOCYSTITIDAE Bather, 1899
O. (LLO–LLO3/CRD) Mar.

First: *Malocystites murchisoni* Billings, 1857, *Canadocystis barrandei* Billings, 1858 and *C. emmonsi* Hudson, 1905, LLO of USA and Canada.
Last: *Wellerocystis kimmswickensis* Foerste, 1920, Upper LLO/Lower CRD, Missouri, USA.

F. BISTOMIACYSTIDAE Sprinkle and Parsley, 1982
O. (LLO3/CRD) Mar.

First and Last: *Bistomiacystis globosa* Sprinkle and Parsley, 1982, Lower Echinoderm Zone, Bromide Formation, Oklahoma, USA.

Plesion (Subclass) CORONATA Jaekel, 1918
O. (L. CRD)–S. (LUD) Mar.

The Coronata have been reviewed recently by Donovan and Paul (1985) and by Brett *et al.* (1983).
First: *Mespilocystites bohemicus* Barrande, 1887, Letna Formation, Trubsko, Czechoslovakia.
Last: *Stephanoblastus mirus* (Barrande, 1887), Kopanina Formation, Lodenice, Czechoslovakia.

Comment: Material identical to *Stephanoblastus angulatus* Conrad, 1842 has been described from the Ludlovian of Kashmir, but this record (Gupta and Webster, 1971) is now considered unreliable.

Class BLASTOIDEA Say, 1825

F. UNNAMED O. (CRD) Mar.

First and Last: *Macurdablastus uniplicatus* Broadhead, 1984, Caradoc, Benbolt Formation, Tennessee, USA.
Comment: Poor preservation of this material precludes ordinal or familial assignment of this species.

Order FISSICULATA Jaekel, 1918

The fissiculate blastoids have been comprehensively reviewed by Breimer and Macurda (1972) and by Macurda (1983). Their classification scheme is adopted here. However, since the spiraculate blastoids are considered to have had a polyphyletic origin from several different fissiculate ancestors, then the fissiculate families in question must necessarily be paraphyletic. Hence the current classification of blastoids cannot be considered to reflect their phylogeny.

Many blastoids were described from Timor, Indonesia, particularly from the Basleo Beds, Sonnebait 'Series'. Although the Basleo faunas have been regarded as Upper Permian in age, recent information (Webster, 1987, 1990) strongly suggests that they are of Artinskian or even Sakmarian age.

F. PHAENOSCHISMATIDAE Etheridge and Carpenter, 1886 S. (WEN)–C. (VRK?) Mar.

First: *Decaschisma pulchellum* (Miller and Dyer, 1878), Waldron Shale, Indiana.
Last: Undescribed phaenoschismatids A and B, Gene Autry Formation, Oklahoma, USA.

F. OROPHOCRINIDAE Jaekel, 1918 D. (EMS)–P. (SAK) Mar.

First: *Brachyschisma ?oostheizeni* Breimer and Macurda, 1972, Bokkeveld Beds, South Africa.
Last: *Anthoblastus stelliformis* Wanner, 1924, Basleo Beds, Timor, Indonesia.

F. NYMPHAEOBLASTIDAE Wanner, 1940 D. (EMS)–P. (ROT) Mar.

First: *Pachyblastus dicki* Breimer and Macurda, 1972, Bockeveld Beds, South Africa and Belen Formation, Bolivia.
Last: *Sphaeroschisma somoholense* Wanner, 1924, Lower Permian, Somohole Beds, Timor, Indonesia.

F. ASTROCRINIDAE Austin and Austin, 1843 C. (ASB–BRI) Mar.

First: *Astrocrinus* sp., Meenymore Formation, Republic of Ireland (Waters and Sevastopulo, 1984).
Last: *Astrocrinus tetragonus* (Austin and Austin, 1843), Charlestown Main Limestone, Scotland, and other formations of the same age in England, UK and the Republic of Ireland.

F. NEOSCHISMATIDAE Wanner, 1940 C. (HAS)–P. (ZEC) Mar.

First: *Hadroblastus blairi* (Miller and Gurley, 1895), Chouteau Limestone, Missouri, USA.
Last: Fragmentary neoschismatid, Oxtrack Formation, Bowen Basin, Queensland, Australia (Breimer and Macurda, 1972).

F. CODASTERIDAE Etheridge and Carpenter, 1886 C. (BRI)–P. (ART) Mar.

First: *Codaster acutus* M'Coy, 1849, Middle Limestone, Yorkshire, and other formations in England, UK and the Republic of Ireland.
Last: *Angioblastus variabilis* Wanner, 1931, *Pteroblastus brevialatus* Wanner, 1931, *P. decemcostis* Wanner, 1931, *P. ferrugineus* Wanner, 1940, *P. gracilis* Wanner 1924, *Nannoblastus cuspidatus* Wanner, 1924, and *N. pyramidalis* Wanner, 1924, Basleo Beds, Timor, Indonesia.

F. CERATOBLASTIDAE Breimer and Macurda, 1972 P. (ART) Mar.

First and Last: *Ceratoblastus nanus* Wanner, 1940, Basleo Beds, Timor, Indonesia.

Order SPIRACULATA Jaekel, 1918

The spiraculates are no longer considered a natural taxon, and may have had a polyphyletic origin from at least seven fissiculate ancestors (Horowitz *et al.*, 1986). The classification adopted here is essentially that used in the *Treatise on Invertebrate Paleontology*, with modifications based on Horowitz *et al.* (1986).

F. TROOSTICRINIDAE Bather, 1899 S. (WEN)–P. (ART) Mar.

First: *Troosticrinus reinwardti* (Troost, 1835), Laurel Limestone, Indiana, USA.
Last: *Rhopaloblastus timoricus* Wanner, 1916, Basleo Beds, Timor, Indonesia.

F. HYPEROBLASTIDAE Fay, 1964 D. (EMS–FAM) Mar.

First: *Conuloblastus malladai* (Etheridge and Carpenter, 1893), Santa Lucia Formation, Spain.
Last: *Petaloblastus ovalis* (Goldfuss, 1821), Übergangskalke, Ratingen, Germany, or *Petaloblastus boletus* (Schmidt, 1930), Isenbugel, Germany.

F. PENTREMITIDAE d'Orbigny, 1851 C. (HAS–CHE/MEL?) Mar.

First: *Strongyloblastus petalus* Fay, 1962, Banff Formation, Canada, or *Strongyloblastus breimeri* Sprinkle and Gutschick, 1990, and *Strongyloblastus* sp., Sprinkle and Gutschick, 1990, Lodgepole Limestone, Montana, USA.
Last: *Pentremites rusticus* Hambach, 1903, Bloyd Formation, Arkansas, and equivalent formations in Oklahoma, USA.

F. AMBOLOSTOMATIDAE Horowitz *et al.*, 1986 C. (ARU) Mar.

First and Last: *Ambolostoma baileyi* Peck, 1930, Brazer Formation, Great Blue Limestone, Utah, USA.

UNNAMED FAMILY Horowitz *et al.*, 1986 C. (HAS) Mar.

First and Last: *'Pentremoblastus' subovalis* Fay and Koenig, 1963, McCraney Limestone, Missouri, USA.

The remaining five families of spiraculate blastoid have yet to be revised and reinterpreted, hence their validity as natural taxonomic groups remains uncertain. The classification used below is that of Fay (1967) as in the *Treatise on Invertebrate Paleontology*.

F. DIPLOBLASTIDAE Fay, 1964
C. (ASB–BRI) Mar.

First: *Diploblastus kirkwoodensis* (Shumard, 1863), St Louis Limestone, Missouri, USA.
Last: *Nodoblastus librovitchi* (Yakovlev, 1941), Namurian, Daibar Mountain, Aktyubinsk Province, Kazakhstan, former USSR.

F. GRANATOCRINIDAE Fay, 1961
C. (HAS–BRI) Mar.

First: *Tanaoblastus roemeri* (Shumard, 1855) and *Poroblastus chouteauensis* (Peck, 1930), Chouteau Limestone, Missouri, or *Tanaoblastus haynesi* Sprinkle and Gutschick, 1990, Lodgepole Limestone, Montana, USA.
Last: *Heteroblastus cumberlandi* Etheridge and Carpenter, 1886, 'Yoredale Series', Northumberland, England, UK.

F. SCHIZOBLASTIDAE Etheridge and Carpenter, 1886 C. (IVO)–P. (ART) Mar.

First: *Schizoblastus aplatus* (Rowley and Hare, 1891), lower Burlington Limestone, USA.
Last: *Deltoblastus elongatus* (Wanner, 1940) and other species of *Deltoblastus*, Basleo and Amarissi Beds, Timor, Indonesia.

F. NUCLEOCRINIDAE Bather, 1899
D. (EIF–GIV) Mar.

First: *Elaeacrinus verneuili* Roemer, 1851 and *E. venustus* (Miller and Gurley, 1895), Onondaga Limestone and equivalents, east and mid-west USA.
Last: *Nucleocrinus elegans* Conrad, 1842, Tully Limestone, New York, USA.

F. ORBITREMITIDAE Bather, 1899
D. (FAM)–P. (ART) Mar.

First: *Doryblastus melonianus* (Schmidt, 1930), Velbert, Germany.
Last: *Orbitremites malaianus* Wanner, 1916, Basleo Beds, Timor, Indonesia.

DIPLOPORITE CYSTOIDS

Paul and Smith (1984) and Paul (1988) consider the diploporite cystoids to be paraphyletic or even polyphyletic. The relationship of the stemless sphaeronitid diploporites to other cystoids is reasonably well understood, but the phylogenetic position of all other diploporites, which encompass both stemmed and stemless forms, remains unknown. They are grouped together here solely for convenience.

Class DIPLOPORITA Müller, 1854

Superfamily SPHAERONITIDA Neumayr, 1889

Plesion (Genus) *LICHENOIDES* Barrande, 1887
€. (STD) Mar.

First and Last: *Lichenoides priscus* Barrande, 1887, upper Middle Cambrian, Jince Formation, Czechoslovakia.
Comment: Paul (1988) considers *Lichenoides* to be sister group to all sphaeronitids.

F. SPHAERONITIDAE Neumayr, 1889
O. (ARG/LLN1)–D. (EIF) Mar.

First: *Sphaeronites minor* Paul and Bockelie, 1983, lower *Asaphus* Marl, Kinnekulle Mountain, Västergötland, Sweden.
Last: *Eucystis flavus* (Barrande, 1887), Azzel Mathi, central Sahara, Morocco. Other representatives of this genus are found in equivalent strata in France, Algeria and Czechoslovakia.
Comment: The Sphaeronitidae are entirely unknown from Silurian strata.

F. HOLOCYSTITIDAE Miller, 1889
O. (RAW)–S. (WEN) Mar.

First: *Brightonicystis gregaria* Paul, 1971, Cystoid Limestone, Cautley District, Yorkshire, England, UK.
Last: *Holocystites alternatus* (Hall, 1864), Racine Dolomite, Wisconsin, USA.

NON-SPHAERONITID DIPLOPORITES

Paul (1988) recognizes three major groups (superfamilies) among the non-sphaeronitid diploporites.

Superfamily PROTOCRINITIDA Bather, 1899

F. PROTOCRINITIDAE Bather, 1899
O. (ARG?/LLN2–LLO3/CRD) Mar.

First: *Protocrinites fragrum* (Eichwald, 1840), Duboviki Formation, Pulkowa and Zarskoje Selo, former USSR. Bockelie (1984, p. 6) states that *Protocrinites* is found as early as the Arenig in Estonia, former USSR.
Last: *Eumorphocystis multiporata* Branson and Peck, 1940, lower Echinoderm Zone, Bromide Formation, Oklahoma, USA.

F. GOMPHOCYSTITIDAE Miller, 1889
O. (LLN)–S. (WEN) Mar.

First: *Pyrocystites pirum* Barrande, 1887, Wosek, Bohemia, Czechoslovakia.
Last: *Gomphocystites bownockeri* Foerste, 1920, Cedarville Dolomite, Ohio, USA.

F. DACTYLOCYSTIDAE Jaekel, 1899
O. (M.–U. CRD) Mar.

First: *Dactylocystis schmidti* Jaekel, 1899, and *Estonocystis antropoffi* Jaekel, 1918, *Hemicosmites* Beds, Reval, Estonia, former USSR.
Last: *Revalocystis mickwitzi* (Jaekel, 1899), Wesenberg Formation, Reval, Estonia, former USSR.

Superfamily ASTEROBLASTIDA Bather, 1900

F. ASTEROBLASTIDAE Bather, 1900
O. (ARG–LLN) Mar.

First: *Asteroblastus sublaevis* Jaekel, 1899, Walchow Formation, Pulkowa, former USSR.
Last: *Metasterocystis micropelta* Jaekel, 1918, lower LLN, Kunda Formation, Estonia and rest of former USSR.

F. MESOCYSTIDAE Bather, 1899
O. (ARG/LLN1) Mar.

First and Last: *Mesocystis pusirefskii* (Hoffman, 1866), Walchow or Kunda Formation, Leningrad, former USSR.

Superfamily ARISTOCYSTITIDA Neumayr, 1889

F. ARISTOCYSTITIDAE Neumayr, 1889
O. (TRE)–S. (WEN) Mar.

First: *Calix dorecki* Sdzuy, 1955, Leimitz-Schiefern, Frankenwald, Germany.
Last: *'Holocystites' gyrinus* Miller and Gurley, Racine Dolomite, Wisconsin, USA.

Class RHOMBIFERA Zittel, 1879

Paul and Smith (1984) and Paul (1988) recognize three distinct groups of rhomb-bearing cystoids. However, they consider that the rhombs in glyptocystitid rhombiferans and in fistuliporite rhombiferans evolved independently, and hence these two groups are not closely related. The relationship of hemicosmitid rhombiferans to other cystoids remains even less clear. The Order Rhombifera, as recognized in the *Treatise*, clearly is polyphyletic, but is retained here for the sake of convenience.

Order FISTULIPORITA Paul, 1968

Plesion (Genus) *NOLICHUCKIA* Sprinkle, 1973
€. (MER) Mar.

First and Last: *Nolichuckia casteri* Sprinkle, 1973, Dresbachian, *Cedaria* Zone, Nolichucky Formation, Virginia and Tennessee, USA.
Comment: Paul and Smith (1984) and Paul (1988) regard *Nolichickia* as primitive sister group to all fistuliporite rhombiferans, derived from *Gogia*.

F. ECHINOSPHAERITIDAE Neumayr, 1889
O. (LLN1–RAW) Mar.

First: *Echinosphaerites aurantium* Gyllenhaal, 1772, Kunda Formation, Reval, Estonia, former USSR. Bockelie (1981) reports that *E. aurantium* may also occur in the Upper Arenig of Estonia, former USSR.
Last: *Echinosphaerites* sp. Swindale Limestone, Knock, Cumbria, England, UK (Paul, 1967).

F. CARYOCYSTITIDAE Jaekel, 1918
O. (ARG?/LLN1–HIR) Mar.

First: *Heliocrinites echinoides* Leuchtenberg, Walchow Formation, Pulkowa and Zarskoje Selo, former USSR. The precise age of this record is uncertain. The first definite record is *Heliocrinites araneus* (von Schlotheim) and *H. balticus* (Eichwald) from the Kunda Formation, Reval, Estonia, former USSR.
Last: *Heliocrinites stellatus* Regnéll, 1945, and *H. variabilis* Regnéll, 1945, Boda Limestone, Dalarna, Sweden.

F. STICHOCYSTIDAE Jaekel, 1918
O. (LLO–M. CRD) Mar.

First: *Stichocystis* sp., Shihtzepu Shale, Kweichou, China (Sun, 1936).
Last: *Stichocystis geometrica* (Angelin, 1878), Kullsberg Limestone, Dalarna, Sweden.

Order DICHOPORITA Jaekel, 1899

Superfamily GLYPTOCYSTITIDA Bather, 1899

Plesion (Genus) *AKADOCRINUS* Sprinkle, 1973
€. (STD) Mar.

First and Last: *Akadocrinus jani* Prokop, 1962 and *A. nuntius* Prokop, 1962, upper Middle Cambrian, Jince Formation, Czechoslovakia.
Comment: Paul (1988) considers *Akadocrinus* to be derived from *Gogia* and to be primitive sister group to all Glyptocystitida.

F. GLYPTOCYSTITIDAE Bather, 1899
O. (LLO3/COS–SOU) Mar.

First: *Glyptocystella loeblichi* (Bassler, 1943), *Hesperocystis deckeri* Sinclair, 1945, *Pirocystella strimplei* Sprinkle, 1982, *P. cooki* Sprinkle, 1982, and *P. bassleri* Sprinkle, 1982, lower Echinoderm Zone, Bromide Formation, Oklahoma, USA.
Last: *Glyptocystites multipora* Billings, 1854, lower Trenton, Kirkfield Formation, central Ontario, Canada.

F. CYSTOBLASTIDAE Jaekel, 1899
O. (LLN/LLO–CRD) Mar.

First: *Cystoblastus leuchtenbergi* Volborth, 1867, *Echinosphaerites* Limestone, Pavlosk, former USSR.
Last: *Cystoblastus kokeni* Jaekel, 1899, Kuckers Shale, Kuckers, Estonia, former USSR.

F. RHOMBIFERIDAE Kesling, 1962
O. (CRD) Mar.

First and Last: *Rhombifera bohemica* Barrande, 1887, CRD, Wraz, Bohemia, Czechoslovakia.

F. CHEIROCRINIDAE Jaekel, 1899
O. (TRE)–S. (GLE/GOR) Mar. (see Fig. 25.2)

First: *Cheirocystella antiqua* Paul, 1972, upper TRE, Fillmore Limestone, Utah, USA.
Last: *Cheirocrinus tertius* Barrande, 1887, upper WEN/lower LUD, Lodenitz and Sedlitz, Bohemia, Czechoslovakia.
Comment: Paul (1968) has described an isolated pectinirhomb plate, possibly a cheirocrinid, from the uppermost Cambrian, Whipple Cave Formation, of Sawmill Canyon, Nevada, USA.

F. ECHINOENCRINITIDAE Bather, 1899
O. (ARG)–S. (GLE) Mar.

First: *Echinoencrinites angulosus* (Pander, 1830), upper ARG, Walchow Formation, Leningrad, former USSR.
Last: *Glansicystites baccata* (Forbes, 1848), Wenlock Limestone, Walsall, Dudley, England, UK (Paul, 1966).

F. PLEUROCYSTITIDAE Neumayr, 1889
O. (LLO_3/COS)–D. (EIF) Mar.

First: Two undescribed species, Ninemile Shale, Nevada, and Fillmore Formation, Utah, USA (Sprinkle, 1990).
Last: *Regulaecystis* sp., Shales at Triangle Point, Torquay, Devon, England, UK (Paul, 1967).

F. CALLOCYSTITIDAE Bernard, 1895
O. (ASH)–D. (FRS) Mar.

First: *Lepadocystis moorei* (Meek, 1871), Elkhorn Formation, Indiana and Ohio, USA.
Last: *Strobilocystis schucherti* Thomas, 1924, Shell Rock Formation, Nora Springs, Iowa, USA.

F. MACROCYSTELLIDAE Bather, 1899
O. (TRE–CRD) Mar.

First: *Macrocystella* cf. *mariae* Callaway, 1877, *Dictyonema flabelliforme* Zone, Moel Llyfnant, Gwynedd, Wales, UK (Paul, 1984).
Last: *Mimocystites bohemicus* Barrande, 1887, Letna Beds, Bohemia, Czechoslovakia.
Comment: Paul (1988) considers *Macrocystella* to be primitive sister group to all glyptocystitids with pectinirhombs. The earliest representative of the *Macrocystella*–glypto-

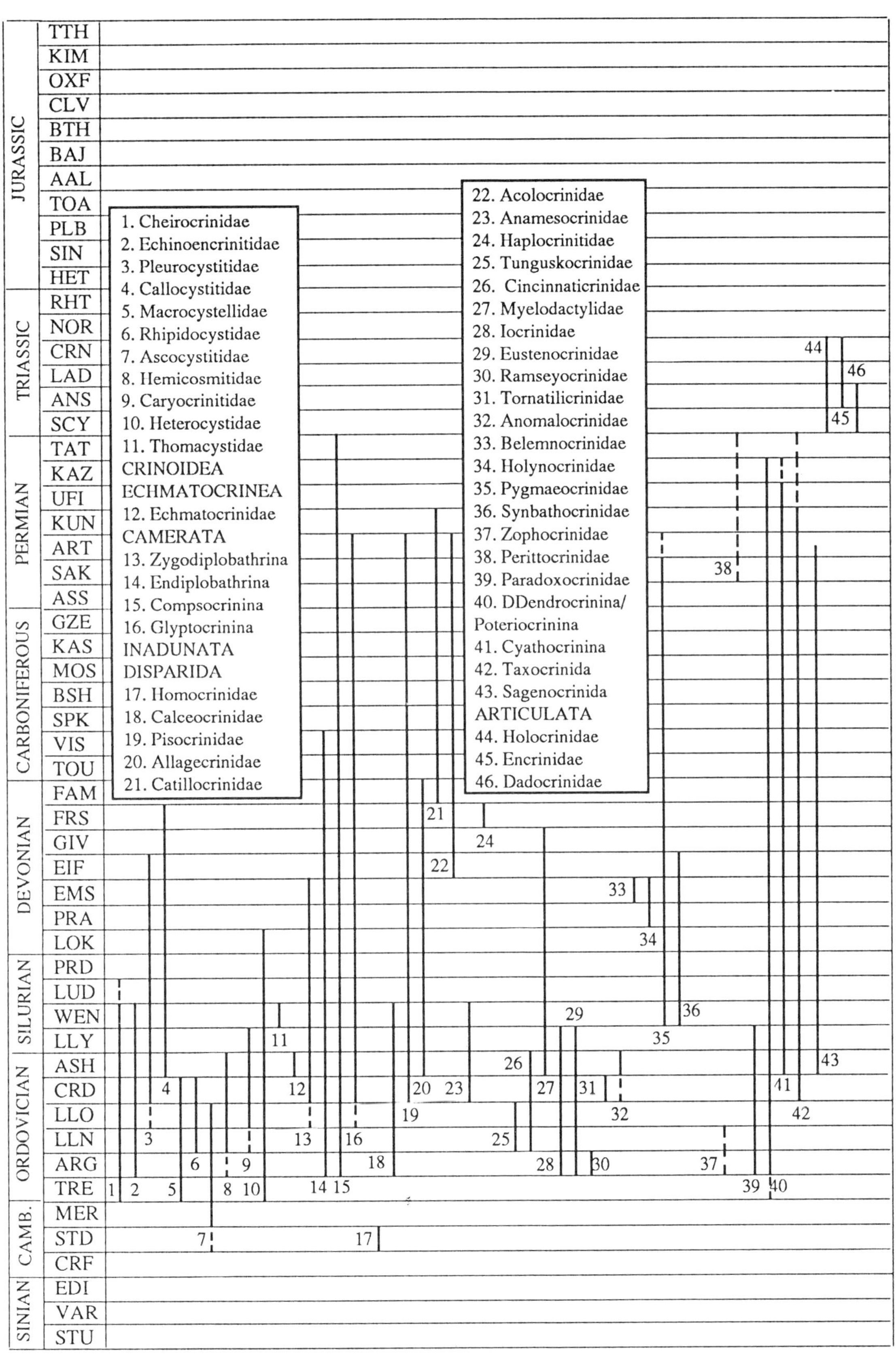
JURASSIC
TTH
KIM
OXF
CLV
BTH
BAJ
AAL
TOA
PLB
SIN
HET
TRIASSIC
RHT
NOR
CRN
LAD
ANS
SCY
PERMIAN
TAT
KAZ
UFI
KUN
ART
SAK
ASS
CARBONIFEROUS
GZE
KAS
MOS
BSH
SPK
VIS
TOU
DEVONIAN
FAM
FRS
GIV
EIF
EMS
PRA
LOK
SILURIAN
PRD
LUD
WEN
LLY
ORDOVICIAN
ASH
CRD
LLO
LLN
ARG
TRE
CAMB.
MER
STD
CRF
SINIAN
EDI
VAR
STU
1. Cheirocrinidae
2. Echinoencrinitidae
3. Pleurocystitidae
4. Callocystitidae
5. Macrocystellidae
6. Rhipidocystidae
7. Ascocystitidae
8. Hemicosmitidae
9. Caryocrinitidae
10. Heterocystidae
11. Thomacystidae
CRINOIDEA
ECHMATOCRINEA
12. Echmatocrinidae
CAMERATA
13. Zygodiplobathrina
14. Endiplobathrina
15. Compsocrinina
16. Glyptocrinina
INADUNATA
DISPARIDA
17. Homocrinidae
18. Calceocrinidae
19. Pisocrinidae
20. Allagecrinidae
21. Catillocrinidae
22. Acolocrinidae
23. Anamesocrinidae
24. Haplocrinitidae
25. Tunguskocrinidae
26. Cincinnaticrinidae
27. Myelodactylidae
28. Iocrinidae
29. Eustenocrinidae
30. Ramseyocrinidae
31. Tornatilicrinidae
32. Anomalocrinidae
33. Belemnocrinidae
34. Holynocrinidae
35. Pygmaeocrinidae
36. Synbathocrinidae
37. Zophocrinidae
38. Perittocrinidae
39. Paradoxocrinidae
40. DDendrocrinina/
Poteriocrinina
41. Cyathocrinina
42. Taxocrinida
43. Sagenocrinida
ARTICULATA
44. Holocrinidae
45. Encrinidae
46. Dadocrinidae

Fig. 25.2

cystitid clade is *Cambrocrinus regularis* Orlowski, 1968, from the Upper Cambrian, Dresbachian, *Olenus* Beds, Holy Cross Mountains, Poland.

F. RHIPIDOCYSTIDAE Jaekel, 1901
O. (LLN–COS) Mar.

First: *Lingulocystis boliviensis* Sprinkle, 1973b, Sella Mendez Province, south Bolivia.
Last: *Rhipidocystis norvegica* Bockelie, 1981, *Coelosphaeridium* Beds, Furuberg Formation, Oslo, Norway.
Comment: Paul (1988) considers the rhipidocystids to represent one of two monophyletic clades, which together constitute the sister group to the *Cambrocrinus*–glyptocystitid clade.

F. ASCOCYSTITIDAE Ubaghs, 1967
€. (STD/MER)–O. (LLN/LLO) Mar.

First: *Pareocrinus ljubzovi* Yakovlev, 1956, upper Middle Cambrian or lower Upper Cambrian, east Siberia, former USSR.
Last: *Ascocystites drabowensis* Barrande, 1887, Middle Ordovician, Bohemia, Czechoslovakia.
Comment: The Ascocystitidae represent the second of the two monophyletic clades mentioned above and are sister group to the Rhipidocystidae.

Superfamily HEMICOSMITIDA Jaekel, 1918

The relationship of hemicosmitids to other cystoids is unclear, although Paul (1988) suggests that their origin may lie among the Coronata.

F. HEMICOSMITIDAE Jaekel, 1918
O. (ARG/LLN1–HIR) Mar.

First: *Hemicosmites* ? sp. A, Orthoceratid Limestone, Vaekkerø, Norway (Bockelie, 1979).
Last: *Hemicosmites sculptus* Bockelie, 1979, 'Stage' 5b 'reef' limestone, Ringerike, Oslo, Norway.

F. CARYOCRINITIDAE Bernard, 1895
O.–D. (LOK) Mar.

First: *Caryocrinites aurora* Bather, 1906, *C. turbo* Bather, 1906, *C. avellana* Bather, 1906, Ordovician, Naungkangyi, Burma.
Last: *Stribalocystites elongatus* Rowley, 1900, Bailey Limestone, Montana, USA.

F. HETEROCYSTIDAE Jaekel, 1918
S. (WEN) Mar.

First and Last: *Heterocystites armatus* Hall, 1851, Shelby Dolomite, Lockport, New York, USA.

F. THOMACYSTIDAE Paul, 1969
O. (RAW) Mar.

First and Last: *Thomacystis tuberculata* Paul, 1969, Rhiwlas Limestone, Bala, Gwynedd, Wales, UK.

Class CRINOIDEA Miller, 1821

The most recent classification of the Crinoidea, as used in the *Treatise on Invertebrate Paleontology*, is far from satisfactory, and cannot be considered to reflect the phylogeny of the group on even the coarsest scale. Of the five presently accepted subclasses, the Echmatocrinea cannot even be regarded with any certainty as crinoids, while the Inadunata is an obviously paraphyletic taxon incorporating the stem groups of several major crinoid clades. The Flexibilia and Articulata appear to represent monophyletic clades, whose origins lie among the 'inadunates', while the Camerata may represent another monophyletic clade perhaps sharing a common ancestry with one of the early 'inadunates', or else derived from them.

Subclass ECHMATOCRINEA Sprinkle and Moore, 1978

Order ECHMATOCRINIDA Sprinkle and Moore, 1978

F. ECHMATOCRINIDAE Sprinkle, 1973
€. (STD) Mar.

First and Last: *Echmatocrinus brachiatus* Sprinkle, 1973, Burgess Shale, British Columbia, Canada.
Comment: The phylogenetic position of this species remains uncertain and it is questionable whether it should even be referred to the Echinodermata.

Subclass CAMERATA Wachsmuth and Springer, 1885

Within the Camerata, it has been assumed that the orders Monobathrida and Diplobathrida are monophyletic, although monophyly is far less probable at subordinal level and below. Camerate suborders are differentiated primarily on the basis of the arrangement of plates in the cup, features which may well have arisen more than once through convergence, rather than on the presence of particular autapomorphies. Hence they are unlikely to represent natural taxa. At still lower taxonomic levels, it is probable that a few, highly distinctive, families represent monophyletic clades but the validity of many others as natural groupings is highly questionable. In consequence of this, the compilation below has been taken only to subordinal level.

Order DIPLOBATHRIDA Moore and Laudon, 1943

Suborder ZYGODIPLOBATHRINA Ubaghs, 1953
O. (LLO3/CRD)–D. (EMS) Mar.

First: *Cleiocrinus bromidensis* Kolata, 1982, lower Echinoderm Zone, Bromide Formation, Oklahoma, USA.
Last: *Spyridiocrinus cheuxi* (Oehlert, 1889), Limestone with *Spirifer paradoxus*, Angers, France.

Suborder EUDIPLOBATHRINA Ubaghs, 1953
O. (ARG)–C. (BRI) Mar.

First: *Proexenocrinus inyoensis* Strimple and McGinnis, 1972, Al Rose Formation, California, USA.
Last: *Rhodocrinites baccatus* Wright, 1939, Lower Limestone Group, Scotland, UK.

Order MONOBATHRIDA Moore and Laudon, 1943

Suborder COMPSOCRININA Ubaghs, 1978
O. (ARG)–P. (TAT) Mar.

First: *Celtocrinus ubaghsi* Donovan and Cope, 1989, Brunel Beds, middle Arenig, Dyfed, South Wales, UK.
Last: *Neocamptocrinus* sp. *nov.*, Liveringa Formation, Hardman Member, Canning Basin, Western Australia (Webster, 1990).

Suborder GLYPTOCRININA Moore, 1952
O. (LLO3/CRD)–P. (ART) Mar.

First: *Abludoglyptocrinus laticostatus* Kolata, 1982, upper Pooleville Member, Bromide Formation, Oklahoma, USA.
Last: *Plesiocrinus pyriformis* Wanner, 1937, *Eutelecrinus poculiformis* Wanner, 1916 and various species of *Platycrinites, Neoplatycrinites* and *Pleurocrinus*. All from Basleo Beds, Timor.
Comment: A late Permian age for the Basleo Beds is now considered uncertain and they are probably of Artinskian or even Sakmarian age (Webster, 1987, 1990).

'Subclass' INADUNATA Wachsmuth and Springer, 1885

The 'subclass' Inadunata is the least clearly resolved of all Palaeozoic crinoid groups. It includes three orders, of which two, the Disparida and Hybocrinida, may be monophyletic. The emended classification of the Disparida used here has enabled stratigraphical ranges of this group to be compiled down to family level, but this has proven impossible for the Hybocrinida, which probably have been subdivided unnecessarily, and the Cladida. Instead these are listed at ordinal level. The Cladida include stem-group representatives of the Articulata and Flexibilia, and hence are demonstrably paraphyletic. At subordinal level, it is also clear that the Dendrocrinina and Poteriocrinina represent grades of development rather than natural groupings, such that the former are paraphyletic and the latter probably polyphyletic. The situation is even worse at lower taxonomic levels. Investigation of the articulate stem group (Simms and Sevastopulo, 1993) has shown that various genera, which cladistic analysis shows to lie close to the common ancestry of post-Palaeozoic crinoids, have been distributed in at least four disparate superfamilies. Furthermore, re-examination of those cladid families which lie close to the articulate stem group (e.g. Ampelocrinidae, Cymbiocrinidae) reveals that fewer than 25% of the constituent genera can justifiably be retained in these families, the remainder being uninterpretable on the data available or, in at least 25% of cases, wrongly assigned altogether. It is evident from even a cursory examination of the relevant part of the *Treatise*, that this problem is not confined to the articulate stem group, but pervades almost the entire classification of Palaeozoic crinoids. Consequently, it is impossible to justify listing the first and last occurrence of cladid crinoids at a taxonomic level any lower than suborder.

Order DISPARIDA Moore and Laudon, 1943

The subdivision of the Disparida follows the *Treatise on Invertebrate Paleontology*, with a few additions and modifications. However, the reader should be aware that this classification leaves a good deal to be desired. The superfamilies recognized in the *Treatise* each either contain only one family or probably are not monophyletic and are, therefore, not used in this compilation. The age of the many crinoids described from the Basleo Beds of Timor raises problems in assessing the ranges of several families of disparids. The Basleo fauna is regarded here as being of Sakmarian or Artinskian age rather than the late Permian age usually assigned to it (see comments under Blastoidea). However, because the Basleo fauna is from widely scattered localities it is possible that it contains some Upper Permian forms. Families based entirely on stem material are not listed.

F. HOMOCRINIDAE Kirk, 1914 O. (LLN)–S. (WEN) Mar.

First: *Ibexocrinus lepton* Lane, 1970, Kanash Shale, Utah, USA.
Last: *Homocrinus parvus* Kirk, 1914, Rochester Shale, New York, USA.

F. CALCEOCRINIDAE Meek and Worthen, 1869 O. (HAR/SOU)–P. (ART) Mar.

First: *Paracremacrinus laticardinalis* Brower, 1977, Mountain Lake Member, Bromide Formation, Oklahoma, USA.
Last: *Epihalysiocrinus tuberculatus* (Yakovlev, 1927), Boetzkay Suite, Urals, former USSR.

F. PISOCRINIDAE Angelin, 1878 O. (RAW)–D. (FAM) Mar.

First: *Eocicerocrinus sevastopuloi* Donovan, 1989, Ladyburn Starfish Bed, Girvan, Scotland, UK.
Last: *Triacrinus granulatus* Münster, 1839, and *T. pyriformis* Münster, 1839, both FAM, Thuringia, Germany.

F. ALLAGECRINIDAE Carpenter and Etheridge, 1881 D. (FAM)–P. (KUN) Mar.

First: *Allagecrinus americanus* Weller, 1930, Louisiana Limestone, Missouri, USA.
Last: *Metallagecrinus palermoensis* Strimple and Sevastopulo, 1982, Sosio Limestone, Sicily, or possibly one of several species of *Metallagecrinus* or *Wrightocrinus* from Timor (?Upper Permian).

F. CATILLOCRINIDAE Wachsmuth and Springer, 1886 D. (EIF)–P. (ART) Mar.

First: *Mycocrinus boletus* Schultze, 1866, *Mycocrinus conicus* Springer, 1923, or *Mycocrinus granulatus* Jaekel, 1895, all EIF, Germany.
Last: *Notiocatillocrinus cephalonus* (Willink, 1978), Wandrawandian Siltstone, New South Wales, Australia, or possibly species of *Isocatillocrinus, Neocatillocrinus, Paracatillocrinus* or *Xenocatillocrinus*, all from Basleo Beds, Timor.
Comment: *Acolocrinus* Kesling and Paul, 1971, and *Agostocrinus* Kesling and Paul, 1971, placed in this family in the *Treatise*, are now assigned to the Acolocrinidae (see below).

F. ACOLOCRINIDAE Brett, 1980 O. (CRD)–S. (WEN) Mar.

First: *Acolocrinus* sp. (Sprinkle, 1982). ? early CRD, Tulip Creek or McLish Formations, Oklahoma, USA, or *Acolocrinus hydraulicus* Kesling and Paul, 1971, Benbolt Formation, Tennessee, and Virginia, USA.
Last: *Paracolocrinus paradoxicus* Brett, 1980, Rochester Shale, New York, USA.

F. ANAMESOCRINIDAE Goldring, 1923 D. (FRS) Mar.

First and Last: *Anamesocrinus lutheri* Goldring, 1923, Laona Sandstone, New York, USA.

F. HAPLOCRINITIDAE Bassler, 1938 D. (EIF–FRS) Mar.

First: *Haplocrinites mespiliformis* Goldfuss, 1831, Germany, or *Haplocrinites* sp. A, Le Menn and Pidal, 1989, Saint Fiacre Formation, Brittany, France.

Last: *Haplocrinites boitardi* (Rouault, 1847), Pineres Formation, Cantabrian Mountains, Spain.
Comment: Records of Silurian and Carboniferous haplocrinitids are regarded as erroneous.

F. TUNGUSKOCRINIDAE Arendt, 1963 O. (LLN/LLO) Mar.

First and Last: *Tunguskocrinus ivanovae* Arendt, 1963, 'Middle' Ordovician, Mangazeyan Stage, Stolbovaya Suite, former USSR.

F. CINCINNATICRINIDAE Warn and Strimple, 1977 O. (LLN–ASH) Mar.

First: *Othneiocrinus priscus* (Lane, 1970), Kanosh Shale, Utah, USA.
Last: *Cincinnaticrinus pentagonus* (Ulrich, 1882), Bull Fork Formation, Ohio, USA.
Comment: Warn and Strimple (1977) introduced the Family Cincinnaticrinidae to replace Heterocrinidae Zittel, 1879, which was recognized in the *Treatise*.

F. MYELODACTYLIDAE S. A. Miller, 1883 O. (HIR)–D. (GIV) Mar.

First: *Musicrinus bodae* Donovan, 1984, Boda Limestone, Dalarna, Sweden.
Last: *Myelodactylus canaliculatus* (Goldfuss, 1826), Kerbelec Formation, Brittany, France, and Skaly Beds, Poland.

F. IOCRINIDAE Moore and Laudon, 1943 O. (ARG)–S. (LLY) Mar.

First: *Iocrinus* sp., lower Arenig, Nine Mile Shale Formation, Nevada, USA (Sprinkle, 1990).
Last: *Pariocrinus heterodactylus* Eckert, 1984, lower LLY, Cabot Head Formation, Ontario, Canada.

F. EUSTENOCRINIDAE Ulrich, 1925 O. (ARG)–S. (LLY) Mar.

First: *Pogonipocrinus antiquus* (Kelly and Ausich, 1978), Lower ARG, Fillmore Formation, Utah, USA.
Last: *Cataractocrinus clementii* Eckert, 1984, lower LLY, Cabot Head Formation, Ontario, Canada.

F. RAMSEYOCRINIDAE Donovan, 1984 O. (ARG) Mar.

First: *Ramseyocrinus cambriensis* (Hicks, 1873), Ogof Hen Formation, Ramsey Island, Wales, UK.
Last: *Ramseyocrinus vizcainoi* Ubaghs, 1983, Montagne Noire, France. Donovan (1988) has suggested that the poorly known *Tetragonocrinus pygmaeus* (Eichwald) should be assigned to this family, in which case it would be the youngest described ramseyocrinid (ARG).

F. TORNATILICRINIDAE Guensberg, 1984 O. (HAR/SOU) Mar.

First and Last: *Tornatilicrinus longicaudis* Guensberg, 1984, Lebanon Limestone Formation, Tennessee, USA.

F. ANOMALOCRINIDAE Wachsmuth and Springer 1886 O. (HAR/SOU–PUS) Mar.

First: *?Anomalocrinus antiquus* Guensberg, 1984, Lebanon Limestone Formation, Tennessee, USA, or *Geraocrinus sculptus* Ulrich, Ottosee Formation, CRD, Tennessee, USA.
Last: *Anomalocrinus incurvus* (Meek and Worthen, 1865), Maysville Formation, Ohio, USA.

F. BELEMNOCRINIDAE S. A. Miller, 1883 C. (?HAS/IVO–IVO/CHD) Mar.

First: *Belemnocrinus sampsoni* S. A. Miller, 1891, Chouteau Formation, Missouri, USA. More probably *B. whiteii* Meek and Worthen, 1866, or *B. pourtalesi* Wachsmuth and Springer, 1877, both from lower Burlington Limestone, Iowa, USA.
Last: *Belemnocrinus typus* White, 1865, or *Whiteocrinus florifer* (Wachsmuth and Springer, 1877), both from upper Burlington Limestone, Iowa, USA.
Comment: This family probably has been assigned incorrectly to the Disparida.

F. HOLYNOCRINIDAE Bouska, 1948 D. (EMS) Mar.

First and Last: *Holynocrinus moorei* Bouska, 1948, and *H. spinifer* Bouska, 1948, both upper EMS, Trebotov Limestone, Czechoslovakia.

F. PERISSOCRINIDAE Strimple, 1963 Invalid.

McIntosh (1979) has shown that this family was based on abnormal cladid crinoids. The systematic position of the one (monotypic) genus of disparid crinoids (*Quiniocrinus* Schmidt, 1942, from the EIF, *Orthrocrinus* Shale, Germany) assigned to this family in the *Treatise* is not known.

F. PYGMAEOCRINIDAE Strimple, 1963 D. (PRA–EMS) Mar.

First: *Pygmaeocrinus kettneri* Bouska, 1947, Lodenice Limestone, Czechoslovakia.
Last: Two undescribed species of *Pygmaeocrinus*, upper EMS, Trebotov Limestone, Czechoslovakia (Prokop, 1987).
Comment: Bouska (1947) refers, without details, to other species of *Pygmaeocrinus* from the Silurian, while Prokop (1987) cites the first occurrence of the genus as Lochkovian (= Gedinnian).

F. SYNBATHOCRINIDAE S. A. Miller, 1889 S. (WEN)–P. (SAK/ART) Mar.

First: *Abyssocrinus antiquus* Strimple, 1952, Henryhouse Formation, Oklahoma, USA.
Last: *Synbathocrinus campanulatus* (Wanner, 1916) and *S. constrictus* (Wanner, 1916), Callythara Formation, Western Australia. Wanner described these taxa from the Basleo Beds of Timor, which were considered to be late Permian in age, but Webster (1987) has described the same species from strata of undoubted ART or SAK age in Australia.
Comment: Prokop (1976) assigned several genera, included here in the Synbathocrinidae, to a new family Ramacrinidae. The family Ramacrinidae is not recognized in this compilation.

F. ZOPHOCRINIDAE S. A. Miller, 1892 S. (WEN)–D. (EIF) Mar.

First: *Zophocrinus howardi* S. A. Miller, 1892, Laurel Limestone, Indiana, USA.
Last: *Tiaracrinus quadrifons* Schultze, 1866, EIF, Germany.

F. PERITTOCRINIDAE Abel, 1920 ?O. (ARG/LLN) Mar.

The provenance of the two species assigned to this family (*Perittocrinus radiatus* (Beyrich) and *Tetracionocrinus transitor* (Jaekel)) is uncertain according to Ubaghs (1971). They are reputed to be from the Kunda Formation (ARG to LLN) of the Baltic region.

F. PARADOXOCRINIDAE Moore and Laudon, 1943 P. (SAK/ART or ZEC) Mar.

First and Last: *Paradoxocrinus patella* Wanner, 1937, Basleo Beds, Timor, Indonesia.
Comment: This family is unlikely to have been assigned correctly to the Disparida.

Order HYBOCRINIDA Jaekel, 1918 O. (ARG)–S. (RHU) Mar.

First: Undescribed hybocrinid, lower Arenig, Fillmore Formation, Utah, USA (Guensberg and Sprinkle, 1990).
Last: *Cornucrinus longicornis* Regnéll, 1948, basal Silurian, *Dalmanitina* Beds, Dalarna, Sweden.

Order CLADIDA Moore and Laudon, 1943

Three suborders are recognized in the *Treatise* within the Cladida; the Dendrocrinina, Poteriocrinina and the Cyathocrinina. The Cyathocrinina probably represent a monophyletic taxon but the Dendrocrinina and Poteriocrinina are clearly grade groups, representing primitive and derived taxa respectively within a continuum, and hence have no natural basis.

Suborders DENDROCRININA Bather, 1899, and POTERIOCRININA Jaekel, 1918 O. (TRE/ARG)–P. (KAZ) Mar.

First: *Aethocrinus moorei* Ubaghs, 1969, Schistes de St Chinian, Hérault, France.
Last: *Nowracrinus ornatus* (Etheridge, 1892), Gerringong Volcanics, New South Wales, Australia.
Comment: Since the Articulata represent, in effect, the crown group of the Cladida, the apparent gap between the last cladids and the first articulates is entirely an artefact of collecting and/or documentation. Crinoids occur in some abundance in certain Tatarian strata, but have remained undescribed.

Suborder CYATHOCRININA Bather, 1899 O. (LLO_3/CRD)–P. (KUN/?ZEC) Mar.

First: *Palaeocrinus hudsoni* Sprinkle, 1982, and *Carabocrinus treadwelli* Sinclair, 1945, lower Echinoderm Zone, Bromide Formation, Oklahoma, USA.
Last: Numerous genera and species are listed in the *Treatise* as from the Basleo Beds of Timor. These strata are now considered to be ART or SAK in age (Webster, 1987) rather than Upper Permian. The latest undoubted cyathocrinitid is *Gissocrinus* ? sp., from the KUN, Catherine Sandstone, central Queensland, Australia.

Subclass FLEXIBILIA Zittel, 1895

Order TAXOCRINIDA Springer, 1913 O. (LLN1)–P. (SAK) Mar.

First: *Archaetaxocrinus burfordi* Lewis, 1981, Oil Creek Formation, southern Oklahoma, and *A. lanei* Lewis, 1981, Kanosh Shale, Utah, USA. These two species are referred tentatively to the Taxocrinida by Lewis (1981), but may well be ancestral to both Taxocrinida and Sagenocrinida.
Last: *Nevadacrinus geniculatus* Lane and Webster, 1966, Bird Spring Formation, Battleship Wash, Nevada, USA.

Order SAGENOCRINIDA Springer, 1913 O. (CRD/ASH)–P. (WOR/CAP) Mar.

First: *Proanisocrinus oswegoensis* (Miller and Gurley, 1894), Maquoketa Group, Oswego, Illinois, USA.
Last: *Trinalicrinus tunisiensis* Lane, 1979, and *Strobocrinus brachiatus* Lane, 1979, Guadalupian, Djebal Tebaga, Tunisia.

Subclass ARTICULATA Miller, 1821

The classification scheme adopted here for the Articulata is based largely upon that of Simms (1988), but has been modified in the light of more recent work. Several major clades (Encrinidae, Isocrinina, Pentacrinitidae with Comatulidia) can be identified within the Articulata, but at lower taxonomic levels the classification used here cannot be considered to reflect relationships between these groups, and is in need of considerable revision. Conflicting classification schemes exist for some groups within the Articulata, but it is not possible to take account of these here.

F. HOLOCRINIDAE Jaekel, 1918 Tr. (SPA–CRN) Mar.

First: *Holocrinus smithi* (Clark, 1915), Mid-Spathian, Thaynes Limestone, Idaho, and equivalent strata in adjacent states, USA.
Last: *Holocrinus lunatus* (Kristen-Tollmann, 1975), lower CRN, Hallstätter Kalk, Taurus Mountains, Turkey.
Comment: The Holocrinidae is a paraphyletic taxon which includes stem members of the encrinids, isocrinids and pentacrinitids. It is retained here, pending further data concerning the precise relationships of these three derived groups.

F. ENCRINIDAE Dujardin and Hupé, 1862 Tr. (ANS–CRN) Mar.

First: *Encrinus terebratularum* Schmid, 1876, lower ANS, Lower Muschelkalk, Gogolin Beds, Silesia, Poland.
Last: *Traumatocrinus caudex* (Dittmar, 1868), middle CRN, lowermost Raibl Beds, Italy.

F. DADOCRINIDAE Lowenstam, 1942 Tr. (SPA–ANS) Mar.

First: *Dadocrinus kunischi* (Wachsmuth and Springer, 1887), upper Spathian, Röt, Silesia, Poland.
Last: *Dadocrinus* sp., middle ANS, lower Muschelkalk, *Terebratula* Beds, Silesia, Poland.

Order ISOCRININA Simms, 1988 Tr. (LAD)–Rec. Mar.

The Order Isocrinina comprises four families, of which two, the Isocrinidae and Cainocrinidae, are demonstrably paraphyletic, the Proisocrinidae is monotypic and has no fossil record, and only the Isselicrinidae is monophyletic. Hence they have not been subdivided here.
First: *Laevigatocrinus laevigatus* (Münster, 1841), *Laevigatocrinus* ? *venustus* (Klipstein, 1845), *Tyrolecrinus subcrenatus* (Münster, 1841), and *'Isocrinus' propinquus* (Münster, 1841), upper LAD, *Pachycardia* Beds, Italy.
Extant

Order COMATULIDINA Clark, 1908

The present classification of the comatulids, as outlined in the *Treatise*, is unsatisfactory and is considered unlikely to reflect phylogenetic relationships within the group. Furthermore, most fossil comatulid material is undiagnostic at higher taxonomic levels and hence stratigraphical range

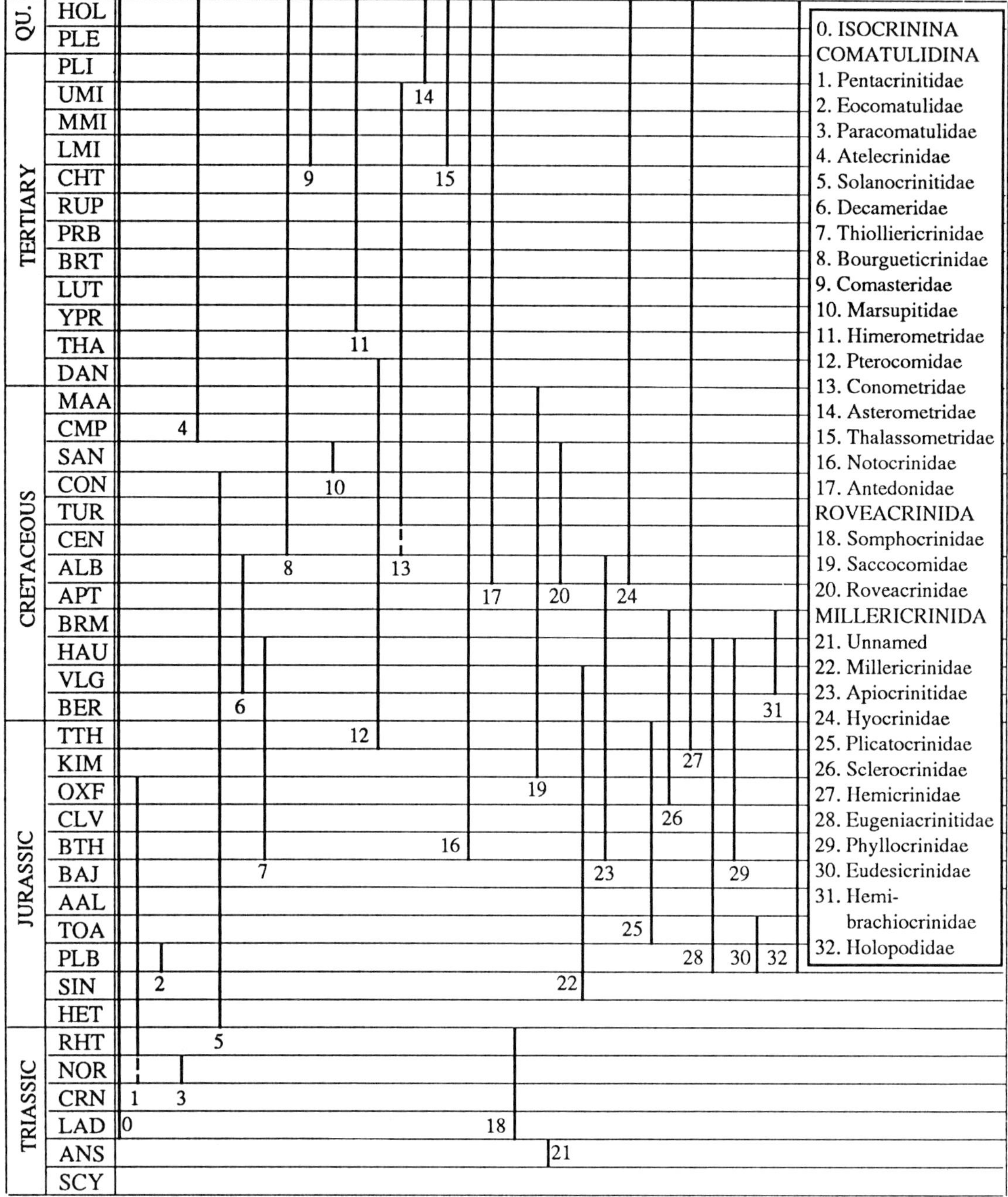

Fig. 25.3

can be ascertained with confidence for only a few distinctive comatulid groups. The classification used here is essentially that of the *Treatise*, with minor revisions where appropriate.

F. PENTACRINITIDAE Gray, 1842 Tr. (NOR/RHT)–J. (OXF) Mar.

First: *Seirocrinus klikushini* Simms, 1989, NOR/RHT, Novosibirsk Islands, former USSR, and *Pentacrinites* sp., Zlambach Marls, Salzkammergut, Austria.
Last: *Pentacrinites* sp., upper Oxfordian, Oolithe Corallienne, France.
Comment: The pentacrinitids are primitive sister group to the comatulids.

Superfamily PARACOMATULACEA Hess, 1951

F. EOCOMATULIDAE Simms, 1988 Tr. (NOR)–J. (PLB)

First: *Eocomatula* sp. *nov.* Simms, in press. Chambara Formation, Shalypayco, central Peru.
Last: *Eocomatula interbrachiatus* (Blake, 1876), upper Pliensbachian, *margaritatus* Zone, Deddington, Oxfordshire, England, UK.

F. PARACOMATULIDAE Hess, 1951 Tr. (NOR)–K. (l.) Mar.

First: *Paracomatula* sp. *nov.* Hagdorn, in press, NOR, New Caledonia.
Last: Undescribed remains, Lower Cretaceous, Hungary (cited by Rasmussen, 1978, T870).

Infra-order COMATULIDIA Clark, 1908

F. ATELECRINIDAE Bather, 1899
K. (CMP)–Rec. Mar.

First: *Jaekelometra gisleni* Rasmussen, 1961, Upper Chalk, Båstad, Sweden. **Extant**

Superfamily SOLANOCRINITACEA Jaekel, 1918

F. SOLANOCRINITIDAE Jaekel, 1918
J. (HET)–K. (CON) Mar.

First: *Palaeocomaster styricus* Kristan-Tollmann, 1988, Allgäu Beds, lower part of Lias alpha, Salzkammergut, Austria.
Last: *Comatulina janenschi* (Sieverts-Doreck, 1958), Middle Chalk, Halberstadt, Germany.

F. DECAMERIDAE Rasmussen, 1978
K. (VLG–ALB) Mar.

First: *Coelometra campichei* (de Loriol, 1879), VLG, Switzerland and France.
Last: *Decameros wortheimi* Peck and Watkins, 1972, Glen Rose Limestone, Texas, USA.

F. THIOLLIERICRINIDAE Clark, 1908
J. (BTH)–K. (HAU) Mar.

First: *Thiolliericrinus ? ooliticus* (M'Coy, 1848), Bradford Clay, Bradford-on-Avon, England, UK.
Last: *Heberticrinus algarbiensis* (de Loriol, 1888), HAU, Algarve, Portugal.
Comment: Klikushin (1987a) considers the Bathonian material not to be attributable to the Thiolliericrinidae. Morphologically and stratigraphically there is little to separate the latest thiolliericrinid from the earliest bourgueticrinid. If a direct relationship between these two groups is proven, which seems highly probable, then the Thiolliericrinidae are a paraphyletic group.

F. BOURGUETICRINIDAE de Loriol, 1882
K. (CEN)–Rec. Mar.

First: *Bourgueticrinus* sp., upper lower TUR, Belbeck Valley, Crimea, former USSR (Klikushin, 1982). **Extant**
Comment: The Family Bourgueticrinidae is here taken to include the families Bourgueticrinidae, Bathycrinidae, Phrynocrinidae and Porphyrocrinidae as listed in the *Treatise*. Their separation into distinct families almost certainly results in the creation of paraphyletic taxa, particularly in the case of the Bourgueticrinidae and Bathycrinidae, while the remaining two families have no fossil record. Klikushin (1987b) assigned an upper Pliensbachian species, *Gutticrinus guttiformis*, to a new family, the Gutticrinidae, within the Bourgueticrinidae. It is probably a cyrtocrinid.

Superfamily COMASTERACEA A. H. Clark, 1908

F. COMASTERIDAE A. H. Clark, 1908
T. (Mio.)–Rec. Mar.

First: *Comaster formae* Noelli, 1901, Miocene, Piemonte, Italy. **Extant**

F. MARSUPITIDAE d'Orbigny, 1852
K. (SAN) Mar.

First: *Uintacrinus socialis* Grinnel, 1876, lower part of *testudinarius* Zone, England, UK.
Last: *Uintacrinus anglicus* Rasmussen, 1961, top of *testudinarius* Zone, England, UK.
Comment: Cladistic analysis indicates that marsupitids were derived from the comasterids, despite the earliest record of a typical comasterid being from the Neogene. This discrepancy is attributable almost certainly to the poorly diagnostic nature of disarticulated material from non-marsupitid comasterids.

Superfamily MARIAMETRACEA A. H. Clark, 1909

F. MARIAMETRIDAE A. H. Clark, 1909
Extant Mar.

F. ZYGOMETRIDAE A. H. Clark, 1908
Extant Mar.

F. EUDIOCRINIDAE A. H. Clark, 1907
Extant Mar.

F. HIMEROMETRIDAE A. H. Clark, 1908
T. (Eoc.)–Rec. Mar.

First: *Himerometra bassleri* Gislén, 1934, Eocene, South Carolina and Louisiana, USA. **Extant**

F. COLOBOMETRIDAE A. H. Clark, 1909
Extant Mar.

Superfamily TROPIOMETRACEA A. H. Clark, 1908

F. TROPIOMETRIDAE A. H. Clark, 1908
Extant Mar.

F. PTEROCOMIDAE Rasmussen, 1978
J. (TTH)–T. (DAN) Mar.

First: *Pterocoma pennata* von Schlotheim, 1813, Solnhofen Limestone, Bavaria, Germany.
Last: *Placometra laticirra* (Carpenter, 1880), upper DAN, Hvalløse, Denmark.

F. CONOMETRIDAE Gislén, 1924
K. (CEN/TUR)–T. (Mio.) Mar.

First: *Amphorometra bellilensis* (Valette, 1935), CEN or TUR, Djebal Bellil, Tunisia.
Last: *Cypelometra iheringi* de Loriol, 1902, Miocene, Argentina, South America.

F. CALOMETRIDAE A. H. Clark, 1911
Extant Mar.

F. PTILOMETRIDAE A. H. Clark, 1914 **Extant** Mar.

F. ASTEROMETRIDAE Gislén, 1924
T. (PLI)–Rec. Mar.

First: Undescribed radials, brachials and cirrals, Pliocene, Indonesia (Sieverts, 1933). **Extant**

F. THALASSOMETRIDAE A. H. Clark, 1908
T. (LMI)–Rec. Mar.

First: *Stenometra pellati* de Loriol, 1897, Lower Miocene, France. **Extant**

F. CHARITOMETRIDAE A. H. Clark, 1909
Extant Mar.

Superfamily NOTOCRINACEA Mortensen, 1918

F. NOTOCRINIDAE J. (BTH)–Rec. Mar.

First: *Semiometra abnormis* Carpenter, 1880, Bradford Clay, Cirencester, England, UK. **Extant**

F. APOROMETRIDAE H. L. Clark, 1938
Extant Mar.

Superfamily ANTEDONACEA Norman, 1865

F. ANTEDONIDAE Norman, 1865
K. (ALB)–Rec. Mar.

First: *Roiometra columbiana* A. H. Clark, 1944, middle ALB, middle Villeta Formation, Cundinamarca, Colombia, South America. **Extant**

F. PENTAMETROCRINIDAE A. H. Clark, 1908
Extant Mar.

Order ROVEACRINIDA Sieverts-Doreck, 1952

Three families are included within the Rovaecrinida, although their relationships to each other and to other articulate groups remain unclear.

F. SOMPHOCRINIDAE Peck, 1978
Tr. (LAD–RHT) Mar.

First: *Osteocrinus spinosus* Kristan-Tollmann, 1970, basal Langobardian, lower LAD, Austria, Turkey and Italy.
Last: *Osteocrinus* sp., RHT, Austria (L. Krystyn, pers. comm.).

F. SACCOCOMIDAE d'Orbigny, 1852
J. (KIM)–K. (MAA) Mar.

First: *Saccocoma* sp., lower Kimmeridge Clay, top of *mutabilis* Zone, southern and eastern England, UK (P. B. Wignall, pers. comm.).
Last: *Applinocrinus* sp., Meerssen Member, Maastricht Formation, upper Maastrichtian, Bergen Terblijt, The Netherlands (Jagt, 1992).

F. ROVEACRINIDAE Sieverts-Doreck, 1952
K. (ALB–SAN) Mar.

First: *Hyalocrinus bulliensis* Estomves, 1984, and *H. magniezi* Estomves, 1984, *Otohoplites raulinianus* Zone, Paris Basin, France.
Last: Undescribed roveacrinid, Meerssen Member, Maastricht Formation, upper Maastrichtian, Maastricht, The Netherlands (Jagt, 1992).

Order MILLERICRINIDA Sieverts-Doreck, 1952

The relationship of the various higher taxa currently included in the Millercrinida is unclear and the group may well prove to be polyphyletic. Many new millericrinid taxa have been described recently from the Lower Jurassic, but the precise position of many of these remains unclear. It is probable that a significant proportion are stem-group representatives of later taxa and hence are difficult to place in the current classification scheme which is based upon Middle Jurassic and later forms. They have, therefore, been excluded from this compilation.

F. UNNAMED Tr. (ANS) Mar.

First and Last: *'Entrochus' silesiacus* Beyrich, 1857, upper ANS, middle Muschelkalk, *Diplopora* Dolomite, Silesia, Poland.
Comment: This species is based on columnals which bear a striking resemblance to those of certain Jurassic millericrinids. They are only tentatively assigned here to the Millericrinida.

Suborder MILLERICRININA Sieverts-Doreck, 1952

Infra-order MILLERICRINIDIA Sieverts-Doreck, 1952

F. MILLERICRINIDAE Jaekel, 1918
J. (SIN)–K. (VLG) Mar.

First: *Shroshaecrinus obliquistratus* Simms, 1989, *densinodulum* Zone, Black Ven Marls, Charmouth, Dorset, England, UK.
Last: *Millericrinus* aff. *neocomiensis* (d'Orbigny, 1850), VLG, Villers-le-Lac, France. The columnals figured by Rasmussen (1961) were assigned by him to *Apiocrinus*, but they bear a much closer resemblance to those of Upper Jurassic species of Millericrinus, to which they are assigned tentatively here.

F. APIOCRINITIDAE d'Orbigny, 1840
J. (BTH)–K. (ALB) Mar.

First: *Apiocrinites parkinsoni* (Schlotheim, 1820), Forest Marble of Dorset, Somerset and Avon, England, UK.
Last: *Apiocrinites gillieroni* (Loriol, 1877), lower ALB, Shenley Limestone, Bedfordshire, England, UK.

Infra-order HYOCRINIDIA Rasmussen, 1978

F. HYOCRINIDAE Carpenter, 1884
K. (ALB)–Rec. Mar.

First: *Taurocrinus tauricus* Klikushin, 1984, *Mortinoceras inflatum* Zone, south-west Crimea, former USSR. **Extant**
Comment: Klikushin (1984, 1987b) has placed a number of Jurassic and Cretaceous genera, together with the Family Cyclocrinidae, in the Hyocrinidia. The taxonomic status of the Cyclocrinidae, based entirely on isolated columnals, is questionable, while most of the Lower Jurassic taxa appear of uncertain phylogenetic position.

Suborder CYRTOCRININA Sieverts-Doreck, 1952

Infra-order CYRTOCRINIDIA Sieverts-Doreck, 1952

F. PLICATOCRINIDAE Zittel, 1879
J. (TOA–TTH) Mar.

First: *Plicatocrinus mayalis* Deslongchamps and *P. deslongchampsi* de Loriol, basal Toarcian, Couche à Leptaena, Calvados, France.
Last: *Plicatocrinus hexagonus* Münster, 1839, TTH, Swabian Alb, Germany.
Comment: *Quenstedticrinus quenstedti* Klikushin, 1987b, from the upper PLB, *margaritatus* Zone, may be closely allied to *Plicatocrinus*.

F. SCLEROCRINIDAE Jaekel, 1918
J. (OXF)–K. (BRM) Mar.

First: *Cyrtocrinus nutans* (Goldfuss, 1826), OXF (?CLV) of France and Germany.
Last: *Sclerocrinus strambergensis* Jaekel, 1891, BRM, Stramberk, Czechoslovakia.

F. HEMICRINIDAE Rasmussen, 1961
J. (TTH)–Rec. Mar.

First: *Hemicrinus tithonicus* Pisera and Dzik, 1979, Red Rogoznik coquina, lower or middle TTH, Rogoznik, Poland. **Extant**

F. EUGENIACRINITIDAE Roemer, 1855
J. (PLB)–K. (HAU) Mar.

First: *Capsicocrinus souti* Delogu and Nicosia, 1984, lower

PLB (or lowermost part of upper PLB), western Anatolia, Turkey.
Last: *Eugeniacrinites gevreysi* (Loriol, 1897), HAU, St Pierre de Cherenne, France.

F. PHYLLOCRINIDAE Jaekel, 1907 J. (BTH)–K. (NEO) Mar.

First: *Phyllocrinus clapsensis* Loriol, 1882, BTH, Bouches-du-Rhône, France.
Last: *Phyllocrinus malbosianus* d'Orbigny, 1850 and several other species of *Phyllocrinus*, Neocomian of France, Switzerland, Romania, Spain, Austria, Italy and southern Germany.

Infra-order HOLOPODINIDIA Arendt, 1974

F. EUDESICRINIDAE Bather, 1899 J. (PLB–TOA) Mar.

First: *Eudesicrinus* cf. *mayalis* (Deslongchamps, 1858), lower Lias, *davoei* Zone, Rottorf, southern Lower Saxony, Germany.
Last: *Eudesicrinus* sp., upper Lias Junction Bed, *bifrons* Zone, Ilminster, Somerset, England, UK.

F. HEMIBRACHIOCRINIDAE Arendt, 1968 K. (VLG–BRM) Mar.

First: *Hemibrachiocrinus manesterensis* Arendt, 1974, *Brachiomonocrinus simplex* Arendt, 1974, *B. subcylindricum* Arendt, 1974 and *Dibrachiocrinus elongatus* Arendt, 1974, lower VLG, Crimea, former USSR.
Last: *Hemibrachiocrinus pumilus* Arendt, 1974, *Brachiomonocrinus exiguus* Arendt, 1974, *Dibrachiocrinus biassalensis* Arendt, 1974 and *D. solovievi* Arendt, 1974, lower BRM, Crimea, former USSR.

F. HOLOPODIDAE Zittel, 1879 J. (PLB)–Rec. Mar.

First: *Cotylederma lineati* Quenstedt, 1852, middle Lias, Württemberg, Germany. **Extant**
Comment: Pisera and Dzik (1979) consider that there is little justification in separating the Jurassic genus *Cotylederma* from the Cretaceous to Recent *Cyathidium*. This view is maintained here, although the possibility that the holopodid morphology arose more than once, through convergence, cannot be dismissed. *Cotylederma oppeli* Terquem, 1855, from the Lias Inféricur (HET) of Lorraine, France, is probably a millericrinid (*s.l.*) holdfast.

Subphylum ELEUTHEROZOA Bell, 1891 (see Fig. 25.4)

Plesion (Genus) *STROMATOCYSTITES* Pompeckj, 1896 €. (CRF–STD) Mar.

First: *Stromatocystites pentangularis* Pompeckj, 1896, and *S. walcotti* Schuchert, 1919, *Olenellus* Beds, Taconian, upper Lower Cambrian, Bonne Bay, Newfoundland.
Last: *Stromatocystites* sp., median Middle Cambrian, *Eccaparadoxides oelandicus* Zone, Sweden (Smith, 1988).

Plesion (Class) EDRIOASTEROIDEA Billings, 1858

The classification of edrioasteroids, including *Stromatocystites*, adopted here is that of Smith (1983, 1985).

Plesion (Genus) *ARKARUA* Gehling, 1987 V. (POU) Mar.

First and Last: *Arkarua adami* Gehling, 1987, Ediacaran Member, Flinders Range, South Australia.
Comment: This problematic fossil is considered by Gehling to represent the earliest known echinoderm, allied to edrioasteroids, a view tentatively accepted by Smith and Jell (1990).

Order EDRIOASTERIDA Bell, 1976

F. TOTIGLOBIDAE Bell and Sprinkle, 1978 €. (STD) Mar.

First and Last: *Totiglobus nimius* Bell and Sprinkle, 1978, lower Middle Cambrian, *Glossopleura* Zone, Chisholm Shale, Nevada, USA.

F. EDRIOASTERIDAE Bather, 1898 €. (STD)–O. (SOU/LON) Mar.

First: *Walcottidiscus typicalis* Bassler, 1935, *Bathyuriscus–Elrathina* Zone, Burgess Shale, British Columbia, Canada.
Last: *Edrioaster bigsbyi* (Billings, 1857), *E. priscus* (Miller and Gurley, 1894), *Edriophus levis* (Bather, 1914) and *E. ? saratogensis* (Ruedemann, 1912), Trenton Limestone, Ontario, Canada, and Michigan, USA.

Order ISOROPHIDA Bell, 1976

Plesion (Genus) *CAMBRASTER* Cabibel *et al.*, 1958 €. (STD) Mar.

First: *Cambraster* sp., lower Middle Cambrian, Coonigan Formation, Australia (Jell *et al.*, 1985).
Last: *Cambraster cannati* (Miquel, 1894), upper Middle Cambrian, *Paradoxides mediterraneus* Zone, Montagne Noire, southern France.

Plesion (Genus) *EDRIODISCUS* Jell *et al.*, 1985 €. (STD) Mar.

First and Last: *Edriodiscus primotica* (Henderson and Shergold, 1971), lower Middle Cambrian, upper Ordian, Yelvertoft Beds, west Queensland, Australia.

F. CYCLOCYSTOIDIDAE Miller, 1882 O. (ARG)–C. (BRI) Mar.

First: *Monocycloides oelandicus* Berg-Madsen, 1987, upper ARG, Volkhov Stage, Langevoja Substage, Hälludden, northern Öland, Sweden.
Last: Undescribed ossicles, Meenymore Formation, Sligo, Republic of Ireland (G. D. Sevastopulo, pers. comm., 1989).

F. AGELACRINITIDAE Chapman, 1860 O. (COS)–C. (KAS/GZE) Mar.

First: *Isorophusella incondita* (Raymond, 1915), Blackriveran, Decorah Formation, *Phylloporina* Bed, Minnesota, USA.
Last: Unnamed material (*Discocystis* cf. *kaskaskiensis* (Hall, 1858)), Virgilian Series, upper part of Madera Formation, Guadalupe Canyon, New Mexico (Bell, 1976).

Order CYATHOCYSTIDA Bockelie and Paul, 1983

F. PYRGOCYSTIDAE Kesling, 1967 O. (COS)–D. (GED/SIG) Mar.

First: *Pyrgocystis sardesoni* Bather, 1915, *Stictopora* Bed, Decorah Formation, St Paul, Minnesota, USA.
Last: *Rhenopyrgus whitei* Holloway and Jell, 1983, Humevale Formation, Kinglake West, central Victoria, Australia.

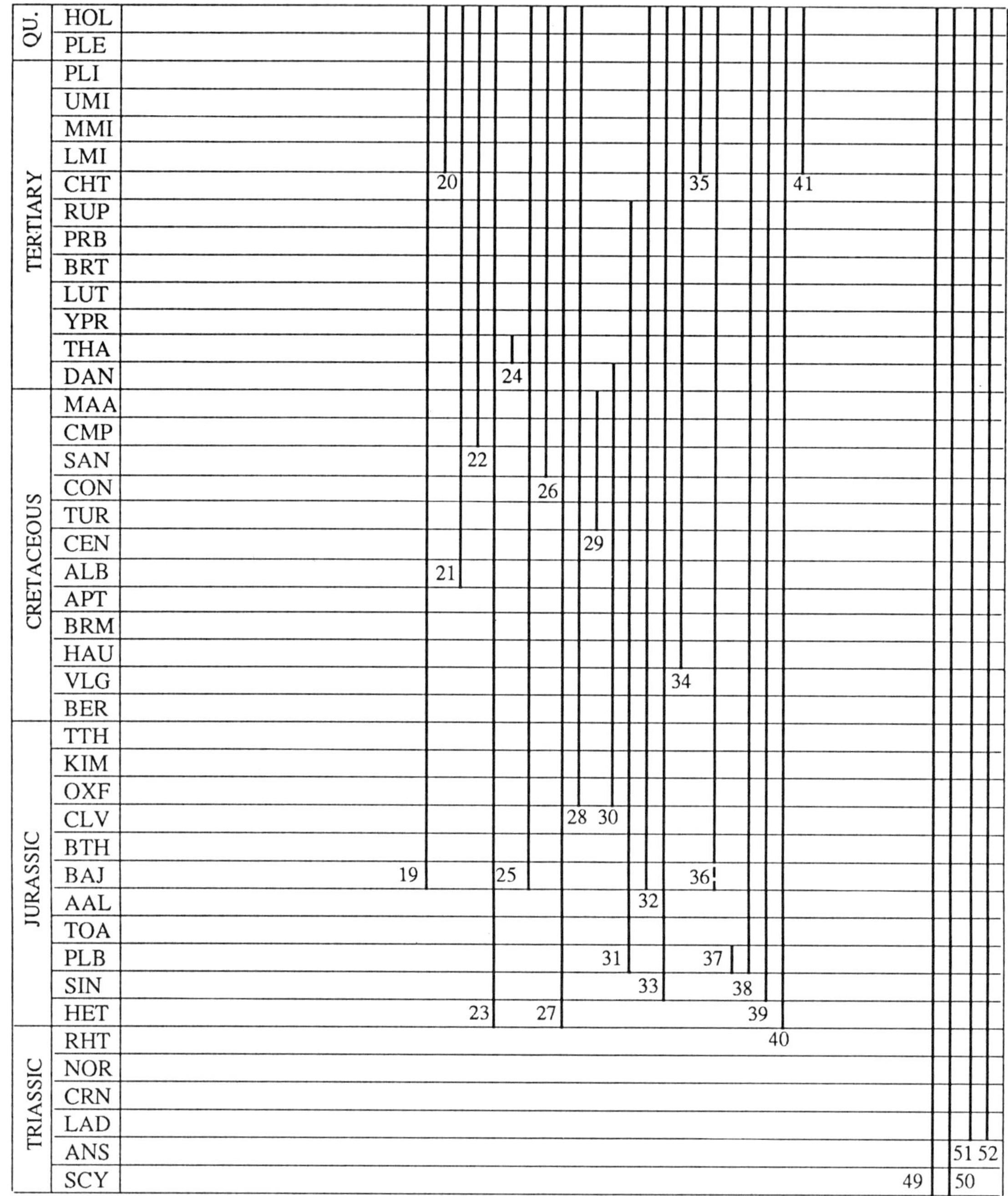

Fig. 25.4

F. CYATHOCYSTIDAE Bather, 1898 O. (LLN–HIR/S. RHU?) Mar.

First: *Cyathotheca corallum*, Kunda Formation, Leningrad, former USSR.

Last: *Cyathotheca suecica* Jaekel, 1927, ?Boda Limestone, Dalarna, Sweden. The precise age of this unique specimen is uncertain, and it may be from the lowermost Silurian (Regnéll, 1945).

Class ASTEROIDEA de Blainville, 1830

The Class Asteroidea is comparatively poorly represented in the fossil record, and although most material is highly fragmentary, published descriptions have been based largely on intact specimens. The classification of post-Palaeozoic asteroids has been the subject of recent revision by Gale (1987) and by Blake (1987), with the former being adopted here. The relationships of Palaeozoic taxa are still rather poorly understood and the compilation below identifies only plesions considered to be monophyletic.

Plesion UNNAMED O. (CRD)–D. (SIG/EMS) Mar.

First: *Schuchertia stellata* (Billings, 1857), Trenton Limestone, Ottawa, Canada.

Last: *Palastericus devonicus* Sturtz, Hünsrückschiefer, Bundenbach, Germany.

Plesion UNNAMED O. (ARG–ASH) Mar.

First: *Uranaster ramseyensis* Hick, Ogof Hên Formation, Movidunian, Ramsey Island, Dyfed, Wales, UK

PERMIAN: TAT, KAZ, UFI, KUN, ART, SAK, ASS
CARBONIFEROUS: GZE, KAS, MOS, BSH, SPK, VIS, TOU
DEVONIAN: FAM, FRS, GIV, EIF, EMS, PRA, LOK
SILURIAN: PRD, LUD, WEN, LLY
ORDOVICIAN: ASH, CRD, LLO, LLN, ARG, TRE
CAMB.: MER, STD, CRF
SINIAN: EDI, VAR, STU

ELEUTHEROZOA
1. Stromatocystites
EDRIOASTEROIDEA
2. Arkarua
3. Totiglobidae
4. Edrioasteridae
5. Cambraster
6. Edriodiscus
7. Cyclocystoididae
8. Agelacrinitidae
9. Pyrgocystidae
10. Cyathocystidae
ASTEROIDEA
11. Unnamed
12. Unnamed
13. Cnemidactis
14. Uractinina
15. Pustulosina
16. Monaster
17. Calliasterella
18. Permaster
NEOASTEROIDEA
19. Astropectinidae
20. Luidiidae
21. Goniopectinidae
22. Porcellanasteridae
23. Benthopectinidae
24. Echinasteridae
25. Goniasteridae
26. Ophidiasteridae
27. Asteropseidae
28. Sphaerasteridae
29. Arthrasteridae
30. Stauranderasteridae
31. Pycinasteridae
32. Poraniidae
33. Asterinidae
34. Chaetasteridae
35. Archasteridae
36. Odontasteridae
37. Tropasteridae
38. Solasteridae
39. Zoroasteridae
40. Asteriidae
41. Brisingidae
42. Bothriocidaridae
HOLOTHUROIDEA
43. Volchoviidae
44. Eucladidae
45. Sollasinidae
46. Rotasaccidae
47. Incertae sedis
48. Palaeocucumariidae
49. Ypsilothuriidae
50. Calclamnidae
51. Exlinellidae, Ludwigiidae & Spinitidae
52. Heterothyonidae

Fig. 25.4.

Last: *Petraster speciosus* (Miller and Dyer, 1878), Richmond Formation, Preble Coundy, Ohio, USA.

Plesion (Genus) *CNEMIDACTIS* Spencer, 1918 O. (RAW) Mar.

First and Last: *Cnemidactis girvanensis* (Schuchert, 1914), Ladyburn Starfish Bed, Drummond Group, Girvan, Scotland, UK.

Plesion URACTININA Spencer and Wright, 1966 O. (LLO)–C. (MOS) Mar.

First: *Urasterella huxleyi* (Billings, 1865), Chazyan, Bed 1, Point Rich, Newfoundland, Canada.
Last: *Urasterella montana* (Sturowsky, 1874), Mosquensis Limestone, Moscow, former USSR.

Plesion PUSTULOSINA Spencer, 1951 O. (CRD)–D. (?EIF) Mar.

First: *Hudsonaster narrawayi* (Hudson, 1912), Black River Formation, Ottawa, Canada.
Last: *Xenaster margaritatus* (Simonovitsch, 1871), Obere Koblenzschichten, Neiderlahnstein, Germany.

Plesion (Genus) *MONASTER* Schuchert, 1915 P. Mar.

First and Last: *Monaster wandageensis* Kesling, and *M. canarvonensis* Kesling, Wandagee Formation, Wandagee, Western Australia.

Plesion (Genus) *CALLIASTERELLA* Schuchert, 1914 C. (MOS–KAS) Mar.

First: *Calliasterella mira* (Traustchold), Mosquensis Limestone, Mjatschkowa, Moscow, former USSR.
Last: *Calliasterella americana* Kesling and Strimple, 1966, LaSalle Limestone, Missourian, Livingston Co., Illinois, USA.

Plesion (Genus) *PERMASTER* Kesling, 1969 P. Mar.

First and Last: *Permaster grandis* Kesling, 1969, Wandagee Formation, Wandagee, Carnarvon Basin, Western Australia.

Subclass NEOASTEROIDEA Gale, 1987

Order PAXILLOSIDA Perrier, 1884

F. ASTROPECTINIDAE Gray, 1840 J. (BAJ)–Rec. Mar.

First: *Tethyaster jurassicus* Blake, 1986, middle Bajocian, Carmel Formation, Teasdale and St George, Utah, USA. **Extant**

F. LUIDIIDAE Verrill, 1889 T. (Mio.)–Rec. Mar.

First: *Luidia* sp., Miocene, California (Blake, 1973). **Extant**

F. GONIOPECTINIDAE Verrill, 1889 K. (ALB)–Rec. Mar.

First: *'Nymphaster' radiatus* Spencer, 1907, Middle Albian, Kent, England, UK. **Extant**

F. CTENODISCIDAE Sladen, 1889 **Extant** Mar.

F. PORCELLANASTERIDAE Sladen, 1883 K. (CMP)–Rec. Mar.

First: *Palaeoctenodiscus campaniurnis* Blake, 1988, CMP, Baja California, USA. **Extant**

Order NOTOMYOTIDA Ludwig, 1910

F. BENTHOPECTINIDAE Verrill, 1894 J. (HET)–Rec. Mar.

First: *Plesiastropecten hallovensis* Blake, upper HET, Hallau, Switzerland. **Extant**

Order SPINULOSIDA Perrier, 1884

F. ECHINASTERIDAE Verrill, 1867 T. (THA) Mar.

First: *Echinaster jacobseni* Rasmussen, upper THA, Fur Formation, Jutland, Denmark. **Extant**

Order VALVATIDA Perrier, 1884

F. GONIASTERIDAE Forbes, 1841 J. (BAJ)–Rec. Mar.

First: *Tylasteria berthandi* (Wright), upper BAJ, upper Hauptrogenstein, Sinznach-Dorf, Switzerland. **Extant**

F. OPHIDIASTERIDAE Verrill, 1867 K. (SAN)–Rec. Mar.

First: *Sladenia fourteaui* de Loriol, 1904, and *Chariaster elegans* de Loriol, 1909, SAN, Abou-Roach, Egypt. **Extant**

F. OREASTERIDAE Fisher, 1911 **Extant** Mar.

F. ASTEROPSEIDAE J. (HET)–Rec. Mar.

First: *Diclidaster gevreyi* de Loriol, 1897, HET, Ardèche, France. **Extant**

F. SPHAERASTERIDAE Schondorf, 1906 J. (OXF)–Rec. Mar.

First: *Sphaeraster annulosus* (Quenstedt), Malm alpha, southern Germany. **Extant**

F. ARTHRASTERIDAE Spencer, 1918 K. (TUR–MAA) Mar.

First: *Arthraster dixoni* Forbes, 1848, Middle Chalk, *T. lata* Zone, Amberley, Sussex, England, UK.
Last: *Arthraster cristatus* Spencer, 1913, White Chalk, Rugen, northern Germany.

F. STAURANDERASTERIDAE Spencer, 1913 J. (?CLV)–T. (DAN) Mar.

First: *Aspidaster delgadoi* de Loriol, 1884, ? CLV, Porcas Valley, near Cinta, Portugal.
Last: *Stauranderaster speculum* Brunnich-Nielsen, 1943, upper DAN, Rejstrup, Denmark.

F. ACANTHASTERIDAE Sladen, 1889 **Extant** Mar.

F. MITHRODIIDAE Viguier, 1879 **Extant** Mar.

F. ASTERODISCIDIDAE Rowe **Extant** Mar.

F. PYCINASTERIDAE Spencer and Wright, 1966 J. (PLB)–T. (RUP) Mar.

?First: *Pycinaster mortenseni* Mercier, 1935, Lower PLB, Crouay, Calvados, Normandy, France.
?Last: *Pycinaster peyroti* Valette, 1925, Stampian, Aquitaine, France.

F. PORANIIDAE Perrier, 1894 J. (BAJ)–Rec. Mar.

?First: *Sphaeriaster jurassicus* Hess, 1972, Upper BAJ, upper Hauptrogenstein, Schinznach-Dorf, Switzerland. **Extant**

F. ASTERINIDAE Gray, 1840 J. (SIN)–Rec. Mar.

First: *Protremaster uniserialis* Smith, 1985, SIN, Antarctica. **Extant**

F. GANERIIDAE Sladen, 1889 **Extant** Mar.

F. CHAETASTERIDAE Sladen, 1889 K. (HAU)–Rec. Mar.

?First: *Chaetasterina gracilis* Hess, 1970, upper HAU, St Blaise, Neuchatel, Switzerland. **Extant**

F. ARCHASTERIDAE Viguier, 1878 T. (Mio.)–Rec. Mar.

?First: *Archaster patersoni* Spencer, 1915, Miocene, Port Elizabeth, South Africa. **Extant**

F. ODONTASTERIDAE Verrill, 1889 J. (BAJ/BTH)–Rec. Mar.

?First: *Odontaster priscus* Fell, 1954, Temaikan, near Onewhero, South Auckland, New Zealand. **Extant**

Order VELATIDA Perrier, 1891

F. TROPASTERIDAE Wright, 1880 J. (PLB) Mar.

First: *Tropidaster pectinatus* Forbes, 1850, lower PLB, *capricornus* Subzone, Mickleton Tunnel, Gloucestershire, England, UK.
Last: *Tropidaster pectinatus* Forbes, 1850, upper PLB, *stokesi* Subzone, Starfish Bed, Eype Cliff, Dorset, England, UK.

F. KORETHRASTERIDAE Danielsson and Koren, 1884 **Extant** Mar.

F. PTERASTERIDAE Perrier, 1884 **Extant** Mar.

F. SOLASTERIDAE Perrier, 1884 J. (PLB)–Rec. Mar.

First: *Solaster murchisoni* (Williamson, 1884), lower PLB, Lower Staithes Formation, *davoei* Zone, Robin Hood's Bay, Yorkshire, England, UK. **Extant**

F. CAYMANOSTELLIDAE Belyaev, 1974 **Extant** Mar.

F. XYLOPLACIDAE Baker *et al.*, 1986 **Extant** Mar.

Order FORCIPULATIDA Perrier, 1884

F. ZOROASTERIDAE Sladen, 1889 J. (SIN)–Rec. Mar.

First: *Terminaster* sp., lower SIN, *bucklandi* or *semicostatum* Zone, Hock Cliff, Gloucestershire, England, UK. **Extant**

F. ASTERIIDAE Gray, 1840 J. (HET)–Rec. Mar.

First: *Germanasterias amplipapularia* Blake, 1990, *angulata* Zone, Gemund, south-east Germany. **Extant**

F. HELISTERIDAE Viguier, 1878 **Extant** Mar.

F. BRISINGIDAE Fisher, 1928 T. (BUR)–Rec. Mar.

First: *Brisingella* sp., upper Burdigalian, Chita Peninsula, Japan. **Extant**

F. FREYELLIDAE Downey, 1986 **Extant** Mar.

INCERTAE SEDIS (Order) BOTHRIOCIDAROIDA Klem, 1904

The systematic position of the three known bothriocidarid genera remains unclear. They have been widely held as primitive echinoids, while Smith (1984b) reinterpreted them as stem Holothuroidea. They are classified here as a separate plesion allied to the stem group of both echinoids and holothurians.

F. BOTHRIOCIDARIDAE Klem, 1904 O. (LLO3/CRD)–S. (LUD) Mar.

First: *Unibothriocidaris bromidensis* Kier, 1982, and *Bothriocidaris kolatai* Kier, 1982, Pooleville Member, Bromide Formation, Oklahoma, and *B. solemi* Kolata, 1975, and *Neobothriocidaris* sp., Platteville Formation, Illinois, USA.
Last: *Neobothriocidaris peculiaris* Paul, 1967, and *N. minor* Paul, 1967, Starfish Bed, middle or upper ASH, Girvan, Scotland, UK. Smith (1984b, p. 168) cites the late Silurian (LUD) as the time of last occurrence.

Class HOLOTHUROIDEA de Blainville, 1834

The fossil record of holothurians consists largely of dissociated sclerites, together with rare occurrences of calcareous ring and/or body fossil material. Most fossil families have been defined on sclerite morphotypes, i.e. as parafamilies. The classification here follows that of Gilliland (1992) in which, where possible, fossil species are referred to the taxonomy of Recent holothurians, i.e. families (F.), and extinct parafamilies synonymous with the level of 'biological family' (P.[F]) below. Many species, however, cannot be assigned satisfactorily and these are discussed under their respective parafamilies (P.).

Plesion (Order) OPHIOCISTIOIDA Sollas, 1899

The Ophiocistioida, globose echinozoa with plated tube feet, a lantern apparatus and 'goniodonts', are included here as probable stem-group holothurians.

F. VOLCHOVIIDAE Hekker, 1938 O. (ARG–LLN1) Mar.

First: *Volchovia mobilis* Hekker, 1938, upper ARG, *Megalaspis* Limestone, Volchov, Leningrad, former USSR.
Last: *Volchovia norvegica* Regnéll, 1948, *Expansus* Shale, Gjeitungholmen, Asker, Norway.

F. EUCLADIDAE Gregory, 1896 S. (GOR) Mar.

First and Last: *Eucladia johnsoni* Woodward, 1869, lower Ludlow Shales, Dudley, England, UK.

F. SOLLASINIDAE Fedotov, 1926 S. (GLE)–D. (EIF) Mar.

First: *Euthemon igerna* Sollas, 1899, Wenlock Limestone, Croft Farm, near Malvern, England, UK.
Last: *Sollasina westfalica* (Richter, 1930), Selscheider Beds, Ebbelinghausen, Westphalia, Germany.

F. ROTASACCIDAE Haude and Langenstrassen, 1976 D. (GIV) Mar.

First and Last: *Rotasaccus dentifer* Haude and Langenstrassen, 1976, upper Wiedenester Schichten, Drolshagen, Öberbergisches Land, Germany.

F. *INCERTAE SEDIS* P. (ASS/SAK) Mar.

Last: Fragmentary goniodonts, Beattie Limestone, Florena Shale Member, Morris County, Kansas, USA (Kornicker and Imbrie, 1958). The wheel *Microantyx permiana* Kornicker and Imbrie, 1958, is found associated with these goniodionts but, although characteristic of ophiocistioids, this morphotype is not restricted to this group and hence these remains cannot be considered conspecific.

Order ARTHROCHIROTIDA Seilacher, 1961

F. PALAEOCUCUMARIIDAE Frizzell and Exline, 1966 D. (LOK/PRA) Mar.

First and Last: *Palaeocucumaria hunsrueckiana* Lehmann, 1958, Hunsrückschiefer, Rhineland, Germany.

Order DACTYLOCHIROTIDA Pawson and Fell, 1965

F. YPSILOTHURIIDAE Heding, 1942 C. (?HAS)–Rec. Mar.

First: *Clavallus spicaudina* (Gutschick *et al.*, 1967), Lodgepole Limestone, Madison Group, Montana, USA. This species could equally be referred to the Paracucumidae (Order Dendrochirotida). The oldest unequivocal species is *Palaeoypsilus liassicus* Gilliland (1992), Blue Lias Formation (SIN), Gloucestershire and Warwickshire, England, UK. **Extant**

Order DENDROCHIROTIDA Grube, 1840

Parafamily CALCLAMNIDAE Frizzell and Exline, 1955 [in part] C. (HAS)–Rec. Mar.

First: *Eocaudina marginata* Langenheim and Epis, 1957, several limestone units including Lodgepole Limestone, Madison Group, Montana, USA. **Extant**
Comment: A variety of 'buttons' and 'button-like plates' of Calclamnidae can be restricted to the Dendrochoritida but not to any one particular family.

Parafamily EXLINELLIDAE Deflandre-Rigaud, 1962, LUDWIGIIDAE Mostler, 1969 and SPINITIDAE Mostler, 1968 Tr. (ANS)–Rec. Mar.

First: *Kuehnites acanthotheelioides* Mostler, 1968 (Spinitidae), Hallstätt Limestone, Anatolia, Turkey. **Extant**

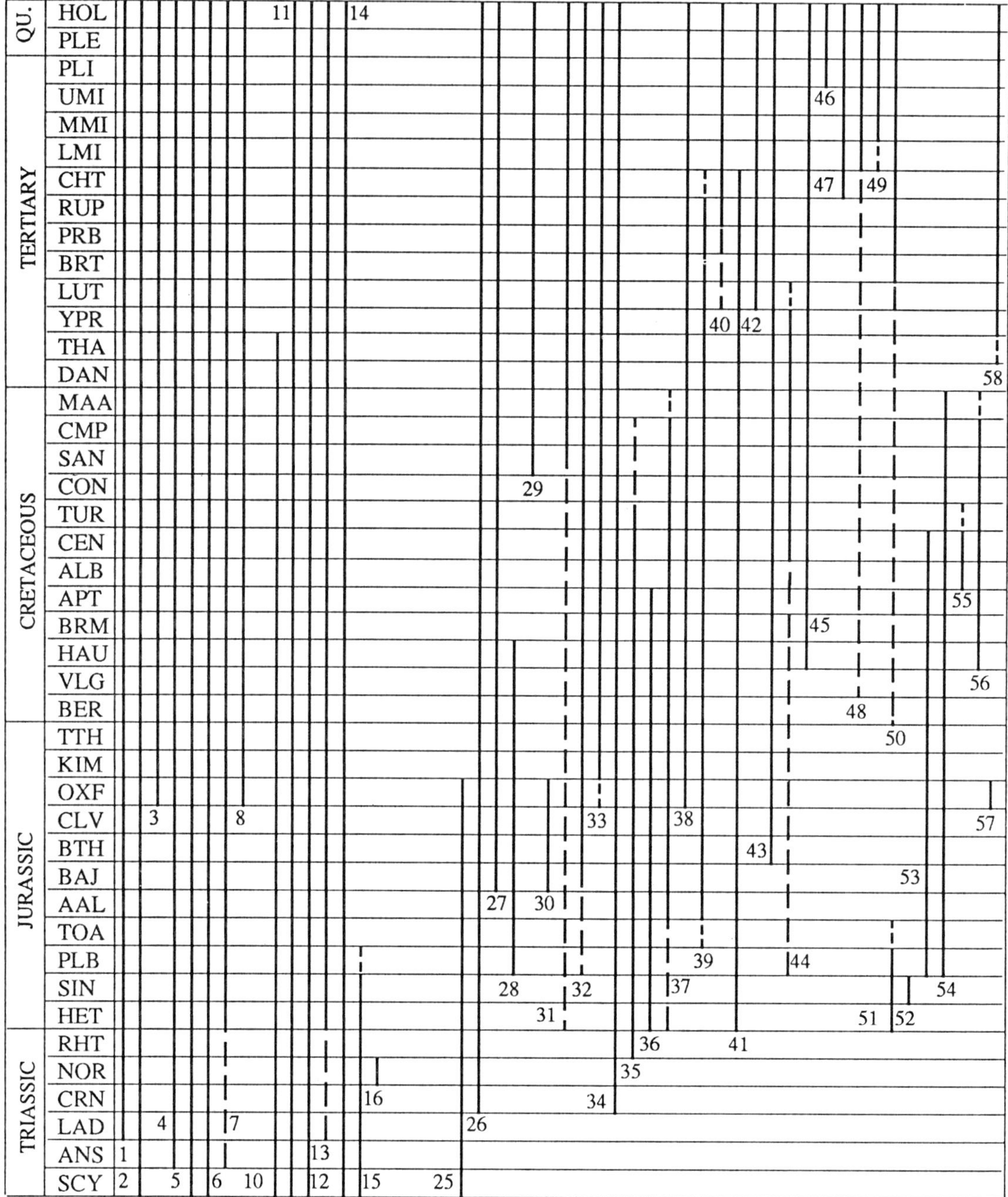

Fig. 25.5

Comment: These three parafamilies include 'basket-like' sclerites similar to those found in several families of the Dendrochirotida.

F. HETEROTHYONIDAE Pawson, 1970 Tr. (LAD)–Rec. Mar.

First: *Strobilothyone rogenti* Smith and Gallemí, 1991, La Riba Reef Formation, Catalonia, Spain (known from many body fossils). **Extant**

F. PSOLIDAE Perrier, 1902 Tr. (LAD)–Rec. Mar. (see Fig. 25.5)

First: *Monilipsolus mirabile* Smith and Gallemí, 1991, La Riba Reef Formation, Catalonia, Spain (known from many body fossils). **Extant**

F. PARACUCUMIDAE Pawson and Fell, 1965 C. (?HAS)–Rec. Mar.

First: *Clavallus spicaudina* (Gutschick *et al.*, 1967), Lodgepole Limestone, Madison Group, Montana, USA (see comments on Ypsilothuriidae above). **Extant**

Order ASPIDOCHIROTIDA Grube, 1840

Early records of the order include two taxa, known from body fossils, whose familial position is uncertain: *Bathysynactites viai* Cherbonnier, 1978, tentatively referred to the Aspidochirotida, from the Muschelkalk of Catalonia, Spain (ANS/LAD), and *Collbatothuria danieli* Smith and Gallemí, 1991, from the La Riba Reef Formation, Catalonia, Spain (LAD).

PERMIAN: TAT, KAZ, UFI, KUN, ART, SAK, ASS
CARBONIFEROUS: GZE, KAS, MOS, BSH, SPK, VIS, TOU
DEVONIAN: FAM, FRS, GIV, EIF, EMS, PRA, LOK
SILURIAN: PRD, LUD, WEN, LLY
ORDOVICIAN: ASH, CRD, LLO, LLN, ARG, TRE
CAMB.: MER, STD, CRF
SINIAN: EDI, VAR, STU

Key for both diagrams

1. Psolidae
2. Paracucumidae
3. Holothuriidae
4. Synallactidae
5. Laetmogonidae
6. Psychropotina
7. Molpadiidae
8. Caudinidae
9. Palaeochiridotidae
10. Achistridae
11. Myriotrochidae
12. Chiridotidae
13. Synaptidae
14. Priscopedatidae
15. Punctatitidae
16. Kozurellidae

ECHINOIDEA
PERISCHOECHINOIDEA

17. Unnamed
18. Echinocystitidae
19. Palaeodiscidae
20. Lepidocentridae
21. Palaechinidae
22. Unnamed
23. Hyattechinidae
24. Archaeocidaridae

CIDAROIDEA

25. Miocidaridae
26. Cidaridae
27. Psychocidaridae
28. Diplocidaridae

EUECHINOIDEA
ECHINOTHURIACEA

29. Echinothuriidae
30. Pelanechinidae

DIADEMATACEA

31. Diadematidae
32. Aspidodiadematidae
33. Micropygidae
34. Pedinidae

ECHINACEA

35. Pseudodiadematidae
36. Hemicidaridae
37. Acrosaleniidae
38. Saleniidae
39. Phymosomatidae
40. Glyptocidaridae
41. Stomechinidae
42. Stomopneustidae
43. Arbaciidae
44. Glyphocyphidae
45. Temnopleuridae
46. Echinidae
47. Echinometridae
48. Parechinidae
49. Strongylocentrotidae
50. Toxopneustidae

IRREGULARIA

51. Eodiadematidae
52. Unnamed

PYGASTEROIDA

53. Pygasteridae

HOLECTYPOIDA

54. Holectypidae
55. Anorthopygidae
56. Discoididae
57. Unnamed
58. Echinoneidae

Fig. 25.5.

F. HOLOTHURIIDAE Ludwig, 1894
J. (OXF)–Rec. Mar.

Tables of Priscopedatidae may be referred to the Holothuriidae (see above). The record here is based upon a morphotype which can be restricted to the family.

First: *Calclamnella elliptica* Deflandre-Rigaud, 1946, and several other species, Oxford Clay, Normandy, France. **Extant**

F. SYNALLACTIDAE Ludwig, 1894
Tr. (ANS)–Rec. Mar.

First: *Priscopedatus triassicus* Mostler, 1968, and several other species. Hallstätt Limestone, Tyrol, Austria. '*P. triassicus*-type' tables are only tentatively included in the Synallactidae. Fossil examples of other synallactid sclerite types are not known. **Extant**

Order **ELASIPODA Théel, 1882**

Suborder **DEIMATINA Hansen, 1975**

Oneirophantites tarragonensis Cherbonnier, 1978, a body fossil from the Muschelkalk (ANS/LAD) of Catalonia, Spain, is accepted as an elasipod but its original subordinal designation is questionable.

F. LAETMOGONIDAE Ekman, 1925
D. (EIF)–Rec. Mar.

First: *Protocaudina kansasensis* Hanna, Dundee Limestone, Ontario, Canada. **Extant**

Suborder **PSYCHROPOTINA Hansen, 1975**
C. (BRI)–Rec. Mar.

First: *Tetravirga fordalensis* Frizzell and Exline, 1955, Carboniferous Limestone Series, Fife, Scotland, UK. **Extant**

Comment: Sclerites of Elpidiidae Théel, 1882, and Psychropotidae Théel, 1882, are similar and fossil taxa cannot be reliably assigned to one or other.

Order **MOLPADIIDA Haekel, 1896**

F. MOLPADIIDAE Muller, 1850
Tr. (ANS?)–Rec. Mar.

First: *Staurocumites bartensteini* Deflandre-Rigaud, 1952, Hallstätt facies, Eastern Alps, Germany. '*Staurocumites*-type' table are only questionably referred to the Molpadiidae. *Priscopedatus* sp. 5 Speckmann, 1968, a '*Priscolongatus*-type' table (Gilliland, 1992a), recorded from the same strata, is also questionably referred to the Molpadiidae. The next oldest record is calcareous ring Type 1 Gilliland (1992a), from the Blue Lias Formation (HET), Warwickshire, England, UK; the sclerite morphospecies from the same 'biological' species as this calcareous ring is interpreted to be *S. bartensteini*. The oldest unequivocal sclerites of the Molpadiidae are several taxa, including *Calcancorella spectabilis* Deflandre-Rigaud, 1959, from the Middle Oligocene, Schleswig-Holstein, Germany. **Extant**
Comment: Sclerites of the Punctatitidae might also be referred to the Molpadiidae.

F. CAUDINIDAE Heding, 1931 J. (OXF)–Rec. Mar.

First: *Pedatopriscus pinguis* (Deflandre-Rigaud, 1946), Oxford Clay, Normandy, France. **Extant**
Comment: Tables of the Priscopedatidae may be referred to the Caudinidae (see above). The record here is based on a morphotype which can be restricted to the family.

Order APODIDA Brandt, 1835

Two early apodid wheel morphospecies cannot be restricted to a particular family (wheels occur in all families except the Achistridae).

Theelia? *hexacneme* Summerson and Campbell, 1958, is recorded from the upper Breathith (upper Pottsville) Formation, Kentucky, USA (BSH). Wheels from the Waulsortian (TOU) of the Republic of Ireland (MacCarthy, pers. comm.) are similar to this species but also resemble ophiocistioid wheels, and hence are not included here. *T. hexacneme* is reported from the Mississippian (PND/ARN) of Slovakia (Kozur *et al.*, 1976), but there are no supporting descriptions or figures.

Thallatocanthus consonus Carini, 1962, occurs in the Wewoka Formation, Oklahoma, USA (MYA). Wheels referred to this species by Gutschick *et al.* (1967) from the Carboniferous (HAS) are referred instead to the Paleochiridotidae.

F.(P.) PALEOCHIRIDOTIDAE Frizzell and Exline, 1955 C. (HAS–KAS) Mar.

First: *Rota martini* Langenheim and Epis, 1957, Escabrosa Limestone, Arizona, USA.
Last: *Paleochiridota plummerae* Croneis, 1932, Keechi Creek Shale, Mineral Wells Formation, Texas, USA.

F.(P.) ACHISTRIDAE Frizzell and Exline, 1955 D. (GIV)–T. (?DAN/THA) Mar.

First: *Porachistrum multiperforata* Beckmann, 1965, *Scutellum*-Schichten, Westfalen, Germany.
Last: *Achistrum issleri* Croneis, Palaeocene, Kutch, India. This record (Soodan, 1972) is unaccompanied by either a figure or description. The next youngest species is *Achistrum monochordata* Hodson *et al.* (1956), Speeton Clay (BRM), Yorkshire, England, UK.

F. MYRIOTROCHIDAE Östergren, 1907 P. (TAT)–Rec. Mar.

First: *Theelia praeacuta* Mostler and Rahimi-Yazd, 1976, Ali Bashi Formation, Ali Bashi Mountains, northern Iran. **Extant**

F. CHIRIDOTIDAE Östergren, 1898 P. (ZEC)–Rec. Mar.

First: *Protheelia geinitziana* (Spandel, 1898), Zechstein, Thüringen, Germany. **Extant**

F. SYNAPTIDAE Burmeister, 1837 Tr. (LAD/CRN?)–Rec. Mar.

First: *Theelia liptovskaensis* Gazdzicki *et al.*, 1978, Korytnica Limestones, Carpathians, Czechoslovakia. This species may belong to the Chiridotidae. The next oldest species is *Theelia synapta* Gilliland (1992), from the Blue Lias Formation (HET) of Dorset and Warwickshire, England, UK. **Extant**

NON-ALIGNED HOLOTHURIAN PARAFAMILIES

Parafamily CALCLAMNIDAE Frizzell and Exline, 1955

This parafamily includes fossil plate morphotypes many of which are not holothurian; those that can be referred to the class are included in the relevant families below.

An equivocally included plate taxon which should be noted, because it may prove to be the oldest recorded holothurian, is *Mercedescaudina triperforata* Schallreuter, 1968, from erratics of the Brick Limestone (Middle Ordovician), Island of Hiddensee, Baltic Sea.

The oldest unequivocal (crown group) holothurian is an undescribed body fossil, complete with calcareous ring and dermal sclerites (Smith and Jell, pers. comm.) from the Upper Silurian (PRD) of Australia. The sclerites, if found isolated, would be assigned to the Calclamnidae.

Parafamily PRISCOPEDATIDAE Frizzell and Exline, 1955 C. (HAS?)–Rec. Mar.

First: *Clavallus spicaudina* Gutschick *et al.*, 1967, a spired plate morphotype (see Ypsilothuriidae below). The oldest recorded table, the main morphotype in the Priscopedatidae, is *Priscopedatus* sp. *nov.* Kozur *et al.*, 1976, from the Carboniferous (PND/ARN?) of Slovakia, Czechoslovakia. This record is not accompanied by either a figure or a description and must be considered tentative.

The next oldest (table) species is *Priscopedatus quinquespinosus* Mostler and Rahimi-Yazd, 1976, from the Permian (TAT), *Arexilevis* Zone Limestones, Ali Bashi Mountains, northern Iran. **Extant**
Comment: An important parafamily of many fossil species of table and spired plate morphotypes which probably includes taxa from four of the six extant orders. A number of different 'groups' can be recognized, some of which can be assigned, with varying degrees of certainty, to biological families (e.g. *Pricopedotus triassicus*-type tables to the Synallactidae). The earliest recorded table species could be accepted questionably as the oldest record of several families of the Dendrochirotida (but not the Heterothyonidae, Paracucumidae or Psolidae), the Aspidochirotida or the Caudinidae (Molpadiida).

Parafamily PUNCTATITIDAE Mostler and Rahimi-Yazd, 1976 P. (TAT)–J. (SIN/PLB) Mar.

First: *Punctatites aequiperforatus* Mostler and Rahimi-Yazd, 1976, *Paratirolites* Beds, Ali Bashi Mountains, northern Iran.
Last: *Punctatites triplex* Mostler, 1972, 'thin-bedded red-brown limestone', Tyrol, Austria.
Comment: The rod-shaped sclerites of this parafamily can be referred to either of the orders Dendrochirotida or Molpadiida (Molpadiidae).

Parafamily (F.) KOZURELLIDAE Mostler, 1972 Tr. (NOR) Mar.

First and Last: *Kozurella formosa* Mostler, 1972, Hallstätt Limestone, Northern Calcareous Alps, Austria.

Class ECHINOIDEA Leske, 1778

Except where indicated, the taxonomic framework adopted here is that of Smith (1984b); authorship and generic plus stratigraphical assignments of species follow Lambert and Thiéry (1909–1925) and Kier and Lawson (1978), with some modification of assignment to genera after Mortensen (1928–1951); generic assignments to Palaeozoic families (Subclass Perischoechinoidea) follow Smith (1984b, table 7.1); generic assignments to post-Palaeozoic families (subclasses Cidaroidea, Euechinoidea) follow Durham *et al.* (1966) or Kier and Lawson (1978) for subsequent ascriptions. Amendments to these assignments are noted only where they affect the recorded family range.

The important and wide-ranging syntheses cited above only partially correct errors and omissions in the extensive earlier (often nineteenth century) literature, and inevitably introduce a few of their own. Taxonomic and stratigraphical details for species cited below are also based largely on the available literature. Consequently, they are not uniformly consistent in concept, and should be interpreted accordingly. For a consistent attempt to distinguish monophyletic from other echinoid families extinct in post-Palaeozoic time, see Smith and Patterson (1988).

For brevity, references cited for years prior to 1967 are not listed in the bibliography where they may be found in Weisbord's (1971) compendium; more recent references are listed only where absent from the key publications of Kier and Lawson (1978) and Smith (1984b). References additional to authorship of taxonomic names are given only where they add significant taxonomic or stratigraphical data; citations and references deemed insignificant by the editors have been excluded.

Class ECHINOIDEA Leske, 1778

Subclass PERISCHOECHINOIDEA M'Coy, 1849

Unnamed stem group O. (RAW)–S. (TEL) Mar.

First: *Aulechinus grayae* Bather and Spencer, 1934, *Ectinechinus lamonti* MacBride and Spencer, 1938, *Eothuria beggi* MacBride and Spencer, 1938; all Drummuck Group, Girvan, Scotland, UK.
Last: *Aptilechinus caledonensis* Kier, 1973, 'Starfish Bed', Gutterford Burn, Pentland Hills, Scotland, UK.
Intervening: None.
Comment: The four genera *Eothuria, Aulechinus, Ectinechinus* and *Aptilechinus*, although not a monophyletic group, comprise the most primitive true echinoids and include the earliest known echinoid species.

Subclass UNNAMED PARAPHYLETIC GROUP

Order ECHINOCYSTITOIDA Jackson, 1912

F. ECHINOCYSTITIDAE Gregory, 1897 S. (GOR)–P. (ASS/SAK) Mar.

First: *Echinocystites pomum* Wyville Thomson, 1861, lower Ludlow, Leintwardine, England, UK; *Gotlandechinus balticus* Regnéll, 1956, Klinteberg Group (lower Ludlow), Gotland, Sweden.
Last: *Xenechinus parvus* Kier, 1958, Niel Ranch Formation (Wolfcampian), west Texas, USA.
Intervening: D1 and C1, known only from single specimens.

Paraphyletic group 'PALAECHINOIDA' Haeckel, 1866

The two monospecific Silurian genera *Myriastiches* and *Koninckocidaris* are placed *incertae sedis* within this group by Smith (1984b). Their eventual family ascription may necessitate increase in the number of families here recognized in the group, or an extension to the stratigraphical range of at least one family.

Suborder PALAEODISCOIDA Smith, 1984

Unnamed stem group (= F. PALAEODISCIDAE Gregory, 1897) S. (GOR) Mar.

First and Last: *Palaeodiscus ferox* Salter, 1857, lower Ludlow, Leintwardine, England. The imperfectly known *P. gothicus* Wyville Thomson, 1861, is from the same locality 'and probably identical with *P. ferox*' (Mortensen, 1928–1951).

F. LEPIDOCENTRIDAE Loven, 1874 D. (EMS?)–P. (TAT?) Mar.

First: *Lepidocentrus lenneanus* Wolburg, 1933, *cultrijugatus* Zone (probably Lower Devonian), Lenne, near Schmallenberg, Germany, and Middle Devonian species of *Lepidocentrus* are poorly known. *Lepidechinoides hunti* Cooper, 1931, Skaneateles Formation, Hamilton Group (Middle Devonian), Ithaca, New York, USA, is the earliest well-documented lepidocentrid.
Last: *Pronechinus anatoliensis* Kier, 1965, Gomaniibrik Formation, Diyarbakir Province, Turkey (?Dzulfian *fide* Philip, *in* Harland (1967)).
Intervening: D3, C.

F. PALAECHINIDAE M'Coy, 1849 D. (LOK/EMS?)–C. (HLK?) Mar.

First: *Porechinus porosus* Dehm, 1961, Bundenbacher Schiefer, Bundenbach, near Kirn (Nahe), Germany.
Last: ?*Melonechinus multiporus* (Norwood and Owen, 1846), St Louis Limestone, St Louis, Missouri, USA; other species of similar date are more poorly known.
Intervening: All other known palaechinids are early Carboniferous in age (Kier, 1965).

Paraphyletic group 'ARCHAEOCIDAROIDA' Smith, 1984

Unnamed stem group D. (FRS?) Mar.

First and Last: *Nortonechinus welleri* Thomas, 1920, Lime Creek Shale, Portland, Iowa, USA (Kier, 1968). Other supposed early archaeocidaroids are known only from spines and isolated plates, so their true ascription remains uncertain. The earliest of these is *Silurocidaris clavata* Regnéll, 1956 from the Upper Silurian, lower Ludlow: Gotland, Sweden.

F. HYATTECHINIDAE Smith, 1984 C. (HAS/IVO–CHD/ARU?) Mar.

First: *Hyattechinus elegans* Jackson, 1929, TOU, Belgium; *H. dixoni* Hawkins, 1935, *H. toreumaticus* Hawkins, 1935, just above Lower Limestone Shales, Z_1 Zone, Pembroke, South Wales, UK, are also TOU.
Last: *Perischodomus fraiponti* Jackson, 1929, lower VIS, Belgium.
Intervening: All representatives of this family have been recorded only from the Lower Carboniferous (lower Mississippian), but precise age relationships are not known.

F. ARCHAEOCIDARIDAE M'Coy, 1849 C. (HAS/IVO–POD?) (P) Mar.

(Family name on ICZN Official List: Direction 41).
First: *Archaeocidaris nerei* (Münster, 1839), TOU, Tournai, Belgium (Jackson, 1929); based on fragmentary material, as are other of the earliest Carboniferous archaeocidaroids. *Lepidocidaris squamosa anglica* Hawkins, 1935, Carboniferous Limestone (?C Zone), Preston, Lancashire, England, UK is ?upper Tournaisian.
Last: *Archaeocidaris immanensis* Kier, 1958, Dewey Limestone, Oklahoma, USA is the only well-preserved Upper Carboniferous archaeocidarid; other Upper Carboniferous archaeocidarids are known only from disassociated test elements. Although Lambert and Thiéry (1909–1925) and Kier and Lawson (1978) together list six supposed archaeocidaroid species from the Permian, Kier (1965) notes that no Permian cidarids with more than two plate columns in each interambulacrum have been found. These species may all be miocidarids.
Intervening: C. (l., u.).

Subclass CIDAROIDEA Claus, 1880

Order CIDAROIDA Claus, 1880

Smith and Wright (1989), in a revised classification and phylogenetic interpretation of this order, recognize only two families (Rhabdocidaridae Tr.?–Rec. and Cidaridae J. (BAJ)–Rec.) additional to the miocidarids, and only 39 assigned genera/subgenera. The less-satisfactory classification of Durham *et al.* (1966) is, however, maintained here, pending assignment of the additional taxa of low species diversity which they recognize but which are excluded from the Smith and Wright revision.

F. MIOCIDARIDAE Durham and Melville, 1957 P. (KAZ)–J. (OXF) (p) Mar.

First: *Miocidaris connorsi* Kier, 1965, Bell Canyon Formation, Texas, USA; *M. keyserlingi* (Geinitz, 1848), Permian, Hungary, and Ford Formation, Zechstein Cycle 1, Sunderland, England, UK; the only two confidently ascribed species of miocidarids reliably recorded from the Palaeozoic (Kier, 1965). *M. cannoni* Jackson, 1912, arguably from the Lower Carboniferous of Colorado, USA, is based on a poorly preserved internal mould of uncertain family and generic ascription.
Last: *Pachycidaris thieryi* Collignon and Lambert, 1928, Upper Jurassic, Europe. (*Miocidaris* itself is known from several Lower Jurassic species).
Intervening: SCY, LAD–RHT.
Comment: A paraphyletic stem group: 'Only those miocidarids with a perignathic girdle of apophyses definitely belong to this clade, the others being advanced stem group echinoids, primitive stem group cidaroids or primitive stem group euechinoids' (Smith and Wright, 1989).

F. CIDARIDAE Gray, 1825 Tr. (CRN)–Rec. Mar.

First: *Leurocidaris montanaro* (Zardini, 1973), St Cassian Beds, northern Italy, a true cidaroid with apophyses but a species and genus difficult to assign to any family with certainty 'because nothing is known of its spines, pedicellariae or peristomial plates' (Kier, 1977). *Polycidaris regularis* (Münster, 1841) is known only from the holotype, whose Triassic origin is doubted by Kier (1977). *Mikrocidaris pentagona* (Münster, 1841), referred to the Cidaridae by Kier (1977) rather than to the Miocidaridae as by Durham *et al.* (1966), is shown to lack apophyses and therefore excluded from the cidaroids by Kier (1984b). Four unnamed species with cidaroid apophyses are, however, known from fragments from the St Cassian Beds. **Extant**
Intervening: Very few gaps between Tr(u.) and Rec.

F. PSYCHOCIDARIDAE Ikeda, 1936 J. (BAJ)–Rec. Mar.

First: *Merocidaris honorinae* (Cotteau, 1880), *Caenocidaris cucumifera* (L. Agassiz, 1840), (Cotteau, 1875–80), *Anisocidaris bajocensis* (Cotteau, 1880), all Middle Jurassic (BAJ) of Europe and probably referable to genus *Balanocidaris sensu* Smith and Wright (1989). *Levicidaris zardinia* Kier, 1977 and *Megaporocidaris mariana* Kier, 1977, both from the St Cassian Beds, northern Italy, were ascribed to the Psychocidaridae by Kier (1977) but later excluded from the cidaroids by Kier (1984b), who observed that they lacked lantern supports (apophyses). **Extant**
Intervening: K.(u.), Pal./Eoc., Mio.

F. DIPLOCIDARIDAE Gregory, 1900 J. (PLB)–K. (HAU) Mar.

First: *Diplocidaris menchikoffi* Lambert, 1937, Lias (upper Domerian), Morocco.
Last: *Diplocidaris bicarinata* Weber, 1934, Hauterivian, Crimea, former USSR.
Intervening: TOA, J.(m., u.), VLG. The family is clearly polyphyletic, according to Smith and Patterson (1988).

Subclass EUECHINOIDEA Bronn, 1860

Infraclass ECHINOTHURIOIDEA Claus, 1880

Cohort ECHINOTHURIACEA Jensen, 1981

Order ECHINOTHURIOIDA Claus, 1880

F. ECHINOTHURIIDAE Thomson, 1872 K. (SAN)–Rec. Mar.

First: *Echinothuria floris* Woodward, 1863, *M. coranguinum* Zone, Upper Chalk, Kent, England, UK. *Araeosoma* (?) *bruennichi* Ravn, 1928, *Asthenosoma* (?) *striatissimum* Ravn, 1928 also occur in the Chalk (upper Senonian) of Denmark, but are known only from spines. Such distinctively 'hoofed' spines are a feature of Recent echinothuriids; Mortensen (1928–1951) notes that 'it is likely that one or other of the spines belongs to *E. floris*'. **Extant**
Intervening: DAN, PLI.

F. PHORMOSOMATIDAE Mortensen, 1934 **Extant** Mar.

F. PELANECHINIDAE Groom, 1887
J. (BAJ–OXF) Mar.

First: *Pelanodiadema oolithicum* Hess, 1973, upper Hauptrogenstein, Schinznach, northern Jura, Switzerland. Congeneric with *Pelanechinus corallinus* according to Smith and Patterson (1988).
Last: *Pelanechinus corallinus* (Wright, 1856), Coralline Oolite, Malton, Yorkshire, and Calne, Wiltshire, England, UK.

Infraclass ACROECHINOIDEA Smith, 1981

Cohort DIADEMATACEA Duncan, 1889

Order DIADEMATOIDA Duncan, 1889

F. DIADEMATIDAE Gray, 1855
J. (HET?)–Rec. Mar.

First: *'Eodiadema' collenoti* (Cotteau, 1882), lower Lias, La Verune, near Joyeuse (Ardèche), Semur, and Saulieu, France (Smith, 1990); probably a diadematoid (Smith and Wright 1990, p. 113) although the Lower Jurassic genus, *Eodiadema*, itself was removed from the Diadematidae *sensu* Durham *et al.* (1966) to the new family Eodiadematidae by Smith (1984b). The only unquestionable fossil diadematoid known from more than isolated spines is *Centrostephanus fragilis* (Wiltshire, *in* Wright, 1882), Upper Chalk (SAN–CMP), England, UK and France (Smith and Wright, 1990). **Extant**
Intervening: There are few Cretaceous and Tertiary records.

F. ASPIDODIADEMATIDAE Duncan, 1889
J. (PLB?)–Rec. Mar.

First: *Eosalenia varusense* (Cotteau, 1881), Lias, France; referred to genus *Eosalenia* together with *E. miranda* Lambert, 1905, Middle Jurassic (BTH), France, and *Pedinothuria barottei* Lambert and Thiéry, 1911, upper BTH, Vesaignes, Haute-Marne, France; new material of *Eosalenia* with spines shows this to be a true aspidodiadematid (Smith, pers. comm.). **Extant**
Intervening: CEN.

F. LISSODIADEMATIDAE Fell, 1966 **Extant** Mar.

Order MICROPYGOIDA Jensen, 1981

F. MICROPYGIDAE Mortensen, 1904
J. (OXF?)–Rec. Mar.

First: *Pedinothuria cidaroides* Gregory, 1897, Weisser Jura, Upper Jurassic (OXF?), Germany, is possibly an early micropygoid (Smith and Wright, 1990). Otherwise this monogeneric family is exclusively Recent, and of deep-water habitat. **Extant**

Order PEDINOIDA Mortensen, 1939

F. PEDINIDAE Pomel, 1883 Tr. (CRN)–Rec. Mar.

First: *Hemipedina(?)incipiens* Bather, 1990, based on a fragment from the Raiblian (upper CRN) of Bakony, Hungary, is considered by Kier (1977) to be intermediate between the cidaroids and the pedinoids. The earliest certain pedinoid is *Hemipedina hudsoni* Kier, 1977, Elphinstone Group (probably Sumra Formation) (NOR), Oman, Arabia. **Extant**
Intervening: A near-continuous record from early Jurassic to Recent.

Cohort ECHINACEA Claus, 1876

F. PSEUDODIADEMATIDAE Pomel, 1883
Tr. (RHT)–K. (CMP?) Mar.

First: *Pseudodiadema silbinense* Stefanini, 1923, Upper Triassic, Selvena, near Sienna, Italy, 'the only Triassic echinoid that with some certainty belongs to the... Echinacea' (Kier, 1977).
Last: *Heterodiadema libycum* (Desor, 1846), widely distributed in CEN of North Africa, Middle East, and elsewhere; possibly TUR, arguably CMP. Cainozoic records of this family are of uncertain or originally incorrect assignment.
Intervening: Good stratigraphical record throughout the Jurassic, early Cretaceous, and especially CEN.
Comment: *Heterodiadema* was transferred from the Hemicidaridae by Smith and Patterson (1988).

Superorder STIRODONTA Jackson, 1912

F. HEMICIDARIDAE Wright, 1857
J. (HET)–K. (APT) Mar.

First: *Pseudodiadema primaevum* Lambert, 1904 = *Hessotiara minor* Lambert, 1904, lower Lias, Le Simon-la-Vineuse (Vendée), France, although taxonomic position uncertain. The supposed hemicidarid *Plesiocidaris florida* (Merian, 1855), lower Lias, Gürbefall, Bernese Alps, Switzerland (Desor and de Loriol, 1868–1872) is a cidaroid (Smith, 1990).
Last: *Hemicidaris prestensis* (Cotteau, 1863), APT. 'CEN records of *Hemicidaris* and *Pseudocidaris* are fragmentary or based on spines and thus inadequate (although they may prove eventually to be correct)' (Smith and Patterson, 1988).

Order CALYCINA Gregory, 1900

(Synomyn of SALENIOIDA Delage and Hérouard, 1903, *fide* Smith and Wright (1990).)

F. ACROSALENIIDAE Gregory, 1900
J. (HET?)–K. (CMP/MAA?) (p) Mar.

First: ?*Acrosalenia chartroni* Lambert, 1904, lower Lias, Revrac, Vendée, France. Kier (1977) does not confirm Lambert and Thiéry's (1909–1925) tentative assignment of *A. balsami* (Stoppani, 1860) (Tr., RHT) to *Acrosalenia*, or pre-HET origins of the family. Smith and Patterson (1988) note that pre-AAL species of supposed acrosaleniids are poorly preserved and of dubious taxonomic assignment. The earliest undoubted acrosaleniid is probably *A. lycetti* Wright, 1851, Crickley Oncolite Member, (AAL), Inferior Oolite Group (Smith, 1984b).
Last: *Eurysalenia minima* Kier, 1966, Pierre Shale, Wyoming, USA; *Polysalenia notabilis* Mortensen, 1932, *P. cottaldi* Mortensen, 1932, both CMP, Scania, Sweden. *Heterosalenia occidentalis* Hawkins, 1923 may extend to MAA (Smith and Wright, 1990).
Intervening: Good stratigraphical record through the middle to late Jurassic and early Cretaceous.
Comment: Recognized as a paraphyletic group by Smith and Patterson (1988); the restricted monophyletic grouping of Smith and Wright (1990) indicates an earlier last appearance.

F. SALENIIDAE L. Agassiz, 1838
J. (OXF)–Rec. Mar.

First: *Salenia taurica* Weber, 1934, Upper Jurassic (Sequanian), Crimea, former USSR. *Poropeltaris sculptopunctata* Quenstedt, 1875, middle OXF, Natheim, Germany, is ascribed (as a subgenus) to *Hyposalenia* by Smith and Wright (1990). **Extant**
Intervening: TTH, NEO (HAU?); thereafter numerous species provide a near-continuous record for the family.

Order PHYMOSOMATOIDA Mortensen, 1904

F. PHYMOSOMATIDAE Pomel, 1883 J. (TOA?)–T. (RUP/CHT?) Mar.

First: ?*'Jacquiertia' minuta* Mortensen and Mercier, 1939, Lias, France. *Leptechinus jutieri* (Cotteau, 1883), Lias (Charmouthian), Mazenay, France, a supposed earlier phymosomatid, is a cidaroid (Smith, pers. comm.).
Last: *Thylechinus sethuramae* Vredenburg, 1922, Oligocene, Burma. *T. pusillus* (Münster, 1826), Oligocene, Germany. *Porosoma* is also known from Oligocene species. Jensen (1981) removed the single genus *Glyptocidaris* from the Phymosomatidae to a new family of its own; hitherto the type species *G. crenularis* A. Agassiz, 1853 had been regarded as the only known Recent phymosomatid.
Intervening: Diverse through the Cretaceous, Palaeocene and Eocene.

F. GLYPTOCIDARIDAE Jensen, 1981 T. (LUT?)–Rec. Mar.

First: *Glyptocidaris heteroporus* (Lambert, 1897), Eocene, Aude, France. Lambert and Thiéry (1909–1925) ascribe two species from the Eocene of Alicante, Spain, to a synonym of *Glyptocidaris*: *Heteractis lloreae* (Cotteau, 1890), *H. vilanovae* (Cotteau, 1890). **Extant**
Intervening: Eoc./Oli., PLI.

F. STOMECHINIDAE Pomel, 1883 J. (HET)–T. (CHT) Mar.

First: *Jeannetia mortenseni* Mercier, 1937, lower Lias, France; *Diplechinus hebbriensis* Lambert, 1931, middle Lias (Domerian), (PLB), Morocco, is more commonly regarded as the earliest stomechinid (Mortensen, 1928–1951).
Last: *Phymotaxis mansfieldi* Cooke, 1941, Suwannee Limestone (Upper Oligocene), Florida, USA.
Intervening: Good stratigraphical record through the Jurassic and Cretaceous.

F. STOMOPNEUSTIDAE Mortensen, 1903 T. (LUT)–Rec. Mar.

First: *Stomopneustes antiquus* Nisiyama, 1966, Eocene, Japan. **Extant**
Intervening: Oli., Mio.

F. ARBACIIDAE Gray, 1855 J. (BTH)–Rec. Mar.

First: *Atopechinus cellensis* Thiéry, 1928, BTH, Celles, Ardèche; also *Acrosaster michaleti* Lambert, *in* Lambert and Thiéry, 1914, also from France. **Extant**
Intervening: Moderately complete record from Jurassic to Recent.

Superorder CAMARODONTA Jackson, 1912

F. GLYPHOCYPHIDAE Duncan, 1889 J. ?(PLB)/K. (HAU)–T. (LUT?) Mar.

First: ?*Glyptodiadema cayluxense* (Cotteau, 1878), Lias (Charmouthian), Caylus, France, but too poorly known for certain ascription (Smith and Patterson, 1988). *Hemidiadema neocomiense* (Cotteau, 1882), Calcaire à Spatangues, Auxerre (Yonne), France, is the earliest definite record of the family.
Last: *Glyphocyphus (Rhabdopleurus) ataxensis* Cotteau, 1886, Middle Eocene, France. (Several other Middle Eocene glyphocyphids are known.)
Intervening: NEO–CEN, Eoc.

Order TEMNOPLEUROIDA Mortensen, 1941

F. TEMNOPLEURIDAE A. Agassiz, 1872 K. (HAU)–Rec. Mar.

First: *Glyptechinus montmolini* (Desor, 1858), Marnes d'Hauterive, Villers-le-Lac (Doubs), France (Cotteau, 1862–1868). **Extant**
Intervening: APT–TUR, Eoc., Mio./PLI.
Comment: Jensen (1981), on the basis of differences in tooth microstructure, indicated that this family might be divided into two or perhaps three families.

Order ECHINOIDA Claus, 1876

F. ECHINIDAE Gray, 1825 T. (ZAN/PIA)–Rec. Mar.

First: *Echinus algirus* Pomel, 1885, Pliocene, Algeria. **Extant**
Intervening: PLE.
Comment: Jensen (1981) restricts this family to the single genus *Echinus*; other formerly ascribed genera are included within the Parechinidae.

F. ECHINOMETRIDAE Gray, 1825 T. (CHT)–Rec. Mar.

First: *'Echinometra' prisca* Cotteau, 1875, although the characteristic plates of the apical disc are not preserved. *Echinometra thomsoni* Haime, 1853, Baluchistan, Pakistan (dated as ?middle Eocene by Philip, *in* Harland (1967)) is shown by Smith (1988b) to be an imperfectly preserved phymosomatoid. The oldest undoubted *Echinometra* is *E. miocenica* de Loriol, 1902 from the Upper Miocene (TOR) of France (Smith, 1988) = *E. mathaei* (Blainville, 1825), and from the Lower Miocene (BUR) according to Negretti *et al.* (1990). **Extant**
Intervening: Mio., PLI, PLE.

F. PARECHINIDAE Mortensen, 1903 (emend. Jensen, 1981) K.? (VLG?)/T. (BUR?)–Rec. Mar.

First: *Psammechinus bernouillensis* Valette, 1908, VLG, Yonne, France and some 10 other Cretaceous species have been ascribed to *Psammechinus* (Lambert and Thiéry, 1909–1925; Kier and Lawson, 1978) plus Eocene species (e.g. *'P.' biarritzensis* Cotteau), but true generic ascription is unknown (Mortensen, 1928–1951). Most may be of *Spaniocyphus*, whose family ascription is uncertain (Durham *et al.*, 1966). True echinids seemingly begin in the Miocene, with many circum-Mediterranean species ascribed to *Psammechinus* or *Stirechinus*, and the South American *Hypechinus patagonensis* (d'Orbigny, 1842) and *Isechinus praecursor* (Ortmann, 1904), both from Patagonia. Smith (1988b) accepts *'Toxopneustes' bouryi* Cotteau, 1883, Upper Miocene, France, as a *Paracentrotus* and probable early parechinid. **Extant**
Intervening: ?Mio., ?PLI.
Comment: Jensen (1981, p. 22, 84) includes, within this family, genera formerly ascribed to the Echinidae, except for the genus *Echinus* itself.

F. STRONGYLOCENTROTIDAE Gregory, 1990
T. (AQT?)–Rec. Mar.

First: *Strongylocentrotus antiquus* Philip, 1965, Longfordian, Australia; *S. franciscanus* (A. Agassiz, 1863), *S. purpuratus* (Stimpson, 1857), both Recent species, are also recorded from the Pliocene of the North American West Coast; the genus is also recorded from the Miocene of Oregon, USA. **Extant**
Intervening: Neogene records of *Strongylocentrotus* from Europe may be based on misidentifications of the parechinid *Paracentrotus* (Mortensen, 1928–1951).

F. TOXOPNEUSTIDAE Troschel, 1872
K.?/T. (BRT?)–Rec. Mar.

First: *Scoliechinus axiologus* Arnold and H. L. Clark, 1927, Cretaceous (?), Leyden region, St James Parish, Jamaica. More confidently dated, supposed early toxopneustids are from the Eocene: *Diplosalenia gosseleti* (Cotteau, 1894), France, and *Lytechinus floralanus* (Cooke, 1941), Ocala Limestone, Alabama and Georgia, USA. **Extant**
Intervening: ?Oli., Mio./PLI.
Comment: Family Parasaleniidae Mortensen, 1903, adopted by Durham *et al.* (1966) was included within the Toxopneustidae by Jensen (1981, p. 84).

Cohort IRREGULARIA Latreille, 1825

F. EODIADEMATIDAE Smith, 1984
J. (HET)–(TOA?) Mar.

First: *Eodiadema bechei* (Wright, 1860), lower Lias, Lyme Regis, England, UK. *E. regulare* (Münster, *in* Wissmann and Münster, 1841), Tr. (CRN), ascribed to *Eodiadema* by Lambert and Thiéry (1909–1925), is reascribed to the cidarid *Polycidaris* by Kier (1977), who doubts its Triassic age.
Last: *Eodiadema pusillum* Lambert, 1990, upper Lias (TOA), *fide* Lambert and Thiéry (1909–1925).
Intervening: SIN, PLB.

Unnamed stem group J. (SIN) Mar.

First and Last: *'Plesiechinus' hawkinsi* Jesionek-Szymanska, 1970, Sunrise Formation, Nevada, USA.
Comment: The monocyclic endocyclic periproct of this species must exclude it from *Plesiechinus* as generally defined. Its perforate genital 5 must, moreover, exclude the species from Superorder Eognathostomata, since loss of genital 5 is a synapomorphy of this clade. It resembles *Atlasaster termieri* Lambert, 1931 (PLB) and *A. jeanneti* Lambert, 1937 (SIN) from the Lower Jurassic of Morocco in apical plate structure, but Plesiechinus lacks the globose test and biserial poriferous zones of Atlasaster.

Superorder EOGNATHOSTOMATA Smith, 1981

Order PYGASTEROIDA Durham and Melville, 1957

F. PYGASTERIDAE Lambert, 1900
J. (PLB)–K. (CEN) Mar.

First: *Plesiechinus reynesi* (Desor, 1868), Lias (*margaritatus* Zone), Cabanous (Aveyron), France.
Last: *Pygaster truncatus* L. Agassiz, 1840, Grès calcarifères (CEN), France.
Intervening: PLB–ALB.

Order HOLECTYPOIDA Duncan, 1899

Suborder HOLECTYPINA Duncan, 1899

F. HOLECTYPIDAE Lambert, 1899
J. (PLB)–K. (MAA) Mar.

First: *Holectypus hians* Lambert, 1933, Lias (lower Domerian), Algeria. *H. conquensis* de Cortazar, 1875, Lias (TOA), Cuenza, Spain, is the only other Lower Jurassic species known.
Last: *Caenholectypus baluchistanensis* (Noetling, 1897), Upper Cretaceous (MAA), Mari Hills, Baluchistan; *C. nachtigali* (Krumbeck, 1906), Jebel Tar, Tripolitania, Libya; *Caenholectypus* is also known from the Shiranish Marls of Kurdistan.
Intervening: TOA, BAJ–OXF, TTH, VLG–CMP.

F. ANORTHOPYGIDAE Wagner and Durham, 1966
K. (ALB)–(TUR?) Mar.

First: *Anorthopygus texanus* Cooke, 1946, Washita Group, Texas, USA.
Last: *Anorthopygus paradoxus* Hawkins, 1935, upper Senonian, Somalia, East Africa.
Intervening: CEN.

F. DISCOIDIDAE Lambert, 1899
K. (HAU)–(CMP/MAA) Mar.

First: *Discholectypus guebhardi* Lambert, 1920, Lower Cretaceous, southern France: both species and genus of uncertain assignment (Rose and Olver, 1985). *Discoides rahbergensis* Jeannet, 1933, Austria, and *D. karakaschi* (Renngarten, 1926), former southern USSR, have also been dated as HAU.
Last: *Metholectypus trechmanni* Hawkins, 1923, Cretaceous, Jamaica; genus of uncertain stratigraphical (CMP/MAA), but probable discoidid assignment (Rose and Olver, 1985). *Lanieria laneiri* (Cotteau, 1881), uppermost Cretaceous, Cuba and Mexico, *L. uvaldana* Cooke, 1953, Texas, USA, and possibly *Discoides menchikoffi* Lambert, 1937, Morocco, are of at least CMP date.
Intervening: APT–TUR.

Suborder ECHINONEINA H. L. Clark, 1925

?Unnamed stem group J. (OXF) Mar.

First and Last: *Pileus pileus* (L. Agassiz, 1847), Upper Jurassic (Rauracian), Europe; a pygasterid according to Durham *et al.* (1966), but an echinoneinid *fide* Smith (1981).

F. ECHINONEIDAE L. Agassiz and Desor, 1847
T. (THA/YPR?)–Rec. Mar.

First: ?*Amblypygus (Paramblypygus) houphoueti* Roman, 1973, Palaeocene (THA), Fresco, Ivory Coast, West Africa; a species of uncertain taxonomic ascription. **Extant**
Intervening: *Echinoneus* itself has a well-defined stratigraphical record from Oligocene to Recent.
Comment: *Amblypygus*, placed 'Family uncertain' by Durham *et al.* (1966), is considered to be a true echinoneid by Rose (1982); its earliest species is *A. dilatatus* L. Agassiz and Desor, 1847, for although typically middle Eocene in age like most *Amblypygus* species, it ranges from the early Eocene (YPR). The supposed Mesozoic echinoneid *Paleoechinoneus* (type *P. hannai* Grant and Hertlein, 1938) is based on an imperfect misidentified specimen of the globatorid *Globator*.

F. CONULIDAE Lambert, 1911
K. (HAU)–(MAA) Mar. (see Fig. 25.6)

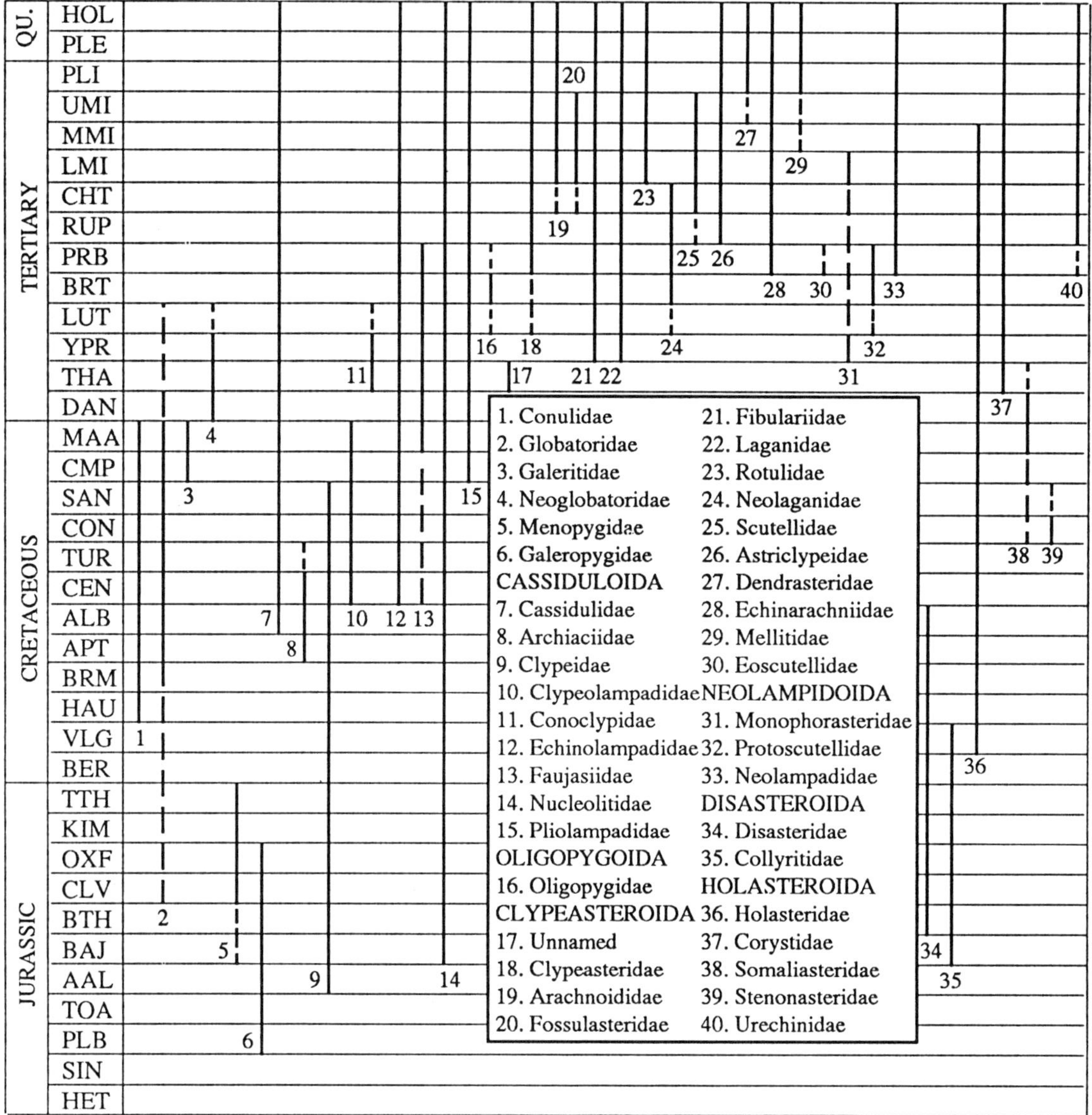

Fig. 25.6

First: *Conulus soubellinsis* Gauthier, 1876, Lower Cretaceous (Neocomian), Algeria; one APT, four ALB species ascribed to the genus by Lambert and Thiéry (1909–1925), Kier and Lawson (1978).
Last: *Conulus chiesai* Airaghi, 1939, *C. sanfilippoi* Checchia-Rispoli, 1930, both Upper Cretaceous (MAA), Libya.
Intervening: APT–CMP.
Comment: Of the four genera assigned by Durham *et al.* (1966) to this family, *Galeraster* is a holasteroid, and *Pygopyrina* and *Globator* should be ascribed to the Globatoridae, leaving only the type genus.

F. GLOBATORIDAE Lambert, 1911
J.? (CLV?)/K. (BRM?)–(MAA)/T. (LUT?) Mar.

First: *?Pygopyrina icaunensis* (Cotteau, 1855), Upper Jurassic (OXF), Yonne, France, ranging from CLV to KIM according to Lambert and Thiéry (1909–1925), (but probably a cassiduloid according to Smith and Patterson, 1988); the few other species ascriptions to *Pygopyrina* are also controversial (Mortensen, 1928–1951).
Last: *Globator ilarionesis* (Dames, 1877), northern Italy, *G. obsoleta* (Bittner, 1880), southern Alps, Austria; both Eocene, but possibly neoglobatorids. Uppermost Cretaceous species include *G. dainellii* Checchia-Rispoli, 1932, Libya, and *G. minuta* (Smiser, 1935), Belgium, both MAA.
Intervening: Mostly Upper Cretaceous, a few Lower Cretaceous.
Comment: Adopted by Endelman (1980) and Rose (1982), but not Durham *et al.* (1966) or Smith (1984b).

F. GALERITIDAE Gray, 1825
K. (CMP)–(MAA) Mar.

First: *Galerites ernsti* Schultz, 1985, Upper Cretaceous Chalk, Lägerdorf, Germany. *Galerites globosus* Roemer, 1841, Upper Cretaceous (CON), north-west Europe, is probably referable to *Echinogalerus* and to a separate lineage (Schultz, 1985).
Last: *Galerites stadensis* (Lambert, 1911), Maastrichtian Chalk of north Germany and Denmark.

F. NEOGLOBATORIDAE Endelman, 1980
T. (DAN)–(LUT?) Mar.

First: *Neoglobator danicus* Endelman, 1980, lower DAN, former USSR. *Neoglobator ovalis* (Smiser, 1935), Upper Cretaceous (MAA), Belgium is recorded only as DAN by van der Ham (1988), and *N. houzeaui* (Cotteau, 1875) occurs below it in the Guelhem Chalk (DAN) at Maastricht, The Netherlands.
Last: *Neoglobator akkajensis* Endelman, 1980, Eocene, former southern USSR. Eocene species currently ascribed to *Globator* should possibly be reascribed to *Neoglobator*.
Intervening: THA, YPR.

Superorder MICROSTOMATA Smith, 1984

F. MENOPYGIDAE Lambert, 1911 J. (BAJ?)–(TTH) Mar.

First: *Menopygus baugieri* (Cotteau, 1873), BAJ–BTH?, France: a poorly known species.
Last: *Infraclypeus thalebensis* Gauthier, 1875, TTH, Algeria (Devriès, 1960b).
Comment: This family was redefined by Rose and Olver (1988) as a plesion of stem Microstomata which span the whole time range BAJ–TTH.

Series NEOGNATHOSTOMATA Smith, 1981

F. GALEROPYGIDAE Lambert, 1911 J. (PLB)–(OXF) Mar.

First: *Galeropygus lacroixi* Lambert, 1925, Lias (Domerian), Géménos, France.
Last: *Laticlypus giganteus* Szörény, 1966, OXF, Hungary. *Stegopygus langeenensis* Devriès and Alcaydé, 1966, CEN, France, is probably referable to the cassiduloid genus *Ochetes* (Smith, pers. comm.).
Intervening: BAJ, BTH, OXF.

Order CASSIDULOIDA Claus, 1880

F. CASSIDULIDAE L. Agassiz and Desor, 1847 K. (ALB)–Rec. Mar.

First: *Ochetes morrisii* (Forbes, 1849), Upper Greensand (ALB), Blackdown, Devon (Kier, 1962b). **Extant**
Intervening: Good stratigraphical record through the Cretaceous and Cenozoic.

F. ARCHIACIIDAE Cotteau and Triger, 1869 K. (APT)–(TUR?) Mar.

First: *Archiacia hungarica* Szörény, 1955, Urgonian–Aptian, Bakony, Hungary.
Last: *Claviaster costatus* Pomel, 1883, TUR?, Algeria; otherwise *Gentilia chouberti* Lambert, 1937, upper CEN, Morocco.
Intervening: ALB–CEN.

F. CLYPEIDAE Lambert, 1898 J. (AAL)–K. (SAN) Mar.

First: *Clypeus michelini* Wright, 1854, *murchisonae* Zone, Crickley Limestone Member, Inferior Oolite Group, Cotswolds, England, UK (Smith, 1984b).
Last: *Pygurus lampassiformis* Tzankov, 1934, Upper Cretaceous (SAN), Bulgaria.
Intervening: Good stratigraphical record through most of the Jurassic and Cretaceous into the Cenomanian.

F. CLYPEOLAMPADIDAE Kier, 1962 K. (CEN)–(MAA) Mar.

First: *Vologesia rhotomagensis* (d'Orbigny, 1856), CEN, France, Belgium and England, UK.
Last: *Clypeolampas ovatus* (Lamarck, 1816), France; also *C. vishnu* Noetling, 1897, *Vologesia helios* (Noetling, 1897), and other MAA species.
Intervening: CMP.

F. CONOCLYPIDAE Zittel, 1879 T. (THA)–(LUT?) Mar.

First: *Conoclypus sindensis* (Duncan and Sladen, 1882), Ranikot Series, Petiani, Pakistan (*C. sanctispiritus* Sánchez Roig, 1949, Cretaceous, Cuba, is of uncertain ascription). *C. leymeriei* is of confirmed late THA age in the northern Pyrenees of France and Spain (Plaziat *et al.*, 1974).
Last: *Conoclypus rostratus* (Duncan and Sladen, 1884), Khirthar Series, hills east of Trak, western Sind, Pakistan. (*C. westraliensis* Crespin, 1944, Middle Miocene, Australia, is a misidentified *Echinolampas* (*Hypsoclypus*), like other Miocene records of this family (Roman, 1965).)
Intervening: Eoc.

F. ECHINOLAMPADIDAE Gray, 1851 K. (CEN)–Rec. Mar.

First: *Arnaudaster gauthieri* Lambert, 1920, Cretaceous (CEN) *Mantelliceras mantelli* Zone, near Fumel, Aquitaine, France. **Extant**
Intervening: TUR–SAN, DAN–PLE.
Comment: The genus Arnaudaster is known from only one species, and that from a single specimen (Kier, 1962b).

F. FAUJASIIDAE Lambert, 1905 K. (CEN?)–T. (PRB) Mar.

First: *Faujasia araripensis* Beurlen, 1966, Brazil, and *F. rancherina* Cooke, 1955, Colombia, are both cited from the Lower Cretaceous (ALB), although Kier (1962b) and Durham *et al.* (1966) do not record the genus before MAA and family before CEN. *Petalobrissus inflatus* (Thomas and Gauthier, 1889), *P. daglensis* (Thomas and Gauthier, 1889), both CEN, Tunisia, are of confirmed ascription and age. *Stigmatopygus malheiroi* de Loriol, 1888, CEN, Angola, lacks recent description.
Last: *Australanthus longianus* (Gregory, 1890), Upper Eocene, Australia.
Intervening: CEN–MAA, Eoc.

F. NUCLEOLITIDAE L. Agassiz and Desor, 1847 J. (BAJ)–Rec. Mar.

First: *Nucleolites latiporus* L. Agassiz, 1839 (= *N. clunicularis* auct.), *N. woodwardii* Wright, 1854, both Inferior Oolite (*parkinsoni* Zone upwards) England, UK and France, *Nucleolites terquemi* L. Agassiz and Desor, 1846, BAJ, France. **Extant**
Intervening: J., K., Eoc., Oli./Mio.
Comment: Includes the Family Apatopygidae of Durham *et al.* (1966).

F. PLIOLAMPADIDAE Kier, 1962 K. (CMP)–Rec. Mar.

First: *Gitolampas tunetana* (Gauthier, 1889), CMP, Midès, Tunisia; possibly *Zuffardia sanfilippoi* (Checchia-Rispoli, 1914), Senonian, Tripoli, Libya. The earlier *Breynella* (? = *Pliolampas*) *baixadoleitensis* Maury, 1934, TUR?, Rio Grande do Norte, Brazil, is of uncertain ascription. Several species of *Gitolampas*, e.g. *G. lamberti* Checchia-Rispoli, 1921, *G. zuffardii* Checchia-Rispoli, 1921, are of undivided Senonian age, and so may be slightly earlier than CMP. **Extant**

Intervening: Good record from the late Cretaceous through the Tertiary to the Recent.

Order OLIGOPYGOIDA Kier, 1967

F. OLIGOPYGIDAE Duncan, 1889 T. (LUT?–PRB?) Mar.

First: ?*Oligopygus phelani* Kier, 1967, Inglis Limestone Formation, Florida, USA. (Most species of *Haimea*, and several of *Oligopygus*, have been recorded from supposed Middle Eocene strata, including many from the Yellow Limestone of Jamaica, so the true first occurrence is likely to be one of these species.)
Last: *O. wetherbyi* de Loriol, 1888, *O. haldemani* (Conrad, 1850); Crystal River Formation, near Ocala, Florida, USA.
Comment: Kier (1967) restricted this family to *Oligopygus* and *Haimea* (including *Bonaireaster* in synonymy), with 12 valid (plus 10 inadequately known) species and 13 valid (plus six inadequately known) species respectively. The age of most of the species relative to each other is unknown. Some are reported from the Middle Eocene, others from the Upper Eocene, but these age determinations are too unreliable to be used. Of all the species, there are only three whose stratigraphical relationships are certain (Kier, 1967, p. 47).

Order CLYPEASTEROIDA A. Agassiz, 1872

Unnamed stem group? T. (THA?) Mar.

First and Last: *Togocyamus seefriedi* (Oppenheim, 1915), Ekekoro Formation, 55 km NW of Lagos, Nigeria; *T. alloiteaui* Roman and Gorodiski, 1959, Senegal; both Palaeocene, of West Africa.
Comment: The genus *Togocyamus*, known from only two species and classified as a fibulariid by Durham *et al.* (1966), has been reinterpreted both as the most primitive clypeasteroid yet known (Smith, 1984b) and as the sole representative of an entirely extinct sister group to Order Clypeasteroida (Mooi, 1990).

Suborder CLYPEASTERINA A. Agassiz, 1872

F. CLYPEASTERIDAE L. Agassiz, 1835 T. (LUT?)–Rec. Mar.

First: *Clypeaster marbellensis* (Boussac, 1911), Auversian, Biarritz, France. Reguant *et al.* (1970) also record *Clypeaster* nov. sp. in the Middle Eocene (upper Biarritzian) of Vic (Barcelona), Spain. **Extant**
Intervening: Excellent stratigraphical record from late Eocene to Recent.

F. ARACHNOIDIDAE Duncan, 1889 T. (CHT?)–Rec.

First: *Fellaster zelandiae* (Gray, 1855), the type and then the only known species was given an Oligocene–Recent range by Durham *et al.* (1966). **Extant**
Intervening: Known from a few Recent and Neogene taxa.

F. FOSSULASTERIDAE Philip and Foster, 1971 T. (CHT?–Mio.) Mar.

First: *Fossulaster halli* Lambert and Thiéry, 1925, and *Willungaster scutellaris* Philip and Foster, 1971, grouped by Philip and Foster (1971) within the new family Fossulasteridae: a taxon overlooked by Smith (1984b); both Upper Oligocene or Lower Miocene (Jankukian or Longfordian), South Australia.
Last: *Scutellinoides patella* (Tate, 1891), Morgan Limestone, Murray River Cliffs, South Australia (Foster and Philip, 1971).
Intervening: Mio.

Suborder SCUTELLINA Haeckel, 1896

Infraorder FIBULARIINA Smith, 1984

F. FIBULARIIDAE Gray, 1825 T. (YPR)–Rec. Mar.

First: *Fibularia jeanneti* (Lambert, 1931), Egypt; *F. planus* (Lambert, 1933), Madagascar, off East Africa; both dated as early Eocene. *F. cyphostomus* (Lambert, *in* Lambert and Jacquet, 1936), Senegal, West Africa, has been redated as middle Eocene (LUT). Records of supposed fibulariids, *Echinocyamus placenta* (Goldfuss, 1826) and *Fibularia subglobosus* (Goldfuss, 1826), from the Upper Chalk of St Pieters, Holland (K., MAA) conditionally accepted by Durham (*in* Durham *et al.*, 1966) and by Philip (*in* Harland, 1967) lack confirmation; *E. kamrupensis* Das Gupta, 1929, upper Senonian, India, is of doubtful taxonomic assignment. Numerous species variously ascribed to *Echinocyamus* and *Fibularia* have an imprecisely known Eocene range, and several may extend back to at least the early Eocene. *Porpitella paleocenica*, (upper THA), northern Pyrenees (France and Spain), is certainly Palaeocene, but *Porpitella* is excluded from the Fibulariidae by Mooi (1989). **Extant**
Intervening: Good stratigraphical record through the Eocene to the Recent.

Infraorder LAGANINA Mortensen, 1948

Superfamily LAGANIDEA A. Agassiz, 1873

F. LAGANIDAE Desor, 1858 T. (YPR)–Rec. Mar.

(Family name on ICZN Official List: Opinion 608.)
First: *Sismondia barabirensis* Lambert, 1931, Lower Eocene, Egypt; many of the 30 species ascribed to *Sismondia* are of imprecise Eocene date. The oldest known *Laganum* is *L. sorigneti* Cotteau, 1890, Eocene, France. **Extant**
Intervening: Good stratigraphical record from Eocene to Recent.

F. ROTULIDAE Gray, 1855 T. (BUR)–Rec. Mar.

First: *Rotuloidea vieirai* Dartevelle, 1953, upper Burdigalian, Angola, south-west Africa. **Extant**
Intervening: Mio., PLI.

F. NEOLAGANIDAE Durham, 1954 T. (LUT/BRT–CHT) Mar.

First: *Durhamella ocalanum* (Cooke, 1942), Lake City Formation (Middle Eocene), Georgia, USA.
Last: *Neorumphia elegans* (Sanchez Roig, 1949), Upper Oligocene, Cuba.
Intervening: Eoc., Oli.

Superfamily SCUTELLIDEA Smith, 1984

F. SCUTELLIDAE Gray, 1825 T. (RUP?–TOR) Mar.

First: *Scutella isseli* Airaghi, 1901, *S. lamberti* Airaghi, 1901, *S. marianii* Airaghi, 1901, all from northern Italy, and three other species of *Scutella* are accepted by Lambert and Thiéry (1909–1925) as of Tongrian (early Oligocene) date, although

generic ascription should perhaps be to *Parascutella* or *Parmulechinus* (*sensu* Durham, 1953). *S. camagueyana* Weisbord, 1934, *S. cubae* Weisbord, 1934, both late Eocene, Cuba, and other 'New World species are unlikely to be *Scutella*' (Mooi, 1989).
Last: *Scutella vindobonensis* Laube *secunda* Schaffer, 1962, middle to upper Tortonian, Austria.
Intervening: Oli.–Mio.

F. ASTRICLYPEIDAE Stefanini, 1911
T. (RUP)–Rec. Mar.

First: *Echinodiscus duffi* (Gregory, 1911), Lower Oligocene, Cyrenaica, Libya (Checchia-Rispoli, 1913). *E. ginauensis* Clegg, 1933, Lower Tertiary, Persian Gulf, and *E. chikuzenensis* Nagao, 1928, Palaeogene, Japan, are less precisely dated. **Extant**
Intervening: Mio.
Comment: *Echinodiscus tiliensis* Wang, 1984, attributed to the Tachien Sandstone, Lower Eocene or Upper Palaeocene, Taiwan, on the appearance of its sandstone matrix, is known only from the holotype, found in a museum collection at the National Taiwan University. The age attribution is inconsistent with present views of the origin not only of the Scutellidea but the entire Clypeasteroida (Mooi, 1989).

F. DENDRASTERIDAE Lambert, 1889
T. (MES?)–Rec. Mar.

First: *Merriamaster lamberti* (Grant and Hertlein, 1938), California, USA is accepted as of middle Miocene age by Kier and Lawson (1978), although Mooi (1989) gives the earliest record for the genus as ?late Miocene, and Durham (1978) excludes both *lamberti* and Miocene records from the genus. **Extant**
Intervening: Mio., PLI.

F. ECHINARACHNIIDAE Lambert, 1914
T. (PRB)–Rec. Mar.

First: *Kewia* sp., Upper Eocene, Oregon, USA (Linder *et al.*, 1988). *K. marquamensis* Linder, Durham and Orr, 1988, Scotts Mills Formation, Oregon, USA is the best documented of the earliest (Upper Oligocene) named species. **Extant**
Intervening: Mio., PLI.

F. MELLITIDAE Stefanini, 1911
T. (LAN/SRV?)–Rec. Mar.

First: *Encope kugleri* Jeannet, 1928, *E. vonderschmitti* Jeannet, 1928, and *E. wiedenmayeri* Jeannet, 1928, all Serie de Capadare, Venezuela, are cited as Middle Miocene by Kier and Lawson (1978). Durham *et al.* (1966) cite the early Miocene as the time of first occurrence for the family. **Extant**
Intervening: PLI.

F. EOSCUTELLIDAE Durham, 1955
T. (PRB?) Mar.

First and Last: *Eoscutella coosensis* (Kew, 1920), Upper Eocene, Oregon, USA. Mortensen (1928–1951) also refers *Scutella vaquerosensis* Kew, 1920 to *Eoscutella*, cited as Miocene in age by Lambert and Thiéry (1909–1925). Parma (1985) has described one, and possibly a second, new species of *Eoscutella* from the Patagonian (Lower Tertiary) of Argentina (Mooi, 1989).
Comment: This monogeneric family (Durham *et al.*, 1966) is not adopted by Smith (1984b, p. 172) in his classification, but a taxon corresponding in range and affinity with *Eoscutella* is distinguished on his (Fig. A.1) diagram which illustrates known stratigraphical ranges of echinoid families.

F. MONOPHORASTERIDAE Lahille, 1896
T. (YPR?–BUR?) Mar.

First: *Iheringiella patagoniensis* (Desor, 1847), from the Lower Miocene of Argentina according to Durham (1955), but Larrain (1984) extends this range into Chile and from the Lower (Middle [?]) Eocene to the Miocene (?) according to Mooi (1989).
Last: *Karlaster pirabensis* Marchesini Santos, 1958, Miocene, Brazil. *Monophoraster darwini* (Desor, 1847), and the two other species ascribed to *Monophoraster*, from Argentina and Chile, are also of imprecise Miocene age, so none of these species may extend beyond the range of *Iheringiella patagoniensis*.

F. PROTOSCUTELLIDAE Durham, 1955
T. (LUT?–PRB) Mar.

First: *Protoscutella mississippiensis* (Twitchell, 1915), Winona Sand Member of Lisbon Formation, Mississippi, and Mount Selman Formation, Texas, USA. Four other species of *Protoscutella* are also known from the Middle Eocene and south-east USA.
Last: *Periarchus (Mortonella) quinquefaria* (Say, 1825), Sandersville Limestone Member of Barnwell Formation, Georgia, USA; a few other species of *Protoscutella*, *Mortonella*, and *Periarchus* are also known from the Upper Eocene of the south-east USA.

Order NEOLAMPADOIDA Philip, 1963

F. NEOLAMPADIDAE Lambert, 1918
T. (PRB)–Rec. Mar.

First: *Pisolampas concinna* Philip, 1963, Tortachilla Limestone (lower Upper Eocene), Aldinga, St Vincent Basin, South Australia. **Extant**
Intervening: LMI.

Series ATELOSTOMATA Zittel, 1879

Order DISASTEROIDA Mintz, 1968

F. DISASTERIDAE Gras, 1848
J. (BTH)–K. (ALB) Mar.

First: *Tithonia sarthacensis* (Cotteau, 1860), BTH, Sarthe, France; known only from the holotype.
Last: *Collyropsis moussoni* (Desor, 1858), Gault (ALB), Swiss Alps and Savoy. Possibly also *Corthya ambayraci* Lambert, 1924; cited plausibly as Cretaceous (ALB) by Lambert and Thiéry (1909–1925), but probably incorrectly as late Eocene in age (BRT) by Kier and Lawson (1978).
Intervening: Moderately good late Jurassic (CLV) through early Cretaceous fossil record.

F. COLLYRITIDAE d'Orbigny, 1853
J. (BAJ)–K. (VLG) Mar.

First: *Cyclolampas kiliani* Lambert, 1909, upper BAJ, Chalet de l'Alpe de Villard d'Arciné (Massif des Ecrins-Isère), France; arguably one ammonite zone older than *Orbignyana ebrayi* Cotteau, 1873 (Jablonski and Bottjer, 1990). *Pygomalus prior* (Desor, 1858), lower Lias, Aargau, Switzerland, and upper Lias, Ardèche, France, a rare, poorly known species, supposedly an earlier collyritid, is now recognized as only an imperfect *C. bicordata* from the upper OXF.

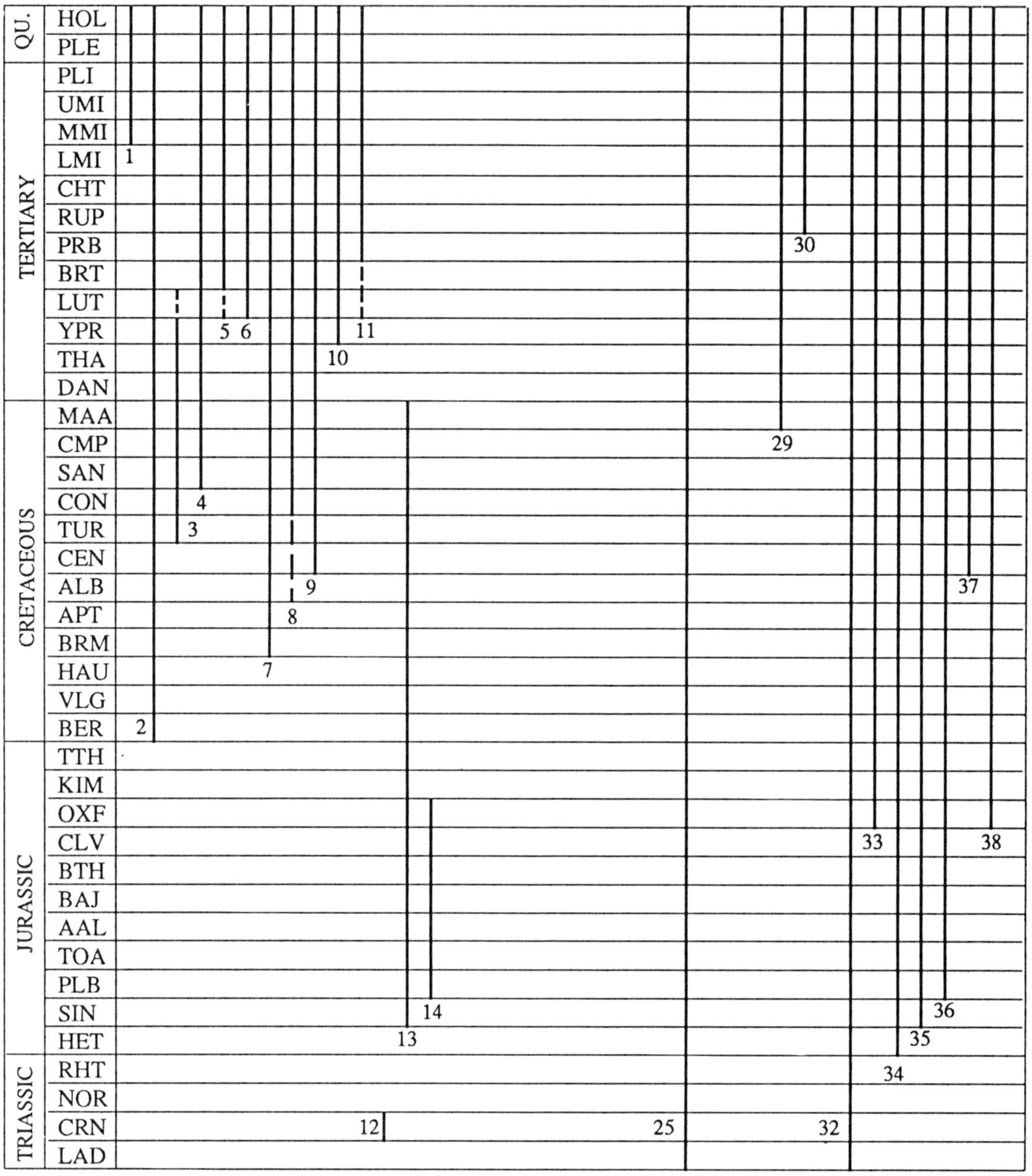

QU.: HOL, PLE
TERTIARY: PLI, UMI, MMI, LMI, CHT, RUP, PRB, BRT, LUT, YPR, THA, DAN
CRETACEOUS: MAA, CMP, SAN, CON, TUR, CEN, ALB, APT, BRM, HAU, VLG, BER
JURASSIC: TTH, KIM, OXF, CLV, BTH, BAJ, AAL, TOA, PLB, SIN, HET
TRIASSIC: RHT, NOR, CRN, LAD

1 2 3 4 5 6 7 8 9 10 11 12 13 14 25 29 30 32 33 34 35 36 37 38

Fig. 25.7

Last: *Cardiopelta oblonga* (d'Orbigny, 1853), Neocomian, Jura, France.
Intervening: J. (m., from BAJ, u.), BER.

Order HOLASTEROIDA Durham and Melville, 1957

F. HOLASTERIDAE Pictet, 1857 K. (VLG)–T. (LAN/SRV) Mar.

First: *H. grasanus* d'Orbigny, 1853, Neocomian, Fontanil, Isère, France; *Holaster valanginensis* Lambert, 1917, Switzerland; *H. cordatus* Dubois, 1836, former USSR; all recorded from the Lower Cretaceous (VLG).
Last: *Toxopatagus italicus* (Manzoni, 1878), Schlier, Bologna, Italy.
Intervening: ALB–Eoc., CHT.

F. CORYSTIDAE Foster and Philip, 1978 T. (THA)–Rec. Mar.

First: *Cardabia bullarensis* Foster and Philip, 1978, Cardabia Group, Carnavon Basin, Western Australia (Middle or Upper Palaeocene). **Extant**
Intervening: PRB–LMI.
Comment: Adopted by David (1988), although not by Smith (1984b).

F. CALYMNIDAE Mortensen, 1907 **Extant** Mar.

F. SOMALIASTERIDAE Wagner and Durham, 1966 K. (CON/MAA?)–T. (DAN/THA) Mar.

First: *Iraniaster morgani* Cotteau and Gauthier, 1895, Upper Cretaceous (Senonian) of Iran; also *I. douvillei* Cotteau and Gauthier, 1895, *I. nodulosus* Gauthier, 1902, of similar origin. *Somaliaster magniventer* Hawkins, 1935, and *S. magniventer* var. *checchiai* Maccagno, 1941, both from East Africa, are of imprecise Cretaceous date.

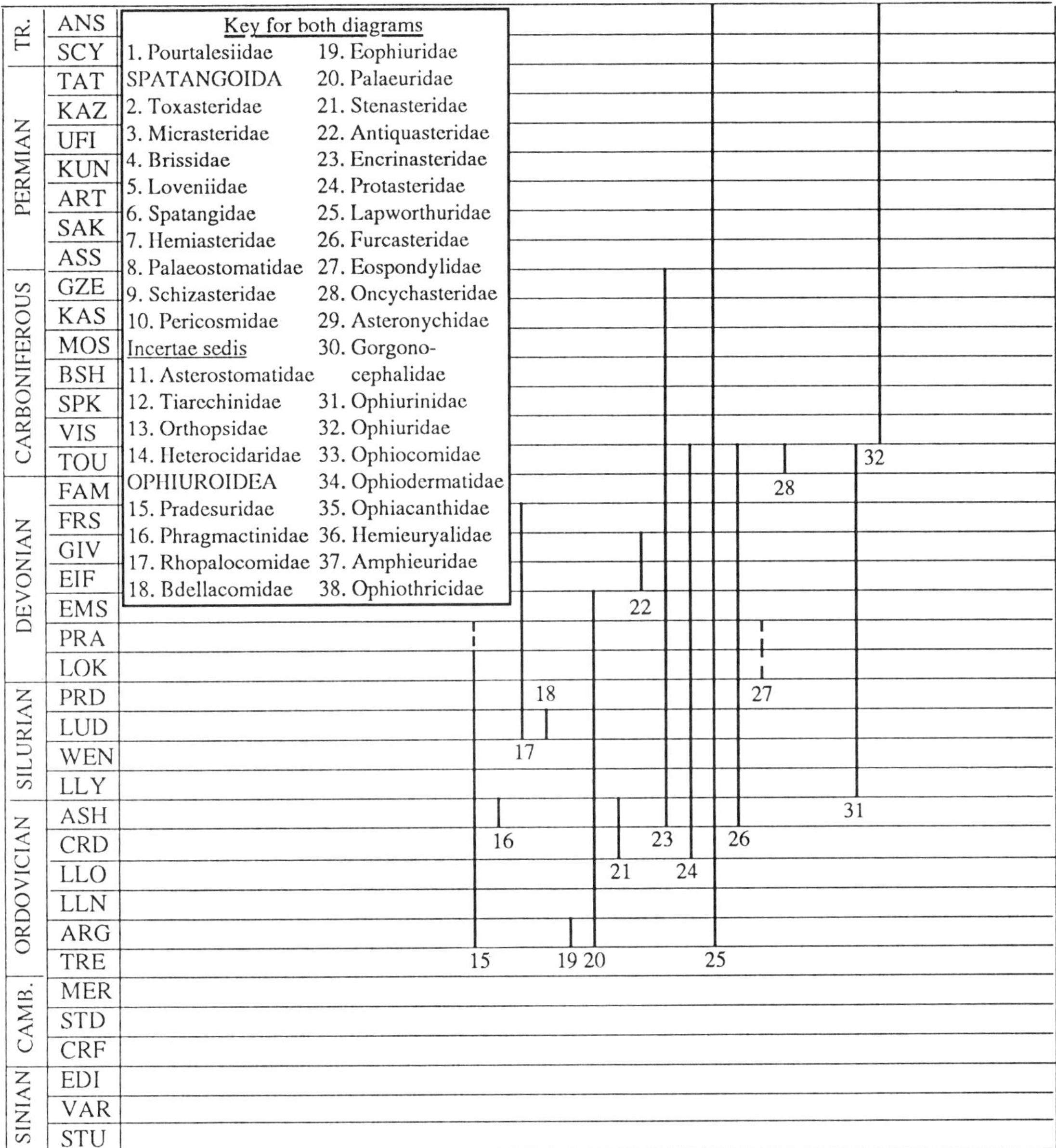

Fig. 25.7.

Last: *Brightonia macfadyeni* Kier, 1957, *Leviechinus gregoryi* (Currie, 1927) both lower Auradu Series, Somalia, East Africa.

F. STENONASTERIDAE Lambert, 1922
K. (CON–SAN?) Mar.

First and Last: *Stenonaster tuberculata* (Defrance, 1816), from Italy, Algeria, Turkey and Tunisia (Zaghbib-Turki, 1987). *S. morgani* Gauthier, 1902, Iran, and *S. douvillei* Lambert, 1928, France, the only other species ascribed to this monogeneric family, are both from imprecisely known (but presumably Upper Cretaceous) stratigraphical horizons.

F. URECHINIDAE Duncan, 1889
T. (PRB?)–Rec. Mar.

First: *Sanchezaster habanensis* Lambert, *in* Sanchez Roig, 1924, Eocene (?Upper Eocene), Habana Province, Cuba (Kier, 1984a). **Extant**
Intervening: Eoc., Mio.

F. POURTALESIIDAE A. Agassiz, 1881
T. (LAN/SRV)–Rec. Mar. (see Fig. 25.7)

First: *Pourtalesia* sp., Tatsukuroiso Mudstone, (Middle Miocene), Honshu, Japan (Kikuchi and Nikaido, 1985). **Extant**
Intervening: No other fossil representatives of this family have been recorded.

Order SPATANGOIDA Claus, 1876

F. TOXASTERIDAE Lambert, 1920
K. (BER)–Rec. Mar.

First: *Toxaster laffittei* Devriès, 1960, Lower Cretaceous (BER), Bel Kheir (Aures), Algeria (although not truly a *Toxaster* according to Fischer, *in* Durham *et al.* (1966), but a taxon intermediate between *Toxaster* and the earlier disasteroid *Collyrites*). *T. rochi* Lambert, 1933, *T. africanus* (Coquand, 1875) also BER, in Tunisia. Several other Lower Cretaceous (VLG–HAU) species ascribed to *Toxaster* are

known, mostly from Europe and North Africa. (Both Mortensen (1928–1951) and Kier and Lawson (1978) erroneously list *Heteraster musandamensis* Lees, 1928 from the Upper Jurassic rather than the Upper Cretaceous.) **Extant**
Intervening: K, DAN, Eoc., Oli.

Suborder MICRASTERINA Fischer, 1966

F. MICRASTERIDAE Lambert, 1920 K. (TUR?)–T. (LUT?) Mar.

First: *Micraster borchardi* Hagenow, 1853, Chalk, Germany (*I. lamarcki* and *I. vancouverensis* Zones) and England, UK (*H. planus* Zone) is the first true *Micraster sensu* Stokes (1975) and therefore first micrasterid *sensu* Durham *et al.* (1966). *M. antiquus* Cotteau, 1887 is of CMP age rather than CEN as listed by Lambert and Thiéry (1909–1925). *M. corbovis* Forbes, 1850, although also *H. planus* Zone, is interpreted as a lineage separate from true *Micraster* by Stokes (1975). *Epiaster michelini* (L. Agassiz, 1847), (upper CEN–TUR of northern Europe) is considered to be ancestral to *Micraster* by Stokes (1975). If genus *Epiaster*, placed in synonymy with the toxasterid *Heteraster* by Durham *et al.* (1966), is maintained and included within the Micrasteridae, the first occurrence of the family would be extended to the early Cretaceous (ALB, ?APT).
Last: *Isopneustes subquadratus* (Desor, 1857), Eocene, Italy.
Intervening: *Micraster* species are numerous and well known throughout the Upper Cretaceous (Stokes, 1975); a few less-common species, notably of *Brissopneustes*, extend the record of the family through the Palaeocene and early Eocene.

F. BRISSIDAE Gray, 1855 K. (SAN)–Rec. Mar.

First: *Diplodetus recklinghausenensis* Schlüter, 1900, Chalk, Germany. (The genus *Plesiaster*, type species *Plesiaster peini* (Coquand, 1862), SAN, Algeria and Tunisia, although classified as a brissid by Durham *et al.* (1966) is considered to be a synonym of *Micraster* by Stokes (1975).) **Extant**
Intervening: Late Cretaceous (mainly CMP); otherwise confined to the Cainozoic (especially post-Palaeocene), with a near-complete stratigraphical record.

F. LOVENIIDAE Lambert, 1905 T. (LUT?)–Rec. Mar.

First: Several Eocene species of *Lovenia* are known, notably *L. carinata* (Cotteau, 1889), Alicante, Spain, *L. gregoryi* (Clark, 1915), USA, *L. lorioli* (Lambert, 1902), Barcelona, Spain, and *L. suessi* Bittner, 1880, Southern Alps, Austria, but precise stratigraphical horizons lack confirmation. **Extant**
Intervening: Good stratigraphical record through the Cainozoic, especially Oligocene and Middle Miocene.

F. SPATANGIDAE Gray, 1825 T. (LUT)–Rec. Mar.

(Family name on ICZN Offical List: Opinion 608.)
First: *Spatangus cosoni* (Sorignet, 1850) France; *Atelospatangus magnus* Szörényi, 1963, Hungary; both Middle Eocene. *S. (Granopatagus) lonchophorus* Meneghini, *in* Desor, 1858, Vincentin, Italy, and *Laevipatagus bigibbus* (Beyrich, 1848) have less-precise ascription within the Eocene. (The record of *Spatangus baixadoleitensis* Maury, 1934, Upper Cretaceous (TUR?), Brazil, accepted by Kier and Lawson (1978) is anomalous; the species is referred to *Hemiaster* by Smith and Bengtson, 1991.) **Extant**
Intervening: Good stratigraphical record through the Cainozoic from the late Eocene.

Suborder HEMIASTERINA Fischer, 1966

F. HEMIASTERIDAE H. L. Clark, 1917 K. (BRM)–Rec. Mar.

First: *Washitaster barremicus* Tanaka and Okubu, 1954, (BRM), Japan; possibly *W.* (?) *macroholcus* Nisiyama, 1950, Neocomian, Japan. **Extant**
Intervening: Very numerous through the Cretaceous (APT–MAA), with a good stratigraphical record through the Cainozoic.

F. PALAEOSTOMATIDAE Loven, 1867 K. (ALB?)–Rec. Mar.

First: *Leiostomaster bosei* Smiser, 1936, Comanchean, Texas, USA. **Extant**
Intervening: Upper Cretaceous, Eoc.

F. AEROPSIDAE Lambert, 1896 **Extant** Mar.

F. SCHIZASTERIDAE Lambert, 1905 K. (CEN)–Rec. Mar.

First: *Linthia dainellii* Stefanini, 1928, Karakorum, Mongolia; *Periaster maugerii* Checchia-Rispoli, 1936, Sicily, central Mediterranean; *Proraster dalli* Clark, 1915, USA; all CEN. **Extant**
Intervening: Mostly confined to the Cainozoic.

F. PERICOSMIDAE Lambert, 1905 T. (YPR)–Rec. Mar.

First: *Pericosmus gregoryi* Currie, 1927, upper Auradu Series (Lower Eocene), Somalia, East Africa. If *Mundaster*, known only from the type species *M. tentugalensis* Soares and Devriès, 1967, upper CEN to lower TUR, Portugal, is included in the family (as by Kier and Lawson, 1978), the range is extended back into the Upper Cretaceous. **Extant**
Intervening: Good fossil record from Eocene to Recent.

Suborder UNKNOWN

F. ASTEROSTOMATIDAE Pictet, 1857 T. (LUT?)–Rec. Mar.

First: *Asterostoma dickersoni* Sanchez Roig, 1949, Cuba, dated as Middle Eocene by Brodermann (1949), although this precise date is not confirmed by Kier (1984a); possibly *Pygospatangus salvae* Cotteau, 1889, Alicante, Spain. These species, and a few attributed to *Antillaster* and *Moronaster*, are of imprecise Eocene age. **Extant**
Intervening: Good late Eocene to Recent stratigraphical record.
Comment: 'Probably a polyphyletic grouping of aberrant members of the Hemiasterina and Micrasterina, which have reduced petals or fascioles or both . . . a taxonomic convenience or necessity rather than a biologically meaningful unit' (Fell, *in* Durham *et al.*, 1966).

UNPLACED REGULAR ECHINOID GROUPS

Order PLESIOCIDAROIDA Duncan, 1889

Subclass uncertain. Considered to be ancestral to the

arbacioids by some authors, but referable to the cidaroids by others (Kier, 1977).

F. TIARECHINIDAE Gregory, 1896
Tr. (CRN) Mar.

First and Last: *Tiarechinus princeps* Neumayr, 1881, St Cassian Beds, northern Italy (Kier, 1977, 1984b).
Comment: Kier (1977) excludes *Lysechinus* from this family: known only from the holotype of *L. incongruens* Gregory, 1896, it is probably a misidentified mollusc. The family is therefore confined to the monospecific *Tiarechinus* 'so different from any other genus that it cannot with any certainty be referred to a suborder' (Kier, 1977).

Order ORTHOPSIDA Mortensen, 1942

Cohort uncertain: Diadematacea or Echinacea

F. ORTHOPSIDAE Duncan, 1889
J. (SIN)–K. (MAA) Mar.

First: *Dubarechinus despujolsi* Lambert, 1937 and *D. termieri* Lambert, 1937, both Lias (upper Domerian), Morocco. *Orthopsis parvituberculata* (Böhm, 1884), Verona, Italy is dated as Triassic (RHT) by Lambert and Thiéry (1909–1925), but as Jurassic (BAJ) by Smith (1990); it is excluded from his review of Triassic echinoids by Kier (1977).
Last: *Orthopsis sanfilippoi* Checchia-Rispoli, 1933, Tripolitania, Libya; *Orthopsis perlata* Noetling, 1897, Mari Hills, Baluchistan, Pakistan; both MAA.
Intervening: BTH–CMP.

Order UNCERTAIN

F. HETEROCIDARIDAE Mortensen, 1934
J. (PLB)–(OXF) Mar.

First: *Heterocidaris bruni* Lambert, *in* Lambert and Thiéry, 1925, Lias (Domerian and Toarcian), near Géménos, France. Also *Cidaris taylorensis*. W. B. Clark, 1893, Hardgrave Sandstone (Lower Jurassic), California, USA, reascribed to *Heterocidaris* by Lambert and Thiéry (1909–1925).
Last: *Heterocidaris dumortieri* Cotteau, 1871, OXF, Ardèche, France.
Intervening: BAJ.
Comment: This monogeneric family is not adopted by Smith (1984b), although *Heterocidaris* is cited as of uncertain order (1984b, p. 20). Durham *et al.* (1966) maintain the family, noting 'The general aspect of the test recalls the Diadematoida, but the cidaroid character of the spines is opposed to such affinity, pointing rather to the Hemicidaroida or even Cidaroida; no precise relationships can be suggested until the lantern structure is known.'

Class OPHIUROIDEA Gray, 1840

Current classifications of the ophiuroids are highly unsatisfactory, being based upon grades of organization rather than phylogenetic relationships. It is probable that most taxa above the level of genus are paraphyletic or even polyphyletic. This is particularly true of Palaeozoic groups, and it is likely that few, if any, represent natural groupings. Furthermore, some of the 'primitive' Lower Palaeozoic ophiuroids almost certainly will be found, upon further investigation, to represent parts of the stem group of both ophiuroids and echinoids. Hence the Ophiuroidea, as currently understood, is a paraphyletic taxon. The situation perhaps is somewhat better for post-Palaeozoic taxa and it is possible that some of these families may represent monophyletic clades. The classification adopted here is largely that of Spencer and Wright (1966), in the *Treatise on Invertebrate Paleontology*, but the monophyly of any particular group within the Ophiuroidea must be regarded as highly uncertain, or indeed improbable, until some attempt has been made to investigate ophiuroid relationships using phylogenetic methods.

Order STENURIDA Spencer, 1951

Suborder PROTURINA Spencer and Wright, 1966

F. PRADESURIDAE Spencer, 1951
O. (ARG)–D. (LOK/PRA) Mar.

First: *Pradesura jacobi* (Thoral, 1935), Schistes de St Chinian, basal Arenig, Hérault, France.
Last: *Sturtzaster spinossima* (Roemer, 1863), Bundenbach Slates, Germany.

F. PHRAGMACTINIDAE Spencer, 1951
O. (RAW) Mar.

First and Last: *Phragmactis grayae* Spencer, 1951, Starfish Bed, Girvan, Scotland, UK.

F. RHOPALOCOMIDAE Spencer and Wright, 1966
S. (GOR)–D. (FRS) Mar.

First: *Rhopalocoma pyrotechnica* Salter, 1857, Lower LUD, Leintwardine, Hereford, and Lake District, England, UK.
Last: *Ptilonaster princeps* Hall, 1868, Ithaca Beds, New York, USA.

F. BDELLACOMIDAE Spencer and Wright, 1966
S. (GOR) Mar.

First and Last: *Bdellacoma vermiformis* Salter, 1857, Lower LUD, Leintwardine, Hereford, England, UK.

Suborder PAROPHIURINA Jaekel, 1923

F. EOPHIURIDAE Schöndorf, 1910
O. (ARG) Mar.

First and Last: *Eophiura bohemica* Schuchert, 1914, upper ARG, Czechoslovakia.

F. PALAEURIDAE Spencer, 1951
O. (ARG)–D. (l.) Mar.

First: *Palaeura neglecta* Schuchert, 1914, upper ARG, Czechoslovakia.
Last: *Medusaster rhenanus* Stürtz, 1890, Dachschiefern, Germany.

F. STENASTERIDAE Schuchert, 1914
O. (COS–RAW) Mar.

First: *Stenaster obtusus* (Forbes, 1848), Black River Limestone, Kentucky and Vermont, USA, Ontario and Ottawa, Canada.
Last: *Stenaster obtusus* (Forbes, 1848), Starfish Bed, Girvan, Scotland, UK.

F. ANTIQUASTERIDAE Kesling, 1971
D. (EIF/GIV) Mar.

First and Last: *Antiquaster magrumi* Kesling, 1971, Silica Formation, north-western Ohio, USA.

Order OEGOPHIURIDA Matsumoto, 1915

Suborder LYSOPHIURINA Gregory, 1896

F. ENCRINASTERIDAE Schuchert, 1914
O. (RAW)–C. (u.) Mar.

First: *Encrinaster grayae* Spencer, 1914, Starfish Bed,

Girvan, Scotland, UK.
Last: *Armathyraster paradoxis* Harper and Morris, 1978, Brush Creek Shale, Glenshaw Formation, Conemaugh Group, Punxsutawney, Pennsylvania, USA.

F. PROTASTERIDAE S. A. Miller, 1889
O. (SOU/HAR)–C. (HAS) Mar.

First: *Protaster salteri* (Salter, 1857), Bala Beds, North Wales, and Spy Wood Grit Shropshire, England, UK.
Last: *Drepanaster scabrosus* (Whidborne, 1896), Pilton Shales, *Cleistopora* Zone, Devon, England, UK.

Suborder ZEUGOPHIURINA Matsumoto, 1929

F. LAPWORTHURIDAE Gregory, 1897
O. (ARG2)–Rec. Mar.

First: *Hallaster* sp., upper Arenig, Czechoslovakia. **Extant**

F. FURCASTERIDAE Stürtz, 1900
O. (RAW)–C. (TOU) Mar.

First: *Furcaster trepidans* Spencer, 1925, Starfish Bed, Girvan, Scotland, UK.
Last: *Tremaster difficilis* Worthen and Miller, 1883, Kinderhookian, Illinois, USA.

Suborder EURYALINA Lamarck, 1816

F. EOSPONDYLIDAE Spencer and Wright, 1966
D. (LOK/PRA) Mar.

First and Last: *Eospondylus primigenius* (Stürtz, 1886) and *Kentrospondylus decadactylus* Lehmann, 1957, Hünsrückschiefer, Rhineland, Germany.

F. ONCYCHASTERIDAE S. A. Miller, 1889
C. (TOU) Mar.

First and Last: *Onychaster barrisi* (Hall, 1861), Lower Carboniferous, Braunton Down, Devon, England, UK, and *O. flexilis* Meek and Worthen, 1868, Lower Carboniferous, Indiana, USA.

F. ASTERONYCHIDAE Miller and Troschel, 1842
K. (MAA)–Rec. Mar.

First: *Asteronyx ornatus* Rasmussen, upper Maastrichtian, Upper Chalk, Denmark. **Extant**

F. ASTEROSCHEMATIDAE Verrill, 1899
Extant Mar.

F. GORGONOCEPHALIDAE Ljungman, 1867
T. (RUP/CHT)–Rec. Mar.

First: Undescribed genus, Oligocene of New Zealand (cited by Spencer and Wright, 1966). **Extant**

F. EURYALIDAE Gray, 1840 **Extant** Mar.

Order OPHIURIDA Müller and Troschel, 1840

Suborder CHILOPHIURINA Matsumoto, 1915

F. OPHIURINIDAE Gregory, 1897
S.–C. (TOU) Mar.

First: *Argentinaster bodenbenderi* Ruedemann, 1916, Silurian, Argentina.
Last: *Silesiaster longivertebralis* Schwarzbach and Zimmermann, 1936, Tournaisian, Germany.

F. OPHIURIDAE Lyman, 1865 C. (CHD)–Rec. Mar.

First: *Aganaster gregarius* (Meek and Worthen, 1869), Keokuk Group, Crawfordsville, Indiana, USA. **Extant**

F. OPHIOLEUCIDAE Matsumoto, 1915
Extant Mar.

F. OPHIOCOMIDAE Ljungman, 1867
J. (OXF)–Rec. Mar.

First: *Ophiocoma ? nereida* (Wright, 1880), Sandsfoot Grit, upper Oxfordian, Weymouth, Dorset, England, UK. **Extant**

F. OPHIONEREIDIDAE Ljungman, 1867
Extant Mar.

F. OPHIODERMATIDAE Ljungman, 1867
J. (HET)–Rec. Mar.

First: *Palaeocoma escheri* (Heer, 1865), Insect Marl, *laqueus* Subzone, Schambelen, Switzerland. **Extant**

Suborder LAEMOPHIURINA Matsumoto, 1915

F. OPHIACANTHIDAE Perrier, 1891
J. (SIN)–Rec. Mar.

First: *Ophiacantha* sp. nov., *bucklandi* or *semicostatum* Zone, Hock Cliff, Fretherne, Gloucestershire, England, UK. **Extant**

F. HEMIEURYALIDAE Verrill, 1899
J. (PLB)–Rec. Mar.

First: *Hemieurylae ? lunaris* Hess, 1962, *davoei* Zone, Aston Magna, Gloucestershire, England, UK. **Extant**

Suborder GNATHOPHIURINA Matsumoto, 1915

F. AMPHILEPIDIDAE Matsumoto, 1915
Extant Mar.

F. OPHIACTIDAE Matsumoto, 1915 **Extant** Mar.

F. AMPHIEURIDAE Ljungman, 1867
K. (CEN)–Rec. Mar.

First: *Nullamphieura felli* Skwarko, 1963, Cenomanian, Australia. **Extant**

F. OPHIOTHRICIDAE Ljungman, 1866
J. (OXF)–Rec. Mar.

First: *Ophiothrix ? royeri* (de Loriol, 1872), Humeralisschichten, upper Oxfordian, Raedersdorf, Switzerland. **Extant**

REFERENCES

Bell, B. M. (1976) A study of North American Edrioasteroida. *Memoir New York State Museum Science Service*, **21**, 1–446.

Blake, D. B. (1973) Ossicle morphology of some recent asteroids and description of some west American fossil asteroids. *University of California Publications in Geological Science*, **104**, 60 pp.

Blake, D. B. (1987) A classification and phylogeny of post-Paleozoic sea stars (Asteroidea: Echinodermata). *Journal of Natural History*, **21**, 481–528.

Bockelie, J. F. (1979) Taxonomy, functional morphology and palaeoecology of the Ordovician cystoid family Hemicosmitidae. *Palaeontology*, **22**, 363–406.

Bockelie, J. F. (1981) The Middle Ordovician of the Oslo Region, Norway. 30. The eocrinoid genera *Cryptocrinites*, *Rhipidocystis* and *Bockia*. *Norsk Geologisk Tidsskrift*, **61**, 123–47.

Bockelie, J. F. (1984) The Diploporita of the Oslo Region, Norway. *Palaeontology*, **27**, 1–68.

Bouska, J. (1947) *Pygmaeocrinus*, new crinoid from the Devonian of Bohemia. *Vestnik Kralovske Ceske Spolecnosti Nauk*, **1946**, 1–4.

Breimer, A. and Macurda, D. B. (1972) The phylogeny of the fissiculate blatoids. *Proceedings Koninklijke Nederlandse Akademie van Wetenschappen*, **26**, 1–390.

Brett, C. E., Frest, T. J., Sprinkle, J. *et al.* (1983) Coronoidea: a new class of blastozoan echinoderms based on taxonomic reevaluation of *Stephanocrinus*. *Journal of Paleontology*, **57**, 627–51.

Cotteau, G. H. (1862–1868) *Paléontologie Française. Description des Animaux Invertébrés: (a) Terrains crétacés*, v. 7, G. Masson, Paris, 894 pp.

David, B. (1988) Origin of the deep-sea holasterid fauna, in *Echinoderm Phylogeny and Evolutionary Biology* (eds C. R. C. Paul and A. B. Smith), Clarendon Press, Oxford, pp. 331–45.

Desor, E. and Loriol, P. de (1868–1872) *Echinologie Helvétique. Terrain jurassique*, Ch. Reinwald, Paris, C. W. Kriedel, Wiesbaden, 441 pp.

Devriès, A. (1960) *Contribution à l'étude de quelques groupes d'échinides fossiles d'Algérie*, Publ. Serv. Carte Geol. Algerie, *Paléont. Mém.*, **3**, 1–276.

Donovan, S. K. (1988) The early evolution of the Crinoidea, in *Echinoderm Phylogeny and Evolutionary Biology* (eds C. R. C. Paul and A. B. Smith). Clarendon Press, Oxford, pp. 235–44.

Donovan, S. K. and Paul, C. R. C. (1985) Coronate echinoderms from the Lower Palaeozoic of Britain. *Palaeontology*, **28**, 527–43.

Durham, J. W. (1955) Classification of clypeasteroid echinoids, *California Univ. Publ. Geol. Sci.*, **31**, 73–198.

Durham, J. W. (1978) Polymorphism in the Pliocene sand dollar *Merriamaster* (Echinoidea). *Journal of Paleontology*, **52**, 275–86.

Durham, J. W., Fell, H. B., Fischer, A. G. *et al.* (1966) Asterozoa – Echinozoa, in *Treatise on Invertebrate Paleontology, Part U, Echinodermata 3* (ed. R. C. Moore), Geological Society of America and University of Kansas, Boulder, Colorado, and Lawrence, Kansas, pp. U211–U640.

Endelman, L. G. (1980) New species of *Neoglobator* (Echinoidea, Holectypoidea) from the Danian–Eocene of the southern USSR. *Byulleten Moskovogo Obstietiestva ispytatelei Prirody: Otdel Geologicheskii*, **55** (3), 93–103 [in Russian].

Fay, R. O. (1967) Phylogeny and evolution [of blastoids], in *Treatise on Invertebrate Paleontology, Part S, Echinodermata 1 (2)* (ed. R. C. Moore), Geological Society of America and University of Kansas, Boulder, Colorado, and Lawrence, Kansas, pp. S392–S396.

Frizzell, D. L. and Exline, H. (1955) Monograph of fossil holothurian sclerites. *Bulletin of the Missouri University School of Mines, Technical Series*, **89**, 204 pp.

Gale, A. S. (1987) Phylogeny and classification of the Asteroidea (Echinodermata). *Zoological Journal of the Linnean Society*, **89**, 107–32.

Gehling, J. G. (1987) Earliest known echinoderm – a new Ediacaran fossil from the Pound Subgroup of South Australia. *Alcheringa*, **11**, 337–45.

Gilliland, P. (1992a) Holothurians in the Blue Lias of southern Britain. *Palaeontology*, **35**, 159–210.

Gilliland, P. M. (1992b) Holothurian faunal changes across the Triassic–Jurassic boundary. *Lethaia*, **25**, 69–84.

Guensberg, T. and Sprinkle, J. (1990) Early Ordovician crinoid-dominated echinoderm fauna from the Fillmore Formation of western Utah. *Geological Society of America, Abstracts with Programs*, **22** (7), A220.

Gupta, V. J. and Webster, G. D. (1971) *Stephanocrinus angulatus* (Crinoidea) from the Silurian of Kashmir. *Palaeontology*, **14**, 262–5.

Gutschick, R. C., Canis, W. F. and Brill, K. G. jnr (1967) Kinderhook (Mississippian) holothurian sclerites from Montana and Missouri. *Journal of Paleontology*, **41**, 1461–80.

Horowitz, A. S., Macurda, D. B. and Waters, J. A. (1986) Polyphyly in the Pentremitidae (Blastoidea, Echinodermata). *Bulletin of the Geological Society of America*, **97**, 156–61.

Jablonski, D. and Bottjer, D. J. (1990) The origin and diversification of major groups: environmental patterns and macroevolutionary lags, in *Major Evolutionary Radiations* (eds P. D. Taylor and G. P. Larwood), Volume 42. Clarendon Press, Oxford, pp. 17–57.

Jackson, R. T. (1929) *Paléozoic Echini of Belgium*, Musée Royal Histoire Nat. Belgique, Mém., **38**, 96 pp.

Jagt, J. W. M. (1992) Campanian–Maastrichtian pelagic crinoids from NE Belgium and SE Netherlands: preliminary observations. *Bulletin de l'Institut Royal des Sciences Naturelles de Belgique, Sciences de la Terre*, **62**, 155–61.

Jell, P. A., Burrett, C. F. and Banks, M. R. (1985) Cambrian and Ordovician echinoderms from eastern Australia. *Alcheringa*, **9**, 183–208.

Jensen, M. (1981) Morphology and classification of Euechinoidea Brouu, 1860. A cladistic analysis. *Videnskabelige Meddelelser fra Dansk Naturhistorisk Forening*, **141**, 7–99.

Kier, P. M. (1962) Revision of the cassiduloid echnoids. *Smithson. Misc. Colls.*, **144**, 1–262.

Kier, P. M. (1965) Evolutionary trends in Paleozoic echinoids. *J. Paleont.*, **39**, 436–65.

Kier, P. M. (1967) Revision of the oligopygoid echinoids. *Smithson. Misc. Colls.*, **152**, 1–147.

Kier, P. M. (1968) Echinoids from the Middle Eocene Lake City Formation of Georgia. *Smithson. Misc. Colls.*, **153**, 1–45.

Kier, P. M. (1977) Triassic echinoids. *Smithsonian Contributions to Paleobiology*, **30**, 1–88.

Kier, P. M. (1984a) Fossil spatangoid echinoids of Cuba. *Smithsonian Contributions to Paleobiology*, **55**, 1–336.

Kier, P. M. (1984b) Echinoids from the Triassic (St Cassian) of Italy, their lantern supports, and a revised phylogeny of Triassic echinoids. *Smithsonian Contributions to Paleobiology*, **56**, 1–4.

Kier, P. M. and Lawson, M. H. (1978) Index to living and fossil echinoids 1924–1970. *Smithsonian Contributions to Paleobiology*, **34**, 1–182.

Kikuchi, Y. and Nikaido, A. (1985) The first occurrence of the abyssal echinoid *Pourtalesia* from the middle Miocene Tatsukuraiso Mudstone in Iberaki Prefecture, northeastern Honshu, Japan. *Annual Report Geoscience, University Tsukuba*, **11**, 32–4.

Klikushin, V. G. (1982) Cretaceous and Paleogene Bourgueticrinina. *Geobios*, **15**, 811–43.

Klikushin, V. G. (1984) Fossil sea lilies of the Suborder Hyocrinina. *Paleontologischeskii Zhurnal*, **3**, 74–85.

Klikushin V. G. (1987a) Thiolliericrinid crinoids from the Lower Cretaceous of Crimea. *Geobios*, **20**, 625–65.

Klikushin, V. G. (1987b) Crinoids from Middle Liassic Rosso ammonitico beds. *Neues Jahrbuch für Geologie und Paläontologie Abhandlungen*, **175**, 235–60.

Kornicker, L. S. and Imbrie, J. (1958) Holothurian sclerites from the Florena Shale (Permian) of Kansas. *Micropaleontology*, **4**, 83–91.

Lambert, J. and Thiéry, P. (1909–1925) *Essai de Nomenclature Raisonée des Echinides*. Librairie Ferriere, Chaumont, 607 pp.

Lane, N. G. (1970) Lower and middle Ordovician crinoids from west-central Utah. *Geological Studies of Brigham Young University*, **17**, 3–17.

Larrain, A. P. (1984) The fossil and recent shallow water irregular echinoids from Chile. Doctoral thesis, University of Southern California, Los Angeles, 235 pp.

Lewis, R. D. (1981) *Archaetaxocrinus*, new genus, the earliest known flexible crinoid (Whiterockian) and its phylogenetic implications. *Journal of Paleontology*, **55**, 227–38.

Linder, R. A., Durham, J. W. and Orr, W. N. (1988) New late Oligocene echinoids from the central western Cascades of Oregon. *Journal of Paleontology*, **62**, 945–8.

Macurda, D. B. (1983) Systematics of the fissiculate Blastoidea. *University of Michigan Museum of Paleontology Papers on Paleontology*, **22**, 1–291.

McIntosh, G. C. (1979) Abnormal specimens of the Middle Devonian crinoid *Bactrocrinites* and their effect on the taxonomy of the genus. *Journal of Paleontology*, **53**, 18–28.

Melville, R. V. and Smith, J. D. D. (eds) (1987) *Official Lists and Indexes of Names and Works in Zoology*. International Trust for Zoological Nomenclature, London, 366 pp.

Mooi, R. (1989) Living and fossil genera of the Clypeasteroida

(Echinoidea: Echinodermata): an illustrated key and annotated checklist. *Smithsonian Contributions to Zoology*, **488**, 1–51.

Mooi, R. (1990) Paedomorphosis, Aristotle's lantern, and the origin of the sand dollars. *Paleobiology*, **16**, 25–48.

Mortensen, T. (1928–1951) *A Monograph of the Echinoidea*, 5 vols, C. A. Reitzel, Copenhagen.

Negretti, B., Philippe, M., Soudet, H. J. *et al.* (1990) *Echinometra miocenica* Loriol, échinide Miocène, synonyme d'*Echinometra mathei* (Blainville), actuel: biogéographie et paléoécologie. *Geobios*, **23**, 445–9.

Parma, S. G. (1985) Eoscutella Grant and Hertlein (Echinodermata, Clypeasteroida) in the Patagoniana (Early Tertiary) of the Province of Santa Cruz, Republica Argentina. *Ameghiniana*, **22**, 35–41 [in Spanish].

Paul, C. R. C. (1966) On the occurrence of *Comarocystites* or *Sinclairocystis* (Paracrinoidea: Comarocystitidae) in the Starfish Bed, Threave Glen, Girvan. *Geological Magazine*, **102**, 474–77.

Paul, C. R. C. (1967) The British Silurian cystoids. *Bulletin of the British Museum (Natural History) Geology Series*, **13**, 297–356.

Paul, C. R. C. (1968) Notes on cystoids. *Geological Magazine*, **105**, 413–20.

Paul, C. R. C. (1984) British Ordovician Cystoids. Part 2. *Palaeontological Society Monograph*, 65–152.

Paul, C. R. C. (1988) The phylogeny of the cystoids, in *Echinoderm Phylogeny and Evolutionary Biology* (eds C. R. C. Paul and A. B. Smith), Clarendon Press, Oxford, pp. 199–213.

Paul, C. R. C. and Smith, A. B. (1984) The early radiation and phylogeny of echinoderms. *Biological Reviews*, **59**, 443–81.

Pisera, A. and Dzik, J. (1979) Tithonian crinoids from Rogoznik (Pieniny Klippen Belt, Poland) and their evolutionary relationships. *Eclogae Geologicae Helvetiae*, **72**, 805–49.

Prokop, R. J. (1976) The Family Ramacrinidae fam. n. (Crinoidea) in the Devonian of Bohemia. *Casopsis pro Mineralogii a Geologii*, **22**, 43–6.

Prokop, R. J. (1987) The stratigraphical distribution of Devonian crinoids in the Barrandian area. *Newsletters in Stratigraphy*, **17**, 101–7.

Rasmussen, H. W. (1961) A monograph on the Cretaceous Crinoidea. *Danske Videnskabernes Selskabs, Biologiske Skrifter*, **12**, 1–428.

Regnéll, G. (1945) *Non-crinoid Pelmatozoa from the Palaeozoic of Sweden*. Meddelanden från Lunds Geologisk-Mineralogiska Institution, no. 108, 255 pp.

Reguant, S., Roman, J. and Villatte, J. (1970) Echinides de l'éocène moyen de la région de Vic (Barcelone). *Comptes Rendues de la Société Géologique de France*, **6**, 209.

Roman, J. (1965) Morphologie et évoloution des *Echinolampas* (échinides cassiduloïdes). *Mémoire du Muséum National d'Histoire Naturelle, Paris, Série C*, **15**, 1–341.

Rose, E. P. F. (1982) Holectypoid echinoids and their classification, in *Echinoderms* (ed. J. M. Lawrence), A. A. Balkema, Rotterdam, pp. 145–52.

Rose, E. P. F. and Oliver, J. B. S. (1985) Slow evolution in the Holectypidae, a family of primitive irregular echinoids, in *Echinodermata* (eds B. F. Keegan and B. D. S. O'Connor), A. A. Balkema, Rotterdam, pp. 81–9.

Rose, E. P. F. and Oliver, J. B. S. (1988) Jurassic echinoids of the family Menopygidae: Implications for the evolutionary interpretation and classification of early Irregularia, in *Echinoderm Biology* (eds R. D. Burke, P. V. Mladenov, P. Lambert *et al.*), A. A. Balkema, Rotterdam, pp. 149–58.

Schultz, M. G. (1985) Die Evolution der Echiniden-Gattung *Galerites* im Campan und Maastricht Norddeutschlands. *Geologische Jahrbuch (A)*, **80**, 3–93,

Sieverts, H. (1933) Jungtertiäre Crinoiden von Seran und Borneo. *Neues Jahrbuch für Mineralogie, Geologie und Paläontologie*, **69**, 145–68.

Simms, M. J. (1988) The phylogeny of post-Palaeozoic crinoids, in *Echinoderm Phylogeny and Evolutionary Biology* (eds C. R. C. Paul and A. B. Smith), Clarendon Press, Oxford, pp. 269–84.

Simms, M. J. (in press) The crinoid fauna of the Chambara Formation, Pucára Group, central Peru. *Palaeontographica*.

Simms, M. J. and Sevastopulo, G. D. (1993) The origin of articulate crinoids. *Palaeontology*, **36**, 91–109.

Smith, A. B. (1981) Implications of lantern morphology for the phylogeny of post-Palaeozoic echinoids. *Palaeont.*, **24**, 779–801.

Smith, A. B. (1983) British Carboniferous Edrioasteroidea. *Bulletin of the British Museum (Natural History) Geology Series*, **37**, 113–38.

Smith, A. B. (1984a) Classification of the Echinodermata. *Palaeontology*, **27**, 431–59.

Smith, A. B. (1984b) *Echinoid Palaeobiology*. Allen and Unwin, London, 190 pp.

Smith, A. B. (1985) Cambrian eleutherozoan echinoderms and the early diversification of edrioasteroids. *Palaeontology*, **28**, 715–56.

Smith, A. B. (1988a) Patterns of diversification and extinction in early Palaeozoic echinoderms. *Palaeontology*, **31**, 799–828.

Smith, A. B. (1988b) Phylogenetic relationship, divergence times, and rates of molecular evolution for camarodont sea urchins. *Molecular Biology and Evolution*, **5**, 345–65.

Smith, A. B. (1990) Echinoid evolution from the Triassic to Lower Liassic. *Cahiers de l'Université Catholique de Lyon, Séries Scientifique*, **3**, 79–117.

Smith, A. and Bengtson, P. (1991) Cretaceous echinoids from North-eastern Brazil. *Fossils and Strata*, **31**, 1–88.

Smith, A. B. and Jell, P. A. (1990) Cambrian edrioasteroids from Australia and the origin of starfishes. *Memoirs of the Queensland Museum*, **28**, 715–78.

Smith, A. B. and Patterson, C. (1988) The influence of taxonomic method on the perception of patterns of evolution. *Evolutionary Biology*, **23**, 127–216.

Smith, A. B. and Wright, C. W. (1989–1990) British Cretaceous Echinoids. *Palaeontographical Society Monograph*, **578** and **583**, 1–198.

Soodan, K. S. (1972) Fossil holothurian sclerites from the Upper Cretaceous and Palaeocene sequences of Kutch, India. *Proceedings of the 59th Indian Science Congress*, **3**, 224–5.

Spencer, W. K. and Wright, C. W. (1966) Asterozoans, in *Treatise on Invertebrate Paleontology, Part S, Echinodermata 1 (2)*, (ed. R. C. Moore), Geological Society of America and University of Kansas, Boulder, Colorado and Lawrence, Kansas, pp. U4–U107.

Sprinkle, J. (1990) New echinoderm fauna from the Ninemile Shale (Lower Ordovician) of central and southern Nevada. *Geological Society of America, Abstracts with Programs*, **22** (7), p. A219.

Sprinkle, J. and Gutschick, R. C. (1990) Early Mississippian blastoids from western Montana. *Bulletin of the Museum of Comparative Zoology*, **152**, 89–166.

Sun, Y. C. (1936) On the occurrence of Aristocystis faunas in China. *Bulletin of the Geological Society of China*, **15**, 477–88.

Ubaghs, G. (1971) Un crïnoide enigmatique Ordovicien: *Perittocrinus* Jaekel. *Neues Jahrbuch für Geologie und Paläontologie Abhandlungen*, **137**, 305–36.

van der Ham, R. W. J. M. (1988) Echinoids from the early Palaeocene (Danian) of the Maastricht area (NE Belgium, SE Netherlands): Preliminary results. *Mededelingen van de Werkgroop voor Tertiaire en Kwartaire Geologie*, **25**, 127–61.

Warn, J. and Strimple, H. L. (1977) The disparid inadunate superfamilies Homocrinacea and Cincinaticrinacea (Echinodermata: Crinoidea), Ordovician–Silurian, North America. *Bulletins of American Paleontology*, **72**, 1–138.

Waters, J. A. and Sevastopulo, G. D. (1984) The stratigraphical distribution and palaeoecology of Irish Lower Carboniferous blastoids. *Irish Journal of Earth Science*, **6**, 137–54.

Webster, G. D. (1987) Permian crinoids from the type-section of the Callytharra Formation, Callytharra Springs, Western Australia. *Alcheringa*, **11**, 95–135.

Webster, G. D. (1990) New Permian crinoids from Australia. *Palaeontology*, **33**, 49–74.

Weisbord, N. E. (1971) Bibliography of Cenozoic Echinoidea including some Mesozoic and Paleozoic titles. *Bulletin of American Paleontology*, **59**, 1–314.

Zaghbib-Turki, D. (1987) *Les Echinides des Crétacé du Tunisie*. D. ès Sc. Thèse, Université de Tunis, 613 pp (unpublished).

26

BASAL DEUTEROSTOMES (CHAETOGNATHS, HEMICHORDATES, CALCICHORDATES, CEPHALOCHORDATES AND TUNICATES)

M. J. Benton

The taxa included in this chapter are all apparently related to graptolites (see Chapter 27), echinoderms (see Chapter 25), and chordates (see Chapters 30–41). Many of these have been called calcichordates, carpoids or stylophorans, and include such forms as the cinctans (homosteleans), solutes (homoiosteleans), cornutes and mitrates. The calcichordates are treated here as a paraphyletic group.

The family-level documentation has been compiled from the publications of Jefferies (1969, 1986, 1990, and MS) and others, with the classification based broadly on the cladistic analyses by Jefferies (1986) and Cripps (1991). Note that this is a recent, cladistically-based classification, which is regarded as controversial by many workers on calcichordates and on modern basal chordates and echinoderms. Basic information on the older taxa is presented in Ubaghs (1967a,b) and Caster (1967). There have been problems in defining 'family-level' groupings: problems typical to all groups, but especially here because of the rate of new discoveries, and because of the fluid nature of the phylogenetic trees. This chapter provides a measure of present, patchy knowledge, and may be tested usefully against future discoveries, and against known earlier times of origin based on cladistic branching points.

Paraphyletic groups are marked with a 'p' in parentheses.

Acknowledgements – I am deeply grateful to Dick Jefferies for his help with this chapter, both in letting me draw from his unpublished MS on the subject, and for checking the details, even though the result does not really reflect his views on the possibility of producing such a taxon list from a cladistic phylogeny. I thank Derek Briggs, Paul Daley and Quentin Bone for other assistance.

Superphylum DEUTEROSTOMIA Grobben, 1908

Phylum CHAETOGNATHA Leuckart, 1854

F. UNNAMED Є. (STD) Mar.

First: Undescribed specimens, Middle Cambrian, Stephen Formation, *Glossopleura* Zone, Mount Stephen, British Columbia, Canada (Briggs and Conway Morris, 1986).
Comments: *Amiskwia* Walcott, 1911, from the Burgess Shale, was described as a chaetognath, although this interpretation was rejected by Conway Morris (1977; see also discussion in Briggs and Conway Morris, 1986). The only other soft-bodied fossil of a chaetognath is *Paucijaculum samamithion*, described by Schram (1973) from the Mazon Creek fauna (Carbondale Formation, Francis Creek Shale (MOS), Peabody Coal Co. Pit 11, Will, Grundy and Kankakee Counties, Illinois, USA). Bieri (1991) asserts that this material contains three species in three genera, but the details are unpublished.

The other matter requiring discussion is the possibility that, if the protoconodonts are not phylogenetically linked via the paraconodonts to the euconodonts, then they may not be chordates. In that case, the striking similarities between protoconodont elements and the grasping spines of chaetognaths (e.g. Szaniawski, 1982) may indicate a relationship (see Chapter 29).

Phylum HEMICHORDATA Bateson, 1885

Details of hemichordate classification and fossil record are taken from Bulman (1970).

Class ENTEROPNEUSTA Gegenbaur, 1870 (see Fig. 26.1)

F. UNCERTAIN J. (SIN)–Rec. Mar.

First: *Megaderaion sinemuriense* Arduini *et al.*, 1981, SIN, Lombardy, Italy (Arduini *et al.*, 1981). **Extant**
Intervening: K. (u.).
Comment: Enteropneusts have a very limited fossil record, with this being one of the few definite fossil examples. Upper Cretaceous examples were noted by Wetzel (1972). So-called enteropneust burrows were described from the Muschelkalk of the Holy Cross Mountains, Poland (see Bulman, 1970, p. V13 for details).

Class PTEROBRANCHIA Lankester, 1877

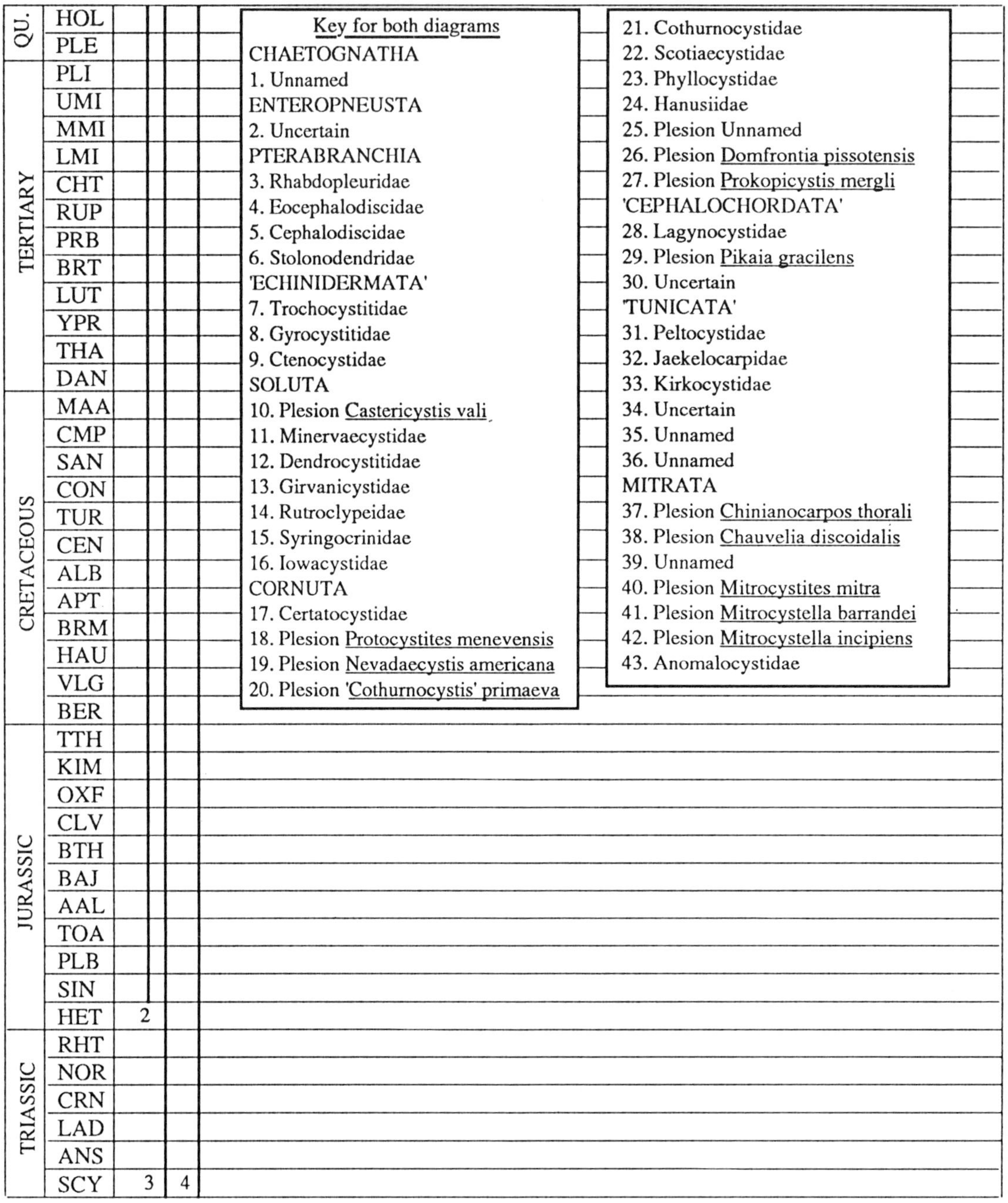

QU.: HOL, PLE
TERTIARY: PLI, UMI, MMI, LMI, CHT, RUP, PRB, BRT, LUT, YPR, THA, DAN
CRETACEOUS: MAA, CMP, SAN, CON, TUR, CEN, ALB, APT, BRM, HAU, VLG, BER
JURASSIC: TTH, KIM, OXF, CLV, BTH, BAJ, AAL, TOA, PLB, SIN, HET 2
TRIASSIC: RHT, NOR, CRN, LAD, ANS, SCY 3 4

Key for both diagrams

CHAETOGNATHA
1. Unnamed
ENTEROPNEUSTA
2. Uncertain
PTERABRANCHIA
3. Rhabdopleuridae
4. Eocephalodiscidae
5. Cephalodiscidae
6. Stolonodendridae
'ECHINIDERMATA'
7. Trochocystitidae
8. Gyrocystitidae
9. Ctenocystidae
SOLUTA
10. Plesion Castericystis vali
11. Minervaecystidae
12. Dendrocystitidae
13. Girvanicystidae
14. Rutroclypeidae
15. Syringocrinidae
16. Iowacystidae
CORNUTA
17. Certatocystidae
18. Plesion Protocystites menevensis
19. Plesion Nevadaecystis americana
20. Plesion 'Cothurnocystis' primaeva
21. Cothurnocystidae
22. Scotiaecystidae
23. Phyllocystidae
24. Hanusiidae
25. Plesion Unnamed
26. Plesion Domfrontia pissotensis
27. Plesion Prokopicystis mergli
'CEPHALOCHORDATA'
28. Lagynocystidae
29. Plesion Pikaia gracilens
30. Uncertain
'TUNICATA'
31. Peltocystidae
32. Jaekelocarpidae
33. Kirkocystidae
34. Uncertain
35. Unnamed
36. Unnamed
MITRATA
37. Plesion Chinianocarpos thorali
38. Plesion Chauvelia discoidalis
39. Unnamed
40. Plesion Mitrocystites mitra
41. Plesion Mitrocystella barrandei
42. Plesion Mitrocystella incipiens
43. Anomalocystidae

Fig. 26.1

Order RHABDOPLEURIDA Fowler, 1892

F. RHABDOPLEURIDAE Harmer, 1905
€. (STD)–Rec. Mar.

First: *Rhabdotubus johanssoni* Bengtson and Urbanek, 1986, *Eccaparadoxides pinus* Zone, Middle Cambrian, Sweden (Bengtson and Urbanek, 1986). **Extant**
Comment: There are several examples of rhabdopleurids scattered through the fossil record, including some from the Jurassic of Poland, and *Rhabdopleura eoceenica* Dighton Thomas and Davis, 1949, from the London Clay (YPR) of southern England, UK.

Order CEPHALODISCIDA Fowler, 1892

F. EOCEPHALODISCIDAE Kozłowski, 1949
O. (TRE) Mar.

First and Last: *Eocephalodiscus polonicus* Kozłowski, 1949, Glauconitic Sandstone, Wysoczki, Poland.

F. CEPHALODISCIDAE Harmer, 1905
O. (TRE/ARG)–Rec. Mar.

First: *Pterobranchites antiquus* Kozłowski, 1967, glacial boulder, Lower Ordovician, Poland. **Extant**
Intervening: LUT/BRT.

Order STOLONOIDEA Kozłowski, 1938

F. STOLONODENDRIDAE Bulman, 1955
O. (TRE) Mar.

First and Last: *Stolonodendrum uniramosum* Kozłowski, 1949, TRE, Poland.
Comment: This group has often been assigned to

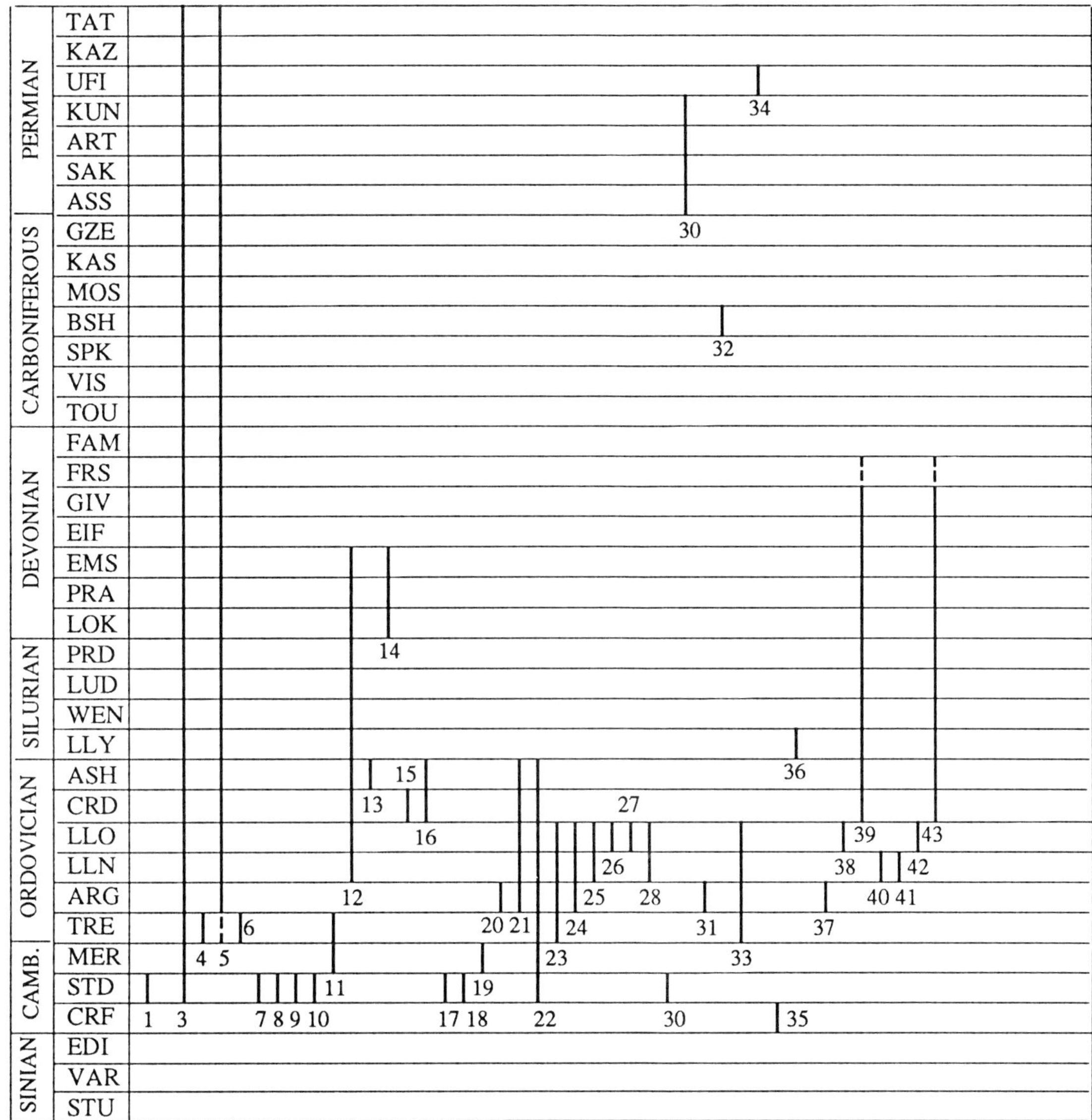

Fig. 26.1.

the Graptolithina. *Melanostrophus fokini* Öpik, 1930, Lower to Upper Ordovician, Europe, may belong here too (Bulman, 1970, p. V53).

Class PLANCTOSPHAEROIDEA van der Horst, 1936

This group has no known fossil record.

Class GRAPTOLITHINA Bronn, 1846

See Chapter 27.

Subsuperphylum DEXIOTHETICA Jefferies, 1979

STEM-GROUP ECHINODERMS

Order CINCTA Jaekel, 1901

F. TROCHOCYSTITIDAE Jaekel, 1901 Є. (STD) Mar.

First and Last: *Trochocystites bohemicus* Barrande, 1887, Middle Cambrian, Bohemia, Czechoslovakia, and ?Bavaria, Germany; *Trochocystoides parvus* Jaekel, 1918, Middle Cambrian, Bohemia, Czechoslovakia (Ubaghs, 1967a).

F. GYROCYSTITIDAE Jaekel, 1918 Є. (STD) Mar.

First and Last: *Gyrocystis barrandei* (Munier-Chalmas and Bergeron, 1889), Middle Cambrian, Morocco; *Decacystis hispanica* Gislén, 1927, Middle Cambrian, Spain.

Class CTENOCYSTOIDEA Robison and Sprinkle, 1969

F. CTENOCYSTIDAE Robison and Sprinkle, 1969 Є. (STD) Mar.

First: *Ctenocystis utahensis* Robison and Sprinkle, 1969, Spence Shale, Wasatch Mountains, Utah, USA (Robison and Sprinkle, 1969).
Last: *Ctenocystis smithi* Ubaghs, 1987, upper Middle Cambrian, Montagne Noire, France (Ubaghs, 1987).

Phylum ECHINODERMATA de Bruguière, 1791 (Fleming, 1828)

See Chapter 25.

Grade SOLUTA Jaekel, 1901 (p)

The solutes are a paraphyletic group, apparently falling on the cladogram between the branch for echinoderms and the branch for chordates. The solute–cornute transition occurs, by tradition, between the Iowacystidae and the Ceratocystidae. The oldest solute known is an unnamed form from the Lower Cambrian Kinzers Formation of Pennsylvania, USA (shown in Paul and Smith, 1984, Fig. 3). It is likely that all known solutes are stem-group chordates.

Plesion *Castericystis vali* Ubaghs and Robison, 1985 Є. (STD) Mar.

First and Last: *Castericystis vali* Ubaghs and Robison, 1985, Marjum Formation, Utah, USA (Ubaghs and Robison, 1985).

?F. MINERVAECYSTIDAE Ubaghs and Caster, 1967 Є. (MER)–O. (TRE) Mar.

First: ?*Minervaecystis*, Trempealeauan, Upper Cambrian, Nevada, USA (Ubaghs, 1969).
Last: *Minervaecystis vidali* (Thoral, 1935), TRE, France.

F. DENDROCYSTITIDAE Bassler, 1938 O. (LLN)–D. (EMS) Mar.

First: *Dendrocystites sedgwicki* (Barrande, 1867), LLN, Bohemia, Czechoslovakia; *Heckericystis kuckersianus* (Hecker, 1940), Middle Ordovician, Estonia, former USSR.
Last: *Dehmicystis globulus* (Dehm, 1934), Bundenbach Shale (Hunsrückian), Rhine region, Germany.
Intervening: LLO, ASH.
Comment: This may well be a paraphyletic assemblage, with *Dehmicystis* probably not a dendrocystitid (P. Daley, pers. comm., 1992).

F. GIRVANICYSTIDAE Caster, 1967 O. (ASH) Mar.

First: *Girvanicystis casteri* Daley, 1992, Sholeshook Limestone (Cautleyan), Pembrokeshire, South Wales, UK (Daley, 1992b).
Last: *Girvanicystis batheri* Caster, 1967, Starfish Bed (upper Rawtheyan), Girvan, Ayrshire, Scotland, UK (Daley, 1992b).

?F. RUTROCLYPEIDAE Gill and Caster, 1960 D. (l.) Mar.

First and Last: *Rutroclypeus junori* Withers, 1933, *R. withersi* Gill and Caster, 1960, and *R. victoriae* Gill and Caster, 1960, all Lower Devonian, Victoria, Australia.

?F. SYRINGOCRINIDAE Parsley and Caster, 1965 O. (CRD) Mar.

First: *Syringocrinus paradoxicus* (Billings, 1859), Sherman Falls Beds, Middle Trenton Group, Quebec, Canada.
Last: *Syringocrinus sinclairi* Parsley and Caster, 1965, Upper Trenton Group, Chateau Richer, Quebec, Canada (Parsley and Caster, 1965).

F. IOWACYSTIDAE Gill and Caster, 1960 O. (CRD–ASH) Mar.

First: *Scalenocystites strimplei* Kolata, 1973, Dunleith Formation, Champlainian Series, Minnesota, USA, *Belemnocystites wetherbyi* Miller and Gurley, 1894, Curdsville Limestone, Champlainian Series, Kentucky, USA, *Myeinocystites natus* Strimple, 1953, Bromide Formation, Champlainian Series, Oklahoma, USA and *M. crossmanni* Kolata *et al.*, 1977, Dunleith Formation, Champlainian Series, Illinois, USA.
Last: *Iowacystis sagittaria* Thomas and Ladd, 1926, Fort Atkinson Formation, Cincinnatian Series, Iowa, USA.
Comment: Includes the F. Belemnocystitidae Parsley, 1967 (Kolata *et al.*, 1977).

Grade CORNUTA Jaekel, 1900 (P)

F. CERATOCYSTIDAE Jaekel, 1900 Є. (STD) Mar.

First and Last: *Ceratocystis perneri* Jaekel, 1900, Skryje Shale, Skryje, Bohemia, Czechoslovakia and *C. vizcainoi* Ubaghs, 1987, Middle Cambrian, Montagne Noire, France (Ubaghs, 1987).

Plesion *Protocystites menevensis* Hicks, 1872 Є. (STD) Mar.

First and Last: *Protocystites menevensis* Hicks, 1872, *Hypagnostus parvifrons* Zone, Menevian, Dyfed, South Wales, UK (Jefferies *et al.*, 1987).

Plesion *Nevadaecystis americana* (Ubaghs, 1963) Є. (MER) Mar.

First and Last: *Nevadaecystis americana* (Ubaghs, 1963), Trempealeauan, Nevada, USA.

Plesion *'Cothurnocystis' primaeva* Thoral, 1935 O. (ARG) Mar.

First and Last: *'Cothurnocystis' primaeva* Thoral, 1935, Lower Ordovician, Montagne Noire, France.

F. COTHURNOCYSTIDAE Bather, 1913 O. (ARG–ASH) Mar.

First: *Cothurnocystis fellinensis* Ubaghs, 1969 and *C. courtessolei* Ubaghs, 1969, both Lower Ordovician, Montagne Noire, France (Ubaghs, 1969).
Last: *Cothurnocystis elizae* Bather, 1913, Starfish Bed, Girvan, Ayrshire, Scotland, UK.
Intervening: None.

F. SCOTIAECYSTIDAE Caster and Ubaghs, *in* Ubaghs, 1967 Є. (STD)–O. (ASH) Mar.

First: *'Cothurnocystis' bifida* Ubaghs, 1988, Wheeler Formation, western Utah, USA (Ubaghs, 1988).
Last: *Scotiaecystis curvata* (Bather, 1913), Starfish Bed, Girvan, Ayrshire, Scotland, UK.
Intervening: ARG, LLO.

F. PHYLLOCYSTIDAE Derstler, 1979 O. (TRE–LLO) Mar.

First: *Prochauvelicystis semispinosa* Daley, 1992a, Arenaceous Beds, Shineton Shales Formation, Cressage, Shropshire, England, UK (Daley, 1992a).
Last: *Milonicystis kerfornei* Chauvel, 1966, Traveusot Formation, Morocco and *Phyllocystis salairica* Dubatolova, *in* Rozova *et al.*, 1985, Chupino Formation, Salaira, former USSR (Chauvel, 1966; Dubatolova, *in* Rozova *et al.*, 1985).
Intervening: ARG, LLN.

F. HANUSIIDAE Cripps, 1991 O. (ARG–LLO) Mar.

First: *Galliaecystis lignieresi* Ubaghs, 1969 and *Progalliaecystis ubaghsi* Cripps, 1989, both basal Ordovician, Montagne Noire, France.
Last: *Hanusia prilepensis* Cripps, 1989, Dobrotivá Forma-

tion, Bohemia, Czechoslovakia (Cripps, 1989, 1991).
Intervening: LLN.

Plesion UNNAMED O. (LLN–LLO) Mar.

First: *Reticulocarpos hanusi* Jefferies and Prokop, 1972, Šárka Shales, Bohemia, Czechoslovakia.
Last: *Beryllia miranda* Cripps, in press, Montagne Noire, France (Cripps, 1991).

Plesion *Domfrontia pissotensis* (Chauvel and Nion, 1977) O. (LLO) Mar.

First and Last: *Domfrontia pissotensis* (Chauvel and Nion, 1977), Middle Ordovician shales, Pissot, Normandy, France.

Plesion *Prokopicystis mergli* Cripps, 1989 O. (LLO) Mar.

First and Last: *Prokopicystis mergli* Cripps, 1989, Dubrotivá Shales, Bohemia, Czechoslovakia (Cripps, 1989).

Phylum CHORDATA Bateson, 1886 (see Chapters 30–41)

STEM-GROUP CEPHALOCHORDATES

F. LAGYNOCYSTIDAE Jaekel, 1918 O. (LLN–LLO) Mar.

First: *Lagynocystis pyramidalis* (Barrande, 1887), Šárka Shales, Bohemia, Czechoslovakia.
Last: *Lagynocystis pyramidalis* (Barrande, 1887), Traveusot Formation, Normandy, France (Chauvel and Nion, 1977).

Plesion *Pikaia gracilens* Walcott, 1911 Є. (STD) Mar.

First and Last: *Pikaia gracilens* Walcott, 1911, Burgess Shale, British Columbia, Canada.
Comment: The phylogenetic position of *Pikaia* is uncertain. It may fall here, or in the stem group of the craniates. It is described (Conway Morris, 1979) as amphioxus-like, and may well be a chordate. If it is a chordate, it may be located in the cladogram as indicated in this listing, or it may be a close sister group of the crown-group craniates (Jefferies, MS). Butterfield (1990) doubts the chordate affinities of *Pikaia*.

Subphylum CEPHALOCHORDATA Owen, 1846

F. UNCERTAIN P. (ROT) Mar.

First and Last: *Palaeobranchiostoma hamatotergum*, Whitehill Formation Ecca Group (Lower Permian), South Africa (Oelofson and Loock, 1981).
Comments: This find is unconfirmed. Other supposed fossil cephalochordates have been discounted (Holland *et al.*, 1967, p. 608).

STEM-GROUP TUNICATES

Suborder PELTOCYSTIDA Jefferies, 1973

F. PELTOCYSTIDAE Ubaghs, 1967 O. (ARG) Mar.

First and Last: *Peltocystis cornuta* Thoral, 1935, lowest ARG, Montagne Noire, France.

F. JAEKELOCARPIDAE Kolata *et al.* 1991 C. (BSH) Mar.

First and Last: *Jaekelocarpus oklahomensis* Kolata *et al.*, 1991, Golf Course Formation, Morrowan, Johnston County, Oklahoma, USA (Kolata *et al.*, 1991).

F. KIRKOCYSTIDAE Caster, 1952 O. (TRE–LLO) Mar.

First: *Guichenocarpos* sp., Shineton Shales, Shropshire, England, UK.
Last: *Guichenocarpos nioni*, Traveusot Formation, Normandy, France.

Subphylum TUNICATA Lamarck, 1816 (UROCHORDATA Bateson, 1885)

Tunicate fossils are rare. They include spicules from the Eocene of France (?Didemnidae) and the problematic *Palaeobotryllus taylori* Müller, 1977 from the Upper Cambrian of Nevada, USA (Jefferies, 1986, pp. 316–17).

Class ASCIDIACEA Blainville, 1824

F. UNCERTAIN P. (UFI) Mar.

First: *Permosoma tunicatum* Jaekel, 1918, lower Zechstein, Sosio-kalk, Palazzo Adriano, Sosio, Sicily, Italy.

Class THALIACEA van der Hoeven, 1850

No known fossil record (see Hoiland *et al.*, 1967, p. 607).

Class APPENDICULARIA Lahille, 1890

Limited fossil record (see Holland *et al.*, 1967, p. 607), other than the appendicularians reported by Zhang (1987) from the Cambrian of China.

F. UNNAMED Є. (CRF) Mar.

First and Last: 'Appendicularians', Lower Cambrian, China (1987).

Order UROCHORDATA *Incertae sedis*

F. UNNAMED S. (LLY) Mar.

First and Last: *Ainiktozoon loganense* Scourfield, 1937, Patrick Burn Formation, Priesthill Group (late LLY in age), Lesmahagow Inlier, Ayrshire, Scotland, UK.
Comments: Originally described as an early chordate, but discounted as a probable arthropod by Holland *et al.* (1967, p. 607). Ritchie (1985), however, regards it as a protochordate, possibly a tunicate.

Subphylum CRANIATA

Order MITRATA Jaekel, 1918 (p)

The mitrates usually include the stem-group cephalochordates and tunicates noted above, as well as the stem-group craniates noted here.

Plesion *Chinianocarpos thorali* Ubaghs, 1961 O. (ARG) Mar.

First and Last: *Chinianocarpos thorali* Ubaghs, 1961, lowest ARG, Montagne Noire, France.

Plesion *Chauvelia discoidalis* Cripps, 1990 O. (LLO) Mar.

First and Last: *Chauvelia discoidalis* Cripps, 1990, Ouine-Inirne Formation, Anti Atlas Group, Zagora, Morocco (Cripps, 1990).

F. UNNAMED O. (CRD)–D. (GIV/FRS) Mar.

First: *Aspidocarpus bohemicus* Ubaghs, 1979, Letná Forma-

tion, lower CRD, Bohemia, Czechoslovakia (Ubaghs, 1979).
Last: *Paranacystis petrii* Caster, 1954, Ponta Grossa Shale, Lower Devonian, Paraná, Brazil.
Intervening: None.

Plesion *Mitrocystites mitra* Barrande, 1887
O. (LLN) Mar.

First and Last: *Mitrocystites mitra* Barrande, 1887, Šárka Shales, Bohemia, Czechoslovakia.

Plesion *Mitrocystella barrandei* Jaekel, 1901
O. (LLN) Mar.

First and Last: *Mitrocystella barrandei* Jaekel, 1901, Šárka Shales, Bohemia, Czechoslovakia.

Plesion *Mitrocystella incipiens* (Barrande, 1887)
O. (LLO) Mar.

First and Last: *Mitrocystella barrandei* Jaekel, 1901, Dobrotivá Beds, Bohemia, Czechoslovakia, and Formation de Traveusot, Redon, Bretagne, France.

F. ANOMALOCYSTIDAE Bassler, 1938
O. (CRD)–D. (GIV/FRS) Mar.

First: *Barrandeocarpus jaekeli* Ubaghs, 1979, Letná Formation, Bohemia, Czechoslovakia.
Last: *Australocystis langei* Caster, 1954, Ponta Grossa Shale, Paraná Basin, Brazil.
Intervening: ASH, LUD–GED, EMS, EIF.

REFERENCES

Arduini, P., Pinna, G. and Teruzzi, G. (1981) *Megaderaion sinemuriense* n. g. n. sp., a new fossil enteropneust of the Sinemurian. *Atti della Società Italiana di Scienze Naturale e del Museo Civico di Storia Naturale di Milano*, **122**, 104–8.

Bengtson, S. and Urbanek, A. (1986) *Rhabdotubus*, a middle Cambrian rhabdopleurid hemichordate. *Lethaia*, **19**, 293–308.

Bieri, R. (1991) Systematics of Chaetognatha, in *The Biology of Chaetognaths* (eds Q. Bone, H. Kapp and A. C. Pierrot-Quilts), Oxford Scientific Publications, pp. 122–36.

Briggs, D. E. G. and Conway Morris, S. (1986) Problematica from the Middle Cambrian Burgess Shale of British Columbia, in *Problematic Fossil Taxa* (eds A. Hoffman and M. H. Nitecki), Oxford University Press, pp. 167–83.

Bulman, O. M. B. (1970) Graptolithina, with sections on Enteropneusta and Pterobranchia, in *Treatise on Invertebrate Paleontology*, Part V, 2nd edn, (eds C. Teichert and R. C. Moore), The Geological Society of America and The University of Kansas, Boulder, Colorado and Lawrence, Kansas, xxxii + 163 pp.

Butterfield, N. J. (1990) Organic preservation of non-mineralizing organisms and the taphonomy of the Burgess Shale. *Paleobiology*, **16**, 272–87.

Caster, K. E. (1967) Homoiostelea, in *Treatise on Invertebrate Paleontology, Part S. Echinodermata 1* (ed. R. C. Moore), The Geological Society of America and The University of Kansas, Boulder, Colorado and Lawrence, Kansas, pp. S581–S627.

Chauvel, J. (1966) *Echinodermes de l'Ordovicien de Maroc*. Cahiers de Paléontologie. CNRS, Paris.

Chauvel, J. and Nion, J. (1977) Echinodermes (Homalozoa: Cornuta et Mitrata) nouveaux pour l'ordovicien du Massif Armoricain et conséquences paléogéographiques. *Geobios*, **10**, 35–49.

Conway Morris, S. (1977) A redescription of the Middle Cambrian worm *Amiskwia sagittiformis* from the Middle Cambrian Burgess Shale of British Columbia. *Paläontologische Zeitschrift*, **51**, 271–87.

Conway Morris, S. (1979) The Burgess Shale (Middle Cambrian) fauna. *Annual Reviews in Ecology and Systematics*, **10**, 327–49.

Cripps, A. P. (1989) A new genus of stem chordate (Cornuta) from the Lower and Middle Ordovician of Czechoslovakia and the origin of bilateral symmetry in the chordates. *Geobios*, **22**, 215–45.

Cripps, A. P. (1990) A new stem cornute from the Ordovician of Morocco and the search for the sister group of the Craniata. *Zoological Journal of the Linnean Society*, **100**, 27–71.

Cripps, A. P. (1991) A cladistic analysis of the cornutes (stem chordates). *Zoological Journal of the Linnean Society*, **102**, 333–66.

Daley, P. E. J. (1992a) Two new cornutes from the Lower Ordovician of Shropshire and southern France. *Paleontology*, **35**, 127–48.

Daley, P. E. J. (1992b) The anatomy of the solute *Girvanicystis batheri* (?Chordata) from the Upper Ordovician of Scotland and a new species of *Girvanicystis* from the Upper Ordovician of South Wales. *Zoological Journal of the Linnean Society*, **105**, 353–75.

Holland, C. H., Rickards, R. B., Skevington, D. *et al.* (1967) Chordata: Hemichordata (including Graptolithina), Pogonophora, Urochordata and Cephalochordata, in *The Fossil Record* (eds W. B. Harland, C. H. Holland, M. R. House *et al.*), Geological Society of London, pp. 601–11.

Jefferies, R. P. S. (1969) *Ceratocystis perneri*–a Middle Cambrian chordate with echinoderm affinities. *Palaeontology*, **12**, 494–535.

Jefferies, R. P. S. (1986) *The Ancestry of the Vertebrates*. British Museum (Natural History), London, 376 pp.

Jefferies, R. P. S. (1990) The solute *Dendrocystoides scoticus* from the Upper Ordovician of Scotland and the ancestry of chordates and echinoderms. *Palaeontology*, **33**, 631–79.

Jefferies, R. P. S., Lewis, M. and Donovan, S. K. (1987) *Protocystites menevensis*–a stem-group chordate (Cornuta) from the Middle Cambrian of South Wales. *Palaeontology*, **30**, 429–84.

Kolata, D. R., Strimple, H. L. and Levorson, C. (1977) Revision of the Ordovician carpoid family Iowacystidae. *Palaeontology*, **20**, 529–57.

Kolata, D. R., Frest, T. J. and Mapes, R. H. (1991) The youngest carpoid: occurrence, affinities, and life mode of a Pennsylvanian (Morrowan) mitrate from Oklahoma. *Journal of Paleontology*, **65**, 844–55.

Oelofson, B. W. and Loock, J. C. (1981) A first record of a fossil cephalochordate? *Palaeontologia Africana*, **24**, 17.

Parsley, R. L. and Caster, K. E. (1965) North American Soluta (Carpoidea, Echinodermata). *Bulletin of American Paleontology*, **49** (221), 109–69.

Paul, C. R. C. and Smith, A. B. (1984) The early radiation and phylogeny of echinoderms. *Biological Review*, **59**, 443–81.

Ritchie, A. (1985) *Ainiktozoon loganense* Scourfield, a protochordate? from the Silurian of Scotland. *Alcheringa*, **9**, 117–42.

Robison, R. A. and Sprinkle, J. (1969) Ctenocystoidea: a new class of primitive echinoderms. *Science*, **166**, 1512–14.

Rozova, A. V., Rozov, S. N. and Dubatolova, Yu. A. (1985) *Stratigrafiya i Fauna Ordovika Severo-Zapadnogo Salaira*. Nauka, Moscow, 176 pp.

Schram, F. R. (1973) Pseudocoelomates and a nemertine from the Illinois Pennsylvanian. *Journal of Paleontology*, **47**, 985–9.

Szaniawski, H. (1982) Chaetognath grasping spines recognized among Cambrian protoconodonts. *Journal of Paleontology*, **56**, 806–10.

Ubaghs, G. (1967a) Homostelea, in *Treatise on Invertebrate Paleontology. Part S. Echinodermata 1* (ed. R. C. Moore), Geological Society of America and University of Kansas Press, Boulder, Colorado and Lawrence, Kansas, pp. S565–81.

Ubaghs, G. (1967b) Stylophora, in *Treatise on Invertebrate Paleontology. Part S. Echinodermata 1* (ed. R. C. Moore), Geological Society of America and University of Kansas Press, Boulder, Colorado and Lawrence, Kansas, pp. S495–565.

Ubaghs, G. (1969) *Les Echinodermes Carpoides de l'Ordovicien Inférieur de la Montagne Noire*. Cahiers de Paléontologie. CNRS, Paris, 112 pp.

Ubaghs, G. (1979) Trois Mitrata (Echinodermata: Stylophora) nouveaux de l'Ordovicien de Tchécoslovaquie. *Paläontologische Zeitschrift*, **53**, 98–119.

Ubaghs, G. (1987) Echinodermes nouveaux du cambrien moyen de la Montagne Noire (France). *Annales de Paléontologie*, **73** (1), 1–22.

Ubaghs, G. (1988) Homalozoan echinoderms of the Wheeler

Formation (Middle Cambrian) of western Utah. *University of Kansas Paleontological Contributions*, **120**, 1–17.

Ubaghs, G. and Robison, R. A. (1985) A new homoiostelean and a new eocrinoid from the Middle Cambrian of Utah. *University of Kansas Paleontological Contributions*, **115**, 1–24.

Wetzel, W. (1972) Flintgefüllte Enteropneusten? Wohnröhren in einen Oberkreide-Geschiebe. *Schriften des Naturwissenschaftliche Vereins von Schleswig-Holstein*, **42**, 104–7.

Zhang, A. (1987) Fossil appendicularians in the early Cambrian. *Scientia Sinica (B)*, **30**, 888–96.

27

GRAPTOLITHINA

R. B. Rickards

The classification of graptolites is currently in a state of flux after many decades of reasonable stability, and the classification adopted here is that of Palmer and Rickards (1991), which broadly hinges on the classification of Fortey and Cooper (1986) and of Mitchell (1987, 1990) with respect to diplograptids. However, I am not entirely convinced by the argument used by Fortey and Cooper in terms of placing the family Anisograptidae in the graptoloids, and hence I retain it in the dendroids, following the Bulman (1970) *Treatise* decision: either decision is arbitrary and accepts the intermediate nature of the family.

Further, I am unable to accept Mitchell's (1987) redefinition of the family Monograptidae to include biserial graptolites occurring as early as the Arenig. By raising to family level some of his constituent subfamilies of the Monograptidae (namely Monograptinae, Dimophograptinae, Glyptograptinae and Eoglyptograptinae), I feel there is less danger of obscuring one of the most spectacular events in graptoloid evolution, namely the development of the uniserial scandent colony. This is especially the case in a volume of this kind which purports to record the true occurrence of major grades and clades. The first edition also recorded the occurrence in time of, for example, the dimorphograptids (partially uniserially scandent) and direct comparison is thus possible of any changes in the record during the past quarter of a century. In this work, I separate the Monograptidae from Mitchell's subfamilies Eoglyptograptinae, Glyptograptinae and Retiolitinae, which I maintain as family-level taxa, as with his Monograptinae and Dimorphograptinae. The Retiolitidae of Bulman (1970) and others are, therefore, split between the Orthograptidae and Abrograptidae (the Ordovician forms) and the Retiolitidae (the Silurian forms; related to the Glyptograptidae), following the views of Mitchell (1987), Bates (1990), and others. Obut and Zaslavskaya (1986), on the other hand, retain all 'retiolitids' in a single group, and recognize numerous families.

Finally, I am in agreement with the work of Mierzejewski (1986) which casts serious doubt on the graptolite affinities of several orders: thus the Stolonoidea are regarded as Pterobranchia, and the orders Dithecoidea and Archaeodendrida are removed from the Hemichordata (see Chapter 26). The families Inocaulidae and Chaunograptidae I provisionally remove to the hydroids (Coelenterata, see Chapter 6), and am in agreement with Mierzejewski (1986) on this.

Acknowledgements – I am extremely grateful to Henry Williams who helped with several areas made problematic as a result of recent reclassifications.

Phylum HEMICHORDATA Bateson, 1885

Class ENTEROPNEUSTA Gegenbaur, 1870

See Chapter 26.

Class PTEROBRANCHIA Lankester, 1877

See Chapter 26.

Class GRAPTOLITHINA Bronn, 1846

Order DENDROIDEA Nicholson, 1872 (see Fig. 27.1)

F. DENDROGRAPTIDAE Roemer, *in* Frech, 1897 €. (CRF)–C. (ARN) Mar.

First: *Callograptus antiquus* Ruedemann, 1947, and *Dendrograptus edwardsi major* Ruedemann, 1947, and *D. hallianus moneymakeri* Ruedemann, 1947, Nolichucky Shale, Tennessee, USA (Ruedemann, 1947); other approximately coeval species of *Dictyonema, Callograptus, Dendrograptus* and *Aspidograptus* from Upper Wilberns Formation, Texas, USA; Deadwood Formation, South Dakota, USA; Trempeleau Formation, Gaspé (Eau Claire), Quebec, Canada (Ruedemann, 1947); and Potsdam Formation, New York, USA. A dendroid graptolite, undescribed, but possibly *Dendrograptus*, has been obtained from the Middle Cambrian *P. davidis* Zone, Comley, England, UK. A *Dictyonema* sp. (Quilty, 1971) probably upper part of Upper Cambrian.

Last: *Dictyonema* sp., E2 Zone, Yorkshire, England, UK (undescribed, but examined by author).

The Fossil Record 2. Edited by M. J. Benton. Published in 1993 by Chapman & Hall, London. ISBN 0 412 39380 8

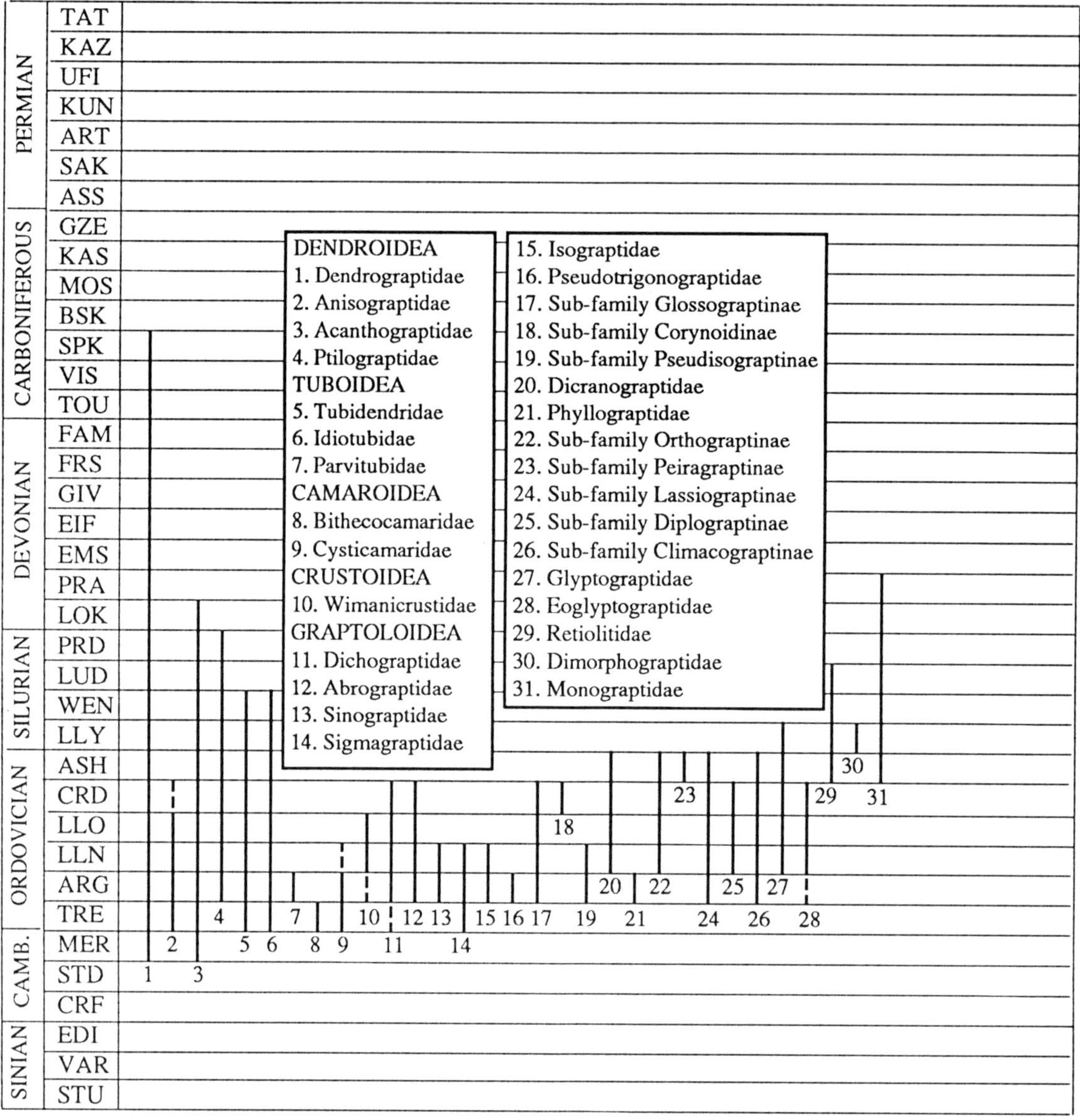

Fig. 27.1

F. ANISOGRAPTIDAE Bulman, 1950
O. (TRE–LLO/CRD) Mar.

First: Undescribed forms referable to *Bryograptus*, mentioned Tjernvik (1958), *D. flabelliforme desmograptoides* Zone, Flagabro, Scania, Sweden.

Last: *Calyxdendrum graptoloides* Kozłowski, 1960, LLO to lower CRD, Poznan and Warsaw, Poland.

Main genera: *Anisograptus* Ruedemann, 1937, *Adelograptus* Bulman, 1941, *Bryograptus* Lapworth, 1880, *Clonograptus* Hall and Nicholson, 1873, *Radiograptus* Bulman, 1950, *Staurograptus* Emmons, 1855, *Triograptus* Monsen, 1925, *Kiaerograptus* Spjeldnaes, 1963, *Calyxdendrum* Kozłowski, 1960.

F. ACANTHOGRAPTIDAE Bulman, 1938
€. (MER)–D. (LOK) Mar.

First: *Acanthograptus priscus* Ruedemann, 1947, Trempealeauan Stage, Jordan Sandstone, Afton, Minnesota and Arbuckle Limestone, Oklahoma, USA (Ruedemann, 1947). *Thallograptus* sp. and *Palaeodictyota* sp. have recently been recorded from the Upper Cambrian by Rickards *et al.* (1990).

Last: *Palaeodictyota rotundatum* Bouček, 1957, *P. textorium* (Počta, 1894), and *P. undulatum* (Počta, 1894), *hercynicus* Zone, Lochkov Limestone, Kasor, Czechoslovakia (Bouček, 1957).

Main genera: *Acanthograptus* Spencer, 1878, *Koremagraptus* Bulman, 1927, *Palaeodictyota* Whitfield, 1902. *Dyadograptus* Obut, 1960 is not recognized here. *Acanthograptus antiquus* Quilty, 1971, recorded by Quilty (1971), is almost certainly referable to the genus *Inocaulis*, which I would place in the hydroids.

F. PTILOGRAPTIDAE Hopkinson, 1875
O. (ARG)–S. (PRD) Mar.

First: Several species of *Ptilograptus* Hall, 1865 from Deepkill Shale and equivalents, North America (Ruedemann, 1947), from San Remo, Australia (Thomas, 1960) and from St David's, Pembrokeshire, Wales, UK.

Last: *Ptilograptus pribyli* Bouček, 1957, Pridoli Reds, Velka Chuchle, Czechoslovakia (Bouček, 1957).

Order TUBOIDEA Kozłowski, 1938

F. TUBIDENDRIDAE Kozłowski, 1949 O. (TRE)–S. (WEN) (= Multitubidae Skevington, 1963) Mar.

First: *Tubidendrum bulmani* Kozłowski, 1949, Poland (Kozłowski, 1949).
Last: *Reticulograptus thorsteinssoni* Bulman and Rickards, 1966, Cornwallis Island, Canadian Arctic.
Constituent genera: *Tubidendrum* Kozłowski, 1949 and *Reticulograptus* Wiman, 1901 (= *Multitubus* Skevington, 1963). Several LUD species placed in *Reticulograptus* (see Bouček, 1957) must now be referred to this genus with reserve, and are not included above.

F. IDIOTUBIDAE Kozłowski, 1949 O. (TRE)–S. (WEN) Mar.

First: *Idiotubus typicalis* Kozłowski, 1949, and several other *Idiotubus* species, *Calycotubus infundibulatus* Kozłowski, 1949, Poland (Kozłowski, 1949).
Last: *Cyclograptus rotadentatus* Spencer, 1883, Niagara Formation, Ontario, Canada (Ruedemann, 1947), and *C. multithecatus* Bouček, 1957, *Cyrtograptus radians* Zone, Czechoslovakia (Bouček, 1957).
Constituent genera: *Idiotubus* Kozłowski, 1949, *Calycotubus* Kozłowski, 1949, *Conitubus* Kozłowski, 1949, *Cyclograptus* Spencer, 1883, *Dendrotubus* Kozłowski, 1949, *Discograptus* Wiman, 1901, *Galeograptus* Wiman, 1901. The genera *Calyptograptus* Spencer, 1878, *Epigraptus* Eisenack, 1941, and *Rhodonograptus* Počta, 1894 are only doubtfully placed in this family.

F. PARVITUBIDAE Skevington, 1963 O. (ARG) Mar.

First and Last: *Parvitubus oelandicus* (Bulman, 1936), Öland, Sweden (Skevington, 1963).
Constituent genus: *Parvitubus* Skevington, 1963.

Order CAMAROIDEA Kozłowski, 1938

F. BITHECOCAMARIDAE Bulman, 1955 O. (TRE) Mar.

First and Last: *Bithecocamara gladiator* Kozłowski, 1949, Poland (Kozłowski, 1949).
Constituent genus: *Bithecocamara* Kozłowski, 1949.

F. CYSTICAMARIDAE Bulman, 1955 O. (TRE–ARG/LLN) Mar.

First: *Cysticamara accollis* Kozłowski, 1949, *Flexicollicamara bryozoaeformis* Kozłowski, 1949, *Graptocamara hyperlinguata* Kozłowski, 1949, *Tubicamara coriacea* Kozłowski, 1949, and *Syringotaenia bystrowi* Obut, 1963, former USSR.
Last: *Graptocamara hyperlinguata* Kozłowski, 1949, Hagudden, Öland, Sweden (Skevington, 1963).
Constituent genera: *Cysticamara* Kozłowski, 1949, *Flexicollicamara* Kozłowski, 1949, *Graptocamara* Kozłowski, 1949, *Tubicamara* Kozłowski, 1949, *Syringotaenia* Obut, 1963.

Order CRUSTOIDEA Kozłowski, 1962

F. WIMANICRUSTIDAE Bulman, 1970 O. (ARG/LLN–LLN/LLO) Mar.

First: *Lapworthicrusta aenigmata* Kozłowski, 1962, Sarbia, Poland (Kozłowski, 1962).
Last: *Bulmanicrusta modesta* Kozłowski, 1962, *Wimanicrusta urbaneki* Kozłowski, 1962, and *W. cistaelingulata* Kozłowski, 1962, Zakroczym, Poland (Kozłowski, 1962).
Comment: All the known crustoids have been obtained from erratic boulders, not all of which have been dated with certainty.

Order GRAPTOLOIDEA Lapworth, 1875

Suborder DICHOGRAPTINA Lapworth, 1873

F. DICHOGRAPTIDAE Lapworth, 1873 O. (?TRE–CRD) Mar.

First: *Didymograptus novus* Berry, 1960, *D. primigenius* Bulman, 1950a, ?*D.* sp. Bulman, 1954, *D.? stoermeri* Erdtmann, 1965, *D. pritchardi* Hall, 1899, *D. taylori* Hall, 1899, *D. norvegicus* Monsen, 1925, *D. klotschichini* Obut, 1961. With the recognition of bithecae on *D. kiaeri* Monsen, 1925, further doubt has been cast on the true graptoloid nature of the Tremadoc 'didymograptids'. In the lowest beds of the Arenig, however, there are several species of *Tetragraptus* and *Didymograptus* (Monsen, 1937; Thomas, 1960), and it is probable that true graptoloids do occur in the Tremadoc.
Last: *Didymograptus superstes* Lapworth, 1880 and *D. serratulus* Hall, 1858, *gracilis* Zone, Glenkiln Shale, Scotland, UK (Elles and Wood, 1901–1918).

F. ABROGRAPTIDAE Mu, 1958 O. (ARG–CRD) Mar.

First: *Dinemagraptus warkae* Kozłowski, 1952, ?*hirundo* Zone, erratic block of Baltic limestone.
Last: *Abrograptus formosus* Mu, 1958, and other genera and species, *gracilis* Zone, Hulo Shale, China. The recently described genus *Metabrograptus* Strachan, 1990 is probably also from the *gracilis* Zone of Scotland, UK (Strachan, 1990).
Comment: The family includes some 'retiolite' genera such as *Reteograptus* Hall, 1859 (Bates, 1990).

F. SINOGRAPTIDAE Mu, 1957 O. (ARG–LLN) Mar.

First: *Holmograptus leptograptoides* (Monsen, 1937), *bifidus* Biozone, mid ARG, North America, Scandinavia, and elsewhere.
Last: *Holmograptus*, *Tylograptus*, *Nicholsonograptus*, *Sinograptus* and others; numerous species, uppermost *tentaculatus* Zone or *decoratus* Zone, middle LLN. NB *Holmograptus* and *Tylograptus* may be congeneric, with *Tylograptus* the junior name.

F. SIGMAGRAPTIDAE Cooper and Fortey, 1982 O. (TRE–LLN) Mar.

First: *Kiaerograptus* (?) cf. *pritchardi* (T. S. Hall, 1914), *Adelograptus antiquus* Subzone, upper TRE, Peel River, Yukon Territory, Canada.
Last: *Goniograptus* spp. Cosmopolitan.

F. ISOGRAPTIDAE Harris, 1933 O. (ARG–LLN) Mar.

First: *Isograptus primulus* Harris, 1933, Chewtonian 1, New Zealand.
Last: *Isograptus caduceus spinifer* Keble and Benson, 1929, *I. ovatus* (T. S. Hall, 1914), and *I. ovatus davidis* Skevington and Jackson, 1976, Australasia and Wales, UK.
Note: *I. primulus* was separated by Fortey and Cooper (1986) as a new family, but not named; for the present, the species is left in the Isograptidae.

F. PSEUDOTRIGONOGRAPTIDAE Fortey and Cooper, 1986 O. (ARG) Mar.

First and Last: *Pseudotrigonograptus ensiformis* Cooper and Fortey, 1982 and *P. minor* Cooper and Fortey, 1982, *hirundo* Zone, numerous localities in Spitsbergen, China, North America, and others (Cooper and Fortey, 1982).
Main genera: Only *Pseudotrigonograptus* Cooper and Fortey, 1982: *Trigonograptus* itself is invalid, and *Tristichograptus* Jackson and Bulman, 1970, is a junior synonym of *Pseudotrigonograptus*.

F. GLOSSOGRAPTIDAE Lapworth, 1873 O. (ARG–CRD) Mar.

Subfamily GLOSSOGRAPTINAE Lapworth, 1873 O. (ARG–CRD) Mar.

First: *Glossograptus acanthus* Elles and Wood, 1901–1918, *Glossograptus* sp. (p) indet., *Cryptograptus antennarius* (Hall, 1865), and *C. hopkinsoni* (Nicholson, 1869), *hirundo* Zone, cosmopolitan (Jackson, 1962; Spjeldnaes, 1953). *Glossograptus* sp. Monsen, 1937, from the *P. densus* (3b γ) and *P. angustifolius elongatus* (3b δ) Zones (? = *nitidus* + *gibberulus* Zones), lower *Didymograptus* Shale, Scandinavia (Monsen, 1937), probably belongs in *Phyllograptus* (Bulman, 1970). Presence of *C. hopkinsoni* in the *gibberulus* Zone, Skiddaw Slates, England, UK, is doubtful (Jackson, 1962).
Last: *Cryptograptus insectiformis* Ruedemann, 1908, top of *pygmaeus* Zone, Oklahoma, USA; *C. tricornis* (Carruthers, 1859), *clingani* Zone, Scotland, UK (Elles and Wood, 1908), Australia, and elsewhere.
Comment: Origin cryptogenetic (Bulman, 1970): from *Isograptus* (Thomas, 1960). Extinction relatively sudden; no known descendants.

Subfamily CORYNOIDINAE Hopkinson and Lapworth, 1875 O. (CRD) Mar.

First: *Corynoides calicularis* Nicholson, 1867, *gracilis* Zone, cosmopolitan.
Last: *Corynoides curtus* Lapworth, 1876, *clingani* Zone, cosmopolitan (Strachan, 1949).
Comment: *Corynites* Kozłowski, 1956, the other genus in the family, is known only from erratic boulders, but is probably also CRD.

Subfamily PSEUDISOGRAPTINAE Cooper and Ni, 1986 O. (ARG–LLN) Mar.

First: *Pseudisograptus hastatus* (Harris, 1933), Castlemainian, *maximus* Zone, middle ARG.
Last: *P. jiangxiensis* (Yu and Fang, 1981) and *P. dumosus* (Harris, 1933), upper Darriwilian, lower LLN, North America, China and Scandinavia.

Suborder VIRGELLINA Fortey and Cooper, 1986

F. DICRANOGRAPTIDAE Lapworth, 1873 (including Nemagraptidae Lapworth, 1873) O. (LLN–ASH) Mar.

First: *Dicellograptus moffatensis* Carruthers, 1858, *bifidus* Zone, Wales, UK.
Last: *Dicellograptus anceps* (Nicholson, 1867), *anceps* Zone, Hartfell Shale, Scotland, UK (Elles and Wood, 1901–1918).

F. PHYLLOGRAPTIDAE Lapworth, 1873 O. (ARG) Mar.

First: *Phyllograptus ilicifolius* Hall, 1858 and *P. typus* Hall, 1858, Bendigonian, lower ARG, North America, Australia, Eurasia.
Last: *Phyllograptus anna* Hall, 1858, *Isograptus* Zone, or *bifidus* Zone, Quebec, Canada and elsewhere.

F. ORTHOGRAPTIDAE Mitchell, 1987 O. (ARG–ASH) Mar.

Subfamily ORTHOGRAPTINAE Mitchell, 1987 O. (LLN–ASH) Mar.

First: *Hustedograptus uplandicus* (Wiman, 1895), widespread.
Last: *Orthograptus abbreviatus* Elles and Wood, 1901–1918, *anceps* Biozone, widespread; *Orthoretiograptus denticulatus* Wang *et al.*, 1987, *anceps* Zone, UK and China.
Comment: Several records of *O. truncatus* and allies from the earliest Silurian are treated sceptically by most graptolithologists, because the proximal-end development is unlike the Ordovician species. The subfamily includes a number of 'retiolite' genera (Bates, 1990).

Subfamily PEIRAGRAPTINAE Jaanusson, 1960 O. (ASH) Mar.

First and Last: *Peiragraptus fallax* Strachan, 1954, Vaureal Formation, Anticosti Island, Canada.
Comment: Family monotypic; probably derived from Diplograptidae by modification of mode of development. Restricted to Canada.

Subfamily LASIOGRAPTINAE Lapworth, 1879 O. (ARG–ASH) Mar.

First: *Hallograptus inutilis* (Hall, 1865), *hirundo* (3bε) Zone, Lower *Didymograptus* Shale, Norway (Spjeldnaes, 1953). NB This is a suspiciously early occurrence.
Last: *Nymphograptus velatus* Elles and Wood, 1901–1918, *anceps* Zone, upper Hartfell Shales, Scotland, UK (Elles and Wood, 1908).
Comment: Family probably derived by excessive thecal differentiation from early Diplograptidae of *austrodentatus* group (Bulman, 1970).

F. DIPLOGRAPTIDAE Lapworth, 1873 O. (LLN–ASH) Mar.

Subfamily DIPLOGRAPTINAE Lapworth, 1873 O. (LLN–CRD) Mar.

First: *Pseudoamplexograptus confertus* (Nicholson, 1868), just above the base of the *artus* Biozone in the UK), widespread. NB Probably several other forms in the same horizon.
Last: *Pseudoamplexograptus* spp., high CRD, various, widespread.
Main genera: *Pseudoamplexograptus* Mitchell, 1987 and *Urbanekograptus* Mitchell, 1987.

Subfamily CLIMACOGRAPTINAE Frech, 1897 O. (ARG–ASH) Mar.

First: *Climacograptus* sp., top of Cowhead Group, upper ARG, Newfoundland, Canada; possibly other forms elsewhere.
Last: *Climacograptus tubuliferus* Lapworth, 1876 and *Pseudoclimacograptus* spp., widespread.
Main genera: *Pseudoclimacograptus* Přibyl, 1947, *s.s.*, *Prolasiograptus* Lee, 1963, *Climacograptus* Hall, 1865, and *Dicaulograptus* Rickards and Bulman, 1965.

F. GLYPTOGRAPTIDAE Mitchell, 1987 (ex Glyptograptinae Mitchell, 1987) O. (LLN)–S. (LLY) Mar.

First: *Glyptograptus* sp., lower LLN, Cuidad Real, Spain (Gutierrez Marco, 1986).

Last: *'Glyptograptus' nebula* Toghill and Strachan, 1970, *griestoniensis* Zone, Telychian, Southern Uplands, Scotland, UK and northern England, UK. An older record is *Pseudoclimacograptus (Metaclimacograptus)* sp. *nov.*, upper *turriculatus* Zone, Armorican Massif, France (Paris *et al.*, 1980).

Main genera: *Glyptograptus* Lapworth, 1873, *Cystograptus* Hundt, 1942, *Pseudoglyptograptus* Bulman and Rickards, 1968, *Metaclimacograptus* Bulman and Rickards, 1968, *Clinoclimacograptus* Bulman and Rickards, 1968.

Comment: Rare biserial specimens from the Ludlow and Devonian are here regarded as insufficiently studied to enable a decision on their evolutionary position, or even their generic attribution.

F. EOGLYPTOGRAPTIDAE Mitchell, 1987 (ex Eoglyptograptinae Mitchell, 1987) O. (?ARG/LLN–CRD) Mar.

First: *Undulograptus* Boucek, 1973, several species, ?ARG, lower LLN, widespread.

Last: *Eoglyptograptus* spp., several (in need of revision).

Main genera: *Eoglyptograptus* Mitchell, 1987 and *Undulograptus* Bouček, 1973.

F. RETIOLITIDAE Lapworth, 1873 *s.s* O. (ASH)–S. (LUD) Mar.

First: *Pseudoretiolites* sp., ?*triangulatus* Zone (Elles and Wood, 1918; Rickards *et al.*, 1977).

Last: *Holoretiolites* sp., *leintwardinensis* Zone, upper LUD, England, UK (Rickards *et al.*, 1977). Slightly older records are *Plectograptus macilentus* (Törnquist, 1887) and *Spinograptus spinosus* (Wood, 1900), *nilssoni* Zone, lower LUD, Bohemia, Czechoslovakia, the latter also from England, UK; and *Agastograptus robustus* Obut and Zaslavskaya, 1986, *nilssoni* Zone, lower LUD, Kaliningrad, former USSR (Obut and Zaslavskaya, 1986).

Comment: *Arachiograptus laqueus* Ross and Berry, 1963, *complanatus* Zone, ASH, Nevada, USA is listed by Bulman (1970, p. V128) as a retiolitine (i.e. retiolitid *sensu stricto*), but the taxon is not well defined.

F. DIMORPHOGRAPTIDAE Elles and Wood, 1908 S. (LLY) Mar.

First: *Akidograptus acuminatus* (Nicholson, 1867) and *A. ascensus* Davies, 1929, *acuminatus* Zone, Birkhill Shale, Scotland, UK.

Last: *Rhaphidograptus toernquisti* (Elles and Wood, 1906), *convolutus* Zone, Birkhill Shale, Scotland, UK.

F. MONOGRAPTIDAE Lapworth, 1873 O. (ASH)–D. (PRA) Mar.

First: *Atavograptus ceryx* (Rickards and Hutt, 1970), *Persculptus* Zone, Lake District, England, UK (Rickards and Hutt, 1970). Similar, although less clearly defined, forms have been found elsewhere at this level, including in China.

Last: *Monograptus yukonensis* Jackson and Lenz, 1962, *M. thomasi* Jaeger, 1970, *M. pacificus* Jaeger, 1970 and *M. atopus* Bouček, 1952; all *yukonensis* Zone, North America, Eurasia, China.

REFERENCES

Bates, D. E. B. (1990) Retiolite nomenclature and relationships. *Journal of the Geological Society, London*, **147**, 717–23.

Bouček, B. (1957) The dendroid graptolites of the Silurian of Bohemia. *Ústředniho Ústavu Geologického Rozpravy*, **23**, 1–105.

Bulman, O. M. B. (1970) Graptolithina, in *Treatise on Invertebrate Paleontology*, 2nd edn (ed. C. Teichert), Geological Society of America and University of Kansas Press, Boulder, Colorado, and Lawrence, Kansas, pp. V1–V163.

Cooper, R. A. and Fortey, R. A. (1982) The Ordovician graptolites of Spitsbergen. *Bulletin of the British Museum (Natural History), Geology Series*, **36**, 157–302.

Elles, G. L. and Wood, E. M. R. (1901–1918) Monograph of British graptolites, parts I–XI. *Palaeontographical Society Monographs*, clxxi + 539 pp.

Fortey, R. A. and Cooper, R. A. (1986) A phylogenetic classification of the graptoloids. *Palaeontology*, **29**, 631–54.

Gutierrez Marco, J. C. (1966) Graptolitos del Ordovicico Español. Doctoral Thesis, Universidad Complutense de Madrid.

Jackson, D. E. (1962) Graptolite zones in the Skiddaw Group in Cumberland, England. *Journal of Paleontology*, **36**, 300–13.

Kozĺowski, R. (1949) Les graptolithes et quelques nouveaux groupes d'animaux du Tremadoc de la Pologne. *Palaeontologia Polonica*, **3**, 1–235.

Kozĺowski, R. (1962) Crustoidea, nouveau groupe de graptolites. *Acta Palaeontologia Polonica*, **7**, 3–52.

Mierzejewski, P. (1986) Ultrastructure, affinities and taxonomy of some Ordovician and Silurian microfossils. *Palaeontologia Polonica*, **47**, 129–220.

Mitchell, C. E. (1987) Evolution and phylogenetic classification of the Diplograptacea. *Palaeontology*, **30**, 353–405.

Mitchell, C. E. (1990) Directional macroevolution of the diplograptacean graptolites: a product of astogenetic heterochrony and directed speciation, in *Major Evolutionary Radiations* (eds P. D. Taylor and G. P. Larwood), Systematics Association Special Volume, 42, pp. 235–64.

Monsen, A. (1937) Die Graptolithenfauna in unteren Didymograptusschiefer (Phyllograptusschiefer) Norwegens. *Norsk Geologisk Tidsskrift*, **16**, 57–266.

Obut, A. M. and Zaslavskaya, N. M. (1986) Families of Retiolitida and their phylogenetic position, in *Palaeoecology and Biostratigraphy of Graptolites* (eds C. P. Hughes and R. B. Rickards), Geological Society Special Publication, 20, London, 207–19.

Palmer, D. and Rickards, R. B. (eds) (1991) *Graptolites: Writing in the Rocks*. Boydell and Brewer, Woodbridge, Suffolk, 167 pp.

Paris, F., Rickards, R. B. and Skevington, D. (1980) Les assemblages de graptolites du Llandovery dans le Synclinorium du Menez-Belais (Massif Armoricain). *Geobios*, **13**, 153–71.

Quilty, P. (1971) Cambrian and Ordovician dendroids and hydroids of Tasmania. *Journal of the Geological Society of Australia*, **17**, 171–89.

Rickards, R. B. and Hutt, J. E. (1970) The earliest monograptid. *Proceedings of the Geological Society of London*, **1664**, 115–19.

Rickards, R. B., Hutt, J. E. and Berry, W. B. N. (1977) Evolution of the Silurian and Devonian graptoloids. *Bulletin of the British Museum (Natural History), Geology*, **28**, 1–120.

Rickards, R. B., Baillie, P. W. and Jago, J. B. (1990) An Upper Cambrian (Idamean) dendroid assemblage from near Smithton, northwestern Tasmania. *Alcheringa*, **14**, 207–32.

Ruedemann, R. (1947) Graptolites of North America. *Geological Society of America Memoir*, **19**, 1–652.

Skevington, D. (1963) Graptolites from the Ontikan Limestones (Ordovician) of Öland, Sweden: 1. Dendroidea, Tuboidea, Camaroidea and Stolonoidea. *Bulletin of the Geological Institution, University of Uppsala*, **42**, 1–62.

Spjeldnaes, N. (1953) The Middle Ordovician of the Oslo Region, Norway. 3. Graptolites dating the beds below the Middle Ordovician. *Norsk Geologisk Tidsskrift*, **31**, 171–84.

Strachan, I. (1949) On the genus *Corynoides* Nicholson. *Geological Magazine*, **86**, 153–60.

Strachan, I. (1990) A new genus of abrograptid graptolite from the Ordovician of Southern Scotland. *Palaeontology*, **33**, 933–6.

Thomas, D. E. (1960) The zonal distribution of Australian graptolites. *Journal and Proceedings of the Royal Society of New South Wales*, **94**, 1–58.

Tjernvik, T. E. (1958) The Tremadocian beds at Flagabro in south-eastern Scania (Sweden). *Geologiska Föreningen i Stockholm, Förhandlingar*, **80**, 260–76.

28

PROBLEMATICA

M. A. Wills and J. J. Sepkoski Jr

A significant number of fossils cannot be placed within existing phyla with any degree of confidence, and are therefore regarded as 'Problematica'. This may be either because the gross body form or 'Bauplan' of the fossil is not known from a recognized phylum (usually in cases where preservation is good), or because material is open to ambiguous interpretation (in which case, assignment to more than one phylum may be possible). Many such forms arise from sites of exceptional preservation ('Konservat-Lagerstätten'), and most of the taxa listed here date from the early Palaeozoic. The Middle Cambrian Burgess Shale of British Columbia provides many of the most celebrated examples, first discussed extensively in a series of monographs by Walcott (e.g. 1908, 1911, 1912, 1918). Faunas of comparable significance have been described more recently at Chenjiang, in the Yunnan Province of China (Jiang, 1980, 1982; Zhang, 1987; Hou, 1987a–c; Hou and Sun, 1988), and in Peary Land, Greenland (Conway Morris *et al.*, 1987).

The response of the taxonomist to such material has been varied. In some cases (e.g. Seilacher, 1984; Gould, 1989), authors postulate the existence of new phyla, often to accommodate single representatives. At the other extreme (e.g. Walcott 1911, 1912, 1918), Problematica can be forced into known phyla, either by compromising the definition of the group, or by biasing an interpretation of the fossil evidence. These assignments are at least partially influenced by the worker's phylum concept *per se*. Few definitions are explicit, and there is certainly no general consensus as to how many 'units of morphological difference' are needed to merit phylum status.

Many hundreds of problematical taxa have been described from the fossil record. This chapter is not a complete listing of all named problematical genera. Only selected taxa erected as families or with particularly distinct or well-described characters are included. An excellent introduction to Problematica in general is given by Bengtson (1986), and to Burgess Problematica in particular by Briggs and Conway Morris (1986).

The chapter is organized into four sections for ease of reference, covering Problematica for which the highest bona fide taxonomic assignments are at the class, order, family and genus levels respectively.

Acknowledgements – Sepkoski acknowledges partial support for his research from the National Aeronautics and Space Administration (USA), under grant NAGW-1693. Wills thanks Dr D. E. G. Briggs for practical assistance with source material. His research was undertaken during the tenure of a University of Bristol Postgraduate Scholarship.

1. PROBLEMATICA ASSIGNED TO UNRELATED CLASSES

Class TRILOBOZOA Fedonkin, 1983

Vendian and Lower Cambrian taxa with unique triradial symmetry. Included groups are from Fedonkin (1985), with the questionable addition of the small calcareous anabaritids, as tentatively suggested by Fedonkin (1986).

F. ALBUMARESIDAE Fedonkin, 1976 V. (u.) Mar. (see Fig. 28.1)

First and Last: *Albumares brunsae* Fedonkin, 1976, Vendian, Vaddaiskaya Series, Onezhskii peschuostrov, former USSR. *Anfesta stankovskii* Fedonkin, 1984, Ust-Pinega Formation, Valdaiskoi Series, Vendian of the White Sea region, former USSR (Fedonkin, 1985).

F. ANABARITIDAE Missarzhevsky, 1974 V. (N–Da)–Є. (ATB) Mar.

First: *Anabarites trisulcatus* Missarzhevsky, 1969, from the widespread Anabarites–Protohertzina Zone (?uppermost Vendian), Siberia, Mongolia, China and Canada (Brasier, 1989). *Anabarites tristichus* Missarzhevsky, 1969, uppermost Precambrian, Siberian Platform, former USSR (Sokolov and Zhuravleva, 1983).

Last: *Anabarites tricarinatus* Missarzhevsky, 1969, Atdabanian, Salanygol, Mongolia. *Tiksitheca korobovi* (Missarzhevsky, 1966), upper Meishucunian, South China (Landing, 1988).

The Fossil Record 2. Edited by M. J. Benton. Published in 1993 by Chapman & Hall, London. ISBN 0 412 39380 8

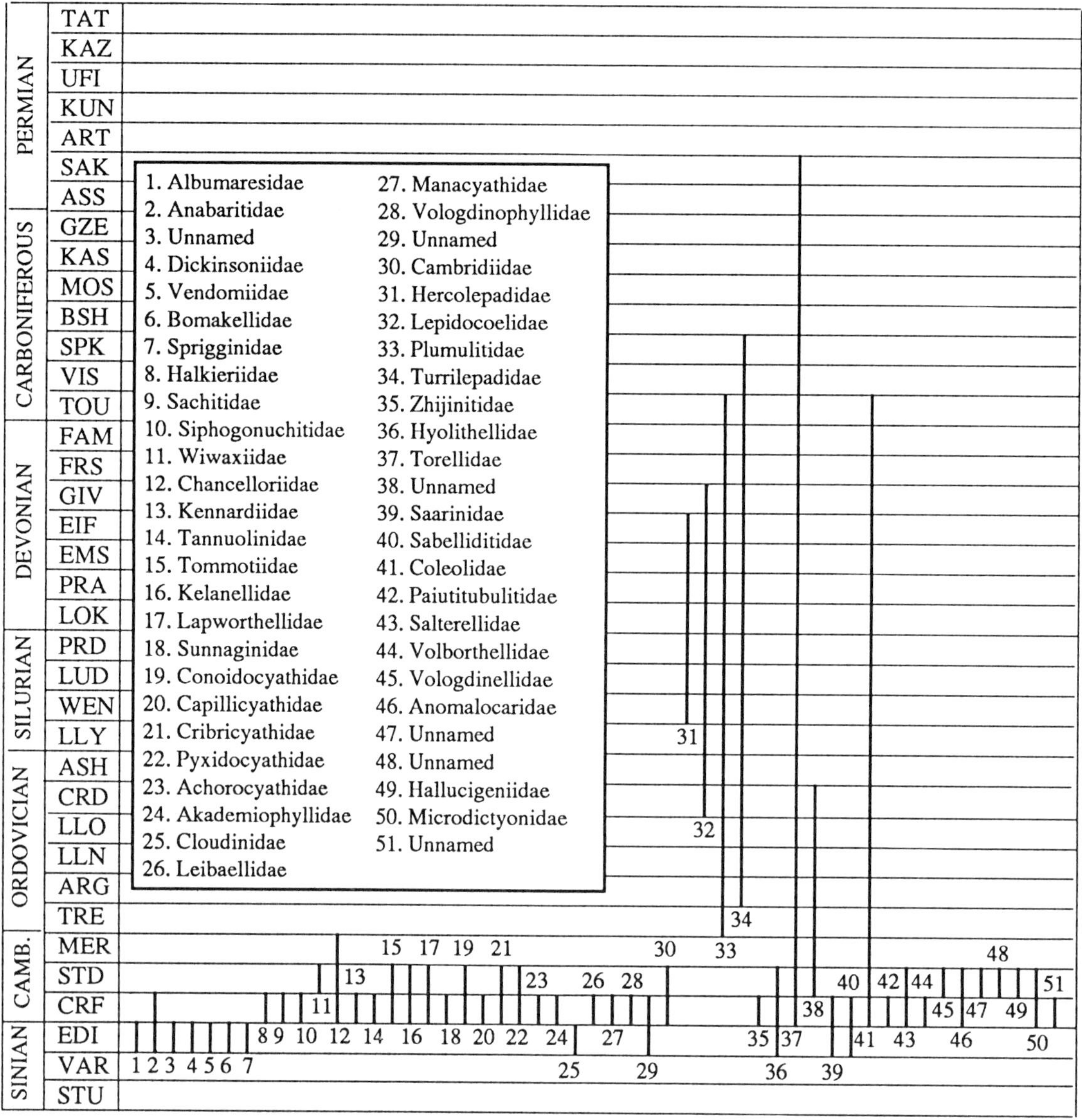

Fig. 28.1

Comment: Valkov and Sysoiev (1970) consider the Angustiochreidae to be synonymous.

F. UNNAMED, 1959 V. (u.) Mar.

First and Last: *Tribrachidium heraldicum* Glaessner, 1959, Vendian, Vaddaiskaya Series, White Sea region, and Pound Quartzite, upper Adelaide Series, Ediacara Hills, South Australia (Fedonkin, 1985).

Class DIPLEUROZOA Harrington and Moore, 1956

F. DICKINSONIIDAE Harrington and Moore, 1955 V. (POU) Mar.

First and Last: *Dickinsonia costata* Sprigg, 1947, and *Dickinsonia lissa* Wade, 1972, Pound Quartzite, upper Adelaide Series, Ediacara Hills, South Australia, and Vendian of the White Sea region. *Palaeoplatoda segmentata* Fedonkin, 1979, Vendian of the White Sea region, former USSR.

Comment: *Dickinsonia* has been variously placed with the polychaetes and the platyhelminthes. Seilacher (1989) treats it as a 'bipolar Vendozoan', thus allying it with the 'Petalonamae' (see Chapter 6, Coelenterata).

Class VENDIAMORPHA Fedonkin, 1985 V. (POU) Mar.

Includes Ediacaran taxa sometimes assigned to the Arthropoda. Seilacher's (1989) treatment as 'Vendozoa' would ally these taxa with the 'Petalonamae' (see Chapter 6, Coelenterata).

F. VENDOMIIDAE Keller, 1976 V. Mar.

First and Last: *Onega stepanovi* Fedonkin, 1976, *Vendomia menneri* Keller, 1976, Vendian of the Russian Platform (Keller and Fedonkin, 1976). *Vendia sokolovi* Keller, 1969, Valdaian Series of the Russian Platform, former USSR (Keller, 1969).

Class PARATRILOBITA Fedonkin, 1985

Includes Ediacaran taxa variously assigned to the Annelida and Arthropoda. Seilacher's (1989) treatment as 'Vendozoa' would ally these taxa with the 'Petalonamae' (see Chapter 6, Coelenterata).

F. BOMAKELLIDAE Fedonkin, 1987 V. (POU) Mar.

First and Last: *Bomakellia kelleri* Fedonkin, 1985, Ust-Pinega Formation, White Sea region, former USSR.

F. SPRIGGINIDAE Glaessner, 1958 V. (POU) Mar.

First and Last: *Spriggina floundersi* Glaessner, 1958, Pound Quartzite, upper Adelaide series, Ediacara Hills, near Ediacara, southern Australia. *Spriggina borealis* Fedonkin, 1979, Vendian of the Russian Platform, Russia.
Comment: Glaessner (1976) places *Spriggina* within the Tomopteridae (Phyllodocemorpha, Polychaeta), (see Chapter 15, Annelida).

Class COELOSCERITOPHORA Bengtson and Missarzhevsky, 1981

Order SACHITIDA He, 1980

F. HALKIERIIDAE Ch. Poulsen, 1967 €. (TOM–ATB) Mar.

First: *Halkieria denlataniformis* Mambetov, 1981, Tommotian Stage, Siberian Platform; *H. sacciformis* (Meshkova, 1969), Tommotian Stage, Siberian Platform, former USSR (Missarzhevsky and Mambetov, 1981).
Last: *Halkieria amorphe* (Meshkova, 1974), *Bercutia cristata* and *Rhombocorniculum cancellatum* Zones, Atdabanian Stage, Mongolia; *H. trianguliformis* Mambetov, 1981, and *H. curvativa* Mambetov, 1981, *Rhombocorniculum cancellatum* Zone, Atdabanian Stage, Siberian Platform, former USSR (Missarzhevsky and Mambetov, 1981).

F. SACHITIDAE Meshkova, 1969 €. (TOM–ATB) Mar.

First: *Sachites proboscideus* Meshkova, 1969, *A. sunnaginicus*, *D. regularis* and *D. lenaicus–T. primegenius* Zones, Tommotian Stage, Kazakhstan, former USSR, and Mongolia (Sokolov and Zhuravleva, 1983).
Last: *Sachites sacciformis* Meshkova, 1969, *A. sunnaginicus*, *D. regularis* and *D. lenaicus–T. primegenius* Zones, Atdabanian Stage, Maly Karatau, southern Kazakhstan, former USSR (Brasier, 1989).

F. SIPHOGONUCHITIDAE Qian, 1977 €. (TOM–LEN) Mar.

First: *Siphogonuchites triangulatus* Qian, 1977, Meischucunian Stage, Zone 1, Shizhonggou, south-western Shaanxi, China.
Last: *Trapezochites huoqiuensis* Xiao and Zhou, 1984, Yutaishan Formation (?Botomian), Anhui, China.

F. WIWAXIIDAE Walcott, 1911 €. (STD) Mar.

First: *Wiwaxia corrugata* (Matthew, 1899), Middle Cambrian Spence Shale, Cataract Canyon, Wellsville Mountains and nearby localities, Brigham City, Utah, USA (Conway Morris, 1989).
Last: *Wiwaxia corrugata* (Matthew, 1899), Phyllopod Bed, Walcott Quarry, Middle Cambrian Burgess Shale and *Ogygopsis* Shale, British Columbia, Canada (Conway Morris, 1985).
Comment: Butterfield (1990) argues that *Wiwaxia* is a polychaete worm, and assigns it to the order Phyllodocida, near the extant Chrysopetalidae and Aphroditidae.

Order CHANCELLORIIDA Walcott, 1920

F. CHANCELLORIIDAE Walcott, 1920 €. (TOM–MER) Mar.

First: *Chancelloria ex gr. lenaica* Zhuravleva and Korde, 1955, *A. sunnaginicus*, *D. regularis* and *D. lenaicus–T. primegenius* Zones, Tommotian and Atdabanian Stages, Siberian Platform, Altaye–Sayanskaya Oblast, Russia, former USSR (Sokolov and Zhuravleva, 1983).
Last: *Chancelloria* sp., lower Dresbachian *Crepicephalus* Zone of Texas and Montana, lower Upper Cambrian, USA (Lochman, 1940); *Chancelloria iranica* Mostler and Mosleh-Yadzi, 1976, Upper Cambrian Mila Formation, Elzburz Mountains, Iran.

Class TOMMOTIIDA Missarzhevsky, 1970

Order MITROSAGOPHORA Bengtson, 1977

Classification is taken from Bengtson (1970, 1977).

?F. KENNARDIIDAE Laurie, 1986 €. (ATB) Mar.

First and Last: *Kennardia reticulata* Laurie, 1986, Todd River Dolomite, Amadeus Basin, central Australia.

F. TANNUOLINIDAE Fonin and Smirnova, 1967 €. (LEN) Mar.

First: *Micrina etheridgei* (Tate, 1892), Todd River Dolomite, Amadeus Basin, central Australia (Laurie, 1986).
Last: *Tannuolina multifora* Fonin and Smirnova, 1967, Shangan Suite, upper reaches of Shivelig-Khem River, Tuva, eastern Tan-u-Ola Range, southern Siberia, former USSR.

F. TOMMOTIIDAE Missarzhevsky, 1970 €. (TOM–STD) Mar.

First: *Camenella garbowskae* Missarzhevsky, 1966, *Tommotia kozlowskii* (Missarzhevsky, 1966), *T. admiranda* (Missarzhevsky, 1966), *T. plana* (Missarzhevsky, 1966), and *T. zonata* (Missarzhevsky, 1969), River Lena, Siberian Platform, former USSR (Zhuravleva, 1970).
Last: *Tommotia plana* (Missarzhevsky, 1966), Toyonian Stage, Siberian Platform, former USSR.

Order UNNAMED

F. KELANELLIDAE Missarzhevsky and Grigor'yeva, 1981 €. (ATB–STD) Mar.

First: *Kelanella altaica* Missarzhevsky, 1966, Kameshki Horizon, Atdabanian, Isha River, Shilovka, Altaj, south-eastern Siberia, former USSR.
Last: *Sonella rostriformis* Missarzhevsky and Grigor'yeva, 1981, lower Amginian, Kazakhstan, former USSR.
Comment: Landing (1984) includes the genera of this family within the Lapworthellidae.

F. LAPWORTHELLIDAE Missarzhevsky, 1966 €. (TOM–STD) Mar.

First: *Lapworthella bella* and *L. tortuosa* Missarzhevsky, 1966, Chekurokovka Village, Lower reaches of River Lena, Siberia, former USSR (Missarzhevsky, 1969; Matthews and Missarzhevsky, 1975).
Last: *Lapworthella lucida* Meshkova, 1969, lower part of Kuranakhskogo Horizon, Siberian Platform; *Lapworthella marginata* Meshkova, 1969, and *L. corniforma* Meshkova, 1969, Kenyadinskii Horizon, Siberian Platform, former USSR.

F. SUNNAGINIDAE Landing, 1984
€. (TOM–ATB) Mar.

First: *Sunnaginia imbricata* Missarzhevsky, 1969, Pestrotsvet Formation (lower Tommotian), Ulakhan-Sulugur, Aldan River, Yakutia, Siberia, former USSR (Sokolov and Zhuravleva, 1983).
Last: *Sunnaginia angulata* Brasier, 1986, Ac3 Beds, Comley Limestone, Comley, Shropshire, England (Brasier, 1986).

Class CRIBRICYATHEA Vologdin, 1961

This class was erected to accommodate a number of calcareous microfossils from the Lower Cambrian of the former USSR, originally assigned to the Phylum Archaeocyatha. All are oblong or isometric, one- or two-walled cups, up to 5 mm long and 1.5 mm wide, although most are smaller (Hill, 1972).

Order CONOIDOCYATHIDA Vologdin, 1964

Commonly conical cups, with a single wall of peripterate structure.

F. CONOIDOCYATHIDAE Vologdin, 1964
€. (LEN–STD) Mar.

First and Last: *Conoidocyathus artus* Vologdin, 1964, Botomian–Lenian Stages, Altay-Sayan; *Conoidocyathus plumosus* Vologdin, 1966, Botomian–Toyonian Stages, Altay-Sayan; *Pubericyathus phialiformis* Vologdin, 1964, Botomian–Toyonian Stages, Kuznetsk Alatau; *Azyricyathus transeptatus* Vologdin, 1964, Lenian, Kuznetsk Alatau, former USSR.

Order CRIBRICYATHIDA Vologdin, 1964

Classification is predominantly taken from Hill (1972). Cribricyathids are all double-walled cups, with a peripterate inner, and striate outer wall.

F. CAPILLICYATHIDAE Vologdin, 1964
€. (LEN) Mar.

First and Last: *Capillicyathus fimbriatus* Vologdin, 1964, Botomian Stage, River Kyzas, west Sayan, former USSR.

F. CRIBRICYATHIDAE Vologdin, 1964
€. (LEN)–€. (STD) Mar.

First: *Apocyathus ovalis* Vologdin, 1964, Botomian Stage, River Abakan, west Sayan; *Lagenicyathus lamellifer* Vologdin, 1964, Botomian Stage, River Abakan, west Sayan; *Thecocyathus tetragonus* Vologdin, 1964, Botomian Stage, River Kyzas, west Sayan, former USSR.
Last: *Cribricyathus longus* Vologdin, 1964, Botomian–Solontsov Stages, Altay-Sayan; *Dolichocyathus effiguratus* Vologdin, 1964, Solontsov Stage, Kuznetsk Alatau; *Lomatiocyathus clathratus* Vologdin, 1964, Solontsov Stage, Altay-Sayan, former USSR.

F. PYXIDOCYATHIDAE Vologdin, 1964
€. (LEN)–E. (STD) Mar.

First: *Pyxidocyathus gracilis* Vologdin, 1964, Botomian Stage, River Kyzas, west Sayan; *Longicyathus pubescens* Vologdin, 1964, Botomian Stage, River Sanashtykgol, west Sayan; *Sunicyathus pulcher* Vologdin, 1964, Botomian Stage, west Sayan; *Turricyathus procerulus* Vologdin, 1964, Botomian Stage, west Sayan, former USSR.
Last: *Pteripterocyathus cirratus* Vologdin, 1964, Solontsov Stage, Kuznetsk Alatau; *Radicicyathus canaliculatus* Vologdin, 1964, Solontsov Stage, Kuznetsk Alatau, former USSR.

Order VOLOGDINOPHYLLIDA Radugin, 1964

These are very small, bilaterally symmetrical, slenderly conical or cylindrical, straight or cornute cups, with one or two walls, the outer of which consists of separate laminar transverse elements, the inner being transversely annulate or monolithic.

F. ACHOROCYATHIDAE Yankauskas, 1965
€. (ATB) Mar.

First and Last: *Acherocyathus perbellus* Yankauskas, 1965, and *Topolinocyathus popovi* Yankauskus, 1965, River Mana, east Sayan, former USSR.

F. AKADEMIOPHYLLIDAE Radugin, 1964
€. (ATB) Mar.

First and Last: *Akademiophyllum conuforme* and *Erphyllum bephylleforme* Radugin, 1964, *Lacerathus cuneatus* Yankauskas, 1969 and *Pterocyathus glaucus* Yankauskas, 1965, River Mana, east Sayan, former USSR.

F. CLOUDINIDAE Hahn and Pflug, 1985
V. (POU) Mar.

First and Last: *Cloudina hartmannae* Germs, 1972, and *C. riemkeae* Germs, 1973, Schwarzkalk Limestone Member (Poundian), Kuibis Formation, Driedoornvlakte, near Schlip, Namibia, and other parts of the world (Grant, 1990).
Comment: Grant (1990) argues that *Cloudina*? *borrelloi* Yochelson and Herrera, 1974, from the upper Lower Cambrian of Argentina probably belongs with the genus *Salterella*.

F. LEIBAELLIDAE Yankauskas, 1965
€. (ATB) Mar.

First and Last: *Leibaella elovica* Yankauskas, 1964, *Dubius uncatus* Yankauskas, 1969, and *Ramifer giratus* Yankaskas, 1965, River Mana, east Sayan, former USSR.

F. MANACYATHIDAE Yankauskas, 1969
€. (ATB) Mar.

First and Last: *Manacyathus mikroporosus* Yankauskas, 1969, River Mana, east Sayan, former USSR (Hill, 1972).

F. VOLOGDINOPHYLLIDAE Radugin, 1964
€. (ATB) Mar.

First and Last: *Vologdinophyllum chachlovi* Radugin, 1962, Atdabanian, Altay-sayan, former USSR. *Cardiophyllum kelleri* Radugin, 1964, *Crispus subdimisdiatus* Yankauskas, 1965, and *Longaevus vitalis* Yankauskas, 1965, Atdabanian, east Sayan, former USSR (Hill, 1972).

F. UNNAMED V. (POU)–€. (TOM) Mar.

First: *Sinotubulites baimatuoensis* Chen *et al.*, 1981, Upper Sinian of Yichang, Hubei Province, China.
Last: *Sinotubulites cienegensis* McMenamin, 1985, La Cienega Formation, unit 1, western Cerro Clemente, Caborca, Mexico.
Comment: Grant (1990) suggests that this genus, along with *Nevadatubulus* and *Wyattia*, are related to, or even congeneric with, *Cloudina*.

Class STENOTHECOIDA Yochelson, 1969

Order CAMBRIDIOIDEA Horny, 1958

F. CAMBRIDIIDAE Horny, 1957
Є. (ATB–STD) Mar.

First: *Stenothecoides* sp., *Dokidocyathus lenaicus* Zone, Salanygol, Mongolia (Brasier, 1989).
Last: *Stenothecoides elongata* (Walcott, 1884), Mount Whyte Formation, Mount Stephen, British Columbia, Canada.

Class MACHAERIDIA Withers, 1926

Classification follows Dzik (1986a) except that the Sachitida and Tommotiida have been removed.

Order HERCOLEPADIDA Dzik, 1986

F. HERCOLEPADIDAE Dzik, 1986
S. (WEN)–D. (EIF) Mar.

First: *Hercolepas signata* (Aurivillus, 1892), Middle Silurian Visby Beds of Gotland, Sweden (Aurivillus, 1892).
Last: *Protobalanus hamiltonensis* Hall and Clarke, 1888, Marcellus Shale, Avon, New York, USA (Van Name, 1926).

Order TURRILEPADIDA Pilsbry, 1916

F. LEPIDOCOELIDAE Withers, 1926
O. (LLN)–D. (GIV) Mar.

First: *Plicacoleus robustus* Dzik, 1986, *E. reclinatus* Zone, Mojcza Limestone, Holy Cross Mountains, Poland, and Baltic area.
Last: *Aulakolepos* sp., Middle Devonian (Pope, 1975).

F. PLUMULITIDAE Jell, 1979
O. (TRE)–C. (TOU) Mar.

First: *Plumulites bohemicus* Barrande, 1872, Wosek and Sta Benigna, Bohemia, Czechoslovakia (Jell, 1979).
Last: Unidentified specimens, Tournaisian, Wierzchowe area, northern Poland, Illustrated by Korejwo (1979) and referred to this family by Dzik (1986a).

F. TURRILEPADIDAE Clarke, 1896
O. (ARG)–C. (SPK) Mar.

First: *Mojezalepas multilamellosa* Dzik, 1986a, Arenig and Llanvirn of the Baltic area, Europe (Dzik, 1986a).
Last: *Turrilepas whitersi* Elias, 1958, and *Clarkeolepis clarkei* Elias, 1958, both upper Mississippian Redoak Hollow Formation, southern Oklahoma, USA.

Class CAMBROCLAVIDA Conway Morris and Chen, 1991

F. ZHIJINITIDAE Qian, 1978 Є. (TOM–LEN) Mar.

First: *Zhijinites longistriatus* Qian, 1978, middle Meishucunian Stage, Yunnan, China.
Last: *Cambroclavus fangxianensis* Qian and Zhang, 1983, Damao Group (Ordian), Damaodong section, Yaxian County, Hainan Island, China.
Comment: The family Cambroclavitidae Mambetov, 1979, is considered to be synonymous (Mambetov and Repina, 1979).

2. PROBLEMATICA ASSIGNED TO UNRELATED ORDERS

Order HYOLITHELMINTHES Fisher, 1962

F. HYOLITHELLIDAE Walcott, 1886
V. (POU)–Є. (STD) Mar.

First: *Hyolithellus* sp., Rovno Horizon, Russian Platform former USSR; *Mobergella holsti* (Moberg, 1892), *M. radiolata* Bengtson, 1968, and *M. turgida* Bengtson, 1968, boulders of Lower Cambrian conglomerate from Venenäs, on the Skäggenäs Peninsula, Baltic region, Norway and Sweden.
Last: *Hyolithellus cuyanus* and *H. mendozanus* Rusconi, 1951, Isidreana Formation (Villavicense Horizon), Mendoza, Argentina.

F. TORELLIDAE Holm, 1893
Є. (TOM)–P. (SAK) Mar.

First: *Torellella curvae* Missarzhevsky, 1966, lower Pestrotsvet Formation, *sunnaginicus* Zone, Dvortsy, Aldan River, Yakutia, Siberia, former USSR.
Last: *Phosphannulus* spp. Müller, Nogami and Lenz, 1974, Hughes Creek Shale Member, Foraker Limestone (Lower Permian, Wolfcampian), road cutting on Interstate 70, Wabaunshee County, Kansas, USA (Welch, 1976).

F. UNNAMED Є. (STD)–O. (CRD) Mar.

First: *Byronia annulata* Matthew, 1899, Stephen Formation, Mount Stephen, British Columbia, Canada.
Last: *Byronia naumovi* Kozłowski, 1967, middle Caradocian erratic boulder, Wyszgrod, Poland (Mierzejewski, 1986).
Comment: Scrutton (1979) and Mierzejewski (1986) consider this to be a scyphozoan.

Order SABELLIDITIDA Sokolov, 1965

F. SAARINIDAE Sokolov, 1965
V. (POU)–Є. (TOM) Mar.

First: *Calyptrina partita* Sokolov, 1965, Yudoma Horizon, uppermost Vendian, River Kotui basin in the north-west of the Anabar uplift (Sokolov, 1968).
Last: *Calyptrina partita* Sokolov, 1965, basal Cambrian, northern former USSR. *Saarina tenera* Sokolov, 1965, lower part of the Baltic Stage, Russian Platform, former USSR (Sokolov, 1968).

F. SABELLIDITIDAE Sokolov, 1965
V. (POU)–Є. (ATB) Mar.

First: *Paleolina evenkiana* Sokolov, 1965, lower part of Platonovka Suite, Upper Vendian, Turukhansk region, Siberian Platform, former USSR (Sokolov, 1968); *Sabellidites* sp., uppermost Vendian, southern Siberia, former USSR.
Last: *Parasabellidites yanachevskyi* Sokolov, 1967, Baltic Stage of Lower Cambrian, Siberian Platform, former USSR (Sokolov, 1968).

Order COLEOLIDA Bouček, 1964

Coleolids were assigned as molluscs under the order Hyolithellida by Sysoiev (1957) and as an order (Coleolida) under the class Tentaculita by Bouček (1964).

F. COLEOLIDAE Fisher, 1962
Є. (TOM)–C. (TOU) Mar.

First: *Coleolella differo* Lendzion, 1972, lower Tommotian, borehole, northern Poland (Brasier, 1989); *Coleolus trigonus* Sysoiev, 1962, and *S. billingsi* (Sysoiev, 1962) lower Pestrotsvet Formation, *sunnaginicus* Zone, Yunikan, Aldan River, Yakutia, Siberia, former USSR.
Last: *Coleolus missouriensis* Howell, 1952, Chouteau Formation, Sedalia, Missouri, USA (Downie *et al.*, 1967).

Order PAIUTIIDA Tynan, 1983

F. PAIUTITUBULITIDAE Tynan, 1983 Є. (ATB) Mar.

First and Last: *Paiutitubulites variabilis* Tynan, 1983, *P. durhami* Tynan, 1983, and *Cambrotubulites trisepta* Tynan, 1983, Lower Cambrian Campito and Poleta Formations, White Mountains, Inyo County, California, USA.

Order VOLBORTHELLIDA Kobayashi, 1937

Yochelson (1977) regards this group as a distinct phylum, 'Agmata', while Glaessner (1976) places them within the Spiomorpha (Polychaeta).

F. SALTERELLIDAE Poulsen, 1932 Є. (LEN–STD) Mar.

First: *Serpulites maccullochi* Murchison, 1959, *Salterella* Grit, Loch Eriboll, Scottish Highlands (Yochelson, 1983); *Salterella acervulosa* Resser and Howell, 1938, Kinzers Formation, near Lancaster, Pennsylvania, USA.
Last: *Salterella* sp., Toquima Range, Nye County, Nevada, USA (Yochelson *et al.*, 1970).

F. VOLBORTHELLIDAE Kiaer, 1916 Є. (ATB–LEN) Mar.

First: *Volborthella tenuis* Schmidt (1888), lower Atdabanian equivalents of northern Norway and northern Poland (Brasier, 1989).
Last: *Volborthella tenuis* Schmidt (1888), *Salterella* Grit, Loch Eriboll, Scottish Highlands, UK (Yochelson, 1983).
Comment: Yochelson (1977) contends that *Volborthella* from the Baltic region, and *Salterella* may be the same organism, only affected by different diagenesis.

F. VOLOGDINELLIDAE Balashov, 1962 Є. (MEN) Mar.

First and Last: *Vologdinella antiquus* (Vologdin, 1930), upper Middle Cambrian, Chingiz Mountains, north-eastern Kazakhstan, former USSR (Tchernycheva, 1965).

Order UNNAMED

D. Collins (pers. comm.) has suggested that the Anomalocaridae, Opabiniidae, *Proboscaris*, and *Hurdia* are related by sharing the apomorphic 'Peytoia' mouth. D. E. G. Briggs (pers. comm.) considers that the Opabiniidae should not be included in the order, since they lack a sphincter-like arrangement of plates around the mouth.

F. ANOMALOCARIDAE Whittington and Briggs, 1985 Є. (ATB–STD) Mar.

First: *Cassubia infercambriensis* (Lendzion, 1975), sub-surface Zawiszany Formation of the basal Cambrian, north-east Poland.
Last: *Anomalocaris nathorsti* (Walcott), Marjum Formation, *Ptychagnostus punctuosus* Zone, House Range, Utah, USA (Briggs and Robison, 1984; Whittington and Briggs, 1985).

F. UNNAMED Є. (MEN) Mar.

First and Last: *Hurdia dentata* Simonentta and Delle Cave, 1975, *H. triangulata* Walcott, 1912, and *H. victoria* Walcott, 1912, Walcott Quarry, mid Middle Cambrian Burgess Shale, Stephen Formation, near Field, British Columbia, Canada.

F. UNNAMED Є. (STD) Mar.

First and Last: *Proboscaris agnosta* Rolfe, 1962, Wheeler Shale, *Ptychagnostus atavus* Zone, House Range, Utah, USA.

Order UNNAMED

Ramsköld and Hou (1991) have argued that the following taxa form a group which is probably related to the onychophoran 'lobopods'.

F. HALLUCIGENIIDAE Conway Morris, 1977 Є. (MEN) Mar.

First and Last: *Hallucigenia sparsa*, Walcott, 1911, Walcott Quarry, Middle Cambrian burgess Shale, near Field, British Columbia, Canada.

F. MICRODICTYONIDAE Chen *et al.*, 1989 Є. (TOM–STD) Mar.

First: *Microdictyon tenuiporatum* Bengtson *et al.*, 1986, Tommotian Stage, River Lena, Yakutia, Siberian Platform, former USSR.
Last: *Microdictyon robisoni* Bengtson *et al.*, 1986, uppermost Swasey Limestone bed, *Ptychagnostus gibbus* Zone (or just below), Middle Cambrian, Topaz Mountain, Drum Mountains, Millard County, Utah, USA.

F. UNNAMED Є. (ATB) Mar.

First and Last: *Luolishania longicruris* Hou and Chen, 1989, Chiungchussu Formation, at Maotianshan, Chengjiang, east Yunnan, China.

3. PROBLEMATICA ASSIGNED TO UNRELATED FAMILIES

F. AMISKWIIDAE Walcott, 1911 Є. (MEN) Mar. (see Fig. 28.2)

First and Last: *Amiskwia sagittiformis* Walcott, 1911, Walcott Quarry, Middle Cambrian Burgess Shale, near Field, British Columbia, Canada.

F. ANCIENTIDAE Ross, 1967 O. (CRD–ASH) Mar.

First: *Ancienta arborea*, *A pomerania* and *A. rossi* Schallreuter, 1981, Middle Ordovician of Europe.
Last: *Ancienta ohioensis* and *A. fortensis* Ross, 1967, Waynesville Formation, Richmond Group, Cincinnation Series (Ashgill), Ohio, USA.

F. CORNULITIDAE O. (CRD)–C. (SPK) Mar.

First: *Cornulites flexuosus* (Hall, 1847) [as *Tentaculites flexuosa* Hall], Trenton Limestone, Trenton Falls, New York, USA (Downie *et al.*, 1967).
Last: *Cornulitella carbonaria* (Young), 'Carboniferous Limestone Group', Ravenscraig Castle, near Kirkcaldy and Whitehouse, near Linlithgow, Scotland, UK (Etheridge, 1880).
Comment: Regarded as an Order in the Class Tentaculita by Bouček (1964).

F. CUPITHECIDAE Duan, 1984 Є. (TOM–LEN) Mar.

First: *Cupitheca manicae* Duan, 1983, Xihaoping Formation (Meichucunian), Shennongjia District, Hubei, China.
Last: *Actinotheca costellata* Xiao and Zhou, 1984, and *A. dolioformis* Xiao and Zhou, 1984, Yutaishan Formation (?Botomian), Anhui, China.

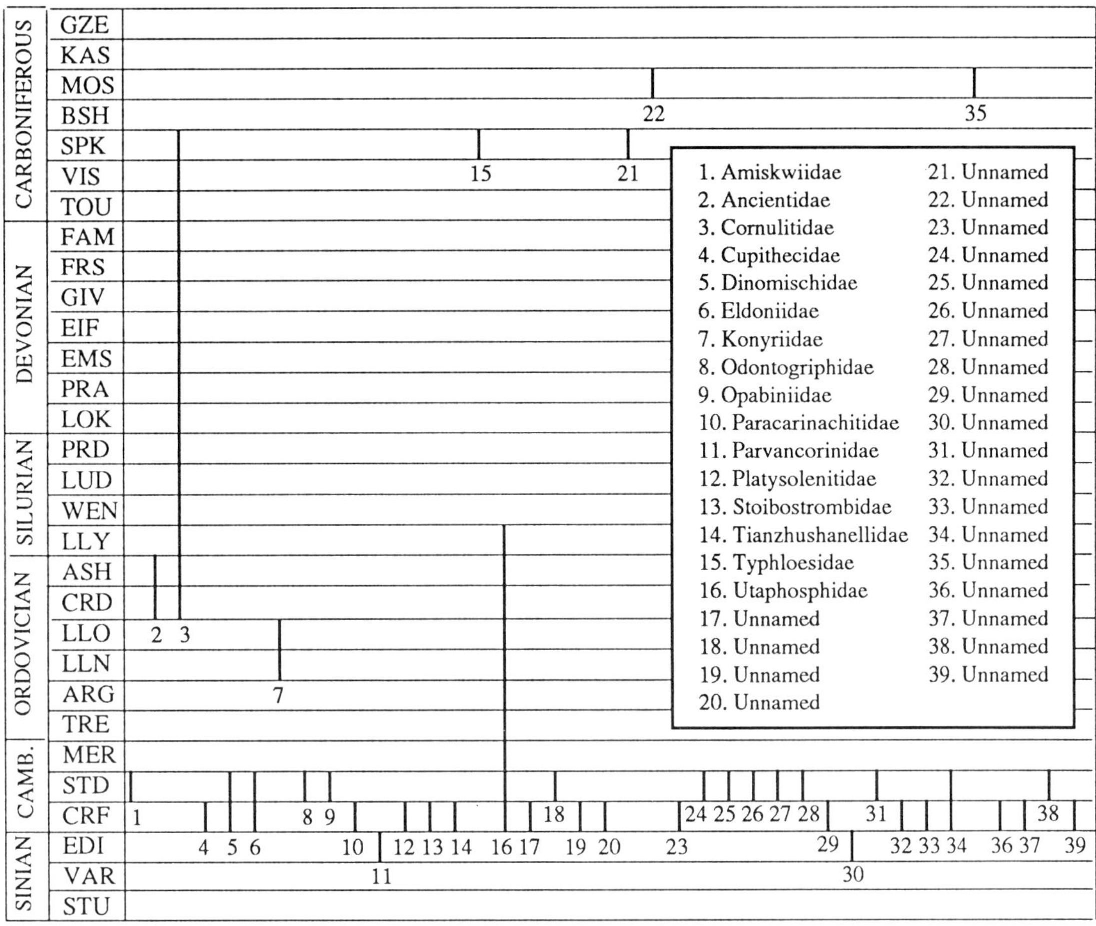

Fig. 28.2

F. DINOMISCHIDAE Conway Morris, 1977 €. (ATB–MEN) Mar.

First: *Dinomischus venustus* Chen *et al.*, 1989, Lower Cambrian, *Nevadella* Zone, Chiungchussu Formation, Yunnan, China.
Last: *Dinomischus isolatus* Conway Morris, 1977, Middle Cambrian Burgess Shale, near Field, British Columbia, Canada.

F. ELDONIIDAE Walcott, 1911 €. (ATB–STD) Mar.

First: *Stellostomites eumorphus* Sun and Hou, 1987 and *Yunnanomedusa eleganta* Sun and Hou, 1987, Lower Cambrian, *Nevadella* Zone, Chiungchussu Formation, Yunnan, China.
Last: *Eldonia ludwigi* Walcott, 1911, middle part of Spence Shale, Antimony Canyon, Wellsville Mountain, USA, and middle of Marjum Formation, House Range, USA (Conway Morris and Robison, 1988).

F. KONYRIIDAE Nazarov and Popov, 1976 O. (LLN–LLO) Mar.

First: *Konyrum varium* Nazarov and Popov, 1976, Middle Table Head Formation, north of Piccadilly, Port au Port Peninsula, western Newfoundland, Canada (Bergström, 1979).
Last: *Konyrum varium* Nazarov and Popov, 1976, upper Fort Peña and lower Woods Hollow Formations of the Marathon area, western Texas, USA (Bergström, 1979).
Comment: S. Bengtson (pers. comm.) suggests that these are hexactinellid sponges.

F. ODONTOGRIPHIDAE Conway Morris, 1976a €. (MEN) Mar.

First and Last: *Odontogriphus omalus* Conway Morris, 1976a, Middle Cambrian Burgess Shale, near Field, British Columbia, Canada (Conway Morris, 1976a).

F. OPABINIIDAE Walcott, 1912 €. (MEN) Mar.

First: *Opabinia regalis* Walcott, 1912, Walcott Quarry, mid Middle Cambrian Burgess Shale, near Field, British Columbia, Canada.

F. PARACARINACHITIDAE Qian, 1988 €. (TOM–ATB) Mar.

First: *Paracarinachites sinensis* Qian and Jiang, 1982, middle Meishucunian Stage, China (Luo *et al.*, 1982).
Last: *Luyanhaochiton spinus* Yu, 1984 (= *P. sinensis*) and *Yangtzechiton elongatus* Yu, 1984, Dengyin Formation, Zhongyicun Member (upper Meishucunian), Jinning, Yunnan, China.
Comment: May be related to the cambroclaves (Qian and Bengtson, 1989; Conway Morris and Chen, 1991).

F. PARVANCORINIDAE Glaessner, 1979 V. (POU) Mar.

First and Last: *Parvancorina minchami* Glaessner, 1959, Pound Quartzite, Ediacara Hills, South Australia (Glaessner and Daily, 1959).

F. PLATYSOLENITIDAE Eichwald, 1969 Є. (TOM–ATB) Mar.

First: *Platysolenites antiquissimus* Eichwald, 1969, upper part of the Wodawa Formation (equivalent to the Rovno), Poland (Brasier, 1989).
Last: *Platysolenites antiquissimus* Eichwald, 1969, lower Purley Shale, Nuneaton, England, UK (Brasier, 1989).
Comment: Føyn and Glaessner (1979) and Loeblich and Tappan (1988) consider this to be a foraminifer.

F. STOIBOSTROMBIDAE Conway Morris and Bengtson, 1990 Є. (ATB) Mar.

First and Last: *Stoibostrombidus crenulatus* Conway Morris and Bengtson, 1990, Ajax Limestone, Mt. Scott Range and other localities, South Australia.

F. TIANZHUSHANELLIDAE Conway Morris, 1990 Є. (TOM–ATB) Mar.

First: *Tianzhushanella ovata* Liu, 1979, upper Dengying Formation (Meishucunian), Yangtze Gorges, Hubei, China.
Last: *Apistoconcha siphonalis* Conway Morris and Bengtson, 1990b, Parara Limestone, Horse Gulley, South Australia.

F. TYPHLOESIDAE Conway Morris, 1990 C. (SPK) Mar.

First and Last: *Typhloesus wellsi* (Melton and Scott, 1973), Bear Gulch Limestone, Montana, USA (Conway Morris, 1990).

F. UTAPHOSPHIDAE Wrona, 1987 Є. (ATB)–S. (LLY) Mar.

First: *Lenargyrion knappologicum* Bengtson, 1977, upper part of the Atdabanian, middle reaches of the River Lena, Yakutia, Siberian Platform, former USSR.
Last: *Hadimopanella silurica* Wang, 1990, Wangjiawan Formation, Telychian, Ningqiang, Shaanxi, China.

F. UNNAMED Є. (ATB–LEN) Mar.

First: *Rushtonites spinosus* Hinz, 1987, lower Comley Limestone, Rushton, Shropshire, England, UK.
Last: *Mongolitubulus squamifer* Missarzhevsky, 1977, Botomian at Prikhubsugul'e, Mongolia.
Comment: Affinity of these two genera has been suggested by S. Bengtson (pers. comm.).

4. DISTINCTIVE PROBLEMATICA NOT ASSIGNED TO TAXA ABOVE THE GENUS RANK

F. UNNAMED Є. (MER) Mar.

First and Last: *Banffia constricta* Walcott, 1911, Walcott Quarry, Middle Cambrian Burgess Shale, near Field, British Columbia, Canada.
Comment: Removed from the Ottoiidae Walcott, 1911, *Nom.* correct Howell, 1962, by Conway Morris, 1977.

F. UNNAMED Є. (ATB) Mar.

First and Last: *Cowiella reticulata* Hinz, 1987, 'Callavia Limestone' (lower Comley Limestone), Comley Quarry, Shropshire, England, UK.

F. UNNAMED 1984 Є. (TOM) Mar.

First and Last: *Cyrtochites pinnoides* Qian, 1983, Dengying Formation (middle Meichucunian), eastern Yunnan, China.

F. UNNAMED C. (SPK) Mar.

First and Last: *Deuteronectanebos papillorum* Schram, 1979, Bear Gulch Limestone, Montana, USA.

F. UNNAMED C. (MOS) Mar.

First and Last: *Etacystis communis* Nitecki and Schram, 1976, Pennsylvanian Essex fauna, Francis Creek Shale, Mazon Creek, northern Illinois, USA.

F. UNNAMED Є. (TOM) Mar.

First: *Fomitchella infundibuliformis*, Missarzhevsky, 1969, lower half of Tommotian, north-western slope of Anbar Massif, rivers Kotui, Eriechka and Fomich, south-east part of Yakutia, Maya River, Siberia; *F. acinaciformis* Missarzhevsky, 1969, Siberia, former USSR.
Last: *Fomitchella infundibuliformis*, Missarzhevsky, 1969, *Dokidocyathus regularis* Zone, upper Tommotian, Siberian Platform, former USSR (Brasier, 1989).
Comment: *Fomitchella infundibuliformis* is regarded by Dzik (1986b) to be the oldest of the true conodonts.

F. UNNAMED Є. (MEN) Mar.

First and Last: *Nectocaris pteryx* Conway Morris, 1976, Walcott Quarry, Middle Cambrian Burgess Shale, near Field, British Columbia, Canada (Conway Morris, 1976b).

F. UNNAMED Є. (MEN) Mar.

First and Last: *Oesia disjuncta* Walcott, 1911, Walcott Quarry, Middle Cambrian Burgess Shale, near Field, British Columbia, Canada.

F. UNNAMED Є. (MER) Mar.

First and Last: *Palaeobotryllus taylori* Müller, 1977, Whipple Cave Formation (Trempealeauan), Sawmill Canyon, near Lund, Nevada, USA.
Comment: Müller (1977) suggested this was an ascidiacean tunicate (see Chapter 26).

F. UNNAMED Є. (STD) Mar.

First and Last: *Pollingeria grandis* Walcott, 1911, Middle Cambrian Burgess Shale, and localities on Fossil Ridge, Mount Field and Mount Stephen, near Field, British Columbia, Canada (Collins *et al.*, 1983).

F. UNNAMED Є. (MEN) Mar.

First and Last: *Portalia mira* Walcott, 1918, Walcott Quarry, Middle Cambrian Burgess Shale, near Field, British Columbia, Canada.

F. UNNAMED Є. (ATB) Mar.

First and Last: *Pyrgites mirabilis* Yu, 1984, Dengying Formation, upper member (upper Meishucunian), Shanxi, China.

F. UNNAMED V. (POU) Mar.

First and Last: *Redkinia spinosa* Sokolov, 1976, borehole at Nepeitsino, Moscow Syneclise, *Redkinia* Horizon, middle Vendian, former USSR.

F. UNNAMED Є. (MEN) Mar.

First and Last: *Redoubtia polypodia* Walcott, 1918, Walcott Quarry, Middle Cambrian Burgess Shale, near Field, British Columbia, Canada.

F. UNNAMED €. (TOM–LEN) Mar.

First: *Rhombocorniculum insolutum* Missarzhevsky, 1981, Tommotian and Atdabanian Stages, Siberian Platform, former USSR (Missarzhevsky and Mambetov, 1981; Sokolov and Zhuravleva, 1983).
Last: *Rhombocorniculum cancellatum* (Cobbold), *Bergeroniells asiaticus* Zone, Botomian, Siberia, former USSR (Missarzhevsky, 1982).
Comment: Assigned to the conodonts by some authors.

F. UNNAMED €. (LEN) Mar.

First and Last: *Stefania longula* Grigor'eva, 1982, Khajrkhan Formation (?Botomian), Salany-Gol, Mongolia (Voronin *et al.*, 1982).

F. UNNAMED €. (LEN–STD) Mar.

First: *Tubulella ?pervetus* (Matthew, 1899), Etcheminian Beds, Botomian, Newfoundland, Canada.
Last: *Tubulella terranovaensis* Howell, 1963, Kellgrew Brook Formation, *Paradoxides davidis* Zone, Murphy's Cove, Little Lawn Harbor, Burin Peninsula, Newfoundland, Canada.

F. UNNAMED C. (MOS) Mar.

First and Last: *Tullimonstrum gregarium* Richardson, 1966, Pennsylvanian Essex fauna, Francis Creek Shale, Mazon Creek, northern Illinois, USA.

F. UNNAMED €. (TOM) Mar.

First and Last: *Tummulduria incomperta* Missarzhevsky, 1969, base of Tommotian, Aldan River, Siberia, former USSR.

F. UNNAMED €. (LEN) Mar.

First and Last: *Westgardia gigantea* Rowland and Carlson, 1983, lower part of Bonnia-Olenellus Zone in upper member of the Puleta Formation, near Westgard Pass in the White-Inyo Mountains of eastern California, USA.
Comment: Preservation very poor.

F. UNNAMED €. (MEN) Mar.

First and Last: *Worthenella cambria* Walcott, 1911, Walcott Quarry, Middle Cambrian Burgess Shale, near Field, British Columbia, Canada.

F. UNNAMED €. (TOM) Mar.

First and Last: *Yunnanodus doleres* Wang and Jiang, 1980, middle Meishucun Stage, Yuhucum Formation, Zhongyicun Member, eastern Yunnan, China (Bengtson, 1983).

REFERENCES

Aurivilus, C. W. S. (1892) Über einige ober-silurische Cirripeden aus Gotland. *Bihang till Kongl. Svenska Vetenskaps-Akademiens Handlingen*, **18,4** (3), 1–24.

Bengtson, S. (1968) The problematic genus *Mobergella* from the Lower Cambrian of the Baltic area. *Lethaia*, **1**, 325–51.

Bengtson, S. (1977) Early Cambrian button-shaped phosphatic microfossils from the Siberian Platform. *Palaeontology*, **20**, 751–62.

Bengtson, S. (1983) The early history of the Conodonta. *Fossils and Strata*, **15**, 5–19.

Bengtson, S. (1986) Introduction: The problem of the Problematica, in *Problematic Fossil Taxa* (eds A. Hoffman and H. Nitecki), Clarendon Press, Oxford, pp. 3–11.

Bengtson, S., Matthews, S. C. and Missarzhevsky, V. V. (1986) The Cambrian netlike fossil *Microdictyon*, in *Problematic Fossil Taxa* (eds A. Hoffman and H. Nitecki), Oxford Monographs on Geology and Geophysics No. 5, Oxford University Press, pp. 97–115.

Bergström, S. M. (1979) First report of the enigmatic Ordovician microfossil *Konyrium* in North America. *Journal of Paleontology*, **53**, 320–7.

Bouček, B. (1964) *The Tentaculites of Bohemia*. Czechoslovak Academy of Science, Prague, 215 pp.

Brasier, M. D. (1986) The succession of small shelly fossils (especially conoidal microfossils) from English Precambrian–Cambrian boundary beds. *Geological Magazine*, **123**, 237–56.

Brasier, M. D. (1989) Towards a biostratigraphy of the earliest skeletal biotas, in *The Precambrian–Cambrian Boundary* (eds J. W. Cowie and M. D. Brasier), Clarendon Press, Oxford, pp. 117–65.

Briggs, D. E. G. and Conway Morris, S. (1986) Problematica from the Middle Cambrian Burgess Shale of British Columbia, in (eds A. Hoffman and H. Nitecki), *Problematic Fossil Taxa*. Clarendon Press, Oxford, pp. 167–83.

Briggs, D. E. G. and Robison, R. A. (1984) Exceptionally preserved nontrilobite arthropods and *Anomalocaris* from the Middle Cambrian of Utah. *University of Kansas Paleontological Contributions*, **111**, 1–23.

Butterfield, N. J. (1990) A reassessment of the enigmatic Burgess Shale fossil *Wiwaxia corrugata* (Matthew) and its relationship to the polychaete *Canadia spinosa* Walcott. *Paleobiology*, **16**, 287–303.

Chen, J-Y., Hou, X.-G. and Lu, H.-Z. (1989) Early Cambrian hock glass-like rare sea animal *Dinomischus* (Entoprocta) and its ecological features. *Acta Palaeontologica Sinica*, **28**, 58–71 [in Chinese with an English abstract].

Chen, M., Chen, Y.-Y. and Qian, Y. (1981) [Some tubular fossils from Sinian–Lower Cambrian boundary sequences, Yangtze Gorges.] *Bulletin of the Tianjin Institute of Geology and Mineral Research. Chinese Academy of Geological Sciences*, **3**, 117–24. [in Chinese].

Collins, D. H., Briggs, D. E. G. and Conway Morris, S. (1983) New Burgess Shale fossil sites reveal Middle Cambrian faunal complex. *Science*, **222**, 163–7.

Conway Morris, S. (1976a) A new Cambrian lophophorate from the Burgess Shale of British Columbia. *Palaeontology*, **19**, 199–222.

Conway Morris, S. (1976b) *Nectocaris pteryx*, a new organism from the Middle Cambrian Burgess Shale of British Columbia. *Neues Jahrbuch für Geologie and Paläontologie, Abhandlungen*, **1976**, 705–13.

Conway Morris, S. (1977) A new entoproct-like organism from the Burgess Shale of British Columbia. *Palaeontology*, **20**, 833–45.

Conway-Morris, S. (1985) The Middle Cambrian metazoan *Wiwaxia corrugata*, from the Burgess Shale and *Ogygopsis* Shale, British Columbia. *Philosophical Transactions of the Royal Society of London, Series B*, **307**, 507–86.

Conway-Morris, S. (1989) The persistence of Burgess Shale-type faunas: implications for the evolution of deeper-water faunas. *Transactions of the Royal Society of Edinburgh: Earth Sciences*, **80**, 271–83.

Conway Morris, S. (1990) *Typhloesus wellsi* (Melton and Scott, 1973), a bizarre metazoan from the Carboniferous of Montana, USA. *Philosophical Transactions of the Royal Society of London, Series B*, **327**, 595–629.

Conway Morris, S. and Bengtson, S. (1990a) Ornamented cones, in *Early Cambrian Fossils from South Australia* (eds S. Bengtson, S. Conway Morris, B. J. Cooper *et al.*), Association of Australasian Palaeontologists Memoir, 9, pp. 145–60.

Conway Morris, S. and Bengtson, S. (1990b) Bivalved organisms of possible brachiopod affinity, in *Early Cambrian Fossils from South Australia* (eds S. Bengtson, S. Conway Morris, B. J. Cooper *et al.*), Association of Australasian Palaeontologists Memoir, 9, pp. 164–86.

Conway Morris, S. and Chen, Meng'e (1991) *Cambroclaves* and paracarinachitids, early skeletal problematica from the Lower Cambrian of South China. *Palaeontology*, **34**, 357–98.

Conway Morris, S. and Robison, R. A. (1988) More soft-bodied animals and algae from the Middle Cambrian of Utah and British Columbia. *University of Kansas Palaeontological Contributions Paper*, **122**, 1–48.

Conway Morris, S., Peel, J. S., Higgins, A. K. *et al.* (1987) A Burgess Shale-like fauna from the Lower Cambrian of North Greenland. *Nature, London*, **326**, 181–3.

Downie, C., Fisher, D. W., Goldring, R. *et al.* (1967) Miscellania, in *The Fossil Record* (eds W. B. Harland *et al.*), Geological Society of London, London, pp. 613–26.

Duan, C.-G. (1984) Small shelly fossils from the Lower Cambrian Xihaoping Formation in the Shennongija District, Hubei Province – hyoliths and fossil skeletons of unknown affinities. *Bulletin of the Tianjin Institute of Geology and Mineral Research. Chinese Academy of Geological Sciences*, **7**, 143–88.

Dzik, J. (1986a) Turrilepadida and other Machaerida, in *Problematic Fossil Taxa* (eds A. Hoffman and H. Nitecki), Clarendon Press, Oxford, pp. 116–34.

Dzik, J. (1986b) Chordate affinities of the conodonts, in *Problematic Fossil Taxa* (eds A. Hoffman and H. Nitecki), Clarendon Press, Oxford, pp. 240–54.

Elias, M. K. (1958) Late Mississippian fauna from the Redoak Hollow Formation of southern Oklahoma. Part 4. Gastropoda, Scaphopoda, Cephalopoda, Ostracoda, Thoracia, and Problematica. *Journal of Paleontology*, **32**, 1–57.

Etheridge, R. Jr (1880) A contribution to the study of British Carboniferous tubicolar annelida. *Geological Magazine*, (2), **7**, 262–4.

Fedonkin, M. A. (1979) Paleoichnology of the Precambrian and Early Cambrian, in *Paleontology of the Precambrian and Early Cambrian* (ed. B. S. Sokolov), Nauka, Leningrad [in Russian], pp. 183–92.

Fedonkin, M. A. (1985) Sistematicheskoe opisanie vendskikh Metazoa, in *Vendskaya Sistema, V.1, Paleontologiya* (eds B. S. Sokolov and A. B. Ivanoskii), Nauka, Moscow.

Fedonkin, M. A. (1986) Precambrian problematic animals: their body plan and phylogeny, in *Problematic Fossil Taxa* (eds A. Hoffman and M. Nitecki), Clarendon Press, Oxford, 267 pp.

Fonin, V. D. and Smirnova, T. N. (1967) New group of problematic early Cambrian organisms and methods of preparing them. *Paleontological Journal*, **1967**, 15–27.

Føyn, S. and Glaessner, M. F. (1979) *Platysolenites* and other animal fossils, and the Precambrian–Cambrian transition in Norway. *Norsk Geologisk Tidsskrift*, **59**, 25–46.

Glaessner, M. F. (1958) New fossil from the base of the Cambrian in South Australia. *Transactions of the Royal Society of South Australia*, **81**, 185–8.

Glaessner, M. F. (1976) Early Phanerozoic annelid worms and their geological and biological significance. *Journal of the Geological Society of London*, **132**, 259–75.

Glaessner, M. F. and Daily, B. (1959) The geology and Late Precambrian fauna of the Ediacara Fossil Reserve. *South Australian Museum, Records*, **13** (13), 369–401.

Gould, S. J. (1989) *Wonderful Life. The Burgess Shale and the Nature of History*. Hutchinson Radius, London, 347 pp.

Grant, S. W. F. (1990) Shell structure and distribution of *Cloudina*, a potential index fossil for the terminal Proterozoic. *American Journal of Science*, **290-A**, 261–94.

Hill, D. (1972) Archaeocyatha, in *Treatise on Invertebrate Paleontology*, Part. E. (ed. C. Teichert), Geological Society of America and University of Kansas Press, Boulder, Colorado, and Lawrence, Kansas, pp. E1–E158.

Hinz, I. (1987) The Lower Cambrian microfauna of Comley and Rushton, Shropshire/England. *Palaeontographica, Abteilung A*, **198**, 41–100.

Hou X.-G. (1987a) Two new arthropods from Lower Cambrian, Chenjiang, Eastern Yunnan. *Acta Palaeontologica Sinica*, **26**, 250–6.

Hou X.-G. (1987b) Three new large arthropods from Lower Cambrian, Chenjiang, Eastern Yunnan. *Acta Palaeontologica Sinica*, **26**, 280–5.

Hou X.-G. (1987c) Early Cambrian large bivalved arthropods from Chenjiang, Eastern Yunnan. *Acta Palaeontologica Sinica*, **26**, 292–8.

Hou X.-G. and Chen J.-Y. (1989) Early Cambrian arthropod–annelid intermediate sea animal *Luolishania* gen. nov. from Cheng-Jiang, Yunnan. *Acta Palaeontologica Sinica*, **28**, 208–13 [in Chinese with an English abstract].

Hou X.-G. and Sun W.-G. (1988) Discovery of Chengjiang Fauna at Meishucun, Jinning, Yunnan. *Acta Palaeontologica Sinica*, **27**, 1–12.

Howell, B. F. (1963) New Cambrian conchostracans from Wyoming and Newfoundland, brachiopods from Vermont, and worm, hydrozoan, and problematicum from Newfoundland. *Journal of Paleontology*, **37**, 264–7.

Jell, P. A. (1979) *Plumulites* and the machaeridian problem. *Alcheringa*, **3**, 253–60.

Jiang Z.-W. (1980) [The Meishucan Stage and fauna of the Jinning County, Yunnan.] *Bulletin of the Chinese Academy of Geological Sciences, Series 1*, **2:1**, 75–92 [in Chinese with an English summary].

Jiang Zhiwen (1982) Homopoda, in *The Sinian–Cambrian Boundary in Eastern Yunnan, China* (eds H.-L. Luo, Z.-W. Jiang, X.-C. Wu *et al.*), People's Publishing House, Yunnan.

Keller, B. M. (1969) Impression of unknown animal from Valdaian Series of Russian Platform, in *The Tommotian Stage and the Cambrian Lower Boundary Problem* (eds A. Yu Rozanov, V. V. Missarzhevsky, N. A. Volkova), Trudy Geologecheskii Institut, Nauka, Moscow, 206 [in Russian. English translation, US Department of the Interior, 1981], pp. 205–6.

Keller, B. M. and Fedonkin, M. A. (1976) New finds of fossils in the Precambrian Valday Series on the River Syuzma. *Akademii Nauka SSSR, Izvestiya Serie Geologiya*, **3**, 38–44 [in Russian].

Korejwo, K. (1979) Biostratigraphy of the Carboniferous sediments from Wierzchowe area (western Pomerania). *Acta Geologica Polonica*, **29**, 457–73.

Landing, E. (1984) Skeleton of lapworthellids and the suprageneric classification of tommotiids (Early and Middle Cambrian phosphatic Problematica). *Journal of Paleontology*, **58**, 1380–98.

Landing, E. (1988) Lower Cambrian of eastern Massachusetts: stratigraphy and small shelly fossils. *Journal of Paleontology*, **62**, 661–95.

Laurie, J. (1986) Phosphatic fauna of the Early Cambrian Todd River Dolomite, Amadeus Basin, central Australia. *Alcheringa*, **10**, 431–54.

Lendzion, K. (1975) Fauna of the *Mobergella* Zone in the Polish Lower Cambrian. *Kwartalnik geologiczny, Warszawa*, **19**, 237–42.

Liu D.-Y. (1979) Earliest Cambrian brachiopods from southwest China *Acta Palaeontologica Sinica*, **18**, 505–11 [in Chinese with English summary].

Lochman, C. (1940) Fauna of the basal Bonneterre Dolomite (Upper Cambrian) of southeastern Missouri. *Journal of Paleontology*, **14**, 1–53.

Loeblich, A. R. and Tappan, H. (1988) *Foraminiferal Genera and Their Classification*, Vol. 1, Van Nostrand Reinhold, New York, 970 pp.

Luo H.-L., Jiang Z.-W., Wu X.-C. *et al.* (1982) *The Sinian Cambrian Boundary in Eastern Yunnan, China*. People's Republic of China. 265 pp. [in Chinese with an English summary].

Mambetov, A. M. and Repina, L. N. (1979) The Lower Cambrian of the Talassk Ala-Too and its correlation with the sections of Karatau Miner and the Siberian Platform. *Trudy Institut Geologiya Geofizika sibersk Otdelenie*, **406**, 98–138.

Matthew, G. F. (1899) Upper Cambrian of Mount Stephen, British Columbia: the trilobites and worms. *Transactions of the Royal Society of Canada, Series 2*, **5**, 39–66.

Matthews, S. C. and Missarzhevsky, V. V. (1975) Small shelly fossils of the late Precambrian and early Cambrian age: a review of recent work. *Journal of the Geological Society of London*, **131**, 289–304.

McMenamin, M. A. S. (1985) Basal Cambrian small shelly fossils from the La Ciénega Formation, northwestern Sonora, Mexico. *Journal of Paleontology*, **59**, 1414–25.

Meshkova, N. P. (1969) K voprosu o paleontologicheskoi kharakteristike nizhnekembriiskikh otlozhenii Sibirskoi Platformi, in *Biostratigrafiya i Paleontologiya Nizhnego Kembriya Evropy i Severnoy Asii* (ed. I. T. Zhuravleva), Nauka, Moscow, pp. 158–74.

Mierzejewski, P. (1986) Ultrastructure, taxonomy and affinities of some Ordovician and Silurian organic microfossils. *Palaeontologia Polonica*, **47**, 129–220.

Missarzhevsky, V. V. (1966) [The first finds of *Lapworthella* in the Lower Cambrian of the Siberian Platform]. *Paleontological Journal*, **1966**, 13–18.

Missarzhevsky, V. V. (1969) [Description of hyolithids, gastropods, hyolithelminthes, camenides and forms of an obscure taxonomic position], in *The Tommotian Stage and the Cambrian Lower Boundary Problem* (eds A. Yu Rozanov, V. V. Missarzhevsky, N. A. Volkova *et al.*), Trudy Geologecheskii Institut, Nauka, Moscow, 206 [in Russian. English translation, US Department of the Interior, 1981], pp. 127–205.

Missarzhevsky, V. V. (1977) [Conodonts(?) and phosphatic problematica from the Cambrian of Mongolia and Siberia], in *Palaeozoic Invertebrata of Mongolia* (eds L. P. Tartarinov *et al.*), Joint Soviet–Mongolian Palaeontological Expedition, Transactions [in Russian], pp. 10–19.

Missarzhevsky, V. V. (1982) [Subdivision and correlation of the Precambrian–Cambrian boundary beds using some groups of the oldest skeletal organisms.] *Byulleten' Moskovskogo Obshchestva Ispytatelei Prirody, Otdelenie Geologii*, **57** (5), 52–67 [in Russian].

Missarzhevsky, V. V. and Grigor'yeva, N. V. (1981) About new representatives of the order Tommotiida. *Paleontological Journal*, **1981** (4), 91–7.

Missarzhevsky, V. V. and Mambetov, A. M. (1981) [*Stratigraphy and Fauna of the Precambrian–Cambrian Boundary Beds of Malyj Karatau.*] Trudy Geologiya-skogo Instituta, AH CCCP 326, 90 pp. [in Russian].

Mostler, H. and Mosleh-Yadzi, A. (1976) Neue Poriferen aus oberkambrischen Gesteinen der Milaformation im Elburzgebirge (Iran). *Geologische und Paläontologische Mitteilungen, Innsbruck*, **5**, 1–36.

Müller, K. J. (1977) *Palaeobotryllus* from the Upper Cambrian of Nevada – a probable ascidian. *Lethaia*, **10**, 107–18.

Nitecki, M. H. and Schram, F. R. (1976) *Etacystis communis*, a fossil of uncertain affinities from the Mazon Creek Fauna (Pennsylvanian of Illinois). *Journal of Paleontology*, **50**, 1157–61.

Pope, J. K. (1975) Evidence for relating the Lepidocoleidae, machaeridian echinoderms, to the mitrate carpoids. *Bulletin of American Paleontology*, **67**, 385–406.

Qian, Y. (1977) Hyolithida and some problematica from the Lower Cambrian Meishucunian Stage in central and southwestern China. *Acta Palaeontologica Sinica*, **16**, 255–75 [in Chinese with English abstract].

Qian, Y. (1978) The Early Cambrian hyolithids in central and southwest China and their stratigraphical significance. *Memoirs of the Nanjing Institute of Geology and Palaeontology*, **11**, 1–38.

Qian, Y. (1983) Several groups of bizarre sclerite fossils from the earliest Cambrian in eastern Yunnan. *Bulletin of the Nanjing Institute of Geology and Palaeontology, Academica Sinica*, **1983**, 85–99.

Qian, Y. and Bengtson, S. (1989) Palaeontology and Biostratigraphy of the Early Cambrian Meishucunian Stage in Yunnan Province, south China. *Fossils and Strata*, **24**, 1–156.

Qian, Y. and Zhang, S.-B. (1983) Small shelly fossils from the Xihaoping Member of the Tongying Formation in Fangxiang county of Hubei Province and their stratigraphical significance. *Acta Palaeontologica Sinica*, **22**, 82–94 [in Chinese with an English summary].

Radugin, K. V. (1962) O rannikh formakh arkheotsiat. *Materialy po geologii Zapadnoy Sibiri*, **63**, 7–10 [in Russian].

Radugin, K. V. (1964) O novoy gruppe drevneyshikh zhivotnykh. *Akademisk Nauk SSSR, Sibirsk Otdelenie Geologiya y Geofizika*, **1964** (1), 145–9 [in Russian].

Ramsköld, L. and Hou, X.-G. (1991) New early Cambrian animal and onychophoran affinities of enigmatic metazoans. *Nature*, **351**, 225–8.

Resser, C. E. and Howell, B. F. (1938) Lower Cambrian Olenellus Zone of the Appalachians. *Bulletin of the Geological Society of America*, **49**, 195–248.

Richardson, E. S. Jr (1966) Wormlike fossil from the Pennsylvanian of Illinois. *Science*, **151**, 75–6.

Ross, J. P. (1967) Fossil problematica from Upper Ordovician, Ohio. *Journal of Paleontology*, **41**, 37–42.

Rowland, S. M. and Carlson, S. J. (1983) *Westgardia gigantia*, a new Lower Cambrian fossil from eastern California. *Journal of Paleontology*, **57**, 1317–20.

Rusconi, C. (1951) Màs trilobitas càmbricos de San Isidro, Cerro Pelado y Canota. *Revista Museo Historia naturale Mendoza*, **5**, 3–30.

Schallreuter, R. (1981) Ordovizische Problematika III. *Ancienta* Ross, 1967, aus Europa. *Paläontologische Zeitschrift*, **55**, 209–18.

Schram, F. R. (1979) Worms of the Mississipian Bear Gulch Limestone of Central Montana, USA. *Transactions of the San Diego Society for Natural History*, **19**, 107–20.

Schmidt, F. (1988) Über eine neuendeckte Unter-Cambrische Fauna in Esthland. *Academie Impèriale de Sciences de St. Pétersbourg, Mémoires, série 7*, **36** (2), 5, 25–6.

Scrutton, C. T. (1979) Early fossil cnidarians, in *The Origin of Major Invertebrate Groups* (ed. M. R. House), Systematics Association Special Volume, 12, Academic Press, New York, pp. 161–208.

Seilacher, A. (1984) Late Precambrian Metazoa: Preservational or Real Extinctions, in *Patterns of Change in Earth Evolution* (eds H. D. Holland and A. F. Trendall), Springer-Verlag, Berlin, pp. 159–68.

Seilacher, A. (1989) Vendozoa: organismic construction in the Proterozoic. *Lethaia*, **22**, 229–39.

Simonetta, A. M. and Delle Cave, L. (1975) The Cambrian non-trilobite arthropods from the Burgess Shale of British Columbia. A study of their comparative morphology, taxonomy and evolutionary significance. *Palaeontographica Italica*, **69**, (n.s. 39), 1–37.

Sokolov, B. S. (1968) Vendian and Early Cambrian Sabelliditida (Pogonophora) of the USSR. *Proceedings of the 23rd International Geological Congress*, pp. 79–86.

Sokolov, B. S. (1976) The organic world of the earth on the way to Phanerozoic differentiation. *Vestnik Akademiya Nauk SSSR*, **1**, 126–43.

Sokolov, B. S. and Zhuravleva, I. T. (1983) Lower Cambrian Stage subdivision of the Siberia. Atlas of fossils. *Transactions of the Institute of Geology and Geophysics*, **558**, 1–216 [in Russian].

Sprigg, R. C. (1947) Early Cambrian (?) jellyfishes from the Flinders Ranges, South Australia. *Transactions of the Royal Society of South Australia*, **71**, 212–24.

Sun, W.-G. and Hou, X.-G. (1987) Early Cambrian worms from Chenjiang, Yunnan, China: *Maotianshania* gen nov. *Acta Palaeontologica Sinica*, **26**, 303–5.

Sysoiev, V. A. (1957) On the morphology, systematic position, and systematics of Hyolithoidea. *Akademiya Nauk. SSSR Leningrad, Doklady*, **116** (2), 1–37.

Sysoiev, V. A. (1962) Cambrian Hyoliths from the Northern Slope of the Aldan Shield. Nauka, Moscow, 66 pp.

Tchernycheva, N. E. (ed.) (1965) *Kembriyskaya Sistem.* Nedra, Moscow.

Tynan, M. C. (1983) Coral-like microfossils from the Lower Cambrian of California. *Journal of Paleontology*, **57**, 1188–211.

Valkov, A. K. and Sysoiev, V. A. (1970) Cambrian angustiochreids of Siberia, in *Stratigrafiya i Paleontologiya Proterozoya i Kembriya Vostoka Sibirskoi Platform. Yakutskoe Knizhnoe Izdatel'stvo* (ed. A. K. Bobrov), Yakutsk [in Russian], pp. 94–100.

Van Name, W. G. (1926) A new specimen of *Protobalanus*, supposed Paleozoic barnacle. *American Museum Novitates*, **227**, 1–6.

Vologdin, A. G. (1964) Kirbritsiaty – novyy klass arkheotsiat. *Doklady Akademii Nauk SSSR*, **157**, 1391–4 [in Russian].

Voronin, A. K., Voronova, L. G., Grigor'yeva, N. V. *et al.* (1982) The Precambrian/Cambrian Boundary in the geosynclinal areas (the reference section of Salany-Gol, MPR). *Trudy sovm. sov.-*

mongol. paleont. Eksped, **18**, 1–141 [in Russian].

Wade, M. (1972) Hydrozoa and Scyphozoa and other medosoids from the Precambrian Ediacara fauna, South Australia. *Palaeontology*, **15**, 197–225.

Walcott, C. D. (1884) On the Cambrian faunas of North America. *Bulletin of the United States Geological Survey*, **10**.

Walcott, C. D. (1908) Mount Stephen rocks and fossils. *Canadian Alpine Journal*, **1**, 232–48.

Walcott, C. D. (1911) Middle Cambrian Merostomata. *Smithsonian Miscellaneous Collections*, **57**, 17–40.

Walcott, C. D. (1912) Middle Cambrian Branchiopoda, Malacostraca, Trilobita and Merostomata. *Smithsonian Miscellaneous Collections*, **57**, 145–228.

Walcott, C. D. (1918) Geological explorations in the Canadian Rockies, in *Explorations and field-work of the Smithsonian Institution in 1917. Smithsonian Miscellaneous Collections*, **63**, 3–20.

Wang C.-Y. (1990) Some Llandovery phosphatic microfossils from south China. *Acta Palaeontologica Sinica*, **29**, 548–56.

Whittington, H. B. and Briggs, D. E. G. (1985) The largest Cambrian animal, *Anomalocaris*, Burgess Shale, British Columbia. *Philosophical Transactions of the Royal Society of London, Series*, **B309**, 569–609.

Welch, J. R. (1976) *Phosphannulus* on Paleozoic crinoid stems. *Journal of Paleontology*, **50**, 218–25.

Xiao L.-G. and Zhou B.-H. (1984) [Early Cambrian Hyolitha from Huainan and Huoqiu County in Anhui Province.] *Professional Papers of Stratigraphy and Palaeontology*, **13**, 141–51 [in Chinese].

Yankauskas, T. V. (1965) Pterotsiatidy-Novyy otryad kribritsiat. *Doklady Akademii Nauk SSSR*, **162**, 438–40 [in Russian].

Yankauskas, T. V. (1969) Pterotsiatidy nizhnego kembriya Krasnoyarskogo kryazha (vostochnyy Sayan), in *Biostratigrafiya i Paleontologiya Nizhnego Kembriya Sibiri i Dal'nego Vostoka* (ed. I. T. Zhuravleva), Nauka, Moscow [in Russian], pp. 114–57.

Yochelson, E. L. (1977) Agmata, a proposed extinct phylum of early Cambrian age. *Journal of Paleontology*, **51**, 437–54.

Yochelson, E. L. (1983) *Salterella* (Early Cambrian; Agmata) from the Scottish Highlands. *Palaeontology*, **26**, 253–66.

Yochelson, E. L., Pierce, J. W. and Taylor, M. E. (1970) *Salterella* from the Cambrian of Central Nevada. *United States Geological Survey Professional Paper*, **643H**, H1–H7.

Yu W. (1984) On merismmoconchids. *Acta Palaeontologica Sinica*, **23**, 432–46.

Zhang W.-T. (1987) Early Cambrian Chengjiang fauna and its trilobites. *Acta Palaeontologica Sinica*, **26**, 232–5.

Zhuravleva, I. T. (1970) Marine faunas and Lower Cambrian stratigraphy. *American Journal of Science*, **269**, 417–45.

29

MISCELLANIA

M. A. Wills

This chapter covers a diverse collection of poorly skeletonized taxa for which there is a minimal or underrepresentative fossil record. Some (e.g. Nematoda) are thought to have been widespread and abundant groups in the past, and their poor records therefore reflect a low fossilization potential. Others (e.g. Onychophora) are also relatively inconspicuous in today's biota, although this, of course, says little of their importance in the past.

Acknowledgements – I thank Drs D. E. G. Briggs and R. J. Aldridge for their guidance and assistance with source material. This chapter was researched during the tenure of a University of Bristol Postgraduate Scholarship.

Phylum CTENOPHORA

See Chapter 6, Coelenterata.

Phylum PLATYHELMINTHES (see Fig. 29.1)

F. UNNAMED V. (EDI) Mar.

First and Last: *Platypholinia pholiata* Fedonkin, 1985, Ust-Pinega Formation, Valdaiskoi Series, Vendian of the White Sea region, former USSR.

F. UNNAMED V. (EDI) Mar.

First and Last: *Vladimissa missarzhevskii* Fedonkin, 1985, Ust-Pinega Formation, Valdaiskoi Series, Vendian of the White Sea region, former USSR.

Class TURBELLARIA

Order TRICLADIDA

F. UNNAMED Q. (PLE) FW

First: Cocoons or fertilized egg capsules found widely distributed in modern lake deposits in the Canadian Arctic, East Africa, Europe, India and the USA (Gray, 1988). **Extant**

Order INCERTAE SEDIS

F. *INCERTAE SEDIS* T. (Mio.)–Rec. Mar.

First: Three unnamed species from three unnamed families, found in calcareous, petroliferous nodules formed in a Miocene lake, Calico Mountains, San Bernardino County, California, USA (Pierce, 1960). **Extant**

Phylum NEMERTEA

F. UNNAMED C. (MOS) Mar.

First: *Archisymplectes rhothon* Schram, 1973, Francis Creek Shale Member, Carbondale Formation, Grundy County, Illinois, USA (Schram, 1973).

Phylum ROTIFERA ?Ehrenberg, 1838

Order MONOGONONTA ?Plate, 1891

F. *INCERTAE SEDIS* T. (LUT)–Rec. Mar.

First: *?Keratella* sp., lower Middle Eocene, North Maslin Sands, near Adelaide, South Australia (Southcott and Lange, 1971). **Extant**

Phylum NEMATOIDA Rudolphi, 1808

The nematodes are known from a number of well-recorded examples, particularly in Tertiary ambers (Conway Morris *et al.*, 1982). Størmer's (1963) description of *Scorpiophagus* in saprobiotic association with a scorpion carcass would place their earliest record in the Lower Carboniferous, although some authors (e.g. Poinar, 1983) dispute this identification. Schram's assignment of forms from the Upper Carboniferous Mazon Creek (Schram, 1973) and Carboniferous Bear Gulch Limestone (Schram, 1979a) to the group are also somewhat equivocal (Conway Morris, 1985).

Class ADENOPHOREA

Order MERMITHIDA

F. MERMITHIDAE Braun, 1883 T. (Eoc)–Rec. Terr.

First: *Heydonius antiquus* Heydon, 1862, found projecting from the anus of a beetle, (*Hesthesis immortua* Heydon) from the Rhine lignite, Eocene of Germany (Taylor, 1935). **Extant**

Intervening: A number of other genera also found in Oligocene Baltic (CHT) amber.

Order ARAEOLAIMIDA

F. PLECTIDAE Chitwood and Chitwood, 1937 T. (CHT)–Rec. Terr.

First: *Oligoplectus succini* von Duisberg, 1862, Oligocene Baltic amber, Könisberg, Germany. **Extant**

Class SECERNENTEA

Order ASCARIDIDA

F. OXYURIDAE Cobbold, 1864 Q. (PLE)–Rec. Terr.

The Fossil Record 2. Edited by M. J. Benton. Published in 1993 by Chapman & Hall, London. ISBN 0 412 39380 8

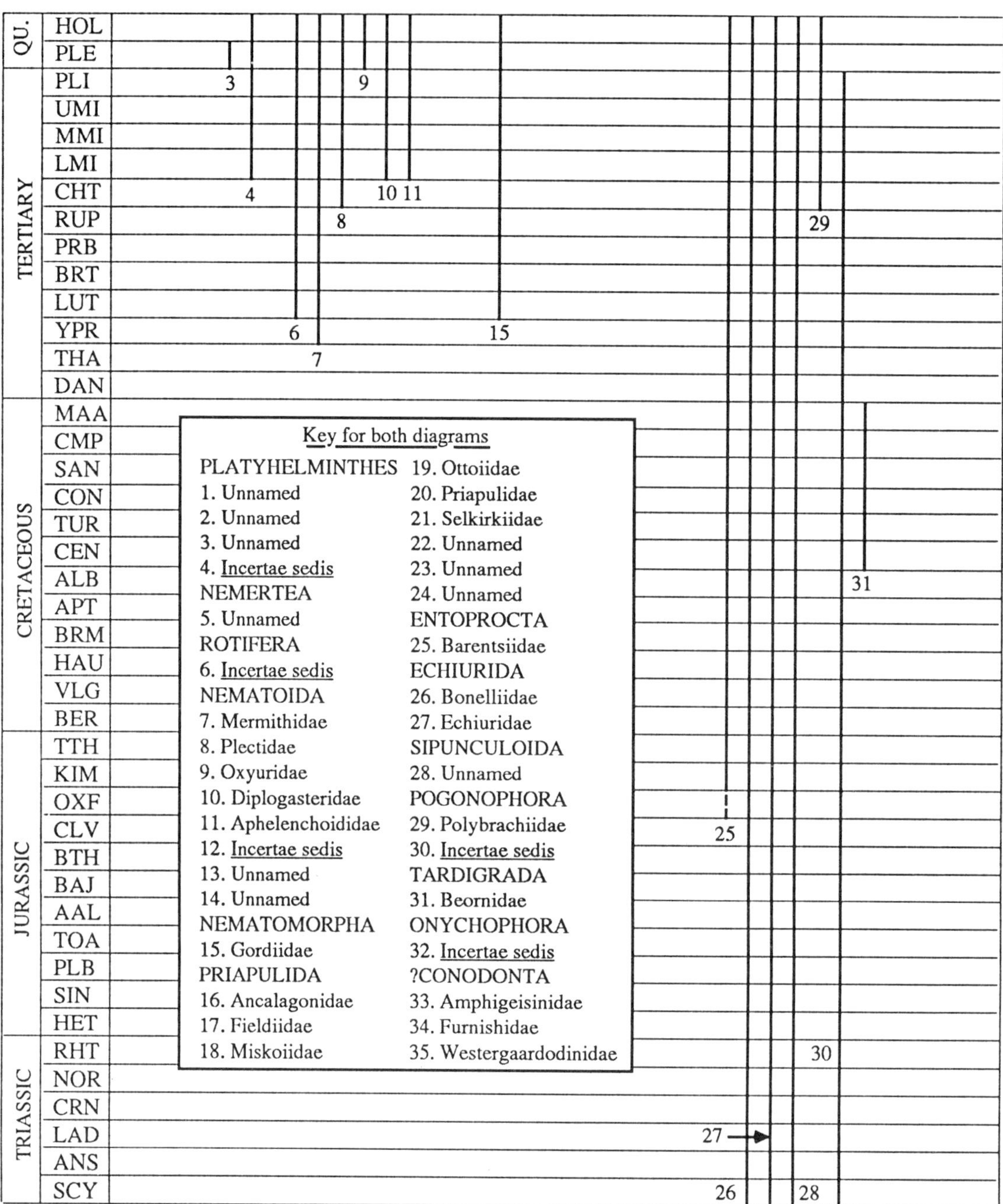

Fig. 29.1

First: *Ascaris obvelata* Rudolphi, 1802, Pleistocene of Siberia, former USSR. **Extant**

Order DIPLOGASTERIDA

F. DIPLOGASTERIDAE T. (?LMI) Terr.

First: Genus and species unknown, Chiapas amber-bearing beds from the Simojovel area, Central Mesa region and parts of the Tabasco Coastal Plain, Mexico (Gray, 1988). **Extant**

Order APHELENCHIDA

F. APHELENCHOIDIDAE T. (?LMI) Terr.

First: Genus and species unknown, Chiapas amber-bearing beds from the Simojovel area, Central Mesa region and parts of the Tabasco Coastal Plain, Mexico (Gray, 1988). **Extant**

Class INCERTAE SEDIS

F. *INCERTAE SEDIS* C. (SPK) Mar.

First and Last: Genus and species undetermined, Lower Carboniferous limestone, Gower Peninsula, South Wales, UK (Wu, 1983).

F. UNNAMED C. (MOS) Terr.

First and Last: *Nemavermes mackeei* Schram, 1973, Essex sub-biota, Mazon Creek, NE Illinois (Schram, 1979b).

F. UNNAMED C. (TOU) Terr.

First and Last: *Scorpiophagus baculiformis* and *Scorpiophagus*

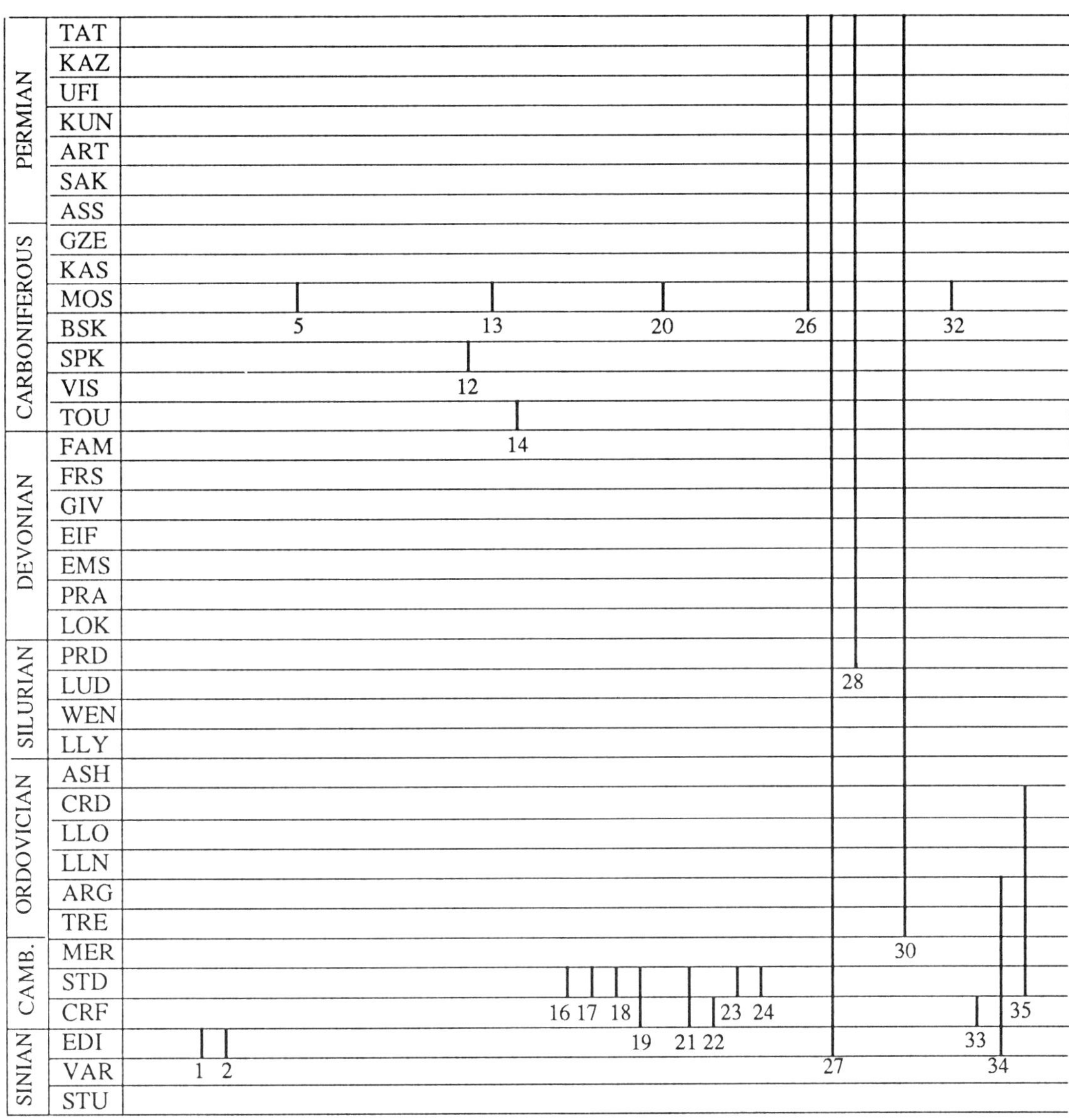

Fig. 29.1.

latus Størmer, 1963, found in saprobiotic association with the carcass of a scorpion, *Giantoscorpio willsi*, grey calcareous shale of the Calciferous Sandstone Series at Langholm, Dumfriesshire, Scotland, UK.

Phylum NEMATOMORPHA Vejovsky, 1886

Order GORDIOIDEA

F. GORDIIDAE May, 1919 T. (LUT)–Rec. Mar.

First: *Gordius tenuifibrosus* Voigt, 1983, Eocene lignite, Geiseltales, Germany. **Extant**

Comment: Sciacchitano (1955) suggests that this species is synonymous with the extant *G. albopunctatus*.

Phylum PRIAPULIDA

Conway Morris (1977) provides an excellent review of this phylum, which includes most of the families listed here. As a group, the priapulids once exceeded the annelids in both abundance and diversity, but are now restricted to marginal environments.

F. ANCALAGONIDAE Conway Morris, 1977 Є. (MEN) Mar.

First and Last: *Ancalagon minor* (Walcott, 1911), Burgess Quarry, Middle Cambrian Burgess Shale, near Field, British Columbia, Canada.

F. FIELDIIDAE Conway Morris, 1977 Є. (MEN) Mar.

First and Last: *Fieldia lanceolata* Walcott, 1912, Phyllopod bed of Burgess Quarry, Middle Cambrian Burgess Shale, near Field, British Columbia, Canada.

F. MISKOIIDAE Walcott, 1911 Є. (MEN) Mar.

First and Last: *Louisiella pedunculata* Walcott, 1911, Burgess Quarry, Middle Cambrian Burgess Shale, near Field, British Columbia, Canada.

F. OTTOIIDAE Walcott, 1911 Є. (TOM–STD) Mar.

First: *Maotianshania cylindrica* Sun and Hou, 1987, Chiungchussu Formation, Mount Maotianshan, Yunnan

Province, China. [*Maotianshania* is a junior pseudonym for *Ottoia*.]
Last: *Ottoia prolifica* Walcott, 1911, Phyllopod bed of Burgess Quarry, Middle Cambrian Burgess Shale, near Field, British Columbia, Canada.

F. PRIAPULIDAE Gosse, 1855 C. (MOS)–Rec. Mar.

First: *Priapulites konecniorum* (Schram, 1973), Moscovian, Francis Creek Shale, Mazon Creek, NE Illinois, USA. **Extant**

F. SELKIRKIIDAE Conway Morris, 1977 €. (LEN–MEN) Mar.

First: *Selkirkia pennsylvanica* Resser and Howell, 1938, Kinzers Formation, near Lancaster, Pennsylvania (Conway Morris, 1977).
Last: *Selkirkia columbia* Walcott, 1911, Middle Cambrian Burgess Shale, near Field, British Columbia, Canada (Conway Morris, 1977).

F. UNNAMED €. (ATB) Mar.

First and Last: *Cricosomia jinningensis* Hou and Sun, 1988, Chiungchussu Formation, Mount Maotianshan, Yunnan Province, China.

F. UNNAMED €. (MEN) Mar.

First and Last: *Lecythioscopa simplex* Walcott, 1931, Phyllopod Bed of Burgess Quarry, Middle Cambrian Burgess Shale, near Field, British Columbia, Canada (Conway Morris, 1977).

F. UNNAMED €. (MEN) Mar.

First and Last: *Scolecofurca rara* Conway Morris, 1977, Raymond's Quarry, Middle Cambrian Burgess Shale, near Field, British Columbia, Canada.

Phylum ENTOPROCTA

The phylum today comprises about 150 species of solitary or colonial, predominantly sessile metazoans. All are marine except members of the genus *Urnatella*. Only one fossil representative is known.

F. BARENTSIIDAE J. (OXF, KIM)–Rec. Mar.

First: *Barentsia* sp., Ampthill or Kimmeridge Clay, South Ferriby, Humberside preserved as bio-immured colonies beneath the oyster *Deltoideum* from the Upper Jurassic of England (Todd and Taylor, 1992). **Extant**

Phylum ECHIURIDA

Despite considerable differences between the adults of the two phyla, a number of authors (e.g. Harms, 1934; Beklemischew, 1958–1960), consider the Echiura to be secondarily reduced annelids, and therefore members of that phylum. The similarities between the two groups are discussed by Clark (1969).

Order ECHIUROINEA Bock, 1942

F. BONELLIIDAE Baird, 1868 (Fischer, 1946) C. (MOS)–Rec. Mar.

First: *Coprinoscolex ellogimus* Jones and Thompson, 1977, Moscovian, Francis Creek Shale, Mazon Creek, NE Illinois, USA. **Extant**

F. ECHIURIDAE Baird, 1868 V. (EDI)–Rec. Mar.

First: *Protechiurus edmonsi* Glaessner, 1979, upper Vendian, Kuibis Quartzite, lower Nama Group, Namibia. **Extant**

Phylum SIPUNCULOIDA Sedgwick, 1898

Skolithos burrows in Palaeozoic rocks of all ages have been regarded as those of sipunculids, but may equally well have been produced by phoronids or any sedentary, vermiform animal.

F. UNNAMED S. (PRD)–Rec. Mar.

First: *Trypanites* sp., uppermost Silurian Bertie Formation, and lowest Devonian Bois Blanc Formation (Emsian), Hagersville to Port Colborne, Southern Ontario, Canada (Pemberton *et al.*, 1980). **Extant**

Phylum POGONOPHORA Johannson, 1937

Fossil tubes interpreted as belonging to pogonophores (e.g. *Sabellidites* and *Saarina*) are abundant in some Upper Precambrian/Lower Cambrian rocks, particularly in eastern Europe and the former USSR (Sokolov, 1972). Differences in ultrastructure between such tubes and those produced by modern Pogonophora have thrown some doubt upon the relationship, and the fossil forms listed are assigned to the phylum on the basis of gross morphology. There are five extant families, in addition to the two listed here with a fossil record.

Order THECANEPHRIA Ivanov, 1955

F. POLYBRACHIIDAE Ivanov, 1952 T. (CHT)–Rec. Mar.

First: *Adekunbiella durhami* Adegoke, 1967, Oligocene Shales of Keasey Formation, west bank of Nehalem River, Oregon, USA. **Extant**

Order THECANEPHRIA or ATHECANEPHRIA?

F. *INCERTAE SEDIS* O. (TRE)–T. (PLI) Mar.

First: *Ivanovites fundibulatus* Kozłowski, 1967, Ordovician erratic rocks, Mochty province of Warsaw, and *Sokolovites poonophoroides* Kozłowski, 1967, Ordovician erratic rocks, Zakroczym-Wyszogród Province, Poland.
Last: *Tasselia* sp. de Heinzelin, 1965, Lower Pliocene concretions in the Antwerp basin, Belgium (Sokolov, 1968).

Phylum TARDIGRADA

F. BEORNIDAE Cooper, 1964 K. (u.) FW?

First and Last: *Beorn leggi* Cooper (1964), amber, Cedar Lake, Manitoba, Canada.

Phylum UNIRAMIA Manton, 1972

Subphylum ONYCHOPHORA Grube, 1853

F. *INCERTAE SEDIS* C. (MOS)–Rec. Mar.

First and Last: *Helenodora inopinata* Thompson and Jones, 1980, Moscovian, Francis Creek Shale, Mazon Creek, NE Illinois, USA. **Extant**
Comment: Assignment to the subphylum level is made only tentatively, since detailed preservation of the extreme anterior and posterior of the animal is lacking. *Aysheaia pedunculata* from the Middle Cambrian Burgess Shale, British Columbia is considered by most authors to represent a sister group to the Onychophora. True

Onychophorans possess jaws and an extension of the body beyond the posterior-most pair of appendages. Ramsköld and Hou (1991) consider *Hallucigenia*, *Microdictyon* and *Luolishania* to form a group which is probably related to the onychophorans. (see Chapter 28, section 2).

Phylum ?CONODONTA Eichenberg, 1930

Class ?CONODONTA Eichenberg, 1930

The true conodonts are dealt with elsewhere (see Chapter 30), where Sweet's (1988) classification, including only the euconodonts, has been adopted. This leaves the protoconodonts and paraconodonts. Müller and Hinz (1991) considered the protoconodonts to constitute a subgroup within the Paraconodontida, as they are similar in their overall direction of growth. The classification in Clarke (1981), which recognizes only the Paraconodontida as a formal taxon, has been adopted here.

Bengtson (1983) interpreted an evolutionary sequence from proto- through para- to euconodonts. This seems quite plausible, since all sclerites are composed of an organic matrix with an inlay of apatitic crystals, and there is remarkable uniformity in the development of structures within the Phylum. There is also a temporal trend from simple to more complicated shapes, consistent with a chemical evolution from early sclerites with a high organic and low phosphatic content to forms with gradually more phosphate and decreased proportions of organic material (Müller and Hinz, 1991).

At present it is unclear whether protoconodonts should necessarily be excluded from the true conodonts (Müller and Hinz, 1991). The term 'protoconodont' was introduced by Bengtson (1976) for slender Cambrian elements with only basal-internal growth increments. This structure was demonstrated for the Middle Cambrian *Gapparodus bisulcatus* (Müller, 1959), and later for the widespread Upper Cambrian *'Prooneotodus' tenuis* Bengtson, 1977, These forms, along with *Amphigeisina danica* (Poulsen, 1966) also show a three-layered wall structure with a thick, laminated middle layer, bounded on either side by thinner lamellae. Szaniawski (1982, 1983) has argued that protoconodonts are the grasping elements of chaetognaths. He made further comparisons between chaetognath teeth and paraconodont elements, suggesting a common ancestry for chaetognaths and the conodonts as a whole (Szaniawski, 1987). Protoconodonts occur from the Precambrian–Cambrian boundary to the lowermost Ordovician.

The genera *Rhombocorniculum* and *Fomitchella* are treated as Problematica not assigned to taxa above the genus level, while *Lapworthella* is included within the class Tommotiida (see Chapter 28, Problematica).

Order PARACONODONTIDA Müller, 1962

There is fairly wide acceptance that the paraconodonts are the ancestors of the true conodonts (euconodonts). Paraconodonts are known from Middle Cambrian to Middle Ordovician times, while the euconodonts appear in the basal Upper Cambrian and become extinct by the Upper Triassic.

Superfamily AMPHIGEISINACEA Miller, 1981

These are non-geniculate elements, distinguished by an unusual three-layered wall. The inner layer lines the basal cavity, the middle layer is thick and mostly apatite, and the outer layer covering the element is thin and mostly or entirely organic.

F. AMPHIGEISINIDAE Miller, 1981
Ꞓ. (LEN) Mar.

First and Last: *Amphigeisina danica* Bengtson, 1976, Lower to Middle Cambrian transition, Sweden, ?England and ?Siberia (Clark, 1981).

Superfamily FURNISHINACEA Miller, 1981

Non-geniculate coniform and unusual coniform multicuspate elements with a two-layered wall. Outer wall is thin and mostly organic, while the inner is thick and composed mostly of apatite (Clark, 1981).

F. FURNISHIDAE Müller and Nogami, 1971
V. (POU)–O. (ARG) Mar.

Elements with growth lamellae discontinuous on the outside of the cusp and on the inside of the basal cavity (Clark, 1981).

First: *Protohertzina anabarica* Missarzhevsky, 1973, Uppermost Precambrian and Tommotian of Siberia and Kazakhstan, former USSR.

Last: *Albiconus postcostatus* Miller, 1980, *Symphysurina* Zone, Lower Ordovician, Utah, Nevada, Oklahoma, Texas, Alberta, USA; *Nogamiconus* spp. Miller, 1980, Asia, Australia and North America; *Proacodus obliquus* Müller, 1959, Nevada, Utah and Wyoming, USA; Sweden and Germany (glacial erratics), China and Siberia; *Problematiconites perforata* Müller, 1959, Nevada, Utah and Wyoming, USA; Sweden and Germany (glacial erratics), Iran and Queensland, Australia; *Phakelodus tenuis* (Müller, 1959), Great Britain; *Prosagittodontus dahlmani* (Müller, 1959), Nevada, Utah and Wyoming, USA; Sweden and Germany (glacial erratics), China, Kazakhstan, Iran.

F. WESTERGAARDODINIDAE Müller, 1959
Ꞓ. (STD)–O. (CRD) Mar.

Aberrant forms with two to five unequally sized projections. The basal cavity is continuous from side to side, and growth lamellae are interrupted on either side or divided into two lateral cavities with growth lamellae round the base (Clark,1981).

First and Last: *Westergaardodina* spp. Müller, 1959, widespread in western and SW. states of USA, glacial erratics of Sweden, Poland and Germany; also Siberia, China, Poland, Iran and Queensland, Australia (Müller, 1962).

REFERENCES

Adegoke, O. S. (1967) A probable pogonophoran from the Early Oligocene of Oregon. *Journal of Paleontology*, **41**, 1090–4.

Beklemischew, W. N. (1958–1960) *Grundlagen der Vergleichenden Anatomie der Wirbellosen*, Vols 1, 2, Deutsche Verlag Wissenschaften, Berlin.

Bengtson, S. (1977) Aspects of problematic fossils in the early Palaeozoic. *Acta Universitatis Upsaliensis. Abstracts of Uppsala Dissertations from the Faculty of Science*, **415**, 1–71.

Bengtson, S. (1983) The early history of the Conodonta. *Fossils and Strata*, **15**, 5–19.

Clarke, D. L. (1981) Classification, *in Treatise on Invertebrate Palaeontology*, Part W (Supplement 2), (ed. R. A. Robison), Geological Society of America and University of Kansas Press, Boulder, Colorado and Lawrence, Kansas.

Clarke, R. B. (1969) Systematics and phylogeny: Annelida, Echiura, Sipuncula, *in Chemical Zoology*, Vol. 4 (eds M. Florkin and B. J. Scheer), Academic Press, New York, pp. 1–68.

Conway Morris, S. (1977) Fossil priapulid worms. *Special Papers in Palaeontology*, **20**, 1–95.

Conway Morris, S. (1985) Non-skeletonized lower invertebrate fossils: a review, in *The Origins and Relationships of Lower Invertebrates* (eds S. Conway Morris, J. D. George and R. Gibson), Systematics Association Special Volume, 28, Clarendon Press, Oxford, pp. 343–59.

Conway Morris, S., Pickerill, R. K. and Harland, T. L. (1982) A possible annelid from the Trenton Limestone (Ordovician) of Quebec, with a review of fossil oligochaetes and other annulate worms. *Canadian Journal of Earth Sciences*, **19**, 2150–7.

Cooper, K. W. (1964) The first fossil tardigrade: *Beorn leggi* Cooper, from cretaceous amber. *Psyche*, **71** (2), 41–8.

Duisberg, H. von (1862) Beitrag zur Bernstein-Fauna. *Königlich Physikalisch und Oekonomisch Gesellschaft Königsberg, Schriften*, **3**, 31–6.

Fedonkin, M. A. (1985) Sistematiueskoe opisanie Vendskich Metazoa, in *Vendskaya Sistema*, Vol. 1 (ed. B. S. Sokolov), Nauka, Moscow [in Russian], pp. 70–106.

Glaessner, M. F. (1979) An echiurid worm from the late Precambrian. *Lethaia*, **12**, 121–4.

Gray, J. (1988) Evolution of the freshwater ecosystem: the fossil record. *Palaeogeography, Palaeoclimatology, Palaeoecology*, **62**, 1–214.

Harms, W. (1934) *Handwörterbuch der Naturwissenschaften*, **ser ii**, **9**, p. 85.

Hou, X.- G. and Sun, W.- G. (1988) Discovery of Chengjiang fauna at Meishucan, Jinning, Yunnan. *Acta Palaeontologica Sinica*, **27**, 1–12 [in Chinese with English abstract].

Jones, D. and Thompson, I. (1977) Echiura from the Pennsylvanian Essex Fauna of northern Illinois. *Lethaia*, **10**, 317–25.

Kozłowski, R. (1967) Sur certains fossiles ordoviciens à test organique. *Acta Palaeontologica Polonica*, **12**, 99–132.

Miller, J. F. (1980) Taxonomic revisions of some upper Cambrian and Lower Ordovician conodonts, with comments on their evolution. *University of Kansas Paleontological Contributions*, **99**, p. 39.

Missarzhevsky, V. V. (1973) Konodontoobraznye organizmy iz pogranichtsykh sloyebv Kembriya i Dokembiya Sibirskoy platformy i Kazakhstana, in *Problemy Paleontologii i Biostratigrafii nizhnego Kembriya i Dalnego Vostoka* (ed. V. V. Missarzhevsky), Trudy Instituta Geologii i Geofiziki, Akademiya Nauk SSSR, Sibirskoe Otdelenie, pp. 53–7.

Müller, K. J. (1959) Kambrische Conodonten. *Zeitschrift des Deutschen Geologischen Gesellschaft*, **111**, 434–85.

Müller, K. J. (1962) Supplement to systematics of conodonts, in *Treatise on Invertebrate Paleontology*. Part W, (ed. C. Teichert), Geological Society of America and University of Kansas Press, Boulder, Colorado, and Lawrence, Kansas, W246–W249.

Müller, K. J. and Hinz, I. (1991) Upper Cambrian conodonts from Sweden. *Fossils and Strata*, **28**, 1–53.

Pemberton, S. G., Kobluk, D. R., Yeo, R. K. *et al.* (1980) The boring *Trypanites* at the Silurian–Devonian disconformity in Southern Ontario. *Journal of Paleontology*, **54**, 1258–66.

Pierce, W. D. (1960) Silicified turbellaria from Calico Mountains nodules. *Bulletin of the South California Academy of Sciences*, **59**, 138–43.

Poinar, G. O. (1983) *The Natural History of Nematodes*. Prentice-Hall, Englewood Cliffs, NJ.

Ramsköld, L. and Hou, X.- G. (1991) New early Cambrian animal and onychophoran affinities of enigmatic metazoans. *Nature*, **351**, 225–8.

Schram, F. R. (1973) Pseudocoelomates and a nemertine from the Illinois Pennsylvanian. *Journal of Paleontology*, **47**, 985–9.

Schram, F. R. (1979a) Worms of the Mississipian Bear Gulch Limestone of Central Montana, USA. *Transactions of the San Diego Natural History Society*, **19**, 107–20.

Schram, F. R. (1979b) The Mazon Creek biotas in the context of a Carboniferous faunal continuum, in *Mazon Creek Fossils* (ed. M. H. Nitecki), Academic Press, New York, pp. 159–90.

Sciacchitano, I. (1955) Su un Gordio fossile. Monitore Zoologico Italiano, **63**, 57–61.

Sokolov, B. S. (1968) Vendian and Early Cambrian Sabelliditida (Pogonophora) of the USSR. *Proceedings of the 23rd International Geological Congress*, pp. 79–86.

Sokolov, B. S. (1972) The Vendian Stage in Earth history. *Proceedings of the 24th International Geological Congress, Montreal, Section 1*, pp. 78–84.

Southcott, R. V. and Lange, R. T. (1971) Acarine and other microfossils from the Maslin Eocene, South Australia. *Records of the South Australian Museum*, **16** (7), 1–21.

Størmer, L. (1963) *Giantoscorpio willsi* a new scorpion from the Lower Carboniferous of Scotland and its associated preying microorganisms. *Skrifter utgitt av Det Norske Videnskaps-Akedemi i Oslo I. Matematika–Naturvisens Klasse, Ny Serie*, **8**, 1–171.

Sun, W. G. and Hou, X. G. (1987) Early Cambrian worms from the Chenjiang, Yunnan, China: *Maotranshania* gen nov. *Acta Palaeontologica Sinica*, **26**, 303–5.

Sweet, W. C. (1988) *The Conodonta*. Clarendon Press, Oxford, 212 pp.

Szaniawski, H. (1982) Chaetognath grasping spines recognized among Cambrian protoconodonts. *Journal of Paleontology*, **56**, 806–10.

Szaniawski, H. (1983) Structure of protoconodont elements. *Fossils and Strata*, **15**, 21–7.

Szaniawski, H. (1987) Preliminary structural comparisons of protoconodont, paraconodont and euconodont elements, in *Paleobiology of Conodonts*. (ed. R. J. Aldridge), Ellis Horwood, Chichester.

Taylor, A. L. (1935) A review of the fossil nematodes. *Proceedings of the Helminthological Society of Washington*, **2**, 47–9.

Thompson, I. and Jones, D. S. (1980) A possible onychophoran from the middle Pennsylvanian Mazon Creek beds of northern Illinois. *Journal of Paleontology*, **54**, 588–96.

Todd, J. A. and Taylor, P. D. (1992) (in press) The first fossil entoproct. *Naturwissenschaften*, **79**.

Voigt, E. (1983) Ein fossiler Saitenwurm (*Gordius tenuifibrosus* n. sp.) aus der eozänen Braunkohle des Geisaltes. *Nova Acta Lepoldina*, **5**, 351–60.

Walcott, C. D. (1911) Middle Cambrian annelids. *Smithsonian Miscellaneous Collections*, **57**, 109–44.

Walcott, C. D. (1912) Middle Cambrian Branchiopoda, Malacostraca, Trilobita and Merostomata. *Smithsonian Miscellaneous Collections*, **57**, 145–228.

Welch, J. R. (1976) *Phosphannulus* on Paleozoic crinoid stems. *Journal of Paleontology*, **50**, 218–25.

Wu X.-T. (1983) Origin and significance of constant-size fenestrae associated with calcispheres from the Lower Carboniferous of the Gower Peninsula, South Wales. *Palaeogeography, Palaeoclimatology, Palaeoecology*, **41**, 139–51.

Animals: Vertebrates

Acanthostega gunnari Jarvik, 1952, skull in right lateral view, from the Britta Dal Formation (Upper Devonian, Famennian) of Stensiöbjerg, Gauss Halvø, East Greenland (specimen MGUH 1300A, Geologisk Museum, Copenhagen). One of the oldest tetrapods. The skull is about 120 mm long. Photograph courtesy of S. E. Bendix-Almgreen, with the permission of the Geologisk Museum, Copenhagen, and J.A. Clack.

30

CONODONTA

R. J. Aldridge and M. P. Smith

Before 1970, conodont classification was entirely utilitarian. Each element type was regarded as a distinct taxon, and suprageneric categories consisted of groupings of elements that showed overall morphological similarities. In the last edition of *The Fossil Record*, the utilitarian classification of Hass (1962) was used, with modifications, by Rhodes (1967), who recognized that it had no phylogenetic significance. Subsequently, classifications have been developed that take account of the multi-element nature of conodont apparatuses and are thus closer to biological reality. The most comprehensive are those of Lindström (1970), Clark (with other authors in the *Treatise on Invertebrate Paleontology*, Robison (ed.) 1981) and Sweet (1988). All of these are exploratory and have limitations (see Fåhræus, 1983, 1984 for a critical discussion of the *Treatise* classification), and none has been based on well-formulated cladistic or other methodological principles.

Despite these reservations we have endeavoured to follow the most recent classification, that of Sweet (1988), as far as possible in this contribution. We have made some modifications to Sweet's scheme and have added families to accommodate additional genera. We emphasize, however, that much of conodont suprageneric classification is still probationary.

The conodont fossil record is very good, and there are few recorded major gaps in familial ranges. At the level of precision documented here, each conodont family is represented in all intervening chronostratigraphical intervals between its recorded first and last occurrences. We have used the stage level for quoting ranges to provide uniformity with other contributions, but have endeavoured to give more precise entries for the first and last occurrences. In most cases, the range limits are quoted to biozonal or sub-biozonal level. For some Middle and Upper Ordovician taxa, published ranges have been given in terms of the graphically correlated composite standard section of Sweet (1984), and are consequently cited here in terms of conodont chronozones; where taxa of this age have not been graphically correlated, Sweet's intervals are used as first-appearance biozones. For chronostratigraphical intervals in which more than one conodont biozonation is available, we have used the most suitable scheme for the particular record; for example, we have employed the North American Midcontinent Province and the North Atlantic Province biozonations as appropriate for Ordovician entries.

Acknowledgements – We appreciate information freely provided by many colleagues.

Phylum CHORDATA Bateson, 1886

Class CONODONTA Eichenberg, 1930
Sensu Clark, 1981

Order PROCONODONTIDA Sweet, 1988

F. CORDYLODONTIDAE Lindström, 1970
Є. (MER)–O. (TRE) Mar. (see Fig. 30.1)

First: *Eoconodontus notchpeakensis* (Miller, 1969), San Saba Member, Wilberns Formation, Threadgill Creek–Lange Ranch section, central Texas, USA; *notchpeakensis* Sub-biozone (Miller *et al.*, 1982).
Last: *Cordylodus angulatus* Pander, 1856, House Formation, Ibex area, Utah, USA; *angulatus* Biozone (Ethington and Clark, 1982).

F. FRYXELLODONTIDAE Miller, 1981
Є. (MER) and/or O. (TRE) Mar.

First: *Fryxellodontus inornatus* Miller, 1969, San Saba Member, Wilberns Formation, Threadgill Creek–Lange Ranch section, central Texas, USA; *inornatus* Sub-biozone (Miller *et al.*, 1982).
Last: *Fryxellodontus lineatus* Miller, 1969, Lava Dam Member, Notch Peak Formation, southern House Range, Utah, USA; *elongatus* Sub-biozone (Miller *et al.*, 1982).

F. PROCONODONTIDAE Lindström, 1970
Є. (MER) Mar.

First: *Proconodontus tenuiserratus* Miller, 1980, Point Peak Member, Wilberns Formation, Threadgill Creek–Lange Ranch section, central Texas, USA; *tenuiserratus* Biozone (Miller *et al.*, 1982).

The Fossil Record 2. Edited by M. J. Benton. Published in 1993 by Chapman & Hall, London. ISBN 0 412 39380 8

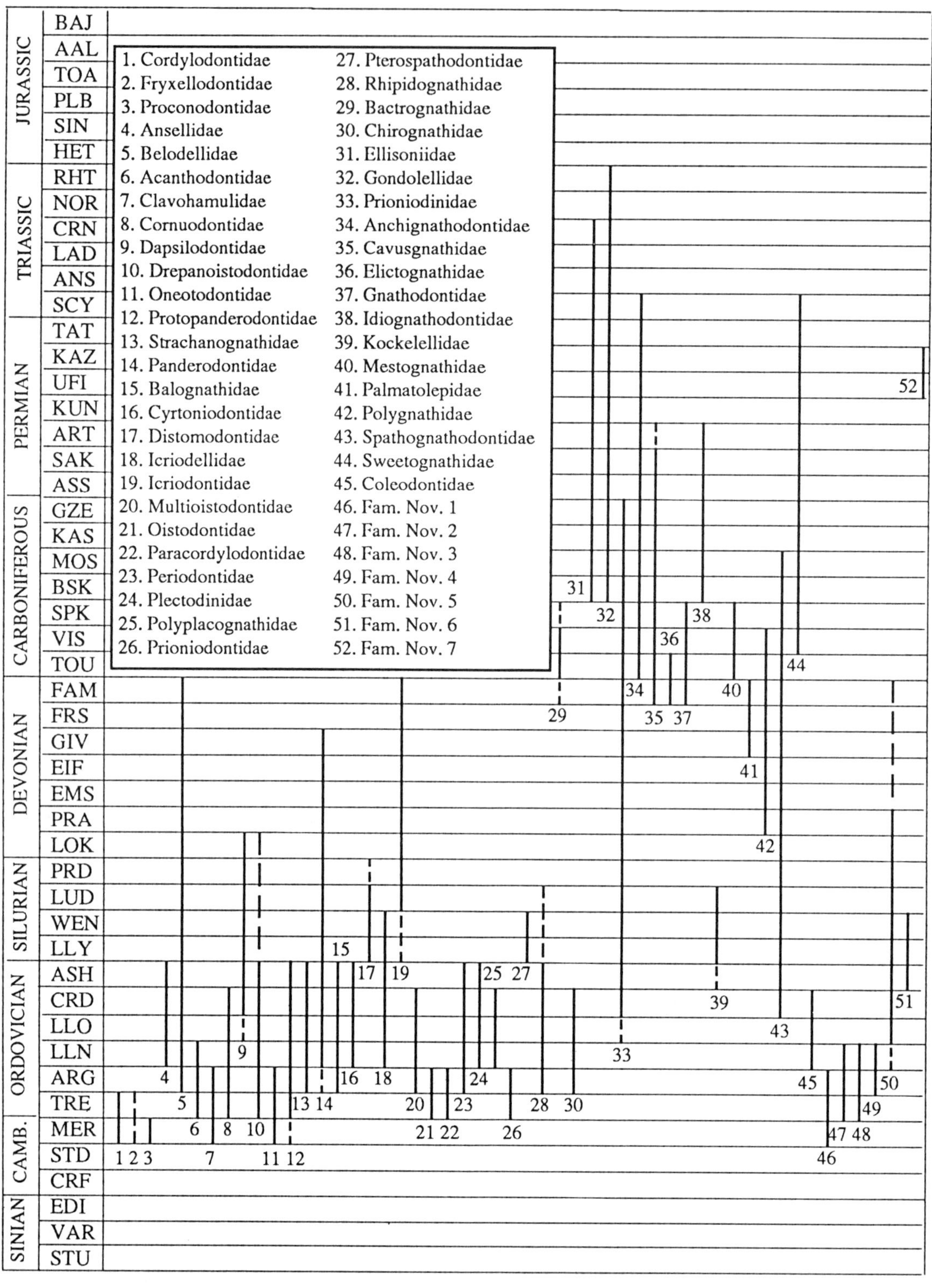

Fig. 30.1

Last: *Procondontus muelleri* Miller, 1969, San Saba Member, Wilberns Formation, Threadgill Creek–Lange Ranch section, central Texas, USA; top of *minutus* Sub-biozone (Miller *et al.*, 1982).

Order BELODELLIDA Sweet, 1988

F. ANSELLIDAE Fåhræus and Hunter, 1985
O. (LLN–ASH) Mar.

In addition to the genera included by Sweet (1988), we follow Fåhræus and Hunter (1985) in including *Goverdina* Fåhræus and Hunter, 1985.

First: *Ansella jemtlandicus* (Löfgren, 1978), Brunflo area, Lake Storsjön, Jämtland, Sweden; top of *flabellum* Sub-biozone (Löfgren, 1978).

Last: *Hamarodus europaeus* (Serpagli, 1967), *mucronata* Beds, Troutbeck Formation, Brow Gill, Cumbria, England,

UK; *ordovicicus* Biozone, Rawtheyan (Orchard, 1980). If *Hamarodus* Viira, 1975 is not included in the Ansellidae, the last is *Ansella nevadensis* (Ethington and Schumacher, 1969); *compressa* Chronozone in the composite standard section of Sweet (1984).

F. BELODELLIDAE Khodalevich and Tschernich, 1973 O. (ARG)–D. (FAM) Mar.

First: *Stolodus stola* (Lindström, 1955), Kalkberget Formation, Flåsjö area, Jämtland, Sweden; base of *elegans* Biozone (Löfgren, 1978).
Last: *Belodella resima* (Philip, 1965), Sadler Limestone (?), Canning Basin, Western Australia (Druce, 1976).

Order PROTOPANDERODONTIDA Sweet, 1988

The Protopanderodontida contains the majority of euconodonts that have their apparatuses composed entirely of coniform elements. Although knowledge of apparatus plans has advanced substantially in recent years, the group is still poorly known in terms of evolutionary relationships.

F. ACANTHODONTIDAE Lindström, 1970 O. (TRE–LLN) Mar.

The family composition adopted here differs from that of Sweet (1988). We refer *Cornuodus* Fåhræus, 1966 and *Scalpellodus* Dzik, 1976 to the Cornuodontidae, but include *Eucharodus* Kennedy, 1980 (synonymized with *Ulrichodina* Furnish, 1938 by Sweet) in the Acanthodontidae. The family remains a particularly poorly understood group in need of further taxonomic work.
First: *Acanthodus lineatus* (Furnish, 1938), House Formation, Ibex area, Utah, USA; *intermedius* Biozone (Ethington and Clark, 1982).
Last: *Parapaltodus flexuosus* (Barnes and Poplawski, 1973), top of Table Cove Formation, Table Point, Newfoundland, Canada; *sulcatus* Sub-biozone (Stouge, 1984).

F. CLAVOHAMULIDAE Lindström, 1970 €. (MER)–O. (ARG) Mar.

Dasytodus Chen and Gong, 1986, *Granatodontus* Chen and Gong, 1986, and *Hispidodontus* Nicoll and Shergold, 1991 are added to the genera included by Sweet (1988); *Serratognathoides* An, 1987 and *Eoserratognathus* An, 1990 also possibly belong.
First: *Granatodontus ani* Chen and Gong, 1986, 'Xiaoyangqiao Lower Section', Dayangcha, Jilin Province, China; *tenuiserratus* Biozone (Chen *et al.*, 1988).
Last: *Serratognathus extensus* Yang, 1983, Liangjiashan Formation, North China; *communis* Biozone (An *et al.*, 1983).

F. CORNUODONTIDAE Stouge, 1984 O. (TRE–CRD) Mar.

We include *Cornuodus* Fåhræus, 1966, *Scalpellodus* Dzik, 1976, and *Macerodus* Fåhræus and Nowlan, 1978. Sweet (1988) did not recognize the family, placing the first two genera in the Acanthodontidae, and omitting *Macerodus* from consideration. Stouge (1984) included additional genera, here placed in other families.
First: *Cornuodus longibasis* (Lindström, 1955), Köpingsklint Formation, Köpingsklint, Öland, Sweden; *deltifer* Biozone (van Wamel, 1974, who described the species as *Protopanderodus longibasis* (Lindström, 1955)).
Last: *Scalpellodus cavus* (Webers, 1966), McGregor Member, Platteville Formation, Olmsted County, Minnesota, USA (Webers, 1966); probably *tenuis* Chronozone.

F. DAPSILODONTIDAE Sweet, 1988 O. (LLO?/CRD)–D. (LOK) Mar.

First: *Besselodus arcticus* Aldridge, 1982, Troedsson Cliff Member, Kap Jackson Formation, Wulff Land, North Greenland; *velicuspis* Biozone (Tull, 1988). Dzik (1983) assigned *Acontiodus semisymmetricus* Hamar, 1966 to *Besselodus*, giving a possible earlier occurrence of: *Besselodus semisymmetricus* (Hamar, 1966), erratic boulder, northern Poland; *robustus* Sub-biozone (Dzik, 1976, 1983).
Last: *Dapsilodus obliquicostatus* (Branson and Mehl, 1933), chert and shale member, Caballos Novaculite, Marathon Uplift, Texas, USA; Lower Devonian (Barrick, 1987).

F. DREPANOISTODONTIDAE Fåhræus and Nowlan, 1978 O. (TRE)–O. (ASH)/D. (LOK) Mar.

We follow Bergström (*in* Robison, 1981) in referring *Nordiodus* Serpagli, 1967 and *Scandodus* Lindström, 1955 to this family in addition to the genera listed by Sweet (1988). *Decoriconus* Cooper, 1975 may well also belong here.
First: *Paltodus deltifer* (Lindström, 1955), Djupvik Formation, Köpingsklint, Öland, Sweden; base of *deltifer* Biozone (van Wamel, 1974, who described the species as *Drepanoistodus inequalis* (Pander, 1856)).
Last: *Drepanoistodus suberectus* (Branson and Mehl, 1933), top of Member 5, Ellis Bay Formation, Anticosti Island, Quebec; *ensifer* Biozone (McCracken and Barnes, 1981). If *Decoriconus* is included in the family, the last occurrence becomes: *Decoriconus fragilis* (Branson and Mehl, 1933), chert and shale member, Caballos Novaculite, Marathon Uplift, Texas, USA; Lower Devonian (Barrick, 1987).

F. ONEOTODONTIDAE Miller, 1981 €. (MER)–O. (ARG) Mar.

To the genera included by Miller (*in* Robison, 1981), we add *Teridontus* Miller, 1980. *Semiacontiodus* Miller, 1969 is transferred to the Protopanderodontidae, and *Pseudopanderodus* Landing, 1979 is a junior synonym of *Panderodus* Ethington, 1959 (Sweet, 1988).
First: *Teridontus nakamurai* (Nogami, 1967), Chatsworth Limestone, Black Mountain, Georgina Basin, western Queensland, Australia; *posterocostatus* Biozone (Nicoll and Shergold, 1991).
Last: *Oneotodus costatus* Ethington and Brand, 1981, Fillmore Formation, Ibex area, Utah, USA; upper part of *communis* Biozone (Ethington and Clark, 1982).

F. PROTOPANDERODONTIDAE Lindström, 1970 €. (MER)/O. (TRE)–O. (ASH)/S. (LLY) Mar.

This is a large family group, probably containing representatives of several lineages, although a common ancestor may prove to exist in *Semiacontiodus nogamii* Miller, 1969. To the genera listed by Sweet (1988), *Anodontus* Stouge and Bagnoli, 1988, *Juanognathus* Serpagli, 1974, *Parapanderodus* Stouge, 1984, *Scolopodus* Pander, 1856 and *Ulrichodina* Furnish, 1938 are added. Following Miller (*in* Robison, 1981), a separate family Oneotodontidae is recognized, in which we include *Monocostodus*, *Oneotodus*, *Teridontus* and *Utahconus*. *Pseudooneotodus* and *Strachanognathus*, question-

ably included in the Protopanderodontidae by Sweet (1988), are assigned to separate families.
First: *Semiacontiodus nogamii* Miller, 1969, Signal Mountain Limestone, Wichita Mountains, Oklahoma, USA; base of *elongatus* Sub-biozone (Miller *et al.*, 1982).
Last: *Staufferella inaligerae* McCracken and Barnes, 1981, base of Member 6, Ellis Bay Formation, Anticosti Island, Quebec, Canada; *nathani* Biozone, Ordovician/Silurian boundary beds (McCracken and Barnes, 1981).

F. STRACHANOGNATHIDAE Bergström, 1981 O. (ARG–ASH) Mar.

First: *Strachanognathus parvus* Rhodes, 1955, Borghamn, Östergötland, Sweden; Volkhovian (Löfgren, 1978).
Last: *Strachanognathus parvus* Rhodes, 1955, *mucronata* Beds, Troutbeck Formation, Brow Gill, Cumbria, England, UK; *ordovicicus* Biozone, top Rawtheyan (Orchard, 1980).

Order PANDERODONTIDA Sweet, 1988

F. PANDERODONTIDAE Lindström, 1970 O. (ARG?/LLN)–D. (GIV) Mar.

To the genera listed by Sweet (1988), *Zanclodus* Nowlan and McCracken, 1988 may be added. The apparatus of *Parapanderodus* Stouge, 1984 differs in composition and architecture from that of *Panderodus* Ethington, 1959 (Smith, 1991). We do not, therefore, follow Stouge (1984) and Sweet (1988) in including *Parapanderodus* in the Panderodontidae, but place it in the Protopanderodontidae. The suggestion of Dzik (1976, 1983) regarding an origin for *Panderodus* in the *Scalpellodus* (Cornuodontidae) lineage warrants further attention.
First: *Panderodus sulcatus* (Fåhræus, 1966), Jämtland, Sweden; upper part of the *gracilis* Sub-biozone (Löfgren, 1978). An unnamed species of *Panderodus* was figured by Serpagli (1974, pl. 24, figs 12, 13) from the *evae* Biozone of the San Juan Formation, Argentina.
Last: *Neopanderodus perlineatus* Ziegler and Lindström, 1971, Dunkle Kalke, Ebbesattels, Germany; *varcus* Biozone (Ziegler, 1965).

Order PRIONIODONTIDA Dzik, 1976

F. BALOGNATHIDAE Hass, 1959 O. (ARG–ASH) Mar.

Pygodus Lamont and Lindström, 1957, *Polonodus* Dzik, 1976, and *Promissum* Kovács-Endrödy, 1986 are added to the genera listed by Sweet (1988).
First: *Baltoniodus crassulus* (Lindström, 1955), Bruddesta Formation, Öland, Sweden (van Wamel, 1974), or *Baltoniodus bohemicus* Dzik, 1984, Klabava Formation, Svatostepánský rybník, Mýto, Czechoslovakia (Dzik, 1984); both *evae* Biozone.
Last: *Gamachignathus ensifer* McCracken *et al.*, 1980 and *Gamachignathus hastatus* McCracken *et al.*, 1980, base of Member 6, Ellis Bay Formation, Anticosti Island, Quebec, Canada; *ensifer* Biozone (McCracken and Barnes, 1981).

F. CYRTONIODONTIDAE Hass, 1959 O. (LLN–ASH) Mar.

First: *Phragmodus harrisi* Bauer, 1989, base of *harrisi* Chronozone (= *P. 'pre-flexuosus'* Chronozone) in the composite standard section of Sweet (1984).
Last: *Phragmodus undatus* Branson and Mehl, 1933, Member 5, Ellis Bay Formation, Anticosti Island, Quebec, Canada; top of *ensifer* Biozone (McCracken and Barnes, 1981).

F. DISTOMODONTIDAE Klapper, 1981 S. (LLY–LUD/PRD) Mar.

First: *Distomodus kentuckyensis* Branson and Branson, 1947, Bronydd Formation, Llandovery, Wales, UK; *kentuckyensis* Biozone, low Rhuddanian (Aldridge and Mohamed, *in* Cocks *et al.*, 1984). Specimens identified as *Distomodus* aff. *D. kentuckyensis* were reported from the lowermost Silurian of the Ellis Bay Formation, Anticosti Island, Quebec, Canada by McCracken and Barnes (1981).
Last: *Dentacodina*? *dubia* (Rhodes, 1953), Long Quarry Formation, Llandovery, Wales, UK (Aldridge, 1975); *eosteinhornensis* Biozone?, latest LUD or earliest PRD.

F. ICRIODELLIDAE Sweet, 1988 O. (LLN)–S. (WEN) Mar.

We do not follow Sweet (1988) in including *Pedavis* Klapper and Philip, 1971, *Sannemannia* Al-Rawi, 1977, or *Streptotaxis* Uyeno and Klapper, 1980 here. Klapper and Bergström (*in* Robison, 1981) referred *Pedavis* and *Sannemannia* to the Icriodontidae, where we would place all three genera.
First: *Icriodella cerata* (Knüpfer, 1967)?, Rzeszòwek slates, Kaczawa Mountains, Sudetes, Poland; *A. variabilis* Biozone (Dzik, 1989).
Last: *Icriodella inconstans* Aldridge, 1972, Brinkmarsh Formation, Tortworth, Bristol, England, UK (Aldridge, 1975); *sagitta rhenana* Biozone, lower Sheinwoodian.

F. ICRIODONTIDAE Müller and Müller, 1957 S. (LLY?/LUD)–D. (FAM) Mar.

We include *Pedavis*, *Sannemannia* and *Streptotaxis* here, rather than follow Sweet (1988) in placing them in the Icriodellidae. *Playfordia* Glenister and Klapper, 1966 may also belong.
First: *Pedavis latialata* (Walliser, 1964), Alticola Kalke, Carnic Alps, Austria; *latialata* Biozone (Walliser, 1964). *Coryssognathus*? sp. nov. Aldridge and Mohamed, 1982, Solvik Formation, Asker, Norway (*kentuckyensis* Biozone, Aeronian?) is a possible progenitor.
Last: *Pelekysgnathus inclinatus* Thomas, 1949, upper *praesulcata* Biozone (Sandberg and Dreesen, 1984).

F. MULTIOISTODONTIDAE Harris, 1964 O. (ARG–CRD) Mar.

Of the genera listed by Sweet (1988), *Leptochirognathus* Branson and Mehl, 1943 is transferred to Fam. *nov.* 3. *Trigonodus* Nieper, 1969, considered a probable junior synonym of *Pteracontiodus* Harris and Harris, 1965 by Sweet (1988), is here accepted as a separate genus.
First: *Trigonodus larapintinensis* (Crespin, 1943), Horn Valley Siltstone, Amadeus Basin, Northern Territory, Australia; *evae* Biozone (Cooper, 1981).
Last: *Pteracontiodus alatus* (Dzik, 1976), Mountain Lake Formation, Oklahoma, USA; *gerdae* Sub-biozone (Dzik, 1983).

F. OISTODONTIDAE Lindström, 1970 O. (TRE–ARG) Mar.

Sweet (1988) divided the Oistodontidae into three subfamilies: Oistodontinae, Tripodontinae and Juanognathinae. There is no diagnostic basis for recognizing the Oistodontinae and Tripodontinae, and their constituent genera,

with the exception of *Protoprioniodus* McTavish, 1973, are here treated simply as members of the Oistodontidae. *Protoprioniodus* is transferred to the Paracordylodontidae. *Juanognathus* Serpagli, 1974, the nominate genus of the Juanognathinae, does not have a prioniodontid apparatus and is transferred, pending a review of that family, to the Protopanderodontidae. *Histiodella* Harris, 1962, also included in the Juanognathinae by Sweet (1988), remains in the Oistodontidae.
First: *Rossodus manitouensis* Repetski and Ethington, 1983, House Limestone, Ibex area, Utah, USA; *manitouensis* Biozone (Ethington and Clark, 1982). Repetski and Ethington (1983) also included *Utahconus tenuis* Miller, 1980 in *Rossodus*; if this is correct then the first occurrence becomes: House Limestone, Ibex area, Utah, USA; *hintzei* Sub-biozone (Miller *et al.*, 1982).
Last: *Histiodella sinuosa* Graves and Ellison, 1941; *friendsvillensis* Chronozone in the composite standard section of Sweet (1984).

F. PARACORDYLODONTIDAE Bergström, 1981 O. (TRE–ARG) Mar.

Bergström (*in* Robison, 1981) erected this family for one genus, *Paracordylodus* Lindström, 1955; Sweet (1988) included neither the family nor the genus in his classification. The concept of the family is expanded here to include *Fahraeusodus* Stouge and Bagnoli, 1988 and *Protoprioniodus* McTavish, 1973, which probably form part of the same lineage.
First: *Paracordylodus gracilis* Lindström, 1955, Köpingsklint Formation, Öland, Sweden; *deltifer* Biozone (van Wamel, 1974).
Last: *Fahraeusodus marathonensis* (Bradshaw, 1969), Kanosh Shale, Ibex area, Utah, USA; *sinuosa* Biozone (Ethington and Clark, 1982).

F. PERIODONTIDAE Lindström, 1970 O. (ARG–ASH) Mar.

First: *Periodon selenopsis* (Serpagli, 1974), upper part of Bed 9, Cow Head Group, Point of Head, Cow Head Peninsula, Newfoundland, Canada; *elegans* Biozone (Stouge and Bagnoli, 1988). *Periodon primus* Stouge and Bagnoli, 1988, from the base of Bed 9 (*proteus* Biozone) is adenticulate and, although possibly ancestral (Löfgren, 1978; Stouge and Bagnoli, 1988), should probably be referred to *Diaphorodus* Kennedy, 1980 (Oistodontidae).
Last: *Periodon grandis* (Ethington, 1959), *velicuspis* Chronozone in the composite standard section of Sweet (1984).

F. PLECTODINIDAE Sweet, 1988 O. (LLN–ASH) Mar.

First: *Plectodina* aff. *flexa* (Rhodes, 1953), Rzeszòwek slates, Karzawa Mountains, Sudetes, Poland; *navis/triangularis* Biozone (Dzik, 1989).
Last: *Plectodina aculeatoides* Sweet, 1979, upper part of *divergens* Chronozone in the composite standard section of Sweet (1984).

F. POLYPLACOGNATHIDAE Bergström, 1981 O. (LLN–CRD) Mar.

We include *Prattognathus* Bergström, 1983 here in addition to the genera listed by Sweet (1988).
First: *Eoplacognathus zgierzensis* Dzik, 1976, −510.35 m, Ohesaare borehole, Estonia, former USSR; *A. variabilis* Biozone (Dzik, 1983).
Last: *Polyplacognathus ramosus* Stauffer, 1935, lower part of *confluens* Chronozone in the composite standard section of Sweet (1984).

F. PRIONIODONTIDAE Bassler, 1925 O. (TRE–ARG) Mar.

First: *Prioniodus gilberti* Stouge and Bagnoli, 1988, uppermost Bed 8, Cow Head Group, Point of Head, Cow Head Peninsula, Newfoundland, Canada; top of *deltifer* Biozone (Stouge and Bagnoli, 1988).
Last: *Oepikodus communis* (Ethington and Clark, 1964), Wah Wah Formation, Ibex area, Utah, USA; *communis* Biozone (Ethington and Clark, 1982). A possible successor species was identified as *Oepikodus*? aff. *minutus* (McTavish, 1973) by Ethington and Clark (1982), from the Kanosh Shale, Ibex area, Utah, USA; *altifrons* Biozone.

F. PTEROSPATHODONTIDAE Cooper, 1977 S. (LLY–WEN) Mar.

Of the genera included by Sweet (1988), *Carniodus* Walliser, 1964 is transferred to Fam. *nov.* 6, and *Johnognathus* Mashkova, 1977, if indeed a distinct genus, probably belongs in the Distomodontidae. The apparatus of *Aulacognathus* Mostler, 1967 resembles that of *Ctenognathodus* Pander, 1856, and the genus is tentatively transferred to the Kockelellidae. The relationships of *Apsidognathus* Walliser, 1964 and *Astropentagnathus* Mostler, 1967 remain to be clarified, but they are retained at present in the Pterospathodontidae. *Pranognathus* Männik and Aldridge, 1989 is also included here, and *Huddlella* Mashkova, 1979 may belong.
First: *Pranognathus siluricus* (Pollock *et al.*, 1971), Neahga Shale, Niagara Falls, Ontario, Canada (Pollock *et al.*, 1971) or *Pranognathus tenuis* (Aldridge, 1972), Solvik Formation, Oslo, Norway (Aldridge and Mohamed, 1982); both species from the upper part of the *kentuckyensis* Biozone, Aeronian.
Last: *Pterospathodus amorphognathoides* Walliser, 1964 from the Trilobiten- und Aulacopleura-Schichten of Cellon, Carnic Alps, Austria; *amorphognathoides* Biozone, lowermost Sheinwoodian (Walliser, 1964). ?*Huddlella johni* Mashkova, 1979, Khakomian Formation, Central Siberia; Homerian (Mashkova, 1979).

F. RHIPIDOGNATHIDAE Lindström, 1970 O. (ARG)–O. (ASH)/?S. (LUD) Mar.

Thrincodus Bauer, 1987 is a possible addition to the genera included by Sweet (1988). *Tasmanognathus* Burrett, 1979 shows some similarities to *Rhipidognathus* Branson *et al.*, 1951, and may also belong.
First: *Bergstroemognathus extensus* (Graves and Ellison, 1941), Bed 9, Cow Head Group, Cow Head Peninsula, Newfoundland, Canada; *proteus* Biozone (Stouge and Bagnoli, 1988), or *B. hubeiensis* An, 1985, Honghuayuan Formation, Yichang, Hubei, China; *diversus* Biozone (An *et al.*, 1985).
Last: *Rhipidognathus symmetricus* Branson *et al.*, 1951, Member 1, Ellis Bay Formation, Anticosti Island, Quebec, Canada; *ensifer* Biozone (McCracken and Barnes, 1981). Possibly 'Apparatus A' Uyeno, 1981, Cape Storm Formation, Goodsir Creek, Cornwallis Island, Canadian Arctic Archipelago; ?*siluricus* Biozone (Uyeno, 1981).

Order PRIONIODINIDA Sweet, 1988

F. BACTROGNATHIDAE Lindström, 1970
?D (FAM)/C. (TOU)–C. (VIS/SPK?) Mar.

In addition to the genera listed by Sweet (1988), *Geniculatus* Hass, 1953 may belong here. *Apatella* Chauff and Klapper, 1978 has a closely similar Pa element to *Bactrognathus* Branson and Mehl, 1941 and may also belong, although the similarities have been interpreted as homeomorphic (Chauff and Klapper, 1978; Chauff, 1985).

First: ?*Apatella ziegleri* Chauff and Klapper, 1978, lower shale member, Sulphur Springs Formation, Pevely, Missouri, USA; Lower *costatus* Biozone (Chauff and Klapper, 1978). Otherwise, *Dollymae saggitula* Hass, 1959, Chappel Limestone, Texas, USA; *isosticha*–Upper *crenulata* Biozone (Hass, 1959, Lane *et al.*, 1980).

Last: *Embsaygnathus asymmetricus* Metcalfe, 1980, Skibeden Shales, Embsay Beck, Skipton, North Yorkshire, England, UK; *commutata* Biozone, Arundian (Metcalfe, 1980). ?*Geniculatus claviger* (Roundy, 1926), Edale Shales, Castleton, Derbyshire, England, UK; *bollandensis* Biozone, Arnsbergian (Higgins, 1975).

F. CHIROGNATHIDAE Branson and Mehl, 1944
O. (ARG–CRD) Mar.

First: *Erraticodon patu* Cooper, 1981, Horn Valley Siltstone, Amadeus Basin, Northern Territory, Australia; *evae* Biozone (Cooper, 1981).

Last: *Erismodus radicans* (Hinde, 1879), *tenuis* Chronozone in the composite standard section of Sweet (1984).

F. ELLISONIIDAE Clark, 1972
C. (BSH)–Tr. (CRN) Mar.

From the genera listed by Sweet (1988), we transfer *Merrillina* Kozur and Mock, 1974 to the Gondolellidae, following its reconstruction by Swift (1986). *Stepanovites* Kozur, 1975 belongs here, if it is a distinct genus.

First: *Ellisonia latilaminata* von Bitter and Merrill, 1983, Eagle Limestone, Mingo County, West Virginia, USA; Morrowan (von Bitter and Merrill, 1983).

Last: *Gladigondolella tethydis* (Huckriede, 1958), Hallstatt Limestone, Feuerkogel, Röthelstein, Steiermark, Austria; *tethydis* Biozone (Huckriede, 1958).

F. GONDOLELLIDAE Lindström, 1970
C. (BSH)–Tr. (RHT) Mar.

To the genera listed by Sweet (1988), we add *Merrillina* Kozur and Mock, 1974 and *Axiothea* Fåhræus and Ryley, 1989; *Chirodella* Hirschmann, 1959 and *Algherella* Bagnoli *et al.*, 1984 should be included if they are indeed distinct genera, and *Parachirognathus* Clark, 1959 and *Icriospathodus* Krahl *et al.*, 1983 may also belong here. If Kozur (1989) is correct about the iterative evolution of gondolellid platforms, then his genera *Celsigondolella* Kozur, 1968, *Parvigondolella* Kozur and Mock, 1972, *Nicoraella* Kozur, 1980, *Mesogondolella* Kozur, 1988, *Pseudogondolella* Kozur, 1988, *Chiosella* Kozur, 1989, *Clarkina* Kozur, 1989, *Mockina* Kozur, 1989, *Norigondolella* Kozur, 1989 and *Scythogondolella* Kozur, 1989 should be included.

First: *Neogondolella clarki* (Koike,1967), Marble Falls Limestone, San Saba County, Texas, USA; Morrowan (von Bitter and Merrill, 1977).

Last: *Axiothea posthernsteini* (Kozur and Mock, 1974), Kössener Schichten, Eiberg, Austria; *posthernsteini* Biozone (Golebiowski, 1986).

F. PRIONIODINIDAE Bassler, 1925
O. (LLO?/CRD)–C. (GZE) Mar.

To the genera listed by Sweet (1988) we tentatively add *Pseudolonchodina* Zhou *et al.*, 1981 (= *Aspelundia* Savage, 1985). True *Apatognathus* Branson and Mehl, 1934 is probably correctly included (Nicoll, 1980; Sweet, 1988).

First: *Oulodus serratus* (Stauffer, 1930), base of the *undatus* Chronozone of the composite standard section of Sweet (1984). Unpublished thesis collections contain a much older species of *Oulodus* from the *sweeti* Biozone of the Thumb Mountain Formation, Canadian Arctic Islands (Nowlan, 1976) and the Morris Bugt Group, North Greenland (Tull, 1988).

Last: *Idioprioniodus typus* Gunnell, 1933, Topeka Limestone, Shawnee Group, Shawnee County, Kansas, USA; Virgilian (von Bitter, 1972, see Merrill and Merrill, 1974).

Order OZARKODINIDA Dzik, 1976

F. ANCHIGNATHODONTIDAE Clark, 1972
C. (TOU)–Tr. (DIE) Mar.

First: *Hindeodus crassidentatus* (Branson and Mehl, 1934), Bushberg Sandstone, Frenchman Creek, Missouri, USA (Branson and Mehl, 1934); *sandbergi* Biozone.

Last: *Hindeodus typicalis* (Sweet, 1970), Khunamuh Formation, Guryul Ravine, Kashmir (Sweet, 1979); *kummeli–cristagalli* Biozone.

F. CAVUSGNATHIDAE Austin and Rhodes, 1981
D. (FAM)–P. (SAK/ART?) Mar.

First: *Patrognathus ourayensis* Sandberg and Ziegler, 1979, Ouray Limestone, San Juan Mountains, Colorado, USA; Middle? or Upper *styriacus* Biozone (Sandberg and Ziegler, 1979).

Last: *Adetognathus paralautus* Orchard, 1984, Harper Ranch Group, Kamloops, British Columbia, Canada; *bisselli* Biozone (Orchard, 1984).

F. ELICTOGNATHIDAE Austin and Rhodes, 1981
D. (FAM)–C. (TOU) Mar.

First: *Alternognathus pseudotrigosus* (Dreesen and Dusar, 1974), Hamoir-Fairon section, near Liege, Belgium; upper *rhomboidea* Biozone (Dreesen and Dusar, 1974).

Last: *Siphonodella isosticha* (Cooper, 1939), pre-Welden Shale, Pontotoc County, Oklahoma (Cooper, 1939); *isosticha*–Upper *crenulata* Biozone.

F. GNATHODONTIDAE Sweet, 1988
D. (FAM)–C. (SPK) Mar.

First: *Protognathodus meischneri* Ziegler, 1969, Hangenberg Schiefer, Stockum, Germany; Lower *praesulcata* Biozone (Ziegler, 1969; Ziegler and Sandberg, 1984).

Last: *Gnathodus bilineatus* Roundy, 1926, Edale Shales, Edale, Derbyshire, England, UK (Higgins, 1975); *bollandensis* Biozone, Arnsbergian.

F. IDIOGNATHODONTIDAE Harris and Hollingsworth, 1933
C. (BSH)–P. (ART) Mar.

First: *Declinognathodus noduliferus* (Ellison and Graves, 1941), Target Limestone, Springer Formation, Carter

County, Oklahoma, USA; *primus* Biozone, Morrowan (Lane and Straka, 1974).
Last: *Streptognathodus elongatus* Gunnell, 1933, topmost Tensleep Sandstone, Mayoworth, Wyoming, USA (Rhodes, 1963); *bisselli/whitei* Biozone.

F. KOCKELELLIDAE Klapper, 1981 ?O. (ASH)/S. (LLY)–S. (LUD) Mar.

In addition to the genera listed by Sweet (1988), this family includes *Ctenognathodus*' Pander, 1856, and possibly *Aulacognathus* Mostler, 1967, and *Tuxekania* Savage, 1985.
First: ?'*Ctenognathodus*' *pseudofissilis* Lindström, 1959, Crug Limestone, Llandeilo, Wales, UK (Lindström, 1959); ?*superbus/ordovicicus* biozonal boundary, Cautleyan (Orchard, 1980). Otherwise *Kockelella manitoulinensis* (Pollock *et al.*, 1971), Manitoulin Dolomite, Manitoulin Island, Ontario, Candada (Pollock *et al.*, 1971); *kentuckyensis* Biozone, ?Rhuddanian.
Last: *Kockelella variabilis* Walliser, 1957 from the Cardiola–Niveau beds of Cellon, Carnic Alps, Austria (Walliser, 1964); *siluricus* Biozone, Ludfordian.

F. MESTOGNATHIDAE Austin and Rhodes, 1981 C. (TOU–SPK) Mar.

First: *Mestognathus groessensi* Belka, 1983, borehole WB-64, Olkusz, Poland; *isosticha*–Upper *crenulata* Biozone (Belka, 1983).
Last: *Mestognathus bipluti* Higgins, 1961, Edale Shales, north Derbyshire, England, UK; *naviculus* Biozone, Pendleian (Higgins, 1975).

F. PALMATOLEPIDAE Sweet, 1988 D. (GIV–FAM) Mar.

First: '*Polygnathus*' *latifossatus* Wirth, 1967, Quinto Real Massif, Pyrenees, Spain (Wirth, 1967); upper *varcus* Subbiozone.
Last: *Palmatolepis gracilis sigmoidalis* Ziegler, 1962, Wocklumeria Limestone, Oberrödinghausen, Rheinisches Schiefergebirge, Germany (Ziegler, 1962); upper *praesulcata* Biozone.

F. POLYGNATHIDAE Bassler, 1925 D. (PRA)–C. (VIS) Mar.

In addition to the genera listed by Sweet (1988), we provisionally include *Rhodalepis* Druce, 1969.
First: *Polygnathus pireneae* Boersma, 1974, Basibe Formation, central Pyrenees, Spain (Boersma, 1974); *pireneae* Biozone.
Last: *Polygnathus bischoffi* Rhodes *et al.*, 1969, Carboniferous Limestone (C Zone), Gower, Wales, UK (Rhodes *et al.*, 1969); *homopunctatus* Biozone.

F. SPATHOGNATHODONTIDAE Hass, 1959 O. (CRD)–C. (MOS) Mar.

Currently, a very broadly conceived family. To the genera listed by Sweet (1988) we add *Kimognathus* Mashkova, 1978, *Homeognathodus* Denkler and Harris, 1988, and *Skeletognathus* Sandberg *et al.*, 1989.
First: ?*Yaoxianognathus? abruptus* (Bergström and Sweet, 1966), *undatus* Chronozone in the composite standard section of Sweet (1984). Otherwise, '*Plectodina*' *tenuis* (Branson and Mehl, 1933), *tenuis* Chronozone in the composite standard section.
Last: *Rhachistognathus minutus declinatus* Baesemann and Lane, 1985, Bird Spring Formation, Arrow Canyon, Nevada, USA; Atokan, ?lower Desmoinesian (Baesemann and Lane, 1985).

F. SWEETOGNATHIDAE Ritter, 1986 C. (VIS)–Tr. (GRI) Mar.

First: *Diplognathodus* spp., Ship Cove Limestone, Codroy, Newfoundland, Canada; '*Diplognathodus*' Biozone (von Bitter and Plint-Geberl, 1982).
Last: *Isarcicella isarcica* (Huckriede, 1958), Werfen Formation, Pufelsbach, Val Gardena, northern Italy; *isarcica* Biozone (Huckriede, 1958).

Order UNKNOWN

F. COLEODONTIDAE Branson and Mehl, 1944 O. (LLN–CRD) Mar.

First: *Coleodus simplex* Branson and Mehl, 1933, Tyner Formation, Cherokee County, Oklahoma, USA; *harrisi* Biozone (Bauer, 1989).
Last: *Coleodus delicatus* Branson and Mehl, 1933, *C. simplex* Branson and Mehl, 1933, *Neocoleodus breviconus* Branson and Mehl, 1933 and *Stereoconus gracilis* Branson and Mehl, 1933, upper sandstone member, Harding Sandstone, Fremont County, Colorado, USA (Sweet, 1955); *undatus* Biozone.

F. **FAM. NOV.** 1 Є. (MER)–O. (ARG) Mar.

Comprises *Cambropustula* Müller and Hinz, 1990, *Polonodus*? *sensu* Stouge and Bagnoli, 1988, and possibly *Nericodus* Lindström, 1955; the group constitutes the oldest euconodont family.
First: *Cambropustula kinnekullensis* Müller and Hinz, 1990, Alum Shale, Kinnekulle, Västergötland, Sweden; *Agnostus pisiformis* trilobite Biozone (Müller and Hinz, 1991).
Last: '*Polonodus*' *corbatoi* (Serpagli, 1974), Bed 11, Cow Head Group, Point of Head, Cow Head Peninsula, Newfoundland, Canada; *evae* Biozone (Stouge and Bagnoli, 1988).

F. **FAM. NOV.** 2 O. (TRE–LLN) Mar.

Contains only *Chosonodina* Müller, 1964; may be in the same order as Fam. *nov.* 3.
First: *Chosonodina herfuthi* Müller, 1964, *C. chirodina* Zhang, 1983, *C. fisheri* Druce and Jones, 1971, and *C. tridentata* Zhang, 1983, Yeli Formation, north China; *angulatus* Biozone (An *et al.*, 1983).
Last: *Chosonodina rigbyi* Ethington and Clark, 1982, Lehman Formation, northern Egan Range, Nevada, USA (Harris *et al.*, 1979); uppermost *harrisi* Biozone or *friendsvillensis* Biozone.

F. **FAM. NOV.** 3 O. (TRE–LLN) Mar.

Includes *Cristodus* Repetski, 1982, *Jumudontus* Cooper, 1981, *Loxodus* Furnish, 1938 and *Leptochirognathus* Branson and Mehl, 1943. Sweet (1988) did not consider the first three genera and placed the last in the Multioistodontidae.
First: *Loxodus bransoni* Furnish, 1938, House Formation, Ibex area, Utah, USA; *manitouensis* Biozone (Ethington and Clark, 1982).
Last: *Leptochirognathus quadratus* Branson and Mehl, 1943, Tulip Creek Formation, Arbuckle Mountains, Oklahoma, USA; *friendsvillensis* Chronozone (Bauer, 1987).

F. FAM. NOV. 4 O. (ARG–LLN) Mar.

Contains only *Dischidognathus* Ethington and Clark, 1982.

First: Unnamed species of *Dischidognathus*, Eleanor River Formation, Canadian Arctic Archipelago; *communis* Biozone (Nowlan, 1976).

Last: *Dischidognathus primus* Ethington and Clark, 1982, Antelope Limestone Formation, Martin Ridge, Monitor Range, Nevada, USA; uppermost *harrisi* Biozone or *friendsvillensis* Biozone (Harris *et al.*, 1979).

F. FAM. NOV. 5 O. (LLN??/LLO)–D. (PRA/FAM?) Mar.

Contains *Pseudooneotodus* Drygant, 1974 and possibly *Fungulodus* Gagiev, 1979 (= *Mitrellataxis* Chauff and Price, 1980).

First: ??*Pseudooneotodus mitratus mitratus* (Moskalenko, 1973), Table Point Formation, Table Head, Newfoundland, Canada; *holodentata* Biozone (Stouge, 1984, single unfigured specimen, lost prior to publication). Otherwise, *P. mitratus*, Gonioceras Bay Member, Kap Jackson Formation, Wulff Land, North Greenland; *sweeti* Biozone (Tull, 1988).

Last: *Pseudooneotodus beckmanni* Bischoff and Sannemann, 1958, Coopers Creek Formation, Gippsland, Victoria, Australia; *sulcatus* Biozone (Philip, 1965, as gen. *et* sp. indet. A). ?*Fungulodus coronella* (Chauff and Price, 1980), Sulphur Springs Formation, Allenton, Missouri, USA; Lower *costatus* Biozone (Chauff and Price, 1980).

F. FAM. NOV. 6 O. (ASH)–S. (WEN) Mar.

Contains *Carniodus* Walliser, 1964 and probably *Eocarniodus* Orchard, 1980.

First: *Eocarniodus gracilis* (Rhodes, 1955), Sally Beck, Cautley, West Yorkshire, England, UK; *superbus* Biozone, upper Pusgillian (Orchard, 1980).

Last: *Carniodus carnulus* Walliser, 1964, Trilobiten- und Aulacopleura-Schichten, Cellon, Carnic Alps, Austria; *amorphognathoides* Biozone, lowermost Sheinwoodian (Walliser, 1964).

F. FAM. NOV. 7 P. (UFI/KAZ) Mar.

Contains only *Caenodontus* Behnken, 1975.

First and Last: *Caenodontus serrulatus* Behnken, 1975, Cherry Canyon and Bell Canyon Formations, Culberson County, Texas, USA; Guadalupian (Behnken, 1975).

REFERENCES

Aldridge, R. J. (1975) The stratigraphic distribution of conodonts in the British Silurian. *Journal of the Geological Society*, **131**, 607–18.

Aldridge, R. J. and Mohamed, I. (1982) Condodont biostratigraphy of the early Silurian of the Oslo Region, in *IUGS Subcommission on Silurian Stratigraphy; Field Meeting Oslo Region 1982* (ed. D. Worsley), Palaeontological Contributions from the University of Oslo, **278**, pp. 109–20.

An, T.-X., Du, G.-Q. and Gao, Q.-Q. (1985) *Ordovician conodonts from Hubei, China*. Publishing House, Beijing, 64 pp. [in Chinese].

An, T.-X., Zhang, F., Xiang, W.-D. *et al.* (1983) *Conodonts of North China and Adjacent Regions*. Science Press, Beijing, 223 pp. [in Chinese].

Baesemann, J. F. and Lane, H. R. (1985) Taxonomy and evolution of the genus *Rhachistognathus* Dunn (Conodonta; late Mississippian to early middle Pennsylvanian). *Courier Forschungsinstitut Senckenberg*, **74**, 93–136.

Barrick, J. E. (1987) Conodont biostratigraphy of the Caballos Novaculite (Early Devonian–Early Mississippian), northwestern Marathon Uplift, west Texas, in *Conodonts: Investigative Techniques and Applications* (ed. R. L. Austin), Ellis Horwood, Chichester, pp. 120–35.

Bauer, J. A. (1987) Conodonts and conodont biostratigraphy of the McLish and Tulip Creek Formations (Middle Ordovician) of South-Central Oklahoma. *Oklahoma Geological Survey Bulletin*, **141**, 1–58.

Bauer, J. A. (1989) Conodont biostratigraphy and paleoecology of Middle Ordovician rocks in eastern Oklahoma. *Journal of Paleontology*, **63**, 92–107.

Behnken, F. H. (1975) Leonardian and Guadalupian (Permian) conodont biostratigrapy in western and southwestern United States. *Journal of Paleontology*, **49**, 284–315.

Belka, Z. (1983) Evolution of the Lower Carboniferous conodont genus *Mestognathus*. *Acta Geologica Polonica*, **33**, 73–84.

Boersma, K. T. (1974) Description of certain Lower Devonian platform conodonts of the Spanish central Pyrenees. *Leidse Geologische Mededelingen*, **49**, 285–301.

Branson, E. B. and Mehl, M. G. (1934) Conodonts from the Bushberg Sandstone and equivalent formations of Missouri. *The University of Missouri Studies*, **8**, 265–99.

Chauff, K. M. (1985) Phylogeny of the multielement conodont genera *Bactrognathus, Doliognathus* and *Staurognathus*. *Journal of Paleontology*, **59**, 299–309.

Chauff, K. M. and Klapper, G. (1978) New conodont genus *Apatella* (Late Devonian), possible homeomorph *Bactrognathus* (Early Carboniferous, Osagean Series), and homeomorphy in conodonts. *Geologica et Palaeontologica*, **12**, 151–64.

Chauff, K. M. and Price, R. C. (1980) *Mitrellataxis*, a new multielement genus of Late Devonian conodont. *Micropaleontology*, **26**, 177–88.

Chen, J.-Y., Qian, Y.-Y., Zhang, J.-M. *et al.* (1988) The recommended Cambrian–Ordovician global boundary stratotype of the Xiaoyangqiao section (Dayangcha, Jilin Province), China. *Geological Magazine*, **125**, 415–44.

Cocks, L. R. M., Woodcock, N. H., Rickards, R. B. *et al.* (1984) The Llandovery Series of the type area. *Bulletin of the British Museum (Natural History), Geology Series*, **38**, 131–82.

Cooper, B. J. (1981) Early Ordovician conodonts from the Horn Valley Siltstone, central Australia. *Palaeontology*, **24**, 147–83.

Cooper, C. L. (1939) Conodonts from a Bushberg–Hannibal horizon in Oklahoma. *Journal of Paleontology*, **13**, 379–422.

Dreesen, R. and Dusar, M. (1974) Refinement of conodont biozonation in the Famennian type area. *International Symposium on Belgian micropalaeontological limits, Namur 1974, Publication* **13**, 36 pp.

Druce, E. C. (1976) Conodont biostratigraphy of the Upper Devonian reef complexes of the Canning Basin, Western Australia. *Bureau of Mineral Resources, Geology and Geophysics, Bulletin*, **158**, 303 pp.

Dzik, J. (1976) Remarks on the evolution of Ordovician conodonts. *Acta Palaeontologica Polonica*, **21**, 395–455.

Dzik, J. (1983) Relationships between Ordovician Baltic and North American Midcontinent conodont faunas. *Fossils and Strata*, **15**, 59–85.

Dzik, J. (1984) Early Ordovician conodonts from the Barrandian and Bohemian-Baltic faunal relationships. *Acta Palaeontologica Polonica*, **28**, 327–68.

Dzik, J. (1989) Conodont evolution in high latitudes of the Ordovician. *Courier Forschungsinstitut Senckenberg*, **117**, 1–28.

Ethington, R. L. and Clark, D. L. (1982) Lower and Middle Ordovician conodonts from the Ibex area, western Millard County, Utah. *Brigham Young University Geology Studies*, **28** (2), 1–155.

Ethington, R. L. and Schumacher, D. (1969) Conodonts of the Copenhagen Formation (Middle Ordovician) in central Nevada. *Journal of Paleontology*, **43**, 440–84.

Fåhræus, L. E. (1983) Phylum Conodonta Pander, 1856 and nomenclatural priority. *Systematic Zoology*, **32**, 455–9.

Fåhræus, L. E. (1984) A critical look at the Treatise family-group classification of Conodonta: an exercise in eclecticism. *Lethaia*, **17**, 293–305.

Fåhræus, L. E. and Hunter, D. R. (1985) Simple-cone conodont taxa from the Cobbs Arm Limestone (Middle Ordovician), New World Island, Newfoundland. *Canadian Journal of Earth Sciences*, **22**, 1171–82.

Golebiowski, R. (1986) Neue Misikellen-Funde (Conodonta) und ihre Bedeutung für die Abgrenzung des Rhät s. str. in den Kössener Schichten. *Sitzungsberichte der mathematisch-naturwissenschaftlichen Klasse der Österreichischen Akademie der Wissenschaften, Wien*, **195**, 53–65.

Harris, A. G., Bergström, S. M., Ethington, R. L. *et al.* (1979) Aspects of Middle and Upper Ordovician conodont biostratigrapy of carbonate facies in Nevada and southeast California and comparison with some Appalachian successions. *Brigham Young University Geology Studies*, **26** (3), 7–33.

Hass, W. H. (1959) Conodonts from the Chappel Limestone of Texas. *U.S. Geological Survey Professional Paper* **294-J**, 365–400.

Hass, W. H. (1962) Conodonts, in *Treatise on Invertebrate Paleontology, Part W, Miscellanea* (ed. R. C. Moore), Geological Society of America and University of Kansas Press, Lawrence, Kansas, W3–W69.

Higgins, A. C. (1975) Conodont zonation of the late Viséan–early Westphalian strata of the south and central Pennines of northern England. *Bulletin of the Geological Survey of Great Britain*, **53**, 1–90.

Huckriede, R. (1958) Die Conodonten der Mediterranean Trias und ihr stratigraphischer Wert. *Paläontologische Zeitschrift*, **32**, 141–75.

Kozur, H. (1989) The taxonomy of the gondolellid conodonts in the Permian and Triassic. *Courier Forschungsinstitut Senckenberg*, **117**, 409–69.

Lane, H. R. and Straka, J. J. II. (1974) Late Mississippian and early Pennsylvanian conodonts, Arkansas and Oklahoma. *Geological Society of America Special Paper* **152**, 1–144.

Lane, H. R., Sandberg, C. A. and Ziegler, W. (1980) Taxonomy and phylogeny of some Lower Carboniferous conodonts and preliminary standard post-*Siphonodella* zonation. *Geologica et Palaeontologica*, **14**, 117–64.

Lindström, M. (1955) Conodonts from the lowermost Ordovician strata of south-central Sweden. *Geologiska Föreningens i Stockholm Förhandlingar*, **76**, 517–604.

Lindström, M. (1959) Conodonts from the Crug Limestone (Ordovician, Wales). *Micropaleontology*, **5**, 427–52.

Lindström, M. (1970) A suprageneric taxonomy of the conodonts. *Lethaia*, **3**, 427–45.

Löfgren, A. (1978) Arenigian and Llanvirnian conodonts from Jämtland, northern Sweden. *Fossils and Strata*, **13**, 129 pp.

Mashkova, T. V. (1979) New Silurian conodonts from central Siberia. *Paleontologicheskii Zhurnal*, **1979** (2), 98–105 [in Russian].

McCracken, A. D. and Barnes, C. R. (1981) Conodont biostratigraphy and paleoecology of the Ellis Bay Formation, Anticosti Island, Quebec, with special reference to Late Ordovician–Early Silurian chronostratigraphy and the systemic boundary. *Geological Survey of Canada Bulletin*, **329**, 51–134.

Merrill, G. K. and Merrill, S. M. (1974) Pennsylvanian nonplatform conodonts, IIa: the dimorphic apparatus of *Idioprioniodus*. *Geologica et Palaeontologica*, **8**, 119–30.

Metcalfe, I. (1980) Conodont zonation and correlation of the Dinantian and early Namurian strata of the Craven Lowlands of northern England. *Institute of Geological Sciences Report* **80/10**, 1–70.

Miller, J. F., Taylor, M. E., Stitt, J. H. *et al.* (1982) Potential Cambrian–Ordovician boundary stratotype sections in the western United States, in *The Cambrian–Ordovician Boundary: Sections, Fossil Distributions and Correlations* (eds M. G. Bassett and W. T. Dean), National Museum of Wales, Cardiff, pp. 155–80.

Müller, K. J. and Hinz, I. (1991) Upper Cambrian conodonts from Sweden. *Fossils and Strata*, **28**, 1–153.

Nicoll, R. S. (1980) The multielement genus *Apatognathus* from the Late Devonian of the Canning Basin, Western Australia. *Alcheringa*, **4**, 133–52.

Nicoll, R. S. and Shergold, J. H. (1991) Revised Late Cambrian (pre-Payntonian–Datsonian) conodont biostratigraphy at Black Mountain, Georgina Basin, western Queensland, Australia. *BMR Journal of Australian Geology and Geophysics*, **12**, 93–118.

Nowlan, G. S. (1976) Late Cambrian to Late Ordovician conodont evolution and biostratigraphy of the Franklinian Miogeosyncline, eastern Canadian Arctic Islands. PhD Thesis (unpublished), University of Waterloo, Ontario.

Orchard, M. J. (1980) Upper Ordovician conodonts from England and Wales. *Geologica et Palaeontologica*, **14**, 9–44.

Orchard, M. J. (1984) Early Permian conodonts from the Harper Ranch Beds, Kamloops area, southern British Columbia. *Current Research, Part B, Geological Survey of Canada Paper*, **84–1B**, 207–15.

Philip, G. M. (1965) Lower Devonian conodonts from the Tyers area, Gippsland, Victoria. *Proceedings of the Royal Society of Victoria*, **79**, 95–117.

Pollock, C. A., Rexroad, C. B. and Nicoll, R. S. (1971) Lower Silurian conodonts from northern Michigan and Ontario. *Journal of Paleontology*, **44**, 743–64.

Repetski, J. E. and Ethington, R. L. (1983) *Rossodus manitouensis* (Conodonta), a new Early Ordovician index fossil. *Journal of Paleontology*, **57**, 684–703.

Rhodes, F. H. T. (1963) Conodonts from the topmost Tensleep Sandstone of the eastern Big Horn Mountains, Wyoming. *Journal of Paleontology* **37**, 401–8.

Rhodes, F. H. T. (1967) Miscellanea; Phylum Uncertain Order Conodontophorida Eichenberg, 1930, in *The Fossil Record* (eds. W. B. Harland *et al.*), Geological Society, London, pp. 613–18.

Rhodes, F. H. T., Austin, R. L. and Druce, E. C. (1969) British Avonian (Carboniferous) conodont faunas, and their value in local and intercontinental correlation. *Bulletin of the British Museum (Natural History) Geology, Supplement* **5**, 313 pp.

Robison, R. A. (ed.) (1981) *Treatise on Invertebrate Paleontology, Part W, Miscellanea, Supplement 2, Conodonta.* Geological Society of America and University of Kansas Press, Boulder, Colorado, and Lawence, Kansas, xxviii + 202 pp.

Sandberg, C. A. and Dreesen, R. (1984) Late Devonian icriodontid biofacies models and alternate shallow-water conodont zonation. *Geological Society of America Special Paper* **196**, 143–78.

Sandberg, C. A. and Ziegler, W. (1979) Taxonomy and biofacies of important conodonts of Late Devonian *styriacus*-Zone, United States and Germany. *Geologica et Palaeontologica*, **13**, 173–212.

Serpagli, E. (1974) Lower Ordovician conodonts from Pre-Cordilleran Argentina (Province of San Juan). *Bollettino della Societa Paleontologica Italiana*, **13**, 17–98.

Smith, M. P. (1991) Early Ordovician conodonts of East and North Greenland. *Meddelelser om Grønland, Geoscience*, **26**, 81 pp.

Stouge, S. (1984) Conodonts of the Middle Ordovician Table Head Formation, western Newfoundland. *Fossils and Strata*, **16**, 145 pp.

Stouge, S. and Bagnoli, G. (1988) Early Ordovician conodonts from Cow Head Peninsula, western Newfoundland. *Palaeontographia Italica*, **75**, 89–179.

Sweet, W. C. (1955) Conodonts of the Harding Formation (Middle Ordovician) of Colorado. *Journal of Paleontology*, **29**, 226–62.

Sweet, W. C. (1979) Late Ordovician conodonts and biostratigraphy of the western Midcontinent Province. *Brigham Young University Geology Studies*, **26** (3), 45–86.

Sweet, W. C. (1984) Graphic correlation of upper Middle and Upper Ordovician rocks, North American Midcontinent Province, U.S.A., in *Aspects of the Ordovician System* (ed. D. L. Bruton), Universitetsforlaget, Oslo, pp. 23–35.

Sweet, W. C. (1988) *The Conodonta: Morphology, Taxonomy, Paleoecology and Evolutionary History of a Long-Extinct Animal Phylum.* Oxford University Press, New York, x + 212 pp.

Swift, A. (1986) The conodont *Merrillina divergens* (Bender & Stoppel) from the Upper Permian of England, in *The English Zechstein and Related Topics* (eds G. M. Harwood and D. B. Smith), Special Publication, Geological Society of London, pp. 55–62.

Tull, S. J. (1988) Conodont micropalaeontology of the Morris Bugt Group (Middle Ordovician–Early Silurian), North Greenland. Unpublishad PhD Thesis, University of Nottingham, England.

Uyeno, T. T. (1981 [1980]) Stratigraphy and conodonts of Upper Silurian and Lower Devonian rocks in the environs of the Boothia Uplift, Canadian Arctic Archipelago. Part II: Systematic study of conodonts. *Geological Survey of Canada Bulletin*, **292**, 39–75.

van Wamel, W. A. (1974) Conodont biostratigraphy of the Upper Cambrian and Lower Ordovician of northwestern Öland, southeastern Sweden. *Utrecht Micropaleontological Bulletins*, **10**, 1–126.

von Bitter, P. H. (1972) Environmental control of conodont distribution in the Shawnee Group (Upper Pennsylvanian) of eastern Kansas. *The University of Kansas Paleontological Contributions*, Article **59**, 1–105.

von Bitter, P. H. and Merrill, G. K. (1977) Neogondolelliform conodonts of Early and Middle Pennsylvanian age. *Royal Ontario Museum Life Sciences Occasional Paper*, **29**, 1–10.

von Bitter, P. H. and Merrill, G. K. (1983) Late Palaeozoic species of *Ellisonia* (Conodontophorida) evolutionary and palaeoecological significance. *Royal Ontario Museum Life Sciences Contributions*, **136**, 1–57.

von Bitter, P. H. and Plint-Geberl, H. A. (1982) Conodont biostratigraphy of the Codroy Group (Lower Carboniferous), southwestern Newfoundland, Canada. *Canadian Journal of Earth Sciences*, **19**, 193–221.

Walliser, O. H. (1964) Conodonten des Silurs. *Abhandlungen des Hessischen Landesamtes für Bodenforschung*, **41**, 1–106.

Webers, G. (1966) The Middle and Upper Ordovician conodont faunas of Minnesota. *Special Publication Series Minnesota Geological Survey*, **SP-4**, 123 pp.

Wirth, M. (1967) Zur Gliederung des höheren Paläozoikums (Givet-Namur) im Gebiet des Quinto Real (Westpyrenäen) mit Hilfe von Conodonten. *Neues Jahrbuch für Geologie und Paläontologie, Abhandlungen*, **127**, 179–244.

Ziegler, W. (1962) Taxionomie und Phylogenie oberdevonischer Conodonten und ihre stratigraphische Bedeutung. *Abhandlungen des Hessischen Landesamtes für Bodenforschung*, **38**, 166 pp.

Ziegler, W. (1965) Zum höchsten Mitteldevon an der Nordflanke des Ebbesattels. *Fortschritte in der Geologie von Rheinland und Westfälen*, **9**, 519–38.

Ziegler, W. (1969) Eine neue Conodontenfauna aus dem höchsten Oberdevon. *Fortschritte in der Geologie von Rheinland und Westfälen*, **17**, 343–60.

Ziegler, W. and Sandberg, C. A. (1984) *Palmatolepis*-based revision of upper part of standard Late Devonian conodont zonation. *Geological Society of America Special Paper*, **196**, 179–94.

31

AGNATHA

L. B. Halstead*

The Agnatha are frequently divided into two groups – the armoured fossil forms and the naked living cyclostomes, the latter until 1968 without a fossil record. Stensiö (1927), however, established the close affinity between the lampreys (Petromyzonida) and cephalaspids (Osteostraci) and anaspids. His contention that the hagfish (Myxini) and pteraspids (Heterostraci) were similarly closely related has not met with such ready acceptance, although it was acknowledged that the lampreys and hags were very distantly related. The Cyclostomata were considered to be an unnatural grouping, resulting from convergent evolution. Work during the 1980s now indicates they are, after all, a natural group, and the classification used here reflects this (Schaeffer and Thomson, 1980; Yalden, 1985). Discoveries of new fossils in China, Australia, and South America in the 1970s and 1980s have added enormously to our understanding, while at the same time necessitating a re-evaluation of the classification. Details of the internal anatomy, coupled with the microstructure of the bony armour, indicate that the Agnatha belong to two distinct classes: the Diplorhina which includes the Heterostraci and Thelodonti, and the Monorhina comprising four subclasses, Osteostraci (cephalaspids), Anaspida, Galeaspida and Cyclostomata (lampreys and hagfish). Some of the new material cannot be assigned to any subclass, and one subclass, the Astraspida, can no longer be linked to either of the established agnathan classes. General references on the ranges and significance of Siluro – Devonian agnathans are Halstead (1966, 1973, 1985a,b) and Novitskaya (1983, 1986).

Acknowledgement The editor thanks Professor David L. Dineley for his considerable help in checking this chapter, and Mrs Jane Hawker for retyping Bev's rather eccentric MS.

Superclass AGNATHA Haeckel, 1895

Class DIPLORHINA Haeckel, 1895 (PTERASPIDOMORPHI Goodrich, 1909)

Subclass HETEROSTRACI Lankester, 1868 (PTERASPIDES) (see Fig. 31.1)

Order ARANDASPIDIFORMES Ritchie and Gilbert-Tomlinson, 1977

F. ARANDASPIDIDAE Ritchie and Gilbert-Tomlinson, 1977 O. (CRD) Mar.

First and Last: *Arandaspis prionotolepis* Ritchie and Gilbert-Tomlinson, 1977, Stairway Sandstone, Northern Territory, Australia; *Saccambaspis janvieri* Gagnier, 1987, Anzaldo Formation, Sacabamba, Bolivia, previously dated as CRD, was then placed in LLN (Gagnier, 1989), but now again in CRD (Elliott *et al.*, 1991).

Order ERIPTYCHIFORMES Tarlo, 1962

F. ERIPTYCHIIDAE Tarlo, 1962 O. (CRD) Mar.

First and Last: *Eryptychius americanus* Walcott, 1892, Harding Sandstone, Canyon City, Colorado, USA; *Eriptychius orvigi* Denison, 1967, Bear Rocks, Bighorn Mountains, Sheridan County, Wyoming, USA.

F. POROPHORASPIDAE **fam. nov.** O. (LLN) Mar.

First and Last: *Porophoraspis crenulata* Ritchie and Gilbert-Tomlinson, 1977, Stairway Sandstone, Northern Territory, Australia.

F. ASERASPIDIDAE **fam. nov.** S. (PRD) FW

First and Last: *Aseraspis canadensis* Dineley and Loeffler, 1976, Delorme Formation, District of Mackenzie, North-west Territories, Canada.

F. LEPIDASPIDIDAE **nov**. S. (PRD) FW

First and Last: *Lepidaspis serrata* Dineley and Loeffler, 1976, Delorme Formation, District of Mackenzie, North-west Territories, Canada.

F. TESSERASPIDIDAE Berg, 1940 S. (PRD)–D. (PRA) FW

First: *Tesseraspis denisoni* Halstead Tarlo, 1965, Beaver River, south-east Yukon, Canada.
Last: *Tesseraspis tessellata* Wills, 1935, Ditton Group, Clee Hills, Shropshire, England, UK.

Order CYATHASPIDIFORMES Kiaer and Heintz, 1935

*This may be one of Bev's last papers: he died tragically, soon after submitting it. [Editor]

The Fossil Record 2. Edited by M. J. Benton. Published in 1993 by Chapman & Hall, London. ISBN 0 412 39380 8

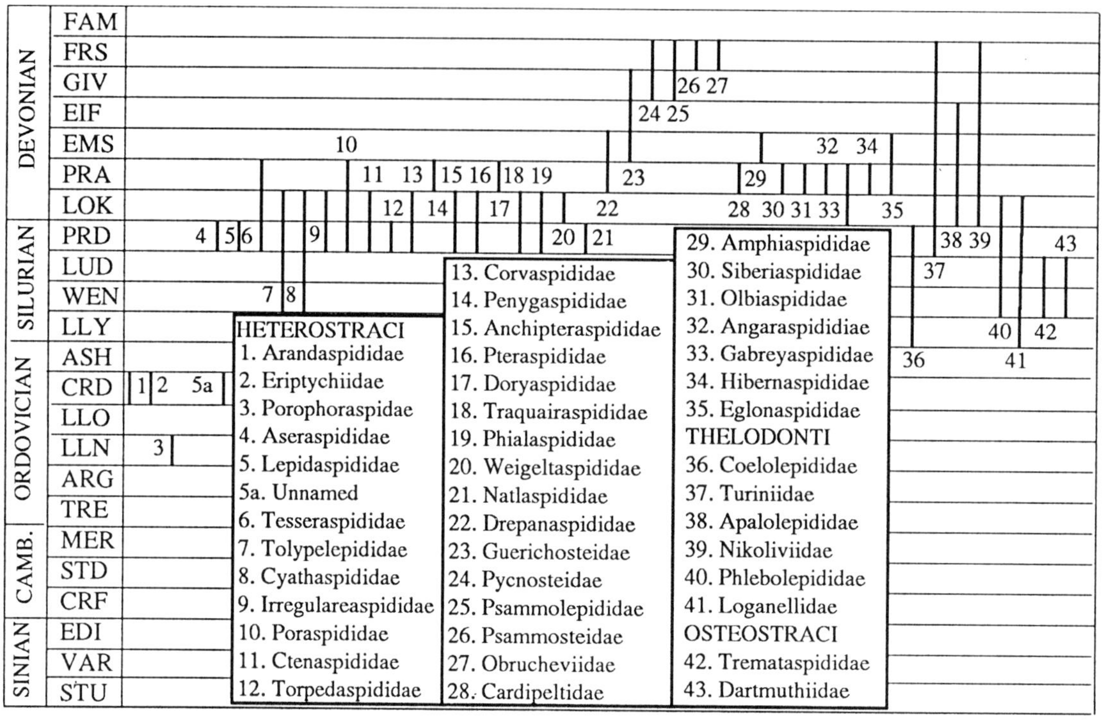

DEVONIAN: FAM, FRS, GIV, EIF, EMS, PRA, LOK
SILURIAN: PRD, LUD, WEN, LLY
ORDOVICIAN: ASH, CRD, LLO, LLN, ARG, TRE
CAMB.: MER, STD, CRF
SINIAN: EDI, VAR, STU

HETEROSTRACI
1. Arandaspididae
2. Eriptychiidae
3. Porophoraspidae
4. Aseraspididae
5. Lepidaspididae
5a. Unnamed
6. Tesseraspididae
7. Tolypelepididae
8. Cyathaspididae
9. Irregulareaspididae
10. Poraspididae
11. Ctenaspididae
12. Torpedaspididae
13. Corvaspididae
14. Penygaspididae
15. Anchipteraspididae
16. Pteraspididae
17. Doryaspididae
18. Traquairaspididae
19. Phialaspididae
20. Weigeltaspididae
21. Natlaspididae
22. Drepanaspididae
23. Guerichosteidae
24. Pycnosteidae
25. Psammolepididae
26. Psammosteidae
27. Obrucheviidae
28. Cardipeltidae
29. Amphiaspididae
30. Siberiaspididae
31. Olbiaspididae
32. Angaraspididiae
33. Gabreyaspididae
34. Hibernaspididae
35. Eglonaspididae
THELODONTI
36. Coelolepididae
37. Turiniidae
38. Apalolepididae
39. Nikoliviidae
40. Phlebolepididae
41. Loganellidae
OSTEOSTRACI
42. Tremataspididae
43. Dartmuthiidae

Fig. 31.1

F. TOLYPELEPIDIDAE Strand, 1934
S. (WEN)–D. (LOK) Mar./FW

First: *Tolypelepis* sp. *nov.*, Allen Bay Formation, Shellabear Creek, Cornwallis Island, North-west Territories, Canada. This material, noted by Thorsteinsson (1958), is still to be described.
Last: *Tolypelepis timanica* Kossovoi and Obruchev, 1962, Eptarmenskaya Beds, Velikaya River, Timan, former USSR.

F. CYATHASPIDIDAE Kiaer, 1932
S. (WEN)–D. (LOK) Mar./FW

First: 'Cyathaspid n. gen. A. sp. B' Denison, 1964, Read Bay Group, Cornwallis Island, North-west Territories, Canada.
Last: *Seretaspis zychi* Stensiö, 1958, Czortkov Series, Lochkovian, Podolia, Ukraine; *Steinaspis miroshnikovi* Obruchev, 1964, lower Zubova Horizon, Norilsk, Nats Okrug, Knasnoyarsky Territory, Siberia, former USSR.

F. IRREGULAREASPIDIDAE Denison, 1964
S. (PRD)–D. (LOK) Mar./FW

First: *Dikenaspis yukonensis* Denison, 1963, Limestone and graptolitic shales, Beaver River, SE Yukon, Canada; *Nahanniaspis mackenziei* Dineley and Loeffler, 1976, Delorme Formation, Mackenzie District, North-west Territories, Canada.
Last: *Irregulareaspis stensioei* Zych, 1931, Czortkow Stage, Lochkovian, Podolia, Ukraine, former USSR; *I. hoeli* (Kiaer, 1932), Red Bay Series, Ben Nevis Formation, Ben Nevis, Spitsbergen.

F. PORASPIDIDAE Kiaer, 1932
S. (PRD)–D. (PRA) Mar./FW

First: *Americaspis americana* (Claypole, 1884), Landisburg Sandstone Member, Wills Creek Formation, Landisburg, Perry County, Pennsylvania, USA; *A. claypolei* Denison, 1964, Longwood Shale, Shin Hollow, Orange County, New York, USA.
Last: *Allocryptaspis elliptica* (Bryant, 1934), Beartooth Butte Formation, Beartooth Butte, Park County, Wyoming, USA; *A. laticostata* Denison, 1960, Holland Quarry, Lucas County, Ohio, USA.
Comment: This family is abundant in Lochkovian in Podolia, Ukraine, Spitsbergen, England and Wales, Canadian Arctic.

F. CTENASPIDIDAE Kiaer, 1930
S. (PRD)–D. (LOK) Mar./FW

First: '*Ctenaspis* n. sp. aff. *dentatus*' Thorsteinsson, 1958, Snowblind Bay Formation (Lochkovian-Pragian), Read Bay, Cornwallis Island, North-west Territories, Canada.
Last: *Ctenaspis dentata* Kiaer, 1930, Ben Nevis Formation, Red Bay Series, Ben Nevis, Spitsbergen; *Ctenaspis obruchevi* Dineley, 1976, Upper Member, Peel Sound Formation (Lochkovian), Prince of Wales Island, North-west Territories, Canada.

F. TORPEDASPIDIDAE **fam. nov.** S. (PRD) FW

First and Last: *Torpedaspis elongata* Broad and Dineley, 1973, Somerset Island and Peel Sound Formations (Pridolian–Lochkovian), Somerset and Prince of Wales Islands, North-west Territories, Canada.

F. CORVASPIDIDAE Dineley, 1953
S. (PRD)–D. (LOK) Mar./FW

First: *Corveolepis arctica* (Loeffler and Dineley, 1976), Peel Sound Formation (Lochkovian), Pressure Point, Somerset Island, North-west Territories, Canada.
Last: *Corvaspis graticulata* Dineley, 1953, Ben Nevis Formation, Red Bay Series, Ben Nevis, Spitsbergen.

Order PTERASPIDIFORMES Berg, 1940

Range data on pteraspidiform heterostracans is derived in part from Blieck (1984).

F. PENYGASPIDIDAE **fam. nov.** D. (PRA) FW

First and Last: *Penygaspis dixoni* (White, 1938), Senni Beds Formation, Breconian, Pen-y-gau Farm, Dyfed, South Wales, UK.

F. ANCHIPTERASPIDIDAE Elliott, 1984
S. (PRD)–D. (LOK) Mar./FW

First: *Ulititaspis aquilonia* Elliott, 1984, Somerset Island Formation (Pridolian), Somerset Island, North-west Territories, Canada.
Last: *Anchipteraspis crenulata* Elliott, 1984, Peel Sound Formation, Somerset Island, North-west Territories, Canada.

F. PTERASPIDIDAE Claypole, 1885
S. (PRD)–D. (EIF) FW

First: *Protopteraspis arctica* Elliott and Dineley, 1983, Upper Somerset Island Formation, Somerset Island, North-west Territories, Canada.
Last: Pteraspidomorphi Føyn and Heintz, 1943, Widje Bay Series, Andredalen, Spitsbergen.

F. DORYASPIDIDAE N. Heintz, 1963
D. (PRA) FW

First and Last: *Doryaspis nathorsti* (Lankester, 1884), Keltiefjellet (Lykta) Division, Wood Bay Series, Spitsbergen.

Order PHIALASPIDIFORMES Berg, 1955

F. TRAQUAIRASPIDIDAE Kiaer, 1932
S. (PRD)–D. (LOK) FW

First: *Traquairaspis campbelli* (Traquair, 1913), '*Dictyocaris* band', Cowie Harbour, Stonehaven, Kincardineshire, Scotland, UK (syn. *Phialaspis pococki* var. *cowiensis* White, 1946).
Last: *Traquairaspis pococki* (White, 1946), *Psammosteus* Limestone Group, Dittonian, Gardener's Bank, Shropshire; Lower Cusop, Herefordshire; Joan's Hole, Worcestershire; Sharpness, Gloucestershire, England, UK.

F. PHIALASPIDIDAE White, 1946
S. (PRD)–D. (LOK) FW

First: *Phialaspis symondsi* (Lankester, 1866), Red Downton Formation, Lye Brook, Morville, Shropshire, England, UK.
Last: *Phialaspis symondsi* (Lankester, 1866), *Pteraspis crouchi* Zone, Ditton Series, Targrove Quarry, Whitbatch, Shropshire, England, UK. *Phialaspis symondsi* is common in the *Psammosteus* Limestone Group, Westhope Hill, Herefordshire; Earnstrey Brook, Shropshire; Holbeache, Worcestershire, England, UK; Crwcws Wood, Brecknockshire; Caldey Island, Pembrokeshire, Wales, UK; and Knoydart Formation, McArras Brook, Antigonish County, Nova Scotia, Canada.

F. WEIGELTASPIDIDAE Brotzen, 1933
D. (LOK) FW

First and Last: *Weigeltaspis alta* Brotzen, 1933, *W. brotzeni* Halstead Tarlo, 1965, Stage I, *Podolaspis lerichei* Zone, Usciezko, Dniestr, Podolia, Ukraine, former USSR; *W. godmani* Halstead Tarlo, 1965, *Pteraspis crouchi* Zone, Ditton Series, Castle Mattock Quarry, Clodock, Herefordshire, England, UK; *W. heintzi* Halstead Tarlo, 1965, Red Bay Series, Ben Nevis Division, Ben Nevis, Spitsbergen.

F. NATLASPIDIDAE **fam. *nov.*** S. (PRD) FW

First and Last: *Natlaspis planicosta* Dineley and Loeffler, 1976, *N. adunata* Dineley and Loeffler, 1976, *Lankesteraspis mackenziensis* (Dineley and Loeffler, 1976), *L. lemniscata* (Dineley and Loeffler, 1976), *L. poolei* (Dineley and Loeffler, 1976), Delorme Formation, Mackenzie District, North-west Territories, Canada.

Order PSAMMOSTEIFORMES Berg, 1940

F. DREPANASPIDIDAE Traquair, 1899
D. (PRA–EMS) Mar./FW

First: *Drepanaspis carteri* (M'Coy, 1851), *Drepanaspis edwardsi* Halstead Tarlo, 1965, *Rhinopteraspis cornubica* Zone, Dartmouth Slates, Lantivit Bay, Polperro, Cornwall, England, UK.
Last: *Drepanaspis lipperti* Gross, 1937, Klerferschichten, Zweifelscheid, Rhineland, Germany.

F. GUERICHOSTEIDAE Halstead Tarlo, 1965
D. (EMS–GIV) FW/Mar.

First: *Guerichosteus kozlowskii* Halstead Tarlo, 1965, *Hariosteus kielanae* Halstead Tarlo, 1965, Placoderm Sandstone, Daleszyce, near Kielce, Holy Cross Mountains, Poland.
Last: *Schizosteus asatkini* Obruchev, 1940, *Pycnosteus palaeformis* Zone, Arukula Horizon, Luga River, near Leningrad, former USSR. *Schizosteus perneri* (Ruzicka, 1929), Chotek Limestone, Horizon g, Holin Hlubocepy, near Prague, Czechoslovakia, is the only psammosteid of EIF age.

F. PYCNOSTEIDAE Tarlo, 1962
D. (GIV–FRS) FW

First: *Pycnolepis splendens* (Eichwald, 1844), *Schizosteus striatus* Zone, Narowa Horizon, River Slawanka, NW of former USSR.
Last: *Ganosteus stellatus* Rohon, 1901, *Psammolepis paradoxa* Zone, Gauja Horizon, Eglina, Leningrad District, former USSR.

F. PSAMMOLEPIDIDAE Tarlo, 1962
D. (GIV–FRS) FW

First: *Psammolepis proia* Mark-Kurik, 1965, *Pycnosteus palaeformis* Zone, Arukula Horizon, Tamme, Estonia, former USSR; *Psammolepis groenlandica* Halstead Tarlo, 1965, Series with *Asterolepis savesoderberghi*, Sydryggen, Canning Land, East Greenland.
Last: *Psammolepis undulata* (Agassiz, 1845), *Psammolepis undulata* Zone, Amata Horizon, Kuke, Latvia, former USSR; Nairn sandstones, Upper Old Red Sandstone, Kingsteps Quarry, Nairn, Scotland, UK.

F. PSAMMOSTEIDAE Traquair, 1896 D. (FRS) FW

First: *Psammosteus praecursor* Obruchev, 1947, *Psammolepis undulata* Zone, Amata Horizon, Yam-Tesov, River Oredesch, NW Russia, former USSR; *Psammosteus markae* Tarlo, 1961, *Psammolepis undulata* Zone, Amata Horizon, Vastseliina, Latvia, former USSR.
Last: *Psammosteus falcatus* Gross, 1942, *Psammosteus falcatus* Zone, Horizon e, Jurenski, River Pededze, Latvia; Scaat Craig Beds, Upper Old Red Sandstone, Scaat Craig, Elgin, Scotland, UK.

F. OBRUCHEVIIDAE Halstead Tarlo, 1965
D. (FRS) FW

First and Last: *Obruchevia heckeri* (Obruchev, 1936), *Psammosteus falcatus* Zone, River Lovat, NW Russia, former

USSR; *Traquairosteus pustulatus* (Traquair, 1897), Scaat Craig Beds, Upper Old Red Sandstone, Scaat Craig, Elgin, Scotland, UK.

Order CARDIPELTIFORMES Tarlo, 1962

F. CARDIPELTIDAE Bryant, 1933 D. (PRA) FW

First and Last: *Cardipeltis wallacii* Branson and Mehl, 1931, Water Canyon Formation, Blacksmith Fork and Cottonwood Canyons, Cache County, Utah, Beartooth Butte Formation, Beartooth Butte, Park County, Wyoming, USA, *C. bryanti* Denison, 1966, and *C. richardsoni* Denison, 1966 from same localities.

Order AMPHIASPIDIFORMES Berg, 1940

F. AMPHIASPIDIDAE Obruchev, 1939 D. (EMS) FW

First and Last: *Amphiaspis argos* Obruchev, 1939, Razvedochinski Horizon, left bank of River Kureyka, tributary of River Yenissei, NW Siberia, former USSR.

F. SIBERIASPIDIDAE Novitskaya, 1968 D. (PRA) FW

First: *Aphataspis kiaeri* Obruchev, 1964, *Putoranaspis prima* Obruchev, 1964, Lower Kureyka Formation, left bank of River Kureyka, NW Siberia, former USSR, *Boothisaspis ovata* Broad, 1973, identified by Broad (1973) as a siberiaspid, Peel Sound Formation (GIV), Prince of Wales Island, North-west Territories, Canada, is here identified as a cyathaspid.
Last: *Argyriaspis tcherkesovae* Novitskaya, 1971, Urumski sloj, Ust-Tareya Series, River Tareya, central Taimyr, NW Siberia, former USSR.

F. OLBIASPIDIDAE Obruchev, 1964 D. (PRA) FW

First: *Kureykaspis salebrosa* Novitskaya, 1968, Lower Kureyka Formation, left bank of River Kureyka, NW Siberia, former USSR. '?Olbiaspididae genus indet.' of Broad (1973), Peel Sound Formation (GIV), Somerset Island, North-west Territories, Canada, is here identified as a cyathaspid.
Last: *Gerronaspis dentata* (Obruchev, 1964), Middle Kureyka Formation, left bank of River Kureyka, NW Siberia: *Gerronaspis elgamarkae* n.sp. (Mark-Kurik, 1974), Rybnaya River Formation, River Ryasakh-Kanga, Kotelney Island, New Siberian Islands, former USSR.

F. ANGARASPIDIDAE **fam. nov.** D. (PRA) FW

First and Last: *Angaraspis urvantzevi* Obruchev, 1964, Middle Kureyka Formation, left bank of River Kureyka, Kureyski Horizon, NW Siberia, former USSR.

F. GABREYASPIDIDAE Novitskaya, 1968 D. (LOK–PRA) FW

First: *Prosarctaspis taimyrica* Novitskaya, 1968, Belokamenski sloj, Ust-Tareya Horizon (Lochkovian–Pragian), River Tareya, central Taimyr, NW Siberia, former USSR.
Last: *Gabreyaspis tarda* Novitskaya, 1968 and *Pelaspis teres* Novitskaya, 1971, Uryum Formation, Ust-Tareyski Horizon (Lochkovian–Pragian), River Tareya, central Taimyr, NW Siberia, former USSR.

F. HIBERNASPIDIDAE Obruchev, 1939 D. (PRA) FW

First and Last: *Hibernaspis macrolepis* Obruchev, 1939 and *Edaphaspis* Novitskaya, 1968, upper Kureyski Horizon, left bank of River Kureyka, NW Siberia, former USSR.

F. EGLONASPIDIDAE Tarlo, 1962 D. (PRA–EMS) FW

First: *Lecaniaspis lata* Novitskaya, 1971, middle Kureyski Horizon, left bank of River Kureyka, NW Siberia, former USSR.
Last: *Pelurgaspis macrorhyncha* Obruchev, 1964, Razvedochinski Horizon, left bank of River Kureyka, near River Nijny, NW Siberia, former USSR.

Subclass THELODONTI (COELOLEPIDES)

Order THELODONTIDA

Range data on thelodontids is based on Turner (1976), and Turner and Dring (1981).

F. COELOLEPIDIDAE Pander, 1856 S. (LLY–PRD) Mar.

First: *Thelodus parvidens* Agassiz, 1839, *Petalocrinus* Limestone (Telychian), Littlehope, Welsh Borderland, England, UK.
Last: *Thelodus parvidens* Agassiz, 1839, Lower Red Downton Group, Herefordshire, England, UK.

F. TURINIIDAE Obruchev, 1964 S. (PRD)–D. (FRS) Mar./FW

First: *Turinia pagei* (Powrie, 1870), upper Red Downton Group, Beaconhill Brook, Herefordshire, England, UK; present throughout Lower–Upper Ditton Group (Lochkovian) in Welsh Borderland, England, UK.
Last: *Australolepis seddoni* Turner and Dring, 1981, Gneudna Formation, Carnarvon Basin, Western Australia; *Turinia pagoda* Wang *et al.*, 1986, Heyuanzhai Formation (GIV), western Yunnan, China.

F. APALOLEPIDIDAE Turner, 1976 D. (LOK–EIF) FW

First: *Apalolepis obruchevi* Karatajute-Talimaa, 1968, Czortkow Stage, Dittonian (Lochkovian), Podolia, Ukraine, former USSR.
Last: *Skamolepis* sp., Turner and Janvier, 1979, Khush-Yeilagh Formation, NE Iran.

F. NIKOLIVIIDAE Karatajute-Talimaa, 1978 D. (LOK–FRS) FW

First: *Nikolivia* sp. Karatajute-Talimaa, 1978, *Pteraspis leathensis* Zone, '*Psammosteus* Limestone Group' (Lochkovian), England and Wales, UK.
Last: Nikoliviid gen. *et* sp. indet., Turner and Dring, 1981, Gneudna Formation, Carnarvon Basin, Western Australia.

Order PHLEBOLEPIDIFORMES Berg, 1940 (syn. KATOPORIDA)

F. PHLEBOLEPIDIDAE Berg, 1940 S. (WEN)–D. (PRA) Mar./FW

First: *Phlebolepis elegans* Pander, 1856, K1 Horizon, Wenlock, Oesel (Saarema), Estonia, former USSR (previously considered LUD).
Last: *Katoporodus grossi* (Karatajute-Talimaa, 1970) and *Gonioporus alatus* (Gross, 1947), *Althaspis leachi* Zone, Ditton Group (Lochkovian–Pragian), Cwm Mill, Abergavenny, Wales, UK.

F. LOGANELLIDAE **fam. nov.** S. (LLY)–D. (LOK) Mar./FW

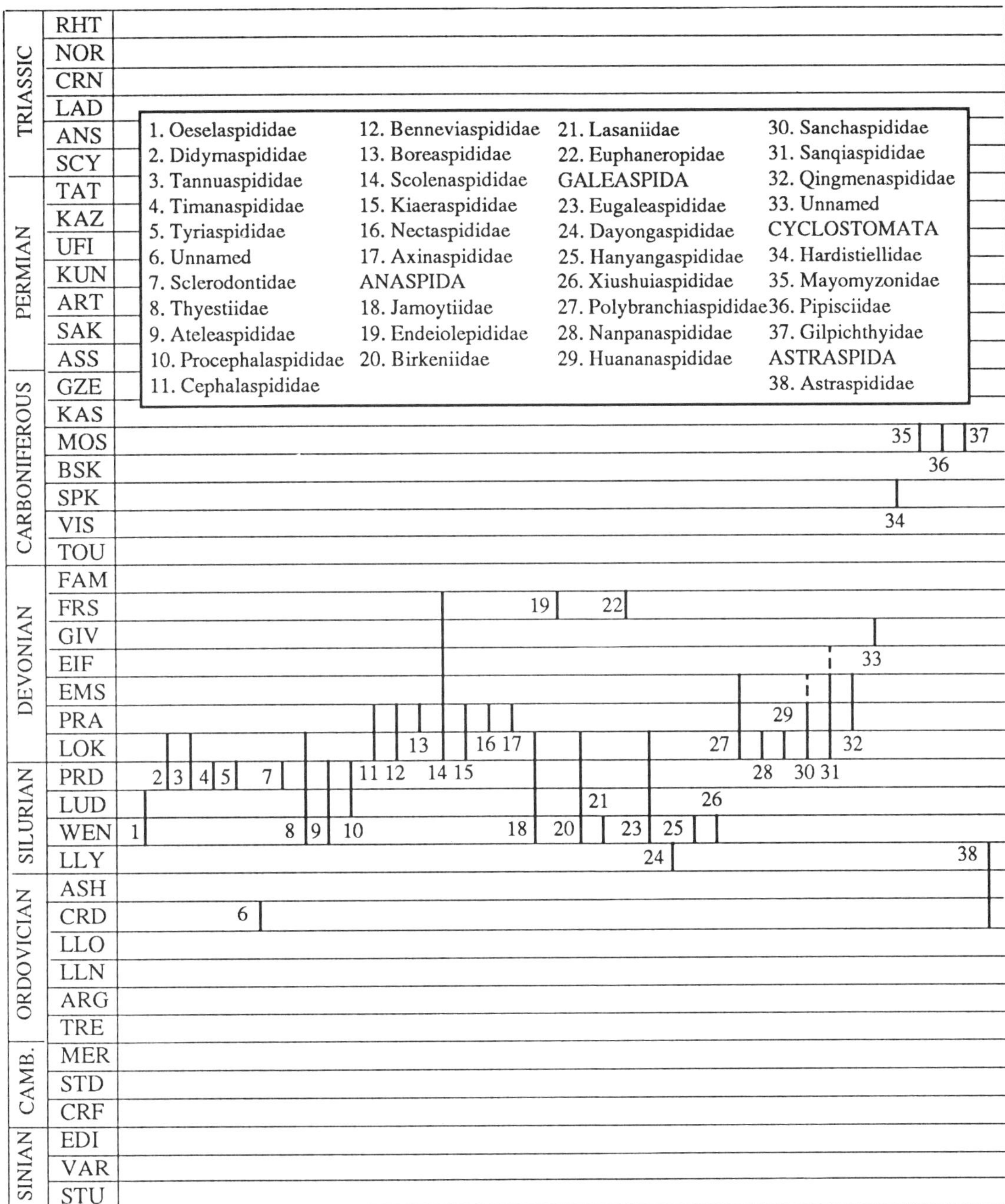

Fig. 31.2

First: *Loganella asiaticus* (Karatajute-Talimaa, 1978), Kizilchrinskie Division, Tuva, Siberia, former USSR; *Loganella scotica* (Traquair, 1898), Venusbank Formation, Hope Quarry, Shropshire, Telychian Wych Beds, Gullet Quarry, Malvern Hills, England, UK.
Last: *Loganella cuneata* (Gross, 1947), Beyrichienkalk erratics, Germany; *Pteraspis leathensis* Zone, '*Psammosteus* Limestone Group', Ross Motorway and Gardiner's Bank, Herefordshire, England, UK.

Class MONORHINA Haeckel, 1895

Subclass OSTEOSTRACI Lankester, 1868 (CEPHALASPIDES)

Order TREMATASPIDIFORMES Halstead Tarlo, 1967

F. TREMATASPIDIDAE Woodward, 1891 S. (WEN–LUD) Mar.

First: *Tremataspis schmidtii* Rohon, 1892, Rootsiküla Formation K_1, Horizon, Oesel (Saaremaa), Estonia, former USSR.
Last: *Tremataspis mammillata* Patten, 1931, *T. milleri* Patten, 1931, and *T. rohoni* Robertson, 1938, Paadla Formation, K_2 Horizon, Oesel (Saaremaa), Estonia, former USSR.

F. DARTMUTHIIDAE Robertson, 1935 S. (WEN–LUD) Mar.

First: *Saaremaaspis mickwitzi* (Rohon, 1892), Rootisküla Formation, K_1 Horizon, Oesel (Saaremaa), Estonia, former USSR.
Last: *Dartmuthia gemmifera* Patten, 1931, Paadla Formation, K_2 Horizon, Oesel (Saaremaa), Estonia, former USSR.

F. OESELASPIDIDAE Robertson, 1935 S. (WEN–LUD) Mar. (see Fig. 31.2)

First: *Oeselaspis pustulata* (Patten, 1931), Rootsiküla Formation, K_1 Horizon, Oesel (Saaremaa), Estonia, former USSR.
Last: *Oeselaspis pustulata* Patten, 1931, Paadla Formation, Oesel (Saaremaa), Estonia, former USSR.

F. DIDYMASPIDIDAE Berg, 1940 S. (PRD)–D. (LOK) Mar./FW

First: *Didymaspis grindrodi* Lankester, 1867, Lower Red Downton Formation, Ledbury and Bush Pitch, Herefordshire, England, UK.
Last: *Didymaspis grindrodi* Lankester, 1867, '*Psammosteus* Limestone Group', Gardener's Bank, Shropshire, Lower Cusop, Herefordshire, England, UK.

F. TANNUASPIDIDAE Obruchev, 1964 S. (PRD)–D. (LOK) Mar./FW

First: *Tuvaspis margaritae* Obruchev, 1956, Samagaltayskaya Group, Kizil, near Lake Khadin, Tuva, former USSR.
Last: *Tannuaspis levenkoi* Obruchev, 1956, Kendeyskaya Group, Kutuk Valley, Tannu-Ola Mountains, Tuva, former USSR.

F. TIMANASPIDIDAE **fam. nov.** S. (PRD) Mar.

First and Last: *Timanaspis kossovoii* Obruchev, 1962, Eptarminskaya Group, Velichkaya River, northern Timan, former USSR.

F. TYRIASPIDIDAE **fam. nov.** S. (PRD) Mar.

First and Last: *Tyriaspis whitei* Heintz, 1967, 'Downtonian' Sandstones, Tyrifjorden, Ringerike, Norway.

F. UNNAMED O. (CRD) Mar.

First and Last: Osteostracan fragments showing cellular dentine (mesodentine) and cellular bone, Harding Sandstone Formation, Canyon City, Colorado, USA (Smith, 1991).

Order SCLERODONTIFORMES

F. SCLERODONTIDAE Fowler, 1947 S. (PRD) Mar.

First and Last: *Sclerodus pustulliferus* Agassiz, 1839, Ludlow Bone Bed, Ludlow, Shropshire; Downton Bridge and Kington, Worcester and Herefordshire; Baggeridge Colliery, Staffordshire, England, UK.

Order THYESTIIFORMES

F. THYESTIIDAE Berg, 1940 S. (WEN)–D. (LOK) Mar.

First: *Thyestes verrucosus* Eichwald, 1854, Rootsiküla Formation, K_1, Horizon, Oesel (Saaremaa), Estonia, former USSR; Hall Formation, Gotland, Sweden.
Last: *Fieldingaspis egertoni* Lankester, 1870, Lower Red Downton Formation (Lochkovian), Ledbury, England, UK.

Order ATELEASPIDIFORMES

F. ATELEASPIDIDAE Traquair, 1899 S. (WEN–PRD) Mar.

First: *Ateleaspis tesselata* Traquair, 1899, Slot Burn, Birkenhead Burn, Lesmahagow Hills, Lanarkshire; Hagshaw Hills, Ayrshire, Scotland, UK; *Witaaspis schrenkii* (Pander, 1856), Rootsiküla Formation, K_1 Horizon, Oesel (Saaremaa), Estonia, former USSR.
Last: *Hemicyclaspis murchisoni* (Egerton, 1857), Lower Red Downton Formation (Pridolian), Ledbury, Herefordshire; Baggeridge Colliery, Staffordshire, England, UK; *Ateleaspis robustus* (Kiaer, 1911) [*Hirella gracilis* (Kiaer, 1911) = immature *A. robustus*], 'Downtonian' Sandstones (?Ludlovian), Rudstangen, Ringerike, Norway.

Order CEPHALASPIDIFORMES Halstead Tarlo, 1967

F. PROCEPHALASPIDIDAE Stensiö, 1958 S. (LUD–PRD) Mar.

First: *Procephalaspis oeselensis* (Robertson, 1939), Paadla Formation, K_2 Horizon, Oesel (Saaremaa), Estonia, former USSR.
Last: *Auchenaspis salteri* Egerton, 1857, Lower Red Downton Formation, Ludlow, Shropshire, England, UK.

F. CEPHALASPIDIDAE Agassiz, 1843 D. (LOK–PRA) FW

First: *Cephalaspis lyelli* Agassiz, 1835, Lower Old Red Sandstone, Glamis, Forfarshire, Scotland, UK.
Last: *Meteoraspis moythomasi* (Wangsjö, 1953) and *M. semicircularis* (Wangsjö, 1952), Stjordalen Division, Wood Bay Formation, Stjordalen Valley, Spitsbergen.

Order BENNEVIASPIDIFORMES

F. BENNEVIASPIDIDAE Denison, 1952 D. (LOK–PRA) FW

First: *Benneviaspis holtedahli* Stensiö, 1927, Ben Nevis Division, Red Bay Formation, Horizon J, L and Cliff at Ben Nevis, Spitsbergen; *B. lankesteri* Stensiö, 1932, *Pteraspis crouchi* Zone, Ditton Series, Cradley, Worcester and Herefordshire, England, UK; *B. anglica* Stensiö, 1932, *Pteraspis crouchi* Zone, Ditton Series, Leverhill and Cradley, Herefordshire, England, UK.
Last: '*Benneviaspis* sp.' (Wängsjö, 1952). Stjordalen Division, Wood Bay Formation, Widje Bay, between Jäderin and Zeipel Valleys, Spitsbergen.

F. BOREASPIDIDAE Stensiö, 1958 D. (PRA) FW

First: '*Boreaspis*' *ceratops* Wängsjö, 1953, '*B*'. *batoides* Wängsjö, 1952, and '*B*'. *ginsbugi* Janvier, 1977, Sigurdfjellet Division, Wood Bay Series, Mt. Sigurd, Wood Bay, Spitsbergen.
Last: *Dicranaspis curtirostris* (Wängsjö, 1952), Lykta Division, Wood Bay Formation, Dickson Bay, Mt. Triplex, Mt. Lykta, Mt. Errol White and *Spatulaspis costata* (Wängsjö, 1952), Lykta Division, Wood Bay Formation, Dickson Bay, Mt. Rebbingen, Mt. Lykton, Spitsbergen.

Order SCOLENASPIDIFORMES

F. SCOLENASPIDIDAE Janvier, 1985 D. (LOK–FRS) FW

First: *Scolenaspis signata* (Wängsjö, 1952), Ben Nevis Formation, Red Bay Group, Mt. Ben Nevis, Red Bay, Spitsbergen.
Last: *Alaspis macrotuberculata* Ørvig, 1957, *A. rosamundae* (Robertson, 1937), and *Escuminaspis laticeps* (Traquair, 1899), Escuminac Formation, Scaumenac Bay (Miguasha Bay), Quebec, Canada.

Order KIAERASPIDIFORMES Halstead Tarlo, 1967

F. KIAERASPIDIDAE Stensiö, 1932 D. (LOK–PRA) FW

First: *Kiaeraspis auchenaspidoides* Stensiö, 1927, Ben Nevis

Division, Red Bay Group, Mt. Ben Nevis, Red Bay, Spitsbergen.
Last: *Norselaspis glacialis* Janvier, 1981, Sigurdfjellet Division, Wood Bay Formation, Sigurdfjellet, Wood Bay, Spitsbergen.

F. NECTASPIDIDAE Stensiö, 1958 D. (PRA) FW

First: *Acrotomaspis instabilis* Wängsjö, 1953, Kapp Kjeldsen Division, Wood Bay Series, Mt. Kronprinz, Wood Bay, Spitsbergen.
Last: *Gustavaspis trinodis* (Wängsjö, 1952), Stjordalen Division, Wood Bay Formation, Mt. Scott Keltie, Wood Bay, Spitsbergen.

F. AXINASPIDIDAE Janvier, 1985 D. (PRA) FW

First: *Axinaspis* sp. 1. (Janvier, 1981), Sigurdfjellet Division, Wood Bay Formation, Sigurdfjellet, Wood Bay, Spitsbergen.
Last: *Axinaspis whitei* Wängsjö, 1952, Kapp Kjeldsen Division, Wood Bay Formation, Mt. Sigurd, Bock Bay, Mt. Kronprinz, Wood Bay, Spitsbergen.

Order JAMOYTIIFORMES

F. JAMOYTIIDAE White, 1946
S. (WEN)–D. (LOK) Mar.

First: *Jamoytius kerwoodi* White, 1946, Patrick Burn Formation, Kip Burn Formation, Lesmahagow Hills, Scotland, UK.
Last: *Jamoytius*-like vertebrates (Janvier and Busch, 1984), Manlius Formation, Paris, New York, USA.

Order ENDEIOLEPIDIFORMES Berg, 1940

F. ENDEIOLEPIDIDAE Stensiö, 1939 D. (FRS) FW

First and Last: *Endeiolepis aneri* Stensiö, 1939, Escuminac Formation, Scaumenac Bay (Miguasha Bay), Quebec, Canada.

Order BIRKENIIFORMES Berg, 1940

F. BIRKENIIDAE Traquair, 1899
S. (WEN)–D. (LOK) Mar./FW

First: *Birkenia elegans* Traquair, 1899, 'Downtonian', Lesmahagow Hills, Lanarkshire, Scotland, UK.
Last: Anaspid scales (Ball *et al.*, 1961), *Phialaspis symondsi* Zone, *Psammosteus* Limestone Group, Targrove Dingle, Whitbatch, Shropshire, England, UK.

F. LASANIIDAE Abel, 1919 S. (WEN) Mar.

First and Last: *Lasanius problematicus* Traquair, 1899, Lesmahagow and Hagshaw Hills, Lanarkshire; *L. armatus* Traquair, 1899, Seggholm, Lanarkshire, Scotland, UK.

F. EUPHANEROPIDAE Woodward, 1900
D. (FRS) FW

First and Last: *Euphanerops longaevus* Woodward, 1900, Escuminac Formation, Escuminac Bay (Miguasha Bay), Quebec, Canada.

Subclass GALEASPIDA Halstead, 1982

Order EUGALEASPIDIFORMES Pan, 1983

F. EUGALEASPIDIDAE Liu, 1980
S. (WEN)–D. (LOK) FW

First: *Sinogaleaspis shankouensis* Pan and Wang, 1980, *S. xikengensis* Pan and Wang, 1980, Xikeng Formation, Xikeng, Xiushui, Jiangxi, *Eugaleaspis zhejiangensis* Pan, 1986, Maoshan Formation, Anji and Sian, Zhe-Jiang and *E. xiushanensis* Liu, 1983, Huixingshao Formation, Shuiyuantou, Xiushan, Szechwan, China.
Last: *Eugaleaspis changi* (Liu, 1965), *Yunnanogaleaspis major* Pan and Wang, 1980, Chifengshan Formation, Qujing, Yunnan, China. No record in S. (LUD, PRD).

Order POLYBRANCHIASPIDIFORMES Liu, 1965

F. DAYONGASPIDIDAE Pan, 1985 S. (LLY) FW

First and Last: *Dayongaspis hunanensis* Pan, 1985, Rongxi Formation, Wentang, Hunan, China.

F. HANYANGASPIDIDAE Pan and Liu, 1975
S. (WEN) FW

First and Last: *Hanyangaspis guodingshanensis* Pan and Liu, 1975, Goudingshan Formation, Wuhan, Hubei and *Latirostraspis chaohuensis* Wang *et al.*, 1980, Fentou Formation, Xiazhucun, Anhui, China (referred to *Hanyangaspis* by Pan, 1986).

F. XIUSHUIASPIDIDAE Pan and Wang, 1983
S. (WEN) FW

First and Last: *Xiushuiaspis jiangxiensis* Pan and Wang, 1983 and *X. ganbiensis* Pan and Wang, 1983, Xikeng, Formation, Xikeng, Xiushui, Jiangxi, China.

F. POLYBRANCHIASPIDIDAE Liu, 1965
D. (LOK–EMS) FW

First: *Polybranchiaspis miandiancunensis* Pan and Wang, 1978 and *Dongfangaspis* sp., Miandiancun Formation Qujing, Yunnan, China.
Last: *Duyunolepis paoyangensis* (Pan and Wang, 1978), Shujiapin Formation, Paoyang, Guizhou, China.

Order NANPANASPIDIFORMES Liu, 1965

F. NANPANASPIDIDAE Liu, 1975 D. (LOK) FW

First and Last: *Nanpanaspis microculus* Liu, 1965, Chifengshan Formation, Qujing, Yunnan, China.

Order HUANANASPIDIFORMES Janvier, 1975

F. HUANANASPIDIDAE Liu, 1973 D. (LOK) FW

First: *Lungmenshanaspis kiangyouensis* Pan and Wang, 1975, *Sinoszechuanaspis yanmenpaensis* (Pan and Wang, 1975), and *S. gracilis* Pan and Wang, 1978, Pingyipu Formation, Yanmenpao, Jiangyou, Szechwan, China.
Last: *Asiaspis expansa* Pan, 1975, Nagaoling Formation, Liujing, Guangxi, China.

F. SANCHASPIDIDAE Pan and Wang, 1981
D. (LOK–PRA/EMS) FW

First: *Antiquisagittaspis cornuta* Liu, 1985, Nagaoling Formation, Hengxian, Guangxi, China.
Last: *Sanchaspis magalarostra* Pan and Wang, 1981, Chifengshan Formation, Sancha, Qujing, Yunnan, China.

F. SANQIASPIDIDAE Liu, 1975
D. (LOK–EMS/EIF) FW

First: *Sanqiaspis zhaotongensis* Liu, 1975, Chifengshan Formation, Zhaotong, Yunnan, *S. rostrata* Liu, 1975, *S. sichuanensis* Pan and Wang, 1978, Pingyipu Formation, Jiangyou, Szechwan, China.

Last: *Wumengshanaspis cuntianensis* Wang and Lan, 1984, Suotoushan Formation, Yiliang, Yunnan, China.

Order QINGMENASPIDIFORMES Pan and Wang, 1981

F. QINGMENASPIDIDAE **fam. nov.** D. (PRA–EMS) FW

First and Last: *Qingmenaspis microculus* Pan and Wang, 1981, Chifengshan Formation, Xishancun, Qujing, Yunnan, China.

F. UNNAMED D. (GIV) Mar.

First and Last: Undescribed remains listed by Pan and Dineley (1988), Yidade Formation, Panxi, Huaning, Yunnan, China.
Comment: Further undescribed 'galeaspid' remains listed by Pan and Dineley (1988) from the Zhongning Formation (FRS), Zingwei, Ningxia, are not here accepted as galeaspid.

Subclass CYCLOSTOMATA

Superorder PETROMYZONIDA

Order PETROMYZONIFORMES Berg, 1940

F. HARDISTIELLIDAE **fam. nov.** C. (SPK) FW

First and Last: *Hardistiella montanensis* Janvier and Lund, 1983, Bear Gulch Member, Heath Formation, Montana, USA.

F. MAYOMYZONIDAE Bardack and Zangerl, 1971 C. (MOS) FW

First and Last: *Mayomyzon pieckoensis* Bardack and Zangerl, 1968, Francis Creek Shale, Carbondale Formation, Illinois, USA.

F. PIPISCIIDAE **fam. nov.** C. (MOS) FW

First and Last: *Pipiscius zangerli* Bardack and Richardson, 1971, Francis Creek Shale, Carbondale Formation, Illinois, USA.

F. GILPICHTHYIDAE **fam. nov.** C. (MOS) FW

First and Last: *Gilpichthys greenei* Bardack and Richardson, 1977, Francis Creek Shale, Carbondale Formation, Illinois, USA.
Comment: *Scaumenella mesacanthus* Graham-Smith, 1935, Escuminac Formation (FRS), Miguasha, Quebec, Canada, tentatively identified as a larval lamprey (Tarlo, 1960), has been shown by Béland and Arsenault (1985) to be the degraded remains of the acanthodian fish *Triazeugacanthus affinis* (Whiteaves, 1887).

Superorder MYXINI

Order MYXINIFORMES Berg, 1940

No known fossil record.

Class INDET.

Subclass ASTRASPIDA

Order ASTRASPIDIFORMES Berg, 1940

F. ASTRASPIDIDAE Eastman, 1917 O. (CRD)–S. (LLY) Mar.

First: *Astraspis desiderata* Walcott, 1982, Harding Formation, Colorado, USA, Gull River Formation, Ontario, Canada, *Pycnaspis splendens* Ørvig, 1958, previously dated as O. (ASH) now placed in O. (CRD) (Ørvig, 1989).
Last: *Tesakoviaspis concentrica* Karatajute-Talimaa, 1978, L. Chunku River, Podkamennaya, Siberia, former USSR.
Comment: The exclusion of the Astraspida from the Heterostraci is based on the possession of enameloid and astraspidin in the former in contrast to dentine and aspidin in the latter (Halstead, 1987, see also Elliott, 1987 and Blieck, 1992).

?AGNATHA *Class* INDET.

Anatolepis heintzi Bockelie and Fortey, 1976, Valhallfonna Formation, O. (ARG–LLN), Ny Friesland, Spitsbergen, attributed to the Heterostraci (figured in Smith and Hall, 1990), has an ornamentation and microstructure unlike any known vertebrate. Its vertebrate assignation is not here accepted (see also Ørvig, 1989). *Anatolepis* cf. *heintzi* Repetski, 1978, Deadwood Formation, €. (MER), Crook County, Wyoming, USA (figured in Benton, 1990) is not here recognized as a vertebrate (see also Ørvig, 1989).

REFERENCES

Ball, H. W., Dineley, D. L. and White, E. I. (1961) The Old Red Sandstone of Brown Clee Hill and the adjacent area. *Bulletin of the British Museum (Natural History), Geology Series*, **5**, 175–310.

Bardack, D. and Richardson, E. S. (1977) New agnathous fishes from the Pennsylvanian of Illinois. *Fieldiana, Geology*, **33**, 489–510.

Béland, P. and Arsenault, M. (1985) Scauménellisation de l'Acanthodii *Triazeugacanthus affinis* (Whiteaves) de la Formation d'Escuminac (Dévonien supérieur de Miguasha, Quebec): révision de *Scaumenella mesacanthi* Graham-Smith. *Canadian Journal of Earth Sciences*, **22**, 514–24.

Benton, M. J. (1990) *Vertebrate Palaeontology*, Unwin Hyman, London, 377 pp.

Blieck, A. (1984) Les Hétérostracés ptéraspidiformes, Agnatha du Silurien-Dévoniens du continent Nord-Atlantique et des blocks avoisinants. *Cahiers de Paléontologie*, 1–199.

Blieck, A. (1992) At the origin of chordates. *Geobios*, **25**, 101–13.

Broad, D. S. (1973) Amphiaspid (Heterostraci) from the Silurian of the Canadian Arctic Archipelago. *Geological Survey of Canada Bulletin*, **222**, 35–50.

Dineley, D. L. and Loeffler, E. J. (1976) Ostracoderm faunas of the Delorme and associated Siluro-Devonian Formations, North West Territories, Canada. *Special Papers in Palaeontology*, **18**, 1–214.

Elliott, D. K. (1987) A reassessment of *Astraspis desiderata*, the oldest North American vertebrate. *Science*, **237**, 190–2.

Elliot, D. K., Blieck, A. R. M. and Gagnier, P.-Y. (1991) Ordovician vertebrates, *in Advances in Ordovician Geology* (eds C. R. Barnes and S. H. Williams), Geological Survey of Canada, Paper 90–9, pp. 93–106.

Gagnier, P.-Y. (1989) The oldest vertebrate: a 470 million-year-old jawless fish, *Saccambaspis janvieri*, from the Ordovician of Bolivia. *National Geographic Research*, **5**, 250–3.

Halstead, L. B. (1973) The heterostracan fishes. *Biological Reviews*, **48**, 279–332.

Halstead, L. B. (1985a) The vertebrate invasion of freshwater. *Philosophical Transactions of the Royal Society of London, Series B*, **309**, 243–58.

Halstead, L. B. (1985b) Extinction and survival of the jawless vertebrates, the Agnatha, in *Extinction and Survival in the Fossil Record* (ed. G. P. Larwood), Clarendon Press, Oxford, pp. 257–67.

Halstead, L. B. (1987) Evolutionary aspects of neural crest-derived skeletogenic cells in the earliest vertebrates, in *Developmental and Evolutionary Aspects of the Neural Crest* (ed. P. F. A. Maderson),

Wiley Interscience, New York, pp. 339–58.

Halstead Tarlo, L. B. (1965) Psammosteiformes (Agnatha) – a review with descriptions of new material from the Lower Devonian of Poland. I. General part. *Palaeontologia Polonica*, **13**, 1–135.

Halstead Tarlo, L. B. (1966) Psammosteiformes (Agnatha) – a review with descriptions of new material from the Lower Devonian of Poland. II. Systematic part. *Palaeontologia Polonica*, **15**, 1–168.

Janvier, P. (1981) *Norselaspis glacialis* n.g., n.sp. et les rélations phylogénétiques entre les Kiaeraspidiens (Osteostraci) du Dévonien inférieur du Spitsberg. *Palaeovertebrata*, **11**, 19–131.

Janvier, P. (1985a) Les Thyestidiens (Osteostraci) du Silurien de Saaremaa (Estonie). *Annales de Paléontologie*, **71**, 83–147, 187–216.

Janvier, P. (1985b) Les Céphalaspides du Spitsberg. *Cahiers de Paléontologie*, 1–244.

Janvier, P. and Lund, R. (1983) *Hardistiella montanensis* n.gen. et sp. (Petromyzontida) from the Lower Carboniferous of Montana, with remarks on the affinities of the lampreys. *Journal of Vertebrate Paleontology*, **2**, 407–13.

Karatajute-Talimaa, V. (1978) *Silurian and Devonian Thelodonts of the USSR and Spitsbergen*. Mokslas Pub., Vilnius, Lithuania, 334 pp.

Märss, T. (1986) Silurian vertebrates of Estonia and West Latvia. *Fossilia Baltica*, **1**, 1–104.

Novitskaya, L. (1971) Les Amphiaspides (Heterostraci) du Dévonien de la Sibérie. *Cahiers de Paléontologie*, 1–130.

Novitskaya, L. (1983) Morphology of ancient jawless fishes. *Trudy Palaeontological Institute of the Academy of Sciences*, **196**, 1–183 [in Russian].

Novitskaya, L. (1986) Ancient jawless fishes of the USSR. Heterostraci: cyathaspids, amphiaspids, pteraspids. *Trudy Palaeontological Institute of the Academy of Sciences*, **219**, 1–160 [in Russian].

Ørvig, T. (1989) Histologic studies of ostracoderms, placoderms and fossil elasmobranchs. 6. Hard tissues of Ordovician vertebrates. *Zoologica Scripta*, **18**, 427–46.

Pan, J. (1986) Note on Silurian vertebrates of China. *Bulletin of the Chinese Academy of Geological Sciences*, **15**, 161–90.

Pan, J. and Dineley, D. L. (1988) A review of early (Silurian and Devonian) vertebrate biogeography and biostratigraphy of China. *Proceedings of the Royal Society of London, Series B*, **235**, 29–61.

Schaeffer, B. and Thomson, K. S. (1980) Reflections on agnathan-gnathostome relationships, in *Aspects of Vertebrate History: Essays in Honor of Edwin Harris Colbert* (ed. L. L. Jacobs), Museum of Northern Arizona Press, Flagstaff, pp. 19–33.

Smith M. M. (1991) Putative skeletal neural crest cells in early Late Ordovician Vertebrates from Colorado. *Science*, **251**, 301–3.

Smith M. M. and Hall, B. K. (1990) Development and evolutionary origins of vertebrate skeletogenic and odontogenic tissues. *Biological Reviews*, **65**, 277–373.

Stensiö, E. A. (1927) The Downtonian and Devonian Vertebrates of Spitsbergen. Part 1. Family Cephalaspidae. *Skrifter on Svalbard og Nordishavet*, **12**, 1–391.

Stensiö, E. A. (1932) *The Cephalaspids of Great Britain*. British Museum (Natural History), London, 220 pp.

Tarlo, L. B. (1960) The invertebrate origins of the vertebrates. *21st International Geological Congress, Copenhagen*, **22**, 113–22.

Thorsteinsson, R. (1958) Cornwallis and Little Cornwallis Islands, District of Franklin, North-west Territories. *Memoirs of the Geological Survey of Canada*, **294**, 1–34.

Turner, S. (1976) Thelodonti (Agnatha). *Fossilium Catalogus*, **122**, 1–35.

Turner, S. and Dring, R. S. (1981) Late Devonian thelodonts (Agnatha) from the Gneudna Formation, Carnavon Basin, Western Australia. *Alcheringa*, **5**, 39–48.

Wängsjö, G. (1952) The Downtonian and Devonian vertebrates of Spitsbergen. IX. Morphologic and systematic studies of the Spitsbergen cephalaspids. *Skrifter Norsk Polarinstitut*, **97**, 1–615.

White, E. I. (1946) The genus *Phialaspis* and the '*Psammosteus* Limestones'. *Quarterly Journal of the Geological Society of London*, **101**, 207–42.

Yalden, D. W. (1985) Feeding mechanisms as evidence for cyclostome monophyly. *Zoological Journal of the Linnean Society*, **84**, 291–300.

32

PLACODERMI

B. G. Gardiner

The placoderms flourished in the Devonian and are almost entirely restricted to this period. However, antiarch remains have been recorded from the marine Upper Silurian of the Guandi Formation in the Qujing District, eastern Yunnan Province, and in western Hunan Province (Pan and Dineley, 1988), while more primitive placoderms (possibly related to palaeacanthaspids) have been recovered from marine Middle Silurian sediments of the Yulongssu Formation, Qujing (M.-M. Chang, pers. comm.). Classification mainly after Denison (1978) and Gardiner (1990).

Order STENSIOELLIDA White, 1952 (see Fig. 32.1)

F. STENSIOELLIDAE Berg, 1940
D. (PRA) Mar.

First and Last: *Stensioella heintzi* Broili, 1933, Hunsrückschiefer, Germany. *Stensioella* is a very poorly known genus (Gröss, 1962) from a single Lower Devonian locality, and can only be associated with placoderms on phenetic grounds (Forey and Gardiner, 1986).

Order PSEUDOPETALICHTHYIDA Denison, 1975

F. PARAPLESIOBATIDAE Berg, 1940
D. (PRA) Mar.

First and Last: *Pseudopetalichthys problematicus* Moy-Thomas, 1939, Hunsrückschiefer, Germany. *Paraplesiobatus heinrichsi* Broili, 1933, and *Nessariostoma granulosum* Broili, 1933, also from the Hunsrückschiefer (Gröss, 1962) are probably both synonymous with *P. problematicus*.

Order PTYCTODONTIDA Gröss, 1952

F. PTYCTODONTIDAE Woodward, 1891
D. (PRA)–C. (TOU) Mar.

First: *Tollodus brevispinus* Mark-Kurik, 1934, shales and limestones of the Siberian Platform, Bysach-Karga and Sokolow Rivers, former USSR.
Last: *Chelyophorus verneuili* Agassiz, 1844, Dankov-Lebedyan Beds, former USSR. (Denison, 1978). *Ptyctodus calceolus* Newberry and Worthen, 1866, from the Bushberg Sandstone, North America (Branson and Mehl, 1938) appears to have been derived from earlier, Devonian strata.
Comment: This order is marine except for one genus, *Rhamphodopsis* which occurs in fresh water, Middle Old Red Sandstone deposits of Scotland, UK.

Order GEMUENDINIDA Gröss, 1963
(= RHENANIDA Broili, 1930)

F. ASTEROSTEIDAE Woodward, 1891
D. (PRA–FRS) Mar.

First: *Gemuendina stuertzi* Traquair, 1903, Hunsrückschiefer, Germany (Gröss, 1963).
Last: *Jagorina pandora* Jaekel, 1921, *Manticoceras* Beds, Germany (Jaekel, 1921; Stensiö, 1925).

Order PALAEACANTHASPIDA Obruchev, 1964

F. PALAEACANTHASPIDIDAE Stensiö, 1944
D. (LOK–PRA) Mar.

First: *Kimaspis tienshanica* Mark-Kurik, 1973, Dzhalpak Formation, southern Tien Shan, Uzbekistan, former USSR *Palaeacanthaspis vasta* Brotzen, 1934, and *Dobrowlania podolica* Stensiö, 1944, Czortkow Stage, Ukraine, former USSR (Stensiö, 1944; Boucot and Pankiwskyja, 1962), *Romundina stellina* Ørvig, 1975, Prince of Wales Island, Canada.
Last: *Breizosteus armoricensis* Goujet, 1980, calcareous shales, Armorique, France (Goujet, 1980).

F. RADOTINIDAE Arambourg, 1958
D. (LOK–PRA) Mar.

First: *Kosoraspis peckai* Gröss, 1959, and *Radotina kosorensis* Gröss, 1950, Radotin Limestone, Czechoslovakia (Gröss, 1958, 1959).
Last: *Radotina tuberculata* Gross, 1958, Taunus Quartzite, Germany (Gröss, 1937, 1958).

F. KOLYMASPIDIDAE Bystrow, 1956
D. (EMS) Mar.

First and Last: *Kolymaspis siberica* Bystrow, 1956, Neliudim Formation, Siberia, former USSR (Denison, 1978).

F. WEEJASPERASPIDAE White, 1978
D. (EMS) Mar.

First and Last: *Murrindalaspis wallacei* Long, 1988, Buchan Group, *Receptaculites* Limestone and *Weejasperaspis gavini* White, 1978, Yarssensis Limestone, Murrumbidgee Series, Australia (Long and Young, 1988).

Order BRINDABELLASPIDA

F. BRINDABELLASPIDAE **fam. nov.**
D. (EMS) Mar.

First and Last: *Brindabellaspis stensioi* Young, 1980, Taemas Limestone (*Receptaculites*, Warroo and Crinoidal Limestone), Australia (Young, 1980).

The Fossil Record 2. Edited by M. J. Benton. Published in 1993 by Chapman & Hall, London. ISBN 0 412 39380 8

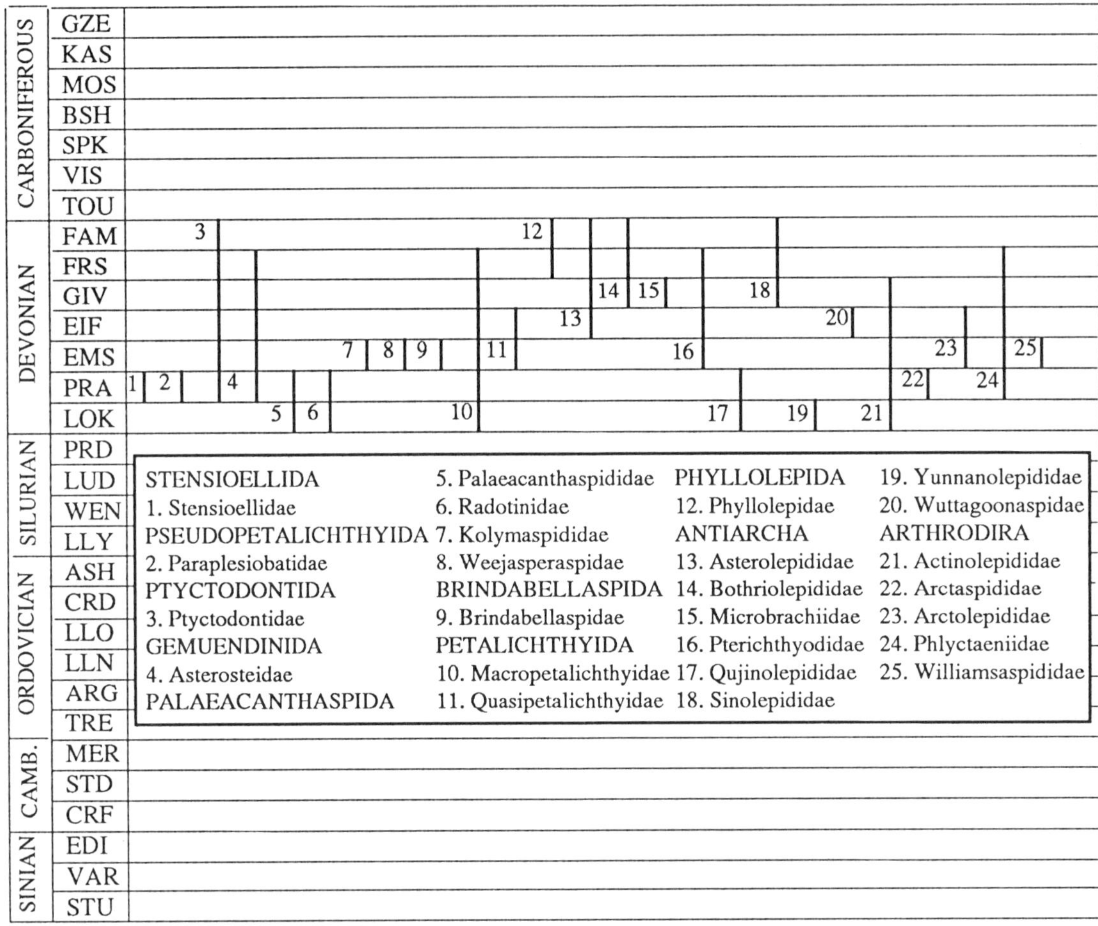

Fig. 32.1

Order PETALICHTHYIDA Jaekel, 1911

F. MACROPETALICHTHYIDAE Eastman, 1898 D. (LOK–FRS) Mar.

First: *Diandongpetalichthys liaojiaoshanensis* P'an and Wang, 1978, Xishancun Member, Quijing, Yunnan, China, and *Xinanpetalichthys shedaowanensis* P'an and Wang, 1978, Xishancun Member, Quijing, Yunnan, China (Pan and Dineley, 1988).
Last: *Epipetalichthys wildungensis* Stensiö, 1925, *Manticoceras* Beds, Germany (Stensiö, 1925).

F. QUASIPETALICHTHYIDAE Liu, 1973 D. (EMS–EIF) Mar.

First: *Neopetalichthys yanmenpaensis* Liu, 1973, Pingyipu Formation, Jiangyou and Sichon, China.
Last: *Latipetalichthys; Quasipetalichthys haikoyensis* Liu, 1973, Haikou Formation, Haikou, Kunming, China (Pan and Dineley, 1988).

Order PHYLLOLEPIDA Stensiö, 1934

F. PHYLLOLEPIDAE Woodward, 1891 D. (FRS–FAM) FW

First: *Austrophyllolepis ritchiei* Long, 1984, Dulcie Sandstone; South Blue Range, Taggerty and Mt. Howitt; Harajica Sandstone, Australia, Victoria Land, Antarctica; *Placolepis budawangensis* Ritchie, Bairdwood, New South Wales, Australia (Young, 1988).
Last: *Phyllolepis* Agassiz, 1844, *Phyllolepis* Series and base of *Remigolepis* Series, East Greenland; Rosebrae Beds and Dura Den Beds, Scotland, UK; also occurs in England, UK, Turkey, Belgium, former USSR, Greenland, North America, Australia and Antarctica (Denison, 1978).

Order ANTIARCHA Cope, 1885

(Earliest (Silurian) members marine, most genera, however, are fresh water – some genera have both marine and freshwater species.)

F. ASTEROLEPIDIDAE Traquair, 1888 D. (EIF–FAM) Mar./FW

First: *Asterolepis estonica* Gröss, 1940, Narova and Aruküla Beds, Baltic States, former USSR.
Last: *Remigolepis* Stensiö, 1931, *Remigolepis* Series, East Greenland (Jarvik, 1961). Also recorded from the Upper Devonian of Australia and North and South China (Young, 1974; Young, MS).

F. BOTHRIOLEPIDIDAE Cope, 1886 D. (GIV–FAM) Mar./FW

First: *Bothriolepis* Eichwald, 1840, Upper Mangzixia Series, Guitou Group, Kwangtung Province; Tiaomachien Series, Hunan and Yunnan Provinces, China (Denison, 1978); Aztec Siltstone, Antarctica (Young, 1988); *Monarolepis verrucosa* (Young, 1988, ex Young and Gorter, 1981), Hatchery Creek Formation, New South Wales, Australia – precise age uncertain (EIF–GIV).

Last: *Bothriolepis* Eichwald, Nadbilovo Stage, former USSR; Alves, Dura Den, Oxendean, Rosebrae and Scaat Craig Beds, Scotland, UK; Upper Ketleri Stage, Latvia, former USSR; *Phyllolepis* Series, East Greenland; Psammites Beds, Belgium. *B. nielseni* Stensiö, 1948, *Remigolepis* Series, East Greenland, is said to be the latest known *Bothriolepis* (Denison, 1978).

F. MICROBRACHIIDAE Gross, 1965
D. (GIV) Mar./FW

First and Last: *Microbrachius dicki* Traquair, 1885, John O'Groats Sandstone, Caithness, Eday Beds, Orkney, Scotland, UK (Denison, 1978); *Hohsienolepis hsintuensis* P'an, Xindu and Hexian Members, China; *Hunanolepis tieni* P'an and Tzeng, 1978, Dahepo, Haikou and Tiaomachin members, China; *Wudinolepis weni* Chang, Yunnan Province, China (Pan and Dineley, 1988).

F. PTERICHTHYODIDAE Cope, 1885
D. (EMS–FRS) Mar./FW

First: *Pterichthyodes* Bleeker, 1859, is reported from New South Wales, Australia (Young, 1974). However, elsewhere it is restricted to the Middle Old Red Sandstone (Middle Devonian) of Scotland, UK.
Last: *Gerdalepis dohmi* (Gröss, 1933) Assize de Masy, Belgium; *Lepadolepis stensioei* (Gröss, 1933), *Mantioceras* Beds, Germany; *Stegeolepis jungata* Malinovskaja, 1973, Taldysay Series, former USSR (Denison, 1978).

F. QUJINOLEPIDIDAE Zhang, 1978
D. (LOK–PRA) Mar.

First: *Qujinolepis gracilis* Zhang, 1978; *Procondylolepis qujingensis* Zhang, 1978; *Zhanjilepis aspratilis* Zhang, 1978, Xitun Member, Cuifengshan Formation, Qujing, China.
Last: *Liujiangolepis suni* Wang, 1987, Xiaoshan and Gruangxi Members, China (Pan and Dineley, 1988).

F. SINOLEPIDIDAE Liu and P'an, 1958
D. (GIV–FAM) Mar.

First: *Xichonolepis qujingensis* P'an and Wang, 1978, Haikou Member, Yunnan, China (closely resembles *Sinolepis*).
Last: *Sinolepis macrocephala* Liu and P'an, 1958, Wutung Series, South China (Pan and Dineley, 1988).

F. YUNNANOLEPIDIDAE Gross, 1965
D. (LOK) Mar.

First and Last: *Yunnanolepis* Liu, 1963, Xishancum, Nagading, Miandiancan, Wudang and Lianhuashan Members, China; *Phymolepis cuifengshanensis* Zhang, 1978, Xhang and Xitun Members, Cuifengshan Formation, Qujing, China (Pan and Dineley, 1988).

Order WUTTAGOONASPIDA Ritchie, 1973

F. WUTTAGOONASPIDAE Ritchie, 1973
D. (EIF) FW

First and Last: *Wuttagoonaspis fletcheri* Ritchie, 1973, Mulga Downs Formation, New South Wales, Australia. This is probably a stem-group arthrodire (Denison, 1978; Forey and Gardiner, 1986).

Order ARTHRODIRA Woodward, 1891

ACTINOLEPIDOIDS Gardiner, 1990

The earliest actinolepids appear in freshwater deposits.

F. ACTINOLEPIDIDAE Gröss, 1940
D. (LOK–GIV) FW/Mar.

First: *Baringaspis dineleyi* Miles, 1973, Peel Sound Formation, Prince of Wales Island, Canada; *Kujdanowiaspis* Stensiö, 1942, Babin Sandstone, Podolia; Dittonian, Great Britain; Old Red Stage 1, Ukraine, former USSR.
Last: *Actinolepis magna* Mark-Kurik, 1973, Burtnieki Beds, Estonia and Latvia, former USSR (Denison, 1978).

PHLYCTAENOIDS Gardiner, 1990

F. ARCTASPIDIDAE Heintz, 1937 D. (PRA) FW

First and Last: *Arctaspis kiaeri* Heintz, 1929, *Dicksonosteus arcticus* Goujet, 1975, Wood Bay Series, Spitsbergen (Heintz, 1929; Goujet, 1975).

F. ARCTOLEPIDIDAE Heintz, 1937
D. (EMS–EIF) FW

First: *Arctolepis decipiens* (Woodward, 1891), Lykta Formation, Wood Bay Series, Spitsbergen.
Last: *Arctolepis* Eastman, 1908, Grey Hoek Formation, Spitsbergen (Heintz, 1929).

F. PHLYCTAENIIDAE Fowler, 1947
D. (PRA–FRS) FW/Mar.

First: *Elegantaspis reticornis* Heintz, 1929, Kapp Kjeldsen Formation, Wood Bay Series, Spitsbergen.
Last: *Neophlyctaenius sherwoodi* (Denison, 1950), Onteora and Katsberg Formation, USA (Denison, 1978).

F. WILLIAMSASPIDIDAE White, 1952
D. (EMS) Mar.

First and Last: *Williamsaspis bedfordi* White, 1952, Murrumbidgee Group, Australia (Denison, 1978). This family appears to be stem-group groenlandaspids.

GROENLANDASPIDS Gardiner, 1990

F. GROENLANDASPIDIDAE Obruchev, 1964
D. (PRA–FAM) Mar./FW (see Fig. 32.2)

First: *Tiaraspis subtilis* (Gröss, 1933), Rheinland, Germany.
Last: *Groenlandaspis mirabilis* Heintz, 1932, *Groenlandaspis* Series, Greenland; Ketorcan Beds, Republic of Ireland; Astec Siltstone, Antarctica. Upper Old Red Sandstone, England, UK; Hunter Siltstone, Australia (Denison, 1978).

HOLONEMATIDS Gardiner, 1990

F. HOLONEMATIDAE Obruchev, 1932
D. (EIF–FRS) Mar./FW

First: *Holonema* Newberry, 1889, Aruküla Beds, Estonia; Narova Beds, former USSR.
Last: *Holonema* Newberry, 1889, Afghanistan, Morocco, Turkey and Iran; Somerset Island, Canada; Redheugh, Scotland, UK; *Aspidichthys* Newberry, 1873, Huron and Olentangy Shale, Ohio, USA, *Manticoceras* Beds, Germany and Poland, Morocco; *Devonema obrucevi* Kulczycki, 1957, *Gyroplacosteus* Obruchev, 1932, Sheldon Beds, Latvia, former USSR; Oberer Plattenkalk, Germany (Denison, 1978).

BUCHANOSTEIDS Gardiner, 1990

F. BUCHANOSTEIDAE White, 1952
D. (PRA–EMS) Mar.

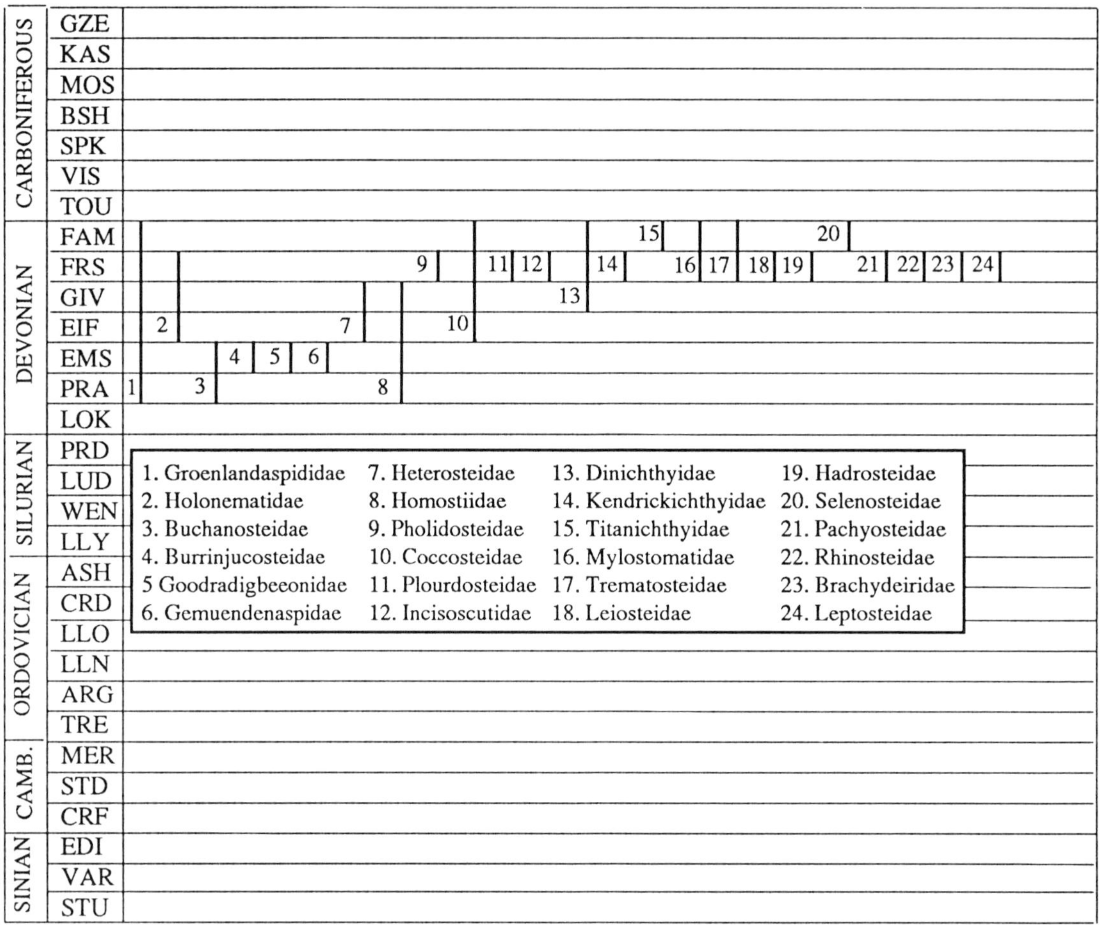

1. Groenlandaspididae
2. Holonematidae
3. Buchanosteidae
4. Burrinjucosteidae
5 Goodradigbeeonidae
6. Gemuendenaspidae
7. Heterosteidae
8. Homostiidae
9. Pholidosteidae
10. Coccosteidae
11. Plourdosteidae
12. Incisoscutidae
13. Dinichthyidae
14. Kendrickichthyidae
15. Titanichthyidae
16. Mylostomatidae
17. Trematosteidae
18. Leiosteidae
19. Hadrosteidae
20. Selenosteidae
21. Pachyosteidae
22. Rhinosteidae
23. Brachydeiridae
24. Leptosteidae

Fig. 32.2

First: *Kweichowlepis sinensis* P'an and Wang, 1975, Wudang Member, Guiyang and Guizhou, China.
Last: *Buchanosteus confertituberculatus* (Chapman, 1916), *Spirifer yassensis* and *Receptaculites* Limestones, Buchan Group, Australia; *Arenipiscis westolli* Young, 1981a, Taeman Limestone, Australia (Young, 1981a,b).

F. BURRINJUCOSTEIDAE White, 1978
D. (EMS) Mar.

First and Last: *Burrinjucosteus asymmetricus* White, 1978, Warro Limestone, Murrumbidgee Series, Australia; *Errolosteus goodradigbeensis* Young, 1981a; *Toombsosteus denisoni* White, 1978, Taemus Limestone, Australia (Young, 1981a,b).

F. GOODRADIGBEEONIDAE White, 1978
D. (EMS) Mar.

First and Last: *Goodradigbeeon australianum* White, 1978, Cavan Bluff Limestone, Murrumbidgee Series, Australia (White, 1978).

F. GEMUENDENASPIDAE Miles, 1962
D. (EMS) Mar.

First and Last: *Gemuendenaspis angusta* (Traquair, 1903), Hunsrückschiefer, Germany. Known from a single specimen which may be a stem heterostiid, homostiid or coccosteoid (Denison, 1978).

HETEROSTIIDS Gardiner, 1990

F. HETEROSTEIDAE Jaekel, 1903
D. (EIF–GIV) Mar./FW

First: *Herasmius granulatus* Ørvig, 1969, Wood Bay Series, Spitsbergen (Ørvig, 1969).
Last: *Heterosteus asmussi* (Agassiz, 1845) Aruküla Beds, Baltic States, former USSR; *Stingocephalus* Beds, Germany; Wijde Bay Formation, Spitsbergen; *Yinosteus major* Wang and Wang, 1984, Haikou Formation, Yunnan, China (Denison, 1978; Pan and Dineley, 1988).

HOMOSTIIDS Gardiner, 1990

F. HOMOSTIIDAE Jaekel, 1903
D. (PRA–GIV) FW

First: *Euleptaspis depressa* (Gröss, 1933), Lower Devonian, Germany; Kapp Kjeldsen Formation, Spitsbergen.
Last: *Homostius milleri* Traquair, 1888, Thurso flagstones, Rousay, Beds, Melby Fish Band, Scotland, UK (Denison, 1978).

PHOLIDOSTEIDS Gardiner, 1990

F. PHOLIDOSTEIDAE Gröss, 1932 D. (FRS) Mar.

First and Last: *Malerosteus gorizdroae* Kulczycki, 1957, Holy Cross Mountains, Poland; *Pholidosteus friedeli* Jaekel, 1907; *Tapinosteus heintzi* Stensiö, 1963, *Manticoceras* Beds, Germany (Denison, 1978).

COCCOSTEOIDS Gardiner, 1990

F. COCCOSTEIDAE Traquair, 1888
D. (EIF–FAM) FW (Predom.)

First: *Coccosteus* Miller, 1841, Caithness flagstones, Scotland, UK; Aruküla and Burtnieki Beds, Baltic States, former USSR; *Livosteus grandis* (Gröss, 1933), Aruküla and Burtnieki Beds, Latvia, former USSR; *Millerosteus orvikui* (Gröss, 1940), Narova and Luga Beds, Baltic States, former USSR. *Protitanichthys fossatus* Eastman, 1907, Delaware Limestone, USA (Denison, 1978).
Last: *Ardennosteus ubaghsi* Lelievre, 1982, Esneux, Belgium (Lelievre, 1982).

F. PLOURDOSTEIDAE Vézina, 1989
D. (FRS) Mar. (Predom.)

First and Last: *Harrytoombsia elegans* Miles and Dennis, 1979; *Kimberleyichthys* (Dennis-Bryan and Miles, 1983); *Torosteus* Gardiner and Miles, 1990, Gogo Formation, Australia; *Panxiosteus* Wang, 1979, Panxi marls, Yunnan, China; *Plourdosteus* Ørvig, 1951, Escuminac Formation, Canada, Holy Cross Mountains, Poland; Pskov Stage and Ganja Beds, former USSR (Denison, 1978; Dennis-Bryan and Miles, 1983).

F. INCISOSCUTIDAE **fam. nov.** D. (FRS) Mar.

First and Last: *Camuropiscis* Dennis and Miles, 1979 *Fallacosteus* Long, 1988; *Incisoscutum mitchiei* Dennis and Miles, 1981; *Latocamurus* Long, 1988, *Tubonasus lennardensis* Dennis and Miles, 1980, all from Gogo Formation, Australia (Long, 1988).

EASTMANOSTEIDS Gardiner, 1990

F. DINICHTHYIDAE Newberry, 1885
D. (GIV–FAM) Mar.

First: *Dunkleosteus* Lehman, 1956, Haikou Formation, Yunnan, China; *Eastmanosteus* Obruchev, 1964, Corniferous and Hamilton Limestone, USA; Akka Formation, Morocco Tabas Series, Iran (Dennis-Bryan, 1987).
Last: *Dunkleosteus*, Huron and Cleveland Shales, USA; Holy Cross Mountains, Poland; Belgium; Morocco; *Heintzichthys gouldii* (Newberry, 1885); *Hussakofia minor* (Newberry, 1878); *Gorgonichthys clarki* Claypole, 1892, Cleveland Shales, USA (Denison, 1978).

KENRICKICHTHYIDS Gardiner, 1990

F. KENDRICKICHTHYIDAE **fam. nov.**
D. (FRS) Mar.

First and Last: *Bullerichthys facidens* Dennis and Miles, 1980; *Bruntonichthys multidens* Dennis and Miles, 1980; *Kendrickichthys cavernosus* Dennis and Miles, 1980, Gogo Formation, Australia (Dennis and Miles, 1980).

TITANICHTHYIDS Gardiner, 1990

F. TITANICHTHYIDAE Dean, 1901
D. (FAM) Mar.

First and Last: *Tafilalichthys lavocati* Lehman, 1954, Morocco; *Titanichthys agassizi* Newberry, 1885, Cleveland Shales, USA; Holy Cross Mountains, Poland; Tafilalet, Morocco (Denison, 1978).

F. MYLOSTOMATIDAE Woodward, 1891
D. (FRS–FAM) Mar.

First: *Dinomylostoma beecheri* Eastman, 1906, Cushaqua and West River shales, Evans Limestone, USA; Morocco; Gogo Formation, Australia.
Last: *Mylostoma* Newberry, 1883, Cleveland Shales, USA (Denison, 1978).

TREMATOSTEIDS Gardiner, 1990

F. TREMATOSTEIDAE Gross, 1932
D. (FRS–FAM) Mar.

First: *Belosteus* Jaekel, 1919; *Brachyosteus* Jaekel, 1927; *Braunosteus* Stensiö, 1959; *Parabelosteus* Miles, 1969; *Trematosteus* Jaekel, 1927; all from the *Manticoceras* Beds (Upper FRS), Germany.
Last: *Bungartius perissus* Dunkle, 1947; *Paramylostoma arcualis* Dunkle and Bungart, 1945, Cleveland Shale, USA (Denison, 1978).

LEIOSTEIDS Gardiner, 1990

F. LEIOSTEIDAE Stensiö, 1963 D. (FRS) Mar.

First and Last: *Erromenosteus* Jaekel, 1919, *Manticoceras* Beds, Germany (Denison, 1978).

HADROSTEIDS Gardiner, 1990

F. HADROSTEIDAE Gröss, 1932 D. (FRS) Mar.

First and Last: *Hadrosteus rapax* Gröss, 1932, *Manticoceras* Beds, Germany (Gröss, 1932).

F. SELENOSTEIDAE Dean, 1901 D. (FAM) Mar.

First and Last: *Selenosteus brevis* (Claypole, 1869); *Gymnotrachelus hydei* Dunkle and Bungart, 1939, Cleveland Shales, USA (Denison, 1978).

PACHYOSTEIDS Gardiner, 1990

F. PACHYOSTEIDAE Gross, 1932 D. (FRS) Mar.

First and Last: *Enseosteus* Jaekel, 1919; *Microsteus* Gröss, 1932; *Pachyosteus bulla* Jaekel, 1903, all from the *Manticoceras* Beds, Germany (Denison, 1978).

RHINOSTEIDS Gardiner, 1990

F. RHINOSTEIDAE Stensiö, 1963 D. (FRS) Mar.

First and Last: *Rhinosteus* Jaekel, 1919, *Manticoceras* Beds, Germany; *Melanosteus occitanus* Lelievre and Goujet, 1987, Serve Formation, France (Lelievre *et al.*, 1987).

BRACHYDEIRIDS Gardiner, 1990

F. BRACHYDEIRIDAE Gröss, 1932 D. (FRS) Mar.

First and Last: *Brachydeirus* Koenen, 1880; *Oxyosteus* Jaekel, 1911, *Synauchenia* Jaekel, 1919, all from the *Manticoceras* Beds, Germany (Denison, 1978).

F. LEPTOSTEIDAE Jaekel, 1911 D. (FRS) Mar.

First and Last: *Leptosteus* Jaekel, 1911, *Manticoceras* Beds, Germany; Java Formation, New York, USA (Denison, 1978).

The informal higher-order terms used above (namely Brachydeirids) refer to the successive levels of organization within the arthrodires, and therefore indicate individual clades.

REFERENCES

Boucot, A. J. and Pankiwskja, K. A. (1962) Llandoverian to Gedinnian stratigraphy of Podolia and adjacent Moldavia. 2

Intenational Arbeit–Silur./Devon.–Grenze, Bonn-Bruxellee 1960; 128 pp.

Branson, E. B. and Mehl, M. G. (1938) Stratigraphy and paleontology of the Lower Mississippian of Missouri. Pt. 2. *University of Missouri Studies*, **13**, 1–242.

Denison, R. (1978) Placodermi, in *Handbook of Paleoichthyology*, Vol. 2, (ed. H. P. Schultze), Fisher, Stuttgart, pp. 1–128.

Dennis-Bryan, K. (1987) A new species of eastmanosteid arthrodire (Pisces: Placodermi) from Gogo, Western Australia, *Zoological Journal of the Linnean Society*, **90**, 1–64.

Dennis-Bryan, K. and Miles, R. S. (1983) Further eubrachythoracid arthrodires from Gogo, Western Australia. *Zoological Journal of the Linnean Society*, **77**, 145–73.

Forey, P. and Gardiner, B. G. (1986) Observations on *Ctenurella* (Pycnodontida) and the classification of placoderm fishes. *Zoological Journal of the Linnean Society*, **86**, 43–74.

Gardiner, B. G. (1990) Placoderm fishes: diversity through time. *Systematics Association, Special Volume*, **42**, 305–19.

Gardiner, B. G. and Miles R. S. (1990) Further primitive eubrachythoracid arthrodires from Gogo, Western Australia. *Zoological Journal of the Linnean Society*, **99**, 159–204.

Goujet, D. (1975) *Dicksonosteus*, un nouvel arthrodire du Dévonien du Spitsberg. Remarques sur le squelette viscéral des Dolichothoraci. *Colloques Internationaux du Centre National de la Recherche Scientifique*, **218**, 81–99.

Goujet, D. (1980) Les poissons, in *Les Schistes et calcaires de l'Armorique (Dévonien inférieur Massif Armoricain). Sédimentologie paléontologie, stratigraphie*. (co-ord. Y. Plusquellec), Mémoires de la Société Géologique et Minéralogique de Bretagne, **23**, pp. 309–15.

Gröss, W. (1932) Die Arthrodira Wildungens. *Geologische und Palaeontologische Abhandlungen*, **19**, 1–61.

Gröss, W. (1937) Die Wirbeltiere des rheinischen Devons 2. *Abhandlungen der Preussischen Geologischen Landesanstalt*, **176**, 1–83.

Gröss, W. (1958) Über die älteste Arthrodiren-Gattung. *Notizblatt Hessischen Landesamtes für Bodenforschung zu Wiesbaden*, **86**, 7–30.

Gröss, W. (1959) Arthrodiren aus dem Obersilur der Prager Mulde. *Palaeontographica, Abteilung A*, **133**, 1–35.

Gröss, W. (1962) Neuuntersuchung der Stensiöellida (Arthrodira, Unterdevon). *Notizblatt des Hessischen Landesamtes für Bodenforschung zu Wiesbaden*, **90**, 48–86.

Gröss, W. (1963) *Gemuendina stuertzi* Traquair, Neuuntersuchung. *Notizblatt des Hessischen Landesamtes für Bodenforschung zu Wiesbaden*, **91**, 36–73.

Heintz, A. (1929) Die downtonischen und devonischen Vertebraten von Spitsbergen. II. Acantheapida. *Skrifter om Svalbard og Ishavet*, **22**, 1–81.

Jaekel, O. (1921) Paläontologische Berichte: Die Stellung der Paläontologie zu einigen Problemen der Biologie und Phylogenie. *Paläontologische Zeitschrift*, **3**, 217–21.

Jarvik, E. (1961) Devonian vertebrates, in *Geology of the Arctic, I*, (ed. G. O. Raasch), University of Toronto Press, pp. 197–204.

Lelievre, H. (1982) *Ardennosteus ubaghsi* n.g., n. sp. Brachythoraci primitif (Vertébrés Placoderme) du Famennin d'Esneux (Belgique). *Annales de la Société Géologique de Belgique*, **105**, 1–7.

Lelievre, H., Feist, R. and Goujet, D. (1987) Les vertébrés Dévoniens de la Montagne Noire (Sud de la France) et leur apport à la phylogénie des Pachyosteomorphes (Placodermes Arthrodires). *Palaeovertébrata*, **17**, 1–26.

Long, J. (1988) A new camuropiscid arthrodire (Pisces: Placodermi) from Gogo, Western Australia. *Zoological Journal of the Linnean Society*, **94**, 233–58.

Long, J. and Young, G. C. (1988) Acanthothoracid remains from the Early Devonian of New South Wales including a complete sclerotic capsule and pelvic girdle. *Memoir of the Association of Australasian Palaeontologists*, **7**, 65–80.

Ørvig, T. (1969) A new brachythoracid arthrodire from the Devonian of Dickson Land, Vestspitsbergen. *Lethaia*, **2**, 261–71.

Pan, J. and Dineley, D. (1988) A review of early (Silurian and Devonian) vertebrate biogeography and biostratigraphy of China. *Proceedings of the Royal Society of London*, **235B**, 29–61.

Stensiö, E. A. (1925) On the head of the macropetalichthyids with certain remarks on the head of the other arthrodires. *Publications of the Field Museum of Natural History, Geological Series*, **232**, 91–197.

Stensiö, E. A. (1944) Contributions to the knowledge of the vertebrate fauna of the Silurian and Devonian of Western Podolia. 2. Notes on two arthrodires from the Downtonian of Podolia. *Arkiv för Zoologi*, **35A**, 1–83.

White, E. I. (1978) The larger arthrodiran fishes from the area of the Burrinjuck Dam, N.S.W. *Transactions of the Zoological Society of London*, **34**, 149–262.

Young, G. C. (1974) Stratigraphic occurrence of some placoderm fishes in the Middle and Late Devonian. *Newsletters of Stratigraphy*, **3**, 243–61.

Young, G. C. (1980) A new Early Devonian placoderm from New South Wales, Australia, with a discussion of placoderm phylogeny. *Palaeontographica, Abteilung A*, **167**, 10–76.

Young, G. C. (1981a) New Early Devonian brachythoracids (placoderm fishes) from the Taemas–Wee Jasper region of New South Wales. *Alcheringa*, **5**, 247–71.

Young, G. C. (1981b) A new fish fauna of Middle Devonian age from the Taemas/Wee Jasper region of New South Wales. *Bulletin of the Bureau of Mineral Resources, Geology and Geophysics, Australia*, **209**, 85–128.

Young, G. C. (1988) New occurrences of phyllolepid placoderms from the Devonian of central Australia. *Bureau of Mineral Resources, Geology and Geophysics, Australia*, **10**, 363–76.

Young, G. C. and Gorter, J. D. (1981) A new fish fauna of Middle Devonian age from the Taemas/Wee Jasper region of New South Wales. *Bulletin of the Bureau of Mineral Resources, Geology and Geophysics, Australia*, **209**, 85–128.

33

ACANTHODII

J. Zidek

The classification below follows Miles (1966), except for the position of the Ischnacanthiformes, which are listed first to allude to the possibility that they may be the most primitive acanthodians (Long, 1986b), and the addition of Culmacanthidae Long, 1983 (Diplacanthoidei). The most recent review of the Acanthodii is Denison (1979), who merged the Euthacanthidae (monogeneric) into the Climatiidae and recognized only one acanthodiform family, the Acanthodidae. However, there is much in favour of upholding the families suppressed by Denison (Euthacanthidae, Mesacanthidae, Cheiracanthidae). Miles and Denison may have erred in including the Diplacanthidae and Gyracanthidae in the Climatiiformes. The histology of diplacanthid scales (Valiukevičius, 1985), and the paucity of our knowledge of the gyracanthids suggest that the two should be regarded as separate orders; however, this is not the place to argue controversial issues, and Miles' arrangement is therefore maintained.

Harland *et al.* (1982) are followed in stage/age assignments, with the exception of the Silurian, where Llandoverian, Wenlockian, Ludlovian and Pridolian are considered stages/ages rather than series/epochs (Jaeger, 1980; Chlupáč *et al.*, 1981).

Class ACANTHODII Owen, 1846

Order ISCHNACANTHIFORMES Berg, 1940

F. ISCHNACANTHIDAE Woodward, 1891 S. (LUD)–C. (MRD) Mar. (see Fig. 33.1)

First: *Gomphonchus sandelensis* (Pander, 1856) and *G. hoppei* (Gross, 1947), both from middle LUD Paadla (K_2) beds of Saaremaa, Estonia, former USSR, and the Hemse Formation of Gotland, Sweden.
Last: *Marsdenius summiti* Wellburn, 1902 and *M. acuta* Wellburn, 1902, both from the Pendleside Limestone (upper Namurian B), Yorkshire, England, UK.
Intervening: PRD–EIF, FRS.
Comment: *Acanthodopsis* Hancock and Atthey, 1868, from the Coal Measures of England and Scotland, is Westphalian B in age (probably VRK), but Long (1986b) has provided a convincing argument for placing this genus in the Acanthodidae.

Order CLIMATIIFORMES Berg, 1940

Suborder CLIMATIOIDEI Miles, 1966

F. CLIMATIIDAE Berg, 1940 S. (LLY)–D. (GIV) Mar./FW

First: *Onchus clintoni* Claypole, 1885, Burnt Ridge Member of Rose Hill Formation, Perry County, Pennsylvania, USA.
Last: *Cheiracanthoides comptus* Wells, 1944, Cürten Formation, Eifel Limestone Synclinorium, Germany (Vieth-Schreiner, 1983).
Intervening: PRD–EIF.
Comment: The Burnt Ridge Member is **upper** LLY. *Onchus graptolitarum* Fritsch, 1907, allegedly roughly contemporaneous with *O. clintoni*, in reality is PRD, as evidenced by the character of the matrix (Chlupáč *et al.*, 1972; Zajíc, 1986). Kříž *et al.* (1986) found *O. graptolitarum* only in the upper PRD. Gross (1951) mentioned *Nostolepis*-like scales from either upper WEN or lower LUD of Bohemia, Czech Republic, and Denison (1979) listed these as *Nostolepis* sp. from the Liteň Beds, which would make them WEN or possibly even LLY. However, these scales are associated with the holotype of *Onchus graptolitarum* and thus are of PRD age. The genera *Onchus* Agassiz, 1837 and *Nostolepis* Pander, 1856 are based on isolated spines and scales, respectively, and the noted spine–scale association is the only instance hitherto known. It would be premature to synonymize *Nostolepis* with *Onchus* on this evidence, but it does allow to transfer *Onchus* from Acanthodii *incertae sedis* (Denison, 1979) to the Climatiidae.

F. EUTHACANTHIDAE Berg, 1940 D. (PRA) FW

First and Last: *Euthacanthus macnicoli* Powrie, 1864, Dundee Formation, Forfarshire, Scotland, UK. This is dated as LOK according to Paton (1976).

Suborder DIPLACANTHOIDEI Miles, 1966

F. DIPLACANTHIDAE Woodward, 1891 D. (EIF–FRS) FW/Mar.(?)

First: *Rhadinacanthus balticus*? Gross, 1973 and *Ptychodictyon rimosum* Gross, 1973, lower part of Middle Narva Regional Substage, Narva Regional Stage, Baltic region.
Last: *Diplacanthus horridus* Woodward, 1892, Escuminac Formation, Quebec, Canada.
Intervening: GIV.
Comment: *Rhadinacanthus* Traquair, 1888 may be syn-

The Fossil Record 2. Edited by M. J. Benton. Published in 1993 by Chapman & Hall, London. ISBN 0 412 39380 8

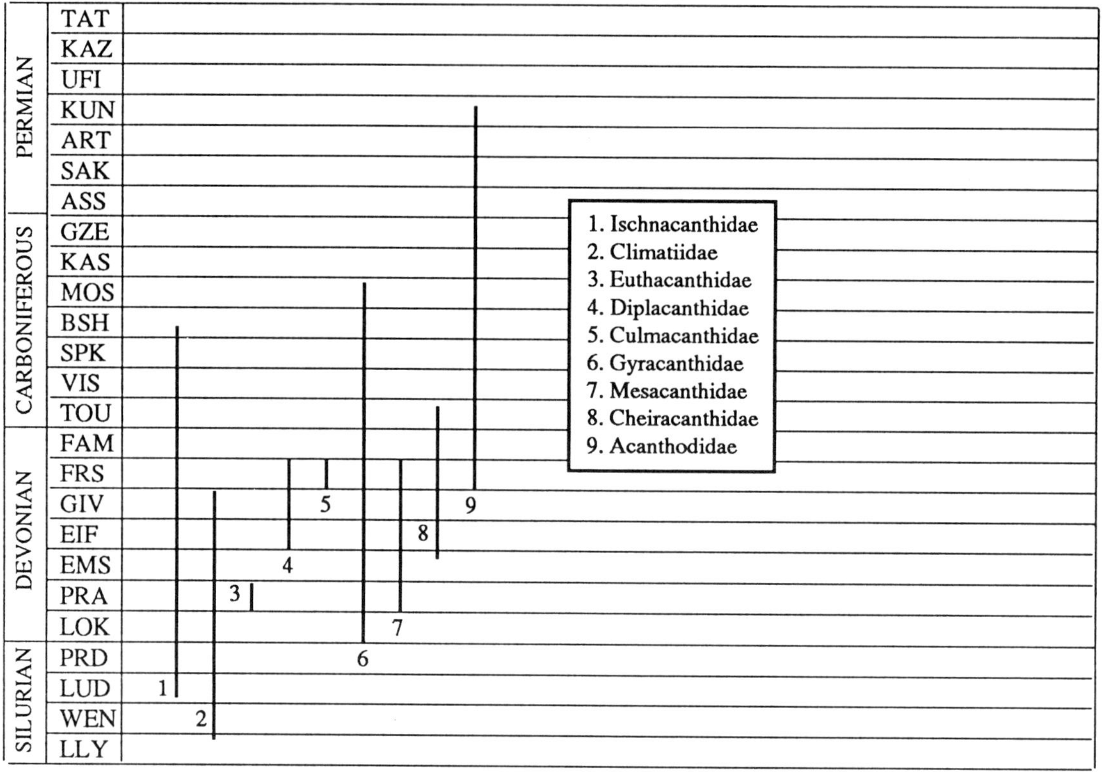

Fig. 33.1

onymous with *Diplacanthus* Agassiz, 1844, but this remains to be ascertained. *Ptychodictyon* Gross, 1973 was placed in Acanthodii *incertae sedis* by Denison (1979), but Valiukevičius (1979, 1985) recognized its diplacanthid scale histology.

Gladiobranchus probaton Bernacsek and Dineley (1977) is a LOK species (Delorme Formation, District of Mackenzie, NWT, Canada) included by Denison (1979) in the Diplacanthidae. However, it is not clear whether this is a diplacanthoid or an ischnacanthiform (Long, 1983), and it is left as *incertae sedis*.

All undisputed diplacanthids but *D. horridus* are EIF and early GIV in age. Valiukevičius (1985) did not correlate the Narva Regional Stage (Narova Beds of Denison, 1979) with the international standard scale, but he observed that the Kernavė Formation of Lithuania, former USSR, which contains conodonts of early EIF age, corresponds with the Upper Narva Regional Substage. This means that his second acanthodian assemblage, which characterizes the lower part of the Middle Narva Regional Substage and contains the first occurrence of *R. balticus*? and *P. rimosum*, must be early EIF in age and thus older than the diplacanthid occurrences in the Middle Old Red Sandstone of Scotland (upper EIF–lower GIV). Even older, earliest EIF, are *Diplacanthus*? sp. nov. 1, *Rhadinacanthus* sp. nov. 1, and *Ptychodictyon* sp. nov. 1 from the Pärnu and Rezekne Regional Stages of the Baltic region, listed but not described by Valiukevičius (1985).

Contrary to the generally accepted interpretation, Schultze (*in* Carroll *et al.*, 1972) proposed a marine depositional environment for the Escuminac Formation.

F. CULMACANTHIDAE Long, 1983 D. (FRS) FW

First and Last: *Culmacanthus stewarti* Long, 1983, lower part of Avon River Group, Mt. Howitt, Victoria, Australia.

F. GYRACANTHIDAE Woodward, 1906 D. (PRA)–C. (VER) FW/Mar.

First: *Gyracanthus*? *convexus* Gross, 1933, Wahnbachschichten and Taunusquarzit, Rheinisches Schiefergebirge, Germany.

Last: *Gyracanthus duplicatus* Dawson, 1868, Joggins Formation, Nova Scotia, Canada.

Intervening: EMS–FAM, TOU–BSH.

Comments: Mader (1986) recorded *Gyracanthus* sp. from the lower LOK Upper Carazo Formation, Palencia Province, northern Spain. *Gyracanthus* sp. from the Mecca Quarry Shale of Carbondale Formation, Kankakee County, Illinois, USA (Baird, 1978), is early Westphalian D, i.e. either the latest POD or the earliest MYA.

Order ACANTHODIFORMES Berg, 1940

F. MESACANTHIDAE Moy-Thomas, 1939 D. (PRA–FRS) FW/Mar.(?)

First: *Mesacanthus mitchelli* (Egerton, 1861), Dundee Formation, Forfarshire, Scotland, UK (dated as LOK by Paton, 1976).

Last: *Triazeugacanthus affinis* (Whiteaves, 1887), Escuminac Formation, Quebec, Canada.

Intervening: EMS–GIV.

F. CHEIRACANTHIDAE Berg, 1940 D. (EMS)–C. (TOU) FW/Lagoonal

First: *Cheiracanthus costellatus* Traquair, 1893, Atholville Beds, New Brunswick, and Battery Point Formation, Quebec, Canada.

Last: *Carycinacanthus lopatini* (Rohon, 1889), Nadaltaiskaia Formation, Minusinsk Depressions, former USSR; and/or *Homalacanthus bergi* (Obruchev, 1962), upper part of Shivelikskaia Formation, Tuva Basin, former USSR.

Intervening: EIF–FAM.

Comments: The Atholville Beds/Battery Point Formation assemblage is usually regarded as early EIF, but *C. costellatus* is known from locality (zone) 1 at D'Aiguillon (Pageau, 1968; Carroll *et al.*, 1972), in the lower part of the Battery Point Formation, which has produced miospores of late EMS age (McGregor, 1973). Should this occurrence prove to be lower EIF, then *C. costellatus* would be roughly contemporaneous with *C. longicostatus* Gross, 1973, *C. brevicostatus* Gross, 1973 and *C. crassus* Valiukevičius, 1985 from the Pärnu and Rezekne Regional Stages in the Baltic region (Valiukevičius, 1985), and probably also with *Isodendracanthus ramiformis* Valiukevičius, 1979 and *Ectopacanthus cristiformis* Valiukevičius, 1979 from the Grey Hoek Formation of Vestspitsbergen.

According to Obruchev (1962), *C. lopatini* is known from the Izykchulskii Fish Horizon in the middle of Bystrianskaia Formation up into the Nadaltaiskaia Formation, and *H. bergi* from Dzharginskaia Formation and the upper part of Shivelikskaia Formation, all belonging in the TOU series. So far as can be discerned from Kotliar *et al.* (1977) and Obruchev's (1962) remarks, the Nadaltaiskaia Formation of the Minusinsk depression corresponds with the upper part of Shivelikskaia Formation in the Tuva Basin, and the two species thus can be said to be roughly contemporaneous, more likely HAS than IVO.

F. ACANTHODIDAE Huxley, 1861 D. (FRS)–P. (SAK) FW/Mar.

First: *Howittacanthus kentoni* Long, 1986a, lower part of Avon River Group, Mt. Howitt, Victoria, Australia.

Last: *Acanthodes luedersensis* (Dalquest *et al.*, 1988), Lueders Formation, uppermost Wichita Group (uppermost Wolfcampian), Baylor County, Texas, USA.

Intervening: TOU, VIS, KIN–ASS.

Comments: *A. luedersensis* was described as a new genus, *Rodriguezichthyes* [*sic*], based chiefly on misinterpretation of the scale histology (Schultze, 1990). The species is said to have been shallow-water marine. *Acanthodes* sp. from the Purcell Sandstone at the base of the Hennessey Group in Tillman County, Oklahoma, USA (Zidek, 1975) is early Leonardian in age, i.e. early ART. The depositional environment is fresh water.

In the Saar–Pfalz region of Germany, *Acanthodes* sp. occurs as high in the Rotliegendes section as in the N4–N5 informal subdivisions of the Nahe Group (Boy, 1987), a level that is considered Saxonian by Boy (1987) and is correlated by him with the Olivětín Member of Broumov Formation in the Intra-Sudeten Basin (Bohemia, Czech Republic), the upper part of Prosečné Formation (Kalná Horizon) in the Krkonoše Piedmont Basin (Bohemia, Czech Republic), and upper Leonardian (KUN) in the USA. Although Czech authors place the named formations in the upper Autunian rather than Saxonian (Zajíc and Štamberg 1986), Boy's (1987) correlation of the N4–N5 levels of the Nahe Group with the upper Leonardian appears to be realistic and is accepted here. The Nahe Group *Acanthodes* thus is KUN, and represents the youngest known occurrence of the family and of the entire class. The depositional environment is pond/lacustrine according to Boy (1987).

REFERENCES

Baird, D. (1978) Studies on Carboniferous freshwater fishes. *American Museum Novitates*, **2641**, 1–22.

Boy, J. A. (1987) Die Tetrapoden-Lokalitäten des saarpfälzischen Rotliegenden (?Ober-Karbon–Unter-Perm; SW-Deutschland) und die Biostratigraphie der Rotliegend-Tetrapoden. *Mainzer geowissenschaftliche Mitteilungen*, **16**, 31–65.

Carroll, R. L., Belt, E. S., Dineley, D. L. *et al.* (1972) *Excursion A59, Vertebrate Paleontology of Eastern Canada*. Guidebook, 24th International Geological Congress, Montreal, Quebec, 1972, 113 pp.

Chlupáč, I., Flügel, H. and Jaeger, H. (1981) Series or stages within Palaeozoic systems? *Newsletter in Stratigraphy*, **10**, 78–91.

Chlupáč, I., Jaeger, H. and Zikmundová, J. (1972) The Silurian–Devonian boundary in the Barrandian. *Bulletin of Canadian Petroleum Geology*, **20**, 104–74.

Dalquest, W. W., Kocurko, M. J. and Grimes, J. V. (1988) Geology and vertebrate paleontology of a Lower Permian deposit on the Brazos River, Baylor County, Texas. With the description of a new genus and species of acanthodian fish. *Tulane Studies in Geology and Paleontology*, **21**, 85–104.

Denison, R. H. (1979) Acanthodii, In *Handbook of Paleoichthyology*, Vol. 5 (ed. H.-P. Schultze), Gustav Fischer, Stuttgart, pp. 1–62.

Gross, W. (1951) Die paläontologische und stratigraphische Bedeutung der Wirbeltierfaunen des Old Reds und der marinen altpaläozoischen Schichten. *Abhandlungen der deutschen Akademie der Wissenschaften zu Berlin, Mathematisch–naturwissenschaftliche Klasse*, **1949**, 1–300.

Harland, W. B., Cox, A. V., Llewellyn, P. G. *et al.* (1982) *A Geologic Time Scale*. Cambridge University Press, 131 pp.

Jaeger, H. (1980) Silurian series and stages: a comment. *Lethaia*, **13**, p. 365.

Kotliar, G. V., Kropacheva, G. S., Verbutskaya, N. G. (eds) (1977) *Stratigraficheskii Slovar SSSR, Karbon, Perm*. 'NEDRA', Leningrad, 535 pp.

Kříž, J., Jaeger, H., Paris, F. *et al.* (1986) Přídolí – the fourth division of the Silurian. *Jahrbuch der geologischen Bundesanstalt*, Vienna, **129**, 291–360.

Long, J. A. (1983) A new diplacanthoid acanthodian from the Late Devonian of Victoria. *Memoirs of the Association of Australasian Palaeontologists*, **1**, 51–65.

Long, J. A. (1986a) A new Late Devonian acanthodian fish from Mt. Howitt, Victoria, Australia, with remarks on acanthodian biogeography. *Proceedings of the Royal Society of Victoria*, **98**, 1–17.

Long, J. A. (1986b) New ischnacanthid acanthodians from the Early Devonian of Australia, with comments on acanthodian interrelationships. *Zoological Journal of the Linnean Society*, **87**, 321–39.

Mader, H. (1986) Schuppen und Zähne von Acanthodiern und Elasmobranchiern aus dem Unter-Devon Spaniens (Pisces). *Göttinger Arbeiten in Geologie und Paläontologie*, **28**, 1–59.

McGregor, D. C. (1973) Lower and Middle Devonian spores of eastern Gaspé, Canada. I. Systematics. *Palaeontographica, Abteilung B*, **142**, 1–74.

Miles, R. S. (1966) The acanthodian fishes of the Devonian Plattenkalk of the Paffrath Trough in the Rhineland. With an appendix containing a classification of the Acanthodii and a revision of the genus *Homalacanthus*. *Arkiv för Zoologi*, **18**, 147–94.

Obruchev, D. V. (1962) Klass Acanthodii. Akantody, In *Biostratigrafia Paleozoia Saiano-Altaiskoi Gornoi Oblasti, Tom III, Verchnii Paleozoi* (ed. L. L. Khalfin), SNIIGGIMS, Novosibirsk, p. 212.

Pageau, M. Y. (1968) Nouvelle faune ichthyologique du Dévonien moyen dans les grès de Gaspé (Québec). I. Géologie et écologie. *Naturaliste Canadien*, **95**, 1459–97.

Paton, R. L. (1976) A catalogue of fossil vertebrates in the Royal Scottish Museum, Edinburgh. *Royal Scottish Museum Information Series, Geology*, **6**, 1–40.

Schultze, H.-P. (1990) A new acanthodian from the Pennsylvanian of Utah, U.S.A., and the distribution of otoliths in gnathostomes.

Journal of Vertebrate Paleontology, **10**, 49–58.
Valiukevičius, J. J. (1979) Cheshui akantod iz eifelskikh otlozhenii Shpitsbergena. *Paleontologicheskii Zhurnal*, **1979** (4), 101–11.
Valiukevičius, J. J. (1985) *Akantody Narovskogo Gorizonta Glavnogo Devonskogo polia*. MOKSLAS, Vilnius, Lithuania, 143 pp.
Vieth-Schreiner, J. (1983) Fisch-Schuppen und -Zähne aus der Eifeler Kalkmulden-Zone (Emsium, Eifelium). *Senckenbergiana Lethaea*, **64**, 129–77.
Zajíc, J. (1986) Stratigraphic position of finds of the acanthodians (Acanthodii) in Czechoslovakia. *Acta Universitatis Carolinae – Geologica*, Špinar Volume **2**, 145–53.
Zajíc, J. and Štamberg, S. (1986) Summary of the Permocarboniferous freshwater fauna of the limnic basins of Bohemia and Moravia. *Acta Musei Reginaehradecensis (A)*, **20**, 61–82.
Zidek, J. (1975) Oklahoma paleoichthyology, part IV: Acanthodii. *Oklahoma Geology Notes*, **35**, 135–46.

34

CHONDRICHTHYES

H. Cappetta, C. Duffin and J. Zidek

The most recent comprehensive treatments of the Palaeozoic and Cainozoic elasmobranchs are respectively by Zangerl (1981) and by Cappetta (1987), whose classifications are followed.

The oldest elasmobranch teeth are *Leonodus carlsi* Mader, 1986 from the lower GED of northern Spain. They have been placed tentatively in the Xenacanthida. The oldest teeth of neoselachian design are *Mcmurdodus whitei* Turner and Young, 1987 from EMS/EIF of western Queensland, Australia. Turner and Young (1987) compared them with *Mcmurdodus featherensis* White, 1968 from GIV/FRS of south Victoria Land, Antarctica, and tentatively placed the genus *Mcmurdodus* White, 1968 in the order Hexanchida. The oldest elasmobranch scales are *Elegestolepis grossi* Karatajute-Talimaa, 1973 from the LDF/PRD or PRD of Tuva, former USSR. They have been likened to edestid scales, but are not allocated to any higher taxon in this review. The oldest elasmobranch(?) spine is *Bulbocanthus rugosus* Bryant, 1932 from the SIG of Wyoming, USA.

Some genera whose familial status is uncertain were not taken into consideration; they belong mainly to the Batomorphii and their number is insignificant compared with the number of genera included in this compilation.

The holocephalans are in desperate need of taxonomic revision. There is a considerable body of literature, based largely upon isolated tooth plates assigned to nominal species. Much synonymy probably exists, especially among Palaeozoic groups. Complete dentitions and articulated skeletons are rare, giving rise to problems with form genera (Smith and Patterson, 1988). The classification scheme for Palaeozoic groups followed here is that of Zangerl (1981), which is also the most recent. The scheme is provisional, in anticipation of the Holocephali volume of the *Handbook of Palaeoichthyology* (in prep.). Owing to the paucity of information, no cladistic analysis is presented here other than that already developed by Zangerl (1981).

Note that the menaspids (WEN ???, TAT; Mar.), formerly based upon *Pilolepis margaritifera* Thorsteinsson, 1973 (earliest Cape Phillips Formation, upper Wenlock of Cornwallis Island, Canadian Archipelago), and *Menaspis armata* Ewald, 1848 from the Kupferschiefer of Mansfeld, Germany, have been recently reassigned to the arthrodires (Ortlam, 1985).

Zangerl (1981) left several higher taxa (one superorder, three orders and one suborder) unnamed in the Subterbranchialia. Only a careful revision of these groups will allow the introduction of new names so, in the present contribution, these taxa remain in open nomenclature.

In the absence of clear official recommendations concerning the spelling of order-group names, and to preserve homogeneity, the order names have been altered, when necessary, to provide a standardized spelling; so, all the order names are formed with the suffix-iformes.

Age determinations from the older literature are sometimes difficult, but the standard nomenclature of Harland *et al.* (1982) has been adopted wherever possible. Some age assignments had to be broadened, especially in certain Carboniferous, Cretaceous and Tertiary records, e.g. Tournaisian, Viséan, Mississippian, Moscovian and Senonian, reflecting the degree of stratigraphical resolution possible from the relevant literature. In a study devoted to Jurassic fishes, Schaeffer and Patterson (1984) consider the fossiliferous localities of the Solnhofen area (Germany) to be Kimmeridgian, when in fact they are lower Tithonian (= Portlandian).

Subclass ELASMOBRANCHII Bonaparte, 1838 (see Fig. 34.1)

Cohort EUSELACHII Hay, 1902

Superfamily CTENACANTHOIDEA Zangerl, 1981

F. CTENACANTHIDAE Dean, 1909
D. (GIV)–C. (ASB) Mar./FW?

First: *Ctenacanthus wrightii* Newberry, 1884 and *C. nodocostatus* Hussakof and Bryant, 1918, both from the Hamilton Group, Erian, New York State, USA.

The Fossil Record 2. Edited by M. J. Benton. Published in 1993 by Chapman & Hall, London. ISBN 0 412 39380 8

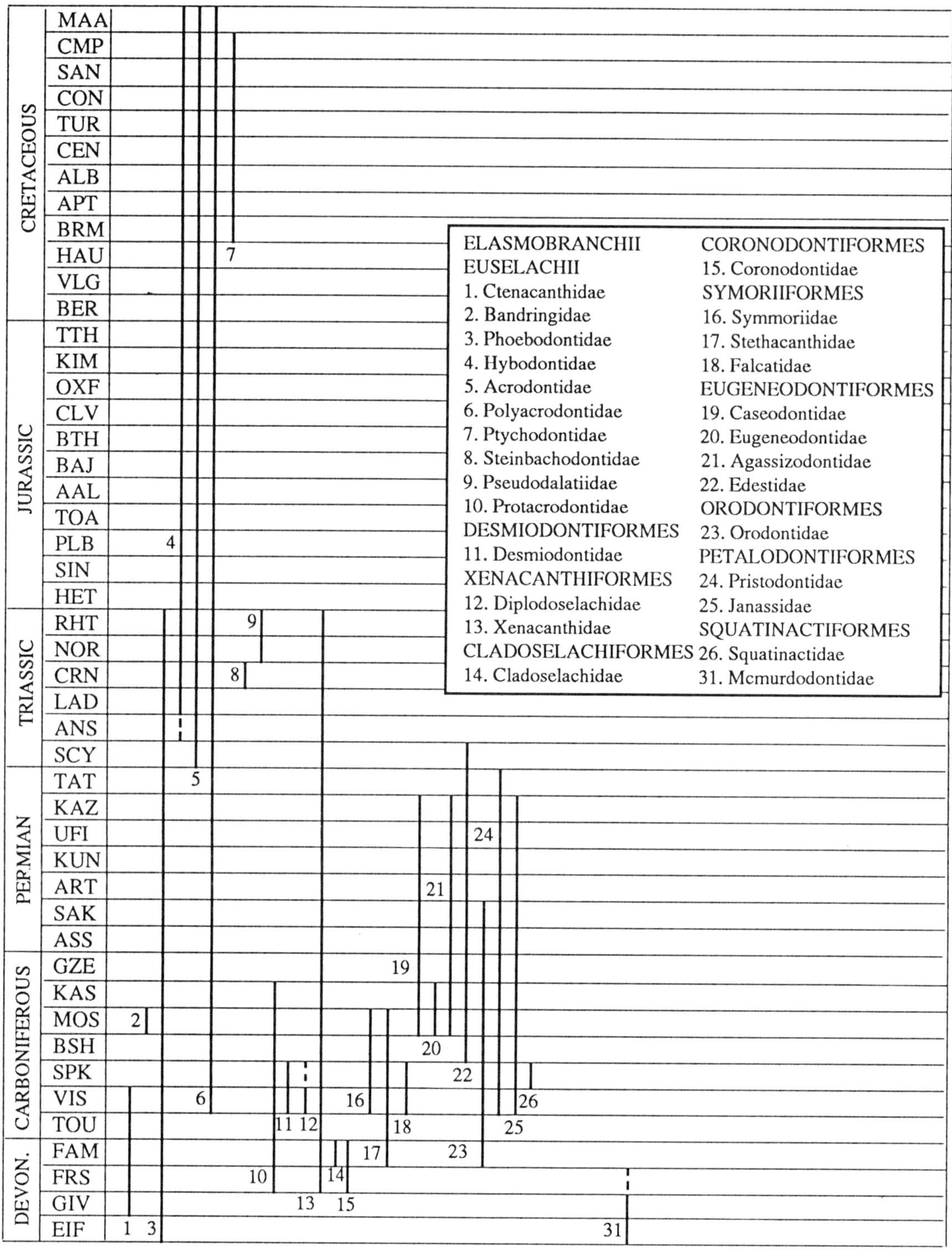

Fig. 34.1

Last: *Ctenacanthus harrisoni* St John and Worthen, 1883, *C. littoni* Newberry, 1889, and *C. pellensis* St John and Worthen, 1883, all from the St Louis Limestone (upper Meramecian) of Illinois, Missouri and Iowa (USA), respectively. [*C. harrisoni* and *C. littoni* may be synonymous with *C. major* according to Maisey (1984), *Ctenacanthus denticulatus* M'Coy, 1855, from Armagh and Drumlish, Northern Ireland, and Shropshire, England, is Asbian and thus correlatable with the upper part of St Louis Limestone.]

Intervening: FRS–HAS, CHD, ARU?, HLK.

Comment: *Ctenacanthus amblyxiphias* Cope, 1891 is known from the Desmoinesian (= upper MOS) of Oklahoma (Zidek, 1976) and the Wolfcampian (= SAK) of Texas (Berman, 1970), but these spines may in fact belong to the petalodont *Megactenopetalus* (cf. Maisey, 1984, pp. 2, 9). Maisey (1981, 1982, 1984) reviewed and revised the genus *Ctenacanthus* and reassigned a number of its species to other genera some of which may not be ctenacanthid or even ctenacanthoid. Therefore, only the species not explicitly

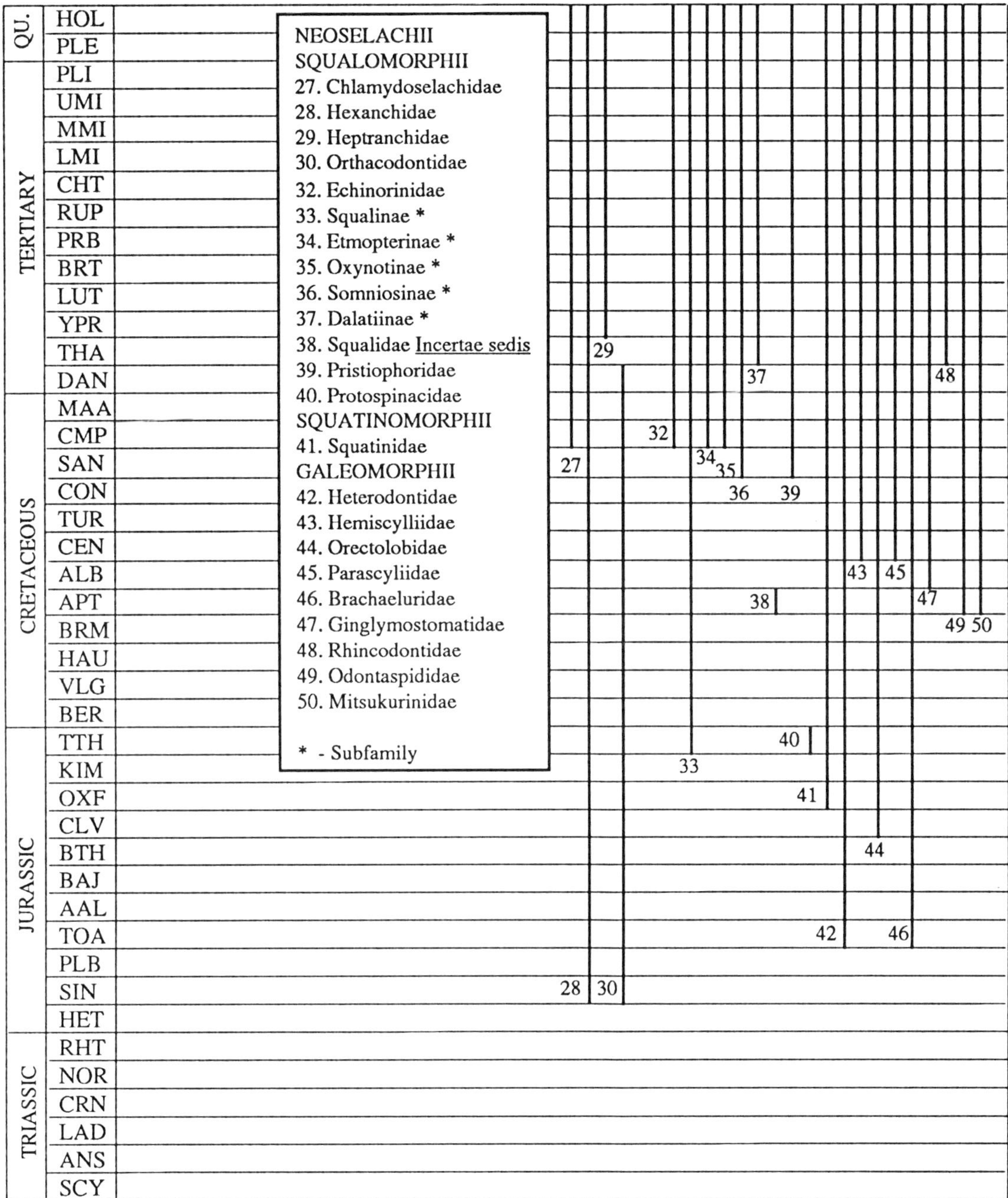

Fig. 34.1

removed from the genus by Maisey (1982, 1984) are considered in this section.

F. BANDRINGIDAE Zangerl, 1969 C. (POD/MYA, MYA) FW/Mar.

First and Last: *Bandringa rayi* Zangerl, 1969 and *B. herdinae* Zangerl, 1979, Essex (predominantly marine, *B. rayi*) as well as Braidwood (predominantly freshwater, *B. herdinae*) faunas of Francis Creek Shale, Carbondale Formation, lower Westphalian D, Mazon Creek, Illinois, USA. *B. rayi* is also known from the Kittanning Formation, middle Westphalian D, Cannelton, Pennsylvania, USA (Baird, 1978). The Cannelton deposit (upper Kittanning Coal) is fresh water.

F. PHOEBODONTIDAE Williams, 1985 D. (EIF)–Tr. (RHT) FW/Mar.

First: *Phoebodus floweri* Wells, 1944, East Liberty and Kiddville Bone Beds, Eifelian, Ohio and Kentucky, USA. **Last:** *Phoebodus brodiei* Woodward, 1893, upper Keuper, Warwick, Warwickshire, England, UK; and *P. keuperensis* Seilacher, 1948, Gipskeuper, Württemberg, Germany. The teeth from the upper Keuper of England may be distinguished from the typical teeth of *Phoebodus* by the absence of secondary cusplets between the principal cusps.
Intervening: GIV–HAS, POD, KLA, NOG, SAK.

F. UNNAMED Tr. (ANS/LAD) Mar.

First and Last: *Acronemus tuberculatus* (Bassanii, 1886), Middle Triassic of the southern Alps (Grenzbituminenzone

of Monte San Giorgio, Tessin, Switzerland). Also in the same beds near Besano, Lombardy, Italy (Rieppel, 1982).
Comment: In Cappetta (1987), this genus seems to be included in the family Phoebodontidae, because the head section was accidentally omitted. In fact, *Acronemus* belongs to a quite different family. On the basis of fin spine structure, *Acronemus* must be assigned to Ctenacanthoidea; yet, the teeth look like those of *Acrodus* (Hybodontoidea), raising the problem of the exact systematic position of the numerous species of *Acrodus* founded on isolated teeth (Rieppel, 1982).

Superfamily HYBODONTOIDEA Zangerl, 1981

The group existed possibly as early as the late Devonian (*'Ctenacanthus' vetustus* Newberry 1873, late FAM, Cleveland Shale Member of Ohio Shale, Ohio, USA), but it is impossible for the moment to group the Palaeozoic forms into families because of our insufficient knowledge (Zangerl, 1981). For the post-Palaeozoic forms, the classification follows Cappetta (1987).

F. HYBODONTIDAE Owen, 1846
Tr. (ANS/LAD)–K. (MAA) Mar./FW

First: *Hybodus plicatilis* Agassiz, 1843, Middle Triassic, Schweuningen, Württemberg, Germany, and Lunéville, north-eastern France. Many other species have been described from the Middle Triassic. In the Fossilium Catalogus (Deecke, 1926) the genus is cited from the SCY of Spitsbergen, but the species in fact belongs to the genus *Polyacrodus*.
Last: *Hybodus* sp., Cappetta and Case, 1975, lower MAA (Mont Laurel Sands) Monmouth Co., New Jersey, USA. The genus is said to occur in the MAA of Niger (Cappetta, 1987, p. 31), but the remains in fact belong to the genus *Asteracanthus*, as previously published (Cappetta, 1972).
Intervening: LAD/ANS, RHT, OXF/KIM, BER/BRM, APT–CEN, CMP–MAA.

F. ACRODONTIDAE Casier, 1959
Tr. (SCY)–K. (MAA) Mar.

First: *Acrodus scaber* Stensiö, 1921, lowest Triassic, Mt. Congress, Spitsbergen.
Last: *Asteracanthus aegyptiacus* Stroemer, 1927, MAA, Mt. Igdaman, Niger (Cappetta, 1972). This species was founded on fin-spines from the CEN of Gebel El Dist, Baharija Oasis, Egypt. Oral teeth from the same locality, corresponding undoubtedly to the fin spines, were recently described as a new genus, *Aegyptobatus*, by Werner (1989) and included, with some doubt, in the new family Distobatidae. The genus *Asteracanthus* (*A. eocaenus* Regan, 1864; see Maisey, 1978) is said to occur in the Palaeocene of southern Australia, but in fact this species must be assigned to the genus *Heterodontus*.
Intervening: LAD/ANS, RHT, HET–TOA, BTH–KIM, VLG, APT–CEN, SAN.

F. POLYACRODONTIDAE Glickman, 1964
C. (VIS)–K. (MAA) Mar./FW

First: *Lissodus wirksworthensis* Duffin, 1985, Cawder Limestones, P1 subzone, Wirksworth, Derbyshire, England, UK (Duffin, 1985).
Last: *Lissodus selachos* (Estes, 1964), Lance Formation, eastern Wyoming, USA.
Intervening: SCY–RHT, BTH, TTH, BER–BRM, ALB–TUR, SAN.

F. PTYCHODONTIDAE Jaekel, 1898
K. (BRM)–K. (CMP) Mar./FW

First: *Hylaeobatis ornatus* (Woodward, 1889), Wealden, Isle of Wight, England, UK.
Last: *Ptychodus mortoni* Agassiz, 1839, 'Grès vert des Etats-Unis', Upper Cretaceous [precise age cannot be established]. This species occurs up into the CMP in England, UK (Dibley, 1911). It is often asserted that the genus *Ptychodus* occurs in the MAA and persists up into the Palaeocene but, as shown by Dibley (1911), this genus disappears during the CMP. Its occurrence in younger levels is the result of reworking or of labelling errors in the collections.
Intervening: ALB–SAN.

F. STEINBACHODONTIDAE Reif, 1980
Tr. (CRN) Mar.

First and Last: *Steinbachodus estheriae* Reif, 1980, lower Carnian, Schwäbisch Hall, southern Germany (Reif, 1980).

F. PSEUDODALATIIDAE Reif, 1978
Tr. (NOR–RHT) Mar.

First and Last: *Pseudodalatias barnstonensis* (Sykes, 1971), Rhaetic, Nottinghamshire, England, UK (Sykes, 1971).
Comment: First described from the Rhaetic of England, this species was later discovered in the Norian (Upper Triassic) of Lombardy, Italy (Tintori, 1980), and in the Upper Triassic of south-eastern France (Cappetta, pers. obs.).

Superfamily PROTACRODONTOIDEA Zangerl, 1981

F. PROTACRODONTIDAE fam. nov.
D. (FRS)–C. (KRE) Mar./Lagoonal

First: *Protacrodus vetustus* Jaekel, 1921, *Manticoceras* Beds, Wildungen, Germany.
Last: *Holmesella equilaterata* Gunnell, 1933, Winterset Limestone of Kansas City Group, Missourian, Missouri, USA.
Intervening: ARN?, CHE, POD, MYA.

Order DESMIODONTIFORMES Zangerl, 1981

F. DESMIODONTIDAE fam. nov.
C. (HLK–ARN) Mar.

First: *Desmiodus tumidus* St John and Worthen, 1875, St Louis Limestone, upper Meramecian, Illinois and Missouri, USA.
Last: *Heteropetalus elegantulus* Lund, 1977, Bear Gulch Limestone Member, Heath Formation, Namurian E_2b, Montana, USA (Lund, 1977).
Comment: Lund (1977) considered *Heteropetalus* to be a petalodont, but Zangerl (1981, pp. 62–3) provisionally included it in the Desmiodontida because of the similarity of its teeth to *Desmiodus*. Lund (1986b, p. 105) stated that *Heteropetalus* 'has little to compare it to other known petalodonts except the general form of the teeth', but most recently he placed it, without any explanation, in the order Hybodontida (Lund, 1990, p. 4).

Order XENACANTHIFORMES Berg, 1940

F. DIPLODOSELACHIDAE Dick, 1981
C. (CHD–HLK/ASB) Lagoonal

First and Last: *Diplodoselache woodi* Dick, 1981, from the base of Lower Oil Shale Group (Granton Sandstones) to the Upper Oil Shale Group (Dunnet Shales), Viséan C_2 S_1 to S_2 or D, Edinburgh area, Scotland, UK (Dick, 1981).

F. XENACANTHIDAE Fritsch, 1889 D. (GIV/FRS)–Tr. (RHT) FW/Lagoonal/Mar.

First: *'Dittodus' priscus* (Eastman, 1899) and *'D.' striatus* (Eastman, 1899), Conodont Bed of Genesee Formation, uppermost Erian/lowermost Senecan, New York State, USA.
Last: *Xenacanthus moorei* (Woodward, 1889), Keuper, Somerset, England, UK (cf. Johnson, 1980, p. 926).
Intervening: FAM, CHD, HLK, ASB, SPK, BSH, POD/MYA, MYA, CHV–KLA, ASS–ART, LAD–NOR.
Comments: A third *'Dittodus'* species from the Genesee Conodont Bed, *D. grabaui* Hussakof and Bryant, 1918, is *Phoebodus* (Phoebodontidae, Ctenacanthoidea). *Anodontacanthus pusillus* Hussakof and Bryant, 1918, also from the Genesee Conodont Bed, is a fragmentary spine which cannot be unequivocally assigned to that genus and, at any rate, the assignment of *Anodontacanthus* to the Xenacanthidae, or even to the Xenacanthida, is equivocal (Zidek, 1978, p. 1075).

Young (1982) described *Antarctilamna prisca* from upper GIV/lower FRS freshwater deposits of south Victoria and New South Wales, and suggested that it is a xenacanth shark immediately related to *Xenacanthus*. However, although the teeth of *A. prisca* are diplodont, their bases appear to be phoebodontid (Ctenacanthoidea) in character, the fin spines are clearly ctenacanthoid, and the scales are taxonomically inconclusive. The isolated teeth identified by Young (1982) as *Xenacanthus* sp. certainly do not belong to that genus and probably are not of xenacanth derivation.

Mader (1986) named isolated teeth from the Lower Devonian (lower LOK–PRA) of northern Spain *Leonodus carlsi*, and assigned them to Xenacanthida *incertae sedis*. Overall, these teeth are similar to *Xenacanthus*, but the basolabial articulating boss, as well as the basolingual margin, show a tendency toward splitting (Mader, 1986, pl. 5, figs 1b, 3, 4), raising the possibility that these teeth are ctenacanthoid.

Poplin and Heyler (1989) proposed a different familial arrangement, recognizing three families, the Diplodoselachidae, Orthacanthidae and Expleuracanthidae. The content of the last two corresponds with the Xenacanthidae of this compilation.

Order CLADOSELACHIFORMES Dean, 1909

F. CLADOSELACHIDAE Dean, 1894 D. (FAM) Mar.

First and Last: *Cladoselache fyleri* (Newberry, 1889) and *Monocladodus clarki* Claypole, 1893, Cleveland Shale Member, Ohio Shale, upper FAM, Ohio, USA.
Comment: Mader (1986) described two types of scales, *Iberolepis aragonensis* and *Lunalepis leonensis* from the Lower Devonian (lower LOK–PRA) of northern Spain, and tentatively assigned them to the Order Cladoselachida.

Order CORONODONTIFORMES Zangerl, 1981

F. CORONODONTIDAE Harris, 1950 D. (GIV/FRS–FAM) Mar./Lagoonal(?)

First: *Coronodus reimanni* Bryant, 1923, Conodont Bed of Genesee Formation, uppermost Erian/lowermost Senecan, New York State, USA.
Last: *Diademodus hydei* Harris, 1950, Cleveland Shale Member of Ohio Shale, upper FAM, Ohio, USA.
Comment: Williams (1985) described *Diademodus*? sp. teeth from the Mecca Quarry Shale of Linton Formation (upper Westphalian C = POD), Indiana, and from the Excello Shale of Carbondale Formation (lower Westphalian D = POD/MYA), Indiana, and noted that a specimen from the Stark Shale of Dennis Formation (upper Westphalian D = MYA), Iowa, may also be assignable to *Diademodus*.

Order SYMMORIIFORMES Zangerl, 1981

F. SYMMORIIDAE Dean, 1909 C. (VIS–POD) Mar./Lagoonal

First: *Denaea fournieri* Pruvost, 1922, Marbres noirs de Denée, lower Viséan (CHD or ARU), Denée area, Belgium.
Last: *Symmorium reniforme* Cope, 1893 and *Denaea meccaensis* Williams, 1985, Excello Shale of Carbondale Formation, lower Westphalian D (= POD/MYA), Indiana, USA (Williams, 1985).
Comment: Mader and Schultze (1987) recorded *Symmorium* sp. from Viséan IIIα3 of the Rheinisches Schiefergebirge, Germany, but since this is in the lower part of the *Goniatites* Zone, it appears to be somewhat younger (HLK) than the occurrence at Denée. Unpublished symmoriid specimens are known from the Virgilian (Stephanian B = KAS–GZE) of New Mexico, USA.

F. STETHACANTHIDAE Lund, 1974 D. (FAM)–C. (MYA) Mar./Lagoonal

First: *Stethacanthus altonensis* (St John and Worthen, 1875), Cleveland Shale Member of Ohio Shale, Ohio, USA (Williams, 1985).
Last: Altamont Limestone Member of Oologah Limestone, Westphalian D (= upper MYA), locality unknown (old specimen, cf. Williams 1985, p. 146); the unit is recognized in south-eastern Kansas, southern Iowa, south-western Missouri, and north-eastern Oklahoma, USA.
Intervening: HAS, IVO, HLK–ARN, POD, MYA.
Comment: *Stethacanthus praecursor* Hussakof and Bryant 1918, from the Conodont Bed of the Genesee Formation (GIV/FRS), New York State, USA, has been declared an 'indeterminate lump' (Lund, 1984, p. 283), but it may well warrant at least the generic assignment (cf. Hussakof and Bryant, 1918, pl. 54, figs 1, 1a, 2).

There is a strong disagreement concerning the species and generic diversity of the family. Zangerl (1981) recognized only *Stethacanthus* and placed all occurrences in *S. altonensis*. Williams (1985) concurred, while recognizing *S. carinatus*, *S. humilis*, and possibly *S. proclivus*, and *S. erectus* as valid species, whereas Lund (1984, 1985a,b, 1986a) recognized a number of species placed in the genera *Stethacanthus* Newberry, 1889, *Orestiacanthus* Lund, 1984, *Falcatus* Lund, 1985 and *Damocles* Lund, 1986.

F. FALCATIDAE Zangerl, 1990 C. (HAS–ARN) Mar./Lagoonal

First: *Falcatus falcatus* (St John and Worthen, 1875), Kinderhookian, Iowa, USA (Lund, 1985a).
Last: *Falcatus falcatus* (St John and Worthen, 1875) and *Damocles serratus* Lund, 1986, both Bear Gulch Limestone, upper Chesterian, Namurian E2b, Montana, USA (Lund, 1985a, 1986a, 1990).
Intervening: HLK/ASB.

Order EUGENEODONTIFORMES Zangerl, 1981

Superfamily CASEODONTOIDEA Zangerl, 1981

F. CASEODONTIDAE Zangerl, 1981 C. (POD)–P. (KAZ) Mar./Lagoonal

First: *Ornithoprion hertwigi* Zangerl, 1966 and *Caseodus eatoni* Zangerl, 1981, Logan Quarry Shale, Staunton Formation, upper Westphalian C, Indiana, USA.
Last: *Fadenia crenulata* Nielsen, 1932 and *Erikodus groenlandicus* (Nielsen, 1932), *Posidonia* Shale Member, Foldvik Creek Formation, East Greenland.
Intervening: MYA.
Comment: Teeth identifiable as *Campodus* sp. (Caseodontoidea *inc. sedis*) are known from the Salem Limestone (middle Meramecian = ARU/HLK) of Missouri, USA (Zangerl, 1981, p. 85).

F. EUGENEODONTIDAE Zangerl, 1981 C. (POD–KRE) Mar./Lagoonal

First: *Gilliodus orvillei* Zangerl, 1981, *G. peyeri* Zangerl, 1981, and *Eugeneodus richardsoni* Zangerl, 1981, Logan Quarry Shale of Staunton Formation, upper Westphalian C, Indiana, USA.
Last: *Bobbodus schaefferi* Zangerl, 1981, Queen Hill Shale of Lecompton Formation, Stephanian A, Nebraska, USA.
Intervening: POD/MYA, MYA.

Superfamily EDESTOIDEA Hay, 1930

F. AGASSIZODONTIDAE Zangerl, 1981 C. (POD/MYA)–P. (KAZ) Mar.

First: *Agassizodus variabilis* (Newberry and Worthen, 1870). The holotype, from Illinois, USA, is presumably early Westphalian D (= POD/MYA) in age, but a new specimen referred to this species by Zangerl (1981, pp. 75, 77), from the Queen Hill Shale of Lecompton Formation, Nebraska, USA, is Stephanian A (= KRE). *Arpagodus rectangulus* Trautschold, 1879, from Myachkova, USSR, is late Westphalian D or Cantabrian (= MYA), and is thus nearly contemporaneous with *A. variabilis*.
Last: *Sarcoprion edax* Nielsen, 1932, *Posidonia* Shale Member of Foldvik Formation, East Greenland.
Intervening: KRE, MYA, ASS/SAK, ART, KUN (incomplete, ages of many specimens uncertain).

F. EDESTIDAE Jaekel, 1899 C. (BSH)–Tr. (GRI) Mar.

First: *Lestrodus newtoni* (Woodward, 1917), upper Millstone Grit, lower BSH, Yorkshire, England, UK.
Last: *Parahelicampodus spaercki* Nielsen, 1952, Wordie Creek Formation, lower SCY, East Greenland. *Helicampodus ealoni* Obruchev, 1965, from Dzhul'fa on the Araks River, Armenia, former USSR, is also SCY, probably roughly contemporaneous with *P. spaercki*.
Intervening: POD, MYA, ART, KUN, KAZ (incomplete, ages of many specimens uncertain).

Order ORODONTIFORMES Zangerl, 1981

F. ORODONTIDAE de Koninck, 1878 D. (FAM)–P. (SAK) Mar./Lagoonal

First: *Hercynolepis meischneri* Gross, 1973, FAM [oberes Oberdevon (to III) of Gross, 1973, p. 102], Harz Mountains, Germany.
Last: *Orodus corrugatus* Romer, 1942, Lueders Formation, uppermost Wolfcampian, Texas, USA (Berman, 1970; Johnson, 1981). Species name preoccupied by *Orodus corrugatus* Newberry and Worthen, 1870.
Intervening: HAS/IVO through ASS; with the exception of POD and MYA, stage names cannot be given due to dating and taxonomic uncertainty.

Order PETALODONTIFORMES Zangerl, 1981

F. PRISTODONTIDAE Woodward, 1889 C. (BRI)–P. (TAT) Mar./Lagoonal

First: *Pristodus falcatus* Davis, 1883, Brigantian, Yorkshire, England, UK.
Last: *Megactenopetalus kaibabanus* David, 1944, Dzulfian (lower TAT), China and Iran (cf. Hansen, 1978).
Intervening: PND, ARN, KRE, NOG, ASS?, KUN–KAZ.

F. JANASSIDAE Jaekel, 1899 C. (ASB)–P. (KAZ) Mar./Lagoonal

First: *Janassa clavata* (M'Coy 1855), Asbian, Armagh, Northern Ireland, UK.
Last: *Janassa kochi* Nielsen, 1932 and *J. unguicula* (Eastman, 1903), *Posidonia* Shale Member, Foldvik Creek Formation, East Greenland.
Intervening: BRI, ARN, CHE–MYA, CHV, KLA, ASS–UFI.
Comment: Zangerl (1981), following Hansen (1978), included *Peripristis* St John, 1870, *Pristodus* Davis, 1883 and *Megactenopetalus* David, 1944 in the Pristodontidae, and left all other petalodont genera without family affiliation. Hansen (1985) revived the family Janassidae (Jaekel's subfamily) to include *Janassa* Muenster, 1839 and *Fissodus* St John and Worthen, 1875, and possibly *Peltodus* Newberry and Worthen, 1870 and *Cholodus* St John and Worthen, 1875, should they prove distinct from the former two. Lund (1986b, 1989) added four more families, the Petalodontidae (*Petalodus* Owen, 1840 and *Polyrhizodus* M'Coy, 1848), Peripristidae (presumably for *Peripristis* St John, 1870, but generic content not specified), Pristodontidae (*Pristodus* Davis, 1883, *Siksika* Lund, 1989, *Petalorhynchus* Newberry and Worthen, 1866, *Peripristis* St John, 1870, and *Megactenopetalus* David, 1944), and Belantseidae (*Belantsea* Lund, 1989, *Netsepoye* Lund, 1989 and *Ctenoptychius* Agassiz, 1838). However, the content of the Petalodontidae is far from clear, the separation of the Peripristidae (Lund, 1986b) has later been retracted (Lund, 1989), and the family Belantseidae appears to be based on specimens that may prove to be a *Ctenoptychius* (*Belantsea montana* Lund, 1989) and possibly a juvenile *Fissodus* (*Netsepoye hawesi* Lund, 1989), i.e. a janassid. To clarify these issues would require a major research undertaking, which is beyond the scope and purpose of this review, and which could easily fail because of the paucity of the existing material. Therefore, only the Pristodontidae and Janassidae are recognized.

Order SQUATINACTIFORMES Zangerl, 1981

F. SQUATINACTIDAE **fam. nov.** C. (ARN) Mar.

First and Last: *Squatinactis caudispinatus* Lund and Zangerl, 1974, Bear Gulch Limestone Member, Heath Formation, Namurian E2b, Montana, USA (Lund and Zangerl, 1974).

Subcohort NEOSELACHII Compagno, 1977

Superorder SQUALOMORPHII Compagno, 1973

Order HEXANCHIFORMES Buen, 1926

Suborder CHLAMYDOSELACHOIDEI Berg, 1958

F. CHLAMYDOSELACHIDAE Garman, 1884 K. (CMP)–Rec. Mar.

First: *Chlamydoselachus thomsoni* Richter and Ward, 1990, Beta Member, Santa Marta Formation, Brandy Bay area, James Ross Island (Antarctic) (Richter and Ward, 1990). An undescribed species also occurs in the Campanian of Angola (Cappetta, 1987). **Extant**
Intervening: LUT, RUP/CHT, BUR, SRV, ZAN.

Suborder HEXANCHOIDEI Garman, 1913

F. HEXANCHIDAE Gray, 1851 J. (SIN)–Rec. Mar.

First: *Hexanchus arzoensis* (Beaumont, 1960), Lotharingian, Tessin, Switzerland. **Extant**
Intervening: OXF, ALB, SAN–YPR, RUP–CHT, TOR, ZAN.

F. HEPTRANCHIDAE Barnard, 1925 T. (YPR)–Rec. Mar.

First: *Heptranchias howellii* (Reed, 1946), Lower Eocene(?), Montmouth Co., New Jersey, USA. The original specimen was found in a boulder probably derived from the Shark River Formation of early Eocene age. This species also occurs in the YPR of Morocco (Cappetta, 1981). **Extant**
Intervening: LUT/BRT, BUR–LAN.

F. ORTHACODONTIDAE Beaumont, 1960 J. (SIN)–T. (DAN) Mar.

First: *Sphenodus helveticus* Beaumont, 1960 [in part], Tessin, Switzerland. The type material is a mixture of teeth of *Sphenodus* and of *Paraorthacodus*.
Last: *Sphenodus lundgreni* (Davis, 1890), Danian of southern Sweden and Denmark. Glickman (1957) made this species the type of the genus *Eychlaodus*.
Intervening: TOA, BAJ–SAN, MAA.

?*Order* HEXANCHIFORMES

?*Suborder* HEXANCHOIDEI

F. MCMURDODONTIDAE White, 1968 D. (EMS/EIF–GIV/FRS) Mar.

First: *Mcmurdodus whitei* Turner and Young, 1987, Lower/Middle Devonian, Cravens Peak Beds, Georgina Basin, western Queensland, Australia (Turner and Young, 1987).
Last: *Mcmurdodus featherensis* White, 1968, Middle/Upper Devonian, south Victoria Land, Antarctica.
Comment: By their morphology, the teeth of *Mcmurdodus* look like hexanchid teeth, mainly the lower symphysial ones. Yet, this resemblance could be the result of convergence. There is a large gap in the fossil record between the last *Mcmurdodus* and the first unquestionable hexanchids (SIN).

Order SQUALIFORMES Goodrich, 1909

F. ECHINORHINIDAE Gill, 1862 K. (CMP)–Rec. Mar.

First: *Echinorhinus* sp., an undescribed species from the CMP of Angola and Morocco (Cappetta, 1987). The occurrence of the family in the CMP of Westphalia (Müller and Schöllmann, 1989) needs to be confirmed by better preserved material. The most ancient described representative of the family is *Gibbechinorhinus lewyi* Cappetta, 1990, lower MAA Oron Syncline, Negev Desert, Israel (Cappetta, 1990). **Extant**
Intervening: MAA–LUT, RUP, BUR–LAN, ZAN.

F. SQUALIDAE Bonaparte, 1834

Subfamily SQUALINAE Bonaparte, 1834 J. (TTH)–Rec. Mar.

First: *Squalogaleus woodwardi* Maisey, 1976, lower TTH, Solnhofen, Germany. Maisey (1976) assigned this genus to the suborder Galeoidea (Lamniformes + Carcharhiniformes *sensu* Compagno, 1973). Cappetta (1987), based mainly on the dental morphology, considered *Squalogaleus* close to the ancestral stock leading to more advanced Squaliformes. In case *Squalogaleus* proves not to be a squaliform, the most ancient representative of the subfamily Squalinae would be *Protosqualus albertsii* Thies, 1981, BRM, Braunschweig, NW Germany (Thies, 1981). **Extant**
Intervening: BRM, ALB, TUR, SAN–LUT, RUP/CHT–TOR, ZAN–PLE.

Subfamily ETMOPTERINAE Fowler, 1934 K. (CMP)–Rec. Mar.

First: *Eoetmopterus supracretaceus* Müller and Schöllmann, 1989, Upper Coesfeder Schichten, Westphalia, NW Germany. A species of *Etmopterus* has been described from the MAA of Hemmoor, NW Germany (*E. hemmoorensis* Herman, 1982) but its dental morphology indicates that it is *Eoetmopterus* Müller and Schollmann, 1989. **Extant**
Intervening: MAA, BUR–LAN, TOR.

Subfamily OXYNOTINAE Rafinesque, 1810 K. (CMP)–Rec. Mar.

First: *Protoxynotus misburgensis* Herman, 1975, upper CMP, Hannover, NW Germany (Herman, 1975). **Extant**
Intervening: AQT/BUR, ZAN.

Subfamily SOMNIOSINAE Jordan, 1888 K. (SAN)–Rec.

First: *Cretascymnus adonis* (Signeux, 1950), upper SAN, Sahel Alma, Lebanon. **Extant**
Intervening: CMP–MAA, CHT, SRV, ZAN, PLE.

Subfamily DALATIINAE Gray, 1851 T. (THA)–Rec. Mar.

First: *Isistius trituratus* (Winkler, 1874), LUT, Woluwe-St-Lambert, Belgium. This species occurs as early as the THA in Morocco (Arambourg, 1952). The genus *Scymnorhinus* (= *Dalatias*) has been recorded from the Palaeocene of former USSR on the basis of a single crown from a lower file (Glickman, 1964), but this occurrence needs to be confirmed by more complete and convincing material.
Intervening: YPR–TOR, ZAN–PIA.

Family ?SQUALIDAE *INCERTAE SEDIS* K. (APT) Mar.

First and Last: *Centropterus lividus* Costa, 1861, APT, Naples area, Italy. The single specimen, a heavily damaged skeleton, shows a second dorsal fin with a spine. Its assignment to the Squalidae is quite provisional.

Order PRISTIOPHORIFORMES Berg, 1958

F. PRISTIOPHORIDAE Bleeker, 1859 K. (SAN)–Rec. Mar.

First: *Pristiophorus tumidens* (Woodward, 1932), upper SAN, Sahel Alma, Lebanon. **Extant**
Intervening: MAA, THA, LUT–PLE.

?***Superorder*** SQUALOMORPHII Compagno, 1973

F. PROTOSPINACIDAE Woodward, 1919 J. (TTH) Mar.

First and Last: *Protospinax annectans* Woodward, 1919, lower TTH, Solnhofen, Germany.
Comment: The systematic position of this genus is still subject to discussion. Maisey (1976) considered *Protospinax* as a junior synonym of *Belemnobatis*, a batoid described by Thiollière (1854) from the KIM of Cérin, France. However, study of the skeletal anatomy of the two genera shows that *Protospinax* is an unquestionable shark and *Belemnobatis* is a batoid. Thies (1983) has grouped together the genera *Protospinax* and *Squalogaleus* Maisey, 1976, regarding the latter as a juvenile specimen of the former. Yet, on the basis of the skeletal calcification, arrangement of the tooth rows, and the dental morphology, Cappetta (1987) considered the two genera as valid. Thies (1983) described as *Protospinax* teeth from the CLV of England and OXF of Germany; these teeth by their morphologies, particularly the root with a completely closed groove, are easily separated from *Protospinax*.

Superorder SQUATINOMORPHII Compagno, 1973

Order SQUATINIFORMES Buen, 1926

F. SQUATINIDAE Bonaparte, 1838 J. (OXF)–Rec. Mar.

First: *Squatina* sp., upper OXF (lower *bimammatum* Zone), Hannover, Germany (Thies, 1983). *S. acanthoderma* Fraas, 1854 is represented by complete specimens in the lower TTH, Nusplingen, southern Germany. **Extant**
Intervening: TTH, VLG, APT, CEN–MAA, THA–PRB, LAN, ZAN.

Superorder GALEOMORPHII Compagno, 1973

Order HETERODONTIFORMES Berg, 1937

F. HETERODONTIDAE Gray, 1851 J. (TOA)–Rec. Mar.

First: *Heterodontus duffini* Thies, 1983, upper TOA (*dispansum* to *aalensis* Zones), near Hannover, Germany. This genus is known by complete skeletons from the lower TTH, *H. falcifer* (Wagner, 1857), Nusplingen, southern Germany. **Extant**
Intervening: AAL, TTH, VLG, APT–TUR, SAN–PRB, AQT/MES, PLE.

Order ORECTOLOBIFORMES Applegate, 1972

F. HEMISCYLLIIDAE Gill, 1862 K. (CEN)–Rec. Mar.

First: *Mesiteia emilae* Kramberger, 1885, CEN, Hakel, Lebanon. It was believed, following the original description, that the type specimen had been collected in the Lower Eocene of Monte Bolca, Italy. Discovery of identical specimens from Lebanon and studies of the microfossils from the matrix of the type specimen have demonstrated that the true source of the type was the Lower Eocene of Italy. **Extant**
Intervening: TUR–CMP, THA–LUT.

F. ORECTOLOBIDAE Jordan and Fowler, 1903 J. (CLV)–Rec. Mar.

First: *Orectoloboides pattersoni* Thies, 1983, Oxford Clay, Bedfordshire, England, UK (Thies, 1983).
Intervening: APT–CEN, CMP–MAA, THA–YPR.

F. PARASCYLLIIDAE Gill, 1862 K. (CEN)–Rec. Mar.

First: *Pararhincodon lehmani* Cappetta, 1980, CEN, Hakel, Lebanon (Cappetta, 1980). **Extant**
Intervening: TUR, YPR.

F. BRACHAELURIDAE Applegate, 1972 J. (TOA)–Rec. Mar.

First: *Palaeobrachaelurus aperizotus* Thies, 1983, upper AAL (*scissum* and *murchisonae* Zones), Hannover, NW Germany. The same species occurs in the upper TOA (*dispansum* to *aalensis* Zones) at the same locality (Thies, 1983). **Extant**
Intervening: AAL, BRM, LUT–BRT.
Comment: In Cappetta (1987), the range of the family is erroneously listed as Barremian to Recent.

F. GINGLYMOSTOMATIDAE Gill, 1862 K. (ALB)–Rec. Mar.

First: *Ginglymostoma lithuanica* Dalinkevicius, 1935, ALB, Sventoji River, Lithuania, former USSR.
Intervening: CEN–TUR, MAA–LUT, LAN.

F. RHINCODONTIDAE Garman, 1913 T. (THA/YPR)–Rec. Mar.

First: *Palaeorhincodon dartevellei* (Arambourg, 1952), YPR, Oued Oussen, Ouled Abdoun Basin, Morocco. This species was originally assigned to the genus *Squatirhina* which is a batoid and not a shark (Cappetta, 1987). Arambourg (1952) recorded *P. dartevellei* from the THA but this probably belongs to another orectolobiform genus of the same family (Cappetta, pers. obs.). **Extant**
Intervening: LUT–BRT, LAN.

Order LAMNIFORMES Berg, 1958

F. ODONTASPIDIDAE Müller and Henle, 1839 K. (APT)–Rec. Mar.

First: *Carcharias striatula* Dalinkevicius, 1935, ALB, Sventoji River, Lithuania, former USSR. This species has then been collected also in the Gargasian, upper APT, Vaucluse, southern France (Cappetta, 1975).
Intervening: ALB–PIA.
Comment: The generic assignment of numerous odontaspidid species has been controversial until recently (Compagno, 1984; Cappetta, 1987; Ward, 1988). All the species previously assigned to the genera *Synodontaspis* White, 1931 and *Eugomphodus* Gill, 1861, must be assigned to the genus *Carcharias* Rafinesque, 1810.

F. MITSUKURINIDAE Jordan, 1898 K. (APT)–Rec. Mar.

First: *Anomotodon principialis* Cappetta, 1975, Gargasian, upper APT, Vaucluse, southern France. **Extant**
Intervening: ALB–MAA, THA–LUT, CHT–SRV, ZAN/PIA.

F. LAMNIDAE Müller and Henle, 1838 T. (THA)–Rec. Mar. (see Fig. 34.2)

First: *Isurus winkleri* (Vincent, 1876), THA (*Pholadomya oblitterata* level), Liège Province, Belgium.
Intervening: YPR–LUT, RUP, AQT–PLE.

F. CRETOXYRHINIDAE Glickman, 1958 K. (VLG)–PRB Mar.

First: *Protolamna infracretacea* Leriche, 1910, upper Neocomian (VLG/HAU), Haute-Marne, NE France. Pictet and Campiche (1858) described *Odontaspis gracilis* Agassiz from the VLG of the Sainte-Croix area, Switzerland; the figured teeth can be assigned to the genus *Protolamna* Cappetta, 1980.
Last: *Cretolamna twiggsensis* Case, 1981, 'Jacksonian', Twiggs County, Georgia, USA. Also in the Qsar-el-Sagha Formation of Fayum, Egypt (Case and Cappetta, 1990).
Intervening: VLG/HAU–SAN, MAA, THA–BRT.

F. OTODONTIDAE Glickman, 1964 T. (THA–PIA) Mar.

First: *Otodus obliquus* Agassiz, 1843, London Clay (YPR), Kent, England, UK. This species appears in the THA: Morocco (Arambourg, 1952), Belgium and the Paris Basin (Leriche, 1906), and Bulgaria (Datchev, 1971).
Last: *Parotodus benedeni* (Le Hon, 1871), Neogene, Belgium. This species persists up into the PLI: Angola (Dartevelle and Casier, 1959), Belgium (Leriche, 1926), England, UK (Woodward, 1894), Italy (Lawley, 1881; Landini, 1977) and North America (Cappetta, pers. obs.). **Extant**
Intervening: YPR–LUT, RUP, AQT–TOR.

F. ALOPIIDAE Bonaparte, 1838 K. (? CEN)–Rec. Mar.

First: *Paranomotodon angustidens* (Reuss, 1845), TUR, Czechoslovakia. This species appears in the CEN: England, UK (Woodward, 1911 [1912]) and Lithuania, former USSR (Dalinkevicius, 1935). *Paranomotodon* is assigned to the Alopiidae because of the great resemblance of its teeth to those of some species of *Alopias* (Cappetta and Case, 1975); however, this resemblance may be the result of convergent evolution. If this genus proves not to be an alopiid, the first unquestionable representative of the family will be *Alopias denticulatus* Cappetta, 1981, lower YPR, Ouled Abdoun Basin, Morocco. **Extant**
Intervening: YPR–BRT, RUP, BUR–SRV.

F. CETORHINIDAE Gill, 1862 T. (RUP)–Rec. Mar.

First: *Cetorhinus parvus* Leriche, 1908, RUP, Belgium. Usually only the gill-rakers are collected and figured; the teeth generally escape attention because of their small size. The family seems to be present in the Eocene of North America (B. Welton *in litteris*), but this needs to be confirmed by published material. **Extant**
Intervening: BUR–SER, ZAN, PLE.

F. ANACORACIDAE Casier, 1947 K. (ALB–MAA) Mar.

First: *Squalicorax australis* (Chapman, 1909), Boulia, Queensland, Australia. The genus is known from the ALB of Angola (Antunes, 1972).
Last: *Pseudocorax affinis* (Münster, *in* Agassiz, 1843); *Paracorax jaekeli* (Woodward, 1895); *Squalicorax pristodontus* (Agassiz, 1843), MAA, Maastricht, The Netherlands; *Squalicorax yangaensis* (Dartevelle and Casier, 1943), Senonian, Yanga Lake, Cabinda, western Africa; MAA, Morocco (Arambourg, 1952); *Squalicorax bassanii* (Gemmellaro, 1920), MAA, Nile Valley and Red Sea phosphate deposits, Egypt; also in the MAA of Morocco as *S. kaupi* (Arambourg, 1952).
Intervening: CEN–CMP.

Order CARCHARHINIFORMES Compagno, 1973

F. SCYLIORHINIDAE Gill, 1862 J. (BTH)–Rec. Mar.

First: New genus and species, BTH, El Mers area, middle Atlas, Morocco (Cappetta, in prep.). The first described species is: *Macrourogaleus hassei* Woodward, 1889, lower TTH, Solnhofen, Germany. **Extant**
Intervening: TTH, VLG, ALB–TUR, SAN–YPR, BUR–TOR, ZAN.

F. TRIAKIDAE Gray, 1851 K. (TUR)–Rec. Mar.

First: *Galeorhinus* sp. [figured as *G. minutissimus* (Arambourg, 1952)], Lublin, Poland (Marcinowski and Radwanski, 1983). The first described species is *Paratriakis bettrechiensis* Herman, 1977, TUR, Bettrechies, northern France (Herman, 1977). **Extant**
Intervening: SAN–LUT, BUR, SRV, ZAN.

F. HEMIGALEIDAE Hasse, 1879 T. (LUT/BRT)–Rec. Mar.

First: *Hemipristis curvatus* Dames, 1883, Middle Eocene, Birket-El-Qurun, Fayum, Egypt. **Extant**
Intervening: PRB, BUR–TOR, PLE.

F. CARCHARHINIDAE Jordan and Evermann, 1896 T. (THA)–Rec. Mar.

First: New genus and species, Ouarzazate Basin, south Morocco (Cappetta *et al.*, 1987). It is probable that *Scyliorhinus africanus* Arambourg, 1952 from the THA of Morocco will be reassigned to the genus *Abdounia*. The first named representatives of the family are: *Abdounia beaugei* (Arambourg, 1952), YPR, Ouled Abdoun Basin, Morocco; *Eogaleus bolcensis* Cappetta, 1975, YPR, Monte Bolca, Italy; *Galeocerdo latidens* (Agassiz, 1843), upper YPR of Morocco (Cappetta, 1987); *Physogaleus secundus* (Winkler, 1874), LUT, Neder-Ockerzeel, Belgium; YPR, Morocco (Arambourg, 1952; Cappetta, 1981, 1987), and Anglo-Franco-Belgian Basin (Leriche, 1905, 1906; Priem, 1908; Casier, 1946, 1966). **Extant**
Intervening: YPR–RUP, AQT–SRV, ZAN, PLE.

F. SPHYRNIDAE Gill, 1872 T. (AQT)–Rec. Mar.

First: *Sphyrna arambourgi* Cappetta, 1970, Langhian, Loupian, southern France (Cappetta, 1970). The teeth from the LUT of Nigeria that White (1926) attributed to this genus in fact belong to the genus *Physogaleus*. An indeterminate species of the genus occurs in the AQT, Caunelle, Montpellier area, southern France (Cappetta, pers. obs.). **Extant**
Intervening: LAN, ZAN, PLE.

Superorder GALEOMORPHII *INCERTAE ORDINIS*

F. PALAEOSPINACIDAE Regan, 1906 Tr. (SCY)–T. (THA) Mar.

First: ?*Palaeospinax* sp. (Thies, 1982), Dienerian, *cristagalli*

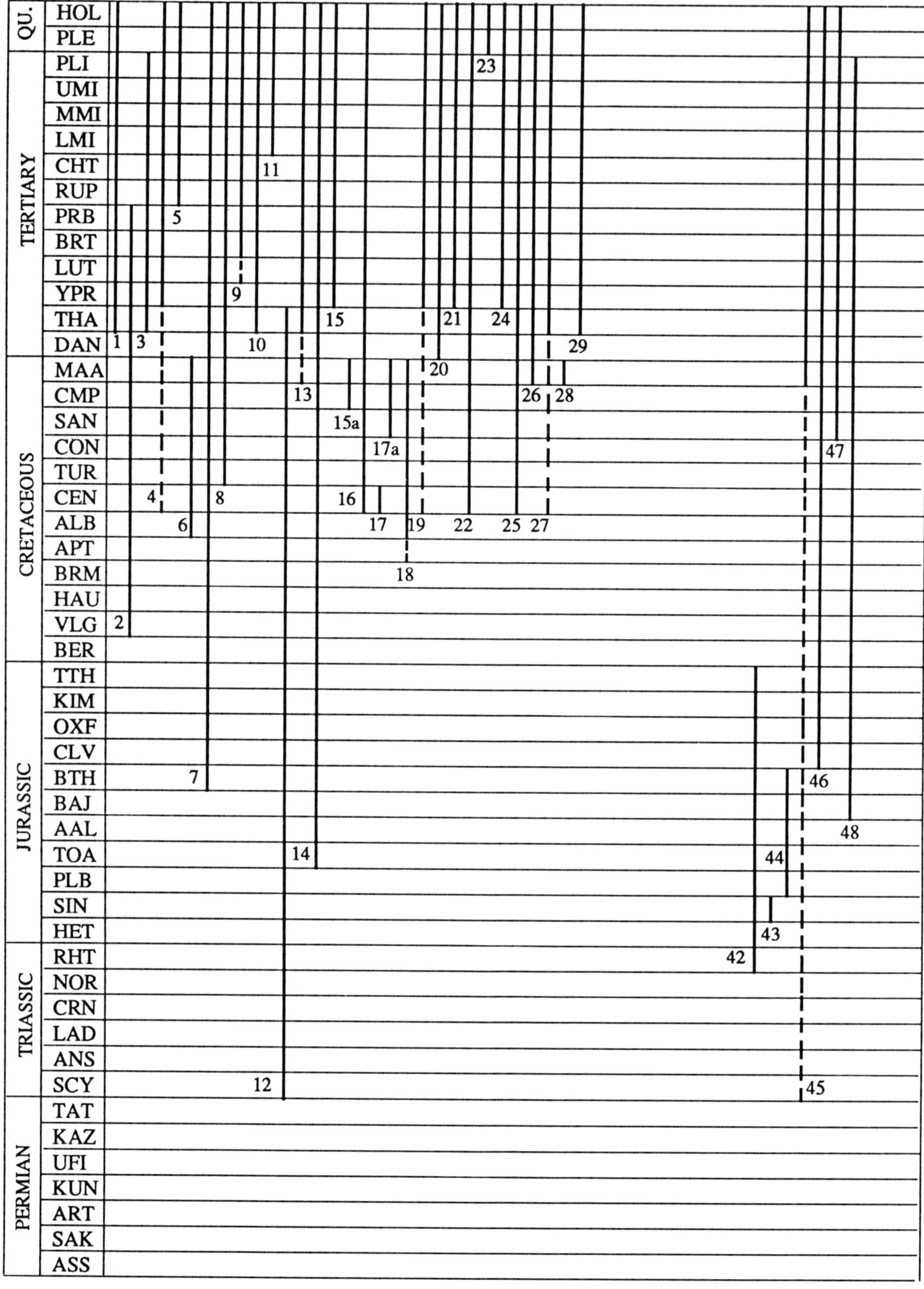

Fig. 34.2

Zone, Kocaeli Peninsula, between Istanbul and Izmit, western Turkey (Thies, 1982). Although incomplete, the tooth figured by Thies really seems to belong to the genus *Palaeospinax* Egerton, 1872.

Last: *Paraorthacodus clarkii* (Eastman, 1901), Aquia Formation, Liverpool Point, Maryland, USA; *P. eocaenus* (Leriche, 1902), THA, Erquelinnes, Belgium. These two species may be synonymous. The last species occurs at Dormaal, Belgium, where it is associated with mammalian remains of early Eocene age; therefore, the teeth of *P. eocaenus* seem to be reworked.

Intervening: RHT, SIN, TOA–AAL, TTH, VLG, APT–TUR, CMP–DAN.

Superorder BATOMORPHII Cappetta, 1980

Order RAJIFORMES Berg, 1940

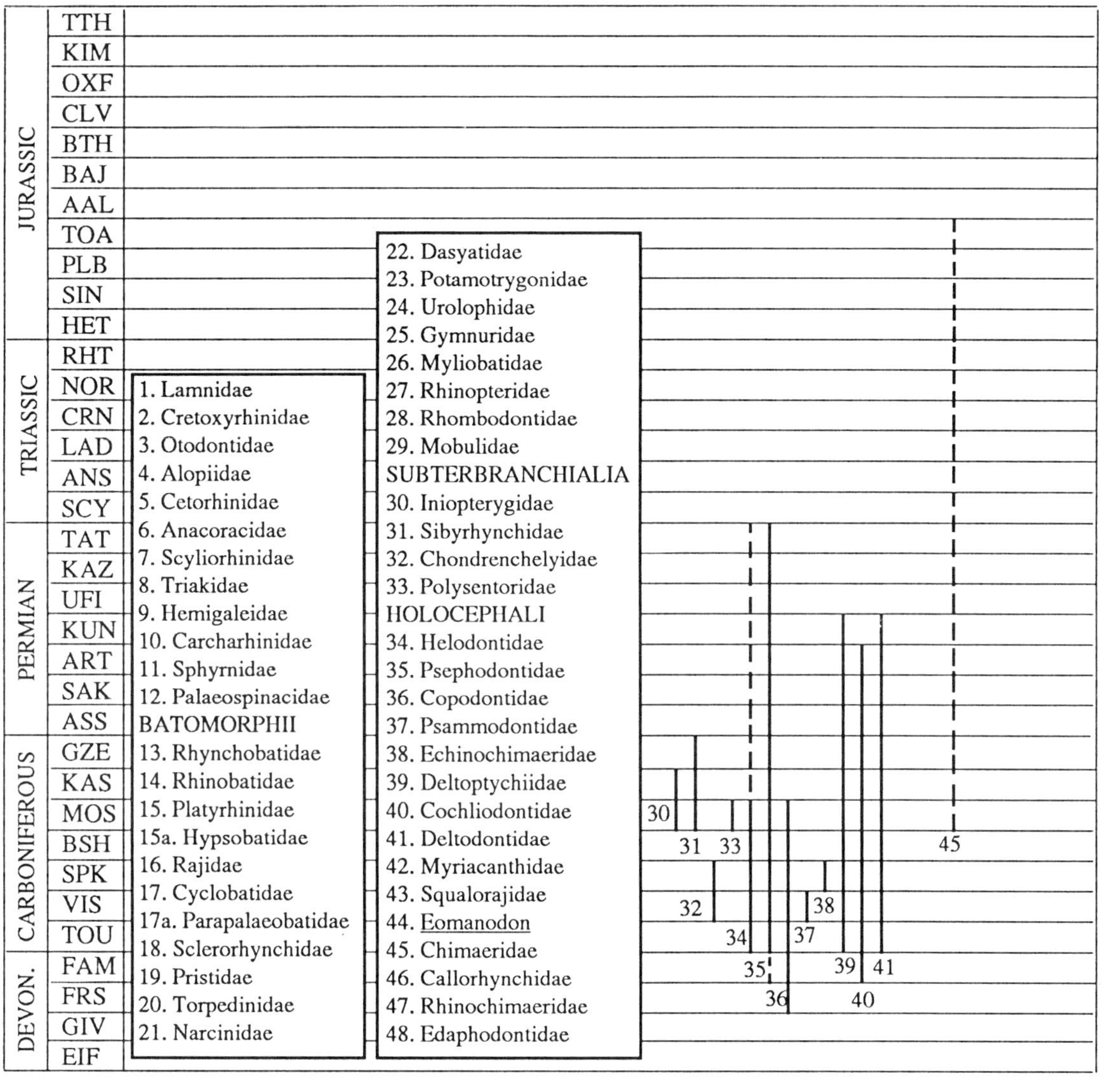

JURASSIC: TTH, KIM, OXF, CLV, BTH, BAJ, AAL, TOA, PLB, SIN, HET
TRIASSIC: RHT, NOR, CRN, LAD, ANS, SCY
PERMIAN: TAT, KAZ, UFI, KUN, ART, SAK, ASS
CARBONIFEROUS: GZE, KAS, MOS, BSH, SPK, VIS, TOU
DEVON.: FAM, FRS, GIV, EIF

1. Lamnidae
2. Cretoxyrhinidae
3. Otodontidae
4. Alopiidae
5. Cetorhinidae
6. Anacoracidae
7. Scyliorhinidae
8. Triakidae
9. Hemigaleidae
10. Carcharhinidae
11. Sphyrnidae
12. Palaeospinacidae
BATOMORPHII
13. Rhynchobatidae
14. Rhinobatidae
15. Platyrhinidae
15a. Hypsobatidae
16. Rajidae
17. Cyclobatidae
17a. Parapalaeobatidae
18. Sclerorhynchidae
19. Pristidae
20. Torpedinidae
21. Narcinidae

22. Dasyatidae
23. Potamotrygonidae
24. Urolophidae
25. Gymnuridae
26. Myliobatidae
27. Rhinopteridae
28. Rhombodontidae
29. Mobulidae
SUBTERBRANCHIALIA
30. Iniopterygidae
31. Sibyrhynchidae
32. Chondrenchelyidae
33. Polysentoridae
HOLOCEPHALI
34. Helodontidae
35. Psephodontidae
36. Copodontidae
37. Psammodontidae
38. Echinochimaeridae
39. Deltoptychiidae
40. Cochliodontidae
41. Deltodontidae
42. Myriacanthidae
43. Squalorajidae
44. Eomanodon
45. Chimaeridae
46. Callorhynchidae
47. Rhinochimaeridae
48. Edaphodontidae

Fig. 34.2

Suborder RHINOBATOIDEI Fowler, 1941

F. RHYNCHOBATIDAE Garman, 1913 ?K. (MAA)/T. (THA)–Rec. Mar.

First: Undescribed species of *Rhynchobatus*, THA, Ouarzazate Basin, south Morocco (Gheerbrant *et al.*, 1993). The first described species is *Rhynchobatus vincenti* Jaekel, 1894, LUT, Woluwe-St-Lambert, Belgium. *Rhynchobatus arganiae* Arambourg, from the Maastrichtian of Morocco, has been recently reassigned to another genus which does not belong to the Rhynchobatidae (Cappetta, 1989). Some undescribed teeth from the MAA of Imin Tanout, Morocco, could represent the most ancient occurrence of the genus *Rhynchobatus*. **Extant**

Intervening: THA–LUT, RUP, AQT–SRV.

F. RHINOBATIDAE Müller and Henle, 1838 J. (TOA)–Rec. Mar.

First: *Jurobatos cappettai* Thies, 1983, upper TOA (*dispansum* to *aalensis* Zones), Hannover, northern Germany (Thies, 1983). **Extant**

Intervening: AAL, BTH, KIM–TTH, VLG, APT, CEN–TUR, SAN–MAA, THA–BRT, BUR–SRV, ZAN.

F. PLATYRHINIDAE Jordan, 1923 T. (YPR)–Rec. Mar.

First: *Platyrhina bolcensis* Molin, 1860 and *Platyrhina egertoni* (Zigno, 1876), Monte Bolca, northern Italy. The teeth of *Platyrhina ypresiensis* Casier, 1946, from the YPR of Belgium, cannot belong to this genus because of their very different morphology. **Extant**

Intervening: None.

F. HYPSOBATIDAE Cappetta, 1992 K. (CMP–MAA) Mar.

First: Undescribed species of *Hypsobatis* Cappetta, 1992, CMP, Negev Desert, Israel (Cappetta, 1992). The first described species is *Hypsobatis weileri* Cappetta, 1992, lower MAA, Level VI, 'Tranchée d'essai' near Benguerir, Ganntour Basin, Morocco. This species occurs also in the lower MAA of Egypt, where it was figured as *Rhombodus* sp. (Weiler, 1930).

Last: *Youssoubatis ganntourensis* Cappetta, 1992, upper MAA, Youssoufia, 'Recette 4, Sillon X', Ganntour Basin, Morocco.

Intervening: CMP–MAA.

Suborder RAJOIDEI Garman, 1913

F. RAJIDAE Bonaparte, 1831 K. (CEN)–Rec. Mar.

First: *Rajorhina expansa* (Davis, 1887), CEN, Hakel, Lebanon; *Mafdetia tibniensis* Werner, 1989, upper CEN, Gebel Dist Member of the Baharija Formation, Baharija Oasis, Egypt (Werner, 1989). **Extant**
Intervening: SAN, THA–YPR, RUP, BUR–TOR, ZAN.

F. CYCLOBATIDAE Cappetta, 1980 K. (CEN) Mar.

First and Last: *Cyclobatis oligodactylus* Egerton, 1844, CEN, Hakel, Lebanon; *C. major* Davis, 1887, CEN, Hakel and Hadjula, Lebanon; *C. tuberculatus* Cappetta, 1980, CEN, Hakel, Lebanon.

Suborder RHINOBATOIDEI or SCLERORHYNCHOIDEI

F. PARAPALAEOBATIDAE Cappetta, 1992 K. (SAN–MAA) Mar.

First: *Parapalaeobates pygmaeus* (Quaas, 1902), MAA, Libyan Desert, Egypt. This species occurs as early as the SAN of Dordogne, south-western France (Landemaine, 1991). This author, on the basis of superficial morphological dental resemblances, attributed erroneously the genus *Parapalaeobates* to the family Heterodontidae. *P. atlanticus* Arambourg, 1952 is well represented in the CMP and lower MAA of Morocco and it remains to be proved that the Egyptian and Moroccan species belong to the same species as suggested by Landemaine (1991).
Last: *P. atlanticus* Arambourg, 1952, MAA, Morocco.
Intervening: SAN–MAA.

Suborder SCLERORHYNCHOIDEI Cappetta, 1980

F. SCLERORHYNCHIDAE Cappetta, 1974 K. (APT–MAA) Mar.

First: *Onchopristis praecursor* Thurmond, 1971, Trinity Group (APT/ALB), Texas, USA (Thurmond, 1971); *O. numidus* (Haug, 1905), upper ALB, Djoua, Algeria.
Last: *Ctenopristis nougareti* Arambourg, 1940, MAA, Ouled Abdoun Basin, Morocco; *Dalpiazia stromeri* Checchia-Rispoli, 1933, MAA, Tripolitania, Libya; *Ganopristis leptodon* Arambourg, 1935, MAA, Ouled Abdoun Basin, Morocco; *Ischirhiza nigeriensis* (Tabaste, 1963), MAA, Mont Igdaman, Niger (several other species of this last genus occur in MAA deposits); *Pucapristis branisi* Schaeffer, 1963, El Molino Formation (MAA), Toro-Toro, Bolivia; *Schizorhiza stromeri* Weiler, 1930, 'Grès de Nubie' (MAA), Egypt.
Intervening: ALB–CMP.

Suborder PRISTIOIDEI Cappetta, 1980

F. PRISTIDAE Bonaparte, 1838 ?K. (CEN)/T. (YPR)–Rec. Mar.

First: *Peyeria libyca* Weiler, 1935, lower CEN, Baharija, Egypt. It is probable that this species is not a pristid; indeed, unquestionable pristids are present only since the early Eocene: *Pristis lathami* Galeotti, 1837, LUT, Melsbroeck, Belgium. This species occurs in the YPR of the Anglo-Franco-Belgian Basin. Rostral teeth of *Anoxypristis* type – i.e. teeth devoid of posterior groove – occur in the lower YPR of the Ouled Abdoun Basin, Morocco (Cappetta, pers. obs.). **Extant**
Intervening: YPR–BRT, AQT, LAN.

Order TORPEDINIFORMES Buen, 1926

Superfamily TORPEDINOIDEA Compagno, 1973

F. TORPEDINIDAE Bonaparte, 1838 T. (DAN)–Rec. Mar.

First: *Eotorpedo zennaroi* Cappetta, 1988, Danian, Imin Tanout, P3 Level, Morocco (Cappetta, 1988). There is a long gap in the fossil record between the last *Eotorpedo* in the YPR and the first *Torpedo* in the LAN.
Intervening: THA–YPR, LAN, ZAN.

Superfamily NARCINOIDEA Compagno, 1973

F. NARCINIDAE Gill, 1862 T. (YPR)–Rec. Mar.

First: *Narcine molini* Jaekel, 1894, YPR, Monte Bolca, northern Italy. Isolated teeth of *Narcine* have been collected in the YPR and LUT of Europe and Africa (Cappetta, 1987, 1988) and then, the genus disappears from the fossil record. Miocene records (Tortonian of Portugal: Jonet, 1968 and Miocene of India: Sahni and Mehrotra, 1981) are based on erroneous determinations (Cappetta, 1987). **Extant**
Intervening: LUT.

Order MYLIOBATIFORMES Compagno, 1973

Superfamily DASYATOIDEA Whitley, 1940

F. DASYATIDAE Jordan, 1888 K. (CEN)–Rec. Mar./FW

First: *Dasyatis* sp. indet., CEN, Texas, USA (Meyer, 1974). The family is very well represented in the fossil record, but its remains are abundant only since the MAA.
Intervening: MAA–BRT, RUP, AQT–SRV, ZAN.

F. POTAMOTRYGONIDAE Garman, 1913 Q. (PLE)–Rec. FW

First: *Potamotrygon africana* Arambourg, 1947, PLE, Lake Rudolph, Ethiopia. **Extant**
Comment: This family occurs only in fresh water in South America. In Africa and Asia, Dasyatoidea occur also in some rivers and lakes but they belong to the genera *Dasyatis* and *Himantura; so the assignment, by Arambourg, of sting rays from the PLE of East Africa to Potamotrygon*, remains questionable. Patterson (1967), following Arambourg (1947), assigned to this family the dermal tubercles described by Larrazet (1886) as *Dynatobatis*, from the Tertiary (probably Neogene) of Rio Parana (Argentina). However, these tubercles could also belong to other batoids (Dasyatidae, Rajidae, etc.), and the fossil occurrence of Potamotrygonidae thus remains to be demonstrated.

F. UROLOPHIDAE Gray, 1851 T. (YPR)–Rec. Mar.

First: *Urolophus crassicauda* Blainville, 1818, YPR, Monte Bolca, northern Italy. Despite the close resemblance between this species and the Recent genus, it is possible, considering the marked differences in the teeth, that the fossil form belongs to a different genus, all the more so because the genus *Urolophus* disappears completely from the fossil record until the PLE of southern California (Fitch, 1964). **Extant**
Intervening: PLE.

F. GYMNURIDAE Fowler, 1934 K. (CEN)–Rec. Mar.

First: *Gymnura laterialata* Werner, 1989, upper CEN, Gebel Dist Member of the Baharija Formation, Baharija Oasis, Egypt (Werner, 1989). **Extant**
Intervening: THA–LUT, RUP, AQT–SRV, ZAN.

Superfamily MYLIOBATOIDEA Compagno, 1973

F. MYLIOBATIDAE Bonaparte, 1838 K. (MAA)–Rec. Mar.

First: *Brachyrhizodus wichitaensis* Romer, 1942, 'Permo-Carboniferous', Godwin Creek, Texas, USA. This species does not come from Palaeozoic deposits, as asserted by Romer; it is not uncommon in the lower MAA deposits of New Jersey, USA (Cappetta and Case, 1975). **Extant**
Intervening: DAN–BRT, RUP, AQT–SRV, ZAN–PIA.

F. RHINOPTERIDAE Jordan and Evermann, 1896 ?K. (CEN)/T. (THA)–Rec. Mar.

First: *Rhinoptera prisca* Woodward, 1907, Palaeocene, Pernambuco Province, Brazil; *Rhinoptera raeburni* White, 1934, THA, Wurno, northern Nigeria. The occurrence of *Rhinoptera* in the upper CEN of Baharija, Egypt (Werner, 1989), rests on the evidence of a single heavily damaged tooth which can probably be assigned to a hybodont.
Intervening: LUT, BUR–SRV.

F. RHOMBODONTIDAE Cappetta, 1987 K. (MAA) Mar.

First and Last: *Rhombodus levis* Cappetta and Case, 1975, Mount Laurel Sands (lower MAA), New Jersey, USA; *R. binkhorsti* Dames, 1881, MAA, Maastricht, Holland; *R. microdon* Arambourg, 1952, MAA, Ouled Abdoun Basin, Morocco; *R. meridionalis* Arambourg, 1952, MAA, Oued Erguita, Morocco. The species *R. bondoni* Arambourg, 1952 (Maastrichtian, Ouled Abdoun Basin, Morocco) previously assigned to the genus *Rhombodus* is now attributed to a new genus inc. fam. (Noubhani and Cappetta, in press). The age of *R. levis*, previously considered to be late CMP, is early MAA in age on the basis on new stratigraphical and micropalaeontological studies (Petters, 1976).

Superfamily MOBULOIDEA Whitley, 1936

F. MOBULIDAE Gill, 1893 T. (THA)–Rec. Mar.

First: *Archaeomanta priemi* Herman, 1979, THA, Ouled Abdoun Basin, Morocco. *Burnhamia* sp., THA, Ouled Abdoun Basin, Morocco (Arambourg, 1952, as *Rhinoptera daviesi*; Cappetta, 1985).
Intervening: YPR–LUT, RUP, AQT–SRV.

Subclass SUBTERBRANCHIALIA Zangerl, 1979

Superorder UNNAMED

Order INIOPTERYGIFORMES Zangerl and Case, 1973

F. INIOPTERYGIDAE Zangerl and Case, 1973 C. (POD–KRE) Mar./Lagoonal

First: *Promexyele peyeri* Zangerl and Case, 1973, Mecca Quarry Shale, Linton Formation (Westphalian C), Vermilion County, Indiana, USA.
Last: *Iniopteryx rushlaui* Zangerl and Case, 1973, Stark Shale, Dennis Formation, Bronson Group, Missouri Series (Westphalian D), Fort Calhoun, Nebraska, USA; *I. tedwhitei* Zangerl and Case, 1973, Wea Shale, Westerville Formation, Kansas City Group (Westphalian D), Papillon, Nebraska, USA. The Westerville Formation belongs in the Kansas City Group which is Stephanian A (= KRE) according to Harland *et al.* (1982) but upper Westphalian D (= MYA) according to Zangerl and Case (1973).
Intervening: MYA.

F. SIBYRHYNCHIDAE Zangerl and Case, 1973 C. (POD–KLA) Mar./Lagoonal

First and Last: The family ranges from the Logan Quarry Shale of Staunton Formation (upper Wesphalian C = POD), Indiana, into the Queen Mill Shale of Lecompton Formation (upper Stephanian B = KLA), Nebraska and Iowa, USA. *Sibyrhynchus denisoni* Zangerl and Case, 1973 is known from both the lowest and highest levels of this interval, *Inioxyele whitei* Zangerl and Case, 1973 occurs only at the highest level, and *Inioptera richardsoni* Zangerl and Case, 1973 ranges from the lowest level up into the Wea Shale of Westerville Formation (KRE).
Comment: Johnson (1981) found sibyrhynchid remains in the middle Wolfcampian (= ASS/SAK) Admiral Formation of Texas, USA. Four undescribed iniopterygian species are known from the Bear Gulch Limestone Member of Heath Formation (Namurian E_2b = ARN) of Montana, USA (Lund, 1990).

Order CHONDRENCHELYIFORMES Moy-Thomas, 1939

F. CHONDRENCHELYIDAE Berg, 1940 C. (ARU–ARN) Mar./Lagoonal

First: *Chondrenchelys problematica* Traquair, 1888, Glencartholm Volcanic Beds, upper Border Group, Viséan C_2S_1, Dumfriesshire, Scotland, UK.
Last: *Harpagofututor volsellorhinus* Lund, 1982, Bear Gulch Limestone Member of Heath Formation, Namurian E_2b, Montana, USA.
Comments: Lund (1982) tentatively placed in this order the genera *Platyxystrodus* Hay, 1899 and *Solenodus* Trautschold, 1874. *Solenodus* is known only from Myachkovo near Moscow, former USSR, and is of MYA age. *Platyxystrodus* is known from England, Northern Ireland, and the USA (Illinois, Iowa), and ranges from HAS to MYA (cf. Lund, 1982, table 1).

Order POLYSENTORIFORMES, *nov.*

F. POLYSENTORIDAE Zangerl, 1979 C. (POD/MYA) Mar./Lagoonal

First and Last: *Polysentor gorbairdi* Zangerl, 1979, Essex (predominantly marine) fauna of Francis Creek Shale, Carbondale Formation, lower Westphalian D, Mazon Creek, Illinois, USA (Zangerl, 1979).

Superorder HOLOCEPHALI Bonaparte, 1832–1841

Order HELODONTIFORMES Patterson, 1965

F. HELODONTIDAE Patterson, 1965 C. (HAS)–P. (TAT) Mar.

First: *Helodus* spp., Louisiana, Missouri (Branson, 1914), and *H. simplex* Agassiz, 1838, also from Missouri, USA (Bryant and Johnson, 1936).
Last: *Helodopsis* spp., Upper *Productus* Limestone, Salt Range, Pakistan (Waagen, 1879).
Comment: Bendix-Almgreen (1975) has shown that Permian '*Helodus*' teeth may well belong to the edestids;

Smith and Patterson (1988, pp. 184, 194) comment that *Helodus* is polyphyletic and used as a form genus for the anterior teeth of various bradyodonts.

Order BRADYODONTIFORMES Smith Woodward, 1921

Suborder UNNAMED

F. PSEPHODONTIDAE Zangerl, 1981 D. (FAM?)–P. (TAT) Mar.

First: *Psephodus* sp., Louisiana, Missouri, USA (Branson, 1914).
Last: *P. indicus* Waagen, 1880 and *P. depressus* Waagen, 1880, upper *Productus* Limestone, Kafir-kot, Salt Range, Pakistan (Waagen, 1879).
Intervening: HAS–BRI.

F. COPODONTIDAE Davis, 1883 D. (FRS)–C. (MOS) Mar.

First: *Acmoniodus clarkei* Hussakof and Bryant, 1918, Genesee Conodont Bed, New York State, USA.
Last: *Copodus* spp., Carboniferous Limestone, UK, USA.
Intervening: HAS–BRI.

F. PSAMMODONTIDAE Koninck, 1878 C. (VIS) Mar.

First and Last: *Psammodus* spp., UK, Belgium, USA, former USSR, etc., and *Lagarodus specularis* (Trautschold, 1874), Moscow, former USSR.

Order CHIMAERIFORMES Berg, 1940

Suborder ECHINOCHIMAEROIDEI Lund, 1977

F. ECHINOCHIMAERIDAE Lund, 1977 C. (ARN) Mar.

First and Last: *Echinochimaera meltoni* Lund, 1977, *E. snyderi* Lund, 1986, *E. kellyi* Lund, 1986, *E. sulphurea* Lund, 1986, *E. indiana* Lund, 1988, and *E. elusiva* Lund, 1988, all Bear Gulch Limestone, Fergus County, Montana, USA (Lund, 1977, 1986b, 1988).

Suborder MENASPOIDEI Patterson, 1965

F. DELTOPTYCHIIDAE Patterson, 1965 C. (TOU–SPK/BSH) Mar.

First and Last: *Deltoptychius armigerus* (Traquair, 1887), upper Calciferous Sandstone Group, Cementstones (C2), Glencartholme, Scotland, UK.
Last: *Deltoptychius armigerus* (Traquair, 1887), Flex Coal, Namurian, Loanhead, Midlothian, Scotland, UK.

F. COCHLIODONTIDAE Owen, 1867 D. (FAM)–P. (ART) Mar.

First: *Thoralodus cabrieri* Lehman, 1953, Cabrières, SE Montagne Noire, France, and *Sandalodus minor* Bryant and Johnson, 1936, Chaffee Formation, Gribble Creek, Fremont County, Colorado, USA.
Last: *Crassidonta subcrenulata* Teichert, 1943, Permian, Wandagee Station, Western Australia, and *C. stuckenbergi* Branson, 1916, Lower Phosphate Member, Phosphoria Formation, Wyoming, USA.
Intervening: HAS–ALP.

F. DELTODONTIDAE Zangerl, 1981 C. (HAS)–P. (?KUN) Mar.

First: *Deltodus* spp., Carboniferous Limestone, UK, USA, former USSR.
Last: *Deltodus mercurii* Newberry, 1876, Kaibab Formation, 'mid-Permian', Arizona and Utah, USA (McKee, 1938).
Intervening: IVO–BRI.

Suborder MYRIACANTHOIDEI Patterson, 1965

F. MYRIACANTHIDAE Woodward, 1889 Tr. (RHT)–J. (TTH) Mar.

First: *Agkistracanthus mitgelensis* Duffin and Furrer, 1981, Kössen Beds Formation, Piz Son Mitgel, Kanton Graubunden, Switzerland (Duffin and Furrer, 1981).
Last: *Chimaeropsis paradoxa* Zittel, 1887, Plattenkalk, Solnhofen Formation, Solnhofen, Bavaria, southern Germany.
Intervening: HET, SIN, TOA.

Suborder SQUALORAJOIDEI Patterson, 1965

F. SQUALORAJIDAE Woodward, 1886 J. (SIN) Mar.

First and Last: *Squaloraja polyspondyla* Agassiz, 1836 and *S. tenuispina* Woodward, 1886, Lower Lias, Lyme Regis, Dorset, England, UK, and Lombardische Kieselkalk Formation (*bucklandi* zone), Osteno, Lombardy, northern Italy (Patterson, 1965; Arduini *et al.*, 1982).

Suborder CHIMAEROIDEI Patterson, 1965

F. UNNAMED J. (PLB–BTH) Mar.

First: *Eomanodon simmsi* Ward and Duffin, 1989 (*incertae familiae*), Middle Lias, *subnodosus* subzone, Gretton, Gloucestershire, England, UK (Ward and Duffin, 1989).
Last: *Ganodus oweni* Agassiz, 1843, *G. rugulosus* Egerton, 1843, and *G. dentatus* Egerton, 1847, Bathonian, UK.

F. CHIMAERIDAE Thienemann, 1828 C. (?MOS)–Rec. Mar.

First: *Similihariotta dabasinskasi* Zangerl, 1979, Francis Creek Shale, Carbondale Formation, Will County, Illinois, USA (Zangerl, 1979). This is a most unlikely determination, being based upon general body form only. The next oldest species are *Elasmodus planus* Leriche, 1929, *E. crassus* (Hebert, 1854), *E. ubaghsi* Leriche, 1929, and *E. greenoughi* Agassiz, 1843 from the MAA of UK and NW Europe. **Extant**
Intervening: MAA–HOL.

F. CALLORHYNCHIDAE Garman, 1901 J. (CLV)–Rec. Mar.

First: *Brachymylus altidens* Woodward, 1892 and *Pachymylus leedsi* Woodward, 1892, Oxford Clay, ?*athleta* zone, Peterborough, Cambridgeshire, England, UK. **Extant**
Intervening: KIM, SEN, THA.

F. RHINOCHIMAERIDAE Garman, 1901 K. (SAN)–Rec. Mar.

First: *Harriotta lehmani* Werdelin, 1986, Sahel Alma, Lebanon (Werdelin, 1986). **Extant**
Intervening: THA, RUP.

F. EDAPHODONTIDAE Owen, 1846 J. (BAJ)–T. (PIA) Mar.

First: *Ischyodus ferrugineus* Riess, 1887, *I. personati*

(Quenstedt, 1852), *I. bifurcati* (Quenstedt, 1887), and *I. aalensis* (Quenstedt, 1852), Brown Jura β, Aalen, Germany.
Last: *Edaphodon pliocenicus* Carraroli, 1897, Piacentino, Italy, and *E. antwerpensis*, Pli., Belgium (Leriche, 1951).
Intervening: BTH, CLV, KIM, ALB–MES.
Comment: *I. cornaliae* Bellotti, 1858, based on a fine spine from the NOR, probably belongs to a hybodont shark.

REFERENCES

Antunes, M. T. (1972) Les squales (Crétacé et Tertiaire): intérêt pour la stratigraphie et sa problématique. *Mémoires du Bureau de Recherches Géologiques et Minières*, **77**, 345–55.

Arambourg, C. (1947) Contributions à l'étude géologique et paléontologique du bassin du Lac Rodolphe et de la basse vallée de l'Omo. Deuxième partie. Paléontologie. *Mission Scientifique de l'Omo*, **1**, 231–562.

Arambourg, C. (1952) Les vertébrés fossiles des gisements de phosphates (Maroc–Algérie–Tunisie). *Services Géologiques du Maroc, Notes et Mémoires*, **92**, 1–372.

Arduini, P., Pinna, G. and Teruzzi, G. (1982) Il giacimento sinemuriano di Osteno in Lombardia e i suoi fossili. In: *Palaeontology, Essentials of Historical Geology* (ed. E. M. Gallitelli), Modena, pp. 495–522.

Baird, D. (1978) Studies on Carboniferous freshwater fishes. *American Museum Novitates*, **2641**, 1–22.

Bendix-Almgreen, S. E. (1975) Fossil fishes from the marine late Paleozoic of Holm Land-Amdrup Land, N.E. Greenland. *Meddelelser om Grønland*, **195**, 1–38.

Berman, D. S. (1970) Vertebrate fossils from the Lueders Formation, Lower Permian of north-central Texas. *University of California, Publications in Geological Sciences*, **86**, 61 pp.

Branson, E. B. (1914) The Devonian fishes of Missouri. *Bulletin of the University of Missouri, Science Series*, **2**, 59–74.

Bryant, W. L. and Johnson, J. H. (1936) Upper Devonian Fish from Colorado. *Journal of Palaeontology*. **10**, 656–9.

Cappetta, H. (1970) Les sélaciens du Miocène de la région de Montpellier. *Palaeovertebrata*, Mémoire Extraordinaire, 139 pp.

Cappetta, H. (1972) Les poissons crétacés et tertiaires du Bassin des Iullemmeden (République du Niger). *Palaeovertebrata*, **5**, 179–251.

Cappetta, H. (1975) Sélaciens et Holocéphale du Gargasien de la région de Gargas (Vaucluse). *Géologie Méditerranéenne*, **2**, 115–35.

Cappetta, H. (1980a) Les sélaciens du Crétacé supérieur du Liban. I: Requins. *Palaeontographica, Abteilung A*, **168**, 69–148.

Cappetta, H. (1980b) Les sélaciens du Crétacé supérieur du Liban. II: Batoides. *Palaeontographica, Abteilung A*, **168**, 149–229.

Cappetta, H. (1981) Additions à la faune de sélaciens fossiles du Maroc. 1: Sur la présence des genres *Heptranchias*, *Alopias* et *Odontorhytis* dans l'Yprésien des Ouled Abdoun. *Geobios*, **14**, 563–75.

Cappetta, H. (1985) Sur une nouvelle espèce de *Burnhamia* (Batomorphii, Mobulidae) dans l'Yprésien des Ouled Abdoun, Maroc. *Tertiary Research*, **7**, 27–33.

Cappetta, H. (1987) Mesozoic and Cenozoic Elasmobranchii, in *Handbook of Paleoichthyology* (ed. H.-P. Schultze), Chondrichthyes II, **3B**, Gustav Fisher Verlag, 193 pp.

Cappetta, H. (1988) Les Torpédiniformes (Neoselachii, Batomorphii) des phosphates du Maroc. Observations sur la denture des genres actuels. *Tertiary Research*, **10**, 21–52.

Cappetta, H. (1989) Sélaciens nouveaux ou peu connus du Crétacé supérieur du Maroc. *Mesozoic Res.*, **2** (1), 11–23.

Cappetta, H. (1990) Echinorhinidae nouveau (Euselachii, Squaliformes) du Crétacé supérieur du Negev (Israel). *Neues Jahrbuch für Geologie und Paläontologie*, **12**, 741–9.

Cappetta, H. (1992) Nouveau Rhinobatoidei (Neoselachii, Rajiformes) à denture specialisée du Maastrichtien du Maroc. Remarques sur l'évolution dentaire des Rajiformes et des Myliobatiformes. *Neues Jahrbuch für Geologie und Paläontologie*, **187** (1), 31–52.

Cappetta, H. and Case, G. R. (1975) Contribution à l'étude des sélaciens du groupe Monmouth (Campanien–Maestrichtien) du New Jersey. *Palaeontographica, Abteilung A*, **151** (1–3), 1–46.

Cappetta, H., Jaeger, J.-J., Sigé, B. *et al.* (1987) Compléments et précisions biostratigraphiques sur la faune paléocène à mammifères et sélaciens du Bassin d'Ouarzazate (Maroc). *Geobios*, **8**, 147–57.

Case, G.R. (1981) Late Eocene selachians from South Central Georgia. *Palaeontographica, Abteilung A*, **176**, 52–79.

Case, G. R. and Cappetta, H. (1990) The Eocene selachian fauna from the Fayum depression of Egypt. *Palaeontographica, Abteilung A*, **212**, 1–30.

Casier, E. (1946) La faune ichthyologique de l'Yprésien de la Belgique. *Mémoires du Musée Royal d'Histoire naturelle de Belgique*, **104**, 267 pp.

Dalinkevicius, J. A. (1935) On the fossil fishes of the Lithuanian Chalk. I. Selachii. *Mémoires de la Faculté des Sciences de l'Université Vytautas le Grand*, **9**, 243–305.

Dartevelle, E. and Casier, E. (1959) Les poissons fossiles du Bas-Congo et des régions voisines. *Annales du Musée Royal du Congo belge, Série A*, (3), **2**, 257–568.

Datchev, D. M. (1971) On the phylogeny and distribution of the genus *Otodus* in Bulgaria. *Sofiiskiia Universitat, Geologo–Geografski Fakultet, Godishnik*, **63**, 11–18 [In Bulgarian.]

Dibley, G. E. (1911) On the teeth of *Ptychodus* and their distribution in the English Chalk. *Quarterly Journal of the Geological Society of London*, **67**, 263–77.

Dick, J. R. F. (1981) *Diplodoselache woodi* gen. et sp. nov., an Early Carboniferous shark from the Midland Valley of Scotland. *Transactions of the Royal Society of Edinburgh: Earth Sciences*, **72**, 99–113.

Duffin, C. J. (1985) Revision of the hybodont selachian genus *Lissodus* Brough (1935). *Palaeontographica, Abteilung A*, **188**, 105–52.

Duffin, C. J. and Furrer, H. (1981) Myriacanthid holocephalan remains from the Rhaetian (Upper Triassic) and Hettangian (Lower Jurassic) of Graubünden (Switzerland). *Eclogae Geologicae Helvetiae*, **74**, 803–29.

Gheerbrant, E., Cappetta, H., Feist, M. *et al.* (1993) La succession des faunes de vertébrés d'âge paléocène supérieur et éocène inférieur dans le bassin d'Ouarzazate, Maroc. Contexte géologique, portée biostratigraphique et paléogéographique. *Newsl. Stratigr.*, **28** (1), 33–58.

Glickman, L. S. (1957) Taxonomic significance of the accessory teeth of sharks of the families Lamnidae and Scapanorhynchidae. *Trudy Geologicheskiu Muzei, Akademiia Nauk SSSR*, **1**, 103–9.

Glickman, L. S. (1964) *Sharks of Paleogene and Their Stratigraphic Significance*. Nauka Press, Moscow and Leningrad, 229 pp.

Gross, W. (1973) Kleinschuppen, Flossenstacheln und Zähne von Fischen aus europäischen und nordamerikanischen Bonebeds des Devons. *Palaeontographica, Abteilung A*, **142**, 51–155.

Hansen, M. C. (1978) A presumed lower dentition and a spine of a Permian petalodontiform chondrichthyan, *Megactenopetalus kaibabanus*. *Journal of Paleontology*, **52**, 55–60.

Hansen, M. C. (1985) Systematic relationships of petalodontiform chondrichthyans. In *Ninth International Congress on Carboniferous Stratigraphy and Geology, Compte Rendu, Vol. 5, Paleontology, Paleoecology, Paleogeography*, Carbondale and Edwardsville, Southern Illinois University Press, pp. 523–41.

Harland, W. B., Cox, A. V., Llewellyn, P. G. *et al.* (1982) *A Geologic Time Scale*. Cambridge University Press, 131 pp.

Herman, J. (1975) Zwei neue Haifischzähne aus der Kreide von Misburg bei Hannover (höheres Campan). *Bericht der Naturhistorische Gesellschaft Hannover*, **119**, 295–302.

Herman, J. (1977) Les sélaciens des terrains néocrétacés et paléocènes de Belgique et des contrées limitrophes. Eléments d'une biostratigraphie inter-continentale. *Mémoires pour servir à l'Explication des Cartes Géologiques et Minières de la Belgique* (1975 published 1977), **15**, 401 pp.

Herman, J. (1979) Additions to the Eocene fish fauna of Belgium. 4. *Archaeomanta*, a new genus from the Belgium and North African Paleogene. *Tertiary Research*, **2**, 61–7.

Herman, J. (1982) Die Selachier-Zähne aus der Maastricht Stufe von Hemmoor, Niederelbe (NW-Deutschland). *Geologische Jahrbuch, A*, **61**, 129–59.

Hussakof, L. and Bryant, W. L. (1918) Catalog of the fossil fishes in the museum of the Buffalo Society of Natural Sciences. *Bulletin of the Buffalo Society of Natural Sciences*, **12**, 346 pp.

Johnson, G. D. (1980) Xenacanthodii (Chondrichthyes) from the Tecovas Formation (Late Triassic) of west Texas. *Journal of Paleontology*, **54**, 923–32.

Johnson, G. D. (1981) Hybodontoidei (Chondrichthyes) from the Wichita–Albany Group (Early Permian) of Texas. *Journal of Vertebrate Paleontology*, **1**, 1–41.

Landemaine, O. (1991) Sélaciens nouveaux du Crétacé supérieur du Sud-Ouest de la France. Quelques apports à la systématique des élasmobranches. *Saga*, **1**, 1–45.

Landini, W. (1977) Revizione degli 'Ittiodontoliti pliocenici' della collezione Lawley. *Palaeontographia Italica*, **70**, 92–134.

Lawley, R. (1881) Studi comparativi sui pesci fossili coi viventi dei generi *Carcharodon*, *Oxyrhina* e *Galeocerdo*. 151 pp., Pisa.

Leriche, M. (1906) Contribution à l'étude des poissons fossiles du Nord de la France et des régions voisines. *Mémoires de la Société Géologique du Nord*, **5**, 430 pp.

Leriche, M. (1926) Les poissons tertiaires de Belgique. IV. Les poissons néogènes. *Mémoires du Musée Royal d'Histoire Naturelle de Belgique*, **32**, 367–472.

Leriche, M. (1951) Les poissons Tertiaires de la Belgique (Supplément). *Mémoires de l'Institut Royal des Sciences Naturelles de Belgique*, **118**, 473–600.

Lund, R. (1977) A new petalodont (Chondrichthyes, Bradyodonti) from the Upper Mississippian of Montana. *Annals of the Carnegie Museum*, **46**, 129–55.

Lund, R. (1982) *Harpagofututor volsellorhinus* new genus and species (Chondrichthyes, Chondrenchelyiformes) from the Namurian Bear Gulch Limestone, *Chondrenchelys problematica* Traquair (Visean), and their sexual dimorphism. *Journal of Paleontology*, **56**, 938–58.

Lund, R. (1984) On the spines of the Stethacanthidae (Chondrichthyes), with a description of a new genus from the Mississippian Bear Gulch Limestone. *Geobios*, **17**, 281–95.

Lund, R. (1985a) The morphology of *Falcatus falcatus* (St John and Worthen), a Mississippian stethacanthid chondrichthyan from the Bear Gulch Limestone of Montana. *Journal of Vertebrate Paleontology*, **5**, 1–19.

Lund, R. (1985b) Stethacanthid elasmobranch remains from the Bear Gulch Limestone (Namurian E_2b) of Montana. *American Museum Novitates*, **2828**, 1–24.

Lund, R. (1986a) On *Damocles serratus*, nov. gen. et sp. (Elasmobranchii: Cladodontida) from the Upper Mississippian Bear Gulch Limestone of Montana. *Journal of Vertebrate Paleontology*, **6**, 12–19.

Lund, R. (1986b) The diversity and relationships of the Holocephali, in *Indo-Pacific Fish Biology: Proceedings of the Second International Conference on Indo-Pacific Fishes*, (ed. T. Uyeno, R. Arai, T. Taniuchi *et al.*), Ichthyological Society of Japan, Tokyo, pp. 97–106.

Lund, R. (1988) New Mississippian Holocephali (Chondrichthyes) and the evolution of the Holocephali. *Mémoires du Museum National d'Histoire Naturelle; Série C, Sciences de la Terre*, **53**, 195–205.

Lund, R. (1989) New petalodonts (Chondrichthyes) from the Upper Mississippian Bear Gulch Limestone (Namurian E_2b) of Montana. *Journal of Vertebrate Paleontology*, **9**, 350–68.

Lund, R. (1990) Chondrichthyan life history styles as revealed by the 320 million years old Mississippian of Montana. *Environmental Biology of Fishes*, **27**, 1–19.

Mader, H. (1986) Schuppen und Zähne von Acanthodiern und Elasmobranchiern aus dem Unter-Devon Spaniens (Pisces). *Göttinger Arbeiten zur Geologie und Paläontologie*, **28**, 59 pp.

Mader, H. and Schultze, H.-P. (1987) Elasmobranchier-Reste aus dem Unterkarbon des Rheinischen Schiefergebirges und des Harzes (W. Deutschland). *Neues Jahrbuch für Geologie und Paläontologie, Abhandlungen*, **175**, 317–46.

Maisey, J. G. (1976) The Jurassic selachian fish *Protospinax* Woodward. *Palaeontology*, **19**, 733–47.

Maisey, J. G. (1978) Growth and form of finspines in hybodont sharks. *Palaeontology*, **21**, 657–66.

Maisey, J. G. (1981) Studies on the Paleozoic selachian genus *Ctenacanthus* Agassiz: n° 1. Historical review and revised diagnosis of *Ctenacanthus*, with a list of referred taxa. *American Museum Novitates*, **2718**, 1–22.

Maisey, J. G. (1982) Studies on the Paleozoic selachian genus *Ctenacanthus* Agassiz: n° 2. *Bythiacanthus* St John and Worthen, *Amelacanthus*, new genus, *Eunemacanthus* St John and Worthen, *Sphenacanthus* Agassiz, and *Wodnika* Münster. *American Museum Novitates*, **2722**, 1–24.

Maisey, J. G. (1984) Studies on the Paleozoic selachian genus *Ctenacanthus* Agassiz: n° 3 Nominal species referred to *Ctenacanthus*. *American Museum Novitates*, **2774**, 1–20.

Marcinowski, R. and Radwanski, A. (1983) The Mid-Cretaceous transgression onto the central Polish Uplands (marginal part of the central European Basin). *Zitteliana*, **10**, 65–95.

McKee (1938) The environment and history of the Toroweap and Kaikab formations of Northern Arizona and Southern Utah. *Publications of the Carnegie Instution of Washington*, **492**, p. 166.

Meyer, R. L. (1974) *Late Cretaceous Elasmobranchs from the Mississippi east Texas embayments of the Gulf Coastal Plain*. Unpublished PhD of the Southern Methodist University, Dallas, 419 pp.

Müller, A. and Schöllmann, L. (1989) Neue Selachian (Neoselachii, Squalomorphii) aus dem Campanium Westfalens (NW-Deutschland). *Neues Jahrbuch für Geologie und Paläontologie, Abhandlungen*, **178**, 1–35.

Noubhani, A. and Cappetta, H. (in press) Révision des Rhombodontidae (Neoselachii, Batomorphii) des bassins à phosphate du Maroc (in press).

Ortlam, D. (1985) Neue Aspekte zur Deutung von *Menaspis armata* Ewald (Kupferschiefer, Zechstein 1, Deutschland) mit Hilfe der stereoskopischen Rontgentechnik. *Geologisches Jahrbuch*, **A81**, 3–57.

Patterson, C. (1965) The phylogeny of the chimaeroids. *Philosophical Transactions of the Royal Society of London*, **B249**, 101–219.

Petters, S. W. (1976) Upper Cretaceous subsurface stratigraphy of Atlantic Coastal Plain of New Jersey. *American Association of Petroleum Geologists Bulletin*, **60**, 87–107.

Pictet, E. J. and Campiche, G. (1858) Description des fossiles du terrain crétacé des environs de Sainte-Croix, 1ère partie. *Matériaux pour la Paléontologie suisse, Série 2*, **2**, 380 pp.

Poplin, C. and Heyler, D. (1989) Evolution et Phylogénie des Xénacanthiformes (= Pleuracanthiformes) (Pisces, Chondrichthyes). *Annales de Paléontologie*, **75**, 187–222.

Reif, W. E. (1980) Tooth enameloid as a taxonomic criterion. 3. A new primitive shark family from the Lower Keuper. *Neues Jahrbuch für Geologie und Paläontologie, Abhandlungen*, **160**, 61–72.

Richter, M. and Ward, D. J. (1990) Fish remains from the Santa Marta Formation (Late Cretaceous) of James Ross Island, Antarctica. *Antarctic Science*, **2**, 67–76.

Schaeffer, B. (1963) Cretaceous fishes from Bolivia, with comments on pristid evolution. *American Museum Novitates*, **2159**, 1–20.

Schaeffer, B. and Patterson, C. (1984) Jurassic fishes from the Western United States, with comments on Jurassic fish distribution. *American Museum Novitates*, **2796**, 1–86.

Smith, A. B. and Patterson, C. (1988) The influence of taxonomic method on the perception of patterns of evolution, in *Evolutionary Biology*, **23**, (eds M. K. Hecht and B. Wallace), Plenum, New York, pp. 127–216.

Sykes, J. H. (1971) A new Dalatid fish from the Rhaetic bone-bed at

Barnstone, Nottinghamshire. *Mercian Geologist*, **4**, 13–22.

Thies, D. (1981) Vier neue Neoselachier-Haiarten aus der NW-deutschen Unterkreide. *Neues Jahrbuch für Geologie und Paläontologie, Monatshefte*, **1981**, 475–86.

Thies, D. (1982) A neoselachian shark tooth from the Lower Triassic of the Kocaeli (= Bithynian) Peninsula, W Turkey. *Neues Jahrbuch für Geologie und Paläontologie, Monatshefte*, **1982**, 272–8.

Thies, D. (1983) Jurazeitliche Neoselachier aus Deutschland und S. England. *Courier Forschungsinstitut Senckenberg*, **58**, 1–116.

Thorsteinsson, R. (1973) Dermal elements of a new Lower Vertebrate from Middle Silurian (Upper Wenlockian) rocks of the Canadian Arctic Archipelago. *Palaeontographica, Abteilung A*, **143**, 51–7.

Thurmond, J. T. (1971) Cartilaginous fishes of the Trinity Group and related rocks (Lower Cretaceous) of North Central Texas. *Southeastern Geology*, **13**, 207–27.

Tintori, A. (1980) Teeth of the selachian genus *Pseudodalatias* (Sykes, 1971) from the Norian (Upper Triassic) of Lombardy. *Rivista Italiana di Paleontologia*, **86**, 19–30.

Turner, S. and Young, G. C. (1987) Shark teeth from the Early–Middle Devonian Cravens Peak Beds, Georgina Basin, Queensland. *Alcheringa*, **11**, 233–44.

Waagen, W. (1879) Salt Range Fossils. 1. *Productus* Limestone Fossils 1) Pisces – Cephalopoda. *Memoirs of the Geological Survey of India, Palaeontologia Indica, Ser.* **13**, 72 pp.

Ward, D. J. (1988) *Hypotodus verticalis* (Agassiz 1843), *Hypotodus robustus* (Leriche 1921) and *Hypotodus heinzelini* (Casier 1967) Chondrichthyes. Lamniformes, junior synonyms of *Carcharias hopei* (Agassiz 1843). *Tertiary Res.*, **10** (1), 1–12.

Ward, D. J. and Duffin, C. J. (1989) Mesozoic Chimaeroids 1. A new chimaeroid from the Early Jurassic of Gloucestershire, England. *Mesozoic Research*, **2**, 45–51.

Weiler, W. (1930) Fischreste aus dem nubischen Sandstein von Mahamid und Edfu und aus den Phosphaten Oberägyptens und der oase Baharije, in *Ergebnisse der Forschungsreisen Prof. E. Stromer's in den Wüsten Ägyptens*. Abhandlungen der bayerischen Akademie der Wissenschaften, Mathematisch-natur Wissenschaftlichen Abtheilung, N.F., pp. 12–42.

Werdelin, L. (1986) A new chimaeroid fish from the Cretaceous of Lebanon. *Geobios*, **19**, 393–7.

Werner, C. (1989) Die Elasmobranchier-Fauna des Gebel Dist Member des Bahariya Formation (Obercenoman) der Oase Bahariya, Ägypten. *Palaeoichthyologica*, **5**, 1–112.

Williams, M. E. (1985) The 'cladodont level' sharks of the Pennsylvanian black shales of central North America. *Palaeontographica, Abteilung A*, **190**, 83–158.

Woodward, A. S. (1894) Note on a tooth of *Oxyrhina* from the Red Crag of Suffolk. *Geological Magazine*, **1** (4), 75–6.

Woodward, A. S. (1911) The fishes of the English chalk. *Palaeontographical Society of London*, **56**, 1–96.

Young, G. C. (1982) Devonian sharks from south-eastern Australia and Antarctica. *Palaeontology*, **25**, 817–43.

Zangerl, R. (1979) New Chondrichthyes from the Mazon Creek Fauna (Pennsylvanian) of Illinois, in *Mazon Creek Fossils* (ed. M. H. Nitecki), Academic Press, New York, pp. 119–500.

Zangerl, R. (1981) Chondrichthyes. I. Paleozoic Elasmobranchii, in *Handbook of Paleoichthyology, Vol. 3A*, Gustav Fischer Verlag, Stuttgart, 115 pp.

Zangerl, R. (1990) Two new stethacanthid sharks (Stethacanthidae, Symmoriida) from the Pennsylvanian of Indiana, U.S.A. *Palaeontographica, Abteilung A*, **213**, 115–41.

Zangerl, R. and Case, G. R. (1973) Iniopterygia, a new order of chondrichthyan fishes from the Pennsylvanian of North America. *Fieldiana, Geology Memoirs*, **6**, 67 pp.

Zidek, J. (1976) Oklahoma paleoichthyology. Part V: Chondrichthyes. *Oklahoma Geology Notes*, **36**, 175–92.

Zidek, J. (1978) New chondrichthyan spines from the late Paleozoic of Oklahoma. *Journal of Paleontology*, **52**, 1070–8.

35

OSTEICHTHYES: BASAL ACTINOPTERYGIANS

B. G. Gardiner

The 'lower' Actinopterygii, until recently called the Chondrostei, include a number of separate lineages or groups that share various primitive actinopterygian characters as well as a variety of derived or specialized ones. Traditionally, the Chondrostei have contained numerous extinct families, assigned to the broadly inclusive Palaeonisciformes plus the living polypterids (Cladistia) and the sturgeons and paddlefishes (Acipenseriformes). It is now apparent that the palaeonisciforms represent a grade group and are hence paraphyletic. The polypterids, in spite of numerous autapomorphies, are related to the primitive palaeonisciforms, while the sturgeons and paddlefishes form a monophyletic group (for which the term Chondrostei is now employed) affiliated with the Triassic saurichthyid fishes.

A great majority of these extinct 'lower' actinopterygian genera are represented by poorly preserved usually compressed specimens, which, more often than not, yield minimal information even for critical parts of the dermal skull pattern (Gardiner, 1984). However, the design of the fins and the squamation may often prove to be diagnostic at least at the family level. The higher classification is based on the dermal skull roof patterns, in particular, the various configurations displayed by the bones on the otic branch of the infra-orbital canal.

Some 270 fossil genera have been described so far and there are six extant genera (*Polypterus, Acipenser, Huso, Scaphirhynchus, Polyodon* and *Psephurus*).

The higher actinopterygians (Neopterygii) are usually subdivided into Holostei and Teleostei (treated in Chapter 36). The teleosts are undoubtedly monophyletic, whereas the Holostei are probably paraphyletic. Accordingly, the grade Holostei is divided into the Division Ginglymodi for the gars, and the Division Halecostomi for the more teleost-like holosteans. The higher actinopterygians are characterized by such features as the elongation of the upper caudal fin-rays, premaxillae with nasal processes, a coronoid process on the mandible, a vertical suspensorium and unpaired fins in which the fin rays have equalled their supports in number. Halecostome characters include an interopercular, uncinate processes on the epibranchials, median neural spines and a mobile maxilla.

One hundred and twenty genera of 'holosteans' (Ginglymodi + Halecostomi, other than Teleostei) have been described, of which two survive to the present day (*Lepisosteus, Amia*).

Class OSTEICHTHYES Huxley, 1880

Subclass ACTINOPTERYGII Klein, 1885

The earliest undoubted actinopterygians are late Devonian in age, although isolated scales have been recorded from the marine Upper Silurian of China (*Naxilepis*) and Europe (*Lophosteus, Andreolepis*), and from Lower Devonian deposits of both Canada (*Dialipina*) and Australia (*Ligulalepis*) (Märss, 1986; Schultze, 1968).

F. UNNAMED S. (LUD) Mar. (see Fig. 35.1)

First and Last: *Andreolepis hedei* Gross, 1968, Hemse Beds, Sweden; Paadla Beds, Baltic; Velikoretskaya Formation, north Timan; Ust-Spokoinaya Formation, Severnaya Zemlya; Long Quarry Beds, South Wales, UK (Märss, 1986).

F. CHEIROLEPIDIDAE Pander, 1860 D. (EIF/GIV–FRS) FW

First: *Cheirolepis trailli* Agassiz, 1835, middle Old Red Sandstone, Nairnshire, Scotland, UK.

Last: *Cheirolepis canadensis* Whiteaves, 1881, Escuminac Bay Formation, Quebec, Canada.

Infraclass CLADISTIA Cope, 1871

F. POLYPTERIDAE Lacépède, 1803 K. (MAA)–Rec. FW

First: *Polypterus* Lacépède, 1803, Niger, Elmolew Formation, Bolivia. **Extant**

Infraclass ACTINOPTERI Cope, 1871

F. MIMIIDAE **fam. *nov.*** D. (FRS) Mar./FW

The Fossil Record 2. Edited by M. J. Benton. Published in 1993 by Chapman & Hall, London. ISBN 0 412 39380 8

QU.: HOL, PLE
TERTIARY: PLI, UMI, MMI, LMI, CHT, RUP, PRB, BRT, LUT, YPR, THA, DAN
CRETACEOUS: MAA, CMP, SAN, CON, TUR, CEN, ALB, APT, BRM, HAU, VLG, BER
JURASSIC: TTH, KIM, OXF, CLV, BTH, BAJ, AAL, TOA, PLB, SIN, HET
TRIASSIC: RHT, NOR, CRN, LAD, ANS, SCY

Key for both diagrams

1. Andreolepis
2. Cheirolepididae
CLADISTIA
3. Polypteridae
ACTINOPTERI
4. Mimiidae
5. Tegeolepididae
6. Stegotrachelidae
7. Osorioichthyidae
8. Kentuckiidae
9. Acrolepididae
10. Ptycholepididae
11. Cosmoptychiidae
12. Willomorichthyidae
13. Rhadinichthyidae
14. Phanerorhynchidae
15. Holuridae
16. Atherstoniidae
17. Dwykiidae
18. Carbovelidae
19. Gonatodidae
20. Strepheoschemidae
21. Cornuboniscidae
22. Amblypteridae
23. Redfieldiidae
24. Haplolepididae
25. Styracopteridae
26. Canobiidae
27. Tarrasiidae
28. Aeduelliidae
29. Platysomidae
30. Amphicentridae
31. Cryphiolepididae
32. Boreolepididae
33. Urosthenidae
34. Brachydegmidae
35. Lawniidae
36. Trissolepididae
37. Commentryidae
DORYPTERIFORMES
38. Dorypteridae

3 9 10 16 23 29 33

Fig. 35.1

First and Last: *Mimia toombsi* Gardiner and Bartram, 1977, Gogo Formation, Australia; *Howqualepis rostridens* Long, 1988, lacustrine shales, Mt. Howitt, Australia.

F. TEGEOLEPIDIDAE Romer, 1945
D. (FAM) Mar.

First and Last: *Tegeolepis clarki* (Newberry, 1888), Cleveland Shale, Ohio, USA.

F. STEGOTRACHELIDAE Gardiner, 1963
D. (EIF/GIV–FAM) Mar./FW

First: *Stegotrachelus* Woodward and White, 1926; *Moythomasia* Gross, 1950, *Orvikuina* Gross, 1953, middle Old Red Sandstone, Scotland, UK; Wildingun, Bergish-Gladbach, Germany, and western Estonia, former USSR.

Last: *Moythomasia*, Ohio and Genesee shales, New York State, USA (Gardiner, 1963).

F. OSORIOICHTHYIDAE Gardiner, 1967
D. (FRS/FAM) Mar.

First and Last: *Osorioichthys marginis* (Casier, 1952), Belgium.

F. KENTUCKIIDAE **fam. *nov.*** C. (TOU/VIS) Mar.

First: *Kentuckia deani* (Eastman, 1905), Waverly Shales, Kentucky, USA.

Last: *Elonichthys robisoni* (Hibbert, 1835), Calciferous Sandstone, Oil Shales, Dunnet Shale, Scotland, UK.

F. ACROLEPIDIDAE Aldinger, 1937
C. (TOU)–Tr. (CRN/NOR) Mar./FW

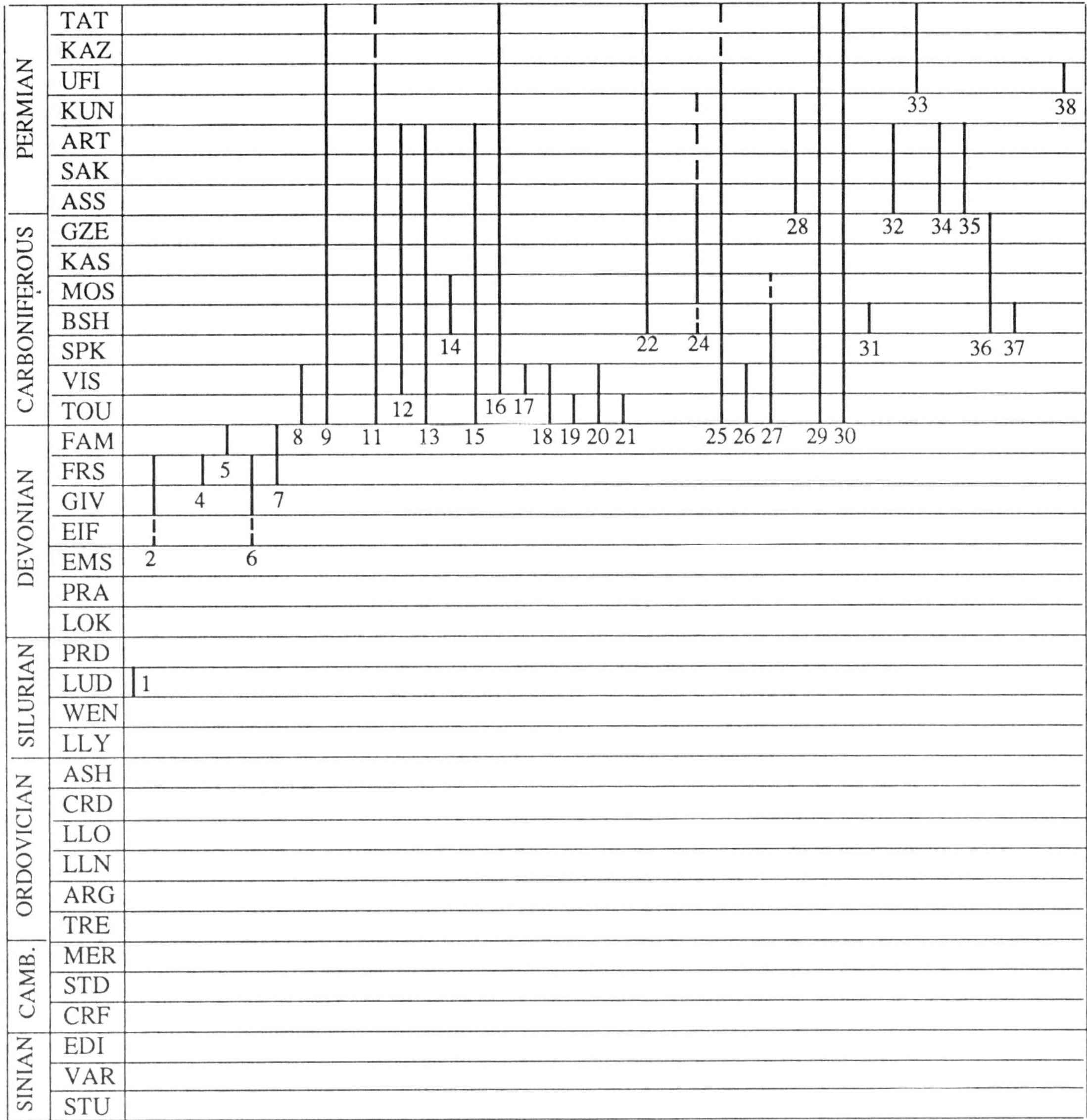

Fig. 35.1

First: *Nematoptychius greenocki* (Traquair, 1866), Calciferous Sandstone, Scotland, UK; *Acrolepis* Agassiz, 1833, several species, Marl Slate, England, UK; Kupferschiefer, Germany.
Last: *Turseodus acutus* Leidy, 1857, Newark Group, Pennsylvania and New Jersey, USA.

F. PTYCHOLEPIDIDAE Brough, 1939 Tr. (SCY)–J. (TOA) Mar./FW

First: *Boreosomus* Stensiö, 1921, several species, Spitsbergen, Madagascar and Greenland.
Last: *Ptycholepis bollensis* Agassiz, 1833, England, Germany and France; *Yuchoulepis szechuanensis* R.-T. Su, 1974, *Chungkingichthys tachuensis* R.-T. Su, 1974, China.

F. COSMOPTYCHIIDAE Gardiner, 1963 C. (TOU)–P. (ZEC) Mar.

First: *Watsonichthys pectinatus* (Traquair, 1877), *Cosmoptychius striatus* (Agassiz, 1835), Calciferous Sandstone, Scotland, UK.
Last: *Inichthys gorelovae* Kazantseva, 1979, *Neuburgella cognominis* Kazantseva, 1981, *Paralogoniscus lautus* Kazantseva, 1981, Jemenei Basin, Kazakhstan, former USSR.

F. WILLOMORICHTHYIDAE Gardiner, 1969 (F. RHABDOLEPIDIDAE Gardiner, 1963) C. (VIS)–P. (ROT) Mar./FW

First: *Willomorichthys striatulus* Gardiner, 1969, upper Witteberg series, South Africa; ?*Bendenius* Traquair, 1878, Belgium.
Last: *Rhabdolepis macropterus* (Bronn, 1829), Lower Permian, Lebach, Germany; *Tienshaniscus longipterus* Lui and Wang, 1978 and *Sinoniscus macrolepis* Lui and Wang, 1978 from the Upper Permian of China doubtfully belong here.

F. RHADINICHTHYIDAE Romer, 1945 C. (TOU)–P. (ROT) Mar./FW

First: *Rhadinichthys* Traquair, 1877, *Cycloptychius* Young, 1866, Calciferous Sandstone, Scotland, UK and elsewhere.
Last: *Illiniichthys cozarti*, *Nozamichthys contorta* Schultze

and Bardack, 1986, Mazon Creek, USA; *Cycloptychius* and *Rhadinichthys* also in the Upper Carboniferous; *Uydenia latifrons* Kazantseva, 1980 and *Eigilia nielseni* Kazantseva, 1981 from the Lower Permian of Kazakhstan also appear to belong to this family.

F. PHANERORHYNCHIDAE Stensiö, 1932 C. (BSK/MOS) FW

First and Last: *Phanerorhynchus armatus* Gill, 1923, Middle Coal Measures, Lancashire, England, UK.

F. HOLURIDAE Moy-Thomas, 1939 C. (TOU)–P. (ROT) FW

First: *Holurus parki* Traquair, 1881, Cementstone Group, Scotland, UK.
Last: *Australichthys longidorsalis* Gardiner, 1969, Upper Witteberg Series, South Africa; *Holuropsis yavorskyi* Berg, 1947, Permian, former USSR and *Palaeonisconotus*, Middle Jurassic, the former, USSR may also belong here.

F. ATHERSTONIIDAE Gardiner, 1969 C. (VIS)–Tr. (SCY) Mar./FW

First: *Aestuarichthys fulcratus* Gardiner, 1969, Upper Witteberg Series, South Africa.
Last: *Atherstonia* Woodward, 1889, ranging from the Upper Permian of Europe to the Lower Triassic of South Africa (Beaufort Beds) and Madagascar.

F. DWYKIIDAE Gardiner, 1969 C. (VIS) Mar.?

First and Last: *Dwykia analensis* Gardiner, 1969, Upper Witteberg Series, South Africa.

F. CARBOVELIDAE Romer, 1945 C. (TOU/VIS) Mar.

First and Last: *Phanerosteon mirabile* Traquair, 1881, Cementstone Group, Scotland, UK.

F. GONATODIDAE Gardiner, 1967 C. (TOU) Mar./FW

First: *Gonatodus punctatus* (Agassiz, 1835), Calciferous Sandstone Series, *Protamblypterus macrolepis* (Traquair, 1877), Blackband Ironstone, Scotland, UK.
Last: *Drydenius molyneuxi* (Traquair, 1877), Deep Mine Ironstone, Staffordshire, England, UK, France, Belgium and Germany; *Brachypareion insperatum, Paradrydenius tinterisi, Paragonatodus magnificus* Kazantseva, 1980, all from Kazakhstan, former USSR.

F. STREPHEOSCHEMIDAE Gardiner, 1985 C. (TOU/VIS) Mar.

First and Last: *Strepheoschema fouldenensis* White, 1927, *Aetheretmon valentiacum* White, 1927, *?Rhadinichthys carinatus* (Agassiz, 1835), *?Bendenius deneensis* (Traquair, 1878), Dinantian of Foulden, Scotland, and Cementstone Group, Northumberland, England, UK.

F. CORNUBONISCIDAE White, 1939 C. (TOU) Mar.

First and Last: *Cornuboniscus budensis* White, 1939, Barren Coal Measures, Bude, Cornwall, England, UK.

F. AMBLYPTERIDAE Romer, 1945 C. (BSH)–P. (TAT) Mar./FW

First: *Paramblypterus decorus* (Egerton, 1850), Coal Measures, France.
Last: *Amblypterus* Agassiz, 1833, *Amblypterina* Berg, 1940, Lebach, Germany, Kargala, former USSR; *Gardinerichthys tewarii* Gupta *et al.*, 1978, Lower Permian, India; *Korutichthys korutensis* Kazantseva, 1980, Lower Upper Permian, former USSR.

F. REDFIELDIIDAE Berg, 1940 Tr. (SCY)–J. (l.) Mar./FW

First: *Sakamenichthys germaini* Nauche, 1959, Madagascar; *Brookvalia* Wade, 1935, *Dictyopyge* Egerton, 1847, Narrabeen Shales, Gosford, New South Wales, Australia; *Atopocephala* Brough, 1934, *Daedalichthys* Brough, 1930, *Helichthys* Broom, 1909, *Cynognathus* Zone, Bekkers Kraal, South Africa.
Last: *Redfieldius* Hay, 1902, *Redfieldius* beds, Newark, USA.

F. HAPLOLEPIDIDAE Westoll, 1944 C. (BSK/MOS)–P. (ASS/KUN) Mar./FW

First: *Haplolepis* Miller, 1892, several species, Linton, Ohio, Mazon Creek, Illinois, USA, Coal Measures, Northumberland and Staffordshire, England, UK.
Last: *Pyritocephalus sculptus* Fric, 1894, 'Cannel' Coal, Nýřany, Czechoslovakia, Mazon Creek, Illinois, USA, Newsham, England, UK.

F. STYRACOPTERIDAE Moy-Thomas, 1939 C. (TOU)–P. (ZEC) Mar.

First: *Styracopterus fulcatus* Traquair, 1890, Cementstone Group, Scotland, UK; *Whiteichthys* Moy-Thomas, 1942, Greenland.
Last: *Chichia gracilis* Liu and Wang, 1978, Upper Permian, China.

F. CANOBIIDAE Aldinger, 1937 C. (TOU–VIS) Mar.

First: *Canobius* Traquair, 1881, *Mesopoma* Traquair, 1890, Cementstone Group, Dumfriesshire, Scotland, UK.
Last: *Sundayichthys elegantulus* Gardiner, 1969, upper Witteberg Series, South Africa; *Charleuxia* Heyler, 1969, from the ASS of France doubtfully belongs to this family.

F. TARRASIIDAE Traquair, 1881 C. (TOU–BAS/MOS) Mar.

First: *Tarrasius problematicus* Traquair, 1881, Cementstone Group, Dumfriesshire, Scotland, UK.
Last: *Palaeophichthys parvulus* Eastman, 1908, Coal Measures, Mazon Creek, Illinois, USA.

F. AEDUELLIIDAE Romer, 1945 P. (ASS/KUN) Mar.

First and Last: *Aeduella blainvillei* (Agassiz, 1833), *Westollia crassus* (Pholig, 1892), *Igornella, Decazella, Burbonella* Heyler, 1969, Autunian of Bourbon-l'Archambault Decazeville and Autun, France. *Palaeothrissum* Blainville, 1818, Marl Slate, England, UK, Germany.

F. PLATYSOMIDAE Young, 1866 C. (TOU)–Tr. (ANS) Mar.

First: *Mesolepis* Young, 1866, *Paramesolepis* Moy-Thomas and Dyne, 1938, *Wardichthys* Traquair, 1875, *Platysomus* Agassiz, 1833, Cementstone Group, Dumfriesshire, Scotland, UK.
Last: *Caruichthys ornatus* Broom, 1913, South Africa;

Dorsolepis virgatus Jörg, 1969, Bunter Sandstone, Karlsruhe, Germany.

F. AMPHICENTRIDAE Moy-Thomas, 1939
C. (TOU)–P. (TAT) Mar.

First: *Cheirodopsis geikiei* Traquair, 1881, Cementstone Group, Dumfriesshire, Scotland, UK.
Last: *Eurynotoides cypriorion* Berg, 1940, Kayala, Ural Basin, former USSR.

F. CRYPHIOLEPIDIDAE Moy-Thomas, 1939
C. (BSH) FW

First and Last: *Cryphiolepis striatus* (Traquair, 1881), Edge Coal Series, Scotland, UK.

F. BOREOLEPIDIDAE Aldinger, 1937
P. (ROT) Mar.

First and Last: *Boreolepis jenseni* Aldinger, 1937, Middle Permian, Greenland.

F. UROSTHENIDAE Woodward, 1931
P. (TAT)–Tr. (LAD) Mar.

First: *Urosthenes latus* Woodward, 1931, Upper Coal Measures, New South Wales, Australia.
Last: *Urosthenes australis* Dana, 1848, Hawkesbury Series, New South Wales, Australia.

F. BRACHYDEGMIDAE Gardiner, 1967
P. (ROT) FW

First and Last: *Brachydegma caelatum* Dunkle, 1939, Red Beds, Texas, USA.

F. LAWNIIDAE Gardiner, 1967 P. (ROT) FW

First and Last: *Lawnia taylorensis* Wilson, 1953, Red Beds, Texas, USA.

F. TRISSOLEPIDIDAE Frič, 1893 C. (u.) FW

First and Last: *Sceletophorus biserialis* Frič, 1894, *Sphaerolepis kounoviensis* Frič, 1877, 'Cannel' coal, Třeošná, Kounová, Záboř and Hředl, Czechoslovakia.

F. COMMENTRYIDAE Gardiner, 1963
C. (BSH) FW

First and Last: *Commentrya traquairi* Sauvage, 1888, Houiller Series, Commentry, France.

Order DORYPTERIFORMES Berg, 1937

F. DORYPTERIDAE Gill, 1926
P. (UFI) FW

First and Last: *Dorypterus hoffmanni* Germar, 1842, ?*Dorypterus althausi* (Münster, 1842), Marl Slate, Durham, England, UK, and Kupferschiefer, Germany; *Dorypterus* sp. Liu and Tseng, 1964, Touling Coal Series, Central China.

Order BOBASATRANIIFORMES Berg, 1937
(see Fig. 35.2)

F. BOBASATRANIIDAE Stensiö, 1932
P. (ZEC)–Tr. (CRN) Mar.

First: *Ebenaqua ritchiei* Campbell and Dug Phuoc, 1983, Rangal Coal Measures, Queensland, Australia.
Last: *Bobasatrania* White, 1932, several species, Greenland, Madagascar and Canada; *Ecrinesomus dixoni* Woodward, 1910, Madagascar; *Polzbergia brochatus* Griffith, 1977, Rheingrabener Schiefer, Austria, may also belong here.

Order SAURICHTHYIFORMES Berg, 1937

F. SAURICHTHYIDAE Goodrich, 1909
(= Belonorynchidae Woodward, 1888)
Tr. (SCY)–J. (TOA) Mar.

First: *Saurichthys* Agassiz, 1834, several species, Muschelkalk, Germany, Raibl, Lombardy, Madagascar, Nepal and Spitsbergen. *Eosaurichthys* Liu and Wel, 1988, is said to come from the Upper Permian of China.
Last: *Saurorhynchus brevirostris* (Woodward, 1895), upper Lias, England and Germany.

Superdivision CHONDROSTEI Müller, 1845

Order ACIPENSERIFORMES Berg, 1940

F. PEIPIAOSTEIDAE Liu and Zhou, 1965
J. (TTH) Mar.

First and Last: *Peipiaosteus pani* Liu and Zhou, 1965, Tsien-shan-tze-kou, Nanling, China.

F. CHONDROSTEIDAE Traquair, 1877
J. (SIN–TOA) Mar./FW

First: *Chondrosteus acipenseroides* Egerton, 1844, *Chondrosteus pachyurus* Egerton, 1858, lower Lias, Lyme Regis and Barrow-on-Soar, Leicestershire, England, UK.
Last: *Strongylosteus hindenburgi* (Pompeckz, 1914), upper Lias Holzmaden, Germany; *Gyrosteus mirabilis* Egerton, 1858, upper Lias, Whitby, Yorkshire, England, UK. *Gyrosteus subdeltoideus* Stinton and Torrens, 1968, is recorded from the BTH.

F. ACIPENSERIDAE Bonaparte, 1831
K. (TUR/CMP)–Rec. Mar./FW.

First: *Paleopsephurus wilsoni* McAlpin, 1947, Hell Creek Beds, Montana; *Acipenser* Linnaeus, 1758, Edmonton Beds, Alberta, Canada. **Extant**

F. POLYODONTIDAE Bonaparte, 1838
T. (YPR)–Rec. Mar./FW

First: *Crassopholis magnicaudata* Cope, 1883, Green River Shales, Wyoming, USA. **Extant**

Superdivision NEOPTERYGII Regan, 1925

Order PALAEONISCIFORMES Goodrich, 1909

F. PALAEONISCIDAE Vogt, 1852
C. (TOU)–J. (SIN) Mar.

First: *Elonichthys serratus* Traquair, 1881, *E. pulcherrimus* Traquair, 1881, Calciferous Sandstone, Scotland, UK.
Last: *Cosmolepis ornatus* (Egerton, 1854), *Cosmolepis egertoni* Egerton, 1858, lower Lias, Lyme Regis, England, UK.

F. CENTROLEPIDIDAE Gardiner, 1960
J. (SIN) Mar.

First and Last: *Centrolepis aspera* Agassiz, 1844, lower Lias, Lyme Regis, England, UK.

F. COCCOCEPHALICHTHYIDAE Romer, 1945
C. (PND/ALP)–P. (ROT) Mar.

First: *Coccocephalichthys wildi* (Watson, 1925), Coal Measures, Lancashire, England, UK.
Last: *Coccocephalichthys tessallatus* Beltan, 1981 = *Monesedeiphus depressus* Beltan, 1990, Uruguay.

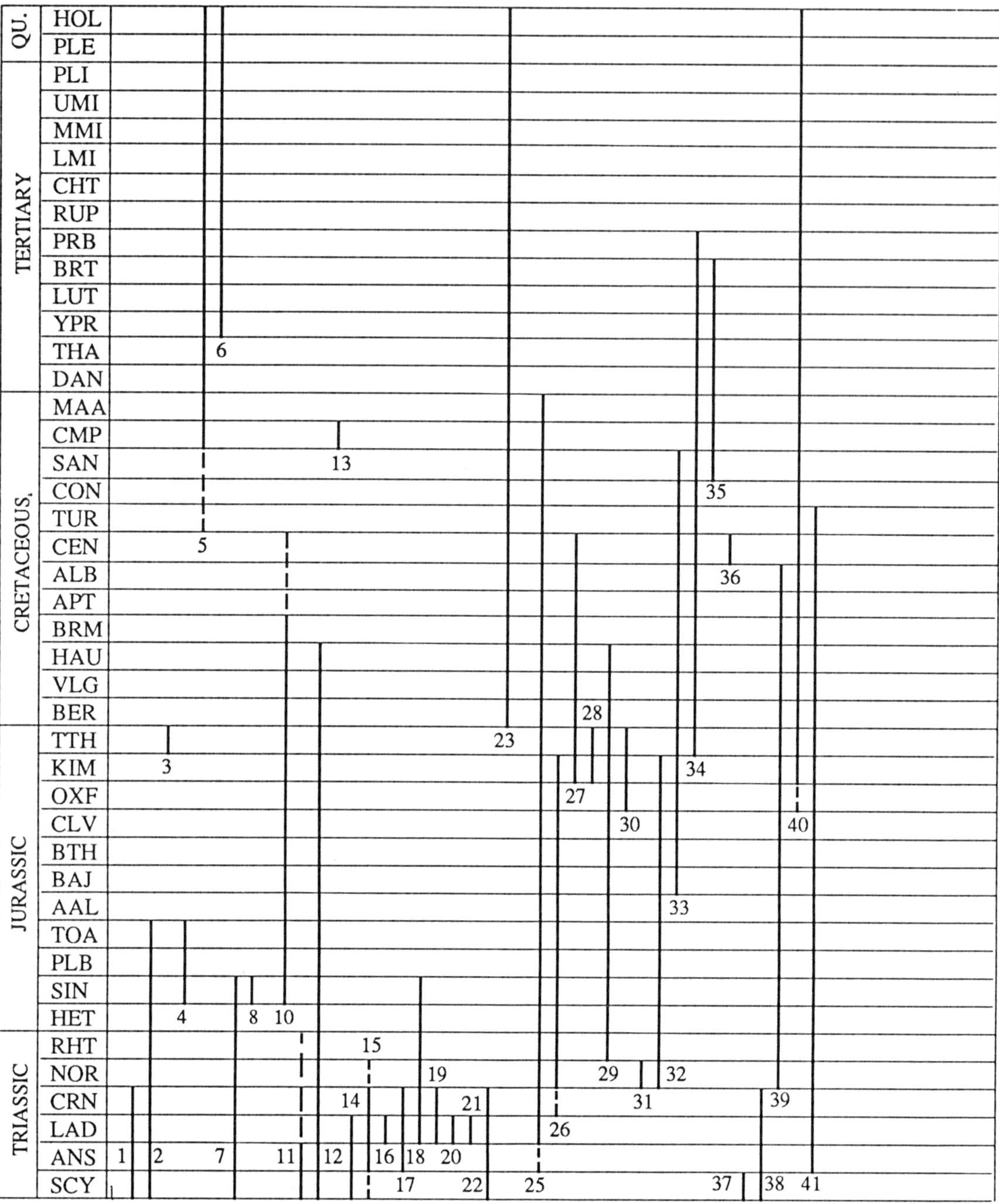

Fig. 35.2

F. COCCOLEPIDIDAE Berg, 1940 J. (SIN)–K. (BRM/CEN) Mar.

First: *Coccolepis liassica* Woodward, 1890, lower Lias, Lyme Regis, England, UK; *Plesiococcolepis humanensis* Wang, 1977, Lower Jurassic, China.

Last: *Sunolepis yumenensis* Liu, 1957, China. *Coccolepis macroptera* Traquair, 1911, Ber, Belgium–*Coccolepis* is widely distributed and is also recorded from the K_1 of Turkestan, former USSR and Australia.

F. SCANILEPIDIDAE Romer, 1945 Tr. (SCY/ANS–RHT) Mar.

First: *Evenkia eunotoptera* Berg, 1941, Tunguska Basin, Siberia, former USSR.

Last: *Scanilepis dubia* (Woodward, 1893), Sweden. *Fukangichthys longidorsalis* Su, 1978, China and *Tanaocrossus kalliokoskii* Schaeffer, 1967, Norian, New Mexico, doubtfully belong in this family,

F. BIRGERIIDAE Aldinger, 1937 Tr. (SCY)–K. (NEO) Mar.

First: *Birgeria* Stensiö, 1919, several species, Europe, Spitsbergen, Madagascar and Greenland.

Last: *Psilichthys selwyni* Hall, 1900, Victoria, Australia.

F. ASAROTIDAE Schaeffer, 1968 K. (CMP) Mar.

First and Last: *Asarotus* Schaeffer, 1968, Niobrara Formation, Kansas, USA.

PERMIAN	TAT
	KAZ
	UFI 1 … 24
	KUN
	ART
	SAK
	ASS
CARBONIFEROUS	GZE
	KAS
	MOS
	BSH
	SPK 9
	VIS
	TOU 7
DEVONIAN	FAM
	FRS
	GIV
	EIF
	EMS
	PRA
	LOK
SILURIAN	PRD
	LUD
	WEN
	LLY
ORDOVICIAN	ASH
	CRD
	LLO
	LLN
	ARG
	TRE
CAMB.	MER
	STD
	CRF
SINIAN	EDI
	VAR
	STU

Key for both diagrams

BOBASTRANIIFORMES	LUGANIIFORMES
1. Bobasatraniidae	21. Luganoiidae
SAURICHTHYIFORMES	22. Thoracopteridae
2. Saurichthyidae	GINGLYMODI
CHONDROSTEI	23. Lepisosteidae
ACIPENSERIFORMES	HALECOSTOMI
3. Peipiaosteidae	24. Acentrophoridae
4. Chondrosteidae	25. Semionotidae
5. Acipenseridae	26. Dapediidae
6. Polyodontidae	27. Oligopleuridae
NEOPTERYGII	28. Ionoscopidae
PALAEONISCIFORMES	29. Macrosemiidae
7. Palaeoniscidae	30. Uarbrichthyidae
8. Centrolepididae	PYCNODONTIFORMES
9. Coccocephalichthyidae	31. Brembodontidae
10. Coccolepididae	32. Mesturidae
11. Scanilepididae	33. Gyrodontidae
12. Birgeriidae	34. Pycnodontidae
13. Asarotidae	35. Nursallidae
PHOLIDO-PLEURIFORMES	36. Coccodontidae
	HALECOMORPHI
14. Pholidopleuridae	PARASEMIONO-TIFORMES
PERLEIDIFORMES	
15. Colobodontidae	37. Parasemionotidae
16. Aetheodontidae	38. Promecosominidae
17. Cleithrolepididae	AMIIFORMES
PELTOPLEURIFORMES	39. Caturidae
18. Peltopleuridae	40. Amiidae
19. Habroichthyidae	41. Ophiopsidae
CEPHALOXENIFORMES	
20. Cephaloxenidae	

Fig. 35.2

Order PHOLIDOPLEURIFORMES Berg, 1937

F. PHOLIDOPLEURIDAE Wade, 1932 Tr. (SCY–LAD) Mar.

First: *Australosomus* Piveteau, 1930, several species, Greenland, Spitsbergen and Madagascar. *Arctosomus sibiricus* Berg, 1941, Induan/Anisian, Tunguska Basin, Siberia, former USSR.
Last: *Pholidopleurus typus* Bronn, 1858, Besano, Italy; *Macroaethes* Wade, 1932, Middle Triassic, Brookvale, Australia.

Order PERLEIDIFORMES Berg, 1937

F. COLOBODONTIDAE Stensiö, 1916 Tr. (SCY/ANS–CRN/NOR) Mar.

First: *Colobodus* Agassiz, 1844, *Perleidus* Deeke, 1911, *Meidiichthys* Borough, 1931, *Tripelta* Wade, 1940, *Chrotichthys* Wade, 1940, *Zeuchthiscus* Wade, 1940, *Pristisomus* Woodward, 1890, *Helmolepis* Nybelin, 1977, *Boreichthys* Selezneva, 1982, *Plesioperleidus* Dazae and Li, 1983, collectively a world-wide distribution.
Last: *Colobodus* Agassiz, 1844, several species, Besano, Italy, and many other European localities.

F. AETHEODONTIDAE Brough, 1939 Tr. (LAD) Mar.

First and Last: *Aetheodontus besanensis* Brough, 1939, Besano, Italy.

F. CLEITHROLEPIDIDAE Wade, 1935 Tr. (ANS–CRN) Mar./FW

First: *Cleithrolepis granulata* Egerton, 1864, *Cleithrolepis alta* Wade, 1935, Narrabeen Shales, Gosford, New South Wales, Australia.
Last: *Dipteronotus cyphus* Egerton, 1854, Worcestershire and Otter Sandstone, Devon, England, UK.

Order PELTOPLEURIFORMES Lehman, 1966

F. PELTOPLEURIDAE Brough, 1939 Tr. (LAD)–J. (SIN) Mar.

First: *Peltopleurus* Kner, 1866, Besano, Italy, and Kueichow, China; *Placopleurus* Brough, 1939, Besano, Italy; *Platysiagum* Egerton, 1872, Besano, Italy.

Last: *Platysiagum sclerocephalus* Egerton, 1872, lower Lias, Lyme Regis, England, UK; *Placopleurus* sp. Italy.

F. HABROICHTHYIDAE Gardiner, 1967 Tr. (LAD–CRN) Mar.

First: *Habroichthys minimus* Brough, 1939, Besano, Italy.
Last: *Nannolepis elegans* Griffith, 1977, Rheingrabener Schiefer, Lunz, Austria.

Order CEPHALOXENIFORMES Lehman, 1966

F. CEPHALOXENIDAE Brough, 1939 Tr. (LAD) Mar.

First and Last: *Cephaloxenus macropterus* Brough, 1939, Besano, Italy.

Order LUGANOIIFORMES Lehman, 1958

F. LUGANOIIDAE Brough, 1939 Tr. (LAD) Mar.

First and Last: *Luganoia* Brough, 1939, several species, *Besania micrognathus* Brough, 1939, Besano, Italy.

F. THORACOPTERIDAE Griffiths, 1977 Tr. (SCY/CRN) Mar.

First and Last: *Thoracopterus* Bronn, 1858, *Gigantopterus* Abel, 1904, Upper Triassic, Carinthie, Austria, and Salerno, Italy.

Division GINGLYMODI, Cope, 1871

F. LEPISOSTEIDAE Cuvier, 1825 K. (l.)–Rec. FW

First: *Paralepidosteus* Arambourg and Joleaud, 1943, Damergou, Niger, Africa; *Lepisosteus* Lacépède, 1903 ranges back into the CMP in North America and Europe and into the Eocene in India. **Extant**

Division HALECOSTOMI Regan, 1923

F. ACENTROPHORIDAE Berg, 1936 P. (ZEC) Mar.

First and Last: *Acentrophorus* Traquair, 1877, Magnesian Limestone, Durham and Northumberland, England, UK.

F. SEMIONOTIDAE Woodward, 1890 Tr. (ANS/LAD)–K. (MAA) Mar./FW

First: *Semionotus* Agassiz, 1832, Europe, North America, and Africa; *Alleiolepis*, *Aphelolepis* Heller, 1953, Europe; *Asialepidotus* Su, 1959, China; *Sinosemionotus* Yuan and Koh, 1936, China; *Allelepidotus* Frič, 1906, Europe; *Paralepidotus* Stolley, 1920, Europe. *Pericentrophorus* Jörg, 1969, Europe, possibly from the Scythian may belong here.
Last: *Lepidotes* Agassiz, 1832, England, France, Germany and Switzerland.

F. DAPEDIIDAE Vogt, 1852 Tr. (CRN/NOR)–J. (KIM) Mar./FW

First: *Dandya* White and Moy-Thomas, 1940, Hallein, Austria; *?Prohalecites* Deeke, 1889, Raibl, Carinthia, Besano, Italy, Polzberg, Austria.
Last: *Heterostrophus* Wagner, 1860, Solnhofen, Germany.

F. OLIGOPLEURIDAE Woodward, 1895 J. (KIM)–K. (CEN) Mar.

First: *Oligopleurus* Thiollière, 1850, Cerin, France.
Last: *Spathiurus* Davis, 1887, Lebanon; *Oshunia* Wentz and Kellner, 1986, APT, Brazil, may belong here.

F. IONOSCOPIDAE Lehman, 1966 J. (KIM–TTH) Mar.

First: *Callopterus* Thiollière, 1858, *Ionoscopus* Thiollière, 1858, Cerin, France, Bavaria, Germany.
Last: *Ionoscopus* Thiollière, 1858, Dalmatia, former Yugoslavia, Portlandian, England, UK.

F. MACROSEMIIDAE Thiollière, 1858 Tr. (RHT)–K. (HAV) Mar.

First: *Legonotus* Egerton, 1854, Rhaetic, Gloucestershire, England, UK, Hallein, Austria.
Last: *Enchelyolepis* Woodward, 1918, Meuse, France; Wealden, England, UK.

F. UARBRICHTHYIDAE Bartram, 1977 J. (u.) FW

First and Last: *Uarbrichthys* Wade, 1941, Talbragar Beds, New South Wales, Australia.

Order PYCNODONTIFORMES Lehman, 1966

F. BREMBODONTIDAE Tintori, 1980 Tr. (NOR) Mar.

First and Last: *Brembodus* Tintori, 1980, *Gibbodon* Tintori, 1980, Zorzino Limestone, Lombardy, Italy.

F. MESTURIDAE Nursall, 1987 Tr. (NOR)–J. (KIM) Mar.

First: *Eomesodon* Woodward, 1918, Barrow-on-Soar, Leicestershire, England, UK.
Last: *Mesturus* Wagner, 1859, Cerin; Bavaria; Kimmeridgian, England, UK.

F. GYRODONTIDAE Berg, 1940 J. (BAJ)–K. (SAN) Mar./FW

First: *Macromesodon* Blake, 1905, Europe; *Gyrodus* Agassiz, 1833, Europe.
Last: *Micropycnodon* Hibbard and Graffam, 1945, Niobrara Formation, Kansas, USA, *Anomoedus* Forir, 1887, Carlisle Shale, Fairport Chalk, Kansas, USA.

F. PYCNODONTIDAE Agassiz, 1833 J. (TTH)–T. (PRB) Mar./FW

First: *Coelodus* Heckel, 1856, England, France, Spain.
Last: *Pycnodus* Agassiz, 1833, Europe; *Palaeobalistum* Blainville, 1818, Monte Bolca, Italy; London Clay, Sheppey, England, UK; Belgium.

F. NURSALLIDAE Blot, 1987 K. (SEN)–T. (BRT) Mar.

First: *Nursallia glutterosum* (Arambourg, 1954), *N. goedei* (Heckel), Europe; *N. flabellatus* (Cope, 1886) is recorded from the SAN, Brazil.
Last: *Nursallia veronae* Blot, 1987, Monte Bolca, Italy.

F. COCCODONTIDAE Berg, 1940 K. (CEN) Mar.

First and Last: *Coccodus* Pictet, 1850; *Trewavasia* White and Moy-Thomas, 1941; *Ichthyoceras* Gayet, 1984, Lebanon.

Subdivision HALECOMORPHI Cope, 1871

Order PARASEMIONOTIFORMES Lehman, 1966

F. PARASEMIONOTIDAE Stensiö, 1932 Tr. (SCY) Mar.

First and Last: *Broughia* Stensiö, 1932; *Helmolepis* Stensiö, 1932, *Ospia* Stensiö, 1932, Greenland; *Jacobulus* Lehman,

1952, *Stensionotus* Lehman, 1952; *Thomasinotus* Lehman, 1952, *Watsonulus* Brough, 1939; *Parasemionotus* Piveteau, 1929, *Lehmanotus, Devillersia, Piveteaunotus* Beltan, 1968, Madagascar.

F. PROMECOSOMINIDAE Wade, 1941 Tr. (SCY–CRN) Mar.

First: *Paracentrophorus madagascariensis* Piveteau, 1941, Madagascar.
Last: *Phaidrosoma lunzensis* Griffith, 1977, Rheingrabener Schiefer, Polzberg, Austria.

Order AMIIFORMES Huxley, 1861

F. CATURIDAE Owen, 1860 (FURIDAE Jordan, 1923) Tr. (NOR)–K. (ALB) Mar./FW

First: *Heterolepidotus* Egerton, 1872, Tyrol, Austria; Perledo, Lombardy, Italy.
Last: *Macrepistius* Cope, 1894, Kansas, USA; *Lophiostomus* Egerton, 1852, from the CEN of Europe may also belong here.

F. AMIIDAE Bonaparte, 1837 J. (OXF/KIM)–Rec. Mar./FW

First: *Amiopsis* Kner, 1863, (= *Megalurus* Agassiz, 1833, *Urocles* Jordan, 1919), *Liodesmus* Wagner, 1859, Europe; *Vidalamia* White and Moy-Thomas, 1941, Lower Cretaceous, Europe; *Sinamia* Stensiö, 1935, *Ikechaoamia* Liu, 1961, Upper Jurassic/Lower Cretaceous, China. *Amia* Linnaeus ranges back into the late Cretaceous Fruitland and Kirtland Formations, New Mexico, USA, and the Palaeogene of Europe and Asia. **Extant**

F. OPHIOPSIDAE Bartram, 1975 Tr. (m.)–K. (TUR) Mar./FW

First: *Ophiopsis attenuata* Wagner, 1859, Perledo, Como, Italy.
Last: *Neorhombolepis excelsus* Woodward, 1888, Wealden, Europe; *Aphanepygus* Bassani, 1879, Lebanon, Lesina, Lago, Italy; Dalmatia, former Yugoslavia.

REFERENCES

Gardiner, B. G. (1963) Certain palaeoniscoid fishes and the evolution of the snout in actinopterygians. *Bulletin of the British Museum (Natural History), Geology Series*, **8**, 255–325.

Gardiner, B. G. (1967) Further notes on palaeoniscoid fishes with a classification of the Chondrostei. *Bulletin of the British Museum (Natural History), Geology Series*, **14**, 143–206.

Gardiner, B. G. (1969) New palaeoniscoid fish from the Witteberg Series of South Africa. *Zoological Journal of the Linnean Society*, **48**, 423–52.

Gardiner, B. G. (1984) The relationships of the palaeoniscoid fishes: a review based on new specimens of *Mimia* and *Moythomasia* from the Upper Devonian of Western Australia. *Bulletin of the British Museum (Natural History), Geology Series*, **37** 173–428.

Heyler, D. (1969) Vertébrés de l'Autunien de France. *Centre National de la Recherche Scientifique, Cahiers de Paléontologie*, **259**, 1–222.

Kazantseva-Selezneva, A. A. (1980) Permian Palaeonisci of Central Siberia. *Paleontologicheski Zhurnal*, **1980** (1), 95–103 [in Russian].

Kazantseva-Selezneva, A. A. (1981) A phylogeny of lower actinopterygians. *Voprosy Ikhtiologii*, **21** (4), 579–94 [in Russian].

Märss, T. (1986) *Silurian Vertebrates of Estonia and West Latvia.* Academy of Sciences Estonian SSR, Institute of Geology, 104 pp. [in Russian].

Schultze, H.-P. (1968) Palaeoniscoide Schuppen aus dem Underdevon Australiens und Kanadas und aus dem Mitteldevon Spitzbergens. *Bulletin of the British Museum (Natural History), Geology Series*, **16**, 341–68.

Stinton, F. C. and Torrens, H. S. (1968) Fish otoliths from the Bathonian of southern England. *Palaeontology*, **11**, 246–58.

Su, Re-Tsao (1974) New Jurassic ptycholepid fishes from Szechuan, S. W. China. *Vertebrata Palasiatica*, **12**, 1–20 [in Chinese, with English summary].

36

OSTEICHTHYES: TELEOSTEI

C. Patterson

The Teleostei, with about 22 000 living species, are by far the largest vertebrate group, accounting for more than half of the total of recognized species of living vertebrates (Nelson, 1984). They are also the vertebrate group to which cladistic analysis was first applied, and, for that reason, teleost systematics has probably progressed further down the road from traditional phenetics towards a phylogenetic system than has the classification of any other vertebrate group. The principal result of cladistic analysis is to uncover and eliminate paraphyletic groups, and therefore the general acceptance of cladistics tends to have more radical effects on the systematics of fossils than on extant taxa. The traditional phenetic method resulted in many extinct and paraphyletic 'ancestral groups' which dissolve under cladistic analysis, either into an assemblage of *incertae sedis* genera, most of them monotypic, or into a series of plesions (Patterson and Rosen, 1977), each the sister taxon of everything distal to it in the cladogram. Of course, we are still very far from a thorough analysis of fossil teleosts along these lines, and deficiencies of preservation convince me that we shall never achieve anything like it. In making this compilation, I have tried to strike a balance between simplicity of presentation and the cladist's expectation that groups shall be monophyletic. Some extinct taxa that I know or guess to be merely phenetic are included; comments on the status of nominal extinct families of teleosts are given in Smith and Patterson (1988).

In general, the arrangement of teleost families used here is that of Nelson (1984), except where subsequent revisions provide better supported classifications of particular subgroups. Nelson's (1984) check-list of extant teleostean families included 409 names. The following list includes 494 families, of which 69 are extinct, leaving 425 extant families. Among the 69 extinct nominal families, 32 (most of them monotypic) are listed as *incertae sedis* at various levels in the hierarchy. About 50 more nominal extinct families are merged or synonymized with other families, living or extinct. A few extinct monotypic nominal families are not listed, because I believe that the fossils are *incertae sedis* at levels in the hierarchy which do not make them the earliest members of the lowest-ranked taxon to which they may be assigned (e.g. Varasichthyidae Arratia, 1981, Chongichthyidae Arratia, 1982). In addition to the 494 families, the compilation includes ten Recent genera of Percoidei which Johnson (1984) listed as *incertae sedis* within that suborder, and because of its importance and extensive fossil record, the family Scombridae is subdivided into nine tribes (following Johnson, 1986 and Schultz, 1987).

The difference of sixteen between Nelson's (1984) 409 extant families and the 425 extant families in this compilation implies that there has been little change (a mere 4% increase in number) in teleost familial classification over the period of about seven years, surely the most active in the history of ichthyology, between the completion of Nelson's list and of this one. In fact, the differences between Nelson's list and this one are much more extensive, involving almost one family in five. Thirty-two families in Nelson's compilation are synonymized or merged with others in this list (because Nelson's book is widely used, those synonymized family names are capitalized for easy recognition), and 48 extant families in this list are absent in Nelson's. Among the 425 extant families listed here, 181 (43%) have no fossil record. Among the 244 extant families that do have a fossil record, that record consists only of otoliths for 58 families (24%).

The notations **O** and **S** mean respectively otolith and skeletal records, where it seems necessary to distinguish the latter from otoliths in a summary of stratigraphical range. Johnson (1984), in his review of Percoidei, uses an asterisk (*) after the family name to indicate percoid families with only one extant genus. That convention is used throughout this compilation (usually following Nelson's (1984) assessment of generic status), and may give those unfamiliar with teleosts a feel for families in which fossils outweigh Recent representatives, or in which a lack of a fossil record

The Fossil Record 2. Edited by M. J. Benton. Published in 1993 by Chapman & Hall, London. ISBN 0 412 39380 8

might be expected. Of the extant families 133 (31%) are monogeneric; among them, 82 (62%) have no fossil record, and of the 51 with a fossil record, 36 (70.5%) are known by skeletal fossils, and 15 (29.5%) have a fossil record only of otoliths. Monogeneric families account for 45% (81 out of 181) of the extant families without known fossils.

The higher classification used here is eclectic but not original, and parts are presented only in outline, as an asymmetrical (with more ranks towards the higher teleosts) hierarchy, principally so that taxa are available for *incertae sedis* fossils. Perciform classification, in particular, is in a state of flux, and I have taken some liberties in seeking to reduce the number of families left in the basal Percoidei. As one example, I have taken Johnson and Fritzsche's (1989) suggestion (based on a shared pattern of the ramus lateralis accessorius nerve) that several 'percoid' families might be related to the stromateoids to mean that those families should be classified with the stromateoids, and I have included them there as Stromateoidei *sensu lato*. Other instances are to be found. A reference placed after the name of a higher taxon means either that the classification of the taxon follows that source, or that the citation contains a review of fossil members. References are generally kept to a minimum.

Three general points on stratigraphy deserve mention here. Firstly, the fish beds at Monte Bolca, Verona, Italy, provide the earliest records for a very large number of teleost higher taxa. Traditionally regarded as Lutetian, Monte Bolca is now generally treated as topmost Ypresian (e.g. Blot, 1980) following the discovery of nannoplankton which place the beds in the zone of *Discoaster sublodoensis* (NP 14). But according to the Harland *et al.* (1990) time-scale used in this volume, NP 14 is lowermost Lutetian, and Monte Bolca records are so entered here. Secondly, the fish beds in the Danatinsk Formation, Kopetdag, SW Turkmenia (former USSR), contain an extensive teleost fauna first described by Danil'chenko (1968). He dated the fishes as late Palaeocene, and that allocation has been accepted in subsequent Russian work on the fauna (e.g. Bannikov, 1985a; Danil'chenko, 1980) and by other palaeoichthyologists. However, Tyler and Bannikov (1992), in describing new tetraodontiforms from the fish beds, point out that in the 1975 compendium on Russian Palaeogene stratigraphy, Solun and Travina (1975) placed the fish beds (which are about 9 m thick) in the middle portion of the Danatinsk Formation, with a suite of foraminifera including *Globorotalia subbotinae* (now called *Morozovella subbotinae*), an index fossil for the lowermost Eocene (basal YPR). Solun and Travina excluded the fish beds from the lower part of the Danatinsk containing *Acarinina subsphaerica* (now called *A. mckannai*), an index fossil for the top Palaeocene. In that 1975 compendium, Danil'chenko himself (1975, p. 436) referred the fishes to the lower Eocene, although elsewhere he subsequently continued to call them Upper Palaeocene. If that Thanetian age for the Danatinsk Formation is accepted, then it marks the earliest record for some fifteen families or higher taxa of teleosts, including Clupeidae, Chanidae, Urosphenidae, Serranidae, Menidae, Caristiidae, Carangidae, Acanthuroidei, Siganidae, Luvaridae, Euzaphlegidae, Scombrini and Sardini. I have decided to accept Solun and Travina's (1975) work as authoritative, and have entered all the Danatinsk fishes as Ypresian. Finally, in translating Russian stratigraphical terms, I have followed the practice of the translators of the *Paleontologicheskii Zhurnal* by taking 'svita' (suite) as 'Formation'.

The authorship of family-group names in teleosts is still an outstanding problem, for no trustworthy compilation exists. In general, I have followed Nolf (1985): the only modern and reasonably complete list that I know; I have commented elsewhere (Patterson, 1987) on the unjust burden thrown on Nolf, a fossil otolith specialist, in attempting to determine the authorship of all teleost family-group names. Nolf had a little help from me, mostly on higher-category names, but he traced the vast majority of the family names and I believe that he did an excellent job. However, a recent authoritative work on eels (Böhlke, 1989) showed that in that group Nolf had the author and/or date wrong for seven out of 25 families (28%). Although I have corrected those and other slips that I have found in Nolf's authorship of families, I have added many more families, and it will not surprise me if my error rate is as high as 28%. I beg indulgence and correction from those better informed.

Acknowledgements – I am particularly grateful to Dirk Nolf, Brussels, and Lance Grande, Chicago, for the time that they have spent on drafts of this work, and for the information that they have given to me. I have also had most valuable advice or information from: G. David Johnson, Lynne Parenti, Victor G. Springer, James C. Tyler and Richard P. Vari in Washington DC; Gordon Howes, Dick Jefferies and Darrell Siebert in London; and Mark Wilson in Edmonton. My sincere thanks to all of them. Doubtless many errors and omissions remain, and I would much appreciate corrections. I hope to maintain an updated and corrected compilation on disc, and would welcome contributions to it and queries about particular groups.

Subclass TELEOSTEI Müller, 1846

F. PACHYCORMIDAE Woodward, 1895 (including Protosphyraenidae Lydekker, 1889) J. (TOA)–K. (CMP) Mar. (see Fig. 36.1)

First: *Euthynotus, Pachycormus, Prosauropsis, Sauropsis, Saurostomus* all in upper Lias, UK, France, Germany.
Last: *Protosphyraena* spp., CMP zones, European Chalk and Demopolis Chalk, Alabama and Mississippi, USA (Stewart, 1988).
Intervening: BAJ, CLV–TTH, ALB–SAN.

F. ASPIDORHYNCHIDAE Bleeker, 1859 (including Vinctiferidae Silva Santos, 1990) J. (BTH)–T. (THA) Mar./FW

First: *Aspidorhynchus crassus* Woodward, 1890, Stonesfield Slate, Oxfordshire, England, UK.
Last: *Belonostomus* sp. Bryant, 1987, Tongue River Formation, Morton County, North Dakota, USA.
Intervening: CLV–MAA.

F. PLEUROPHOLIDAE Saint-Seine, 1949 (including Ligulellidae Saint-Seine, 1955, Majokiidae Saint-Seine, 1955) J. (KIM)–K. (APT) Mar./FW

First: *Pleuropholis thiollieri* Sauvage, 1883, Cerin, Ain, France.
Last: *Pleuropholis decastroi* Bravi, 1991, Calcari ad ittioliti, Pietraroja, Campania, Italy.
Intervening: TTH–BRM.

F. PHOLIDOPHORIDAE Woodward, 1890 (including Archaeomenidae Boulenger, 1904) Tr. (CRN)–K. (APT) Mar./FW

First: *Pholidophoretes salvus* Griffith, 1977, Reingrabener Schiefer, Austria.
Last: *Neopholidophoropsis serrata* Taverne, 1981, *Pholidophoristion ornatus* (Agassiz, 1844), *P. spaethi* Taverne, 1981, Töck, Heligoland; *Wadeichthys oxyops* Waldman, 1971, Koonwarra, Victoria, Australia (?APT/ALB) (Smith and Patterson, 1988).
Intervening: NOR–BER.

F. ICHTHYOKENTEMIDAE Griffith and Patterson, 1963 (? = Catervariolidae Saint-Seine, 1955, Galkiniidae Yakovlev, 1962) Tr. (CRN)–J. (TTH) Mar./FW

First: *Elpistoichthys pectinatus* Griffith, 1977, Reingrabener Schiefer, Austria.
Last: *Ceramurus macrocephalus* Egerton, 1854, *Ichthyokentema purbeckensis* (Davis, 1887), lower Purbeck, Dorset, England, UK.
Intervening: J (m.).

F. SIYUICHTHYIDAE Su, 1985 K. (l.) FW

First and Last: Five genera, Shengchinkow Formation, Xinjiang, China (Su, 1985).

F. LEPTOLEPIDIDAE Lydekker, 1889 (including Koonwarriidae Waldman, 1971) J. (SIN)–K. (APT/ALB) Mar./FW

First: *Proleptolepis* spp. (Nybelin, 1975), *obtusum* subzone, Lower Lias, Dorset, England, UK; undetermined fishes, Quebrada Vaquillas Altas, Chile (Arratia, 1987).
Last: *Koonwarria manifrons* Waldman, 1971, *Leptolepis koonwarri* Waldman, 1971, Koonwarra, Victoria, Australia.
Intervening: PLB–HAU.

F. ALLOTHRISSOPIDAE Patterson and Rosen, 1977 J. (KIM–TTH) Mar.

First: *Allothrissops regleyi* (Thiollière, 1854), Cerin, Ain, France.
Last: *Allothrissops salmoneus* (Blainville, 1818), *A. mesogaster* (Agassiz, 1833), Solnhofen, etc., Bavaria, Germany.

F. ICHTHYODECTIDAE Crook, 1892 J. (KIM)–K. (CMP) Mar.

First: *Thrissops formosus* Agassiz, 1833, *T. subovatus* Münster, 1833, Cerin, Ain, France.
Last: *Gillicus, Ichthyodectes, Xiphactinus*, Niobrara Formation, Kansas, USA.
Intervening: TTH–SAN.

F. SAURODONTIDAE Cope, 1871 K. (CEN–MAA) Mar.

First: *Saurodon intermedius* (Newton, 1878), Lower Chalk, Kent, England, UK.
Last: *Saurocephalus lanciformis* Harlan, 1824, Navesink Formation, New Jersey, USA; Maastricht, The Netherlands; Gramame Formation, Pernambuco, Brazil (Silva Santos and de Figueiredo, 1987).
Intervening: TUR–CMP.

Supercohort OSTEOGLOSSOMORPHA Greenwood *et al.*, 1966

Order OSTEOGLOSSIFORMES Regan, 1909

Osteoglossiformes *incertae sedis* (F. HUASHIIDAE Chang and Chou, 1977) J. (u.)/K. (l.) FW

First and Last: *Kuntulunia longipterus* Liu *et al.*, 1982, Inner Mongolia and Ningxia, China, in beds thought to be 'somewhere near the Jurassic–Cretaceous boundary' (Chang and Chow, 1986).

Suborder NOTOPTEROIDEI Jordan, 1923

F. HIODONTIDAE* Valenciennes, 1846 (including Lycopteridae Cockerell, 1925) J. (u.)/K. (l.)–Rec. FW

First: *Lycoptera* spp. and other genera, widespread in China (Liaoning, Hebei, Jilin, Gansu, Shanxi, Shandong, Ningxia), Mongolia and south-eastern former USSR, in beds thought to be 'somewhere near the Jurassic–Cretaceous boundary' (Chang and Chow, 1986). **Extant**
Intervening: K. (m.)–RUP.

?F. OSTARIOSTOMIDAE Schaeffer, 1949 K. (MAA)/T. (DAN) FW

First and Last: *Ostariostoma wilseyi* Schaeffer, 1949, Livingston Formation, Madison County, Montana, USA (Grande and Cavender, 1991).

F. NOTOPTERIDAE Bleeker, 1859 **O** T. (?DAN)–Rec. FW

First: **O** *'Notopteridarum' nolfi* Rana, 1988, Intertrappeans between Deccan Trap Flows no. 4 and 5, Rangapur, Hyderabad, India. **S** ?Eoc., *Notopterus notopterus* (Pallas,

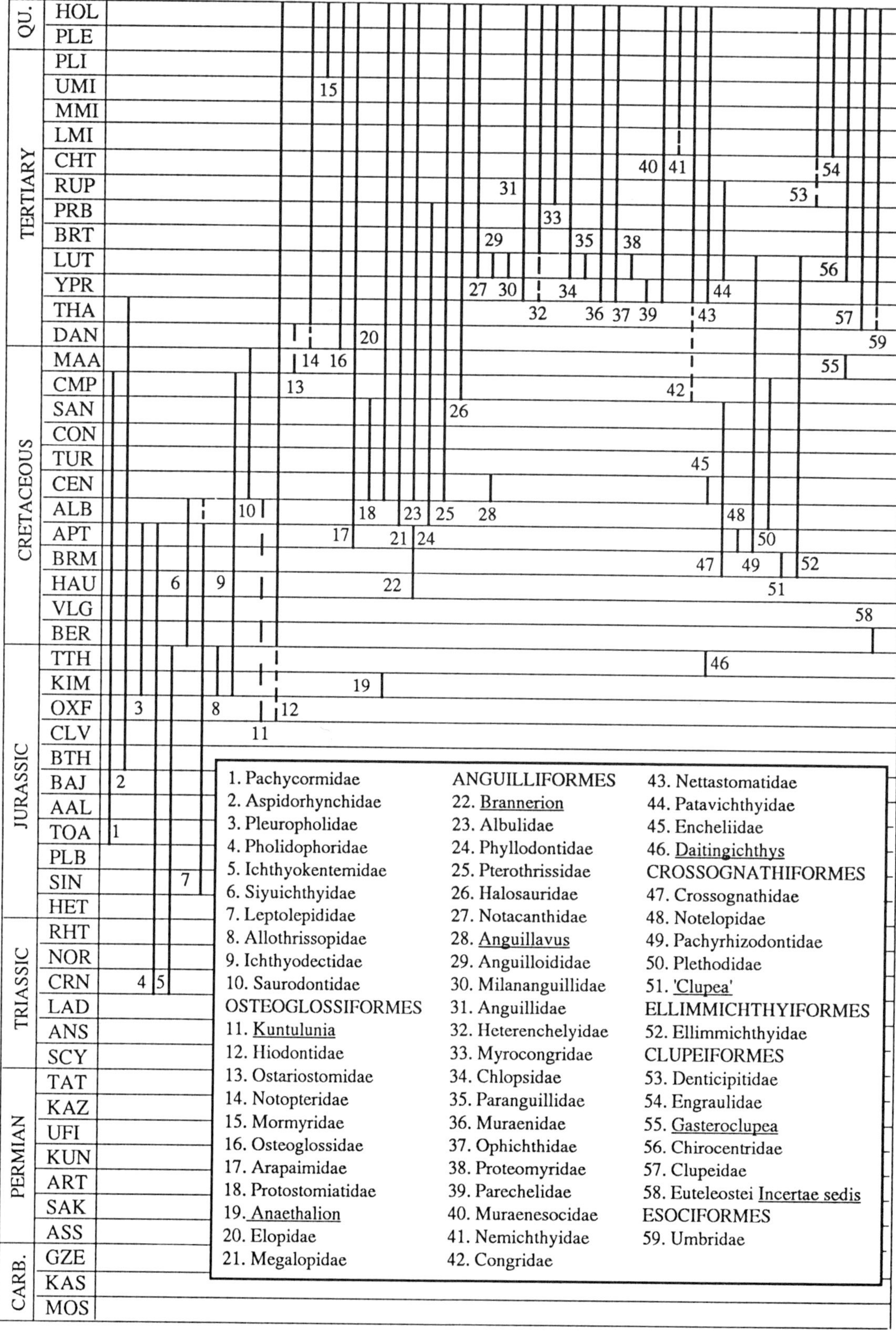

Fig. 36.1

1769, ?including *N. primaevus* Günther, 1876), Padang, Sumatra. The Eocene age of the fish shales at Padang is still dubious. **Extant**

Intervening: ?Eoc.

F. MORMYRIDAE Bonaparte, 1832 T. (PLI)–Rec. FW

First: *Hyperopisus* sp., Wadi el Natrun, Egypt (Greenwood, 1974). **Extant**

Intervening: PLE.

F. GYMNARCHIDAE* Bleeker, 1859 **Extant** FW

Suborder OSTEOGLOSSOIDEI Regan, 1909

F. OSTEOGLOSSIDAE Bonaparte, 1832 (including PANTODONTIDAE Peters, 1876) T. (DAN)–Rec. Mar./FW

First: Scales assigned to *Musperia*, *Phareoides* and Osteoglossidae *incertae sedis*, Intertrappean beds, central India; *Brychaetus caheni* Taverne, 1969, Montian, Landana, Cabinda, Angola. *Phareodus* is also reported from the Lameta Formation (?MAA) and equivalent infratrappean strata in India by Sahni and Bajpai (1988) and Prasad (1989). **O** *'Osteoglossidarum' deccanensis* Rana, 1988, *'O.' intertrappus* Rana, 1988, Intertrappeans between Deccan Trap Flows no. 4 and 5, Rangapur, Hyderabad, India. **Extant**
Intervening: THA–BRT.

F. ARAPAIMIDAE Bleeker, 1859 K. (APT)–Rec. FW

First: *Laeliichthys ancestralis* Silva Santos, 1985, Areado Formation, Minas Gerais, Brazil. **Extant**
Intervening: CEN.

Supercohort ELOPOCEPHALA Patterson and Rosen, 1977

ELOPOCEPHALA *incertae sedis*

F. PROTOSTOMIATIDAE Arambourg, 1954 K. (CEN–SAN) Mar.

First: *Protostomias maroccanus* Arambourg, 1954, Jebel Tselfat, Morocco; Cinto Euganeo, Padova, Italy.
Last: *Pronotacanthus sahelalmae* (Davis, 1887), Sahel Alma, Lebanon. This family was assigned to the Stomiiformes by Arambourg (1954), and may belong there (L. Taverne, pers. comm.), but no stomiiform apomorphies are yet recorded.

Cohort ELOPOMORPHA Greenwood *et al.*, 1966

ELOPOMORPHA *incertae sedis* (including Anaethalionidae Gaudant, 1968) J. (KIM) Mar.

First: *Anaethalion* spp., Cerin, Ain, France. The genus also occurs in TTH and VLG.

F. ELOPIDAE* Bonaparte, 1832 K. (CEN)–Rec. Mar.

First: *Davichthys gardneri* Forey, 1973, Hakel, Lebanon; *D. lacostei* (Arambourg, 1954), Jebel Tselfat, Morocco. **Extant**
Intervening: SAN, Eoc. (l.–m.), ?UMI.

F. MEGALOPIDAE* Jordan and Gilbert, 1882 (including Sedenhorstiidae Goody, 1969) K. (ALB)–Rec. Mar.

First: *Elopoides tomassoni* Wenz, 1965, Vallentigny, Aube, France; **O** *'Megalopidarum' bicrenulatus* (Stinton, 1973), Gault, Kent, England, UK. **Extant**
Intervening: CEN, CMP, YPR, UMI.

Order ANGUILLIFORMES Regan, 1909

Suborder ALBULOIDEI Jordan, 1923

ALBULOIDEI *incertae sedis* (including Osmeroididae Forey, 1973) **O** K. (HAU) **S** K. (APT)

First: **O** *'Albuloideorum' ventralis* Weiler, 1971, lower HAU, Engelbostel, Hannover, Germany; **S** *Brannerion vestitum* Jordan and Branner, 1908, Santana Formation, Ceara, Brazil.

F. ALBULIDAE* Bleeker, 1859 K. (CEN)–Rec. Mar.

First: *Lebonichthys lewisi* (Davis, 1887), Hakel, Lebanon. **Extant**
Intervening: SNT, CMP, **O** THA–LMI.

F. PHYLLODONTIDAE Jordan, 1923 K. (ALB)–T. (PRB) Mar.

First: *Casierius heckelii* (Costa, 1864), Albian of Italy, France, UK and Texas (Estes, 1969).
Last: *Paralbula stromeri* (Weiler, 1929), Qasr-el-Sagha Formation, Egypt.
Intervening: CMP–BRT (Estes and Hiatt, 1978).

F. PTEROTHRISSIDAE* Gill, 1893 K. (CEN)–Rec. Mar.

First: *Hajulia multidens* Woodward, 1942, Hakel and Hajula, Lebanon. **Extant**
Intervening: SAN, CMP, **O** THA–PLI.

F. HALOSAURIDAE Günther, 1868 K. (CMP)–Rec. Mar.

First: *Echidnocephalus troscheli* Marck, 1858, Sendenhorst, Westphalia, Germany. **Extant**
Intervening: UMI.

F. NOTACANTHIDAE Rafinesque, 1810 (including LIPOGENYIDAE Goode and Bean, 1894) **O** T. (LUT)–Rec. Mar.

First: **O** *Notacanthus* sp. Nolf and Lapierre, 1979, Calcaire Grossier, Thiverval, Yvelines, France. **Extant**

Suborder ANGUILLOIDEI Regan, 1909 (arrangement of extant families follows Böhlke, 1989)

ANGUILLOIDEI *incertae sedis* K. (CEN) Mar.

First and Last: *Anguillavus quadripinnis* Hay, 1903, Hakel and Hajula, Lebanon; *Urenchelys germanus* Hay, 1903, Hajula, Lebanon; *U. hakelensis* (Davis, 1887), Hakel, Lebanon; ?*Enchelurus anglicus* Woodward, 1901, *H. subglobosus* zone, Lower Chalk, SE England, UK.
Comment: These fishes, conventionally allocated to the (paraphyletic) extinct families Anguillavidae Hay, 1903 and Urenchelyidae Regan, 1912, are primitive eels. Robins (1989) examined specimens of *Anguillavus* and *Urenchelys* and asserted that 'there is no basis to align them with the Anguilliformes' (i.e. they are not eels) but I believe that they are true eels (Patterson, in press). Anguilloid otoliths are recorded from the Severn Formation (MAA), Maryland, by Huddleston and Savoie (1983).

Superfamily ANGUILLOIDEA Rafinesque, 1810

ANGUILLOIDEA *incertae sedis*

F. ANGUILLOIDIDAE Blot, 1978
T. (LUT) Mar.

First and Last: *Anguilloides branchiostegalis* (Eastman, 1905), *Veronanguilla ruffoi* Blot, 1978, Monte Bolca, Verona, Italy.

F. MILANANGUILLIDAE Blot, 1978
T. (LUT) Mar.

First and Last: *Milananguilla lehmani* Blot, 1978, Monte Bolca, Verona, Italy.
Comment: These two families (Anguilloididae, Milananguillidae) are said by Blot (1984) to be close to the ancestral stock of anguilloids.

F. ANGUILLIDAE* Rafinesque, 1810
T. (YPR)–Rec. Mar./FW

First: O *Anguilla annosa* Stinton, 1975, upper London Clay Formation, SE England, UK; first skeletal remains are LUT, *Eoanguilla leptoptera* (Agassiz, 1835), Monte Bolca, Verona, Italy. **Extant**
Intervening: BRT–PLI.

F. MORINGUIDAE Gill, 1885
Extant Mar.

F. HETERENCHELYIDAE* Regan, 1912
O T. (?YPR)–Rec. Mar.

First: O *'Heterenchelyidarum' richardsi* Nolf, 1988, one otolith from the Argile de Gan, Gan, Pyrénées-Atlantiques, France, probably belongs to this species, which is otherwise PRB. **Extant**
Intervening: O BRT, PRB, Mio.–PLI.

Superfamily MURAENOIDEA Risso, 1826

F. MYROCONGRIDAE* Gill, 1890
?T. (RUP)–Rec. Mar.

First: ?*Myroconger roustami* Arambourg, 1967, Elam, Luristan, Iran. **Extant**

F. CHLOPSIDAE Rafinesque, 1810
(= XENOCONGRIDAE Regan, 1912)
T. (LUT)–Rec. Mar.

First: *Whitapodus breviculus* (Agassiz, 1839), Monte Bolca, Verona, Italy. **Extant**

F. PARANGUILLIDAE Blot, 1980
T. (LUT) Mar.

First and Last: *Paranguilla tigrina* (Agassiz, 1839), *Dalpiaziella brevicauda* Cadrobbi, 1962, Monte Bolca, Verona, Italy. The family is said by Blot (1984) to be close to the ancestral stock of muraenids.

F. MURAENIDAE Risso, 1826
T. (YPR)–Rec. Mar.

First: *Eomuraena sagittidens* Casier, 1967, 'Untereozän 2', Katharinenhof, Fehmarn, Schleswig-Holstein, Germany. **Extant**
Intervening: O LUT, UMI.

Superfamily CONGROIDEA Kaup, 1856

F. SYNAPHOBRANCHIDAE Johnson, 1862
(including Simenchelyidae Gill, 1879,
Dysommatidae Regan, 1912,
Nettodariidae Whitley, 1951,
Dyssomminidae Böhlke and Hubbs, 1951)
Extant Mar.

F. OPHICHTHIDAE Duméril, 1806
(including Neenchelyidae Bamber, 1915,
Aoteidae Phillips, 1926) T. (YPR)–Rec. Mar.

First: *Micromyrus fehmarnensis* Casier, 1967, *Palaeomyrus franzi* Casier, 1967, 'Untereozän 2', Katharinenhof, Fehmarn, Schleswig-Holstein, Germany; *Rhynchorhinus branchialis* Woodward, 1901, London Clay Formation, SE England, UK. **Extant**
Intervening: LUT, O, Mio.–PLI.

F. PROTEOMYRIDAE Blot, 1980
T. (LUT) Mar.

First and Last: *Proteomyrus ventralis* (Agassiz, 1839), Monte Bolca, Verona, Italy. Said by Blot (1984) to be intermediate between Ophichthidae and Congridae.

F. PARECHELIDAE Casier, 1967
T. (YPR) Mar.

First and Last: *Parechelus prangei* Casier, 1967, 'Untereozän 2', Katharinenhof, Fehmarn, Schleswig-Holstein, Germany. Said to be intermediate between Ophichthidae and Muraenidae.

F. COLOCONGRIDAE* Smith, 1971
Extant Mar.

F. DERICHTHYIDAE Gill, 1884 (including
Nessorhamphidae Schmidt, 1931) **Extant** Mar.

F. MURAENESOCIDAE Günther, 1870
O T. (YPR)–Rec. Mar.

First: O *Muraenesox cymbium* Stinton, 1966, London Clay Formation, SE England, UK, and Argile des Flandres, Belgium. **Extant**
Intervening: O LUT.

F. NEMICHTHYIDAE Günther, 1870
T. (?BUR)–Rec. Mar.

First: ?*Mastygocercus vermiformis* de Beaufort (1926), Patanuang Asu E, Maros, SW Sulawesi; a single specimen, said by de Beaufort to have a filamentous tail like the Recent nemichthyid *Cercomitus* (=*Nemichthys scolopaceus*). No other fossil nemichthyid is known. **Extant**

F. CONGRIDAE Kaup, 1856 (including
MACROCEPHENCHELYIDAE Fowler, 1934)
O ?K. (CMP)–Rec. Mar.

First: O Two juvenile congrid otoliths, Tupelo Tongue of the Coffee Sand Formation, Chapelville, Mississippi, USA (Nolf and Dockery, 1990). The next oldest records are otoliths of *Conger*, *Gnathophis* and *Hildebrandia* from the Ypresian of Aquitaine and Belgium, and of *'Congridarum'* from the Ypresian of England, Belgium, Aquitaine, Mississippi and New Zealand. **Extant**
Intervening: YPR–PLE.

F. NETTASTOMATIDAE Jordan and Davis, 1891
O T (YPR)–Rec. Mar.

First: O *Hoplunnis ariejansseni* Nolf, 1988, Argile de Gan, Gan, Pyrénées-Atlantiques, France. **Extant**
Intervening: O PLI.

F. PATAVICHTHYIDAE Blot, 1980 T. (LUT–RUP) Mar.

First: *Patavichthys bolcensis* (Bassani, 1898), Monte Bolca, Verona, Italy.
Last: *Proserrivomer mecquenemi* Arambourg, 1967, Elam, Luristan, Iran. This genus, previously placed in the succeeding family, was transferred to Patavichthyidae by Blot (1984).

F. SERRIVOMERIDAE Roule and Bertin, 1929 **Extant** Mar.

Superfamily SACCOPHARYNGOIDEA Bleeker, 1859

SACCOPHARYNGOIDEI *incertae sedis* (F. ENCHELIIDAE Hay, 1903) K. (CEN) Mar.

First and Last: *Enchelion montium* Hay, 1903, Hakel, Lebanon (Patterson, in press).

FF. CYEMATIDAE Regan, 1912, SACCOPHARYNGIDAE* Bleeker, 1859, EURYPHARYNGIDAE* Gill, 1883, MONOGNATHIDAE* Bertin, 1936 **Extant** Mar.

Cohort CLUPEOCEPHALA Patterson and Rosen, 1977

CLUPEOCEPHALA *incertae sedis* J. (TTH)

First: *Daitingichthys tischlingeri* Arratia, 1987, Malm Zeta 3, Daiting, Bavaria, Germany.

Order CROSSOGNATHIFORMES Taverne, 1989

Suborder CROSSOGNATHOIDEI Taverne, 1989

F. CROSSOGNATHIDAE Woodward, 1901 (= Apsopelicidae Cragin, 1901, Syllaemidae Romer, 1966) K. (BRM–CMP) Mar.

First: *Crossognathus sabaudianus* Pictet, 1858, Voirons, Switzerland; Hilsthön, Hildesheim, Hannover, Germany.
Last: *Apsopelix anglicus* (Dixon, 1850), Pierre Shale, western interior USA; Fox Hills Sandstone, Colorado, USA.
Intervening: ALB–SAN.
Comment: Taverne (1989), who has revised the crossognathids, regards them as occurring first in the lower Aptian (Töck, Heligoland), but does not comment on the stratigraphy of the localities in Switzerland (Voirons) and Hannover (Hildesheim) from which *Crossognathus* was first recorded (Woodward, 1901). These localities are traditionally regarded as Neocomian, but the ammonites associated with the Voirons fishes imply Barremian age.

Suborder PACHYRHIZODONTOIDEI Forey, 1977

F. NOTELOPIDAE Forey, 1977 K. (APT) ?Mar.

First and Last: *Notelops brama* (Agassiz, 1841), Santana Formation, Serra do Araripe, Ceará, Brazil.

F. PACHYRHIZODONTIDAE Cope, 1872 (including Greenwoodellidae Taverne, 1973 and probably Araripichthyidae Silva Santos, 1985) K. (APT)–T. (LUT) Mar.

First: *Rhacolepis buccalis* Agassiz, 1841, ?*Araripichthys castilhoi* Silva Santos, 1985, Santana Formation, Serra do Araripe, Ceara, Brazil; *Greenwoodella tockensis* Taverne and Ross, 1973, Töck, Heligoland.
Last: *Platinx macropterus* (de Blainville, 1818), Monte Bolca, Verona, Italy.
Intervening: ALB–CMP, THA, YPR.

Suborder TSELFATIOIDEI Bertin and Arambourg, 1958

F. PLETHODIDAE Loomis, 1900 (including Bananogmiidae Applegate, 1970, Niobraridae Jordan, 1924, Protobramidae Le Danois and Le Danois, 1964, Thryptodontidae Jordan, 1923, Tselfatiidae Arambourg, 1944) K. (ALB–CMP) Mar.

First: *Plethodus expansus* Dixon, 1850, Gault, SE England, UK.
Last: *Bananogmius* spp., *Moorevillia hardi* Applegate, 1970, Mooreville Formation, Alabama, USA.
Intervening: CEN–SAN.

Subcohort CLUPEOMORPHA Greenwood *et al.*, 1966 (Grande, 1985)

CLUPEOMORPHA *incertae sedis* K. (BRM) Mar./FW

First: *'Clupea' antiqua* Pictet, 1858, *'C'. voironensis* Pictet, 1858, Voirons, Switzerland.
Comments: These beds, conventionally dated as Neocomian, contain a cephalopod fauna implying Barremian age. *'Diplomystus'* spp. said to be of Neocomian age include *'D'. goodi* Eastman, 1912, from the middle part of the Cocobeach Series in Gabon and Rio Muni, and *'D'. primotinus* Uyeno, 1979 and *'D.' kokuraensis* Uyeno, 1979 from the Wakamiya Formation, Kitakyushu, Fukuoka, Japan. The first of these is here included in *Ellimmichthys* (below); the two Japanese species occur in poorly dated freshwater sediments, and there is no good reason to believe that they are earlier than Barremian. *Paraclupea chetungensis* Du, 1950 is from freshwater deposits in SE China dated as late Jurassic or early Cretaceous, but correlated (by Chang and Chow, 1986, using fishes) with the Cocobeach Series of West Africa and the Ilhas Formation of Brazil. *Paraclupea* may be an ellimmichthyid (below), but no apomorphies indicating that are recorded.

Order ELLIMMICHTHYIFORMES Grande, 1982

F. ELLIMMICHTHYIDAE Grande, 1982 (= Diplomystidae Patterson, 1970, ? = Paraclupeidae Chang and Chou, 1977) K. (BRM)–T. (LUT) Mar./FW

First: *Ellimmichthys longicostatus* (Cope, 1886), Ilhas Formation, Bahia, Brazil and *'Diplomystus' goodi* Eastman, 1912 (here taken to belong in *Ellimmichthys*), Cocobeach Series, Rio Muni. These freshwater beds are conventionally dated as Neocomian, but could well be Barremian or even Aptian (e.g. de Klasz and Gageonnet, 1965; Brito, 1979).
Last: *Diplomystus dentatus* Cope, 1877, Green River

Formation, Wyoming, USA; *Diplomystus shengliensis* Zhang *et al.*, 1985, Shahejiezu Formation, Henan, China.
Intervening: ?APT, CEN, SAN, YPR.

Order CLUPEIFORMES Bleeker, 1859

Suborder DENTICIPITOIDEI Greenwood *et al.*, 1966

F. DENTICIPITIDAE* Clausen, 1959
T. (Oli./Mio.)–Rec. FW

First: *Palaeodenticeps tanganikae* Greenwood, 1960, lacustrine shales, Singida, Tanzania. **Extant**

Suborder CLUPEOIDEI Bleeker, 1859

Superfamily ENGRAULOIDEA Gill, 1861

F. COILIIDAE* Jordan and Seale, 1925
Extant Mar./FW

F. ENGRAULIDAE Gill, 1861
O T. (BUR) **S** T. (UMI/PLI (l.))–Rec. Mar./FW

First: **S** *Engraulis tethensis* Grande and Nelson, 1985, Mesaoria Group, Lyssi, Cyprus; **O** *Anchoa nitida* Schwarzhans, 1980, Altonian, Otago, New Zealand. **Extant**
Intervening: **O** PLI, PLE.

Superfamily PRISTIGASTEROIDEA Jordan and Evermann, 1896

PRISTIGASTEROIDEA *incertae sedis*
K. (MAA) Mar./FW

First: *Gasteroclupea branisai* Signeux, 1964, El Molino Formation, Cayara, Bolivia.

F. PELLONIDAE Gill, 1861 **Extant** Mar./FW

F. PRISTIGASTERIDAE Jordan and Evermann, 1896
Extant Mar./FW

Superfamily CLUPEOIDEA Cuvier, 1817

F. CHIROCENTRIDAE* Bleeker, 1859
O T. (LUT)–Rec. Mar.

First: **O** *Chirocentrus exilis* Stinton, 1977, Selsey Formation, southern England, UK. **Extant**

F. CLUPEIDAE Cuvier, 1817
T. (THA)–Rec. Mar./FW

First: *Knightia vetusta* Grande, 1982, Tongue River Formation, Bay Horse, Montana, USA.
Intervening: YPR–PLI.

Subcohort EUTELEOSTEI Greenwood *et al.*, 1967

EUTELEOSTEI *incertae sedis* K. (BER) FW

First: *Pattersonella formosa* (Traquair, 1911), 'Wealden', Bernissart, Belgium (BER); *Wenzichthys congolensis* (Arambourg and Schneegans, 1935), Cocobeach Series, Gabon (?BRM) (both genera included in F. Pattersonellidae Taverne, 1982); **O** *'Euteleosteorum' lobatus* (Weiler, 1954), 'Wealden 2', boreholes at Menslage and Aldorf, NW Germany.

Order ESOCIFORMES Bleeker, 1859

F. UMBRIDAE Bleeker, 1859 (inc. Palaeoesocidae Berg, 1936) T. (THA)/(YPR)–Rec. FW

First: *Boltyshia brevicauda* Sytchevskaya and Danil'chenko, 1975, *B. truncata* Sytchevskaya, 1976, 'sapropelitic layer', Bolt'yishka, Ukraine, former USSR. **Extant**
Intervening: YPR–UMI, PLE.

F. ESOCIDAE* Cuvier, 1817
K. (CMP)–Rec. FW (see Fig. 36.2)

First: *Estesesox foxi* Wilson *et al.*, 1992, *Oldmanesox canadensis* Wilson *et al.*, 1992, Judith River Formation and Milk River Formation, Alberta, Canada.
Intervening: MAA–PLE.

Division OSTARIOPHYSI Sagemehl, 1885

Series ANOTOPHYSI Rosen and Greenwood, 1970

Order GONORYNCHIFORMES Regan, 1909

GONORYNCHIFORMES *incertae sedis* K. (BER)

First: *Aethalionopsis robustus* (Traquair, 1911), Bernissart, Belgium.

F. CHANIDAE* Günther, 1868
(including Halecopsidae Casier, 1946)
T. (YPR)–Rec. Mar.

First: *Chanos torosus* Danil'chenko, 1968, Danatinsk Formation, Kopetdag, Turkmenia, former USSR; *Halecopsis insignis* (Delvaux and Ortlieb, 1887), London Clay Formation, southern England, UK; Argile des Flandres, Hainaut, Belgium; Nord, France; and equivalents in NW Germany. **Extant**
Intervening: YPR, LUT, Oli.

F. GONORYNCHIDAE Bleeker, 1859
(including Cromeriidae Boulenger, 1904, Grasseichthyidae Géry, 1964, Judeichthyidae Gayet, 1985, KNERIIDAE Günther, 1868, PHRACTOLAEMIDAE Boulenger, 1904) K. (CEN)–Rec. Mar./FW

First: *Charitosomus hakelensis* (Davis, 1887), Hakel and Hajula, Lebanon; *C. hermani* Taverne, 1976, Kipala, Kwango, Zaïre; *Judeichthys haasi* Gayet, 1985, *Ramallichthys orientalis* Gayet, 1982, Ramallah, Israel. **Extant**
Intervening: SAN, CMP, THA–CHT.

Series OTOPHYSI Garstang, 1931

?OTOPHYSI *incertae sedis*
(including Salminopsidae Gayet, 1985) K. (CEN)

First and Last: *Lusitanichthys characiformis* Gayet, 1981, Laveiras, Portugal; *Salminops ibericus* Gayet, 1985, Caranguejeira, Portugal.
Comment: The status of these two genera is controversial; if they are not otophysans, the first otophysans are the MAA catfishes cited below.

Order CYPRINIFORMES Bleeker, 1859

F. CYPRINIDAE Cuvier, 1817 T. (YPR)–Rec. FW

First: *Parabarbus* sp. Sytchevskaya (1986), Obailinskaya Formation, River Chaibulak, Kazakhstan, former USSR. **Extant**
Comment: The earliest cyprinids are difficult to document precisely; because of their freshwater habitat, the precise age of the many Asian Eocene records (e.g. Sytchevskaya, 1986; Chang and Chow, 1986; Zhou, 1990) is not known

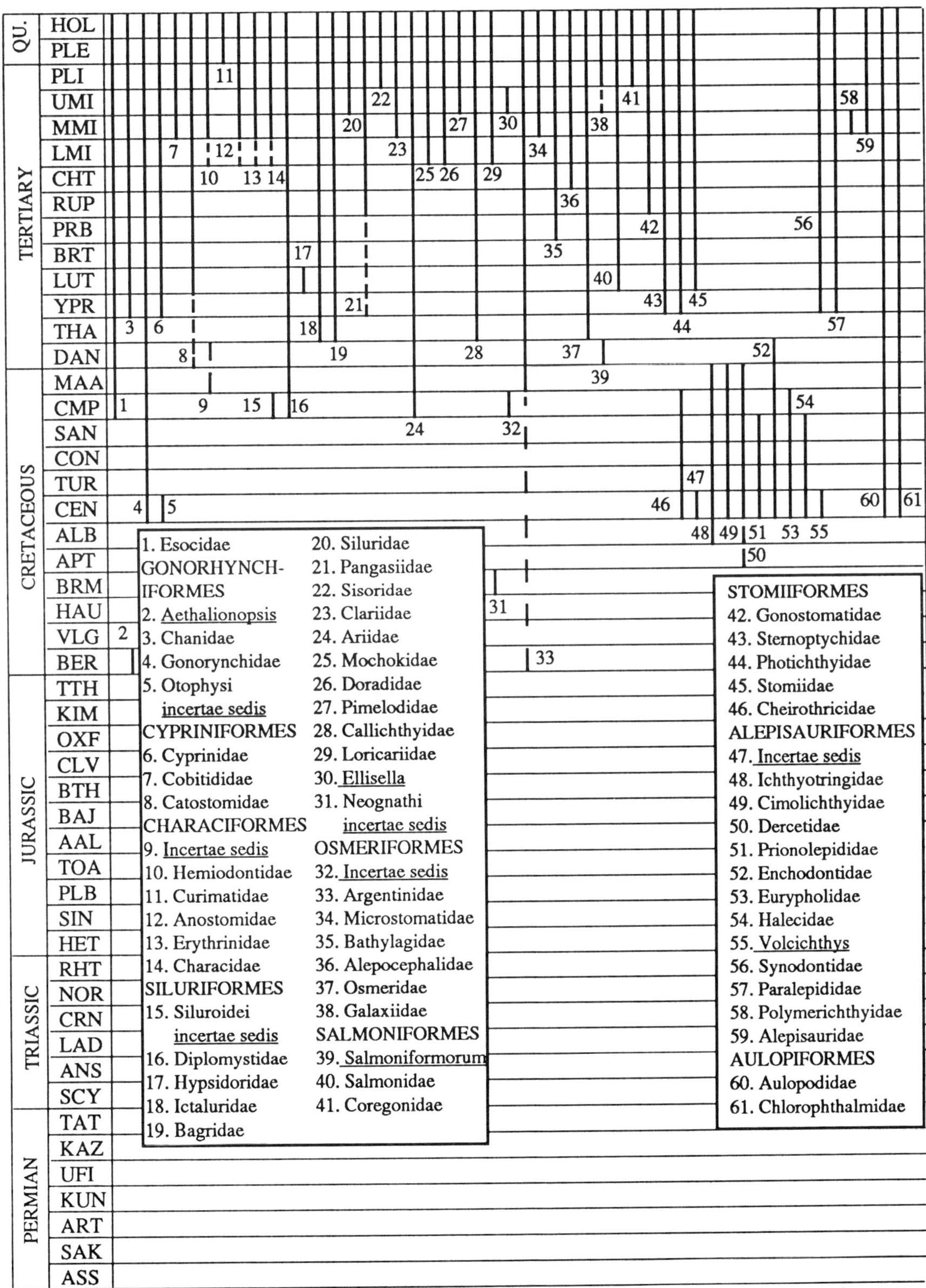

Fig. 36.2

precisely. In agreement with Fink *et al.* (1984), I do not accept *Molinichthys inopinatus* Gayet, 1982, from the El Molino Formation (MAA/DAN), Bolivia, as a cyprinid or cyprinoid. *Blicca croydonensis* White, 1931, from the THA Woolwich and Reading Formation, SE England, UK, was cited as the earliest cyprinid in the first edition of this compilation (Patterson, 1967), but there is no evidence that the single specimen is cyprinid or even otophysan.
Intervening: LUT–PLE.

F. PSILORHYNCHIDAE* Hora, 1925 **Extant** FW

F. HOMALOPTERIDAE Bleeker, 1859 **Extant** FW

F. COBITIDIDAE Swainson, 1839
T. (MMI)–Rec. FW

First: *Cobitis centrochir* Agassiz, 1835, Öhningen, Bavaria, Germany; *C.* cf. *taenia* Linné, 1758, western Kazakhstan,

former USSR. *Noemacheilus musceli* Pauca, 1929, RUP, Romania, is an argentinid. **Extant**
Intervening: UMI–PLI.

F. GYRINOCHEILIDAE* Boulenger, 1904
Extant FW

F. CATOSTOMIDAE Heckel, 1843
T. (?DAN)–Rec. FW

First: Cleithra like those of *Amyzon*, Paskapoo Formation (Tiffanian), Alberta, Canada (Wilson, 1980); the next oldest records are LUT: *Amyzon gosiutensis* Grande *et al.*, 1982, Laney Shale member, Green River Formation, Wyoming, USA, and *A. brevipinne* Cope, 1893 and *A. aggregatum* Wilson, 1977, respectively from the Allenby Formation and Horsefly River Beds of British Columbia, Canada; *Amyzon zaissanicus* (Sytchevskaya, 1983) and Catostomidae indet., Obailinskaya Formation (YPR/LUT), River Ul'ken-Ulast'y, Kazakhstan, former USSR.
Intervening: LUT–PLE.

Order CHARACIFORMES Goodrich, 1909

CHARACIFORMES *incertae sedis*
K. (MAA)/T. (DAN) FW

First: Teeth referred to Characidae and Erythrinidae from the El Molino Formation, Bolivia (de Muizon *et al.*, 1983), and similar teeth from the Vilquechico Formation, Peru (pers. observation). These formations have been described as MAA but their dating is controversial (Mourier *et al.*, 1988; Van Valen, 1988; Marshall, 1989; Cappetta, 1990). Characiformes *incertae sedis* are also known from the Palaeocene of Morocco (THA, Cappetta, 1990) and Niger (undescribed BMNH material).

F. CITHARINIDAE Günther, 1864 **Extant** FW

F. HEMIODONTIDAE Bleeker, 1859
T. (LMI)–Rec. FW

First: *Parodon* sp. Roberts, 1975, Loyola Formation, Cuenca Basin, Ecuador. **Extant**

F. CURIMATIDAE Bleeker, 1859
T. (PLE)–Rec. FW

First: *Curimata mosesi* Travassos and Silva Santos, 1955, Tremembé, São Paulo, Brazil. **Extant**

F. ANOSTOMIDAE Günther, 1864
T. (LMI)–Rec. FW

First: Anostomidae *incertae sedis*, *Leporinus* sp. Roberts, 1975, Loyola Formation, Cuenca Basin, Ecuador. **Extant**

F. ERYTHRINIDAE Gill, 1858
T. (LMI)–Rec. FW

First: Erythrinidae *incertae sedis*, *Hoplias* sp. Roberts, 1975, Loyola Formation, Cuenca Basin, Ecuador. **Extant**

F. LEBIASINIDAE Eigenmann, 1910 **Extant** FW

F. GASTEROPELECIDAE Bleeker, 1859
Extant FW

F. CTENOLUCIIDAE Schultz, 1944
(= Xiphostomidae Boulenger, 1904) **Extant** FW

F. HEPSETIDAE* Fowler, 1958 **Extant** FW

F. CHARACIDAE Bleeker, 1859
T. (LMI)–Rec. FW

First: Characidae *incertae sedis* (Roberts, 1975), Loyola Formation, Cuenca Basin, Ecuador; *Sindacharax lepersonnei* Greenwood and Howes, 1975, Kabiga Formation and Mohari and Karugamania beds, Zaïre; cf. *Alestes*, Lamitina beds, Mt. Elgon, Uganda and Daban Formation, Biyo Gora, Somalia (Van Couvering, 1977).
Intervening: MMI–PLE.

Order SILURIFORMES Cuvier, 1817

Suborder SILUROIDEI Cuvier, 1817

SILUROIDEI *incertae sedis*
K. (CMP) Mar./FW

First: Undetermined fin-spines (Cione, 1987), Los Alamitos Formation, Rio Negro, Argentina. Early South American siluroids also occur in the Bauru Group, Brazil (Gayet and Brito, 1989, SEN), Coli Toro Formation (lower MAA), Argentina, and in the Yacoraite Formation (Argentina), the Vilquechico Formation (Peru) and the El Molino Formation (Bolivia), which are probably correlated (Mourier *et al.*, 1988; Van Valen, 1988), and are late MAA or DAN in age. Skulls from the El Molino have been assigned to three extinct monotypic genera, one in a monotypic F. Andinichthyidae Gayet, 1988 (Gayet, 1990). In Africa, MAA siluroids occur at In Beceten, Niger (BMNH material).

F. DIPLOMYSTIDAE Eigenmann, 1890
K. (CMP)–Rec. FW

First: 'cf. Diplomystidae' Cione 1987, fin-spines from Los Alamitos Formation, Rio Negro, Argentina. **Extant**

F. HYPSIDORIDAE Grande, 1987 T. (LUT) FW

First and Last: *Hypsidoris farsonensis* Lundberg and Case, 1970, Laney Shale Member, Green River Formation, Wyoming, USA.

F. ICTALURIDAE Gill, 1861
(= Am[e]iuridae Günther, 1864)
T. (THA)–Rec. FW

First: ?*Astephus* sp. Lundberg, 1975, Polecat Bench Formation, Big Horn County, Wyoming, USA. **Extant**
Intervening: YPR–PLE.

F. BAGRIDAE Bleeker, 1858 T. (THA)–Rec. FW

First: *Nigerium gadense* White, 1935, *Eomacrones wilsoni* (White, 1935), Gada, Nigeria (two genera tentatively assigned to Bagridae by White). **Extant**
Intervening: LUT–RUP, PLI, PLE.

F. CRANOGLANIDIDAE* Myers, 1931
Extant FW

F. SILURIDAE Cuvier, 1817 T. (TOR)–Rec. FW

First: *Silurus* sp., Howenegg-im-Hegau, Baden, Germany. **Extant**
Intervening: PLI.

F. SCHILBEIDAE Bleeker, 1859 **Extant** FW

F. PANGASIIDAE Bleeker, 1859
T. (?Eoc.)–Rec. FW

First: *Pangasius indicus* von der Marck (1876), Padang, Sumatra. The Eocene age of the fish shales at Padang is still dubious. **Extant**

F. AMBLYCIPITIDAE Regan, 1911 **Extant** FW

F. AMPHILIDAE Regan, 1911 **Extant** FW

F. PARAKYSIDAE* Roberts, 1989 **Extant** FW

F. AKYSIDAE Günther, 1864 **Extant** FW

F. SISORIDAE Bleeker, 1859
(= Bagariidae Günther, 1864)
T. (PLI)–Rec. FW

First: *Bagarius yarrelli* (Hamilton-Buchanan, 1822), Siwaliks, Nahan, India. **Extant**

F. CLARIIDAE Günther, 1864
T. (MMI)–Rec. FW

First: cf. *Clarias* and Clariidae *incertae sedis*, Ngorora Formation, Tugen Hills and Kirimun Beds, Kirimun, Kenya (Van Couvering, 1977).
Intervening: UMI–PLE.

F. HETEROPNEUSTIDAE* Hora, 1936
Extant FW

F. CHACIDAE* Bleeker, 1859 **Extant** FW

F. OLYRIDAE* Hora, 1936 **Extant** FW

F. MALAPTERURIDAE* Bleeker, 1859
Extant FW

F. ARIIDAE Günther, 1864
K. (CMP)–Rec. Mar./FW

First: 'cf. Ariidae' Cione, 1987, fin-spines from Los Alamitos Formation, Rio Negro, Argentina; **O** juvenile otolith, Tupelo Tongue of the Coffee Sand Formation, Chapelville, Mississippi, USA (Nolf and Dockery, 1990).
Intervening: MAA–PLE.

F. PLOTOSIDAE Bleeker, 1859
Extant Mar./FW

F. MOCHOKIDAE Regan, 1911
T. (LMI)–Rec. FW

First: *Synodontis* sp., BUR, Moghara, Egypt; LMI, Rusinga Island, Lake Victoria, Kenya.
Intervening: MMI–PLE.

F. DORADIDAE Günther, 1864
T. (Mio.)–Rec. FW

First: Doradidae indet. Cione, 1986, Ituzaingo Formation, Parana, Argentina.

F. AUCHENIPTERIDAE Eigenmann and Eigenmann, 1890 **Extant** FW

F. PIMELODIDAE Bleeker, 1859
T. (UMI)–Rec. FW

First: *Phractocephalus hemiliopterus* (Bloch and Schneider, 1801), Urumaco Formation, Falcon State, Venezuela; *Pimelodus* sp., Cione, 1986, Ituzaingo Formation, Parana, Argentina. **Extant**
Intervening: PLI–PLE.

FF. AGENEIOSIDAE Bleeker, 1858,
HELOGENIDAE Regan, 1911,
CETOPSIDAE Bleeker, 1858,
HYPOPHTHALMIDAE* Cope, 1871
ASPREDINIDAE Bleeker, 1859,
NEMATOGENYIDAE* Günther, 1864
TRICHOMYCTERIDAE Bleeker, 1858
Extant FW

F. CALLICHTHYIDAE Bleeker, 1859
T. (THA)–Rec. FW

First: *Corydoras revelatus* Cockerell, 1925, Maiz Gordo Formation, NW Argentina. **Extant**
Intervening: PLE.

F. SCOLOPLACIDAE* Bailey and Baskin, 1976
Extant FW

F. LORICARIIDAE Bonaparte, 1831
T. (Mio.)–Rec. FW

First: *Loricaria*?, *Hypostomus* sp., Cione, 1986, Ituzaingo Formation, Parana, Argentina. **Extant**

F. ASTROBLEPIDAE* Eigenmann, 1922
Extant FW

Suborder GYMNOTOIDEI Gill, 1872

GYMNOTOIDEI *incertae sedis*
T. (UMI) FW

First: *Ellisella kirschbaumi* Gayet and Meunier, 1991, Yecua Formation, Rio Moile, Bolivia.

FF. STERNOPYGIDAE Cope, 1871
RHAMPHICHTHYIDAE Regan, 1911,
HYPOPOMIDAE Magio-Leccia, 1978,
APTERONOTIDAE Jordan, 1923,
GYMNOTIDAE* Bonaparte, 1831,
ELECTROPHORIDAE* Gill, 1872
Extant FW

Division NEOGNATHI Rosen, 1973

NEOGNATHI *incertae sedis*
(with PU1 + U1 fused) K. (?BRM) FW

First: *Casieroides yamangaensis* (Casier, 1961), *Chardonius longicaudatus* (Casier, 1961), *Pseudoleptolepis minor* (Casier, 1961), Couches de la Loia, Yamangi Moke, Zaïre.

Order OSMERIFORMES

OSMERIFORMES *incertae sedis*
O K. (CMP)

First: **O** juvenile otoliths from Beds B and E of the Tupelo Tongue of the Coffee Sand Formation, Chapelville, Mississippi, USA (Nolf and Dockery, 1990).

Suborder ARGENTINOIDEI (Begle, 1992)

Superfamily ARGENTINOIDEA

F. ARGENTINIDAE Bonaparte, 1823
?K. (BER)–Rec. Mar.

First: *Nybelinoides brevis* (Traquair, 1911), Bernissart (FW), Belgium, described by Taverne (1982) as an argentinid. The next oldest records of this otherwise marine family are: **O** undetermined argentinid otoliths, Severn Formation (MAA), Maryland, USA (Huddleston and Savoie, 1983), and *Argentina extenuata* Stinton, 1966, London Clay Formation (YPR), SE England, UK; **S** *Glossanodon* (*Proargentina*) *nebulosa* (Danil'chenko, 1960), Dabaxanskaya Formation (LUT/BRT), Georgia, former USSR. **Extant**

F. MICROSTOMATIDAE Gill, 1861
O T. (SRV)–Rec. Mar.

First: **O** *Nansenia* sp. Nolf and Steurbaut, 1983, Tonen der Ziegelei Sunder, Twistringen, Bremen, Germany.
Intervening: **O** TOR–PIA.

F. BATHYLAGIDAE Gill, 1884
(including OPISTHOPROCTIDAE Roule, 1915)
O T (PRB)–Rec. Mar.

First: **O** *Opisthoproctus weitzmani* Nolf, 1988, Marnes de Brihande, Chalosse, France.
Intervening: MMI–UMI, **O** PLI.

Superfamily ALEPOCEPHALOIDEA

F. ALEPOCEPHALIDAE Valenciennes, 1846
(including Bathylaconidae Parr, 1948,
Bathyprionidae Marshall, 1966,
Leptochilichthyidae Marshall, 1966,
Platytroctidae Roule, 1919 = Searsiidae
Parr, 1951) T. (CHT)–Rec. Mar.

First: *Carpathichthys polonicus* Jerzmanska, 1979, Zone IPM 5, Menilite Beds, Korzeniec, Carpathians, Poland.
Intervening: **O** MMI, PLI.

Suborder OSMEROIDEI Weitzman, 1967

Superfamily OSMEROIDEA Regan, 1913

F. OSMERIDAE Regan, 1913
(including PLECOGLOSSIDAE Bleeker, 1859)
T. (THA)–Rec. Mar./FW

First: *Speirsaenigma lindoei* Wilson and Williams, 1991, Paskapoo Formation, Joffre Bridge, Alberta, Canada; **O** *'Osmeridarum' tricrenulatus* (Stinton, 1965), Thanet Formation, Pegwell Bay, Kent, England, UK. **Extant**
Intervening: **S** RUP–CHT, PLE; **O** LUT–RUP, LMI, PLI–PLE.

Superfamily GALAXIOIDEA Bonaparte, 1832

F. LEPIDOGALAXIIDAE* Frankenberg, 1968
Extant FW

F. SALANGIDAE Bleeker, 1859
(including SUNDASALANGIDAE Roberts, 1981)
Extant Mar./FW

F. RETROPINNIDAE McCulloch, 1927
Extant Mar./FW

F. GALAXIIDAE Bonaparte, 1832
T. (?PLI)–Rec. Mar./FW

First: *Galaxias brevipinnis* Günther, 1866, Kaikorai Valley, Dunedin, New Zealand, and *G. vulgaris* Stokell, Foulden Hills, Middlemarch, New Zealand, both from diatomaceous shales of 'Taranakian (Upper Miocene) to Waitotoran (Pliocene) age' (McDowall, 1976). **Extant**

Order SALMONIFORMES Bleeker, 1859

SALMONIFORMES *incertae sedis*
O T. (?DAN) FW

First: **O** *'Salmoniformorum' rectangulus* Rana, 1988, Intertrappeans between Deccan Trap Flows no. 4 and 5, Rangapur, Hyderabad, India. According to D. Nolf (pers. comm.) the allocation of these otoliths to Salmoniformes lacks real support.

F. SALMONIDAE Rafinesque, 1815
T. (LUT)–Rec. Mar./FW

First: *Eosalmo driftwoodensis* Wilson, 1977, Driftwood Creek Beds, British Columbia. *Helgolandichthys schmidi* Taverne, 1981, from the Aptian Töck, Heligoland, was described as a salmonid *sensu lato*, and would be the first if correctly placed. **Extant**
Intervening: UMI–PLE.

F. COREGONIDAE Gill, 1892
T. (PIA)–Rec. FW

First: *Prosopium prolixus* Smith, 1975, Glenns Ferry Formation, Idaho, USA. **Extant**
Intervening: PLE.

Subdivision NEOTELEOSTEI Nelson, 1969

Order STOMIIFORMES Regan, 1909

F. GONOSTOMATIDAE Gill, 1893
T. (RUP)–Rec. Mar.

First: *Scopeloides glarisianus* (Agassiz, 1839), Switzerland, Poland, Czechoslovakia, Romania, Caucasus (former USSR). **Extant**
Intervening: MMI–PLI.

F. STERNOPTYCHIDAE Duméril, 1806
O T. (YPR)–Rec. Mar.

First: **O** *Valenciennellus* sp. Nolf, 1988, Argile de Gan, Gan, Pyrénées-Atlantiques, France; **S** Eoc. (m), *Polyipnoides levis* Danil'chenko, 1962, Dabaxanskaya Formation, Georgia, former USSR. **Extant**
Intervening: LUT–PLI.

F. PHOTICHTHYIDAE Weitzman, 1974
O T. (YPR)–Rec. Mar.

First: **O** *Polymetme dartagnan* Nolf, 1988, *'Photichthyidarum'* sp. Nolf, 1988, Argile de Gan, Gan, Pyrénées-Atlantiques, France; **S** Eoc. (m), *Vinciguerria distincta* Danil'chenko, 1962, Dabaxanskaya Formation, Georgia. *Idrissia jubae* Arambourg, 1954 and *Paravinciguerria praecursor* Arambourg, 1954, both from K. (CEN), Jebel Tselfat, Morocco, the latter also from K. (CEN), Cinto Euganeo, Padova, Italy, were described as gonostomatids, the latter as most similar to *Vinciguerria*. Weitzman (1967) doubted that the first of these was 'even a stomiatoid', but thought that the second might be related to *Vinciguerria*. *Idrissia* is also reported from the Oli. of Poland. **Extant**
Intervening: LUT/BRT–PLI.

F. STOMIIDAE Bleeker, 1859
(including Astronesthidae Gill, 1882,
Chauliodontidae Bleeker, 1859,
Idiacanthidae Gill, 1892,
Malacosteidae Gill, 1892,
Melanostomiidae Parr, 1927,
following Fink, 1985) T. (LUT/BRT)–Rec. Mar.

First: *Astronesthes praevius* Danil'chenko, 1962, Dabaxanskaya Formation, Georgia, former USSR. **Extant**
Intervening: UMI.

Section EURYPTERYGII Rosen, 1973

EURYPTERYGII *incertae sedis*

F. CHEIROTHRICIDAE Woodward, 1901 K. (CEN–CMP) Mar.

First: *Exocoetoides minor* Davis, 1887, *Telepholis tenuis* (Davis, 1887), Hakel, Lebanon.
Last: *Telepholis acrocephalus* von der Marck, 1868, *Cheirothrix guestphalicus* (Schlüter, 1858), Baumberg, Westphalia, Germany.
Intervening: SAN.

Order ALEPISAURIFORMES Regan, 1911

ALEPISAURIFORMES *incertae sedis* K. (CEN) Mar.

First: *Rharbichthys ferox* Arambourg, 1954, Jebel Tselfat, Morocco; Cinto Euganeo, Padova, Italy.

Rosen (1973) regarded all the genera making up the following three extinct suborders (including seven extinct families) as Alepisauriformes *incertae sedis*. They are listed here much as grouped by Goody (1969).

Suborder ICHTHYOTRINGOIDEI Goody, 1969

F. ICHTHYOTRINGIDAE Jordan, 1905 (including Apateopholidae Goody, 1969) K. (ALB–MAA) Mar.

First: *Apateodus glyphodus* (Blake, 1863), Gault, SE England, UK.
Last: *Apateodus corneti* (Forir, 1887), Lower Maestricht Chalk, The Netherlands.
Intervening: TUR–CMP.

F. CIMOLICHTHYIDAE Goody, 1969 K. (CEN–MAA) Mar.

First: *Cimolichthys levesiensis* Leidy, 1857, *H. subglobosus* Zone, Chalk, SE England, UK.
Last: *Cimolichthys manzadinensis* Dartevelle and Casier, 1949, Calcaire de Manzada, Zaïre.
Intervening: TUR–CMP.

F. DERCETIDAE Pictet, 1850 (including Stratodontidae Cope, 1872) K. (?APT–MAA) Mar.

First: *Benthesikyme* sp. Taverne, 1981, Töck, Heligoland, based on a single fragmentary braincase; *Dercetis* (*Benthesikyme*) is otherwise unknown before TUR. If the APT Töck specimen is not a dercetid, then the first would be K. (CEN), *Rhynchodercetis* spp. from Lebanon, Israel, Morocco and the former Yugoslavia, and *Dercetoides venator* Chalifa, 1989 from Israel.
Last: *Stratodus apicalis* Cope, 1872, MAA phosphates, Morocco and Israel.
Intervening: TUR–CMP.

F. PRIONOLEPIDIDAE Goody, 1969 K. (CEN–?SAN) Mar.

First: *Prionolepis cataphractus* (Pictet and Humbert, 1866), Hakel and Hajula, Lebanon.
Last: ?*Leptecodon rectus* Williston, 1899, Smoky Hill Chalk Member, Niobrara Formation, Kansas.
Intervening: TUR.

Suborder ENCHODONTOIDEI Berg, 1937

F. ENCHODONTIDAE Lydekker, 1889 K. (CEN)–T. (DAN) Mar./?FW

First: *Enchodus* spp., Lower Chalk, SE England, UK; Hakel and Hajula, Lebanon; Jebel Tselfat, Morocco; Cinto Euganeo, Padova, Italy; Woodbine Formation, Texas, USA; *Parenchodus longipterygius* Raab and Chalifa, 1987, Kefar Shaul Formation, Israel.
Last: *Enchodus* spp., Montian Phosphates, Morocco; Intertrappeans, Gitti Khadan, Madhya Pradesh, India.
Intervening: TUR–MAA.

F. EURYPHOLIDAE Goody, 1969 K. (CEN–CMP) Mar.

First: *Eurypholis boissieri* Pictet, 1850, Hakel and Hajula, Lebanon; *Saurorhamphus freyeri* Heckel, 1850, Scisti di Comeno, Comeno, NW Italy (these beds are conventionally assigned to CEN, but may range up to SEN; Medizza and Sorbini, 1980).
Last: *Eurypholis japonicus* Uyeno and Minakawa, 1983, Izumi Group, Matsuyama, Ehime, Japan (*Saurorhamphus freyeri* may be younger, above).
Intervening: TUR.

Suborder HALECOIDEI Goody, 1969

F. HALECIDAE Agassiz, 1834 K. (CEN–SAN) Mar.

First: *Halec eupterygius* (Dixon, 1850), *H. subglobosus* Zone, Chalk, SE England, UK, *Phylactocephalus microlepis* Davis, 1887, Hakel and Hajula, Lebanon, *Hemisaurida hakelensis* Goody, 1969, Hakel, Lebanon.
Last: *Halec eupterygius* (Dixon, 1850), *M. coranguinum* Zone, Chalk, SE England, UK.
Intervening: TUR, CON.

Suborder SYNODONTOIDEI Gill, 1872

SYNODONTOIDEI *incertae sedis* K. (CEN?)

First and Last: *Volcichthys dainellii* D'Erasmo, 1946, Volci, Comeno, NW Italy (these beds may range in age from CEN to TUR or higher, see under Eurypholidae above).

F. PSEUDOTRICHONOTIDAE* Yoshino and Araga, 1975 **Extant** Mar.

F. SYNODONTIDAE Gill, 1872 (including Harpadontidae McCulloch, 1929) T. (YPR)–Rec. Mar.

First: **S** *Argillichthys toombsi* Casier, 1966, London Clay Formation, Sheppey, Kent, England, UK; **O** *Saurida davisi* (Frost, 1925), London Clay Formation, SE England, UK, *Saurida* sp. Nolf, 1988, Argile de Gan, Gan, Pyrénées-Atlantiques, France. The MAA **O** '*Synodontidarum*' *pseudoperca* Nolf and Dockery, 1990, Coffee Sand Formation, Mississippi, USA, described as a synodontid, is a chlorophthalmid (D. Nolf, pers. comm.).
Intervening: **O** YPR–PRB, Mio., PLI **S**. UMI–PLI.

F. GIGANTURIDAE Brauer, 1906 **Extant** Mar.

Suborder ALEPISAUROIDEI Regan, 1911

F. PARALEPIDIDAE Gill, 1872 **O** T. (YPR)–Rec. Mar.

First: **O** *Lestidiops ypresiensis* Nolf, 1988, Argile de Gan, Gan, Pyrénées-Atlantiques, France; **S** LUT, *Holosteus*

esocinus Agassiz, 1839, Monte Bolca, Verona, Italy. A '*Holosteus*'-like fish is also recorded from the THA Fur Formation (Mo Clay) of NW Denmark (Bonde, 1987). **Extant**
Intervening: LUT, RUP, UMI, PLI.

F. POLYMERICHTHYIDAE Uyeno, 1967
T. (MMI) Mar.

First and Last: *Polymerichthys nagurai* Uyeno, 1967, Tubozawa Formation, Aichi, Japan.

F. ANOTOPTERIDAE* Fowler, 1936 **Extant** Mar.

F. EVERMANNELLIDAE Fowler, 1901
Extant Mar.

Comment: The Pliocene otoliths from Nice, France, reported as *Evermannella* by Nolf (1985) do not belong here but to the zeiform *Zenion* (D. Nolf, pers. comm.).

F. OMOSUDIDAE* Regan, 1911 **Extant** Mar.

F. ALEPISAURIDAE* Bonaparte, 1832
T. (MMI)–Rec. Mar.

First: *Alepisaurus paraonai* D'Erasmo, 1924, Rosignano, Piemonte, Italy. **Extant**

Order AULOPIFORMES Rosen, 1973

F. AULOPODIDAE* Cope, 1872
K. (CEN)–Rec. Mar.

First: *Nematonotus bottae* (Pictet and Humbert, 1866), Hakel and Hajula, Lebanon; *N. longispinus* Hay, 1903, Hajula, Lebanon. **Extant**
Intervening: YPR.

F. CHLOROPHTHALMIDAE Jordan, 1923
K. (CEN)–Rec. Mar.

First: *Acrognathus dodgei* Hay, 1903, Hakel and Hajula, Lebanon. **Extant**
Intervening: TUR–SAN, **O** MAA, PRB, CHT–MMI, PLI.

F. IPNOPIDAE Gill, 1883 (including Bathypteroidae Jordan, 1923) **Extant** Mar.

F. SCOPELARCHIDAE Regan, 1911
O T. (CHT)–Rec. Mar. (see Fig. 36.3)

First: **O** *Scopelarchus nolfi* Steurbaut, 1982, Argile de Saint-Etienne-d'Orthe, Aquitaine, France. **Extant**
Intervening: **O** PLI.

F. NOTOSUDIDAE Parr, 1928
(including Scopelosauridae Marshall, 1966)
O T. (BRT)–Rec. Mar.

First: **O** *Scopelosaurus brevirostris* Schwarzhans, 1980, Kaiatan, McCullough's Bridge, Canterbury, New Zealand.

Subsection CTENOSQUAMATA Rosen, 1973

CTENOSQUAMATA *incertae sedis*
(including Aulolepidae Patterson, 1964,
Ctenothrissidae Woodward, 1901,
Pattersonichthyidae Gaudant, 1978,
Pateropercidae Gaudant, 1978)
K. (CEN) Mar.

First: Several genera, Hakel and Hajula, Lebanon, and *H. subglobosus* Zone, Chalk, SE England, UK.

Order MYCTOPHIFORMES Regan, 1911

F. MYCTOPHIDAE Gill, 1893
K. (CMP)–Rec. Mar.

First: *Sardinius cordieri* (Agassiz, 1839), Baumberg and Sendenhorst, Westphalia, Germany. **Extant**
Intervening: **O** DAN/THA–PLE.

F. NEOSCOPELIDAE Parr, 1928
(?including Sardinioididae Goody, 1969)
?K. (CEN)–Rec. Mar.

First: ?*Sardinioides attenuatus* Woodward, 1901, *S. minimus* (Agassiz, 1839), *S. pontivagus* (Hay, 1903), Hakel and Hajula, Lebanon. If *Sardinioides* is not a neoscopelid, the only fossil is an undescribed CHT otolith from Aquitaine, France (D. Nolf, pers. comm.). **Extant**
Intervening: TUR–CMP, **O** CHT.

Sept ACANTHOMORPHA Rosen, 1973

ACANTHOMORPHA *incertae sedis*

F. ASINEOPIDAE Cope, 1884
?K. (CMP)–T. (LUT) Mar./FW

First: *Nardoichthys francisci* Sorbini and Bannikov, 1991, Calcare di Mellissano, Nardo, Italy; this genus is described as a percoid, but in my view is most similar to *Asineops*.
Last: *Asineops squamifrons* Cope, 1870, Green River Formation, Wyoming, USA.
Intervening: THA, YPR.

F. PHARMACICHTHYIDAE Patterson, 1964
K. (CEN) Mar.

First and Last: *Pharmacichthys* spp., Hakel, Lebanon, and Ramallah, Israel (Gayet, 1980a).

The following two nominal extinct families are conventionally placed in the Perciformes, either in the suborder Scombroidei (e.g. Danil'chenko, 1964; Blot, 1980) or the suborder Xiphioidei (e.g. Fierstine and Applegate, 1974). Johnson (1986) has shown that xiphioids (billfishes) are a subgroup of Scombridae, and has characterized Scombroidei and Scombridae by a number of apomorphies, none of which is yet demonstrated in blochiids or palaeorhynchids. Palaeorhynchids may be scombroids; blochiids are almost certainly unrelated to them.

F. BLOCHIIDAE Bleeker, 1859
K. (CEN)–T. (BRT) Mar.

First: *Cylindracanthus libanicus* (Woodward, 1942), Hajula, Lebanon; *C. cretaceus* (Dixon, 1850), zone of *Holaster subglobosus*, Chalk, Sussex, England, UK.
Last: *Cylindracanthus rectus* (Agassiz, 1844), Barton Clays, Barton Formation, southern England, UK.
Intervening: CMP–LUT.

F. PALAEORHYNCHIDAE Günther, 1880
T. (YPR–CHT) Mar.

First: *Enniskillenus radiatus* Casier, 1966, London Clay Formation, Sheppey, Kent, England, UK. There may be a palaeorhynchid in the THA Fur Formation (Mo Clay) of NW Denmark (Bonde, 1987).
Last: *Pseudotetrapturus luteus* Danil'chenko, 1960, Abadzekhski Horizon, Maikop Series, Caucasus, former USSR.

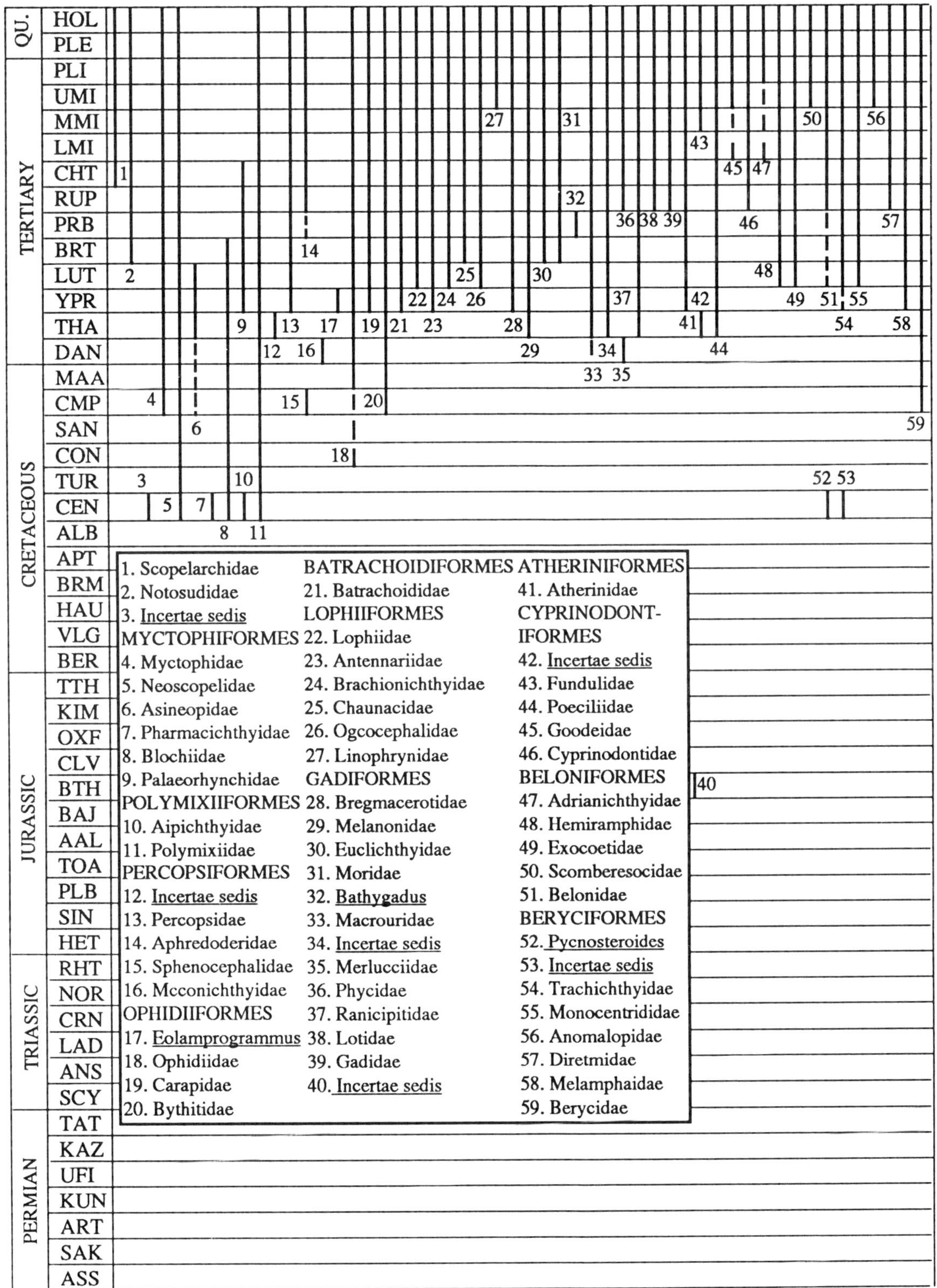

Fig. 36.3

Intervening: LUT–RUP.

Order POLYMIXIIFORMES Patterson, 1964

F. AIPICHTHYIDAE Patterson, 1964 (including Aipichthyoididae Gayet, 1980) K. (CEN) Mar.

First and Last: *Aipichthys* spp., Hakel and Hajula, Lebanon; Lower Chalk, Sussex, England, UK; *Aipichthyoides* spp., Ramallah, Israel (Gayet, 1980a,b).

F. POLYMIXIIDAE* Poey, 1868 (including Berycopsidae Regan, 1911, ?Dalmatichthyidae Radovcic, 1975, Omosomopsidae Gaudant, 1978) K. (CEN)–Rec. Mar.

First: *Berycopsis elegans* Dixon, 1850, *Homonotichthys* spp., Lower Chalk, Kent and Sussex, England, UK; *Omosoma tselfatensis* Gaudant, 1978, *Omosomopsis simum* (Arambourg, 1954), Jebel Tselfat, Morocco. **Extant**
Intervening: TUR, SAN–CMP, **O** MAA, THA, PRB, Mio.

Superorder PARACANTHOPTERYGII Greenwood *et al.*, 1966

Order PERCOPSIFORMES Berg, 1937

PERCOPSIFORMES *incertae sedis* T. (THA) FW

First: cf. *Amphiplaga* (Wilson, 1980), Paskapoo Formation, Alberta, Canada.

F. PERCOPSIDAE* Agassiz, 1846 (including Erismatopteridae Jordan, 1905) T. (YPR)–Rec. FW

First: *Amphiplaga brachyptera* Cope, 1877, Fossil Butte Member, Green River Formation, Wyoming, USA. **Extant**
Intervening: LUT.

F. APHREDODERIDAE* Bonaparte, 1832 T. (?PRB/RUP)–Rec. FW

First: *Trichophanes hians* Cope, 1872, Osino Shales, Osino, Nevada, USA (Cavender, 1986). **Extant**
Intervening: ?CHT.

F. AMBLYOPSIDAE Bonaparte, 1832 **Extant** FW

ANACANTHINI Müller, 1846

ANACANTHINI *incertae sedis*

F. SPHENOCEPHALIDAE Patterson, 1964 K. (CMP) Mar.

First and Last: *Sphenocephalus fissicaudus* Agassiz, 1839, Baumberg, Westphalia, Germany.

F. MCCONICHTHYIDAE Grande, 1988 T. (DAN) FW

First and Last: *Macconichthys longipinnis* Grande, 1988, Tullock Formation, McCone County, Montana, USA.

Order OPHIDIIFORMES Berg, 1937

OPHIDIIFORMES *incertae sedis* T. (YPR) Mar.

First: *Eolamprogrammus senectus* Danil'chenko, 1968, Danatinsk Formation, Kopetdag, Turkmenia, former USSR.

F. OPHIDIIDAE Rafinesque, 1810 O K. (SEN)–Rec. Mar.

First: **O** *'Sirembinorum' bavaricus* (Koken, 1891), Senonian, Siegsdorf, Bavaria, Germany; **S** *Ampheristus toliapicus* König, 1825, London Clay Formation, SE England, UK. **Extant**
Intervening: **O** DAN/THA–PLE; **S** ?LUT, RUP–MMI.

F. CARAPIDAE Jordan and Fowler, 1902 (including Pyramodontidae Smith, 1955) O T. (YPR)–Rec. Mar.

First: **O** *Onuxodon kiriakoffi* Nolf, 1980, Argile de Gan, Gan, Pyrénées-Atlantiques, France. **Extant**
Intervening: **O** LUT–PLI.

F. BYTHITIDAE Gill, 1861 O K. (CMP)–Rec. Mar.

First: **O** *'Dinematichthyinorum' crepidatus* (Voigt, 1926), erratic boulder, Cöthen, Anhalt, Germany. **Extant**
Intervening: **O** THA–PLI.

F. APHYONIDAE Jordan and Evermann, 1898 **Extant** Mar.

Order BATRACHOIDIFORMES Goodrich, 1909

F. BATRACHOIDIDAE Jordan and Evermann, 1898 O T. (YPR)–Rec. Mar.

First: **O** *'Batrachoididarum' trapezoidalis* Nolf, 1988, Argile de Gan, Gan, Pyrénées-Atlantiques, France; **S** *Batrachoides didactylus* Bloch and Schneider, 1801, MES, Oran, Algeria. **Extant**
Intervening: **O** RUP/CHT–PLI.

Order LOPHIIFORMES Garman, 1899

F. LOPHIIDAE Rafinesque, 1810 T. (LUT)–Rec. Mar.

First: *'Lophius' brachysomus* Agassiz, 1835, Monte Bolca, Verona, Italy (Blot, 1980). **Extant**
Intervening: BRT–PLI.

F. ANTENNARIIDAE Gill, 1863 O T. (YPR)–Rec. Mar.

First: **O** *Antennarius euglyphus* (Stinton, 1966), London Clay Formation, Sheppey, Kent, England, UK; Argile de Gan, Gan, Pyrénées-Atlantiques, France. **Extant**
Intervening: **O** LUT–PRB.

F. LOPHICHTHYIDAE* Boeseman, 1964 **Extant** Mar.

F. BRACHIONICHTHYIDAE* Gill, 1863 T. (LUT)–Rec. Mar.

First: *Histionotophorus bassanii* (de Zigno, 1887), Monte Bolca, Verona, Italy. **Extant**

F. CHAUNACIDAE* Gill, 1863 O T. (BRT)–Rec. Mar.

First: **O** *Chaunax semiangulatus* Stinton, 1978, Barton Formation, Hampshire, England, UK. **Extant**
Intervening: **O** PRB, UMI.

F. OGCOCEPHALIDAE Gill, 1893 O T. (LUT)–Rec. Mar.

First: **O** *Ogcocephalus cirrhosus* Stinton, 1978, Selsey Formation, Selsey, Sussex, England, UK; **S** an undescribed genus from Monte Bolca, Verona, Italy, is mentioned by Blot (1980). **Extant**
Intervening: **O** PRB, Mio.

FF. CAULOPHRYNIDAE Regan, 1912, CERATIIDAE Gill, 1864, GIGANTACTINIDAE Boulenger, 1904, NEOCERATIIDAE* Regan, 1926 **Extant** Mar.

F. LINOPHRYNIDAE Regan, 1926 T. (UMI)–Rec. Mar.

First: *Acentrophryne longidens* Regan, 1926, Puente Formation, Hacienda Heights, Los Angeles County, California, USA. **Extant**

FF. ONEIRODIDAE Gill, 1878, THAUMATICHTHYIDAE Smith and Radcliffe, 1912, CENTROPHRYNIDAE* Bertelsen, 1951, DICERATIIDAE Regan and Trewavas, 1932, HIMANTOLOPHIDAE* Gill, 1861, MELANOCETIDAE* Gill, 1878 **Extant** Mar.

Order GADIFORMES Goodrich, 1909 (Cohen, 1989)

F. MURAENOLEPIDIDAE* Regan, 1903 **Extant** Mar.

F. BREGMACEROTIDAE* Gill, 1872 T. (YPR)–Rec. Mar.

First: **O** *'Bregmacerotidarum' rappei* Nolf, 1988, Argile de Gan, Gan, Pyrénées-Atlantiques, France. **Extant**
Intervening: **S** LUT–PLI.

F. MELANONIDAE* Marshall, 1965 **O** T. (THA)–Rec. Mar.

First: **O** *Melanonus ellesmerensis* Schwarzhans, 1985, Eureka Sound Formation, Ellesmere Island, NWT, Canada. **Extant**
Intervening: **O** RUP–PLI.

F. EUCLICHTHYIDAE* Cohen, 1984 **O** T. (BRT)–Rec. Mar.

First: **O** *Euclichthys eocenicus* Schwarzhans, 1980, Kaiatan, McCullough's Bridge, Canterbury, New Zealand. **Extant**
Intervening: **O** BUR.

F. STEINDACHNERIIDAE* Marshall and Cohen, 1973 **Extant** Mar.

F. MORIDAE Goode and Bean, 1896 **O** T. (BRT)–Rec. Mar.

First: **O** *Tripterophycis immutatus* Schwarzhans, 1980, Kaiatan, McCullough's Bridge, Canterbury, New Zealand; Aldingian, South Australia; *T. elongatissimus* Schwarzhans, 1985, Aldingian, South Australia. *Eophycis*, a genus known by skeletons in the RUP of Poland, and which is probably a morid, may also occur in the THA Fur Formation of NW Jutland, Denmark (Bonde, 1987, p. 36).
Intervening: **O** CHT–PLI.

Suborder MACROUROIDEI Garman, 1899

MACROUROIDEI *incertae sedis* **O** T. (PRB)

First: **O** *Bathygadus mauli* Nolf, 1988, Couches de Cauneille, Chalosse, France.

F. MACROURIDAE Bonaparte, 1838 **O** T. (DAN/THA)–Rec. Mar.

First: **O** *Nezumia lindsayi* Schwarzhans, 1985, Dartmoor Formation, Mt. Gambier, South Australia. **Extant**
Intervening: **O** UMI–PLI.

Suborder GADOIDEI Goodrich, 1909

GADOIDEI *incertae sedis* T. (DAN) Mar.

First: 'Protocodus' Cohen, 1984, undescribed fish from Kangilia Formation, Nugssuaq, West Greenland.

F. MERLUCCIIDAE Adams, 1864 T. (THA)–Rec. Mar.

First: *Rhinocephalus* sp., Fur Formation (Mo Clay), NW Jutland, Denmark. **Extant**
Intervening: YPR–PLI.

F. PHYCIDAE Swainson, 1838 **O** T. (RUP)–Rec. Mar.

First: **O** *Phycis praecongatus* Schwarzhans, 1977, Hückelhoven, Nordrhein-Westphalen, Germany. **Extant**
Intervening: **O** CHT–PLI.

F. RANICIPITIDAE* Gill, 1872 **O** T. (THA)–Rec. Mar.

First: **O** *Raniceps hermani* Nolf, 1978, Formation de Heers, Orp-le-Grand, Marte, Belgium; **S** *Onobrosmius sagus* (Fedotov, 1974), CHT, Voskovogorskii Horizon, northern Caucasus, former USSR. **Extant**
Intervening: **O** RUP–UMI.

F. LOTIDAE Bonaparte, 1838 T. (RUP)–Rec. Mar./FW

First: **O** *'Lotidarum' marinus* Gaemers, 1985, *Nucula* Clay, Heide-Boskant, Lubbeek, Belgium; **S** *Palaeomolva monstrata* Fedotov, 1974, CHT, Zurakamentskii Horizon, Caucasus, former USSR. **Extant**
Intervening: CHT–PLE.

F. GADIDAE Rafinesque, 1810 **O** T. (RUP)–Rec. Mar.

First: **O** *'Parvicolliolus' minutulus* (Gaemers, 1978), Berg Sand, *Nucula* Clay, Belgium, Ratum Formation, The Netherlands; *Gadiculus altus* (Gaemers and van Hinsbergh, 1978), Brinkherune Formation, The Netherlands, Boom Clay, Belgium; **S** first in UMI. **Extant**
Intervening: CHT–PLE.

Superorder ACANTHOPTERYGII Gouan, 1770

ACANTHOPTERYGII *incertae sedis* ??J. (BTH)

First: **O** *'Acanthopterygiorum' circularis* (Stinton, 1968) and *'Acanthopterygiorum' dorsetensis* (Stinton, 1968), otoliths from the Bradford Clay, Bradford on Avon, Wiltshire, and Fuller's Earth Clay, Langton Herring, Dorset, UK. Originally assigned (without good reason) to pycnodonts, these otoliths are listed by Nolf (1985) under the heading given here. If they are acanthopterygian, they precede the earliest skeletal record of the group by some 60 Ma.

Series ATHERINOMORPHA Greenwood *et al.*, 1966 (Rosen and Parenti, 1981)

Order ATHERINIFORMES Rosen, 1964

F. ATHERINIDAE Risso, 1826 **O** T. (YPR)–Rec. Mar./FW

First: **O** *'Atherinidarum'* sp. Nolf, 1988, Argile de Gan, Gan, Pyrénées-Atlantiques, France; **S** *Palaeoatherina rhodanica* Gaudant, 1976, Mormoiron, Vaucluse, France, *P. vardinis* (Sauvage, 1883), Bassin d'Alés, Gard, France, *P. formosa* Gaudant, 1984, Orgnac-l'Aven, Ardèche, France, all PRB.
Intervening: **O** LUT–BRT, Mio.–PLE.

F. BEDOTIIDAE Jordan and Hubbs, 1919 **Extant** FW

F. ISONIDAE Rosen, 1964 **Extant** Mar.

F. MELANOTAENIIDAE Weber and de Beaufort, 1922 **Extant** FW

F. TELMATHERINIDAE* Rosen and Parenti, 1981 **Extant** Mar.

F. PHALLOSTETHIDAE Regan, 1916 (including NEOSTETHIDAE Aurich, 1937) **Extant** Mar./FW

F. DENTATHERINIDAE* Patten and Ivantsoff, 1983 **Extant** Mar.

Order CYPRINODONTIFORMES Jordan, 1923 (Parenti, 1981)

CYPRINODONTIFORMES *incertae sedis* ?T. (THA)

First: *Cyprinodon ?primulus* Cockerell, 1936 was based on isolated scales from the THA Maiz Gordo Formation, NW Argentina; Cockerell stated that the scales might be poeciliid. **O** *'Cyprinodontoideorum' ornatissimus* Nolf, 1988, YPR (Argile de Gan, Gan, Pyrénées-Atlantiques, France), is probably a percoid (*Mene*) rather than cyprinodontiform (D. Nolf, pers. comm.).

Suborder APLOCHEILOIDEI Parenti, 1981

F. APLOCHEILIDAE Bleeker, 1859 **Extant** FW

F. RIVULIDAE Myers, 1925 **Extant** FW

Suborder CYPRINODONTOIDEI Jordan, 1923

F. PROFUNDULIDAE* Hoedeman and Bronner, 1951 **Extant** FW

F. FUNDULIDAE Jordan and Gilbert, 1882 T. (LAN)–Rec. FW/Mar.

First: *Fundulus lariversi* Lugaski, 1977, Siebert Tuff, Nevada, USA. **Extant**
Intervening: UMI–PLE.

F. VALENCIIDAE* Parenti, 1981 **Extant** FW

F. ANABLEPIDAE Bonaparte, 1837 (including JENYNSIIDAE Garman, 1895) **Extant** FW/Mar.

F. POECILIIDAE Bonaparte, 1837 T. (THA)–Rec. FW

First: Cione (1986) records 'Poeciliidae indet.' from the Maiz Gordo Formation (THA), Lumbrera Formation (YPR) and Rio Sali Formation (Mio.) in NW Argentina, but no descriptions have been published. Parenti (1981, p. 507) cited 'Upper Tertiary fossil poeciliids . . . from Brazil' and tells me that this was based on information from D. E. Rosen; the record has not been traced. The only other record is an undescribed PLE poeciliid, Rancho La Brisca, Sonora, Mexico (M. L. Smith, 1981).
Intervening: YPR, Mio., PLE.

F. GOODEIDAE Jordan, 1923 T. (Mio.)–Rec. FW

First: *Tapatia occidentalis* Alvarez and Arriola, 1972, Santa Rosa, Jalisco, Mexico. **Extant**
Intervening: PLI, PLE.

F. CYPRINODONTIDAE Agassiz, 1834 T. (RUP)–Rec. FW/Mar.

First: *Prolebias aymardi* (Sauvage, 1869), Ronzon, Le Puy-en-Velay, Haute-Loire, France; *P. rhenanus* Gaudant, 1981, Streifige Mergel, Alsace, France, and Baden, Germany; *P. catalaunicus* Gaudant, 1982, Sarreal, Tarragona, Spain. *Cyprinodon ?primulus* Cockerell, 1936, based on isolated scales from the THA Maiz Gordo Formation, Santa Barbara, Jujuy, Argentina, is here treated as Cyprinodontiformes *incertae sedis*. **Extant**
Intervening: CHT–MES.

Order BELONIFORMES Berg, 1937

Suborder ADRIANICHTHYOIDEI Rosen and Parenti, 1981

F. ADRIANICHTHYIDAE Weber, 1913 (including HORAICHTHYIDAE Kuhlkarni, 1942, ORYZIATIDAE Rosen, 1964) ?T. (Ng.)–Rec. FW/Mar.

First: ? *Lithopoecilus brouweri* de Beaufort, 1934, Gimpoe Basin, central Sulawesi. Age unknown, but presumably Neogene, and said by de Beaufort to be intermediate between *Oryzias* and adrianichthyids. **Extant**

Suborder EXOCOETOIDEI Regan, 1911

F. HEMIRAMPHIDAE Gill, 1861 T. (LUT)–Rec. Mar./FW

First: **O** *Hemiramphus ovalis* Stinton, 1978, Selsey Formation, southern England, UK; **S** *?H. edwardsi* Bassani, 1876, Monte Bolca, Verona, Italy. **Extant**
Intervening: BRT–PLI.

F. EXOCOETIDAE Risso, 1826 T. (LUT)–Rec. Mar.

First: *Rhamphexocoetus volans* Bannikov *et al.*, 1985, Monte Bolca, Verona, Italy. **Extant**

F. SCOMBERESOCIDAE Richardson, 1846 T. (MES)–Rec. Mar.

First: *Scomberesox licatae* Sauvage, 1880, Licata, Sicily, and Oran, Algeria; *S. edwardsi* Jordan and Gilbert, 1919, California, USA. **Extant**

F. BELONIDAE Bonaparte, 1837 T. (LUT/RUP)–Rec. Mar./FW

First: *?Xiphopterus falcatus* (Volta, 1796), Monte Bolca, Verona, Italy, may be a belonid (Blot, 1980); if not, the **first** records are RUP, *Belone harmati* Weiler, 1933, Eger, Hungary, *B. menelitica* Pauca, 1938, Tarlesti, Romania. **Extant**
Intervening: CHT–PLI.

Series PERCOMORPHA Rosen, 1973

Order BERYCIFORMES Regan, 1909

BERYCIFORMES *incertae sedis* (including Digoriidae Bannikov and Danil'chenko, 1985, Dinopterygiidae Jordan, 1923, Pycnosteroididae Patterson, 1964) K. (CEN) Mar.

First: *Pycnosteroides levispinosus* (Hay, 1903), Hajula, Lebanon.

Suborder TRACHICHTHYOIDEI Moore, 1990

TRACHICHTHYOIDEI *incertae sedis* (including Hoplopterygiidae Jordan, 1923, Lissoberycidae Gayet, 1980) K. (CEN)

First: Several genera, Hakel and Hajula, Lebanon, Ramallah, Israel, Lower Chalk, SE England, UK; **O** otoliths assigned to Trachichthyidae are recorded from CMP and MAA of USA, but they are treated here as trachichthyoids.

Superfamily TRACHICHTHYOIDEA Parr, 1933

F. TRACHICHTHYIDAE Bleeker, 1859 **O** T. (?YPR)–Rec. Mar.

First: **O** *Hoplostethus densus* Stinton, 1978, London Clay Formation, Essex, England, UK. According to D. Nolf (pers. comm.) this may represent the berycid *Centroberyx*; if so the first trachichthyid is LUT, *Gephyroberyx hexagonalis* (Leriche, 1905), Bruxelles Formation, Belgium. See above (Trachichthyoidei) for K otoliths. **S** *Gephyroberyx robustus* (Bogatshov, 1933), RUP, Caucasus, former USSR.
Intervening: **O** LUT–PLI.

F. MONOCENTRIDIDAE Bleeker, 1859 **O** T. (LUT)–Rec. Mar.

First: **O** *Monocentris erectus* (Schwarzhans, 1980), Bortonian, Waihao River, Canterbury, New Zealand. **Extant**
Intervening: **O** PRB–CHT.

F. ANOMALOPIDAE Gill, 1885 **O** T. (TOR)–Rec. Mar.

First: **O** *Kryptophaneron* sp. Nolf and Steurbaut, 1983, Tortonian, Montegibbio, Italy. **Extant**
Intervening: **O** Pli.

F. DIRETMIDAE* Goode and Bean, 1896 **O** T. (RUP)–Rec. Mar.

First: **O** *Diretmus* sp. Nolf and Steurbaut, 1987, Antognola Formation, Pizzocorno, NW Italy. **Extant**

Superfamily STEPHANOBERYCOIDEA Gill, 1884

F. STEPHANOBERYCIDAE Gill, 1884 **Extant** Mar.

F. ANOPLOGASTRIDAE* Gill, 1892 **Extant** Mar.

F. GIBBERICHTHYIDAE* Parr, 1893 **Extant** Mar.

F. MELAMPHAIDAE Gill, 1892 **O** T. (YPR)–Rec. Mar.

First: **O** *Melamphaes* spp. 1 and 2 Nolf, 1988, Argile de Gan, Gan, Pyrénées-Atlantiques, France. **Extant**
Intervening: **S** UMI, **O** RUP, UMI, PLI.

FF. RONDELETIIDAE* Goode and Bean, 1892, BARBOURISIIDAE* Parr, 1945, CETOMIMIDAE Goode and Bean, 1892, HISPIDOBERYCIDAE* Kotlyar, 1981, MIRAPINNIDAE* Bertelsen and Marshall, 1956, EUTAENIOPHORIDAE Bertelsen and Marshall, 1958, MEGALOMYCTERIDAE Myers and Freihofer, 1966 **Extant** Mar.

Suborder BERYCOIDEI Regan, 1909

F. BERYCIDAE Gill, 1862 **O** K. (CMP)–Rec. Mar.

First: **O** *'Berycidarum' senoniensis* (Voigt, 1926), Cöthen, Anhalt, Germany. **Extant**
Intervening: **O** THA–Oli.

Suborder HOLOCENTROIDEI new

HOLOCENTROIDEI *incertae sedis* (including Caproberycidae Patterson, 1967, Alloberycidae Gayet, 1982, Stichocentridae Gayet, 1982) K. (CEN) (see Fig. 36.4)

First: *Stichocentrus* spp., Hajula, Lebanon; *Stichoberyx polydesmus* (Arambourg, 1954), Jebel Tselfat, Morocco; *Trachichthyoides ornatus* Woodward, 1902, zone of *Holaster subglobosus*, Chalk, Kent, England, UK. **O** CMP, juvenile holocentrid, cf. *Sargocentron* (= *Adioryx*) Nolf and Dockery, 1990, Tupelo Tongue of the Coffee Sand Formation, Chapelville, Mississippi, USA.

F. HOLOCENTRIDAE Richardson, 1846 **O** T. (YPR)–Rec. Mar.

First: **O** *Holocentrus sheppeyensis* (Frost, 1934), London Clay Formation, southern England, UK; *Sargocentron amplus* (Schwarzhans, 1980), Mangaorapan/Heretaungan, Waihao River, Canterbury, New Zealand; *S.* sp. (Nolf, 1988), Argile de Gan, Gan, Pyrénées-Atlantiques, France. **Extant**
Intervening: **O** LUT, MMI–PLI., **S** PLI.

F. MYRIPRISTIDAE Nelson, 1955 T. (YPR)–Rec. Mar.

First: **O** *'Myripristinarum' sinuatus* (Stinton, 1978), London Clay Formation, southern England, UK; *'M.'* sp. Nolf, 1988, Argile de Gan, Gan, Pyrénées-Atlantiques, France; **S** *Naupygus bucklandi* Casier, 1966, London Clay Formation, Sheppey, Kent, England, UK. **Extant**
Intervening: LUT–UMI.

Order LAMPRIDIFORMES Regan, 1909

LAMPRIDIFORMES *incertae sedis* T. (LUT) Mar. (see Fig. 36.4)

First: *Bajaichthys elegans* Sorbini and Bottura, 1988, Monte Bolca, Verona, Italy.

F. LAMPRIDIDAE* Gill, 1862 T. (UMI)–Rec. Mar.

First: *Lampris zatima* (Jordan and Gilbert, 1919), Monterey Formation, Lompoc, California, USA. **Extant**

F. VELIFERIDAE* Bleeker, 1859 T. (DAN)–Rec. Mar.

First: *Bathysoma lutkeni* Davis, 1890, Danian, Limhamn, southern Sweden. **Extant**
Intervening: THA, LUT, RUP.

F. LOPHOTIDAE* Bleeker, 1859 T. (LUT/BRT)–Rec. Mar.

First: *Eolophotes lenis* (Danil'chenko, 1962), Dabaxanskaya Formation, Georgia, former USSR. **Extant**
Intervening: RUP.

FF. RADIICEPHALIDAE* Osorio, 1917, TRACHIPTERIDAE Swainson, 1839, REGALECIDAE* Gill, 1885, STYLEPHORIDAE* Swainson, 1839 **Extant** Mar.

F. ATELEOPODIDAE Kaup, 1858 **O** T. (SRV)–Rec. Mar.

First: **O** *Ateleopus* sp. Huyghebaert and Nolf, 1979, Sables

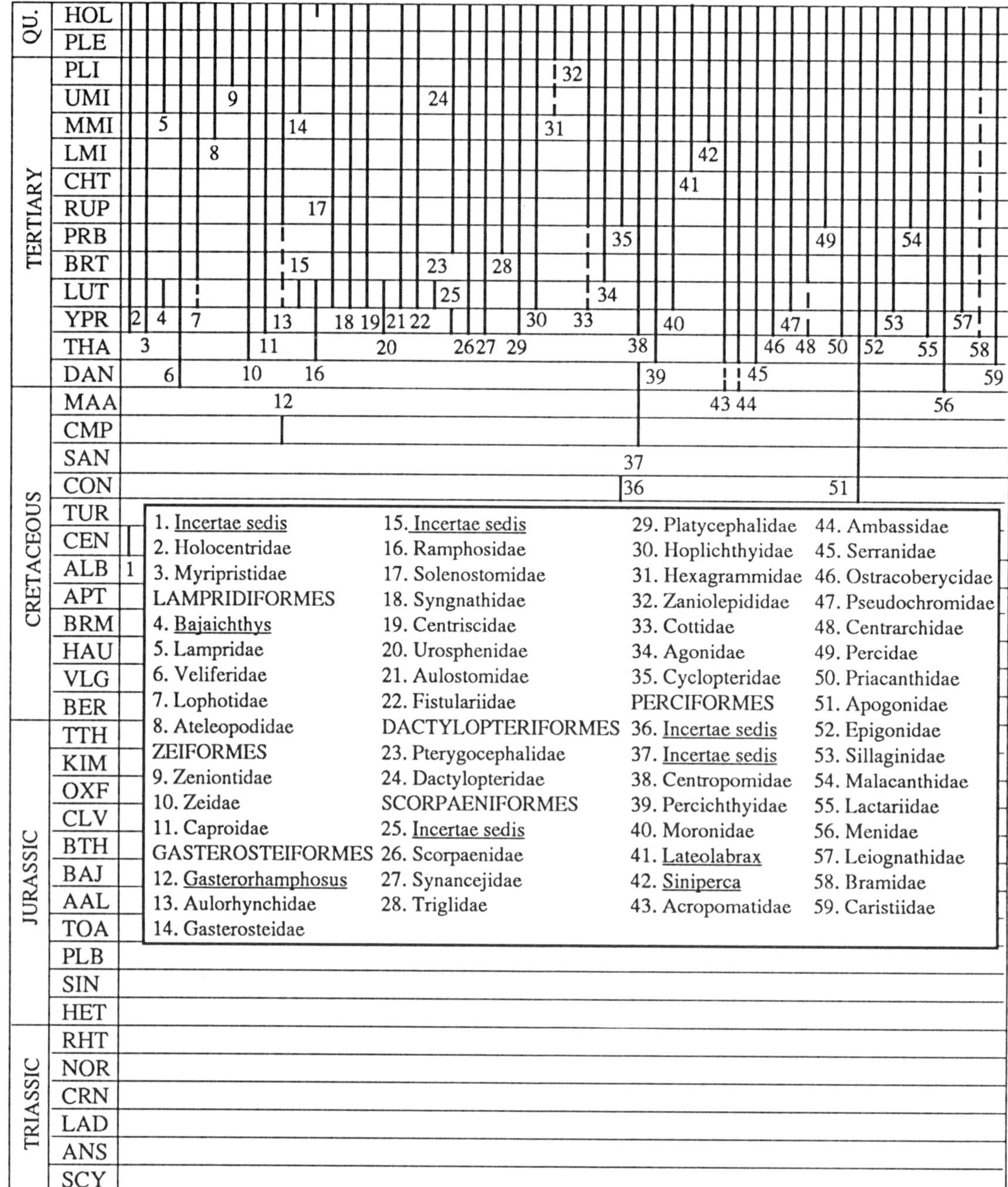

Fig. 36.4

de Zonderschot, Heist-op-den-Berg, Belgium. **Extant**
Intervening: **O** PLI.

Order ZEIFORMES Regan, 1909

F. PARAZENIDAE* Greenwood *et al.*, 1966 **Extant** Mar.

F. MACRUROCYTTIDAE* Myers, 1960 **Extant** Mar.

F. ZENIONTIDAE Myers, 1960 **O** T. (ZAN)–Rec. Mar.

First: **O** *Zenion hololepis* Goode and Bean, 1896, Zanclian, SE France. **Extant**

F. ZEIDAE Bonaparte, 1831 T. (THA)–Rec. Mar.

First: Undescribed ?zeid, Fur Formation (Mo Clay), NW Jutland, Denmark. *Palaeocyttus princeps* Gaudant, 1978, K. (CEN), Laveiras, Portugal, was described as a zeid, but differs substantially from other members of the family; it may be a zeiform *incertae sedis*. **Extant**
Intervening: Oli., UMI.

F. OREOSOMATIDAE Bleeker, 1859 **Extant** Mar.

F. GRAMMICOLEPIDIDAE Poey, 1873 **Extant** Mar.

F. CAPROIDAE Lowe, 1844 T. (YPR)–Rec. Mar.

First: **O** *Antigonia* sp. Nolf, 1988, Argile de Gan, Gan, Pyrénées-Atlantiques, France; **S** *Antigonia veronensis* Sorbini, 1983, Monte Bolca (LUT), Verona, Italy. *Microcapros libanicus* Gayet, 1980, K. (CEN), Hajula, Lebanon, was described as a caproid, but no features relating it to this

group are cited. An undescribed THA *'Antigonia'* from the Fur Formation, Denmark, is illustrated by Bonde and Christensen (1991). **Extant**
Intervening: Oli., Mio.

Order GASTEROSTEIFORMES Goodrich, 1909 (Pietsch, 1978)

GASTEROSTEIFORMES *incertae sedis* K. (CMP) Mar.

First: *Gasterorhamphosus zuppichinii* Sorbini, 1981, Calcare di Mellissano, Nardo, Lecce, Italy.

Suborder GASTEROSTEOIDEI Goodrich, 1909

F. HYPOPTYCHIDAE* Jordan, 1923 **Extant** Mar.

F. AULORHYNCHIDAE Gill, 1861 (? including Protosyngnathidae Boulenger, 1902) ?T. (Eoc.)–Rec. Mar./FW

First: *Protosyngnathus sumatrensis* von der Marck, 1876, Padang, Sumatra. The Eocene age of the fish shales at Padang is still dubious. If *Protosyngnathus* is not an aulorhynchid, the family has no fossil record, since *Protaulopsis bolcensis* Woodward, 1901, from Monte Bolca, Verona, Italy (LUT) – the only other fossil assigned to the family – belongs elsewhere according to Blot (1980). **Extant**

F. GASTEROSTEIDAE Bonaparte, 1832 T. (UMI)–Rec. Mar./FW

First: *Gasterosteus aculeatus* Linné, 1758, Monterey Formation, Lompoc, California, USA; *Pungitius hexacanthus* (Schtylko, 1934), Pavlodarskaya Formation, western Siberia, former USSR.
Intervening: PLI, PLE.

Suborder SYNGNATHOIDEI Regan, 1909

SYNGNATHOIDEI *incertae sedis* T. (LUT) Mar.

First: *Aulorhamphus bolcensis* (Steindachner, 1863), *Pseudosyngnathus opisthopterus* (Agassiz, 1833), Monte Bolca, Verona, Italy.

Superfamily PEGASOIDEA Bonaparte, 1832

F. INDOSTOMATIDAE* Prasad and Mukerji, 1929 **Extant** FW

F. PEGASIDAE* Bonaparte, 1832 **Extant** Mar.

F. RAMPHOSIDAE Gill, 1884 T. (THA–LUT) Mar.

First: *Ramphosus rosenkrantzi* Nielsen, 1960, Fur Formation (Mo Clay), NW Jutland, Denmark.
Last: *R.* spp. (Blot, 1980), Monte Bolca, Verona, Italy.

Superfamily SYNGNATHOIDEA Bonaparte, 1832

F. SOLENOSTOMIDAE* Kaup, 1853 T. (LUT)–Rec. Mar.

First: Three genera (Blot, 1980), Monte Bolca, Verona, Italy. **Extant**

F. SYNGNATHIDAE Bonaparte, 1832 T. (LUT)–Rec. Mar./FW

First: *'Syngnathus'* spp., Monte Bolca, Verona, Italy. **Extant**
Intervening: RUP–PLI.

Superfamily MACRORHAMPHOSOIDEA Bleeker, 1859

F. MACRORHAMPHOSIDAE Bleeker, 1859 **Extant** Mar.

Two monotypic fossil genera, *Gasterorhamphosus*, K. (CMP), Calcare di Mellissano, Nardo, Italy, and *Protorhamphosus*, T. (YPR), Danatinsk Formation, Turkmenia, former USSR, have been placed in this family, but they are here treated as Gasterosteiformes *incertae sedis*.

F. CENTRISCIDAE Rafinesque, 1826 (including Paraeoliscidae Blot, 1980) T.(LUT)–Rec. Mar.

First: Three genera (Blot, 1980), Monte Bolca, Verona, Italy. **Extant**
Intervening: RUP–LMI, PLI.

Superfamily AULOSTOMOIDEA Regan, 1909

F. UROSPHENIDAE Gill, 1884 T. (YPR–LUT) Mar.

First: *Urosphenopsis sagitta* Danil'chenko, 1968, Danatinsk Formation, Kopetdag, Turkmenia, former USSR.
Last: *Urosphen dubia* (de Blainville, 1818), Monte Bolca, Verona, Italy.

F. AULOSTOMIDAE* Latreille, 1825 T. (LUT)–Rec. Mar.

First: Four genera (Blot, 1980), Monte Bolca, Verona, Italy. **Extant**
Intervening: RUP.

F. FISTULARIIDAE* Bonaparte, 1832 (including Parasynarcualidae Blot, 1980, Fistularioididae Blot, 1980) T. (LUT)–Rec. Mar.

First: *Fistularioides veronensis* Blot, 1980, *F. phyllolepis* Blot, 1980, *Parasynarcualis longirostris* (de Blainville, 1818), ?*Aulostomoides tyleri* Blot, 1980, Monte Bolca, Verona, Italy. **Extant**
Intervening: RUP, UMI.

Order DACTYLOPTERIFORMES Regan, 1914

?F. PTERYGOCEPHALIDAE Hubbs, 1952 T. (LUT) Mar.

First and Last: *Pterygocephalus paradoxus* Agassiz, 1839, Monte Bolca, Verona, Italy. This fish may be related to dactylopterids (Blot, 1984b).

F. DACTYLOPTERIDAE Gill, 1885 (= Cephalacanthidae Lacépède, 1802) T. (PRB)–Rec. Mar.

First: *Prevolitans faedoensis* Gayet and Barbin, 1985, PRB, River Faedo, northern Italy. **Extant**
Intervening: PLI.

Order SCORPAENIFORMES Garman, 1899

Suborder SCORPAENOIDEI Garman, 1899

SCORPAENOIDEI *incertae sedis* O T. (YPR) Mar.

First: O *'Scorpaenoideorum' prominens* (Stinton, 1978), London Clay Formation, southern England, UK.

F. SCORPAENIDAE Risso, 1826
O T. (YPR)–Rec. Mar.

First: **O** *'Scorpaenidarum' acutus* (Frost, 1934), London Clay Formation, southern England, UK. No scorpaenid skeletons are reported earlier than MMI, apart from an isolated LUT opercular from the Subathu Formation, Jammu and Kashmir, referred to *Scorpaena* by Khare (1976). **Extant**
Intervening: **O** BRT–PLE.

F. SYNANCEJIDAE Adams, 1854 ?T. (YPR)–Rec. Mar.

First: *Eosynanceja brabantica* Casier, 1966, Argile des Flandres, Quenast, Brabant, Belgium; known only by a premaxilla, a quadrate, and a few vertebrae. No other fossils recorded. **Extant**

FF. CARACANTHIDAE* Gill, 1889, APLOACTINIDAE Regan, 1913, PATAECIDAE Gill, 1872, CONGIOPODIDAE Gill, 1889 **Extant** Mar.

F. TRIGLIDAE Risso, 1826 **O** T. (PRB)–Rec. Mar.

First: **O** *'Lepidotrigla' cadenati* Steurbaut, 1984, Couches de Cauneille (PRB), Cauneille, Chalosse, France; *'Triglidarum' cor* (Koken, 1888), Jacksonian, Jackson, Mississippi, USA. Earliest skeletal remains RUP. **Extant**
Intervening: RUP–PLE.

Suborder PLATYCEPHALOIDEI Greenwood *et al.*, 1966

F. PLATYCEPHALIDAE Bleeker, 1859
O T. (YPR)–Rec. Mar.

First: **O** *Platycephalus janeti* (Priem, 1911), Argile de Gan, Gan, Pyrénées-Atlantiques, France. **Extant**
Intervening: **O** LUT–MMI.

F. HOPLICHTHYIDAE* Gill, 1889
O T. (LUT)–Rec. Mar.

First: **O** *'Hoplichthyidarum' pulcher* Schwarzhans, 1980, Bortonian, Waihao River, Canterbury, New Zealand. **Extant**
Intervening: **O** BRT, MMI, PLI.

Suborder ANOPLOPOMATOIDEI Nelson, 1984

F. ANOPLOPOMATIDAE Gill, 1863 **Extant** Mar.

Suborder HEXAGRAMMOIDEI Greenwood *et al.*, 1966

F. HEXAGRAMMIDAE Gill, 1889
T. (UMI/PLE)–Rec. Mar.

First: *Achrestogrammus achrestus* (Jordan and Gilbert, 1919), Monterey Formation, Lompoc, California, USA. Placed in this family with a query by David (1946). **Extant**
Intervening: **O** PLE.

F. ZANIOLEPIDIDAE* Jordan, 1923
O T. (PLE)–Rec. Mar.

First: **O** *Zaniolepis latipinnis* Girard, 1857, Palos Verdes Sand, Playa del Rey, Los Angeles, California, USA. **Extant**

Suborder COTTOIDEI Bleeker, 1859

F. NORMANICHTHYIDAE* Clark, 1937
Extant Mar.

F. EREUNIIDAE Yabe, 1981 **Extant** Mar.

F. COTTIDAE Bonaparte, 1832 (including Icelidae Jordan, 1923) T. (?LUT/RUP)–Rec. Mar./FW

First: ?*Eocottus veronensis* (Volta, 1796), Monte Bolca, Verona, Italy. This form is conventionally placed in the Cottidae (e.g. Woodward, 1901; Blot, 1980), but needs revision. Regan (1913) thought that it was a goby, and J. R. Norman (MS) placed it in the Eleotrididae, where it was placed in the first edition of this compilation (Patterson, 1967). If *Eocottus* is not a cottid, the first record is *Cottus cervicornis* Storms, 1894, Argile de Boom, etc. (RUP), Belgium. **Extant**
Intervening: UMI–PLE.

F. COTTOCOMEPHORIDAE Berg, 1907
Extant FW

F. COMEPHORIDAE* Bleeker, 1859 **Extant** FW

F. PSYCHROLUTIDAE Günther, 1861
Extant Mar.

F. AGONIDAE Swainson, 1839
O T. (BRT)–Rec. Mar.

First: **O** *Podothecus costulatus* Stinton, 1978, Barton Formation, Dorset, England, UK. **Extant**
Intervening: **O** RUP, UMI, PLI.

F. CYCLOPTERIDAE Bonaparte, 1832
O T. (RUP)–Rec. Mar.

First: **O** *'Liparis' minusculus* Nolf, 1977, Argile de Boom, Kruibeke, Belgium. **Extant**
Intervening: **S** PLE.

F. BATHYLUTICHTHYIDAE* Balushkin and Voskoboynikova, 1990 **Extant** Mar.

Order PERCIFORMES Bleeker, 1859

PERCIFORMES *incertae sedis*
O K. (CON) Mar.

First: **O** *'Perciformorum' transitus* (Sieber and Weinfurter, 1967), Tiefe Gosau, Ennstaler Alpen, Austria (CON); *'Perciformorum' cepoloides* Nolf and Dockery, 1990, Tupelo Tongue of Coffee Sand Formation, Lee County, Mississippi, USA (CMP).

Suborder PERCOIDEI Bleeker, 1859

The percoids include the major remaining problems in teleost systematics. Johnson (1984) reviewed the Recent percoids from a phylogenetic viewpoint, and estimated that there are 80 families (26 of them monotypic) and 12 *incertae sedis* genera. Johnson's arrangement is followed here, listing the *incertae sedis* genera as if they were families, adjacent to the families in which they are conventionally placed.

PERCOIDEI *incertae sedis*
O K (CMP), **S** T (DAN)

First: **O** *'Percoideorum' pseudochanda* Nolf and Dockery, 1990, Tupelo Tongue of Coffee Sand Formation, Lee

County, Mississippi, USA; **S** *Proserranus lundensis* (Davis, 1890), Danian, Limhamn, southern Sweden; *Eoserranus hislopi* Woodward, 1908, Lameta Formation (? DAN), Dongargaon, Madhya Pradesh, India. On *Platacodon* (K (MAA)) see below under Sciaenidae.

F. CENTROPOMIDAE Poey, 1868
T. (YPR)–Rec. Mar./FW

First: **O** *'Centropomidarum' annectens* Stinton, 1978, *'C.' excavatus* Stinton, 1966, London Clay Formation, southern England, UK; **S** *Eolates gracilis* (Agassiz, 1833), Monte Bolca, Verona, Italy; undescribed BMNH material, ?YPR, Tilemsi Valley, Mali. **Extant**
Intervening: PRB, CHT–PLI.

F. PERCICHTHYIDAE Jordan and Eigenmann, 1890 (*sensu* Johnson, 1984, including GADOPSIDAE Günther, 1862, Maccullochellidae McCulloch, 1929, Macquariidae Munro, 1961, Perciliidae Jordan, 1923, Plectroplitidae Munro, 1961) T. (THA)–Rec. FW

First: *Percichthys lonquimayensis* Chang and Arratia, 1978, *P. sandovali* Arratia, 1982, Lonquimay Mountains, Chile; *Properca angusta* (Agassiz, 1834), Menat, Puy-de-Dome, France. **Extant**
Intervening: YPR–PLI.

F. MORONIDAE Fowler, 1907
O T. (LUT)–Rec. Mar.

First: **O** *Morone eschmeyeri* Nolf and Lapierre, 1979, Calcaire Grossier, Paris Basin, France; **S** ?*Palaeoperca proxima* Micklich, 1985, Messel Formation (LUT), Darmstadt, Germany, described as a percichthyid, was referred to Moronidae by Gaudant (1988); otherwise *Morone vogdtii* (Bogatshov, 1942), LMI, Taman, former USSR.
Intervening: **O** PRB, LMI–PLI.

Genus *Hapalogenys* Richardson, 1844, *Hemilutjanus* Bleeker, 1876, *Howella* Ogilby, 1899 **Extant** Mar.

Genus *Lateolabrax* Bleeker, 1857
?T. (AQT)–Rec. FW

First: ?*Avitolabrax denticulatus* Takai, 1942, Siramizu Formation, Joban Coalfield, Hukusima, Japan, said to be 'ancestral' to *Lateolabrax*. **Extant**

Genus *Polyprion* Oken, 1817 **Extant** Mar.

Genus *Siniperca* Gill, 1862 T. (MMI)–Rec. FW

First: cf. *Siniperca* Chang and Chow, 1986, Shandong and Zhejiang Provinces, China. **Extant**
Intervening: PLI.

Genus *Stereolepis* Ayres, 1859 **Extant** Mar.

F. ACROPOMATIDAE Gill, 1891
O T. (DAN/THA)–Rec. Mar.

First: **O** *Acropoma antiqua* Schwarzhans, 1985, Dartmoor Formation, Mt. Gambier, South Australia; **S** *A. lepidotus* (Agassiz, 1833), LUT, Monte Bolca, Verona, Italy. **Extant**
Intervening: **O** YPR–PLE.

F. AMBASSIDAE Boulenger, 1904 (= Chandidae)
O T. (DAN/THA)–Rec. Mar./FW

First: **O** *Dapalis* sp. Schwarzhans, 1985, Dartmoor Formation, Mt. Gambier, South Australia; *D.* sp. Rana, 1988, Intertrappeans, Rangapur, Hyderabad, India; **S** *D. ventralis* (Agassiz, 1836), PRB, Montmartre, Paris, and Mormoiron, Vaucluse, France. **Extant**
Intervening: YPR–MMI.

F. SERRANIDAE Swainson, 1839 (including GRAMMISTIDAE Gill, 1892, Pseudogrammatidae Greenwood *et al.*, 1966)
T. (THA)–Rec. Mar./FW

First: *Tretoperca vestita* Sytchevskaya, 1986, Cherkassy, Ukraine, former USSR; **O** *'Serranidarum' serranoides* (Stinton, 1965), Woolwich and Reading Formation, SE England, UK. **Extant**
Intervening: YPR–UMI.

F. GIGANTHIIDAE* Katayama, 1960 **Extant** Mar.

F. OSTRACOBERYCIDAE* Fowler, 1934
T. (YPR)–Rec. Mar.

First: **O** *Ostracoberyx pattersoni* Nolf, 1988, Argile de Gan, Gan, Pyrénées-Atlantiques, France. **Extant**

F. CALLANTHIIDAE Katayama, 1959
Extant Mar.

F. CENTROGENYSIDAE* Weber and de Beaufort, 1931 **Extant** Mar.

F. DINOPERCIDAE Heemstra and Hecht, 1986
Extant Mar.

Genus *Symphysanodon* Bleeker, 1878 **Extant** Mar.

F. PSEUDOCHROMIDAE Müller and Troschel, 1849 (including Anisochromidae Smith, 1954, Congrogadidae Gill, 1872, Pseudoplesiopidae Bleeker, 1875
O T. (LUT)–Rec. Mar.

First: **O** *Haliophis colletti* Nolf and Lapierre, 1979, Calcaire Grossier and Auvers Formation, Paris Basin, France. **Extant**

F. GRAMMIDAE Böhlke, 1960 **Extant** Mar.

F. PLESIOPIDAE Günther, 1861 (including ACANTHOCLINIDAE Günther, 1861)
Extant Mar.

F. BANJOSIDAE* Jordan, 1923 **Extant** Mar.

F. ELASSOMIDAE* Jordan and Gilbert, 1882
Extant FW

F. CENTRARCHIDAE Bleeker, 1859
T. (Eoc.)–Rec. FW

First: Undescribed centrarchids, NW Montana, USA (Cavender, 1986). **Extant**
Intervening: RUP, LMI–PLE.

F. PERCIDAE Cuvier, 1817 **O** T. (RUP)–Rec. FW

First: **O** *Perca hassiaca* Weiler, 1961, Melanienton, Hesse, Germany; **S** not recorded before UMI.
Intervening: **O** LMI; **S** PLI, PLE.

F. PRIACANTHIDAE Gill, 1872
O T. (YPR)–Rec. Mar.

First: **O** *Pristigenys bella* Stinton, 1980, London Clay Formation, Essex, England, UK; *P.* cf. *caduca* Nolf, 1973, Argile

de Gan, Gan, Pyrénées-Atlantiques, France; **S** *Pristigeny substriatus* (de Blainville, 1818), Monte Bolca, Verona, Italy (LUT). **Extant**
Intervening: LUT, RUP–UMI.

F. APOGONIDAE Jordan and Gilbert, 1882 **O** K. (CON)–Rec. Mar.

First: **O** *'Apogonidarum' weinbergeri* Sieber and Weinfurter, 1967, Tiefe Gosau, Ennstaler Alpen, Austria (CON); Apogonidae (Nolf and Dockery, 1990), Tupelo Tongue of Coffee Sand Formation, Lee County, Mississippi, USA (CMP); **S** *Apogon spinosus* Agassiz, 1836, Monte Bolca, Verona, Italy (LUT). **Extant**
Intervening: **O** MAA–PLI.

F. EPIGONIDAE Fraser, 1972 **O** T. (YPR)–Rec. Mar.

First: **O** *Epigonus polli* Nolf, 1988, *'Epigoninarum' malamphoides* Nolf, 1988, 'aff. *Scombrosphyraena' ganensis* Nolf, 1988 (*Scombrosphyraena* = *Sphyraenops*, Johnson, 1984), Argile de Gan, Gan, Pyrénées-Atlantiques, France. **Extant**
Intervening: **O** PRB, LMI, UMI, PLI.

F. DINOLESTIDAE* Whitley, 1948 **Extant** Mar.

F. SILLAGINIDAE Richardson, 1846 **O** T. (LUT)–Rec. Mar.

First: **O** *Sillago* sp. Schwarzhans, 1980, Bortonian, Waihao River, Canterbury, New Zealand (two eroded juvenile specimens); the next records are RUP, *Sillago hassovicus* (Koken, 1891), Meeressande, Waldböckelheim, Mainz Basin, Germany, and *S. lamberti* (Priem, 1906), Sables d'Ormoy, Ormoy, Paris Basin, France. **Extant**
Intervening: **O** LMI, MMI, PLI.

F. MALACANTHIDAE Günther, 1861 (including Branchiostegidae Jordan, 1923) **O** T. (RUP)–Rec. Mar.

First: **O** *'Malacanthidarum' cadenati* Steurbaut, 1984, Sables d'Yrieu, St-Martin-de-Seignanx, Landes, France. These otoliths were originally described as 'aff. *Lepidotrigla*'; according to D. Nolf (pers. comm.) a number of other **O** records of supposed triglids are malacanthids. **S** Mio. (m), *Lopholatilus* sp., Calvert Formation, Maryland and Virginia, USA. **Extant**
Intervening: **O** CHT–MMI, **S** MMI, MES.

F. LACTARIIDAE* Jordan, 1923 **O** T. (YPR)–Rec. Mar.

First: **O** *Lactarius* sp. Nolf, 1988, Argile de Gan, Gan, Pyrénées-Atlantiques, France. **Extant**
Intervening: **O** LUT, CHT–LMI.

F. MENIDAE* Gill, 1885 T. (DAN)–Rec. Mar.

First: *Mene phosphaticus* Astre, 1927, Montian, Tunis. **Extant**
Intervening: THA–RUP.

F. LEIOGNATHIDAE Jordan, 1923 **O** T. (LUT)–Rec. Mar.

First: **O** *'Leiognathidarum' bercherensis* Nolf and Lapierre, 1979, 'aff. *Gazza' pentagonalis* Nolf and Lapierre, 1979, Calcaire Grossier, Paris Basin, France. **Extant**
Intervening: **O** RUP–CHT.

F. BRAMIDAE Lowe, 1836 T. (?YPR/PLI)–Rec. Mar.

First: *Bramoides brieni* Casier, 1966 and *Goniocranion arambourgi* Casier, 1966, London Clay Formation, SE England, UK, were both assigned to Bramidae by Casier, but the attribution is not well supported. *Brama* sp. Sorbini, 1988, PLI, Marecchia River, NE Italy, is the only other record of the family. **Extant**

F. CARISTIIDAE Gill and Smith, 1905 (including Exelliidae Blot, 1969, = Semiophoridae Jordan, 1923) T. (THA)–Rec. Mar.

First: *Exellia* sp., Fur Formation (Mo Clay), NW Denmark (Bonde, 1987). Blot (1969) placed his Exelliidae closest to Caristiidae. **Extant**
Intervening: YPR, LUT, UMI.

F. EMMELICHTHYIDAE Jordan and Evermann, 1898 **O** T. (LUT)–Rec. Mar. (see Fig. 36.5)

First: **O** *Emmelichthys* sp. Nolf and Lapierre, 1977, Sables du Bois-Gouet, Loire-Atlantique, France. **Extant**

F. LUTJANIDAE Gill, 1884 **O** T. (LUT)–Rec. Mar. (see Fig. 36.5)

First: **O** *Apsilus latus* Stinton, 1980, Selsey Formation, southern England, UK; **S** *Hypsocephalus atlanticus* Swift and Ellwood, 1972, Crystal River Formation (PRB), Jackson County, Florida, USA. **Extant**
Intervening: RUP.

F. CAESIONIDAE Klunzinger, 1870 **O** T. (LUT)–Rec. Mar.

First: **O** *'Caesio' bourdoti* (Priem, 1906), Calcaire Grossier, Paris Basin, France. **Extant**

F. PARASCORPIDIDAE* Smith, 1949 **Extant** Mar.

Genus *Caesioscorpis* Whitley, 1945 **Extant** Mar.

F. LOBOTIDAE* Gill, 1883 **Extant** Mar.

Genus *Datnioides* Bleeker, 1853 **Extant** FW/Mar.

F. GERREIDAE Bleeker, 1859 T. (YPR)–Rec. Mar./FW

First: **O** *Gerres latidens* Stinton, 1980, Wittering Formation, southern England, UK; *'Gerreidarum' aquitanicus* Nolf, 1988, Argile de Gan, Gan, Pyrénées-Atlantiques, France. **Extant**
Intervening: **O** LUT, PRB.

F. HAEMULIDAE Richardson, 1848 (= Pomadasyidae Regan, 1913) **O** T. (THA)–Rec. Mar./FW

First: **O** *Isacia remensis* (Leriche, 1908), Sables de Chalons-sur-Vesle, Marne, France; Oldhaven Formation, SE England, UK; *'Pomadasyidarum' gullentopsi* Nolf, 1978, Landen Formation, Wansin, Belgium; **S** *?Pomadasys furcatus* (Agassiz, 1839), Monte Bolca, Verona, Italy (LUT). **Extant**
Intervening: **O** YPR–PLE.

F. INERMIIDAE Jordan, 1923 **Extant** Mar.

F. SPARIDAE Bonaparte, 1832 T. (?THA/YPR)–Rec. Mar.

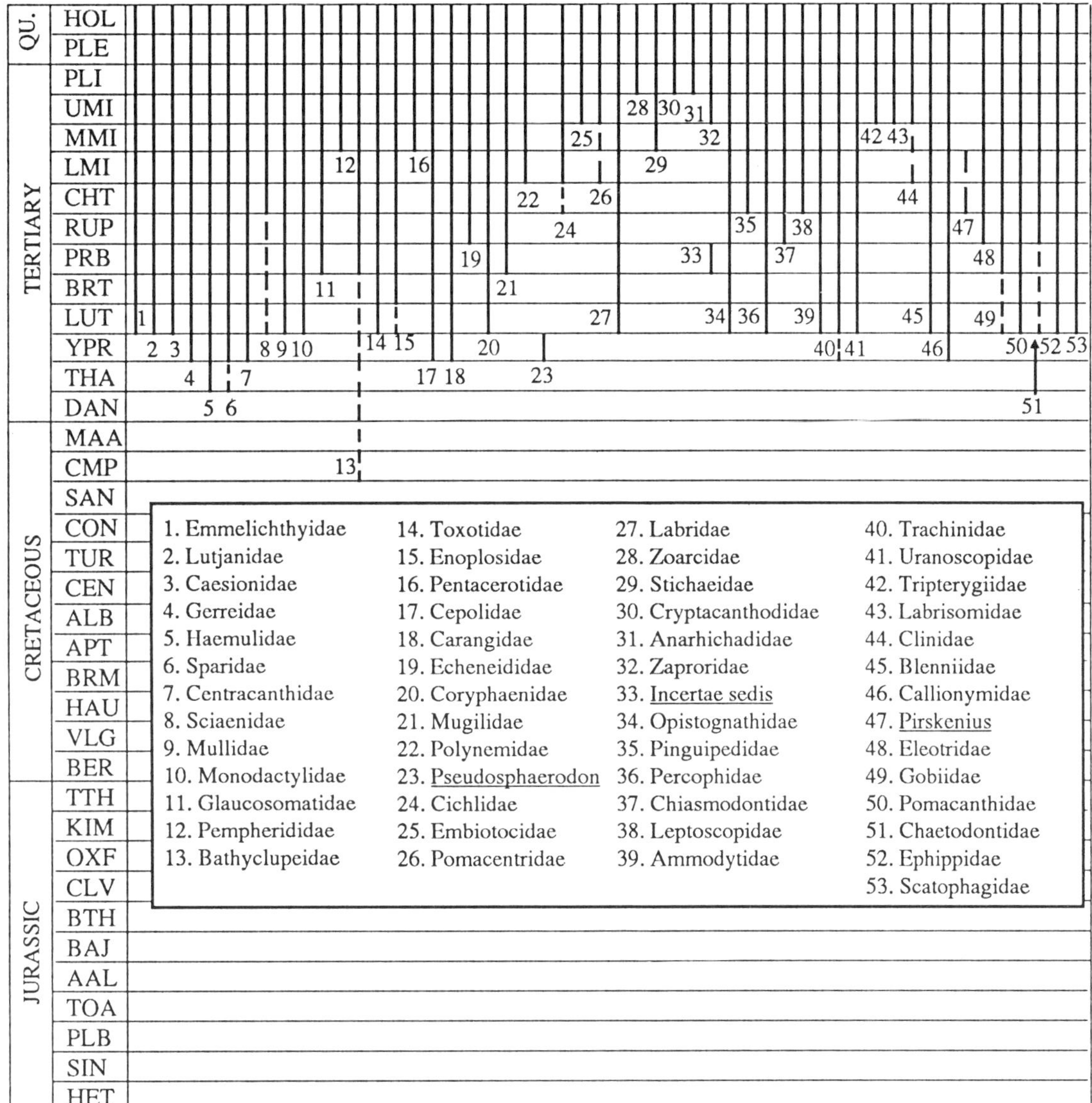

Fig. 36.5

First: *Sparus* sp. Arambourg, 1952, isolated teeth, THA phosphates, Ouled Abdoun and Ganntour, Morocco; YPR sparids include *Sciaenurus bowerbanki* Agassiz, 1845, *Podocephalus* spp. Casier, 1966, **O** *Dentex pentagonalis* Stinton, 1957 and *Pagrus bognorensis* Stinton, 1966, all from the London Clay Formation, southern England, UK, and **O** 'Sparidarum' sp. Steurba and Nolf, 1990, from the Ieper Formation, Belgium. **Extant**
Intervening: YPR–PLE.

F. CENTRACANTHIDAE Gill, 1891 (= Maenidae Cuvier, 1829) **O** T. (YPR)–Rec. Mar.

First: **O** *Centracanthus cahuzaci* Nolf, 1988, Argile de Gan, Gan, Pyrénées-Atlantiques, France. **Extant**
Intervening: **O** LUT, LMI–MMI.

F. LETHRINIDAE Regan, 1913 **Extant** Mar.

F. NEMIPTERIDAE Regan, 1913 (including Pentapodidae Smith, 1941) **Extant** Mar.

F. SCIAENIDAE Cuvier, 1829 T. (?LUT/CHT)–Rec. Mar./FW

First: **O** *'Sciaenidarum' claybornensis* Koken, 1888, Claiborne, Mississippi, USA. Sciaenids are otherwise unknown before the CHT: **S** *Larimus ignotus* (Smirnov, 1936), Maikop, Caucasus, former USSR; **O** *'Sciaenidarum' eporrectus* Koken, 1888, Vicksburg, Newton, Mississippi, USA; *Platacodon nanus* Marsh, 1889, Lance Formation (MAA), Wyoming, USA, contains isolated pharyngeals (not dentaries: see Wilson *et al.*, 1988) referred to Sciaenidae by Estes (1964), but compared with Cichlidae by Cavender (1986). According to M. V. H. Wilson (pers. comm.), they are not sciaenid, may not be acanthomorph, and could be ostariophysan. **Extant**
Intervening: CHT–PLE.

F. MULLIDAE Cuvier and Valenciennes, 1828 T. (LUT)–Rec. Mar.

First: Undescribed mullid (Blot, 1980), Monte Bolca, Verona, Italy. **Extant**
Intervening: LMI–PLI.

F. MONODACTYLIDAE* Jordan, 1923 T. (LUT)–Rec. Mar./FW

First: *Pasaichthys pleuronectiformis* Blot, 1969, *Psettopsis* *subarcuatus* (de Blainville, 1818), Monte Bolca, Verona, Italy. **Extant**

F. GLAUCOSOMATIDAE* Jordan, 1923 T. (PRB)–Rec. Mar.

First: **O** *Glaucosoma* sp. Nolf, 1988, Saint-Estéphe Formation, Blaye, Gironde, France. **Extant**

F. PEMPHERIDIDAE Gill, 1862 **O** T. (MMI)–Rec. Mar.

First: **O** *Pempheris fornicata* (Stinton, 1963), Balcombian, Mornington, Victoria, Australia. Huddleston and Savoie (1983) illustrated an otolith from the Severn Formation (MAA), Prince George County, Maryland, USA, as 'cf. Pempheridae', but the attribution is here treated as questionable. **Extant**
Intervening: **O** UMI.

F. LEPTOBRAMIDAE* Ogilby, 1913 **Extant** Mar./FW

F. BATHYCLUPEIDAE* Goode and Bean, 1896 **O** ?K. (CMP)–Rec. Mar.

First: **O** 'aff. *Bathyclupea*' sp. Nolf and Dockery, 1990, Tupelo Tongue of Coffee Sand Formation, Lee County, Mississippi, USA. **Extant**
Intervening: **O** PRB, LMI.

F. TOXOTIDAE* Bleeker, 1859 T. (LUT)–Rec. Mar./FW

First: **O** *Toxotes wheeleri* Nolf and Lapierre, 1979, Calcaire Grossier, Paris Basin, France; **S** undescribed toxotid (Blot, 1980), Monte Bolca, Verona, Italy. **Extant**

F. CORACINIDAE* Smith, 1949 **Extant** Mar.

F. ENOPLOSIDAE* Regan, 1913 ?T. (LUT)–Rec. Mar.

First: *Enoplosus pygopterus* Agassiz, 1836, Monte Bolca, Verona, Italy (species in need of revision). **Extant**

F. PENTACEROTIDAE Gill, 1885 (including Histiopteridae Gill, 1891) T. (MMI)–Rec. Mar.

First: *Pentaceros sakhalinicus* Gretchina, 1975, Agnevzkaya Formation, Sakhalin, former USSR. **Extant**

F. NANDIDAE Bleeker, 1859 (including Badidae Barlow *et al.*, 1968, Pristolepididae Regan, 1913) **Extant** FW

Hora (1939) assigned isolated scales from Intertrappean beds (?DAN), Deothan, Madhya Pradesh, India, to the Nandidae, comparing one type with *Nandus* and another with *Pristolepis*, but I doubt the reliability of these determinations. Other Tertiary scales from India have been assigned to various genera of Nandidae by Nair (1945) and Khare (1976).

Superfamily CIRRHITOIDEA Gray, 1846

FF. CIRRHITIDAE Gray, 1846, CHIRONEMIDAE Jordan, 1923, APLODACTYLIDAE Bleeker, 1859, CHEILODACTYLIDAE Bleeker, 1859, LATRIDIDAE Jordan, 1923 **Extant** Mar.

Superfamily CEPOLOIDEA Bonaparte, 1832

F. OWSTONIIDAE Jordan, 1923 **Extant** Mar.

F. CEPOLIDAE Bonaparte, 1832 **O** T. (YPR)–Rec. Mar.

First: **O** *Cepola robusta* Nolf, 1988, Argile de Gan, Gan, Pyrénées-Atlantiques, France. **Extant**
Intervening: **O** PRB, RUP, LMI, PLI.

Superfamily CARANGOIDEA Rafinesque, 1815 (*sensu* Johnson, 1984, where group is not named)

F. NEMATISTIIDAE* Gill, 1862 **Extant** Mar.

F. CARANGIDAE Rafinesque, 1815 (including Apolectidae Jordan, 1923, Formionidae Berg, 1940) T. (YPR)–Rec. Mar.

First: *Trachicaranx tersus* Danil'chenko, 1968, *Archaeus oblongus* Danil'chenko, 1968, Danatinsk Formation, Kopetdag, Turkmenia, former USSR; *Teratichthys antiquitatis* König, 1825, *Eothynnus salmoneus* Woodward, 1901, London Clay Formation, SE England, UK. **Extant**
Intervening: LUT–PLE.

F. ECHENEIDIDAE Rafinesque, 1810 T. (RUP)–Rec. Mar.

First: *Opisthomyzon glaronensis* (Wettstein, 1886), Glarus, Switzerland; *Echeneis carpathica* Szajnocha, 1926, Menilite Beds, Carpathians, Poland. **Extant**
Intervening: LMI.

F. CORYPHAENIDAE Lowe, 1839 (including Rachycentridae Gill, 1895, ?Ductoridae Blot, 1969) T. (LUT)–Rec. Mar.

First: **O** *Rachycentron* sp. Nolf, 1985, 'Middle Eocene of Belgium'; **S** *Ductor vestenae* (Volta, 1796), Monte Bolca, Verona, Italy, only member of the Ductoridae, is said by Blot (1969) to be close to Rachycentridae. **Extant**
Intervening: ?UMI.

Suborder MUGILOIDEI Regan, 1909

F. MUGILIDAE Cuvier, 1829 T. (PRB)–Rec. Mar./FW

First: **O** *'Mugilidarum' debilis* Koken, 1888, Jackson River, Mississippi; **S** *Mugil princeps* Agassiz, 1843, Menilite Beds, Poland and the former USSR (RUP). **Extant**
Intervening: RUP–PLI.

Suborder POLYNEMOIDEI Regan, 1909

F. POLYNEMIDAE Cuvier, 1828 T. (BUR)–Rec. Mar./FW

First: **O** *'Polynemidarum' gaemersi* Steurbaut, 1984, Faluns de Pont Pourquey, Saucats, Gironde, France. Khare (1976) referred a fragment of preopercular from the LUT Subathu Formation, Jammu and Kashmir, to *Polydactulus*, but the attribution is doubtful. **Extant**
Intervening: MMI, PLI.

Suborder LABROIDEI Bleeker, 1859

? LABROIDEI *incertae sedis* T. (YPR) FW

First: *Pseudosphaerodon antiquus* Casier, 1966, London Clay Formation, Sheppey, Kent, England, UK. The monotypic Tortonesidae Sorbini *et al.*, 1991 (*Tortonesia esilis* Sorbini, 1983, LUT, Monte Bolca, Verona, Italy) belongs here.

F. CICHLIDAE Gill, 1872 T. (?CHT)–Rec. FW

First: *Macfadyena dabanensis* Van Couvering, 1982, indet. Cichlidae Van Couvering, 1982, Middle Daban Series, Berbera, Somalia ('questionably Oligocene,' Van Couvering, 1982). *Aequidens saltensis* Bardack, 1961 and *Acaronia longirostrum* Bardack, 1961 are from the Anta Formation in NW Argentina, which is CHT or Mio.
Intervening: LMI–PLI.

F. EMBIOTOCIDAE Günther, 1862 T. (UMI)–Rec. Mar./FW

First: *Eriquius plectrodes* Jordan, 1924, Monterey Formation, Lompoc, California, USA. **Extant**
Intervening: PLI–PLE.

F. POMACENTRIDAE Girard, 1858 T. (?LMI)–Rec. Mar.

First: *Izuus nakamurai* Tokunaga and Saito, 1938, Yugasima Group, Izu, Japan. This fish was compared by Tokunaga and Saito with Recent pomacentrids and with *Odonteus* Agassiz, from Monte Bolca, Verona, Italy (LUT), and conventionally assigned to Pomacentridae. Blot (1980), who had begun revising *Odonteus*, repeated Arambourg's (1927) comment that the genus has nothing to do with Pomacentridae. The only other fossil pomacentrid is MES, *Chromis savornini* Arambourg, 1927, Raz-el-Ain, Oran, Algeria. **Extant**

F. LABRIDAE Cuvier, 1817 (including ODACIDAE Gill, 1885, SCARIDAE Rafinesque, 1810, following Kaufman and Liem, 1982) T. (LUT)–Rec. Mar.

First: *Eocoris bloti* Bannikov and Sorbini, 1991, *Phyllopharyngodon longipinnis* Bellwood, 1991, *Eolabroides szajnochae* (de Zigno, 1887), *Gillidia antiqua* (Agassiz, 1835), *Labrus? valenciennesi* Agassiz, 1835, all from Monte Bolca, Verona, Italy, the last three taxa being in need of revision; *Symphodus eocenicus* (White, 1926), Ameki, Nigeria. **Extant**
Intervening: PRB–RUP, MMI–PLE.

Suborder ZOARCOIDEI Garman, 1899

F. BATHYMASTERIDAE Jordan, 1885 **Extant** Mar.

F. ZOARCIDAE Cuvier, 1829 O T. (PLI)–Rec. Mar.

First: O *Lycodopsis pacifica* (Collett, 1879), unspecified PLI, California, USA. **Extant**
Intervening: O PLE.

F. STICHAEIDAE Gill, 1872 T. (MMI)–Rec. Mar.

First: *Ascoldia agnevica* Grechina, 1980, *Ernogrammus litoralis* Grechina, 1980, Agnevo Formation, Agnevo, Sakhalin, former USSR. **Extant**

F. CRYPTACANTHODIDAE Gill, 1861 O. T. (PLI)–Rec. Mar.

First: O *Lyconectes aleutensis* Gilbert, 1895, unspecified PLI, California, USA. **Extant**
Intervening: O PLE.

F. PHOLIDIDAE Gill, 1893 **Extant** Mar.

F. ANARHICHADIDAE Gill, 1865 T. (PLI)–Rec. Mar.

First: *Anarhichas lupus* L. 1758, Coralline Crag, Suffolk, England, UK. **Extant**
Intervening: PLE.

F. PTILICHTHYIDAE* Gill, 1885 **Extant** Mar.

F. ZAPRORIDAE* Jordan, 1896 T. (UMI)–Rec. Mar.

First: *Araeosteus rothi* Jordan and Gilbert, 1920, Modelo and Monterey Formations, southern California, USA. **Extant**

F. SCYTALINIDAE* Jordan and Evermann, 1898 **Extant** Mar.

Suborder NOTOTHENIOIDEI Greenwood *et al.*, 1966

FF. BOVICHTHYIDAE Gill, 1861, NOTOTHENIIDAE Gill, 1861, HARPAGIFERIDAE* Gill, 1861, ARTEDIDRACONIDAE Andriyashev, 1967, BATHYDRACONIDAE Regan, 1914, CHANNICHTHYIDAE Gill, 1861 **Extant** Mar.

Suborder TRACHINOIDEI Bertin and Arambourg, 1958 *s.l.* (Nelson, 1984)

TRACHINOIDEI *incertae sedis* O T. (PRB) Mar.

First: O *'Trachinoideorum' schwarzhansi* Nolf, 1988, Marnes de Brihande and Couches de Cauneille, Landes, France.

F. OPISTOGNATHIDAE Gill, 1872 O T. (LUT)–Rec. Mar.

First: O *'Opistognathidarum' bloti* Nolf and Lapierre, 1979, Calcaire Grossier, Paris Basin, France. **Extant**
Intervening: O LMI.

F. NOTOGRAPTIDAE* Regan, 1912 **Extant** Mar.

F. TRICHODONTIDAE Bleeker, 1859 **Extant** Mar.

Suborder TRACHINOIDEI Bertin and Arambourg, 1958 *s.s.* (Pietsch and Zabetian, 1990)

F. CHEIMARRICHTHYIDAE* Regan, 1913 **Extant** FW/Mar.

F. PINGUIPEDIDAE Günther, 1860 (including MUGILOIDIDAE Jordan, 1923, Parapercidae Gill, 1892) O T. (CHT)–Rec. Mar.

First: O *Parapercis finlayi* Frost, 1924, *P. fatuus* Schwarzhans, 1980, Duntroonian, Otago, New Zealand. **Extant**
Intervening: O LMI–UMI.

F. PERCOPHIDAE Adams, 1854 O T. (LUT)–Rec. Mar.

First: O *'Hemerocoetinarum' apertus* (Schwarzhans, 1980), Bortonian, Waihao River, Canterbury, New Zealand. **Extant**
Intervening: O PRB–LMI, PLI.

F. TRICHONOTIDAE Regan, 1913 **Extant** Mar.

F. CREEDIIDAE Regan, 1913 **Extant** Mar.

F. CHIASMODONTIDAE Gill, 1882 T. (RUP)–Rec. Mar.

First: *Pseudoscopelus grossheimi* (Danil'chenko, 1960), lower Khadum, Caucasus, former USSR. **Extant**
Intervening: O PLI.

F. CHAMPSODONTIDAE* Jordan, 1923
Extant Mar.

F. LEPTOSCOPIDAE Gill, 1872
O T. (CHT)–Rec. Mar.

First: O *Leptoscopus progressus* Schwarzhans, 1980, Waitakian, Otago, New Zealand. **Extant**
Intervening: O LMI–UMI.

F. AMMODYTIDAE Bonaparte, 1832
T. (LUT)–Rec. Mar.

First: O *Ammodytes vasseuri* Nolf and Lapierre, 1977, Sables de Bois-Gouet, Loire-Atlantique, France. **Extant**
Intervening: PRB–MMI, PLI.

F. TRACHINIDAE* Risso, 1826
T. (?YPR)–Rec. Mar.

First: *Trachinus* sp. Casier, 1946, Sables de Forest, Forest-lez-Bruxelles, Belgium (isolated opercular). There is more convincing evidence of LUT trachinids: *Callipteryx speciosus* Agassiz, 1838, *C. recticaudus* Agassiz, 1838, Monte Bolca, Verona, Italy; **O** *Trachinus* sp. Nolf and Lapierre, 1979, Calcaire Grossier, Paris Basin, France. **Extant**
Intervening: LUT–PLI.

F. URANOSCOPIDAE Bleeker, 1859
O T. (LUT)–Rec. Mar.

First: O *Uranoscopus ignavus* Schwarzhans, 1980, Bortonian, Waihao River, Canterbury, New Zealand; *'Uranoscopidarum' orbis* Nolf, 1991, Khirthar Formation, Punjab, Pakistan. **Extant**
Intervening: O MMI, PLI.

Suborder BLENNIOIDEI Bleeker, 1859

? F. PHOLIDICHTHYIDAE* Jordan and Evermann, 1898 **Extant** Mar.

F. TRIPTERYGIIDAE Hubbs, 1952
T. (MES)–Rec. Mar.

First: *Tripterygion pronasus* Arambourg, 1927, Oran, Algeria. **Extant**

F. DACTYLOSCOPIDAE Gill, 1872 **Extant** Mar.

F. LABRISOMIDAE Hubbs, 1952
T. (MES)–Rec. Mar.

First: *Labrisomus pronuchipinnis* Arambourg, 1927, Oran, Algeria. **Extant**

F. CLINIDAE Gill, 1885 **O** T. (Mio.)–Rec. Mar.

First: Undescribed otolith, Hispaniola (Nolf, 1985). **Extant**

F. CHAENOPSIDAE Gill, 1856 **Extant** Mar.

F. BLENNIIDAE Rafinesque, 1810
T. (LUT)–Rec. Mar./FW

First: O *'Blenniidarum' blondeaui* Nolf and Lapierre, 1979, Calcaire Grossier, Paris Basin, France; **S** ?*Oncolepis isseli* Bassani, 1898, Monte Bolca, Verona, Italy, otherwise TOR, *Blennius fossilis* Kramberger, 1891, Dolje, Croatia, former Yugoslavia. **Extant**
Intervening: PRB–RUP, UMI, PLE.

Suborder ICOSTEOIDEI Berg, 1937

F. ICOSTEIDAE* Jordan and Gilbert, 1882
Extant Mar.

Suborder CALLIONYMOIDEI Berg, 1937

F. CALLIONYMIDAE Bonaparte, 1832
T. (YPR)–Rec. Mar./FW

First: *Callionymus eocaenus* Casier, 1946, Sables de Forest, Forest-lez-Bruxelles, Belgium (isolated preoperculars); otoliths of *Callionymus* in Calcaire Grossier (LUT), Paris Basin, France. **Extant**
Intervening: LUT, RUP, UMI.

F. DRACONETTIDAE Jordan and Fowler, 1903
Extant Mar.

Suborder GOBIESOCOIDEI Berg, 1937

F. GOBIESOCIDAE Bleeker, 1859 (including ALABETIDAE Gill, 1906) **Extant** Mar./FW

Suborder GOBIOIDEI Jordan and Evermann, 1896

GOBIOIDEI *incertae sedis* (F. PIRSKENIIDAE Obrhelova, 1961) T. (CHT/LMI) Mar./FW

First and Last: *Pirskenius diatomaceus* Obrhelova, 1961, diatomite, Hrazeny, northern Bohemia, Czechoslovakia.

F. RHYACICHTHYIDAE* Jordan, 1923 **Extant** FW

F. ELEOTRIDAE Gill, 1861 **O** T. (RUP)–Rec. Mar./FW

First: O *'Eleotridarum'* sp. Steurbaut, 1984, Faluns de Gaas, Landes, France. **Extant**
Intervening: O LMI.

F. GOBIIDAE Bonaparte, 1832
T. (?LUT/PRB)–Rec. Mar./FW

First: *'Gobius' microcephalus* Agassiz, 1839, Monte Bolca, Verona, Italy, doubtfully a gobiid or even a gobioid. One gobiid otolith is reported by Nolf and Bajpai (in press) from the Harudi Formation (LUT), Katchchh, India, and two gobiid otoliths from the Naggulan Formation (BRT), Java. The first reliably identified skeletal gobiids are PRB: *Pomatoschistus* (?) cf. *bleicheri* (Sauvage, 1883), Fishbourne Member of Headon Hill Formation, Isle of Wight, England, UK (Gaudant and Quayle, 1988); also PRB is **O** *'Gobiidarum'* sp. Steurbaut, 1984, Sables d'Yrieu, Landes, France. **Extant**
Intervening: RUP–PLE.

F. XENISTHMIDAE Miller, 1973 **Extant** Mar.

F. SCHINDLERIIDAE* Giltay, 1934 (Johnson and Brothers, 1990) **Extant** Mar.

F. GOBIOIDIDAE Jordan, 1923 **Extant** Mar./FW

F. TRYPAUCHENIDAE Günther, 1861
Extant Mar./FW

F. KRAEMERIIDAE Whitley, 1935 **Extant** Mar./FW

F. MICRODESMIDAE Regan, 1912 **Extant** Mar.

Suborder KURTOIDEI Regan, 1909

F. KURTIDAE* Bleeker, 1859 **Extant** FW/Mar.

Suborder ACANTHUROIDEI Berg, 1937 (*sensu lato*, Tyler *et al.*, 1989)

F. DREPANIDAE* Kaup, 1860 **Extant** Mar.

F. POMACANTHIDAE Jordan and Evermann, 1898 **O** T. (LUT)–Rec. Mar.

First: **O** *Pomacanthus fitchi* Nolf, 1973, Calcaire Grossier, Paris Basin, France. **Extant**

F. CHAETODONTIDAE Bonaparte, 1832 (?including Pygaeidae Jordan, 1904) T. (?LUT/RUP)–Rec. Mar.

First: *Pygaeus* Agassiz, 1838 (about six nominal species), *Malacopygaeus* Leriche, 1906 (one species) and *Parapygaeus* Pellegrin, 1907 (one species), all from Monte Bolca, Verona, Italy, are conventionally assigned to Chaetodontidae, but all are in need of revision. *Chaetodon penniger* Bogatshov, 1964, RUP, Moroskin, Transcaucasia, former USSR, is otherwise the first chaetodontid. **Extant**
Intervening: RUP–PLI.

F. EPHIPPIDAE Bleeker, 1859 (including Chaetodipteridae Bleeker, 1877, Platacidae Bleeker, 1877, RHINOPRENIDAE Munro, 1964) T. (LUT)–Rec. Mar.

First: *Archaephippus asper* (Volta, 1796), *Eoplatax papilio* (Volta, 1796), Monte Bolca, Verona, Italy. **Extant**
Intervening: ?RUP, PLI.

F. SCATOPHAGIDAE Gill, 1883 T. (LUT)–Rec. Mar.

First: *Scatophagus frontalis* Agassiz, 1835, Monte Bolca, Verona, Italy. **Extant**
Intervening: RUP.

Suborder ACANTHUROIDEI *s.s.* (Tyler *et al.*, 1989)

ACANTHUROIDEI *incertae sedis* (F. Kushlukiidae Danil'chenko, 1968) T. (YPR) (see Fig. 36.6)

First and Last: *Kushlukia permira* Danil'chenko, 1968, Danatinsk Formation, Kopetdag, Turkmenia, former USSR; *Kushlukia* may also occur in the YPR Fuller's Earth, Barmer, SW Rajasthan, India (J. C. Tyler, pers. comm.).

F. SIGANIDAE Richardson, 1836 T. (YPR)–Rec. Mar.

First: *Siganopygaeus rarus* Danil'chenko, 1968, Danatinsk Formation, Kopetdag, Turkmenia, former USSR. **Extant**
Intervening: LUT, RUP.

F. LUVARIDAE* Gill, 1885 (including Beerichthyidae Casier, 1966) T. (YPR)–Rec. Mar.

First: *Proluvarus necopinatus* Danil'chenko, 1968, Danatinsk Formation, Kopetdag, Turkmenia, former USSR; *Beerichthys ingens* Casier, 1966, London Clay Formation, SE England, UK. *Eoluvarus bondei* Sahni and Choudhary, 1977, YPR, Barmer, SW Rajasthan, India, was accepted as a luvarid by Tyler *et al.* (1989), but Tyler (pers. comm.), after examining the type and only specimen, believes that it belongs not in this family but with *Exellia*, which is here included in the Caristiidae. Tyler *et al.* (1989) doubted that *Luvarus preimperialis* Arambourg, 1956, RUP, Elam, Luristan, Iran, is a luvarid, and Tyler (pers. comm.) has since examined the type material and confirmed that it is unrelated to *Luvarus*. **Extant**

F. ACANTHURIDAE Bleeker, 1859 (including Caprovesposidae Bannikov and Fedotov, 1984, Teuthidae Latreille, 1825, Zanclidae Gill, 1885) T. (LUT)–Rec. Mar.

First: Eight genera (Blot and Tyler, 1991), Monte Bolca, Verona, Italy, and one unnamed acronurus-like specimen (Bannikov and Tyler, 1992), Dabakham Formation, Tbilisi, Georgia, former USSR. **Extant**
Intervening: Eoc. (u), RUP, MMI; these records are of *Caprovesposus*, long thought to be a zeiform, which Bannikov and Tyler (1992) show to be acronurus-stage larvae of acanthurids.

Suborder SCOMBROIDEI Bleeker, 1859 *s.l.* (Johnson, 1986)

? F. EUZAPHLEGIDAE Danil'chenko, 1960 (= Zaphlegidae Jordan, 1920) T. (YPR–UMI) Mar.

First: *Palimphyes palaeocenicus* Danil'chenko, 1968, Danatinsk Formation, Kopetdag, Turkmenia, former USSR.
Last: Three genera (David, 1943), Modelo and Monterey Formations, southern California, USA.
Intervening: LUT/BRT, RUP–LMI.
Comment: This nominal family, discussed by David (1943: 'nearly related to the Gempylidae') and Danil'chenko (1960: 'intermediate between Scombridae and Gempylidae'), seems to lack most of the characters of Scombroidei *s.s.* and of Gempylidae listed by Johnson (1986). There is no indication that it is monophyletic.

F. SCOMBROPIDAE* Gill, 1891 **O** T. (YPR)–Rec. Mar.

First: **O** *Scombrops* sp. Nolf, 1988, Argile de Gan, Gan, Pyrénées-Atlantiques, France. **Extant**

F. SCOMBROLABRACIDAE* Roule, 1922 **Extant** Mar.

F. POMATOMIDAE* Gill, 1865 T. (?LUT/LMI)–Rec. Mar.

First: ?*Carangopsis brevis* (de Blainville, 1818), *C. dorsalis* Agassiz, 1844, Monte Bolca, Verona, Italy, placed 'provisionally' in Pomatomidae by Blot (1969), and compared with the only other fossil pomatomids known at that time, *Lophar miocenicus* Jordan and Gilbert, 1919 and *Pseudoseriola gillilandi* David, 1943, both UMI, southern California. Bannikov and Fedotov (1989) refer *Lednevia oligocenica* (Smirnov, 1936), LMI, Crimea and Caucasus, to this family. **Extant**

Suborder SCOMBROIDEI *s.s.* (Johnson, 1986)

F. SPHYRAENIDAE* Rafinesque, 1815 T. (YPR)–Rec. Mar.

First: *Sphyraena* spp., London Clay Formation, southern England, UK; Ieper Formation, Belgium; Phosphates, Morocco (isolated teeth). **Extant**
Intervening: LUT–PLE.

F. GEMPYLIDAE Gill, 1862 (including TRICHIURIDAE Rafinesque, 1810) T. (DAN)–Rec. Mar.

First: *Eutrichiurides orpiensis* (Leriche, 1906), Montian

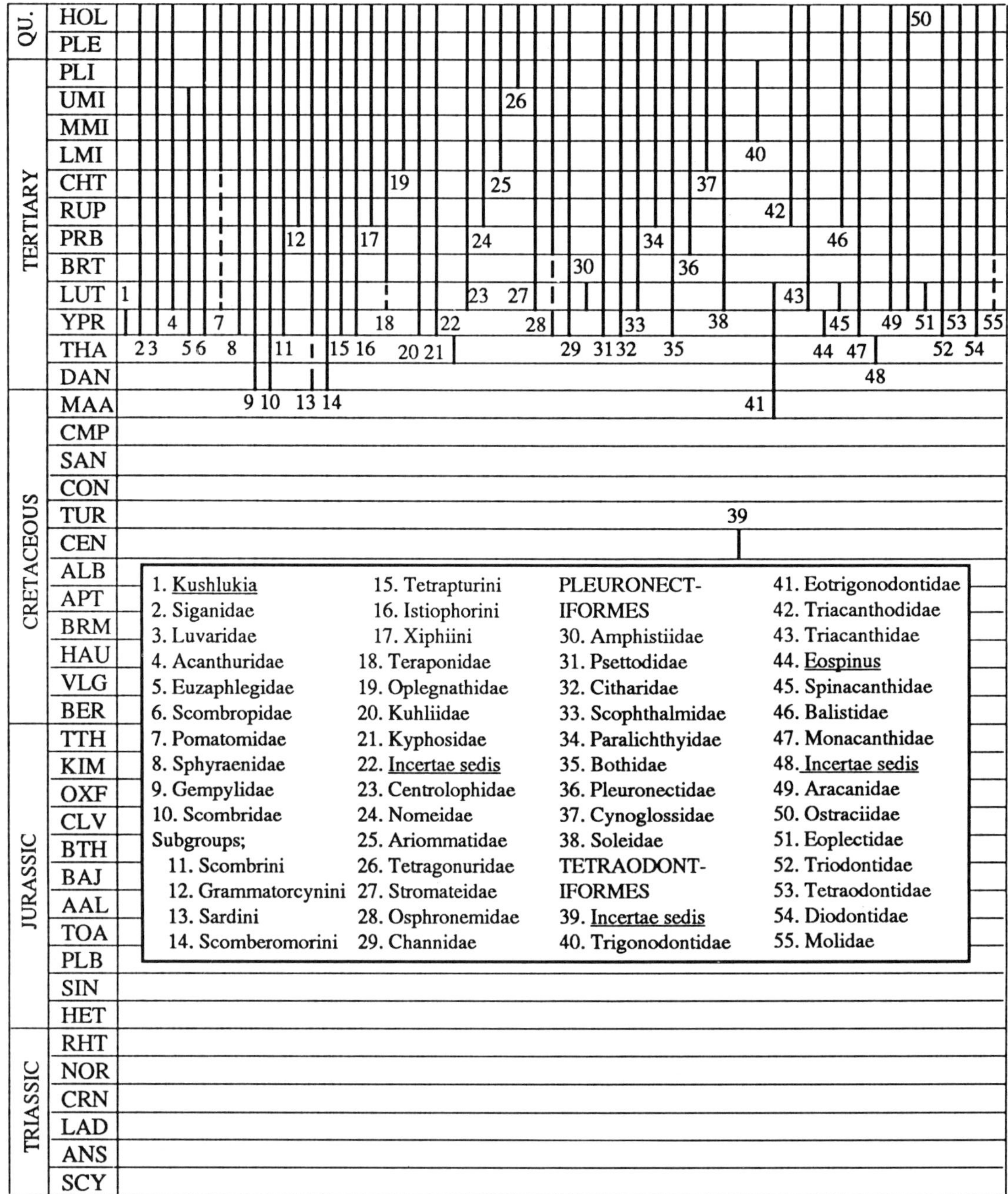

Fig. 36.6

Phosphates, Morocco; *E. africanus* Dartevelle and Casier, 1949, Landana, Cabinda, Angola (isolated teeth). **Extant Intervening:** THA–PLI.

F. SCOMBRIDAE Rafinesque, 1815 (including Istiophoridae Lütken, 1875, Xiphiidae Swainson, 1839, Xiphiorhynchidae Regan, 1909) T. (DAN)–Rec. Mar. (Bannikov, 1985a)

Tribe GASTEROCHISMATINI Gill, 1892 **Extant** Mar.

Tribe SCOMBRINI Rafinesque, 1815 T. (YPR)–Rec. Mar.

First: *Scombrosarda turkmenica* Danil'chenko, 1968, Danatinsk Formation, Kopetdag, Turkmenia, former USSR; *S. decipiens* (Casier, 1966), London Clay Formation, SE England, UK.
Intervening: LUT–PLI.

Tribe GRAMMATORCYNINI* Johnson, 1986 T. (RUP)–Rec. Mar.

First: *Grammatorcynus scomberoides* Arambourg, 1967, Elam, Luristan, Iran. **Extant**
Intervening: MMI.

Tribe SARDINI Starks, 1910 T. (?DAN)–Rec. Mar.

First: *Sarda palaeocaena* (Leriche, 1909), Landana, Cabinda, Angola (isolated teeth); earliest complete fish is YPR, *Palaeothunnus parvidentatus* Danil'chenko, 1968, Danatinsk

Formation, Kopetdag, Turkmenia, former USSR. **Extant**
Intervening: YPR–PLE.

Tribe SCOMBEROMORINI* Starks, 1910
T. (DAN)–Rec. Mar.

First: *Landanichthys lusitanicus* Dartevelle and Casier, 1949, *L. moutai* Dartevelle and Casier, 1949, *Sphyraenodus multidentatus* Dartevelle and Casier, 1959, Landana, Cabinda, Angola. **Extant**
Intervening: THA–UMI.

Tribe ACANTHOCYBIINI* Starks, 1910
Extant Mar.

Tribe TETRAPTURINI Smith, 1956
T. (YPR)–Rec. Mar. (Schultz, 1987)

First: *Hemirhabdorhynchus depressus* Casier, 1946, *H. ypresiensis* Casier, 1946, Sables de Forest, Forest-lez-Bruxelles, Belgium; *Aglyptorhynchus* spp., London Clay Formation, SE England, UK; and Sables de Forest, Belgium. **Extant**
Intervening: LUT–PLI.

Tribe ISTIOPHORINI Lütken, 1875
(including Xiphiorhynchidae Regan, 1909)
T. (YPR)–Rec. Mar. (Schultz, 1987)

First: *Xiphiorhynchus priscus* (Agassiz, 1839), London Clay Formation, SE England, UK, and Sables de Forest, Belgium; *X. parvus* Casier, 1966, London Clay Formation, SE England, UK. **Extant**
Intervening: LUT–PLE.

Tribe XIPHIINI* Swainson, 1839 T. (RUP)–Rec.
Mar. (Schultz, 1987)

First: *Xiphias rupeliensis* Leriche, 1909, Argile de Boom, Belgium, and Böhlener Schichten, Leipzig, Germany. **Extant**
Intervening: MMI–PLI.

Suborder STROMATEOIDEI Regan, 1929 *s.l.*
(Johnson and Fritzsche, 1989)

F. ARRIPIDAE* Regan, 1913 **Extant** Mar.

F. TERAPONIDAE Richardson, 1856
?T. (LUT)–Rec. Mar./FW

First: ?*Pelates quindecimalis* Agassiz, 1834, Monte Bolca, Verona, Italy. This is the only record of a fossil teraponid, a single specimen (assigned to a Recent genus) which requires study. **Extant**

F. OPLEGNATHIDAE Bleeker, 1859
T. (LMI)–Rec. Mar.

First: *Oplegnathus* sp., Batesford, Victoria, Australia. **Extant**
Intervening: UMI, PLI.

F. KUHLIIDAE* Jordan and Evermann, 1896
O T. (YPR)–Rec. Mar./FW

First: **O** *Kuhlia lepta* Stinton, 1980, Wittering Formation, southern England, UK; *K.* sp. Nolf, 1988, Argile de Gan, Gan, Pyrénées-Atlantiques, France. **Extant**
Intervening: **O** LUT, PRB, CHT, LMI.

F. KYPHOSIDAE Gill, 1893 (including Girellidae Gill, 1862, Labracoglossidae Regan, 1913, Microcanthidae Fowler, 1953, Scorpididae Günther, 1860, *Neoscorpis* Smith, 1931)
O T. (YPR)–Rec. Mar.

First: **O** *'Kyphosidarum' goujeti* Nolf, 1988, Argile de Gan, Gan, Pyrénées-Atlantiques, France. **Extant**

Suborder STROMATEOIDEI Regan, 1929 *s.s.*
(Bannikov, 1985b)

? STROMATEOIDEI *incertae sedis*
T. (THA) Mar.

First: Undescribed fishes, questionably referred to the suborder (Bonde, 1987), Fur Formation (Mo Clay), NW Denmark.

F. AMARSIPIDAE* Haedrich, 1969 **Extant** Mar.

F. CENTROLOPHIDAE Gill, 1861
O T. (LUT)–Rec. Mar.

First: **O** *Schedophilus confinis* (Nolf, 1973), Lede Formation, Belgium, and Selsey Formation, southern England, UK. **Extant**
Intervening: **O** BRT, CHT–LMI.

F. NOMEIDAE Günther, 1860
T. (RUP)–Rec. Mar.

First: *Psenicubiceps alatus* Danil'chenko, 1980, Pshekha Horizon, northern Caucasus, former USSR. **Extant**
Intervening: PLI.

F. ARIOMMATIDAE* Haedrich, 1967
O T. (LMI)–Rec. Mar.

First: **O** *'Ariomma'* sp. Steurbaut, 1984, Marnes de Saubrigues, Landes, France. **Extant**
Intervening: **O** MMI.

F. TETRAGONURIDAE* Risso, 1826
T. (PIA)–Rec. Mar.

First: *Tetragonurus* sp. Sorbini, 1988, Marrechia River, NW Italy. **Extant**

F. STROMATEIDAE Rafinesque, 1810
T. (LUT)–Rec. Mar.

First: **O** *Stromateus brailloni* Nolf, 1975, Auvers Formation, Paris Basin, France. **Extant**
Intervening: RUP, LMI.

Suborder ANABANTOIDEI Regan, 1909

F. ANABANTIDAE Richardson, 1836 **Extant** FW

F. BELONTIIDAE Liem, 1963
(= Polyacanthidae Gill, 1893) **Extant** FW

Comment: Hora (1939) assigned a scale from Interappean beds (?DAN), Deothan, Madhya Pradesh, India, to the extant *Macropodus*, but I doubt the determination. Nair (1945) and Khare (1976) also assigned isolated Tertiary scales from India to *Macropodus*.

F. HELOSTOMATIDAE* Gill, 1872 **Extant** FW

F. OSPHRONEMIDAE* Bleeker, 1859
T. (?Eoc.)–Rec. FW

First: *Osphronemus goramy* Lacépède, 1802, Padang,

Sumatra. The Eocene age of the fish shales at Padang is still dubious. **Extant**

Suborder LUCIOCEPHALOIDEI Berg, 1937

F. LUCIOCEPHALIDAE* Bleeker, 1859 **Extant** FW

Suborder OPHICEPHALOIDEI Bleeker, 1859 (= Channiformes Greenwood *et al.*, 1966)

F. CHANNIDAE* Berg, 1940 (= Ophicephalidae Bonaparte, 1831) T. (YPR)–Rec. FW

First: **S** *Eochanna chorlakkiensis* Roe, 1991, Kuldana Formation, Chorlakki, Kohat, Pakistan. **Extant**
Intervening: **S** LUT, MMI–PLE, **O** PRB, LMI.

Order SYNBRANCHIFORMES Regan, 1909

FF. SYNBRANCHIDAE Swainson, 1838, CHAUDHURIIDAE Anandale, 1918, MASTACEMBELIDAE Günther, 1861 **Extant** FW

Order PLEURONECTIFORMES Bleeker, 1859

? PLEURONECTIFORMES *incertae sedis*

F. AMPHISTIIDAE Boulenger, 1902 T. (LUT) Mar.

First and Last: *Amphistium paradoxum* Agassiz, 1834, Monte Bolca, Verona, Italy; *A. altum* (Agassiz, 1842), Calcaire Grossier, Paris Basin, France.

Suborder PSETTODOIDEI Regan, 1909

F. PSETTODIDAE* Regan, 1910 (inc. Joleaudichthyidae Chabanaud, 1937) **O** T. (YPR)–Rec. Mar.

First: **O** *Psettodes bavayi* Nolf, 1988, *P.* sp. Nolf, 1988, Argile de Gan, Gan, Pyrénées-Atlantiques, France, and *P.* sp. Nolf and Steurbaut, 1990, Ieper Formation, Belgium; **S** *Joleaudichthys sadeki* Chabanaud, 1937, lower Mokattam, Jebel Tourah, Egypt. **Extant**
Intervening: LUT/BRT, LMI.

Suborder PLEURONECTOIDEI Bleeker, 1859

F. CITHARIDAE Hubbs, 1945 **O** T. (YPR)–Rec. Mar.

First: **O** *Citharus circularis* (Stinton, 1966), London Clay Formation, southern England, UK and Argile de Gan, Gan, Pyrénées-Atlantiques, France. **Extant**
Intervening: **O** PRB–PLI.

F. SCOPHTHALMIDAE Jordan, 1923 T. (LUT)–Rec. Mar.

First: *Imhoffius lutetianus* Chabanaud, 1940, Calcaire Grossier, Paris Basin, France. **Extant**
Intervening: **O** CHT–MMI.

F. PARALICHTHYIDAE Regan, 1910 **O** T. (RUP)–Rec. Mar.

First: **O** *Monolene prudhommae* Steurbaut, 1984, Sables d'Yrieu, Landes, France; **S** UMI, Algeria and California, USA. **Extant**
Intervening: UMI–PLE.

F. BOTHIDAE Regan, 1910 T. (YPR)–Rec. Mar.

First: **O** *'Bothidarum' lapierrei* Nolf, 1988, Argile de Gan, Gan, Pyrénées-Atlantiques, France; **S** LUT *Eobothus minimus* (Agassiz, 1839), Monte Bolca, Verona, Italy; *E. vialovi* Berg, 1914, Alaiskiy Horizon, SW Uzbekistan, former USSR. *Eobothus singhi* Sahni and Choudhary, 1972, YPR, Barmer, SW Rajasthan, India, was very inadequately described and may not be pleuronectiform. **Extant**
Intervening: LUT–PLI.

F. PLEURONECTIDAE Rafinesque, 1810 **O** T. (PRB)–Rec. Mar./FW

First: **O** *'Pleuronectidarum'* sp. Nolf, 1988, Couches de Cauneille, Landes, France; **S** *Platichthys parvulus* (Smirnov, 1936), CHT, northern Caucasus, former USSR.
Intervening: RUP–PLE.

Suborder SOLEOIDEI Norman, 1931

F. CYNOGLOSSIDAE Jordan and Goss, 1889 **O** T. (BUR)–Rec. Mar./FW

First: **O** *Paraplagusia roseni* Nolf and Cappetta, 1980, Montpeyroux, Hérault, France; *P. alta* (Lafond-Grellety, 1982), Gironde and Landes, France. **Extant**
Intervening: **O** LMI–UMI, PLE.

F. SOLEIDAE Bonaparte, 1833 (including Eobuglossidae Chabanaud, 1937) T. (LUT)–Rec. Mar./FW

First: *Eobuglossus eocenicus* (Woodward, 1910), *Turahbuglossus cuvillieri* Chabanaud, 1937, lower Mokattam, Jebel Tourah, Egypt. **Extant**
Intervening: BRT–PLI.

Order TETRAODONTIFORMES Regan, 1929 (Tyler, 1980, Winterbottom, 1974)

Tyler (1980) and Winterbottom (1974) give very different interpretations of the systematics of Recent tetraodontiforms and of several fossil genera. Winterbottom's scheme is followed here, since it is explicitly cladistic and since, although his work was published six years before Tyler's, he had access to Tyler's interpretation of the fossils, and to a large part of Tyler's manuscript.

TETRAODONTIFORMES *incertae sedis* K. (CEN) Mar.

First: *Plectocretacicus clarae* Sorbini, 1979, Hakel, Lebanon.

? F. TRIGONODONTIDAE Arambourg, 1927 T. (MMI–PLI) Mar.

First: *Trigonodon oweni* Sismonda, 1849, various Middle Miocene localities in Europe.
Last: *T. oweni*, Pliocene, Tuscany, Italy.

? F. EOTRIGONODONTIDAE White, 1935 K. (MAA)–T. (LUT) Mar.

First: *Stephanodus libycus* (Dames, 1883), MAA, Morocco, Tunisia, Libya, Nigeria, Niger, Zaïre, Israel, Jordan, Iraq; according to Jain and Sahni (1983) this species also occurs in the Mount Laurel Sand of New Jersey, USA, which is late CMP in age, but it is not clear whether the material comprises only pharyngeal teeth (*'Ancistrodon'*), which have a wider stratigraphical distribution than the jaw teeth (*Stephanodus*). The Lameta Formation, Maharashtra, India, probably MAA, contains jaw teeth named *Indotrigonodon ovatus* Jain and Sahni, 1983, *Eotrigonodon wardhaensis* Jain and Sahni, 1983, and *Pisdurodon spatulatus* Jain and Sahni,

1983. *Kankatodus cappettai* Kumar and Loyal, 1987, MAA, Niger.
Last: *Eotrigonodon* spp., LUT, Belgium, France, Egypt.
Intervening: DAN–YPR.
Comment: Eotrigonodontids, although conventionally treated as tetraodontiforms, are probably not teleosts but pycnodonts, a group whose stratigraphical range includes that of eotrigonodontids.

Suborder TRIACANTHOIDEI Winterbottom, 1974

F. TRIACANTHODIDAE Gill, 1862 T. (RUP)–Rec. Mar.

First: *Cryptobalistes brevis* (Rath), Glarus, Switzerland. Tyler (1980) places *Cryptobalistes* in the succeeding family, and includes four monotypic LUT genera from Monte Bolca, Verona, Italy, in this one. **Extant**

F. TRIACANTHIDAE Bleeker, 1859 T. (LUT)–Rec. Mar.

First: *Protacanthodes ombonii* (de Zigno, 1884), Monte Bolca, Verona, Italy. **Extant**
Intervening: RUP, Mio.

Suborder TETRAODONTOIDEI Berg, 1937

Infra-order BALISTOIDEO Berg, 1937

BALISTOIDEO *incertae sedis* T. (YPR)

First and Last: *Eospinus daniltshenkoi* Tyler and Bannikov, 1992, Danatinsk Formation, Kopetdag, Turkmenia, former USSR.

F. SPINACANTHIDAE Tyler, 1968 T. (LUT) Mar.

First and Last: *Spinacanthus cuneiformis* (de Blainville, 1818) and *Protobalistum imperiale* (Massalongo, 1857), Monte Bolca, Verona, Italy.

Superfamily BALISTOIDEA Rafinesque, 1810

F. BALISTIDAE Rafinesque, 1810 T. (RUP)–Rec. Mar.

First: *Balistomorphus ovalis* (Agassiz, 1842), *B. spinosus* (Agassiz, 1842), *B. orbiculatus* (Heer, 1865), Glarus, Switzerland; *Oligobalistes robustus* Danil'chenko, 1960, Khadum Horizon, Maikop Series, Caucasus, former USSR. **Extant**
Intervening: MMI–PLI.

F. MONACANTHIDAE Nardo, 1844 **O** T. (YPR)–Rec. Mar.

First: **O** *Amanses sulcifer* Stinton, 1966, London Clay Formation, southern England, UK; **S** PLI, *Alutera* sp. Sorbini, 1988, Marecchia River, NE Italy. **Extant**
Intervening: PLI.

Superfamily OSTRACIOIDEA Rafinesque, 1810

OSTRACIOIDEA *incertae sedis* T. (THA) Mar.

First: *'Ostracion' meretrix* Daimeries, 1891, Aquia Formation, Virginia, USA (dermal plates), also in Woolwich and Reading Formation, SE England, UK. *Ostracion* is reported by Gayet *et al.* (1984) from the ?DAN, Intertrappean beds, Gitti Khadan, Madhya Pradesh, India, and from K. (MAA), In Beceten, Niger, but both records require confirmation.

F. ARACANIDAE Hollard, 1860 T. (LUT)–Rec. Mar.

First: *Proaracana dubia* (de Blainville, 1818), Monte Bolca, Verona, Italy. **Extant**
Intervening: CHT.

F. OSTRACIIDAE Rafinesque, 1810 T. (LUT)–Rec. Mar.

First: *Eolactoria sorbinii* Tyler, 1975, Monte Bolca, Verona, Italy. **Extant**
Intervening: Dermal plates assigned to *Ostracion* occur in THA–BRT, RUP, but are better treated as Ostracioidea *incertae sedis* (above).

Infra-order TETRAODONTOIDEO Berg, 1937

F. EOPLECTIDAE Tyler, 1973 (including Zignoichthyidae Winterbottom, 1974) T. (LUT) Mar.

First and Last: *Eoplectus bloti* Tyler, 1973 and *Zignoichthys oblongus* (Zigno, 1874), Monte Bolca, Verona, Italy.

Superfamily TRIODONTOIDEA Bleeker, 1865

F. TRIODONTIDAE* Bleeker, 1859 T. (YPR)–Rec. Mar.

First: *Triodon antiquus* Leriche, 1905, Ieper Formation, Belgium, and London Clay Formation, southern England, UK. Tyler (1980) pointed out that the isolated jaws on which this species is based could be equally well from primitive diodontids as from *Triodon*, but a recently collected London Clay skull indicates that the fish is a *Triodon*. **Extant**
Intervening: LUT–BRT.

Superfamily TETRAODONTOIDEA Bonaparte, 1832

F. TETRAODONTIDAE Bonaparte, 1832 T. (LUT)–Rec. Mar./FW

First: *Eotetraodon pygmaeus* (de Zigno, 1887), Monte Bolca, Verona, Italy. **Extant**
Intervening: MMI–PLI.

F. DIODONTIDAE Bibron, 1855 (including Eodiodontidae Tavani, 1955) T. (YPR)–Rec. Mar.

First: *Diodon* sp. Kumar and Loyal, 1987, lower Subathu Formation, Solan, Himachal Pradesh, India; *Kyrtogymnodon* sp. Weems and Horman, 1983, lower Nanjemoy Formation, Popes Creek, Maryland, USA (both records isolated tooth plates, not surely diodontid). The first intact skeletons are LUT, *Prodiodon erinaceus* (Agassiz, 1844), *P. tenuispinus* (Agassiz, 1833), Monte Bolca, Verona, Italy. **Extant**
Intervening: LUT–PLI.

Superfamily MOLOIDEA Ranzani, 1837

F. MOLIDAE Ranzani, 1837 T. (LUT/PRB)–Rec. Mar.

First: *Eoranzania* Tyler and Bannikov (in press), isolated jaws from the Kumskii Formation, Pshekha River, northern Caucasus, former USSR. **Extant**
Intervening: ?CHT (Woodward, 1901, p. 576), LAN–PLI.

REFERENCES

Arambourg, C. (1927) Les poissons fossiles d'Oran. *Matériaux pour la Carte Géologique de l'Algérie*, (1), **6**, 1–298.

Arambourg, C. (1954) Les poissons crétacés du Jebel Tselfat. *Notes et Mémoires du Service Géologique du Maroc*, **118**, 1–188.

Arratia, G. (1987) Jurassic fishes from Chile and critical comments, in *Biogeografia de los Sistemas Regionales del Jurasico y Cretacico en America del Sur*, **1**, Mendoza (eds W. Volkheimer and E. A. Musacchio), pp. 257–86.

Bannikov, A. F. (1985a) Fossil scombrids of the USSR. *Trudy Paleontologicheskogo Instituta, Akademiya Nauk SSSR*, **210**, 1–111 [in Russian].

Bannikov, A. F. (1985b) Fossil stromateoid fishes (Teleostei) of the Caucasus. *Paleontologicheskii Zhurnal*, **1985**, 77–83 [in Russian].

Bannikov, A. F. and Fedotov, V. F. (1989) On systematic position of perciforms of genera *Lednevia* and *Abadzekhia* from Maikopian of south of USSR. *Byulleten' Moskovskogo Obshchestva Ispytatelei Prirody (Geol.)*, **64**, 85–91 [in Russian].

Bannikov, A. F. and Tyler, J. C. (1992) *Caprovesposus* from the Oligocene of Russia: the pelagic acronurus presettlement stage of a surgeonfish (Teleostei: Acanthuridae). *Proceedings of the Biological Society of Washington*, **105**, 810–20.

Begle, D. P. (1992) Monophyly and relationships of the argentinoid fishes. *Copeia*, **1992**, 350–66.

Blot, J. (1969) Les poissons fossiles du Monte Bolca classés jusqu'ici dans les familles des Carangidae, Menidae, Ephippidae, Scatophagidae. *Studi e Richerche sui giacimenti Terziari di Bolca*, **1**, 1–525.

Blot, J. (1980) La faune ichthyologique des gisements du Monte Bolca (Province de Vérone, Italie). *Bulletin du Muséum National d'Histoire Naturelle, Paris*, (4), **2**, C, 339–96.

Blot, J. (1984a) Les Apodes fossiles du Monte Bolca. II. *Studi e Ricerche sui giacimenti Terziari di Bolca*, **4**, 61–238.

Blot, J. (1984b) Famille des Pterygocephalidae Blot, 1980. *Studi e Richerche sui giacimenti Terziari di Bolca*, **4**, 265–99.

Blot, J. and Tyler, J. C. (1991) New genera and species of fossil surgeon fishes and their relatives (Acanthuroidei, Teleostei) from the Eocene of Monte Bolca, Italy, with application of the Blot formula to both Recent and fossil forms. *Studi e Ricerche sui giacimenti Terziari di Bolca*, **6**, 13–92.

Böhlke, E. B. (ed.) (1989) *Orders Anguilliformes and Saccopharyngiformes (Fishes of the Western North Atlantic Part 9* Vol. 1), Sears Foundation for Marine Research, New Haven, Conn., xvii + 655 pp.

Bonde, N. (1987) *Moler – its Origin and its Fossils especially Fishes*. Skamol, Nykøbing Mors, Denmark, 52 pp.

Bonde, N. and Christensen, E. F. (1991) Det første Danekrae – en ganske lille fisk. *VARV*, **1991**, 1, 12–14.

Brito, I. M. (1979) *Bacias Sedimentares e Formacoes Pos-paleozoicas do Brasil*. Interciencia, Rio de Janeiro, 179 pp.

Cappetta, H. (1990) Tertiary dinosaurs in South America? A reply to some of Van Valen's assertions. *Historical Biology*, **3**, 265–8.

Cavender, T. M. (1986) Review of the fossil history of North American freshwater fishes, in *The Zoogeography of North American Freshwater Fishes* (eds C. H. Hocutt and E. O. Wiley), John Wiley and Sons, New York, pp. 699–724.

Chang, M.-M. and Chow, C.-C. (1986) Stratigraphic and geographic distribution of the late Mesozoic and Cenozoic fishes of China, in *Indo-Pacific Fish Biology* (eds T. Uyeno, R. Arai, T. Taniuchi *et al.*), Ichthyological Society of Japan, Tokyo, pp. 523–39.

Cione, A. L. (1986) Los peces continentales del Cenozoico de Argentina. *Actas del Congreso Argentino de Paleontologia y Bioestratigrafia*, **4**, 2, 101–6.

Cione, A. L. (1987) The late Cretaceous fauna of Los Alamitos, Patagonia, Argentina. Part II – the fishes. *Revista del Museo Argentino de Ciencias Naturales 'Bernardino Rivadivia' e Instituto Nacional de Investigacion de las Ciencias Naturales*, **3**, 111–20.

Cohen, D. M. (ed.) (1989) Papers on the systematics of gadiform fishes. *Science Series, Natural History Museum of Los Angeles County*, **32**, 1–262.

Danil'chenko, P. G. (1960) Bony fishes of the Maikop deposits of the Caucasus. *Trudy Paleontologicheskogo Instituta, Akademiya Nauk SSSR*, **78**, 1–208 [in Russian].

Danil'chenko, P. G. (1964) Teleostei, in *Osnovy Paleontologii*, **11**, Nauka, Moscow, pp. 396–484 [in Russian].

Danil'chenko, P. G. (1968) Ryby verkhnego paleotsena Turkmenii, in *Ocherko po Filogenii i Sistematike Iskopaemykh ryb i Bezcheliustnykh*. Nauka, Moscow, pp. 113–56 [in Russian].

Danil'chenko, P. G. (1975) Kostistye ryby, in *Stratigrafiya SSSR, Paleogenovaya Sistema* (eds V. A. Grossgeim and I. A. Korobkov), Nedra, Moskva, 524 pp. [in Russian].

Danil'chenko, P. G. (1980) Osnovnye kompleksy ikhtiofauny kajnozojckikh morej Tetisa. *Trudy Paleontologicheskogo Instituta*, **178A**, 175–83 [in Russian].

David, L. R. (1943) Miocene fishes of southern California. *Special Papers, Geological Society of America*, **43**, 1–193.

De Beaufort, L. F. (1926) On a collection of marine fishes from the Miocene of South Celebes. *Jaarboek van het Mijnwezen in Nederlandsch Oost-Indië*, **1925**, 113–48.

de Muizon, C., Gayet, M., Lavenu, A. *et al.* (1983) Late Cretaceous vertebrates, including mammals, from Tiupampa, southcentral Bolivia. *Geobios*, **16**, 747–53.

Estes, R. (1969) Studies on fossil phyllodont fishes: *Casierius*, a new genus of albulid from the Cretaceous of Europe and North America. *Eclogae Geologicae Helveticae*, **62**, 751–5.

Estes, R. and Hiatt, R. (1978) Studies on fossil phyllodont fishes: a new species of *Phyllodus* (Elopiformes, Albuloidea) from the late Cretaceous of Montana. *Paleobios*, **28**, 1–10.

Fierstine, H. L. and Applegate, S. P. (1974) *Xiphiorhynchus kimblalocki*, a new billfish from the Eocene of Mississippi, with remarks on the systematics of xiphioid fishes. *Bulletin of the Southern California Academy of Sciences*, **73**, 14–22.

Fink, S. V., Greenwood, P. H. and Fink, W. L. (1984) A critique of recent work on fossil ostariophysan fishes. *Copeia*, **1984**, 1033–41.

Fink, W. L. (1985) Phylogenetic interrelationships of the stomiid fishes (Teleostei: Stomiiformes). *Miscellaneous Publications of the Museum of Zoology, University of Michigan*, **171**, 1–127.

Gaudant, J. (1988) L'ichthyofaune éocène de Messel et du Geiseltal (Allemagne): Essai d'approche paléobiogéographique. *Courier Forschungsinstitut Senckenberg*, **107**, 355–67.

Gaudant, J. and Quayle, W. J. (1988) New palaeontological studies on the Chapelcorner Fish Bed (Upper Eocene, Isle of Wight). *Bulletin of the British Museum (Natural History) Geology Series*, **44**, 15–39.

Gayet, M. (1980a) Contribution à l'étude anatomique et systématique des poissons cénomaniens du Liban, anciennement placés dans les acanthoptérygiens. *Mémoires du Muséum National d'Histoire Naturelle, Paris* (n.s.), C, **44**, 1–149.

Gayet, M. (1980b) Recherches sur l'ichthyofaune cénomanienne des Monts de Judée: les 'acanthoptérygiens'. *Annales de Paléontologie (Vertébrés)*, **66**, 75–128.

Gayet, M. (1990) Nouveaux siluriformes du Maastrichtien de Tiupampa (Bolivie). *Comptes Rendus Hebdomadaire des Séances de l'Académie des Sciences, Paris*, (II), **310**, 867–72.

Gayet, M. and Brito, P. M. (1989) Ichthyofaune nouvelle du Crétacé supérieur du Groupe Bauru (états de Sao Paulo et Minas Gerais, Brésil). *Geobios*, **22**, 841–7.

Gayet, M., Rage, J.-C. and Rana, R. S. (1984) Nouvelles ichthyofaune et herpétofaune de Gitti Khadan, le plus ancien gisement connu du Deccan (Crétacé/Paléocène) à microvertébrés. *Mémoires de la Société Géologique de France, N.S.*, **147**, 55–65.

Goody, P. C. (1969) The relationships of certain Upper Cretaceous teleosts with special reference to the myctophoids. *Bulletin of the British Museum (Natural History) Geology Series, Supplement*, **7**, 1–255.

Grande, L. (1985) Recent and fossil clupeomorph fishes with materials for revision of the subgroups of clupeoids. *Bulletin of the American Museum of Natural History*, **181**, 231–372.

Grande, L. and Cavender, T. M. (1991) Description and phylogenetic reassessment of the monotypic Ostariostomidae (Teleostei). *Journal of Vertebrate Paleontology*, **11**, 405–16.

Greenwood, P. H. (1974) Review of the Cenozoic freshwater fish faunas in Africa. *Annals of the Geological Survey of Egypt*, **4**, 211–32.

Harland, W. B., Armstrong, R. L., Cox, A. V. *et al.* (1990) *A Geologic Time Scale*. Cambridge University Press, Cambridge, xvi + 263 pp.

Hora, S. L. (1939) On some fossil fish-scales from the Intertrappean beds at Deothan and Kheri, Central Province. *Records of the Geological Survey of India*, **73**, 267–94.

Huddleston, R. W. and Savoie, K. M. (1983) Teleostean otoliths from the late Cretaceous (Maestrichtian) Severn Formation of Maryland. *Proceedings of the Biological Society of Washington*, **96**, 658–63.

Jain, S. L. and Sahni, A. (1983) Some Upper Cretaceous vertebrates from central India and their palaeogeographic implications. *Special Publications, Palaeontological Society of India*, **2**, 66–83.

Johnson, C. D. (1986) Scombroid phylogeny: an alternative hypothesis. *Bulletin of Marine Science*, **39**, 1–41.

Johnson, G. D. (1984) Percoidei: development and relationships, in *Ontogeny and Systematics of Fishes* (eds H. G. Moser, W. J. Richards, D. M. Cohen *et al.*), Special Publication No. 1, American Society of Ichthyologists and Herpetologists, pp. 464–98.

Johnson, G. D. and Brothers, E. B. (1990) *Schindleria*, a paedomorphic goby. *Abstracts, Annual Meeting, American Society of Ichthyologists and Herpetologists*, **1990**, p. 105.

Johnson, G. D. and Fritzsche, R. A. (1989) *Graus nigra*, an omnivorous girellid, with a comparative osteology and comments on relationships of the Girellidae (Pisces: Perciformes). *Proceedings of the Academy of Natural Sciences of Philadelphia*, **141**, 1–27.

Kaufman, L. S. and Liem, K. F. (1982) Fishes of the suborder Labroidei (Pisces: Perciformes): phylogeny, ecology, and evolutionary significance. *Breviora*, **472**, 1–19.

Khare, S. K. (1976) Eocene fishes and turtles from the Subathu Formation, Beragua Coal Mine, Jammu and Kashmir. *Journal of the Palaeontological Society of India*, **18**, 36–43.

de Klasz, I. and Gageonnet, R. (1965) Biostratigraphie du bassin gabonais. *Mémoires du Bureau de Recherches Géologiques et Minières*, **32**, 277–303.

Marshall, L. G. (1989) The K–T boundary in South America: on which side is Tiupampa? *National Geographic Research*, **5**, 268–70.

McDowall, R. M. (1976) Notes on some *Galaxias* fossils from the Pliocene of New Zealand. *Journal of the Royal Society of New Zealand*, **6**, 17–22.

Medizza, F. and Sorbini, L. (1980) Il giacimento di Comeno (Carso), in *I Vertebrati Fossili Italiani* (ed. L. Sorbini, A. B. Bosi, L. Altichieri *et al.*), La Grafica, Verona, pp. 115–17.

Mourier, T., Bengtson, P., Bonhomme, M. *et al.* (1988) The Upper Cretaceous–Lower Tertiary marine to continental transition in the Bagua basin, northern Peru. *Newsletters on Stratigraphy*, **19**, 143–77.

Nair, K. K. (1945) On some fossil fish-scales from the lower Nimradic System of Kandana, Salt Range, Punjab. *Proceedings of the National Institute of Sciences of India*, **11**, 122–32.

Nelson, J. S. (1984) *Fishes of the World*, 2nd edn. John Wiley and Sons, New York, xv + 523 pp.

Nolf, D. (1985) Otolithi Piscium. *Handbook of Paleoichthyology*, **10**. Gustav Fischer Verlag, Stuttgart, 145 pp.

Nolf, D. (1988) Les otolithes de téléostéens éocènes d'Aquitaine et leur intérêt stratigraphique. *Mémoires, Académie Royale de Belgique, Classe des Sciences*, 4° (2), **19**, 2, 1–147.

Nolf, D. and Bajpai, S. (in press) Marine Middle Eocene fish otoliths from India and Java. *Bulletin, Institut Royal des Sciences Naturelles de Belgique* (Sciences de la Terre), in press.

Nolf, D. and Dockery, D. T. (1990) Fish otoliths from the Coffee Sand (Campanian) of northeastern Mississippi. *Mississippi Geology*, **10** (3), 1–14.

Parenti, L. R. (1981) A phylogenetic and biogeographic analysis of cyprinodontiform fishes (Teleostei, Atherinomorpha). *Bulletin of the American Museum of Natural History*, **168**, 335–557.

Patterson, C. (1967) Teleostei, in *The Fossil Record* (eds W. B. Harland *et al.*), Geological Society of London, London, pp. 654–66.

Patterson, C. (1987) Otoliths come of age. *Journal of Vertebrate Paleontology*, **7**, 346–8.

Patterson, C. (in press) Comments on Cretaceous eels. *Copeia*.

Patterson, C. and Rosen, D. E. (1977) Review of ichthyodectiform and other Mesozoic teleost fishes and the theory and practice of classifying fossils. *Bulletin of the American Museum of Natural History*, **158**, 81–172.

Pietsch, T. W. (1978) Evolutionary relationships of the sea moths (Teleostei: Pegasidae) with a classification of gasterosteiform families. *Copeia*, **1978**, 517–29.

Pietsch, T. W. and Zabetian, C. R. (1990) Osteology and interrelationships of the sand lances (Teleostei: Ammodytidae). *Copeia*, **1990**, 78–100.

Prasad, G. V. R. (1989) Vertebrate fauna from the Infra- and Intertrappean Beds of Andhra Pradesh: age implications. *Journal Geological Society of India*, **34**, 161–73.

Regan, C. T. (1913) The osteology and classification of the teleostean fishes of the Order Scleroparei. *Annals and Magazine of Natural History*, (8), **13**, 169–84.

Roberts, T. R. (1975) Characoid fish teeth from Miocene deposits in the Cuenca Basin, Ecuador. *Journal of Zoology, London*, **175**, 259–71.

Robins, C. R. (1989) The phylogenetic relationships of the anguilliform fishes, in *Orders Anguilliformes and Saccopharyngiformes* (*Fishes of the Western North Atlantic*, Volume 1), (ed. E. B. Böhlke), Sears Foundation for Marine Research, New Haven, Conn., pp. 9–23.

Rosen, D. E. (1973) Interrelationships of higher euteleostean fishes, in *Interrelationships of Fishes* (eds P. H. Greenwood, R. S. Miles and C. Patterson), Academic Press, London, pp. 397–513.

Rosen, D. E. and Parenti, L. R. (1981) Relationships of *Oryzias*, and the groups of atherinomorph fishes. *American Museum Novitates*, **2719**, 1–25.

Sahni, A. and Bajpai, S. (1988) Cretaceous–Tertiary boundary events: the fossil vertebrate, palaeomagnetic and radiometric evidence from Peninsular India. *Journal Geological Society of India*, **32**, 382–96.

Schultz, O. (1987) Taxonomische Neugruppierung der Überfamilie Xiphioidea (Pisces, Osteichthyes). *Annalen des Naturhistorischen Museums, Wien*, **89A**, 95–202.

Silva Santos, R. and de Figueiredo, F. J. (1987) Sobre um Saurocephalidae da Formaçao Gramame (Camada da Fosfato), Estada da Pernambuco. *Anais do Congresso Brasiliero de Paleontologia*, **10**, 7–19.

Smith, A. B. and Patterson, C. (1988) The influence of taxonomic method on the perception of patterns of evolution. *Evolutionary Biology*, **23**, 127–216.

Smith, M. L. (1981) Late Cenozoic fishes in the warm deserts of North America: a reinterpretation of desert adaptations, in *Fishes in North American Deserts* (eds R. J. Naiman and D. L. Soltz), John Wiley and Sons, New York, pp. 11–38.

Solun, V. I. and Travina, T. F. (1975) Yuzhnaya chast' Turanskoj plity (Tsentralnye Karakumy, Tuarkyr), Kopetdag i Badkhyz, in *Stratigrafiya SSSR, Paleogenovaya Sistema* (eds V. A. Grossgeim and I. A. Korobkov), Nedra, Moskva, pp. 231–59.

Stewart, J. D. (1988) The stratigraphic distribution of late Cretaceous *Protosphyraena* in Kansas and Alabama. *Fort Hays Studies*, (3), **10**, 80–94.

Su, D. (1985) On late Mesozoic fish fauna from Xinjiang (Sinkiang), China. *Memoirs, Institute of Vertebrate Palaeontology and Palaeoanthropology*, **17**, 61–136 [in Chinese, with English summary].

Sytchevskaya, E. K. (1986) Palaeogene freshwater fish fauna of the USSR and Mongolia. *Trudy Sovmestnaya Sovetsko-Mongol'skaya Paleontologicheskaya Ekspeditsiya*, **29**, 1–157 [in Russian, with English summary].

Taverne, L. (1982) Sur *Pattersonella formosa* (Traquair, R. H., 1911) et *Nybelinoides brevis* (Traquair, R. H., 1911), téléostéens salmoniformes argentinoides du Wealdien inférieur de Bernissart,

Belgique, précédemment attribués au genre *Letolepis* Agassiz, L., 1832. *Bullétin de l'Institut Royal des Sciences Naturelles de Belgique, Sciences de la Terre*, **54**, 3, 1–27.

Taverne, L. (1989) *Crossognathus* Pictet, 1858 du Crétacé inférieur de l'Europe et systématique, paléozoogéographie et biologie des Crossognathiformes nov. ord. du Crétacé et du Tértiaire. *Palaeontographica, Abteilung A*, **207**, 79–105.

Tyler, J. C. (1980) Osteology, phylogeny, and higher classification of the fishes of the order Plectognathi. *NOAA Technical Report, NFMS Circular*, **434**, 1–422.

Tyler, J. C. and Bannikov, A. F. (1992) A remarkable new genus of tetraodontiform fish with features of both balistids and ostraciids from the Eocene of Turkmenistan. *Smithsonian Contributions to Paleobiology*, **72**, 1–14.

Tyler, J. C., Johnson, G. D., Nakamura, I. *et al.* (1989) Morphology of *Luvarus imperialis* (Luvaridae), with a phylogenetic analysis of the Acanthuroidei (Pisces). *Smithsonian Contributions to Zoology*, **485**, 1–78.

Van Couvering, J. A. H. (1977) Early records of freshwater fishes in Africa. *Copeia*, **1977**, 163–6.

Van Couvering, J. A. H. (1982) Fossil cichlid fish of Africa. *Special Paper in Palaeontology*, **29**, 1–103.

Van Valen, L. (1988) Paleocene dinosaurs or Cretaceous ungulates in South America? *Evolutionary Monographs*, **10**, 1–79.

Weitzman, S. H. (1967) The origin of the stomiatoid fishes with comments on the classification of salmoniform fishes. *Copeia*, **1967**, 507–40.

Wilson, M. V. H. (1980) Oldest known *Esox* (Pisces: Esocidae), part of a new Paleocene teleost fauna from western Canada. *Canadian Journal of Earth Sciences*, **17**, 307–12.

Wilson, M. V. H., Brinkman, D. B. and Neuman, A. G. (1988) A Cretaceous fossil record for the Esocidae. *Abstracts Annual Meeting, American Society of Ichthyologists and Herpetologists*, **68**, p. 191.

Winterbottom, R. (1974) The familial phylogeny of the Tetraodontiformes (Acanthopterygii: Pisces) as evidenced by their comparative myology. *Smithsonian Contributions to Zoology*, **155**, i–iv + 1–201.

Woodward, A. S. (1901) *Catalogue of the Fossil Fishes in the British Museum (Natural History)*, Vol. 4. British Museum (Natural History), London, xxxix + 636 pp.

Zhou, J. (1990) The Cyprinidae fossils from Middle Miocene of Shanwang Basin. *Vertebrata Palasiatica*, **28**, 95–127 [in Chinese with English summary].

37

OSTEICHTHYES: SARCOPTERYGII

H.-P. Schultze

Actinopterygians and sarcopterygians are the two divisions of the osteichthyans. The sarcopterygians are a monophyletic taxon that include piscine representatives (Dipnoiformes, Onychodontida, Actinistia and Rhipidistia) and Tetrapoda (Schultze, 1987). The Tetrapoda, together with the Panderichthyida and the Osteolepidida, form the Choanata, a subdivision of the infraclass Rhipidistia. This classification reflects the position of the Tetrapoda and their closest relatives in a phylogenetic system. Of course, herpetologists, ornithologists and mammalogists ascribe to their respective groups a higher taxonomic rank than the phylogenetic position of these groups would require. Thus, it is customary to retain the divisions found in the traditional Linnean classification. However, I prefer not to elevate the rank of the piscine sarcopterygians based on their early branching points (Hennig, 1966). The piscine sarcopterygians form a relatively small group compared to the tetrapods.

In this chapter, only the piscine sarcopterygians are considered. The sequence of higher taxa (infraclass and order) follows Schultze (1987) with a sequential change in the position of Onychodontida and Actinistia. The Dipnoi are united with their sister group Diabolepidida *nov.* order in the infraclass Dipnoiformes. The Dipnoiformes do not include *Youngolepis* as proposed by Cloutier (1993). *Youngolepis* is the sister taxon to *Powichthys* within the Youngolepiformes. Within the Dipnoi, the families listed are monophyletic according to Marshall's 1987 strict consensus tree (his fig. 5); some families contain only one genus (Conchopomatidae, Uronemidae). Suborders of Dipnoi as proposed by Vorobyeva and Obruchev (1964) are all paraphyletic except for the Ceratodontoidei.

At present, an acceptable division of the Actinistia into families and orders is not available. Cladistic analyses (Forey, 1981, 1984, 1988; Cloutier, 1991a,b) give a sequence of genera without grouping into higher categories. Here I base the assignment of genera on Cloutier's analysis, so that the 49 nominal genera are divided into closely related groups. The families Miguashaiidae *nov.*, Diplocercidae, Hadronectoridae, Whiteiidae *nov.*, Mawsoniidae *nov.* and Latimeriidae are monophyletic. The other three families, Rhabdodermatidae, Laugiidae and Coelacanthidae, are paraphyletic, each including a section of sequentially arranged genera on the cladogram of Cloutier (1991a). Contrary to common usage, the family Laugiidae is not restricted to *Laugia*; here it also includes *Rhabdoderma madagascariensis, Synaptotylus, Coccoderma* and *Coelacanthus granulatus*. Above family level, there are no monophyletic units except for the suborder Latimerioidei; all other orders are paraphyletic with respect to their descendants.

The infraclass Rhipidistia comprises five orders and the tetrapods, and of the five orders three contain only one family. Representatives of the Youngolepiformes, Holoptychiida and Rhizodontida possess two external nasal openings, in contrast to Osteolepidida and Panderichthyida with one external and one internal opening, a true choana. Therefore, only the last two orders, together with the Tetrapoda, can be classified as Choanata. The division of Osteolepidida into families follows Worobjewa (1975) with the exception of two families raised to the rank of order (Rhizodontida and Panderichthyida) and for Lamprotolepididae, a family based on one species of one genus. The family Canowindridae has recently been placed within the Osteolepidida (Young *et al.*, 1992).

Acknowledgements – The author is indebted to R. Cloutier for review and criticism of this MS.

Class SARCOPTERYGII Romer, 1955

Infraclass DIPNOIFORMES Cloutier, 1993 (see Fig. 37.1)

Order DIABOLEPIDIDA *nov.*

F. DIABOLEPIDIDAE **fam. nov.** D. (LOK) Mar.

First and Last: *Diabolepis speratus* (Chang and Yu, 1984), Xitun Member, Cuifengshan Formation, Yunnan, China (Chang and Yu, 1984).

The Fossil Record 2. Edited by M. J. Benton. Published in 1993 by Chapman & Hall, London. ISBN 0 412 39380 8

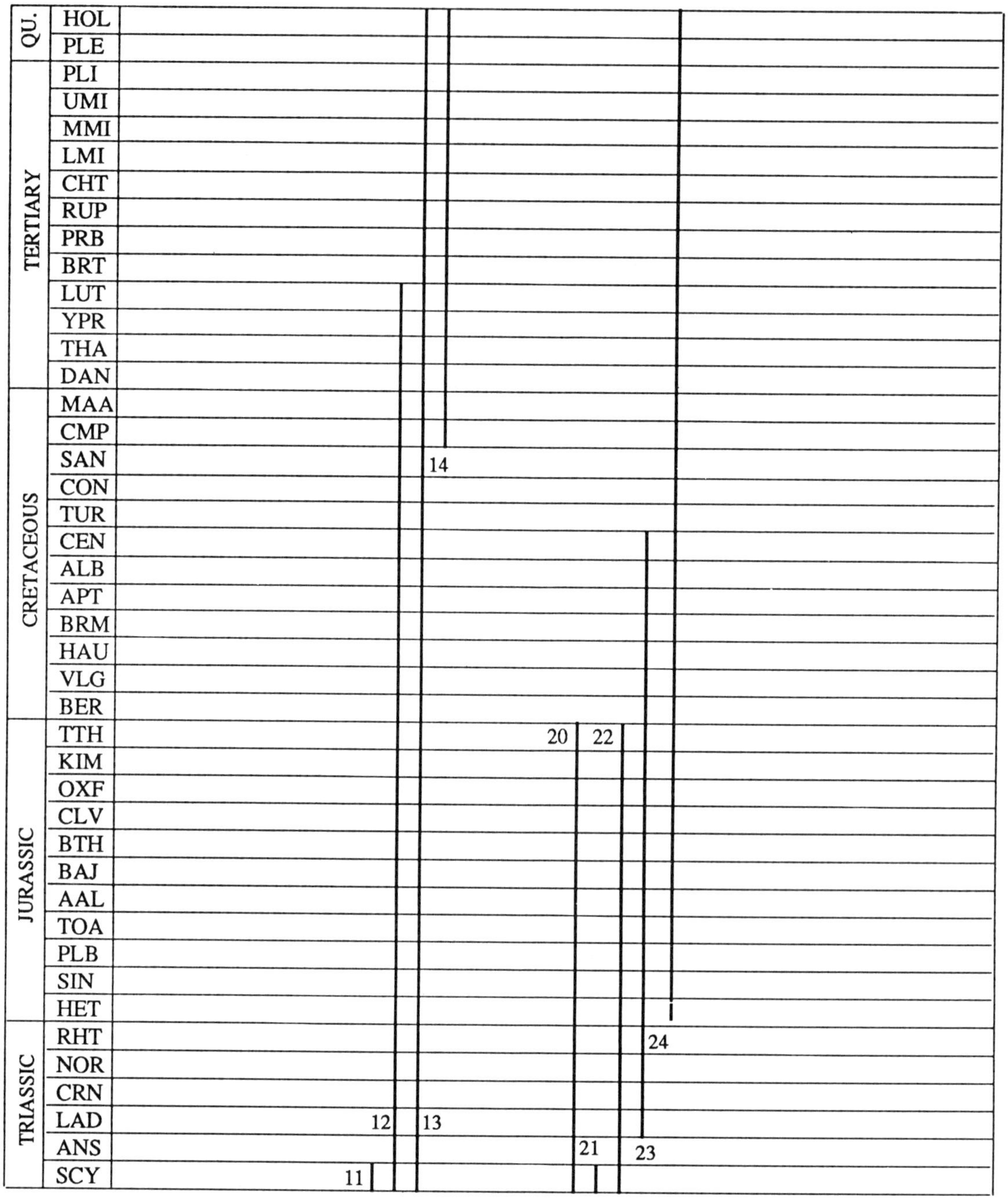

Fig. 37.1

Order DIPNOI Müller, 1844

The dipnoans can be traced back to the Gedinnian, based on unpublished records from southern China.

F. URANOLOPHIDAE Miles, 1977 D. (PRA) Mar.

First and Last: *Uranolophus wyomingensis* Denison, 1968, Water Canyon and Beartooth Butte Formation, Utah, Idaho and Wyoming, USA (Denison, 1968).

F. DIPNORHYNCHIDAE Berg, 1940 D. (PRA, EMS–?EIF) Mar.

First: *Speonysedrion lehmanni* (Westoll, 1949), Hunsrückschiefer, Pragian, Rhineland-Palatinate, Germany (Lehmann and Westoll, 1952).

Last: *Dipnorhynchus suessmilchi* (Etheridge, 1906), Warroo Limestone, Taemas region, New South Wales, Australia (Thomson and Campbell, 1971).

Suborder DIPTEROIDEI Vorobyeva and Obruchev, 1964

F. DIPTERIDAE Owen, 1846 D. (PRA–FAM) Mar./?FW

First: *Dipterus* sp., Water Canyon Formation, Idaho, USA (Denison, 1968).
Last: *Conchodus jerofejewi* Pander, 1858, Zagare Formation, Estonia, former USSR (Blieck *et al.*, 1988).
Intervening: EIF–FRS.

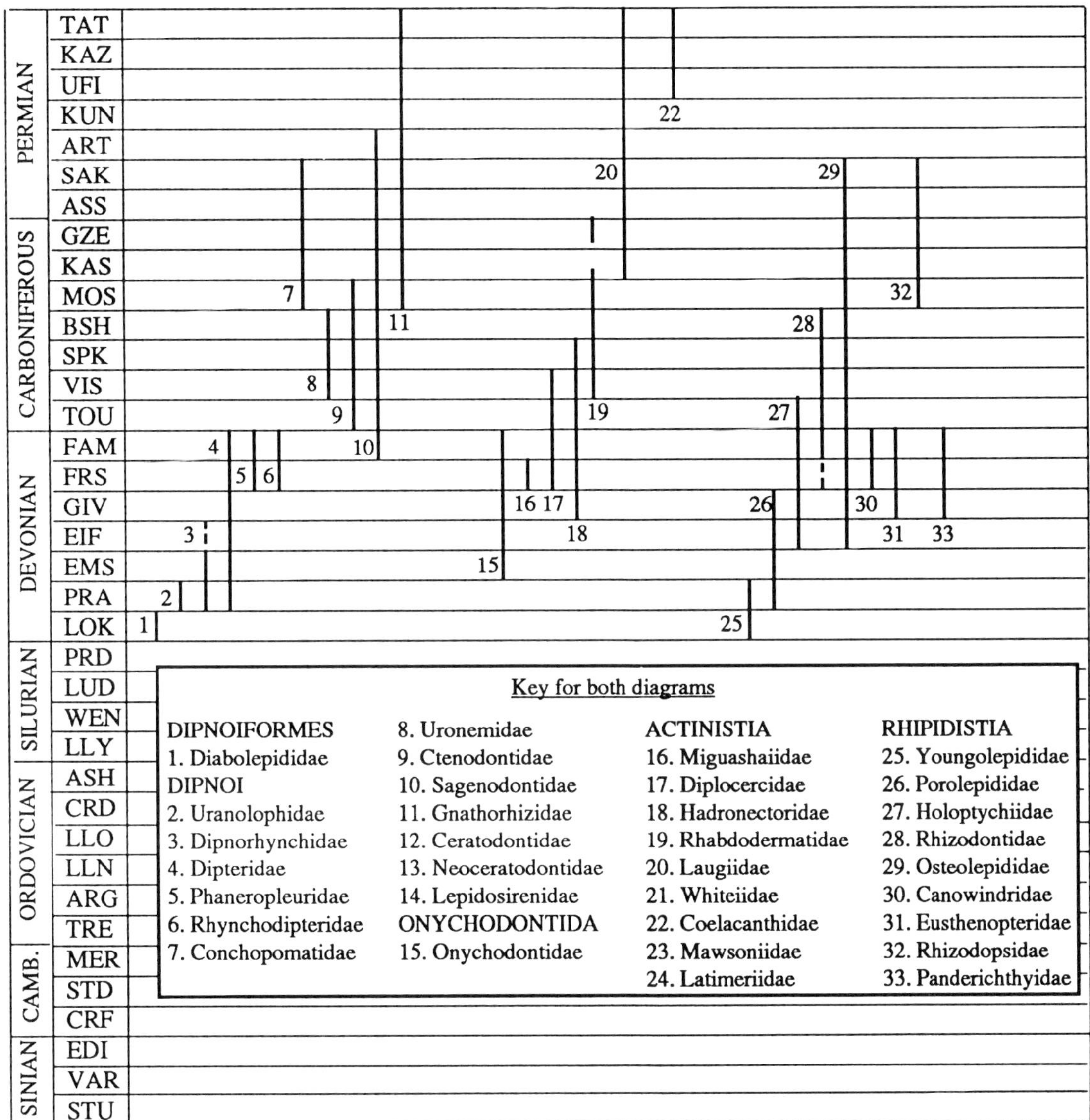

Fig. 37.1

F. PHANEROPLEURIDAE Huxley, 1861
D. (FRS, FAM) FW/Mar.

First: *Scaumenacia curta* (Whiteaves, 1881), Escuminac Formation, Quebec, Canada (Ørvig, 1957).
Last: *Oervigia nordica* Lehman, 1959, *Remigolepis* Series, East Greenland (Bendix-Almgreen, 1976).

F. RHYNCHODIPTERIDAE Berg, 1940
D. (FRS, FAM) FW/?Mar.

First: *Fleurantia denticulata* Graham-Smith and Westoll, 1937, Escuminac Formation, Quebec, Canada (Ørvig, 1957).
Last: *Soederberghia groenlandica* Lehman, 1959 and *Jarvikia arctica* Lehman, 1959, *Remigolepis* Series, East Greenland (Bendix-Almgreen, 1976).

F. CONCHOPOMATIDAE Berg, 1940
C. (MOS)–P. (SAK) FW/Mar.

First: *Conchopoma edesi* Denison, 1969, Carbondale Formation, Middle Pennsylvanian, Illinois, USA (Denison, 1969).
Last: *Conchopoma gadiforme* Kner, 1868, Lebach Group, Lower Rotliegend, Palatinate, Germany (Heidtke, 1986).

Suborder CTENODONTOIDEI Vorobyeva and Obruchev, 1964

F. URONEMIDAE Traquair, 1890 C. (VIS, BSH) FW/Mar.

First: *Ganopristodus lobatus* (Agassiz, 1844), Burdiehouse Limestone, Midlothian, Scotland, UK (Henrichsen, 1972).
Last: *Ganopristodus splendens* Traquair, 1881, Burghlee Ironstone, Namurian E_1, Midlothian, Scotland, UK (Henrichsen, 1972).
Intervening: SPK.

F. CTENODONTIDAE Woodward, 1891
C. (TOU–MOS) FW

First: *Ctenodus romeri* Thomson, 1965, lower Calciferous Sandstone Measures, Berwickshire, Scotland, UK (Thomson, 1965).

Last: *Ctenodus murchisoni* Ward, 1890, Phalen Coal, Sydney coalfield, Nova Scotia, Canada (Baird, 1978).
Intervening: VIS–BSH.

F. SAGENODONTIDAE Romer, 1966 D. (FAM)–P. (ART) FW/Mar.

First: *Proceratodus wagneri* (Newberry, 1889), Cleveland Shale, Cleveland, Ohio, USA (Romer and Smith, 1934).
Last: *Sagenodus periprion* (Cope, 1878), Vale Formation, Hennessey Group, Oklahoma, USA (Simpson, 1979).
Intervening: TOU–ART.

F. GNATHORHIZIDAE Miles, 1977 C. (MOS)–Tr. (OLK) FW/Mar.

First: *Palaeophichthys parvulus* Eastman, 1908, Francis Creek Shales, Carbondale Formation, Illinois, USA (Eastman, 1908).
Last: *Gnathorhiza bogdensis*, *G. otschevi* and *G. triassica* Minikh, 1977, Baskunchek Series, Mezen River Basin, former USSR (Minikh, 1977).
Intervening: KAS–ART, TAT, IND.

Suborder CERATODONTOIDEI Vorobyeva and Obruchev, 1964

F. CERATODONTIDAE Gill, 1872 Tr. (DIE)–T. (LUT) FW/Mar.

First: *Paraceratodus germaini* Lehman *et al.*, 1959, Middle Sakamena Group, Madagascar (Schaeffer and Mangus, 1976).
Last: *Ceratodus humei* Priem, 1914, Lutetian, In Farghas, North Africa (Martin, 1984).
Intervening: ANS–RHT, J.(l), CLV, TTH, NEO–CEN, CON, CMP, MAA, Pal.

F. NEOCERATODONTIDAE Miles, 1977 Tr. (SCY)–Rec. FW/Mar.

First: *Epiceratodus* sp., Lower Triassic, Orenburg Region (southern Urals), (Vorobyeva and Obruchev, 1964). **Extant**
Intervening: ALB, Mio., PLE.

F. LEPIDOSIRENIDAE Bonaparte, 1841 K. (CMP)–Rec. FW

First: *Protopterus ? regulatus* Schaal, 1984, Mut Formation, Campan, Egypt (Schaal, 1984). **Extant**
Intervening: MAA–PLE.

Infraclass ONYCHODONTIDA Andrews, 1973

F. ONYCHODONTIDAE Woodward, 1891 D. (EMS)–(FAM) Mar.

First: Struniiform crossopterygian, gen. *et* sp. indet. *Spirifer yassensis* Limestone, New South Wales, Australia (Ørvig, 1969). Earlier records refer probably to acanthodian teeth.
Last: *Onychodus dellei* Gross, 1942, Ketleri Beds, Latvia (Blieck *et al.*, 1988). Later records from the Carboniferous (C(u): Huene, 1943: lower Westphalian, Silesia; or Schultze, 1973: Desmoinesian (MOS), North America) may be teeth of rhizodonts (*Strepsodus*).
Intervening: EIF–FRS.

Infraclass ACTINISTIA Cope, 1871

Suborder DIPLOCERCIDOIDEI Berg, 1937

F. MIGUASHAIIDAE *nov.* D. (FRS) Mar.

First and Last: *Miguashaia bureaui* Schultze, 1973, Escuminac Formation, Quebec, Canada (Schultze, 1973).

F. DIPLOCERCIDAE Stensiö, 1922 D. (FRS)–C. (VIS) Mar.

First: *Diplocercides heiligenstockiensis* (Jessen, 1966), Oberer Plattenkalk (early FRS), Germany (Jessen, 1966).
Last: *Diplocercides davisi* (Moy-Thomas, 1937), Viséan P_1, Republic of Ireland and Scotland, UK (Moy-Thomas, 1937).

Suborder HADRONECTOROIDEI Lund and Lund, 1984

F. HADRONECTORIDAE Lund and Lund, 1984 D. (GIV)–C. (SPK) Mar.

First: *Euporosteus eifelianus* Jaekel, 1927, Crinoidenmergel, Germany (Gross, 1950).
Last: *Hadronector donbairdi* Lund and Lund, 1984, and *Polyosteorhynchus simplex* Lund and Lund, 1984, Bear Gulch Limestone, Montana, USA (Lund and Lund, 1984).
Intervening: FAM.

F. RHABDODERMATIDAE Berg, 1958 C. (VIS–MOS/?GZE) Mar./FW

First: *Coelacanthopsis curta* Traquair, 1901, and *Rhabdoderma ardrossense* Moy-Thomas, 1937, Calciferous Sandstone Series, Scotland, UK (Henrichsen, 1972).
Last: *Rhabdoderma elegans* (Newberry, 1856), Allegheny Group, Ohio, USA (Hook and Baird, 1986). May extend into GZE, if Coelacanthidae gen. *et* sp. indet. (Zidek, 1975) from the Wild Cow Formation, New Mexico, USA, and Coelacanthidae indet. (Schultze and Chorn, 1989) from the Bern Limestone Formation, Wabaunsee Group, Kansas, USA, belong to the family.
Intervening: SPK.

Suborder COELACANTHOIDEI Berg, 1937

F. LAUGIIDAE Berg, 1940 C. (KAS)–J. (TTH) Mar.

First: *Synaptotylus newelli* (Hibbard, 1933), Stanton Limestone, Lansing Group, Kansas, USA (Echols, 1963).
Last: *Coccoderma suevicum* Quenstedt, 1858, Lithographic Limestone, Bavaria, Germany (Reis, 1888).
Intervening: IND: GRI, DIE, KIM.

F. WHITEIIDAE **fam. nov.** Tr. (DIE) Mar.

First and Last: *Whiteia woodwardi* Moy-Thomas, 1935, Middle Sakamena Group, Madagascar (Schaeffer and Mangus, 1976).

F. COELACANTHIDAE Agassiz, 1843 P. (UFI)–J. (TTH) Mar.

First: *Coelacanthus granulatus* Agassiz, 1839, Kupferschiefer, Germany and England, UK (Schaumberg, 1978).
Last: *Heptanema willemoesi* (Vetter, 1881), Lithographic Limestone, Bavaria, Germany (Reis, 1888).
Intervening: KAZ, TAT, OLK, ANS, LAD, ?CRN, RHT.

Suborder LATIMERIOIDEI **subord. nov.**

F. MAWSONIIDAE **fam. nov.** Tr. (LAD)–K. (CEN) FW/Mar.

First: *Alcoveria brevis* Beltan, 1972, Upper Muschelkalk, Spain (Beltan, 1972).

Last: *Mawsonia libyca* Weiler, 1935, Baharija Formation, Egypt (Schaal, 1984).
Intervening: RHT, HET, OXF–ALB.

F. LATIMERIIDAE Berg, 1940 J. (? HET)–Rec. Mar.

First: *Holophagus barroviensis* (Woodward, 1890), lower Lias, England, UK (Woodward, 1890). **Extant**
Intervening: SIN, TTH, CEN, TUR, CMP.

Infraclass RHIPIDISTIA Cope, 1889

Order YOUNGOLEPIFORMES Gardiner, 1984

F. YOUNGOLEPIDIDAE Gardiner, 1984 D. (LOK–PRA) Mar.

First and Last: *Youngolepis praecursor* Zhang and Yu, 1981, Xitun Member of Cuifengshan Series, Yunnan, China (Chang, 1982); and *Powichthys thorsteinssoni* Jessen, 1975, Drake Bay Formation, Prince of Wales, Canadian Arctic (Jessen, 1980).

Order HOLOPTYCHIIDA Andrews, 1967

F. POROLEPIDIDAE Berg, 1940 D. (PRA–GIV) Mar./?FW

First: *Porolepis siegenensis* Gross, 1936, Siegenian, Rhineland, Germany, and *Porolepis spitsbergensis* Jarvik, 1937, *P. elongata* Jarvik, 1937 and *P. brevis* Jarvik, 1937, Wood Bay Series, Spitsbergen (Ørvig, 1969).
Last: *Porolepis posnaniensis* (Kade, 1858), Aruküla Beds, Estonia, former USSR (Blieck *et al.*, 1988).
Intervening: EMS, EIF.

F. HOLOPTYCHIIDAE Owen, 1860 D. (EIF)–C. (TOU) FW/Mar.

First: *Glyptolepis leptopterus* Agassiz, 1844, Tynet Burn Fish Beds, Morayshire, Scotland, UK (Westoll, 1951).
Last: *Holoptychius* sp., *Groenlandaspis* Series, East Greenland (Bendix-Almgreen, 1976).
Intervening: GIV, FRS.

Order RHIZODONTIDA Andrews and Westoll, 1970

F. RHIZODONTIDAE Traquair, 1881 D. (?FRS, FAM)–C. (BSH) Mar./FW

First: *Sauripterus taylori* Hall, 1843, Catskill Formation, Blossburgh, Pennsylvania, USA (Andrews and Westoll, 1970). *Propycnoctenion nephroides* Vorobyeva and Obrucheva, 1977, Kochajaska Beds (FRS), Kazahkstan, may belong here (Vorobyeva and Obrucheva, 1977).
Last: *Strepsodus sauroides* (Binney, 1941), upper Middle Coal Measures, Newsham, Northumberland, England, UK (Andrews and Westoll, 1970).
Intervening: TOU, VIS.

Order OSTEOLEPIDIDA Boulenger, 1901

F. OSTEOLEPIDIDAE Cope, 1889 D. (EIF)–P. (SAK) Mar./FW

First: *Thursius macrolepidotes* (Sedgwick and Murchison, 1829), Wick Beds, Caithness, Scotland, UK (Jarvik, 1948).
Last: *Ectosteorhachis nitidus* Cope, 1880, Belle Plains Formation, Wichita Group, Texas (Romer, 1958).
Intervening: GIV–FAM, VIS–MOS, GZE.

F. CANOWINDRIDAE Young *et al.*, 1992 D. (FRS, FAM) FW/?Mar.

First: *Beelarongia patrichae* Long, 1987, Avon River Group, Mount Howitt, Victoria, Australia (Long, 1987).
Last: *Canowindra grossi* Thomson, 1973, Mandagery Sandstone, Canowindra, New South Wales, Australia (Long, 1985).

F. EUSTHENOPTERIDAE Berg, 1955 D. (GIV/FAM) FW/Mar.

First: *Tristichopterus alatus* Egerton, 1861, John O'Groat's Sandstone, Caithness, Scotland, UK (Andrews and Westoll, 1970).
Last: *Eusthenodon waengsjoei* Jarvik, 1952, *Remigolepis* Series, East Greenland (Jarvik, 1952).
Intervening: FRS.

F. RHIZODOPSIDAE Berg, 1940 C. (MOS)–P. (SAK) FW/Mar.

First: *Rhizodopsis sauroides* (Williamson, 1837), Coal Measures, UK (Andrews and Westoll, 1970).
Last: Rhizodopsidae gen. *et* sp. indet, Lebach group, Lower Rotliegend, Palatine, Germany (Schultze and Heidtke, 1986).

Order PANDERICHTHYIDA Vorobyeva, 1989

F. PANDERICHTHYIDAE Vorobyeva and Lyarskaya, 1968 D. (GIV–FAM) Mar./?FW

First: *Panderichthys rhombolepis* (Gross, 1930), Gauja Beds, Baltic and Leningrad region (Blieck *et al.*, 1988).
Last: *Panderichthys bystrowi* Gross, 1941, Ketleri Formation, Latvia, former USSR (Blieck *et al.*, 1988).
Intervening: FRS.

REFERENCES

Andrews, S. M. and Westoll, T. S. (1970) The postcranial skeleton of rhipidistian fishes excluding *Eusthenopteron*. *Transactions of the Royal Society of Edinburgh*, **68**, 391–489.

Baird, D. (1978) Studies on Carboniferous freshwater fishes. *American Museum Novitates*, **2641**, 1–22.

Beltan, L. L. (1972) La faune ichthyologique du Muschelkalk de la Catalogne. *Memorias de la Real Academia de Ciencias y Artes de Barcelona*, **41**, 281–325.

Bendix-Almgreen, S. E. (1976) Palaeovertebrate faunas of Greenland, in *Geology of Greenland* (eds A. Escher and W. S. Watt), Groenlands Geologiske Undersoegelse, Odense, pp. 536–73.

Blieck, A., Mark-Kurik, E. and Märss, T. (1988) Biostratigraphical correlations between Siluro-Devonian invertebrate-dominated and vertebrate-dominated sequences: the East Baltic example, in *Paleontology, Paleoecology and Biostratigraphy* (*Devonian of the World. Volume III*) (eds N. J. McMillan, A. F. Embry and D. J. Glass), Canadian Society of Petroleum Geologists, Calgary, Alberta, pp. 579–87.

Chang, M.-M. (1982) *The Braincase of* Youngolepis, *a Lower Devonian Crossopterygian from Yunnan, South-western China*, 113 pp. Doctoral Dissertation, Department of Geology, University of Stockholm.

Chang, M.-M. and Yu, X. (1984) Structure and phylogenetic significance of *Diabolichthys speratus* gen. et sp. nov., a new dipnoan-like form from the Lower Devonian of Eastern Yunnan, China. *Proceedings of the Linnean Society of New South Wales*, **107**, 171–84.

Cloutier, R. (1991a) Patterns, trends, and rates of evolution within the Actinistia. *Environmental Biology of Fishes*, **32**, 23–58.

Cloutier, R. (1991b) Interrelationships of Palaeozoic actinistians:

patterns and trends, in *Early Vertebrates and Related Problems in Evolutionary Biology*, Science Press, Beijing, China, pp. 379–438.

Cloutier, R. (1993) Phylogenetic status, basal taxa, and interrelationships of lower sarcopterygian groups. *Zoological Journal of the Linnean Society*, London.

Denison, R. H. (1968) Early Devonian lungfishes from Wyoming, Utah, and Idaho. *Fieldiana, Geology*, **17**, 353–413.

Denison, R. H. (1969) New Pennsylvanian lungfishes from Illinois. *Fieldiana, Geology*, **12**, 193–211.

Eastman, C. R. (1908) Devonian Fishes of Iowa. *Iowa Geology Survey*, **18**, *Annual Report*, **1907**, 29–386.

Echols, J. (1963) A new genus of Pennsylvanian fish (Crossopterygii, Coelacanthiformes) from Kansas. *University of Kansas, Museum of Natural History Publications*, **12**, 475–501.

Forey, P. L. (1981) The coelacanth *Rhabdoderma* in the Carboniferous of the British Isles. *Palaeontology*, **24**, 203–29.

Forey, P. L. (1984) The coelacanth as a living fossil, in *Living Fossils* (eds N. Eldredge and S. M. Stanley), Springer, New York, pp. 166–9.

Forey, P. L. (1988) Golden jubilee for the coelacanth *Latimeria chalumnae*. *Nature* (*London*), **336**, 727–32.

Gross, W. (1950) Die paläontologische und stratigraphische Bedeutung der Wirbeltierfaunen des Old Reds und der marinen, altpaläozoischen Schichten. *Abhandlungen der Deutschen Akademie der Wissenschaften in Berlin, Mathematisch-naturwissenschaftliche Klasse, Jahrgang*, **1949**, 1–130.

Heidtke, U. (1986) Über Neufunde von *Conchopoma gadiforme* KNER (Dipnoi: Pisces). *Paläontologische Zeitschrift*, **60**, 299–312.

Hennig, W. (1966) *Phylogenetic Systematics*. University of Illinois Press, Urbana, Illinois, 263 pp.

Henrichsen, I. G. C. (1972) *A Catalogue of Fossil Vertebrates in the Royal Scottish Museum, Edinburgh. Part Three/Actinistia and Dipnoi*. Royal Scottish Museum Information Series, Geology, 3, v + 26 pp.

Hook, R. W. and Baird, D. (1986) The Diamond Coal Mine of Linton, Ohio and its Pennsylvanian-age vertebrates. *Journal of Vertebrate Paleontology*, **6**, 174–90.

Huene, E. von (1943) Ein *Onychodus* aus dem oberschlesischen Kohlengebirge. *Berichte der Reichsstelle für Bodenforschung in Berlin, Zweigstelle Wien*, **1943**, 123–31.

Jarvik, E. (1948) On the morphology and taxonomy of the Middle Devonian osteolepid fishes of Scotland. *Kungliga Svenska VetenskapsAkademiens Handlingar*, (3)**25**, 1–301.

Jarvik, E. (1952) On the fish-like tail in the ichthyostegid stegocephalian and a new crossopterygian from the Upper Devonian of East Greenland. *Meddelser om Grønland*, **114**, 1–90.

Jessen, H. (1966) Die Crossopterygier des Oberen Plattenkalkes (Devon) der Bergisch-Gladbach–Paffrather Mulde (Rheinisches Schiefergebirge) unter Berücksichtigung von amerikanischem und europäischem *Onychodus*-Material. *Arkiv för Zoologi, Stockholm*, (2)**18**, 305–89.

Jessen, H. (1980) Lower Devonian Porolepiformes from the Canadian Arctic with special reference to *Powichthys thorsteinssoni* JESSEN. *Palaeontographica, Abteilung A*, **167**, 180–214.

Kemp, A. and Molnar, R. E. (1981) *Neoceratodus forsteri* from the Lower Cretaceous of New South Wales, Australia. *Journal of Paleontology*, **55**, 211–17.

Lehmann, W. and Westoll, T. S. (1952) A primitive dipnoan fish from the Lower Devonian of Germany. *Proceedings of the Royal Society, London*, (B)**140**, 403–21.

Long, J. (1985) New information on the head and shoulder girdle of *Canowindra grossi* Thomson, from the Late Devonian Mandagery Sandstone, New South Wales. *Records of the Australian Museum*, **37**, 91–9.

Long, J. (1987) An unusual osteolepiform fish from the Late Devonian of Victoria, Australia. *Palaeontology*, **30**, 839–52.

Lund, R. and Lund, W. L. (1984) New genera and species of coelacanths from the Bear Gulch Limestone (Lower Carboniferous) of Montana (USA). *Geobios*, **17**, 237–44.

Marshall, C. R. (1987) Lungfish: Phylogeny and Parsimony, in *The Biology and Evolution of Lungfishes* (eds W. E. Bemis, W. W. Burggren and N. E. Kemp), Alan R. Liss, Inc., New York, pp. 151–62.

Martin, M. (1984) Deux Lepidosirenidae (Dipnoi) crétacés du Sahara, *Protopterus humei* (PRIEM) et *Protopterus protopteroides* (TABASTE). *Paläontologische Zeitschrift*, **58**, 265–77.

Minikh, M. G. (1977) *Triasovyye Dvoyakodyshashchiye Ryby Vostoka Yevropeyskoy Chasti SSSR* [Triassic Dipnoan Fishes from the East of the European USSR]. Saratov University Press, Saratov, USSR, 62 pp.

Moy-Thomas, J. A. (1937) The Carboniferous coelacanth fishes of Great Britain and Ireland. *Proceedings of the Zoological Society of London, (B)*, **3**, 383–415.

Ørvig, T. (1957) Remarks on the vertebrate fauna of the Lower Upper Devonian of Escuminac Bay, P.Q., Canada, with special reference to the Porolepiform Crossopterygians. *Arkiv för Zoologi, Stockholm*, (2)**10**, 367–426.

Ørvig, T. (1969) Vertebrates from the Wood Bay Group and the position of the Emsian–Eifelian boundary in the Devonian of Vestspitsbergen. *Lethaia*, **2**, 273–83.

Reis, O. M. (1888) Die Coelacanthinen, mit besonderer Berücksichtigung der im Weissen Jura Bayerns vorkommenden Gattungen. *Palaeontographica*, **35**, 1–96.

Romer, A. S. (1958) The Texas Permian Redbeds and their vertebrate fauna, in *Studies on Fossil Vertebrates* (ed. T. S. Westoll), Univ. London, Athlone Press, London, pp. 157–79.

Säve-Söderbergh, G. (1937) On *Rhynchodipterus elginensis* n. g., n. sp., representing a new group of Dipnoan-like *Choanata* from the Upper Devonian of East Greenland and Scotland. *Arkiv för Zoologi, Stockholm*, **29B**, 1–8.

Schaal, S. (1984) Oberkretazische Osteichthyes (Knochenfische) aus dem Bereich von Bahariya und Kharga, Ägypten, und ihre Aussagen zur Palökologie und Stratigraphie. *Berliner Geowissenschaftliche Abhandlungen, Reihe A*, **53**, 79 pp.

Schaeffer, B. and Mangus, M. (1976) An Early Triassic fish assemblage from British Columbia. *Bulletin of the American Museum of Natural History*, **156**, 515–63.

Schaumberg, G. (1978) Neubeschreibung von *Coelacanthus granulatus* Agassiz (Actinistia, Pisces) aus dem Kupferschiefer von Richelsdorf (Perm, W.-Deutschland). *Paläontologische Zeitschrift*, **52**, 169–97.

Schultze, H.-P. (1973) Crossopterygier mit heterozerker Schwanzflosse aus dem Oberdevon Kanadas, nebst einer Beschreibung von Onychodontida – Resten aus dem Mitteldevon Spaniens und aus dem Karbon der USA. *Palaeontographica, Abteilung A*, **143**, 188–208.

Schultze, H.-P. (1987) Dipnoans as Sarcopterygians, in *The Biology and Evolution of Lungfishes* (eds W. E. Bemis, W. W. Burggren and N. E. Kemp), Alan R. Liss, Inc., New York, pp. 39–74.

Schultze, H.-P. and Chorn, J. (1989) The Upper Pennsylvanian vertebrate fauna of Hamilton, Kansas, in *Regional Geology and Paleontology of Upper Paleozoic Hamilton Quarry Area in Southeastern Kansas*. Kansas Geological Survey Guidebook Series, **6** (for 1988), pp. 147–54.

Schultze, H.-P. and Heidtke, U. (1986) Rhizodopside Rhipidistia (Pisces) aus dem Perm der Pfalz (W.-Deutschland). *Neues Jahrbuch für Geologie und Paläontologie, Monatshefte*, **1986**, 165–70.

Simpson, L. C. (1979) Upper Gearyan and Lower Leonardian terrestrial vertebrate faunas of Oklahoma. *Oklahoma Geological Notes*, **39**, 3–21.

Thomson, K. S. (1965) On the relationships of certain Carboniferous Dipnoi; with descriptions of four new forms. *Proceedings of the Royal Society of Edinburgh, (B)*, **69**, 221–45.

Thomson, K. S. and Campbell, K. S. W. (1971) The structure and relationships of the primitive Devonian lungfish – *Dipnorhynchus sussmilchi* (Etheridge). *Peabody Museum of Natural History, Bulletin*, **38**, 1–109.

Vorobyeva, E. I. (1989) [Panderichthyida – new order of Paleozoic crossopterygian fishes (Rhipidistia)]. *Doklady Akademija Nauk SSSR*, Moscow, **306** (1), 188–9.

Vorobyeva, E. I. and Obruchev, D. W. (1964) Sarcopterygii, in *Osnovy Paleontologii. Agnatha, Pisces* (ed. D. Obruchev), Akad. Nauk SSR, Moscow, pp. 268–322.

Vorobyeva, E. I. and Obrucheva, Y. D. (1977) [Rhizodont crossopterygian fish (fam. Rhizodontidae) from middle Paleozoic deposits of the Asiatic part of the USSR], in *Ocherki po Filogenii i Sistematike Iskopayemykh Ryb i Beschelyustnykh* (ed. V. V. Menner), Izdtelysto Nauka, Moscow, pp. 89–97.

Westoll, T. S. (1951) The vertebrate-bearing strata of Scotland. *18th International Geological Congress, Report*, Part **XI**, 5–21.

Woodward, A. S. (1890) Notes on some ganoid fishes from the English Lower Lias. *Annals and Magazine of Natural History, London*, (6)**5**, 430–6.

Worobjewa, E. J. (1975) Formenvielfalt und Verwandtschaftsbeziehungen der Osteolepidida (Crossopterygii, Pisces). *Paläontologische Zeitschrift*, **49**, 44–5.

Young, G., Long, J. and Ritchie, A. (1992) Crossopterygian fishes from the Devonian of Antarctica: systematics, relationships and biogeographic significance. *Records of the Australian Museum*, suppl., **14**, 1–77.

Zidek, J. (1975) Some fishes of the Wild Cow Formation (Pennsylvanian), Manzanita Mountains, New Mexico. *New Mexico Bureau of Mines and Mineral Resources, Circular*, **135**, 22 pp.

38

AMPHIBIAN-GRADE TETRAPODA

A. R. Milner

The term 'Amphibia' is generally used for all tetrapods between the fish grade and the amniote clade of vertebrate organization. In this review I have divided these forms into four groups:

1. the Stem Tetrapoda, comprising the forms believed to be branches off the common stem of all living tetrapods;
2. the Aïstopoda, a group of early tetrapods, the relationships of which are enigmatic;
3. the Amphibia *sensu stricto*, comprising the living Lissamphibia and a series of stem groups referred to as Stem Amphibia;
4. the Stem Amniota.

This follows the general pattern put forward by Smithson (1985b) and Panchen and Smithson (1988), but differs in that those authors did not recognize any groups as stem tetrapods whereas I have placed two genera and four families in this category.

STRATIGRAPHICAL NOTES

There does not seem to be a precise correlation of the Moscow Basin late Carboniferous ages used by Harland *et al.* (1982) with the Westphalian/Stephanian stages widely used for western Europe. The following approximate correlations are used for the Moscow Basin ages which are asterisked (*) throughout the text.

NOG*	Stephanian C
KLA*	upper Stephanian B
DOR*	middle Stephanian B
CHV*	lower Stephanian B
KRE*	Stephanian A
MYA*	middle/upper Westphalian D
POD*	upper Westphalian C/lowermost Westphalian D
KSK*	uppermost Westphalian B/lower Westphalian C
VRK*	lower/middle Westphalian B
MEL*	upper Westphalian A
CHE*	lower Westphalian A

The stratigraphical terminology for the formations within the lower Texas Red Bed sequence follows that of Hentz (1989) and Hook (1989). Stratigraphical correlation of the European Lower Permian with the Pre-Ural standard is currently in a state of flux. In this work, most of the Autunian/lower Rotliegendes is equated with the Asselian, while the uppermost lower Rotliegendes is treated as indeterminate Asselian/Sakmarian. The Tambach and Wadern Schichten are treated as indeterminate Sakmarian/Artinskian.

STEM TETRAPODA

STEM TETRAPODA *incertae sedis*
D. (FRS/FAM) FW? (Fig. 38.1)

First and Last: *Metaxygnathus denticulus* Campbell and Bell, 1977, Cloghnan Shale, Forbes, New South Wales, Australia (Campbell and Bell, 1977).
Comment: This specimen is a mandible which has not been placed in a family but is diagnostically tetrapod (Clack, 1988), and therefore merits inclusion as probably the earliest recorded tetrapod fragment.

F. ICHTHYOSTEGIDAE Säve-Söderbergh, 1932
D. (FAM) FW

First: *Ichthyostega* spp., Aina Dal Formation, East Greenland.
Last: *Ichthyostega stensioi* Säve-Söderbergh, 1932 and *Ichthyostegopsis wimani* Säve-Söderbergh, 1932, Britta Dal Formation, East Greenland (Clack, 1988).

F. ACANTHOSTEGIDAE Jarvik, 1952
D. (FAM) FW

First: *Acanthostega gunnari* Jarvik, 1952, undescribed referred material from Aina Dal Formation, East Greenland (J. A. Clack, pers. comm.).
Last: *Acanthostega gunnari* Jarvik, 1952, Britta Dal Formation, East Greenland (Clack, 1988).

F. UNNAMED D. (FAM) FW/Terr.

First and Last: *Tulerpeton curtum* Lebedev, 1984, Suvorov, Tula Province, former USSR (Lebedev, 1984).
Comments: This specimen has not yet formed the basis of

The Fossil Record 2. Edited by M. J. Benton. Published in 1993 by Chapman & Hall, London. ISBN 0 412 39380 8

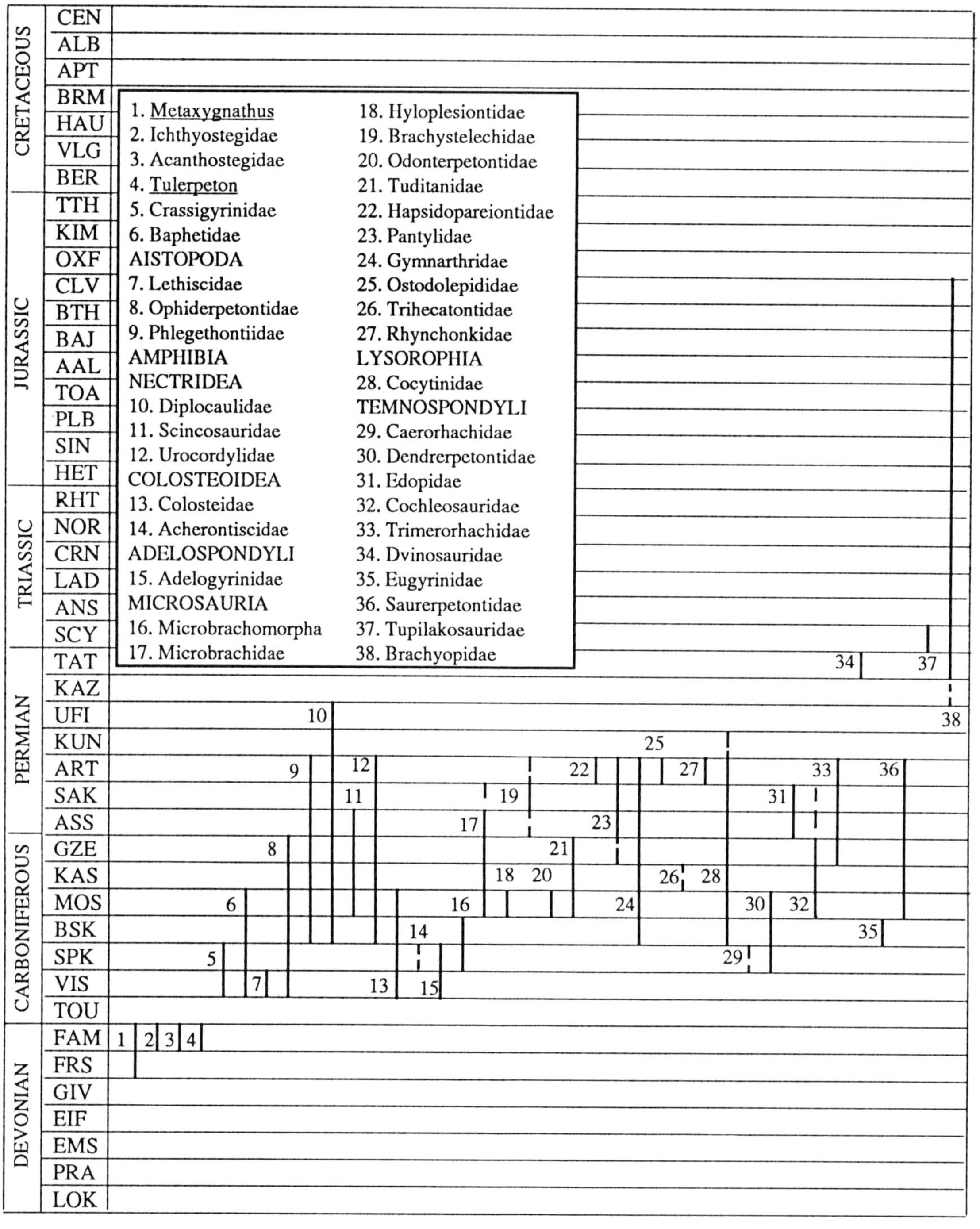
CRETACEOUS
JURASSIC
TRIASSIC
PERMIAN
CARBONIFEROUS
DEVONIAN
CEN
ALB
APT
BRM
HAU
VLG
BER
TTH
KIM
OXF
CLV
BTH
BAJ
AAL
TOA
PLB
SIN
HET
RHT
NOR
CRN
LAD
ANS
SCY
TAT
KAZ
UFI
KUN
ART
SAK
ASS
GZE
KAS
MOS
BSK
SPK
VIS
TOU
FAM
FRS
GIV
EIF
EMS
PRA
LOK
1. Metaxygnathus
2. Ichthyostegidae
3. Acanthostegidae
4. Tulerpeton
5. Crassigyrinidae
6. Baphetidae
AISTOPODA
7. Lethiscidae
8. Ophiderpetontidae
9. Phlegethontiidae
AMPHIBIA
NECTRIDEA
10. Diplocaulidae
11. Scincosauridae
12. Urocordylidae
COLOSTEOIDEA
13. Colosteidae
14. Acherontiscidae
ADELOSPONDYLI
15. Adelogyrinidae
MICROSAURIA
16. Microbrachomorpha
17. Microbrachidae
18. Hyloplesiontidae
19. Brachystelechidae
20. Odonterpetontidae
21. Tuditanidae
22. Hapsidopareiontidae
23. Pantylidae
24. Gymnarthridae
25. Ostodolepididae
26. Trihecatontidae
27. Rhynchonkidae
LYSOROPHIA
28. Cocytinidae
TEMNOSPONDYLI
29. Caerorhachidae
30. Dendrerpetontidae
31. Edopidae
32. Cochleosauridae
33. Trimerorhachidae
34. Dvinosauridae
35. Eugyrinidae
36. Saurerpetontidae
37. Tupilakosauridae
38. Brachyopidae

Fig. 38.1

a separate family but is clearly distinct from all other early tetrapods, and will ultimately merit a new family.

F. CRASSIGYRINIDAE Watson, 1929 C. (BRI–PND) FW

First: *Crassigyrinus scoticus* (Lydekker, 1890), Lower Limestone Group, Gilmerton, Scotland, UK.
Last: *Crassigyrinus scoticus* (Lydekker, 1890), referred material from Limestone Coal Group, Cowdenbeath, Fife, Scotland, UK (Panchen, 1985).

F. BAPHETIDAE Cope, 1875 (= LOXOMMATIDAE Lydekker, 1889) C. (BRI–MYA*) FW

First: *Loxomma allmanni* Huxley, 1862, Lower Limestone Group, Gilmerton, Scotland, UK.
Last: *Baphetes bohemicus* (Fritsch, 1885), Gaskohle, Kladno Formation, Nýřany, Czechoslovakia. (Beaumont, 1977).
Intervening: PND, CHE*, VRK*–KSK*.

TETRAPODA *INCERTAE SEDIS*

Order AÏSTOPODA Miall, 1875

The most recent general review of the aïstopods is that of Baird, 1964.

F. LETHISCIDAE Wellstead, 1982 C. (HLK) Terr./FW

First and Last: *Lethiscus stocki* Wellstead, 1982, Wardie Shales, lower Oil Shale Group, Wardie, Edinburgh, Scotland, UK.

F. OPHIDERPETONTIDAE Schwarz, 1908
C. (BRI–KLA*) Terr./FW

First: Unnamed material, East Kirkton Limestone, Bathgate, Scotland, UK (A. C. Milner, pers. comm.).
Last: *Ophiderpeton vicinum* Fritsch, 1880, Kounov Series, Slaný Formation, Kounov, Czechoslovakia.
Intervening: MEL*, KSK*–MYA*, KRE/CHV*.

F. PHLEGETHONTIIDAE Cope, 1875
C. (MEL*)–P. (ART) Terr.

First: *Phlegethontia phanerapha* Thayer, 1985, Black Prince Limestone, Swisshelm Mountains, Arizona, USA.
Last: *Sillerpeton permianum* Lund, 1978, upper Garber Formation, Fort Sill Fissures, Oklahoma, USA.
Intervening: POD*–MYA*.
Comments: A slightly older possible phlegethontiid is '*Dolichosoma*' *emersoni* Wright and Huxley, 1866, Jarrow Coal, Kilkenny, Republic of Ireland (= CHE*). This material is very poorly preserved and though often reported as a phlegethontiid, might equally be a juvenile ophiderpetontid (K. Bossy, cited by Lund, 1978, p. 54).

AMPHIBIA

The term 'Amphibia' is here taken to include the clade Lissamphibia and all those extinct amphibian groups which are more closely related to the lissamphibians than to the amniotes.

'STEM AMPHIBIA'

Order NECTRIDEA Miall, 1875

The major source for data on nectrideans is A. C. Milner (1980).

F. DIPLOCAULIDAE Cope, 1881
C. (CHE*)–P. (UFI) FW

First: *Keraterpeton galvani* Wright and Huxley, 1866, Jarrow Coal, Kilkenny, Republic of Ireland.
Last: *Diplocaulus parvus* Olson, 1972, Chickasha Formation, Blaine County, Oklahoma, USA.
Intervening: KSK*–MYA*, KRE/CHV*, ASS–KUN.

F. SCINCOSAURIDAE Jaekel, 1909
C. (MYA*)–P. (ASS) Terr.

First: Scincosaurid specimen mentioned by A. C. Milner (1980), Allegheny Group, Linton, Ohio, USA.
Last: *Sauravus cambrayi* Thevenin, 1910, Les Télots, Autun, France.
Intervening: CHV/KLA*.

F. UROCORDYLIDAE Lydekker, 1889
C. (CHE*)–P. (ART) FW

First: *Urocordylus wandesfordii* and *Lepterpeton dobbsii* Wright and Huxley, 1866, Jarrow Coal, Kilkenny, Republic of Ireland.
Last: *Crossotelos annulatus* Case, 1902, Wellington Formation, Orlando, Oklahoma, USA.
Intervening: MEL*–MYA*, CHV/KLA*, SAK.

Order COLOSTEOIDEA Tatarinov, 1964 *sensu* Hook, 1983

The most recent general source for data on the colosteoids is Hook (1983).

F. COLOSTEIDAE Cope, 1875
C. (ASB–MYA*) FW

First: *Pholidogaster pisciformis* Huxley, 1862 (referred specimen originally described as *Otocratia modesta* Watson, 1929; see Panchen, 1975), upper Oil Shale Group, Burdiehouse, Scotland, UK.
Last: *Colosteus scutellatus* (Newberry) Cope, 1871, Allegheny Group, Linton, Ohio, USA.
Intervening: BRI–PND, CHE*.

F. ACHERONTISCIDAE Carroll, 1969
C. (?PND) FW

First and Last: *Acherontiscus caledoniae* Carroll, 1969, ?Limestone Coal Group, ?Loanhead, Scotland, UK (locality and horizon uncertain).

Order ADELOSPONDYLI Watson, 1929

The most recent review of the adelospondyls is that of Andrews and Carroll (1991).

F. ADELOGYRINIDAE Werner, 1931
C. (ASB–PND) FW

First: *Palaeomolgophis scoticus* Brough and Brough, 1967, Pumpherston Oil Shale, Broxburn, Scotland, UK.
Last: *Adelogyrinus* sp., Shale above South Parrot Coal Seam, Niddrie, Edinburgh, Scotland, UK (Smithson, 1985a).
Intervening: None.

Order MICROSAURIA Dawson, 1863

The most recent review of the microsaurs is that of Carroll and Gaskill (1978).

Suborder MICROBRACHOMORPHA Carroll and Gaskill, 1978

MICROBRACHOMORPHA *incertae sedis*
C. (ALP/KIN) FW

First: *Utaherpeton franklini* Carroll *et al.*, 1991, Manning Canyon Shale Formation, Utah County, Utah, USA.

F. MICROBRACHIDAE Fritsch, 1883
C. (MYA*)–P. (ASS/SAK) FW

First: *Microbrachis pelikani* Fritsch, 1876, Gasköhle, Kladno Formation, Nýřany, Czechoslovakia.
Last: *Paramicrobrachis fritschi* Kuhn, 1959, Lauterecken-Odernheim Formation, Glan Group, Lebach, Saarpfalz region, Germany.
Intervening: None.
Comments: A specimen from the Jarrow Coal, Kilkenny, Republic of Ireland (= CHE*) described as ?microbrachid by Carroll and Gaskill (1978) is a poorly preserved member of the Cocytinidae.

F. HYLOPLESIONTIDAE Carroll and Gaskill, 1978
C. (MYA*) Terr./FW

First: *Hyloplesion longicostatum* Fritsch, 1876, Gasköhle, Kladno Formation (= upper MYA*), Nýřany, Czechoslovakia.
Last: *Hyloplesion longicostatum* Fritsch, 1876, Plattelkohle (= uppermost MYA*), Třemošná, Czechoslovakia.

F. BRACHYSTELECHIDAE Carroll and Gaskill, 1978 P. (ASS/SAK–?ART) Terr.

First: *Batropetes fritschi* (Geinitz and Deichmuller),

Niederhässlich-Schweinsdorf Beds, Niederhässlich, Germany (Carroll, 1991).
Last: *Quasicaecilia texana* Carroll, 1990, ?Arroyo Formation, ?Baylor County, Texas, USA (locality data uncertain).

F. ODONTERPETONTIDAE Carroll and Gaskill, 1978 C. (MYA*) FW

First and Last: *Odonterpeton triangulare*, Moodie, 1909, Allegheny Group, Linton, Ohio, USA.

Suborder TUDITANOMORPHA Carroll and Gaskill, 1978

F. TUDITANIDAE Cope, 1875 C. (VRK*-KLA*) Terr.

First: *Asaphestera intermedia* (Dawson, 1894), Joggins Formation, Joggins, Nova Scotia.
Last: *Boii crassidens* (Fritsch, 1876), Kounov Series, Slaný Formation, Kounov, Czechoslovakia.
Intervening: MYA*.

F. HAPSIDOPAREIONTIDAE Daly, 1973 P. (ART) Terr.

First: *Hapsidopareion lepton* Daly, 1973, Hennessy Formation, South Grandfield, Oklahoma, USA.
Last: *Llistrofus pricei* Carroll and Gaskill, 1978, upper Garber Formation, Fort Sill Fissures, Oklahoma, USA.
Comments: Following Schultze and Foreman (1981), *Ricnodon* is not included in the Hapsidopareiontidae but is considered to be indeterminate.

F. PANTYLIDAE Case, 1911 C./P. (NOG*/ASS)-P. (ART) Terr.

First: *Stegotretus agyrus* Berman *et al.*, 1988, Cutler Formation, New Mexico, USA.
Last: *Pantylus cordatus* Cope, 1881, Waggoner Ranch Formation, Mitchell Creek, Texas, USA.
Intervening: SAK.
Comments: Following Schultze and Foreman (1981) and Berman *et al.* (1988), *Trachystegos megalodon* Carroll, 1966, from the Joggins Formation (= VRK*), Joggins, Nova Scotia, Canada, is not included in the Pantylidae, but is taken to be indeterminate.

F. GYMNARTHRIDAE Case, 1910 C. (MEL*)-P. (ART) Terr.

First: *Elfridia bulbidens* Thayer, 1985, Black Prince Limestone, Swisshelm Mountains, Arizona, USA.
Last: *Euryodus primus* Olson, 1939, Arroyo Formation, Brushy Creek, Texas, USA.
Intervening: VRK*, MYA*, KLA*, SAK.

F. OSTODOLEPIDIDAE Romer, 1945 P. (ART) Terr.

First and Last: *Ostodolepis brevispinatus* Williston, 1913 and *Pelodosotis elongatum* Carroll and Gaskill, 1978, Arroyo Formation, Coffee Creek, Texas, USA.

F. TRIHECATONTIDAE Vaughn, 1972 C. (KRE/CHV*) Terr.

First and Last: *Trihecaton howardinus* Vaughn, 1972, Sangre de Cristo Formation, Colorado, USA.

F. RHYNCHONKIDAE Zanon, 1988 P. (ART) Terr.

First and Last: *Rhynchonkos stovalli* (Olson, 1970), Hennessy Formation, South Grandfield, Oklahoma, USA.

Order LYSOROPHIA Romer, 1930

The most recent review of the lysorophians is that of Wellstead (1991).

F. COCYTINIDAE Cope, 1875 (= LYSOROPHIDAE Williston, 1908); C. (CHE*)-P. (ART/KUN) FW.

First: Indeterminate cocytinid (Wellstead, 1991), Jarrow Coal, Kilkenny, Republic of Ireland.
Last: *Brachydectes elongatus* Wellstead, 1991, referred material (as *Lysorophus* cf. *L. tricarinatus* Cope), Choza Formation, Texas, USA (Olson, 1956).
Intervening: KSK*-MYA*, KRE/CHV*, ASS-ART.

Grade TEMNOSPONDYLI Zittel, 1888 (P.)

There is no comprehensive work on temnospondyls subsequent to that of Romer (1947), and much of the following data is compiled by the author from primary sources, and is a summary of the families used in the *Temnospondyli* volume of the *Handbuch der Palaeoherpetologie* (Milner, in prep.). Four useful reviews summarizing earlier literature are Cosgriff (1984: Lower Triassic assemblages), Kalandadze *et al.* (1968: Russian Permo-Triassic temnospondyls), Milner (1987: Westphalian assemblages) and Werneburg (1989: Lower Permian of Europe).

STEM TEMNOSPONDYLS

F. CAERORHACHIDAE Carroll, 1988 C. (?PND) Terr.

First and Last: *Caerorhachis bairdi* Holmes and Carroll, 1977, ?Limestone Coal Group, ?Loanhead, Scotland, UK (locality and horizon uncertain).

F. DENDRERPETONTIDAE Fritsch, 1889 C. (BRI-VRK*) Terr.

First: Unnamed material, East Kirkton Limestone, Bathgate, Scotland, UK (Milner *et al.*, 1986).
Last: *Dendrerpeton acadianum* Owen, 1853, Joggins Formation, Joggins, Nova Scotia, Canada.
Intervening: CHE*.

F. EDOPIDAE Romer, 1945 P. (ASS) FW/Terr.

First: *Edops craigi* Romer, 1936, Markley Formation, Texas, USA.
Last: *Edops craigi* Romer, 1936, Archer City Formation, Texas, USA.

F. COCHLEOSAURIDAE Broili, 1923 C. (VRK*-C./P. NOG*/SAK) FW

First: Unnamed vomers from Joggins Formation, Joggins, Nova Scotia, Canada (Carroll, 1967, fig. 7A, as *Dendrerpeton*).
Last: *Chenoprosopus milleri* Mehl, 1913, Cutler Formation, Arroyo de Agua, New Mexico, USA.
Intervening: MYA*.

THE TRIMERORHACHOID-BRACHYOPOID GROUP

F. TRIMERORHACHIDAE Cope, 1882 C. (KLA*)-P. (ART) FW

First: *Dawsonerpeton polydens* (Fritsch, 1879), Kounov Series, Slaný Formation, Kounov, Czechoslovakia, is probably a trimerorhachid as suggested by Romer (1945) and is currently being restudied by Dr M. Maňourová (Prague). The first certain trimerorhachid is: *Lafonius lehmani* Berman, 1973, Wild Cow Formation (= NOG*), Manzanita Mountains, New Mexico, USA.
Last: *Trimerorhachis rogersi* Olson, 1955, Choza Formation, Foard County, Texas, USA.
Intervening: NOG*–SAK.

F. DVINOSAURIDAE Watson, 1919 P. (TAT) FW

First and Last: All dvinosaurids are from the Zone IV (= TAT) of the former USSR (Shishkin, *in* Kalandadze *et al.*, 1968). They are as follows: *Dvinosaurus primus* Amalitsky, 1921, Kotlas, Arkhangel Province; *Dvinosaurus egregius* Shishkin, 1968, Vyazniki, Vladimir Province; and *Dvinosaurus purlensis* Shishkin, 1968, Purla, Gorkiy Province.

F. EUGYRINIDAE Watson, 1940 C. (MEL*) FW

First and Last: *Eugyrinus wildi* (Woodward, 1891), Lower Coal Measures, Trawden, Lancashire, England, UK.

F. SAURERPETONTIDAE Chase, 1965 C. (POD*)–P. (ART) FW

First: *Saurerpeton obtusum* (Cope, 1868), referred specimen from Carbondale Formation, Mazon Creek, Illinois, USA (Milner, 1982).
Last: *Isodectes megalops* Cope, 1896, referred specimen (*Acheloma casei* of Broili, 1913) from Arroyo Formation, Coffee Creek, Texas (Baird, *in* Welles and Estes, 1969).
Intervening: MYA*.

F. TUPILAKOSAURIDAE Kuhn, 1960 Tr. (DIE) ?Mar.

First and Last: *Tupilakosaurus heilmani* Nielsen, 1954, Wordy Creek Formation, Kap Stosch, East Greenland, and *Tupilakosaurus wetlugensis* Shishkin, 1961, Vetluga Series, Spasskoye, Vetluga Basin, former USSR.

F. BRACHYOPIDAE Lydekker, 1885 P. (KAZ/TAT)–J. (CLV) FW

First: *Bothriceps major* Woodward, 1909, Lithgow Coal Measures, Airly, New South Wales, Australia.
Last: *Ferganobatrachus riabinini* Nessov, 1990, Balabansay Formation, Tashkumyr, Kirgizistan, former USSR.
Intervening: GRI–ANS, NOR, BAJ/BTH.
Comments: Nessov (1990) described *Ferganobatrachus* as a 'capitosauroid' but the holotype clavicle appears to be brachyopid. Shishkin (1991) has described a brachyopid, *Gobiops desertus*, from the Upper Jurassic (stage uncertain) of Shara-Teg, Mongolia.

STEM STEREOSPONDYLS

F. ACTINODONTIDAE Lydekker, 1885 P. (ASS–UFI) FW (see Fig. 38.2)

First: *Sclerocephalus bavaricus* (Branca, 1886), Altenglan Formation, Kusel Group, Ohmbach, Saarpfalz region, Germany.
Last: *Syndiodosuchus tetricus* Konzhukova, 1956, Inta Formation, Komi, former USSR.
Intervening: SAK.

F. INTASUCHIDAE Konzhukova, 1953 P. (ASS/SAK–UFI) FW

First: *Cheliderpeton vranyi* Fritsch, 1877, Broumov Formation, Olivětín, Czechoslovakia.
Last: *Intasuchus silvicola* Konzhukova, 1956, Inta Formation, Komi, former USSR.
Intervening: None.

F. ARCHEGOSAURIDAE Lydekker, 1885 P. (ASS/SAK–TAT) FW

First: *Archegosaurus decheni* Goldfuss, 1847, Lauterecken-Odernheim Formation, Glan Group, Lebach, Saarpfalz region, Germany.
Last: *Platyoposaurus vjuschkovi* Gubin, 1989, lower Tatarian, Malaya Kinel' River, Orenburg Region, former USSR; and *Platyoposaurus* sp. (as *Platyops* sp.) Barberena and Daemon (1974), Rio do Rasto Formation, Parana State, Brazil.
Intervening: KAZ.

CROWN-GROUP STEREOSPONDYLS

F. RHINESUCHIDAE Watson, 1919 P. (KAZ–KAZ/TAT) FW

First: *Rhinesuchus whaitsi* Broom, 1908, *Tapinocephalus* Zone, Cape Province, South Africa.
Last: *Rhinesuchus africanus* (Lydekker, 1890), *R. capensis* Haughton, 1925, *R. broomianus* Huene, 1931 and *R. rubidgei* Broom, 1948, all *Cistecephalus* Zone, Cape Province, South Africa.

F. URANOCENTRODONTIDAE Romer, 1947 P. (TAT)–Tr. (GRI) FW

First: *Laccocephalus insperatus* Watson, 1919 and *Laccosaurus watsoni* Haughton, 1925, *Daptocephalus* Zone, South Africa.
Last: *Uranocentrodon senekalensis* (van Hoepen, 1911), *Lystrosaurus* Zone, Orange Free State, South Africa.

F. SCLEROTHORACIDAE Romer, 1947 Tr. (SPA) Terr.

First and Last: *Sclerothorax hypselonotus* Huene, 1932, middle Bunter, Queck, Oberhessen, Germany.

F. WETLUGASAURIDAE Säve-Söderbergh, 1935 Tr. (DIE–SPA) FW

First: *Wetlugasaurus groenlandicus* Säve-Söderbergh, 1935, Wordy Creek Formation, Kap Stosch, East Greenland.
Last: *Wetlugasaurus kzilsajensis* Ochev, 1966, Zone VI, Andreyevka, Orenburg Province, former USSR.
Intervening: None.
Comment: Most species of *Wetlugasaurus* occur in the Vetluga Series of the former USSR which is variously correlated with the Dienerian and the Spathian (Cosgriff, 1984, p. 38). The species quoted above are those of least equivocal stratigraphical position.

F. CAPITOSAURIDAE Watson, 1919 Tr. (GRI/DIE–NOR) FW

First: *Parotosuchus rewanensis* Warren, 1980, *P. gunganj* Warren, 1980, and *P. aliciae* Warren and Hutchinson, 1988b, Arcadia Formation, Queensland, Australia (= GRI/DIE); and *Parotosuchus madagascariensis* (Lehman, 1961),

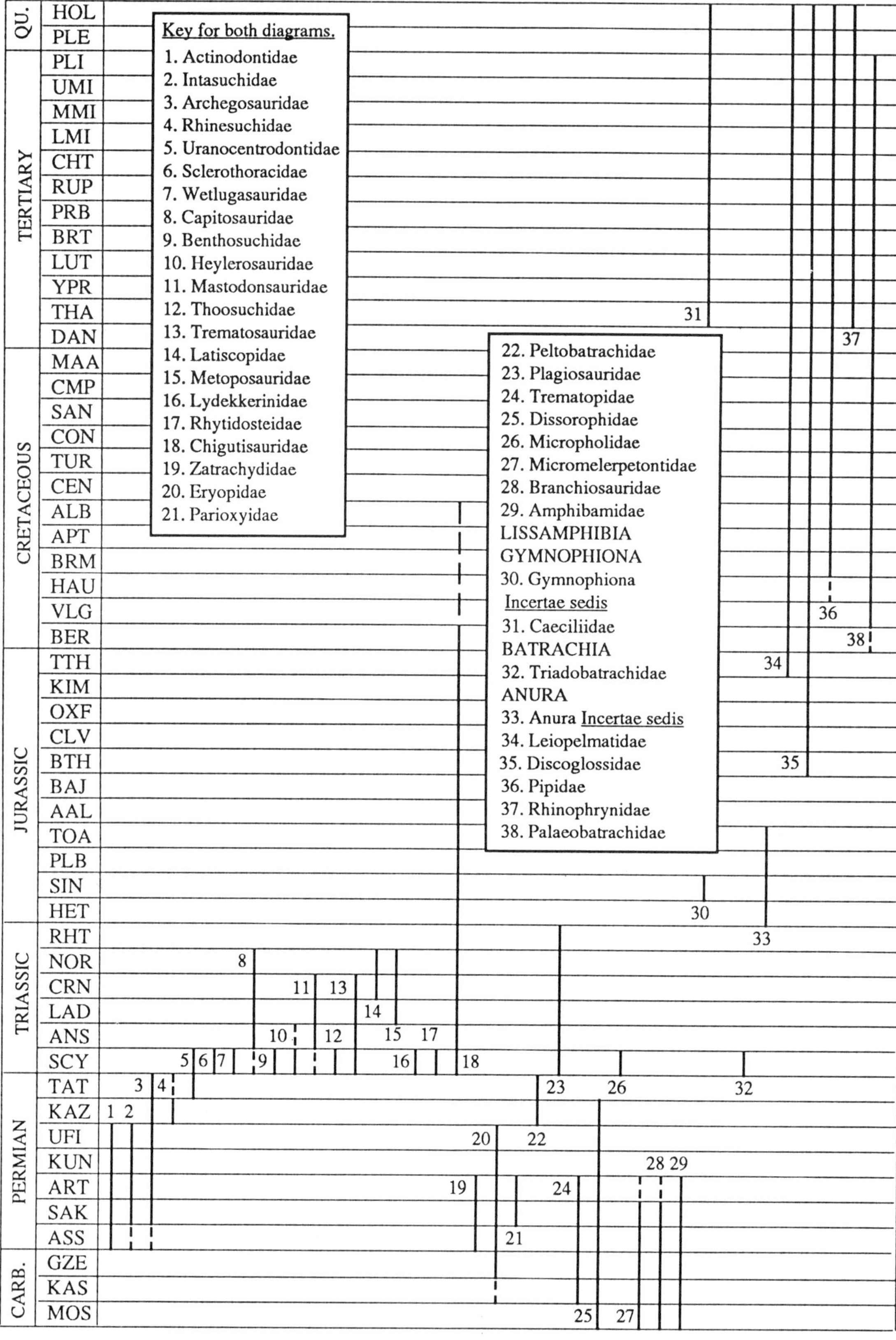

Fig. 38.2

Sakamena Formation, Madagascar (= DIE) (Warren and Hutchinson, 1988a).

Last: *Cyclotosaurus carinidens* (Jaekel, 1914) (including *Hercynosaurus carinidens* Jaekel, 1914, *Hemprichisaurus keuperianus* Kuhn, 1939 and *Cyclotosaurus hemprichi* Kuhn, 1942), Knollenmergel, Halberstadt, Germany. This is the latest certain capitosaurid material.

Intervening: SMI–CRN.

Comments: Nessov (1988) reported capitosaurid? material from the Jurassic (CLV) of former Soviet Central Asia, and later described it as *Ferganobatrachus riabinini* (Nessov, 1990). The material is certainly temnospondyl but is not critically diagnostic. The type clavicle appears to belong to a brachyopid.

F. BENTHOSUCHIDAE Efremov, 1940 Tr. (DIE–SPA) FW

First: All species of *Benthosuchus* derive from Vetluga Series horizons in the European part of the former USSR. They include: *B. sushkini* (Efremov, 1937), Sharzhenga River, Vologda Province; *B. bashkiricus* Ochev, 1967, Muraptalovo, Bashkirian, former USSR; *B. korobkovi* Ivakhnenko, 1972, Tikhvinskoye, Yaroslavl Province; *B. uralensis* (Ochev, 1966), Blumental Site, Orenburg Province.
Last: *Benthosphenus lozovskii* Shishkin, 1979, Zone VI?, Russkiy Island near Vladivostock, former USSR.
Intervening: None.
Comment: As used here, the family Benthosuchidae (restricted to *Benthosuchus* and *Benthosphenus*) is probably paraphyletic with respect to the Heylerosauridae, Mastodonsauridae, Thoosuchidae, Trematosauridae, Latiscopidae and Metoposauridae.

F. HEYLEROSAURIDAE Shishkin, 1980 Tr. (SMI–SPA/ANS) FW

First: *Odenwaldia heidelbergensis* Morales and Kamphausen, 1984, upper Konglomerat Horizon, Middle Buntsandstein, near Heidelberg, Germany.
Last: *Eocyclotosaurus woschmidti* Ortlam, 1970, Lower Röt, upper Buntsandstein, Rötfelden, Schwarzwald, Germany.
Comment: As used here, the Heylerosauridae is probably paraphyletic with respect to the Mastodonsauridae. This family was revised, as the Heylerosaurinae, by Kamphausen (1989).

F. MASTODONSAURIDAE Lydekker, 1885 Tr. (SPA/ANS–CRN) FW

First: *Mastodonsaurus cappelensis* Wepfer, 1923, upper Buntsandstein, Kappel, Baden-Württemberg, Germany.
Last: *Mastodonsaurus keuperinus* Fraas, 1889, Schilfsandstein, Stuttgart, Germany.
Intervening: LAD.

F. THOOSUCHIDAE Getmanov, 1982 Tr. (DIE/SPA) FW

First and Last: All thoosuchids derive from the Zone V horizons of European part of the former USSR. They include: *Thoosuchus jakovlevi* (Riabinin, 1926), Vetluga Series, Yaroslavl Province and *Trematotegmen otschevi* Getmanov, 1982, Kyzyl-Say Formation, Orenburg Province.
Comments: As used here, the Thoosuchidae is probably paraphyletic with respect to the Trematosauridae. The same stratigraphical ambiguities apply here as for the Wetlugasauridae (q.v.).

F. TREMATOSAURIDAE Watson, 1919 Tr. (GRI–CRN) Mar./FW

First: *Gonioglyptus longirostris* Huxley, 1865, *Glyptognathus fragilis* Lydekker, 1882, and *Panchetosaurus panchetensis* Tripathi, 1969, are all fragments of long-snouted trematosaurid from the Panchet Formation, Bengal, India (Cosgriff, 1984).
Last: *Hyperokynodon keuperinus* Plieninger, 1852, Schilfsandstein, Heilbronn, Baden-Württemberg, Germany (Hellrung, 1987).
Intervening: DIE–SPA.

F. LATISCOPIDAE Wilson, 1948 Tr. (CRN–NOR) FW

First: *Almasaurus habbazi* Dutuit, 1972, t.5 Beds, Argana Formation, Alma, Morocco.
Last: *Latiscopus disjunctus* Wilson, 1948, upper Dockum Group, Texas, USA.

F. METOPOSAURIDAE Watson, 1919 Tr. (LAD–NOR) FW

First: *Trigonosternum latum* Schmidt, 1931, Lettenkeuper, Kolleda, Germany. This taxon is a *nomen vanum*, but is nevertheless a metoposaurid. Also from the Lettenkeuper is an undescribed skull from Eschenau, Baden-Württemburg, Germany (Morales, 1988).
Last: *Metoposaurus stuttgartensis* Fraas, 1913, Lehrbergstufe, Stuttgart-Sonnenberg, Germany.
Intervening: CRN.
Comments: Murry (1987) suggested that metoposaurid material from the Redonda Formation, Apache Canyon, New Mexico, USA, was post-Carnian and might prove to be Hettangian. Hunt and Lucas (1990) have argued that this site is late Triassic (Norian) in age.

F. LYDEKKERINIDAE Watson, 1919 Tr. (GRI–SMI) FW

First: *Lydekkerina huxleyi* (Lydekker, 1889), and *Limnoiketes paludinatans* Parrington, 1948, both *Lystrosaurus* Zone, Orange Free State, South Africa.
Last: *Chomatobatrachus halei* Cosgriff, 1974, Knocklofty Formation, Tasmania, Australia.
Intervening: DIE.
Comments: As used here and elsewhere, the Lydekkerinidae (comprising *Lydekkerina, Limnoiketes, Luzocephalus* and *Chomatobatrachus*) is probably paraphyletic with respect to the Rhytidosteidae.

F. RHYTIDOSTEIDAE Huene, 1920 Tr. (GRI–SPA) FW/Mar./Terr.

First: *Pneumatostega potamica* Cosgriff and Zawiskie, 1979, *Lystrosaurus* Zone, Cape Province, South Africa; and *Indobrachyops panchetensis* Huene and Sahni, 1958, Panchet Formation, Bengal, India.
Last: *Laidleria gracilis* Kitching, 1957, *Cynognathus* Zone, Cape Province, South Africa.
Intervening: DIE–SMI.

F. CHIGUTISAURIDAE Rusconi, 1951 Tr. (GRI/DIE)–K. (BER/ALB) FW

First: *Keratobrachyops australis* Warren, 1981, Arcadia Formation, Queensland, Australia.
Last: Unnamed material, Strzelecki Formation, Victoria, Australia (Jupp and Warren, 1986).
Intervening: CRN, NOR, PLB/TOA.

THE ERYOPID–DISSOROPHOID GROUP

F. ZATRACHYDIDAE Cope, 1882
P. (ASS–ART) FW

First: *Acanthostomatops vorax* (Credner, 1883), Niederhässlich-Schweinsdorf Beds, Niederhässlich, Germany.
Last: *Zatrachys* sp., Arroyo Formation, Coffee Creek, Texas, USA (Murry and Johnson, 1987).
Intervening: SAK.
Comments: Most North American zatrachydids are either ART or of uncertain stratigraphical position. The late record is one of relatively secure stratigraphical position.

F. ERYOPIDAE Cope, 1882
C. (KRE/NOG*)–P. (UFI) FW/Terr.

First: *Eryops avinoffi* (Romer, 1952), Conemaugh Group, Pennsylvania, USA.
Last: *Clamorosaurus nocturnus* Gubin, 1983, Sheshminskiy Horizon, and *C. borealis* Gubin, 1983, Inta Formation, Komi, former USSR.
Intervening: ASS–ART.

F. PARIOXYIDAE Moustafa, 1955
P. (SAK–ART) Terr.

First: *Parioxys bolli* Carroll, 1964, Archer City Formation, Archer City Bone bed, Texas, USA.
Last: *Parioxys ferricolus* Cope, 1878, Petrolia Formation, Texas, USA.

F. PELTOBATRACHIDAE Kuhn, 1960
P. (KAZ/TAT) Terr.

First and Last: *Peltobatrachus pustulatus* Panchen, 1959, Kawinga Formation, Ruhuhu Valley, Tanzania.

F. PLAGIOSAURIDAE Jaekel, 1914
Tr. (GRI/DIE–RHT) FW

First: *Plagiobatrachus australis* Warren, 1985, Arcadia Formation, Queensland, Australia.
Last: *Gerrothorax rhaeticus*, Nilsson, 1934, Rhaetic, Bjuv, Scania, Sweden.
Intervening: SPA–NOR.

F. TREMATOPIDAE Williston, 1910
C. (DOR*)–P. (ART) Terr.

First: *Actiobates peabodyi* Eaton, 1973, Stanton Formation, Garnett, Kansas, USA (Berman *et al.*, 1987a).
Last: *Acheloma cumminsi* Cope, 1882, Arroyo Formation, Coffee Creek, Texas, USA.
Intervening: NOG*–SAK.

F. DISSOROPHIDAE Boulenger, 1902
C. (MYA*)–P. (KAZ) Terr.

First: *Stegops newberryi* (Cope, 1875), Allegheny Group, Linton, Ohio, USA (Hook and Baird, 1986). Although widely cited as an early zatrachydid, undescribed material shows this to be an armoured dissorophid (R. Hook, pers. comm). The earliest dissorophid described as such is: *Astreptorhachis ohioensis* Vaughn, 1971, Conemaugh Group (= KRE/NOG*), Wayne Township, Ohio, USA.
Last: *Kamacops acervalis* Gubin, 1980, Belebey Formation, Perm and Orenburg Provinces, former USSR.
Intervening: KRE/NOG*–UFI.

F. MICROPHOLIDAE Watson, 1919
Tr. (GRI) Terr./FW

First and Last: *Micropholis stowi* Huxley, 1859, *Lystrosaurus* Zone, Cape Province, South Africa; and *Lapillopsis nanus* Warren and Hutchinson, 1990, Arcadia Formation, Queensland, Australia.

F. MICROMELERPETONTIDAE Boy, 1972
C. (MYA*)–P. (SAK/ART) FW

First: *Limnogyrinus elegans* (Fritsch, 1881), Gaskohle, Kladno Formation, Nýřany, Czechoslovakia.
Last: *Eimerosaurus* (*'Tersomius'*) *graumanni* (Boy, 1980), lower Wadern Beds, Nahe Group, Sobernheim, Saarpfalz Region, Germany.
Intervening: KLA/NOG*–SAK.

F. BRANCHIOSAURIDAE Fritsch, 1883
C. (MYA*)–P. (SAK/ART) FW

First: *Branchiosaurus salamandroides* Fritsch, 1876, Gaskohle, Kladno Formation, Nýřany, Czechoslovakia.
Last: *Melanerpeton* sp. (as *Branchiosaurus* sp.), Prosečné Formation, Horní Kalná, Czechoslovakia (Maňourová 1981).
Intervening: KLA*–SAK.

F. AMPHIBAMIDAE Moodie, 1910
C. (POD*)–P. (ART) Terr./FW

First: *Amphibamus grandiceps* Cope, 1865, Carbondale Formation, Mazon Creek, Illinois, USA.
Last: *Doleserpeton annectens* Bolt, 1969, Upper Garber Formation, Fort Sill fissures, Oklahoma, USA.
Intervening: MYA*, ASS–SAK.

Crown division LISSAMPHIBIA Haeckel, 1866

The most recent major summary of data on lissamphibians is that of Duellman and Trueb (1986) and the recent families used in the following section are those used in that work. The higher-group taxonomy and the indeterminate position of the Albanerpetontidae follows Milner (1988).

Superorder GYMNOPHIONA Rafinesque, 1814

GYMNOPHIONA *incertae sedis* J. (SIN)

First: Undescribed material, family undesignated, Kayenta Formation, Arizona, USA (Jenkins and Walsh, 1990).
Comments: The earliest described specimen is an unnamed vertebra, indeterminate at family level, unnamed formation (=MAA), Tiupampa, Bolivia (Rage, 1986).

F. RHINOTREMATIDAE Nussbaum, 1977
Extant Terr.

F. ICHTHYOPHIDAE Taylor, 1968 **Extant** Terr.

F. URAEOTYPHLIDAE Nussbaum, 1979
Extant Terr.

F. SCOLECOMORPHIDAE Taylor, 1969
Extant Terr.

F. CAECILIIDAE Gray, 1825 T. (THA)–Rec. Terr.

First: *Apodops pricei* Estes and Wake, 1972, Itaborai Fissures, Rio de Janeiro State, Brazil (Estes, 1981). **Extant**
Intervening: None.

F. TYPHLONECTIDAE Taylor, 1968 **Extant** FW

Superorder BATRACHIA Brongniart, 1799

Order SALIENTIA Laurenti, 1768

There is no recent review of all fossil frogs. The latest major source of data on Mesozoic and some Cenozoic fossil frogs is Estes and Reig (1973).

F. TRIADOBATRACHIDAE Kuhn, 1962
Tr. (DIE) FW.

First and Last: *Triadobatrachus massinoti* (Piveteau, 1936), marine shales, Betsieka, Madagascar (Rage and Roček, 1989).

Crown order ANURA Rafinesque, 1815

F. ANURA *incertae sedis* J. (HET/TOA)

First: *Vieraella herbstii* Reig, 1961, Roca Blanca Formation, Santa Cruz Province, Argentina.
Comment: This is the earliest fossil with the suite of anuran skeletal characters, and merits inclusion for this reason even though it is not placed in a family. Estes and Reig (1973) placed it in the Leiopelmatidae (under the synonym Ascaphidae) and most subsequent authors have followed them. However, that position was entirely based on primitive characters and *Vieraella* could equally be a 'stem frog' with no immediate relationship to any living family. For this reason it is left as Anura *incertae sedis*.

Suborder DISCOGLOSSOIDEI Sokol, 1977

F. LEIOPELMATIDAE Mivart, 1869
J. (TTH)–Rec. FW

First: *Notobatrachus degiustoi* Reig, 1955, La Matilde Formation, Santa Cruz Province, Argentina. **Extant**
Intervening: None.

F. DISCOGLOSSIDAE Guenther, 1859
J. (BTH)–Rec. FW/Terr.

First: *Eodiscoglossus oxoniensis* Evans *et al.*, 1990, Forest Marble Formation, Kirtlington, Oxfordshire, England, UK. **Extant**
Intervening: TTH, BER/VAL, BRM/APT, MAA–DAN, PRB–RUP, AQT–SRV, ZAN–PLE.

Suborder RANOIDEI Sokol, 1977

Superfamily PIPOIDEA Fitzinger, 1843

F. PIPIDAE Gray, 1825 K. (HAU?)–Rec. FW

First: *Shomronella jordanica* Estes *et al.*, 1978, Tayasir Formation, Shomron, Israel. **Extant**
Intervening: BRM, SAN/CMP, THA, RUP, BUR/LAN.

F. RHINOPHRYNIDAE Guenther, 1859
T. (THA)–Rec. Terr.

First: *Eorhinophrynus* sp., Fort Union Formation, Princeton Quarry, Wyoming, USA (Estes, 1975). **Extant**
Intervening: LUT, RUP, PLE.

F. PALAEOBATRACHIDAE Cope, 1865
K. (BER/VAL)–T. (PIA) FW

First: *Neusibatrachus wilferti* Seiffert, 1972, Montsech, Lerida Province, Spain.
Last: *Pliobatrachus langhae* Fejervary, 1917, from Betfia, Romania, and referred material from Rebielice Krolewskie and Weze, Poland.
Intervening: MAA, THA, LUT, PRB–RUP, AQT, BUR/SRV, ZAN.

Superfamily PELOBATOIDEA Bolkay, 1919

F. PELOBATIDAE Bonaparte, 1850
J. (TTH)–Rec. Terr. (see Fig. 38.3)

First: Undetermined pelobatid, Morrison Formation, Como Bluff, Wyoming, USA (Evans and Milner, in press).
Intervening: ALB, CON–SAN, MAA, LUT, PRB–RUP, AQT, BUR/SRV, ZAN–PLE. **Extant**
Comment: The 'pelobatid' material reported from the Liassic Kota Formation of India (Yadagiri, 1986) is not diagnostically anuran.

F. PELODYTIDAE Bonaparte, 1850
T. (PRB)–Rec. Terr.

First: cf. *Pelodytes*, Phosphorites de Quercy, La Bretou, France (Rage, 1988). **Extant**
Intervening: SRV/TOR, PLE.
Comment: *Propelodytes wagneri* Weitzel, 1938, Geiseltal, Germany (= LUT), was described as the earliest pelodytid. Duellman and Trueb (1986) still cite it as such, but other workers have questioned its determinacy (e.g. Rage, 1988).

Superfamily HYLOIDEA Wied, 1856

F. MYOBATRACHIDAE Schlegel, 1850
T. (BUR/SRV)–Rec. FW/Terr.

First: *Limnodynastes archeri* Tyler, 1982, Etadunna Formation, mid-Miocene, South Australia, Australia. **Extant**
Intervening: PLE.
Comment: *Indobatrachus pusillus* Noble, 1931, from the Lower Eocene of India was described as a myobatrachid and is often cited as such, but has most recently been reinterpreted as a leptodactylid (Špinar and Hodrová, 1985).

F. HELEOPHRYNIDAE Noble, 1931 **Extant** FW

F. SOOGLOSSIDAE Noble, 1931 **Extant** Terr.

F. LEPTODACTYLIDAE Werner, 1896
K. (SAN/MAA)–Rec. FW/Terr.

First: Undescribed material reported from the Marila Formation, Minas Gerais State, Brazil (Baez, 1985). **Extant**
Intervening: THA, YPR/LUT, PRB–AQT, BUR/SRV, ZAN, PLE.

F. BUFONIDAE Gray, 1825
T. (THA)–Rec. Terr./FW

First: *Bufo* sp. Undescribed material from the Itaborai Fissures, Rio de Janeiro Province, Brazil (Estes and Reig, 1973). **Extant**
Intervening: AQT–TOR, ZAN–PLE.

F. BRACHYCEPHALIDAE Guenther, 1859
Extant Terr.

F. RHINODERMATIDAE Bonaparte, 1850
Extant Terr.

F. PSEUDIDAE Fitzinger, 1843 **Extant** FW/Terr.

F. HYLIDAE Gray, 1825 T. (THA)–Rec. Terr./FW

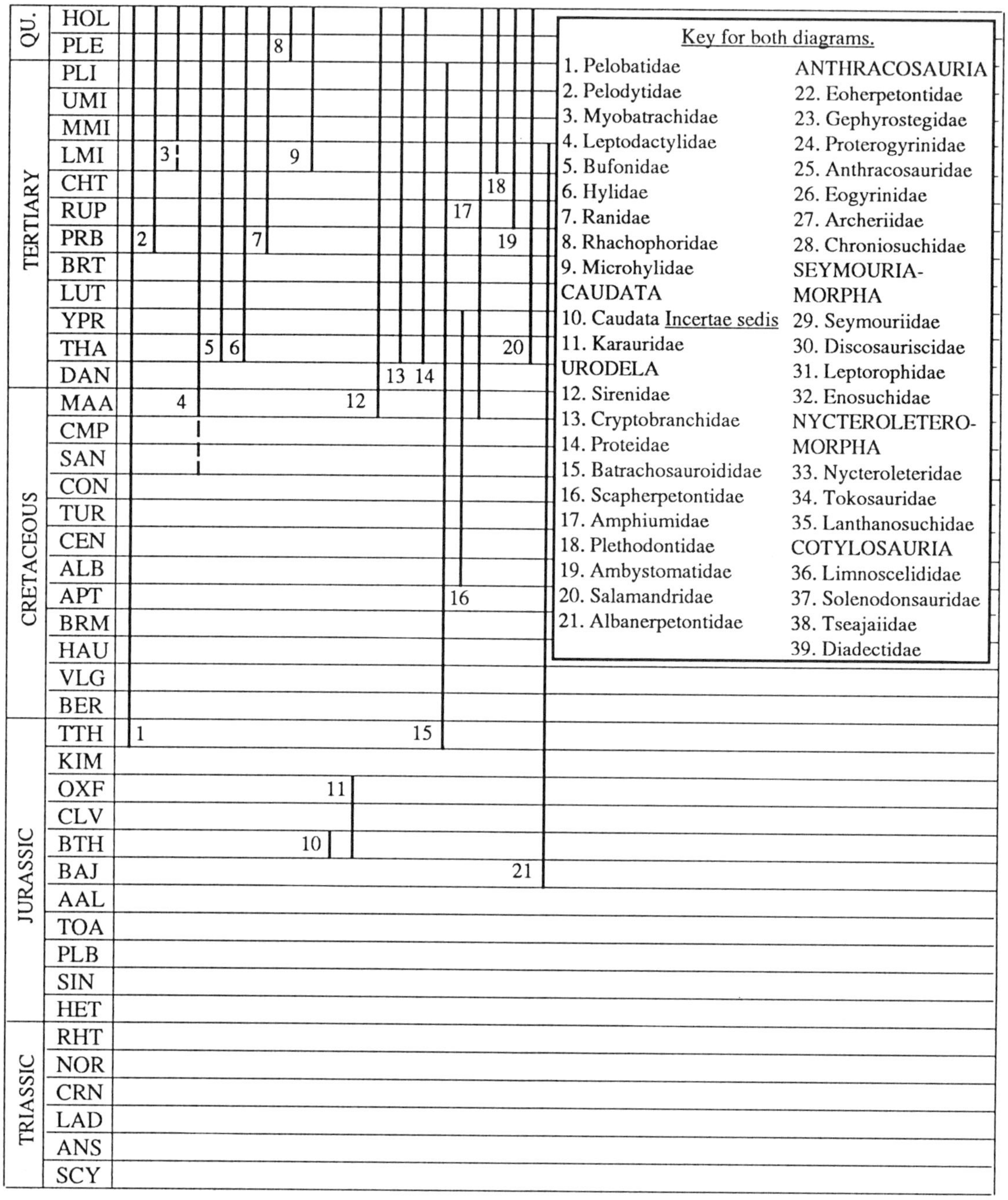
QU.
HOL
PLE
TERTIARY
PLI
UMI
MMI
LMI
CHT
RUP
PRB
BRT
LUT
YPR
THA
DAN
CRETACEOUS
MAA
CMP
SAN
CON
TUR
CEN
ALB
APT
BRM
HAU
VLG
BER
JURASSIC
TTH
KIM
OXF
CLV
BTH
BAJ
AAL
TOA
PLB
SIN
HET
TRIASSIC
RHT
NOR
CRN
LAD
ANS
SCY
Key for both diagrams.
1. Pelobatidae
2. Pelodytidae
3. Myobatrachidae
4. Leptodactylidae
5. Bufonidae
6. Hylidae
7. Ranidae
8. Rhachophoridae
9. Microhylidae
CAUDATA
10. Caudata Incertae sedis
11. Karauridae
URODELA
12. Sirenidae
13. Cryptobranchidae
14. Proteidae
15. Batrachosauroididae
16. Scapherpetontidae
17. Amphiumidae
18. Plethodontidae
19. Ambystomatidae
20. Salamandridae
21. Albanerpetontidae
ANTHRACOSAURIA
22. Eoherpetontidae
23. Gephyrostegidae
24. Proterogyrinidae
25. Anthracosauridae
26. Eogyrinidae
27. Archeriidae
28. Chroniosuchidae
SEYMOURIA-
MORPHA
29. Seymouriidae
30. Discosauriscidae
31. Leptorophidae
32. Enosuchidae
NYCTEROLETERO-
MORPHA
33. Nycteroleteridae
34. Tokosauridae
35. Lanthanosuchidae
COTYLOSAURIA
36. Limnoscelididae
37. Solenodonsauridae
38. Tseajaiidae
39. Diadectidae

Fig. 38.3

First: Undescribed material from the Itaborai Fissures, Rio de Janeiro Province, Brazil (Estes and Reig, 1973). **Extant**
Intervening: RUP, AQT, ZAN, PLE.

F. CENTROLENIDAE Taylor, 1951 **Extant** Terr.

F. DENDROBATIDAE Cope, 1865 **Extant** Terr.

Superfamily RANOIDEA Fitzinger, 1826

F. RANIDAE Gray, 1825 T. (PRB)–Rec. FW/Terr.

First: Unnamed ranid material from Robiac-equivalent horizon, Grisolles, France (Rage, 1984). **Extant**
Intervening: RUP, AQT–BUR/LAN, ZAN–PLE.

F. HYPEROLIIDAE Laurent, 1943 **Extant** Terr.

F. RHACHOPHORIDAE Hoffman, 1932
Q. (PLE)–Rec. Terr.

First: *Rhachophorus* cf. *R. japonicus* and *Rhachophorus* cf. *R. viridis*, Pleistocene, Ryukyu Islands, Japan (Hasegawa, 1980). **Extant**

Superfamily MICROHYLOIDEA Duellman, 1975

F. MICROHYLIDAE Guenther, 1859
T. (AQT)–Rec. Terr.

First: *Gastrophryne* cf. *G. carolinensis*, Thomas Farm deposits, Gilchrist County, Florida, USA (Holman, 1965). **Extant**
Intervening: PLE.

Order CAUDATA Oppel, 1811

The most recent major review of fossil caudates is Estes (1981). Later work by Nessov is summarized in Nessov (1988).

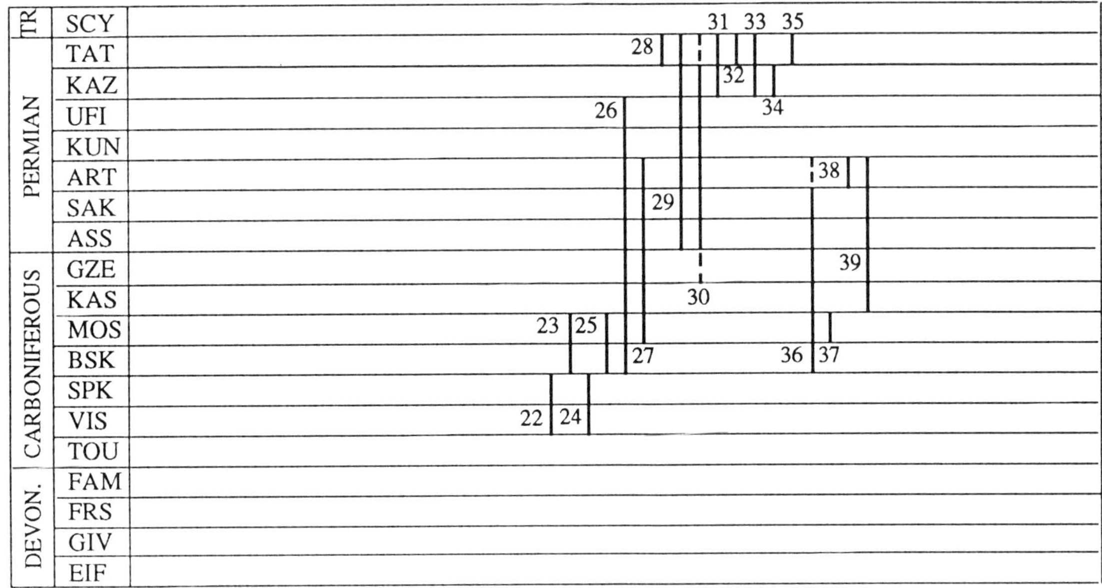

Fig. 38.3

F. CAUDATA *incertae sedis* J. (BTH) FW

First: *Marmorerpeton kermacki* and *M. freemani* Evans *et al.*, 1988, Forest Marble Formation, Kirtlington, Oxfordshire, England, UK.

F. KARAURIDAE Ivakhnenko, 1978 J. (BTH–KIM) FW?

First: *Kokartus honorarius* Nessov, 1988, black and red shales, Kizylsu River, Kirghizia, former USSR.
Last: *Karaurus sharovi* Ivakhnenko, 1978, Karabastau Formation, Kazakhstan, former USSR.
Intervening: None.

Crown order URODELA Latreille, 1825

F. SIRENIDAE Gray, 1825 K. (CMP)–Rec. FW

First: *Habrosaurus dilatus* Gilmore, 1928, referred material from the Judith River Formation, Montana and the 'Mesaverde' Formation, Wyoming, USA (Estes, 1981). **Extant**
Intervening: MAA–THA, LUT, AQT–LAN, MES/ZAN, PLE.
Comment: The 'sirenid' material reported from the Liassic Kota Formation of India (Yadagiri, 1986) is not diagnostically urodelan.

F. HYNOBIIDAE Cope, 1859 **Extant** FW

F. CRYPTOBRANCHIDAE Fitzinger, 1826 T. (THA)–Rec. FW

First: *Aviturus exsecratus* Gubin, 1991, and *Ulanurus fractus* Gubin, 1991, from Naran Bulak Formation, Naran Bulak, Gobi Desert, Mongolia; and '*Cryptobranchus*' *saskatchewensis* Naylor, 1981, Upper Ravenscrag Formation, Saskatchewan, Canada (Gubin, 1991). **Extant**
Intervening: YPR, PRI–CHT, BUR–PLE.

F. PROTEIDAE Gray, 1825 T. (THA)–Rec. FW

First: *Necturus krausei* Naylor, 1978, upper Ravenscrag Formation, Saskatchewan, Canada. **Extant**
Intervening: LAN/SRV, PIA–PLE.

F. BATRACHOSAUROIDIDAE Auffenberg, 1958 J. (TTH)–T. (ZAN) FW

First: Unnamed material from the Freshwater Member, Purbeck Limestone Formation, Langton Matravers, Dorset, England, UK (Ensom *et al.*, 1991).
Last: *Peratosauroides problematica* Naylor, 1981, 'San Pablo' Formation, California, USA.
Intervening: CON, CMP–LUT, AQT–LAN.
Comment: The earliest fully described batrachosauroidid is *Mynbulakia surgayi* Nessov, 1981, Middle Taikarshin Beds (= CON), Dzharakhuduk, former USSR.

F. DICAMPTODONTIDAE Tihen, 1958 **Extant** FW

Ambystomichnus montanensis (Gilmore, 1928) from the Palaeocene of Montana is frequently cited as a dicamptodont footprint trail, but I have excluded it as not susceptible to further phylogenetic investigation. Estes (1981) has included a series of four fossil genera in the Dicamptodontidae, namely *Bargmannia* Herre, 1955, *Chrysotriton* Estes, 1981, *Geyeriella* Herre, 1950 and *Woltersdorfiella* Herre, 1950. All of them have a vertebral structure similar to that of *Dicamptodon*, but this is a gradistic resemblance and not a cladistic one.

F. SCAPHERPETONTIDAE Auffenberg and Goin, 1959 K. (ALB)–T. (YPR) FW

First: *Horezmia gracile* Nessov, 1981, Lower/Middle Chodzhakul Formation, Karakalbakia, Uzbekistan, former USSR.
Last: *Piceoerpeton* cf. *P. willwoodense* (Meszoely, 1967), Eureka Sound Formation, Ellesmere Island, Canada (Estes and Hutchinson, 1978).
Intervening: CEN–THA.

F. AMPHIUMIDAE Gray, 1825 K. (MAA)–Rec. FW

First: *Proamphiuma cretacea* Estes, 1969, Hell Creek Formation, Montana, USA. **Extant**
Intervening: THA, BUR/LAN, PLE.

F. PLETHODONTIDAE Gray, 1850
T. (AQT)–Rec. Terr./FW

First: *Aneides* sp. and *Plethodon* sp. Tihen and Wake (1981), Cabbage Patch Formation, Montana, USA. **Extant**
Intervening: MES, PIA–PLE.

F. AMBYSTOMATIDAE Hallowell, 1856
T. (RUP)–Rec. FW/Terr.

First: *Ambystoma tiheni* Holman, 1968, Cypress Hills Formation, Saskatchewan, Canada. **Extant**
Intervening: BUR–TOR, ZAN–PLE.

F. SALAMANDRIDAE Gray, 1825
T. (THA)–Rec. FW/Terr.

First: *Koaliella genzeli* Herre, 1950, Walbeck, Germany, and material referred to *K. genzeli* from Cernay, France (Estes *et al.*, 1967). **Extant**
Intervening: YPR–LUT, PRB–LAN, TOR/MES, PIA–PLE.

LISSAMPHIBIA *incertae sedis*

F. ALBANERPETONTIDAE Fox and Naylor, 1982
J. (BAJ)–T. (BUR/LAN) FW

First: Atlas centrum, referred to *Albanerpeton megacephalus*, Aveyron, France (Estes, 1981).
Last: *Albanerpeton inexpectatum* Estes and Hoffstetter, 1976, fissures, La Grive St Alban, France.
Intervening: BTH–OXF, VLG/HAU–CEN, CON, CMP–DAN/THA

STEM AMNIOTA

Order ANTHRACOSAURIA Säve-Söderbergh, 1935

Recent reviews of anthracosaurian families include Smithson (1985b: eoherpetontids), Smithson (1986: proterogyrinids), Clack (1987a: anthracosaurids), Clack (1987b: eogyrinids), Holmes (1989: archeriids), and Tverdokhlebova (1972: chroniosuchids).

F. EOHERPETONTIDAE Panchen, 1980
C. (BRI–PND) Terr.

First: *Eoherpeton watsoni* Panchen, 1975, Gilmerton Ironstone, Gilmerton, Edinburgh, Scotland, UK.
Last: *Eoherpeton watsoni* Panchen, 1975, Burghlee Ironstone, Loanhead, Edinburgh, and Dora Bone Bed, Cowdenbeath, Fife, Scotland, UK.

F. GEPHYROSTEGIDAE Jaekel, 1909
C. (MRD–MYA*) Terr.

First: *Bruktererpeton fiebigi* Boy and Bandel, 1973, Hagener Beds, Wuppertal, Germany.
Last: *Gephyrostegus bohemicus* Jaekel, 1902, referred material from Plattelkohle, Třemošná, Czechoslovakia. This material is currently in the literature under the names *Hemichthys problematica* Fritsch, 1895, and *Nummulosaurus kolbii* Fritsch, 1901.
Intervening: None.

F. PROTEROGYRINIDAE Romer, 1970
C. (BRI–PND) FW

First: *Proterogyrinus scheeli* Romer, 1970, Bluefield Formation, Greer, West Virginia, USA.
Last: *Proterogyrinus pancheni* Smithson, 1986, Burghlee Ironstone, Loanhead, Edinburgh, and Dora Bone Bed, Cowdenbeath, Fife, Scotland, UK.

F. ANTHRACOSAURIDAE Cope, 1875
C. (MEL*–KSK*) Terr.

First: *Anthracosaurus russelli* Huxley, 1863, referred material from Top Busty Seam, Washington, Co. Durham, England, UK.
Last: *Anthracosaurus russelli* Huxley, 1863, referred material from Low Main Seam, upper *Modiolaris* Zone, Newsham, Northumberland, England, UK.
Intervening: VRK*.

F. EOGYRINIDAE Watson, 1929
C. (CHE*)–P. (UFI) FW

First: *Pholiderpeton scutigerum* Huxley, 1869, Black Coal Bed, Toftshaw, Yorkshire, England, UK.
Last: *Aversor dmitrievi* Gubin, 1985, Sheshma Horizon, Komi, former USSR.
Intervening: MEL*–KSK*, MYA*, CHV/DOR*, ASS.
Comment: Several earlier poor embolomere specimens have been attributed to the Eogyrinidae but none is diagnostically eogyrinid. The earliest of these is *'Pholiderpeton' bretonense* Romer, 1958, Point Edward Formation, Nova Scotia, Canada.

F. ARCHERIIDAE Kuhn, 1965
C. (MYA*)–P. (ART) FW

First: Undescribed archeriid, Linton, Ohio, USA (reported but not described by Hook and Baird, 1986: p. 183).
Last: *Archeria crassidisca* (Cope, 1884), referred material from middle Garber Formation, Cotton County, Oklahoma, USA (Holmes, 1989).
Intervening: ASS–SAK.

F. CHRONIOSUCHIDAE Viushkov, 1957
P. (TAT) FW/Terr.

First and Last: All known chroniosuchids are from upper TAT horizons in the former USSR. The genera and species are: *Chroniosuchus mirabilis* Viushkov, 1957, Gorkiy and Orenburg Provinces; *C. paradoxus* Viushkov, 1957, Pronkino, Orenburg Province; *C. licharevi* (Riabinin, 1962), Sokolki, Arkhangel Province, and *Chroniosaurus dongusensis* Tverdokhlebova, 1972, Donguz River, Orenburg Province.

Order SEYMOURIAMORPHA Watson, 1917

Information on Russian material is mainly derived from Ivakhnenko, 1987.

Suborder SEYMOURIIDA Tatarinov, 1971

F. SEYMOURIIDAE Williston, 1911
P. (ASS–TAT) Terr./FW

First: *Seymouria sanjuanensis* Vaughn, 1966, referred material from Cutler Formation, New Mexico, USA, equivalent to Archer City Formation of Texas (Berman *et al.*, 1987b).
Last: A wide range of seymouriids *sensu* Ivakhnenko (1987) (i.e. including the kotlassiids) occur in formations of

Zone IV Permian (= TAT) in pre-Ural former USSR. These include: *Kotlassia prima* Amalitsky, 1921, Arkhangel Province; and *Karpinskiosaurus secundus* (Amalitsky) Sushkin, 1925, Arkhangel Province.
Intervening: SAK–KAZ.

F. DISCOSAURISCIDAE Romer, 1947 C./P. (???)–P. (KAZ/TAT) FW

First: *Utegenia schpinari* Kuznetzov and Ivakhnenko, 1981, Alma-Ata, Kazakhstan, former USSR, may be late Carboniferous or early Permian in age.
Last: *Discosauriscus netschaevi* (Riabinin, 1911), Orenburg Province, former USSR.
Intervening: ASS/SAK–ART.
Comment: Many discosauriscids derive from basins in Central Asia which are poorly correlated with those of Europe.

Suborder LEPTOROPHIDA Ivaknenko, 1987

F. LEPTOROPHIDAE Ivakhnenko, 1987 P. (KAZ–TAT) FW/Terr?

First: *Leptoropha talonophora* (Tchudinov, 1955), Shikovo-Cherki, Kirov Province, former USSR.
Last: *Raphanodon ultimus* (Tchudinov and Viushkov, 1956), Pronkino, Orenburg Province; and *R. tverdochlebovae* Ivakhnenko, 1987, Donguz VI, Orenburg Province, former USSR.

F. ENOSUCHIDAE Konzhukova, 1955 P. (TAT) FW

First and Last: *Enosuchus breviceps* Konzhukova, 1955, Isheevo, Tatarian, former USSR.

Order NYCTEROLETEROMORPHA Ivakhnenko, 1987

Information on this order is mainly derived from Ivakhnenko (1987).

Suborder NYCTEROLETERIDA Tatarinov, 1972

F. NYCTEROLETERIDAE Romer, 1956 P. (KAZ–TAT) Terr.

First: *Nycteroleter bashkyricus* Efremov, 1940, Belebey, Bashkirian, former USSR; and *N. kassini* Tchudinov, 1955, Shikhovo-Cherki, Kirov Province, former USSR.
Last: *Nycteroleter ineptus* Efremov, 1938 and *Macroleter poezicus* Tverdokhlebova and Ivakhnenko, 1984, both from Mezen River Basin, Arkhangel Province, former USSR.

F. TOKOSAURIDAE Tverdokhlebova and Ivakhnenko, 1984 P. (KAZ) Terr.

First and Last: *Tokosaurus perforatus* Tverdokhlebova and Ivakhnenko, 1984, Krimskiy, Orenburg Province, former USSR.

Suborder LANTHANOSUCHIDA Tatarinov, 1972

F. LANTHANOSUCHIDAE Efremov, 1946 P. (TAT) FW?

First: *Lanthaniscus efremovi* Ivakhnenko, 1980, Mezen River, Arkhangel Province, former USSR (= early TAT).
Last: *Lanthanosuchus watsoni* Efremov, 1946, Isheevo, Tatarian, former USSR; and *L. lukjanovae* Ivakhnenko, 1980, Novo Nikol'skoye, Orenburg Province, former USSR (both = late TAT in age).

Order COTYLOSAURIA Cope, 1880

There is no recent taxonomic review of this group as a whole, but papers incorporating reference to earlier work are Langston (1966: later limnoscelidids) and Moss (1972: tseajaiids, diadectids and solenodonsaurids).

F. LIMNOSCELIDIDAE Williston, 1911 C. (CHE*)–P. (SAK/ART) Terr.

First: *Romeriscus periallus* Baird and Carroll, 1967, Port Hood Formation, Cape Linzee, Nova Scotia, Canada.
Last: Undescribed limnoscelidid, Tambacher Formation, Upper Rotliegendes, Bromacker, Thuringia, Germany (Martens, pers. comm.).
Intervening: MYA*, KRE/CHV*, NOG*/ASS.

F. SOLENODONSAURIDAE Broili, 1924 C. (MYA*) Terr.

First and Last: *Solenodonsaurus janenschi* Broili, 1924, Gaskohle, Kladno Formation, Nýřany, Czechoslovakia.

F. TSEAJAIIDAE Vaughn, 1964 P. (ART) Terr.

First and Last: *Tseajaia campi* Vaughn, 1964, Organ Rock Shale, Utah, USA.

F. DIADECTIDAE Cope, 1880 C. (KRE/CHV*)–P. (ART) Terr.

First: *Desmatodon hollandi* Case, 1908, Conemaugh Group, Pitcairn, Pennsylvania, USA; and *Desmatodon hesperis* Vaughn, 1969, Sangre de Cristo Formation, Colorado, USA.
Last: *Diadectes* sp. Vale Formation, Taylor County, Texas (Olson and Mead, 1982).
Intervening: ASS–SAK.

REFERENCES

Andrews, S. M. and Carroll, R. L. (1991) The Order Adelospondyli: Carboniferous lepospondyl amphibians. *Transactions of the Royal Society of Edinburgh; Earth Sciences*, **82**, 239–75.

Baez, A. M. (1985) Anuro leptodactilido en el Cretacico superior (Grupo Bauru) de Brasil. *Ameghiniana*, **22**, 75–9.

Baird, D. (1964) The aïstopod amphibians surveyed. *Breviora*, **206**, 1–17.

Baird, D. and Carroll, R. L. (1967) *Romeriscus*, the oldest known reptile. *Science*, **157**, 56–9.

Barberena, M. C. and Daemon, R. F. (1974) A primeira ocorrência de Amphibia (Labyrinthodonta) na Formação Rio de Rasto; implicações geocronológicas e estratigráficas. *Congresso Brasileiro de Geologia, Anais*, **28**, 251–61.

Beaumont, E. H. (1977) Cranial morphology of the Loxommatidae (Amphibia: Labyrinthodontia). *Philosophical Transactions of the Royal Society of London, Series B*, **280**, 29–101.

Berman, D. S. (1973) A trimerorhachid amphibian from the Upper Pennsylvanian of New Mexico. *Journal of Paleontology*, **47**, 932–45.

Berman, D. S., Eberth, D. A. and Brinkman, D. B. (1988) *Stegotretus agyrus* a new genus and species of microsaur (amphibian) from the Permo-Pennsylvanian of New Mexico. *Annals of the Carnegie Museum*, **57**, 293–323.

Berman, D. S., Reisz, R. R. and Eberth, D. A. (1987a) A new genus and species of trematopid amphibian from the late Pennsylvanian of north-central New Mexico. *Journal of Vertebrate Paleontology*, **7**, 252–69.

Berman, D. S., Reisz, R. R. and Eberth, D. A. (1987b) *Seymouria sanjuanensis* (Amphibia: Batrachosauria) from the Lower Permian Cutler Formation of north-central New Mexico and the occurrence of sexual dimorphism in that genus questioned. *Canadian*

Journal of Earth Sciences, **24**, 1769–84.

Bolt, J. R. (1969) Lissamphibian origins: possible protolissamphibian from the Lower Permian of Oklahoma. *Science*, **166**, 888–91.

Campbell, K. W. S. and Bell, M. W. (1977) A primitive amphibian from the Late Devonian of New South Wales. *Alcheringa*, **1**, 369–81.

Carroll, R. L. (1967) Labyrinthodonts from the Joggins Formation. *Journal of Paleontology*, **41**, 111–42.

Carroll, R. L. (1990) A tiny microsaur from the Lower Permian of Texas: size constraints in Palaeozoic tetrapods. *Palaeontology*, **33**, 893–909.

Carroll, R. L. (1991) *Batropetes* from the Lower Permian of Europe – a microsaur, not a reptile. *Journal of Vertebrate Paleontology*, **11**, 229–42.

Carroll, R. L. and Gaskill, P. (1978) The order Microsauria. *Memoir of the American Philosophical Society*, **126**, 1–211.

Carroll, R. L., Bybee, P. and Tidwell, W. D. (1991) The oldest microsaur. *Journal of Paleontology*, **65**, 314–22.

Clack, J. A. (1987a) Two new specimens of *Anthracosaurus* (Amphibia: Anthracosauria) from the Northumberland Coal Measures. *Palaeontology*, **30**, 15–26.

Clack, J. A. (1987b) *Pholiderpeton scutigerum* Huxley, an amphibian from the Yorkshire Coal Measures. *Philosophical Transactions of the Royal Society of London, Series B*, **318**, 1–107.

Clack, J. A. (1988) New material of the early tetrapod *Acanthostega* from the Upper Devonian of East Greenland. *Palaeontology*, **31**, 699–724.

Cosgriff, J. W. (1984) The temnospondyl labyrinthodonts of the earliest Triassic. *Journal of Vertebrate Paleontology*, **4**, 30–46.

Duellman, W. E. and Trueb, L. (1986) *Biology of Amphibians*. McGraw-Hill, New York, 670 pp.

Dutuit, J.-M. (1972) Un nouveau genre de stégocéphale du Trias supérieur marocain: *Almasaurus habbazi*. *Bulletin du Muséum national d'Histoire naturelle Serie 3*, **72**, 73–81.

Ensom, P. C., Evans, S. E. and Milner, A. R. (1991) Amphibians and reptiles from the Purbeck Limestone Formation (Upper Jurassic) of Desert, in *Fifth Symposium on Mesozoic Terrestrial Ecosystems*. Contributions from the Paleontological Museum, University of Oslo, **364**, 19–20.

Estes, R. (1975) Lower Vertebrates from the Fort Union Formation, late Paleocene, Big Horn Basin, Wyoming. *Herpetologica*, **31**, 365–85.

Estes, R. (1981) Gymnophiona, Caudata, in *Handbuch der Paläoherpetologie*, Vol. 2 (ed. P. Wellnhofer), Fischer Verlag, Stuttgart, pp. 1–115.

Estes, R. and Hutchinson, J. H. (1978) Eocene lower vertebrates from Ellesmere Island, Canadian Arctic Archipelago. *Palaeogeography, Palaeoclimatology, Palaeoecology*, **30**, 325–75.

Estes, R. and Reig, O. A. (1973) The early fossil record of frogs: a review of the evidence, in *Evolutionary Biology of the Anurans: Contemporary Research on Major Problems*, (ed. J. L. Vial), Univ. Missouri Press, Columbia, pp. 1–63.

Estes, R., Špinar, Z. V. and Nevo, E. (1978) Early Cretaceous pipid tadpoles from Israel (Amphibia: Anura). *Herpetologica*, **34**, 374–93.

Evans, S. E. and Milner, A. R. (in press) Frogs and salamanders from the Upper Jurassic Morrison Formation (Quarry Nine, Como Bluff) of North America. *Journal of Vertebrate Paleontology*.

Evans, S. E., Milner, A. R. and Mussett, F. (1988) The earliest known salamanders (Amphibia: Caudata): a record from the Middle Jurassic of England. *Geobios*, **21**, 539–52.

Evans, S. E., Milner, A. R. and Mussett, F. (1990) A discoglossid frog (Amphibia: Anura) from the Middle Jurassic of England. *Palaeontology*, **33**, 299–311.

Fox, R. C. and Naylor, B. G. (1982) A reconsideration of the relationships of the fossil amphibian *Albanerpeton*. *Canadian Journal of Earth Sciences*, **19**, 118–28.

Gubin, Y. M. (1980) New Permian dissorophids of the Ural forelands. *Paleontologicheskii Zhurnal*, **1980** (3), 82–90 [in Russian]. *Paleontological Journal*, **14** (3), 88–96 [in English].

Gubin, Y. M. (1983) The first eryopids from the Permian of the east European platform. *Paleontologicheskiy Zhurnal*, **1983** (4), 110–15 [in Russian]. *Paleontological Journal*, **17** (4), 105–10 [in English].

Gubin, Y. M. (1985) The first anthracosaur from the Permian of the East European Platform. *Paleontologicheskii Zhurnal*, **1985** (3), 118–22 [in Russian]. *Paleontological Journal*, **19** (3), 105–8 [in English].

Gubin, Y. M. (1989) The systematic position of the labyrinthodonts from the Malaya Kinel' locality (Orenberg Region). *Paleontologicheskii Zhurnal*, **1989** (3), 116–20 [in Russian]. *Paleontological Journal*, **23** (3), 115–19 [in English].

Gubin, Y. M. (1991) Paleocene salamanders from Southern Mongolia. *Paleontologicheskiy Zhurnal*, **1991** (1); 96–106 [in Russian]. *Paleontological Journal*, **25** (1), 91–102 [in English].

Harland, W. B., Cox, A. V., Llewellyn, P. G. *et al.* (1982) *A Geologic Time Scale*. Cambridge University Press, 131 pp.

Hasegawa, Y. (1980) Vertebrates of the late Pleistocene–Holocene of the Ryukyu Islands. *Quaternary Research (Japan Association for Quaternary Research)*, **18**(4), 263–7.

Hellrung, H. (1987) Revision von *Hyperokynodon keuperinus* Plieninger (Amphibia: Temnospondyli) aus dem Schilfsandstein von Heilbronn (Baden-Württemberg). *Stuttgarter Beiträge zur Naturkunde Serie B, Geologie und Paläontologie*, **136**, 1–28.

Hentz, T. F. (1989) Permo-Carboniferous lithostratigraphy of the vertebrate-bearing Bowie and Wichita Groups, north-central Texas, in *Permo-Carboniferous Paleontology, Lithostratigraphy, and Depositional Environments of North-Central Texas* (ed. R. W. Hook), Field Trip Guide Book No. 2, 49th Annual Meeting of the Society of Vertebrate Paleontology, Austin, Texas, pp. 1–21.

Holmes, R. (1989) The skull and axial skeleton of the Lower Permian anthracosauroid amphibian *Archeria crassidisca* Cope. *Paläontographica, Abteilung A*, **207**, 161–206.

Holmes, R. and Carroll, R. L. (1977) A temnospondyl amphibian from Mississippian of Scotland. *Bulletin of the Museum of Comparative Zoology, Harvard University*, **147**, 489–511.

Hook, R. W. (1983) *Colosteus scutellatus* (Newberry), a primitive temnospondyl amphibian from the Middle Pennsylvanian of Linton, Ohio. *American Museum Novitates*, **2770**, 1–41.

Hook, R. W. (1989) Stratigraphic distribution of tetrapods in the Bowie and Wichita Groups, Permo-Carboniferous of north-central Texas, in *Permo-Carboniferous Paleontology, Lithostratigraphy, and Depositional Environments of North-Central Texas* (ed. R. W. Hook), Field Trip Guide Book No. 2, 49th Annual Meeting of the Society of Vertebrate Paleontology. Austin, Texas, pp. 47–53.

Hook, R. W. and Baird, D. (1986) The Diamond Coal Mine of Linton, Ohio, and its Pennsylvanian-age vertebrates. *Journal of Vertebrate Paleontology*, **6**, 174–90.

Hunt, A. P. and Lucas, S. G. (1990) The status of 'Jurassic' metoposaurs in the American southwest. *Stegocephalian Newsletter*, **1**, 16–17.

Ivakhnenko, M. F. (1972) A new benthosuchid from the Lower Triassic of the Upper Volga region. *Paleontologicheskii Zhurnal*, **1972** (4), 93–9 [in Russian]. *Paleontological Journal*, **6** (4), 532–6 [in English].

Ivakhnenko, M. F. (1978) Urodelans from the Triassic and Jurassic of Soviet Central Asia. *Paleontologicheskii Zhurnal*, **1978** (3), 84–9 [in Russian]. *Paleontological Journal*, **12** (3), 362–8 [in English].

Ivakhnenko, M. F. (1987) [Permian parareptiles of the USSR.] *Trudy Paleontologicheskogo Instituta, Akademiya Nauk SSSR*, **223**, 1–159 [in Russian].

Jenkins, F. A. Jr and Walsh, D. M. (1990) During the Jurassic, caecilians had limbs. *Journal of Vertebrate Paleontology*, **10** (3 supplement), p. 29A.

Jupp, R. and Warren, A. A. (1986) The mandibles of the Triassic temnospondyl amphibians. *Alcheringa*, **10**, 99–124.

Kalandadze, N. N., Ochev, V. G., Tatarinov, L. P. *et al.* (1968) [A catalogue of the Permian and Triassic tetrapods of the USSR], in [*Upper Palaeozoic and Mesozoic Amphibians and Reptiles of the USSR*.) Nauka Press, Moscow [in Russian], pp. 72–7.

Kamphausen, D. (1989) Der Schadel von *Eocyclotosaurus woschmidti* Ortlam (Amphibia, Stegocephalia) aus dem Oberen Buntsandstein (Trias) des Schwarzwaldes (SW-Deutschland). *Stuttgarter Beiträge zur Naturkunde Serie B, Geologie und Paläontologie*, **149**, 1–65.

Langston, W. Jr (1966) *Limnosceloides brachycoles* (Reptilia: Captorhinomorpha), a new species from the Lower Permian of New Mexico. *Journal of Paleontology*, **40**, 690–5.

Lebedev, O. A. (1984) The first find of a Devonian tetrapod in the USSR. *Doklady Akademiy Nauk SSSR*, **278**, 1470–3 [in Russian].

Lund, R. (1978) Anatomy and relationships of the family Phlegethontiidae (Amphibia, Aïstopoda). *Annals of the Carnegie Museum*, **47**, 53–79.

Maňourová, M. (1981) Nový nález Branchiosaurů na lokalitě Horní Kalná u Vrchlabí. *Časopis Národniho Muzea v Praze, Řada Přírodovédna*, **150**, 169–72.

Milner, A. C. (1980) A review of the Nectridea (Amphibia), in *The Terrestrial Environment and the Origin of Land Vertebrates* (ed. A. L. Panchen), Academic Press, London, pp. 377–405.

Milner, A. R. (1987) The Westphalian tetrapod fauna; some aspects of its geography and ecology. *Journal of the Geological Society of London*, **144**, 495–506.

Milner, A. R. (1988) The relationships and origin of living amphibians, in *The Phylogeny and Classification of the Tetrapods*, Vol. 1 (ed. M. J. Benton), Clarendon Press, Oxford, pp. 59–102.

Milner, A. R., Smithson, T. R., Milner, A. C. *et al.* (1986) The search for early tetrapods. *Modern Geology*, **10**, 1–28.

Morales, M. (1988) New metoposaurid and capitosauroid labyrinthodonts from the Triassic of Germany and the Soviet Union. *Journal of Vertebrate Paleontology*, **8** (3, Supplement), p. 23A.

Moss, J. L. (1972) The morphology and phylogenetic relationships of the Lower Permian tetrapod *Tseajaia campi* Vaughn (Amphibia: Seymouriamorpha). *University of California Publications in Geological Sciences*, **98**, 1–63.

Murry, P. A. (1987) Notes on the stratigraphy and paleontology of the Upper Triassic Dockum Group. *Journal of the Arizona–Nevada Academy of Sciences*, **22**, 73–84.

Murry, P. A. and Johnson, G. D. (1987) Clear Fork vertebrates and environments from the Lower Permian of north-central Texas. *Texas Journal of Science*, **39**, 253–66.

Nessov, L. A. (1988) Late Mesozoic amphibians and lizards of Soviet Middle Asia. *Acta Zoologica Cracoviensis*, **31**, 475–86.

Nessov, L. A. (1990) [The latest labyrinthodonts (Amphibia, Labyrinthodontia) and other relict groups of vertebrates from northern Fergana.] *Paleontologicheskii Zhurnal*, **1990** (3), 82–90 [in Russian].

Olson, E. C. (1955) Fauna of the Upper Vale and Choza: 10. *Trimerorhachis*: including a revision of pre-Vale species. *Fieldiana: Geology*, **10**, 225–74.

Olson, E. C. (1956) Fauna of the Upper Vale and Choza: 11. *Lysorophus*: Vale and Choza. *Diplocaulus*, *Cacops* and Eryopidae: Choza. *Fieldiana: Geology*, **10**, 313–22.

Olson, E. C. and Mead, J. G. (1982) The Vale Formation (Lower Permian): its vertebrates and paleoecology. *Bulletin of the Texas Memorial Museum*, **29**, 1–46.

Panchen, A. L. (1959) A new armoured amphibian from the Upper Permian of East Africa. *Philosophical Transactions of the Royal Society of London, Series B*, **242**, 207–81.

Panchen, A. L. (1975) A new genus and species of anthracosaur amphibian from the Lower Carboniferous of Scotland and the status of *Pholidogaster pisciformis* Huxley. *Phil. Trans. Royal Society of London, Series B*, **269**, 581–640.

Panchen, A. L. (1985) On the amphibian *Crassigyrinus scoticus* Watson from the Carboniferous of Scotland. *Philosophical Transactions of the Royal Society of London*, **309**, 505–68.

Panchen, A. L. and Smithson, T. R. (1988) The relationships of early tetrapods, in *The Phylogeny and Classification of the Tetrapods*, Vol. 1 (ed. M. J. Benton), Clarendon Press, Oxford, pp. 1–32.

Rage, J.-C. (1984) Are the Ranidae (Anura, Amphibia) known prior to the Oligocene? *Amphibia–Reptilia*, **5**, 281–8.

Rage, J.-C. (1986) Le plus ancien Amphibien apode (Gymnophiona) fossile. Remarques sur la répartition et l'histoire paléobiogéographique des Gymnophiones. *Comptes Rendus de l'Académie des Sciences, Série II*, **302**, 1033–6.

Rage, J.-C. (1988) Le Gisement du Bretou (Phosphorites du Quercy, Tarne-et-Garonne, France) et sa faune de vertébrés de l'Eocène supérieur. 1. Amphibiens et reptiles. *Palaeontographica, Abteilung A*, **205**, 3–27.

Rage, J.-C. and Roček, Z. (1989) Redescription of *Triadobatrachus massinoti* (Piveteau, 1936) an anuran amphibian from the early Triassic. *Palaeontographica, Abteilung A*, **206**, 1–16.

Romer, A. S. (1947) Review of the Labyrinthodontia. *Bulletin of the Museum of Comparative Zoology, Harvard College*, **99**, 1–138.

Schultze, H.-P. and Foreman, B. (1981) A gymnarthrid microsaur from the Lower Permian of Kansas with a review of the tuditanomorph microsaurs (Amphibia). *Occasional Papers of the Museum of Natural History, University of Kansas*, **91**, 1–25.

Shishkin, M. A. (1991) A labyrinthodont from the Jurassic of Mongolia. *Paleontologicheskii Zhurnal*, **1991** (1), 81–95 [in Russian].

Smithson, T. R. (1985a) Scottish Carboniferous amphibian localities. *Scottish Journal of Geology*, **21**, 123–42.

Smithson, T. R. (1985b) The morphology and relationships of the Carboniferous amphibian *Eoherpeton watsoni* Panchen. *Zoological Journal of the Linnean Society*, **85**, 317–410.

Smithson, T. R. (1986) A new anthracosaur amphibian from the Carboniferous of Scotland. *Palaeontology*, **29**, 603–28.

Špinar, Z. V. and Hodrová, M. (1985) New knowledge of the genus *Indobatrachus* (Anura) from the Lower Eocene of India. *Amphibia–Reptilia*, **6**, 363–76.

Thayer, D. W. (1985) New Pennsylvanian lepospondyl amphibians from the Swisshelm mountains, Arizona. *Journal of Paleontology*, **59**, 684–700.

Tverdokhlebova, G. I. (1972) A new batrachosaurian genus from the Upper Permian of Southern Cisuralia. *Paleontologicheskii Zhurnal*, **1972**, 95–103 [in Russian]. *Paleontological Journal*, **6**, 84–90 [in English].

Tyler, M. J. (1982) Tertiary frogs from South Australia. *Alcheringa*, **6**, 101–3.

Vaughn, P. P. (1971) A *Platyhistrix*-like amphibian with fused vertebrae, from the Upper Pennsylvanian of Ohio. *Journal of Paleontology*, **45**, 464–9.

Warren, A. A. (1980) *Parotosuchus* from the early Triassic of Queensland and Western Australia. *Alcheringa*, **4**, 25–36.

Warren, A. A. (1985) Triassic Australian plagiosauroid. *Journal of Paleontology*, **59**, 236–41.

Warren, A. A. and Hutchinson, M. N. (1988a) The Madagascan capitosaurs. *Bulletin du Muséum National d'Histoire naturelle, Section C*, **10** (4), 23–30.

Warren, A. A. and Hutchinson, M. N. (1988b) A new capitosaurid amphibian from the early Triassic of Queensland, and the ontogeny of the capitosaur skull. *Palaeontology*, **31**, 857–76.

Warren, A. A. and Hutchinson, M. N. (1990) *Lapillopsis*, a new genus of temnospondyl amphibians from the early Triassic of Queensland. *Alcheringa*, **14**, 149–58.

Welles, S. P. and Estes, R. (1969) *Hadrokkosaurus bradyi* from the Upper Moenkopi Formation of Arizona, with a review of the brachyopid labyrinthodonts. *University of California Publications in Geological Sciences*, **84**, 1–56.

Wellstead, C. F. (1982) A Lower Carboniferous aistopod amphibian from Scotland. *Palaeontology*, **25**, 193–208.

Wellstead, C. F. (1991) Taxonomic revision of the Lysorophia, Permo-Carboniferous lepospondyl amphibians. *Bulletin of the American Museum of Natural History*, **209**, 1–90.

Werneburg, R. (1989) Labyrinthodontier (Amphibia) aus dem Oberkarbon und Unterperm Mitteleuropas – systematik, phylogenie und biostratigraphie. *Freiberger Forschungshefte, Serie C*, **436**, 7–57.

Yadagiri, P. (1986) Lower Jurassic lower vertebrates from Kota Formation, Pranhita–Godavari Valley, India. *Journal of the Palaeontological Society of India*, **31**, 89–96.

39

REPTILIA

M. J. Benton

The stratigraphical assignments of many of the terrestrial deposits containing fossil reptiles were difficult because of doubtful correlations with the marine standard stages summarized in Harland *et al.* (1982, 1990). This is particularly true for the Carboniferous and Lower Permian. Assignments of the Carboniferous units are based on Carroll (1984) and Milner (1987), and the Lower Permian formations of the south-western United States were dated according to summary tables in Olson and Vaughn (1970), Hentz (1989) and Hook (1989). The stage-level ages of many terrestrial Mesozoic formations were obtained from Weishampel's (1990) compilation on dinosaurian localities, and many Cainozoic ages were based on Savage and Russell's (1983) compilation of mammalian faunas. Note that the Guimarota locality in Portugal is accepted as Oxfordian in age on the basis of several lines of evidence, rather than the oft-quoted Kimmeridgian (see Evans, 1989), although the question is not settled.

Faunal zones in South Africa have been revised recently. The scheme used here (Rubidge, 1992) is:

Eodicynodon–Tapinocaninus Assemblage Zone (= lower part of the *Tapinocephalus* Zone, and lower portion of the Dinocephalian Assemblage Zone of Keyser and Smith (1979)).
Tapinocephalus–Bradysaurus Assemblage Zone (= middle part of the *Tapinocephalus* Zone, and upper portion of the Dinocephalian Assemblage Zone of Keyser and Smith (1979)).
Pristerognathus–Diictodon Assemblage Zone (= 'upper' *Tapinocephalus* Zone).
Tropidostoma–Endothiodon Assemblage Zone (= *Endothiodon* Zone, or *Tropidostoma microtrema* Assemblage Zone of Keyser and Smith (1979)).
Aulacephalodon–Cistecephalus Assemblage Zone (= *Cistecephalus* Zone, or *Aulacephalodon baini* Assemblage Zone of Keyser and Smith (1979)).
Dicynodon–Theriognathus Assemblage Zone (= *Daptocephalus* Zone, or *Dicynodon lacerticeps* Assemblage Zone of Keyser and Smith (1979)).
Lystrosaurus–Procolophon Assemblage Zone (= *Lystrosaurus* Zone).
Cynognathus–Diademodon Assemblage Zone (= *Cynognathus* Zone, or *Kannemeyeria* Assemblage Zone of Keyser and Smith (1979)).

Paraphyletic taxa are indicated by (p).

Acknowledgements – I thank Chris Bennett, Eric Buffetaut, Bob Carroll, Jim Clark, Susan Evans, Gene Gaffney, Howard Hutchison, Gillian King, Theagarajen Lingham-Soliar, Judy Massare, Chris McGowan, Peter Meylan, Mike Parrish, Jean-Claude Rage, Olivier Rieppel, Patrick Spencer, Glenn Storrs, Hans Sues, Bob Sullivan, Mike Taylor, David Unwin, Peter Wellnhofer and Frank Westphal for comments and corrections on parts of the text. This contribution was compiled as part of a project on the tetrapod fossil record, funded by the Leverhulme Trust.

Series AMNIOTA

AMNIOTA *incertae sedis*

F. UNNAMED C. (VIS) Terr. (see Fig. 39.1)

First and Last: *Westlothiana lizziae* Smithson and Rolfe, 1991, East Kirkton Limestone, Brigantian, West Lothian, Scotland, UK.
Comment: This specimen is said to be the oldest reptile, but the preliminary description (Smithson, 1989) did not indicate a familial assignment.

F. BOLOSAURIDAE Cope, 1878 P. (ART–KAZ) Terr.

First: *Bolosaurus striatus* Cope, 1878, lower Wichita Beds, Texas, USA.
Last: *Davletkulia gigantea* Ivakhnenko, 1990, upper KAZ, Davletkulovo settlement, right bank of Yaman-Yushatyr' River, Bashkirian, former USSR (Ivakhnenko, 1990).

F. ACLEISTORHINIDAE Daly, 1969 P. (ART) Terr.

The Fossil Record 2. Edited by M. J. Benton. Published in 1993 by Chapman & Hall, London. ISBN 0 412 39380 8

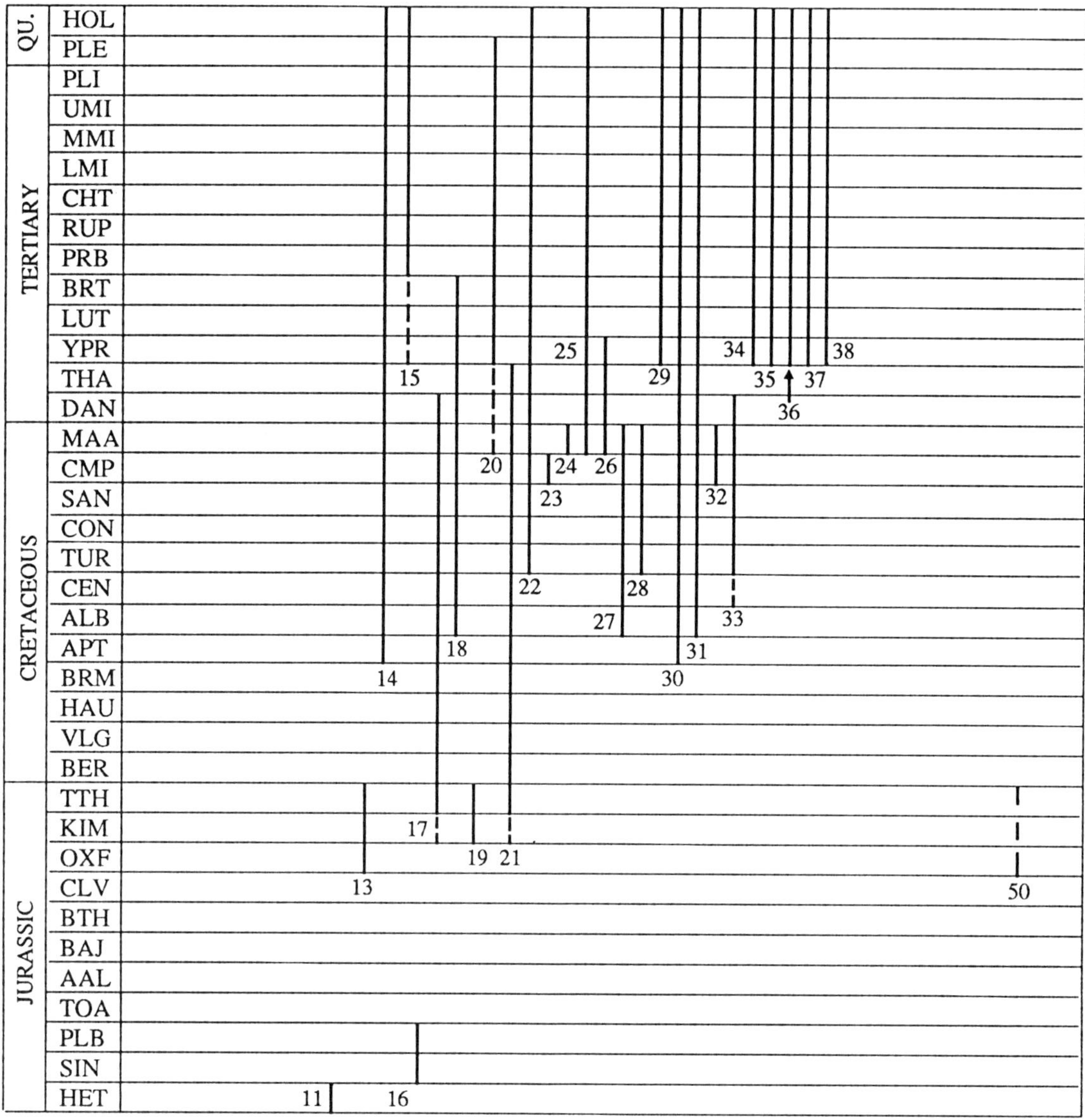

Fig. 39.1

First and Last: *Acleistorhinus pteroticus* Daly, 1969, Garber Formation, Oklahoma, USA.

F. EUNOTOSAURIDAE Romer, 1956 P. (UFI) Terr.

First and Last: *Eunotosaurus africanus* Seeley, 1892, Abrahamskraal Formation, *Tapinocephalus-Bradysaurus* Assemblage Zone, Beaufort West, Karoo Basin, South Africa.

Class REPTILIA Laurenti, 1768 (p)

Subclass ANAPSIDA Williston, 1917

The Procolophonidae have been proposed (Reisz and Laurin, 1991) as the closest known sister group of the Testudines. The Nyctiphruretidae are generally reckoned to be close relatives of the Procolophonidae, so they are placed here. The Captorhinidae were earlier (Gaffney and McKenna, 1979) proposed as turtle relatives, but Reisz and Laurin (1991) regard them as more distant than the procolophonids.

General information on the stratigraphical distribution of early anapsids was obtained from Kuhn (1969) and Anderson and Cruickshank (1978). The Mesosauridae, Millerettidae, Procolophonidae and Pareiasauridae were tentatively included in a new clade 'Parareptilia' by Gauthier et al. (1988). However, the Procolophonidae have been removed to the Anapsida, as a close sister group of the Testudines (Reisz and Laurin, 1991), and the Pareiasauridae and others may follow suit (P. S. Spencer, pers. comm., 1993).

F. MESOSAURIDAE Baur, 1889 P. (ART) Mar.

First and Last: *Mesosaurus tenuidens* Gervais, 1865, White Band, Ecca Group, South Africa, White Band equivalent, south-western Africa, and Irati Formation, Passa Dois Group, Paraná Basin, Brazil.

F. MILLERETTIDAE Romer, 1956 P. (KAZ–TAT) Terr.

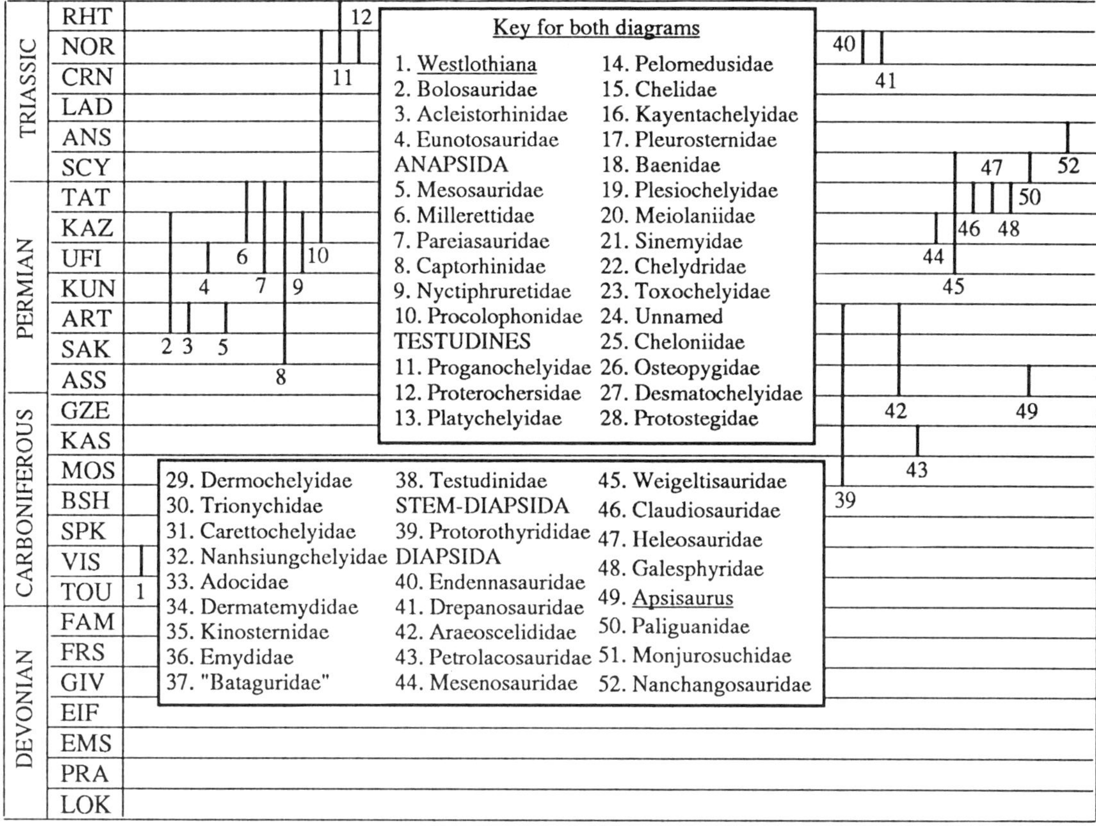

Fig. 39.1

First: *Broomia perplexa* Watson, 1914, *Tapinocephalus–Bradysaurus* Assemblage Zone, Karoo Basin, South Africa.
Last: *Milleretta rubidgei* Broom, 1938, and three or four other possible species, *Aulacephalodon–Cistecephalus* Assemblage Zone, Karoo Basin, South Africa.

F. PAREIASAURIDAE Cope, 1896
P. (UFI–TAT) Terr.

First: ?*Rhiphaeosaurus tricuspidens* Efremov, 1940, *Leptoropha novojilovi* Chudinov, 1955 and *Parabradysaurus udmurticus* Efremov, 1954, all from Zone II, Bashkir Republic and Kirov Province, former USSR.
Last: *Pareiasaurus serridens* Owen, 1876, *Dicynodon–Theriognathus* Assemblage Zone of the Karoo Basin, South Africa, and several other species from that formation, and equivalents, in the former USSR, China and Scotland, UK.
Intervening: KAZ.

F. CAPTORHINIDAE Case, 1911
P. (SAK–TAT) Terr.

First: *Romeria primus* Clark and Carroll, 1973, Moran Formation, Wichita Group, Archer County, Texas, USA (Clark and Carroll, 1973).
Last: *Moradisaurus grandis* Ricqlès and Taquet, 1982, Moradi Formation, Niger, and *Protocaptorhinus* sp., Middle Madumabisa Mudstones, Middle Zambezi Basin, Zimbabwe (Gaffney and McKenna, 1979).
Intervening: ART–KAZ.

F. NYCTIPHRURETIDAE Efremov, 1938
P. (UFI–KAZ) Terr.

First: *Nyctiphruretus acudens* Efremov, 1938, Zone III, Mesen district, former USSR.
Last: *Barasaurus besairiei* Piveteau, 1955, ?uppermost Permian, Madagascar.
Comment: It has been assumed that nyctiphruretids and procolophonids are related, but the material is poorly known. *Nyctiphruretus* may be related to procolophonids, but *Barasaurus* shows no clear affinities (P. S. Spencer, pers. comm., 1992).

F. PROCOLOPHONIDAE Cope, 1889
P. (KAZ)–Tr. (NOR) Terr.

First: *Owenetta rubidgei* Broom, 1939, *Aulacephalodon–Cistecephalus* Assemblage Zone, South Africa.
Last: *Hypsognathus fenneri* Gilmore, 1928, upper Passaic Formation, New Jersey and Pennsylvania, USA.
Intervening: ANS–CRN.
Comment: *Sphodrosaurus pennsylvanicus* Colbert, 1960, also upper Passaic Formation, New Jersey and Pennsylvania, USA, seems to be a diapsid, while the RHT or latest NOR 'procolophonoid' described by Cuny (1991) from the St Nicolas de Port locality in France is incorrectly identified (P. S. Spencer, pers. comm., 1992).

Order TESTUDINES Batsch, 1788

The classification of turtles used here is based on that of Gaffney and Meylan (1988), and information on stratigraphical distributions comes from Młynarski (1976), de Broin (1988) and Gaffney (1990). Authors of familial names are based on those authors, with corrections from Bour and Dubois (1984).

F. PROGANOCHELYIDAE Baur, 1888
Tr. (NOR)–J. (HET) Terr./FW

First: *Proganochelys quenstedtii* Baur, 1887, Mittlere and Obere Stubensandstein, Baden-Württemberg and Halberstadt, Germany.
Last: Unnamed proganochelyid, upper Elliot Formation (Red Beds), Orange Free State, South Africa (Gaffney, 1986).
Comment: The age of *P. ruchae* is assumed to be equivalent to the German formations, but that is not certain.

Suborder PLEURODIRA Cope, 1868

F. PROTEROCHERSIDAE Nopcsa, 1928
Tr. (NOR) Terr./FW

First and Last: *Proterochersis robusta* E. Fraas, 1913, Untere Stubensandstein, Baden-Württemberg, Germany.

F. PLATYCHELYIDAE Bräm, 1965
J. (OXF–TTH) FW/Terr.

First: cf. *Platychelys* sp., Oxfordian (formerly, Kimmeridgian), Guimarota Mine, Leiria, Portugal.
Last: *Platychelys oberndorferi* A. Wagner, 1853, Solothurn, Switzerland.
Intervening: KIM.

F. PELOMEDUSIDAE Cope, 1868
K. (APT)–Rec. FW/Mar.

First: *Araripemys barretoi* Price, 1975, Santana Formation, Ceará, Brazil. **Extant**
Intervening: ALB–PLE (de Broin, 1988).
Comment: Following Gaffney and Meylan (1988), the podocnemines and bothremydines are included here in the Pelomedusidae. Other authors, however, maintain the Podocnemididae Baur, 1888 (ALB–Rec) and Bothremydidae Baur, 1891 (ALB–YPR) as separate families (Antunes and de Broin, 1988; de Broin, 1988).

F. CHELIDAE Lindholm, 1929 (Gray, 1825)
T. (Eoc.)–Rec. FW

First: Unnamed form, Eocene, Tasmania, Australia (Gaffney, 1992). **Extant**
Intervening: ?THA, PRB, CHT, LAN–PLE (de Broin, 1988).
Comment: The oldest records of chelids given by de Broin (1988) are based on indeterminate material of uncertain age. The next oldest chelids noted by de Broin (1988, p. 136) are several species from the Campanian (Upper Cretaceous) and Palaeocene of Chubut Province, Argentina, but Gaffney (pers. comm.) regards these as pelomedusids. De Broin (1988, p. 138) also notes the chelid *Hydromedusa* sp. Wagler 1830, from the Upper Eocene of Chubut Province, Argentina. Gaffney (1975, 1990) indicates that the oldest chelid is Eocene.

Suborder CRYPTODIRA Cope, 1868

Infra-order STEM SELMACRYPTODIRES

F. KAYENTACHELYIDAE Gaffney *et al.*, 1987
J. (SIN/PLB) Terr./FW

First and Last: *Kayentachelys aprix* Gaffney *et al.*, 1987, Kayenta Formation, Coconino County, Arizona, USA (Gaffney *et al.*, 1987).

Infra-order SELMACRYPTODIRA Gaffney *et al.*, 1987

Superfamily PLEUROSTERNOIDEA Romer, 1956

F. PLEUROSTERNIDAE Cope, 1868
J. (KIM/TTH)–T. (DAN) Mar./FW

First: *Glyptops plicatulus* (Cope, 1877), Morrison Formation, Colorado, USA (Gaffney, 1979).
Last: *Compsemys victa* Leidy, 1859, Torrejonian, San Juan Basin, New Mexico, USA (Hutchison, 1987).
Intervening: BER, MAA.

Superfamily BAENOIDEA Williams, 1950

F. BAENIDAE Cope, 1882 K. (ALB)–T. (BRT) FW

First: *Trinitichelys hiatti* Gaffney, 1972, Trinity Sand, Trinity Group, Montague County, Texas, USA.
Last: *Chisternon undatum* (Leidy, 1871) and *Baena arenosa* Leidy, 1870, both with types from the Bridger Formation of Wyoming and Utah, but latest records from the Uinta Formation of Utah, USA (Gaffney, 1972).
Intervening: CMP–LUT.

Infra-order STEM POLYCRYPTODIRES

F. PLESIOCHELYIDAE Rütimeyer, 1873
J. (KIM–TTH) Mar.

First: *Plesiochelys etalloni* (Pictet and Humbert, 1857), Kimmeridgian, Solothurn, and other localities, Switzerland (Gaffney, 1975).
Last: *Portlandemys mcdowelli* Gaffney, 1975, Portland Stone, Dorset, England, UK (Gaffney, 1975).

F. MEIOLANIIDAE Lydekker, 1887
K. (MAA?)–Q. (PLE) Terr.

First: *Niolamia argentina* Ameghino, 1899, Upper Cretaceous? (Młynarski, 1976, p. 119), 'Pre-Oligocene, Post-Jurassic' (Gaffney, 1981, p. 20), Patagonia, Argentina.
Last: *Meiolania platyceps* Owen, 1881, *M. mackayi* Anderson, 1925, and *M. oweni* Woodward, 1888, Pleistocene of Lord Howe Island, Walpole Island (New Caledonia), and Queensland and New South Wales, Australia, respectively (Gaffney, 1981).
Intervening: YPR?, MMI.

Infra-order POLYCRYPTODIRA Gaffney and Meylan, 1988

F. SINEMYIDAE Wiman, 1930
J. (KIM?)–T. (THA) FW

First: *Sinemys lens* Wiman, 1930, Upper Jurassic, China.
Last: *Protochelydra zangerli* Erickson, 1973, Tongue River Formation, Billings County, North Dakota, USA (Erickson, 1973).
Intervening: TTH, APT, ALB, CMP, MAA (?) (Ckhikvadzé, 1988).

Superfamily CHELYDROIDEA Gaffney and Meylan, 1988

F. CHELYDRIDAE Gray, 1831 K. (TUR)–Rec. FW

First: Unnamed forms, Turonian, North America (Hutchison and Archibald, 1986). **Extant**
Intervening: CON–PLE.

Superfamily CHELONIOIDEA Baur, 1889

F. TOXOCHELYIDAE Baur, 1895 K. (CMP) Mar.

First and Last: *Toxochelys latiremis* Cope, 1873, Niobrara Formation, Kansas, USA.

F. UNNAMED K. (MAA) Mar.

First and Last: *Ctenochelys tenuitesta* Zangerl, 1953 and *C. acris* Zangerl, 1953, Selma Formation, Alabama, USA.

F. CHELONIIDAE Oppel, 1811 T. (MAA)–Rec. Mar.

First: *Dollochelys ('Toxochelys') atlantica* (Zangerl, 1953), Hornerstown Formation, Gloucester County, New Jersey, USA. **Extant**
Intervening: DAN–PLE.

F. OSTEOPYGIDAE Zangerl, 1953
K. (MAA)–T. (YPR) Mar.

First: *Osteopygis emarginatus* Cope, 1868, Hornerstown Formation, Gloucester County, New Jersey, USA.
Last: *Erquelinnesia gosseleti* (Dollo, 1886), Erquelinnes Sands, Upper Landenian and Sparnacian, Belgium.
Intervening: ?DAN.

F. DESMATOCHELYIDAE Gaffney, 1990
K. (ALB–MAA) Mar.

First: *Notochelone costata* (Owen, 1882), Toolebuc Formation, Flinders River, Queensland, Australia (Gaffney, 1981).
Last: *Desmatochelys lowi* Williston, 1898, Benton Group, Nebraska and South Dakota, USA (Zangerl and Sloan, 1960).
Intervening: CMP.

F. PROTOSTEGIDAE Cope, 1889
K. (TUR–MAA) Mar.

First: *Protostega eaglefordensis* Zangerl, 1953, Eagle Ford Shale, McLennan County, Texas, USA (Zangerl, 1953).
Last: *Pneumatoarthrus peloreus* Cope, 1870, ?Hornerstown Formation, Monmouth County, New Jersey, USA (Baird, 1978).
Intervening: SAN, CMP.

F. DERMOCHELYIDAE Baur, 1888
T. (YPR)–Rec. Mar.

First: *Eosphargis gigas* (Owen, 1861), London Clay, Kent, England, UK. **Extant**
Intervening: LUT-PLE.

Superfamily TRIONYCHOIDEA Gray, 1870

F. TRIONYCHIDAE Fitzinger, 1826
K. (APT)–Rec. FW/Terr.

First: Oldest trionychids, Aptian/Albian of Inner Mongolia, China (Nessov, 1988, pp. 9–10). **Extant**
Intervening: CEN, CMP–PLE.

F. CARETTOCHELYIDAE Boulenger, 1887
K. (ALB)–Rec. FW/Terr.

First: *Kizylkumemys* sp., lower or middle Chodzhakul Formation, Chodzhakul Lake, Kazakhstan, former USSR (Nessov, 1985). **Extant**
Intervening: CEN, YPR–PRB, PLE.
Comment: *Kizylkumemys schultzi* Nessov, 1977, was described from the upper part of the Chodzhakul Formation, dated as CEN, but Nessov (1985) notes older indeterminate specimens of this genus from older parts of the same formation.

F. NANHSIUNGCHELYIDAE Yeh, 1966
K. (CMP–MAA) Terr./FW

First: *Nanhsiungchelys wuchingensis* Yeh, 1966, Nanxiong Formation, Guandong, China, and *Zangerlia testudinimorpha* Młynarski, 1972, lower Nemegt Formation, Nemegt Basin, Gobi Desert, Mongolia, both late CMP to MAA in age (Meylan and Gaffney, 1989; Weishampel, 1990).
Last: *Basilemys sinuosa* Riggs, 1906, Hell Creek Formation, Montana, USA.

F. ADOCIDAE Cope, 1870
K. (CEN/TUR)–T. (DAN) FW/Terr.

First: *Adocus amtgai* Narmandakh, 1985, upper Bainshireinskaya Formation, Amtgay, eastern Gobi, Mongolia (Narmandakh, 1985).
Last: *Adocus onerosus* Gilmore, 1919, Nacimiento Formation, Torrejonian, San Juan Basin, New Mexico, USA.
Intervening: CMP, MAA.

F. DERMATEMYDIDAE Gray, 1870
T. (YPR)–Rec. Terr./FW

First: *Baptemys tricarinata* Hay, 1908, Wind River Formation, Wasatchian, Wyoming, USA (Hutchison, 1980). **Extant**
Intervening: LUT, BRT, LMI.

F. KINOSTERNIDAE Gray, 1869
T. (YPR)–Rec. Terr./FW

First: *Baltemys staurogastros* Hutchison, 1991, Willwood Formation, Wasatchian, Wyoming, USA. **Extant**
Intervening: PLI, PLE.
Comments: Earlier records of supposed MAA kinosternids are given by Hutchison and Archibald (1986).

Superfamily TESTUDINOIDEA Baur, 1893

F. EMYDIDAE Gray, 1825
T. (YPR)–Rec. Terr./FW

First: *Chrysemys bicarinata* (Bell, 1849) and *C. testudiniformis* (Owen, 1844), London Clay, Kent, England, UK. **Extant**
Intervening: LUT–PLE.
Comment: Hutchison (pers. comm., 1991) notes that these European forms could be batagurids, and that all Eocene records of *Chrysemys* are in question.

F. 'BATAGURIDAE' Gray, 1869
T. (YPR)–Rec. Terr./FW

First: *Echmatemys testudinea* (Cope, 1872), Wasatch Formation, Wyoming, USA. **Extant**
Intervening: LUT–PLE.

F. TESTUDINIDAE Batsch, 1788
T. (YPR)–Rec. Terr.

First: *Hadrianus majusculus* Hay, 1904, Willwood and Wasatch Formations, Wasatchian, Wyoming, USA. **Extant**

Intervening: LUT–PLE.

STEM DIAPSIDA

F. PROTOROTHYRIDIDAE Price, 1937
C. (VRK)–P. (ART) Terr.

First: *Hylonomus lyelli* Dawson, 1860, Cumberland Group, Joggins, Nova Scotia, Canada.
Last: Unnamed protorothyridid, Arroyo Formation, Clear Fork Group, Fort Sill, Oklahoma, USA (Reisz, 1980).
Intervening: POD, MYA, ASS.

Subclass DIAPSIDA Osborn, 1903

The classification of early diapsids is based on Benton (1985), Evans (1988) and Laurin (1991). Stratigraphical ranges are taken from papers cited by those authors, as well as Kuhn (1969) and Anderson and Cruickshank (1978), as well as more recent references cited.

DIAPSIDA *incertae sedis*

F. ENDENNASAURIDAE Carroll, 1987
Tr. (NOR) Terr.

First and Last: *Endennasaurus acutirostris* Renesto, 1984, Calcare di Zorzino, Bergamo, Italy (Renesto, 1984).

F. DREPANOSAURIDAE Carroll, 1987
Tr. (NOR) Terr.

First and Last: *Drepanosaurus unguicaudatus* Pinna, 1980, Calcare di Zorzino, Bergamo, Italy (Pinna, 1980).

Order ARAEOSCELIDIA Williston, 1913

F. ARAEOSCELIDIDAE Williston, 1910
P. (ASS–ART) Terr.

First: *Zarcasaurus tanyderus* Brinkman *et al.*, 1984, Cutler Formation, Rio Arriba County, New Mexico, USA (Brinkman *et al.*, 1984).
Last: *Araeoscelis gracilis* Williston, 1910, Arroyo Formation, Clear Fork Group, Baylor County, Texas, USA.

F. PETROLACOSAURIDAE Peabody, 1952
C. (KAS) Terr.

First and Last: *Petrolacosaurus kansensis* Lane, 1945, Stanton Formation, Lansing Group, Garnett, Kansas, USA.

STEM-GROUP NEODIAPSIDA

F. MESENOSAURIDAE Romer, 1956
P. (KAZ) Terr.

First and Last: *Mesenosaurus romeri* Efremov, 1940, Zone II, Mezen' River, Archangel Province, former USSR.

F. WEIGELTISAURIDAE Romer, 1933
P. (UFI)–Tr. (SCY) Terr.

First: *Weigeltisaurus jaekeli* (Weigelt, 1930), Kupferschiefer, Hesse, Germany; Marl Slate, Durham, England, UK.
Last: *Wapitisaurus problematicus* Brinkman, 1988, Vega-Phroso Member, Sulphur Mountain Formation, British Columbia, Canada (Brinkman, 1988).
Intervening: KAZ.

F. CLAUDIOSAURIDAE Carroll, 1981
P. (TAT) FW

First and Last: *Claudiosaurus germaini* Carroll, 1981, upper part of the Lower Sakamena Formation, Leoposa, Madagascar.

?F. HELEOSAURIDAE Haughton, 1924
P. (TAT) Terr.

First and Last: *Heleosaurus scholtzi* Broom, 1907, *Aulacephalodon–Cistecephalus* Assemblage Zone, Victoria West, Karoo Basin, South Africa.

?F. GALESPHYRIDAE Currie, 1981 P. (TAT) Terr.

First and Last: *Galesphyrus capensis* Broom, 1914, *Aulacephalodon–Cistecephalus* Assemblage Zone, Cape Province, South Africa.

F. UNNAMED P. (ASS) Terr.

First and Last: *Apsisaurus witteri* Laurin, 1991, Archer City bone bed, Archer City Formation, Wichita Group, lower Wolfcampian, Archer County, Texas, USA (Laurin, 1991).

Infraclass NEODIAPSIDA Benton, 1985

NEODIAPSIDA *incertae sedis*

F. PALIGUANIDAE Broom, 1926 Tr. (SCY) Terr.

First and Last: *Paliguana whitei* Broom, 1903, ?*Lystrosaurus–Procolophon* Assemblage Zone, Tarkastad, Karoo Basin, South Africa.

?F. MONJUROSUCHIDAE Endo, 1940 J. (u) Terr.

First and Last: *Monjurosuchus splendens* Endo, 1940, Chiufotang Formation, Lingyung Basin, Manchuria, China.

F. NANCHANGOSAURIDAE Wang, 1959
Tr. (ANS) Mar.

First and Last: *Nanchangosaurus suni* Wang, 1959 and *Hupehsuchus nanchangensis* Young and Dong, 1972, Jialingjiang Formation and Daye Limestone, Hubei Province, China (Carroll and Dong, 1991).

Division YOUNGINIFORMES Romer, 1933

F. UNNAMED P. (TAT) Terr. (see Fig. 39.2)

First and Last: *Acerosodontosaurus piveteaui* Currie, 1980, lower Sakamena Formation, Sakamena River Valley, Madagascar.

F. YOUNGINIDAE Broom, 1914 P. (TAT) Terr.

First and Last: *Youngina capensis* Broom, 1914, *Dicynodon–Theriognathus* Assemblage Zone, New Bethesda, Karoo Basin, South Africa.

F. TANGASAURIDAE Camp, 1945
P. (TAT)–Tr. (SCY) FW

First: *Hovasaurus boulei* Piveteau, 1926 and *Thadeosaurus colcanapi* Carroll, 1981, Lower Sakamena Formation, Benenitra to Ranohira region, Madagascar; *Tangasaurus mennelli* Haughton, 1924, 'Upper Permian', Tanga, Tanzania.
Last: *Kenyasaurus mariakaniensis* Harris and Carroll, 1977, Maji ya Chumvi Beds, lower Middle Duruma Sandstone Series, Mariakani, Kenya.

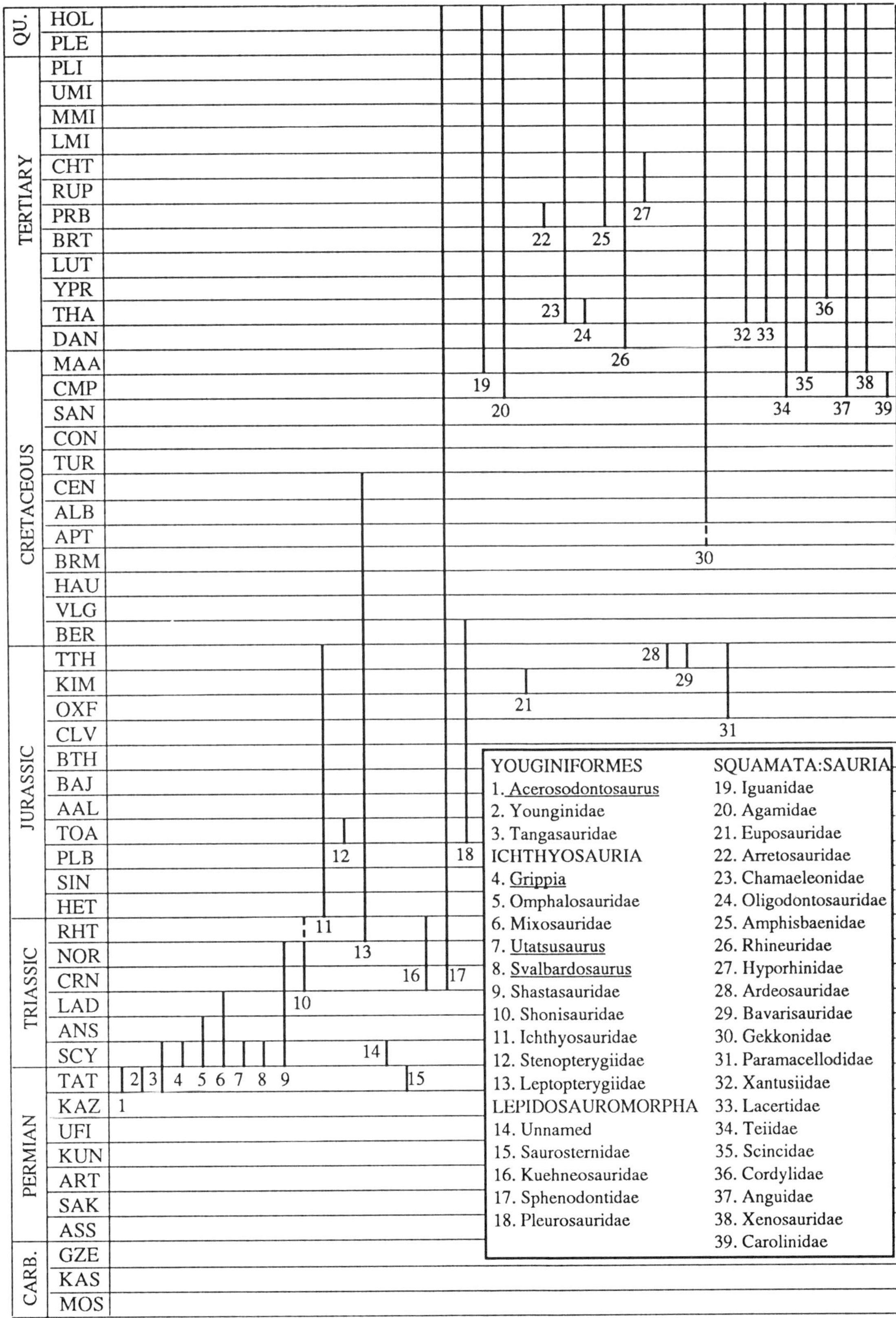

Fig. 39.2

Order ICHTHYOSAURIA de Blainville, 1835

Ichthyosaur classification and stratigraphical distributions are taken from Mazin (1982, 1988) and Massare and Callaway (1990), but there has been no recent comprehensive phylogenetic analysis of the group, and the families are rather fluid in composition. Massare and Callaway (1990) argue that the ichthyosaurs are closely related to the younginiforms.

F. UNNAMED Tr. (SCY) Mar.

First and Last: *Grippia longirostris* Wiman, 1928, Sticky Keep Formation (Spathian), Svalbard, Spitsbergen.

F. OMPHALOSAURIDAE Merriam, 1906 Tr. (SCY–ANS) Mar.

First: *Omphalosaurus nettarhynchus* Mazin and Bucher, 1987, Prida Formation (Spathian), Humboldt Range, Nevada, USA.
Last: *Omphalosaurus nevadanus* Merriam, 1906, Prida Formation, West Humboldt Range, Nevada, USA.

F. MIXOSAURIDAE Baur, 1887 Tr. (SCY–LAD) Mar.

First: *Mixosaurus* cf. *M. nordenskioeldii* (Hulke, 1873), Sulphur Mountain Formation, Wapiti Lake, British Columbia, Canada.
Last: *Mixosaurus nordenskioeldii* (Hulke, 1873), Tschermakfjellet Formation, Svalbard, Spitsbergen; *Mixosaurus* sp., upper Muschelkalk, Bavaria and Baden-Württemberg, Germany.
Intervening: ANS.

F. UNNAMED Tr. (SCY) Mar.

First and Last: *Utatsusaurus hataii* Shikama, Kamei and Murata, 1978, Osawa Formation, northern Honshu, Japan.

F. UNNAMED Tr. (SCY) Mar.

First and Last: *Svalbardosaurus crassidens* Mazin, 1981, Sticky Keep Formation, Svalbard, Spitsbergen.

F. SHASTASAURIDAE Merriam, 1902 Tr. (SCY–NOR) Mar.

First: *Cymbospondylus* sp., Thaynes Formation, Idaho, USA (Massare and Callaway, 1992).
Last: *Shastasaurus* cf. *S. osmonti* Merriam, 1902, Martin Bridge Formation, Wallowa Mountains, Oregon, USA.
Intervening: ANS–CRN.
Comment: Mazin (1988, p. 54) notes possible RHT shastasaurids from the Germanic Basin.

F. SHONISAURIDAE Camp, 1980 Tr. (CRN–NOR/RHT) Mar.

First: *Shonisaurus popularis* Camp, 1976, *S. mulleri* Camp, 1976, and *S. silberlingi* Camp, 1976, Luning Formation, Nye County, Nevada, USA.
Last: *Shonisaurus* sp., Kössen Formation, Switzerland.

F. ICHTHYOSAURIDAE Bonaparte, 1841 J. (HET–TTH) Mar.

First: *Ichthyosaurus communis* Conybeare, 1821, lower Lias (*Psiloceras planorbis* Zone), Somerset, England, UK.
Last: ?*Ophthalmosaurus* sp., Purbeck Beds, Dorset, England, UK.
Intervening: SIN.
Comment: McGowan (1978) notes a humerus of *Ichthyosaurus* sp. from the Lower Cretaceous of North-west Territories, Canada, but its exact age is uncertain.

F. STENOPTERYGIIDAE Kuhn, 1934 J. (TOA) Mar.

First: *Stenopterygius quadriscissus* (Quenstedt, 1858), and six other species, Posidonienschiefer (*Dactylioceras tenuicostatum* and *Harpoceras falciferum* Zones), Baden-Württemberg, Germany.
Last: *Stenopterygius acutirostris* (Owen, 1840), Alum Shales Formation (*Hildoceras bifrons* Zone), Yorkshire, England, UK.

F. LEPTOPTERYGIIDAE Kuhn, 1934 Tr. (RHT)–K. (CEN) Mar.

First: *Leptopterygius tenuirostris* Conybeare, 1822, Kössen Formation, Switzerland (McGowan, 1989).
Last: ??*Platypterygius* sp., lower SAN, Western Australia (Wade, 1990). If these are derived from older rocks, then *Platypterygius americanus* (Nace, 1939), Mowry Shales, Wyoming, USA, *P. kipijanoffi* (Romer, 1968), Sewerisch Sandstone, former USSR, and *P. campylodon* (Carter, 1846), upper Greensand and Lower Chalk, Cambridgeshire and Kent, England, UK (all CEN), are the youngest.
Intervening: HET, SIN, BTH, CLV, KIM, TTH, HAU–CEN.
Comment: Supposedly later ichthyosaurs, one from the New Egypt Formation of New Jersey, USA (late MAA in age), and one from the Bearpaw Shale of Saskatchewan, Canada (late CMP), turn out to be isolated bones of plesiosaurs (Baird, 1984). Ventura (1984) notes a Miocene ichthyosaur from Malta, but is doubtful of its true provenance!

Division LEPIDOSAUROMORPHA Benton, 1983

F. UNNAMED Tr. (SCY) Terr.

First and Last: *Palaeagama vielhaueri* Broom, 1926, ?*Lystrosaurus–Procolophon* Assemblage Zone, Mount Frere district, Karoo Basin, South Africa.

STEM-GROUP LEPIDOSAURIA

?F. SAUROSTERNIDAE Haughton, 1924 P. (TAT) Terr.

First and Last: *Saurosternon bainii* Huxley, 1868, *Cistecephalus* or *Dicynodon–Theriognathus* Assemblage Zone, Sneeuwberg, Karoo Basin, South Africa.

F. KUEHNEOSAURIDAE Romer, 1966 Tr. (CRN–RHT) Terr.

First: *Icarosaurus siefkeri* Colbert, 1966, Lockatong Formation (upper Carnian), North Bergen, New Jersey, USA; and '?kuehneosaur jaw fragments', Petrified Forest Member (uppermost Carnian), Chinle Formation, St Johns, Arizona, USA (Murry, 1987).
Last: *Kuehneosaurus latus* Robinson, 1962, Pant-y-ffynon Quarry, Glamorgan, Wales (Crush, 1983).
Comment: Pant-y-ffynon Quarry is dated as RHT. The type material of *K. latus* comes from Emborough Quarry, Somerset, England, UK, whose age is probably NOR, but this is not certain. Later supposed kuehneosaurs, or close relatives, such as *Cteniogenys antiquus* Gilmore, 1928 from the Upper Jurassic and *Litakis gilmorei* Estes, 1964 from the Upper Cretaceous (Estes, 1983) are very doubtful. *Cteniogenys* has been reclassified as a choristodere (Evans, 1989).

Superorder LEPIDOSAURIA Haeckel, 1866 (Duméril and Bibron, 1839)

Order SPHENODONTIA Williston, 1925

F. SPHENODONTIDAE Cope, 1870 Tr. (CRN)–Rec. Terr. (p)

First: 'sphenodontian cf. *Planocephalosaurus*', Turkey

Branch Formation, ?lower CRN, Virginia, USA (Sues and Olsen, 1990); *Brachyrhinodon taylori* Huene, 1912, Lossiemouth Sandstone Formation, ?Upper CRN, Elgin, Scotland, UK. **Extant**
Intervening: NOR, KIM, TTH, APT.
Comment: *Brachyrhinodon* is probably the oldest confirmed sphenodontid. Other upper Upper CRN examples have been reported from Arizona, New Mexico and Texas (Murry, 1986, 1987). Other Upper Triassic taxa from England, Germany, Zimbabwe, and the USA (Fraser and Benton, 1989) are probably NOR in age. Older supposed sphenodontids, such as *Palacrodon* from the Lower Triassic of South Africa, and *Anisodontosaurus* from the Middle Triassic of Arizona, may be procolophonids (Murry, 1987; Evans, 1988). *Elachistosuchus* is an archosauromorph (Evans, 1988). The family Sphenodontidae, as presented here, is paraphyletic because of the exclusion of the Pleurosauridae. *Sapheosaurus* is included here within the Sphenodontidae (Evans, 1988; Fraser and Benton, 1989) and is not given in a separate family. In addition, *Gephyrosaurus bridensis* Evans, 1980, from a fissure fill, Glamorgan, South Wales, UK (HET/SIN), is included within Sphenodontidae, and is not regarded as the representative of a separate family (Fraser and Benton, 1989).

F. PLEUROSAURIDAE Lydekker, 1888 J. (TOA)–K. (BER) Mar.

First: *Palaeopleurosaurus posidoniae* Carroll, 1985, Posidonienschiefer, Baden-Württemberg, Germany (Carroll, 1985).
Last: *Pleurosaurus ginsburgi* Fabre, 1974, Gisement des Bessons, Var, France.
Intervening: KIM, TTH.

Order SQUAMATA Oppel, 1811

Cladistic analyses of squamates (Evans, 1984; Estes *et al.*, 1988; Rieppel, 1988) show that the snakes (Serpentes) are a monophyletic group nested among the squamates. Hence the lizards (Sauria) form a paraphyletic group, which is retained here. The location of Serpentes among the 'lizard' groups is uncertain. The classification and stratigraphical distribution of families of Squamata are based on Estes (1983), Estes *et al.* (1988), and Rage (1984).

Suborder SAURIA McCartney, 1802 (p)

Infra-order IGUANIA Cuvier, 1817 (Cope, 1864)

F. IGUANIDAE Gray, 1827 K. (MAA)–Rec. Terr.

First: *Pristiguana brasiliensis* Estes and Price, 1973, Baurú Formation, Minas Gerais, Brazil. **Extant**
Intervening: THA–HOL.

F. AGAMIDAE Gray, 1827 K. (CMP)–Rec. Terr.

First: *Mimeosaurus crassus* Gilmore, 1943, Djadokhta Formation, Mongolia. **Extant**
Intervening: THA–PRB, UMI–HOL.

F. EUPOSAURIDAE Camp, 1923 J. (KIM) Terr.

First and Last: *Euposaurus thiollierei* Lortet, 1892, *E. cirinensis* Lortet, 1892, and *E. lorteti* Hoffstetter, 1964, all from Calcaire lithographique, Cerin (Ain), France.

?F. ARRETOSAURIDAE Gilmore, 1943 T. (PRB) Terr.

First and Last: *Arretosaurus ornatus* Gilmore, 1943, Ulan Gochu Formation, Shara Murun, Mongolia.

F. CHAMAELEONIDAE Gray, 1825 T. (THA)–Rec. Terr.

First: *Anquingosaurus brevicephalus* Hou, 1976, Wang-Hu-Dun Series, Qian-Shan District, Anhui, China. **Extant**
Intervening: LMI–UMI, PLE, HOL.

Infra-order SCLEROGLOSSA Estes, de Queiroz and Gauthier, 1988

SCLEROGLOSSA *incertae sedis*

Included here are the Dibamidae, Amphisbaenia and Serpentes (listing follows all the 'lizard' groups), according to Estes *et al.* (1988).

F. DIBAMIDAE Boulenger, 1884 **Extant** Terr.

Parvorder AMPHISBAENIA Gray, 1844

F. OLIGODONTOSAURIDAE Estes, 1975 T. (THA) Terr.

First and Last: *Oligodontosaurus wyomingensis* Gilmore, 1942, Fort Union Formation, Park County, Wyoming, and *Oligodontosaurus* sp., Tongue River Formation, Carter County, Montana and Bison Basin deposits, Fremont County, Wyoming, USA.
Comment: Possible amphisbaenians have been reported from the Upper Cretaceous (?MAA) of Spain (Astibia *et al.*, 1990).

F. AMPHISBAENIDAE Gray, 1865 T. (PRB)–Rec. Terr.

First: *Omoiotyphlops edwardsi* (de Rochebrune, 1884), Phosphorites de Quercy, France. **Extant**
Intervening: LMI, UMI.

F. RHINEURIDAE Vanzolini, 1951 T. (DAN)–Rec. Terr.

First: *Plesiorhineura tsentasi* Sullivan, 1985, upper part of Nacimiento Formation (Torrejonian), Torreon Wash, New Mexico, USA (Sullivan, 1985). **Extant**
Intervening: THA–LUT, RUP–UMI, PLE.

F. HYPORHINIDAE Baur, 1893 T. (RUP–CHT) Terr.

First: *Hyporhina tertia* Berman, 1972, White River Formation, Fremont County, Wyoming, USA.
Last: *Hyporhina antiqua* Baur, 1893, White River Formation, Washington County, South Dakota, USA.

F. BIPEDIDAE Taylor, 1951 **Extant** Terr.

F. TROGONOPHIDAE Gray, 1865 **Extant** Terr.

Parvorder GEKKOTA Cuvier, 1817

F. ARDEOSAURIDAE Camp, 1923 J. (TTH) Terr.

First and Last: *Ardeosaurus brevipes* Meyer, 1855, *A. digitalellus* Grier, 1914, and *Eichstaettosaurus schroederi* Broili, 1938, Solnhofener Schichten, Eichstätt, Germany.

Comment: Estes (1983) includes *Yabeinosaurus tenuis* Endo and Shikama, 1942, Tsaotzushan Formation, Manchuria, and *Y. youngi* Hoffstetter, 1964, Ketzutung, Liaoning, China, in this family. These occurrences are dated merely as 'Upper Jurassic', so may extend the range of the family.

F. BAVARISAURIDAE Kuhn, 1961 J. (TTH) Terr.

First and Last: *Bavarisaurus macrodactylus* Wagner, 1852 and *Palaeolacerta bavarica* Cocude-Michel, 1961, Solnhofener Schichten, Bavaria, Germany.

F. GEKKONIDAE Bonaparte, 1831 K. (APT/ALB)–Rec. Terr.

First: *Hoburogekko suchanovi* Alifanov, 1989, APT/ALB, Ubur-Khangay aymak, Mongolia (Alifanov, 1989). **Extant**
Intervening: MAA, THA, YPR, PRB, RUP, LMI–HOL.

F. PYGOPODIDAE Gray, 1845 **Extant** Terr.

Parvorder SCINCOMORPHA Camp, 1923

F. PARAMACELLODIDAE Estes, 1983 J. (OXF–TTH) Terr.

First: *Becklesius hoffstetteri* (Seiffert, 1973), *Saurillodon proraformis* (Seiffert, 1973), and *S. henkeli* (Seiffert, 1973), Guimarota Lignite Mine, Leiria, Portugal.
Last: *Paramacellodus oweni* Hoffstetter, 1967, *Pseudosaurillus becklesi* Hoffstetter, 1967 and *Saurillus obtusus* Owen, 1854, Purbeck Beds, Dorset, England, UK.
Intervening: KIM.

F. XANTUSIIDAE Baird, 1859 T. (THA)–Rec. Terr.

First: *Palaeoxantusia fera* Hecht, 1956, Tongue River Formation, Carter County, Montana, and Fort Union Formation, Carbon County, Wyoming, USA. **Extant**
Intervening: LUT–RUP, UMI, PLE, HOL.
Comment: *Eoxanta lacertifrons* Borsuk-Bialynicka, 1988, from the red beds of Khermeen Tsav (?middle Campanian) of the Gobi Desert, Mongolia, is classified as the sister group of the Xantusiidae (Borsuk-Bialynicka, 1988).

F. LACERTIDAE Gray, 1825 T. (THA)–Rec. Terr.

First: *Plesiolacerta? paleocenicus* (Kuhn, 1940) and *Pseudeumeces? wahlbeckensis* (Kuhn, 1940), upper Palaeocene, Wahlbeck, Germany. **Extant**
Intervening: YPR–HOL.
Comment: These two earliest species are based on material now lost, and hence diagnosis is uncertain (Estes, 1983). The next oldest lacertid is a specimen of *Plesiolacerta lydekkeri* Hoffstetter, 1942, from the Lower Eocene of Dormaal, Belgium.

F. TEIIDAE Gray, 1827 K. (CMP)–Rec. Terr.

First: *?Polyglyphanodon sternbergi* Gilmore, 1940, El Gallo Formation, Baja California, Mexico (?middle CMP). **Extant**
Intervening: MAA, THA, CHT, LMI–HOL.
Comment: Numerous other teiids are reported from the CMP and MAA of North America and Mongolia, such as *Adamisaurus, Chamops, Cherminosaurus, Darchanosaurus, Erdenetesaurus, Haptosphenus, Leptochamops, Macrocephalosaurus, Meniscognathus, Paraglyphanodon* and *Peneteius*, but the exact ages of certain of the Mongolian formations in particular are uncertain.

F. GYMNOPHTHALMIDAE Merrem, 1820 **Extant** Terr.

F. SCINCIDAE Gray, 1825 K. (MAA)–Rec. Terr.

First: *Contogenys sloani* Estes, 1969, Hell Creek Formation, McCone County, Montana, and *Sauriscus cooki* Estes, 1964, Lance Formation, Niobrara County, Wyoming, USA. **Extant**
Intervening: DAN, THA, PRB, PUR–HOL.

F. CORDYLIDAE Gray, 1837 T. (YPR)–Rec. Terr.

First: *Pseudolacerta* sp., Lower Eocene, Dormaal, Belgium. **Extant**
Intervening: PRB, LMI, HOL.

Parvorder ANGUIMORPHA Fürbringer, 1900

F. ANGUIDAE Gray, 1825 K. (CMP)–Rec. Terr.

First: *Odaxosaurus piger* Gilmore, 1928, Mesaverde Formation, Natrona County, Wyoming and Judith River Formation, Chouteau County, Montana, and 'cf. *Gerrhonotus*', Fruitland Formation, San Juan County, New Mexico, USA. **Extant**
Intervening: MAA–HOL.

F. XENOSAURIDAE Cope, 1866 K. (MAA)–Rec. Terr.

First: *Exostinus lancensis* Gilmore, 1928, Lance Formation, Niobrara County, Wyoming, and Hell Creek Formation, McCone County, Montana, USA. **Extant**
Intervening: DAN–YPR, CHT.

F. CAROLINIDAE Borsuk-Bialynicka, 1985 K. (CMP) Terr.

First and Last: *Carusia intermedia* (Borsuk-Bialynicka, 1985), Red Beds of Khermeen Tsav, Omnogov, Mongolia and *Shinisauroides latipalatum* Borsuk-Bialynicka, 1985, Barun Goyot Formation, Omnogov, Mongolia.

F. DORSETISAURIDAE Hoffstetter, 1967 J. (OXF–TTH) Terr. (see Fig. 39.3)

First: *Dorsetisaurus purbeckensis* Hoffstetter, 1967, Guimarota Lignite Mine, Leiria, Portugal (described as *Introrsisaurus pollicidens* Seiffert, 1973).
Last: *Dorsetisaurus purbeckensis* Hoffstetter, 1967 and *D. hebetidens* Hoffstetter, 1967, Purbeck Beds, Dorset, England, UK.

F. NECROSAURIDAE Hoffstetter, 1943 K. (MAA)–T. (CHT) Terr.

First: *Parviderma inexacta* Borsuk-Bialynicka, 1984, Barun Goyot Formation, ?middle CMP, Khulsan, Nemegt Basin, Mongolia (Borsuk-Bialynicka, 1984).
Last: *Necrosaurus eucarinatus* (Kuhn, 1940), 'Middle' Oligocene, Europe (Augé, 1986).
Intervening: MAA–RUP.

F. HELODERMATIDAE Gray, 1837 K. (MAA)–Rec. Terr.

First: *Paraderma bogerti* Estes, 1964, Lance Formation, Niobrara County, Wyoming, USA. **Extant**
Intervening: THA, PRB–LMI, HOL.

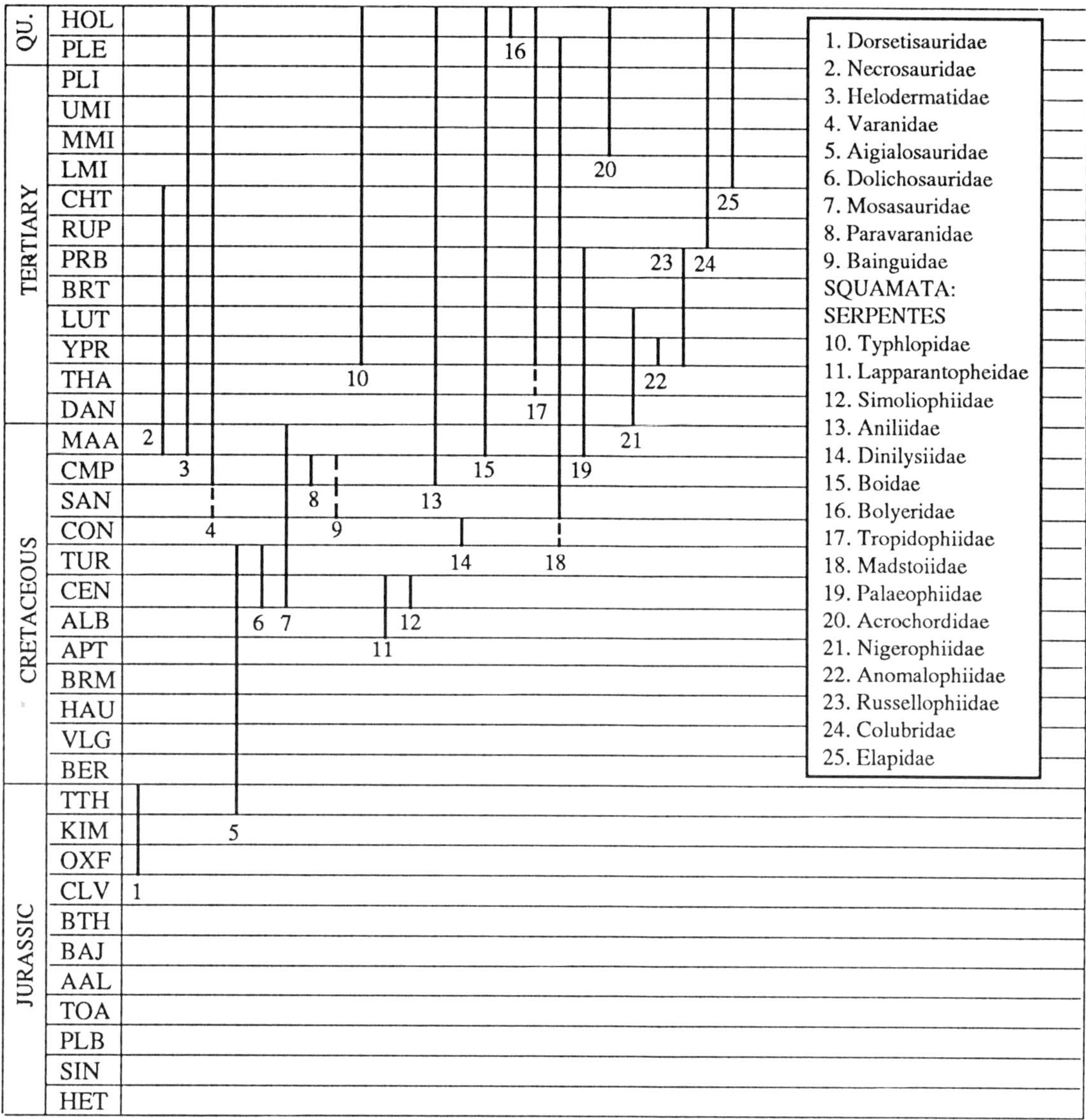

Fig. 39.3

F. VARANIDAE Gray, 1827 K. (SAN/CMP)–Rec. Terr. (includes LANTHANOTIDAE Steindachner, 1978)

First: *Telmasaurus grangeri* Gilmore, 1943, Djadokhta Formation, ?upper SAN and/or ?lower CMP, Bayn Dzak, Gobi Desert, Mongolia (Borsuk-Bialynicka, 1984). **Extant**
Intervening: CMP–HOL.

F. AIGIALOSAURIDAE Gorjanovic-Kramberger, 1892 ?J. (TTH)–K. (TUR) Mar.

First: *Proaigialosaurus huenei* Kuhn, 1958, Solnhofener Schichten, Eichstättt, Bavaria, Germany.
Last: *Aigialosaurus dalmaticus* Kramberger, 1892, *Opetiosaurus buccichi* Kornhuber, 1901, and *Carsosaurus marchesetti* Kornhuber, 1893, Fischschiefer, Lesina and Comeno, Dalmatia, former Yugoslavia (Russell, 1967).
Intervening: CEN.
Comment: *Proaigialosaurus* is based on limited material, and it is not clear whether it is an aigialosaurid or not (Carroll and Debraga, 1992). *Carsosaurus* lacks a skull, and is also of uncertain familial assignment (Carroll and Debraga, 1992).

F. DOLICHOSAURIDAE Gorjanovic-Kramberger, 1892 K. (CEN–TUR) Mar.

First: *Dolichosaurus longicollis* Owen, 1850, Lower Chalk, Kent and Sussex, England, UK (Russell, 1967); *Coniasaurus crassidens* Owen, 1850, Lower Chalk, Sussex, England and Eagle Ford Group, Texas, USA (Bell *et al.*, 1982).
Last: *Acteosaurus tommasini* Meyer, 1860, *Pontosaurus lesinensis* Kornhuber, 1873, and *Eidolosaurus* Nopcsa, 1923, Fischschiefer, Lesina and Comeno, Dalmatia, former Yugoslavia (CEN–TUR; Russell, 1967).

F. MOSASAURIDAE Gervais, 1853 K. (CEN–MAA) Mar.

First: Mosasaur jaws, Middle Chalk, Cuxton, Kent (Russell, 1967); undescribed specimens, Eagle Ford Formation (CEN/TUR), Texas, USA.

Last: *Mosasaurus hoffmanni* Mantell, 1829, Maastricht Calcarenite, upper Gulpen Formation, Maastricht, The Netherlands; *Leiodon sectorius* Cope, 1871, Tuffeau of Maastricht, Maastricht Formation, Maastricht, The Netherlands; *Carinodens fraasi* (Dollo, 1913) and *Plioplatecarpus marshi* Dollo, 1882, both Craie Grossierea Silex Gris, Maastricht Formation, Limburg, The Netherlands; *Goronyosaurus nigeriensis* (Swinton, 1930 *pars*) and *Mosasaurus* sp., Dukamaje Formation, Sokoto, NW Nigeria and Niger (T. Lingham-Soliar, pers. comm., 1992); *Mosasaurus dekayi* Bronn, 1838, *M. maximus* Cope, 1869, *Liodon sectorius* Cope, 1871, and *Plioplatecarpus depressus* (Cope, 1869), all 'Navesink Formation and younger Cretaceous' (Russell, 1967), New Jersey, USA; *Prognathodon rapax* (Hay, 1902) and *Halisaurus platyspondylus* Marsh, 1869, New Egypt Formation, New Jersey, USA; all upper MAA (Russell, 1967).
Intervening: TUR–CMP.

SAURIA *incertae sedis*

F. PARAVARANIDAE Borsuk-Bialynicka, 1984
K. (CMP) Terr.

First and Last: *Paravaranus angustifrons* Borsuk-Bialynicka, 1984, Barun Goyot Formation, ?middle CMP, Khulsan, Nemegt Basin, Mongolia (Borsuk-Bialynicka, 1984).

F. BAINGUIDAE Borsuk-Bialynicka, 1984
K. (SAN/CMP) Terr.

First and Last: *Bainguis parvus* Borsuk-Bialynicka, 1984, Djadokhta Formation, ?upper SAN and/or ?lower CMP, Bayn Dzak, Gobi Desert, Mongolia (Borsuk-Bialynicka, 1984).

Suborder SERPENTES Linnaeus, 1758

The classification, and distributions, are taken from Rage (1984), unless otherwise stated.

Suborder SCOLECOPHIDIA Duméril and Bibron, 1844

F. ?TYPHLOPIDAE Gray, 1825
T. (YPR)–Rec. Terr.

First: Scolecophidia indet., Lower Eocene, Dormaal, Belgium. **Extant**
Intervening: PRB, MMI, HOL.

Suborder ALETHINOPHIDIA Hoffstetter, 1955

Superfamily SIMOLIOPHEOIDEA Nopcsa, 1925

F. LAPPARANTOPHIIDAE Hoffstetter, 1968
K. (ALB/CEN–CEN) Terr.

First: *Lapparantophis defrennei* Hoffstetter, 1960, 'Continental intercalaire', In Akhamil, Algeria.
Last: *Pouitella pervetus* Rage, 1988, lower or middle CEN, Maine-et-Loire, France (Rage, 1988).

F. SIMOLIOPHIIDAE Nopcsa, 1925
K. (CEN) Mar.

First and Last: *Simoliophis rochebrunei* Sauvage, 1880, CEN, south-western France, Portugal; *Simoliophis* sp., CEN, Egypt.

Superfamily ANILIOIDEA Fitzinger, 1826

F. ANILIIDAE Fitzinger, 1826
K. (CMP)–Rec. Terr.

First: *Coniophis cosgriffi* Armstrong-Ziegler, 1978, Fruitland Formation, New Mexico, USA. **Extant**
Intervening: MAA, LUT, PRB, MMI, PLI.

F. UROPELTIDAE Müller, 1832 **Extant** Terr.

Superfamily BOOIDEA Hoffstetter, 1955

F. DINILYSIIDAE Romer, 1956 K. (CON) Terr.

First and Last: *Dinilysia patagonica* Woodward, 1901, Rio Colorado Formation, Neuquén Group, Neuquén, Argentina.
Comment: The Rio Colorado Formation has been dated as tentatively CON by Bonaparte (1991).

F. XENOPELTIDAE Bonaparte, 1845 **Extant** Terr.

F. BOIDAE Gray, 1825 K. (MAA)–Rec. Terr.

First: Indeterminate boid, Hell Creek Formation, Montana, USA, and from equivalent deposits in Portugal and India (J.-C. Rage, pers. comm., 1991). **Extant**
Intervening: DAN–PLE.

F. BOLYERIIDAE Hoffstetter, 1946
Q. (HOL)–Rec. Terr.

First: Subfossil *Casarea*, Mauritius. **Extant**

F. TROPIDOPHIIDAE Cope, 1894
T. (THA?)–Rec. Terr.

First: 'Tropidophiid', Palaeocene, South America. **Extant**
Intervening: YPR.

F. MADSTOIIDAE Hoffstetter, 1961
K. (CON/SAN)–Q. (PLE) Terr.

First: *Madstoia* aff. *M. madagascariensis*, lower Senonian, Niger.
Last: *Wonambi naracoortensis* Smith, 1976, upper PLE, South Australia.
Intervening: CMP, YPR, PRB.

F. PALAEOPHIIDAE Lydekker, 1888
K. (MAA)–T. (PRB) FW

First: *Palaeophis* sp., MAA, Morocco.
Last: *Pterosphenus schucherti* Lucas, 1889, Jackson Formation, Alabama, USA; *P. schweinfurthi* (Andrews, 1901), Qasr el Sagha Formation, Fayûm Basin, Egypt; and *P. sheppardi* Hoffstetter, 1958, Seca Formation, Ancon, Ecuador.
Intervening: DAN–BRT.

Superfamily ACROCHORDOIDEA Bonaparte, 1838

F. ACROCHORDIDAE Bonaparte, 1838
T. (MMI)–Rec. FW/Mar.

First: *Acrochordus dehmi* Hoffstetter, 1964, Chinji Formation, Siwalik Group, Chhoinja, Pakistan. **Extant**

F. NIGEROPHIIDAE Rage, 1975
T. (DAN–LUT) FW/Mar.

First: *Nigerophis mirus* Rage, 1975, Palaeocene, Krebb de Sessao, Niger.
Last: *Woutersophis novus* Rage, 1980, Bruxellian, Brussels, Belgium.

Superfamily COLUBROIDEA Fitzinger, 1826

F. ANOMALOPHIIDAE Oppel, 1811 T. (YPR) FW

First and Last: *Anomalophis bolcensis* (Massalongo, 1859), Monte Bolca, Veneto, Italy.

F. RUSSELLOPHIIDAE Rage, 1978 T. (YPR–PRB) FW

First: *Russellophis* sp., Lower Eocene, Dormaal, Belgium.
Last: Russellophiid indet, Phosphorites de Quercy, France.

F. COLUBRIDAE Gray, 1825 T. (RUP)–Rec. Terr.

First: *Coluber cadurci* Rage, 1974 and *Natrix mlynarskii* Rage, 1988, Phosphorites de Quercy, upper Suevian, Lot, France. **Extant**
Intervening: CHT–PLE.

F. ELAPIDAE Boie, 1827 T. (LMI)–Rec. FW/Mar.

First: *Naja romani* (Hoffstetter, 1939), Orleanian, France. **Extant**
Intervening: MMI, PLI.

F. VIPERIDAE Gray, 1825 T. (LMI)–Rec. Terr. (see Fig. 39.4)

First: Viperid, Agenian, France. **Extant**
Intervening: MMI, UMI–PLE.

Division ARCHOSAUROMORPHA Huene, 1946

Order CHORISTODERA Cope, 1876

Relationships of the group, and division into families, are based on the cladogram of Evans (1990).

F. PACHYSTROPHEIDAE Kuhn, 1961 Tr. (RHT) FW/Mar.

First and Last: *Pachystropheus rhaeticus* E. von Huene, 1935, Rhaetic Bone Bed, southern England, UK, Germany.

F. UNNAMED J. (BTH–KIM) FW

First: *Cteniogenys antiquus* Gilmore, 1928, Chipping Norton Formation, lower BTH, Gloucestershire, England, UK (Metcalf *et al.*, 1993); Forest Marble Formation, upper BTH, Oxfordshire, England, UK (Evans, 1989).
Last: *Cteniogenys antiquus* Gilmore, 1928, Morrison Formation, Wyoming, USA.
Intervening: OXF.
Comments: The familial assignment of these early choristoderes has not been confirmed, and relationships to the pachystropheids and to later champsosaurs are unclear at present. Earlier choristoderes, perhaps belonging to this group have been noted from the Kayenta Formation (SIN/PLB: J. Clark, pers. comm., 1991).

F. CHAMPSOSAURIDAE Cope, 1876 (1884?) K. (APT/ALB)–T. (RUP) FW

First: *Tchoiria namsarai* Efimov, 1985, Lower Cretaceous, Khamaril-Khural, Mongolia.
Last: *Lazarussuchus inexpectatus* Hecht, 1992, Stampian Limestone, Armissan quarry, Aude, France (Hecht, 1992).
Intervening: CMP–YPR.

F. SIMOEDOSAURIDAE Lemoine, 1884 K. (ALB/CEN)–T. (YPR?) FW

First: *Ikechosaurus sunailinae* Sigogneau-Russell, 1981, Middle Cretaceous, Otok District, Inner Mongolia (Sigogneau-Russell, 1981).
Last: *Simoedosaurus* sp., Clarkforkian, Park County, Wyoming, USA (Uppermost Palaeocene or Lowermost Eocene).
Intervening: CMP–THA.

Order RHYNCHOSAURIA Osborn, 1903

F. RHYNCHOSAURIDAE Huxley, 1887 (Cope, 1870) Tr. (SCY–CRN) Terr.

First: *Howesia browni* Broom, 1905 and *Mesosuchus browni* Watson, 1912, *Cynognathus–Diademodon* Assemblage Zone, Karoo Basin, South Africa.
Last: *Hyperodapedon gordoni* Huxley, 1859, Lossiemouth Sandstone Formation, Morayshire, Scotland, UK; *Scaphonyx sanjuanensis* Sill, 1970, Ischigualasto Formation, San Juan, Argentina; *Scaphonyx sulcognathus* Azevedo and Schultz, 1988, Caturrita Formation, Rio Grande do Sul, Brazil; *Otischalkia elderae* Hunt and Lucas, 1991, Dockum Group, Texas, USA; undescribed rhynchosaur, Wolfville Formation, Nova Scotia, Canada (Hunt and Lucas, 1991b).
Intervening: ANS, LAD.
Comment: Carroll (1976) suggested that *Noteosuchus colletti* (Watson, 1912) from the *Lystrosaurus–Procolophon* Assemblage Zone of South Africa was the oldest rhynchosaur, but it lacks diagnostic characters of the group (Benton, 1985). Other upper Carnian rhynchosaurs are known, but these are dated as 'early' late Carnian by Hunt and Lucas (1991b), while the 'Lasts' listed above are given as 'late' late Carnian.

Order THALATTOSAURIA Merriam, 1904

F. ASKEPTOSAURIDAE Kuhn, 1952 Tr. (ANS/LAD) Mar.

First: *Askeptosaurus italicus* Nopcsa, 1925, Grenzbitumenzone, Monte San Giorgio, Kt. Tessin, Switzerland.
Last: Thalattosaurid indet., Pardonet Formation, British Columbia, Canada (Storrs, 1991b).

F. THALATTOSAURIDAE Merriam, 1904 Tr. (CRN–NOR) Mar.

First: *Thalattosaurus alexandrae* Merriam, 1904, and *Nectosaurus balius* Merriam, 1905, Hosselkus Limestone, California, USA.
Last: Thalattosaurid indet., Pardonet Formation, British Columbia, Canada (Storrs, 1991b).

F. CLARAZIIDAE Peyer, 1936 Tr. (ANS/LAD) Mar.

First and Last: *Clarazia schinzi* Peyer, 1936, and *Hescheleria ruebeli* Peyer, 1936, Grenzbitumenzone, Monte San Giorgio, Kt. Tessin, Switzerland.

F. TRILOPHOSAURIDAE Gregory, 1945 Tr. (CRN–RHT) Terr.

First: *Trilophosaurus buettneri* Case, 1928, lower Dockum Group, Crosby County, Texas, USA.
Last: *Tricuspisaurus thomasi* Robinson, 1957, Upper Triassic (?NOR/RHT), Ruthin Quarry fissure, Glamorgan, Wales, UK (Fraser, 1986).
Intervening: NOR.

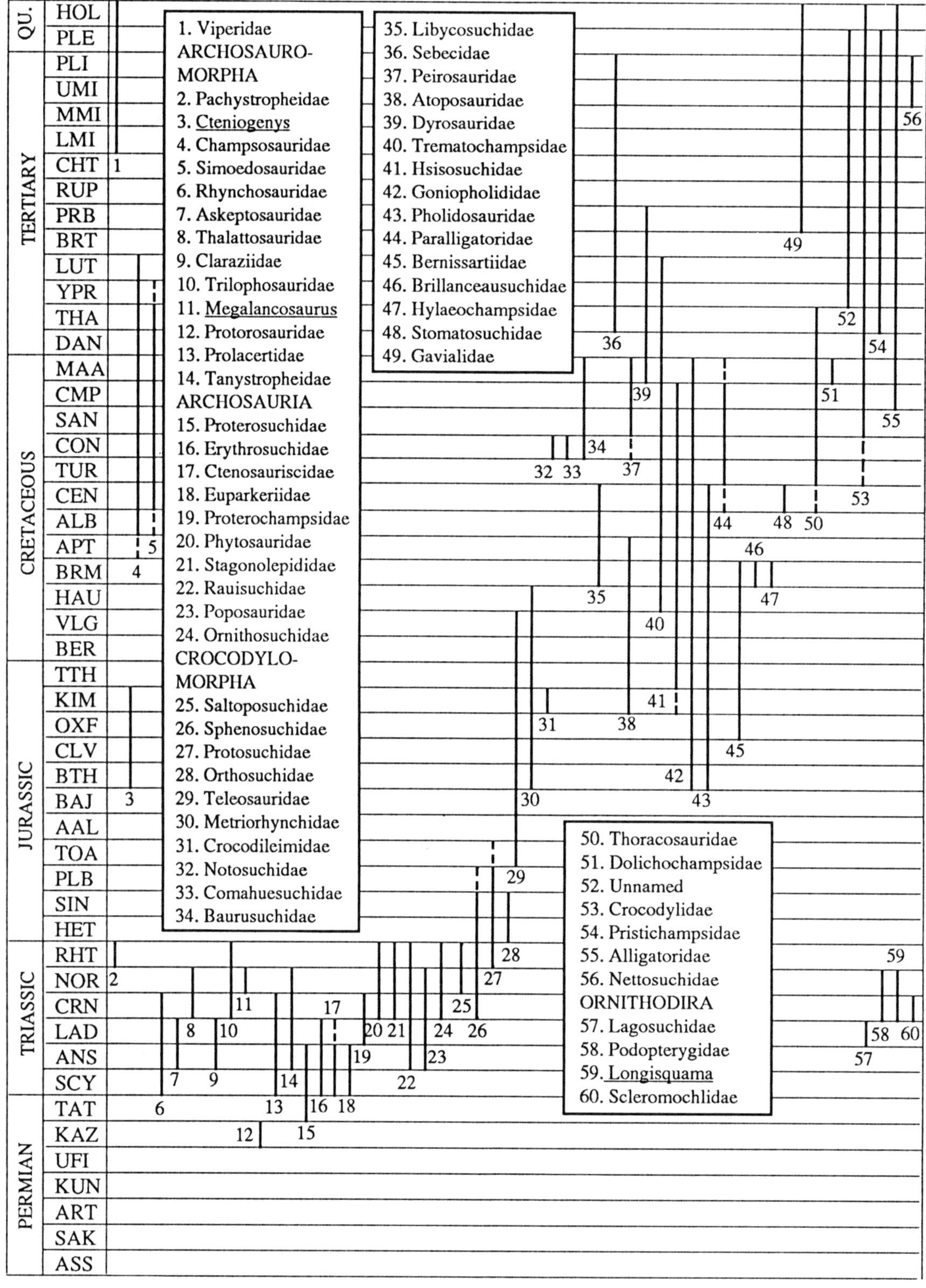

Fig. 39.4

Comment: Earlier supposed trilophosaurids, such as *Doniceps*, *Anisodontosaurus*, and *Gomphiosaurus*, as well as *Toxolophosaurus* from the Lower Cretaceous, are probably not trilophosaurids (Murry, 1987). It is unclear whether *Tricuspisaurus* and *Variodens*, both from the English–Welsh fissures, are trilophosaurids or not.

F. UNNAMED Tr. (NOR) Terr.

First and Last: *Megalancosaurus preonensis* Calzavara *et al.*, 1980, Dolomia di Forni, Udine, Italy.

Order PROLACERTIFORMES Camp, 1945

F. PROTOROSAURIDAE Baur, 1889 (Cope, 1871) P. (KAZ) Terr.

First and Last: *Protorosaurus speneri* Meyer, 1832, Kupferschiefer, Thuringia, Germany.

F. PROLACERTIDAE Parrington, 1935 Tr. (SCY–CRN) Terr.

First: *Prolacerta broomi* Parrington, 1935, *Lystrosaurus–Procolophon* Assemblage Zone, Karoo Basin, South Africa, and Fremouw Formation, Antarctica.
Last: *Malerisaurus robinsonae* Chatterjee, 1980, Maleri Formation, Andhra Pradesh, India; and *M. langstoni* Chatterjee, 1986, Tecovas Member, lower Dockum Group, Howard County, Texas, USA.
Intervening: ANS, LAD.

F. TANYSTROPHEIDAE Romer, 1945 (Gervais, 1859) Tr. (ANS–NOR) Terr./Mar.

First: *'Tanystropheus' conspicuus* Huene, 1931, Obere Buntsandstein, southern Germany.
Last: *Tanystropheus fossai* Wild, 1980, Argillite di Riva di Solto, Val Brembana, Italy.
Intervening: LAD, CRN.

Subdivision ARCHOSAURIA Cope, 1869

The classification of basal archosaurs ('thecodontians') is based on Benton and Clark (1988), Sereno and Arcucci (1990), and Sereno (1991b), and stratigraphical distributions are taken from Charig (1976), and more recent references noted.

F. PROTEROSUCHIDAE Huene, 1908 P. (TAT)–Tr. (ANS) Terr./FW

First: *Archosaurus rossicus* Tatarinov, 1960, Zone IV, Vladimir region, former USSR.
Last: *Chasmatosaurus ultimus* Young, 1964, Er-Ma-Ying Series, SE Shansi Province, China.
Intervening: SCY.

F. ERYTHROSUCHIDAE Watson, 1917 Tr. (SCY–LAD) Terr.

First: *Fugusuchus hejiapensis* Cheng, 1980, He Shang-gou Formation, Fuku County, Shanxi Province, North China, and *Garjainia prima* Ochev, 1958, Yarenskian Horizon, upper part of Stage V, Orenburg region, southern Urals, former USSR, both mid to late Scythian in age.
Last: *Cuyosuchus huenei* Reig, 1961, Cacheuta Formation, Mendoza Province, Argentina.
Intervening: ANS.

F. CTENOSAURISCIDAE Kuhn, 1964 Tr. (SCY–ANS/LAD) Terr.

First: *Ctenosauriscus koeneni* (Huene, 1902), Mittlere Buntsandstein, Göttingen, Germany.
Last: *Lotosaurus adentus* Zhang, 1975, Batung Formation, Hunan, China.
Comment: These two taxa of (?) archosaurs share long dorsal neural spines, but their systematic position is uncertain. It is not clear whether they are related to each other or not.

F. EUPARKERIIDAE Huene, 1920 Tr. (SCY–ANS) Terr.

First: *Euparkeria capensis* Broom, 1913, *Cynognathus–Diademodon* Assemblage Zone, Karoo Basin, South Africa.
Last: *Halazhaisuchus giaoensis* Wu, 1982, Lower Er-Ma-Ying Formation, Shansi Province, China; *Turfanosuchus dabanensis* Young, 1973, Er-Ma-Ying equivalent, Sinkiang, China; and *T. shageduensis* Wu, 1982, Lower Er-Ma-Ying Formation, Inner Mongolia, China.
Comment: *Turfanosuchus* may be a pseudosuchian (J. M. Parrish, pers. comm., 1991).

F. PROTEROCHAMPSIDAE Sill, 1967 Tr. (LAD–CRN) Terr./FW

First: *Chanaresuchus bonapartei* Romer, 1971 and *Gualosuchus reigi* Romer, 1971, Chañares Formation, La Rioja Province, Argentina.
Last: *Proterochampsa barrionuevoi* Reig, 1959, Ischigualasto Formation, San Juan Province, Argentina.

Infradivision CRUROTARSI Sereno and Arcucci, 1990

F. PHYTOSAURIDAE Lydekker, 1888 Tr. (CRN–RHT) FW/Terr.

First: *'Rutiodon* sp.', Pekin Formation, middle CRN, North Carolina, USA (Olsen *et al.*, 1989).
Last: *Rutiodon* sp., Rhät, Switzerland, North Germany; 'phytosaurs', upper Passaic Formation, New Jersey, Upper New Haven Arkose, Connecticut, USA.
Intervening: NOR.
Comment: Apparently older phytosaurs, *Mesorhinosuchus fraasi* (Jaekel, 1910) from the Mittlere Buntsandstein (SCY) of Bernburg, Germany, and others from the Muschelkalk of Germany (ANS–LAD) are all doubtful records (Westphal, 1976). There are numerous upper CRN phytosaurs, *Paleorhinus bransoni* Williston, 1904, Popo Agie Formation, Fremont County, Wyoming, USA, and other species of *Paleorhinus* from Arizona and Texas, USA, Morocco, Germany, Austria and India (Hunt and Lucas, 1991a).

F. STAGONOLEPIDIDAE Lydekker, 1887 Tr. (CRN–RHT) Terr.

First: cf. *Typothorax*, Pekin Formation, middle CRN, Chatham County, North Carolina, USA (Olsen *et al.*, 1989); *Longosuchus meadei* (Sawin, 1947), lower Dockum Group, Howard County, Texas; Salitral Member, Chinle Formation, Rio Arriba County, New Mexico, USA (Hunt and Lucas, 1990).
Last: *Neoaetosauroides engaeus* Bonaparte, 1969, upper Los Colorados Formation, La Rioja, Argentina; aetosaur elements, Penarth Group ('Rhaetian'), SW England (Fraser, 1988).
Intervening: NOR.
Comment: There are numerous late CRN stagonolepidids, *Stagonolepis robertsoni* Agassiz, 1844, Lossiemouth Sandstone Formation, Morayshire, Scotland, UK; *Aetosauroides scagliai* Casamiquela, 1960 and *Argentinosuchus bonapartei* Casamiquela, 1960, Ischigualasto Formation, San Juan, Argentina; *Desmatosuchus haplocerus* (Cope, 1892), lower units of the Chinle Formation and Dockum Group, New Mexico and Texas, USA; unnamed aetosaur, Wolfville Formation, Nova Scotia, Canada.

F. RAUISUCHIDAE Huene, 1942 Tr. (ANS–RHT) Terr.

First: *Wangisuchus tzeyii* Young, 1964 and *Fenhosuchus cristatus* Young, 1964, Er-Ma-Ying Series, Shansi, China; *Vjushkovisaurus berdjanensis* Ochev, 1982, Donguz Series, Orenburg region, former USSR; *Stagonosuchus major* (Haughton, 1932) and '*Mandasuchus*', upper bone bed of the Manda Formation, Ruhuhu region, Tanzania; 'rauisuchid', Yerrapalli Formation, India.
Last: *Fasolasuchus tenax* Bonaparte, 1978, upper Los Colorados Formation, La Rioja, Argentina.
Intervening: LAD–NOR.
Comment: Some of the early 'rauisuchids' are of uncertain affinities, particularly *Fenhosuchus*, *Vjushkovisaurus*, and *Wangisuchus* (Bonaparte, 1984; Benton, 1986). Parrish (pers. comm., 1991) regards this family as non-monophyletic.

F. POPOSAURIDAE Nopcsa, 1928
Tr. (ANS–NOR) Terr.

First: *Bromsgroveia walkeri* Galton, 1985, Bromsgrove Sandstone Formation, Warwick, England, UK.
Last: Poposaurid, upper Redonda Formation, ?late NOR, New Mexico, USA.
Intervening: LAD, CRN.
Comment: If the 'last' record is not confirmed, there are several lower and middle NOR poposaurids, *Teratosaurus suevicus* Meyer, 1861, Mittlere Stubensandstein, Baden-Württemberg, Germany; *Postosuchus kirkpatricki* Chatterjee, 1985, upper Dockum Group, Texas, USA.

F. ORNITHOSUCHIDAE Huene, 1908
Tr. (CRN–RHT) Terr.

First: *Ornithosuchus longidens* Newton, 1894, Lossiemouth Sandstone Formation, Morayshire, Scotland, UK, and *Venaticosuchus rusconii* Bonaparte, 1971, Ischigualasto Formation, La Rioja, Argentina.
Last: *Riojasuchus tenuiceps* Bonaparte, 1969, upper Los Colorados Formation, La Rioja, Argentina.
Intervening: None.

Superorder CROCODYLOMORPHA Walker, 1968

The classification of crocodilomorphs is based on Clark, *in* Benton and Clark (1988), and stratigraphical distributions are taken from Steel (1973), Buffetaut (1982) and references cited below.

F. SALTOPOSUCHIDAE Crush, 1984
Tr. (NOR–RHT) Terr.

First: *Saltoposuchus connectens* Huene, 1921, Mittlere Stubensandstein, Württemberg, Germany.
Last: *Terrestrisuchus gracilis* Crush, 1984, Ruthin Quarry, Glamorgan, Wales, UK.

F. SPHENOSUCHIDAE Huene, 1922
Tr. (CRN)–J. (SIN/PLB) Terr. (p)

First: *Hesperosuchus agilis* Colbert, 1952, lower Chinle Group, Cameron, Arizona, USA.
Last: *Kayentasuchus* sp., Kayenta Formation, Arizona, USA (Clark, 1993).
Intervening: RHT, HET.
Comment: *Hallopus victor* (Marsh, 1877) is a crocodilomorph which may belong to this clade (Clark, *in* Benton and Clark, 1988). It is probably from the lower Ralston Creek Formation (CLV) of Freemont County, Colorado, USA (Norell and Storrs, 1989).

Order CROCODYLIA Gmelin, 1788

F. PROTOSUCHIDAE Brown, 1934
Tr. (RHT)–J. (PLB/TOA) Terr.

First: *Hemiprotosuchus leali* Bonaparte, 1969, upper Los Colorados Formation, La Rioja, Argentina.
Last: Unnamed form, Kayenta Formation (SIN/PLB), Arizona, USA (Clark, *in* Benton and Clark, 1988), or *Stegomosuchus longipes* Lull, 1953, upper Portland Group (PLB/TOA), Connecticut, USA.
Intervening: HET, SIN.
Comment: The range of Protosuchidae could be much greater if one includes *Dyoplax arenaceus* Fraas, 1867, Schilfsandstein, Germany (CRN), as Walker (1961) suggests, and *Edentosuchus tienshanensis* Young, 1973, Wuerho, China (VLG/ALB), as Clark (*in* Benton and Clark, 1988) suggests.

F. ORTHOSUCHIDAE Whetstone and Whybrow, 1983 J. (HET/SIN) FW/Terr.

First and Last: *Orthosuchus stormbergi* Nash, 1968, upper Elliot Formation, Lesotho, South Africa.

Suborder MESOEUCROCODYLIA Whetstone and Whybrow, 1983

Infra-order THALATTOSUCHIA Fraas, 1901

F. TELEOSAURIDAE Geoffroy, 1831
J. (TOA)–K. (VLG) Mar.

First: *Steneosaurus bollensis* (Jaeger, 1828), *S. brevior* Blake, 1876, and *S. gracilirostris* Westphal, 1961, Whitby Mudstone Formation, Yorkshire, England, UK; Posidonienschiefer, Baden-Württemberg, Germany.
Last: 'Teleosaurid', Valanginian, Bouches-du-Rhône, France.
Intervening: AAL–TTH.
Comment: Huene and Mauberge (1954) reported teleosaurid vertebrae from the Lotharingian (HET/SIN) of Lorraine, France.

F. METRIORHYNCHIDAE Fitzinger, 1843
J. (BTH)–K. (HAU) Mar./FW

First: *Teleidosaurus calvadosi* (J. A. Eudes-Deslongchamps, 1866), *T. gaudryi* Collot, 1905, and *T. bathonicus* (Mercier, 1933), Bathonian, Normandy and Burgundy, France.
Last: *Dakosaurus maximus* (Plieninger, 1846), Hauterivian, Provence, France.
Intervening: CLV–TTH.

?F. CROCODILEIMIDAE Buffetaut, 1979
J. (KIM) Mar.

First and Last: *Crocodileimus robustus* Jourdan, 1871, Calcaires lithographiques, Cerin, Ain, France.

Infra-order METAMESOSUCHIA Clark, *in* Benton and Clark, 1988

F. NOTOSUCHIDAE Dollo, 1914 K. (CON) Terr.

First and Last: *Uruguaysuchus aznaresi* Rusconi, 1933, Guichon Formation, Paysandú, Uruguay; *Notosuchus terrestris* Woodward, 1896, Rió Colorado Formation, Neuquén, Argentina.
Comment: The Rió Neuquén Formation is dated tentatively as CON by Bonaparte (1991).

F. COMAHUESUCHIDAE Bonaparte, 1991
K. (CON) Terr.

First and Last: *Comahuesuchus brachybuccalis* Bonaparte, 1991, Rió Colorado Formation, Neuquén, Argentina (Bonaparte, 1991).

F. BAURUSUCHIDAE Price, 1945 K. (CON–MAA) Terr.

First: *Cynodontosuchus rothi* Woodward, 1896, Rió Colorado Formation, Neuquén, Argentina.
Last: *Baurusuchus pachecoi* Price, 1945, Baurú Formation, Sao Paolo, Brazil.

F. LIBYCOSUCHIDAE Stromer, 1914 K. (BRM–CEN) Terr.

First: 'Libycosuchid', BRM, Niger.
Last: *Libycosuchus brevirostris* Stromer, 1914, Cenomanian, Baharija, Egypt.
Intervening: ALB.

F. SEBECIDAE Simpson, 1937 T. (THA–PLI) Terr.

First: *Sebecus* sp., Upper Palaeocene, Itaboraí, Brazil.
Last: cf. *Sebecus*, Pliocene, Australia.
Intervening: YPR, PRB–CHT, MMI.

F. PEIROSAURIDAE Gasparini, 1982 K. (?CON–MAA) Terr.

First: *Peirosaurus tormini* Price, 1955 and *Lomasuchus palpebrosus* Gasparini, Chiappe, and Fernandez, 1991, Rió Colorado Formation, Neuquén Province, Argentina (Gasparini *et al.*, 1991).
Last: *Peirosaurus tormini* Price, 1955, Baurú Formation, State of Minais Gerais, Brazil.

Parvorder NEOSUCHIA Clark, *in* Benton and Clark, 1988

F. ATOPOSAURIDAE Gervais, 1871 J. (KIM)–K. (APT) Terr.

First: *Alligatorium meyeri* Gervais, 1871, Calcaire lithographique, Cerin, Ain, France.
Last: Atoposaurid, BRM/APT, Spain.
Intervening: TTH, BRM.
Comment: The atoposaurids are essentially an Upper Jurassic group, but extend into the Lower Cretaceous if *Theriosuchus* is included in the family (Buffetaut, 1982; Clark, *in* Benton and Clark, 1988).

F. DYROSAURIDAE de Stefano, 1903 K. (MAA)–T. (PRB) FW/Mar.

First: *Sokotosuchus ianwilsoni* Halstead, 1975, Dukamaje Formation, Sokoto, Nigeria; *Hyposaurus rogersii* Owen, 1849, Hornerstown Formation, New Jersey, USA; *H. derbianus* Cope, 1886, MAA, Pernambuco Province, Brazil.
Last: 'Dyrosaurid', Upper Eocene, Burma.
Intervening: DAN–BRT.
Comment: An older possible dyrosaurid is reported from the CEN of Nazaré, Portugal (Buffetaut, 1982).

F. TREMATOCHAMPSIDAE Buffetaut, 1974 K. (HAU)–T. (LUT) FW/Terr.

First: *Amargasuchus minor* Chiappe, 1988, La Amarga Formation, Neuquén Province, Argentina (Chiappe, 1988).
Last: *Bergisuchus dietrichbergi* Kuhn, 1968, Messel lignite, Hesse, Germany (Buffetaut, 1988).
Intervening: ?CON, CMP, MAA, YPR, BRT.

F. HSISOSUCHIDAE Young and Chow, 1953 J. (?KIM)–K. (CMP) Terr.

First: *Hsisosuchus chungkingensis* Young and Chow, 1953, Upper Jurassic, Szechwan, China.
Last: ?*Doratodon carcharidens* (Bunzel, 1871), Gosau Formation, Austria.
Intervening: None.

F. GONIOPHOLIDIDAE Cope, 1875 J. (BTH)–K. (MAA) FW/Terr.

First: 'Goniopholids', Ostracod Limestone, Skye, Scotland, UK (Savage, 1984), Chipping Norton, White Limestone, and Forest Marble Formations, Gloucestershire and Oxfordshire, England, UK (Metcalf *et al.*, 1993).
Last: *'Goniopholis' kirtlandicus* Wiman, 1931, MAA, New Mexico, USA.
Intervening: KIM–ALB, TUR, SAN.

F. PHOLIDOSAURIDAE Eastman, 1902 J. (BTH)–K. (CEN) FW/Terr.

First: *Anglosuchus geoffroyi* (Owen, 1884), *A. laticeps* (Owen, 1884), White Limestone Formation, Oxfordshire, England, UK.
Last: *Teleorhinus mesabiensis* Erickson, 1969, CEN, Iron Range, Minnesota, USA.
Intervening: KIM–ALB.

F. PARALLIGATORIDAE Konjukova, 1954 K. (CEN/TUR–CMP/MAA)

First: *Shamosuchus major* (Efimov, 1981), *S. ulgicus* (Efimov, 1981), Baynshirenskaya Svita, Mongolia (Efimov, 1981).
Last: *Shamosuchus ancestralis* (Konjukova, 1954), Nemegt Formation, Omnogov, Mongolia.
Intervening: SAN.

F. BERNISSARTIIDAE Dollo, 1883 J. (OXF)–K. (BRM) Terr./FW

First: *Bernissartia* sp., Guimarota, Leiria, Portugal.
Last: *Bernissartia* sp., Wealden, Isle of Wight, England, UK.
Intervening: VLG, HAU.

F. BRILLANCEAUSUCHIDAE Michard *et al.*, 1990 K. (BRM?) Terr./FW

First and Last: *Brillanceausuchus babouriensis* Michard *et al.*, 1990, from the Lower Cretaceous of Babouri-Figuil Basin, north Cameroon (Michard *et al.*, 1990).

Suborder EUSUCHIA Huxley, 1875

F. HYLAEOCHAMPSIDAE Andrews, 1913 K. (BRM) Terr./FW

First and Last: *Hylaeochampsa vectiana* Owen, 1874, Wealden, Isle of Wight, England, UK.

F. STOMATOSUCHIDAE Stromer, 1925 K. (CEN) FW

First and Last: *Stomatosuchus inermis* Stromer, 1925, ?*Aegyptosuchus peyeri* Stromer, 1933, Baharija Formation, Marsa Matruh, Egypt.

F. GAVIALIDAE Cuvier, 1807 T. (PRB)–Rec. FW/Terr.

First: *Eogavialis africanus* (Andrews, 1905), Upper Eocene,

Fayum, Egypt. **Extant**
Intervening: LUT–HOL.

F. THORACOSAURIDAE Cope, 1871
K. (CEN/TUR)–T. (THA) FW

First: *Thoracosaurus cherifiensis* Lavocat, 1955, southern Morocco.
Last: *Thoracosaurus macrorhynchus* (Blainville, 1839–1864), Bourgogne, Marne, France.
Intervening: CMP–DAN.
Comment: *T. cherifiensis* may be a pholidosaurid (E. Buffetaut, pers. comm., 1991), in which case the oldest *Thoracosaurus* species are MAA in age (Hornerstown Formation, New Jersey; Ripley Formation, Mississippi, USA).

F. DOLICHOCHAMPSIDAE Gasparini and Buffetaut, 1980 K. (MAA) FW

First and Last: *Dolichochampsa minima* Gasparini and Buffetaut, 1980, Yacoraite Formation, Salta Province, Argentina and El Molino Formation, southern Bolivia (Buffetaut, 1987).

F. UNNAMED T. (YPR–PLE) FW

First: ?*Eosuchus lerichei* Dollo, 1907, Ypresian, Jeumont, northern France.
Last: Unnamed form, Quaternary?, Murua Island, Solomon Sea (Molnar, 1982).
Intervening: LUT, LMI, UMI, PLI.

F. CROCODYLIDAE Cuvier, 1807
K. (TUR/SAN)–Rec. FW/Terr.

First: ?*Tadzhikosuchus macrodentis* Efimov, 1982, Yalovachskaya Formation, Tadzhikistan (Efimov, 1982). **Extant**
Intervening: CMP–HOL.

F. PRISTICHAMPSIDAE Kuhn, 1968
T. (THA–PLE) Terr.

First: *Planocrania datangensis* Li, 1976, Nonshan Formation, Guangdong, China; *P. hengdongensis* Li, 1984, Palaeocene (?) red beds, Hunan, China; *Wanosuchus atresus* Zhang, 1981, Palaeocene (?), Anhui, China.
Last: *Quinkana fortirostrum* Molnar, 1981, cave deposits, north Queensland, Australia (Molnar, 1981).
Intervening: YPR–BRT, PLI.

F. ALLIGATORIDAE Gray, 1844 (Cuvier, 1807)
K. (CMP)–Rec. FW/Terr.

First: *Albertochampsa langstoni* Erickson, 1972, Judith River Formation, Alberta, Canada; *Bottosaurus perrugosus* Cope, 1874, Belly River Formation, Alberta, Canada. **Extant**
Intervening: MAA–HOL.

F. NETTOSUCHIDAE Langston, 1965
T. (UMI–PLI) FW

First: *Mourasuchus atopus* (Langston, 1965), Honda Beds, La Venta, Huila, Colombia.
Last: *Mourasuchus amazonensis* Price, 1964, Acre State, Brazil.

Infradivision ORNITHODIRA Gauthier, 1986

F. LAGOSUCHIDAE Arcucci, 1987
Tr. (LAD) Terr.

First and Last: *Lagosuchus talampayensis* Romer, 1971, *Lagerpeton chanarensis* Romer, 1971, and *Pseudolagosuchus major* Arcucci, 1987, all Chañares Formation, La Rioja, Argentina (Arcucci, 1987).

F. PODOPTERYGIDAE Sharov, 1971
Tr. (CRN/NOR) Terr.

First and Last: *Sharovipteryx mirabilis* (Sharov, 1971), Madyigenskaya Svita, Fergana, Kirgizia, former USSR.

F. UNNAMED Tr. (CRN/NOR) Terr.

First and Last: *Longisquama insignis* Sharov, 1970, Madyigenskaya Svita, Fergana, Kirgizia, former USSR (Haubold and Buffetaut, 1987).

F. SCLEROMOCHLIDAE Huene, 1914
Tr. (CRN) Terr.

First and Last: *Scleromochlus taylori* Woodward, 1907, Lossiemouth Sandstone Formation, Morayshire, Scotland, UK.

Order PTEROSAURIA Owen, 1840 (Kaup, 1834)
(see Fig. 39.5)

The classification of pterosaurs is based on Wellnhofer (1978, 1991), Howse (1986), Bennett (1989), and Unwin (1991), with distribution data from Wellnhofer (1978, 1991), and other references noted.

Suborder RHAMPHORHYNCHOIDEA
Plieninger,1901 (p)
F. UNNAMED Tr. (NOR) Terr.

First and Last: *Preondactylus buffarini* Wild, 1983, lower middle part of the 'Dolomia Principale', Udine Province, Italy (Wild, 1983).

F. DIMORPHODONTIDAE Seeley, 1870
Tr. (NOR)–J. (SIN) Terr.

First: *Peteinosaurus zambellii* Wild, 1978, upper half of the Calcare di Zorzino, Bergamo, Italy (Wild, 1978).
Last: *Dimorphodon macronyx* (Buckland, 1829), upper Blue Lias, Dorset, England, UK.
Intervening: None.

F. EUDIMORPHODONTIDAE Wellnhofer, 1978
Tr. (NOR) Terr.

First and Last: *Eudimorphodon ranzii* Zambelli, 1973, upper half of the Calcare di Zorzino, Bergamo, Italy.

F. ANUROGNATHIDAE Kuhn, 1937
J. (TTH) Terr.

First and Last: *Anurognathus ammoni* Döderlein, 1923, Solnhofener Schichten, Bavaria, Germany; and ?*Batrachognathus volans* Riabinin, 1948, Upper Jurassic, Karatau Mountains, Michailokva, Kazakhstan, former USSR.

F. UNNAMED J. (OXF–TTH) Terr.

First: *Nesodactylus hesperius* Colbert, 1969, OXF, Province Pinar del Rio, west Cuba.
Last: *Comodactylus ostromi* Galton, 1981, upper part of Morrison Formation, Wyoming, USA.
Intervening: KIM?

F. RHAMPHORHYNCHIDAE Seeley, 1870
J. (TOA–TTH) Terr.

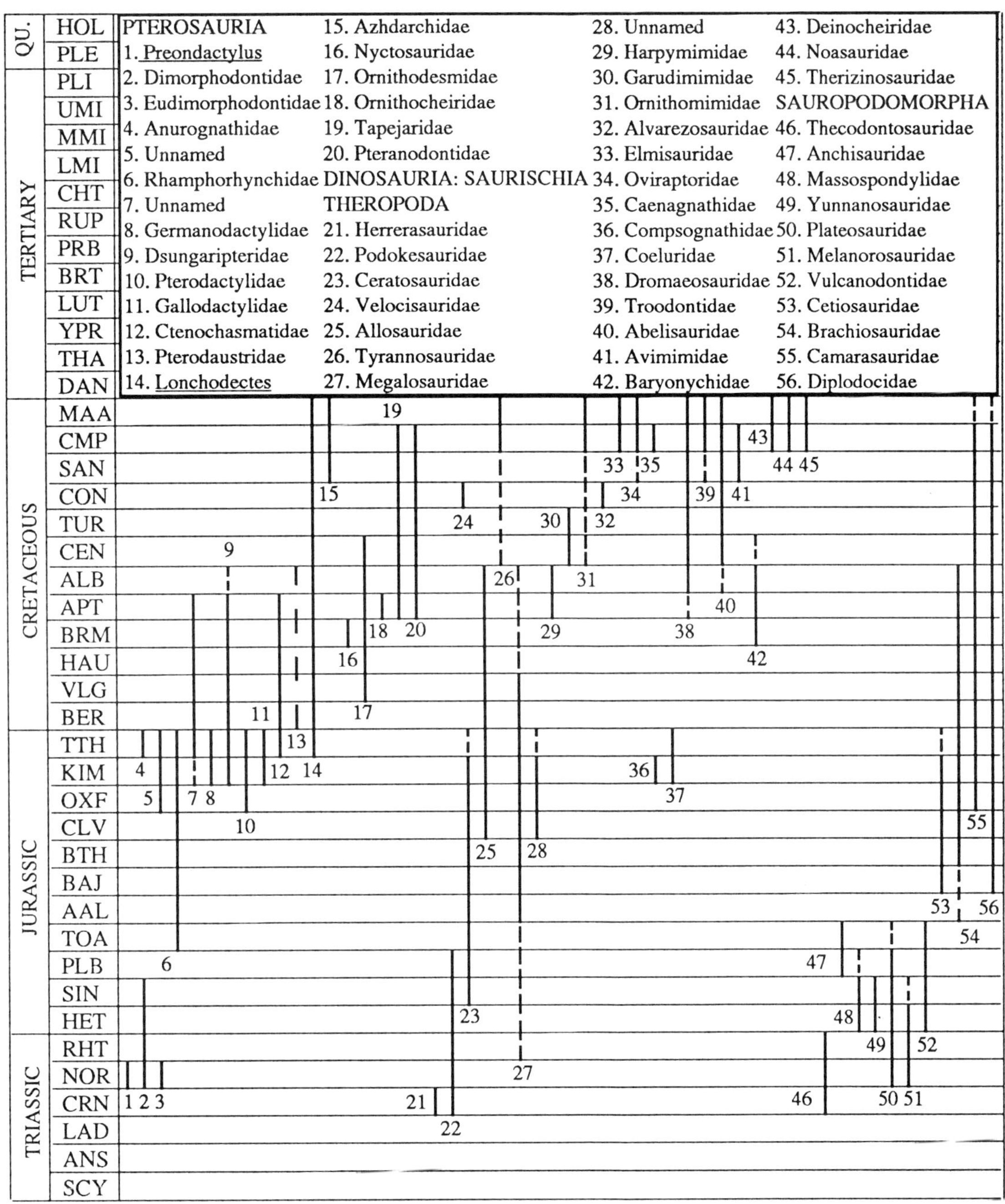

Fig. 39.5

First: *Parapsicephalus purdoni* (Newton, 1888), upper Lias, Yorkshire, England, UK; *Dorygnathus banthensis* (Theodori, 1930), upper Lias, Germany.
Last: *Rhamphorhynchus longicaudus* (Münster, 1839), *R. intermedius* Koh, 1937, *R. muensteri* (Goldfuss, 1831), *R. gemmingi* Meyer, 1846, *R. longiceps* Woodward, 1902, *Scaphognathus crassirostris* (Goldfuss, 1831), and *Odontorhynchus aculeatus* Stolley, 1936 (?*nom. nud.*), Solnhofener Schichten, Bavaria, Germany.
Intervening: BTH–KIM.

Suborder PTERODACTYLOIDEA Plieninger, 1901

Superfamily DSUNGARIPTEROIDEA Young, 1964

F. UNNAMED J. (KIM/TTH)–K. (APT) Terr.

First: *Dermodactylus montanus* (Marsh, 1878), Morrison Formation, Wyoming, USA.
Last: *Araripedactylus dehmi* Wellnhofer, 1977, Santana Formation, Estado do Ceará, Brazil.
Intervening: None.

F. GERMANODACTYLIDAE Young, 1964 J. (KIM–TTH) Terr.

First: *Germanodactylus* sp., Kimmeridge Clay, Dorset, England, UK (Unwin, 1987).
Last: *Germanodactylus cristatus* (Wiman, 1925), Solnhofener Schichten, Bavaria; *G. rhamphastinus* (Wagner, 1851), Mörnsheimer Schichten, Bavaria, Germany.

F. DSUNGARIPTERIDAE Young, 1964 J. (KIM)–K. (APT/ALB) Terr.

First: ?*Dsungaripterus brancai* (Reck, 1931), Tendaguru Beds, Tendaguru, Tanzania (Galton, 1980).
Last: *Dsungaripterus weii* Young, 1964 and *Noripterus complicidens* Young, 1973, Tugulu Group, Junggar Basin, Xinjiang, China; 'dsungaripterid', Qingshan Formation, Shandong, China.
Intervening: BER–HAU.
Comment: *Dsungaripterus brancai* is not noted as a dsungaripterid by Bennett (1989), so that the oldest certain member of the family is *Dsungaripterus parvus* Bakhurina, 1982, from the Tsagantsabskaya Svita, Khovd, Mongolia (VLG).

Superfamily UNNAMED

F. PTERODACTYLIDAE Bonaparte, 1838 (p) J. (OXF–TTH) Terr.

First: *Pterodactylus* sp. (teeth), Guimarota lignite mine, Leiria, Portugal.
Last: *Pterodactylus antiquus* (Soemmerring, 1812), *P. kochi* (Wagner, 1837), *P. micronyx* Meyer, 1856, *P. elegans* Wagner, 1861, and *P. longicollum* Meyer, 1854, Solnhofener Schichten, Bavaria, Germany.
Comment: Evans and Milner (1991) note 'pterodactylid' teeth from the Forest Marble, upper BTH, of Oxfordshire, England, UK.

F. GALLODACTYLIDAE Fabre, 1974 J. (KIM–TTH) Terr.

First and Last: *Gallodactylus suevicus* (Quenstedt, 1855), Solnhofener Schichten, Bavaria, Germany and *G. canjuersensis* Fabre, 1974, 'Portlandien', Var, France.

F. CTENOCHASMATIDAE Nopcsa, 1928 J. (TTH)–K. (APT) Terr.

First: *Ctenochasma roemeri* Meyer, 1852, Purbeck, Hannover, Germany; *C. gracile* Oppel, 1862 and *Gnathosaurus subulatus* Meyer, 1834, Solnhofener Schichten, Bavaria, Germany; *Ctenochasma* sp., Portlandien inférieur, Haute Marne, France; *Huanhepterus quingyangensis* Dong, 1982, ?Tithonian, Ordos Basin, China.
Last: *Cearadactylus atrox* Leonardi and Borgomanero, 1985, Santana Formation, Estado do Ceára, Brazil.
Comment: *Aidachar paludalis* Nessov, 1981, Taykarshinskaya Member (TUR/SAN), Uzbekistan, former USSR, turns out to be based on the remains of a teleost fish (P. Wellnhofer, pers. comm., 1991). *Cearadactylus* is included here according to Unwin's (1991) cladistic analysis, although Wellnhofer (1991) places it in a separate family.
Intervening: None.

F. PTERODAUSTRIDAE Bonaparte, 1971 K. (l.) Terr.

First and Last: *Pterodaustro guinazui* Bonaparte, 1970, Lagarcito Formation, San Luis, Argentina.

F. UNNAMED K. (VLG–TUR) Terr.

First: ?*Lonchodectes sagittirostris* (Owen, 1874), Hastings Beds, Sussex, England, UK.
Last: *Lonchodectes compressirostris* (Owen, 1851), Middle Chalk, Kent, England, UK.
Intervening: ALB, CEN.
Comment: This family is based on Unwin's (1991) work.

F. AZHDARCHIDAE Nessov, 1984 J. (TTH)–K. (MAA) Terr.

First: *Doratorhynchus validus* Seeley, 1875, Purbeck, Dorset, England, UK.
Last: *Quetzalcoatlus northropi* Lawson, 1975, Tornillo Group, Brewster County, Texas, USA; *Quetzalcoatlus* sp., Lance Formation, Wyoming, USA; MAA, New Jersey, USA; *Arambourgiana philadelphiae* (Arambourg, 1959), Maastrichtian, Amman, Jordan (Nessov, 1984; Bennett, 1989).
Intervening: ALB–CON, CMP.

Superfamily ORNITHOCHEIROIDEA Seeley, 1891

F. NYCTOSAURIDAE Williston, 1903 K. (SAN–MAA) Terr.

First: *Nyctosaurus gracilis* Marsh, 1876, Smoky Hill Chalk Member, upper Niobrara Formation, western Kansas, USA.
Last: *Nyctosaurus lamegoi* (Price, 1953), Gramame Formation, Paraíba, Brazil.
Intervening: None.

F. ORNITHODESMIDAE Hooley, 1913 K. (BRM) Terr.

First and Last: *Ornithodesmus latidens* Seeley, 1901, Wealden, Isle of Wight, England, UK.
Comment: *O. cluniculus* (Hooley, 1913), also from the Wealden of the Isle of Wight, is based on a dinosaur sacrum (A. R. Milner, in prep.).

F. ORNITHOCHEIRIDAE Seeley, 1870 K. (VLG–CEN) Terr.

First: *Coloborhynchus clavirostris* (Owen, 1874), Hastings Sand, Sussex, England, UK.
Last: *Anhanguera cuvieri* (Bowerbank, 1851), Lower Chalk, Kent, England, UK.
Intervening: APT, ALB.
Comment: This family is based on Unwin's (1991) analysis, and includes the Anhangueridae Campos and Kellner, 1985.

F. TAPEJARIDAE Kellner, 1990 K. (APT) Terr.

First and Last: *Tapejara wellnhoferi* Kellner, 1990 and *Tupuxuara longicristatus* Kellner and Campos, 1989, Santana Formation, Estado do Ceára, Brazil.

F. PTERANODONTIDAE Marsh, 1876 K. (APT–CMP) Terr.

First: *Ornithostoma sedgwicki* Seeley, 1891, Cambridge Greensand, Cambridgeshire, England, UK.
Last: *Pteranodon longiceps* Marsh, 1876 and *P. sternbergi* Harksen, 1966, Smoky Hill Chalk Member, upper Niobrara Formation and Pierre Shale, Kansas, South Dakota, and Wyoming, USA.
Intervening: CEN–SAN.
Comment: The Pteranodontidae, according to Bennett (1989) should be expanded to include pterosaurs assigned by Wellnhofer (1978, 1991), and others, to the families Criorhynchidae, Ornithocheiridae and Ornithodesmidae. However, the ornithodesmids and ornithocheirids are retained as separate families here, according to Unwin's (1991) work. Species of *Pteranodon* other than those named are regarded as synonyms or *nomina dubia* by C. Bennett (pers. comm., 1991).

Superorder DINOSAURIA Owen, 1842 (P)

The classification of dinosaurs, and stratigraphical distributions of families are taken from Weishampel *et al.* (1990).

Order SAURISCHIA Seeley, 1877

Suborder THEROPODA Marsh, 1881

F. HERRERASAURIDAE Benedetto, 1973 Tr. (CRN) Terr.

First and Last: *Staurikosaurus pricei* Colbert, 1970, *Scaphonyx Assemblage Zone*, Santa Maria Formation, Rio Grande do Sul, Argentina; *Herrerasaurus ischigualastensis* Reig, 1963, and *Ischisaurus cattoi* Reig, 1963, Ischigualasto Formation, San Juan, Argentina.

Infra-order CERATOSAURIA Gauthier, 1986

F. PODOKESAURIDAE Huene, 1914 Tr. (CRN)–J. (PLB) Terr.

First: *Coelophysis bauri* (Cope, 1889), lower part of Petrified Forest Member, Chinle Formation, Arizona, USA.
Last: *Syntarsus kayentakatae* Rowe, 1989, Kayenta Formation, Willow Springs, Arizona, USA.
Intervening: NOR, HET, SIN.
Comment: The famous *Coelophysis* quarry at Ghost Ranch, New Mexico, USA, is in the upper part of the Petrified Forest Member, dated lower NOR.

F. CERATOSAURIDAE Marsh, 1884 (P) J. (SIN–KIM/TTH) Terr.

First: *Sarcosaurus woodi* Andrews, 1921, Lias, Leicestershire, England, UK.
Last: *Ceratosaurus nasicornis* Marsh, 1884, Morrison Formation, Canyon City, Colorado, USA.
Intervening: PLB?

F. VELOCISAURIDAE Bonaparte, 1991 K. (CON) Terr.

First and Last: *Velocisaurus unicus* Bonaparte, 1991, lower part of Rió Colorado Formation Neuquén Province, Argentina (Bonaparte, 1991).

Infra-order CARNOSAURIA Huene, 1920

F. ALLOSAURIDAE Marsh, 1879 J. (CLV)–K. (ALB) Terr.

First: *Piatnitzkysaurus floresi* Bonaparte, 1979, Cañadon Asfalto Formation, Chubut, Argentina.
Last: *Chilantaisaurus marotuensis* Hu, 1964, unnamed unit, Nei Mongol Zizhiqu, China.
Intervening: OXF–HAU, APT

F. TYRANNOSAURIDAE Osborn, 1905 K. (?CEN–MAA) Terr.

First: *Alectrosaurus olseni* Gilmore, 1933, Iren Dabasu Formation, Nei Mongol Zizhqu, China.
Last: *Tyrannosaurus rex* Osborn, 1905, Hell Creek Formation, Montana, and numerous other upper Maastrichtian formations in the mid-west of Canada and USA.
Intervening: CMP.

F. MEGALOSAURIDAE Huxley, 1869 Tr. (RHT)?–K. (VLG/ALB) Terr.

First: *Megalosaurus cambrensis* (Newton, 1899), Rhaetic, Glamorgan, Wales, UK.
Last: *Kelmayisaurus petrolicus* Dong, 1973, Lianmugin Formation, Xinjiang Uygur Zizhqu, China.
Intervening: AAL–BTH, OXF.
Comment: The family Megalosauridae is not accepted by Molnar *et al.* (1990), although they suggest that *Megalosaurus, Magnasaurus,* and *Kelmayisaurus* may be related. There is little evidence that *M. cambrensis* is a true megalosaur. If not, the earliest records of *Megalosaurus* are AAL and BAJ.

F. UNNAMED J. (CLV–KIM/TTH) Terr.

First: *Eustreptospondylus oxoniensis* Walker, 1964, Oxford Clay, Oxfordshire and Buckinghamshire, England, UK.
Last: *Torvosaurus tanneri* Galton and Jensen, 1979, Morrison Formation, Colorado, USA.
Intervening: None.
Comment: This family is hinted at by Molnar *et al.* (1990, p. 209), in suggesting a phyletic link between *Eustreptospondylus, Torvosaurus* and *Yangchuanosaurus*.

Infra-order ORNITHOMIMOSAURIA Barsbold, 1976

F. HARPYMIMIDAE Barsbold and Perle, 1984 K. (APT/ALB) Terr.

First and Last: *Harpymimus okladnikovi* Barsbold and Perle, 1984, Shinekhudukskaya Svita, Dundgov, Mongolia.

F. GARUDIMIMIDAE Barsbold, 1981 K. (CEN/TUR) Terr.

First and Last: *Garudimimus brevipes* Barsbold, 1981, Baynshirenskaya Svita, Omnogov, Mongolia.

F. ORNITHOMIMIDAE Marsh, 1890 K. (?CEN–MAA) Terr.

First: *Archaeornithomimus asiaticus* (Gilmore, 1933), Iren Dabasu Formation, Nei Mongol Zizhqu, China.
Last: *Ornithomimus velox* Marsh, 1890, Denver Formation, Colorado; Kaiparowits Formation, Utah, USA.
Intervening: CMP.
Comment: An older 'ornithomimid' has been reported from the Sebayashi Formation (BRM/APT) of Japan (Manabe and Hasegawa, 1991), but its exact affinities are unclear.

STEM-GROUP MANIRAPTORANS

?F. ALVAREZOSAURIDAE Bonaparte, 1991 K. (CON) Terr.

First and Last: *Alvarezosaurus calvoi* Bonaparte, 1991, Bajo de la Carpa Member, Rió Colorado Formation, Neuquén Province, Argentina (Bonaparte, 1991).

F. ELMISAURIDAE Osmólska, 1981 K. (CMP–MAA) Terr.

First: *Chirostenotes pergracilis* Gilmore, 1924 and *Elmisaurus elegans* (Parks, 1933), Judith River Formation, Alberta, Canada.
Last: *Elmisaurus rarus* Osmólska, 1981, Nemegt Formation, Omnogov, Mongolia.

F. OVIRAPTORIDAE Barsbold, 1976 K. (SAN/CMP–MAA) Terr.

First: *Oviraptor philoceratops* Osborn, 1924, Djadochta Formation, Omnogov, Mongolia.

Last: *Oviraptor mongoliensis* Barsbold, 1986, Nemegt Formation, Omnogov, Mongolia.

F. CAENAGNATHIDAE Sternberg, 1940 K. (CMP) Terr.

First and Last: *Caenagnathus collinsi* Sternberg, 1940 and *C. sternbergi* Cracraft, 1971, Judith River Formation, Alberta, Canada.

Infra-order MANIRAPTORA Gauthier, 1986

F. COMPSOGNATHIDAE Cope, 1871 J. (KIM) Terr.

First and Last: *Compsognathus longipes* Wagner, 1861, Solnhofener Schichten, Bavaria, Germany, and Lithographic Limestone, Canjuer, Var, France.

F. COELURIDAE Marsh, 1881 J. (KIM/TTH)

First and Last: *Coelurus fragilis* Marsh, 1879, Morrison Formation, Wyoming and Utah, USA.

F. DROMAEOSAURIDAE Matthew and Brown, 1922 K. (APT/ALB–MAA) Terr.

First: *Deinonychus antirrhopus* Ostrom, 1969, Cloverly Formation, Wyoming and Montana, USA.
Last: *Adasaurus mongoliensis* Barsbold, 1983, Nemegtskaya Svita, Bayankhongor, Mongolia.
Intervening: ?SAN, CMP.
Comment: Older (?) dromaeosaurid teeth have been noted from the Kitadani Formation (HAU/APT) of Japan (Manabe and Hasegawa, 1991). Even older dromaeosaurid/troodontid-type teeth have been noted from the BTH of the Cotswolds of England, UK (Metcalf *et al.*, 1993).

F. TROODONTIDAE Gilmore, 1924 K. (SAN/CMP–MAA) Terr.

First: *Saurornithoides mongoliensis* Osborn, 1924, Djadochta Formation, Omnogov, Mongolia.
Last: *Troodon formosus* Leidy, 1856, Hell Creek Formation, Montana, and Lance Formation, Wyoming, USA.

THEROPODA *incertae sedis*

F. ABELISAURIDAE Bonaparte and Novas, 1985 K. (ALB/CEN–MAA) Terr.

First: *Carnotaurus sastrei* Bonaparte, 1985, Gorro Frigio Formation, Chubut, Argentina.
Last: *Abelisaurus comahuensis* Bonaparte and Novas, 1985, Allen Formation, lower MAA, Rió Negro, Argentina; 'abelisaurid', Rognacien, Provence, France.
Intervening: CMP.

F. AVIMIMIDAE Kurzanov, 1981 K. (SAN/CMP) Terr.

First and Last: *Avimimus portentosus* Kurzanov, 1981, 'Barungoyotskaya' Svita, Omnogov, and 'Djadochtinskaya' Svita, Ovorkhangai, Mongolia.

F. BARYONYCHIDAE Charig and Milner, 1986 K. (BRM–ALB/CEN) Terr.

First: *Baryonyx walkeri* Charig and Milner, 1986, Weald Clay, Surrey, England, UK.
Last: '*Spinosaurus* cf. *aegypticus*', continental red beds, Hammada du Guir, Morocco (Buffetaut, 1989).

F. DEINOCHEIRIDAE Osmólska and Roniewicz, 1970 K. (CMP/MAA) Terr.

First and Last: *Deinocheirus mirificus* Osmólska and Roniewicz, 1970, Nemegt Formation, Omnogov, Mongolia.

F. NOASAURIDAE Bonaparte and Powell, 1980 K. (CMP/MAA) Terr.

First and Last: *Noasaurus leali* Bonaparte and Powell, 1980, Lecho Formation, El Brete, Salta, Argentina.

F. THERIZINOSAURIDAE Maleev, 1954 K. (CMP/MAA) Terr.

First and Last: *Therizinosaurus cheloniformis* Maleev, 1954, Nemegt Formation, Omnogov, and White Beds of Khermeen Tsav, Bayankhongor, Mongolia.

Suborder SAUROPODOMORPHA Huene, 1932

Infra-order PROSAUROPODA Huene, 1920 (p)

F. THECODONTOSAURIDAE Huene, 1908 Tr. (CRN–RHT) Terr.

First: *Azendhosaurus laaroussi* Dutuit, 1972, Argana Formation, Marrakech, Morocco.
Last: *Thecodontosaurus antiquus* Riley and Stutchbury, 1836, Magnesian Conglomerate, Avon, England, UK; fissure fillings, Glamorgan, Wales, UK.
Intervening: NOR.

F. ANCHISAURIDAE Marsh, 1885 J. (PLB/TOA) Terr.

First and Last: *Anchisaurus polyzelus* (Hitchcock, 1865), Upper Portland Formation, Connecticut and Massachusetts, USA.

F. MASSOSPONDYLIDAE Huene, 1914 J. (HET–SIN/PLB) Terr.

First: *Massospondylus carinatus* Owen, 1854, upper Elliot Formation, Clarens Formation, and Bushveld Sandstone, South Africa; Forest Sandstone, Zimbabwe; upper Elliot Formation, Lesotho.
Last: *Massospondylus* sp., Kayenta Formation, Arizona, USA.

F. YUNNANOSAURIDAE Young, 1942 J. (HET/SIN) Terr.

First and Last: *Yunnanosaurus huangi* Young, 1942, upper Lower Lufeng Series, Yunnan, China.

F. PLATEOSAURIDAE Marsh, 1895 Tr. (NOR)–J. (PLB/TOA) Terr.

First: *Sellosaurus gracilis* Huene, 1907–1908, Untere and Mittlere Stubensandstein, Baden-Württemberg, Germany.
Last: *Ammosaurus major* (Marsh, 1891), upper Portland Formation, Connecticut; Navajo Sandstone, Arizona, USA.
Intervening: RHT.

F. MELANOROSAURIDAE Huene, 1929 Tr. (NOR–HET/SIN) Terr.

First: *Euskelosaurus browni* Huxley, 1866, lower Elliot Formation and Bushveld Sandstone, South Africa; lower Elliot Formation, Lesotho; Mpandi Formation, Zimbabwe; *Melanorosaurus readi* Haughton, 1924, lower Elliot Formation, South Africa.

Last: *Lufengosaurus huenei* Young, 1941, upper Lower Lufeng Series, Yunnan, China.
Intervening: RHT.
Comment: Dong *et al.* (1983) note cf. *Lufengosaurus* from the Zhenzhunchong Formation, Szechwan, China, dated as
TOA/BAJ (Weishampel, 1990).

Infra-order SAUROPODA Marsh, 1878

F. VULCANODONTIDAE Cooper, 1984 (p?) J. (HET–TOA) Terr.

First: *Vulcanodon karibaensis* Raath, 1972, *Vulcanodon* Beds, Mashonaland North, Zimbabwe.
Last: *Ohmdenosaurus liasicus* Wild, 1978, Posidonienschiefer, Baden-Württemberg, Germany.
Intervening: PLB?

F. CETIOSAURIDAE Lydekker, 1888 (p) J. (BAJ–KIM/TTH) Terr.

First: *Cetiosaurus medius* Owen, 1842, Inferior Oolite, West Yorkshire, England, UK; *Amygdalodon patagonicus* Cabrera, 1947, Cerro Carnerero Formation, Chubut, Argentina; ?*Rhoetosaurus brownei* Longman, 1925, ?Injune Creek Beds, Queensland, Australia.
Last: *Haplocanthosaurus priscus* (Hatcher, 1903) and *H. delfsi* McIntosh and Williams, 1988, Morrison Formation, Colorado and Wyoming, USA.
Intervening: BTH, CLV.

F. BRACHIOSAURIDAE Riggs, 1904 J. (?AAL/BTH)–K. (ALB) Terr.

First: 'Brachiosaurid', Northamptonshire Sand Formation, Northamptonshire, England, UK (Cope *et al.*, 1980).
Last: *Brachiosaurus nougaredi* Lapparent, 1960, 'Continental Intercalaire', Wargla, Algeria; *Chubutisaurus insignis* Corro, 1974, Gorro Frigio Formation, Chubut, Argentina.
Intervening: BTH, CLV, KIM, TTH, VAL–APT.
Comment: If the Northamptonshire brachiosaurid is not confirmed, definite BTH examples include: *Bothriospondylus robustus* Owen, 1875, Forest Marble, Wiltshire, England, UK; *B. madagascariensis* Lydekker, 1895 and *Lapparentosaurus madagascariensis* Bonaparte, 1979, Isalo Formation, Majunga, Madagascar.

F. CAMARASAURIDAE Cope, 1877 J. (OXF)–K. (CMP/MAA) Terr.

First: *Tienshanosaurus chitalensis* Young, 1937, Shishugou Formation, Xinjiang, China.
Last: *Opisthocoelicaudia skarzynskii* Borsuk-Bialynicka, 1977, Nemegt Formation, Omnogov, Mongolia.
Intervening: KIM, TTH, HAU–BRM.

F. DIPLODOCIDAE Marsh, 1884 J. (BAJ)–K. (CMP/MAA) Terr.

First: *Cetiosauriscus longus* (Owen, 1842), Inferior Oolite, West Yorkshire, England, UK.
Last: *Nemegtosaurus mongoliensis* Nowinski, 1971, Nemegt Formation, Omnogov, Mongolia.
Intervening: BTH, CLV, KIM, TTH, ALB, ?SAN.

F. TITANOSAURIDAE Lydekker, 1885 J. (KIM)–K. (MAA) Terr. (Fig. 39.6)

First: *Tornieria robusta* (Fraas, 1908), upper Tendaguru Beds, Mtwara, Tanzania.
Last: *Alamosaurus sanjuanensis* Gilmore, 1922, Javelina Formation, upper MAA, Texas, USA; *Magyarosaurus dacus* (Nopcsa, 1915), *M. transsylvanicus* Huene, 1932, and *M. hungaricus* Huene, 1932, Sinpetru Beds, upper MAA, Hunedoara, Romania.
Intervening: VAL–BRM, ALB, TUR–CMP.
Comment: Titanosaurids are known from numerous CMP/MAA formations in Asia, South America, North America and Europe, but most are not dated as late MAA.

SAURISCHIA *incertae sedis*

F. SEGNOSAURIDAE Perle, 1979 K. (CEN/TUR–CMP) Terr.

First: *Enigmosaurus mongoliensis* Barsbold and Perle, 1983, *Erlikosaurus andrewsi* Perle, 1980, and *Segnosaurus galbinensis* Perle, 1979, Baynshirenskaya Svita, Omnogov and Dornogov, Mongolia.
Last: *Nanshiungosaurus brevispinus* Dong, 1979, Nanxiong Formation, Guandong, China.
Intervening: SAN, CMP.

Order ORNITHISCHIA Seeley, 1887

F. PISANOSAURIDAE Casamiquela, 1967 Tr. (CRN) Terr.

First and Last: *Pisanosaurus merti* Casamiquela, 1967, Ischigualasto Formation, La Rioja Province, Argentina.

F. FABROSAURIDAE Galton, 1972 J. (HET/SIN) Terr.

First and Last: *Lesothosaurus diagnosticus* Galton, 1978, Upper Elliot Formation, Mafeting District, Lesotho.
Comment: Other supposed fabrosaurids such as *Technosaurus* and *Revueltosaurus* (CRN), *Scutellosaurus* (HET), *Fabrosaurus*, *Tawasaurus*, and *Fulengia* (HET/SIN), *Xiaosaurus* (BTH), *Alocodon* and *Trimucrodon* (OXF), *Nanosaurus* (KIM), and *Echinodon* (BER) are not regarded as fabrosaurids, but merely Ornithischia *incertae sedis*, or thyreophorans (e.g. *Scutellosaurus*), or prosauropods (e.g. *Fulengia, Tawasaurus*, ?*Technosaurus*) (Weishampel and Witmer, 1990; Sereno, 1991a).

Suborder THYREOPHORA Nopcsa, 1915

F. SCELIDOSAURIDAE Huxley, 1869 (?p) J. (SIN–TTH?) Terr.

First: *Scelidosaurus harrissoni* Owen, 1861, lower Lias, Dorset, England, UK.
Last: *Echinodon becklesi* Owen, 1861, middle Purbeck Beds, Dorset, England, UK.
Intervening: ?PLB.
Comment: The family Scelidosauridae is equated here with the 'basal Thyreophora' of Coombs *et al.* (1990). If *Echinodon* is not a 'basal thyreophoran', the family range becomes SIN–PLB?, with *Scutellosaurus lawleri* Colbert, 1981, as the youngest member.

Infra-order STEGOSAURIA Marsh, 1877

F. HUAYANGOSAURIDAE Dong *et al.*, 1982 J. (HET/PLB–BTH/CLV) Terr.

First: *Tatisaurus oehleri* Simmons, 1965, Dark Red Beds of the Lower Lufeng Group, Yunnan, China (Dong, 1990).
Last: *Huayangosaurus taibaii* Dong *et al.*, 1982, Xiashaximiao Formation, Szechwan, China.

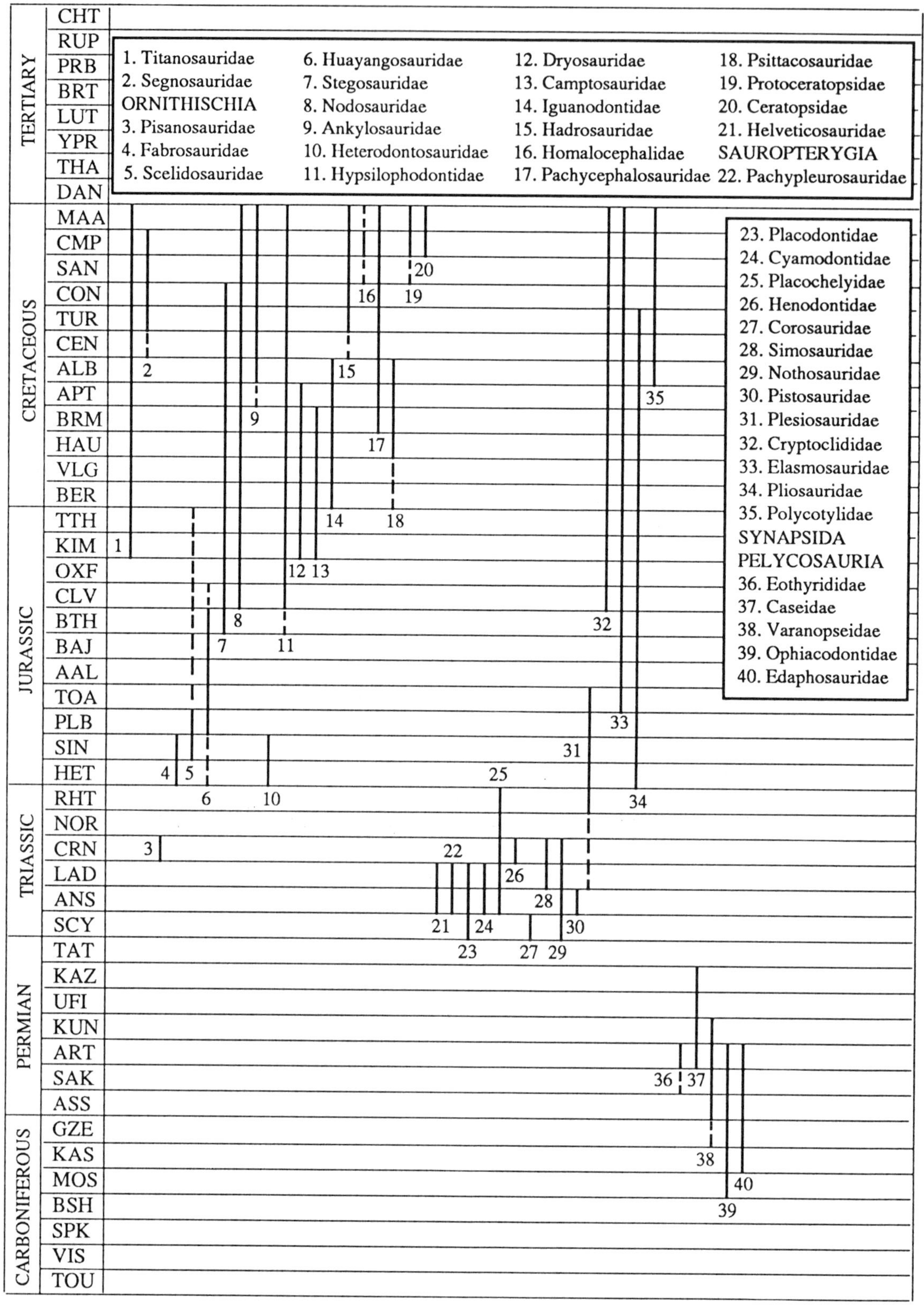

Fig. 39.6

F. STEGOSAURIDAE Marsh, 1880
J. (BTH)–K. (CON) Terr.

First: Unnamed stegosaur, Chipping Norton Formation, lower BTH, Gloucestershire, England, UK (Metcalf *et al.*, 1993); and from other BTH localities in Gloucestershire and Oxfordshire, England, UK (Evans and Milner, 1991).

Last: *Dravidosaurus blanfordi* Yadagiri and Ayyasami, 1979, Trichinopoly Group, Tamil Nadu, India.
Intervening: CLV–ALB.

Infra-order ANKYLOSAURIA Osborn, 1923

F. NODOSAURIDAE Marsh, 1890
J. (CLV)–K. (MAA) Terr.

First: *Sarcolestes leedsi* Lydekker, 1893, lower Oxford Clay, Cambridgeshire, England, UK.
Last: *'Struthiosaurus transilvanicus'* Nopcsa, 1915, Sinpetru Beds, Hunedoara, Romania; Gosau Formation, Niederösterreich, Austria; *Denversaurus schlessmani* Bakker, 1988, Lance Formation, South Dakota, USA.
Intervening: KIM, VLG–CEN, CMP.

F. ANKYLOSAURIDAE Brown, 1908 K. (APT/ALB–MAA) Terr.

First: *Shamosaurus scutatus* Tumanova, 1983, Khukhtekskaya Svita, Dornogov, Mongolia.
Last: *Ankylosaurus magniventris* Brown, 1908, Hell Creek Formation, Montana, USA; Lance Formation, Wyoming, USA; Scollard Formation, Alberta, Canada.
Intervening: CEN–CMP.

Suborder CERAPODA Sereno, 1986

Infra-order ORNITHOPODA Marsh, 1881

F. HETERODONTOSAURIDAE Romer, 1966 J. (HET/SIN–SIN) Terr.

First and Last: *Lycorhinus angustidens* Haughton, 1924, *Lanasaurus scalpridens* Gow, 1975, and *Abrictosaurus consors* (Thulborn, 1975), upper Elliot Formation, South Africa and/or Lesotho.
Last: *Heterodontosaurus tucki* Crompton and Charig, 1962, Clarens Formation, Cape Province, South Africa.

F. HYPSILOPHODONTIDAE Dollo, 1882 J. (BTH/CLV)–K. (MAA) Terr.

First: *Yandusaurus honheensis* He, 1979, Xiashaximiao Formation, Szechwan, China.
Last: *Thescelosaurus neglectus* Gilmore, 1913, Lance Formation, Wyoming, USA; Hell Creek Formation, Montana and South Dakota, USA; Scollard Formation, Alberta, Canada; ?*T. garbanii* Morris, 1976, Hell Creek Formation, Montana, USA.
Intervening: KIM, TTH, BRM–ALB, CMP.

F. DRYOSAURIDAE Milner and Norman, 1984 J. (KIM)–K. (APT) Terr.

First: *Dryosaurus lettowvorbecki* (Virchow, 1919), Tendaguru Beds, Mtwara, Tanzania.
Last: *Valdosaurus nigeriensis* Galton and Taquet, 1982, El Rhaz Formation, Agadez, Niger.
Intervening: TTH–BRM.

F. CAMPTOSAURIDAE Marsh, 1885 J. (KIM)–K. (BRM) Terr.

First: *Camptosaurus prestwichi* (Hulke, 1880), Kimmeridge Clay, Oxfordshire, England, UK.
Last: ?*Camptosaurus depressus* Gilmore, 1909, Lakota Formation, South Dakota, USA.
Intervening: TTH.

F. IGUANODONTIDAE Huxley, 1869 K. (BER–ALB) Terr.

First: *Iguanodon hoggi* Owen, 1874, upper Purbeck Beds, Dorset, England, UK.
Last: *'Iguanodon' orientalis* Rozhdestvensky, 1952, Khukhtekskaya Svita, Dundgov, Ovorkhangai; Shinekhudukskaya Svita, Dundgov, Mongolia.
Intervening: VLG–APT.

F. HADROSAURIDAE Cope, 1869 K. (?CEN–MAA) Terr.

First: *Gilmoreosaurus mongoliensis* (Gilmore, 1933) and *Bactrosaurus johnsoni* Gilmore, 1933, Iren Dabasu Formation, Nei Mongol Zizhiqu, China. The Iren Dabasu Formation is dated variously as CEN or MAA. If the latter, the oldest hadrosaurid is *Aralosaurus tubiferus* Rozhdestvensky, 1968, Beleutinskaya Svita, Kazakhstan, former USSR (?TUR/SAN).
Last: *Edmontosaurus regalis* Lambe, 1917, *E. annectens* (Marsh, 1892), *E. saskatchewanensis* (Sternberg, 1926), and *'Anatosaurus' copei* Lull and Wright, 1942, Scollard Formation, Alberta, Canada; Frenchman Formation, Saskatchewan, Canada; Hell Creek Formation, Montana, North Dakota, South Dakota, USA; Lance Formation, South Dakota, Wyoming, USA; Laramie Formation, Colorado, USA.
Intervening: TUR–CMP.

Infra-order PACHYCEPHALOSAURIA Maryanska and Osmólska, 1974

F. HOMALOCEPHALIDAE Dong, 1974 K. (?SAN–?MAA) Terr.

First: *Goyocephale latimorei* Perle *et al.*, 1982, unnamed unit, Ovorkhangai, Mongolia.
Last: *Homalocephale calathocercos* Maryanska and Osmólska, 1974, Nemegt Formation, Omnogov, Mongolia.
Intervening: CMP.

F. PACHYCEPHALOSAURIDAE Sternberg, 1945 K. (BRM–MAA) Terr.

First: *Yaverlandia bitholus* Galton, 1971, Wealden Marls, Isle of Wight, England, UK.
Last: *Pachycephalosaurus wyomingensis* Brown and Schlaikjer, 1943, Lance Formation, Wyoming, USA; Hell Creek Formation, South Dakota, Montana, USA; *Stegoceras edmontonense* (Brown and Schlaikjer, 1943), Hell Creek Formation, Montana, USA; *Stygimoloch spinifer* Galton and Sues, 1983, Hell Creek Formation, Montana, USA; Lance Formation, Wyoming, USA.
Intervening: CMP.

Infra-order CERATOPSIA Marsh, 1890

F. PSITTACOSAURIDAE Osborn, 1923 K. (BER/HAU–ALB) Terr.

First: *Psittacosaurus mongoliensis* Osborn, 1923, Shestakovskaya Svita, Gorno-Altayaskaya Autonomous Region, former USSR, and several formations in Mongolia and China.
Last: *Psittacosaurus guyangensis* Cheng, 1983, and *P. osborni* Young, 1931, Lisangou Formation, Nei Mongol Zizhiqu, China; and possibly other species of *Psittacosaurus* from China (Sereno, 1990, p. 589) in rocks dated as APT–ALB.
Intervening: APT.
Comment: An older 'psittacosaurid' has been reported from the Kitadani Formation (BRM) of Japan (Manabe and Hasegawa, 1991).

F. PROTOCERATOPSIDAE Granger and Gregory, 1923 K. (SAN/CMP–MAA) Terr.

First: *Protoceratops andrewsii* Granger and Gregory, 1923, Beds of Toogreeg and Beds of Alag Teg, Omnogov, Mongolia.

Last: *Leptoceratops gracilis* Brown, 1914, Scollard Formation, Alberta, Canada; Lance Formation, Wyoming, USA.

F. CERATOPSIDAE Marsh, 1890 K. (CMP–MAA) Terr.

First: *Chasmosaurus mariscalensis* Lehman, 1989, Aguja Formation, Texas, USA.
Last: *Torosaurus latus* Marsh, 1891 and *Triceratops horridus* Marsh, 1889, Lance Formation, Wyoming, USA; Evanston Formation, Wyoming, USA; Hell Creek Formation, Montana, South Dakota, USA; Laramie Formation, Colorado, USA; Javelina Formation, Texas, USA; Scollard Formation, Alberta, Canada; Frenchman Formation, Saskatchewan, Canada.

ARCHOSAUROMORPHA *incertae sedis*

F. HELVETICOSAURIDAE Peyer and Kuhn-Schnyder, 1955 Tr. (ANS/LAD) Mar.

First and Last: *Helveticosaurus zollingeri* Peyer, 1943, Grenzbitumen Horizon, Monte San Giorgio, Kanton Tessin, Switzerland.

NEODIAPSIDA *incertae sedis*

Superorder SAUROPTERYGIA Owen, 1860

Data on sauropterygian classification come from Sues (1987), Tschanz (1989) and Storrs (1991), and on distributions from Kuhn (1971), Mazin (1988) and Storrs (1991).

Order PACHYPLEUROSAURIA Sanz, 1980

F. PACHYPLEUROSAURIDAE Nopcsa, 1928 Tr. (ANS–LAD) Mar.

First: *Dactylosaurus gracilis* Gürich, 1884, lower Muschelkalk, Silesia, Poland; ?*Anarosaurus multidentatus* Huene, 1958, base of the Muschelkalk, Lechtaler Alpen, Germany; *Keichousaurus yuanensis* Young, 1965, basal Anisian, Kweichou, China.
Last: *Neusticosaurus pusillus* (Fraas, 1881), terminal Muschelkalk, Baden-Württemberg, Germany; *Psilotrachelosaurus toeplitschi* Nopcsa, 1928, upper Muschelkalk, Töplitsch, Germany.

Order NOTHOSAURIFORMES Storrs, 1991

Suborder PLACODONTIA Zittel, 1887–1890 (Owen, 1859)

F. PLACODONTIDAE Meyer, 1863 Tr. (SCY–LAD) Mar.

First: ?*Placodus impressus* Agassiz, 1839, upper Buntsandstein, Pfalz, Germany. A reputed upper SCY record from Makhtech Ramon, Israel is not a placodontian, but is rather a temnospondyl (Zanon, 1991).
Last: *Placodus gigas* Agassiz, 1833, Tonplatten, upper Muschelkalk, Bayreuth, Germany.
Intervening: ANS.

F. CYAMODONTIDAE Nopcsa, 1923 Tr. (ANS–LAD) Mar.

First: *Cyamodus tarnowitzensis* Gürich, 1884, lower Muschelkalk, Silesia, Poland.
Last: *Cyamodus rostratus* (Münster, 1830), upper Muschelkalk, Bayreuth, Germany.

F. PLACOCHELYIDAE Jaekel, 1907 Tr. (ANS–RHT) Mar.

First: *Saurosphargis volzi* Frech, 1903, Wellenkalk, lower Muschelkalk, Silesia, Poland.
Last: *Psephoderma alpinum* Meyer, 1858 and *P. raeticum* (Schubert-Klempnauer, 1975), Rät, Bavaria, Germany; *P. anglicum* Meyer, 1867, Rhaetic, Avon, England, UK.
Intervening: LAD, CRN, ?NOR.

F. HENODONTIDAE Huene, 1936 Tr. (CRN) Mar./FW

First and Last: *Henodus chelyops* Huene, 1936, Gipskeuper, Baden-Württemberg, Germany.

Suborder EUSAUROPTERYGIA Tschanz, 1989

Infra-order NOTHOSAURIA Seeley, 1882 (p)

F. COROSAURIDAE Kuhn, 1964 Tr. (SCY) Mar.

First and Last: *Corosaurus alcovensis* Case, 1936, Alcova Limestone, Chugwater Group, Natrona County, Wyoming, USA (Storrs, 1991a).

F. SIMOSAURIDAE Gervais, 1859 Tr. (LAD–CRN) Mar.

First: *Simosaurus gaillardoti* Meyer, 1842, upper Muschelkalk, France, Germany.
Last: *Simosaurus guilelmi* Meyer, 1855, Lettenkohle, Hoheneck, Germany.

F. NOTHOSAURIDAE Baur, 1889 Tr. (SCY–CRN) Mar.

First: ?*Nothosaurus mirabilis* Münster, 1834, Obere Buntsandstein, Germany; ?*Kwangsisaurus orientalis* Young, 1959, Lower Triassic, Kwangsi, China.
Last: *Nothosaurus edingerae* Schultze, 1970, Gipskeuper, Bayreuth, Germany.
Intervening: ANS, LAD.

Infra-order PLESIOSAURIA Blainville, 1835

Plesiosaur classification is based on Brown (1981) and Storrs and Langston (1993), and distributional data come from Persson (1963), Brown (1981), Mazin (1988) and Storrs and Langston (1993).

F. PISTOSAURIDAE Baur, 1887 Tr. (ANS) Mar.

First and Last: *Pistosaurus longaevus* Meyer, 1847–1855, upper Muschelkalk, Bayreuth, Bavaria, Germany.

F. PLESIOSAURIDAE Gray, 1825 Tr. (LAD??/RHT)–J. (TOA) Mar.

First: *Plesiosaurus priscus* Huene, 1902, Lettenkohle of Bibersfeld, Germany, is based on plesiosaur-like vertebrae (Mazin, 1988, p. 119). Other plesiosaurian vertebrae are known from the Middle Triassic German Muschelkalk and the Ladinian of the former USSR (Storrs, 1991a, p. 81). If these are not plesiosaurid, then the oldest remains are *Plesiosaurus costatus* Owen, 1840, and other species, from the Rhaetic of Avon, Leicestershire, and Nottinghamshire, England, UK; Morayshire, Scotland, UK; Autun, France; and Baden-Württemberg, Germany.
Last: *Plesiosaurus brachypterygius* Huene, 1923 and *P. guilelmiimperatoris* Dames, 1895, Lias-ε, Posidonienschiefer, Baden-Württemberg, Germany.
Intervening: HET–PLB.

F. CRYPTOCLEIDIDAE Williston, 1925
J. (CLV)–K. (MAA) Mar.

First: *Cryptoclidus eurymerus* (Phillips, 1871), Oxford Clay, Bedfordshire and Cambridgeshire, England, UK; *C. richardsoni* (Lydekker, 1889), Oxford Clay, Dorset, England, UK.
Last: *Aristonectes parvidens* Cabrera, 1941, Cañadon del Loro, Chubut, Argentina; *Turneria seymourensis* Chatterjee and Small, 1989, Lopez de Bertodano Formation, Seymour Island, Antarctica (Chatterjee and Small, 1989).
Intervening: KIM.

F. ELASMOSAURIDAE Cope, 1869
J. (TOA)–K. (MAA) Mar.

First: *Microcleidus macropterus* (Seeley, 1865) and *M. homalospondylus* (Owen, 1840), Alum Shale Member, Yorkshire, England, UK.
Last: *Mauisaurus haasti* Hector, 1874, Haumurian, South Island, New Zealand (Welles and Gregg, 1971); elasmosaurid vertebrae, Nacatoch Formation, Texas, USA (Storrs and Langston, 1993); elasmosaurid vertebrae and tooth, Nekum Chalk and Emael Chalk, Maastricht Formation, Limburg, The Netherlands (Mulder, 1990); elasmosaurid, Lopez de Bertodano Formation, Seymour Island, Antarctica (Chatterjee and Small, 1989).
Intervening: OXF–TTH, ALB–CMP.

F. PLIOSAURIDAE Seeley, 1874
J. (HET)–K. (TUR) Mar.

First: ?*Eurycleidus arcuatus* (Owen, 1840), and others, lower Lias, Zone of *Psiloceras planorbis*, Dorset, England, UK.
Last: *Polyptychodon hudsoni* Welles and Slaughter, 1963, Arcadia Park Formation, Eagle Ford Group, Dallas County, Texas, USA (Storrs and Langston, 1993).
Intervening: SIN–TOA, BTH–TTH, APT–CEN.
Comment: If *Eurycleidus arcuatus* and relatives are not pliosaurids, then the oldest confirmed examples are CLV and OXF from Europe.

F. POLYCOTYLIDAE Williston, 1908
K. (ALB–MAA) Mar.

First: *Trinacomerum* sp., Kiamichi Formation, Denton County, Texas, USA (Storrs and Langston, 1993).
Last: *Polycotylus* sp., Haumurian, South Island, New Zealand (Welles and Gregg, 1971), Fox Hills Formation, New Mexico, USA.

Subclass SYNAPSIDA Osborn, 1903(p)

Order PELYCOSAURIA Cope, 1878(p)

All pelycosaur records were obtained from Reisz (1986), unless otherwise stated. The ages of the terrestrial Lower Permian tetrapod-bearing formations of the United States are hard to correlate with the type Russian marine sections, so there is some uncertainty over the dating of many pelycosaur records (Olson and Vaughn, 1970).

F. EOTHYRIDIDAE Romer and Price, 1940
P. (SAK?–ART) Terr.

First: *Oedaleops campi* Langston, 1965, Abo/Cutler Formation, Cutler Group, Rio Arriba County, New Mexico, USA.
Last: *Eothyris parkeyi* Romer, 1937, Belle Plains Formation, Wichita Group, Archer County, Texas, USA.

F. CASEIDAE Williston, 1912
P. (ART–KAZ) Terr.

First: *Casea broilii* Williston, 1910, uppermost Arroyo Formation or lowermost Vale Formation, Clear Fork Group, Baylor County, Texas, USA.
Last: *Ennatosaurus tecton* Efremov, 1956, Zone II, Kazanian, Pinega River, former USSR.
Intervening: KUN.

F. VARANOPSEIDAE Romer and Price, 1940
C./P. (NOG/ASS)–P. (KUN) Terr.

First: *Aerosaurus greenleeorum* Romer, 1937, Abo/Cutler Formation, Cutler Group, Rio Arriba County, New Mexico, USA.
Last: *Varanodon agilis* Olson, 1965, Chickasha Formation, equivalent of the middle Flowerpot Formation, Blaine County, Oklahoma, USA.
Intervening: SAK, ART.

F. OPHIACODONTIDAE Nopcsa, 1923
C. (MYA)–P. (ART) Terr.

First: *Archaeothyris florensis* Reisz, 1972, Morien Group, Florence, Nova Scotia, Canada.
Last: *Ophiacodon major* Romer and Price, 1940, Clyde Formation, Clear Fork Group, Baylor County, Texas, USA.
Intervening: KRE/CHV, KLA–SAK.
Comment: A possible ophiacodontid, *Varanosaurus acutirostris* Broili, 1904, from the Arroyo Formation, Clear Fork Group of Texas, would be the youngest representative of that family (ART) if correctly determined. However, Reisz (1986, p. 85) regards it as 'Pelycosauria *incertae sedis*'.

F. EDAPHOSAURIDAE Case, 1907
C. (KRE/CHV)–P. (ART) Terr.

First: *Edaphosaurus ? raymondi* (Case, 1908), Round Knob Formation, Conemaugh Group, Pitcairn, Pennsylvania, USA.
Last: *Edaphosaurus pogonias* Cope, 1882, Arroyo Formation, Clear Fork Group, Baylor County, Texas, USA.
Intervening: KLA/NOG–SAK.

F. SPHENACODONTIDAE Williston, 1912
C. (KRE)–P. (UFI) Terr. (Fig. 39.7)

First: *Haptodus garnettensis* Currie, 1977, Stanton Formation, Lansing Group, Garnett, Kansas, USA.
Last: *Dimetrodon angelensis* Olson, 1962, upper San Angelo Formation, Pease River Group, Knox County, Texas, USA.
Intervening: ASS–KUN.

Order THERAPSIDA Broom, 1905

Therapsid classification is based broadly on Kemp (1982) and Hopson and Barghusen (1986). Distributional data and more detailed family designations are based on Sigogneau-Russell (1989) for Phthinosuchia, Biarmosuchia, Eotitanosuchia and Gorgonopsia, King (1988) for Dinocephalia and Dicynodontia, Tatarinov (1974), Mendrez-Carroll (1975) and Kemp (1982), and Hopson and Barghusen (1986) for the classification of Therocephalia, and Hopson and Kitching (1972), Tatarinov (1974), Battail (1982), Kemp (1982) and Hopson and Barghusen (1986) for the classification of Cynodontia. Additional distributional data came from Haughton and Brink (1954), Kitching (1977), Anderson and Cruickshank (1978), and Bonaparte (1978).

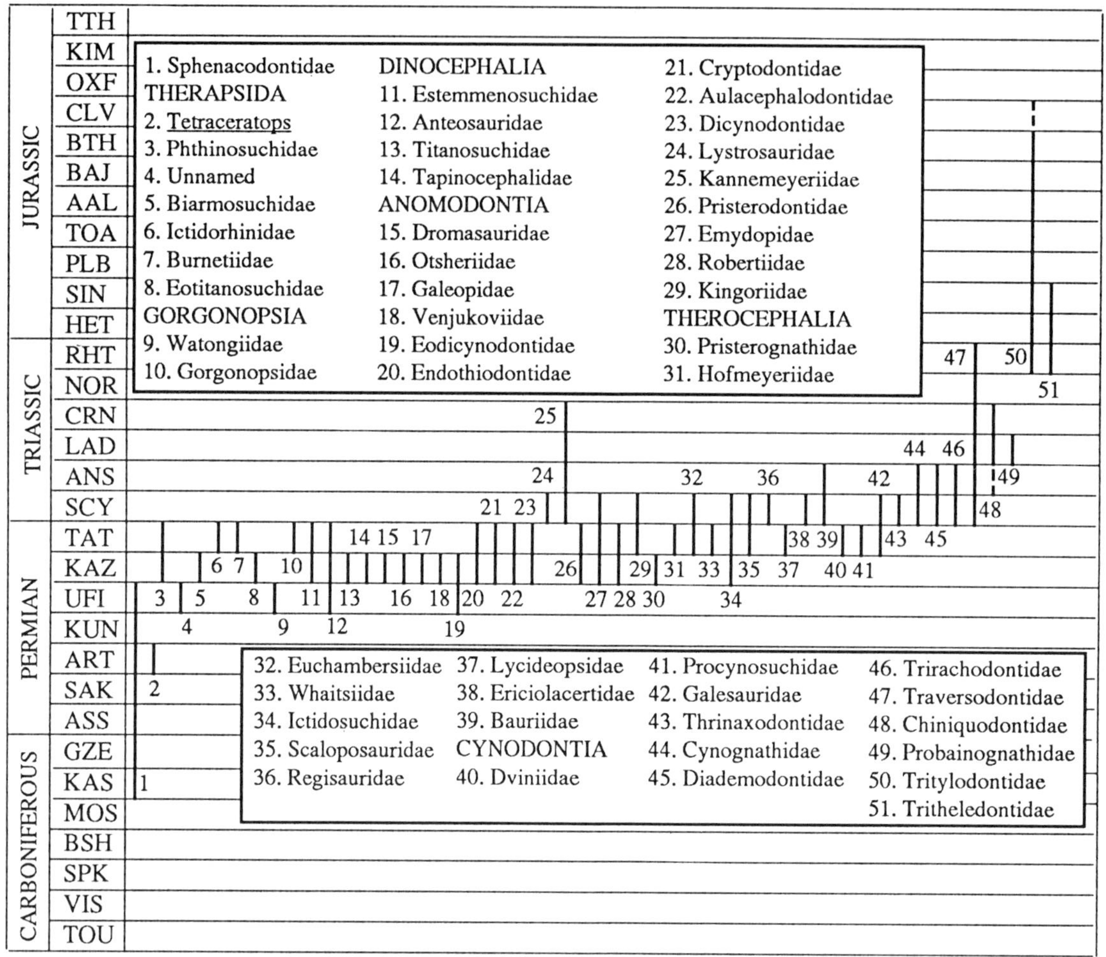

Fig. 39.7

F. UNNAMED P. (ART) Terr.

First and Last: *Tetraceratops insignis* Matthew, 1908, basal Clear Fork Group, Baylor County, Texas, USA (Laurin and Reisz, 1990).

Suborder PHTHINOSUCHIA Romer, 1961

F. PHTHINOSUCHIDAE Efremov, 1954 P. (KAZ–TAT) Terr.

First: *Phthinosaurus borissiaki* Efremov, 1940, Uralian Cupric Sandstones (Ezhovo), western Cisuraly, former USSR.
Last: *Phthinosuchus discors* Efremov, 1954, Uralian Cupric Sandstones (Isheevo), western Cisuraly, former USSR.

F. UNNAMED P. (UFI) Terr.

First: ?*Knoxosaurus niteckii* Olson, 1962, upper San Angelo Formation, Knox County, Texas, USA.
Last: *Steppesaurus gurleyi* Olson and Beerbower, 1953, lower Flower Pot Formation, Hardeman County, Texas, USA.

Suborder BIARMOSUCHIA Hopson and Barghusen, 1986

F. BIARMOSUCHIDAE Olson, 1962 P. (KAZ) Terr.

First and Last: *Biarmosuchus tener* Tchudinov, 1960, Ezhovo, Perm Province, former USSR.

F. ICTIDORHINIDAE Broom, 1932 P. (TAT) Terr. (including Hipposauridae)

First: *Hipposaurus boonstrai* Haughton, 1929 and *H. brinki* Sigogneau, 1970, *Eodicynodon–Tapinocaninus* Assemblage Zone, South Africa.
Last: *Ictidorhinus martinsi* Broom, 1913 and *Rubidgina angusticeps* Broom, 1942, *Dicynodon–Theriognathus* Assemblage Zone, Graaff-Reinet, South Africa.

F. BURNETIIDAE Broom, 1923 P. (TAT) Terr.

First and Last: *Proburnetia viatkensis* Tatarinov, 1968, upper TAT, Kirov Province, Kotelnitch, former USSR; *Burnetia mirabilis* Broom, 1923, *Dicynodon–Theriognathus* Assemblage Zone, Graaff-Reinet, South Africa.

Suborder EOTITANOSUCHIA Boonstra, 1963

F. EOTITANOSUCHIDAE Tchudinov, 1960 P. (KAZ) Terr.

First and Last: *Eotitanosuchus olsoni* Tchudinov, 1960, Ezhovo, Ocher Province, former USSR.

Suborder GORGONOPSIA Seeley, 1895

F. WATONGIIDAE Sigogneau-Russell, 1989 P. (UFI) Terr.

First and Last: ?*Watongia meieri* Olson, 1974, Chickasha Formation, Blaine County, Oklahoma, USA.

F. GORGONOPSIDAE Lydekker, 1890 P. (TAT) Terr.

First: *Broomisaurus planiceps* (Broom, 1913), *Eoarctops vanderbyli* Haughton, 1929, *Galesuchus gracilis* Haughton, 1925, and *Scylacognathus parvus* Broom, 1913, *Eodicynodon–Tapinocaninus* Assemblage Zone, South Africa.
Last: *Aelurosaurus wilmanae* Broom, 1940, *Aloposaurus ? tenuis* (Brink and Kitching, 1953), *Arctognathus curvimola* (Owen, 1876), *Arctops watsoni* Brink and Kitching, 1953, *Broomicephalus laticeps* (Broom, 1940), *Clelandina scheepersi* (Brink and Kitching, 1953), *Cyonosaurus rubidgei* (Broom, 1947), *Dinogorgon pricei* (Broom and George, 1950), *Leontocephalus cadlei* Broom, 1940, *Lycaenops angusticeps* (Broom, 1913), *Paragalerhinus rubidgei* (Broom, 1936), *Prorubidgea maccabei* Broom, 1940, *Rubidgea platyrhina* Brink and Kitching, 1953, and *Sycosaurus laticeps* Haughton, 1924, *Dicynodon–Theriognathus* Assemblage Zone, South Africa; *Inostrancevia alexandri* Amalitsky, 1922 and *Pravoslavlevia parva* (Pravoslavlev, 1927), Northern Dvina Horizon, Archangelsk Province, former USSR; ?*Niuksenitia sukhonensis* Tatarinov, 1977, Vologda Province, former USSR.

Infra-order DINOCEPHALIA Seeley, 1894

F. ESTEMMENOSUCHIDAE Tchudinov, 1960 P. (KAZ–TAT) Terr.

First: *Estemmenosuchus uralensis* Tchudinov, 1960, *E. mirabilis* Tchudinov, 1968, *Anoplosuchus tenuirostris* Tchudinov, 1968, and *Zopherosuchus luceus* Tchudinov, 1983, upper KAZ, Ezhovo, former USSR.
Last: *Molybdopygus arcanus* Tchudinov, 1964, lower TAT, Kirov, former USSR.

F. ANTEOSAURIDAE Boonstra, 1954 P. (UFI–TAT) Terr. (= Brithopidae)

First: *Eosyodon hudsoni* Olson, 1962, San Angelo Formation, Knox County, Texas, USA.
Last: *Notosyodon guvesi* Tchudinov, 1968, *Syodon biarmicum* Kutorga, 1838, *Titanophoneus potens* Efremov, 1938, *Doliosauriscus yanshinovi* (Orlov, 1958), *Deuterosaurus biarmicus* Eichwald, 1860, and *Admetophoneus kargalensis* Efremov, 1954, lower TAT, southern Cisurals region, former USSR; *Anteosaurus magnificus* Watson, 1921 and *Paranteosaurus primus* Boonstra, 1954, *Tapinocephalus–Bradysaurus* Assemblage Zone, South Africa.
Intervening: KAZ.

F. TITANOSUCHIDAE Boonstra, 1972 P. (KAZ) Terr.

First and Last: *Jonkeria truculenta* Van Hoepen, 1916 and six other species, and *Titanosuchus ferox* Owen, 1879, *Tapinocephalus–Bradysaurus* Assemblage Zone, South Africa.

F. TAPINOCEPHALIDAE Lydekker, 1890 P. (KAZ) Terr.

First: *Ulemosaurus svigagensis* Riabinin, 1938, Zone II, Isheevo, Tatar Republic former USSR; *Tapinocaninus pamelae* Rubidge, 1988, *Eodicynodon–Tapinocaninus* Assemblage Zone, South Africa.
Last: *Struthiocephalus whaitsi* Haughton, 1915, and 17 other species, *Tapinocephalus–Bradysaurus* Assemblage Zone, South Africa.

Infra-order ANOMODONTIA (DICYNODONTIA) Owen, 1859

F. DROMASAURIDAE Abel, 1919 P. (KAZ) Terr.

First: *Patronomodon nyaphulii* Rubidge and Hopson, 1991, *Eodicynodon–Tapinocaninus* Assemblage Zone, South Africa (Rubidge and Hopson, 1991).
Last: *Galepus jouberti* Broom, 1910, *Aulacephalodon–Cistecephalus* Assemblage Zone, South Africa.

F. OTSHERIIDAE Tchudinov, 1960 P. (KAZ) Terr.

First and Last: *Otsheria netsvetajevi* Tchudinov, 1960, Zone II, KAZ, Ezhovo, Ural region, former USSR.

F. GALEOPIDAE Broom, 1912 P. (KAZ) Terr.

First and Last: *Galeops whaitsi* Broom, 1912, *Tapinocephalus–Bradysaurus* Assemblage Zone, South Africa.

F. VENJUKOVIIDAE Efremov, 1940 P. (KAZ) Terr.

First: *Venjukovia prima* Amalitsky, 1922, Copper Sandstones, Zone II, KAZ, Ural region, former USSR.
Last: *Venjukovia invisa* Efremov, 1938, Copper Sandstones, Zone II, KAZ, Ural region, former USSR.

F. EODICYNODONTIDAE Cluver and King, 1983 P. (UFI/KAZ) Terr.

First and Last: *Eodicynodon oosthuizeni* Barry, 1974, *Eodicynodon–Tapinocaninus* Assemblage Zone, Cape Province, South Africa.

F. ENDOTHIODONTIDAE Owen, 1876 P. (KAZ–TAT) Terr.

First: *Chelydontops altidentalis* Cluver, 1975, *Tapinocephalus–Bradysaurus* Assemblage Zone, Cape Province, South Africa.
Last: *Endothiodon* sp., *Aulacephalodon–Cistecephalus* Assemblage Zone, South Africa. Equivalents in Brazil, India and Zambia.

F. CRYPTODONTIDAE Owen, 1859 P. (KAZ–TAT) Terr.

First: *Tropidostoma microtrema* (Seeley, 1889), *Cteniosaurus platyceps* Broom, 1935, *Rhachiocephalus magnus* (Owen, 1876), *Oudenodon bainii* Owen, 1860, and other species of these genera, *Tropidostoma–Endothiodon* Assemblage Zone, Beaufort West, South Africa.
Last: *Oudenodon baini* (Owen, 1860), *Dicynodon–Theriognathus* Assemblage Zone, South Africa. Equivalents from Zambia.

F. AULACEPHALODONTIDAE Cluver and King, 1983 P. (KAZ–TAT) Terr.

First: *Aulacephalodon baini* (Owen, 1845) and *Pelanomodon rubidgei* Broom, 1938, *Aulacephalodon–Cistecephalus* Assemblage Zone, South Africa, and equivalents in Zambia and Tanzania respectively.
Last: *Geikia elginensis* Newton, 1893, Cutties Hillock Sandstone Formation, Morayshire, Scotland, UK.

F. DICYNODONTIDAE Owen, 1859 P. (KAZ–TAT) Terr.

First: *Dicynodon acutirostris* Broom, 1935, and other species of this genus, *Tropidostoma–Endothiodon* Assemblage Zone, South Africa.

Last: *Dicynodon traquairi* (Newton, 1893), Cutties Hillock Sandstone, Morayshire, Scotland, UK.
Comment: Supposedly older species of *Dicynodon*, from the *Pristerognathus–Diictodon* Assemblage Zone of South Africa, are hard to substantiate (King, 1988).

F. LYSTROSAURIDAE Broom, 1903 Tr. (SCY) Terr.

First and Last: *Lystrosaurus murrayi* (Huxley, 1859), *Lystrosaurus–Procolophon* Assemblage Zone, Cape Province, South Africa; and 12 other species from this zone, and supposedly equivalent zones in Antarctica, the former USSR, China, India and Laos.

F. KANNEMEYERIIDAE Huene, 1948 Tr. (SCY–CRN) Terr.

First: *Kannemeyeria simocephalus* (Weithofer, 1888), Lower Etjo Beds, south-west Africa; *K. wilsoni* Broom, 1937, *Cynognathus–Diademodon* Assemblage Zone, South Africa; *K. argentinensis* Bonaparte, 1966, Puesto Viejo Formation, Mendoza Province, Argentina; *Vinceria andina* Bonaparte, 1967, Cerro de Las Cabras Formation, Mendoza Province, Argentina.
Last: *Jachaleria colorata* Bonaparte, 1971, boundary between Ischigualasto Formation and lower Los Colorados Formation, La Rioja Province, Argentina.
Intervening: ANS, LAD.

F. PRISTERODONTIDAE King, 1988 P. (KAZ–TAT) Terr.

First: *Pristerodon merwevillensis* (Broili and Schroeder, 1935), *Tapinocephalus–Bradysaurus* Assemblage Zone, South Africa.
Last: *Pristerodon mackayi* Huxley, 1868, and 13 other species of that genus, *Aulacephalodon–Cistecephalus* Assemblage Zone, South Africa.

F. EMYDOPIDAE Cluver and King, 1983 P. (KAZ)–Tr. (SCY) Terr.

First: *Emydops* sp., *Tapinocephalus–Bradysaurus* Assemblage Zone, South Africa.
Last: *Myosaurus gracilis* Haughton, 1917, *Lystrosaurus–Procolophon* Assemblage Zone, South Africa.
Intervening: TAT.

F. ROBERTIIDAE Cluver and King, 1983 P. (KAZ–TAT) Terr.

First: *Robertia broomiana* Boonstra, 1948, *Diictodon joubertii* (Broom, 1905), and four other species of the latter genus, *Tapinocephalus–Bradysaurus* Assemblage Zone, South Africa.
Last: *Diictodon nanus* (Broom, 1936), *Dicynodon–Theriognathus* Assemblage Zone, South Africa.

F. KINGORIIDAE King, 1988 P. (TAT)–Tr. (SCY) Terr.

First: *Kingoria nowacki* (Huene, 1942), Kawinga Formation, Kongori, Tanzania; *K. recurvidens* (Owen, 1876), and four other species of that genus, and *Dicynodontoides parringtoni* Broom, 1940, *Aulacephalodon–Cistecephalus* Assemblage Zone, South Africa.
Last: *Kombuisia frerensis* Hotton, 1974, *Cynognathus–Diademodon* Assemblage Zone, Cape Province, South Africa.

Suborder THEROCEPHALIA Broom, 1903

F. PRISTEROGNATHIDAE Broom, 1906 P. (KAZ) Terr.

First: *Porosteognathus efremovi* Vjuschkov, 1952, Zone I, Urals region, former USSR; 'pristerognathid', upper Ecca Group, Cape Province, South Africa (Rubidge *et al.*, 1983).
Last: *Pristerognathus polyodon* Seeley, 1895, and other species, upper *Pristerognathus–Diictodon* Assemblage Zone, South Africa.

F. HOFMEYERIIDAE Hopson and Barghusen, 1986 P. (TAT) Terr.

First and Last: *Hofmeyria atavus* Broom, 1935 and *Ictidostoma hemburyi* (Broom, 1912), ?*Aulacephalodon–Cistecephalus* Assemblage Zone, South Africa (Brink, 1960).

F. EUCHAMBERSIIDAE Boonstra, 1934 P. (TAT)–Tr. (SCY) Terr. (= Moschorhinidae; Annatherapsidae; Akidnognathidae)

First: *Euchambersia mirabilis* Broom, 1931, *Aulacephalodon–Cistecephalus* Assemblage Zone; *Annatherapsidus petri* (Amalitzky, 1922), Zone IV, Urals region, former USSR.
Last: *Moschorhinus kitchingi* Broom, 1920, *Lystrosaurus–Procolophon* Assemblage Zone, South Africa.

F. WHAITSIIDAE Haughton, 1918 P. (TAT) Terr.

First: *Whaitsia* sp. *Aulacephalodon–Cistecephalus* Assemblage Zone, Graaf-Reinet, South Africa.
Last: *Whaitsia platyceps* Haughton, 1918, *Dicynodon–Theriognathus* Assemblage Zone, South Africa and equivalent, Ruhuhu Valley, Tanzania; *Moschowhaitsia vjushkovi* Tatarinov, 1963, Zone IV, TAT, Urals region, former USSR.

F. ICTIDOSUCHIDAE Broom, 1903 P. (KAZ)–Tr. (SCY) Terr.

First: *Icticephalus polycynodon* Broom, 1915, *Tapinocephalus–Bradysaurus* Assemblage Zone, South Africa.
Last: *Olivieria parringtoni* Brink, 1965, *Lystrosaurus–Procolophon* Assemblage Zone, Orange Free State, South Africa. Note that Colbert and Kitching (1981) regard *Olivieria* as a juvenile *Moschorhinus*.

F. SCALOPOSAURIDAE Broom, 1914 P. (TAT)–Tr. (SCY) Terr.

First: *Scaloposaurus constrictus* Owen, 1876 and *Nanictocephalus richardi* Broom, 1940, *Aulacephalodon–Cistecephalus* Assemblage Zone, South Africa (Mendrez-Carroll, 1979).
Last: *Pedaeosaurus parvus* Colbert and Kitching, 1981, Fremouw Formation, Antarctica.
Comment: Mendrez-Carroll (1979) and Hopson and Barghusen (1986) noted that the *Lystrosaurus–Procolophon* Assemblage Zone examples of scaloposaurids belong to other taxa, but Colbert and Kitching (1981) described *Pedaeosaurus* as a scaloposaurid.

F. REGISAURIDAE Hopson and Barghusen, 1986 Tr. (SCY) Terr.

First and Last: *Regisaurus jacobi* Mendrez, 1972, *Lystrosaurus–Procolophon* Assemblage Zone, Cape Province, South Africa.

F. LYCIDEOPSIDAE Broom, 1931 P. (TAT) Terr.

First and Last: *Lycideops longiceps* Broom, 1931,

Aulacephalodon–Cistecephalus Assemblage Zone, South Africa.

F. ERICIOLACERTIDAE Watson and Romer, 1956 Tr. (SCY) Terr.

First and Last: *Ericiolacerta parva* Watson, 1931, *Lystrosaurus–Procolophon* Assemblage Zone, Orange Free State, South Africa; Fremouw Formation, Antarctica (Colbert and Kitching, 1981).

F. BAURIIDAE Broom, 1911 Tr. (SCY–ANS) Terr.

First: *Bauria cynops* Broom, 1909, *Sesamodon browni* Broom, 1932, *Cynognathus–Diademodon* Assemblage Zone, South Africa.
Last: *Dongusaurus schepetovi* Vjuschkov, 1964 and *Nothogomphodon danilovi* Tatarinov, 1974, Donguz Series, Urals region, former USSR; *Herpetogale marsupialis* Keyser and Brink, 1979, Omingonde Formation, Etjo Mountain, south-west Africa (Keyser and Brink, 1979).

Suborder CYNODONTIA Owen, 1860

F. DVINIIDAE Tatarinov, 1968 P. (TAT) Terr.

First and Last: *Dvinia prima* Amalitzky, 1922, Zone IV, Urals region, former USSR.

F. PROCYNOSUCHIDAE Broom, 1937 P. (TAT)

First and Last: *Procynosuchus delaharpeae* Broom, 1937, ?*Aulacephalodon–Cistecephalus* and *Dicynodon–Theriognathus* Assemblage Zones, South Africa; Madumabisa Mudstones, Luangwa Valley, Zambia; *Procynosuchus* sp., Randkalk, northern Hessen, Germany (Sues and Boy, 1988).

F. GALESAURIDAE Lydekker, 1890 P. (TAT)–Tr. (SCY) Terr.

First: *Cynosaurus suppostus* (Owen, 1876), *Dicynodon–Theriognathus* Assemblage Zone, South Africa.
Last: *Galesaurus planiceps* Owen, 1859, *Lystrosaurus–Procolophon* Assemblage Zone, South Africa.

F. THRINAXODONTIDAE Seeley, 1894 Tr. (SCY) Terr.

First and Last: *Thrinaxodon liorhinus* Seeley, 1894, *Lystrosaurus–Procolophon* Assemblage Zone, South Africa; Fremouw Formation, Antarctica (Colbert and Kitching, 1977).

F. CYNOGNATHIDAE Seeley, 1895 Tr. (SCY–ANS) Terr.

First: *Cynognathus crateronotus* Seeley, 1895, *Cynognathus–Diademodon* Assemblage Zone, South Africa; *C. minor* Bonaparte, 1967, Puesto Viejo Formation, Argentina (Bonaparte, 1978).
Last: *Cynognathus* sp., Omingonde Mudstone Formation, south-west Africa.

F. DIADEMODONTIDAE Haughton, 1925 Tr. (SCY–ANS) Terr.

First: *Diademodon tetragonus* Seeley, 1894, *D. grossarthi* (Broili and Schröder, 1935) and *D. mastacus* Seeley, 1894, *Cynognathus–Diademodon* Assemblage Zone, South Africa.
Last: *Diademodon rhodesiensis* Brink, 1963, Ntawere Formation, Luangwa Valley, Zambia.

F. TRIRACHODONTIDAE Crompton, 1955 Tr. (SCY–ANS) Terr.

First: *Trirachodon berryi* Seeley, 1894, *Cynognathus–Diademodon* Assemblage Zone, South Africa.
Last: *Cricodon metabolus* Crompton, 1955, Manda Formation, Ruhuhu Valley, Tanzania.

F. TRAVERSODONTIDAE Huene, 1936 Tr. (SCY–RHT) Terr.

First: *Pascualgnathus polanskii* Bonaparte, 1966, Puesto Viejo Formation, and *Andescynodon mendozensis* Bonaparte, 1967 and *Rusconiodon mignonei* Bonaparte, 1972, Rio Mendoza Formation, Mendoza Province, Argentina (Bonaparte, 1978).
Last: *Microscalenodon nanus* Hahn *et al.* 1988, lower RHT Bone Bed, Gaume, southern Belgium (Hahn *et al.*, 1988).
Intervening: ANS–NOR.

F. CHINIQUODONTIDAE Huene, 1948 Tr. (?ANS–CRN) Terr.

First: *Aleodon brachyramphus* Crompton, 1955, Manda Formation, Ruhuhu Valley, Tanzania. If this is not a chiniquodontid (Battail, 1982; Kemp, 1982, p. 208), the oldest representatives are *Probelesodon lewisi* Romer, 1969 and *Chiniquodon* sp. from the Chañares Formation, La Rioja Province, Argentina (LAD).
Last: *Chiniquodon theotonicus* Huene, 1936, Santa Maria Formation, Estado Rio Grande do Sul, Brazil.
Intervening: LAD.
Comment: Hahn *et al.* (1987) report a chiniquodontid tooth, *Lepagia gaumensis* Hahn *et al.*, 1987, from the lower RHT Bone Bed of Gaume, southern Belgium.

F. PROBAINOGNATHIDAE Romer, 1973 Tr. (LAD) Terr.

First and Last: *Probainognathus jenseni* Romer, 1970, lower beds of Ischichuca Formation, La Rioja Province, Argentina.

F. TRITYLODONTIDAE Cope, 1884 Tr. (RHT)–J. (BTH/CLV) Terr.

First: 'cf. *Tritylodon*', upper beds of Los Colorados Formation, La Rioja Province, Argentina (Bonaparte, 1978).
Last: *Bienotheroides wanhsienensis* Young, 1982, upper Xiashaximiao Formation, Szechwan, China (Sues, 1986).
Intervening: HET–PLB, BTH.

F. TRITHELEDONTIDAE Broom, 1912 Tr. (RHT)–J. (SIN) Terr.

First: *Chaliminia musteloides* Bonaparte, 1980, Los Colorados Formation, La Rioja Province, Argentina.
Last: *Pachygenelus monus* Watson, 1913, Clarens Formation, South Africa, Lesotho.
Intervening: HET.
Comment: *Therioherpeton cargnini* Bonaparte and Barberena, 1975, Santa Maria Formation, Parana Basin, Brazil (CRN), was described as the oldest tritheledontid, but Shubin *et al.* (1991) argue that this assignment is incorrect.

REFERENCES

Alifanov, V. R. (1989) The oldest gecko (Lacertilia, Gekkonidae) from the Lower Cretaceous of Mongolia. *Palaeontological Journal*, **23**, 128–31.

Anderson, J. M. and Cruickshank, A. R. I. (1978) The biostratigraphy of the Permian and Triassic. Part 5. A review of the classification and distribution of Permo-Triassic tetrapods. *Palaeontologia Africana*, **21**, 15–44.

Antunes, M. T. and Broin, F. de (1988) Le Crétacé terminal de la province Beira Litoral, Portugal: remarques stratigraphiques et écologiques; étude complémentaire de *Rosasia soutoi* (Chelonii, Bothremydidae). *Ciencias da Terra*, **9**, 141–52.

Arcucci, A. (1987) Un nuevo Lagosuchidae (Thecodontia–Pseudosuchia) de la fauna de Los Chañares (Edad Reptil Chañarense, Triasico Medio), La Rioja, Argentina. *Ameghiniana*, **24**, 89–94.

Astibia, H., Buffetaut, E., Buscalioni, A. D. *et al.* (1990) The fossil vertebrates from Lano (Basque Country, Spain); new evidence on the composition and affinities of the Late Cretaceous continental faunas of Europe. *Terra Nova*, **2**, 460–6.

Augé, M. (1986) *Les Lacertiliens (Reptiles, Squamata) de l'Eocène supérieur et de l'Oligocène ouest européens.* PhD, Université de Paris, VI, 218 pp.

Baird, D. (1978) *Pneumatoarthrus* Cope, 1879, not a dinosaur but a sea-turtle. *Proceedings of the National Academy of Natural Sciences of Philadelphia*, **129**, 71–81.

Baird, D. (1984) No ichthyosaurs in the Upper Cretaceous of New Jersey . . . or Saskatchewan. *The Mosasaur*, **2**, 129–33.

Battail, B. (1982) Essai de phylogénie des cynodontes (Reptilia, Therapsida). *Geobios, Mémoire Spécial*, **6**, 157–67.

Bell, B. A., Murry, P. A. and Osten, L. W. (1982) *Coniasaurus* Owen, 1850 from North America. *Journal of Paleontology*, **56**, 520–4.

Bennett, S. C. (1989) A pteranodontid pterosaur from the Early Cretaceous of Peru, with comments on the relationships of Cretaceous pterosaurs. *Journal of Paleontology*, **63**, 669–77.

Benton, M. J. (1985) Classification and phylogeny of the diapsid reptiles. *Zoological Journal of the Linnean Society*, **84**, 97–164.

Benton, M. J. (1986) The late Triassic reptile *Teratosaurus* – a rauisuchian, not a dinosaur. *Palaeontology*, **29**, 293–301.

Benton, M. J. and Clark, J. (1988) Archosaur phylogeny and the relationships of the Crocodylia, in *Amphibians, Reptiles, Birds* (*The Phylogeny and Classification of the Tetrapods*: Vol. 1), (ed. M. J. Benton), Systematics Association Special Volume 35A, Clarendon Press, Oxford, 377 pp.

Bonaparte, J. F. (1978) El Mesozoico de América del Sur y sus tetrápodos. *Opera Lilloana*, **26**, 1–596.

Bonaparte, J. F. (1984) Locomotion in rauisuchid thecodonts. *Journal of Vertebrate Paleontology*, **3**, 210–18.

Bonaparte, J. F. (1991) Los vertebrados fosiles de la Formacion Rio Colorado, de la ciudad de Neuquén y Cercanias, Cretacico Superior, Argentina. *Revista del Museo Argentino de Ciencias Naturales 'Bernardino Rivadavia' e Instituto Nacional de Investigacion de las Ciencias Naturales, Paleontologia*, **4**, 15–123.

Borsuk-Bialynicka, M. (1984) Anguimorphans and related lizards from the Late Cretaceous of the Gobi Desert, Mongolia. *Palaeontologia Polonica*, **46**, 5–105.

Borsuk-Bialynicka, M. (1988) Carolinidae, a new family of xenosaurid-like lizards from the Upper Cretaceous of Mongolia. *Acta Palaeontologica Polonica*, **30**, 151–76.

Bour, R. and Dubois, A. (1984) Nomenclature ordinale et familiale des tortues (Reptilia). *Studia Palaeocheloniologica*, **1**, 77–86.

Brink, A. S. (1960) A new type of primitive cynodont. *Palaeontologia Africana*, **7**, 119–54.

Brinkman, D. (1988) A weigeltisaurid reptile from the Lower Triassic of British Columbia. *Palaeontology*, **31**, 951–5.

Brinkman, D., Berman, D. S. and Eberth, D. A. (1984) A new araeoscelid reptile, *Zarcasaurus tanyderus*, from the Cutler Formation (Lower Permian) of north-central New Mexico. *New Mexico Geology*, **1984(5)**, 34–9.

Broin, F. de (1988) Les tortues et le Gondwana. Examen des rapports entre le fractionnement du Gondwana et la dispersion géographique des tortues pleurodires à partir du Crétacé. *Studia Palaeocheloniologica*, **2**, 103–42.

Brown, D. S. (1981) The English Upper Jurassic Plesiosauroidea (Reptilia) and a review of the phylogeny and classification of the Plesiosauria. *Bulletin of the British Museum (Natural History), Geology Series*, **35**, 253–347.

Buffetaut, E. (1982) Radiation évolutive, paléoecologie et biogéographie des crocodiliens mésosuchiens. *Mémoires de la Société Géologique de France*, **142**, 1–88.

Buffetaut, E. (1987) Occurrence of the crocodilian *Dolichochampsa minima* (Eusuchia, Dolichochampsidae) in the El Molino Formation of Bolivia. *Bulletin de la Société Belge de Géologie*, **96**, 195–9.

Buffetaut, E. (1988) The ziphodont mesosuchian crocodile from Messel: a reassessment. *Courier Forschungsinstitut Senckenberg*, **107**, 211–21.

Buffetaut, E. (1989) New remains of the enigmatic dinosaur *Spinosaurus* from the Cretaceous of Morocco and the affinities between *Spinosaurus* and *Baryonyx*. *Neues Jahrbuch für Geologie und Paläontologie, Monatshefte*, **1989**, 79–87.

Carroll, R. L. (1976) *Noteosuchus* – the oldest known rhynchosaur. *Annals of the South African Museum*, **72**, 37–57.

Carroll, R. L. (1984) Problems in the use of terrestrial vertebrates for zoning the Carboniferous, in *Neuvième Congrès International de Stratigraphie et de Géologie du Carbonifère, Compte Rendu*, Vol. 2, *Biostratigraphy* (eds P. K. Sutherland and W. L. Sutherland), Southern Illinois University Press, Carbondale and Edwardsville, pp. 135–47.

Carroll, R. L. (1985) A pleurosaur from the Lower Jurassic and the taxonomic position of the Sphenodontida. *Palaeontographica, Abteilung A*, **189**, 1–28.

Carroll, R. L. and Debraga, M. (1992) Aigialosaurs: Mid-Cretaceous varanoid lizards. *Journal of Vertebrate Paleontology*, **12**, 66–86.

Carroll, R. L. and Dong Z.-M. (1991) *Hupehsuchus*, an enigmatic aquatic reptile from the Triassic of China, and the problem of establishing relationships. *Philosophical Transactions of the Royal Society of London, Series B*, **331**, 131–53.

Charig, A. J. (1976) Order Thecodontia Owen 1859, in *Handbuch der Paläoherpetologie* (ed. O. Kuhn), Gustav Fischer, Stuttgart, pp. 7–10.

Chatterjee, S. and Small, B. J. (1989) New plesiosaurs from the Upper Cretaceous of Antarctica, in *Origins and Evolution of the Antarctic Biota* (ed. J. A. Crame), Geological Society Special Publications, London, pp. 197–215.

Chiappe, L. M. (1988) A new trematochampsid crocodile from the Early Cretaceous of north-western Patagonia, Argentina and its palaeobiogeographical and phylogenetic implications. *Cretaceous Research*, **1988**, 379–89.

Ckhikvadzé, V. M. (1988) Sur la classification et les caractères de certaines tortues fossiles d'Asie rares et peu etudiées. *Studia Palaeocheloniologica*, **2**, 55–86.

Clark, J. (1993) Cranial anatomy of *Protosuchus richardsoni* (Brown) and two new protosuchids, and the relationships of the 'Protosuchia' (Archosauria: Crocodylomorpha). *Bulletin of the American Museum of Natural History*, in press.

Clark, J. and Carroll, R. L. (1973) Romeriid reptiles from the Lower Permian. *Bulletin of the Museum of Comparative Zoology*, **144**, 353–407.

Colbert, E. H. and Kitching, J. W. (1977) Triassic cynodont reptiles from Antarctica. *American Museum Novitates*, **2611**, 1–30.

Colbert, E. H. and Kitching, J. W. (1981) Scaloposaurian reptiles from the Triassic of Antarctica. *American Museum Novitates*, 1–22.

Coombs, W. P. Jr, Weishampel, D. B. and Witmer, L. M. (1990) Basal Thyreophora, in *The Dinosauria* (eds D. B. Weishampel, P. Dodson and H. Osmólska), University of California Press, Berkeley, pp. 427–34.

Cope, J. C. W., Duff, K. L., Parsons, C. F. *et al.* (1980) Jurassic. Part 2. Middle and Upper Jurassic. *Special Report of the Geological Society of London*, **15**, 1–109.

Crush, P. J. (1983) Fossil Tetrapods from the Pant-y-Ffynon fissure site, South Wales. PhD, University of London.

Cuny, G. (1991) Nouvelles données sur la faune et l'age de Saint Nicholas de Port. *Revue de Paléobiologie*, **10**, 69–78.

Dong Z.-M. (1990) Stegosaurs of Asia, in *Dinosaur Systematics: Perspectives and Approaches* (ed. K. Carpenter and P. J. Currie), Cambridge University Press, Cambridge, pp. 255–68.

Dong, Z.-M., Zhou, S. and Zhang, Y. (1983) [The dinosaurian remains from Sichuan Basin, China.] *Palaeontologica Sinica, Series*

C, **23**, 139–45 [in Chinese with English summary].

Efimov, M. B. (1981) [New paralligatorids from the Lower Cretaceous of Mongolia.] *Trudy Sovmestnaya Sovetsko-Mongolkskaya Paleontologicheskoi Ekspeditsii*, **15**, 26–8.

Efimov, M. B. (1982) A two-fanged crocodile from the Upper Cretaceous in Tadzhikistan. *Paleontological Journal*, **1982(4)**, 103–5.

Erickson, B. R. (1973) A new chelydrid turtle *Protochelydra zangerli* from the Late Paleocene of North Dakota. *Scientific Publications of the Science Museum of Minnesota, New Series*, **2(2)**, 1–16.

Estes, R. (1983) Sauria terrestria, Amphisbaenia. *Handbuch der Paläoherpetologie*, **10A**, 1–249.

Estes, R., Queiroz, K. de and Gauthier, J. (1988) Phylogenetic relationships within Squamata, in *Phylogenetic Relationships of the Lizard Families. Essays Commemorating Charles L. Camp* (eds R. Estes and G. Pregill), Stanford University Press, Stanford, pp. 119–281.

Evans, S. E. (1984) The classification of the Lepidosauria. *Zoological Journal of the Linnean Society*, **82**, 87–100.

Evans, S. E. (1988) The early history and relationships of the Diapsida, in *The Phylogeny and Classification of the Tetrapods* (ed. M. J. Benton), Systematics Association Special Volume, 35A, Clarendon Press, Oxford, pp. 221–60.

Evans, S. E. (1989) New material of *Cteniogenys* (Reptilia: Diapsida; Jurassic) and a reassessment of the phylogenetic position of the genus. *Neues Jahrbuch für Geologie und Paleotologie, Monatshefte*, **1989**, 577–89.

Evans, S. E. (1990) The skull of *Cteniogenys* (Reptilia: Archosauromorpha) from the Middle Jurassic of Oxfordshire. *Zoological Journal of the Linnean Society*, **99**, 205–37.

Evans, S. E. and Milner, A. R. (1991) Middle Jurassic microvertebrate faunas from the British Isles, in *Fifth Symposium on Mesozoic Terrestrial Ecosystems and Biota* (eds Z. Kielan-Jaworowska, N. Heintz and H. A. Nakrem), Contributions from the Paleontological Museum, University of Oslo, 364. University of Oslo, Oslo, 72 pp.

Fraser, N. C. (1986) Terrestrial vertebrates at the Triassic–Jurassic boundary in south west Britain. *Modern Geology*, **10**, 147–57.

Fraser, N. C. (1988) Rare tetrapod remains from the Late Triassic fissure infillings of Cromhall Quarry, Avon. *Palaeontology*, **31**, 567–76.

Fraser, N. C. and Benton, M. J. (1989) The Triassic reptiles *Brachyrhinodon and Polysphenodon* and the relationships of the spenodontids. *Zoological Journal of the Linnean Society*, **96**, 413–45.

Gaffney, E. S. (1972) The systematics of the North American family Baenidae (Reptilia, Cryptodira). *Bulletin of the American Museum of Natural History*, **147**, 241–320.

Gaffney, E. S. (1975) A phylogeny and classification of the higher categories of turtles. *Bulletin of the American Museum of Natural History*, **155**, 387–436.

Gaffney, E. S. (1979) The Jurassic turtles of North America. *Bulletin of the American Museum of Natural History*, **162**, 91–136.

Gaffney, E. S. (1981) A review of the fossil turtles of Australia. *American Museum Novitates*, **2720**, 1–38.

Gaffney, E. S. (1986) Triassic and Early Jurassic turtles, in *The Beginning of the Age of Dinosaurs* (ed. K. Padian), Cambridge University Press, Cambridge, pp. 183–7.

Gaffney, E. S. (1990) The comparative osteology of the Triassic turtle *Proganochelys*. *Bulletin of the American Museum of Natural History*, **194**, 1–263.

Gaffney, E. S. (1992) An introduction to turtles, with a review of the turtles of Australia, in *Vertebrate Evolution in Australia* (eds M. Archer and S. Hand), in press.

Gaffney, E. S. and McKenna, M. C. (1979) A Late Permian captorhinid from Rhodesia. *American Museum Novitates*, **2688**, 1–15.

Gaffney, E. S. and Meylan, P. A. (1988) A phylogeny of turtles, in *Amphibians, Reptiles, Birds (The Phylogeny and Classification of the Tetrapods: Vol. 1)*, (ed. M. J. Benton), Systematics Association Special Volume, 35A, Clarendon Press, Oxford, pp. 157–219.

Gaffney, E. S., Hutchison, J. H., Jenkins, F. A. *et al.* (1987) Modern turtle origins: the oldest known cryptodire. *Science*, **237**, 289–91.

Galton, P. M. (1980) Avian-like tibiotarsi of pterodactyloids (Reptilia: Pterosauria) from the Middle and Upper Jurassic. *Paläontologische Zeitschrift*, **54**, 331–42.

Gasparini, Z., Chiappe, L. M. and Fernandez, M. (1991) A new Senonian peirosaurid (Crocodylomorpha) from Argentina and a synopsis of the South American Cretaceous crocodilians. *Journal of Vertebrate Paleontology*, **11**, 316–33.

Gauthier, J. A., Kluge, A. G. and Rowe, T. (1988) The early evolution of the Amniota, in *Amphibians, Reptiles, Birds (The Phylogeny and Classification of the Tetrapods: Vol. 1)*, (ed. M. J. Benton), Systematics Association Special Volume, 35A, Clarendon Press, Oxford, pp. 103–55.

Hahn, G., Lepage, J.-C. and Wouters, G. (1988) Traversodontiden-Zähne (Cynodontia) aus der Ober-Trias von Gaume (Süd-Belgien). *Bullétin de l'Institut Royal des Sciences naturelles de Belgique*, **58**, 177–86.

Hahn, G., Wild, R. and Wouters, G. (1987) Cynodontier-Zähne aus der Ober-Trias von Gaume (S-Belgien). *Mémoires pour servir à l'Explication des Cartes Géologiques et Minières de la Belgique*, **24**, 1–33.

Harland, W. B., Armstrong, R. L., Cox, A. V. *et al.* (1990) *A Geologic Time Scale 1989*. Cambridge University Press, Cambridge, 263 pp.

Harland, W. B., Cox, A. V., Llewellyn, P. G. *et al.* (1982) *A Geologic Time Scale*. Cambridge University Press, Cambridge, 131 pp.

Haubold, H. and Buffetaut, E. (1987) Une nouvelle interprétation de *Longisquama insignis*, reptile énigmatique du Trias supérieur d'Asie centrale. *Comptes rendus de l'Académie des Sciences, Paris*, **305 (II)**, 65–70.

Haughton, S. H. and Brink, A. S. (1954) A bibliographic list of Reptilia from the Karroo beds of Africa. *Palaeontologia Africana*, **2**, 1–187.

Hecht, M. K. (1992) A new choristodere (Reptilia, Diapseda) from the Oligocene of France: an example of the Lazarus effect. *Geobios*, **25**, 115–31.

Hentz, T. F. (1989) Permo-Carboniferous lithostratigraphy of the vertebrate-bearing Bowie and Wichita Groups, north-central Texas, in *Permo-Carboniferous Vertebrate Paleontology, Lithostratigraphy, and Depositional Environments of North-Central Texas* (ed. R. W. Hook), Field Trip Guidebook No. 2, 49th Annual Meeting of the Society of Vertebrate Paleontology. Society of Vertebrate Paleontology, Austin, Texas, pp. 1–21.

Hook, R. W. (1989) Stratigraphic distribution of tetrapods in the Bowie and Wichita groups, Permo-Carboniferous of North-Central Texas, in *Permo-Carboniferous Vertebrate Paleontology, Lithostratigraphy and Depositional Environments of North-Central Texas* (ed. R. W. Hook), Field Trip Guidebook No. 2, 49th Annual Meeting of the Society of Vertebrate Paleontology. Society of Vertebrate Paleontology, Austin, Texas, pp. 47–53.

Hopson, J. A. and Barghusen, H. R. (1986) An analysis of therapsid relationships, in *The Ecology and Biology of Mammal-like Reptiles* (eds N. Hotton III, P. D. MacLean and J. J. Roth *et al.*), Smithsonian Institution Press, Washington, DC, pp. 83–106.

Hopson, J. A. and Kitching, J. W. (1972) A revised classification of cynodonts (Reptilia: Therapsida). *Palaeontologia Africana*, **14**, 71–85.

Howse, S. C. B. (1986) On the cervical vertebrae of the Pterodactyloidea (Reptilia: Archosauria). *Zoological Journal of the Linnean Society*, **88**, 307–28.

Huene, F. von and Mauberge, P. L. (1954) Sur quelques restes de sauriens du Rhétien et du Jurassique lorrains. *Bullétin de la Société Géologique de France, Série 6*, **4**, 105–9.

Hunt, A. P. and Lucas, S. G. (1990) Re-evaluation of *'Typothorax' meadei*, a Late Triassic aetosaur from the United States. *Paläontologische Zeitschrift*, **64**, 317–28.

Hunt, A. P. and Lucas, S. G. (1991a) The *Paleorhinus* Biochron and the correlation of the nonmarine Upper Triassic of Pangaea. *Palaeontology*, **34**, 487–501.

Hunt, A. P. and Lucas, S. G. (1991b) A new rhynchosaur from the Upper Triassic of west Texas, U.S.A., and the biochronology of

Late Triassic rhynchosaurs. *Palaeontology*, **34**, 927–38.

Hutchison, J. H. (1980) Turtle stratigraphy of the Willwood Formation, Wyoming: preliminary results. *University of Michigan Papers in Paleontology*, **24**, 115–18.

Hutchison, J. H. (1987) New cranial material of *Compsemys* (Testudines) and its systematic implications. *Journal of Vertebrate Paleontology*, **7** (supplement to no. 3), p. 19A.

Hutchison, J. H. and Archibald, J. D. (1986) Diversity of turtles across the Cretaceous/Tertiary boundary in northeastern Montana. *Palaeogeography, Palaeoclimatology, Palaeoecology*, **55**, 1–22.

Ivakhnenko, M. F. (1990) Elements of the Early Permian tetrapod faunal assemblages of Eastern Europe. *Palaeontological Journal*, **1990**, 104–12.

Kemp, T. S. (1982) *Mammal-like Reptiles and the Origin of Mammals*. Academic Press, London, 363 pp.

Keyser, A. W. and Brink, A. S. (1979) A new bauriamorph (*Herpetogale marsupialis*) from the Omingonde Formation (Middle Triassic) of South West Africa. *Annals of the Geological Survey of South Africa*, **12**, 91–105.

Keyser, A. W. and Smith, R. M. H. (1979) Vertebrate biozonation of the Beaufort Group with special reference to the western Karoo basin. *Annals of the Geological Survey (South Africa)*, **12**, 1–36.

King, G. M. (1988) Anomodontia. *Handbuch der Paläoherpetologie*, **17C**, 1–174.

Kitching, J. W. (1977) The distribution of the Karroo vertebrate faunas. *Memoirs of the Bernard Price Institute for Palaeontological Research*, **1**, 1–131.

Kuhn, O. (1969) Proganosauria, Bolosauria, Placodontia, Araeoscelidia, Trilophosauria, Weigeltisauria, Millerosauria, Rhynchocephalia, Protorosauria. *Handbuch der Paläoherpetologie*, **9**, 1–74.

Kuhn, O. (1971) *Die Saurier der Deutschen Trias*. Gebrüder Geiselberger, Altötting, 105 pp.

Laurin, M. (1991) The osteology of a Lower Permian eosuchian from Texas and a review of diapsid phylogeny. *Zoological Journal of the Linnean Society*, **101**, 59–95.

Laurin, M. and Reisz, R. R. (1990) *Tetraceratops* is the oldest known therapsid. *Nature, London*, **345**, 249–50.

Manabe, M. and Hasegawa, Y. (1991) The Cretaceous dinosaur fauna of Japan, in *Fifth Symposium on Mesozoic Terrestrial Ecosystems and Biota* (eds Z. Kielan-Jaworowska, N. Heintz and H. A. Nakrem), Contributions from the Paleontological Museum, University of Oslo, 364. University of Oslo, Oslo, pp. 41–2.

Massare, J. A. and Callaway, J. M. (1990) The affinities and ecology of Triassic ichthyosaurs. *Geological Society of America Bulletin*, **102**, 406–16.

Massare, J. A. and Callaway, J. M. (1992) *Cymbospondylus* (Ichthyosauria, Shastasauridae) from the Early Triassic Thaynes Formation of southeastern Idaho, in press.

Mazin, J.-M. (1982) Repartition stratigraphique et géographique des Mixosauria (Ichthyopterygia), provincialité marine au Trias moyen, in *Symposium Paléontologique Georges Cuvier* (eds E. Buffetaut, J.-M. Mazin and E. Salmon), Actes du Symposium Paléontologique G. Cuvier, Montbéliard, pp. 375–86.

Mazin, J.-M. (1988) Paléobiogéographie des Reptiles Marins du Trias. Phylogenie, Systématique, Écologie et Implications Paléobiogéographiques. PhD, Académie de Paris. Université Pierre et Marie Curie, 313 pp.

McGowan, C. (1978) An isolated ichthyosaur coracoid from the Upper Cretaceous Mowry Formation of Wyoming. *Canadian Journal of Earth Sciences*, **15**, 169–71.

McGowan, C. (1989) *Leptopterygius tenuirostris* and other long-snouted ichthyosaurs from the English Lower Lias. *Palaeontology*, **32**, 409–27.

Mendrez, C. H. (1975) On the skull of *Regisaurus jacobi*, a new genus and species of Bauriamorpha Watson and Romer, 1956 (= Scaloposauria Boonstra, 1953), from the *Lystrosaurus* Zone of South Africa, in *Studies in Vertebrate Evolution* (eds K. A. Joysey and T. S. Kemp), Oliver & Boyd, Edinburgh, pp. 191–212.

Mendrez-Carroll, C. H. (1979) Nouvelle étude du crâne du type de *Scaloposaurus constrictus* Owen, 1876, spécimen jeune, Therocephalia, Scaloposauria, Scaloposauridae, de la zone à *Cistecephalus* (Permien supérieur) d'Afrique australe. *Bullétin du Museum Nationale d'Histoire Naturelle de Paris, 4ème. Série*, **1C**, 155–201.

Metcalf, S. J., Vaughan, R. F., Benton, M. J. *et al.* (1993) A new Bathonian (Middle Jurassic) microvertebrate site, within the Chipping Norton Limestone Formation at Hornsleasow Quarry, Gloucestershire. *Proceedings of the Geologists' Association*, **103**, 321–42.

Meylan, P. A. and Gaffney, E. S. (1989) The skeletal morphology of the Cretaceous cryptodiran turtle, *Adocus*, and the relationships of the Trionychoidea. *American Museum Novitates*, **2941**, 1–60.

Michard, J.-G., Broin, F. de, Brunet, M. *et al.* (1990) Le plus ancien crocodilien néosuchien spécialisé à caractèters 'eusuchiens' du continent africain (Crétacé inférieur, Cameroun). *Comptes rendus de l'Académie de Sciences, Paris*, **311 (II)**, 365–8.

Milner, A. R. (1987) The Westphalian tetrapod fauna; some aspects of its geography and ecology. *Journal of the Geological Society of London*, **144**, 495–506.

Młynarski, M. (1976) Testudines. *Handbuch der Paläoherpetologie*, **7**, 1–130.

Molnar, R. E. (1981) Pleistocene ziphodont crocodilians of Queensland. *Records of the Australian Museum*, **33**, 803–34.

Molnar, R. E. (1982) A longirostrine crocodilian from Murua (Woodlark), Solomon Sea. *Memoirs of the Queensland Museum*, **20**, 675–85.

Molnar, R. E., Kurzanov, S. M. and Dong, Z. (1990) Carnosauria, in *The Dinosauria* (eds D. B. Weishampel, P. Dodson and H. Osmólska), University of California Press, Berkeley, pp. 169–209.

Mulder, E. W. A. (1990) Ein Elasmosaurierzahn aus der oberen Kreide des St. Pieterberges bei Maastricht, Süd-Limburg, Niederlande. *Paläontologische Zeitschrift*, **64**, 145–51.

Murry, P. A. (1986) Vertebrate paleontology of the Dockum Group, western Texas and eastern New Mexico, in *The Beginning of the Age of Dinosaurs* (ed. K. Padian), Cambridge University Press, Cambridge, pp. 109–37.

Murry, P. A. (1987) New reptiles from the Upper Triassic Chinle Formation of Arizona. *Journal of Paleontology*, **61**, 773–86.

Narmandakh, P. (1985) A new chelonian species of the genus *Adocus* from the Upper Cretaceous of Mongolia. *Paleontological Journal*, **1985(2)**, 81–8.

Nessov, L. A. (1984) Upper Cretaceous pterosaurs and birds from Central Asia. *Palaeontological Journal*, **1984**, 47–57.

Nessov, L. A. (1985) Data on late Mesozoic turtles from the USSR. *Studia Palaeocheloniologica*, **1**, 215–24.

Nessov, L. A. (1988) On some Mesozoic turtles of the Soviet Union, Mongolia and China, with comments on systematics. *Studia Palaeocheloniologica*, **2**, 87–102.

Norell, M. A. and Storrs, G. W. (1989) Catalogue and review of the type fossil crocodilians in the Yale Peabody Museum. *Postilla*, **203**, 1–28.

Olsen, P. E., Schlische, R. W. and Gore, P. J. W. (1989) *Tectonic, Depositional, and Paleoecological History of Early Mesozoic Rift Basins, Eastern North America*. American Geophysical Union, Washington, DC, 174 pp.

Olson, E. C. and Vaughn, P. P. (1970) The changes of terrestrial vertebrates and climates during the Permian of North America. *Forma et Functio*, **3**, 113–38.

Persson, P. O. (1963) A revision of the classification of the Plesiosauria with a synopsis of the stratigraphical and geographical distribution of the group. *Lunds Universitets Årsskrifter, N. F.*, **59**, 1–60.

Pinna, G. (1980) *Drepanosaurus unguicaudatus*, nuovo genere e nuovo specie di Lepidosaurio del Trias Alpino (Reptilia). *Atti della Societá Italiana di Scienze Naturale e del Museo Civico di Storia Naturale di Milano*, **121**, 181–92.

Rage, J.-C. (1984) Serpentes. *Handbuch der Paläoherpetologie*, **11**,

1–80.
Rage, J.-C. (1988) Un serpent primitif (Reptilia, Squamata) dans le Cénomanien (base du Crétacé supérieur). *Comptes rendus de l'Academie de Sciences, Paris*, **307(II)**, 1027–32.
Reisz, R. R. (1980) A protorothyridid captorhinomorph reptile from the Lower Permian of Oklahoma. *Life Sciences Contributions, Royal Ontario Museum*, **121**, 1–16.
Reisz, R. R. (1986) Pelycosauria. *Handbuch der Paläoherpetologie*, **17A**, 1–102.
Reisz, R. R. and Laurin, M. (1991) *Owenetta* and the origin of turtles. *Nature, London*, **349**, 324–6.
Renesto, S. (1984) A new lepidosaur (Reptilia) from the Norian beds of the Bergamo Prealps (preliminary note). *Rivista Italiano di Paleontologia e Stratigrafia*, **90**, 165–76.
Rieppel, O. (1988) The classification of the Squamata, in *Amphibians, Reptiles, Birds (The Phylogeny and Classification of the Tetrapods: Vol. 1)*, (ed. M. J. Benton), Systematics Association Special Volume, 35A, Clarendon Press, Oxford, pp. 261–93.
Rubidge, B. S. (1992) *Biostratigraphy of the Karoo Sequence*. South African Commission for Stratigraphy, Cape Town, South Africa, in press.
Rubidge, B. S. and Hopson, J. A. (1991) A new anomodont therapsid from South Africa and its bearing on the ancestry of Dicynodontia. *South African Journal of Science*, **86**, 165–6.
Rubidge, B. S., Kitching, J. W. and Van den Heever, J. A. (1983) First record of a therocephalian (Therapsida: Pristerognathidae) from the Ecca of South Africa. *Navorsinge van die Nasionale Museum Bloemfontein*, **4**, 229–35.
Russell, D. A. (1967) Systematics and morphology of American mosasaurs (Reptilia, Sauria). *Bulletin of the Peabody Museum of Natural History*, **23**, 1–237.
Savage, D. E. and Russell, D. E. (1983) *Mammalian Paleofaunas of the World*. Addison-Wesley, Reading, Massachusetts, 432 pp.
Savage, R. J. G. (1984) Mid-Jurassic mammals from Scotland, in *Third Symposium on Mesozoic Terrestrial Ecosystems, Tübingen* (eds W.-E. Reif and F. Westphal), Attempto, Tübingen, pp. 211–13.
Sereno, P. C. (1990) Psittacosauridae, in *The Dinosauria* (ed. D. B. Weishampel, P. Dodson, and H. Osmólska), University of California Press, Berkeley, pp. 579–92.
Sereno, P. C. (1991a) *Lesothosaurus*, 'fabrosaurids', and the early evolution of Ornithischia. *Journal of Vertebrate Paleontology*, **11**, 168–97.
Sereno, P. C. (1991b) Basal archosaurs: phylogenetic relationships and functional implications. *Journal of Vertebrate Paleontology, Supplement*, **11**, 1–53.
Sereno, P. C. and Arcucci, A. B. (1990) The monophyly of crurotarsal archosaurs and the origin of bird and crocodile ankle joints. *Neues Jahrbuch für Geologie und Paläontologie, Abhandlungen*, **180**, 21–52.
Shubin, N. H., Crompton, A. W., Sues, H.-D. *et al.* (1991) New fossil evidence on the sister-group of mammals and early Mesozoic faunal distributions. *Science*, **251**, 1063–5.
Sigogneau-Russell, D. (1981) Présence d'un nouveau champsosauridé dans le Crétacé supérieur de Chine. *Comptes rendus de l'Académie de Sciences, Paris*, **292(II)**, 541–4.
Sigogneau-Russell, D. (1989) Theriodontia I. Phthinosuchia, Biarmosuchia, Eotitanosuchia, Gorgonopsia. *Handbuch der Paläoherpetologie*, **17B/I**, 1–127.
Smithson, T. R. (1989) The earliest known reptile. *Nature, London*, **342**, 676–8.
Steel, R. (1973) Crocodylia. *Handbuch der Paläoherpetologie*, **16**, 1–116.
Storrs, G. W. (1991a) Anatomy and relationships of *Corosaurus alcovensis* (Diapsida: Sauropterygia) from the Triassic Alcova Limestone of Wyoming. *Bulletin of the Peabody Museum of Natural History*, **44**, 1–151.
Storrs, G. W. (1991b) Note on a second occurrence of thalattosaur remains (Reptilia: Neodiapsida) in British Columbia. *Canadian Journal of Earth Sciences*, **28**, 2065–8.
Storrs, G. W. and Langston, Jr, W. (1993) The Plesiosauria (Diapsida: Sauropterygia) in Texas; a review with description of new material. *Bulletin of the Texas Memorial Museum*, in press.
Sues, H.-D. (1986) Relationships and biostratigraphic significance of the Tritylodontidae (Synapsida) from the Kayenta Formation of northeastern Arizona, in *The Beginning of the Age of Dinosaurs* (ed. K. Padian), Cambridge University Press, Cambridge, pp. 279–84.
Sues, H.-D. (1987) Postcranial skeleton of *Pistosaurus* and interrelationships of the Sauropterygia (Diapsida). *Zoological Journal of the Linnean Society*, **90**, 109–31.
Sues, H.-D. and Boy, J. A. (1988) A procynosuchid cynodont from central Europe. *Nature*, **331**, 523–4.
Sues, H.-D. and Olsen, P. E. (1990) Triassic vertebrates of Gondwanan aspect from the Richmond Basin of Virginia. *Science*, **249**, 1020–3.
Sullivan, R. M. (1985) A new middle Paleocene (Torrejonian) rhineurid amphisbaenian, *Plesiorhineura tsentasi* new genus, new species, from the San Juan Basin, New Mexico. *Journal of Paleontology*, **59**, 1481–5.
Tatarinov, L. P. (1974) Teriodonti SSSR. *Trudy Paleontologicheskogo Instituta, Akademiya Nauk SSSR*, **143**, 1–250 (in Russian).
Tschanz, K. (1989) *Lariosaurus buzzii* n. sp. from the Middle Triassic of Monte San Giorgio (Switzerland), with comments on the classification of nothosaurs. *Palaeontographica, Abteilung A*, **208**, 153–79.
Unwin, D. M. (1987) A new pterosaur from the Kimmeridge Clay of Kimmeridge, Dorset. *Proceedings of the Dorset Natural History and Archaeological Society*, **109**, 150–3.
Unwin, D. M. (1991) The Morphology, Systematics, and Evolutionary History of Pterosaurs from the Cretaceous Cambridge Greensand of England. PhD, University of Reading.
Ventura, C. S. (1984) The fossil herpetofauna of the Maltese Islands: a review. *Naturalista siciliana*, **8**, 93–106.
Wade, M. (1990) A review of the Australian Cretaceous longipinnate ichthyosaur *Platypterygius* (Ichthyosauria, Ichthyopterygia). *Memoirs of the Queensland Museum*, **28**, 115–37.
Weishampel, D. B. (1990) Dinosaurian distribution, in *The Dinosauria* (eds D. B. Weishampel, P. Dodson and H. Osmólska), University of California Press, Berkeley, pp. 63–139.
Weishampel, D. B. and Witmer, L. M. (1990) *Lesothosaurus*, *Pisanosaurus*, and *Technosaurus*, in *The Dinosauria* (eds D. B. Weishampel, P. Dodson and H. Osmólska), University of California Press, Berkeley, pp. 416–25.
Weishampel, D. B., Dodson, P. and Osmólska, H. (eds) (1990) *The Dinosauria*. University of California Press, Berkeley, 733 pp.
Welles, S. P. and Gregg, D. R. (1971) Late Cretaceous marine reptiles of new Zealand. *Records of the Canterbury Museum*, **9**, 1–111.
Wellnhofer, P. (1978) Pterosauria. *Handbuch der Paläoherpetologie*, **19**, 1–82.
Wellnhofer, P. (1991) *The Illustrated Encyclopedia of Pterosaurs*. Salamander Books Ltd, London, 192 pp.
Westphal, F. (1976) Phytosauria, in *Handbuch der Paläoherpetologie 13* (ed. O. Kuhn), Gustav Fischer, Stuttgart, pp. 79–120.
Wild, R. (1978) Die Flugsaurier (Reptilia, Pterosauria) aus der Oberen Trias von Cene bei Bergamo. *Bollettino della Società Paleontologia Italiana*, **17**, 176–256.
Wild, R. (1983) A new pterosaur (Reptilia, Pterosauria) from the Upper Triassic (Norian) of Friuli, Italy. *Gortania – Atti del Museo Friulano di Storia Naturale*, **5**, 45–62.
Zangerl, R. (1953) The vertebrate fauna of the Selma Formation of Alabama. Part 3. The turtles of the family Protostegidae. Part 4. The turtles of the family Toxochelyidae. *Fieldiana, Geological Memoirs*, **3**, 61–277.
Zangerl, R. and Sloan, R. E. (1960) A new specimen of *Desmatochelys lowi* Williston. A primitive cheloniid sea turtle from the Cretaceous of South Dakota. *Fieldiana, Geology*, **14**, 7–40.
Zanon, R. T. (1991) *Negevodus ramonensis* Mazin, 1986, reintepreted as a temnospondyl, not a placodont. *Journal of Vertebrate Paleontology*, **11**, 515–18.

40

AVES

D. M. Unwin

Avian palaeontology has experienced a tremendous growth in popularity during the last two decades, as clearly shown by a simple index based on rate of publication (Unwin, 1988), which demonstrates that activity is now at an all-time high. It is also reflected in the data presented here; the temporal ranges of fewer than 70 of the 200 or so families listed by Fisher in 1967 remain unchanged, and during the intervening period, some 33 new families, four new orders and a new subclass have been proposed.

Archaeopteryx remains the earliest bird and, with the recent discovery of a sixth specimen (Wellnhofer, 1988), one of the best-known forms from the Mesozoic. Molnar's recent review (1985) of contenders for the title of earliest-known bird, mostly represented by very fragmentary remains, concluded that none could definitely be assigned to Aves. Two putative Jurassic avians, *Laopteryx* Marsh, 1881 and *Palaeopteryx* Jensen, 1981, have recently been reidentified as pterosaurian (Jensen, 1981; Ostrom, 1986; Jensen and Padian, 1989). *Praeornis sharovi* (Rautian, 1978) is almost certainly not a plant frond as has been suggested (Bock, 1986; Nessov, 1992), but its true identity (feather, prefeather, scale or perhaps none of these) remains unclear, as does the identity of the animal upon which it was borne.

The systematic layout of this section largely follows traditional lines, and only where a clear consensus of opinion seems to have emerged, have taxa been relocated. Practically all references prior to 1985 can be found in Brodkorb's *Catalogue of Fossil Birds* (1963, 1964, 1967, 1971, 1978) or Olson's recent review of the avian fossil record (1985). Only references not included in these works are given. Please note the additional abbreviation: Lit. = littoral.

Avian systematics continues to be riven by disputes, particularly with respect to the composition and phylogenetic arrangement of families. This stems, in part, from the variety of systematic methods employed by avian palaeontologists (Cracraft, 1988), and while some approaches may yield more durable results than others, none can fairly claim to have a monopoly on objectivity, or, for that matter, 'the truth'. In fact, the imposition of a single system may have a stultifying rather than invigorating effect on avian systematics. The interminable taxonomic wrangles have led to a continual shuffling of taxa between families. The data presented here represent the latest state of play, but many changes in composition of families and their temporal range are to be expected.

Acknowledgements – Many thanks to all those who contributed to this work in its formative stages. Special thanks go to Herculano Alvarenga, Zygmunt Bochenski, Walter Boles, Luis Chiappe, Jacques Cheneval, Clive Coy, Juan Cuello, Paul Davis, Karlheinrich Fischer, Colin Harrison, Lian Hou, Peter Houde, Hildegarde Howard, Evgeny Kurochkin, Alexander Karkhu, Larry Martin, Joseph McKee, Jiří Mlíkovský, Cecile Mourer-Chauviré, Storrs Olson, Stephan Peters, Pat Rich, David Steadman, J. Stuart, Claudia Tambussi, Eduardo Tonni and Cyril Walker, who drew attention to numerous errors and frequently provided much, as yet unpublished, information. I am particularly indebted to Mike Benton for inviting me to undertake this project, the late Bev Halstead for providing much advice and encouragement, and most especially to Natalia Bakhurina for her never-failing support and tireless enthusiasm.

Class AVES

Order ARCHAEOPTERYGIFORMES Fürbringer, 1888

F. ARCHAEOPTERYGIDAE Huxley, 1872
J. (TTH)–K. (BER?) Terr. (see Fig. 40.1)

First: *Archaeopteryx lithographica* Meyer, 1861, Solnhofener Schichten, Altmühl-Alb, Bavaria, Germany.
Last: *Archaeopteryx* sp., Cornet, Padurea Crailui Mountains, Bihor, Romania (Kessler, 1984)? Until this identification is confirmed, *A. lithographica* remains the only certain record for this family.

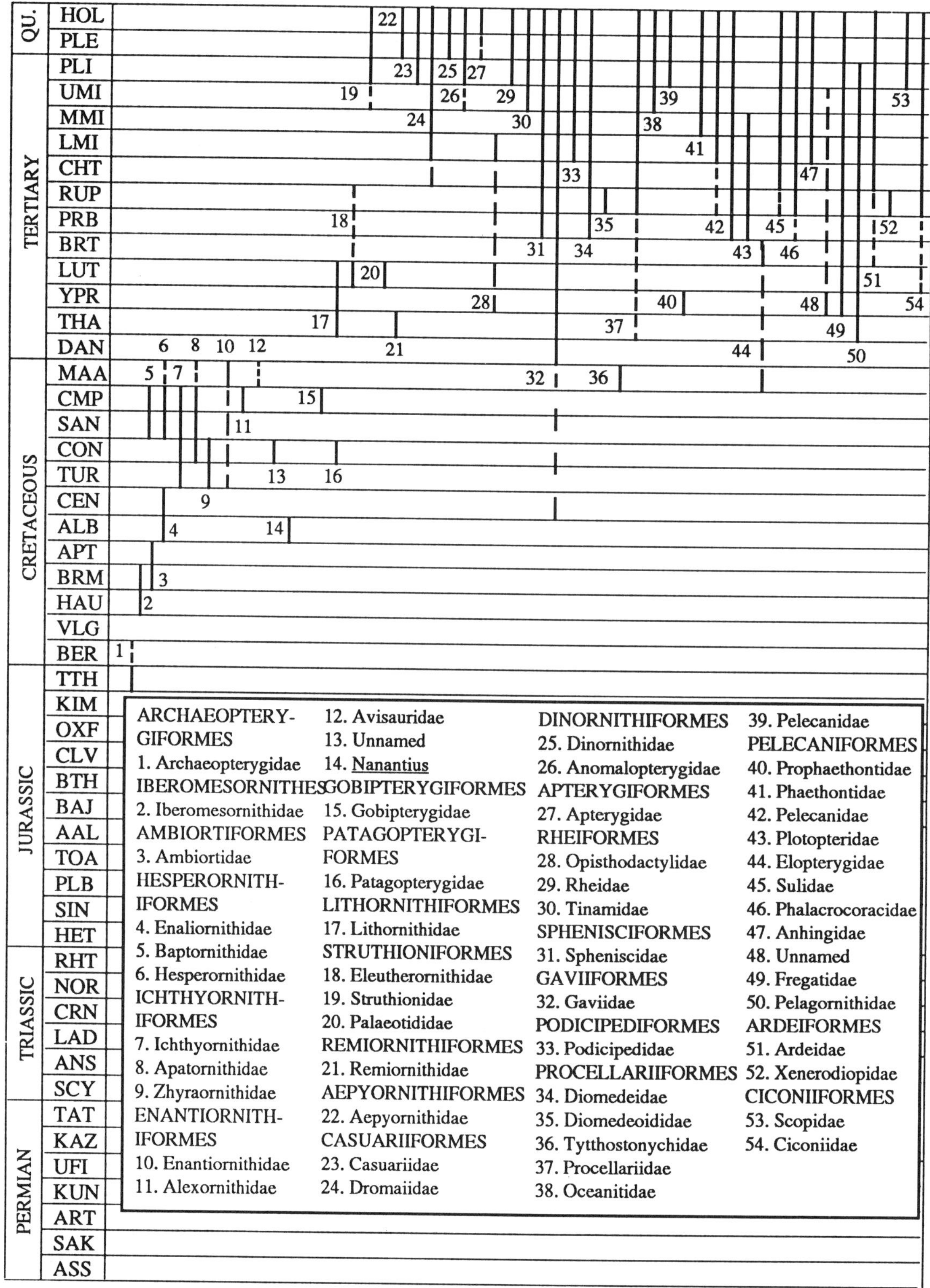

Fig. 40.1

Order IBEROMESORNITHIFORMES Sanz and Bonaparte, 1992

F. IBEROMESORNITHIDAE Sanz and Bonaparte, 1992 K. (HAU/BRM) Terr.

First and Last: *Iberomesornis romerali* Sanz and Bonaparte (1992), La Huérgina Formation, Las Hoyas, La Cierva Township, Cuenca Province, Spain.

Order AMBIORTIFORMES Kurochkin, 1982

F. AMBIORTIDAE Kurochkin, 1982 K. (BRM/APT) Terr.

First and Last: *Ambiortus dementjevi* Kurochkin, 1982, Unduruhinskaya Formation, Khurilt-Ulan-Bulak, Bajan-Khongor aimak, People's Republic of Mongolia. Further remains from the same formation at Holbotu, also in the

Bajan-Khongor aimak, currently referred to as *'Holbotia ponomarenkoi'* and thought to be pterosaur (Kurochkin, 1991), probably belong in the Ambiortidae, and possibly to *Ambiortus*.

Order HESPERORNITHIFORMES Fürbringer, 1888

F. ENALIORNITHIDAE Fürbringer, 1888 K. (ALB–CEN) Mar.

First: *Enaliornis barretti* Seeley, 1864 and *E. sedgwicki* Seeley, 1864, Cambridge Greensand, Cambridge, England, UK.
Last: *'Enaliornis* like' tarso-metatarsus, Greenhorn Formation, Kansas, USA (Martin, 1983).

F. BAPTORNITHIDAE American Ornithologists' Union, 1910 K. (SAN–CMP) Mar.

First: *Baptornis advenus* Marsh, 1877, Smoky Hill Chalk, Niobrara Formation, Kansas, USA.
Last: *Baptornis* sp., Judith River Formation, Saskatchewan Landing, Saskatchewan, Canada (Tokaryk and Harington, in press).
Comment: *Neogaeornis,* earlier thought to be a baptornithid, has recently been transferred to the Gaviidae (Olson, 1992).

F. HESPERORNITHIDAE Marsh, 1872 K. (SAN–MAA?) Mar.

First: *Hesperornis regalis* Marsh, 1872, *H. crassipes* Marsh, 1876, *H. gracilis* Marsh, 1876 and *Parahesperornis alexi* Martin, 1983, Smoky Hill Chalk, Niobrara Formation, Kansas, USA. Note, the three species of *Hesperornis* are probably conspecific (Martin, 1980).
Last: Hesperornithid remains have been recovered from the Kanguk Formation (MAA), of Bylot Island, north-west of Baffin Island, North-west Territories, Canada (Currie, pers. comm., 1989), but until this is confirmed, the youngest reliable record is *Hesperornis* cf. *H. regalis*, from the Foremost Formation (middle CMP), South Saskatchewan River, Alberta, Canada (Fox, 1974).

Order ICHTHYORNITHIFORMES Marsh, 1873

F. ICHTHYORNITHIDAE Marsh, 1873 K. (TUR–CMP) Lit./Mar.

First: *Ichthyornis* sp., Vimy Member, Kaskapau Formation, Watino, Alberta, Canada (Fox, 1984). Note, teeth with some similarity to those of *Ichthyornis* have been recovered from upper Lower Cretaceous (APT/ALB) sediments of Hoboor, Central Mongolia (Kurochkin, 1988).
Last: *Ichthyornis*, Pembina Formation, Pierre Shale, Manitoba, Canada (Martin and Stewart, 1982) and the Mooreville Chalk, of Alabama and Mississippi, USA (Nicholls and Russell, 1990).
Intervening: CON, SAN.

F. APATORNITHIDAE Fürbringer, 1888 K. (CON–MAA?) Lit./Mar.

First: *Apatornis celer* (Marsh, 1873), Smoky Hill Chalk, Niobrara Formation, Kansas, USA.
Last: *Palintropus retusus* (Marsh, 1892), Lance Formation, Lance Creek, Niobrara County, Wyoming, USA. Studies currently in progress suggest that this form is indeed an apatornithid (C. Coy, pers. comm., 1990; Tyrrell Museum of Palaeontology, 1988). Otherwise, the earliest reliable record consists of undescribed remains from the Judith River Formation (CMP), Dinosaur Provincial Park, Alberta, Canada (C. Coy, pers. comm., 1990). Note, there may be some grounds for supposing the Apatornithidae to belong within the Charadriiformes (Olson, 1985).

F. ZHYRAORNITHIDAE Nessov, 1984 K. (TUR/CON) Lit.

First and Last: *Zhyraornis kashkarovi* Nessov, 1984, Bissekty Formation, Dzhyrakhuduk water well, Bukhara region, central Kyzylkum desert, Uzbekistan, former USSR.
Comment: There is some doubt as to the relationship of this family with ichthyornithiformes (Nessov, 1992).

Order ENANTIORNITHIFORMES Walker, 1981

F. ENANTIORNITHIDAE Nessov and Borkin, 1983 K. (MAA) Terr.

First and Last: *Enantiornis leali* Walker, 1981, Lecho Formation, El Brete, Salta Province, Argentina. *Sazavis prisca* Nessov, 1989, based on very fragmentary remains from the Taykarshin Beds (TUR/CON) of Dzhyrakhuduk, Bukhara region, Uzbekistan, former USSR, might possibly represent an earlier record for this family (Nessov, 1984; Nessov and Yarkov, 1989).

F. ALEXORNITHIDAE Brodkorb, 1976 K. (CMP) Terr.

First and Last: *Alexornis antecedens* Brodkorb, 1976, Bocana Roja Formation, Baja California del Norte, Mexico.

F. AVISAURIDAE Brett-Surman and Paul, 1985 K. (MAA?) Terr.

First and Last: *Avisaurus archibaldi* Brett-Surman and Paul, 1985, Hell Creek Formation, Garfield County, Montana, USA. Described as a new family of theropods, recent authors (e.g. Cracraft, 1986; Chiappe and Calvo, 1989a) regard the Avisauridae as probably avian.

F. UNNAMED Chiappe and Calvo, 1989a K. (CON) Terr.

First and Last: Undescribed remains from the Rio Colorado Formation, Neuquén, Neuquén Province, Argentina (Chiappe and Calvo, 1989a). These remains may be shown eventually to belong within the Enantiornithidae (Chiappe, 1991).

F. UNNAMED Molnar, 1986 K. (ALB) Terr.

First and Last: *Nanantius eos* Molnar, 1986, Toolebuc Formation, Warra Station, Queensland, Australia.

Order GOBIPTERYGIFORMES Elzanowski, 1974

F. GOBIPTERYGIDAE Elzanowski, 1974 K. (CMP) Terr.

First and Last: *Gobipteryx minuta* Elzanowski, 1974, Barun Goyot Formation, Khulsan, Nemegt Basin, Mongolia.

Order PATAGOPTERYGIFORMES Alvarenga and Bonaparte, 1992

F. PATAGOPTERYGIDAE Alvarenga and Bonaparte, 1992 K. (CON) Terr.

First and Last: *Patagopteryx deferrariisi* Alvarenga and Bonaparte, 1992, Rio Colorado Formation, Neuquén,

Neuquén Province, Patagonia, Argentina (see Chiappe and Calvo, 1989b; Bonaparte, 1991).

Order LITHORNITHIFORMES Houde, 1988

F. LITHORNITHIDAE Houde, 1988 T. (THA–LUT) Terr.

First: *Lithornis celetius* Houde, 1988, Fort Union Formation, Bangtail Quarry, Park County, Montana, USA.
Last: Lithornithidae genus indet., Bridgerian sediments, Henry's Fork, Bridger Basin, Wyoming, USA (Houde, 1988).
Intervening: YPR.

Order STRUTHIONIFORMES Latham, 1790

F. ELEUTHERORNITHIDAE Wetmore, 1951 T. (LUT–RUP?) Terr.

First: *Eleutherornis helveticus* Schaub, 1940, Egerkingen gamma, Bohnerz, Switzerland. Fischer (1987) has reidentified *Saurornis matthesi* Fischer, 1967 from Neumark West, Saxony, Germany, as a perissodactyl.
Last: *Proceriavis martini* Harrison and Walker, 1979a, Hampstead Beds, Yarmouth, Isle of Wight, England, UK. *Proceriavis* is only tentatively included within the Eleutherornithidae (Harrison and Walker, 1979a), thus *Eleutherornis* may represent the sole occurrence of this family.

F. STRUTHIONIDAE Vigors, 1825 T. (UMI)–Rec. Terr.

First: A pedal phalanx and egg-shell fragments (!) tentatively ascribed to *Struthio brachydactylus* Burchak-Abramovich, 1939, lower to middle Upper Miocene, Anatolia, Turkey (Sauer, 1979). The earliest reliable record is *Struthio orlovi* Kurochkin and Lungu, 1970, from the Upper Miocene (MES), Varnitsa, Bendersky Region, Moldavia, former USSR.
Intervening: ZAN–HOL.

F. PALAEOTIDIDAE Houde and Haubold, 1987 T. (LUT) Terr.

First: *Palaeotis weigelti* Lambrecht, 1928, lowermost LUT, Messel Oil Shale, Messel Quarry, Hessen, Germany (Houde and Haubold, 1987; Peters, 1988).
Last: *Palaeotis weigelti* Lambrecht, 1928, Braunkohle des Geiseltales, Saxony, Germany (Houde, 1986).

Order REMIORNITHIFORMES Martin, 1992

F. REMIORNITHIDAE Martin, 1992 T. (THA) Terr.

First and Last: *Remiornis heberti* Lemoine, 1881. Cernay les Reims, Dept. Haut-Rhin and Mont du Berru, France.

Order AEPYORNITHIFORMES Newton, 1884

F. AEPYORNITHIDAE Bonaparte, 1853 Q. (PLE/HOL) Terr.

First and Last: *Mullerornis betsilei* Milne-Edwards and Grandidier, 1894, *M. agilis* Milne-Edwards and Grandidier, 1894, *M. rudis* Milne-Edwards and Grandidier, 1894, *Aepyornis maximus* Geoffroy Saint-Hilaire, 1851, *A. medius* Milne-Edwards and Grandidier, 1866, *A. hildebrandti* Burckhardt, 1893, and *A. gracilis* Monnier, 1913, Madagascar. The Upper Eocene/Lower Oligocene genera *Eremopezus* Andrews, 1904 and *Stromeria* Lambrecht, 1929 'cannot be positively diagnosed as ratites, much less as aepyornithids' (Olson, 1985, p. 104; see also Rasmussen *et al.*, 1987).

Order CASUARIIFORMES Sclater, 1880

F. CASUARIIDAE Kaup, 1847 T. (PLI)–Rec. Terr.

First: *Casuarius* sp., Awe fauna, Bulolo, south-east New Guinea (Plane, 1967).
Intervening: PLE.

F. DROMAIIDAE Gray, 1870 T. (CHT–LMI)–Rec. Terr.

First: *Emuarius gidju* (Patterson and Rich, 1987), Leaf Locality, Wipajiri Formation, Lake Ngapakaldi, Lake Eyre Basin, South Australia (Boles, 1992). **Extant**
Intervening: MMI, UMI–HOL.

Order DINORNITHIFORMES Gadow, 1893

F. DINORNITHIDAE Bonaparte, 1853 Q. (PLE/HOL) Terr.

First and Last: *Dinornis novaezealandiae* Owen, 1843, *D. ingens* Owen, 1844, *D. giganteus* Owen, 1844, *D. gazella* Oliver, 1949, and *D. hercules* Oliver, 1949 from North Island, New Zealand and *D. torosus* Hutton, 1891, *D. robustus* Owen, 1846 and *D. maximus* Owen, 1867, South Island, New Zealand.

F. ANOMALOPTERYGIDAE Oliver, 1930 T. (MES/ZAN)–Rec. Terr.

First: *Anomalopteryx antiquus* Hutton, 1892, Timaru, South Island, New Zealand.
Last: *Megalapteryx didinus* Owen, 1883, Takahe Valley, South Island, New Zealand; the last remaining anomalopterygid became extinct about 200 years ago (Scarlett, 1974).
Intervening: PLE, HOL.

Order APTERYGIFORMES Haeckel, 1866

F. APTERYGIDAE Gray, 1840 Q. (PLE?)–Rec. Terr.

First: *Pseudapteryx gracilis* Lydekker, 1891, possibly Upper Pleistocene but probably Holocene, New Zealand. **Extant**
Intervening: HOL.

Order RHEIFORMES Forbes, 1884

F. OPISTHODACTYLIDAE Ameghino, 1895 T. (THA?–AQT) Terr.

First: ?*Diogenornis fragilis* Alvarenga, 1983, Basal Limestone of Itaboraí, Cabuçu district, Rio de Janeiro State, Brazil.
Last: *Opisthodactylus patagonicus* Ameghino, 1891, Santa Cruz Formation, Santa Cruz Province, Argentina. This genus may represent both the first and last occurrence of this family, as Alvarenga (1983) only tentatively assigns *Diogenornis* to the Opisthodactylidae.

F. RHEIDAE Bonaparte, 1853 T. (PIA)–Rec. Terr.

First: *Heterorhea dabbenei* Rovereto, 1914, Monte Hermoso Formation, Buenos Aires Province, Argentina. **Extant**
Intervening: PLE, HOL.

Order TINAMIFORMES Huxley, 1872

F. Tinamidae Gray, 1840 T. (MES)–Rec. Terr.

First: *Eudromia* sp. indet. Epecuén Formation, Salinas Grandes de Hidalgo, La Pampa Province, Argentina (Tambussi, 1987). Remains of a tinamou from the Santacruzian Beds (LMI) of Patagonia have recently been identified by R. Chandler, but remain unpublished (L. Chiappe, pers. comm., 1989). **Extant**
Intervening: ZAN–HOL.

Order SPHENISCIFORMES Sharpe, 1891

F. SPHENISCIDAE Bonaparte, 1831 T. (PRB)–Rec. Mar.

First: *Pachydyptes simpsoni* Jenkins, 1974, Blanche Point Marls (lower PRB), Blanche Point, South Australia. **Extant**
Intervening: RUP–HOL.

Order GAVIIFORMES Wetmore and Miller, 1926

F. GAVIIDAE Allen, 1897 K. (u.?)–Rec. FW/Mar.

First: *Neogaeornis wetzeli* Lambrecht, 1929, Quirquina Formation, San Vincente bay, Conceptión Province, Chile. Originally placed within the Baptornithidae, Olson (1992) has reidentified this bird as a diver. In addition, the partial skeleton of another diver has recently been recovered from the Lopez de Bartodano Formation, Seymour Island, Antarctica (Chatterjee, 1989). **Extant**
Intervening: PRB, AQT–HOL.

Order PODICIPEDIFORMES (Fürbringer, 1888)

F. PODICIPEDIDAE Bonaparte, 1831 T. (BUR)–Rec. FW/Lit.

First: *Miobaptus walteri* Svec, 1982, Dolnice, Cheb Basin, Czechoslovakia. **Extant**
Intervening: TOR–HOL.

Order PROCELLARIIFORMES Fürbringer, 1888

F. DIOMEDEIDAE Gray, 1840 T. (PRB)–Rec. Mar.

First: Diomedeidae gen. *et* sp. indet., La Meseta Formation, Seymour Island, Antarctica (Tambussi and Tonni, 1988). **Extant**
Intervening: CHT–HOL.

F. DIOMEDEOIDIDAE Fischer, 1985 T. (RUP) Mar.

First and Last: *Diomedeoides minimus* Fischer, 1985, phosphorite nodule horizon, Braunkohlentagebau, Espenhain, south of Leipzig, Germany.

F. TYTTHOSTONYCHIDAE Olson and Parris, 1987 K. (MAA) Mar.

First and Last: *Tytthostonyx glauconiticus* Olson and Parris, 1987, Hornerstown Formation, Sewell, Gloucester County, New Jersey, USA.

F. PROCELLARIIDAE Boie, 1826 T. (THA??)–Rec. Mar.

First: *Eopuffinus kazachstanensis* Nessov, 1986, Upper Palaeocene beds of Zhylga, Tsimkent District, Kazakhstan, former USSR. Until the validity of this taxon, based on a fragmentary paired frontal (!), can be established, the oldest reliable record is *Puffinus raemdonckii* (Van Beneden, 1871), Rupelian Sand (RUP), Edeghem, Antwerp, Belgium (Brodkorb, 1962). **Extant**
Intervening: CHT–HOL.

F. OCEANITIDAE Salvin, 1896 T. (UMI)–Rec. Mar.

First: *Oceanodroma hubbsi* Miller, 1951, Capistrano Formation, Capistrano Beach, Orange County, California, USA. **Extant**
Intervening: ZAN, PLE, HOL.

F. PELECANOIDIDAE Gray, 1871 T. (ZAN)–Rec. Mar.

First: *Pelecanoides cymatotrypetes* Olson, 1984, Quartzose Sand Member, Varswater Formation, Langebaanweg, Cape Province, South Africa. **Extant**
Intervening: PLE, HOL.

Order PELECANIFORMES Sharpe, 1891

F. PROPHAETHONTIDAE Harrison and Walker, 1976 T. (YPR) Mar.

First and Last: *Prophaethon shrubsolei* Andrews, 1899, London Clay, Isle of Sheppey, Kent, England, UK.

F. PHAETHONTIDAE Bonaparte, 1853 T. (MMI)–Rec. Mar.

First: Undescribed remains, Calvert Formation, Maryland, USA (Olson, 1985). **Extant**
Intervening: HOL.

F. PELECANIDAE Vigors, 1825 T. (RUP?)–Rec. FW/Mar.

First: *Miopelecanus gracilis* (Milne-Edwards, 1863), Labeur, Commune de Vaumas, Allier, France, Recent stratigraphical studies suggest that the beds which yielded *Miopelecanus* may only be early Miocene (AQT) in age (Cheneval, 1984). **Extant**
Intervening: AQT–HOL.

F. PLOTOPTERIDAE Howard, 1969 T. (PRB–LLA) Mar.

First: *Phocavis maritimus* Goedert, 1988, Keasey Formation, Vernonia, Washington County, Oregon, USA.
Last: Unnamed form, Central Japan (Hasegawa *et al.*, 1977; Olson, 1985).
Intervening: RUP, CHT?, LMI.

F. ELOPTERYGIDAE Lambrecht, 1933 K. (MAA?)–T. (MEO?) Mar./FW

First: *Elopteryx nopcsai* Andrews, 1913, Transylvanian freshwater limestone, Hatszeg, Romania. Considerable doubt has been expressed about the nature of remains ascribed to *Elopteryx*, including suggestions that they are non-avian (Olson, 1985); an avian skull has now apparently been discovered (Kessler, 1987), although whether it can be legitimately assigned to *Elopteryx* remains to be demonstrated. According to Mlíkovský (1987) the holotype mandible is avian, but possibly belongs within the Sulidae.
Last: *Eostega lebedinsky* Lambrecht, 1929, 'coarse chalk', near Cluj–Napoca (formerly Kolozsvar), Romania. While *Eostega* may be pelecaniform, there is some doubt as to whether it belongs here (Olson, 1985). The only remaining

genus within the Elopterygidae, *Argillornis* Owen, 1878, from the London Clay (YPR), Isle of Sheppey, Kent, England, UK, belongs in the Pelagornithidae (Harrison and Walker, 1976).

F. SULIDAE Reichenbach, 1849 T. (RUP?)–Rec. Mar.

First: *Sula ronzoni* Milne-Edwards, 1867, calcareous marl of Ronzon, Auvergne, France. This material is difficult to interpret and cannot definitely be assigned to the Sulidae (Olson, 1985); it may even belong in the Phalacrocoracidae (Harrison, 1975). The earliest reliable record is *Empheresula avernensis* (Milne-Edwards, 1867–1871), Calcaire de Gannat (CHT), Allier, France (Cheneval, 1984), but see also comments on *Elopteryx* above. **Extant**
Intervening: AQT–HOL.

F. PHALACROCORACIDAE Bonaparte, 1854 T. (PRB/RUP)–Rec. Lit./FW

First: Unnamed genus, Phosphorites du Quercy, France (Mourer-Chauviré, 1982). Note, *Piscator tenuirostris*, a putative cormorant from the Upper Eocene (PRB) of Hordle, Hampshire (Harrison and Walker, 1976), was referred to Aves *Incertae Sedis* by Brodkorb (1978). **Extant**
Intervening: RUP–HOL.

F. ANHINGIDAE Ridgway, 1887 T. (LMI)–Rec. Lit./FW

First: *Anhinga subvolans* (Brodkorb, 1956), Thomas Farm, Gilchrist County, Florida, USA (Becker, 1986). **Extant**
Intervening: LLA–HOL.

F. UNNAMED Rich *et al.*, 1986 T. (Eoc./Mio.) FW/Terr.

First and Last: *Protoplotus beauforti* Lambrecht, 1930, freshwater fish beds, whose age may be anywhere between Eocene and Miocene (Rich *et al.*, 1986), Sipang, Sumatra.

F. FREGATIDAE Garrod, 1874 T. (YPR)–Rec. Mar./FW

First: *Limnofregata azygosternon* Olson, 1977, Green River Formation, Wyoming, USA. **Extant**
Intervening: PLE, HOL.

F. PELAGORNITHIDAE Fürbringer, 1888 T. (THA–PIA?) Mar.

First: *Pseudodontornis tenuirostris* Harrison, 1985, Oldhaven Beds, Herne Bay, Kent, England, UK.
Last: *Pseudodontornis stirtoni* Howard and Warter, 1969, Greta Siltstone, Motunae Beach, South Island, New Zealand. As *P. stirtoni* is of uncertain age, pelagornithid material from the Middle Pliocene of Waihi Beach, Hawera, North Island, New Zealand, may represent the youngest record for this family (McKee, 1985).
Intervening: YPR–ZAN.
Comment: Following Olson (1985), I assume here that Pelagornithidae = Odontopterygidae, Cyphornithidae, Pseudodontornithidae and possibly Dasornithidae (see also Goedert, 1989).

Order ARDEIFORMES Wagler, 1830

F. ARDEIDAE Vigors, 1825 T. (BRT/CHT)–Rec. FW/Terr.

First: *Proardea amissa* (Milne-Edwards, 1892), Phosphorites du Quercy, France. **Extant**
Intervening: RUP–HOL.

F. XENERODIOPIDAE Rasmussen, Olson and Simons, 1987 T. (RUP) FW/Terr.

First and Last: *Xenerodiops mycter* Rasmussen *et al.*, 1987, upper sequence of the Jebel Qatrani Formation, Quarry M, Fayum Province, Egypt.

Order CICONIIFORMES Bonaparte, 1854

F. SCOPIDAE Bonaparte, 1853 T. (ZAN)–Rec. Terr.

First: *Scopus xenopus* Olson, 1984, Varswater Formation, eastern Quarry, Langebaanweg, Cape Province, South Africa. **Extant**

F. CICONIIDAE (Gray, 1840) T. (MEO?)–Rec. FW/Terr.

First: *Eociconia sangequanensis* Hou, 1989, Yi-Xi-Bai-La Formation, Sangequan, Xinjiang, China. While this is probably the oldest known stork, until confirmed, *Palaeoephippiorhynchus dietrichi* Lambrecht, 1930, from the Jebel Qatrani Formation (RUP), north of Qasr Quarun and possibly also from Quarry M in the upper sequence of the Jebel Qatrani Formation, Fayum Province, Egypt, are the earliest reliable records for this family (Rasmussen *et al.*, 1987). **Extant**
Intervening: AQT–HOL.

F. BALAENICIPITIDAE Bonaparte, 1853 T. (RUP)–Rec. FW/Terr. (Fig. 40.2)

First: *Goliathia andrewsi* Lambrecht, 1930, lower sequence of Jebel Qatrani Formation, Fayum Province, Egypt. **Extant**
Intervening: UMI.

F. PLATALEIDAE Bonaparte, 1838 T. (LUT)–Rec. FW/Lit.

First: *Rhynchaeites messelensis* Wittich, 1899, Messel Oil Shales, Messel Quarry, Hessen, Germany (Peters, 1983). **Extant**
Intervening: PRB, AQT–ZAN, PIA, PLE, HOL.

Order CATHARTIFORMES Coues, 1884

F. VULTURIDAE Illiger, 1811 T. (BRT)–Rec. Terr.

First: *Diatropornis ellioti* (Milne-Edwards, 1892), Le Bretou, Phosphorites du Quercy, France (Mourer-Chauviré, 1988a). *Eocathartes robustus* Lambrecht, 1935, from the Braunkohle des Geiseltales (LUT), Halle, Germany, is almost certainly not a vulture (Houde, quoted in Olson, 1985). **Extant**
Intervening: RUP, BUR, MMI, TOR, ZAN–HOL.

F. TERATORNITHIDAE Miller, 1909 T. (MES)–Q. (PLE) Terr.

First: *Argentavis magnificens* Campbell and Tonni, 1980, Epecuén Formation, Salinas Grandes de Hidalgo, La Pampa, Buenos Aires Province, Argentina.
Last: *Teratornis merriami* Miller, 1909, and *Cathartornis gracilis* Miller, 1910, Rancho La Brea, Los Angeles County, California and *Teratornis incredibilis* Howard, 1952, Smith Creek Cave, White Pine County, Nevada, USA.
Intervening: ZAN.

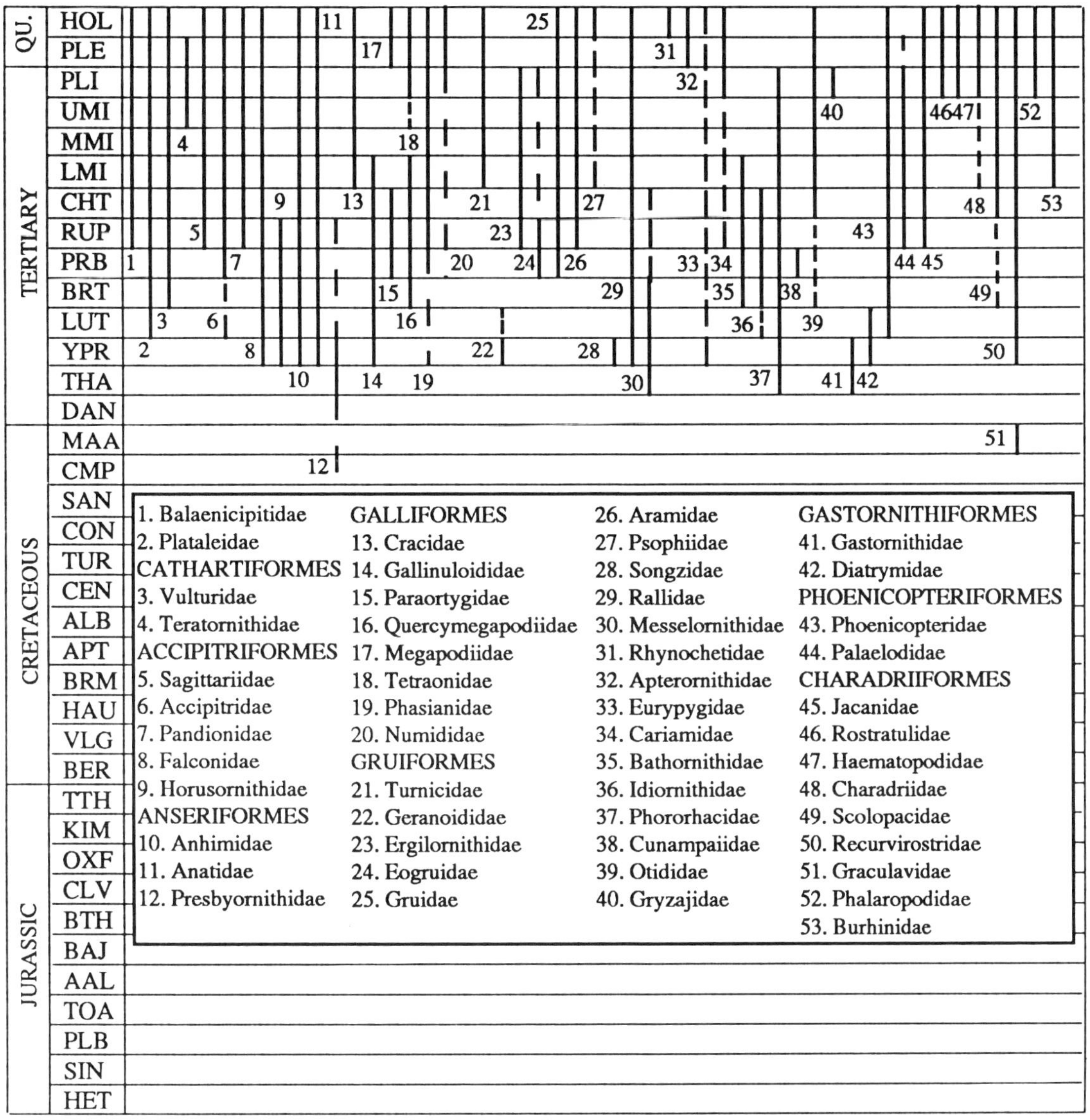

QU. HOL PLE
TERTIARY PLI UMI MMI LMI CHT RUP PRB BRT LUT YPR THA DAN
CRETACEOUS MAA CMP SAN CON TUR CEN ALB APT BRM HAU VLG BER
JURASSIC TTH KIM OXF CLV BTH BAJ AAL TOA PLB SIN HET

1. Balaenicipitidae
2. Plataleidae
CATHARTIFORMES
3. Vulturidae
4. Teratornithidae
ACCIPITRIFORMES
5. Sagittariidae
6. Accipitridae
7. Pandionidae
8. Falconidae
9. Horusornithidae
ANSERIFORMES
10. Anhimidae
11. Anatidae
12. Presbyornithidae
GALLIFORMES
13. Cracidae
14. Gallinuloididae
15. Paraortygidae
16. Quercymegapodiidae
17. Megapodiidae
18. Tetraonidae
19. Phasianidae
20. Numididae
GRUIFORMES
21. Turnicidae
22. Geranoididae
23. Ergilornithidae
24. Eogruidae
25. Gruidae
26. Aramidae
27. Psophiidae
28. Songzidae
29. Rallidae
30. Messelornithidae
31. Rhynochetidae
32. Apterornithidae
33. Eurypygidae
34. Cariamidae
35. Bathornithidae
36. Idiornithidae
37. Phororhacidae
38. Cunampaiidae
39. Otididae
40. Gryzajidae
GASTORNITHIFORMES
41. Gastornithidae
42. Diatrymidae
PHOENICOPTERIFORMES
43. Phoenicopteridae
44. Palaelodidae
CHARADRIIFORMES
45. Jacanidae
46. Rostratulidae
47. Haematopodidae
48. Charadriidae
49. Scolopacidae
50. Recurvirostridae
51. Graculavidae
52. Phalaropodidae
53. Burhinidae

Fig. 40.2

Order ACCIPITRIFORMES Vieillot, 1816

F. SAGITTARIIDAE Finsch and Hartlaub, 1870 T. (RUP)–Rec. Terr.

First: *Pelargopappus schlosseri* (Gaillard, 1908), Phosphorites du Quercy, France (Mourer-Chauviré and Cheneval, 1983). **Extant**
Intervening: CHT, AQT.

F. ACCIPITRIDAE (Vieilliot, 1816) T. (LUT/PRB)–Rec. Terr.

First: *Milvoides kempi* Harrison and Walker, 1979b, upper Bracklesham Beds (probably LUT), Lee on Solent, Hampshire, England, UK. A small raptor belonging to the Accipitridae has recently been identified in the Messel Oil Shale (LUT) Messel Quarry, Hessen, Germany (Peters, 1991). **Extant**
Intervening: RUP–HOL.

F. PANDIONIDAE Sclater and Salvin, 1873 T. (RUP)–Rec. FW/Terr.

First: Genus and species indet., aff. *Pandion*, upper sequence of the Jebel Qatrani Formation, Quarry M, Fayum Province, Egypt (Rasmussen *et al.*, 1987). **Extant**
Intervening: MMI, TOR, ZAN–HOL.

F. FALCONIDAE Vigors, 1824 T. (YPR)–Rec. Terr.

First: *Stintonornis mitchelli* Harrison, 1984a, London Clay, Isle of Sheppey, Kent, England, UK. *Parvulivenator watteli* Harrison, 1982, also from the London Clay, is slightly older, but is only provisionally assigned to the Falconidae (Harrison, 1982). **Extant**
Intervening: LUT, PRB/RUP, LMI, MMI, ZAN–HOL.

F. HORUSORNITHIDAE Mourer-Chauviré, 1991 T. (YPR–RUP) Terr.

First: *Horusornis vianeyliaudae* Mourer-Chauviré, 1991, La Bouffie, Phosphorites du Quercy, France.
Last: Unnamed species, Lower Oligocene, USA (Mourer-Chauviré, 1991).

Order ANSERIFORMES Wagler, 1831

F. ANHIMIDAE Stejneger, 1885 T. (YPR)–Rec. FW

First: Remains of an anhimid have recently been recovered from the Lower Eocene (Tyrberg, pers. comm., 1992). **Extant**
Intervening: ?RUP, HOL.

F. ANATIDAE Vigors, 1825 T. (EOC)–Rec. Mar./FW

First: Undetermined remains from Burgin Xian, Xinjiang and Erenhot, Inner Mongolia, China (Rich *et al.*, 1986), and possibly *Palaeopapia eous* from the Lower Headon Beds (PRB) of Hordle, Hampshire, England, UK (Harrison and Walker, 1976). **Extant**
Intervening: RUP–HOL.

F. PRESBYORNITHIDAE Wetmore, 1926 K. (CMP?)–T. (RUP?) FW/Lit.

First: Presbyornithidae gen. *et* sp. indet. from the Barun-Goyot Formation, Udan-Sair, Southern Mongolia (Kurochkin, 1988). Based on a single, fragmentary, tarsometatarsus this record is difficult to confirm. The oldest certain record is *Presbyornis* sp., Palaeocene (THA) of Utah, USA (Olson, 1985).
Last: *Headonornis hantoniensis* (Lydekker, 1891) Hamstead Beds, Hamstead, and Bembridge Marls, Burnt Wood, Isle of Wight, England, UK (Harrison and Walker, 1979a). These might represent the youngest records for this family, but until confirmed, the youngest certain records are *Presbyornis pervetus* Wetmore, 1926, and *Presbyornis antiquus* (Howard, 1955), Green River Formation (YPR), Uintah County, Utah, USA, and Casamayor Formation (YPR), Cañadón Vaca, Chubut Province, Argentina (Tonni and Tambussi, 1986).

Order GALLIFORMES Temminck, 1820

F. CRACIDAE Vigors, 1825 T. (AQT)–Rec. Terr.

First: *Boreortalis laesslei* Brodkorb, 1954, Thomas Farm Beds, Thomas Farm, Gilchrist County, Florida, USA.
Intervening: MMI, UMI, ZAN, PLE, HOL.

F. GALLINULOIDIDAE Lucas, 1900 T. (YPR–AQT) Terr.

First: *Gallinuloides wyomingensis* Eastman, 1900, Green River Shales, Lincoln County, and Bridger Formation, Uinta County, Wyoming, USA.
Last: *Taoperdix gallica* (Milne-Edwards, 1869), *T. brevipes* (Milne-Edwards, 1869) and *T. phasianoides* (Milne-Edwards, 1869), Langy and St Gérand-le-Puy, Dept. Allier, France.
Intervening: PRB, RUP.

F. PARAORTYGIDAE Mourer-Chauviré, 1992 T. (PRB–CHT) Terr.

First: *Paraortyx lorteti* Gaillard, 1908, La Bouffie, Phosphorites du Quercy, France.
Last: *Pirortyx major* (Gaillard, 1938), Pech du Fraysse, Phosphorites du Quercy, France.
Intervening: RUP.

F. QUERCYMEGAPODIIDAE Mourer-Chauviré, 1992 T. (BRT–AQT) Terr.

First: *Quercymegapodius brodkorbi* Mourer-Chauviré (1992), Lavergne, France.
Last: Unnamed genus, Saint-Gérand-le-Puy, France, (C. Mourer-Chauviré, pers. comm., 1989).
Intervening: PRB.
Comment: Crowe and Short (1992) would place both this family and the Paraortygidae within the Gallinuloididae.

F. MEGAPODIIDAE Swainson, 1837 Q. (PLE)–Rec. Terr.

First: *Progoura gallinacea* De Vis, 1889, Upper Pleistocene, Darling Downs Beds, Southern Queensland and caves in eastern New South Wales, *Progoura naracoortensis* Van Tets, 1974, Pleistocene, south-east South Australia (Van Tets, 1974), and remains of *Megapodius*, Pleistocene cave deposits, south-eastern Australia (Rich and Van Tets, 1982). **Extant**
Intervening: HOL.

F. PHASIANIDAE Vigors, 1825 T. (PRB/RUP)–Rec. Terr.

First: *Nanortyx inexpectatus* Weigel, 1963, Cypress Hills Formation, Calf Creek, Eastend, Saskatchewan, Canada (possibly referable to the uppermost Upper Eocene, Mourer-Chauviré, 1992). *Argillipes aurorum* from the London Clay (YPR) of Kent, England, UK (Harrison and Walker, 1977) might possibly represent an earlier record. **Extant**
Intervening: AQT–PLE, HOL.

F. NUMIDIDAE Reichenbach, 1850 T. (PRB?)–Rec. Terr.

First: *Telecrex grangeri* Wetmore, 1934, Irdin Manha Formation, Shara Murun Region, Suiyuan Province, Inner Mongolia, China. This is only tentatively assigned to the Numididae (Olson, 1974; Mourer-Chauviré,1992a), thus the oldest certain guinea-fowl is *Numida meleagris* (Linnaeus, 1758), from six, putative, Upper Pleistocene sites in Hungary, Czechoslovakia and Germany (Brodkorb, 1964). **Extant**
Intervening: PLE, HOL.

Order GRUIFORMES Bonaparte, 1854

F. TURNICIDAE Gray, 1840 T. (Mio.)–Rec. Terr.

First: Undescribed remains from the Miocene of Virginia, USA (Olson, 1989). **Extant**
Intervening: ZAN, PLE, HOL.
Comment: Recent authors (Mourer-Chauviré, 1981; Olson and Steadman, 1981; Olson, 1985; Hesse, 1988b) would exclude the button quails from the Gruiformes.

F. GERANOIDIDAE Wetmore, 1933 T. (YPR–LUT?) Terr.

First: *Geranoides jepseni* Wetmore, 1933, *Paragrus prentici* (Loomis, 1906), *P. shufeldti* Cracraft, 1969, *Palaeophasianus meleagroides* Shufeldt, 1913, *P. incompletus* Cracraft, 1969 and *Eogeranoides campivagus* Cracraft, 1969, Willwood Formation, Bighorn County, Wyoming, USA.
Last: *Geranodornis aenigma* Cracraft, 1969, Uinta County, Wyoming, USA. Olson (1985) expresses considerable doubt over the identity of this form, thus the Geranoididae may be confined to the Ypresian.

F. ERGILORNITHIDAE Kozlova, 1960 T. (RUP–PIA) Terr.

First: *Ergilornis rapidus* Kozlova, 1960, and *E. minor* Kozlova, 1960, Ergilyeen Dzo and Khoer Dzan, south-east Gobi Desert, People's Republic of Mongolia.

Last: *Amphipelargus dzabghanensis* Kurochkin, 1985, Hirgis Nur Suite (Upper Middle Pliocene), Dzagso Hirhan, western part of People's Republic of Mongolia.
Intervening: LMI, UMI, ZAN.
Comments: Some authors (e.g. Mlíkovský, 1985) consider the Ergilornithidae and Eogruidae to be a single family.

F. EOGRUIDAE Wetmore, 1934
T. (PRB–UMI?) Terr.

First: *Eogrus turanicus* (Bendukidze, 1971), Obayla Formation, Kalmakpai River, eastern Kazakhstan, former USSR, *Eogrus aeola* Wetmore, 1934, Irdin Manha, Shara Murun region, Suiyan Province, Inner Mongolia, China, and similar remains from Upper Eocene beds near Iren Dabasu, Inner Mongolia, China (Kurochkin, 1981).
Last: *Eogrus wetmorei* Brodkorb, 1967, Tung Gur Formation, Iren Dabasu, Inner Mongolia, China. Until the identity of this form is confirmed, the youngest record for this family is *Sonogrus gregalis* Kurochkin, 1981, Lower Oligocene (RUP), Khoer Dzan, People's Republic of Mongolia.
Comment: The Geranoididae, Ergilornithidae and Eogruidae quite possibly belong in the Struthioniformes (e.g. Olson, 1985).

F. GRUIDAE Vigors, 1925 T. (PRB)–Rec. FW/Terr.

First: *Geranopsis hastingsiae* Lydekker, 1891, Hordle Beds, Hordle, Hampshire, England, UK. **Extant**
Intervening: CHT, AQT, TOR–HOL.

F. ARAMIDAE Bonaparte, 1849
T. (RUP)–Rec. Terr.

First: *Badistornis aramus* Wetmore, 1940, Brule Formation, Washington County, South Dakota, USA. **Extant**
Intervening: ZAN, PLE, HOL.

F. PSOPHIIDAE Bonaparte, 1831
T. (LMI?)–Rec. Terr.

First: *Anisolornis excavatus* Ameghino, 1891, Santa Cruz Formation, Southern Patagonia, Argentina. *Anisolornis* is only very tentatively identified as a trumpeter (Olson, 1985), otherwise they have no fossil record. **Extant**

F. SONGZIDAE Hou, 1990 T. (YPR) Terr.

First and Last: *Songzia heidangkouensis* Hou, 1990, Yangxi Formation, Heidangkou, Songzi County, Hubei Province, China.

F. RALLIDAE T. (YPR)–Rec. Terr.

First: *Parvirallus bassetti* Harrison, 1984a, *P. medius* Harrison, 1984a and *P. gassoni* Harrison, 1984a, London Clay, Isle of Sheppey, Kent, England, UK. **Extant**
Intervening: LUT/BRT?–TOR, ZAN–HOL.

F. HELIORNITHIDAE Gray, 1849 **Extant** Terr.

F. MESITORNITHIDAE Wetmore, 1960
Extant Terr.

F. MESSELORNITHIDAE Hesse, 1988
T. (THA–BRT/CHT) Terr.

First: Remains from Mont Berru, France (Mourer-Chauviré, 1992b).
Last: Undescribed species from the BRT–CHT of France and North America (Hesse, 1988a).
Intervening: LUT.

F. EURYPYGIDAE Bonaparte, 1849
T. (YPR?)–Rec. Terr.

First: Remains from the Lower Eocene of Wyoming, USA, are tentatively ascribed to this family (Olson, 1989). Otherwise, sun bitterns have no fossil record. **Extant**

F. RHYNOCHETIDAE Newton, 1868
Q. (HOL)–Rec. Terr.

First: *Rhynochetos orarius* Balouet and Olson, 1989, Holocene deposits, Pindai Cave, Nepoui Peninsula, New Caledonia. **Extant**

F. APTERORNITHIDAE Olson, 1985
Q. (PLE–HOL) Terr.

First: *Apterornis otidiformis* (Owen, 1848), mennacenite beds at Te Rangatapu, Waingongoro and Wainganui, North Island, New Zealand.
Last: *Apterornis otidiformis* (Owen, 1848), Pyramid Valley Swamp and fourteen other sites (Brodkorb, 1964), South Island, New Zealand.

F. CARIAMIDAE Bonaparte, 1853
T. (RUP?)–Rec. Terr.

First: *Riacama caliginea* Ameghino, 1899, Deseado Formation, Santa Cruz Province, Argentina. As *Riacama* may not be a cariamid (Mourer-Chauviré, 1981), the oldest certain record is *Chunga incerta* from the lower Middle Pliocene (PIA), Monte Hermoso Formation, Buenos Aires Province, Argentina (Tonni, 1974). **Extant**
Intervening: PLE, HOL.

F. BATHORNITHIDAE Wetmore, 1927
T. (BRT–LMI) Terr.

First: '*Neocathartes grallator*' (Wetmore, 1944), Washakie Formation, Sweetwater County, Wyoming, USA (Olson, 1985).
Last: *Bathornis fricki* Cracraft, 1968, Willow Creek, Converse County, Wyoming, USA.
Intervening: PRB, RUP, CHT.
Comment: Both this family and the Idiornithidae probably belong within the Cariamidae (Mourer-Chauviré, 1981; Hesse, 1988b).

F. IDIORNITHIDAE Brodkorb, 1965
T. (LUT?–CHT) Terr.

First: Remains from the Messel Oil Shale, Messel, Germany (Peters, 1988) possibly represent the earliest idiornithids. Until confirmed, the oldest are *Elaphrocnemus alfhildae* (Shufeldt, 1915), Washakie Formation (BAR), Sweetwater County, Wyoming, USA and *Idiornis gaillardi* (Cracraft, 1973) Le Bretou (BAR), Phosphorites du Quercy, France (Mourer-Chauviré, 1988a).
Last: *Idiornis cursor* (Milne-Edwards, 1891), *I. gracilis* (Milne-Edwards, 1891), *I. itardiensis* Mourer-Chauviré, 1983, and *Elaphrocnemus crex* Milne-Edwards, 1891, Phosphorites du Quercy, France (Mourer-Chauviré, 1983).
Intervening: PRB, RUP.

F. PHORORHACIDAE Brodkorb, 1963
T.(THA–PIA) Terr.

First: *Paleopsilopterus itaboraiensis* Alvarenga, 1985, Basal Limestone of the Povoado de San José, Cabaçu District, Itaboraí, Brazil.
Last: *Titanis walleri* Brodkorb, 1963, Santa Fé River,

boundaries of Gilchrist and Columbia Counties, Florida, USA, recently redated from Pleistocene to late Pliocene (Olson, 1985).
Intervening: YPR, LUT, PRB–ZAN.

F. CUNAMPAIIDAE Rusconi, 1946 T. (PRB) Terr.

First and Last: *Cunampaia simplex* Rusconi, 1946, Divisadero Largo Formation, Las Heras, Mendoza Province, Argentina.

F. OTIDIDAE (Gray, 1840) T. (BRT/CHT)–Rec. Terr.

First: Undetermined remains, Phosphorites du Quercy, France (Mourer-Chauviré, 1982). **Extant**
Intervening: MMI, UMI–HOL.
Comment: There are some grounds for believing that the bustards belong within the Charadriiformes, near to the Glareolidae (see Olson, 1985).

F. GRYZAJIDAE Brodkorb, 1967 T. (PIA) Terr.

First and Last: *Gryzaja odessana* Zubareva, 1939, Kotlovina, southern Ukraine and Etulya, Vulkaneshty District, Moldavia, former USSR (Bochenski and Kurochkin, 1987).
Comment: This family probably belongs within the Otididae (see Olson, 1985, p. 180).

Order GASTORNITHIFORMES Stejneger, 1885

F. GASTORNITHIDAE Fürbringer, 1888 T. (THA–YPR) Terr.

First: *Gastornis parisiensis* Hébert, 1855, Conglomerate de Meudon, Seine-et-Oise; and Sables de Rilly and Conglomerate de Cernay, Marne, France; a fissure fill, Walbeck, Haldensleben County, Magdeburg District, Saxony, Germany; Upper Palaeocene beds of Meusin near Mons, Hainault, Belgium; and Bottom Bed, lower Woolwich Beds, Croydon, England, UK, and *G. russelli* Martin, 1992, Berru in Reims, France.
Last: *Zhongyuanus xichuanensis* Hou, 1980, Yu-Huang-Ding Series, Xichuan district, Hunan Province, China.

F. DIATRYMIDAE Shufeldt, 1913 T. (YPR–LUT) Terr.

First: *Diatryma gigantea* Cope, 1876, Wasatchian of Park County, and Graybull Member, Willwood Formation, South Elk Greek, Big Horn County, Wyoming, USA.
Last: *Diatryma geiselensis* Fischer, 1978, Braunkohle des Geiseltales, Neumark West, Saxony, Germany, and Messel Oil Shale, Messel Quarry, Hessen, Germany (Fischer, 1978).
Comment: There is a growing consensus (e.g. Olson, 1985; Andors, 1992; Martin, 1992) for combining Gastornithidae with Diatrymidae.

Order PHOENICOPTERIFORMES Fürbringer, 1888

F. PHOENICOPTERIDAE Bonaparte, 1831 T. (LUT)–Rec. FW/Terr.

First: *Juncitarsus merkeli* Peters, 1987, Messel Oil Shales, Messel Quarry, Hessen, Germany. **Extant**
Intervening: PRB?, RUP, AQT, MMI–ZAN, PIA, PLE, HOL.

F. PALAELODIDAE (Stejneger, 1885) T. (RUP–ZAN) FW/Terr.

First: Genus indet. aff. *Palaelodus* species 1 and 2, upper sequence of Jebel Quatrani Formation, Quarry M, Fayum Province, Egypt (Rasmussen *et al.*, 1987).
Last: *Megapaloelodus opsigonus* Brodkorb, 1961, Juntura Formation, Malheur County, Oregon, USA (Brodkorb, 1961). This family might range into the Pleistocene (Baird and Rich, quoted in Cheneval and Escuillié, 1992).
Intervening: AQT, MMI, UMI.

Order CHARADRIIFORMES Huxley, 1867

F. JACANIDAE Stejneger, 1885 T. (RUP)–Rec. FW/Terr.

First: *Nupharanassa tolutaria* Rasmussen *et al.*, 1987, lower sequence of the Jebel Qatrani Formation, Quarry E, Fayum Province, Egypt. **Extant**
Intervening: TOR, PIA, PLE, HOL.

F. ROSTRATULIDAE Ridgway, 1919 T. (ZAN)–Rec. FW/Terr.

First: 'New species of painted snipe', Langebaanweg, south-western Cape Province, South Africa (Olson and Eller, 1989). **Extant**

F. HAEMATOPODIDAE (Gray, 1840) T. (ZAN)–Rec. Lit./Terr.

First: *Haematopus* sp., Lee Creek, North Carolina, USA (Olson, 1985) and *Haematopus sulcatus* (Brodkorb, 1955), Bone Valley Mining District, Polk County, Florida, USA. **Extant**
Intervening: PIA–HOL.

F. CHARADRIIDAE Vigors, 1825 T. (BUR?)–Rec. Lit./Terr.

First: Very fragmentary charadriid remains have been described from a siderolithic fissure fill, Vieux-Collonges, Dept. Rhône, France (Ballmann, 1972), but until they can be verified the earliest certain record is *Charadrius* sp., Rębielice Królewskie I, southern Poland (Jánossy, 1974), now dated as late Pliocene (PIA) following recent stratigraphical revision (Z. Bochenski, pers. comm., 1989; J. Mlíkovský, pers. comm., 1990). **Extant**
Intervening: PLE, HOL.

F. SCOLOPACIDAE Vigors, 1825 T. (BRT/CHT)–Rec. Lit./Terr.

First: *Totanus edwardsi* Gaillard, 1908, Phosphorites du Quercy, France. **Extant**
Intervening: RUP, AQT, MMI–HOL.

F. RECURVIROSTRIDAE Bonaparte, 1854 T. (YPR)–Rec. Lit./Terr.

First: *Fluviatilavis antunesi* Harrison, 1983, Lower Mondego Region, Silveirinha, central Portugal. **Extant**
Intervening: PRB, MMI, PIA, PLE, HOL.

F. GRACULAVIDAE Fürbringer, 1888 K. (MAA) Mar./Lit.

First and Last: *Graculavus velox* Marsh, 1872, *Telmatornis priscus* Marsh, 1870, *Anatalavis rex* Olson and Parris, 1987, *Laornis edwardsianus* Marsh, 1870, *Palaeotringa littoralis* Marsh, 1870 and *P. vagans*, Marsh, 1872, Hornerstown or Navesink Formation, New Jersey, USA (Olson and Parris, 1987).

F. PHALAROPODIDAE Bonaparte, 1831 T. (PIA)–Rec. Mar./Lit.

First: *Phalaropus eleonorae* Kurochkin, 1985, Chikoyskaya Suite, Beregovaya, Buichurski Region, Buryatia, former USSR. **Extant**
Intervening: PLE.

F. DROMADIDAE (Gray, 1840) **Extant** Lit./Terr.

F. BURHINIDAE Matthews, 1913
T. (LMI)–Rec. Terr.

First: *Burhinus lucorum* Bickart, 1982, Middle Sheep Creek Formation, Thomson Quarry, Sioux City, Nebraska, USA. **Extant**
Intervening: PLI, PLE, HOL.

F. GLAREOLIDAE Brehm, 1831 T. (LMI)–Rec. Terr.

First: *Paractiornis perpusillus* Wetmore, 1930, Harrison Formation, Sioux County, Nebraska, USA. **Extant**
Intervening: MMI, PLE.

F. THINOCORIDAE Gray, 1845
Q. (PLE)–Rec. Terr. (see Fig. 40.3)

First: *Thinocorus koepckeae* Campbell, 1979 and *T. rumicivorus* Eschscholtz, 1829, Talara, Piura, Peru (Cuello, 1987). **Extant**

F. PEDIONOMIDAE **Extant** Terr.

F. CHIONIDIDAE Bonaparte, 1832 **Extant** Terr.

F. STERCORARIIDAE (Gray, 1871)
T. (AQT?)–Rec. Lit./Mar.

First: *Larus desnoyersii* Milne-Edwards, 1863, Saint-Gérand-le-Puy, Allier, France. Probably a skua, but needs to be confirmed (Olson, 1985). The oldest certain record is *Stercorarius* sp., Calvert Formation (MMI), Maryland, USA (Olson, 1985). **Extant**
Intervening: ZAN, PLE, HOL.

F. LARIDAE Vigors, 1825 T. (PRB/RUP)–Rec. Lit./Mar.

First: Undetermined remains, Phosphorites du Quercy, France (Mourer-Chauviré, 1982). **Extant**
Intervening: RUP?, AQT, UMI, ZAN, PLE, HOL.

F. RHYNCHOPIDAE Bonaparte, 1838 **Extant** Lit./FW

F. ALCIDAE Vigors, 1825 T. (PRB?)–Rec. Mar.

First: *Hydrotherikornis oregonus* Miller, 1931, Sunset Bay, Coos County, Oregon, USA? According to R. Chandler (pers. comm., 1989) *Hydrotherikornis* is probably not an auk, thus *Petralca austriaca* (Mlíkovský, 1987), from the Schieferton (part of the Puchkirchener Schichtengruppe) (CHT) of Traun near Linz, Austria, is the oldest reliable record. **Extant**
Intervening: MMI, UMI–HOL.

Order COLUMBIFORMES Latham, 1790

F. PTEROCLIDAE Bonaparte, 1831
T. (PRB/CHT)–Rec. Terr.

First: *Archaeoganga validus* Milne-Edwards, 1892, *A. larvatus* Milne-Edwards, 1892, and *A. pinguis* Mourer-Chauviré, 1992d, Phosphorites du Quercy, France. **Extant**
Intervening: AQT, ZAN–PLE.

F. COLUMBIDAE (Illiger, 1811)
T. (AQT)–Rec. Terr.

First: *Gerandia calcaria* (Milne-Edwards, 1871), Aquitainian beds between St-Gerand-le-Puy and Langy, Allier, France. **Extant**
Intervening: UMI–HOL.

F. RAPHIDAE Wetmore, 1930
Q. (PLE/HOL) Terr.

First and Last: *Raphus cucullatus* (Linnaeus, 1758), Mauritius, *R. solitaria* (Selys-Lonchamps, 1846), Reunion Island and *Pezophaps solitaria* (Gmelin, 1789), Rodriguez.

Order PSITTACIFORMES Wagler, 1830

F. PSITTACIDAE (Illiger, 1811)
T. (YPR?)–Rec. Terr.

First: *Palaeopsittacus georgei* Harrison, 1982, London Clay, Walton-on-the-Naze, Essex, England, UK. Some doubts have been expressed about this identification (Olson, 1985), thus the oldest reliable record is *Conuropsis fratercula* Wetmore, 1927, Upper Miocene, Snake Creek Quarries, Sioux County, Nebraska, USA. **Extant**
Intervening: LUT/BRT? PRB, AQT, MMI–HOL.

Order MUSOPHAGIFORMES Seebohm, 1890

F. MUSOPHAGIDAE Bonaparte, 1831
T. (THA?)–Rec. Terr.

First: An isolated sternum from the Moler Diatomite, Fur Formation, Limfjorden West, Denmark has been tentatively identified as that of a turaco (Bonde, 1987). The earliest certain record is genus and species indet., aff. *Crinifer*, upper sequence of the Jebel Qatrani Formation (RUP), Quarry M, Fayum Province, Egypt (Rasmussen *et al.*, 1987). **Extant**
Intervening: CHT, BUR, TOR.

Order CUCULIFORMES Wagler, 1830

F. CUCULIDAE Vigors, 1825 T. (BRT/CHT)–Rec. Terr.

First: *Dynamopterus velox* Milne-Edwards, 1892, Phosphorites du Quercy, France. **Extant**
Intervening: RUP, LMI, ZAN, PLE, HOL.

F. PARVICUCULIDAE Harrison, 1982
T. (YPR) Terr.

First: *Procuculus minutus* Harrison and Walker, 1977, lower Fish-tooth Bed, Division B of London Clay, Bognor Regis, Sussex, England, UK.
Last: *Parvicuculus minor* Harrison and Walker, 1977, Division D of London Clay, Burnham-on-Crouch, Essex, England, UK.

F. OPISTHOCOMIDAE Gray, 1840
T. (MMI)–Rec. Terr.

First: *Hoazinoides magdalenae* Miller, 1953, La Venta Formation, Magdalena Valley, Huila, Colombia. **Extant**

F. FORATIDAE Olson, 1992 T. (YPR) Terr.

First and Last: *Foro panarium* Olson, 1992, Fossil Butte Member of Green River Formation, Thompson Quarry, Kemmerer, Lincoln County, Wyoming, USA.

Order STRIGIFORMES Wagler, 1830

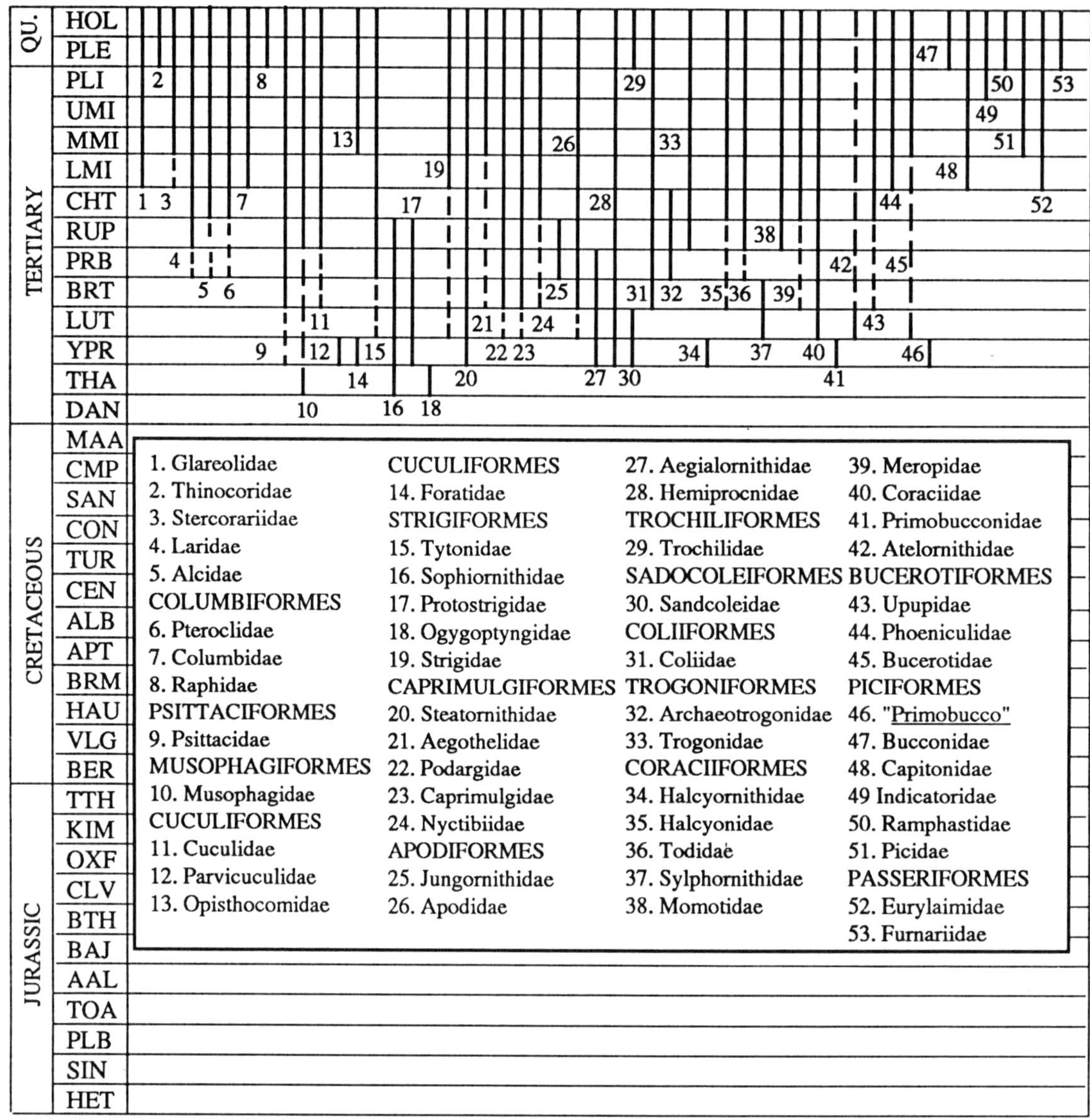

Fig. 40.3

F. TYTONIDAE Ridgway, 1914 T. (LUT?)–Rec. Terr.

First: *Necrobyas rossignoli* Milne-Edwards, 1892, Gisement de Perrière and *Nocturnavis incerta* (Milne-Edwards, 1892), Gisement d'Escamps, Phosphorites du Quercy (PRB), France. **Extant**
Intervening: RUP–AQT, MMI–MES, PIA–HOL.

F. SOPHIORNITHIDAE Mourer-Chauviré, 1987 T. (THA–RUP) Terr.

First: Unnamed genus, Mont Berru, France (Mourer-Chauviré, 1987).
Last: *Sophiornis quercynus* Mourer-Chauviré, 1987, Belgarric 1, Phosphorites du Quercy, France.

F. PROTOSTRIGIDAE Wetmore, 1933 T. (YPR–RUP) Terr.

First: *Eostrix mimica* (Wetmore, 1938), Knight Formation, Washakie County, Wyoming, USA and possibly *Eostrix vincenti* Harrison, 1980, London Clay, England, UK.
Last: *Oligostrix rupelensis* Fischer, 1983, Middle Oligocene, Weisselsterbecken, near Leipzig, Germany.
Intervening: LUT/BRT, PRB.

F. OGYGOPTYNGIDAE Rich and Bohaska, 1981 T. (THA) Terr.

First and Last: *Ogygoptynx wetmorei* Rich and Bohaska, 1976, San Juan River, south-west Colorado, USA.

F. STRIGIDAE Vigors, 1825 T. (LUT?)–Rec. Terr.

First: *Eoglaucidium pallas* Fischer, 1987, Braunkohle des Geiseltales, Neumark-Sud, near Halle, Saxony, Germany. J. Mlíkovský (pers. comm., 1990) seriously doubts the identification of *Eoglaucidium* as a true owl. *Palaeoglaux artophoron* Peters, 1992, Messel Oil Shale (LUT), Messel Quarry, Hessen, Germany and *Palaeoglaux perrierensis* Mourer-Chauviré, 1987, Gisement de Perrière, Phosphorites du Quercy (PRB), France, may be strigids, but until confirmed the next oldest record is *'Bubo' poirrieri* Milne-Edwards, 1863 (AQT), Saint-Gérand-le-Puy, Dept. Allier, France (Mourer-Chauviré, 1987). **Extant**
Intervening: AQT–HOL.

Order CAPRIMULGIFORMES Ridgway, 1881

F. STEATORNITHIDAE (Gray, 1846) T. (YPR)–Rec. Terr.

First: *Prefica nivea* Olson, 1987, Green River Formation, Lincoln City, Wyoming, USA. **Extant**
Intervening: CHT?.

F. AEGOTHELIDAE Bonaparte, 1853 T. (BRT/CHT?)–Rec. Terr.

First: Fragment of a sternum from the Phosphorites du Quercy, France, tentatively assigned to the Aegothelidae by Mourer-Chauviré (1982). The next oldest record is *Quipollornis koniberi* Rich and McEvey, 1977, from a caldera lake deposit (LLA), Warrumbungle Mountains, eastern New South Wales, Australia. **Extant**
Intervening: PLE, HOL.

F. PODARGIDAE (Gray, 1840) T. (LUT?)–Rec. Terr.

First: Undescribed remains from the Messel Oil Shale, Messel Quarry, Hessen, Germany (Peters, 1988). The earliest certain record is *Quercypodargus olsoni* Mourer-Chauviré, 1988b, Phosphorites du Quercy (BRT), France. **Extant**

F. CAPRIMULGIDAE Vigors, 1825 T. (LUT?)–Rec. Terr.

First: Undescribed remains from the Messel Oil Shale, Messel Quarry, Hessen, Germany (Peters, 1988). The earliest certain record is *Ventivorus ragei* Mourer-Chauviré, 1988b, Le Bretou, Phosphorites du Quercy (BRT), France. **Extant**
Intervening: PLE, HOL.

F. NYCTIBIIDAE (Bonaparte, 1853) T. (BRT/CHT)–Rec. Terr.

First: *Euronyctibius kurochkini* Mourer-Chauviré, 1988b, Phosphorites du Quercy, France. **Extant**
Intervening: PLE, HOL.

Order APODIFORMES Peters, 1940

F. JUNGORNITHIDAE Karkhu, 1988 T. (PRB–RUP) Terr.

First: *Palescyvus escampensis* Karkhu, 1988, Escampes III, Dept. Lot, France.
Last: *Jungornis tesselatus* Karkhu, 1988, Maykop Formation, Abadzekhskaya Station, Adygey, northern Caucasus, former USSR.

F. APODIDAE Hartert, 1897 T. (LUT/BRT)–Rec. Terr.

First: *Scaniacypselus wardi* Harrison, 1984b, Røsnaes Clay, Ølst, Jutland, Denmark. **Extant**
Intervening: PRB, AQT, MMI–HOL.

F. AEGIALORNITHIDAE Lydekker, 1891 T. (YPR–PRB) Terr.

First: *Primapus lacki* Harrison and Walker, 1974, Fish-Tooth Beds, London Clay, Bognor Regis, Sussex, and London Clay, Warren Point, Isle of Sheppey, Kent, England, UK.
Last: *Aegialornis broweri* Collins, 1976, Sainte-Néboule, Phosphorites du Quercy, France (C. Mourer-Chauviré, pers. comm., 1989).
Intervening: LUT.

F. HEMIPROCNIDAE Oberholser, 1906 T. (YPR)–Rec. Terr.

First: *Eocypselus vincenti* Harrison, 1984b, London Clay, Walton-on-the-Naze, Essex, England, UK. **Extant**
Intervening: BRT.
Comment: Some authors (e.g. Mlíkovský, 1985) would combine Aegialornithidae with Hemiprocnidae.

Order TROCHILIFORMES Wagler, 1830

F. TROCHILIDAE Vigors, 1825 Q. (PLE)–Rec. Terr.

First: *Chlorostilbon bracei* (Lawrence, 1897) and *C. ricordii* (Gervais, 1835), New Providence Island, Bahamas (Graves and Olson, 1987); *Anthracothorax dominicus* (Linnaeus, 1758), Cerro San Francisco, Dominican Republic, and *Clytolaema rubricauda* (Boddaert, 1783), Lapa de Escrivania, Minas Geraës Province, Brazil (Brodkorb, 1978). **Extant**
Intervening: HOL.

Order SANDCOLEIFORMES Houde and Olson, 1992

F. SANDCOLEIDAE Houde and Olson, 1992 T. (YPR–LUT) Terr.

First: *Sandcoleus copiosus* Houde and Olson, 1992, Sand Coulee Beds, Willwood Formation, Clark's Ford Basin, Clark Quadrangle, Park County, Wyoming, USA.
Last: *Eobucco brodkorbi* Feduccia and Martin, 1976, 56 km north of Green River, Sweetwater County, Wyoming, USA.

Order COLIIFORMES Murie, 1872

F. COLIIDAE Swainson, 1837 T. (BRT)–Rec. Terr.

First: *Primocolius sigei* Mourer-Chauviré, 1988a, Le Bretou, Phosphorites du Quercy, France. **Extant**
Intervening: PRB, AQT, MMI–ZAN.

Order TROGONIFORMES American Ornithologists' Union, 1886

F. ARCHAEOTROGONIDAE Mourer-Chauviré, 1980 T. (PRB–CHT) Terr.

First: *Archaeotrogon venustus* Milne-Edwards, 1892, Mouillac, Tarn-et-Garonne, France (Mourer-Chauviré, 1982).
Last: *Archaeotrogon venustus* Milne-Edwards, 1892, *A. cayluxensis* Gaillard, 1908, *A. zitteli* Gaillard, 1908 and *A. hoffsteteri* Mourer-Chauviré, 1980, Phosphorites du Quercy, Bach (Lot), Caylux and Mouillac (Tarn-et-Garonne), France.
Intervening: RUP.

F. TROGONIDAE Swainson, 1831 T. (RUP)–Rec. Terr.

First: Unnamed specimen originally referred to *Protornis glarniensis* Meyer, 1854, Glarner Fischschiefer, Sernftal, Glarus Canton, Switzerland (Olson, 1976). **Extant**
Intervening: AQT, PLE.

Order CORACIIFORMES Forbes, 1884

F. HALCYORNITHIDAE Harrison and Walker, 1972 T. (YPR) Terr.

First and Last: *Halcyornis toliapicus* Köenig, 1825, London Clay, Isle of Sheppey, Kent, England, UK.

F. HALCYONIDAE Vigors, 1825
T. (BRT/CHT)–Rec. Terr.

First: Undetermined remains, Phosphorites du Quercy, France (Mourer-Chauviré, 1982). **Extant**
Intervening: ZAN, PLE, HOL.

F. TODIDAE Vigors, 1825
T. (PRB/RUP)–Rec. Terr.

First: *Palaeotodus escampsiensis* Mourer-Chauviré, 1985, Gisement d'Escamps, Phosphorites du Quercy, France. **Extant**
Intervening: RUP, CHT, PLE.

F. SYLPHORNITHIDAE Mourer-Chauviré, 1988
T. (LUT–BRT) Terr.

First: Undescribed remains from the Messel Oil Shale, Messel Quarry, Hessen, Germany (Peters, 1991).
Last: *Sylphornis bretouensis* Mourer-Chauviré, 1988b, Le Bretou, Phosphorites du Quercy, France.

F. MOMOTIDAE (Gray, 1840)
T. (RUP)–Rec. Terr.

First: *Protornis glarniensis* Meyer, 1854, Glarner Fischschiefer, Sernftal, Glarus Canton, Switzerland (Olson, 1976). **Extant**
Intervening: UMI, PLE, HOL.

F. MEROPIDAE Vigors, 1825
T. (BRT/CHT)–Rec. Terr.

First: Undetermined remains, Phosphorites du Quercy, France (Mourer-Chauviré, 1982). **Extant**
Intervening: MMI, PLE.

F. CORACIIDAE Vigors, 1825
T. (LUT)–Rec. Terr.

First: Undescribed remains from the Messel Oil Shale, Messel Quarry, Hessen, Germany (Peters, 1988). **Extant**
Intervening: BRT/CHT, PRB, PLE, HOL.

F. PRIMOBUCCONIDAE Feduccia and Martin, 1976
T. (YPR) Terr.

First and Last: *Primobucco mcgrewi* Brodkorb, 1970, Green River Formation, Fossil, Lincoln County, Wyoming, USA.

F. ATELORNITHIDAE Bonaparte, 1854
T. (LUT?)–Rec. Terr.

First: Undescribed remains from the Messel Oil Shale, Messel Quarry, Hessen, Germany (Peters, 1988). Otherwise, this family has no fossil record. **Extant**

F. LEPTOSOMIDAE Bonaparte, 1850 **Extant** Terr.

Order BUCEROTIFORMES Fürbringer, 1888

F. UPUPIDAE Bonaparte, 1831
T. (BRT/CHT)–Rec. Terr.

First: Undetermined remains, Phosphorites du Quercy, France (Mourer-Chauviré, 1982). **Extant**
Intervening: PLE, HOL.

F. PHOENICULIDAE Sclater, 1824
T. (AQT)–Rec. Terr.

First: *Limnatornis paludicola* Milne-Edwards, 1871, St-Gerand-le-Puy, Allier, France. **Extant**
Intervening: BUR.

F. BUCEROTIDAE Vigors, 1825
T. (LUT?)–Rec. Terr.

First: Candidates include *Geiseloceros robustus* Lambrecht, 1935 (LUT), possibly a vulture (Olson, 1985), but almost certainly not a hornbill (Mlíkovský, pers. comm., 1990); *Cryptornis antiquus* Gervais, 1852 (PRB), possibly a coraciid (Harrison, 1979) and *Homalopus picoides* Milne-Edwards, 1871 (MMI), which might be a phoeniculid but is almost certainly not a hornbill (Olson, 1985). The oldest reliable record for this family is *Bucorvus brailloni* Brunet, 1971, Middle Miocene of Beni Mellal, north of Atlas Mountains, Morocco (Olson, 1985). **Extant**

Order PICIFORMES Meyer and Wolf, 1810

F. GALBULIDAE Bonaparte, 1831 **Extant** Terr.

F. UNNAMED Houde and Olson, 1989
T. (YPR) Terr.

First and Last: *'Primobucco' olsoni* (Feduccia and Martin, 1976), Green River Formation, Nugget, Lincoln County, Wyoming and *'Neanis' kistneri* (Fedducia, 1973), Green River Formation, Sweetwater County, Wyoming, USA (Houde and Olson, 1989).

F. BUCCONIDAE Boie, 1826 Q. (PLE)–Rec. Terr.

First: *Nystalus chacuru* (Vieillot, 1816) and *Malacoptila striata* (Spix, 1824) Lapa de Escrivania, Minas Geraës Province, Brazil. **Extant**

F. CAPITONIDAE Bonaparte, 1840
T. (BUR)–Rec. Terr.

First: *Capitonides europeus* Ballmann, 1969, Wintershof West, near Eichstätt, Bavaria, Germany. **Extant**
Intervening: SER/TOR, PLE.

F. INDICATORIDAE Swainson, 1837
T. (ZAN)–Rec. Terr.

First: Undescribed remains, Varswater Formation, Langebaanweg, South Africa (Olson, 1985). **Extant**

F. RAMPHASTIDAE Vigors, 1825
Q. (PLE)–Rec. Terr.

First: *Ramphastos dicolurus* Linnaeus, 1758, and *R. toco* Müller, 1846, Lapa de Escrivania, Minas Geraës Province, Brazil. **Extant**

F. PICIDAE Vigors, 1825 T. (MMI)–Rec. Terr.

First: Unnamed genus, Middle Miocene (Upper Barstovian) of New Mexico, USA (Olson, 1985). **Extant**
Intervening: LMI, ZAN, PIA, PLE, HOL.

Order PASSERIFORMES Linnaeus, 1766

F. EURYLAIMIDAE (Swainson, 1837)
T. (BUR)–Rec. Terr.

First: Eurylaimid gen. *et* sp. indet., Wintershof West, near Eichstätt, Bavaria, Germany (Ballmann, 1966). **Extant**

F. FURNARIIDAE (Gray, 1840)
Q. (PLE)–Rec. Terr.

First: *Cinclodes major* Tonni, 1977, Miramar Formation, Mar del Plata, Buenos Aires Province, Argentina. **Extant**
Intervening: HOL.

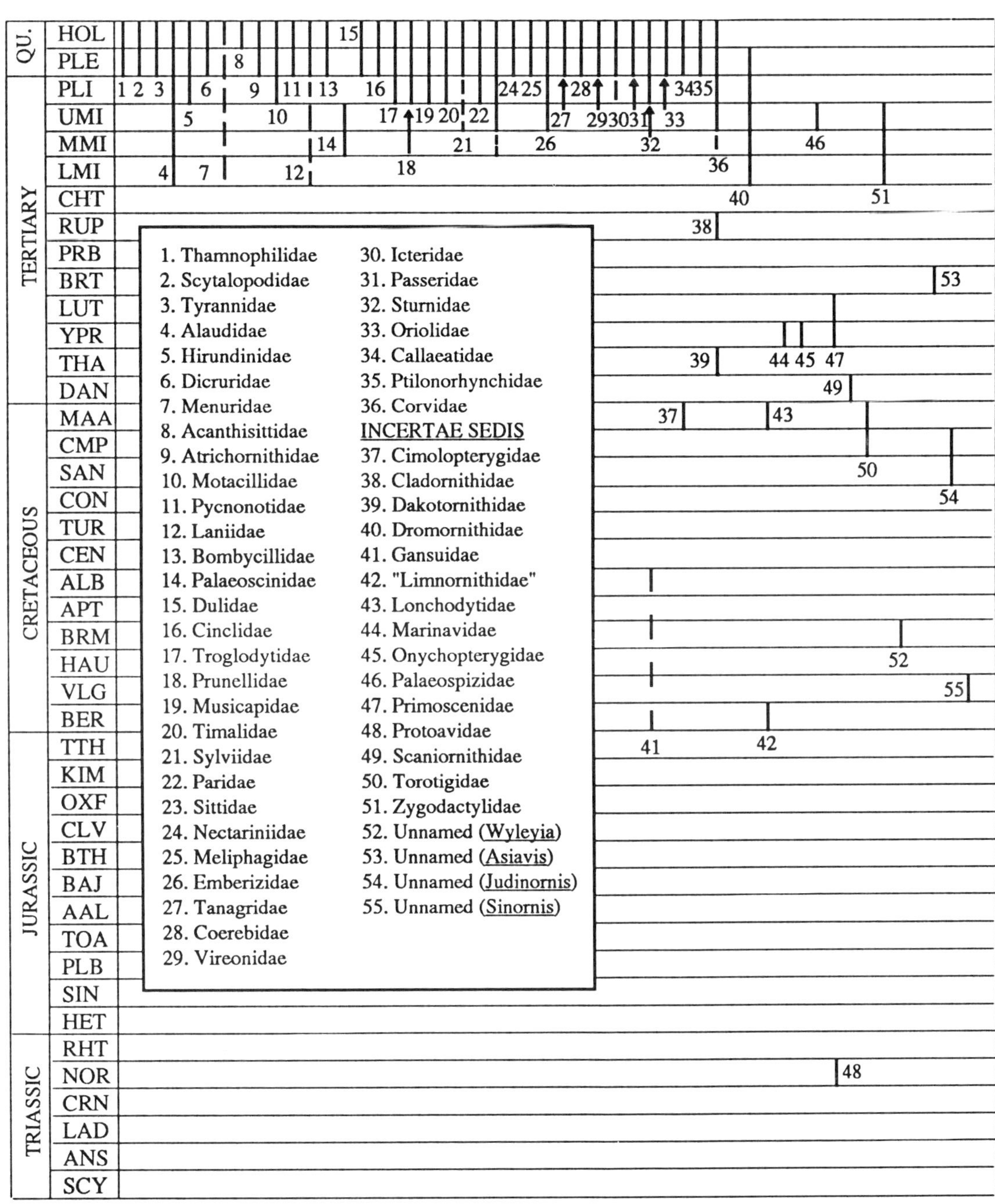

Fig. 40.4

F. THAMNOPHILIDAE Vigors, 1825
Q. (PLE)–Rec. Terr. (Fig. 40.4)

First: *Chamaeza brevicauda* (Vieillot, 1818), Upper Pleistocene, Lapa da Escrivania, Minas Geraës, Brazil. **Extant**

F. SCYTALOPODIDAE (Müller, 1846)
Q. (PLE)–Rec. Terr.

First: ?*Scytalopus* sp., Upper Pleistocene, Cueva de Los Fosiles, north-east of Camagüey, Camaguey Province, Cuba (Olson and Kurochkin, 1987). **Extant**
Intervening: ?HOL.

F. PITTIDAE Bonaparte, 1850 **Extant** Terr.

F. PHILEPITTIDAE (Sharpe, 1870) **Extant** Terr.

F. TYRANNIDAE (Vigors, 1825)
Q. (PLE)–Rec. Terr.

First: *Tyrannus tyrannus* (Linnaeus, 1758), Reddick Beds (Middle Pleistocene) Reddick, Florida, USA. Eleven other neospecies are listed from the Pleistocene (Brodkorb, 1978). **Extant**
Intervening: HOL.

F. OXYRUNCIDAE Ridgway, 1906 **Extant** Terr.

F. PIPRIDAE Vigors, 1825 **Extant** Terr.

F. QUERULIDAE (Swainson, 1837) **Extant** Terr.

F. PHYTOTOMIDAE (Swainson, 1837)
Extant Terr.

F. ALAUDIDAE (Vigors, 1825)
T. (BUR/LAN1)–Rec. Terr.

First: Undetermined genus, siderolithic fissure fill, Vieux-Collonges, Dept. Rhône, France (Ballmann, 1972). **Extant**
Intervening: ZAN, PLE, HOL.

F. HIRUNDINIDAE Vigors, 1825
T. (PIA)–Rec. Terr.

First: *Hirundo aprica* Feduccia, 1967, Rexroad Formation, Fox Canyon, Mead County, Kansas, USA. **Extant**
Intervening: PLE, HOL.

F. DICRURIDAE (Vigors, 1825)
Q. (PLE)–Rec. Terr.

First: *Dicrurus macrocercus* (Vieillot, 1816), localities 1 and 15, Zhoukoudian area, Beijing, China (Rich *et al.*, 1986). **Extant**

F. MENURIDAE (Gray, 1847)
T. (LMI?)–Rec. Terr.

First: Undescribed remains from Riversleigh, north-west Queensland, Australia, are tentatively assigned to this family (W. Boles, pers. comm., 1989), otherwise lyrebirds have no fossil record. **Extant**

F. ACANTHISITTIDAE (Sundevall, 1872)
Q. (HOL)–Rec. Terr.

First: *Dendroscansor decurvirostris* Millener and Worthy, 1991, Upper Quaternary (HOL) of New Zealand. **Extant**

F. ATRICHORNITHIDAE Stejneger, 1855
Q. (PLE)–Rec. Terr.

First: Undescribed remains from the Pleistocene of Australia (P. Rich, pers. comm., 1989). **Extant**

F. MOTACILLIDAE Vigors, 1825
T. (ZAN)–Rec. Terr.

First: *Anthus seductus* Kurochkin, 1985, Lower Pavlodarski Suite, Pavlodar, eastern Kazakhstan, former USSR. **Extant**
Intervening: PIA–HOL.

F. CAMPEPHAGIDAE (Vigors, 1825) **Extant** Terr.

F. PYCNONOTIDAE (Gray, 1840)
Q. (PLE)–Rec. Terr.

First: *Pycnonotus barbatus* (Desfontaines, 1789), 'Ubeidiya Formation, 'Ubeidiya, SW bank of Sea of Galilee, Israel (Tchernov, 1980). **Extant**

F. LANIIDAE (Swainson, 1824)
T. (AQT?)–Rec. Terr.

First: *Lanius miocaenus* Milne-Edwards, 1871, Langy, Dept. Allier, France. Milne-Edwards' original identification of these remains is now strongly doubted (Olson, 1985); the earliest reliable records are *Lanius excubitor* Vieillot, 1808 and *Lanius* sp. from the 'Ubeidiya Formation, 'Ubeidiya, SW bank of Sea of Galilee, Israel, *Lanius minor* Gmelin, 1788, from Layer 24, Petralona, Greece and *Lanius* cf. *minor* from locality 2, Betfia, Romania, all of which are early Middle Pleistocene in age (Tyrberg, pers comm., 1992). **Extant**
Intervening: HOL.

F. VANGIDAE Shelley, 1896 **Extant** Terr.

F. BOMBYCILLIDAE (Swainson, 1832)
Q. (PLE)–Rec. Terr.

First: *Bombycilla garrulus* (Linnaeus, 1758), Upper Middle Pleistocene of Orgnac 3, France (Mourer-Chauviré, 1975). **Extant**
Intervening: HOL.

F. PALAEOSCINIDAE Howard, 1957
T. (MMI/UMI) Terr.

First and last: *Palaeoscinis turdirostris* Howard, 1957, Monterey Formation, Tepusquet Creek, Santa Barbara County, California, USA.

F. DULIDAE (Sclater, 1862) Q. (PLE)–Rec. Terr.

First: *Dulus dominicus* (Linnaeus, 1758), Cerro de Francisco, Dominican Republic (Bernstein, 1965). **Extant**

F. CINCLIDAE (Cabanis, 1847)
Q. (PLE)–Rec. Terr.

First: *Cinclus cinclus* (Linnaeus, 1758), layer G2, Hunas, Germany (Janossy, 1983), and Grotte du Lazaret and La Fage, France (Mourer-Chauviré, 1975). **Extant**

F. TROGLODYTIDAE (Swainson, 1832)
T. (PIA)–Rec. Terr.

First: *Troglodytes troglodytes* (Linnaeus, 1758), Pedrera de S'Onix, Mallorca (Mourer-Chauviré *et al.*, 1977). **Extant**
Intervening: PLE, HOL.

F. PRUNELLIDAE Richmond, 1908
T. (PIA)–Rec. Terr.

First: *Prunella* cf. *modularis*, Pedrera de S'Onix, Mallorca (Mourer-Chauviré *et al.*, 1977). **Extant**
Intervening: PLE, HOL.

F. MUSICAPIDAE Vigors, 1825
T. (PIA)–Rec. Terr.

First: *Oenanthe infima* Kurochkin, 1985, Hirgis Nur Suite (Middle Pliocene), Shar Gain Gobi, western part of People's Republic of Mongolia. **Extant**
Intervening: PLE, HOL.

F. TIMALIIDAE Delacour, 1946
T. (PIA)–Rec. Terr.

First: *Turdoides borealis*, Osztramos 1, Hungary (Jánossy, 1979). **Extant**

F. SYLVIIDAE (Vigors, 1825)
T. (UMI?)–Rec. Terr.

First: Genus indet., Gargano Peninsula, Italy (Ballmann, 1973). Ballmann only tentatively identified these remains as sylviid; the oldest certain records are *Sylvia* cf. *atricapilla*, Pedrera de S'Onix (PIA), Mallorca (Mourer-Chauviré *et al.*, 1977), and *Hippolais* sp., Csarnota 2 (PIA), Hungary (Jánossy, 1979). **Extant**
Intervening: PLE, HOL.

F. PARIDAE Boie, 1826 T. (PIA)–Rec. Terr.

First: *Aegithalos* cf. *caudatus*, *Parus* cf. *ater* and *P.* cf. *major*, Pedrera de S'Onix, Mallorca (Mourer-Chauviré *et al.*, 1977). **Extant**
Intervening: PLE, HOL.

F. SITTIDAE Bonaparte, 1831
T. (MMI/UMI)–Rec. Terr.

First: *Sitta*, La-Grive-Saint-Alban, France (Ballmann, 1973; see also Olson, 1985). **Extant**
Intervening: ZAN–HOL.

F. PARDALOTIDAE Bonaparte, 1850 **Extant** Terr.

F. NECTARINIIDAE Vigors, 1825
Q. (PLE)–Rec. Terr.

First: *Nectarinia osea* Oustalet, 1904, Upper Pleistocene, Hayonim Cave, Israel (Tchernov, 1979). **Extant**

F. ZOSTEROPIDAE Bonaparte, 1853 **Extant** Terr.

F. MELIPHAGIDAE Vigors, 1825
Q. (PLE)–Rec. Terr.

First: *Prosthemadera novaeseelandiae* (Gmelin, 1788), Pyramid Valley Swamp and Martinborough Cave V, South Island, New Zealand. *Anthornis melanura* (Sparrman), Awakino–Mohoenui area, North Island, New Zealand (Medway, 1971). **Extant**
Intervening: HOL.

F. EMBERIZIDAE Vigors, 1831
T. (MES)–Rec. Terr.

First: *Ammodramus hatcheri* (Shufeldt, 1913), Long Island, Phillips County, Kansas, USA (Steadman, 1982). **Extant**
Intervening: ZAN–HOL.

F. TANAGRIDAE (Vigors, 1825)
Q. (PLE)–Rec. Terr.

First: *Pyrrhuloxia cardinalis* (Linnaeus, 1758), Arredondo Clay member, Wicomico Formation (Middle Pleistocene), Arredondo, Alachua County, Florida, USA (Fisher, 1967). **Extant**
Intervening: HOL.

F. COEREBIDAE (Gray, 1840)
Q. (PLE)–Rec. Terr.

First: *Geothlypis trichas* (Linnaeus, 1766), Reddick Beds (Middle Pleistocene), Reddick, Marion County, Florida, USA (Brodkorb, 1957). **Extant**
Intervening: HOL.

F. VIREONIDAE Swainson, 1837
Q. (PLE)–Rec. Terr.

First: *Cyclarhis gujanensis* (Gmelin, 1789), Lapa da Escrivania (Upper Pleistocene), Lagoa Santa, Minas Geraës Province, Brazil; *Vireo calidris* (Linnaeus, 1758) and *V. nanus* (Lawrence, 1875), Cerro de San Francisco, Dominican Republic; *V. griseus* (Boddaert, 1783), Haile, Florida, USA (Brodkorb, 1978). **Extant**
Intervening: HOL.

F. ICTERIDAE Vigors, 1825 T. (ZAN?)–Rec. Terr.

First: *'Colinus eatoni'* Shufeldt, 1915, ?Oglalla Formation, Fort Wallace, western Kansas (Brodkorb, 1978). There is some doubt as to the age of this material (Brodkorb, 1978) thus the oldest certain records are *Pandanaris floridana* Brodkorb, 1957, Reddick Beds (middle Upper Pleistocene), Reddick, Marion County and *Cremaster tytthus* Brodkorb, 1959, Arredondo, Alachua County, Florida, USA. **Extant**
Intervening: HOL.

F. PASSERIDAE (Illiger, 1811)
Q. (PLE)–Rec. Terr.

First: *Passer domesticus* (Linnaeus, 1758), Grotte L'Escale, France (Mourer-Chauviré, 1975). **Extant**
Intervening: HOL.

F. STURNIDAE Vigors, 1825 T. (PIA)–Rec. Terr.

First: Genus indet., Chikoyskaya Suite, Selenginski Aimak, Shamar, northern part of People's Republic of Mongolia (Kurochkin, 1985). **Extant**
Intervening: PLE, HOL.

F. ORIOLIDAE (Vigors, 1825)
Q. (PLE)–Rec. Terr.

First: *Oriolus* sp., 'Ubeidiya Formation, 'Ubeidiya, SW bank of Sea of Galilee, Israel (Tchernov, 1980). Remains of orioles have been found recently in the Upper Pliocene (PIA) of Varshets, west Balkan Range, Bulgaria (Boev, 1992). **Extant**
Intervening: HOL.

F. CALLAEATIDAE (Gray, 1841)
Q. (PLE)–Rec. Terr.

First: *Callaeas cinerea* (Gmelin, 1788) and *Creadion carunculatum* (Gmelin, 1789), Pyramid Valley Swamp, South Island, New Zealand (Scarlett, 1955). **Extant**
Intervening: HOL.

F. ARTAMIDAE Blyth, 1849 **Extant** Terr.

F. PTILONORHYNCHIDAE Gray, 1846
Q. (PLE)–Rec. Terr.

First: *Turnagra capensis* (Sparrman, 1787) (neosp.), Awakino-Mohoenui area, North Island, and possibly Pyramid Valley Swamp, South Island, New Zealand (Scarlett, 1955; Medway, 1971). **Extant**

F. PARADISAEIDAE Vieillot, 1825 **Extant** Terr.

F. CORVIDAE Vigors, 1825 T. (MMI?)–Rec. Terr.

First: Corvidae gen. *et* sp. indet., Shanwang Formation, Linqu, Shandong, China (Yeh and Sun, 1989). This record has not yet been confirmed, and *Miocorvus larteti* (Milne-Edwards, 1871), from the Upper Miocene of Sansan, Dept. Gers, France, also needs to be checked (Brodkorb, 1978). The earliest certain record is *Miocitta galbreathi* Brodkorb, 1972 (UMI) Kennesaw local fauna, Pawnee Creek Formation, Logan County, Colorado, USA. **Extant**
Intervening: PIA–HOL.

INCERTAE SEDIS

F. CIMOLOPTERYGIDAE Brodkorb, 1963
K. (MAA) Terr.

First and Last: *Cimolopteryx rara* Marsh, 1889, *C. minima* Brodkorb, 1963 and *C. maxima* Brodkorb, 1963, *Palintropus retusus* (Marsh, 1892), (see entry for Apatornithidae) and *Ceramornis major* Brodkorb, 1963, Lance Formation, Lance Creek, Niobrara County, Wyoming, USA; *Cimolopteryx* sp., Frenchman Formation, Frenchman Valley, south of Shaunavon, Saskatchewan, Canada (Tokaryk and James, 1989).

F. CLADORNITHIDAE Ameghino, 1895
T. (RUP) Lit./?Terr.

First and Last: *Cladornis pachypus* Ameghino, 1895, Deseado Formation, Rio Deseado, Santa Cruz Province, Argentina.

F. DAKOTORNITHIDAE Erickson, 1975
T. (THA) Terr.

First and Last: *Dakotornis cooperi* Erickson, 1975, Tongue River Formation, Wannagan Creek, Billings County, North Dakota, USA.

F. DROMORNITHIDAE Fürbringer, 1888
T. (LMI)–Q. (PLE) Terr.

First: *Barawertornis tedfordi* Rich, 1979, Riversleigh Homestead, north-western Queensland, Australia. *Bullockornis planei* Rich, 1979, has now also been recovered from this locality (W. Boles, pers, comm., 1989).
Last: *Genyornis newtoni* Stirling and Zeitz, 1896, Upper Pleistocene beds of Lake Callabonna, Normanville; Baldina Creek near Burra; Parroo River and Mount Gambier, South Australia, and Mudgee, New South Wales, Australia (Rich, 1979).
Intervening: MMI, UMI, PLI.

F. GANSUIDAE Hou and Liu, 1983
K. (l.) Terr.

First and Last: *Gansus yumenensis* Hou and Liu, 1983, Xiagou Formation, Yumen, Gansu Province, China.

F. 'LIMNORNITHIDAE' Kessler and Jurcsák, 1984
K. (BER) Terr.

First and Last: *Eurolimnornis corneti* Kessler and Jurcsák, 1984, Cornet, Padurea Crailui Mountains, Bihor, Romania.

F. LONCHODYTIDAE Brodkorb, 1963
K. (MAA) Terr.

First and Last: *Lonchodytes estesi* Brodkorb, 1963 and *L. pterygius* Brodkorb, 1963, Lance Formation, Lance Creek, Niobrara County, Wyoming, USA.

F. MARINAVIDAE Harrison and Walker, 1977
T. (YPR) Lit./Mar.

First and Last: *Marinavis longirostris* Harrison and Walker, 1977, Blackheath Beds, Abbey Wood, Kent, England, UK. Considerable doubt has been cast on the validity of this taxon (Brodkorb, 1978; Steadman, 1981).

F. ONYCHOPTERYGIDAE Cracraft, 1971
T. (YPR) Terr.

First and Last: *Onychopteryx simpsoni* Cracraft, 1971, Casamayor Formation, Cañadón Hondo near Paso Niemann, Chubut, Argentina.

F. PALAEOSPIZIDAE Wetmore, 1925
T. (UMI) Terr.

First and Last: *Palaeospiza bella* Allen, 1878, Florissant Lake Beds, Florissant, Teller County, Colorado, USA.

F. PRIMOSCENIDAE Harrison and Walker, 1977
T. (YPR–LUT) Terr.

First: *Primoscens minutus* Harrison and Walker, 1977, lower Fish-Tooth Beds, London Clay, Bognor Regis, West Sussex, England, UK.
Last: Undescribed remains from the Messel Oil Shale, Messel Quarry, Hessen, Germany (Peters, 1991).

F. PROTOAVIDAE Chatterjee, 1991
Tr. (NOR) Terr.

First and Last: *Protoavis texenis* Chatterjee, 1991, Dockum Formation, Post, west Texas, USA. The identification of the remains of *Protoavis* as avian is highly controversial (Beardsley, 1986; Ostrom, 1987) and will only be resolved by further study (Ostrom, 1991).

F. SCANIORNITHIDAE Lambrecht, 1933
T. (DAN) Mar.

First and Last: *Scaniornis lundgreni* Dames, 1890, Saltholm Chalk, Annetorp Quarry, Limhamn, Sweden.

F. TOROTIGIDAE Brodkorb, 1963
K. (CMP–MAA) Lit./Terr.

First: *Parascaniornis stensioi* Lambrecht, 1933, Shell Fragment Limestone, Ivö, Sweden.
Last: *Torotix clemensi* Brodkorb, 1964, Lance Formation, Lance Creek, Niobrara County, Wyoming, USA.

F. ZYGODACTYLIDAE Brodkorb, 1971
T. (BUR–TOR) Terr.

First: *Zygodactylus ignotus* Ballmann, 1967, Wintershof West, near Eichstätt, Bavaria, Germany.
Last: *Zygodactylus grivensis* Ballmann, 1969 and possibly also *Z. gaudryi* (Depéret, 1887), La-Grive-Saint-Alban, Dept. Isère, France (Brodkorb, 1971).
Intervening: MMI?

F. UNNAMED Harrison and Walker, 1973
K. (BRM) Terr.

First and Last: *Wyleyia valdensis* Harrison and Walker, 1973, Weald Clay, Henfield, Sussex, England, UK. Opinions differ as to whether this is a reptile or bird (Brodkorb, 1978; Elzanowski, 1983; Olson, 1985). It may possibly be enantiornithid (Walker, pers. comm., 1988).

F. UNNAMED Nessov, 1986 T. (BRT) Terr.

First and Last: *Asiavis phosphatica* Nessov, 1986, uppermost Middle Eocene beds of Tashkura, Central Kizylkum, Uzbekistan, former USSR. Represented by a single incomplete wing bone, this bird supposedly belongs in the Gruiformes.

F. UNNAMED Nessov and Borkin, 1983
K. (CMP/MAA) Terr.

First and Last: *Judinornis nogontzavensis* Nessov and Borkin, 1983, Nemegt Formation, Nogon-Tsav, southern Mongolia.

F. UNNAMED Rao and Sereno, 1990
K. (VAL?) Terr.

First and Last: *Sinornis santensis* Sereno and Rao, 1992, lake sediments, Jiufuotang Formation, Chaoyoung County, Liaoning Province, north-east China (Rao and Sereno, 1990). Further remains representing two new genera have been recently recovered from the Jiufuotang Formation, which is now generally accepted to be early Cretaceous in age (Zhou *et al.*, 1992).

REFERENCES

Alvarenga, H. M. F. (1985) Um novo Psilopteridae (Aves: Gruiformes) dos sedimentos Terciaros de Itaborai, Rio De Janeiro, Brasil. *VIII Congresso Brasiliano Paleontologia: 1983 MME–DNMP Séries. Geologia* 27, Paleontologia/Estratigrafia, **2**, pp. 17–20.

Alvarenga, H. M. F. and Bonaparte, J. F. (1992) A new flightless landbird from the Cretaceous of Patagonia, in *Papers in Avian*

Paleontology Honoring Pierce Brodkorb (ed. K. E. Campbell), *Science Series, Natural History Museum of Los Angeles County*, **36**, Los Angeles, California, pp. 51–64.

Andors, A. V. (1992) Reappraisal of the Eocene ground bird Diatryma (Aves: Anseromorphae), in *Papers in Avian Paleontology Honoring Pierce Brodkorb* (ed. K. E. Campbell), *Science Series, Natural History Museum of Los Angeles County*, **36**, Los Angeles, California, pp. 109–25.

Balouet, J. C. and Olson, S. L. (1989) Fossil birds from Late Quaternary deposits in New Caledonia. *Smithsonian Contributions to Zoology*, **469**, 1–38.

Beardsley, T. (1986) Fossil bird shakes evolutionary hypothesis. *Nature*, **322**, p. 677.

Becker, J. J. (1986) Reidentification of *'Phalacrocorax' subvolans* Brodkorb as the earliest record of Anhingidae. *Auk*, **103**, 807–8.

Bochenski, Z. and Kurochkin, E. N. (1987) Pliocene bustards (Aves: Otididae and Gryzajidae) of Moldavia and S. Ukraine. *Documents de la Laboratorie de Géologie, Lyon*, **99**, 173–87.

Bock, W. J. (1986) The arboreal origin of avian flight. *Memoirs of the Californian Academy of Sciences*, **8**, 57–72.

Boev, Z. (1992) Upper Pliocene birds from Varshets (West Balkan Range, Bulgaria) in *Abstracts of the 3rd Symposium of the Society of Avian Paleontology and Evolution*, Frankfurt-am-Main, Germany.

Boles, W. E. (1992) Revision of *Dromaius gidju* Patterson and Rich 1987 from Riversleigh, Northwestern Queensland, Australia, with a reassessment of its generic position, in *Papers in Avian Paleontology Honoring Pierce Brodkorb* (ed. K. E. Campbell), *Science Series, Natural History Museum Los Angeles County*, **36**, Los Angeles, California, pp. 195–208.

Bonaparte, J. F. (1991) Los vertebrados fosiles de la Formacion Rio Colorado, de la Ciudad de Neuquén y Cercanias, Cretacio superior, Argentina. *Revista del Museo Argentino de Ciencias Naturales 'Bernadino Rivadiva' E. Instituto Nacionales de Investigación de las Ciencias Naturales*, **4** (3), 17–123.

Bonde, N. (1987) *Moler – its Origin and its Fossils especially Fishes*. Skamal Skarrehage Molervaerk A/S, Nykobing Mors, 52 pp.

Brett-Surman, M. K. and Paul, G. S. (1985) A new family of bird-like dinosaurs linking Laurasia and Gondwanaland. *Journal of Vertebrate Paleontology*, **5**, 133–8.

Brodkorb, P. (1963) Catalogue of fossil birds: Part 1 (Archaeopterygiformes through Ardeiformes). *Bulletin of the Florida State Museum*, **7**, 179–293.

Brodkorb, P. (1964) Catalogue of fossil birds: Part 2 (Anseriformes through Galliformes). *Bulletin of the Florida State Museum*, **8**, 195–335.

Brodkorb, P. (1967) Catalogue of fossil birds: Part 3 (Ralliformes, Ichthyornithiformes, Charadriiformes). *Bulletin of the Florida State Museum*, **11**, 99–220.

Brodkorb, P. (1971) Catalogue of fossil birds: Part 4 (Columbiformes through Piciformes). *Bulletin of the Florida State Museum*, **15**, 163–266.

Brodkorb, P. (1978) Catalogue of fossil birds: Part 5 (Passeriformes). *Bulletin of the Florida State Museum*, **23**, 139–228.

Chatterjee, S. (1991) Cranial anatomy and relationships of a new Triassic bird from Texas. *Philosophical Transactions of the Royal Society, Series B*, **332**, 277–342.

Cheneval, J. (1984) Les oiseaux aquatiques (Gaviformes a Anseriformes) du gisement Aquitanien de Saint-Gérand-le-Puy (Allier, France): révision systématique. *Palaeovertebrata*, **14**, 33–115.

Cheneval, J. and Escuillié, F. (1992) New data concerning *Palaelodus ambiguus* (Aves: Phoenicopteriformes: Palaelodidae): ecological and evolutionary interpretations, in *Papers in Avian Paleontology Honoring Pierce Brodkorb* (ed. K. E. Campbell), *Science Series, Natural History Museum of Los Angeles County*, **36**, Los Angeles, California, pp. 209–24.

Chiappe, L. M. (1991) Cretaceous birds of Latin America. *Cretaceous Research*, **1991** (12), 55–63.

Chiappe, L. M. and Calvo, J. O. (1989a) El primer Enantiornithes (Aves) del Cretacio de Patagonia. *VI Jornados Argentino Paleontologia Vertebrados*, La Plata, May 1989, San Juan, Argentina, pp. 19–21.

Chiappe, L. M. and Calvo, J. O. (1989b) Nuevos hallazgos de Aves en el Cretacio de Patagonia. *Ciencia e Investigacion, Paleontologia*, **43**, 20–4.

Chure, D. (1991) (ed.) *The Morrison Times*, No. 1, Spring, 1991.

Cracraft, J. (1986) The origin and early diversification of birds. *Paleobiology*, **12**, 383–99.

Cracraft, J. (1988) The major clades of birds, in *Amphibians, Reptiles, Birds (The Phylogeny and Classification of the Tetrapods*: Vol. 1) (ed. M. J. Benton), Systematics Association Special Volume 35A, Clarendon Press, Oxford, pp. 333–61.

Crowe, T. M. and Short, L. L. (1992) A new gallinaceous bird from the Oligocene of Nebraska, with comments on the phylogenetic position of the Gallinuloididae, in *Papers in Avian Paleontology Honoring Pierce Brodkorb* (ed. K. E. Campbell), *Science Series, Natural History Museum of Los Angeles County*, **36**, Los Angeles, California, pp. 179–85.

Cuello, J. P. (1987) Lista de las aves fósiles de la región neotropical y de las islas antillanas. *Paula-Coutiana, Porto Alegre*, **2**, 3–79.

Fischer, K. (1978) Neue Reste des Riesenlaufvogels *Diatryma* aus dem Eozän des Geiseltales bei Halle (DDR). *Mitteilungen der Zoologischen Museum, Berlin*, **54**, 133–44.

Fischer, K. (1983) *Oligostrix rupelensis* n. gen., n. sp., eine neue Ureule (Protostrigidae, Strigiformes, Aves) aus dem marinen Mitteloligozän des Weisselsterbecken bei Leipzig (DDR). *Zeitschrift für Geologische Wissenschaften der DDR*, **11**, 483–7.

Fischer, K. (1987) Eulenreste (*Eoglaucidium pallas* nov. gen., nov. sp., Strigiformes, Aves) aus der mitteleozän Braunkohle des Geiseltales bei Halle (DDR). *Mitteilungen der Zoologischen Museum, Berlin*, **63**, 137–42.

Goedert, J. L. (1988) A new late Eocene species of Plotopteridae (Aves: Pelecaniformes) from northwestern Oregon. *Proceedings of the California Academy of Sciences*, **45**, 97–102.

Goedert, J. L. (1989) Giant Late Eocene marine birds (Pelecaniformes: Pelagornithdae) from northwestern Oregon. *Journal of Paleontology*, **63**, 939–44.

Graves, G. R. and Olson, S. L. (1987) *Chlorostilbon bracei* Lawrence, an extinct species of hummingbird from New Providence Island, Bahamas. *Auk*, **104**, 296–302.

Harrison, C. J. O. (1982) A new tiny raptor from the Lower Eocene of England. *Ardea*, **70**, 77–80.

Harrison, C. J. O. (1983) A new wader, Recurvirostridae (Charadriiformes) from the early Eocene of Portugal. *Ciencias da Terra (UNL)*, **7**, 9–16.

Harrison, C. J. O. (1984a) Further additions to the fossil birds of Sheppey: a new Falconid and three small rails. *Tertiary Research*, **5**, 179–87.

Harrison, C. J. O. (1984b) A revision of the fossil swifts (Vertebrata, Aves, Suborder Apodi), with descriptions of three new genera and two new species. *Mededelingen van de Werkgroep voor Tertiaire en Kwartaire Geologie*, **21**, 157–77.

Harrison, C. J. O. (1985) A bony-toothed bird (Odontopterygiformes) from the Palaeocene of England. *Tertiary Research*, **7**, 23–5.

Harrison, C. J. O. and Walker, C. A. (1979a) Birds of the British Lower Oligocene. *Tertiary Research Special Paper*, **5**, 29–42.

Harrison, C. J. O. and Walker, C. A. (1979b) Birds of the British Middle Eocene. *Tertiary Research, Special Paper*, **5**, 19–26.

Hesse, A. (1988a) Die Messelornithidae – eine neue Familie der Kranichartigen (Aves: Gruiformes: Rhynocheti) aus dem Tertiär Europas und Nordamerikas. *Journal für Ornithologie*, **129**, 83–95.

Hesse, A. (1988b) Taxonomie der Ordnung Gruiformes (Aves) nach osteologischen und morphologischen kriterien unter besonderer berücksichtigung der Messelornithidae Hesse, 1988. *Courier Forschungsinstitut Senckenberg*, **107**, 235–47.

Hou, L. (1989) A Middle Eocene bird from Sangequan, Xinjiang. *Vertebrata Palasiatica*, **27**, 65–70.

Hou, L. and Liu, Z. (1983) A new fossil bird from the Lower

Cretaceous of Gansu and the early evolution of birds. *Scientia Sinica (B)*, **27**, 1296–302.

Houde, P. (1986) Ostrich ancestors found in the Northern Hemisphere suggest new hypothesis of ratite origins. *Nature*, **324**, 563–5.

Houde, P. (1988) Paleognathous birds from the Early Tertiary of the Northern Hemisphere. *Publications of the Nuttall Ornithological Club*, **22**, 1–148.

Houde, P. and Haubold, H. (1987) *Palaeotis weigelti* restudied: a small Middle Eocene Ostrich (Aves: Struthioniformes). *Palaeovertebrata*, **17**, 27–42.

Houde, P. and Olson, S. L. (1989) Small arboreal nonpasserine birds from the Early Tertiary of Western North America. *Acta XIX Congressus Internationalis Ornithologici*, 2, Symposium 35, The Early Radiation of Birds. University of Ottawa Press, Ottawa, pp. 2030–6.

Houde, P. and Olson, S. L. (1992) A radiation of coly-like birds from the Eocene of North America (Aves: Sandcoleiformes New Order), in *Papers in Avian Paleontology Honoring Pierce Brodkorb* (ed. K. E. Campbell), *Science Series, Natural History Museum of Los Angeles County*, **36**, Los Angeles, California, pp. 137–60.

Jánossy, D. (1979) Plio-Pleistocene bird remains from the Carpathian Basin IV. Anseriformes, Gruiformes, Charadriiformes, Passerifomes. *Aquila*, **85**, 11–39.

Jánossy, D. (1983) Die Jungmittelpleistozäne Vogelfauna von Hunas (Hartmannshof), in *Die Höhlenruine Hunas bei Hartmannshof (Landkreis Nurnberger Land)* (eds F. Heller *et al.*), Quartär-Bibliothek, **4**, 265–88.

Jensen, J. A. (1981) A new oldest bird? *Anima*, **1981**, 33–9 [in Japanese].

Jensen, J. A. and Padian, K. (1989) Small dinosaurs and pterosaurs from the Uncompahgre Fauna (Brushy Basin Member, Morrison Formation: ?Tithonian), Late Jurassic, Western Colorado. *Journal of Paleontology*, **63**, 364–73,

Karkhu, A. A. (1988) A new family of Swift-like birds from the Paleogene of Europe. *Paleontologicheskii Zhurnal*, **1988** (3), 78–88.

Kessler, E. (1984) Lower Cretaceous birds from Cornet, Roumania, in *Third Symposium on Mesozoic Terrestrial Ecosystems* (eds W. E. Reif and F. Westphal), Tübingen 1984. Attempto Verlag, Tübingen, 259 pp.

Kessler, E. (1987) New contributions to the knowledge about the Lower and Upper Cretaceous birds from Roumania, in *Fourth Symposium on Mesozoic Terrestrial Ecosystems* (eds P. J. Currie and E. H. Koster), Occasional Paper of the Tyrrell Museum of Palaeontology, **3**, pp. 133–5.

Kessler, E. and Jurcsák, T. (1984) Fossil bird remains in the bauxite from Cornet (Roumania, Bihor County). *Travaux de la Muséum d'Histoire Naturelle Grigore Antipa (Bucarest)*, **25**, 393–401.

Kurochkin, E. N. (1985) Birds from the Pliocene of Central Asia. *Transactions of the Joint Soviet–Mongolian Palaeontological Expedition*, **26**, 1–120 [in Russian].

Kurochkin, E. N. (1988) Cretaceous birds from Mongolia and their significance for the study of bird phylogeny. *Transactions of the Joint Soviet–Mongolian Palaeontological Expedition*, **34**, 33–42 [in Russian].

Kurochkin, E. N. (1991) *Protoavis, Ambiortus* and other palaeornithological rarities. *Priroda*, **1991** (12), 43–53 [in Russian].

Martin, L. D. (1980) Foot-propelled diving birds of the Mesozoic. *Act. Congr. Int. Ornithol.*, **17**, 1237–42.

Martin L. D. (1992) The status of the late Paleocene birds *Gastornis* and *Remiornis*, in *Papers in Avian Palaeontology Honoring Pierce Brodkorb* (ed. K. E. Campbell), *Science Series, Natural History Museum of Los Angeles County*, **36**, Los Angeles, California, pp. 97–108.

McKee, J. W. A. (1985) A pseudodontorn (Pelecaniformes: Pelagornithidae) from the middle Pliocene of Hawera, Taranaki, New Zealand. *New Zealand Journal of Zoology*, **12**, 181–4.

Millener, P. R. and Worthy, T. H. (1991) Contribution to New Zealand's late Quaternary avifauna. II: *Dendroscansor decurvirostris*, a new genus and species of wren (Aves: Acantisittidae). *Journal of the Royal Society of New Zealand*, **21** (2), 179–200.

Mlíkovský, J. (1985) Towards a new classification of birds, in *Acta XVIII Congressus Internationalis Ornithologici* (eds V. D. Ilyichev and V. M. Gavrilov), Nauka Press, Moscow, pp. 1145–6.

Mlíkovský, J. (1987) A new alcid (Aves: Alcidae) from the Upper Oligocene of Austria. *Annalen des Naturhistorischen Museums in Wien A*, **88a**, 131–47 [with a note by J. Kovar].

Molnar, R. E. (1985) Alternatives to *Archaeopteryx*: a survey of proposed early or ancestral birds, in *The Beginnings of Birds*. Proceedings of the International Archaeopteryx Conference 1984 (eds M. K. Hecht, J. H. Ostrom and G. Viohl), Freunde des Jura-Museums Eichstätt, Willibaldsburg, 382 pp.

Molnar, R. E. (1986) An enantiornithine bird from the Lower Cretaceous of Queensland, Australia. *Nature*, **322**, 736–8.

Mourer-Chauviré, C. (1985) Les Todidae (Aves, Coraciiformes) des Phosphorites du Quercy (France). *Proceedings of the Koninklijke Nederlandse Akademie van Wetenschappen, B*, **88**, 407–14.

Mourer-Chauviré C. (1987) Les Strigiformes (Aves) des Phosphorites du Quercy (France): Systematique, Biostratigraphie et Paleobiogeographie. *Documents de la Laboratoire de Géologie, Lyon*, **99**, 89–135.

Mourer-Chauviré C. (1988a) Le Gisement du Bretou (Phosphorites du Quercy, Tarn-et-Garonne, France) et sa faune de vertébrés de L'Éocene Supérieur. II. Oiseaux. *Paleontographica Abteilung A*, **205**, 29–50.

Mourer-Chauviré C. (1988b) Les Caprimulgiformes et les Coraciiformes de L'Eocène et de L'Oligocène des phosphorites du Quercy et description de deux genres nouveaux de Podargidae et Nyctibiidae. *Acta XIX Congressus Internationalis Ornithologici*, 2, Symposium 35, The Early Radiation of Birds. University of Ottawa Press, Ottawa, pp. 2047–55.

Mourer-Chauviré, C. (1991) Les Horusornithidae *nov. fam.*, Accipitriformes (Aves) a articulation intertarsienne hyperflexible de l'Éocene du Quercy. *Geobios*, **1991**, 13, 183–92.

Mourer-Chauviré, C. (1992a) The Galliformes (Aves) of the Phosphorites du Quercy (France): Systematics and biostratigraphy, in *Papers in Avian Paleontology Honoring Pierce Brodkorb* (ed. K. E. Campbell), *Science Series, Natural History Museum of Los Angeles County*, **36**, Los Angeles, California, pp. 67–95.

Mourer-Chauviré, C. (1992b) The Messelornithidae (Aves: Gruiformes) from the Paleogene of France, in *Abstracts of the 3rd Symposium of the Society of Avian Paleontology and Evolution*, Frankfurt-am-Main, Germany.

Mourer-Chauviré, C. (1992c) Un ganga primitif (Aves, Columbiformes, Pteroclidae) de très grande taille dans le Paléogène des Phosphorites du Quercy (France) *Comptes Rendus Académie des Sciences, Paris*, **314**, *Série II*, 229–35.

Mourer-Chauviré, C., Moya, S. and Adrover, R. (1977) Les oiseaux des gisements Quaternaires de Majorque. *Nouvelles Archives du Museum Naturelle de Lyon, Supplement*, **15**, 61–4.

Nessov, L. A. (1984) Pterosaurs and birds from the late Cretaceous of Central Asia, *Palaeontologicheskii Zhurnal*, **1984** (1), 47–57.

Nessov, L. A. (1986) First discovery of the Late Cretaceous bird *Ichthyornis* in the Old World and some other bird bones from the Cretaceous and Palaeogene of Soviet Central Asia, in *Ecological and Faunistic Investigations of Birds* (ed. R. L. Potapov), *Proceedings of the Zoological Institute, Leningrad*, **147**, pp. 31–8 [in Russian].

Nessov, L. A. (1992) Mesozoic and Palaeogene birds of the USSR and their palaeoenvironments, in *Papers in Avian Paleontology Honoring Pierce Brodkorb* (ed. K. E. Campbell), *Science Series, Natural History Museum of Los Angeles County*, **36**, Los Angeles, California, pp. 465–78.

Nessov, L. A. and Yarkov, A. A. (1989) New Cretaceous–Palaeogene birds of the USSR and some remarks on the origin and evolution of the Class Aves, in *Faunistic and Ecological Studies of Eurasian Birds* (ed. R. L. Potapov), *Proceedings of the Zoological Institute, USSR Academy of Sciences*, **197**, pp. 78–97 [in Russian].

Nicholls, L. S. and Russell, A. P. (1990) Paleobiogeography of the Cretaceous Western Interior Seaway of North America: the vertebrate evidence. *Palaeogeography, Palaeoclimatology, Palaeoecology*, **79**, 149–69.

Olson, S. L. (1985) The fossil record of birds, in *Avian Biology*, Vol. 8 (eds D. S. Farner, J. R. King and K. C. Parkes), Academic Press, New York, pp. 79–256.

Olson, S. L. (1987) An Early Eocene oilbird from the Green River Formation of Wyoming (Caprimulgiformes: Steatornithidae). *Documents de la Laboratoire de Géologie de Lyon*, **99**, 57–69.

Olson, S. L. (1989) Aspects of global avifaunal dynamics during the Cenozoic. *Acta XIX Congressus Internationalis Ornithologici* **2**, Symposium 35, The Early Radiation of Birds. University of Ottawa Press, Ottawa, pp. 2023–9.

Olson, S. L. (1992a) *Neogaeornis wetzeli* Lambrecht, a Cretaceous loon from Chile (Aves: Gaviidae). *Journal of Vertebrate Paleontology*, **12** (1), 122–4.

Olson, S. L. (1992b) A new family of primitive landbirds from the Lower Eocene Green River Formation of Wyoming, in *Papers in Avian Paleontology Honoring Pierce Brodkorb* (ed. K. E. Campbell), *Science Series, Natural History Museum of Los Angeles County*, **36**, Los Angeles, California, pp. 127–36.

Olson, S. L. and Eller, K. G. (1989) A new species of painted snipe (Charadriiformes: Rostratulidae) from the Early Pliocene at Langebaanweg, southwestern Cape Province, South Africa. *Ostrich*, **60**, 118–21.

Olson, S. L. and Kurochkin, E. N. (1987) Fossil evidence of a tapaculo in the Quaternary of Cuba (Aves: Passeriformes: Scytalopodidae). *Proceedings of the Biological Society of Washington*, **100**, 353–7.

Olson, S. L. and Parris, D. C. (1987) The Cretaceous birds of New Jersey. *Smithsonian Contributions to Paleobiology*, **63**, 1–22.

Ostrom, J. H. (1986) The Jurassic 'bird' *Laopteryx priscus* re-examined. *Contributions to Geology, University of Wyoming, Special Paper* **3**, 11–19.

Ostrom, J. H. (1987) *Protoavis*, a Triassic bird? *Archaeopteryx*, **5**, 113–14.

Ostrom, J. H. (1991) The bird in the bush. *Nature*, **353**, p. 212.

Patterson, C. and Rich, P. V. (1987) The fossil history of the emus, *Dromaius* (Aves: Dromaiinae). *Records of the South Australian Museum*, **21**, 85–117.

Peters, S. (1987) *Juncitarsus merkeli* n. sp. stützt die Ableitung der Flamingos von Regenpfeifervögeln (Aves: Charadriiformes: Phoenicopteridae). *Courier Forschungs-Institut Senckenberg*, **97**, 141–55.

Peters, S. (1988) Die Messel Vogel – eine Landvogelfauna, in *Messel – Ein Schaufenster in die Geschichte der Erde und des Lebens* (eds S. Schaal and W. Ziegler), Vetlag Waldemar Kramer, Frankfurt am Main, pp. 137–51.

Peters, S. (1991) Zoogeographical relationships of the Eocene avifauna from Messel (Germany). *Acta XX Congressus Internationalis Ornithologici*, **1**, Symposium 6, The Methodology of Reconstructing the Past. New Zealand Ornithological Congress Trust Board, Christchurch, New Zealand, pp. 572–7.

Peters, S. (1992) A new species of owl (Aves: Strigiformes) from the middle Eocene Messel Oil Shale, in *Papers in Avian Paleontology Honoring Pierce Brodkorb* (ed. K. E. Campbell), *Science Series, Natural History Museum of Los Angeles County*, **36**, Los Angeles, California, pp. 161–9.

Plane, M. D. (1967) Stratigraphy and vertebrate fauna of the Otibanda Formation, New Guinea. *Bulletin of the Bureau of Mineral Resources, Australia*, **86**, 1–64.

Rao, C. and Sereno, P. (1990) Early evolution of the avian skeleton: new evidence from the Lower Cretaceous of China. *Journal of Vertebrate Palaeontology*, **10** (supplement to 3), 38A–39A.

Rasmussen, D. T., Olson, S. L. and Simons, E. L. (1987) Fossil birds from the Oligocene Jebel Qatrani Formation, Fayum Province, Egypt. *Smithsonian Contributions to Paleobiology*, **62**, 1–20.

Rautian, A. S. (1978) A unique bird feather from Jurassic lake sediments of the Karatau mountains. *Paleontologicheskii Zhurnal*, **1978** (4), 106–14 [in Russian].

Rich, P. V., Hou, L. H., Keiichi, O. *et al.* (1986) A review of the fossil birds of China, Japan and southeast Asia. *Geobios*, **19**, 755–72.

Sanz, J. L. and Bonaparte, J. F. (1992) A new order of birds (class Aves) from the Lower Cretaceous of Spain, in *Papers in Avian Paleontology Honoring Pierce Brodkorb* (ed. K. E. Campbell), *Science Series, Natural History Museum of Los Angeles County*, **36**, Los Angeles, California, pp. 39–49.

Sereno, P. C. and Rao, C. (1992) Early evolution of avian flight and perching: new evidence from the Lower Cretaceous of China. *Science*, **255**, 845–8.

Tambussi, C. P. (1987) Catalogo critico de los Tinamidae (Aves: Tinamiformes) fosiles de la Republica Argentina. *Ameghiniana*, **24**, 241–4.

Tambussi, C. P. and Tonni, E. P. (1988) Un Diomedeidae (Aves: Procellariiformes) del Eoceno Tardio de Antartida. *V Jornados Argentino Paleontologia Vertebrados*, May 1988, La Plata, Argentina, p. 34.

Tchernov, E. (1979) Quaternary fauna, in *The Quaternary of Israel* (ed. A. Horowitz), New York, pp. 259–90.

Tchernov, E. (1980) *The Pleistocene birds of 'Ubeidiya, Jordan Valley*. The Israel Academy of Science and Humanities, Jerusalem.

Tokaryk, T. T. and Harington, C. R. (in press) *Baptornis* sp. (Aves: Hesperornithiformes) from the Judith River Formation (Campanian) of Saskatchewan, Canada. *Journal of Paleontology*.

Tokaryk, T. T. and James, P. C. (1989) *Cimolopteryx* sp. (Aves, Charadriiformes) from the Frenchman Formation (Maastrichtian), Saskatchewan. *Canadian Journal of Earth Sciences*, **26**, 2729–30.

Tonni, E. P. and Tambussi, C. P. (1986) Las Aves del Cenozoico de la Republica Argentina. En: Simposio 'Evolución de los Vertebrados Cenozoicos'. *IV Congreso Argentino de Paleontologia y Biostratigrafia, Actas*, **2**, 131–42.

Tyrrell Museum of Paleontology (1988) Palaeoecology of Upper Cretaceous Judith River Formation at Dinosaur Provincial Park, Alberta, Canada. *Occasional Paper of the Tyrrell Museum of Paleontology*, **7**, 1–37.

Unwin, D. M. (1988) Extinction and survival in birds, in *Extinction and Survival in the Fossil Record* (ed. G. P. Larwood), Systematics Association Special Paper **34**. Clarendon Press, Oxford, pp. 295–318.

Wellnhofer, P. (1988) A new specimen of *Archaeopteryx*. *Science*, **240**, 1790–2.

Yeh, X. and Sun, B. (1989) Fossil rail and crow from Linqu, Shandong. *Zoological Research*, **10** (3), 177–84.

Zhou, Z., Jin, F. and Zhang, J. (1992) Preliminary report on a Mesozoic bird from Liaoning, China. *Chinese Science, Bulletin*, **37** (16), 1365–8.

41

MAMMALIA

R. K. Stucky and M. C. McKenna*

Since the original publication of the chapter on the Mammalia (Butler *et al.*, 1967), a number of critical references on the phylogenetic relations and stratigraphical distributions of mammals have been published. The classification adopted here is based on McKenna (1975), and a current revision of the classification of mammals now in progress by McKenna *et al.* (MS). Alternative classifications of families and orders are indicated in the references cited for each group. Recent reviews of the relationships of higher taxa include Novacek (1990), Novacek *et al.* (1988), and references cited therein. Classification is principally the work of MCM and data on ranges and localities were collated by RKS.

Correlations to stage level often are not possible for many taxa of mammals because of difficulties in direct correlation with standard marine sequences. Exceptions to this include ranges for mammals from the Mesozoic and those from Europe. For the many Cainozoic records, occurrences are recorded according to lower (l.), middle (m.) and upper (u.) epochal units, which, at current levels of mammalian biostratigraphical resolution, are generally equivalent from continent to continent. General correlations follow Savage and Russell (1983), with the following exceptions: South American faunal correlation is based on the work of MacFadden (1990) and Pascual and Jaureguizar (1990), and North American correlations follow works in Woodburne (1987), except that:

1. the Palaeocene-Eocene boundary is here considered to coincide with the Clarkforkian–Wasatchian boundary (Krishtalka *et al.*, 1987; this boundary may lie within the Wasatchian, Lucas, 1989; Wing, 1984; Rea *et al.*, 1990; Beard and Tabrum, 1991);
2. Chadronian faunas are considered to be late Eocene in age, equivalent in part to the Priabonian (Swisher and Prothero, 1990).

Major references from which first and last occurrences of mammalian families are derived, include the temporal and geographic summaries listed below. Data regarding the lithostratigraphical unit are often lacking for many of the reported taxa. In many cases, either the original locality name is reported or the 'faunal' unit from which the taxon occurs is listed. This especially applies to the European and South American record. Mesozoic mammals: Clemens *et al.* (1979); Lillegraven *et al.* (1979); Cainozoic mammals: Thenius (1959); de Paula Couto (1979); Savage and Russell (1983); Dawson and Krishtalka (1984); Pleistocene mammals: Kurtén and Anderson (1980); Anderson (1984); Martin and Klein (1984); Africa: Simons (1968); Maglio and Cooke (1978); Antarctica: Marshall *et al.* (1990); Asia: Li and Ting (1983); Russell and Zhai (1987); Australia: Woodburne *et al.* (1985); Archer *et al.* (1989); Marshall *et al.* (1990); Europe: Hooker and Insole (1980); Russell *et al.* (1982); Savage and Russell (1983); Remy *et al.* (1987); Schmidt-Kittler (1987b and references therein); Heissig (1987); North America: Webb (1984); Woodburne (1987 and references therein); Archibald *et al.* (1987); Lillegraven and McKenna (1986); Stucky (1990;, 1992); and South America: Ameghino (1906); Pascual *et al.* (1966); de Paula Couto (1979); Cifelli (1982); Marshall *et al.* (1983); Marshall *et al.* (1990).

Acknowledgements – The authors thank Mary R. Dawson (rodents), Ken Carpenter (Mesozoic mammals), Richard H. Tedford (carnivores and Australian mammals), Peter de Toledo (South American mammals), Rosendo Pascual (South American mammals), Michael Woodburne (Australian mammals) and Peter Robinson (African and North American mammals) for information regarding various groups of fossil mammals. Special thanks are extended to Marge Pries for corrections and additions to the manuscript.

*Order of authorship determined by the flip of a coin.

The Fossil Record 2. Edited by M. J. Benton. Published in 1993 by Chapman & Hall, London. ISBN 0 412 39380 8

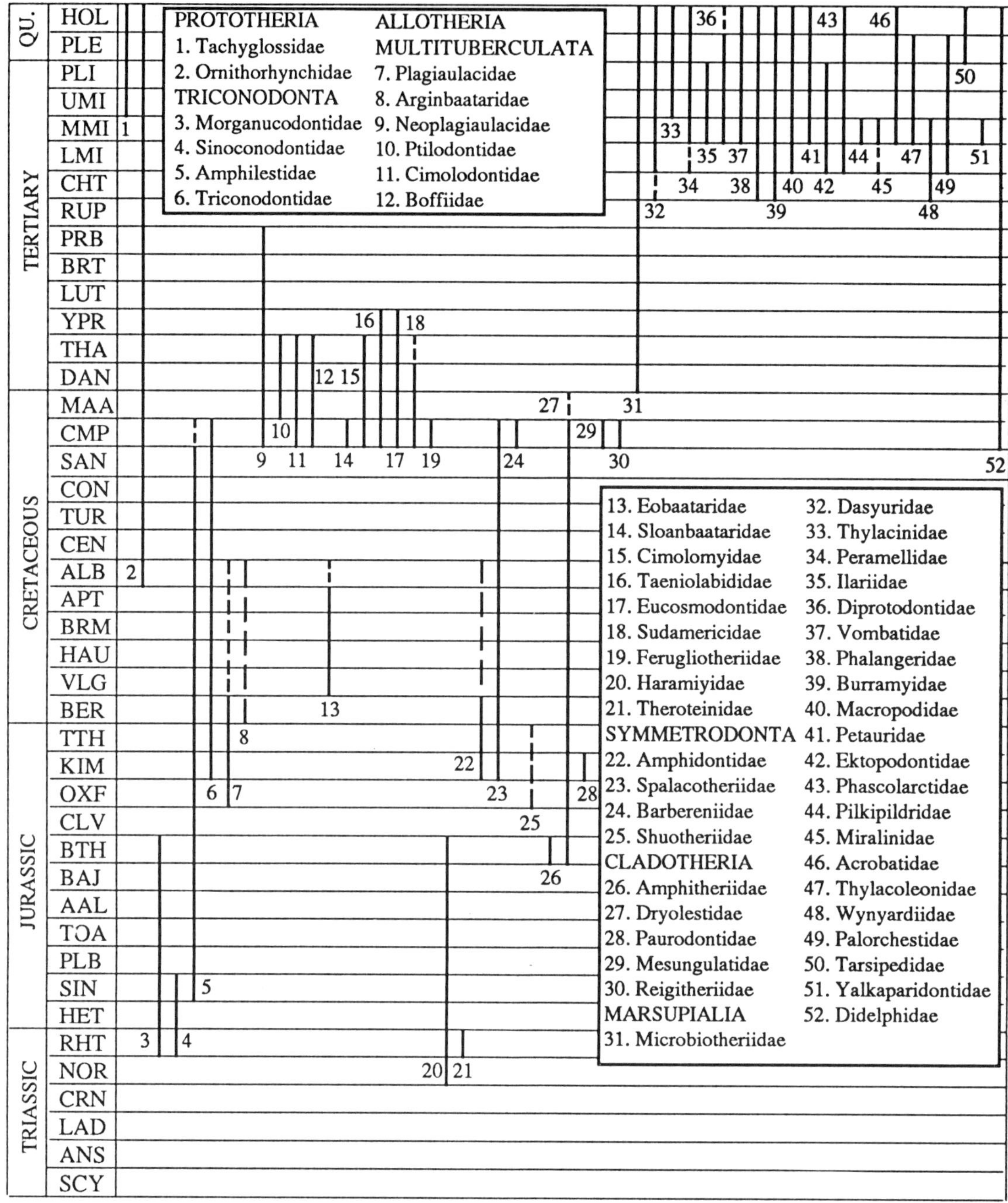

Fig. 41.1

Class MAMMALIA Linnaeus, 1758

Subclass PROTOTHERIA Gill, 1872, new rank, McKenna, this volume

Order TACHYGLOSSA Gill, 1872, new rank, McKenna, this volume

F. TACHYGLOSSIDAE Gill, 1872
T. (UMI)–Rec. Terr. (Fig. 41.1)

First: *Zaglossus robusta* Dun, 1895, Gulgong, Australia. **Extant**

Order PLATYPODA Gill, 1872, new rank, McKenna, this volume

F. ORNITHORHYNCHIDAE Burnett, 1830
K. (ALB)–Rec. Terr.

First: *Steropodon* Archer *et al.*, 1985, Griman Creek Formation, Australia. **Extant**
Intervening: Mio.

Subclass TRICONODONTA Osborn, 1888, new rank, McKenna, this volume

Order MORGANUCODONTA Kermack *et al.*, 1973, new rank, McKenna, this volume

F. MORGANUCODONTIDAE Kühne, 1958
Tr. (RHT)–J. (BTH) Terr.

First: *Eozostrodon* Parrington, 1941, Somerset, England, UK; ?Asia. *Morganucodon* Kühne, 1949, Glamorgan, Wales, UK; *Helvetiodon* Clemens, 1980, Hallau, Switzerland; *Brachyzostrodon* Sigogneau-Russell, 1983, Europe.
Last: *Wareolestes rex* Freeman, 1979, Forest Marble Formation, Oxfordshire, England, UK.

Order EUTRICONODONTA Kermack *et al.*, 1973, new rank, McKenna, this volume

F. SINOCONODONTIDAE Mills, 1971 Tr. (RHT)–J. (SIN) Terr.

First and Last: *Sinoconodon* Patterson and Olson, 1961, and *Lufengoconodon* Yang, 1982, both lower Lufeng Formation, Yunnan, China.
Last: *Megazostrodon* Crompton and Jenkins, 1968, Stormberg Group, South Africa.

F. AMPHILESTIDAE Osborn, 1888 J. (SIN)–K. (?CMP) Terr.

First: *Dinnetherium* Jenkins *et al.*, 1983, Kayenta Formation, Arizona, USA.
Last: *Guchinodon* Trofimov, 1978, and *Gobiconodon* Trofimov, 1978, both Mongolia.

F. TRICONODONTIDAE Marsh, 1887 J. (KIM)–K. (CMP) Terr.

First: *Triconodon* Owen, 1859, Lulworth Beds, Dorset, England, UK; *Priacodon* Marsh, 1887, Morrison Formation, Wyoming, USA; *Trioracodon* Simpson, 1928, Lulworth Beds, Dorset, England, UK; Morrison Formation, Wyoming, USA.
Last: *Alticonodon lindoei* Fox, 1970, Milk River Formation, Alberta, Canada; *Austrotriconodon* Bonaparte, 1986, Los Alamitos Formation, Argentina.

Subclass ALLOTHERIA Marsh, 1880

Infraclass MULTITUBERCULATA Cope, 1884, new rank, McKenna, this volume

Order PLAGIAULACIDA McKenna, 1971, new rank, McKenna, this volume

F. PLAGIAULACIDAE Gill, 1872 J. (OXF)–K. (BER/APT/ALB) Terr.

First: *Pseudobolodon* Hahn, 1977, *Paulchoffatia* Kühne, 1961, *Guimarotodon* Hahn, 1969, and *Henkelodon* Hahn, 1977, all Guimarota, Portugal; *Plagiaulax* Falconer, 1857, and *Bolodon* Owen, 1871, both Lulworth Beds, Dorset, England, UK; *Psalodon* Simpson, 1926, and *Ctenacodon* Marsh, 1879, both Morrison Formation, Wyoming, USA.
Last: *Paulchoffatia* Kühne, 1961, and *Bolodon* Owen, 1871, both Galve, Spain. A plagiaulacid has been reported from the Trinity Sands, Texas, USA (Clemens *et al.*, 1979).

Order CIMOLODONTA McKenna, 1975, new rank, McKenna, this volume

F. ARGINBAATARIDAE Hahn and Hahn, 1983 K. (l) Terr.

First and Last: *Arginbaatar* Trofimov, 1980, Khovboor Beds, Mongolia.

F. NEOPLAGIAULACIDAE Ameghino, 1890 K. (CMP)–T. (PRB) Terr.

First: *Cimexomys* Sloan and Van Valen, 1965, and *Mesodma* Jepsen, 1940, both Milk River Formation, Alberta, Canada.
Last: ?*Ectypodus* Matthew and Granger, 1921, White River Formation, Wyoming and Nebraska, USA; Calf Creek, Saskatchewan, Canada (Krishtalka *et al.*, 1982).

F. PTILODONTIDAE Gregory and Simpson, 1926 K. (MAA)–T. (THA) Terr.

First: *Kimbetohia* Simpson, 1936, Kirtland and Fruitland Formations, New Mexico, USA.
Last: *Prochetodon* Jepsen, 1940, Willwood Formation, Wyoming, USA (Krause, 1987).

F. CIMOLODONTIDAE Marsh, 1889 K. (CMP)–T. (THA) Terr.

First: *Cimolodon* Marsh, 1889, Milk River Formation, Alberta, Canada.
Last: *Anconodon* Jepsen, 1940, Paskapoo Formation, Alberta, Canada; *Liotomus* Cope, 1884, MP6, Cernay, France (Godinot, 1987).

F. BOFFIIDAE Hahn and Hahn, 1983 T. (THA) Terr.

First and Last: *Boffius* Vianey-Liaud, 1973, MP1–5, Hainin, Belgium (Godinot, 1987).

F. EOBAATARIDAE Kielan-Jaworowska *et al.*, 1987 K. (VLG–APT/ALB) Terr.

First: *Loxaulax* Simpson, 1928, Wadhurst Clay, Sussex, England, UK.
Last: *Eobaatar* Kielan-Jaworowska *et al.*, 1987, Khovboor Beds, People's Republic of Mongolia.

F. SLOANBAATARIDAE Kielan-Jaworowska, 1974 K. (CMP) Terr.

First and Last: *Sloanbaatar* Kielan-Jaworowska, 1974, Djadokhta Formation, Mongolia.

F. CIMOLOMYIDAE Marsh, 1889 K. (CMP)–T. (THA) Terr.

First: *Cimolomys* Marsh, 1889, Oldman Formation, Alberta, Canada; *Meniscoessus* Cope, 1882, Milk River Formation, Alberta, Canada.
Last: *Hainina* Vianey-Liaud, 1979, MP6, Cernay, France.

F. TAENIOLABIDIDAE Granger and Simpson, 1929 K. (CMP)–T. (YPR) Terr.

First: *Kamptobaatar* Kielan-Jaworowska, 1970, and *Catopsalis* Cope, 1882, both Djadokhta Formation, Mongolia.
Last: *Prionessus* Matthew and Granger, 1925, Bayan Ulan Formation, Inner Mongolia; *Lambdopsalis* Chow and Qi, 1978, and *Sphenopsalis* Matthew *et al.*, 1928, both Nomogen Formation, Inner Mongolia.

F. EUCOSMODONTIDAE Jepsen, 1940 K. (CMP)–T. (YPR) Terr.

First: *Tugrigbaatar* Kielan-Jaworowska and Dashzeveg, 1978, Toogreeg, Mongolia; *Chulsanbataar* Kielan-Jaworowska, 1974, and *Nemegtbataar* Kielan-Jaworowska, 1974, both Barun Goyot Formation, Mongolia; *Bulganbataar* Kielan-Jaworowska, 1974, Djadokhta Formation, Mongolia.
Last: *Neoliotomus ultimus* Jepsen, 1930, Willwood Formation, Wyoming, USA.

F. SUDAMERICIDAE Scillato-Yane and Pascual, 1984 K. (CMP)–T. (DAN/THA) Terr.

First: *Gondwanatherium* Bonaparte, 1986, Los Alamitos Formation, Argentina.
Last: *Sudamerica* Scillato-Yane and Pascual, 1984, Banco Negro, Argentina.

F. FERUGLIOTHERIIDAE Bonaparte, 1986 K. (CMP) Terr.

First and Last: *Feruglio therium* Bonaparte, 1986, Los Alamitos Formation, Argentina.

Infraclass HARAMYOIDEA Hahn, 1973, new rank, McKenna, this volume

F. HARAMIYIDAE Simpson, 1947 Tr. (NOR)–J. (BTH) Terr.

First and Last: *Haramiya* Simpson, 1947, Holwell Quarry, Somerset, England, UK. See also Lillegraven *et al.* (1979).

ALLOTHERIA *incertae sedis*

F. THEROTEINIDAE Sigogneau-Russell *et al.*, 1986 Tr. (RHT) Terr.

First and Last: *Theroteinus* Sigogneau-Russell *et al.*, 1986, Saint-Nicolas-de-Port, France.

Subclass TRECHNOTHERIA McKenna, 1975, new rank, McKenna, this volume

Infraclass SYMMETRODONTA Simpson, 1925, new rank, McKenna, this volume

Order AMPHIDONTOIDEA Prothero, 1981, new rank, McKenna, this volume

F. AMPHIDONTIDAE Simpson, 1925 J. (KIM)–K. (l.) Terr.

First: *Amphidon* Simpson, 1925, Morrison Formation, Wyoming, USA.
Last: *Manchurodon* Yabe and Shikama, 1938, Sakusiyo Coal, Manchuria. *Gobiodon* Trofimov, 1980, Khovboor Beds, Mongolia.

Order SPALACOTHERIOIDEA Prothero, 1981, new rank, McKenna, this volume

F. SPALACOTHERIIDAE Marsh, 1887 J. (KIM)–K. (CMP) Terr.

First: *Spalacotherium* Owen, 1854, Lulworth Beds, England, UK.
Last: *Symmetrodontoides* Fox, 1976, Milk River Formation, Alberta, Canada; ?*Brandonia* Bonaparte, 1990, Los Alamitos Formation, Argentina.

F. BARBERENIIDAE Bonaparte, 1990 K. (CMP) Terr.

First and Last: *Barberenia* Bonaparte, 1990, and *Quirogatherium* Bonaparte, 1990, both Los Alamitos Formation, Argentina.

SYMMETRODONTA *incertae sedis*

F. SHUOTHERIIDAE Chow and Rich, 1982 J. (u.) Terr.

First and Last: *Shuotherium* Chow and Rich, 1982, Chongging Group, northern Sichuan, China.

Infraclass CLADOTHERIA McKenna, 1975, new rank, McKenna, this volume

Legion DRYOLESTOIDEA Butler, 1939, new rank, McKenna, this volume

Order AMPHITHERIIDA Prothero, 1981, new rank, McKenna, this volume

F. AMPHITHERIIDAE Owen, 1846 J. (BTH) Terr.

First and Last: *Amphitherium* De Blainville, 1838, Stonesfield Slate, Oxfordshire, England, UK

Order DRYOLESTIDA Prothero, 1981, new rank, McKenna, this volume

F. DRYOLESTIDAE Marsh, 1879 J. (BTH)–K. (CMP/MAA) Terr.

First: Upper Bathonian, Forest Marble, Oxfordshire, England, UK (Freeman, 1976).
Last: *Leonardus* Bonaparte, 1990, and *Groebertherium* Bonaparte, 1986, both Los Alamitos Formation, Argentina. Lillegraven and McKenna (1986) report a dryolestid from CMP or MAA strata of the Mesa Verde Formation, Wyoming, USA.

F. PAURODONTIDAE Marsh, 1887 J. (KIM) Terr.

First and Last: *Paurodon* Marsh, 1887, *Archaeotrigon* Simpson, 1927, *Tathiodon* Simpson, 1927, *Pelicopsis* Simpson, 1927, *Araeodon* Simpson, 1937, and *Foxraptor* Bakker, 1990, all Morrison Formation, Wyoming, USA.

F. MESUNGULATIDAE Bonaparte, 1986 K. (CMP) Terr.

First and Last: *Mesungulatum* Bonaparte and Soria, 1985, Los Alamitos Formation, Argentina.

F. REIGITHERIIDAE Bonaparte, 1990 K. (CMP) Terr.

First and Last: *Reigitherium* Bonaparte, 1990, Los Alamitos Formation, Argentina.

Legion THERIA Parker and Haswell, 1897, new rank, McKenna, this volume

Supercohort MARSUPIALIA Illiger, 1811, new rank, McKenna, this volume

An alternative classification of the Marsupialia is that of Marshall *et al.* (1990). The earliest known marsupials may be from the Straight Cliffs Formation, Utah, of Turonian age (Cifelli, 1990). The ages of marsupials from the Etadunna Formation and Carl Creek Limestone are currently not precisely known (compare Woodburne *et al.*, 1985; Archer *et al.*, 1989). According to Woodburne (pers. comm. to RKS, 1991), marsupials from the Etadunna Formation are probably early Miocene in age, and those from the Carl Creek Limestone are probably middle Miocene.

Cohort AUSTRALIDELPHIA Szalay, 1982

Magnorder MICROBIOTHERIA Ameghino, 1889, new rank, McKenna, this volume

F. MICROBIOTHERIIDAE Ameghino, 1887 T. (DAN)–Rec. Terr.

First: *Khasia* Marshall and de Muizon, 1988, El Molino Formation, Bolivia. **Extant**

Magnorder EOMETATHERIA Simpson, 1970, new rank, McKenna, this volume

Superorder DASYUROMORPHIA Gill, 1872, new rank, McKenna, this volume

F. DASYURIDAE Goldfuss, 1820 T. (CHT/?LMI)–Rec. Terr.

First: ?Upper Oligocene: Tedford *et al.* (1975) report a dasyurid from the Geilston Travertine, Tasmania. Lower Miocene: *Ankotarinja* Archer, 1976, *Dasylurinjia* Archer, 1982, *Keeuna* Archer, 1976, all from Etadunna Formation, South Australia. **Extant**

F. THYLACINIDAE Bonaparte, 1838 T. (UMI)–Rec. Terr.

First: *Thylacinus* Temminck, 1827, New Guinea; Waite Formation, Northern Territory, Australia. **Extant**

Superorder SYNDACTYLI Gill, 1871, new rank, McKenna, this volume

Order PERAMELINA Gray, 1825, new rank, McKenna, this volume

F. PERAMELIDAE Gray, 1825 T. (?LMI/MMI/PLI)–Rec. Terr.

First: ?*Perameles* Geoffrey, 1804, Allingham Formation, Queensland; Wipajiri Formation, South Australia. Woodburne *et al.* (1985) report a taxon near *Microperorctes* Stein, 1932 from the Etadunna Formation, South Australia. **Extant**

Order DIPROTODONTIA Owen, 1866

F. ILARIIDAE Tedford and Woodburne, 1987 T. (MMI–PLI) Terr.

First: *Ilaria* Tedford and Woodburne, 1987 and *Kuterintja* Pledge, 1987, both Australia.
Last: *Koobor* Archer, 1976, Chinklea Sands, Queensland, Australia.

F. DIPROTODONTIDAE Gill, 1872 T. (MMI)–Q. (PLE/HOL) Terr.

First: *Bematherium* Tedford, 1967, and *Neohelos* Stirton, 1967, both Carl Creek Limestone, Queensland, Australia; *Raemeotherium* Rich *et al.*, 1978, Namba Formation, South Australia.
Last: *Zygomaturus* Owen, 1858, Tasmania; *Nototherium* Owen, 1845, Australia; *Diprotodon* Owen, 1838, King Island, Tasmania; Calabonna, Australia (possibly Rec.; Butler *et al.*, 1967); *Stenomerus* DeVis, 1907, Tasmania, Australia.

F. VOMBATIDAE Burnett, 1930 T. (MMI)–Rec. Terr.

First: *Rhizophascolonus* Stirton *et al.*, 1967, Wipajiri Formation, South Australia. **Extant**

F. PHALANGERIDAE Thomas, 1888 T. (CHT)–Rec. Terr.

First: Tedford *et al.* (1975) report a phalangerid from the Geilston Travertine, Tasmania. **Extant**

F. BURRAMYIDAE Broom, 1898 T. (CHT)–Rec. Terr.

First: Tedford *et al.* (1975) report a possible burramyid from the Geilston Travertine, Tasmania. **Extant**

F. MACROPODIDAE Gray, 1821 T. (LMI)–Rec. Terr.

First: *Nambaroo* Flannery and Rich, 1986, Namba Formation, South Australia; *Ekaltadeta* Archer and Flannery, 1985, Australia. *Purtia* Case, 1984, Etadunna Formation, South Australia. **Extant**

F. PETAURIDAE Gill, 1872 T. (MMI)–Rec. Terr.

First: *Marlu* Woodburne *et al.*, 1987, Etadunna Formation, South Australia; *Paljara* Woodburne *et al.*, 1987, and *Pildra* Woodburne *et al.*, 1987, both from Carl Creek Limestone, Queensland, Australia, Also, ?Lower or Middle Miocene petaurids are reported from the Etadunna Formation, South Australia (Woodburne *et al.*, 1985), and Carl Creek Limestone, Queensland (Archer *et al.*, 1989). **Extant**

F. EKTOPODONTIDAE Stirton *et al.*, 1967 T. (LMI–PLI) Terr.

First: *Chunia* Woodburne and Clemens, 1985, Etadunna Formation, South Australia.
Last: *Darcius* Rich, 1985, Grange Burn Formation, lower PLI, Victoria, Australia.

F. PHASCOLARCTIDAE Owen, 1839 T. (LMI)–Rec. Terr.

First: *Madakoala* Woodburne *et al.*, 1987, and *Perikoala* Stirton, 1957, both from Etadunna Formation, South Australia. **Extant**

F. PILKIPILDRIDAE Archer *et al.*, 1987 T. (MMI) Terr.

First and Last: ?*Djilgaringa* Archer *et al.*, 1987, Carl Creek Limestone, Queensland. *Pilkipildra* Archer *et al.*, 1987, Australia.

F. MIRALINIDAE Woodburne *et al.*, 1987 T. (?LMI)–T. (MMI) Terr.

First and Last: *Miralina* Woodburne *et al.*, 1987, Etadunna Formation, South Australia.

F. ACROBATIDAE Aplin and Archer, 1987 T. (MMI)–Rec. Terr.

First: cf. *Acrobates* Desmarest, 1817, Carl Creek Limestone, Queensland, Australia (Archer *et al.*, 1989). **Extant**

F. THYLACOLEONIDAE Gill, 1872 T. (MMI)–Q. (PLE) Terr.

First: *Prisciléo* Rauscher, 1987, Australia; *Wakaleo* Clemens and Plane, 1974, Wipajiri Formation, South Australia; Camfield Beds, Northern Territory; Carl Creek Limestone, Queensland, Australia.
Last: *Thylacoleo* Owen, 1858, Australia; Tasmania.

F. WYNYARDIIDAE Osgood, 1921 T. (CHT–MMI) Terr.

First: *Wynyardia* Spencer, 1900, Table Cape, Tasmania, Australia.
Last: *Muramura* Pledge, 1987, Etadunna Formation, South Australia; *Namilamedeta* Rich and Archer, 1979, Namba Formation, South Australia.

F. PALORCHESTIDAE Tate, 1948 T. (LMI)–Q. (PLE) Terr.

First: *Ngapakaldia* Stirton, 1967, and *Pitikantia* Stirton, 1967, both Etadunna Formation, South Australia.

Last: *Palorchestes* Owen, 1873, Upper PLE, Tasmania, Australia.

F. TARSIPEDIDAE Gervais and Verreaux, 1842 Q. (PLE)–Rec. Terr.

First: *Tarsipes* Gervais and Verreaux, 1842, Australia. **Extant**

Cohort EOMETATHERIA *incertae sedis*

Order NOTORYCTEMORPHIA Kirsch, 1977

F. NOTORYCTIDAE Ogilby, 1892 **Extant** Terr.

Order YALKAPARIDONTIA Archer *et al.*, 1988

F. YALKAPARIDONTIDAE Archer *et al.*, 1988 T. (MMI)–Terr.

First and Last: *Yalkaparidon* Archer *et al.*, 1988, Carl Creek Limestone, Queensland, Australia.

Cohort AMERIDELPHIA Szalay, 1982

Order DIDELPHIPHORMES Szalay, 1982

F. DIDELPHIDAE Gray, 1821 K. (CMP)–Rec. Terr.

First: *Albertatherium* Fox, 1971, and *Alphadon* Simpson, 1927, both Milk River Formation, Alberta, Canada; *Protalphadon* Cifelli, 1990, Wahweap Formation, Utah, USA; *Incadelphys* Marshall and de Muizon, 1988, *Mizquedelphys* Marshall and de Muizon, 1988, *Pucadelphys* Marshall and de Muizon, 1988, *Jaskhadelphys* Marshall and de Muizon, 1988, and *Tiulordia* Marshall and de Muizon, 1988, all El Molino Formation, Bolivia. **Extant**

F. SPARASSOCYNIDAE Reig, 1958 T. (UMI–PLI) Terr. (Fig. 41.2)

First and Last: *Sparassocynus* Mercerat, 1898, Huayquerian, UMI–Upper PLI, Argentina.

Order PAUCITUBERCULATA Ameghino, 1894

F. CAENOLESTIDAE Trouessart, 1898 T. (CHT/LMI)–Rec. Terr.

First: *Pseudhalmarhiphus* Ameghino, 1903, *Acdestis* Ameghino, 1887, *Palaeothentes* Ameghino, 1887, all Deseadan, Argentina. **Extant**

F. POLYDOLOPIDAE Ameghino, 1897 T. (THA–RUP) Terr.

First: *Epidolops* de Paula Couto, 1952, Itaboraí, Brazil.
Last: *Antarctodolops* Woodburne and Zinsmeister, 1984, and *Eurydolops* Woodburne and Chaney, 1988, both Seymour Island, Antarctica. Pascual and Bond (1981) report Oligocene polydolopids from Salla, Bolivia.

F. PREPIDOLOPIDAE Pascual, 1980 T. (YPR–PRB/RUP) Terr.

First and Last: *Prepidolops* Pascual, 1980, Lumbrera and Pozuelos formations, Argentina.

F. BONAPARTHERIIDAE Pascual, 1980 T. (YPR) Terr.

First and Last: *Bonapartherium* Pascual, 1980, Casamayoran, Argentina.

F. ARGYROLAGIDAE Ameghino, 1904 T. (CHT/LMI–PLI) Terr.

First: *Proargyrolagus* Wolff, 1984, Deseadan, Argentina.
Last: *Argyrolagus* Ameghino, 1904, Chapadamalalan, Upper PLI, Argentina.

F. GROEBERIIDAE Patterson, 1952 T. (RUP) Terr.

First and Last: *Groeberia* Patterson, 1952, Divisadero Largo Formation, Argentina.

F. PATAGONIIDAE Pascual and Carlini, 1987 T. (LMI) Terr.

First and Last: *Patagonia* Pascual and Carlini, 1987, Patagonia, Argentina.

F. GLASBIIDAE Clemens, 1966 K. (MAA) Terr.

First and Last: *Glasbius* Clemens, 1966, Lance Creek Formation, Wyoming, USA.

F. CAROLOAMEGHINIIDAE Ameghino, 1901 T. (DAN–YPR) Terr.

First: *Roberthoffstatteria* Marshall *et al.*, 1983, El Molino Formation, Bolivia.
Last: *Caroloameghinia* Ameghino, 1901, Argentina.

Order SPARASSODONTA Ameghino, 1894

F. BORHYAENIDAE Ameghino, 1894 T. (THA–PLI) Terr.

First: *Allqokirus* Marshall and de Muizon, 1988, El Molino Formation, Bolivia.
Last: *Eutemnodus* Burmeister, 1885, *Parahyaenodon* Ameghino, 1904, and *Notocynus* Mercerat, 1891, all Montehermosan, Lower PLI, Argentina; *Thylacosmilus* Riggs, 1933, Chapadmalalan, Argentina (Pascual *et al.*, 1966).

F. KOLLPANIIDAE Marshall *et al.*, 1990 T. (DAN–THA) Terr.

First: *Kollpania* Marshall and de Muizon, 1988, El Molino Formation, Bolivia.
Last: *Zeusdelphys* Marshall, 1987, Itaboraí, Brazil.

F. HONDADELPHIDAE Marshall *et al.*, 1990 T. (MMI) Terr.

First and Last: *Hondadelphys* Marshall, 1976, Santacrucian, Honda Group, Columbia.

AMERIDELPHIA *incertae sedis*

Superfamily PROTODIDELPHINAE Marshall, 1987 T. (THA) Terr.

First and Last: *Bobbschaefferia* Paula Couto, 1970, *Guggenheimia* Paula Couto, 1952, *Procaroloameghinia* Marshall, 1982, *Protodelphis* Paula Couto, 1952, *Robertbutleria* Marshall, 1987, all Itaboraí, Brazil.

MARSUPIALIA *incertae sedis*

F. PEDIOMYIDAE Simpson, 1927 K. (CMP–MAA) Terr.

First: *Aquiladelphis* Fox, 1971, *Iqualadelphis* Fox, 1987, and *Pediomys* Marsh, 1889, all Milk River Formation, Alberta, Canada.

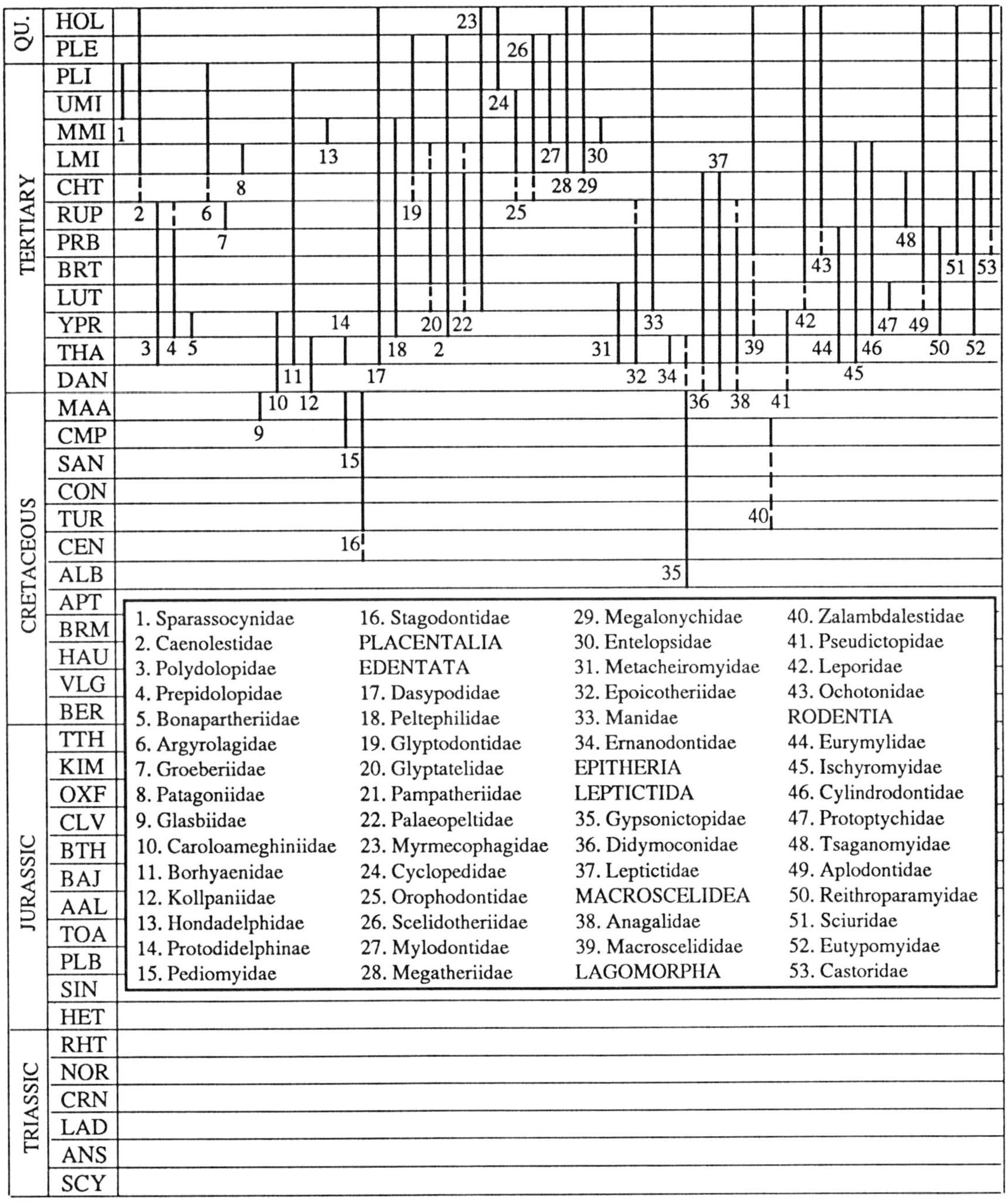

QU.
HOL
PLE
TERTIARY
PLI
UMI
MMI
LMI
CHT
RUP
PRB
BRT
LUT
YPR
THA
DAN
CRETACEOUS
MAA
CMP
SAN
CON
TUR
CEN
ALB
APT
BRM
HAU
VLG
BER
JURASSIC
TTH
KIM
OXF
CLV
BTH
BAJ
AAL
TOA
PLB
SIN
HET
TRIASSIC
RHT
NOR
CRN
LAD
ANS
SCY
1. Sparassocynidae
2. Caenolestidae
3. Polydolopidae
4. Prepidolopidae
5. Bonapartheriidae
6. Argyrolagidae
7. Groeberiidae
8. Patagoniidae
9. Glasbiidae
10. Caroloameghiniidae
11. Borhyaenidae
12. Kollpaniidae
13. Hondadelphidae
14. Protodidelphinae
15. Pediomyidae
16. Stagodontidae
PLACENTALIA
EDENTATA
17. Dasypodidae
18. Peltephilidae
19. Glyptodontidae
20. Glyptatelidae
21. Pampatheriidae
22. Palaeopeltidae
23. Myrmecophagidae
24. Cyclopedidae
25. Orophodontidae
26. Scelidotheriidae
27. Mylodontidae
28. Megatheriidae
29. Megalonychidae
30. Entelopsidae
31. Metacheiromyidae
32. Epoicotheriidae
33. Manidae
34. Ernanodontidae
EPITHERIA
LEPTICTIDA
35. Gypsonictopidae
36. Didymoconidae
37. Leptictidae
MACROSCELIDEA
38. Anagalidae
39. Macroscelididae
LAGOMORPHA
40. Zalambdalestidae
41. Pseudictopidae
42. Leporidae
43. Ochotonidae
RODENTIA
44. Eurymylidae
45. Ischyromyidae
46. Cylindrodontidae
47. Protoptychidae
48. Tsaganomyidae
49. Aplodontidae
50. Reithroparamyidae
51. Sciuridae
52. Eutypomyidae
53. Castoridae

Fig. 41.2

Last: *Pediomys* Marsh, 1889, Kirtland and Fruitland Formations, New Mexico, USA; Scollard Formation, Alberta, Canada; Frenchman Formation, Saskatchewan, Canada; Hell Creek Formation, Montana, Lance Creek Formation, Wyoming, and North Horn Formation, Utah, USA.

F. STAGODONTIDAE Marsh, 1889
K. (???TUR/CEN–MAA) Terr.

First: *Pariodens* Cifelli and Eaton, 1987, Dakota Formation, Utah, USA. Cifelli (1990) questionably refers a premolar to the stagodontids from the Straight Cliffs Formation, Utah, USA, of Turonian age.

Last: *Didelphodon* Marsh, 1889, Lance Creek Formation, Wyoming and Hell Creek Formation, Montana, USA; Scollard Member, Alberta, Canada.

Supercohort PLACENTALIA Owen, 1837

Cohort EDENTATA Cuvier, 1798

Order CINGULATA Illiger, 1811

F. DASYPODIDAE Gray, 1921
T. (THA)–Rec. Terr.

First: *Prostegotherium* Ameghino, 1902, Itaboraí, Brazil (Marshall *et al.*, 1983). **Extant**

F. PELTEPHILIDAE Ameghino, 1894
T. (YPR–MMI) Terr.

First: Marshall *et al.* (1983) report Peltephilinae indet. from the Lower Eocene Casamayoran, Argentina.

Last: *Epipeltephilus* Ameghino, 1904, Chasicoan, Argentina (Pascual *et al.*, 1966).

F. GLYPTODONTIDAE Burmeister, 1879
T. (CHT/LMI)–Q. (PLE) Terr.

First: Propalaehoplophorinae Ameghino, 1891, reported by Marshall *et al.* (1983) from Deseadan, Argentina.
Last: *Hoplophorus* Lund, 1838, Minas Geraes, Brazil; Buenos Aires, Argentina; *Lomaphorus* Ameghino, 1889, Buenos Aires, Argentina; Brazil; *Plaxhaplous* Ameghino, 1884, El Paso de la Virgen, Argentina; *Doedicurus* Burmeister, 1874, Brazil; Buenos Aires, Argentina; *Glyptodon* Owen, 1839, Brazil; Buenos Aires, Argentina; Venezuela; Uruguay; *Heteroglyptodon* Roselli, 1976, Uruguay; *Glyptotherium* Osborn, 1903, Texas, USA and Mexico; *Sclerocaluptus* Ameghino, 1891, Santa Isabel, Argentina; *Neothoracophorus* Ameghino, 1889, Buenos Aires, Argentina, all Upper PLE (Pascual *et al.*, 1966; Kurtén and Anderson, 1980).

F. GLYPTATELIDAE Castellanos, 1932
T. (LUT/BRT–CHT/LMI) Terr.

First: *Glyptatelus* Ameghino, 1897, Mustersan, Argentina.
Last: *Glyptatelus* Ameghino, 1897, and *Clypeotherium* Scillato-Yane, 1977, both Deseadan, Argentina.

F. PAMPATHERIIDAE Paula Couto, 1954
T. (YPR)–Q. (PLE) Terr.

First: *Machlydotherium* Ameghino, 1902, Casamayoran, Argentina.
Last: Upper PLE: *Pampatherium* Ameghino, 1875, Buenos Aires, Argentina; Toca de Esperanca, Bahia, and Arroio Touro Passo, Brazil. Lower PLE: *Holmesina* Simpson, 1930, Uruguay; Coleman 2A, Florida; Texas; Oklahoma; Kansas, USA (Kurtén and Anderson, 1980).

F. PALAEOPELTIDAE Ameghino, 1895
T. (LUT/BRT–CHT/LMI) Terr.

First and Last: *Palaeopeltis* Ameghino, 1895, Mustersan and Deseadan, Argentina.

Order PILOSA Flower, 1883

F. MYRMECOPHAGIDAE Gray, 1825
T. (LUT)–Rec. Terr.

First: *Eurotamandua* Storch, 1981, MP11, Messel, Germany. **Extant**

F. CYCLOPEDIDAE Pocock, 1924
T. (PLI)–Rec. Terr.

First: *Palaeomyrmidon* Rovereto, 1914, Tunuyan Formation, Lower PLI, Argentina. **Extant**

F. OROPHODONTIDAE Ameghino, 1895
T. (CHT/LMI–UMI) Terr.

First: *Orophodon* Ameghino, 1895, and *Octodontotherium* Ameghino, 1895, both Deseadan, Argentina.
Last: *Octomylodon* Ameghino, 1904, Chasicoan, Argentina.

F. SCELIDOTHERIIDAE Ameghino, 1899
T. (CHT/LMI)–Q. (PLE) Terr.

First: *Chubutherium* Cattoi, 1962, Deseadan, Argentina (Marshall *et al.*, 1983).
Last: *Scelidotherium* Owen, 1839, Salado-Indio Rico l.f., Upper PLE, Argentina; Chile; Uruguay; Bolivia; Peru; Brazil.

F. MYLODONTIDAE Gill, 1872
T. (MMI)–Q. (PLE) Terr.

First: *Glossotheriopsis* Scillato-Yane, 1976, Colloncurense, Rio Negro, Argentina.
Last: *Glossotherium* Owen, 1839 (including *Paramylodon* Brown, 1903) South and pan-North America (see Anderson, 1984); *Leston* Gervais, 1865, Buenos Aires, Argentina; *Lestodontidion* Roselli, 1976, Uruguay, all Upper PLE.

F. BRADYPODIDAE Gray, 1821 **Extant** Terr.

F. MEGATHERIIDAE Gray, 1821
T. (LMI)–Rec. Terr.

First: *Proprepotherium* Ameghino, 1904, Colhuehuapian, Argentina. **Extant**

F. MEGALONYCHIDAE Ameghino, 1889
T. (LMI)–Rec. Terr.

First: *Proschismotherium* Ameghino, 1887, Colhuehuapian, Argentina. **Extant**

F. ENTELOPSIDAE Ameghino, 1889
T. (MMI) Terr.

First and Last: *Entelops* Ameghino, 1887, and *Delotherium* Ameghino, 1889, both Santacrucian, Argentina.

Order PHOLIDOTA Weber, 1904

F. METACHEIROMYIDAE Wortman, 1903
T. (THA–LUT) Terr.

First: *Propalaeanodon* Rose, 1979, Polecat Bench Formation, Wyoming, USA.
Last: *Metacheiromys* Wortman, 1903, Bridger Formation, Wyoming, USA.

F. EPOICOTHERIIDAE Simpson, 1927
T. (THA–PRB/RUP) Terr.

First: *Amelotabes* Rose, 1978, Polecat Bench Formation, Wyoming, USA.
Last: *Xenocranium* Colbert, 1942, White River Formation, Wyoming. *Epoicotherium* Simpson, 1927, White River Formation, Wyoming; Renova Formation, Montana, USA. See Heissig (1982) for possible Lower Oligocene occurrence in Germany.

F. MANIDAE Gray, 1821 T. (LUT)–Rec. Terr.

First: *Eomanis* Storch, 1978, MP11, Messel, Germany (Hooker, 1987). **Extant**

EDENTATA *incertae sedis*

F. ERNANODONTIDAE Ding, 1979
T. (THA) Terr.

First and Last: *Asiabradypus* Nesov, 1987, Dzhilga, Kazakhstan, former USSR. *Ernanodon* Ding, 1979, Nungshan Formation, Kwantung, China.

Cohort EPITHERIA McKenna, 1975

Superorder LEPTICTIDA McKenna, 1975

F. GYPSONICTOPIDAE Van Valen, 1967
K. (ALB)–K. (MAA)/T. (THA) Terr.

First: *Prokennalestes* Kielan-Jaworowska and Dashzeveg, 1989, Khovboor Beds, Mongolia.
Last: Middle Paleocene: ?*Stilpnodon* Simpson, 1935, Lebo Formation, Montana, USA (assignment questionable, MCM). Upper Cretaceous: *Gypsonictops* Simpson, 1927, Scollard Formation, Alberta, Canada; Hell Creek Formation, Montana; Lance Creek Formation, Wyoming, USA.

F. DIDYMOCONIDAE Kretzoi, 1943 T. (DAN/THA–CHT) Terr.

First: *Zeuctherium* Tang and Yan, 1976, Wang-hu-dun Formation, Anhui, China.
Last: *Didymoconus* Matthew and Granger, 1924, Kansu, China.

F. LEPTICTIDAE Gill, 1872 T. (DAN–CHT) Terr.

First: *Prodiacodon* Matthew, 1929, Tullock Formation, Montana, USA.
Last: *Leptictis* Leidy, 1868, White River Formation, South Dakota and Colorado, USA.

Superorder PREPTOTHERIA McKenna, 1975, new rank, McKenna

Grandorder ANAGALIDA Szalay and McKenna, 1971

Order MACROSCELIDEA Butler, 1956

F. ANAGALIDAE Simpson, 1931 T. (DAN/THA–PRB/RUP) Terr.

First: *Anaptogale* Xu, 1976, *Chianshania* Xu, 1976, and *Wanogale* Xu, 1976, all Wang-hu-dun Formation, Anhui, China; *Linnania* Chow *et al.*, 1973, Shang-hu Formation, Guongdong, China; *Stenanagale* Wang, 1975, Zaoshi Formation, Hunan, China.
Last: *Anagale* Simpson, 1931, Ulan Gochu Formation, Inner Mongolia, China; *Anagalopsis* Bohlin, 1951, Shih-ehr-ma-cheng Loc., Gan-su, China.

F. MACROSCELIDIDAE Bonaparte, 1838 T. (YPR/PRB)–Rec. Terr.

First: Lower Eocene: *Chambius* Hartenberger, 1986, Kasserine Platform, Tunisia (questionable assignment). Upper Eocene: *Metoldobotes* Schlosser, 1910, Jebei Qatrani Formation, Egypt. **Extant**

Order LAGOMORPHA Brandt, 1855

F. ZALAMBDALESTIDAE Gregory and Simpson, 1926 K. (?TUR/?CON/CMP–CMP) Terr.

First and Last: *Zalambdalestes* Gregory and Simpson, 1926, Djadokhta Formation, Mongolia; *Barunlestes* Kielan-Jaworowska, 1975, Barun Guyot Formation, Nemegt, Mongolia. See Nesov (1985) for possible Turonian or Coniacian record of *Zalambdalestes* from Uzbekistan, former USSR.

F. PSEUDICTOPIDAE Sulimski, 1969 T. (DAN/THA–YPR) Terr.

First: *Paranictops* Qiu, 1977, *Anictops* Qiu, 1977, and *Cartictops* Ding and Tong, 1979, all Wang-hu-dun Formation, Anhui, China.
Last: Lower Eocene: *Pseudictops* Matthew *et al.*, 1929, Bayan Ulan Formation, Inner Mongolia, China. Lower Eocene or Upper Palaeocene: *Mingotherium* Schoch, 1985, Black Mingo Formation, South Carolina, USA.

F. LEPORIDAE Fischer, 1817 T. (LUT/BRT)–Rec. Terr.

First: *Lushilagus* Li, 1965, Lushih Formation, Honan, China; *Procaprolagus* Gureev, 1960, Swift Current Creek, Saskatchewan, Canada (Storer, 1984); *Mytonolagus* Burke, 1938, Uinta Formation, Utah; Wagon Bed Formation, Wyoming, USA. **Extant**
Comments: North American taxa are late Uintan in age.

F. OCHOTONIDAE Thomas, 1897 T. (PRB/RUP)–Rec. Terr.

First: *Desmatolagus* Matthew and Granger, 1923, Ulan Gochu Formation, Hsanda Gol Formation, Inner Mongolia, China. **Extant**

Order RODENTIA Bowdich, 1821

Papers in Luckett and Hartenberger (1985) provide alternative classifications and additional information on rodents.

Suborder SCIUROMORPHA Brandt, 1855

F. EURYMYLIDAE Matthew *et al.*, 1929 T. (THA–PRB) Terr.

First: *Heomys* Li, 1977, and *Mimotona lii* Dashzeveg and Russell, 1988, both Wang-hu-dun Formation, Anhui, China.
Last: *Gomphos* Shevyreva, 1975, Narun-Bulak Beds, Mongolia; *Rhombomylus* Zhai, 1978, Shisanjiangfang Formation, Turpan Basin, China.
Comment: The eurymylids have been reviewed by Dashzeveg and Russell (1988).

F. ISCHYROMYIDAE Alston, 1876 (including PARAMYIDAE Miller and Gidley, 1918; see Black, 1971) T. (THA–LMI) Terr.

First: *Acritoparamys* Korth, 1984, Eagle Coal Mine, Montana; Fort Union Formation, Wyoming, USA (Ivy, 1990); ?*Paramys* Leidy, 1871, Togwotee, Wyoming, USA; MP7, Meudon, France (Russell *et al.*, 1988); *Microparamys* and '*Pseudoparamys*', both MP7, Meudon, France (Russell *et al.*, 1988).
Last: *Paracitellus* Dehm, 1950, Europe.

F. CYLINDRODONTIDAE Miller and Gidley, 1918 T. (YPR–LMI) Terr.

First: *Dawsonomys* Gazin, 1961, Wasatch Formation, Wyoming, USA.
Last: *Downsimys* Flynn *et al.*, 1986, Asia.

F. PROTOPTYCHIDAE Schlosser, 1911 T. (LUT) Terr.

First and Last: *Protoptychus* Scott, 1895, Washakie Formation, Wyoming; Uinta Formation, Utah, USA.

F. TSAGANOMYIDAE Matthew and Granger, 1923 T. (RUP–CHT) Terr.

First and Last: *Tsaganomys* Matthew and Granger, 1923, Hsanda Gol, Mongolia; *Beatomus* Shevyreva, 1972, Ergil Obo, Mongolia; *Sepulkomys* Shevyreva, 1972, Takal Gol, Mongolia.

F. APLODONTIDAE Brandt, 1855 T. (LUT/BRT)–Rec. Terr.

First: *Spurimus* Black, 1971, Wagon Bed Formation, Wyoming, USA. **Extant**

F. REITHROPARAMYIDAE Wood, 1962 T. (YPR–PRB) Terr.

First and Last: *Reithroparamys* Matthew, 1920, Western Interior, North America.

F. SCIURIDAE Gray, 1821 T. (PRB)–Rec. Terr.

First: *Oligospermophilus* Korth, 1987, White River Formation, Nebraska. Earliest European record is *Palaeosciurus*

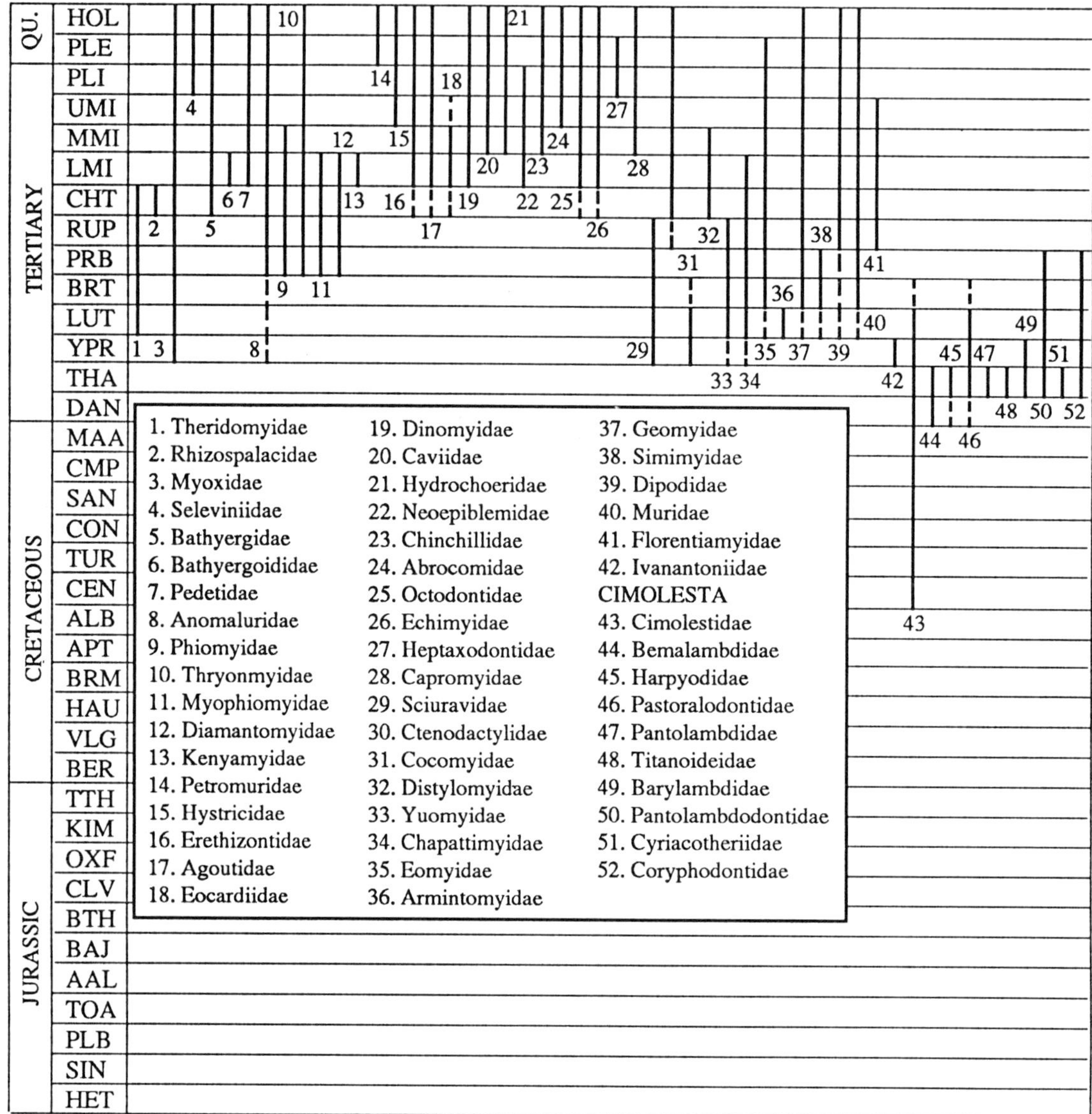

Fig. 41.3

Vianey-Liaud, 1974, MP21, Mas de Got, Quercy, France. **Extant**

F. EUTYPOMYIDAE Miller and Gidley, 1918 T. (YPR–CHT) Terr.

First: *Mattimys* Korth, 1984, Wind River Formation, Wyoming, USA.
Last: *Eutypomys* Matthew, 1905, Sharps Formation, South Dakota, USA.

F. CASTORIDAE Hemprich, 1820 T. (PRB/RUP)–Rec. FW/Terr.

First: Upper Eocene: *Agnotocastor* Stirton, 1935, White River Formation, Wyoming and Colorado, USA. Lower Oligocene: *Steneofiber* Geoffroy, 1833, MP21, Mohren, Germany; Lower Hamstead Beds, England, UK; Lignite Sandstone Formation, Thrace, Turkey (Unay-Bayraktar, 1989). **Extant**

F. THERIDOMYIDAE Alston, 1876 T. (LUT–CHT) Terr. (see Fig. 41.3)

First: *Protadelomys* Hartenberger, 1968, MP13, Bouxwiller and Saint Martin, France (Hartenberger, 1990).
Last: *Columbomys* Thaler, 1962, MP30, level of Coderet, France (Brunet and Vianey-Liaud, 1987).

F. RHIZOSPALACIDAE Thaler, 1966 T. (CHT) Terr.

First and Last: *Rhizospalax* Miller and Gidley, 1919, MP29, Rickenbach, Switzerland; MP30, Coderet, France (Brunet and Vianey-Liaud, 1987).

F. MYOXIDAE Gray, 1821 T. (YPR)–Rec. Terr.

First: *Eogliravus* Hartenberger, 1971, MP13, level of Geiseltal, Germany (Franzen, 1987). **Extant**

F. SELEVINIIDAE Bashanov and Belosludov, 1939 ?T. (PLI)–Rec. Terr.

First: *Plioselevinia* Sulimski, 1962, Poland. **Extant**

Suborder HYSTRICOMORPHA Brandt, 1855

F. BATHYERGIDAE Waterhouse, 1841 T. (CHT)–Rec. Terr.

First: *Morosomys* Shevyreva, 1972, Middle Oligocene, Ergil Obo, Mongolia. **Extant**

F. BATHYERGOIDIDAE Lavocat, 1973 T. (LMI) Terr.

First and Last: *Bathyergoides* Stromer, 1926, south-west and east Africa.

F. PEDETIDAE Gray, 1821 T. (AQT)–Rec. Terr.

First: *Megapedetes* MacInnes, 1957, Negev, Israel. **Extant**

F. ANOMALURIDAE Gervais, 1849 T. (?YPR/PRB)–Rec. Terr.

First: *Nementchamys* Jaeger *et al.*, 1985, Nementcha Mountains, Algeria. An indeterminate Lower Eocene anomalurid is recorded from the Kasserine Platform, Tunisia, by Hartenberger *et al.* (1985). **Extant**

F. PHIOMYIDAE Wood, 1955 T. (PRB–MMI) Terr.

First: *Phiomys* Osborn, 1908, Jebel Qatrani Formation, Egypt.
Last: *Phiomys* Osborn, 1908, Songhor, Koru, Kenya. *Andrewsimys* Lavocat, 1973, Songhor, Kenya. Sen (1977) reports an unnamed phiomyid from the Middle Miocene of Anatolia.

F. THRYONOMYIDAE Pocock, 1922 T. (PRB)–Rec. Terr.

First: *Gaudeamus* Wood, 1968, lower Jebel Qatrani Formation, Egypt. *Paraphiomys* Andrews, 1914, upper Jebel Qatrani Formation, Egypt. **Extant**

F. MYOPHIOMYIDAE Lavocat, 1973 T. (PRB–LMI) Terr.

First: *Phiocricetomys* Wood, 1968, Jebel Qatrani Formation, Egypt.
Last: *Myophiomys* Lavocat, 1973, Rusinga and Songhor, Kenya.

F. DIAMANTOMYIDAE Schaub, 1958 T. (PRB–LMI) Terr.

First: *Metaphiomys* Osborn, 1908, Jebel Qatrani Formation, Egypt.
Last: *Diamantomys* Stromer, 1922, Rusinga and Songhor, Kenya.

F. KENYAMYIDAE Lavocat, 1973 T. (LMI) Terr.

First and Last: *Simonimys* Lavocat, 1973, Songhor, East Africa. *Kenyamys* Lavocat, 1973, Rusinga, Kenya.

F. PETROMURIDAE Tullberg, 1899 Q. (PLE)–Rec. Terr.

First: *Petromus* Smith, 1831, Taung, South Africa. **Extant**

F. HYSTRICIDAE Fischer von Waldheim, 1817 T. (UMI)–Rec. Terr.

First: First known in the Upper Miocene of Egypt, Hungary, and India (Chaline and Mein, 1979; Wood, 1985); *Miohystrix* Kretzoi, 1951, Europe; *Sivacanthion* Colbert, 1933, Asia. **Extant**

F. ERETHIZONTIDAE Bonaparte, 1845 T. (CHT/LMI)–Rec. Terr.

First: *Protosteriomys* Wood and Patterson, 1959, Deseadan, Argentina. **Extant**

F. AGOUTIDAE Gray, 1821 T. (CHT/LMI)–Rec. Terr.

First: *Incamys* Hoffstetter and Lavocat, 1970, and *Branisamys* Hoffstetter and Lavocat, 1970, both Salla Beds, Bolivia; *Cephalomys* Ameghino, 1897, *Scotamys* Loomis, 1914, and *Litodontomys* Loomis, 1914, all Deseadan, Argentina. **Extant**

F. EOCARDIIDAE Ameghino, 1891 T. (CHT/LMI–MMI/UMI) Terr.

First: *Chubutomys* Wood and Patterson, 1959, *Asteromys* Ameghino, 1897, and *Phanomys* Ameghino, 1887, all Deseadan, Argentina.
Last: *Neophanomys* Rovereto, 1914, Friasian, Argentina.

F. DINOMYIDAE Toschel, 1874 T. (LMI)–Rec. Terr.

First: *Scleromys* Ameghino, 1887, Santacrucian, Argentina. **Extant**

F. CAVIIDAE Fischer de Waldheim, 1817 T. (MMI)–Rec. Terr.

First: *Neoprocavia* Ameghino, 1889, Barrancos del Parana, Argentina (Ameghino, 1906); *Prodolichotis* Kraglievich, 1932, Uruguay. **Extant**

F. HYDROCHOERIDAE Gray, 1825 T. (MMI)–Rec. Terr.

First: *Procardiatherium* Ameghino, 1885, and *Cardiatherium* Ameghino, 1883, both Chasicoan, Argentina (Bondesio, 1978). **Extant**

F. NEOEPIBLEMIDAE Kraglievich, 1926 T. (LMI–PLI) Terr.

First: *Perimys* Ameghino, 1887, Colhuehuapian, Argentina.
Last: *Neoepiblema* Ameghino, 1889, Parana, Argentina; *Phoberomys* (= *Dabbenea*) Kraglievich, 1926, Archipeliego de las Antillas, Argentina (Bondesio and Villanueva Bocquentin, 1987).

F. CHINCHILLIDAE Bennett, 1833 T. (MMI)–Rec. Terr.

First: *Pliolagostomus* Ameghino, 1887, and *Prolagostomus* Ameghino, 1887, both Santacrucian, Argentina. **Extant**

F. ABROCOMIDAE Miller and Gidley, 1918 T. (UMI)–Rec. Terr.

First: *Protabrocoma* Kraglievich, 1927, Huayquerian, Argentina. **Extant**

F. OCTODONTIDAE Waterhouse, 1839 T. (CHT/LMI)–Rec. Terr.

First: *Platypittamys* Wood, 1949, Deseadan, Argentina. *Migraveramus* Patterson and Wood, 1982, Salla Beds, Bolivia. **Extant**

F. ECHIMYIDAE Gray, 1825 T. (CHT/LMI)–Rec. Terr.

First: *Sallamys* Hoffstetter and Lavocat, 1970, Salla Beds, Bolivia; *Deseadomys* Wood and Patterson, 1959, and *Xylechimys* Patterson and Pascual, 1968, both Deseadan, Argentina. **Extant**

F. HEPTAXODONTIDAE Anthony, 1917
T. (PLI)–Q. (PLE) Terr.

First: *Pentastylomys* Kraglievich, 1926, and *Tetrastylomys* Kraglievich, 1926, both Archipeliego de las Antillas, Argentina.
Last: *Clidomys* Anthony, 1920, Jamaica; *Amblyrhiza* Cope, 1868, Lesser Antilles.

F. CAPROMYIDAE Smith, 1842
T. (MMI)–Rec. Terr.

First: Unnamed capromyids are reported from Santacrucian strata of Honda, Bolivia (MacFadden and Wolff, 1981). **Extant**

Suborder MYOMORPHA Brandt, 1855

F. SCIURAVIDAE Miller and Gidley, 1918
T. (YPR–RUP) Terr.

First: *Knightomys* Gazin, 1961, Wasatch and Willwood Formation, Wyoming, USA.
Last: A sciuravid is known from the White River Formation of the Douglass area, Wyoming, USA (D. Kron, pers. comm. to RKS).

F. CTENODACTYLIDAE Gervais, 1853
T. (RUP/CHT)–Rec. Terr.

First: *Woodomys* Shevyreva, 1971, Tschelkar-Teniz, Kazakhstan; *Karakoromys* Matthew and Granger, 1923, Asia. **Extant**

F. COCOMYIDAE De Bruijn *et al.*, 1982
T. (YPR–LUT/BRT) Terr.

First: *Cocomys* Dawson *et al.*, 1984, Ling-cha Formation, Hunan, China; *Bumbanomys* Shevyreva, 1989, *Adolomys* Shevyreva, 1989 and *Tsagankushumys* Shevyreva, 1989, all Tsagan Khush, Mongolia.
Last: *Tsinglingomys* Li, 1963, Lushi Formation, Honan, China; *Tamquammys* Shevyreva, 1971, Zaisan Basin, Kazakhstan, former USSR.

F. DISTYLOMYIDAE Wang, 1988
T. (CHT–MMI) Terr.

First: *Prodistylomys* Wang and Qi, 1989, Suosuoquan Formation, China.
Last: *Distylomys* Wang, 1988, Tung Gur, China.

F. YUOMYIDAE Dawson, Li and Qi, 1984
T. (?YPR–RUP) Terr.

First: *Advenimus* Dawson, 1964, Yu-huang-ding Formation, Sichuan, China; Mongolia; *Petrokozlovia* Shevyreva, 1972, Mongolia and Kazakhstan, former USSR.
Last: *Dianomys* Wang, 1984, Caijiachong Formation, Yunan, China.

F. CHAPATTIMYIDAE Hussain *et al.*, 1978
T. (?YPR/LUT/BRT–LMI) Terr.

First: *Chapattimys* Hussain *et al.*, 1978, Kuldana Formation, Pakistan; Sabutha Formation, India; *Birbalomys* Sahni and Khare, 1973, Kuldana Formation, Pakistan; *Shkhikvadzomys* Shevyreva, 1989, *Bolosomys* Shevyreva, 1989, and *Esasempomys* Shevyreva, 1989, all Tsagan Khush, Mongolia. Hartenberger (1990) suggests that Chapattimyids occur in the Middle Eocene of Spain.
Last: *Lindsaya* Flynn *et al.*, 1989, *Lophibaluchia* Flynn *et al.*, 1989, *Baluchimys* Flynn *et al.*, 1989, *Hodsahibia* Flynn *et al.*, 1989, and *Fallomus* Flynn *et al.*, 1989, all Bugti Beds, Pakistan.

F. EOMYIDAE Winge, 1887
T. (LUT/BRT)–Q. (PLE) Terr.

First: ?*Namatomys* Black, 1965, Cypress Hills Formation, Saskatchewan, Canada; Ventura County, California, USA; *Protadjidaumo* Burke, 1934, Cypress Hills Formation, Saskatchewan, Canada; Duchesne River Formation, Utah, USA; *Yoderimys* Wood, 1955, and *Omegodus* (= *Adjidaumo*) Pomel, 1853, both Cypress Hills Formation, Saskatchewan, Canada (Storer, 1987, 1988).
Last: *Estramomys* Janossy, 1969, Osztramos, Hungary. Possibly also *Meteomys* Kretzoi, 1952, Europe.

F. ARMINTOMYIDAE Dawson *et al.*, 1990
T. (LUT) Terr.

First and Last: *Armintomys* Dawson *et al.*, 1990, Wind River
Formation, Wyoming, USA.

F. GEOMYIDAE Bonaparte, 1845
T. (LUT/BRT)–Rec. Terr.

First: *Heliscomys* Cope, 1873, Wagon Bed Formation, Wyoming, USA (M. R. Dawson, pers. comm. to MCM, 1987); Cypress Hills Formation, Saskatchewan, Canada (Storer, 1988). **Extant**

F. SIMIMYIDAE Wood, 1980
T. (LUT/BRT–PRB) Terr.

First: *Simimys* Wilson, 1935, ?Chambers Formation, Texas; Mission Valley Formation, California, USA.
Last: *Nonomys* Emry and Dawson, 1973, Vieja Group, Texas; White River Formation, Wyoming, USA (Emry, 1981).

F. DIPODIDAE Fischer von Waldheim, 1817
T. (LUT/RUP)–Rec. Terr.

First: Middle Eocene: *Elymys* Emry and Korth, 1989, Sheep Pass Formation, Nevada, USA (questionably referred to zapodids). Lower Oligocene: *Plesiosminthus* Viret, 1926, Caijiachong Formation, Yunnan, China. **Extant**

F. MURIDAE Illiger, 1811
T. (LUT/BRT)–Rec. Terr.

First: *Eumys* Leidy, 1856, Cypress Hills Formation, Saskatchewan, Canada (Storer, 1988). **Extant**

F. FLORENTIAMYIDAE Wood, 1936
T. (RUP–UMI) Terr.

First: *Ecclesimus* Korth, 1989, White River Formation, Cedar Creek Member, Colorado, USA.
Last: *Harrymys* Munthe, 1988, Split Rock Formation, Wyoming, USA.

F. IVANANTONIIDAE Shevyreva, 1989
T. (YPR) Terr.

First and Last: *Ivanantonia* Shevyreva, 1989, Asia.

Grandorder FERAE Linnaeus, 1758

Order CIMOLESTA McKenna, 1975

Suborder DIDELPHODONTA McKenna, 1975

F. CIMOLESTIDAE Marsh, 1889
K. (CEN)–T. (LUT/BRT) Terr.

First: *Otlestes* Nesov, 1985, Uzbekistan, former USSR. Possibly also *Deccanolestes* Prasad and Sahni, 1988, Andhra Pradesh, India.
Last: *Didelphodus* Cope, 1882, Swift Current Creek, Saskatchewan, Canada (Storer, 1984).

Suborder PANTODONTA Cope, 1873

F. BEMALAMBDIDAE Chow *et al.*, 1973 T. (DAN/THA) Terr.

First: *Bemalambda* Chow *et al.*, 1973 and *Hypsilolambda* Wang, 1975, both Zao-shi Formation, Hunan, China.
Last: *Bemalambda* Chow *et al.*, 1973, Wang-hu-dun Formation, Anhui, China.

F. HARPYODIDAE Wang, 1979 T. (DAN/THA–THA) Terr.

First: *Harpyodus* Chiu and Li, 1977, Wang-hu-dun Formation, Anhui, China.
Last: *Harpyodus* Chiu and Li, 1977, Chi-jiang Formation, Kiangsi, China.

F. PASTORALODONTIDAE Chow and Qi, 1978 T. (DAN/THA–LUT/BRT) Terr.

First: *Altilambda* Chow and Wang, 1978, Wang-hu-dung Formation, Anhui, China.
Last: *Pastoralodon* Chow and Qi, 1978, Nomogen, Inner Mongolia, China; Bayan Ulan Formation, Inner Mongolia, China.

F. PANTOLAMBDIDAE Cope, 1883 T. (THA) Terr.

First: *Pantolambda* Cope, 1882, Nacimiento Formation, New Mexico; Polecat Bench Formation, Wyoming; Lebo and Mellville formations, Montana, USA.
Last: *Caenolambda* Gazin, 1956, Polecat Bench Formation, Wyoming, USA; Alberta, Canada.

F. TITANOIDEIDAE Patterson, 1934 T. (THA) Terr.

First and Last: *Titanoides* Gidley, 1917, Western Interior, North America (Archibald *et al.*, 1987).

F. BARYLAMBDIDAE Patterson, 1937 T. (THA–YPR) Terr.

First: *Barylambda* Patterson, 1937, Blacks Peak Formation, Texas, USA.
Last: *Barylambda* Patterson, 1937, Wind River Formation, Wyoming, USA.
Comments: Last record is based on unpublished materials in the collections of Carnegie Museum of Natural History from the Lysite Member, Wind River Formation, Wyoming, USA.

F. PANTOLAMBDODONTIDAE Granger and Gregory, 1934 T. (THA–PRB) Terr.

First: *Archaeolambda* Flerov, 1952, Naran Bulak, Mongolia.
Last: *Pantolambdodon* Granger and Gregory, 1934, Irdin Manha Formation, Inner Mongolia, China.

F. CYRIACOTHERIIDAE Rose and Krause, 1982 T. (THA) Terr.

First and Last: *Cyriacotherium* Rose and Krause, 1982, Paskapoo Formation, Alberta, Canada; Willwood and Polecat Bench Formations, Wyoming; Togwotee, Wyoming, USA.

F. CORYPHODONTIDAE Marsh, 1876 T. (THA–PRB) Terr.

First: *Coryphodon* Owen, 1845, Wasatch Formation, Colorado, USA; Polecat Bench and Willwood Formations, Wyoming, USA; London Clay, Blackheath and Suffolk Pebble beds, England, UK; Meudon, Argile à Lignites and Argiles Plastiques, France; Vertian and Orp-le-Grand, Belgium; Dabu Formation, Xinjiang, China; Yuli Formation, Shanxi, China; White Beds, Narun-Bulak Beds, Mongolia (Lucas, 1989).
Last: *Eudinoceras* Osborn, 1924, Shandong, China.

F. WANGLIIDAE Van Valen, 1988 T. (DAN–THA) Terr. (see Fig. 41.4)

First: *Alcidedorbignya* de Muizon and Marshall, 1987, El Molino Formation, Bolivia.
Last: *Wanglia* Van Valen, 1988, Jiang-xi, China.

Suborder PANTOLESTA McKenna, 1975

F. PANTOLESTIDAE Cope, 1884 T. (THA–CHT) Terr.

First: *Propalaeosinopa* Simpson, 1927, Tongue River Formation, Montana.
Last: *Kochictis* Kretzoi, 1943, Egerer, Hungary (Thenius, 1959).

F. PAROXYCLAENIDAE Weitzel, 1933 T. (YPR–PRB) Terr.

First: *Spaniella* Crusafont-Pairo and Russell, 1967, Spain; *Merialus* Russell and Godinot, 1988, MP7, Palette, France (Godinot, 1987).
Last: *Paroxyclaenus* Teilhard de Chardin, 1922, Quercy, France; *Euhookeria* Russell and Godinot, 1988, Headon Beds, England, UK.

F. PTOLEMAIIDAE Osborn, 1908 T. (PRB) Terr.

First and Last: *Ptolemaia* Osborn, 1908, *Qarunavus* Simons and Gingerich, 1974, and *Cleopatrodon* Bown and Simons, 1987, all Jebel Qatrani Formation, Egypt.

Suborder TAENIODONTA Cope, 1876

F. STYLINODONTIDAE Marsh, 1875 T. (THA–LUT) Terr.

First: *Onychodectes* Cope, 1888, Nacimiento Formation, New Mexico; North Horn Formation, Utah; *Wortmania* Hay, 1899, Nacimiento Formation, New Mexico, USA.
Last: *Stylinodon* Marsh, 1874, Uinta Formation, Utah, USA.

Suborder APATOTHERIA Scott and Jepsen, 1936

F. APATEMYIDAE Matthew, 1909 T. (THA–RUP) Terr.

First: *Jepsenella* Simpson, 1940, Fort Union and Polecat Bench Formations, Wyoming; Lebo Formation, Montana, USA.
Last: *Sinclairella* Jepsen, 1934, White River Formation, South Dakota, USA.

Suborder TILLODONTIA Marsh, 1875, new rank, McKenna

F. TILLOTHERIIDAE Marsh, 1875 T. (DAN/THA–PRB) Terr.

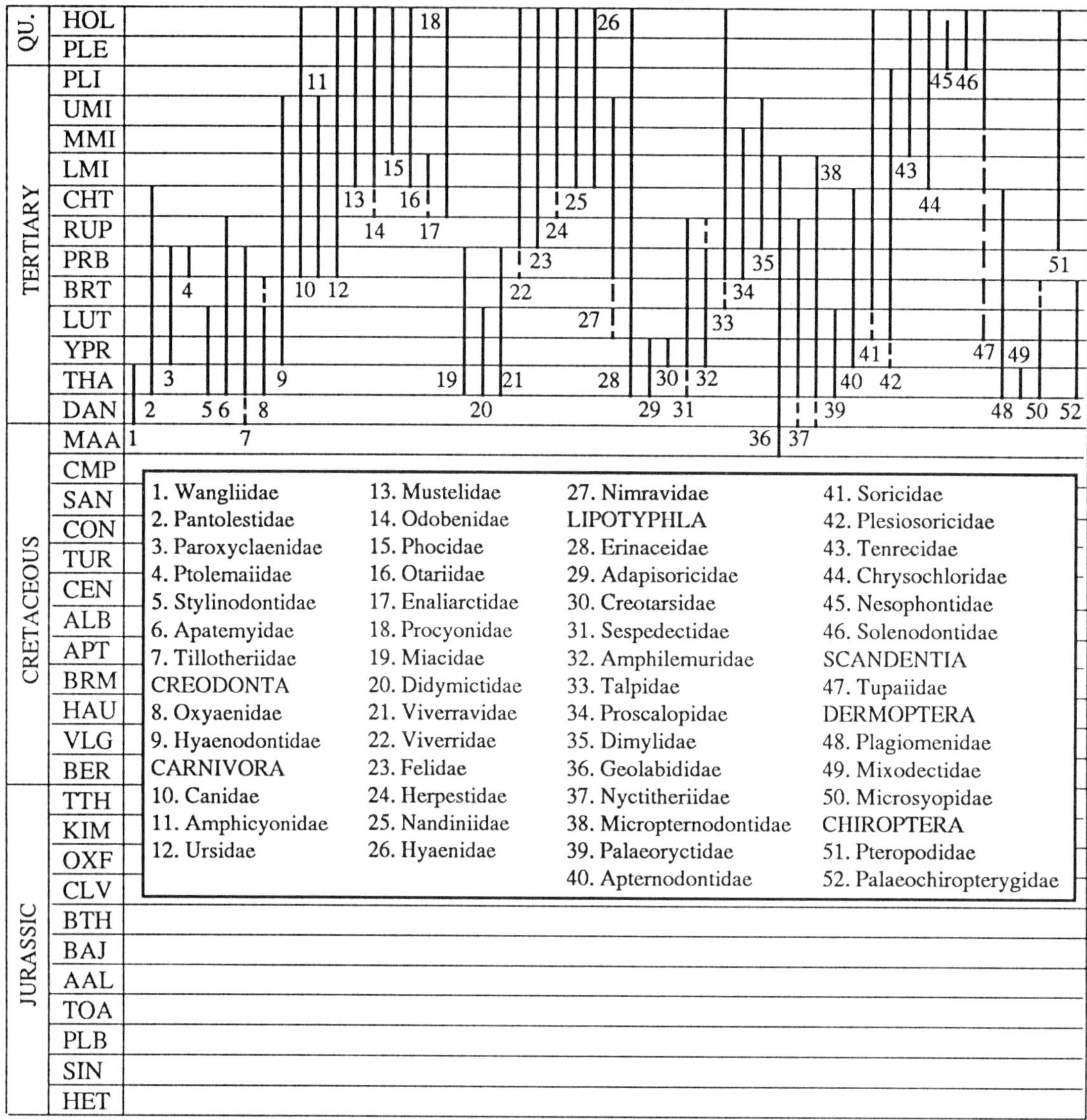

Fig. 41.4

First: *Lofochaius* Chow *et al.*, 1973, Shang-hu Formation, Kwantung, China; *Anchilestes* Chiu and Li, 1977, Wang-hu-din Formation, Anhui, China.
Last: *Adapidium* Young, 1937, He-ti Formation, Shansi, China.

Order CREODONTA Cope, 1875

F. OXYAENIDAE Cope, 1877
T. (THA–LUT/BRT) Terr.

First: *Tytthaena* Gingerich, 1980, Jepsen Quarry, Polecat Bench Formation, Wyoming, USA.
Last: *Sarkastodon* Granger, 1938, Irdin Manha Formation, Inner Mongolia, China.

F. HYAENODONTIDAE Leidy, 1869
T. (YPR–UMI) Terr.

First: *Parvagula* Lange-Badre, 1987, MP7, Palette, France; *Arfia* Van Valen, 1965, MP7, Dormaal, Belgium; MP7, Rians and Meudon, France; MP7, Suffolk Pebble Beds, England, UK; Willwood and Wasatch Formations, Wyoming, USA; *Paratritemnodon* Rao, 1973, Kuldana Formation, Pakistan; *Proviverra* Rutimeyer, 1862, MP7, Level of Dormaal, Europe; *Prototomus* Cope, 1874, Willwood Formation, Wyoming, USA; MP10, Grauves, France. *Prolimnocyon* Matthew, 1915, Willwood Formation, Wyoming; MP7, Dormaal, Belgium; *Acarictis* Gingerich and Deutsch, 1987, and *Galecyon* Gingerich and Deutsch, 1987, both Willwood Formation, Wyoming, USA.
Last: *Pterodon* de Blainville, 1839, Nagri Beds, India.
Comments: A specimen of middle Palaeocene age of *Prolimnocyon* has been reported from Swain Quarry, Fort Union Formation, Wyoming, USA (Archibald *et al.*, 1987). According to Ivy (pers. comm. to RKS), these specimens are not from a creodont. Gheerbrant (1987) has reported a proviverrine hyaenodont from the Upper Paleocene of l'Adrar Mgorn, Morocco.

Order CARNIVORA Bowditch, 1821

F. CANIDAE Fischer von Waldheim, 1817
T. (PRB)–Rec. Terr.

First: *Hesperocyon* Scott, 1890, White River Formation, Wyoming, USA. Storer (1988) has a queried identification of this genus from the Cypress Hills Formation, Saskatchewan, Canada (Duchesnean). **Extant**

F. AMPHICYONIDAE Trouessart, 1885 T. (PRB–UMI) Terr.

First: *Daphoenus* Leidy, 1853, Wagon Bed Formation, Wyoming; Chambers Formation, Texas, USA; *Cynodictis* Bravard and Pomel, 1850, MP18, La Debruge, France.
Last: *Ischyrocyon* Matthew, 1904, Clarendon, Texas, USA; *Agnotherium* Kaup, 1832, Eppelsheim, Germany; Africa; *Arctamphicyon* Pilgrim, 1932, Europe; *Pseudarctos* Schlosser, 1899, Tutzing and Hader, Germany.

F. URSIDAE Fischer von Waldheim, 1817 T. (PRB)–Rec. Terr.

First: *Plesiocyon* Schlosser, 1887, Quercy, France; *Cephalogale* Jourdan, 1862, Quercy, France; Bose Basin, Guanxi, China; ?*Parictis* Scott, 1893 (including *Campylocynodon* Chaffee, 1954), Cypress Hills Formation, Saskatchewan, Canada (Storer, 1987). **Extant**

F. MUSTELIDAE Fishcher de Waldheim, 1817 T. (AQT)–Rec. Terr/Mar./FW

First: *Miomephitis* Dehm, 1950, *Mionictis* Matthew, 1924, *Paralutra* Roman and Viret, 1934, and *Martes* Pinel, 1792, all Europe; *Miomustela* Hall, 1930, *Leptarctus* Leidy, 1856, *Dinogale* Cook and MacDonald, 1962, and *Craterogale* Gazin, 1936, all North America; *Kenyalutra* Schmidt-Kittler, 1987a,b, MN4–5, Songhor and Rusinga, Kenya; *Luogale* Schmidt-Kittler, 1987a,b, MN4–5, Rusinga, Kenya. **Extant**
Comments: A number of 'paleomustelid' taxa have been assigned to the mustelids in the past. These taxa, such as *Aelurocyon* Peterson, 1906, may or may not be closely related to modern mustelids (R. H. Tedford, pers. comm. to RKS, 1991).

F. ODOBENIDAE Allen, 1880 T. (?CHT/LMI)–Rec. Mar.

First: *Pinnarctidion* Barnes, 1979, North America. **Extant**

F. PHOCIDAE Gray, 1821 T. (MMI)–Rec. Mar.

First: *Monachopsis* Kretzoi, 1941, Europe; *Monotherium* Van Beneden, 1876, Antwerp Basin, Belgium; St Marys Formation, Maryland, USA; *Properiptychus* Ameghino, 1897, Argentina; *Palmidophoca* Ginsburg and Janvier, 1975, Europe; *Praepusa* Kretzoi, 1941, Crimea, former USSR; *Phocanella* Van Beneden, 1876, Belgium; *Pontophoca* Kretzoi, 1941, Kishenev, former USSR; *Leptophoca* True, 1906, Maryland, USA. **Extant**

F. OTARIIDAE Gray, 1825 T. (LMI)–Rec. Mar.

First: *Pteronarctos* Barnes, 1989, Pacific North America. **Extant**

F. ENALIARCTIDAE Mitchell and Tedford, 1973 T. (?CHT–LMI) Mar.

First and Last: *Enaliarctos* Mitchell and Tedford, 1973, Jewett Formation, California, USA (Berta and Ray, 1990). **Extant**

F. PROCYONIDAE Gray, 1825 T. (CHT)–Rec. Terr.

First: *Amphictis* Pomel, 1853, *Plesictis* Pomel, 1846, and *Plesiogale* Pomel, 1847, all Quercy, France; *Pseudobassaris* Pohle, 1917, Quercy, France; North America. **Extant**

F. MIACIDAE Cope, 1880 T. (THA–PRB) Terr.

First: *Uintacyon rudis* Matthew, 1915, Willwood Formation, Wyoming, USA.
Last: *Miacis* Cope, 1872, MP17, level of Fons 4, Europe (Legendre, 1987); Little Egypt and Blue Cliff horizons, Chambers Formation, Texas, USA (Wilson, 1986).

F. DIDYMICTIDAE Flynn and Galiano, 1982 T. (THA–LUT) Terr.

First and Last: *Protictis* Matthew, 1937, Western Interior, North America (Flynn and Galiano, 1982).

F. VIVERRAVIDAE Wortman and Matthew, 1899 T. (THA–PRB) Terr.

First: *Simpsonictis* MacIntyre, 1962, Fort Union and Polecat Bench Formations, Wyoming; Lebo Formation, Montana, USA.
Last: *Viverravus* Marsh, 1872, Europe; *Tapocyon* Stock, 1934, and *Plesiomiacis* Stock, 1935, both Sespe Formation, California, USA; *Procynodictis* Wortman and Matthew, 1899, Quercy, France.

F. VIVERRIDAE Gray, 1821 T. (PRB/RUP)–Rec. Terr.

First: *Stenoplesictis* Filhol, 1880, MP22, Villebramar, France; *Stenogale* Schlosser, 1887, MP22, Quercy, France; *Palaeoprionodon* Filhol, 1880, Quercy, France; Osborne Formation, England, UK; Tatal Gol, Mongolia (Russell and Zhai, 1987). **Extant**
Comments: Family also reported from the Oligocene of Hsanda Gol, Mongolia (Schmidt-Kittler, 1987b; Hunt, 1989).

F. FELIDAE Fischer von Waldheim, 1817 T. (RUP)–Rec. Terr.

First: *Proailurus* Filhol, 1879, MP21, Aubrelong and Somailles, France (Hunt, 1989). **Extant**

F. HERPESTIDAE Bonaparte, 1845 T. (CHT/AQT)–Rec. Terr.

First: *Herpestes* Illiger, 1911 (including *Leptoplesictis* Forsyth Major, 1903), MN4, France (Hunt, 1989). **Extant**

F. NANDINIIDAE Pocock, 1929 T. (LMI)–Rec. Terr.

First: Nandiniids are first reported from the Lower Miocene of Songhor, Kenya (Hunt, 1989). **Extant**

F. HYAENIDAE Gray, 1821 T. (AQT)–Rec. Terr.

First: *Herpestides* de Beaumont, 1967, MN2, St Gerand, France (Hunt, 1989). **Extant**

F. NIMRAVIDAE Cope, 1880 T. (LUT/BRT/PRB–UMI) Terr.

First: *Hoplophoneus* Cope, 1874, and *Dinictis* Leidy, 1854, both White River Formation, North America. ?Middle Eocene: cf. *Eusmilus* Gervais, 1876, Lushi Formation, Honan, China.
Last: *Barbourofelis* Schultz *et al.*, 1970, Kimball Formation, Nebraska, USA; Asia. *Sansanosmilus* Kretzoi, 1929, Sansan, France.

Grandorder LIPOTYPHLA Haeckel, 1866, new rank, McKenna

Order ERINACEOMORPHA Gregory, 1910

See Novacek *et al.* (1985) for an alternative classification of the Erinaceomorpha.

F. ERINACEIDAE Fischer von Waldheim, 1817 T. (THA)–Rec. Terr.

First: *Litolestes* Jepsen, 1930, Polecat Bench Formation, Wyoming, USA; Tongue River Formation, Montana, USA; Paskapoo Formation, Alberta, Canada. **Extant**

F. ADAPISORICIDAE Schlosser, 1887 T. (THA–YPR) Terr.

First and Last: *Adapisorex* Lemoine, 1883, MP6, Cernay, France.
Last: *Neomatronella* Russell *et al.*, 1975, MP8+9, Mutigny and Avenay, France (Godinot, 1987).

F. CREOTARSIDAE Hay, 1930 T. (YPR) Terr.

First and Last: *Creotarsus* Matthew, 1918, Willwood Formation, Wyoming, USA.

F. SESPEDECTIDAE Novacek, 1985 T. (?THA–RUP) Terr.

First: *Scenopagus* McKenna and Simpson, 1959, Willwood Formation, Wyoming, USA.
Last: *Ankylodon* Patterson and McGrew, 1937, White River Formation, Wyoming, USA.

F. AMPHILEMURIDAE Heller, 1935 T. (YPR–PRB/RUP) Terr.

First: *Macrocranion* Weitzel, 1949, Willwood Formation, Wyoming, USA; MP7, Palette, France; *Dormaalius* Quinet, 1964, MP7, Dormaal, Belgium.
Last: *Gesneropithex* Hurzeler, 1946, MP18, Gosgen (Solothurn), Switzerland. A 'dormaliid' has been reported from the lower Oligocene Caijiachong Formation, China, by Wang (1992).

Order SORICOMORPHA Gregory, 1910

F. TALPIDAE Fischer de Waldheim, 1817 T. (BRT/PRB)–Rec. Terr.

First: *Eotalpa* Sige *et al.*, 1977, MP17, Headon, England, UK. **Extant**

F. PROSCALOPIDAE Reed, 1961 T. (PRB–MMI) Terr.

First: *Proscalops* Matthew, 1901, White River Formation, South Dakota and Nebraska; Renova Formation, Montana, USA; *Oligoscalops* Reed, 1961, White River Formation, North America.
Last: *Mesoscalops* Reed, 1960, Deep River Formation, Montana; Split Rock Formation, Wyoming; Batesland and Rosebud Formations, South Dakota, USA (Barnosky, 1981).

F. DIMYLIDAE Schlosser, 1887 T. (RUP–UMI) Terr.

First: *Exoedaenodus* Hurzeler, 1944, Quercy, France.
Last: *Plesiodimylus* Gaillard, 1897, Europe; *Cordylodon* von Meyer, 1859, Ulm, Germany.

F. GEOLABIDIDAE McKenna, 1960 K. (MAA)–T. (LMI) Terr.

First: *Batodon* Marsh, 1892, Lance Creek Formation, Wyoming, USA.
Last: *Centetodon* Marsh, 1872, Gering Formation, Nebraska; Monroe Creek Formation, South Dakota, USA (Lillegraven *et al.*, 1982).

F. NYCTITHERIIDAE Simpson, 1928 T. (DAN/THA–RUP) Terr.

First: *Leptacodon* Matthew and Granger, 1921, Tullock Formation, Montana, USA.
Last: *Darbonetus* Crochet, 1974, MP23, Itardies, France.

F. MICROPTERNODONTIDAE Stirton and Rensberger, 1964 T. (DAN/THA–LMI) Terr.

First: *Prosarcodon* McKenna *et al.*, 1984, Fangou Formation, Shaanxi Province, China.
Last: *Micropternodus* Matthew, 1903, John Day Formation, Oregon, USA.

F. PALAEORYCTIDAE Winge, 1917 T. (THA–LUT) Terr.

First: *Palaeoryctes* Matthew, 1913, Nacimiento Formation, New Mexico; Fort Union Formation, Wyoming, USA.
Last: Unnamed taxon, Friars Formation, California, USA (Novacek, 1976).

F. APTERNODONTIDAE Matthew, 1910 T. (YPR–CHT?) Terr.

First: *Parapternodus* Bown and Schankler, 1982, Willwood Formation, Wyoming, USA.
Last: *Apternodus* Matthew, 1903, White River Formation, Wyoming, USA.

F. SORICIDAE Fischer von Waldheim, 1917 T. (LUT/BRT)–Rec. Terr.

First: *Domnina* Cope, 1873, loc. 5, Wagon Bed Formation, Wyoming, USA (Krishtalka and Setoguchi, 1977). **Extant**

F. PLESIOSORICIDAE Winge, 1917 T. (YPR/LUT–PLI) Terr.

First: *Pakilestes* Russell and Gingerich, 1981, Kuldana Formation, Pakistan.
Last: *Meterix* Hall, 1929, ?Upper PLI, Nevada, USA.

F. TENRECIDAE Gray, 1821 T. (MMI)–Rec. Terr.

First: *Erythrozootes* Butler and Hopwood, 1957, and *Protenrec* Bulter and Hopwood, 1957, both Napak and Songhor, Kenya; *Parageogale* Butler, 1984, Hiwegi Formation, Kenya. **Extant**

F. CHRYSOCHLORIDAE Gray, 1825 T. (LMI)–Rec. Terr.

First: *Prochrysochloris* Butler and Hopwood, 1957, Songhor, Kenya. **Extant**

F. NESOPHONTIDAE Anthony, 1916 Q. (PLE–HOL) Terr.

First and Last: *Nesophontes* Anthony, 1916, West Indies.

F. SOLENODONTIDAE Gill, 1872 Q. (PLE)–Rec. Terr.

First: *Solenodon* Brandt, 1833, Cuba; Hispaniola. **Extant**

Grandorder ARCHONTA Gregory, 1910

Order SCANDENTIA Wagner, 1855

F. TUPAIIDAE Gray, 1825 T. (?LUT/BRT/UMI)–Rec. Terr.

First: Eocene: *Eodendrogale* Tong, 1988, He-tao-yuan Formation, Henon, China; Miocene: *Prodendrogale* Qiu, 1986, Lufeng, China. **Extant**

Order DERMOPTERA Illiger, 1811

Fossils of dermopterans are unknown from Miocene to Recent.

F. PLAGIOMENIDAE Matthew, 1918 T. (THA–CHT) Terr.

First: *Elpidophorus* Simpson, 1927, Lebo Formation, Montana, USA.
Last: *Ekgmowechashala* MacDonald, 1963, Sharps Formation, South Dakota, and John Day Formation, Oregon, USA, *fide* McKenna (1990).

F. MIXODECTIDAE Cope, 1883 T. (THA) Terr.

First: *Dracontolestes* Gazin, 1941, North Horn Formation, Utah, USA.
Last: *Mixodectes* Cope, 1883, Nacimiento Formation, New Mexico; Fort Union Formation, Wyoming, USA.

F. GALEOPITHECIDAE Gray, 1821 **Extant** Terr.

F. MICROSYOPIDAE Osborn, 1892 T. (THA–LUT/BRT) Terr.

First: *Berruvius* Russell, 1964, MP6, Berru and Cernay, France; *Navajovius* Matthew and Granger, 1921, Nacimiento Formation, Colorado; Black Peaks Formation, Texas; Fort Union and Evanston Formations, Wyoming, USA.
Last: *Craseops* Stock, 1935, Sespe Formation, California, USA.
Comments: See Gunnell (1989) for the most recent revision of this group.

Order CHIROPTERA Blumenbach, 1779

Despite the occurrence of many bat families during the Eocene, the fossil record of chiropterans is incomplete and discontinuous. Sigé and Legendre (1983) provide a summary of the fossil record of bats in Europe.

F. PTEROPODIDAE Gray, 1821 T. (RUP)–Rec. Volant

First: *Archaeopteropus* Meschinelli, 1903, Monteviale, Italy. **Extant**

F. PALAEOCHIROPTERYGIDAE Revilliod, 1917 T. (THA–BRT) Volant

First: *Archaeonycteris* Revilliod, 1917, and *Palaeochiropteryx* Revilliod, 1917, both MP7, Rians, France; MP7, Dormaal, Belgium.
Last: Both genea MP11, Messel, Germany.

F. ICARONYCTERIDAE Jepsen, 1966 T. (THA–LUT) Volant

First: *Icaronycteris* Jepsen, 1966, MP7, Dormaal, Belgium; MP7, ?Meudon, France.
Last: *Icaronycteris* Jepsen, 1966, Wind River Formation, Wyoming, USA (Stucky, 1984).

F. EMBALLONURIDAE Gervais, 1855 T. (LUT/BRT)–Rec. Volant

First: *Vespertiliavus* Schlosser, 1887, Ludian, Quercy, France. **Extant**

F. RHINOPOMATIDAE Bonaparte, 1838 **Extant** Volant

F. CRASEONYCTERIDAE Hill, 1974 **Extant** Volant

F. RHINOLOPHIDAE Gray, 1825 T. (LUT)–Rec. Volant

First: *Hipposideros* Gray, 1821; Rhinolophids are reported from MP16, Robiac, France; Grissolles, France; Chamblon, Switzerland; MP14, Egerkingen, Switzerland; and MP16, Creechbarrow, England, UK. **Extant**

F. MEGADERMATIDAE Allen, 1864 T. (PRB)–Rec. Volant

First: *Necromantis* Weithofer, 1887, Quercy, France. **Extant**

F. NYCTERIDAE Van der Hoeven, 1855 T. (CHT)–Rec. Volant

First: ?Nycteridae, MP29, Verneuil (Allier), France. **Extant**

F. PHYLLOSTOMIDAE Gray, 1825 T. (MMI)–Rec. Volant

First: *Notonycteris* Savage, 1951, La Venta, Honda Group, Colombia. **Extant**

F. MORMOOPIDAE de Saussure, 1860 Q. (PLE)–Rec. Volant

First: *Pteronotus* Gray, 1838, Lower PLE, Central America (Webb and Perrigo, 1984). **Extant**

F. NOCTILIONIDAE Gray, 1821 Q. (PLE)–Rec. Volant

First: *Noctilio* Linnaeus, 1766, Upper PLE, West Indies. **Extant**

F. VESPERTILIONIDAE Gray, 1821 T. (BRT)–Rec. Volant

First: *Stehlinia* Revilliod, 1919, MP14, Chamblon and Egerkingen, Switzerland. **Extant**

F. MOLOSSIDAE Gervais, 1855 T. (LUT/BRT)–Rec. Volant

First: *Wellia* Storer, 1984, Swift Current Creek, Saskatchewan, Canada (Legendre, 1985). **Extant**

F. PHILISIDAE Sigé, 1985 T. (PRB) Volant

First and Last: *Philisis* Sigé, 1985, Jebel Qatrani Formation, Egypt.

F. NATALIDAE Gray, 1866 Q. (PLE)–Rec. Volant

First: *Natalus* Gray, 1838, Cuba. **Extant**

F. MYZOPODIDAE Thomas, 1904 Q. (PLE)–Rec. Volant

First: *Myzopoda* Milne Edwards and Grandidier, 1878, Lower PLE, Africa. **Extant**

F. FURIPTERIDAE Gray, 1866 **Extant** Volant

F. THYROPTERIDAE Miller, 1907 **Extant** Volant

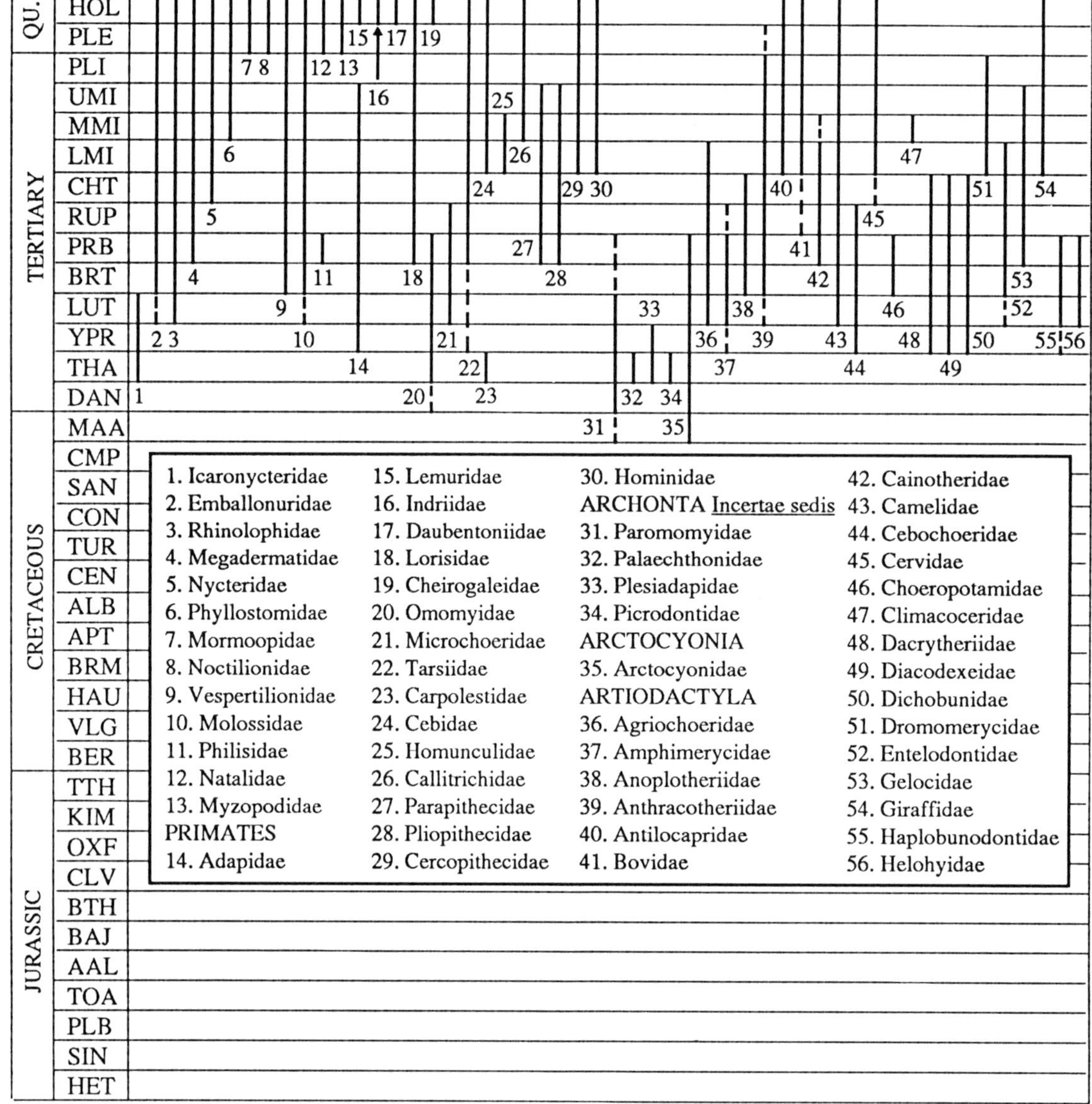

Fig. 41.5

F. MYSTACINIDAE Dobson, 1875 **Extant** Volant

Order PRIMATES Linnaeus, 1758

Here only the 'euprimates' are considered. As discussed below, a variety of phylogenetic arrangements for Primates have been proposed that have included the paromomyids, plesiadapids, microsyopids and their allies. Paromomyids, plesiadapids, palaechthonids, and picrodontids are included below in Archonta, *incertae sedis*. At practically all phylogenetic levels, opinions vary as to the included taxa. For different interpretations and reviews of the phylogenetic relationships among Primates, consult Szalay and Delson (1979), Grine *et al.* (1987), Szalay *et al.* (1987), Wible and Covert (1987), Simons (1989), Fleagle and Rosenberger (1990), Martin (1990) and Beard *et al.* (1991).

F. ADAPIDAE Trouessart, 1879
T. (YPR–UMI) Terr.

First: *Donrussellia* Szalay, 1976, MP7, Palette, France; *Cantius* Simons, 1962, MP7, Dormaal, Belgium; Willwood Formation, Wyoming, USA; MP7, Meudon, France (Russell *et al.*, 1988).

Last: *Sivaladapis* Gingerich and Sahni, 1979, and *Indraloris* Lewis, 1933, both Nagri Beds, India; *Sinoadapis* Wu and Pan, 1985, Lufeng Formation, Yunnan, China.

F. LEMURIDAE Gray, 1821 Q. (HOL)–Rec. Terr.

Comments: Subfossil lemurids from Madagascar include species of *Varecia* Gray, 1863 and *Megaladapis* Major, 1894. **Extant**

F. INDRIIDAE Burnett, 1828 Q. (HOL)–Rec. Terr.

Comments: Subfossil indriids from Madagascar include species of *Mesopropithecus* Standing, 1905, *Archaeolemur* Filhol, 1895, *Hadropithecus* Lorenz, 1899, *Thaumastolemur* Filhol, 1895, and *Archaeoindris* Standing, 1908. **Extant**

F. DAUBENTONIIDAE Gray, 1863
Q. (HOL)–Rec. Terr.

Comments: *Daubentonia* Geoffrey, 1895, subfossil, Madagascar. **Extant**

F. LORISIDAE Gray, 1821 T. (PRB)–Rec. Terr.

First: Lorisidae, gen. indet., Jebel Qatrani Formation, Egypt (Simons *et al.*, 1986). **Extant**

F. CHEIROGALEIDAE Gray, 1872 Q. (HOL)–Rec. Terr.

Comments: Szalay and Delson (1979) report subfossil specimens of living taxa from Madagascar. **Extant**

F. OMOMYIDAE Trouessart, 1879 T. (DAN/THA–PRB) Terr.

First: Upper Paleocene: *Decoredon* Xu, 1977, Wang-hudung Formation, Anhui, China, Asia. Upper Palaeocene: *Altiatlasius* Sige *et al.*, 1990, Adrar Mgorn, Morocco (Gingerich, 1990).
Last: *Macrotarsius* Clark, 1941, Renova Formation, Montana, USA; *Rooneyia* Wilson, 1966, Chambers Formation, Texas, USA.
Comments: Recent evidence on the basicranial anatomy calls into question the monophyly of the omomyids (*sensu* Szalay, 1976). Two subfamilies are generally recognized, the Anaptomorphinae Cope, 1883 and the Omomyinae Trouessart, 1879. The former appears to be monophyletic (Bown and Rose, 1987), whereas the latter now appears to be paraphyletic as it has classically been characterized by a number of workers. *Shoshonius* Granger, 1910 possesses several derived characteristics which suggest that it shares a more recent common ancestry with living *Tarsius* than other omomyids in which the relevant anatomy is known (Beard *et al.*, 1991; Martin, 1991). If tarsiids are the sister group of Anthropoidea, then the earliest anthropoids are earlier than late early Eocene in age (YPR).

F. MICROCHOERIDAE Lydekker, 1887 T. (LUT–RUP) Terr.

First: *Nannopithex* Stehlin, 1916, MP13, Bouxwiller, France.
Last: *Microchoerus* Wood, 1844, MP19, San Cugat, Spain; MP19, Mormont and Entreroches, Switzerland; *Pseudoloris* Stehlin, 1916, MP19, San Cugat, Spain (Szalay and Delson, 1979).

F. TARSIIDAE Gray, 1825 T. (?YPR/PRB)–Rec. Terr.

First: *Afrotarsius* Simons and Bown, 1985, Jebel Qatrani Formation, Egypt. ?Lower Eocene: *Shoshonius* Granger, 1910, Wind River Formation, Wyoming, USA (Beard *et al.*, 1991). **Extant**

F. CARPOLESTIDAE Simpson, 1935 T. (THA) Terr.

First: *Elphidotarsius* Gidley, 1923, Polecat Bench Formation, Wyoming; Lebo and Tongue River Formations, Montana, USA.
Last: *Carpolestes* Simpson, 1928, Willwood Formation, Wyoming, USA.

F. CEBIDAE Bonaparte, 1831 T. (LMI)–Rec. Terr.

First: *Tremacebus* Hershkovitz, 1974, Sacanana, Colhuehuapian, Argentina; *Dolichocebus* Kraglievich, 1951, Gaiman, Colhuehuapian, Argentina. **Extant**
Comments: *Branisella* Hoffstetter, 1969, Salla, Bolivia, may be a cebid or callitrichid according to the classification used here (Rosenberger *et al.*, 1990).

F. HOMUNCULIDAE Ameghino, 1894 T. (LMI–MMI) Terr.

First and Last: *Homunculus* Ameghino, 1891, Santa Cruz Beds, Santacrucian, Argentina.

F. CALLITRICHIDAE Gray, 1821 T. (MMI)–Rec. Terr.

First: *Micodon* Setoguchi and Rosenberger, 1985, *Mohanamico* Luchterhand *et al.*, 1986, both La Venta, Friasian, Columbia. **Extant**

F. PARAPITHECIDAE Schlosser, 1911 T. (PRB–UMI) Terr.

First: *Oatrania* Simons and Kay, 1983, *Apidium* Osborn, 1908 and *Parapithecus* Schlosser, 1911, all Jebel Qatrani Formation, Egypt.
Last: *Oreopithecus* Gervais, 1872, numerous localities in Italy, and possibly from the former USSR.

F. PLIOPITHECIDAE Zapfe, 1960 T. (PRB–UMI) Terr.

First: *Catopithecus* Simons, 1989, and *Proteopithecus* Simons, 1989, both Jebel Qatrani Formation, Egypt.
Last: *Pliopithecus* Gervais, 1849, Rudibanya, Hungary.

F. CERCOPITHECIDAE Gray, 1821 T. (LMI)–Rec. Terr.

First: *Victoriapithecus* von Koenigswald, 1969, Maboko and Ombo, Kenya; *Prohylobates* Fourtau, 1918, Jebel Zelten, Libya; Wadi Moghara, Egypt. **Extant**

F. HOMINIDAE Gray, 1825 T. (LMI)–Rec. Terr.

First: *Dryopithecus* Lartet, 1856, numerous localities, principally in East Africa (Szalay and Delson, 1979). **Extant**

ARCHONTA *incertae sedis*

There is considerable controversy concerning the ordinal affinities of the families included here, many of which have classically been included among the primates. New evidence on the skull and postcranial skeleton of paromomyids and plesiadapids (Szalay and Drawhorn, 1980; Beard, 1990; Kay *et al.*, 1990) suggests that these families are related to galeopithecids rather than primates among living mammals. For this paper, these families are tentatively assigned to the Archonta, *incertae sedis*. Picrodontids and palaechthonids are included here as well. For varying recent interpretations of the phylogenetic affinities of these taxa see Szalay and Delson (1979), Szalay *et al.* (1987), MacPhee *et al.* (1988), Gunnell (1989), Beard (1990), and Kay *et al.* (1990).

F. PAROMOMYIDAE Simpson, 1940 K. (MAA)/T. (DAN)–T. (LUT/BRT–PRB) Terr.

First: ?Cretaceous and/or Lower Palaeocene: *Purgatorius* Van Valen and Sloan, 1965, Hell Creek Formation, Montana, USA. Palaeocene: *Paromomys* Gidley, 1923, North Horn Formation, Utah; Nacimiento Formation, New Mexico, USA.
Last: *Ignacius* Matthew and Granger, 1921, Wagon Bed Formation, Wyoming, USA; *Phenacolemur* Matthew, 1915, Cypress Hills Formation, Saskatchewan, Canada (Storer, 1990).
Comments: Gunnell (1989) places *Purgatorius* in the monophyletic family Purgatoriidae Gunnell, 1986. Beard (1990), and Kay *et al.* (1990) refer *Ignacius* to Dermoptera.

F. PALAECHTHONIDAE Szalay, 1969 T. (THA) Terr.

First: ?*Plesiolestes* Jepsen, 1930, North Horn Formation, Utah, USA.

Last: *Palaechthon* Gidley, 1923, and *Palenochtha* Simpson, 1935, both Fort Union Formation, Wyoming, USA; *Plesiolestes* Jepsen, 1930, New Mexico; Utah; Fort Union and Evanston Formations, Wyoming; Montana, USA.

F. PLESIADAPIDAE Trouessart, 1897
T. (THA–YPR) Terr.

First: *Pronothodectes* Gidley, 1923, Evanston, Fort Union and Polecat Bench Formations, Wyoming; Tongue River and Lebo formations, Montana, USA; Saunders Creek Formation, Alberta, Canada.
Last: *Plesiadapis* Gervais, 1887, Willwood Formation, Wyoming, USA; MP7, Meudon, France; *Platychoerops* Charlesworth, 1854, MP7, Dormaal, Belgium; MP7, Pourcy, France (Szalay and Delson, 1979).
Comments: Kay *et al.* (1990) suggest that this family is closely allied with Dermoptera.

F. PICRODONTIDAE Simpson, 1937
T. (THA) Terr.

First: *Draconodus* Tomida, 1982, North Horn Formation, Utah, USA.
Last: *Zanycteris* Matthew, 1917, Animas Formation, Colorado, USA; *Picrodus* Douglass, 1908, Fort Union Formation, Wyoming, USA.

Grandorder UNGULATA Linnaeus, 1766

Mirorder EPARCTOCYONA McKenna, 1975

Order ARCTOCYONIA Van Valen, 1969

F. ARCTOCYONIDAE Gervais, 1870
K. (MAA)–T. (PRB) Terr.

First: *Protungulatum* Sloan and Van Valen, 1965, Frenchman Formation, Saskatchewan, Canada.
Last: *Lantianius* Chow, 1964, Shensi, China.

Order ARTIODACTYLA Owen, 1848

F. AGRIOCHOERIDAE Leidy, 1869
T. (LUT–LMI) Terr.

First: *Protoreodon* Scott and Osborn, 1887, Washakie Formation, Colorado (unpublished data, University of Colorado Museum, RKS); Friars Formation, California, USA (Golz and Lillegraven, 1977).
Last: *Agriochoerus* Leidy, 1850, Sharps Formation, South Dakota, USA.

F. AMPHIMERYCIDAE Stehlin, 1910
T. (?YPR–?RUP) Terr.

First: *Amphimeryx* Pomel, 1848, MP19, Montmartre, France.
Last: *Pseudamphimeryx* Stehlin, 1910, Bembridge Marls, England, UK (Hooker and Insole, 1980); *Amphimeryx* Pomel, 1848, Ronzon, France (Sudre, 1978).

F. ANOPLOTHERIIDAE Gray, 1821
T. (BRT–CHT) Terr.

First: *Robiacina* Sudre, 1969, Robiac, France; *Robiatherium* Sudre, 1988, MP16, Robiac Nord, France; *Dacrytherium* Filhol, 1876, MP16, Creechbarrow Limestone, England, UK.
Last: *Ephelcomenus* Hurzeler, 1938, ?MP24, Cadibona, Italy.

F. ANTHRACOTHERIIDAE Leidy, 1869
T. (LUT/BRT–PLI)/Q. (PLE?) Terr.

First: *Anthracokeryx* Pilgrim and Cotter, 1916, Ganda Kas, Pakistan; Burma; *Siamotherium* Suteethorn *et al.*, 1988, Thailand; *Probrachyodus* Xu and Chiu, 1962, Dongjun Formation, Guanxi, China; *Brachyodus* Deperet, 1895, Lumeiyi Formation, Yunnan, China; *Bakalovia* Nikolov and Heissig, 1985, and *Anthracosenex* Zdansky, 1930, both Heti Formation, Henan, China; *Prominatherium* Teller, 1884, former Yugoslavia; Romania; *Ulausuodon* Hu, 1963, Sharon Murun Formation, Mongolia.
Last: *Merycopotamus* Falconer and Cautley, 1845, Lower PLI or ?PLE, Sahabi, Libya.

F. ANTILOCAPRIDAE Gray, 1866
T. (LMI)–Rec. Terr.

First: *Merycodus* Leidy, 1854, Split Rock Formation, Wyoming; Sheep Creek Formation, Nebraska, USA. **Extant**

F. BOVIDAE Gray, 1821
T. (?RUP/BUR)–Rec. Terr.

First: Oligocene: *Hanhaicerus* Huang, 1985, Ulantatal, Mongolia; ?*Palaeohypsodontus* Trofimov, 1957, Hsanda Gol Formation, Mongolia. Lower Miocene: *Gazella* de Blainville, 1816, Gebel Zelten, Libya; Negev, Israel; *Eotragus* Pilgrim, 1939, Gebel Zelten, Libya; Artenay, France (Ginzburg, 1968); ?Asia; *Oioceras* Gaillard, 1902, Saudi Arabia; *Sinopalaeoceros* Chen, 1988, Xiejiaean, China; *Protragocerus* Deperet, 1887, Gebel Zelten, Libya. **Extant**

F. CAINOTHERIIDAE Cope, 1881
T. (PRB)–T. (LMI/MMI) Terr.

First: *Oxacron* Filhol, 1884, and *Plesiomeryx* Gervais, 1873, both MP17–19, Mouillac, France; *Paroxacron* Hurzeler, 1936, Mormount, DeBruge, France.
Last: *Cainotherium* Bravard, 1828, Bunol, Spain.

F. CAMELIDAE Gray, 1821 T. (LUT)–Rec. Terr.

First: *Poebrodon kayi* Gazin, 1955, Uinta Formation, Utah; Santiago Formation, California, USA (Golz and Lillegraven, 1977). **Extant**

F. CEBOCHOERIDAE Lydekker, 1883
T. (YPR–RUP) Terr.

First: *Cebochoerus* Gervais, 1848, MP10, Los Saleres, Spain.
Last: *Acotherulum* Gervais, 1850, MP21, Soumailles, France.

F. CERVIDAE Goldfuss, 1820
T. (CHT/LMI)–Rec. Terr.

First: Lower Oligocene: Cervid, indet. has been reported from Chaganbulage, China (Wang, 1992); *Dicrocerus* Lartet, 1837, Sansan, France; Tung Gur Formation, China; *Acteocemas* Ginsburg, 1985, Chilleurs-aux-Bois, France; *Stephanocemas* Colbert, 1936, Europe. **Extant**

F. CHOEROPOTAMIDAE Owen, 1845
T. (BRT–PRB) Terr.

First and Last: *Choeropotamus* Cuvier, 1821, Malperie and Quercy, France. Possible early record from MP16, Creechbarrow Limestone, England, UK.

F. CLIMACOCERIDAE Hamilton, 1978
T. (MMI) Terr.

First and Last: *Nyanzameryx* Thomas, 1984, and *Climacoceras* MacInnes, 1936, both Africa.

F. DACRYTHERIIDAE Deperet, 1917 T. (YPR–CHT) Terr.

First: *Cuisitherium* Sudre *et al.*, 1983, MP8, Avenay, France.
Last: *Tapirulus* Gervais, 1850, MP18, Ranet and Aubrelong, France; *Dacrytherium* Filhol, 1876, MP18, level of La Debruge.

F. DIACODEXEIDAE Gazin, 1955 T. (YPR–CHT) Terr.

First: *Diacodexis* Cope, 1882, Willwood Formation, Wyoming, USA; MP-7, Dormaal, Belgium; MP7, Palette, France; MP7, Silveirinha, Portugal; Kuldana Formation, Pakistan.
Last: *Leptochoerus* Leidy, 1856, Sharps Formation, South Dakota, USA.
Comments: *Diacodexis* is the most primitive artiodactyl and may be paraphyletic as it is currently defined (Sudre *et al.*, 1983; Krishtalka and Stucky, 1985; Stucky and Krishtalka, 1990).

F. DICHOBUNIDAE Turner, 1849 T. (YPR–CHT) Terr.

First: *Aumelasia* Sudre, 1980, and *Protodichobune* Lemoine, 1879, both from MP10, Grauves and other localities, France.
Last: *Metriotherium* Filhol, 1882, MP26, level of Mas de Pauffie, France.
Comments: The dichobunids formerly included taxa here assigned to the Diacodexeidae and Homacodontidae (Sudre *et al.*, 1983). Here they are considered an endemic Eurasian group.

F. DROMOMERYCIDAE Frick, 1937 T. (BUR–PLI) Terr.

First: *Asiagenes* Vislobokova, 1983, Mongolia; *Barbouromeryx* Frick, 1937, Antelope Creek, Nebraska; *Bouromeryx* Frick, 1937, Nebraska, USA; *Pediomeryx* Stirton, 1936, Coffee Ranch, Texas, USA (Webb, 1983).
Last: *Cranioceras* Matthew, 1918, North America.

F. ENTELODONTIDAE Lydekker, 1883 T. (LUT/BRT–LMI) Terr.

First: *Brachyhyops* Colbert, 1937, Porvenir, Texas; Duchesne River Formation, Utah; Wagon Bed Formation, Wyoming, USA; *Eoentelodon* Chow, 1958, Lumeiyi Formation, Yunnan, China; *Entelodon* Aymard, 1847, Villebramar, Rozon, Soumalles, France.
Last: *Daeodon* Cope, 1878, Runningwater Formation, Nebraska, USA.
Comments: An unnamed taxon, formerly referred to *Parahyus* (see West and Dawson, 1976) from the lower Uintan (Shoshonean) of the Washakie Formation, Colorado, USA, may represent the earliest entelodontid.

F. GELOCIDAE Schlosser, 1886 T. (PRB–UMI) Terr.

First: *Lophiomeryx* Pomel, 1854, Sevkhul, Asia; *Phaneromeryx* Schlosser, 1886, Quercy, France.
Last: *Pseudoceras* Frick, 1937, Gracias Formation, Honduras; Round Mountain Quarry, New Mexico, USA (Webb and Perrigo, 1984).

F. GIRAFFIDAE Gray, 1821 T. (LMI)–Rec. Terr.

First: *Palaeotragus* Gaudry, 1861, Rusinga, Kenya; Gebel Zelten, Libya. **Extant**

F. HAPLOBUNODONTIDAE Sudre, 1978 T. (YPR/LUT–PRB) Terr.

First: *Rhagatherium* Pictet and Humbert, 1855, MP10, ?Mas de Gimel, France; MP13, Geiseltal, Germany.
Last: *Amphirhagatherium* Deperet, 1908, MP20, level of Saint Capraise, France (Sudre, 1978; Legendre, 1987).

F. HELOHYIDAE Marsh, 1877 T. (LUT–PRB) Terr.

First: *Helohyus* Marsh, 1872, Bridger and Aycross Formations, Wyoming, USA; *Gobiohyus* Matthew and Granger, 1925, Lushi Formation, Honan, China.
Last: *Gobiohyus* Matthew and Granger, 1925, Irdin Manha Formation, Mongolia, Lu-mei-yi Formation, Yunnan and He-ti Formation, Shansi, China.
Comments: Report of a specimen of *Helohyus* from the Wind River Formation (Stucky, 1984) was in error. The specimen represents a deciduous tooth fragment of the titanothere *Eotitanops*.

F. HIPPOPOTAMIDAE Gray, 1821 T. (MMI)–Rec. Terr. (Fig. 41.6)

First: *Kenyapotamus* Pickford, 1983, Fort Ternan, Kenya. **Extant**

F. HOMACODONTIDAE Sinclair, 1891 T. (YPR–PRB) Terr.

First: *Hexacodus* Gazin, 1952, Wasatch and Wind River Formations, Wyoming, USA.
Last: *Bunomeryx* Wortman, 1898, *Mesomeryx* Peterson, 1919, *Pentacemylus* Peterson, 1931, and *Mytonomeryx* Gazin, 1955, all Uinta Formation, Utah, USA; *Tapochoerus* McKenna, 1958, Sespe Formation, California, USA.

F. HOPLITOMERYCIDAE Leinders, 1983 T. (UMI) Terr.

First and Last: *Hoplitomeryx matthei* Leinders, 1983, Gargano, Italy.

F. HYPERTRAGULIDAE Cope, 1879 T. (PRB–MMI) Terr.

First: *Parvitragulus* Emry, 1978, Devil's Graveyard and Capote Mountain Formations, Texas; Vieja Group, Texas; White River Formation, Wyoming, USA; *Hypertragulus* Cope, 1874, Prietos Formation, Mexico; Devil's Graveyard Formation, Texas, USA (Wilson, 1986); *Miomeryx* Matthew and Granger, 1925, Sevkhul Beds, Asia.
Last: *Andegameryx* Ginsburg, 1971, Pontigne, France.

F. LEPTOMERYCIDAE Zittel, 1893 T. (LUT/BRT–MMI) Terr.

First: *Hendryomeryx* Black, 1978, Wagon Bed Formation, Wyoming; ?Devil's Graveyard Formation, Texas, USA.
Last: *Pseudoparablastomeryx* Frick, 1937, Olcott Formation, Nebraska; Trinity River Pit, Texas, USA (Taylor and Webb, 1976; Tedford, 1981).

F. MIXTOTHERIIDAE Lydekker, 1883 T. (YPR–PRB) Terr.

First and Last: *Mixtotherium* Filhol, 1880, Les Badies, Spain; Quercy, France; Creechbarrow Limestone, England, UK.

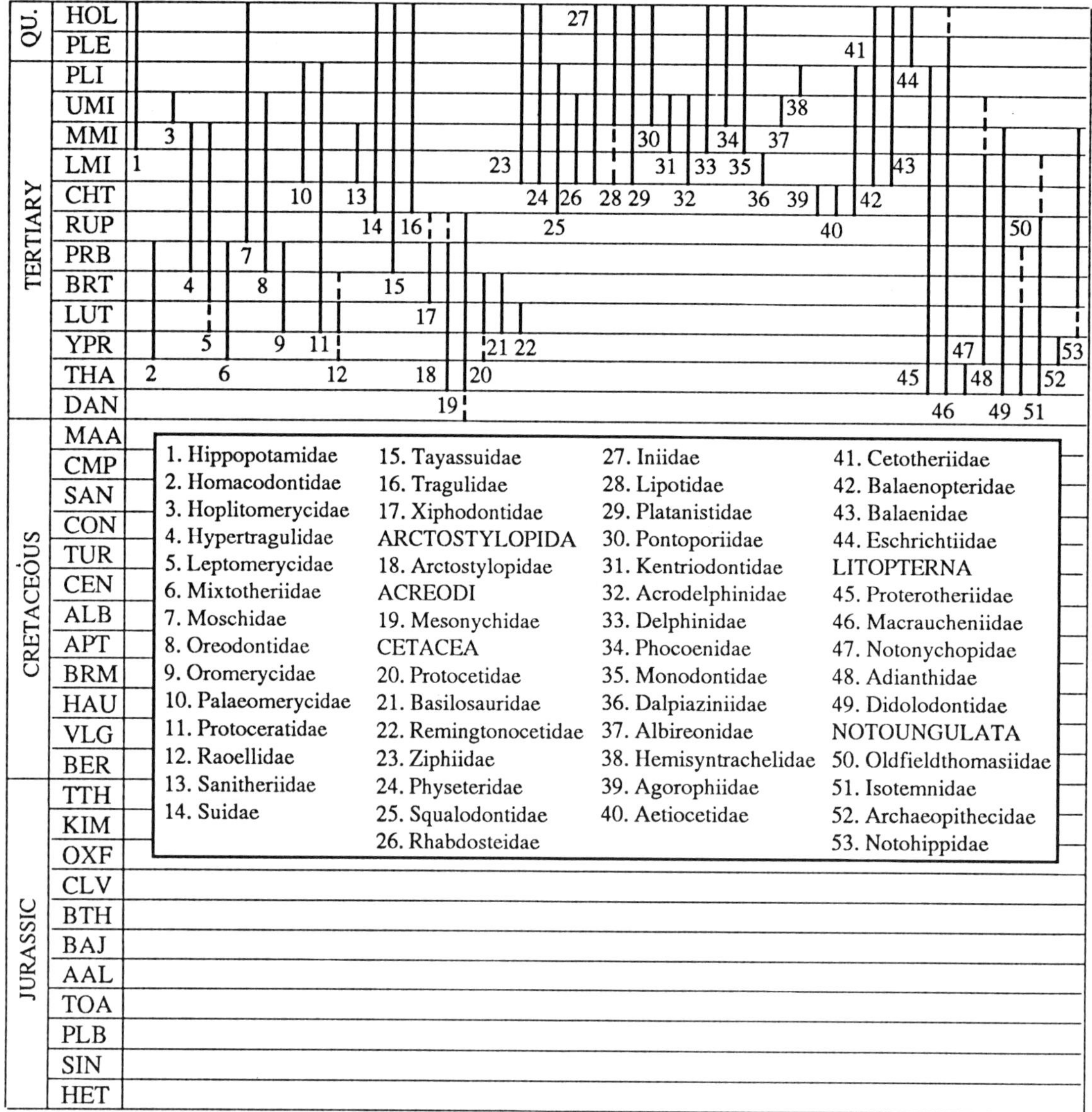

Fig. 41.6

F. MOSCHIDAE Gray, 1821 T. (RUP)–Rec. Terr.

First: *Dremotherium* Geoffrey, 1833, MP28, Pech Desse and Pech du Fraysse, France. **Extant**

F. OREODONTIDAE Leidy, 1869 T. (PRB–UMI) Terr.

First: *Aclistomycter* Wilson, 1971, Devil's Graveyard Formation, Texas, USA.

Last: *Ustatochoerus* Schultz and Falkenbach, 1941, Snake Creek Formation, Nebraska, USA.

F. OROMERYCIDAE Gazin, 1955 T. (LUT–PRB) Terr.

First: *Camelodon arapahovius* Granger, 1910, Wagon Bed Formation, Wyoming, USA; *Oromeryx* Marsh, 1894, Uinta Formation, Utah, USA; *Protylopus* Wortman, 1898, Uinta Formation, Utah, USA; Mission Valley Formation, California, USA; *Merycobunodon* Golz, 1976, Friars Formation, California, USA.

Last: *Eotylopus* Matthew, 1910, White River Formation, Wyoming, USA; *Montanatylopus* Prothero, 1986, Renova Formation, Montana; Rubio Peak Formation, New Mexico, USA (Lucas, 1986).

F. PALAEOMERYCIDAE Lydekker, 1883 T. (BUR–PLI) Terr.

First: *Palaeomeryx* Meyer, 1834, Gebel Zelten, Libya; Artenay, France; *Oriomeryx* Ginsburg, 1985, MN2, Saint-Gerard-le-Puy, France; *Propalaeomeryx* Lydekker, 1883, Nari Series, Baluchistan.

Last: *Palaeomeryx* Meyer, 1834, ?Lower PLI, Switzerland (Hamilton, 1973).

F. PROTOCERATIDAE Marsh, 1891 T. (LUT–PLI) Terr.

First: *Leptoreodon* Wortman, 1898, Uinta Formation, Utah; Friars Formation, California, USA (Golz and Lillegraven, 1977).

Last: *Kyptoceras amatorum* Webb, 1981, Bone Valley Formation, Lower PLI, Florida, USA.

F. RAOELLIDAE Sahni *et al.*, 1981 T. (?YPR–LUT/BRT) Terr.

First: *Khirtharia* Pilgrim, 1940, Kuldana Formation, Pakistan; *Indohyus* Rao, 1971, Kuldana Formation, Pakistan; *Kunmunella* Sahni and Khare, 1971, and *Raoella* Sahni and Khare, 1971, both Asia.
Last: Above genera and *Metkatius* Kumar and Sahni, 1985, Sabuthu Group, Pakistan (Thewissen *et al.*, 1987).

F. SANITHERIIDAE Simpson, 1945 T. (LMI–MMI) Terr.

First: *Diamantohyus* Stromer, 1922, Africa and Asia.
Last: *Sanitherium* Meyer, 1866, Africa; Koshialgarh, India; Bruck an der Mur, Germany; Greece (Thenius, 1959).

F. SUIDAE Gray, 1821 T. (CHT)–Rec. Terr.

First: *Palaeochoerus* Pomel, 1847, MP29, Montiviale, Italy. An earlier record may be *Hemichoerus* Filhol, 1882, Quercy, France. **Extant**

F. TAYASSUIDAE Palmer, 1897 T. (PRB)–Rec. Terr.

First: *Perchoerus* Leidy, 1869, White River Formation, South Dakota, USA; *Doliochoerus* Filhol, 1882, Argiles de Marbly, St Andre, Pech du Fraysse, Antoingt, France (Thenius, 1959). **Extant**

F. TRAGULIDAE Milne Edwards, 1864 T. (CHT)–Rec. Terr.

First: *Cryptomeryx* Schlosser, 1886, MP22, Mas de Got, France; *Iberomeryx* Gabunia, 1964, Benara, Georgia, former USSR (Sudre, 1984). **Extant**

F. XIPHODONTIDAE Flower, 1884 T. (BRT–?RUP) Terr.

First: *Dichodon* Owen, 1848, MP13–14, Egerkingen, Switzerland (Franzen, 1987).
Last: *Xiphodon* Cuvier, 1822, MP20, Saint Capraise, France.

Order ARCTOSTYLOPIDA Cifelli *et al.*, 1989

F. ARCTOSTYLOPIDAE Schlosser, 1923 T. (THA–PRB/?RUP) Terr.

First: *Asiostylops* Zheng, 1979, Chi-jiang Formation, Kiangsi, China; *Gashatostylops* Cifelli *et al.*, 1989, and *Palaeostylops* Matthew and Granger, 1925, both Gashato, Bayan Ulan, Naran Bulak and Nomogen Formations, Mongolia; *Sinostylops* Tang and Yan, 1976, Dou-mu Formation, Shung-ta-si Group, Anhui, China; *Bothriostylops* Zheng and Huang, 1986, Chijiang Formation, Jiang-xi Province, China; *Kazachostylops* Nesov, 1987, Pretashkent Svita, Kazakhstan, former USSR; *Arctostylops* Matthew, 1915, Polecat Bench Formation, Wyoming, USA.
Last: *Anatolostylops* Zhai, 1978, Lian-kan Formation, Sinkiang, China.

Mirorder CETE Linnaeus, 1758

Order ACREODI Matthew, 1909

F. MESONYCHIDAE Cope, 1875 T. (DAN/THA–RUP) Terr.

First: ?*Hukoutherium* Chow *et al.*, 1973, *Yantanglestes* Ideker and Yan, 1980, and *Dissacusium* Chow *et al.*, 1977, all Shang-hu Formation, Kwantung, China; *Ankalagan* Van Valen, 1980, Nacimiento Formation, New Mexico, USA.
Last: *Mongolestes* Szalay and Gould, 1966, Ulan Gochu, Mongolia.

Order CETACEA Brisson, 1762

Suborder ARCHAEOCETI Flower, 1883

F. PROTOCETIDAE Stromer, 1908 T. (YPR/LUT–BRT) Mar.

First: *Pakicetus* Gingerich and Russell, 1981, *Gandakasia* Dehm and Oettingen-Spielberg, 1958, *Ichthyolestes* Dehm and Oettingen-Spielberg, 1958, all Kuldana Formation, Pakistan. These genera may be Lutetian in age and, as such, first occurrences would also include *Indocetus* Sahni and Mishra, 1975, Berwali Series, India; *Protocetus* Fraas, 1904, Mokattum Formation, Egypt; India; *Pappocetus* Andrews, 1920, Ameki Formation, Nigeria.
Last: *Eocetus* Fraas, 1904, Birket-el-Qurum Formation, Egypt.

F. BASILOSAURIDAE Cope, 1868 T. (LUT–BRT) Mar.

First: *Prozeuglodon* Andrews, 1906, Wadi-Rayan, Egypt; *Zygorhiza* True, 1908, Creechbarrow Limestone, England, UK; *Basilosaurus* Harlan, 1834, Europe.
Last: *Basilosaurus* Harlan, 1834, Barton Clay, England, UK; *Zygorhiza* True, 1908, Jackson Formation, Georgia; Harleyville Formation, South Carolina, USA; Barton Clay, England, UK; *Dorudon* Gibbes, 1845, Kaiatan, New Zealand; Qasr el Sagha, Egypt.

Suborder ODONTOCETI Flower, 1867

F. REMINGTONOCETIDAE Kumar and Sahni, 1986 T. (LUT) Mar.

First and Last: *Andrewsiphius* Sahni and Mishra, 1975, Berwali Series, India (Russell and Zhai, 1987); *Remingtonocetus* Kumar and Sahni, 1986, India.

F. ZIPHIIDAE Gray, 1850 T. (BUR)–Rec. Mar.

First: *Squalodelphis* Dal Piaz, 1916, Belluno, Italy; *Notocetus* Moreno, 1892, Chubut, Argentina; Ziphiidae undet. Turkana Grit, Kenya (Barnes and Mitchell, 1978); *Medocinia* de Muizon, 1988, Gironde, France. **Extant**

F. PHYSETERIDAE Gray, 1821 T. (LMI)–Rec. Mar.

First: *Miokogia* Pilleri, 1986, and *Physeter* Linnaeus, 1758, both Europe; *Orycterocetus* Leidy, 1853, Pontigne and Pont Boutard, France; *Diaphorocetus* Ameghino, 1894, Chubut, Argentina; *Scaldicetus* Du Bus, 1867, Belluno, Italy. **Extant**

F. SQUALODONTIDAE Brandt, 1873 T. (CHT–PLI) Mar.

First: *Patriocetus* Abel, 1913, Linz, Belgium; *Microcetus* Kellogg, 1923, Asia; Europe; New Zealand; *Metasqualodon* Hall, 1911, and *Parasqualodon* Hall, 1911, both Junjukian, Australia; *Kelloggia* Mchedlidze, 1976, Azerbaijan, former USSR; *Eosqualodon* Rothausen, 1968, Europe; *Agriocetus* Abel, 1913, Linz, Belgium; *Squalodon* Grateloup, 1840, Jewett Sand, California, USA; Linz, Belgium; Ashiyu Formation, Japan; Duntroonian, New Zealand.
Last: *Squalodon* Grateloup, 1840, Europe.

F. RHABDOSTEIDAE Gill, 1871 T. (LMI–UMI) Mar.

First: *Macrodelphinus* Wilson, 1935, Jewett Sand, California, USA; *Phocaenopsis* Huxley, 1859, Blue Clay of

Parimoa, New Zealand; *Argyrocetus* Lydekker, 1894, Jewett Sand, California, USA; Chubut, Argentina.
Last: *Ziphiodelphis* Dal Piaz, 1912, Belluno, Italy; *Eurhinodelphis* Du Bus, 1867, Antwerp, Belgium; Pontigne, France.

F. INIIDAE Gray, 1846 T. (LMI)–Rec. FW/Mar.

First: *Proinia* True, 1910, Patagonian, Argentina. Iniid also from Visiano, Italy. **Extant**

F. LIPOTIDAE Zhou *et al.*, 1979
T. (MIO.)–Rec. FW

First: *Prolipotes* Zhou *et al.*, 1984, Guipong Co., Guangxi, China. **Extant**

F. PLATANISTIDAE Gray, 1846
T. (LMI)–Rec. FW.

First: *Allodelphis* Wilson, 1935, Pyramid Hill, California, USA. **Extant**

F. PONTOPORIIDAE Gray, 1870
T. (UMI)–Rec. FW/Mar.

First: *Brachydelphis* de Muizon, 1988, Cerro la Bruja, Peru. **Extant**

F. KENTRIODONTIDAE Slijper, 1936
T. (MMI–UMI) Mar.

First: *Kentriodon* Kellogg, 1927, *Liolithax* Kellogg, 1931, and *Delphinodon* Leidy, 1869, all Calvert Formation, Maryland, USA; *Leptodelphis* Kirpichnikov, 1954, and *Sarmatodelphis* Kirpichnikov, 1954, both Sarmat, Moldavia, former USSR; *Microphocaena* Kudrin and Tatarinov, 1965, Ternpol, Ukraine, former USSR; *Kampholophus* Rensberger, 1969, Monterey Formation, California, USA; *Atocetus* de Muizon, 1988, Pisco Formation, Peru.
Last: *Liolithax* Kellogg, 1931, Santa Margarita Formation, California, USA.

F. ACRODELPHINIDAE Abel, 1905
T. (LMI–UMI) Mar.

First: *Chamsodelphis* Gervais, 1848, Gironde, France; *Eoplatanista* Dal Piaz, 1916, Belluno, Italy; *Schizodelphis* Gervais, 1861, Italy; Moghara Formation, Egypt.
Last: *Acrodelphis* Abel, 1900, Europe; *Schizodelphis* Gervais, 1861, France.

F. DELPHINIDAE Gray, 1821
T. (MMI)–Rec. Mar.

First: *Tursiops* Gervais, 1855, Europe; *Orcinus* Fitzinger, 1860, Cava de Verrua, Italy. **Extant**

F. PHOCOENIDAE Gray, 1825
T. (UMI)–Rec. Mar.

First: *Salumiphocaena* Barnes, 1985, Monterey Formation, California, USA; *Piscolithax* de Muizon, 1983, Almejas Formation, California, USA; Pisco Formation, Peru; *Lomacetus* de Muizon, 1986, and *Australithax* de Muizon, 1988, both Pisco Formation, Peru. **Extant**

F. MONODONTIDAE Gray, 1821
T. (MMI)–Rec. Mar.

First: *Delphinapterus* Lacepede, 1804, Asia. **Extant**

F. DALPIAZINIIDAE de Muizon, 1988
T. (BUR) Terr.

First and Last: *Dalpiazinia* de Muizon, 1988, Molasses de Bologne, Italy.

F. ALBIREONIDAE Barnes, 1984 T. (UMI) Mar.

First and Last: *Albireo* Barnes, 1984, North America.

F. HEMISYNTRACHELIDAE Slijper, 1936
T. (PLI) Mar.

First and Last: *Hemisyntrachelus* Brandt, 1873, Lower PLI, Italy.
Comments: An upper Miocene hemisyntrachelid is reported from California by Savage and Barnes (1972).

F. AGOROPHIIDAE Abel, 1913 T. (CHT) Mar.

First and Last: *Agorophius* Cope, 1895, Ashley Marl, South Carolina, USA.

Suborder MYSTICETI Flower, 1864

F. AETIOCETIDAE Emlong, 1966 T. (CHT) Mar.

First and Last: *Aetiocetus* Emlong, 1966, North America.

F. CETOTHERIIDAE Brandt, 1872
T. (CHT–PLI) Mar.

First: *Cetotheriopsis* Brandt, 1871, Linz, Austria.
Last: *Rhegnopsis* Cope, 1896, City Point, Virginia, USA; *Cephalotropis* Cope, 1896, Yorktown Formation, Maryland, USA, both Upper PLI.

F. BALAENOPTERIDAE Gray, 1864
T. (LMI)–Rec. Mar.

First: *Plesiocetus* Van Beneden, 1859, Rio Negro Formation, Argentina; ?Europe. **Extant**

F. BALAENIDAE Gray, 1821 T. (LMI)–Rec. Mar.

First: *Morenocetus* Cabrera, 1926, Patagonian, Argentina. **Extant**

F. ESCHRICHTIIDAE Ellerman and Morrison-Scott, 1951 Q. (PLE)–Rec. Mar.

First: *Eschrichtius* Gray, 1864, Upper PLE, North America. **Extant**

Mirorder MERIDIUNGULATA McKenna, 1975

Order LITOPTERNA Ameghino, 1889

F. PROTEROTHERIIDAE Ameghino, 1887
T. (THA–PLI) Terr.

First: *Paranisolambda* Cifelli, 1983, and *Anisolambda* Paula Couto, 1952, both Itaboraí, Brazil.
Last: *Brachytherium* Ameghino, 1883, Chapadmalalan, Upper PLI, Argentina; *Licaphrium* Ameghino, 1887, San José Formation, Uruguay.

F. MACRAUCHENIIDAE Gill, 1872
T. (THA)–Q. (PLE/HOL) Terr.

First: *Victorlemoinea* Ameghino, 1901, Itaboraí, Brazil.
Last: *Macrauchenia* Owen, 1838, Buenos Aires, Argentina; Bolivia; Uruguay; Brazil; *Windhausenia* Kraglievich, 1930, South America (Anderson, 1984); *Xenorhinotherium* Cartella and Lessa, 1988, Bahia, Brazil, all Upper PLE or HOL.

F. NOTONYCHOPIDAE Soria, 1989
T. (THA) Terr.

First and Last: *Notonychops* Soria, 1989, Rio Loro Formation, Riochican, Tucuman, Argentina.

F. ADIANTHIDAE Ameghino, 1891
T. (YPR–Mio.) Terr.

First: *Indalecia* Bond and Vucetich, 1983, *Proectocion* Ameghino, 1904, and *Adiantoides* Simpson and Minoprio, 1949, all Casamyoran, Argentina.
Last: *Adianthus* Ameghino, 1891, Santacrucian, Argentina.

F. DIDOLODONTIDAE Scott, 1913
T. (THA–MMI) Terr.

First: *Paulacoutoia* Cifelli, 1983, *Lamegoia* Paula Couto, 1952, *Asmithwoodwardia* Ameghino, 1901, all Itaboraí, Brazil.
Last: *Megadolodus* McKenna, 1956, La Venta, Columbia.

Order NOTOUNGULATA Roth, 1903

F. OLDFIELDTHOMASIIDAE Simpson, 1945
T. (THA–LUT/BRT/PRB) Terr.

First: *Colbertia* de Paula Couto, 1952, and *Itaboraitherium* Paula Couto, 1970, both from Itaboraí, Brazil.
Last: *Tsamnichoria* Simpson, 1936, Mustersan, Argentina.
Comments: Cifelli (1982) indicates that several taxa are known from the Divisadero Largo Formation, Argentina.

F. ISOTEMNIDAE Ameghino, 1897
T. (THA–CHT/LMI) Terr.

First: *Isotemnus* Ameghino, 1897, Rio Chico, Argentina.
Last: *Trimerostephanos* Ameghino, 1895, *Pleurocoelodon* Ameghino, 1895, *Trigonolophodon* Roth, 1903, and *Lophocoelus* Ameghino, 1904, all Deseadan, Argentina.

F. ARCHAEOPITHECIDAE Ameghino, 1897
T. (YPR) Terr.

First and Last: *Archaeopithecus* Ameghino, 1897, and *Acropithecus* Ameghino, 1904, both Casamayoran, Argentina. According to Cifelli (1982), these taxa may be Riochican in age as well.

F. NOTOHIPPIDAE Ameghino, 1894
T. (LUT/BRT–MMI) Terr.

First: *Eomorphippus* Ameghino, 1901, Mustersan, Argentina.
Last: *Notohippus* Ameghino, 1891, Santacrucian, Brazil.

F. LEONTINIIDAE Ameghino, 1895
T. (CHT/LMI–MMI) Terr. (see Fig. 41.7)

First: *Ancylocoelus* Ameghino, 1895, *Leontinia* Ameghino, 1895, and *Scarrittia* Simpson, 1934, all Deseadan, Argentina. Cifelli (1982) reports an unnamed taxon from the Mustersan, Argentina.
Last: *Huilatherium* Villarroel and Guerro Diaz, 1985, La Venta, Columbia.

F. HOMALODOTHERIIDAE Gregory, 1910
T. (RUP/LMI–UMI) Terr.

First: *Asmodeus* Ameghino, 1895, Deseadan, Argentina.
Last: *Chasicotherium* Cabrera and Kraglievich, 1931, Chasicoan, Argentina.

F. TOXODONTIDAE Gervais, 1847
T. (RUP/LMI)–Q. (PLE) Terr.

First: *Proadinotherium* Ameghino, 1895, Deseadan, Argentina; *Posnanskytherium* Lazarte, 1943, Bolivia.
Last: *Mixotoxodon* van Frank, 1957, Hormiguero, El Salvador; Yeroconte and Orillas del Humuya, Hondurus; *Toxodon* Owen, 1840, Lujan Formation, Argentina; Arroio Touro Passo, Brazil (Pascual *et al.*, 1966; Anderson, 1984), all Upper PLE.

F. INTERATHERIIDAE Ameghino, 1887
T. (?THA–PLI) Terr.

First: *Transpithecus* Ameghino, 1901, Rio Chico, Argentina.
Last: *Protypotherium* Ameghino, 1882, Chasicoan, Argentina.

F. CAMPANORCIDAE Bond *et al.*, 1984
T. (Eoc.) Terr.

First and Last: *Campanorca* Bond *et al.*, 1984, South America.

F. MESOTHERIIDAE Alston, 1876
T. (PRB)–T. (CHT/LMI)/Q. (PLE) Terr.

First: Trachytheriinae Ameghino, 1894, Divisadero Largo Formation, Argentina (Simpson *et al.*, 1962). Cifelli (1982) suggests that mesotheriids are present in the Mustersan, Argentina.
Last: *Mesotherium* Serres, 1867, Deseadan, Argentina; ?Pulchense (Lower PLE), Argentina (Pascual *et al.*, 1966).

F. ARCHAEOHYRACIDAE Ameghino, 1897
T. (?THA)–T. (CHT/LMI) Terr.

First: *Eohyrax* Ameghino, 1901, Rio Chico, Argentina.
Last: *Archaeohyrax* Ameghino, 1897, Deseadan, Argentina.

F. HEGETOTHERIIDAE Ameghino, 1894
T. (RUP)–Q. (PLE) Terr.

First: *Ethegotherium* Simpson *et al.*, 1962, Divisadero Largo Formation, Argentina.
Last: *Paedotherium* Burmeister, 1888, Argentina.

F. HENRICOSBORNIIDAE Ameghino, 1901
T. (THA)–T. (PRB/RUP) Terr.

First: *Othnielmarshia* Ameghino, 1901, Itaboraí, Brazil.
Last: *Othnielmarshia* Ameghino, 1901, *Henricosbornia* Ameghino, 1901, and *Peripantostylops* Ameghino, 1904, all Casamayoran, Argentina. Cifelli (1982) reports an henricosborniid from the Deseadan of Argentina.

F. NOTOSTYLOPIDAE Ameghino, 1897
T. (YPR–LUT/BRT) Terr.

First: *Boreastylops* Vucetich, 1980, Lumbrera Formation, Argentina; *Edvardtrouessartia* Ameghino, 1901, *Homalostylops* Ameghino, 1901, and *Notostylops* Ameghino, 1897, all Casamayoran, Argentina.
Last: *Otronia* Roth, 1901, Mustersan, Argentina.

Order ASTRAPOTHERIA Lydekker, 1894

F. TRIGONOSTYLOPIDAE Ameghino, 1901
T. (THA)–LUT/BRT) Terr.

First: *Tetragonostylops* de Paula Couto, 1963, Itaboraí, Brazil.
Last: *Trigonostylops* Ameghino, 1897, Mustersan, Argentina (Soria and Bond, 1984).

F. ASTRAPOTHERIIDAE Ameghino, 1887
T. (YPR–MMI) Terr.

First: *Albertogaudrya* Ameghino, 1901, and *Scaglia* Simpson, 1957, both Casamayoran, Argentina.
Last: *Astrapotherium* Burmeister, 1879, Santacrucian,

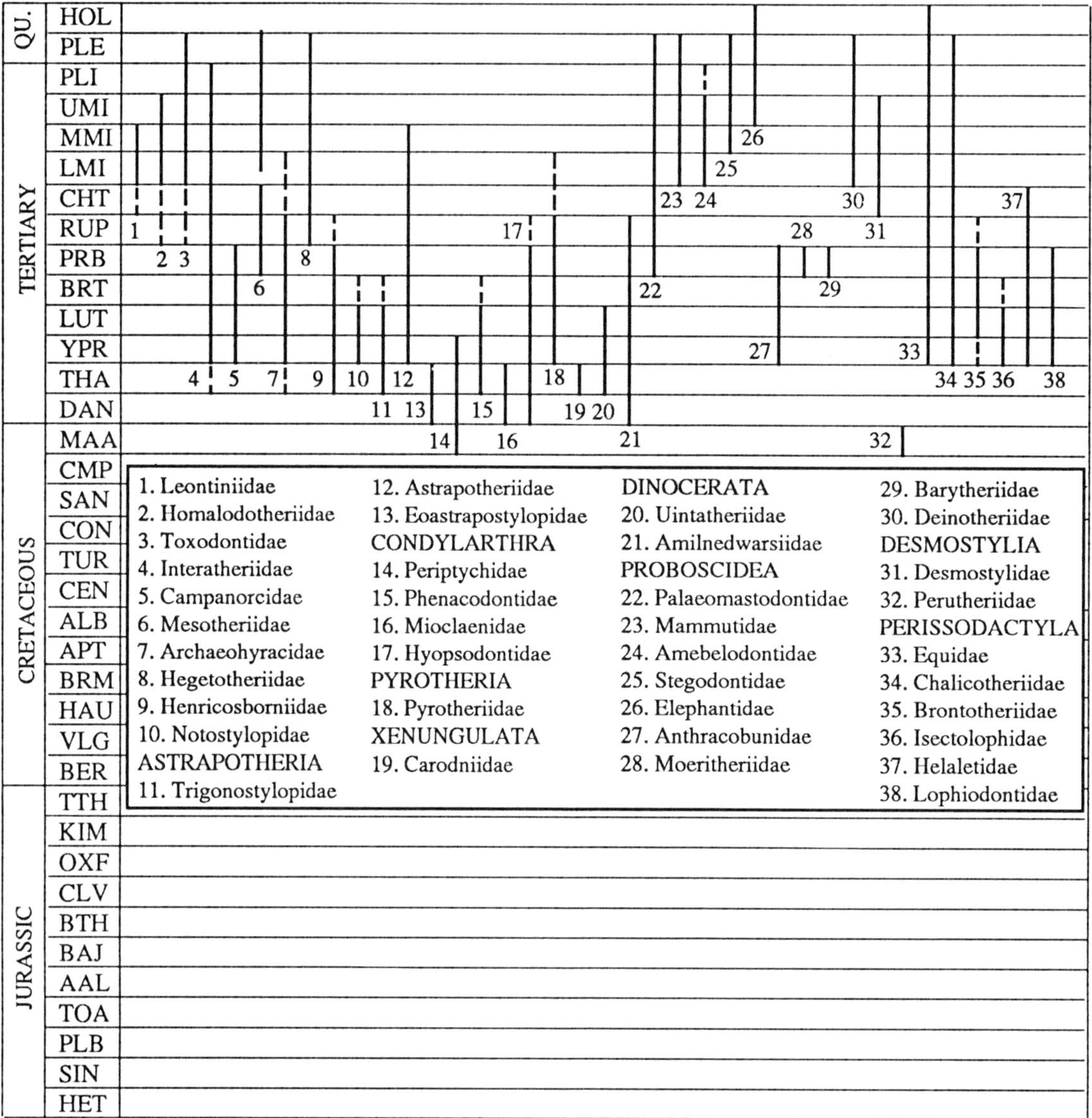

Fig. 41.7

Argentina; *Monoeidodon* Roth, 1898, Rio Collon Cura, Argentina.

F. EOASTRAPOSTYLOPIDAE Soria and Powell, 1981 T. (DAN/THA) Terr.

First and Last: *Eoastrapostylops* Soria and Powell, 1981, Rio Loro Formation, Argentina.

Order CONDYLARTHRA Cope, 1881

F. PERIPTYCHIDAE Cope, 1882 K. (MAA)–T. (YPR) Terr.

First: *Mimatuta* Van Valen, 1978, Hell Creek Formation, Montana, USA.
Last: *Lessnessina* Hooker, 1979, MP8+9, Blackheath Beds, England, UK.

F. PHENACODONTIDAE Cope, 1881 T. (THA–LUT/BRT) Terr.

First: *Tetraclaenodon* Scott, 1892, Nacimiento Formation, New Mexico; Goler Formation, California, USA.
Last: *Phenacodus* Cope, 1873, Wagon Bed and Bridger Formations, Wyoming, USA.
Comments: Most recent review by Thewessin (1990).

F. MIOCLAENIDAE Osborn and Earle, 1895 T. (DAN–THA) Terr.

First: *Bubogonia* Johnson, 1984, Ravenscrag Formation, Saskatchewan, Canada; *Tiznatzinia* Simpson, 1936, North America; *Choeroclaenus* Simpson, 1937, and *Promioclaenus* Trouessart, 1904, both from Nacimiento Formation, New Mexico, USA; *Tiuclaenus* de Muizon and Marshall, 1987, and *Molinodus* de Muizon and Marshall, 1987, both El Molino Formation, Bolivia. Also possibly Lower Palaeocene: *Palasiodon* Tong *et al.*, 1976, Shang-hu Formation, Kwantung, China.
Last: *Orthaspidotherium* Lemoine, 1885, and *Pleuraspidotherium* Lemoine, 1878, both MP6, Cernay, France; ?*Protoselene* Matthew, 1897, Blacks Peak Formation, Texas, Togwotee, Wyoming, USA.

F. HYOPSODONTIDAE Trouessart, 1879 T. (DAN)–T. (PRB/RUP) Terr.

First: *Litomylus* Simpson, 1935, and *Haplaletes* Simpson, 1935, both Western Interior, North America (Archibald *et al.*, 1987).
Last: *Hyopsodus* Leidy, 1870, Uinta Formation, Utah; Wagon Bed Formation, Wyoming, USA; *Epapheliscus* Van Valen, 1966, Europe; *Heptaconodon* Zdansky, 1930, Shantung, China.

Order PYROTHERIA Ameghino, 1895

F. PYROTHERIIDAE Ameghino, 1889
T. (YPR–CHT/LMI) Terr.

First: *Proticia* Patterson, 1977, north-west Venezuela; *Propyrotherium* Ameghino, 1901, Mustersan, Argentina.
Last: *Pyrotherium* Ameghino, 1888, Deseadan, Argentina.

Order XENUNGULATA Paula Couto, 1952

F. CARODNIIDAE Paula Couto, 1952
T. (THA) Terr.

First: *Carodnia* Simpson, 1935, Itaboraí, Brazil.
Last: *Carodnia* Simpson, 1935, Rio Chico, Argentina.

Order DINOCERATA Marsh, 1873

F. UINTATHERIIDAE Flower, 1876
T. (THA–LUT) Terr.

First: *Prodinoceras* Matthew *et al.*, 1929, Gashato Formation, Mongolia; Polecat Bench Formation, Wyoming, USA.
Last: *Tetheopsis* Cope, 1885, Washakie Formation, Wyoming, USA; *Uintatherium* Leidy, 1872, ?Uinta Formation, Utah, USA; *Eobasileus* Cope, 1872, Washakie Formation, Wyoming and Colorado, USA; Uinta Formation, Utah, USA; *Gobiatherium* Osborn and Granger, 1932, Irdin Manha Formation, Mongolia.

Order MERIDIUNGULATA *incertae sedis*

F. AMILNEDWARSIIDAE Soria, 1984
T. (DAN–RUP) Terr.

First: *Andinodus* de Muizon and Marshall, 1987, El Molino Formation, Argentina.
Last: *Acamana* Simpson *et al.*, 1962, Divisadero Largo, Argentina.

Mirorder TETHYTHERIA McKenna, 1975

Order PROBOSCIDEA Illiger, 1811

Tassy (1990) and Domning *et al.* (1986) provide information on the relationships among proboscideans, sirenians and desmostylians.

F. PALAEOMASTODONTIDAE Andrews, 1906
T. (PRB)–Q. (PLE) Terr.

First: *Phiomia* Andres and Beadnell, 1902, Jebel Qatrani Formation, Egypt; *Palaeomastodon* Andrews, 1901, Dor el Talha, Libya, Qasr el Sagha and Jebel Qatrani Formations, Egypt.
Last: *Cuvieronius* Osborn, 1923, Upper PLE, North and Central America; Chile.

F. MAMMUTIDAE Hay, 1922
T. (LMI)–Q. (PLE) Terr.

First: *Zygolophodon* Vacek, 1877, Djilancik, Kazakhstan, former USSR; Europe; *Mammut* Blumenbach, 1799, Lower Miocene, Africa and Europe.
Last: *Mammut* Blumenbach, 1799, Upper Pleistocene, North America.

F. AMEBELODONTIDAE Barbour, 1927
T. (LMI)–T. (UMI/PLI) Terr.

First: *Archaeobelodon* Tassy, 1984, Africa and Europe; *Platybelodon* Borissiak, 1928, Loperot, Kenya.
Last: *Amebelodon* Barbour, 1927 and *Platybelodon* Borissiak, 1928, both ?Lower PLI, North America.

F. STEGODONTIDAE Osborn, 1918
T. (MMI)–Q. (PLE) Terr.

First: *Stegolophodon* Schlesinger, 1917, India.
Last: *Stegodon* Falconer, 1857, Upper PLE, India; Sulawesi; Timor; Flores (Anderson, 1984).

F. ELEPHANTIDAE Gray, 1821
T. (UMI)–Rec. Terr.

First: *Primelephas* Maglio, 1970, and *Stegotetrabelodon* Petrocchi, 1941, both North and East Africa. **Extant**

F. ANTHRACOBUNIDAE Wells and Gingerich, 1983
T. (YPR/LUT/BRT–YPR/PRB) Terr.

First and Last: *Anthracobune* Pilgrim, 1940, Asia; *Ishatherium* Sahni and Kumar, 1980, Subatha Formation, India; *Lammidhania* Gingerich, 1977, Kuldana Formation, Pakistan.
Comments: *Minchenella* Zhang, 1980 from the Upper Palaeocene Nonshan Formation, Gungdang, China may be included here (Domning *et al.*, 1986).

F. MOERITHERIIDAE Andrews, 1906
T. (PRB) Terr.

First and Last: *Moeritherium* Andrews, 1901, Qasr el Sagha, Egypt; Libya; In Tafidet, Gao, Mali; M'Bodione Dadera, Senegal.

F. BARYTHERIIDAE Andrews, 1906
T. (PRB) Terr.

First and Last: *Barytherium* Andrews, 1901, Qasr el Sagha and Dor el Talha, Egypt.

F. DEINOTHERIIDAE Bonaparte, 1845
T. (BUR)–Q. (PLE) Terr.

First: *Prodeinotherium* Ehik, 1930, Korunga Beds, Kenya, Asia, Europe; *Deinotherium* Kaup, 1835, Asia.
Last: *Deinotherium* Kaup, 1835, Bed II, Middle PLE, Olduvai Gorge, Kenya; Hadar, Ethiopia (Butler *et al.*, 1967; Anderson, 1984).

Order DESMOSTYLIA Reinhart, 1953 (Fig. 41.7)

F. DESMOSTYLIDAE Osborn, 1905
T. (CHT–UMI) Terr.

First: *Cornwallius* Hay, 1923, Sooke Formation, British Columbia, Canada; *Behemotops* Domning *et al.*, 1986, Japan; Pysht Formation, Washington, USA.
Last: *Paleoparadoxia* Reinhart, 1959, Santa Margarita Formation, California, USA; ?Asia.

UNGULATA *incertae sedis*

F. PERUTHERIIDAE Van Valen, 1978
K. (MAA) Terr.

First and Last: *Perutherium* Thaler, 1967, Vilquechico, Peru.

Order PERISSODACTYLA Owen, 1848

See articles in Prothero and Schoch (1989) for data on Perissodactyla and alternative classifications.

F. EQUIDAE Gray, 1821 T. (YPR)–Rec. Terr.

First: *Hyracotherium* Owen, 1840, MP7, Pourcy, France; Willwood Formation, Wyoming, USA; Baja California, Mexico. **Extant**

F. CHALICOTHERIIDAE Gill, 1872
T. (YPR)–Q. (PLE) Terr.

First: *Paleomoropus* Radinsky, 1964, Willwood Formation, Wyoming, USA; *Lophiaspis* Deperet, 1910, MP7, Palette, France.
Last: *Nestoritherium* Kaup, 1859, Sangiran, Java (Hooijer, 1964); *Ancylotherium* Gaudry, 1863, Olduvai, Kenya; Makapansgat, South Africa (Anderson, 1984), Lower or ?Middle PLE.

F. BRONTOTHERIIDAE Marsh, 1873
T. (??YPR/LUT)–PRB/RUP) Terr.

First: ??Lower Eocene: *Xenicohippus* Bown and Kihm, 1981, Willwood Formation, Wyoming, USA; Huerfano Formation, Colorado, USA; *Lambdotherium* Cope, 1880, Wind River Formation, Wyoming, USA; ?*Eotitanops dayi* Dehm and Oettingen-Spielberg, 1958, Kuldana Formation, Pakistan. First typical brontothere represented by *Eotitanops* from the Wind River Formation, Wyoming; Huerfano Formation, Colorado, USA (Stucky, 1984).
Last: Brontotheriidae, indet., White River Formation, Wyoming, USA; *Pygmaetitan* Miao, 1982, Shinao Formation, China; *Embolotherium* Osborn, 1929, Houldjin Formation, Inner Mongolia, China; *Titanodectes* Granger and Gregory, 1943, Ulan Gochu, Shara Murun, Urtyn Obo, Mongolia.
Comments: The last brontothere from North America may be either Upper Eocene or Lower Oligocene in age. A skeleton is known from the Brule Member of the White River Formation, a unit that is considered to be early Oligocene in age. The skeleton was found above Ash 5 in the Douglass area, Wyoming, USA, which has a date of 33.7 Ma (Evanoff, 1990).

F. ISECTOLOPHIDAE Peterson, 1919
T. (YPR)–T. (LUT/BRT) Terr.

First: *Homogalax* Hay, 1899, Wu-tu Formation, Shantung, China; Willwood Formation, Wyoming, USA.
Last: *Isectolophus* Scott and Osborn, 1887, Uinta Formation, Utah, USA.

F. HELALETIDAE Osborn, 1892
T. (YPR–CHT) Terr.

First: *Heptodon* Cope, 1882, Wu-tu Formation, Shantung, China; Willwood Formation, Wyoming; Wind River Formation, Wyoming; San José Formation, New Mexico, USA; *Cymbalophus* Hooker, 1984, MP7, Suffolk Pebble Beds, England, UK; MP7, Palette, France; ?Asia (Hooker, 1984).
Last: *Colodon* Marsh, 1890, Renova Formation, Montana; White River Formation, Wyoming, USA.

F. LOPHIODONTIDAE Gill, 1872
T. (YPR–PRB) Terr.

First: *Lophiodon* Cuvier, 1822, MP7, Epernay, France.
Last: *Lophiodon* Cuvier, 1822, MP16, Hengitsbury, England, UK.

F. TAPIRIDAE Gray, 1825 T. (RUP)–Rec.
Terr. (Fig. 41.8)

First: *Protapirus* Filhol, 1877, Quercy, France; MP21, Mohren, Germany (Heissig, 1987); ?Aral Svita, Kazakhstan, former USSR (Russell and Zhai, 1987). **Extant**

F. LOPHIALETIDAE Matthew and Granger, 1925
T. (LUT/BRT–PRB) Terr.

First: *Parabreviodon* Reshetov, 1975, Mongolia; *Kalakotia* Ranga Rao, 1972, Kalakot, India; Sabutha, India; *Eoletes* Biryukov, 1974, Kolpak Svita, Kazakhstan, former USSR (Russell and Zhai, 1987); *Breviodon* Radinsky, 1965, Mongolia; *Schlosseria* Matthew and Granger, 1926, Arshanto Formation, Inner Mongolia, China.
Last: *Simplaletes* Qi, 1980, Bailuyuan, China (Wang, 1992); *Breviodon* Radinsky, 1965, Mongolia; *Schlosseria* Matthew and Granger, 1926, Mongolia.

F. DEPERETELLIDAE Radinsky, 1965
T. (LUT/BRT–RUP) Terr.

First: *Pachylophus* Tong and Lei, 1984, Hetaoyuan Formation, Henan, China; *Teleolophus* Matthew and Granger, 1925, Arshanto Formation, Inner Mongolia, China.
Last: *Teleolophus* Matthew and Granger, 1925, Ergilin Dzo Formation, Mongolia; Chaganbulage, China (Dashzeveg, 1992; Wang, 1992); *Haagella* Heissig, 1978, MP21, Mohren, Germany.

F. HYRACODONTIDAE Cope, 1879
T. (LUT–LMI) Terr.

First: *Hyrachyus* Leidy, 1871, Wind River Formation, Wyoming; Huerfano Formation, Colorado, USA; MP10, Grauves, France; *Yimengia* Wang, 1988, and *Rhodopagus* Radinsky, 1965, from Guan-zhuang Formation, Shantung, China; *Forstercooperia* Wood, 1939, Arshanto Formation, Inner Mongolia, China.
Last: *Peraceratherium* Forster Cooper, 1911, Agispe-Petrovslova, former USSR.

F. RHINOCEROTIDAE Gray, 1825
T. (LUT/BRT)–Rec. Terr.

First: *Amynodon* Marsh, 1877, and *Epitriplopus* Wood, 1927, both Uinta Formation, Utah, USA; *Prohyracodon* Koch, 1897, Europe; *Euryodon* Xu *et al.*, 1979, Da-cang-fang Formation, Honan, China; *Lushiamynodon* Chow and Xu, 1965, Lushi Formation, Honan, China. **Extant**

F. PALAEOTHERIIDAE Bonaparte, 1850
T. (YPR–CHT) Terr.

First: *Propachynolophus* Lemoine, 1891, MP10, Grauves, France.
Last: *Palaeotherium* Cuvier, 1804, MP 21, Soumaille, France; Bembridge Marl, England, UK; *Plagiolophus* Lemoine, 1891, MP 22, Villebramar, France; Bembridge Marl, England, UK.

Order TUBULIDENTATA Huxley, 1872

F. ORYCTEROPODIDAE Gray, 1821
T. (RUP)–Rec. Terr.

First: *Leptomanis* Filhol, 1893, Quercy, France (Thewissen, 1985). **Extant**

Order HYRACOIDEA Huxley, 1869, new rank, McKenna, this volume

Rasmussen (1989) has most recently revised the hyracoids.

F. PLIOHYRACIDAE Osborn, 1899
T. (?YPR/LUT/BRT–Q. (PLE) Terr.

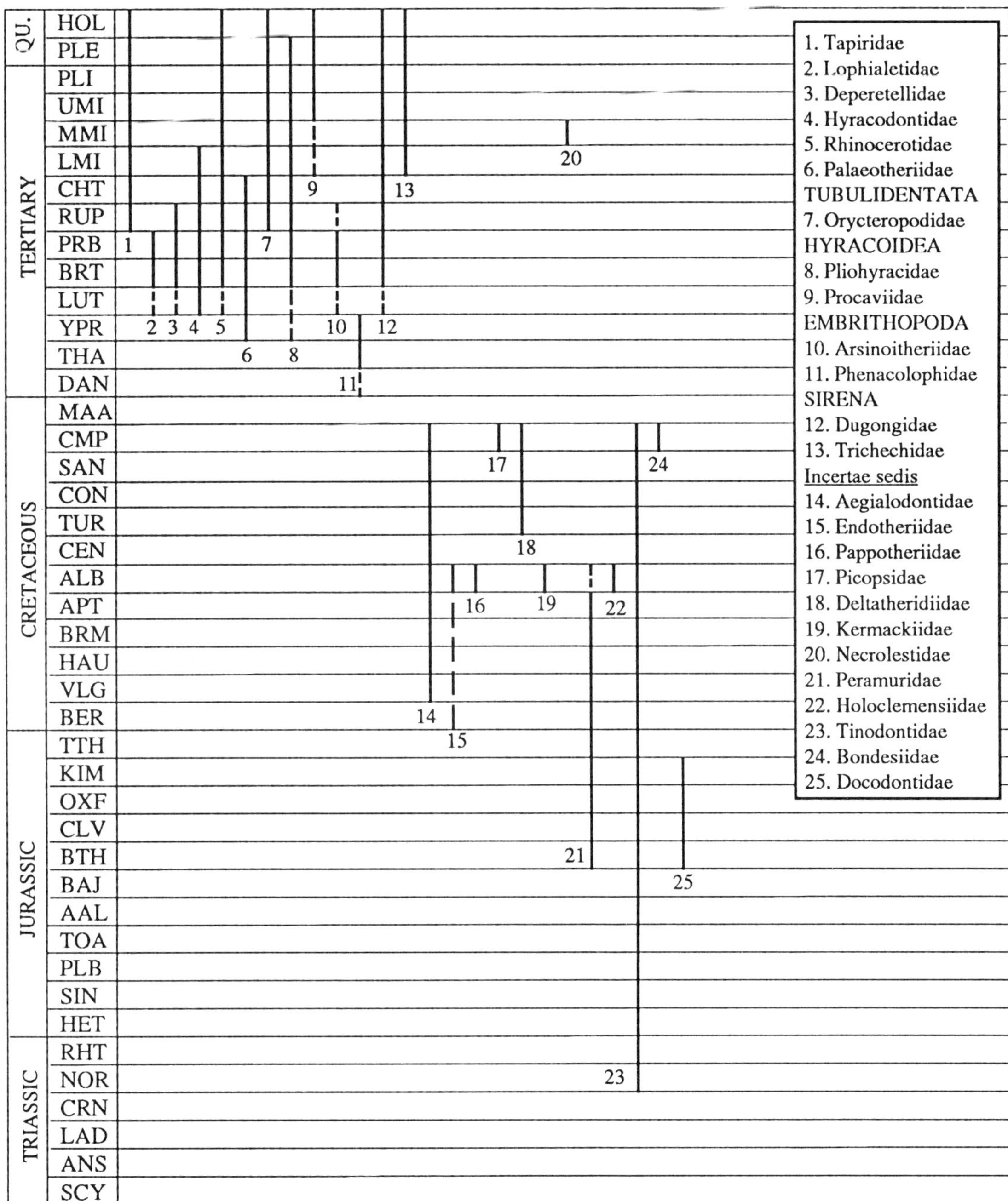

Fig. 41.8

First: ?Lower Eocene: *Titanohyrax* Matsumoto, 1921, Gour Lazib, Algeria.
Last: *Postschizotherium* von Koenigswald, 1932, Lower PLE, Choukoutien, China.

F. PROCAVIIDAE Thomas, 1892
T. (LMI/UMI)–Rec. Terr.

First: A procaviid is known from the Upper Miocene of Nakali, Kenya; *Prohyrax* Stromer, 1926, may be a procaviid (Rasmussen, 1989). **Extant**

Order EMBRITHOPODA Andrews, 1906

F. ARSINOITHERIIDAE Andrews, 1904
T. (LUT/BRT–PRB/RUP) Terr.

First: *Palaeoamasia* Ozansoy, 1966, Turkey.
Last: *Crivadiatherium* Radulesco *et al.*, 1976, Romania.

F. PHENACOLOPHIDAE South China Red Bed Team, 1977, ?T. (DAN/THA–YPR) Terr.

First and Last: *Phenacolophus* Matthew and Granger, 1925, Kashat Svita, Mongolia; Phenacolophidae, gen. *et* sp. *nov.*, Nung-shan Formation, Kwantung, China.

EPITHERIA *incertae sedis*

Order SIRENIA Illiger, 1811

F. DUGONGIDAE Gray, 1821
T. (LUT/BRT)–Rec. FW

First: *Eotheroides* Palmer, 1899, Mokattam Beds and Qasr el Sagha, Egypt; ?*Protosiren* Abel, 1904, Mokattam Beds, Egypt; Blaye Limestone, France; Asia (Butler *et al.*, 1967); *Prorastomus* Owen, 1855, Yellow Limestone Group, Jamaica (Butler *et al.*, 1967). **Extant**

F. TRICHECHIDAE Gray, 1825
T. (LMI)–Rec. FW

First: *Sirenotherium* de Paula Couto, 1967, Argentina.
Extant

THERIA *incertae sedis*

F. AEGIALODONTIDAE Kermack *et al.*, 1968
K. (VLG–CMP) Terr.

First and Last: *Aegialodon* Kermack *et al.*, 1965, Cliff End, England, UK; Asia.
Last: *?Zygiocuspis* Cifelli, 1990, Wahweap Formation, Utah, USA (tentatively assigned here).

F. ENDOTHERIIDAE Shikama, 1947 K. (?l) Terr.

First and Last: *Endotherium* Shikama, 1947, Fuhsin Coal Field, Liaoning, China.

F. PAPPOTHERIIDAE Slaughter, 1965
K. (ALB) Terr.

First and Last: *Pappotherium* Slaughter, 1965, and *Slaughteria* Butler, 1978, Trinity Sands, Texas, USA; *Bobolestes* Nesov, 1985, Uzbekistan, former USSR.

F. PICOPSIDAE Fox, 1980 K. (CMP) Terr.

First and Last: *Picopsis* Fox, 1980, Milk River Formation, Alberta, Canada.

F. DELTATHERIDIIDAE Gregory and Simpson, 1926
K. (TUR–CMP) Terr.

First: *Sulestes* Nesov, 1985, Uzbekistan, former USSR.
Last: *Deltatheridium* Gregory and Simpson, 1926, Djadohkta Formation, Mongolia.

F. KERMACKIIDAE Butler, 1978 K. (ALB) Terr.

First and Last: *Kermackia* Slaughter, 1971, and *Trinititherium* Butler, 1978, both Trinity Sands, Texas, USA.

F. NECROLESTIDAE Ameghino, 1894
T. (MMI) Terr.

First and Last: *Necrolestes* Ameghino, 1891, Santacrucian, Argentina.

F. PERAMURIDAE Kretzoi, 1946
J. (BTH)–K. (?ALB) Terr.

First: *Palaeoxonodon* Freeman, 1976, Forest Marble, England, UK.
Last: *Arguimus* Dashzeveg, 1979, Asia.

F. HOLOCLEMENSIIDAE Aplin and Archer, 1987
K. (ALB) Terr.

First and Last: *Holoclemensia* Slaughter, 1968, Paluxy Formation, Texas, USA.

MAMMALIA *incertae sedis*

F. TINODONTIDAE Marsh, 1887
(including Kuehneotheriidae)
Tr. (NOR)–K. (CMP) Terr.

First: *Kuehneotherium* Kermack *et al.*, 1968, Bridgend, South Wales, UK.
Last: *Mictodon* Fox, 1984, Milk River Formation, Alberta, Canada; *Bondesius* Bonaparte, 1990, El Molino Formation, Argentina.

F. BONDESIIDAE Bonaparte, 1990
K. (CMP) Terr.

First and Last: *Bondesia* Bonaparte, 1990, Los Alamitos Formation, Argentina.

Order DOCODONTA Kretzoi, 1946

F. DOCODONTIDAE Simpson, 1947
J. (BTH–KIM) Terr.

First: *Borealestes* Waldman and Savage, 1972, Ostracod Limestone, Scotland, UK; *Simpsonodon* Kermack, 1987, Forest Marble, England, UK.
Last: *Haldanodon* Kühne and Krusat, 1972, Guimarota, Portugal; *Docodon* Marsh, 1881, Morrison Formation, Colorado and Wyoming, USA.

REFERENCES

Ameghino, F. (1906) *Les Formations Sédimentaires du Crétacé Supérieur et du Tertiaire Patagonie: Avec un Parallele entre leurs Faunas Mammalogiques et Celles de l'Ancien Continent*. Juan A. Alsina, Buenos Aires, 567 pp.

Anderson, E. (1984) Who's who in the Pleistocene: a mammalian bestiary, in *Quaternary Extinctions: A Prehistoric Revolution* (eds P. S. Martin and R. G. Klein), University of Arizona Press, Tucson, pp. 40–89.

Archer, M., Godthelp, H., Hand, S. J. *et al.* (1989) Fossil mammals of Riversleigh, Northwestern Queensland: Preliminary overview of biostratigraphy, correlation and environmental change. *Australian Zoologist*, **25**, 29–65.

Archibald, J. D., Clemens, W. A. and Gingerich, P. D. (1987) First North American land mammal ages of the Cenozoic Era, in *Cenozoic Mammals of North America: Geochronology and Biostratigraphy* (ed. M. O. Woodburne), University of California Press, Berkeley, pp. 24–76.

Barnes, L. G. and Mitchell, E. (1978) Cetacea, in *Evolution of African Mammals* (eds J. J. Maglio and H. B. S. Cooke), Harvard University Press, Cambridge, pp. 582–602.

Barnosky, A. D. (1981) A skeleton of *Mesoscalops* (Mammalia, Insectivora) from the Miocene Deep River Formation, Montana, and a review of the proscalopid moles: evolutionary, functional, and stratigraphic relations. *Journal of Vertebrate Paleontology*, **1**, 285–339.

Beard, K. C. (1990) Gliding behaviour and palaeoecology of the alleged primate family Paromomyidae (Mammalia, Dermoptera). *Nature*, **345**, 340–1.

Beard, K. C. and Tabrum, A. R. (1991) The first early Eocene mammal from eastern North America: an omomyid primate from the Bashi Formation, Lauderdale County, Mississippi. *Mississippi Geology*, **11** (2), 1–6.

Beard, K. C., Krishtalka, L. and Stucky, R. K. (1991) First skulls of the early Eocene primate *Shoshonius cooperi* and the anthropoid–tarsier dichotomy. *Nature*, **349**, 64–7.

Berta, A. and Ray, C. E. (1990) Skeletal morphology and locomotor capabilities of the archaic pinniped *Enaliarctos mealsi*. *Journal of Vertebrate Paleontology*, **10**, 141–57.

Black, C. C. (1971) Paleontology and Geology of the Badwater Creek area, central Wyoming. Part 7. Rodents of the family Ischyromyidae. *Annals of Carnegie Museum*, **43**, 179–213.

Bonaparte, J. F. (1990) New Late Cretaceous mammals from the Los Alamitos Formation, northern Patagonia. *National Geographic Research*, **6**, 63–93.

Bonaparte, J. F., Krause, D. W. and Kielan-Jaworowska, Z. (1989) *Ferugliotherium windhauseni* Bonaparte, the first known multituberculate from Gondwanaland. *Journal of Vertebrate Paleontology*, **10**, p. 14A.

Bondesio, P. (1978) Nuevos restos de Cardiatheriinae (Rodentia, Hydrochoeridae) en el territorio Argentino; inferencias paleoambientales; nota preliminar. *Ameghiniana*, **15**, 229–34.

Bondesio, P. and Villenueva Bocquentin, J. (1987) Novedos restas de Neoepiblemidae (Rodentia, Hystricognathi) del Mioceno tardio de Venezuela, inferences paleoabvientales. *Ameghiniana*, **25**, 31–7.

Bown, T. M. and Rose, K. D. (1987) Patterns of dental evolution in early Eocene anaptomorphine Primates (Omomyidae) from the Bighorn Basin, Wyoming. *Paleontological Society Memoir*, **23**, 1–162.

Brunet, M. and Vianey-Liaud, M. (1987) Mammalian reference levels MP 21–30. *Münchner geowissenschaftlichen Abhandlungen, Munich*, (A), **10**, 30–1.

Butler, P. M., Clemens, W. A., Graham, S. F. *et al.* (1967) Mammalia, in *The Fossil Record* (eds W. B. Harland, C. H. Holland, M. R. House *et al.*), The Geological Society of London, London, pp. 763–87.

Chaline, J. and Mein, P. (1979) *Les Rongeurs et l'Évolution*. Doin Editeurs, Paris, 235 pp.

Cifelli, R. L. (1982) South American ungulate evolution and extinction, in *The Great American Biotic Interchange* (eds F. G. Stehli and S. D. Webb), Plenum Press, New York, pp. 249–66.

Cifelli, R. L. (1990) Cretaceous mammals of southern Utah. III. Therian mammals from the Turonian (early Late Cretaceous). *Journal of Vertebrate Paleontology*, **10**, 332–45.

Clemens, W. A., Lillegraven, J. A., Lindsay, E. H. *et al.* (1979) Where, when and what – a survey of known Mesozoic mammal distribution, in *Mesozoic Mammals: the First Two-thirds of Mammalian History* (eds J. A. Kielan-Jaworowska and W. A. Clemens), University of California Press, Berkeley, pp. 7–58.

Dashzeveg, D. and Russell, D. E. (1988) Palaeocene and Eocene Mixodontia (Mammalia, Glires) of Mongolia and China. *Palaeontology*, **31**, 129–64.

Dawson, M. R. and Krishtalka, L. (1984) Fossil history of the families of recent mammals, in *Orders and Families of Recent Mammals of the World* (eds S. Anderson and J. K. Jones), John Wiley and Sons, New York, pp. 11–57.

Domning, D. P., Ray, C. E. and McKenna, M. C. (1986) Two new Oligocene desmostylians and a discussion of tethytherian systematics. *Smithsonian Contributions to Paleobiology*, **59**, 1–56.

Emry, R. J. (1981) New material of the Oligocene muroid rodent *Nonomys*, and its bearing on muroid origins. *American Museum Novitates*, **2712**, 1–14.

Evanoff, E. (1990) Late Eocene and Early Oligocene Paleoclimates as Indicated by the Sedimentology and Nonmarine Molluscs of the White River Formation near Douglass, Wyoming. Unpublished PhD dissertation, University of Colorado, Boulder, 440 pp.

Fleagle, J. G. and Rosenberger, A. L. (1990) (eds) *The Platyrrhine Fossil Record*. Academic Press, London, 254 pp.

Flynn, J. J. and Galiano, H. (1982) Phylogeny of early Tertiary Carnivora, with a description of a new species of *Protictis* from the middle Eocene of northwestern Wyoming. *American Museum Novitates*, **2725**, 1–64.

Franzen, J. L. (1987) Mammalian reference levels MP 10–13. *Münchner Geowissenschaftlichen Abhandlungen, Munich*, (A), **10**, 24–5.

Freeman, E. F. (1976) Mammal teeth from the Forest Marble (middle Jurassic) of Oxfordshire, England. *Science*, **194**, 1053–5.

Gheerbrant, E. (1987) Les vertébrés continentaux de l'Adrar Mgorn (Maroc, Paleocene): une dispersion de mammifères transtéthysienne aux environs de la limite Mesozoique–Cénozoique. *Geodynamica Acta, Paris*, **1**, 233–46.

Gingerich, P. D. (1990) African dawn for primates. *Nature*, **346**, p. 411.

Gingerich, P. D. and Deutsch, H. A. (1989) Systematics and evolution of early Eocene Hyaenodontidae (Mammalia, Creodonta) in the Clarks Fork Basin, Wyoming. *Contributions from the University of Michigan, Museum of Paleontology*, **27**, 327–91.

Godinot, M. (1987) Mammalian reference levels MP 1–10. *Münchner geowissenschaftenlichen Abhandlungen, Munich*, (A), **10**, 21–3.

Golz, D. J. and Lillegraven, J. A. (1977) Summary of known occurrences of terrestrial vertebrates from Eocene strata of southern California. *Contributions to Geology, Laramie*, **15**, 43–60.

Grine, F. E., Fleagle, J. G. and Martin, L. B. (eds) (1987) *Primate Phylogeny*. Academic Press, London, 146 pp.

Gunnell, G. F. (1989) Evolutionary history of Microsyopoidea (Mammalia, ?Primates) and the relationship between Plesiadapiformes and Primates. *University of Michigan, Papers on Paleontology*, **27**, 1–157.

Hamilton, W. R. (1973) The lower Miocene ruminants of Gebel Zelten, Libya. *Bulletin of the British Museum (Natural History), Geology Series*, **21**, 75–150.

Hartenberger, J. (1990) L'origine des Theridomyoidea (Mammalia, Rodentia): données nouvelles et hypothèses. *Comptes Rendus de l'Académie des Sciences, Paris*, **311**, 1017–23.

Hartenberger, J., Martinez, C. and Said, A. B. (1985) Découverte de mammifères d'âge Eocene inférieur in Tunisie Centrale. *Comptes Rendus de l'Académie des Sciences*, **301**, Série 2, p. 649.

Heissig, K. (1982) Ein Edentate aus dem Oligózän Süddeutschlands. *Mitteilungen der Bayerische Staatssammlung für Paläontologie und Historische Geologie*, **22**, 91–6.

Heissig, K. (1987) Changes in the rodent and ungulate fauna in the Oligocene fissure fillings of Germany. *Münchner Geowissenschaftlichen Abhandlungen*, (A), **10**, 101–8.

Hooijer, D. A. (1964) New records of mammals from the Middle Pleistocene of Sangiran, central Java. *Zoologische Mededlingen, Leiden*, **40**, 73–88.

Hooker, J. J. (1984) A primitive ceratomorph (Perissodactyla, Mammalia) from the early Tertiary of Europe. *Zoological Journal of the Linnean Society*, **82**, 229–44.

Hooker, J. J. (1987) Mammalian reference levels MP 14–16. *Münchner geowissenschaftlichen Abhandlungen, Munich*, (A), **10**, 26–7.

Hooker, J. J. and Insole, A. N. (1980) The distribution of mammals in the English Palaeogene. *Tertiary Research*, **3**, 31–45.

Hunt, R. M. Jr (1989) Evolution of the aeluroid Carnivora: Significance of the ventral promontorial process of the petrosal, and the origin of basicranial patterns in the living families. *American Museum Novitates*, **2930**, 1–32.

Ivy, L. D. (1990) Systematics of late Paleocene and early Eocene Rodentia (Mammalia) from the Clarks Fork Basin, Wyoming. *Contributions from the Museum of Paleontology, University of Michigan*, **28**, 21–70.

Kay, R. F., Thorington, R. W. and Houde, P. (1990) Eocene plesiadapiform shows affinities with flying lemurs, not primates. *Nature*, **345**, 342–5.

Kielan-Jaworoska, Z., Dashzeveg, D. and Trofimov, B. A. (1987) Early Cretaceous multituberculates from Mongolia and a comparison with late Jurassic forms. *Acta Paleontologica Polonica*, **32**, 3–47.

Krause, D. W. (1987) Systematic revision of the genus *Prochetodon* (Ptilodontidae, Multituberculata) from the late Paleocene and early Eocene of western North America. *Contributions from the Museum of Paleontology, The University of Michigan*, **27**, 221–36.

Krishtalka, L. and Setoguchi, T. (1977) Paleontology and geology of the Badwater Creek area, central Wyoming. Part 13. The late Eocene Insectivora and Dermoptera. *Annals of Carnegie Museum*, **46**, 71–99.

Krishtalka, L. and Stucky, R. K. (1985) Revision of the Wind River faunas, early Eocene of central Wyoming. Part 7. Revision of *Diacodexis* (Mammalia, Artiodactyla). *Annals of Carnegie Museum*, **54**, 413–86.

Krishtalka, L., Emry, R. J., Storer, J. E. *et al.* (1982) Oligocene multituberculates (Mammalia: Allotheria): Youngest known record. *Journal of Paleontology*, **56**, 791–4.

Krishtalka, L., Stucky, R. K., West, R. M. *et al.* (1987) Eocene (Wasatchian through Duchesnean) chronology of North America, in *Cenozoic Biochronology of North America* (ed. M. O. Woodburne), University of California Press, Berkeley, 336 pp.

Kurtén, B. and Anderson, E. (1980) *Pleistocene Mammals of North America*. Columbia University Press, New York, 442 pp.

Legendre, S. (1985) Molossides (Mammalia, Chiroptera) cénozoïques de l'ancien et du nouveau monde; statut systématique; intégration phylogénique des données. *Neues Jahrbuch für Geologie und Paläontologie, Abhandlungen*, **170**, 205–27.

Legendre, S. (1987) Mammalian reference levels MP 17–20. *Münchner geowissenschaftlichen Abhandlungen, Munich*, (A), **10**, 28–9.

Li, C. and Ting, S. (1983) The Paleogene mammals of China. *Bulletin of Carnegie Museum of Natural History*, **21**, 1–93.

Lillegraven, J. A. and McKenna, M. C. (1986) Fossil mammals from the 'Mesaverde' formation (Late Cretaceous, Judithian) of the Bighorn and Wind River basins, Wyoming, with definitions of late Cretaceous North American land-mammal 'ages'. *American Museum Novitates*, **2840**, 1–68.

Lillegraven, J. A., Kielan-Jaworowska, Z. and Clemens, W. A. (eds) (1979) *Mesozoic Mammals: the first two-thirds of Mammalian History*. University of California Press, Berkeley, 311 pp.

Lillegraven, J. A., McKenna, M. C. and Krishtalka, L. (1982) Evolutionary relationships of middle Eocene and younger species of *Centetodon* (Mammalia, Insectivora, Geolabididae) with a description of the dentition of *Ankylodon* (Adapisoricidae). *University of Wyoming Publications*, **45**, 1–115.

Lucas, S. G. (1986) The first Oligocene mammal from New Mexico. *Journal of Paleontology*, **60**, 1274–6.

Lucas, S. G. (1989) Fossil mammals and Paleocene–Eocene boundary in Europe, North America and Asia. *28th International Geological Congress, Abstracts*, **2**, p. 335.

Luckett, W. P. and Hartenberger, J.-L. (1985) *Evolutionary Relationships Among Rodents: A Multidisciplinary Analysis*. NATO ASI Series A, Life Sciences, Vol. 92, Plenum Press, 721 pp.

MacFadden, B. J. (1990) Chronology of Cenozoic primate localities in South America. *Journal of Human Evolution*, **19**, 7–21.

MacFadden, B. J. and Wolff, R. G. (1981) Geological investigations of late Cenozoic vertebrate-bearing deposits in southern Bolivia, in *Anais do II Congresso Latino-Americano de Paleontologia*, **2**, 765–78.

MacPhee, R. D. E., Novacek, M. J. and Storch, G. (1988) Basicranial morphology of early Tertiary erinaceomorphs and the origin of Primates. *American Museum Novitates*, **2921**, 1–42.

Maglio, V. J. and Cooke, H. B. S. (eds) (1978) *Evolution of African Mammals*. Harvard, Cambridge, 641 pp.

Marshall, L. G., Hoffstetter, R. and Pascual, R. (1983) Mammals and stratigraphy: Geochronology of the continental mammal-bearing Tertiary of South America. *Palaeovertebrata, Memoire Extraordinaire*, **1983**, 1–93.

Marshall, L. G., Case, J. A. and Woodburne, M. O. (1990) Phylogenetic relationships of the families of marsupials. *Current Mammalogy*, **2**, 433–505.

Martin, P. S. and Klein, R. G. (eds) (1984) *Quaternary Extinctions: A Prehistoric Revolution*. University of Arizona Press, Tucson, 892 pp.

Martin, R. D. (1990) *Primate Origins and Evolution*. Chapman and Hall, London, 804 pp.

Martin, R. D. (1991) New fossils and primate origins. *Nature*, **349**, 19–20.

McKenna, M. C. (1975) Toward a phylogenetic classification of the Mammalia, in *Phylogeny of the Primates: a Multidisciplinary Approach* (eds W. P. Luckett and F. S. Szalay), Plenum Publishing, New York, 483 pp.

McKenna, M. C. (1990) Plagiomenids (Mammalia: ?Dermoptera) from the Oligocene of Oregon, Montana, and South Dakota, and middle Eocene of northwestern Wyoming. *Geological Society of America, Special Paper*, **243**, 211–34.

Nesov, L. A. (1985) [Rare bony fish, terrestrial lizards and mammals from estuarine and littoral zones of the Cretaceous of the Kyzyl-Kum Desert.] *Ezhegodnik Vsesoiuznoe Paleontologicheskoe Obshchestvo*, **28**, 199–219 [in Russian].

Novacek, M. J. (1976) Insectivora and Proteutheria of the later Eocene (Uintan) of San Diego County, California. *Natural History Museum of Los Angeles County, Contributions in Science*, **283**, 1–52.

Novacek, M. J. (1990) Morphology, paleontology, and the higher clades of mammals. *Current Mammalogy*, **2**, 507–43.

Novacek, M. J., Bown, T. M. and Schankler, D. (1985) On the classification of the early Tertiary Erinaceomorpha (Insectivora, Mammalia). *American Museum Novitates*, **2813**, 1–22.

Novacek, M. J., Wyss, A. R. and McKenna, M. C. (1988) The major groups of eutherian mammals, in *Mammals (The Phylogeny and Classification of the Tetrapods, Vol. 2)* (ed. M. J. Benton), Systematics Association Special Volume, **35B**. Clarendon Press, Oxford, 329 pp.

Pascual, R. and Bond, M. (1981) Epidolopinae, subfam. nov. de los Polydolopidae (Marsupialia, Plodolopoidea). *Anais II Congresso Latino-Americano Paleontologia, Porto Alegre, Abril, 1981*, 480–8.

Pascual, R. and Jaureguizar, E. O. (1990) Evolving climates and mammal faunas in Cenozoic South America. *Journal of Human Evolution*, **19**, 23–60.

Pascual, R., Cattoi, N. V., Francis, J. C. *et al.* (1966) *Paleontografia Bonaerense, Fasciculo IV. Vertebrata*. Comision de Investigacion Cientifica, La Plata, 201 pp.

Paula Couto, C. de (1979) *Tratado de Paleomastozoologia*. Academia Brasileira de Ciencias, Rio de Janiero, 590 pp.

Prothero, D. R. and Schoch, R. M. (eds) (1989) *The Evolution of the Perissodactyla*. Clarendon Press, New York, 537 pp.

Rasmussen, D. T. (1989) The evolution of the Hyracoidea: a review of the fossil evidence, in *The Evolution of the Perissodactyla* (eds D. R. Prothero and R. M. Schoch), Clarendon Press, New York, pp. 57–78.

Rea, D. K., Zachos, J. C., Owen, R. M. *et al.* (1990) Global change at the Paleocene–Eocene boundary: climatic and evolutionary consequences of tectonic events. *Palaeogeography, Palaeoclimatology and Palaeoecology*, **79**, 117–28.

Remy, J. A., Crochet, J.-Y., Sigé, B. *et al.* (1987) Biochronologie des phosphorites du Quercy: mise à jour des listes fauniques et nouveaux gisements de mammiféres fossiles. *Münchner Geowissenschaftlichen Abhandlungen A*, **10**, 169–88.

Rosenberger, A. L., Setoguchi, T. and Shigehara, N. (1990) The fossil record of callitrichine primates. *Journal of Human Evolution*, **19**, 209–36.

Russell, D. E. and Zhai, R. (1987) The Paleogene of Asia: mammals and stratigraphy. *Mémoires du Museum Nationale de l'Histoire Naturelle, (C) Sciences de la Terre*, **52**, 1–488.

Russell, D. E., Galoyer, A. and Gingerich, P. D. (1988) Nouveaux Vertébrés sparnaciens du Conglomérat de Meudon à Meudon, France. *Comptes Rendus de l'Academie des Sciences, Paris*, **307**, 429–33.

Russell, D. E., Hartenberger, J., Pomerol, C. *et al.* (1982) Mammals and stratigraphy: the Paleogene of Europe. *Palaeovertebrata, Mémoire extraordinaire*, **1982**, 1–77.

Savage, D. E. and Barnes, L. G. (1972) Miocene vertebrate geochronology of the west coast of North America, in *Proceedings of the Pacific Coast Miocene Biostratigraphic Symposium* (ed. E. H. Stinemeyer), Society of Economic Paleontologists and Mineralogists, pp. 124–45.

Savage, D. E. and Russell, D. E. (1983) *Mammalian Paleofaunas of the World*. Addison-Wesley, London, 432 pp.

Schmidt-Kittler, N. (1987a) The Carnivora (Fissipedia) from the lower Miocene of East Africa. *Palaeontographica*, **197**, 85–126.

Schmidt-Kittler, N. (ed.) (1987b) European reference levels and correlation tables. *Münchner geowissenschaftlichen Abhandlungen A*, **10**, 13–31.

Sen, S. (1977) First study of a Pliocene rodent fauna from Anatolia. *Turkey, Mineral Research and Exploration Institute, Bulletin*, **89**, 84–9.

Sigé, B. and Legendre, S. (1983) L'histoire des Peuplements de Chiropterès du Bassin Méditerranéen: L'apport comparé des remplissages karstiques et des dépôts fluvio-lacustres. *Memoires Biospeléologiques*, **10**, 209–25.

Simons, E. L. (1968) African Oligocene mammals: introduction, history of study, and faunal succession. *Peabody Museum of Natural History, Yale University, Bulletin*, **28**, 1–21.

Simons, E. L. (1989) Description of two genera and species of late Eocene Anthropoidea from Egypt. *Proceedings of the National Academy of Sciences, USA*, **86**, 9956–60.

Simons, E. L., Bown, T. M. and Rasmussen, D. T. (1986) Discovery of two additional prosimian primate families (Omomyidae, Lorisidae) in the African Oligocene. *Journal of Human Evolution*, **15**, 431–7.

Simpson, G. G., Minoprio, J. L. and Patterson, B. (1962) The mammalian fauna of the Divisadero Largo formation, Mendoza, Argentina. *Bulletin of the Museum of Comparative Zoology, Harvard*, **127**, 239–93.

Soria, M. and Bond, M. F. (1984) Adiciones al conocimiento de *Trigonostylops* Ameghino, 1897 (Mammalia; Astrapotheria; Trigonostylopidae). *Ameghiniana*, **19**, 155–68.

Storer, J. E. (1984) Mammals of the Swift Current Creek local fauna (Eocene: Uintan), Saskatchewan. *Natural History Contributions, Saskatchewan Museum of Natural History*, **7**, 1–158.

Storer, J. E. (1987) Dental evolution and radiation of Eocene and early Oligocene Eomyidae (Mammalia, Rodentia) of North America, with new material from the Duchesnean of Saskatchewan. *Dakoterra*, **3**, 108–17.

Storer, J. E. (1988) The rodents of the Lac Pelletier lower fauna, late Eocene, (Duchesnean) of Saskatchewan. *Journal of Vertebrate Paleontology*, **8**, 84–101.

Storer, J. E. (1990) Primates of the Lac Pelletier lower fauna (Eocene: Duchesnean), Saskatchewan. *Canadian Journal of Earth Sciences*, **27**, 520–4.

Stucky, R. K. (1984) The Wasatchian–Bridgerian land mammal age boundary (early to middle Eocene) in western North America. *Annals of Carnegie Museum*, **53**, 347–82.

Stucky, R. K. (1990) Evolution of land mammal diversity in North America during the Cenozoic. *Current Mammalogy*, **2**, 375–432.

Stucky, R. K. (1992) Mammalian faunas in North America of Bridgerian to early Arikareean 'ages' (Eocene and Oligocene), in *Eocene–Oligocene Climatic and Biotic Events* (eds D. Prothero and W. Berggren), Princeton University Press, Princeton, pp. 464–93.

Stucky, R. K. and Krishtalka, L. (1990) Revision of the Wind River Faunas, early Eocene of central Wyoming. Part 10. *Bunophorus* (Mammalia, Artiodactyla). *Annals of Carnegie Museum*, **59**, 149–71.

Sudre, J. (1978) Les Artiodactyles de l'Éocène moyen et supérieur d'Europe occidentale. *Mémoires et Travaux de l'Institut de Montpellier de l'École Pratique des Hautes Études*, **7**, 1–299.

Sudre, J. (1984) *Cryptomeryx* Schlosser, 1886, Tragulide de L'Oligocene D'Europe; relations du genre et considérations sur l'origine des ruminants. *Palaeovertebrata*, **14(1)**, 1–31.

Sudre, J., Russell, D. E., Louis, P. *et al.* (1983) Les Artiodactyles de l'Éocène inférieur d'Europe. *Bulletin du Museum Nationale d'Histoire Naturelle, Paris*, **5**, 281–333, 339–65.

Swisher, C. C. and Prothero, D. R. (1990) Single-crystal Ar^{40}/Ar^{39} dating of the Eocene–Oligocene transition in North America. *Science*, **249**, 760–2.

Szalay, F. S. (1976) Systematics of the Omomyidae (Tarsiiformes, Primates): taxonomy, phylogeny and adaptations. *Bulletin of the American Museum of Natural History*, **156**, 157–450.

Szalay, F. S. and Delson, E. (1979) *Evolutionary History of the Primates*. Academic Press, New York, 580 pp.

Szalay, F. S. and Drawhorn, G. (1980) Evolution and diversification of the Archonta in an arboreal Millieu, in *Comparative Biology and Evolutionary Relationships of Tree Shrews* (ed. P. Luckett), Plenum Press, New York, pp. 133–69.

Szalay, F. S., Rosenberger, A. L. and Dagosto, M. (1987) Diagnosis and differentiation of the Order Primates. *Yearbook of Physical Anthropology*, **30**, 75–106.

Tassy, P. (1990) Phylogénie et classification des Proboscidea (Mammalia): historique et actualité. *Annales de Paléontologie*, **76**, 159–224.

Taylor, B. E. and Webb, S. D. (1976) Miocene Leptomerycidae (Artiodactyla, Ruminantia) and their relationships. *American Museum Novitates*, **2596**, 1–22.

Tedford, R. H. (1981) Mammalian biochronology of the late Cenozoic basins of New Mexico. *Geological Society of America Bulletin*, **92**, 1008–22.

Tedford, R. H., Banks, M. R., Kemp, N. R. *et al.* (1975) Recognition of the oldest known fossil marsupials from Australia. *Nature*, **255**, 141–2.

Thenius, E. (1959) *Tertiar: Zweiter Teil, Wirbeltierfaunen, Handbuch der Stratigraphischen Geologie*. III Band 2. Teil. Ferdinand Enke Verlag, Stuttgart, 328 pp.

Thewissen, J. G. M. (1985) Cephalic evidence for the affinities of Tubulidentata. *Mammalia*, **49**, 257–84.

Thewissen, J. G. M. (1990) Evolution of Paleocene and Eocene Phenacodontidae (Mammalia, Condylarthra). *University of Michigan, Papers on Paleontology*, **29**, 1–107.

Thewissen, J. G. M., Gingerich, P. D. and Russell, D. E. (1987) Artiodactyla and Perissodactyla (Mammalia) from the early–middle Eocene Kuldana Formation of Kohat (Pakistan). *Contributions from the Museum of Paleontology, The University of Michigan*, **27**, 247–74.

Unay-Bayraktar, E. (1989) *Rodents from the Middle Oligocene of Turkish Thrace*. Loonzetteri, Abe, Hoogeren, Utrecht, 75 pp.

Wang, B. (1992) The Chinese Oligocene – preliminary review of the mammalian localities and local faunas, in *Eocene–Oligocene Climatic and Biotic Evolution*, Princeton University Press, Princeton, pp. 529–47.

Webb, S. D. (1983) A new species of *Pediomeryx* from the late Miocene of Florida, and its relationship within the subfamily Cranioceratinae (Ruminantia: Dromomerycidae). *J. Mammalogy*, **64** (2), 261–76.

Webb, S. D. (1984) Ten million years of mammal extinctions in North America, in *Quaternary Extinctions: A Prehistoric Revolution* (eds P. S. Martin and R. G. Klein), University of Arizona Press, Tucson, 892 pp.

Webb, S. D. and Perrigo, S. C. (1984) Late Cenozoic vertebrates from Honduras and El Salvador. *Journal of Vertebrate Paleontology*, **4**, 237–54.

West, R. M. and Dawson, M. R. (1975) Eocene fossil Mammalia from the Sand Wash Basin, northwestern Moffat County, Colorado. *Annals of Carnegie Museum*, **45**, 231–53.

Wible, J. R. and Covert, H. H. (1987) Primates: cladistic diagnosis and relationships. *Journal of Human Evolution*, **16**, 1–21.

Wilson, J. A. (1986) Stratigraphic occurrence and correlation of early Tertiary vertebrate faunas, Trans-Pecos Texas: Agua Fria-Green Valley areas. *Journal of Vertebrate Paleontology*, **6**, 350–73.

Wing, S. L. (1984) A new basis for recognizing the Paleocene/Eocene boundary in Western Interior North America. *Science*, **226**, 439–41.

Wood, A. E. (1985) The relationships, origin and dispersal of the hystricognathous rodents, in *Evolutionary Relationships among Rodents: a Multidisciplinary Approach* (eds W. P. Luckett and J. Hartenberger), Plenum Press, New York, pp. 475–514.

Woodburne, M. O. (1987) *Cenozoic Mammals of North America: Geochronology and Biostratigraphy*. University of California Press, Berkeley, 336 pp.

Woodburne, M. O., Tedford, R. H., Archer, M. *et al.* (1985) Biochronology of the continental mammal record of Australia and New Guinea. *Special Publication, South Australian Department of Mines and Energy*, **5**, 347–63.

Plants

Cooksonia pertoni Lang, 1937, branching axis and terminal sporangia, from the Cloncannon Formation (Gleedon, Wenlock, Middle Silurian) of County Tipperary, Ireland. The oldest vascular plant. The whole specimen is about 700 μm long. Photograph courtesy of D. E. Edwards.

42

BRYOPHYTA

D. Edwards

'Mosses and liverworts have long challenged taxonomists to devise a classification which will be at once reasonably natural and clearly workable' (Watson, 1964, p. 13). Challenges imposed by their exceedingly inadequate fossil record seem even more insuperable because of the lack of critical information relating to anatomy, sexual reproductive characters or sporophytes. Such uncertainties are reflected in the large numbers of fossils called *Muscites*, *Thallites* and *Hepaticites* in the Palaeozoic and Mesozoic, and also in this analysis, where records during these times are possible only at the ordinal level for hepatics and at subclass level for mosses.

In contrast, the vast majority of bryophytes, particularly mosses, recorded from post-Palaeogene sediments are assigned to extant taxa, and thus to families. In that estimates of moss families range from the forties to over eighty (e.g. Crosby and Magill, 1977), and records in the Neogene (e.g. Miller, 1984) and Quaternary (see Dickson, 1973) are very extensive, treatment at the family level has not been attempted here. This compilation relies heavily on reviews published in *Manual of Bryology* (1984) by Krassilov and Schuster for the Palaeozoic and Mesozoic, and Miller for the Tertiary.

Kingdom PLANTAE

Division BRYOPHYTA Schimper, 1879

Class HEPATICOPSIDA Rothmaler, 1951

Of the generally accepted major hepatic orders, fossil records are lacking for the Takakiales, Calobryales and Monocleales (Schuster, 1981). The taxonomic position of the Lower Devonian *Sporogonites* remains uncertain. Hepatic assignment is based on unbranched axes apparently attached to a thalloid gametophyte (Andrews, 1960). Three-dimensional anatomy – essential for the elimination of the possibility of vascular status – has not yet been recorded. Schuster (1966) placed it in its own order, the Sporogonitales. If indeed a hepatic, it would be the oldest representative of the group.

Order METZGERIALES Schuster, ex Schljakov
D. (FRS)–Rec. Terr. (Fig. 42.1)

First: *Pallaviciniites devonicus* (Hueber) Schuster, 1966, New York State, USA. **Extant**
Comments: Schuster (1966) related this taxon to the extant Pallaviciniaceae with closest similarities to *Pallavicinia* and *Symphyogyna*. New combinations for Lower Carboniferous taxa, namely *Blasites lobatus* (Walton) Schuster, 1966, *Treubiites kidstonii* (Walton) Schuster, 1966 and *Metzgeriothallus metzgerioides* (Walton) Schuster, 1966 are considered representatives of three further families within the Metzgeriales (Krassilov and Schuster, 1984), but these were not named. The Metzgeriales is the only order thought with any confidence to be represented in the Palaeozoic. Earliest fossil assignable to a family is *Riccardia palmata* (Aneuraceae) from Tertiary amber (Jovet-Ast, 1967).

Order MARCHANTIALES Engler, 1892
Tr. (?CRN)–Rec. Terr.

First: *Eomarchantites cyathodoides* (Townrow) Schuster (*in* Krassilov and Schuster, 1984). Molteno Formation, Upper Triassic, Natal, South Africa. **Extant**
Comments: Krassilov and Schuster (1984) found no evidence for this group in the Palaeozoic, but document a number of genera from the Triassic into the Eocene. Most are called *Marchantites* or *Thallites*. See also list in Schuster (1966) and Jovet-Ast (1967). The earliest record of *Marchantia* itself is *M. lignitica* (Ward) Brown, 1962 from the Palaeocene, Fort Union Formation, Yellowstone River, USA.

Order JUNGERMANNIALES Halle, 1913
T. (PRB)–Rec. Terr.

First: *Schizolepidella gracilis* Halle, 1913, Graham Land, Antarctica. **Extant**
Comment: A most remarkable Upper Eocene assemblage of Jungermanniales preserved in Baltic amber has been revised recently by Grolle (1980a,b, 1981a,b). It contains the first records of the families Jungermanniaceae (*s.s.*), Cephaloziellaceae, Lepidoziaceae, Radulaceae, Frullaniaceae and Lejeuneaceae. Most of the representatives are assigned to modern genera. It seems likely that they grew as epiphytes on the trees producing the amber.

Order SPHAEROCARPALES Cavers, 1910
Tr. (RHT)–Rec. Terr.

First: *Naiadita lanceolata* Buckman, 1850 emend. Harris 1938 (Harris, 1939), Cotham Beds, Avon, England, UK.
Comment: Sporophytic characters place this species in

The Fossil Record 2. Edited by M. J. Benton. Published in 1993 by Chapman & Hall, London. ISBN 0 412 39380 8

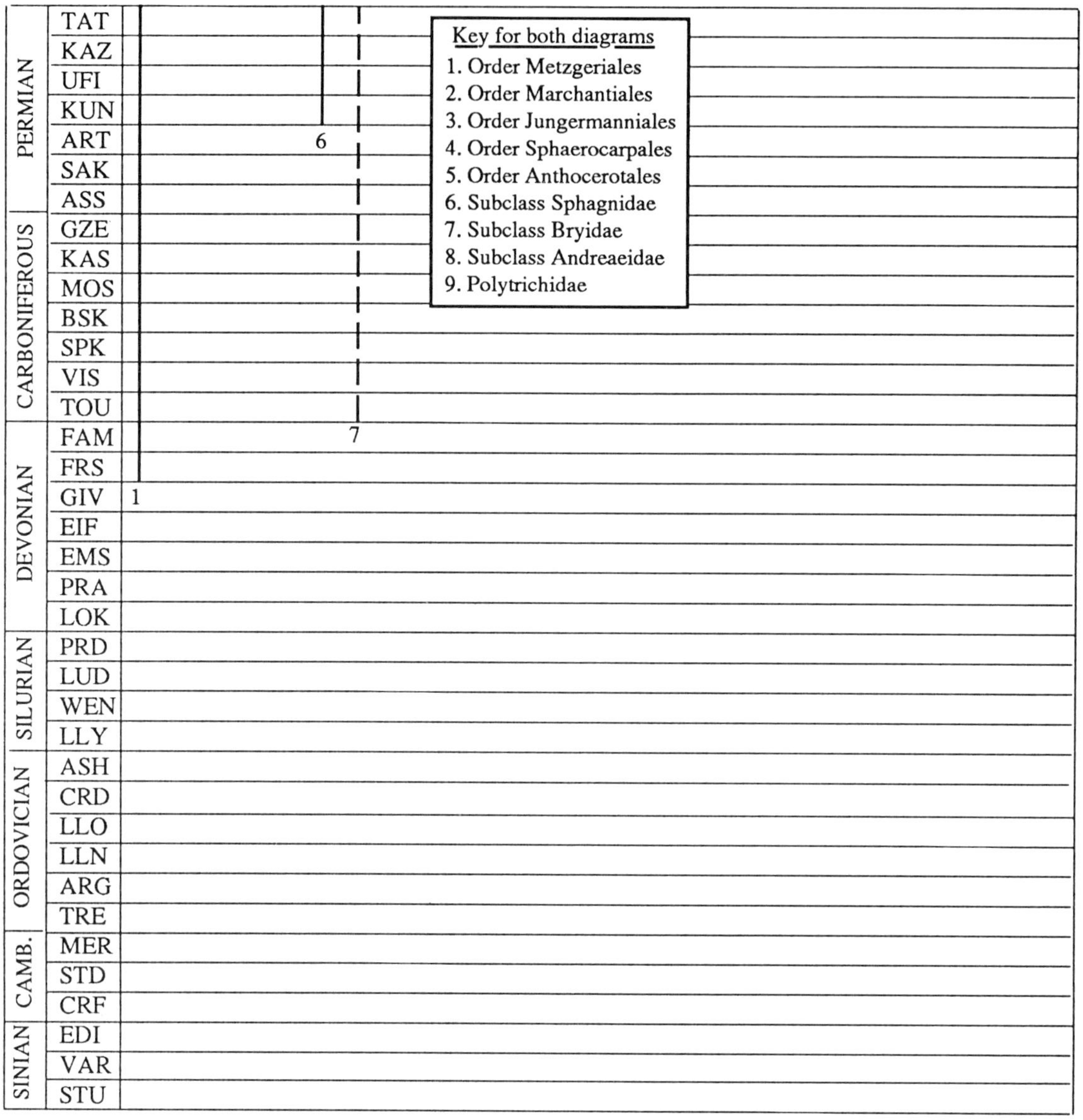

Fig. 42.1

Key for both diagrams
1. Order Metzgeriales
2. Order Marchantiales
3. Order Jungermanniales
4. Order Sphaerocarpales
5. Order Anthocerotales
6. Subclass Sphagnidae
7. Subclass Bryidae
8. Subclass Andreaeidae
9. Polytrichidae

PERMIAN: TAT, KAZ, UFI, KUN, ART, SAK, ASS
CARBONIFEROUS: GZE, KAS, MOS, BSK, SPK, VIS, TOU
DEVONIAN: FAM, FRS, GIV, EIF, EMS, PRA, LOK
SILURIAN: PRD, LUD, WEN, LLY
ORDOVICIAN: ASH, CRD, LLO, LLN, ARG, TRE
CAMB.: MER, STD, CRF
SINIAN: EDI, VAR, STU

1 6 7

the Sphaerocarpales, although its gametophyte being erect, radial and isophyllous is calobryalean (see discussion *in* Krassilov and Schuster, 1984). There are no Cainozoic representatives (Miller, 1984).

Class ANTHOCEROTOPSIDA Rothmaler, 1951

Order ANTHOCEROTALES Muller, 1940
K. (MAA)–Rec. Terr.

First: *Phaeoceros* Jarzen (1979: spore forms A–C) Frenchman Formation, Saskatchewan, Canada; Selma Group, Alabama, USA.
Comment: Earliest records are based on spores (Jarzen, 1979). The first sporophytes are reported (as petrifactions) in the uppermost Cretaceous or Palaeocene Deccan Intertrappan beds of Mahgaonkalan, India (Chitaley and Yawale, 1980).

Class BRYOPSIDA Rothmaler, 1951

Unequivocal mosses occur in the Carboniferous, but are not reliably assigned to higher taxa. The earliest is *Muscites plumatus* from the Lower Carboniferous of southern England, tentatively compared with *Grimmia* by Thomas (1972). Upper Carboniferous records include *M. polytrichaceus* Renault and Zeiller and *M. bertrandii* Lignier. Of the five extant subclasses, the Buxbaumiidae has no known fossils.

Subclass SPHAGNIDAE Engler *et al.*, 1954
P. (KUN)–Rec. Terr.

First: *Protosphagnum nervatum, Vorcutannularia plicata, Junjagia,* KUN, UF1, KA2 and TAT Stages, Angaraland, former USSR (Neuberg, 1956).
Comment: Neuberg (1956) placed these three genera in a new order, the Protosphagnales, but Krassilov and Schuster (1984) believe that these and the nine genera that she assigned to the Bryales, constitute a single natural group with some similarities to the Isobryales. Ignatov (1990) discussed the relationships of the three protosphagnaleans to each other and to the Sphagnales, and in erecting a new genus *Palaeosphagnum* from Upper Permian strata of the Russian Platform with closest affinity to *Vorcutannularia,* he was convinced of the existence of the

Period	Stage	Range no.
QU.	HOL	
	PLE	
TERTIARY	PLI	8
	UMI	
	MMI	
	LMI	
	CHT	
	RUP	
	PRB	
	BRT	3, 9
	LUT	
	YPR	
	THA	
	DAN	
CRETACEOUS	MAA	
	CMP	5
	SAN	
	CON	
	TUR	
	CEN	
	ALB	
	APT	
	BRM	
	HAU	
	VLG	
	BER	
JURASSIC	TTH	
	KIM	
	OXF	
	CLV	
	BTH	
	BAJ	
	AAL	
	TOA	
	PLB	
	SIN	
	HET	
TRIASSIC	RHT	
	NOR	4
	CRN	
	LAD	2
	ANS	
	SCY	1, 6, 7

Fig. 42.1

extant order at the end of the Palaeozoic, although he excluded Neuberg's *Protosphagnum* and *Junjagia*. Although there is a record of Sphagnales in the Lower Jurassic of Bavaria (Reissinger, 1950), Miller (1984) does not consider the Sphagnidae to be represented in pre-Cainozoic rocks, and so *Sphagnum*? from a Lower Tertiary/?Palaeogene/ lignite in northern Ellesmere Island becomes the earliest undisputed record of the group.

Subclass BRYIDAE Engler *et al.*, 1954
?C. (l./u.)/P. (TAT)–Rec. Terr.

Comment: *Muscites plumatus* Thomas (Lower Carboniferous) and *M. bertrandii* (Upper Carboniferous) Lignier are probable Palaeozoic representatives, but cannot be assigned to higher taxa. As mentioned above, most of Neuberg's genera from the Lower Permian and Upper Permian of the Kuznetsk, Pechora and Tunguska Basins, former USSR, may belong here (see *The Fossil Record*, 1967), but although they resemble Mniaceae in certain respects, they show a unique mode of leaf attachment. Ignatov (1990) described nine new taxa from the Upper Permian (upper TAT) Viatsky Suite from the Vologda Province in the northern part of the Russian Platform. On the basis of cell shape and dimensions, on leaf margin characters and on the structure of costa, he compared the fragmentary vegetative fossils with extant genera and families, and in conclusion considered that they included representatives of the extant orders Dicranales, Pottiales, Funariales, Leucodontales and Hypnales. A possible Mesozoic record of this very large subclass is *M. guescelini* Townrow (1959) from the Middle Triassic of Natal, South Africa, thought to be close to the Leucodontaceae (Isobryales). The oldest specimens to be assigned to extant taxa, and hence with confidence to families, are species of *Calliergon* and *Drepanocladus* (Amblystegiaceae in the Hypnobryales) from a Lower Tertiary (Palaeogene) lignite on northern Ellesmere Island, Arctic Canada (Kuc, 1973). Most Palaeogene fossils, although clearly belonging to the Bryidae are too badly preserved or incomplete for more precise placement and are named under form genera. In contrast the great majority in Europe and North America are assigned to extant genera. A comprehensive survey is given in Miller (1984).

Subclass ANDREAEIDAE Engler *et al.*, 1954
Q. (PLE)–Rec. Terr.

First: *Andreaea rothii* Weber and Mohr, Riss glacial deposit, Zamszany Poland (Szafran, 1952). **Extant**

Subclass POLYTRICHIDAE Engler *et al.*, 1954
T. (PRB)–Rec. Terr.

First: *Polytrichum subseptentrionale* Goeppert, 1853, *Pogonatum suburnigerum* (Goeppert and Menge) Dixon, 1927, *Catharinea subundulata* (Goeppert) Dixon, 1927. Upper Eocene/Tertiary Baltic amber.
Comment: The Tertiary record is particularly poor and most records (e.g. *Polytrichites*) are doubtful, because they are based on few or no structural details. There is no compelling evidence that Upper Carboniferous *Muscites polytrichaceus* Renault and Zeiller belongs to this subclass.

REFERENCES

Andrews, H. N. (1960) Notes on Belgian specimens of *Sporogonites*. *The Palaeobotanist*, **7**, 85–9.

Chitaley, S. and Yawale, N. R. (1980) *Notothylites nirulai* gen. et sp. nov., a petrified sporogonium from the Deccan Intertrappen beds of Mahgaonkalan, M.P. India. *Botanique*, **9**, 111–18.

Crosby, M. R. and Magill, R. E. (1977) *A Dictionary of Mosses*. Missouri Botanical Garden, St Louis, 43 pp.

Dickson, J. H. (1973) *Bryophytes of the Pleistocene. The British Record and its Chorological and Ecological Implications*. Cambridge University Press, lx + 246 pp.

Grolle, R. (1980a) Lebermoose in Bernstein 1. *Feddes Reprium*, **91**, 183–90.

Grolle, R. (1980b) Lebermoose in Bernstein 2. *Feddes Reprium*, **91**, 401–7.

Grolle, R. (1981a) *Nipponolejeunea* fossil in Europe. *Journal of the Hattori Botanical Laboratory*, **50**, 143–57.

Grolle, R. (1981b) Was ist *Lejeunea schumannii* Caspary aus dem baltischen Bernstein? *Occasional Papers of the Farlow Herbarium, Harvard University*, **16**, 101–10.

Harris, T. M. (1939) *Naiadita*, a fossil bryophyte with reproductive organs. *Annals of Bryology*, **12**, 57–70.

Ignatov, M. S. (1990) Upper Permian mosses from the Russian Platform. *Palaeontographica, Abteilung B*, **217**, 147–89.

Jarzen, D. M. (1979) Spore morphology of some Anthoceroteae and the occurrence of *Phaeoceros* spores in the Cretaceous of North America. *Pollen et Spores*, **21**, 211–31.

Jovet-Ast, S. (1967) Bryophyta, in *Traité de Paléobotanique*, Vol. II (ed. E. Boureau), Masson et Cie, Paris, pp. 17–186.

Krassilov, V. A. and Schuster, R. M. (1984) Paleozoic and Mesozoic Fossils, in *New Manual of Bryology* (ed. R. M. Schuster), Hattori Botanical Laboratory, Nichinan, pp. 1173–93.

Kuc, M. (1973) Plant macrofossils in Tertiary coal and amber from northern Lake Hazen, Ellesmere Island, N.W.T. *Geological Survey of Canada, Paper*, **73-1(B)**, p. 143.

Miller, N. G. (1984) Tertiary and Quaternary fossils, in *New Manual of Bryology* (ed. R. M. Schuster), Hattori Botanical Laboratory, Nichinan, pp. 1194–232.

Neuberg, M. F. (1956) The discovery of mosses in the Permian deposits of the USSR. *Doklady Akademii Nauk SSSR*, **107**, 321–4 [in Russian].

Reissinger, A. (1950) Die 'Pollenanalyse' ausgedehnt auf alle Sedimentgesteine der geologischen Vergangenheit. *Palaeontographica, Abteilung B*, **90**, 99–126.

Schuster, R. M. (1966) *The Hepaticae and Anthocerotae of North America*, Vol. 1, Columbia University Press, New York, 802 pp.

Schuster, R. M. (1981) Paleoecology, origin, distribution through time, and evolution of Hepaticae and Antherotae, in *Paleobotany, Paleoecology and Evolution* (ed. K. J. Niklas), Praeger, New York, pp. 129–91.

Szafran, B. (1952) Pleistocene mosses from Poland and the adjacent eastern territories. *Panstwowa Instytut Geologiczny*, **68**, 5–38.

Thomas, B. A. (1972) A probable moss from the Lower Carboniferous of the Forest of Dean, Gloucestershire. *Annals of Botany*, **36**, 155–61.

Townrow, J. A. (1959) Two Triassic bryophytes from South Africa. *Journal of South African Botany*, **25**, 1–22.

Watson, E. V. (1964) *The Structure and Life of Bryophytes*. Hutchinson, London, 192 pp.

43

PTERIDOPHYTA

C. J. Cleal

In this analysis, I take the Pteridophyta to include all vascular plants which do not bear seeds. It has been argued that the pteridophytes are polyphyletic, but the consensus now seems to be that they probably originated from a single algal ancestor (Stewart, 1983). It is quite clearly a paraphyletic group, however, since the gymnosperms and angiosperms are an offshoot of one of its classes, probably the Progymnospermopsida.

The main pteridophyte classes have become essentially standardized in recent years (Lycopsida, Equisetopsida, Filicopsida and Progymnospermopsida). Details of the palaeontology of these classes can be found in the standard palaeobotany textbooks (Taylor, 1981; Stewart, 1983; Meyen, 1987; Thomas and Spicer, 1987).

The position with the earliest pteridophytes is less stable. At the time of publication of the first edition of *The Fossil Record*, they were divided into three families of one class, the Psilophytopsida. Banks (1968, 1975) subsequently proposed to raise these families in rank to subdivisions (Rhyniophytina, Zosterophyllophytina, Trimerophytina). Recent work, including critical anatomical investigations, has confirmed the validity of the Zosterophyllophytina, but the limits and content of the Rhyniophytina and Trimerophytina require re-evaluation. Also, new higher taxa need to be erected for plant fossils that have been found to have novel combinations of characters, such as *Aglaophyton major* (Kidston and Gwynne-Vaughan) D. S. Edwards, 1986, which has a rhyniophyte morphology but moss-like conducting tissues. As far as possible, I have attempted to use Banks' classification, except that I do not use the subdivisional taxa.

There is no recent *Treatise*-like analysis of the fossil record of plants and it has thus been impossible to use the so-called 'hierarchical' approach adopted elsewhere in this volume. This explains what may seem an excessively long list of references given at the end of this chapter. The protologue of a particular taxon is only referenced at the end when the types are the oldest and/or youngest known specimens for the family. In the Zosterophyllaceae, for instance, the reference to *Z. myretonianum* Penhallow, 1892, is not listed, as the oldest specimens were described by Edwards (1975), while that to *S. ornata* Hueber, 1971, is given as the types include the youngest known specimens.

Acknowledgements – Thanks go to Josephine Camus (Natural History Museum, London) for advice on fern taxonomic nomenclature, and to Richard Bateman (Smithsonian Institution) for providing unpublished data. I am also grateful to Dianne Edwards (University of Wales, Cardiff), Barry Thomas (National Museum of Wales), Judy Skog (George Mason University, Virginia) and Chris Hill (Natural History Museum, London) for reading early drafts of the manuscript.

Class RHYNIOPSIDA Banks, 1975 (see Fig. 43.1)

Lack of information of xylem from the oldest (*Cooksonia*) and more complex presumed members (e.g. *Renalia hueberi*) hampers delineation of this class (Edwards and Edwards, 1986). Most important from an evolutionary standpoint are the small, isotomously branching forms (e.g. *Pertonella*, *Salopella*), usually considered to be the earliest 'higher' (i.e. pteridophytic) plants, but in which xylem has not yet been demonstrated. Edwards and Edwards (1986) have used the term 'rhyniophytoid' for such fossils and, although it is clearly not a formal taxon, is included in the following analysis.

Rhyniophytoid fossils *sensu* Edwards and Edwards, 1986 S. (GLE)–D. (FRS) Terr.

First: *Cooksonia pertoni* Lang, 1937, *C. hemisphaerica* Lang, 1937, and *C. cambrensis* Edwards, 1979, Cloncannon Formation, County Tipperary, Republic of Ireland (Edwards *et al.*, 1983).
Last: *Taeniocrada lesquereuxii* White, 1903, Chemung Formation, Pennsylvania, USA.
Intervening: LDF–EMS.

The Fossil Record 2. Edited by M. J. Benton. Published in 1993 by Chapman & Hall, London. ISBN 0 412 39380 8

QU.	HOL				
	PLE				
TERTIARY	PLI				
	UMI				
	MMI				
	LMI				
	CHT				
	RUP				
	PRB				
	BRT				
	LUT				
	YPR				
	THA				
	DAN				
CRETACEOUS	MAA				
	CMP				
	SAN				
	CON				
	TUR				
	CEN				
	ALB				
	APT				
	BRM				
	HAU				
	VLG				
	BER				
JURASSIC	TTH				
	KIM				
	OXF				
	CLV				
	BTH				
	BAJ				
	AAL				
	TOA				
	PLB	10	11	24	35
	SIN				
	HET				

<u>Key for both diagrams</u>
RHYNIOPSIDA
1. Rhyniophytoids
2. Rhyniaceae
ZOSTEROPHYLLOPSIDA
3. Zosterophyllaceae
HORNEOPHYTOPSIDA
4. Horneophytaceae
TRIMEROPHYTOPSIDA
5. Trimerophytaceae
<u>Incertae sedis</u>
6. Barinophytaceae
LYCOPSIDA
7. Drepanophycaceae
8. Protolepidodendraceae
9. Eleutherophyllaceae
10. Lycopodiaceae
11. Selaginellaceae
12. Cyclostigmaceae
13. Flemingitaceae
14. Sigillariostrobaceae
15. Lepidocarpaceae
16. Spenceritaceae
17. Diaphorodendraceae
18. Caudatocarpaceae
19. Pinakodendraceae
20. Sporangiostrobaceae
21. Cyclodendraceae
22. Pleuromeiaceae
23. Miadesmiaceae
24. Isoetaceae
25. Chaloneriacea
26. Takhtajanodoxaceae
EQUISETOPSIDA
27. Pseudoborniaceae
28. Bowmanitaceae
29. Eviostachyaceae
30. Cheirostrobaceae
31. Archaeocalamitaceae
32. Calamostacnyaceae
33. Tchernoviaceae
34. Gondwanostachyaceae
35. Equisetaceae
36. Echinostachyaceae
FILICOPSIDA
37. Cladoxylaceae
38. Ibykaceae

Fig. 43.1

Order RHYNIALES Banks, 1975

F. RHYNIACEAE Kidston and Lang, 1920
D. (LOK–EMS) Terr.

First: *Cooksonia pertoni* Lang, Ditton Series (*micrornatus-newportensis* spore biozone), Welsh Borderland, Shropshire, England, UK (Edwards, Davies and Axe, 1992). Older specimens of *Cooksonia* have not been demonstrated as vascular and hence are included in the rhyniophytoids.
Last: *Hsüa robusta* Li, 1982, Xujiachong Formation, Yunnan, China.
Comments: *Salopella australis* and *Hedeia* sp. from the LDF? of Victoria, Australia (Tims and Chambers, 1984) may also belong here, but are preserved as impressions showing no evidence of vascular tissue.

Class ZOSTEROPHYLLOPSIDA Banks, 1975

Order ZOSTEROPHYLLALES Banks, 1975

Niklas and Banks (1990) emphasize two distinct growth patterns in the fertile axes of the Zosterophyllales, exemplified by *Zosterophyllum* and *Gosslingia*. This is probably evidence that two families exist in the order. However, this taxonomic distinction has yet to be established formally.

F. ZOSTEROPHYLLACEAE Kräusel, 1938
D. (LOK–FRS) Terr.

First: *Zosterophyllum myretonianum* Penhallow, 1892, Dundee Formation, Tayside Region, Scotland, UK (Edwards, 1975).
Last: *Sawdonia ornata* Hueber, 1971, Onteora Formation, New York, USA.
Intervening: PRA–EIF.
Comments: Tims and Chambers (1984) record putative zosterophylls from the LDF? of Victoria, Australia, but they have not yet been described.

Class HORNEOPHYTOPSIDA Němejc, 1960

Order HORNEOPHYTALES Němejc, 1960

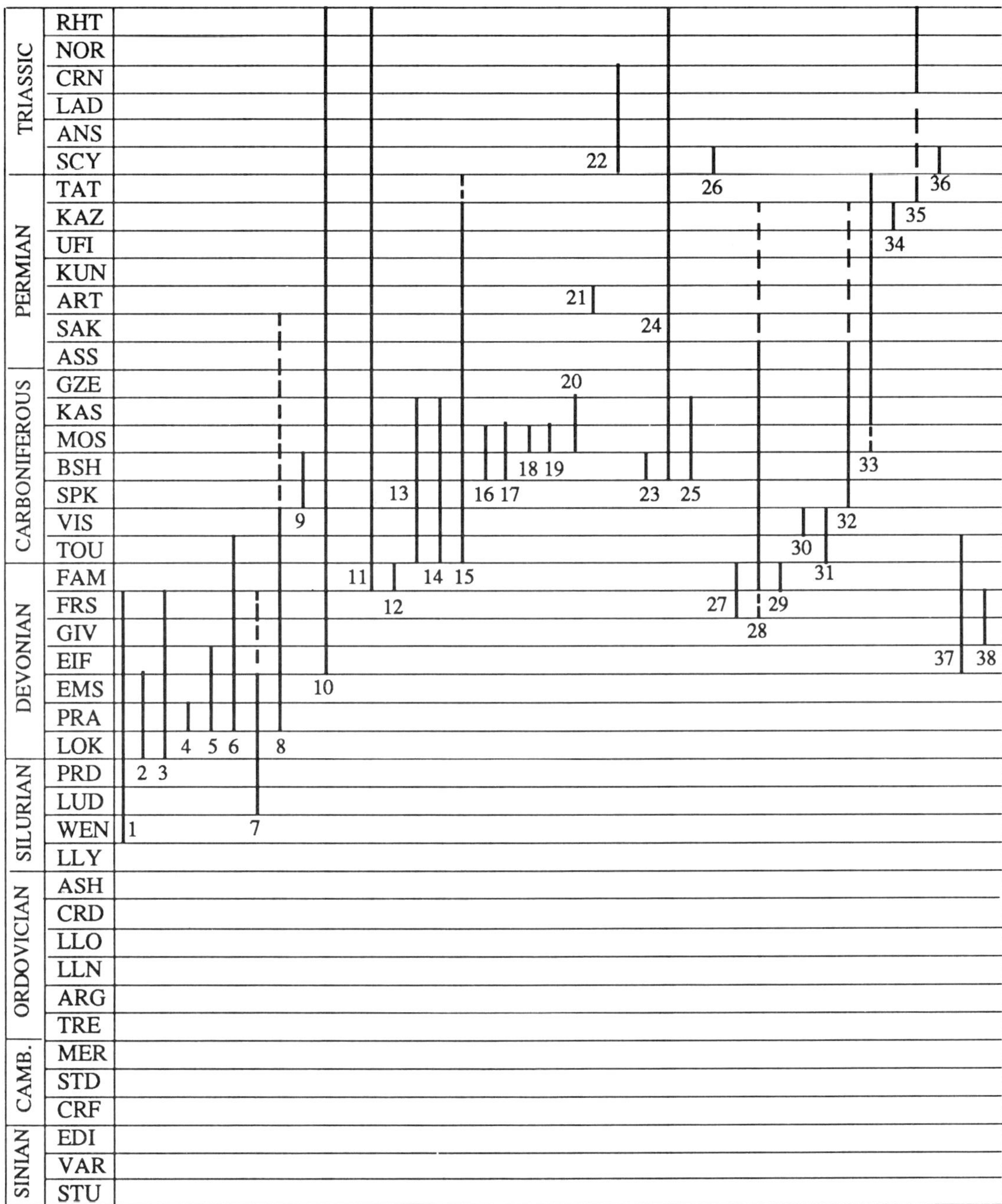

Fig. 43.1

F. HORNEOPHYTACEAE Němejc, 1960
D. (PRA) Terr.

First and Last: *Horneophyton lignieri* (Kidston and Lang) Barghoorn and Darrah, 1938, Rhynie Chert, Grampian Region, Scotland, UK (El-Saadawy and Lacey, 1979).

Class TRIMEROPHYTOPSIDA Banks, 1975

Order TRIMEROPHYTALES Banks, 1975

F. TRIMEROPHYTACEAE Banks, 1975
D. (PRA–EIF) Terr.

First: ?*Dawsonites* sp., Senni Beds, Powys, Wales, UK (Croft and Lang, 1942). These are fragments of trusses of sporangia, clearly belonging to *Psilophyton*, but with insufficient characters to assign it to a species. The next oldest, *Psilophyton burnotense* (?=*P. goldschmidtii*) is of controversial affinity. The oldest recognized species are *Psilophyton forbesii*, *P. crenulatum* and *P. princeps* from the basal EMS of Belgium (Gerrienne, 1983) and Canada (Gensel and Andrews, 1984).

Last: *Psilophyton dapsile* Kaspar *et al.*, 1974 and *Pertica quadrifaria* Kaspar and Andrews, 1972, Trout Valley Formation, Maine, USA (Kaspar *et al.*, 1988).

Intervening: PRA–EMS.

INCERTAE SEDIS

Order BARINOPHYTALES Høeg, 1967

F. BARINOPHYTACEAE Høeg, 1967
D. (PRA)–C. (IVO) Terr.

First: *Krithodeophyton croftii* Edwards, 1968, Senni Beds, Powys, Wales, UK.
Last: *Protobarinophyton* sp., Price Formation, Virginia, USA (Scheckler, 1984).
Intervening: PRA–EMS, FAM.
Comments: This family is clearly a natural taxon, which also includes *Pectinophyton*, *Protobarinophyton* and *Barinostrobus*. It appears to fall broadly within the concept of the Pteridophyta, as used here. However, it does not appear to be closely related to any of the other pteridophyte classes referred to in this chapter (Brauer, 1981).

Class LYCOPSIDA Scott, 1909

The classification is based on Thomas and Brack-Hanes (1984), in which only fertile material is definitely included in families. However, mention is also made of isolated sterile specimens (e.g. stems) which may extend the range of particular families (e.g. *Lepidodenron* for the Flemingitaceae).

Order DREPANOPHYCALES Pichi-Sermolli, 1958

F. DREPANOPHYCACEAE Kräusel and Weyland, 1949 S. (LUD)–D. (FRS?) Terr.

First: *Baragwanathia longifolia* Lang and Cookson, 1935, Yea Formation, Victoria, Australia (Garratt, 1978).
Last: ?*Drepanophycus spinaeformis* Göppert, 1850, New York State, USA (Stubblefield and Banks, 1978). This is based on sterile axes. The youngest fertile specimens are *D. spinaeformis* from the EMS of the Rhineland, Germany (Kräusel and Weyland, 1930).
Intervening: LOK, PRA, GIV?
Comments: Rayner (1984) places this family in a separate class (Drepanophycopsida), but we have here followed the traditional view and retained it within the lycopsids.

Order PROTOLEPIDODENDRALES Pichi-Sermolli, 1958

F. PROTOLEPIDODENDRACEAE Kräusel and Weyland, 1949 D. (PRA)–P. (SAK?) Terr.

First: *Sugambrophyton pilgeri* Schmidt, 1954, Hamberg Formation, Westphalia, Germany.
Last: *Brasilodendron pedroanum* Chaloner *et al.*, 1979, Itararé Formation, Paraná Basin, Brazil.
Intervening: EMS–CHD, KSK?–ASS?
Comments: In the palaeo-equatorial regions, the family is restricted to the Devonian and Lower Carboniferous, but in southern palaeolatitudes it extends up into the Permian. Sterile stems occur as high as KAZ? (Lemoigne and Brown, 1980).

F. ELEUTHEROPHYLLACEAE Kräusel and Weyland, 1949 C. (CHO–MRD) Terr.

First: *Eleutherophyllum waldenburgense* (Stur) Zimmermann, 1930, Wałbrzych (formerly Waldenburg) Beds, Lower Silesia, Poland.
Last: *Eleutherophyllum hamatum* Josten, 1983, Hagener Formation, Ruhr, Germany.

Order LYCOPODIALES Potonié, 1899

Skog (1986) has recently proposed a new lycopodialean family, the Tanydoraceae, based on the Cretaceous *Onychiopsis psilotoides* (Stokes and Webb) Ward, 1905, traditionally regarded as fern foliage (e.g. Watson, 1969). However, structurally preserved specimens of the latter species described by Friis and Pedersen (1990) have confirmed that it is a fern, probably of the Dicksoniaceae.

F. LYCOPODIACEAE Mirbel, 1802 D. (EIF)–Rec. Terr.

First: *Lycopodites oosensis* Kräusel and Weyland, 1937, Ooser Plattenkalk, Eifel, Germany. **Extant**
Intervening: HLK, POD–MYA, RHT–BER.
Comments: Much of the quoted intervening range is based on sterile axes.

Order SELAGINELLALES Potonié, 1899

F. SELAGINELLACEAE Willkes, 1861 D. (FAM)–Rec. Terr.

First: *Barsostrobus famennensis* Fairon-Demaret, 1977, Evieux Formation, Barse, Belgium. **Extant**
Intervening: HLK–ASB, MEL–DOR, CRN–RHT, BER.

Order LEPIDOCARPALES Thomas and Brack-Hanes, 1984

F. CYCLOSTIGMACEAE Kräusel and Weyland, 1949 D. (FAM) Terr.

First and Last: *Cyclostigma kiltorkense* Haughton, 1855, Kiltorcan Beds, Republic of Ireland (Chaloner, 1968).
Comments: This is the only species for which the fertile structures are known. Sterile specimens from the Pennant Formation of South Wales, UK (MYA), have been described as ?*Cyclostigma cambricum* Crookall, 1964.

F. FLEMINGITACEAE Thomas and Brack-Hanes, 1984 C. (IVO–DOR) Terr.

First: *Flemingites allantonensis* (Chaloner) Brack-Hanes and Thomas, 1983, Cementstone Group, Borders Region, Scotland, UK (Chaloner, 1953).
Last: *Flemingites major* (Germar) Brack-Hanes and Thomas, 1983, Blanzy, France (Zeiller, 1906).
Intervening: CHD–KRE.
Comments: This corresponds to the Lepidodendraceae of many authors, and occurs most commonly in the Namurian and Westphalian (KIN–MYA). Records of *Lepidodendron* stems range up to the Upper Permian in China (e.g. Zhao *et al.*, 1980), but they lack fertile structures.

F. SIGILLARIOSTROBACEAE Thomas and Brack-Hanes, 1984 C. (IVO–KLA) Terr.

First: *Mazocarpon pettycurense* Benson, 1918, Cementstone Group, Borders Region, Scotland, UK (Long, 1968a).
Last: *Sigillariostrobus serreatus* Teixeira, 1956, São Pedro da Corva Mine, Douro Basin, Portugal.
Intervening: ASB, VRK–DOR.
Comments: Sterile *Sigillaria* stems have been reported as high as the Upper Permian (e.g. Feng *et al.*, 1977).

F. LEPIDOCARPACEAE Schopf, 1941 C. (IVO)–P. (TAT?) Terr.

First: *Achlamydocarpon scoticum* Long, 1968a, Cementstone Group, Borders Region, Scotland, UK.
Last: *Achlamydocarpon sinensis* Tian and Guo, 1987, Wangjiazhai, Guizhou, China. The chronostratigraphical position of these specimens is stated only as 'Upper Permian'.
Intervening: VRK–MYA.

F. SPENCERITACEAE Thomas and Brack-Hanes, 1984 C. (YEA–MYA) Terr.

First: *Spencerites membranaceus* Kubart, 1910, Koksflöz, Ostrava-Karviná Coalfield, Czechoslovakia.
Last: *Spencerites moorei* (Cridland) Leisman, 1962, Herrin Coal, Illinois, USA (Leisman and Stidd, 1967).
Intervening: MEL, KSK.

F. DIAPHORODENDRACEAE DiMichele and Bateman, 1992 C. (YEA–MYA) Terr.

First: *Diaphorodendron vasculare* (Binney) DiMichele, 1985, Hauptflöz, Ruhr Coalfield, Germany (Phillips, 1980).
Last: *Diaphorodendron scleroticum* (Pannell) DiMichele, 1985, Herrin Coal, Illinois, USA (Phillips, 1980).
Intervening: VRK–POD.

F. CAUDATOCARPACEAE Thomas and Brack-Hanes, 1984 C. (VRK–POD) Terr.

First: *Caudatocarpus braidwoodense* (Arnold) Brack-Hanes 1981, Copland Seam, Illinois, USA.
Last: *C. braidwoodense*, Carbondale Formation, Illinois, USA (Arnold, 1938).
Intervening: KSK.

F. PINAKODENDRACEAE Chaloner, 1967 C. (VRK) Terr.

First and Last: *Pinakodendron ohmannii* Weiss, 1893, Veine de l'Olive, Mariemont, Belgium; also Veine 9 Paumes, Anzin, France (Kidston, 1911; Rousseau, 1933).
Comments: Mathieu (1937) records this species from Vendée, France (?PND) but does not state if the specimens are fertile.

F. SPORANGIOSTROBACEAE Thomas and Brack-Hanes, 1984 C. (KSK–DOR) Terr.

First: *Sporangiostrobus rugosus* Bode, 1928, and *S. orzeschensis* Bode, 1928, Orzesze Formation, Upper Silesia, Poland.
Last: *Sporangiostrobus feistmantelii* (Feistmantel) Němejc, 1931, Seam III, Puertollano, Spain (Wagner, 1983).
Intervening: POD–MYA.

F. CYCLODENDRACEAE Thomas and Brack-Hanes, 1984 P. (?ART) Terr.

First and Last: *Cyclodendron leslii* (Seward) Kräusel, 1928, Ecca Group, Hammanskraal, Transvaal, South Africa (Rayner, 1985).
Comments: Sterile stems of this species range from the middle Ecca to lower Beaufort groups (?SAK–?KUN) (Rayner, 1985).

F. PLEUROMEIACEAE Potonié, 1902 Tr. (GRI?–CRN) Terr.

First: *Pleuromeia jiaochengensis* Wang and Wang, 1982, Luijiakou Formation, Shanxi, China.
Last: *Annalepsis zeilleri* Fliche, 1905, Lettenkohle, Lorraine, France (Grauvogel-Stamm and Duringer, 1983).
Intervening: DIE–SPA.
Comments: The chronostratigraphical position of many of the localities yielding this family is in doubt (Wang and Wang, 1982). Other than *Annalepsis*, however, the family appears to be restricted mainly to the SCY.

Order MIADESMIALES Chaloner, 1967

F. MIADESMIACEAE Chaloner, 1967 C. (MEL) Terr.

First and Last: *Miadesmia membranacea* C. E. Bertrand, 1895, Union Seam, Lancashire, England, UK.

Order ISOETALES Engler, 1924

F. ISOETACEAE Dumortier, 1829 C. (MEL)–Rec. Terr.

First: *Bothrodendrostrobus mundus* Hirmer, 1927, Union Seam, Lancashire, England, UK (Stubblefield and Rothwell, 1981). **Extant**
Intervening: Tr, VLG–HOL.

F. CHALONERIACEAE Pigg and Rothwell, 1983 C. (MEL–CHV) Terr.

First: *Polysporia mirabilis* Newberry, 1873, No. 1 Coal, Tallmadge, Ohio, USA (Crookall, 1966).
Last: *Chaloneria cormosa* Pigg and Rothwell, 1983, Duquense Seam, Ohio, USA.
Intervening: VRK–POD.

F. TAKHTAJANODOXACEAE Thomas and Brack-Hanes, 1984 Tr. (SCY) Terr.

First and Last: *Takhtajanodoxa mirabilis* Snigirevskaya, 1980, Tutonchansk 'Suite', eastern Siberia, former USSR.
Comments: The exact chronostratigraphical position of these strata within the Lower Triassic is uncertain.

Class EQUISETOPSIDA Takhtajan ex Němejc, 1963

This is the Sphenopsida of some authors. The classification is based essentially on Meyen (1987), and relies on the recognition of fertile structures. In contrast to Meyen, however, the Pseudoborniales is retained as a distinct order; the Eviostachyaceae and Cheirostrobaceae are maintained as separate families within the Bowmanitales; the family Calamostachyaceae is included in the Equisetales, rather than being separated into its own order; and the Archaeocalamitaceae (Asterocalamitaceae auct.) is reinstated for the distinctive group of Lower Carboniferous equisetalean taxa.

Order PSEUDOBORNIALES Nathorst, 1902

F. PSEUDOBORNIACEAE Nathorst, 1902 D. (FRS–FAM) Terr.

First: *Pseudobornia ursina* Nathorst, 1902, Bear Island, Arctic (Schweitzer, 1967).
Last: *P. ursina* Nathorst, 1902, Alaska, USA (Mamay, 1962).

Order BOWMANITALES Meyen, 1978

F. BOWMANITACEAE Meyen, 1978 D. (FRS?)–P. (KAZ?) Terr.

First: ?*Sphenophyllum subtenerrimum* Nathorst, 1920, Bear Island, Arctic. This is based on sterile foliage. The oldest strobili appear to be *Bowmanites tumbana* Remy and Spassov, 1959 (associated with *S. subtenerrimum* foliage), Tumba, Bulgaria (FAM).
Last: ?*Sphenophyllum sinocoreanum* Yabe, Lungtan Formation, Jiangxi, China (Mei, 1984). This is based on sterile foliage. The youngest strobili are *Bowmanites simonii* Remy, 1961, upper Autunian, Thuringia, Germany (ASS).
Intervening: IVO–NOG.

F. EVIOSTACHYACEAE Boureau, 1964
D. (FAM) Terr.

First and Last: *Eviostachya hoegii* Stockmans, 1948, Evieux Formation, Belgium (Leclercq, 1957).

F. CHEIROSTROBACEAE Scott, 1907
C. (ASB) Terr.

First and Last: *Cheirostrobus pettycurensis* D. H. Scott, 1898, Pettycur Limestone, Fife Region, Scotland, UK.

Order EQUISETALES Trevisan, 1876

F. ARCHAEOCALAMITACEAE Stur, 1875
C. (IVO–ASB) Terr.

First: *Archaeocalamites* sp. and Fructification-type A of Scott *et al.* (1985), Loch Humphrey Burn, Strathclyde Region, Scotland, UK.
Last: *Protocalamostachys pettycurensis* Chaphekar, 1963, Pettycur Limestone, Fife Region, Scotland, UK.
Intervening: ARU–HLK.
Comments: Distinguishing this family from the Calamostachyaceae is not always easy, particularly with sterile material. Archaeocalamites stems have been found ranging from FAM (Ischenko, 1965) to ARN (Novik, 1968). There is also an unpublished record of stems and foliage from the Clyde Formation (?ART), Texas, USA (R. Bateman, pers. comm.).

F. CALAMOSTACHYACEAE Meyen, 1987
C. (ALP)–P. (KAZ?) Terr.

First: *Calamostachys polystachya* (Sternberg) Weiss, 1876, Zone de Sippenaken, Gives-Groynne, Belgium (Stockmans and Willière, 1953).
Last: ?*Calamostachys* (?) sp., Ural Mountains, former USSR (Vakhrameev *et al.*, 1978). Better documented are *Calamostachys grandeuryi* (Renault) Jongmans, 1911; Autun, France (ASS) (Renault, 1876) and *C. dumasii* (Zeiller) Jongmans, 1911, Sobernheim, Nahe, Germany (ASS) (Kerp, 1984).
Intervening: KIN–NOG.

F. TCHERNOVIACEAE Meyen, 1969
C. (MYA?)–P. (TAT) Terr.

First: ?*Phyllotheca deliquescens* (Göppert) Schmalhausen, Keregetassker Formation, Pribalchasje, Kazakhstan, former USSR (Radchenko, 1967). This record is based on sterile foliage of equivocal chronostratigraphical position. Better documented is the fructification *Tchernovia ungensis* Gorelova, 1962, Alykaevskii Formation (KAS), Kuznetsk, former USSR (Meyen, 1982).
Last: *Sendersonia matura* Meyen and Menshikova, 1983, Tailuganskii Formation, Kuznetsk, former USSR.
Intervening: NOG?–KAZ.
Comments: The Angaran sphenopsids have been reclassified by Meyen (1971) and Meyen and Menshikova (1983) (see also Meyen, 1982, for a review). There is some evidence that the family may range up to the Lower Jurassic based on the distribution of foliage such as *Neokoretrophyllites*. As pointed out by Meyen (1971), however, assigning such foliage to a particular family without evidence of the fructifications can be suspect, and so this expanded stratigraphical range is not quoted here.

F. GONDWANOSTACHYACEAE Meyen, 1969
P. (KAZ) Terr.

First and Last: *Gondwanostachys australis* (Brongniart) Meyen, 1969, Newcastle Group, New South Wales, Australia (Townrow, 1956).
Comments: Townrow also mentions foliage from the Raniganj Formation of India, the Beaufort Formation of South Africa, and from Madagascar, Argentina and the Falkland Islands. These are all probably of about the same age.

F. EQUISETACEAE Richard ex De Candolle, 1805
P. (TAT?)–Rec. Terr.

First: ?*Neocalamites superpermicus* Kon'no, 1973, Toyoma Formation, Kitakami, Japan. This record is based on sterile stems. The oldest fertile structures are *Neocalamostachys pedunculatus* Kon'no, 1962, Momonoki Formation, Kitakami, Japan (CRN). **Extant**
Intervening: SPA?–LAD?, RHT–HOL.

F. ECHINOSTACHYACEAE Grauvogel-Stamm, 1978 Tr. (SMI?) Terr.

First and Last: *Echinostachys oblonga* Brongniart, 1828, *E. cylindrica* Schimper and Mougeot, 1844, and *Schizoneura paradoxa* Schimper and Mougeot, 1844, *Voltzia* Sandstone, Vosges, France (Grauvogel-Stamm, 1978).

Class FILICOPSIDA Pichi-Sermolli, 1958

This class has also been called the Pteropsida and Polypodiopsida. There is little agreement on the classification of the group, or even its taxonomic rank. The scheme used here is based on Meyen (1987), with modifications based on Danzé (1956), Galtier and Scott (1979), Brousmiche (1983) and Scott and Galtier (1985).

Order CLADOXYLALES Hirmer, 1923

F. CLADOXYLACEAE Unger, 1856
D. (EIF)–C. (IVO) Terr.

First: *Hyenia elegans* Kräusel and Weyland, 1926, and *Calamophyton primaevum* Kräusel and Weyland, 1926, Lindlar, Rhineland, Germany (Schweitzer, 1972, 1973).
Last: *Cladoxylon waltonii* Long, 1968b, Cementstone Group, Borders Region, UK.
Intervening: GIV–HAS.
Comments: *Protohyenia janovii* Ananiev, 1957, Torgachine, western Siberia, former USSR (EMS) has been included by some authors in this family (e.g. Meyen, 1987), but was described from specimens belonging to more than one species.

Order IBYKALES Skog and Banks, 1973

F. IBYKACEAE Skog and Banks, 1973
D. (GIV–FRS) Terr.

First: *Ibyka amphikoma* Skog and Banks, 1973, New York State, USA.
Last: *Asteropteris noveboracensis* Dawson, 1881, Portage Group, Milo, New York State, USA (Bertrand, 1913).
Comments: The taxonomic position of this order is still in doubt, and some authors have assigned it to the Equisetopsida.

Order COENOPTERIDALES Zimmermann, 1930

F. RHACOPHYTACEAE Barnard and Long, 1975
D. (EIF)–C. (HAS) Terr. (see Fig. 43.2)

First: *Protocephalopteris praecox* (Høeg) Ananiev, 1960, Saian-Altai Mountains, Siberia, former USSR.
Last: *Rhacophyton* sp., Lower Limestone Shales, Avon Gorge, England, UK (Utting and Neves, 1970).
Intervening: GIV, FAM.
Comments: The Avon Gorge assemblage has yet to be named or described in detail. However, CJC has examined the material (in the collection of R. H. Wagner), and it includes fertile structures (see also Wagner, 1984, p. 113).

F. ZYGOPTERIDACEAE Bertrand, 1909 D. (FAM?)–C. (BRI) Terr.

First: ?*Clepsydropsis campbellii* Read, 1936, New Albany Shale, Kentucky, USA. This species is based on sterile foliage. The oldest fructification appears to be Fructification-type D from Loch Humphrey Burn, Strathclyde Region, Scotland, UK (IVO?) (A. C. Scott *et al.*, 1985).
Last: *Musatea globata* Galtier, 1968, Esnost and Roannais, Loire, France. These fertile structures are associated with fronds *Diplolabis roemeri* Solms-Laubach, 1892, *Metaclepsydropsis duplex* Williamson, 1874, and *Dineuron pteroides* Renault, 1896 (A. C. Scott *et al.*, 1984).
Intervening: CHD?–HLK?, ASB.
Comments: This family has traditionally been interpreted in a broad sense, for a group of Upper Devonian to Lower Permian fern fossils, typically including stems of the zygopterid type. Following evidence provided by Galtier and Scott (1979), however, we treat it in a more restricted sense, including only those species which are known to have radially arranged sessile sori of elongate exannulate sporangia with inwards-facing dehiscence. Most of the other zygopterids can be assigned to the stratigraphically higher families Corynepteridaceae and Biscalithecaceae (below).

F. STAUROPTERIDACEAE Hirmer, 1927 D. (FAM)–C. (VRK) Terr.

First: *Gillespiea randolphensis* Erwin and Rothwell, 1989, Hampshire Formation, West Virginia, USA.
Last: *Stauropteris biseriata* Cichan and Taylor, 1982, upper Path Fork Coal, Kentucky, USA.
Intervening: IVO, ASB, MEL.
Comments: Erwin and Rothwell (1989) argue that this is an artificial family, encompassing plant fossils of simple morphology and anatomy, but derived from several independent lineages.

F. CORYNEPTERIDACEAE Cleal, fam. nov. C. (IVO?–CHV) Terr.

First: ?Fructification-type E of Scott *et al.* (1985), Loch Humphrey Burn, Strathclyde Region, Scotland, UK. This record is based on isolated sporangia, which have not yet been fully described. Stockmans and Willière (1955) describe sterile foliage possibly belonging to this family as *Alloiopteris quercifolia* (Göppert) Potonié, 1897, Zone de Bioul, Belgium (CHO). The oldest record of fertile foliage is *Corynepteris stellata* Baily, 1860, ?upper Namurian, County Limerick, Republic of Ireland (MRD?), (Galtier and Scott, 1979).
Last: *Corynepteris angustissima* (Sternberg) Němejc, Illingen Seam, Saarland, Germany (Brousmiche, 1983).
Intervening: ALP–KRE.
Comments: The circumscription and diagnosis of this family was suggested by Galtier and Scott (1979), but they failed to give it a formal name; I therefore propose the new name here, Corynepteridaceae. It includes those species, previously assigned to the Zygopteridaceae, where sessile sori are attached to laminate pinnules. Each sorus consists of elongate sporangia (1–2.5 mm long) with a V-shaped annulus. The type is the form genus *Corynepteris*. Reports of corynepterid fronds from higher horizons (Broutin, 1981; Wagner and Sousa, 1983) need to be verified.

F. BISCALITHECACEAE Cleal, fam. nov. C. (MYA)–P. (ASS) Terr.

First: *Biscalitheca musata* Mamay, 1957, Baker Coal Member, Kentucky, USA (Phillips, 1980).
Last: *Biscalitheca dubius* (Renault) Galtier, 1978, Autunian, Champ des Borgis, Autun, France.
Intervening: KRE–NOG.
Comments: This is the second of the Upper Carboniferous–Permian groups of zygopterid-like ferns which Galtier and Scott (1979) suggested probably belonged to a separate family, but for which no name was provided. According to Galtier and Scott, it includes those species with long-stalked sori attached directly to the rachis (rather than to the pinnules, as in the Corynepteridaceae). Also, the sporangia are longer (3–4 mm) and have two distinct longitudinal annuli. The type is *Biscalitheca* Mamay, 1957. Other form genera included are *Nemejcopteris* Barthel, 1968, and *Schizostachys* Grand'Eury, 1877.

Order BOTRYOPTERIDALES Meyen, 1987

F. PSALIXOCHLAENACEAE Holmes, 1981 C. (IVO?–CHV?) Terr.

First: ?*Psalixochlaena berwickense* Long, 1976, Cementstone Group, Borders Region, Scotland, UK. This species is based on vegetative axes without fructifications. The youngest known species with fertile structures is *Psalixochlaena cylindrica* (Williamson) Holden, 1960, Union Seam (MEL), Lancashire, England, UK (Holmes, 1981).
Last: ?*Hymenophyllites quadridactylites* (Gutbier) Kidston, 1923, Wahlschied Seam, Saarland, Germany (Brousmiche, 1983). This record is based on Meyen's (1987) suggestion that *Hymenophyllites* (Zeiller) Kidston, 1923, may belong to the Psalixochlaenaceae. The youngest recorded unequivocal member of the family is *Norwoodia angustum* Rothwell, 1976, Cabaniss Formation (POD), Kansas, USA (Good and Rothwell, 1988). Good (1981) argued that *Botryopteris antiqua* Kidston, 1908, belongs to this family, but it differs in details of the vascular anatomy and position of attachment of the sporangia.

F. TEDELEACEAE Eggert and Taylor, 1966 C. (IVO?)–P. (ASS) Terr.

First: ?Fructification-type C of A. C. Scott *et al.* (1985), Loch Humphrey Burn, Strathclyde Region, Scotland, UK. This specimen has yet to be described fully or named.
Last: ?*Senftenbergia saxonica* Barthel, 1976, Döhlener Formation, Döhlener Basin, Germany. *Tubicaulis solenites* Cotta, 1832, Leukersdorfer Formation, Erzgebirge, Germany (Barthel, 1976).
Intervening: ASB–NOG.
Comments: The quoted range assumes that permineralized fructifications known as *Tedelea* Eggert and Taylor, 1966, can be correlated with the compression fructifications *Senftenbergia* Corda, 1845 (Meyen, 1987).

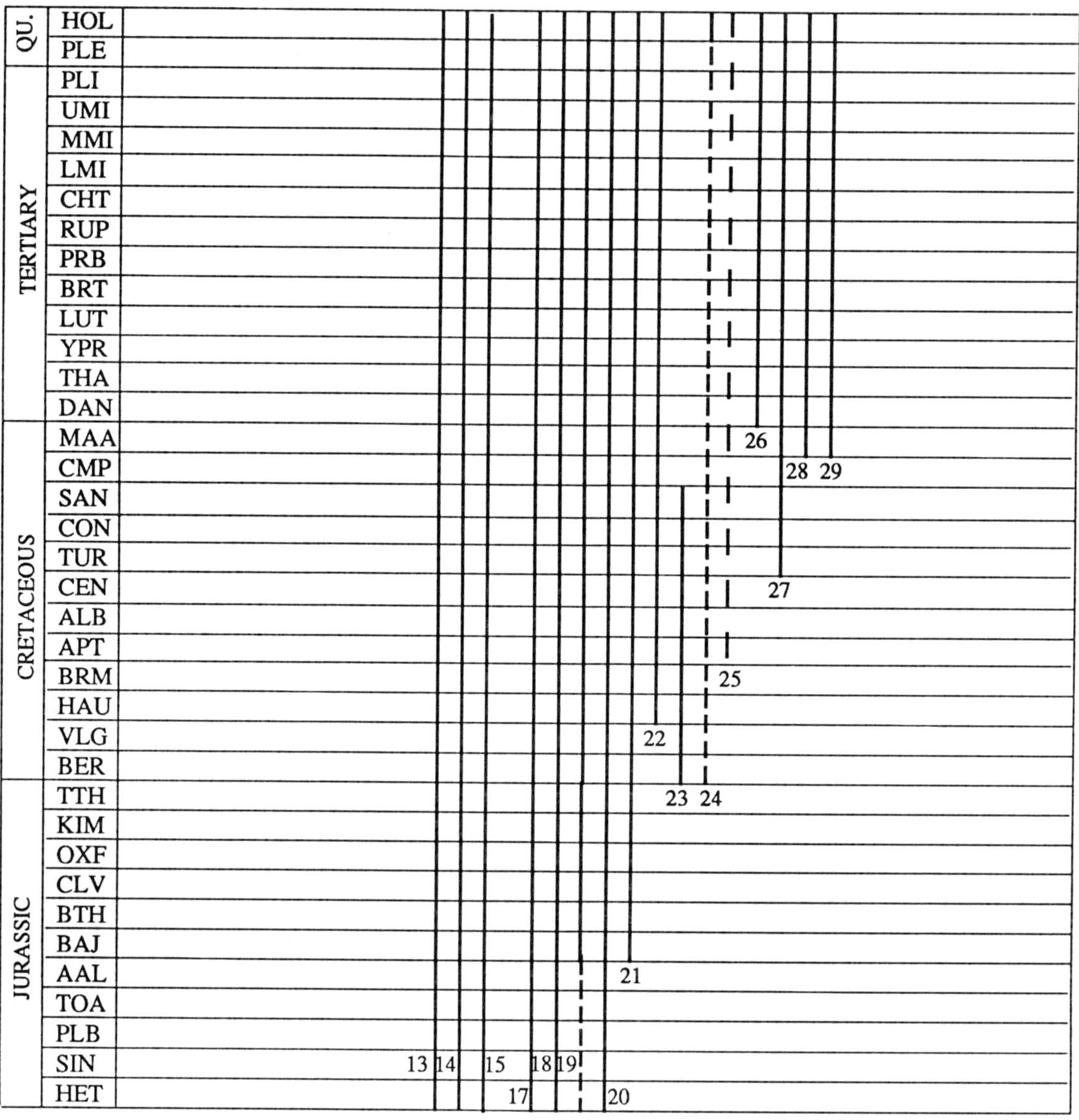

Fig. 43.2

F. BOTRYOPTERIDACEAE Renault, 1883
C. (IVO)–P. (ASS?) Terr.

First: *Botryopteris* cf. *antiqua* Kidston, 1908, and Fructification-type F, Loch Humphrey Burn, Strathclyde Region, Scotland, UK (A. C. Scott *et al.*, 1985). This fructification has yet to be fully described, but the structure of the annulus differs slightly from typical *Botryopteris antiqua*, as described by Galtier (1967). However, its assignment to the family appears to be confirmed by the anatomy of the associated sterile axes.

Last: ?*Botryopteris burgkensis* (Sterzel) Barthel, 1976, tonstein above Seam 2, Döhlener Formation, Döhlener Basin, Germany. This record is based on adpressions, and Galtier (1986) has expressed doubt as to their attribution. Better documented is *Botryopteris forensis* Renault, 1875, Rive de Gier Formation, Grand'Croix, France (CHV), (Galtier, 1971).

Intervening: ASB, MEL, POD–CHV.

Comments: The generic taxonomy of this family probably needs to be revised (e.g. Good, 1981), but the species are united at the rank of family by the vascular anatomy and sporangial structure.

F. SERMEYACEAE Eggert and Delevoryas, 1967
C. (VRK)–P. (KAZ) Terr.

First: *Oligocarpia brongniartii* Stur, 1883, Barnsley Seam, South Yorkshire, England, UK (Kidston, 1923).

Last: *Oligocarpia permiana* Fefilova, 1973, Seidinskaya 'Suite', Pechora Basin, former USSR.

Intervening: KSK–DOR.

Comments: The quoted range relies on the correlation between the permineralized *Sermeya* Eggert and Delevoryas, 1967, and the adpression *Oligocarpia* Göppert, 1841 (Meyen, 1987), although some still regard the latter as in the Gleicheniaceae (e.g. Galtier and Scott, 1985; Yao and Taylor, 1988). Triassic records of *Oligocarpia* from Lunz (LAD?), (Vakhrameev *et al.*, 1978) need to be verified. The range of permineralizations is MYA–CHV.

Order URNATOPTERIDALES Danzé, 1956

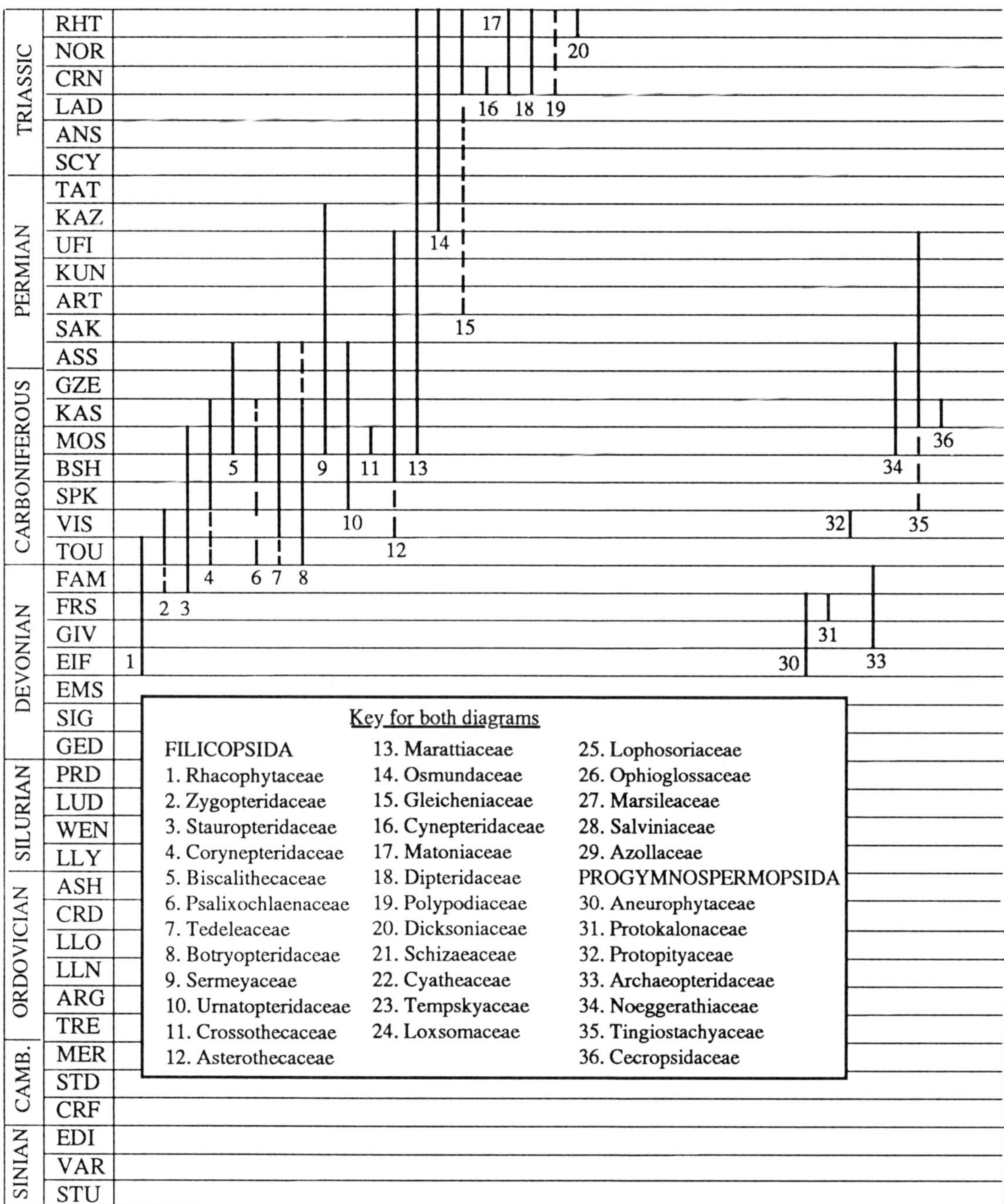

Fig. 43.2

F. URNATOPTERIDACEAE Danzé, 1956
C. (ALP)–P. (ASS) Terr.

First: *Renaultia gracilis* (Brongniart) Zeiller, 1883, Malonne Formation, Lontzen, Belgium (Stockmans and Willière, 1953).
Last: *Renaultia lebachensis* (Weiss) Brousmiche, 1983, Lower Rotliegend, Saarland, Germany (Kerp and Fichter, 1985).
Intervening: KIN–NOG.

Order CROSSOTHECALES Danzé, 1956

F. CROSSOTHECACEAE Danzé, 1956
C. (VRK–MYA)

First: *Crossotheca schatzlarensis* Stur, 1885, Fettkohle, Dortmund, Germany (Gothan, 1935).
Last: *Crossotheca crepinii* Zeiller, 1883, Steinbesch Formation, Lorraine, France (Brousmiche, 1983).
Intervening: KSK–POD.

Order MARATTIALES Engler and Prantl, 1902

There has been considerable disagreement as to the classification of the Marattiales. The view advanced by Hill and Camus (1986) has been adopted here, and just two families recognized. The oldest representative of the order appears to be *Burnitheca pusilla* Meyer-Berthaud and Galtier, 1986, Loch Humphrey Burn, Strathclyde Region, Scotland, UK (IVO?). However, it is based on an isolated sporangial cluster, which does not show the features necessary to assign it to a family.

F. ASTEROTHECACEAE Sporne, 1962
C. (BRI?)–P. (UFI)

First: ??*Megaphyton protuberans* Lesquereux, 1866, lower Chesterian, Illinois, USA (Pfefferkorn, 1976). The oldest permineralized axes are *Psaronius simplicicaulis* DiMichele and Phillips, 1977, Morrowan Group, Illinois, USA (YEA?). The oldest fructifications are *Scolecopteris minor* var. *parvifolia* Mamay, 1950, Upper Foot Seam, Lancashire, England, UK (MEL), (Phillips, 1980).
Last: *Gemellitheca saudica* Wagner *et al.*, 1985, Unayzah, Saudi Arabia. The taxonomic positions of *Asterotheca (?) pluseriata* Fefilova, 1973, Pechora Basin, former USSR (KAZ); *A. merianii* (Brongniart) Stur, 1885, from Lunz and Spitsbergen (CRN); and *A. cottonii* Zeiller, 1903, from Vietnam (RHT), (Vakhrameev *et al.*, 1978) are uncertain.
Intervening: VRK–ASS.
Comments: This is the same as the Psaroniaceae of many authors. However, the Asterothecaceae is a better name, being based on *Asterotheca* Presl, 1845, a form genus of sporangia (*Psaronius* is a form genus for anatomically preserved stems).

F. MARATTIACEAE Berchtold and Presl, 1820
C. (MYA)–Rec. Terr.

First: *'Radstockia' kidstonii* Taylor, 1967, Mazon Creek, Illinois, USA (a new form-genus name is required for this species – Brousmiche, 1983). The marginally earlier (MYA) *Danaeites saraepontanus* Stur, 1885, Beust Seam, Saarland, Germany (Corsin, 1951) may also belong here (Hill *et al.*, 1985). **Extant**
Intervening: ASS?, UFI?, RHT–HOL.
Comments: The macrofossil record of this family is discussed by Hill *et al.* (1985) and Hill and Camus (1986).

Order OSMUNDALES Zimmermann, 1959

F. OSMUNDACEAE Berchtold and Presl, 1820
P. (KAZ)–Rec. Terr.

First: *Zalesskaya gracilis* (Eichwald) Kidston and Gwynne-Vaughan, 1908, *Z. diploxylon* Kidston and Gwynne-Vaughan, 1908, *Thamnopteris schlechtendalii* (Eichwald) Brongniart, 1849, *Bathypteris rhomboidea* (Kutorga) Eichwald, 1860, and *Anomorrhoea fischeri* Eichwald, 1860, all from a Kazanian sandstone, Orenburg, former USSR (Kidston and Gwynne-Vaughan, 1908, 1909). *Palaeosmunda williamsii* Gould, 1970, and *P. playfordii* Gould, 1970, Bowen Basin, Queensland, Australia, are quoted as Upper Permian, but it is not stated whether they are older than the Orenburg species. **Extant**
Intervening: SMI?, CRN–HOL.
Comments: Meyen (1987) has suggested that some of the Permian species may represent an independent family within the Osmundales, but this has yet to be formalized.

Order FILICALES Engler and Prantl, 1902

F. GLEICHENIACEAE (Brown) Presl, 1825
P. (ART?)–Rec. Terr.

First: ?*Chansitheca kidstonii* Halle, 1927, lower Shihhotse Formation, Shanxi, China. This has gleicheniacean-like sporangia, but atypical vegetative organs. A less equivocal member of the family is *Wingatea plumosa* (Daugherty) Ash, 1969, Chinle Formation, New Mexico and Arizona, USA (CRN). **Extant**
Intervening: UFI?, PLB, BAJ, BER–HOL.
Comments: The Palaeozoic form genus *Oligocarpia* Göppert, 1841, has been often assigned to the Gleicheniaceae (e.g. Abbott, 1954), but is now more usually referred to the Sermeyaceae (Botryopteridales). The Gleicheniaceae is rare in the fossil record below the Cretaceous (Arnold, 1964).

F. CYNEPTERIDACEAE Ash, 1969
Tr. (CRN) Terr.

First and Last: *Cynepteris lasiphora* Ash, 1969, Chinle Formation, New Mexico and Arizona, USA.

F. MATONIACEAE Presl, 1847
Tr. (CRN)–Rec. Terr.

First: *Phlebopteris smithii* (Daugherty) Arnold, 1956 and *P. utensis* Arnold, 1956, Chinle Formation, Utah, Arizona and New Mexico, USA. **Extant**
Intervening: RHT–HOL.
Comments: Records of *Phlebopteris* Brongniart, 1828, from the Korvuncansker plant fossil assemblage (DIE?), Tunguska Basin, Siberia, former USSR (Vakhrameev *et al.*, 1978) need to be verified.

F. DIPTERIDACEAE Seward and Dale, 1901
Tr. (CRN)–Rec. Terr.

First: *Clathropteris walkeri* Daugherty, 1941, and *Apachea arizonica* Daugherty, 1941, Chinle Formation, New Mexico and Arizona, USA (Ash, 1969). Other, less well-documented records from CRN localities, such as Lunz, are reviewed by Vakhrameev *et al.* (1978). **Extant**
Intervening: NOR–HOL.
Comments: The sudden appearance of this family in such abundance in the Upper Triassic is quite striking, but there is no direct evidence on its origins.

F. POLYPODIACEAE Brown, 1810
Tr. (CRN?)–Rec. Terr.

First: ?*Polypodites cladophleboides* Brick, 1952, Kurasasajsker Group, Ural Mountains, former USSR. This appears to be based on sterile foliage. The oldest fertile fronds are *Aspidites thomasii* Harris, 1961, lower Deltaic Group, Yorkshire, England, UK (BAJ). **Extant**
Intervening: BER–HOL.
Comments: This family is treated in the generalized sense used in most palaeontological studies. It includes those groups assigned by some botanists (e.g. Sporne, 1975) to the Dennstaediaceae and Adiantaceae.

F. DICKSONIACEAE Bower, 1908
Tr. (RHT)–Rec. Terr.

First: *Coniopteris hymenophylloides* (Brongniart) Seward, 1900, 'Lower Series', Karmozd-Zirab, Iran (Klipper, 1964). **Extant**
Intervening: HET–BER, CMP–HOL.
Comments: See comments on Cyatheaceae.

F. SCHIZAEACEAE Kaulfuss, 1827
J. (BAJ)–Rec. Terr.

First: *Stachypteris spicans* Pomel, Scalby Formation, North Yorkshire, England, UK (Harris, 1961). **Extant**
Intervening: BTH–HOL.
Comments: The Palaeozoic form genus *Senftenbergia* Corda, 1845, has been widely included in the Schizaeaceae (e.g. Radforth, 1939), but is now more usually referred to the Tedelaceae (Botryopteridales).

F. CYATHEACEAE Kaulfuss, 1827 K. (HAU)–Rec. Terr.

First: *Cibotiocaulis tateiwae* Ogura, 1927 and *Cyathocaulis naktongensis* Ogura, 1927, North Kyong Sang Province, Korea. Nishida (1989) has recorded a permineralized stem from 'the Upper Jurassic or Lower Cretaceous' of Tasmania, which might belong to this family, but the evidence has not yet been published. **Extant**
Intervening: BRM–APT, TUR–SAN, DAN.
Comments: Most of the macrofossil evidence for this family comes from stem permineralizations. It is possible that at least some of the records usually attributed to the Dicksoniaceae may in fact be more properly placed here.

F. TEMPSKYACEAE Read and Brown, 1937 K. (BER–SAN) Terr.

First: *Tempskya schimperi* Corda, 1845, Ashdown Sand Formation, Sussex, England, UK (Seward, 1894).
Last: *Tempskya cretacea* Hosius and von der Marck, 1880, Haltern Quartzite, Westfalia, Germany.
Intervening: APT–ALB, CEN.
Comments: There is little direct evidence as to the fronds attached to the *Tempskya* Corda, 1845, stems, and therefore of the fructifications. The taxonomic rank and position of this taxon is thus uncertain. Records from the Tertiary (Kidston and Gwynne-Vaughan, 1911; Chandler, 1968) appear to be based on reworked fragments of permineralized stems.

F. LOXSOMACEAE Presl, 1875 K. (BER?)–Rec. Terr.

First: ?*Loxsomopteris anasilla* Skog, 1976, College Park, Maryland, USA. **Extant**
Comments: This is the only reported representative of the family in the macrofossil record, and is based on a fusainized rhizome.

F. LOPHOSORIACEAE Pichi Sermolli, 1970 K. (APT?)–Rec. Terr.

First: ?*Lophosoriorhachis japonica* Nishida, 1982, Miyako 'Series', Choshi, Japan. **Extant**
Comments: This is the only known macrofossil attributed to the family, and shows no evidence of fructifications. The palaeontological background to the family is discussed by Kurmann and Taylor (1987).

Order OPHIOGLOSSALES Engler and Prantl, 1902

F. OPHIOGLOSSACEAE (Brown) Agardh, 1822 T. (DAN)–Rec. Terr.

First: *Botrychium wightonii* Rothwell and Stockey, 1989, Genesee, Alberta, Canada. **Extant**
Comments: This is the only unequivocal representative of the order in the macrofossil record.

Order MARSILEALES Zimmermann, 1959

F. MARSILEACEAE Mirbel, 1802 K. (TUR)–Rec. FW

First: *Marsilea vera* Jarmolenko, 1935, clay beds of Kysyl-djar Hill, Kara-Tau, Urals, former USSR. **Extant**
Intervening: MAA–HOL.
Comments: The fossil record of this family is based mainly on isolated megaspores. A possible relationship with the Schizaeaceae has been argued by some authors.

Order SALVINIALES Zimmermann, 1959

The fossil record of this order is reviewed by Jain (1971) and Collinson (1980). Its taxonomy is far from settled, some authors recognizing just one family (Salviniaceae) while others recognize three (the two quoted below, plus the Aziniaceae).

F. SALVINIACEAE Dumortier, 1829 K. (MAA)–Rec. FW

First: *Salvinia stewartii* Jain, 1971, Edmonton Formation, Genesee, Alberta, Canada.
Intervening: DAN–HOL.

F. AZOLLACEAE Wettstein, 1903 K. (MAA)–Rec. FW

First: *Azolla extincta* Jain, 1971, and *A. distincta* Snead, 1969, Edmonton Formation, Genesee, Alberta, Canada (Jain, 1971).
Intervening: DAN–HOL.

Class PROGYMNOSPERMOPSIDA Beck, 1960

The following classification has been synthesized mainly from Barnard and Long (1975) and Meyen (1987).

Order ANEUROPHYTALES Kräusel and Weyland, 1941

F. ANEUROPHYTACEAE Kräusel and Weyland, 1941 D. (EIF–FRS) Terr.

First: *Protopteridium thomsonii* (Dawson) Kräusel and Weyland, 1933, Sandwick Fish Bed, Orkney, Scotland, UK (Lang, 1926).
Last: *Aneurophyton germanicum* Kräusel and Weyland, 1923, and *Tetraxylopteris schmidtii* Beck, 1957, Delaware River Formation, New York State, USA (Bonamo and Banks, 1967; Serlin and Banks, 1978).
Intervening: GIV.
Comments: The taxonomy and stratigraphical distribution of this family are extensively reviewed by Scheckler and Banks (1971).

F. PROTOKALONACEAE Barnard and Long, 1975 D. (FRS) Terr.

First and Last: *Protokalon petryi* Scheckler and Banks, 1971, Oneonta Formation, New York State, USA.
Comments: This species was separated from the Aneurophytaceae by Barnard and Long (1975) because of the differentiation of the protosteles in the stems and branches. However, its fertile structures have not yet been described.

F. PROTOPITYACEAE Banks, 1968 C. (ARU–ASB?) Terr.

First: *Protopitys scotica* Walton, 1957, Loch Humphrey Burn, Strathclyde Region, Scotland, UK (Smith, 1962).
Last: *Protopitys buchiana* Göppert, 1850, Falkenberg, Germany.
Comments: This is often treated as a separate progymnosperm order, but only differs significantly from the rest of the Aneurophytales by being heterosporous. A. C. Scott *et al.*, (1984) list it as a pteridosperm, but this is not supported by evidence given by Smith (1962). The chronostratigraphical position of the Falkenberg assemblage is uncertain due to the lack of recent studies (A. C. Scott *et al.*,

1984). It is estimated here as ASB?, as it compares well with assemblages of this age from France and the UK.

Order ARCHAEOPTERIDALES Zimmermann, 1930

F. ARCHAEOPTERIDACEAE Zimmermann, 1930 D. (GIV–FAM) Terr.

First: *Svalbardia scotica* Chaloner, 1972, North Gavel Beds, Fair Isle, Shetland, Scotland, UK (Allen and Marshall, 1986).
Last: *Archaeopteris hibernica* Forbes, 1853, from Kiltorcan, Republic of Ireland (Chaloner, 1968), and Borders Region, Scotland, UK (Miller, 1857), and Bear Island, Arctic (Kaiser, 1970), all HAS?
Intervening: FRS–FAM.
Comments: Meyen (1987) has suggested that *Federkurtzia* Archangelsky, 1981 and *Botrychiopsis* Kurtz, 1895 from the Lower Permian of Gondwanaland might be affiliated with this family, but the evidence is as yet equivocal.

Order NOEGGERATHIALES Darrah, 1939

F. NOEGGERATHIACEAE Darrah, 1939 C. (VRK)–P. (ASS) Terr.

First: *Rhacophyton (? Noeggerathia) westermannii* Gothan, 1931, Kohlscheider Group, Aachen, Germany. A fragment of foliage without fertile structures. The oldest fertile structures are *Discinites jongmansii* Hirmer, 1941, Limburg, The Netherlands (VRK).
Last: *Noeggerathia zamitoides* Sterzel, 1918, Middle Rotliegend, Saxony, Germany (Hirmer, 1941).
Intervening: KSK–POD, KLA–NOG.
Comments: The affinities of this family are still somewhat in doubt, but there is growing consensus that it is progymnospermous (e.g. Meyen, 1987). The taxonomy and distribution are reviewed by Hirmer (1941).

F. TINGIOSTACHYACEAE Gao and Thomas, 1987 C. (KIN?)–P. (TAT?) Terr.

First: ??*Tingia placida* Tidwell, 1967, Manning Canyon Shale, Utah, USA. This is based on sterile fragments, which occur outside of the normal palaeogeographical range of the form genus (i.e. the Cathaysia palaeocontinent). *Tingia* sp., Tongshan Formation, Kaiping, China? (Stockmans and Mathieu, 1939) are the oldest specimens of foliage from Cathaysia (VRK?). *Tingiostachya* sp., Taiyuan Formation, Kaiping, China (Stockmans and Mathieu, 1957) are the oldest strobili, associated with *Tingia trilobata* Stockmans are Mathieu, 1939, and *T. partita* Halle, 1927, foliage (CHV?).
Last: ?*Tingia subcarbonica* Kon'no *et al.*, 1971, Lingin 'Flora', Johore, Malaysia and *T. yichuanensis* Feng *et al.*, 1977, Henan, China. These are based on sterile foliage fragments. The youngest known strobili are *Tingiostachya tetralocularis* Kon'no, 1929, lower Shihezi Formation, Shanxi, China (Gao and Thomas, 1987), (UFI? – not Lower Permian, as stated by Gao and Thomas).
Intervening: ROT.
Comments: This family has been reviewed by Gao and Thomas (1987), who concluded that it does not belong to the Noeggerathiales, and may not even be a progymnosperm. However, it has many features in common with this order, both in the form of the strobili and the foliage, and so the traditional view as to its taxonomic position has been followed here.

Order CECROPSIDALES Stubblefield and Rothwell, 1969

F. CECROPSIDACEAE Stubblefield and Rothwell, 1969 C. (DOR) Terr.

First and Last: *Cecropsis luculentum* Stubblefield and Rothwell, 1989, Duquesne Seam, Ohio, USA.

REFERENCES

Abbott, M. L. (1954) Revision of the Paleozoic fern genus *Oligocarpia*. *Palaeontographica, Abteilung B*, **96**, 39–65.

Allen, K. C. and Marshall, J. E. A. (1986) *Svalbardia* and the 'corduroy' plant from the Devonian of the Shetland Islands, Scotland. *Special Papers in Palaeontology*, **35**, 7–20.

Ananiev, A. R. (1957) New fossil plants from the Lower Devonian deposits near the village of Torgashino in the southeastern part of Western Siberia. *Botanicheskii Zhurnal*, **42**, 691–702 [in Russian].

Ananiev, A. R. (1960) Study of the Middle Devonian flora of the Saian–Altai mountain region. *Botanicheskii Zhurnal*, **45**, 649–66 [in Russian].

Arnold, C. A. (1938) Note on a lepidophyte strobilus containing large spores, from Braidwood, Illinois. *American Midland Naturalist*, **20**, 709–12.

Arnold, C. A. (1956) Fossil ferns of the Matoniaceae from North America. *Journal of the Palaeontological Society of India*, **1**, 118–21.

Arnold, C. A. (1964) Mesozoic and Tertiary fern evolution and distribution. *Bulletin of the Torrey Botanical Club*, **21**, 58–66.

Ash, S. R. (1969) Ferns from the Chinle Formation (Upper Triassic) in the Fort Wingate area, New Mexico. *United States Geological Survey Professional Paper*, **613**, D1–D52.

Banks, H. P. (1968) The early history of land plants, in *Evolution and Environment* (ed. E. Drake), Yale University Press, New Haven and London, pp. 73–107.

Banks, H. P. (1975) Reclassification of Psilophyta. *Taxon*, **24**, 401–13.

Barnard, P. D. W. and Long, A. G. (1975) *Triradioxylon*–a new genus of Lower Carboniferous petrified stems and petioles together with a review of the classification of early Pterophytina. *Transactions of the Royal Society of Edinburgh*, **69**, 231–50.

Barthel, M. (1976) Die Rotliegendflora Sachsens. *Abhandlungen des Staatlichen Museums für Mineralogie und Geologie zu Dresden*, **24**, 1–190.

Bertrand, C. E. (1895) Sur une nouvelle centradesmide de l'Époque houillère. *Association française avancement sciences, Comptes rendus 23e session*, **1894**, 588–93.

Bertrand, P. (1913) Étude du stipe de l'*Asteropteris noveboracensis*. *Compte rendu 12e Congrès géologie du Canada*, 909–24.

Bode, H. (1928) Über eine merkwurdige Pteridophytenfruktifikation aus dem oberschlesischen Carbon. *Preussische geologische Landesanstalt Jahrbuch*, **49**, 245–7.

Bonamo, P. M. and Banks, H. P. (1967) *Tetraxylopteris schmidtii*: its fertile parts and its relationship within the Aneurophytales. *American Journal of Botany*, **54**, 755–68.

Brack-Hanes, S. D. (1981) On a lycopsid cone with winged spores. *Botanical Gazette*, **142**, 294–304.

Brauer, D. F. (1981) Heterosporous, barinophytacean plants from the Upper Devonian of North America and a discussion of the possible affinities of the Barinophytaceae. *Review of Palaeobotany and Palynology*, **33**, 347–62.

Brick, M. I. (1952) *The Fossil Flora and the Stratigraphy of the Lower Mesozoic Strata in the Middle Course of the Ilek in west Kazakhstan*. Gosgeolizdat, Moscow [in Russian].

Brousmiche, C. (1983) Les fougères sphénoptéridiennes du Bassin Houiller Sarro-Lorrain. *Société Géologique du Nord Publication*, **10**, 1–480.

Broutin, J. (1981) Étude Paléobotanique et Palynologique du Passage Carbonifère–Permien dans le Bassins Continentaux du Sud-Est de la Zone d'Ossa–Morena (Environs de Guadalcanal.

Espagne du Sud). Unpublished Doctoral Thesis, Université Pierre et Marie Curie, Paris.

Chaloner, W. G. (1953) On the megaspores of four species of *Lepidostrobus*. *Annals of Botany*, **17**, 264–73.

Chaloner, W. G. (1968) The cone of *Cyclostigma kiltorkense* Haughton, from the Upper Devonian of Ireland. *Botanical Journal of the Linnean Society*, **61**, 25–36.

Chaloner, W. G., Leistikow, K. U. and Hill, A. (1979) *Brasilodendron* gen. nov. and *B. pedroanum* (Carruthers) comb. nov., a Permian lycopod from Brazil. *Review of Palaeobotany and Palynology*, **28**, 117–36.

Chandler, M. E. S. (1968) A new *Tempskya* from Kent. *Bulletin of the British Museum (Natural History) Geology Series*, **15**, 169–79.

Chaphekar, M. (1963) Some calamitean plants from the Lower Carboniferous of Scotland. *Palaeontology*, **6**, 408–29.

Cichan, M. A. and Taylor, T. N. (1982) Structurally preserved plants from southeastern Kentucky: *Stauropteris biseriata* sp. nov. *American Journal of Botany*, **69**, 1491–6.

Collinson, M. E. (1980) A new multiple-floated *Azolla* from the Eocene of Britain with a brief review of the genus. *Palaeontology*, **23**, 213–29.

Corsin, P. (1951) Bassin houiller de la Sarre et de la Lorraine. I. Flore fossile. 4. Pécoptéridées. *Études des Gîtes Minéraux de la France*, 176–370.

Croft, W. N. and Lang, W. H. (1942) The Lower Devonian flora of the Senni Beds of Monmouthshire and Breconshire. *Philosophical Transactions of the Royal Society of London, B*, **231**, 131–64.

Crookall, R. (1964) Fossil plants of the Carboniferous rocks of Great Britain [Second section]. Part 3. *Memoirs of the Geological Survey of Great Britain, Palaeontology*, **4**, 217–354.

Crookall, R. (1966) Fossil plants of the Carboniferous rocks of Great Britain [Second section]. Part 4. *Memoirs of the Geological Survey of Great Britain, Palaeontology*, **4**, 355–572.

Danzé, J. (1956) Contribution à l'étude des *Sphenopteris*. Les fougères sphénoptéridiennes du bassin houiller du Nord de la France. *Études Géologiques pour l'Atlas de Topographie Souterrain*, **1** (2), 1–568.

Dimichele, W. A. and Bateman, R. M. (1992) Diaphorodendraceae, fam. *nov.* (Lycopsida: Carboniferous): systematics and evolutionary relationships of *Diaphorodendron* and *Synchysidendron*, gen. nov. *American Journal of Botany*, **79**, 605–17.

Dimichele, W. A. and Phillips, T. L. (1977) Monocyclic *Psaronius* from the lower Pennsylvanian of the Illinois Basin. *Canadian Journal of Botany*, **55**, 2514–24.

Edwards, D. (1968) A new plant from the Lower Old Red Sandstone of South Wales. *Palaeontology*, **11**, 683–90.

Edwards, D. (1975) Some observations on the fertile plants of *Zosterophyllum myretonianum* Penhallow from the Lower Old Red Sandstone of Scotland. *Transactions of the Royal Society of Edinburgh*, **69**, 251–65.

Edwards, D. and Edwards, D. S. (1986) A reconsideration of the Rhyniophytina, in *Systematic and Taxonomic Approaches in Palaeobotany* (eds R. A. Spicer and B. A. Thomas), Systematics Association Special Volume, No. 31. Clarendon Press, Oxford, pp. 199–220.

Edwards, D., Davies, K. L. and Axe, L. M. (1992) A vascular conducting strand in the early land plant *Cooksonia*. *Nature*, **357**, 683–5.

Edwards, D., Feehan, J. and Smith, D. G. (1983) A late Wenlock flora from Co. Tipperary, Ireland. *Botanical Journal of the Linnean Society*, **86**, 19–36.

Edwards, D. S. (1986) *Aglaophyton major*, a non-vascular plant from the Devonian Rhynie Chert. *Botanical Journal of the Linnean Society*, **93**, 173–204.

El-Saadawy, W. and Lacey, W. S. (1979) The sporangia of *Horneophyton lignieri* (Kidston and Lang) Barghoorn and Darrah. *Review of Palaeobotany and Palynology*, **28**, 137–44.

Erwin, D. M. and Rothwell, G. W. (1989) *Gillespiea randolphensis* gen. et sp. nov. (Stauropteridales), from the Upper Devonian of West Virginia. *Canadian Journal of Botany*, **67**, 3063–77.

Fairon-Demaret, M. (1977) A new lycophyte cone from the Upper Devonian of Belgium. *Palaeontographica, Abteilung B*, **162**, 51–63.

Fefilova, L. A. (1973) *Permian Pteropsida from the North of the Fore-Ural Basin*. Nauka, Leningrad [in Russian].

Feng S., Chen G., Xi Y. *et al.* (1977) Plants, in *Fossil Atlas of Central–South China, 2*, Geological Publishing House, Hupei Institute of Geological Sciences, Beijing, pp. 622–74 [in Chinese].

Friis, E. M. and Pedersen, K. R. (1990) Structure of the Lower Cretaceous fern *Onychiopsis psilotoides* from Bornholm, Denmark. *Review of Palaeobotany and Palynology*, **66**, 47–63.

Galtier, J. (1967) Les sporanges de *Botryopteris antiqua* Kidston. *Compte rendu des sénaces de l'Académie des Sciences, Paris, Série D*, **265**, 897–900.

Galtier, J. (1968) Un nouveau type de fructification filicinéenne du Carbonifère inférieur. *Comptes Rendus des Séances de l'Académie des Sciences, Paris, Série D*, **266**, 1004–7.

Galtier, J. (1971) La fructification de *Botryopteris forensis* Renault (Coenoptéridales de Stéphanien Français): précisions sur les sporanges et les spores. *Naturalia monspeliana, Série Botanique*, **22**, 145–55.

Galtier, J. (1978) Précisions sur 《Zygopteris lacattei》 & 《Botryopteris dubius》, fougères très rares de l'Autunien d'Autun. *Bulletin de la Société d'Histoire Naturelle d'Autun*, **88**, 17–25.

Galtier, J. (1986) Taxonomic problems due to preservation: comparing compression and permineralized taxa, in *Systematic and Taxonomic Approaches in Palaeobotany* (eds R. A. Spicer and B. A. Thomas), The Systematics Association Special Volume, 31, Clarendon Press, Oxford, pp. 1–16.

Galtier, J. and Scott, A. C. (1979) Studies of Palaeozoic ferns: on the genus *Corynepteris*. A redescription of the type and some other European species. *Palaeontographica, Abteilung B*, **170**, 81–125.

Galtier, J. and Scott, A. C. (1985) Diversification of early ferns. *Proceedings of the Royal Society of Edinburgh B*, **86**, 289–301.

Gao Z. and Thomas, B. A. (1987) A re-evaluation of the plants *Tingia* and *Tingiostachya* from the Permian of Taiyuan, China. *Palaeontology*, **30**, 815–28.

Garratt, M. J. (1978) New evidence for a Silurian (Ludlow) age for the earliest *Baragwanathia* flora. *Alcheringia*, **2**, 217–24.

Gensel, P. G. and Andrews, H. N. (1984) *Plant Life in the Devonian*. Praeger, New York.

Gerrienne, P. (1983) Les plantes emsiennes de Marchin (Vallée du Hoyoux, Belgique). *Annales de la Societé Géologique de Belgique*, **106**, 19–35.

Good, C. W. (1981) A petrified fern sporangium from the British Carboniferous. *Palaeontology*, **24**, 483–92.

Good, C. W. and Rothwell, G. W. (1988) A reinterpretation of the Paleozoic fern *Norwoodia angustum*. *Review of Palaeobotany and Palynology*, **56**, 199–204.

Göppert, H. R. (1850) *Monographie der Fossilen Coniferen*. Leiden.

Gothan, W. (1931) Die Steinkohlenflora der westlichen paralischen Carbonreviere Deutschlands. 2. *Arbeiten aus dem Institut für Paläobotanik und Petrographie der Brennsteine*, **1**, 49–96.

Gothan, W. (1935) Die Steinkohlenflora der westlichen paralischen Steinkohlenreviere Deutschlands. 3. *Abhandlung der Preussischen Geologischen Landesanstalt, Neue Folge*, **167**, 8–58.

Gould, R. E. (1970) *Palaeosmunda*, a new genus of siphonostelic osmundaceous trunks from the Upper Permian of Queensland. *Palaeontology*, **13**, 10–28.

Grauvogel-Stamm, L. (1978) La flore du grès à *Voltzia* (Buntsandstein supérieur) des Vosges du Nord (France). Morphologie, anatomie, interprétations phylogénique et paléogéographique. *Université L. Pasteur de Strasbourg, Institute Géologique, Mémoires*, **50**, 1–225.

Grauvogel-Stamm, L. and Duringer, P. (1983) *Annalepsis zeilleri* Fliche 1910 emend., un organe reproducteur de Lycophyte de la Lettenkohle de l'Est de la France. *Geologische Rundschau*, **72**, 23–51.

Halle, T. G. (1927) Palaeozoic plants from central Shansi. *Palaeontologica Sinica, Series A*, **2**, 1–316.

Harris, T. M. (1961) *The Yorkshire Jurassic Flora. I. Thallophyta–Pteridophyta*. British Museum (Natural History), London.

Hill, C. R. and Camus, J. M. (1986) Evolutionary cladistics of marattialean ferns. *Bulletin of the British Museum (Natural History) Botany*, **14**, 219–300.

Hill, C. R., Wagner, R. H. and El-Khayal, A. A. (1985) *Qasimia* gen. nov., an early *Marattia*-like fern from the Permian of Saudi Arabia. *Scripta Geologica*, **79**, 1–50.

Hirmer, M. (1941) *Noeggerathia*, neuendeckte verwandte Formen und ihre Stellung im System der Farne. *Biologia Generalis*, **15**, 134–71.

Holmes, J. C. (1981) The Carboniferous fern *Psalixochlaena cylindrica* as found in Westphalian A coal balls from England. Part II. The frond and fertile parts. *Palaeontographica, Abteilung B*, **176**, 147–73.

Hosius, A. and Marck, W. von der (1880) Flora der westfälischen Kreideformation. *Palaeontographica*, **26**, 127–241.

Hueber, F. M. (1971) *Sawdonia ornata*: a new name for *Psilophyton princeps* var. *ornata*. *Taxon*, **16**, 641–2.

Ischenko, T. A. (1965) *Devonian Floras of Greater Donbass*. Akademia Nauk, Ukraine SSR, Kiev [in Russian].

Jain, R. K. (1971) Pre-Tertiary records of Salviniaceae. *American Journal of Botany*, **58**, 487–96.

Jarmolenko, A. V. (1935) The Upper Cretaceous flora of northwest Kara-Tau. *Trudy Sredneaziatskogo Gosudarsd-venego Universitet*, Series 8–b Botany, **28**, 1–36 [in Russian].

Josten, K.-H. (1983) Die fossilen Floren im Namur des Ruhrkarbons. *Fortschritte in der Geologie von Rheinland und Westfalen*, **31**, 1–327.

Kaiser, H. (1970) Die Oberdevon-flora der Bäreninsel 3. Microflora des höheren Oberdevons und des Unterkarbons. *Palaeontographica, Abteilung B*, **129**, 72–124.

Kaspar, A. E., Gensel, P. G., Forbes, W. H. *et al.* (1988) Plant paleontology in the State of Maine – a review. *Maine Geological Survey. Studies in Maine Geology*, **1**, 109–28.

Kerp, J. H. F. (1984) Aspects of Permian palaeobotany and palynology. V. On the nature of *Asterophyllites dumasii* Zeiller, its correlation with *Calamites gigas* Brongniart and the problem concerning its sterile foliage. *Review of Palaeobotany and Palynology*, **41**, 301–17.

Kerp, J. H. F. and Fichter, J. (1985) Die Makrofloren des saarpfälzischen Rotliegenden (?Ober-Karbon–Unter Perm; SW-Deutschland). *Mainzer geowissenschaftliche Mitteilungen*, **14**, 159–286.

Kidston, R. (1911) Les végétaux houillers recueillis dans le Hinaut Belge. *Mémoires du Musée Royal d'Histoire Naturelle de Belgique*, **4**, 1–282.

Kidston, R. (1923) Fossil plants of the Carboniferous rocks of Great Britain. Parts 1–4. *Memoirs of the Geological Survey of Great Britain, Palaeontology*, **2**, 1–375.

Kidston, R. and Gwynne-Vaughan, D. T. (1908) On the fossil Osmundaceae. Part II. *Transactions of the Royal Society of Edinburgh*, **46**, 213–32.

Kidston, R. and Gwynne-Vaughan, D. T. (1909) On the fossil Osmundaceae. Part III. *Transactions of the Royal Society of Edinburgh*, **46**, 651–67.

Kidston, R. and Gwynne-Vaughan, D. T. (1911) On a new species of *Tempskya* from Russia. *Verhandlung der Russisch-Kaiserlichen Mineralogischen Gesellschaft zu St. Petersburg*, **48**, 1–20.

Klipper, K. (1964) Über eine Rät-Lias-Flora aus dem nördlichen Abfall des Alburs-Gebirges in Nordiran. Teil 1: Bryophyta und Pteridophyta. *Palaeontographica, Abteilung B*, **114**, 1–78.

Kon'no, E. (1962) Some species of *Neocalamites* and *Equisetites* in Japan and Korea. *Science Reports of the Tohoku University Sendai, 2nd Series Geology*, **5**, 21–7.

Kon'no, E. (1973) New species of *Pleuromeia* and *Neocalamites* from the Upper Scythian beds in the Kitakami Massif, Japan – with a brief note on some equisetacean plants from the Upper Permian bed in the Kitakami Massif. *Science Reports of the Tohoku University Sendai, 2nd Series, Geology*, **43**, 99–115.

Kon'no, E., Asama, K. and Rajah, S. S. (1971) The later Permian Lingin flora from the Gunong Blumut Area, Johore, Malaysia, in *Geology and Palaeontology of Southeast Asia*, 9, (eds T. Kobayashi and O. Toriyama), University Press, Tokyo, pp. 1–86.

Kräusel, R. and Weyland, H. (1930) Die Flora des deutschen Unterdevons. *Abhandlung preussische geologische Landesanstalt, Neue Folge*, **131**, 1–92.

Kräusel, R. and Weyland, H. (1937) Pflanzenreste aus dem Devon. X. Zwei Pflanzenfunde im Oberdevon der Eifel. *Senckenbergiana*, **19**, 338–55.

Kubart, B. (1910) Untersuchungen über die Flora des Ostrau-Karwiner Kohlenbeckens. I. Die Spore von *Spencerites membranaceous* nov. spec. *Denkschrift der Akademie Wissenschaft in Wien, Mathematisch-Naturwissenschaft Klasse*, **85**, 83–9.

Kurmann, M. H. and Taylor, T. N. (1987) Sporoderm ultrastructure of *Lophosoria* and *Cyatheacidites (Filicopsida)*: systematic and evolutionary implications. *Plant Systematics and Evolution*, **157**, 85–94.

Lang, W. H. (1926) Contributions to the study of the Old Red Sandstone flora of Scotland. III. On *Hostimella (Ptilophyton) Thomsoni*, and its inclusion in a new genus, *Milleria*. IV. On a specimen of *Protolepidodendron* from the Middle Old Red Sandstone of Caithness. V. On the identification of the large 'stems' in the Carmylie Beds of the Lower Old Red Sandstone as *Nematophyton*. *Transactions of the Royal Society of Edinburgh*, **54**, 785–99.

Leclercq, S. (1957) Étude d'une fructification de Sphenopside à structure conservée du dévonien supérieur. *Mémoires d'Académie royaume Belge, Classe Science*, **4-2-14-3**, 1–39.

Leisman, G. A. and Stidd, B. A. (1967) Further occurrences of *Spencerites* from the middle Pennsylvanian of Kansas and Illinois. *American Journal of Botany*, **54**, 316–23.

Lemoigne, Y. and Brown, J. T. (1980) Sur une flore à Glossopteridopsida et Lycopsida de Namibie (Sud-Oest africain). *Geobios*, **13**, 541–53.

Li C.-S. (1982) *Hsüa robusta*, a new land plant from the Lower Devonian of Yunnan, China. *Acta Phytotaxonomica Sinica*, **8**, 331–42.

Long, A. G. (1968a) Some specimens of *Mazocarpon, Achlamydocarpon* and *Cystosporites* from the Lower Carboniferous rocks of Berwickshire. *Transactions of the Royal Society of Edinburgh*, **67**, 359–72.

Long, A. G. (1968b) Some specimens of *Cladoxylon* from the Calciferous Sandstone Series of Berwickshire. *Transactions of the Royal Society of Edinburgh*, **68**, 45–61.

Long, A. G. (1976) *Psalixochlaena berwickense* sp. nov., a Lower Carboniferous fern from Berwickshire. *Transactions of the Royal Society of Edinburgh*, **69**, 513–21.

Mamay, S. H. (1962) Occurrence of *Pseudbornia* Nathorst in Alaska. *Palaeobotanist*, **11**, 19–22.

Mathieu, G. (1937) Recherches Géologiques sur les Terrains Paléozoïques de la Région Vendéene. Unpublished Thesis, University of Lille.

Mei M. (1984) *The Analysis of the Floras of Permian Coal-bearing Strata in Fujian, Jiangxi and Sichuan Provinces*. Graduate School, China Institute of Mining, Bejing (prepared for the 2nd Conference of the International Organization of Palaeobotany, Edmonton, 1984).

Meyen, S. V. (1969) New data on relationship between Angara and Gondwana Late Paleozoic floras, in *Gondwana Stratigraphy, IUGS Symposium, Buenos Aires*, 1967, pp. 141–57.

Meyen, S. V. (1971) *Phyllotheca*-like plants from the Upper Palaeozoic flora of Angaraland. *Palaeontographica, Abteilung B*, **133**, 1–33.

Meyen, S. V. (1982) The Carboniferous and Permian floras of Angaraland (a synthesis). *Biological Memoirs*, **7**, 1–109.

Meyen, S. V. (1987) *Fundamentals of Palaeobotany*. Chapman and Hall, London.

Meyen, S. V. and Menshikova, L. V. (1983) Systematics of the Upper Palaeozoic articulate family Tchernoviaceae. *Botanicheskii Zhurnal*, **68**, 721–9 [in Russian].

Meyer-Berthaud, B. and Galtier, J. (1986) Une nouvelle fructification du Carbonifère inférieur d'Ecosse: *Burnitheca*, Filicinée ou Ptéridospermale? *Comptes rendus des séances de l'Académie des Sciences, Paris, Série II*, **303**, 1263–8.

Miller, H. (1857) *The Testimony of the Rocks*. Nimmo, Edinburgh.

Nathorst, A. G. (1902) Zur oberdevonischen Flora der Bäreninsel. *Kungl. Svenska Vetenskapsakademiens Handlingar*, **36**, 1–60.

Niklas, K. J. and Banks, H. P. (1990) A reevaluation of the Zosterophyllophytina with comments on the origin of lycopods. *American Journal of Botany*, **77**, 274–83.

Nishida, H. (1982) *Lophosoriorachis japonica* n. gen. et sp., from the lower Cretaceous of Choshi, Chiba Prefecture, Japan. *Palaeontographica, Abteilung B*, **181**, 118–22.

Nishida, H. (1989) Structure and affinities of the petrified plants from the Cretaceous of Japan and Saghalien., V. Tree fern stems from Hokkaido, *Paracyathocaulis ogurae* gen. et comb. nov. and *Cyathocaulis yezopteroides* sp. nov. *Botanical Magazine, Tokyo*, **102**, 255–82.

Novik, K. O. (1968) *Flora from the Lower Carboniferous of the Donets Basin and its Western Continuation*. Akademia Nauk Ukraine SSR, Kiev [in Russian].

Ogura, Y. (1927) On the structure and affinities of some fossil tree-ferns from Japan. *Journal of the Faculty of Science, Imperial University, Tokyo, Section 3 Botany*, **1**, 351–80.

Pfefferkorn, H. W. (1976) Pennsylvanian tree fern compressions *Caulopteris*, *Megaphyton* and *Artisophyton* gen. nov. in Illinois. *Illinois State Geological Survey Circular*, **492**, 11–31.

Phillips, T. L. (1980) Stratigraphic and geographic occurrences of permineralized coal-swamp plants – Upper Carboniferous of North America and Europe, in *Biostratigraphy of Fossil Plants* (eds D. L. Dilcher and T. N. Taylor), Dowden, Hutchinson and Ross, Stroudsburg, pp. 25–92.

Pigg, K. B. and Rothwell, G. W. (1983) *Chaloneria* gen. nov.; heterosporous lycophytes from the Pennsylvanian of North America. *Botanical Gazette*, **144**, 132–47.

Radchenko, M. I. (1967) *The Carboniferous Flora of Southeast Kazakhstan*. 'Nauka' Kazakhstan SSR, Alma Alta [in Russian].

Radforth, N. W. (1939) Further contributions to our knowledge of the fossil Schizaeaceae; genus *Senftenbergia*. *Transactions of the Royal Society of Edinburgh*, **59**, 745–61.

Rayner, R. J. (1984) New finds of *Drepanophycus spinaeformis* Göppert from the Lower Devonian of Scotland. *Transactions of the Royal Society of Edinburgh: Earth Science*, **75**, 353–63.

Rayner, R. J. (1985) The Permian lycopod *Cyclodendron leslii* from South Africa. *Palaeontology*, **28**, 111–20.

Read, C. B. (1936) A Devonian flora from Kentucky. *Journal of Paleontology*, **10**, 215–27.

Remy, R. (1961) Beiträge zur Flora des Autunien. III. *Monatsbericht der Deutschen Akademie der Wissenschaften zu Berlin*, **3**, 331–6.

Remy, W. and Spassov, C. (1959) Der paläobotanische Nachweiss von Oberdevon in Bulgarien. *Monatsbericht der Deutschen Akademie der Wissenschaften zu Berlin*, **1**, 384–6.

Renault, B. (1876) Végétaux silicifies d'Autun et du Saint-Étienne. Nouvelles recherches sur la structure des *Sphenophyllum* et leur affinities botaniques. *Annals des Sciences Naturelle, Série 6 Botanique*, **4**, 277–311.

Rothwell, G. W. and Stockey, R. A. (1989) Fossil Ophioglossales in the Paleocene of western North America. *American Journal of Botany*, **76**, 637–44.

Rousseau, A. (1933) Contribution à l'étude de *Pinakodendron ohmanni* Weiss. *Mémoires du Musée Royal d'Histoire Naturelle de Belgique*, **59**, 1–32.

Scheckler, S. E. (1984) Persistence of the Devonian plant group Barinophytaceae into the basal Carboniferous of Virginia, U.S.A. *Compte Rendu 9e Congrès International de Stratigraphie et de Géologie du Carbonifère* (Urbana 1979), **2**, 223–8.

Scheckler, S. E. and Banks, H. P. (1971) *Protokalon* a new genus of progymnosperms from the Devonian of New York State and its bearing on phylogenetic trends in the group. *American Journal of Botany*, **58**, 874–84.

Schmidt, W. (1954) Pflanzenreste aus der Tonschiefer-Gruppe (unteres Siegen) des Sigerlandes – 1, *Sugambrophyton pilgeri* n.g., n.sp., eine Protolepidodendraceae aus den Hamberg-Schichten. *Palaeontographica, Abteilung B*, **97**, 1–46.

Schweitzer, H.-J. (1967) Die Oberdevon-flora der Bärinsel, 1: *Pseudobornia ursina* Nathorst. *Palaeontographica, Abteilung B*, **120**, 116–37.

Schweitzer, H.-J. (1972) Die Mitteldevon-Flora von Lindlar (Rheinland). 3. Filicinae – *Hyenia elegans* Kräusel & Weyland. *Palaeontographica, Abteilung B*, **137**, 154–75.

Schweitzer, H.-J. (1973) Die Mitteldevon-Flora von Lindlar (Rheinland). 4. Filicinae – *Calamophyton primaevum* Kräusel & Weyland, *Palaeontographica, Abteilung B*, **140**, 117–50.

Scott, A. C. and Galtier, J. (1985) Distribution and ecology of early ferns. *Proceedings of the Royal Society of Edinburgh B*, **86**, 289–301.

Scott, A. C., Galtier, J. and Clayton, G. (1984) Distribution of anatomically-preserved floras in the Lower Carboniferous in western Europe. *Transactions of the Royal Society of Edinburgh: Earth Sciences*, **75**, 311–40.

Scott, A. C., Galtier, J. and Clayton, G. (1985) A new late Tournaisian (Lower Carboniferous) flora from the Kilpatrick Hills, Scotland. *Review of Palaeobotany and Palynology*, **44**, 81–99.

Scott, D. H. (1898) On the structure and affinities of fossil plants from the Palaeozoic rocks. Pt 1. On *Cheirostrobus*. *Philosophical Transactions of the Royal Society of London, B*, **189**, 1–34.

Serlin, B. S. and Banks, H. P. (1978) Morphology and anatomy of *Aneurophyton*, a progymnosperm from the Late Devonian of New York State. *Palaeontographica Americana*, **51**, 343–59.

Seward, A. C. (1894) *The Wealden Flora, Part I. Thallophyta–Pteridophyta*. British Museum (Natural History), London.

Skog, J. E. (1976) *Loxsomopteris anasilla*, a new fossil fern rhizome from the Cretaceous of Maryland. *American Fern Journal*, **66**, 8–14.

Skog, J. E. (1986) The supposed fern *Onychiopsis psilotoides* from the English Wealden (Lower Cretaceous) reinterpeted as a lycopod. *Canadian Journal of Botany*, **64**, 1453–66.

Skog, J. E. and Banks, H. P. (1973) *Ibykia amphikoma*, gen. et sp. n., a new protoarticulate precursor from the late Middle Devonian of New York State. *American Journal of Botany*, **60**, 366–80.

Smith, D. L. (1962) Three fructifications from the Scottish Lower Carboniferous. *Palaeontology*, **5**, 225–37.

Snigirevskaya, N. S. (1980) A new fossil genus of Isoetopsida from the early Triassic of east Siberia. *Botanicheskii Zhurnal*, **65**, 95–6 [in Russian].

Sporne, K. R. (1975) *The Morphology of the Pteridophytes*, 4th edn, Hutchinson, London.

Stewart, W. N. (1983) *Paleobotany and the Evolution of Plants*. University Press, Cambridge.

Stockmans, F. and Mathieu, F.-F. (1939) *La Flore paléozoïque du Bassin houiller de Kaiping (Chine)*. Musée Royal d'Histoire Naturelle de Belgique, Brussels.

Stockmans, F. and Mathieu, F.-F. (1957) La flore paléozoïque du bassin houiller de Kaiping (Chine) (Deuxième partie). *Association pour l'Étude de la Paléontologie et de la Stratigraphie Houillères*, **32**, 1–89.

Stockmans, F. and Willière, Y. (1953) Végétaux namuriens de la Belgique. *Association pour l'Étude de la Paléontologie et de la Stratigraphie Houillères*, **13**, 1–382.

Stockmans, F. and Williére, Y. (1955) Végétaux Namuriens de la Belgique. Assise de Chokier, Zone de Bioul. *Association pour l'Étude de la Paléontologie et de la Stratigraphie Houillères*, **23**, 1–35.

Stubblefield, S. P. and Banks, H. P. (1978) The cuticle of *Drepanophycus spinaeformis*, a long-ranging Devonian lycopod from New York and eastern Canada. *American Journal of Botany*, **65**, 100–18.

Stubblefield, S. P. and Rothwell, G. W. (1981) Embryology and reproductive biology of *Bothrodendrostrobus mundus* (Lycopsida). *American Journal of Botany*, **68**, 625–34.

Stubblefield, S. P. and Rothwell, G. W. (1989) *Cecropsis luculentum* gen. et sp. nov.: evidence for heterosporous progymnosperms in the Upper Pennsylvanian of North America. *American Journal of*

Botany, **76**, 1415–28.

Taylor, T. N. (1967) On the structure and phylogenetic relationships of the fern *Radstockia* Kidston. *Palaeontology*, **10**, 43–6.

Taylor, T. N. (1981) *Paleobotany. An Introduction to Fossil Plant Biology*. McGraw-Hill, New York.

Teixeira, C. (1956) Une nouvelle espèce de *Sigillariostrobus* du Stéphanien portugais. *Boletim do Museu e Laboratório Mineralógico e Geológico da Faculdade de Ciências da Universidade de Lisboa, 7a Série*, **24**, 3–5.

Thomas, B. A. and Brack-Hanes, S. D. (1984) A new approach to family groupings in the lycophytes. *Taxon*, **33**, 247–55.

Thomas, B. A. and Spicer, R. A. (1987) *The Evolution and Palaeobiology of Land Plants*. Croom Helm, London.

Tian B.-L. and Guo Y.-T. (1987) A new species of lepidodendroid fructification – *Achlamydocarpon sinensis*. *Acta Botanica Sinica*, **29**, 218–22 [in Chinese].

Tidwell, W. D. (1967) Flora of the Manning Canyon Shale. Part I: A lowermost Pennsylvanian flora from the Manning Canyon Shale, Utah, and its stratigraphic significance. *Brigham Young University Geology Studies*, **14**, 3–66.

Tims, J. D. and Chambers, T. C. (1984) Rhyniophytina and Trimerophytina from the early land flora of Victoria, Australia. *Palaeontology*, **27**, 265–79.

Townrow, J. A. (1956) On some species of *Phyllotheca*. *Journal and Proceedings of the Royal Society of New South Wales*, **89**, 39–63.

Utting, J. and Neves, R. (1970) Palynology of the Lower Limestone Shale Group (Basal Carboniferous Limestone series) and Portishead Beds (Upper Old Red Sandstone). In *Colloque sur la Stratigraphie du Carbonifère. Liège Université, Congrès et Colloques*, **55**, (eds M. Streel and R. H. Wagner), pp. 411–27.

Vakhrameev, V. A., Dobruskina, I. A., Meyen, S. V. *et al*. (1978) *Paläozoische und Mesozoische Floren Eurasiens und die Phytogeographie dieser Zeit*. Gustav Fischer, Jena.

Wagner, R. H. (1983) Upper Stephanian stratigraphy and palaeontology of the Puertollano Basin, Cuid Real, Spain. *Anals Faculdade de Ciências, Porto, Supplement*, **64**, 171–231.

Wagner, R. H. (1984) Megafloral zones of the Carboniferous. *Compte rendu 9e Congrès International de Stratigraphie et de Géologie du Carbonifère* (Washington and Urbana 1979), **2**, 109–34.

Wagner, R. H. and Sousa, M. J. L. (1983) The Carboniferous megafloras of Portugal – a revision of identifications and discussion of stratigraphic ages. *Memórias dos Serviços Geológicos de Portugal*, **29**, 127–52.

Wagner, R. H., Hill, C. R. and El-Khayal, A. A. (1985) *Gemellitheca* gen. nov., a fertile pecopterid fern from the upper Permian of the Middle East. *Scripta Geologica*, **79**, 51–74.

Wang Z. and Wang L. (1982) A new species of the lycopsid *Pleuromeia* from the Early Triassic of Shanxi, China, and its ecology. *Palaeontology*, **25**, 215–25.

Ward, L. (1905) Status of the Mesozoic floras of the United States. *United States Geological Survey Monograph*, **48**.

Watson, J. (1969) A revision of the English Wealden flora, 1 Charales – Ginkgoales. *Bulletin of the British Museum (Natural History) Geology Series*, **17**, 209–54.

White, D. (1903) Description of a fossil alga from the Chemung of New York with remarks on the genus *Haliserites* Stbg. *Report of the New York State Museum of Natural History*, **1901**, 593–605.

Yao Z. and Taylor, T. N. (1988) On a new gleicheniaceous fern from the Permian of South China. *Review of Palaeobotany and Palynology*, **54**, 121–34.

Zeiller, R. (1906) Bassin houiller et permien de Blanzy et du Creusot. Flore fossile. *Étude Gîtes Mineraux France*.

Zhao X., Mo Z., Zhang S. *et al*. (1980) Late Permian flora from W. Guizhou and E. Yunnan, in *Stratigraphy and Palaeontology of Upper Permian Coal Measures in W. Guizhou and E. Yunnan*. Scientific Press, Beijing [in Chinese], pp. 70–122.

Zimmermann, F. (1930) Zur Kenntnis von *Eleutherophyllum mirabile* (Sternberg) Stur (*'Equisetites' mirabilis* Sternbg.). *Arbeit der Institut Paläobotanik und Petrographie der Brennsteine*, **2**, 83–102.

44

GYMNOSPERMOPHYTA

C. J. Cleal

Since the publication of the 1967 edition of *The Fossil Record*, there have been radical changes in gymnosperm taxonomy. For instance, some traditional taxa, such as the Pteridospermopsida, are now thought to be polyphyletic 'grade-groups', and are not given formal taxonomic status, while families established mainly on foliar characters (e.g. Callipteridiaceae Corsin, 1960) are not normally accepted as useful. However, trying to find a coherent alternative classification is far from easy. In many ways the most useful scheme is that of Meyen (1984, 1987), if only because the taxa are formally named and circumscribed. It has, however, been subjected to severe criticism on a variety of fronts (e.g. Beck, 1985; Miller, 1985; Rothwell, 1985; for a reply, see Meyen, 1986), but no alternative formal taxonomy has been proposed. Cladistic analyses (e.g. Hill and Crane, 1982; Crane, 1985; Doyle and Donoghue, 1986) have produced results which partly contradict Meyen's view, although they also disagree to varying extents with each other because of differences in methodology and in the characters analysed. Perhaps the most significant results of these cladistic analyses are:

1. that the seed-bearing plants are a monophyletic group, contrary to some views expressed previously (e.g. Arnold, 1948);
2. that the Cycadales and Benettitales are not closely related;
3. that the angiosperms seem to be most closely related to the Bennettitales and Gnetales.

In the following analysis, Meyen's scheme has formed the core of the classification adopted, but partly modified to make it compatible with some of the results of the cladistic studies, mentioned above. For instance, the Ginkgoaceae is removed from the group of so-called Mesozoic pteridosperms, and returned to its traditional position in the Pinopsida. Such a compromise scheme will undoubtedly attract more criticism than praise, but it nevertheless provides a reasonably coherent classification on which to base the following analysis.

In virtually all cases, the quoted ranges are based on macrofossil evidence alone. This is mainly because identifying families in the dispersed pollen/spore record is at best very difficult. This factor, combined with the extreme partiality of the plant macrofossil record, must be borne in mind when trying to draw floristic inferences (e.g. diversity analysis) from the results.

For simplicity, the term 'seed' has been used in the following chapter, to refer both to pre-fertilization ovules and post-fertilization seeds.

Acknowledgements – Thanks go to Barry Thomas (Natural Museum of Wales), Joan Watson (University of Manchester) and Chris Hill (Natural History Museum, London) for reading the manuscript and making many helpful suggestions.

Class LAGENOSTOMOPSIDA Cleal, **cl.** ***nov.*** (see Fig. 44.1)

Meyen (1987) included the Lagenostomales within the Cycadopsida, but analyses by both Crane (1985) and Doyle and Donoghue (1986) point to its being a quite distinct and primitive group of gymnosperms. Therefore it has been placed here in a separate class, the Lagenostomopsida. Following ICBN Article 16, it is automatically typified by the form genus *Lagenostoma* Oliver and Scott, 1904. The diagnosis is identical to that of its only order, Lagenostomales.

Order LAGENOSTOMALES Seward, 1917

The most comprehensive attempt at a family classification of the order is that based on seed structure by Long (1975). This has to be modified by excluding, for instance, Eurystomaceae Long, 1975, which in part belongs to the Calamopityales, and Callamospermaceae Long, 1975, which is synonymous with the Callistophytaceae (Callistophytales).

F. ELKINSIACEAE Rothwell *et al.*, 1989

D. (FAM)–C. (ASB?) Terr.

The Fossil Record 2. Edited by M. J. Benton. Published in 1993 by Chapman & Hall, London. ISBN 0 412 39380 8

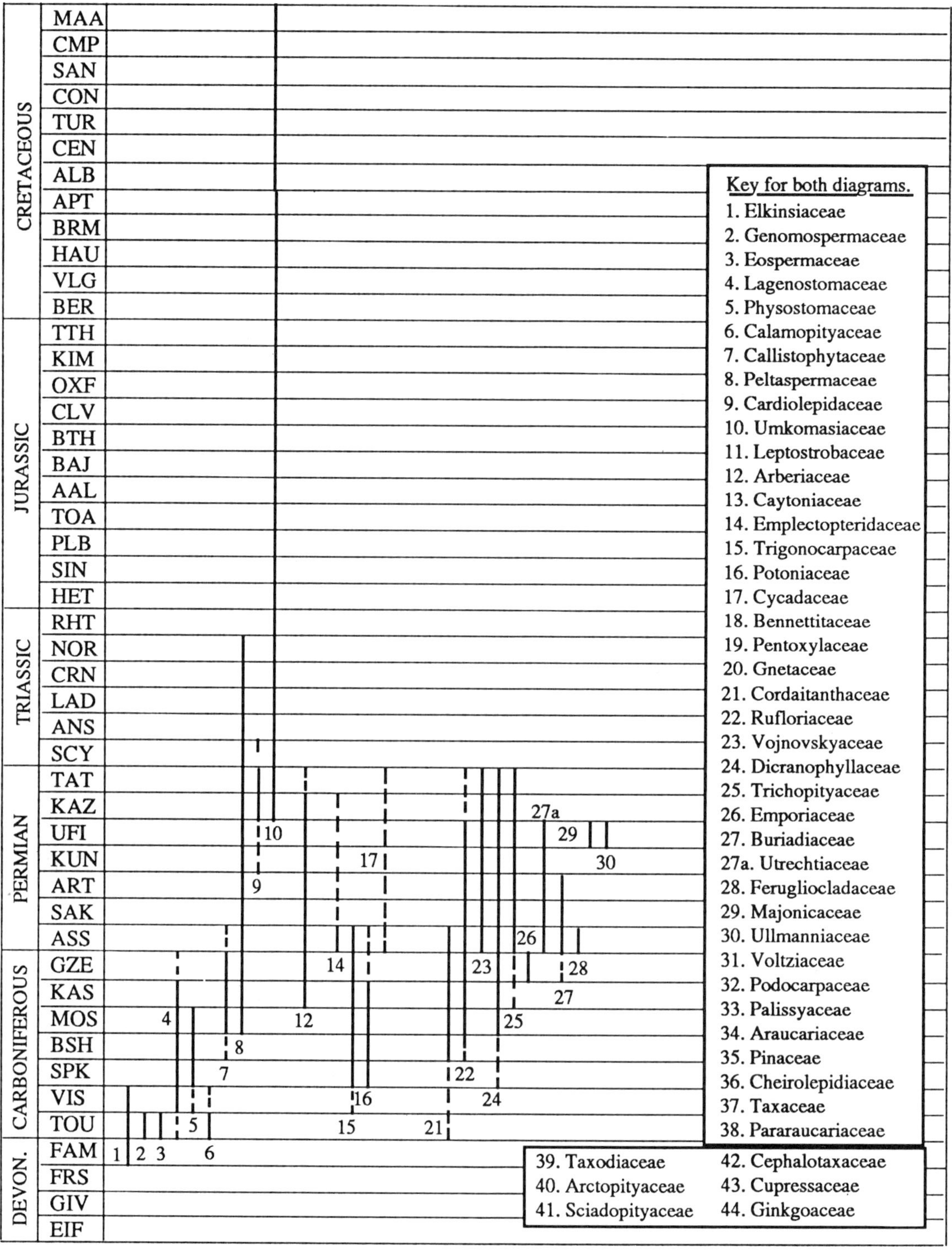

Fig. 44.1

First: *Elkinsia polymorpha* Rothwell *et al.*, 1989, Hampshire Formation, West Virginia, USA.

Last: ??*Megatheca thomasii* Andrews, 1940, Oil Shale Group, Lothian Region, Scotland, UK. This is a compression of a cupule very similar in form to the permineralized *Calathospermum* Walton, 1940, but there is no evidence of the seeds that it contained. The youngest permineralization is *Calathospermum scoticum* Walton 1949b, Loch Humphrey Burn (HLK?), Strathclyde Region, Scotland, UK.

Intervening: HAS–ARU.

Comments: Rothwell *et al.* (1989) do not diagnose this family, and include only one form genus (*Elkinsia*). However, they imply a close relationship with a number of other Upper Devonian and Lower Carboniferous fossils, and it is in this wider context that the family is interpreted here. It refers to fossil plants with lagenostomalean seeds borne in multi-ovulate cupules. The seeds have an integument that is free from the nucellus above the plinth and does not form a distinct micropyle. The distal part of the nucellus forms a wide lagenostome with a prominent central plug. Included are seeds of the form genera *Hydra-*

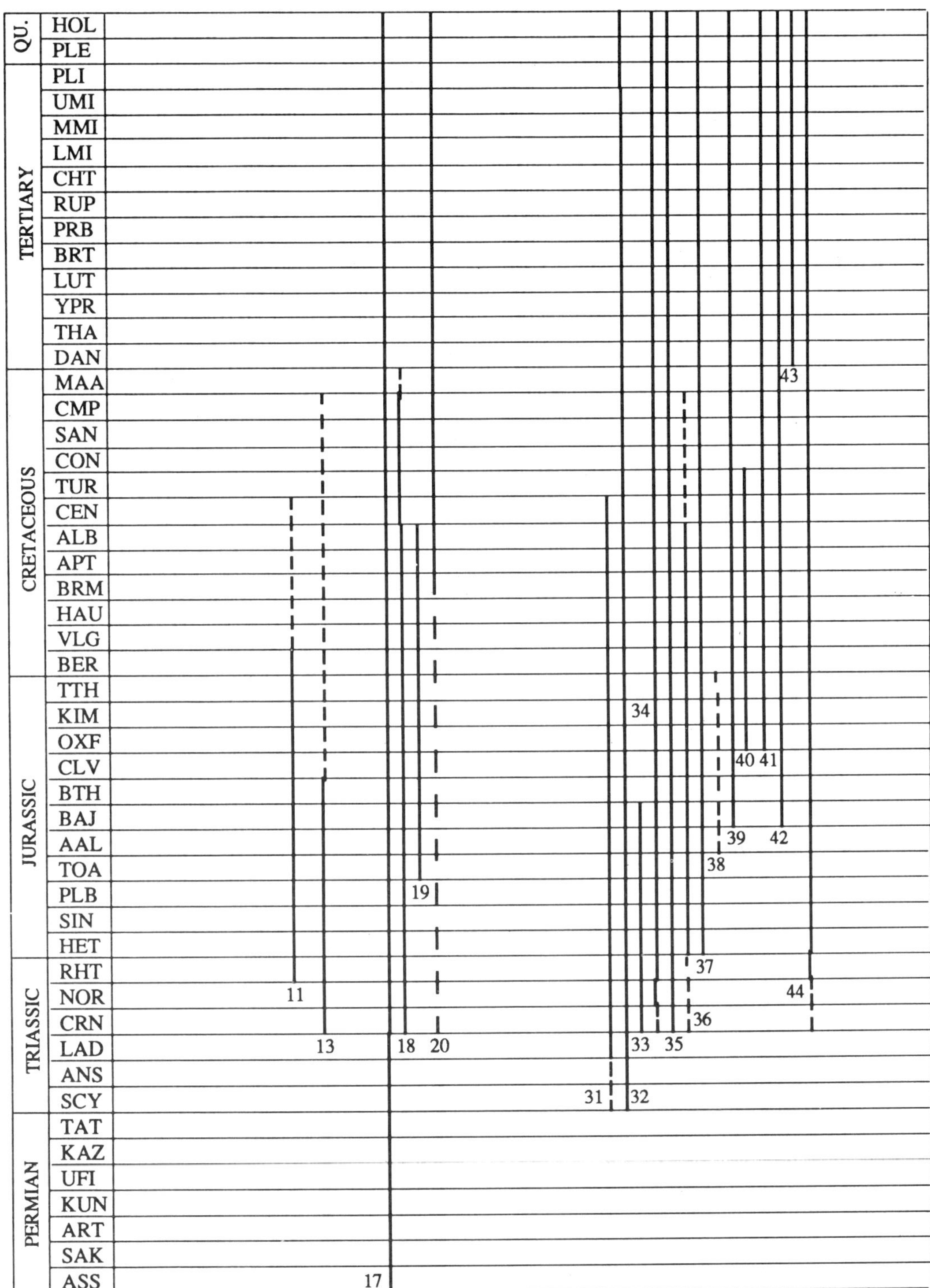

Fig. 44.1

sperma Long, 1961, *Stamnostoma* Long, 1960, and *Salpingostoma* Gordon, 1941.

F. GENOMOSPERMACEAE Long, 1975 C. (IVO) Terr.

First and Last: *Genomosperma kidstonii* (Calder) Long, 1959 and *G. latens* Long, 1959, Cementstone Group, Borders Region, Scotland, UK.

Comments: Rothwell *et al.* (1989) compare *Genomosperma* with *Elkinsia* (see previous family). It is assigned to a separate family here because it has:

1. a totally free nucellus;
2. a distinctive short, thin lagenostome;
3. was not borne in a cupule.

However, it is recognized that more information is needed on *Genomosperma* Long, 1959, before this family distinction can be confirmed. *Geminitheca scotica* Smith, 1959 from Loch Humphrey Burn, Strathclyde Region, Scotland, UK (HLK?) also has a free nucellus, but was borne in a multi-ovulate cupule and has quite a distinct lagenostome (the taxonomic position of this species is uncertain). The

Genomospermaceae is widely regarded as the archetypal primitive gymnosperm family, although it occurs in the fossil record somewhat higher than the Elkinsiaceae.

F. EOSPERMACEAE Long, 1975 C. (IVO) Terr.

First and Last: *Eosperma edromense* Long, 1966, *Deltasperma fouldense* Long, 1961a, *Eccroustosperma langtonense* Long, 1961b and *Camptosperma berniciense*, Long 1961b, Cementstone Group, Borders Region, Scotland, UK; and *Eosperma oxroadense* Barnard, 1959, Garleton Hills Volcanic Formation, Lothian Region, Scotland, UK.

Comments: This family is represented by a series of isolated, apparently non-cupulate seeds that have a highly distinctive, tapered lagensostome, with a conical central plug. Unlike other lagenostomaleans, they are platyspermic, and show varying degrees of curvature, from flat (*Eosperma* Barnard, 1959) to fully campylotropous (*Camptosperma* Long, 1961). As they have not been found attached to fronds or stems, their position within the Lagenostomales is provisional.

F. LAGENOSTOMACEAE Seward, 1917 C. (IVO?–KLA?) Terr.

First: ??cf. *Sphaerostoma* sp., Garleton Hills Volcanic Formation, Lothian Region, Scotland, UK (R. M. Bateman, pers. comm., 1989). Better documented is *Sphaerostoma ovale* Benson, 1914, Pettycur Limestone (ASB), Fife, Scotland, UK.

Last: ?*Eusphenopteris rotundiloba* Němejc, 1937, Kounov, central Bohemia, Czechoslovakia. This is based on the observation that some lagenostomacean permineralized stems bore eusphenopterid foliage (Shadle and Stidd, 1975). Adpression foliage known as *Pseudomariopteris* Danzé-Corsin, 1953, which ranges up into the Lower Permian, may have lagenostomacean affinities, but this has yet to be confirmed. The youngest permineralized fructifications are the seeds *Gnetopsis elliptica* Renault and Zeiller, 1884, and *Conostoma* sp., Rive de Gier Formation (CHV), Grand'Croix, France (Phillips, 1980).

Intervening: BRI–KRE, DOR?

Comments: This family represents those lagenostomaleans with seeds having a fully developed micropyle. Long (1975) also included *Stamnostoma* Long, 1960, but the lagenostome of these seeds is surrounded by an integumental collar, rather than a true micropyle. The family contains a wide diversity of seeds, including non-cupulate seeds (*Conostoma* Williamson, 1876), and seeds borne in uniovulate (*Lagenostoma* Williamson, 1876) and multi-ovulate (*Gnetopsis* Renault and Zeiller, 1884) cupules. It may thus be further divided as the relationships between the seeds to whole-plant reconstructions become better established.

F. PHYSOSTOMACEAE Long, 1975 C. (ASB?–POD) Terr.

First: ??*Physostoma* sp., Pettycur Limestone, Fife, Scotland, UK (Gordon, 1910). This is an unillustrated record of an allegedly poorly preserved specimen. Better documented is *Physostoma elegans* (Williamson) Oliver, 1909, Upper Foot Seam (MEL), Lancashire, England, UK.

Last: *Physostoma calcaratum* Leisman, 1964, Cabaniss Subgroup, Kansas, USA.

Comments: This family is in many ways similar to the Elkinsiaceae, except that the seeds were not borne in a cupule. Also, the seeds have a much smaller, thinner lagenostome, which appears to lack a central plug.

Class UNNAMED

Most recent analyses (Crane, 1985; Doyle and Donoghue, 1986; Meyen, 1987) agree that the Mesozoic pteridosperms cluster together with the glossopterids. Meyen referred to them as the Ginkgoopsida, since he included the extant Ginkgoales, but the cladistic analyses of Crane, and Doyle and Donoghue support the traditional view that the ginkgos cluster with the conifers. The latter analyses could be used to argue for a single class to encompass conifers, ginkgos, Mesozoic pteridosperms and glossopterids (among others). However, the conifer/ginkgo complex has been retained here as a separate class, and consequently an alternative class name is required for the remainder.

Order CALAMOPITYALES Taylor, 1981

F. CALAMOPITYACEAE Solms-Laubach, 1896 C. (HAS–BRI?) Terr.

First: *Calamopitys americana* Scott and Jeffrey, 1914, *C. foerstii* Read, 1936, *Stenomyelon muratum* Read, 1936, *Diichnia kentuckiensis* Read, 1936, and *Bostonia perplexa* Stein and Beck, 1978, Falling Run Member, New Albany Shales, Kentucky, USA (Read, 1937; Stein and Beck, 1978). The chronostratigraphical position of these fossils has been changed since the first edition of *The Fossil Record*, following Stein and Beck (1978). An unnamed seed from the Lydiennes Formation (HAS), Hérault, France (Galtier and Rowe, 1989) is the oldest fructification which may belong to the family.

Last: ?*Spathulopteris ettingshausenii* (Feistmantel) Kidston, 1923, and *S. clavigera* (Kidston) Walton, 1928, Upper Black Limestone, Clwyd, Wales, UK (Walton, 1931). This is based on foliage adpression fossils. The youngest fructification which may belong here is *Alcicornopteris hallei* Walton, 1949a, Loch Humphrey Burn (HLK?), Strathclyde Region, Scotland, UK. The youngest stem permineralizations are *Stenomyelon tuedianum* Kidston ex Kidston and Gwynne-Vaughan, 1912, *S. heterangioides* Long, 1964, and *S. primaevum* Long, 1964, Cementstone Group (IVO), Borders Region, Scotland, UK (Long, 1964).

Comments: The taxonomy of this order is problematic because it is not certain what fructifications were attached to calamopityalean stems and vegetation. There is good circumstantial evidence that seeds belonging to the Eurystomaceae *pro. parte* Long, 1975 (*Lyrasperma* Long, 1960, *Eurystoma* Long, 1960, *Dolichosperma* Long, 1961) may belong here, and it was partly because of this that Meyen (1987) suggested a link with the 'Mesozoic pteridosperms'. However, evidence of organic connection between these seeds and the calamopityalean plants has never been demonstrated, and neither is there unequivocal proof that the seeds were non-cupulate: both key points if Meyen's classification is to be accepted (Beck, 1985).

Order CALLISTOPHYTALES Rothwell, 1981

F. CALLISTOPHYTACEAE Stidd and Hall, 1970 C. (MEL?)–P. (ASS?) Terr.

First: ??*Callistophyton* sp., First Coal, Lancashire, England, UK (Phillips, 1980). This is a questionable example of a fusainized stem. Better documented is the permineralized pollen organ *Idanothekion glandulosum* Millay and Eggert, 1970, from the Buffalowville Coal (KSK), Indiana, USA (Phillips, 1980).

Last: ?*Dicksonites beyrichii* (Weiss) Doubinger, 1956, Lower Rotliegend, Saarland, Germany (Kerp and Fichter, 1985). This assumes that *Dicksonites* is the adpression analogue of permineralized *Callistophyton* Delevoryas and Morgan, 1954 fronds. The youngest permineralizations are *Callistophyton* sp., Redstone Coal (KLA), Ohio, or Pittsburgh Coal, West Virginia, USA (Phillips, 1980).
Intervening: POD–DOR, NOG?

Order PELTASPERMALES Němejc, 1968

F. PELTASPERMACEAE Thomas ex Harris, 1937
C. (KLA)–Tr. (RHT) Terr.

First: *Autunia conferta* (Sternberg) Kerp and *Lodevia nicklesii* (Zeiller) Haubold and Kerp, Faisceau de Beaubrun, St Étienne, France (Bouroz and Doubinger, 1977).
Last: *Peltaspermum ottonis* (Göppert) Poort and Kerp, 1990, Mine Formation, Scania, Sweden.
Intervening: NOG–NOR.
Comments: This family is as interpreted by Kerp and Haubold (1988).

F. CARDIOLEPIDACEAE Meyen, 1977
P. (KUN?–TAT) Terr.

First: ?*Phylladoderma chalyshevii* Fefilova and Smoller ex Meyen, 1983, Lekvorkutsk 'Suite', Pechora Coalfield, former USSR. This record is of foliage fragments with cuticles preserved. Better documented are the fructifications *Nucicarpus piniformis* Neuburg, 1965, and *Cardiolepis piniformis* Neuburg, 1965, Seidinsk 'Suite' (UFI), Pechora Coalfield, former USSR (Meyen, 1983, 1988).
Last: *Amphorispermum* sp., *Phylladoderma tatarica* Meyen, 1986 and *Doliostomia krassilovii* Meyen, 1986, Titov, Russian Platform, former USSR (Gomankov and Meyen, 1986).
Intervening: KAZ.
Comments: Meyen (1977) originally assigned this family to the conifers (Pinopsida), but subsequently transferred it to the peltasperms (Meyen, 1984).

F. UMKOMASIACEAE Meyen, 1984
P. (KAZ)–K. (APT) Terr.

First: *Rhaphidopteris praecursor* Meyen, 1979, Pritimanie Deposits, Komi, former USSR. This record is based on foliage, which is similar to the Mesozoic form genus *Pachypteris*. It is also associated with pollen (*Alisporites*) similar to that found in Jurassic pollen organs belonging to the family (e.g. *Pteroma*), and so has been accepted here as definite evidence of the family.
Last: *Ruflorina sierra* Archangelsky, 1963 and *Ktalenia circularis* Archangelsky, 1963, Ticó, Santa Cruz Province, Argentina (Taylor and Archangelsky, 1985).
Intervening: SMI–HET, AAL, BTH, KIM–HAU.
Comments: This is the Corystospermaceae of some authors. Its taxonomic position is uncertain, and it is arguable that it belongs to the Glossopteridales or even its own distinct order.

Order LEPTOSTROBALES Meyen, 1984

F. LEPTOSTROBACEAE Meyen, 1984
Tr. (RHT)–K. (CEN?) Terr.

First: *Leptostrobus longus* Harris, 1935, Scoresby Sound, Greenland; and *Irania hermaphroditica* Schweitzer, 1977, Iran.
Last: ?*Czekanowskia* ex group *rigida* Heer, 1876 and *Phoenicopsis steenstrupii* Seward, Koèvunjsker Formation, Anadyr River and Arkagalinsker and Armanjsker formations, Kolyma River, eastern Siberia, former USSR (Vakhrameev, 1966; Samylina, 1973). These records are based on adpressions of foliage. The youngest fructifications are of *Leptostrobus laxiflorus* Heer, 1876, Ilinureksker Formation (BER), Tyl River, eastern Siberia, former USSR (Vakhrameev, *in* Vakhrameev *et al.*, 1978).
Intervening: HET–TTH, VLG?–ALB?.
Comments: This has also been called the Czekanowskiaceae. Many authors have regarded it as being closely related to the Ginkgoaceae, but it is now thought to be closer to the Peltaspermales and Callistophytales (Meyen, 1984, 1987; Crane, 1985).

Order ARBERIALES Meyen, 1984

F. ARBERIACEAE Meyen, 1984
C. (KLA?)–Tr. (DIE?) Terr.

First: *Arberia umbellata* Surange and Lele, 1955, Talchir Formation, India. The detailed correlations between the various Gondwana sequences and how they relate to the internationally recognized chronostratigraphy are still uncertain. The KLA estimate for the position of the Talchir Formation must therefore be regarded as provisional. Furthermore, it may be pre-dated by *Ottokaria* sp., Itararé Subgroup, Paraná Basin, Brazil (Rösler, 1978).
Last: ?*Glossopteris papillosa* Srivastava, 1969 and *G. sennii* Srivastava, 1969, Nidpur, Madhya Pradesh, India. This record is based on foliage. Records of Middle Triassic glossopterid foliage from India (as reviewed by Vakhrameev *et al.*, 1978) are not accepted by Chandra and Surange (1979). *Mexiglossa varis* Delevoryas and Person, 1975, Zorrillo and Simón formations (BAJ?), Oaxaca, Mexico, has been interpreted as a late example of glossopterid foliage, but is associated with the pollen-bearing organ *Perezlaria oaxacensis* Delevoryas and Gould, 1971, which appears to have caytoniacean affinities. The youngest fructifications appear to be *Plumsteadia natalensis* Lacey *et al.*, 1975, *P. gibbosa* (Benecke) Anderson and Anderson, 1985, *Estcourtia vandijkii* Anderson and Anderson, 1985, *E. bergvillensis* Anderson and Anderson, 1985, *Rigbya arberoides* Lacey *et al.*, 1975, *Lidgettonia africana* Thomas, 1958, *L. mooriverensis* Anderson and Anderson, 1985, *L. inhluzanensis* Anderson and Anderson, 1985, *L. lidgettonioides* (Lacey *et al.*, 1975) Anderson and Anderson, 1985, and *L. elegans* (Lacey *et al.*, 1975) Anderson and Anderson, 1985, Estcourt Formation (TAT?), NE Karoo Basin, South Africa (Anderson and Anderson, 1985).
Intervening: NOG–KAZ.
Comments: This is the Glossopteridaceae of some authors.

F. CAYTONIACEAE Thomas, 1925
Tr. (CRN?)–K. (CMP?) Terr.

First: ?*Pramelreuthia halberfelneri* Krasser, 1909, Lettenkohle, Lunz, Austria (Kräusel, 1949). The affinities of this pollen organ have been queried by Harris (1964), although it should be noted that the typically caytoniacean foliage, *Sagenopteris* sp., has been reported from about the same stratigraphical horizon at Raibl, Austria (Stur, 1885). Better documented, however, are *Caytonanthus kochii* Harris, 1932, *Amphorispermum ellipticum* Harris, 1932, *A. rotundum* Harris, 1932, *A. major* Harris, 1932, and *Sagenopteris hallei* Harris, 1932, Scoresby Sound (RHT), Greenland (Harris, 1932a,b).
Last: ??*Sagenopteris variabilis* Velenovsky, Barykovsker

Formation, Ugol'naja Basin, eastern Siberia, former USSR (Vakhrameev, 1966). This record is based on adpressions of foliage. The youngest fructifications are *Caytonia nathorstii* (Thomas) Harris, 1940, Scalby Formation (BTH), North Yorkshire, England, UK (Harris, 1964).
Intervening: RHT–BTH, OXF?, BER?, BRM?, SAN?
Comments: Meyen (1987) places this family in a separate order. However, both Crane (1985) and Doyle and Donoghue (1986) found that it clusters with the Arberiaceae.

Order GIGANTONOMIALES Meyen, 1987

F. EMPLECTOPTERIDACEAE Wagner, 1967 P. (ASS–KAZ?) Terr.

First: *Emplectopteris triangularis* Halle, 1927, Shansi Formation, Taiyuan Coalfield, China (Halle, 1932).
Last: ?*Gigantopteris taiyuanenesis* (Asama) Meyen, 1987, upper part of Upper Shihhotse Formation, Taiyuan Coalfield, China (Asama, 1962). This is based on leaf fossils. The youngest fructifications are *Gigantonomia fukiensis* (Yabe and Oishi) Li and Yao, 1983, and *Gigantotheca paradoxa* Li and Yao, 1983, lower part of upper Shihhotse Formation (UFI), Taiyuan Coalfield, China.
Intervening: SAK?–KUN?
Comments: Wagner (1967) based this family on the '*Emplectopteris* Series' of Asama (1962). It also corresponds to the Gigantopteridaceae of Li and Yao (1983). Meyen (1987) has argued that the latter name has to be rejected, because some *Gigantopteris* Schenk, 1883 foliage belongs to the Peltaspermales, and instead proposed the name Gigantonomiaceae. However, it is essentially identical to the Emplectopteridaceae, except in the degree of dissection of the leaf; the seeds of *Emplectopteris* Halle, 1927 are similar to *Gigantonomia* Li and Yao, 1983 in both form and position of attachment to the frond (Halle, 1932). Thus there seems little reason for separating the former into a separate family. The family was not included in the analyses of Crane (1985), or Doyle and Donoghue (1986), and it has been placed in its current taxonomic position following Meyen (1987). It is mainly restricted to Cathaysian assemblages. Very similar foliage is found in the Lower Permian of North America (e.g. Read and Mamay, 1964), but has yielded no evidence of fructifications or cuticles and Meyen (1987) assigned them to the Peltaspermales. The first edition of *The Fossil Record* included *Lescuropteris* Schimper, 1869, from the Upper Carboniferous of Europe and North America, but this almost certainly belongs to the Trigonocarpales.

Class CYCADOPSIDA Barnard and Long, 1975

This class is treated in essentially the same way as Meyen (1987), except that the Lagenostomales and Bennettitales are excluded and placed in separate classes.

Order TRIGONOCARPALES Seward, 1917

F. TRIGONOCARPACEAE Seward, 1917 C. (BRI?)–P. (ASS) Terr.

First: ?*Neuropteris antecedens* Stur, 1875, and *Holcospermum ellipsoideum* (Göppert) Walton, 1931, upper Black Limestone, Clwyd, Wales, UK (Walton, 1931). This record is based on adpressions yielding little anatomical information. Better documented are *Rhynchosperma quinnii* Taylor and Eggert, 1967b, *Quaestoria amplecta* Mapes and Rothwell, 1980, and *Medullosa* sp. Taylor and Eggert, 1967a, lower Fayetteville Formation (PND), Arkansas, USA.
Last: ?*Odontopteris subcrenulata* (Rost) Zeiller, 1888, lower Shihhotse Formation, Shanxi, China (Halle, 1927). This is based on adpressions of foliage and can be dated no more accurately than Early Permian (Wagner *et al.*, 1983). More accurately dated are the stem and rachis permineralizations *Medullosa stellata* Cotta, 1832, *M. leuckartii* Göppert and Sterzel, 1881, *M. porosa* Cotta, 1832, *M. solmsii* Schenk, 1889 and *Myeloxylon elegans* (Cotta) Brongniart, 1849, associated with the foliage adpressions *Neurocallipteris planchardii* (Zeiller) Cleal *et al.*, 1990, and *Alethopteris schneideri* Sterzel, 1981, Leukersdorf Formation (ASS), Erzgebirge, Germany (Barthel, 1976).
Intervening: ARN–NOG, SAK?–KUN?
Comments: This is the Medullosaceae of many authors. Attempts to subdivide the Trigonocarpales at the family level have not been satisfactory. For instance, the scheme proposed by Corsin (1960), and used in the first edition of *The Fossil Record*, is too heavily based on frond architecture for it be 'natural'. Only the Potonieaceae appears to be sufficiently well circumscribed (below). The Trigonocarpaceae is therefore used here to encompass the entire Trigonocarpales, exclusive of the Potonieaceae.

F. POTONIEACEAE Halle, 1933 C. (PND?)–P. (ASS?) Terr.

First: ?*Paripteris gigantea* (Sternberg) Gothan, 1941, Tseishui Formation, Hunan, China (Yang *et al.*, in Wagner *et al.*, 1983). This refers to foliage adpressions from the upper part of the Lower Carboniferous, in the Chinese classification (its assignment to the PND is tentative). The oldest fructifications are *Hexagonocarpus modestae* (Bertrand) Seward, 1917 and *Potoniea adiantiformis* Zeiller, 1899, associated with *Paripteris gigantea* (Sternberg) Gothan foliage, Zone de Baulet (MRD), Belgium (Stockmans and Willière, 1953).
Last: ?*Linopteris gangamopteroides* (de Stefani) Wagner, 1958 (syn. *L. brongniartii* Zeiller, 1888, *non* Gutbier, 1835), Surmoulin and Millery Formations, Autun, France (Bouroz and Doubinger, 1977). This is based on foliage adpressions. The youngest fructifications are permineralized seeds *Hexapterospermum pachypterum* Brongniart, 1881 and *H. stenopterum* Brongniart, 1881, Rive de Gier Formation (CHV), Grand'Croix, France (Phillips, 1980).
Intervening: CHO–KRE, KLA?–NOG?
Comments: This is the only well-circumscribed family within the Trigonocarpales, and is based on the connection between *Sutcliffia* Scott, 1906 stems, *Potoniea* Zeiller, 1899 pollen organs, *Hexagonocarpus* Renault, 1896, *Hexapterospermum* Brongniart, 1874 seeds and *Paripteris* Gothan, 1941 and *Linopteris* Presl, 1838 fronds (Bertrand, 1930; Stidd *et al.*, 1975; Stidd, 1978). It differs from other trigonocarpaleans in producing trilete, rather than monolete, pre-pollen.

Order CYCADALES Engler, 1892

The Dirhopalostachyaceae Krassilov, 1975 was proposed as an extinct family of cycads (in its broad sense), which showed a number of characters believed to be 'proangiospermous' (Krassilov, 1975, 1977). However, Doludenko and Kostina, *in* Meyen (1987) have reported evidence suggesting that Krassilov's fossils are in fact early pinaceaen conifers. The family is thus not included in this analysis.

F. CYCADACEAE Persoon, 1807
C. (KIN??)–Rec. Terr.

First: ??*Lesleya cheimarosa* Leary and Pfefferkorn, 1977, channel-fill immediately below Abbott Formation, Illinois, USA (Leary, 1990). This is a leaf with seeds attached to the axis, found in an extrabasinal ('upland') deposit. It was compared with various other putative early cycads, such as *Archaeocycas whitei* Mamay, 1973, Belle Plains Formation (ASS), Texas, USA, *Phasmatocycas kansana* Mamay, 1973, Wellington Formation (ASS), Kansas, USA, and *Sobernheimia jonkeri* Kerp, 1983, Waderner Group (ASS), Nahe, Germany (Mamay, 1973, 1976; Kerp, 1983; Gillespie and Pfefferkorn, 1986). Crane (1985) initially expressed doubt as to the cycadalean affinities of these Lower Permian fossils, as they had platyspermic seeds, but he later (Crane, 1988) accepted that this was not necessarily a bar to them belonging to that order. Nevertheless they are quite different morphologically from cycads in the normally accepted sense, and it is perhaps wiser to regard them as cycad precursors, rather than true cycads. The oldest unequivocal cycad remains are *Crossozamia chinensis* (Zhu and Du) Gao and Thomas, 1989, *C. minor* Gao and Thomas, 1989, *C. spadicia* Gao and Thomas, 1989, *C. cucullata* (Halle) Gao and Thomas, 1989, *Tianbaolinia circinalis* Gao and Thomas, 1989, *Yuania chinensis* Zhu and Du, 1981 and *Taeniopteris taiyuanensis* Halle, 1927, Taiyuan Formation, Shanxi, China (Gao and Thomas, 1989). At least part of the Taiyuan Formation is Upper Carboniferous (Li and Yao, 1985), but Gao and Thomas claim that the cycadacean fossils originated from the Lower Permian (ASS). **Extant**

Intervening: ASS, CRN, RHT–HOL.

Comments: Various other families have been mentioned within the Cycadales, including the Zamiaceae, Stangeriaceae and Nilssoniaceae. However, there has been no coherent attempt to analyse the distribution of these families in the fossil record. Such an analysis is beyond the scope of this report, and so the view of Harris (1961) has been followed, and these groups regarded as subfamilies within the Cycadaceae. Taylor (1969) suggested that *Lasiostrobus polysaccii* Taylor, 1969 from the Calhoun Coal, Illinois, might be a Carboniferous cycad pollen cone, but later work (Taylor, 1970) cast doubt on this.

Class GNETOPSIDA Engler, 1954

This class is taken here to have a wider circumscription than in most previous studies. It refers to the gymnospermous orders included within the angiosperm/bennettite/pentoxylid/gnetalean clade by Crane (1985) and Doyle and Donoghue (1986).

Order BENNETTITALES Engler, 1892

F. BENNETTITACEAE Engler, 1892
Tr. (CRN)–K. (MAA?) Terr.

First: *Haitingeria krasseri* (Schuster) Krasser, 1919, *Leguminanthus siliquosus* (Leuthardt) Kräusel and Schaarschmidt, 1966, *Westersheimia pramelreuthensis* Krasser, 1919, *Sturianthus langeri* Kräusel, 1949 and *Bennetticarpus wettsteinii* (Krasser) Kräusel, 1949, Lettenkohle, Lunz, Austria; *Leuthardtia ovalis* Kräusel and Schaarschmidt, 1966, Lettenkohle, Neuewelt, Switzerland (Crane, 1986, 1988).

Last: ??*Pterophyllum* sp., Augustovka River near Boshniakovo, Sakhalin Range, Siberia, former USSR (Krassilov, 1978). This is based on foliage without preserved cuticles. Better documented is *Monanthesia magnifica* Wieland ex Delevoryas, 1959, Mesaverde Formation (CMP), New Mexico, USA, which includes fructifications.

Intervening: RHT–ALB.

Comments: The traditional division of the Bennettitales into the Williamsoniaceae and Cycadeoidaceae (e.g. Taylor, 1981; Stewart, 1983) is regarded by Crane (1985, 1988) as dubious and is not recognized here. Some early authors regarded the Bennettitales as closely related to the Cycadales, due to similarities in the foliage and general habit of the plants (e.g. Chamberlain, 1935). Harris (1969) argued, however, that the phylogenetic relationship is only remote, although he gave no opinion as to the position of the Bennettitales. Recent analyses suggest that the Bennettitales cluster together with the Pentoxylales and Gnetales (Crane, 1985, 1988; Doyle and Donoghue, 1986).

Order PENTOXYLALES Pilger and Melchior, 1954

F. PENTOXYLACEAE Pilger and Melchior, 1954
J. (TOA)–K. (ALB) Terr.

First: *Taeniopteris spatulata* McClelland, 1850 and *Carnoconites* sp., Talbragar Fish Beds, New South Wales, Australia (White, 1981).

Last: *Pentoxylon sahnii* Srivastava, 1944, *Nipanioxylon guptai* Srivastava, 1944, *Nipaniophyllum raoi* Sahni, 1948, *Carnoconites compactum* Srivastava, 1944, *C. rajmahalensis* (Wieland) Bose *et al.*, 1984, and *Sahnia nipaniensis* Vishnu-Mittre, 1953, Rajmahal Formation, Bihar, India (Rao, 1976, 1981; Bose *et al.*, 1984). There has been considerable disagreement as to the chronostratigraphical position of the Rajmahal Formation. It is taken here to be ALB, based on K–Ar dating by McDougall and McElhinny (1970).

Intervening: TTH, HAU, VLG.

Comments: The distribution of this family has been reviewed by Drinnan and Chambers (1985), who conclude that it was a significant element in the mid-Mesozoic floras of Gondwanaland.

Order GNETALES Engler, 1892

F. GNETACEAE Lindley, 1834
Tr. (CRN?)–Rec. Terr.

First: ?*Dechellyia gormanii* Ash, 1972 and *Masculostrobus clathratus* Ash, 1972, Chinle Formation, Arizona, USA. The taxonomic position of these enigmatic fossils is still somewhat in doubt. Better documented is *Drewria potomacensis* Crane and Upchurch, 1987, Potomac Group (APT), Virginia, USA. **Extant**

Intervening: APT.

Comments: Some authors divide the order into three families, one for each of the extant genera. The meagre fossil record of the order is discussed by Crane (1988).

Class PINOPSIDA Meyen, 1984

This is the Coniferopsida of many authors. The classification adopted here is essentially that of Meyen (1987), except that the Ginkgoales have been incorporated. The close relationship of the ginkgos to the conifers is reflected in most traditional classifications (e.g. Chamberlain, 1935) and is supported by cladistic analysis (Crane, 1985; Doyle and Donghue, 1986). Some expansion of the classification of Palaeozoic conifers has also been included, following Clement-Westerhof (1988) and Mapes and Rothwell (1991).

Order CORDAITANTHALES Meyen, 1984

F. CORDAITANTHACEAE Meyen, 1984 C. (IVO?)–P. (ASS) Terr.

First: ?*Mitrospermum bulbosum* Long, 1977, Cementstone Group, Borders Region, Scotland, UK. This is based on an isolated seed, which has many features in common with younger seeds assigned to this family, but there is no evidence of the rest of the plant that produced it. Better documented is *Cordaitanthus pitcairniae* (Lindley and Hutton) Renault, 1881 and *Cordaites palmaeformis* (Göppert) Weiss, 1871, Assise de Chokier (KIN), Belgium (Stockmans and Willière, 1954).
Last: cf. *Cordaitanthus gemmifer* Grand'Eury, 1877 and cf. *C. baccifer* Grand'Eury, 1877, Döhlener Formation, Döhlener Basin, Germany (Barthel, 1976).
Intervening: MRD–NOG.
Comments: This family is known only from palaeo-equatorial assemblages. There are records of *Cordaites* Unger, 1850 foliage as high as the Upper Permian in Cathaysia (e.g. Mei, 1984) but, in view of the difficulty of separating the foliage of this family from that of other cordaitanthalean families, they are not referred to in the above-quoted range.

F. RUFLORIACEAE Meyen, 1982 C. (KIN?)–P. (TAT?) Terr.

First: ?*Rufloria subangusta* (Zalessky) Meyen, 1963, Kaezovsky 'Suite', Kuznetsk, former USSR (Meyen, 1982). This record is based on foliage adpressions for which there is no epidermal evidence. The oldest fructifications are *Krylovia sibirica* Chachlov, 1938, Mazurovsky 'Suite' (VRK?), Kemerov and Tom'-Usinsk regions, Kuznetsk, former USSR (Gorelova, *in* Gorelova *et al.*, 1973).
Last: ?*Rufloria mitinaensis* (Gorelova) Meyen, 1963, *R. brevifolia* (Gorelova) Meyen, 1966, Tailugansky 'Suite', Leninsko–Erunakovsky Region, Kuznetsk, former USSR (Gorelova, *in* Gorelova *et al.*, 1973). These records are based on foliage adpressions. The youngest fructification is *Gaussia scutellata* Neuberg, 1934, Uskatsky 'Suite' (KAZ), Prokop'evsko–Kiselevsky and Bunguro–Chumyshsky regions, Kuznetsk, former USSR (Gorelova, *in* Gorelova *et al.*, 1973).
Intervening: MRD?–MEL?, KSK–UFI.
Comments: See comments on the Vojnovskyaceae.

F. VOJNOVSKYACEAE Meyen, 1982 P. (ASS)–P. (TAT) Terr.

First: *Vojnovskaya usjatensis* Gorelova, *in* Gorelova *et al.*, 1973, Promezhutochny 'Suite', Kemerov Region, Kuznetsk, former USSR.
Last: *Kuznetskia planiuscula* Meyen, 1982, Tailugansky 'Suite', Kuznetsk, former USSR (Meyen, 1982).
Intervening: SAK–KAZ.
Comments: The assignment of this family and the Rufloriaceae to the Cordaitanthales is following Meyen (1987, 1988), although some doubt has been expressed by Miller (1985) and Rothwell (1988). *Cordaites* Unger, 1850, leaves, which might belong to the Vojnovskyaceae, are reported as low as the Mazurovsky 'Suite' (VRK?) in Angaraland (Vakhrameev *et al.*, 1978). However, their identity has not been confirmed by epidermal evidence, and they are not associated with fructifications.

Order DICRANOPHYLLALES Němejc emend. Archangelsky and Cúneo, 1990

F. DICRANOPHYLLACEAE Němejc emend. Archangelsky and Cúneo, 1990 C. (ALP?)–P. (TAT) Terr.

First: ?*Dicranophyllum richirii* Renier, 1907, Zone de Malonne, Baudour, Belgium (Stockmans and Willière, 1953). This is based on sterile foliage. The oldest fructification is *Dicranophyllum gallicum* Grand'Eury, 1877, Commentry (KLA?), France (Renault, 1890).
Last: *Mostochkia* sp. with associated fructifications, Kama River, near Ustinov, west of Ural Mountains, former USSR (Meyen and Smoller, 1986); also *Slivkovia petschorensis* Meyen, 1969, Pechora 'Series', Pechora, former USSR.
Intervening: KSK, KLA–ASS, KUN–KAZ.
Comments: Meyen and Smoller (1986) characterize this family mainly on features of the foliage, including epidermal structure. There is very limited evidence available on their fructifications, however, and it is far from certain that it is a homogeneous group.

F. TRICHOPITYACEAE Florin emend. Archangelsky and Cúneo, 1990 C. (DOR?)–P. (KUN) Terr.

First: ?*Dichophyllum moorei* Ellias ex Andrews, 1941, Rock Lake Shale and Ireland Sandstone, Kansas, USA (Cridland and Morris, 1963). This is based on foliage, which Meyen (1987) included in this family. Better documented is *Trichopitys heteromorpha* Florin, 1949, Lydiennes Formation, Hérault, France (ASS).
Last: *Biarmopteris pulchra* Zalessky, 1939 and *Mauerites gracilis* Zalessky, 1939, Sylva River, Middle Fore-Urals, former USSR (Meyen, 1982).
Comments: This family has been variously assigned to the Ginkgoales (Andrews, 1941; Florin, 1949) and Peltaspermales (Meyen, 1984). Its presently accepted position in the Dicranophyllales follows Archangelsky and Cúneo (1990).

Order PINALES Meyen, 1984

Fragments of conifer foliage have been widely documented in the Upper Carboniferous, the earliest being *Swillingtonia denticulata* Scott and Chaloner, 1983, Middle Coal Measures, West Yorkshire, England, UK, (VRK: see Lyons and Darrah, 1989 for a review). However, they rarely have fructifications attached, and so cannot be assigned to families.

F. EMPORIACEAE Mapes and Rothwell, 1991 C. (KLA?) Terr.

First and Last: *Emporia lockardii* (Mapes and Rothwell) Mapes and Rothwell, 1991, Shawnee Group, Hamilton Quarry, Kansas, USA.

F. BURIADIACEAE Pant, 1977 C. (KLA?)–P. (ART) Terr.

First: *Buriadia heterophylla* (Feistmantel) Seward and Sahni, 1920, Itararé Subgroup, Paraná Basin, Brazil (Rocha Campos and Archangelsky, *in* Wagner *et al.*, 1985). The chronostratigraphical position of these fossils is not certain, but seems more likely to be Upper Carboniferous than Lower Permian.
Last: *Walchia* sp. (syn. *B. heterophylla* – see Florin, 1940), upper Sadong Formation, Pyongyang Coalfield, Korea (Kawasaki, 1934).
Intervening: ASS.

F. UTRECHTIACEAE Mapes and Rothwell, 1991 P. ASS–UFI Terr.

First: *Utrechtia floriniformis* Mapes and Rothwell, 1991, Lower Rotliegend, Germany.
Last: *Ortiseia leonardii* Florin, 1964, Val Gardena Formation, Dolomites and Vicentinian Alps, Italy (Clement-Westerhof, 1984).
Intervening: SAK–KUN.
Comments: This family is approximately equivalent to the Walchiaceae *sensu* Clement-Westerhof (1984) and Kerp *et al.* (1990). However, Mapes and Rothwell (1991) gave the family a more rigorous definition based mainly on ovulate cone structure, which required a change of name.

F. FERUGLIOCLADACEAE Archangelsky and Cúneo, 1987 P. (ASS) Terr.

First and Last: *Ferugliocladus riojanum* Archangelsky and Cúneo, 1987, *F. patagonicus* (Feruglio) Archangelsky and Cúneo, 1987 and *Ugartecladus genoensis* Archangelsky and Cúneo, 1987, Rio Genoa Group, Central Patagonian Basin, Argentina.

F. MAJONICACEAE Clement-Westerhof, 1987 P. (UFI) Terr.

First and Last: *Majonica palina* Clement-Westerhof, 1987, *Dolmitia cittertiae* Clement-Westerhof, 1987 and *Pseudovoltzia sjerpii* Clement-Westerhof, 1987, Val Gardena Formation, Dolomites and Vicentinian Alps, Italy. Also *Pseudovoltzia liebeana* (Geinitz) Florin, 1927, Kupferschiefer, Lower Rhine, Germany (Schweitzer, 1963); and Marl Slate, Cumbria and Durham, England, UK (Stoneley, 1958).

F. ULLMANNIACEAE Zimmermann, 1959 P. (UFI) Terr.

First and Last: *Ullmannia bronnii* Göppert, 1850 and *U. frumentaria* Göppert, 1850, Kupferschiefer, Lower Rhine, Germany (Schweitzer, 1963); and Marl Slate, Cumbria and Durham, UK (Stoneley, 1958).
Comments: The natural status of this family has still to be confirmed (Clement-Westerhof, 1988). *Ullmannia* Göppert, 1850 has also been reported from older (SAK?) and slightly younger (KAZ) Angaran assemblages (summarized by Vakhrameev *et al.*, 1978), but their relationship to the European species (and thus to the family) is unclear.

F. VOLTZIACEAE Florin, 1951 Tr. (SMI?)–K. (CEN) Terr.

First: *Aethophyllum speciosum* Schimper, 1869 and *Voltzia heterophylla* Brongniart, 1835, Upper Buntsandstein, Germany (Mägdefrau, 1956).
Last: *Protodammara speciosa* Hollick and Jeffrey, 1909 and *Dectylolepis cyrptomerioides* Hollick and Jeffery, 1909, Raritan Formation, Staten Island, USA.
Intervening: LAD–BTH.
Comments: Many authors have included within this family a wide variety of Palaeozoic and Mesozoic conifers (e.g. Taylor, 1981; Stewart, 1983). However, Clement-Westerhof (1988) has separated out most of the Palaeozoic taxa into the Majonicaceae and Ullmanniaceae. Whether the remaining Mesozoic members form a 'natural' group, and their relationship to the Taxodiaceae, remain uncertain (Miller, 1982). It includes the Cycadocarpidiaceae, listed separately in the first edition of *The Fossil Record*, and the Swedenborgiaceae Zimmermann, 1959.

F. PODOCARPACEAE Endlicher, 1847 Tr. (DIE?)–Rec. Terr.

First: ?*Voltzia* cf. *heterophylla* Carpentier, 1935 *non* Brongniart, 1835 (syn. *Rissikia media* (Tennison Woods) Townrow, 1967), Sakoa Group, Madagascar. This record is based on foliage. The oldest frutifications are *R. media* (Tennison Woods) Townrow, 1967, Burnera Waterfall, Natal, South Africa (SMI?). **Extant**
Intervening: NOR, BAJ, BER?, CMP–HOL.

F. PALISSYACEAE Florin, 1958 Tr. (CRN)–J. (BAJ) Terr.

First: *Stachyotaxus lipoldii* (Stur) Kräusel, 1952, Lettenkohle, Lunz, Austria; and *S. sahnii* Kräusel, 1952, Lettenkohle, Neuewelt, Switzerland.
Last: *Palissya* sp., Saltwick Formation, North Yorkshire, England, UK (Hill and van Konijnenburg van Cittert, 1973). This material has not been described in detail, but is reported to include a female cone.
Intervening: RHT–HET.
Comments: The validity of this family has recently been questioned (Meyen, 1984; Miller, 1985) but, in the absence of any formal taxonomic changes, the traditional concept has been maintained.

F. ARAUCARIACEAE Henkel and Hochstetter, 1865 Tr. (CRN?)–Rec. Terr.

First: *Araucarites parsorensis* Lele, 1955 and *A. indica* Lele, 1962, Parsora Formation, South Rewa, India. The exact chronostratigraphical position of these fossils is uncertain, but appears to be younger than TAT but older than NOR (Vakhrameev *et al.*, 1978). **Extant**
Intervening: RHT–HOL.
Comments: The fossil record of this family is reviewed by Stockey (1982).

F. PINACEAE Lindley, 1836 Tr. (CRN)–Rec. Terr.

First: *Compsostrobus neotericus* Delevoryas and Hope, 1973 and *Millerostrobus pekinensis* Taylor *et al.*, 1987, Pekin Formation, North Carolina, USA. **Extant**
Intervening: RHT–HET, BAJ–BTH, BER–HOL.
Comments: *Compsostrobus* Delevoryas and Hope, 1973, was originally placed in a separate family (Compsostrobaceae Delevoryas and Hope, 1973), but most authors now include it in the Pinaceae. The pollen cone *Millerostrobus* Taylor *et al.*, 1987, has features indicative of both the Pinaceae and the Podocarpaceae.

F. CHEIROLEPIDIACEAE Takhtajan, 1963 Tr. (CRN?)–K. (CMP?) Terr.

First: ??*Brachyphyllum hegewaldia* Ash, 1973 and *Pagiophyllum* spp., Chinle Formation, Arizona, USA (Watson, 1988). These refer to foliage with cheirolepidacean-like cuticles. Better documented are records from the RHT of *Hirmeriella muensteri* (Schenk) Jung, 1968, Frankonia, Germany, and from the Keuper Conglomerate, Cowbridge, South Glamorgan, Wales, UK (Harris, 1957; Jung, 1968); also *H. airelensis* Muir and van Konijnenburg-van Cittert, 1970, Airel, France.
Last: ??*Frenelopsis hoheneggeri* (Ettingshausen) Carpentier, 1937, Sainte Baume, France. This is based on incompletely described and poorly preserved specimens. Better documented is *Frenelopsis oligostomata* Romariz, 1946, Esgueira (TUR), Beira Littorale, Portugal (Alvin, 1977).

Intervening: RHT–ALB.

Comments: The confused nomenclature for this family has been discussed and stabilized by Watson (1982, 1988). Its circumscription has become considerably enlarged to encompass taxa originally included in the Cupressaceae (e.g. in the first edition of *The Fossil Record*).

F. TAXACEAE Gray, 1821 J. (HET)–Rec. Terr.

First: *Palaeotaxus rediviva* Nathorst, 1908, Upper Coal Bed, Skrombergia Colliery, Scania, Sweden (Florin, 1958). **Extant**

Intervening: TOA-BHT, BER, CMP–HOL.

Comments: This family is sometimes placed in its own order (Taxales).

F. PARARAUCARIACEAE Stockey, 1977 J. (?m./u.) Terr.

First and Last: *Pararaucaria patagonica* Wieland, 1929, Cerro Cuadrado fossil forest, Patagonia, Argentina (Stockey, 1977).

Comments: The exact chronostratigraphical position of these fossils cannot be fixed beyond Middle to Upper Jurassic. They are similar to both the Pinaceae and Taxodiaceae, and suggest a close phylogenetic relationship between these families.

F. TAXODIACEAE Warming, 1890 J. (BAJ)–Rec. Terr.

First: *Elatides thomasii* Harris, 1979, Saltwick Formation, Yorkshire, England, UK. **Extant**

Intervening: BTH–HOL.

F. ARCTOPITYACEAE Manum and Bose, 1989 J. (OXF)–K. (TUR) Terr.

First: *Sciadopityoides persulcata* (Johansson) Sveshnikova, 1981 and *S. nathorstii* (Halle) Sveshnikova, 1981, Ramsa Formation, Andøya, Norway (Manum, 1987).

Last: *Sciadopityoides uralensis* (Dorofeev and Sveshnikova) Sveshnikova, 1981, Ural Mountains, former USSR.

Intervening: KIM–ALB.

Comments: This family was proposed for a variety of leafy shoots from mainly Arctic regions, which were traditionally included in the Sciadopityaceae (Manum and Bose, 1989). However, full details of the proposal have still to be published. No fructifications are known.

F. SCIADOPITYACEAE Seward, 1919 J. (OXF)–Rec. Terr.

First: *Sciadopitys macrophylla* (Florin) Manum, 1987, *S. lagerheimii* (Johansson) Manum, 1987 and *Sciadopitys*-like cone scales, Ramsa Formation, Andøya, Norway (Bose, 1955; Manum, 1987). **Extant**

Intervening: DAN–HOL.

Comments: Virtually all macrofossils assigned to this family are foliage fragments, although there is good palynological evidence for the existence of this family in the Lower Tertiary (Manum and Bose, 1988). Many authors have included *Sciadopitys* Siebold and Zuccarini, 1870 and its allies within the Taxodiaceae (e.g. Taylor, 1981). However, Christophel (1976) has argued cogently for the separation of the Sciadopityaceae, a view which has been adopted here. Florin (1958) assigned foliage fragments (*Elatocladus* sp.) from the Scalby Formation (BAJ), North Yorkshire, England, UK, to this family, an opinion which has been disputed by Harris (1979). The inclusion of *Sciadopitys scania* Florin, 1922 from the Rhaetic (RHT) of Sweden has also been rejected (Bose, 1955).

F. CEPHALOTAXACEAE Neger, 1907 J. (BAJ?)–Rec. Terr.

First: ?*Elatocladus zamioides* (Leckenby) Seward, 1919, Cloughton Formation, North Yorkshire, England, UK (Harris, 1979). This record is based on foliage alone. **Extant**

Intervening: BTH–HOL.

Comments: Foliage regarded as typical of this family occurs reasonably commonly in the fossil record, but records of fructifications are equivocal. For instance, some of the seeds identified as *Cephalotaxospermum* Berry, 1910 might in fact belong to the Podocarpaceae or Taxaceae (Miller, 1977).

F. CUPRESSACEAE Bartling, 1830 T. (DAN)–Rec. Terr.

First: *Cupressinocladus interruptus* (Newberry) Schweitzer, 1974, Volcanic Tuff, Smoky Tower, Alberta, Canada (Christophel, 1976). **Extant**

Comments: Many authors have extended the range of this family back to the Triassic. However, this is based on fossils which, at least in part, probably belong to the Cheirolepidiaceae (Watson, 1977, 1988).

Order GINKGOALES Engler, 1897

F. GINKGOACEAE Engler, 1897 Tr. (CRN?)–Rec. Terr.

First: *Ginkgoites lunzensis* (Stur) Florin, 1936, Lettenkohle, Lunz, Austria (Kräusel, 1943). This record is based on foliage adpressions. The oldest fertile material is *Allicospermum xystrum* Harris, 1935, Scoresby Sound, Greenland (RHT). **Extant**

Intervening: RHT, HET, BAJ–BTH, BER, CEN–HOL.

Comments: Meyen (1987) classified the Ginkgoales with the so-called 'Mesozoic pteridosperms'. However, Crane (1985) and Doyle and Donoghue (1986) show that they are more closely aligned with the Pinales. Many of the putative ginkgoaleans from the Palaeozoic mentioned in the first edition of *The Fossil Record* probably belong to the Peltaspermopsida (Meyen, 1987). Attempts to classify the Ginkgoales into families have tended to rely on vegetative structures and have not been widely followed.

REFERENCES

Alvin, K. L. (1977) The conifers *Frenelopsis* and *Manica* in the Cretaceous of Portugal. *Palaeontology*, **20**, 387–404.

Anderson, J. M. and Anderson, H. M. (1985) *Palaeoflora of Southern Africa. Prodromus of South African Megafloras Devonian to Lower Cretaceous*. Balkema, Rotterdam.

Andrews, H. N. (1940) A new cupule from the Lower Carboniferous of Scotland. *Bulletin of the Torrey Botanical Club*, **67**, 595–601.

Andrews, H. N. (1941) *Dichophyllum moorei* and certain associated seeds. *Annals of the Missouri Botanical Garden*, **28**, 375–84.

Archangelsky, S. and Cúneo, R. (1987) Ferugliocladaceae, a new conifer family from the Permian of Gondwana. *Review of Palaeobotany and Palynology*, **51**, 3–30.

Archangelsky, S. and Cúneo, R. (1990) *Polyspermophyllum*, a new Permian gymnosperm from Argentina, with considerations about the Dicranophyllales. *Review of Palaeobotany and Palynology*, **63**, 117–35.

Arnold, C. A. (1948) Classification of gymnosperms from the viewpoint of paleobotany. *Botanical Gazette*, **110**, 2–12.

Asama, K. (1962) Evolution of Shansi flora and origin of simple leaf. *Science Reports, Tohoku University, 2nd Series (Geology) Special Volume*, **5**, 247–74.

Ash, S. R. (1972) Late Triassic plants from the Chinle Formation in northeastern Arizona. *Palaeontology*, **15**, 598–618.

Barnard, P. D. W. (1959) On *Eosperma oxroadense* gen. et sp. nov. a Lower Carboniferous seed from East Lothian. *Annals of Botany, New Series*, **90**, 284–96.

Barthel, M. (1976) Die Rotliegendflora Sachsens. *Abhandlungen des Staatlichen Museums für Mineralogie und Geologie zu Dresden*, **24**, 1–190.

Beck, C. B. (1985) Gymnosperm phylogeny – a commentary on the views of S. V. Meyen. *Botanical Review*, **51**, 273–94.

Benson, M. J. (1914) *Sphaerostoma ovale* (*Conostoma ovale* et *intermedium* Williamson), a Lower Carboniferous ovule from Pettycur, Fifeshire, Scotland. *Transactions of the Royal Society of Edinburgh*, **50**, 1–16.

Bertrand, P. (1930) Bassin houiller de la Sarre et de la Lorraine. I. Flore fossile. 1. Neuroptéridées. *Étude Gîtes Mineraux France*, 1–58.

Bose, M. N. (1955) *Sciadopitytes variabilis* n. sp. from the Arctic of Canada. *Norsk Geologisk Tidsskrift*, **35**, 53–68.

Bose, M. N., Pal, P. K. and Harris, T. M. (1984) *Carnoconites rajmahalensis* (Wieland) comb. nov. from the Jurassic of Rajmahal Hills, India. *Palaeobotanist*, **32**, 368–9.

Bouroz, A. and Doubinger, J. (1977) Report on the Stephanian–Autunian boundary and on the contents of Upper Stephanian and Autunian in their stratotypes, in *Symposium on Carboniferous Stratigraphy* (eds V. Holub and R. H. Wagner), Geological Survey, Prague, pp. 147–69.

Carpentier, A. (1935) Étude paléobotanique sur le groupe de la Sakoa et le groupe de la Sakmena (Madagascar). *Annales géologiques du Service des Mines, Madagascar*, **5**, 7–32.

Carpentier, A. (1937) Remarques sur des empreintes de *Frenelopsis* trouvées dans le Campanien inférieur de la Sainte Baume. *Annals de la Musée Histoire naturelle Marseille*, **28**, 5–14.

Chamberlain, C. J. (1935) *Gymnosperms Structure and Evolution*. University Press, Chicago.

Chandra, S. and Surange, K. R. (1979) Revision of the Indian species of *Glossopteris*. *Birbal Sahni Institute of Palaeobotany, Monograph*, **2**, 1–291.

Christophel, D. C. (1976) Fossil floras of Smoky Tower locality, Alberta, Canada. *Palaeontographica, Abteilung B*, **157**, 1–43.

Clement-Westerhof, J. A. (1984) Aspects of Permian palaeobotany and palynology. 4. The conifer *Ortiseia* Florin from the Val Gardena Formation of the Dolomites and the Vicentinian Alps (Italy) with special reference to a revised concept of the *Walchiaceae* (Göppert) Schimper. *Review of Palaeobotany and Palynology*, **41**, 51–66.

Clement-Westerhof, J. A. (1987) Aspects of Permian palaeobotany and palynology. 7. The *Majonicaceae*, a new family of Late Permian conifers. *Review of Palaeobotany and Palynology*, **52**, 375–402.

Clement-Westerhof, J. A. (1988) Morphology and phylogeny of Paleozoic conifers, in *Origin and Evolution of Gymnosperms* (ed. C. B. Beck), Columbia University Press, New York, pp. 298–337.

Corsin, P. (1960) Classification des Ptéridophytes et des Ptéridospermophytes du Carbonifère. *Bulletin de la Société Géologique de France, 7e Série*, **2**, 566–72.

Crane, P. R. (1985) Phylogenetic analysis of seed plants and the origin of angiosperms. *Annals of the Missouri Botanical Garden*, **72**, 716–93.

Crane, P. R. (1986) The morphology and relationships of Bennettitales, in *Systematic and Taxonomic Approaches in Palaeobotany* (eds R. A. Spicer and B. A. Thomas), Systematics Association Special Volume 31, pp. 163–75.

Crane, P. R. (1988) Major clades and relationships in the 'Higher' gymnosperms, in *Origin and Evolution of Gymnosperms* (ed. C. B. Beck), Columbia University Press, New York, pp. 218–72.

Crane, P. R. and Upchurch, G. R. (1987) *Drewia potomacensis* gen. et sp. nov., an early Cretaceous member of Gnetales from the Potomac Group of Virginia. *American Journal of Botany*, **74**, 1722–36.

Cridland, A. A. and Morris, J. E. (1963) *Taeniopteris*, *Walchia* and *Dichophyllum* in the Pennsylvanian System of Kansas. *University of Kansas Science Bulletin*, **44**, 71–85.

Delevoryas, T. (1959) Investigations of North American cycadeoides: *Monanthesia*. *American Journal of Botany*, **46**, 657–66.

Delevoryas, T. and Gould, R. E. (1971) An unusual fossil fructification from the Jurassic of Oaxaca, Mexico. *American Journal of Botany*, **58**, 616–20.

Delevoryas, T. and Hope, R. C. (1973) Fertile coniferophyte remains from the Late Triassic Deep River Basin, North Carolina. *American Journal of Botany*, **60**, 810–18.

Delevoryas, T. and Person, C. P. (1975) *Mexiglossa varia* gen. et sp. nov., a new genus of glossopterid leaves from the Jurassic of Oaxaca, Mexico. *Palaeontographica, Abteilung B*, **154**, 114–20.

Doyle, J. A. and Donoghue, M. J. (1986) Relationships of angiosperms and Gnetales: a numerical cladistic approach, in *Systematic and Taxonomic Approaches in Palaeobotany* (eds R. A. Spicer and B. A. Thomas), Systematics Association Special Volume 31, pp. 177–98.

Drinnan, A. N. and Chambers, T. C. (1985) A reassessment of *Taeniopteris daintreei* from the Victoria Early Cretaceous: a member of the Pentoxylales and a significant Gondwanaland plant. *Australian Journal of Botany*, **33**, 89–100.

Florin, R. (1922) On the geological history of the Sciadopitineae. *Svensk Botanisk Tidskrift*, **16**, 260–70.

Florin, R. (1940) Die Koniferen des Oberkarbons und des unteren Perms. Teil 5. *Palaeontographica, Abteilung B*, **85**, 243–364.

Florin, R. (1949) The morphology of *Trichopitys heteromorpha* Saporta, a seed plant of Palaeozoic age, and the evolution of the female flowers in the Ginkgoinae. *Acta Horti Bergiana*, **15**, 79–109.

Florin, R. (1958) On the Jurassic taxads and conifers from north-western Europe and eastern Greenland. *Acta Horti Bergiana*, **17**, 257–402.

Galtier, J. and Rowe, N. P. (1989) A primitive seed-like structure and its implications for early gymnosperm evolution. *Nature*, **340**, 225–7.

Gao Z. and Thomas, B. A. (1989) A review of fossil cycad megasporophylls, with new evidence of *Crossozamia* Pomel and its associated leaves from the Lower Permian of Taiyuan, China. *Review of Palaeobotany and Palynology*, **60**, 205–23.

Gillespie, W. H. and Pfefferkorn, H. W. (1986) Taeniopterid lamina on *Phasmatocycas* megasporophylls (Cycadales) from the Lower Permian of Kansas, U.S.A. *Review of Palaeobotany and Palynology*, **49**, 99–116.

Gomankov, A. V. and Meyen, S. V. (1986) *Tatarina Flora (Composition and Distribution in the Late Permian of Eurasia)*. USSR Academy of Sciences (Transactions of the Order of the Red Banner of Labour Geological Institute 401) [in Russian].

Gordon, W. T. (1910) On a new species of *Physostoma* from the Lower Carboniferous rocks of Pettycur (Fife). *Proceedings of the Cambridge Philosophical Society*, **15**, 395–7.

Gorelova, S. G., Men'shikova, L. V. and Khalfin, L. L. (1973) Phytostratigraphy and plant taxonomy in the Upper Palaeozoic coal-bearing deposits of the Kuznetsk Basin. *Trudy Sibirskogo Nauchno-Issledovatel'skogo, Geologii, Geofiziki i Mineral'nogo*, **140**, 1–169 [in Russian].

Halle, T. G. (1927) Palaeozoic plants from central Shansi. *Palaeontologica Sinica*, **A2** (1), 1–316.

Halle, T. G. (1932) On the seeds of the pteridosperm *Emplectopteris triangularis*. *Geological Society of China Bulletin*, **11**, 301–6.

Harris, T. M. (1932a) The fossil flora of Scoresby Sound, East Greenland. Part 2. Description of seed plants *incertae sedis* together with a discussion of certain cycadophyte cuticles. *Meddelelser om Grønland*, **85** (3), 1–114.

Harris, T. M. (1932b) The fossil flora of Scoresby Sound, East Greenland. Part 3. Caytoniales and Bennettitales. *Meddelelser om Grønland*, **85** (5), 1–133.

Harris, T. M. (1935) The fossil flora of Scoresby Sound, East

Greenland. Part 4. Ginkgoales, Coniferales, Lycopodiales and isolated fructifications. *Meddelelser om Grønland*, **112** (1), 1–176.

Harris, T. M. (1957) A Liasso-Rhaetic flora in south Wales. *Proceedings of the Royal Society of London, Series B*, **147**, 289–308.

Harris, T. M. (1961) The fossil cycads. *Palaeontology*, **4**, 313–23.

Harris, T. M. (1964) *The Yorkshire Jurassic Flora II Caytoniales, Cycadales and Pteridosperms*. British Museum (Natural History), London.

Harris, T. M. (1969) *The Yorkshire Jurassic Flora. III. Bennettitales*. British Musem (Natural History), London.

Harris, T. M. (1979) *The Yorkshire Jurassic Flora. V. Coniferales*. British Museum (Natural History), London.

Hill, C. R. and Crane, P. R. (1982) Evolutionary cladistics and the origin of angiosperms, in *Problems of Phylogenetic Reconstruction* (eds K. A. Joysey and A. E. Friday), Systematics Association Special Volume 21, pp. 269–361.

Hill, C. R. and Konijnenburg-van Cittert, J. H. A. van (1973) Species of plant fossils collected from the Middle Jurassic plant bed at Hasty Bank, Yorkshire. *The Naturalist*, **1973**, 59–63.

Hollick, A. and Jeffrey, E. C. (1909) Studies on Cretaceous coniferous remains from Kreisherville, New York. *Memoirs of the New York Botanical Gardens*, **3**, 1–76.

Jung, W. (1968) *Hirmerella munsteri* (Schenk) Jung nov. comb., eine bedeutsame Konifere des Mesozoikums. *Palaeontographica, Abteilung B*, **122**, 55–93.

Kawasaki, S. (1934) The flora of the Heian System. Part 2. *Bulletin of the Geological Survey (Korea) Chosen*, **6**, 46–311.

Kerp, J. H. F. (1983) Aspects of Permian palaeobotany and palynology. 1. *Sobernheimia jonkeri* nov. gen., nov. sp., a new fossil plant of cycadalean affinity from the Waderner Gruppe of Sobernheim. *Review of Palaeobotany and Palynology*, **38**, 173–83.

Kerp, J. H. F. and Fichter, J. (1985) Die Makrofloren des saarpfälzischen Rotliegenden (?Ober-Karbon – Unter Perm; SW-Deutschland). *Mainzer geowissenschaftliche Mitteilungen*, **14**, 159–286.

Kerp, J. H. F. and Haubold, H. (1988) Aspects of Permian palaeobotany and palynology. VIII. On the reclassification of the west- and central-European species of the form-genus *Callipteris* Brongniart 1849. *Review of Palaeobotany and Palynology*, **54**, 135–50.

Kerp, J. H. F., Poort, R. J., Swinkels, H. A. J. M. *et al.* (1990) Aspects of Permian palaeobotany and palynology. IX. Conifer-dominated Rotliegend floras from the Saar–Nahe Basin (?Late Carboniferous–Early Permian; SW-Germany) with special reference to the reproductive biology of early conifers. *Review of Palaeobotany and Palynology*, **62**, 205–48.

Krassilov, V. A. (1975) Dirhopalostachyaceae – a new family of proangiosperms and its bearing on the problem of angiosperm ancestry. *Palaeontographica, Abteilung B*, **153**, 100–10.

Krassilov, V. A. (1977) The origin of angiosperms. *Botanical Review*, **43**, 143–76.

Krassilov, V. A. (1978) Late Cretaceous gymnosperms from Sakhalin and the terminal Cretaceous event. *Palaeontology*, **21**, 893–905.

Kräusel, R. (1943) Die Ginkgophyten der Trias vom Lunz in Nieder-Österreich und von Neuewelt bei Basel. *Palaeontographica, Abteilung B*, **87**, 59–93.

Kräusel, R. (1949) Koniferen und andere Gymnospermen aus der Trias von Lunz, Nieder-Österreich. *Palaeontographica, Abteilung B*, **89**, 35–82.

Kräusel, R. (1952) *Stachyotaxus sahnii* n. sp., eine Konifere aus der Trias von Neuewelt bei Basel. *Palaeobotanist*, **1**, 285–8.

Leary, R. L. (1990) Possibly Early Pennsylvanian ancestor of the Cycadales. *Science*, **249**, 1152–4.

Leisman, G. A. (1964) *Physostoma calcaratum* sp. nov., a tentacled seed from the middle Pennsylvanian of Kansas. *American Journal of Botany*, **51**, 1069–75.

Lele, K. M. (1955) Plant fossils from Parsora in the South Rewa Gondwana basin, India. *Palaeobotanist*, **4**, 23–34.

Lele, K. M. (1962) Studies in the Indian Middle Gondwana flora 2. Plant fossils from the South Rewa Gondwana basin. *Palaeobotanist*, **10**, 69–82.

Li X. and Yao Z. (1983) Fructifications of gigantopterids from South China. *Palaeontographica, Abteilung B*, **185**, 11–26.

Li X. and Yao Z. (1985) Carboniferous and Permian floral provinces in East Asia. *Compte rendu 9e Congrès International de Stratigraphie et Géologie du Carbonifère*, (Washington and Urbana, 1979), **5**, 95–101.

Long, A. G. (1959) On the structure of *Calymmatotheca kidstonii* Calder (emended) and *Genomosperma latens* gen. et sp. nov. from the Calciferous Sandstone Series of Berwickshire. *Transactions of the Royal Society of Edinburgh*, **64**, 29–44.

Long, A. G. (1961a) On the structure of *Deltasperma fouldense* gen. et sp. nov., and *Camptosperma berniciense* gen. et sp. nov., petrified seeds from the Calciferous Sandstone Series of Berwickshire. *Transactions of the Royal Society of Edinburgh*, **64**, 281–95.

Long, A. G. (1961b) Some pteridosperm seeds from the Calciferous Sandstone Series of Berwickshire. *Transactions of the Royal Society of Edinburgh*, **64**, 401–19.

Long, A. G. (1964) Some specimens of *Stenomyelon* and *Kalymma* from the Calciferous Sandstone Series of Berwickshire. *Transactions of the Royal Society of Edinburgh*, **65**, 435–46.

Long, A. G. (1966) Some Lower Carboniferous fructifications from Berwickshire, together with a theoretical account of the evolution of ovules, cupules and carpels. *Transactions of the Royal Society of Edinburgh*, **66**, 345–75.

Long, A. G. (1975) Further observations on some Lower Carboniferous seeds and cupules. *Transactions of the Royal Society of Edinburgh*, **69**, 267–93.

Long, A. G. (1977) Observations on Carboniferous seeds of *Mitrospermum*, *Conostoma* and *Lagenostoma*. *Transactions of the Royal Society of Edinburgh*, **70**, 37–61.

Lyons, P. C. and Darrah, W. C. (1989) Earliest conifers of North America: upland and/or paleoclimatic indicators? *Palaios*, **4**, 480–6.

Mägdefrau, K. (1956) *Paläobiologie der Pflanzen*. Gustav Fischer, Jena.

Mamay, S. H. (1973) *Archaeocycas* and *Phasmatocycas* – new genera of Permian cycads. *Journal of Research, US Geological Survey*, **1**, 687–9.

Mamay, S. H. (1976) Paleozoic origin of cycads. *Professional Paper US Geological Survey*, **934**, 1–48.

Manum, S. B. (1987) Mesozoic *Sciadopitys*-like leaves with observations on four species from the Jurassic of Andøya, northern Norway, and emendation of *Sciadopityoides* Sveshnikova. *Review of Palaeobotany and Palynology*, **51**, 145–68.

Manum, S. B. and Bose, M. N. (1988) Sciadopityaceae – en gammel bartrefamilie belyst ved norske fossiler. *Blyttia*, **46**, 189–94.

Manum, S. B. and Bose, M. N. (1989) A new and prominent conifer family in the Arctic Lower Cretaceous revealed by cuticle studies (Abstract). *Contributions from the Paleontological Museum, University of Oslo*, **359**, 19–20.

Mapes, G. and Rothwell, G. W. (1980) *Quaestoria amplecta* gen. et sp. nov., a structurally simple medullosan stem from the Upper Mississippian of Arkansas. *American Journal of Botany*, **67**, 636–47.

Mapes, G. and Rothwell, G. W. (1991) Structure and relationships of primitive conifers. *Neues Jahrbuch für Geologie und Paläontologie, Abhandlungen*, **183**, 269–87.

McDougall, I. and McElhinny, M. W. (1970) The Rajmahal Traps of India – K–Ar ages and palaeomagnetism. *Earth and Planetary Science Letters*, **9**, 371–8.

Mei M. (1984) *The Analysis of the Floras of Permian Coal-bearing Strata in Fujian, Jiangxi and Sichuan Provinces*. Graduate School, China Institute of Mining, Beijing (prepared for the 2nd Conference of the International Organization of Palaeobotany, Edmonton, 1984).

Meyen, S. V. (1969) New genera *Entsovia* and *Slivkovia*, from the Permian of the Russian Platform and Cisuralia. *Palaeontologiskii Zhurnal*, **4**, 93–100 [in Russian].

Meyen, S. V. (1977) Cardiolepidaceae – a new coniferalean family from the Upper Permian of north Eurasia. *Palaeontologiskii Zhurnal*, **3**, 130–40 [in Russian].

Meyen, S. V. (1979) Permian predecessors of the Mesozoic pteridosperms in western Angaraland. *Review of Palaeobotany and Palynology*, **28**, 191–201.

Meyen, S. V. (1982) The Carboniferous and Permian floras of Angaraland (a synthesis). *Biological Memoirs*, **7**, 1–109.

Meyen, S. V. (ed.) (1983) *Palaeontological Atlas for the Permian Strata of the Pechora Coal-Basin*. Institute of Geology, USSR Academy of Sciences, Leningrad [in Russian].

Meyen, S. V. (1984) Basic features of gymnosperm systematics and phylogeny as evidenced by the fossil record. *Botanical Review*, **50**, 1–111.

Meyen, S. V. (1986) Gymnosperm systematics and phylogeny: a reply to commentaries by C. B. Beck, C. N. Miller and G. W. Rothwell. *Botanical Review*, **52**, 300–20.

Meyen, S. V. (1987) *Fundamentals of Palaeobotany*. Chapman and Hall, London.

Meyen, S. V. (1988) Gymnosperms of the Angara flora, in *Origin and Evolution of Gymnosperms* (ed. C. B. Beck), Columbia University Press, New York, pp. 338–81.

Meyen, S. V. and Smoller, H. G. (1986) The genus *Mostotchkia* Chachlov (Upper Palaeozoic of Angaraland) and its bearing on the characteristics of the order Dicranophyllales (Pinopsida). *Review of Palaeobotany and Palynology*, **47**, 205–23.

Miller, C. N. (1977) Mesozoic conifers. *Botanical Review*, **43**, 217–80.

Miller, C. N. (1982) Current status of Paleozoic and Mesozoic conifers. *Review of Palaeobotany and Palynology*, **37**, 99–114.

Miller, C. N. (1985) A critical review of S. V. Meyen's 'Basic features of gymnosperm systematics and phylogeny as evidenced by the fossil record'. *Botanical Review*, **51**, 295–318.

Muir, M. and Konijnenburg-van Cittert, J. H. A. van (1970) A Rhaeto-Liassic flora from Airel, northern France. *Palaeontology*, **13**, 433–42.

Němejc, F. (1937) The Sphenopterides stated in the Permocarboniferous of Central Bohemia (a preliminary report, II. part). *Věstník Kralovske Ceske spolec Nauk*, 1–14.

Oliver, F. W. (1909) On *Physostoma elegans*, Williamson, an archaeic type of seed from the Palaeozoic rocks. *Annals of Botany*, **23**, 73–116.

Phillips, T. L. (1980) Stratigraphic and geographic occurrences of permineralized coal-swamp plants – Upper Carboniferous of North America and Europe, in *Biostratigraphy of Fossil Plants* (eds D. L. Dilcher and T. N. Taylor), Dowden, Hutchinson and Ross, Stroudsburg, pp. 25–92.

Poort, R. J. and Kerp, J. H. F. (1990) Aspects of Permian palaeobotany and palynology. XI. On the recognition of true peltasperms in the Upper Permian of western and central Europe and a reclassification of species formerly included in *Peltaspermum*. *Review of Palaeobotany and Palynology*, **63**, 197–225.

Rao, A. R. (1976) Problems in the Pentoxyleae. *Palaeobotanist*, **25**, 393–6.

Rao, A. R. (1981) The affinities of the Pentoxyleae. *Palaeobotanist*, **28/29**, 207–9.

Read, C. B. (1937) The flora of the New Albany Shale Part 2. The Calamopityaceae and their relationships. *United States Geological Survey, Professional Paper*, **186-E**, 81–91.

Read, C. B. and Mamay, S. H. (1964) Upper Paleozoic floral zones and floral provinces of the United States. *United States Geological Survey, Professional Paper*, **454-K**, 1–35.

Renault, B. (1890) Étude sur le terrain houiller de Commentry. Livre 2è: flore fossile, 2è partie. *Bulletin de la Societé Industrie Minière*, **4**, 381–712.

Rösler, O. (1978) The Brazilian eogondwanic floral succession. *Boletin Instituto Geociências Universitad São Paulo*, **9**, 91–5.

Rothwell, G. W. (1985) The role of comparative morphology and anatomy in interpreting the systematics of fossil gymnosperms. *Botanical Review*, **51**, 319–27.

Rothwell, G. W. (1988) Cordaitales, in *Origin and Evolution of Gymnosperms* (ed. C. B. Beck), Columbia University Press, New York, pp. 273–97.

Rothwell, G. W., Scheckler, S. E. and Gillespie, W. H. (1989) *Elkinsia* gen. nov., a Late Devonian gymnosperm with cupulate ovules. *Botanical Gazette*, **150**, 170–89.

Samylina, V. A. (1973) The correlation of continental upper Cretaceous deposits in northeastern USSR through palaeobotanical evidence. *Sovetskya Geologia*, **8** [in Russian].

Schweitzer, H.-J. (1963) Der weibliche Zapfen von *Pseudovoltzia liebeana* und seine Bedeutung für die Phylogenie der Koniferen. *Palaeontographica, Abteilung B*, **113**, 1–29.

Schweitzer, H.-J. (1977) Die Räto-Jurassischen Floren des Iran und Afghanistans. 4. Die Rätische Zwitterblutte *Irania hermaphroditica* nov. spec. und ihre Bedeutung für die Phylogenie der Angiospermen. *Palaeontographica, Abteilung B*, **161**, 98–145.

Scott, A. C. and Chaloner, W. G. (1983) The earliest fossil conifer from the Westphalian B of Yorkshire. *Proceedings of the Royal Society of London, Series B*, **220**, 163–82.

Shadle, G. L. and Stidd, B. M. (1975) The frond of *Heterangium*. *American Journal of Botany*, **62**, 67–75.

Smith, D. L. (1959) *Geminitheca scotica* gen. et sp. nov., a pteridosperm from the Lower Carboniferous of Dumbartonshire. *Annals of Botany, New Series*, **23**, 477–91.

Srivastava, S. C. (1969) Two new species of *Glossopteris* from the Triassic of Nidpur, Madhya Pradesh, India, in *J. Sen Memorial Volume*, Calcutta, pp. 299–303.

Stein, W. E. and Beck, C. B. (1978) *Bostonia perplexa* gen. et sp. nov., a calamopityan axis from the New Albany Shale of Kentucky. *American Journal of Botany*, **65**, 459–65.

Stewart, W. N. (1983) *Paleobotany and the Evolution of Plants*. University Press, Cambridge.

Stidd, B. M. (1978) An anatomically preserved *Potoniea* with *in situ* spores from the Pennsylvanian of Illinois. *American Journal of Botany*, **65**, 677–83.

Stidd, B. M., Oestry, L. L. and Phillips, T. L. (1975) On the frond of *Sutcliffia insignis* var. *tuberculata*. *Review of Palaeobotany and Palynology*, **20**, 55–66.

Stockey, R. A. (1977) Reproductive biology of the Cerro Cuadrado (Jurassic) fossil conifers: *Pararaucaria patagonica*. *American Journal of Botany*, **64**, 733–44.

Stockey, R. A. (1981) Some comments on the origin and evolution of conifers. *Canadian Journal of Botany*, **59**, 1932–40.

Stockey, R. A. (1982) The Araucariaceae: an evolutionary perspective. *Review of Palaeobotany and Palynology*, **37**, 133–54.

Stockmans, F. (1954) Flores Namuriennes de la Belgique: Incertitudes et hypothèses de travail, in *Volume Victor van Straelen*. Institut Royal Science Naturelle de Belgique, Brussels, pp. 117–32.

Stockmans, F. and Willière, Y. (1953) Végétaux namuriens de la Belgique. *Association pour l'Étude de Paléontologie et de Stratigraphie Houillères*, **13**, 1–382.

Stoneley, H. M. M. (1958) The Upper Permian flora of England. *Bulletin of the British Museum (Natural History), Geology*, **3**, 293–337.

Stur, D. (1885) Die obertriasische Flora der Lunzer-Schichten und des bituminösen Schiefers von Raibl. *Sitzungsberichte der Mathematisch–Naturwissenschaftlichen Classe der Kaiserlichen Akademie der Wissenschaften*, I, **91**, 93–103.

Surange, K. R. and Lele, K. M. (1955) Studies in the *Glossopteris* flora of India. 3. Plant fossils from Talchir Needle shales from Giridih coalfield. *Palaeobotanist*, **4**, 153–7.

Sveshnikova, I. N. (1981) The new fossil genus *Sciadopityoides* (Pinopsida). *Botaniskii Zhurnal*, **66**, 1721–9 [in Russian].

Taylor, T. N. (1969) Cycads: evidence from the Upper Pennsylvanian. *Science*, **164**, 294–5.

Taylor, T. N. (1970) *Lasiostrobus* gen n., a staminate strobilus of gymnospermous affinity from the Pennsylvanian of North America. *American Journal of Botany*, **57**, 670–90.

Taylor, T. N. (1981) *Paleobotany. An Introduction to Fossil Plant Biology*. McGraw-Hill, New York.

Taylor, T. N. and Archangelsky, S. (1985) The Cretaceous pteridosperms *Ruflorina* and *Ktalenia* and implications on cupule and carpel evolution. *American Journal of Botany*, **72**, 1842–53.

Taylor, T. N. and Eggert, D. A. (1967a) Petrified plants from the Upper Mississippian (Chester Series) of Arkansas. *Transactions of the American Microscopical Society*, **86**, 412–16.

Taylor, T. N. and Eggert, D. A. (1967b) Petrified plants from the

Upper Mississippian of North America. I: The seed *Rhynchosperma* gen. n. *American Journal of Botany*, **54**, 984–92.

Taylor, T. N., Delevoryas, T. and Hope, R. C. (1987) Pollen cones from the Late Triassic of North America and implications on conifer evolution. *Review of Palaeobotany and Palynology*, **53**, 141–9.

Townrow, J. A. (1967) On *Rissikia* and *Mataia* podocarpaceous conifers from the Lower Mesozoic of southern lands. *Papers and Proceedings of the Royal Society of Tasmania*, **101**, 103–36.

Vakhrameev, V. A. (1966) The upper Cretaceous floras of USSR coastal regions of the Pacific Ocean, the peculiarities of its composition and its stratigraphical significance. *Izvestia Akademia Nauk SSSR*, (Geological Series) **3** [in Russian].

Vakhrameev, V. A., Dobruskina, I. A., Meyen, S. V. *et al.* (1978) *Paläozoische und Mesozoische Floren Eurasiens und die Phytogeographie Dieser Zeit*. Gustav Fischer, Jena.

Wagner, R. H. (1967) Two new family names in the class Pteridospermopsida. *Proceedings of the Geological Society of London*, **1640**, 150–1.

Wagner, R. H., Winkler Prins, C.F. and Granados, L. F. (1983) *The Carboniferous of the World. I. China, Korea, Japan and S.E. Asia*. Instituto Geológicos y Minero de España, Madrid (IUGS Publication No. 16).

Wagner, R. H., Winkler Prins, C. F. and Granados, L. F. (eds) (1985) *The Carboniferous of the World. II Australia, Indian Subcontinent, South Africa, South America, and North Africa*. Instituto Geológicos y Minero de España, Madrid (IUGS Publication No. 20).

Walton, J. (1931) Contributions to the knowledge of Lower Carboniferous plants. – Part III. *Philosophical Transactions of the Royal Society of London, B*, **219**, 347–79.

Walton, J. (1949a) A petrified example of *Alcicornopteris* (*A. Hallei* sp. nov.) from the Lower Carboniferous of Dumbartonshire. *Annals of Botany, New Series*, **13**, 445–52.

Walton, J. (1949b) *Calathospermum scoticum* – an ovuliferous fructification of Lower Carboniferous age from Dumbartonshire. *Transactions of the Royal Society of Edinburgh*, **61**, 719–28.

Watson, J. (1977) Some Lower Cretaceous conifers of the Cheirolepidiaceae from the U.S.A. and England. *Palaeontology*, **20**, 715–49.

Watson, J. (1982) The Cheirolepidiaceae: a short review. *Phyta, Studies on Living and Fossil Plants, Pant Commemorative Volume*, pp. 265–73.

Watson, J. (1988) The Cheirolepidiaceae, in *Origin and Evolution of Gymnosperms* (ed. C. B. Beck), Columbia University Press, New York, pp. 382–447.

White, M. E. (1981) Revision of the Talbragar Fish Bed flora (Jurassic) of New South Wales. *Records of the Australian Museum*, **33**, 695–721.

45

MAGNOLIOPHYTA ('ANGIOSPERMAE')

M. E. Collinson, M. C. Boulter and P. L. Holmes

We consider that the angiosperms are a special case, requiring a format which differs from other chapters in this volume. Our reasons are explained below.

Acknowledgements – We particularly thank D. Mai, L. Stuchlik, B. H. Tiffney and J. A. Wolfe for helpful comments and for drawing our attention to relevant references. Helpful discussion and information was also provided by P. R. Crane, E. M. Friis, C. Gee, D. Greenwood, R. S. Hill, N. F. Hughes, Z. Kvacek and S. R. Manchester. MEC would also like to acknowledge the generous provision of reprints by many colleagues, without which this work could not have been undertaken. This work was carried out while MEC was in receipt of a Royal Society 1983 University Research Fellowship, which is gratefully acknowledged.

COMMENTS ON THE 1967 LIST

The presentation of stratigraphical ranges of angiosperm families in the first edition of *The Fossil Record* (Chesters *et al.*, 1967) was based on the wisdom of those three authors. It provided valuable suggestions about the likely current concensus for about 150 angiosperm families. The influence of the late Marjorie Chandler was not insignificant in the judgements (N. F. Hughes, pers. comm.). There was no attempt to justify the taxonomic composition of the family names, or to caution the users of uncertainties involving taxonomy, migration and evolution. Many users of these data may have accepted the records unaware of the subjective evidence upon which they were based.

PROBLEMS OF THE ANGIOSPERM RECORD

We are anxious to avoid these pitfalls and to stress the limitations of what we understand from present knowledge. Most importantly, we are anxious that the data presented here are not seen as accurate expressions of the stratigraphical ranges of any angiosperm taxa; although the most well-researched examples (noted under 'comments') should indicate minimum ages. The differences between the first and second editions of the angiosperm chapter in *The Fossil Record* may seem to be a regression, especially in comparison to the way in which other fossil groups are treated in this volume. This is because we believe the angiosperms require special treatment.

Angiosperms are generally represented in the fossil record by individual organs like pollen, flowers, fruits, seeds, leaves and wood. Each organ provides a different set of characters and the extent to which these are diagnostic of modern taxa (or can be used to diagnose extinct taxa) varies considerably, both within and between organs. These problems have been exacerbated in the past by a common tendency to include fossils in modern taxa based on superficial similarity rather than upon in-depth analysis. Although the latter is now the rule rather than the exception, many older determinations have not yet been revised. The diversity of many modern angiosperm taxa renders these revisions time-consuming and often beyond the scope of an individual or even a small group of researchers. Angiosperm organs also vary in their preservation potential which is further influenced by the habitat, growth habit and functional biology of the plant on which they are borne. The Cretaceous (Lidgard and Crane, 1990) and early Tertiary (Collinson, 1990) are periods of rapid diversification with novel character combinations on individual organs and reconstructed plants. These partly influence, and are influenced by, co-adaptation with insects and mammals. Combined, these phenomena make the angiosperms an exceptionally complex taxon for which to present a synthesis of family groupings. A stratigraphical range chart for angiosperm families is thus unobtainable at present, and has not been produced to accompany this chapter.

PROCEDURE FOR THE CURRENT LIST

We base our primary occurrence data on two defined authoritative sources; Muller (1981, with occasional reference to Muller, 1985) for pollen, and Holmes *et al.* (1991) for the fossil remains of other plant organs. The age quoted in these documents for the fossil genera within each family is listed below. For the pollen this age ('Muller First') is the first occurrence recognized by Muller, while for other plant organs, it is the age of the type species quoted in the Plant Fossil Record database version 1.0 (Holmes *et al.*, 1991) for each genus ('PFR First').

Comments are given below each of these entries. In most cases for PFR these do not relate to the specific PFR record but are a general indication of the early fossil record of the family as familiar to one of us (MEC) from the current literature. (For further details, see section below 'Standards and principles for comments'.)

These data sources are known to have many limitations and weaknesses. The PFR (Holmes *et al.*, 1991) records only the first appearances of type species of fossil angiosperm genera within the families represented and only includes

The Fossil Record 2. Edited by M. J. Benton. Published in 1993 by Chapman & Hall, London. ISBN 0 412 39380 8

fossil genera described up to the early 1980s. The PFR also includes at least 213 generic names for angiosperms which were not assigned to families by the original authors and thus could not be considered here. Both PFR and Muller (1981, 1985) include untested and therefore doubtful indications of family affinity.

The taxonomic system utilized is believed to be most widely accepted, that of Cronquist (1981) and, where possible, the time ranges are set against the matching stratigraphical scheme of Harland *et al.* (1990). All the established angiosperm families (except Priscaceae) have modern representatives and so we only record the first appearances of the generic types. Extinct groups, e.g. Czekanowskiales and Dirhopalostachyaceae, which are considered to be 'proangiosperms' (see Krassilov, pp. 7–10, *in* Douglas and Christophel, 1990) have been excluded (covered in Chapter 44, this volume).

EXTINCT ANGIOSPERM FAMILIES

Some palaeobotanical researchers try to fit fossil genera into a modern family, even when they may actually fall between two such families. Others give only an ordinal, class or division assignment for these 'intermediates'. Workers are reluctant to form a new intermediary family for extinct angiosperms; some are even reluctant to entertain the concept that high-ranking groups during the early times of 'angiosperm' evolution may have been very different from the modern families and orders. At its most extreme, this part of our argument means that the Magnoliophyta, a group based on modern plants, is a name that has less and less meaning back in time; the earliest plants of the lineage, taken on their own, might have had a very different definition and probably should be named accordingly and differently.

These difficult taxonomic problems were made easier through the Montreal–Leningrad editions of the ICBN by the concept of the 'organ-genus'. This confusingly named taxon allowed 'some genera of fossil plants not to be assigned to a family'. The organ-genus taxon was removed from the Code at the Melbourne congress, leaving fossil plant genera to be either modern genera or form genera. The latter 'may or may not be assigned to a family'. The Code does not distinguish between taxa of family rank and higher, being based on modern or on fossil material. Nevertheless, we are aware of only one extinct angiosperm taxon at a higher rank than that of genus.

The major reason that hardly any extinct angiosperm families or orders have been established by palaeobotanists is that plant fossils are found as separate organs. To diagnose any large group, extinct or living, requires evidence from as many organs as possible. Usually the evidence of such associations is lacking for fossil angiosperms. Exceptions are for the modern families Betulaceae, Cercidiphyllaceae, Fagaceae, Platanaceae, Juglandaceae, Ulmaceae and to some extent Trochodendraceae, Salicaceae, Aceraceae and the extinct Priscaceae (comments herein and review by Collinson, 1990). This detailed knowledge is largely the result of intensive research over the last ten or so years. More frequently, there is detail available for individual organs such as pollen, leaves, woods and fructifications, but their links are rarely clear. It is generally considered inadequate to establish families from the evidence of just one organ.

The palynological evidence, taken on its own, does show evidence of some large groups of extinct angiosperms. Muller (1985) summarized the widely held view that the normapolles and triprojectate pollen forms are from plants of two distinct and extinct families. Also, small, clavate pollen from the Cretaceous may be from a third, even older, group of family rank. Several palaeobotanists are working hard to show details of other organs that were associated with these fossils: other extinct angiosperm families may soon be described. In this present work the palynological record of these three most ancient groups is recognized within the Juglandales (see 'Comments', Juglandaceae), the Santalales (see 'Comments', Loranthaceae) and the Laurales (see 'Comments', Chloranthaceae) respectively.

Of course, these limitations in palaeobotanical taxonomy are highlighted when a compendium such as this is prepared. If nothing else, a major result of this chapter is to demonstrate these theoretical constraints in a much more practical way using the fossils themselves. The primary data presented here, although authoritative, give a very clouded view of angiosperm evolution, which can only be partially improved by our 'Comments' additions. This may encourage more consideration of the high-ranking extinct angiosperm taxa that are, so far, largely not established (but see Priscaceae Retallack and Dilcher, 1981; Kvaček, 1992), and which might guide a scientific re-assessment of Cretaceous and early Tertiary angiosperm evolution. We hope that this chapter will initiate production of a more intelligent and informed set of data in the (hopefully) computerised third edition of *The Fossil Record*.

STANDARDS AND PRINCIPLES FOR OUR COMMENTS

For each of the family listings we have extracted the earliest occurrence of the family which is cited in the PFR (Plant Fossil Record) database (Holmes *et al.*, 1991). Where several examples are cited with a broad age range (e.g. Tertiary) but one is more focused (e.g. Eocene), we have taken that more accurate example. This means that an earlier record may have been excluded, but this cannot be corrected without a complete revision of all the stratigraphical citations in PFR. We have excluded tentative family assignments, except where these were the only examples cited in PFR. One of us (MEC) has commented on the megafossil record (following each PFR entry) and MCB has commented on the pollen records (Muller, 1981) after each Muller entry. The use of the words 'pending' and 'rejected' is explained by Muller (1981).

Comments on megafossils are given only when they are based on additional information not included within the entire PFR, and they reflect only personal knowledge rather than a substantive search of the entire literature on fossil angiosperms. While such a detailed search would have been preferable, it is impossible within the time-scale of two months available for this project, ever since the PFR database became available for use. Furthermore, comments on all the specific PFR citations are impossible at this time as they would require major world travel; consultation with many specialists world-wide and numerous searches of museum collections for old and forgotten specimens, some of which might well have deteriorated or even have been lost. The comments are restricted to the pre-Pleistocene fossil record. We hope that this publication will stimulate the necessary revisions both of the PFR database and of the evidence cited in our comments.

Comments on megafossils follow the philosophy of Collinson (1986b) whereby an isolated organ can be assigned to a modern genus, and hence family, provided that it exhibits a suite of diagnostic characters which fall within the range of variation of the organ in the modern representatives of the family and which are unique to that family. Obviously, additional information from multiple organs, especially in organic connection, strengthens the proposed relationship (Collinson, 1990). It is widely accepted that many early determinations of leaves to modern genera and families were incorrect (see discussion in Taylor, 1990, p. 281). In the Cretaceous, recent studies of both leaves and woods have shown that these organs often possess some characteristics of several modern families or have a generalized form perhaps for an order or subclass (e.g. Page, 1981; Crabtree, 1987; Wheeler *et al.*, 1987; Upchurch and Dilcher, 1990). For these reasons, we have not accepted records from older literature unless they have been accepted as part of a more recent study.

F. ACANTHACEAE de Jussieu, 1789

PFR: No record.
Comments: A single seed from the uppermost Eocene of England was identified to modern *Acanthus* by Reid and Chandler (1926). The original has deteriorated and no new specimens have been found. This record should be considered as unconfirmed. We are not aware of any well-substantiated megafossil record for the family.
Muller First: *Multiareolites formosus* (form taxon) Pares Regali *et al.*, 1974a, Lower Miocene, Brazil. *Multimarginites vanderhammeni* (form taxon) Germeraad *et al.*, 1968, Lower Miocene, northern South America.
Comments: Unchallenged. See also list in Taylor (1990) for possible records from the Middle Eocene to Lower Oligocene of North America.

F. ACERACEAE de Jussieu, 1789

PFR First: *Negundoides acutifolia* Lesquereux, 1868, Leaf. Cretaceous.
Comments: The modern genus *Acer* is recorded from the latest Palaeocene onwards and the related *'Acer' arcticum* complex occurs in the upper Maastrichtian (Wolfe and Tanai, 1987). These records are based on fruits and associated foliage.
Muller First: *Acer campestre* type Piel, 1971. Oligocene, British Columbia, Canada.
Comments: Unchallenged.

F. ACTINIDIACEAE Hutchinson, 1926

PFR First: *Actinidioxylon princeps* (R. Ludwig) W. R. Müller-Stoll and E. Mädel-Angeliewa, 18 August 1969. Wood. Tertiary, Pliocene, Germany: Westerwald, Montbaur, Dernbach.
Comments: Seeds like those of modern *Saurauia* were recorded from the Maastrichtian onwards in Europe (Knobloch and Mai, 1986). Seeds like those of modern *Actinidia* are known from the Upper Eocene onwards in Europe (Friis, 1985a). Leaves like those of *Saurauia* occur in the Middle Eocene of North America (Taylor, 1990).
Pollen: No record.

F. AGAVACEAE Endlicher, 1841

PFR First: *Asteliaephyllum italicum* Squinabol, 1892. Leaf. Tertiary, Italy: Santa Guistina.
Comments: Tidwell and Parker (1990) described stems of *Protoyucca* (an arborescent monocotyledon with secondary growth), but although they noted similarity to modern *Yucca*, they did not make a family assignment for the fossils. We know of no well-substantiated megafossil record of the family (Daghlian, 1981).
Muller First: *Phormium* Couper, 1960. Upper Eocene, New Zealand.
Comments: Too early; no supporting evidence available of this modern genus in the Eocene.

F. AKANIACEAE Stapf, 1912

PFR: No record.
Comments: Romero and Hickey (1976) described a single leaf impression from the Palaeocene of Argentina as *Akania americana* Romero and Hickey, which they included in the family Akaniaceae, although expressing caution in the discussion regarding the determination to the family based on such limited material.
Pollen: No record.

F. ALANGIACEAE de Candolle, 1828

PFR First: *Alangioxylon scalariforme* N. Awasthi, August 1969. Wood. Middle Tertiary. India, Madras, South Arcot District, Murattandichavadi.
Comments: Leaves and fruits like those of modern *Alangium* are recorded from the Middle Eocene of North America (Taylor, 1990) and the Lower (Chandler, 1964), Middle (Mai, 1976), and Upper (Mai and Walther, 1985) Eocene of Europe.
Muller First: *Alangiopollis eocaenicus* (form taxon). Krutzsch 1969a. Lower Eocene, Germany.
Comments: Unchallenged.

F. ALISMATACEAE Ventenat, 1799

PFR First: *Alismaphyllum victor-masoni* (Ward) E. W. Berry, 1911. Leaf? Lower Cretaceous, USA, Virginia, White House Bluff.
Comments: Erwin and Stockey (1989) assigned a permineralized petiole from the Middle Eocene of Canada to the family. Collinson (1983) documented fruits from the uppermost Eocene/Lower Oligocene of England, UK, but the majority of the records are Oligocene or younger (Mai, 1985a; Collinson, 1988a; Erwin and Stockey, 1989).
Muller First: Pending.

F. AMARANTHACEAE de Jussieu, 1789

PFR: No record.
Comments: Negru (1979) recorded seeds like those of modern *Amaranthus* in the uppermost Miocene (Pontian) of Moldavia, former USSR. Friis (pers. comm., 1991) has unpublished seeds from the Santonian/Campanian of Sweden.
Muller First: see Chenopodiaceae.
Comments: It is not possible to distinguish the pollen of these two families.

F. ANACARDIACEAE Lindley, 1830

PFR First: *Edenoxylon parviareolatum* Kruse, July 1954. Wood. Tertiary, L. Eocene. USA, Wyoming, Farson.
Comments: Fruits are diverse in the Lower Eocene of England (Chandler, 1964) and fruits, leaves and woods are reported from the Middle Eocene of North America (Taylor,

1990). Mai (1987a) lists an Upper Palaeocene fruit, and Knobloch and Mai (1986, p. 170) refer to Upper Cretaceous foliage from several areas, but all these pre-Eocene records are in need of revision. Romero (1986a) cites one Coniacian record followed by a Lower Tertiary diversification of the family in South America.
Muller First: *Rhus* type = *Tricolporopollenites cingulum* (form taxon). Gruas-Cavagnetto, 1976a, Palaeocene, France.
Comments: Possible.

F. ANNONACEAE de Jussieu, 1789

PFR First: *Xylopiaecarpum eocaenicum* Râsky, 1956. Fruit. Tertiary, Lower Eocene, Hungary: Tokod.
Comments: Seeds are widespread in the Eocene; the earliest record is from the Maastrichtian of Nigeria and one species is recorded in the Middle Palaeocene of Pakistan (Tiffney and McClammer, pp. 13–20, *in* Collinson, 1988c). Leaves are reported in the Middle Eocene of North America (Taylor, 1990).
Muller First: *Foveomorphomonocolpites humbertoides* (form taxon). Sole de Porta, 1971: Maastrichtian, Colombia.
Comments: Unchallenged.

F. APIACEAE Lindley, 1846

PFR First: *Peucedanites spectabilis* Heer, post-October 1859. Fruit. Tertiary: Miocene, Switzerland: Oeningen.
PFR First: *Pimpinellites zizioides* Unger, 1839. Fruit. Tertiary, Miocene, former Yugoslavia: Croatia: Radoboj. *Umbelliferospermum latahense* E. W. Berry, 18 April 1929. Fruit. Tertiary, Miocene, USA, Washington, Spokane brickyard.
Comments: Gregor (1982) described fruits named *Umbelliferopsis* from the European Middle Miocene; most other records are Pliocene (Mai, 1985a; Szafer, 1954).
Muller First: *Hydrocotyle* type. Gruas-Cavagnetto and Cerceau-Larrival 1978. Lower Eocene. France. *Bupleurum* type. Gruas-Cavagnetto and Cerceau-Larrival, 1978. Lower Eocene, France.
Comments: Unchallenged.

F. APOCYNACEAE de Jussieu, 1789

PFR First: *Aspidospermoxylon uniseriatum* Kruse, 1954. Wood. Tertiary, L Eocene. USA, Wyoming, Hays Ranch, 16 miles east of Farson.
Comments: The fossil record for this family is sporadic, but we see no reason to reject fruits and seeds recorded in the Lower Eocene of England, UK (Chandler, 1964) and foliage in the Middle Eocene of Germany (Wilde, 1989). Cretaceous woods included in the family by Taylor (1990) were described only as 'woods with characteristics of Apocynaceae' by Wheeler *et al.* (1987) who also noted similarity to Simaroubaceae and other families.
Muller First: *Alyxia* type Muller, 1968, Palaeocene, Borneo. *Diporites iskaszentgyorgii* (form taxon). Gruas-Cavagnetto, 1978, Lower Eocene, France.
Comments: Well-supported by other records of pollen and megafossils.

F. APONOGETONACEAE Agardh, 1858

PFR: No record.
Comments: Zhilin (1974a,b, 1989) and Pneva (1988) have described leaves like those of modern *Aponogeton* from the Oligocene of the former USSR.
Pollen: No record.

F. AQUIFOLIACEAE Bartling, 1830

PFR First: *Ilicoxylon* Greguss, 1943. Wood. Tertiary, Miocene, Hungary: Tokaj-Eperjesi Mountains.
Comments: Fruits like those of modern *Ilex* are recorded from the Maastrichtian onwards in Europe (Knobloch and Mai, 1986; Mai, 1987a) although there is only one doubtful record from the well-known Tertiary floras of England, UK (Chandler, 1964). Two Eocene and one Upper Palaeocene records of *Ilex* leaves are cited by Taylor (1990). The latter may need revision.
Muller First: *Ilexpollenites* (form genus). H. A. Martin, 1977. Turonian, Australia.
Comments: Well-supported for family.

F. ARACEAE de Jussieu, 1789

PFR First: *Limnophyllum primaevum* Hosius and Marck, April 1880. Leaf. Cretaceous, Senonian, Germany, Westfalen.
Comments: Fruits and seeds occur in the Middle Eocene of North America and seeds have a widespread occurrence in younger strata (Cevallos-Ferriz and Stockey, 1988a). An Upper Palaeocene seed form was listed by Collinson (1986a). Leaves are recorded from the Middle Eocene of Europe (Wilde, 1989) and Lower and Middle Eocene of North America (Taylor, 1990).
Muller First: *Spathiphyllum* type: Graham, 1976. Upper Miocene, Mexico. Leopold, 1969; Miocene, Palau, West Pacific.
Comments: Unchallenged.

F. ARALIACEAE de Jussieu, 1789

PFR First: *Araliopsoides breviloba* Berry, 1916. Leaf. Upper Cretaceous, USA: Maryland, Cecil County, Bull Mountain.
Comments: Knobloch and Mai (1986) recorded fruits like those of modern *Aralia* and *Acanthopanax* in the Maastrichtian of Europe, while fruits like modern *Aralia*, *Pentapanax* and *Schefflera* are recorded from the Upper Eocene onwards (Mai and Walther, 1985; Palamarev, pp. 97–106, *in* Collinson, 1988c). Araliaceae leaves (especially like modern *Dendropanax* and *Oreopanax*) are listed from the Palaeocene and Middle Eocene onwards in North America (Taylor, 1990). Leaves from the Middle Eocene of Europe, reported by Wilde (1989), were not determined to modern genera.
Muller First: *Tricolporopollenites armatus* (form taxon). Gruas-Cavagnetto and Bui, 1976. Upper Palaeocene, France.
Comments: Unchallenged.

F. ARECACEAE Schultz-Schultzenstein, 1832

PFR First: *Eolirion primigenium* Schenk, June, 1869. Leaf. Cretaceous, Urgonian, Austria.
Comments: Pinnate palm foliage is recorded from the Lower Campanian of North America (Crabtree, 1987), and Daghlian (1981) refers to possible Santonian examples. Daghlian (1981) also reviews subsequent records of fruits, stems and foliage. Palm flowers, fruits and leaves are recorded in association in the Middle Eocene of Germany (Schaarschmidt and Wilde, 1986). The record of fruits like those of modern *Nypa* (Nypoideae; Nypaceae of some authors) extends to the Lower Palaeocene in Africa, although these fruits are more widespread from the Early Eocene onwards (see review by Gee, pp. 315–19, *in* Knobloch and Kvacek, 1990).
Muller First: Nypoideae *Nypa* = *Spinizonocolpites baculatus*

(form taxon). Germeraad *et al.*, 1968, etc. South America. Jardine and Magloire, 1963, etc. Africa, Venkatachala, 1974, India, Muller, 1968, Borneo, Maastrichtian.
Comments: See review by Gee (pp. 315–19, *in* Knobloch and Kvacek, 1990). Maastrichtian, South America, Africa and Malaysia.
Muller First: Arecoideae *Retimonocolpites pluribaculatus* (form taxon) Salard-Cheboldaeff, 1978, Maastrichtian, Africa. Cocoideae (Cronquist: Cocosoideae) *Trichotomosulcites antiquus* (form taxon) Krutzsch and Lenk, 1969, Germany, Maastrichtian. Lepidocaryoideae *Monocolpites franciscoi* (form taxon) Muller, 1970, Palaeocene, South America. *Dicolpopollis malesianus* and *D. elegans* (form taxa) Muller, 1979, Palaeocene, Borneo. Caryotoideae *Arenga* Muller, 1979, Lower Miocene, Borneo.
Comments: All unchallenged.

F. ARISTOLOCHIACEAE de Jussieu, 1789

PFR First: *Aristolochioxylon prakashii* Kulkarni and Patil, 1977. Wood. Lower Tertiary, Eocene? India: Nawargon, Wardha District, Maharashtra.
Comments: Leaves assigned to modern *Aristolochia* are listed by Taylor (1990) from North America and reported in Takhtajan (1974) from Africa; the earliest are Middle Eocene, but all may need revision. Wolfe (pers. comm., 1991) considers that Middle Eocene examples from North America are probably reliable.
Pollen: No record.

F. ASCLEPIADACEAE Brown, 1810

PFR First: *Asclepiadites laterita* MacGinitie, 15 November 1941. Leaf. Tertiary, Eocene, USA: California, Nevada County, You Bet (*sic.*).
Comments: Seeds from the uppermost Eocene of England were assigned to modern *Phyllanthera* and *Tylophora* by Reid and Chandler (1926). The former was represented by six specimens, the latter by a single specimen. The originals are in poor condition, and no new material has been found. Palamarev (1968) assigned a Miocene seed from Bulgaria to the family. We consider all these records to be unconfirmed, and we know of no well-substantiated megafossil record for this family.
Muller First: *Polyporotetradites laevigatus* (form taxon) Salard-Cheboldaeff, 1978, Oligocene, Cameroon.
Comments: Unchallenged.

F. ASTERACEAE Dumortier, 1822

PFR First: *Hieracites salyorum* Saporta, 1861. Leaf: Tertiary: Eocene, France: Aix-en-Provence.
Comments: A Lower Eocene fruit described by Chandler (1978, p. 30) was tentatively compared with this family. The fruit record seems otherwise to be restricted to the Miocene onwards in Europe (e.g. Mai, 1985a; Szafer, 1954).
Muller First: *Tricolporopollenites microechinatus* (form taxon) Hochuli, 1978, Oligocene, Austria. *Tubulifloridites antipodica* (form taxon) Kemp and Harris, 1975, Oligocene, Indian Ocean. *Tubiflorae* type Krutzsch, 1970d, Oligocene, Germany; Leopold and Macginitie, 1972, Oligocene, USA.
Comments: More ultrastructural examination required.

F. BALANOPHORACEAE L. C. and A. Richard, 1822

PFR: No record.
Comments: We known of no well-substantiated megafossil record for this family.
Muller First: *Balanopollis minutus* Salard-Cheboldaeff, 1978. Lower Miocene, Cameroon.
Comments: Unchallenged.

F. BERBERIDACEAE de Jussieu, 1789

PFR First: *Winchellia triphylla* Lesquereux, October 1893. Leaf. Cretaceous, USA: Yellowstone River, near mouth of Powder River.
Comments: Leaves like those of modern *Mahonia* are listed from the Middle Eocene and commonly from the Oligocene of North America (Taylor, 1990). Lower and Upper Oligocene examples (the former *Mahonia*, the latter *Mahonia* and *Berberis*) were accepted by Wolfe and Schorn (1989) and Manchester and Meyer (1987). The single Maastrichtian record, of a leaf assigned to the family listed by Taylor (1990), may need reassessment. Wolfe (pers. comm., 1991) considers the Middle Eocene *Mahonia* as the earliest reliable records. Tertiary records (Oligocene onwards) from the former USSR are documented in Takhtajan (1974). Fruits like those of modern *Achlys* were described from the Lower Palaeocene of Europe by Mai (1987a).
Muller First: Records rejected; no proven evidence.

F. BETULACEAE Gray, 1821

PFR First: *Betuloxylon oligocenicum* Kaiser, 1880. Stem. Tertiary, Oligocene; *Coryloxylon nemejcii* U. Prakash *et al.*, Mai, 1971. Wood. Tertiary: Oligocene, Czechoslovakia: northern Bohemia, Doupovske Mountains, Kadan.
Comments: Crane (chapter 6, vol. 2, *in* Crane and Blackmore, 1989) reviewed the fossil record. He recognized modern *Alnus* and modern *Betula* (Betuleae) as well defined by the Middle Eocene, based upon multiple organ evidence. The earliest reproductive structures of *Alnus* are late Palaeocene in age and foliage from the Maastrichtian onwards may represent Betuleae. Coryleae are represented in the Upper Palaeocene by nuts like those of *Corylus*, and by the extinct genus *Palaeocarpinus* with associated foliage.
Muller First: *Alnipollenites eminens* (form taxon) Miki, 1977, Santonian, Japan. *Betulaceoipollenites* (form genus) Jarzen and Norris, 1975, Santonian, Canada.
Comments: May be too early for the family due to confusion with extinct groups yielding porate pollen.

F. BIGNONIACEAE de Jussieu, 1789

PFR First: *Bignonicapsula formosa* E. W. Berry, 1930. Tertiary, L Eocene, USA: Tennessee: Madison County; Denmark.
Comments: One leaf, assigned to modern *Chilopsis*, from the Upper Oligocene of North America is listed by Taylor (1990), but this was rejected as indeterminate by Wolfe and Schorn (1989). However, Wolfe and Schorn (1989) did list a leaf assigned to modern *Catalpa* from the Upper Oligocene of North America (for details see Wolfe and Schorn, 1990). Gregor (1982) lists *Catalpa* leaves from the European Miocene. Seeds from the Uppermost Eocene of England were assigned to modern *Catalpa*, *Radermachera* and *Incarvillea*, each based on a single specimen. The originals are in poor condition and no new material has been found (Reid and Chandler, 1926). We consider these seed records to be unconfirmed.
Muller First: *Albertipollenites araneosus* (form taxon) Frederiksen, 1973, Middle Eocene, south-east USA.
Comments: Unchallenged.

F. BIXACEAE Link, 1831

PFR First: *Scolopioidea palaeocenica* M. Langeron, 1899. Leaf. Tertiary: Eocene, France: Sêzanne.
Comments: We know of no well-substantiated megafossil record for this family.
Pollen: No record.

F. BOMBACACEAE Kunth, 1822

PFR First: *Bombacites formosus* E. W. Berry, 1916. Leaf. Tertiary: Lower Eocene, USA: Tennessee, Henry County, Puryear.
Comments: Many of the original identifications made by Berry are incorrect (see discussion in Taylor, 1990, p. 281). No leaves are listed for this family from North America by Taylor (1990). The Maastrichtian wood, named *Parabombacaceoxylon*, listed by Taylor (1990) as a member of this family, was specifically stated by Wheeler *et al.* (1987, p. 90) to be 'not intended to imply affinities with only the Bombacaceae' in spite of the name. Supposed *Bombax* leaves from the Tertiary of Australia were rejected by Hill (1988). Woods assigned to *Bombacoxylon* occur in the Oligocene onwards of Africa and France. These may have affinities with Bombacaceae (Wheeler *et al.*, 1987, Boureau *et al.*, 1983).
Muller First: *Bombax* type Wolfe, 1975. Maastrichtian, south-east USA.
Comments: Possible for family.

F. BORAGINACEAE de Jussieu, 1789

PFR First: *Davisella ehretioides* E. M. Reid and M. E. J. Chandler, 25 November 1933. Fruit. Tertiary: Eocene, UK: England, Middlesex: Harefield.
Comments: This Lower Eocene fruit is now included in modern *Ehretia* (see Chandler, 1964). Gregor (1978) documented fruits like those of modern *Argusia* from the Oligocene to Pliocene of Europe and the former USSR. Boraginaceae are diverse in the Late Miocene of North America (Thomasson, 1987a) and have been recorded in the Neogene and possible Palaeocene of India (Mathur and Mathur, 1985).
Muller First: *Tournefortia bicolor* type Graham and Jarzen, 1969, Oligocene, Puerto Rico.
Comments: Unchallenged.

F. BRASSICACEAE Burnett, 1835

PFR First: *Isatides microcarpa* Saporta, 1889. Fruit. Tertiary: Eocene, France: Aix-en-Provence.
Comments: Fossil fruits from the Oligocene of Wyoming were considered equivocal by Taylor (1990). Dorofeev (1957) and Mädler (1939) described Pliocene seeds like those of modern *Bunias*.
Muller First: Cruciferae. Naud and Suc, 1975. Upper Miocene. France.
Comments: Possible.

F. BROMELIACEAE de Jussieu, 1789

PFR First: *Bromelites dolinskii* J. Schmalhausen, 1883. Stem. Tertiary, Eocene, former USSR near Kiev.
Comments: One possible leaf from the mid-Tertiary of South America is listed by Taylor (1990). We consider all these records as unconfirmed.
Pollen: No record.
Comments: Graham (1987) lists pollen of *Tillandsia* type from the Upper Eocene of Central America.

F. BRUNELLIACEAE Engler, *in* Engler and Prantl, 1897

PFR: No record.
Comments: We know of no claims for a megafossil record for this family.
Muller First: Pending.

F. BURMANNIACEAE Blume, 1827

PFR: No record.
Comments: We know of no claims for a megafossil record for this family.
Muller First: Pollen record rejected.

F. BURSERACEAE Kunth, 1824

PFR First: *Bursericarpum angulatum* E. Reid and M. E. J. Chandler, 1933, Fruit, Tertiary: Eocene, England: Kent: Sheppey. *Burserites fayettensis* E. W. Berry, 1924. Leaf. Tertiary: Eocene, Venezuela: Betijoque, State of Trujillo. *Palaeobursera bognorensis* M. E. J. Chandler, 1961. Fruit. Tertiary: Eocene, England: Sussex: Bognor. *Protocommiphora europaea* E. M. Reid and M. E. J. Chandler, 25 November 1933. Fruit. Tertiary: Eocene, England: Kent: Sheppey.
Comments: These fruit records reflect the diversity of the family in the Lower Eocene of England. Leaves and fruits occur in the Middle Eocene (one tentative assignment in the Upper Palaeocene) of North America listed by Taylor (1990).
Muller First: Pending.

F. BUTOMACEAE Richard, 1815

PFR First: *Butomites cretaceus* Velenovsky, 1889. Leaf. Upper Cretaceous, Czechoslovakia: Bohemia: Vidovle prope Jinonice.
Comments: We know of no well-substantiated leaf fossils for this family. Seeds like those of modern *Butomus* occur in the Oligocene onwards in Europe (Mai, 1985a). Kovach and Dilcher (1988) noted suggestions of similarity between Cretaceous seed-like cuticles named *Costatheca* and *Butomaceae* seeds, but they rejected an assignment to the family without futher comparative surveys and in the absence of an intervening fossil record. Seeds of this form, borne on wetland plants, have high fossilization and recovery potential, so intervening examples should not have passed unreported.
Pollen: No record.

F. BUXACEAE Dumortier, 1822

PFR: No record.
Comments: Mai and Walther (1985) described seeds like those of modern *Pachysandra* (as *P. ascidiiformis* Mai) from the Upper Eocene as the first megafossil record in Europe. Kvacek *et al.* (1982) reviewed the fossil record of leaves and fruits like those of modern *Buxus* from the Early Miocene onwards in Eurasia. Uemura (1979) described leaves like those of modern *Buxus* from the Miocene of Japan. Taylor (1990) lists no megafossils of the family in North America. Flowers very similar to those in modern Buxaceae have been recorded in the Albian of North America (Drinnan *et al.*, 1991).
Muller First: *Erdtmanipollis* (form genus) Jarzen and Norris, 1975. Campanian, Canada. *Grootipollis calvoerdensis* (form taxon) Krutzsch and Lenk, 1969. Campanian–Danian, Germany.
Comments: Unchallenged.

F. CABOMBACEAE Richard, 1828

PFR First: *Braseniella nymphaeoides* (P. I. Dorofeev) P. I. Dorofeev, April–June 1973. Seed. Tertiary: Oligocene, former USSR: Belojarka, bank of Tavdy River: Oblast: Sverdlovskaya and Tyumenskaya. *Dusembaya turgaica* (P. I. Dorofeev) P. I. Dorofeev, April–June 1973. Seed. Tertiary: Oligocene, former USSR: Kazakhstan: Ojusembaj.
Comments: The fossil history is reviewed by Mai (1985a), Cevallos-Ferriz and Stockey (1989), and Collinson (1980). Several fossil seed forms (including those cited above) are intermediate in character between genera now segregated in Cabombaceae and Nymphaeaceae *sensu stricto*. However, seeds identical with those of modern *Brasenia* (Cabombaceae) occur in the Upper Eocene of England, UK (Collinson, 1980 and in prep.).
Pollen: No record.

F. CACTACEAE de Jussieu, 1789

PFR First: *Eopuntia douglassii* Chaney, 7 November 1944. Tertiary, Middle Eocene.
Comments: The material of *Eopuntia* was re-examined by MacGinitie (1969, p. 91) who stated that 'the objects cannot be assigned to the Cactaceae'. They were thought to be possible monocotyledonous tubers. We know of no well-substantiated megafossil record for this family.
Pollen: No record.

F. CAESALPINIACEAE R. Brown, *in* Flinders, 1814

PFR First: *Gleditschiacanthus alsaticus* C. Lakowitz, 1895. Tertiary, Oligocene. France, Alsace, Brunstatt. *Afzelioxylon furoni* J. Koeniguer, 1973, Devonian, Chad: L'oasis de Kirdimi.
Comments: In their revisions of fossil legumes, Herendeen and Dilcher (1990b, 1991) document pods and leaves of several genera in this family (as Caesalpinioideae) from the Eocene, especially Middle Eocene, of North America. They note that most records of the family in the older literature still await revision.
Muller First: *Sindorapollis* (form genus) Krutzsch, 1969b, Maastrichtian, Siberia, former USSR.
Comments: Unchallenged.

F. CALLITRICHACEAE Link, 1821

PFR: No record.
Comments: Fruits are recorded from the Pliocene in Europe (Mai, 1985a).
Muller First: Pending.

F. CALYCANTHACEAE Lindley, 1819

PFR: No record.
Comments: Fruits like those of modern *Calycanthus* were recorded by Mai (1987b) from the Middle Miocene of Germany.
Pollen: No record.

F. CAMPANULACEAE de Jussieu, 1789

PFR: No record.
Comments: Seeds are recorded in the Upper Miocene of Poland (Łancucka-Środoniowa, 1979).
Muller First: Pending.

F. CANELLACEAE Martius, 1832

PFR: No record.
Comments: We know of no claims for a megafossil record of this family.
Muller First: Pending.

F. CANNABACEAE Endlicher, 1837

PFR: No record.
Comments: The fossil record was reviewed by Collinson (Chapter 18, vol. 2, *in* Crane and Blackmore, 1989) who accepted fruits like those of modern *Humulus* (and a related extinct genus) from the Oligocene of the former USSR and Bulgaria.
Pollen: No record.

F. CANNACEAE de Jussieu, 1789

PFR First: *Cannaites intertrappea* B. S. Trivedi and C. L. Verma, 1971. Root. Tertiary: Eocene, India: Mohgaon Kalan, Chhindwara.
Comments: Leaves previously assigned to Cannaceae were considered to be 'not well understood' by Daghlian (1981).
Pollen: No record.

F. CAPPARACEAE de Jussieu, 1789

PFR First: *Capparites cynphylloides*, 1919. Leaf. Upper Cretaceous, USA: Alabama, Fayette County, Shirleys Mill.
Comments: Fruits and seeds were recorded by Chandler (1964) in the Lower Eocene of England, UK. This is an isolated record for the family, but we know of no reason to reject it. Taylor (1990) noted a wood assigned to the family from the Middle Eocene of North America. We know of no well-substantiated leaf fossils for the family.
Muller First: van Campo, 1976, Upper Miocene, Spain.
Comments: Unchallenged.

F. CAPRIFOLIACEAE de Jussieu, 1789

PFR: No record.
Comments: Seeds identical with those of modern *Sambucus* occur from the Early Eocene onwards (Chandler, 1964) and fruits of *Abelia* occur in the uppermost Eocene/Lower, Oligocene (Crane, *in* Collinson, 1988c, pp. 21–30) of England, UK. *Sambucus* also occurs in the Middle Eocene of Messel, Germany (Collinson, 1988b). Several Palaeocene records for modern *Viburnum* listed by Taylor (1990) may require revision. According to Hickey (1977) some represent 'only plausible generic assignments retained pending further study'.
Muller First: *Tricolporopollenites viburnoides* (form taxon) Gruas-Cavagnetto, 1978. Middle Eocene, France.
Comments: The associated megafossil record does support the family identity.

F. CARYOCARACEAE Szyszylowicz, *in* Engler and Prantl, 1893

PFR: No record.
Comments: We know of no claims for megafossils of this family.
Muller First: *Retisyncolporites angularis* (form taxon). Gonzalez-Guzman, 1967; Muller, 1970, Middle Eocene, Venezuela.
Comments: Very characteristic parasyncolpate pollen.

F. CARYOPHYLLACEAE de Jussieu, 1789

PFR First: *Hantsia pulchra* (M. E. J. Chandler) M. E. J. Chandler, July 1960. Seed. Tertiary: Eocene: Bartonian, England, UK: Hampshire: Hordle.

Comments: There are several records of this and another species in the Middle and lower Upper Eocene of England, UK (Chandler, 1964). These are isolated records, but we know of no reason to reject them. Van der Burgh (1987) records seeds like those of modern *Stellaria* from the Upper Miocene of Europe.
Muller First: *Caryophyllidites polyoratus* (form taxon) Muller, 1970. Oligocene. New Zealand.
Comments: Unchallenged.

F. CASUARINACEAE Brown, *in* Flinders, 1814

PFR First: *Casuaroxylon anglica* Goeppert and Stache, 15 May 1855. Wood? No age is given for this genus in PFR, there is one Upper Triassic taxon assigned tentatively to the Casuarinaceae.
Comments: The family is well documented in Australasia by reproductive structures and leafy stems. The earliest examples belong to modern *Gymnostoma*, and this is common and well defined in the Middle Eocene onwards (Scriven and Christophel, *in* Douglas and Christophel, 1990, pp. 137–47); the record from Mt. Hotham cited by these authors is probably Palaeocene (Greenwood, pers. comm., 1991). The fossil record is also summarized by Johnson and Wilson (Chapter 9, vol. 2, *in* Crane and Blackmore, 1989) who note one megafossil of *Gymnostoma* in Patagonia. This is stated (as *Casuarina*) by Romero (1986a) to be Palaeocene.
Muller First: *Haloragacidites* (= *Triorites*) *harrisii* (form taxon) Mildenhall, 1980. Lower Palaeocene. New Zealand.
Comments: Well supported in the Southern Hemisphere.

F. CELASTRACEAE Brown, *in* Flinders, 1814

PFR First: *Sugoia opposita* Samylina, 1976. Leaf. Cretaceous, former USSR: Magadan District, Omsukchan.
Comments: Fruits and seeds are recorded in the Lower Eocene of England, UK (Chandler, 1964) and leaves are listed from the Middle Eocene of North America (Taylor, 1990).
Muller First: *Microtropis* type. Lobreau-Callen and Caratini, 1973. Lower Oligocene. France. *Peritassa* type. Lobreau-Callen and Caratini, 1973, Oligocene, France. *Campylostemon* type = *Triporotetradites campylostemonoides* (form taxon) Salard-Cheboldaeff, 1974b, Oligocene, Cameroon. *Hippocratea volubilis* type = *Polyadopollenites macroreticulatus* (form taxon) Salard-Cheboldaeff, 1973. Oligocene, Cameroon.
Comments: Unchallenged.

F. CENTROLEPIDACEAE Endlicher, 1836

PFR First: *Podostachys bureauana* Marion, 1872, Tertiary, France, Ronzoon.
Comments: Unconfirmed. We know of no well-substantiated megafossil record of this family.
Muller First: *Milfordia incerta* (form taxon) Krutzsch, 1970a. Palaeocene Germany.
Comments: Possible; indistinguishable from Restionaceae.

F. CERATOPHYLLACEAE Gray, 1821

PFR: No record.
Comments: Herendeen *et al.* (1990) review the fossil record in North America and document the earliest fruits *Ceratophyllum furcatispinum* Herendeen *et al.* in the Upper Palaeocene of North America. Mai (1985a) reviewed the European record which extends from the Oligocene onwards.
Pollen: No record.

F. CERCIDIPHYLLACEAE Engler, 1909

PFR First: *Cercidiphylloxylon kadanense* U. Prakash *et al.* (*'kadanende'*) May, 1971. Wood. Tertiary, Oligocene. Czechoslovakia: north Bohemia: Kadan. Doupovske Hory Mountains.
Comments: Extinct partially reconstructed whole plants (the *Joffrea* and *Nyssidium* 'plants' being most thoroughly reconstructed), are widespread in uppermost Cretaceous and Palaeocene floras of the mid to high latitudes of the Northern Hemisphere (see review in Friis and Crane, *in* Crane and Blackmore, 1989). These were included in Cercidiphyllaceae by Crane and Stockey (1985, p. 363; 1986) and Crane *et al.* (1990) and they are more closely related to *Cercidiphyllum* than to any other extant genus (Crane and Stockey, 1985, Friis and Crane, *in* Crane and Blackmore, 1989). Fossil plants similar to modern *Cercidiphyllum* do not occur until the Oligocene (Crane and Stockey, 1985).
Muller First: *Cercidiphyllites brevicolpatus* (form taxon). Jarzen and Norris, 1975, Campanian, Canada.
Comments: Unchallenged.

F. CHENOPODIACEAE Ventenat, 1799

PFR First: *Salicornites massalongoi* Principi, 1926. Tertiary, Oligocene. Italy, Chiavon.
Comments: Seeds like those of modern *Chenopodium* were described from the Middle and Upper Miocene of Europe (Gregor, 1982; Van der Burgh, 1987).
Muller First: *Polyporina cribraria* (form taxon) Srivastava, 1969b, Maastrichtian, Canada.
Comments: Difficult to distinguish from *Liquidambar* type and some *Amaranthaceae* pollen and some dinoflagellate cysts.

F. CHLORANTHACEAE Brown ex Lindley, 1821

PFR: No record.
Comments: Androecia of Chloranthaceae were described by Crane *et al.* (1989) from the Upper Cretaceous (Upper Santonian/Lower Campanian) of Sweden. These are very similar to modern *Chloranthus* but contain spiraperturate pollen not found in the family today although similar in exine structure to grains of modern *Sarcandra* and some species of *Chloranthus*. Crane *et al.* (1989) also describe a single Lower Cretaceous chloranthoid androecium which contains tricoplate pollen, but they do not assign it to Chloranthaceae. They note that several Lower Cretaceous dispersed pollen forms (especially *Clavatipollenites* and *Asteropollis*), are considered to be very similar to grains of modern *Chloranthaceae* (*Ascarina* and *Hedyosmum*, respectively) and that abundance of these pollen indicates wind-pollinated parent plants; in contrast to the two fossil androecia which suggest insect pollination. Leaf fossils from the Lower Cretaceous resemble Chloranthaceae but are not sufficiently diagnostic to assign to the family (Crane *et al.*, 1989). Crepet *et al.* (1991) draw attention to differences between Lower Cretaceous chloranthoid flowers and those of modern Chloranthaceae.
Muller First: *Clavatipollenites–Ascarina* complex, in Muller, 1981, Aptian, Central Africa, South America, North and Central America, Europe, Australia.
Comments: Well-supported records may be assignable to an extinct and related family.

F. CHRYSOBALANACEAE Brown, *in* Tuckey, 1818

PFR: No record.
Comments: Fruits tentatively assigned to modern *Parinari* were recorded in the Pliocene of Colombia (Wijninga and Kuhry, 1990).
Pollen: No record.

F. CISTACEAE de Jussieu, 1789

PFR First: *Cistinocarpum roemeri* Conwentz, 1886. Fruit. Lower Tertiary, Germany: West Prussia.
Comments: Gottwald (1992) assigned an Upper Eocene wood to this family.
Muller First: *Cistacearumpollenites* (form genus) Konzalova, 1976a. Lower Miocene, Czechoslovakia.
Comments: Possible.

F. CLETHRACEAE Klotzsch, 1851

PFR First: *Clethraecarpum asepalum* Menzel, 1914. Fruit. Tertiary (Braunkohle), Germany: Herzogenrath.
Comments: Fruits and seeds of modern *Clethra* are recorded by Friis (1985a) from the Middle Miocene of Denmark. Friis (1985a) rejected earlier records then known, and noted that the above PFR record had been reassigned to *Visnea* (Theaceae). Knobloch and Mai (1986) assigned Maastrichtian fruits named *Discoclethra* to the Clethraceae. However, the material was noted to show features of Clethraceae, Celastraceae, Hamamelidaceae, Cunoniaceae and Ericaceae, and thus we do not consider that it should be taken as a record of Clethraceae. (See also our discussion under Cyrillaceae.)
Muller First: *Clethra* type Chmura, 1973. Maastrichtian, California, USA.
Comments: Too early; difficult to distinguish tricolporate pollen and from that of the Cyrillaceae. According to Taylor (1990), pollen of Cyrillaceae cannot be distinguished from that of Clethraceae. Seeds assigned to Cyrillaceae have a record more consistent with that of these pollen types.

F. CLUSIACEAE Lindley, 1826

PFR First: *Mammaeites francheti* Fliche, 1896. Seed. Cretaceous: Cenomanian, France: Sainte-Menehould: Chaudefontaine.
Comments: Seeds like those of modern *Hypericum* occur in the Oligocene onwards in Europe (Friis, 1985a).
Muller First: *Kielmeyerapollenites eocenicus* (form taxon) Sah and Kar, 1974. Lower Eocene, India.
Comments: Unchallenged.

F. COMBRETACEAE Brown, 1810

PFR First: *Combretiphyllum acuminatum* Menzel, 1909. Leaf. Cretaceous: Senonian, Cameroon: Balangi.
Comments: Fruits once assigned to *Terminalia*, the only megafossils of the family reported by Taylor (1990) from the Upper Oligocene of North America, were considered to be in need of revision (Manchester and Meyer, 1987) being similar to fruits of several modern families. Fruits very similar to those of modern *Combretum* and *Terminalia* are abundant in the Lower Miocene of Kenya (Chesters, 1957; Collinson, in prep.). Woods assigned to Combretaceae are common in the Oligocene and Neogene of Africa and south-east Asia (Bande and Prakash, 1986, Boureau *et al.*, 1983). Gregor (1978) documented fruits like those of modern *Quisqualis* from the Middle Miocene onwards in Europe.
Muller First: *Heterocolpites laevigatus* (form taxon) Salard-Cheboldaeff, 1978. Upper Eocene, Cameroon.
Comments: Unchallenged.

F. COMMELINACEAE Brown, 1810

PFR First: *Commelinacites dichorisandroides* R. Caspary, 1880 (post-4 June). Flower. Tertiary, Germany: Prussia.
Comments: Fruits and leaves assigned to modern *Pollia* (*P. tugenensis* Jacobs and Kabuye) were described from the upper Middle Miocene of Kenya (Jacobs and Kabuye, 1989).
Pollen: No record.

F. CONNARACEAE Brown, *in* Tuckey, 1818

PFR First: *Connaracanthium roureoides* Conwentz, 1886. Inflorescence. Lower Tertiary, Germany: West Prussia.
Comments: Unconfirmed. Fruits like those of modern *Cnestis* are recorded from the Lower Miocene of Kenya (Chesters, 1957).
Pollen: No record.

F. CONVOLVULACEAE de Jussieu, 1789

PFR First: *Palaeoipomoea fukuiensis* Matsuo Mai, 1956. Leaf. Tertiary: Middle Miocene, Japan, Fukui Prefecture.
Comments: We know of no well-substantiated megafossil record for this family. See discussion under Verbenaceae for comments on *Porana* cited by Chesters *et al.* (1967) as Convolvulaceae.
Muller First: *Calystegiapollis microechinatus* (form taxon) Salard-Cheboldaeff, 1975a, Lower Eocene, Cameroon.
Comments: Unchallenged.

F. CORIARIACEAE de Candolle, 1824

PFR: No record.
Comments: The fossil record was reviewed by Gregor (1980) who described new material of seeds assigned to *Coriaria* from the Lower Miocene of Germany.
Muller First: *Coriaria*. Van Campo, *in* Muller, 1981. Upper Miocene, Spain.
Comments: Unchallenged.

F. CORNACEAE Dumortier, 1829

PFR First: *Cornoxylon erraticum* Conwentz, October–December, 1884. Wood. Cretaceous: lower Senonian, Germany: Braunschweig: Helmstedt.
Comments: Knobloch and Mai (1986) recorded fruits of four extinct genera of Mastixioideae (as Mastixiaceae) from the Maastrichtian of Europe, and the fossil record is widespread and diverse from then onwards. The fossil record of Cornoideae was reviewed in detail by Eyde (1988), who accepted fruits of blue-line *Cornus* in the Middle Eocene and of red-line *Cornus* in the Lower Eocene. Eyde (1988) noted problems in identifying fossil leaves and woods to the family, but accepted leaves from the Lower Eocene as 'vestiges of early dogwoods' (p. 269). These and other leaves are listed as *Cornus* by Taylor (1990).
Muller First: Pending (the pending status is reaffirmed by Eyde (1988, pp. 269–70) who discussed problems in identifying pollen to this family with certainty).

F. CRASSULACEAE de Candolle, *in* Lamarck and de Candolle, 1805

PFR: No record.
Comments: According to Friis and Skarby (1982) all mega-

fossils assigned to this family are questionable and require revision.
Pollen: No record.

F. CRYPTERONIACEAE de Candolle, 1868

PFR: No record.
Comments: We know of no claims for megafossils of this family.
Muller First: *Dactylocladus* type. Anderson and Muller, 1975. Upper Miocene, Borneo.
Comments: There is no evidence to distinguish from pollen of Combretaceae and Melastomataceae.

F. CUCURBITACEAE de Jussieu, 1789

PFR First: *Cucumites* sp. Bowerbank, post-24 March 1840. Fruit. Tertiary: Eocene, UK, England: Kent, Sheppey.
Comments: Seeds of several species of the form genus *Cucurbitospermum* occur in the Lower Eocene of England (Chandler, 1964). Undescribed seeds also occur in the uppermost Palaeocene of England (Collinson, 1986a).
Muller First: *Hexacolpites echinatus* (form taxon) Salard-Cheboldaeff, 1978, Oligocene, Cameroon.
Comments: Unchallenged.

F. CUNONIACEAE Brown, *in* Flinders, 1814

PFR First: *Cunonioxylon* sp. E. Hofmann, April 1952. Wood. Tertiary, Upper Oligocene, Eastern Alps: Prambachkirchen.
Comments: Leaves similar to those of modern *Weinmannia* and *Cunonia* are reported from the Oligocene of Tasmania (Hill, *in* Douglas and Christophel, 1990, pp. 31–42). Palaeocene leaves from Europe and Greenland have also been assigned to *Weinmannia* based upon morphological and epidermal characters, but these should be regarded with caution in view of the largely Southern Hemisphere distribution of the modern genus (Friis and Skarby, 1982; Friis, 1990). Wolfe (pers. comm., 1991) does not accept any Northern Hemisphere records for the family. Lower Eocene leaves from North America, identified as *Lamanonia* by Hickey (1977), are now known to be *Platycarya* of the Juglandaceae (see citations in Manchester, 1987). Gottwald (1992) assigned an Upper Eocene wood to this family.
Muller First: *Weinmannia* type Mildenhall, 1980. Oligocene, New Zealand.
Comments: Unchallenged.

F. CYCLANTHACEAE Dumortier, 1829

PFR First: *Cyclanthodendron sahnii* (Rode) Sahni and Surange, 22 July 1944. Stem. Tertiary: Eocene, Deccan: India: Chhindwara: Mohgaon Kalan.
Comments: This taxon was revised by Biradar and Bonde (*in* Douglas and Christophel, 1990, pp. 51–7), who concluded that this was an underground rhizomatous stem which continued into an aerial, woody pseudostem (named *Musocaulon*), bearing strap-like leaves with sheathing based (of which petioles when found isolated had been named *Heliconiates*). Anatomical evidence suggested that fruits named *Tricoccites* were borne on this plant. The reconstructed plant was considered to combine characters of Musaceae and Strelitziaceae, and hence cannot be assigned to Cyclanthaceae. Collinson (1988b) reported fruiting cycles very similar to those of modern *Cyclanthus*, in the Middle Eocene of Messel, but did not assign these to Cyclanthaceae.
Pollen: No record.

F. CYMODOCEACEAE N. Taylor, 1909

PFR: No record.
Comments: Daghlian (1981) rejected previously published records of the family. Recently described Upper Middle Eocene material from Florida (USA) includes leaves and rhizomes of species assigned to modern *Thalassodendron* and *Cymodocea* (and leaves tentatively assigned to *Halodule*), representing a seagrass bed in which characteristic epibionts and an associated fauna are also preserved (Ivany *et al.*, 1990).
Pollen: No record.
Comments: The pollen in the modern representatives is without exine (Cronquist, 1981) and is very unlikely to be preserved.

F. CYPERACEAE de Jussieu, 1789

PFR First: *Caricopsis laxa* V. A. Samylina, March 1960. Leaf. L Cretaceous, former USSR: north-east Siberia: Kolyma Basin.
Comments: The family is well represented by a diversity of fruits from the Middle Palaeocene onwards in Europe (Chandler, 1964; Mai and Walther, 1985; Mai, 1987). Fruiting heads (of *Caricoidea*) similar to those of modern Mapaniodeae are found in the Middle Eocene of Germany (Collinson, 1988b). We know of no well-substantiated leaf fossils of the family.
Muller First: *Cyperaceapollis* (form genus) Krutzsch, 1970a, Middle Eocene, Germany.
Comments: Well supported.

F. CYRILLACEAE Endlicher, 1841

PFR First: *Cyrilloxylon europaeum* J. van der Burgh, July 1964. Wood. Tertiary: Miocene, The Netherlands.
Comments: Fruits and seeds, assigned to three extinct genera and forms referable to modern *Cyrilla*, are recorded from the Maastrichtian to Miocene in Europe (Knobloch and Mai, 1986; Mai and Walther, 1978, 1985). The extinct *Epacridicarpum* Chandler occurs commonly in the Maastrichtian to Miocene strata, however, Friis (1985a) questioned its inclusion in Cyrillaceae, suggesting that some species might belong in Ericaceae. Fruits of *Valvaecarpus* Knobloch and Mai were stated to show features of Cyrillaceae and Clethraceae. Friis (1985b) described Upper Cretaceous (Santonian/Campanian) flowers, fruits and seeds of Ericales which showed similarities with members of the Cyrillaceae, Clethraceae, Ericaceae, Epacridaceae and Diapensaceae (perhaps being most similar to the latter). In view of this complex character mixing, we consider all records of these families provisional unless the fossils can be matched exactly in one modern genus (see also comments on Clethraceae). Fruits like those of modern *Cyrilla* are only known from one Upper Eocene locality (Mai and Walther, 1985).
Muller First: *Clethra* type Chmura, 1973. Maastrichtian, California, USA.
Comments: Too early; difficult to distinguish tricolpate pollen and especially Clethraceae (see coments on Clethraceae).

F. DAPHNIPHYLLACEAE Muell.-Arg. *in* de Candolle, 1869

PFR: No record.
Comments: Leaves of *Daphniphyllum protomacropodum* Murai were described by Uemura (1988) from the Upper

Miocene of Japan.
Pollen: No record.

F. DATISCACEAE Lindley, 1830

PFR First: *Tetrameleoxylon prenudiflora* R. N. Lakhanpal and J. K. Verma, April 1966. Wood. Tertiary, probably Lower Eocene. India, M.P., Chhindwara, Mohgaon Kalan.
Comments: We are unable to comment on this wood record. We know of no other megafossil record for this family.
Pollen: No record.

F. DIAPENSIACEAE Lindley, 1836

PFR: No record.
Comments: No megafossils can be assigned to the family but many characteristics of the family are seen in Upper Cretaceous flowers, fruits and seeds (see discussion under Cyrillaceae).
Pollen: No record.

F. DICHAPTELACEAE Baillon, *in* Martius, 1886

PFR: No record.
Comments: We know of no claims for megafossils of this family.
Muller First: *Tapura ivorense* type. Medus, 1975b, Miocene, Senegal.
Comments: Well-supported evidence of characteristic pollen morphology but no megafossil data.

F. DIDYMELACEAE Leandri, 1937

PFR: No record.
Comments: We know of no claims for megafossils of this family.
Muller First: *Didymeles* = *Schizocolpus marlinensis*. Harris, 1974, Palaeocene, Indian Ocean and New Zealand.
Comments: Well supported by pollen morphology, but there is no support from the megafossil record.

F. DILLENIACEAE Salisbury, 1807

PFR First: *Dillenites microdentatus* (Hollick) Berry, 1916. Leaf. Tertiary, L Eocene. USA, Louisiana, Red River Parish, Coushatta.
Comments: Seeds like those of modern *Tetracera* and *Hibbertia* were recorded in the Lower Eocene of England, UK (Chandler, 1964). Dilleniid leaves are common in Upper Cretaceous floras but they are not diagnostic of modern Dilleniaceae (e.g. Crabtree, 1987; Taylor, 1990).
Muller First: *Curatella* type Wymstra, 1971. Middle Miocene, Guyana.
Comments: Unchallenged.

F. DIOSCOREACEAE Brown, 1810

PFR First: *Dioscoroides lyelli* (A. Watelet) P. H. Fritel, 15 October 1904. Leaf. Tertiary, Eocene, France: Bassin de Paris: Belleu. *Prototamus paucinervis* M. Langeron, 1899. Leaf. Tertiary: Eocene, France: Sêzanne.
Comments: Daghlian (1981) considered all records of the family questionable and in need of revision. A Lower Oligocene record from Hungary (Andreánszky, 1959) was apparently not included in Daghlian's work, but we can find no reason to treat this as more reliable.
Muller First: Pending.

F. DIPSACACEAE de Jussieu, 1789

PFR: No record.
Comments: Rare fruits of *Scabiosa* cf. *columbaria* L. were described by Szafer (1954) from the Pliocene of Poland.
Muller First: *Scabiosaepollenites magnus* (form taxon). Nagy, 1969, Middle Miocene, Hungary.
Comments: Unchallenged.

F. DIPTEROCARPACEAE Blume, 1825

PFR First: *Woburnia porosa* M. C. Stopes, 9 July 1912. Wood. Cretaceous: Lower Aptian, England, UK: Bedfordshire, Woburn.
Comments: The material on which this PFR record is based is considered to be of questionable derivation and has no attached matrix from which this might be reassessed. Wolfe *et al.* (1975, p. 819) felt that further consideration of *Woburnia* was valueless. We agree that this record must not be cited as evidence for Cretaceous Dipterocarpaceae. Wolfe (1977) described leaves of *Parashorea* from the Middle Eocene of North America, and noted the rarity of the family in the Palaeogene. This observation is confirmed by the tabulated review of fossil Dipterocarpaceae leaves provided by Lakhanpal and Guleria (1986) who recognized *Dipterocarpus* leaves in the Neogene of India. Woods assigned to *Dipterocarpoxylon* are common in the Neogene of south-east Asia (Bande and Prakash, 1986).
Muller First: *Dipterocarpus* type. Muller, 1970. Oligocene, Borneo.
Comments: Unchallenged.

F. DROSERACEAE Salisbury, 1808

PFR: No record.
Comments: Seeds like those of modern *Aldrovanda* are documented from the Upper Eocene onwards in Europe (Mai, 1985a; Chandler, 1964). Seeds from the Maastrichtian of Europe were assigned to *Palaeoaldrovanda* by Knobloch and Mai (1986). This early record should be treated with caution in view of the absence of intervening examples.
Muller First: *Saxonipollis saxonicus* (form taxon). Krutzsch, 1970b. Lower Eocene, Germany.
Comments: Unchallenged.

F. EBENACEAE Gürke, *in* Engler and Prantl, 1891

PFR First: *Diospyrocarpum* (L. Criê) N. Vaudois-Mieja, 11 February 1980. Fruit. Tertiary: Eocene, France: Sarthe, near Man, St Pavace.
PFR First: *Diospyropsis* E. P. Korovin, 1956. Fructification. Eocene, former USSR: Turkmenistan, Badhyz, Er-Orlan-Duz Lake.
Comments: Basinger and Christophel (1985) described Ebenaceae flowers and leaves from the Middle Eocene of Australia.
Muller First: *Tricolporopollenites milonii* (form taxon). Ollivier-Pierre, 1979 = *Diospyros* type Muller, 1981. Lower Eocene, France.
Comments: Supported.

F. ELAEAGNACEAE de Jussieu, 1789

PFR First: *Elaeagnites campanulatus* O. Heer, 1870. Flower. Tertiary: Miocene, Norway: Spitsbergen: Cape Staratschin.
Comments: Taylor (1990) lists two records of leaves assigned to modern *Shepherdia* in the Oligocene of North America. One was rejected as indeterminate by Wolfe and Schorn (1989) and Wolfe (pers. comm., 1991) does not accept the other. Wolfe (1964) described leaves of *Elaeagnus* from the Middle Miocene of North America. These should

now be included in *Shepherdia* and represent the earliest record of the family in North America (Wolfe, pers. comm., 1991).

Muller First: *Slowakipollis hippophaeoides* (form taxon). Krutzsch, 1962, Oligocene, Germany.

Comments: Possible. *Bohlensipollis hohli* has now been recorded from many Oligocene sediments in north-west Europe (Vinken, 1988), and according to lists in Taylor (1990), from the Middle and Upper Eocene of North America. It has demisyncolpate apertures, not present in modern *Elaeagnus* pollen, and may be from extinct groups of this family.

F. ELAEOCARPACEAE de Candolle, 1824

PFR First: *Sloaneaecarpum eocenicum* K. Râsky, 1962. Fruit. Upper Eocene, Hungary: Budapest-Obuda.

Comments: Rozefelds (*in* Douglas and Christophel, 1990, pp. 123–6) described Lower Oligocene fruits like those of modern *Elaeocarpus*, and Christophel and Greenwood (1987) recorded leaves similar to those of modern *Sloanea* and *Elaeocarpus* from the Middle Eocene – both in Australia. Chandler (1964) recorded fruits like those of modern *Echinocarpus* in the Lower Eocene of England, UK.

Muller First: *Elaeocarpus* type. Mildenhall, 1980. Oligocene. New Zealand.

Comments: Unchallenged.

F. ELATINACEAE Dumortier, 1829

PFR: No record.

Comments: Seeds like those of modern *Elatine* were documented by Mai (1985a) from the Miocene of Europe.

Pollen: No record.

F. EMPETRACEAE Gray, 1821

PFR: No record.

Comments: Van der Burgh (1987) recorded fruits like those of modern *Corema* from the Upper Miocene of the Lower Rhenish Basin, Europe.

Muller First: No record.

Comments: *Empetrum* sp. Boulter, 1971, Upper Miocene, England, UK.

F. EPACRIDACEAE Brown, 1810

PFR First: *Dulaurensia pulchra* E. M. Reid, 1930. Fruit. Eocene, France: Saint Trudy: Finistère.

Comments: The genus *Epacridicarpum* Chandler has been included in Cyrillaceae. We know of no well-substantiated megafossils of this family. (See discussion under Cyrillaceae.)

Pollen: No record.

F. ERICACEAE de Jussieu, 1789

PFR First: *Kalmiophyllum marcodurense* Kräusel and Weyland, May 1959. Leaf? Upper Oligocene, Germany: Rhineland: Duren.

Comments: Seeds like those of modern *Rhododendron* occur in the Upper Palaeocene of England, UK (Collinson and Crane, 1978) and fruits and seeds like those of modern *Leucothoe*, and also representatives of two extinct genera, were recorded in the Maastrichtian of Europe (Knobloch and Mai, 1986). The family was diverse in the Middle Miocene of Europe (Friis, 1985a). Taylor (1990) lists leaves assigned to *Kalmia* in the Palaeocene of North America but these may need revision. (See also comments under Cyrillaceae.)

Muller First: *Calluna* type. Menke, 1976. Pliocene, Germany.

Comments: Too late. *Erecipites* tetrads are found commonly throughout the European Tertiary and also occur in the Maastrichtian of Wyoming and California (Drugg, 1967; Farabee and Canright, 1986).

F. ERIOCAULACEAE Desvaux, 1828

PFR: No record.

Comments: Chesters *et al.* (1967) cited a North American Palaeocene record of *Eriocaulon*, but this was not listed by Taylor (1990). We know of no well-substantiated megafossil record of this family.

Pollen: No record.

F. ERYTHROXYLACEAE Kunth, *in* Humbolt, Bonpland and Kunth, 1822

PFR: No record.

Comments: Chesters *et al.* (1967) cited a Miocene record of *Erythroxylon* and Cronquist (1981) refers to fossils attributed to the family from the Eocene of Argentina. We know of no well-substantiated megafossil record of this family.

F. EUCOMMIACEAE Engler, 1909

PFR: No record.

Comments: Rare leaves and fruits assigned to modern *Eucommia* are listed by Taylor (1990) from the Upper Eocene and Oligocene of North America. Fruits are recorded in the Upper Miocene of Europe (Szafer, 1961). *Eucommia* is also recorded in the early Miocene of the former USSR (Zhilin, 1989).

Muller First: *Tricolporopollenites parmularius* (form genus). Krutzsch, 1970b, Lower Palaeocene, Germany.

Comments: Unchallenged. Smooth tricolporate pollen is variable and commonly found throughout the Tertiary, and gives poor evidence of affinity on its own.

F. EUCRYPHIACEAE Endlicher, 1841

PFR: No record.

Comments: Several species of *Eucryphia* leaves are recorded from the Upper Palaeocene and Eocene of south-eastern Australia (Hill, 1991b).

Pollen: No record.

F. EUPHORBIACEAE de Jussieu, 1789

PFR First: *Securinegoxylon biseriatum* E. Mädel, 15 August 1962. Wood. Cretaceous: Senonian, South Africa: eastern Pondoland, Umzamba.

Comments: Dilcher and Manchester (*in* Collinson, 1988c, pp. 45–58) considered Cretaceous records inconclusive. They document flowers and fruits of tribe Hippomaneae in the Middle Eocene. Lower Eocene fruits are found in England, UK (Chandler, 1964) and leaves referred to as 'euphorbs' are recorded by Wolfe and Upchurch (1986) in the Upper Maastrichtian of North America. Tanai (1990) also reviewed the fossil history of the family and treated the PFR wood record as euphorbioid. He documented leaves of four modern genera from the Middle and Upper Eocene of Japan.

Muller First: *Malvacipollis* diversus (form taxon). H. A. Martin, 1974, Upper Palaeocene, Australia. *Crotonopollis burdwanensis* (form taxon), Baksi *et al.*, 1979, Palaeocene, India. *Retitricolporites irregularis* (form taxon), Muller, 1970. Lower Eocene, Venezuela.

Comments: Unchallenged.

F. EUPOMATIACEAE Endlicher, 1841

PFR: No record.
Comments: We know of no megafossil record of this family.
Muller First: Pending.

F. EUPTELEACEAE Wilhelm, 1910

PFR: No record.
Comments: Leaves assigned to *Euptelea orientalis* (Sanborn) Wolfe were documented by Wolfe (1977, p. 89) from the Upper Eocene of North America. A wood from the Middle Eocene of Oregon is the only record of the family listed by Taylor (1990). This wood record seems acceptable, although no associated fruits or foliage have been found in Oregon (Manchester, pers. comm., 1991).
Muller First: *Eupteleapollis* (form taxon). Krutzsch, 1966c, Upper Eocene, Germany.
Comments: Unchallenged.

F. FABACEAE Lindley, 1836

PFR First: *Palaeocassia augustifolia* Ettingshausen, 1867. Leaf. Cretaceous: Cenomanian, Germany, Saxony.
Comments: Fruits like those of modern *Diplotropis* are recorded (under Leguminosae, Papillionoideae) in the Middle Eocene and papillionoid flowers from the Palaeocene/Eocene both in North America (Herendeen and Dilcher, 1990b).
Muller First: *Indigofera* and *Astragalus*, Beucher, 1975. Pliocene, Sahara.
Comments: Unchallenged.

F. FAGACEAE Dumortier, 1829

PFR First: *Nothofagoxylon scalariforme* W. Gothan, 1908. Wood. Upper Cretaceous, Seymour Island.
Comments: Crepet (*in* Crane and Blackmore, 1989) considered castaneoid and trigonobalanoid inflorescences with pollen, fruits and associated foliage in the Palaeocene/Eocene of North America as the earliest unequivocal Fagaceae. He suggested an uppermost Cretaceous origin for the family based upon this simultaneous appearance of multiple organ fossils comparable with both modern tribes. Fagaceae fossils in Europe from the Eocene onwards are reviewed by Kvacek and Walther (1989). *Nothofagus* (Nothofagaceae of some authors) fossils were reviewed by Hill (1991a) who noted that Cretaceous and many Lower Tertiary fossils need to be revised. Evidence from fossil cupules and leaves shows that all extant subgenera were established as distinct entities (with no intermediates) by the Oligocene. Fossil wood, named *Nothofagoxylon*, from the Upper Cretaceous of South America may be the earliest megafossil of the family (noted in Hill, 1991a; Romero, 1986b).
Muller First: *Nothofagidites senectus* (form taxon). Dettman and Playford, 1969, Santonian, Australia. *Tricolporopollenites* sp. (form genus). Rouse *et al.*, 1971. Santonian/Campanian, Canada. *Castanea*, Zagwijn *in* Muller, 1981. Lower Campanian, Netherlands.
Comments: Unchallenged without *in situ* evidence for this small and varied tricolp(or)ate morphology.

F. FLACOURTIACEAE de Candolle, 1824

PFR First: *Aphloioxylon* Mathiesen, 1937. Leaf. Palaeocene, Argentina: Patagonia: between Chubut and Santa Cruz.
Comments: Wolfe and Upchurch (1986) refer to 'flacourt' foliage in the Upper Maastrichtian and Taylor (1990) lists one Maastrichtian leaf fossil. Chandler (1964) recorded fruits in the Lower Eocene of England, UK. Leaves, fruits and seeds like those of modern *Idesia* are reported from the Lower Eocene onwards in North America (Taylor, 1990).
Muller First: *Casearia* type. Graham and Jarzen, 1969. Oligocene, Puerto Rico.
Comments: Unchallenged. Muller rejects comparisons with *Pistillipollenites macgregorii* (form taxon).

F. FLAGELLARIACEAE Dumortier, 1829

PFR: No record.
Comments: We are not aware of any claims for megafossils of this family.
Muller First: *Flagellaria* type. Anderson and Muller (1975). Upper Miocene, Borneo.
Comments: Unchallenged.

F. FOUQUIRIACEAE de Candolle, 1828

PFR: No record.
Comments: We are not aware of any claims for megafossils of this family.
Muller First: Earlier records rejected.

F. FUMARIACEAE de Candolle, 1821

PFR: No record.
Comments: Dorofeev (1964) listed a seed of *Corydalis* sp. from the uppermost Miocene and Mai (pers. comm., 1991) has seeds like those of modern *Corydalis* in the Miocene of Germany.
Pollen: No record.

F. GARRYACEAE Lindley, 1834

PFR: No record.
Comments: Cronquist (1981) refers to leaves, seeds and inflorescences like those of modern *Garrya* from the Miocene of the western United States. Wolfe (1964) described the leaves from the Middle Miocene of North America and he considers these to be the oldest examples (Wolfe, pers. comm., 1991).
Pollen: No record.

F. GENTIANACEAE de Jussieu, 1789

PFR: No record.
Comments: Flowers from the Lower Eocene of North America were assigned to this family (Crepet and Daghlian, 1981). These contain *Pistillipollenites* pollen but this is also found in other flowers which are not Gentianaceae (Stockey and Manchester, 1988). The inclusion of all records of *Pistillipollenites* in this family (e.g. Taylor, 1990), implying a Maastrichtian to Eocene fossil record, is therefore incorrect.
Muller First: Pending.

F. GERANIACEAE de Jussieu, 1789

PFR: No record.
Comments: A possible Upper Pliocene *Geranium* or *Erodium* is described by Strauss (1969). We know of no well-substantiated megafossils of this family.
Muller First: *Geranium phaeum* type. Bortenschlager, 1967. Upper Miocene. Spain. *Erodium* type Van Campo, 1976, Upper Miocene, Spain.
Comments: Unsupported Spanish records are based on only one and 14 specimens respectively with no evidence from other organs.

F. GOODENIACEAE Brown, 1810

PFR: No record.
Comments: We know of no claims for megafossils of this family.
Muller First: *Poluspissusites digitatus.* (form taxon) Salard-Cheboldaeff, 1978. Oligocene, Cameroon.
Comments: Unchallenged.

F. GROSSULARIACEAE de Candolle, *in* Lamarck and de Candolle, 1805

PFR: No record.
Comments: Leaves like those of modern *Quintinia* are recorded (as Escalloniaceae) from the Middle Eocene of Australia (Christophel *et al.*, 1987; Christophel and Greenwood, 1988). Friis (1990) described Upper Cretaceous (Santonian/Campanian) flowers of Saxifragales (with associated leaves) which showed close correlation of characters with modern Escalloniaceae, especially *Quintinia*, but also showed similarity with other woody Saxifragales. Several records of leaves like those of modern *Ribes* are listed by Taylor (1990) from Upper Eocene onwards in North America. Upper Oligocene examples are accepted by Wolfe and Schorn (1989). Mai (1985b) recorded fruits like those of modern *Itea* (Iteaceae of some authors) from the Miocene and Pliocene in Europe. A flower from the Baltic amber is similar to Iteaceae (Friis and Skarby, 1982). Leaves like those of modern *Itea* are recorded in the Middle Eocene of North America (Taylor, 1990).
Muller First: Under Grossulariaceae: no record. Under Escalloniaceae: *Quintinia* type Mildenhall, 1980. Upper Eocene, New Zealand. Under Brexiaceae: *Tetracolporites ixerboides* (form taxon) Couper, 1960. Oligocene, New Zealand. Under Iteaceae: *Iteapollenites angustiporatus* Gruas-Cavagnetto, 1976a, Upper Eocene, England, UK.
Comments: All unchallenged.

F. GUNNERACEAE Meissner, 1841

PFR: No record.
Comments: No megafossil record is known to us for *Gunnera*, the only genus Cronquist includes in the family.
Muller First: *Retitricolpites microreticulatus* (form taxon). Brenner, 1968. Turonian, Peru.
Comments: Unchallenged.

F. HALORAGACEAE Brown, *in* Flinders, 1814

PFR First: *Hippuridella stacheana* W. N. Edwards, August 1932.Tertiary, L Eocene, former Yugoslavia: Central Istria: Pisino: Gorge of the Foiba.
Comments: Fruits like those of modern *Proserpinaca* and *Myriophyllum* are recorded from the Oligocene onwards in Europe (Mai, 1985a).
Muller First: *Myriophyllum* sp. Gruas-Cavagnetto and Praglowski, 1977, Palaeocene, France.
Comments: Unchallenged.

F. HAMAMELIDACEAE Brown, *in* Abel, 1818

PFR First: *Hamamelidoxylon renaulti* O. Lignier, 1907 (post-4 January). Wood. Cretaceous: Cenomanian, France: near Vimoutiers.
Comments: Seeds are recorded from the Upper Cretaceous onwards (Chandler, 1964; Knobloch and Mai, 1986; Mai, 1987a) and flowers very similar to those of modern *Hamamelis*, subfamily Hamamelidoideae, occur in the Santonian/Campanian of Sweden; subfamily Liquidambaroideae may also be distinct in the Upper Cretaceous (Friis and Crane, *in* Crane and Blackmore, 1989; Endress and Friis, 1991). Leaves of Hamamelidaceae also occur in the Campanian (Crabtree, 1987).
Muller First: *Liquidambar* type. Muller, 1970, Palaeocene, south Europe.
Comments: Unchallenged.

F. HELICONIACEAE Nakai, 1941 (see comments under MUSACEAE)

F. HERNANDIACEAE Blume, 1826

PFR: No record.
Comments: Berry (1937) described fruits of *Gyrocarpus* from the Upper Miocene of Venezuela. We know of no other megafossils from this family, and consider that this record requires confirmation.
Pollen: No record.

F. HIPPOCASTANACEAE de Candolle, 1824

PFR: No record.
Comments: Fruits like those of modern *Aesculus* are recorded in the Upper Miocene and Pliocene of Europe (Szafer, 1961) with tentative records in the Middle Miocene (Gregor, 1982). Upper Eocene leaf records from North America were considered questionable by Taylor (1990).
Muller First: *Aesculus*. Smiley *et al.*, 1975, Miocene, Idaho, USA.
Comments: Unchallenged.

F. HIPPOCRATEACEAE de Jussieu, 1811

PFR: No record.
Comments: We know of no megafossil record for this family.
Muller First: See Celastraceae.

F. HIPPURIDACEAE Link, 1821

PFR: No record.
Comments: Fruits like those of modern *Hippuris* are recorded from the Oligocene of Europe (Mai, 1985a).
Pollen: No record.

F. HUGONIACEAE Arnott, 1834

PFR: No record.
Comments: We know of no megafossil record for this family.
Muller First: under Ctenolophonaceae: *Ctenolophonidites costatus* (form taxon). Muller, 1970, Maastrichtian, Nigeria.
Comments: Unchallenged.

F. HUMIRIACEAE de Jussieu, *in* St-Hilaire, 1829

PFR: No record.
Comments: Romero (1986a) refers to work by Berry documenting fruit casts and leaf impressions of *Humiria* in the Maastrichtian and Lower Eocene of Colombia. The material was originally thought to be Miocene, but the sections were redated palynologically. These records should probably be considered as unconfirmed and requiring re-examination. Endocarps like those of modern *Sarcoglottis* and others tentatively assigned to modern *Humiriastrum* were reported in the Pliocene of Colombia (Wijninga and Kuhry, 1990).
Muller First: *Humiria* type, van de Hammen *et al.*, 1973, Pliocene, Colombia.
Comments: Unchallenged.

F. HYDRANGEACEAE Dumortier, 1829

PFR: No record.
Comments: Flowers of *Hydrangea* are documented from the Middle Eocene of North America (Manchester and Meyer, 1987; Manchester, *in* Knobloch and Kvacek, 1990, pp. 183–7) and the Middle Eocene to Pliocene of Europe (Mai, 1985b). Fruits and seeds like those of modern *Hydrangea, Dichroa* and *Schizophragma* are also reported from the Middle Oligocene and later in Europe (Mai, 1985b). Wolfe and Schorn (1989) accepted the determination of leaves of modern *Jamesia* in the Upper Oligocene of North America. Various Cretaceous and Palaeocene leaf records, assigned to *Hydrangea* and *Philadelphus*, should probably be considered in need of revision (Taylor, 1990; Friis and Skarby, 1982; Mai, 1985b). Wolfe (pers. comm., 1991) considers that there are no reliable pre-early Eocene records.
Pollen: No record.

F. HYDROCHARITACEAE de Jussieu, 1789

PFR First: *Hydrocharites obcordatus* C. O. Weber, December 1855. Leaf. Miocene, Germany: Rhenish Prussia: Rott.
Comments: The family is represented by seeds of *Stratiotes* from the Upper Palaeocene upwards in England, UK (Collinson, 1986a, 1990) and by foliage of other freshwater genera in the Middle and Upper Eocene of Europe (Wilde, 1989; Mai and Walther, 1978, 1985). Seagrass beds from the Upper Middle Eocene of North America (see Cymodoceaceae) include representatives of modern *Thalassia* (Ivany *et al.*, 1990).
Pollen: No record.

F. ICACINACEAE Miers, 1851

PFR First: *Calatoloides eocenicum* E. W. Berry, April 1922. Fruit. Tertiary: Eocene, USA: Texas: Freestone County. *Icacinicarya platycarpa* Reid and Chandler, 25 November 1933. Fruit. Tertiary: Eocene, UK: England: Kent: Sheppey. *Palaeophytocrene foveolata* Reid and Chandler, 25 November 1933. Fruit. Tertiary: Eocene, UK: England: Kent: Sheppey. *Sphaeriodes ventricosa* (Bowerbank) Reid and Chandler, 25 November 1933. Fruit. Tertiary: Eocene, UK: England: Kent: Sheppey. *Stizocarya communis* Reid and Chandler, 25 November 1933. Fruit. Tertiary: Eocene, UK: England: Kent: Sheppey.
Comments: Eocene fruits of the family are diverse, but rarer examples occur in the Upper Palaeocene (Crane *et al.*, 1990; Collinson, 1986a; Mai, 1987a) and possibly Upper Cretaceous (Tanai, 1990) of Europe, Africa and North America. Leaves are known in the Lower and Middle Eocene of North America (Taylor, 1990), Middle and Upper Eocene of Japan (Tanai, 1990) and tentatively in the Middle Eocene of Europe (Wilde, 1989). Maastrichtian woods named *Icacinoxylon* listed by Taylor (1990) are not conclusively included in the family (e.g. Wheeler *et al.*, 1987).
Muller First: *Compositoipollenites rhizophorus*. Kedves, 1970. Lower Palaeocene, Germany.
Comments: Well supported.

F. ILLICIACEAE Smith, 1947

PFR: No record.
Comments: Leaves like those of modern *Illicium* are recorded from the Middle Eocene of Germany (Wilde, 1989) and North America (Taylor, 1990). Fruits are documented in the Oligocene of North America (Tiffney and Barghoorn, 1979).
Muller First: *Illicium* Chmura, 1973. Maastrichtian, California, USA.
Comments: No supporting evidence.

F. IRIDACEAE de Jussieu, 1789

PFR First: *Irites alaskana* Lesquereux, 17 May 1888. Leaf. Lower Cretaceous, USA: Alaska, Cape Lisbourne.
Comments: We know of no well-substantiated early megafossil record for this family, although Pleistocene seeds are known from Europe and Japan (Miki, 1961; Mai, 1985a).
Muller First: Pending.

F. JUGLANDACEAE Richard ex Kunth, 1824

PFR First: *Juglandiphyllum integrifolium* W. M. Fontaine, September–December 1889. Leaf. Lower Cretaceous, USA: Virginia, White House Bluff.
Comments: The major diversification of this family occurred in the Palaeocene and is documented from a range of organs, including inflorescences, fruits and foliage (Manchester, 1987; Stone *in* Crane and Blackmore, 1989). Cretaceous flowers (Friis, 1983), some of which contain *Normapolles* pollen (*Manningia* containing *Trudopollis* and *Caryanthus* containing *Plicapollis*), are sometimes assigned to the family (Knobloch and Mai, 1986). These are best considered either as Juglandales or intermediates between Juglandales and Myricales. They differ from Juglandaceae in the small size of the fruits, bisexual flowers and undivided locule (Friis and Crane, *in* Crane and Blackmore, 1989).
Muller First: *Momipites fragilis* type. Nichols, *in* Muller, 1981. Lower Campanian, USA.
Comments: Current megafossil evidence (above) suggests that Muller's four pollen groups (*Engelhardia–Alfaroa, Carya, Platycarya* and *Juglans–Pterocarya*) developed in the Tertiary from earlier groups closely related to the Juglandales and the Myricales, which produced Nichols' *Momipites* and the extinct Normapolles.

F. JULIANIACEAE Hemsley, 1906

PFR: No record.
Comments: Wolfe (pers. comm., 1991) considers that the leaf material which he described in 1964 as a species of *Schinus* may in fact belong in this family.
Pollen: No record.

F. JUNCACEAE de Jussieu, 1789

PFR: No record.
Comments: Seeds like those of modern *Juncus* are recorded in the Upper Eocene/Lower Oligocene of England, UK (Collinson, 1983) and from the Miocene onwards elsewhere in Europe (Mai, 1985a). Kovach and Dilcher (1988) noted the suggestions that Cretaceous seed-like cuticles named *Spermatites* might have some relationship with Juncaceae, but considered these suggestions to be unconfirmed in the absence of detailed comparative studies and in view of the absence of intervening records. Seeds like those of *Juncus*, if borne on wetland species, would have high fossilization and recovery potential, and intervening examples should not have passed unreported.
Pollen: No record.

F. JUNCAGINACEAE Richard, 1808

PFR First: *Lamprocarpites nitidus* O. Heer, 1882. Fruit. Upper Cretaceous, Greenland: Upernivik.
Comments: Neither Mai (1985a) nor Collinson (1988a)

recorded this family in their reviews of fossils of wetland plants.
Pollen: No record.

F. LACTORIDACEAE Engler, *in* Engler and Prantl, 1888

PFR First: No record.
Comments: We know of no claims for megafossils of this family.
Pollen: *Lactoripollenites africanus* (form taxon) Zavada and Benson, 1987. Cretaceous: Turonian to Campanian, South Africa.
Comments: The first report of this family in the fossil record, is from Turonian–Campanian sediments off the south-west coast of South Africa. The permanent tetrahedral tetrads consist of anasulcate pollen which is primitive in structure and ancestry.

F. LAMIACEAE Lindley, 1836

PFR First: *Ajuginucula smithii* E. M. Reid and Chandler, 1926. Fruit. Tertiary: Oligocene, UK: England: Isle of Wight.
Comments: Mai (1985a) and Łancucka-Środoniowa (1979) record fruits like those of modern *Lycopus* from the Oligocene onwards in Europe.
Muller First: *Salvia* Emboden, 1964. Upper Miocene, USA: Alaska.
Comments: Unchallenged.

F. LARDIZABALACEAE Decaisne, 1838

PFR First: *Lardizabaloxylon lardizabaloides* Schonfeld, 1954. Wood? Tertiary, Argentina: Patagonia.
Comments: We are unable to comment on this wood record. Mai (1980) recorded seeds like those of modern *Decaisnea* from the Oligocene of Germany.
Muller First: Previous record rejected.

F. LAURACEAE de Jussieu, 1789

PFR First: *Daphnophyllum fraasii* Heer, 1869. Leaf? Cretaceous: Cenomanian, Czechoslovakia: Moravia, Moletein.
Comments: Drinnan *et al.* (1990) record flowers from the lower Cenomanian of North America. They summarize briefly the subsequent widespread early Tertiary record of leaves, woods, fruits and flowers of the family. Drinnan *et al.* (1990) noted some similarity between the new material and material previously described as *Prisca* Retallack and Dilcher (1981), but preservational differences prevented a full comparison. Kvacek (1992) considered a suite of European Upper Cretaceous leaves and reproductive structures, along with the North American *Prisca*, to represent an extinct family, Priscaceae, closely related to Lauraceae within the Laurales.
Muller First: *Inaperturopollenites palaeogenicus* (form taxon). Gruas-Cavagnetto, 1978. Upper Palaeocene, France.
Comments: Pollen of this family has little sporopollenin in the exine and thus low preservation potential (Drinnan *et al.*, 1990; Lidgard and Crane, 1990). The flowers noted above failed to yield pollen in spite of their excellent preservation.

F. LECYTHIDACEAE Poiteau, 1825

PFR First: *Barringtonioxylon deccanense* Shallom, 1960. Wood. Tertiary: Palaeocene, Deccan, India: Nagpur.
Comments: Other Tertiary woods assigned to the family are described from south-east Asia (Bande and Prakash, 1986). Seeds from the South American Upper Oligocene were described as *Lecythidospermum* by Pons (1983), who reviewed the fossil record from the area, including a fruit from the Maastrichtian assigned to the family. One Middle Eocene leaf form, requiring confirmation, is noted by Taylor (1990) from North America.
Muller First: *Marginipollis concinnus* (form taxon). Venkatachala and Rawat, 1972. Lower Eocene, India.
Comments: Unchallenged.

F. LEEACEAE Dumortier, 1829

PFR: No record.
Comments: We know of no megafossils assigned to *Leea*, in this monotypic family often included in Vitaceae.
Muller First: Previous record rejected.

F. LEITNERIACEAE Bentham, *in* Bentham and Hooker, 1880

PFR: No record.
Comments: Chesters *et al.* (1967) cited a Miocene record, and Cronquist (1981) refers to Miocene and possible Oligocene records of the single modern genus *Leitneria*.

F. LEMNACEAE Gray, 1821

PFR First: *Lemnospermum pistiforme* V. P. Nikitin, 1976. Seed. Tertiary, Pliocene? or Miocene, former USSR, western Siberia, Omsk District: Andreevka: Mamomtova Gora.
Comments: Seeds like those of modern *Lemna* are recorded by Dorofeev (1988), Mai (1985a), and Mai and Walther (1978) from the Oligocene onwards in Europe and Asia. (*Lemnospermum* was placed *incertae sedis* by Mai and Walther, 1978.) Leaves assigned to modern *Spirodella* are reported from the Palaeocene and Eocene of Canada (Taylor, 1990).
Pollen: No record.

F. LENTIBULARIACEAE Richard, *in* Poiteau and Turpin, 1808

PFR: No record.
Comments: *Utricularia* is known only in the Quaternary (Mai, 1985a; Collinson, 1988a).
Muller First: *Urticularia minor* type. Graham, 1976, Upper Miocene, Mexico; Miocene, Senegal.
Comments: Unchallenged.

F. LILIACEAE de Jussieu, 1789

PFR First: *Cretovarium* M. Stopes and Fujii, March 1910. Upper Cretaceous, Japan: Hokkaido.
Comments: We know of no well-substantiated megafossil record for this family (Daghlian, 1981).
Muller First: *Astelia*. Mildenhall, 1980. Upper Eocene. New Zealand.
Comments: Unchallenged.

F. LIMNANTHACEAE Brown, 1833

PFR: No record.
Comments: One tentative and unconfirmed leaf fossil from the Oligocene of North America is noted by Taylor (1990).

F. LINACEAE Gray, 1821

PFR First: *Wetherellia variabilis* Bowerbank, post-24 March

1840. Fruit. Tertiary: Eocene, UK: England: Kent: Sheppey.
Comments: This genus probably belongs in Euphorbiaceae (Dilcher and Manchester, *in* Collinson, 1988c, pp. 45–58). We are not aware of other megafossils except *Decaplatyspermum* Reid and Chandler which was only tentatively assigned to the family (Chandler, 1964).
Muller First: *Linum* type. Van Campo, 1976. Upper Miocene, Spain.
Comments: Unchallenged.

F. LOASACEAE Dumortier, 1822

PFR: No record.
Comments: Chesters *et al.* (1967) cited a Miocene record of *Mentzelia*. Wolfe (pers. comm., 1991) considers that this record needs re-examination. We know of no well-substantiated megafossils of this family.

F. LOGANIACEAE Martius, 1827

PFR: No record.
Comments: Mai (1968) assigned fruits and seeds named *Saxifragaceaecarpum* Menzel from the Middle and Upper Miocene of Europe to this family.

F. LORANTHACEAE de Jussieu, 1808

PFR First: *Loranthacites succineus* Conwentz, 1886. Stem. Lower Tertiary, Germany: West Prussia.
Comments: Leaves of the family occur in the Middle and Upper Eocene of Germany (Wilde, 1989; Mai and Walther, 1978) and also in the Middle Eocene of Anglesea, Australia (D. C. Christophel, pers. comm., 1991).
Muller First: *Gothanipollis* (form genus). Krutzsch, 1970d, 20 form species recognized! Lower Eocene, Germany. *Spinulaepollis arceuthobioides* (form taxon), Krutzsch, 1970d. Lower Eocene. Germany.
Comments: These two taxa by Krutzsch are now supported with megafossil evidence of the family from Germany. Muller (1985) places the *Aquilapollenites* complex in the Santalales 'as an extinct family, with closest affinity to Loranthaceae, its exact phylogenetic relationship within the order remains to be discovered.' However, Muller refrained from actually creating or naming the extinct family. Muller's views were stimulated by the work of Wiggins (1982) who suggested that *Expressipollis striatus* from the Campanian of Alaska can be assigned to the Loranthaceae, and this species could form part of a transitional series.

F. LYTHRACEAE Jaume St-Hilaire, 1805

PFR First: *Palaeolythrum bournense* M. E. J. Chandler, July 1960. Seed. Tertiary: Eocene, Bartonian, UK, England: Hampshire, Bournemouth.
Comments: The family is diverse in the Eocene with several Lower Eocene fruits from England (Chandler, 1964). Seeds similar to those of modern *Decodon* occur in the Upper Palaeocene in England (Collinson, 1986a) and are common in the Eocene in Europe and North America (Eyde, 1972; Cevallos-Ferriz and Stockey, 1988b).
Muller First: *Verrutricolporites rotundiporis* (form taxon). Muller, 1970. Upper Eocene, northern South America.
Comments: Unchallenged.

F. MAGNOLIACEAE de Jussieu, 1789

PFR First: *Liriodendropsis simplex* Newberry, December 1895. Leaf. Cretaceous, USA: New Jersey: Woodbridge. *Magnoliaephyllum alternans* (Heer) Seward, 1926, May 1896. Leaf. Cretaceous, former Yugoslavia: Moravia: Kundstadt. *Magnoliaestrobus gilmouri* A. C. Seward and V. Conway, 31 December 1935. Fruit. Cretaceous, West Greenland.
Comments: Seeds from the Upper Cretaceous and Palaeocene of Europe are assigned to the family (Knobloch and Mai, 1986; Mai, 1987a). Flowers named *Archaeanthus* from the Cenomanian of North America are said to be 'probably closely related to extant Magnoliaceae' (Crane *et al.*, 1989). Magnolialean foliage is common in the Upper Cretaceous, but this often possesses generalized features, combining the characters of two or more modern families (e.g. Crabtree, 1987; Upchurch and Dilcher, 1990). Cevallos-Ferriz and Stockey (1990a) summarize the fossil record and note that by the Eocene widespread seeds provide clear evidence for several modern genera. *Magnolia*-like leaves are listed by Taylor (1990) from the Palaeocene onwards in North America. Upper Cretaceous to Eocene woods can be compared with the family, but are not identical to woods of any extant genera (Cevallos-Ferriz and Stockey, 1990a).
Muller First: *Magnolipollis graciliexinus* and *M. megafiguratus* (form taxa). Krutzsch, 1970a, Middle Eocene, Germany.
Comments: Supported by megafossil evidence.

F. MALPIGHIACEAE de Jussieu, 1789

PFR First: *Banisteriophyllum australiense* Ettingshausen, 1887. Leaf. Tertiary: Eocene, Australia: Tingha.
Comments: Specialized, bee-pollinated flowers of the family (named *Eoglandulosa*) occur in the Middle Eocene of North America (Taylor and Crepet, 1987). Fossil fruits like those of modern *Tetrapteris* have recently been revised, and most North American records removed from this genus and family (Manchester, 1991).
Muller First: *Perisyncolporites pokornyi* (form taxon). Pares Regali *et al.*, 1974a. Middle Eocene, Brazil.
Comments: Unchallenged.

F. MALVACEAE de Jussieu, 1789

PFR First: *Hibiscoxylon niloticum* Kräusel, 1939. Wood. Cretaceous, upper Senonian, Egypt.
Comments: We know of no recent revision of megafossils of this family. No megafossils are cited for North America by Taylor (1990). However, Rasky (1956) reported *Kydia* from the Upper Eocene/Lower Oligocene of Hungary and Iljinskaya (1986) reported Eocene leaves of the family in the former USSR. Seeds of *Kosteletzkya* were recorded in the Miocene of the former USSR by Dorofeev (1959, 1988).
Muller First: *Echiperiporites estelae* (form taxon). Muller, 1970. Upper Eocene, Venezuela.
Comments: Unchallenged.

F. MARANTACEAE Petersen, *in* Engler and Prentl, 1888

PFR: No record.
Comments: Although a possible record is cited by Chesters *et al.* (1967), Daghlian (1981) did not accept a fossil record for the family.
Pollen: No record.

F. MARCGRAVIACEAE Choisy, *in* de Candolle, 1824

PFR First: *Ruyschioxylon sumatrense* H. Hoffmann, 1884. Wood. Tertiary, Indonesia, Sumatra.
Comments: We are unable to comment on this wood

record. We know of no other megafossil record for this family.
Muller First: Pending.

F. MELASTOMATACEAE de Jussieu, 1789

PFR First: *Melastomites druidum* Unger, January–April 1850. Leaf. Tertiary: Eocene, Austria: Styria, Sotzka.
Comments: Seeds occur in the Miocene of Europe (Dorofeev, 1988; Collinson and Pingen, 1992; material previously misidentified as Portulacaceae). Hickey (1977) described Lower Eocene leaves as *Acrovena*, noting that all features could be matched only in this family. One other record is given for this genus in Taylor (1990). Other leaf records either require revision or have been rejected (Pingen and Collinson, in prep.).
Muller First: Pending.

F. MELIACEAE de Jussieu, 1789

PFR First: *Paratrichilioxylon russellii* J.-C. Koeniguer (*'russelli'*), 1971. Wood. Tertiary: Palaeocene, Niger: Sessao.
Comments: Fruits occur in the Lower Eocene of England, UK (Chandler, 1964). Leaves similar to *Cedrela* are listed by Taylor (1990) from the Middle Eocene onwards in North America; in one Lower Oligocene example, leaves and seeds occur in association (Manchester and Meyer, 1987). Fruits like those of *Guarea* are also listed by Taylor (1990) from the Maastrichtian and Palaeocene. These may need revision.
Muller First: *Psilastephanocolporites grandis* (form taxon). Salard-Cheboldaeff, 1978, Oligocene, Cameroon: Guarea. Graham and Jarzen, 1969, Oligocene, Puerto Rico.
Comments: Possible.

F. MENISPERMACEAE de Jussieu, 1789

PFR First: *Menispermophyllum celakovskii* Velenovsky, 1900. Leaf. Cretaceous: Cenomanian, Czechoslovakia: Praha: Chuchle prope.
Comments: Endocarps diagnostic of the family are recorded in the Maastrichtian and Palaeocene of Europe (Collinson, 1986a; Knobloch and Mai, 1986; Mai, 1987a), although they are much more diverse in the Lower and Middle Eocene (Chandler, 1964; Collinson, 1988b), where leaves are also recorded (Wilde, 1989; Takhtajan, 1974). Endocarps are also diverse in the Middle Eocene of North America (Manchester *in* Knobloch and Kvacek, 1990, pp. 183–8), and one form is recorded in the Upper Palaeocene (Crane *et al.*, 1990). Taylor (1990) lists numerous Eocene leaf records with four Palaeocene examples. Crabtree (1987) states that an early Campanian leaf is 'entirely consistent with the leaf morphology of extant Menispermaceae'.
Pollen: No record.

F. MENYANTHACEAE Dumortier, 1829

PFR: No record.
Comments: Seeds like those of modern *Menyanthes* are recorded from the Oligocene onwards in Europe (Mai, 1985a).
Pollen: No record.

F. MIMOSACEAE Brown *in* Flinders, 1814

PFR First: *Acaciaphyllites grevilleoides* E. W. Berry, 1914. Leaf. Upper Cretaceous, USA: South Carolina: Chesterfield County, Middendorf.
Comments: Herendeen and Dilcher (1990c) documented leaves and fruits in association with *Eomimosoidea* inflorescences, and suggested these were all derived from a single plant species which shared several features with extant *Dinizia*. Isolated flowers from the Palaeocene/Eocene boundary were also assigned to this family (as Mimosoideae), although they shared certain characters with Caesalpinioideae (Crepet and Taylor, 1986).
Muller First: *Brevicolporites guinetii* (form taxon) Guinet and Salard-Cheboldaeff, 1975. Middle Eocene, Cameroon. *Eomimosoidea plumosa* (form taxon) Crepet and Dilcher, 1977. Middle Eocene, Tennessee, USA.
Comments: Mutually supported by PFR and Muller. Records from the Lower Eocene are reviewed in Herendeen and Dilcher (1990c).

F. MONIMIACEAE de Jussieu, 1809

PFR First: *Hedycaryoxylon subaffine* (Vater) Suss, 29 August 1960. Wood. Cretaceous: Lower Senonian, Germany: Helmstedt near Braunschweig.
PFR First: *Protohedycarya ilicoides* (O. Heer) L. Rüffle, January 1965. Leaf. Cretaceous: lower Senonian, Germany, Quedlinburg.
Comments: We find no evidence to confirm these early records, although Knappe and Ruffle (1975) also described European Santonian leaves. Leaves with some similarity to those of modern *Doryphora* occur in the Middle Eocene of Anglesea, Australia (Christophel and Greenwood, pers. comm., 1991). Gottwald (1992) assigned Upper Eocene Woods to two genera of Monimiaceae. Leaves like those of modern *Atherosperma* were reported from the Pliocene/Pleistocene of Australia (Hill and Macphail, 1985) which is now considered to be Lower Middle Pleistocene (Hill, pers. comm., 1991).
Muller First: *Laurelia* type. Muller, 1970, Oligocene, New Zealand.
Comments: Muller, 1970, pending: may be too early.

F. MORACEAE Link, 1831

PFR First: *Arthmiocarpus hesperus* (Wieland) Delevoryas, 1964. Fruit. Upper Cretaceous, USA, South Dakota, Grand River valley, slightly west of its junction with Cottonwood Creek (Wieland, 1908).
PFR First: *Combretiphyllum acuminatum* P. Menzel, 1909. Leaf. Cretaceous Senonian, Cameroon, Balangi. *Debeya serrata* F. A. W. Miquel, 1853. Leaf. Upper Cretaceous: Senonian, The Netherlands: Limburg, Kunrade.
Comments: The fossil record was reviewed by Collinson (*in* Crane and Blackmore, 1989), who accepted several genera (including forms like those of modern *Ficus* and *Morus*) based on endocarps and achenes from the Lower Eocene of Eurasia. All earlier records were considered in need of critical revision and many leaf fossils and putative fleshy fruits, previously assigned to the family, were rejected.
Muller First: *Ficus* type, Potter, 1976. Middle Eocene, Tennessee, USA.
Comments: Unchallenged.

F. MORINGACEAE Dumortier, 1829

PFR: No record.
Comments: We know of no claims for megafossils of this family.
Muller First: Pollen record rejected.

F. MUSACEAE de Jussieu, 1789

PFR First: *Haastia speciosa* Ettingshausen, 1887. Leaf.

Upper Cretaceous, New Zealand: Pakawau, Nelson. This use of the genus is nomenclaturally invalid as it is a junior homonym of *Haastia* J. D. Hooker, 1864.
Comments: Large *Musa*-like leaves, including *Musophyllum* and *Musocaulon* (see comments on Cyclanthaceae, and Boyd, 1990) should not be considered to belong in Musaceae, but may combine characters of Heliconiaceae and Strelitziaceae with those of Musaceae, or simply may not be diagnostic of any single family. One record of *Musa* fruits from the Eocene of the Deccan was noted by Daghlian (1981) and Manchester (*in* Knobloch and Kvacek, 1990, pp. 183–8) lists *Musa* as a component of the Middle Eocene Clarno flora.
Pollen: No record.

F. MYRICACEAE Blume, 1829

PFR First: *Myricophyllum longepetiolatum* C. von Ettingshausen, 1893 (post-13 April). Leaf. Cretaceous, Australia: Queensland, between Warragh and Oxley. This use of the genus is nomenclaturally invalid as it is a junior homonym of *Myricophyllum* G. Saporta, 1862, assigned to the Proteaceae.
Comments: Hill (1988) listed all Australian leaf fossils of Myricaceae as incorrectly identified. Macdonald (*in* Crane and Blackmore, 1989) reviewed the literature on fossil history for this family and concluded that megafossils diagnostic of Myricaceae could not be identified earlier than the Eocene. In Europe endocarps like those of *Myrica* occur in the Lower Eocene but are much more abundant from the Upper Eocene onwards (Chandler, 1964; Mai and Walther, 1978, 1985, Friis, 1985a). *Comptonia* foliage and endocarps also occur in the Oligocene onwards of Europe (Mai and Walther, 1978) and Middle Eocene leaves of *Comptonia* are reported in North America (Taylor, 1990). (See also discussion under Juglandaceae.)
Muller First: aff. *Triatriopollenites* sp. (form genus) Doyle and Robbins, 1977, Santonian, eastern USA.
Comments: May be too early: poor support from megafossils.

F. MYRISTICACEAE Brown, 1810

PFR First: *Myristicoxylon princeps* E. Boureau, 10 October 1950. Wood. Tertiary: Palaeocene: Danian. Sydan: Sahara, Asselar.
Comments: This family has a scant fossil record. Wolfe (1977) described leaves from the Middle Eocene of North America and considers that these are probably correctly assigned to the family (Wolfe, pers. comm., 1991). Fruits (*Myristicacarpum*) from the Middle Miocene of Europe were described by Gregor (1978). We are unable to comment on the wood record.
Muller First: *Echimonocolpites major* (form taxon) du Chene, 1978a. Upper Eocene, Nigeria.
Comments: Unchallenged.

F. MYROTHAMNACEAE Niedenzu, *in* Engler and Prentl, 1891

PFR: No record.
Comments: We know of no claims for megafossils of this family.
Muller First: Pending.

F. MYRSINACEAE Brown, 1810

PFR First: *Myrsinopsis succinea* Conwentz, 1886. Flower. Tertiary, Germany: western Prussia. *Pleiomerites reticulatus* Ettingshausen, 1868. Leaf. Tertiary, Czechoslovakia: Bohemia: Kutschlin. *Pleiomeropsis rottensis* Weyland, July 1938. Inflorescence. Tertiary, Germany: Siebengebirge: Rott.
Comments: We have found no recent work on putative megafossils of this family. Fruits like those of modern *Ardisia* are recorded in the Lower Eocene of England, UK (Chandler, 1964).
Muller First: *Suttonia* type. Couper, 1960. Oligocene, New Zealand.
Comments: Unchallenged.

F. MYRTACEAE de Jussieu, 1789

PFR First: *Myrtophyllum* Turczaninow, 1869. Leaf. Cretaceous, Cenomanian, Czechoslovakia: Moravia: Moletein.
Comments: Fruits of Myrtaceae (Myrtoideae–?*Psidium*) were described by Crane *et al.* (1990) from the Upper Palaeocene of North America, while leaves assigned to modern *Eugenia* are listed by Taylor (1990) from the Middle Eocene. *Myrtus* seeds were described by Friis (1985a) from the Lower Miocene of Denmark. Leaves of *Rhodomyrtophyllum* are common in the Middle and Upper Eocene of Europe (Mai and Walther, 1985; Wilde, 1989; Walther, *in* Knobloch and Kvacek, 1990, pp. 149–58). Myrtaceae foliage is common in the Middle Eocene of Australia (Christophel and Greenwood 1987, 1988; Christophel *et al.*, 1987). Leaves and fruits of *Eucalyptus* have also recently been reported from the Middle Eocene Nelly Creek flora (Greenwood, pers. comm., 1991).
Muller First: *Myrtaceidites lisamae* (= *Syncolporites lisamae*) (form taxa). Boltenhagen, 1976a. Santonian, Gabon.
Comments: Unchallenged.

F. NAJADACEAE de Jussieu, 1789

PFR First: *Cymodoceites parisiensis* (Brongniart) Bureau, 25 January–30 January 1886. Tertiary: Eocene, France: Loire-Inférieure, Arthon.
Comments: Seeds like those of modern *Najas* are common from the Oligocene onwards in Europe (Mai, 1985a; Collinson, 1988a).
Pollen: No record.

F. NELUMBONACEAE Dumortier, 1828

PRE: No record.
Comments: Leaves assigned to *Nelumbo* occur in the Lower Eocene of North America and *Nelumbo* is recorded from the Eocene onwards in Europe (listed by Mai, 1985a; Taylor, 1990). Nymphaealean leaves occur in the Upper Cretaceous (e.g. Crabtree, 1987), but these are generally not determinable to a modern family.
Muller First: *Buravicolpites venustus* (form taxon) Bratzeva, 1976, Eocene, Siberia, former USSR.
Comments: Unchallenged.

F. NEPENTHACEAE Dumortier, 1829

PFR: No record.
Comments: We know of no claims for megafossils of this family.
Muller First: *Nepenthes*. Anderson and Muller, 1975. Lower Miocene, Borneo.
Comments: Unchallenged.

F. NYCTAGINACEAE de Jussieu, 1789

PFR First: *Nyctaginites ellipticus* E. W. Berry, 1938. Leaf. Tertiary, Argentina: Rio Pichileufu.

Comments: Berger (1954) referred to a Miocene record for *Abronia*, and Wolfe and Upchurch (1986) cited a possible Palaeocene record of the family. However, we know of no well-substantiated megafossils for this family.
Muller First: *Lymingtonia* (form genus) Muller, 1970. Lower Eocene, England, UK.
Comments: Supported.

F. NYMPHAEACEAE Salisbury, 1805

PFR First: *Braseniopsis venulosa* G. Saporta, June 1893. Leaf. Cretaceous: Albian, Portugal: Buarcos.
Comments: Nymphaealean leaves are common in the Upper Cretaceous, but generally cannot be assigned to a modern family (e.g. Crabtree, 1987). Palaeocene leaves are listed by Taylor (1990), but their exact family affinities are equally unclear. Seeds (*Sabrenia* and *?Palaeonymphaea*) occur in the Upper Palaeocene (Collinson, 1980, 1986a), but many more forms occur from the Middle Eocene onwards in Europe, with one record in North America (Collinson, 1980; Mai, 1985a; Cevallos-Ferriz and Stockey, 1989). Collinson (1980) considered that certain extinct genera like *Sabrenia* were intermediate in characters between Cabombaceae and Nymphaeaceae *s. s.*, whereas Cevallos-Ferriz and Stockey (1989) concluded that these two families were probably distinguishable by the Middle Eocene.
Muller First: *Zonosulcites scollardensis* and *Z. parvus* (form taxa) Srivastava, 1969b, Maastrichtian, Canada.
Comments: Unchallenged.

F. NYSSACEAE Dumortier, 1829

PFR First: *Nyssoidea eocenica* M. E. J. Chandler, July–December 1962. Fruit. Tertiary: Eocene, UK, England: Dorset: Bournemouth. *Palaeonyssa multilocularis* E. M. Reid and M. E. J. Chandler, 25 November 1933. Fruit. Tertiary: Eocene, UK, England: Kent, Sheppey. *Protonyssa bilocularis* Reid and Chandler, 25 November 1933. Fruit. Tertiary: Eocene, UK, England: Kent: Sheppey.
Comments: These Lower Eocene fruits of the family are highly diagnostic. *Protonyssa* was later included in *Nyssa* and Mai and Walther (1978) reported the *N. bilocularis* in the Lower, Middle and Upper Eocene of Europe. Middle Oligocene fruits are reported from North America (Taylor, 1990).
Muller First: *Tricolporopollenites kruschi* (form taxon). Haseldonckx, 1973, Palaeocene, Spain; Gruas-Cavagnetto, 1978, Palaeocene, France.
Comments: Possible.

F. OCHNACEAE de Candolle, 1811

PFR: No record.
Comments: Chesters *et al.* (1967) cited a North American Eocene record of *Ouratea*, but this was not listed by Taylor (1990).
Muller First: Pending.

F. OLACACEAE Mirble ex de Candolle, 1824

PFR: No record.
Comments: The record for this family is sporadic. Fruits like those of modern *Olax* and *Erythropallum* occur in the Lower Eocene of England, UK (Chandler, 1964). Leaves assigned to *Schoepfia* occur in the Middle Eocene of North America (Taylor, 1990). The *Olax* fruit listed by Taylor (1990) was not listed by Manchester (*in* Knobloch and Kvacek, 1990, pp. 183–8) in his summary of the Clarno flora, and Manchester (pers. comm., 1991) has confirmed that Olacaceae are not represented in the Clarno Flora; the supposed *Olax* fruits being *Musa* seeds.
Muller First: *Anacolosidites cretacicus* (form taxon). Krutzsch and Lenk, 1969, Maastrichtian, Germany. *Anacolosidites* sp., Chlonova, 1974, Maastrichtian, Siberia, former USSR. *A. baculatus*. Miki, 1977, Maastrichtian, Japan. *A. meyerorum*. Chmura, 1973, Maastrichtian, California, USA.
Comments: Well supported palynologically.

F. OLEACEAE Hoffmannsegg and Link, 1813–1820

PFR First: *Oleiphyllum boreale* Conwentz, 1886. Leaf. Lower Tertiary, Germany: Western Prussia.
Comments: Many examples of leaves and fruits assigned to *Fraxinus* are listed by Taylor (1990) from the Palaeocene onwards in North America. However, we are not aware of any recent revision of this record. Wolfe (pers. comm., 1991) accepts Middle Eocene and later records. Endocarps like those of modern *Olea* are recorded in the Upper Eocene of England, UK (Chandler, 1964), and seeds like those of modern *Jasminium* occur in the Miocene and Pliocene of Europe (Szafer, 1961).
Muller First: *Retitricolporites oleoides* (form taxon). Roche and Schuler, 1976. Oligocene, Belgium.
Comments: Unchallenged.

F. ONAGRACEAE de Jussieu, 1789

PFR First: *Palaeeucharidium cellulare* E. M. Reid and M. E. J. Chandler, 25 November 1933. Fruit. Tertiary: Eocene, UK, England: Kent: Sheppey, Minister.
Comments: This is an isolated record, but we know of no reason to reject it. Seeds like those of modern *Ludwigia* are common from the Middle Oligocene onwards, especially in the Miocene, of Europe (Friis, 1985a; Mai, 1985a).
Muller First: *Trivestibulopollenites* sp. (form genus). Chmura, 1973, Maastrichtian, California, USA. *Jussitriporites* sp. (form genus). Pares Regali *et al.*, 1974. Maastrichtian, Brazil.
Comments: Unchallenged.

F. OPILIACEAE Valeton; 1886

PFR First: *Opilioxylon nigerinum* J.-C. Koeniguer, 1970. Wood. Upper Cretaceous. Southern Sahara, Eastern Niger, Mt. Kanak.
Comments: We are unable to comment on this wood record. We know of no other megafossil record for this family.
Pollen: No record.

F. ORCHIDACEAE de Jussieu, 1789

PFR First: *Palaeorchis rhyzoma* A. Massalongo, post-22 August 1858. Stem, rhizome and leaf. Tertiary: Eocene, Italy: Monte Bolca: Veronese. *Protorchis monorchis* Massalongo, post-22 August 1858. Tertiary: Eocene, Italy: Monte Bolca.
Comments: Schmid and Schmid (1973) concluded that the family had no reliable fossil record, but considered a Pliocene example (Strauss, 1969) as most plausible. Friis (1985a) described a Lower Miocene seed which was tentatively compared with Orchidaceae.
Muller First: Pending.

F. OROBANCHACEAE Ventenat, 1799

PFR: No record.

Comments: We know of no claims for megafossils of this family.
Muller First: Cistanche type. Beucher, 1975, Pliocene, Sahara.
Comments: Unchallenged.

F. OXALIDACEAE Brown, *in* Tuckey, 1817

PFR First: *Oxalidites brachysepalus*. R. Caspary, 1887. Fruit. Tertiary, Germany: Baltic Prussia: Samland.
Comments: Seeds like those of modern *Oxalis* were recorded by Mai and Walther (1988) in the Pliocene of Germany. Leaves (*Averrhoites* Hickey) described by Hickey (1977) were morphologically similar to leaves of this family, but this was not considered to indicate a close relationship.
Pollen: No record.

F. PAEONIACEAE Rudolphi, 1830 (see RANUNCULACEAE)

F. PANDANACEAE Brown, 1810

PFR First: *Pandanocarpum oolithicum* (Carruthers) Zigno, 1873. Fruit. Jurassic, UK, England: Kingsthorpe near Northampton. *Podocarya bucklandi* Goeppert, *in* Bronn, 1848, p. 1023, 1836 (post-30 May). Fruit. Jurassic, UK, England: Dorset: Charmouth.
Comments: Daghlian (1981) stated that the Pandanaceae have no reliable fossil record, although it is not clear if he considered and rejected all possible records such as the Oligocene to Miocene material from Europe summarized in Weyland (1957). Lakhanpal *et al.* (1984) assigned Eocene leaves to the family. We consider that the fossil record of this family requires a thorough, critical revision.
Muller First: *Pandanus tectoria* type Jarzen, 1978b. Maastrichtian, Canada.
Comments: Unchallenged.

F. PAPAVERACEAE de Jussieu, 1789

PFR: No record.
Comments: A single seed from the Uppermost Eocene of England was assigned to modern *Papaver* by Reid and Chandler (1926). The original has deteriorated and no new specimens have been found. Dorofeev (1969) recorded seeds like those of modern *Macleaya* from the Middle Miocene. We consider these isolated records to be unconfirmed. Flowers (*Princetonia* Stockey) with psilate pentacolpate pollen, described by Stockey (1987) from the Middle Eocene of North America show similarities with Flacourtiaceae and Papaveraceae, particularly with Eschscholziaeae of the Papaveraceae. However, the similarities are not sufficient to include the flowers in the modern family, especially in the absence of fossil evidence for other parts of the plant.
Muller First: Pollen record rejected.

F. PASSIFLORACEAE de Jussieu ex Kunth, 1817

PFR First: *Passifloraephyllum kraeuseli* K. Râsky, 29 August 1960. Leaf. Tertiary, Upper Eocene, Hungary: Budapest-Obuda.
Comments: Seeds like those of modern *Passiflora* were described from the Middle Miocene of Europe by Mai (1964) and Gregor (1978, 1982). We know of no well-substantiated leaf record of this family.
Muller First: Pending.

F. PEDALIACEAE Brown, 1810

PFR: No record.
Comments: Fruits of *Trapella* (Trapellaceae of many authors) occur in the Miocene onwards in Europe (Mai, 1985a; Collinson, 1988a).

F. PELLICIERACEAE Beauvisage, 1920

PFR: No record.
Comments: We know of no claims for megafossils of this family.
Muller First: *Verrutricolporites crassus* (form taxon) Fuchs, 1970. Upper Eocene, northern South America.
Comments: Unchallenged.

F. PENTAPHYLLACEAE Engler and Prantl, 1897

PFR: No record.
Comments: Fruits from the Upper Cretaceous of Europe were described as two species of a new genus *Allericarpus* Knobloch and Mai; others were included in an extinct species of the modern genus *Pentaphylax* (Knobloch and Mai, 1986). We know of no other fossil record for this family and thus consider that these isolated records should be treated with caution.
Pollen: No record.

F. PHYTOLACACEAE Brown, *in* Tuckey, 1819

PFR First: *Stachycarpus eocenica* Meunier, 1898. Fruit. Tertiary: Eocene, France: Bêthune, Beuvry. This use of the genus is nomenclaturally invalid as it is a junior homonym of *Stachycarpus* (Endlicher) Van Tieghem, 1891.
Comments: This PFR record was only tentatively assigned to the family. We know of no well-substantiated megafossils of the family.
Pollen: No record.

F. PIPERACEAE Agardh, 1825

PFR First: *Piperites miquelianus* Goeppert, 1854. Leaf. Tertiary, Java: Dorfe Tandjung.
Comments: Seeds like those of modern *Peperomia* were recorded by Dorofeev (1988) from the Miocene of Siberia and Tambov, former USSR.
Pollen: No record.

F. PITTOSPORACEAE Brown, *in* Flinders, 1814

PFR First: *Billardierites longistylus* R. Caspary, 1881 (post-27 May). Flower. Tertiary: Miocene, Germany: Prussia, Samland.
Comments: This was the only megafossil noted by Friis and Skarby (1982) as having been assigned to the family. The determination should be considered unconfirmed.
Muller First: *Pittosporum* Beucher, 1975. Pliocene. Sahara.
Comments: Unchallenged.

F. PLANTAGINACEAE de Jussieu, 1789

PFR First: No record.
Comments: Seeds like those of modern *Plantago* are recorded in the Pliocene of Poland (Szafer, 1954).
Muller First: *Plantago coronopus* type Van Campo, 1976, Upper Miocene, Spain. *P. major* type, Van Campo, 1976, Upper Miocene, Spain. *P. lanceolata* type, Van Campo, 1976, Upper Miocene, Spain. *Plantaginacearumpollis* (form genus), Upper Miocene: Gray, 1964, USA; Naud and Suc, 1975, France; Krutzsch, 1966c, Germany.
Comments: Unchallenged.

F. PLATANACEAE Dumortier, 1829

PFR First: *Pseudoaspidiophyllum platanoides* Hollick, 1930. Leaf. Upper Cretaceous. USA: Alaska: Yukon River, one and a half miles below Seventymile Creek.
PFR First: *Pseudoprotophyllum marginatum* Hollick, 1930. Leaf. Upper Cretaceous, USA: Alaska: Yukon, 6 miles above Nahochatilton.
Comments: Friis *et al.* (1988) document pistillate and staminate inflorescences in the Albian (upper Lower Cretaceous) of North America and Santonian to Campanian of North America and Sweden. An extensive fossil record occurs throughout the Tertiary, including inflorescences, infructescences and associated foliage, and reconstructed whole plants (see Pigg and Stockey, 1991 and references cited in Friis *et al.*, 1988).
Muller First: *Platanus occidentaloides* Frederiksen, 1980. Upper Eocene, south-east USA.
Comments: *Platanus* pollen is known in the Upper Palaeocene of Scotland, UK (Boulter and Kvacek, 1989). Pollen is well preserved in the flowers quoted above.

F. PLUMBAGINACEAE de Jussieu, 1789

PFR: No record.
Comments: We know of no claims for megafossils of this family.
Muller First: *Aegialitis* type. Morley, 1977. Middle Miocene. Borneo.
Comments: Unchallenged.

F. POACEAE Barnhart, 1895

PFR First: *Yorkia gramineoides* L. F. Ward, 1900. Leaf. Triassic, York Haven, York County, Pennsylvania, USA.
Comments: Whole plants including spikelets and inflorescences of Poaceae have been recorded recently from the Palaeocene/Eocene boundary in North America (Crepet and Feldman, 1991). These represent the earliest unequivocal grasses. Other acceptable records, including occurrences of modern grassland grass genera (Lower Miocene onwards) and examples of C4 and C3 photosynthetic pathways confirmed from anatomy and isotopic signatures (Upper Miocene and Pliocene), are summarized by Thomasson (1987b) and Tidwell and Nambudiri (1990).
Muller First: *Monoporites annulatus* (form taxon) Pares Regali *et al.*, 1974a, Palaeocene, Brazil. Salard-Cheboldaeff, 1978, Palaeocene, Cameroon. Adegoke *et al.*, 1978, Palaeocene, Nigeria. *Graminidites* (form genus) Harris, 1965b, Palaeocene, Queensland, Australia.
Comments: Unchallenged.

F. PODOSTEMACEAE Richard ex Agardh, 1822

PFR First: *Nitophyllites zaisanica* I. A. Ilinskaja, 1963 (post-12 Jan.). Leaf. Tertiary: Upper Eocene, USSR: Zaisan Basin. This generic name is listed as nomenclaturally invalid in PFR; however, it still serves as a record for this family.
Comments: Szafer (1952) described leaves and flowers from the Pliocene of Poland and a probable example from the lower most Miocene of Germany is discussed by Mosbrugger (*in* van Koenigswald, 1989).
Pollen: No record.

F. POLEMONIACEAE de Jussieu, 1829

PFR: No record.
Comments: We know of no claims for megafossils of this family.
Muller First: *Gilia filififormis* type Muller, 1970. Upper Miocene, USA
Comments: Unchallenged.

F. POLYGALACEAE Brown, *in* Flinders, 1814

PFR: No record.
Comments: We know of no published megafossils of this family, but Wolfe (pers. comm., 1991) has an unpublished leaf of *Securidaca* from the Lower Miocene of Oregon.
Muller First: *Psilastephanocolporites fissilis* (form taxon). Doubinger and Chotin, 1975, Palaeocene, Chile.
Comments: Unchallenged.

F. POLYGONACEAE de Jussieu, 1789

PFR First: *Coccolobites cretaceus* E. W. Berry, 1916. Leaf. Upper Cretaceous, USA, Maryland, Cecil County, Grove Point. This use of the genus is nomenclaturally invalid as it is a junior homonym of *Coccolobites* Visiani, 1858, also assigned to the family Polygonaceae but of Eocene age.
Comments: Fruits like those of modern *Rumex* and *Polygonum* occur in floras from the Lower Miocene and younger strata in Europe and Asia (Friis, 1985a; Dorofeev, 1988). We know of no well-substantiated leaf fossils for the family.
Muller First: *Persicarioipollis* (form genus) Krutzsch, 1970d, Palaeocene, Germany.
Comments: There is some support in Europe.

F. PONTEDERIACEAE Kunth, 1816

PFR: No record.
Comments: Seeds like those of modern *Monochoria* and seeds and leaves like those of modern *Eichornia* are recorded from the Upper Eocene onwards in Europe (Mai and Walther, 1978, 1985; Wilde, 1989).
Muller First: Pollen record rejected.

F. PORTULACACEAE de Jussieu, 1789

PFR: No record.
Comments: Upper Tertiary seed records have been reidentified as Melastomataceae (see comments on Melastomataceae).
Muller First: *Portulaca* cf. *oleracea*. Van Campo, 1976. Lower Miocene, Spain.
Comments: Unchallenged.

F. POSIDONIACEAE Lotsy, 1911

PFR: No record.
Comments: Daghlian (1981) considered that this family did not have a megafossil record inspite of several claims in the literature.
Pollen: No record.

F. POTAMOGETONACEAE Dumortier, 1829

PFR First: *Potamogetophyllum vernonense* Fontaine, 1905. Leaf. L Cretaceous. USA: Virginia: Mount Vernon.
Comments: Fruits like those of modern *Potamogeton* are widespread from the lower Middle Eocene onwards in Europe and Asia (Collinson, 1982, 1988a; Mai, 1985a) and leaves assigned to *Potamogeton* are listed by Taylor (1990) from the Middle Eocene of North America. Fruits named *Limnocarpus* (and related fossil genera), often assigned to this family, are considered intermediate between those of this family and Ruppiaceae (Collinson, 1982). (See comments on Ruppiaceae.)
Muller First: *Potamogeton*. Upper Miocene: Van Campo,

1976. Spain, Naud and Suc, 1975, France, Menke, 1976; von der Brelie, 1977, Germany.
Comments: Unchallenged.

F. PRIMULACEAE Ventenat, 1799

PFR: No record.
Comments: Seeds like those of modern *Lysimachia* are recorded from the Upper Oligocene onwards in Europe and Asia (Friis, 1985a).
Muller First: *Samolus valerandi* type. Van Campo, 1976. Upper Miocene, Spain.
Comments: Unchallenged.

F. PRISCACEAE Retallack and Dilcher, 1981 Upper Cretaceous only

PFR: No record.
First: *Prisca reynoldsii* Retallack and Dilcher, 1981. Fruiting axis with associated leaves. Lower Cenomanian, Dakota Formation, central Kansas, USA.
Comments: Kvacek (1992) proposes that other material of leaves and reproductive structures from the European Upper Cretaceous should be included in the family. The family is closely related to Lauraceae, within Laurales (Kvacek, 1992) and the reproductive structures are similar to the lower Cenomanian material from Maryland included in Lauraceae by Drinnan *et al.* (1990). (See comments on Lauraceae.)
Pollen: No record. No pollen is preserved in any of the reproductive structures mentioned by the above authors.

F. PROTEACEAE de Jussieu, 1789

PFR First: *Proteaephyllum reniforme* Fontaine, 1889. Leaf. Lower Cretaceous, USA: Virginia: Fredericksburg.
PFR First: *Rogersia* sp. 1889. L Cretaceous, Potomac Group. USA.
Comments: Proteaceae are now well documented in the Middle Eocene onwards in Australia. Records include foliage similar to modern *Banksia* and *Dryandra* of tribe Banksieae (*Banksieaephyllum* Cookson and Duigan, 1950 – Hill and Christophel, 1988); foliage similar to modern *Lomatia* (Carpenter and Hill, 1988) and inflorescences of *Musgraveinanthus alcoensis* Christophel (1984) also of tribe Banksieae. Leaves and fruits assigned to *Lomatia* from the Middle Eocene onwards in North America are listed by Taylor (1990), but these require confirmation.
Muller First: *Knightia* aff. *excelsa*. Couper, 1960, Upper Senonian, New Zealand. *Propylipollis* (*Proteacidites*) *dehaani* (form taxon). Germeraad *et al.*, 1968, Maastrichtian, North America and Central Africa. *Proteacidites* sp. Chmura, 1973, Maastrichtian, California, USA.
Comments: There is no evidence from megafossils for such an early origin.

F. PUNICACEAE Horaninow, 1834

PFR First: *Punicites hesperidum* C. O. Weber, December 1855. Flower. Tertiary, Germany, Rott. This genus was only tentatively assigned to this family.
Comments: Friis (1985a) rejected the earlier assignment by Gregor of *Carpolithes natans* Niktin ex Dorofeev to modern *Punica* citing differences such as the absence of a heteropyle and presence of a raphal chamber on the opposite side of the seed to the germination valve in the fossils. The species occurs in the Middle Oligocene to Upper Miocene in Europe. Gregor (1978) recorded fruits and seeds of another species, which he assigned to *Punica*, from the Middle Miocene of Germany.
Pollen: No record.

F. RANUNCULACEAE de Jussieu, 1789

PFR First: *Paeoniaecarpum hungaricum* G. Andreânszky, 1961. Fruit. Tertiary: Miocene: Sarmatian, Hungary: Szelecsi Valley.
Comments: The PFR record may perhaps belong in Paeoniaceae *sensu* Cronquist (1981). Fruits of *Myosurus* were recorded in the Upper Oligocene of Germany and England, UK (Mai and Walther, 1978). Fruits like those of *Ranunculus* are recorded from the Oligocene onwards in Europe and former USSR (Mai, 1985a; Łancucka-Srodoniowa, 1979; Takhtajan, 1974).
Muller First: *Punctioratipollis ludwigi* (form taxon). Krutzsch, 1966c, Lower Miocene, Germany.
Comments: According to megafossils this record is too late.

F. RESEDACEAE Gray, 1821

PFR: No record.
Comments: We know of no claims for megafossils of this family.
Muller First: *Reseda*. Beucher, 1975. Pliocene, Sahara.
Comments: Unchallenged.

F. RESTIONACEAE Brown, 1810

PFR First: *Restiacites pleiocaulis* Saporta, 1861. Tertiary: Eocene, France, Provence.
Comments: This PFR record was only tentatively assigned to the family. No megafossil record was accepted for this family by Daghlian (1981).
Muller First: *Graminidites* sp. (form genus). Jardine and Magloire, 1963, Maastrichtian, Senegal.
Comments: See *Milfordia* within Centrolepidaceae; possibly from an extinct family.

F. RHAMNACEAE de Jussieu, 1789

PFR First: *Eorhamnidium* sp. Berry, 1919. Leaf. Upper Cretaceous.
PFR First: *Rhamnoxylon* sp. Chitaley and Kate, 1972. Wood. Upper Cretaceous, India: Madhyar Pradesh: Chhindwara: Mohgaon-Kalan.
Comments: Leaves from the Middle Eocene onwards are listed for North America by Taylor (1990). Wolfe (pers. comm., 1991) considers that Maastrichtian and Palaeocene leaves from North America may be assigned to the family. Fruits and seeds like those of modern *Paliurus* occur in the Middle Miocene of Denmark (Friis, 1985a) and like those of modern *Zizyphus* in the Lower Miocene of Kenya (Chesters, 1957).
Muller First: *Rhamnus brandonensis*. Traverse, 1955. Oligocene. Vermont, USA. *Tricolporopollenites haanradensis* (form taxon). Petrescu, 1973, Oligocene, Romania.
Comments: Unchallenged.

F. RHIZOPHORACEAE Brown, *in* Flinders, 1814

PFR First: *Palaeobruguiera elongata* M. E. J. Chandler, January 1961. Fruit. Tertiary: Eocene, UK: England: Hampton: Swale Cliff.
Comments: Viviparous embryos of *Ceriops* have also been reported from the Lower Eocene of England (Wilkinson, 1981). Leaves assigned to *Kandelia* from the Middle Eocene of North America are listed by Taylor (1990).

Muller First: *Zonocostites ramonae* (form taxon). Germeraad *et al.*, 1968. Upper Eocene, Caribbean. Pares Ragali *et al.*, 1974a, Upper Eocene, Brazil. Venkatachala and Rawat, 1972, Upper Eocene, India.
Comments: Unchallenged.

F. RHOIPTELEACEAE Handel-Mazetti, 1932

PFR: No record.
Comments: Fruits assigned to *Rhoiptelea pontwallensis* (Vangerow) Knobloch and Mai, 1986 were recorded from the Maastrichtian of Europe. We know of no other megafossil record for this family and, in view of the complexity of Upper Cretaceous Juglandales/Myricales (discussed under 'Juglandaceae'), feel that this record should be considered as unconfirmed.
Muller First: Pending.

F. ROSACEAE de Jussieu, 1789

PFR First: *Crataegites borealis* V. A. Samylina, March 1960. Leaf. L Cretaceous, former USSR: north-east Siberia: Kolyma Basin.
Comments: Numerous examples are listed by Taylor (1990) from North America, which include several well-substantiated Middle Eocene leaves (of Spiraeoideae and Rosoideae) and one flower. One citation is from the Maastrichtian, and leaves and fruits named *Prunus* (Prunoideae) are listed in the Palaeocene. Woods of Prunoideae are known from the Lower Eocene onwards (Cevallos-Ferriz and Stockey, 1990c). The record of *Prunus* fruits in Europe (from the Eocene onwards) was reviewed in detail by Mai (1984). Foliage of Rosaceae is recorded in the Middle Eocene of Europe (Wilde, 1989).
Muller First: *Psilatricolporites undulatus* (form taxon). Salard-Cheboldaeff, 1978. Oligocene, Cameroon.
Comments: Unchallenged.

F. RUBIACEAE de Jussieu, 1789

PFR First: *Ixorophyllum anceps* H. T. Geyler. Leaf. Tertiary: Eocene, Borneo: Labuan.
PFR First: *Paleorubiaceophyllum eocenicum* (E. W. Berry) J. L. Roth and D. L. Dilcher, 10 December 1979. Leaf. Tertiary: Eocene, USA.
PFR First: *Rubiaceocarpum markgrafii* Kräusel (*'markgrafi'*), post-May 1939. Seed. Tertiary: Eocene, Egypt.
Comments: Taylor (1990) lists foliage from the Middle Eocene onwards in North America. Vaudois-Mieja (1976) documented fruits from the Lower Eocene of France. *Cephalanthus* fruits are widespread from the Upper Eocene onwards in Europe (Friis, 1985a; Mai and Walther, 1985).
Muller First: *Triporotetradites nachterstedtensis* (form taxon). Krutzsch, 1970b. Upper Eocene, Germany.
Comments: Unchallenged.

F. RUPPIACEAE Hutchinson, 1934

PFR: No record.
Comments: Fruits of the extinct *Limnocarpus* and related genera were considered to be intermediate in character between Ruppiaceae and Potamogetonaceae (Collinson, 1982). These occur from the Upper Palaeocene onwards in Europe and former USSR. Fruits identical with those of modern *Ruppia* were only recognized in the Pleistocene by Collinson (1982), who considered that Miocene fossil fruits (*Midravalva* Collinson, *Medardus* Collinson and a grouping of 'fossil *Ruppia*') were closest to modern *Ruppia* but somewhat intermediate between that genus and Potamogetonaceae.
Muller First: Pending.

F. RUTACEAE de Jussieu, 1789

PFR First: *Evodioxylon oweni* (Carruthers) A. Chiarugi, 20 June–15 July 1933. Wood. Cretaceous, and Tertiary: Miocene. Australia; southern Italian East Africa (Somaliland): Scec-Gure.
Comments: Rutaceae are well documented from seeds from the Eocene onwards in North America, Europe and Asia (e.g. Collinson and Gregor, pp. 67–80, *in* Collinson, 1988c; Gregor, 1989), and seeds were also reported from the Maastrichtian (Knobloch and Mai, 1986) and Palaeocene (Mai, 1987a). Leaves occur in the Middle Eocene of Europe (Wilde, 1989) and North America (Taylor, 1990).
Muller First: *Phellodendron* type. Zagwijn, 1960. Pliocene, The Netherlands.
Comments: Possible.

F. SABIACEAE Blume, 1851

PFR First: *Sabiocaulis sakuraii* M. C. Stopes and K. Fujii, 14 February 1910. Stem, Upper Cretaceous, Japan, Hokkaido.
Comments: Endocarps like those of modern *Sabia* and *Meliosma*, as well as an extinct form *Insitiocarpus* Knobloch and Mai, were recorded by Knobloch and Mai (1986) in the Maastrichtian of Europe. Palaeocene examples are reported by Crane *et al.* (1990) and Mai (1987a) from North America and Europe, respectively. Leaves assigned to *Meliosma* are listed by Taylor (1990) from the Palaeocene onwards in North America.
Pollen: No record.

F. SALICACEAE Mirbel, 1815

PFR First: *Credneria integerrina* Zenker, 1833. Leaf. Upper Cretaceous, Germany: Blankenburg.
PFR First: *Dryoxylon jenense* Schleiden, 1853. Wood. Middle Triassic, Germany: near Jena.
Comments: Well-substantiated Salicaceae occur in the uppermost Palaeocene (*Populus*) and Lower Eocene (*Salix*) of North America. Middle Eocene examples of both genera are based on connected foliage and reproductive organs (Manchester *et al.*, 1986).
Muller First: *Salix*. Graham and Jarzen, 1969, Oligocene, Puerto Rico.
Comments: Unchallenged.

F. SANTALACEAE Brown, 1810

PFR First: *Thesianthium inclusum* Conwentz, 1886. Flower. Lower Tertiary, Germany: West Prussia.
Comments: Mai (1976) described seeds assigned to modern *Santalum* from the Middle Eocene of Germany. Chesters *et al.* (1967) cited a North American Cenomanian leaf, but this is not listed by Taylor (1990).
Muller First: *Santalumidites cainozoicus* (form taxon). Cookson and Pike, 1954, Eocene, Australia.
Comments: Unchallenged.

F. SAPINDACEAE de Jussieu, 1789

PFR First: *Sapindopsis cordata* Fontaine, 1889. Leaf fragment. L Cretaceous.
Comments: The fossil record for Sapindaceae is rather sporadic. Seeds (assigned to *Sapindospermum* Reid and Chandler) occur in Europe in the Maastrichtian (Knobloch and Mai, 1986). A variety of fruit and seed forms and twigs

occur in the Lower Eocene of England, UK (Chandler, 1964, Wilkinson, *in* Collinson, 1988c, pp. 81–6). In a review of the fossil record for Sapindaceae, Erwin and Stockey (1990) do not accept any Cretaceous megafossils from North America, but note various Palaeocene and Eocene leaf forms (see also list in Taylor, 1990). Erwin and Stockey (1990) described flowers from the Middle Eocene of Canada as *Wehrwolfea* Erwin and Stockey which they assign to Sapindaceae. Wolfe and Tanai (1987) discuss the extinct fruit *Bohlenia*, which is somewhat intermediate between Aceraceae and Sapindaceae. Fruits of *Pteleaecarpus*, which range from the Eocene to Miocene in Europe, Asia and North America, were assigned to Sapindaceae (Buzek *et al.*, 1989). However, this assignment was incorrect (Manchester, pers. comm., 1991).
Muller First: *Tricolporites* sp. (form genus). Kemp, 1976. Middle Eocene, central Australia.
Comments: Tricolporate pollens are notoriously difficult to name and to assign to modern taxa.

F. SAPOTACEAE de Jussieu, 1789

PFR First: *Sapotacites sideroxyloides* Ettingshausen, 1853. Leaf. Tertiary: Eocene, Austria: Tyrol: Haering. *Isonandrophyllum* sp. 1887. Leaf. Tertiary: Eocene, Indonesia: Borneo: Labuan. *Sapoticarpum rotundatum* 25 November 1933. Fruit. Tertiary: Eocene, UK, England: Kent: Sheppey. *Sapotispermum sheppeyense* 25 November 1933. Seed. Tertiary: Eocene, UK, England: Kent: Sheppey.
Comments: The fossil record of the Sapotaceae is scant, but we have no reason to reject these Lower Eocene fruit and seed records. Foliage also occurs in the Middle Eocene of Europe (Wilde, 1989).
Muller First: *Sapotaceoidaepollenites robustus* (form taxon). Muller, 1970, Senonian, Borneo.
Comments: Unchallenged.

F. SARGENTODOXACEAE Stapf, ex Hutchinson, 1926

PFR: No record.
Comments: Tiffney (pers. comm., 1992) has recognized seeds of this family in the Oligocene Brandon lignite, Vermont, USA. Details of the locality may be found in Tiffney and Barghoorn (1979).
Pollen: No record.

F. SAURURACEAE Meyer, 1827

PFR First: *Saururopsis niponensis* M. Stopes and Fujii, March 1910. Stem. Upper Cretaceous, Japan: Hokkaido. This use of the genus is nomenclaturally invalid as it is a junior homonym of *Saururopsis* Turczaninow, 1848.
Comments: Seeds like those of modern *Saururus* range from the Upper Eocene to the Pliocene in Europe (Friis, 1985a).
Pollen: No record.

F. SAXIFRAGACEAE de Jussieu, 1789

PFR First: *Saxifragispermum spinosissimum* E. M. Reid and M. E. J. Chandler, 25 November 1933. Fruit. Tertiary: Eocene, England: Kent: Sheppey.
Comments: This fruit was reassigned to Flacourtiaceae (Chandler, 1964). *Saxifragaceaecarpum* was reassignd to Loganiaceae (Mai, 1968). Several Upper Cretaceous Saxifragalean flowers have been reported (Friis, 1990), but these combine features of modern Saxifragaceae with those of Hydrangeaceae and Escalloniaceae (= Grossulariaceae *sensu* Cronquist), perhaps being most similar to the latter. Leaves listed as Saxifragaceae in the Upper Oligocene of North America (Taylor, 1990, p. 302) are Hydrangeaceae *sensu* Cronquist (*Philadelphus*, *Jamesia* and *Fendlera*). Flowers from the Baltic amber previously included in Saxifragaceae should not be included in the family, due to lack of information on the androecium (Friis and Skarby, 1982). Seeds like those of modern *Chrysosplenium* and *Mitella* have been recorded in the Pliocene of Europe (noted in Friis and Skarby, 1982).
Pollen: No record.

F. SCHEUCHZERIACEAE Rudolphi, 1830

PFR: No record.
Comments: Chesters *et al.* (1967) cite a Pliocene record. We know of no well-substantiated megafossil record for this family.
Pollen: No record.

F. SCHISANDRACEAE Blume, 1830

PFR: No record.
Comments: Leaves like those of modern *Schisandra* are recorded in the Upper Eocene of Europe (Mai and Walther, 1985; Wilde, 1989) and fruits of *Schisandra* are listed by Manchester (*in* Knobloch and Kvacek, 1990, pp. 183–8) from the Middle Eocene of North America.
Muller First: *Schisandra* type. Chmura, 1973, Maastrichtian, California, USA.
Comments: Unchallenged.

F. SCROPHULARIACEAE de Jussieu, 1789

PFR First: *Paulownioxylon hondoense* S. Watari, 1948. Wood. Tertiary, Lower Miocene, Japan, Hanenisi, Simane Prefecture, Kute Toun, Anno District.
Comments: Seeds like those of modern *Limosella* are recorded from the Upper Eocene onwards in Europe (Mai and Walther, 1985). Seeds of *Gratiola* are recorded from the Upper Miocene of Poland (Łancucka-Srodoniowa, 1979).
Muller First: Pending.

F. SIMAROUBACEAE de Candolle, 1811

PFR First: *Ailanthoxylon indicum* Prakash, 1959, Wood, Tertiary, Deccan Intertrappean: India: Chhindwara District: Madhya Pradesh: Mohgaon: Kalan.
Comments: Fruits like those of modern *Ailanthus* occur from the Middle Eocene onwards in North America (Taylor, 1990) and they are also recorded in the Middle Eocene of Europe (Collinson, 1988b).
Muller First: Pending.

F. SMILACACEAE Ventenat, 1799

PFR First: *Majanthemophyllum petiolatum* C. O. Weber, December 1851. Leaf. Tertiary: Oligocene, Germany: Quegstein.
Comments: Several leaf forms with morphology and cuticular details similar to those of modern *Smilax* are assigned to Smilacaceae (as Smilacoideae) from the Middle Eocene of Messel, Germany; other examples occur in the Oligocene and Miocene in Europe (Mai and Walther, 1978; Wilde, 1989). One leaf of this type was noted in the Oligocene of North America (Taylor, 1990).
Muller First: *Smilax*. Graham, 1976. Upper Miocene, Mexico.
Comments: Unchallenged.

F. SOLANACEAE de Jussieu, 1789

PFR First: *Cantisolanum daturoides* E. M. Reid and M. E. J. Chandler, 25 November 1933. Fruit. Tertiary: Eocene, UK, England: Kent: Sheppey.
Comments: This is an isolated record, but we have no reason to reject it. Seeds like those of modern *Physalis* are recorded in the Middle Miocene and Pliocene of Europe (Szafer, 1961) and seeds like those of modern *Solanum* are reported from the European Middle to Upper Miocene (Gregor, 1982; Van der Burgh, 1987).
Pollen: No record.

F. SONNERATIACEAE Engler and Gilg, 1924

PFR First: *Sonneratiorhizos raoii* S. D. Chitaley, August 1969. Wood and/or root. Tertiary: Palaeocene? India: Madhyar Pradesh, Chhindwara District, Mohgaon, Kalan.
Comments: Mehrotra (1988) described a wood from the Lower Tertiary, Deccan Intertrappean Beds as a species of *Sonneratioxylon* and critically reviewed other records.
Muller First: *Florschuetzia levipoli* (form taxon). Muller, 1970, Lower Miocene, Borneo.
Comments: Unchallenged.

F. SPARGANIACEAE Rudolphi, 1830

PFR First: *Sparganiocarpus terminalis* Velenovsky and Viniklâr, 1929. Fruit. Cretaceous: Cenomanian, Czechoslovakia: Silvenec.
Comments: Fruits like those of modern *Sparganium* are known from the Upper Eocene onwards in Europe (Chandler, 1964; Mai and Walther, 1978; Mai, 1985a; Friis, 1985a).
Muller First: *Sparganiaceaepollenites* spp. (form genus). Krutzsch, 1970a, Palaeocene, Germany. Schumacker-Lambry and Roche, 1973a, Palaeocene, Belgium. Gruas-Cavagnetto, 1976a. Palaeocene, France and England, UK; *Sparganium globipites* Wilson and Webster, 1946, Palaeocene, Montana, USA.
Comments: These monoporate pollens may be from an extinct group with ecological, facies or systematic connections to the Typhaceae, Centrolepidaceae and Restionaceae.

F. STACHYURACEAE Agardh, 1858

PFR: No record.
Comments: Seeds like those of modern *Stachyurus* were described from the Middle Miocene of Germany (Mai, 1964).
Pollen: No record.

F. STAPHYLEACEAE Lindley, 1829

PFR: No record.
Comments: Fruits like those of modern *Tapiscia* occur in the Lower and Middle Eocene of Europe and North America (Manchester, *in* Collinson, 1988c, pp. 59–66). Leaves, seeds and woods assigned to *Staphylea* and *Turpinia* are listed by Taylor (1990) from the Upper Palaeocene onwards in North America.
Muller First: *Staphylea*. Zagwijn, 1967, Pliocene, The Netherlands.
Comments: Unchallenged.

F. STEMONACEAE Engler, *in* Engler and Prantl, 1887

PFR: No record.
Comments: Numerous species of *Spirellea* Knobloch and Mai were described from the Upper Cretaceous of Europe (Knobloch and Mai, 1986). Two of these (Maastrichtian to Lower Palaeocene) were tentatively assigned to Stemonaceae by Mai (1987a). This is an isolated fossil record of the family and requires confirmation.
Pollen: No record.

F. STERCULIACEAE Bartling, 1830

PFR First: *Chattawaya paliformis* S. R. Manchester, 16 January 1980. Wood. Tertiary: Middle Eocene, USA: Oregon, Nut Beds. *Triplochitioxylon oregonensis* S. R. Manchester, 10 July 1979. Wood. Tertiary: Middle Eocene, USA: Oregon.
Comments: These woods are critically documented Middle Eocene examples of the family. Leaves, some similar to modern *Sterculia*, others to modern *Dombeya*, are recorded from the Lower, Middle and Upper Eocene of Europe and North America (Wilde, 1989; Taylor, 1990). The Palaeocene leaves listed by Taylor (1990) as '(*Pterospermites*) *Penosophyllum*' were not considered as members of the Sterculiaceae by Manchester (1980).
Muller First: *Reevesiapollenites* spp. (form genus), Krutzsch, 1970f, Palaeocene, Germany.
Comments: Unchallenged.

F. STRELITZIACEAE Hutchinson, 1934

PFR: No record.
Comments: Zhilin (1974b) assigned an Upper Oligocene/Lower Miocene leaf to *Strelitzia* sp., but this was not mentioned by Daghlian (1981). Zhilin (1989, p. 258) included other Upper Oligocene leaves in *Strelitzia*. (See also comments under 'Musaceae' and 'Cyclanthaceae'.)
Pollen: No record.

F. STYRACACEAE Dumortier, 1829

PFR: No record.
Comments: Leaves assigned to *Styrax* are listed by Taylor (1990) from the Middle Eocene of North America, but these may require revision. Fruits like those of modern *Pterostyrax* were recorded by Mai and Walther (1985) in the Upper Eocene of Europe, and fruits like those of modern *Rehderodendron* have been recorded in the Lower Eocene of England, UK, and France (Vaudois-Mieja, 1983).
Pollen: No record.

F. SURIANACEAE Arnott, *in* Wight and Arnott, 1834

PFR: No record.
Comments: Taylor (1990) noted a wood from the Eocene of North America assigned to this family, but indicated that the record required confirmation or reinvestigation.
Pollen: No record.

F. SYMPLOCACEAE Desfontaines, 1820

PFR First: *Durania ehrenbergi* F. Kirchheimer, 31 December 1936. Fruit. Tertiary: Oligocene, Germany: Knonzendorf.
Comments: Fruits like those of modern *Symplocos* occur in the Lower Eocene onwards in Europe (Mai and Walther, 1985; Chandler, 1964) and leaves like those of *Symplocos* occur in the Middle Eocene of Europe (Wilde, 1989) and North America (Taylor, 1990).
Muller First: *Triporopollenites andersonii* and *T. scabroporus* (form taxa) Chmura, 1973. Maastrichtian, California, USA.
Comments: Too early.

F. TACCACEAE Dumortier, 1829

PFR: No record.
Comments: Seeds like those of modern *Tacca* were recorded by Gregor (1983) from the Oligocene of Czechoslovakia.
Pollen: No record.

F. TAMARICACEAE Link, 1821

PFR First: *Tamaricoxylon africanum* (Kräusel) Boureau, 1951. Quaternary, Somaliland.
Comments: Gokhtuni and Takhtajan (1988) have recorded shoots like those of modern *Tamarix* from the uppermost Miocene of the former USSR.
Pollen: No record.

F. TETRACENTRACEAE Van Tieghem, 1900

PFR: No record.
Comments: Leaves assigned to *Tetracentron* (the only modern genus in the family) are recorded from the Upper Cretaceous and Lower Tertiary of North America, Europe and the former USSR (Takhtajan, 1974; Wolfe, 1977). However, some of these leaves are associated with (or very similar to those associated with) the partially reconstructed plants assigned to Cercidiphyllaceae and Trochodendraceae (see discussion on these families). Confirmation of a fossil record for Tetracentraceae must await complete revision of all the leaf fossils and recovery of appropriate reproductive material.
Muller First: See comments on Trochodendraceae.

F. THEACEAE Don, 1825

PFR First: *Schimoxylon dachelense* (R. Kräusel) K. Kramer (*'dachalense'*), 29 March 1974. Wood. Cretaceous: Senonian, Egypt: Oase Dachel.
Comments: In their review of fossil Theaceae, Grote and Dilcher (1989) considered all North American Cretaceous and Palaeocene leaf records as needing revision. They cited unpublished *Gordonia*-like fruits from the Middle Eocene and described new Middle Eocene fruit material of subfamily Camellioideae. Seeds like those of modern subfamilies Ternstroemioideae and Camellioideae are widespread in Europe from the Upper Cretaceous onwards (summary in Grote and Dilcher, 1989).
Muller First: *Tricolporopollenites srivastavai* (form taxon) Gruas-Cavagnetto, 1978; Ollivier-Pierre, 1979. Lower Eocene, France.
Comments: Unchallenged.

F. THEOPHRASTACEAE Link, 1829

PFR First: *Clavijopsis staubi* G. Schindehütte, 1907. Leaf. Tertiary. Germany: Homberg, Hessen.
Comments: We know of no well-substantiated megafossils of this family.
Pollen: No record.

F. THYMELAEACEAE de Jussieu, 1789

PFR First: *Daphnites goepperti* Ettingshausen, post-7 February 1867. Leaf. Cretaceous, Austria: Aigen. This use of the genus is considered by some to be nomenclaturally invalid as a junior homonym of *Daphnitis* K. P. J. Sprengel, 1824.
Comments: Taylor (1990) lists an unpublished flower of this family from the Upper Palaeocene of North America. Mai and Walther (1978, 1985) recorded fruits like those of modern *Aquilaria* from the Lower Eocene to Middle Oligocene and seeds of *Thymelaeaspermum* from the Middle Eocene to Miocene in Europe.
Muller First: *Pseudospinaepollis pseudospinosus* (form taxon) Krutzsch, 1966c, Germany; Gruas-Cavagnetto, 1976a England, UK.
Comments: Unchallenged.

F. TILIACEAE de Jussieu, 1789

PFR First: *Etheridgea subglobosa* C. von Ettingshausen, 1893 (post-13 April). Fruit. Cretaceous, Australia: Queensland: between Warragh and Oxley. *Tiliaephyllum dubium* Newberry, 1895. Leaf. Cretaceous, USA: New Jersey.
Comments: Leaves like those of modern *Tilia* and *Willisia* are recorded from the Middle Eocene of North America (Taylor, 1990). Several fruits are recorded in the Lower Eocene of Europe (Chandler, 1964; Vaudois-Mieja, *in* Collinson, 1988c, pp. 31–44).
Muller First: *Discoidites borneensis* (form taxon) Muller, 1970. Palaeocene, Borneo. *Intratriporopollenites pseudinstructus* (form taxon) Mai, 1961. Lower Palaeocene, Germany. Frederiksen, 1979, Lower Palaeocene, USA.
Comments: Unchallenged.

F. TRAPACEAE Dumortier, 1828

PFR First: *Prototrapa douglassi* V. N. Vasilev, 19 May–30 June 1967, Fruit, Cretaceous. Australia, Victoria.
Comments: Mai (1985a, fig. 485) gives an elegant diagrammatic representation of the fossil record of this family, which is represented by fruits assigned to *Hemitrapa* in the Upper Oligocene and to modern *Trapa* in the Miocene onwards in Europe. Fruits or seeds superficially similar (and previously assigned to *Hemitrapa*) from the Lower Cretaceous of Australia were rejected as not angiospermous (Drinnan and Chambers, 1986).
Muller First: *Sporotrapoidites illingensis* (form taxon) Konzalova, 1976a, Lower Miocene, Czechoslovakia.
Comments: Unchallenged.

F. TROCHODENDRACEAE Prantl, *in* Engler and Prantl, 1888

PFR First: *Trochodendroides rhomboideus* (Lesquereux) E. W. Berry, 23 March 1922. Leaf. Upper Cretaceous, Arthur's Bluff, Texas, USA.
Comments: The extinct, partially reconstructed *Nordenskioldia*-plant, represented by associated infructescences and leaves (Crane *et al.*, 1990, 1991) is assigned to this family. It was widespread from the Palaeocene onwards in the Northern Hemisphere and ranges into the Miocene (Crane *et al.*, 1991; Manchester *et al.*, 1991). *Trochodendron* is first known in the Miocene (Manchester *et al.*, 1991). Upper Cretaceous material may also represent *Nordenskioldia* (Friis and Crane, *in* Crane and Blackmore, 1989), but this has not been confirmed. (See also comments on Tetracentraceae.)
Muller First: Rejected; no record.

F. TRIMENIACEAE Gibbs, 1917

PFR: No record.
Comments: We know of no claims for megafossils of this family.
Muller First: Pending.

F. TURNERACEAE de Candolle, 1828

PFR: No record.

Comments: We know of no claims for megafossils of this family.
Muller First: Pending.

F. TYPHACEAE de Jussieu, 1789

PFR First: *Typhaephyllum scammonii* U. Prakash and E. Boureau, 1970. Tertiary, Upper Miocene, USA, Washington, Vantage.
Comments: Fruits and seeds like those of modern *Typha* are widespread in the Upper Cretaceous onwards in Europe (Mai, 1985a; Knobloch and Mai, 1986; Collinson, 1988a). Grande (1984, p. 265) figures a striking example of a complete spike of *Typha* sp., probably in the fruiting condition, from the Lower Eocene of North America.
Muller First: *Typha latifolipites* Wilson and Webster, 1946. Palaeocene, Montana, USA.
Comments: Pollen may be difficult to distinguish from that of the Sparganiaceae.

F. ULMACEAE Mirbel, 1815

PFR First: *Celtidophyllum praeaustrale* F. Krasser, 1911. Leaf. Cretaceous, Czechoslovakia: Moravia, Kunstadt.
Comments: Manchester (*in* Crane and Blackmore, 1989) reviewed the fossil history of the family and, while accepting Upper Cretaceous (Santonian–Campanian) and Palaeocene foliage as examples of Ulmoideae, noted that no well-substantiated fruits of the subfamily were known until the Eocene. In the Eocene, leafy shoots with attached fruits are known for *Ulmus*, *Zelkova* and an extinct genus *Cedrelospermum*. The fossil record of Celtoideae includes foliage, fruit and flowers, with the earliest examples being fruits from the Maastrichtian of Europe and Palaeocene of North America.
Muller First: *Triorites minutipora* (form taxon) Muller, 1968. Turonian, Sarawak.
Comments: Unchallenged.

F. URTICACEAE de Jussieu, 1789

PFR First: *Urticicarpum scutellum* E. M. Reid and M. E. J. Chandler, 25 November 1933. Fruit. Tertiary: Eocene, UK, England: Kent: Sheppey: Minster.
Comments: The PFR record was only tentatively assigned to the family. Collinson (*in* Crane and Blackmore, 1989) reviewed the fossil history and was unable to confirm any records prior to the Upper Eocene. Cretaceous fruits assigned to the family were dissimilar to modern examples. Friis and Crane (*in* Crane and Blackmore, 1989) suggested that some of these might be similar to *Caryanthus* a genus of Cretaceous Juglandales/Myricales. Fruits like those of modern *Pilea* and *Laportea* occur in the Oligocene onwards in Europe and Asia.
Muller First: Pending.

F. VALERIANACEAE Batsch, 1802

PFR First: *Valerianellites capitatus* G. Saporta, 1862. Inflorescence. Tertiary, Aix en Provence, France.
Comments: The PFR record was only tentatively assigned to the family. Fruits like those of modern *Patrinia* occur in the Miocene to Pliocene of Europe and the former USSR (Łańcucka-Środoniowa, 1979; Dorofeev, 1988) and Mai (1985a) lists *Valeriana* from the Miocene onwards in Europe.
Muller First: Pending.

F. VERBENACEAE Jaume St-Hilaire, 1805

PFR First: *Premnophyllum trigonum* J. Velenovsky, 20 July 1884. Leaf. Upper Cretaceous, Czechoslovakia: Bohemia: Vyserovic.
Comments: Taylor (1990) lists several occurrences of a flower said to be similar to *Holmskioldia*, but notes that revision is required. Manchester and Meyer (1987) note that flowers given this name are the same as those of *Florissantia physalis* Knowlton (and others termed *Porana speirii*). They possess a large, fused, five-lobed perianth, five-carpelled ovary with a single style. They are considered to represent an extinct plant, possibly within Malvales, and are thus unlikely to represent Verbenaceae (Lamiales). The only other fossils listed for this family by Taylor (1990) are Middle Eocene leaves assigned to modern *Clerodendrum*. Bande (1986) recorded wood like that of modern *Gmelina* from the Lower Tertiary Deccan Intertrappean Beds of India.
Muller First: *Avicennia* type Leopold, 1969. Lower Miocene, Marshall Islands, West Pacific.
Comments: Unchallenged.

F. VIOLACEAE Batsch, 1802

PFR First: No record.
Comments: Several species of *Viola* are represented by seeds in the Oligocene to Pliocene of Europe (Łańcucka-Środoniowa, 1979).
Muller First: Pending.

F. VISCACEAE Miers, 1851

PFR: No record.
Comments: Selmeier (1975) described *Viscoxylon haustoria* from within *Pinus* wood from the Upper Miocene of Germany. Although originally assigned to Loranthaceae, a close comparison was indicated with *Viscum* which is placed in Viscaceae *sensu* Cronquist (1981). Flowers and fruits like those of modern *Arceuthobium* occur in the Upper Miocene of Poland (Łańcucka-Środoniowa, 1980).
Pollen: No record.

F. VITACEAE de Jussieu, 1789

PFR First: *Ieoleoia dorofeevii* V. A. Samylina, 1976 (post-22 September). Fruit, Seed. Cretaceous: Albian, former USSR: Magadan District: Sugoi River Basin: Auci Brook: Omsukchan.
Comments: Vitaceae are well represented by seeds from the Upper Palaeocene onwards in Europe (Chandler, 1964; Mai and Walther, 1978, 1985; Collinson, 1986a; Mai, 1987a). Foliage is recorded from the Middle Eocene (Wilde, 1989) and Upper Eocene to Oligocene (Mai and Walther, 1985) of Europe. Notably, the strongly diagnostic seeds are not recorded in the Upper Cretaceous material of Knobloch and Mai (1986). Seeds are also well represented in the Middle Eocene of North America (Manchester, *in* Knobloch and Kvacek, 1990, pp. 183–8; Cevallos-Ferriz and Stockey, 1990b) and foliage is listed by Taylor (1990) from the Palaeocene onwards in North America.
Muller First: *Tricolporopollenites marcodurensis* (form taxon) Gruas-Cavagnetto, 1978, Oligocene, France. Petrescu, 1973. Oligocene, Romania. Traverse, 1955, Oligocene, Vermont, USA. *Vitis forestdalensis* Traverse, 1955. Oligocene, Vermont, USA.
Comments: Unchallenged.

F. VOCHYSIACEAE St-Hilaire, 1820

PFR First: *Qualeoxylon itaquaquecetubense* K. Suguio and

D. Mussa, 1978. Wood. Quaternary, Brazil: Sao Paulo: Itaquaquecetuba.
Comments: We know of no earlier claims for megafossils of this family.
Pollen: No record.

F. WINTERACEAE Lindley, 1820

PFR: No record.
Comments: The one example of fossil leaves included in Winteraceae (Taktajan, 1974) had been revised and included in Magnoliaceae by Kirchheimer (1957). Hill and Macphail (1985) described *Tasmannia* seeds from the Plio-Pleistocene of Australia, but this material is now considered to be early middle Pleistocene in age (Hill, pers. comm., 1991). Hill (pers. comm., 1991) may have Winteraceae seeds in the Australian Oligocene. Gottwald (1992) assigned an Upper Eocene wood to this family.
Muller First: *Pseudowinterapollis* = (*Gephyrapollenites*) *wahooensis* Mildenhall and Crosbie, 1979; Stover and Partridge, 1973. Maastrichtian, Australasia.
Comments: Doyle *et al.* (1990) described ulcerate pollen tetrads from the Upper Barremian–Lower Aptian of Gabon as *Walkeripollis gabonensis*, which resemble pollen of this family.

F. XYRIDACEAE Agardh, 1823

PFR: No record.
Comments: Mai (1985a) lists *Xyris* in the Miocene of Europe.
Pollen: No record.

F. ZANNICHELLIACEAE Dumortier, 1829

PFR: No record.
Comments: Fruits like those of modern *Zannichellia* are recorded from the Miocene and Pliocene in Europe (Szafer, 1961).
Pollen: No record.

F. ZINGIBERACEAE Lindley, 1835

PFR First: *Spirematospermum wetzleri* Heer, 1925. Fruit. Tertiary, Upper Eocene, UK: England: Hordle, Hampshire.
Comments: *Spirematospermum* is now known from the Santonian/Campanian of North America and Maastrichtian of Europe (Goth, 1986; Knobloch and Mai, 1986; Friis, *in* Collinson, 1988c, pp. 7–12). This seed record is supported by leaves of *Zingiberopsis* also in the Upper Cretaceous of North America (Friis, *in* Collinson, 1988c, pp. 7–12; Taylor, 1990). Subsequent seed records occur throughout the Tertiary (Friis, *in* Collinson, 1988c, pp. 7–12) and leaves at least in the Middle and Upper Eocene (Wilde, 1989; Taylor, 1990).
Pollen: No record.

F. ZOSTERACEAE Dumortier, 1829

PFR: No record.
Comments: Daghlian (1981) considered all megafossil records to be unreliable.
Pollen: No record.

F. ZYGOPHYLLACEAE Brown *in* Flinders, 1814

PFR First: *Bubulcia globifera* A. B. Massalongo, post-22 August 1858, Tertiary: Eocene, Italy: Monte Bolca.
Comments: We know of no well-substantiated megafossils for this family.
Muller First: Pending.
Comments: See Oxalidaceae.

REFERENCES

Andreánszky, G. (1959) Contribution a la conaissance de la flore de l'Oligocene Inferieur de la Hongrie et une essai sur reconstruction de la Flore Contemporiare. *Acta Bontanica de l'Académie Scientifique de Hungarie*, **5**, 1–36.

Bande, M. B. (1986) Fossil wood of *Gmeliña* Linn. (Verbenaceae) from the Decean Intertrappean beds of Nawargaon with comments on the nomenclature of Tertiary woods. *The Palaeobotanist*, **35**, 165–70.

Bande, M. B. and Prakash, U. (1986) The Tertiary flora of southeast Asia with remarks on its palaeoenvironment and phytogeography of the Indo-Malayan region. *Review of Palaeobotany and Palynology*, **49**, 203–33.

Basinger, J. F. and Christophel, D. C. (1985) Fossil flowers and leaves of the Ebenaceae from the Eocene of southern Australia. *Canadian Journal of Botany*, **63**, 1825–43.

Berger, W. (1954) Pflanzenreste aus dem Miozänen Ton von Weingraben bei Drasmarkt (Mittelburgenland). II. *Sitzumgsberichte der Mathematisch–Naturwissenschaftlichen Klasse der Bayerischen Akademie der Wissenschaften zu München, Abteiling 1*, **162**, 17–24.

Berry, E. W. (1937) Gyrocarpus and other fossil plants from the Cumarebo field in Venezuela. *Journal of the Washington Academy of Sciences*, **27**, 501–6.

Boulter, M. C. (1971) A palynological study of two Neogene plant beds in Derbyshire. *Bulletin of the British Museum of Natural History (Geology)*, **19**, 360–410.

Boulter, M. C. and Kvacek, Z. (1989) The Palaeocene flora of the Isle of Mull. *Special Papers in Palaeontology*, **42**, 1–149.

Boureau, E., Cheboldaeff-Salard, M., Koeniguer, J.-C. *et al.* (1983) Evolution des flores et de la végétation Tertiaires en Afrique, au nord de l'Equateur. *Bothalia*, **14**, 355–67.

Boyd, A. (1990) The Thyra O Flora: toward an understanding of the climate and vegetation during the Early Tertiary in the High Arctic. *Review of Palaeobotany and Palynology*, **62**, 189–203.

Bůžek, C., Kvaček, Z. and Manchester, S. R. (1989) Sapindaceous affinities of *Pteleaecarpum* fruits from the Tertiary of Eurasia and North America. *Botanical Gazette*, **150**, 477–89.

Carpenter, R. J. and Hill, R. S. (1988) Early Tertiary *Lomatia* (Proteaceae) megafossils from Tasmania, Australia. *Review of Palaeobotany and Palynology*, **56**, 141–50.

Cevallos-Ferriz, S. and Stockey, R. A. (1988a) Permineralized fruits and seeds from the Princeton Chert (Middle Eocene) of British Columbia: Araceae. *American Journal of Botany*, **75**, 1099–113.

Cevallos-Ferriz, S. and Stockey, R. A. (1988b) Permineralized fruits and seeds from the Princeton Chert (Middle Eocene) of British Columbia: Lythraceae. *Canadian Journal of Botany*, **66**, 303–12.

Cevallos-Ferriz, S. and Stockey, R. A. (1989) Permineralized fruits and seeds from the Princeton Chert (Middle Eocene) of British Columbia: Nymphaeaceae. *Botanical Gazette*, **150**, 207–17.

Cevallos-Ferriz, S. and Stockey, R. A. (1990a) Vegetative remains of the Magnoliaceae from the Princeton Chert (Middle Eocene) of British Columbia. *Canadian Journal of Botany*, **68**, 1327–39.

Cevallos-Ferriz, S. and Stockey, R. A. (1990b) Permineralised fruits and seeds from the Princeton chert (Middle Eocene) of British Columbia: Vitaceae. *Canadian Journal of Botany*, **68**, 288–95.

Cevallos-Ferriz, S. and Stockey, R. A. (1990c) Vegetative remains of the Rosaceae from the Princeton Chert (Middle Eocene) of British Columbia. *International Association of Wood Anatomists Bulletin*, **11**, 261–80.

Chandler, M. E. J. (1964) *The Lower Tertiary Floras of Southern England. IV A Summary and Survey of Findings in the Light of Recent Botanical Observations.* British Museum (Natural History), London, xii + 151 pp.

Chesters, K. I. M. (1957) The Miocene flora of Rusinga Island, Lake Victoria, Kenya. *Palaeontographica, Abteilung B*, **101**, 30–71.

Chesters, K. I. M., Gnauck, F. R. and Hughes, N. F. (1967) Angiospermae, in *The Fossil Record* (ed W. B. Harland, C. H. Holland, M. R. House), Geological Society, London, pp. 269–88.

Christophel, D. C. (1984) Early Teriary Proteaceae: The first floral evidence of Musgraveinae. *Australian Journal of Botany*, **32**, 177–86.

Christophel, D. C. and Greenwood, D. R. (1987) A megafossil flora from the Eocene of Golden Grove, South Australia. *Transactions of the Royal Society of South Australia*, **111**, 155–62.

Christophel, D. C. and Greenwood, D. R. (1988) A comparison of Australian tropical rainforest and Tertiary fossil leaf beds. *Proceedings of the Ecological Society of Australia*, **15**, 139–48.

Christophel, D. C., Harris, W. K. and Syber, A. K. (1987) The Eocene flora of the Anglesea Locality, Victoria. *Alcheringa*, **11**, 303–23.

Collinson, M. E. (1980) Recent and Tertiary seeds of the Nymphaeaceae *sensu lato* with a revision of *Brasenia ovula* (Brong.) Reid and Chandler. *Annals of Botany*, **46**, 603–32.

Collinson, M. E. (1982) A reassessment of fossil Potamogetoneae fruits with description of new material from Saudi Arabia. *Tertiary Research*, **4**, 83–104.

Collinson, M. E. (1983) Palaeofloristic assemblages and palaeoecology of the Lower Oligocene Bembridge Marls, Hamstead Ledge, Isle of Wight. *Botanical Journal of the Linnean Society*, **86**, 177–225.

Collinson, M. E. (1986a) The Felpham Flora: – a preliminary report. *Tertiary Research*, **8**, 29–32.

Collinson, M. E. (1986b) Use of modern generic names for plant fossils, in *Systematic and Taxonomic Approaches in Palaeobotany* (eds R. A. Spicer and B. A. Thomas), The Systematics Association Special Volume, 31, Clarendon Press, Oxford, pp. 91–104.

Collinson, M. E. (1988a) Freshwater macrophytes in palaeolimnology. *Palaeogeography, Palaeoclimatology, Palaeoecology*, **62**, 317–42.

Collinson, M. E. (1988b) The special significance of the Middle Eocene fruit and seed flora from Messel, West Germany. *Courier Forschungsinstitut Senckenberg*, **107**, 187–97.

Collinson, M. E. (ed.) (1988c) Plants and their palaeoecology: examples from the last 80 million years. *Tertiary Research*, **9**, 1–235.

Collinson, M. E. (1990) Plant evolution and ecology during the early Cainozoic diversification. *Advances in Botanical Research*, **17**, 1–98.

Collinson, M. E. and Crane, P. R. (1978) *Rhododendron* seeds from the Palaeocene of southern England. *Botanical Journal of the Linnean Society*, **76**, 195–205.

Collinson, M. E. and Pingen, M. (1992) Seeds of the Melastomataceae from the Miocene of Central Europe, in *Palaeovegetational Development in Europe and Regions Relevant to its Palaeofloristic Evolution* (ed. J. Kovar-Eder). Museum of Natural History, Vienna, pp. 129–39.

Crabtree, D. R. (1987) Angiosperms of the northern Rocky Mountains: Albian to Campanian (Cretaceous) macrofossil floras. *Annals of the Missouri Botanic Garden*, **74**, 707–47.

Crane, P. R. and Blackmore, S. (eds) (1989) *Evolution, Systematics, and Fossil History of the Hamamelidae*. Vol. 1: *Introduction and 'Lower' Hamamelidae*; Vol. 2: *'Higher' Hamamelidae*. Systematics Association Special Volume 40A, 40B, Clarendon Press, Oxford, Vol. 1 (40A) 302 pp., Vol. 2 (40B) 353 pp.

Crane, P. R. and Stockey, R. A. (1985) Growth and reproductive biology of *Joffrea speirsii* gen. et sp. nov., a *Cercidiphyllum*-like plant from the Late Palaeocene of Alberta, Canada. *Canadian Journal of Botany*, **63**, 340–64.

Crane, P. R. and Stockey, R. A. (1986) Morphology and development of pistillate inflorescences in extant and fossil Cercidiphyllaceae. *Annals of the Missouri Botanical Garden*, **73**, 382–93.

Crane, P. R., Friis, E. M. and Pedersen, K. R. (1989) Reproductive structure and function in Cretaceous Chloranthaceae. *Plant Systematics and Evolution*, **165**, 211–26.

Crane, P. R., Manchester, S. R. and Dilcher, D. L. (1990) A preliminary survey of fossil leaves and well-preserved reproductive structures from the Sentinel Butte Formation (Palaeocene) near Almont, North Dakota. *Fieldiana, Geology, New Series*, **20**, 1–63.

Crane, P. R., Manchester, S. R. and Dilcher, D. L. (1991) Reproductive and vegetative structure of *Nordenskioldia* (Trochodendraceae), a vesseless dicotyledon from the Early Tertiary of the Northern Hemisphere. *American Journal of Botany*, **78**, 1311–34.

Crepet, W. L. and Daghlian, C. P. (1981) Lower Eocene and Paleocene Gentianaceae: floral and palynological evidence. *Science*, **214**, 75–7.

Crepet, W. L. and Feldman, G. D. (1991) The earliest remains of grasses in the fossil record. *American Journal of Botany*, **78**, 1010–14.

Crepet, W. L. and Taylor, D. W. (1986) Primitive mimosoid flowers from the Paleocene–Eocene and their systematic and evolutionary implications. *American Journal of Botany*, **73**, 548–63.

Crepet, W. L., Friis, E. M. and Nixon, K. C. (1991) Fossil evidence for the evolution of biotric pollination. *Philosophical Transactions of the Royal Society of London*, B **333**, 187–95.

Cronquist, A. (1981) *An Integrated System of Classification of Flowering Plants*, Columbia University Press, New York, xviii + 1262 pp.

Daghlian, C. P. (1981) A review of the fossil record of monocotyledons. *The Botanical Review*, **47**, 517–55.

Dorofeev, P. I. (1957) O pliotsenovai flore nagavskikh glin na donu. *Doklady Akademia Nauk SSSR*, **117**, 124–6.

Dorofeev, P. I. (1959) Materialy k Poznaniyu Miotsenovoi Flory Rostovskoi Oblasti. *Problemy Botaniki*, **4**, 143–89.

Dorofeev, P. I. (1964) Sarmatskaya Flora g. Apsheronska. *Doklady Akademia Nauk SSSR*, **156**, 82–4.

Dorofeev, P. I. (1969) *The Miocene Flora of the Mamontova Mountains*. Nauka, Leningrad, 124 pp.

Dorofeev, P. I. (1988) *Miocene Floras of the Tambov District*. Posthumous work edited by F. Ju. Velichkevich, Akad. Nauk, Leningrad, 196 pp.

Douglas, J. G. and Christophel, D. C. (eds) (1990) *Proceedings of the 3rd International Organisation of Palaeobotany Conference*, Melbourne 1988. A–Z Printers, International Organisation of Palaeobotany, Place of publication not stated. (? Melbourne, Mount Waverley, or Adelaide), 154 pp.

Doyle, J. A., Hotton, C. A. and Ward, J. V. (1990) Early Cretaceous tetrads, zonasulcate pollen and Winteraceae. I. Taxonomy, morphology and ultrastructure; II. Cladistic analysis and implications. *American Journal of Botany*, **77**, 1544–57, 1558–68.

Drinnan, A, N. and Chambers, T. C. (1986) Flora of the Lower Cretaceous Koonwarra Fossil Bed (Korumbarra Group), South Gippsland, Victoria, in *Plants and Invertebrates from the Lower Cretaceous Koonwarra Fossil Bed, South Gippsland, Victoria* (eds P. A. Jell and J. Roberts), Association of Australian Palaeontologists, Sydney, Memoir, **3**, pp. 1–205.

Drinnan, A. N., Crane, P. R., Friis, E. M. *et al.* (1990) Lauraceous flowers from the Potomac Group (Mid-Cretaceous) of eastern North America. *Botanical Gazette*, **151**, 370–84.

Drinnan, A. N., Crane, P. R., Friis, E. M. *et al.* (1991) Angiosperm flowers and tricolpate pollen of Buxaceous affinity from the Potomac Group (Mid-Cretaceous) of Eastern North America. *American Journal of Botany*, **78**, 153–76.

Drugg, W. S. (1967) Palynology of the Upper Moreno Formation (Late Cretaceous–Palaeocene), Escarpado Canyon, California. *Palaeontographica, Abteilung B*, **120**, 1–71.

Endress, P. K. and Friis E. M. (1991) *Archamamelis*, hammamelidalean Flowers from the Upper Cretaceous of Sweden. *Plant Systematics and Evolution*, **175**, 101–14.

Erwin, D. M. and Stockey, R. A. (1989) Permineralized monocotyledons from the Middle Eocene Princeton chert (Allenby Formation) of British Columbia: Alismataceae. *Canadian Journal of Botany*, **67**, 2636–45.

Erwin, D. M. and Stockey, R. A. *et al.* (1990) Sapindaceous flowers from the Middle Eocene of British Columbia, Canada. *Canadian Journal of Botany*, **68**, 2025–34.

Eyde, R. H. (1972) Note on geologic histories of flowering plants. *Brittonia*, **24**, 111–16.

Eyde, R. H. (1988) Comprehending *Cornus*: puzzles and progress in the systematics of the Dogwoods. *The Botanical Review*, **54**, 233–351.

Farabee, M. J. and Canright, J. E. (1986) Stratigraphic palynology of the lower part of the Lance Formation (Maestrichtian) of

Wyoming. *Palaeontographica, Abteilung B*, **199**, 1–89.

Friis, E. M. (1983) Upper Cretaceous (Senonian) floral structures of Juglandalean affinity containing Normapolles pollen. *Review of Palaeobotany and Palynology*, **39**, 161–88.

Friis, E. M. (1985a) Angiosperm fruits and seeds from the Middle Miocene of Jutland, Denmark. *Biologiske Skrifter*, **24**, 1–165.

Friis, E. M. (1985b) *Actinocalyx* gen. nov., sympetalous angiosperm flowers from the Upper Cretaceous of southern Sweden. *Review of Palaeobotany and Palynology*, **45**, 171–83.

Friis, E. M. (1990) *Silvianthemum suecicum* gen. et sp. nov., a new saxifragalean flower from the Late Cretaceous of Sweden. *Biologiske Skrifter*, **36**, 1–35.

Friis, E. M. and Skarby, A. (1982) *Scandianthus* gen. nov., angiosperm flowers of Saxifragalean affinity from the upper Cretaceous of southern Sweden. *Annals of Botany*, **50**, 569–83.

Friis, E. M., Crane, P. R. and Pedersen, K. J. (1988) Reproductive structures of Cretaceous Platanaceae. *Biologiske Skrifter*, **31**, 1–55.

Gokhtuni, N. G. and Takhtajan, A. L. (1988) Additional data on the Late Sarmatian plants from the Nakhichevan satiferous deposits. *Botanicheskii Zhurnal*, **73**, 1708–10.

Goth, K. (1986) Erster Nachweis von Spirematospermum-Samen aus der Oberkreide von Kossen in Tirol. *Courier Forschungsinstitut Senckenberg*, **86**, 171–5.

Gottwald, H. (1992) Woods from marine sands of the late Eocene near Helmstedt (Lower Saxony/Germany). *Palaeontographica, Abteilung B*, **225**, 27–103.

Graham, A. (1987) Tropical American Tertiary floras and paleoenvironments: Mexico, Costa Rica and Panama. *American Journal of Botany*, **74**, 1519–31.

Grande, L. (1984) Paleontology of the Green River Formation, with a review of the fish fauna. *Bulletin of the Geological Survey of Wyoming*, **63**, 1–333.

Gregor, H-J. (1978) Die Miozänen Frucht- und Samen-Floren der oberpfalzer Braunkohle I. Funde aus den sandigen Zwichenmitteln. *Palaeontographica, Abteilung B*, **167**, 8–103.

Gregor, H-J. (1980) Seeds of the genus *Coriaria* Linné (Coriariaceae) in the European Neogene. *Tertiary Research*, **3**, 61–9.

Gregor, H-J. (1982) *Die Jungtertiaren Floren Suddeutschlands*. Ferdinand Enke, Stuttgart, 278 pp.

Gregor, H-J. (1983) Erstnachweis der Gattung *Tacca* Forst 1776 (Taccaceae) im Europaischen Alttertiar. *Documenta Naturae*, **6**, 27–31.

Gregor, H-J. (1989) Aspects of the fossil record and phylogeny of the family Rutaceae. *Plant Systematics and Evolution*, **162**, 251–65.

Grote, P. J. and Dilcher, D. L. (1989) Investigations of angiosperms from the Eocene of North America: a new genus of Theaceae based on fruit and seed remains. *Botanical Gazette*, **150**, 190–206.

Harland, W. B., Armstrong, R. L., Cox, A. V. *et al.* (1990) *A Geologic Timescale 1989*. Cambridge University Press, 263 pp.

Herendeen, P. S. and Dilcher, D. L. (1990a) *Diplotropis* (Leguminosae, Papilionoideae) from the Middle Eocene of Southeastern North America. *Systematic Botany*, **15**, 526–33.

Herendeen, P. S. and Dilcher, D. L. (1990b) Reproductive and vegetative evidence for the occurrence of *Crudia* (Leguminosae, Caesalpinoideae) in the Eocene of Southeastern North America. *Botanical Gazette*, **151**, 402–13.

Herendeen, P. S. and Dilcher, D. L. (1990c) Fossil mimosoid legumes from the Eocene and Oligocene of southeastern North America. *Review of Palaeobotany and Palynology*, **62**, 339–61.

Herendeen, P. S. and Dilcher, D. L. (1991) *Caesalpinia* subgenus *Mezoneuron* (Leguminosae, Caesalpinoideae) from the Tertiary of North America. *American Journal of Botany*, **78**, 1–12.

Herendeen, P. S., Les, D. H. and Dilcher, D. L. (1990) Fossil *Ceratophyllum* (Ceratophyllaceae) from the Tertiary of North America. *American Journal of Botany*, **77**, 7–16.

Hickey, L. J. (1977) Stratigraphy and Paleobotany of the Golden Valley Formation (Early Tertiary) of Western North Dakota. *Geological Society of America Memoir*, **150**, 1–183.

Hill, R. S. (1988) Australian Tertiary angiosperm and gymnosperm leaf remains – an updated catalogue. *Alcheringa*, **12**, 207–19.

Hill, R. S. (1991a) Tertiary *Nothofagus* (Fagaceae) macrofossils from Tasmania and Antarctica and their bearing on the evolution of the genus. *Botanical Journal of the Linnean Society*, **105**, 73–112.

Hill, R. S. (1991b) Leaves of *Eucryphia* (Eucryphiaceae) from Tertiary sediments in South-eastern Australia. *Australian Systematic Botany*, **4**, 481–97.

Hill, R. S. and Christophel, D. C. (1988) Tertiary leaves of the tribe Banksieae (Proteaceae) from south-eastern Australia. *Botanical Journal of the Linnean Society*, **97**, 205–27.

Hill, R. S. and Macphail, M. K. (1985) A fossil flora from rafted Plio-Pleistocene mudstones at Regatta Point, Tasmania. *Australian Journal of Botany*, **33**, 497–517.

Holmes, P. L. (ed.-in-chief) (1991) *The Plant Fossil Record Database, Version 1.0*, International Organization of Palaeobotany, London, Magnetic Publication.

Iljinskaya, I. A. (1986) Izmienienie flory Zajsankoy vpadiny a konca mela po Miocen. *Problemy Paleobotaniki*. Nauka, Leningrad.

Ivany, L. C., Portell, R. W. and Jones, D. S. (1990) Animal–plant relationships and paleobiogeography of an Eocene seagrass community from Florida. *Palaios*, **5**, 244–58.

Jacobs, B. F. and Kabuye, C. H. S. (1989) An extinct species of *Pollia* Thunberg (Commelinaceae) from the Miocene Ngorora Formation, Kenya. *Review of Palaeobotany and Palynology*, **59**, 67–76.

Kirchheimer, F. (1957) *Die Laubgewachse der Braunkohlenzeit*. Willhelm Knapp, Halle (Saale), 783 pp.

Knappe, H. and Ruffle, L. (1975) Neue Monimiaceen-Blatter im Santon des Subherzyn und ihre phytogeographischen Beziehungen zur Flora des ehemaligen Gondwana-Kontinentz. *Wissenschaftliche Zeitschrift der Humboldt-Universitat zu Berlin. Mathematisch–Naturwissenschaftiche Reihe*, **24**, 493–9.

Knobloch, E. and Kvaček, Z. (eds) (1990) *Proceedings of the Symposium Paleofloristic and Paleoclimatic Changes in the Cretaceous and Tertiary*. Geological Survey Publisher, Prague, 322 pp.

Knobloch, E. and Mai, D. H. (1986) Monographie der Fruchte und Samen in der Kreide von Mitteleuropa. *Rozpravy Ustredniho Ustavu Geologickeho Praha*, **47**, 1–219.

Koenigswald, W. van (1989) *Fossillagerstatte Rott bei Hennef am Siebengebirge*. Rheinlandia Verlag, Siegburg, 82 pp.

Kovach, W. L. and Dilcher, D. L. (1988) Megaspores and other dispersed plant remains from the Dakota Formation (Cenomanian) of Kansas, U.S.A. *Palynology*, **12**, 89–119.

Kvaček, Z. (1992) Lauralean angiosperms in the Cretaceous. *Courier Forschungsinstitut Senckenberg*, **147**, 345–67.

Kvaček, Z. and Walther, H. (1989) Paleobotanical studies in Fagaceae of the European Tertiary. *Plant Systematics and Evolution*, **162**, 213–29.

Kvaček, Z., Bůžek, C. and Holy, F. (1982) Review of *Buxus* fossils and a new large-leaved species from the Miocene of Central Europe. *Review of Palaeobotany and Palynology*, **37**, 361–94.

Lakhanpal, R. N. and Guleria, J. S. (1986) Fossil leaves of *Dipterocarpus* from the Lower Siwalik beds near Jawalamukhi, Himachal Pradesh. *The Palaeobotanist*, **35**, 258–62.

Lakhanpal, R. N., Guleria, J. S. and Awasthi, N. (1984) The fossil flora of Kachchh. III. Tertiary megafossils. *The Palaeobotanist*, **33**, 228–319.

Łańcucka-Środoniowa, M. (1979) Macroscopic plant remains from the freshwater Miocene of the Nowy Sacz Basin (West Carpathians, Poland). *Acta Palaeobotanica*, **20**, 3–117.

Łańcucka-Środoniowa, M. (1980) Macroscopic remains of the dwarf mistletoe *Arceuthobium* Bieb. (Loranthaceae) in the Neogene of Poland. *Acta Palaeobotanica*, **21**, 61–6.

Lidgard, S. and Crane, P. R. (1990) Angiosperm diversification and Cretaceous floristic trends: a comparison of palynofloras and leaf macrofloras. *Paleobiology*, **16**, 77–93.

MacGinitie, H. D. (1969) The Eocene Green River flora of Northwestern Colorado and Northeastern Utah. *University of California Publications in Geological Sciences*, **83**, 1–140.

Mädler, K. (1939) Die Pliozane Flora von Frankfurt am Main. *Abhandlungen der Senckenbergischen Naturforschenden Gesellschaft*, **446**, 1–202.

Mai, D. H. (1964) Die Mastixioideen-Floren in Tertiar der Oberlausitz. *Paläontologische Abhandlungen, Abteilung B*, **2**, 1–192.

Mai, D. H. (1968) Zwei ausgestorbene Gattungen im Tertiar Europas und ihre florengeschichtliche Bedeutung. *Palaeontographica, Abteilung B*, **123**, 184–99.

Mai, D. H. (1976) Fossile Fruchte und Samen aus dem Mitteleozän des Geiseltales. *Abhandlungen des Zentralen Geologischen Instituts, Paläontologische Abhandlungen*, **26**, 93–149.

Mai, D. H. (1980) Zur Bedeutung von Relikten in der Florengeschichte, in *100 Jahre Arboretum 1879–1979* (ed. W. Vent), Berlin, pp. 281–307.

Mai, D. H. (1984) Karpologische Untersuchungen der Steinkerne fossiler und rezenter Amygdalaceae (Rosales). *Feddes Repertorium*, **95**, 299–322.

Mai, D. H. (1985a) Entwicklung der Wasser- und Sumpfpflanzen-Gesellschaften Europas von der Kreide bis ins Quartar. *Flora*, **176**, 449–511.

Mai, D. H. (1985b) Beiträge zur Geschichte einiger holziger Saxifragales-Gattungen. *Gleditschia*, **13**, 75–88.

Mai, D. H. (1987a) Neue Fruchte und Samen aus Palaozanen Ablagerungen Mitteleuropas. *Feddes Repertorium*, **98**, 197–229.

Mai, D. H. (1987b) Neue Arten nach Früchten und Samen aus dem Tertiär von Nordwestsachsen und der Lausitz. *Feddes Repertorium* **98**, 105–26.

Mai, D. H. and Walther, H. (1978) Die Floren der Haselbacher Serie im Weisselster-Becken (Bezirk Leipzig, DDR). *Abhandlungen des Staatliches Museums für Mineralogie und Geologie zu Dresden*, **28**, 1–200.

Mai, D. H. and Walther, H. (1985) Die obereozänen Floren des Weisselster-Beckens und seiner Randgebiete. *Abhandlungen des Staatliches Museums fur Mineralogie und Geologie zu Dresden*, **33**, 1–260.

Mai, D. H. and Walther, H. (1988) Die Pliozänen Floren von Thuringen/Deutsche Demokratische Republik. *Quartärpaläontologie*, **7**, 55–297.

Manchester, S. R. (1980) *Chattawaya* (Sterculiaceae): a new genus of wood from the Eocene of Oregon and its implications for the xylem evolution of the extant genus *Pterospermum*. *American Journal of Botany*, **67**, 59–67.

Manchester, S. R. (1981) Fossil plants of the Eocene Clarno Nut Beds. *Oregon Geology*, **43**, 75–81.

Manchester, S. R. (1987) The fossil history of the Juglandaceae. *Monographs in Systematic Botany from the Missouri Botanical Garden*, **21**, 1–137.

Manchester, S. R. (1991) *Cruciptera*, a new genus of juglandaceous winged fruit from the Eocene and Oligocene of western North America. *Systematic Botany*, **16**, 715–25.

Manchester, S. R. and Meyer, H. W. (1987) Oligocene fossil plants of the John Day Formation, Fossil, Oregon. *Oregon Geology*, **49**, 115–27.

Manchester, S. R., Crane, P. R. and Dilcher, D. L. (1991) *Nordenskioldia* and *Trochodendron* (Trochodendraceae) from the Miocene of Idaho and Washington, U.S.A. *Botanical Gazette*, **152**, 346–58.

Manchester, S. R., Dilcher, D. L. and Tidwell, W. D. (1986) Interconnected reproductive structures and vegetative remains of *Populus* (Salicaceae) from the Middle Eocene Green River Formation, northeastern Utah. *American Journal of Botany*, **73**, 156–60.

Mathur, A. K. and Mathur, U. B. (1985) Boraginaceae (angiosperm) seeds and their bearing on the age of Lameta Beds of Gujarat. *Current Science*, **54**, 1070–1.

Mehrotra, R. C. (1981) Fossil wood of *Sonneratia* from the Deccan Intertrappean Beds of Mandla District, Madhya Pradesh. *Geophytology*, **18**, 129–34.

Miki, S. (1961) Aquatic floral remains in Japan. *Journal of Biology of the Osaka City University*, **12**, 91–121.

Muller, J. (1981) Fossil pollen records of extant angiosperms. *The Botanical Review*, **47**, 1–146.

Muller, J. (1985) Significance of fossil pollen for angiosperm history. *Annals of the Missouri Botanical Garden*, **71**, 419–43.

Negru, A. G. (1979) *The Early Pontian Flora from the Region Between Dnestr and Prutsk*. Akad. Nauk. Moldavian S.S.R. Botanic Garden, Stiinca, Kishinev, 110 pp. [in Russian].

Page, V. M. (1981) Dicotyledonous wood from the Upper Cretaceous of Central California III, Conclusions. *Journal of the Arnold Arboretum*, **62**, 437–55.

Palamarev, E. (1968) Karplogische Reste aus dem Miozäne Nordbulgariens. *Palaeontographica, Abteilung B*, **123**, 200–12.

Pigg, K. B. and Stockey, R. A. (1991) Platanaceous plants from the Paleocene of Alberta, Canada. *Review of Palaeobotany and Palynology*, **70**, 125–46.

Pingen, M. and Collinson, M. E. (in prep.) The fossil history of the Melastomataceae.

Pneva, G. P. (1988) Novy tretichny vid roda Aponogeton (Aponogetonaceae) iz Kazakhstana in Karakalpakii. *Botanicheskii Zhurnal*, **73**, 1597–9.

Pons, D. (1983) *Lecythidospermum bolivarensis* (Berry) nov. comb., graine de fossile Lecythidaceae (Angiosperme) du Tertiare de la Colombie. *Annales de Paléontologie*, **69**, 1–12.

Rasky, K. (1956) Fossilis Novenyek a Budapest Kornyeki 'Budai' Margaosszietbol. *Foldtani Koziony*, **86**, 167–9.

Reid, E. M. and Chandler, M. E. J. (1926) *The Bembridge Flora (Catalogue of Cainozoic Plants in the Department of Geology*: Vol. 1), British Museum (Natural History), London, 206 pp.

Retallack, G. and Dilcher, D. L. (1981) Early angiosperm reproduction: *Prisca reynoldsii*, gen. et sp. nov. from Mid-Cretaceous coastal deposits in Kansas, U.S.A. *Palaeontographica, Abteilung B*, **179**, 103–37.

Romero, E. J. (1986a) Paleogene phytogeography and climatology of south America. *Annals of the Missouri Botanical Gardens*, **73**, 449–61.

Romero, E. J. (1986b) Fossil evidence regarding the evolution of *Nothofagus* Blume. *Annals of the Missouri Botanical Gardens*, **73**, 276–83.

Romero, E. J. and Hickey, L. J. (1976) A fossil leaf of *Akaniaceae* from Paleocene beds in Argentina. *Bulletin of the Torrey Botanical Club*, **103**, 126–31.

Schaarschmidt, F. and Wilde, V. (1986) Palmenbluten und-blatter aus dem Eozän von Messel. *Courier Forschungsinstitut Senckenberg*, **86**, 177–202.

Schmid, R. and Schmid, M. J. (1973) Fossils attributed to the Orchidaceae. *American Orchid Society Bulletin*, **42**, 17–27.

Selmeier, A. (1975) *Viscoxylon pini* novum genus et nova species, Mistel-Senker in einem verkieselten *Pinus*-Holz aus jungtertiaren Sedimenten von Falkenberg, Oberpfalz (Bayern). *Bericht der Bayerischen Botanischen Gesellschaft zur Erforschung der Heimischen Flora, Munchen*, **46**, 93–109.

Stockey, R. A. (1987) A permineralised flower from the Middle Eocene of British Columbia. *American Journal of Botany*, **74**, 1878–87.

Stockey, R. A. and Manchester, S. R. (1988) A fossil flower with *in situ Pistillipollenites* from the Eocene of British Columbia. *Canadian Journal of Botany*, **66**, 313–18.

Strauss, A. (1969) Beitrage zur Kenntnis der Pliozanflora von Willershausen VII. Die angiospermen Fruchte und Samen. *Argumenta Palaeobotanica*, **3**, 163–97.

Szafer, W. (1952) A member of the family Podostemaceae in the Tertiary of the West Carpathian Mountains. *Acta Societas Botanica Polonica*, **21**, 747–69 [in Polish, English summary].

Szafer, W. (1954) *Pliocene Flora from the Vicinity of Czorsztyn (West Carpathians) and its Relationship to the Pleistocene*. Instytut Geologiczny Prace, XI, Wydawnictwa Geologiczne, Warsaw, 238 pp.

Szafer, W. (1961) *Miocene Flora from Stare Gliwice in Upper Silesia*. Instytut Geologiczny Prace, XXXIII, Wydawnictwa Geologiczne, Warsaw, 205 pp.

Takhtajan, A. (ed.) (1974) *Magnoliaceae–Eucommiaceae (Magnoliophyta Fossilia URSS*, Vol. 1), Nauka, Leningrad [in Russian], 188 pp.

Tanai, T. (1990) Euphorbiaceae and Icacinaceae from the Paleogene of Hokkaido, Japan. *Bulletin of the National Science Museum Tokyo, Series C*, **16**, 91–118.

Taylor, D. W. (1990) Paleobiogeographic relationships of angio-

sperms from the Cretaceous and Early Tertiary of the North American area. *The Botanical Review*, **56**, 279–416.
Taylor, D. W. and Crepet, W. L. (1987) Fossil floral evidence of Malpighiaceae and an early plant pollinator relationship. *American Journal of Botany*, **74**, 274–86.
Thomasson, J. R. (1987a) Late Miocene plants from northeastern Nebraska. *Journal of Paleontology*, **61**, 1065–79.
Thomasson, J. R. (1987b) Fossil grasses: 1820–1986 and beyond, in *Grass Systematics and Evolution* (eds T. R. Soderstrom, K. W. Hilu, C. S. Campbell *et al.*), Smithsonian Institution Press, Washington, DC, pp. 159–71.
Tidwell, W. D. and Nambudiri, E. M. V. (1990) *Tomlinsonia stichkania* sp. nov., a permineralized grass from the Pliocene to (?)Pleistocene China Ranch Beds in Sperry Wash, California. *Botanical Gazette*, **151**, 263–74.
Tidwell, W. D. and Parker, L. R. (1990) *Protoyucca shadishii* gen. et sp. nov., an arborescent monocotyledon with secondary growth from the Middle Miocene of Northwestern Nevada, U.S.A. *Review of Palaeobotany and Palynology*, **62**, 79–95.
Tiffney, B. H. and Barghoorn, E. S. (1979) Flora of the Brandon lignite. IV. Illiciaceae. *American Journal of Botany*, **66**, 321–9.
Uemura, K. (1979) Leaf compressions of *Buxus* from the Upper Miocene of Japan. *Bulletin of the National Science Museum*, (C) **5**, 1–8.
Uemura, K. (1988) *Late Miocene floras in Northeast Honshu, Japan*. National Science Museum, Tokyo, 174 pp.
Upchurch, G. R. and Dilcher, D. L. (1990) Cenomanian angiosperm leaf megafossils, Dakota Formation, Rose Creek Locality, Jefferson County, Southeastern Nebraska. *United States Geological Survey Bulletin*, **1915**, 1–52.
Van der Burgh, J. (1987) Miocene floras in the Lower Rhenish Basin and their ecological interpretation. *Review of Palaeobotany and Palynology*, **52**, 299–366.
Vaudois-Mieja, N. (1976) Sur deux fruits fossiles de Rubiaceae provenant des Grès á Sabals de l'Anjou. *Actes du 97è Congres National des Sociétés de Savantes, Nantes 1972, Section des Sciences*, **4**, 167–83.
Vaudois-Mieja, N. (1983) Extension paléogeographique en Europe de l'actuel genre asiatique *Rehderodendron* Hu (Styracacées). *Comptes Rendus de l'Académie de Sciences de Paris, Série II*, **296**, 125–30.
Vinken, R. (compiler) (1988) The Northwest European Tertiary Basin. *Geologische Jahrbuch*, **100**, 1–500.
Weyland, H. (1957) Kritische Untersuchungen zur kuticularan algre tertiärer Blatter III. Monocotylen der rheinischen Braunkohle. *Palaeontographica B*, **103**, 34–74.
Wheeler, E. F., Lee, M. and Matten, L. C. (1987) Dicotyledonous woods from the Upper Cretaceous of southern Illinois. *Botanical Journal of the Linnean Society*, **95**, 77–100.
Wiggins, V. D. (1982) *Expressipollis striatus* n. sp. to *Anacolosidites striatus* n. sp. An Upper Cretaceous example of suggested pollen aperture evolution. *Grana*, **21**, 39–49.
Wijninga, V. M. and Kuhry, P. (1990) A Pliocene florule from the Subachoque valley (Cordillera Oriental, Columbia). *Review of Palaeobotany and Palynology*, **62**, 249–90.
Wilde, V. (1989) Untersuchungen zur Systematik der Blattreste aus dem Mitteleozän der Grube Messel bei Darmstadt (Hessen, Bundesrepublik Deutschland). *Courier Forschungsinstitut Senckenberg*, **115**, 1–213.
Wilkinson, H. P. (1981) The anatomy of the hypocotyls of *Ceriops* Arnott (Rhizophoraceae), Recent and fossil. *Botanical Journal of the Linnean Society*, **82**, 139–64.
Wolfe, J. A. (1964) The Miocene floras from Fingerrock Wash Southwestern Nevada. *Professional Paper of the US Geological Survey*, **454-N**, N1–N36.
Wolfe, J. A. (1977) Paleogene floras from the Gulf of Alaska Region. *Professional Paper of the United States Geological Survey*, **997**, 1–108.
Wolfe, J. A. and Schorn, H. E. (1989) Paleoecologic, paleoclimatic, and evolutionary significance of the Oligocene Creede flora, Colorado. *Paleobiology*, **15**, 180–98.
Wolfe, J. A. and Schorn, H. E. (1990) Taxonomic revision of the Spermatopsida of the Oligocene Creede flora, southern Colorado. *US Geological Survey Bulletin*, **1923**, 1–40.
Wolfe, J. A. and Tanai, T. (1987) Systematics, phylogeny, and distribution of *Acer* (maples) in the Cenozoic of western North America. *Journal of the Faculty of Science, Hokkaido University*, **22**, 1–246.
Wolfe, J. A. and Upchurch, G. R. (1986) Vegetation, climatic and floral changes at the Cretaceous–Tertiary boundary, *Nature, London*, **324**, 148–51.
Wolfe, J. A., Doyle, J. A. and Page, V. M. (1976) The bases of angiosperm phylogeny: paleobotany. *Annals of the Missouri Botanical Garden*, **62**, 801–24.
Zavada, M. S. and Benson, J. M. (1987) First fossil evidence for the primitive angiosperm family Lactoridaceae. *American Journal of Botany*, **74**, 1590–4.
Zhilin, S. G. (1974a) The first Tertiary species of the genus Aponogeton (Aponogetonaceae). *Botanicheskii Zhurnal*, **59**, 1203–6.
Zhilin, S. G. (1974b) *Tretichny flory Ustyurta*. *Nauka*, Leningrad.
Zhilin, S. G. (1989) History of the development of the temperate forest flora in Kazakhstan, U.S.S.R. from the Oligocene to the early Miocene. *The Botanical Review*, **55**, 205–330.

INDEX